ANNUAL REVIEW OF BIOCHEMISTRY

EDITORIAL COMMITTEE (1985)

JOHN N. ABELSON
GÜNTER BLOBEL
PAUL D. BOYER
JAMES E. DARNELL, JR.
IGOR B. DAWID
EUGENE KENNEDY
J. MURRAY LUCK
ALTON MEISTER
CHARLES C. RICHARDSON
JAMES C. WANG

Responsible for the organization of Volume 54
(Editorial Committee, 1983)

JOHN N. ABELSON
PAUL D. BOYER
JAMES E. DARNELL, JR.
EUGENE KENNEDY
M. DANIEL LANE
J. MURRAY LUCK
ALTON MEISTER
CHARLES C. RICHARDSON
ESMOND E. SNELL
HERBERT TABOR
JAMES C. WANG
MARLENE DELUCA (GUEST)
FRANK M. HUENNEKENS (GUEST)
HERBERT STERN (GUEST)
CELIA TABOR (GUEST)

Production Editor	WENDY A. CAMPBELL
Indexing Coordinator	MARY A. GLASS
Subject Indexer	STEVEN A. SORENSEN

ANNUAL REVIEW OF BIOCHEMISTRY

VOLUME 54, 1985

CHARLES C. RICHARDSON, *Editor*

Harvard Medical School

PAUL D. BOYER, *Associate Editor*

University of California, Los Angeles

IGOR B. DAWID, *Associate Editor*

National Institutes of Health

ALTON MEISTER, *Associate Editor*

Cornell University Medical College

ANNUAL REVIEWS INC.   4139 EL CAMINO WAY   PALO ALTO, CALIFORNIA 94306 USA

ANNUAL REVIEWS INC.
Palo Alto, California, USA

International Standard Serial Number: 0066–4309
International Standard Book Number: 0–8243–0854-9
Library of Congress Catalog Card Number: 50–13143

Typesetting by Kachina Typesetting Inc., Tempe, Arizona; John Olson, President
Typesetting coordinator, Janis Hoffman

PRINTED AND BOUND IN THE UNITED STATES OF AMERICA

Annual Review of Biochemistry
Volume 54, 1985

CONTENTS

SOME RELATED ARTICLES IN OTHER *ANNUAL REVIEWS*

From the *Annual Review of Cell Biology*, Volume 1 (1985)

Traffic Through the Golgi Complex, Marilyn Farquhar
Receptor-Mediated Endocytosis: Concepts Emerging from the LDL Receptor System,
Joseph L. Goldstein, Michael S. Brown, Richard G. W. Anderson, David W.
Russell, and Wolfgang J. Schneider
Biogenesis of Membrane Proteins, George E. Palade
Myosin, James A. Spudich
Nonmuscle Actin-Binding Proteins, T. P. Stossel, C. Chaponnier, R. Ezzell, J. H.
Hartwig, P. Janmey, D. Kwiatkowski, S. Lind, D. Smith, F. S. Southwick, H. L.
Yin, K. S. Zaner
Chromosome Segregation in Mitosis and Meiosis, Andrew Murray and Jack Szostak
Biosynthesis of Paroxisomes, Paul Lazarow and Yukio Fujiki

From the *Annual Review of Genetics*, Volume 18 (1984)

*The Mutation and Polymorphism of the Human β-Globin Gene and its Surrounding
DNA,* Stuart H. Orkin and Haig H. Kazazian, Jr.
The Evolutionary Implications of Mobile Genetic Elements, Michael Syvanen
The Molecular Genetics of Cellular Oncogenes, Harold E. Varmus

From the *Annual Review of Immunology*, Volume 3 (1985)

*Immunobiology of Myasthenia Gravis, Experimental Autoimmune Myasthenia Gra-
vis, and Lambert-Eaton Syndrome,* Jon Lindstrom
*T-Lymphocyte Recognition of Antigen in Association with Gene Products of the Major
Histocompatibility Complex,* Ronald H. Schwartz
Molecular Genetics of T-Cell Receptor Beta Chain, Mark Davis

From the *Annual Review of Medicine*, Volume 36 (1985)

The Molecular Mechanism of Insulin Action, C. Ronald Kahn

From the *Annual Review of Microbiology*, Volume 39 (1985)

The Biochemistry and Industrial Potential of Spirulina, Orio Ciferri and Orsola
Tiboni
Oncogenes: Their Role in Neoplastic Transformation, L. Ratner, S. F. Josephs, and
Flossie Wong-Staal
The DNA Translocating Vertex of dsDNA Bacteriophage, Christopher Bazinet and
Jonathan King

For the convenience of readers, a detachable order form/envelope is bound into the back of this volume.

Ann. Rev. Biochem. 1985. 54:1–41

THEN AND NOW

Harland G. Wood

Department of Biochemistry, Case Western Reserve University, School of Medicine, Cleveland, Ohio 44106

CONTENTS

0066-4154/85/0701-0001$02.00

SCIENCE AS A HOBBY

Scientists are among the fortunate few who earn their livelihood by pursuit of a hobby. This hobby sometimes consumes their every thought but usually it provides a deeply satisfying life. Some, I'm sure, and perhaps with reason, will not agree with this statement. In my case, in my seventy-seventh year, biochemistry is still exciting, just as it was 51 years ago when I published my first scientific paper. Throughout those years, with few exceptions, I have let my own curiosity guide my research efforts without much regard for where they would eventually lead or what practical value they might have. Some of the curiosity detours have been fruitless, but these frustrations have been surpassed by the satisfaction from experiments that went nicely or resulted in unexpected findings. Sharing these experiences with coworkers has made life in the laboratory very fulfilling.

The *Then* of the title of this Prefatory Chapter is September 2, 1907, the date I was born in Delavan, a small village in south central Minnesota. The *Now* is this day in my life of retirement. I receive no salary from the University, but through the kindness of Richard W. Hanson, our Department Chairman, I am provided with ample laboratory space, and through grants from NIH and NSF, I have a fine group of coworkers.

Many highly successful scientists desert the laboratory bench early in their careers and thereafter direct the research of their coworkers. My own goal has been to remain personally active in the laboratory as long as I am involved in science. Although I was Chairman of the Department for 19 years, I did laboratory experiments whenever possible. Most importantly, I always took my sabbatical leave and devoted that time fully to research. The sabbatical provided a good excuse not to accept new appointments to committees and to resign from all current committees. Following my return, it usually took a year or so before I was again caught up in committees. By this device, I had times when I was free from administrative work. During my sabbatical year, full responsibility for the department was turned over to Merton F. Utter, who served as Acting Chairman. I left my coworkers in the laboratory to shift for themselves; ours was a congenial department and they could get advice if they had the initiative. Usually, all went well in my absence; if not, I thought, so be it. I considered it essential to be free of red tape for a time and get rejuvenated by laboratory experiments. Fitzi Lynen thought that if I had spent full time directing the work of my coworkers, more would have been accomplished. Perhaps, but I don't agree. I think collaborators work harder when they have a major role in designing their experiments. Furthermore, part of our task is training, and this method leads to independent thinkers and doers.

MY FOREFATHERS

My middle name is Goff, which is my mother's family name. That family traces back to William Goffe, who was born in 1619 and was known as the "fugitive regicide." William Goffe was one of the appointed judges whose verdict caused the monarch, Charles I, to be beheaded. Charles II came to the throne in 1660 and he offered a reward of one hundred pounds for the capture of William Goffe. For the rest of his life, Goffe was a fugitive from officers of the crown and lived a shadowy and perilous life in the American colonies.

His brother had a son named John Goffe, who settled in the colonies. He was the first with that name in America. There were numerous descendents named John Goff. A book concerning these ancestors, *The John Goffe's Legacy* by George Woodbury, was published in 1955 by W. W. Norton and Company.

The records from the Wood side of the family are far less complete. Many who disagreed with the doctrines of the Church of England were forced to leave after the Act of Uniformity in 1662, and they settled in Holland. Our early ancestors may have been from this group. It is known that my grandfather, Peter Wood, spoke Dutch until he attended school.

My grandparents on both sides were farmers. In 1866, my grandfather and grandmother, Peter and Emily Wood, traveled by ox cart to Minnesota with 10 children. The number later increased to 15, with my father, William, being the youngest. They homesteaded a farm in the wide open prairie near Delavan, Minnesota. During the first winter, the family of 12 lived in a 20 × 14 foot sod house. When I was young, I worked on that farm, which was then owned by Uncle Leonard Wood; by then, there was a comfortable home.

My grandfather, John A. Goff, homesteaded land near Mapleton, Minnesota, about 30 miles from Delavan. His first wife died and he married Ann Augusta Tenney in 1874. They had five children, my mother, Inez, being the second born. They were prosperous farmers and in 1901 built a home with five bedrooms and a newfangled bathroom with a flush toilet. I was much impressed; we did not have such luxury on our small 20-acre farm near Mankato, Minnesota. Later, they constructed a round barn, 70 feet in diameter, which became one of the showpieces of the region. It was a joy to play in its lofty haymow.

In his youth, my grandfather watched the simultaneous hanging in Mankato of some 38 Sioux Indians who had been rounded up following a series of massacres of early settlers. I saw paintings of this event in the windows of the numerous saloons on Front Street in Mankato; in addition, a big granite monument marked the location of the hanging. These have all disappeared; people now are not so certain it was an event of which to be proud.

PARENTS AND MY YOUTH

Following their marriage, my parents, William and Inez, were teachers for a year in one-room schools of adjoining school districts. Mother had graduated from high school and Father had an eighth-grade education. Only about 10% of the students finished high school in those days. The following year, Father entered the real estate business, in which he continued to work for the remainder of his life. We moved to Mankato, Minnesota in 1908, when I was one year old, and since then, that town has remained the focus for our closely knit family. My father acquired a former 14-room resort hotel and 10 acres of land on Lake Washington, which is about 12 miles from Mankato. This beautiful site on a hill overlooking the lake was our summer home to which the sons and daughter returned during high school and college vacations. During the Depression, we set up a public swimming beach which the family ran. We all joined in building a bath house on pillars extending out over the lake. The beach business helped Mother and Dad with the expense of feeding the summer gathering which, by then, included some wives and grandchildren. To this day, the Wood clan, which sometimes numbers 20 to 30 people, gather at Lake Washington for 2 or 3 weeks of reunion.

When I was 4 years old, we moved from the city to a 20-acre farm near Mankato. The house was on a hill with a fine view. It has now been converted to the Valley View Cemetery, and my folks are buried there. The farm was about a mile from the grade school to which we walked and from which we returned home at noon for lunch and to water the stock. Then, we walked or ran back for afternoon classes. It was good training for track meets.

I spent two years in kindergarten and two years in the first grade. I was frail as a child, and I think the extra time was good for me. It allowed me to mature and do well in sports. The two years in the first grade were due to a drive to Raymondville, Texas during the winter of 1914. My father sold land there to groups of people that the real estate company sent to Texas by train from the north. We were a family of eight and made the trip in a Model T Ford. My sister and I had to sit on boxes between the front and rear seats and Mother held Wilbur, who was not yet a year old, on her lap. This crossing of the United States was a real adventure. I recall this as my first encounter with intolerance. The roads were often bad. One evening, our Ford became stuck in the mud. When Father asked for help at a nearby mansion, his northern accent was recognized and the dogs were set on him. A cowboy passed by and tried to pull us out with his lariat attached to the saddle on his horse. This failing, he told us of a black family a mile or so down the road. We tramped there through the mud, and they provided for us as best they could. We all spent an uncomfortable night crammed in crosswise on a bed.

Our family did not have a strong professional or academic tradition, yet all of the children completed college and then continued to higher degrees. Two of the sons earned PhD degrees, a third, a PhD/M.D., a fourth, an M.D., and the fifth, an LL.D. Sister Louise would have received her PhD if she hadn't lost her nearly completed thesis in a car accident. I have often wondered what our folks did that gave their children this orientation. They did teach us to work hard and be self-reliant, but I don't recall that they ever directly encouraged us to strive for advanced degrees. Because we were good athletes, we were sought out by colleges. Chester, the oldest, went to Huron College. All others graduated from Macalester. In those days, athletes were not highly paid. I was promised a job for my board and I was awarded tuition for one year on the basis of superior grades in high school.

LIFE AT MACALESTER COLLEGE

Macalester, a Presbyterian denominational college, had an enrollment of about 800 students. Classes in religion were required, and once a week there was chapel at which attendance was taken. The professors were friendly and congenial. I majored in science but found some of the courses on the history of religion interesting. It was at Macalester that I met Mildred Davis, and we were married during our junior year, on September 14, 1929. It is now hard to believe that I was asked to see the President to obtain permission to continue in college following our marriage. His major request was that we not entertain unmarried couples in our living quarters, which was a room we rented in a private home. How times have changed; now they have coed dorms at Macalester.

The stock market crashed in 1929 and the terrible Depression was under way. I had very little financial aid from my parents, but we managed quite well. My grades improved because I became serious about the future. I worked for my board in the dormitory making salads, corrected exam papers in chemistry, and represented a clothing store on campus. I was told if I continued athletics, I would be paid by the hour if I taped up the players before the games. In addition, an alumnus offered $50.00 toward tuition which, at that time, was $87.50 per semester. That was attractive and I continued to participate in football, swimming, and track. Milly worked in the library and helped correct exam papers. She had her meals in the girls' dormitory and I in the men's dormitory. We both graduated in 1931 and with no debts.

I majored in chemistry. Jobs were scarce during the Depression and I decided I needed a higher degree to meet the competition. I, therefore, set about applying for fellowships in chemistry. The biology professor, O. T. Walters, was very helpful with my applications. I had taken his course in bacteriology, and it was he who suggested that I apply for a fellowship in bacteriology at Iowa State College. I did this but without much enthusiasm because I thought chemistry offered the best opportunities. The Iowa application was the only one I made in bacteriology and the only one that was successful. It was a fortunate accident that I applied there. The application was reviewed by C. H. Werkman, who had been trained as an immunologist but at that time was shifting his field of interest. He was influenced by the writings of A. J. Kluyver, a Dutch microbiologist at Delft, where there was a long tradition of study of the physiology of bacteria and particularly the chemistry of the fermentations. Therefore, Werkman wanted a student in chemistry. Among the applications, mine no doubt stood out since most were from biology majors. In Werkman's laboratory, the emphasis was on intermediary metabolism of bacteria, which, in turn, led to consideration of the studies of Warburg, Meyerhof, Theorell, Cori, Lipmann, Keilin, and Kluyver, who were the giants of the then burgeoning field of enzymology and intermediary metabolism. My applications in chemistry were to departments that were emphasizing organic synthesis and training for jobs in industry. If I had received one of those fellowships, my scientific orientation would almost certainly have been different. Small incidents change one's life.

DISCOVERY OF UTILIZATION OF CO$_2$ AND MY PhD THESIS

My fellowship paid $50.00 a month for 9 months and included tuition exemption. The rent for our two-room attic apartment was $32.50. Later, I received $60.00 a month for 9 months and we moved to an apartment house where I fired furnace and the rent was $15.00 a month. Ours was a Spartan life; we had no car and little money for amusements. Our first daughter, Donna, was born during these graduate student days. The Depression was very severe, and there was no social security to help the unemployed. Many formerly well-to-do were without work and in desperate straits. I felt very fortunate to have some income and a chance to further my education. During the summers, we went to Lake Washington in Mankato and returned with lots of home-canned vegetables and fruit as provisions for the coming winter.

My assignment for my PhD research was to investigate the fermentation by propionic acid bacteria. I have previously reviewed in more detail some of the material that will be considered here (1, 2, 3). When I started these studies in 1931, the Embden-Meyerhof Pathway, the Krebs Cycle, the Pentose Cycle, the Calvin Cycle, and in fact all the presently familiar pathways of metabolism were unknown. There were no commercial sources of enzymes, ATP, DPN, DPNH, etc; nothing was known about the enzymes of carbohydrate metabolism in bacteria; in fact, crude extracts of bacteria that would ferment carbohydrates had not been obtained. Paper and gel chromatography were unknown and there were no isotopic tracers. In those days, study of bacterial metabolism was done by determining the products of the fermentations. The accuracy of these determinations was estimated using carbon and oxidation reduction balances. If the determinations were accurate and complete, 100% of the carbon of the fermented substrate would be recovered in the products; if the recovery was incomplete, it indicated some product had escaped detection and should be sought for. In anaerobic fermentation, i.e. when oxygen is not the electron acceptor, the formation of an oxidized product is accompanied by the formation of a reduced product from the substrate. Thus, the ratio of oxidized products to reduced products (O/R) should equal 1.

My assignment didn't seem to offer much promise. Van Niel (4) had published a comprehensive thesis in 1928 describing the isolation of numerous species of propionic acid bacteria and reporting the products from glucose, glycerol, and lactate fermentations. Werkman offered no explanation of the purpose of my studies or what he expected me to find that was new. He just said, "Before you start, read everything you can find about these bacteria." I did as I was told. I now know the prospect wasn't so dismal; usually there is more to be learned about any subject of research. In fact, even today my group is investigating enzymes from these bacteria, and questions concerning their mechanism of metabolism still remain.

My reading showed that propionate, acetate, CO_2, and also occasionally succinate had been found as products of the fermentation. Van Niel had observed formation of succinate, but he concluded that it was formed from aspartate present in the crude extract (prepared from yeast) that he included in his medium as a source of nitrogen and essential growth factors. The vitamin and amino acid requirements for growth of bacteria were unknown at that time. I set as my first goal to determine whether or not glucose was the source of succinate. I carefully determined the products. From the carbon and oxidation reduction balances, it became clear that succinate was formed from the glucose. Both the carbon and O/R balances were good if succinate was included; otherwise they were not.

Without any specific reason, I then investigated fermentations of glycerol. There I made what I consider to be the most significant contribution of my scientific career. The results showed that CO_2 is used as a substrate by heterotrophic bacteria. Heterotrophs require organic compounds for growth, in contrast to autotrophs, which can grow with CO_2 as the sole source of carbon. The products from glycerol were found to be propionate, succinate, and acetate, but surprisingly, no CO_2. Instead, the CO_2, formed from the $CaCO_3$ used to neutralize the acids, was actually utilized as a substrate (5). That the determinations were not in error was evident from the carbon and oxidation reduction balances. As noted below, I did not recognize this fact at first.

The reason I think this discovery was a very significant contribution is that it destroyed an erroneous dogma held by biochemists. It was a firmly intrenched tenet that CO_2 is an inert end product of the metabolism of all living forms except the specially adapted chemosynthetic and photosynthetic autotrophs. Prior to this discovery, no mechanism was considered for heterotrophs which involved combination of CO_2 with another organic compound. Thus, such considerations as the formation of oxalacetate by combination of CO_2 with pyruvate or phosphoenolpyruvate were excluded. This limitation alone made it impossible to explain pathways of carbohydrate metabolism. Today, of course, we know that CO_2 is an essential building block of metabolism and is used not only in carbohydrate metabolism but also in the metabolism of fatty acids, nucleic acids, and amino acids.

The discovery of CO_2 utilization by heterotrophs was a true case of serendipity. It is these types of experiments that are truly exciting. If one finds what is planned, more often than not, it only confirms what is already known. I believed so strongly that CO_2 was not used by heterotrophs that I at first considered there had been some error in the experiments. As a consequence, I did not mention the utilization of CO_2 in my PhD thesis. The thesis was being typed and I was at my desk trying to decide if I could include something about the glycerol fermentations. Suddenly, the idea struck home. What if there hadn't been an error and the missing CO_2 from the carbonate was used as a substrate? I quickly calculated the O/R balances on the basis that the CO_2 had been used and was converted to succinate, propionate, or acetate. The O/R balances were perfect. I knew then and there that CO_2 was used by these bacteria. I rushed to Professor Werkman and told him I wanted to rewrite the thesis. Clearly, if CO_2 was utilized, what I had written about the mechanism was wrong. Professor Werkman said, "The thesis is all typed except the bibliography; we don't want to type it again." Perhaps he wasn't convinced. I am told on good authority that when I presented these results at a microbiology meeting in 1935 (6), a person next to him leaned over and said, "I don't believe a word of it." Werkman then replied, "I don't either." If he didn't believe it then, he certainly did later and was very proud of this discovery. My delay in recognizing this discovery immediately is a good example of how hard it is to approach a question with an open mind.

POSTDOCTORAL FELLOWSHIP AT THE UNIVERSITY OF WISCONSIN

The University of Wisconsin has always been a major center of microbiology. I applied for a National Research Council Postdoctoral Fellowship, which I received with W. H. Peterson. This fellowship provided $2400 as a stipend and a $200 supplement to cover expenses for birth of a child. We did not overlook the supplement; our second daughter was born in Madison. There Ed Tatum was biding time waiting for a job or a postdoctoral position. Later, he received a fellowship with Kögel in the Netherlands; following that, he joined George Beadle in studying the genetics of *Neurospora,* which led to the Nobel Prize. Wisconsin then was a center for nutritional studies, and Tatum and I investigated the growth factor requirements of propionibacteria. We (7) showed for the first time that vitamin B_1 is required for the growth of a microorganism, the propionic acid bacteria. My experience at the University of Wisconsin broadened my outlook on science. I had many stimulating discussions with, among others, Marvin Johnson, Perry Wilson, and Wayne Wooley; I also met Esmond Snell, who was then a graduate student.

Professor Werkman offered me a position as an Assistant Research Professor. I had written to Otto Meyerhof hoping to spend the second year of the fellowship in his laboratory. Jobs were few and far between so I accepted Werkman's offer. I have often wondered how a year with Meyerhof would have influenced my career. Almost certainly I would have received useful training in enzymology there. I remained at Iowa State University for seven years.

LIFE AT IOWA STATE UNIVERSITY

An Introduction to the Stable Isotope of Carbon, ^{13}C

There was an International Congress of Microbiology in New York City in 1939. There I first learned about ^{11}C and of its availability in Berkeley. At that time, we had indirect evidence that the CO_2 is fixed in succinate. It was clear with $^{11}CO_2$ that this could be proved conclusively. I was told if I could obtain the succinate in about 5 hr, it would be possible to do the experiment. The timing was necessary because ^{11}C has a half-life of 20.5 min. I started work immediately and found if I used a thick suspension of washed cells I could ferment glycerol rapidly. Then, I acidified the mixture and mixed it with plaster of paris, thus obtaining a dry powder. The succinic acid was extracted by pouring ether through the powder, and the succinate was isolated as the calcium salt in good yield in an overall time of less than 4 hr. I told Professor Werkman of my success and that I planned to drive to Berkeley that summer at my own expense. I was amazed when he said, "No, you can't go." He never offered an explanation, but my guess is that he thought we would lose control of the problem since we didn't have access to ^{11}C.

I doubt if Professor Werkman deserves any credit, but as events turned out, it was a fortunate decision. That summer, at Lake Washington, I told my brother Earl about this incident. He was studying for his PhD and M.D. degrees at the University of Minnesota, and he knew about studies that were being initiated with ^{13}C, the stable isotope of carbon. He told me ^{13}C was being concentrated in a thermal diffusion column at the University of Minnesota; he was certain I could collaborate with a young physicist named Alfred Nier, who measured the ^{13}C with a mass spectrometer. This all proved to be true. Using ^{13}C gave one the great advantage of not needing to rush through the experiment. Professor Werkman agreed to this collaboration, and we soon proved that $^{13}CO_2$ is fixed in the succinate. Furthermore, since there was no limit on time, we degraded the succinate and showed the ^{13}C was exclusively in the carboxyl groups (8). This result was very satisfying since it was in accord with our proposal that CO_2 combined with pyruvate forms oxalacetate which is then reduced to succinate.

Utilization of CO_2 by Liver and the Symmetry of Citrate

I was of the opinion that the full significance of the utilization of CO_2 would never be fully appreciated until it was shown to occur in animals. However, Professor Werkman would not grant permission to do these experiments. He was of the opinion that bacteriologists should confine their experiments to microorganisms. He finally did relent when Evans (9) published results that made it almost certain CO_2 was fixed by higher forms of life. He showed that in the presence of malonate, pigeon liver converts pyruvate to C_4-dicarboxylic acids, α-ketoglutaric acid, and CO_2. It seemed to me that the pyruvate was being converted to oxalacetate by fixation of CO_2; and oxalacetate in turn was used to synthesize citrate, which was oxidized to α-ketoglutarate via the Krebs Cycle. Accordingly, the fixed $^{13}CO_2$ would be found in only one of the primary carboxyls of the citrate. However, I made a serious error in setting up the experiment. I thought that citrate was a symmetrical molecule and that therefore the aconitase would not differentiate between the two primary carboxyls; the resulting isocitrate would then have the fixed $^{13}CO_2$ distributed on the average in both primary carboxyl groups. If so, the resulting α-ketoglutarate would also be labeled in both carboxyl groups, and the succinate formed from it would likewise contain fixed CO_2. I isolated the α-ketoglutarate as the 2,4-dinitrophenylhydrazone. For the ^{13}C analysis, it was necessary to convert the compounds to CO_2 for introduction into the mass spectrometer. If I oxidized the hydrazone to CO_2, only one of the eleven carbons would contain ^{13}C; the ^{13}C would thus be diluted and make accurate assay of the excess ^{13}C by mass spectrometry difficult. Instead, the hydrazone was oxidized with permanganate to succinate, which was isolated and then oxidized to CO_2. We found the succinate contained no excess ^{13}C. I was both astounded and disappointed, concluding that CO_2 must not be fixed by liver. We all know now that my reasoning was wrong. Aconitase does handle citrate asymmetrically. By this error in reasoning, we missed the opportunity to be the first to show fixation of CO_2 by higher forms of life.

A few months later, I was amazed when I read that Evans & Slotin (10), using $^{11}CO_2$, had found ^{11}C in the α-ketoglutarate that they isolated as the hydrazone. Thus, they were the first to demonstrate fixation of CO_2 by liver. After much thought, I concluded the only reasonable explanation was that the fixed CO_2 was exclusively in the carboxyl adjacent to the keto group of α-ketoglutarate and it was lost when we oxidized to succinate. Further study by Evans & Slotin (11) and by us (12) showed this to be true. We then both concluded that citrate per se was not in the cycle; if present, it had to be a derivative, such as phosphocitrate, which would make it asymmetric (13). This conclusion was accepted by biochemists for eight years, until Ogston (14) pointed out the possibility that enzymes could differentiate between identical groups of compounds that do not contain an asymmetric carbon. It was then soon shown that this is true for citrate.

Construction of a Thermal Diffusion Column

We had been collaborating with Alfred Nier in all of these studies, and he had been very helpful; but it was clear that having our own facility for ^{13}C would make us much more efficient. We therefore set about assembling a mass spectrometer and a thermal diffusion column. This was a tremendous undertaking for a group of microbiologists. The thermal diffusion column extended 5 stories in an elevator shaft from the basement to the attic of the science building. It consisted of an iron pipe which was heated electrically by an inner nichrome wire covered with 3-inch-long porcelain insulators over the entire 60 feet of wire. The insulators were necessary to prevent the wire from touching the iron pipe and shorting out. The iron pipe constituted the inner wall of the space in which methane gas was heated to about 300°C. Because of the heat, the 12-foot lengths of pipe had to be silver soldered together using an oxygen torch. The outer cold wall was of brass tubing. Steel pins were brazed to the inner iron pipe and small lavite spacers were placed on these. The pipe was slipped into the brass tubing and the pins held the pipe in a central position. Surrounding the brass tubing was an outer brass tubing which formed the jacket through which cold water passed to provide a cold surface to the gas space. At the very top was a 45-gallon tank containing a reservoir of normal methane. At the bottom was a one-liter receiver in which the enriched ^{13}C methane was collected. All the connections had to be gas tight since the whole column was to be degassed under the vacuum.

This task was accomplished with the help of graduate students alone, among whom Lester Krampitz played a major role. There were many problems; perhaps the most tragic, yet most amusing in retrospect, occurred after the column had been in operation for a short time. One day we found the column warped and distorted—it seemed the water must have been shut off briefly. It required a good bit of sweat and tears to repair the column. With the door of the laboratory open, I could hear the water running from the column. Whenever the flow slowed down, I rushed out and pulled the switch. Finally, it occurred to me that this seemed to happen when the bell rang and the home economics classes let out. This suggested an experiment. We found that when two toilets were flushed simultaneously, all was well, but if three were set off at once, the water pressure fell so low that it would not push the water up the five stories of our column. We suggested to Professor Werkman that he destroy one of the women's toilets but instead, he obtained a separate water line to our column. The thermal diffusion column then worked well, giving us about 13% ^{13}C in the methane.

Construction of a Mass Spectrometer

The construction of the mass spectrometer was not as spectacular, but it presented many problems. It required a very high vacuum in which the gas to be analyzed was ionized by an electron beam. High voltage accelerated the ionized gas through a strong magnet, which caused the paths of the ionized gases to curve. The various gases were then separated by centrifugal force acting on their different masses. By variation of the voltage, a selected mass could be focused on a collector plate, which then generated a current of about 10^{-10} A and was measured using an amplifying circuit.

The mass spectrometer tube (with its ion source and collector plates) and the glass mercury diffusion pumps to generate the high vacuum required for free flow of the ionized gas were fabricated at the University of Minnesota. The magnet was made in our shops at Iowa State University. Our job was to assemble the equipment; Al Nier was to come for the shakedown tests. We had to learn about the electrical circuits and, of course, do a substantial amount of glassblowing including attachment of the large mercury pump to the spectrometer tube and then mounting the equipment. We were disappointed to learn that Nier couldn't come for the tests and that he would send his graduate student, Ed Ney. When Ney couldn't come during the Christmas holiday of 1941, Nier sent an undergraduate named McClure. Krampitz and I worked with him night and day, but we could get no evidence that there was an ion beam coming to the collector plate. Our instructions had been to evacuate the housing containing the electrometer tube of the amplifying circuit, but we had hesitated to do this because we suspected it had not been constructed properly in our machine shop. Finally, in desperation, on Christmas Eve we hooked it to the vacuum. To our horror, the cover collapsed and broke the electrometer tube. We closed down and I drove all night to Minneapolis to spend Christmas with my wife and children, who were with Milly's parents.

While in Minneapolis, I went to see Nier. Whereas previously I had been freely admitted to his laboratories, now I wasn't allowed near them. Furthermore, Nier said, "There are not any electrometer tubes available; they have been requisitioned by the government." I was amazed by the change and only much later learned why the laboratories were off-limits and why Nier and Ney were too busy to come to Iowa. They were engaged in secret work for the

Manhattan Project. Using some very large mass spectrometers which they had constructed, they were separating the first ^{235}U for a test of its utility as an atomic bomb.

I returned to Ames much depressed, thinking all our work had been in vain, but Krampitz said, "Chin up, man" and was off to the physics department. Within the hour, he was back with both an electrometer tube and the physicist's doubts that we would ever get it to work. He said, "The circuit is so sensitive, it goes wild every time a fly lands on the hot wires of the electrocuting fly trap they use at the School of Veterinary Medicine." Yet we knew we had seen it work in Nier's laboratory.

Our failure with McClure served one good purpose; in the process, we had learned a good bit about the spectrometer. We tried everything we could think of; finally, by elimination, we decided the problem had to lie with the magnet. The coils of the magnet were covered with black glossy paper. To avoid tearing this, we had pushed it down just enough to see the direction of the winding. Now we removed some of the paper and found, of all things, that the machinist had tied the wire by making a few turns in the opposite direction. We had been bending the ions down instead of up all this time! We switched the leads to the magnet and produced a glorious swing of the galvanometer in response to the ion beam. Our problems were not over, but now we were in business.

Departure from Iowa State University

Werkman was an enigma. I have often wondered how he managed to set up a laboratory, which after about 15 years gained a fine reputation in bacterial metabolism. He did select students who worked very hard. There were few distractions in the small town of Ames, and little money was available for frivolity. Werkman provided little direction for his lab's research projects, though fruitful discussions took place among his students and assistants. We learned that the best way to initiate an experiment was to mention it casually to Werkman and then drop the subject. Within a short time, Werkman would usually suggest the same experiment.

Mine had been an enjoyable and productive time (11 years) with Werkman. By then, he had a strong group; I, Lester Krampitz, and Merton F. Utter were full-time research assistants, augmented by a good group of graduate students. Quite abruptly, however, I found it best to leave this productive group. A problem became apparent after Milly and I bought a house. When I told Werkman this news, he said, "Why did you do that, do you think you can stay here forever?" This was shocking because he had indicated previously that I could stay and expect promotions. In fact, I had just turned down a position at the University of Minnesota with a promotion to Associate Professorship and a higher salary. I had thought it would be foolish to leave Iowa State. Nowhere else in the world was there a microbiology department with facilities to do tracer studies with ^{13}C and with a comparable fine group geared to use this tool. In the ensuing discussion with Werkman, however, it became obvious that there would be no opportunity for future independence. I immediately accepted the position at the University of Minnesota. I have never quite forgiven Werkman for wrecking this opportunity for some truly excellent research and science. Thereafter, people left when they had the chance, and the productivity of Werkman's laboratory fell rapidly.

LIFE AT THE UNIVERSITY OF MINNESOTA, POLIOMYELITIS, LABELING OF GLYCOGEN

The change was abrupt. Suddenly, in the department of physiology at the medical school, I was infecting cotton rats and monkeys with polio virus and learning how to work with brain tissue. The University had received a large interdepartmental grant from the National Foundation for Infantile Paralysis, and I was paid from it to investigate the effect of poliomyelitis on the enzymes of carbohydrate metabolism in nervous tissue. After about a year, Merton Utter joined me in these studies. It had been reported (15a, 15b) that when cotton rats are infected with the Lansing strain of polio virus, glycolysis in the brain is inhibited. Our large series of experiments showed that the infection had no significant effect on glycolysis (16a, 16b).

At the University of Minnesota, I had a marvelous opportunity to work with physiologists and with ^{13}C to test whether the metabolic pathways that had been put together by biochemists occurred in vivo in animals. At that time, some questioned whether the results obtained with chopped tissue and enzymes had any relevance to the metabolism of intact animals. Nier's laboratories were deserted; the physicists were all elsewhere working on the Manhattan Project; however, a mass spectrometer was made available to us.

We started by studying the conversion of $[^{13}C]NaHCO_3$ to glycogen. In another fine example of serendipity, animal utilization of CO_2 in vivo had first been demonstrated at Harvard in 1941 (17). Solomon et al were doing experiments on the conversion of $[^{11}C]lactate$ to liver glycogen in rats. Since everything had to be done quickly, they wanted a control to show that their purified glycogen was free of extraneous ^{11}C. They considered that $^{11}CO_2$ would be inert and used $[^{11}C]NaHCO_3$ as a control. Hastings stated at a conference in 1971 (18), "By administering cold lactic acid and ^{11}C-bicarbonate, we expected to find no radioactivity in the liver glycogen. Instead, the glycogen was full of radioactivity. . . . It is extremely difficult for anyone sitting around this table to comprehend how surprising this result was to us." Apparently, they were unaware of our work on bacterial metabolism or considered that the metabolism of heterotrophic bacteria had little relevance to what might occur in animals.

In our experiments, we proposed to use the distribution of the label in the six carbons of the glucose unit of the glycogen as an indicator of the metabolic pathways. From the Krebs Cycle and the Embden-Meyerhof scheme, which by then had been mapped out, it could be predicted where the tracer should be located. The question was, did the distribution fit prediction? With $[^{13}C]NaHCO_3$, the prediction was that pyruvate would be converted to oxalacetate by fixation of $^{13}CO_2$. Then, by reversible conversion of the oxalacetate to fumarate via malate dehydrogenase and fumarase, the oxalacetate would acquire ^{13}C in both its carboxyl carbons. From this oxalacetate $[1-^{13}C]$-phosphoenolpyruvate would be formed and by gluconeogenesis via the Embden-Meyerhof pathway would yield glycogen-containing glucose units with the excess ^{13}C exclusively in carbons 3 and 4.

The physiologists, Lifson and Lorber, handled the feeding of fasted rats by stomach tube and the injection of the $[^{13}C]NaHCO_3$ intraperitoneally. After 3.5 hr, the rats were sacrificed and the glycogen isolated from the liver. My job was to degrade the glucose of the glycogen. For that, I put my microbiological experience to good use. The sugar was fermented with *Lactobacillus casei* which converts glucose almost quantitatively to lactate. If the fermentation proceeded by the glycolytic pathway as predicted, the 3 and 4 carbons of the glucose would be in the carboxyl group of lactate. The lactate was oxidized with permanganate to acetaldehyde and CO_2; I thus obtained carbons 3 and 4 in the CO_2 and 1, 2, 5, 6 in the acetaldehyde. These experiments were a check on both the metabolism of the bacteria and of the rats.

The results came out just as predicted (19). We were all somewhat surprised that during the numerous reactions occurring in the rat and the bacteria, there had not been scrambling of the ^{13}C in the products. We confirmed the results by a complete chemical degradation of the methylglucoside prepared from the glucose.

The procedure was later modified by fermenting the glucose with *Leuconostoc mesenteroides*. By this fermentation and chemical degradations, each of the six carbons of the glucose could be obtained separately (20). This type of study was extended to studies of other labeled compounds (21).

Buchanan et al (22) had concluded that the carbon of acetate is converted to glycogen only after it is oxidized to CO_2. We found the label from $[1-^{13}C]$-acetate entered carbons 3 and 4 of glucose but with $[2-^{14}C]$acetate, the ^{13}C was predominantly in carbons 1, 2, 5, and 6 (23). It thus was evident that acetate was converted to glycogen by pathways other than by fixation of CO_2. The carbon of acetate enters intermediates of the Krebs Cycle and thereby intermediates of glycolysis and thus into the glycogen. This occurs even though two CO_2 are formed per acetate in the cycle and there is no net synthesis of glucose from acetate.

LIFE AT CASE WESTERN RESERVE
UNIVERSITY (CWRU)

In 1946, I accepted a position as the Chairman of the Department of Biochemistry at the Medical School of Case Western Reserve University. I have been there ever since. We moved with reluctance. Minnesota was our home state. We enjoyed frequent visits with our parents, and I enjoyed hunting and fishing with my father and brothers. One of the conditions that I put to Dean Joseph Wearn in my interview was that I be permitted to return to Minnesota every November for the annual deer hunt. He said, "If you don't tell on me, I won't tell on you." Dean Wearn was an avid hunter. He retired to his Castle Hill Plantation near Yemasse, South Carolina, and died this year at age 91. Three years ago, some of his former associates from the medical school gathered at his estate, and Lester Krampitz and I went quail hunting with him. He was still sharp with his double-barrel shotgun.

It is remarkable that I was selected for the position of Chairman of Biochemistry. I had never taken or taught a course in biochemistry, and I was a graduate in microbiology from a school of agriculture. However, my training was in intermediary metabolism which, at that time, was at the forefront of biochemistry. My lack of schooling in traditional medical biochemistry was perhaps an advantage. Soon changes in the medical curriculum at CWRU were to be considered; I had no entrenched ideas that had to be altered to meet this change.

I had the opportunity to set up a department with an entirely new staff. Merton Utter, Victor Lorber, and Warwick Sakami came with me from Minnesota; the others were Lester Krampitz, John Muntz, Tom Singer, and Robert Greenberg. A year and a half later, Lester Krampitz became Chairman of the Department of Microbiology at the Medical School. After studying the laboratory courses of a number of schools, we decided they did not reflect modern biochemistry and set about designing some experiments. We developed an experiment with rat brain which showed that ATP and DPN are required. Using minced heart muscle, the effect of malonate and cyanide on respiration of succinate was demonstrated. The students did an experiment in which $[1\text{-}^{14}C]$-acetate and $[2\text{-}^{14}C]$acetate were fed to rats and the glycogen was isolated and degraded. It illustrated the flow of carbon through the Embden-Meyerhof Pathway and Krebs Cycle. For this experiment, we had to obtain permission from the Atomic Energy Commission. Lester Krampitz set up an experiment on alloxan diabetes in rats, in which the blood sugar and acetone bodies were followed. The department was a beehive of activity in getting the research laboratories under way and designing these laboratory experiments for the course to be given in the spring semester of that first year.

The Integrated Medical Curriculum

A readiness at the Medical School of CWRU to change the way medicine was taught was evident the day I arrived. It culminated in a new curriculum, initiated in 1952. In brief, subjects were organized under the supervision of multidisciplinary committees rather than taught in isolation by the separate departments. The first year was called Phase I, and it dealt with normal structure, function, growth, and development. By 1956 (24), there were 5 subject committees in Phase I: 1. Cell Biology; 2. Tissue Biology and Neuro Muscular Physiology; 3. Cardiovascular and Respiratory Physiology; 4. Metabolism, and 5. Endocrines. There were biochemists, physiologists, anatomists or histologists, and a clinical representative on all committees; microbiologists and representatives from other departments served on some. The committees planned and presented the lectures and laboratories. Anatomy, physiology, histology, biochemistry, and microbiology experiments were all done in a multidisciplinary laboratory in which each student had his own writing desk as his home base.

To improve overall correlation, there was a coordinator of Phase I. The coordinator reviewed the entire plans with the chairman of each of the subject committees. The overall plan was then submitted to the chairmen of the involved departments for their approval and for approval of the proposed teaching personnel. There were five examinations, one by each subject committee, and a final comprehensive examination at the end of the year.

Each student was required to enroll in a project. The projects consisted of laboratory work, library research, or a study of clinical or other subjects. The staff of both the preclinical and clinical departments submitted projects, which were screened by a committee. A day each week was set aside in the second half-year of Phase I and in the first half of the second year in Phase II for projects. A thesis on the subject was required in the fourth year.

Free time was built in—a day and a half each week throughout the first two and a half years of Phase I and Phase II in which the student was at liberty to choose his own activities. This free time was to encourage the student to carry on his own self-education. Some put in part of this time on their project.

Clinical instruction was started in the first year: two hours on Thursday morning and one hour Saturday morning. A physician served as a preceptor of a group of eight students. Each student was assigned a normal pregnant mother. The student visited her home in his free time to follow the mother's course, and he observed the delivery. The discussions with the preceptor were integrated with the observation of the mother and child.

Phase II dealt with abnormal structure, function, growth, and disease. It was taught by suitable subject committees under an organization similar to that of Phase I. It lasted for one and a half years. Phase III was clinical in the final one and a half years.

This curriculum became known as the Western Reserve Experiment and it has been critically reviewed (25). It has been modified over the years, but teaching by subject committees continues to this day.

Clearly, this change was not accomplished without intensive debate, but it was done in a democratic manner with full faculty vote on all aspects.

I was the chairman of the committee to organize Phase I and thus was in the center of the debate. I favored the change because I had observed the students were inundated with exams; when a department felt that the students were neglecting its course, it increased the number of exams. Some subjects were covered by several departments and were overtaught; other subjects were hardly touched. Contradictory facts were sometimes presented by the various departments.

The planning committee of Phase I had set as its goal a starting date of September, 1952. It was evident that the faculty would need a fairly clear idea of how a subject committee might function in order to judge how the change could be accomplished. Towards that end, I set up a committee to present a plan in detail for correlation of the subject material of renal anatomy, histology, biochemistry, and physiology. The plan was well received by the faculty. It listed the individual lectures, laboratory experiments, and correlation conferences; included were clinical correlation conferences headed by the clinical representative. In the meantime, plans for the multidisciplinary laboratory were made by a committee headed by Lester Krampitz. Among others, there were committees on projects, free time, and clinical science. By the spring of 1952, the plans for Phase I were in hand, but some thought the inauguration should be delayed.

Dean Wearn agreed to call a meeting of the general faculty to vote on this issue. The assembly was packed, with standing room only. Some individuals were designated to speak for the delay, others for starting in 1952. The opposition held that there was no certainty that a change in the curriculum would produce better students and that the present plans didn't provide such evidence. My reply was that this was an experiment in medical education and that in experiments, one doesn't know what the results will be. I proposed we do the experiment and then act on the basis of the results. In fact, no matter how long the delay, we never would know the results unless we started the experiment. The vote was overwhelming in favor of starting in 1952.

The First Sabbatical (1955), Enzymes from the Propionic Acid Bacteria, Lactose Synthesis

New Zealand was the choice for the first half of my sabbatical. I had heard about the hunting and fishing and the beauty of that country. I knew that

Norman Edson, the Chairman of the Department of Biochemistry at the University of Dunedin, had an active group that was preparing enzymes from bacteria. I knew I wanted to have uninterrupted time for research and be far away from routine telephone calls. New Zealand met this last requirement. We had shown that the fermentations of glucose by propionic acid bacteria were not straightforward as with *Lactobacillus casei* and *Leuconostoc mesenteroides*. Instead, from $[1-^{14}C]$, $[2-^{14}C]$, $[3,4-^{14}C]$, or $[6-^{14}C]$ labeled glucose, the ^{14}C was randomized into every carbon of the propionate, succinate, and acetate (26). I wanted to know whether we could obtain a soluble enzyme system from the propionic acid bacteria that would produce similar results. I expected that such an enzyme system would prove useful in determining how this strange shuffling of the carbons occurred.

We went by ship with two of our daughters, Louise and Beverly (our third daughter, Donna, was married), to Auckland with stops at Hawaii and Fiji, and then by car through the beautiful north and south islands to Dunedin. There, we had a trim little cottage on one of the hills that surround Dunedin, with a beautiful view of the city and the Pacific. The cottage was heated by a coal burning stove in the dining room and a fireplace in the living room. On cold days, we often gathered in the dining room around the stove. I saw more of the family during that time than I had in years. We found New Zealand rustic, with spectacular mountains, lakes, and streams. Fords crossed the shallows of small streams on the side roads, and sometimes after a rain, the roads were impassable until the streams went down. The country's people were friendly.

All went well in the laboratory. They had the equipment for radioactive assays, but owing to uncertainty about safety regulations for radioactive isotopes it hadn't been used. My stay was to be less than six months. I told Professor Edson, who was quite conservative, that I could drink the small amount of $[1-^{14}C]$-glucose I had brought with me without harm and that I'd take a chance on the regulations. I often took long weekends to enjoy the country with my family but otherwise worked night and day in the laboratory. Working nights was not the custom there, but before long, the younger staff and graduate students were showing up. The glassblower fabricated equipment that would have taken ages to obtain on order in that remote country. We soon had the radioactive instruments running and the equipment necessary for chemical degradation of the products of fermentation set up. In the meantime, Professor Edson assigned an undergraduate student named Dick Kulka to me and we determined how to prepare extracts from *Propionibacterium shermanii* and *Proprionibacterium arabinosum*. Dick Kulka subsequently went to England and got his PhD in Kreb's laboratory. Before I left, we had shown we could prepare an enzyme system that would ferment $[1-^{14}C]$glucose and that the ^{14}C was randomized into every carbon of the products. Furthermore, there was net fixation of CO_2 (27). These studies laid the groundwork for our future studies of enzymes of the propionic acid bacteria. For a comprehensive review of how the tracer becomes randomized in the fermentation, see reference (3).

Our stay was too short in New Zealand and we hated to leave. From there, we flew on around the world with stops for sightseeing in Singapore, Bangkok, Calcutta, New Delhi, the Taj Mahal, Beirut, and then to Brussels for my first attendance at an International Congress of Biochemistry. I had made arrangements to spend the last six months of my sabbatical in Copenhagen in the laboratory of Herman Kalckar working in enzymology. On arrival, I was surprised to learn that Kalckar had left for the United States. Those who know Herman well know that his very active and creative mind jumps from one point to another while ordinary mundane things are often forgotten. I therefore changed the research plans and studied the synthesis of lactose by cows.

The background of these studies is the following. In 1951, I had visited George Popjak's laboratory in London. He had told me about experiments in which he injected [1-^{14}C]acetate intravenously into goats. He had isolated the fats from the milk and investigated the distribution of ^{14}C in the fatty acids. The lactose samples from the milk of these experiments had been isolated and were on a shelf in his laboratory. It seemed to me that it would be interesting to determine whether or not the isotope patterns were the same in the glucose and galactose of the lactose. Popjak listened with interest and said, "You're welcome to the lactose; take it with you." Georges Peeters of the Veterinary School of Medicine of Ghent had perfused an isolated mammary gland of a cow with blood containing [1-^{14}C]acetate. Popjak also had the lactose from this experiment and gave it to me.

Later, when Per Schambye of the Royal Veterinary and Agricultural College of Copenhagen came to my laboratory, he degraded the glucose and galactose of the lactose samples. In the lactose from in vivo experiments with goats, he found rather small differences in the distribution of ^{14}C in the two hexoses. To our surprise, however, he found that the galactose of lactose from the in vitro perfusion experiment was heavily labeled but the glucose contained very little ^{14}C. This result seemed strange, particularly in view of the fact that Dimant, Smith & Lardy (28) had found when they perfused a cow's udder with blood containing [1-^{14}C]glucose that the lactose contained nearly equal amounts of ^{14}C in the glucose and galactose.

The question was, why the difference with [1-^{14}C]acetate and [1-^{14}C]glucose? Since Per Schambye and I were both in Copenhagen, this presented a fine opportunity to verify Per Schambye's earlier results, which we had never published. He had done his degradations using *Lactobacillus casei,* and the hexoses had not been purified extensively. We decided to repeat the perfusion experiments very carefully. Now we would use *Leuconostoc mesenteroides* to determine the ^{14}C in each carbon of the hexoses. The results were conclusive. They showed clearly that more than 90% of the ^{14}C of the lactose was in the galactose moiety (29).

We also degraded lactose in which [^{14}C]NaHCO, [1-^{14}C]acetate, [2-^{14}C]acetate, and [1-^{14}C]glucose had been injected intravenously in cows. The results were the same as with the goats. The total ^{14}C in the glucose and galactose moieties was almost equal, although with each labeled compound there were small differences in the distribution of ^{14}C in the two hexoses (30).

The question then arose, does the difference in results obtained by perfusion and those obtained in vivo arise because of some aberration occurring during the perfusion? At that time, mid-December, I was scheduled to make a trip to London and then to Munich, and during the flight to London, it occurred to me that the circulation of blood to the two halves of the udder of a cow might be separated; if so, we could check our results in vivo. I wrote to Schambye from England to ask the surgeons whether there was such separation. If so, I wanted to know if it would be feasible to reach the desired artery for an injection. If it were feasible, I wanted him to set up the experiment. Rapid action was needed since I was to return to the United States in January.

It turned out that right and left pudic arteries flow to the right and left sides of the udder. We injected [1-^{14}C]acetate at a constant rate for 10 min in the left pudic artery and then at intervals thereafter; milk from the two halves of the udder were collected separately. Thus, the [1-^{14}C]acetate entered directly into the left half of the udder, but it reached the right side via the general circulation following passage through the left udder. Therefore, we predicted that the lactose from the milk of the right side should be labeled like that from the cows that had received intravenous injections, whereas that from the milk of the left side should be labeled like that of a perfusion experiment. The results were as predicted. For example, at 100 min, the lactose from the milk of the right, non-injected side had almost equal ^{14}C activity in the glucose and galactose, whereas that from the left, injected side, which received the [1-^{14}C]acetate directly, contained almost 9 times more ^{14}C in the galactose than the glucose (31). Clearly, the results obtained in the perfusion experiments were representative of what occurs in vivo in the udder, per se.

The question then was, what explains these facts? We concluded that free glucose is the precursor of the glucose moiety of lactose. We reasoned as follows: In the udder, the [^{14}C]acetate would be metabolized via the Krebs Cycle and thereby its ^{14}C would enter oxalacetate from which phosphoenol-pyruvate would be formed. It, in turn, via the glycolytic pathway, would give rise to ^{14}C-labeled hexose phosphates and thus to [^{14}C]glucose-1-P. About 5 liters of whole blood were used in the perfusion experiments. If some of the highly labeled hexose phosphate made in the udder was hydrolyzed to free glucose and passed into the blood, its ^{14}C would be diluted by the large quantity of unlabeled glucose in the perfusion fluid. Thus the free glucose would have a low specific activity compared to that of the hexose phosphates in the udder. We proposed the lactose is synthesized by the following reactions:

$$UTP + [^{14}C]glucose\text{-}1\text{-}P \rightleftharpoons UDP\ [^{14}C]glucose + PP_i \quad 1.$$

$$UDP\ [^{14}C]glucose \rightleftharpoons UDP\ [^{14}C]galactose \quad 2.$$

$$UDP\ [^{14}C]galactose + glucose \rightleftharpoons [^{14}C]lactose + UDP \quad 3.$$

Accordingly, since the free glucose of the perfusion has a low ^{14}C activity, the radioactivity of the glucose moiety would be low (Reaction 3). Since the $[^{14}C]glucose\text{-}1\text{-}P$ formed in the udder has a much higher ^{14}C activity and is the precursor of the UDP glucose (Reaction 1), which in turn is converted to UDP galactose (Reaction 2) and then to the galactose moiety (Reaction 3), the galactose moiety would have a much higher ^{14}C activity than the glucose moiety.

The situation is quite different when $[^{14}C]glucose$ is the labeled substrate. Reis & Barry (32) and Kleiber et al (33) have shown using labeled glucose that four fifths of the carbon of lactose arises from plasma glucose. Thus, the majority of the hexose phosphate will be formed by phosphorylation of the $[^{14}C]glucose$ from the perfusion. As a consequence, the hexose phosphates and thereby the UDP galactose as well as the free glucose will be highly labeled. These facts explain why the glucose and galactose moieties of lactose were nearly equally labeled in the experiments of Dimant et al (28) with [1-$^{14}C]glucose$.

The same explanation applies to the equal labeling when $[^{14}C]acetate$ is injected intravenously in vivo. The acetate flows to the liver, where ^{14}C is incorporated into hexose phosphate precursors via the Krebs Cycle; from these, $[^{14}C]glucose$ is formed and enters the blood stream. Since $[^{14}C]glucose$ is the major source of carbon of lactose, both the glucose and galactose are labeled about equally.

The procedure of direct injection in one pudic artery was used in numerous other experiments done later at the University of Illinois in collaboration with Gaurth Hansen. In the experiment at Copenhagen, the cow was anesthetized with ether and was lying down. This was an awkward position for milking. At Illinois, the veterinary surgeon, Harry Hardenbrook, injected procaine in the epidural space in the lumbar region of the spinal column, causing a regional anesthesia covering one flank of the cow. The cow remained standing; an incision was made on the flank through which Hardenbrook, by touch, found the pudic artery and inserted a syringe needle. The labeled compound was administered via this needle. These were large-scale experiments, often involving ten or more people whom Hansen recruited. Some drew blood from the jugular vein from which glucose and CO_2 were collected, others milked the cow on the right and left sides, others recorded the time and events, and some just looked on, enjoying the scene but ready to step in if needed.

There were many accidents that in retrospect are amusing. On one occasion, when all was in readiness and the crowd was waiting, I was carefully dissolving some labeled glucose (which was very expensive) in a small volumetric flask. I flicked the flask with my finger to stir the contents. There was a flaw in the flask and to my horror, the bottom dropped off. Fortunately, I was working over a blotter at a desk. I quickly picked the blotter up and cut out the wet portion. While everyone waited, Gaurth Hansen and I shredded the paper, washed it on a suction funnel, and then flash evaporated the solution. We finished the experiments late in the evening. It cost me the price of meals and drinks for all the crew. On another occasion, Hardenbrook's assistant missed the epidural space and injected the procaine directly into the spinal column. The cow sat down and that experiment terminated then and there until the next day.

Space does not permit discussion of the experiments done in collaboration with Gaurth Hansen; see reference (34) for a review. From these studies, it became clear that there is an active pentose cycle in the mammary gland. This led to a series of papers written in collaboration with Joe Katz and Bernie Landau on methods for calculation of the relative amount of glucose metabolized via the Embden-Meyerhof pathway and the pentose cycle. These are reviewed in reference (35) and there is still controversy about the pathway (36a, 36b, 36c).

Gander et al (37) had presented evidence from experiments with enzyme preparations from cow's udder that lactose is formed from UDP galactose and glucose-1-P. If this were the case, since glucose-1-P would be the precursor of UDP glucose, the hexose moieties of lactose would have had similar labeling. Not until 1962 was it shown in vitro (38) that free glucose is the precursor. I'm rather proud of the fact that we were able to show this earlier in vivo.

Inorganic Pyrophosphate (PP$_i$) and Inorganic Polyphosphate (Poly P$_n$) as a Source of Energy in Place of ATP

Reference (39) is an extensive review of the enzymes using PP$_i$. These studies began in 1960 and have persisted to this day. We weren't looking for the utilization of PP$_i$; we were looking for the enzyme that catalyzes the utilization of CO_2 by propionic acid bacteria. The outcome was the discovery of utilization of PP$_i$. Rune Stjernholm and I undertook the study and expected to find one of the previously identified enzymes such as P-enolpyruvate carboxykinase, pyruvate carboxylase, the malate enzyme, or perhaps propionyl-CoA carboxylase. With a crude extract of *Propionibacterium shermanii*, it was soon found that oxalacetate was formed from $^{14}CO_2$ and P-enolpyruvate. We then tried to purify the enzyme by ammonium sulfate precipitation, but the precipitate was inactive. Then we found if we boiled some of the crude enzyme and added it to the ammonium sulfate precipitate, it was active. We then thought the boiled extract contained a coenzyme that had been removed by the precipitation. We tried addition of all the cofactors we could find as a replacement for the boiled extract, but all failed. Then we ashed the boiled extract and took the residue up in water. It then activated the ammonium sulfate precipitate. The factor was almost certainly a metal, but no combination of metals replaced the ash. By then, we were at a loss about what to do next. A graduate student, Patrick Siu, was looking for a project so we decided to let him try his hand at the problem. I was astonished but pleased when he returned in a few days and said he had prepared an ammonium sulfate precipitate that was active without additions. A close examination of his protocols and ours showed that he had taken the precipitate up in phosphate buffer and we in Tris buffer. It then became obvious that phosphate was the essential factor, and it was soon proved (40a, 40b) that the reaction is

$$P_i + \text{P-enolpyruvate} + CO_2 \rightleftharpoons \text{oxalacetate} + PP_i.$$

The enzyme was named phosphoenolpyruvate carboxytransphosphorylase. The reaction is reversible. P-enolpyruvate was formed from oxalacetate and PP_i. Thus, PP_i was serving as a source of high-energy phosphate.

The discovery of this enzyme turned our attention to the possibility that the enzyme might be coupled with pyruvate carboxylase to form P-enolpyruvate from pyruvate as follows:

$$\text{Pyruvate} + CO_2 + \text{ATP} \rightleftharpoons \text{oxalacetate} + \text{ADP} + P_i \qquad 4.$$

$$\text{Oxalacetate} + PP_i \rightleftharpoons \text{P-enolpyruvate} + CO_2 + P_i \qquad 5.$$

$$\text{Sum: Pyruvate} + \text{ATP} + PP_i \rightleftharpoons \text{P-enolpyruvate} + \text{ADP} + 2P_i \quad 6.$$

We quite naturally considered this possibility since Keech and Utter were just down the hall from us and they had shown that P-enolpyruvate is formed from pyruvate in liver by a similar series of reactions, except that in Reaction 5 GTP replaces PP_i and the reaction is catalyzed by P-enolpyruvate carboxykinase (41). In this case, two high-energy phosphates from ATP and GTP are utilized to form P-enolpyruvate. If the above reactions occurred in propionibacteria, PP_i would be replacing GTP as a source of energy.

Herbert Evans took up this study for his Ph.D. thesis. He didn't find the above sequence but he did find another enzyme that utilized PP_i, which again illustrates the fact that what you seek is not always what you find. At first Evans thought he had found pyruvate carboxylase. With pyruvate, ATP, $^{14}CO_2$, and the crude extract of *P. shermanii,* he found oxalacetate was formed. But when he fractionated with ammonium sulfate, the enzyme was inactive. But this time I knew the right question to ask. It turned out he was using Tris buffer to dissolve the precipitate. When he used phosphate buffer, the ammonium sulfate fraction was active.

Evans then showed that the oxalacetate was being formed by the following sequence of reactions in which Reaction 5 is catalyzed by carboxytransphosphorylase:

$$\text{Pyruvate} + \text{ATP} + P_i \rightleftharpoons \text{P-enolpyruvate} + \text{AMP} + PP_i \qquad 7.$$

$$\text{P-enolpyruvate} + CO_2 + P_i \rightleftharpoons \text{oxalacetate} + PP_i \qquad 5.$$

$$\text{Sum: Pyruvate} + \text{ATP} + CO_2 + 2P_i \rightleftharpoons \text{oxalacetate} + \text{AMP} + 2PP_i \qquad 8.$$

The new enzyme that catalyzed Reaction 7, was called pyruvate,phosphate dikinase—a dikinase because both pyruvate and phosphate were phosphorylated by ATP (42a, 42b). In the same year, this enzyme was found in plants by Hatch & Slack (43), in *E. histolytica* by Reeves (44), and also in *Bacteroides symbiosus* (45).

From the sequence of Reactions 7, 5, 8, it became clear that the propionic acid bacteria and liver deal with oxalacetate and pyruvate quite differently. In liver, the oxalacetate is made directly from pyruvate by pyruvate carboxylase, and then P-enolpyruvate is formed from it. In propionibacteria, the sequence is reversed. P-enolpyruvate is formed directly from pyruvate, and oxalacetate is formed from it. Comparative biochemistry has shown that many reactions are identical in bacteria, plants, and animals; but occasionally there are differences.

Pyruvate,phosphate dikinase is a very intriguing enzyme, and our investigations have continued with it to the present. The catalysis involves three partial reactions:

$$\text{Enzyme} + \text{ATP} \rightleftharpoons \text{enzyme-PP} + \text{AMP} \qquad 9.$$

$$\text{Enzyme-PP} + P_i \rightleftharpoons \text{enzyme-P} + PP_i \qquad 10.$$

$$\text{Enzyme-P} + \text{pyruvate} \rightleftharpoons \text{enzyme} + \text{P-enolpyruvate} \qquad 11.$$

$$\text{Sum: ATP} + P_i + \text{pyruvate} \rightleftharpoons \text{P-enolpyruvate} + \text{AMP} + PP_i \qquad 7.$$

Each partial reaction is catalyzed at a functionally distinct site. There is a histidyl group of the enzyme which accepts the β,γ-pyrophosphate group from the ATP in Partial Reaction 9. The histidyl group then serves as the carrier of the pyrophosphoryl group to the P_i site where the γ-phosphate is transferred to the P_i forming PP_i, Partial Reaction 10. Then the β-phosphoryl group is transferred via the histidyl group to the pyruvate site forming P-enolpyruvate. The evidence for this sequence is as follows: Tri Uni Uni Ping Pong kinetics are observed as predicted for a mechanism involving three forms of the enzyme. The phosphoryl group has been shown to be an amide linkage with the 3'N of the histidyl group. The amino acid sequence around this pivotal histidyl group has been determined (46). Bromopyruvate reacts with a cysteinyl group at the pyruvate site with complete inhibition of Partial Reaction 11, but there is no inhibition of Partial Reaction 9 and only a small inhibition of Partial Reaction 10 (47). The 2,'3'-dialdehyde of AMP reacts at the ATP,AMP site inhibiting Partial Reaction 9 but not Partial Reaction 11, and there is a small inhibition of Partial Reaction 10 (48). The enzyme has been modified with pyridoxal phosphate, which reacts with a lysine residue and inhibits all three partial reactions to approximately the same extent; this indicates it is affecting the reactivity of the pivotal histidyl residue (49). Presently, the amino acid sequence around the modified sites is being determined. Eventually, we hope to determine the three-dimensional structure of this enzyme and to gain some insight concerning how the histidyl group serves as a carrier between the three subsites.

The discovery in propionibacteria of a third enzyme that catalyzes utilization of PP_i has practically clinched the evidence that PP_i is a source of energy in such organisms. This enzyme, pyrophosphate phosphofructokinase, catalyzes the following reaction:

$$PP_i + \text{fructose-6-P} \rightleftharpoons P_i + \text{fructose-1,6-diP} \qquad 12.$$

This reaction was first observed by Reeves et al (50) in *Entamoeba histolytica*. He informed us of his finding and we soon found the enzyme in *P. shermanii* (51). There is little or no ATP phosphofructokinase in *P. shermanii*. Since the propionic acid bacteria ferment glucose by the Embden-Meyerhof pathway and the ATP phosphofructokinase is absent, it is clear that these bacteria use PP_i in place of ATP to phosphorylate fructose-6-P.

PP_i is formed in many biosynthetic reactions, and the dogma has held that the PP_i is rapidly hydrolyzed to provide the driving force for these endothermic reactions. Clearly, all that is required for this purpose is to maintain a low concentration of PP_i. This could be accomplished either by utilizing the PP_i or by wastefully hydrolyzing it. The evidence is conclusive in the case of *E. histolytica* and *P. shermanii* that the PP_i is used as a source of energy in place of

ATP (see references 52 and 53). Furthermore, there is considerable evidence that PP_i may be used by plants. It has recently been shown that fructose-2,6-diP is an activator of PP_i phosphofructokinase of plants; with it present, the enzyme has been demonstrated in many plants (54).

Recently, we have taken up the study of the role of inorganic polyphosphate in the propionic acid bacteria. These linear anhydrides of phosphoric acid, which vary in length from a few phosphates to a few hundred, occur in practically all forms of life—bacteria, fungi, algae, protozoa, insects, plants, and animals. Their role has been a subject of interest for many years [see Kulaev (55) for an extensive review]. Our attention was focused on the possible role of polyphosphate in the metabolism of propionibacteria by Kulaev et al (56). They reported that these bacteria contain:

Polyphosphate kinase
$$\text{ATP} + \text{Poly } P_n \rightleftharpoons \text{ADP} + \text{Poly } P_{n+1} \qquad\qquad 13.$$

Polyphosphate 3-phosphoglycerate kinase
$$\text{1,3-diphosphoglycerate} + \text{Poly } P_n \rightleftharpoons \text{3-phosphoglycerate} + \text{Poly } P_{n+1} \quad 14.$$

Polyphosphate glucokinase
$$\text{Glucose} + \text{Poly } P_n \rightleftharpoons \text{glucose-6-P} + \text{Poly } P_{n-1} \qquad\qquad 15.$$

Of these three enzymes, we have found polyphosphate kinase and polyphosphate glucokinase present in propionibacteria but have not detected polyphosphate 3-phosphoglycerate kinase. The ATP 3-phosphoglycerate kinase is very active. The polyphosphate glucokinase is four or more times as active in some species of propionibacteria as the ATP glucokinase (57). We find the polyphosphate kinase almost completely inactive unless histones, protamine, or polylysine are added. The enzyme precipitates in the presence of these basic compounds; and the activity and the synthesized polyphosphate are confined to the precipitate (58). Far more study will be required to determine what role polyphosphates have in the metabolism of the propionic acid bacteria.

It is an intriguing possibility that inorganic polyphosphates as well as pyrophosphates were used as a source of energy by the earliest forms of life. Inorganic anhydrides of phosphoric acid were almost certainly present on the earth's crust before there were living forms on earth.

The Second Sabbatical (1962), Transcarboxylase and the Carboxylation Site

For my second sabbatical, I decided to go to Fitzi Lynen's laboratory in Munich, Germany. We took the long route, visiting New Zealand again and vacationing in that beautiful country. Lynen et al (59) had made the surprising discovery that the biotin enzyme, β-methylcrotonyl-CoA carboxylase, with ATP and CO_2, carboxylates free biotin. They showed that the product is

1'N-carboxybiotin and proposed that the same type of reaction occurs to the biotin of the enzyme which then serves as the CO_2 carrier in the carboxylation of β-methylcrotonyl-CoA. We had isolated from the propionic acid bacteria the biotin enzyme, transcarboxylase, which catalyzes the reaction by which propionate is formed in the fermentation. The long background to this discovery has been reviewed (60).

Transcarboxylase is the only known enzyme catalyzing transfer of a carboxyl group from one compound to another. With it, the carboxylating agent is methyl-malonyl-CoA rather than ATP and CO_2 and accordingly, the sequence is as follows in which TC is transcarboxylase.

$$CH_3CH(COO^-)COSCoA + TC\ biotin \rightleftharpoons CH_3CH_2COSCoA + TC\ biotin \cdot COO^- \quad 16.$$

$$TC\ biotin \cdot COO^- + CH_3COCOO^- \rightleftharpoons TC\ biotin + {}^-OOCCH_2COCOO^- \quad 17.$$

$$Sum:\ CH_3CH(COO^-)COSCoA + CH_3COCOO^- \rightleftharpoons CH_3CH_2COSCoA + {}^-OOCCH_2COCOO^- \quad 18.$$

The questions I planned to investigate during the sabbatical were: (*a*) Are there two partial reactions? Is the enzyme first carboxylated by methylmalonyl-CoA and then the carboxyl transferred in a separate step to pyruvate? (*b*) Is the carboxyl covalently linked to the 1'N of the biotin in Reaction 16? (*c*) What is the linkage of the biotin to the enzyme?

Lynen's laboratory was flourishing. At that time, there were 10 postdoctorals from the United States in the laboratory including Esmond Snell, Dan Lane, and Richard Himes. There were many joyous *Nachsitzungen* and uproarious parties including the *Oktoberfest* and *Fasching*. At Christmas time, the laboratory closed down for two weeks. There was very little heat in the laboratories and it was impossible to work. Many went skiing, so we adapted and went to Lech, Austria. There, we took lessons as our first introduction to skiing and enjoy it to this day.

Lynen assigned Hans Lochmüller, an M.D. postdoctoral, to work with me and a technician, Christl Reipertinger. She could not speak English and I had hoped to learn German from her, but Hans and Christl wanted to learn English. I took German lessons. To force me to use English, they used a *Bäyerisch* dialect impossible for me to understand. With all the Americans around, they picked up English and I very little German.

While in Lynen's laboratory, reports filtered through from the United States that Wakil & Waite (61a) were unable to confirm Lynen et al (59). Using acetyl-CoA carboxylase, they carboxylated the enzyme with ${}^{14}CO_2$ and ATP and reported the fixed ${}^{14}CO_2$ was incorporated into an acid-stable linkage which, on hydrolysis with 4 N H_2SO_4 at 121°C, yielded biotin with ${}^{14}C$

exclusively in the uriedo carbon. Waite & Wakil (61a, 61b) concluded that the fixed $^{14}CO_2$ is not on the 1'N of the biotin and that the transfer involves the uriedo carbon of the biotin. This controversy added zest to our studies.

The work went well. At the same time, Lane was doing similar studies in Lynen's laboratory with propionyl-CoA carboxylase (62) and Knappe et al (63) in Heidelberg with β-methylcrotonyl-CoA carboxylase. Transcarboxylase was purified and then carboxylated with [3-^{14}C]methylmalonyl-CoA. The ^{14}C-labeled enzyme was separated from the excess [3-^{14}C]methylmalonyl-CoA by filtration on Sephadex G-50. We found the ^{14}C was rapidly removed when the carboxylated enzyme was heated (64); thus, our results differed from those of Waite & Wakil (61b). When incubated with pyruvate, the ^{14}C from the carboxylated enzyme was transferred yielding oxalacetate. Thus, the two partial reactions were demonstrated and my first question was answered.

We stabilized the carboxylated enzyme by esterification with diazomethane and then treated it exhaustively with pronase. By comparison with authentic chemically synthesized material, the product was shown to be 1'N-carbomethoxybiocytin. The 1'N-carbomethoxybiocytin was then hydrolyzed with biotinidase and the product identified as 1'N-carbomethoxybiotin. Thus, the biotin was shown to be in amide linkage to the ε amino group of lysine of the enzyme and the carboxylation on the 1'N (64). My second and third questions were answered. Similar results were obtained by Lane & Lynen (62) and by Knappe et al (63).

At a meeting of the Federated Societies of Biological Sciences, Waite presented a ten-minute talk which I followed with our presentation. There was a surprisingly large crowd in attendance. Wakil later published that they were unable to repeat their results. Many, when they make a mistake, never do admit it.

Other Enzymes of the Propionic Acid Bacteria and Further Studies on Transcarboxylase

We purified many enzymes involved in the propionic acid fermentation including the CoA transferase, phosphotransacetylase, acetyl kinase, and malate dehydrogenase (65). In addition, methylmalonyl-CoA racemase and the B_{12} coenzyme–dependent methylmalonyl isomerase (mutase) were investigated in some detail, including the mechanisms of these very interesting reactions. The results are reviewed in reference (66).

Studies of transcarboxylase have continued to this day; references may be found in reviews (67) and (68). The enzyme has a complex structure, which is illustrated schematically in Figure 1. The following are some pertinent facts about the enzyme. The subunits have been isolated and the active enzyme reconstituted. The isolated central subunit specifically catalyzes Reaction 16 with the isolated biotinyl subunit serving as the carboxyl acceptor; the outside

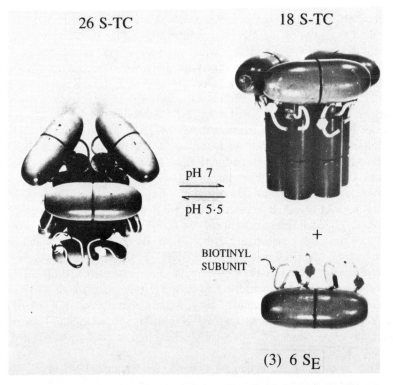

Figure 1   Diagrammatic structure of transcarboxylase. The cylindrical central subunit is made up of six identical peptides of $M_r = 6.0 \times 10^4$. The six outside subunits are dimeric of $M_r = 1.2 \times 10^5$ and each are attached to the central subunit by two biotinyl subunits of $M_r = 1.2 \times 10^4$ (biotin shown as a hexagon). The intact enzyme, $M_r = 1.2 \times 10^6$ with an $s_{20,w} = 26$ S dissociates at neutral pH to the 18 S form with loss of three outside subunits from one face of the cylindrical subunit with the biotinyl subunits still attached and this form is designated the $6S_E$ subunit.

subunit specifically catalyzes Reaction 17 with the carboxylated biotinyl sub-unit serving as the carboxyl donor. By photoaffinity labeling with paraazi-dobenzoyl-CoA, it has been shown that there are 12 CoA ester sites per central subunit, i.e. 2 per polypeptide. The 12 biotinyl subunits function as the carboxyl carriers between the 12 CoA ester sites on the central subunits and the 12 keto acid sites on the 6 dimeric outside subunits. It has recently been shown by ultracentrifugation, glycerol gradient centrifugation, and electron micro-scopy, that sequence 2–26 of the biotinyl subunit promotes binding of the outside subunits to the central subunits (69). Of this sequence, a portion of 2–14 binds to the central subunit and a portion of 15–26 to the outside subunit; there is probably a portion in between that is not bound, which may account for the variable distance observed between the outside and central subunits by electron microscopy. The biotin of the biotinyl subunit is at residue 89. Undoubtedly, there is a specific sequence for the binding surrounding the biotin to orient it so that it can serve as a carboxyl carrier between the central and outer subunits.

David Samols (of this department) has cloned the gene for the biotinyl subunit of transcarboxylase, which will be very useful in future studies. We have crystallized both the 26 S transcarboxylase and the central subunit. We expect to use them for X-ray crystallography and electron microscopy of thin crystals to determine eventually the three-dimensional structure of this intriguing enzyme.

The Third Sabbatical (1969), Synthesis of Acetate from CO_2

For my third sabbatical, I decided to go to Lars Ljungdahl's laboratory at the University of Georgia. The object was to continue investigations we had begun while he was a graduate student and postdoctoral in my laboratory and then an assistant professor in the Department. We were studying the mechanism by which *Clostridium thermoaceticum* ferments glucose with the formation of nearly 3 mol of acetate per mol of glucose; for references see reviews (70) and (71).

The background is as follows: These anaerobic, thermophilic bacteria were isolated in 1942 and found to produce almost a quantitative yield of acetate from glucose. In one of the first experiments done with ^{14}C, Barker & Kamen (72) showed that $^{14}CO_2$ is converted to acetate with ^{14}C in both positions. They proposed the following reactions as a representation of the overall conversion:

$$C_6H_{12}O_6 + 2H_2O \rightleftharpoons 2CH_3COOH + 8H + 2CO_2 \qquad 19.$$

$$8H + 2\,^{14}CO_2 \rightleftharpoons {}^{14}CH_3{}^{14}COOH + 2H_2O \qquad 20.$$

Accordingly, this fermentation would involve a total synthesis of acetate from CO_2. However, it remained possible the labeled acetate was a mixture of $[2-^{14}C]CH_3COOH$ and $[1-^{14}C]CH_3COOH$. I therefore conducted fermentations in the presence of $^{13}CO_2$ and, after converting the acetate to ethylene, determined the masses by mass spectrometry. The results showed that about one third of the acetate was formed entirely from CO_2 (73). We then embarked on an investigation of this synthesis with the thought it might provide an insight into autotrophic pathways. The Calvin pathway of photosynthesis had not yet been discovered. Our attempts to determine the pathway met with little success.

Then, in 1964, Poston, Kuratomi & Stadtman (74) made a discovery that opened a new vista. They showed with extracts of *C. thermoaceticum* fermenting pyruvate to which Co-$^{14}CH_3$ vitamin B_{12} was added that $[2-^{14}C]CH_3COOH$ was formed. Vitamin B_{12} is a tetrapyrol with a coordinated cobalt and the

^{14}C-methyl group was ligated to the cobalt. Vitamin B_{12} is one of many compounds classified as corrinoids. Earl Stadtman has told me that they knew *C. thermoaceticum* contained plentiful corrinoids and considered they must be there for a reason. Intrinsic factor, which binds B_{12}, inhibited the fixation of $^{14}CO_2$. This observation prompted the tests with ^{14}C-labeled methyl B_{12}.

We therefore exposed *C. thermoaceticum* cells that were fermenting glucose to $^{14}CO_2$ for 15 sec and then isolated the corrinoids from the cells. Of these, only the Co-methyl corrinoids contained ^{14}C, and the ^{14}C activity was very much greater than that of the acetate, in accord with a precursor–product relationship (75). Subsequently, we proposed (76) a pathway in which CO_2 is reduced to formate, which combines with tetrahydrofolate. The formyltetrahydrofolate is reduced to methyltetrahydrofolate (CH_3THF). Then the methyl is transferred to the cobalt of a corrinoid with which CO_2 somehow combines, forming acetate. This proposal was based in part on the fact that formate had been found to be a better precursor of the methyl of acetate than CO_2 and in part on the known mechanism of formation of the methyl of methionine from formate via a corrinoid enzyme.

In the studies at the University of Georgia we attempted to verify this proposed mechanism. There were two obvious questions: (*a*) Is CH_3THF a precursor of the methyl of acetate? (*b*) Is a corrinoid enzyme involved?

Using an enzyme preparation from *C. thermoaceticum* with pyruvate as the substrate, we found that CH_3THF was as good a source of the methyl of acetate as methyl B_{12} (77). We then did studies with propyl iodide. It was known from the studies of methionine synthesis that corrinoid enzymes are inactivated by alkylation of the cobalt and that the alkyl group can be removed by photolysis with accompanying reactivation of the enzyme. We used this procedure and found the enzyme preparation was inactivated by propyl iodide and reactivated by light. Since the reactivation by light is quite specific, this gave strong support for the occurrence of a corrinoid enzyme in the system.

One surprise developed. We found that the carboxyl group of pyruvate is converted to the carboxyl group of acetate without conversion to CO_2 (78). However, the enzyme cleaving pyruvate catalyzes a rapid exchange of CO_2 with the carboxyl group of pyruvate. Thus CO_2 can indirectly enter the carboxyl of acetate via this exchange.

The Synthesis of Acetate from Methyltetrahydrofolate and CO or CO_2 and H_2 by C. thermoaceticum and Relation to Autotrophism

Long ago, we expressed the view that the mechanism of acetate synthesis present in *C. thermoaceticum* might represent a pathway for autotrophism (79a, 79b). There are numerous anaerobic bacteria, which are acetogenic and grow with Co or CO_2 and H_2 as the source of carbon and energy; see Zeikus (80) for a review and references. These bacteria do not use the Calvin Cycle or other known pathways for growth on C_1 compounds but use a pathway in which acetyl-CoA is used for the anabolic processes. We now know that these bacteria utilize the pathway shown in Figure 2 for autotrophic growth. After these many years, it is very gratifying to me that this newly recognized pathway has been derived from studies with *C. thermoaceticum*.

Four enzymes, phosphotransacetylase, methyltransferase, pyruvate ferredoxin oxidoreductase, and ferredoxin were purified from *C. thermoaceticum*, and a fifth fraction F_3 was obtained. The five components were found to catalyze the conversion of pyruvate, CoASH, phosphate, and CH_3THF to acetylphosphate (81). If the phosphotransacetylase was omitted, the product was acetyl-CoA. Then, an unexpected break occurred. Diekert & Thauer (82) had found that *C. thermoaceticum* contains CO dehydrogenase which catalyzes the following reactions in which $[C_1]$ is an unidentified intermediate.

$$CO + H_2O \rightleftharpoons [C_1] \rightleftharpoons CO_2 + 2H^+ + 2e \qquad 21.$$

Tests of our fraction F_3 showed that it contained CO dehydrogenase (83). To our amazement, it was then found that pyruvate could be replaced by CO in the

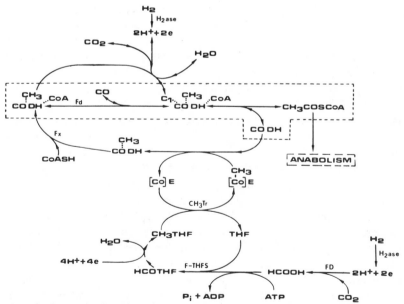

Figure 2 Proposed pathway by which anaerobic bacteria that are acetogenic grow on CO or CO_2 and H_2 as the source of carbon and energy. CODH is CO dehydrogenase, [Co]E is corrinoid enzyme, CH_3Tr is methyltransferase, F_x is an essential enzyme with an unidentified function, F-THFS is formyltetrahydrofolate synthetase, FD is formic dehydrogenase, and THF is tetrahydrofolate.

synthesis of acetyl-CoA and only the methyltransferase and fraction F_3 were required (84). The reaction is as follows where CH_3Tr is methyltransferase.

$$CO + CH_3THF + CoASH \xrightarrow{CH_3Tr,\ ATP,\ F_3} CH_3COSCoA + THF \qquad 22.$$

Fraction F_3 not only contains CO dehydrogenase, it also contains a corrinoid enzyme and an unidentified enzyme that is essential for the synthesis. The CO dehydrogenase has been purified by Ragsdale et al (85) and Diekert & Ritter (86).

We next showed that acetyl-CoA is synthesized from CH_3THF, CO_2, H_2, and CoA. Drake (87) had demonstrated that *C. thermoaceticum* contains hydrogenase which catalyzes the following reaction.

$$H_2 \rightleftharpoons 2H^+ + 2e \qquad 23.$$

We have found that fraction F_3 contains hydrogenase and demonstrated the following in which Fd is ferredoxin (88).

$$CO_2 + H_2 + CoASH + CH_3THF \xrightarrow{CH_3Tr,\ F_3,\ Fd,\ ATP} CH_3COSCoA + THF + H_2O \qquad 24.$$

The corrinoid enzyme has now been purified (89) and it has been shown with the purified methyltransferase to serve as a methyl acceptor from CH_3THF. Recently, a fourth enzyme has been isolated (90) and with it, it is now possible to synthesize acetyl-CoA from CH_3THF, CO, and CoASH using a "clean" system consisting of the methyltransferase, the corrinoid enzyme, CO dehydrogenase, and this fourth enzyme of which the function still remains unknown.

Very recently (S. W. Ragsdale, H. G. Wood, *J. Biol. Chem.*, In press), there has been a most unexpected development. It has been found when [1-^{14}C]-acetyl-CoA is incubated with unlabeled CO and CO dehydrogenase (no other enzymes are required) that the CO acquires ^{14}C and the acetyl-CoA loses ^{14}C, but there is no net decrease in the amount of acetyl-CoA (see the dashed line enclosure of Figure 2). For this to occur, the bonds between C-2 and C-1 and between C-1 and the sulfur must be cleaved and the C-1 carbonyl group must be converted to ^{14}CO and then ^{14}CO must mix with the ^{12}CO in the solution and gas phase. Consequently, when the acetyl-CoA is resynthesized from this CO with diluted ^{14}C, it has a much lower radioactivity and the net result is loss of ^{14}C from the acetyl-CoA. Clearly, in the reverse of the exchange, CO dehydrogenase is catalyzing the synthesis of acetyl-CoASH from CO, CoASH, and a methyl group. Thus its catalytic capacity is much greater than that required for simply catalyzing Reaction 21. Furthermore, the CO dehydrogenase must have binding sites for the methyl and CoASH groups (as indicated in Figure 2) since neither a methyl acceptor nor CoASH is required for the exchange. Electron spin resonance has provided further evidence of this binding since the signals due to binding of CO to the Ni of the CO dehydrogenase are altered in the presence of acetyl-CoA or CoASH.

We have now modified our previous concepts (71, 84) of the pathway and have assigned to CO dehydrogenase the central role of catalyzing the final steps of the synthesis of acetyl-CoA. The overall pathway of synthesis of acetyl-CoA from CO_2 and H_2 or CO can now be described in some detail. With CO_2, hydrogen via hydrogenase, supplies the electrons for the reductive processes. As shown, starting at the lower right of Figure 2, the CO_2 is reduced to formate by formic dehydrogenase. The formate in turn is converted to formyltetrahydrofolate by formyltetrahydrofolate synthetase. This is followed by a series of reductive steps which yield methyltetrahydrofolate. A large amount of research has been done that demonstrates convincingly the portion of the pathway from CO_2 to formate, to formyltetrahydrofolate and to CH_3THF (70, 71, 91). The methyltransferase catalyzes the transfer of the methyl to the corrinoid enzyme (89) from which the methyl is transferred to the CO dehydrogenase yielding CH_3-CODH. Protein F_x (90) may catalyze the addition of CoASH to the CH_3-CODH forming the CH_3-, CoA-CODH complex. As shown at the top of Figure 2, H_2 via hydrogenase supplies the electrons for reduction of the second CO_2, which is followed by formation of the CH_3-, CoA-, C_1-CODH complex, which in turn is converted to acetyl-CoA with regeneration of CODH. When CO is the source of carbon, it is the direct source of C_1 for formation of the CH_3-, CoA-, C_1,-CODH complex, and CO via CO dehydrogenase (Reaction 21) serves as the source of electrons in place of H_2. The acetyl-CoA then serves as the source of carbon for the anabolic processes.

Plants derive their energy from low potential electrons by photolysis of water whereas the acetogenic anaerobic bacteria derive their energy from low-potential electrons generated from CO or H_2. All the anaerobic acetogenic bacteria contain CO dehydrogenase and corrinoids which are key components of this system. It is noteworthy that Kerby & Zeikus (92) have now shown that *C. thermoaceticum* can be grown on CO or CO_2 with H_2 as the source of carbon and energy. Thus, *C. thermoaceticum* is a facultative anaerobe. This finding came as a bit of a surprise, but it is logical since *C. thermoaceticum* has the necessary enzymes for such growth.

There are a large number of metallo enzymes involved in this pathway. There is cobalt in the corrinoid enzyme; nickel, zinc, and iron in the CO dehydrogenase; selenium, tungsten and iron in the formate dehydrogenase; and iron in the ferredoxin. It seems likely that metal catalysis occurred in prebiological time on the earth's crust. CO, CO_2, H_2, and phosphate were almost certainly present on the earth's crust. Perhaps the earliest forms of life adopted metallocatalysis for synthesis of acetyl-CoA and acetylphosphate and for use in their growth.

The Fourth Sabbatical (1978), Biotination of Apotranscarboxylase and Its Subunits

The first six months I spent in the laboratory of Robert Becker at Oregon State University. I had three purposes: (a) to learn whether or not we liked Oregon as a place to live (we have a home on the Applegate River near Grants Pass to which we will move when I stop laboratory work); (b) to consider the possibility of moving my laboratory to Oregon State University, if the then-to-be-appointed Chairman of Biochemistry at Case Western Reserve University preferred that I leave; and (c) to work on the sequence of the biotinyl subunit of transcarboxylase. This work was nearing completion and was finished while I was at Oregon (93). We found we loved Oregon. Although facilities at Oregon State University were great, I stayed at Case Western Reserve University.

For the second six months, I went to Lynen's laboratory at the Max-Planck Institute in Martinsreid near Munich. Fitzi was nearing retirement; he was deeply involved as Vice-President of the Max-Planck Society and President of the Alexander von Humboldt Foundation. I was the only postdoctoral from the United States, and the pace in the laboratory and in the country as a whole was far more leisurely than it had been when I was with Lynen in 1962.

The purpose of the studies was to compare the rates of biotination of intact apotranscarboxylase and its subunits. The biotination is posttranslational by a synthetase. From ATP and biotin this synthetase forms adenylbiotin, which then reacts with a specific lysine of the enzyme. Does the biotination occur after the assembly of the apoenzyme from its aposubunits or at the aposubunit stage? All previous studies had been done with intact apoenzymes.

The propionic acid bacteria were grown in a medium containing a minimal amount of biotin. In such a medium they form the transcarboxylase lacking biotin. The 26 S form of apotranscarboxylase was then isolated and dissociated at pH 7 to obtain the outside subunit with the apobiotinyl subunit still attached (see Figure 1). This aposubunit was in turn dissociated at pH 9 into the outside subunit and the apobiotinyl subunit; the latter was then isolated. In addition, the synthetase for the attachment of the biotin to the apoproteins was isolated and purified.

Lynen was much interested in this project and assigned first one and later a second technician to me. We found that the intact apoenzyme and the two types of aposubunits were biotinated by the synthetase at the same rate (94). Thus, these results provided no evidence on the order of events in the intact cell, except to indicate all three were possibilities.

Lynen never lived to see these studies published. He had surgery in June for an aneurysm in his aorta. The last time I was to see my good friend alive was when I left him at the hospital on July 1, 1979, prior to returning to the States. He was to have assumed the office of President of the International Union of Biochemistry on July 7th at the International Congress in Toronto. As President-Elect, he knew I would be acting in his place at the Congress. He wished me luck with IUB, good experiments, and a good life. He died August 6, 1979, from complications following the surgery.

RESEARCH PANELS, COMMITTEES, OFFICES

I find it curious that, with no training in medicine, I have been asked to serve on many committees dealing with clinical research. My first research panel was as Chairman of the Isotope Panel of the Committee on Growth, 1947–1948, which later became the American Cancer Society. The next was on the initial Research Committee of the American Heart Association beginning in 1948. I was the only basic scientist on the committee and pushed hard to have a policy adopted that basic research be supported even when it bears no apparent relationship to cardiovascular function. My view was that some percentage of the funds should be set aside to support all types of basic research. My belief was that no one really knew where the next essential information would be found; it might be by studying $E.$ $coli$ or a fly. There was long debate, and in the end, 15% was set aside for this purpose. I seldom, if ever, had to call for its specific use for basic research. My clinical colleagues almost always rated these applications highly because the methods and objectives were clearly evident, in contrast to some of the more clinical applications. The fact that basic research was to be supported became well known and was a factor in our receipt of good applications from a number of basic scientists. Ours was the first committee to adopt the policy of supporting Career Investigators. Some very fine basic scientists have been supported by this program. I then served on the AEC Advisory Committee for Biology and Medicine; the Advisory Council for Life Insurance Medical Research; and the American Cancer Society Advisory Board. In 1973, I was appointed on a Study Section for NIH that dealt primarily with applications from biochemists and with a committee made up largely of people with PhDs.

My first call to be an editor was for the *Journal of Biological Chemistry*, beginning in 1949. There were 20 members on the Editorial Board. Ed Tatum, Joe Fruton, Es Snell, and I were the only young members on the Board. The appointment was for five years, but it was a tradition for the editors to stay on for the remainder of their scientific careers. I was to review papers in which isotopes were used, be the subject formation of bone, transport, or intermediary metabolism. These were the early days of isotopic studies and many investigators were naive about their use. In trying to be helpful and fair, some of my critiques were quite long. Carl Cori as the seconder of one of my decisions once wrote, "What are you trying to do, run a correspondence course?"

I thought there should be more turnover of the Board and decided at the end of my five years that I would not accept another appointment. The annual meetings of the Board were at very fine dinners with plenty of liquid refreshments, followed by a business meeting at which, among other things, the number of papers reviewed by each member of the board was announced and how many papers he had accepted and declined. This was supposed to be an indication of who worked the hardest, who was too soft, and who was a good tough editor. I announced at the meeting that I was stepping down. The reaction by some of the older members was very heated, almost as if I were a traitor. I persisted in my resolve. Finally, after a continued attack, I became angry and said, "Listen, if all you guys died tomorrow, a good Board could be picked the next day to replace you." Some years later, the appointment for five years was made subject to one immediate renewal; a second renewal could not be made before a lapse of a year. The membership of the Editorial Board now is about 200. It is no longer the cozy group of old.

I served on the President's Scientific Advisory Committee under the last part of President Johnson's term and first part of President Nixon's. It was an interesting experience and I think it served a good purpose even though the administration as a rule followed only those suggestions that suited their purposes. Kennedy used the committee's advice extensively, Johnson was a bit suspicious of the motives of scientists, and Nixon did away with the committee.

I have spent a good bit of effort in recent years on the International Union of Biochemistry, first as a Council Member, then as General Secretary from 1970–1973 and beginning in 1979, as Acting President during the term Feodor Lynen was to have been in office, and now as President until 1985. When I became General Secretary, I felt IUB needed to be vitalized and needed good leadership to accomplish this. While I held the office, E. C. "Bill" Slater was made Treasurer and Osamu Hayaishi, President. William "Bill" Whelan succeeded me as General Secretary. These and later officers have done much towards accomplishing the goal of IUB of promoting the discipline of biochemistry worldwide. I take some pride in aiding in this expanded role of IUB.

CONCLUDING COMMENTS

One wonders whether the next 50 years will witness a revolution in biochemistry comparable to that of the past 50 years. These days, I listen to research seminars that go far beyond the wildest dreams of my graduate student days. Nowadays, the amino acid sequence of an enzyme is derived from the nucleotides of a sequence of a clone of the gene in a matter of days or weeks. When I was a graduate student, we never dreamed it would be possible to sequence an enzyme, and not so long ago, the process took years, not weeks. It wasn't known that a gene is made up of DNA, to say nothing of generating mutations or of manipulating the gene for technological advance. We no doubt are at the threshold of many new discoveries, but scientists of today seem more sophisticated and accustomed to rapid change. I wonder if the new discoveries will ever carry the same impact, thrill, and amazement as those during the past 51 years.

Literature Cited

1. Wood, H. G. 1972. In *The Molecular Bases of Biological Transport*, ed. J. F. Woessner, Jr., F. Huiging, pp. 1–54. Miami Winter Symposia. New York: Academic
2. Wood, H. G. 1981. *Curr. Top. in Cell. Regul.* 18:255–87
3. Wood, H. G. 1982. In *Of Oxygen, Fuels and Living Matter*, Part 2, ed. G. Semenza, pp. 173–250. New York: Wiley
4. van Niel, C. B. 1928. *The Propionic Acid Bacteria*, Haarlem N. V. Uitgeverijzaak, J. W. Boisevain and Co.
5. Wood, H. G., Werkman, C. H. 1936. *Biochem. J.* 30:48–53
6. Wood, H. G., Werkman, C. H. 1935. *J. Bacteriol.* 30:332 (Abstr.)
7. Tatum, E. L., Wood, H. G., Peterson, W. H. 1936. *Biochem. J.* 30:1898–1904
8. Wood, H. G., Werkman, C. H., Hemingway, A., Nier, A. O. 1941. *J. Biol. Chem.* 139:373–81
9. Evans, E. A. Jr. 1940. *Biochem. J.* 34:829–37
10. Evans, E. A. Jr., Slotin, L. 1940. *J. Biol. Chem.* 136:301–02
11. Evans, E. A. Jr., Slotin, L. 1941. *J. Biol. Chem.* 141:439–50
12. Wood, H. G., Werkman, C. H., Hemingway, A., Nier, A. O. 1941. *J. Biol. Chem.* 141:483–84
13. Wood, H. G., Werkman, C. H., Hemingway, A., Nier, A. O. 1942. *J. Biol. Chem.* 142:31–45
14. Ogston, A. G. 1948. *Nature* 162:963
15a. Racker, E., Kabat, H. 1942. *J. Exp. Med.* 76:579–85
15b. Nickle, M., Kabat, H. 1944. *J. Exp. Med.* 80:247–55
16a. Wood, H. G., Rusoff, I. I., Reiner, J. M. 1945. *J. Exp. Med.* 81:151–59
16b. Utter, M. F., Reiner, J. M., Wood, H. G. 1945. *J. Exp. Med.* 82:217–26
17. Solomon, A. K., Vennesland, B., Klemper, F. W., Buchanan, J. M., Hastings, A. B. 1941. *J. Biol. Chem.* 140:171–82
18. Hastings, A. B. 1971. In *Proceedings of the Conference on the History of Biochemistry and Molecular Biology*. Brookline, MA: Am. Acad. Arts Sci., p. 21–23
19. Wood, H. G., Lifson, N., Lorber, V. 1945. *J. Biol. Chem.* 159:475–89
20. Bernstein, I. A., Wood, H. G. 1957. *Meth. Enzymol.* 4:561–84
21. Wood, H. G. 1948. *Symp. Quant. Biol.* 13:201–10
22. Buchanan, J. M., Hastings, A. B., Nesbett, F. B. 1943. *J. Biol. Chem.* 150: 413–25
23. Lifson, N., Lorber, V., Sakami, W., Wood, H. G. 1948. *J. Biol. Chem.* 176:1263–84
24. Wood, H. G. 1956. *Fed. Proc.* 15:865–70
25. Williams, G. 1980. *Western Reserve's Experiment in Medical Education and its Outcome*. London: Oxford Univ. Press
26. Wood, H. G., Stjernholm, R., Leaver, F. W. 1955. *J. Bacteriol.* 70:510–20
27. Wood, H. G., Kulka, R. G., Edson, N. L. 1956. *Biochem. J.* 63:177–82
28. Dimant, E., Smith, V. R., Lardy, H. A. 1953. *J. Biol. Chem.* 201:85–91
29. Wood, H. G., Schambye, P., Peeters, G. J. 1957. *J. Biol. Chem.* 226:1023–34
30. Schambye, P., Wood, H. G., Kleiber, M. 1957. *J. Biol. Chem.* 226:1011–21
31. Wood, H. G., Siu, P., Schambye, P. 1957. *Arch. Biochem. Biophys.* 69:390–404
32. Reiss, O. K., Barry, J. M. 1953. *Biochem. J.* 55:783–85
33. Kleiber, M., Black, A., Brown, M. A., Baxter, C. F., Luick, J. R., Stadman, F. H. 1955. *Biochem. Biophys. Acta* 17: 252–60
34. Wood, H. G. 1958. In *Polysaccharides in Biology*. Trans. Third Conf., ed. G. F. Springer, pp. 173–95. Madison, NJ: Madison Printing Co.
35. Wood, H. G., Katz, J., Landau, B. R. 1963. *Biochem. Z.* 338:809–47
36a. Landau, B. R., Wood, H. G. 1983. *Trends Biochem. Sci.* 7:292–96
36b. Williams, J. F., Krishman, K. A., Longenecker, J. P. 1983. *Trends Biochem. Sci.* 7:275–77
36c. Landau, B. R., Wood, H. G. 1983. *Trends Biochem. Sci.* 8:312–13
37. Gander, J. E., Peterson, W. E., Boyer, P. D. 1956. *Arch. Biochem. Biophys.* 60:259–61
38. Watkins, W. M., Hazzid, W. Z. 1962. *J. Biol. Chem.* 237:1432–40

39. Wood, H. G., O'Brien, W. E., Michaels, G. 1977. *Adv. Enzymol.* 45: 85–155
40a. Siu, P. M. L., Wood, H. G., Stjernholm, R. L. 1961. *J. Biol. Chem.* 236:PC21
40b. Siu, P. L. M., Wood, H. G. 1962. *J. Biol. Chem.* 237:3044–51
41. Keech, D. B., Utter, M. F. 1963. *J. Biol. Chem.* 238:2609–14
42a. Evans, H. J., Wood, H. G. 1968. *Proc. Natl. Acad. Sci. USA* 61:1448–53
42b. Evans, H. J., Wood, H. G. 1971. *Biochemistry* 10:721–29
43. Hatch, M. D., Slack, C. R. 1968. *Biochem. J.* 106:141–46
44. Reeves, R. E. 1968. *J. Biol. Chem.* 243:3203–4
45. Reeves, R. E., Manzies, R. H., Hsu, D. S. 1968. *J. Biol. Chem.* 243:5486–91
46. Goss, N. H., Evans, C. T., Wood, H. G. 1980. *Biochemistry* 19:5805–9
47. Yoshida, H., Wood, H. G. 1978. *J. Biol. Chem.* 253:7650–55
48. Evans, C. T., Goss, N. H., Wood, H. G. 1980. *Biochemistry* 19:5809–14
49. Phillips, N. F. B., Goss, N. H., Wood, H. G. 1983. *Biochemistry* 22:2518–23
50. Reeves, R. E., South, D. J., Blytt, H. J., Warren, L. G. 1974. *J. Biol. Chem.* 249:7737–41
51. O'Brien, W. E., Bowien, S., Wood, H. G. 1975. *J. Biol. Chem.* 250:8690–95
52. Reeves, R. E. 1976. *Trends Biochem. Sci.* 1:53–55
53. Wood, H. G. 1977. *Fed. Proc.* 38:2197–2205
54. Carnal, N. W., Black, C. C. 1983. *Plant Physiol.* 71:150–54
55. Kulaev, I. S. 1980. *The Biochemistry of Inorganic Polyphosphates.* New York: Wiley. (English transl.)
56. Kulaev, I. S., Vorob'eva, L. I., Konovalvo, L. V., Bobyk, M. A., Konoshenko, G. I., Uryson, S. O. 1973. *Biokimiya* 38:712 (English transl. in *Biochem. Russ.* 38:595–99)
57. Wood, H. G., Goss, N. H. 1984. *Proc. Natl. Acad. Sci. USA.* In press
58. Robinson, N. A., Goss, N. H., Wood, H. G. 1984. *Biochem. Int.* 8:757–69
59. Lynen, F., Knappe, J., Lorch, E., Jutting, G., Ringleman, E. 1959. *Angew. Chem.* 71:481–86
60. Wood, H. G. 1972. In *The Enzymes,* ed. P. D. Boyer, 6:83–115. New York: Academic
61a. Wakil, S., Waite, M. 1962. *Biochem. Biophys. Res. Commun.* 9:18–24
61b. Waite, M., Wakil, S. J. 1963. *J. Biol. Chem.* 238:81–90
62. Lane, M. D., Lynen, F. 1963. *Proc. Natl. Acad. Sci. USA* 49:379–85
63. Knappe, J., Wenger, B., Wiegand, U. 1973. *Biochem. Z.* 337:232–46
64. Wood, H. G., Lochmüller, H., Riepertinger, C., Lynen, F. 1963. *Biochem. Z.* 337:247–66
65. Allen, S. H. G., Kellermeyer, R. W., Stjernholm, R. L., Wood, H. G. 1964. *J. Bacteriol.* 87:171–87
66. Wood, H. G., Kellermeyer, R. W., Stjernholm, R., Wood, H. G. 1964. *Ann. NY Acad. Sci.* 112:661–79
67. Wood, H. G., Zwolinski, G. K. 1976. *CRC Crit. Rev. Biochem.* 4:47–122
68. Wood, H. G. 1979. *CRC Crit. Rev. Biochem.* 7:143–60
69. Kumar, G. K., Bahler, C. R., Wood, H. G., Merrifield, R. B. 1982. *J. Biol. Chem.* 257:13828–34
70. Ljungdahl, L. G., Wood, H. G. 1982. In *Vitamin B₁₂,* ed. D. Dolphin, pp. 166–202. New York: Wiley
71. Wood, H. G. 1982. *Proceedings Biochem. Symp.,* ed. E. Snell, pp. 29–56. Palo Alto, Calif.: Annual Reviews Inc.
72. Barker, H. A., Kamen, M. D. 1945. *Proc. Natl. Acad. Sci. USA* 31:219–25
73. Wood, H. G. 1952. *J. Biol. Chem.* 194:905–31
74. Poston, J. M., Kuratomi, K., Stadtman, E. 1964. *Ann. NY Acad. Sci.* 112:804–6
75. Ljungdahl, L., Irion, E., Wood, H. G. 1965. *Biochemistry* 4:2771–80
76. Ljungdahl, L., Irion, E., Wood, H. G. 1966. *Fed. Proc.* 25:1642–48
77. Ghambeer, R. K., Wood, H. G., Schulman, M., Ljungdahl, L. 1971. *Arch. Biochem.* 143:471–84
78. Schulman, M., Ghambeer, R. K., Ljungdahl, L. G., Wood, H. G. 1973. *J. Biol. Chem.* 248:6255–61
79a. Wood, H. G., Stjernholm, R. L. 1962. In *The Bacteria,* eds. I. C. Gunsalus, R. Y. Stanier, 3:41–117. New York: Academic
79b. Wood, H. G., Utter, M. F. 1965. In *Essays in Biochemistry,* ed. P. Campbell, 1:1–21. London: Oxford Univ. Press
80. Zeikus, J. G. 1983. *Adv. Microbiol. Physiol.* 24:215–99
81. Drake, H. L., Hu, S.-I., Wood, H. G. 1981. *J. Biol. Chem.* 256:11137–44
82. Diekert, G. B., Thauer, R. K. 1978. *J. Bacteriol.* 136:597–606
83. Drake, H. L., Hu, S.-I., Wood, H. G. 1980. *J. Biol. Chem.* 255:7174–88
84. Hu, S.-I., Drake, H. L., Wood, H. G. 1982. *J. Bacteriol.* 149:440–48
85. Ragsdale, S. W., Clark, J. E., Ljung-

dahl, L. G., Lundie, L. L., Drake, H. L. 1983. *J. Biol. Chem.* 258:2364–69
86. Diekert, G., Ritter, M. 1983. *FEBS Lett.* 151:41–44
87. Drake, H. L. 1980. *J. Bacteriol.* 150:702–9
88. Pezacka, E., Wood, H. G. 1984. *Arch. Microbiol.* 137:63–69
89. Hu, S.-I., Pezacka, E., Wood, H. G. 1984. *J. Biol. Chem.* 259:8892–97
90. Pezacka, E., Wood, H. G. 1984. *Proc. Natl. Acad. Sci. USA* In press

91. Clark, J. E., Ljungdahl, L. G. 1984. *J. Biol. Chem.* 259:10845–49
92. Kerby, R., Zeikus, J. G. 1983. *Curr. Microbiol.* 8:27–30
93. Maloy, W. L., Bowien, B. U., Zwolinski, G. K., Kumar, K. G., Wood, H. G. et al. 1979. *J. Biol. Chem.* 254:11615–22
94. Wood, H. G., Harmon, F. R., Wühr, B., Hübner, K., Lynen, F. 1980. *J. Biol. Chem.* 255:7397–409

Ann. Rev. Biochem. 1985. 54:43–71

TIME-RESOLVED FLUORESCENCE OF PROTEINS

Joseph M. Beechem and Ludwig Brand

Department of Biology, The Johns Hopkins University, Baltimore, Maryland 21218

CONTENTS

PERSPECTIVES AND SUMMARY

Time-resolved fluorescence spectroscopy monitors events that occur during the lifetime of the excited singlet state. This time scale ranges from a few picoseconds to hundreds of nanoseconds. Relevant biological events that occur in this time domain include the rotation of whole proteins, of segments of polypeptide chains, and of individual amino acid side chains. In addition, the excited chromophore can interact with solvent molecules or neighboring amino acid residues.

43

0066-4154/85/0701-0043$02.00

The classical review by Weber in 1953 (1) represents a turning point in the application of fluorescence spectroscopy to studies of proteins. The use of fluorescence polarization to study rotational diffusion was discussed in detail, and it was suggested that the intrinsic fluorescence of proteins should originate from the aromatic amino acids and appear in the ultraviolet region of the spectrum. This was soon followed by the observation of fluorescence due to proteins and the aromatic amino acids by Bowman et al (2), Shore & Pardee (3), Duggan & Udenfriend (4), Konev (5), and Vladimirov & Konev (6). Weber (1) also introduced the concept of an extrinsic fluorescence probe by covalently attaching DNS (1-dimethylamino naphthalene-5-sulfonyl chloride) to proteins. This fluorophore has desirable spectroscopic characteristics, which makes it a useful probe for measuring the rotational motion of macromolecules.

Fluorescence studies of proteins can be done in three ways: with extrinsic covalent probes, with noncovalent (adsorbed) extrinsic probes, or by making use of the intrinsic fluorescence of proteins. The first approach has the advantage that a probe may be designed to have the most desirable spectroscopic properties. A difficulty is that one has to study a modified protein whose functional characteristics may differ from the native molecule. In addition, it is desirable, but not trivial, to prepare a protein conjugate modified at a selected and unique site. The present review will be limited to the third approach: time-resolved studies of the intrinsic fluorescence of proteins.

It is now well established that changes in the quantum yields and emission maxima of intrinsic fluorescence can be used to monitor conformational changes of proteins and their interaction with ligands. See (7–12) for typical examples and (13–16) for reviews. An understanding of the photophysics of the aromatic amino acids is necessary for the interpretation of the fluorescence of proteins. This topic has been reviewed (14, 17–20). We will summarize the current status of work in this area, with an emphasis on time-resolved studies.

Steady-state measurements provide an intensity weighted average of the underlying decay processes. Therefore, steady-state signals are often proportional, not to the most populated state, but simply to the state that emits the most light. Time-resolved studies, however, can provide detailed information concerning the population distribution of molecular species in the excited state. Steady-state emission spectra can often be time resolved into component parts representing distinct molecular species. Similarly, time-resolved anisotropy measurements can uncover differently rotating species.

GENERAL CONSIDERATIONS

Some pure fluorophores show monoexponential decay kinetics when they are dissolved in noninteracting solvents. The equation for this is:

$$I(t) = I_0 \exp[-t/\tau].$$
1.

In the case of biomolecules, multiexponential decay kinetics, with the equation

$$I(t) = \sum_i \alpha_i \exp[-t/\tau_i], \qquad\qquad 2.$$

or nonexponential decay kinetics (for example kinetics whose equation contains a $\sqrt{t}$ term) are more common. It is useful to consider possible origins of complex fluorescence decay in proteins.

Clearly a mixture of fluorophores could give rise to multiexponential decay, with the number of decay constants equal to the number of components in the mixture. In this context, a single chromophore in different microenvironments may be considered as a mixture of fluorophores. It might thus be possible to assign two decay constants found in a protein containing two tryptophans to the individual tryptophan residues. Likewise if a protein has only one tryptophan residue but exists in two conformations, biexponential decay kinetics might reflect these two states of the protein and could be used to investigate their interconversion. Microheterogeneity can also be more subtle. As will be discussed in detail below, tryptophan zwitterion itself shows biexponential decay, and one interpretation of this finding suggests that tryptophan exists in microheterogeneous (rotameric) forms.

Multi- or non-exponential decay behavior may also arise from a pure homogeneous fluorophore undergoing an excited-state reaction. These include processes such as exciplex formation, excited-state solvent relaxation, configurational changes (such as *cis-trans* isomerization), excited-state proton transfer, and energy transfer, just to mention a few. A system that is homogeneous in the ground state may exhibit emission from two or more excited states. If the kinetics of an excited-state reaction are rapid compared to the time-scale of measurement, complex decay may still be treated in terms of microheterogeneity. If the excited-state reaction occurs on the same time scale as the time-resolved measurement, complex decay may be found even if the system is homogeneous in the ground state. In this case, the observed decay reflects the dynamic approach to the excited-state equilibrium. A hallmark of this type of excited-state reaction is a negative preexponential at a wavelength at which emission due to a species created in the excited state predominates. Unfortunately, this negative amplitude is often masked owing to spectral overlap or because the quantum emission of a species created in the excited state is small compared to the emission of the directly excited fluorophore.

FLUORESCENCE DECAY OF TRYPTOPHAN AND DERIVATIVES

There are many problems involved in interpretation of complex decay of tryptophan and its derivatives, both in solution and in proteins. The problem arises because the measured fluorescence lifetime is a function of many different rate parameters:

$$\tau_m = 1/(\tau_n^{-1} + \Sigma \, k_{nr}), \hspace{4cm} 3.$$

where τ_n is the fluorescence lifetime that would be measured in the absence of any nonradiative processes and $\Sigma \, k_{nr}$ is the sum over all nonradiative rate constants.

In many molecules there are only a limited number of nonradiative pathways, so that changes in the observed lifetime can be interpreted in terms of these processes. However, in indole and related compounds, a great number of nonradiative excited state processes have been documented. A brief description of some of the processes which compete with, and/or complicate, the interpretation of the fluorescence decay from tryptophan and related compounds will be described. For a recent review see Creed (18).

The Photophysics of Indole

The near ultraviolet absorption of light by indole and related compounds involves two overlapping transitions: ^1L_a and ^1L_b (21). Complications arising from absorption into these two transitions are twofold. Interpretation of the decay of the polarized light from indole-containing molecules at very short times will be complicated by equilibration between the differently polarized ^1L_a and ^1L_b transitions. The theory describing this process has recently been presented by Cross et al (22). Also, the ^1L_a transition, due to its large excited-state dipole, undergoes solvent relaxation to a much greater extent than the ^1L_b transition (23). This solvent relaxation has recently been described by Meech et al as actually being a ^1L_a/charge transfer complex (24). Dual fluorescence from the ^1L_a and ^1L_b was once invoked to describe the biexponential decay of tryptophan at neutral pH (25), but single exponential behavior of many related compounds has mitigated against this hypothesis. Solvent relaxation however, has been shown to occur in N-acetyl-tryptophan amide (NATA)(26), and there is no reason to doubt that in certain proteins this may be a complicating factor in the interpretation of fluorescent lifetimes as a function of emission wavelength (27).

Indole and related compounds have been found to form both excited state and ground state complexes with many different polar compounds (28–33). These complexes may or may not be fluorescent, and their presence can result in shifts in the absorption/emission spectra and multiple fluorescence lifetimes. These ground-state and excited-state complexes may each have very different nonradiative rate constants (28), such that interpretation of the actual measured lifetime may be complicated.

Indole-containing compounds have been shown to undergo electron ejection (34–40). This may be enhanced by the coupling of the ^1L_a Franck-Condon state, to a solvated Rydberg state (41). The ionization threshold for both tryptophan and indole is much smaller in polar solvents than in nonpolar solvents (36). Therefore, tryptophans located near the surface of a protein will be more apt to undergo electron ejection than those buried in the hydrophobic interior. This photoejection has been reported to occur from a prefluorescent state (36, 38). However, recent studies (42) suggest the lowest excited singlet state as the precursor to monophotonic photoionization. The temperature and excitation wavelength dependence of the measured lifetime in tryptophyl compounds probably arises from changes in the photoionization rate constants.

Charge transfer reactions have been shown to occur in indole-containing compounds (43, 44). These reactions involve electron orbital interactions between an electron donor and an electron acceptor (24, 46, 47). This donor-acceptor complex may have its own distinct absorption and emission spectrum. For indole-containing compounds, the charge transfer complex has been described with indole as the donor, and the carbonyl (43) and ammonium (45) as the acceptor. The indole-carbonyl charge transfer complex rapidly undergoes nonradiative decay to the ground state. Charge transfer efficiencies in tryptophyl compounds should therefore increase as the electrophilicity of the acceptor increases. Trends of this type have indeed been found in the study of Werner & Forster (48).

Intersystem crossing is another nonradiative process occurring in these compounds (34, 38, 49, 50). The rate constant for intersystem crossing as well as the τ_N^{-1} appear to be temperature independent processes (51) in water. Under normal solution conditions, no phosphorescence is observed due to competing nonradiative decay processes which occur on a much faster time scale. For a very complete description of these triplet and photoionizing intermediates see Bent & Hayon (34) and Klein et al (37).

Effect of pH on Tryptophan Fluorescence

The effect of pH on the fluorescence of tryptophan and related compounds has been investigated in detail (52–57). Protonation of the amino and the carboxyl group decreases the fluorescence yield and the titration of these functional groups can be monitored by the fluorescence from the indole. Edelhoch et al (56) studied a series of compounds, $NH_2(CH_2)_nCO\text{-Trp}$, where $n = 1-6$. It was found that both the charged and neutral forms of the amine could act as quenchers. The distance of the amine from the indole determines which quenches the most. Proximate functional groups in proteins may thus have unexpected effects on the tryptophan fluorescence.

De Lauder & Wahl (58) investigated the pH dependence of the fluorescence decay of tryptophan. The average lifetime followed the relative quantum yield-pH curve as expected. Decay times of about 1, 3, and 9 ns within the pH range of 2 to 10 could be explained in terms of the mixtures of the different ionic species present in solution (protonated, zwitterionic, and ionic species). It appeared that each species had a characteristic lifetime and contributed to the total decay in proportion to its molar fraction.

These studies suggested that the decay of the tryptophan zwitterion could be described in terms of a single decay constant. Several other workers have also reported monoexponential decay for this species. For example, Alpert et al (59) carried out a detailed investigation of the fluorescence decay of dilute solutions of L-tryptophan. Both the phase shift technique and the method of single photon counting with synchrotron radiation as a light source were used to obtain the decay data with excitation in the range of 220 to 320 nm. A single exponential decay of 3.1 ns, independent of excitation wavelength, was reported. More recently, Georghiou et al (60) used a boxcar averager method to obtain the decay of tryptophan in Tris buffer at pH 7. These workers report a monoexponential decay of about 2.6 ns which increases slightly with emission wavelength.

The Biexponential Decay of Tryptophan (Rotamers)

The work of Szabo and his coworkers (25, 61, 62), and from many other laboratories (40, 45, 63, 64), provides compelling evidence that a biexponential with decay times near 3.1 and 0.5 ns is required to describe the fluorescence of the tryptophan zwitterion (see Table 1 for a representative comparison). The emission spectra associated with the long and short decay time differ. The 3.1 ns decay is associated with an emission maximum of 350 nm, while the 0.5 ns component has an emission maximum of 335 nm.

In order to interpret the origin of the double exponential decay of the tryptophan zwitterion, it is important to realize that indole, and several indole

Table 1 Decay constants for tryptophan in aqueous solution

pH	T(C)	tau(1)	tau(2)	tau(3)	α_1	α_2	α_3	λ(nm)	Ref.
7.0	20	3.1	0.53	—	.67	.33	—	330	62
10.5	20	4.55	—	10.1	.44	—	.56	330	62
5.0	20	3.1	0.53	—	.67	.33	—	330	62
7.0	20	3.2	0.62	—	.78	.22	—	>320	40, 64
11.0	18	—	—	9.1	—	—	1	>320	40, 64
3.0	18	3.0	0.29	—	.73	.27	—	>320	40, 64
7.1	20	3.0	—	—	1	—	—	>290	58
10.6	20	—	—	8.9	—	—	1	>290	58
3.5	20	3.0	1.2	—	.94	.06	—	>290	58
7.0	20	3.2	0.6	9.1	.94	.06	.01	>305	75
10.5	20	3.1	0.5	9.0	.03	.01	.96	>305	75
3.0	20	2.6	0.5	—	.92	.08	—	>305	75
7.0	20	3.2	0.85	—	.7	.3	—	355	63
7.0	20	2.6	—	—	1	—	—	350	60
7.0	20	3.1	—	—	1	—	—	340	59
7.0	20	3.1	0.55	—	.725	.275	—	330	146

derivatives such as 3-methyl indole, exhibit monoexponential decay behavior in aqueous solution. Likewise N-acetyltryptophanamide shows monoexponential decay. Szabo & Rayner (62) suggested that the dual emission of neutral tryptophan originates from different rotamers of the alanyl side chain of tryptophan. A similar interpretation was proposed by Wahl and his coworkers to explain the fluorescence decay of tryptophanyl (65) and tyrosyl (66) diketopiperazines. Newman projections of three rotamers of tryptophan along the C^α–C^β bond are shown in Figure 1.

The existence of preferred populations of C^α–C^β bond rotamers has been demonstrated with the aid of NMR spectroscopy (67, 68). Dezube et al (67) indicate that at least six conformers should be considered (see Figure 1); g−(A), t(B), and g+(C) each have a possible perpendicular or antiperpendicular configuration of the indole ring (the A, B, C refer to the nomenclature commonly used in fluorescence studies). They conclude that for tryptophan in solution, the g−(A) configuration predominates and that the ring is perpendicular to the C^α–C^β bond. At the same time, the NMR data which includes coupling constants, nuclear Overhauser effects, and lanthanide perturbations require that other conformations contribute at least 20% and possibly 40% of the population. X-ray structural analysis of crystals of tryptophan hydrochloride (69) suggests that in the crystal, the g+(C) conformer with the ring in the perpendicular orientation predominates. It is of interest that Hochstrasser (70) has measured the fluorescence decay of a crystal of tryptophan, finding the same biexponential decay observed in solution. Janin et al (71) indicate that the principal side-chain configuration of tryptophan in protein crystals is 57% g+(C), 31% t(B), 15% g−(A). Bhat et al (72) state that tryptophan in proteins

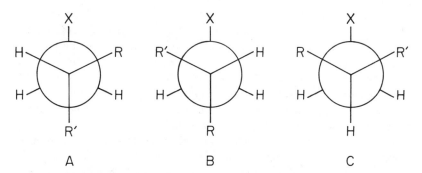

Figure 1   g−(A),t(B),g+(C) conformers about the C^α–C^β bond. For tryptophyl derivatives: X = indole; For NATA: R = NHCOCH3, R' = CONH2, AA: R = CONH2, R' = CONH2; For EE: R = COOEt, R' = COOET, AE: R = CONH2, R' = COOEt; For tyrosyl derivatives: X = phenol; For Series 1 glycyl dipeptides: R NHCOCH2NH3, R' = COO−; For Series 2 glycyl dipeptides: R = NH3, R' = CONHCH2COO−

can exist in several conformations, but the g−(A)(perpendicular) is quite common.

Szabo & Rayner (62) argue that rotation about the $C^\beta–C^\gamma$ bond will not affect the spectral assignment in terms of rotamers g−(A), t(B), and g+(C). They suggest two alternative assignments for the emitting species of the tryptophan zwitterion (see Figure 1). Their discussion is based in part on a consideration of the general form of the potential energy surfaces for the interconversion of the three rotamers in the ground state and the excited state. It is reasoned that g−(A) is the most stable, followed by g+(C) and t(B). In the excited state, g−(A) will be stabilized by interaction of the positively charged ammonium group with the large dipole of the excited indole nucleus. In contrast, the interaction between the excited indole and the carboxylate substituent is expected to be less favorable due to electrostatic repulsion, and rotamer t(B) would be destabilized in the excited state. It is thus suggested that g−(A) should emit at longer wavelengths than the other rotamers and this assigns the 3 ns decay with an emission maximum near 350 nm. This is also in accord with the fact that this spectral species is present in the highest proportion. Szabo & Rayner (62) also suggest that the 335 nm (0.5 ns)–emitting species is due to g+(C). This is in accord with the fact that g+(C) has two quenching groups: the ammonium and the carboxylate near the indole ring. In the g−(A) rotamer, only the ammonium is in this position. These workers do not rule out an alternative proposal that rotamer t(B) has an emission maximum at 335 nm and a 0.5 ns decay, while rotamers g−(A) and g+(C) have emission maxima near 350 nm and a 3 ns decay time.

Chang et al (45) and Petrich et al (64) carried out an extensive study of the time-resolved fluorescence of tryptophan and derivatives under a variety of conditions. These workers (64) agree that the different rotamers can explain the deviation of the tryptophan zwitterion from monoexponential decay behavior. The assignment of rotamers to the specific emitting species differs from that discussed above. Fleming and his coworkers (64) emphasize possible problems involved with a quantitative comparison between ground-state rotamer concentrations as determined by NMR and results obtained by fluorescence. NMR experiments are typically done with 100 times more concentrated solutions than those used for fluorescence studies and solute-solute and solute-solvent interactions may influence relative rotamer populations (73, 74).

It is well established, from studies of the pH dependence of the fluorescence of tryptophan and derivatives, that in those compounds having an amino group near the indole ring, the fluorescence is more quenched when the amino group is protonated. The amount of quenching depends on the distance of the protonated amine from the indole ring. Chang et al (45) favor the idea that this quenching process is a charge transfer. These workers suggest that the carboxylate group is not an efficient electron acceptor unless an adjacent group such as the protonated amine is present to decrease its electron density. This reasoning must be reconciled with the fact that quenching by the protonated amino group can be seen with gly-trp, gly-gly-trp and alanyl-trp. Petrich et al (64) have investigated the decay of several tryptophan derivatives including the three novel compounds: AA, (3-indolylmethyl)-malonamide; EE, (diethyl (3-indolylmethyl)-malonate; and AE, (ethyl (3-indolylmethyl)-malonamate (Figure 1). The assignment of each rotameric form to the emitting species has been made on the assumption that the main nonradiative decay process competing with fluorescence is charge transfer. The rate of charge transfer depends on the distance between the donor and acceptor and the ionization potential of the donor relative to the electron affinity of the acceptor. The 0.5 ns (335 nm) emitting species is assigned to rotamers g−(A) and g+(C) in which the protonated amino group is equidistant from the indole nitrogen (see Figure 1). The 3 ns (350 nm) component is assigned to rotamer t(B) in which the protonated amino group is further from the indole ring. The monoexponential decay (3 ns) exhibited by NATA is explained by the fact that both groups adjacent to the indole ring have peptidelike carbonyls and all three rotamers have the carbonyl carbon equidistant from the indole nitrogen. In the compounds AA and EE, mentioned above, the amide and ester groups are symmetrically placed about the indole ring. They both show monoexponential decay. The ester group has a higher electron affinity than the amide group which causes EE to have the shorter decay time. AE exhibits biexponential decay and the longer decay component is attributed to g−(A).

In a study by Gudgin et al (75, 76), the fluorescence characteristics of tryptophan in H_2O, D_2O, and various solvents were examined. In protic solvents at neutral pH, where tryptophan is free to exchange the carboxyl, amino, and ring nitrogen protons, the lifetime of tryptophan is increased by a factor of approximately two. This isotope effect was thought to reflect a proton transfer from the ammonium group to the 2 position on the indole ring (40, 77, 78). It was suggested that immediately after excitation, some percentage of the available conformers were in exactly the proper configuration to undergo proton transfer. Other side-chain conformers would need to diffuse into the proper position to undergo proton transfer. The decay law expected for this type of process would involve a $\sqrt{t}$ term. Beddard et al (78) proposed that this $\sqrt{t}$ term could be the cause of the observed double exponential character of the decay of the tryptophan zwitterion at neutral pH. Gudgin (76) however, further examined the isotope effect in the aprotic solvent DMSO, in order to determine whether this was a molecular, or a solvent effect. It was found that the lifetimes of deuterated tryptophan in DMSO were identical with nondeuterated tryptophan. This data suggests that solvation plays an important role in the observed D_2O effect.

Saito et al (79) have described an interesting series of experiments which suggest that the ammonium group of tryptophan quenches fluorescence by proton transfer at the 4-position of the indole. Tryptophan was irradiated with ultraviolet light in D_2O and the incorporation of D_2O into the 4-position of indole was followed directly by ^1H NMR. The quantum yield of deuterium incorporation was estimated to be 0.14. The yield was not decreased with saturating concentrations of N_2O, which suggests that electron ejection was not involved in the photosubstitution reaction. The converse experiment showed that irradiation of C-4–deuterated tryptophan resulted in formation of non-deuterated tryptophan. The reaction could be inhibited by potassium iodide (a quencher), which indicates that this was an excited-state process. These sub-stitution reactions only occur at a pH where the α amino group is protonated. It is proposed that the protonation at the C-4 position, with formation of a diprotonated intermediate, could have a role in the nonradiative decay of excited singlet tryptophan at neutral pH.

This work provides direct evidence for proton transfer from the charged amino group of the tryptophan zwitterion to the indole ring. It remains to be determined how the protonated intermediate at the indole 4-position (79) or possibly the 2-position (40, 78) influences the decay kinetics.

FLUORESCENCE DECAY OF TYROSINE

Gauduchon & Wahl (66) examined the time-dependent fluorescent decays of tyrosine, and a series of derivatives with substituents on both the amino (series 1 derivatives) and carboxyl groups (series 2 derivatives). At pH = 5.5, 20°C, the fluorescence decay of tyrosine, tyramine, p-cresol, and all series 1 deriva-tives were found to decay as a simple monoexponential. It was found that all carboxy-substituted tyrosines (series 2) show biexponential decay of the fluorescence intensity. This could not be attributed to different ionic species because at this pH only a single ionic species was present. Similar to tryp-tophan, a rotamer model is proposed.

It had been shown by Cowgill (80) that the carbonyl from both peptide and amide groups could quench the fluorescence of phenol. The work of Tournon et al (81) suggested that charge transfer between the aromatic ring and the carbonyl was responsible for this quenching. Therefore, the rotamer with the shortest lifetime (most quenched) would be expected to have carbonyl groups closest to the aromatic ring.

The possible rotamers of tyrosine are shown in Figure 1. NMR data (82, 83) on tyrosine and phenylalanine indicate that rotamer A is favored >50%. In series 1 dipeptides it is argued that electrostatic repulsion between the carboxy-late and the electron cloud of the aromatic ring must decrease the ground state populations of rotamers B and C. In zwitterions containing the glycyl residue, the conformation favoring the carbonyl peptide backbone (with its partial negative charge) interacting with the ammonium group should be preferred. Therefore for series 1 compounds only the preferential rotamer A allows close contact between the carbonyl and phenol group. In series 2 compounds, it is only the unfavored rotamer C that allows carbonyl quenching. This rotamer, furthermore, forces the carbonyl group into close contact with the fluorophore, since rotation about the tyrosyl residue is sterically hindered. This is not the case for rotamer A of series 1. Therefore intramolecular collisional quenching by the carbonyl of rotamer A series 1 must be less than rotamer C of series 2. This also predicts that the differences in deactivation rates (and hence lifetimes) of series 1 compounds should be less than that in series 2 compounds. From this logic the biexponential decay of series 2 compounds are assigned as: the short lifetime component of 0.5 ns at pH 5.5 to rotamer C, the longer lived 1.6 ns component to rotamers A and B. Given this assignment, the relative amplitude of the shorter component should be less than that of the longer lived component. This is not found to be the case, and Gauduchon & Wahl conclude that interconversion of the rotamers on the same time scale as fluorescence emission is affecting the relative amplitudes associated with each lifetime. The monoexponential character of the series 1 derivatives are explained by assuming that either the rate of exchange between the rotamers is much faster than the deactivation process, or that the deactivation processes and therefore the lifetimes of the different rotamers are too close to resolve.

We can conclude that the photophysics of tryptophan and tyrosine is complex and not yet completely understood. A variety of mechanisms may influence the decay kinetics. At the same time, it is clear that detailed dynamic information on the atomic level is available from time-resolved fluorescence measurements.

The interpretation of time-resolved fluorescence studies of proteins must rely on our understanding of the photophysics of tryptophan and small tryptophyl and tyrosyl peptides. It is evident from the discussion above that not only the interpretations but even some of the results obtained with these derivatives remain to be resolved. In spite of the fact that the photophysics of these simple model compounds have not been completely explained, considerable progress has been made in studies of the fluorescence decay of proteins.

TIME-RESOLVED INTRINSIC PROTEIN FLUORESCENCE

In an early paper, Chen et al (84) showed that the mean lifetime of a number of proteins was in the range of 2–5 ns. De Lauder & Wahl (85) found that the single tryptophan containing Human Serum Albumin (HSA) has biexponential decay. In a very comprehensive survey by Grinvald & Steinberg (86), a whole series of single- and multi-tryptophan–containing proteins was examined. It was found that all of these proteins, with the exception of apoazurin, showed at least biexponential decay characteristics. Table 2 summarizes the tryptophan decay times in single and multitryptophan proteins. In an attempt to relate some physically meaningful quantity to these multiple lifetime components, Wahl pioneered the analysis of fluorescence decay curves as a function of excitation/emission wavelength (87, 88). This allowed the resolution of the steady state emission spectra into components associated with each particular lifetime. This type of association aids in the identification of specific tryptophyl residues in proteins. The relationship between the steady-state spectrum and a time-resolved spectra is:

$$I_i(\lambda) = I_{ss}(\lambda)[\alpha_i(\lambda) \, \tau_i \, / \, (\sum_i \alpha_i(\lambda) \, \tau_i)] \qquad\qquad 4.$$

where I_{ss} is the total steady-state spectrum, I_i is the spectrum associated with the ith component, and the α_i are the preexponential terms at wavelength λ_i. For an alternative implementation of this decay association technique see Knutson et al (89).

Two-Tryptophan Proteins

LAC REPRESSOR   A typical spectrum generated from data obtained from *Escherichia coli* lac repressor protein is shown in Figure 2. Lac repressor is a two-tryptophan–containing protein which has been found to decay as a biexponential with lifetimes of 3.8 and 9.8 ns (88). Spectrum 1 is associated with the 3.8 ns lifetime, and spectrum 2 is associated with the 9.8 ns lifetime. Sommer (90) had previously investigated the steady state fluorescence of a series of mutant lac repressor molecules lacking either tryptophan 190 or 209. It was found that the mutant lacking tryptophan 209 had an emission spectrum with a maximum of 325 nm. The mutant lacking tryptophan 190 had emission maximum of 338 nm. By comparison of the time-resolved emission spectra with Sommer's data, Brochon et al were able to assign tryptophan 190 to spectrum 1 and tryptophan 209 to spectrum 2. Therefore, at least in the case of this protein, the complex decay of the emission of tryptophan can be accounted for simply as the sum of the emission rates of the two individual tryptophan residues.

ALCOHOL DEHYDROGENASE   Horse liver alcohol dehydrogenase (HLADH) has long enjoyed favor as a protein with which to evaluate new methods and techniques in protein chemistry. One of the first applications of fluorescence was the classical investigation of the binding of nicotinamide adenine dinu-cleotide to HLADH by Boyer & Theorell (91). This approach led to a very extensive literature utilizing fluorescence to measure the interaction of coen-zymes with dehydrogenases. It was subsequently found that binding of the pyridine nucleotides leads to quenching of the intrinsic protein fluorescence and can be used to measure binding.

Table 2 Intrinsic fluorescence decay of single and multiple tryptophan–containing proteins

Protein	#trps	tau(1)	tau(2)	tau(3)	<tau>[a]	Ref.
apoazurin	1	4.9				101
nucleaseB	1	5.1				140
melittin	1	3.1				60
basic myelin	1	4.7	1.97		2.8	140
holoAzurin	1	4.8	0.18		1.1	101
glucagon	1	3.3	1.1		2.77	165
adenocorticotropin	1	5.1	2.0		3.79	63
nuclease	1	5.7	2.0		5.2	86
ribonuclease T1	1	3.3	1.5		3.0	86
subtilisin Carlsberg	1	4.9	1.6		3.4	86
glucagon	1	3.6	1.1		2.8	86
human serum albumin	1	7.8	3.3		4.8	85, 166
myelin coat	1	4.0	1.6		2.8	86
naja naja toxin	1	2.4	0.8		1.1	86
phospholipase A2	1	7.2	2.9	0.96	3.25	167
apolipoprotein C-I	1	6.8	2.9	0.4	2.1	141
lumazine protein	1	6.9	1.9		4.45	168
tuna met-myoglobin	1	2.17	0.132	0.031	0.08	118
tuna apo-myoglobin	1	2.81	0.33		2.04	118
enzyme-I	2	7.5	2.7		5.8	143
horse liver alc. dehydrogen.	2	7.0	3.8		4.66	89, 93
E. coli maltose receptor	2	9.8	4.4		N/A[b]	169
lac repressor	2	9.8	3.8		8.12	88
bovine serum albumin	2	7.1	2.7		6.3	86
yeast 3-P-glycerate kinase	2	7.0	3.1	0.6	2.42	99
ferredoxin	2	6.9	0.55		1.18	170
sperm whale myoglobin	2	2.68	0.106	0.014	0.13	118
sperm whale met-myoblobin	2	2.22	0.135	0.016	0.10	118
sperm whale apo-myoglobin	2	2.68	0.627		1.82	118
human deoxy hemoglobin A	3	3.6	0.04		0.21	126
human oxy hemoglobin A	3	2.6	0.03		0.22	126
human deoxy hemoglobin	3	5.4	1.9	0.09	N/A	126
human oxy hemoglobin	3	4.9	1.8	0.09	N/A	126
gramicidin	4	7.8	2.5		6.2	171
beef liver glut. dehydrogen.	3	6.25	2.1		4.53	172
papain	5	7.1	3.7	1.1	4.1	86
human lysozyme	6	1.2	0.4		0.7	86
porcine pepsinogen	5	7.3	1.6		6.2	86
chicken pepsin	5	7.1	1.4		6.1	86
chicken pepsinogin	5	7.3	0.69		7.5	173
lactogen hormone	N/A	4.9	1.5		2.1	86
carbonic anhydrase	5–7	5.6	0.8		5.0	86
myosin subfragment 1	5	8.7	4.7	0.8	3.77	111
phosphotidyl-choline transp.	5	5.7	2.3	0.8	1.66	174
lactate dehydrogenase	6	8.0	4.0	1.0	3.83	112
hen egg white lysozyme	6	2.6	0.9		1.9	86, 103, 104

[a] <tau>: average lifetime defined as per (153).
[b] N/A: not available from reference.

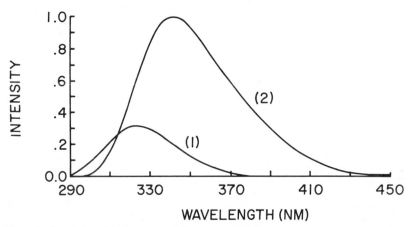

Figure 2   Resolution of the *lac* repressor fluorescence spectrum into two components associated with decay times of 3.8 ns (curve 1) and 9.8 ns (curve 2). Ref. (88).

The intrinsic fluorescence of HLADH was shown to decrease, together with measurable spectral shifts, when the protein was denatured (7). It was concluded from the denaturation experiments that the enzyme had at least two "types" of tryptophans, one type quenched by denaturation and a remainder not affected. It is now established (92) that HLADH is a dimer made up of identical subunits and that each subunit has two tryptophan residues. One of the tryptophan residues (Trp-314) is buried in a cage of hydrophobic residues at the subunit interface. The other is well exposed to solvent (92–96).

Several workers have measured the fluorescence decay of HLADH by both phase and pulse methods (97, 98). Biexponential decay kinetics are observed with decay times close to 3.8 and 7 ns. The mean lifetime increases with increasing emission wavelength (27). Ross et al (93) showed that this can be explained by an increase of the fractional amplitude associated with the longer decay time with increasing wavelength. The 7 ns decay was quenched by potassium iodide and was assigned to the exposed tryptophan residue (Trp 15). The short decay was not affected by potassium iodide. The 3.8 ns decay was selectively quenched in the ternary complex between enzyme-NAD+ and pyrazole. It was suggested that this could be explained by resonance energy transfer between the "inner" Trp-314 and the coenzyme. It was shown that the fluorescence emission of HLADH could be decay associated into two spectral components with Trp-15 (exposed) having an emission maximum at 337 nm and Trp-314 (buried) at 324 nm. These results are in excellent agreement with spectral resolution achieved by Abdallah et al (94) by selective quenching with potassium iodide. Knutson et al (89) decay associated the spectral contributions of the two tryptophan residues by a different procedure and obtained identical results. In addition it was found that while only the buried Trp-314 undergoes dynamic quenching, the emission of both tryptophans decreases. The exposed Trp-15 shows a decrease in fluorescence intensity and a small red shift in the emission maximum upon complex formation. HLADH thus represents another example of a two tryptophan protein exhibiting biexponential decay, where it is reasonable to assign one decay to each of the two specific tryptophan residues. Ross et al (93) suggest that the apparent monoexponentiality of Trp-15 and 314 can be explained by the argument that both residues are in highly restricted environments. The indole rings would thus experience a sufficiently narrow range of interactions with the neighboring side groups such that no kinetically distinct second environments exist.

3-PHOSPHOGLYCERATE KINASE   Yeast 3-phosphoglycerate kinase (PGK) represents an interesting example of a two-tryptophan–containing protein that decays as a triple exponential. By simultaneous analysis of a series of decay curves obtained under differing excitation/emission wavelengths, Privat et al (99) were able to resolve the decay of PGK into three lifetimes of 0.6, 3.1, and 7.0 ns. The spectrum associated with the short lifetime component had an emission maximum of 325 nm and represented 13% of the total intensity at pH = 3.9 and 50% of the intensity at pH = 7.2. The spectra associated with both the 3.1 and 7.0 ns component were identical, and had maxima at 335 nm. Even though there are three distinct decay times (which cannot each represent a different tryptophan residue), there are only *two* distinct spectra, which can be attributed to each of the two tryptophans. The static quenching of the fluorescence intensity of the spectra associated with the medium and long lifetimes yielded a pK of 4.7. The data suggest that a carboxylate moiety from either an aspartic or glutamic acid is involved. It is suggested that one of the lifetimes associated with the blue shifted spectra results from ground state complexing of the carboxylate with the indolic N-H group. The other lifetime associated with this spectrum is suggested to be due to dynamic quenching of the indole moiety with the acid form of the carboxyl group, possibly through a charge transfer mechanism.

In the cases discussed so far, the resolution of the complex decay of tryptophan in proteins has been assigned to individual tryptophan residues. The question arises, why do most of the single tryptophan–containing proteins also show multiexponential decay? As has been discussed in the previous section on amino-acid photophysics, complex decay from a single tryptophan can arise from a number of causes. In addition to all these processes, single tryptophan residues in proteins may be subjected to multiple environments corresponding to different protein conformations. For this case, the relative amplitudes associated with the different lifetimes components can be related to the relative populations of the various protein conformations.

Single Tryptophan Proteins

AZURIN   The blue copper protein azurin has a single tryptophan residue. Although decay of the fluorescence intensity is biexponential, examination of apozurin (azurin without the bound copper) shows monoexponential decay kinetics (100, 101). Therefore, the addition of the copper ligand must create a microheterogenous environment for the tryptophan that does not exist in the apoenzyme. The tryptophan in azurin could be modeled well by the compound 3-methyl indole in methylcyclohexane, which indicates that this tryptophan was buried deep within the hydophobic core of the protein. Examination of fluorescence decay as a function of emission wavelength showed that the spectra associated with the two lifetimes are identical. Careful comparison of both the absorption and emission spectra of azurin and apoazurin also showed no significant changes in shape. Szabo et al (101) propose that the two lifetimes originate from two conformers of the copper binding site which interact differently with the single tryptophan residue. The existence of two copper site conformers is supported by several other lines of investigation. From the ratio of the preexponential components associated with each lifetime, an equilibrium constant between the two forms of copper ligand is estimated to be 0.15–0.26 with the tetrahedral geometry (associated with the 0.18 ns component) predominating.

Multiple Tryptophan–Containing Proteins

The fluorescence decay of proteins containing many tryptophans can often be analyzed in terms of classes of emitting species. Emitters in a single class have the same relative lifetime and spectral distribution, so their emission can be treated as a single decay parameter or as a distribution of closely spaced decays (102).

LYSOZYME   Hen egg-white lysozyme, a well-characterized enzyme, contains six tryptophans (residues 28, 62, 63, 108, 111, and 123). Chen et al (84) reported the average lifetime of lysozyme to be 2.0 ns. Later Yashinsky (103) resolved this lifetime into a double exponential of 2.6 and 0.9 ns. Formoso et al (104), in a very comprehensive study, assigned these lifetime components to specific classes of tryptophans in lysozyme.

Trp-62 was shown to be essential for enzyme activity, and could be selectively reacted with N-bromosuccinimide (NBS) (105) to form oxindole. Hartdegen et al (106) provided the methodology to selectively react Trp-108 with NBS. Steady-state work on these derivatives showed that approximately 80% of the fluorescence in the native protein comes from Trp-62 and Trp-108 (107). Raising the pH from 5 to 8 greatly increases the fluorescence contribution of Trp-108. This is because of the removal of quenching due to protonated Glu-35 which is in van der Waals contact with Trp-108, and has a pK of 6.0. Fluorescence studies on the Glu-35 ester did not show this behavior.

Time-resolved fluorescence studies with these derivatives (104) were able to resolve these tryptophan lifetime components into three classes. Trp 28, 63, 111 and 123 all decay with a lifetime of 0.5 ns.; Trp 62 has a lifetime of 2.5 ns; and Trp-108 a lifetime of 1.7 ns. Integration over the recovered decay parameters showed that Trp-62 is the largest contributor to lysozyme emission. The lifetime of Trp-108 decreases in the presence of Trp-62, which indicates that there is energy transfer from Trp-108 to Trp-62. The rate of energy transfer from Trp-108 to Trp-62, calculated from the lifetimes obtained from Trp-108 lysozyme with and without Trp-62, was approximately 10^{-8} sec^{-1}. Upon the addition of potassium iodide only Trp-62 is quenched, indicating that it is exposed to solvent. Binding of the substrate, tri-(N-acetylglucosamine) resulted in complex changes in both the lifetimes and the number of primary emitters in lysozyme.

The fluorescence of lysozyme that contain all six tryptophan residues can be resolved into three distinct classes, which are determined by different exposure to solvent, proximity to neighboring quenching groups, and energy transfer to other tryptophans.

MYOSIN   Myosin subfragment 1 contains five tryptophan residues. ATP has been shown to react with this subfragment (108). This reaction is accompanied by a shift in the tryptophan region of the absorption spectra (109) and a large increase in the fluorescence emission (110).

Torgerson (111) examined the intrinsic decay of the fluorescence of this subfragment as a function of emission wavelength and found that the data could be analyzed in terms of three lifetime components of 0.7, 4.5, and 8.8 ns. The emission maxima and percent intensity associated with these lifetimes are 320 nm (9%), 330 nm (45%), and 345 nm (46%) respectively. Upon addition of ATP, the spectral component associated with the medium-lived component increases by 30%, while the long lifetime component increases 9%. The short-lived component does not change. Upon addition of the quencher, acrylamide, only the longer lifetime component is affected. While the information available with this protein is still limited, it is clear that even with a protein containing five tryptophans, classes of tryptophans each with an associated spectra can be resolved. These classes of tryptophans behave very differently upon addition of substrate and quencher.

LACTATE DEHYDROGENASE   Torikata et al (112) investigated the time-resolved fluorescence of pig heart lactic dehydrogenase, a tetrameric enzyme which contains six tryptophan residues per subunit. The fluorescence was resolved into three classes based on decay times, quenching by potassium iodide, and quenching by bound NADH.

Time-resolved studies provide a nonperturbing probe that permits simultaneous monitoring of multiple protein domains. These domains may be subdivided according to their accessibility to various types of quenchers, i.e. according to whether they are exposed to the external solvent or buried within the core of the protein. Through examination of the emission spectra, one can deduce the polarity of the local environment. The magnitude of the fluorescence lifetime can be a sensitive monitor of neighboring group interactions. As an added benefit, all of these measurements can be performed with physiological solution conditions.

HEMOGLOBIN AND MYOGLOBIN   Warburg & Negelein showed in 1928 (113) that energy transfer could occur from the aromatic amino acids to the heme in a heme protein. Weber & Teale (114) found that the tryptophan fluorescence yield of hemoglobin is very low due to resonance energy transfer to the heme residues. Even though the fluorescence intensity is only about one percent of that found with the apoproteins, the fluorescence of hemoglobin and myoglobin has attracted considerable interest. Initial time-resolved studies indicated that at least one component of the fluorescence decay had a lifetime of 1–4 ns, which is in the same time range found for proteins that do not contain a heme moiety (115, 116). Energy transfer depends not only on the distance between donor and acceptor but also on their relative orientation. Since detectable fluctuations in the protein matrix occur on the nanosecond time scale (117), a small fraction of tryptophans may in fact not show any energy transfer. The workers in this area are well aware of the implicit danger of working with fluorescence intensities as low as those observed with the heme proteins. Even studies with pure fluorophores (under conditions where 99% of their emission is quenched) are plagued by emission due to contaminants. This problem has been addressed in several ways. Alpert et al (116) have done measurements on samples purified with care. They have further used bis-anilino-napthalene sulfonate, which shows a large fluorescence enhancement when it binds to the heme pocket of apoglobins, to conclude that their preparations contain less than 0.5% apohemoglobin. In a study of the fluorescence decay of myoglobin, Hochstrasser & Negus (118), carried out fluorescence decay studies in the presence of excess hemin so that the concentration of apomyoglobin would be insignificant at the equilibrium.

The fluorescence of aplysia and sperm whale apomyoglobins were studied by Anderson et al (119). They found an average lifetime in the range of 3.9 ns and the spectra of the aplysia apomyoglobin revealed two emission maxima at about 330 and 355 nm. The two tryptophanyl residues of mammalian apomyoglobin (120) also show microheterogeneity with emission maxima at 321 and 333 nm for Trp-14 and Trp-7 respectively. In contrast, tuna apomyoglobin, (missing Trp-7) shows only one emission at 321 nm. The tuna myoglobin has also been used to advantage in fluorescence decay studies of myoglobins and apomyoglobins. Hochstrasser & Negus (118) have calculated the theoretical reciprocal energy transfer rate constants for Trp-7 and Trp-14 for sperm whale myoglobin and obtained values of 125 ps and 31 ps. In the case of myoglobin carbon monoxide complex (MbCO) they obtained values of 121 ps and 41 ps. The fluorescence decay of myoglobin could be analyzed in terms of a triple exponential. Decay times of 106 ps and 14 ps were obtained in good agreement with the predicted values. In addition, a small fraction of a 2.7 ns component was observed, which was not predicted. The MbCO gave decays of 132 ps and 26 ps again in good agreement with the predicted values. A small fraction of a 2.8 ns component was also found in this case. In the case of tuna metMb (only Trp 14), the predominant decay time found was 31 ps. The interpretation of these findings was that the lifetimes in the range 15 to 30 ps are associated with Trp-14, while the decays in the range of 106 to 136 ps correspond to Trp-7. It is suggested that the small fraction of decay component in the 2 ns range observed with all the apomyoglobins tested arises from some nonheme-related source.

The fluorescence characteristics of human hemoglobins have also been examined in some detail. Hirsch et al (121, 122) observed tryptophan emission from HbA, which they attributed to beta Trp-37 because of the much diminished fluorescence from Hb Rothschild, which lacks this trytophan residue. It was found that this fluorescence is sensitive to the R-T transition. In further studies (123) it was found that fluorescence of hemoglobin dimers is different from that of the tetramer and this was attributed to increased exposed surface area of beta Trp-37. Itoh et al and Mizukoshi et al (124, 125) found that the fluorescence intensity and also the fluorescence lifetime of HbA change during the R-T transition. No significant change in fluorescence was found with hemoglobin Kempsey, which does not exhibit the R-T transition. There is a logical relation between changes in the long decay component of hemoglobin emission and in known R-T transitions and lack of fluorescence changes in mutant hemoglobins that do not show this transition. This relation supports the idea that in these cases, even though the fluorescence is of low intensity, it may have its origin in the native hemoglobin itself.

Szabo et al (126) have measured the fluorescence decay of several different forms of human hemoglobin (HbO, HbCO, Hb[deoxy], and Hb[Met]). Each derivative exhibited triple exponential decay kinetics with decay time ranges of 70–90 ps, 1.8–1.9 ns, an 4.9–5.4 ns. The fractional contribution between the components depended on the hemoglobin derivative. The results were rationalized by assigning the three decay components to three different average conformations of the hemoprotein. It is assumed that the relative amounts of each conformer varies with the hemoglobin derivative. The short decay component represents a conformation in which the heme and the tryptophan residues are aligned so that there is an efficient energy transfer to the heme. The other lifetimes represent protein conformers in which the interaction of the tryptophan with its neighboring environment are different. It should be noted that here, in contrast to several of the other studies cited, multiexponential decay behavior is attributed to different conformations of the protein rather than an assignment to different tryptophan residues in proteins of one conformation.

TIME-RESOLVED ANISOTROPY DECAY STUDIES

The beginning portion of this article has described the analysis of time-resolved decay of the total intensity. As we have seen, this quantity is directly related to the photophysical properties of the molecule. If one examines the time decay of the polarized components of the light intensity, one obtains a signal which is related to both the photophysical and rotational properties of the molecule.

The measured quantities in a fluorescence polarization experiment are the intensity of light measured parallel and perpendicular to the polarization of the exciting beam. These quantities are related to relevant physical parameters by the following expressions:

$$I_{||}(t) = 1/3[S(t)(1 + 2\ r(t))], \text{ and} \qquad\qquad 5.$$

$$I_{\perp}(t) = 1/3[S(t)(1 - r(t))], \qquad\qquad\qquad 6.$$

where the decay $S(t) = \Sigma_i\ \alpha_i\ \exp[-t/\tau_i]$ and the anisotropy decay $r(t) = \Sigma_j\ \beta_j\ \exp[-t/\phi_j]$. The α_i are wavelength-dependent weighting factors for the total intensity of the emission and the β_j are functions of the excitation wavelength and related to the angle between the absorption and emission dipoles of the molecules and the principle diffusion axes of the molecule. ϕ is the rotational correlation time of the particle and for spheres (using Stokes-Einstein relationship) is equal to: $\eta V/kT$, with η the viscosity, V the volume, k Boltzmann's constant, and T the absolute temperature. In the literature, polarization, not anisotropy, values are often reported. Polarization is defined as: $P = (I_{||} - I_{\perp})/(I_{||} + I_{\perp})$. Anisotropy is defined as: $R = (I_{||} - I_{\perp}) / (I_{||} + 2I_{\perp})$. For steady-state measurements, either quantity may be used. However, for time-resolved measurements, polarization does not decay as a simple sum of exponentials, whereas the anisotropy does. Therefore anisotropy is exclusively used with time-resolved studies. For relevant techniques concerning analysis of this data see (127–129) and the section at the end of this contribution.

Because of the relatively short lifetime of tryptophan in proteins, and the small amount of light that can be obtained from standard flashlamps at 295 nm, little work has been done utilizing the intrinsic fluorescence to measure rotational correlation times. Extrinsic probes have normally been utilized to measure the motions of whole proteins (1) and protein segments (130, 131), owing to both the longer lifetimes and greater quantum yields associated with these dyes. The arrival of lasers and synchrotrons as a light source has both improved the timing resolution and eliminated the very weak signals that are often obtained with proteins. Therefore, using the intrinsic fluorescence from proteins, one should now be able to follow motions occurring from 50 ps to 150 ns or longer.

The Motions of Tryptophan in Proteins

At one time proteins were considered to have static structures. Enzymes were considered to interact with their substrates according to a "lock and key" model. This was followed by the "induced fit" concept (132), which gave some flexibility to the protein and helped to explain control over enzyme function. More recently, the field of molecular dynamics has provided evidence that proteins are in a constant state of flux (133, 134). Fluorescence studies have provided direct experimental evidence for these dynamic motions.

Lakowicz & Weber (135) studied the quenching of tryptophan fluorescence in 14 different proteins and found that no tryptophan residues were inaccessible to oxygen during the lifetime of the excited state (~4 ns). The diffusion rates for oxygen in proteins was found to differ from the diffusion rate in water by only 20–50%. This type of data supports the concept of fast dynamic motions in proteins. Lakowicz et al (136) extended this concept to the study of steady state anisotropy and found that tryptophan residues were able to undergo angular displacements of 0–37° in subnanosecond time ranges. Molecular dynamics calculations had predicted motions of this magnitude to occur both with tyrosine (137, 138), and tryptophan (139), in time scales ranging from 0.2–10 picoseconds.

Munro et al (140), using synchrotron radiation as a light source, were able to directly measure the rotational relaxation times of tryptophan in a number of single tryptophan–containing proteins. It was found that the tryptophan in nuclease B had little rotational freedom, only rotating with the protein as a whole, with a correlation time of 9.9 ns. In contrast, tryptophan in myelin basic protein was found to undergo very rapid rotational relaxation, with correlation time of 0.09 and 1.3 ns. A correlation time of 6.9 ns would have been predicted for rotation of the protein as a whole. The single tryptophan in human serum albumin at 8°C showed a single rotational correlation time of 31.4 ns. At 43°C however, two rotational correlation times are present, one of 0.14 ns (representing local tryptophan motion) and another of 14 ns (representing rotation of the entire protein). Thus human serum albumin at 43°C exists in a conformation where the local mobility of the single tryptophan is greatly increased.

The single tryptophan in holazurin is buried deep within the hydrophobic interior of the protein. This tryptophan was found to exhibit very fast motion of 0.51 ns and a slower component of 11.8 ns. The rotational correlation time of 0.51 ns can be interpreted as tryptophan wobbling in a cone of semi-angle 34 degrees. Thus it appears that for tryptophan in azurin, the core of the protein is not static at all on the nanosecond time scale, and in fact is closer to being a "fluid."

The decay of the anisotropy of the single tryptophan in a synthetic analog of adrenocorticotropin hormone, ACTH(1 = >24) tetracosapeptide, has been reported by Ross et al (63). It was found that both the decay of the total intensity, and the emission anisotropy could be well represented as double exponentials. The spectra associated with each of the total intensity lifetimes were found to be the same. The independence of the amplitudes and time constants as a function of emission wavelength does not support the explanation that the biexponential decay is due to a simple excited-state reaction such as solvent relaxation or exciplex formation. It is proposed that the tryptophan at position 9 is interacting with the neighboring guanidinium at position 8, leading to this complex decay kinetics. The complex decay of the emission anisotropy is interpreted as independent rotation of the tryptophan, representing the short rotational correlation time of 0.92 ns, and a 4.5 ns component representing the complex rotation of ACTH as a whole. Since the tryptophan in ACTH undergoes subnanosecond rotation, the complex decay kinetics found in the total intensity could arise due to a "sampling" of a variety of side chain conformations during the lifetime of the excited state.

Tran et al (9) have examined the decay of the anisotropy of the single tryptophan in the 29 amino acid–long glucagon. It was also found that the decay of the total intensity and the anisotropy could be interpreted as a biexponential. The decay of the anisotropy was found to be 0.38 ns and 1.8 ns at 8°C. Again these results are interpreted as local rotation of the individual tryptophan residue coupled with the rotation of glucagon as a whole.

Jonas et al (141) examined the lifetime and rotational properties of the single tryptophan in apolipoprotein, the major protein in high density lipoproteins. It was found that the decay of the total intensity was best represented as a three exponential decay of 0.4, 2.9, and 6.7 ns. Lux et al (141a) also note complex decay of a related apoproteolipid. Two rotational correlation times were found: a short 0.2–0.4 ns component representing fast rotation of the tryptophan residue, and a slower component arising from the rotation of the protein as a whole. This slower component could be used to determine the aggregation state of the molecule with a 3.6 ns component representing the monomer, a 7.2 ns component the oligomer, and a 44 ns component the complete protein-lipid complex.

Subnanosecond motion of tryptophans in proteins has now been well documented. It has also been shown that the tryptophan(s) in some proteins may show little or no motion on this time scale (see Table 3). Studies of this type are finally bringing the theoretical conclusions obtained from molecular dynamics together with experimental data. Further information on these fast motions (or lack of them) will certainly be appearing in the literature.

Table 3 Time-resolved anisotropy studies of various single- and multi-tryptophan containing proteins

Protein	#trps	β_1	ϕ_1	β_2	ϕ_2	r_0	T(°C)	Ref.
nuclease B	1	.177	9.9			.177	20	140
myelin basic protein	1	.15	0.09	.196	1.26	.286	20	140
human serum albumin	1	.19	31.4			.19	8	140
human serum albumin	1	.06	0.14	.17	14.	.232	43	140
human serum albumin	1	.07	0.15	.16	3.	.23	47	148
human serum albumin	1	.03	6.0	.15	23.0	.18	31	166
melittin	1	.185	1.1			.185	25	60
glucagon	1	.094	0.42	.08	1.67	.174	19	9
endonuclease	1	.07	8.9			.07	20	142
apolipoprotein C-I	1	.12	0.2	.09	3.6	.21	15	141
adenocorticotropin	1	.13	0.92	.05	4.5	.18	4	63
lac repressor	1	.10	9.5			.10	23	175
lumazine protein	1	.13	1.2	.02	15.5	.15	4	168
bovine serum albumin	2	.05	6.0	.09	29	.14	21	166
enzyme 1	2	.17	66.			.17	4	143
horse liver alc. dehy.	2	.21	56.			.21	10	93

Motion of the Entire Protein

We have concentrated on the use of fluorescence anisotropy studies to reveal subnanosecond motion within proteins. Tryptophan anisotropy studies can also be used to obtain motion of the protein as a whole. This motion occurs on the nanosecond time scale and can be used to monitor various association reactions, and the effect of substrate on the conformation of proteins.

In a very early study of Brochon et al (142), the intrinsic fluorescence of tryptophan was used to calculate the rotational correlation time of low molecular weight (149 amino acids) endonuclease. The single tryptophan of this molecule was found to rotate with the protein as a whole, giving rise to a single rotational correlation time of 8.9 ns at 21°C. It was found that the addition of substrate did not affect the rotational correlation time, but it did selectively quench the fluorescence of the tyrosine residues. The measured correlation time corresponds to a molecular volume of 42 nm^3. The difference between this radius and that calculated for an anhydrous sphere of molecular weight 16,800 was found to be 0.4 nm, the proper magnitude for the thickness of a single water molecular layer.

Georghiou et al (60) utilized the intrinsic fluorescence of the single tryptophan in melittin, the major component of bee venom, to monitor its rotational properties. It was found that melittin in Tris buffer at pH = 7 showed a single intensity lifetime of 3.1 ns, with a small wavelength dependence. However, addition of 1 mM phosphate resulted in a blue shift in the emission spectrum, and biexponential decay in the intensity of 0.6 and 2.5 ns. Analysis of the time-dependent emission anisotropy indicated that the addition of phosphate was accompanied by an association reaction. The rotational correlation time in the absence of phosphate was 1.1 ns, and with phosphate 3.7 ns. This change corresponds to a monomer-tetramer interaction.

Neyroz et al (143) utilized the intrinsic fluorescence of the two-tryptophan–containing enzyme I of the bacterial phosphotransferase system to monitor a monomer-dimer reaction. The decay of the total intensity was found to be a biexponential of 2.7 and 7.5 ns at 2°C and 4.4 and 7.7 ns at 20°C. At low temperatures, this protein yielded a single rotational correlation time of 66 ns. At higher temperatures, a rotational correlation time of 102 ns was found. Use of the Stokes-Einstein relationship yielded molecular radii of 33 Å at low temperatures (where the protein exists as a monomer) and 46 Å at high temperatures (dimer). Independent measurements of this monomer-dimer reaction were done using gel exclusion chromotography (143a) and it was found that the monomer radius was 32.9 Å and the dimer radius was 46.2 Å. The very good agreement of these two quite different techniques supports the idea that reasonable results can be obtained from time-resolved fluorescence studies even when complex total intensity decay behavior is observed.

INSTRUMENTATION AND DATA ANALYSIS

Spectroscopy in general and fluorescence spectroscopy in particular is capable of providing detailed static and dynamic structural information about protein molecules and their interaction with each other and with ligands. While steady-state measurements have been successfully used for several years, time-resolved studies of the total intensity and the emission anisotropy are clearly required to unravel the complexity introduced by microheterogeneity, dynamic interactions, internal motions, and segmental flexibility.

Advances in instrumentation as well as in procedures for data analysis are likely to have a major impact on time-resolved fluorescence studies of proteins. In regard to pulse fluorometry, the availability of high repetition picosecond dye lasers as a light source (61, 118, 144, 145, 146) represents a significant improvement over the flash lamps used in the past. With the use of time-correlated single photon counting as the detection system, the picosecond laser allows the rapid collection of high density data in both a short time and with picosecond timing resolution. The introduction of micro-channel plate detectors (147) may further increase the fast timing resolution. While initial applications of the streak camera as a detector were disappointing, advances in this technology may hold promise for experimental systems that require time resolution of a few picoseconds (118, 145, 148).

There have also been significant advances in regard to the phase-shift technique. Gratton et al (149, 150, 151) and Lakowicz (152) have described instruments that work over a continuous frequency range, providing data that can be analyzed in terms of complex decay equations.

The importance of data analysis in time-resolved fluorescence studies cannot be overemphasized. Perhaps it is fortunate that fluorescence decay curves, obtained by pulse techniques, are convolution products of the impulse response and the excitation function. For this reason, it was recognized in the very early days of nanosecond fluorometry that linearization of data by means of logarithmic transformation was not an appropriate method for distinguishing mechanisms or obtaining decay constants. The search for reasonable ways for "deconvolving" the data led to considerable interest in the analysis of fluorescence decay data. We now know that convolution is not the most important problem at all.

The determination of decay parameters from functions that are sums of exponentials is not a well-conditioned problem. The investigator is faced with the task of distinguishing between models and quantitatively recovering rate parameters. This task is complicated by the presence of a variety of systematic errors in the data; convolution is one example.

Most workers in the field (78) use the method of nonlinear least squares (153, 154) for data analysis. Isenberg and his colleagues (155, 156) advocate a transform approach: the method of moments. Data analysis of fluorescence decay data has been reviewed elsewhere (17) and will not be covered in detail here. We will, however, point out some recent advances in the analysis of both total intensity and emission anisotropy decay data.

It is not uncommon for the time-resolved anisotropy or total intensity decay of a protein to be defined by a function containing six or eight independent parameters. To evaluate such a system (or a more complex case), it is desirable to perform experiments under a variety of conditions. For example, the excitation or emission wavelength, polarizer orientation, pH, or temperature might be varied. Simultaneous analysis of multiple fluorescence decay experiments with the use of transform methods (87, 157), or by nonlinear least squares (158, 159, 160) facilitates the ability to distinguish between different mechanisms and to resolve decay constants in a complex system. The nonlinear least squares approach (159, 160) is completely general and allows one to associate any given parameter in a single experiment with any given parameter in another experiment. Therefore instead of sequentially analyzing two-dimensional slices of a fluorescence decay surface, one can analyze the entire multidimensional experimental surface for an internally consistent set of decay parameters. Model testing is imposed directly on the data instead of on the recovered parameters from individual experiments.

For total intensity data, the choice of models is relatively limited: sums of exponentials or square root of time decay laws. The decay of the anisotropy is inherently more difficult to interpret. The form of the decay of the anisotropy in isotropic systems for any assymetric body (161–164) is a sum of five exponentials. For the case of ellipsoids of revolution, this reduces to a sum of three exponentials. Ideally, from anisotropy decay measurements, one could recover all three rotational correlation times and therefore determine the exact hydrodynamic shape of the protein examined. As we have seen however, tryptophan in proteins may show rotational freedom independent of the protein; therefore, the anisotropy decay must have additional term(s) for this local motion.

Improved instruments and more sophisticated computational methods are now available. It can be predicted that the next review on time-resolved studies of proteins will describe detailed mechanistic studies of proteins not only in solution but also within biological membranes.

ACKNOWLEDGMENTS

Work in our laboratory is supported by NIH grant No. GM11632. We thank Julie Kang for help with the manuscript and Nancy Beechem for graphics. J. M. B. supported by NIH training grant HD07103. We thank Drs. Jay Knutson and Marcel Ameloot for helpful discussions.

Literature Cited

1. Weber, G. 1953. In *Advances in Protein Chemistry*, ed. M. L. Anson, K. Bailey, J. T. Edsall, 8:415–59. New York: Academic
2. Bowman, R. L., Caufield, P. A., Udenfriend, S. 1955. *Science* 122:32–33
3. Shore, V. G., Pardee, A. B. 1956. *Arch. Biochem. Biophys.* 60:100–7
4. Duggan, D., Udenfriend, S. 1956. *J. Biol. Chem.* 223:313–19
5. Konev, S. V. 1957 *Some features of photochemical transformations in biological systems.* PhD thesis. Moscow (In Russian)
6. Vladimirov, Yu. A., Konev, S. V. 1960. *Biofizika* 5:385–88
7. Brand, L., Everse, J., Kaplan, N. O. 1962. *Biochemistry* 1:423–34
8. Schechter, A. N., Chen, R. F., Anfinsen, C. B. 1970. *Science* 167:886–87
9. Tran, C. D., Beddard, G. S., Osborne, A. D. 1982. *Biochim. Biophys. Acta* 709:256–64
10. Burstein, E. A., Permyakov, E. A., Yashin, V. A., Burkhanov, S. A., Finazzi-Agro, A. 1977. *Biochim. Biophys. Acta* 491:155–59
11. Komiyama, T., Miwa, M. 1980. *J. Biochem.* 87:1029–36
12. Maliwal, B. P., Lakowicz, J. R. 1984. *Biophys. Chem.* 19:337–44
13. Chen, R. F. 1967. In *Fluorescence: Theory, Instrumentation, and Practice*, ed. G. G. Guilbault, pp. 443–509. New York: Dekker
14. Lakowicz, J. R. 1983. *Principles of Fluorescence Spectroscopy.* New York: Plenum
15. Longworth, J. W. 1983. See Ref. 17, pp. 651–725
16. Kronman, M. J. 1976. In *Biochemical Fluorescence: Concepts*, ed. R. F. Chen, H. Edelhoch, 2:487–514. New York: Dekker
17. Cundall, R. B., Dale, R. E., eds. 1983. *Time-Resolved Fluorescence Spectroscopy in Biochemistry and Biology.* New York: Plenum
18. Creed, D. 1984. *Photochem. Photobiol.* 39:537–62
19. Creed, D. 1984. *Photochem. Photobiol.* 39:563–75
20. Weinryb, I., Steiner, R. F. 1971. In *Excited States of Proteins and Nucleic Acids*, ed. R. F. Steiner, I. Weinryb, pp. 277–318. New York: Plenum
21. Valeur, B., Weber, G. 1977. *Photochem. Photobiol.* 25:441–44

22. Cross, A. J., Waldeck, D. H., Fleming, G. R. 1983. *J. Chem. Phys.* 78:6455–67
23. Strickland, E. H., Horwitz, J., Billups, C. 1970. *Biochemistry* 9:4914
24. Meech, S. R., Phillips, D., Lee, A. G. 1983. *Chem. Phys.* 80:317–28
25. Rayner, D. M., Szabo, A. G. 1978. *Can. J. Chem.* 56:743–45
26. Lakowicz, J. R., Balter, A. 1982. *Photochem. Photobiol.* 36:125–32
27. Lakowicz, J. R., Cherek, H. 1980. *J. Biol. Chem.* 255:831–34
28. Hopkins, T. R., Lumry, R. 1972. *Photochem. Photobiol.* 15:555–56
29. Tatischeff, I., Klein, R., Zemb, T., Duquesne, M. 1978. *Chem. Phys. Lett.* 54:394–98
30. Lasser, N., Feitelson, J., Lumry, R. 1977. *Isr. J. Chem.* 16:330–34
31. Hershberger, M. V., Lumry, R., Verrall, R. 1981. *Photochem. Photobiol.* 33: 609–17
32. Skalski, B., Rayner, D. M., Szabo, A. G. 1980. *Chem. Phys. Lett.* 70:587–90
33. Gonzalo, I., Escudero, J. L., Lopez-Campillo, A. 1984. *J. Lumin* 29:55–64
34. Bent, D. V., Hayon, E. 1975. *J. Am. Chem. Soc.* 97:2612–19
35. Feitelson, J. 1971. *Photochem. Photobiol.* 13:87–96
36. Amouyal, E., Bernas, A., Grand, D., Mialocq, J. C. 1982. *Faraday Discuss. Chem. Soc.* 1982(74):147–59
37. Klein, R., Tatischeff, I., Bazin, M., Santus, R. 1981. *J. Phys. Chem.* 85:670–77
38. Grossweiner, L. I., Brendzel, A. M., Blum, A. 1981. *Chem. Phys.* 57:147–55
39. Pigault, C., Hasselman, C., Laustriat, G. 1982. *J. Phys. Chem.* 86:1755–57
40. Robbins, R. J., Fleming, G. R., Beddard, G. S., Robinson, G. W., Thistlethwaite, P. J., Woolfe, G. J. 1980. *J. Am. Chem. Soc.* 102:6271–79
41. Lami, H. 1977. *J. Chem. Phys.* 67:3274–81
42. Bazin, M., Patterson, L. K., Santus, R. 1983. *J. Phys. Chem.* 87:189–90
43. Ricci, R. W., Nesta, J. M. 1976. *J. Phys. Chem.* 80:974–80
44. Truong, T. B. 1980. *J. Phys. Chem.* 84:960–64
45. Chang, M. C., Petrich, J. E., McDonald, D. B., Fleming, G. R. 1983. *J. Am. Chem. Soc.* 105:3819–24
46. Mulliken, R. S. 1950. *J. Am. Chem. Soc.* 72:600–8
47. Mulliken, R. S. 1952. *J. Am. Chem. Soc.* 74:811–24
48. Werner, T. C., Forster, L. S. 1979. *Photochem. Photobiol.* 29:905–14
49. Santus, R., Grossweiner, L. I. 1972. *Photochem. Photobiol.* 15:101–5
50. Bryant, F. 1975. *J. Phys. Chem.* 79:2711
51. Kirby, E. P., Steiner, R. F. 1970. *J. Phys. Chem.* 74:4480–90
52. White, A. 1959. *Biochem. J.* 71:217–20
53. Cowgill, R. W. 1963. *Arch. Biochem. Biophys.* 100:36–44
54. Konev, S. V. 1967. *Fluorescence and Phosphorescence of Proteins and Nucleic Acids.* New York: Plenum
55. VanderDonckt, E. 1969. *Bull. Soc. Chim. Belg.* 78:69–75
56. Edelhoch, H., Brand, L., Wilchek, M. 1967. *Biochemistry* 6:547–59
57. Cowgill, R. W. 1963. *Biochim. Biophys. Acta* 75:272–73
58. De Lauder, W. B., Wahl, P. 1970. *Biochemistry* 9:2750–54
59. Alpert, B., Jameson, D. M., Lopez-Delgado, R., Schooley, R. 1979. *Photochem. Photobiol.* 30:479–81
60. Georghiou, S., Thompson, M., Mukhopadhyay, A. K. 1982. *Biochim. Biophys. Acta* 688:441–52
61. Szabo, A. G., Rayner, D. M. 1980. *Biochem. Biophys. Res. Commun.* 94: 909–15
62. Szabo, A. G., Rayner, D. M. 1980. *J. Am. Chem. Soc.* 102:554–63
63. Ross, J. B., Rousslang, K. W., Brand, L. 1981. *Biochemistry* 20:4361–69
64. Petrich, J. W., Chang, M. C., McDonald, D. B., Fleming, G. R. 1983. *J. Am. Chem. Soc.* 105:3824–32
65. Donzel, B., Gauduchon, P., Wahl, P. 1974. *J. Am. Chem. Soc.* 96:801–8
66. Gauduchon, P., Wahl, P. 1978. *Biophys. Chem.* 8:87–104
67. Dezube, B., Dobson, C. M., Teague, C. E. 1981. *J. Chem. Soc. Perkin Trans. 2* 1981:730–35
68. Takigawa, T., Ashida, T., Sasada, Y., Kakudo, M. 1966. *Bull. Chem. Soc. Jpn.* 39:2369
69. Newmark, R. A., Miller, M. A. 1971. *J. Phys. Chem.* 75:505–8
70. Hochstrasser, R. M. Personal communication
71. Janin, J., Wodak, S., Levitt, M., Maigret, B. 1978. *J. Mol. Biol.* 125:357–86
72. Bhat, T. N., Sasisekharan, V., Vijayan, M. 1979. *Int. J. Pept. Protein Res.* 13:170–84
73. Kobayashi, J., Higashijima, T., Sekido, S., Miyazawa, T. 1981. *Int. J. Pept. Protein Res.* 17:486–94
74. Cavanaugh, J. R. 1970. *J. Am. Chem. Soc.* 92:1488–93
75. Gudgin, E., Lopez-Delgado, R., Ware, W. R. 1981. *Can. J. Chem.* 59:1037–44
76. Gudgin, E., Lopez-Delgado, R., Ware, W. R. 1983. *J. Phys. Chem.* 87:1559–65
77. Ricci, R. W. 1970. *Photochem. Photobiol.* 12:67–75

78. Beddard, G. S., Fleming, G. R., Porter, G., Robbins, R. J. 1980. *Philos. Trans. R. Soc. London A* 298:321–34
79. Saito, I., Sugiyama, H., Yamamoto, A., Muramatsu, S., Matsuura, T. 1984. *J. Am. Chem. Soc.* 106:4286–87
80. Cowgill, R. W. 1970. *Biochim. Biophys. Acta* 200:18–25
81. Tournon, J., Kuntz, E., El-Bayoumi, M. A. 1972. *Photochem. Photobiol.* 16:425–33
82. Kainosho, M., Ajisaka, K., Kamisaku, M., Murai, A. 1975. *Biochem. Biophys. Res. Commun.* 64:425–32
83. Hansen, P. E., Feeney, J., Roberts, G. C. K. 1975. *J. Magn. Reson.* 17:249–61
84. Chen, R. F., Vurek, G. G., Alexander, N. 1967. *Science* 156:949–51
85. De Lauder, W. B., Wahl, P. 1971. *Biochem. Biophys. Res. Commun.* 42:398–404
86. Grinvald, A., Steinberg, I. Z. 1976. *Biochim. Biophys. Acta* 427:663–78
87. Wahl, P., Auchet, J. C. 1972. *Biochim. Biophys. Acta* 285:99–117
88. Brochon, J. C., Wahl, P., Charlier, M., Maurizot, J. C., Helene, C. 1977. *Biochem. Biophys. Res. Commun.* 79:1261–71
89. Knutson, J. R., Walbridge, D. G., Brand, L. 1982. *Biochemistry* 21:4671–79
90. Sommer, H., Lu, P., Miller, J. H. 1976. *J. Biol. Chem.* 251:3774–79
91. Boyer, P. D., Theorell, H. 1956. *Acta Chem. Scand. Ser. A* 10:447–50
92. Branden, C. I., Jornvall, H., Eklund, H., Furugren, B. 1975. *Enzymes* 11:103–90
93. Ross, J. B. A., Schmidt, C. J., Brand, L. 1981. *Biochemistry* 20:4369–77
94. Abdallah, M. A., Biellmann, J. F., Wiget, P., Joppich-Kuhn, R., Luisi, P. L. 1978. *Eur. J. Biochem.* 89:397–405
95. Eklund, H., Nordstrom, B., Zeppezauer, E., Soderland, G., Boiwe, T., et al. 1976. *J. Mol. Biol.* 102:27–59
96. Eftink, M. R., Selvidge, L. A. 1982. *Biochemistry* 21:117–25
97. Eftink, M. R., Jameson, D. M. 1982. *Biochemistry* 21:4443–49
98. Beechem, J. M., Knutson, J. R., Ross, J. B. A., Turner, B. W., Brand, L. 1983. *Biochemistry* 22:6054–58
99. Privat, J. P., Wahl, P., Auchet, J. C. 1980. *Biophys. Chem.* 11:239–48
100. Grinvald, A., Schlessinger, J., Pecht, I., Steinberg, I. Z. 1975. *Biochemistry* 14:1921–29
101. Szabo, A. G., Stepanik, T. M., Wayner, D. M., Young, N. M. 1983. *Biophys. J.* 411:233–44
102. Provencher, S. W., Dovi, V. G. 1979. *J. Biochem. Biophys. Methods* 1:313–18
103. Yashinsky, G. Y. 1972. *FEBS Lett.* 26:123–26
104. Formoso, C., Forster, L. S. 1975. *J. Biol. Chem.* 250:3738–45
105. Katsuya, H., Imoto, T., Gunki, F., Funatsu, M. 1965. *J. Biochem.* 58:227–35
106. Hartdegan, F. J., Rupley, J. A. 1964. *Biochim. Biophys. Acta* 92:625–27
107. Imoto, T., Forster, L. S., Rupley, J. A., Tanaka, F. 1971. *Proc. Natl. Acad. Sci. USA* 69:1151–55
108. Lowey, S., Luck, S. M. 1969. *Biochemistry* 8:3195–99
109. Morita, F. 1967. *J. Biol. Chem.* 242:4501–6
110. Werber, M. M., Szent-Györgyi, A. G., Fasman, G. D. 1972. *Biochemistry* 11:2872–82
111. Torgerson, P. M. 1984. *Biochemistry* 23:3002–7
112. Torikata, T., Forster, L. S., O'Neal, C. C. Jr., Rupley, J. A. 1979. *Biochemistry* 18:385–90
113. Warburg, O., Negelein, E. 1928. *Biochem. Z.* 193:339–46
114. Weber, G., Teale, F. J. W. 1959. *Discuss. Faraday Soc.* 27:134–41
115. Alpert, B., Lopez-Delgado, R. 1976. *Nature* 263:445–46
116. Alpert, B., Jameson, D. M., Weber, G. 1980. *Photochem. Photobiol.* 31:1–4
117. Lakowicz, J. R., Weber, G. 1973. *Biochemistry* 12:4171–79
118. Hochstrasser, R. M., Negus, D. K. 1984. *Proc. Natl. Acad. Sci. USA* 81:4399–403
119. Anderson, S. R., Brunori, M., Weber, G. 1970. *Biochemistry* 9:4723–29
120. Irace, G., Balestrieri, C., Parlato, G., Servillo, L., Colonna, G. 1981. *Biochemistry* 20:792–99
121. Hirsch, R. E., Zukin, R. S., Nagel, R. L. 1980. *Biochem. Biophys. Res. Commun.* 93:432–39
122. Hirsch, R. E., Nagel, R. L. 1981. *J. Biol. Chem.* 256:1080–83
123. Hirsch, R. E., Squires, N. A., Discepola, C., Nagel, R. L. 1983. *Biochem. Biophys. Res. Commun.* 116:712–18
124. Itoh, M., Mizukoshi, H., Fuke, K., Matsukawa, S., Mawatari, K., et al. 1981. *Biochem. Biophys. Res. Commun.* 100:1259–65
125. Mizukoshi, H., Itoh, M., Matsukawa, S., Mawatari, K., Yoneyama, Y. 1982. *Biochim. Biophys. Acta* 700:143–47
126. Szabo, A. G., Krajcarski, D., Zuker, M., Alpert, B. 1984. *Chem. Phys. Lett.* 108:145–49
127. Wahl, P. 1979. *Biophys. Chem.* 10:91–104
128. Cross, A. J., Fleming, G. R. 1984. *Biophys. J.* 46:45–56

129. Badea, M. G., Brand, L. 1979. *Methods Enzymol.* 61:378–425
130. Yguerabide, J., Epstein, H. F., Stryer, L. 1970. *J. Mol. Biol.* 51:573–90
131. Hanson, D. C., Yguerabide, J., Schumaker, V. N. 1981. *Biochemistry* 20: 6842–52
132. Koshland, D. E. Jr., Neet, K. E. 1968. *Ann. Rev. Biochem.* 37:359–410
133. Karplus, M., McCammon, J. A. 1981. *Crit. Rev. Biochem.* 9:293–349
134. McCammon, J. A., Karplus, M. 1983. *Acc. Chem. Res.* 16:187–93
135. Lakowicz, J. R., Weber, G. 1973. *Biochemistry* 12:4171–79
136. Lakowicz, J. R., Maliwal, B. P., Cherek, H., Balter, A. 1983. *Biochemistry* 22:1741–52
137. McCammon, J. A., Wolynes, P. G., Karplus, M. 1979. *Biochemistry* 18:928–41
138. Levy, R. M., Szabo, A. 1982. *J. Am. Chem. Soc.* 104:2073–75
139. Ichiye, T., Karplus, M. 1983. *Biochemistry* 22:2884–93
140. Munro, I., Pecht, I., Stryer, L. 1979. *Proc. Natl. Acad. Sci. USA* 76:56–60
141. Jonas, A., Privat, J., Wahl, P., Osborne, J. C. Jr. 1982. *Biochemistry* 21:6205–11
141a. Lux, B., Helynck, G., Trifilieff, E., Luu, B., Gerard, D. 1984. *Biophys. Chem.* 19:345–53
142. Brochon, J. C., Wahl, P., Auchet, J. C. 1974. *Eur. J. Biochem.* 41:577–83
143. Neyroz, P., Beechem, J. M., Brand, L., Roseman, S. 1984. *Photochem. Photobiol.* 39:41S
143a. Kukuruzinska, M. A., Turner, B. W., Ackers, G. K., Roseman, S. 1984. *J. Biol. Chem.* 259:11679–81
144. Small, E. W., Libertini, L. J., Isenberg, I. 1984. *Rev. Sci. Instrum.* 55:879–85
145. Doukas, A. G., Buchert, J., Alfano, R. R. 1982. In *Biology Probed by Ultrafast Laser Spectroscopy*, ed. R. R. Alfano, New York: Academic
146. Van Resandt, R. W. W., Vogel, R. H., Provencher, S. W. 1982. *Rev. Sci. Instrum.* 53:1392–97
147. Shapiro, S. L., ed. 1977. *Topics in Applied Physics*, 18:1–15. Heidelberg: Springer-Verlag
148. Nordlund, T. M., Podolski, D. A. 1983. *Photochem. Photobiol.* 38:665–69
149. Gratton, E., Limkeman, M. 1983. *Biophys. J.* 44:315–24
150. Jameson, D. M., Gratton, E., Hall, R. D. 1984. *Appl. Spectrosc. Rev.* 20:55–106
151. Gratton, E. I., Jameson, D. M., Hall, R. D. 1984. *Ann. Rev. Biophys. Bioeng.* 11:105–24
152. Lakowicz, J. R., Maliwal, B. P. 1984. *Biophys. Chem.* In press
153. Grinvald, A., Steinberg, I. Z. 1974. *Anal. Biochem.* 59:583–98
154. Knight, A. E. W., Selinger, B. K. 1971. *Spectrochim. Acta Part A* 27:1223–34
155. Isenberg, I., Small, E. W. 1982. *J. Chem. Phys.* 77:2799–805
156. Isenberg, I. 1983. *Biophys. J.* 43:141–48
157. Eisenfeld, J., Ford, C. C. 1979. *Biophys. J.* 26:73–84
158. Knorr, F. J., Harris, J. M. 1981. *Anal. Chem.* 53:272–76
159. Knutson, J. R., Beechem, J. M., Brand, L. 1983. *Chem. Phys. Lett.* 102:501–7
160. Beechem, J. M., Knutson, J. R., Ross, J. B. A., Turner, B. W., Brand, L. 1983. *Biochemistry* 22:6054–58
161. Tao, T. 1969. *Biopolymers* 8:609–32
162. Chuang, T. J., Eisenthal, K. B. 1972. *J. Chem. Phys.* 57:5094–5097
163. Ehrenberg, M., Rigler, R. 1972. *Chem. Phys. Lett.* 14:539–44
164. Small, E. W., Isenberg, I. 1977. *Biopolymers* 16:1907–28
165. Cockle, S. A., Szabo, A. G. 1981. *Phochem. Photobiol.* 34:23–27
166. Van Hoek, A., Vervoort, J., Visser, A. J. W. G. 1983. *J. Biochem. Biophys. Methods* 7:243–54
167. Ludescher, R. D., Volwerk, J. J., de Haas, G. H., Hudson, B. S. 1984. *Biochemistry.* Submitted
168. Lee, J., O'Kane, D. J., Visser, A. J. W. G. 1984. *Biochemistry.* In press
169. Zukin, R. S. 1979. *Biochemistry* 18: 2139–45
170. Gafni, A., Werber, M. M. 1979. *Arch. Biochem. Biophys.* 196:363–70
171. Masotti, L., Cavatorta, P., Spisni, A., Casali, E., Sartor, G., et al. 1983. In *Structure and Function of Membrane Proteins*, ed. E. Quagliariello, F. Palmieri, pp. 3–10. Amsterdam: Elsevier
172. Brochon, J. C., Wahl, P., Jallon, J. M., Iwatsubo, M. 1976. *Biochemistry* 15: 3259–65
173. Grinvald, A., Steinberg, I. Z. 1974. *Biochemistry* 13:5170–78
174. Berkhout, T. A., Visser, A. J. W. G., Wirtz, K. W. A. 1984. *Biochemistry* 23:1505–13
175. Bandyopadhyay, P. K., Wu, F. Y. H., Wu, C. W. 1981. *J. Mol. Biol.* 145:375–404

Ann. Rev. Biochem. 1985. 54:73–100

ADP-RIBOSYLATION

Kunihiro Ueda

Department of Medical Chemistry, Kyoto University Faculty of Medicine, Kyoto 606, Japan

Osamu Hayaishi

Osaka Medical College, Takatsuki, Osaka 569, Japan

CONTENTS

PERSPECTIVES AND SUMMARY

ADP-ribosylation encompasses a group of posttranslational modifications of proteins that utilize NAD as a donor of their modification group (1). Since the discovery of poly(ADP-ribose) synthesis in animal tissues in 1966 (2–4), the ADP-ribosylation family has expanded into almost all forms of life and almost all compartments of the cell. ADP-ribosylation may now be considered as one

73

0066-4154/85/0701-0073$02.00

of the most general ways by which living organisms modify their protein structures and functions (5, 6). ADP-ribosylation reactions are classified into two major groups: mono(ADP-ribosyl)ation and poly(ADP-ribosyl)ation. These two groups are different not solely in the length of the ADP-ribose chain but also in the chemical nature of the ADP-ribosyl protein bond (N-glycoside vs O-glycoside), the character of enzyme (many microbial toxins vs eukaryotic enzymes), and the site of reaction (cytoplasm and cell membrane vs nucleus). The biological significance of each reaction is better understood for the members of mono(ADP-ribosyl)ation group than those of poly(ADP-ribosyl)ation group. Recently, however, data have been rapidly accumulating that suggest close relationships between poly(ADP-ribosyl)ation and many important biological events such as DNA repair, the cell cycle, cellular differentiation, and oncogenesis. The molecular mechanisms of the relationships are being unveiled (5).

Because of the rapid progress in this field of research, earlier reviews (1, 7–13) may no longer be informative enough for biologists or biochemists. Recent reviews (13–20) and the proceedings of the last two international symposia (21, 22) are of value as references to specific topics treated therein. This article was undertaken to summarize important findings in wider aspects of ADP-ribosylation reported since our last review in this series (1) and to illuminate possible biological significance of some of the latest observations. The bibliography was filed as of June, 1984. For details of reports published before 1982, readers are referred to a monograph edited by the authors (23).

MONO(ADP-RIBOSYL)ATION

Various mono(ADP-ribosyl)ation reactions so far reported are listed in Table 1. It is evident that many of the ADP-ribosyltransferases are microbial toxins with targets (acceptors) in eukaryotic cells. Recently, it has been found that the same acceptors as used by toxins may be modified by the enzymes of host-cell origin (40, 41). Furthermore, it has been shown that mono(ADP-ribosyl)ation predominates by far (>10 fold) over poly(ADP-ribosyl)ation in mammalian cells, particularly in extranuclear compartments (14, 49). These findings point to the importance of mono(ADP-ribosyl)ation in eukaryotic cells under physiological conditions. According to the amino acids that accommodate ADP-ribosylation, mono(ADP-ribosyl)ation reactions are classified into four subtypes: diphthamide-, arginine-, and asparagine-specific enzymatic ADP-ribosylations, and lysine-specific nonenzymatic ADP-ribosyl adduct (Schiff base) formation (50). It is noteworthy that, in spite of the difference in acceptor amino acids, all mono(ADP-ribosyl)ation reactions involve side-chain nitrogen atoms and produce N-glycosides (Figure 1). This is in contrast to nuclear poly(ADP-ribosyl)ation, which starts with mono(ADP-ribosyl)ation on carboxyl groups to form O-glycosides (Figure 1). N- and O-glycosides are most conveniently, though less definitely, distinguished by differential sensitivities to neutral NH$_2$OH (51). A series of analyses carried out by Hilz and coworkers (14, 52, 53) revealed the existence of both NH$_2$OH-sensitive and resistant mono(ADP-ribosyl) protein conjugates and their independent changes in various tissues and cells. These results argued against a tacitly accepted notion that, in eukaryotes, mono(ADP-ribosyl) proteins simply provide the initiation sites for elongation or represent the products of poly(ADP-ribosyl) protein degradation, and strongly argued that different classes of mono(ADP-ribosyl) proteins have different origins, fates, and probably functions (14).

Diphthamide-Specific ADP-Ribosylation of Elongation Factor 2

ADP-ribosyltransferases documented to utilize diphthamide as the acceptor amino acid are diphtheria toxin (24, 54), *Pseudomonas aeruginosa* exotoxin A (25), and a recently discovered mammalian cytosolic enzyme (40). The finding that the forward and reverse reactions were catalyzed interchangeably by these three enzymes (40, 55) indicated that these enzymes modify the same site of the same acceptor.

DIPHTHERIA TOXIN   Mono(ADP-ribosyl)ation of the specific acceptor, elongation factor 2 (EF-2), by diphtheria toxin was first demonstrated by Honjo et al (24) and confirmed soon afterwards by others (56, 57). EF-2 is a protein

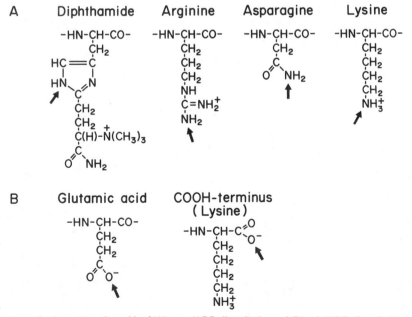

Figure 1   Acceptor amino acids of (A) mono(ADP-ribosyl)ation and (B) poly(ADP-ribosyl)ation.

Table 1 Natural distribution of various mono(ADP-ribosyl)ation reactions

Enzyme	Acceptor Protein	Amino acid	Reference
Prokaryotes			
Diphtheria toxin	elongation factor 2	diphthamide	24
Pseudomonas aeruginosa toxin A	elongation factor 2	diphthamide	25
T4 phage *mod* protein	*E. coli* RNA polymerase α subunit and other proteins	arginine	26
T4 phage *alt* protein	*E. coli* RNA polymerase α subunit and other proteins	arginine (?)	27
Cholera toxin	N$_s$ protein of adenylate cyclase complex	arginine (?)	28, 29
	microtubule proteins	arginine (?)	30, 31
	transducin	arginine	32
Escherichia coli enterotoxin LT	N$_s$ protein of adenylate cyclase complex (?)	arginine (?)	33, 34
Pertussis toxin (Islet-activating protein)	N$_i$ protein of adenylate cyclase complex	asparagine (?)	35
	transducin	asparagine	36
Pseudomonas aeruginosa exoenzyme S	elongation factor 1 (?) and associated proteins	unknown	37
N$_4$ phage	*E. coli* proteins	unknown	38
Escherichia coli (non-infected) enzyme	*E. coli* proteins	unknown	39
Eukaryotes			
Mammalian cytosolic enzyme	elongation factor 2	diphthamide	40
Avian erythrocyte enzyme	N$_s$ protein (?) and other soluble proteins	arginine (?)	41
Mammalian cytoplasmic enzyme	soluble proteins	arginine (?)	42
Mammalian membrane enzyme	N$_s$ protein and other membrane proteins	arginine (?)	43, 44, 45
Mammalian mitochondrial enzyme	mitochondrial inner membrane protein	arginine (?)	46, 47
Avian nuclear enzyme	nuclear proteins	arginine (?)	48

component required for polypeptide chain elongation on ribosomes in eukaryotic cells. ADP-ribosylation accompanies inactivation of its translocase activity, thereby inhibiting protein synthesis and ultimately causing cell death. EF-2 from a wide variety of eukaryotes ranging from mammals to yeast (58), and also certain archaebacteria (59), proved to be active as substrate. Diphtheria toxin is produced by *Corynebacterium diphtheriae* lysogenic for corynephage β carrying the *tox* gene. After secretion as a single polypeptide, the toxin is cleaved by co-secreted proteases to produce disulfide-linked A (active ADP-ribosyltransferase) and B (binding subunit) fragments. For these earlier findings, readers are referred to previous reviews (1, 8, 60–63). Recently, the primary structures of the toxin (535 amino acids, mol wt = 58,342) as well as fragment A and B were determined (64, 65). Functional structures of diphtheria toxin have been analyzed in detail by Collier and associates (63). The cellular receptor for diphtheria toxin (66) and the mechanism by which fragment B enables the A entry (67) are in controversy.

The target amino acid in EF-2 was identified by Bodley and coworkers (54) as a modified histidine, 2-[3-carboxy-amido-3-(trimethylammonio)propyl]-histidine, which is also termed "diphthamide" (Figure 1). Oppenheimer & Bodley (68) showed that the ADP-ribosylation accompanied anomeric inversion from β (NAD) to α (ADP-ribosyl) at C-1 of the ribose. The complex structure of diphthamide proved to be generated via a series of elaborate posttranslational modifications (69).

PSEUDOMONAS TOXIN   Exotoxin A produced by *P. aeruginosa* was found by Iglewski & Kabat (25) to catalyze exactly the same reaction as diphtheria toxin does, although the two toxins are quite different in many aspects (70, 71). Although exotoxin A is secreted as an enzymatically inactive and toxic single polypeptide (613 amino acids; mol wt = 66,583) (71), functionally distinct "A" (enzymatically active) and "B" (receptor binding) domains could be distinguished in vitro (70). Neither the in vivo processing mechanism nor the structure of specific receptor(s) for exotoxin A has yet been elucidated (70). Available data support the hypothesis that exotoxin A may play a role in certain types of *P. aeruginosa* infections (70).

MAMMALIAN CYTOSOLIC ENZYMES   Recently, an endogenous ADP-ribosyltransferase that catalyzed the reaction apparently identical with that of diphtheria toxin or *P. aeruginosa* exotoxin A was discovered by Lee & Iglewski (40) in polyoma virus-transformed baby hamster kidney cells. It is noteworthy that this enzyme activity was inhibited by a charcoal-adsorbed cytoplasmic extract as well as by histamine (cf 72). A similar enzyme activity was isolated from beef liver (72a). Because such an enzyme activity must be under stringent control or it would cause cell death, it seems plausible that the interaction of the cellular ADP-ribosyltransferase and its inhibitor is a rather ubiquitous mechanism of controlling protein synthesis at the level of functional EF-2 (40).

Arginine-Specific ADP-Ribosylation and Guanine Nucleotide-Binding Proteins

Three kinds of ADP-ribosyltransferases modifying arginine residues are known: phage-encoded proteins (73), bacterial enterotoxins (10, 11, 15, 74, 75), and vertebrate enzymes (76) (Table 1).

PHAGE PROTEINS   The *alt* protein of coliphage T4 (27) and the *mod* protein induced in *Escherichia coli* by T4 infection (26, 77) were previously reviewed (1). ADP-ribosyltransferase isolated from noninfected *E. coli* cells (39) modified various proteins of *E. coli*. The reported chemical stability of the ADP-ribosyl protein bond produced by this enzyme suggested its N-glycosidic nature.

ENTEROTOXINS   Cholera toxin (choleragen) (28, 29) and *E. coli* heat-labile enterotoxin LT (23) have been shown to possess ADP-ribosyltransferase activities. These two toxins resemble each other strongly in many aspects, including subunit structure (AB_5); molecular weights of whole toxin (82,000/84,000), subunits (A, 25,000–28,000; B, ~11,500) and active fragment A_1 (22,000/24,000); amino acid sequence and genomic organization; cellular binding site (ganglioside G_{M_1}); a clinical symptom (watery diarrhea/"traveller's diarrhea"); and biological actions (activation of adenylate cyclase) (11, 75, 78). Furthermore, the mechanism of the cyclase activation was found to be identical, i.e. through ADP-ribosylation of the α subunit (mol wt = 42,000) of a guanine nucleotide regulatory protein, N (also termed G or G/F), in the adenylate cyclase complex (28, 29, 33, 79). The N protein is a component mediating signal transmission from a ligand (hormone)/receptor complex to the cyclase catalytic unit at the expense of GTP hydrolysis (80, 81). ADP-ribosylation of the α subunit resulted in inactivation of its GTPase activity, which converted N to a permanently activated (GTP-bound) state (11, 74, 82, 83). Besides this N, which is currently known as N_s (stimulatory), another N, N_i (inhibitory), was recently found to be involved in cyclase regulation (83). This new regulatory protein proved to be ADP-ribosylated by another toxin, pertussis toxin (35), as will be described below. N_s (mol wt $\approx$ 80,000) was recently purified to apparent homogeneity from rabbit liver and turkey erythrocytes, and shown to consist of three dissimilar subunits, α, β, and γ (83). Activation of N_s by GTP or ADP-ribosylation was effected by dissociation of α and β (inhibitor of α) subunits (84). ADP-ribosylation of N_s by cholera toxin required GTP and a cytosolic macromolecule (85), presumably, for induction of a conformational change of N_s (86). A protein apparently related to the cytosolic factor was recently purified from plasma membranes (87).

Another acceptor recently identified is transducin (32). This retinal protein is also a guanine nucleotide-binding regulatory protein that mediates signals, i.e. activation of a cyclic GMP-specific phosphodiesterase by photolyzed rhodopsin (88, 89). Transducin was purified to apparent homogeneity, and shown to be structurally and functionally homologous to N_s and N_i (32, 88–91). In fact, the β subunit appeared to be identical among the three regulatory proteins (92). Furthermore, the α subunit of transducin was ADP-ribosylated, as is N_s, by cholera toxin, and its GTPase activity was inactivated (32). Analysis of a tryptic digest of ADP-ribosylated transducin indicated that the ADP-ribose was linked to the guanidinium group of arginine in the peptide Ser-Arg-Val-Lys (93).

The ADP-ribosyltransferase activity of cholera toxin (and *E. coli* enterotoxin LT) was shown to be associated with an NAD glycohydrolase activity (94, 95). Furthermore, it was shown to utilize low-molecular-weight guanidino compounds such as arginine, arginine methyl ester, agmatine, or guanidine as acceptors (95, 96). The configuration of ADP-ribosyl arginine was identified as α regarding C-1 of ribose (97). Cholera toxin has also been shown to catalyze, at least in vitro, ADP-ribosylation of a variety of proteins such as histone H1, protamine (98), cytoskeletal proteins (tubulin, microtubule-associated proteins, intermediate filament proteins) (30, 31, 99), and glycopeptide hormones (thyrotropin, lutropin, follitropin, human chorionic gonadotropin, corticotropin) (98). A possibility remained that some of the ADP-ribosylations might occur nonenzymatically using free ADP-ribose released by the toxin-linked NAD glycohydrolase activity (50, 74).

VERTEBRATE ENZYMES   The presence of arginine-specific ADP-ribosyltransferases in vertebrates was first demonstrated by Moss & Vaughan (41). The enzyme, found in turkey erythrocytes, modified arginine and its derivatives, and was capable of activating the adenylate cyclase of rat brain in the presence of NAD. The ADP-ribosyltransferase was purified to apparent homogeneity (100). This enzyme existed in two different forms, protomeric (fully active) and oligomeric (relatively inactive). Activation was promoted by chaotropic salts, histones (76), lysolecithin and certain nonionic detergents. Although NAD was clearly the preferred substrate, the enzyme could use NADP (76). The enzyme activity was inhibited by nicotinamide, thymidine, and theophylline (100), as is poly(ADP-ribose) synthetase (101). The ADP-ribosyl histone (or agmatine) bond produced by this enzyme was relatively resistant to both alkali and neutral NH_2OH (102). Similar ADP-ribosyltransferases were thereafter found in rat liver (42), human and rabbit erythrocytes (76), and chicken liver nuclei (48). These enzymes exhibited kinetic and physical properties not unlike those of the turkey erythrocyte enzyme (76). Furthermore, membrane fractions of various mammalian cells were found to possess endogenous ADP-ribosyltransferases (43–45, 103–105).

Findings of these arginine-specific ADP-ribosyltransferases in vertebrates carried two major implications; one was the possible roles in signal transmission, and the other concerned heterogeneity of ADP-ribosyl protein bonds. The findings of stimulation of rat liver adenylate cyclase by the turkey erythrocyte enzyme (41), of partial homology of amino acid sequences between several glycopeptide hormones and cholera toxin (106), of stimulation of ADP-ribosylation by specific hormones (103), and of ADP-ribosylation of apparently the same membrane protein, the α subunit of N_s, by endogenous enzymes and cholera toxin (43, 44, 103; cf 105, 107), all supported the view that the vertebrate enzymes might function as physiological counterparts of cholera toxin or *E. coli* enterotoxin LT. The relatively high stability of the ADP-ribosyl guanidino bond (102) as opposed to the NH_2OH-sensitive and alkali-sensitive ADP-ribosyl carboxylate bond (51, 108) could account for a majority of NH_2OH-resistant and/or alkali-resistant conjugates (14, 52). Recently, a new assay method for arginine-specific ADP-ribosyltransferases was invented using guanylhydrazones and high-performance liquid chromatography (109).

Asparagine-Specific ADP-Ribosylation and Guanine Nucleotide-Binding Proteins

Islet-activating protein, one of the pertussis toxins, is the only known asparagine-specific ADP-ribosyltransferase (35, 36). This protein was isolated from the culture medium of *Bordetelle pertussis* by Ui and associates (110). The purified islet-activating protein potentiated markedly a secretagogue-induced insulin release at a dose of 1 μg in rat or at 1–10 pg/ml in pancreatic islets cultures (111). The enhancement of insulin release was caused by accumulation of cyclic AMP produced by activated adenylate cyclase (111–113). The effect was observed also with rat glioma C6 cells (114). In the glioma system, islet-activating protein was found to inhibit a specific and low-K_m GTPase activity (114). The finding of NAD requirement for the adenylate cyclase activation, as in the case of cholera toxin (28, 29), led to the discovery of ADP-ribosylation of a specific protein (mol wt = 41,000) by islet-activating protein (35, 115–117). As described in the preceding section, the activity of adenylate cyclase in many types of cells is controlled by two guanine nucleotide regulatory proteins, N_s (stimulatory) and N_i (inhibitory). N_i was recently purified from rabbit liver by Gilman and associates (83), and was shown to be composed of three dissimilar subunits, α, β, and γ. The ADP-ribosyl acceptor for islet-activating protein was identified as this α subunit (83). ADP-ribosylation of N_i, like that of N_s, resulted in inhibition of ligand-stimulated GTP hydrolysis (116). The ADP-ribosylation of N_i, however, unlike that of N_2, promoted association, rather than dissociation, of α and β subunits. Islet-activating protein (mol wt = 117,000) consists of five dissimilar subunits (118). The biggest subunit, S-1, is an ADP-ribosyltransferase.

The target amino acid for this transferase was identified with transducin, another acceptor of islet-activating protein-catalyzed ADP-ribosylation (36). As already mentioned, an arginine residue in the α subunit of transducin is ADP-ribosylated by cholera toxin (32). Gilman and coworkers found that the same subunit of transducin was ADP-ribosylated by islet-activating protein, but at a different site (36). Analysis of a tryptic peptide of ADP-ribosylated transducin revealed its primary structure, Glu-Asn-Leu-Lys-Asn(ADP-ribose)-Gly-Leu-Phe (36), which means that ADP-ribosylation occurred on an asparagine residue and formed an N-glycoside. Recently, partial homology of amino acid sequences of the above tryptic peptide of transducin and of the *ras* gene product has been noted (83).

ADP-Ribose·Protein Adducts (Schiff Bases)

Nonenzymatic formation of an adduct of ADP-ribose (or ribose 5-phosphate) and various proteins through Schiff bases at lysine residues was first demonstrated by Kun and associates (50). Ueda et al (119) confirmed the adduct formation and made use of the ADP-ribose·histone H1 adduct as a primer for elongation catalyzed by poly(ADP-ribose) synthetase. The presence of very potent NAD glycohydrolase activities that generate ADP-ribose in membranes of mitochondria (120, 121), and similar efficiencies of NAD and ADP-ribose as the precursor of covalent modification (47) or in suppression of mitochondrial DNA polymerase (120), suggested that the adduct formation may play a significant role in modification of mitochondrial proteins under certain conditions. (See also the next section.)

Mono(ADP-Ribosyl)ation in Mitochondria and Calcium Transport

Analysis of the distribution of subtypes of ADP-ribosyl proteins by Hilz and coworkers (52, 122) showed that most of the mono(ADP-ribosyl) proteins of NH_2OH-resistant form were located in mitochondria in rat liver cells. An ADP-ribosyltransferase activity that appeared to be responsible for these conjugates was found by Adamietz et al (46) in isolated rat liver mitochondria. This enzyme was insensitive to thymidine but was inhibited by arginine methyl ester or other basic compounds. An enzyme activity related, most probably, to this mitochondrial activity was found by Richter and associates (47) in rat liver mitochondria. A prominent acceptor protein (protomer mol wt≈30,000) was identified in the inner mitochondrial membrane (123). Although the enzyme that catalyzed this modification has not been identified yet, the protein modification was shown to be closely related to a hydroperoxide-induced irreversible oxidation of pyridine nucleotides, their hydrolysis and calcium release; all these four phenomena were inhibited by ATP and also by nicotinamide (47, 124). It was also shown that ADP-ribosyl groups bound to submitochondrial particles were rapidly turning over. These results were consistent with the view of a regulatory role of mono(ADP-ribosyl)ation by mitochondrial NAD glycohydrolase (121) in the calcium release mechanism. The identity or relationship of this ADP-ribosylation and "M-band" oligo(ADP-ribosyl)ation (120) has not yet been determined.

POLY(ADP-RIBOSYL)ATION

Various poly(ADP-ribosyl)ation reactions so far known are summarized in Table 2. The fact that a majority (usually >95%) of poly(ADP-ribosyl)ating activity resides in the nucleus in a variety of cells was established by subcellular fractionation and immunohistochemistry (101, 150).

Nuclear Poly(ADP-Ribosyl)ation

Figure 2 illustrates schematically the outline of poly(ADP-ribosyl) protein metabolism. Proposed structures of sugar "X" in the ADP-ribosyl protein lyase product are also presented.

STRUCTURE OF POLY(ADP-RIBOSE)   The primary structure of poly(ADP-ribose) has been well understood (1, 7, 13, 152). Notable features of the polymer, in contrast to DNA and RNA, are the ribose $(1''{\rightarrow}2')$ribose-phosphate-phosphate backbone and the occurrence of branching. When 2'-deoxyNAD was used as a substrate for poly(ADP-ribose) synthesis, a ribose$(1''{\rightarrow}3')$ribose linkage was produced (153–155). The structure of branched portion was determined as ribose $(1'''{\rightarrow}2'')$ribose$(1''{\rightarrow}2')$ribose by Miwa and coworkers (152, 156, 157). The anomeric configuration of all ribose→ribose bonds was determined as α (157–160). The branched structure of poly(ADP-ribose) was recently observed with electron microscopy (161, 162). The frequency of branching was estimated as once per 30–50 ADP-ribose residues (157, 163). The polymer isolated from nature also had branches (164, 165). A higher-order structure of poly(ADP-ribose) was earlier suggested by separation of two forms, H (heavy) and L (light), with different sedimentation velocities, buoyant densities, solubilities in salt solutions, and susceptibilities to enzyme actions (7, 152). A double-strand model was constructed by Gill (166). Recently, Minaga & Kun (167, 168) showed the dependency of hypochromicity and/or circular dichroism θ values on chain lengths of oligo(ADP-ribose), on temperatures, and on salts, and postulated a helical conformation of long-chain poly(ADP-ribose).

ISOLATION AND QUANTITATION OF POLY- AND OLIGO(ADP-RIBOSE) Chromatography on hydroxyapatite (169–171) or DEAE-cellulose (172), gel filtration (167), and electrophoresis (169, 171, 173) have been established as the methods for isolation of poly- and oligo(ADP-ribose). The natural contents of poly(ADP-ribose) were estimated by a number of methods (49, 174). Earlier

Table 2 Natural distribution of various poly(ADP-ribosyl)ation reactions

Enzyme	Acceptor		Reference
	Protein	Amino acid	
Nuclear enzyme			
Poly(ADP-ribose) synthetase	core histones (H2B)	glutamate	125–128
	linker histone (HI)	glutamate, COOH-terminus	125, 126 129, 130
	Mg^{2+}, Ca^{2+}-dependent endonuclease	unknown	131
	Poly(ADP-ribose) synthetase	unknown	132
	RNA polymerase α	unknown	133
	A24 protein	unknown	134
	high-mobility group proteins	unknown	135, 136
	actin	unknown	137
	RNases	unknown	138
	SV40 T antigen	unknown	139
	heterogeneous nuclear RNA-associated protein	unknown	140
	nuclear matrix proteins	unknown	18
	adenovirus T antigen and core proteins	unknown	141
	polyoma virus minichromosome proteins	unknown	142
	topoisomerase I	unknown	143, 144
	stress-induced protein	unknown	145
Nucleolar enzyme	nucleolar proteins	unknown	146
Extranuclear enzyme			
Microsomal enzyme	histones	unknown	147
Mitochondrial enzyme	"M-band" proteins	unknown	148, 120
Reovirus enzyme	capsid proteins	unknown	149

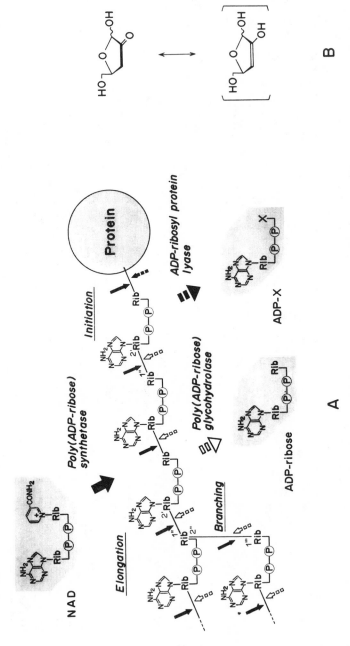

Figure 2 (A) Biosynthesis and degradation of poly(ADP-ribosyl) protein. (B) Structures of sugar X in the split product of ADP-ribosyl protein lyase; (upper) 3-deoxy-D-*glycero*-pentos-2-ulose, (lower) 3'-deoxypent-2-enofuranose (151).

methods employed isotope dilution (175), enzyme coupling (176), or fluorometry (177) after enzymatic and/or chemical degradation, or radioimmunoassays using anti-poly(ADP-ribose) antibodies (178–180). The methods were improved by using antibodies specific to AMP or *iso*ADP-ribose in radioimmunoassays (181–183), and also by affinity chromatography on boronate (122, 184–187) and high-performance liquid chromatography (165, 185–187). Sensitivity was increased by chemical derivatization to fluorescent (185, 186) or ^{3}H-labeling with KB^3H_4 reduction after $NaIO_4$ oxidation (165). The currently accepted value of the natural content in most tissues is 5–60 pmols (about 3–30 ng) per mg of DNA (188). Distribution of the polymer in tissues and cells was estimated by an indirect (150, 189) or direct immunofluorescence method (190) using anti-poly(ADP-ribose) antibodies.

ACCEPTORS OF POLY(ADP-RIBOSE) AND ADP-RIBOSYL PROTEIN LINKAGE   A number of nuclear proteins have been identified as acceptors for poly(ADP-ribosyl)ation (191) (Table 2). Boronate affinity chromatography was established as the method for isolation of ADP-ribosylated proteins (184, 192–194). Although histones have been well characterized as the acceptors (51, 108, 126, 184, 191), it has recently been noticed that histones, especially nucleosomal core histones in vivo, are modified primarily by monomers or short oligomers rather than by long polymers (18, 125, 128, 130, 188, 195). One of the main nonhistone acceptors proved to be poly(ADP-ribose) synthetase under various conditions (196–199).

Based on the sensitivities to alkali and neutral NH_2OH, a major portion of poly(ADP-ribosyl) protein bonds had been suggested to be carboxyl esters (51, 108). This type of bond was identified with two kinds of acceptors, histone H1 and H2B of rat liver (108). H1 accepted ADP-ribose at four sites, Glu-2, Glu-14, Glu-116, and COOH-terminal Lys-213 (129, 130), and H2B at a single site, Glu-2 (127, 128). In all these cases, the ADP-ribosyl residue was attached to a carboxyl group, either γ- (Glu) or α-carboxyl (COOH-terminal Lys), and an ester bond (O-glycoside) was formed (Figure 1). Besides the NH_2OH-sensitive ester bonds, a minor portion of poly(ADP-ribosyl) proteins was shown to exist in NH_2OH-resistant (and/or alkali-resistant) form(s) (191, 195, 200) (see the section on Mono(ADP-ribosyl)ation, above). In view of the fact that the ADP-ribose chain attached chemically or enzymatically (with arginine-specific enzyme) to basic proteins could be elongated by poly(ADP-ribose) synthetase (48, 119), most of the NH_2OH-resistant poly(ADP-ribosyl) proteins appeared to be initiated by ADP-ribosyltransferases (40, 41, 48).

BIOLOGICAL FUNCTIONS OF POLY(ADP-RIBOSYL)ATION   Besides the implications in DNA repair and cell differentiation, which will be discussed separately, a number of possible biological roles have been postulated for poly(ADP-ribosyl)ation.

A role in NAD metabolism has been well recognized (201). The pyridine nucleotide cycle prevailing in eukaryotic cells (NAD → nicotinamide → NMN → NAD) (201, 202) and the experiments on enucleated cells (203) indicated the importance of nuclear splitting of nicotinamide-ribose linkage in cellular NAD metabolism. Among several enzymes responsible for this splitting, ADP-ribosyltransferases were noted in view of a good correlation of the NAD level and the extent of mono(ADP-ribosyl)ation of proteins in various hepatic tissues (204). However, under certain conditions like DNA damage, the marked NAD breakdown could be attributed to increased poly(ADP-ribose) synthesis (185, 205–208).

A close relationship between poly(ADP-ribosyl)ation and chromatin structure has been demonstrated by Smulson and coworkers (209). The results indicated preferential interaction of poly(ADP-ribose) synthetase and nucleosomes with a periodicity of 8–10, which corresponds to the size of a full superhelical turn (209). Thus, promotion of polynucleosome aggregation by poly(ADP-ribosyl)ation was suggested (210). Evidence was obtained for a role of "histone H1 dimer"—a structure composed of two H1 molecules linked by poly(ADP-ribose) of a chain length of 15 (211)—in the nucleosome aggregation (212). In marked contrast to these findings, Poirier et al (213) recently reported relaxation, rather than condensation, of chromatin with poly(ADP-ribosyl)ation. It is of interest that this effect also appeared to be caused by hyper(ADP-ribosyl)ated histone H1 that resembled "H1 dimer" (214). Another discrepancy was disclosed by Holtlund et al (215), who suggested a function of poly(ADP-ribosyl)ated nonhistone proteins, rather than histones, in the mitotic events. Recently, a close relationship between poly(ADP-ribosyl)ation and phosphorylation of histone H1 was suggested (216; cf 217).

In relation to chromatin condensation/decondensation, a marked change in poly(ADP-ribosyl)ation during the cell cycle has been noted (218). Although the results obtained in various systems were not always consistent (9), a peak of polymer content and/or synthetase activity was observed around G_2 phase in many types of cells (1, 9, 179, 218–220). Analyses by Kidwell and coworkers (218) revealed that various inhibitors of poly(ADP-ribose) synthetase caused G_2 arrest or, at least, transient growth arrest, while reagents causing G_2 arrest induced a marked increase in poly(ADP-ribose) level with or without detectable DNA strand breaks.

Koide and coworkers (221, 222) observed that preincubation with NAD suppressed subsequent DNA synthesis in isolated nuclei. This effect was largely explained by the inhibition of DNA fragmentation by poly(ADP-ribosyl)ation of Ca^{2+}, Mg^{2+}-dependent endonuclease (223), by poly(ADP-ribose)-sensitive Mg^{2+}-dependent endonuclease (224), or both. A remarkable increase in poly(ADP-ribose) synthesis induced by suppression of DNA synthesis under various conditions (220) was also attributed to the increase in DNA strand breaks. The idea that rapidly proliferating cells exhibit relatively higher activities of poly(ADP-ribosyl)ation than quiescent cells (1) has been further supported by observations on SV40-transformed cells (225), mitogen-stimulated lymphocytes (226–228), regenerating rat liver (187), corticoid-treated chick embryo liver (229), rat cardiac cells responsive to circulatory stress (230), and fertilizing sea urchin egg (222) (cf 53). Several clues correlating poly(ADP-ribosyl)ation and DNA replication have been obtained, such as poly(ADP-ribosyl)ation of Ap_4A (a ligand of DNA polymerase α) (231), poly(ADP-ribosyl)ation of DNA polymerase α and β (232), and requirement of poly(ADP-ribosyl)ation for ligation of Okazaki fragments (233).

The biological significance of the inhibition of nuclear protease activity by poly(ADP-ribose) (234) is not known. Anti-poly(ADP-ribose) antibodies found in sera of systemic lupus erythematosus patients (235, 236) have been noted from diagnostic and etiological viewpoints (237).

Poly(ADP-Ribose) Synthetase and Glycohydrolase

POLY(ADP-RIBOSE) SYNTHETASE (POLYMERASE)   Multiple functions performed by this versatile protein (101) include initiation of ADP-ribose chains on protein (carboxyl ester formation) (238, 239), elongation (119, 238) and branching of poly(ADP-ribose) (glycosidic bond formation) (17, 163, 240) (Figure 2A), automodification (acceptance of ADP-ribose) (132, 163), and NAD hydrolysis (241). This enzyme (or enzyme activity) has been found in almost all eukaryotic cells so far examined (101), including animals, plants, slime molds, *Tetrahymena,* plasmodium (malaria parasites) (242), and yeast (243). The only exception is the mature granulocyte (189, 244). The nuclear localization of this enzyme was recently confirmed by immunohistochemistry (189, 245). The enzyme has been purified extensively from many tissues (101), such as rat, calf, cow, pig, and lamb thymuses (246–251); mouse testis (252); Ehrlich ascites tumor cells (253); HeLa cells (197); and human tonsils (254). All these enzymes exhibited very similar properties (101), e.g. mol wt (110,000–130,000), pI (9–10), almost absolute DNA dependency, activation by polycations (Mg^{2+}, histones, polyamines), SH–reducing agent requirement, automodification, and an inherent NAD glycohydrolase activity. The most notable property, DNA dependency, was investigated in some detail (249, 255–257); double-strandedness, the presence of nicks or strand ends, and a length of >10 base pairs were identified as structural requirements for activation. The effect of polycations was related to the presence of the polyanion, DNA (101, 239, 258). Initiation of ADP-ribose chains on exogenous proteins such as histones was shown to require appropriate ratios of protein to DNA (239) and ionic milieus (259). Elongation proved to proceed partly processively (239) by a terminal addition mechanism (119). Automodification was shown to take place at about 15 sites on the enzyme molecule (239).

This multiple modification accounted for decreased affinity for the activator DNA (161, 258) and thus partial inactivation upon extensive auto-poly-(ADP-ribosyl)ation (239). Based on analyses of variously automodified enzymes, Ferro & Olivera (259) proposed an idea of a "repulsion point," while Zahradka & Ebisuzaki (258) proposed a "shuttle mechanism." The architecture of the enzyme was recently revealed by Shizuta and coworkers (260); it consists of three functional domains for substrate binding, ADP-ribosylation (automodification), and DNA binding. Proteolytic processing of poly(ADP-ribose) synthetase was investigated by others (199, 261).

INHIBITORS OF POLY(ADP-RIBOSE) SYNTHETASE   Dozens of compounds have so far been reported to inhibit poly(ADP-ribose) synthetase in vitro (101). Among them, the most frequently employed in in vivo studies are nicotinamide, benzamide (262), and their derivatives. Although their effects are fairly specific on poly(ADP-ribose) synthetase, the following side actions have been noticed, and the possibility of misinterpretation of in vivo results because of these actions was pointed out (263). Nicotinamide, administered in vivo, proved not always inhibitory (218) but often stimulatory to ADP-ribosylation (204). It is a substrate for N-methyltransferase and phosphoribosyltransferase, and thus has a potential to disturb the metabolism of S-adenosylmethionine or 5-phosphoribosyl-1-pyrophosphate under certain conditions. 5-Methylnicotinamide proved to inhibit nicotinamide N-methyltransferase. Benzamide, an analog of nicotinamide, was reported to weakly inhibit the activities of cyclic AMP phosphodiesterase, carboxypeptidase A, and chymotrypsin (264). Benzamide and derivatives proved to be potent inhibitors of nicotinamide N-methyltransferase (264), and, in certain cells, to variably inhibit de novo synthesis of DNA (262).

POLY(ADP-RIBOSE) GLYCOHYDROLASE   Degradation of poly(ADP-ribosyl) protein is carried out by consecutive actions of poly(ADP-ribose) glycohydrolase (152, 265–267) and ADP-ribosyl protein lyase (151) (Figure 2A). The branched portion was also split by the glycohydrolase (268). Recently, Tavassoli et al (269) purified the glycohydrolase from pig thymus 12,300 fold and estimated its molecular weight as 60,000.

ADP-Ribosyl Protein Lyase and Glutamyl Ribose 5-Phosphate Storage Disease

Removal of the ADP-ribosyl moiety from acceptor protein is catalyzed by ADP-ribosyl protein lyase (151) [initially termed "ADP-ribosyl histone splitting enzyme" (270)] (Figure 2A). The lyase was purified from rat liver to apparent homogeneity (mol wt $\approx 80,000$) by Oka et al (151). The enzyme split the bond between ADP-ribose and various acceptors such as histone H1, H2B, NH_2-terminal pentapeptide of H2B, or nonhistone proteins (a mixture). The enzyme activity was highly specific to mono(ADP-ribosyl) carboxylate esters. The split product was inferred to be 5'-ADP-3''-deoxypent-2''-enofuranose, based on the identified structure, 3-deoxy-D-*glycero*-pentos-2-ulose, of the terminal, ribose-derived sugar (151, 271) and its possible tautomerism (Figure 2*B*). The splitting reaction thus proved to proceed by elimination, and not hydrolysis, at C-3'' of the ribose.

The importance of this enzyme in poly(ADP-ribosyl) protein metabolism is apparent from its key role in the rate-limiting step, i.e. the removal of primary ADP-ribosyl groups from acceptor proteins, in the overall turnover of poly(ADP-ribosyl) groups in the cell (256, 272). A more explicit indication for the importance of this enzyme was recently obtained by Williams and associates (273). They found a thitherto unknown substance accumulating in lysosomes of various tissues, notably brain and kidney, of an eight-year-old boy who succumbed after a six-year course of progressive neurologic deterioration and renal failure (274). Isolation and analyses of the accumulated substance revealed its structure as glutamyl ribose 5-phosphate (273). Judging from this structure representing the linkage region in ADP-ribosyl histones, the primary defect in this patient appeared to be a genetic abnormality of ADP-ribosyl protein lyase. Family history indicated an X-linked recessive inheritance. This is the first instance of genetic defects in poly(ADP-ribosyl)ation metabolism.

Poly(ADP-Ribosyl)ation and DNA Repair

An indispensable role of poly(ADP-ribosyl)ation in DNA excision repair has been suggested by many lines of evidence (275, 276). The initial observation was that various DNA damages brought about a marked increase in poly(ADP-ribosyl)ation concomitant with a remarkable decrease in the cellular NAD level (185, 205, 206, 277, 278). Subsequently, various inhibitors of the synthetase were shown to effectively abolish the NAD depletion (207, 208). Analyses by Shall and coworkers further revealed that the synthetase inhibitors not only retarded the repair of damaged DNA (208; also 279), but also increased the cytotoxicity of DNA-damaging agents (208, 280). The effect of lowering the NAD level by nutritional nicotinamide deprivation resembled that of the synthetase inhibitors (208). The step impaired by inhibitors of poly(ADP-ribose) synthetase proved to be the ligation step (208, 279, 281, 282); neither incision nor excision step was affected, and the DNA repair synthesis rather increased with the synthetase inhibitors (275, 276, 283). Postreplication repair was not affected (275).

The enhancement of poly(ADP-ribosyl)ation by DNA damage was ascribed to the known activation of poly(ADP-ribose) synthetase by DNA strand ends

(101, 256, 257, 284). Studies on mutant cells (xeroderma pigmentosum and ataxia telangiectasia) supported this view (285, 286). As for the mechanism by which poly(ADP-ribosyl)ation supported the ligation of damaged DNA, two possibilities have been proposed; Creissen & Shall (287) suggested direct ADP-ribosylation of DNA ligase II, while we (288) postulated the reversal of inhibitory effect of histones on DNA ligase activity by automodified synthetase. In support of the latter model, the synthetase was identified as a main acceptor for poly(ADP-ribosyl)ation during DNA repair (196, 289). Enhancement of sister chromatid exchange by the synthetase inhibitors (275) appeared to be related to the persistence of DNA strand breaks. The diabetogenicity of streptozotocin and alloxan was ascribed to their DNA-damaging actions in islet β-cells (290–292). Abnormally low levels of NAD or poly(ADP-ribosyl)ation activity were found in some cases of Cockayne syndrome (293), Fanconi's anemia (294, 295), or Bloom's syndrome (296). Some of poly(ADP-ribose) synthetase inhibitors were found, as expected from the effect of interfering with DNA repair, to enhance therapeutic effects of antitumor drugs (297–299), while, under different conditions, to promote chemical oncogenesis (292, 300). Recently, Kun and associates (301) reported cell cycle–dependent intervention by benzamide of carcinogen-induced neoplastic transformation (cf 263). Apparent contradiction to the general scheme described above was the repair of UV-irradiated (281, 302) or intercalated DNA (303), and the repair in heat-shocked *Drosophila melanogaster* cells (304).

Poly(ADP-Ribosyl)ation and Cell Differentiation

A close correlation between chondrocytic differentiation of embryonal chick limb mesenchymal cells and a suppressed poly(ADP-ribosyl)ating activity (305) was confirmed by Caplan and associates (306) assaying the chemical content of poly(ADP-ribose), and by Nishio et al (307) using inhibitors of poly(ADP-ribose) synthetase. A decrease in poly(ADP-ribosyl)ation during differentiation has been observed in many other types of cells, such as intestinal epithelial cells (308), granulocytes (244), and differentiation-induced mouse preadipocytes (309), murine teratocarcinoma cells (310), murine erythroleukemia (Friend) cells (311; cf 312), and human leukemic promyelocytes HL-60 (313). Furthermore, terminal differentiation of the last three cells was successfully induced by the synthetase inhibitors (243, 310, 311, 314). However, apparently contradictory observations were also reported; Farzaneh et al (315) observed an increase, rather than a decrease, in poly(ADP-ribosyl)ation during differentiation of cultured chick embryonal myoblasts due, probably, to an increase in DNA strand breaks, and further showed inhibition of the differentiation by the synthetase inhibitors. Johnstone & Williams (316) also demonstrated the inhibitory effect of the synthetase inhibitors on mitogen-stimulated human peripheral lymphocytes, probably through inhibition of rejoining of single-strand breaks of DNA. Similar inhibition was observed by Ittel et al

(317), but the result was interpreted to show inhibition primarily of cell proliferation. Although the reason for the discrepancy remains to be investigated, it was noted that the change in poly(ADP-ribosyl)ation preceded, rather than accompanied, the phenotypic changes in many cases (309, 310, 317).

Poly(ADP-Ribosyl)ation and RNA Synthesis/Processing/Degradation

The finding of inhibition of RNA synthesis by nicotinamide in hypertrophic rat kidney (318) triggered the work that ultimately led to the discovery of poly(ADP-ribose) synthesis (2). The theme, the metabolic relationship between pyridine nucleotides and other nucleotides, was revived recently in light of poly(ADP-ribosyl)ation enhancement in DNA-damaged cells; general suppression of macromolecular (DNA, RNA and protein) syntheses (283, 290, 319) after DNA damage was explained by depletion of NAD and ATP pools (283). It seems that the exhaustion of cellular NAD by enhanced poly(ADP-ribosyl)ation impaired operation of ATP generating systems and thus affected many energy-dependent processes including RNA synthesis.

A more direct relationship between poly(ADP-ribosyl)ation and RNA synthesis was suggested by Müller & Zahn (133), who found poly(ADP-ribosyl)ation of DNA-dependent RNA polymerase I of quail oviduct and its decrease after progesterone administration. In analogy to mono(ADP-ribosyl)ation of E. coli RNA polymerase by T4 phage, modulation of transcriptional specificity by poly(ADP-ribosyl)ation was postulated. The mechanism of transcriptional control is currently a most intriguing problem in molecular biology. Among many factors isolated and suggested for this control, a factor termed TFIIC was shown to eliminate random (nick-induced) transcription by RNA polymerase II in in vitro systems (320). Purification and analyses of the factor by Roeder and associates (320) indicated its identity with poly(ADP-ribose) synthetase. The synthetase appeared to restrict the transcription initiation at promoter-specified sites by binding to nick sites of DNA template (320). Recently, the "cleaning" activity of the synthetase was shown to reside in a restricted area termed the DNA-binding domain of the enzyme molecule (321).

Another possible link between poly(ADP-ribosyl)ation and transcriptional control has been suggested by modification of specific groups of chromosomal proteins; poly(ADP-ribosyl)ation of protein A24 (134) and high-mobility group (HMG) proteins 14 and 17 (135, 136, 214, 322) might affect their proposed functions in ribosomal gene expression (323) or in transcriptionally active nucleosomes (324), respectively. A close correlation between suppression of poly(ADP-ribosyl)ation of HMG 14/17 and glucocorticoid-regulated mammary tumor virus RNA synthesis in cultured mouse cells (325) appeared to support this view. A proposal of preferential distribution of poly(ADP-ribose) synthetase in transcriptionally active chromatin (326) has been substantiated (327, 328).

Besides the step of transcription, the processing of transcripts has also been suggested to be related to poly(ADP-ribosyl)ation. Costantini & Johnson (323) found disproportionate accumulation of 18S and 28S rRNAs in normal rat kidney cells after treatment with picolinic acid or 5-methylnicotinamide, a poly(ADP-ribose) synthetase inhibitor. The imbalance appeared to result from an unusual instability of 28S rRNA and/or 32S precursor RNA. Poly(ADP-ribosyl)ation of a variety of RNases (138) and heterogeneous nuclear RNA (hnRNA)-associated proteins (140) might be related to this phenomenon.

Extranuclear Poly- and Oligo(ADP-Ribosyl)ation

The activity found in the microsomal fraction of HeLa cells (147) resembled that of nuclear poly(ADP-ribose) synthetase, and represented, most probably, nascent molecules of the synthetase staying on ribosomes. The enzyme activity found in the inner membrane–matrix complex of rat liver mitochondria (329) also resembled nuclear poly(ADP-ribose) synthetase. The enzyme activity found by Kun and associates (148) in rat liver mitochondria, by contrast, differed from the nuclear enzyme in the insensitivity to thymidine and DNA. Recent analyses (120) revealed its localization in "M-band" or a mitochondrial inner membrane/DNA/RNA complex. Evidence was obtained suggesting a role of the oligo(ADP-ribosyl)ation in regulation of mitochondrial DNA replication. Two other enzyme activities similar to nuclear poly(ADP-ribose) synthetase were reported, one in mitochondria of *Xenopus laevis* oocytes (330) and the other in the postmitochondrial fraction of baby hamster kidney cells (331). The poly(ADP-ribosyl)ating activity found in the microsomal fraction of glucocorticoid-treated chick embryo liver cells (332) has not yet been well characterized.

Viral Oligo- and Poly(ADP-Ribosyl)ation

Several animal viruses have been shown to contain oligo- or poly(ADP-ribosyl)ated proteins and/or ADP-ribosylating enzyme activities. The first example, reovirus, was reported by Carter et al (149). They showed that a group of outer-capsid proteins μ_1, μ_{1_c}, and viii, were modified by oligo(adenylate) (149) and oligo(ADP-ribose) of about 1.5 average chain length (149–333). ADP-ribosylating activity was also found in purified virions (333). Inhibitor studies indicated that oligo(ADP-ribosyl)ation might play a role in viral replication (333). Poly(ADP-ribosyl)ation of large T antigen of simian virus 40 (SV40) was found by Goldman et al (139). Another oncogenic DNA virus, adenovirus 5, was shown by Goding et al (141) to have acceptors, core proteins V and VII, for ADP-ribosylation by nuclear extracts prepared from infected cells. In this case, 3-aminobenzamide, an inhibitor of poly(ADP-ribose) synthetase, had no effect on viral replication. Recently, Prieto-Soto et al (142) reported that histones H2A and H2B as well as several nonhistone proteins associated with polyoma virus minichromosomes were poly(ADP-ribosyl)ated by endogenous poly(ADP-ribose) synthetase. Analysis of replicative intermediates indicated a possible role of poly(ADP-ribosyl)ation in polyoma replication or transcription.

CONCLUDING REMARKS

During the last several years, an increasing number of researchers from various research fields joined in studying ADP-ribosylation and made remarkable contributions to elucidation of its biological roles. Demonstration by pharmacologists of structural and functional similarities of several guanine nucleotide-binding proteins that serve as ADP-ribose acceptors (35, 83, 89), a dispute by radiobiologists about a role of poly(ADP-ribosyl)ation in DNA repair (22, 263), and the discovery by pediatricians of glutamyl ribose 5-phosphate storage disease (273) are examples of such achievements. Further collaboration among researchers with different backgrounds will advance our knowledge of ADP-ribosylation reactions. Currently, a major focus of interests in mono(ADP-ribosyl)ation pertains to the enzymes in eukaryotes that mimic the actions of bacterial toxins (40, 41). Such enzymes should participate in the physiological regulation of, for example, protein synthesis or signal transmission. Precise identification of natural acceptors for these enzymes as well as for their controls will shed new light on cellular physiology. A continuing interest in poly(ADP-ribosyl)ation is coupled with its still unclear biological significance. A major difficulty of obtaining direct and definitive evidence for the roles of poly(ADP-ribosyl)ation has stemmed from the complexity of the systems used, and also from a lack of expedient inhibitors of poly(ADP-ribose)-metabolizing enzymes. A breakthrough would, therefore, be expected by reconstituting a model system in vitro from well-defined components, and by finding or developing specific, nonmetabolizable, and cell-permeable inhibitors of relevant enzymes. Our recent success in demonstrating DNA ligase activation by auto-poly(ADP-ribosyl)ated synthetase in an in vitro system (288), and the valuable contribution of benzamide derivatives introduced by Purnell & Whish (262) to studies of DNA repair (208, 280) and cell differentiation (310, 314, 315) encouraged our endeavor in this line. In addition, the recent discovery of glutamyl ribose 5-phosphate storage disease (273) appears to provide us with a new kind of tool for in vivo studies.

NOTE ADDED IN PROOF: The amino acid sequence of the peptide fragment of transducin α subunit, as recently deduced from a DNA sequence analysis of a cDNA clone, has aspartic acid and cysteine in place of asparagine at a putative ADP-ribosylation site by islet-activating protein, suggesting that ADP-ribosylation takes place either at the aspartic acid residue directly (O-glycoside) or after posttranslational amidation (N-glycoside) or at the cysteine residue (S-glycoside). Hurley, J. B., Simon, M. I., Teplow, D. B., Rabishaw, J. D., Gilman, A. G. 1984. *Science* 226:860–62

94 UEDA & HAYAISHI

Literature Cited

1. Hayaishi, O., Ueda, K. 1977. *Ann. Rev. Biochem.* 46:95–116
2. Chambon, P., Weill, J. D., Doly, J., Strosser, M. T., Mandel, P. 1966. *Biochem. Biophys. Res. Commun.* 25:638–43
3. Sugimura, T., Fujimura, S., Hasegawa, S., Kawamura, Y. 1967. *Biochim. Biophys. Acta* 138:438–41
4. Nishizuka, Y., Ueda, K., Nakazawa, K., Hayaishi, O. 1967. *J. Biol. Chem.* 242:3164–71
5. Hayaishi, O., Ueda, K. 1982. See Ref. 23, pp. 3–16
6. Wold, F. 1981. *Ann. Rev. Biochem.* 50:783–814
7. Sugimura, T. 1973. *Prog. Nucleic Acid Res. Mol. Biol.* 13:127–51
8. Honjo, T., Hayaishi, O. 1973. *Curr. Top. Cell. Regul.* 7:87–127
9. Hilz, H., Stone, P. 1976. *Rev. Physiol. Biochem. Pharmacol.* 76:1–58
10. Gill, D. M. 1977. *Adv. Cyclic Nucleotide Res.* 8:85–118
11. Moss, J., Vaughan, M. 1979. *Ann. Rev. Biochem.* 48:581–600
12. Purnell, M. R., Stone, P. R., Whish, W. J. D. 1980. *Biochem. Soc. Trans.* 8:215–27
13. Sugimura, T., Miwa, M., Saito, H., Kanai, M., Ikejima, K., et al 1980. *Adv. Enzyme Regul.* 18:195–220
14. Hilz, H. 1981. *Hoppe-Seyler's Z. Physiol. Chem.* 362:1415–25
15. Vaughan, M., Moss, J. 1981. *Curr. Top. Cell. Regul.* 20:205–46
16. Mandel, P., Okazaki, H., Niedergang, C. 1982. *Prog. Nucleic Acid Res. Mol. Biol.* 27:1–51
17. Ueda, K., Ogata, N., Kawaichi, M., Inada, S., Hayaishi, O. 1982. *Curr. Top. Cell. Regul.* 21:175–87
18. Ueda, K., Kawaichi, M., Ogata, N., Hayaishi, O. 1983. In *Nucleic Acid Research: Future Development*, ed. K. Mizobuchi, I. Watanabe, J. D. Watson, pp. 143–64. New York: Academic
19. Pekala, P. H., Moss, J. 1983. *Curr. Top. Cell. Regul.* 22:1–49
20. Hayaishi, O., Ueda, K., Oka, J., Kawaichi, M., Komura, H., Nakanishi, K. 1984. *Curr. Top. Cell. Regul.* In press
21. Smulson, M. E., Sugimura, T., eds. 1980. *Novel ADP-Ribosylations of Regulatory Enzymes and Proteins. Development in Cell Biology*, Vol. 6. New York/Amsterdam/Oxford: Elsevier/North-Holland. 452 pp.
22. Miwa, M., Hayaishi, O., Shall, S., Smulson, M., Sugimura, T., eds. 1983.

ADP-Ribosylation, DNA Repair and Cancer. Proc. Int. Symp. Princess Takamatsu Cancer Res. Fund, 13th, Tokyo, 1982. Tokyo: Jap. Sci. Soc. Press. 338 pp.
23. Hayaishi, O., Ueda, K., eds. 1982. *ADP-Ribosylation Reactions: Biology and Medicine.* New York: Academic. 698 pp.
24. Honjo, T., Nishizuka, Y., Hayaishi, O., Kato, I. 1968. *J. Biol. Chem.* 243:3553–55
25. Iglewski, B. H., Kabat, D. 1975. *Proc. Natl. Acad. Sci. USA* 72:2284–88
26. Goff, C. G. 1974. *J. Biol. Chem.* 249:6181–90
27. Rohrer, H., Zillig, W., Mailhammer, R. 1975. *Eur. J. Biochem.* 60:227–38
28. Cassel, D., Pfeuffer, T. 1978. *Proc. Natl. Acad. Sci. USA* 75:2669–73
29. Gill, D. M., Meren, R. 1978. *Proc. Natl. Acad. Sci. USA* 75:3050–54
30. Amir-Zaltsman, Y., Ezra, E., Scherson, T., Zutra, A., Littauer, U. Z., Salomon, Y. 1982. *EMBO J.* 1:181–86
31. Hawkins, D. J., Browning, E. T. 1982. *Biochemistry* 21:4474–79
32. Abood, M. E., Hurley, J. B., Pappone, M.-C., Bourne, H. R., Stryer, L. 1982. *J. Biol. Chem.* 257:10540–43
33. Moss, J., Richardson, S. H. 1978. *J. Clin. Invest.* 62:281–85
34. Tait, R. M., Booth, B. R., Lambert, P. A. 1980. *Biochem. Biophys. Res. Commun.* 96:1024–31
35. Katada, T., Ui, M. 1982. *Proc. Natl. Acad. Sci. USA* 79:3129–33
36. Manning, D. R., Fraser, B. A., Kahn, R. A., Gilman, A. G. 1984. *J. Biol. Chem.* 259:749–56
37. Iglewski, B. H., Sadoff, J., Bjorn, M. J., Maxwell, E. 1978. *Proc. Natl. Acad. Sci. USA* 75:3211–15
38. Pesce, A., Casoli, C., Schito, G. C. 1976. *Nature* 262:412–14
39. Skórko, R., Kur, J. 1981. *Eur. J. Biochem.* 116:317–22
40. Lee, H., Iglewski, W. J. 1984. *Proc. Natl. Acad. Sci. USA* 81:2703–7
41. Moss, J., Vaughan, M. 1978. *Proc. Natl. Acad. Sci. USA* 75:3621–24
42. Moss, J., Stanley, S. J. 1981. *J. Biol. Chem.* 526:7830–33
43. Beckner, S., Blecher, M. 1981. *Biochim. Biophys. Acta* 673:477–86
44. De Wolf, M. J. S., Vitti, P., Ambesi-Impiombato, F. S., Kohn, L. D. 1981. *J. Biol. Chem.* 256:12287–96
45. Walaas, O., Horn, R. S., Lystad, E., Adler, A. 1981. *FEBS Lett.* 128:133–36
46. Adamietz, P., Wielckens, K., Brede-

horst, R., Lengyel, H., Hilz, H. 1981. *Biochem. Biophys. Res. Commun.* 101:96–103

47. Richter, C., Winterhalter, K. H., Baumhüter, S., Lötscher, H.-R., Moser, B. 1983. *Proc. Natl. Acad. Sci. USA* 80:3188–92

48. Tanigawa, Y., Tsuchiya, M., Imai, Y., Shimoyama, M. 1984. *J. Biol. Chem.* 259:2022–29

49. Hilz, H., Wielckens, K., Bredehorst, R. 1982. See Ref. 23, pp. 305–21

50. Kun, E., Chang, A. C. Y., Sharma, M. L., Ferro, A. M., Nitecki, D. 1976. *Proc. Natl. Acad. Sci. USA* 73:3131–35

51. Nishizuka, Y., Ueda, K., Yoshihara, K., Yamamura, H., Takeda, M., Hayaishi, O. 1969. *Cold Spring Harbor Symp. Quant. Biol.* 34:781–86

52. Hilz, H., Bredehorst, R., Adamietz, P., Wielckens, K. 1982. See Ref. 23, pp. 207–19

53. Bredehorst, R., Wielckens, K., Adamietz, P., Steinhagen-Thiessen, E., Hilz, H. 1981. *Eur. J. Biochem.* 120:267–74

54. Van Ness, B. G., Howard, J. B., Bodley, J. W. 1980. *J. Biol. Chem.* 255:10710–16

55. Iglewski, B. H., Liu, P. V., Kabat, D. 1977. *Infect. Immun.* 15:138–44

56. Gill, D. M., Pappenheimer, A. M. Jr., Brown, R., Kurnick, J. J. 1969. *J. Exp. Med.* 129:1–21

57. Goor, R. S., Maxwell, E. S. 1970. *J. Biol. Chem.* 245:616–23

58. Van Ness, B. G., Howard, J., Bodley, J. W. 1978. *J. Biol. Chem.* 253:8667–90

59. Kessel, M., Klink, F. 1980. *Nature* 287:250–51

60. Gill, D. M., Pappenheimer, A. M. Jr., Uchida, T. 1973. *Fed. Proc.* 32:1508–15

61. Collier, R. J. 1975. *Bacteriol. Rev.* 39:54–85

62. Pappenheimer, A. M. Jr. 1977. *Ann. Rev. Biochem.* 46:69–94

63. Collier, R. J. 1982. See Ref. 23, pp. 575–92

64. Greenfield, L., Bjorn, M. J., Horn, G., Fong, D., Buck, G. A., et al. 1983. *Proc. Natl. Acad. Sci. USA* 80:6853–57

65. Ratti, R., Rappuoli, R., Giannini, G. 1983. *Nucleic Acids Res.* 11:6589–95

66. Proia, R. L., Hart, D. A., Holmes, R. K., Holmes, K. V., Eidels, L. 1979. *Proc. Natl. Acad. Sci. USA* 76:685–89

67. Kajan, B. L., Finkelstein, A., Colombini, M. 1981. *Proc. Natl. Acad. Sci. USA* 78:4950–54

68. Oppenheimer, N. J., Bodley, J. W. 1981. *J. Biol. Chem.* 256:8579–81

69. Dunlop, P. C., Bodley, J. W. 1983. *J. Biol. Chem.* 258:4754–58

70. Thomson, M. R., Iglewski, B. H. 1982. See Ref. 23, pp. 661–74

71. Gray, G. L., Smith, D. H., Baldridge, J. S., Harkins, R. N., Vasil, M. L., et al. 1984. *Proc. Natl. Acad. Sci. USA* 81:2645–49

72. Gill, D. M., Dinius, L. L. 1973. *J. Biol. Chem.* 248:654–58

72a. Iglewski, W. J., Lee, H., Muller, P. 1984. *FEBS Lett.* 173:113–18

73. Skórko, R. 1982. See Ref. 23, pp. 647–59

74. Gill, D. M. 1982. See Ref. 23, pp. 593–621

75. Moss, J., Vaughan, M. 1982. See Ref. 23, pp. 623–36

76. Moss, J., Vaughan, M. 1982. See Ref. 23, pp. 637–45

77. Mailhammer, R., Yang, H. L., Reiness, G., Zubay, G. 1975. *Proc. Natl. Acad. Sci. USA* 72:4928–32

78. Spicer, E. K., Kavanaugh, W. M., Dallas, W. S., Falkow, S., Konigsberg, W. H., Schafer, D. E. 1981. *Proc. Natl. Acad. Sci. USA* 78:50–54

79. Johnson, G. L., Kaslow, H. R., Bourne, H. R. 1978. *J. Biol. Chem.* 253:7120–23

80. Cassel, D., Selinger, Z. 1976. *Biochim. Biophys. Acta* 452:538–51

81. Ross, E. M., Gilman, A. G. 1980. *Ann. Rev. Biochem.* 49:533–64

82. Cassel, D., Selinger, Z. 1977. *Proc. Natl. Acad. Sci. USA* 74:3307–11

83. Gilman, A. G. 1984. *Cell* 36:577–79

84. Kahn, R. A., Gilman, A. G. 1984. *J. Biol. Chem.* 259:6235–40

85. Gill, D. M. 1976. *J. Infect. Dis.* 133:S55–63

86. Enomoto, K., Gill, D. M. 1980. *J. Biol. Chem.* 255:1252–58

87. Kahn, R. A., Gilman, A. G. 1984. *J. Biol. Chem.* 259:6228–34

88. Stryer, L., Hurley, J. B., Fung, B. K.-K. 1981. *Curr. Top. Memb. Transp.* 15:93–108

89. Stryer, L., Hurley, J. B., Fung, B. K.-K. 1981. *Trends Biochem. Sci.* 6:245–47

90. Bitensky, M. W., Wheeler, M. A., Rasenick, M. M., Yamazaki, A., Stein, P. J., et al. 1982. *Proc. Natl. Acad. Sci. USA* 79:3408–12

91. Houslay, M. D. 1984. *Trends Biochem. Sci.* 9:39–40

92. Manning, D. R., Gilman, A. G. 1983. *J. Biol. Chem.* 258:7059–63

93. Van Dop, C., Tsubokawa, M., Bourne, H. R., Ramachandran, J. 1984. *J. Biol. Chem.* 259:696–98

94. Moss, J., Manganiello, V. C., Vaughan, M. 1976. *Proc. Natl. Acad. Sci. USA* 73:4424–27

95. Moss, J., Garrison, S., Oppenheimer, N.

J., Richardson, S. M. 1979. *J. Biol. Chem.* 254:6270–72
96. Moss, J., Vaughan, M. 1977. *J. Biol. Chem.* 252:2455–57
97. Oppenheimer, N. J. 1978. *J. Biol. Chem.* 253:4907–10
98. Trepel, J. B., Chuang, D.-M., Neff, N. H. 1981. *J. Neurochem.* 36:538–43
99. Kaslow, H. R., Groppi, V. E., Abood, M. E., Bourne, H. R. 1981. *J. Cell Biol.* 91:410–13
100. Moss, J., Stanley, S. J., Watkins, P. A. 1980. *J. Biol. Chem.* 255:5838–40
101. Ueda, K., Kawaichi, M., Hayaishi, O. 1982. See Ref. 23, pp. 117–55
102. Moss, J., Yost, D. A., Stanley, S. J. 1983. *J. Biol. Chem.* 258:6466–70
103. Vitti, P., De Wolf, M. J. S., Acquaviva, A. M., Epstein, M., Kohn, L. D. 1982. *Proc. Natl. Acad. Sci. USA* 79:1525–29
104. Bernofsky, C., Amamoo, D. G. 1984. *Biochem. Biophys. Res. Commun.* 118: 663–68
105. Lester, H. A., Steer, M. L., Michaelson, D. M. 1982. *J. Neurochem.* 38:1080–86
106. Ledley, F. D., Mullin, B. R., Lee, G., Aloj, S. M., Fishman, P. H., et al. 1976. *Biochem. Biophys. Res. Commun.* 69: 852–59
107. Rebois, R. V., Beckner, S. K., Brady, R. O., Fishman, P. H. 1983. *Proc. Natl. Acad. Sci. USA* 80:1275–79
108. Burzio, L. O. 1982. See Ref. 23, pp. 103–16
109. Soman, G., Tomer, K. B., Graves, D. J. 1983. *Anal. Biochem.* 134:101–10
110. Yajima, M., Hosoda, K., Kanbayashi, Y., Nakamura, T., Nogimori, K., et al. 1978. *J. Biochem.* 83:295–303
111. Katada, T., Ui, M. 1979. *Endocrinology* 104:1822–27
112. Katada, T., Ui, M. 1980. *J. Biol. Chem.* 255:9580–88
113. Katada, T., Ui, M. 1979. *J. Biol. Chem.* 254:469–79
114. Katada, T., Amano, T., Ui, M. 1982. *J. Biol. Chem.* 257:3739–46
115. Katada, T., Ui, M. 1982. *J. Biol. Chem.* 257:7210–16
116. Burns, D. L., Hewlett, E. L., Moss, J., Vaughan, M. 1983. *J. Biol. Chem.* 258:1435–38
117. Codina, J., Hildebrandt, J., Iyengar, R., Birnbaumer, L., Sekura, R. D., Manclark, C. R. 1983. *Proc. Natl. Acad. Sci. USA* 80:4276–80
118. Tamura, M., Nogimori, K., Murai, S., Yajima, M., Ito, K., et al. 1982. *Biochemistry* 21:5516–22
119. Ueda, K., Kawaichi, M., Okayama, H., Hayaishi, O. 1979. *J. Biol. Chem.* 254:679–87
120. Kun, E., Kirsten, E. 1982. See Ref. 23, pp. 193–205

121. Moser, B., Winterhalter, K. H., Richter, C. 1983. *Arch. Biochem. Biophys.* 224:358–64
122. Wielckens, K., Bredehorst, R., Adamietz, P., Hilz, H. 1981. *Eur. J. Biochem.* 117:69–74
123. Lötscher, H.-R., Winterhalter, K. H., Carafoli, E., Richter, C. 1980. *J. Biol. Chem.* 255:9325–30
124. Hofstetter, W., Mühlebach, T., Lötscher, H.-R., Winterhalter, K. H., Richter, C. 1981. *Eur. J. Biochem.* 117:361–67
125. Nishizuka, Y., Ueda, K., Honjo, T., Hayaishi, O. 1968. *J. Biol. Chem.* 243:3465–67
126. Ueda, K., Omachi, A., Kawaichi, M., Hayaishi, O. 1975. *Proc. Natl. Acad. Sci. USA* 72:205–9
127. Burzio, L. O., Riquelme, P. T., Koide, S. S. 1979. *J. Biol. Chem.* 254:3029–37
128. Ogata, N., Ueda, K., Hayaishi, O. 1980. *J. Biol. Chem.* 255:7610–15
129. Riquelme, P. T., Burzio, L. O., Koide, S. S. 1979. *J. Biol. Chem.* 254:3018–28
130. Ogata, N., Ueda, K., Kagamiyama, H., Hayaishi, O. 1980. *J. Biol. Chem.* 255:7616–20
131. Yoshihara, K., Tanigawa, Y., Koide, S. S. 1974. *Biochem. Biophys. Res. Commun.* 59:658–65
132. Yoshihara, K., Hashida, T., Yoshihara, H., Tanaka, Y., Ohgushi, H. 1977. *Biochem. Biophys. Res. Commun.* 78: 1281–88
133. Müller, W. E. G., Zahn, R. K. 1976. *Mol. Cell. Biochem.* 12:147–59
134. Okayama, H., Hayaishi, O. 1978. *Biochem. Biophys. Res. Commun.* 84: 755–62
135. Kawaichi, M., Ueda, K., Hayaishi, O. 1978. *Seikagaku* 50:920 (Abstr.) (In Japanese)
136. Reeves, R., Chang, D., Chung, S.-C. 1981. *Proc. Natl. Acad. Sci. USA* 78: 6704–8
137. Kun, E., Romaschin, A. D., Blaisdell, R. J., Jackowski, G. 1981. In *Metabolic Interconversion of Enzymes 1980,* ed. H. Holzer, pp. 280–93. Berlin/New York: Springer-Verlag
138. Leone, E., Farina, B., Faraone Mennella, M. R., Mauro, A. 1981. See Ref. 137, pp. 294–302
139. Goldman, N., Brown, M., Khoury, G. 1981. *Cell* 24:567–72
140. Kostka, G., Schweiger, A. 1982. *Biochim. Biophys. Acta* 969:139–44
141. Goding, C. R., Shaw, C. H., Blair, G. E., Russell, W. C. 1983. *J. Gen. Virol.* 64:477–83
142. Prieto-Soto, A., Gourlie, B., Miwa, M., Pigiet, V., Sugimura, T., et al. 1983. *J. Virol.* 45:600–6

143. Jongstra-Bilen, J., Ittel, M.-E., Niedergang, C., Vosberg, H.-P., Mandel, P. 1983. *Eur. J. Biochem.* 136:391–96
144. Ferro, A. M., Higgins, N. P., Olivera, B. M. 1983. *J. Biol. Chem.* 258:6000–3
145. Carlsson, L., Lazarides, E. 1983. *Proc. Natl. Acad. Sci. USA* 80:4664–68
146. Kawashima, K., Izawa, M. 1981. *J. Biochem.* 89:1889–1901
147. Roberts, J. H., Stark, P., Giri, C. P., Smulson, M. 1975. *Arch. Biochem. Biophys.* 171:305–15
148. Kun, E., Zimber, P. H., Chang, A. C. Y., Puschendorf, B., Grunicke, H. 1975. *Proc. Natl. Acad. Sci. USA* 72:1436–40
149. Carter, C. A., Lin, B. Y., Metlay, M. 1980. *J. Biol. Chem.* 255:6479–85
150. Ikai, K., Ueda, K., Hayaishi, O. 1980. *J. Histochem. Cytochem.* 28:670–76
151. Oka, J., Ueda, K., Hayaishi, O., Komura, H., Nakanishi, K. 1984. *J. Biol. Chem.* 259:986–95
152. Miwa, M., Sugimura, T. 1982. See Ref. 23, pp. 43–63
153. Suhadolnik, R. J., Lennon, M. B., Uematsu, T., Monahan, J. E., Baur, R. 1977. *J. Biol. Chem.* 252:4125–33
154. Suhadolnik, R. J., Baur, R., Lichtenwalner, D. M., Uematsu, T., Roberts, J. H., et al. 1977. *J. Biol. Chem.* 252:4134–44
155. Suhadolnik, R. J. 1982. See Ref. 23, pp. 65–75
156. Miwa, M., Saikawa, N., Yamaizumi, Z., Nishimura, S., Sugimura, T. 1979. *Proc. Natl. Acad. Sci. USA* 76:595–99
157. Miwa, M., Ishihara, M., Takishima, S., Takasuka, N., Maeda, M., et al. 1981. *J. Biol. Chem.* 256:2916–21
158. Miwa, M., Saitô, H., Sakura, H., Saikawa, N., Watanabe, F., et al. 1977. *Nucleic Acids Res.* 4:3977–4005
159. Ferro, A. M., Oppenheimer, N. J. 1978. *Proc. Natl. Acad. Sci. USA* 75:809–13
160. Inagaki, F., Miyazawa, T., Miwa, M., Saitô, H., Sugimura, T. 1978. *Biochem. Biophys. Res. Commun.* 85:415–20
161. de Murcia, G., Jongstra-Bilen, J., Ittel, M.-E., Mandel, P., Delain, E. 1983. *EMBO J.* 2:543–48
162. Hayashi, K., Tanaka, M., Shimada, T., Miwa, M., Sugimura, T. 1983. *Biochem. Biophys. Res. Commun.* 112: 102–7
163. Kawaichi, M., Ueda, K., Hayaishi, O. 1981. *J. Biol. Chem.* 256:9483–89
164. Juarez-Salinas, H., Levi, V., Jacobson, E. L., Jacobson, M. K. 1982. *J. Biol. Chem.* 257:607–9
165. Kanai, M., Miwa, M., Kuchino, Y., Sugimura, T. 1982. *J. Biol. Chem.* 257:6217–23
166. Gill, D. M. 1975. In *Poly(ADP-ribose),*

Int. Symp., Bethesda, 1974, ed. M. Harris. Fogarty Int. Cent. Proc. No. 2, pp. 85–99. Washington DC: GPO
167. Minaga, T., Kun, E. 1983. *J. Biol. Chem.* 258:725–30
168. Minaga, T., Kun, E. 1983. *J. Biol. Chem.* 258:5726–30
169. Tanaka, M., Saikawa, N., Yamaizumi, Z., Nishimura, S., Sugimura, T. 1978. *Nucleic Acids Res.* 5:3183–94
170. Kanai, Y., Kawamitsu, H., Tanaka, M., Matsushima, T., Miwa, M. 1980. *J. Biochem.* 88:917–20
171. Tanaka, M., Miwa, M., Hayashi, K., Kubota, K., Matsushima, T., Sugimura, T. 1977. *Biochemistry* 7:1485–89
172. Kawaichi, M., Oka, J., Ueda, K., Hayaishi, O. 1981. *Biochem. Biophys. Res. Commun.* 101:672–79
173. Adamietz, P., Bredehorst, R., Hilz, H. 1978. *Biochem. Biophys. Res. Commun.* 81:1377–83
174. Niedergang, C., Mandel, P. 1982. See Ref. 23, pp. 287–303
175. Stone, P. R., Bredehorst, R., Kittler, M., Lengyel, H., Hilz, H. 1976. *Hoppe-Seyler's Z. Physiol. Chem.* 357:51–56
176. Goebel, M., Stone, P., Lengyel, H., Hilz, H. 1977. *Hoppe-Seyler's Z. Physiol. Chem.* 358:13–21
177. Niedergang, C., Okazaki, H., Mandel, P. 1978. *Anal. Biochem.* 88:20–28
178. Kanai, Y., Miwa, M., Matsushima, T., Sugimura, T. 1974. *Biochem. Biophys. Res. Commun.* 59:300–6
179. Kidwell, W. R., Mage, M. G. 1976. *Biochemistry* 15:1213–17
180. Sakura, H., Miwa, M., Tanaka, M., Kanai, Y., Shimada, T., et al. 1977. *Nucleic Acids Res.* 4:2903–15
181. Ikejima, M., Sakura, H., Miwa, M., Kanai, Y., Sezawa, K., Sugimura, T. 1980. See Ref. 21, pp. 165–72
182. Bredehorst, R., Ferro, A. M., Hilz, H. 1978. *Eur. J. Biochem.* 82:115–21
183. Bredehorst, R., Schlüter, M.-M., Hilz, H. 1981. *Biochim. Biophys. Acta* 652: 16–28
184. Okayama, H., Ueda, K., Hayaishi, O. 1978. *Proc. Natl. Acad. Sci. USA* 75:1111–15
185. Juarez-Salinas, H., Sims, J. L., Jacobson, M. K. 1979. *Nature* 282:740–41
186. Sims, J. L., Juarez-Salinas, H., Jacobson, M. K. 1980. *Anal. Biochem.* 106:296–306
187. Romaschin, A. D., Kun, E. 1981. *Biochem. Biophys. Res. Commun.* 102: 952–57
188. Minaga, T., Romaschin, A. D., Kirsten, E., Kun, E. 1979. *J. Biol. Chem.* 254: 9663–68
189. Ikai, K., Ueda, K., Hayaishi, O. 1982. See Ref. 23, pp. 339–60

190. Kanai, Y., Tanuma, S., Sugimura, T. 1981. *Proc. Natl. Acad. Sci. USA* 78: 2801–4
191. Adamietz, P. 1982. See Ref. 23, pp. 77–101
192. Ademietz, P., Klapproth, K., Hilz, H. 1979. *Biochem. Biophys. Res. Commun.* 91:1232–38
193. Braeuer, H.-C., Adamietz, P., Nellessen, U., Hilz, H. 1981. *Eur. J. Biochem.* 114:63–68
194. Romaschin, A. D., Kirsten, E., Jackowski, G., Kun, E. 1981. *J. Biol. Chem.* 256:7800–5
195. Adamietz, P., Bredehorst, R., Hilz, H. 1978. *Eur. J. Biochem.* 91:317–26
196. Ogata, N., Kawaichi, M., Ueda, K., Hayaishi, O. 1980. *Biochem. Int.* 1:229–36
197. Jump, D. B., Smulson, M. 1980. *Biochemistry* 19:1031–37
198. Ogata, N., Ueda, K., Kawaichi, M., Hayaishi, O. 1981. *J. Biol. Chem.* 256: 4135–37
199. Surowy, C. S., Berger, N. A. 1983. *J. Biol. Chem.* 258:579–83
200. Adamietz, P., Hilz, H. 1976. *Hoppe-Seyler's Z. Physiol. Chem.* 357:527–34
201. Olivera, B. M., Ferro, A. M. 1982. See Ref. 23, pp. 19–40
202. Hillyard, D., Rechsteiner, M., Ramos, P. M., Imperial, J. S., Cruz, L. J., Olivera, B. M. 1981. *J. Biol. Chem.* 256:8491–97
203. Rechsteiner, M., Catanzarite, V. 1974. *J. Cell. Physiol.* 84:409–22
204. Bredehorst, R., Lengyel, H., Hilz, H., Stärk, D., Siebert, G. 1980. *Hoppe-Seyler's Z. Physiol. Chem.* 361:559–62
205. Whish, W. L. D., Davies, M. I., Shall, S. 1975. *Biochem. Biophys. Res. Commun.* 65:722–30
206. Smulson, M. E., Schein, P., Mullins, D. W. Jr., Sudhaker, S. 1977. *Cancer Res.* 37:3006–12
207. Skidmore, C. J., Davies, M. I., Goodwin, P. M., Halldorsson, H., Lewis, P. J., et al. 1979. *Eur. J. Biochem.* 101: 135–42
208. Durkacz, B. W., Omidiji, O., Gray, D., Shall, S. 1980. *Nature* 283:593–96
209. Butt, T., Smulson, M. 1982. See Ref. 23, pp. 173–91
210. Butt, T. R., Jump, D. B., Smulson, M. E. 1979. *Proc. Natl. Acad. Sci. USA* 76:1628–32
211. Stone, P. R., Lorimer, W. S. III, Kidwell, W. R. 1977. *Eur. J. Biochem.* 81:9–18
212. Nolan, N. L., Butt, T. R., Wong, M., Lambrianidou, A., Smulson, M. E. 1980. *Eur. J. Biochem.* 113:15–25

213. Poirier, G. G., De Murcia, G., Jongstra-Bilen, J., Niedergang, C., Mandel, P. 1982. *Proc. Natl. Acad. Sci. USA* 79: 3423–27
214. Poirier, G. G., Niedergang, C., Champagne, M., Mazen, A., Mandel, P. 1982. *Eur. J. Biochem.* 127:437–42
215. Holtlund, J., Kristensen, T., Østvold, A. C., Laland, S. G. 1983. *Eur. J. Biochem.* 130:47–51
216. Wong, M., Miwa, M., Sugimura, T., Smulson, M. 1983. *Biochemistry* 22: 2384–89
217. Tanigawa, Y., Tsuchiya, M., Imai, Y., Shimoyama, M. 1983. *Biochem. Biophys. Res. Commun.* 113:135–41
218. Kidwell, W. R., Nolan, N., Stone, P. 1982. See Ref. 23, pp. 373–88
219. Tanuma, S., Enomoto, T., Yamada, M. 1978. *Exp. Cell Res.* 117:421–30
220. Berger, N. A., Weber, G., Kaichi, A. S., Petzold, S. J. 1978. *Biochim. Biophys. Acta* 519:105–17
221. Burzio, L. O., Koide, S. S. 1970. *Biochem. Biophys. Res. Commun.* 40: 1013–20
222. Koide, S. S. 1982. See Ref. 23, pp. 361–71
223. Tanaka, Y., Yoshihara, K., Itaya, A., Kamiya, T., Koide, S. S. 1984. *J. Biol. Chem.* 259:6579–85
224. Tanigawa, Y., Shimoyama, M. 1983. *J. Biol. Chem.* 258:9184–91
225. Miwa, M., Oda, K., Segawa, K., Tanaka, M., Irie, S., et al. 1977. *Arch. Biochem. Biophys.* 181:313–21
226. Berger, N. A., Adams, J. W., Sikorski, G. W., Petzold, S. J., Shearer, W. T. 1978. *J. Clin. Invest.* 62:111–18
227. Rochette-Egly, C., Ittel, M. E., Bilen, J., Mandel, P. 1980. *FEBS Lett.* 120:7–11
228. Perrella, F. W. 1982. See Ref. 23, pp. 451–64
229. Kitamura, A., Tanigawa, Y., Yamamoto, T., Kawamura, M., Doi, S., Shimoyama, M. 1979. *Biochem. Biophys. Res. Commun.* 87:725–33
230. Jackowski, G., Heymann, M. A., Rudolph, A. M., Kun, E. 1982. *Experientia* 38:1068–69
231. Yoshihara, K., Tanaka, Y. 1981. *J. Biol. Chem.* 256:6756–61
232. Yoshihara, K., Itaya, A. 1984. *Seikagaku* 56: In press
233. Lönn, U., Lönn, S. 1984. *Proc. Natl. Acad. Sci. USA* 81:674 (Abstr.) (In Japanese)
234. Inagaki, T., Miura, K., Murachi, T. 1980. *J. Biol. Chem.* 255:7746–50
235. Kanai, Y., Kawaminami, Y., Miwa, M., Matsushima, T., Sugimura, T. et al. 1977. *Nature* 265:175–77
236. Okolie, E. E., Shall, S. 1979. *Clin. Exp. Immunol.* 36:151–77

237. Kanai, Y., Sugimura, T. 1982. See Ref. 23, pp. 533–46
238. Ueda, K., Hayaishi, O., Kawaichi, M., Ogata, N., Ikai, K., et al. 1979. In *Modulation of Protein Function*, ed. D. E. Atkinson, C. F. Fox, pp. 47–64. New York: Academic
239. Kawaichi, M., Ueda, K., Hayaishi, O. 1980. *J. Biol. Chem.* 255:816–19
240. Ueda, K., Kawaichi, M., Oka, J., Hayaishi, O. 1980. See Ref. 21, pp. 47–57
241. Kawaichi, M., Ueda, K., Hayaishi, O. 1981. *Seikagaku* 53:877 (Abstr.) (In Japanese)
242. Okolie, E. E., Onyezili, N. I. 1983. *Biochem. J.* 209:687–93
243. Miwa, M., Iijima, H., Kondo, T., Kata, M., Kawamitsu, H., et al. 1983. *Seikagaku* 55:547 (Abstr.) (In Japanese)
244. Ikai, K., Ueda, K., Fukushima, M., Nakamura, T., Hayaishi, O. 1980. *Proc. Natl. Acad. Sci. USA* 77:3682–85
245. Ikai, K., Ueda, K. 1983. *J. Histochem. Cytochem.* 31:1261–64
246. Okayama, H., Ueda, K., Hayaishi, O. 1980. *Methods Enzymol.* 66:154–58
247. Niedergang, C., Okazaki, H., Mandel, P. 1979. *Eur. J. Biochem.* 102:43–57
248. Ito, S., Shizuta, Y., Hayaishi, O. 1979. *J. Biol. Chem.* 254:3647–51
249. Yoshihara, K., Hashida, T., Tanaka, Y., Ohgushi, H., Yoshihara, K., Kamiya, T. 1978. *J. Biol. Chem.* 253:6459–66
250. Tsopanakis, C., Leeson, E., Tsopanakis, A., Shall, S. 1978. *Eur. J. Biochem.* 90:337–45
251. Petzold, S. J., Booth, B. A., Leimbach, G. A., Berger, N. A. 1981. *Biochemistry* 20:7075–81
252. Agemori, M., Kagamiyama, H., Nishikimi, M., Shizuta, Y. 1982. *Arch. Biochem. Biophys.* 215:621–27
253. Kristensen, T., Holtlund, J. 1978. *Eur. J. Biochem.* 88:495–501
254. Carter, S. G., Berger, N. A. 1982. *Biochemistry* 21:5475–81
255. Ohgushi, H., Yoshihara, K., Kamiya, T. 1980. *J. Biol. Chem.* 255:6205–11
256. Benjamin, R. C., Gill, D. M. 1980. *J. Biol. Chem.* 255:10493–501
257. Yoshihara, K., Kamiya, T. 1982. See Ref. 23, pp. 157–171
258. Zahradka, P., Ebisuzaki, K. 1982. *Eur. J. Biochem.* 127:579–85
259. Ferro, A. M., Olivera, B. M. 1982. *J. Biol. Chem.* 257:7808–13
260. Kameshita, I., Matsuda, Z., Taniguchi, T., Shizuta, Y. 1984. *J. Biol. Chem.* 259:4770–76
261. Holtlund, J., Jemtland, R., Kristensen, T. 1983. *Eur. J. Biochem.* 130:309–14
262. Purnell, M. R., Whish, W. J. D. 1980. *Biochem. J.* 185:775–77
263. Cleaver, J. E., Bodell, W. J., Borek, C., Morgan, W. F., Schwartz, J. L. 1983. See Ref. 22, pp. 195–207
264. Johnson, G. S. 1981. *Biochem. Int.* 2:611–17
265. Ueda, K., Narumiya, S., Miyakawa, N., Hayaishi, O. 1972. *Biochem. Biophys. Res. Commun.* 46:516–23
266. Miwa, M., Tanaka, M., Matsushima, T., Sugimura, T. 1974. *J. Biol. Chem.* 249:3475–82
267. Miwa, M., Nakatsugawa, K., Hara, K., Matsushima, T., Sugimura, T. 1975. *Arch. Biochem. Biophys.* 167:54–60
268. Miwa, M., Kato, M., Iijima, H., Tanaka, Y., Kondo, T., et al. 1983. See Ref. 22, pp. 27–37
269. Tavassoli, M., Tavassoli, M. H., Shall, S. 1983. *Eur. J. Biochem.* 135:449–53
270. Okayama, H., Honda, M., Hayaishi, O. 1978. *Proc. Natl. Acad. Sci. USA* 75:2254–57
271. Komura, H., Iwashita, T., Naoki, H., Nakanishi, K., Oka, J., et al. 1983. *J. Am. Chem. Soc.* 105:5164–65
272. Wielckens, K., Schmidt, A., George, E., Bredehorst, R., Hilz, H. 1982. *J. Biol. Chem.* 257:12872–77
273. Williams, J. C., Chambers, J. P., Liehr, J. G. 1984. *J. Biol. Chem.* 259:1037–42
274. Williams, J. C., Butler, I. J., Rosenberg, H. S., Verani, R., Scott, C. I., Conley, S. B. 1984. *N. Engl. J. Med.* 311:152–55
275. Shall, S. 1982. See Ref. 23, pp. 477–520
276. Shall, S. 1984. *Adv. Radiat. Biol.* In press
277. Smulson, M., Stark, P., Gazzoli, M., Roberts, J. 1975. *Exp. Cell Res.* 90:175–82
278. Berger, N. A., Sikorski, G. W., Petzold, S. J., Kurohara, K. K. 1979. *J. Clin. Invest.* 63:1164–71
279. Hayaishi, O., Kawaichi, M., Ogata, N., Ueda, K. 1981. See Ref. 137, pp. 4–9
280. Nduka, N., Skidmore, C. J., Shall, S. 1980. *Eur. J. Biochem.* 105:525–30
281. James, M. R., Lehmann, A. R. 1982. *Biochemistry* 21:4007–13
282. Morgan, W. F., Cleaver, J. E. 1983. *Cancer Res.* 43:3104–7
283. Sims, J. L., Berger, S. J., Berger, N. A. 1983. *Biochemistry* 22:5188–94
284. Miller, E. G. 1975. *Biochim. Biophys. Acta* 395:191–200
285. Berger, N. A., Sikorski, G. W., Petzold,S. J., Kurohara, K. K. 1980. *Biochemistry* 19:289–93
286. Edwards, M. J., Taylor, A. M. R. 1980. *Nature* 287:745–49
287. Creissen, D., Shall, S. 1982. *Nature* 296:271–72

288. Ohashi, Y., Ueda, K., Kawaichi, M., Hayaishi, O. 1983. Proc. Natl. Acad. Sci. USA 80:3604–7

289. Kreimeyer, A., Wielckens, K., Adamietz, P., Hilz, H. 1984. J. Biol. Chem. 259:890–96

290. Okamoto, H. 1981. Mol. Cell Biochem. 37:43–61

291. Yamamoto, H., Uchigata, Y., Okamoto, H. 1981. Nature 294:284–86

292. Okamoto, H., Yamamoto, H. 1983. See Ref. 22, pp. 297–308

293. Fujiwara, Y., Goto, K., Kano, Y. 1982. Exp. Cell Res. 139:207–15

294. Berger, N. A., Berger, S. J., Catino, D. M. 1982. Nature 299:271–73

295. Klocker, H., Auer, B., Hirsch-Kauffmann, M., Altmann, H., Burtscher, H. J., Schweiger, M. 1983. EMBO J. 2:303–7

296. Kanai, M., Miwa, M., Kondo, T., Tanaka, Y., Nakayasu, N., Sugimura, T. 1981. Proc. Jpn. Mol. Biol. Meet., 4th, Kanazawa, p. 37. (Abstr.) (In Japanese)

297. Berger, N. A., Catino, D. M., Vietti, T. J. 1982. Cancer Res. 42:4382–86

298. Kawamitsu, H., Miwa, M., Tanaka, Y., Sakamoto, H., Terada, M., et al. 1982. J. Pharm. Dyn. 5:900–4

299. Sakamoto, H., Kawamitsu, H., Miwa, M., Terada, M., Sugimura, T. 1983. J. Antibiot. 36:296–300

300. Takahashi, S., Ohnishi, T., Denda, A., Konishi, Y. 1982. Chem.-Biol. Interact. 39:363–68

301. Kun, E., Kirsten, E., Milo, G. E., Kurian, P., Kumari, H. L. 1983. Proc. Natl. Acad. Sci. USA 80:7219–23

302. Althaus, F. R., Lawrence, S. D., Sattler, G. L., Pitot, H. C. 1982. J. Biol. Chem. 257:5528–35

303. Zwelling, L. A., Kerrigan, D., Pommier, Y., Michaels, S., Steren, A., Kohn, K. W. 1982. J. Biol. Chem. 257:8957–63

304. Nolan, N. L., Kidwell, W. R. 1982. Radiat. Res. 90:187–203

305. Caplan, A. I., Rosenberg, M. J. 1975. Proc. Natl. Acad. Sci. USA 72:1852–57

306. Caplan, A. I., Niedergang, C., Okazaki, H., Mandel, P. 1979. Dev. Biol. 72:102–9

307. Nishio, A., Nakanishi, S., Doull, J., Uyeki, E. M. 1983. Biochem. Biophys. Res. Commun. 111:750–59

308. Porteous, J. W., Furneaux, H. M., Pearson, C. K., Lake, C. M., Morrison, A. 1979. Biochem. J. 180:455–63

309. Pekala, P., Lane, M. D., Watkins, P. A., Moss, J. 1981. J. Biol. Chem. 256:4871–76

310. Ohashi, Y., Ueda, K., Hayaishi, O., Ikai, K., Niwa, O. 1984. Proc. Natl. Acad. Sci. USA 81:7132–36

311. Morioka, K., Tanaka, K., Nokuo, T., Ishizawa, M., Ono, T. 1979. Gann 70:37–46

312. Rastl, E., Swetly, P. 1978. J. Biol. Chem. 253:4333–40

313. Kanai, M., Miwa, M., Kondo, T., Tanaka, Y., Nakayasu, M., Sugimura, T. 1982. Biochem. Biophys. Res. Commun. 105:404–11

314. Terada, M., Fujiki, H., Marks, P., Sugimura, T. 1979. Proc. Natl. Acad. Sci. USA 76:6411–14

315. Farzaneh, F., Zalin, R., Brill, D., Shall, S. 1982. Nature 300:362–66

316. Johnstone, A. P., Williams, G. T. 1982. Nature 300:368–70

317. Ittel, M. E., Jongstra-Bilen, J., Rochette-Egly, C., Mandel, P. 1983. Biochem. Biophys. Res. Commun. 116:428–34

318. Revel, M., Mandel, P. 1982. Cancer Res. 22:456–62

319. Taniguchi, T., Agemori, M., Kameshita, I., Nishikimi, M., Shizuta, Y. 1982. J. Biol. Chem. 257:4027–30

320. Slattery, E., Dignam, J. D., Matsui, T., Roeder, R. G. 1983. J. Biol. Chem. 258:5955–59

321. Ohtsuki, M., Sekimizu, K., Agemori, M., Shizuta, Y., Natori, S. 1984. FEBS Lett. 168:275–77

322. Tanuma, S., Johnson, G. S. 1983. J. Biol. Chem. 258:4067–70

323. Constantini, M. G., Johnson, G. S. 1981. Exp. Cell Res. 132:443–51

324. Weisbrod, S., Groudine, M., Weintraub, H. 1980. Cell 19:289–301

325. Tanuma, S., Johnson, L. D., Johnson, G. S. 1983. J. Biol. Chem. 258:15371–75

326. Mullins, D. W. Jr., Giri, C. P., Smulson, M. 1977. Biochemistry 16:506–13

327. Levy-Wilson, B. 1981. Arch. Biochem. Biophys. 208:528–34

328. Hough, C. J., Smulson, M. E. 1984. Biochemistry. 23:5016–23

329. Burzio, L. O., Saez, L., Cornejo, R. 1981. Biochem. Biophys. Res. Commun. 103:369–75

330. Burzio, L. O., Luke, M., Koide, S. S. 1979. Fed. Proc. 38:618 (Abstr.)

331. Furneaux, H. M., Pearson, C. K. 1977. Biochem. Soc. Trans. 5:743–45

332. Kitamura, A., Tanigawa, Y., Doi, S., Kawakami, K., Shimoyama, M. 1980. Arch. Biochem. Biophys. 204:455–63

333. Carter, C. A., Pozzatti, R. O., Lin, B. Y. 1982. See Ref. 23, pp. 221–39

Ann. Rev. Biochem. 1985. 54:101–134
Copyright © 1985 by Annual Reviews Inc. All rights reserved

GENETIC ANALYSIS OF PROTEIN EXPORT IN *ESCHERICHIA COLI* K12

S. A. Benson[1]

Laboratory of Genetics and Recombinant DNA, LBI-Basic Research Program, NCI-Frederick Cancer Research Facility, Frederick, Maryland 21701

M. N. Hall

Department of Biochemistry and Biophysics, University of California, San Francisco, California 94143

T. J. Silhavy[1]

Laboratory of Genetics and Recombinant DNA, LBI-Basic Research Program, NCI-Frederick Cancer Research Facility, Frederick, Maryland 21701

CONTENTS

[1]Present address: Department of Molecular Biology, Princeton University, Princeton, New Jersey 08544

101

0066-4154/85/0701-0101$02.00

PERSPECTIVES AND SUMMARY

All cells produce proteins that are localized (exported) from the cytoplasm into or through various membrane structures. The processes involved in the vectorial movement of proteins into and across membranes are selective and efficient, and considerable research effort has been directed towards understanding the underlying molecular mechanisms. One striking outcome of this work is the realization that both procaryotic and eucaryotic cells appear to use very similar mechanisms. That the processes have been conserved throughout evolution is illustrated by the fact that bacteria will export various eucaryotic secretory proteins (1, 2) and, conversely, eucaryotic cells will secrete certain bacterial proteins (3).

Through the application of electron microscopy and biochemistry, investigators working with eucaryotic systems laid the ground work that has shaped our present view of protein localization. This work has been extensively reviewed (4–5a). On the other hand, investigators working with bacteria have led the field with respect to genetic analysis. Bacteria such as *Escherichia coli* are haploid, are easily manipulated genetically, and grow rapidly. The ease and sophistication of the genetic approaches available with this organism far surpass those of any other system.

With respect to protein localization, *E. coli* and other gram-negative bacteria have three noncytoplasmic locations to which proteins are exported. These are the inner (cytoplasmic) and outer membranes and an aqueous compartment, termed the periplasm, that is bounded by these two structures. True protein secretion, i.e., the efficient transport and release of proteins into the growth media, is uncommon and this process appears to require the presence of extrachromosomal genetic elements (6). Most genetic studies have focused on proteins destined for either the periplasm or the outer membrane. These proteins, with one exception, are all made initially in precursor form with an NH_2-terminal signal sequence that is proteolytically removed during the export process. The one exception, phospholipase A, has a signal sequence but it is not processed (7). Available evidence indicates that both periplasmic and outer membrane proteins are routed through a common export pathway (8).

The genetic analysis of protein export in *E. coli* is a relatively young area of research. The first bacterial signal sequence was discovered only seven years ago (9, 10) and the first signal sequence mutation was reported one year later (11). Since then, the number of signal sequence mutations has increased to >60, and at least 18 different chromosomal genes that may specify components of the cellular export machinery have been identified. This wealth of data is due, in large part, to the application of modern genetic technologies such as gene fusion and recombinant DNA, and it demonstrates the complexity of this important cellular process.

It is not yet possible to interpret clearly all of the results obtained from the analysis of export-defective mutants. Nevertheless, considerable progress has been made. We have learned that the genetic information specifying correct cellular localization is contained within the structural gene. Some of this information resides in the signal sequence; other information is located in relatively small discrete regions corresponding to mature protein sequences. A mechanism to couple translation and export has been uncovered that appears to be at least formally analogous to an equivalent coupling mechanism used by eucaryotic cells (5a). Several heretofore unknown proteins have been identified and shown to function in protein export. Moreover, we can begin to assign probable functions and to predict direct physical interactions between members of this complex, multicomponent system.

In this review we have limited our discussion to various genetic approaches that have yielded information about protein export in *E. coli*. A number of excellent reviews detailing biochemical studies in procaryotes and genetic analysis in other organisms, such as yeast, have been published (12–19).

SELECTIONS FOR EXPORT-DEFECTIVE MUTANTS

The most straightforward genetic approach for the study of protein localization is the isolation of export-defective mutants. Characterization of such mutants should define the cellular components involved, identify regions of export information within a given gene, and provide insights into the molecular mechanism(s) of the process. These mutants in turn would provide useful materials for in vitro biochemical reconstructions. This approach, while conceptually easy, is in reality, difficult technically. The difficulty arises from the fact that phenotypically there is no easy way to differentiate between mutations that prevent export of a given protein and mutations that prevent function or synthesis of that protein. Without a selectable or scorable phenotype that is specific for an export defect, one is forced to utilize biochemical screens such as cytoplasmic accumulation of the precursor form of the protein. In addition to being labor intensive, this approach historically has not been successful. Of the thousands of mutations which cause a null phenotype for a variety of genes specifying exported proteins, very few are known to affect the localization process.

Mutations That Define Intragenic Export Information

A systematic search for export-defective mutations in the gene specifying the periplasmic β-lactamase *(bla)* was conducted by Koshland & Botstein (20). Hundreds of mutations mapping throughout the gene were isolated and characterized, and none were found that prevent export of the protein from the cytoplasm. Two of the mutations lie in the region of the gene encoding the signal sequence, but both are chain-terminating nonsense mutations. Nonsense mutations at the 3' end of the gene cause production of a truncated polypeptide that is exported from the cytoplasm but is not released in H_2O-soluble form in the periplasm (21). Whether this reflects a normal step in the export pathway or simply aggregation of the abnormal molecule remains to be determined.

Similar results with nonsense mutations have been obtained in other systems. For example, two *malE* (which specifies the periplasmic maltose-binding protein) nonsense mutations have been studied in detail (22). One produces a peptide 90% complete that is exported and processed normally. The other produces a peptide only 30% complete that is exported from the cytoplasm but is not released in H_2O-soluble form in the periplasm.

The first signal sequence missense mutation was obtained by Wu & Lin (11). This mutation changes gly at position 14 of the lipoprotein (Lpp) signal sequence to asp (Table 2). It was obtained following a complicated "suicide" selection and it was recognized because it blocks the characteristic NH_2-terminal diglyceride modification and this blocks processing by the lipoprotein signal peptidase (SPase II, see below), an essential enzyme specific for lipoproteins. Although the mutation prevents signal sequence processing, it has little effect on export to the outer membrane.

The significance of the results described in this section can be summarized as follows: First, export information must lie in the 5' portion of the structural gene. The 3' portion and the COOH terminus of the protein are not essential. Second, specific protein conformations do not appear to be required for protein export from the cytoplasm. Finally, proteolytic removal of the signal sequence is not an obligatory step in protein export. These conclusions will be further strengthened by results presented below.

Pleiotropic Export-Defective Mutants

Several approaches have been used to isolate mutants that exhibit a pleiotropic export defect and thus obtain a genetic "handle" on cellular components of the machinery of protein export. Wanner et al (23) identified *perA* (75 min, Table 1) as a mutation that decreased *phoA* (periplasmic alkaline phosphatase) but not *lacZ* (cytoplasmic β-galactosidase) expression. Although this mutation has pleiotropic effects, it is unclear whether it affects secretion or regulation, or perhaps both. This is especially pertinent in this case since the mutation appears to lie in a nonessential gene *(envZ)* that is involved in the transcriptional

Table 1 Genetic loci that may specify components of the cellular export machinery

Gene	Map position (min)	Gene product or mutant phenotypes[a]	Reference
expA	22	Decreases appearance of certain proteins in the envelope	25
lep	55	Gene for signal peptidase I; essential gene	85, 86
lsp	0.5	Gene for signal peptidase II; essential gene	91, 92
perA (envZ)	75	Decreases appearance of certain proteins in the envelope; transcriptional regulation of porin proteins	23
prlA (secY)	72	General suppressor of signal sequence mutations; different alleles suppress *secA*[ts]	55, 60
prlB (rbsB)	84	Suppresses *lamB* signal sequence mutations via a bypass mechanism	12, 55
prlC	68	Suppressor of *lamB* and *malE* signal sequence mutations	12, 55
prlD	2.5	Suppressor of certain *malE* and *lamB* signal sequence mutations	59
prlE	8.5	Reduces export of LamB, MalE, and PhoA; slight cs phenotype	(see text)
prlF	70	Relieves Mal[s] of *lamB-lacZ* and *malE-lacZ* fusions; slight cs phenotype	44
secA	2.5	Blocks export of envelope proteins; am and ts conditionally lethal; am decreases translation of exported proteins	47, 48
secB	80.5	Blocks export of certain envelope proteins; not an essential gene.	51
secC (rpsO)	69	Gene for ribosomal protein S15; suppresses *secA*[ts]; decreases translation of exported proteins; cs conditionally lethal.	61
ssaD	10	Suppresses *secA*[ts]; decreases translation of MalE; cs conditionally lethal	(see text)
ssaE	50	Suppresses *secA*[ts]; decreases translation of MalE; cs conditionally lethal	(see text)
ssaF	83	Suppresses *secA*[ts]; decreases translation of MalE; cs conditionally lethal	(see text)
ssaG	41	Suppresses *secA*[ts]; decreases translation of MalE; cs conditionally lethal	(see text)
secH	94	Suppresses *secA*[ts]; decreases translation of MalE; cs conditionally lethal	(see text)
secY (prlA)	72	Blocks export of envelope proteins; am and ts conditionally lethal	26, 27
ssyA	54	Suppresses *secY*[ts], slight cs phenotype; slows protein synthesis at low growth temperature	62

[a] am = amber; cs = cold-sensitive; ts = temperature-sensitive

Table 2 Mutational Alterations of Signal Sequences[a]

Protein	Charged Segment	Hydrophobic Segment

PhoA

```
                      MET LYS|GLN SER THR ILE ALA ALA LEU ALA LEU PRO LEU LEU PHE THR PRO VAL THR LYS ALA↓arg
                                              GLN
                                                  GLU
                                          HIS
                                                              ARG
                                                              GLN
                                              ///////
```

MalE

```
MET LYS ILE LYS THR GLY ALA ARG|ILE LEU ALA LEU SER ALA LEU THR THR MET MET PHE SER ALA SER ALA LEU ALA ̇lys
                            PRO
                            ARG
                                GLU
                                PRO
                                SER
                                            GLU
                                            PRO
                                                        LYS
                                                        ARG
                                                            ARG
                                //////////////     //////////////
```

Bla

```
MET SER ILE GLN HIS PHE ARG|VAL ALA LEU ILE PRO PHE PHE ALA ALA PHE CYS LEU PRO VAL PHE ALA ̇his
                        VAL
                    PRO CYS ARG
                                                                SER
                                                                LEU
                                                ALA PHE LEU PHE
                                LEU ARG HIS PHE ALA PHE LEU PHE
```

Lpp

```
                                                                                            R
                                                                                            |
MET LYS ALA THR LYS|LEU VAL LEU GLY ALA VAL ILE LEU GLY SER THR LEU LEU ALA GLY↓cys
LYS ASP
//// ALA
```

GLU ASP

LamB   MET MET ILE THR LEU ARG LYS | LEU PRO LEU ALA VAL ALA ALA GLY VAL MET SER ALA GLN ALA MET ALA $\downarrow$ val
                              SER   ASP   ASP
                                                               ASP   VAL
                                                                    GLU   ☐
                                                                         GLU   ALA
                                                                              ARG   ALA
                                                                              ASP
                                                                              ARG
                                                                              LYS

VAL ☐

OmpF   MET MET LYS ARG | ASN ILE VAL ILE LEU ALA VAL LEU LEU VAL ALA GLY THR ALA ASN ALA $\downarrow$ ala
            AM   MET SER LEU LEU   THR   THR
                                                      LEU LEU   THR VAL   VAL
                                                   ☐ LEU   THR VAL SER SER ARG TYR CYS LYS ARG

ᵃ The amino acids of the signal sequences for the following periplasmic proteins; PhoA (93), MalE (43), and Bla (94); and the outer membrane proteins Lpp (10), LamB (95), and OmpF (96, 97) are shown. The (↓) above each of the sequences denotes the processing site. The R group above the cys residue in LPP designates the position of the thioetherdiglyceride modification that is present in the mature protein. Specific mutational changes within the signal sequence are shown below each of the wild-type sequences. Deletions are shown as boxes. See the text for a complete description of the phenotypes conferred by each of the mutations. AM designates an amber mutation.

regulation of the major outer membrane porin proteins, OmpF and OmpC (24).

In a similar manner, Dassa & Boquet (25) have identified a mutation, *expA,* (22 min, Table 1) that pleiotropically decreases expression of the genes specifying 10 periplasmic proteins and at least one outer membrane protein. Here again, it is not clear whether the mutation affects synthesis or export.

Using local mutagenesis and a brute-force biochemical screen for precursor accumulation at nonpermissive growth temperature, Ito and coworkers identified a gene, *secY,* whose product appears to be required for protein export (26, 27). This hypothesis is strengthened by the finding that *secY* and *prlA* are allelic. As described in detail below, mutations in this gene exhibit a number of export-related phenotypes.

The results described above raise a difficult problem that will recur. In the absence of a reliable biochemical assay for protein translocation in vitro, it is difficult to demonstrate if a given mutation affects protein export directly or indirectly or whether some other step in gene expression has been altered.

APPLICATION OF GENE FUSION TECHNOLOGY

Beckwith and coworkers exploited the sophistication of *lac* genetics and the unusual properties of the cytoplasmic enzyme β-galactosidase to develop in vivo techniques that allow fusion of *lacZ* to any gene in *E. coli* [for review, see (28)]. Such fusions specify a hybrid protein composed of an NH_2-terminal sequence from the target gene product and a large, functional COOH-terminal portion of β-galactosidase. By constructing a series of fusions differing only in the amount of target gene DNA contained in the hybrid gene and determining the cellular location of the hybrid protein that is produced, investigators have been able to define the location of intragenic export information.

In vivo gene fusion technology has been applied extensively to study the localization of two periplasmic proteins, MalE and PhoA, and the outer membrane protein LamB, the receptor for bacteriophage λ. To a lesser extent, the technique has also been used to study export of the inner membrane maltose transport protein MalF and the outer membrane porin protein OmpF. Results with these fusions are quite consistent [for review, see (12, 13, 15)]. First, fusions constructed so as to contain a substantial portion of a gene specifying a noncytoplasmic protein produce a hybrid protein that is exported, at least to some degree, from the cytoplasm. This result demonstrates that export information is contained within the structural gene, and, consistent with results presented above, it indicates that the information must lie at a position corresponding to the NH_2-terminal end of the protein. The 3' end of the gene, and thus the COOH terminus of the protein, is not required. Second, fusions that contain only a small portion of a gene specifying an exported protein produce a hybrid protein that remains in the cytoplasm. With *lamB,* fusions have been constructed that specify a hybrid protein that contains the complete signal sequence plus as many as 27 amino acids of the mature product. Despite the presence of the signal sequence, these hybrid proteins remain in the cytoplasm (29, 30). This finding suggests that, at least for LamB, the signal sequence is not sufficient to cause export from the cytoplasm.

Finally, β-galactosidase sequences can be exported to the inner or outer membrane but not to the periplasm. This suggests that, although export of periplasmic and other membrane proteins may be similar in the early stages (8), the export pathways diverge before completion. Genes that specify membrane proteins may contain additional export information.

Recombinant DNA technology has extended the gene fusion approach to a variety of cloned genes. Fusions have constructed not only to β-galactosidase, but also to a variety of other exported and nonexported proteins. These include gene fusions between the *bla* gene and the eucaryotic pre-proinsulin gene (31), *bla-phoE, phoE-ompF* [*phoE* specifies an outer membrane porin protein, (32)], *lamB-phoA, phoA-bla* (33), *bla-lpp, lacZ-lpp* (34, M. Inouye, personal communication), and *ompF-lpp* (35). In addition, the *ompA* gene (which specifies a major outer membrane protein) has been fused to the gene specifying tetracycline resistance and to the major antigenic determinant of foot and mouth disease (36). All of these studies indicate that signal sequences are interchangeable, and do not appear to encode the sorting information that directs an envelope protein to its proper location. Moreover, these signal sequences can function from an internal site in the protein.

SELECTIONS BASED ON PROPERTIES CONFERRED BY *lacZ* GENE FUSIONS

A bonus in the use of gene fusions to genetically dissect protein export in *E. coli* stems from the unusual phenotypes often exhibited by certain fusion strains (37). These phenotypes result from the cell's attempt to export the hybrid protein, the bulk of which is the carboxy-terminal β-galactosidase moiety. One such phenotype, termed overproduction lethality, is characteristic of fusion strains that produce a hybrid protein that is incorrectly or inefficiently localized. In such cases, high level expression of the hybrid gene often leads to a situation in which the export machinery appears to be jammed. Presumably, this occurs because sequences within the β-galactosidase portion of the hybrid protein are incompatible with the export system. A direct consequence of this jamming is that normally exported proteins are prevented from reaching their destination. This results in their accumulation as precursor forms in the cytosol. Eventually, the cells die, presumably because essential envelope proteins are prevented from reaching the correct envelope location.

In the case of *malE-lacZ* and *lamB-lacZ* gene fusions, this overproduction lethality can be observed by adding maltose to the growth media. Maltose induces high level expression of the operons which encode these fusions. Accordingly, this phenotype has been termed maltose sensitivity (Mal^s). One can easily select for loss of the Mal^s phenotype by growth on media containing maltose. Many of these maltose resistant (Mal^r) mutants have genetic alterations that prevent synthesis of the hybrid protein. However, a small proportion of the Mal^r mutants carry export-defective mutations that relieve the Mal^s phenotype but do not prevent synthesis of the hybrid protein. Such mutants can be readily identified by their LacZ$^+$ phenotype.

A variation on this overproduction lethality phenomenon has been observed with *ompF-lacZ* gene fusions (E. Sodergren, personal communication). When this hybrid gene is present in a specialized λ-transducing phage it grows normally on wild-type strains. However, its efficiency of plating on *ompR$_2$* [a mutation causing constitutive, high-level expression of *ompF*, (38)] strains is decreased by 5 logs. Apparently, this results from the high-level synthesis of a poorly exported hybrid protein.

A second unusual and genetically useful phenotype results when the efficiency of export of a LacZ hybrid protein into the membrane structure is high. In such an environment the hybrid protein exhibits reduced enzymatic activity, probably due to the inability of the hybrid protein to form an active conformation within the membrane environment. Consequently these strains are unable to grow on lactose (Lac$^-$). Thus, by selecting for growth on lactose, (Lac$^+$) mutations that prevent or reduce export of the hybrid protein can be obtained.

Intragenic Mutations that Relieve Overproduction Lethality

LAMB   Starting with a Mal^s *lamB-lacZ* fusion, 42-1, three types of signal sequence mutations have been isolated by selecting for Mal^r. One type blocks localization of the hybrid protein and wild-type LamB protein by over 95%. All of the point mutations conferring this phenotype alter one of four residues at positions 14, 15, 16, or 19 of the signal sequence (Table 2). Each of these mutations have been isolated repeatedly (39, 40). It is believed that these four residues constitute a recognition site involved in an early step of the export pathway. All of the alterations share the property that they introduce a charged residue into the hydrophobic core of the signal sequence. The Mal^r selection also yields various deletion mutations. All of the deletions, with one exception, remove one or more of these critical residues. The one exception is the internal signal sequence deletion S78 that removes only 12 bp of DNA. This mutation appears to block export by altering the conformation of the signal sequence (see below).

The second type of signal sequence mutations obtained in the Mal^r selection utilizing the *lamB-lacZ* fusion, 42-1, are alterations that eliminate maltose sensitivity, and block export of the hybrid protein, but which have very little effect when recombined into the wild-type *lamB* gene (41). These mutations map at position 17 of the signal sequence, and convert the gly residue to either an acidic or a basic residue (arg or asp, Table 2). These mutations strengthen the hypothesis that the residues at position 14, 15, 16, and 19 constitute a site, and that mutations at these residues block export by altering this site. The mutations that alter residue 17 demonstrate that simply placing a charged residue in this region of the signal sequence is not, by itself, sufficient to block export. In this regard, these mutations are similar to the gly to asp mutation in the Lpp signal sequence (see above).

A final class of *lamB* signal sequence mutations isolated using the Mal^r selection are alterations in the NH$_2$-terminal hydrophilic region of the signal sequence. By selecting for Mal^r and then screening for mutants in which levels of β-galactosidase activity were lowered, a mutation was identified that appears to affect the coupling of export and translation (42). This mutation changes the arg codon at residue 6 of the signal sequence to ser. It reduces expression of wild-type *lamB* and *lamB-lacZ* hybrid genes that specify exported hybrid proteins by ~ 80%. This block in expression is not observed when the mutation is present on smaller *lamB-lacZ* fusions that specify nonexported hybrid proteins. A similar type of mutation within the NH$_2$-terminal hydrophilic region has been generated in vitro in the cloned *lpp* gene (see below).

MALE   Selections for Mal^r in strains carrying a large Mal^s *malE-lacZ* fusion yields both base pair substitutions at positions 10, 14, 16, 18, and 19, and deletion mutations that alter the hydrophobic core (43, 15). These alterations, when placed cis to a wild-type *malE* gene, impose a strong block on the export of MalE. All of the base pair substitution mutations introduce a charged residue into the hydrophobic core with the exception of the mutation 10-1, which substitutes a pro residue for a leu residue at position 10. It is believed this change behaves in a manner analogous to the *lamB*S78 deletion in that it imposes a block in localization by preventing the hydrophobic core of the signal sequence from forming an α-helical conformation. Again, it would appear that a subset of critical residues (14, 16, 18, and 19) within the hydrophobic core form a recognition sequence, perhaps in a manner analogous to residue 14, 15, 16, and 19 of LamB. Repeatedly, it is this subset of changes that are found when one selects Mal^r.

If one starts with a smaller *malE-lacZ* fusion that confers a less severe Mal^s phenotype, then one obtains a different subset of *malE* signal sequence mutations by selecting Mal^r (15). Of four mutations obtained this way, three are substitutions of one uncharged amino acid for another at positions 11 and 14, and the remaining mutation introduces a charged amino acid at position 10. All of these mutations have a less dramatic effect on export of the hybrid protein and of the wild-type MalE protein than do mutations selected using the larger, more maltose-sensitive *malE-lacZ* fusion. Only one mutation selected this way (pro at position 14) confers Mal^r when placed cis to the larger *malE* fusion. Thus, not unexpectedly, different *malE* fusions yielded different export-defective mutations in the Mal^r selection, and, as with *lamB,* more than one type of signal sequence mutation is observed.

Unlinked Mutations That Relieve Overproduction Lethality

If the Mal^s phenotype is the result of β-galactosidase sequences jamming export sites and consequently preventing the localization of essential envelope proteins, then it should be possible to alter the export sites by mutation and relieve the jamming. Using a modified Mal^r selection, several mutations in a previously undefined genetic locus have been identified (44). These *prlF* mutations relieve the Mal^s phenotype of both *lamB-lacZ* and *malE-lacZ* fusions. In such strains the hybrid proteins no longer jam the export sites, as evidenced by the fact that no precursor forms of other exported proteins can be detected accumulating in the cells. In the mutant strains, the hybrid proteins are efficiently incorporated into the envelope. This occurs because in the absence of jamming, the hybrid protein does not block its own export. As a consequence of this more efficient export, LacZ activity is reduced 50–100 fold. The *prlF1* mutation maps at 71 minutes and confers a slight but easily detectable cold-sensitive growth phenotype that is independent of gene fusions. Whether this mutation alters a component of the export machinery directly, causes an increase in the amounts of some limiting component, or whether it relieves Mal^s by some indirect mechanism remains to be determined. In any event, analysis of *prlF* should shed light on the nature of the export-related Mal^s phenotype.

Intragenic Mutations That Alter the Location of Membrane-Associated Hybrid Proteins

LAMB   Strains carrying the largest *lamB-lacZ* fusion, 42-12, are Lac$^-$ owing to the high efficiency ($\sim$ 90%) of hybrid protein export (12). Selections for growth on minimal media supplemented with lactose yields both signal sequence point mutations and deletions. The deletions all remove at least part of the hydrophobic core and extend into the mature portion of the protein. The point mutations are of two types. The first are point mutations at positions 15 and 19. These are identical to ones isolated in the Mal^r Lac$^+$ selection using the Mal^s 42-1 fusion. The second class contains novel signal sequence mutations at positions 12 and 13 (Table 2). Both of these mutations confer a Mal^r Lac$^+$

phenotype. When they are crossed cis to the wild-type gene, the block in protein export is less than that observed with mutations that alter residues 14, 15, 16 or 19, but it is greater than that observed with mutations that alter position 17. This was determined by measuring the ratio of precursor to mature *lamB* present in a total cellular extract (S. Benson, personal communication).

Over 300 independent Lac$^+$ mutants of the 42-12 fusion have been obtained by selecting for Lac$^+$ on lactose tetrazolium agar, a complex medium (30). These mutants are of two types, Lac$^+$ Mal^r and Lac$^+$ Mal^s. The Lac$^+$ Mal^r are the predominant class. All of the mutants appear to have resulted from deletion events that remove both *lamB* and *lacZ* sequences. Consequently, these deletions create a number of new *lamB-lacZ* fusions. At least two new regions specifying intragenic export information present in the early part of *lamB* have been defined by analysis of these fusions. These regions are required in addition to the signal sequence. Defined in terms of amino acids of mature protein, the regions are as follows. The first region lies between amino acids 27 and 39, and it appears to be required to direct insertion of the nascent chain into the export site. If this region is deleted, the fusion does not confer a Mal^s phenotype and the hybrid protein remains in the cytoplasm. Preliminary results suggest that this region may participate in the coupling of translation and export (see below). It should be noted that although the region is defined in terms of amino acids of mature protein, the export information may be read from the mRNA. A second region between 39 and 49 appears to direct sorting to the outer membrane. If this region is absent the cell attempts to export the hybrid protein, but it does not reach the outer membrane. This hypothesis has been strengthened by the finding of Nikaido & Wu (45). They have shown that the outer membrane proteins OmpA, OmpF, and LamB all contain a region of amino acids that are unique to outer membrane proteins and suggest that this may be the tag the cell uses to sort these proteins to the outer membrane. For LamB this region overlaps the 39-49 region identified as a sorting signal using gene fusions.

PHOA   Fusions between alkaline phosphatase *(phoA)* and β-galactosidase *(lacZ)* that are efficiently localized to the envelope are phenotypically Lac$^-$. By selecting for Lac$^+$ it has been possible to isolate mutations that alter the *phoA* signal sequence. Five base pair substitutions at positions 8, 9, 10, and 14 and one small deletion that removes codons 8 and 9 have been characterized at the level of the DNA sequence [(46), S. Michaelis, J. Hunt, J. Beckwith, personal communication]. Here again two classes of mutations are observed. Three of the mutations block export to >95%; glu at position 9, his at position 10, and arg at position 14. In contrast, gln at position 8 and gln at position 14 have a lesser effect, 84% and 70% respectively.

Unlinked Mutations That Alter the Location of Membrane-Associated Hybrid Proteins

As described above, *malE-lacZ*, *lamB-lacZ*, and *phoA-lacZ* fusions that specify a hybrid protein that is efficiently incorporated into a membrane confer a Lac⁻ phenotype. The properties of these strains suggest that only a slight degree of hybrid protein internalization would provide sufficient β-galactosidase activity to allow growth on lactose. This is the case, and this observation has been exploited to obtain mutations that define two genetic loci in *E. coli* (*secA* and *secB*) whose gene products appear to be involved in protein export. In a similar fashion the largest *lamB-lacZ* fusion, 42-12, has been useful in the selection of mutations which also appear to pleiotropically effect protein export. These mutations define a single genetic locus *prlE*.

SECA   The original *secA* mutation was isolated using the Lac⁻ *malE-lacZ* fusion 72-47 and is temperature sensitive (*secA*ᵗˢ). Strains containing this mutation grow normally at 30°C; at 42°C, growth stops and cells form filaments and accumulate precursor forms of several different exported proteins (47). However, not all exported proteins are affected by *secA*ᵗˢ. Some proteins are localized normally at the nonpermissive temperature. The mutation is not strictly conditional; some precursors can be detected at permissive temperatures. The *secA* gene maps at 2.5 min on the *E. coli* chromosome. Using a *secA* transducing phage and UV-irradiated cells, Oliver & Beckwith (48) have shown that the gene product is a protein of 92,000 daltons. Fusions, (*secA-lacZ*), have been constructed, and the hybrid protein has been used to obtain antisera directed against SecA. In addition, a *secA* amber mutation has been isolated. With these reagents, evidence has been obtained indicating that the SecA protein is essential for growth and that it is regulated in response to the secretion requirements of the cell; i.e., if the export machinery is jammed with a hybrid protein, or if export is blocked by a *secA*ᵗˢ mutation, expression of *secA* is derepressed at least 10 fold (49).

Synthesis of the SecA protein can be controlled by constructing a *secA*ᵃᵐ *supF*ᵗˢ double mutant. In this strain, at the permissive temperature (30°C), the *secA*ᵃᵐ is suppressed by the *supF*ᵗˢ allele; at the nonpermissive temperature, the suppressor function is inactivated and this results in the production of an inactive truncated SecA protein. Using this system, Oliver & Beckwith (49) have shown that at the nonpermissive temperature synthesis of certain, but not all, exported proteins is blocked. For example, synthesis of the wild-type MalE protein is blocked at the nonpermissive temperature. Certain *malE* signal sequence mutations (50) as well as *prlA* alleles (see below), relieve this translational block. These results suggest that SecA may function in the coupling of translation and export.

SEC*B*   The *secB* mutation was identified using the same selection that yielded *secA*. It maps at 80.5 min on the *E. coli* chromosome. In contrast to *secA* mutations, the original *secB* mutation is not a temperature-sensitive lethal (51). Indeed, *secB*::Tn*5* insertion mutations have now been constructed, and it is clear from results with this mutation that *secB* is not an essential gene. In the absence of the *secB* gene product, the export of only a class of envelope proteins is blocked. Exported gene products affected by the *secB* mutation include LamB, MalE, and OmpF. Other envelope proteins, such as ribose-binding protein or PhoA, do not require SecB for normal localization (C. Kumamoto, J. Beckwith, personal communication).

Combinations of various *secB* and *secA* alleles produce a synergistic effect in that the block on protein export is more severe than that of either mutant alone or than the sum of the two effects (51). The various *secB, secA* combinations exhibit allele specificity; i.e., certain combinations yield a stronger export defect than others. This allele specificity suggests that the *secA* and *secB* gene products interact with each other. The function of SecB, however, remains to be determined.

PRL*E*   The *prlE* mutations were isolated using the Lac⁻ *lamB-lacZ* fusion 42-1. These mutations, which map at 8.5 min on the *E. coli* chromosome, cause accumulation of the precursor forms of the envelope proteins MalE, LamB, and PhoA (S. Benson, D. Kiino, C. Gardel, personal communication). They confer a slight growth defect at lower growth temperatures (30°C) that is independent of any gene fusion. A similar mutation has been isolated starting with a Lac⁻ *phoA-lacZ* fusion (C. Gardel, J. Hunt, S. Michaelis, J. Beckwith, personal communication). The fact that two different selections for export-defective mutations yielded *prlE* strengthens the proposal that the gene specifies a component of the protein localization machinery. However, more work is required.

UNUSUAL EXPORT-RELATED PHENOTYPES

Careful biochemical analysis of protein export in strains harboring strong *malE* and *lamB* signal sequence mutations has uncovered a second phenotype caused by these alterations (52). In addition to imposing a severe block in the export of MalE or LamB, respectively, they interfere with the export of other envelope proteins including MalE, LamB, OmpA, and PhoA, but not Lpp, causing their transient accumulation in precursor forms (V. Bankaitis, personal communication). The interference phenotype is observed only during high level synthesis of the export-defective protein. Relief of the export defect by either intragenic or extragenic suppressor mutations concomitantly relieves the interference phenotype. Strains carrying deletions that remove all but the first six amino acids of either the MalE or LamB signal sequence show the interference phenotype, and it is also observed when strong export defective mutations (either point mutations or deletions) are present on large *lamB-lacZ* or *malE-lacZ* fusions. In the latter case, the fusion must contain at least 567 bp of *malE*. This suggests that information in the mature portion of the MalE protein is required in addition to a defective signal sequence. This region has been defined by analyzing various internal *malE* deletions (15, 53). It lies between amino acids 160 and 189 of the mature protein and has been termed the internal export site (IES). The function of IES is not known.

ALLELE SPECIFIC SUPPRESSORS

As previously illustrated, one approach to isolating export-defective mutations is to select or screen for mutations that pleiotropically alter the export of one or more envelope proteins. An alternative approach is to select for the restoration [i.e., pseudoreversion, for review see (54)] of a phenotype in a mutant which carries an export defect. This restoration can be the export of a specific protein or the loss of a conditionally lethal phenotype that results from a pleiotropic defect in protein export. In either selection, the identification of suppressor mutations in a genetic locus other than the one that contains the original mutation provides evidence for the direct physical interaction between the gene products of the two loci. Such mutations would define components of the export machinery. The suppressor approach can also yield intragenic alterations that restore a phenotype. These mutations in themselves can provide information about protein structure or interactions between specific protein domains.

Suppressors of Signal Sequence Mutations

In order for *E. coli* to grow on maltodextrins (Dex$^+$) with a chain length of 4 glucose residues or more, a functional LamB protein is required. Consequently, mutations that block the localization of LamB confer a Dex$^-$ phenotype. Utilizing strains where the export of LamB is blocked due to a signal sequence mutation, it has been possible to identify second site mutations that suppress the Dex$^-$ phenotype. Using this selection, three different second-site suppressor mutations, termed *prlA*, -B, and -C, have been identified (55).

PRLA   Certain alleles of *prlA* phenotypically suppress export-defective signal sequence mutations in *lamB, malE, phoA,* and *ompF;* this restores apparently normal export and processing [(46, 55, 56); E. Sodergren, personal communication]. They do not cause any growth defect, nor do they appear to alter normal protein export. Genetic mapping places the *prlA* gene at 72 min on the *E. coli* chromosome, at the extreme distal end of the *spc* operon (57). This operon specifies 10 ribosomal proteins in addition to PrlA. DNA sequence analysis of *prlA4* (J. Shultz, personal communication) and the *secY*^{ts} allele of *prlA* (27) demonstrate that the mutations alter a large open reading frame at the end of the *spc* operon, capable of encoding a 443-amino-acid protein (58). In vitro construction of an inframe fusion of this open reading frame to *lacZ* confirms that this open reading frame is transcribed and translated in vivo (57). However, the PrlA protein has not yet been identified, despite repeated efforts to express this protein in a variety of expression systems, including minicells and maxicells. This may be a consequence of the unusual physical structure of this protein. The predicted amino acid sequence contains two very large hydrophobic regions and an overall preponderance of basic residues (pI>10).

The function of PrlA remains to be determined. The observed allele specificity between different *prlA* alleles and the various *lamB* and *malE* signal sequence mutations suggests a direct interaction between the signal sequence and the PrlA protein. The pleiotropic defects in the export of both periplasmic and outer membrane proteins observed with $secY^{ts}$ allele of *prlA* confirm that the *prlA* gene product is required for protein export (26, 27). In addition, the identification of *prlA* alleles that suppress the $secA^{ts}$ mutation argues that PrlA may interact directly with SecA (see below).

PRL*B*   The single *prlB* allele is a small deletion within the gene *(rbsB)* specifying the periplasmic ribose-binding protein (J. Garwin, S. Emr, personal communication). It suppresses only lamB signal sequence mutations, and unlike *prlA* it does not restore processing. It is unclear why this mutation suppresses the export defect of *lamB* signal sequence mutations. One possibility is that in *prlB* strains the LamB protein is localized via a mechanism that bypasses the normal export pathway.

PRL*C*   To date only two *prlC* alleles have been isolated, in contrast to over 100 *prlA* alleles. These suppressors, like *prlA,* suppress both *lamB* and *malE* signal sequence mutations, restore normal processing, and do not result in any growth defect. The *prlC* alleles map at 68 min on the *E. coli* chromosome (J. Shultz, N. Trun, personal communication). The function of this gene product is unknown.

PRL*D*   The strongest *malE* signal sequence mutations block function and confer a Mal⁻ (unable to grow on maltose) phenotype. By selecting for a Mal⁺ phenotype, Bankaitis & Bassford (59) have isolated extragenic suppressor mutations; all but one of these, *prlD,* map to the *prlA* locus. The *prlD* suppressor maps very close to the *secA* locus on the *E. coli* chromosome. Complementation studies have shown that these two loci are independent. The *prlD1* mutation allows weak suppression of the *malE* signal sequence deletion mutation Δ12-18 and some but not all of the *malE* signal sequence point mutations. In addition, it has been shown to weakly suppress at least one of the *lamB* signal sequence mutations. It has an antagonistic effect on the export of other *lamB* signal sequence mutants. It, like the other suppressor mutations, does not effect cell growth or the export of wild-type envelope proteins. However, when the *prlD1* mutation is combined with certain *prlA* alleles, the resulting double mutants exhibit a severe growth defect with the accumulation of the precuror forms of MalE, LamB, and OmpA. This observation suggests that *prlA* and *prlD* might interact.

Suppressors of secA^{ts}

The search for suppressors of the *secA*^{ts} phenotype has yielded mutations that map to a number of genetic loci, *prlA, secC, ssaD, -E, -F, -G,* and *-H*. All of these suppressors confer a cold-sensitive phenotype.

The *prlA1012* allele, unlike previously isolated *prlA* mutations, does not restore export of *lamB* or *phoA* signal sequence mutations. The mutation suppresses all five *secA*^{ts} alleles but does not suppress the *secA*^{am} (60). These results indicate that SecA and PrlA interact.

Two *secC* cold-sensitive mutations have been identified, and mapped to 68.5 min on the *E. coli* chromosome (61). They appear to be alterations of the *rpsO* gene that specifies the ribosomal protein S15 (S. Ferro-Novick, personal communication). Each of the *secC* alleles suppresses the *secA*^{ts} defect for growth and protein export, and each exhibits allele specificity. The range of *secA*^{ts} alleles that are suppressed by the two *secC*^{cs} mutants overlap but are not identical. Both *secC* alleles confer a cold-sensitive growth phenotype in the absence of the *secA*^{ts} mutation. In each case, at the nonpermissive temperature, translation of periplasmic and outer membrane proteins is blocked, whereas synthesis of nonexported proteins remains unaffected. The block in synthesis of MalE at the nonpermissive temperature is overcome by mutations that alter the hydrophobic region of the MalE signal sequence. These results suggest that the *secC* gene product is involved in the coupling of translation and export.

The remaining suppressors of *secA*^{ts}, from *ssaD* to *-H,* are similar to *secC* in that they decrease synthesis of MalE in *secA*$^+$ strains at low growth temperatures. They have not as yet been extensively characterized (D. Oliver, personal communication) (Table 2).

Suppressors of secY^{ts} (prlA)

Selections starting with the *secY*^{ts} *(prlA)* mutation have yielded a single mutation, *ssyA,* that suppresses the *secY* defect and confers a cold-sensitive phenotype (62). The *ssyA*^{cs} mutation maps between min 54 and 55 on the *E. coli* chromosome. The mutation does not cause the accumulation of precursors of exported proteins nor does it specifically inhibit their synthesis at either 42°C or 30°C, either in the presence or absence of *secY*^{ts}. However, these strains do exhibit poor growth at 30°C. This may be related to a general decrease in the rate of protein synthesis at low growth temperature. Since *ssyA* exhibits allele specificity, it is unlikely to suppress the *secY*^{ts} phenotype via a bypass mechanism; however, the function of the gene product is not yet known.

Intragenic Suppressors of Signal Sequence Mutations

BLA   Starting with two different independent frameshift mutations that alter the signal sequence of the periplasmic enzyme β-lactamase (Bla), Koshland et al (21) were able to isolate two signal sequence mutations that affect processing but not export. The first of these mutations changes the pro at position 20 to a ser; the second alters amino acids 18–21, which changes the wild-type sequence (cys-leu-pro-val) to ala-phe-leu-phe. It should be noted that both of these changes, although they do not block export from the cytoplasm, appear to prevent the release of the protein into the periplasmic space. This selection also yielded a class of signal sequence mutations that drastically reduces export across the inner membrane. The first of these changes residues 14–21 (phe-ala-ala-phe-cys-leu-pro-val) of the signal sequence to leu-arg-his-phe-ala-phe-leu-phe. The second alters residue 7–9 (arg-val-ala) to pro-cys-arg (Table 2).

OMPF   The use of frameshift mutations as a tool for obtaining signal sequence mutations has also been exploited in the *ompF* system. Starting with an *ompF-lacZ* gene fusion carrying a frameshift mutation that alters the signal sequence, a variety of signal sequence mutations that restore a correct reading frame have been isolated (E. Sodergren, personal communication). All eight mutants appear to cause a tight block in the export of the wild-type OmpF protein because they confer a K20 resistant (phage K20 uses OmpF as receptor) phenotype. The nature of the alterations to the signal sequence range from a single amino acid deletion to those that alter 9 of the 22 residues in the OmpF signal sequence, and they suggest that pro at position 12 and leu at position 14 are especially important residues (Table 2).

MALE   Bankaitis et al (63) have selected second-site pseudo-revertants of the 18 bp *malE* signal sequence deletion mutation Δ12-18. This mutation removes three of the four residues (14, 16, 18) that are believed to be critical for function and it reduces the length of the hydrophobic core. Six second-site mutations have been analyzed (Table 3). Five alter residues within the signal sequence, allowing export of the mutant protein. Three of these restore export to wild-type levels. The other two have less dramatic effects. In each case the length of the hydrophobic core is increased in size. The observation that either of two single amino acids substitutions at position 8 (arg→leu or arg→cys) can restore export to near wild-type levels suggests that an important requirement for a functional signal is hydrophobic axis length (64). The sixth and by far the weakest second-site mutation maps at a position corresponding to amino acid 19 of mature MalE. It is unclear why this change should partially restore export; however, this result provides further evidence that sequences within the mature portion of the protein are involved in the export process.

Table 3 Intragenetic Suppressors of Signal Sequence Mutations[a]

```
LamB (WT)   MET MET ILE THR LEU ARG LYS LEU PRO LEU VAL ALA VAL ALA ALA     GLY VAL MET SER ALA GLN ALA MET ALA↓val
     (S78)  MET MET ILE THR LEU ARG LYS LEU PRO /////// ALA VAL ALA ALA     GLY VAL MET SER ALA GLN ALA MET ALA↓val
   (S78r1)  MET MET ILE THR LEU ARG LYS LEU PRO /////// ALA VAL ALA ALA CYS GLY VAL MET SER ALA GLN ALA MET ALA↓val
   (S78r2)  MET MET ILE THR LEU ARG LYS LEU LEU /////// ALA VAL ALA ALA     GLY VAL MET SER ALA GLN ALA MET ALA↓val

                                                                                          R
                                                                                          |
Lpp (WT)    MET LYS ALA THR LYS LEU VAL LEU GLY ALA VAL ILE LEU GLY SER THR LEU LEU ALA GLY↓cys

                                                                    |
   (mlpA)   MET LYS ALA THR LYS LEU VAL LEU GLY ALA VAL ILE LEU ASP SER THR LEU LEU ALA GLY↓cys

                                                                    R*
                                                                    |
  (14R21)   MET LYS ALA THR LYS LEU VAL LEU GLY ALA VAL ILE LEU ASN SER THR LEU LEU ALA GLY↓cys


MalE (WT)   MET LYS ILE LYS THR GLY ALA ARG ILE LEU ALA LEU SER ALA LEU THR THR MET MET PHE SER ALA SER ALA LEU ALA↓lys
            ile glu gly lys leu val ile trp ile asn gly asp lys gly tyr asn gly leu ala glu
```

```
MalE (12-18)    MET LYS ILE LYS THR GLY ALA ARG ILE LEU ALA  ///////////  MET PHE SER ALA SER ALA LEU ALA↓lys

MalE (12-18r2)  MET LYS ILE LYS THR GLY ALA LEU ILE LEU ALA  ///////////  MET PHE SER ALA SER ALA LEU ALA↓lys

MalE (12-18r3)  MET LYS ILE LYS THR GLY ALA CYS ILE LEU ALA  ///////////  MET PHE SER ALA SER ALA LEU ALA↓lys

MalE (12-18r4)  MET LYS ILE LYS THR GLY ALA ARG ILE LEU LEU ALA  [   ]    MET PHE SER ALA SER ALA LEU ALA↓lys

MalE (12-18r5)  MET LYS ILE LYS THR GLY ALA ARG ILE LEU VAL  ///////////  MET PHE SER ALA SER ALA LEU ALA↓lys

MalE (12-18r1)  MET LYS ILE LYS THR GLY ALA ARG ILE LEU ALA  ///////////  MET LEU ALA MET PHE SER ALA SER ALA SER↓lys

MalE (12-18r6)  MET LYS ILE LYS THR GLY ALA ARG ILE LEU ALA  ///////////  MET PHE SER ALA SER ALA LEU ALA↓lys
                ile glu glu gly lys leu val ile trp ile asn gly asp lys gly tyr asn VAL leu ala glu
```

aThe amino acids of the wild type (WT) signal sequences for the exported proteins LamB, Lpp, and MalE are shown. A segment (amino acids 1-22) of mature MalE is also shown (lower case letters). The (↓) above the line indicates the processing site. The R group above the cys in Lpp denotes the thioetherdiglyceride modification; its absence in the mlpA allele indicates lack of this modification and the R* represents a reduction in the amount of this modification. The signal sequence of mutants that served as the starting allele in selection of intragenic suppressor mutations are shown below the WT sequences. The signal sequences of the second site suppressors are shown below the mutant sequences. Deletions are represented by boxes, and single amino acid insertions or substitutions are underlined. For MalE, the suppressor mutation at position 19 of the mature protein is also shown. See the text for a complete description of these mutants.

LAMB   Starting with a strain carrying the 12 bp deletion, S78, that does not alter any of the residues believed to comprise a critical recognition site and selecting for Dex$^+$, it has been possible to obtain second site mutations within the *lamB* signal sequence that restore function (Table 3). DNA sequence analysis and predictive rules for polypeptide structure suggested that two independent second-site mutations restore export by allowing the region containing the critical residues at positions 14, 15, and 16 to assume an α-helical conformation (65). Physical measurements using synthetic signal sequence polypeptides have confirmed this hypothesis (66). These results indicate that conformation is important for signal sequence function.

The arg to ser alteration at position six of the LamB signal sequence has also been used in the selection of second site pseudorevertants. This mutation decreases translation of *lamB* and *lamB-lacZ* fusions that specify exported hybrid proteins (42). Starting with a *lamB-lacZ* fusion that is Lac$^-$ owing to the presence of this alteration and selecting Lac$^+$, a number of second site alterations within the *lamB* gene have been identified (M. Hall, S. Benson, personal communication). These alterations are of three types; pseudorevertants that restore an arg residue at position six; signal sequence mutations that alter one or more of the amino acids within the hydrophobic core; and a novel deletion mutation that removes residues 10 to 99 of the mature LamB protein. Whether these mutations relieve the translational block imposed by the alteration at position six, or whether they confer a Lac$^+$ phenotype by internalizing and thus activating the hybrid protein, remains to be determined. Nevertheless, this novel deletion provides further evidence that *lamB* sequences corresponding to the NH$_2$-terminal portion of the mature protein contain important export information.

LPP   The *mlpA* allele of *lpp* results in the substitution of asp for gly at position 14 of the Lpp signal sequence (11). As described above, this mutation confers resistance of globomycin by preventing modification and consequently processing of the peptide. Spontaneous revertants are isolated by selecting for resistance to EDTA. One pseudorevertant changes the mutant asp codon to asn (Table 3) (67). This alteration restores modification and allows subsequent processing. In this pseudorevertant the rate of modification is reduced relative to the wild-type *lpp* while the rate of processing both in vivo and in vitro is more rapid. Neither the original mutation nor the pseudorevertant appear to significantly alter the localization of Lpp.

IN VITRO MUTAGENESIS

The advent of recombinant DNA technology provided several additional means to produce mutations that alter the export process. One approach is to generate in vitro deletions that remove specific segments of the coding sequence; a second approach is to generate insertions by the addition of small oligonucleotide linkers into the coding sequence, and a third approach is to construct desired mutations using site-specific oligonucleotide-directed mutagenesis.

OMPA   Henning and coworkers have used recombinant DNA to produce a series of 3' *ompA* deletions that specify truncated OmpA polypeptides (68, 69). These studies show that while the mature protein contains 325 amino acids, only the first 193 are necessary for export and stable incorporation into the outer membrane. Surprisingly, this 193-amino-acid fragment is functional and complements all of the *ompA* phenotypes except EDTA sensitivity. In contrast, a slightly shorter peptide of 133 amino acids is nonfunctional and very unstable. Recently, Henning et al (36) have succeeded in the in vitro construction of fusions between the *ompA* gene and a variety of short open reading frames in the plasmid PUC8, in the tetracycline gene of pBR322, and in the *Vp 1* gene of the foot and mouth virus. Hybrid proteins that contain the first 163 amino acids of OmpA are very unstable, and apparently not exported. Similar fusions that contain 193 amino acids of *ompA* are stable and exported. This is consistent with the hypothesis that mature sequences are required for normal export.

F1 GENE III   The gene III product (pIII) of the filamentous phage f1 is an integral membrane protein of 406 amino acids. This protein spans the bilayer once and contains a 23-amino-acid hydrophobic sequence at the COOH terminus that appears to serve as a membrane anchor. Support for this comes from the following constructions (70, 71). Deletions that remove 210 amino acids from the COOH terminus yield a truncated peptide that is secreted into the periplasm. Deletions that remove the 23 hydrophobic amino acids also have this effect. In contrast, deletions that remove the basic residue at the carboxy-terminal boundary and deletions that remove fewer than seven amino acids from the hydrophobic segment have no effect on localization. Longer deletions into the hydrophobic domain gradually destabilize the membrane attachment. Replacement of this domain with a similar region from the gene VIII product of the f1 phage restores a membrane localization. Conversely, replacement of this region with a segment of the λc1 gene product that lacks a definable hydrophobic domain does not (J. Boeke, personal communication).

These results indicate that all of the signals required for export from the cytoplasm are contained in the first half of the coding sequence, and they suggest that the carboxy-terminal membrane anchor sequence may behave in a fashion analogous to the stop transfer sequence observed with a variety of eucaryotic proteins (5).

LAMB   In the *lamB* system, a different general approach has been used to direct mutagenesis to specific regions of the gene. Starting with a cloned *lamB-lacZ* gene fusion, a series of deletions within the coding sequence of the mature protein was generated (72). These deletions, ranging in size from ~12 to 400 base pairs, are of two types. The majority (8 out of 10) do not alter localization of either the hybrid protein, or LamB itself. Taken as a group, these deletions remove the region from approximately amino acid 70 to 220 of the mature protein [Mature LamB protein contains 421 amino acids, (73)]. Since none of these deletions affect localization, it would appear that this region does not contain essential information required for export of LamB to the outer membrane. Furthermore, a number of these deletions remove significant portions of the protein (over 100 amino acids). Since the internally truncated protein is localized, this provides further evidence that overall protein structure is not a driving factor in the localization of LamB.

The remaining two deletions define a region that is required for efficient LamB-LacZ hybrid protein export to the outer membrane. This region includes amino acids 220–241 of LamB and the amino acids specified by the *lamB-lacZ* fusion joint. Whether this region also functions in wild-type LamB export, or whether it is an artifact caused by the abnormal DNA joint, is not yet clear. Evidence obtained by linker mutagenesis of *lamB* supporting the latter explanation has been presented by Bouges-Bocquet et al (74).

LPP   Site-specific oligonucleotide-directed mutagenesis has been applied to both the hydrophilic and hydrophobic regions of the Lpp signal sequence. Although this approach has great potential, the results were somewhat disappointing in that most of these mutations have only marginal effects. Of four separate directed alterations to the hydrophilic region of the Lpp signal sequence, only the most dramatic change has any significant effect on export of the protein. This alteration changes the lys-ala at position 2 and 3 to glu-asp. This change introduces two negatively charged residues and alters the charge normally present in this region from a $+1$ to a -2. These alterations, like the alteration in the hydrophilic part of the *lamB* signal sequence, appear to affect the coupling between export and translation (75, 76).

Site directed mutagenesis of residues in the hydrophobic region has yielded the following observations. Substitutions of val for gly at positions 9 and 14 either singly or in combination do not affect export; likewise deletion of the gly at position 9 has no effect. Deletion of the gly residue at position 14 slows modification and processing of the peptide. This effect can be suppressed by simultaneously deleting the gly at position 9 (77). Substitution of an ala residue for the ser at position 15 results in transient accumulation of membrane-bound unmodified pro-Lpp. This same substitution at position 16 (thr) has no affect. The double mutant ala15-ala16 exhibits a greater accumulation of the membrane-bound precursor that the ala15 alteration alone (78). The results with *lpp* demonstrate that the signal sequence can tolerate numerous changes and still function in a normal or near normal fashion.

OMPF   Recently, site-directed mutagenesis has been applied to the *ompF* signal sequence with an important modification (R. Zagursky, M. Berman, personal communication). The lys codon at position three was converted to an amber codon. By using the appropriate suppressor it has been possible to substitute a variety of different amino acids at this position (E. Sodergren, personal communication). These results indicate that substitution of tyr for the lys at this position results in a functional signal sequence, whereas substitution of a ser does not. Since neither tyr nor ser reinstates a charged residue, it follows that it is structure rather than charge that is critical for this region. Whether or not these changes affect OmpF synthesis in a manner analogous to similar *lamB* or *lpp* mutations remains to be determined.

CLONING OF THE SIGNAL PEPTIDASE GENES

One striking and diagnostic feature of all known periplasmic and outer membrane proteins in *E. coli* is the presence of an NH_2-terminal signal sequence. In most cases, this signal sequence is proteolytically removed during the export process. Processing is not required for the export from the cytoplasm, at least not in any of the cases examined to date. However, to be processed the protein must cross the inner membrane; this requirement results from the extracytosolic location of the signal peptidase activities. Accordingly, processing has been and continues to be a bench mark by which export is measured. *E. coli* has been shown to have at least two distinct signal peptidases; signal peptidase I (SPase I or leader peptidase) and signal peptidase II (SPase II or lipoprotein signal peptidase) (17).

The gene specifying SPase I *(lep)* has been cloned, the DNA sequence determined, and the gene product purified (79–84). The enzyme (36,000 daltons) has been reported to process the precursor form of a variety of periplasmic and outer membrane proteins. The *lep* gene has been mapped (55 min) and evidence has been presented to suggest that it is an essential gene in *E. coli* (85, 86).

Signal peptidase II processes only glyceride-modified precursor proteins such as Lpp and it is sensitive to the inhibitory effect of the antibiotic globomycin. The gene encoding this activity *(lsp)* has been cloned and the DNA sequence determined by two independent groups (87–92). The *lsp* gene has been mapped at 0.5 min. Its essential nature has been demonstrated by the isolation of a conditionally lethal mutation and by the cell's sensitivity to globomycin. The enzyme has a monomer subunit molecular weight of approximately 18,000. It appears to be an integral inner membrane protein with four hydrophobic regions. The two SPases do not share sequence homology at the DNA or protein levels, and neither enzyme appears to have a cleavable NH_2-terminal signal sequence (17). How these proteins are assembled into the membrane is not yet clear.

THE SIGNAL HYPOTHESIS AND *E. COLI*

A commonly accepted model for the early steps in protein localization is the signal hypothesis (5a, 98, 99). This model is an oversimplification and it does not adequately explain all of the experimental data. Nevertheless, it is striking that a number of predictions made from a model proposed primarily on the basis of biochemical studies with eucaryotic cells are supported by genetic data obtained with bacteria.

In Figure 1 we present an updated and slightly modified version of the original signal hypothesis. According to this model, translation of mRNA for exported proteins is initiated by soluble ribosomes in the cytoplasm. Translation proceeds until the signal sequence (---- in the mRNA, •••• in the protein in Figure 1) emerges from the large ribosomal subunit. At this point the signal sequence is recognized by a component of the cellular export machinery, signal recognition particle (SRP). SRP, which is composed of 6 different proteins and 7S RNA, binds the signal sequence and the ribosome and arrests translation (100–102). This block is maintained until the polysome diffuses to the membrane (the rough endoplasmic reticular membrane is eucaryotes, the inner membrane in procaryotes) where it interacts with another component of the export machinery docking protein (103, 104) or SRP receptor (105). This interaction releases the translation arrest and synthesis is resumed. The signal sequence loops (106) into the bilayer and a competent export site is formed. This site is thought to be composed of a protein(s) yet undiscovered. SRP and docking protein are not part of this site and they are released at this point to function again (107). As synthesis proceeds, the nascent chain is transferred vectorially across the membrane. For secreted proteins in eucaryotic cells or for periplasmic proteins in gram-negative bacteria this vectorial transfer continues until the protein is discharged on the opposite side of the membrane. Membrane proteins are thought to contain additional export information such as stop transfer sequences (membrane anchor sequences) that stop vectorial transfer without stopping translation, which leaves the molecule imbedded in the membrane (5, 5a).

Cellular Components of the Export Machinery

Genetic analysis in *E. coli* has revealed a number of genes whose products may be components of the cellular export machinery. In many cases the evidence

Periplasm

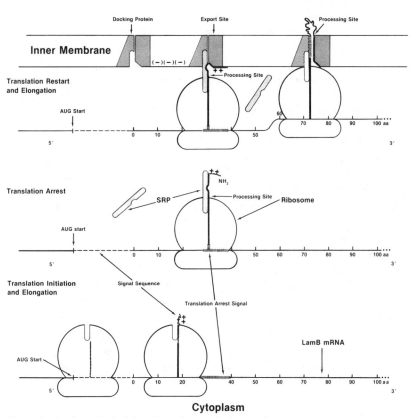

Cytoplasm

Figure 1   A schematic depicting the early steps in the protein export. See text for details.

supporting gene product involvement is weak and in certain cases, i.e., *prlB* or *perA,* available data argues to the contrary. However, in at least two cases, *prlA (secY)* and *secA,* the evidence is very compelling. Mutations in these genes exhibit several different export-related phenotypes and the gene products are essential for cell growth and viability. Moreover, these two gene products probably interact directly. It is important to note that the null phenotype for each of these two genes, i.e., the phenotype in the absence of the gene product, is quite different. For *prlA (secY),* the null phenotype is cytoplasmic precursor accumulation. For *secA,* it is a specific defect in the translation of the mRNA specifying exported proteins. These results suggest the existence of a mechanism that couples export and translation. Further support for this suggestion comes from the analysis of intragenic export-defective mutants (see below) and from the characterization of allele specific suppressors of *secA*[ts], most notably *secC (rpsO)*. Here the essential gene product, the ribosomal protein S15, is known to be involved in translation.

The model shown in Figure 1 proposes at least nine important cellular components of the export machinery: seven for SRP, one for docking protein, and one for the export site. Also, of course, it makes a very clear prediction concerning the mechanism that couples export and translation. We would expect to find mutations that block export by preventing the translation arrest or by inactivating the membrane export site. Also, we would expect mutations that prevent release of the translation arrest. The first and second type of mutations would lead to cytoplasmic precursor accumulation. The third would cause a decrease in net synthesis. Viewed in these terms, PrlA could be a component of the procaryotic equivalent of SRP (pSRP) required to impose the translation arrest or it could be a component of the membrane export site. In either case, the genetic data indicate that it interacts directly with the signal sequence. SecA, on the other hand, would be required to relieve the translation arrest imposed by pSRP. Accordingly, it could be a component of pSRP or it could be the procaryotic equivalent of docking protein. In any event, an interaction with the ribosomal protein S15 would appear to be required to relieve the translation arrest.

Reports have appeared suggesting that SecA and 6S RNA may be components of pSRP (108, 109). These have not been confirmed and recent results demonstrate that 6S RNA is required neither for protein export nor for cellular viability (C. Lee, J. Beckwith, personal communication). Accordingly, there is no direct evidence to support the existence of a prokaryotic equivalent of either SRP or docking protein. Whether these will be discovered or whether bacteria use a somewhat different mechanism to couple synthesis and export remains to be seen.

Signal Sequence Mutations

If we consider only base substitution mutations that change one amino acid in the signal sequence to another, we can identify at least three classes of signal sequence mutations. The class I mutants cause a severe export block (>95%) and thus confer a null phenotype. Generally these mutations are confined to a specific subset of amino acids within the signal sequence and they introduce a charge into the hydrophobic core. Comparison of class I mutations in different signal sequences does not reveal a consistently reproducible pattern. Class II signal sequence mutations are leaky and often confer no easily detectable phenotype. They block export to varying degrees and may vary the rate of export or processing or both. Class III signal sequence mutations are quite distinct. Rather than causing a pronounced export defect, they decrease translation thus providing further evidence supporting the existence of a mechanism to couple synthesis and export. These mutations appear to be confined to the codons that specify the basic amino acids that define the border between the NH_2-terminal hydrophilic portion and the hydrophobic core of the signal sequence. The simplest explanation for this confusing complexity is to propose that the signal sequence performs multiple functions. Certain mutations may affect one function while others may affect another or perhaps be multiply defective.

The signal hypothesis does indeed propose multiple functions for the signal sequence. It should interact with SRP and the membrane and perhaps docking protein and/or the export site as well. Class I mutations are probably multiply defective. Two lines of evidence suggest that these mutations may prevent recognition by pSRP. First, they do not significantly reduce net synthesis. Second, all of the class I signal sequence mutations tested so far suppress the translation defect caused by certain *secA* and *secC* alleles. This is satisfying logically because a mutation that prevents release of the translation arrest should have no effect on synthesis if a second mutation prevents imposition of the translation arrest in the first place. The class I mutations probably also inhibit the interaction between the signal sequence and the membrane because they introduce a charged amino acid and disrupt what has been termed the hydrophobic axis length (64). The best evidence supporting this suggestion comes from the analysis of second-site signal sequence mutations that phenotypically suppress class I mutations that alter the MalE signal sequence (see above). All of these second-site mutations restore an acceptable hydrophobic axis length.

Class II signal sequence mutations are difficult to explain. They may weakly affect one function of the signal sequences or another. It is not yet possible to comment meaningfully on these mutations.

Class III signal sequence mutations alter a basic residue near the NH_2 terminus of the signal sequence and could decrease net synthesis by inhibiting release of the translation arrest imposed by pSRP. If this is true and if class I signal sequence mutations prevent recognition of the signal sequence by pSRP as described above, then class I signal sequence mutations should overcome the inhibition of translation that is caused by class III mutations. As described in a previous section, a selection has been devised to search for second-site pseudorevertants of a class III signal sequence mutation in *lamB*. This selection does yield class I signal sequence mutations in accordance with predictions made here. However, until these double mutants are characterized more carefully, this result should be viewed as tentative.

According to the model shown in Figure 1, the signal sequence itself does not pass through the bilayer. Rather it loops into the membrane, and the NH_2 terminus of the signal sequence never leaves the cytoplasm. Convincing evidence supporting this topology has been presented by Inouye and coworkers (34). As predicted, attaching up to 145 random amino acids (specified by *lacZ*) to the NH_2 terminus of the signal sequence does not block export of Lpp to the outer membrane.

OTHER INTRAGENIC EXPORT INFORMATION

Evidence indicating the presence of intragenic information required in addition to the signal sequence for correct protein localization has been presented in several different experimental systems. In the case of *lamB*, two additional export signals have been identified and defined rather precisely. These signals lie in the regions of *lamB* corresponding to amino acids 27–39 and 39–49 of the mature product. The first appears to be required along with the signal sequence for protein export from the cytoplasm. The second appears to be a sorting signal to direct proper routing of LamB to the outer membrane.

The signal hypothesis makes no specific predictions about any intragenic information other than the signal sequence required for early steps in the export process. However, it is intriguing that when SRP binds the signal sequence and the ribosome and stops translation, we would expect the ribosome to stall somewhere in the vicinity of codon 30; in other words, in the region 27–39. We expect this because the signal sequence must emerge from the large ribosomal subunit to be recognized by SRP, and the ribosome protects ~ 28 amino acids. These observations raise the possibility that certain amino acids in the 27–39 region or perhaps more likely, certain sequences in the corresponding mRNA may be required for the translation arrest. Support for this possibility comes from the following observation: The *lamB* mutation that alters one of the basic residues near the NH_2-terminal end of the signal sequence (class III) decreases synthesis of *LamB* and all exported LamB-LacZ hybrid proteins. Indeed, it decreases translation of all *lamB-lacZ* fusions that specify hybrid proteins containing the signal sequence and 39 or more amino acids of LamB. In contrast, the mutation has no effect on fusions that specify a hybrid protein containing the signal sequence and 27 or fewer amino acids of LamB (S. Benson, M. Hall, personal communication). This is striking because the cell attempts to export LamB-LacZ hybrid proteins containing the signal sequence and 39 or more amino acids of LamB while hybrid proteins containing the signal sequence and 27 amino acids or less remain in the cytoplasm. This correlation indicates a mechanism to couple synthesis and export and identifies the 27–39 region as one containing important information required for this mechanism.

Computer searches for amino acid sequence homology have defined a region common to all outer membrane proteins, and it was suggested that this region may identify for the cell a protein destined for this particular location; i.e., a sorting sequence. As described earlier, results with *lamB-lacZ* fusions that identified the 39–49 region of LamB as a sorting signal are consistent with this suggestion. The mechanism of sorting, however, remains unknown.

In *malE* (and perhaps *lamB* as well, see above), evidence suggesting yet another signal, termed internal export site (IES), has been presented. It appears to be located between amino acids 160 and 189 and when MalE export is prevented, the presence of this site in the cytoplasm inhibits localization of several other envelope proteins. The function of IES is not clear. Recently, Randall (110) has presented evidence to suggest that translocation across the inner membrane does not occur until a substantial portion of the protein has been synthesized. These results may reflect a requirement for IES and they imply a more complicated mechanism of protein transport through the inner membrane than that implied in Figure 1.

Early versions of the signal hypothesis predicted that membrane proteins would contain additional export information, stop transfer sequences or membrane anchors, that would serve to stop translocation of the protein across the membrane without stopping synthesis. This would serve to imbed the protein in the membrane bilayer. Such membrane anchors have been identified for several eucaryotic viral glycoproteins [for review see (5)] and for the procaryotic phage protein pIII. As expected, if these sequences are deleted, the complete protein is translocated across the membrane.

POSTTRANSLATIONAL EXPORT

For years, the question of whether protein export occurred during translation (cotranslational) or whether it occurred posttranslationally has been shrouded in controversy. Results presented with eucaryotic systems and genetic evidence in bacteria were most consistent with a cotranslational mode. Nevertheless, a substantial amount of biochemical evidence exists in bacteria supporting a posttranslational mode. Space and the scope of this review do not permit a detailed discussion of this data. However, with respect to bacterial systems, several general statements can be made. First, the energy required for export is not derived from translation. Energy appears to be provided by the membrane potential (16, 111, 112). Second, certain inner membrane proteins, because of their physical properties, may insert into the membrane spontaneously as predicted by the membrane trigger hypothesis of Wickner (113). With few exceptions, these proteins have not been studied genetically. Finally, although many proteins may be localized normally by a cotranslational mechanism, under appropriate in vitro conditions or in certain mutants, export may occur by a posttranslational mechanism. Whether this posttranslational export represents a completely different mechanism or whether certain components of the normally cotranslational machinery are still required remains to be determined.

ACKNOWLEDGMENTS

Research sponsored by the National Cancer Institute, DHHS, under Contract No. NO1-CO-23909 with Litton Bionetics, Inc. The contents of this publication do not necessarily reflect the views or policies of the Department of Health and Human Services, nor does mention of trade names, commercial products, or organizations imply endorsement by the U.S. Government.

Literature Cited

1. Fraser, T. H., Bruce, B. J. 1978. *Proc. Natl. Acad. Sci. USA* 75:5936–40
2. Talmadge, K., Stahl, S., Gilbert, W. 1980. *Proc. Natl. Acad. Sci. USA* 77:3369–73
3. Wiedman, M., Huth, A., Rapoport, T. A. 1984. *Nature* 309:637–39
4. Palade, G. E. 1975. *Science* 189:347–58
5. Sabatini, D. D., Kreibich, G., Morimoto, T., Adesnik, M. 1982. *J. Cell Biol.* 92:1–22
5a. Walter, P., Gilmore, R., Blobel, G. 1984. *Cell* 38:5–8
6. Goebel, W., Hedgpeth, J. 1982. *J. Bacteriol.* 151:1290–98
7. de Geus, P., Verhei, H. M., Riegman, N. H., Hoekstra, W. P. M., de Haas, G. H. 1984. *EMBO J.* 3:1799–1802
8. Ito, K., Bassford, P. J. Jr., Beckwith, J. 1981. *Cell* 24:707–17
9. Inouye, H., Beckwith, J. 1977. *Proc. Natl. Acad. Sci. USA* 74:1440–44
10. Inouye, S., Wang, S., Sekizawa, J., Halegoua, S., Inouye, M. 1977. *Proc. Natl. Acad. Sci. USA* 74:1004–8
11. Lin, J. J.-C., Kanazawa, H., Ozols, J., Wu, H. C. 1978. *Proc. Natl. Acad. Sci. USA* 75:4891–95
12. Silhavy, T. J., Benson, S. A., Emr, S. D. 1983. *Microbiol. Rev.* 47:313–44
13. Michaelis, S., Beckwith, J. 1982. *Ann. Rev. Microbiol.* 36:435–65
14. Osborn, M. J., Wu, H. C.-P. 1980. *Ann. Rev. Microbiol.* 34:369–422
15. Bankaitis, V. A., Ryan, J. P., Rasmussen, B. A., Bassford, P. J. Jr. 1984. *Current Topics in Membranes and Transport*, ed. P. A. Knauf, J. Cook. New York: Academic. In press
16. Randall, L. L., Hardy, S. J. S. 1984. *Modern Cell Biology*, ed. S. Birgit. New York: Liss. In press
17. Wu, H. 1984. *J. Cell Biol.* In press
18. Schatz, G., Butow, R. A. 1983. *Cell* 32:316–18
19. Scheckman, R. 1982. *Trends Biochem. Sci.* 7:243–46
20. Koshland, D., Botstein, D. 1980. *Cell* 20:749–60
21. Koshland, D., Sauer, R. T., Botstein, D. 1982. *Cell* 30:903–14
22. Ito, K., Beckwith, J. R. 1981. *Cell* 25:143–50
23. Wanner, B. L., Sarthy, A., Beckwith, J. 1979. *J. Bacteriol.* 140:229–39
24. Garrett, S., Taylor, R. K., Silhavy, T. J. 1983. *J. Bacteriol.* 156:62–69
25. Dassa, E., Boquet, P. L. 1981. *Mol. Gen. Genet.* 181:192–200
26. Ito, K., Wittekind, W., Nomura, M., Miura, A., Shiba, K., et al. 1983. *Cell* 32:789–97
27. Shiba, K., Ito, K., Yura, T., Cerretti, D. P. 1984. *EMBO J* 3:631–35
28. Weinstock, G. M., Berman, M. L., Silhavy, T. J. 1983. *Expression of Cloned Genes in Procaryotic and Eucaryotic Vectors,* ed. T. S. Papas, M. Rosenberg, J. Chirikjian, pp. 28–64. New York: Elsevier
29. Moreno, F., Fowler, A. V., Hall, M., Silhavy, T. J., Zabin, I., Schwartz, M. 1980. *Nature* 286:356–59
30. Benson, S. A., Bremer, E., Silhavy, T. J. 1984. *Proc. Natl. Acad. Sci. USA* 81:3830–34
31. Talmadge, K., Brosius, J., Gilbert, W. 1981. *Nature* 294:176–78
32. Tommassen, J., van Tol, H., Lugtenberg, B. 1983. *EMBO J* 2:1275–79
33. Wright, A., Hoffman, C., Fishman, Y. 1983. *J. Cell. Biochem.* 7B:346 (Abstr.)
34. Ghrayeb, J., Inouye, M. 1984. *J. Biol. Chem.* 259:In press
35. Yu, F., Furukawa, H., Nakamura, K., Mizushima, S. 1984. *J. Biol. Chem.* 259:6013–18
36. Henning, U., Cole, S. T., Bremer, E., Hindennach, I., Schaller, H. 1983. *Eur. J. Biochem.* 136:233–40
37. Beckwith, J., Silhavy, T. J. 1984. *Methods Enzymol.* 97:3–11
38. Hall, M. N., Silhavy, T. J. 1981. *J. Mol. Biol.* 151:1–15
39. Emr, S. D., Silhavy, T. J. 1980. *J. Mol. Biol.* 141:63–90
40. Emr, S. D., Hedgpeth, J., Clement, J. M., Silhavy, T. J., Hofnung, M. 1980. *Nature* 285:82–85
41. Emr, S. D., Silhavy, T. J. 1982. *J. Cell Biol.* 95:689–96

42. Hall, M. N., Gabay, J., Schwartz, M. 1983. *EMBO J*. 2:15–19
43. Bedouelle, H., Bassford, P. J., Fowler, A. V., Zabin, I., Beckwith, J., et al. 1980. *Nature* 285:78–81
44. Kiino, D. R., Silhavy, T. J. 1984. *J. Bacteriol*. 158:878–83
45. Nikaido, H., Wu, H. 1984. *Proc. Natl. Acad. Sci. USA* 81:1048–52
46. Michaelis, S., Inouye, H., Oliver, D., Beckwith, J. 1983. *J. Bacteriol*. 154:366–74
47. Oliver, D. B., Beckwith, J. 1981. *Cell* 25:765–72
48. Oliver, D. B., Beckwith, J. 1982. *J. Bacteriol*. 150:686–91
49. Oliver, D. B., Beckwith, J. 1982. *Cell* 30:311–19
50. Kumamoto, C., Oliver, D. B., Beckwith, J. 1984. *Nature* 308:863–64
51. Kumamoto, C., Beckwith, J. 1984. *J. Bacteriol*. 154:253–60
52. Bankaitis, V. A., Bassford, P. J. 1984. *J. Biol. Chem*. In press
53. Bankaitis, V. A., Bassford, P. J. 1984. *J. Bacteriol*. In press
54. Botstein, D., Maurer, R. 1982. *Ann. Rev. Genet*. 16:61–85
55. Emr, S. D., Hanley-Way, S., Silhavy, T. J. 1981. *Cell* 23:79–88
56. Emr, S. D., Bassford, P. J. Jr. 1982. *J. Biol. Chem*. 257:2852–60
57. Shultz, J., Silhavy, T. J., Berman, M. L., Niil, N., Emr, S. D. 1982. *Cell* 31:227–35
58. Cerretti, D. P., Dean, D., Davis, G. R., Bedwell, D. M., Nomura, M. 1983. *Nucleic Acids Res*. 11:2599–616
59. Bankaitis, V. A., Bassford, P. J. 1984. *J. Bacteriol*. In press
60. Brickman, E. R., Oliver, D. B., Garwin, J. L., Kumamoto, C., Beckwith, J. 1984. *Mol. Gen. Genet*. In press
61. Ferro-Novick, S., Honma, M., Beckwith, J. 1984. *Cell* 30:211–17
62. Shiba, K., Ito, K., Yura, T. 1984. *J. Bacteriol*. In press
63. Bankaitis, V. A., Rasmussen, B. A., Bassford, P. J. Jr. 1984. *Cell* 37:243–52
64. Bedouelle, H., Hofnung, M. 1981. *Membrane Transport and Neuroreceptors*, ed. D. Oxender, pp. 309–43. New York: Liss
65. Emr, S. D., Silhavy, T. J. 1983. *Proc. Natl. Acad. Sci. USA* 80:4599–603
66. Briggs, M., Gierasch, L. M. 1984. *Biochemistry*. 23:3111–14
67. Tokunaga, H., Wu, H. C. 1984. *J. Biol. Chem*. 259:6098–104
68. Bremer, E., Beck, E., Hindennach, I., Sonntag, I., Henning, U. 1980. *Mol. Gen. Genet*. 179:13–20
69. Bremer, E., Cole, S. T., Hindennach, I.,

Henning, U., Beck, E., et al. 1982. *Eur. J. Biochem*. 122:223–31
70. Boeke, J. D., Model, P. 1982. *Proc. Natl. Acad. Sci. USA* 79:5200–4
71. Davis, N. G., Boeke, J. D., Model, P. 1984. *J. Mol. Biol*. In press
72. Benson, S., Silhavy, T. 1983. *Cell* 32:1325–35
73. Clement, J. M., Hofnung, M. 1981. *Cell* 27:507–14
74. Bouges-Bocquet, B., Villarroya, H., Hofnung, M. 1984. *J. Cell Biol*. In press
75. Inouye, S., Soberon, X., Franceschini, T., Nakamura, K., Itakura, K., et al. 1982. *Proc. Natl. Acad. Sci. USA* 79:3438–41
76. Vlasuk, G. P., Inouye, S., Ito, H., Itakura, K., Inouye, M. 1983. *J. Biol. Chem*. 258:7141–48
77. Inouye, S., Vlasuk, G. P., Hsiung, H., Inouye, M. 1984. *J. Biol. Chem*. 259:3729–33
78. Vlasuk, G. P., Inouye, S., Inouye, M. 1984. *J. Biol. Chem*. 259:6195–200
79. Chang, C. N., Blobel, G., Model, P. 1978. *Proc. Natl. Acad. Sci. USA* 75:361–65
80. Mandel, G., Wickner, W. 1979. *Proc. Natl. Acad. Sci. USA* 76:236–40
81. Date, T., Wickner, W. 1981. *Proc. Natl. Acad. Sci USA* 78:6106–10
82. Wolfe, P. B., Silver, P., Wickner, W. 1982. *J. Biol. Chem*. 257:7898–902
83. Zwizinski, C., Date, T., Wickner, W. 1981. *J. Biol. Chem*. 256:3593–97
84. Zwizinski, C., Wickner, W. 1980. *J. Biol. Chem*. 255:7973–77
85. Date, T. 1983. *J. Bacteriol*. 154:76–83
86. Silver, P., Wickner, W. 1983. *J. Bacteriol*. 154:569–72
87. Tokunaga, M., Loranger, J. M., Wu, H. C. 1983. *J. Biol. Chem*. 258:12102–5
88. Innis, M. A., Tokunaga, M., Williams, M. E., Loranger, J. M., Chang, S., et al. 1984. *Proc. Natl. Acad. Sci. USA* 81:3708–12
89. Yamagata, H., Daishima, K., Mizushima, S. 1983. *FEBS Lett*. 158:301–8
90. Yamagata, H., Ippolito, C., Inukui, M., Inouye, M. 1982. *J. Bacteriol*. 152:1163–68
91. Regue, M., Remenick, J., Tokunaga, M., Mackie, G. A., Wu, H. C. 1984. *J. Bacteriol*. 158:632–35
92. Yamagata, H., Taguchi, N., Daishima, K., Mizushima, S. 1983. *Mol. Gen. Genet*. 192:10–14
93. Inouye, H., Barnes, W., Beckwith, J. 1982. *J. Bacteriol*. 149:434–39
94. Sutcliffe, J. G. 1978. *Proc. Natl. Acad. Sci. USA* 75:3737–41
95. Hedgpeth, J., Clement, J. M., Marchal,

C., Perrin, D., Hofnung, M. 1980. *Proc. Natl. Acad. Sci. USA* 77:2621–25

96. Berman, M. L., Jackson, D. E., Fowler, A., Zabin, I., Christensen, L., et al. 1984. *Gene. Anal. Tech.* 1:143–51
97. Mutoh, N., Inokuchi, K., Mizushima, M. 1982. *FEBS Lett.* 137:171–74
98. Blobel, G., Dobberstein, B. 1975. *J. Cell Biol.* 67:835–51
99. Blobel, G., Dobberstein, B. 1975. *J. Cell Biol.* 67:852–62
100. Walter, P., Blobel, G. 1980. *Proc. Natl. Acad. Sci. USA* 77:7112–16
101. Walter, P., Blobel, G. 1981. *J. Cell Biol.* 91:557–61
102. Walter, P., Blobel, G. 1982. *Nature* 299:691–98
103. Meyer, D. I., Dobberstein, B. 1980. *J. Cell Biol.* 87:503–8

104. Meyer, D. I., Krause, E., Dobberstein, B. 1982. *Nature* 297:647–50
105. Gilmore, R., Walter, P., Blobel, G. 1982. *J. Cell Biol.* 95:470–77
106. Inouye, M., Halegoua, S. 1979. *Crit. Rev. Biochem.* 7:339–71
107. Gilmore, R., Blobel, G. 1983. *Cell* 35:677–85
108. Robertson, M. 1984. *Nature* 307:594–95
109. Walter, P., Blobel, G. 1983. *Cell* 34:525–33
110. Randall, L. L. 1983. *Cell* 33:231–40
111. Enequist, H. G., Hirst, T. R., Harayama, S., Hardy, S. J. S., Randall, L. L. 1981. *Eur. J. Biochem.* 116:227–33
112. Daniels, C. J., Bole, D. G., Quay, S. C., Oxender, D. L. 1981. *Proc. Natl. Acad. Sci. USA* 78:5396–400
113. Wickner, W. 1979. *Ann. Rev. Biochem.* 48:23–45

Ann. Rev. Biochem. 1985. 54:135–169

CELL ADHESION AND THE MOLECULAR PROCESSES OF MORPHOGENESIS

Gerald M. Edelman

Laboratory of Molecular and Developmental Biology, The Rockefeller University, New York, NY 10021

CONTENTS

PERSPECTIVES AND SUMMARY

The development of new assays and the application of modern techniques have permitted a new biochemical assault on the classical problem of cell adhesion as it relates to morphogenesis. The purpose of this review is to outline the

135

0066-4154/85/0701-0135$02.00

problem of morphogenesis in genetic and evolutionary terms, briefly considering the role of primary processes of development and then singling out cell adhesion as a key regulatory primary process definable at the molecular level. A description of multistage immunological adhesion assays is followed by an account of some of the molecular properties of cell adhesion molecules, their general tissue distributions, and their spatiotemporal variations both in the course of embryogenesis and of histogenesis. Brain histogenesis is discussed as a key example.

Three cell adhesion molecules have been detected and isolated so far in a variety of vertebrate species: N-CAM (neural cell adhesion molecule), Ng-CAM (neuron-glia cell adhesion molecule) and L-CAM (liver cell adhesion molecule). They are all cell surface glycoproteins of relatively high molecular weight that differ in their structures. N-CAM, unlike the other two, has an unusual carbohydrate moiety containing large amounts of polysialic acid in α-2-8 linkage. N- and L-CAMs are so-called primary CAMs defined as such by the fact that they are expressed at very early stages of embryogenesis in derivatives of more than one germ layer. Ng-CAM is a secondary CAM derived from a single germ layer (neuroectoderm) and is seen first only in postmitotic neurons and at the beginning of tract formation in the nervous system.

Studies on the basic polypeptide chain structures of the three known CAMs show them to be different. Most is known about the structure-function relationships of N-CAM. N-CAM binding is directly to N-CAM on an opposing cell (homophilic binding) and is mediated by a binding region in the NH_2-terminal third of the molecule. The sialic acid–rich portion of the molecule is in an adjoining domain and the COOH-terminal domain is responsible for the molecule's plasma membrane association as an integral protein. N-CAM and Ng-CAM do not require Ca^{2+} either for their conformational integrity or binding, but L-CAM does.

All of the CAMs undergo remarkable sequences of expression, changes in concentration or prevalence, and changes in localization at the cell surface (so-called local cell surface modulation), and they appear in a definite spatiotemporal order during development and histogenesis. The order as established so far is: 1. L-CAM, 2. N-CAM (in early embryogenesis, both molecules show prevalence modulation) followed by 3. Ng-CAM appearance in the nervous system, and by 4. chemical modulation of N-CAM in the perinatal period leading to loss of about two-thirds of its polysialic acid (the so-called embryo to adult or E-A conversion). Physicochemical studies on rates of binding of N-CAM inserted into reconstituted lipid vesicles indicate that alterations in surface concentration and in E-A conversion independently lead to large changes in binding rates. This provides a basis for suggesting that, in vivo, surface modulations reflect definite functional changes. In accord with this view, perturbation experiments with specific antibodies to CAMs show definite alterations in tissue structure in vitro and in vivo.

The accumulated data on the function, modulation, and spatiotemporal sequences of CAMs can be incorporated into an evolutionary hypothesis relating the genetic specification of the CAMs to their regulatory role in morphogenesis and histogenesis. This hypothesis and the availability of new assays provide a basis for relating the role of CAMs to that of other central molecules in morphogenesis—the substrate adhesion molecules (laminin, fibronectin, collagen, etc) and cell junctional molecules, constituting such structures as desmosomes, and gap and tight junctions. The data indicate that no single static factor such as CAM specificity can account for histogenesis. Although CAMs cannot be sufficient for the establishment of form, the biochemical and cellular evidence suggests that they are absolutely necessary, and that they play pivotal roles in constraining other primary processes, particularly morphogenetic movements.

INTRODUCTION

To explain tissue structure and animal form in molecular terms is one of the most challenging and complex problems in modern biochemistry. In attempting to formulate this problem, one must distinguish between explanation in principle and exhaustive molecular descriptions. The latter can be achieved by correlation of cell biological and biochemical data; the former requires an understanding of evolutionary, genetic, and ontogenetic mechanisms. In these terms, there can be no molecular histology without a molecular embryology and it may therefore be useful to state at the outset of this review what the major problems (1, 2) of developmental biology are. The first might be called the developmental genetic problem: How can a one-dimensional genetic code specify a three-dimensional organism? The second is the evolutionary problem: How can the mechanisms of ontogeny be reconciled with rapid changes in animal form over short periods of evolutionary time? The third relates to growth and scaling: What determines the size of an organism and the relative scale of its organs and appendages?

In a strict sense, this review shall be concerned with one aspect of the first problem: how cell adhesion molecules (CAMs) serve to regulate various primary processes of development in time, particularly that of cell movement. But before going into the details of this relatively new subject, it is useful to develop a general framework for thinking about molecular aspects of morphogenesis. Morphogenesis is the result of a number of primary processes: cell division, cell movement, cell adhesion, cell differentiation, and cell death. The extent and timing of these processes changes dramatically in vertebrate development at different stages and in different embryonic locations. Each process represents a complex of molecular events often comprising epigenetic sequences—changes which must occur at particular times and places before other changes can occur. This review points up several important features of development. 1. It occurs at several levels, molecular, cellular, and multicellular. 2. It is under both genetic and epigenetic control. 3. A key to the relationship between the biochemical level and the various other levels lies in the position of cells and in the sequences of their mutual encounter. 4. Because of this central need for positional specification, the solution to the developmental genetic problem must be in part mechanochemical i.e. the sequential expression of certain gene products must coordinate with their function in regulating cell movement, cell proximity, and cell contact, for all of these factors condition the epigenetic sequences that result in orderly embryonic induction and cell differentiation at the right times and places.

Evidence is accumulating that cell adhesion molecules are candidates for such a role in mechanochemical regulation (3). This evidence indicates that CAMs are necessary but not sufficient to account for form, and that while they have different binding specificities, it is the broad range of dynamic selectivity embodied in a number of cellular processes that underlies their function in determining cellular positions during development. A cell does not recognize a molecular address directly, but rather undergoes a spatiotemporal sequence of events involving a hierarchy of complex biochemical regulatory pathways and occasional irreversible events.

This view of the developmental genetic problem has both consequences and requirements. One consequence is that it would be fruitless to search for myriads of cell surface markers that might pre-specify the position and relationships of each cell in a complex tissue. One requirement is that subtle control mechanisms must exist to regulate the specificity, selectivity, and sequences of embryogenetic events. Clearly, a set of control mechanisms must exist at the level of gene expression, particularly in regulatory genes that control the expression of molecules mediating cell adhesion. Another set must be concerned with the synthesis and export to the cell surface of these CAMs. A third, related to the second but not identical to it, concerns the modulation of the cell surface in such a way that cells sense their environment and coordinate the switching of their various primary processes, including adhesion.

Before turning to CAMs and the actual evidence that they undergo modulation, consider the various forms that cell surface modulation might take. Global modulation (4) is an alteration of the entire cell surface affecting the mobility of the majority of cell surface molecules regardless of their specificity, particularly those that are attached to the cytoskeleton. Global modulation can be induced in vitro by nonspecific local cross linkage of a small percentage of cell surface molecules, depends upon the state of microtubules, and can alter entry into the cell division cycle. So far, it has not been proven to have a role in vivo, but its demonstration in vitro was one of the first indicators of the existence of structural and functional transmembrane linkage.

Local cell surface modulation (4, 5) refers to changes in the surface density, distribution (polar or uniform), and chemical state of a given cell surface molecule of a definite specificity (Figure 1). By its very definition, the functional consequences of local modulation depend upon several variables that must be considered separately for each particular cell surface receptor. As reviewed below, there is now a body of evidence indicating that the various forms of local modulation occur for cell adhesion molecules (CAMs) both in vitro and in vivo.

With these preliminaries out of the way, we are in a position to consider the structure, function and interaction of cell adhesion molecules in various epochs of development and stages of histogenesis. Most of the issues dealt with here will concern vertebrate species. No attempt will be made to review the phenomenology of cell adhesion on which there is a vast literature [for bibliography, see e.g. (6–10)]. A number of specific aspects of CAMs have been discussed in previous reviews (5, 11, 12); the present excursus may be considered an update with some shift in emphasis in order to highlight the chemistry of adhesion.

THE CHEMICAL AND PHYSICAL NATURE OF CELL ADHESION

Even when considered on strictly chemical terms as a transaction amongst molecules, adhesion is a complex phenomenon. A major advance in simplifying the subject came when it was demonstrated unequivocally that it is a property of defined surface molecules (13, 14) and not merely a physicochemical ensemble property (10) of the cell surface. Even so, a number of distinctions must be made to avoid confusion between the specificities and selectivities expressed at both cellular and molecular levels (Figure 2). Binding

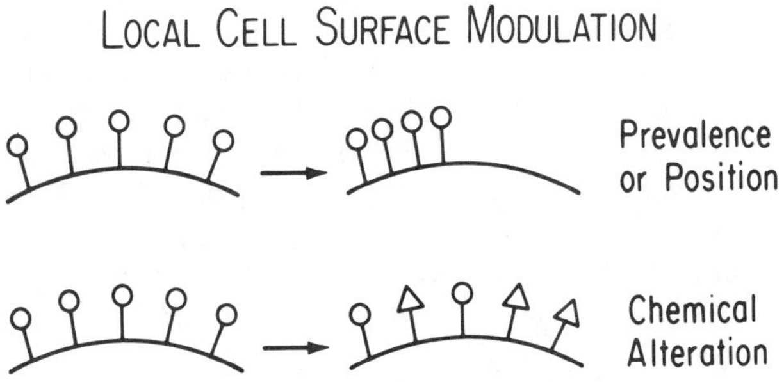

LOCAL CELL SURFACE MODULATION

Prevalence
or Position

Chemical
Alteration

Figure 1   Schematic representation of local cell-surface modulation. The various elements represent a specific glycoprotein (for example N-CAM) on the cell surface. The upper sequence shows modulation by alteration of both the prevalence of a particular molecule and its distribution on the cell surface. The lower sequence shows modulations by chemical modification resulting in the appearance of new or related forms (triangles) of the molecule with altered activities. Local modulation is distinct from global modulation, which refers to alterations in the whole membrane that affect a variety of different receptors independent of their specificity [see (4)].

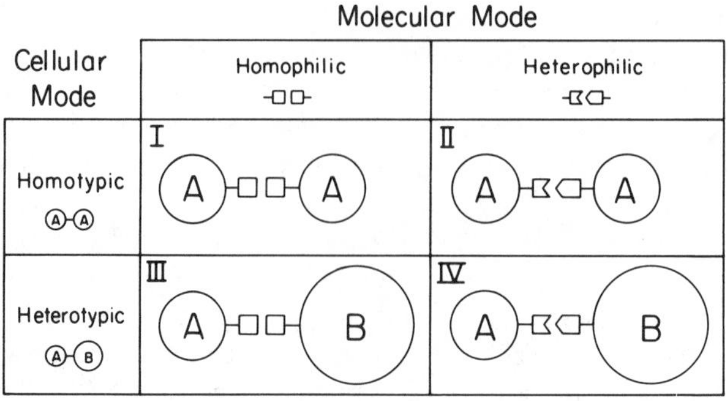

Molecular Mode

Figure 2 Schematic representation of different modes of cell-cell adhesion at the cellular and molecular levels. The circles represent cells; similar cells are represented by circles (panels I and II) of the same size, different cells are represented by circles of different sizes (panels III and IV). Cell adhesion molecules are represented by squares (panels I and II) and by the five-sided figures (panels II and IV).

between like cells is homotypic; that between unlike cells is heterotypic. Adhesion between the same adhesion molecules on opposing cell surfaces is called homophilic; that between unlike molecules is heterophilic. Heterotypic adhesion can occur by a homophilic mechanism as we shall see later for nerve-muscle interactions. This is a sufficient justification for the nomenclatural distinction but it is not the only one; it also suggests that selective mechanisms beyond mere molecular specificity are at work.

Operationally, the most significant advances in identifying and isolating CAMs came from the development of short-term in vitro assays (13) for quantitation of direct adhesion. These assays involve several subtleties which we shall consider here; the importance of the operational details must not blur the distinction that should be made between the results of any given assay and the list of criteria that must be fulfilled to qualify as a cell adhesion molecule. This list is open-ended but to date includes the following: 1. capacity to mediate cell-cell adhesion in a short-term adhesion assay under defined conditions of temperature and shear, 2. demonstration that the putative CAM is present on the cell surface of cells from the tissue or organ in question prior to and after dissociation of the tissue, 3. isolation of the CAM, 4. characterization of its binding mechanism, 5. demonstration that removal of the CAM or immunological perturbation of its binding in vitro and finally in vivo disrupts a morphogenetic process dependent upon cellular adhesion, 6. evidence of distribution of a particular CAM in tissues in a pattern consistent with its postulated mode of adhesion.

So far the only CAM to have met all these criteria is N-CAM, the neural cell adhesion molecule. Nonetheless, the other adhesion molecules meet the majority of criteria and further studies promise to bring them to the status of N-CAM. A minimum condition is the fulfillment of the first three criteria. In many instances, molecules identified by monoclonal antibodies or other means and lacking a function have been subsequently identified as CAMs (15–21) by their close resemblance to other proteins (5, 13, 14) that have fulfilled the minimal condition.

Of the various adhesion assays, an indirect immunologically based assay (13) consisting of several steps has provided the most successful general approach; it has served as a paradigm for a number of identifications of adhesion molecules in different laboratories (15–27). The assay procedure uses a short time period (30–60 min) in which binding between cells or between cells and membrane vesicles is measured quantitatively. Fab' fragments from the Igs of polyspecific antisera to the cell surface are then used to inhibit the in vitro adhesion. In order to obtain highly specific antibodies and their corresponding antigens (putative CAMs), various fractions of molecules released from the cell surface are checked for their capacity to neutralize the blocking capacity of the Fab' fragments. Active neutralizing fractions can then be used either to reimmunize for polyspecific antibodies that are more highly specific for the CAM or to prepare a variety of monoclonal antibodies.

Antibodies of proven specificity can be used to identify the intact CAM on the cell membrane, to detect the CAM distribution in various tissues during development, or to perturb CAM function in vitro and in vivo. Monoclonal antibodies have been used to establish the identity of the neutralizing antigens, deplete the antigen from the original neutralizing extract, and in some cases inhibit adhesion in the assay [for example, see (28, 29)]. Once a CAM is identified, the molecule or its soluble fragments may be used to test for binding to its cell of origin (30, 31); a particularly strong test is the ability to introduce the purified CAM into synthetic lipid vesicles and show binding of the vesicles similar to that seen for those cells of origin (32).

As indicated, all of the assays must be quantitative under defined conditions and must avoid the pitfalls of contaminating nonspecific molecules inducing adhesion or partial destruction of the CAMs (5). It must be emphasized that the assays are not conducted under conditions of thermodynamic equilibrium. Nonetheless, kinetic assays using lipid vesicles have been derived that provide great insight into CAM binding efficacy and inasmuch as CAM binding in vivo is likely to be a kinetically constrained event, these assays are particularly illuminating. It is obviously important to understand the thermodynamics (33) of CAM binding; current measurements on intact molecules are being carried out with the molecules embedded in lipid bilayers to avoid artificial interactions of membrane-associated hydrophobic portions of CAMs that would otherwise give spurious results.

STRUCTURE, BINDING MECHANISMS, AND BIOCHEMISTRY OF CAMs

Although no description of a CAM at the chemical level is likely to give a direct insight into its role in morphogenesis, a knowledge of the protein and carbohydrate structure, binding mechanisms, and metabolism of the CAM is nonetheless essential to any explanation of that physiological role. Three CAMs of different specificities have been isolated in quantities sufficient to explore structure at a protein chemical level: N-CAM, Ng-CAM, and L-CAM (5, 11, 23, 34–36). Their isolation in milligram quantities is attributable to the choice of the chick embryo as a primary source—so far, one-half million embryos have been used to obtain sufficient amounts. But it should be stressed that, for certain explorations, other species are more useful. Fortunately, as shall be discussed below, CAM structures are closely similar in vertebrates; this both facilitates biochemical comparison and justifies cautious interpretive extrapolations.

In this section, I shall discuss each CAM in terms of its polypeptide chain structure, associated carbohydrate, phosphorylation and sulfation, and membrane attachment. Proposed minimal models (exclusive of valence) will be considered in terms of the role of each CAM as an integral or extrinsic protein; its domain structure and antigenic sites; its attachment sites for carbohydrate, peptide maps and known sequences; and its binding mechanisms. Inasmuch as most is known about N-CAM, this molecule will be considered in greatest detail.

N-CAM

N-CAM is the first vertebrate CAM to be identified and described in structural and functional terms (5, 29, 31). Like all CAMs so far isolated, it is a glycoprotein found at the cell surface and synthesized by the cell to which it is attached. Two main kinds of N-CAM polypeptides with M_r 160,000 and 130,000 have been identified (29, 31). The evidence suggests that the major difference in these chains is in their COOH-terminal regions. It is not known whether these chains are associated at the cell surface or whether each forms a separate subvariety on single cells, which would imply that they exist on several cell populations. In any case, both forms are found in intracellular fractions and also at the cell surface and they have very similar peptide maps and identical NH_2-terminal sequences. It is therefore likely that, if one is a posttranslational modification of the other, the modification occurs intracellularly. Recent studies indicate that two species of N-CAM mRNA exist (37); they may in fact code for the two polypeptides. In view of the gross similarity of the two polypeptides, most of the discussion will be restricted to the longer chain.

The carbohydrate attached to N-CAM consists of neutral and amino sugars in "normal" amounts but the molecule contains an unusually large amount of sialic acid (29, 31, 38–41). It has been proposed that this exists as polysialic acid (31, 38) and the linkage has been confirmed to be α-2-8 (40). A comparison of N-CAM from embryonic (E) tissues and adult (A) tissues has revealed a large difference in sialic acid content although the polypeptide chain structure appears similar or identical at these two epochs of development time (38). This developmental change in sialic acid content, which is known as E-A conversion, confers different binding efficacies on the molecule (32), and occurs in a characteristic tissue distribution pattern (42) as discussed in later sections.

The E form of N-CAM migrates on SDS PAGE as a microheterogeneous broad band ranging from apparent M_r 180,000 to M_r 250,000 on SDS PAGE (29, 38). Although, on average, each N-CAM contains about 130–150 residues of sialic acid, this finding suggests that variations in the amount on each molecule may contribute to the microheterogeneity in gels. In contrast, the A forms migrate as three sharp bands (38, 42) of apparent M_r 180,000, 140,000, and 120,000 (this last being a minor component which may correspond to the lower–molecular weight polypeptide chain of N-CAM). Treatment with neuraminidase alters the migration of both E and A forms and converts them to closely spaced doublet bands migrating respectively in the region of M_r 170,000 and 140,000 (29, 38). Treatment of embryonic N-CAM with endoglycosidase F results in its conversion to bands of M_r 160,000 and 130,000, and synthesis of N-CAM by cells treated with tunicamycin produces bands that co-migrate in the same region (31). This suggests that the carbohydrate is N-linked; evidence on the number of attachment sites will be discussed later.

A number of prosthetic groups can be found on N-CAM (43, 44): When embryonic chicken brain tissue was cultured in media containing $^{35}SO_4$ or $^{32}PO_4$; ^{35}S or ^{32}P was incorporated into N-CAM (43). The ^{35}S label was located in Asn-linked carbohydrates on both glycopeptides (M_r 170,000 and 140,000), but not in the sialic acid. The ^{32}P label was detected in phosphoamino acids in the carboxyl-terminal third of both polypeptides, but the ratio of phosphoserine to phosphothreonine differed in the two polypeptide components. More recent studies indicate that labeled palmitate can be incorporated into the COOH-terminal region (B. Sorkin, G. M. Edelman, B. A. Cunningham, unpublished observation). The function of these various prosthetic groups is unknown but they may provide additional sites for functional control or modulation.

N-CAM appears to be an integral membrane protein that can be removed from membrane fractions only by detergent extraction (29, 32). Recent studies using a monoclonal antibody directed against the COOH-terminal region and fluorescent antibody tracing on permeabilized cells confirm this conclusion (45). Incubation of N-CAM at 37°C results in spontaneous cleavage of the NH_2-terminal region (Fragment 1 or Fr 1, M_r 65,000), suggesting the presence of an exposed region of polypeptide chain subject to limited proteolysis (29). Treatment of membrane fractions containing N-CAM with V8 protease results in cleavage of a fragment of larger size (Fr 2, M_r 108,000) which includes the NH_2-terminal region corresponding to Fr 1 (31). This suggests the possibility that the polypeptide stretches subject to limited proteolysis delimit separate structural domains, a fact consistent with electron micrographic examinations of the molecule (46).

Several monoclonal antibodies (Mabs) reactive with each of these regions have been obtained. Clone 1 reacts with Fr 1, and clone 2 with Fr 2 but not with Fr 1 (29, 31). Taken together with the findings on the COOH-terminal region mentioned above (44), Mabs exist for determinants on each of the major CAM regions. The molecule is highly immunogenic and after it was identified in terms of its function, antibodies that react with N-CAM were identified as such by several groups that had been searching for antigens in neural tissue. The synaptosomal protein D_2 (15, 16), and the NS-4 (17), BSP-2 (18–20), and 224-1A6-A1 (21) antigens all appear to be N-CAM.

Treatment of N-CAM with CNBr yields approximately up to six major and seven minor CNBr fragments (29, 41). As many as five of these fragments can be attributed to the NH_2-terminal region corresponding to Fr 1. A recent study (41) has succeeded in isolating one major CNBr fragment corresponding for the most part to the difference region between Fr 1 and Fr 2. A detailed study of the CHO content and particularly the sialic acid of this major CNBr sialopeptide from E and A forms of N-CAM has established that both forms have three N-linked attachment sites for carbohydrate. The major difference between the forms is in the sialic acid content; this is also present as polysialic acid in A forms although they have only one-third as much sialic acid as the E form. These findings establish that E-A conversion is likely to involve solely the linkage of sialic acid, and they focus attention upon cellular sialidases or sialyl transferases as the means by which E-A conversion occurs.

An analysis of the binding of N-CAM and Fr 1 to cells and to N-CAM (31), and of the binding behavior of N-CAM in reconstituted lipid vesicles (32), has suggested that the *trans* binding between apposed cells is N-CAM to N-CAM, i.e. it is homophilic. A detailed discussion of this mechanism is given below; here it is sufficient to note that these data suggest that the binding region is located in the NH_2 portion corresponding to Fr 1 and that removal of sialic acid still allows binding to occur.

On the basis of all of these facts, a polypeptide chain model of N-CAM can be constructed (Figure 3). Functions can be mapped onto this linear model (31) which, however, does not take valence into account. Recent electron micrographic studies (46) show that N-CAM can form triskelions and higher order structures, raising the possibility of association of multiple N-CAM chains on the same membrane (*cis* binding). While this would lead to multivalent structures if it occurred in vivo, it should be noted that the multimeric forms have only been seen so far on grids, not on membranes. More persuasive is the direct electron micrographic visualization (46) of structures consistent with domains, with intervening bends in the polypeptide chain that may correspond roughly to sites of cleavage Fr 1 and Fr 2.

DETAILED ANALYSIS OF THE HOMOPHILIC BINDING MECHANISM   One of the most important objectives of chemical studies of CAMs is to define their binding mechanisms in kinetic and thermodynamic terms. Careful measure-

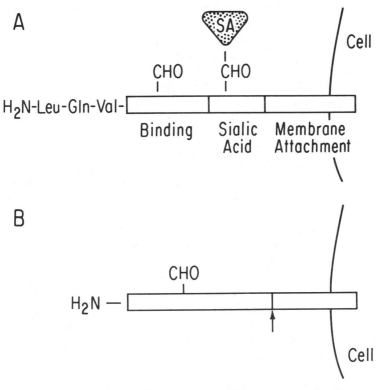

Figure 3   Comparison of the linear structures of N-CAM and L-CAM. (*A*) Three structural and functional regions of the N-CAM deduced from studies of the intact molecule and a series of fragments. The NH_2-terminal region includes a specific binding domain and carbohydrate (CHO) but little, if any, sialic acid. The neighboring region is very rich in sialic acid (SA), present mainly as polysialic acid. There are three attachment sites for carbohydrate in this region; the COOH-terminal region is associated with the plasma membrane. (*B*) Linear structure of L-CAM obtained by comparing the intact molecule, (M_r 124,000) released by detergent extraction with a fragment (M_r 81,000) released by trypsin (see arrow). The carbohydrate (CHO) is known to be attached at four major sites.

ments (32) of the rates of aggregation of purified N-CAM reconstituted into synthetic lipid vesicles have provided insights into the factors that govern homophilic binding. These studies not only confirm the location of the N-CAM binding region but allow an analysis of the influences of various other portions of the molecule on binding as well as of various extrinsic physical and chemical factors. Moreover, native brain vesicles of similar size containing N-CAM together with other membrane proteins may be compared to these artificial vesicles and thus one may assess the effects of these other proteins upon N-CAM binding behavior.

Both Fr 1 and Fr 2 inhibit vesicle binding, confirming that the binding region (31) is in the NH_2-terminal region of the molecule. As shown in the model (Figure 3), although this region has a small amount of carbohydrate, it lacks the major sialic acid moieties which thus do not take part directly in binding. The important question nonetheless arises whether E-A conversion influences binding efficacy. Vesicle experiments (32) clearly indicate that reduction of sialic acid content from the levels seen in E forms to those seen in A forms results in a four- to fivefold increase in the apparent second-order rate of appearance of particles above a detection threshold. Native brain vesicles (E or A) at comparable levels show comparable rates, suggesting that neither the isolation procedure to obtain pure CAMs nor the presence of other membrane proteins influences the results. Removal of all sialic acid from vesicles containing N-CAM in the A form has no further effect on binding. A recent qualitative study (47) has given results consistent with these more rigorous kinetic findings.

These observations suggest that the chemical form of local cell surface modulation, which is represented by E-A conversion, has important functional effects, increasing binding efficacy as indicated by increased binding rates. Although determination of the exact basis for the perturbation of binding by polysialic acid must await additional structural studies, the result is compatible with the charge perturbation model of homophilic binding (5). According to this model, the sialic acid may (a) perturb by direct charge repulsion in *trans* binding (b) alter the conformation of the neighboring NH_2-terminal domain (c) change the hinge angle of this domain in a heterogeneous fashion in N-CAMs on one cell so that the various angles of approach for *trans* binding would lead to unfavorable strain upon the plasma membrane should binding occur. This last possibility points up our ignorance of the distribution of microheterogeneous E forms at the cellular level. Does each cell make all forms or are there differences among different cell populations even before E-A conversion occurs?

The findings of three attachment sites for sugar (41) in the middle domain, of residual polysialic acid in A forms, and of failure to influence rates upon removal of sialic acid from the A form (32) suggest that sialic acid attached at different sites may have different functions. One function may relate to *trans* homophilic binding; the residual functions may relate to charge repulsion between bound cells as well as between mobile N-CAMs *cis* in the membrane.

Another major question concerns the effects of different amounts of N-CAM at the cell surface. Here the experimental results (32) are unequivocal: a twofold increase in the amount of N-CAM associated with the reconstituted vesicles results in greater than 30-fold increases in binding rates. While the interpretation of this result also must await more information, this fifth-order dependence of rates upon prevalence suggests the hypothesis that, *at equilibrium*, the rate change may be the result of higher valence states achieved by *cis* interactions of N-CAMs when they are present at higher densities at the cell surface. It is relevant to these considerations that N-CAM has been shown to be mobile in the plane of the plasma membrane with an apparent diffusion constant of 6×10^{-10} cm^2/sec (48). In any case, the results suggest that major functional effects are to be expected by prevalence modulation of N-CAM.

Ng-CAM, A Second Neuronal CAM

By means of a carefully designed modification of the original (13) indirect immunological assay, a second neuronal CAM responsible for neuron-glial adhesion has been identified (34). This molecule (Ng-CAM, neuron-glia CAM) is responsible for a heterotypic interaction and, inasmuch as it is not found on glial cells in any quantity, it operates by a putative heterophilic binding mechanism. The identification of this molecule was achieved by conducting the assay in the presence of anti–N-CAM antibodies to block homophilic neuron-neuron interaction and thus probe for inhibition of neuron-glial interactions.

A variety of monoclonal antibodies and several polyclonal antibodies were found to inhibit these interactions in the chick (34, 35); more recently comparable antibodies have been raised against mouse Ng-CAM (M. Grumet, G. M. Edelman, unpublished observation). Specific anti–Ng-CAM antibodies recognized a complex of polypeptide chains: a major component with M_r 135,000 and two minor components with M_r 200,000 and 80,000. The two components of higher M_r had similar peptide maps (36). Comparison of cross-reactivities among these components with a panel of Mabs showed that the M_r 135,000 and M_r 80,000 components were both antigenically related to the M_r 200,000 component but not to each other (36). Whether the two smaller components are posttranslational cleavage products of the largest component or are separately synthesized remains to be determined, but all three polypeptides appear to be present at the neuronal cell surface. It is pertinent that Ng-CAM synthesized in brain tissue in the presence of ^{32}P phosphate incorporated the label into the M_r 200,000 and the M_r 80,000 component but not into the M_r 135,000 component (36). Inasmuch as these various components are visualized in electrophoretic gels after dissociation by NaDodSo$_4$, it is not clear whether they are all associated at the cell surface or whether some occur as separate proteins on the neuron. Although it is not proven that Ng-CAM components are intrinsic proteins, they cannot be removed from the cell membrane by agents other than detergents.

Ng-CAMs are glycoproteins but lack the polysialic acid that makes N-CAM so unusual (36). Like the neuron-neuron binding mediated by N-CAM, neuron-glia binding mediated by Ng-CAM is Ca^{2+}-independent. N-CAM and Ng-CAM appear to differ grossly in their major chemical characteristics. After removal of sialic acid by neuraminidase, none of the polypeptides of the two neuronal CAMs (both of which may appear at the surface of the same individual neuron) show similar molecular weights (35, 36). Moreover, peptide maps of the two proteins are strikingly different (35). Despite these differences, two independently derived monoclonal antibodies cross-react with both molecules. Recent studies (36) confirm that one of these Mabs recognizes common determinants in the carbohydrate moieties other than those in sialic acid.

Until the putative counterpart ligand (so-called Gn-CAM) is isolated from glial cells, it is not possible precisely to define the mechanism of heterophilic binding involving Ng-CAM. In the meantime, several important facts about Ng-CAM and N-CAM binding have emerged (36). These facts suggest that the two molecules together may influence neuron-neuron binding and may be summarized as follows: Anti–N-CAM has no effect on neuron-glial binding. Monoclonal and polyclonal anti–Ng-CAMs block neuron-glia interactions. Interestingly, polyclonal anti–Ng-CAM (but not a monoclonal anti–Ng-CAM) inhibits both neuron-neuron binding and aggregation of neuron membrane vesicles. The combination of anti–N-CAM and polyclonal anti–Ng-CAM blocks such aggregation even more effectively; this can be demonstrated in the presence of the monoclonal anti–Ng-CAM indicating that this combined effect is not the result of heterotypic interactions with glial cells or glial membranes in the preparations.

These findings raise the possibility that Ng-CAM 1. may interact with N-CAM *cis* thus allowing anti–Ng-CAM to inhibit the homophilic *trans* binding mediated by N-CAM by steric means, 2. may actually bind to N-CAM from another cell *trans* or, 3. may bind to yet another *trans* ligand on neurons. While the situation has not yet been clarified, there are some observations that favor the first two of these possibilities: Ng-CAM and N-CAM have been found to interact in solution as shown by migration in gradients and by competition experiments using labeled molecules to bind to neuronal surfaces (36; S. Hoffman, G. M. Edelman, unpublished observation). Resolution of the issue should come from mixed vesicle experiments. In the meanwhile, the intriguing possibility exists that *cis* interacting Ng-CAM would be removed from poten-tial glial interactions by the binding of its N-CAM partner to N-CAM *trans* on neurons. The reciprocal situation would hold for Ng-CAM binding to glia, removing N-CAM from *trans* binding and providing a cross-modulation mechanism that might be useful in selectivity at the cellular level.

Despite this potential complexity, it is clear that N-CAM and Ng-CAM have different roles and specificities in cell adhesion. This is particularly obvious in considering their sequence of appearance in the embryo and their germ layer derivatives. While N-CAM is a primary CAM appearing well before neural induction and found on both ectodermal and mesodermal derivatives (48, 50), Ng-CAM appears first on postmitotic neurons and is seen only on neuroectodermal derivatives (50a). We shall review these distributions in detail in a later section.

L-CAM, The CAM Of Early And Late Nonneural Epithelia

In the early experiments using the indirect assays, a search for nonneural CAMs was deliberately started to provide a control from other tissues. After early reports on putative CAMs from liver (22, 24, 51) appeared, a number of laboratories identified what now appear to be related nonneural CAMs. These molecules L-CAM (23), uvomorulin (25), cadherin (26), and cell CAM 120/80 (27), appear to be homologous forms of a single CAM present from the earliest stages of embryogenesis and remaining in the adult on a great variety of nonneural epithelial cells in the adult.

Because chick L-CAM is the only one of these molecules that has received chemical characterization in detail (23, 53), and because its analysis has rationalized the relationships among the others, which were at first either detected as fragments or were incompletely defined, we shall refer to all of these molecules as L-CAMs. L-CAM was first definitively identified by means of an M_r 81,000 fragment released from liver membranes by treatment with trypsin in the presence of Ca^{2+} (23). Early studies (54–59) suggested the importance of Ca^{2+} in binding mechanisms; it is now clear that the ion is essential for conformational integrity of the molecule. The intact form of the molecule which has an M_r of 124,000 was then unequivocally identified (23) on the hepatocyte surface by means of specific antibodies to the fragment. L-CAM is a large glycoprotein, lacks polysialic acid, depends upon Ca^{2+} for its active binding conformation, and is rapidly degraded by proteolysis (23) in the absence of this ion.

L-CAM is immunologically unrelated to N-CAM and Ng-CAM and does not bind to either of these molecules. The binding mechanism of L-CAM has not been established definitively as homophilic but a variety of studies using antibodies to perturb various binding functions, histotypic aggregation, or compaction mechanisms (25, 60–63) have nonetheless shed light on its function. A molecule, cell CAM 120/80, identified in human mammary carcinoma cells (27) and an apparently related set of molecules implicated in the compaction of mouse cleavage stage embryos [uvomorulin (25) and cadherin (26)] have similar molecular characteristics to L-CAM. All of these molecules have trypsin fragments similar to those of L-CAM and depend for their conformation and binding upon Ca^{2+}. Antibodies to uvomorulin and cadherin strongly cross-react (26); antibodies to all three can block compaction. It has recently been shown that antibodies to cadherin block liver cell aggregation (56) and prevent the formation of the inner cell mass in 8- to 16-cell stage embryos (57). The original demonstration (22) and subsequent more refined experiments (23) on L-CAM show that antibodies to this molecule not only block liver cell adhesion but also prevent formation of characteristic histotypic liver cell colonies in culture. All of these chemical and cell biological findings leave little doubt that the chick, mouse, and human CAMs are homologs of L-CAM. As we shall see later, this conclusion is corroborated by studies of their tissue distribution.

A definitive chemical analysis of chick L-CAM has begun (53) and provides a basis for a linear model of the molecule. The NH_2 terminus extends away from the membrane (Figure 3), and at the membrane there is a site susceptible to limited proteolytic cleavage even in the presence of Ca^{2+}. The proteolysis yields an M_r 81,000 component (Ft 1) containing the NH_2 terminus, and a COOH terminus associated with the cell membrane. It is not yet established whether L-CAM is an integral membrane protein. Experiments on uvomorulin in various solvent-detergent mixtures and extractions with EDTA suggest it is extrinsic (64). While experiments on its phase distributions in similar mixtures are in agreement with these results, L-CAM cannot be readily extracted from liver cells by EDTA without detergents (53). Experiments on incorporation of ^{32}P inorganic phosphate (53) show that Ft 1 contains little or no ^{32}P but that the more COOH-terminal region is phosphorylated at threonine and serine residues.

A detailed analysis (53) for sugar sites on this glycoprotein succeeded in identifying four N-linked attachment sites, one of which was for mannose-rich carbohydrate as indicated by differential susceptibility to endoglycosidases F and H. Ft 1 has all of these N-linked carbohydrate groups. As mentioned previously, no unusual polysialic acid is present.

Although not enough information has accumulated to warrant discussion in a separate section, there are two other arenas of great structural and chemical interest bearing upon the CAMs: identification and isolation of their gene sequences by cloning techniques, and studies of their metabolic turnover, transport to the cell surface, and posttranslational modification.

Recent studies (37) have successfully isolated two cDNA probes for sequences homologous to messenger RNA capable of synthesizing chick N-CAM. Both plasmids hybridized to two discrete species of mRNA six to seven kilobases long present in poly(A)$^+$ RNA from embryonic chick brain. These mRNA's possibly correspond to the two known polypeptide chains. So far, however, DNA corresponding to the protein sequence remains to be isolated complete and subjected to sequence analysis. It is significant that Southern blotting experiments detected only one hybridizing fragment in chicken genomic DNA digested with several restriction enzymes, suggesting that sequences corresponding to those within the region of N-CAM mRNA are present, possibly only once in the genome.

In view of the manifest effects on binding of changes in the surface density or chemical modulation of CAMs, it is important to understand the turnover, transport, and posttranslational modification of these molecules. A recent study (65) has identified precursors in the synthesis of protein D2 which has been shown to be N-CAM. Of particular interest is the question of whether E-A conversion occurs by means of the action of surface sialidases, blockade of an intracellular sialyltransferase, or both. A decision requires the estimation of the turnover rate of the molecule. An in vitro system using chick neural elements and pulse chase experiments to follow E-A conversion has recently been developed (D. Friedlander, R. Brackenbury, C.-M. Chuong, G. M. Edelman, unpublished results). It was found that the A form succeeding the E form in this system was newly synthesized, suggesting turnover. In the absence of detailed population analyses, however, this result cannot yet be definitively interpreted at the level of single cells; the fact that it could be obtained in the presence of cytosine arabinoside which blocks cell proliferation rules out one variable, but that of differential cell death remains to be explored.

SPECIES DISTRIBUTION AND EVOLUTIONARY CONSERVATION OF CAM FUNCTION

As hinted at in previous sections, there is accumulating evidence of strong evolutionary conservation of CAM structure and function. Although this has not been explored in detail until recently, both N-CAM and L-CAM have been found in a variety of species (23, 27, 28, 66–68). Ng-CAM has been found in the chick and mouse, and cross-reactivity of chick antibodies with frog brain extracts suggests it is also present in amphibia. E-A conversion of N-CAM also is seen in frogs (C. M. Chuong, G. M. Edelman, unpublished observations).

A more recent study of N-CAM (68) has extended previous observations to include a wide variety of classes and species down to elasmobranchs. The data suggest that N-CAM will be found in all vertebrates and perhaps in all descendents of protochordates. The reactivity of these N-CAM homologs with Mab 15G8, which appears to have specificity for the polysialic acid moiety of N-CAM (28), also suggests that this structure is conserved. It remains to be seen whether Ng-CAM and L-CAM will be found in all these species, but it appears likely.

More striking even than the structural resemblances and antigenic cross-reactivity across species lines is the conservation of the specificity and function of the N-CAM binding region in homophilic binding. By adapting the kinetic vesicle-binding assay to detect cross-species homophilic vesicle aggregation and comparing it to intraspecies aggregation, the question of conservation of specificity could be tested (68). The startling result is that cross-species aggregation occurred at the same rates as intraspecies aggregation among frog, chick, and mouse brain vesicles. In the frog-mouse pairing, little antigenic cross-reactivity was found between specific monoclonal antibodies to the respective species. This allowed a further test in support of the hypothesis of homophilic binding: both antifrog and antimouse Mabs were required fully to inhibit cross-species and intraspecies aggregation in the same assay. In accord with the idea that E-A conversion changes only the efficacy but not the specificity of binding, various cross-species E and A pairings gave consonant results with those found within a species.

All of these findings suggest that the overall structure and particularly that of the binding region of N-CAM have been evolutionarily conserved. Inasmuch as binding is homophilic and the construction of proteins from L amino acids prevents mirror symmetry, the N-CAM binding region must contain at least two sites or subregions. Deleterious mutations within either in the absence of complementing mutations would cause loss of CAM binding and be strongly selected against. It is important to note that these observations do not indicate that CAMs from different species are nonspecific but rather that the mechanism by which they contribute to form and histogenesis must be dynamic and hierarchical, involving regulatory mechanisms as well as protein binding specificity. Whether all of the known CAMs of different specificity have had a common evolutionary precursor is unknown and a decision shall have to await more detailed sequence information. So far, no strong homologies have been seen, but the data are incomplete.

TISSUE DISTRIBUTION OF CAMs AND CONSTRUCTION OF CAM FATE MAPS

With this structural, functional, and evolutionary background, we are in a position to relate some of the findings on the cellular and spatiotemporal appearance of CAMs in different epochs of development from early embryogenesis to the adult. In the present section, we will consider the gross distributions of the two primary CAMs, N-CAM and L-CAM, and note the first appearance of the secondary Ng-CAM. We will then consider the significance of the order and locations of appearance of these molecules in early embryogenesis by relating these variables to fate maps which describe the invariances maintained by cells in the various germ layers as they give rise to various key structures of the embryo. In the next section, we will discuss the histogenesis of the nervous system in greater detail, using this system as an example to describe the changing relationships between two different CAMs as development proceeds in a complex tissue.

As shown in Table 1, the two primary CAMs appear at least as early as the blastoderm stage in the chick, then diverge in location; subsequently, they are found concordantly in adult tissue derivatives of the germ layers in which they were first found (49, 50, 69). N-CAM is seen in ectoderm and mesoderm, and L-CAM is seen in all three germ layers (27, 69); neither is found uniformly in every derivative of these layers but rather their distribution is a dynamic one. Their persistence into adult life indicates that the primary CAMs serve functions in all epochs of development. Although there is little reason to doubt that their binding specificities persist (as we shall see from certain in vivo perturbation experiments), what is likely to change is the dynamic mechanism by which they undergo cell surface modulation in different tissues at different times.

In considering the morphogenetic questions raised at the beginning of this review, it is illuminating to review the spatiotemporal course of appearance of N- and L-CAM in the chick embryo in more detail. As shown in a series of immunohistochemical studies of the CAMs (49, 50, 69), they are more or less uniformly distributed on the blastoderm of the chick. (L-CAM has actually been found at the four-cell stage in mouse embryos (63); so far, the distribution of primary CAMs at the very early cleavage stages in chicks has eluded investigation). At the blastoderm stage, just as gastrulation proceeds, the epiblast and hypoblast stain with fluorescent antibodies to both CAMs (49, 50). Later on, as the so-called primitive streak develops, the middle-layer (mesoblast) cells that are migrating though the streak between the other two layers do not stain. Somewhat after this point in time, a remarkable transition occurs: L-CAM appears on Henson's node at the cephalic end of the primitive streak and remains with it for its existence; cells that will become the neural plate lose evidence of L-CAM and stain very strongly with antibodies to N-CAM. Reciprocally, at the border of the ectoderm and neuroectoderm, the ectodermal cells stain only with antibodies to L-CAM as do the endodermal cells beneath. This emerging pattern of CAM segregation and border formation accompanies the key event of primary or neural induction in which neural and nonneural tissues are distinguished and the axis of the animal is realized. It is important to stress that the segregation is not accomplished by cellular migration but by differences in relative CAM expression at particular points in the cellular sheet.

Embryologists use fate maps to record and relate the origins of various differentiated structures from cells in particular positions in the different germ layers (2). Fate maps summarize what will become of cells in each embryonic region in terms of the structures to which they will give rise for a defined period of time or epoch. A fate map is only a summary of events that occur in such a

Table 1 Distribution of chick L-CAM and N-CAM in three epochs

0–3 day Embryo	5–13 day Embryo	Adult
L-CAM		
Ectoderm		
Upper layer	Epidermis	Skin: Stratum
Epiblast	Extraembryonic	germinativum
Presumptive	ectoderm	
epidermis		
Placodes		
Mesoderm		
Wolffian duct	Wolffian duct	Epithelium of:
	Ureter	Kidney
	Most meso- and	Oviduct
	metanephric	
	epithelium	
Endoderm		
Endophyll	Epithelium of:	Epithelium of:
Hypoblast	Esophagus	Tongue
Gut primordium	Proventriculus	Esophagus
and buddings	Gizzard	Proventriculus
	Instestine	Gizzard
	Liver	Intestine
	Pancreas	Liver
	Lung	Pancreas
	Thymus	Lung
	Bursa	Thymus
	Thyroid	Thyroid
	Parathyroid	Parathyroid
	Extraembryonic	Bursa
	endoderm	
N-CAM		
Ectoderm		
Upper layer	Nervous system	Nervous system
Epiblast		
Neural plate		
Placodes		
Mesoderm		
Notochord	Striated muscle	Footnote a
Somites	Adrenal cortex	Cardiac muscle
Dermomyotome	Gonad cortex	Testis
Somato- and	Some mesonephric and	
splanchno-	metanephric epithelia	
pleural	Somato- and splanchno-	
mesoderm	pleural elements	
Heart	Heart	
Mesonephric		
primordium		

[a]It is not yet known whether adult striated muscle contains N-CAM.

time period and a series of such maps is usually necessary to cover several epochs, particularly if one wishes to follow the whole line of development to maturity. Fate maps are similar from individual to individual but differ in precision depending both upon the time chosen and the degree to which individual cells in a species give rise only to one exact line of structures. It is important to realize that a future four-dimensional distribution (time and three spatial dimensions) of cells is being mapped back onto a two-dimensional surface, for example, that consisting of the sheet of blastoderm progenitor cells. This results in an epochal and topological map rather than a topographic one in real time; i.e. exact details of structure are sacrificed to reveal connectedness and neighboring relationships. It is these relationships that are particularly important in understanding embryonic induction events, which result from the successive interactions of cells of different history brought together by various kinds of morphogenetic movements.

By examining the descendents of blastodermal cells in the chick fate map (Figure 4A) as revised by Vakaet (70) for their expression of different kinds of CAM, a composite CAM fate map (Figure 4B) has been constructed (50). This map has a number of striking features: 1. the calcium-independent N-CAM regions that will give rise to the neural plate, to the notochord, and to certain

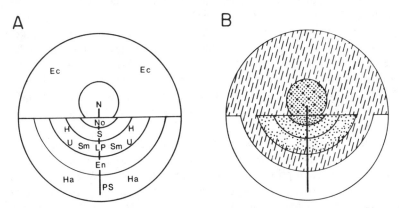

Figure 4 Composite CAM fate map. (A) Fate map of the blastodisc [L. Vakaet (70), and unpublished observations] showing areas of cells that will give rise to differentiated tissues (indicated by letters; see below for designations). (B) Map of cells that will express CAMs. The distribution of N-CAM (stippled) or L-CAM (slashed) in tissues at 5–14 days (stage 26–40) as determined by immunofluorescence staining is mapped back onto the blastodisc fate map. Cells that will give rise to the urinary tract (U) express both L-CAM and N-CAM. Smooth muscle (Sm) and hemangioblastic (Ha) tissues express neither N-CAM nor L-CAM; areas giving rise to these tissues are blank on this map. Ng-CAM is represented on neuroectoderm by (•) symbols; it is seen only after three days. The vertical bar represents the primitive streak (PS). Other abbreviations: Ec = intraembryonic and extraembryonic ectoderm; En = endoderm; H = heart; LP = lacteral plate (splanchnosomatopleural mesoderm); N = nervous system; No = prechordal and chordamesoderm; S = somite.

parts of the lateral plate mesoderm are completely surrounded by a contiguous, simply connected ring of regions that will express the calcium-dependent L-CAM—regions that together comprise the rest of the ectoderm and the endoderm. 2. There is a cephalocaudal coarse gradient of N-CAM antibody staining which appears to be most intense in the region of the neural plate. In underlying regions such as the notochord, the staining is less intense and the pattern is dynamic as revealed by sections taken at successive times—at first, there is no staining for N-CAM, then the notochord (a mesodermal structure) stains intensely, and in later stages the stain disappears. 3. A similar dynamic pattern of N-CAM appearance is seen in the somites just as they segregate into segments. Moreover, the early sequence—(L-CAM staining plus N-CAM staining followed by loss of L-CAM staining)—is seen in placodes, the epithelial structures that will form ganglia. Both CAMs are also seen at the apical ectodermal ridge of the limb bud, a key structure necessary for formation of the appendages. 4. A singular feature of the map can be seen in the kidney elements which come from the mesoderm: both L-CAM and N-CAM appear and disappear in sequences corresponding to the reciprocal embryonic induction of the so-called mesonephric mesenchyme by the tubular structure called the Wolffian duct. L-CAM first appears on the Wolffian duct, then N-CAM appears on mesonephric tubules as they organize, then this is replaced by L-CAM as these tubules extend. 5. Certain regions of the lateral plate mesoderm that give rise to smooth muscle, and regions of the map containing the hemangioblastic precursors, do not stain at all with either CAM. This raises the possibility that at least one other primary CAM mediates the adhesive interactions of these areas in early development; it is too early to say how many others will be found. 6. Finally, as will be discussed in detail in the next section, the secondary Ng-CAM appears only in derivatives of the neuroectoderm (Figure 4B) and is not seen in very early events (50a).

Three large generalizations emerge from these observations on the early embryo. 1. CAMs undergo dynamic changes in distribution, sequence, and amount at regions of primary and secondary induction. 2. Wherever epithelia are converted to so-called mesenchyme (loosely associated cells, often undergoing motion), CAMs appear to be lost from the cell surface. 3. After differentiation of a tissue (e.g. the nervous system), new CAMs of different specificity can appear. These observations not only indicate that prevalence modulation at the cell surface occurs for the primary CAMs, they also indicate that the modulation occurs in orderly sequences at specific places and times and in specific relation to the formation of germ layers and their early derivatives. These conclusions are central to attempts to understand the evolutionary and ontogenetic significance of the CAMs, a subject that will be considered in the final part of this review.

It is particularly important to note that these sequential and parallel events of CAM expression precede most cell differentiation events that occur in the various induced tissues and organs. What happens as this differentiation begins to occur? To answer this question, it is useful to consider the later changes and additional modulation mechanisms undergone by a primary CAM as it is expressed in later histogenesis. We will do this extensively by reviewing what is known about N-CAM and the secondary Ng-CAM during the process of neurogenesis.

CAMs IN THE BRAIN: SPECIFICITY AND SELECTIVITY THROUGH ADDITIONAL MECHANISMS OF MODULATION

During development, the brain and nervous system undergo enormous changes in which high numbers of cells interact through the primary processes to yield extraordinarily complicated structures and maps (71). This example of exquisite histogenesis, unparalleled in its complexity, serves as an excellent vehicle to illustrate a succession of mechanisms of cell surface modulation of CAMs. After the appearance of N-CAM exclusively on the neural plate and after neurulation (formation of the neural tube), the molecule is seen ubiquitously (although not exclusively, see Table 1) in neural derivatives. However, it does undergo sharp changes in amount at the cell surface during certain cellular rearrangements. A striking example of this is the prevalence modulation of N-CAM in chick neural crest cells, the cells responsible for forming much of the peripheral nervous system among other structures (49). When crest cells first appear at the top of the neural tube, they stain for N-CAM. As soon as they begin their migration, however, they lose this staining and fibronectin (a substrate adhesion molecule or SAM) appears in their path (72). When they reach their destination and just before they form ganglia, N-CAM reappears on their surface. As in the case of mesoblast cells described earlier, this classical example of the conversion of a sheet of cells to a mesenchyme consisting of loosely attached mobile cells is clearly correlated with the occurrence of strong CAM modulation. It is not yet known, however, whether the N-CAM on the neural crest cell surface is masked during migration or whether it is turned over and actually lost.

It is well known that many neuronal migrations within the structures deriving from neural tube require interactions with glial cells (73). N-CAM does not appear to mediate adhesion of neurons to glia as does Ng-CAM. It is striking that Ng-CAM is absent from the surface of neurons and neuroepithelial precursors as they divide and before neuronal migration and fiber extension occur. Ng-CAM first appears at the surface of early postmitotic neurons in the chick (50a). By three days in the embryo, the first neurons to be detected by anti–Ng-CAM antibodies are found in the ventral neural tube and as these precursors of motor neurons emit fibers to the periphery, the fibers are particularly well stained. Neural crest cells and early ganglion rudiments do not show

Ng-CAM at first; the first neurons in the peripheral nervous system to be stained appear in the cranial ganglia at four days. Progressively, all neurons and neurites in the periphery display Ng-CAM. By 4 ½ days, fiber networks in the CNS (rhombencephalon and telencephalon) show the molecule and by six days, the ventral horn and dorsal white matter of the spinal cord display it. By 13 days, one may observe remarkable changes in mapped structures such as the optic tectum and the cerebellum: fibers penetrating from the optic nerve into the tectum show intense anti–Ng-CAM staining and the migratory granule cells and their processes in the cerebellum also stain. In striking contrast to N-CAM (74), the distribution of which was more even, Ng-CAM showed marked changes (36) from layer to layer: it was absent in the proliferative zone, and present in the premigratory zone, internal granular layer, fiber tracts, and in highest amounts in the molecular layer. After birth, the relative intensity of anti–Ng-CAM staining was markedly decreased whereas that of N-CAM remained more or less the same. Most strikingly, many central nervous system neurons show Ng-CAM associated with axons rather than with cell bodies, a clear-cut example of polarity modulation.

At this account shows, unlike N-CAM, Ng-CAM is expressed only on neurons of the CNS and PNS in the later epochs of development concerned with neural histogenesis. Ng-CAM is thus a secondary CAM, a specific differentiation product of neuroectoderm. Ng-CAM is found on developing neurons at about the same time that neurofilaments appear, a time at which glial cells are still undergoing differentiation from neuroepithelial precursors and its appearance and binding function suggest that it has a role in neuronal migration (73, 75) on these glial cells.

At all these times and places, N-CAM is also present on neurons and their processes; the impression is that it is more ubiquitously distributed in the various brain structures than Ng-CAM in early stages. This does not imply that N-CAM is maintained in constant amounts at the cell surface in all regions. Indeed, if the modulation hypothesis is correct, one should see evidence during tract formation not only of continuing changes in surface prevalence of N-CAM such as those seen in the early fate map, but also in other forms of modulation. This is in fact the case, as shown by changes in N-CAM levels and by the advent of E-A conversion which appears in later stages of neural development around hatching in the chick and in rodents (42). The entire expression sequence (50a) of the two neuronal CAMs is given in Table 2.

An analysis of E-A conversion in different gross brain regions shows strong differences in pattern in each region (42). This suggests that the rate of conversion, its time of initiation, or its degree in different cell types may differ in these histologically different regions. Conversion is an epigenetic event. Although its mechanism is not known, the structural studies reviewed earlier imply that it results either from enzymatic cleavage of the sialic acid from

Table 2 Expression sequence of two neuronal CAMS in the developing chick

N-CAM (1°)		Ng-CAM (2°)

```
                                          AGE
                                     DAYS│STAGE

                  BLASTODERM           0 ┬  1

NON-NEURAL DERIVATIVES   NEURAL DERIVATIVES
NOTOCHORD, SOMITE;       NEURAL PLATE;
LATERAL PLATE MESODERM;  NEURAL TUBE;
MYOCARDIUM;              NEURAL CREST;      2 ┼ 13
MYOBLAST;                DISAPPEARS FROM
TRANSIENT APPEARNCE ON     MIGRATING NEURAL
  PLACODES AND             CREST CELLS;
  MESONEPHRIC             REAPPEARS IN
TUBULES;                   AGGREGATING
MYOTUBES;                  NEURAL CREST    4 ┼ 23    VENTRAL NEURAL TUBE;
                           CELLS                     CILIARY GANGLIA;
                                                     DORSAL ROOT GANGLIA;
STRIATED MUSCLE;                                     SPINAL CORD: WHITE COLUMN;
                                                     RETINA: GANGLION CELL FIBERS;
                                                     TECTUM: FIBER LAYER UNDER PIA;
                                           6 ┼ 29    SYMPATHETIC GANGLIA;

                                                     SPINAL CORD: GRAY MATTER;
                                                     TECTUM: STRATUM OPTICUM AND
                                                       ALBUM CENTRALE;

                                           8 ┼ 34


                                          10 ┼ 36

                                                     CEREBELLUM: FIBER TRACTS;

                                          12 ┼ 38    CEREBELLUM: MOLECULAR LAYER;


                                          14 ┼ 40


                                          16 ┼ 42

                                                     CEREBELLUM: MIGRATING GRANULE CELLS;

                                          18 ┼ 44

                 E
                 ↓                       BIRTH ┼ 46
                 A

N-CAM ON STRIATED    REMAINS ON          ADULT       DIMINISHES IN SPINAL CORD
  MUSCLE CONCENTRATED  ALL SUBSE-                     WHITE MATTER, CEREBELLAR FIBER
  AT THE END PLATE;    QUENT NER-                     TRACTS AND TECTAL STRATUM OPTICUM
CARDIAC MUSCLE;        VOUS TISSUES.                  AND ALBUM CENTRALE.
TESTIS; OVARY;
OVIDUCT.
```

surface N-CAM, or from turnover of E forms followed by replacement by A forms of N-CAM that have lesser amounts of sialic acid. In the latter case, an intracellular enzyme or sialyl transferase responsible for linking sialic acid to N-CAM would be implicated. Whatever the mechanism turns out to be, the significant observation is that, during organogenesis and particularly during the perinatal period, grossly different amounts of E and A forms are present in different neural regions. As indicated by the influence of conversion on the binding kinetics of N-CAM, this would be expected to change the binding efficiencies of various cells in these regions in different ways during histogenesis and thus alter neural structure during cell differentiation and movement.

A failure in the normal scheduling of E-A conversion has been seen in genetic mutants of mice with cerebellar connectional defects, the so-called *staggerer* mice (76). *Staggerer* is expressed only in homozygous animals; cerebella of these animals show faults in the formation of synapses between parallel fibers and Purkinje cells, which do not appear to have normal tertiary dendritic branches. After failure to make these synapses, the granule cells die in great quantities; the consequence of these anomalies is an ataxic animal with a small and disordered cerebellum destined to die without further care at about one month after birth.

E-A conversion was found to be greatly delayed in the cerebellum of *staggerer* homozygotes, but the N-CAM polypeptide appeared to be normal. On the other hand, *reeler* and *weaver* (other cerebellar mutants) and normal littermates of *staggerer* had normal schedules of E-A conversion (76). While these findings do not provide an explanation for the cause of disease in *staggerer* mice, they suggest that one consequence of the genetic defect is a failure in the activity or in the expression of enzymes responsible for E-A conversion. It may well be that this leads in turn to a failure to terminate certain key cellular binding events with a consequent failure in coordination of the various aspects of neural process formation, migration, and synapse formation.

This brief survey of neurogenesis shows that N-CAM is used after early neural induction, undergoes variations at the cell surface particularly in migrating cells such as neural crest cells, and is joined later on the neuronal surface by Ng-CAM. Thus two CAMs can be present on a single cell (35), but at the same time undergo different modulation mechanisms. In the CNS, Ng-CAM shows polarity modulation on the same cells on which N-CAM is more or less uniformly distributed. Despite the power of modulation mechanisms, we thus know that there are certain histogenetic circumstances such as glial interactions and neurulation that require CAMs of different specificity. We conclude that it is the combination of both specificity and modulation that are important in morphogenetic change. The need during morphogenesis for CAMs of different specificities is clearly context-dependent and within a given tissue, it depends upon timing, modulation, concurrent presence of different cell types and interactions with the other primary processes of development. An examination of the histogenetic expression sequence for the two CAMs indicates that there must be at least three kinds of regulatory signals responsible for modulation: 1. signals for N-CAM and Ng-CAM expression; 2. signals for alteration of surface expression; 3. signals for E-A conversion.

One particularly convincing demonstration of the action of some of these signals is given by the alteration of N-CAM expression at the surface of virally transformed cells derived from primary cultures of chick brain cells (77) and cerebellar cell lines of the rat (78). The evidence on chick primary cells indicates that, after transformation by RSV (79), there is a series of differentiation events accompanied by loss of N-CAM at the cell surface, loss of adhesion, increased mobility, and changes of cell shape and expression of 34kD $pp60^{src}$ substrate. All of these changes occur in neuroepithelial cells that were definitely N-CAM positive before transformation. In rat cell lines (80) infected with temperature-sensitive mutants, this cycle of events can be reversed at least once by return to the nonpermissive temperature (78). The cells aggregate with return of surface N-CAM and regain their original morphology; subsequent attempts to revise the sequence fail after one cycle. These findings dramatically illustrate the role of the control of surface N-CAM in maintaining cell-cell interactions and have obvious implications for a possible permissive role of N-CAM modulation in metastasis.

The observations reviewed in this section are consistent with the hypothesis that N-CAM and Ng-CAM play major roles in establishing and maintaining a variety of neural structures. It is likely that the coordinated cell surface modulations of these two CAMs play key roles in neuronal migration and mapping, in fasciculation and its reversal, in fiber bundle extension (both CAMs are found on growth cones) and, in the case of Ng-CAM, in neuron-glia binding. The succession of N-CAM and Ng-CAM modulations takes on potential functional significance when viewed in the light of the previously described kinetic experiments showing the high order dependence of N-CAM binding rates upon surface density and upon polysialic acid decrement.

Perhaps even more compelling evidence that CAMs actually function in tissues comes from certain perturbation experiments using anti–N-CAM and anti–Ng-CAM antibodies in vitro and in vivo. Anti–N-CAM antibodies can perturb neurite fasciculation of dorsal root ganglia in vitro (81, 82), orderly layering of chick retina in organ culture (83), and neuron-myotube interactions (84, 85). By means of an agar implantation technique, it has been shown that anti–N-CAM antibodies implanted in the tectum of the frog *Xenopus laevis* cause alteration in both the order and precision of the retinotectal map (67). Of particular interest was the observation that, at longer times, when the in vivo antibody levels were decreased, restoration of such maps occurred. This is consistent with the dynamism expected for primary processes regulated by modulation events.

Perturbation experiments have also been carried out in vitro with anti–Ng-CAM [(36); C.-M. Chuong, G. M. Edelman, unpublished observation] and anti–L₁ antibodies (86, 87) on tissue slices of mouse cerebellum removed just as the granule cell population in the external granular layer was moving on glial fibers to deeper layers. In controls, such movements of postmitotic neurons occurred; in those slices treated with the anti–Ng-CAM antibodies, the majority of migrating cells was arrested. These findings and structural studies (35, 36) suggest that, although L1 has not been shown to mediate neuron-glia binding, it is in fact mouse Ng-CAM.

How many different molecular specificities are likely to be required in forming fiber tracts and maps in the CNS is at present not known (12). Until an exact causal analysis is made of the role of CAMs in neural mapping in vivo, interpretation must be guarded. Nonetheless, a number of observations suggest that whatever new molecules are found as markers, they will not have the function of cellular addresses directly responsible for pattern. These observations can be listed as follows. 1. It is necessary and sufficient to block only N-CAM binding to block neuron-neuron binding and to disrupt neural patterns in a variety of tissues. Neural CAMs change binding efficacy by modulation, and it does not seem necessary that they represent a very large family (i.e. greater than 100) of different molecular specificities. Indeed, the same N-CAM that plays a key role in early neurogenesis is ubiquitous on differentiated neurons and plays a key role in later histogenesis. 2. Neural patterning depends critically upon glial interactions as well as upon neural interactions, particularly for neural migration; just before the appearance of glia, Ng-CAM shows modulation in its expression rather than a host of different specificities. While Ng-CAM may be involved indirectly in neuron-neuron binding (74), its major function is to mediate neuron-glia binding. 3. A number of primary processes of development contribute to neural patterns in a dynamic fashion that is responsive in part to CAM expression. It should be emphasized in this context that the graded nature of surface modulation events at the molecular level can

---→

Figure 5   CAM expression and control. (*A*) Schematic diagram showing the temporal sequence of expression of CAMs. After an initial differentiation event, N-CAM and L-CAM diverge in cellular distribution and are then modulated in prevalence (⇅) within various regions of inductions or actually disappear (○) when mesenchyme appears or cell migration occurs. Note that placodes which have both CAMs echo the events seen for neural induction. Just before appearance of glia, a secondary set CAM (Ng-CAM) emerges; unlike the other two CAMs, this CAM would not be found in the map shown in Figure 4 before 3.5 days. In the perinatal period, a series of epigenetic modulations occurs: E-A conversion for N-CAM and polar redistribution for L-CAM. The diagrammed events are based mainly on work on the chick. (*B*) Regulatory effects of CAM expression and modulation. While the major effect is on the adhesion process, CAM expression and

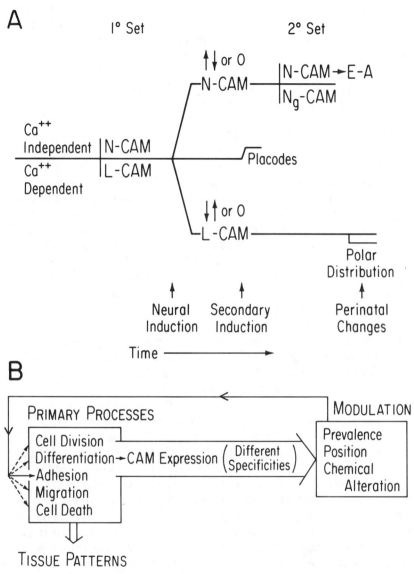

modulation may alter the sequence and extent of each of the other primary processes either indirectly or (as suggested by the dotted arrows) directly. These effects may lead in turn to different tissue patterns. The regulator hypothesis states that activation of a set of regulatory genes controlling a small number of CAM structural genes alters cell-cell adhesion and therefore morphogenetic cell movements. This controls embryonic induction and activation of other regulatory genes, particularly those concerned with cytodifferentiation. Regulation of CAM genes is thus considered to be prior to and relatively independent of cytodifferentiation schedules until later histogenesis.

lead to a very large number of cell binding states. This puts the onus for the establishment of regular form upon developmental genetic and epigenetic control events selected for during evolution, and brings us back to the key issues mentioned at the beginning of this review.

GENE CONTROL AND CAM EXPRESSION: MORPHOGENESIS RECONSIDERED

We are now in a position to look at the whole CAM expression sequence (Figure 5) and consider it in terms of how molecules specified by a one-dimensional genetic code, such as the CAMs, can help to specify a three-dimensional animal (what I have termed the developmental genetic problem in the introduction to this review). Clearly this issue must remain sketchy and hypothetical until much more knowledge is obtained about old and new CAMs, their genes, and particularly the control mechanisms governing their expression. Nevertheless, a useful working hypothesis (the regulator hypothesis) emerges even from the scanty data now in hand (3). The main virtue of this hypothesis is that it encompasses the molecular, cellular, genetic, and evolutionary levels contributing to form and histogenesis in an attempt to integrate them into a cogent and consistent whole.

Before considering this hypothesis, three points must be stressed. 1. As shown in Figure 5, the different CAMs are expressed in particular places at particular times in a regular expression sequence. 2. This sequence nonetheless does not deviate from the topological order of the composite fate map (Figure 4) despite the complex topography of organ formation. 3. CAMs appear at times much earlier than most tissue-specific differentiation products but these same CAMs are also expressed as various cell lineages continue to differentiate in organogenesis. Taken together with the facts reviewed above, these conclusions suggest that both the expression of genes for different CAMs in particular sequences during embryonic development and the mechanisms of transport and epigenetic modulation of CAMs at cell surfaces are fundamental in morphogenesis. The sequence shown in Figure 5 must be confirmed by actual determinations of the precise times of CAM gene expression. Nonetheless, the biochemical data, the expression sequence, and the composite fate map (Figure 4) all provide enough evidence on the ordered spatiotemporal expression and function of the CAMs to formulate a testable hypothesis on their role in connecting genotype to phenotype.

Following the original formulation (3), the key assumptions and conclusions of the CAM regulator hypothesis may be summarized as follows.

1. CAMs play a central role in morphogenesis by acting through adhesion as steersmen or regulators for other primary processes, particularly morphogenetic movements. Because of the obvious linking of cells by CAMs, CAM expression must play a direct role in the control of motion which is the result of

the play between cellular motility, tension in tissue sheets, and adhesion. CAMs exercise their role as regulators by means of local cell surface modulation.

2. Genes for CAMs are expressed in schedules that are prior to and relatively independent of those for particular networks of cytodifferentiation in different organs. This appears to be reflected in the fact that a CAM in the composite fate map (Figure 4) is expressed early and remains in structures (Table 1) that span classical map boundaries not only within a germ layer (e.g. somites, heart, kidney) but also across germ layers (all three for L-CAM; ectoderm and mesoderm for N-CAM).

3. The control of CAM structural genes by regulatory genes is reflected in the control of morphogenetic movements, and both processes together are responsible for the body plan as seen in fate maps. In the chick, the origin of this plan is reflected in a topological order (Figure 4): a simply connected central region expressing N-CAM surrounded by a contiguous simply connected ring of cells expressed L-CAM.

4. Morphogenetic movements are resultants of the inherent motility of cells, tissue sheet folding, and CAM expression as they are coordinated with the expression of substrate adhesion molecules (SAMs) and substrates such as fibronectin (9, 21). These movements, which are regulated in part by CAM modulation, are responsible for bringing cells of different history together to result in various embryonic inductions including neural induction.

5. Natural selection acts to eliminate inappropriate movements by selecting against mutant organisms with failures of movement or those that express CAM genes in sequences that fail to regulate movements appropriately, leading to failures in embryonic induction. On the other hand, any variant combinations of movements and covariant alteration in timing of CAM gene expression (resulting from variation in regulatory genes) that together lead to appropriate inductive sequences will in general be evolutionarily selected. This covariant set of changes would allow for great alteration in the details of fate maps from species to species but at the same time would tend to conserve the basic body plan.

6. Small changes in CAM regulatory genes that do not abrogate this principle of selection could nevertheless lead to large changes in form in relatively short evolutionary times.

It has been pointed out (3) that this model is parsimonious—the total number of genes involved need not be large. Because of the wide dynamic range of cell surface modulation effects (4, 5) and their temporal permutions, the developmental and evolutionary effects of variation in the schedules of expression of a rather small number of genes related to cell adhesion could be momentous. The predictions of the regulator hypothesis and the contrary results that would falsify the hypothesis have been reviewed elsewhere (3), but in the present context, one prediction is particularly germane: If the topology of CAM expression in neural induction is not the same in various animal species that have similar body plans but different morphogenetic movements, then the hypothesis must be abandoned. For example, the frog should show a similar composite CAM fate map topology to that of the chicken (Figure 4) despite the great differences in early morphogenetic movements in the two species. If true, this forces the prediction that the temporal expression of CAM regulatory genes in ontogeny would have different schedules in the two species. This fits with notions of evolutionary change based on heterochrony (1, 88).

The regulator hypothesis focuses attention upon a central determinant issue of both early and late morphogenesis: What regulates the expression of CAM genes themselves and how are the other primary processes controlled? It remains an open question whether the action of CAM regulatory genes is triggered by morphogens or is triggered by feedback from CAM interactions, spatial asymmetries, or by global cell surface modulation induced by binding of substrate adhesion molecules (SAMs). It is not known whether CAM gene expression and the ensuing cell-cell adhesion events that irreversibly alter cell movements can actually feed back directly to affect other differentiation or division events. It does appear likely that, at the least, differential CAM expression will have indirect effects which change the proportional contribution of each primary process to morphogenesis (Figure 5). One impression already emerges from this dynamic, multilevel picture: The diversity of control loops is likely to be very large depending upon the actual number of CAMs of different specificity and the number of modulation mechanisms employed by different differentiated cells. Nonetheless, the underlying picture of a relatively small number of gene products (CAMs) used in different modes is consistent with the known plasticity of development (1, 2) and with relatively rapid evolutionary change (88) and, in this sense, it is satisfyingly simple.

THE CELL IN CONTACT: INTERACTIVE AND COMPLEMENTARY MOLECULAR SYSTEMS FOR FUTURE STUDY

As already shown in Table 1, the CAMs appear in adult tissues in distributions that follow from those seen in embryonic life. But in most cases, they constitute a smaller proportion of tissues, appear only in certain limited tissue locations, and in some cases (such as the pancreas), they are found distributed on cells in a polar fashion. The sparser CAM distribution is perhaps not surprising, because adult tissues contain increased amounts of derivatives of connective tissue and show evidence for increased cellular interactions with SAMs, [molecules that include collagen, laminin, fibronectin, glycosoaminoglycans, etc (89, 90)]; moreover cells of adult tissues have formed specialized junctions (91) of various types by means of cell junctional molecules (CJMs). The central nervous system is an exception: tracts which form a major part of the membrane surface in the nervous system do not have major amounts of CJMs and SAMs, and while SAMs and CJMs are present in neural tissue, CAMs predominate.

It is a reasonable surmise that, while the major determinants of tissue specificity derive from the selectivity of CAM modulation, the different SAMs and CJMs also play very important parts in morphogenesis. The known functions of SAMs are related to intermodulation with CAMs and cell migration [as seen for example in the neural crest cells (49, 50, 72)], to tissue partition, and to the development of hard tissues (89, 90). The known functions of CJMs are cell connection and communication as seen in gap junctions or in the sealing of the surfaces of epithelial sheets (91). Based on the biochemical evidence accumulated so far, CAMs, SAMs, and CJMs seem to be completely separate families of molecules with disparate but conjugate functions.

The main point is that no view of the molecular basis of morphogenesis would be complete without considering the potential interactions of molecules in all of these groups. Such interactions are already revealed clearly in the case of neural crest cells as mentioned above and they can be studied for gap junctions in the earliest embryos. Furthermore, SAMs have been implicated in certain embryonic induction events (89) and it is reasonable to conjecture that this may be related to their support of cell migration. It remains to be seen how the temporal expression of all three functional molecular families of cell binding molecules is regulated in particular tissues during embryogenesis. One intriguing possibility is that SAMs such as fibronectin and laminin may induce global cell surface modulation (4) which in turn may alter CAM prevalence, which is a local modulation event. If this turns out to be the case, one of the key missing elements in the molecular understanding of the control of morphogenesis would be at hand. In any event, there is ample reason for a concerted effort to determine the sequences of appearance at various morphogenetic stages of all three classes of molecules and to search further for direct or indirect interactions amongst them. This represents a formidable but potentially rewarding challenge to biochemists interested in unraveling the complexities of morphogenesis.

Literature Cited

1. Raff, R. A., Kaufman, T. C. 1983. *Embryos, Genes, and Evolution.* New York: Macmillan
2. Slack, J. M. W. 1983. *From Egg to Embryo.* New York: Cambridge Univ. Press
3. Edelman, G. M. 1984. *Proc. Natl. Acad. Sci. USA* 81:1460–64
4. Edelman, G. M. 1976. *Science* 192:218–26
5. Edelman, G. M. 1983. *Science* 219:450–57
6. Frazier, W., Glaser, L. 1979. *Ann. Rev. Biochem.* 48:491–523
7. Lilien, J., Balsamo, J., McDonough, J., Hermolin, J., Cook, J., et al. 1979. In *Surfaces of Normal and Malignant Cells,* ed. R. O. Hynes, p. 389. New York: Wiley
8. Moscona, A. A. 1974. In *The Cell Sur-*

face in Development, ed. A. A. Mosco-na, p. 67. New York: Wiley

9. Steinberg, M. S. 1970. J. Exp. Zool. 173:395–433

10. Curtis, A. S. G. 1967. The Cell Surface: Its Molecular Role in Morphogenesis. New York: Academic

11. Edelman, G. M. 1984. Ann. Rev. Neurosci. 7:339–77

12. Edelman, G. M. 1984. Trends Neurosci. 7:78–84

13. Brackenbury, R., Thiery, J.-P., Rutishauser, U., Edelman, G. M. 1977. J. Biol. Chem. 252:6835–40

14. Thiery, J.-P., Brackenbury, R., Rutishauser, U., Edelman, G. M. 1977. J. Biol. Chem. 252:6841–45

15. Jorgensen, O. S., Bock, E. 1974. J. Neurochem. 23:879–80

16. Jorgensen, O. S., Delouvée, A., Thiery, J.-P., Edelman, G. M. 1980. FEBS Lett. 111:39–42

17. Schachner, M., Wortham, K. A., Carter, L. D., Chaffee J. K. 1975. Dev. Biol. 44:313–25

18. Hirn, M., Pierres, M., Deagostini-Bazin, H., Hirsch, M., Goridis, C. 1981. Brain Res. 214:433–39

19. Hirn, M., Pierres, M., Deagostini-Bazin, H., Hirsch, M.-R., Goridis, C., et al. 1982. Neuroscience 7:239–50

20. Goridis, C., Deagostini-Bazin, H., Hirn, M., Hirsch, M.-R., Rougon, G., et al. 1983. Cold Spring Harbor Symp. Quant. Biol. 48:527–37

21. Lemmon, V., Staros, E. B., Perry, H. E., Gottlieb, D. I. 1982. Dev. Brain Res. 3:349–60

22. Bertolotti, R., Rutishauser, U., Edelman, G. M. 1980. Proc. Natl. Acad. Sci. USA 77:4831–35

23. Gallin, W. J., Edelman, G. M., Cunningham, B. A. 1983. Proc. Natl. Acad. Sci. USA 80:1038–42

24. Nielsen, L. D., Pitts, M., Grady, S. R., McGuire, E. J. 1981. Dev. Biol. 86:315–26

25. Hyafil, F., Morello, D., Babinet, C., Jacob, F. 1980. Cell 21:927–34

26. Yoshida-Noro, C., Suzuki, N., Takeichi, M. 1984. Dev. Biol. 101:19–27

27. Damsky, C. H., Richa, J., Solter, D., Knudsen, K., Buck, C. A. 1983. Cell 34:455–66

28. Chuong, C.-M., McClain, D. A., Streit, P., Edelman, G. M. 1982. Proc. Natl. Acad. Sci. USA 79:4234–38

29. Hoffman, S., Sorkin, B. C., White, P. C., Brackenbury, R., Mailhammer, R., et al. 1982. J. Biol. Chem. 257:7720–29

30. Rutishauser, U., Hoffman, S., Edelman, G. M. 1982. Proc. Natl. Acad. Sci. USA 79:685–89

31. Cunningham, B. A., Hoffman, S., Rutishauser, U., Hemperly, J. J., Edelman, G. M. 1983. Proc. Natl. Acad. Sci. USA 80:3116–20

32. Hoffman, S., Edelman, G. M. 1983. Proc. Natl. Acad. Sci. USA 80:5762–66

33. Bell, G. I. 1978. Science 200:618–27

34. Grumet, M., Edelman, G. M. 1984. J. Cell Biol. 98:1746–56

35. Grumet, M., Hoffman, S., Edelman, G. M. 1984. Proc. Natl. Acad. Sci. USA 81:267–271

36. Grumet, M., Hoffman, S., Chuong, C.-M., Edelman, G. M. 1984. Proc. Natl. Acad. Sci. USA. 81:7989–93

37. Murray, B. A., Hemperly, J. J., Gallin, W. J., MacGregor, J. S., Edelman, G. M., et al. 1984. Proc. Natl. Acad. Sci. USA 81:5584–88

38. Rothbard, J. B., Brackenbury, R., Cunningham, B. A., Edelman, G. M. 1982. J. Biol. Chem. 257:11064–69

39. Rougon, G., Deagostini-Bazin, H., Hirn, M., Goridis, C. 1982. EMBO J. 1:1239–44

40. Finne, J., Finne, U., Deagostini-Bazin, H., Goridis, C. 1983. Biochem. Biophys. Res. Commun. 112:482–87

41. Crossin, K. L., Edelman, G. M., Cunningham, B. A. 1984. J. Cell Biol. 99:1848–55

42. Chuong, C.-M., Edelman, G. M. 1984. J. Neurosci. 4:2354–68

43. Sorkin, B. C., Hoffman, S., Edelman, G. M., Cunningham, B. A. 1984. Science 225:1476–78

44. Annunziata, P., Regan, C., Balazs, R. 1983. Brain Res. 284:261–73

45. Gennarini, G., Rougon, G., Deagostini-Bazin, H., Hirn, M., Goridis, C. 1984. Eur. J. Biochem. 142:57–64

46. Edelman, G. M., Hoffman, S., Chuong, C.-M., Thiery, J.-P., Brackenbury, R., et al. 1983. Cold Spring Harbor Symp. Quant. Biol. 48:515–26

47. Sadoul, K., Hirn, M., Deagostini-Bazin, H., Rougon, G., Goridis, C. 1983. Nature 304:347–49

48. Gall, W. E., Edelman, G. M. 1981. Science 213:903–5

49. Thiery, J.-P., Duband, J.-L., Rutishauser, U., Edelman, G. M. 1982. Proc. Natl. Acad. Sci. USA 79:6737–41

50. Edelman, G. M., Gallin, W. J., Delouvée, A., Cunningham, B. A., Thiery, J.-P. 1983. Proc. Natl. Acad. Sci. USA 80:4384–88

50a. Thiery, J.-P., Delouvée, A., Grumet, M., Edelman, G. M. 1984. J. Cell Biol. In press

51. Ocklind, C., Obrink, B. 1982. *J. Biol. Chem.* 257:6788–95
52. Ocklind, C., Forsum, U., Obrink, B. 1983. *J. Cell Biol.* 96:1168–71
53. Cunningham, B. A., Leutzinger, Y., Gallin, W. J., Sorkin, B. C., Edelman, G. M. 1984. *Proc. Natl. Acad. Sci. USA* 81:5787–91
54. Takeichi, M. 1977. *J. Cell Biol.* 75:464–74
55. Urushihara, H., Ozaki, H. S., Takeichi, M. 1979. *Dev. Biol.* 70:206–16
56. Urushihara, H., Takeichi, M. 1980. *Cell* 20:363–71
57. Steinberg, M. S., Armstrong, P. B., Granger, R. E. 1973. *J. Membr. Biol.* 13:97–128
58. Brackenbury, R., Rutishauser, U., Edelman, G. M. 1981. *Proc. Natl. Acad. Sci. USA* 78:387–91
59. Grunwald, G. B., Pratt, R. S., Lilien, J. 1982. *J. Cell Sci.* 55:69–83
60. Hyafil, F., Babinet, C., Jacob, F. 1981. *Cell* 26:447–54
61. Yoshida, C., Takeichi, M. 1982. *Cell* 28:217–24
62. Ogou, S. I., Yoshida-Noro, C., Takeichi, M. 1983. *J. Cell Biol.* 97:944–48
63. Shirayoshi, Y., Okada, T. S., Takeichi, M. 1983. *Cell* 35:631–38
64. Peyrieras, N., Hyafil, F., Louvard, D., Ploegh, H. L., Jacob, F. 1983. *Proc. Natl. Acad. Sci. USA* 80:6274–77
65. Lyles, J. M., Norrild, B., Bock, E. 1984. *J. Cell Biol.* 98:2077–81
66. McClain, D. A., Edelman, G. M. 1982. *Proc. Natl. Acad. Sci. USA* 79:6380–84
67. Fraser, S. E., Murray, B. A., Chuong, C.-M., Edelman, G. M. 1984. *Proc. Natl. Acad. Sci. USA* 81:4222–26
68. Hoffman, S., Chuong, C.-M., Edelman, G. M. 1984. *Proc. Natl. Acad. Sci. USA* 81:6881–85
69. Thiery, J.-P., Delouvée, A., Gallin, W. J., Cunningham, B. A., Edelman, G. M. 1984. *Dev. Biol.* 102:61–78
70. Vakaet, L. 1984. In *Chimeras in Developmental Biology,* ed. N. M. LeDourain, A. McLaren. New York: Academic. In press
71. Cowan, W. M. 1978. *Int. Rev. Physiol. Neurophysiol. III* 17:149–91
72. Thiery, J.-P. 1983. In *Fibronectin,* ed. D. Mosher, E. Ruoslahti. New York: Academic. In press
73. Rakic, P. 1981. In *The Organization of the Cerebral Cortex,* ed. F. O. Schmitt, F. G. Worden, G. Adelman, S. G. Dennis, p. 7. Cambridge: MIT Press
74. Edelman, G. M. 1984. *Sci. Am.* 250:118–29
75. Rakic, P. 1972. *J. Comp. Neurol.* 141:283–312
76. Edelman, G. M., Chuong, C.-M. 1982. *Proc. Natl. Acad. Sci. USA* 79:7036–40
77. Brackenbury, R., Greenberg, M. E., Edelman, G. M. 1984. *J. Cell Biol.* 99:1944–54
78. Greenberg, M. E., Brackenbury, R., Edelman, G. M. 1984. *Proc. Natl. Acad. Sci. USA* 81:969–73
79. Calothy, G., Poirier, F., Dambrine, G., Mignatti, P., Combes, P., Pessac, B. 1980. *Cold Spring Harbor Symp. Quant. Biol.* 44:983–90
80. Giotta, G. J., Heitzmann, J., Cohn, M. 1980. *Brain Res.* 202:445–58
81. Rutishauser, U., Gall, W. E., Edelman, G. M. 1978. *J. Cell Biol.* 79:382–93
82. Rutishauser, U., Edelman, G. M. 1980. *J. Cell Biol.* 87:370–78
83. Buskirk, D., Thiery, J.-P., Rutishauser, U., Edelman, G. M. 1980. *Nature* 285:486–89
84. Grumet, M., Rutishauser, U., Edelman, G. M. 1982. *Nature* 295:693–95
85. Rutishauser, U., Grumet, M., Edelman, G. M. 1983. *J. Cell Biol.* 97:145–52
86. Rathjen, F. G., Schachner, M. 1984. *EMBO J.* 3:1–10
87. Lindner, J., Rathjen, F. G., Schachner, M. 1983. *Nature* 305:427–30
88. Gould, S. J. 1977. *Ontogeny and Phylogeny.* Boston: Belknap
89. Hay, E. D. 1981. *Cell Biology of Extracellular Matrix,* ed. E. D. Hay. New York: Plenum
90. Trelstad, R. L., ed. 1984. *The Role of Extracellular Matrix in Development.* New York: Liss
91. Gilula, N. B. 1978. In *Intercellular Junctions and Synapses: Receptors and Recognition,* ed. J. Feldman, N. B. Gilula, J. D. Pitts, Ser. B, 36:3. London: Chapman & Hall

Ann. Rev. Biochem. 1985. 54:171–204

MECHANISM AND CONTROL OF TRANSCRIPTION INITIATION IN PROKARYOTES

William R. McClure

Department of Biological Sciences,Carnegie-Mellon University,Pittsburgh, Pennsylvania, 15213

TABLE OF CONTENTS

PERSPECTIVES AND SUMMARY

RNA polymerase initiates RNA synthesis at sites called promoters. These sites are segments of DNA that are defined by both genetic and biochemical criteria. The biochemical investigation of transcription initiation and its control is aided immensely by the other biological criteria that exist as a test for significance. In

0066-4154/85/0701-0171$02.00

particular, mechanisms of regulation that are proposed on the basis of in vitro experimentation can be compared directly to genetic and to other in vivo results as a criterion for proof of a particular pathway and its control. For transcription initiation as with other biological control processes, the relation of structure to function is central to an understanding of the biological processes. For RNA polymerase, structural information is rather meager. However, structural information on promoters is more detailed. Sequence homologies and the location of promoter mutations have shown that two separate regions within the bacterial promoter participate in the RNA polymerase initiation reaction. The two regions are located at approximately 35 and at 10 base pairs upstream from the start point of RNA synthesis. Thus, the current view we have of transcription initiation and its control is decidedly one-sided and, for better or worse, RNA polymerase is frequently considered to be a relatively constant entity in transcription experiments. This review will emphasize recent biochemical evidence on the mechanism and control of RNA chain initiation and indicate how this work compares to in vivo evidence.

The investigation of transcription initiation has for many years depended upon a rather simple model involving the three overall steps shown schematically below:

$$R + P \underset{K_B}{\rightleftharpoons} RP_c \xrightarrow{k_f} RP_o \xrightarrow{NTP^s} \to \to \cdots \to RNA \qquad 1.$$

1) binding   2) isomerization   3) promoter clearance

This scheme was suggested by the work of Zillig and coworkers (1) and involves the initial binding of RNA polymerase and the promoter with a binding constant, K_B, to form an inactive intermediate termed the closed complex (RP_c). The closed complex subsequently isomerizes with rate constant k_f to form the transcriptionally active open complex (RP_o)[1]. Both of these initial steps involve noncovalent interactions. The energy available in nucleoside triphosphates is not utilized until the subsequent binding of the template directed–nucleoside triphosphates occurs and RNA chain elongation commences.

The mechanism shown in Scheme 1 can be considered from several points of view. This review will emphasize mechanisms as they relate to physiological

---

[1]The names and symbols used in Equation 1 are adapted from previously published descriptions of the two functional steps involved in the formation of a transcriptionally active RNA polymerase-promoter complex. The designations "closed" and "open" complex are from Chamberlin (2). K_B has been referred to as K_I and K_{A^*}; k_f is called k_2 elsewhere. The rationale for using these latter changes is that the steps identified by K_B and k_f are likely to remain valid descriptions of the two principal steps in open complex formation, whereas detailed mechanistic work is likely to show that both K_B and k_f can be dissected into separate steps (e.g. K_1, K_2, etc or k_2, k_3, etc) (See THE PROMOTER section).

control. In particular, we will describe the pathway(s) of open complex formation, and the importance of the binding and isomerization steps and how they, in turn, relate to DNA sequence and to DNA structure. I will then examine how the action of ancillary proteins such as repressors and activators is superimposed upon the basic recognition by the enzyme for the promoter. I will consider the effects of closely-spaced promoters on the control of transcription initiation. Finally, the effects of DNA supercoiling and altered conformation within the promoter will be examined.

A striking physiological observation that motivates much of the biochemical investigation is the fact that RNA chain initiation frequencies occur over a dynamic range of about 10^4. To a first approximation, genes exist in one copy per cell and only one enzyme is responsible for transcribing most of them. However, some genes are transcribed every few seconds while others are transcribed once per generation or less (3). The term *promoter strength* refers to the relative rate of synthesis of full length RNA product from a given promoter, and initiation frequency expresses the same idea in absolute units of reciprocal time (e.g. 10 chains/min, once/generation, etc). The discussion of biochemical kinetics in this review is intended to emphasize the contribution of initiation frequency to gene expression and to provide a quantitative basis for comparing in vitro results with in vivo observations.

The control of transcription has been reviewed recently in other publications. I note here the review on molecular aspects of transcription that appeared last year by von Hippel and coworkers (4). *Promoters: Structure and Function* (5) contains much interesting and up-to-date information, and the Cold Spring Harbor volume still contains important and timely information on RNA polymerases (6). The 1982 Cold Spring Harbor Symposium covered many related aspects of DNA structure and gene expression (7). Additional reviews on RNA polymerase and promoters can be consulted (8–10). Finally, the review by Raibaud & Schwartz (11) is concerned with genetic experiments on the positive control of transcription initiation and is complementary to the topics that are discussed here.

RNA POLYMERASES

The enzyme isolated from *Escherichia coli* is the best studied RNA polymerase from prokaryotic organisms. The enzyme is a complex structure with subunits (and molecular weights) β' (155,613), β (150,618), σ (70,263), and α (36,512). The synthesis of the enzyme is under intricate transcriptional and translational control [see (12, 13) for reviews]. The DNA sequence of the genes for β', β, and σ has been determined (14–16). The structure of the enzyme is understood only at rather low resolution from studies using neutron and X-ray small angle scattering (17). It is well established that the core enzyme $\beta\beta'\alpha_2$

(termed E) contains all of the catalytic machinery required for the synthesis of an RNA chain; the combination of σ subunit and core is called holoenzyme (Eσ). Eσ can bind specifically to promoter sites and initiate RNA chains correctly. The enzyme is present in rather large quantities in *E. coli*, approximately 3,000 molecules per cell (18). It is relatively straightforward to purify the enzyme and to separate it from interfering activities using the procedures originally worked out by Burgess and coworkers (19, 20). Some physical properties of the core-σ binding have been investigated (21). Sedimentation analysis has shown that holoenzyme undergoes dimerization at ionic strengths less than 0.1 M. At about the same salt concentrations the core enzyme aggregates extensively (22).

Following the discovery of σ subunit (23), the hypothesis was proposed that there would be a group of σ factors within *E. coli*, each specifying initiation selectivity for specific classes of genes. For many years it was thought that the "σ hypothesis" applied only to *Bacillus subtilis* and to its phages. The recent discovery by Grossman et al (24) that another σ factor exists in *E. coli* suggests that the σ hypothesis as originally proposed may yet have a wider significance for *E. coli* than was previously imagined. Kassavetis & Geiduschek have found that *E. coli* core and a positive control protein (gp 55) of bacteriophage T4 alone can direct the initiation of the late T4 genes in vitro (24a).

The investigation of RNA synthesis in *B. subtilis* has shown that there are at least five σ factors present. One of these is sporulation-induced and is responsible in part for transcriptional specificity during the early stages of the sporulation developmental program. [see (25, 26) for review]. Pero and coworkers and Geiduschek and coworkers have shown that SP01, a *B. subtilis* phage, encodes two σ factors that are responsible for the temporal expression of genes from that bacteriophage (27). The *B. subtilis* core enzyme is similar to the *E. coli* core and provides the catalytic machinery for all of these Eσ combinations. The vegetative *B. subtilis* cell contains one principal σ subunit termed σ^{55}, and three additional minor species, σ^{37}, σ^{32}, and σ^{28} (28–30). (The superscripts correspond to their molecular weights in kilodaltons.) The promoters recognized by $E\sigma^{55}$ show many of the conserved sequence features of *E. coli* $E\sigma^{70}$ (31). From the sequences of several promoters corresponding to other Eσ classes, it now appears that recognition of both the −35 and −10 regions is unique for each holoenzyme. The isolation of several mutations in an $E\sigma^{37}$ promoter (32) promises to reveal in greater detail the basis of promoter selectivity by the various Eσ combinations in *B. subtilis*.

The bacteriophage-induced RNA polymerases display a very high degree of specificity for the promoters on the corresponding bacteriophage DNA. [see (33) for review.] In contrast to the bacterial host enzymes, these RNA polymerases recognize a contiguous sequence of DNA bases near the transcription start point. Evidence for a melted region of DNA near the startsite, and for some movement of the enzyme within the separated strands, was found for the T7-induced RNA polymerase (33a). Diverse strategies are exhibited by the various bacteriophages in using their specific RNA polymerases. As shown in Table 1, the diversity is reflected by the fact that some of the phages use the host enzyme early and then encode a phage specific enzyme for the middle and late transcriptional units. Other phages use a modified host enzyme throughout development. One of the more novel strategies is used by the coliphage N4, which injects with its DNA one or two molecules of the N4-specific RNA polymerase (39). This enzyme transcribes the early region of N4 DNA. One of the genes transcribed by the virion enzyme codes for a second enzyme complex that transcribes the middle transcriptional units (40, 41), and finally late in infection the host RNA polymerase, perhaps in a modified form, is responsible for N4 late transcription.

RNA polymerases from archaebacterial organisms such as thermophiles and halophiles resemble more closely the enzymes from eukaryotes than those from the eubacteria (42). Zillig and coworkers have reported the purification and partial functional characterization of several of these enzymes and their subunits (43–45).

THE PROMOTER

Sequence Analyses

The DNA sequences of more than 100 promoters from *E. coli* have now been reported (46). Homologies between promoter regions were noted (47, 48) soon after DNA sequencing was possible and a so-called consensus sequence was later formulated (49, 50). The current consensus sequence is shown in Figure 1. In each region, there are three bases that are highly conserved: the TTG at -35 and the TA---T near -10. For convenience, we refer to the -35 and -10 hexamers as the principal features in each region. However, it is clear that conserved and weakly conserved sequence homologies exist on either side of

Table 1 Participation of bacteriophage-induced RNA polymerases in the patterns of temporal development.

	RNA Polymerase(s) Used			
Bacteriophage[a]	Early	Middle	Late	References[b]
1. T7, T3, SP6, gh1	host		phage	33–37
2. PBS2	modified host?	—	phage	38
3. N4	virion	phage	modified host?	40

[a]The bacteriophages listed in the first column are classified on the basis of their distinctive use of different RNA polymerases during the three stages of viral development.
[b]The references indicated contain citations to the earlier work on each enzyme.

the hexamers. The surprising finding was that the consensus sequence based only on a compilation of many diverse promoters appears to be maximal in promoter strength. That conclusion is based largely on the fact that the location and identity of promoter mutations show that alterations that decrease homology to the consensus sequence are down mutations and mutations that increase homology to the consensus are up promoter mutations. The sequences of bona fide promoters in *E. coli* do not always contain striking homologies to the consensus sequence. One contributing factor to the lack of homology is likely to be related to the range of initiation frequencies encountered in the promoters. The notion that the consensus promoter is likely to be nearly maximal in activity is also supported by the construction of the semisynthetic promoter termed TAC in which the -35 and -10 hexamers are precisely homologous to the consensus sequence (51–54). This promoter and several variants have been found to promote RNA chain initiation at very high frequencies both in vivo and in vitro (54, 55). The fact that no wild-type *E. coli* promoter has been found that completely matches the consensus promoter even in the highly conserved hexamer regions suggests that promoter function is ordinarily optimized in vivo and not maximized.

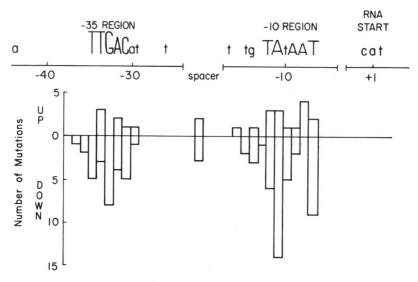

Figure 1 E. Coli promoter consensus sequence and distribution of mutations within known promoters.

The sequence homologies at positions within the promoter are shown in three classes. The highly conserved bases (>75%; 9 S.D.) are large capital letters. The conserved bases (>50%; 5 S.D.) are small capital letters. The weakly conserved bases (>40%; 3 S.D.) are lower case letters. The distance between the -35 and -10 regions (spacer) is ordinarily 17 ± 1 base pairs. Below each position, including the spacer, the number and the nature of known promoter mutations are shown as bars. Adapted from Hawley & McClure (46).

If we accept the proposition that "consensus-is-best" for promoter function, four rules can be useful. First, all known *E. coli* promoters that use Eσ[70] have at least two of the three most highly conserved bases in the −10 region. Second, all promoters have at least one of the most highly conserved TTG residues in the −35 region. Third, those promoters with poor homology to the consensus in the −35 region are frequently positively controlled by dissociable activators (11). Finally, the promoters used by *E. coli* Eσ[32] during the heat shock response have similar −35 region sequences, but very different −10 region sequences (56) (i.e. they break rule one). In the future, the utility of these rules might be realized in biochemical searches for missing activators (poor −35 homology) or for additional σ subunits (poor −10 homology). As discussed by Raibaud & Schwartz (11), both classes of positive regulators would appear very similar in genetic experiments.

DNA Conformations

The conformation of DNA within the promoter region and longer range interactions have been suggested for many years to contribute to the overall promoter function in a fashion that would be superimposed upon the sequence-specific and electrostatic binding that occurs at the promoter. See (57) for discussion. There is now abundant evidence from a variety of sources that the B form DNA double helix is actually a family of structures and moreover that it can be converted to rather different families of structures with suitable treatments (58–61). From the standpoint of transcription initiation and its control, it is appropriate to note the magnitude and the rates of the interconversions that can occur in the DNA double helix and to relate some of these, if possible, to transcription initiation.

The time scale of some of the perturbations that are known to occur in the DNA double helix is shown in Figure 2. The range of known perturbations spans several orders of magnitude. The most rapid changes shown (nsec) are the angular alterations due to twisting and bending of the helix (62, 63). Near the melting temperature, intramolecular helix-coil transitions occur in the μsec time range (64, 65). DNA replication proceeds at about 1000 nucleotides/sec; RNA chain elongation occurs at about 50 nucleotides/sec (67). Most of the rate limiting steps in transcription initiation fall in the region of slower transformations in the enzyme DNA complex. The kinetics of proton exchange as measured by magnetic resonance and isotope exchange techniques (68–70) could relate to structural changes occurring either immediately before or during DNA melting near the transcriptional start site. The NMR studies have also shown a more flexible structure associated with specific sequences such as TATAAT (68) and GTG (70). At present we do not know which of the above perturbations, if any, contributes directly to the major steps in transcription initiation. Nor do we know which ones might provide fine tuning or modulation of these steps. The purpose of this section is to highlight classes of structural alterations in the DNA helix that might be relevant to transcription initiation when the latter process is dissected in molecular detail.

In addition to the time scale for the perturbations, it is also important to consider the magnitude of the structural changes that can occur. Clearly the change in helical sense from right-handed to left-handed is a dramatic structural change in the DNA (59). Somewhat more modest in structural alteration but possibly of significance is the transformation betweeen B and A forms (61). Both of these transformations in DNA are cooperative and that means that several base pairs are affected simultaneously, so that nearest neighbors are perturbed to some extent. A recently reported alteration in DNA solution structure for which kinetic evidence is not yet available are localized bends or kinks. (72–74). In one case it was found that the binding of catabolite gene activator protein (CAP) induced a bend in the DNA (72, 73). In the second case it was argued that the bend in the DNA was a static property of the DNA molecule (74). Since some bent DNA species can easily be detected by their abnormal gel mobilities, we should learn soon how widespread this altered structure is in transcriptional control regions. The earlier suggestions that cruciforms or segments of Z DNA might function in vivo as recognition signals in DNA now appear unlikely. Both of these dramatic changes in DNA require high superhelical densities and the structures are formed rather slowly (75, 75a, 76).

Overlapping Promoter Geometries

The location of promoters per se may play a regulatory role in transcription initiation [see also (77) and REGULATION OF INITIATION section]. Of the 112 promoters compiled by Hawley & McClure, more than one-third are close

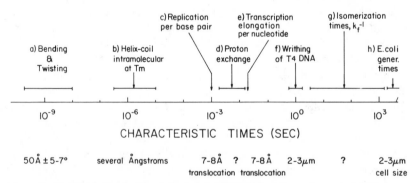

Figure 2 The time scale of selected physical and biological processes that perturb DNA structures.

Each of the perturbations listed is placed at a point or within a region on the logarithmically scaled time axis. The magnitudes of the structural changes, where known, are indicated below the time axis. References for each of these processes are: a.(62, 63); b.(64, 65); c.(67); d.(68, 71); e.(67); f.(70); g.(see text).

to other promoters as shown schematically in Figure 3. The geometry associated with these closely spaced promoters suggests three classes. First, tandem orientations exist where the two (or more) promoters are oriented in the same direction and transcribe the same gene or operon. The second class comprises the divergently oriented promoters in which two RNA polymerases can bind within a common (control) region and transcribe in opposite directions into separate genes or operons. Third, a less frequently encountered class are the convergent promoters where RNA polymerases actually oppose one another and transcribe both strands of the DNA over a common interval. In many of the cases shown in Figure 3 the close proximity has been shown to be an important part of the regulatory function displayed by these promoters. If the two (or more) promoters are so close that RNA polymerase can bind to only one site, interference will occur. In the simplest case, the resulting competition in promoter utilization is predicted to correspond to the ratio of K_B values for each promoter. In principle, promoter proximity could also result in an enhancement of transcription from one or both sites. Some evidence that is consistent with a positive role for two binding sites has been reported for the tRNA^{Tyr} (78) and *ilv*GEDA (C. Adams, W. Hatfield, personal communication) promoters. Closeness per se does not imply an interaction; for example, the three strong

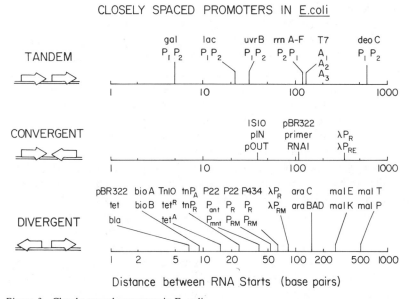

Figure 3 Closely spaced promoters in E. coli.

Of the promoters compiled by Hawley & McClure (46), those pairs that are close to one another are displayed in three groups according to their orientation. For each geometry, RNA polymerase is shown schematically as an arrow.

tandem promoters on phage T7 have not been shown to interact as yet. At a considerably greater distance ($>$ 10 kb) the λ P_L promoter was shown in vivo to interfere with a tandemly oriented downstream *gal* promoter (79). The "promoter occlusion" of *gal* initiation depended on active transcription from the strong λ P_L promoter. The converse of promoter occlusion, transcriptional activation of downstream sites, has not been reported, but it was suggested as a formal possibility several years ago (80). The relationship of promoter geometry to the control of transcription initiation will be considered in more detail with examples in the following sections.

RNA POLYMERASE - PROMOTER INTERACTIONS

Assays for RNA Polymerase-Promoter Complexes

The following section compares the advantages and the limitations of the diverse and commonly used assays that examine RNA polymerase-promoter interactions. The principal assays that have been used historically are continually under development and there is some unevenness in the quality of the published work. We will consider here only briefly the physical basis for each of the assays and the types of information that can be obtained from them. The filter binding technique is based on the fact that protein-DNA complexes bind to nitrocellulose filters, whereas free DNA is not retained. As a consequence, radioactively-labeled DNA fragments, complexed to RNA polymerase, show up as retained radioactivity in this type of assay. This approach has been refined and optimized by Record and coworkers (81, 82). Several artifacts in the older work are now rather well understood, including the binding of holoenzyme to the ends of DNA (83) and at other sites that do not correspond to promoters within a particular DNA segment. Nevertheless, under ideal conditions and with proper controls, the filter binding technique can reveal the physical presence of polymerase on a DNA fragment. This information can be used both to measure the rates and the extents of binding (84, 85).

A more versatile technique that yields the same physical information is the so-called footprinting technique (86, 87) which is based on the protection of a region within a DNA fragment from limited digestion by DNaseI. The advantage over the filter binding technique is that the footprinting assay provides a direct measure of both the extent and the location of complex formation, and moreover, provides information in cases where additional control proteins are present. The analysis of more complex promoter geometries has been aided by using *E. coli* exonuclease III digestion of the enzyme-DNA complexes (88). In this case the limits of protection can be used to assess fractional binding at two or more overlapping sites (89, 90). The results of more detailed protection experiments that have been performed on open complexes were reviewed by von Hippel et al (4).

The two functional assays that are used to detect the formation of the active complex are based on the abortive initiation reaction and on direct transcriptional analysis. In the abortive initiation reaction the rates of the steady-state reaction are proportional to the concentration of open complex. Ordinarily, radioactive products are quantified, but a newly developed detection assay uses fluorescence (91). The assay can be used with sensitivity and convenience to measure the extent and rate of open complex formation (92, 93). It is nevertheless an indirect assay for that complex, since maximum activities must be shown independently to correspond to complete promoter occupancy. Transcription assays have the advantage that they rely upon direct RNA product analysis, and thus the location of the promoter within a particular DNA fragment can be assessed. The latter assay suffers in part because of the variety of problems that may be encountered in doing transcriptional analyses in vitro, such as premature termination, artifactual pausing, etc. Stoichiometric amounts of RNA synthesis per template molecule are not the rule, although strong promoters under optimized conditions can result in the expected numbers of RNA chains (94, 95).

If we use Scatchard's criteria (96) for useful binding information, namely, "How many? How tightly? Where? Why?" it can be seen that at present only a combination of the techniques discussed above can provide all of the desired information. Indeed, one should expect that in the future, combinations of techniques such as footprinting and the activity-based methods will be used to investigate promoters.

The separate determination of K_B and k_f relies on the enzyme concentration dependence of the kinetics of open complex formation. At saturating enzyme concentrations the rate of open complex formation approaches the maximal value of k_f. (Typical values are 10^{-3}–10^{-1} s^{-1}.) At an enzyme concentration equal to K_B^{-1} ($K_B[R] = 1$), the rate is half maximal. (K_B values are 10^6 M^{-1}–$10^9 M^{-1}$.) Abortive initiation, transcription, and filter-binding have all been used as assays for following these kinetics.

Competition assays can also be used in conjunction with any of the detection techniques discussed above. Properly performed, a competition assay can provide a measure of the relative second order rates ($K_B \times k_f$) of open complex formation (55). The concentrations of both promoters in these experiments must be held in large excess relative to the active enzyme concentration used. The simplicity of performing these assays is offset by the inability to determine K_B and k_f separately. Several studies of promoter hierarchies based on competition assays have been reported (55, 97–99).

Finally, since detailed mechanistic studies on promoters require an estimate of the concentration of enzyme required to half saturate RP_C, the activity of $E\sigma$ preparations must be determined. Estimates based on RNA synthesis templated by calf thymus DNA or on the appearance of the protein on SDS polyacrylamide gels are insufficient at best. We have found that transcription assays using the T7 D111 template (94) agree with promoter titration assays (100). The two assays are sensitive to somewhat different activity defects in Eσ, and both can be used to determine whether a stoichiometric complement of active σ is present. Because RNA polymerase is known to bind at sites other than those scored by the assay, the activities reported are slightly lower estimates. Nevertheless, most workers in this field usually tolerate enzyme activities of 40–60%. Following a suggestion from E. Stadtman, we have recently found that partial exclusion of oxygen (autoclaved buffers stored under N_2 and N_2 purging of crude extracts) during the early steps of the Burgess & Jendrisak procedure (19, 20) resulted in Eσ activities of 60–75%. A full accounting of the additional factors responsible for enzyme preparations with less than 100% activity remains to be determined.

Pathways to Open Complexes

The simple reaction scheme, Equation 1, shown in the first section suffices to describe the major steps involved in transcription initiation. Nevertheless, that scheme is undoubtedly oversimplified in the same sense that the Michaelis-Menten mechanism is a simplification of the detailed physical steps that occur in any enzyme-catalyzed reaction. In the following we will consider separately the three principal phases of transcription initiation and consider how the subdivision of those three separate steps might yield additional information. Finally we will consider whether any of the complexities suggested for this overall pathway in transcription initiation alter or confirm the simple view that two principal steps suffice to describe the control of transcription initiation.

First, we examine the binding phase. In the following scheme three steps are included prior to the principal rate-limiting isomerization to the open complex:

$$R + D \underset{}{\overset{1}{\rightleftharpoons}} RD \underset{}{\overset{2}{\rightleftharpoons}} RP_c \underset{}{\overset{3}{\rightleftharpoons}} RP'_c \underset{k_f}{\rightarrow} RP_o \qquad \text{2a.}$$

In this scheme we will ignore the reversibility of open complex formation. Operationally, this condition can be met by using experimental conditions that favor the overall equilibrium. In most cases these conditions also result in very long open complex lifetimes (lasting hours). In the first step of the reaction scheme shown above, the enzyme is shown binding to nonspecific DNA to form an RD complex. This portion of the pathway has been treated in detail by von Hippel et al (4). The conversion to a specifically bound complex, RP_C, would in his model correspond to linear diffusion of the enzyme along the DNA until a specific site has been encountered. Some evidence that such a diffusion occurs has been provided by Wu and coworkers by showing that the enzyme, in association with T7 DNA, can be covalently cross-linked to the DNA by short pulses of UV light (101). The other two steps included in the binding reaction come from the work of Chamberlin and coworkers (102, 103) and from a model proposed by Hawley & McClure (104). Basically these two isomerizations may involve the interconversion between one closed complex and another. In both cases, the additional equilibrium constant was proposed as a rationalization for complexities in a kinetic scheme. The important point to emphasize here is not the number and identity of these binding steps, but rather that, in the current view, all of these steps are expected to be rapidly in equilibrium with the free enzyme and the free DNA. The rapid equilibrium assumed for simplicity in using Equation 1 must be verified experimentally. In most cases, the results of template challenge experiments show this approximation to be valid (93, 104); the times required for dissociation of RP_C were faster than manual mixing and sampling times.

An important mechanistic issue is whether the second step, the diffusion along nonspecific DNA, can ever limit the overall reaction rate. In the simplest scheme, the third step would not be limiting unless diffusion to the promoter occurred more slowly than all of the subsequent steps including the isomerization to the open complex. There is no direct evidence that that is ever the case for $E.\ coli$ RNA polymerase binding to promoters. Indirect evidence reported by Bujard et al (105) was based on very high association rate constants for a bacteriophage T5 promoter. A reinvestigation of these kinetics suggests that the observed rate constant $(3 \times 10^8\ M^{-1}\ s^{-1})$ no longer demands a contribution of linear diffusion to the pathway [(106), H. Bujard, personal communication]. Thus, the experiments that show that the enzyme can slide do not establish that it must do so as a rate enhancement mechanism.

We will now consider the isomerization steps separately:

$$R + P \underset{K_B}{\rightleftharpoons} RP_c \overset{4}{\rightleftharpoons} RP_i \overset{5}{\rightleftharpoons} RP_o \qquad \text{2b.}$$

In this scheme, all of the binding steps are packaged together into K_B, as shown in Equation 1, and we examine a proposal from Buc & McClure (107) and from Roe et al (85) that an additional intermediate occurs after the rate limiting isomerization reaction. In this case, we include the reverse steps to accommodate the scheme that was proposed. These studies employed the use of low temperatures or high ionic strengths to achieve a distribution of complexes between the closed and open complex that were shown to be inactive but that were likely to be significant intermediates on the pathway. It is important to emphasize that the additional intermediate in the isomerization reaction was observable only at extremes of temperature and ionic strength. Under conditions that are close to physiological, the last step proposed in this scheme was very rapid and may correspond to the unwinding of the DNA helix at the promoter.

The rate-limiting isomerization (k_4 in Scheme 2) occurs before the rapid ($k_5 \geq 1 \ sec^{-1}$) DNA melting step at the UV5 promoter (107). Therefore, the structural change in DNA associated with the spectrum of k_f values shown in Figure 2 was not specified. It is known that formation of the open complex ultimately results in: (a) a topological unwinding of the DNA of 540° (108); (b) exposure to chemical reagents of ≥ 10 base pairs of DNA localized near the startsite (109, 110); and (c) hyperchomicity of DNA bases (111–113). What is unknown are the nature and magnitude of the structural changes in RNA polymerase or DNA that immediately precede these dramatic alterations in the template.

In considering the pathway to open complex formation, then, we could fuse the above general sequential pathway and get a total of five separate steps. In such a general mechanistic pathway the forward rate constant (k_f) could correspond to any of the steps between 2 and 5, and the binding phase could contain any of the steps between 1 and 4. It is a reasonable mechanistic question to ask whether promoters can in fact be that variable in function. The answer is very likely to be yes. The changes observed in the rate limiting step for two promoters under different reaction conditions suggests that different steps could be rate limiting for different promoters under a standard set of conditions. The analysis of promoter mutations has provided that type of information. Further investigation should show whether a similar sequential pathway to open complex formation is a general characteristic of RNA polymerase-promoter association. If we consider the extremes of promoter strength the focus returns to the role of RD complexes. They might enhance the rate of association of the strongest promoters if these complexes turn out to be on the main pathway; they might inhibit the extent of association of the weakest promoters if these complexes are shown to overlap with the promoter site.

Promoter Clearance

Finally we will consider the promoter clearance portion of the initiation process separately:

$$R + P \underset{K_B}{\rightleftharpoons} RP_c \xrightarrow{k_f} RP_o \overset{NTP^1}{\longleftrightarrow} RP_o^1 \overset{NTP^2}{\longleftrightarrow} RP_o^2 \overset{NTP^3}{\longleftrightarrow} \cdots \downarrow_{\sigma} \quad \begin{matrix} \text{core-RNA} \\ \text{template} \end{matrix} \quad 2c.$$

Here we show the Scheme 1 formulation for the binding and isomerization steps, but deal with the individual steps that follow the formation of the first phosphodiester bond. Operationally, we will define the end of the initiation process as the release of σ subunit (see below). As shown in the scheme above, several steps are associated with the binding of nucleoside triphosphates and the formation of phosphodiester bonds. Following the formation of the first phosphodiester bond, three other contributions to initiation frequency must be considered under the general topic of the promoter clearance time. RNA polymerase is a large enzyme. When bound in an open complex it protects DNA from enzymatic digestion over a region of approximately 65 base pairs (4). In order to achieve maximal initiation frequency the promoter must be regenerated by translocation of the elongating transcription complex out of the region so that another RNA polymerase can bind, isomerize, and initiate an RNA chain. Several factors have been shown to participate in the duration of the promoter clearance time. First, it is known from a variety of in vitro studies that the abortive synthesis of oligonucleotides can occur at several promoters, even in the presence of all 4 nucleoside triphosphates. The cycling reaction has been quantified for the *lac*UV5 and Tn5 promoters (114, 115). In the UV5 case it is likely that abortive synthesis contributes significantly in vitro to the overall frequency of long chain RNA synthesis (115). A second factor is RNA polymerase pausing near the startsite, even under conditions where abortive cycling of oligonucleotides is not observed. In some cases, for example, the $P_{R'}$ promoter of λ, the pause can be significant (116). One example of such a pause occurs at a position 16 base pairs downstream from the start of $\lambda P_{R'}$ transcription (J. Roberts, personal communication). A third mechanistic event that operationally serves to define the end of the initiation phase is the release of σ subunit. The cycling of σ in vitro was originally demonstrated by Travers & Burgess (117). Hansen & McClure showed that σ was released following transcription of 8 or 9 bases on the synthetic poly[d(AT)] template (118). All three of these factors could contribute to the overall time required to clear the promoter.

A minimum clearance time for the promoter of 1–2 sec can be calculated using the maximum rate of elongation (30–50 nucleotides per second). Thus, the maximum initiation frequency in vitro and in vivo is about one chain/sec. The transition between the open complex and transcribing complex has been examined with a photo affinity reagent at the 5' end of the RNA (119) and with chemical protection experiments (120). Both studies revealed changes in the enzyme-RNA or enzyme-DNA contacts as transcription proceeded. An additional feature of the initiation reaction noted by several groups is that with the formation of a certain number of phosphodiester bonds, the ternary complex thus formed becomes resistant to a shift in reaction conditions, such as to high salt or low temperature, that would otherwise result in the dissociation of binary complexes.

A tidy explanation for all of these features of the promoter clearance process might be that the abortive cycling and pausing cease following the dissociation of sigma subunit to result in a stable transcribing complex. However, experimental results demonstrating a diversity among these features are not consistent with the simplest model. A stable complex is observed after the formation of 3 phosphodiester bonds on the T7A1 promoter (121). Pausing occurs on the $P_{R'}$ promoter at $+16$ (116). Abortive cycling ceases after 6 or 9 nucleotides at the *gal* promoters (122) and after 9 nucleotides at the *lac*UV5 promoter (123). Sigma subunit has been shown to dissociate after 8 or 9 nucleotides on the poly[d(A-T)] template (118). The diversity seen in these events during promoter clearance may yet reflect some underlying mechanistic simplicity in this phase of the reaction. For example, at some point (or points) all of the specific contacts within the promoter must be relinquished by RNA polymerase during its conversion to an elongating ternary complex. Do base pairs in the -35 or -10 regions affect this conversion? Do the conserved base pairs around the startsite (that are mutationally silent) participate? In short, do the examples of frequent abortive initiation and pausing correspond to a DNA sequence–encoded time delay in RNA synthesis? If not, then these processes are simply a biochemical nuisance. A more appealing explanation is that initiation frequency may be affected after the formation of open complexes in some cases (115).

Perhaps the most important characteristic of the initation reaction that remains to be established is the precise point (or points) of σ release on a natural promoter. A convincing demonstration and analysis of this event on promoters would be a benchmark in the study of the promoter clearance time. It might also reveal additional loci for the control of RNA synthesis. Until this aspect of the reaction is established for one or more promoters, the diverse contributions of the processes described above will remain speculative. A particularly challenging goal would be to assess the importance of some of the promoter clearance events to in vivo initiation frequency. A direct measurement of abortive initiation in vivo would be difficult inasmuch as the products formed are likely to be degraded by cellular nucleases. The deletion of a strong in vitro pause site in the *rrn*B operon had no effect on RNA synthesis in vivo (124). Additional genetic challenges of biochemical models should help clarify the contribution of promoter clearance to promoter strength.

Having dealt with known complexities in this overall pathway and admitting that the specific properties of particular promoters could add more diversity, it is appropriate to ask whether the first two steps in Scheme 1 actually suffice to describe the overall mechanism that controls the initiation of RNA synthesis. At two different levels we can answer that question with opposite answers. First, at a functional level, two steps probably do suffice because there is a good correlation between the magnitude of K_B and k_f with in vivo initiation frequency (125). The correlation has also led to new questions about in vivo control. However, at a mechanistic level the two steps certainly do not suffice to describe the pathway to open complex formation. For example, there may be extremes of promoter strength in which the rapid equilibrium assumption fails. In that case the strongest promoters may be of the hit-and-stick class, i.e. once associated with a promoter (RP_C) the RNA polymerase always isomerizes before dissociating.

We have seen above that the mechanistic complexities in open complex formation corresponding to several individual steps in each phase of the reaction can, under most experimental conditions, be grouped together into a rapid equilibrium portion termed binding (K_B) and an essentially irreversible portion called isomerization (k_f). We could now ask whether the sequences or regions within the promoter contribute discretely to each of these two principal steps in open complex formation. In fact, rather soon after two regions of homology were identified for the promoter, Gilbert (126) proposed that the −35 region was involved in recognition (K_B) and that the −10 region was involved in the "melting-in" of the polymerase at the promoter in the vicinity of the startsite (k_f). This bipartite model for structure function correlation is appealing in its simplicity and was a useful formulation because it could readily be challenged with experimental measurements of K_B and k_f.

It now seems clear that the individual base pairs within each region of the promoter can have varying effects on both K_B and k_f (see Figure 4A). Thus, in its simplest form, the bipartite model is probably an oversimplification. An alternative explanation is that the interpretation of in vitro measurements of K_B and k_f is flawed and results in complexities that do not really reflect the simple recognition of polymerase for the promoter in the way postulated in the bipartite model. An argument favoring the validity of the in vitro measurements is the following. First, although single base pair alterations can affect both steps, a generalization that holds for the in vitro analysis of promoter mutants is that no down mutations have been found that increase K_B or increase k_f. The converse is generally true for up promoter mutations (see Figure 4A). In principle if the mechanism of recognition were sufficiently complex, mutants of the two classes described might occur. For example, a mutation that caused very tight binding and which in turn resulted in very slow isomerization as a consequence, would be predicted to be a down mutation. An explanation for these findings is that most single base pair changes within the promoter simply remove a favorable contact or introduce an unfavorable one. Second, as discussed in the REGULATION OF INITIATION section, the influence of activators and repressors on promoter strength also suggests a simplicity of interaction as described by effects on K_B and k_f.

Finally, as with other enzyme mechanism studies, it is to be expected that the magnitude of K_B and k_f determined in function-based assays will be an important criterion by which to judge the significance of the amounts and rates of interconversion of postulated intermediates on the open complex pathway.

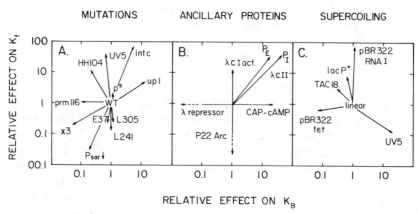

Figure 4 Selectivity maps for three effectors of promoter strength.

For each panel, the relative changes in k_f are shown on the ordinate and changes in K_B are shown on the abscissa. Both scales are logarithmic.

A. A representative set of the relative effects observed for promoter mutations is shown by arrows that indicate the direction and magnitude of the promoter strength changes compared to the wild-type promoter. All down mutations are found in the lower left quadrant; most up mutations are found in the upper right quadrant. The in vitro characterization of these and other promoter mutations can be found in (93, 104, 127–134).

B. The effects of ancillary proteins on promoter strength are displayed as described for panel *A*, except that the reference values correspond to the unactivated or unrepressed promoter. In this case, activators are found only in the upper right and repressors are found in the lower left. References to the experimental results are in Table 2 and the text (REGULATION OF INITIATION section).

C. The effects of supercoiling on promoter strength are displayed as described for panel *A* except that the reference values correspond to K_B and k_f on linear templates. In this case, various wild-type and mutant promoters are found to be affected by changes in combinations of K_B and k_f that are not observed with mutation or ancillary proteins. The supercoiling studies appeared in (55, 129, 172, 173).

At this time it is premature to judge in any detail the agreement between the more detailed mechanistic proposals and the overall measurements of functional constants. The simplest current hypothesis is that the formation of the open complex completes what is likely to be the most significant portion of the pathway that determines the frequency of RNA chain initiation.

Promoter Strength Prediction

The measurements of K_B and k_f performed in vitro can also be tested for significance by attempts to use them to predict promoter strength in vitro based on DNA sequence alone. The first attempt to do so was published by Mulligan et al (135). The evaluation procedure relied upon statistical values related to the occurrences of base pairs within the promoter as compiled by Hawley & McClure (46). Several groups have described computer programs that employ similar statistical bases for locating and evaluating promoter sites within DNA sequences (136–139). None of the other groups attempted to relate the statistical score to in vitro measurements of promoter function. As an illustration of the limitation in using the statistical bases for predictive schemes of promoter strength, we note that the algorithm employed by Staden (139) used a rather different procedure for constructing the weighting matrix for the evaluation of promoter sequences than did Mulligan et al (135). Nevertheless that algorithm when applied to the same data base published by Mulligan et al gave essentially the same goodness of fit; namely, the product of $K_B k_f$ was predicted to within a factor of 3–4 over four orders of magnitude (M. Mulligan, unpublished).

In general, all of these statistical approaches are expected to make reasonable predictions for promoter strength because the major determinants of promoter activity are found in the most highly conserved bases. Detailed mechanisms for the interaction of the weakly conserved bases and the possibility that unfavorable interactions are not properly weighted would seem to doom to failure any statistics-based weighting matrix approach. What is required is a set of experimental data that accurately quantifies the contributions of individual base pairs to promoter strength. The incorporation into the weighting procedures of the experiment-based values could be tested on additional sequences. The work of Youderian et al on the P22 P_{ant} promoter provides a set of more than 20 different mutations in a single promoter (140). Preliminary promoter strength measurements have been reported (141), and additional characterization is currently under way.

A possible complexity that may arise in the attempt to refine these prediction schemes using experimental data can be discussed under the general topic of the "additivity rule." Like the bipartite model for structure function in promoters, the additivity rule suffers from oversimplicity. The advantage is that the simple predictions made are readily testable. In its simplest form, the additivity rule states that the individual promoter strength contributions of base pairs, spacer length, DNA conformation, and electrostatic binding within the promoter can be added together (in a free energy sense) to obtain the total promoter strength ($K_B \times k_f$). The additivity rule can be illustrated with the following numerical example. Alterations in the three most highly conserved bases of the -35 and -10 regions can cause decreases in promoter strength of 100-fold or more. Mutations in weakly conserved bases can result in 10-fold decreases. The additivity rule predicts that the pairwise combination of such mutations will result in 1000-fold decreases. An up mutation is predicted to compensate for a down mutation in a similar fashion. An appropriate challenge to this rule would be the finding that the effect of two individual base pair changes within a promoter do not combine in the predicted fashion to define promoter strength. As yet, no successful challenges exist. The effects of two down mutations in the *lac* promoter on the promoter strength of two up promoter mutations are in good agreement with the second prediction of the rule (128). Additional tests are required before the utility of the additivity rule can be assessed. A second feature of this formulation is that certain alterations in the promoter almost certainly introduce interfering groups in addition to removing favoravle ones as previously suggested (46, 49). This conclusion follows from the above example in that nonconsensus base pairs at all six highly conserved bases could decrease promoter strength by $> 10^{12}$! Since a reduction in RNA polymerase binding to DNA of this magnitude is impossible (the nonspecific binding to DNA is $\sim 10^4$–10^5 M^{-1}), there are clear limits to the applicability of the additivity rule. Unless the correlation of DNA sequence and in vitro function is entirely fortuitous, the current results suggest that the main determinants of promoter strength in *E. coli* promoters do contribute roughly in an additive fashion. Experimental challenges of this idea are expected to refine the weighting procedures used in predicting promoter strength even if the challenges turn out to be successful in overthrowing the simplest form of the additivity rule.

REGULATION OF INITIATION

In the preceding sections most of the discussion has focused on the contribution of individual base pairs within the promoter to the mechanism and regulation of initiation. In this section the next level of complexity that is superimposed upon the contributions of DNA sequence to promoter strength will be examined, namely, the action of ancillary proteins such as repressors and activators and also the effects of DNA structure and DNA supercoiling.

Repression

Repressors of transcription initiation act by binding to specific sites (operators) near the promoter. There is considerable diversity in the location of operators (142). In fact, they have been found in regions as far upstream as -60 and centered as far downstream as $+12$. In many well-known cases the operators overlap the promoters, not only sterically but also in their DNA sequence. In fact the steric blocking mechanism for repression has become a part of the lore of transcription repression. That this mechanism is actually seen and contributes significantly to initiation frequency was recently demonstrated by Hawley et al (143) where the λ cI repressor was found to inhibit the rate of λ P_R open complex formation, specifically, by decreasing K_B. No significant effect on the isomerization rate was detected in those studies. These experimental results were the first to demonstrate the kinetic component of the steric blocking mechanism. In addition, the finding added support to the value of separating the discrete steps in initiation into the initial binding and isomerization steps. It is possible that some repressors may also affect k_f. Preliminary results suggest that this may be the case for Arc protein repression of the P22 P_{ant} promoter

(S.-M. Liao, W. R. McClure, unpublished). The recent reports that two separated operators in *gal* (144, 144a) and in *ara* (145) are required for full repression suggest that the formation of a DNA loop in the promoter region could be responsible for repression. The effects on K_B and k_f might be less important if the putative loops abolish access to the promoter site.

In addition to the diversity of operator locations within the promoter, it is significant that there is a wide spectrum of repressor-operator affinities. Some of the tightest binding repressors, e.g. the *E. coli lac* (146) and P22 *mnt* (147) repressors, bind essentially stoichiometrically to their operators under physiological conditions with a dissociation constant of $\sim 10^{-11}$M. The *c*I and *cro* repressors of bacteriophage λ have dissociation constants of about 10^{-8}–10^{-9} M (148). The Arc repressor of bacteriophage P22 is the weakest of these repressors with a dissociation constant of about 10^{-7} M (A.D. Vershon, R. T. Sauer, personal communication). Thus, the tightest binding repressors can result in an absolute shutoff of chain initiation, at least on a physiological time scale. In other cases, the weaker binding repressors may contribute simply to a down modulation. The advantage to the cell or to the bacteriophage, in the latter case, is that the reduced expression from a particular promoter could then be affected by other cellular or phage products.

Activation

It has become difficult in recent years to discuss activators of transcription separately from repressors of transcription inasmuch as many repressors are also activators. Table 2 compiles the characteristics of some of the activators that have been characterized in vitro. Others are known to have the bifunctional role based on genetic evidence (11). It is not surprising that activators can also be repressors as they share the same DNA binding specificities and characteristics as the repressor proteins. The location of activator binding sites distinguishes activators from repressors. In particular, all of the known activator binding sites are located near or upstream from the −35 region. The above classification of activators and their binding sites is perhaps more convincing for the class of activators that are known to activate transcription initiation directly by an effect on one of the initiation constants.

As shown in Table 2, the λ repressor activates P_{RM} by a direct 11-fold increase in k_f, the isomerization step. In contrast CAP-cAMP activates *lac*P$^+$ transcription by an increase in K_B. The λ *c*II protein is known to activate three promoters on bacteriophage λ (151, 152). The P_E and P_I promoters are activated substantially by increases in both K_B and k_f. These effects are all of the direct sort, as determined in vitro by assays of promoter strength in the presence and absence of the activator. An important feature in several activation mechanisms is the finding that indirect activation can occur and that its effects can be significant. To explain indirect activation I return to the topic of

Table 2 Activation characteristics of positive control proteins

Promoter	Activator	Location of activator site[a]	Extents of activation in vitro			Activation in vivo	References
			K_B	k_2	Indirect		
1. λP_{RM}	λcI Protein	-42 bp	None	$11\times$	None	$8-10\times$	104, 131
	λcI (pc-2)	-42	None	$2-3\times$	None	$1.5\times$	149
2. $lacP_1$	CAP-cAMP	-61	$\geq 20\times$	None	$3-10\times$	$25-50\times$	129
3. λP_{RE}			$15\times$	$40\times$	None	$>100\times$	133, 134, 151
λP_I	λcII protein	-33	$100\times$	$100\times$	None	$>100\times$	b
λP_{aQ}					None	—	152, b
4. P22 P_{mnt}	Mnt protein	-45	None	None	$10\times$	—	c

[a]The location at the center of each binding site is given in base pairs (bp) from the startpoint of the promoters listed in column one.
[b]Hoopes, B. C., & McClure, W. R. unpublished.
[c]Liao, S.-M., & McClure, W. R. unpublished.

promoter location (see THE PROMOTER section) where it was shown that closely-spaced promoters can interfere with one another in binding RNA polymerase. Under these conditions the expression from both promoters is reduced from what would be observed in the absence of the competing promoter. Therefore it is not surprising to find that a protein repressor that inactivates one competing promoter will be predicted to be an indirect activator of the second promoter. An example that has recently been investigated suggests that the Mnt protein by virtue of its repression of the P22 P_{ant} promoter results in an indirect activation of the divergently oriented P_{mnt} promoter (S.-M. Liao, W. R. McClure, unpublished). No effects on K_B or k_f have been detected. In the case of the *lac* operon promoter, contributions from direct effects on K_B and indirect effects due to repression of competing polymerase binding sites have been reported (90, 129, 150). In general the magnitude of indirect activation effects on transcription are predicted to be less than those already demonstrated for direct activation.

The biochemical effects of ancillary proteins are shown in the selectivity map of Fig. 4B. Two important conclusions are apparent in this representation. First, the same simplicity of function seen with mutations is observed, i.e. activators only increase K_B or k_f or both—none have been found that increase one constant while decreasing the other. The converse rule applies to repressors, although fewer examples have been examined. Second, for most of the systems studied, activation and repression are superimposed in a simple way upon the effect of promoter mutations. For example, CAP-cAMP increased K_B for the wild-type *lac* promoter and for two up promoter mutants (129); λ *c*I increased k_f for the wild-type λ P_{RM} promoter and did the same for an up mutant (104) and several down mutants (131).

The structures of three ancillary proteins have been determined using X-ray crystallography [see (153) for review]. Models for the specific complexes between DNA and λ*c*I (154), λ*cro* (155), and CAP-cAMP (156) have been proposed. All three proteins appear to bind right-handed DNA, and all three have a common secondary structural motif that is argued to be directly involved in sequence recognition (157, 158). A possibility for DNA conformational changes in the three complexes has been suggested, but the nature and magnitude of the alterations cannot be specified yet. The only example of a protein-DNA complex of known structure is the *Eco*RI endonuclease - dodecamer substrate structure (159); Rosenberg and coworkers find three well-defined kinks in an otherwise B-DNA double helix.

The structure of the λ *c*I DNA binding domain has already been important in understanding the mechanism of direct activation at λ P_{RM}. Even though the structure of RNA polymerase is unknown, the combination of structural (154), genetic and physiological (160, 161), and biochemical (149, 162) evidence argues strongly for a direct contact between *c*I protein and RNA polymerase at

λ P_{RM}. Finally, N. Irwin and M. Ptashne (personal communication) have found that mutations in the gene for the β subunit of RNA polymerase can partially suppress the defect of a mutant activator (the bacteriophage P22 analog of λ cI). These results provide additional evidence for protein-protein contacts in the activation of P_{RM}.

A significant biochemical difference between activation and repression may be important in the switch-like control exerted by both classes of effectors. At saturating concentrations of activator, the increase in promoter strength reaches a plateau (e.g. 11-fold for cI activation of λ P_{RM}); at saturating concentrations of cI repressor, P_R promoter strength is reduced to zero. Thus, positive control can effect switches between basal expression and some predetermined maximum, whereas negative control can operate between zero and basal expression. The magnitude of the changes can be comparable for both forms of regulation. Finally, in systems that are subject to several cellular signals, it is not surprising to find both negative and positive control mechanisms in place.

Promoter Structure and Conformation

Recent evidence suggests that regulation can also occur as a result of structural and conformational changes in the DNA template. Methylation at GATC sites by the *E. coli dam* gene product is required for the expression of the *mom* gene in bacteriophage Mu (163, 164). In contrast, the same methylation system inhibits transcription initiation at the Tn10 transposase promoter in vivo (D. Roberts, N. Kleckner, personal communication) and in vitro (B. C. Hoopes & W. R. McClure, unpublished). Similar results have been obtained for the Tn5 transposase promoter (J. C.-P. Yin, W. S. Reznikoff, personal communication). The methylation of the Tn10 and Tn5 promoters occurs in the -10 region and presumably blocks an essential recognition site for RNA polymerase. A conformational change in DNA has been correlated to normal transcription of the *Salmonella typhimurium* tRNA^{His} gene; (L. Bossi, D. M. Smith personal communication) have found that a three base pair deletion at -70 reduced transcription in vivo and in vitro. They also find that a DNA fragment bearing the wild-type promoter migrates abnormally on acrylamide gels suggesting a bend or a kink in the DNA (see THE PROMOTER section). The corresponding DNA fragment from the mutant strain shows normal mobility on gels. Deletion(-fusions) of DNA 50–80 base pairs upstream of the *E. coli* tRNA^{Tyr} (165) and *rrn*B P_1 (R. Gourse, M. Nomura, personal communication) promoters has also been found to decrease promoter strength. Gourse and Nomura have also found that DNA fragments containing *rrn* B P_1 migrate abnormally in acrylamide gels. The contribution from a positive control protein in these cases has not yet been ruled out. Experiments performed in vitro can now be used to measure the direct effects of the altered DNA conformation(s).

DNA Supercoiling

DNA supercoiling results in a torsional tension within the DNA double helix; it has been shown both in vivo and in vitro to have direct effects on certain promoters. Surprisingly, it is found that DNA supercoiling can decrease as well as increase gene expression and that linear bacteriophage templates are affected (166, 167). The in vivo observations of alterations in gene expression are subject to the caveat that DNA supercoiling could be indirectly affecting initiation frequency by favoring or disfavoring the binding of an ancillary protein. A possible example of this type of indirect effect of supercoiling is the finding by Sanzey that most CAP-cAMP activated promoters are sensitive to drugs that inhibit DNA gyrase (168). Thus, CAP-cAMP binding and activation may be enhanced by supercoiling or the CAP-activated promoters may be stronger on supercoiled templates, or both effects may be operating. In vitro experiments are required to distinguish these possibilities. Menzel & Gellert (169) have studied the inhibitory effect of supercoiling on the gyrA and B promoters, both in vitro and in vivo. They were able to conclude that supercoiling directly decreased the promoter strength.

There is a significant paradox in thinking about the effect of DNA supercoiling on RNA polymerase–promoter interactions. As detailed above, we know that RNA polymerase when bound in an open complex has unwound the DNA by ~15 base pairs. In a strict thermodynamic sense the negative superhelical torsion within the DNA double helix should favor the equilibrium of a process involved in the unwinding of DNA. It is not clear, however, that all of the steps on the pathway to open complex formation would necessarily be favored by DNA supercoiling, and that is in fact observed. The selectivity map representation of the effect of supercoiling on promoter strength is shown in Figure 4C. Although fewer in vitro studies have been performed on supercoiled templates than on linear DNA, it is already apparent that supercoiling results in considerably more diversity in the patterns of promoter strength than do mutations or ancillary proteins. I note especially the observation that supercoiling can simultaneously increase k_f and decrease K_B and vice versa. Neither of these combinations was found with mutants or ancillary proteins. A second surprising finding is that supercoiling can affect mutant promoters (up and down) in dramatically different ways. For example, the lacUV5 promoter shows a much higher K_B on the supercoiled template, but k_f is decreased approximately 7 fold (129). The wild-type promoter shows a simple 10-fold increase in k_f on supercoiled templates. The paradox stated above extends to in vivo observations where the UV5 promoter was found to be about 50% more active under conditions where gyrase activity was inhibited (168). The agreement between the in vivo and in vitro effects of supercoiling at this promoter adds further support to the description of promoter strength being defined principally by K_B and k_f.

It is puzzling that an energetically favorable contribution in the form of superhelical tension within the DNA helix does not in all cases increase promoter strength. This topic remains a very interesting problem for further investigation, both in vivo and in vitro. Perhaps what is needed is the investigation of the rates and extents of open complex formation on superhelical templates of varying torsional tension. The study of promoter function with DNAs of varying superhelical densities would serve two purposes. First, as normally isolated, plasmids are more underwound ($\sigma = -0.06$) than their in vivo counterparts ($\sigma = -0.03$) (170). Thus, most experiments have been performed using DNAs with higher than physiological superhelical tension. Second, if a discrete step in open complex formation is affected by supercoiling, the prediction, based on energetics, is that it should follow a quadratic dependence on superhelical density (σ). In short, studies of this type are expected to provide both functional and mechanistic information of value.

RNA Polymerase in vivo

The description of RNA chain initiation frequency in terms of two principal steps carries with it an important prediction about the relative activities of various promoters in vivo. In particular, the association of free enzyme and promoter to form a closed complex follows a hyperbolic binding curve with a half saturation value corresponding to a particular RNA polymerase concentration. As we have seen, the in vitro measurements of K_B can be useful for the description of the activators, repressors, and mutations. We now turn to the question of what relevance the K_B values (10^6 M^{-1}–10^9 M^{-1}) have in vivo. The chart shown in Figure 5 depicts the state of the RNA polymerase molecules within an exponentially growing $E.\ coli$ cell. As shown, about one half of the 3,000 molecules are actively transcribing at any one time (171). The other half are partitioned among several locations available for the enzyme. A rough estimate of the average K_B values (10^7 M^{-1}–10^8 M^{-1}) found for wild-type promoters suggests that approximately half of the nontranscribing molecules may be specifically bound to promoters as closed complexes with the other half likely to be nonspecifically bound to various sites within the cell including coding regions in the DNA, tRNAs, other cell constituents, and the like. The free concentration of holoenzyme through which all of these species flow, would in this view be a rather small fraction of the total—approximately 1% [cf (174)]. The attractive feature of this model is that it places a number on the effective concentration or activity of RNA polymerase in vivo. The actual number and distribution of the molecules shown in Figure 5 must be characterized at this time as speculative, but it does draw attention to the fact that the free enzyme concentration within the cell is very likely to play an important role in regulation of transcription initiaton. A similar role for nonspecific binding of lac repressor to DNA has been proposed by von Hippel et al (174a).

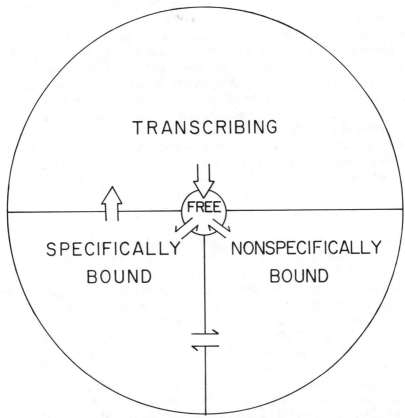

Figure 5   Hypothetical distribution of RNA polymerase molecules within an *E. coli* cell.

In exponentially growing *E. coli* cells about half of the RNA polymerase is actively engaged in RNA synthesis (171). Most of the remaining fraction is conjectured to be distributed between specific promoter-bound complexes (RP_C) and nonspecific complexes (see text). A very small fraction (~1%) is hypothesized to be free (125). The free fraction is considered to be in (rapid) equilibrium ($\rightleftharpoons$) with the other nontranscribing pools of enzyme. The pathway followed during RNA synthesis is shown as ($\Rightarrow$). For comparison to in vitro experiments, 30 molecules in a volume equal to that of an *E. coli* cell corresponds to 30 nM. The limitations of this model and other caveats are discussed in the text.

The argument presented above is essentially the same as similar discussions of the effects of substrate concentration on the activities of enzymes in intermediary metabolism. In those cases it is frequently found that the substrate concentration or activity in vivo is similar to the apparent Km values for the substrate found in vitro and where exceptions exist they are frequently found to be related to interesting control properties. In this case the ligand that saturates the small number of sites within the chromosome is in fact RNA polymerase itself. The 1–2,000 promoters in *E. coli* could in this enzymological view be described as cofactors for the RNA synthesis reaction. An unattractive feature of the model shown in Figure 5 is that it might be very difficult to test experimentally. In fact, because of the rapid equilibria between specifically bound, nonspecifically bound, and free holoenzyme within the cell, the free or active fraction is very low and is effectively buffered by all of the other cellular constituents. Thus, a straightforward test of this model by an attempt to perturb the free holoenzyme concentration within the cell is likely to reveal the properties of the buffering that are suggested to exist. For the time being the model is proposed simply to focus attention on this important point of RNA chain synthesis with the hope that some imaginative tests of the model can be formulated.

ADDITIONAL ROLES FOR RNA POLYMERASE

In this section I consider some regulatory and other functions of RNA polymerase that have been shown to contribute to macromolecular metabolism in *E. coli*. In most cases the biochemical investigations of these more complicated processes are not as detailed as the simple in vitro studies of promoter strength and promoter regulation summarized above. As such they may represent opportunities for fruitful research in the future.

The initiation of DNA synthesis requires an RNA primer. In some cases RNA polymerase synthesizes the primer [see (175) for review]. Two examples illustrate the novel participation of RNA polymerase. First, H. Schaller and coworkers (176) showed that RNA polymerase bound to a specific region of the single-stranded DNA from phage fd in the presence of *E. coli* single-stranded binding protein. The interaction ultimately resulted in the specific initiation and termination (after ~ 25 nucleotides) of the RNA primer. The segment of DNA "recognized" by RNA polymerase looks nothing like a consensus promoter or even a poor match to the consensus! Second, R. McMacken and coworkers (personal communication) have shown that initiation of λ DNA replication in vitro requires activation by a transcribing RNA polymerase which initiates transcription at the P_R promoter. Both of these in vitro results correspond faithfully to in vivo observations.

Stable RNA synthesis in *E. coli* comprises about one half of the catalytic activity of RNA polymerase during exponential growth [(171), see (177) for review]. The tandem ribosomal RNA promoters and the tRNA promoters are subject to several controls related to growth rate. M. Cashel and coworkers have dissected some of these contributions using separated promoters and a combination of in vivo and in vitro experimental approaches (178–180). As noted above in the section on Promoter Structure and Conformation some of these promoter regions may contain altered DNA structures. Additional biochemical and genetic experiments are expected to reveal the factors that result in the high initiation frequencies and controls that are observed *in vivo*.

Rifampicin blocks RNA synthesis at steps soon after the formation of the first phosphodiester bond (181–183). Rifampicin resistance in bacteria maps almost exclusively to loci within *rpo*B, the gene for the β subunit of RNA polymerase. There is a wealth of both specific and anecdotal evidence that Rif^R mutations also result in alterations of RNA polymerase function. Examples of the pleiotropy observed are: *B. subtilis,* sporulation and stable RNA synthesis defects (184); *E. coli,* increase in uridine phosphorylase expression (185); *E. coli,* alteration in stable RNA synthesis (186); *S. typhimurium,* alteration in *pyr*A expression (187); *E. coli,* increase in *ilv* expression (188). C. Yanofsky and coworkers have investigated the properties of Rif^R RNA polymerases that either increase or decrease transcriptional pausing and termination at the *trp* attenuator (189). The in vitro properties reflect the in vivo pleitropic effects. Since many aspects of RNA synthesis can be altered in the Rif^R strains, in vitro experiments are required to decide whether the mutant enzyme is altered in initiation, elongation, or termination. The suggestions based on in vivo evidence alone that some of the Rif^R mutants display altered promoter specificity are therefore premature. Altered promoter recognition specificity in RNA polymerase would not be surprising in some of these cases. A mutation that confers an altered sequence recognition specificity to a DNA binding protein is not unprecedented; this type of mutation was deliberately sought and found in the *mnt* repressor (147).

Multivalent control of the expression of certain genes and operons in *E. coli* has been investigated using genetics primarily. The results show that some promoters are controlled by two or more factors. Each factor can be separately involved in the control of an entire class of genes (regulons). Some examples are the factors specific for (*a*) assimilation of carbon (190), nitrogen (191), phosphorus (192, 193), and sulfur (194); (*b*) anerobic gene expression (195); (*c*) DNA damage, i.e., the SOS response (196–198); (*d*) the reduction of Hg^{2+} for detoxification (199); and (*e*) the heat shock response (200). Multiple promoters, repressors, positive control proteins, and small molecular effectors have all been shown (or suggested) to participate separately and in combination in the regulation of certain promoters within these classes. In one case, the heat shock response, a new *E. coli* σ factor has been identified as the positive control factor (24). An interesting feature in all of these systems is that priorities of control are displayed. The biochemical dissection of (overlapping) sites, control proteins and other features of these complex systems should provide a better understanding of the regulons and the functional interaction between regulons.

Future directions for the less complex studies of in vitro promoter strength can be seen by noting two limitations in our knowledge. First, the relationship of promoter sequence and structure to in vitro promoter strength is incomplete. Part of the task here is to obtain rules for the interaction that will be of predictive value. Second, the measurements of in vitro promoter strength are also being related to in vivo initiation frequency and its control. In this case as well we have seen that our understanding is rudimentary. Finally, it is not unrealistic to expect that in the future in vivo initiation frequency will be understood at a level that would allow its prediction based on the promoter sequence and structure alone.

ACKNOWLEDGMENTS

I am grateful to Kathryn Galligan for her expert assistance in preparing the manuscript. I thank many colleagues for communicating results and comments prior to publication. I thank my coworkers for critical comments on preliminary versions of the paper. Our research on RNA polymerase is supported by the NIH (GM 30375).

Literature Cited

1. Walter, G., Zillig, W., Palm, P., Fuchs, E. 1967. *Eur. J. Biochem.* 3:194–201
2. Chamberlin, M. J. 1974. *Ann. Rev. Biochem.* 43:721–75
3. Hahn, W. E., Pettijohn, D. E., Van Ness, J. 1977. *Science* 197:582–85
4. von Hippel, P. H., Bear, D. G., Morgan, W. D., McSwiggen, J. A. 1984. *Ann. Rev. Biochem.* 53:389–446
5. Rodriguez, R. L., Chamberlin, M. J., eds. 1982. *Promoters: Structure and Function.* New York: Praeger
6. Losick, R., Chamberlin, M. J., eds. 1976. *RNA Polymerase.* New York: Cold Spring Harbor Lab.
7. Cold Spring Harbor Symp. Quant. Biol. 1983. *Structures of DNA,* Vol. 47. New York: Cold Spring Harbor Lab.
8. Chamberlin, M. J. 1982. *Enzymes* 15:61–86
9. Chamberlin, M. J., Kingston, R., Gilman, M., Wiggs, J., deVera, A. 1983. *Methods Enzymol.* 101:540–69
10. Wu, C.-W., Tweedy, N. 1982. *Mol. Cell. Biochem.* 47:129–49
11. Raibaud, O., Schwartz, M. 1984. *Ann. Rev. Genet.* 18:178–206
12. Yura, T., Ishihama, A. 1979. *Ann. Rev. Genet.* 13:59–97
13. Spangler, R., Zubay, G. 1983. In *Multifunctional Proteins: Catalysis Structure and Regulation,* ed. J. F. Kane, pp. 73–85. Boca Raton, Fla: CRC
14. Ovchinnikov, Y. A., Monastyrskaya, G. S., Gubanov, V. V., Guryev, S. O., Chertov, O. Y., et al 1981. *Eur. J. Biochem.* 116:621–29
15. Ovchinnikov, Y. A., Monastyrskaya, G.
    S., Gubanov, V. V., Guryev, S. O., Chertov, O. Y., et al 1981. *Dok. Biochem.* 261:385–90
16. Burton, Z. F., Burgess, R. R., Lin, J., Moore, D., Holder, S., Gross, C. A. 1981. *Nucleic Acids Res.* 9:2889–903
17. Meisenberger, O., Henmann, H., Pilz, I. 1980. *FEBS Lett.* 122:117–20
18. Burgess, R. R. 1976. In *RNA Polymerase,* ed. R. Losick, M. Chamberlin, pp. 69–100. New York: Cold Spring Harbor Lab.
19. Burgess, R. R., Jendrisak, J. J. 1975. *Biochemistry* 14:4634–38
20. Lowe, P. A., Hager, D. A., Burgess, R. R. 1979. *Biochemistry* 18:1344–52
21. Wu, F. Y.-H., Yarbrough, L. R., Wu, C.-W. 1976. *Biochemistry* 15:3254–58
22. Shaner, S. L., Piatt, D. M., Wensley, C. G., Yu, H., Burgess, R. R., Record, M. T. Jr. 1982. *Biochemistry* 21:5539–51
23. Burgess, R. R., Travers, A. A., Dunn, J. J., Bautz, E. K. F. 1969. *Nature* 221:43–46
24. Grossman, A. D., Erickson, J. W., Gross, C. A. 1984. *Cell* 38:383–90
24a. Kassavetis, G. A., Geiduschek, E. P. 1984. *Proc. Natl. Acad. Sci. USA* 81:5101–5
25. Losick, R., Youngman, P. 1984. In *Microbial Development,* ed. R. Losick, L. Shapiro, pp. 63–88. New York: Cold Spring Harbor Lab.
26. Doi, R. H. 1982. *Arch. Biochem. Biophys.* 214:772–81
27. Losick, R., Pero, J. 1981. *Cell* 25:582–84

27a. Chelm, B. K., Duffy, J. J., Geiduschek, E. P. 1982. *J. Biol. Chem.* 257:6501–8
28. Gilman, M. Z., Chamberlin, M. J. 1983. *Cell* 35:285–93
29. Johnson, W. C., Moran, C. P. Jr., Losick, R. 1983. *Nature* 302:800–4
30. Wong, S.-L., Doi, R. H. 1982. *J. Biol. Chem.* 257:11932–36
31. Moran, C. P. Jr., Lang, N., LeGrice, S. F. J., Lee, G., Stephens, M., et al. 1982. *Mol. Gen. Genet.* 186:339–46
32. Tatti, K. M., Moran, C. P. Jr. 1984. *J. Mol. Biol.* 175:285–97
33. Chamberlin, M. J., Ryan, T. 1982. *Enzymes* 15:87–108
33a. Osterman, H. L., Coleman, J. E. 1981. *Biochemistry* 20:4884–92
34. Bailey, J. N., Klement, J. F., McAllister, W. T. 1983. *Proc. Natl. Acad. Sci. USA* 80:2814–18
35. Chapman, K. A., Wells, R. D. 1982. *Nucleic Acids Res.* 10:6331–40
36. Kassavetis, G. A., Butler, E. T., Roulland, D., Chamberlin, M. J. 1982. *J. Biol. Chem.* 257:5779–88
37. Towle, H. C., Jolly, J. F., Boezi, J. A. 1975. *J. Biol. Chem.* 250:1723–33
38. Clark, S., Losick, R., Pero, J. 1974. *Nature* 252:21–24
39. Falco, S. C., Zehring, W., Rothman-Denes, L. B. 1980. *J. Biol. Chem.* 255:4339–47
40. Zehring, W. A., Rothman-Denes, L. B. 1983. *J. Biol. Chem.* 258:8074–80
41. Zehring, W. A., Falco, S. C., Malone, C., Rothman-Denes, L. B. 1983. *Virology* 126:678–87
42. Huet, J., Schnabel, R., Sentenac, A., Zillig, W. 1983. *EMBO J.* 2:1291–94
43. Schnabel, R., Zillig, W., Schnabel, H. 1982. *Eur. J. Biochem.* 129:473–77
44. Madon, J., Zillig, W. 1983. *Eur. J. Biochem.* 133:471–74
45. Madon, J., Leser, U., Zillig, W. 1983. *Eur. J. Biochem.* 135:279–83
46. Hawley, D. K., McClure, W. R. 1983. *Nucleic Acids Res.* 11:2237–55
47. Pribnow, D. 1975. *J. Mol. Biol.* 99:419–43
48. Schaller, H., Gray, C., Herman, K. 1975. *Proc. Natl. Acad. Sci. USA* 72:737–41
49. Rosenberg, M., Court, D. 1979. *Ann. Rev. Genet.* 13:319–53
50. Siebenlist, U., Simpson, R. B., Gilbert, W. 1980. *Cell* 20:269–81
51. Amann, E., Brosius, J., Ptashne, M. 1983. *Gene* 25:167–78
52. DeBoer, H. A., Comstock, L. J., Vasser, M. 1983. *Proc. Natl. Acad. Sci. USA* 80:21–25
53. Rossi, J. J., Soberon, X., Marumoto, Y., McMahon, J., Itakura, K. 1983. *Proc.*

54. Brosius, J., Erfle, M., Storella, J. 1985. *J. Biol. Chem.* In press
55. Mulligan, M. E., Brosius, J., McClure, W. R. 1984. *J. Biol. Chem.* In press
56. Cowing, D., Bardwell, C. A., Craig, E. A., Woolford, C., Hendrix, R., Gross, C. A. 1985. *Proc. Natl. Acad. Sci. USA.* In press
57. Crothers, D. M., Fried, M. 1983. *Cold Spring Harbor Symp. Quant. Biol.* 47:263–69
58. Dickerson, R. E., Drew, H. R., Conner, B. N., Kopka, M. L., Pjura, P. E. 1983. *Cold Spring Harbor Symp. Quant. Biol.* 47:13–24
59. Wang, A. H.-J., Fujii, S., van Boom, J. H., Rich, A. 1983. *Cold Spring Harbor Symp. Quant. Biol.* 47:33–44
60. Wang, J. C., Peck, L. J., Becherer, K. 1983. *Cold Spring Harbor Symp. Quant. Biol.* 47:85–91
61. Ivanov, V. I., Minchenkova, L. E., Minyat, E. E., Schyolkina, A. K. 1983. *Cold Spring Harbor Symp. Quant. Biol.* 47:243–50
62. Barkley, M. D., Zimm, B. H. 1979. *J. Chem. Phys.* 70:2991–3007
63. Hogan, M., Wang, J., Austin, R. H., Monitto, C. L., Hershkowitz, S. 1982. *Proc. Natl. Acad. Sci. USA* 79:3518–22
64. Pörshke, D., Eigen, M. 1971. *J. Mol. Biol.* 62:361–81
65. Craig, M. E., Crothers, D. M., Doty, P. 1971. *J. Mol. Biol.* 62:383–401
66. Deleted in proof
67. Kornberg, A. 1980. *DNA Replication*, p. 233. New York: Freeman
68. Patel, D. J., Kozlowski, S. A., Bhatt, R. 1983. *Proc. Natl. Acad. Sci. USA* 80:3908–12
69. Englander, S. W., Kallenbach, N. R. 1984. *Q. Rev. Biophys.* 16:521–655
70. Lu, P., Cheung, S., Arndt, K. 1983. *J. Biomol. Struct. Dynam.* 1:509–21
71. Vanagida, M., Hiraoka, Y., Katsura, I. 1983. *Cold Spring Harbor Symp. Quant. Biol.* 47:177–87
72. Kolb, A., Spassky, A., Chapon, C., Blazy, B., Buc, H. 1983. *Nucleic Acids Res.* 11:7833–52
73. Wu, H.-M., Crothers, D. 1984. *Nature* 308:509–13
74. Hagerman, P. J. 1984. *Proc. Natl. Acad. Sci. USA* 81:4632–36
75. Courey, A. J., Wang, J. C. 1983. *Cell* 33:817–29
75a. Gellert, M., O'Dea, M. H., Mizuuchi, K. 1983. *Proc. Natl. Acad. Sci. USA* 80:5545–49
76. Stirdivant, S. M., Klysik, J., Wells, R. D. 1982. *J. Biol. Chem.* 257:10159–65
77. McClure, W. R., Hawley, D. K. 1983. *Mobility and Recognition in Cell Biolo-*

gy, ed. H. Sund, C. Veeger, pp. 317–33. New York: de Gruyter

78. Travers, A. A., Lamond, A. I., Mace, H. A. F., Berman, M. L. 1983. *Cell* 35:265–73

79. Adhya, S., Gottesman, M. 1982. *Cell* 29:939–44

80. Herskowitz, I., Signer, E. 1970. *Cold Harbor Symp. Quant. Biol.* 35:355–68

81. Shaner, S. L., Melancon, P., Lee, K. S., Burgess, R. R., Record, M. T. Jr. 1983. *Cold Spring Harbor Symp. Quant. Biol.* 47:463–72

82. Melancon, P., Burgess, R. R., Record, M. T. Jr. 1982. *Biochemistry* 21:4318–31

83. Melancon, P., Burgess, R. R., Record, M. T. JR. 1983. *Biochemistry* 22:5169–76

84. Hinkle, D. C., Chamberlin, M. J. 1972. *J. Mol. Biol.* 70:157–95

85. Roe, J.-H., Burgess, R. R., Record, M. T. Jr. 1984. *J. Mol. Biol.* 176:495–521

86. Galas, D., Schmitz, A. 1978. *Nucleic Acids Res.* 5:3157–70

87. Johnson, A. D., Meyer, B. J., Ptashne, M. 1979. *Proc. Natl. Acad. Sci. USA* 76:5061–65

88. Shalloway, D., Kleinberger, T., Livingston, D. M. 1980. *Cell* 20:411–22

89. Chan, D. T., Lebowitz, J. 1983. *Nucleic Acids Res.* 11:1099–1116

90. Peterson, M. L., Reznikoff, W. S. 1985. *J. Mol. Biol.* Submitted for publication

91. Bertrand-Burggraf, E., LeFevre, J. F., Daune, M. 1984. *Nucleic Acids Res.* 12:1697–1706

92. McClure, W. R. 1980. *Proc. Natl. Acad. Sci. USA* 77:5634–38

93. Hawley, D. K., McClure, W. R. 1980. *Proc. Natl. Acad. Sci. USA* 77:6381–85

94. Chamberlin, M. J., Nierman, W. C., Wiggs, J., Neff, N. 1979. *J. Biol. Chem.* 254:10061–69

95. Stefano, J. E., Gralla, J. D. 1980. *J. Biol. Chem.* 255:10423–30

96. Scatchard, G. 1949. *Ann. NY Acad. Sci.* 51:660–72

97. Kajitani, M., Ishihama, A. 1983. *Nucleic Acids Res.* 11:671–86

98. Kajitani, M., Ishihama, A. 1983. *Nucleic Acids Res.* 11:3873–88

99. von Gabain, A., Bujard, H. 1979. *Proc. Natl. Acad. Sci. USA* 76:189–93

100. Cech, C. L., McClure, W. R. 1980. *Biochemistry* 19:2440–47

101. Park, C.-S., Wu, F. Y.-H., Wu, C.-W. 1982. *J. Biol. Chem.* 257:6950–56

102. Kadesch, T. R., Rosenberg, S., Chamberlin, M. J. 1982. *J. Mol. Biol.* 155:1–29

103. Rosenberg, S., Kadesch, T. R., Chamberlin, M. J. 1982. *J. Mol. Biol.* 155:31–51

104. Hawley, D. K., McClure, W. R. 1982. *J. Mol. Biol.* 157:493–525

105. Bujard, H., Niemann, A., Breunig, K., Roisch, U., Dressel, A., et al. 1982. See Ref. 5, pp. 121–40

106. Brunner, M. 1983. Diplomarbeit, Univ. Heidelberg, FRG

107. Buc, H., McClure, W. R. 1985. *Biochemistry.* In press

108. Gamper, H. B., Hearst, J. E. 1982. *Cell* 29:81–90

109. Siebenlist, U. 1979. *Nature* 279:651–52

110. Kirkegaard, K., Buc, H., Spassky, A., Wang, J. C. 1983. *Proc. Natl. Acad. Sci. USA* 80:2544–48

111. Hsieh, T.-S., Wang, J. C. 1978. *Nucleic Acids Res.* 5:3337–45

112. Riesbig, R. R., Woody, A.-Y. M., Woody, R. W. 1979. *J. Biol. Chem.* 254:11208–17

113. Mulligan, M. E., McClure, W. R. 1985. *Biochemistry.* Submitted for publication

114. Munson, L. M., Reznikoff, W. S. 1981. *Biochemistry* 20:2081–85

115. Stefano, J. E., Gralla, J. 1979. *Biochemistry* 18:1063–67

116. Dahlberg, J. E., Blattner, F. R. 1973. In *2nd I. C. N. Symp. Mol. Biol.,* ed. C. F. Fox, W. S. Robinson, p. 533. New York: Academic

117. Travers, A. A., Burgess, R. R. 1969. *Nature* 222:537–40

118. Hansen, U. M., McClure, W. R. 1980. *J. Biol. Chem.* 255:9564–70

119. Hanna, M. M., Meares, C. F. 1983. *Biochemistry* 22:3546–51

120. Spassky, A., Kirkegaard, K., Buc, H. 1985. *Biochemistry.* In press

121. Kinsella, L., Hsu, C.-V., Schulz, W., Dennis, D. 1982. *Biochemistry* 21:2719–23

122. DiLauro, R., Taniguchi, T., Musso, R., de Crombrugghe, B. 1979. *Nature* 279:494–500

123. Carpousis, A. J., Gralla, J. D. 1980. *Biochemistry* 19:3245–53

124. Gourse, R. L., Stark, M. J. R., Dahlberg, A. E. 1983. *Cell* 32:1347–54

125. McClure, W. R. 1983. In *Biochemistry of Metabolic Processes,* ed. D. L. F. Lennon, F. W. Stratman, R. N. Zahlten, pp. 207–17. New York: Elsevier

126. Gilbert, W. 1976. See Ref. 18, pp. 193–205

127. Stefano, J. E., Gralla, J. D. 1982. *Proc. Natl. Acad. Sci. USA* 70:1069–72

128. Stefano, J. E., Gralla, J. D. 1982. *J. Biol. Chem.* 257:13924–29

129. Malan, T. P., Kolb, A., Buc, H., McClure, W. R. 1984. *J. Mol. Biol.* In press

130. Liao, S.-M., Chiang, C. H., Wu, T.-H., Susskind, M. M., McClure, W. R. 1985. Submitted for publication

131. Shih, M.-C., Gussin, G. N. 1983. *Proc. Natl. Acad. Sci. USA* 80:496–500
132. Simons, R. W., Hoopes, B. C., McClure, W. R., Kleckner, N. 1983. *Cell* 34:673–82
133. Shih, M.-C., Gussin, G. N. 1983. *Cell* 34:941–49
134. Shih, M.-C., Gussin, G. N. 1984. *J. Mol. Biol.* 172:489–506
135. Mulligan, M. E., Hawley, D. K., Entriken, R., McClure, W. R. 1984. *Nucleic Acids Res.* 12:789–800
136. Harr, R., Haggström, M., Gustafsson, P. 1983. *Nucleic Acids Res.* 11:2943–57
137. Stormo, G., Schneider, T., Gold, L., Ehrenfeucht, A. 1982. *Nucleic Acids Res.* 10:2997–3011
138. Artem'ev, I. V., Vasil'ev, G. V., Gurevich, A. I. 1983. *Bioorg. Khim.* 9:1544–57
139. Staden, R. 1984. *Nucleic Acids Res.* 12:505–19
140. Youderian, P., Bouvier, S., Susskind, M. M. 1982. *Cell.* 30:843–53
141. McClure, W. R., Hawley, D. K., Youderian, P., Susskind, M. M. 1983. *Cold Spring Harbor Symp. Quant. Biol.* 47:477–81
142. Miller, J. H., Reznikoff, W. S., eds. 1978. *The Operon.* New York: Cold Spring Harbor Lab.
143. Hawley, D. K., Johnson, A. D., McClure, W. R. 1984. *J. Biol. Chem.* Submitted for publication
144. Orosz, L., Adyha, S. 1983. *Cell* 32:783–88
144a. Majumdar, A., Adhya, S. 1984 *Proc. Natl. Acad. Sci. USA* 81:6100–4
145. Dunn, T., Hahn, S., Ogden, S., Schleif, R. 1984. *Proc. Natl. Acad. Sci. USA* 81:5017–20
146. Barkley, M. D., Bourgeois, S. 1978. See Ref. 142, pp. 177–220
147. Youderian, P., Vershon, A., Bouvier, S., Sauer, R. T., Susskind, M. M. 1983. *Cell* 35:777–83
148. Johnson, A. D., Meyer, B. J., Ptashne, M. 1979. *Proc. Natl. Acad. Sci. USA* 76:5061–65
149. Hawley, D. K., McClure, W. R. 1983. *Cell* 32:327–33
150. Malan, T. P., McClure, W. R. 1984. *Cell* 39:93–97
151. Ho, Y. S., Wulff, D. L., Rosenberg, M. 1983. *Nature* 304:703–8
152. Hoopes, B. C., McClure, W. R. 1985. *Proc. Natl. Acad. Sci. USA.* In press
153. Pabo, C. O., Sauer, R. T. 1984. *Ann. Rev. Biochem.* 53:293–321
154. Pabo, C. O., Lewis, M. 1982. *Nature* 298:443–47
155. Ohlendorf, D. H., Anderson, W. F., Fisher, R. G., Takeda, Y., Matthews, B. W. 1982. *Nature* 298:718–23
156. Steitz, T. A., Weber, I. T., Matthew, J. B. 1983. *Cold Spring Harbor Symp. Quant. Biol.* 47:419–26
157. Steitz, T. A., Ohlendorf, D. H., Anderson, W. F., Takeda, Y. 1982. *Proc. Natl. Acad. Sci. USA* 79:3097–100
158. Sauer, R. T., Yocum, R. R., Doolittle, R. F., Lewis, M., Pabo, C. O. 1982. *Nature* 298:447–51
159. Frederick, C. A., Grable, J., Melia, M., Samudzi, C., Jen-Jacobson, L., et al. 1984. *Nature* 309:327–31
160. Guarente, L., Nye, J. S., Hochschild, A., Ptashne, M. 1982. *Proc. Natl. Acad. Sci. USA* 79:2236–39
161. Hochschild, A., Irwin, N., Ptashne, M. 1983. *Cell.* 32:319–25
162. McClure, W. R., Hawley, D. K. 1984. In *Microbiology 1984,* ed. L. Leive, D. Schlessinger, pp. 93–97. Washington, DC: Am. Soc. Microbiol.
163. Kahmann, R. 1983. *Cold Spring Harbor Symp. Quant. Biol.* 47:639–46
164. Hattman, S., Goradia, M., Monaghan, C., Bukhari, A. I. 1983. *Cold Spring Harbor Symp. Quant. Biol.* 47:647–53
165. Lamond, A. I., Travers, A. A. 1983. *Nature* 305:248–50
166. Gellert, M. 1981. *Ann. Rev. Biochem.* 50:879–910
167. Overbye, K. M., Basu, S. K., Margolin, P. 1983. *Cold Spring Harbor Symp. Quant. Biol.* 47:785–91
168. Sanzey, B. 1979. *J. Bacteriol.* 138:40–47
169. Menzel, R., Gellert, M. 1983. *Cell* 34:105–113
170. Sinden, R. R., Carlson, J. O., Pettijohn, D. E. 1980. *Cell* 21:773–83
171. Ingraham, J. L., Maaløe, O., Neidhardt, F. C. 1983. *Growth of the Bacterial Cell,* pp. 370–71. Sunderland, MA: Sinauer Assoc.
172. Bertrand-Burggraf, E., Schnarr, M., Lefevre, J. F., Daune, M. 1984. *Nucleic Acids Res.* 12:7741–52
173. Wood, D. C., Lebowitz, J. 1984. *J. Biol. Chem.* 259:11184–87
174. Crooks, J. H., Ullman, M., Zoller, M., Levy, S. R. 1983. *Plasmid* 10:66–72
174a. von Hippel, P. H., Revzin, A., Gross, C., Wang, A. C. 1974. *Proc. Natl. Acad. Sci. USA* 71:4808–12
175. Nossal, N. G. 1983. *Ann. Rev. Biochem.* 53:581–615
176. Geider, K., Beck, E., Schaller, H. 1978. *Proc. Natl. Acad. Sci. USA* 75:645–49
177. Nomura, M., Gourse, R., Baughman, G. 1984. *Ann. Rev. Biochem.* 53:75–117
178. Sarmientos, P., Sylvester, J. E., Contente, S., Cashel, M. 1983. *Cell* 32:1337–46

179. Glaser, G., Sarmientos, P., Cashel, M. 1983. *Nature* 302:74–76
180. Sarmientos, P., Cashel, M. 1983. *Proc. Natl. Acad. Sci. USA* 80:7010–13
181. Sippel, A., Hartmann, G. 1968. *Biochim. Biophys. Acta* 157:218–19
182. McClure, W. R., Cech, C. L. 1978. *J. Biol. Chem.* 253:8949–56
183. Schulz, W., Zillig, W. 1981. *Nucleic Acids Res.* 9:6889–906
184. Losick, R., Sonenshein, A. L., Shorenstein, R. G., Hussey, C. 1970. *Cold Spring Harbor Symp. Quant. Biol.* 35:443–50
185. Astvatsaturyan, M. Z., Mironov, A. S., Sukhodolets, V. V. 1983. *Genetika* 19:1070–74
186. Little, R., Ryals, J., Bremer, H. 1983. *J. Bacteriol.* 154:787–92
187. Neuhard, J., Jensen, K. F., Stauning, E., 1982. *EMBO J.* 1:1141–45
188. Gordeev, V. K., Turkov, M. I. 1983. *Genetika* 19:1426–32
189. Fisher, R. F., Yanofsky, C. 1983. *J. Biol. Chem.* 258:8146–50
190. de Crombrugghe, B., Busby, S., Buc, H. 1984. *Science* 224:831–38
191. Ow, D. W., Ausbel, F. M. 1983. *Nature* 301:307–13
192. Shinagawa, H., Makino, K., Nakata, A. 1983. *J. Mol. Biol.* 168:477–88
193. Wanner, B. L. 1983. *J. Mol. Biol.* 166:283–308
194. Jagura-Bardzy, G., Kredich, N. 1983. *J. Bacteriol.* 155:578–85
195. Shaw, D. J., Rice, D. W., Guest, J. R. 1983. *J. Mol. Biol.* 166:241–47
196. Walker, G. C., Kenyon, C. J., Bagg, A., Elledge, S. J., Perry, K. L., Shanabruch, W. G. 1982. *Basic Life Sci.* 20:43–63
197. Little, J. W., Mount, D. W. 1982. *Cell* 29:11–13
198. Little, J. W. 1983. *J. Mol. Biol.* 167:791–808
199. Ni'bhriain, N. N., Silver S., Foster, T. J. 1983. *J. Bacteriol.* 155:690–703
200. Neidhardt, F. C., Van Bogelen, R. A., Lan, E. T. 1983. *J. Bacteriol.* 153:597–603

Ann. Rev. Biochem. 1985. 54:205–35

RECEPTORS AND PHOSPHOINOSITIDE-GENERATED SECOND MESSENGERS[1]

Lowell E. Hokin

Department of Pharmacology, University of Wisconsin Medical School, Madison, Wisconsin 53706

CONTENTS

[1]Abbreviations used: PI, phosphatidylinositol; PIP, phosphatidylinositol-4-phosphate; PIP_2, phosphatidylinositol-4,5-bisphosphate; ER, endoplasmic reticulum; IP, inositol-1-phosphate; Ca^{2+}_c, cytosolic Ca^{2+}; IP_3, inositol-1,4,5-trisphosphate; DG, diacylglycerol; SAG, 1-stearoyl-2-arachidonoyl-*sn*-glycerol; C-kinase, protein kinase C; OAG, 1-oleoyl-2-acetyl-*sn*-glycerol; CCK/PZ, cholecystokinin/pancreozymin; ACh, acetylcholine; PA, phosphatidic acid; IP_2, inositol-1,4-bisphosphate; CDP-DG, cytidine diphosphodiacylglycerol; TPA, 12-tetradecanoylphorbol-13-acetate.

0066-4154/85/0701-0205$02.00

PERSPECTIVES AND SUMMARY

In animal cells, there are three myoinositol-containing phosphatides: phosphatidylinositol[1-(3-*sn*-phosphatidyl)-D-myoinositol] (PI), phosphatidylinositol-4-phosphate or diphosphoinositide (PIP), and phosphatidylinositol-4,5-bisphosphate or triphosphoinositide (PIP$_2$) [see (1–6)]. They are highly active metabolically. They usually account for less than 10% of the total phospholipid in animal cells, PI accounting for over 90% of this. Most of the polyphosphoinositides and a small amount of the total PI are in the plasma membrane. Most of the PI is in the endoplasmic reticulum (ER). Although it has been known for three decades that agonist activation of membrane receptors stimulates PI turnover (at latest count over 25 such receptors), it has only been within the past few years that a coherent picture of the physiological function of phosphoinositides has begun to emerge. This involves agonist-stimulated phosphodiesteratic (phospholipase C) cleavage of the phosphoinositides, which generates second-messenger products (Figures 1 and 2). Recently, attention has been focused on the rapid breakdown of polyphosphoinositides in the plasma membrane and the liberation of inositol phosphates into the cytosol. The accumulation of inositol phosphates is amplified in the presence of lithium (Li$^+$), which inhibits inositol-1-phosphate (IP) phosphatase. The breakdown of polyphosphoinositides precedes the loss of PI, which occurs mainly in the ER. There is debate as to whether the stimulated loss of PI is exclusively via phosphorylation in the plasma membrane to regenerate polyphosphoinositides or partially by direct action of phospholipase C on PI in the ER.

Although it was suggested ten years ago that stimulated breakdown of PI may serve to "open Ca^{2+} gates" in the plasma membrane, it has only been within the past year that a role for phosphoinositides in elevating cytosolic Ca^{2+} (Ca$^{2+}_c$) has been established. This has been made possible by permeabilization of cells with detergents or by hypotonicity to permit entry of added inositol-1,4,5-trisphosphate (IP$_3$), an immediate breakdown product of PIP$_2$. Addition of less than micromolar amounts of IP$_3$ causes a rapid release into the cytosol of "trigger Ca^{2+}," apparently from the ER, and this effect is not mimicked by other inositol phosphates. Work within the past few years has also shown that the other phosphodiesteratic cleavage product of all of the phosphoinositides, diacylglycerol (DG), which is mainly of the 1-stearoyl-2-arachidonoyl-*sn*-glycerol (SAG) variety, activates a protein kinase distinct from cAMP- or cGMP-activated protein kinase. In addition to a requirement for DG, this kinase, termed protein kinase C (C-kinase), requires phospholipids

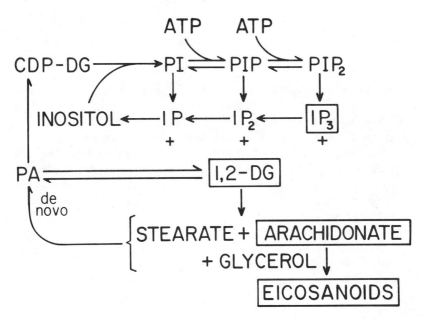

Figure 1 The phosphoinositide cycle. The substances enclosed in squares or rectangles are second messengers. Abbreviations are given in the text.

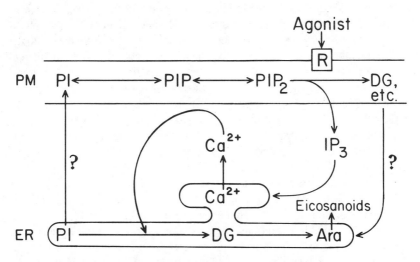

Figure 2 Possible interactions between the plasma membrane and the endoplasmic reticulum. Abbreviations: PM, plasma membrane; R, receptor; Ara, arachidonate. Other abbreviations are given in the text.

(e.g., phosphatidylserine) and Ca^{2+} for maximum activity. Its affinity for Ca^{2+} is dramatically increased by DG. It appears that tumor promoter phorbol esters, as well as 1-oleoyl-2-acetyl-*sn*-glycerol (OAG), can activate at the DG site of C-kinase in intact cells; this has facilitated probing of C-kinase function. The IP_3 branch and the DG branch of the phosphoinositide signal cascade appear to act synergistically to phosphorylate proteins, the former by elevating Ca_c^{2+} to activate calmodulin-dependent protein phosphorylation and the latter to activate C-kinase, which phosphorylates a different set of proteins. The elevated Ca_c^{2+} may also activate reactions not dependent on calmodulin. The third physiologically active breakdown product of phosphoinositides in many but perhaps not all cells is arachidonate, which is the obligatory substrate for synthesis of eicosanoids which mediate or modulate a wide variety of physiological (autocrine) functions. There is some evidence that arachidonate itself may directly activate certain processes, particularly release of cGMP.

There is a considerable body of evidence, although primarily of a circumstantial nature, that phosphoinositide turnover is involved in proliferation. Studies are beginning to emerge suggesting that phosphoinositides may be involved in oncogene action. This suggested function is intriguing but unproven.

Many aspects of the phosphoinositide cycle require further investigation. How does receptor occupancy activate attack by cytosolic phospholipase C on phosphoinositides, which are in membranes? Is the disappearance of as much as 70% of the total cellular PI in some cells due exclusively to phosphorylation to form polyphosphoinositides in the plasma membrane, or is part of the PI attacked directly in the ER and possibly the plasma membrane by phospholipase C? The former mechanism would require transfer of PI from the ER to the plasma membrane and the transfer of phosphoinositide-cycle intermediate(s) back to the ER. Very little is known about the function of PI (or other lipid) transfer proteins vis-a-vis the phosphoinositide response. The direct breakdown of PI in the ER would require activation of phospholipase C attack on PI in the ER by a mediator, presumably elevated Ca_c^{2+}, triggered by release of IP_3 from the plasma membrane (Figure 2). Since the enzymes for PI synthesis as well as cyclooxygenase are in the ER and phospholipase C is in the cytosol, a release in the ER of arachidonate from PI via DG would obviate the necessity for hypothetical transport proteins for carrying lipoidal substances between the ER and the plasma membrane. Another fundamental question is what is the molecular mechanism whereby IP_3 releases Ca^{2+} from the ER?

With a few notable exceptions, the mechanism of arachidonate release has not been clarified. In what tissues does this release occur by sequential action of phospholipase C and DG lipase, and in what tissues is it by action of phospholipase A_2, or by both phospholipases? What is the role of substantial amounts of free arachidonate released on stimulation of phosphoinositide turnover in some tissues? Is the elevation of cGMP which accompanies the phosphoinositide response in many tissues triggered by elevations of free arachidonate, and what role does Ca^{2+}_c play in this process?

Further work is required to elucidate the role of phosphoinositides in proliferation and oncogene action. So far, the number of oncogenes that have been shown to encode proteins involved in or related to the phosphoinositide cycle is very limited. Additional oncogenes as well as their encoded proteins should be studied. The importance of phosphorylation of PI and PIP as a result of oncogene action must be placed in perspective with the known phosphorylation of tyrosine on proteins by oncogene products.

INTRODUCTION

In recent years, there has been widespread interest in the agonist-stimulated breakdown and turnover of myoinositol-containing phospholipids. Although these lipids comprise less than 10% of the total cellular phospholipid, their stimulated turnover accounts for a considerably higher percentage of total phospholipid turnover. This phenomenon has not been reviewed before in this series. A review of this subject now in *Annual Review of Biochemistry* is timely since there have been major developments within the past few years.[1] Diverse aspects of the phosphoinositide response are beginning to come together to make a relatively cohesive story, although much more work needs to be done. It appears that the phosphoinositide effect is a multiregulatory mechanism involving the release of various moieties of the parent molecules, which serve as second messengers for myriad changes in the cell elicited by receptor activation.

Numerous reviews of this subject have been published dating as far back as 1956 and as recently as July, 1984 [for selected reviews, see (1–28)]. Space limitations prevent a detailed review of the numerous cell types in which the receptor mechanisms and the role of phosphoinositides in these mechanisms have been studied. In a review in 1982, Michell (24) stated that over 25 receptor types had so far been associated with a phosphoinositide effect. Examples of receptors coupled to phosphoinositide metabolism are muscarinic-cholinergic, α_1 adrenergic, serotononinergic, dopaminergic, H_1-histaminergic, and peptidergic [e.g., cholecystokinin/pancreozymin (CCK/PZ) and congeners, thyroid-stimulating hormone, thyrotrophin-releasing hormone, corticotrophin-releasing hormone, angiotensin II, V_1-vasopressin, bradykinin, substance P, thrombin, and fMet-Leu-Phe (in neutrophils)]. Fairly detailed accounts of the receptor mechanisms with particular emphasis on the relationship between phosphoinositide metabolism and Ca^{2+} mobilization as well as some informa-

[1]This review covers material up to July 1, 1984.

tion on the physiological responses in various target tissues can be found in a recent full issue of *Cell Calcium* devoted to this subject (Vol. 3, pp. 285–451, 1982). These tissues include platelets (29), mast cells and neutrophils (30), adrenal medulla (31), smooth muscle (32), mammalian salivary gland (33), insect salivary gland (34), liver (35), central nervous system (36), dividing and differentiating cells (37), and target tissues for steroidogenic hormones (38).

This review has been written from an historical perspective. Within the space limitations, primary sources will be referred to. The emphasis in this review reflects to some extent the author's personal involvement in the field in the early years.

EARLY OBSERVATIONS

Discovery of the "Phospholipid Effect"

The discovery of the "phospholipid effect" arose out of an accidental observation over thirty years ago. The details of this discovery have been reviewed (7). Essentially, the incorporation of $^{32}P_i$ into the "RNA" fraction of Schmidt & Thannhauser (39) increased considerably (40) on cholinergic stimulation of enzyme secretion (41) in pigeon pancreas slices. It eventually turned out that a hydrolytic product(s) of phospholipids contaminating the "RNA" fraction was responsible for the stimulation (42), although it was not initially known that it was a phosphoinositide product that was responsible. When the phospholipids were looked at directly, they were found to show a remarkable increase (approximately tenfold) in incorporation of $^{32}P_i$ on stimulation with acetylcholine (ACh) or carbamylcholine (43, 44).

In the initial studies, the incorporation of $^{32}P_i$ into total phospholipid was followed since methods were not available for separation of individual phospholipids from small samples of tissue. The stimulated ^{32}P incorporation appeared to be confined to the phospholipids since there was no stimulation of ^{32}P incorporation into other phosphorus-containing fractions (43). The stimulation was also blocked by the muscarinic cholinergic antagonist, atropine. In a subsequent study (44), it was found that enzyme secretion and ^{32}P incorporation into phospholipids did not parallel each other as the ACh concentration was increased. For example, enzyme secretion was 80% of maximal at 10^{-7} M ACh, but incorporation of ^{32}P into phospholipids was only 15% of maximal. At 10^{-6} M ACh, enzyme secretion was maximal, but incorporation of ^{32}P into phospholipids was only 65% of maximal. These observations indicated that the incorporation of ^{32}P into phospholipids and enzyme secretion were not tightly coupled. That the stimulated incorporation of ^{32}P into phospholipids was not a unique effect of cholinergic drugs unrelated to stimulated enzyme secretion was indicated by the observation that CCK/PZ, which is also a physiological secretogogue for enzyme secretion but which binds to a separate receptor, gave identical results to those with ACh (45), except that its actions were not blocked by atropine. In the intervening years, a similar dissociation between phosphoinositide effects and physiological responses at increasing agonist concentrations has been observed (16, 46, 47). It appears that the phospholipid effect corresponds more closely to agonist binding to receptors and that there are spare receptors in some tissues (16).

When methods became available for separation of the deacylated products of the individual phospholipids (48) and somewhat later the intact phospholipids (49) from small samples of tissue, it was possible to determine which phospholipids showed the phospholipid response. In pigeon pancreas slices, the incorporation of ^{32}P into PI was increased about 15 fold and that into phosphatidic acid (PA) about threefold (50, 51). The incorporation of [^{3}H]myoinositol into PI was also stimulated in pancreas (51) and brain (52), although in the former this stimulation was not as great as that of ^{32}P. The incorporation of ^{32}P and the respective head-groups were also stimulated in phosphatidylcholine and phosphatidylethanolamine (50) but to a considerably lesser extent than that in PI and PA. These smaller effects in the two phospholipids which comprise the major fraction of the total phospholipid phosphorus but which usually turn over much more slowly than PI and PA have not received much attention; they may be related to increases in steady-state levels of their precursor DG associated with increased phosphoinositide breakdown (see below).

That some of the stimulated incorporation of ^{32}P and [^{3}H]myoinositol into PI and of ^{32}P into PA in pancreas was due to increased de novo synthesis of these phosphatides was shown by the observation that ACh increased the incorporation of [^{14}C]glycerol into PI fivefold and into PA threefold (51). Stimulated de novo synthesis of PI in exocrine pancreas has been confirmed and extended (53, 54). These observations are in line with recent studies which show a substantial release of free fatty acids and glycerol from PI on stimulation of enzyme secretion in mouse pancreatic minilobules (55). These moieties can only re-enter lipids by the sequential action of glycerol kinase and sn-glycero-3-phosphate transacylase to form PA, which is the branch point for the synthesis of other phospholipids (56–58). This increased synthesis of PI and PA from their building blocks as a result of their total degradation to these moieties is probably different from the stimulated net synthesis of phosphoinositides and other phospholipids by trophic hormones [see (25, 27)].

At the time that the phospholipid effect in pancreas was first reported (43), ACh was also shown to stimulate the incorporation of ^{32}P into phospholipids in guinea pig brain cortex slices, albeit much less than that in pancreas. Since that time, this effect in slices of synaptic tissue and in cell-free preparations containing synaptosomes has been studied extensively [see (3, 4, 12, 36, 59, 60)]. Very high concentrations of ACh (10^{-2} M) were required to achieve maximal effects in brain cortex slices (52, 61). This appeared to be due to very poor permeation of ACh into the slices since maximal effects in cell-free preparations of brain were seen with concentrations of ACh three orders of magnitude lower (12, 62, 63). Synaptosomes, which are really "minicells," comprise the only "cell-free" system in which stimulated incorporation of precursors into PI and PA has been shown.

In addition to the effects in pancreas and brain cortex already discussed, phospholipid effects were seen in submaxillary and parotid glands in response to cholinergic or adrenergic stimulation (64, 65), ACTH secretion in pituitary in response to corticotrophin-releasing hormone (66), epinephrine secretion in adrenal medulla in response to ACh (67), NaCl secretion in the avian salt gland in response to ACh (68), some but not all parts of the CNS in response to ACh and norepinephrine (69, 70), sympathetic ganglia in response to ACh (69, 71), human sweat glands in response to ACh (72, 73), pigeon esophageal mucosa in response to ACh (65), toad and turtle bladder in response to ACh (74), leukocytes on stimulation of phagocytosis with starch or polystyrene particles (75–77) and thyroid in response to thyroid-stimulating hormone (78, 79).

Interconversions between Phosphatidylinositol and Phosphatidic Acid in the Avian Salt Gland in Response to Acetylcholine and Atropine

Acetylcholine is a muscarinic agonist for stimulated NaCl secretion in the avian salt gland (80). If avian salt gland slices were incubated with $^{32}P_i$ in the absence and presence of ACh, the incorporation of ^{32}P into PA and PI was increased 15 fold and threefold, respectively, by ACh (9, 10, 68). Detailed kinetic studies of the turnover of PA and PI under various conditions revealed several interesting features. If ACh was added to salt gland slices preincubated with $^{32}P_i$, a fraction of PA amounting to no more than 20% of the total PA became rapidly labeled and continued to turn over (9, 10, 81). Under these conditions, in contrast to the addition of ACh and $^{32}P_i$ at zero time, there was only a small increase in radioactivity in PI due to ACh (10), presumably as a result of simultaneous breakdown of prelabeled PI and its increased turnover (see below). If the action of ACh was blocked with atropine, the renewing fraction of PA rapidly disappeared and was accompanied by a rapid spurt in radioactivity in PI (10). The rise in radioactivity in PI was quite close to the loss in radioactivity in PA. A similar rise in radioactivity in PI could be seen with [^3H]myoinositol. If the blocking action of atropine was overridden by addition of a concentration of ACh 33 times higher than the initial ACh concentration, the labeled PI which had been formed on adding atropine disappeared and PA gained the label lost from PI.

If albatross salt gland slices were incubated without and with ACh, the average PI-phosphorus in the stimulated tissue was 20% lower than that in the control, but this difference did not quite reach statistical significance (9). However, later calculations showed that if the percentage difference between controls and ACh-stimulated values were taken for each animal and these differences averaged, there was a 40% decrease in PI-phosphorus which was statistically significant [see (15)].

If avian salt gland slices were prelabeled with [^{14}C]glycerol and then stimulated with ACh, there was no increased incorporation of radioactivity into PA. This favored DG kinase rather than sn-glycero-3-phosphate transacylase as the enzyme responsible for the increased incorporation of ^{32}P into PA. On the basis of these observations in the avian salt gland, a scheme was proposed, called the "phosphatidylinositol-phosphatidic acid" cycle (10) (Figure 1), in which on stimulation with ACh, PI breaks down to DG, catalyzed by PI phosphodiesterase (phospholipase C) (82, 83), and DG is phosphorylated by ATP to form PA (84). On removal of ACh, PA is converted back to PI by the sequential actions of CTP-PA cytidyl transferase and PI synthase (85, 86). From the data presented earlier, it is obvious that all steps in the cycle are occurring under constant stimulation. It should be pointed out that the results with [^{14}C]glycerol incorporation in the avian salt gland were somewhat different from those with pigeon pancreas, where stimulation with ACh produced appreciable increases in incorporation of [^{14}C]glycerol into PI and PA due to de novo synthesis (51, 55). This suggests that in the avian salt gland, unlike the pancreas, there is little breakdown of DG to fatty acids and glycerol. Interconversions between PA and PI similar to those in salt gland were later shown for the exocrine mouse pancreas [see (15)], where the secretory process and products are quite different from those in the avian salt gland; it appears, however, that in both systems secretion is coupled to receptor occupancy via elevations in Ca^{2+}_c(87–89).

Cellular Site of the Phosphatidylinositol Effect

If pigeon pancreas slices were incubated without and with ACh, followed by homogenization and differential centrifugation, about 80% of the ACh-induced increment in ^{32}P incorporation or [^{14}C]glycerol incorporation in PI was in the microsomal fraction (90). In the pancreas, this fraction is derived from the smooth-surfaced and rough-surfaced ER (91) and presumably the plasma membrane. The remaining increment was distributed in the nuclear + zymogen granule fraction (contaminated by whole cells), mitochondrial fraction (contaminated by microsomes), and soluble fraction (contaminated by microsomes). The stimulated ^{32}P incorporation in the avian salt gland was also microsomal, derived primarily from invaginated plasma membrane (68). A radioautographic technique for measuring the incorporation of [^{3}H]myoinositol into PI was developed (92, 93), and grain counts showed the stimulation of [^{3}H]myoinositol incorporation into PI in pancreas to be about equally distributed between the basophilic (site of rough-surfaced ER) and nonbasophilic (site of smooth-surfaced ER) cytoplasm (93). Radioautography showed that in sympathetic ganglia (92) and the avian salt gland (D. Gerber, L. E. Hokin, unpublished data), the site of stimulation of [^{3}H]myoinositol incorporation was also cytoplasmic. Since differential centrifugation had already identified the microsomal fraction as being the site of stimulation, the combined radioautographic and differential centrifugation studies strongly suggested that the site of stimulated [^{3}H]myoinositol incorporation into PI was in the ER and excluded the plasma membrane as a major site of stimulation. Similar radioautographic results have recently been obtained with photoreceptor cells of the retina in response to light (94, 95).

Partial Dissociation of the Phosphatidylinositol Effect from Secretion by Omission of Ca^{2+}

By the middle 1960s, it was already known that Ca^{2+} played an important role in agonist-stimulated secretion in the adrenal medulla (96), the submaxillary gland (97), and neurohypophysis (98). It was found that if pigeon pancreas slices were incubated without Ca^{2+}, amylase secretion in response to ACh was inhibited 98%, but the stimulated ^{32}P incorporation into PI, PA, and phosphatidylethanolamine was inhibited only 38%, 41%, and −8%, respectively (99). These results again showed loose coupling between the phospholipid effect and enzyme secretion and suggested that the phospholipid effect was concerned with some step in the "overall process of excitation and secretion" other than exocytotic enzyme secretion per se. These studies also showed that the phospholipid effect was partially independent of at least external Ca^{2+}.

Early Studies on Polyphosphoinositide Turnover

As early as 1962, the polyphosphoinositides were shown to turn over rapidly in brain, and it was suggested that they may play an important physiological role (100). They were present on radiochromatograms of lipid extracts from brain particulate fractions undergoing oxidative phosphorylation in the presence of ^{32}P$_i$, and the radioactivity in "Spot C" (later identified as PIP) (101) was 30% lower when the fractions were incubated for one hour with ACh (62). Later, it was shown that there was substantial radioactivity in polyphosphoinositides in a wide variety of tissues incubated with ^{32}P$_i$ (101, 102). If sea gull salt gland slices were incubated with ^{32}P$_i$ for 80 min, followed by addition of ACh and incubated for an additional 20 min, the radioactivity in PIP and PIP$_2$ was reduced to 67.5 ± 14.7 and 77.0 ± 10.2% of control slices, respectively (101).

The presence of PIP and PIP$_2$, which were rapidly labeled in their monoesterified phosphate positions by incubation of erythrocyte membranes with [$\gamma-^{32}$P]ATP, was also shown (103). Addition of exogenous PI to the erythrocyte membranes stimulated the incorporation of ^{32}P from [$\gamma-^{32}$P]ATP into PIP, indicating the presence of a PI kinase. Phosphatidylinositol kinase was also demonstrated about the same time in brain (104, 105). Phosphatidylinositol kinase was reported to be present in plasma membranes other than that of the erythrocyte (106, 107). There have been several recent reviews devoted primarily to the chemistry and metabolism of polyphosphoinositides (2–6).

THE MODERN ERA

Agonist-Stimulated Breakdown of Phosphatidylinositol

In 1974, Hokin-Neaverson (14, 108) observed a decrease in mass of PI and a roughly equivalent increase in mass of PA on stimulation of mouse pancreas with ACh or CCK/PZ, and Jones & Michell (109) noted a decrease in mass of PI on stimulation of the parotid with ACh or epinephrine. This confirmed and extended the studies on the kinetics of ^{32}P and [^{3}H]myoinositol labeling as well as the chemical measurements in the avian salt gland made a decade earlier (9, 10). Until the last couple of years, work in this field was aimed almost exclusively at agonist-stimulated PI breakdown and metabolism, with very little attention being paid to the polyphosphoinositides. No attempt will be made here to review in detail the extensive literature on agonist-stimulated "breakdown" and turnover of PI during the last twenty years. In his seminal review in 1975 (1), Michell presented an exhaustive list of tissues which showed a PI effect in response to various stimuli. More recent lists have been published (4, 16, 18, 25, 27). In the past couple of years, studies on PI effects have been extended or observed for the first time in adipocytes (21, 110), adrenal cortical cells (111, 112), exocrine pancreas (21, 55, 113), blood vessels (114–116), ileal smooth muscle (117), nervous tissue (118–122), pineal cells (123), atria (124–126), pancreatic islets (127–132), anterior pituitary cells (133–135), retina (in response to light) (94, 95, 136–138), platelets (139–143), thyroid (21, 144, 145), neutrophils (146–148), peritoneal macrophages (149), mast cells (150), and ovarian granulosa cells (151).

Agonist-Stimulated Breakdown of Polyphosphoinositides

During the past five years or so, there has been an explosive increase in work aimed at elucidation of the biochemical mechanism and physiological role of agonist-induced phosphoinositide responses. Although it was known for a long time that ACh stimulated the breakdown of polyphosphoinositides in "cell-free" preparations of brain (synaptosomes) (62, 152–156) and avian salt gland slices (101), or ACh or norepinephrine, in iris smooth muscle (157), this phenomenon has not received much attention until quite recently. It was also shown by Durell and his collaborators in the late 1960s (59, 60) and by Abdel-Latif and his associates in the late 1970s (157, 158) that IP, inositol-1,4-bisphosphate (IP$_2$), and IP$_3$ were formed on stimulation of synaptosomes by

ACh or of iris smooth muscle by ACh or norepinephrine. This suggested that all of the phosphoinositides were cleaved by a phospholipase C action.

There was a rekindling of interest in the polyphosphoinositides when it was found that Ca^{2+}-mobilizing agonists such as vasopressin in the liver (19, 35, 159) caused a very rapid breakdown in polyphosphoinositides, particularly PIP_2. Unlike the situation in iris smooth muscle (157, 158), the breakdown of polyphosphoinositides in liver was only partially dependent on Ca^{2+} and was not induced by A23187. [Recent studies with vas deferens suggest that the stimulatory effects of A23187 on polyphosphoinositide turnover in smooth muscle may be due to release of agonist ACh rather than rises in Ca_c^{2+} (160).] In most cases, but perhaps not all (161), the loss of polyphosphoinositides was an earlier event than the loss of PI, occurring in some systems in a matter of seconds. On the basis of these observations, it was suggested (19) that the primary event was a breakdown of polyphosphoinositides and that the decreases in levels of PI were secondary to phosphorylation of PI to replenish polyphosphoinositides. The rapid breakdown of polyphosphoinositides was confirmed for liver (162, 163) and has also been demonstrated in parotid gland (33, 164, 165), platelets (166–175) [see, however, (176)], insect salivary gland (177, 178), exocrine pancreas (179, 180), neutrophils (181, 182), macrophages (183), Friend erythroleukemic cells (184), leukocytes (185), cloned rat pituitary cells (186–189), atria (190), islets (191), and ileal smooth muscle (192).

Although it is generally assumed that the agonist-induced breakdown of polyphosphoinositides occurs at the plasma membrane, this has so far been clearly demonstrated only in the liver (193, 194).

The common belief held by many that the agonist-induced loss in PI is due to phosphorylation to form polyphosphoinositides rather than a direct action of phospholipase C on PI is based on the observation that in many systems the "breakdown" of PI appears to require energy since it is blocked by agents which inhibit formation of ATP; e.g., in whole mouse pancreas (14), platelets (141), rat liver (195), and parotid (165, 196). In the parotid, the formation of IP was also inhibited, and it was suggested that it was derived from IP_2 and IP_3 by phosphatase action (165) (see below). A requirement for energy does not prove that PI loss is via phosphorylation to polyphosphoinositides since there are other ways in which energy may be involved. Two laboratories have reported a small stimulation of PI loss by vasopressin (197) or epinephrine (198) in plasma membranes of liver without an added energy source, although deoxycholate was also required in the one study (197) and cytosol in the other (198). The requirement for cytosol is not surprising since it contains phospholipase C. Another argument used to support the phosphorylation of PI as the mechanism for its removal is the earlier disappearance of polyphosphoinositides than PI, and the production of IP_2 and IP_3 in excess of the amount of polyphosphoinositides broken down. Also, in many cells under constant stimulation the levels of polyphosphoinositides recover to basal levels while the level of PI remains depressed, suggesting continual conversion of PI to polyphosphoinositides. Although these studies suggest that some of the PI loss is due to phosphorylation, they do not rule out a later disappearance of some of the PI by hydrolysis. There is evidence that under certain conditions there may be a primary breakdown of PI to IP and DG in the exocrine pancreas. If the maximally responsive pool of PI was prelabeled under "stimulating conditions" (15), i.e., first stimulating the tissue with a cholinergic drug in the presence of [^{3}H]myoinositol, followed by quenching with atropine, which converts all of the rapidly renewing PA to PI, as much as 70% of the total cellular PI broke down on a second stimulation with a CCK/PZ analog (55). This technique is similar to that used to label a unique pool of PI in the avian salt gland (see EARLY OBSERVATIONS). Under these conditions, the second stimulation led to a release of IP and very little release of IP_2 or IP_3 in pancreatic acinar cells (199) or minilobules (J. F. Dixon, L. E. Hokin, unpublished data). Also, 10^{-4} M 2,4-dinitrophenol caused a 74% drop in ATP levels but did not inhibit PI breakdown. If the prelabeling was done under "nonstimulating conditions," the formation of IP, IP_2, and IP_3 was observed [(200); J. F. Dixon, L. E. Hokin, unpublished data], and there was a 36% inhibition of IP_3 formation by 10^{-4} M 2,4-dinitrophenol. Other studies showing release of IP_2 and IP_3 also did not utilize prelabeling under "stimulating conditions." These observations suggest that there may be a pool of PI prelabeled uniquely under "stimulating conditions" which is broken down directly by phospholipase C action (Figure 2).

In any event, it is clear that PI is the main source of DG and arachidonate (see below), whatever the mechanism of PI loss is, since there is far more DG and arachidonate released in pancreatic minilobules, for example, than can be accounted for by the breakdown of polyphosphoinositides (55).

Accumulation of Inositol Phosphates

In the past couple of years, considerable attention has been focused on the rapid release of water-soluble inositol phosphates following stimulation of various tissues with their physiological agonists. This has been made possible by the interesting property of the antimanic drug, Li^+, to inhibit IP phosphatase (199, 201, 202). This leads to an accumulation of IP during phosphoinositide turnover and a corresponding decrease in free myoinositol (199, 202–205). The amplification of the accumulation of inositol phosphates in the presence of Li^+ is a sensitive tool for the measurement of phosphoinositide degradation (36, 121, 199, 202). Attention was focused initially on the accumulation of IP (199, 202, 206–209) and more recently also on the accumulation of IP_2 and IP_3 (3, 165, 177, 186, 187, 201, 210–213). In several systems, the rapid release of IP_3 coincides with the fall in level of PIP_2 [see, for example, (167)]. Kinetic analysis indicates that at very short time periods [e.g., 5 sec in insect salivary gland (177)], there is an increased formation of IP_3 and IP_2, but increased formation of IP occurs later. Similar results have been obtained in pituitary cells in response to thyrotrophin-releasing hormone (186, 187, 211), in parotid acinar cells in response to methacholine (213), in pancreatic acinar cells in response to carbamylcholine (200), and in hepatocytes in response to vasopressin (214). These results are consonant with the earlier disappearances of polyphosphoinositides and suggest the earliest effect of agonists is to stimulate the release of IP_3 and IP_2, although the rapid formation of IP_2 may result from action of a phosphomonoesterase on IP_3.

Enzymes Involved in Agonist-Induced Phosphoinositide Turnover and their Sites

All of the reactions in the phosphoinositide cycle depicted in Figure 1 have been demonstrated. Several reviews dealing partially or exclusively with these enzymes have appeared in the past few years [see (2, 4–6, 215–217)].

The phosphoinositide phosphodiesterases are Ca^{2+} dependent (1, 82, 83, 215, 216, 218, 219) (see below). The data strongly suggest but do not establish unequivocally that there is one enzyme that hydrolyzes both PIP and PIP_2 that is a separate enzyme from the one that hydrolyzes PI (2, 218), although there appear to be several isozymes. The phosphoinositide phosphodiesterases are generally cytosolic [see, however, (220)]. Since the phosphoinositide phosphodiesterase(s) are presumably the initiating catalyst(s) for the phosphoinositide cycle, their mechanism of control is extremely important. Virtually nothing is known about this. Subtle changes in membrane structure in response to agonists rendering the phosphoinositide substrates accessible to their cytosolic enzymes may be the mechanism for their control (218).

Phosphatidylinositol kinase and PIP kinase are in the erythrocyte membrane (103, 221, 222) and in the cell membrane and lysosomes in other tissues [see (2, 4–6, 215–217)]. In the erythrocyte membrane, the kinases are active only from the cytoplasmic facing side (221).

There are phosphomonoesterases in the erythrocyte membrane that degrade PIP_2 to PIP (223) and PIP to PI (224). There is an enzyme in the erythrocyte membrane that degrades IP_3 to IP_2 (225), and there is evidence in the insect salivary gland for a similar enzyme that degrades IP_2 to IP (210). Inositol-1-phosphate is degraded to free myoinositol by an IP phosphatase.

Some (6) place the breakdown and resynthesis of all of the phosphoinositides at the plasma membrane. This ignores the fact that the site of agonist-stimulated synthesis of the bulk of the PI is the ER (90, 92–95, 226) (see EARLY OBSERVATIONS). Michell (1) was aware of this in his 1975 review where he suggested that newly synthesized PI was moved to the plasma membrane from the ER by PI transfer proteins [see (227)]. The difficulty with this point is that there is very little information concerning the transfer of PI to the plasma membrane where polyphosphoinositides and possibly PI are presumably acted on by phospholipase C. There is no firm evidence that PI transfer proteins are present in many of the tissues where phosphoinositide responses occur. Also, there is no transfer of newly formed PI from the ER to other fractions (with the possible exception of the plasma membrane, which was not separated from the ER) in stimulated guinea pig pancreas in vivo for up to 12 hr (226). One or more of the fat-soluble intermediates in the PI cycle [DG, PA, cytidine diphospho-diacylglycerol (CDP-DG), or free fatty acids] must be recycled back to the ER. No such recycling or carriers for any of these lipids has been demonstrated. It has been reported that agonist-stimulated breakdown of PI in the exocrine pancreas also occurs in the ER rather than the plasma membrane (15, 228). The breakdown of PI in pancreatic minilobules (70% of total PI) cannot all be accounted for by breakdown in the plasma membrane (55) without replenishment from the ER involving PI transfer proteins, since the PI in the plasma membrane accounts for less than 10% of total PI. An independent breakdown of PI in the ER may serve to generate DG and arachidonate at later times or at higher agonist concentrations than the generation of IP_2 and IP_3 from PIP and PIP_2, respectively, at the plasma membrane (Figure 2). The signal for this breakdown may be elevated levels of Ca_c^{2+}(28, 229) (see Figure 2 and below).

Action of Lithium on Phosphoinositide Metabolism and its Possible Therapeutic Implications

Allison & Stewart (203) noted a decrease in levels of free myoinositol in the brains of rats treated with Li^+. It was subsequently shown that the levels of IP were raised (204, 205). Inositol-1-phosphate is derived by synthesis from glucose (230) as well as from phosphoinositide breakdown. By inhibiting IP phosphatase (see above), Li^+ inhibits the synthesis of the phosphoinositides by limiting the availability of myoinositol for the enzyme PI synthase. The only other source of myoinositol is the active transport from extracellular fluid (231), and entry by this mechanism is quite limited in brain (232, 233) and in peripheral nerves (234). Thus, nervous tissue would be particularly susceptible to limitations in supply of free myoinositol by inhibition of IP phosphatase with a decline in levels of phosphoinositides as a consequence. This could very well be the mechanism of the antimanic action of Li^+ (199, 202, 205). In fact, the concentration of Li^+ required for half-maximal accumulation of IP in cholinergically stimulated pancreatic acini is quite close to the therapeutic blood levels of Li^+ (199). Lithium may thus inhibit synaptic transmission by inhibition of those mechanisms involving receptors linked to phosphoinositide metabolism in brain, e.g., muscarinic-cholinergic, α-adrenergic, serotoninergic, histaminergic, and peptidergic (36, 202).

Phosphoinositide Degradation Products as Second Messengers

There is now compelling evidence that the immediate phosphodiesteratic cleavage products of PIP_2, i.e., IP_3 and DG, are second messengers. Diacylglycerol may also be derived from PIP and, at later times, PI. These compounds fulfill many of the criteria of second messengers in much the same way as cAMP, i.e., they are formed very rapidly (within a few seconds), they are rapidly degraded by specific enzymes, and they act at very low concentrations. Diacylglycerol is further degraded via monoacylglycerol (28) to stearate, arachidonate, and glycerol in platelets [see (29, 235)] and the exocrine pancreas (55), although there is very little accumulation of MG in the latter. Arachidonate, in some instances through formation of its eicosanoid metabolites, mediates or modulates many physiological responses, and there is evidence that it may be involved in the elevation of cGMP which accompanies phosphoinositide responses in many cell types [see (1, 6)]. Recent evidence suggests that unesterified arachidonate may activate several responses without being converted to eicosanoids.

INOSITOL-1,4,5-TRISPHOSPHATE AS A SECOND MESSENGER FOR RELEASE OF CALCIUM FROM THE ENDOPLASMIC RETICULUM   It has been known since the time of Ringer over a century ago (236) that Ca^{2+} is essential for contraction of the heart. Beginning with the classical work of Katz on ACh release from nerve terminals and of Douglas on stimulus-secretion coupling [see (87, 237–240)], it was becoming apparent that many agonist-evoked responses involved Ca^{2+} as a second messenger. Namely, the physiological response required extracellular Ca^{2+}, and increases in intracellular Ca^{2+} derived from intracellular stores or from extracellular Ca^{2+} were seen. In surveying a large variety of tissues showing the PI response, Michell (1, 16) noted that the stimulus-response mechanism in these tissues involved Ca^{2+} as a second messenger. He felt that the independence or only partial dependence of the PI effect on Ca^{2+} in most of the tissues which had been studied at that time, as well as the stimulation by Ca^{2+} ionophores of the physiological response but not of PI breakdown, favored PI breakdown as antecedent to elevated cell Ca^{2+}. He suggested that PI breakdown opened "Ca^{2+} gates" in the plasma membrane. Additional support for the "Ca^{2+} gating" hypothesis was obtained in the blowfly salivary gland (241, 242), where stimulation with 5-hydroxytryptamine caused PI loss and entry of Ca^{2+} into the epithelial cells. Supramaximal stimulation with 5-hydroxytryptamine caused a loss of the small pool of responsive PI and a fall in Ca^{2+} transport (desensitization). Incubation of washed glands with myoinositol restored both PI sensitivity to 5-hydroxytryptamine and Ca^{2+} transport.

Until recently, there has been considerable controversy vis-a-vis the "Ca^{2+} gating" hypothesis [see (243–245)]. This was based on the fact that in some tissues the phosphoinositide response, usually that of PI, is Ca^{2+} dependent [see (4, 21, 25, 26)]. Also, Ca^{2+} ionophores such as A23187 stimulate phosphoinositide breakdown in some cases, again usually that of PI, suggesting that phosphoinositide breakdown in these situations might be secondary to rises in Ca^{2+}_c. The argument was further complicated by the fact that the phosphoinositide phosphodiesterases are Ca^{2+} dependent. What seems to be emerging is that under physiological conditions of ionic strength and pH the Ca^{2+} concentrations required for phosphodiesteratic cleavage of at least PIP_2 are at or below resting Ca^{2+}_c ($\sim$0.1 μM). Thus, Ca^{2+}_c is not a rate-limiting factor for PIP_2 cleavage [see (19, 28, 229, 240, 246)], although it may be rate limiting in some cases of PI and PIP cleavage (28, 229) (Figure 2). Although the details of Michell's initial proposal for involvement of PI in "Ca^{2+} gating" at the plasma membrane [which was later moved to an intracellular Ca^{2+} storage site (24)] were incorrect, the concept that one function of accelerated phosphoinositide metabolism is to raise Ca^{2+}_c, which is an essential link in many stimulus-response mechanisms (87, 89, 240, 247–250), appears to have been substantiated.

It appears that in those tissues in which a phosphoinositide effect occurs, the immediate source of Ca^{2+}_c is intracellular (87, 239, 240, 249, 251), and apparently the ER (252–254) (Figure 2). This intracellular pool of Ca^{2+}, which is released into the cytosol within a matter of seconds, is often referred to as "trigger Ca^{2+}," and its release is followed only later by an influx of extracellular Ca^{2+} (240, 249), which maintains the response at later times and also tends to replenish the "trigger" pool which is initially depleted.

It is obvious that a messenger is required to transfer information from the receptor at the plasma membrane to the site of release of the "trigger Ca^{2+}." When Berridge (177) found a very rapid liberation of IP_3 on stimulation of blowfly salivary gland with 5-hydroxytryptamine, he suggested that IP_3 may be a second messenger for "Ca^{2+} gating." Within the past year, it has been shown that if IP_3 is added to "permeabilized" (255, 256) pancreas (257), hepatocytes (214, 258, 259), or arterial smooth muscle cells (260), there is a rapid liberation of Ca^{2+} measured by release of $^{45}Ca^{2+}$ (258, 260) or by a Ca^{2+} electrode (257), or by the fluorescent intracellular Ca^{2+} probe quin2 (214, 259). The IP_3 concentration required for half-maximal Ca^{2+} release is in the 0.1–1.0 μM range (257–260). There is no release with IP_2, IP, or inositol-1,2-cyclic phosphate (257). Inositol-1,4,5-trisphosphate has been shown to release Ca^{2+} when added directly to microsome fractions of exocrine pancreas (252), rat insulinoma cells (253), and liver (254). Inositol-1,4,5-trisphosphate did not release Ca^{2+} from mitochondrial fractions (252–254).

Whether phosphoinositide metabolism plays a role in the rapid Ca^{2+} channels of the plasma membrane [see (261)] is not clear. At this point, there is not much evidence for this idea, although there have been numerous suggestions that PA may serve as a Ca^{2+} ionophore in the plasma membrane [see (16, 18, 26, 33, 262)]. Results with PA as an ionophore in liposomes have been conflicting (263, 264).

ACTIVATION OF PROTEIN KINASE C   The other primary product of phospholipase C action on phosphoinositides is DG. The fatty acid composition of the DG moiety of PI is predominantly of the SAG variety (265–268). Geison et al (268) showed about a decade ago that PA became enriched with the SAG moiety on stimulation of PI breakdown in whole mouse pancreas, thus more closely approaching the fatty acid composition of PI. This was further confirmation of the proposal that the PA formed on agonist stimulation was derived from PI. There was also an increased formation of DG (269), but very little of this appeared to be SAG (270). The rapid formation of DG, demonstrated either by an increase in mass of DG or an increase in radioactivity in DG after prelabeling of phosphoinositides with appropriate radioactive precursors, usually arachidonate, has been seen more recently on stimulation of phosphoinositide breakdown in platelets (23, 29, 217, 271–274), pancreas (55), mast cells (275), 3T3 cells (276), pituitary cells (134, 186, 211), and liver (163). In platelets (274), aortic epithelial cells (276), and pituicytes (134), it was shown that the content of stearoyl and arachidonoyl residues in DG more closely approached that of PI on stimulation, suggesting a conversion of PI to DG in these systems. The very rapid formation of PA within seconds on stimulation has been presumed to be due to release of its precursor DG from phosphoinositides.

Beginning in 1979, Nishizuka and coworkers demonstrated a protein kinase which required Ca^{2+}, phospholipid (e.g., phosphatidylserine), and DG for maximum activity, termed protein kinase C [see (277, 278)]. The affinity for Ca^{2+} was enormously increased by DG so that in the presence of the latter, maximal enzyme activity was seen at basal cytosolic concentrations of Ca^{2+}. C-kinase has so far been found in every mammalian tissue and is present in various phyla (278). It has been suggested that C-kinase may be cytosolic and that the fraction found in particulate fractions is loosely adsorbed, although in some cases detergents such as 0.2–0.4% Triton X-100 are required to remove it (278). That the DG released during agonist-induced phosphoinositide breakdown in physiologically responsive cells is likely to activate C-kinase is strongly suggested by the stimulation of this enzyme by addition of OAG to intact cells.

An interesting recent development is the activation of C-kinase by tumor promoter phorbol esters such as 12-tetradecanoylphorbol-13-acetate (TPA) (279–281). Phorbol esters apparently do this by mimicking DG in much the same way as many drugs mimic agonists at receptors. This bears out the prediction of Rohrscheider & Boutwell (282) over a decade ago that phorbol esters might act by binding to some physiological receptor. Diacylglycerol also competitively inhibits phorbol ester binding to a brain cytosolic phorbol ester aporeceptor (283). There is some evidence that part of the physiological regulation of C-kinase may involve binding of soluble enzyme to the plasma membrane since it has been shown that TPA causes a rapid decrease in cytosolic C-kinase and a corresponding increase in the amount bound to the plasma membrane [see (240, 277)]. Transfer of C-kinase from the cytosol to the plasma membrane would bring it into contact with its DG and phospholipid activators. Phorbol ester activation of C-kinase has been used as a selective and sensitive probe of the DG-activated C-kinase pathway in intact cells.

It is not known if or how activation of C-kinase by phorbol esters may be involved in tumor promotion. One possibility relates to the stimulation of superoxide generation in human neutrophils by phorbol esters or OAG and the inhibition of this stimulation by retinal, a known inhibitor of C-kinase (284). C-kinase is also capable of phosphorylating binding proteins for retinol and retinoic acid, and it was suggested that this may play a role in the ability of retinoids to function as antipromoters in chemically induced tumorigenesis, as well as control of physiological aspects of retinoid action (285).

The physiological effects of DG activation of C-kinase is presumably via phosphoryation of specific proteins [see (277, 278)] that are not phosphorylated by other protein kinases and that are presumed to be involved in the physiological response, especially secretion and proliferation. There may be some overlap in the proteins phosphorylated by the various kinases, however (286).

The steady-state level of DG for activation of C-kinase would be a function of its production by phosphoinositides and its phosphorylation to PA, or its degradation to its building blocks [see, for example, (55)].

FORMATION OF ARACHIDONATE AND METABOLITES   In platelets (28, 29, 235) and pancreatic minilobules (55), there is agonist-evoked liberation of unesterified arachidonate presumably via DG-lipase action on DG liberated from the phosphoinositides. In pancreatic minilobules, there is elevation of SAG and a release of substantial amounts [up to 50% of the total PI breakdown (see below)] of stearate, arachidonate, and glycerol on stimulation of secretion with the secretogogue caerulein (55). Unlike the situation in platelets, there is no appreciable accumulation of monoacylglycerol. In this model system, the DG-lipase inhibitor, RHC 80267 (287), reduces the liberation of free stearate, arachidonate, and glycerol, and elevates the steady-state level of SAG (55). These observations in the pancreas provide strong evidence that stimulation of phosphoinositide breakdown generates arachidonate via the sequential action of phospholipase C and DG lipase. Although 10 μM RHC 80267 almost completely inhibits DG lipase in platelet microsomes, it has not been possible to demonstrate a clear-cut increase in steady-state level of DG and reduced formation of arachidonate in agonist-stimulated platelets in the presence of RHC 80267 (288, 289). A reduced formation of monoacylglycerol was, however, seen with 10 μM RHC 80267 (288). At very high concentrations (250 μM) that were not specific for DG lipase (290), RHC 80267 inhibited agonist-stimulated production of arachidonate in platelets.

Another pathway that appears to liberate arachidonate in platelets is a phospholipase A_2-mediated breakdown of phosphatidylcholine, phosphatidyl-ethanolamine, and PI (29, 291–296). The affinity of this enzyme for Ca^{2+} is much lower (higher K_d) than that of phospholipase C, suggesting that phospholipase A_2 action on phospholipids is likely to be stimulated by agonist-evoked rises in Ca^{2+}_c (from ~0.1 μM to ~1.0 μM). Thus arachidonate release may respond to the transient rises in Ca^{2+} evoked by IP_3 [see (241)]. A phospholipase A_2 action on PA has also been suggested (139, 297), but it has been shown more recently that half of the arachidonate was released on stimulation of platelets with thrombin before any rise in PA (298).

The obligate intermediate for formation of prostaglandins and other eicosa-noids is arachidonate (299). Eicosanoids mediate or modulate numerous phys-iological responses [see (300–302)], including many which also involve stimu-lated phosphoinositide metabolism [see (18, 28, 87, 294)].

In the pancreas, the elevation in PI-derived arachidonate on agonist-stimulated enzyme secretion (55) is several orders of magnitude higher than the stimulated formation of PGE_2 and $PGF_{2\alpha}$ (55, 303, 304) and an unidentified eicosanoid that is present in higher amounts than PGE_2 or $PGF_{2\alpha}$ (M. C. Sekar, J. F. Dixon, L. E. Hokin, unpublished data). A similar relationship, although less striking, was noted on stimulation of perfused heart or kidney with bradykinin (305). Although the free arachidonate level is much higher than that of either PGE_2 or $PGF_{2\alpha}$ in caerulein-stimulated pancreas, the concentrations of prostaglandins appear to be regulated by the arachidonate levels (55). The role of a substantial release of unmetabolized arachidonate on stimulation of phosphoinositide breakdown is not clear. It is unlikely that the only function for the large increases in arachidonate is to regulate levels of eicosanoids which are formed in much lower amounts than that of arachidonate in some tissues. Recently, Kolesnick et al (306) showed that 3 μM arachidonate added exoge-nously to cloned pituitary GH_3 cells stimulated $^{45}Ca^{2+}$ efflux and prolactin secretion. Eicosatetraynoic acid and indomethacin, inhibitors of the lipoxyge-nase and cyclooxygenase pathways for arachidonate metabolism, respectively, did not inhibit either of these effects, suggesting a direct effect of arachidonate on $^{45}Ca^{2+}$ efflux and prolactin secretion. Two earlier studies (307, 308) had shown secretogogue-stimulated release of free arachidonate in thyrotropic pituitary cells but apparently not GH_3 (mammotropic) cells using the [^{14}C]arachidonate prelabeling technique, which is a relatively insensitive method of measurement as compared to radioimmunoassay and [^{3}H]-

arachidonate prelabeling (55). Arachidonate itself has been shown to mobilize Ca^{2+} from isolated organelles such as mitochondria (309) and sarcoplasmic reticulum (310), and it has been suggested that arachidonate may be an intracellular mediator for Ca^{2+} mobilization from mitochondria in stimulated hepatocytes (311). In addition to platelets, pancreas, and pituitary, a release of unesterified arachidonate in other tissues which show a phosphoinositide response has been demonstrated (18, 87, 294, 312).

In his 1975 review, Michell (1) noted that in a large number of tissues which showed phosphoinositide responses to specific agonists, the levels of cGMP were also elevated but the levels of cAMP were usually not [see also (6, 17)]. The physiological significance of cGMP is not as clear as that of cAMP [see (87, 313)]. The activation of guanylate cyclase by arachidonate and other fatty acids (87, 313–315) suggests that cGMP levels may be regulated by arachidonate levels. Arachidonate-generated cGMP may act as a negative feedback inhibitor in some instances. For example, CCK/PZ binding to pancreatic acini, which respond to this hormone with increased phosphoinositide turnover and arachidonate release (55), is inhibited by cGMP (316). It has generally been thought that the guanylate cyclase system requires Ca^{2+} since Ca^{2+} deprivation prevents rises in cGMP (87) and rises in cGMP can be elicited by the Ca^{2+} ionophore A23187 (87). It is a common view [see (6, 20)] that the Ca^{2+} dependence for cGMP formation may be due to a Ca^{2+} dependency for the release of arachidonate. This would be particularly true if the arachidonate release were due to the action of phospholipase A_2. There is also evidence that DG lipase may require Ca^{2+} (312). Very recently, however, data were presented suggesting that neurotransmitter receptors that utilize phosphoinositide turnover mediate cGMP formation by involvement of arachidonate and lipoxygenase without rises in intracellular Ca^{2+} (317).

Synergistic Interactions between Diacylglycerol and Inositol-1,4,5-Trisphosphate Branches of the Phosphoinositide Cascade

Several investigators [see (6, 240)] have pointed out the versatility in control afforded by a bifurcation in a signal pathway such as that resulting from the phosphodiesteratic cleavage of PIP_2. One pathway phosphorylates proteins by activation of C-kinase by DG, and the other pathway phosphorylates proteins by activation of calmodulin by the elevated Ca_c^{2+} signalled by IP_3. Some systems responding to increased Ca_c^{2+} may not depend on calmodulin; for example, Ca^{2+} activation of phospholipase A_2 (318). Additional versatility in control would be afforded by independent phosphodiesteratic cleavage of the three phosphoinositides. In the case of breakdown of PI or PIP, there would be formation of DG but no IP_3. This would emphasize the DG and arachidonate pathways. Very recently, it was shown (229) that in GH_3 pituitary cells, unlike the situation with stimulation by thyrotrophin-releasing hormone (134), Ca^{2+} ionophore increased primary formation of IP_2 and IP, as well as DG, arachidonate, and PA, but not IP_3, suggesting that PIP_2 hydrolysis resulting from thyrotrophin-releasing hormone is independent of rises in Ca^{2+}_c but the hydrolysis of PI and PIP may depend on rises in Ca^{2+}_c brought about by thyrotrophin-releasing hormone-stimulated PIP_2 hydrolysis (Figure 2). This sequence has also been suggested for platelets (28).

The synergism between the two branches has been demonstrated by elevating Ca^{2+}_c with ionophores such as A23187 and by stimulating C-kinase with phorbol esters or OAG. In most cases, neither agent alone produces maximal physiological response seen on stimulation with agonist but the two together do. Synergism between TPA or other phorbol esters and A23187 has been seen in platelets (319, 320), lymphocytes (321), hepatocytes (322), adrenal glomerulosa cells (323), neutrophils (324), the exocrine pancreas (325), mast cells (326), adrenocortical cells (327), and pancreatic islets (328).

Involvement of Phosphoinositides in Cell Proliferation

This subject has recently been reviewed (6, 37). Fisher & Mueller (329, 330) showed more than 15 years ago that when a mixed population of human leukocytes was stimulated with phytohemagglutinin, which is a mitogen for T-lymphocytes, there was an increased incorporation of ^{32}P into PI and PA within 3 min, whereas the metabolism of other phospholipids was not stimulated for the first 30 min. Most of the other stimulatory effects of phytohemagglutinin did not occur for hours or even days. More recently, changes in phosphoinositide metabolism on stimulation of pure populations of lymphocytes with mitogens (331, 332) and on activation of proliferation in cells in tissue culture [see (37)] have been reported. An increase in polyphosphoinositide content was recently shown on fertilization of sea urchin eggs (333).

Growth factors such as epidermal growth factor and the platelet-derived growth factor have also been shown to stimulate phosphoinositide metabolism (276, 334). Stimulation of cell proliferation with the platelet-derived growth factor was accompanied by an accumulation of DG, monoacylglycerol, and arachidonate in much the same way as in platelets and exocrine pancreas (276) (see above). The dose-response curves for binding of the epidermal growth factor and the increase in PI metabolism are similar (334).

The involvement of Ca^{2+} in initiation of proliferation was suggested by the observation in lymphocytes that A23187 could mimic the effects of phytohemagglutinin (335–337). Several laboratories have shown at least a transient elevation in Ca^{2+}_c at some stage during the cell cycle [see (37, 338, 339)]. The increase in PI metabolism appears to occur during the G1 and S phases of the cell cycle (340), which are claimed to be the stages where control by Ca^{2+} is crucial (339).

Studies with the quin2 have shown a rapid increase in Ca_c^{2+} on stimulation of Swiss 3T3 cells with the platelet-derived growth factor (341). Removing extracellular Ca^{2+} did not prevent the rise in Ca_c^{2+}, suggesting that the rise was due to release from intracellular stores (342), possibly by the action of IP_3 second messenger.

One of the earliest actions of growth factors on quiescent cells is the activation of a plasma-membrane-bound, amiloride-inhibitable, neutral Na^+/H^+ antiporter, leading to cytoplasmic alkalinization [see (343)]. Intracellular pH has been postulated as a possible mitogenic signal. Recently, the growth of mutants of Chinese hamster lung fibroblasts lacking the Na^+/H^+ antiporter was shown to be pH conditional (pH 8.0–8.3) in bicarbonate-free medium, in contrast to wild-type cells which grow over a wide range of pH's (6.6–8.2) (344). The importance of the Na^+/H^+ antiporter vis-a-vis the phosphoinositide cycle is the observation that phorbol esters (and by inference phosphoinositide-derived DG) can increase pH in 3T3 cells (345).

Arachidonate-activated cAMP formation also appears to be involved in proliferation (346). The activation of adenyl cyclase by arachidonate in this instance is likely to be via conversion to a prostaglandin since the formation of cAMP is blocked by indomethacin. Although the evidence suggesting an involvement of phosphoinositides in proliferation is largely circumstantial, the data are quite compelling.

Involvement of Phosphoinositides in Oncogene Action

Recent studies have suggested that several steps in the phosphoinositide cycle may be involved in oncogene action. Oncogenes, which appear to be involved in regulation of cell growth, show very close homology with transforming genes of retroviruses. Activation of cellular oncogenes can lead to oncogenic transformation identical to that induced by their viral counterparts. The protein products of the *ros* and *src* genes phosphorylate tyrosine on proteins. Recent developments have shown that the protein product of the *ros* gene from avian sarcoma virus UR2 phosphorylates PI (347) to form PIP and that from the *src* gene from Rous sarcoma virus phosphorylates both PI and PIP (348). In the case of the *ros* gene, the concentrations of the phosphodiesteratic cleavage products of PIP and PIP_2, i.e., IP_2 and IP_3, respectively, were increased in cells transformed with UR2. In the case of the *src* gene, within 20 min after induction of transformation by lowering the temperature from 41°C to 35°C, the incorporation of $^{32}P_i$ into PIP, PIP_2, and PA was increased. These studies taken together suggest that the protein products of the *ros* and *src* genes may function by phosphorylating phosphoinositides so as to increase levels of PIP_2 and PIP and thus their two second messenger products, IP_3 and DG. One problem with this theory is that the catalytic efficiency (V_{max}/K_m) of the phosphoinositide kinase encoded by the *src* gene is only one-hundredth to one-thousandth of that of the protein-bound-tyrosine kinase encoded by the same gene (348).

Further evidence that relates oncogenes to phosphoinositides is the finding that the *v-sis* gene codes for a protein that is almost identical to the platelet-derived growth factor (349, 350), which stimulates phosphoinositide metabolism. Also, the v-*erb*-B gene codes for a protein that has a very similar structure to a segment of the receptor for the epidermal growth factor (351), which also stimulates phosphoinositide metabolism (334). At this stage of investigation, the involvement of phosphoinositide metabolism in oncogene action is intriguing but not proven.

Acknowledgements

The author is grateful to M. Chandra Sekar and John F. Dixon for valuable discussions concerning the manuscript and to Karen E. Wipperfurth for her great dedication and skill in its preparation. Research related to this review was supported by grants HL16318 and DA03699 from the National Institutes of Health.

Literature Cited

1. Michell, R. H. 1975. *Biochim. Biophys. Acta* 415:81–147
2. Downes, C. P., Michell, R. H. 1982. *Cell Calcium* 3:467–502
3. Agranoff, B. W. 1983. *Life Sci.* 32:2047–54
4. Abdel-Latif, A. A. 1983. In *Handbook of Neurochemistry*, ed. A. A. Lajtha, 3:91–131. New York: Plenum. 2nd ed.
5. Hawthorne, J. N. 1983. *Biosci. Rep.* 3:887–904
6. Berridge, M. J. 1984. *Biochem. J.* 220:345–60
7. Hokin, L. E., Hokin, M. R. 1956. *Can. J. Biochem. Physiol.* 34:349–58
8. Hokin, L. E., Hokin, M. R. 1962. In *Ciba Foundation Symposium on the Exocrine Pancreas*, ed. A. V. S. de Reuck, M. P. Cameron, pp. 186–204. London: Churchill
9. Hokin, L. E., Hokin, M. R. 1963. *Fed. Proc.* 22:8–18
10. Hokin, M. R., Hokin, L. E. 1965. In *Metabolism and Physiological Significance of Lipids*, ed. R. M. C. Dawson, D. N. Rhodes, pp. 423–34. New York: Wiley
11. Hokin, L. E. 1968. *Int. Rev. Cytol.* 23:187–208
12. Hokin, L. E. 1969. In *The Structure and Function of Nervous Tissue*, ed. G. H. Bourne, 3:161–84. New York: Academic
13. Hokin, L. E., Hokin, M. R. 1969. *Ann. NY Acad. Sci.* 165:695–709
14. Hokin, M. R. 1974. In *Secretory Mechanisms of Exocrine Glands*, ed. N. A. Thom, O. N. Petersen, pp. 101–12. Copenhagen: Munksgaard
15. Hokin-Neaverson, M. 1977. In *Function and Biosynthesis of Lipids*, ed. N. G. Bazan, R. R. Brenner, N. M. Giusto, pp. 429–66. New York: Plenum
16. Michell, R. H., Jafferji, S. S., Jones, L. M. 1977. *Adv. Exp. Med.* 83:447–64
17. Berridge, M. J. 1981. *Mol. Cell. Endocrinol.* 24:115–40
18. Putney, J. W. Jr. 1981. *Life Sci.* 29:1183–94
19. Michell, R. H., Kirk, C. J., Jones, L. M., Downes, C., Creba, J. A. 1981. *Philos. Trans. R. Soc. London Ser. B* 296:123–37
20. Berridge, M. J. 1982. In *Calcium and Cell Function*, ed. W. Y. Cheung, 3:1–36. New York: Academic
21. Fain, J. N. 1982. In *Hormone Receptors*, ed. L. D. Kohn, pp. 237–76. New York: Wiley
22. Hawthorne, J. N. 1982. In *New Comprehensive Biochemistry*, ed. J. N. Hawthorne, G. B. Ansell, 4:263–78. Amsterdam: Elsevier
23. Irvine, R. F., Dawson, R. M. C., Freinkel, N. 1982. *Contemp. Metab.* 2:301–42
24. Michell, R. H. 1982. *Cell Calcium* 3:285–94
25. Farese, R. V. 1983. *Metabolism* 32:628–41

26. Putney, J. W. Jr. 1983. In *Adrenoceptors and Catecholamine Action*, Part B, ed. G. Kunos, pp. 51–64. New York: Wiley
27. Farese, R. V. 1984. *Mol. Cell. Endocrinol.* 35:1–14
28. Majerus, P. W., Neufeld, E. J., Wilson, D. B. 1984. *Cell* 37:701–3
29. Rittenhouse, S. E. 1982. *Cell Calcium* 3:311–22
30. Cockcroft, S. 1982. *Cell Calcium* 3:337–49
31. Hawthorne, J. N., Swilem, A. F. 1982. *Cell Calcium* 3:351–58
32. Takenawa, T. 1982. *Cell Calcium* 3:359–68
33. Putney, J. W. Jr. 1982. *Cell Calcium* 3:369–83
34. Berridge, M. J. 1982. *Cell Calcium* 3:385–87
35. Kirk, C. J. 1982. *Cell Calcium* 3:399–411
36. Downes, C. P. 1982. *Cell Calcium* 3:413–28
37. Michell, R. H. 1982. *Cell Calcium* 3:429–40
38. Farese, R. V. 1982. *Cell Calcium* 3:441–50
39. Schmidt, G., Thannhauser, S. J. 1945. *J. Biol. Chem.* 161:83–89
40. Hokin, L. E. 1952. *Biochim. Biophys. Acta* 8:225–26
41. Hokin, L. E. 1951. *J. Biol. Chem.* 48:320–26
42. Hokin, L. E., Hokin, M. R. 1954. *Biochim. Biophys. Acta* 13:401–12
43. Hokin, M. R., Hokin, L. E. 1953. *J. Biol. Chem.* 203:967–77
44. Hokin, M. R., Hokin, L. E. 1954. *J. Biol. Chem.* 209:549–58
45. Hokin, L. E., Hokin, M. R. 1956. *J. Physiol.* 132:442–53
46. Hokin, M. R. 1968. *Arch. Biochem. Biophys.* 124:271–79
47. Hokin, M. R. 1968. *Arch. Biochem. Biophys.* 124:280–84
48. Dawson, R. M. C. 1954. *Biochim. Biophys. Acta* 14:374–79
49. Marinetti, G. V., Stotz, E. 1956. *Biochim. Biophys. Acta* 21:168–70
50. Hokin, L. E., Hokin, M. R. 1955. *Biochim. Biophys. Acta* 18:102–10
51. Hokin, L. E., Hokin, M. R. 1958. *J. Biol. Chem.* 233:805–10
52. Hokin, L. E., Hokin, M. R. 1958. *J. Biol. Chem.* 233:818–21
53. Calderon, P., Furnelle, J., Cristophe, J. 1979. *Biochim. Biophys. Acta* 574:391–403
54. Calderon, P., Furnelle, J., Cristophe, J. 1979. *Biochim. Biophys. Acta* 574:404–13
55. Dixon, J. F., Hokin, L. E. 1984. *J. Biol. Chem.* 259:14418–25
56. Kornberg, A., Pricer, W. E. Jr. 1953. *J. Biol. Chem.* 204:329–43
57. Kornberg, A., Pricer, W. E. Jr. 1953. *J. Biol. Chem.* 204:345–57
58. Kennedy, E. P. 1957. *Ann. Rev. Biochem.* 26:119–48
59. Durell, J., Garland, J. T., Friedel, R. O. 1969. *Science* 165:862–86
60. Durell, J., Garland, J. T. 1969. *Ann. NY Acad. Sci.* 165:743–54
61. Hokin, L. E., Hokin, M. R. 1955. *Biochim. Biophys. Acta* 16:229–37
62. Hokin, L. E., Hokin, M. R. 1958. *J. Biol. Chem.* 233:822–26
63. Redman, C. M., Hokin, L. E. 1964. *J. Neurochem.* 11:155–63
64. Hokin, L. E., Sherwin, A. L. 1957. *J. Physiol.* 135:18–29
65. Eggman, L. D., Hokin, L. E. 1960. *J. Biol. Chem.* 235:2569–71
66. Hokin, M. R., Hokin, L. E., Saffran, M., Schally, A. V., Zimmermann, B. U. 1958. *J. Biol. Chem.* 233:811–13
67. Hokin, M. R., Hokin, L. E. 1958. *J. Biol. Chem.* 233:814–17
68. Hokin, L. E., Hokin, M. R. 1960. *J. Gen. Physiol.* 44:61–85
69. Hokin, M. R., Hokin, L. E., Shelp, W. S. 1960. *J. Gen. Physiol.* 44:217–26
70. Hokin, M. R. 1969. *J. Neurochem.* 16:127–34
71. Hokin, L. E. 1966. *J. Neurochem.* 13:179–84
72. Hokin, L. E., Hokin, M. R., Lobeck, C. C. 1963. *J. Clin. Invest.* 42:1232–37
73. Gerber, D. E., Lobeck, C. C., Hokin, L. E. 1971. *Biochem. Med.* 5:116–34
74. Hestrin-Lerner, S., Hokin, L. E. 1964. *Am. J. Physiol.* 206:136–42
75. Sbarra, A. J., Karnovsky, M. L. 1960. *J. Biol. Chem.* 235:2224–29
76. Karnovsky, M. L., Wallach, D. F. H. 1961. *J. Biol. Chem.* 236:1895–1901
77. Sastry, P. S., Hokin, L. E. 1966. *J. Biol. Chem.* 241:3354–61
78. Freinkel, N. 1957. *Endocrinology* 61:448–60
79. Scott, T. W., Jay, S. M., Freinkel, N. 1966. *Endocrinology* 79:591–600
80. Schmidt-Nielsen, K. 1960. *Circulation* 21:955–67
81. Hokin, M. R., Hokin, L. E. 1967. *J. Gen. Physiol.* 50:793–811
82. Dawson, R. M. C. 1959. *Biochim. Biophys. Acta* 33:68–77
83. Kemp, P., Hubscher, G., Hawthorne, J. N. 1961. *Biochem. J.* 79:193–200
84. Hokin, M. R., Hokin, L. E. 1959. *J. Biol. Chem.* 234:1381–90
85. Agranoff, B. W., Bradley, R. M., Brady, R. O. 1958. *J. Biol. Chem.* 233:1077–83

86. Paulus, H., Kennedy, E. P. 1960. *J. Biol. Chem.* 235:1303–11
87. Rubin, R. P. 1982. *Calcium and Cellular Secretion.* New York: Plenum
88. Peaker, M. 1976. In *Comparative Physiology: Water, Ions, and Fluid Mechanics,* ed. K. Schmidt-Nielsen, L. Bolis, S. H. P. Maddrell, pp. 207–12. Cambridge: Cambridge Univ. Press
89. Petersen, O. H. 1982. *Br. Med. Bull.* 38:297–302
90. Redman, C. M., Hokin, L. E. 1959. *J. Biophys. Biochem. Cytol.* 6:207–14
91. Palade, G. E., Siekevitz, P., Caro, L. G. 1962. See Ref. 8, pp. 23–55
92. Hokin, L. E. 1965. *Proc. Natl. Acad. Sci. USA* 53:1369–76
93. Hokin, L. E., Huebner, D. 1967. *J. Cell Biol.* 33:521–30
94. Schmidt, S. Y. 1983. *J. Cell Biol.* 97:832–37
95. Anderson, R. E., Maude, M. B., Kelleher, P. A., Rayborn, M. E., Hollyfield, J. G. 1983. *J. Neurochem.* 41:764–71
96. Douglas, W. W., Rubin, R. P. 1961. *J. Physiol.* 159:40–57
97. Douglas, W. W., Poisner, M. 1963. *J. Physiol.* 165:528–41
98. Douglas, W. W., Poisner, M. 1964. *J. Physiol.* 172:1–18
99. Hokin, L. E. 1966. *Biochim. Biophys. Acta* 115:219–21
100. Brockerhoff, H., Ballou, C. E. 1962. *J. Biol. Chem.* 237:49–52
101. Santiago-Calvo, E., Mule, S., Redman, C. M., Hokin, M. R., Hokin, L. E. 1964. *Biochim. Biophys. Acta* 84:550–62
102. Lee, T. C., Huggins, C. G. 1968. *Arch. Biochem. Biophys.* 126:206–13
103. Hokin, L. E., Hokin, M. R. 1964. *Biochim. Biophys. Acta* 84:563–75
104. Colodzin, M., Kennedy, E. P. 1965. *J. Biol. Chem.* 240:3771–80
105. Kai, M., White, G. L., Hawthorne, J. N. 1966. *Biochem. J.* 101:328–37
106. Michell, R. H., Hawthorne, J. N. 1965. *Biochem. Biophys. Res. Commun.* 21:333–38
107. Harwood, J. L., Hawthorne, J. N. 1969. *Biochim. Biophys. Acta* 171:75–88
108. Hokin-Neaverson, M. R. 1974. *Biochem. Biophys. Res. Commun.* 58:763–68
109. Jones, L. M., Michell, R. H. 1974. *Biochem. J.* 142:583–90
110. Mohell, N., Wallace, M., Fain, J. N. 1984. *Mol. Pharmacol.* 25:64–69
111. Balla, T., Enyedi, P., Hunyady, L., Spat, A. 1984. *FEBS Lett.* 171:179–82
112. Elliott, M. E., Goodfriend, T. L., Farese, R. V. 1983. *Life Sci.* 33:1771–78
113. Tennes, K. A., Roberts, M. L. 1982. *Biochim. Biophys. Acta* 719:238–43
114. Villalobos-Molina, R., Uc, M., Hong, E., Garcia-Sainz, J. A. 1982. *J. Pharmacol. Exp. Ther.* 222:258–59
115. Villalobos-Molina, R., Garcia-Sainz, J. A. 1983. *Eur. J. Pharmacol.* 90:457–59
116. Zeleznikar, R. J. Jr., Quist, E. E., Drewes, L. R. 1983. *Mol. Pharmacol.* 24:163–67
117. Sekar, M. C., Roufogalis, B. D. 1984. *Cell Calcium* 5:191–204
118. Fisher, S. K., Klinger, P. D., Agranoff, B. W. 1983. *J. Biol. Chem.* 258:7358–63
119. Harris, R. A., Fenner, D., Leslie, S. W. 1983. *Life Sci.* 32:2661–66
120. Cohen, N. M., Schmidt, D. M., McGlennen, R. C., Klein, W. L. 1983. *J. Neurochem.* 40:547–54
121. Brown, E., Kendall, D. A., Nahorski, S. R. 1984. *J. Neurochem.* 42:1379–87
122. Kendall, D. A., Nahorski, S. R. 1984. *J. Neurochem.* 42:1388–94
123. Hauser, G., Smith, T. L. 1981. *Neurochem. Res.* 6:1067–79
124. Quist, E. E. 1982. *Biochem. Pharmacol.* 31:3130–33
125. Brown, S. L., Brown, J. H. 1983. *Mol. Pharmacol.* 24:351–56
126. Brown, J. H., Masters, S. B. 1984. *Fed. Proc.* 43:2613–17
127. Freinkel, N., El Younsi, E., Dawson, R. M. C. 1975. *Eur. J. Biochem.* 59:245–52
128. Clements, R. S. Jr., Rhoten, W. B. 1976. *J. Clin. Invest.* 57:684–91
129. Tanigawa, K., Kuzawa, H., Okamoto, M., Imura, H. 1983. *Biochem. Biophys. Res. Commun.* 112:419–24
130. Axen, K. V., Schubart, U. K., Blake, A. D., Fleischer, N. 1983. *J. Clin. Invest.* 72:13–21
131. Schrey, M. P., Montague, W. 1983. *Biochem. J.* 216:433–41
132. Best, L., Malaisse, W. J. 1983. *Biochem. Biophys. Res. Commun.* 116:9–16
133. Sutton, C. A., Martin, T. F. J. 1982. *Endocrinology* 110:1273–80
134. Rebecchi, M. J., Kolesnick, R. N., Gershengorn, M. C. 1983. *J. Biol. Chem.* 258:227–34
135. Canonico, P. L., Cronin, M. J., Thorner, M. O., MacLeod, R. M. 1983. *Am. J. Physiol.* 245:E587–90
136. Anderson, R. E., Hollyfield, J. G. 1981. *Biochim. Biophys. Acta* 665:619–22
137. Schmidt, S. Y. 1983. *J. Neurochem.* 40:1630–38
138. Schmidt, S. Y. 1983. *J. Biol. Chem.* 258:6863–68
139. Lapetina, E. G. 1982. *J. Biol. Chem.* 257:7314–17

140. Shukla, S. D., Hanahan, D. J. 1982. *Biochem. Biophys. Res. Commun.* 106:697–703
141. Holmsen, H., Kaplan, K. L., Dangelmaier, C. A. 1982. *Biochem. J.* 208:9–18
142. Mahadevappa, V. G., Holub, B. J. 1983. *J. Biol. Chem.* 258:5337–39
143. Siess, W., Siegel, F. L., Lapetina, E. G. 1983. *J. Biol. Chem.* 258:11236–42
144. Gerard, C., Haye, B., Jacquemin, C., Mauchamp, J. 1982. *Biochim. Biophys. Acta* 710:359–69
145. Uzumaki, H., Muraki, T., Kato, R. 1982. *Biochem. Pharmacol.* 31:2237–41
146. Bennett, J. P., Cockcroft, S., Caswell, A. H., Gomperts, B. D. 1982. *Biochem. J.* 208:801–8
147. Yano, K., Hattori, H., Imai, A., Nozawa, Y. 1983. *Biochim. Biophys. Acta* 752:137–44
148. Takenawa, T., Homma, Y., Nagai, Y. 1983. *J. Immunol.* 130:2849–55
149. Matsubara, T., Hirohata, K. 1983. *Exp. Cell Biol.* 51:77–82
150. Ishizuka, Y., Nozawa, Y. 1984. *Biochem. Biophys. Res. Commun.* 117:710–17
151. Naor, Z., Zilberstein, M., Zakut, H., Dekel, N. 1984. *Life Sci.* 35:389–98
152. Durell, J., Sodd, M. A. 1966. *J. Neurochem.* 13:487–91
153. Yagihara, Y., Hawthorne, J. N. 1972. *J. Neurochem.* 19:355–67
154. Schacht, J., Agranoff, B. W. 1972. *J. Biol. Chem.* 247:771–77
155. Griffin, H. D., Hawthorne, J. N. 1979. *Biochem. J.* 176:541–52
156. Fisher, S. K., Agranoff, B. W. 1981. *J. Neurochem.* 37:968–77
157. Abdel-Latif, A. A., Akhtar, R. A., Hawthorne, J. N. 1977. *Biochem. J.* 162:61–73
158. Akhtar, R. A., Abdel-Latif, A. A. 1980. *Biochem. J.* 192:783–91
159. Creba, J. A., Downes, C. P., Hawkins, P. T., Brewster, G., Michell, R. H., Kirk, C. J. 1983. *Biochem. J.* 212:733–47
160. Warenycia, M. W., Vohra, M. M. 1983. *Can. J. Physiol. Pharmacol.* 61:97–101
161. Litosch, I., Lin, S.-H., Fain, J. N. 1983. *J. Biol. Chem.* 258:13727–32
162. Rhodes, D., Prpic, V., Exton, J. H., Blackmore, P. F. 1983. *J. Biol. Chem.* 258:2770–73
163. Thomas, A. P., Marks, J. S., Coll, K. E., Williamson, J. R. 1983. *J. Biol. Chem.* 258:5716–25
164. Weiss, S. J., McKinney, J. S., Putney, J. W. Jr. 1982. *Biochem. J.* 206:555–60
165. Downes, C. P., Wusteman, M. M. 1983. *Biochem. J.* 216:633–40
166. Billah, M. M., Lapetina, E. G. 1982. *J. Biol. Chem.* 257:12705–8
167. Agranoff, B. W., Murthy, P., Seguin, E. B. 1983. *J. Biol. Chem.* 258:2076–78
168. Mauco, G., Chap, H., Douste-Blazy, L. 1983. *FEBS Lett.* 153:361–65
169. Imai, A., Nakashima, S., Nozawa, Y. 1983. *Biochem. Biophys. Res. Commun.* 110:108–15
170. Leung, N. L., Vickers, J. D., Kinlough-Rathbone, R. L., Reimers, H.-J., Mustard, J. F. 1983. *Biochem. Biophys. Res. Commun.* 113:483–90
171. Billah, M. M., Lapetina, E. G. 1983. *Proc. Natl. Acad. Sci. USA* 80:965–68
172. Rendu, F., Marche, P., MacLouf, J., Girard, A., Levy-Tolendano, S. 1983. *Biochem. Biophys. Res. Commun.* 116:513–19
173. Simon, M.-F., Chap, H., Douste-Blazy, L. 1984. *FEBS Lett.* 170:43–48
174. Shukla, S. D., Hanahan, D. J. 1984. *Arch. Biochem. Biophys.* 227:626–29
175. Graff, G., Nahas, N., Nikolopoulou, M., Natarajan, V., Schmid, H. H. O. 1984. *Arch. Biochem. Biophys.* 228:299–308
176. Vickers, J. D., Kinlough-Rathbone, R. L., Mustard, J. F. 1984. *Biochem. J.* 219:25–31
177. Berridge, M. J. 1983. *Biochem. J.* 212:849–58
178. Litosch, I., Lee, H. S., Fain, J. N. 1984. *Am. J. Physiol.* 246:C141–47
179. Putney, J. W. Jr., Burgess, G. M., Halenda, S. P., McKinney, J. S., Rubin, R. P. 1983. *Biochem. J.* 212:483–88
180. Orchard, J. L., Davis, J. S., Larson, R. E., Farese, R. V. 1984. *Biochem. J.* 217:281–87
181. Volpi, M., Yassin, R., Naccache, P. H., Sha'afi, R. I. 1983. *Biochem. Biophys. Res. Commun.* 112:957–64
182. Yano, K., Nakashima, S., Nozawa, Y. 1983. *FEBS Lett.* 161:296–300
183. Emilsson, A., Sundler, R. 1984. *J. Biol. Chem.* 259:3111–16
184. Rawyler, A. J., Roelofsen, B., Wirtz, K. W. A., Op den Kamp, J. A. F. 1982. *FEBS Lett.* 148:140–44
185. Hirayama, T., Kato, T. 1983. *FEBS Lett.* 157:46–50
186. Martin, T. F. J. 1983. *J. Biol. Chem.* 258:14816–22
187. Rebecchi, M. J., Gershengorn, M. C. 1983. *Biochem. J.* 216:299–308
188. Schlegel, W., Roduit, C., Zahnd, G. R. 1984. *FEBS Lett.* 168:54–60
189. Macphee, C. H., Drummond, A. H. 1984. *Mol. Pharmacol.* 25:193–200

190. Brown, J. H., Brown, S. L. 1984. *J. Biol. Chem.* 259:3777–81
191. Dunlop, M. E., Larkins, R. G. 1984. *J. Biol. Chem.* 259:8407–11
192. Sekar, M. C., Roufogalis, B. D. 1984. *Biochem. J.* 223:527–31
193. Seyfred, M. A., Wells, W. W. 1984. *J. Biol. Chem.* 259:7659–65
194. Seyfred, M. A., Wells, W. W. 1984. *J. Biol. Chem.* 259:7666–72
195. Prpic, V., Blackmore, P. F., Exton, J. H. 1982. *J. Biol. Chem.* 257:11323–31
196. Poggioli, J., Weiss, S. J., McKinney, J. S., Putney, J. W. Jr. 1983. *Mol. Pharmacol.* 23:71–77
197. Wallace, M. A., Randazzo, P., Li, S.-Y., Fain, J. N. 1982. *Endocrinology* 3:341–43
198. Harrington, C. A., Eichberg, J. 1983. *J. Biol. Chem.* 258:2087–90
199. Hokin-Neaverson, M., Sadeghian, K. 1984. *J. Biol. Chem.* 259:4346–52
200. Rubin, R. P., Godfrey, P. P., Chapman, D. A., Putney, J. W. Jr. 1984. *Biochem. J.* 219:655–59
201. Hallcher, L. M., Sherman, W. R. 1980. *J. Biol. Chem.* 255:10896–901
202. Berridge, M. J., Downes, C. P., Hanley, M. R. 1982. *Biochem. J.* 206:587–95
203. Allison, J. H., Stewart, M. A. 1971. *Nature New Biol.* 233:267–68
204. Allison, J. H., Blisner, M. E., Holland, W. H., Hipps, P. P., Sherman, W. R. 1976. *Biochem. Biophys. Res. Commun.* 71:664–70
205. Sherman, W. R., Leavitt, A. L., Honchar, M. P., Hallcher, L. M., Phillips, B. E. 1981. *J. Neurochem.* 36:1947–51
206. Daum, P. R., Downes, C. P., Young, J. M. 1983. *Eur. J. Pharmacol.* 87:497–98
207. Watson, S. P., Downes, C. P. 1983. *Eur. J. Pharmacol.* 93:245–53
208. Daum, P. R., Downes, C. P., Young, J. M. 1984. *J. Neurochem.* 43:25–32
209. Mantyh, P. W., Pinnock, R. C., Downes, C. P., Goedert, M., Hunt, S. P. 1984. *Nature* 309:795–97
210. Berridge, M. J., Dawson, R. M. C., Downes, C. P., Heslop, J. P., Irvine, R. F. 1983. *Biochem. J.* 212:473–82
211. Drummond, A. H., Bushfield, M., Macphee, C. H. 1984. *Mol. Pharmacol.* 25:201–8
212. Vicentini, L. M., Meldolesi, J. 1984. *Biochem. Biophys. Res. Commun.* 121:538–44
213. Aub, D. L., Putney, J. W. Jr. 1984. *Life Sci.* 34:1347–55
214. Thomas, A. P., Alexander, J., Williamson, J. R. 1984. *J. Biol. Chem.* 259:5574–84
215. Shukla, S. D. 1982. *Life Sci.* 30:1323–35
216. Irvine, R. F. 1982. *Cell Calcium* 3:295–309
217. Irvine, R. F. 1982. *Biochem. J.* 204:3–16
218. Thompson, W., Dawson, R. M. C. 1964. *Biochem. J.* 91:233–36
219. Billah, M. M., Lapetina, E. G. 1982. *Biochem. Biophys. Res. Commun.* 109:217–22
220. Abdel-Latif, A. A., Luke, B., Smith, J. P. 1980. *Biochim. Biophys. Acta* 614:425–34
221. Garrett, R. J., Redman, C. M. 1975. *Biochim. Biophys. Acta* 382:58–64
222. Quist, E. E. 1982. *Arch. Biochem. Biophys.* 219:58–64
223. Koutouzov, S., Marche, P. 1982. *FEBS Lett.* 144:16–20
224. Mack, S. E., Palmer, F. B. St. C. 1984. *J. Lipid Res.* 25:75–85
225. Downes, C. P., Mussat, M. C., Michell, R. H. 1982. *Biochem. J.* 203:169–77
226. Gerber, D., Davies, M., Hokin, L. E. 1973. *J. Cell Biol.* 56:736–45
227. Wirtz, K. W. A. 1982. In *Lipid-Protein Interactions,* ed. P. C. Jost, O. H. Griffith, 1:151–231. New York: Wiley
228. Harris, D. W., Hokin-Neaverson, M. 1977. *Fed. Proc.* 36:639
229. Kolesnick, R. N., Gershengorn, M. C. 1984. *J. Biol. Chem.* 259:9514–19
230. Eisenberg, P. Jr. 1978. In *Cyclitols and Phosphoinositides,* ed. W. W. Wells, F. Eisenberg Jr., pp. 269–78. New York: Academic
231. Johnstone, R. M., Sung, C.-P. 1967. *Biochim. Biophys. Acta* 135:1052–55
232. Margolis, R. U., Press, R., Altszuler, N., Stewart, M. A. 1971. *Brain Res.* 28:535–39
233. Spector, R., Lorenzo, A. V. 1975. *Am. J. Physiol.* 228:1510–18
234. Gillon, K. R. W., Hawthorne, J. N. 1983. *Biochem. J.* 210:775–81
235. Majerus, P. W. 1983. *Clin. Invest.* 72:1521–25
236. Ringer, S. 1883. *J. Physiol.* 4:29–42
237. Katz, B. 1969. *The Release of Neural Transmitter Substances,* pp. 1–60. Springfield: Thomas
238. Douglas, W. W. 1974. In *Secretory Mechanisms of Exocrine Glands,* ed. N. A. Thom, O. H. Petersen, pp. 116–36. Copenhagen: Munksgaard
239. Putney, J. W. Jr. 1978. *Pharmacol. Rev.* 30:209–45
240. Rasmussen, H., Barrett, P. Q. 1984. *Physiol. Rev.* 64:938–84
241. Fain, J. N., Berridge, M. J. 1979. *Biochem. J.* 178:45–58
242. Fain, J. N., Berridge, M. J. 1979. *Biochem. J.* 180:655–61

243. Hawthorne, J. N. 1982. *Nature* 295:281–82
244. Cockcroft, S. 1981. *Trends Pharmacol. Sci.* 2:340–42
245. Michell, R. H. 1982. *Nature* 296:492–93
246. Irvine, R. F., Letcher, A. J., Dawson, R. M. C. 1984. *Biochem. J.* 218:177–85
247. Putney, J. W. Jr. 1979. *Pharmacol. Rev.* 30:209–45
248. Petersen, O. H., Maruyama, Y. 1984. *Nature* 307:693–96
249. Schulz, I., Stolze, H. H. 1980. *Ann. Rev. Physiol.* 42:127–56
250. Williams, J. A. 1980. *Am. J. Physiol.* 238:G269–79
251. Schulz, I. 1980. *Am. J. Physiol.* 239:G335–44
252. Streb, H., Bayerdorffer, E., Haase, W., Irvine, R. F., Schulz, I. 1984. *J. Membr. Biol.* 81:241–53
253. Prentki, M., Biden, T. J., Janjic, D., Irvine, R. F., Berridge, M. J., Wollheim, C. B. 1984. *Nature* 309:562–64
254. Dawson, A. P., Irvine, R. F. 1984. *Biochem. Biophys. Res. Commun.* 120:858–64
255. Wilson, S. P., Kirshner, N. 1983. *J. Biol. Chem.* 258:4994–5000
256. Streb, H., Schulz, I. 1983. *Am. J. Physiol.* 245:G347–57
257. Streb, H., Irvine, R. F., Berridge, M. J., Schulz, I. 1983. *Nature* 306:67–69
258. Burgess, G. M., Godfrey, P. P., McKinney, J. S., Berridge, M. J., Irvine, R. F., Putney, J. W. Jr. 1984. *Nature* 309:63–66
259. Joseph, S. K., Thomas, A. P., Williams, R. J., Irvine, R. F., Williamson, J. R. 1984. *J. Biol. Chem.* 259:3077–81
260. Suematsu, E., Hirata, M., Hashimoto, T., Kuriyama, H. 1984. *Biochem. Biophys. Res. Commun.* 120:481–85
261. Reuter, H. 1973. *Prog. Biophys. Mol. Biol.* 26:1–43
262. Lapetina, E. G. 1983. *Life Sci.* 32:2069–82
263. Serhan, C. N., Anderson, P., Goodman, E., Dunham, P., Weissmann, G. 1981. *J. Biol. Chem.* 256:2736–41
264. Holmes, R. P., Yoss, N. L. 1983. *Nature* 305:637–38
265. Keenen, R. W., Hokin, L. E. 1964. *Biochim. Biophys. Acta* 84:458–60
266. Holub, B. J., Kuksis, A. 1971. *J. Lipid Res.* 12:699–705
267. Baker, R. R., Thompson, W. 1972. *Biochim. Biophys. Acta* 270:489–503
268. Geison, R. L., Banschbach, M. W., Sadeghian, K., Hokin-Neaverson, M. 1976. *Biochem. Biophys. Res. Commun.* 68:343–49
269. Banschbach, M. W., Geison, R. L., Hokin-Neaverson, M. 1974. *Biochem. Biophys. Res. Commun.* 58:714–18
270. Banschbach, M. W., Geison, R. L., Hokin-Neaverson, M. 1981. *Biochim. Biophys. Acta* 663:34–45
271. Rittenhouse-Simmons, S. 1979. *J. Clin. Invest.* 63:580–87
272. Prescott, S. M., Majerus, P. W. 1983. *J. Biol. Chem.* 258:764–69
273. Dawson, R. M. C., Irvine, R. F. 1978. *Adv. Prostaglandin Thromboxane Res.* 3:47–54
274. Bell, R. L., Kennerly, D. A., Stanford, N., Majerus, P. W. 1979. *Proc. Natl. Acad. Sci. USA* 76:3238–41
275. Igarashi, Y., Kondo, Y. 1980. *Biochem. Biophys. Res. Commun.* 97:759–65
276. Habenicht, A. J. R., Glomset, J. A., King, W. C., Nist, C., Mitchell, C. D., Ross, R. 1981. *J. Biol. Chem.* 256:12329–35
277. Nishizuka, Y. 1984. *Nature* 308:693–98
278. Kuo, J. F., Schatzman, R. C., Turner, R. C., Mazzei, G. J. 1984. *Mol. Cell. Endocrinol.* 35:65–73
279. Castagna, M., Takai, Y., Kaibuchi, K., Sano, K., Kikkawa, U., Nishizuka, Y. 1982. *J. Biol. Chem.* 257:7847–51
280. Niedel, J. E., Kuhn, L. J., Vandenbark, G. R. 1983. *Proc. Natl. Acad. Sci. USA* 80:36–40
281. Sando, J. J., Young, M. C. 1983. *Proc. Natl. Acad. Sci. USA* 80:2642–46
282. Rohrscheider, A. V., Boutwell, R. K. 1973. *Nature New Biol.* 243:212–13
283. Sharkey, N. A., Leach, K. L., Blumberg, P. M. 1984. *Proc. Natl. Acad. Sci. USA* 81:607–10
284. Fujita, I., Irita, K., Takeshige, K., Minakami, S. 1984. *Biochem. Biophys. Res. Commun.* 120:318–24
285. Cope, F. O., Staller, J. M., Mahsem, R. A., Boutwell, R. K. 1984. *Biochem. Biophys. Res. Commun.* 120:593–601
286. Thams, P., Capito, K., Hedeskov, C. J. 1984. *Biochem. J.* 221:247–53
287. Sutherland, C. A., Amin, D. 1982. *J. Biol. Chem.* 257:14006–10
288. Chau, L. Y., Tai, H. H. 1983. *Biochem. Biophys. Res. Commun.* 113:241–47
289. Bross, T. E., Prescott, S. M., Majerus, P. W. 1983. *Biochem. Biophys. Res. Commun.* 116:68–74
290. Oglesby, T. D., Gorman, R. R. 1984. *Biochim. Biophys. Acta* 793:269–77
291. Van den Bosch, H. 1980. *Biochim. Biophys. Acta* 604:191–246
292. Billah, M. M., Lapetina, E. G., Cuatrecasas, P. 1980. *J. Biol. Chem.* 255:10227–31
293. Broekman, M. J., Ward, J. W., Marcus, A. J. 1980. *J. Clin. Invest.* 66:275–83

294. Rubin, R. P. 1982. *Fed. Proc.* 41:2181–87
295. Bills, T. K., Smith, J. B., Silver, M. J. 1977. *J. Clin. Invest.* 60:1–6
296. Mahadevappa, V. G., Holub, B. J. 1984. *J. Biol. Chem.* 259:9369–73
297. Billah, M. M., Lapetina, E. G., Cuatrecasas, P. 1981. *J. Biol. Chem.* 256:5399–403
298. Neufeld, E. J., Majerus, P. W. 1983. *J. Biol. Chem.* 258:2461–67
299. Samuelsson, B. 1981. *Harvey Lect.* 75:1–40
300. Moncada, S., Vane, J. R. 1978. *Pharm. Rev.* 30:293–331
301. Hammarstrom, S. 1982. *Arch. Biochem. Biophys.* 214:431–45
302. Wolfe, L. S. 1982. *J. Neurochem.* 38:1–14
303. Banschbach, M. W., Hokin-Neaverson, M. 1980. *FEBS Lett.* 117:131–33
304. Bauduin, H., Galand, N., Boeyhaems, J. M. 1981. *Prostaglandins* 22:35–51
305. Isakson, P. C., Raz, A., Denny, S. E., Wyche, A., Needleman, P. 1977. *Prostaglandins* 14:853–71
306. Kolesnick, R. N., Musacchio, I., Thaw, C., Gershengorn, M. C. 1984. *Am. J. Physiol.* 246:E458–62
307. Kolesnick, R. N., Musacchio, I., Thaw, C., Gershengorn, M. C. 1984. *Endocrinology* 114:671–76
308. Naor, Z., Catt, K. J. 1981. *J. Biol. Chem.* 256:2226–29
309. Roman, I., Gmaj, P., Nowicka, C., Angielski, S. 1979. *Eur. J. Biochem.* 102:615–23
310. Cheah, A. M. 1981. *Biochim. Biophys. Acta* 648:113–19
311. Whiting, J. A., Barritt, G. J. 1982. *Biochem. J.* 206:121–29
312. Litosch, I., Saito, Y., Fain, J. N. 1982. *Am. J. Physiol.* 243:C222–26
313. Goldberg, N. D., Haddox, M. K. 1977. *Ann. Rev. Biochem.* 46:823–96
314. Peach, M. J. 1981. *Biochem. Pharmacol.* 30:2745–51
315. Gerzer, R., Hamet, P., Ross, A. H., Lawson, J. A., Hardman, J. G. 1983. *J. Pharmacol. Exp. Ther.* 226:180–86
316. Peikin, S. R., Costenbader, C. L., Gardner, J. D. 1979. *J. Biol. Chem.* 254:5321–27
317. Snider, R. M., McKinney, M., Forray, C., Richelson, E. 1984. *Proc. Natl. Acad. Sci. USA* 81:3905–9
318. Withnell, M. R., Brown, T. J., Diocee, B. K. 1984. *Biochem. Biophys. Res. Commun.* 121:507–13
319. Kaibuchi, K., Takai, Y., Sawamura, M., Hoshijima, M., Fujikura, T., Nishizuka, Y. 1983. *J. Biol. Chem.* 258:6701–4
320. Rink, T. J., Sanchez, A., Hallam, T. J. 1983. *Nature* 305:317–19
321. Mastro, A. M., Smith, M. C. 1983. *J. Cell Physiol.* 116:51–56
322. Roach, P. J., Goldman, M. 1983. *Proc. Natl. Acad. Sci. USA* 80:7170–72
323. Kojima, I., Lippes, H., Kojima, K., Rasmussen, H. 1983. *Biochem. Biophys. Res. Commun.* 116:555–62
324. Sha'afi, R. I., White, J. R., Molski, T. F. P., Shefcyk, J., Volpi, M., et al. 1983. *Biochem. Biophys. Res. Commun.* 114:638–45
325. de Pont, J. J. H. H. M., Fleuren-Jakobs, A. M. M. 1984. *FEBS Lett.* 170:64–68
326. Katakami, Y., Kaibuchi, K., Sawamura, M., Takai, Y., Nishizuka, Y. 1984. *Biochem. Biophys. Res. Commun.* 121:573–78
327. Culty, M., Vilgrain, I., Chambaz, E. M. 1984. *Biochem. Biophys. Res. Commun.* 121:499–505
328. Zawalich, W., Brown, C., Rasmussen, H. 1983. *Biochem. Biophys. Res. Commun.* 117:448–55
329. Fisher, D. B., Mueller, G. C. 1968. *Proc. Natl. Acad. Sci. USA* 60:1396–1402
330. Fisher, D. B., Mueller, G. C. 1971. *Biochim. Biophys. Acta* 248:434–48
331. Hui, D. Y., Harmony, J. A. K. 1980. *Biochem. J.* 192:91–98
332. Hasegawa-Sasaki, H., Sasaki, T. 1982. *J. Biochem.* 91:463–68
333. Turner, P. R., Sheetz, M. P., Jaffe, L. A. 1984. *Nature* 310:414–15
334. Sawyer, S. T., Cohen, S. 1981. *Biochemistry* 20:6280–86
335. Allan, D., Michell, R. H. 1977. *Biochem. J.* 164:389–97
336. Crumpton, M. J., Allan, D., Auger, J., Green, N. M., Maino, V. C. 1975. *Philos. Trans. R. Soc. Ser. B* 272:123–80
337. Greene, W. C., Parker, C. M., Parker, C. W. 1976. *Cell. Immunol.* 25:74–89
338. Berridge, M. J. 1975. *Adv. Cyclic Nucleotide Res.* 6:1–98
339. Whitfield, J. F., Boynton, A. L., MacManus, J. P., Rixon, R. H., Sikorska, M., et al. 1981. *Ann. NY Acad. Sci.* 339:216–40
340. Dubois, C., Rampini, C. 1978. *Biochemie* 60:1307–13
341. Moolenaar, W. H., Tertoolen, L. G. J., de Laat, S. W. 1984. *J. Biol. Chem.* 259:7563–69
342. Lopez-Rivas, A., Rozengurt, E. 1983. *Biochem. Biophys. Res. Commun.* 114:240–47
343. Pouyssegur, J., Chambard, J. C., Franchi, A., L'Allemain, G., Paris, S., Van Obberghen-Schilling, E. 1983. In *Hor-*

monally Defined Media, ed. G. Fischer, R. Wieser, pp. 88–102. Heidelberg: Springer-Verlag

344. Pouyssegur, J., Sardet, C., Franchi, A., L'Allemain, G., Paris, S. 1984. *Proc. Natl. Acad. Sci. USA* 81:4833–37

345. Burns, C. P., Rozengurt, E. 1983. *Biochem. Biophys. Res. Commun.* 116: 931–38

346. Rozengurt, E., Stroobant, P., Waterfield, M. D., Deuel, T. F., Keehan, M. 1983. *Cell* 34:265–72

347. Macara, I. G., Marinetti, G. V., Balduz-zi, P. C. 1984. *Proc. Natl. Acad. Sci. USA* 81:2728–32

348. Sugimoto, Y., Whitman, M., Cantley, L. C., Erikson, R. L. 1984. *Proc. Natl. Acad. Sci. USA* 81:2117–21

349. Doolittle, R. F., Hunkapiller, M. W., Hood, L. E., Devare, S. G., Robbins, K. S., et al. 1983. *Science* 221:275–77

350. Waterfield, M. D., Scrace, G. T., Whittle, N., Stroobant, P., Johnsson, A., et al. 1983. *Nature* 304:35–39

351. Downward, J., Yarden, Y., Mayes, E., Scrace, G., Totty, N., et al. 1984. *Nature* 307:521–27

Ann. Rev. Biochem. 1985. 54:237–71

THIOREDOXIN

Arne Holmgren

Department of Physiological Chemistry, Karolinska Institutet, S-104 01 Stockholm, Sweden

CONTENTS

PERSPECTIVES AND SUMMARY

Thioredoxin is a small, ubiquitous protein with two redox-active half-cystine residues in an exposed active center, having the amino acid sequence: -Cys-Gly-Pro-Cys-. The protein exists either in reduced form (thioredoxin-$(SH)_2$) with a dithiol, or in oxidized form (thioredoxin-S_2), when the half-cystine

237

0066-4154/85/0701-0237$02.00

residues form an intramolecular disulfide bridge. Thioredoxin participates in redox reactions through the reversible oxidation of its active center dithiol, to a disulfide, and catalyzes dithiol-disulfide exchange reactions. Thioredoxin-S_2 is generally reduced by NADPH and flavoprotein thioredoxin reductase; thioredoxin-$(SH)_2$ is a powerful protein disulfide reductase that participates in many thiol-dependent, cellular-reductive processes.

Thioredoxin has now been known for 20 years, since its original purification from *Escherichia coli* in 1964 (1). It was then isolated in vitro as a hydrogen donor for ribonucleotide reductase, an essential enzyme that forms the deoxyribonucleotide precursors of DNA from the corresponding ribonucleotides and NADPH. Research on the structure and function of thioredoxin has since then turned out to be a fruitful pursuit; many surprising discoveries have been made of how this heat-stable protein, of unique structure, functions in many important biological phenomena. Some of the many uses of thioredoxin include: acting as a substrate for reductive enzymes, functioning as a protein disulfide oxido-reductase, being a regulatory factor for enzymes or receptors, and serving as a subunit of a virus DNA polymerase or as an essential component for assembly of small viruses. These multiple roles of thioredoxin have led to a growing interest in the molecule and its associated enzymes among researchers in many fields: DNA and protein synthesis, protein structure and folding, evolution, enzyme mechanisms, photosynthesis, endocrinology, virology, and cancer.

Thioredoxin is present in many different procaryotes and eucaryotes, and appears to be truly ubiquitous in all living cells. Bacteria, yeast, plant, and animal cells all contain thioredoxins with M_r around 12,000 that have evolved from a common ancestor.

Thioredoxin-S_2 from *E. coli* is the best characterized thioredoxin, with the amino acid sequence of its 108 amino acid residues known, and a crystallographically determined three-dimensional structure to 2.8 Å resolution. Also, the gene for thioredoxin, *trxA*, located at 84.7 min, has been sequenced. In thioredoxin-S_2, the active center disulfide is located in a unique protrusion of the three-dimensional structure between a β-pleated sheet and an α-helix. The thioredoxin structure contains a single twisted β-sheet composed of five strands flanked by four α-helical segments. From the homology of primary structures, it seems that all cellular thioredoxins may have this common fold.

Thioredoxin serves as a model for understanding of protein evolution, folding, and the dynamic behavior of proteins in solution. The structure of thioredoxin-$(SH)_2$ is not known in detail yet, but evidence for a localized conformational change on reduction of thioredoxin-S_2 has been obtained from spectroscopic results.

The thioredoxin system (thioredoxin and NADPH-thioredoxin reductase) is today implicated in a wide variety of biochemical systems. These include hydrogen donor functions for enzymes such as ribonucleotide reductase, methionine sulfoxide reductases, and sulfate reductase. Thioredoxin is also a general protein disulfide reductase, implicated in degradation of for example insulin and in regulation of enzyme activity by "thiol redox control." In phage T7-infected *E. coli* thioredoxin-$(SH)_2$ is an essential subunit of the viral DNA polymerase. Recent evidence shows that thioredoxin-$(SH)_2$ is also essential for the assembly of the single-stranded DNA phage f1. A localization of thioredoxin in the periphery of *E. coli* cells may be important for this function. In chloroplasts of plants, two forms of thioredoxin (f and m) regulate the activity of photosynthetic enzymes like fructose-1,6-bisphosphatase and malate dehydrogenase by light. This process is coupled to reduction of thioredoxin by electrons from ferredoxin by a special enzyme, ferredoxin-thioredoxin reductase, generally present in oxygenic photosynthetic cells. In mammalian cells, regulation of the activity of the glucocorticoid receptor and initiation of protein synthesis in reticulocytes by the thioredoxin system have been demonstrated.

The analysis of *E. coli*–viable mutants lacking thioredoxin but with fully active NADPH-dependent deoxyribonucleotide synthesis led to the discovery of glutathione (GSH) and glutaredoxin as hydrogen donor system for ribonucleotide reductase. Glutaredoxin is also a small, ubiquitous protein with a different dithiol-active center that catalyzes GSH-disulfide transhydrogenase reactions; in contrast to thioredoxin it is specific for the monothiol GSH and is not reduced by thioredoxin reductase. The discovery of glutaredoxin has called into question the role of thioredoxin as physiological hydrogen donor for ribonucleotide reductase. However, it has also led to the realization that thioredoxin and glutaredoxin may have specific but overlapping functions in many reduction systems that are still incompletely understood.

A thioredoxin induced by bacteriophage T4 is structurally and functionally similar to *E. coli* glutaredoxin but has also the capacity to be reduced by thioredoxin reductase; thereby it may support T4 DNA synthesis maximally via a T4-induced ribonucleotide reductase. In primary structure T4 thioredoxin (87 residues) is not homologous to *E. coli* thioredoxin; however, the two molecules have similar three-dimensional structures.

Thioredoxin systems in mammalian cells are still largely uncharacterized. Thioredoxin is widely distributed in tissues and organs, irrespective of DNA synthesis, and is partly membrane associated. Its capacity to catalyze protein dithiol-disulfide oxidoreductions suggests many still unknown specialized functions. These may, in particular, relate to formation of protein disulfides, secretory processes, regulation of enzyme activity, and messenger functions in hormone action and receptor mechanisms.

Reviews covering thioredoxin (2–4) and a conference volume on thioredoxins, with emphasis on photosynthetic regulation (5), have been published. Several reviews on ribonucleotide reductase cover essential aspects of thioredoxin (6–8).

INTRODUCTION

This review summarizes current knowledge about thioredoxin and related topics, with emphasis on the best-characterized systems. The literature on thioredoxin covers a fascinating variety of biochemical systems in which thioredoxin is implicated in functional roles. As introduction to a more detailed description of thioredoxin, a historical perspective is useful.

Thioredoxin was first isolated in 1964 and characterized as the hydrogen donor for the enzymatic synthesis of deoxyribonucleotides by ribonucleotide reductase in $E.$ $coli$ (1). The name thioredoxin was assigned to the small protein since its function was dependent on the cyclic reduction-oxidation of a single S-S-group (1). The S-S-bond of oxidized thioredoxin (thioredoxin-S_2) is reduced by NADPH and a specific enzyme called thioredoxin reductase (9):

$$\text{Thioredoxin-}S_2 + \text{NADPH} + H^+ \xrightarrow{\text{thioredoxin reductase}} \text{thioredoxin-}(SH)_2 + NADP^+$$

Reduced thioredoxin (thioredoxin-$(SH)_2$) serves as an efficient in vitro hydrogen donor for reduction of ribonucleotides (rNDP) to deoxyribonucleotides (dNDP) by ribonucleotide reductase of $E.$ $coli$ and many other species (7).

$$\text{Thioredoxin-}(SH)_2 + \text{rNDP} \xrightarrow{\text{ribonucleotide reductase}} \text{thioredoxin-}S_2 + \text{dNDP} + H_2O$$

Actually, thioredoxin was first described in 1960, under different names, as hydrogen donor for enzymes in yeast-reducing methionine sulfoxide (10) and sulfate (11) by NADPH. The identity was confirmed by isolation of thioredoxin and thioredoxin reductase from yeast (12, 13) in 1970, and demonstration of activity with the methionine sulfoxide and sulfate reductases.

The primary structure of $E.$ $coli$ thioredoxin was determined in 1968 (14). This structure showed that the active center contained two vicinal Cys-residues in the sequence -Trp-Cys-Gly-Pro-Cys-Lys- within a single polypeptide chain of 108 amino acid residues. In 1971, the active center of yeast thioredoxin was shown to have the same sequence, demonstrating that thioredoxins were evolutionarily-related proteins (15). Thioredoxin-S_2 from $E.$ $coli$ was crystallized in 1970 (16), and its three-dimensional structure was solved by X-ray crystallographic methods to 2.8 Å resolution in 1975 (17). Thioredoxin-S_2 is a "male" protein, with the S-S-bridge of the active center located in a protruding part of the three-dimensional structure. The protein consists of a central core of five strands of β-pleated sheet surrounded by four α-helixes.

Studies of bacteriophage systems then greatly changed the picture of a simple structure-function relationship for thioredoxins. Phage T4 infection of *E. coli* induces both a T4-encoded ribonucleotide reductase (18) and a T4 thioredoxin (19). The latter was defined by its capacity to be a specific hydrogen donor for T4 ribonucleotide reductase; this enzyme shows no activity with *E. coli* thioredoxin-$(SH)_2$ (19). T4 thioredoxin is reduced by NADPH and the host *E. coli* thioredoxin reductase (19, 20). The primary structure of T4 thioredoxin (87 residues) was determined in 1972 (21) and strangely enough exhibited no sequence homology with *E. coli* thioredoxin. However, crystallization of T4 thioredoxin-S_2 and determination of its three-dimensional structure in 1978 (22) showed that T4 thioredoxin and *E. coli* thioredoxin-S_2 indeed have a common fold and similar three-dimensional structures.

Mutants of *E. coli* having altered thioredoxins became available from unrelated studies of *E. coli* cells that had lost the ability to support growth of bacteriophage T7 (23). Studies on bacteriophage T7 infection of one class of such mutants (*tsn*C) resulted in the discovery that thioredoxin is an essential subunit of phage T7 DNA polymerase (24). Analysis of *tsn*C mutants showed no detectable thioredoxin in one mutant (*E. coli* 7004) (25), which had unimpaired NADPH-dependent ribonucleotide reduction (26). This finding resulted in the discovery of glutaredoxin as a GSH-dependent hydrogen donor system for ribonucleotide reductase (26), and revealed that thioredoxin is not essential for DNA synthesis and growth of *E. coli*. The thioredoxin gene *(trxA)* is located at around 84 min on the *E. coli* K12 genetic map (27).

The function of the thioredoxin system as a general protein disulfide reductase was established with several proteins, in particular insulin (28). Both thioredoxin-$(SH)_2$ and thioredoxin-S_2 show unusually high reaction rates in protein dithiol-disulfide exchange reactions, when compared with artificial dithiols such as dithiothreitol (29). Thioredoxin thus catalyzes the reduction of insulin disulfides by dithiothreitol, but also with dihydrolipoic acid (29).

Research on the regulation of photosynthetic enzymes in plant chloroplasts by light had resulted in the isolation of two regulatory proteins called ARP_b and ARP_a (30). In 1977 it was shown that ARP_b was a chloroplast thioredoxin (31) and ARP_b from rat liver was thioredoxin (32). This discovery greatly stimulated research on regulatory roles for thioredoxin as a redox messenger, and also led to the discovery of ferredoxin-thioredoxin reductase specific for oxygenic photosynthetic cells (33).

Studies of mammalian thioredoxin systems, initially as hydrogen donors for ribonucleotide reductase, resulted in isolation of pure thioredoxin from Novikoff hepatoma (34), calf liver (35), and rat liver (36). All these proteins contain extra structural sulfhydryl groups that may have regulatory capacity upon oxidation (36, 37). In addition, thioredoxin reductase in mammalian cells has a wider substrate specificity (36). Localization of rat thioredoxin in tissues and cells by immunohistochemical techniques (38) has shown a wide distribution, particularly in liver, epithelial, nerve cells, and different secretory cells, and a partial association with membranes. Recently, thioredoxin has been identified as the endogenous activator of the rat glucocorticoid receptor (39, 40) to a steroidbinding state.

A final piece of information in the present thioredoxin story comes from the recent identification of thioredoxin as the *fip* gene product of *E. coli* (42a) required for filamentous phage assembly (41–43).

THIOREDOXIN FROM *ESCHERICHIA COLI*

Isolation and Cellular Localization

ISOLATION   *E. coli* B thioredoxin was originally purified from a neutralized pH 5.0 supernatant by heat treatment to 85°C, followed by chromatography on DEAE-cellulose and filtration through Sephadex G-50 (1, 44). The use of a second, isocratic, DEAE-cellulose step at pH 4.79 removes minor impurities (44). Both thioredoxin and thioredoxin reductase, as well as glutathione reductase and lipoamide dehydrogenase, may be purified from the same *E. coli* cells (45). Thioredoxin has been isolated by immunoadsorbent chromatography (46); this has been combined with 2',5'-ADP-Sepharose affinity chromatography to isolate thioredoxin reductase and glutathione reductase (47).

The best source of *E. coli* thioredoxin is a recently developed K12 strain, SK 3981 (48). This strain contains a derivative of the plasmid pBR 325 (pBHK8) into which a 3 Kb *Pvu* II fragment of *E. coli* DNA containing the thioredoxin gene has been cloned. Growth of SK 3981 into stationary phase results in the overproduction of thioredoxin 100 to 200-fold compared to wild-type cells or 10^6 copies of thioredoxin/cell (48, 25). Thioredoxin purified from SK 3981 (48) is identical to the *E. coli* B thioredoxin (49).

Thioredoxin-S_2 is an acidic (I.P. 4.5) protein. At pH values below 4.5, it aggregates to form dimers and, after lyophilization, to form insoluble multimers: a process that is reversible at pH 8 or above (50). The dimers are probably the same as observed in the crystal structure (17). Thioredoxin-$(SH)_2$ is even more prone to reversible aggregation at pH values below 5.5 (50). No evidence for dimers in the cell, or a function for dimeric thioredoxin, has yet been found (25). Thioredoxin-$(SH)_2$ is readily obtained by NADPH and thioredoxin reductase (9), or by a small molar excess of dithiothreitol (29). Thioredoxin-$(SH)_2$ is oxidized by O_2 in air at neutral pH; faster at higher pH values. In the presence of EDTA and under N_2, thioredoxin-$(SH)_2$ is stable (1).

The isolation of phosphothioredoxin, containing a labile phosphothiol linkage, has been reported (51, 52). Based on the alkylation of the thiols in thioredoxin by iodoacetic acid (51, 52), the phosphate was thought to be located on Cys-32. However, only one thiol group is alkylated by iodoacetic acid at pH 7.5 in native thioredoxin-$(SH)_2$, and this may simulate phosphorylation of the other (53). As much as 94% of the thioredoxin of E. coli was reported to be recovered in the phosphorylated form (52). Later reports have suggested 8% phosphorylation (54). Other studies have questioned the existence of significant phosphothioredoxin in E. coli (53, 55), and no function for thiolphosphorylated thioredoxin is so far known.

CELLULAR LOCALIZATION   Thioredoxin normally behaves as a soluble protein after cell disruption (1, 25, 56). A peripheral localization at the inner membrane in E. coli is indicated from the quantitative release of thioredoxin by osmotic shock (56), and its partial association with a membrane fraction prepared after gentle lysis (56, 57). Specifically, thioredoxin may be located at adhesion sites (56); these sites (200–400 per cell in log phase) represent multifunctional fusions of the inner and other membranes (58). Direct demonstration of a peripheral localization of thioredoxin, close to the membrane in E. coli cells, has been obtained by immunoelectron microscopy (59). However, in some cells, a localization in the nucleoid region was also observed (59). The significance of this is not known.

Attempts to analyze the in vivo oxidation state of thioredoxin in E. coli suggest that substantial fractions of thioredoxin-S_2 (30–40%) exist in log phase cells (55).

Structural Aspects

E. coli thioredoxin is a single polypeptide chain of 108 amino acid residues (M_r 11,700) (14, 49, 60, 61) (See Figure 1). The active center residues, Cys-32 and Cys-35, are joined to a disulfide bridge (or a 14-membered disulfide ring) in thioredoxin-S_2. Many acidic residues are located in the N-terminal third of the molecule, whereas the C-terminal two thirds contain many hydrophobic residues.

The crystallization of thioredoxin-S_2 required the addition of Cu^{2+} ions, and the use of ethanol or 2-methyl-2,4-pentanediol as a precipitant (16). Crystals were only obtained in a narrow pH range, close to the isoelectric point (pH 4.5); attempts to obtain crystals of either thioredoxin-S_2 or thioredoxin-$(SH)_2$ by other methods have so far been unsuccessful [(16); A. Holmgren, unpublished

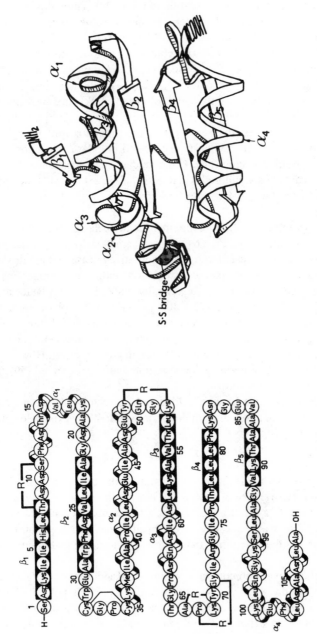

Figure 1 A. The amino acid sequence of thioredoxin-S$_2$ from *E. coli* (14, 49) and the secondary structure elements (17). R = reverse turn. *B.* Schematic drawing of three-dimensional structure of *E. coli* thioredoxin-S$_2$ (17).

results]. The structure of thioredoxin has been solved to 4.5 Å (62) and 2.8 Å resolution (17, 63) by X-ray crystallographic techniques and is shown in Figure 1.

The essential features of the structure are: (a) The molecule is built from a central core of five strands of β-pleated sheet (β_1–β_5), three parallel and two antiparallel, surrounded by four α-helices (α_1–α_4). Clearly defined reversed turns (R) are observed in four places. Including reverse turns, the thioredoxin structure has more than 75% of the residues involved in well-defined secondary structure. This unusually high figure for a protein of M_r 12,000, explains the exceptional stability of thioredoxin. (b) The two Cys-residues involved in forming the active center disulfide bridge form a protrusion between the middle strand of the pleated sheet and one of the helixes. Thus, thioredoxin-S_2 is the first example of a protein with its active center located on a protrusion rather than in a cleft. (c) The C-terminal third of the thioredoxin molecule forms a folding unit consisting of two strands of antiparallel pleated sheet joined by a helix. (d) The redox-active sulfur atoms of the S-S-bridge are accessible from one side of the structure (63). A molecular surface area on this side of the redox-active S-S-bridge is flat and hydrophobic (63); this area is formed by residues Gly-33, Pro-34, and Ile-75, Pro-76 plus Val-91, Gly-92, Ala-93. This area might be involved in binding thioredoxin to other protein molecules (63). On the other, more shielded side, there are a number of charged residues including Glu-30, Lys-36, and Lys-57 plus an aromatic residue, Trp-28. These residues are conserved in other known thioredoxins (63). The residues on the shielded side may participate in the redox reactions. Particularly Lys-36 may form a thiol-base ion pair upon reduction of thioredoxin-S_2 (53).

The basic structure of thioredoxin-S_2 thus consists of a single twisted β-sheet core composed of five strands, flanked by four α-helices integrated in a supersecondary structure. A similar structure, called the NAD-binding domain, has been observed to occur in a large number of proteins with diverse functions such as flavodoxin and dehydrogenases (64, 65).

The thioredoxin-S_2 crystals contain a dimer of the protein, and the function of the Cu^{2+} ions is to link molecules into layers (62). The β_5 strand and helix α_4 are involved in homologous interactions in the dimers (17).

Thioredoxin-S_2 is thus well characterized (17, 63), but the detailed three-dimensional structure of thioredoxin-$(SH)_2$ is still missing, since it has not been possible to crystallize this form.

Structure-Function Relationships

Evidence of a localized conformational change upon reduction of thioredoxin-S_2 has been obtained by fluorescence spectroscopy (66, 67). Reduction gives a 2.5-fold increase in the strongly quenched tryptophan emission of thioredoxin-S_2; the ORD or CD spectra are essentially unchanged (66, 67). The tryptophan emission of thioredoxin-S_2 is strongly quenched over the pH range 2–10 (68). In contrast, the fluorescence of thioredoxin-$(SH)_2$ is strongly pH dependent, with a maximum at pH 5 (68) and an apparent titration curve for a quenching group with a pK of 6.35 (67). This quenching group is supposed to be the deprotonated form of the sulfhydryl group of Cys-32, which has an estimated pH value of 6.7 judged from chemical modification by iodoacetate (53). The modification of cysteine residues in thioredoxin-$(SH)_2$, by alkylation using iodoacetamide or iodoacetic acid, leads to labeling of only one cysteine residue (Cys-32) below pH 8, and the reaction is strongly pH dependent (53). Both cysteine residues are alkylated, when the protein is denatured in 4.5 M guanidine-HCl, showing that the folded structure is responsible for the reactivity of thioredoxin-$(SH)_2$ (53). Presumably the lowered pH value of Cys-32 results from interaction with the ε-amino group of Lys-36, which may stabilize the thiolate and enhance its reactivity, and from the vicinity of Cys-35 (pK > 9.0) (53). Both thioredoxin-S_2 and thioredoxin-$(SH)_2$ show the same exceptional heat stability (unchanged structure after hours at +80°C) (67).

The conformational change in the vicinity of the S-S-bridge in thioredoxin-S_2 upon reduction affects the position of the tryptophan residues, as seen by NMR (50). It is likely that the main central supersecondary structure of thioredoxin is identical in both the oxidized and reduced protein. Further NMR studies will be of interest, since a limited total number of resonances are seen to change as a result of reduction of thioredoxin-S_2 in proton NMR spectra at 500 MHz. (A. Ehrenberg, A. Holmgren, to be published).

Oxidation of tryptophan residues by N-bromosuccinimide has been used to selectively modify the exposed Trp-31 (17, 69, 70) to a nonfluorescent, oxindolealanine residue. Reduction of this thioredoxin-S_2 species gives a sixfold increase in tryptophan fluorescence, originating from Trp-28 (69, 70). Thus, in thioredoxin-S_2, both tryptophan residues are probably quenched by the disulfide bond (68). Reduction of thioredoxin-S_2 changes the microenvironment around Trp-28, so that its fluorescence quantum yield will increase sixfold in thioredoxin-$(SH)_2$ (70). Trp-31 appears to have a strongly quenched fluorescence, in both oxidized and reduced thioredoxin.

Reversible chemical modification of Trp-28 and Trp-31 with formic acid–HCl results in an inactive aggregated thioredoxin; active protein is regenerated by alkali treatment (71). Also oxidation of both Trp-residues by N-bromosuccinimide inactivates thioredoxin (69); whereas modification of only Trp-31 leads to partial activity of thioredoxin as a substrate for thioredoxin reductase. Thus, Trp-28 appears essential as part of a surface required for interaction with other proteins (63).

Gly-92 has been found to be essential for the function of thioredoxin-$(SH)_2$ as a subunit of phage T7 DNA polymerase (72, 73); the altered thioredoxin isolated from E. coli 7007 has Gly-92 changed to an Asp residue (73). The altered thioredoxin reacts differently with thioredoxin reductase (K_m increases threefold and V_{max} decreases sevenfold) (73). It has about 20% activity in its reduced form with ribonucleotide reductase and lower activity in a coupled system for reduction of insulin (73). By radioimmunoassay, the antigen activity of the altered thioredoxin is very low, whereas its behavior in immunoprecipitation is identical to that of thioredoxin, reflecting the more stringent requirement of a competition antibody assay (73), and a major antigenic determinant in the Gly-92 region (73).

Several well-characterized systems for noncovalent reconstitution of thioredoxin from peptide fragments exist (74–79). Thioredoxin has single residues of methionine (Met-37) and arginine (Arg-73)(14)(Figure 1). Cleavage with CNBr results in two structureless fragments: thioredoxin-C-(1–37) and thioredoxin-C-(38–108) (44). The thioredoxin C-(1–37) containing the active site S-S-bridge is enzymatically inactive, but by mixing with thioredoxin-C-(38–108), it reconstitutes noncovalently to an active form called thioredoxin-C' (74). The thioredoxin-C' is a weak complex (K_d ~2 × 10^{-6} M), rapidly formed (t½ 1–2 min), and it shows about 50% activity with thioredoxin reductase in comparison with native thioredoxin-S_2 (75). In immunoprecipitation, it shows full activity as an antigen (75). The reduced form of thioredoxin-C' is not active as a hydrogen donor for ribonucleotide reductase or as an insulin disulfide reductase (75), suggesting that Met-37 and its peptide bond are essential for the conformational change taking place in the oxidation of thioredoxin-$(SH)_2$. Thioredoxin-C-(1–37) is an inhibitor of thioredoxin reductase in the presence of NADPH and 5.5'-dithiobis(2-nitrobenzoic acid) (DTNB), suggesting nucleated folding of the fragment to form a binding site for thioredoxin reductase (77). Such folding is induced at low pH values (78).

Cleavage of thioredoxin at Arg-73 by trypsin, after reversible blocking of the lysine residues by citraconic anhydride, generates two structureless and inactive fragments: thioredoxin-T-(1–73) and thioredoxin-T-(74–108) (75, 76). The latter fragment corresponds to the C-terminal folding unit of thioredoxin (17). Mixing the two fragments generates a complex called thioredoxin-T', with full immunoprecipitation activity with antibodies against thioredoxin (76). The thioredoxin-T' complex is slowly formed, has high stability (K_D ~10^{-8} M), and can be crystallized (76). It shows low activity with thioredoxin reductase (1–2%, major K_m increase) and has activity (3–5%) with ribonucleotide reductase (77). Spectroscopic studies of thioredoxin-T-(1–73) show that it is induced to fold to a structure similar to thioredoxin-S_2 at low pH (67). The results with thioredoxin-T' suggest that the area around Arg-73 is of critical importance for the interaction with both thioredoxin reductase and ribonucleotide reductase.

A third form of reconstituted thioredoxin has been isolated from E. coli cells as a protease-nicked molecule where the Pro-64 bond has been cleaved (79).

The resulting peptides, 1–64 and 65–108, apparently form a tight complex with almost full activity with thioredoxin reductase (79).

Folding Studies

Thioredoxin is heat-resistant but unfolded by high concentrations of guanidine-HCl or urea, as seen by tryptophan fluorescence measurements (68). The reaction is reversible and has been studied in depth by spectral and hydrodynamic measurements (80, 81). Tryptophan fluorescence measurements of unfolding of thioredoxin-S_2 by 4 M guanidine-HCl reveal a single kinetic phase with a relaxation time of 7 s (80). In contrast, refolding is complex and slow with evidence for intermediate states. Measurements indicate three kinetic phases having relaxation times of 0.54 s, 14 s, and 500 s, accounting for 12%, 10%, and 78% of the change (80). The activation enthalphy of the slowest phase is 22 kcal/mol, and it has characteristics of obligatory peptide isomerization of Pro-76 to the cis-isomer prior to refolding (80). In the three-dimensional structure of thioredoxin, Pro-76 has the unusual cis-conformation (17, 63). Pro-76 is very close to the disulfide bond in thioredoxin-S_2 (17). Refolding in urea demonstrates accumulation of a globular refolded intermediate (81).

Model peptides of the active site of thioredoxin have been synthesized and their spectral properties compared with thioredoxin (82, 83). These linear cyclic disulfides appear to have different conformations (83), implying that the folded structure of thioredoxin-S_2 will induce strain in the disulfide, which enhances its reactivity.

Genetic Aspects

The thioredoxin gene in E. coli is denoted trxA (27). No simple selection procedure for thioredoxin mutants existed before the discovery of its essential role in T7 DNA replication (24). However, the isolation of mutants in the trxA gene is readily made with T7 phage (23, 24). Recently, thioredoxin has been shown to be the product of the fip gene of E. coli, a gene whose product is essential for filamentous phage replication (41, 42, 42a); phage M13 containing a functional trxA gene can grow on thioredoxin-negative mutants of E. coli (42a, 43).

The first thioredoxin mutants (tsnC) were isolated in a B-strain (23). The 7004 strain contains no thioredoxin as measured in radioimmunoassays, immunoprecipitation, immunodiffusion, and various enzyme assays (25). No altered growth properties are found except a lag before exponential growth begins (25). This lag can be partially eliminated by cystine (25). A lower growth yield is also observed (25). The nature of the 7004 mutation is not known, but it does not appear to be an amber mutation (J. Fuchs, personal communication). A modified T7 selection scheme (23) was used to isolate a thioredoxin-defective mutant in a K-strain of E. coli (24). The mutant, E. coli

JM 109, F⁻ *polA1, rha⁻, lacZ(am), thyA⁻, str*^R, *trxA*⁻ has no altered growth properties (24). *TrxA* was mapped at 84 min on the *E. coli* chromosome near *metE* (24). Recent results suggest a chromosomal alteration (a duplication) near the *trxA* gene and that *trxA* maps closer to *ilv* than *metE* (42, 43).

A 3 Kb *pvu* II fragment of *E. coli* DNA with the thioredoxin gene has been ligated into pBR325, giving the plasmid pBHK8 (48). The nucleotide sequence of the thioredoxin gene from pBHK8 has recently been determined (49). The coding region of 324 base pairs is initiated by an AUG-codon and ends with an ochre (UAA) codon, and is preceded by a putative promotor region 70 base pairs upstream of the translation initiation codon (47). Thus, the synthesis of thioredoxin is controlled by its own promotor and the protein has no signal sequence (49). From results with *rho* gene (84, 85) and *fip* (41, 42), now known to be thioredoxin (42a), the gene order: *rep - trxA - rho - cyA* can be established. The thioredoxin gene is 330 nucleotides upstream of *rho* in *E. coli* K12 (49) at 84.7 min (42a).

The plasmid PLC 36-14 and others in the Clarke-Carbon library of *E. coli* DNA in a ColEl vector (86) also contain the thioredoxin gene that is expressed as seen in two-dimensional gels (87).

Site-directed mutagenesis experiments with the thioredoxin gene should now allow detailed studies of structure-function relationship of the protein.

THIOREDOXIN-DEPENDENT ENZYME REACTIONS

Thioredoxin Reductase

Thioredoxin reductase is a FAD-containing enzyme, originally found in *E. coli* (9) and purified to homogeneity (88, 45). The enzyme has structural and functional similarities to glutathione reductase and lipoamide dehydrogenase, two other pyridine nucleotide disulfide oxido-reductases; for review see (4, 89).

E. coli thioredoxin reductase has an M_r of 70,000 and consists of two probably identical subunits linked by noncovalent bonds (90, 91). Each subunit has one firmly bound FAD molecule and a redox-active disulfide in its active center. The sequence of this disulfide is -Cys-Ala-Thr-Cys- (91, 92). It is thus similar to thioredoxin in having a 14-membered ring but different from glutathione reductase or lipoamide dehydrogenase (4, 89). The enzyme is highly specific for NADPH, with a K_m value of 1.2 μM at 25°C (89). NADP⁺ is a competitive inhibitor of the enzyme, with a K_i of 15 μM (89). The enzyme is also highly specific for *E. coli* thioredoxin-S₂ with a K_m value of 2.8 μM at 25°C (89). The turnover number is about 2,000 min⁻¹ per FAD (89). Kinetic evidence favors a Ping-Pong mechanism (89).

The enzyme mechanism of the thioredoxin system (Figure 2) involves the reversible transfer of electrons from NADPH via a system of redox-active

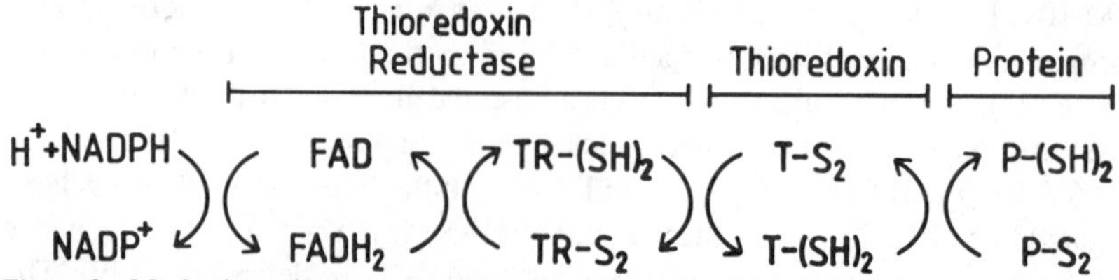

Thioredoxin
Reductase   Thioredoxin   Protein

H$^+$+NADPH ⟍ ⟋ FAD ⟍ ⟋ TR-(SH)$_2$ ⟍ ⟋ T-S$_2$ ⟍ ⟋ P-(SH)$_2$

NADP$^+$ ⟋ ⟍ FADH$_2$ ⟋ ⟍ TR-S$_2$ ⟋ ⟍ T-(SH)$_2$ ⟋ ⟍ P-S$_2$

Figure 2 Mechanism of protein disulfide reduction catalyzed by the thioredoxin system.

disulfides. It is thought that electrons flow from NADPH to the FAD, from the FAD to the redox-active disulfide in the enzyme, and then to thioredoxin (4, 89). In a four-electron reduction of thioredoxin reductase, there is a gradual bleaching of the flavin absorbance throughout the titration (93). This is consistent with the E_m values of the FAD and disulfide couples of thioredoxin reductase that are approximately equal (94). The active center also contains a base that lowers the pK value of one of the thiols in the reduced enzyme by forming a thiol base ion pair (95).

The thioredoxin reductase reaction is readily reversible, and NADPH may be formed from thioredoxin-(SH)$_2$ and NADP$^+$ (9). The equilibrium constant of [NADP$^+$] × [T-(SH)$_2$]/[T-S$_2$] × [NADPH] falls by a factor of 10 for each unit rise in pH between 7 and 9. At pH 7, a value of 48 for the equilibrium constant has been determined (9).

A mutant of *E. coli* with undetectable thioredoxin reductase in vitro has been isolated and partly characterized (96). The mutation in thioredoxin reductase *(trxB)* maps at 20–21 min on the *E. coli* genetic map (97). The mutant grows normally on rich medium and has no defective ribonucleotide reduction (96); it was selected in a *met$^-$* strain on the basis of its slow growth on methionine sulfoxide, reflecting a partial requirement of the thioredoxin system for methionine sulfoxide reductase in *E. coli* (97). The other system is supposed to be GSH and glutaredoxin (97). Thioredoxin may be reduced by other systems as well, perhaps by exchange with dihydrolipoamide groups of the pyruvate dehydrogenase complex (29).

Ribonucleotide Reductase

Ribonucleotide reductase is an essential component of most living cells since it is responsible for the synthesis of the four deoxyribonucleotides required for DNA synthesis (7). The enzyme actually catalyzes the first unique step in DNA synthesis and its synthesis is tightly coupled to this process (7). Ribonucleotide reductases are complex allosterically regulated enzymes of two classes containing either adenosylcobalamin *(Lactobacillus Leichmanni)* or Fe^{3+} and a tyrosine-free radical (*E. coli* and mammalian cells) (7). Extensive reviews on ribonucleotide reductase have been published (6–8).

The *E. coli* enzyme is the best characterized (7). It is encoded by two recently sequenced linked genes, *nrdA* and *nrdB* (98). The enzyme consists of two nonidentical subunits, protein B1 and protein B2, both required for activity (7). The B1 subunit (M_r 194,000) is a dimer of two polypeptides with 774 residues (98); it binds substrate and allosteric affectors and contains a redox-active dithiol (7). The B2 subunit (M_r 86,000) is also a dimer of two polypeptides with 375 residues (98); it contains iron and the tyrosine-free radical (7).

The enzyme requires a dithiol as hydrogen donor (1, 7), and in vitro dithiothreitol or dihydrolipoic acid in mM concentration to be active (7). *E. coli* thioredoxin-$(SH)_2$ is active in μM concentrations (7). The mechanism of the enzyme involves a redox-active disulfide as the immediate electron acceptor of electrons from thioredoxin-$(SH)_2$ (99). The structure of this disulfide in the *E. coli* enzyme has been suggested to be -Cys-Cys-Gly-Lys-Cys-Arg-; this sequence is located in the C-terminal part of the enzyme (98). This would imply a disulfide structure in the B1 subunit forming a 14-membered disulfide ring similar to the active centers of both thioredoxin (14) and thioredoxin reductase (91, 92).

The discovery of the glutaredoxin system (NADPH, glutathione reductase, GSH, and glutaredoxin) (26, 6) showed that thioredoxin is not essential for ribonucleotide reduction, and called into question its role in this process. Glutaredoxin is about 10 times more active than thioredoxin as a dithiol substrate for ribonucleotide reductase (100, 101). Studies on the activity of in vitro inactivated oxidized preparations of ribonucleotide reductase show that thioredoxin may both activate the enzyme, by reducing structural disulfide(s), and be hydrogen donor in the enzyme reaction (101). Glutaredoxin lacks the general protein disulfide reductase activity and shows no activity with such inactive oxidized forms of ribonucleotide reductase (101). Thus, thioredoxin may regulate the activity of ribonucleotide reductase by "thiol redox control" (6).

No mutant in the glutaredoxin gene of *E. coli* has yet been identified. Obviously, mutants in both thioredoxin and glutaredoxin genes will be required before we can assess their relative contributions as hydrogen donors for ribonucleotide reductase, in different species and physiological situations. Evidence against a general role for thioredoxin as hydrogen donor for ribonucleotide reductase has been found in *Corynebacterium nephridii* and rabbit bone marrow, where the pure homologous thioredoxins fail to serve as hydrogen donors for their respective ribonucleotide reductases (102, 103). However, calf thymus ribonucleotide reductase shows activity with both the homologous glutaredoxin and thioredoxin (104, 105).

Sulfoxide and Sulfate Reductases

SULFOXIDE REDUCTASES Enzymes in this group catalyze the following overall reduction of a sulfoxide (RSO) by NADPH:

$$R\text{-}SO + NADPH + H^+ \rightarrow R\text{-}S + NADP^+ + H_2O.$$

The first enzyme of this group was a methionine sulfoxide (Met-SO) reductase activity detected in yeast, requiring three protein fractions (10). They are now known as a methione sulfoxide reductase, thioredoxin, and thioredoxin reductase (12, 13). Also, *E. coli* has a very active system for reduction of free Met-SO, that is dependent on the thioredoxin or glutaredoxin system (106–108). A different enzyme also catalyzes the reduction of peptide or protein bound Met-SO residues (109). This enzyme has been found in *E. coli,* rat tissues, HeLa cells, *Euglena gracilis,* and spinach (109). Finally, a third type of reaction is the reduction of the sulfoxide in sulindac, a sulfoxide with antiinflammatory activity as the sulfide, by rat liver (110), or rat kidney tissues (111). All these enzymes appear to require thioredoxin-$(SH)_2$ as hydrogen donor (106–111). The protein-bound Met-SO reductase may be of considerable importance to protect cells against toxic actions of oxidizing agents. It is thus known that, for example, α-1-antitrypsin, ribosomal protein L12, or Met-enkephalin are inactivated when they contain Met-SO residues (112). The thioredoxin-dependent sulfoxide reductases may thus be a repair system for proteins of importance in, for example, toxicity or aging. In fact, it has been suggested that a defect of the thioredoxin system may account for lens cataract (113). In the development of cataract, extensive oxidation of Met and Cys residues in lens proteins occurs, but the lens methionine sulfoxide reductase is unchanged in activity (113) when assayed with dithiothreitol as hydrogen donor.

SULFATE REDUCTASE   Assimilatory reduction of SO_4^{2-} in plants and microorganisms, but not in mammals, serves a source of sulfur in amino acids like cysteine and methionine (114). One pathway of assimilatory reduction is via the PAPS (adenosine 3'-phosphate-5'-phosphosulfate) reduction first described in yeast (11) and there found to be linked to the thioredoxin system as a hydrogen donor from NADPH (12). A similar PAPS reduction system yielding SO_3^{2-} (sulfite) and involving the thioredoxin system has been found in *E. coli* (115). The first report (115) suggesting an intermediate of thioredoxin with SO_3^{2-} bound (thioredoxin as a carrier of sulfite or sulfate) is probably not correct (116), and this compound has never been isolated. The PAPS reductase (M_r 58,000), which is a dimer, probably contains a redox-active disulfide in its active center (116) that is reduced by thioredoxin-$(SH)_2$ in analogy with the mechanism for ribonucleotide reductase (7). Also the *E. coli* glutaredoxin system is active with the PAPS reductase as a hydrogen donor (116), but the relative contributions of the GSH-dependent and thioredoxin-dependent pathways are not known.

THIOREDOXIN AS A PROTEIN DISULFIDE OXIDO-REDUCTASE

All hydrogen donor functions for thioredoxin are now consistent with reduction of an active center disulfide in a specific substrate reductase. However, the thioredoxin system is also an unspecific protein disulfide reductase (see Figure 2) observed in yeast (10), *E. coli* (9), and mammalian cells (35). The NADPH-dependent reduction of the disulfide in DTNB (1, 9) or a protein such as insulin (35, 36) serves as a convenient assay of thioredoxin and thioredoxin reductase.

Measurements of the reaction rates of thioredoxin in insulin reduction (28, 29) led to the realization that thioredoxin is an enzyme that catalyzes dithiol-disulfide oxidoreductions. As seen from Table 1, thioredoxin-S_2 is reduced by dithiothreitol with a second order rate constant, which is about two orders of magnitude greater than when dithiothreitol reacts with other disulfides (117). Furthermore, thioredoxin-$(SH)_2$ reacts with insulin disulfides about four orders of magnitude faster than dithiothreitol does (28, 29). Thus thioredoxin catalyzes reactions of the type:

$$R\text{-}(SH)_2 + \text{protein-}S_2 \rightarrow R\text{-}S_2 + \text{protein-}(SH)_2,$$

where R may be dithiothreitol, dihydrolipoamide, or a dithiol in a protein-like ribonuclease (A. Holmgren, unpublished). Thioredoxin may thus be regarded as an enzyme operating in a Ping-Pong type of reaction (28, 29).

What makes thioredoxin so efficient in disulfide oxidoreductions? So far this is largely unknown. Clearly the rates of thiol-disulfide exchange reactions are strongly influenced by electrostatic effects, since only the thiolate ion is known to participate in these reactions (118). The pK-value of Cys-32 in thioredoxin has been calculated to be 6.7 (53). Furthermore, both thiols in thioredoxin-$(SH)_2$ are vicinal, making the effective concentration very high (117). The microenvironment of the dithiol/disulfide in thioredoxin, particularly the posi-

Table 1   Rate constants ($M^{-1} s^{-1}$) for disulfide reductions at pH 7 and 25°C (28, 29, 117).

Disulfide	Dithiol	
	Thioredoxin-$(SH)_2$	Dithiothreitol
Cystamine	100	10
Cystine	1,000	3
GSSG	570	—
Insulin	100,000	5
Thioredoxin-S_2		1,600

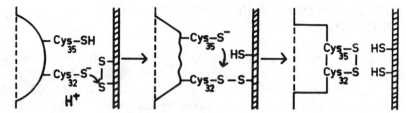

Figure 3   Suggested mechanism of thioredoxin-(SH)₂–catalyzed protein disulfide reduction.

tive change of Lys-36, may be responsible for the enhanced reactivity of Cys-32 by forming a thiol-anion base pair (53). Another suggestion, as yet not experimentally proven, is that thioredoxin-(SH)$_2$ forms a noncovalent catalytic complex with other proteins prior to a mixed disulfide intermediate. A mechanism is presented in Figure 3.

The redox potential calculated for *E. coli* thioredoxin is -0.26 V at pH 7.0 (9, 14). A value of -0.24 V has been reported for yeast thioredoxin (12). The corresponding values for dithiothreitol and lipoic acid are -0.33 V and -0.29 V; NADPH is -0.31 V (4). Since -0.03 V is equivalent to a factor of 10 in equilibrium constant for a 2-electron reduction (118), thioredoxin is essentially quantitatively reduced by dithiothreitol at pH 7.0 (53).

The thioredoxin system offers a convenient procedure for the controlled reduction of exposed protein disulfides. The reaction can be followed from the oxidation of NADPH spectrophotometrically at 340 nm (119). A number of proteins have reducible disulfide bonds; these include insulin (28), human choriogonadotropin (120), fibrinogen (121), and blood coagulation factors (122), particularly factor VIII (123) and factor X (124). Thioredoxin-(SH)$_2$ is a specific probe for certain exposed protein disulfides (119). Large differences have been observed. Thus, in human serum albumin, no disulfide of 17 is reduced, whereas in a factor VIII complex, 75 disulfides of 100 are rapidly reduced (123). For insulin the reduction of the exposed A7-B7 disulfide is controlled by Zn^{2+} ions; Zn^{2+} is a powerful inhibitor of the thioredoxin system and also apparently stabilizes the insulin structure (A. Holmgren, unpublished results). Uncontrolled degradation of insulin by thioredoxin in the β-cells of the pancreas may have implications for the etiology of diabetes (A. Holmgren, unpublished results).

THIOREDOXIN IN BACTERIOPHAGE INFECTION

The infection of *E. coli* by bacteriophages T4, T7, and f1 results in host-virus interactions all involving thioredoxin but in widely different aspects of the virus life cycle. Studies of these systems have greatly widened the perspective of thioredoxin functions.

Bacteriophage T4

Infection of *E. coli* by phage T4 results in induction of a T4-coded ribonu-cleotide reductase and a T4 thioredoxin (18, 19). The genes for the reductase subunits *nrd A* and *nrd B* (125) and T4 thioredoxin *nrd C* (126) are not linked on the T4 map. T4 thioredoxin-S_2 is reduced by NADPH and the host thioredoxin reductase (19). T4 thioredoxin-$(SH)_2$ is hydrogen donor for the T4 ribonu-cleotide reductase (K_m 0.7 μM) (19); this enzyme shows no activity with *E. coli* thioredoxin-$(SH)_2$ (19), and T4 thioredoxin-$(SH)_2$ is not active with *E. coli* ribonucleotide reductase (19). T4 ribonucleotide reductase is similar to the *E. coli* enzyme in subunit structure, iron-radical content, and requirement for a dithiol (7). However, it has a different allosteric regulation and is not turned off by accumulation of deoxyribonucleoside triphosphates (7).

Immunologically, T4 thioredoxin and *E. coli* thioredoxin show no cross-reactivity (127); this has been used for simultaneous purification of both T4 and *E. coli* thioredoxins by immunoadsorbent chromatography using two consecu-tive columns containing specific antibodies (46). T4 thioredoxin contains 87 amino acid residues and its primary structure (Figure 4) shows no homology with *E. coli* thioredoxin (21). T4 thioredoxin-S_2 also contains a single oxida-tion reduction active disulfide with the sequence: -Cys_{14}-Val-Tyr-Cys_{17}-. Crystallization of T4 thioredoxin-S_2 required the addition of Cd^{2+} ions (22). The three-dimensional structure of phage T4 thioredoxin-S_2 has been solved by X-ray crystallography to 2.8 Å resolution (22, 63). The molecule is clearly homologous in folding to *E. coli* thioredoxin-S_2 (Figure 4) (22) and has two simple folding units: a βαβ unit from the N-terminal end (residues 1–36) and a ββα unit from the C-terminal end (residues 67–87). The redox-active S-S-bridge is part of a protrusion of the molecule as in *E. coli* thioredoxin (22, 17). The most obvious difference from *E. coli* thioredoxin is that the first 22 residues of this molecule, including $β_1$ and $α_1$ (Figure 1) are absent in the T4 thioredoxin molecule (Figure 4), which starts directly with a β-strand ($β_2$). The redox-active S-S-bridge is at the N-terminal end of helix $α_2$ in both structures. Alignment of homologous residues in structural elements has been used to define a common fold containing 68 Cα-atoms with a root mean square difference of 2.6 Å (63). A molecular surface on one side of the S-S-bridge is more accessible. This side is hydrophobic and includes residues 15–16, 65–66, and 76–78 in T4 thioredoxin (63). *E. coli* thioredoxin has a similar surface, probably representing the area in both molecules involved in binding to thiore-doxin reductase (63). T4 thioredoxin and *E. coli* thioredoxin thus are examples of molecules with similar folding but unrelated primary structures.

The discovery of the GSH-glutaredoxin system in *E. coli* (26) resulted in similar experiments with T4 thioredoxin, which also catalyzes GSH-dependent ribonucleotide reduction and thus has glutaredoxin activity (128). Further-more, glutaredoxin from *E. coli* is a hydrogen donor for T4 ribonucleotide

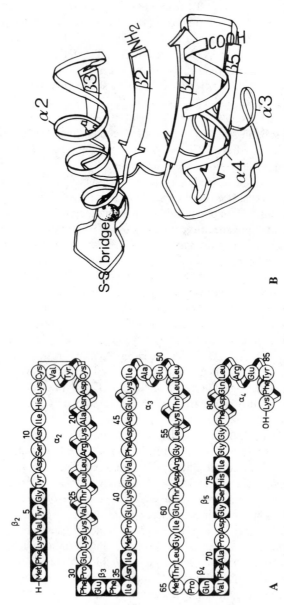

Figure 4   A. Amino acid sequence of T4 thioredoxin-S$_2$ (21) and the secondary structure elements (22). R = reverse turn. *B*. Schematic drawing of three-dimensional structure of T4 thioredoxin-S$_2$ (22). See also Figure 1.

reductase (128). T4 thioredoxin also has GSH-disulfide transhydrogenase activity similar to glutaredoxin (128). The primary structures of T4 thioredoxin and glutaredoxin are clearly homologous (129). Allowing for deletions in the glutaredoxin sequence or insertions in the T4 thioredoxin sequence at four places, there are identical residues at 25 positions of 77 compared (32% identity) (129). Thus, the idea that T4 thioredoxin is actually closely related to glutaredoxin (128, 101), has received strong support from the primary structures, from the enzyme activities, and from a proposed three-dimensional structure for glutaredoxin (63). So far, the two glutaredoxins analyzed, *E. coli* and calf thymus glutaredoxin, exhibit the active center sequence -Cys-Pro-Tyr-Cys- (129, 130), distinguishing them from thioredoxins with -Cys-Gly-Pro-Cys- (14). T4 thioredoxin differs from a glutaredoxin by its capacity to also react with *E. coli* thioredoxin reductase. Thereby it may aid T4 DNA synthesis maximally by utilizing two pathways from NADPH to its ribonucleotide reductase (131).

T4 thioredoxin may operate in GSH-disulfide transhydrogenase reactions via a mixed disulfide with GSH rather than by a dithiol form (128). A conformational change increasing the tyrosine fluorescence occurs on reduction of T4 thioredoxin-S_2. As for *E. coli* thioredoxin-S_2, the disulfide bridge has been proposed as the quencher of fluorescence (131). The redox potential of T4 thioredoxin at pH 7.0 and 25°C is -0.23 V (20) and thus higher compared to a value of -0.26 V for the *E. coli* protein (14). This enables *E. coli* thioredoxin reductase to catalyze reduction of T4 thioredoxin-S_2 by the bacterial thioredoxin $(SH)_2$ (131), a reaction independent of NADPH involving the redox-active disulfide in the enzyme. Together with the specificity of T4 thioredoxin for its homologous ribonucleotide reductase (19), this may lead to preferential use of the T4-induced system after phage infection (131).

In summary, T4 thioredoxin seems to represent a versatile adaptation to the requirements of the virus replication. The large T4 chromosome (166 Kb) requires an increased rate of synthesis of deoxyribonucleotide precursors for DNA replication. If T4 ribonucleotide reductase cannot function with *E. coli* thioredoxin, then the low amounts of glutaredoxin in *E. coli* are not sufficient for T4 ribonucleotide reductase. Modification of a gene for "glutaredoxin" to include reactivity with thioredoxin reductase will enable the phage to utilize all reducing power for T4 directed synthesis. There is evidence from dNTP pool data (132) that the *E. coli* ribonucleotide reductase is not active during T4 infection. This is probably because *E. coli* thioredoxin is oxidized and not active as a hydrogen donor for the enzyme. Furthermore, oxidized *E. coli* thioredoxin in turn may inactivate the *E. coli* ribonucleotide reductase by oxidizing sulfhydryl groups on the B1 subunit (6, 101). *E. coli* glutaredoxin shows no activity with such oxidized preparations of the reductase (101).

Induction of new thioredoxins and ribonucleotide reductases has also been demonstrated after infection with bacteriophage T6 and T5, but not by T7 or λ (133). The T6-induced thioredoxin and ribonucleotide reductase cross-reacts functionally and immunologically with the T4-induced proteins (133). T5-thioredoxin-S_2 is reduced by *E. coli* thioredoxin reductase but is immunologically different from *E. coli* thioredoxin (133). T5 ribonucleotide reductase also utilizes *E. coli* thioredoxin as hydrogen donor (133).

Bacteriophage T7

Phage T7 contains a linear duplex DNA molecule with 39,936 base pairs of known sequence (134), coding for 50 close-packed genes of which expression of 38 has been observed (134). The DNA replication of the phage is largely independent of host functions and has provided a very useful model for analysis of mechanisms of DNA replication (135–138). Prototrophic mutants of *E. coli* unable to support the intracellular growth of phage T7 (*tsn*C) were isolated in 1974 (23). T7 infection of *tsn*C mutants is characterized by normal cell killing and shut off of host functions as well as by normal patterns of phage transcription and protein synthesis; the block is at the level of DNA replication (23). Examination of T7-infected *E. coli* *tsn*C mutants showed no T7 DNA polymerase activity (139). However, extracts from *E. coli* *tsn*C$^+$ cells could restore the activity, and this resided in a heat-stable protein with M_r 12,000 (*tsn*C protein)(139). Purification of T7 DNA polymerase from *E. coli* *tsn*C$^+$ cells showed that the phage T7 DNA polymerase specified by the gene 5 is an enzyme composed of virus and host-specified components in a 1:1 stoichiometry (140). Later work identified the *tsn*C protein as identical to *E. coli* thioredoxin (24). Thus, T7 DNA polymerase is a 1:1 complex of the gene 5 protein (M_r 79,000 from amino sequence predicted by nucleotide sequence data) and thioredoxin M_r 11,700 (141).

Analysis of *E. coli* *tsn*C mutants revealed two classes (25): *E. coli* strains 7004 and 7007. *E. coli* 7004 contains no detectable level of thioredoxin by radioimmunoassays or enzymatic measurements (25), and may be a deletion mutant. *E. coli* 7007 contains reduced levels of enzymatic activity and expresses normal levels of a missense protein with a Gly_{92} → Asp exchange (73).

What is the role of thioredoxin in T7 DNA polymerase? So far no molecular explanation for its presence has been found. Isolation of gene 5 protein either from infected *trxA*-cells (141) or denatured T7 DNA polymerase (142) shows that it retains the single-stranded 3'→5' exonuclease activity of T7 DNA polymerase. However, the 3' → 5' double-stranded exonuclease of T7 DNA polymerase requires thioredoxin (142, 143). Thus, thioredoxin gives the phage-coded subunit the ability to function on double-stranded nucleic acids.

T7 DNA polymerase has been isolated in pure form also with antithioredoxin immunoadsorbent affinity chromatography (72). Binding of the enzyme through the thioredoxin subunit is strong, and active enzyme is eluted by a pulse of alkaline buffer (72). After a final phosphocellulose chromatography, homogenous enzyme is obtained (144).

In T7 infection of *E. coli* B cells, about 5% of the thioredoxin molecules are bound to the gene 5 protein (140, 72). Only thioredoxin-$(SH)_2$ is active in reconstituting T7 DNA polymerase activity or forming a complex with the gene 5 protein; thioredoxin-S_2 shows no activity or competition (145).

How is thioredoxin bound to the gene 5 protein? So far, information concerning this question is limited. A highly specific interaction of complimentary surfaces appears required. Modification of SH-groups in thioredoxin by iodoacetic acid, iodoacetamide, N-ethylmaleimide, or performic acid all destroy its activity (139, 145). Furthermore, only thioredoxin-$(SH)_2$ from *E. coli* has been found to activate gene 5 protein, whereas reduced thioredoxin m and f from spinach chloroplasts, T4 thioredoxin, glutaredoxin, or mammalian thioredoxins all fail to induce activity with the pure gene 5 protein (146). One report suggests that chloroplast thioredoxins f and m give activity (147). These results were obtained with a crude gene 5 protein preparation and it is not known why the discrepancy in results was obtained.

Are the SH-groups of thioredoxin involved in the mechanism of T7 DNA polymerase? There is no obvious requirement for a redox cycle, since under anaerobic conditions the enzyme is fully active in the absence of dithiothreitol (145, 146). Furthermore, the binding of thioredoxin in the enzyme appears to partly shield the SH-groups, since they show no normal reactivity with the disulfides of insulin (I. Slaby, A. Holmgren, unpublished), iodoacetic acid or N-ethylmaleimide (145, 146).

Filamentous Phages f1 and M13.

An *E. coli* mutant, which did not support the growth of filamentous phage f1, defined a new bacterial gene called *fip* (for filamentous phage production) (41). The *fip* gene is located near 84.7 min on the *E. coli* genetic map, just upstream of the *rho* gene and transcribed in the same direction as *rho* (42). Minicells containing *fip*$^+$ phage or plasmids synthesize a protein M_r 12,000 of cytoplasmic localization (42). The *fip* protein was recently identified as *E. coli* thioredoxin (42a). Furthermore, assays of thioredoxin in *fip* strains (43) showed changed activity (43). Neither f1 nor M13 can grow on *E. coli* trxA$^-$ mutants (42a, 43).

Bacteriophage f1 belong to a group of viruses, including M13 and fd, that have filamentous morphology and that are specific for the F$^+$ and Hfr strains of *E. coli* (148). These viruses exhibit unusual mechanisms of replication and maturation and in many respects resemble endosymbiotic animal viruses rather than typical lytic bacteriophages (148). Thus, the viruses are secreted into the medium as the host continues to grow and divide. Release of mature virus particles from the cell involves continuous extrusion through the host membrane (148). The DNA component of the mature virus is a single-stranded covalently closed molecule (M_r around 2×10^6). The complete nucleotide sequences of f1 (6 407 nucleotides) and of M13 and fd are very similar (149). A large untranslated intergenic region (nucleotides 5501–6005) of the filamentous coliphage genome has been of great importance with the discovery that foreign DNA can be incorporated in vitro into this part of the sequence without affecting the maturation and assembly of virus particles (150). This observation has led to the development of derivatives of M13 as cloning vectors of particular value for rapid nucleotide sequencing (151, 152).

The circular DNA of f1 is enclosed in a tubular sheet (~860 nm × ~7 nm) composed of about 2700 copies of the major coat protein and several copies of four minor capsid proteins in unique orientations at the ends (148, 153). The assembly process of f1 is intimately associated with the bacterial cytoplasmic membrane. The major coat protein (p VIII) and one minor coat protein are synthesized with signal sequences and are transmembrane proteins before they appear in phage (154). Other phage f1 proteins are also membrane associated (148, 153). The original *fip*1 mutant (A95) was temperature sensitive and permitted growth of f1 at 34°C but not at 41.5°C where the block was absolute (41). Phage f1 injection, DNA replication, and gene expression are not altered in *fip*1 mutant hosts which grow normally and show no phenotype (41). Several f1 gene I mutants, with morphogenic function, were unusually sensitive to the *fip* I mutation, whereas mutations in the other eight genes were of no effect (41). Spontaneous mutants capable of normal growth on *fip* cells at the nonpermissive temperature can be isolated *(gfip)* (41, 42). These phages contain a mutation at nucleotide 3619 (A → T). Gene I encodes an Asn at this position (142 in gene I); *gfip* specifies Tyr. Interestingly, M13 gene I protein contains His at this position. This is the only difference between these two closely related phages in the gene I region. M13, grows albeit poorly, on *fip* strains at 41.5°C. Thus, all data suggest that in f1 and M13, gene I product and thioredoxin interact to promote filamentous phage production. Nothing is known of the specific role of gene I protein in phage assembly. Very small amounts are made in vitro and it has not been observed in vivo.

Recent results (42a) show that the original *fip*1 mutant (41, 42) has a Pro-34 → Ser substitution in thioredoxin. Furthermore, thioredoxin reductase is also involved in phage assembly (42a), since *trxB* strains give lower efficiency of phage production. A hypothetical model, involving an engine for phage assembly driven by disulfide exchange reactions between the gene I protein and the thioredoxin system has been presented (42a). The peripheral localization of thioredoxin in *E. coli* at the membrane (56, 59) might thus be of importance for the virus assembly function.

THIOREDOXIN IN PHOTOSYNTHETIC CELLS

Regulation of the activity of photosynthetic enzymes in vitro by thioredoxin was discovered in 1977 when a small chloroplast protein originally called ARP_b (assimilation regulatory protein) (30) was shown to be replaceable with *E. coli* thioredoxin (31). Furthermore, ARP_b protein from rabbit liver was shown to be thioredoxin (32).

Sunlight provides the energy for photosynthesis in green tissues of plants (33). Light energy trapped in chloroplasts by chlorophylls is converted into ATP and NADPH, which are used to assimilate carbon and produce other cellular components (33). Light also has important regulatory functions (155), such as controlling the activity of specific chloroplast stromal enzymes (3, 33). It acts as a switch, changing metabolism from oxidative in the dark to photosynthetic in the light by modulating the catalytic activity of selective enzymes of degradative and biosynthetic pathways. This prevents "futile cycles": the simultaneous operation of starch biosynthesis and breakdown (33). Different mechanisms have been proposed to explain the regulation by light, such as increased stromal pH (pH $7{\rightarrow}8$), increased Mg^{2+} concentration ($1 \rightarrow 3$ mM), changes in effector concentrations (e.g. increased ATP and NADPH), and specific thioredoxin-mediated reductive activation of enzymes (33). The last mechanism provides an important link between the light-absorbing pigment system and several key enzymes. The components involved are ferredoxin, ferredoxin-thioredoxin reductase (originally called ARP_a) (30, 31), and several chloroplast thioredoxins (3, 33) (Figure 5).

In the light, electrons from chlorophyll are transferred to ferredoxin (Fd), a small strongly reducing (E_0 -0.42 V) iron-sulfur protein (156) and then via the enzyme ferredoxin-thioredoxin reductase (31) to thioredoxin according to the following reactions:

$$\begin{array}{c} \text{light} \\ \text{thylakoids} \\ 4\ Fd_{ox} + 2\ H_2O \xrightarrow{\hspace{2cm}} 4\ Fd_{red} + O_2 + 4H^+ \end{array}$$

$$\begin{array}{c} \text{ferredoxin-} \\ \text{thioredoxin} \\ \text{reductase} \\ 2\ Fd_{red} + \text{thioredoxin-}S_2 + 2\ H^+ \xrightarrow{\hspace{2cm}} 2\ Fd_{ox} + \text{thioredoxin-}(SH)_2 \end{array}$$

In vitro these two reactions can be replaced by dithiothreitol to reduce thioredoxin-S_2 (3).

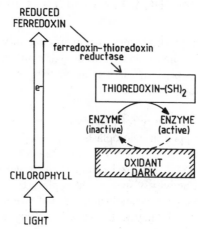

Figure 5 Ferredoxin-thioredoxin system for regulation of photosynthetic enzymes by light (3).

Reduced thioredoxins can activate many enzymes. These include four enzymes of the reductive pentose phosphate cycle (Calvin cycle), namely: fructose-1,6-bisphosphatase (FBPase) (31), sedoheptulose-1,7-bisphosphatase (157), NADPH-glyceraldehyde-3-phosphate dehydrogenase (158), and phosphoribulokinase (159). Photosynthetically reduced thioreoxin is also activating NADP-malate-dehydrogenase (160, 161), phenylalanine ammonia lyase (162), and ATPase activity associated with chloroplast coupling factor (163, 164). Activation of ribulose bisphosphate carboxylase by thioredoxin has also been reported (165).

All plant tissues analyzed have two or more thioredoxin species (3, 166) in contrast to nonphototropic bacteria and animals that only have one. The thioredoxin pattern of plant tissues has been subject to several investigations with partly differing results with regard to the number of isoproteins, their molecular weight, and specificity of enzyme activation (166–168). The best characterized plant thioredoxin system is from spinach chloroplasts (168) where two types of thioredoxin exist: the f-type, which activates fructose-1,6-bisphosphatase exclusively, and the m-type, which activates NADP-malate dehydrogenase (but also fructose-1,6-bisphosphatase with different kinetics) (168). Thioredoxin f and m have different amino acid compositions and terminal sequences (168). Thioredoxin m is further present as two very similar isomers, mb and mc; the only difference is an additional N-terminal lysine residue in mb (168). All these thioredoxins have M_r of about 12,000 (168). Recently, the first partial amino acid sequences of chloroplast thioredoxins showed (169) that thioredoxin f and m_c have an identical active center, -Cys-Gly-Pro-Cys-, which is the same as in *E. coli* thioredoxin (14). Both these proteins are thus thioredoxins and share a common ancestor with *E. coli* thioredoxin (169). Thioredoxin m shows considerable homology with *E. coli* thioredoxin, which is also consistent with their similar spectrum of activities (168, 169). Thioredoxin f, on the other hand, has little homology in the area surrounding the redox-active S-S-bridge, and may have undergone evolution to have a modified enzyme specificity (169).

The regulation of fructose-1,6-bisphosphatase by thioredoxin f is influenced by effectors (sugar phosphates, Mg^{2+}, Ca^{2+}) and has been studied by separating the activation phase from the enzyme catalytic reaction (170). The molecular change in fructose-1,6-bisphosphatase in activation and deactivation is unknown; possibly it is not a net reduction, but a thiol-disulfide exchange (33). Some evidence for a noncovalent complex between thioredoxin f and fructose-1,6-bisphosphatase has been presented (171).

Plant seeds, wheat or soybean, have been shown to contain three thioredoxins (I–III) of M_r 12,000 (172), separable by chromatography on DEAE-cellulose. The soybean isothioredoxins all activate spinach malate dehydrogenase but not fructose-1,6-bisphosphatase and are hydrogen donors for *E. coli* ribonucleotide reductase (172). An NADPH-thioredoxin reductase (M_r 70,000) exists in seeds (172), but this enzyme disappears upon greening of ethiolings (173). The role of the multiple thioredoxins in seeds is not yet established.

Green tissues have ferredoxin-thioredoxin reductase (FTR) (33). The properties of this enzyme are still a matter of discussion. FTR has been isolated as a single subunit protein of M_r 38,000 with no chromophores and essentially no metals (174). In contrast, another report (175) describes FTR purified to homogeneity as a protein of M_r 38,000 containing as a chromophore, a nonheme iron sulfur complex (2–4 Fe and S per molecule). This FTR was composed of nonequal subunits; two of M_r 12,000 and one of M_r 14,000 (175). Further studies of chloroplast FTR will be required to settle this question and define the active center of the enzyme. One report describes still another form of ferredoxin-thioredoxin reductase (176). The anaerobic sacharolytic fermenting bacterium *Clostridium pasteurianum* living in soil unexposed to light utilizes a M_r 53,000 (2 M_r 30,000 subunits) FTR with a typical FAD spectrum similar to thioredoxin reductase of *E. coli* (176).

Thioredoxin has also been demonstrated in, for example, green algae *(Scenedesmus obliquus)* (177) and cyanobacteria (178, 179). Thioredoxin purified to homogeneity from *Anabaena sp.* (179) has the same active center as *E. coli* thioredoxin and functions as hydrogen donor for the B_{12}-dependent ribonucleotide reductase of *Anabaena* and also for *E. coli* ribonucleotide reductase (179). In cyanobacteria, activation of glutamine synthetase, a key enzyme in nitrogen assimilation by thioredoxin, has been suggested (180). A similar activation has been proposed in *Chlorella* (181). Deactivation of glucose-6-phosphate dehydrogenase by thioredoxin-$(SH)_2$ has been demonstrated in cyanobacteria (182, 183).

In *Rhodopseudomonas spheroides,* an activation of 5-amino-levulinic acid (ALA)-synthetase by thioredoxin-$(SH)_2$ has been demonstrated (184, 185). This nonsulfur purple bacterium can grow anaerobically in light and aerobically in both light and dark. The ALA-synthetase is the first enzyme in bacteriochlorophyll synthesis, which is regulated by O_2 probably via thioreoxin (184, 185). *Rhodopseudomonas sp.* contains both a thioredoxin and an NADPH-thioredoxin reductase, and these proteins show about 10% enzymatic cross-reactivity with the corresponding *E. coli* thioredoxin system (185).

THIOREDOXIN SYSTEMS IN MAMMALIAN CELLS

Mammalian cells contain a classical thioredoxin system with one thioredoxin and a thioredoxin reductase. Thioredoxin has been purified to homogeneity from rat Novikoff hepatoma (34), calf liver (35), rat liver (36), rabbit bone marrow (103), and human placenta (A. Ehrnberg, A. Holmgren, unpublished). These thioredoxins all have molecular weights of about 12,000 and are able to serve as hydrogen donors for *E. coli* ribonucleotide reductase and also reduce insulin disulfides. With the exception of rabbit bone marrow (103), they also serve as hydrogen donors for their homologous ribonucleotide reductases. Partial amino acid sequences of mammalian thioredoxins (4, 36, 186) show that these proteins have the same invariant active center sequence (-Trp-Cys-Gly-Pro-Cys-Lys-) and are clearly homologous to *E. coli* thioredoxin (14); they should thus have the same fold as *E. coli* thioredoxin. However, there are certain differences with respect to the *E. coli* thioredoxin system: (*a*) All mammalian thioredoxins have, in addition to the active center disulfide, two additional structural half-cystine residues in the C-terminal half of the molecule (34–36). Oxidation of these two half-cystine residues to an intramolecular disulfide leads to aggregation of the protein and loss of its activity as a substrate for thioredoxin reductase (36, 37). Thus, purification of mammalian thioredoxins is only possible in the reduced form, since the oxidized protein displays multiple peaks in chromatography systems, falsely suggesting the occurrence of multiple species (34, 35). The obvious potential of the structural cysteine residues is to regulate the activity of the thioredoxin system (37). Thus, mammalian thioredoxins have self-regulatory capacity, since thioredoxin-$(SH)_2$ will reduce oxidized molecules (37). (*b*) Thioredoxin reductase from mammalian cells is considerably larger: M_r 116,000 (subunits of 58,000) (36, 187–189). It is specific for NADPH but has a much less stringent disulfide substrate specificity and will reduce thioredoxin from *E. coli* (K_m increases 14 fold) but also other substances like DTNB and menadione or alloxan (36).

The distribution of thioredoxin in organs and subcellular fractions from a week-old calf has been studied by radioimmunoassay (190). Different subcellular fractions contained thioredoxin; this included the plasma membrane fraction where a fraction of immunoreactive thioredoxin was found to be associated with membrane structures (190). Using antisera prepared against pure rat liver thioredoxin and thioredoxin reductase, immunohistochemical studies in adult rats show these two proteins to be widely distributed in tissues and organs (191). However, variations were seen between cell types; epithelial cells, neuronal cells, and secretory cells exocrine and endocrine showed high immunoreactivity whereas mesenchymal cells showed low activity (191). Surface lining epithelial cells and keratinizing cells showed high activity. The immunofluorescence was localized in the cytoplasm of cells with enrichment at the plasma membrane or subplasma membrane zone (191). Variations in secretory cells were seen in relation to feeding and starvation and metabolic activity (191). From these results, thioredoxin functions in relation to protein synthesis, intracellular transport, and different forms of secretion may be inferred (191).

Analysis of mouse pancreas revealed the presence of the thioredoxin system both in exocrine and endocrine cells (192; H. A. Hansson, A. Holmgren, B. Rozell, I.-B. Täljedal, unpublished). The β-cells of pancreatic islets contained thioredoxin and thioredoxin reductase surrounding the secretory granules containing insulin. In the brain, all the big neuronal cells and axons contained high levels of thioredoxin (191). The control of the expression of thioredoxin in mammalian cells is unknown, but not related to that of ribonucleotide reductase in normal and regenerating liver (193) or liver tumor cells (194).

So far, relatively little is known about specific functions of the thioredoxin system in mammalian cells, except for hydrogen donor functions for reductive enzymes. Regulation of CMP kinase (195, 196), initiation of protein synthesis in rabbit reticulocytes (197), and regulation of the glucocorticoid receptor activity (39, 40) have been demonstrated. A function of the thioredoxin system in alloxan toxicity and diabetogenic action has been suggested (198). The involvement of the thioredoxin system in protein disulfide formation and secretion has yet not been studied experimentally, although it may play an important role in these reactions (2, 199, 200). Particularly the possibility of ordered oxidation of thiols by the specificity of thioredoxin may be of importance. As yet, no studies of the possible involvement of thioredoxin in the life cycle of mammalian viruses have been reported.

EVOLUTIONARY ASPECTS

The present knowledge of thioredoxin suggests that this protein is ubiquitous in living cells and has the unique active center sequence -Cys-Gly-Pro-Cys- in a polypetide chain of about 100 residues ($M_r \sim 12,000$). From an evolutionary point of view, thioredoxin is a very old protein, present in procaryotes like cyanobacteria (178, 179). So far, the complete amino acid sequences of *E. coli* thioredoxin (14, 49) and *Corynebacterium nephridii* (102) have been reported to be about 50% homolgous. Determination of the complete amino acid and gene sequences for other thioredoxins should be of interest in tracing evolutionary relationships between species. The advantage of thioredoxin in such studies is its small size, and a function dependent only upon amino acid residues.

Development of cellular control mechanisms may be reflected both in the structure of thioredoxin, and in its mode of reduction and cellular localization. An example of this is thioredoxin f in spinach chloroplast (3, 168, 169). Regulation of protein functions via reversible oxidation of sulfhydyl groups to disulfides ("thiol redox control") (2) may be of considerable importance, perhaps comparable to protein phosphorylation. It will be important to determine the relative functions of thioredoxin and GSH-dependent systems (201, 202) in these processes in future studies.

Many catalytically active disulfides have the structure -Cys-XXX-YYY-Cys- (14, 90, 91, 98). This includes glutaredoxin from *E. coli* (129) and calf thymus (130). Thioredoxin from *E. coli* and glutaredoxin have essentially unrelated amino acid sequences (129) but may have similar three-dimensional structures (63). It is not known whether this is convergent evolution for functional reasons or reflects divergence from a distant evolutionary precursor.

CONCLUDING REMARKS

Despite extensive and in-depth work in many systems, rigorous genetic proofs of essential functions for thioredoxin only exist regarding its role as a subunit of phage T7 DNA polymerase (24) and its role in assembly of the single-stranded DNA phage f1 (42a). Even these functions of thioredoxin are so far little understood. Research to date leads to the following definition of thioredoxin: a well-characterized ubiquitous protein of unique structure, which exists in oxidized and reduced forms, and which has the capacity to catalyze dithiol-disulfide oxido-reductions of proteins.

The thioredoxin field has witnessed several surprising new insights, when a small heat-stable unknown protein, operationally defined, turned out to be identical with thioredoxin. The analogy of thioredoxin-$(SH)_2$ as a physiological, specific cellular equivalent of dithiothreitol is appealing. Also, the knowledge that thioredoxin, with its reversible reactions with NADPH/NADP$^+$ or dithiols/disulfides, may not only be a reductant, but also an oxidizing agent depending on its microenvironment, may help to identify new functions. It seems likely that evolution of complex organisms has included potentially useful aspects of the structure and function of thioredoxin, in particular for regulating integrated parts of metabolism and cellular differentiation. This leaves the impression that much work remains to be done to obtain a final picture of the functional roles of thioredoxin.

ACKNOWLEDGMENTS

Many colleagues generously provided preprints. The work from the author's laboratory was supported by the Swedish Cancer Society (961), the Swedish Medical Research Council (13X-3529), and the Knut and Alice Wallenberg Foundation.

Literature Cited

1. Laurent, T. C., Moore, E. C., Reichard, P. 1964. *J. Biol. Chem.* 239:3436–44
2. Holmgren, A. 1981. *Trends Biochem. Sci.* 6:26–29
3. Buchanan, B. B., Wolosiuk, R. A., Schürmann, P. 1979. *Trends Biochem. Sci.* 4:93–96
4. Holmgren, A. 1980. *Dehydrogenases Requiring Nicotinamide Coenzymes,* ed. J. Jeffery, pp. 149–80. Basel: Birkhauser Verlag
5. Gadal, P., ed. 1983. *Thioredoxins Structure and Function. Proc. Colloq. Int. CNRS-NASA, Berkeley, Calif., 1981,* pp. 1–288. Paris: CNRS
6. Holmgren, A. 1981. *Curr. Top. Cell. Regul.* 19:47–76
7. Thelander, L., Reichard, P. 1979. *Ann. Rev. Biochem.* 48:133–58
8. Lammers, M., Follmann, H. 1983. *Struct. Bonding* 54:27–91
9. Moore, E. C., Reichard, P., Thelander, L. 1964. *J. Biol. Chem.* 239:3445–52
10. Black, S., Harte, E. M., Hudson, B., Wartofsky, L. 1960. *J. Biol. Chem.* 235:2910–16
11. Wilson, L. G., Asahi, T., Bandurski, R. S. 1961. *J. Biol. Chem.* 236:1822–29
12. Gonzalez Porqué, P., Baldesten, A., Reichard, P. 1970. *J. Biol. Chem.* 245:2363–70
13. Gonzalez Porqué, P., Baldesten, A., Reichard, P. 1970. *J. Biol. Chem.* 245:2371–74
14. Holmgren, A. 1968. *Eur. J. Biochem.* 6:475–84
15. Hall, D. E., Baldesten, A., Holmgren, A., Reichard, P. 1971. *Eur. J. Biochem.* 23:328–35
16. Holmgren, A., Söderberg, B.-O. 1970. *J. Mol. Biol.* 54:387–90
17. Holmgren, A., Söderberg, B.-O., Eklund, H., Brändén, C.-I. 1975. *Proc. Natl. Acad. Sci. USA* 72:2305–9
18. Berglund, O., Karlström, O., Reichard,

P. 1969. *Proc. Natl. Acad. Sci. USA* 62:829–35
19. Berglund, O. 1969. *J. Biol. Chem.* 244:6306–8
20. Berglund, O., Sjöberg, B.-M. 1970. *J. Biol. Chem.* 245:6030–35
21. Sjöberg, B.-M., Holmgren, A. 1972. *J. Biol. Chem.* 247:8063–68
22. Söderberg, B.-O., Sjöberg, B.-M., Sonnerstam, U., Brändén, C.-I. 1978. *Proc. Natl. Acad. Sci. USA* 75:5827–30
23. Chamberlin, M. 1974. *J. Virol.* 14:509–16
24. Mark, D. F., Richardson, C. C. 1976. *Proc. Natl. Acad. Sci. USA* 73:780–84
25. Holmgren, A., Ohlsson, I., Grankvist, M.-L. 1978. *J. Biol. Chem.* 253:430–36
26. Holmgren, A. 1976. *Proc. Natl. Acad. Sci. USA* 73:2275–79
27. Mark, D. F., Chase, J. W., Richardson, C. C. 1977. *Mol. Gen. Genet.* 155:145–52
28. Holmgren, A. 1979. *J. Biol. Chem.* 254:9113–19
29. Holmgren, A. 1979. *J. Biol. Chem.* 254:9627–32
30. Schürmann, P., Wolosiuk, R. A., Breazeale, V. D., Buchanan, B. B. 1976. *Nature* 263:257–58
31. Wolosiuk, R. A., Buchanan, B. B. 1977. *Nature* 266:565–67
32. Holmgren, A., Buchanan, B. B., Wolosiuk, R. A. 1977. *FEBS Lett.* 82:351–54
33. Buchanan, B. B. 1980. *Ann. Rev. Plant Physiol.* 31:341–74
34. Herrmann, C. E., Moore, E. C. 1973. *J. Biol. Chem.* 248:1219–23
35. Engström, N. E., Holmgren, A., Larsson, A., Söderhäll, S. 1974. *J. Biol. Chem.* 249:205–10
36. Luthman, M., Holmgren, A. 1982. *Biochemistry* 21:6628–33
37. Holmgren, A. 1977. *J. Biol. Chem.* 252:4600–6
38. Rozell, B., Hansson, H.-A., Luthman,

M., Holmgren, A. 1985. *EMBO J*. In press

39. Grippo, J. F., Tienrungroj, W., Dahmer, M. K., Housley, P. R., Pratt, W. B. 1983. *J. Biol. Chem.* 258:13658–64

40. Grippo, J. F., Holmgren, A., Pratt, W. B. 1985. *J. Biol. Chem.* In press

41. Russel, M., Model, P. 1983. *J. Bacteriol.* 154:1064–76

42. Russel, M., Model, P. 1984. *J. Bacteriol.* 154:526–32

42a. Russel, M., Model, P. 1984. *Proc. Natl. Acad. Sci. USA*. In press

43. Lim, C.-J., Haller, B., Fuchs, J. A. 1985. *J. Bacteriol.* In press

44. Holmgren, A., Reichard, P. 1967. *Eur. J. Biochem.* 2:187–96

45. Williams, C. H., Lanetti, G., Arscott, L. D., McAllister, J. K. 1967. *J. Biol. Chem.* 242:5226–31

46. Sjöberg, B.-M., Holmgren, A. 1973. *Biochim. Biophys. Acta* 315:176–80

47. Pigiet, V. P., Conley, R. R. 1977. *J. Biol. Chem.* 252:6367–72

48. Lunn, C. A., Kathju, S., Wallace, B. J., Kushner, S. R., Pigiet, V. 1984. *J. Biol. Chem.* In press

49. Höög, J.-O., von Bahr-Lindström, H., Josephson, S., Wallace, B. J., Kushner, S. R., et al. 1984. *Bioscience Reports*. 4:917–23

50. Holmgren, A., Roberts, G. 1976. *FEBS Lett.* 71:261–65

51. Pigiet, V., Conley, R. R. 1978. *J. Biol. Chem.* 253:1910–20

52. Conley, R. R., Pigiet, V. 1978. *J. Biol. Chem.* 253:5568–72

53. Kallis, G.-B., Holmgren, A. 1980. *J. Biol. Chem.* 255:10261–65

54. Lunn, C. A., Skolnick, E. B., Pigiet, V. P. 1983. See Ref. 5, pp. 49–58

55. Fagerstedt, M., Holmgren, A. 1982. *J. Biol. Chem.* 257:6926–30

56. Lunn, C. A., Pigiet, V. P. 1982. *J. Biol. Chem.* 257:11424–30

57. Lunn, C. A., Pigiet, V. P. 1979. *J. Biol. Chem.* 254:5008–14

58. Bayer, M. E. 1979. In *Bacterial Outer Membranes: Biogenesis and Function*, ed. M. Inouye, pp. 167–202. New York: Wiley

59. Nygren, H., Rozell, B., Holmgren, A., Hansson, H.-A. 1981. *FEBS Lett.* 133:145–50

60. Holmgren, A. 1967. *Eur. J. Biochem.* 5:359–65

61. Holmgren, A. 1967. *Eur. J. Biochem.* 6:467–74

62. Söderberg, B.-O., Holmgren, A., Brändén, C.-I. 1974. *J. Mol. Biol.* 90:143–52

63. Eklund, H., Cambillau, C., Sjöberg,

B.-M., Holmgren, A., Jörnvall, H., et al. 1984. *EMBO J*. 3:1443–49

64. Rao, S. T., Rossmann, M. G. 1973. *J. Mol. Biol.* 76:241–56

65. Ohlsson, I., Nordström, B., Brändén, C.-I. 1974. *J. Mol. Biol.* 89:339–54

66. Stryer, L., Holmgren, A., Reichard, P. 1967. *Biochemistry* 6:1016–20

67. Reutimann, H., Straub, B., Luisi, P.-L., Holmgren, A. 1981. *J. Biol. Chem.* 256:6796–803

68. Holmgren, A. 1972. *J. Biol. Chem.* 247:1992–98

69. Holmgren, A. 1973. *J. Biol. Chem.* 248:4106–11

70. Holmgren, A. 1981. *Biochemistry* 20:3204–7

71. Holmgren, A. 1972. *Eur. J. Biochem.* 26:528–34

72. Nordström, B., Randahl, H., Slaby, I., Holmgren, A. 1981. *J. Biol. Chem.* 256:3112–17

73. Holmgren, A., Kallis, G.-B., Nordström, B. 1981. *J. Biol. Chem.* 256:3118–24

74. Holmgren, A. 1972. *FEBS Lett.* 24:351–54

75. Slaby, I., Holmgren, A. 1975. *J. Biol. Chem.* 250:1340–47

76. Slaby, I., Holmgren, A. 1979. *Biochemistry* 18:5584–91

77. Holmgren, A., Slaby, I. 1979. *Biochemistry* 18:5591–99

78. Reutimann, H., Luisi, P.-L., Holmgren, A. 1983. *Biopolymers* 22:107–11

79. McEvoy, M., Lantz, C., Lunn, C. A., Pigiet, V. 1981. *J. Biol. Chem.* 256:6646–50

80. Kelley, R., Stellwagen, E. 1984. *Biochemistry*, In press

81. Kelley, R., Stellwagen, E. 1985. *Biochemistry*, In press

82. Kishore, R. K., Mathew, M. K., Balaram, P. 1983. *FEBS Lett.* 159:221–24

83. Ravi, A., Balaram, P. 1983. *Biochim. Biophys. Acta* 745:301–9

84. Brown, S., Albrechtsen, B., Pedersen, S., Klemm, P. 1982. *J. Mol. Biol.* 162:283–98

85. Pinkham, J. L., Platt, T. 1983. *Nucleic Acid. Res.* 11:3531–45

86. Clarke, L., Carbon, J. 1976. *Cell* 9:91–99

87. Neidhardt, F. C., Vaughn, V., Phillips, T. A. 1984. *Microbiol. Rev.* 47:231–84

88. Thelander, L. 1967. *J. Biol. Chem.* 852–59

89. Williams, C. H. Jr. 1976. *Enzymes* 13:89–173

90. Thelander, L. 1968. *Eur. J. Biochem.* 4:407–22

91. Ronchi, S., Williams, C. H. Jr. 1972. *J. Biol. Chem.* 247:2083–86
92. Thelander, L. 1970. *J. Biol. Chem.* 245:6026–29
93. Zanetti, G., Williams, C. H. Jr. 1967. *J. Biol. Chem.* 242:5232–36
94. O'Donnell, M. E., Williams, C. H. Jr. 1983. *J. Biol. Chem.* 258:13795–805
95. O'Donnell, M. E., Williams, C. H. Jr. 1984. *J. Biol. Chem.* 259:2243–51
96. Fuchs, J. 1977. *J. Bacteriol.* 129:967–72
97. Haller, B. L., Fuchs, J. A. 1984. *J. Bacteriol.* In press
98. Carlson, J., Fuchs, J. A., Messing, J. 1984. *Proc. Natl. Acad. Sci. USA.* In press
99. Thelander, L. 1974. *J. Biol. Chem.* 249:4858–62
100. Holmgren, A. 1979. *J. Biol. Chem.* 254:3664–71
101. Holmgren, A. 1979. *J. Biol. Chem.* 254:3672–78
102. Meng, M., Hogenkamp, H. P. C. 1981. *J. Biol. Chem.* 256:9174–82
103. Hopper, S., Iurlano, D. 1983. *J. Biol. Chem.* 258:13453–57
104. Luthman, M., Eriksson, S., Holmgren, A., Thelander, L. 1979. *Proc. Natl. Acad. Sci. USA* 76:2158–62
105. Luthman, M., Holmgren, A. 1982. *J. Biol. Chem.* 257:6686–90
106. Ejiri, S.-I., Weissbach, H., Brot, N. 1979. *J. Bacteriol.* 139:161–64
107. Ejiri, S.-I., Weissbach, H., Brot, N. 1980. *Anal. Biochem.* 102:393–98
108. Carlsson, J., Fuchs, J. A. 1983. See Ref. 5, pp. 111–17
109. Brot, N., Weissbach, L., Werth, J., Weissbach, H. 1981. *Proc. Natl. Acad. Sci. USA* 78:2155–58
110. Anders, M. W., Ratnayake, J. H., Hanna, P. E., Fuchs, J. A. 1980. *Biochem. Biophys. Res. Commun.* 97:846–51
111. Anders, M. W., Ratnayake, J. H., Hanna, P. E., Fuchs, J. A. 1981. *Drug Metab. Dispos.* In press
112. Abrams, W. R., Weinbaum, G., Weissbach, L., Weissbach, H., Brot, N. 1981. *Proc. Natl. Acad. Sci. USA* 78:7483–86
113. Spector, A., Scotto, R., Weissbach, H., Brot, N. 1982. *Biochem. Biophys. Res. Commun.* 108:429–34
114. Cooper, A. J. L. 1983. *Ann. Rev. Biochem.* 52:187–222
115. Tsang, M. L.-S., Schiff, J. A. 1976. *J. Bacteriol.* 125:923–33
116. Tsang, M. L.-S. 1983. See Ref. 5, pp. 103–10
117. Creighton, T. E. 1978. *Prog. Biophys. Mol. Biol.* 33:231–97
118. Jocelyn, P. C. 1972. *Biochemistry of the SH Group,* pp. 1–404. London: Academic
119. Holmgren, A. 1984. *Methods Enzymol.* 107:295–300
120. Holmgren, A., Morgan, F. J. 1976. *Eur. J. Biochem.* 70:377–83
121. Blombäck, B., Blombäck, M., Finkbeiner, W., Holmgren, A., Kowalska-Loth, B., Olovson, G. 1974. *Thromb. Res.* 4:55–75
122. Savidge, G., Carlebjörk, G., Thorell, L., Hessel, B., Holmgren, A., Blombäck, B. 1979. *Thromb. Res.* 16:587–99
123. Hessel, B., Jörnvall, H., Thorell, L., Söderman, S., Larsson, K., et al. 1984. *Thromb. Res.* 35:637–51
124. Sugo, T., Björk, I., Holmgren, A., Stenflo, J. 1984. *J. Biol. Chem.* 259:5705–10
125. Yeh, Y. C., Dubovi, E. J., Tessman, I. 1969. *Virology* 37:615–23
126. Yeh, Y. C., Tessman, I. 1972. *Virology* 47:767–72
127. Holmgren, A., Sjöberg, B.-M. 1972. *J. Biol. Chem.* 247:4160–64
128. Holmgren, A. 1978. *J. Biol. Chem.* 253:7424–30
129. Höög, J.-O., Jörnvall, H., Holmgren, A., Carlquist, M., Persson, M. 1983. *Eur. J. Biochem.* 136:223–32
130. Klintrot, I.-M., Höög, J.-O., Jörnvall, H., Holmgren, A., Luthman, M. 1984. *Eur. J. Biochem.* In press
131. Berglund, O., Holmgren, A. 1975. *J. Biol. Chem.* 250:2778–82
132. Mathews, C. K., Allen, J. R. 1983. In *Bacteriophage T4,* ed. C. K. Mathews, E. M. Kutter, G. Mosig, P. B. Berget, pp. 59–70. Washington, DC: Am. Soc. Microbiol.
133. Eriksson, S., Berglund, O. 1974. *Eur. J. Biochem.* 46:271–78
134. Dunn, J. J., Studier, F. W. 1983. *J. Mol. Biol.* 166:477–535
135. Richardson, C. C. 1983. In *Developments in Molecular Virology,* ed. Y. Becker, The Hague: Nijkoff, 2:163–204
136. Richardson, C. C. 1983. *Cell* 33:315–17
137. Lehman, I. R. 1981. *Enzymes* 14:51–65
138. Nossal, N. G. 1983. *Ann. Rev. Biochem.* 52:581–615
139. Modrich, P., Richardson, C. C. 1975. *J. Biol. Chem.* 250:5508–14
140. Modrich, P., Richardson, C. C. 1975. *J. Biol. Chem.* 250:5515–22
141. Hori, K., Mark, D. F., Richardson, C. C. 1979. *J. Biol. Chem.* 254:11591–97
142. Adler, S., Modrich, P. 1979. *J. Biol. Chem.* 254:11605–14
143. Hori, K., Mark, D. F., Richardson, C. C. 1979. *J. Biol. Chem.* 254:11598–604

144. Randahl, H., Slaby, I., Holmgren, A. 1982. *Eur. J. Biochem.* 128:445–49
145. Adler, S., Modrich, P. 1983. *J. Biol. Chem.* 258:6956–62
146. Randahl, H., Holmgren, A. 1985. *J. Biol. Chem.* In press
147. Harth, G. H., Geider, K., Schürmann, P., Tsugita, A. 1981. *FEBS Lett.* 136:37–40
148. Denhardt, D. T. 1975. *Crit. Rev. Microbiol.* 4:161–223
149. Hill, D. F., Petersen, G. B. 1982. *J. Virol.* 44:32–46
150. Messing, J. B., Gronenborn, B., Müller-Hill, B., Hofschneider, P. H. 1977. *Proc. Natl. Acad. Sci. USA* 74:3642–46
151. Sanger, F., Coulson, A. R., Barrell, B. G., Smith, A. J. H., Roe, B. A. 1980. *J. Mol. Biol.* 143:161–78
152. Messing, J. 1983. *Methods Enzymol.* 101:20–78
153. Lopez, J., Webster, R. E. 1983. *Virology* 127:177–93
154. Russel, M., Model, P. 1982. *Cell* 28:177–84
155. Pedersen, T. A., Kirk, M., Bassham, J. M. 1966. *Plant. Physiol.* 19:219–31
156. Sweeney, W. V., Rabinowitz, J. C. 1980. *Ann. Rev. Biochem.* 49:139–61
157. Breazeale, V. G., Buchanan, B. B., Wolosiuk, R. A. 1978. *Z. Naturforsch. Teil C* 33:521–28
158. Wolosiuk, R. A., Buchanan, B. B. 1978. *Plant Physiol.* 61:669–71
159. Wolosiuk, R. A., Buchanan, B. B. 1978. *Arch. Biochem. Biophys.* 189:97–101
160. Jacquot, J. P., Vidal, J., Gadal, P. 1976. *FEBS Lett.* 71:223–27
161. Wolosiuk, R. A., Buchanan, B. B., Crawford, N. A. 1977. *FEBS Lett.* 81:253–58
162. Nishizawa, A. N., Wolosiuk, R. A., Buchanan, B. B. 1979. *Planta* 145:7–12
163. McKinney, D. W., Buchanan, B. B., Wolosiuk, R. A. 1979. *Biochem. Biophys. Res. Commun.* 86:1178–84
164. Mills, J. D., Mitchell, P., Schürmann, P. 1980. *FEBS Lett.* 112:173–77
165. Wu, G-Y., Den, Y-F., Wu, X-Y. 1980. *Acta Bot. Sinica* 22:241–46
166. Jacquot, J. P., Vidal, J., Gadal, P., Schürmann, P. 1978. *FEBS Lett.* 96:243–46
167. Wolosiuk, R. A., Crawford, N. A., Yee, B. C., Buchanan, B. B. 1979. *J. Biol. Chem.* 254:1627–32
168. Schürmann, P., Maeda, K., Tsugita, A. 1981. *Eur. J. Biochem.* 116:37–45
169. Tsugita, A., Maeda, K., Schürmann, P. 1983. *Biochem. Biophys. Res. Commun.* 115:1–7
170. Hertig, C. M., Wolosiuk, R. A. 1983. *J. Biol. Chem.* 258:984–89
171. Pla, A., Lopez-Gorge, J. 1981. *Biochim. Biophys. Acta* 636:113–18
172. Berstermann, A., Vogt, K., Follmann, H. 1983. *Eur. J. Biochem.* 131:339–44
173. Sushe, G., Wagner, W., Follmann, H. 1979. *Z. Naturforsch. Teil C* 34:214–21
174. de la Torre, A., Lara, C., Wolosiuk, R. A., Buchanan, B. B. 1979. *FEBS Lett.* 107:141–145
175. Schürmann, P. 1981. *Proc. 5th Int. Congr. Photosynthesis. Regulation of Carbon Metabolism,* ed. G. Akogunoglou, 4:273–80. Philadelphia: Balaban Int. Sci. Serv.
176. Hammel, K. E., Cornwell, K. L., Buchanan, B. B. 1983. *Proc. Natl. Acad. Sci. USA* 80:3681–85
177. Wagner, W., Follmann, H. 1977. *Biochem. Biophys. Res. Commun.* 77:1044–51
178. Schmidt, A., Christen, U. 1979. *Z. Naturforsch. Teil C.* 34:1272–74
179. Gleason, F., Holmgren, A. 1981. *J. Biol. Chem.* 256:8306–9
180. Papen, H., Bothe, H. 1984. *FEMS Microbiol. Lett.* 23:41–46
181. Schmidt, A. 1981. *Z. Naturforsch. Teil C* 36:396–99
182. Cossar, J. D., Rowell, P., Stewart, W. D. P. 1984. *J. Gen. Microbiol.* 130:991–98
183. Udvardy, J., Borbely, G., Juhasz, A., Farkas, G. L. 1984. *J. Bacteriol.* 157:681–83
184. Clement-Metral, J. D. 1979. *FEBS Lett.* 101:116–120
185. Clement-Metral, J. D., Holmgren, A. 1983. See Ref. 5, pp. 59–68
186. Guevara, J. Jr., Moore, E. C., Ward, D. N. 1983. See Ref. 5, pp. 79–83
187. Chen, C.-C., BornsMcCall, B. L., Moore, E. C. 1977. *Prep. Biochem.* 7:165–77
188. Chen, C.-C., Moore, E. C., BornsMcCall, B. L. 1978. *Cancer Res.* 38:1885–88
189. Tsang, M.-L., Weatherbee, J. A. 1981. *Proc. Natl. Acad. Sci. USA* 78:7478–82
190. Holmgren, A., Luthman, M. 1978. *Biochemistry* 17:4071–77
191. Rozell, B., Hansson, H.-A., Luthman, M., Holmgren, A. 1985. *Eur. J. Cell Biol.* In press
192. Grankvist, K., Holmgren, A., Luthman, M., Täljedal, I.-B. 1982. *Biochem. Biophys. Res. Commun.* 107:1412–18
193. Larsson, A. 1973. *Eur. J. Biochem.* 35:346–49
194. Elford, H. L., Freese, M., Passamani,

E., Morris, H. P. 1970. *J. Biol. Chem.* 245:5228–33

195. Maness, P., Orengo, A. 1975. *Biochemistry* 14:1484–89

196. Kobayashi, S., Kanayama, K. 1977. *Biochem. Biophys. Res. Commun.* 74:1249–55

197. Hunt, T., Herbert, P., Campbell, E. A., Delidakis, C., Jackson, R. J. 1983. *Eur. J. Biochem.* 131:303–311

198. Holmgren, A., Lyckeborg, C. 1980. *Proc. Natl. Acad. Sci. USA* 77:5149–52

199. Freedman, R. B. 1979. *FEBS Lett.* 97:201–210

200. Freedman, R. B., Hillson, D. A. 1980. In *The Enzymology of Post-translational Modification of Proteins,* ed. R. B. Freedman, H. C. Hawkins, pp. 157–212. London: Academic

201. Mannervik, B., Axelsson, K., Sundewall, A.-C., Holmgren, A. 1983. *Biochem. J.* 213:519–23

202. Meister, A., Anderson, M. E. 1983. *Ann. Rev. Biochem.* 52:711–60

Ann. Rev. Biochem. 1985. 54:273–304
Copyright © 1985 by Annual Reviews Inc. All rights reserved

THE MEMBRANE SKELETON OF HUMAN ERYTHROCYTES AND ITS IMPLICATIONS FOR MORE COMPLEX CELLS

Vann Bennett

Department of Cell Biology and Anatomy, Johns Hopkins University School of Medicine, Baltimore, Maryland 21205

CONTENTS

PERSPECTIVES AND SUMMARY

Membrane-cytoskeletal associations have been discussed in general terms for over a decade as an important feature of many phenomena in cell biology. For example, linkages between membrane-spanning integral proteins and cytoplasmic structural proteins are involved in attachment of cells to surfaces, cell

273

motility, movement of intracellular organelles, and organization of membrane proteins on cell surfaces (1–7). The human erythrocyte has provided a simple, experimentally accessible system for study of membrane-cytoskeletal interactions at a biochemical level. Advances in understanding the red cell membrane have provided the first detailed "map" of the protein linkages involved in attachment of actin to a plasma membrane. Actin filaments are associated in a ternary complex with band 4.1 and spectrin. Spectrin is a flexible, rod-shaped molecule 200 nm in length and forms a meshwork underlying the cytoplasmic surface of the plasma membrane. Spectrin is attached to the membrane bilayer by association with ankyrin. Ankyrin, in turn, is associated with the cytoplasmic domain of the anion transporter, which is an integral membrane protein. Each of the linking proteins have been purified and their associations with neighboring proteins described quantitatively.

In recent developments, spectrin and the anion channel protein have been partially sequenced, and the domain structure of ankyrin, and band 4.1 has been resolved. An additional association between spectrin and the membrane by linkage of band 4.1 to glycoproteins has been suggested. Erythrocyte forms of myosin and tropomyosin have been purified and offer new possibilities for cytoskeletal control of erythrocyte shape.

The mechanism of interaction of actin with erythrocyte membranes has direct relevance to more complex and biologically more interesting cells since analogs of erythrocyte membrane proteins have been discovered in many other tissues. Proteins closely related to spectrin, ankyrin, and band 4.1 have been purified from membranes of brain and other tissues, and immunoreactive forms of the anion channel have been detected. Tissue spectrin is present underneath caps of surface proteins in lymphocytes, postsynaptic densities, and at nodes of Ranvier, and is in the right place to participate in interactions with membrane receptors and ion channels. The list of protein associations of tissue forms of spectrin and ankyrin is not limited to the familiar proteins of erythrocytes. Ankyrin is a tubulin-binding protein, while spectrin interacts in some way with microtubules, intermediate filaments, and a protein that is a major substrate for the sarc kinase. Band 4.1 in brain is identical to a protein named synapsin, which is localized on secretory vesicles in nerve terminals and is a major substrate for cyclic AMP and Ca/calmodulin-regulated protein kinases. The lesson emerging from these studies is that analogs of erythrocyte membrane proteins will have diverse roles in integrating various structural proteins of the cytoplasmic matrix with plasma membranes and intracellular membrane systems.

Understanding the organization of erythrocyte membrane proteins also has helped explain the biochemical basis for certain inherited hemolytic anemias in humans and mice. Hereditary elliptocytosis is associated in some cases with a lower efficiency of spectrin tetramer assembly from dimers, and several families with this disorder have a deficiency in band 4.1. Hereditary spherocytosis in mice is a severe anemia associated with partial to nearly complete lack of spectrin. In humans, spherocytosis also is associated with deficient spectrin, with a close correlation between severity of the disease and the extent of spectrin deficiency.

This chapter will cover the organization and biochemistry of membrane-cytoskeletal proteins in normal and abnormal erythrocytes, and also will review the rapidly growing literature regarding closely related forms of erythrocyte membrane proteins in other tissues. Other recent reviews in these areas also are available (8–17).

STRUCTURAL PROTEINS OF HUMAN ERYTHROCYTE MEMBRANES

The erythrocyte is an unusually durable cell, surviving thousands of passes through the circulation during its 120 day life. The basis for the resilience of this cell is the mechanical properties of the plasma membrane, since the usual cytoplasmic structural proteins and organelles have been lost in differentiation. Direct measurements of deformation of erythrocytes have demonstrated that the membrane behaves like a semisolid with elastic properties which are not observed with simple lipid vesicles (18–20). It is this ability of erythrocyte membranes to store energy during brief periods of deformation, such as occur in the circulation, and then return to their original shape that provides these cells with their longevity.

The structural component of the erythrocyte responsible for the elastic properties is a membrane-associated assembly of proteins commonly (and somewhat inaccurately) referred to as the cytoskeleton or membrane skeleton. This structure was discovered by the simple maneuver of extracting erythrocyte ghosts with a nonionic detergent that removed phospholipids and integral proteins (21). A meshwork of proteins remained after solubilization of the traditional membrane components, and this meshwork retained the same shape as the original ghosts. The cytoskeleton has been visualized as a two-dimensional meshwork of filaments about 100–140 nm in length (22–24). The principal protein components remaining in the cytoskeleton after high salt extraction are spectrin, band 4.1, erythrocyte actin, and band 4.9 (25, 26) (nomenclature based on mobility of proteins on SDS-gels; see Table 1 for a list of membrane proteins).

Several lines of evidence indicate the importance of the cytoskeleton for maintaining the integrity of the erythrocyte membrane. Extraction of cytoskeletal components spectrin and actin from ghosts at low ionic strength is accompanied by disintegration of ghosts into small vesicles. Strains of anemic mice have been developed (27, 28) that have substantial reduction in the

Table 1 Major Proteins in Human Red Cell Membranes

Protein	Subunit M_r	Probable Assembly State	Approximate Copies/Cell
Peripheral Proteins			
Spectrin	α = 260,000	$(\alpha,\beta)_2$ tetramer	10^5 tetramers
	β = 225,000		
Ankyrin	215,000	Monomer	10^5
Band 4.1[a]	78,000	—	2×10^5
Band 4.2[a]	72,000	—	2×10^5
Band 4.9[a]	45,000	—	5×10^4
Actin[b]	43,000	Oligomer of 12–17 Subunits	5×10^5
Glyceraldehyde 3-Phosphodehydrogenase	35,000	Tetramer	5×10^5
Band 7[a]	29,000	—	5×10^5
Band 8[a]	23,000	—	10^5
Tropomyosin	29,000	Dimer	7×10^4 dimers
Integral Proteins			
Band 3[a]	89,000	Dimer/Tetramer	10^6
Glycophorin A[b]	31,000	Dimer	4×10^5
Glycophorin B[c]	23,000	—	$\sim 10^5$
Glycophorin C[c] (Glycoconnectin)	29,000	—$\sim 10^5$	

[a]Nomenclature based on electrophoretic mobility (31).
[b]Data from (112).
[c]Nomenclature of (249); estimated assuming 60% carbohydrate for all 3 glycophorins (9).

amount of spectrin in their erythrocytes (29, 30). Erythrocytes from these animals are extremely fragile, and in the severely affected strains the erythrocytes lyse soon after entering the circulation. Finally, humans with hereditary hemolytic anemias have abnormally shaped or fragile erythrocytes and have been discovered to have defects in amounts or function of cytoskeletal proteins (see section below).

Many of the major structural proteins of erythrocyte membranes have been purified and studied in detail. The properties of these proteins will be discussed below.

Integral Proteins

The erythrocyte membrane contains two major proteins, band 3 and glycophorin A, that span the phospholipid bilayer with distinct domains expressed on outer and cytoplasmic surfaces (31–33). Band 3 [for current reviews, see (32, 34)] is a glycoprotein present in about 10^6 copies per cell and contains an anion transport channel (35). An important feature of band 3, in terms of the cytoskeleton, is that this membrane protein contains a large water-soluble domain of $M_r \sim 43,000$ which is localized on the inner surface of the membrane (36–38). This cytoplasmic domain of band 3 can be purified easily in quantities of 10–20 mg from digests of inside-out vesicles (37–39). The domain is associated with glycolytic enzymes, aldolase, phosphofructokinase and glyceraldehyde-3-phosphodehydrogenase as well as with a $M_r = 72,000$ polypeptide (band 4.2) of unknown function (32, 34). The binding site for the glycolytic enzymes is localized within 75 residues of the amino terminus (40, 41). The cytoplasmic region of band 3 also interacts with hemoglobin (42, 43). The band 3 binding site of hemoglobin has been identified as the 2,3 DPG site expressed in deoxyhemoglobin (44). The cytoplasmic domain of band 3 is attached to ankyrin which in turn is linked to spectrin and the cytoskeleton (26, 37, 45, 46) (see below). Band 3 is a stable homodimer in detergent extracts as well as erythrocyte ghosts (47–49). Band 3 dimers associate to form tetramers or higher oligomers in the membrane (50), and in solution (51, 52). The first 200 residues containing the amino terminus of the cytoplasmic domain have been sequenced (53). The residues at the amino terminus are extremely acidic, with 18 amino acids out of 33 being aspartic or glutamic acid, and no basic amino acids. The carboxyl terminal two-thirds of the fragment, in contrast, is relatively basic. Band 3 contains a single sequence, at least in this region, and thus is not a mixture of closely related polypeptides.

Glycophorin A (PAS bands 1 and 2) is the major sialic acid-containing glycoprotein, and contains some blood group antigens and binding sites for lectins and viruses (33). Glycophorin also may contain a recognition site that allows invasion of erythrocytes by the malarial parasite *Plasmodium falciparum* (54, 55), although band 3 also functions in this process (56). Glycophorin A has been sequenced (33) and the portion of the polypeptide chain exposed on the cytoplasmic membrane surface has been determined (57). Unlike band 3, the cytoplasmic domain of glycophorin is small ($M_r = 3,000$). Glycophorin incorporated into liposomes binds to band 4.1, presumably at this site (58). Glycophorin A is a stable dimer, even in the presence of sodium dodecyl sulphate. Glycophorin is thought to be associated with band 3 in erythrocyte membranes based on the indirect evidence that addition of antibody against glycophorin slows the rotational diffusion rate of band 3 (59). The linkage between band 3 and glycophorin is not maintained after these proteins have been solubilized in nonionic detergents (48). Band 3 and glycophorin are major constituents of the intramembrane particles visualized by freeze-fracture electron microscopy (60, 61). The function of glycophorin A is not clear, since a family has been studied that lacks glycophorin A entirely, and yet has normal erythrocytes (62). These cells still have minor sialoglycoproteins (glycophorins B and C) which could compensate for lack of glycophorin A.

Membrane Skeletal Proteins

About 50% of erythrocyte membrane proteins are water-soluble proteins local-ized on the cytoplasmic surface of the membrane. Some of these proteins participate as structural components of the membrane skeleton and will be discussed here.

SPECTRIN   Spectrin (10^5 tetramers/cell), is the major protein of the membrane skeleton. Spectrin was discovered in low salt extracts of erythrocyte ghosts (63) and can easily be purified in amounts of 50–100 milligrams (39). Spectrin, as viewed by electron microscopy, is a flexible, rod-shaped molecule about 100 nm in length composed of two parallel polypeptide chains of M_r = 260,000 (alpha chain) and 225,000 (beta chain) (64). The subunits have been isolated by chromatography in urea (65, 66) or after SDS-denaturation (67). Spectrin heterodimers self-associate at one end of the molecule to form tetramers of 200 nm in length. It has been proposed (64, 65, 68) that in the tetramer, each alpha chain is associated by head-head linkage with a beta chain, so that no alpha-alpha or beta-beta contacts occur. The tetramer is most likely the major form of spectrin in ghosts based on chemical cross-linking experiments (69), presence of spectrin tetramer in membrane extracts prepared under mild conditions (70), and characterization of the dimer-tetramer equilibrium (71). It has also been proposed that higher oligomers of spectrin such as hexamers, octomers, etc may also be present in membranes and account for the polymerization of spectrin into a meshwork (72, 73).

Spectrin has specific binding sites for a number of proteins. Spectrin binds to F-actin by lateral association (74–77) and this association is promoted by band 4.1 (4.1a and 4.1b) (75–79). Spectrin also binds directly to band 4.1 in the absence of actin with a K_D of 10^{-7} M and a stoichiometry of two 4.1 molecules per spectrin dimer (68). The alpha and beta subunits of spectrin each have a binding site for band 4.1, thus explaining the 2:1 stoichiometry (80). The measurements of binding of spectrin and band 4.1 have been reproduced in other laboratories (81, 82). The binding sites for actin and band 4.1 have been localized to the same region at the tails of spectrin tetramers by electron microscopy (68, 76, 83). It is likely that all three proteins are associated in a ternary complex at this site (84, 85). An equilibrium constant of 10^{-12} M^2 has been estimated for the spectrin-actin-band 4.1 complex (85). Spectrin also has a binding site (one per dimer) for ankyrin (see below) and these sites are localized symmetrically at the midregion of tetramers about 40 nm from each other (83).

In addition to interactions with structural proteins, spectrin binds to calmod-ulin, although with low affinity ($K_D \sim 5 \mu$M) (86). The association of spectrin and calmodulin was dependent on calcium, but was measured in 6 M urea. In another study, binding of spectrin dimer to calmodulin was detected under mild conditions (87). The physiological significance of the weak spectrin-calmodulin interaction is not known, but it should be kept in mind that calmodulin is present in cytosol at micromolar levels (88). Furthermore, spectrin from other tissues (see below) also binds calmodulin, although with higher affinity.

The domains of spectrin responsible for some of the protein associations have been identified (89–91). An 80,000 M_r polypeptide from the amino-terminal end of the alpha chain contains a binding site for the beta subunit that is required for formation of tetramers. The binding site for ankyrin is localized on the beta subunit (65, 67), and a 50,000 M_r fragment of the beta chain has been purified that contains the ankyrin-binding site (90). Binding sites for lateral association between alpha and beta subunits have also been identified and mapped (90). Binding of spectrin to actin requires both alpha and beta subunits since isolated subunits were inactive (65, 80).

A partial amino acid sequence of spectrin alpha and beta subunits has been determined by protein sequencing (92–94). This daunting task has yielded several important insights into the structure and possible evolutionary origin of spectrin. Most of the spectrin alpha subunit is composed of homologous segments with a 106–amino acid repeating unit. The beta subunit also has a 106–amino acid repeating unit, and these are homologous to those of the alpha chain. The two subunits are arranged antiparallel with respect to each other, with the amino terminus of the alpha chain and carboxyl terminus of the beta chain at the tetramer-forming end, and the corresponding termini at the opposite end. Spectrin thus may have evolved by a series of gene duplications and gene fusion steps beginning with a 106–amino acid primordial protein (94). The alpha and beta subunits appear to have initially evolved together, and then diverged to the point that they now associate laterally with each other but do not form homodimers. Spectrin has properties in common with other rod-shaped actin binding proteins such as filamin, alpha-actinin, and microtubule-associated protein 2 (MAP2). It will be of interest to determine if these proteins also have a repeating unit of approximately 12,000 daltons and are homologous to spectrin. Sequence homology between spectrin and MAP2 is likely, since these proteins share antigenic determinants (95).

Spectrin, like many proteins in the erythrocyte membrane, is phosphorylated as isolated from ghost membranes. The four sites of phosphorylation have been mapped to the carboxyl-terminal end of the beta subunit (96). The phosphory-lated end of the beta chain is the region of the molecule where end-to-end association with alpha chain occurs in tetramer assembly, and also is close to the ankyrin binding site. The functional importance of beta chain phosphoryla-tion is not known, since removal of phosphates by phosphatase or protease has no obvious effect on the dimer-tetramer-equilibrium (71), binding of spectrin to F-actin (74), or binding of spectrin to ankyrin (97).

ANKYRIN   Ankyrin is a monomeric phosphoprotein of $M_r = 215,000$ that is present in about 10^5 copies per cell (98) and links spectrin to the cytoplasmic domain of band 3 (see below). Ankyrin has been purified from high salt extracts of inverted vesicles (83), and from detergent-extracted cytoskeletons (39, 99). The physical properties of ankyrin have been characterized ($R_s = 58$ Å, $S_{20,w} = 6.9s$; f/fo = 1.46; somewhat hydrophobic although soluble in the absence of detergents). The spectrin-binding site is in an $M_r = 72,000$ chymotryptic fragment (100, 101) while the band 3–binding site is localized in an $M_r = 90,000$ domain (102). More extensive digestions with trypsin have resolved the spectrin-binding region to a polypeptide of $M_r = 55,000$, and the band 3–binding site to a basic domain of $M_r = 82,000$ (46, 46a). The phosphorylation sites are localized in the spectrin-binding domain (46, 101), although phosphorylation has little effect on spectrin binding (V. Bennett, unpublished data). Ankyrin binds to microtubules with an apparent K_D of 2 μM, and also binds to unpolymerized tubulin dimers (102–104). The tubulin-binding domain of ankyrin is in the same 90,000 M_r fragment that contains the band 3–binding site (102). The tubulin-binding activity of ankyrin is not utilized in mature erythrocytes which lack tubulin. However, tubulin is present as a membrane-associated marginal band in mammalian erythroblasts and primitive circulating fetal erythrocytes (105) as well as mature nucleated erythrocytes of all vertebrates except mammals. Ankyrin in these erythrocytes and ankyrin in brain (102) may mediate linkage of tubules to the plasma membrane. Ankyrin shares antigenic determinants with microtubule-associated protein 1 in brain (103) and a microtubule-associated protein in the mitotic spindle (103, 106), and thus may be a member of a family of tubulin-binding proteins.

ACTIN   Erythrocyte actin was first discovered in 1962 (107) and has been studied more recently in detail (108–110). Purified erythrocyte actin is capable of polymerizing to form 7 nm filaments, activates myosin ATPase, and has all of the properties of other cellular actins. In spite of these similarities to other actins, erythrocyte actin has not been visualized as extended filaments in cells or ghosts even though the concentration of 270 μg/ml is well above its critical concentration for polymerization in solution. Actin is thought, instead, to exist as oligomers of 12–17 subunits (111, 112). Short actin filaments have been visualized in erythrocyte membrane skeletons by electron microscopy (112a). The slow-growing end of membrane actin filaments may be capped since this end does not participate in monomer exchange or support growth of actin filaments in the presence of cytochalasin (111–113). The basis for stable actin oligomers probably involves accessory actin-binding proteins. Band 4.9, for example, which is another cytoskeletal protein, binds to actin and may fragment and/or bundle existing actin filaments (114). Tropomyosin has also been discovered in erythrocyte membranes (115, 116) and the association of this protein could stabilize short filaments, or regulate association of actin with proteins such as spectrin or band 4.9.

Another unexplained feature of erythrocyte actin is that it is present as the beta isoform instead of a mixture of isoelectric variants as in other cells (117). Since erythroblasts presumably have beta and gamma actins before terminal differentiation, this suggests that some feature of beta-actin strongly favors its association with the membrane in erythrocytes relative to other isoforms. It will be important in future studies of actin interactions in erythrocytes to use erythrocyte actin. Erythrocyte actin can be purified in milligram amounts (116), and as a pure beta isoform, also will be useful for studies of beta-actin function in other cells.

BAND 4.1   Band 4.1 includes two polypeptides of M_r = 78,000 and 80,000 closely related in sequence (118, 119). Band 4.1 has been purified from high salt extracts of inverted vesicles (83), and is a monomer in solution (119a). Band 4.1 binds to spectrin at the same region as actin at the tails of spectrin tetramers (see above). Band 4.1 also interacts directly with actin and severs F-actin in vitro (120). Band 4.1 has a membrane-binding site in addition to spectrin. The identity of the 4.1 binding protein is under investigation and two candidates are glycophorin A (58) or another glycoprotein named glycoconnectin (121) (see below).

Structural analysis of 4.1 indicates that this protein has an acidic domain of M_r = 48,000 and a basic region of M_r = 30,000 that contains the spectrin-binding domain (119). Band 4.1 is phosphorylated by a cyclic AMP-dependent protein kinase at site(s) in the acidic region (119), and by a phorbol ester-activated protein kinase (122). As is generally the case at this time for other erythrocyte phosphoproteins, no function for phosphorylation is known.

TROPOMYOSIN   Tropomyosin is a component of erythrocyte ghosts isolated in the presence of 1 mM $MgCl_2$ (115, 116). Erythrocyte tropomyosin has been purified using procedures developed for other tropomyosins and is identical to nonmuscle tropomyosins in terms of its properties that include (a) amino acid composition, (b) two subunits of M_r = 29,000 and 27,000 that are associated as dimers, (c) physical properties (R_s = 59 Å, $S_{20,w}$ = 2.5 S, frictional ratio of 2.07, indicating a highly asymmetric shape, (d) ability to bind to F-actin at a stoichiometry of 1 tropomyosin dimer per 6–7 actin monomers. Erythrocyte tropomyosin binds to actin at lower concentrations of Mg^{2+} (2 mM) than other nonmuscle tropomyosins, and also exhibits a highly cooperative binding curve (Hill coefficient = 2.8). In this respect, the erythrocyte protein more closely resembles muscle tropomyosin which is capable of head-tail self-association in addition to the lateral binding to actin filaments. Erythrocyte tropomyosin is present at about 60,000 copies per cell, which is sufficient to coat about 80% of the actin in red cell ghosts. The functions of red cell tropomyosin are open to conjecture at this point, but likely activities include stabilization of actin filaments as is observed with tropomyosin and actin in vitro (123–125), and regulation of interactions of actin with spectrin (116), band 4.1, and possibly other unidentified proteins. About 80% of tropomyosin dissociates from erythrocyte ghosts prepared in the usual manner with no magnesium ion. It will therefore be important in certain studies, such as those involving cell shape and stability of erythrocyte membrane skeletons, to use magnesium-prepared ghosts.

MYOSIN   Myosin has been discussed as a possible erythrocyte component since 1961 when ATP-dependent shape changes were observed in human red cells (126–129). Myosin has been purified from erythrocyte cytosol (129, 130), and demonstrated to be an authentic vertebrate myosin with the following properties: (a) polypeptide composition of a heavy chain of $M_r = 210,000$ and light chains of $M_r = 25,000$ and $19,500$, (b) appearance by electron microscopy as an extended molecule with a rod-shaped tail 150 nm in length and two globular heads, (c) ability to form bipolar filaments at low ionic strength, (d) characteristic smooth muscle myosin ATPase activities (Ca^{2+}-stimulated, no activity with Mg^{2+}). The Mg^{2+}-ATPase of erythrocyte myosin was not stimulated by actin, which has also been observed with nonphosphorylated forms of other nonmuscle myosins (131). However, phosphorylation of erythrocyte myosin $M_r = 19,500$ light chain induces some actin-activated ATPase activity (132). Erythrocyte myosin heavy chain is closely related in peptide maps to the heavy chain of platelet myosin. However, the M_r of erythrocyte myosin light chains are distinct form those of platelet myosin, and resemble instead the light chains of skeletal or cardiac myosin. Erythrocyte myosin is present in about 6,000 copies per cell, and is evident both in the cytosol and as a membrane-associated form. The physiological significance of erythrocyte myosin remains to be established, but it is possible that the speculations of Nakao and colleagues (126) concerning a role of myosin in ATP-dependent shape changes will be correct.

ORGANIZATION OF THE ERYTHROCYTE MEMBRANE SKELETON

The membrane skeleton can be viewed as being constructed around spectrin. Spectrin tetramers are involved in two independent classes of protein associations that are both essential for the final structure: (a) formation of the two-dimensional meshwork underlying the lipid bilayer, by associations with actin oligomers, band 4.1, and possibly other spectrin molecules, (b) linkage of the spectrin/actin meshwork to integral membrane proteins via association with ankyrin and possibly band 4.1.

Assembly of a meshwork requires some type of polymerization reaction leading to branched structures. A frequently discussed arrangement of proteins (76, 111, 112, 112a) involves a basic structural unit composed of actin filaments, containing 12–17 actin monomers and accessory proteins such as band 4.9 and tropomyosin, which are attached at the ends of multiple spectrin tetramer/band 4.1 complexes (see Figure 1). These actin-spectrin complexes can then assemble to form a branching polymeric structure by interconnections between the free ends of spectrin tetramers and other actin oligomers. It has also been proposed that self-assembly of spectrin into hexamers and higher-order oligomers may provide an additional mechanism for polymerization (72). Actin oligomers from erythrocyte membranes initiate polymerization of G-actin by a cytochalasin inhibitable reaction and thus these actin filaments have free fast-growing ends (111–113). In cells such as developing erythrocytes and other tissues with related forms of spectrin (see below), the spectrin-actin meshwork could have extended actin filaments cross-linked along their length by spectrin or other actin-binding proteins to form a three-dimensional structure. Presumably, during erythrocyte maturation, actin in excess of the final 5 $\times 10^5$ copies per cell is lost, perhaps as the consequence of a band 4.1–severing activity (120), and the red cell is left with a shell of short actin filaments adjacent to the membrane.

Morphological experiments provided initial evidence that spectrin is linked

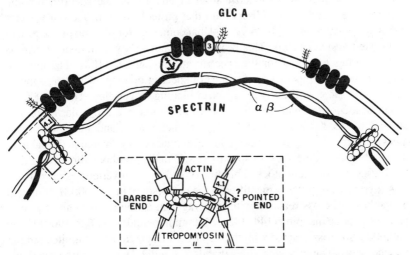

Figure 1 Schematic model of the human erythrocyte membrane skeleton. The barbed end of the actin filament is the fast-growing end while the pointed end is the slow-growing end.

in some way to integral membrane proteins (133–136). The interaction of spectrin with erythrocyte membranes has been analyzed directly by measuring reassociation of purified, radiolabeled spectrin with inside-out vesicles depleted of spectrin and actin (67, 100, 101, 137, 138). Spectrin initially was metabolically labeled with $^{32}P_i$ (label incorporates into the beta subunit in unstimulated cells) in order to maintain as closely as possible the native state of the protein (137, 139). Identical measurements have also been obtained by radiolabeling with ^{125}I (67), reductive formylation (138), and with Bolton Hunter reagent (39, 68, 140). Binding of radiolabeled spectrin heterodimer occurred with inverted vesicles, but not right-side-out membranes, and exhibited features consistent with a specific, protein-protein interaction (12, 67, 137). Quantitatively similar results have been reported using spectrin tetramer (141). Spectrin bound to inverted membranes with a K_D of 10^{-7} to 10^{-8} M to a single class of sites present in about 10^5 copies per cell, which is equivalent to the number of spectrin tetramers in erythrocyte ghosts. Ability of spectrin to reassociate with vesicles required some tertiary structure in spectrin since binding activity was destroyed by thermal denaturation in a highly cooperative manner between 49 and 51°C. It is of interest that this is the same temperature range where erythrocytes disintegrate and where spectrin exhibits major changes in its ORD behavior (142, 143). Binding of spectrin to extracted erythrocyte lipids also occurred but was of doubtful significance, since the binding was not abolished by heat-denaturation of spectrin, and was actually increased by this treatment (137).

The studies that led to identification of ankyrin as the spectrin binding protein were based on the observation that proteolysis of inside-out vesicles released a polypeptide of $M_r = 72,000$ that competed for binding of spectrin to undigested vesicles (100). This proteolytic fragment by a number of criteria had properties expected for the spectrin attachment site (100). The purified fragment was a competitive antagonist of spectrin binding, with a K_i approximately the same as the K_D for spectrin binding to vesicles. The 72,000 M_r fragment associated with spectrin dimer in solution in a 1:1 molar ratio. The amount of 72,000 M_r fragment in ghost membranes and intact cells was measured at $\sim 10^5$ copies per cell by radioimmunoassay, which is the same as the number of binding sites for spectrin (98). Finally, antibody directed against the 72,000 M_r fragment blocked binding of spectrin to inverted vesicles (101).

Ankyrin (band 2.1) was identified as the precursor of the 72,000 M_r fragment on the basis of cross-reaction with anti-fragment antibody (101) and by comparative peptide mapping (144, 145). Minor polypeptides of $M_r = 190,000$ and 160,000 also cross-reacted with the 72,000 M_r fragment and exhibited similar peptide maps. These polypeptides (bands 2.2 and 2.3) are present in variable amounts in human membranes, can be generated by proteolysis of pure band 2.1, and most likely are derived from ankyrin. Purified ankyrin binds to spectrin in solution with high affinity ($K_D = 5 \times 10^{-8}$ M) at the same site as the 72,000 M_r fragment, and forms a complex with a stoichiometry of 1 mol of ankyrin per mol of spectrin heterodimer (68, 101). Association of ankyrin with spectrin has been visualized by electron microscopy (68, 83) and occurs at a site on spectrin about 20 nm from the head of the molecule where spectrin heterodimers join to form a tetramer.

Ankyrin links spectrin to the membrane by direct association with the cytoplasmic domain of band 3, the anion transport protein (26, 37, 45). The first evidence of linkage between ankyrin and band 3 was the observation that antiankyrin Ig immunoprecipitated band 3 and ankyrin in a 1:1 molar ratio from detergent extracts of spectrin-depleted vesicles (26). More direct evidence for association between these proteins is that the cytoplasmic domain of band 3 binds to ankyrin in solution with a K_D of 5×10^{-9} M and in a 1:1 molar ratio (37, 46). Radiolabeled ankyrin reassociates with inverted vesicles depleted of ankyrin as well as most of the peripheral membrane proteins (37, 45). Band 3 is the binding site for ankyrin in inverted vesicles since the reassociation is abolished by anti–band 3 fragment IgG, by selective proteolytic cleavage of band 3, and by the cytoplasmic fragment of band 3. Furthermore, ankyrin binds to vesicles and to liposomes reconstituted with band 3 in a nearly identical manner.

Measurements of ankyrin reassociation with membranes and liposomes have demonstrated convincingly that ankyrin binds directly to band 3. However, these studies have also raised some new questions. Band 3 is present in about 10-fold the number of ankyrin molecules. It is unlikely that ankyrin is binding to a subpopulation of band 3 since the population of band 3 unassociated with ankyrin isolated by extraction of cytoskeletons (which retain ankyrin-linked band 3) can still bind to ankyrin either when reconstituted into liposomes (45), or in solution (99). Furthermore, ankyrin-linked band 3 has been purified as an oligomeric complex with ankyrin, and found to be nearly identical to the free population of band 3 by two-dimensional peptide mapping (146). The excess band 3 over ankyrin probably can be explained by the fact that these band 3 proteins are necessary to meet anion transport requirements of mature erythrocytes. A further complexity of the ankyrin-membrane interaction is that Scatchard plots of binding data were curvilinear. A possible explanation for this data is that ankyrin binds to band 3 preferentially in certain oligomeric states, such as a tetramer rather than a monomer or dimer.

The linkage of spectrin to ankyrin and of ankyrin to band 3 provides one mechanism for a membrane-cytoskeletal attachment. Additional interactions also probably occur between the membrane and spectrin or other cytoskeletal proteins. Band 4.1, for example, remains associated with membranes after removal of spectrin and actin, and requires high ionic strength for extraction. Measurements of association of radiolabeled band 4.1 with 4.1-depleted membranes indicate the existence of a protein attachment site for 4.1 (58, 147). The membrane binding site for 4.1 has been suggested to be glycophorin A based on binding of 4.1 to glycophorin A liposomes (58). However, another sialogly-coprotein termed glyconnectin has also been proposed as a 4.1 binding protein (121). Glyconnectin remains in detergent-extracted cytoskeletons, suggesting an association with some protein component of the cytoskeletons. Evidence that glyconnectin is associated with band 4.1 is based on the observation that membranes deficient in band 4.1 (from a patient with hereditary elliptocytosis) do not retain glyconnectin in the cytoskeleton.

A peculiar feature of the proposed 4.1-glycophorin A linkage is that gly-cophorin A does not remain associated with 4.1 in the cytoskeleton following solubilization of membranes with detergent. Furthermore, erythrocytes lacking glycophorin A entirely still have normal amounts of 4.1 and normal membrane properties. The identity of the binding site for 4.1 thus is not entirely clear at this point.

Band 4.1 clearly is not involved in the high-affinity binding of spectrin measured in reassociation assays, since extraction of 4.1 does not alter spectrin binding (79, 101). Pure 4.1 did not inhibit spectrin binding to vesicles (83). Furthermore, patients completely lacking erythrocyte 4.1 have unaltered amounts of spectrin (148). It is possible that a 4.1-mediated spectrin-membrane interaction is subject to regulation, and is of low affinity in resting cells. However, a 4.1 linkage could be involved in phenomena such as lectin-induced changes in cell morphology and membrane rigidity (149, 150).

Actin is another cytoskeletal component that conceivably could associate directly with the membrane. However, the conclusion of several studies is that actin does not associate with membranes depleted of spectrin and band 4.1 (77, 79, 151, 152). Band 4.9 is another potential candidate for mediating a mem-brane linkage, but its membrane association has not yet been examined.

IMPLICATIONS FOR THE ERYTHROCYTE MEMBRANE

The spectrin-ankyrin-band 3 complex provides a mechanism for linking the membrane bilayer through integral membrane proteins to the underlying spec-trin/4.1/actin meshwork. How are these proteins arranged in the native mem-brane? Band 3 is most likely a tetramer in the membrane, and, in principal, band 3 tetramers could be attached to up to four ankyrin molecules. It is more likely that on the average, each band 3 tetramer is attached to one ankyrin, since no evidence of positive cooperativity was observed in measurements of ankyrin binding to membranes (37, 45). If ankyrin is distributed randomly among band 3 tetramers, then the average nearest-neighbor distance for ankyrin-band 3 complexes is about 35 nm (assuming 10^5 ankyrin/160^2 μM of membrane). Each spectrin tetramer could bind to ankyrin by only one of its two potential sites, or alternatively half of the spectrin tetramers could bind two ankyrin molecules while the other half are not attached to ankyrin. Evidence supporting binding of one ankyrin to each spectrin tetramer is that spectrin tetramer binds to vesicles in the same number of copies as spectrin dimer (141). Thus in the membrane most spectrin tetramers are probably bound to one ankyrin-band 3 complex. (See Figure 1.)

The fact that only a fraction of band 3 molecules is linked to the cytoskeleton is in agreement with measurements of the rotational diffusion of band 3, which indicated an unhindered local environment and little interaction with spectrin (153). More detailed measurements of band 3 rotational diffusion have resolved a population of molecules with decreased mobility which can be freed from restraints by cleavage of the cytoplasmic domain, or by extraction of ankyrin and band 4.1 with high salt (154). About 40% of band 3 (4×10^5/cell) is immobilized in this way, which is consistent with ankyrin linkage to 10^5 band 3 tetramers.

Association of integral proteins to the cytoskeleton would be expected to have a substantial effect on their lateral mobility. Measurements of lateral mobility of band 3 and glycophorin confirm that these proteins are restricted in their lateral movement (155, 156). Several types of experiments indicate that linkage to ankyrin and spectrin restricts mobility of band 3 and glycophorin. One approach has been to examine mobility of band 3 in membranes from erythrocytes deficient in spectrin (157). Band 3 (covalently labeled with a fluorescent probe) exhibited a 50-fold higher rate of diffusion in spectrin-deficient membranes, as measured by the technique of fluorescence photobleach-recovery. Another approach has been to expose erythrocyte ghosts to conditions known to promote dissociation of spectrin from ankyrin (e.g. low ionic strength). Such incubations increased the diffusion constant for band 3 up to 50 fold, and also increased substantially the mobile fraction of band 3 from 10% to 80% (158). Selective dissociation of spectrin from ankyrin can be achieved by use of the 72,000 M_r ankyrin fragment to compete with membrane bound ankyrin for spectrin sites in ghosts. Exposure of ghosts to this spectrin-binding fragment increased the rate of diffusion of band 3 as well as the proportion of band 3 molecules that were mobile (159, 160).

Only a fraction of band 3 proteins interact with high affinity with spectrin, and yet over 90% of these proteins are restricted in lateral movement (158). It is possible that immobilization results from lower affinity interaction of band 3 with spectrin or other cytoskeletal proteins, or that band 3 is simply trapped in the cytoskeletal meshwork. Another possibility, which may occur in parallel, is that band 3 tetramers associate with each other with low affinity, which would decrease the diffusion constant owing to increased effective size of the aggregate, and would immobilize those proteins associated with an ankyrin-linked band 3.

UNANSWERED QUESTIONS

The structural model in Figure 1 is likely to be far from complete. For example, the placement and function is unknown of relatively common membrane proteins such as polypeptides in the M_r range of 44,000–72,000 (band "4.5"), and the tightly membrane-bound component of band 7. It is also likely that additional actin-binding proteins remain to be discovered, which will participate in maintenance of stable actin filaments, and regulate interactions of actin with proteins such as erythrocyte myosin. The number of these actin-related proteins may be on the order of the number of actin filaments, i.e. 3×10^4 per cell, and thus may not be obvious bands on SDS-gels.

Another potentially important and unexplored group of proteins are the calmodulin-binding proteins. Calmodulin, present in 5×10^5 copies per cell, is known to associate with and regulate the Ca^{2+}-transporter (500 copies per cell) (162, 163). Calmodulin also binds to spectrin in a calcium-dependent manner (86, 87) although functional consequences of this interaction remain to be established. Two other calmodulin-binding proteins of unknown activity present in $\sim 1,000$ copies per cell have been identified that are distinct from the Ca-transporter (164). Recently, a calmodulin-binding protein associated with the spectrin/actin meshwork has been isolated (165). This protein has two subunits of M_r 110,000 and 100,000, and is present in about 3×10^4 copies per cell, but its function(s) are not known.

In addition to having only a partial inventory of the protein components of the membrane, we also are limited in our knowledge of two important classes of protein-protein associations: (a) those interactions of low affinity (i.e., K_D greater than 10^{-6} M) which are unstable under dilute conditions and (b) interactions involving three or more protein components. Membrane proteins are highly concentrated in their native state, in a thin shell at most 100 nm thick. Spectrin, for example, is about 10^{-4} M on the membrane. Unfortunately, spectrin and other proteins are difficult to handle by conventional methods at physiological concentrations, and binding assays have not yet been possible. Ternary and higher-order complexes also probably occur in the densely populated region of the membrane skeleton but these are much more difficult to discover than binary interactions and will provide a future experimental challenge.

Regulation of membrane properties by modulation of protein-protein associations is an intriguing and presently unrealized possibility. Erythrocytes and erythrocyte ghosts undergo a shape change from discocytes (disc-shaped cells) to echinocytes (cells rounded up with spicules on their surface). The shape change is induced in cells by depletion of ATP, and elevation of intracellular Ca^{2+} (126, 166–168). ATP-dependent shape changes also occur in erythrocyte ghosts (169–171). Ghosts also are capable of endocytosis in the presence of

ATP (172). A protein interaction probably plays a role in the shape change, since shape changes are blocked by mild digestion of ghosts with trypsin that cleaves only ankyrin (173). Phosphorylation of spectrin or other structural proteins is a postulated (169, 170) but unsubstantiated regulatory mechanism to explain the ATP-dependent shape change. At this point, all of the erythrocyte structural proteins including actin (174) are known to be phosphorylated, but no functional consequences of any phosphorylation reaction are known (71, 74, 97). The role of phosphorylation of spectrin and ATP-dependent shape changes in erythrocytes has been reevaluated, with the conclusion that spectrin phosphorylation is not directly involved in shape change (171). In evaluating these negative results, it is important to keep in mind that phosphorylation may have subtle effects on protein interactions such as a small change in affinity.

The presence of erythrocyte tropomyosin (115, 116), myosin (129, 130, 132), and possibly troponin I (161) suggests another mechanism to regulate cell shape by an acto-myosin contractile event regulated either by Ca^{2+}; tropomyosin and troponins, as in skeletal muscle; or by a Ca^{2+} calmodulin, myosin light chain phosphorylation mechanism as occurs in smooth muscle (131). Considerable work needs to be done to evaluate a function for myosin beginning with characterization of how myosin is associated with the membrane, and direct evaluation of the role of these proteins in shape change.

MEMBRANE PROTEINS IN ABNORMAL ERYTHROCYTES

Recent studies with hereditary hemolytic anemias in humans and mice have introduced the beginnings of genetics in the area of membrane skeletal proteins and already have demonstrated clearly that abnormal or deficient membrane proteins lead to major defects in erythrocyte shape and mechanical stability. Hereditary spherocytosis (HS) and elliptocytosis (HE) are relatively common ($\sim$ 1:5000) disorders of the erythrocyte membrane leading to fragile and/or abnormally shaped cells (8, 9, 17). Erythrocytes are accumulated in the spleen in both cases, and the anemia can be essentially cured by splenectomy. Hereditary spherocytosis is in about 80% of the cases transmitted as an autosomal dominant trait, while hereditary elliptocytosis is almost entirely autosomal dominant. These diseases were recognized initially on the basis of abnormally appearing erythrocytes associated with varying degrees of anemia, that could not be explained by altered hemoglobin, glycolytic enzymes, or autoimmune features. It is likely in both disorders that the molecular lesions will be heterogeneous. A distinctive feature of the spherocytosis group is that these cells are more susceptible to hypotonic lysis, presumably due to a decreased surface area/volume ratio.

Several families with the dominant form of HS have defective binding of spectrin and band 4.1 (81, 82). The capacity of spectrin for band 4.1 is reduced by about 50%, with an unaltered affinity in the remaining sites. Pyropoikilocytosis is a rare recessively inherited variant of hereditary elliptocytosis, and is characterized by erythrocytes that fragment and assume bizarre shapes after warming to 45–46°C, which is about 4° less than the disintegration temperature for normal erythrocyte (143). Spectrin isolated from these cells denatures at the same temperature of 45–46°, based on change in the ORD spectrum, while normal spectrin denatures at 49–51°. Spectrin from patients with pyropoikilocytosis also has a lowered affinity for the dimer-tetramer equilibrium (175, 176). It is likely that a change in spectrin amino acid sequence is responsible for these features since an altered tryptic cleavage has been detected in an 80,000 M_r domain of the alpha chain (the portion of the molecule involved in dimer self-association with the adjacent beta chain). A similar defect in dimer self-association has also been observed in some patients with mild hereditary elliptocytosis (177–179). It has not been established that the altered affinity for tetramer formation measured in vitro actually causes the disease. It is possible that the abnormal spectrin is less stable in erythrocytes, and is degraded or precipitated.

High-affinity ankyrin binding sites are reduced by 50% in erythrocyte membranes in two families with an elliptocytosis-type anemia (140). The cytoplasmic domain of band 3 purified from the defective membranes binds to ankyrin normally, however. These erythrocytes may have an altered arrangement of band 3 molecules rather than a defect in the binding site itself.

Spectrin binding sites and ability of patient spectrin to bind to normal membranes were unaltered in the erythrocytes deficient in high-affinity ankyrin binding, and also were normal in several cases of spherocytosis (140, 180). These results do not exclude defects in spectrin binding in some families, however. It is also possible that spectrin-binding defects play a role in hemolysis in other forms of anemia. The binding of ankyrin to spectrin is sensitive to alkylation of a sulfhydral group on ankyrin (100). It is possible that metabolic disorders, such as glucose-6-phosphate dehydrogenase deficiency, which lead to lowered levels of reduced glutathione, could result in oxidation of ankyrin sulfhydral groups. The ability of ankyrin to bind to spectrin would be compromised, and possibly result in increased erythrocyte fragility.

The defects discussed up to this point have involved altered function of structural proteins. Another mechanism for disease is the absence of decreased quantity of these proteins due to abnormal synthesis or instability of the product. The first examples of deficient proteins came from strains of mutant mice that were developed at the Jackson Laboratory (27, 28). SDS-gels of ghost membranes from these strains revealed a striking decrease in the quantity of spectrin, and the extent of deficiency correlated with severity of the anemia (29, 181).

Decreased spectrin results from different mechanisms, depending on the strain of mouse (182). One mutant (nb/nb) produced no stable ankyrin, but did synthesize normal amounts of spectrin. Thus spectrin deficiency in nb/nb mice is secondary to lack of ankyrin. These mice still have some membrane-associated spectrin, indicating some spectrin/actin meshwork assembly and/or membrane associations of spectrin can occur in the absence of ankyrin. Another strain (ja/ja) lacks ability to synthesize the beta subunit of spectrin, although spectrin alpha chain mRNA was present in elevated amounts. Lack of the beta subunit, which has the ankyrin binding site, thus prevents association of spectrin with the membrane and prevents assembly of the spectrin/actin meshwork since actin-binding requires both subunits (65, 80). Excess alpha chain that is not assembled into alpha/beta dimers is apparently unstable and is degraded, since the alpha subunit did not accumulate. The results with ja/ja mice complement studies of spectrin synthesis in chicken erythrocytes (183, 184) that demonstrate synthesis of alpha chain in excess of beta chain, and rapid degradation of free alpha chain. Two strains of mice have mutations at a locus in chromosome 1 that contains the structural gene for the alpha chain (182). One mutant (sph^{ha}/sph^{ha}) synthesizes large amounts of mRNA coding for an unstable alpha chain that is rapidly degraded. The other mutant (sph/sph) synthesizes no detectable alpha chain mRNA and apparently has a defect in transcription, some step in mRNA processing, and/or in stability of the mRNA. The heterozygote parents, in both cases, are nearly normal, indicating one normal gene, at least in a favorable genetic background, can compensate for these defects. The mutant mice have already provided important insights in synthesis and assembly of the spectrin-membrane skeleton. With the advent of cDNA probes and DNA sequencing in this area, these mutants and similar defects in humans (see below) may provide a molecular understanding of spectrin synthesis as the thalasemmias have with hemoglobin.

Examples of markedly deficient protein 4.1 and spectrin have also been reported for human hemolytic anemias. A form of elliptocytosis with abnormal erythrocyte morphology and increased fragility has been associated with lack of band 4.1 (121, 148). The parents had 50% of the normal complement of band 4.1, and were asymptomatic with mildly abnormal erythrocytes. The affected individuals had a complete absence of band 4.1. It is not clear whether band 4.1 is lacking due to ineffective synthesis, abnormal degradation, decreased association with membranes, or if band 4.1 is present but at a different molecular weight. An unresolved discrepancy is that the parents lacking 50% of band 4.1 are essentially normal while patients lacking 40% of spectrin binding sites for band 4.1 have a significant anemia.

Patients with a severe recessive form of spherocytosis have been reported that have a 50% reduction in spectrin polypeptides on SDS-gels (180), and are similar to the sph^{ha}/sph^{ha} mice. Mild spectrin deficiency of 10–40% assayed by radioimmunoassay has recently been demonstrated to be a general feature of dominant spherocytosis (185). The amount of reduction in spectrin in patients with spherocytosis correlated closely with severity of the disease and with the reduction in membrane stability measured by the osmotic fragility assay, while other membrane proteins were essentially unchanged. Patients with a 10–20% spectrin deficiency have a mild, compensated anemia that may be subclinical, while patients with 20–40% reduction in spectrin are moderately to severely anemic, usually require splenectomy, and are detected as children. The molecular basis for spectrin deficiencies is likely to be diverse and, by analogy with the mutant mice, fall into two major categories: (*a*) unstable spectrin due to a defect in assembly or intrinsic susceptibility of spectrin to protease, (*b*) lowered synthesis of spectrin due to problems with mRNA synthesis or stability. Although the detailed mechanisms remain to be determined, deficiency of spectrin provides a simple explanation for the major common feature of spherocytosis which is a decreased membrane surface area/volume ratio leading to spherical rather than disc-shaped cells.

ERYTHROCYTE PROTEINS IN OTHER CELL TYPES

Recent studies indicate that proteins closely related to the major structural proteins of erythrocyte membranes will be present and organized in a similar way in many other types of cells. As a general rule, these proteins in other tissues are products of different genes from the erythroid homologs, and have substantial differences in antigenic sites and peptide maps, even though the major functional features have been conserved. This divergence is not surprising since erythrocytes have been evolving as a distinct cell type for at least several hundred million years beginning with early chordates. The relationship between erythroid and nonerythroid forms of proteins is probably similar to that of the different intermediate filament proteins. Desmin, vimentin, keratins, and neurofilament proteins have distinct antigenic sites, and different peptide maps. Nevertheless, these proteins in their respective tissues all form intermediate filaments, and have substantial homology in the amino acid sequence within a large conserved domain (186). The fact that erythrocyte proteins and their tissue analogs are not exactly the same led initially to some confusion and still offers problems with nomenclature. In general, out of kindness to students, a reasonable policy is to minimize new names for proteins and retain the erythrocyte name unless the erythrocyte analog turns out to be a well-characterized protein in its own right. Erythrocyte proteins that have been discovered and purified from other tissues include, at this time, spectrin and ankyrin, and represent essentially new proteins. Band 4.1 has also been found in other tissues, and in brain 4.1 is identical to a previously characterized protein named synapsin (187). These proteins will be discussed below. Immunoreactive forms of band 3 have also been identified (188–190) but these proteins have not yet been purified or characterized.

Spectrin

Within a period of about a year many laboratories independently discovered a membrane-associated protein that now is known to be closely related to spectrin. A high molecular weight protein with subunits of ~265,000 (alpha) and 260,000 M_r (beta) that constituted 3% of the total membrane protein and associated with actin filaments was isolated from brain and intestine (191–196). The alpha subunit of this protein also binds to calmodulin in a Ca^{2+}-dependent manner, as determined by overlays of SDS-gels and adsorption to calmodulin affinity columns (191, 195, 197, 198). The protein cross-reacted with erythrocyte spectrin and was localized by immunofluorescence to the plasma membrane of many types of cells (193, 199–201). The brain protein and a similar actin-binding protein from intestinal epithelial cells have been visualized by low angle rotary shadowing as flexible rods about 200 nm in length, which are similar to erythrocyte spectrin (194–196, 202, 203).

The brain protein was identified as a form of spectrin based on the following properties shared with mammalian erythrocyte spectrin: (a) ability to bind to ankyrin sites on erythrocyte membranes; (b) similar structure of a tetramer with the morphology of a 200 nM flexible rod; (c) common antigenic sites in both alpha and beta subunits (201, 202). An additional important feature in common between the brain protein and erythrocyte spectrin is the fact that functional hybrid molecules can be formed with the alpha subunit of brain and the beta subunit from erythrocyte spectrin (203). The subunits of brain spectrin are most likely arranged the same way as those of erythrocyte spectrin, with laterally associated alpha and beta dimers attached by head-to-head linkage of each alpha chain with a beta chain (203). Furthermore, the amino acid compositions of brain and erythrocyte spectrin are remarkably similar (196, 203). Brain spectrin also binds to erythrocyte band 4.1, and its association with actin is promoted by band 4.1 (204, 205). The binding sites of brain spectrin for brain ankyrin have been localized to the beta subunit, and have been visualized by electron microscopy to positions in the midregion of tetramers very similar to the erythrocyte system (102, 206). The brain protein will be referred to here as brain spectrin based on these similarities to erythrocyte spectrin. It has also been named calmodulin-binding protein I (191), brain actin-binding protein (192), and fodrin (193) before the close relationship to erythrocyte spectrin was established. Another name is calspectin, which has been suggested because of calmodulin-binding activity of tissue spectrins (207). However spectrin from avian erythrocytes binds calmodulin as well as tissue spectrins (198), and mammalian erythrocyte spectrin also binds calmodulin with lower affinity (86, 87).

A major problem that delayed recognition of nonerythroid spectrin is the limited cross-reactivity between mammalian erythrocyte and tissue spectrins. These proteins also have distinct peptide maps (202, 208) and thus are products of different genes. Another problem is that the spectrin family does not have a single, definitive feature such as actin-activated ATPase activity of the myosins. Mammalian erythrocyte spectrin also clearly differs from brain spectrin with respect to certain functions. The alpha chain of tissue and avian erythrocyte spectrin binds calmodulin in overlays of SDS-gels while the alpha subunit of mammalian erythrocyte spectrin does not bind calmodulin under these conditions (198, 209), presumably due to a lower affinity. Another difference is the dimer-tetramer equilibrium. Human erythrocyte spectrin tetramers dissociate into dimers above 30°C at concentrations less than 0.1 μM (71), while brain spectrin tetramers are quite stable under these conditions. Finally, the subunits of erythrocyte spectrin appear on the average somewhat more loosely associated than those of brain spectrin as visualized by rotary shadowing with platinum.

It is likely that mammalian erythrocyte spectrins are the most divergent members of the spectrin family. Avian erythrocyte spectrin appears much closer to tissue spectrins in terms of cross-reactivity and calmodulin-binding, even when compared with mammalian tissues (200, 209). Thus the divergence of mammalian erythrocyte spectrin occurred relatively recently, most likely during evolution of nonnucleated erythrocytes. It will be important to obtain sequence data on tissue spectrins to determine if these proteins have a similar 106-residue repeating unit as observed for erythrocyte spectrin (92–94), and to evaluate the extent of homology between members of the spectrin family.

Spectrin proteins have been purified and characterized in several tissues other than brain, including the intestinal brush border, where the protein has been referred to as TW 260/240 (194, 210), and HeLa cells (211). Immunoreactive polypeptides of $M_r \sim 260,000$ cross-reacting with brain or erythrocyte spectrin have been characterized by antigenicity and peptide maps in a number of tissues (195, 208). The alpha subunit of brain spectrin (i.e., the subunit that hybridizes with erythrocyte spectrin beta subunit and binds calmodulin) is generally well conserved, by the criterion of peptide mapping among different tissues including the intestine. The beta subunit (i.e. the ankyrin-binding subunit) also exhibits similar but not identical peptide maps among tissues with the exception of the TW 260/240 and muscle spectrin. Little is known at a biochemical level regarding the variant subunits in muscle and terminal web spectrins. It will be of interest to determine if these variants have an ankyrin-binding site, and if they can hybridize with the alpha subunit of brain spectrin.

The $M_r = 260,000$ subunit of TW 260/240 does not cross-react with brain spectrin beta subunit, and, in contrast to other spectrins, TW 260/240 is not localized at the plasma membrane (212, 213). Intestinal epithelial cells have a conventional spectrin that is localized along the membrane, appears initially during development, and coexists with TW 260/240 in differentiated intestinal cells. During embryogenesis, with formation of specialized terminal web, TW 260/240 appears (208) and interconnects actin filament bundles in the terminal web (212, 213). Muscle spectrin also has a divergent beta subunit, which surprisingly is closely related antigenically to erythroid beta subunit (208, 214, 215). As is the case with TW 260/240, in embryonic intestinal cells the muscle spectrin is not present in primitive myoblasts, but appears later during the process of maturation and accompanies other differentiated features of muscle such as the Z-band and fusion of cells into extended myotubes (215). TW 260/240 and muscle spectrin thus have special functions and their expression is developmentally regulated.

The list of protein associations of brain spectrin is not limited to familiar proteins of the erythrocyte membrane. Spectrin has been reported to associate with tau proteins from microtubules (216) and to bundle microtubules (217). Brain spectrin also modulates actin-activated ATPase of myosin (192) and the effects of spectrin are dependent on calcium (218). Antibodies against spectrin microinjected into living cells precipitate intermediate filaments, indicating some linkage of intermediate filament proteins with spectrin (219). Spectrin binds to a $M_r = 36,000$ polypeptide that is a major substrate for sarc tyrosine-specific protein kinase (220). The $M_r = 36,000$ protein was purified from intestinal microvilli and demonstrated to bind to spectrin as well as actin in the presence of calcium (220). The $M_r = 36,000$ protein has been localized by immunofluorescence to the plasma membrane in a pattern identical with that of spectrin (221, 222). The binding of an oncogene-regulated actin-binding protein to spectrin suggests a possible role for spectrin-related structures in the changes in cell shape and actin filament organization that occur in virally transformed cells.

Spectrin is a major binding site for calmodulin, but the function of this association is not known. One possibility is that spectrin simply provides a reservoir of calmodulin close to the membrane. The binding site of spectrin for calmodulin has been localized to two sites at the midregion of spectrin tetramers (223). The calmodulin binding site of spectrin is close to the ankyrin binding sites and the region where dimers associate to form tetramers. Regulation of these protein associations by calmodulin remains to be investigated.

Ankyrin

The presence of immunoreactive analogs of erythrocyte ankyrin in other cells was first indicated by radioimmunoassay (98). Proteins cross-reacting with ankyrin include two classes: (a) microtubule-associated proteins (103) and (b) a membrane-associated form (102, 206, 224). The microtubule-associated proteins share with ankyrin ability to bind to microtubules, but not other activities (102). The microtubule-associated proteins thus may share an ancestoral relationship with ankyrin, but clearly have distinct functions. The membrane-associated proteins that cross-react with ankyrin, in contrast, are very similar to erythrocyte ankyrin in terms of function and domain structure, and have been purified from brain.

Brain ankyrin includes two polypeptides of $M_r = 220,000$ and $210,000$ that have identical peptide maps and are tightly associated with membranes. Brain ankyrin (102) and a $M_r = 72,000$ fragment of brain ankyrin (206) have been purified by steps that include affinity chromatography on spectrin columns. Brain ankyrin is a monomeric protein based on the molecular weight calculated from physical properties (102). Brain ankyrin binds to brain spectrin at two sites in the midregion of tetramers about 80 nm from each end. Binding of brain ankyrin and spectrin occurred with substantial positive cooperativity, and was half-maximal at 50 nM ankyrin. Brain ankyrin also has a binding site for the cytoplasmic domain of erythrocyte band 3, and for microtubules. The domain of brain ankyrin involved in binding to spectrin was identified as a polypeptide of $M_r = 72,000$, while the binding sites for band 3 and tubulin was localized in a fragment of $M_r = 95,000$. Erythrocyte ankyrin has a similar domain structure except that the polypeptide with binding sites for band 3 and tubulin was somewhat smaller ($M_r = 90,000$). Peptide maps of $M_r = 72,000$ fragments of brain ankyrin and erythrocyte ankyrin had no peptides in common, indicating these proteins are products of different genes.

Brain ankyrin is present in about 100 pmol/mg membrane protein, while spectrin tetramers are present at 30 pmol/mg. Ankyrin is thus a likely candidate as a membrane-attachment site for spectrin in other tissues as well as erythrocytes. The function of ankyrin in excess of spectrin may be related to the microtubule-binding activity. It is of interest that a membrane-associated population of tubulin has been noted in brain (225), and that membrane-binding sites for tubulin have been measured (226). Ankyrin does not promote tubulin polymerization (102), but may mediate attachment of microtubules to membranes.

Important clues to the activity of ankyrin may come from determination of its cellular localization. Ankyrin has been localized by immunofluorescence in skeletal muscle in a pattern coincident with muscle spectrin (224), supporting an association between these proteins. It will be of interest to determine the distribution of ankyrin and spectrin at an ultrastructural level. It will also be important to localize ankyrin and spectrin in a cell that also has microtubules, using an antibody that does not cross-react with microtubule-associated proteins.

Ankyrin in avian erythrocytes is phosphorylated by protein kinases regulated by cAMP and by Ca^{2+}/calmodulin, and was named goblin since these studies were performed with turkey erythrocytes (184, 227). It is likely ankyrin in brain and other tissues also will be phosphoproteins. Another modification of ankyrin that may have regulatory significance is proteolytic processing by a calcium-regulated protease. Ankyrin in brain is especially sensitive to proteolysis during purification, and is cleaved to fragments of M_r = 190,000, then to two dissimilar polypeptides of M_r = 95,000. The M_r = 190,000 form of ankyrin binds preferentially to band 3-affinity columns, is not as easily extracted from membranes as the higher M_r polypeptides, and may have a higher affinity for band 3–like proteins (unpublished data).

The membrane-attachment site(s) for tissue ankyrin remain to be determined. However, a band 3–like protein is likely since brain ankyrin binds to erythrocyte band 3 (102), and since immunoreactive forms of band 3 are present in nonerythroid cells (188–190).

Band 4.1 (Synapsin)

Polypeptides of M_r ~ 80,000 that cross-react with erythrocyte 4.1 are present in fibroblasts and were localized by immunofluorescence to stress-fibers (228). Immunoreactive forms of 4.1 have also been reported in brain (229) and other tissues (230). A spectrin-binding protein that cross-reacts with 4.1 has been purified from brain membranes where it is a doublet of polypeptides of nearly identical peptide maps with apparent M_r = 73,000 and 75,000 and is present in a 2:1 ratio with spectrin tetramer (187). As is the case with brain spectrin and ankyrin, brain 4.1 has distinct peptide maps from its erythrocyte counterpart and is the product of a different gene. Brain 4.1 is identical in M_r on SDS gels, physical properties, immunoreactivity, and peptide maps to a previously characterized protein named synapsin (187). Synapsin is membrane-associated protein that is a major substrate for cyclic AMP and Ca^{2+}/calmodulin dependent protein kinases (231, 232). Synapsin also is phosphorylated in vivo and in vitro following stimulation by hormones, drugs, and nerve impulses (233–235). Synapsin is localized at nerve termini throughout the peripheral and central nervous system (236). Synapsin is a major component of synaptic vesicles, and comprises 6% of the total synaptic vesicle membrane protein (237, 238).

Synapsin thus is a likely candidate to participate in secretion, although detailed activities of this protein have not been studied. The identity of synapsin with an analog on erythrocyte membrane skeletal protein indicates that synapsin is a structural protein with protein associations that may be modulated by phosphorylation. It is already known that synapsin binds to spectrin from brain and erythrocytes (187) and that synapsin interacts with microtubules (239). Synapsin, by analogy with erythrocyte 4.1, may have a membrane-attachment site in addition to spectrin, and thus may mediate certain spectrin-membrane interactions. Spectrin, for example, could link synapsin-coated secretory vesicles to sites of secretion on the plasma membrane and/or couple secretory vesicles to an actomyosin system during axonal transport.

Synapsin, by criteria of immunoreactivity, is present only in the nervous system. However, as is the case with erythrocyte proteins, results with antibodies should not be taken too seriously. 4.1 is present in fibroblasts, and it is likely members of the synapsin family will be widespread in other tissues as well.

POSSIBLE FUNCTIONS OF ERYTHROCYTE PROTEINS IN OTHER CELL TYPES

Tissue forms of erythrocyte proteins have only recently been discovered, and it is too early to know the role(s) of these proteins. Inferences about function can be made based on analogies with erythrocytes. It should be emphasized, however, that neurons and other complex cells are not simply large red cells, and that almost certainly the erythrocyte membrane will be an incomplete model for other cells. The spectrin-actin meshwork supports and stabilizes the lipid bilayer of erythrocytes, and may have an analogous function in certain cells. For example, simple squamous cells such as endothelial cells, alveolar cells and lens epithelial cells have extended plasma membrane surfaces, and minimal cytoplasm. Spectrin in these cells may function as a membrane-supporter. It is pertinent that membranes from lung and lens have large amounts of spectrin (203). In more typical cells such as HeLa cells, spectrin probably is not required to stabilize the membrane since these cells are unaffected by aggregation of spectrin following microinjection of antibody (219).

One likely function of spectrin and ankyrin will be to mediate association of actin, microtubules and possibly intermediate filaments to the plasma membrane. Spectrin is well established as an actin-binding protein, but also may interact with microtubules (217), as well as intermediate filaments (219), while ankyrin binds to microtubules (102, 103). An example of a membrane-cytoskeletal linkage is in striated muscle, where myofibril membrane attachment is thought to result from a spectrin-containing structure, resolved by immunofluorescence (240). This structure has been called a costamere because it occurs in regular rib-like bands that run transversely across muscle cells. Costameres are in register with I bands, and are interconnected by longitudinal fibers. Costameres were initially discovered using antibodies against vinculin (241) and have subsequently been observed to contain the gamma isoform of actin and intermediate filament proteins (240). Ankyrin also colocalizes to costameres and is a reasonable candidate to mediate membrane linkage of the spectrin component (224).

Spectrin is not present in adhesion plaques in cultured cells, and is excluded from the region of stress fibers (200, 201). Similarly, spectrin does not attach actin filaments to Z-bands in striated muscle. Thus spectrin is only one of several mechanisms for attachment of actin filaments to membranes.

The spectrin-ankyrin-band 3 linkage provides a mechanism for association of integral membrane proteins with cytoplasmic structural proteins. Association of spectrin with integral proteins in other cells could permit localization of membrane proteins to specific regions of the cell surface, such as in brain where receptors for neurotransmitters are known to be concentrated at synapses. Spectrin is a major component of postsynaptic densities (242), and is present in elevated amounts in synaptic endings (243). Spectrin also is concentrated at nodes of Ranvier (243), which are highly enriched in ion channels. Spectrin may be associated with cell surface proteins in lymphocytes, since capping of surface components is accompanied by redistribution of spectrin underneath the clustered membrane proteins (244–246). Spectrin in brain and lymphocytes is thus in the right location to participate in localization of membrane proteins, although biochemical evidence is not yet available for such linkages.

Spectrin may also be involved in active intracellular movement of proteins and organelles. In brain, Willard and his colleagues have studied anterograde axonal transport of spectrin and found it moves at different rates with several different groups of proteins (247, 248). Other proteins, in contrast, appear to move in discrete groupings. Spectrin thus may link at least some proteins and vesicles to an actomyosin system. In the intestinal brush border, the terminal web protein TW 260/240 could be involved in membrane traffic since this protein has been visualized in association with membrane vesicles that presumably move between the brush border and basal regions of the cells (212). Synapsin (band 4.1) may mediate attachment of spectrin to intracellular vesicles, and a possible role of these proteins in secretion has been discussed above.

CONCLUSION

The work discussed in this review has led to a detailed understanding of membrane-protein interactions in the human erythrocyte, which is of clinical significance and has direct relevance to other cell types. The strategy in erythrocyte membrane studies of sequential removal of one protein followed by its purification, reassociation, and identification of its binding site also may have application in elucidating details of other complex membrane structures. For example, desmosomes, postsynaptic densities, Z-discs, and adhesion plaques most likely represent stable assemblies of protein components which are ultimately attached to the membrane, and would be amenable to such an approach. It is probable, based on the erythrocyte analogy, that end-on attachment of filament-forming proteins such as actin or intermediate filament proteins to membranes will require several intermediate proteins which may themselves be associated to form an extensive, lateral structure.

ACKNOWLEDGMENTS

Research by the author has been funded in part by grants from the National
Institutes of Health (Research Career Development Award and research grants
RO1 AM 29808 and GM 33996) and the Muscular Dystrophy Association.
Valuable assistance by Jonathan Davis is gratefully acknowledged. The manu-
script was prepared by Arlene Daniel.

Literature Cited

1. Singer, S. J. 1974. *Ann. Rev. Biochem.*
   43:805–33
2. Nicolson, G. L. 1974. *Biochim. Biophys.
   Acta* 457:57–108
3. Edelman, G. M. 1976. *Science* 192:218–
   26
4. Bourguignon, L. Y. W., Singer, S. J.
   1977. *Proc. Natl. Acad. Sci. USA* 74:
   5031–35
5. Tank, D. W., Wu, E. S., Webb, W. W.
   1982. *J. Cell Biol.* 92:207–12
6. Weihing, R. R. 1979. *Methods. Achiev.
   Exp. Pathol.* 8:42–109
7. Geiger, B. 1983. *Biochim. Biophys. Acta*
   737:305–42
8. Lux, S. E. 1979. *Semin. Hematol.* 16:
   21–51
9. Lux, S. E. 1983. In *The Metabolic Basis
   of Inherited Disease,* ed. Stanbury,
   Wyngaarden, Fredrickson, Goldstein,
   Brown, pp. 1573–605. New York:
   McGraw-Hill
10. Branton, D., Cohen, C. M., Tyler, J.
    1981. *Cell* 24:24–32
11. Bennett, V. 1982. *J. Cell. Biochem.* 18:
    49–65
12. Bennett, V. 1983. In *Cell Membranes:
    Methods and Reviews,* ed. E. Elson, W.
    Frazier, L. Glaser. New York: Plenum.
    2:149–195
13. Haest, C. W. 1982. *Biochim. Biophys.
    Acta* 694:331–52
14. Cohen, C. M. 1983. *Semin. Hematol.*
    20:141–58
15. Goodman, S. R., Shiffer, K. 1983. *Am.
    J. Physiol.* 244:121–46
16. Marchesi, V. T. 1983. *Blood* 61:1–11
17. Palek, J., Lux, S. E. 1983. *Semin.
    Hematol.* 20:189–224
18. Rand, R. P., Burton, A. C. 1964. *Bio-
    phys. J.* 4:115–35
19. Evans, E. A., Hochmuth, R. M. 1978.
    Curr. Top. Membr. Transp. 10:1–64
20. Kwok, R., Evans, E. 1981. *Biophys. J.*
    35:637–52
21. Yu, J., Fischman, D. A., Steck, T. L.
    1973. *J. Supramol. Struct.* 1:233–48
22. Hainfield, J. F., Steck, T. L. 1977. *J.
    Supramol. Struct.* 6:301–17
23. Nermut, M. L. 1981. *Eur. J. Cell Biol.*
    25:265–71
24. Mohandas, N., Wyatt, J., Mal, S., Ros-
    si, M. E., Shohet, S. B. 1982. *J. Biol.
    Chem.* 257:6537–43
25. Sheetz, M. P. 1979. *Biochim. Biophys.
    Acta* 557:122–34
26. Bennett, V., Stenbuck, P. J. 1979. *Na-
    ture* 280:468–73
27. Bernstein, S. E. 1980. *Lab. Anim. Sci.*
    30:197–205
28. Russell, E. S. 1979. *Adv. Genet.* 20:357–
    459
29. Greenquist, A. C., Shohet, S. B., Bern-
    stein, S. E. 1978. *Blood* 51:1149–55
30. Shohet, S. B. 1979. *J. Clin. Invest.*
    64:483–93
31. Steck, T. L. 1974. *J. Cell Biol.* 62:1–19
32. Steck, T. L. 1978. *J. Supramol. Struct.*
    8:311–24
33. Marchesi, V. T., Furthmayr, H., Tomi-
    ta, M. 1976. *Ann. Rev. Biochem.* 45:
    667–98
34. Macara, I. G., Cantley, L. C. 1983. *Cell
    Membr. Methods Rev.* 1:41–87
35. Cabantchik, Z. I., Knauf, P. A., Roth-
    stein, A. 1978. *Biochim. Biophys. Acta*
    515:239–302
36. Steck, T. L., Ramos, B., Strapazon, E.
    1976. *Biochemistry* 15:1154–61
37. Bennett, V., Stenbuck, P. J. 1980. *J.
    Biol. Chem.* 255:6424–32
38. Appell, K. C., Low, P. S. 1981. *J. Biol.
    Chem.* 256:11104–11
39. Bennett, V. 1983. *Methods Enzymol.*
    96:313–24
40. Tsai, I. H., Murthy, S. N., Steck, T. L.
    1982. *J. Biol. Chem.* 257:1438–42
41. Murthy, S. N., Liu, T., Kaul, R. K.,
    Kohler, H., Steck, T. L. 1981. *J. Biol.
    Chem.* 256:11203–8
42. Salhany, J. M., Shaklai, N. 1979.
    Biochemistry 18:893–99
43. Salhany, J. M., Cordes, K. A., Gaines,
    E. D. 1980. *Biochemistry* 19:1447–54
44. Walder, J. A., Chatterjee, R., Steck, T.
    L., Low, P. S., Musso, G. F., et al.
    1984. *J. Biol. Chem.* 259:10238–46
45. Hargreaves, W. R., Giedd, K. N., Verk-

leij, A., Branton, D. 1980. *J. Biol. Chem.* 255:11965–72
46. Weaver, D. C., Pasternack, G. R., Marchesi, V. T. 1984. *J. Biol. Chem.* 259:6170–75
46a. Wallin, R., Culp, E., Coleman, D., Goodman, S. R. 1984. *Proc. Natl. Acad. Sci. USA* 81:4095–99
47. Steck, T. L. 1972. *J. Mol. Biol.* 66:295–305
48. Yu, J., Steck, T. L. 1975. *J. Biol. Chem.* 250:9176–84
49. Nigg, E., Cherry, R. J. 1979. *Nature* 277:493–94
50. Nigg, E. A., Cherry, R. J. 1979. *Biochemistry* 18:3457–65
51. Dorst, H., Schubert, D. 1979. *Hoppe-Seyler's Z. Physiol. Chem.* 360:1605–18
52. Nakashima, H., Nakagawa, Y., Makino, S. 1981. *Biochim. Biophys. Acta* 643:509–18
53. Kaul, R. K., Murthy, S. N., Reddy, A. G., Steck, T. L., Kohler, H. 1983. *J. Biol. Chem.* 258:7981–90
54. Perkins, M. 1981. *J. Cell Biol.* 90:563–67
55. Pasval, G., Wainscoat, J. S., Weatherall, D. J. 1982. *Nature* 297:64–66
56. Okoye, V., Bennett, V. 1985. *Science.* 227:169–171
57. Cotmore, S. F., Furthmayr, H., Marchesi, V. T. 1977. *J. Mol. Biol.* 113:539–53
58. Anderson, R. A., Lovrien, R. E. 1984. *Nature* 307:655–58
59. Nigg, E. A., Bron, C., Giradet, M., Cherry, R. J. 1980. *Biochemistry* 19:1887–93
60. Pinto da Silva, P., Nicolson, G. L. 1974. *Biochim. Biophys. Acta* 363:311–19
61. Yu, J., Branton, D. 1976. *Proc. Natl. Acad. Sci. USA* 73:3891–95
62. Dahr, W., Uhlenbruch, G., Wagstaff, W., Leikola, J. 1976. *J. Immunogenet.* 3:383–93
63. Marchesi, V. T., Steers, E. 1968. *Science* 159:203–4
64. Shotton, D. M., Burke, B. E., Branton, D. 1979. *J. Mol. Biol.* 131:303–29
65. Calvert, R., Bennett, P., Gratzer, W. 1980. *Eur. J. Biochem.* 107:355–61
66. Yoshino, H., Marchesi, V. T. 1984. *J. Biol. Chem.* 259:4496–500
67. Litman, D., Hsu, C. J., Marchesi, V. T. 1980. *J. Cell Sci.* 42:1–22
68. Tyler, J. M., Reinhardt, B. N., Branton, D. 1980. *J. Biol. Chem.* 255:7034–39
69. Ji, T. H., Kiehm, D. J., Middaugh, G. R. 1980. *J. Biol. Chem.* 255:2990–93
70. Ralston, G. B. 1978. *J. Supramol. Struct.* 8:361–73

71. Ungewickell, E., Gratzer, W. 1978. *Eur. J. Biochem.* 88:379–85
72. Morrow, J. S., Marchesi, V. T. 1981. *J. Cell Biol.* 88:463–68
73. Liu, S.-C., Windisch, P., Kim, S., Palek, J. 1984. *Cell* 37:587–94
74. Brenner, S. L., Korn, E. D. 1979. *J. Biol. Chem.* 254:8620–27
75. Fowler, V., Taylor, D. L. 1980. *J. Cell Biol.* 85:361–76
76. Cohen, C. M., Tyler, J. M., Branton, D. 1980. *Cell* 21:875–83
77. Fowler, V. M., Luna, E. J., Hargreaves, W. R., Taylor, D. L., Branton, D. 1981. *J. Cell Biol.* 88:388–95
78. Ungewickell, E., Bennett, P. M., Calvert, R., Ohanian, V., Gratzer, W. B. 1979. *Nature* 280:811–14
79. Cohen, C. M., Foley, S. F. 1982. *Biochim. Biophys. Acta* 688:691–701
80. Cohen, C. M., Langley, C. 1984. *Biochemistry.* 23:4488–95
81. Wolfe, L. C., John, K. M., Falcone, J. C., Byrne, A. M., Lux, S. E. 1982. *N. Engl. J. Med.* 307:1367–74
82. Goodman, S. R., Shiffer, K. A., Casoria, L. A., Eyster, M. E. 1982. *Blood* 60:772–84
83. Tyler, J. M., Hargreaves, W. R., Branton, D. 1979. *Proc. Natl. Acad. Sci. USA* 76:5192–96
84. Cohen, C. M., Foley, S. F. 1984. *Biochemistry.* In press
85. Ohanian, V., Wolfe, L., John, K., Pinder, J., Lux, S. E., Gratzer, W. B. 1984. *Biochemistry* 23:4416–20
86. Sobue, K., Muramoto, Y., Fujita, M., Kakiuchi, S. 1981. *Biochem. Biophys. Res. Commun.* 100:1063–70
87. Berglund, A., Backman, L., Shanbhag, V. P. 1984. *FEBS Lett.* 172:109–12
88. Jarrett, H. W., Penniston, J. T. 1978. *J. Biol. Chem.* 253:4676–82
89. Speicher, D. W., Morrow, J. S., Knowles, W. J., Marchesi, V. T. 1980. *Proc. Natl. Acad. Sci. USA* 77:5673–77
90. Morrow, J. S., Speicher, D. W., Knowles, W. J., Hsu, C. J., Marchesi, V. T. 1980. *Proc. Natl. Acad. Sci. USA* 77:6592–96
91. Speicher, D. W., Morrow, J. S., Knowles, W. J., Marchesi, V. T. 1982. *J. Biol. Chem.* 257:9093–101
92. Speicher, D. W., Davis, G., Marchesi, V. T. 1983. *J. Biol. Chem.* 258:14938–47
93. Speicher, D. W., Davis, G., Yurchenco, P. D., Marchesi, V. T. 1983. *J. Biol. Chem.* 258:14931–37
94. Speicher, D. W., Marchesi, V. T. 1984. *Nature* 311:177–80

95. Davis, J., Bennett, V. 1982. *J. Biol. Chem.* 257:5816–20
96. Harris, H. W., Lux, S. E. 1980. *J. Biol. Chem.* 255:11512–20
97. Anderson, J. M., Tyler, J. M. 1980. *J. Biol. Chem.* 255:1259–65
98. Bennett, V. 1979. *Nature* 281:597–99
99. Bennett, V., Stenbuck, P. J. 1980. *J. Biol. Chem.* 255:2540–48
100. Bennett, V. 1978. *J. Biol. Chem.* 253:2292–99
101. Bennett, V., Stenbuck, P. J. 1979. *J. Biol. Chem.* 254:2533–41
102. Davis, J., Bennett, V. 1984. *J. Biol. Chem.* 259:13550–59
103. Bennett, V., Davis, J. 1981. *Proc. Natl. Acad. Sci. USA* 78:7550–54
104. Bennett, V., Davis, J. 1982. *Cold Spring Harbor Symp. Quant. Biol.* 46:647–57
105. Van Deurs, B., Behnke, O. 1973. *Z. Anat. Entwicklungsgesch.* 143:43–47
106. Izant, J. G., Weatherbee, J. A., McIntosh, J. R. 1982. *Nature* 295:248–50
107. Ohnishi, T. 1962. *J. Biochem.* 52:307–8
108. Tilney, L. G., Detmers, P. 1975. *J. Cell Biol.* 66:508–20
109. Sheetz, M. P., Painter, R. G., Singer, S. J. 1976. *Biochemistry* 15:4486–92
110. Nakashima, K., Beutler, E. 1979. *Proc. Natl. Acad. Sci. USA* 76:935–38
111. Brenner, S. L., Korn, E. D. 1980. *J. Biol. Chem.* 255:1670–76
112. Pinder, J. C., Gratzer, W. B. 1983. *J. Cell Biol.* 96:768–75
112a. Shen, B. W., Josephs, R., Steck, T. L. 1984. *J. Cell Biol.* 99:810–21
113. Lin, D. C., Lin, S. 1979. *Proc. Natl. Acad. Sci. USA* 76:2345–49
114. Siegel, D. L., Branton, D. 1982. *J. Cell Biol.* 95:265a
115. Fowler, V. M., Bennett, V. 1984. *J. Biol. Chem.* 259:5978–89
116. Fowler, V. M., Bennett, V. 1984. In *Erythrocyte Membranes 3: Recent Clinical and Experimental Advances*, pp. 57–71. New York: Liss
117. Pinder, J. C., Ungewickell, E., Bray, D., Gratzer, W. B. 1978. *J. Supramol. Struct.* 8:439–45
118. Goodman, S. R., Yu, J., Whitfield, C. F., Culp, E. N., Pasnak, E. J. 1982. *J. Biol. Chem.* 257:4564–69
119. Leto, T. L., Marchesi, V. T. 1984. *J. Biol. Chem.* 259:4603–8
119a. Ohanian, V., Gratzer, W. 1984. *Eur. J. Biochem.* 144:375–79
120. Pinder, J., Ohanian, V., Gratzer, W. B. 1984. *FEBS Lett.* 169:161–64
121. Mueller, T. J., Morrison, M. 1981. In *Erythrocyte Membranes 2: Recent Clinical and Experimental Advances*, ed.

Kruckeberg, Eaton, Brewer, pp. 95–112. New York: Liss
122. Ling, E., Sapirstein, V. 1984. *Biochem. Biophys. Res. Commun.* 120:291–98
123. Wegner, A. 1982. *J. Mol. Biol.* 161:217–27
124. Bernstein, B. W., Bamburg, J. R. 1982. *Cell Motility* 2:1–8
125. Fattoum, A., Hartwig, J. H., Stossel, T. P. 1983. *Biochemistry* 22:1187–93
126. Nakao, M., Nakao, T., Yamazol, S., Yoshikawa, H. 1961. *J. Biochem.* 49:487–92
127. Rosenthal, A. S., Krezenor, F. M., Moses, H. L. 1970. *Biochim. Biophys. Acta* 196:254–62
128. Schrier, S. L., Hardy, B., Junga, I., Ma, L. 1981. *Blood* 58:953–62
129. Kirkpatrick, F. H., Sweeney, M. L. 1980. *Fed. Proc.* 39:2049a
130. Fowler, V. M., Davis, J. Q., Bennett, V. 1985. *J. Cell Biol.* 100:47–55
131. Adelstein, R. S., Eisenberg, E. 1980. *Ann. Rev. Biochem.* 49:921–56
132. Wong, A. L., Kiehart, D., Pollard, T. D. 1985. *J. Biol. Chem.* 260:46–49
133. Nicolson, G. L., Painter, R. G. 1973. *J. Cell Biol.* 59:395–406
134. Elgsaeter, A., Branton, D. 1974. *J. Cell Biol.* 63:1018–30
135. Elgsaeter, A., Shotton, D. M., Branton, D. 1976. *Biochim. Biophys. Acta* 426:101–22
136. Shotton, D. M., Thompson, K., Wofsy, L., Branton, D. 1978. *J. Cell Biol.* 76:512–31
137. Bennett, V., Branton, D. 1977. *J. Biol. Chem.* 252:2753–63
138. Baskin, G. S., Langdon, R. G. 1981. *J. Biol. Chem.* 256:5428–35
139. Bennett, V. 1977. *Life Sci.* 21:433–40
140. Agre, P., Orringer, E., Chui, D., Bennett, V. 1981. *J. Clin. Invest.* 68:1566–76
141. Goodman, S. R., Weidner, S. A. 1980. *J. Biol. Chem.* 255:8082–86
142. Brandts, J. F., Erickson, L., Lysko, K., Schwartz, A. T., Taverna, R. D. 1977. *Biochemistry* 16:3450–54
143. Chang, K., Williamson, J. R., Zarkowsky, H. S. 1979. *J. Clin. Invest.* 64:326–28
144. Yu, J., Goodman, S. R. 1979. *Proc. Natl. Acad. Sci. USA* 76:2340–44
145. Luna, E. J., Kidd, G. H., Branton, D. 1979. *J. Biol. Chem.* 254:2526–32
146. Bennett, V. 1982. *Biochim. Biophys. Acta* 689:475–84
147. Shiffer, K., Goodman, S. R. 1984. *Proc. Natl. Acad. Sci. USA* 81:4044–408
148. Tchernia, G., Mohandas, N., Shohet, S. B. 1981. *J. Clin. Invest.* 68:454–60

149. Lovrien, R. E., Anderson, R. A. 1980. *J. Cell Biol.* 85:534–48
150. Evans, E., Leung, A. 1984. *J. Cell Biol.* 98:1201–8
151. Cohen, C. M., Branton, D. 1979. *Nature* 279:163–65
152. Cohen, C. M., Foley, S. F. 1980. *J. Cell Biol.* 86:694–98
153. Cherry, R. J., Burkli, A., Busslinger, M., Schneider, G., Parish, G. R. 1976. *Nature* 263:389–93
154. Nigg, E. A., Cherry, R. J. 1980. *Proc. Natl. Acad. Sci. USA* 77:4702–6
155. Peters, R., Peters, J., Tews, K. H., Bahr, W. 1974. *Biochim. Biophys. Acta* 367: 282–94
156. Fowler, V., Branton, D. 1977. *Nature* 268:23–26
157. Sheetz, M. P., Schindler, M., Koppel, D. 1980. *Nature* 285:510–12
158. Golan, D. E., Veatch, W. 1980. *Proc. Natl. Acad. Sci. USA* 77:2537–41
159. Fowler, V., Bennett, V. 1978. *J. Supramol. Struct.* 8:215–21
160. Golan, D. E., Veatch, W. R. 1982. *Biophys. J.* 37:177a
161. Maimon, J., Puszkin, S. 1978. *J. Supramol. Struct.* 9:131–41
162. Niggli, V., Penniston, J. T., Caraldi, E. 1979. *J. Biol. Chem.* 254:9955–58
163. Hinds, T. R., Andreasen, T. J. 1982. *J. Biol. Chem.* 256:7877–82
164. Agre, P., Gardner, K., Bennett, V. 1983. *J. Biol. Chem.* 258:6258–65
165. Gardner, K., Bennett, V. 1985. *J. Biol. Chem.* In preparation
166. Weed, R. I., LaCelle, P. L., Merrill, E. 1969. *J. Clin. Invest.* 48:795–809
167. Palek, J., Curby, W. A., Lionetti, F. J. 1970. *Am. J. Physiol.* 220:19–26
168. Kirkpatrick, F. H., Hillman, D. G., LaCelle, P. L. 1975. *Experientia* 31: 653–54
169. Sheetz, M. P., Singer, S. J. 1977. *J. Cell Biol.* 73:638–46
170. Birchmeier, W., Singer, S. J. 1977. *J. Cell Biol.* 73:647–59
171. Patel, V. P., Fairbanks, G. 1981. *J. Cell Biol.* 88:430–40
172. Schrier, S. L., Bensch, K. G., Johnson, M., Junga, I. 1975. *J. Clin. Invest.* 56:8–22
173. Jinbu, V., Sato, S., Nakao, T., Nakao, M., Tsukita, S., et al. 1984. *Biochim. Biophys. Acta* 773:237–45
174. Boivin, P. M., Gabarg, M., Dhermy, D., Galand, C. 1981. *Biochim. Biophys. Acta* 647:1–6
175. Liu, S. C., Palek, J., Prchal, J., Castleberry, R. P. 1981. *J. Clin. Invest.* 68:597–605
176. Knowles, W. J., Morrow, J. S., Speicher, D. W., Zarkowsky, H. S., Mohandas, N., et al. 1983. *J. Clin. Invest.* 71:1867–77
177. Liu, S. C., Palek, J., Prchal, J. T. 1982. *Proc. Natl. Acad. Sci. USA* 79:2072–76
178. Coetzer, T., Zail, S. 1982. *Blood* 59: 900–5
179. Evans, J. P., Baines, A. J., Hann, I. M., Al-Hakin, J., Knowles, S. M., Hoffbrand, A. V. 1983. *Br. J. Haematol.* 54:163–72
180. Agre, P., Orringer, E., Bennett, V. 1982. *N. Engl. J. Med.* 306:1155–61
181. Lux, S. E., Pease, B., Tomasselli, M., John, K., Bernstein, S. E. 1979. In *Normal and Abnormal Red Cell Membranes*, ed. S. E. Lux, V. T. Marchesi, C. F. Fox, pp. 463–69. New York: Liss
182. Bodine, D., Birkenmier, C., Barker, J. 1984. *Cell* 37:721–30
183. Blikstad, I., Nelson, W. J., Moon, R. T., Lazarides, E. 1983. *Cell* 32:1081–91
184. Moon, R. T., Lazarides, E. 1984. *J. Cell Biol.* 98:1899–1904
185. Agre, P., Cassella, J., Zinkham, W., McMillan, C., Bennett, V. 1985. *Nature.* In press
186. Weber, K., Geisler, N. 1984. In *Cancer Cells I, The Transformed Phenotype,* ed. A. J. Levine, G. F. Vande Woude, W. C. Topp, J. D. Watson, pp. 153–59
187. Baines, A., Bennett, V. 1985. *Nature.* Submitted for publication
188. Bennett, V., Davis, J., Fowler, W. 1982. *Phil. Trans. R. Soc. London Ser. B* 299:301–12
189. Kay, M. M. B., Tracey, C. M., Goodman, J. R., Cone, J. C., Bassel, P. S. 1983. *Proc. Natl. Acad. Sci. USA* 80: 6882–86
190. Drenckhahn, D., Kinke, K., Schawer, U., Appell, K., Low, P. 1984. *Eur. J. Cell Biol.* 34:144–50
191. Davies, P. J. A., Klee, C. B. 1981. *Biochem. Inter.* 3:203–12
192. Shimo-Oka, T., Watanabe, Y. 1981. *J. Biochem.* 90:1297–1307
193. Levine, J., Willard, M. 1981. *J. Cell Biol.* 90:631–43
194. Glenney, J. R., Glenney, P., Osborn, M., Weber, K. 1982. *Cell* 28:843–54
195. Glenney, J. R., Glenney, P., Weber, K. 1982. *Proc. Natl. Acad. Sci. USA* 79:4002–5
196. Glenney, J. R., Glenney, P., Weber, K. 1982. *J. Biol. Chem.* 257:9781–87
197. Kakiuchi, S., Sobue, K., Fujita, M. 1981. *FEBS Lett.* 132:144–48
198. Palfrey, H. C., Schiebler, W., Greengard, P. 1982. *Proc. Natl. Acad. Sci. USA* 79:3780–84
199. Goodman, S. R., Zagon, I. S., Kulikowski, R. R. 1981. *Proc. Natl. Acad. Sci. USA* 78:7570–74

200. Repasky, E., Granger, B., Lazarides, E. 1982. *Cell* 29:821–33
201. Burridge, K., Kelly, T., Mangeat, P. 1982. *J. Cell Biol.* 95:478–86
202. Bennett, V., Davis, J., Fowler, W. 1982. *Nature* 299:126–31
203. Davis, J., Bennett, V. 1983. *J. Biol. Chem.* 258:7757–66
204. Burns, N. R., Ohanian, V., Gratzer, W. B. 1983. *FEBS Lett.* 153:165–68
205. Lin, D. C., Flanagan, M. D., Lin, S. 1983. *Cell Motility.* 3:375–82
206. Davis, J. Q., Bennett, V. 1984. *J. Biol. Chem.* 259:1874–81
207. Sobue, K., Kanda, K., Adachi, J., Kakiuchi, S. 1982. *Biomed. Res.* 3:561–70
208. Glenney, J. R., Glenney, P. 1983. *Cell* 34:503–12
209. Bartelt, D. C., Carlin, R. K., Scheele, G. A., Cohen, W. D. 1982. *J. Cell Biol.* 95:278–84
210. Pearl, M., Fishkind, D., Mooseker, M., Kenne, D., Keller, T. 1984. *J. Cell Biol.* 98:66–78
211. Mangeat, P. H., Burridge, K. 1983. *Cell Motility* 3:657–69
212. Hirokawa, N., Cheng, R., Willard, M. 1983. *Cell* 32:953–65
213. Glenney, J. R., Glenney, P., Weber, K. 1983. *J. Cell Biol.* 96:1491–96
214. Nelson, W. J., Lazarides, E. 1983. *Proc. Natl. Acad. Sci. USA* 80:363–67
215. Nelson, W. J., Lazarides, E. 1983. *Nature* 304:364–68
216. Carlier, M. F., Simon, C., Zassoly, R., Pradel, G. A. 1984. *Biochimie* 66:305
217. Ishikawa, M., Murofushi, H., Sakai, H. 1983. *J. Biochem.* 94:1209–18
218. Wagner, P. 1984. *J. Biol. Chem.* 259:6306–10
219. Mangeat, P. H., Burridge, K. 1984. *J. Cell Biol.* 98:1363–77
220. Gerke, V., Weber, K. 1984. *EMBO J.* 3:117
221. Greenberg, M. E., Edelman, G. M. 1983. *Cell* 33:767–79
222. Lehto, V. P., Virtanen, I., Paasivuo, R., Ralston, R., Alitalo, K. 1983. *EMBO J.* 2:1701–5
223. Tsukita, S., Tsukita, S., Ishikawa, H., Kurokawa, M., Morimoto, K., et al. 1983. *J. Cell Biol.* 97:574–78
224. Nelson, W. J., Lazarides, E. 1984. *Proc. Natl. Acad. Sci. USA* 81:3292–96
225. Bhattacharyya, B., Wolff, J. 1975. *J. Biol. Chem.* 250:7639–46
226. Bernier-Valentin, F., Aunis, D., Rousset, B. 1983. *J. Cell Biol.* 97:209–16
227. Beam, K. G., Alper, S. L., Palade, G. E., Greengard, P. 1979. *J. Cell Biol.* 83:1–15
228. Cohen, C. M., Foley, S. F., Korsgren, C. 1982. *Nature* 299:648–50
229. Granger, B. L., Lazarides, E. 1984. *Cell* 37:595–607
230. Goodman, S. R., Casoria, L., Coleman, D., Zagon, I. S. 1984. *Science* 224:1433–35
231. Ueda, T., Greengard, P. 1977. *J. Biol. Chem.* 252:5155–63
232. Huttner, W. B., DeGennaro, L. J., Greengard, P. 1981. *J. Biol. Chem.* 256:1482–88
233. Nestler, E. J., Greengard, P. 1980. *Proc. Natl. Acad. Sci. USA* 77:7479–83
234. Dolphin, A. C., Greengard, P. 1981. *Nature* 289:76–79
235. Nestler, E. J., Greengard, P. 1982. *Nature* 296:452–54
236. DeCamilli, P., Cameron, R., Greengard, P. 1983. *J. Cell Biol.* 96:1337–54
237. DeCamilli, P., Harris, M., Huttner, W. B., Greengard, P. 1983. *J. Cell Biol.* 96:1355–73
238. Huttner, W. B., Schiebler, W., Greengard, P., DeCamilli, P. 1983. *J. Cell Biol.* 96:1374–85
239. Baines, A., Bennett, V. 1985. Unpublished data
240. Craig, S. W., Pardo, J. V. 1983. *Cell Motility* 3:449–62
241. Pardo, J. V., Siliciano, J. D., Craig, S. W. 1983. *Proc. Natl. Acad. Sci. USA* 80:1008–12
242. Carlin, R. C., Bartelt, D. C., Siekevitz, P. 1983. *J. Cell Biol.* 96:443–48
243. Koenig, E., Repasky, E. A. 1984. *J. Neurosci.* In press
244. Levine, J., Willard, M. 1983. *Proc. Natl. Acad. Sci. USA* 80:191–95
245. Nelson, W. J., Colaco, C., Lazarides, E. 1983. *Proc. Natl. Acad. Sci. USA* 80:1626–30
246. Repasky, E. A., Symer, D. E., Bankert, R. B. 1984. *J. Cell Biol.* 99:350–55
247. Lorenz, T., Willard, M. 1978. *Proc. Natl. Acad. Sci. USA* 75:505–9
248. Levine, J., Willard, M. 1980. *Brain Res.* 194:137–54
249. Furthmayr, H. 1979. See Ref. 181, p. 195

Ann. Rev. Biochem. 1985. 54:305–29

ROLE OF REVERSIBLE OXIDATION-REDUCTION OF ENZYME THIOLS-DISULFIDES IN METABOLIC REGULATION

D. M. Ziegler

Clayton Foundation Biochemical Institute, The University of Texas at Austin, Austin, Texas 78712

CONTENTS

0066-4154/85/0701-0305$02.00

PERSPECTIVES AND SUMMARY

In principle, any enzyme bearing an accessible thiol essential for activity is capable of forming protein-mixed disulfides or intramolecular disulfides by reacting with small disulfides. Formation of mixed disulfides or intramolecular disulfides can increase or decrease catalytic activity and examples of both are known. Furthermore, the extent of enzyme S-thiolation[1] would depend on the thiol : disulfide redox potential as well as the nature of the small disulfide and the microenvironment around the accessible protein thiol. These parameters are at least potentially capable of conferring the specificity required for a biological control mechanism through signals transmitted by changes in the thiol : disulfide redox potential as a function of different metabolic states. However, the evidence for regulation of metabolic processes by oxidation and reduction of enzyme thiols is compelling only for oxygenic photosynthetic systems. In chloroplasts there is little question that, in the light, electrons from chlorophyll are transferred to thioredoxin via ferredoxin and the reduced thioredoxin then modulates the activities of key enzymes.[2]

This review will focus on the far more controversial role of reversible enzyme S-thiolation in the regulation of metabolic reactions in vertebrates. The most persuasive arguments for metabolic regulation by this mechanism in animal tissues are usually based on observations that enzymes catalyzing opposing reactions in some closely regulated steps are activated and inactivated in a reciprocal fashion upon incubation with disulfides or thiols. Studies on metabolic changes in cell cultures, perfused organs, and whole animals by agents that produce large changes in cellular thiol to disulfide balance are consistent with the changes in enzyme activity observed in vitro.

On the other hand, crucial evidence for changes in the redox state of cellular proteins in response to physiological stimuli (e.g. hormones) that normally modulate metabolic reactions is virtually nonexistent. Techniques for assessing the thiol : disulfide status or even the existence, let alone changes, in the concentration of protein-mixed disulfides in intact tissues are not adequate to measure the small changes that may occur. Problems in preventing spontaneous oxidation of thiols or reduction of disulfides during cell rupture and analysis are formidable and better methods than those currently available will have to be developed before small changes in the cellular concentrations of S-thiolated enzymes can be adequately assessed experimentally.

---

[1]This term, recently introduced by J. A. Thomas, is an accurate chemical description for the oxidation of protein thiols by disulfides and will be used occasionally in this chapter to denote formation of protein-mixed disulfides by thiol : disulfide exchange with low-molecular-weight disulfides. Since it is a concise description of these reactions its use in future publications in this area is strongly recommended.

[2]Recent reviews (1, 2) have described this aspect of enzyme regulation in chloroplasts in detail and references to the plant work will be limited to that of possible relevance to vertebrates.

Large changes in cellular redox states do not occur normally in most tissues. However the oxidative burst during phagocytosis in leucocytes is an exception and the activation of collagenase by thiol oxidation in these cells is apparently physiologically significant. With this one exception there is still no direct evidence that oxidation or reduction of enzyme thiols is a significant general mechanism regulating metabolic reactions in animals.

REGULATION BY THIOL : DISULFIDE EXCHANGE—AN OVERVIEW

As pointed out by Guzman Baron (3) over 30 years ago, the activity of some enzymes can be modified by the formation of protein disulfides. Covalent modification can result from either the formation of protein-mixed disulfides (equation 1)

$$\text{EnzSH} + \text{RSSR} \rightleftharpoons \text{EnzSSR} + \text{RSH} \qquad 1.$$

or through an intramolecular disulfide (equation 2)

$$\text{Enz(SH)}_2 + \text{RSSR} \rightleftharpoons \text{EnzS}_2 + 2\text{RSH} \qquad 2.$$

These reactions are chemically similar to those involved in the formation of disulfide bonds that stabilize the tertiary structure of proteins (4, 5, 6). They differ from the latter only in that in the native enzyme, disulfides involved in regulation must be accessible to reduction by external thiols since regulation by this mechanism in response to changes in cellular redox state can occur only if the reaction is freely reversible.

The rate of thiol : disulfide exchange between protein thiols and low-molecular-weight disulfides is markedly affected by reaction conditions and is usually quite slow below pH 8.0. The exchange reaction is catalyzed by enzymes present in the cytosol (7) and membrane fractions (8–10), which suggests that it is probably not rate limiting in vivo. The reaction of disulfides with protein thiols is also markedly affected by the nature of the disulfide (11–13). Charge rather than size of the disulfide appears to be the more important variable. Studies with physiological disulfides are usually restricted to the disulfides of glutathione, cysteine, cysteamine, and coenzyme A, although some studies have been carried out with peptide hormones that contain a disulfide bond (14). The oxidation of protein thiols by disulfides is generally nonspecific, and this lack of specificity remains one of the major problems regarding physiological significance of these reactions.

Any of a number of thiols are capable of reducing accessible protein disulfides in vitro, but the high cellular concentration of glutathione (GSH) suggests that this tripeptide, either directly or indirectly, is a major cellular protein disulfide reductant. This and other important biological functions of glutathione have been summarized in proceedings of recent meetings (15–18) and in reviews (19, 20).

Studies on potential regulation in animals of metabolic reactions by thiol : disulfide exchange have focused on four different but interrelated aspects of this problem: (*a*) metabolic changes in animals or intact cells by oxidants that modify the cellular thiol : disulfide redox balance; (*b*) reversible activation or inactivation of enzymes in vitro by thiols and disulfides; (*c*) the possible existence and concentration of low-molecular-weight disulfides and protein mixed disulfides in vivo; and (*d*) enzymatic mechanisms capable of maintaining the cellular thiol : disulfide redox balance in vivo. The studies in each of these areas that are frequently cited in support of regulation by this mechanism will be examined in the following sections.

IN VIVO STUDIES

Tissue thiol : disulfide balance can be perturbed by treating animals with readily absorbed disulfides (21, 22) or with agents that lead to rapid cellular oxidation of thiols. The latter includes direct chemical oxidation of GSH by diamide (23), and enzymatic peroxidation by administered peroxides (24–27), by agents that generate peroxide in situ (25, 28–31), or by compounds that induce enzymatic thiol oxidation by peroxide-independent pathways (32, 33). While the oxidation of thiols by external oxidants affects a large number of complex metabolic processes, only those for which there is some evidence that the changes involve thiol : disulfide exchange with enzymes will be considered. For effects of thiol oxidants on membrane integrity (20, 34), drug toxicity (35–37), calcium homeostasis (38, 39), and tubulin polymerization and mitosis (40), the reader is referred to the references cited. Most studies carried out on intact animals are also not included since many of the metabolic changes are probably due to epinephrine release following the administration of the nearly lethal doses of foreign compounds required to produce significant changes in cellular thiol : disulfide states.

There are, however, a few studies that suggest that changes in carbohydrate metabolism can be directly attributed to large modifications in tissue thiol : disulfide balance. For instance, early work on the mechanism of action of radioprotective agents demonstrated that cysteamine and cystamine produce dramatic alterations in carbohydrate metabolism in all species tested (41–44). Cysteamine and cystamine are metabolically equivalent in the intact animal since they are rapidly equilibrated by known enzyme pathways (33, 45). Both cysteamine and cystamine administered to rats lead to extensive formation of protein-mixed disulfides in serum (46) and in virtually all organs examined (21). Eldjarn & Pihl (11) conclude that the ability to form mixed disulfides is an intrinsic property of sulfur-containing radioprotective agents, and that the effects of these aminothiols on liver glycogen are usually attributed to oxidation of hepatic protein thiols. Liver glycogen can drop to less than 10% of control values within two hours in animals treated with 100–200 mg cystamine or cysteamine/kg body weight. The glycogenolytic activity of these agents is still observed in adrenalectomized and pancreatectomized animals (44), which suggests that the effects are not mediated by hormones.

In addition to their glycogenolytic activity, cystamine and related radio-protective agents inhibit utilization of glucose by human erythrocytes (47), ascites tumor cells (48), and by tissue slices of various rat organs (49), presumably by formation of mixed disulfides between glycolytic enzymes and added aminothiol.

Sies et al (26) have also shown that rat liver perfused with hemoglobin-free media containing peroxides stimulates glucose efflux. Subsequent studies from the same laboratory show that t-butylhydroperoxide perfused into rat liver for 10–12 minutes increases cellular GSSG from 18 to 65 nmol/g tissue (25), and longer perfusion times produce even greater changes in the ratio of GSSG to GSH (50). Whether glucose efflux and oxidation of cellular thiols are directly related is unknown, but the results suggest that a more oxidative environment stimulates glycogen hydrolysis and inhibits glucose oxidation by the glycolytic pathway.

Activation of the phosphogluconate pathway in erythrocytes (51, 52), leuco-cytes (53), isolated adipocytes (54–57), rat heart (58), and lung tissue (59–62) by thiol oxidants has been described. Oxidative stress induced by the herbicide paraquat also inhibits the synthesis of fatty acids in the lung (62, 63). Paraquat is known to generate hydrogen peroxide by cyclic reduction by NADPH and reoxidation by oxygen (64, 65), and paraquat-dependent peroxidation of GSH in mice (66) and in rat livers perfused with this herbicide has been demonstrated (29, 30). However, Keeling et al (62) failed to detect changes in either GSH or GSSG in rat lung tissue two hours after the administration of paraquat, but they did find a linear correlation between lung paraquat and protein-mixed di-sulfides. Inhibition of acetate incorporation into lipid and stimulation of $^{14}CO_2$ production from [1-^{14}C] glucose correlated in a linear fashion with the concen-tration of protein-mixed disulfides. The observed metabolic changes were apparently not due to changes in the ratio of $NADP^+$ to NADPH since Keeling & Smith (67) found that paraquat did not affect steady state concentrations of these nucleotides in the lung, although this herbicide does increase the ratio of $NADP^+$ to NADPH in rat liver (30).

Changes in carbohydrate metabolism produced by external agents that oxi-dize cellular thiols either directly or indirectly as described above are frequently cited to support regulation of these processes by thiol : disulfide exchange. Studies on intact animals or tissues must be interpreted with caution since external oxidants can affect cell surface hormone receptors and mimic hormone action (40, 68). The effects on carbohydrate metabolism in intact tissues are, however, largely consistent with in vitro studies on effects of oxidation and reduction on activities of regulatory enzymes in these pathways.

ENZYMES MODIFIED BY THIOL : DISULFIDE EXCHANGE

Glycogen Phosphorylase Phosphatase and Glycogen Synthase

Glycogen deposition and utilization are controlled by multiple positive and negative effectors and by a complex multicyclic cascade that modifies activity of key enzymes by phosphorylation and dephosphorylation in response to hormonal stimuli [see refs. (69–71) for recent reviews]. Since glycogen synthesis and glycogenolysis are primarily regulated by phosphorylation-dephosphorylation of glycogen synthase and glycogen phosphorylase, changes in activity induced by oxidation-reduction of the associated kinases or phosphatase would also affect glycogen deposition and utilization. Although the effects of disulfides on phosphorylase kinase are not known, phosphorylase phosphatase (protein phosphatase-1) is rapidly inactivated by small disulfides (72–74.) The catalytic subunit of rabbit liver phosphorylase phosphatase is inactivated by GSSG, cystine, homocystine, or cystamine (73). Binding studies with radiolabeled GSSG indicate that inactivation paralleled incorporation of label into protein by thiol : disulfide exchange with one of the two cysteinyl residues of the phosphatase. Treatment of the labeled enzyme with GSH restored activity and was accompanied by loss of the protein-bound radioactivity. The activity of phosphatase treated with cystamine, however, was not reversed by thiols. The irreversible inhibition of phosphorylase phosphatase by cystamine may be the biochemical basis for rapid glycogen loss observed in animals treated with this disulfide (41–44). This interpretation is also consistent with the report of Duyckaerts et al (75) that liver phosphorylase *a* increases in rats treated with cystamine.

While the inactivation of the phosphatase alone could explain the glycogenolytic effects of thiol oxidants, inhibition of glycogen synthase by disulfides would further augment glycogen depletion. Inactivation of rat liver glycogen synthase D by thiol : disulfide exchange with GSSG has been described (76, 77). Four of the eight sulfhydryl groups in the glycogen synthase-glycogen complex are accessible to GSSG. Progressive oxidation of the cysteinyl residues leads to loss of activity and dissociation of the enzyme from glycogen granules. The process can be reversed by addition of dithiothreitol or GSH along with glucose-6-phosphate.

Bovine heart muscle glycogen-free synthase I is also inactivated upon oxidation of peptide cysteinyl residues by GSSG (78, 79). The disulfide-mediated inactivation parallels incorporation of radiolabeled disulfide into protein, and activity is restored by thiols. Inactivation of the synthase by GSSG is blocked by UDP-glucose or glycogen and retarded by glucose-6-phosphate, which indicates that substrates and modifiers affect accessibility of the sensitive protein thiol to GSSG. While these studies show that glycogen synthase is reversibly inactivated by S-thiolation, the physiological significance is not clear. The ratio of GSSG to GSH required to produce significant changes in activity of the liver enzyme (77) is quite high and appears to be well outside normal concentrations present in this tissue.

Phosphofructokinase and Fructose 1,6-Bisphosphatase

Regulation of enzymes catalyzing the interconversion of fructose-6-phosphate and fructose 1,6-bisphosphate is instrumental in switching carbohydrate metabolism between glycolysis and gluconeogenesis, and both phosphofructokinase and fructose 1,6-bisphosphatase are subject to multiple forms of control (80–82). Possible regulation of these enzymes by thiol : disulfide exchange with cellular disulfides has also been suggested and is based largely on in vitro studies with purified enzymes.

Phosphofructokinase from heart muscle (83) and skeletal muscle (84) are reversibly inhibited by GSSG and other small disulfides, and regulation of the skeletal muscle enzyme by thiol : disulfide exchange has been examined in considerable detail by Gilbert (84). In addition to GSSG, the enzyme is inactivated upon incubation with coenzyme A disulfide or CoA-glutathione mixed disulfide at micromolar concentrations. The activity of phosphofructokinase equilibrated with redox buffers containing varying ratios of GSH and GSSG increased about threefold as the ratio of GSH to GSSG increased from 5 to 30. In the presence of ATP, phosphofructokinase is less sensitive to inactivation by GSSG, and significant (greater than 40%) inhibition occurs only at GSH to GSSG ratios below 5. Therefore, significant modulation of phosphofructokinase by this mechanism in vivo is feasible only if changes in the cellular thiol : disulfide redox potential occur over this range.

In contrast to phosphofructokinase, preparations of purified liver and kidney fructose 1,6-bisphosphatase are activated by small disulfides (85–89). Formation of the protein-mixed disulfide restores activity at neutral pH resulting from proteolytic modifications of the native enzyme during isolation from liver (90, 91). Activation of the phosphatase by disulfides requires Mn^{2+} ions but replacing Mn^{2+} with Mg^{2+} leads to a significant decrease in catalytic activity (85, 89). The activation is associated with modification of apparently only one cysteinyl residue in the enzyme (85), and the changes in the catalytic activity are completely reversed upon incubation of the oxidized enzyme with thiols. The enzyme is activated by incubation with cystamine, CoA disulfide, homocystine, and coenzyme A-glutathione mixed disulfide, but not by GSSG.

While the physiological significance of work carried out on an enzyme modified by proteolysis is open to question, a more recent report (92) shows that a more native form of the liver enzyme is also activated by cystamine. Furthermore, the cystamine-activated enzyme is more resistant to proteolysis by cathepsin M. Pontremoli et al (92) suggest that both factors could contribute to the regulation of fructose 1,6-bisphosphatase by thiol : disulfide exchange with cellular disulfides.

Moser et al (93) isolated two forms of fructose 1,6-bisphosphatase from rat liver (apparently identical in molecular weight) that differed only in the oxidation state of the protein thiols. The K_m for fructose 1,6-bisphosphate is virtually identical for both forms of the enzyme, but the more oxidized enzyme is less sensitive to inhibition by AMP. Reduction of the more oxidized form restored sensitivity to AMP inhibition, which suggests that modification of accessible cysteinyl residues by oxidation and reduction can modulate inhibition of this enzyme by the allosteric effector AMP. From the data presented it appears that conversion of the reduced to the oxidized form of the enzyme occurred during the final stages of purification in the absence of dithiothreitol. To what extent, if any, the oxidized form exists in the intact hepatocytes is unknown.

Hexokinase and Glucose-6-Phosphatase

In the liver, cytosolic hexokinases and the membrane-bound glucose-6-phosphatase play a key role in maintaining blood glucose within narrow limits. As Stadtman (94) points out, close regulation of both enzymes is essential to prevent uncontrolled hydrolysis of ATP. The pronounced inhibition of hexokinases by glucose-6-phosphate is undoubtedly the most important mechanism for kinetic regulation of the major isozymes of hexokinase (95).

In addition to regulation by substrate and product a few studies suggest that hexokinases are inactivated by disulfides and reactivated by thiols. Hexokinase purified to homogeneity from bovine brain is extremely sensitive to inactivation by 5,5'-dithiobis(2-nitrobenzoate)(DTNB), presumably by formation of protein-mixed disulfides (96). Oxidation of no more than 2 of the 11 to 12 cysteinyl residues leads to complete inactivation. Full activity is recovered upon addition of excess thiol. The rate of inactivation by DTNB is retarded by glucose and completely blocked by glucose-6-phosphate, which suggests that conformational changes induced by substrate and product influence accessibility of the protein thiols to external oxidants. Inhibition of hexokinase by cystamine and related radioprotective disulfides in rat liver homogenates (97) and other tissue preparations (49) has also been described.

During glucose efflux from the liver the membrane-bound glucose-6-phosphatase must be activated. The extreme lability of this enzyme upon extraction from the membrane has hindered attempts to define mechanisms controlling activity. Although the importance of the phospholipid environment in regulation has been suggested (98), other mechanisms controlling activity of this enzyme are largely unknown. Direct regulation by thiol : disulfide exchange with cellular disulfides is not likely because the enzyme apparently lacks thiols accessible to sulfhydryl blocking agents (99). However, Burchell & Burchell (100) find that GSSG stabilizes glucose-6-phosphatase activity of solubilized preparations. The action of the disulfide is apparently indirect since GSSG only prevents inactivation and cannot restore activity of inactive preparations. Although the mechanism of action of GSSG in stabilizing activity of glucose-6-phosphatase in crude extracts is not known, the work of Burchell & Burchell does show that GSSG can influence activity of this enzyme; perhaps indirectly through inhibition of a protein phosphatase. Increased activity of hepatic glucose-6-phosphatase in the presence of high disulfide concentration is also implied by the rise in blood glucose observed in adrenalectomized rats treated with cystamine (44) and with the efflux of glucose in liver perfused with peroxides (26).

Pyruvate Kinase

Pyruvate kinase catalyzes the last step in glycolysis, and activity of this enzyme in the liver is closely regulated by positive and negative effectors and by phosphorylation-dephosphorylation [see (101) for review]. Two reports suggest that the activity of liver pyruvate kinase L is regulated in vitro by thiol : disulfide exchange with physiological disulfides (102, 103).

Preincubation of rat liver pyruvate kinase with GSSG decreases catalytic activity and this inhibition is completely reversed upon incubating the oxidized enzyme with GSH or mercaptoethanol (102). Modification of this enzyme, presumably by formation of enzyme-mixed disulfides, increases the K_m for phosphoenolpyruvate and the K_a for the allosteric effector, fructose 1,6-bisphosphate. A subsequent report by Mannervik & Axelsson (103) shows that oxidation and reduction of liver pyruvate kinase by GSSG and GSH, respectively, is extremely fast in the presence of the cytosolic thioltransferase. Concentrations of thioltransferase less than one-tenth that present in hepatocytes catalyzes the complete reduction of the oxidized form of pyruvate kinase by GSH in less than one minute.

The physiological significance of the observed reversible inactivation of pyruvate kinase by disulfides is difficult to evaluate because interconversion of the oxidized and reduced forms has not been examined in redox buffers that approach the intracellular ratio of GSSG to GSH. Furthermore, the report by Ballard & Hopgood (104) that phosphoenolpyruvate carboxykinase is inactivated by cystine would argue against regulation of this step in glycolysis and gluconeogenesis by simple changes in cellular thiol : disulfide redox state because high disulfide concentration would inhibit interconversion of phosphoenolpyruvate and pyruvate in both directions. However, the report by

Ballard & Hopgood gave few details on the inactivation of phosphoenolpyruvate carboxykinase by cystine. Furthermore, they state that GSSG did not inactivate the enzyme, which suggests a degree of specificity not usually observed in enzymes modified by thiol : disulfide exchange.

To what extent other enzymes involved in glycolysis and gluconeogenesis are affected by thiol : disulfide exchange with small disulfides is largely unknown. The reported inhibition of glyceraldehyde-3-phosphate dehydrogenase by GSSG (105) was not confirmed by Pihl & Lange (106). The latter investigators could not detect formation of mixed disulfide or changes in activity of the rabbit muscle enzyme treated with GSSG or cystamine, and this enzyme common to both pathways is apparently not affected by changes in the thiol : disulfide redox balance of the media.

Glucose-6-Phosphate Dehydrogenase

The first step in the phosphogluconate pathway in liver, catalyzed by glucose-6-phosphate dehydrogenase, is generally considered rate limiting, and the rate of the reaction is primarily controlled by the ratio of $NADP^+$ to NADPH (107, 108). The enzyme is inhibited by NADPH, and the inhibition approaches 100% at ratios of NADPH to $NADP^+$ greater than 9 (108). Since the ratio of NADPH to $NADP^+$ in mammalian tissues is nearly 100 to 1 (109), Eggleston & Krebs (110) suggest that other mechanisms for regulation of glucose-6-phosphate dehydrogenase must be present in vivo. After an extensive search for a cellular constituent that could relieve the NADPH inhibition of this enzyme, they conclude that of the over 100 metabolites tested, only GSSG is effective at concentrations that approach cellular concentrations. Activation was not due to oxidation of NADPH by GSSG since inhibition of glutathione reductase did not prevent activation. Activation required GSSG and some other factor in liver cytosol because dialysis of liver extracts or purification of the enzyme abolished the effect of GSSG. This factor was not identified, but it was present in all tissues examined except lactating mammary gland. Although this work suggests that GSSG can stimulate glucose-6-phosphate dehydrogenase, the mechanism of activation is obscure.

The stimulation of glucose-6-phosphate dehydrogenase in intact mammalian tissues by agents that oxidize GSH to GSSG (52–55, 57, 59–61) is consistent with the in vitro work but must be interpreted with considerable caution. The reduction of GSSG by NADPH generates $NADP^+$, and the increased availability of the oxidized nucleotide alone could account for the effects observed in intact tissues.

3-Hydroxy-3-Methylglutaryl-CoA Reductase

The reduction of 3-hydroxy-3-methylglutaryl-coenzyme A by NADPH catalyzed by HMG-CoA reductase is the committed step in the biosynthesis of cholesterol (111). Regulation of this reaction by allosteric effectors and by phosphorylation-dephosphorylation of HMG-CoA reductase is described in detail in a recent review (112). The possible regulation of this enzyme by changes in the oxidation state of protein thiols has also been suggested (113–116). Both the yeast (113) and the rat liver enzymes (114) are extremely sensitive to inactivation by disulfides, especially by CoA disulfide. The yeast enzyme is rapidly inactivated upon addition of a slight molar excess of CoA disulfide, and full activity is restored upon incubation with dithiothreitol. The enzyme is also inactivated by GSSG and DTNB, although somewhat higher concentrations are required to achieve complete inactivation. The studies of Dotan & Schecter (114) on the rat liver microsomal enzyme suggest that oxidation of HMG-CoA reductase leads to conformational changes that retard binding of substrate because the reduced but not the oxidized form of the enzyme binds to β-hydroxymethylglutaryl-CoA affinity columns. Analysis of kinetic data obtained with enzyme preparations reduced with increasing concentrations of thiols suggests that the conformational changes are quite complex, and forms of enzyme intermediate between the fully oxidized and reduced protein may be present (115). In addition to formation of enzyme-mixed disulfides and intramolecular disulfides (115), the formation of intermolecular disulfides of HMG-CoA reductase has also been proposed (117). Although these studies indicate that regulation of HMG-CoA reductase by oxidation and reduction is possible, further studies are necessary before the physiological significance of regulation by this mechanism can be evaluated.

Acetyl-Coenzyme A Hydrolase and Serotonin N-Acetyltransferase

The concentration of melatonin (N-acetyl-5-methoxytryptamine) in the pineal gland undergoes large changes during a 24-hour period (118). The concentration, low during the light hours, increases dramatically in the dark and the changes in concentration correlate with the fluctuation in activities of serotonin-N-acetyltransferase (119, 120) and acetyl-CoA hydrolase (120, 121). The transferase is inhibited during the day and activated at night, whereas the hydrolase is most active during the day. The activities of these enzymes appear to be regulated by several factors (119), but Klein and his associates (119, 122) have demonstrated that both are subject to regulation by thiol : disulfide exchange with a number of physiological disulfides.

The pineal N-acetyltransferase is inactivated by preincubation with cystamine and a variety of disulfide-containing polypeptides, of which insulin is the most effective (14, 119, 122). Inactivation by disulfides is reversed by thiols suggesting that enzyme S-thiolation is responsible for inactivation. In contrast to the transferase, acetyl-CoA hydrolase is activated by disulfides both in pineal homogenates (123, 124) and in cultures of pineal cells (125). Of the disulfides tested cystamine is the most effective. The hydrolase is also activated by somatostatin and vasopressin but not by insulin (124).

The differential effects of disulfides and thiols on two key enzymes directly or indirectly involved in the biosynthesis of melatonin certainly suggest that regulation of these enzymes by thiol : disulfide exchange in vivo is possible. Namboodiri et al (14) speculate that a specific disulfide-containing peptide that fits the microenvironment around the sensitive enzyme thiol could confer a high degree of specificity. However, the existence of such a regulatory peptide has not been demonstrated and, as for virtually all enzymes modified by disulfides, the lack of specificity remains one of the major obstacles to regulation of enzymes by this mechanism in vivo.

Protein Kinases

The initiation of protein synthesis in eukaryotes is regulated in part by phosphorylation and dephosphorylation of eIF-2 [see Ref. (126) for recent review]. The formation of the initiation complex in rabbit reticulocyte lysates is extremely sensitive to GSSG (127) and the expression of GSSG inhibition is affected by phosphorylated sugars (128). GSSG-mediated inhibition of protein synthesis appears quite complex, but the studies of Ernst et al (129) suggest that the inhibition may be attributed to activation of a cyclic AMP-independent protein kinase that phosphorylates the alpha subunit of eIF-2. This has not been unequivocally established, and the potential regulation of protein synthesis initiation by GSSG will require further studies on more purified preparations of the associated protein kinases and/or phosphatases.

The possible regulation of bovine kidney pyruvate dehydrogenase kinase by thiol : disulfide exchange has also been described (130). The endogenous kinase activity of purified pyruvate dehydrogenase multienzyme complex is markedly inhibited by some synthetic disulfides and the inhibition is reversed by thiols. None of the physiological disulfides tested were effective inhibitors of the kinase. While specific regulation of this kinase by a natural disulfide is conceivable, the existence of such a regulatory disulfide has not been demonstrated.

Guanylate Cyclase

The guanylate cyclase activity of mammalian tissues is markedly affected by a variety of different oxidants, and the possible role of endogenous peroxides or endoperoxides in the regulation of this enzyme has been reviewed (131). Liver (132) and lung (133) guanylate cyclase are also subject to potential regulation by thiol : disulfide exchange with cellular disulfides and thiols. For example, homogeneous preparations of the rat lung enzyme are inactivated upon incubation with cystamine, cystine, and CoA disulfide but not by GSSG, and the inhibition is reversed by thiols. The bovine liver enzyme is also reversibly inactivated by DTNB, cystamine and, to a lesser extent, by GSSG (132). The inhibition is apparently due to the formation of S-thiolated enzyme because radiolabeled cystine incorporated into the protein is liberated by thiols. Enzyme thiols are also involved in activation of guanylate cyclase activity by nitric oxide and related compounds (134, 135) since GSH or dithiothreitol block activation by nitric oxide. The molecular changes produced by nitric oxide are not known in detail, but guanylate cyclase treated with this agent becomes quite sensitive to inactivation by GSSG (136). These observations suggest that nitric oxide produces conformational changes that expose protein thiols not accessible to GSSG in the untreated enzyme. Although potential concerted or sequential regulation of guanylate cyclase by cellular peroxides and GSSG is possible, the physiological significance of these processes in modulating activity of this enzyme is largely unknown.

The possible regulation of adenylate cyclase by thiol : disulfide exchange has also been reported. The adenylate cyclase in rat brain homogenates is apparently inhibited (137), whereas activity in rat spleen is activated (138), by disulfides. Both studies conclude that disulfides act directly on adenylate cyclase, but further work on more purified preparations appears necessary before this conclusion can be accepted.

Leucocyte Collagenase

Collagenases in vertebrate tissues are usually present in latent forms that can be activated by limited proteolysis or by thiol-blocking agents [see Ref. (139) for recent review]. Activation of collagenase in human polymorphonuclear leucocytes by thiol : disulfide exchange with GSSG, cystamine, or peptides containing accessible disulfides leads to a decrease in molecular weight of the latent collagenase (140) similar to that produced by limited proteolysis. Although the exact mechanism for liberation of the inhibitory peptide is not known, the available evidence suggests that the peptide is attached to collagenase by a disulfide bond. Apparently, oxidation of an adjacent thiol by GSSG initiates intramolecular thiol : disulfide exchange with other cysteinyl residues that leads to reduction of the disulfide bridge joining the inhibitor to collagenase. The reactions are freely reversible, and changes in the ratio of GSH to GSSG cycle the enzyme between the active and inactive forms (141).

Activation of collagenase by this mechanism appears to be physiologically significant. During phagocytosis, leucocytes generate large amounts of hydrogen peroxide (142) and peroxidation of GSH catalyzed by glutathione peroxidase and myeloperoxidase (141) produces GSSG in quantities (143) more than sufficient to activate the latent collagenase.

CONCENTRATION OF TISSUE DISULFIDES

The physiological significance of biochemical studies on regulation of enzymes by reversible S-thiolation depends ultimately on whether disulfides (relative to thiols) are normally present in the concentration range required to modulate activity of enzymes. In the broadest sense, the thiol : disulfide redox state would include total cellular thiols and disulfides that are accessible to cyclic metabolic oxidation and reduction. Since such an all-inclusive model cannot be tested experimentally, most studies have been restricted to measuring GSH, GSSG, and protein-mixed disulfides. It is generally assumed that the various small disulfides and protein-mixed disulfides are rapidly equilibrated in vivo and changes in the ratio of any one redox pair (e.g. GSSG/GSH) reflect changes in the overall thiol : disulfide balance. The validity of this assumption is not known and it should also be pointed out that, because of experimental limitations, measurements on tissue homogenates ignore possible subcellular compartments or unknown redox couples that are not in rapid equilibration with GSH and GSSG.

GSSG

Although tissue GSH can change as a function of age (144) or nutritional state (145), the concentration of GSH is very high and constant relative to GSSG. Therefore, calculations of [GSH]/[GSSG] ratios in intact tissues depend almost entirely on the accuracy of the GSSG determination. There are a number of sensitive and accurate methods for measuring GSH and GSSG and the advantages and limitations of the various procedures have been reviewed (19). The problems of measuring tissue GSSG levels are not in the analysis itself, but lie primarily in preventing artifacts introduced during cell breakage and preparation of the extracts. Although it is generally recognized that thiols are readily oxidized in neutral or alkaline solutions, their facile oxidation by peroxides in acid solutions is not fully appreciated. Oxygenated tissues may contain some endogenous peroxides, and peroxides, as well as other reactive oxygen species, can also be formed upon denaturation of oxygenated hemoproteins. There is, therefore, no assurance that nonenzymatic oxidation of GSH can ever be completely blocked, even in frozen tissue homogenized in strong acid solutions. Because of these and other problems, literature values for tissue GSSG levels must be carefully evaluated and any measurements of tissue GSSG claiming reasonable accuracy must employ methods for preventing oxidation of GSH to GSSG during cell breakage and analysis.

The better methods disrupt the tissue in media containing metal chelators and a thiol-alkylating agent (usually N-ethylmaleimide) (146–148). The concentration of GSSG determined under these conditions is usually less than in procedures where one or both agents are omitted. For example, Vina et al (149) and Harisch & Mahmoud (150) report values for GSSG in rat liver that are 15–20 times higher than the concentration found under more carefully controlled conditions (25, 30, 151, 152). The ratio of GSH to GSSG in perfused rat liver reported by Sies and his associates (30, 50, 153) is no less than 300 to 1. This appears to be the most accurate value currently available for this tissue.

The ratio of GSH to GSSG in rat heart may be somewhat lower, with reported values ranging from 30 (150) to 100 (147, 153). The latter values were obtained on perfused heart essentially by the procedure described by Wendell (147) and appear to be the more accurate. Even these values may be too low because Wendell found that the concentration of GSSG was not significantly altered in hearts perfused with oxygen-free media. Because anaerobiosis is known to affect the [NADP$^+$]/[NADPH] ratio (154), the failure to observe changes in [GSH]/[GSSG] is difficult to explain. There is either a small pool of GSSG not accessible to glutathione reductase in the intact tissue or some GSSG is inevitably formed during tissue disruption and analysis. Further studies are required to resolve these problems. However, from the best values currently available it appears that the ratio of [GSH]/[GSSG] normally present in tissues is far higher than that required to modulate the activity of enzymes in vitro.

Large changes in [GSH]/[GSSG] ratios reported in tissues treated with exogenous thiol oxidants appear to reflect intracellular changes (26, 30, 50, 62, 153). Severe oxidative stress can apparently bring the ratio of [GSH]/[GSSG] and of glutathione-coenzyme A mixed disulfide (155) into the range required to affect the enzymes subject to regulation by reversible S-thiolation. Under these nonphysiological conditions, metabolic regulation by this mechanism is possible.

Protein-Mixed Disulfides

The presumed natural occurrence of a significant pool of protein-mixed disulfides has been reported for ascites tumor cells (156, 157) and a variety of normal mammalian tissues (25, 50, 62, 158–163). These reports are all based on the observation that cellular proteins precipitated with strong acids liberate acid-soluble thiols upon reduction with sodium borohydride or dithiothreitol. The older studies assumed that these acid-soluble thiols were largely GSH and cysteine. More recent studies, however, show that GSH accounts for only a very small fraction of the acid-soluble thiol that is apparently liberated by reduction (30, 50, 157). Modig et al (157) suggest that protein fragments liberated by borohydride account for most of the thiol present in these fractions. After a careful evaluation of the published methods for measuring cellular protein-mixed disulfide, Poulsen (unpublished work this laboratory) also concludes that virtually all of the apparent acid-soluble thiol released upon reduction is due to peptides and unprecipitated protein. It would appear that the reported high concentration of protein-mixed disulfide in normal liver, lung, or heart are experimental artifacts. While protein-mixed disulfides are undoubtedly present, especially in tissues under oxidative stress (25, 50, 51), they appear to be present at levels too low to measure accurately by current published methods.

More recently, Thomas and his associates at Iowa State University (personal communication) have used a different approach in attempting to detect protein-mixed disulfides present within cultured neonatal heart cells. The cells were incubated for short periods with cycloheximide and radiolabeled cysteine. The soluble proteins were extracted, separated electrophoretically, and radioactivity released from individual proteins by dithiothreitol was measured. Intracellular labeling appears quite selective and relatively few (about 20) proteins extracted in the presence of NEM contain detectable radioactivity. However, even the four most heavily labeled proteins contain less than one mole of labeled cysteine per mole of protein subunit. This apparent high degree of selectivity and the resulting low concentration of labeled proteins would explain why protein-mixed disulfides have been so difficult to measure in whole tissue homogenates. This technique for separating and following changes in S-thiolation of specific proteins appears promising. If these preliminary results can be verified, this would be the first unambiguous evidence that S-thiolation of specific proteins occurs normally in mammalian tissues.

ENZYMATIC DISULFIDE REDUCTION AND THIOL OXIDATION

Regulation of enzymes by thiol : disulfide exchange in response to changes in the thiol : disulfide potential must rely on enzyme systems capable of maintaining and regulating this potential. Although enzyme systems that respond to physiological stimuli in mammals have not been described, some of the more likely candidates for such systems are considered below.

Disulfide Reduction

Lipoamide dehydrogenase, glutathione reductase, and thioredoxin reductase are the only well-defined enzymes capable of catalyzing net reduction of disulfides. The reactions catalyzed by the latter two enzymes undoubtedly provide most of the disulfide reductive capacity present in mammalian tissues.

Glutathione reductase catalyzes the reduction of GSSG by NADPH (equation 3). GSH, in turn, can reduce by transhydrogenation (equation 4) a wide variety of disulfides. The latter reactions are catalyzed by cytosolic thioltransferases (7, 164), glutaredoxin (165), and by membrane-bound thiol : disulfide oxidoreductases (protein-disulfide isomerases) (10, 166, 167).

$$GSSG + NADPH + H^+ \rightarrow 2GSH + NADP^+ \qquad\qquad 3.$$

$$2GSH + R_1SSR_2 \rightleftharpoons GSSG + R_1SH + R_2SH \qquad\qquad 4.$$

The substrate specificities of the enzymes catalyzing thiol : disulfide exchange (equation 4) between GSH and other disulfides including protein-mixed disulfides are rather broad. The membrane-bound thioltransferases are primarily involved in catalyzing the isomerization of thiols and disulfides in proteins and are undoubtedly essential catalysts in the formation of intramolecular disulfide bonds during protein biosynthesis (166, 167). They also catalyze reversible formation of protein-mixed disulfides and their participation in the possible modification of enzymes by thiol : disulfide exchange appears likely.

The cytosolic thioltransferase and glutaredoxin appear to be the principal catalysts for the reduction of small disulfides by GSH. Thioltransferase (164) and glutaredoxin (165) purified from rat liver and bovine thymus respectively are similar in molecular weight (about 11,000) and in their GSH-transhydrogenase activities. They are, however, distinctly different peptides and differ significantly in catalytic properties.

Glutaredoxin only catalyzes GSH-dependent transhydrogenation of disulfides, whereas thioltransferase catalyzes transhydrogenation of a large variety of different thiols and disulfides as well as reversible formation of protein-mixed disulfides (168, 169). The broad substrate specificity and high catalytic capacity suggest that, at least in liver, the cytosolic thioltransferase plays a central role in rapidly transferring reducing equivalents from GSH to metabolic networks throughout the cell.

However, transhydrogenation reactions do not change the ratio of thiol to disulfide and the reductive capacity of GSH-dependent transhydrogenations depend ultimately on the reduction of GSSG by NADPH. Regulation of disulfide reduction by this route could occur only if modulation of glutathione reductase takes place in vivo. Within the cell, however, it is generally assumed that the velocity of GSSG reduction is dependent only on the availability of GSSG and NADPH, and physiological effectors of glutathione reductase have not been reported.

The reduction of thioredoxin disulfide by NADPH (equation 5),

$$ThS_2 + NADPH + H^+ \rightarrow Th(SH)_2 + NADP^+ \qquad 5.$$

catalyzed by thioredoxin reductase, may also be an important route for reduction of cellular disulfides. Most studies on thioredoxin function in mammals have focused on its role in the biosynthesis of deoxyribonucleotides (170). However, the extensive studies of Buchanan and his associates (1, 2, 171) have shown that in chloroplasts the reduction of enzyme disulfides by thioredoxin is an important mechanism regulating activity of light-activated biosynthetic pathways. Furthermore, plants contain multiple forms of thioredoxins that exhibit selective reduction of specific enzymes (2), illustrating the importance of thioredoxins in the regulation of plant enzymes.

In contrast to these extensive studies in plants, the role of thioredoxin in the reduction (and potential regulation) of mammalian enzymes has been largely neglected. Mannervik et al (172) have, however, shown that liver thioredoxin coupled with its reductase catalyzes reduction by NADPH of insulin and various disulfides including protein-mixed disulfides. These investigators also compared the reductive capacity of the thioredoxin system in rat liver with that of the GSH-thioltransferase system. Although both systems catalyze reduction of disulfides by NADPH, some significant differences are apparent. The thioltransferase system is more efficient in catalyzing reduction of small disulfides, but peptides with exposed disulfides are more effectively reduced by thioredoxin. These studies suggest that the reduction of protein disulfides by thioredoxin may be more important than generally recognized, and in light of their function in plants further work on regulation of mammalian enzymes by thiol : disulfide exchange with thioredoxin is certainly warranted.

Thiol Oxidation

The autooxidation of thiols is very slow at physiological pH in metal-free solutions (173) and nonenzymatic oxidation of thiols within intact cells is probably insignificant. Disulfide required for synthesis of protein disulfide bonds, whether intramolecular or mixed disulfides, must be generated enzymatically. Enzyme-catalyzed reactions capable of net generation of disulfides within cells are quite limited and only a few have been well characterized.[3]

The oxidation of GSH by H_2O_2 (equation 6), catalyzed by glutathione peroxidase, is usually considered the major route for enzymatic generation of cellular disulfide.

$$2GSH + H_2O_2 \rightarrow GSSG + 2H_2O \qquad\qquad 6.$$

H_2O_2 is formed extensively in tissues (24), and peroxide produced in the cytosol is removed almost exclusively through reduction by GSH (174). The general properties of glutathione peroxidase are discussed at length in a recent review (175).

Isaacs & Binkley (181) proposed that the velocity of GSH peroxidation in vivo is controlled indirectly by cyclic AMP, in view of their observation that catalase activity in rat liver varies diurnally and decreases when rats are treated with dibutyryl cyclic AMP. They speculate that the decrease in catalase activity directs more H_2O_2 into the GSH peroxidase pathway, thereby increasing the rate of GSSG generation and subsequent S-thiolation of regulatory enzymes. Based on the assumption that the change in S-thiolation of enzymes is one of the normal mechanisms regulating glycolysis and gluconeogenesis, their model predicts that both catalase and glutathione peroxidase are essential for processing hormonal signals transmitted by cyclic AMP. However, other studies

---

[3]A number of relatively nonspecific thiol oxidases are also present in extracellular fluids (176, 177) and on the external surface of plasma membranes in some tissues (178–180). The properties and probable function of these thiol oxidases have been reviewed (19). From their location, it appears that they are accessible only to extracellular thiols which makes their participation in the generation of intracellular thiols unlikely. However, translocation of disulfide from extracellular fluids cannot be excluded.

suggest that neither enzyme has an essential role in regulating metabolic reactions in vivo. For instance, catalase activity in acatalasemics is extremely low (182), but these individuals do not exhibit any striking metabolic abnormalities. Furthermore, the glutathione peroxidase activity in tissues from rats maintained on selenium-deficient diets drops to less than 1–2% of control values (27, 183, 184). Yet the virtual absence of this activity does not produce any marked changes in carbohydrate or lipid metabolism (183). Thus it is highly unlikely that catalase or glutathione peroxidase function as postulated (181) in regulating metabolic reactions by controlling enzyme S-thiolation in response to cyclic AMP. The physiological significance of the cyclic AMP induced loss in liver catalase (181) is difficult to interpret at this time.

In addition to peroxidation of GSH, the desulfuration of 3-mercaptopyruvate, catalyzed by the cytoplasmic mercaptopyruvate transsulfurase, is a potential major source of cellular disulfide. The properties of this enzyme and nature of products formed with different acceptors are described in detail in a recent review (185) and only the reaction with thiol acceptors will be considered here.

In the presence of thiols, the desulfuration of 3-mercaptopyruvate yields pyruvate and the persulfide (equation 7).

$$\text{3-mercaptopyruvate} + \text{RSH} \rightarrow \text{Pyruvate} + \text{RSSH} \qquad 7.$$

Persulfides readily transhydrogenate with GSH (or other thiols) generating the disulfide and H_2S (186). The velocity of disulfide formation by this route is apparently determined by the rate of formation of 3-mercaptopyruvate from transamination of cysteine (185). Therefore, changes in the velocity of transamination relative to other routes for the catabolism of cysteine could potentially regulate the generation of disulfide by desulfuration of 3-mercaptopyruvate. Although mechanisms that control cysteine catabolism by alternate routes are not known, the wide tissue distribution of the transsulfurase (186, 187) suggests that desulfuration of 3-mercaptopyruvate may be a major source of cellular disulfide.

The oxidation of cysteamine to cystamine (equation 8) catalyzed by the membrane-bound flavin-containing monooxygenase is the only other reasonably well-defined enzymatic route for generation of intracellular disulfides.

$$2\ H_2NCH_2CH_2SH + NADPH + H^+ + O_2 \rightarrow (H_2NCH_2CH_2S-)_2 +$$
$$NADP^+ + 2\ H_2O \qquad\qquad 8.$$

This monooxygenase is separated from glutathione reductase by compartmentalization, differs in substrate specificity, and is potentially capable of maintaining significant levels of disulfide in the immediate vicinity of cellular membranes. However, questions as to the specificity of the enzyme for cysteamine and the availability of this aminothiol in mammalian tissues have not been fully resolved.

The flavin-containing monooxygenase, purified to homogeneity from hog liver microsomes (188) and more recently from rat liver (189) catalyzes the oxygenation of a wide variety of foreign compounds bearing nucleophilic nitrogen or sulfur functional groups. The role of this enzyme in the metabolism of xenobiotics has been extensively reviewed (190, 191). The catalytic mechanism of the monooxygenase is known in considerable detail (192, 193). Substrates are oxygenated by an intermediate 4a-flavin hydroperoxide form of the enzyme. Activation of substrate is not required for oxygen transfer and any nucleophile that can bind to the active site is oxidized. However, access to the catalytic site is largely restricted to nonionic or nucleophiles bearing no more than one cationic group. Of a wide variety of possible physiological thiols tested, only cysteamine is oxidized at a significant rate (194). Furthermore cysteamine, in contrast to foreign thiols (which are usually sequentially oxygenated to higher oxidation states), is oxidized only to the disulfide, cystamine. At physiological pH cystamine exists as the dication and dications are, without exception, excluded from the active site (188, 190). This property of the enzyme confers a high degree of specificity and insures that this enzyme catalyzes the oxidation of cysteamine only to cystamine. Although these studies suggest that oxidation of cysteamine to cystamine could be a potential source of cellular disulfide, the velocity of the reaction in vivo would be severely limited by the low concentration of cysteamine.

Although it is now generally accepted that cysteamine is a physiological intermediate in the biosynthesis of taurine from cysteine (195), the concentration of cysteamine in mammalian tissue is very low. Ida et al (196) estimate that free cysteamine never exceeds 0.2 μM in rat tissues, although the concentration of free plus protein-cysteamine mixed disulfides may be somewhat higher (194). However, the concentration of cysteamine is apparently much less than that of the flavin-containing monooxygenase in all tissues examined. For example, the concentration of the monooxygenase in hog liver is close to 50 μM (197). Although the concentration of monooxygenase is less in other tissue, it is still an order of magnitude higher than that of cysteamine (194). If cysteamine is the physiological substrate, the velocity of the reaction in vivo must be limited by the concentration of substrate rather than enzyme, and the generation of cellular disulfide by this route could be controlled by the rate of cysteamine biosynthesis. However, mechanisms regulating the rate of cysteamine synthesis have not been determined.

CONCLUDING REMARKS

Regulation of metabolic pathways in vertebrates by reversible S-thiolation of enzymes is an attractive but unproven hypothesis. Better methods than those currently in use must be developed before the existence of S-thiolated enzymes or changes in their concentration in response to normal physiological signals can be adequately evaluated.

The general lack of specificity of disulfide oxidants or thiol reductants in biochemical studies on modulation of enzyme activity by thiol : disulfide exchange remains a major unsolved problem. Closer attention to the role of thioredoxin, glutaredoxin, and/or thioltransferases might reveal that these reactions are more specific than generally recognized. Further work on the number and specificities of cytosolic and membrane-bound enzymes catalyzing thiol : disulfide transhydrogenation could be quite productive.

Protein disulfides, apparently formed within the cell, are present in large concentration in extracellular fluids and marked changes in serum albumin-GSH mixed disulfide as a function of development has been clearly demonstrated (198). Yet little is known regarding mechanisms that generate and maintain significant levels of disulfide in the strong reducing environment present in mammalian cells. Further work on the nature of enzymes maintaining cellular thiol : disulfide balance may eventually reveal how this potential is modulated in vivo. The development of inhibitors specific for different thiol oxidases and disulfide reductases would facilitate such studies.

ACKNOWLEDGMENTS

I am indebted to my associates L. L. Poulsen and K. M. Green for their many helpful suggestions and I also want to thank Ms. Peggy Dunn for typing the manuscript.

Literature Cited

1. Buchanan, B. B. 1980. *Ann. Rev. Plant Physiol.* 31:341–74
2. Buchanan, B. B. 1983. See Ref. 15, pp. 231–42
3. Guzman Baron, E. S. 1951. *Adv. Enzymol.* 11:201–66
4. Anfinsen, C. B., Scheraga, H. A. 1975. *Adv. Protein Chem.* 29:205–300
5. Creighton, T. E. 1978. *Prog. Biophys. Mol. Biol.* 33:231–97
6. Wetlaufer, D. B., Ristow, S. 1973. *Ann. Rev. Biochem.* 42:135–58
7. Mannervik, B., Axelsson, K. 1975. *Biochem. J.* 149:785–88
8. Goldberger, R. F., Epstein, C. J., Anfinsen, C. B. 1963. *J. Biol. Chem.* 238:628–35
9. Venetianer, P., Straub, F. B. 1963. *Biochim. Biophys. Acta* 67:166–68
10. Freedman, R. B., Hawkins, H. C. 1977. *Biochem. Soc. Trans.* 5:348–57
11. Eldjarn, L., Pihl, A. 1960. In *Mechanisms in Radiobiology,* ed. M. Errera, A. Forssberg, 2:231–96. New York: Academic
12. Wilson, J. M., Wu, D., Motiu-DeGrood, R., Hupe, D. J. 1980. *J. Am. Chem. Soc.* 102:359–63
13. Whitesides, G. M., Lilbern, J. E., Szajewski, R. P. 1977. *J. Org. Chem.* 42:332–38
14. Namboodiri, M. A. A., Favilla, J., Klein, D. C. 1981. *Science* 213:571–75
15. Larsson, A., Orrenius, S., Holmgren, A., Mannervik, B., eds. 1983. *Functions of Glutathione.* New York: Raven. 403 pp.
16. Sies, H., Wendel, A., eds. 1978. *Functions of Glutathione in Liver and Kidney,* pp. 1–212. Berlin/Heidelberg/New York: Springer-Verlag
17. Arias, I. M., Jakoby, W. B., eds. 1976. *Glutathione: Metabolism and Function.* Kroc. Found. Ser. 6:1–382. New York: Raven
18. Sakamoto, Y., Higashi, T., Tateishi, N.,

eds. 1983. *Glutathione: Storage, Transport and Turnover in Mammals.* Tokyo: Japan Sci. Soc. Press and Utrecht: VNU Press. 202 pp.

19. Meister, A., Anderson, M. E. 1983. *Ann. Rev. Biochem.* 52:711–60
20. Kosower, N. S., Kosower, E. M. 1978. *Int. Rev. Cytol.* 54:109–60
21. Mondovi, B., Tentori, L., De Marco, C., Cavallini, D. 1962. *Int. J. Rad. Biol.* 4:371–78
22. Eldjarn, L., Nygaard, O. 1954. *Arch. Int. Physiol.* 62:476–86
23. Kosower, N. S., Kosower, E. M., Wertheim, B., Correa, W. 1969. *Biochem. Biophys. Res. Commun.* 37: 593–96
24. Chance, B., Sies, H., Boveris, A. 1979. *Physiol. Rev.* 59:527–605
25. Akerboom, T. P. M., Bilzer, M., Sies, H. 1982. *J. Biol. Chem.* 257:4248–52
26. Sies, H., Gerstenecker, C., Menzel, H., Flohe, L. 1972. *FEBS Lett.* 27:171–75
27. Burk, R. F., Nishiki, K., Lawrence, R. A., Chance, B. 1978. *J. Biol. Chem.* 253:43–46
28. Jones, D. P., Eklow, L., Thor, H., Orrenius, S. 1981. *Arch. Biochem. Biophys.* 210:505–16
29. Brigelius, R., Anwer, M. S. 1981. *Res. Commun. Chem. Pathol. Pharmacol.* 31:493–502
30. Brigelius, R., Lenzen, R., Sies, H. 1982. *Biochem. Pharmacol.* 31:1637–41
31. Farrington, J. A., Ebert, M., Land, E. J., Fletcher, K. 1973. *Biochim. Biophys. Acta* 314:372–81
32. Kreiter, P., Ziegler, D. M., Hill, K. E., Burk, R. F. 1984. *Mol. Pharmacol.* 26:122–27
33. Lauterburg, B. H., Mitchell, J. R. 1981. *Hepatology* 1:525
34. Kosower, N. S., Kosower, E. M. 1983. See Ref. 15, pp. 307–15
35. Mitchell, J. R., Nelson, S. D., Thorgeirsson, S. S., McMurtry, R. J., Dybing, E. 1976. In *Progress in Liver Disease,* ed. H. Popper, F. Schaffner, 5:259–76. New York: Grune & Stratton
36. Lauterburg, B. H., Smith, C. V., Mitchell, J. R. 1984. In *Drug Metabolism and Drug Toxicity,* ed. J. R. Mitchell, M. G. Horning, pp. 321–30. New York: Raven
37. Sies, H., Wahllander, A., Waydhas, C., Soboll, S., Haberle, D. 1980. *Adv. Enzyme Regul.* 18:303–20
38. Orrenius, S., Jewell, S. A., Bellomo, G., Thor, H., Jones, D. P., et al. 1983. See Ref. 15, pp. 261–71
39. Sies, H., Graf, P., Estrela, J. M. 1981. *Proc. Natl. Acad. Sci. USA* 78:3358–62
40. Rebhun, L. I., Miller, M., Schnaitman,

T. C., Jayasree, N., Mellon, M. 1976. *J. Supramol. Struct.* 5:199–219
41. Bacq, Z. M., Fischer, P. 1953. *Arch. Int. Physiol.* 61:417–18
42. Fischer, P. 1954. *Arch. Int. Physiol.* 62:134–36
43. Duyckaerts, C., Bleiman, C., Winand-Devigne, J., Liebecq, C. 1969. *Arch. Int. Physiol. Biochim.* 77:374–76
44. Sokal, J. E., Sarcione, E. J., Gerszi, K. E. 1959. *Am. J. Physiol.* 196:261–64
45. Ziegler, D. M., Poulsen, L. L., Richerson, R. B. 1983. In *Radioprotectors and Anticarcinogens,* ed. O. F. Nygaard, M. G. Simic, pp. 191–202. New York: Academic
46. Eldjarn, L., Pihl, A. 1956. *J. Biol. Chem.* 223:341–52
47. Eldjarn, L., Bremer, J. 1962. *Biochem. J.* 84:286–91
48. Ciccarone, P., Milani, R. 1964. *Biochem. Pharmacol.* 13:183–90
49. Nesbakken, R., Eldjarn, L. 1963. *Biochem. J.* 87:526–32
50. Sies, H., Brigelius, R., Akerboom, T. P. M. 1983. See Ref. 15, pp. 51–64
51. Jacob, H. S., Jandl, J. H. 1966. *J. Biol. Chem.* 241:4243–50
52. Puente, J., Lupu, M., Sapag-Hagar, M. 1981. *IRCS Med. Sci.-Biochem.* 9:813–14
53. Reed, P. W. 1969. *J. Biol. Chem.* 244:2459–64
54. May, J. M. 1981. *Arch. Biochem. Biophys.* 207:117–27
55. Carter, J. R., Martin, D. B. 1969. *Biochim. Biophys. Acta* 177:521–26
56. Czech, M. P., Lawrence, J. C. Jr., Lynn, W. S. 1974. *J. Biol. Chem.* 249:1001–6
57. May, J. M., de Haen, C. 1979. *J. Biol. Chem.* 254:2214–20
58. Zimmer, H.-G., Bunger, R., Koschine, M., Steinkopff, G. 1981. *J. Mol. Cell. Cardiol.* 13:531–35
59. Fisher, H. K., Clements, J. A., Tierney, D. F., Wright, R. R. 1976. *Am. J. Physiol.* 228:1217–23
60. Rose, M. S., Smith, L. L., Wyatt, I. 1976. *Biochem. Pharmacol.* 25:1763–67
61. Bassett, D. J. P., Fischer, A. B. 1978. *Am. J. Physiol.* 234:E653–59
62. Keeling, P. L., Smith, L. L., Aldridge, W. N. 1982. *Biochim. Biophys. Acta* 716:249–57
63. Smith, L. L., Rose, M. S. 1977. In *Biochemical Mechanisms of Paraquat Toxicity,* ed. A. P. Autor, pp. 187–99. New York: Academic
64. Gage, J. C. 1968. *Biochem. J.* 109:757–61
65. Bus, J. S., Cagen, S. Z., Olgaard, M.,

Gibson, J. E. 1976. *Toxicol. Appl. Pharmacol.* 35:501–13
66. Gibson, J. E., Cagen, S. Z. 1977. See Ref. 63, p. 117
67. Keeling, P. L., Smith, L. L. 1980. *Pharmacologist* 22:198 (Abstr. 213)
68. Czech, M. P. 1977. *Ann. Rev. Biochem.* 46:359–84
69. Cohen, P., Klee, C. B., Picton, C., Shenolikar, S. 1980. *Ann. NY Acad. Sci.* 356:151–61
70. Cohen, P. 1978. *Curr. Top. Cell. Regul.* 14:117–95
71. Hers, H. G. 1976. *Ann. Rev. Biochem.* 45:167–89
72. Gratecos, D., Detwiler, T. C., Hurd, S., Fischer, E. H. 1977. *Biochemistry* 16:4812–17
73. Shimazu, T., Tokutake, S., Usami, M. 1978. *J. Biol. Chem.* 253:7376–82
74. Usami, M., Matsushita, H., Shimazu, T. 1980. *J. Biol. Chem.* 255:1928–31
75. Duyckaerts, C., Gilles, L., Liebecq, C. 1971. *J. Physiol. Paris* 63:207A
76. Ernest, M. J., Kim, K.-H. 1973. *J. Biol. Chem.* 248:1550–55
77. Ernest, M. J., Kim, K.-H. 1974. *J. Biol. Chem.* 249:5011–18
78. Lau, K.-H. W., Thomas, J. A. 1983. *J. Biol. Chem.* 258:2321–26
79. Thomas, J. A., Lau, K.-H. W., Mellgren, R. L. 1981. *Cold Spring Harbor Conf. Cell Proliferation* 8:401–11
80. Uyeda, K., Furuya, E., Richards, C. S., Yokoyama, M. 1982. *Mol. Cell. Biochem.* 48:97–120
81. Hers, H.-G., Van Schaftingen, E. 1982. *Biochem. J.* 206:1–12
82. Pilkis, S. J., El-Maghrabi, M. R., McGrane, M., Pilkis, J., Fox, E., Claus, T. H. 1982. *Mol. Cell. Endocrinol.* 25:245–66
83. Froede, H. C., Geraci, G., Mansour, T. E. 1968. *J. Biol. Chem.* 243:6021–29
84. Gilbert, H. F. 1982. *J. Biol. Chem.* 257:12086–91
85. Pontremoli, S., Traniello, S., Enser, M., Shapiro, S., Horecker, B. L. 1967. *Proc. Natl. Acad. Sci. USA* 58:286–93
86. Nakashima, K., Pontremoli, S., Horecker, B. L. 1969. *Proc. Natl. Acad. Sci. USA* 64:947–51
87. Pontremoli, S., Horecker, B. L. 1970. *Curr. Top. Cell. Regul.* 2:173–99
88. Nakashima, K., Horecker, B. L. 1970. *Arch. Biochem. Biophys.* 141:579–87
89. Horecker, B. L., Pontremoli, S., Traniello, S., Nakashima, K., Rosenberg, J. 1974. In *Metabolic Interconversion of Enzymes,* ed. E. H. Fischer, E. G. Krebs, H. Neurath, E. R. Stadtman, pp. 271–84.

Berlin/Heidelberg/New York: Springer-Verlag
90. Nakashima, K., Horecker, B. L. 1971. *Arch. Biochem. Biophys.* 146:153–60
91. Traniello, S., Pontremoli, S., Tashima, Y., Horecker, B. L. 1971. *Arch. Biochem. Biophys.* 146:161–66
92. Pontremoli, S., Melloni, E., Michetti, M., Salamino, F., Sparatore, B., Horecker, B. L. 1982. *Arch. Biochem. Biophys.* 213:731–33
93. Moser, U. K., Althaus-Salzmann, M., Van Dop, C., Lardy, H. A. 1982. *J. Biol. Chem.* 257:4552–56
94. Stadtman, E. R. 1966. *Adv. Enzymol.* 28:41–154
95. Colowick, S. P. 1976. *Enzymes* 9:1–48
96. Redkar, V. D., Kenkare, U. W. 1972. *J. Biol. Chem.* 247:7576–84
97. Lelievre, P., Betz, E. H. 1960. *CR Soc. Biol.* 154:466–68
98. Zakim, D. 1970. *J. Biol. Chem.* 245:4953–61
99. Garland, R. C., Cori, C. F., Chang, H. W. 1976. *Mol. Cell. Biochem.* 12:23–31
100. Burchell, A., Burchell, B. 1980. *FEBS Lett.* 118:180–84
101. Engstrom, L. 1978. *Curr. Top. Cell. Regul.* 13:29–51
102. van Berkel, J. C., Koster, J. F., Hulsmann, W. C. 1973. *Biochim. Biophys. Acta* 293:118–24
103. Mannervik, B., Axelsson, K. 1980. *Biochem. J.* 190:125–30
104. Ballard, F. J., Hopgood, M. F. 1976. *Biochem. J.* 154:717–24
105. Velick, S. F. 1955. In *Methods in Enzymology,* ed. S. P. Colowick, N. O. Kaplan, I:401–6. New York: Academic
106. Pihl, A., Lange, R. 1962. *J. Biol. Chem.* 237:1356–62
107. Gumaa, K. A., McLean, P., Greenbaum, A. L. 1971. *Essays Biochem.* 7:39–86
108. Bonsignore, A., De Flora, A. 1972. *Curr. Top. Cell. Regul.* 6:21–56
109. Veech, R. L., Eggleston, L. V., Krebs, H. A. 1969. *Biochem. J.* 115:609–19
110. Eggleston, L. V., Krebs, H. A. 1974. *Biochem. J.* 138:425–35
111. Rodwell, V. W., McNamara, D. J., Shapiro, D. J. 1973. *Adv. Enzymol.* 38:373–412
112. Beg, Z. H., Brewer, H. B. 1981. *Curr. Top. Cell. Regul.* 20:139–84
113. Gilbert, H. F., Stewart, M. D. 1981. *J. Biol. Chem.* 256:1782–85
114. Dotan, I., Shechter, I. 1982. *Biochim. Biophys. Acta* 713:427–34
115. Roitelman, J., Shechter, I. 1984. *J. Biol. Chem.* 259:870–77

116. Kawachi, T., Rudney, H. 1970. *Biochemistry* 9:1700–5
117. Ness, G. C., Smith, M., Phillips, C. E., McCreery, M. J. 1984. *Fed. Proc.* 43:1786 (Abstr. 2161)
118. Klein, D. C., Weller, J. L. 1970. *Science* 169:1093–95
119. Klein, D. C., Weller, J. L., Auerbach, D. A., Namboodiri, M. A. A. 1980. In *Enzymes and Neurotransmitters in Mental Disease*, ed. E. Usdin, T. L., Sourkes, M. B. H. Youdim, Chap. VI.7, pp. 603–26. Chichester: Wiley
120. Binkley, S., MacBride, S. E., Klein, D. C., Ralph, C. L. 1973. *Science* 181:273–75
121. Klein, D. C., Weller, J. L. 1970. *Science* 169:1093–95
122. Namboodiri, M. A. A., Weller, J. L., Klein, D. C. 1980. *J. Biol. Chem.* 255:6032–35
123. Namboodiri, M. A. A., Favilla, J. T., Klein, D. C. 1979. *Biochem. Biophys. Res. Commun.* 91:1166–73
124. Namboodiri, M. A. A., Favilla, J. T., Klein, D. C. 1982. *J. Biol. Chem.* 257:10030–32
125. Namboodiri, M. A. A., Weller, J. L., Klein, D. C. 1980. *Biochem. Biophys. Res. Commun.* 96:188–95
126. Hardesty, B., Kramer, G., Wollny, E., Fullilove, S., Zardeneta, G., et al. 1983. In *Protein Synthesis*, ed. A. K. Abraham, T. S. Eikhom, I. F. Pryme, pp. 429–46. Clifton, NJ: Humana
127. Kosower, N. S., Vanderhoff, G. A., Kosower, E. M. 1972. *Biochim. Biophys. Acta* 272:623–37
128. Wu, J. M. 1981. *FEBS Lett.* 133:107–11
129. Ernst, V., Levin, D. H., London, I. M. 1978. *Proc. Natl. Acad. Sci. USA* 75:4110–14
130. Pettit, F. H., Humphreys, J., Reed, L. J. 1982. *Proc. Natl. Acad. Sci. USA* 79:3945–48
131. Goldberg, N. D., Haddox, M. K. 1977. *Ann. Rev. Biochem.* 46:823–96
132. Tsai, S.-C., Adamik, R., Manganiello, V. C., Vaughan, M. 1981. *Biochem. Biophys. Res. Commun.* 100:637–43
133. Brandwein, H. J., Lewicki, J. A., Murad, F. 1981. *J. Biol. Chem.* 256:2958–62
134. Braughler, J. M., Mittal, C. K., Murad, F. 1979. *J. Biol. Chem.* 254:12450–54
135. Craven, P. A., DeRubertis, F. R. 1978. *J. Biol. Chem.* 253:8433–43
136. Braughler, J. M. 1983. *Biochem. Pharmacol.* 32:811–18
137. Baba, A., Lee, E., Matsuda, T., Kihara, T., Iwata, H. 1978. *Biochem. Biophys. Res. Commun.* 85:1204–10

138. Soltysiak-Pawluczuk, D., Bitny-Szlachto, S. 1978. *FEBS Lett.* 96:173–74
139. Woolley, D. E., Evanson, J. M. 1980. *Collagenases in Normal and Pathological Connective Tissues.* Chichester: Wiley. 292 pp.
140. Macartney, H. W., Tschesche, H. 1980. *FEBS Lett.* 119:327–32
141. Tschesche, H., Macartney, H. W. 1981 *Eur. J. Biochem.* 120:183–90
142. Baehner, R. L., Gilman, N., Karnovsky, M. L. 1970. *J. Clin. Invest.* 49:692–700
143. Voetman, A. A., Loos, J. A., Roos, D. 1980. *Blood* 55:741–47
144. Hazelton, G. A., Lang, C. A. 1980. *Biochem. J.* 188:25–30
145. Tateishi, N., Sakamoto, Y. 1983. See Ref. 18, pp. 13–38
146. Guntherberg, H., Rost, J. 1966. *Anal. Biochem.* 15:205–210
147. Wendell, P. L. 1970. *Biochem. J.* 117:661–65
148. Akerboom, T. P. M., Sies, H. 1981. *Methods Enzymol.* 77:373–82
149. Vina, J., Hems, R., Krebs, H. A. 1978. *Biochem. J.* 170:627–30
150. Harisch, G., Mahmoud, M. F. 1980. *Hoppe-Seyler's Z. Physiol. Chem.* 361:1859–62
151. Moron, M. S., DePierre, J. W., Mannervik, B. 1979. *Biochim. Biophys. Acta* 582:67–78
152. Adams, J. D. Jr., Lauterburg, B. H., Mitchell, J. R. 1983. *J. Pharmacol. Exp. Ther.* 227:749–54
153. Ishikawa, T., Sies, H. 1984. *J. Biol. Chem.* 259:3838–43
154. Williamson, J. R., Kreisberg, R. A. 1965. *Biochim. Biophys. Acta* 97:347–349
155. Crane, D., Haussinger, D., Sies, H. 1982. *Eur. J. Biochem.* 127:575–78
156. Modig, H. G. 1968. *Biochem. Pharmacol.* 17:177–86
157. Modig, H. G., Edgren, M., Revesz, L. 1971. *J. Radiol. Biol.* 22:257–68
158. Harrap, K. R., Jackson, R. C., Riches, P. G., Smith, C. A., Hill, B. T. 1973. *Biochim. Biophys. Acta* 310:104–110
159. Isaacs, J., Binkley, F. 1977. *Biochim. Biophys. Acta* 497:192–204
160. Habeeb, A. F. S. A. 1973. *Anal. Biochem.* 56:60–65
161. Harding, J. J. 1970. *Biochem. J.* 117:957–60
162. Ziegler, D. M., Duffel, M. W., Poulsen, L. L. 1980. *Ciba Found. Symp.* 72:191–204. New York: Excerta Media
163. Meredith, M. J. 1983. *Anal. Biochem.* 131:504–9
164. Axelsson, K., Eriksson, S., Mannervik, B. 1978. *Biochemistry* 17:2978–84

165. Luthman, M., Holmgren, A. 1982. *J. Biol. Chem.* 257:6686–90
166. Freedman, R. B., Brockway, B. E., Forster, S. J., Lambert, N., Mills, E. N. C., et al. 1983. See Ref. 15, pp. 273–83
167. Morin, J. E., Axelsson, K., Dixon, J. E. 1983. See Ref. 15, pp. 285–95
168. Axelsson, K., Mannervik, B. 1980. *Biochim. Biophys. Acta* 613:324–36
169. Mannervik, B. 1980. In *Enzymatic Basis of Detoxication*, ed. W. B. Jakoby, 2:229–44. New York: Academic
170. Thelander, L., Reichard, P. 1979. *Ann. Rev. Biochem.* 48:133–58
171. Buchanan, B. B., Wolosiuk, R. A., Schurmann, P. 1979. *Trends Biochem. Sci.* 4:93–96
172. Mannervik, B., Axelsson, K., Sundewall, A. C., Holmgren, A. 1983. *Biochem. J.* 213:519–23
173. Misra, H. P. 1974. *J. Biol. Chem.* 249:2151–55
174. Jones, D. P., Eklow, L., Thor, H., Orrenius, S. 1981. *Arch. Biochem. Biophys.* 210:505–16
175. Wendel, A. 1980. See Ref. 169, 1:333–53
176. Janolino, V. G., Swaisgood, H. E. 1975. *J. Biol. Chem.* 250:2532–38
177. Chang, T. S. K., Morton, B. 1975. *Biochem. Biophys. Res. Commun.* 66:309–15
178. Ormstad, K., Lastbom, T., Orrenius, S. 1981. *FEBS Lett.* 130:239–43
179. Grafstrom, R., Stead, A. H., Orrenius, S. 1980. *Eur. J. Biochem.* 106:571–77
180. Takamori, K., Thorpe, J. M., Goldsmith, L. A. 1980. *Biochim. Biophys. Acta* 615:309–23
181. Isaacs, J. T., Binkley, F. 1977. *Biochim. Biophys. Acta* 498:29–38
182. Aebi, H. F., Wyss, S. R. 1978. In *The Metabolic Basis of Inherited Disease*, ed.

J. B. Stanbury, J. B. Wyngaarden, D. S. Fredrickson, pp. 1792–807. New York: McGraw-Hill
183. Hafeman, D. G., Sunde, R. A., Hoekstra, W. G. 1974. *J. Nutr.* 104:580–87
184. Burk, R. F. 1983. *Ann. Rev. Nutr.* 3:53–70
185. Cooper, A. J. L. 1983. *Ann. Rev. Biochem.* 52:187–222
186. Meister, A., Fraser, P. E., Tice, S. V. 1954. *J. Biol. Chem.* 206:561–75
187. Van Den Hammer, C. J. A., Morell, A. G., Scheinberg, I. M. 1967. *J. Biol. Chem.* 242:2514–16
188. Ziegler, D. M., Mitchell, C. H. 1972. *Arch. Biochem. Biophys.* 150:116–25
189. Kimura, T., Kodama, M., Nagata, C. 1983. *Biochem. Biophys. Res. Commun.* 110:640–45
190. Ziegler, D. M. 1984. In *Drug. Metabolism and Drug Toxicity*, ed. J. R. Mitchell, M. G. Horning, pp. 33–53. New York: Raven
191. Ziegler, D. M. 1980. See Ref. 169, 1:201–28
192. Poulsen, L. L., Ziegler, D. M. 1979. *J. Biol. Chem.* 254:6449–55
193. Beaty, N. B., Ballou, D. P. 1981. *J. Biol. Chem.* 256:4619–25
194. Ziegler, D. M., Poulsen, L. L., York, B. M. 1983. See Ref. 15, pp. 297–305
195. Federici, G., Ricci, G., Santoro, L., Atonucci, A., Cavallini, D. 1979. In *Natural Sulfur Compounds*, ed. D. Cavallini, G. E. Gaull, V. Zappia, pp. 187–94. New York/London: Plenum
196. Ida, S., Tanaka, Y., Ohkuma, S., Kuriyama, K. 1984. *Anal. Biochem.* 136:352–56
197. Dannan, G. A., Guengerich, F. P. 1982. *Mol. Pharmacol.* 22:787–94
198. Bump, E. A., Reed, D. J. 1977. *In Vitro* 13:115–18

Ann. Rev. Biochem. 1985. 54:331–65
Copyright © 1985 by Annual Reviews Inc. All rights reserved

MOLECULAR BIOLOGY AND GENETICS OF TUBULIN

Don W. Cleveland and Kevin F. Sullivan

Department of Biological Chemistry, The Johns Hopkins University School of Medicine, Baltimore, Maryland 21205

CONTENTS

0066-4154/85/0701-0331$02.00

PERSPECTIVES AND SUMMARY

Some 18 years ago, tubulin was originally identified as the soluble protein found in most eukaryotic cells that tightly bound the antimitotic drug colchicine (1, 2). Quickly thereafter, this same protein was shown to be a major component of ciliary microtubules (3–6) and the principal protein subunit of microtubules in the cell cytoplasm (7). Although such eukaryotic microtubules had been described morphologically by electron microscopists in the early 1960s as tubular fibrous elements that were present in an impressively wide series of arrays in plant cells, animal cells, and protozoa, it was these seminal biochemical observations coupled with the later discovery of conditions that supported in vitro assembly and disassembly of tubulin into microtubules (8, 9) which opened the way for a careful analysis of the tubulin polypeptide, its polymerization properties, and its companion microtubule associated proteins (MAPs).

It will not be our purpose here to extensively review this early work. Several reviews summarizing what has been learned have previously appeared [e.g., refs. (10–13)] in addition to the authoritative volume of Dustin (14) and the book edited by Roberts and Hyams (15). For a more in depth account of individual aspects of the molecular biology of tubulin we suggest ref. (16).

What has become abundantly clear in the last 18 years is that tubulin polymerizes in cells into a diverse number of microtubule arrays whose assembly and functional properties are defined both by specific programs of cellular differentiation and by cell cycle determinants. For example, microtubules represent the principal structural components of mitotic and meiotic spindles, of eukaryotic cilia and flagella, and of the elongated processes characteristic of neuronal cells. Moreover, they participate in several aspects of intracellular transport {pigment movement in chromatophores, for example [reviewed in ref. (17)]}, in maintenance of various cell surface properties {such as receptor capping [reviewed in ref. (18)]}, and, in concert with actin filaments and 8–10 nm intermediate filaments, they establish overall cell shape and internal cytoplasmic architecture [beautifully illustrated by refs. (19–21) and reviewed in ref. (22)].

This diversity of microtubule function, coupled with demonstration of differential stability of different microtubule classes to depolymerization induced by drugs or other agents (23, 24), quickly fueled speculation that different microtubule classes might be assembled from different tubulin subunits, thereby establishing functionally distinct microtubules. Initially, this hypothesis was tested using a variety of biochemical methods. Most recently, newer approaches utilizing molecular biology and molecular genetics have focused on the question of how microtubule specificity and function, both temporally and spatially, is established in eukaryotes.

We now know, as had long been predicted, that the tubulins are encoded by multigene families which direct the synthesis of peptides which are both highly conserved and heterogeneous. In vertebrates these different tubulins are expressed in complex developmental patterns which all but rule out simple kinds of tissue specific expression of individual polypeptides. Indeed, the complex dominance/epistatic relationships among mutant tubulins in both lower and higher eukaryotic cells imply that multiple tubulin gene products interact with each other in cells, although surprisingly little direct data have yet emerged. Additional structural variation among tubulin gene products is achieved by posttranslational modifications which are subject to strict developmental and morphogenetic controls.

While the data from some lower eukaryotic species have demonstrated that single gene products are sufficient for construction of all essential microtubules, results from higher species clearly suggest that multiple gene sequences are required. This requirement may arise from either of two sources. From the finding that an individual β tubulin polypeptide functions in assembly of all microtubules in spermiogenesis, Raff and coworkers (197) have presented a compelling argument that multiple tubulin genes may be functionally equivalent, representing duplicated genes which have evolved into different genetic regulatory environments. On the other hand, the sum of the presently available sequence data indicates interspecies conservation of putative isotype specific variable region sequences, thus offering strong molecular evidence for the alternative hypothesis of different, functionally specialized tubulin gene products. It seems likely that both hypotheses will prove correct in some instances.

Finally, the study of tubulin regulation has emerged as an important model system in which to dissect the fundamental mechanisms of eukaryotic gene regulation. Cells regulate tubulin gene expression at several levels: (a) selection of which of multiple genes to express, (b) coordination of the synthesis of α and β tubulins, and (c) control of the level of tubulin expression during the cell cycle and during development and differentiation. Regulation both at the level of mRNA stability as well as at the level of transcription has been found to contribute to the proper orchestration of tubulin synthesis.

We have not endeavored in the present review to provide an exhaustive citation of the complete literature, but rather have attempted to discuss those experiments that have documented new discoveries or redirected the field toward new avenues of investigation. In so doing, we have had to make numerous judgments, some of which will no doubt prove to be in error. For such errors, both of commission and omission, we ask our colleagues' and readers' indulgence. At the least, we hope that we have succeeded in clarifying which important questions have been answered, which ones have solutions that are now on the horizon, and which questions have yet to be experimentally posed.

HETEROGENEITY OF TUBULIN POLYPEPTIDES: AN HISTORICAL CHRONICAL

Based on the apparent functional and biochemical heterogeneity of microtubules, many investigators have attempted to address the question of whether cells utilize functionally specialized tubulins in different microtubules. In fact, upon observing biochemical differences among different microtubule systems within developing spermatids of the cranefly, Behnke & Forer (23) first posed this question prior to the identification of tubulin. Unfortunately, two factors slowed definitive work in this area for many years. One was the demonstration, by a variety of techniques, {including electrophoresis [e.g., refs. (25–27)], immunological analysis (28, 29), pharmacological analysis (30, 31), protein sequencing (32), and in vitro assembly studies [(e.g., ref. 33)]} that tubulins are remarkably conserved during evolution. This apparent invariance gave rise to a widely held impression that 'tubulin is tubulin is tubulin' (34). A second and perhaps more significant factor was that analytical techniques capable of distinguishing subtle differences within a group of proteins did not become routinely available until fairly recently. Thus, many reports of tubulin heterogeneity were based on data that can most charitably be described as less than convincing, leading many in the field to overlook the potential significance of the question.

Tubulin heterogeneity is, however, quite real and, as we shall show below, is manifest at both the genetic and posttranslational levels in a complex pattern that correlates with the development and differentiation of a great variety of eukaryotic cells. It is thus important and appropriate to review the significant findings made at the protein level, since it was these efforts which defined the problems that are now approachable by more well-defined experimental techniques.

Following the initial discovery that tubulin in the native form exists as a dimer (4), it soon became apparent that highly purified tubulin contains two different polypeptides present in equimolar quantities (25, 35–38). Bryan & Wilson (35) demonstrated that the two chains, which they named α and β tubulin, had different amino acid compositions and proposed that native tubulin was a heterodimer. The heterodimer model has since been supported by chemical crosslinking studies (39), optical diffraction images of zinc-induced tubulin sheets (40, 41), by the demonstration of both chains in equimolar quantities in all tubulins [see ref. (41a)], and by coordinate synthetic regulation (34, 42, 43).

A number of reports in the early 1970s suggested that the heterogeneity of tubulin extended beyond the well-documented α/β heterogeneity. Witman et al (44) presented convincing evidence showing that the outer doublet tubulins of *Chlamydomonas* flagella resolved into 5 bands by isoelectric focusing. Subsequently, electrophoretic heterogeneity was documented in the α tubulin chains of sea urchin mitotic and ciliary but not flagellar tubulins (45). In a frequently cited study, Fulton and coworkers (46, 47) advanced a 'multi-tubulin hypothesis' which proposed the existence of multiple genes encoding functionally distinct tubulin subunits. This hypothesis was based on their preparation of an antiserum that was reported to react only with flagellar tubulin and not with the abundant cytoplasmic tubulin found in the amoeboid stage of *Naegleria gruberii*, thus apparently demonstrating that the two tubulins were chemically distinct. Without direct data, the two tubulins were postulated to be the products of different genes. In fact, in light of the significant overestimation of the quantity of tubulin in the amoeboid form of *Naegleria* (48), their conclusions were based on an immunological reagent for which, in hindsight, there is no compelling proof of specificity. Nevertheless, their hypothesis presaged subsequent definitive studies.

Based on electrophoresis, peptide mapping, and amino acid analysis, the biochemical search for multiple tubulins continued as Stephens reported differences not only between cytoplasmic and flagellar tubulins but between the central pair, A- and B-subfiber tubulins of flagellar and ciliary microtubules of the sea urchin (49–51). These were interpreted as indicative of primary sequence differences among tubulins within individual organelles of a single species. It is now not clear whether this conclusion was entirely warranted. The possibility of contamination of tubulin samples by nontubulin proteins that possess molecular weights similar to α and β tubulin [as have been shown to be stably associated with sea urchin doublet microtubules (52)] cannot be rigorously excluded. Furthermore, it is quite possible that the majority of ciliary tubulin derives from maternal cytoplasmic tubulin in the sea urchin (53). These difficulties of interpretation, even with the carefully implemented experiments of Stephens, highlight the difficulties of demonstrating primary sequence differences by indirect procedures.

The most extensive microheterogeneity in tubulin has been documented in vertebrate brain. Early reports demonstrated that tubulins isolated from brain tissue as well as cultured mouse neuroblastoma and glioma cells could be resolved into 5–9 components that differed in isoelectric point (54–58). Improved isoelectric focusing techniques have recently allowed more definitive resolution of brain tubulins: 7 α species and 10–14 β species that focus from pH 5.3 to 5.8 have been documented in several species (60–65). Purified isoelectric variants yield single bands of the original isoelectric point upon refocusing (60), thus demonstrating that isoelectric heterogeneity is intrinsic to individual tubulin species.

During vertebrate neural development, this surprisingly complex tubulin pool undergoes significant modulation, as initially demonstrated at low resolution (56, 57, 66) and subsequently confirmed at higher resolution in several laboratories (61–64). In the rat, the β tubulin pool increases from an apparent 3 species in embryos to 14 species in adults; α tubulin on the other hand has been resolved into 7 species that are present throughout development but change dramatically in relative abundance, accumulating in the more acidic variants during maturation [e.g., ref. (62)]. No further changes have been seen upon aging in the adult (61). A similar modulation is also observed in the chick, although in this instance the main developmental differences involve alteration in the relative quantities of species that are present throughout development (64).

Moreover, the extensive heterogeneity of brain tubulin is not due simply to the superposition of tubulins derived from the many cell types of the brain since individual neurons in culture possess complex tubulin pools (67). However, different cells or cell populations within the brain appear to possess different complements of tubulins. Several groups have reported differences in the composition of tubulin protein or tubulin obtained by in vitro translation of mRNA from different regions of the brain (68, 69) or different cell lines in culture (70, 71). In fact, the demonstration of heterogeneity in rat brain tubulins translated in vitro by Marotta and coworkers (72) was perhaps the first definitive indication that tubulins are encoded by multiple genes in vertebrates.

Microheterogeneity is not restricted to flagellar and brain tubulins. Multiple tubulins have been detected in *Drosophila* (73, 74), brine shrimp (75), *Physarum* (76), the unicellular algae *Chlamydomonas* and *Polytomella* (77, 78), the trypanosomatid *Crithidia fasciculata* (79), the fungus *Aspergillus* (80), and essentially all vertebrate tubulin sources examined (81–85).

COMPLEXITY OF TUBULIN GENE SEQUENCES

Although the initial studies of tubulin heterogeneity often failed to receive widespread acceptance because they could not address the source of the observed heterogeneity, these experiments when taken in toto, clearly indicated that both α and β tubulins exist in multiple forms in many, if not all, eukaryotic organisms. With the application of molecular cloning techniques, it became possible to investigate the origin of tubulin polypeptide heterogeneity by analysis of the genes that encode each subunit.

The earliest construction and identification of cloned sequences complimentary to α or β tubulin mRNAs came from the work of Cleveland et al (87). Starting from mRNA isolated from embryonic chick brain which was known to be heavily enriched in tubulin RNAs (88), these authors obtained cDNA clones carrying copies of a chick brain α and a chick brain β tubulin mRNA. Using these cloned sequences as hybridization probes, they were able to establish (as had long been anticipated) that tubulin DNA sequences were highly conserved throughout metazoans. Moreover, these data demonstrated the existence of multiple (and apparently unlinked) gene segments for α and β tubulin in a variety of organisms.

Soon after this initial report, cloned tubulin sequences were isolated from *Drosophila* (90–92) *Chlamydomonas* (43, 93, 94) sea urchin (95), rat (96, 97), and human (98, 99). In each case, the initial experiments (confirmed by later work) again suggested that each species investigated contained small multigene families of α and β tubulin sequences which (for the most part) were unlinked to each other in the genome.

From these and later studies, a firm picture of tubulin sequences within eukaryotic genomes has now emerged. This is summarized in Table 1 in which the complexity, number of known functional genes, and organization within the genome for tubulin gene sequences in a variety of species has been catalogued. The assumption implicit in identification of tubulin genes in each organism is that all tubulin sequences are sufficiently highly conserved to be detected by hybridization to heterologous sequences. As we shall see later, this is, in fact, not always the case.

Tubulin Genes in Unicellular Organisms

Beginning with the simplest example, the budding yeast *Sacchromyces cervisiae* contains a single intronless β tubulin gene, initially identified by its weak hybridization to a cloned chicken β tubulin (130). No additional β genes could be detected even by hybridization to the isolated yeast sequence. On the other hand, two α tubulin genes have been identified in a fission yeast, *Schizosaccharomyces pombe* (129). Although both genes are functional, one is intronless while the other contains a single intron interrupting amino acid codon 19.

Like yeast, a comparatively simple tubulin gene family is also present in the unicellular algae *Chlamydomonas* which contains two α and two β tubulin genes (43, 93). Each of these four sequences represents a functional gene (43, 131). In contrast, the amoeboflagellate *Naegleria* appears to contain as many as 8 α tubulin genes (127). Moreover, unlike yeast and *Chlamydomonas*, no significant hybridization between *Naegleria* and vertebrate tubulin sequences has been detected.

Curiously, while tubulin gene sequences are dispersed in essentially all metazoans investigated [with the exception of some sequence clustering in sea

Table 1 Tubulin gene sequences in various species

Species		No. known sequences per haploid genome	No. known functional genes	Known intron positions (residue no.)	Organization within the genome	References
Human	α	15–20	2	—	Dispersed	99,100
	β	15–20	3	19/20, 56, 93	Dispersed	98, 101–103
Rat	α	10–15	2	1/2, 76, 125/126	Dispersed	96, 97, 104
	β	10–15	3		Dispersed	105, 106, 106a
Mouse	α	10–15	2	—	Dispersed	107
Chicken	α	4–5	2	—	Dispersed	87, 108
	β	6–8	6	19/20, 56, 93	Dispersed	109–112
Sea urchin	α	10–12	2	—	Some clustered α's	95, 114, 115
	β	11–13	3	—	Some clustered β's	95, 114, 115
Drosophila	α	4	4	—	Dispersed	34, 74, 90–92, 116, 117
	β	4	4	—	Dispersed	74, 90, 117
Trypanosomes	α	13–17	all (?)	—	Tandemly duplicated as α/β pairs	118
	β	13–17	all (?)	—		118
Leishmania	α	7–15	all (?)	intronless	Tandemly duplicated α's	119–121
	β	8–9	all (?)	intronless	Tandemly duplicated β's	119–121
Physarum	α	4	2	—	Dispersed	76, 123, 124
	β	3	3	—	Dispersed	76, 122–124
Aspergillus	α	—	2	—		125, 126
	β	—	2	—		125, 126
Naegleria	α	8		—		127
Tetrahymena	β	1	1	—	Single gene	128
Chlamydomonas	α	2	2	—	Dispersed	43, 93
	β	2	2	8/9, 57, 131	Dispersed	43, 93
Yeast	α[a]	2	2	1 intronless; 1 at aa 19	Dispersed	129
	β[b]	1	1	intronless	Single gene	130

[a] *Schizosaccharomyces pombe.*
[b] *Saccharomyces cerevisiae.*

urchin (114)], the arrangement of tubulin genes in unicellular parasites of metazoans has established that clustered, tandemly duplicated arrays of α and β tubulin sequences do exist in these organisms. Moreover, as if to demonstrate that all conceivable possibilities are represented, the clustering may be of all α tubulin genes or all β tubulin genes (119–121) or as tandemly linked α/β pairs (118).

Tubulin Genes in Drosophila

For the most extensively studied system, *Drosophila,* the combined data from four groups have demonstrated the presence of four α and four β tubulin gene sequences, each of which is represented once per haploid genome (34, 90–92, 117). In situ hybridization against polytene chromosomes provided the original demonstration that, in contrast to most previously studied multigene families (e.g., the globins, histones, or immunoglobulins), the *Drosophila* tubulin genes are not linked but are instead dispersed throughout the genome (90–92). As discussed in detail below, all eight *Drosophila* tubulins are functional genes.

Tubulin Genes in Mammals: Authentic Genes Amid a Sea of Pseudogenes

In contrast to the relatively simple gene families in many lower eukaryotes, the matter of tubulin gene sequences in vertebrate genomes is another issue. As initially reported (87), mammalian genomes contain up to 20 sequences that carry strong homology to both the N-terminal and C-terminal coding sequences for either α tubulin or β tubulin. The pioneering work of Cowan and his coworkers (102, 132) in the analysis of the human tubulin gene families has shown that many, if not most, of these sequences are pseudogenes. In the case of the human β tubulin family, for example, of twelve genomic segments analyzed thus far (101–103, 132; N. Cowan, personal communication), only three represent functional genes. All of the remaining nine sequences are pseudogenes which contain multiple deletions and/or in-frame translation termination codons within the exon sequences.

The nature of these tubulin pseudogenes is an intriguing story in itself. Only one of the nine human β sequences is a 'traditional' pseudogene that contains intervening sequences. The other eight are of a novel and unexpected class characterized by (*a*) the lack of all intervening sequences, (*b*) the presence of a long coded poly-A tract at the 3' end, and (*c*) the presence of a 10–15 base pair direct repeat in both the 5' and 3' flanking genomic DNA. These findings clearly imply that these pseudogenes arose by a reverse transcription event in which a mature tubulin mRNA was copied into DNA and inserted at a staggered chromosomal break (132). This class of pseudogene has now been seen in a variety of other genes including a rat α tubulin (104) and will probably prove to be a common eukaryotic phenomonon.

Given the complexity introduced by these pseudogenes, the determination of functional tubulin gene sequences in mammals is a truly formidable task. Nonetheless, by the initial isolation of cDNA clones, three human β tubulin genes (101, 102; N. Cowan, personal communication), three rat β tubulin genes (105, 106), two human α tubulin genes (100), and two rat α tubulin genes (96, 97) have now been identified to be functional. It is not known how many additional α or β tubulin genes in either genome represent authentic genes.

Tubulin Genes in the Chicken

Unlike other vertebrate genomes investigated, the chicken contains relatively few sequences with strong homology to α tubulin or β tubulin (87). Among the β genes, the four gene segments that contain sequence homology detectable by hybridization to the extreme N-terminal and C-terminal coding sequences were initially isolated (110). In addition, 2–3 more divergent segments were later identified (112), one of which has been cloned. When viewed together with the discovery at the protein level of an erythrocyte β tubulin polypeptide (84, 85) which may not derive from any of these 6–7 gene segments (J. Havercroft, D. Cleveland, D. Murphy, unpublished) and with the inferred presence of an homologous, constitutively expressed sister gene to β4 (111, 113) these data now suggest that as many as 7–9 different β genes may exist. Moreover, if additional genes as divergent in sequence as the erythrocyte β tubulin are present, they might also be overlooked by the hybridization assay. At present, at least five chicken β tubulin genes are known to encode authentic polypeptides (84, 85, 110–113).

MULTIPLE TUBULIN GENES ENCODE POLYPEPTIDES THAT ARE BOTH CONSERVED AND HETEROGENEOUS

With the isolation of the initial cloned sequences for α or β tubulin (87), it became possible to rapidly deduce the encoded polypeptides through DNA sequence analysis. Thus, Valenzuela et al (89) obtained the sequence of a chick β tubulin (the first complete tubulin sequence available from any source) and 92% of the polypeptide sequence for a chick brain α tubulin subunit. Comparison of these sequences demonstrated a 40% homology between the α and β polypeptides, thus confirming the widely held belief of a common evolutionary origin of the two subunits.

Shortly thereafter, data obtained both by direct protein sequencing (133, 134) and by DNA sequencing rapidly expanded the number of known tubulin sequences. Although the complete repertoire of tubulin sequences within a single species is only available from the simplest organisms, several significant principles have already begun to emerge from comparative sequence data.

Based partly upon the conserved biochemical properties of tubulins from different species and partly upon comparisons derived from fragmentary N-terminal sequence data (32), it has been widely believed that the tubulins are among the most highly conserved proteins yet studied. Initially, this prejudice was seemingly confirmed when comparison of the primary sequences of a chicken brain (89) and porcine brain β tubulin polypeptide (134) revealed a conservation of >99%. In addition, although protein sequencing identified microheterogeneity among α tubulins (133) and β tubulins (134) in porcine brain, this heterogeneity was detected at only a handful of residue positions, suggesting that the differences between different tubulin polypeptides were quite subtle.

However, subsequent data, first from Cowan's laboratory and then from Cleveland's and Farmer's, have made it clear that this high degree of sequence conservation is both correct and very probably misleading, at least for β tubulins. To illustrate this finding, the sequences for all presently available β tubulin polypeptides have been assembled in Figure 1. Although considerable sequence conservation is evident, a surprising degree of sequence heterogeneity exists among different expressed tubulins within a single species. For example, comparison of one partial (53%) and four essentially complete chicken β tubulin sequences demonstrates divergence in as many as 8.7% (40 of 449) of the residue positions. This level of divergence is greater than that seen in any interspecies comparison of available metazoan β tubulin sequences. Moreover, the deduced polypeptides encoded by the 5 chicken β genes specify four different isotypic classes differentiable by primary sequence. Similar isotypic differences have been noted in sequence data from two human β tubulins [Dβ1 and Dβ2 in Figure 1 (101, 103)] and in partial sequence data from three developmentally regulated rat brain β tubulins [rat 1, rat 2, and rat 3 in Figure 1 (106a)].

Regional Variability in Metazoan β Tubulin Polypeptides Suggests Distinct Isotypic Classes

The sequence heterogeneity among the known metazoan tubulin polypeptides is not, however, randomly distributed throughout the polypeptide, but rather is localized into clusters of heterogeneity and clusters of extreme conservation. The most striking example of sequence conservation lies between residues 401 and 425. This region is identical in all currently known β tubulins including yeast. Evidence for the most pronounced variable cluster, which is confined to the extreme C-terminal residues beyond amino acid position 430, first emerged from comparison of two human β tubulin isotypes (101). Surprisingly, this region is heterogeneous in length as well as in sequence. In addition, as we have

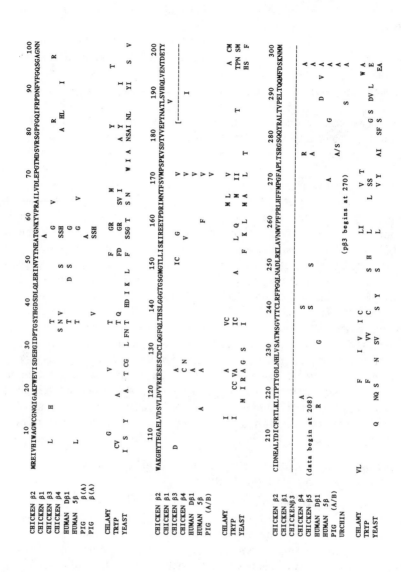

```
                   310        320        330        340        350        360        370        380        390        400
CHICKEN  β1/β2  AACDPRHGRYLTVAAIFRGRMSMKEVDEQMLNVQNKNSSYFVEWIPNNVKTAVCDIPPRGLKMSATFIGNSTAIQELFKRISEQFTAMFRRKAFLHWYTG
CHICKEN  β3     [------]       V
CHICKEN  β4                    TV              AI  S        V      S
CHICKEN  β5                    TV  P           AI           V      AS                                      S          F
HUMAN    Dβ1                   V               MD                  AV
HUMAN    5β                    V          S  S                     AV
PIG      (A/B)                 V
URCHIN         (pβ2 begins at 316)              (pβ1 begins at 349)

                                                                                     M V
CHLAMY          C  A     AS L    T          I                   SS         K         M R VG  L    A
TRYP            Q  A     LSL     T    KV V   EDE HK   S   D   LF I SS   Q   SVA Q  D A  A  S   VGD  S    K        S
YEAST           A  N     F       KV V                                      K   AV  N C

                   410        420        430        440
CHICKEN  β1/β2  EGMDEMEFTEAESNMNDLVSEYQQYQDATADEQGE*FEEEGEEDEAB
CHICKEN  β3                            E E   *    A   EAEB
CHICKEN  β4                            E E  *MY DD   ESEQGAKB
CHICKEN  β5                     E      NDGE A  DDE   INEB
HUMAN    Dβ1                           E EED* G  A  *  B
HUMAN    5β                            •     A   EV  B
PIG      (A/B)                                   A   B
RAT   RBT.1    (data begin at 431)         E     EG   B
RAT   RBT.2    (data begin at 431)     E • •   A   EV  B
RAT   RBT.3    (data begin at 431)     E EED* G  A  *  B
URCHIN   β1                            E E  • D  EG   EAAB
URCHIN   β2/β3                         E E  • D  EG*  EAAB

CHLAMY                            S E E  • G E  AB
TRYP                              IE E  • D  EYB
YEAST                      E   VEDDE *VD N DFGFAPONQDEP1TENFEB
```

Figure 1 Polypeptide sequences of all known β tubulins. The one letter amino acid code has been utilized to display the known β tubulin polypeptide sequences. Except for chicken β1/β2 for which the complete sequence is shown, only amino acid positions which differ from β1/β2 are indicated for the other polypeptides. The letter B represents the translation termination codon position. The sequences were taken from: chicken β1 - K. Sullivan, J. Lau, and D. Cleveland, unpublished; chicken β2 - ref. (89); chicken β3 - ref. (112); chicken β4 - ref. (113); chicken β5 - ref. (112); human Dβ1 - ref. (101); human 5β - ref. (103); Pig A and B - ref. (134); Rat1, rat2 and rat 3 - ref. (106a); Chlamy (*Chlamydomonas*) - ref. (131); Tryp (Trypanosome) - N. Agabian, personal communication; yeast - ref. (130); urchin β1–β3 (*S. Purpuratus*) - ref. (114).

recently noted (113), a second markedly divergent cluster is found between residues 33 and 59.

Moreover, the present data demonstrate that certain sequences, within the clusters of amino acid divergence, have been highly conserved during evolution. For example, the C-terminal variable regions of the dominant neural β tubulins in chicken [β2 (89)], pig [variant A (134)], and rat [rat 1 (106a)] are essentially identical. Further, the apparent constitutively expressed β tubulins from rat [rat 3 (106a)] and human [Dβ2 (101)] are also *identical* to each other in this region. A similar situation exists for an additional pair of rat [rat 2 (106a)] and human [Dβ1 (101)] β tubulin subunits. Such conservation of variable region sequences in functionally related tubulins of different species suggests that sequence divergence in these regions is not selectively neutral, but rather that specific sequences have been evolutionarily maintained by positive selective pressure. Consideration of these findings has lead us to suggest (113) that these variable domains may define specific β tubulin isotypes.

For metazoan α tubulins, although Ponstingl et al (133) initially identified four different α tubulins that differed subtly in sequence in porcine brain, no substantial sequence heterogeneity has yet been seen among two human (100), two rat (96, 97), and one chicken α tubulin (89). Whether isotypic classes are present for α tubulins is therefore still an open question.

Tubulin Polypeptides in Lower Eukaryotes

For tubulin sequences in unicellular eukaryotes, the situation is somewhat different. In the yeast *Sacchromyces cerevisiae,* only a single β tubulin gene is present; for *Chlamydomonas* as well, although two different β tubulin genes are expressed (43, 131), both encode identical polypeptides (131). These polypeptides, although clearly homologous to higher eukaryotic β tubulins, display amino acid homology to metazoan brain sequences of only 89%. Similarly, the yeast β tubulin sequence shows only 70% homology and is 12 amino acids longer than the dominant vertebrate brain subunit (130). In contrast, the two α tubulin polypeptides of the yeast *Schizosaccharomyces pombe* are surprisingly divergent, retaining not only a low 76% homology to known mammalian α tubulins but also showing only 85% homology to each other. Curiously, the major regions of divergence between the two yeast α isotypes are concentrated between residues 444–455 and 31–52, strikingly reminiscent of the positions of the clusters of highest divergence in the metazoan β tubulins [see Figure 1 and ref. (113)].

MULTIPLE TUBULIN GENES ARE EXPRESSED IN COMPLEX DEVELOPMENTAL PROGRAMS

It appears certain from the data currently available (Table 1) that the expression of different α and β tubulin genes (and at least in some cases, different polypeptides—see above and Figure 1) is a general property of metazoan species. This is also the case for some unicellular eukaryotes. Collectively, these findings raise the interesting and important question of whether there exist developmentally regulated programs that specify the timing of expression of individual tubulin genes. The answer for a variety of genomes is an unambiguous yes.

Differential gene expression of a tubulin gene family was first documented in *Drosophila*. Initially, by identification of a mutation in a β tubulin polypeptide whose phenotype was male sterility, Raff and coworkers (73) documented the presence of a β tubulin polypeptide which was expressed in testis but not ovary. Later work from this group failed to detect expression of this 'testis specific' polypeptide (distinguishable from other β subunits by isolectric focusing and encoded at position 85D on the third chromosome) in embryonic development or in other adult tissues (74). [This spermatocyte specificity may not, however, be absolute since Natzle & McCarthy have detected transcripts apparently encoded by this gene in early embryonic development (117)]. Data derived both at the polypeptide level (74) and at the RNA level (117) showed one of the remaining four β genes (at chromosome position 56D) to be expressed ubiquitously, while a third gene (at position 60C) is expressed during midembryogenesis. The fourth β tubulin gene is also ubiquitously expressed at low levels, although the corresponding polypeptide has not been clearly identified. Similarly, for *Drosophila* α tubulins, two genes (at chromosome positions 84B and 84D) are ubiquitously expressed (92, 116, 117). One (at position 85E) encodes two RNA transcripts, the larger of which is present in larvae and the smaller of which is present in pupae and adult males and may be testis specific (116). The final gene (at position 67C) is a more divergent gene (90, 91, 116, 135) which encodes a maternal α tubulin RNA present in ovaries and eggs (116, 117).

Examples of differential gene/polypeptide expression in other genomes have also been reported. For example, in the most extensively studied vertebrate, the chicken, four β tubulin mRNAs are each encoded by a different gene (110, 111). One transcript (β4') is constitutively expressed, one (β2) represents the dominant neural β tubulin, one (β3) is overwhelmingly expressed in testis, and the fourth (β1) is expressed in significant amounts only in a few tissue/cell types (e.g., developing skeletal muscle). Although markedly preferential expression is often seen, no β tubulin subunit (with the possible exception of a divergent erythrocyte β tubulin (84) encoded by an as yet unidentified additional fifth gene) is truly cell type specific (111). Moreover, no cell or tissue type investigated expresses only one β tubulin gene, although many express three or more.

Similarly, in addition to a constitutively expressed gene, Farmer and coworkers have demonstrated two rat β tubulin genes to be expressed primarily in neural tissue and to be differentially regulated during brain development (105,

106, 106a). One gene encodes a β tubulin utilized early in cerebellar development, whereas the second gene is expressed only at later stages. An analogous situation may also be present in human α tubulins since a constitutively expressed and a brain specific gene have been identified (100).

Finally, among lower eukaryotes differential expression of α and β tubulin polypeptides has been documented by Dove and coworkers throughout the life cycle of the slime mold *Physarum* (76, 122, 136). At least three β tubulin genes are present in this organism; one is specific to myxamoebae; one is specific to the plasmodial form; and one is found in both stages of the life cycle (122). Similarly, one α tubulin subunit is known to be specific to plasmodia. These data document minimum contributions of differential tubulin gene expression since assignment of 3–4 additional α tubulin variants has not yet been achieved. It remains a surprise that more genes are expressed in the plasmodial form even though this form apparently possesses only mitotic microtubules (137).

POSTTRANSLATIONAL MODIFICATION OF TUBULIN: FINE TUNING OF FUNCTIONAL SPECIFICITY?

In addition to isotypic heterogeneity among tubulin polypeptides which arises directly from genetic contributions, posttranslational modifications have also been described which appear to be related to specific biological functions of tubulin. We have chosen to include a discussion of these phenomona because it is our feeling that collectively the data demonstrate the importance of chemical variation among tubulins in different cellular structures and underscore the emerging realization that tubulin is a structurally dynamic protein.

Tyrosination of α Tubulin

α Tubulin undergoes a unique posttranslational modification involving the enzymatic removal and addition of tyrosine at the carboxy-terminus of the protein. Originally identified by Rodriguez and coworkers as an RNA-independent incorporation of tyrosine and phenylalanine into protein in rat brain extracts (138), these workers identified α tubulin as the unique tyrosine acceptor protein (139, 140). Subsequently, identification of a carboxy-terminal tyrosine codon in α tubulin mRNAs from several species (89, 96, 97, 100) brought about the surprising realization that the primary modification of this putatively cyclic posttranslational modification process is the removal of the encoded tyrosine and that the specific retyrosination reaction restores the α tubulin subunit to the original state.

Both enzymes of this remarkable modification cycle have been partially purified and characterized. Tubulin:tyrosine ligase (TTLase) is a 35 kd protein which catalyzes the ATP-dependent formation of a peptide linkage of tyrosine to the carboxy-terminal glutamic acid residue of α tubulin (141, 142). TTLase has been detected in all vertebrate tissues examined (143–145), but particularly high activity is found in embryonic brain and muscle tissue where it undergoes strong developmental modulation apparently correlated with the morphological differentiation of asymetric cells (146, 147).

On the other side of the modification scheme, a tyrosyl-tubulin specific carboxypeptidase (TTCPase) has been detected in brain tissues. The enzyme has an apparent molecular weight of 90 kd (148) and shows a high degree of specificity for the carboxy-terminal tyrosine of α tubulin (148–150).

α Tubulin isolated from brain has approximately 0.3 mol C-terminal tyrosine per mole tubulin dimer and approximately 20–30% of brain tubulin can serve as a substrate for TTLase in vitro while the remaining 40–50% of brain tubulin is apparently incapable of being tyrosinated (133, 140, 144, 151). The nature of the difference between substrate and nonsubstrate α tubulin is not known although the possibility that each class arises from different genes is an attractive one. Analysis of microtubule assembly in vitro using maximally tyrosinated or detyrosinated tubulin has revealed little or no difference in the assembly properties of the two forms of tubulin (144, 150, 152).

In vitro studies using partially purified enzymes have shown that TTLase probably acts primarily on unpolymerized tubulin (152), while TTCPase acts preferentially on microtubules (150). The possibility that this scenario may be operative in vivo is supported by the findings of Rodriguez & Borisy who demonstrated that soluble α tubulin contains about twice as much C-terminal tyrosine as does α tubulin polymerized into microtubules when the two fractions are isolated under microtubule stabilizing conditions (153). With this in mind, detection of tyrosination by pulse labeling in vivo (144, 154, 155) and the demonstration that the rapid turnover of the C-terminal tyrosine in cultured muscle cells is inhibited by microtubule depolymerizing agents (156) indicate that this reaction cycle may reflect the dynamic equilibrium between tubulin dimers and microtubules in living cells.

Two powerful new tools for investigating this unique modification cycle have recently become available. The first of these is a monoclonal antibody, YL1/2 (157), which has been shown to recognize only the tyrosinated form of α tubulin (158, 159). The second reagent is a pair of polyclonal antibodies prepared against synthetic peptides corresponding to the C-terminus of tyrosinated or detyrosinated α tubulin, respectively (159a). Surprisingly, each of the antipeptide sera is completely specific for either tyrosinated or detyrosinated α tubulin polypeptides. Using indirect immunofluorescence, the YL1/2 antibody has been found to bind to most if not all cytoplasmic microtubules in PTK2 cells (158, 159). However, more recent work with the antipeptide antibodies has demonstrated the presence in monkey cells of a limited subset of interphase microtubules in which α tubulin is not tyrosinated (159a). Moreover, in these cells detyrosinated α tubulin appears to be completely absent from astral microtubules of metaphase spindles. In addition, double immunofluorescence experiments have revealed the presence of tyrosinated and nontyrosinated α tubulin in the same microtubule, indicating that there are not completely separate populations of fully tyrosinated or detyrosinated microtubules (159a). However, no gradient of tyrosinated or detyrosinated α tubulin has been observed within a single microtubule with either YL1/2 (158, 159) or the polyclonal antibodies (159a).

Microinjection of the tyrosination specific monoclonal YL1/2 into living cells results in a concentration-dependent bundling and reorganization of microtubules correlated with inhibition of saltatory motion, cell movement, mitosis, and dispersion of the Golgi complex (160). These pleiotropic effects probably result from the extensive binding of YL1/2 to cytoplasmic microtubules and do not reflect specific functions of tyrosinated tubulin per se.

A remarkable and exciting specificity of the tyrosination reaction has been found in the central nervous system by Cumming and coworkers (161, 162). Immunocytochemical staining of different regions of the brain reveals a specific lack of YL1/2 staining in the small, unmyelinated axonal fibers in the molecular layer of the cerebellum and the stratum radium of the hippocampus, even though dendrites and glial cells present in these areas do stain with YL1/2 and even though all cell processes stain with other anti-α tubulin antibodies (161). Analysis of development in the cerebellar cortex, however, has demonstrated YL1/2 reactivity in immature granule cell parallel fiber axons of the molecular layer at early stages in development (10 days postnatal); subsequently, YL1/2 reactivity of the parallel fibers disappears in a progressive wave from the lower regions of the molecular layer toward the external granular layer (162). The developmental time course and anatomical distribution of apparent detyrosination correlates remarkably well with the time course of parallel fiber maturation, organization and synaptogenesis in the molecular layer.

The possibility that tyrosination plays a crucial role in cytoskeletal reorganization during neural differentiation is certain to generate intensive experimental scrutiny in the future.

Axonal Tubulins

In addition to tyrosination, a number of studies indicate that tubulin in the axons of neurons may be specifically modified in other ways, either upon entry into the axon or during axonal transport. Hoffman & Lasek (163) originally demonstrated that after injection of radiolabeled amino acids into ganglia containing neuronal cell bodies, proteins subject to axonal transport move as discrete waves down the nerve tract, with tubulin and neurofilament protein constituting the slowest component (SCa) of the several classes of transported proteins. Thus, by exploiting these transport properties, axonal tubulin may be specifically labeled in vivo and by observing the pulse labeled protein in nerves at different distances from the site of injection, it is possible to observe the fate of transported tubulin. Using this approach, Brown et al (164) reported that axonal α tubulin is specifically depleted during axonal transport in the mouse optic tract. Similar studies in rat optic tract (165) and guinea pig sympathetic ganglia (166) noted not only a decreased α/β tubulin ratio throughout the nerve, but also demonstrated the presence of a more acidic α tubulin species, termed α3, which migrates faster than the bulk of brain α tubulin.

The data reported in these studies are, on the whole, convincing that there exists an axon-specific tubulin composition different from that seen in whole brain. It is difficult, however, to determine precisely what these differences are and mechanistically how they arise. This problem is compounded by differences in the various gel systems used to resolve the tubulins and in differences in neuronal tissues and experimental protocols utilized. Resolution of these difficulties in the future will be necessary for developing an understanding of regional or subcellular differences within axonal microtubule systems.

An important step in this direction has already been made by Brady et al (167). These investigators have coupled the axonal labeling paradigm with a functional fractionation of axonal microtubule proteins. After an injection/sacrifice interval appropriate for specific visualization of axonal tubulin, proteins of the particulate fraction obtained after cold and calcium extraction of the dissected optic system were analyzed by electrophoresis and peptide mapping (167). An amazingly large fraction (50%) of the total nerve α and β tubulin was not extractable (depolymerizable) by cold, calcium, or dilution as assayed by one dimensional gel electrophoresis. Similar findings were reported by Black et al (168). Moreover, Brady et al (167) have found that when proteins were analyzed by two dimensional gel electrophoresis, the α tubulin polypeptide(s) were not detectable. Nonequilibrium electrofocusing established that the α tubulin present in the cold and calcium stable fraction migrated with an extremely basic isoelectric point such that at equilibrium it migrated off the basic end of the gel. Clearly, the α tubulin subunits of this cold stable, particulate fraction are highly (covalently) modified, presumably as a consequence of an important functional requirement.

Flagellar Tubulins

Further evidence for functionally specific utilization of chemically distinct tubulins comes from analysis of flagellar morphogenesis in unicellular organisms. In the best studied system, *Chlamydomonas,* flagella can be induced to regenerate after amputation. Although flagellar regeneration depends in part on the presence of a pool of cytoplasmic precursors and in part on new tubulin synthesis, the major flagellar α tubulin, α3, has a mobility different from preexisting cytoplasmic α tubulin, α1 (169). Since α tubulin polypeptides translated in vitro from mRNA isolated during flagellar regeneration migrate as α1, these data suggest that cytoplasmic α tubulin is posttranslationally modified prior to incorporation into flagella (169). This suggestion has now been supported by pulse-chase experiments with regenerating flagella of *Chlamydomonas* (77, 170) and *Polytomella* (77) and by posttranslational labeling of *Chlamydomonas* with [3H]-acetate (171). These experiments have demonstrated that although no α3 is detectable in the cytoplasm of cells either before or after deflagellation, labeled α1 polypeptides can be chased into the α3 position in the absence of new protein synthesis.

Perhaps one of the most significant aspects of this work is the suggestion, based on inhibition of α modification by colchicine (170), that the flagellar modification reaction is closely coupled, both mechanistically and spatially, to flagellar microtubule assembly.

A similar posttranslational modification of flagellar α tubulin has been described in the trypanosome, *Crithidia fasciculata* (79). Curiously, the antibody YL1/2 which is specific for tryosinated α tubulin has been shown to stain only the flagellar microtubules of *Trypanosoma rhodensiense* (172), indirectly suggesting that flagellar α tubulin modification in this organism might arise in part from tyrosination.

Other Posttranslational Modifications of Tubulin

Among other posttranslational modifications of tubulin that have been identified, brain β tubulin has been reported to contain covalently bound phosphate (173, 174), but reports of the presence of specific tubulin kinases [e.g., ref. (175)] must be tempered by the realization that tubulin may be an adventitious substrate for such enzymes. In particular, the oncogene sarc (176), which phosphorylates tyrosine residues, utilizes tubulin as one of the best known in vitro substrates, even though no in vivo action of sarc on tubulin has ever been documented (A. Levinson, T. Hunter, personal communications). In addition, studies of protein phosphorylation in cultured cells have not identified tubulin as a phosphoprotein [e.g. ref. (177)]. Early reports of tubulin glycosylation [e.g. ref. (178)] have not been followed by more convincing studies, nor have in vitro studies reporting the cholera toxin–catalyzed ADP-ribosylation of tubulin (179) offered compelling arguments for modification in vivo.

Nonetheless, it is clear that tubulin is a substrate for posttranslational modifications and future studies of this structurally dynamic protein may reveal additional, biologically significant postsynthetic modifications.

GENETICS OF MICROTUBULE SYSTEMS

From the foregoing data it has become clear that the tubulins constitute a family of different proteins whose expression and modification are closely regulated during development and differentiation. But to what functional end are these programs of gene expression and polypeptide modification utilized? An initial answer to this important question has come from genetic approaches in a variety of organisms.

Microtubule Genetics in Aspergillus

The dissection of tubulin function using genetic approaches was first described for the mold *Aspergillus* in a series of papers from Morris and Oakley and their colleagues. These investigators analyzed a class of mutations that confer resistance or supersensitivity to the antimicrotubule drug benomyl. Drug resistance was mapped to three loci, one of which *(benA)* was shown to encode a β tubulin polypeptide (80). Exploiting the finding that some of the benomyl resistant *benA* alleles were temperature-sensitive for growth, these investigators also identified an α tubulin locus by ability of mutations at that locus to suppress the temperature sensitive *benA* phenotype (125). The requirement for microtubule function in both nuclear division and migration of nuclei along the germ tube of germinating spores was established by demonstration that both processes are resistant or supersensitive to benomyl in strains carrying β tubulin mutants whose phenotype is benomyl resistance or supersensitivity, respectively (180). Furthermore, a particularly interesting β tubulin allele *(benA33)* which confers both benomyl resistance and temperature sensitivity for growth was shown to block nuclear division and nuclear movement at the restrictive temperature not by the expected failure of microtubule assembly, but rather by a temperature-dependent hyperstabilization of microtubules (181–183). As had long been predicted (184), these data have established that disassembly of spindle microtubules is an essential aspect of the mitotic mechanism.

Microtubule Biochemistry and Genetics in Yeast

Historically, the investigation of microtubule function in yeast was slowed by the widely held belief that yeast contained remarkably few microtubules of restricted function and that in any case study of these few microtubules would not be particularly illuminating for analysis of microtubule systems in more complicated eukaryotes. Indeed, initial attempts simply to identify tubulin polypeptides in yeast were not particularly convincing (185, 186), indirectly confirming the belief that tubulin was a very minor component of the yeast cell.

This initial pessimism has now been largely overturned by the successful purification and characterization of yeast tubulin by Kilmartin (187), by an extensive analysis of spindle and cytoplasmic microtubules both by electron [e.g., ref. (188–190)] and light microscopic (191–194) techniques, and by successful identification of several enlightening tubulin mutants.

Using the elegant technique of gene disruption in which a functional chromosomal gene is inactivated by homologous recombination with a cloned gene fragment, Neff et al (130) initially demonstrated to no one's surprise that the single yeast β tubulin gene is essential for growth and that lesions in this gene block the cell division cycle at mitosis. Moreover, as had long been anticipated from the obvious requirement for microtubules in the construction of spindles, tubulin mutations have been shown (129) to be responsible for two of the previously identified cell division cycle (cdc) mutants in which cell growth is arrested at a specific point in the cell cycle. The first of these mutants, *nda2*, is a cold-sensitive mutation in which the block is at a late stage of nuclear division. Toda et al (129) initially cloned the wild-type (NDA2) locus by complementation of the mutant with exogenously introduced yeast DNA fragments carried on a plasmid vector. Two different complementing genes were obtained. By tetrad analysis one of these (α1, corresponding to the first yeast α tubulin sequence in Table 1) was shown to be the actual NDA2 wild-type gene. The second complementing sequence was not linked to NDA2 and represented the second α tubulin gene in this yeast (α2, the second yeast α tubulin sequence in Table 1). This complementation is somewhat unexpected since the endogenous (wild-type) α2 gene resident in the *nda2* mutant line does not support growth. This may be the result of increased expression of α2 from the complementing plasmid construct. A second cdc mutant, *nda3*, encodes β tubulin (Y. Hiraoka, T. Toda, M. Yanagida, unpublished). Although the cell cycle blockage of these mutants is well documented, it is not yet known whether there are concomitant morphological alterations in cytoplasmic microtubules. From other work, however, it seems clear that the events of bud emergence and bud enlargement in budding yeast strains do not require functional cytoplasmic microtubules (191–193, 195).

Microtubule Genetics in Physarum

An unexpected finding concerning the role of multiple tubulin genes in establishment of microtubule function has come from the work of Dove and coworkers (76, 122–124) who have investigated benzimidazole-resistant mutants in *Physarum*. Of four loci which can confer drug resistance in haploid myxamoebae, two (*benA* and *benD*) have been shown to encode β tubulin polypeptides (122, 123). Thus, mutation in either of these β tubulins expressed in haploid myxamoebae is sufficient to confer resistance. However, nonhaploid myxamoebae which are heterozygous for a resistant and a wild-type (sensitive) allele at *benD* are sensitive. At the molecular level these results are consistent with the hypothesis that *Physarum* microtubules are copolymers of the two β tubulins and that only above a threshold level of mutant subunits do the microtubules become drug resistant. Alternatively, it is possible that individual microtubules are comprised of single β tubulin isotypes thereby creating *benA*-specific and *benD*-specific microtubules (123). In this view either of these parallel microtubule sets is sufficient for myxamoebal microtubule function. Only when both microtubule classes contain sensitive alleles (for example in nonhaploids heterozygous for drug resistance at either or both alleles), is the overall phenotype sensitive. This provocative scenario predicts that only resistant β tubulins are present in spindle and cytoskeletal microtubules in cells grown in the presence of benzimidazoles, although this has not yet been directly tested.

Microtubule Genetics in Drosophila

With the discovery in *Drosophila* of a dominant β tubulin mutation whose phenotype was male sterility, it became possible to begin to investigate in detail the role of differential gene expression in construction of microtubules destined for different cellular functions. Initially, Raff and coworkers (73) identified a β tubulin subunit [named β2 and encoded at 85D (196)] which was expressed only in testis and was the sole β tubulin component of mature motile sperm (73). From these data, it seemed plausible that this subunit might be specifically tailored for construction of the sperm flagella axonemal microtubules (the only axonemal microtubules in the fly). However, ultrastructural analysis showed that the mutant sperm were defective not only in axonemal assembly, but also in meiosis, suggesting that the β2 subunit is multifunctional. Unfortunately, since the original mutant is dominant, the possibility that the variant subunit interferes in processes where the wild-type subunit does not normally function could not initially be eliminated.

The subsequent isolation of recessive mutations in the β2 locus has now clearly demonstrated that the β2 subunit is multifunctional (197). Two classes of mutations have been obtained. The first class, distinguished by production of a β2 polypeptide that is unstable, yields a greatly reduced pool of both α and β tubulin subunits and a complete blockage of meiosis, nuclear shaping, and axenomal assembly in mutant spermatocytes (197–199). The second class, characterized by a stable β2 subunit, consists of a set of mutants each with a distinct phenotypic pattern of microtubule disruption (34, 199). Since by definition, a recessive mutation does not disrupt the wild type function, it seems clear that the recessive β2 polypeptides do not interfere with the wild-type subunits and that β2 subunits are a normal component in the various sperm microtubules.

Tubulin Function Ascertained by Mammalian Somatic Cell Genetics

For genetic analysis of microtubules in mammalian cells, techniques of somatic cell genetics have been employed (200–209). Chinese hamster ovary (CHO) cells have been employed because of their rapid doubling time, simple and relatively stable karyotype, and the ease with which mutants can be obtained in culture. By selection for resistance to antimicrotubule drugs following mutagenesis, cell lines have been obtained which have increased resistance to colcemid (200, 203), griesiofulvin (203), and taxol (204, 206). In some instances the lesion has been localized to a tubulin polypeptide. For example, colcemid-resistant mutant *Cmd4* displays a β tubulin subunit which is more basic than the wild-type both in vivo and in in vitro translations of mRNA derived from the mutant cell line (203). Moreover, this mutant [as well as several others (208)] confers a temperature-sensitive phenotype which reverts concomitantly with a loss of colcemid resistance and loss of the altered β tubulin subunit. Similarly, a colcemid-resistant α tubulin has been identified by the mutant polypeptide's altered isoelectric point (201) and by a lower affinity of this mutant cell tubulin for the drug (200, 201). Although not rigorously addressable, these mutations appear to be dominant in the sense that in all cases apparently normal subunits are also present. This apparent dominance may be misleading, however, since several different α and β tubulin genes are almost certainly expressed in these cells and the high frequency of obtaining mutations suggests that at least some of the parent cells are hemizygous for individual tubulin genes. Indeed, the weakness of the somatic cell approach continues to be the paucity of data that identify the number and cell cycle–dependent expression of functional tubulin gene sequences in mammalian cells.

Nonetheless, what has been learned about microtubule function with these mutants? Initially, indirect immunofluorescence experiments using three different mutant cell lines that are temperature sensitive for growth, demonstrated that the cytoplasmic arrays of interphase microtubules were normal in each cell line at both permissive and nonpermissive temperatures. Time lapse video studies indicated, however, that mitosis at the nonpermissive temperature was markedly abnormal, resulting in a mitotic delay, defective cytokinesis, multinucleation, and ultimately, cell death (208). Clearly, interphase and mitotic microtubules differ in some important functional requirement for α and β tubulin polypeptides. Similarly, an intriguing mutant line *(Tax-18)* has been isolated by Cabral (206) on the basis of its resistance to taxol, a drug which strongly induces inappropriate microtubule polymerization in wild-type cells (210, 211). When grown in the absence of taxol, interphase microtubules of the mutant cell line again appear normal but the cells cannot form a proper spindle apparatus and cytokinesis is inhibited (207). Some pole-to-kinetochore microtubules are assembled but interpolar microtubules are not. The precise lesion has not yet been identified, although Cabral (206) has speculated that the mutation may lie in a nontubulin microtubule associated protein. This is an exciting possibility, since it will begin to allow a genetic assessment of in vivo interactions among microtubule proteins.

Microtubule Function Deduced from Analysis of Mutations in Nontubulin Proteins.

The genetics of microtubule function by analysis of nontubulin microtubule associated proteins has been lucratively addressed in two lower eukaryotic systems. In the first of these, work primarily from Luck and Piperno and their colleagues has analyzed mutations that affect the structural components of the flagellum of *Chlamydomonas*. Using mutations that affect flagellar structure or motility, these investigators have identified numerous proteins that comprise substructures within the flagellum and have deduced the order in which these proteins are assembled. Moreover, their work has addressed the mechanism of flagellar motility. For example, they have shown that the radial spokes that connect flagellar outer doublet microtubules to the inner pair are not in fact essential to flagellar motility (212). This considerable amount of work has recently been reviewed in detail by Luck (213).

A second system in which genetics has provided insights concerning microtubule function and nucleation has been in the nematode *Caenorhabditis elegans* (214–216). Although microtubules have almost universally been found to be comprised of 13 protofilaments, the microtubules of this nematode have only 11 protofilaments except in the six touch-receptor neurons which have 15 protofilaments. These 15 protofilament microtubules, which apparently mediate touch sensitivity, are crosslinked in rigid, hexagonally packed bundles. However, mutations in the gene *mec-7* result phenotypically in the loss of touch sensitivity and ultrastructurally in the replacement of the original microtubule array with a less regular set of un-crosslinked 11 protofilament microtubules. Although the lesion involved has not been identified at the protein level, indirect evidence suggests that it does not lie in a tubulin subunit, but rather in a product responsible for specifying the protofilament number of touch cell microtubules. If so, this mutation may well provide a genetic tool for identifying the mechanism of microtubule nucleation in vivo.

TUBULIN: AN ATTRACTIVE MODEL FOR ANALYSIS OF EUKARYOTIC GENE REGULATION

Underlying many of the initial investigations of tubulin gene organization and function and of the corresponding tubulin genetics is the companion goal of determining how cells establish and monitor the appropriate level of tubulin synthesis and content. From the preceeding sections, it is obvious that the requisite regulatory mechanisms must control both the selection of genes that are to be expressed and the specification of the appropriate quantitative level of expression. Although no data have yet begun to address the molecular mechanisms that underlie tissue specific programs of expression, several different experimental systems have yielded provocative findings concerning the mechanisms utilized for establishing the proper level of tubulin gene expression.

Apparent Autoregulatory Control of Tubulin Synthesis in Cultured Animal Cells

A priori it seems obvious that eukaryotic cells can and do regulate their tubulin content. But precisely how such control is achieved is not so apparent. Indeed, such regulation could be achieved either at the level of polymer or the level of the α/β subunit or both. In a pioneering experiment which addressed this question, Ben Ze'ev et al (217) proposed that cultured mammalian cells modulate tubulin gene expression through a feedback control mechanism linked to the level of unpolymerized tubulin subunits. The impetus for this hypothesis sprang from their finding that treatment of an established line of mouse fibroblasts (3T6) with colchicine {a drug which causes microtubule depolymerization [e.g., ref. (218)] and an approximate twofold increase in the level of unpolymerized subunits (219, 220)} results in a specific repression of new tubulin polypeptide synthesis. Since no such repression was seen with the antimicrotubule drug vinblastine [which also induces microtubule depolymerization but results in precipitation of the unpolymerized subunits into paracrystals (221, 222)], it was proposed that the regulation of tubulin expression was achieved by a mechanism that monitors the pool of unpolymerized subunits, rather than the amount of polymer.

Subsequent work from Kirschner's laboratory (42) established that the kinetics of the cellular response are rapid (half time of repression approximately 1 hr) and that this regulation can be demonstrated in a wide variety of primary or established cell types from diverse species. Moreover, consistent with the autoregulatory hypothesis, these studies demonstrated that treatment of cells with nocodozale [which like colchicine increases the pool of unpolymerized subunits (223)] results in the rapid suppression of tubulin synthesis, whereas treatment with taxol, which dramatically decreases the pool of unassembled subunits (210, 211), induces a mild increase in synthesis.

Unfortunately, interpretation of these initial experiments in terms of an autoregulatory control mechanism rested entirely on the presumptive effects of the various antimicrotubule drugs. This caveat was not trivial, as the drugs induce gross morphological alterations and the absolute specificities of action could not be assumed with certainty. However, this ambiguity was clarified by microinjection of purified tubulin into cells to artificially elevate the intracellular content of tubulin in the absence of drug treatments and morphological changes (224). With this approach, it was found that tubulin synthesis is rapidly and specifically suppressed by injection of an amount of tubulin roughly equivalent to 25–50% of the amount initially present in the cell. Although in these experiments it was not possible to determine the fate of the injected tubulin (i.e., unassembled or assembled), companion experiments in which the injected subunits were prebound to colchicine strongly suggested that in fact it is the subunit form of tubulin which is monitored.

What has been deduced concerning the molecular mechanism responsible for this specific modulation of tubulin synthesis in response to changes in the apparent pool of subunits? Initially, using cloned sequences to detect α or β tubulin mRNAs, it was shown that the loss of tubulin synthesis coincided quantitatively with the loss of cytoplasmic tubulin mRNAs (42). Moreover, in those cells in which multiple tubulin genes are expressed, it was found that the RNAs from each gene were lost concomitantly. Although indirect evidence suggested that the ultimate control might be at the level of tubulin RNA transcription (225), this possibility now seems less likely because the level of tubulin RNA transcripts elongated in isolated nuclei by in vivo initiated RNA polymerase molecules is indistinguishable from nuclei prepared from control or colchicine-treated cells (226).

Programmed Synthesis of Tubulin During Flagellar Assembly

Because of the ease with which mechanical shearing or alterations in growth conditions can be utilized to induce flagellar amputation or resorption (or in the case of *Naegleria* initial flagellar outgrowth), the stimulation of ciliary or flagellar protein synthesis during the growth of these organelles has increasingly been used as an attractive model system to study the specific induction and regulated expression of a specific set of eukaryotic genes. Investigations have included analyses of patterns of protein induction determined with in vivo labeling in sea urchins (227), *Tetrahymena* (228), and *Polytomella* (229). Not surprisingly, in vitro translation of RNA from *Tetrahymena* (230), *Naegleria* (231), and sea urchins (232) has shown that for each of these organisms increased synthesis of flagellar proteins (of which the tubulins are the most prominent) is due to increased levels of the corresponding mRNAs encoding flagellar proteins.

Without question, however, the organism that has received the most study and for which the flagellar induction process has been identified in greatest detail is the unicellular flagellate *Chlamydomonas*. Removal of flagella results in the rapid regeneration of new, nearly full length flagella within 90 min (233, 234). Although sufficient flagellar protein reserves exist in the cytoplasm to allow regeneration of flagella that are ⅓ to ½ of full length (234), increased translatable tubulin mRNAs were initially found by in vitro translation of isolated polyribosomes or purified RNA to accumulate as early as 8 min postdeflagellation (169, 234). This accumulation peaks between 45 and 90 min and slowly declines to the basal level by 180 min. Early work indicated that a simple model in which depletion of the substantial tubulin subunit pool during flagellar regrowth following deflagellation or resorption could not be the signal for induction since stimulation of tubulin synthesis was observed even when regrowth was blocked with the drug inhibitors isobutylmethylxanthine or colchicine (169, 235, 236).

The isolation of cloned sequences for the *Chlamydomonas* tubulins in Weeks' and Rosenbaum's laboratories afforded a more detailed look at the mechanism underlying the induction process. Both groups demonstrated that increased levels of tubulin synthesis corresponded quantitatively with an increased level of tubulin mRNAs (93, 94). Moreover, the mRNAs derived from both α tubulin and both β tubulin genes were seen to be induced coordinately (43), although the careful quantitation experiments of Schloss et al (237) have argued for slightly different accumulation kinetics.

Given the dogmatic bias endowed by most previously studied eukaryotic genes in which regulation of expression has been shown to derive principally from transcriptional mechanisms, work from the Rosenbaum laboratory has produced the surprising finding that modulation in transcription rates is only part of the regulatory story. Nuclei isolated from vegetative cells 20 min after deflagellation were found to transcribe between 4 and 10 times as many tubulin RNAs as control nuclei, thus indicating an apparent transcriptional enhancement as a result of deflagellation (238). Later work using in vivo pulse labeling with $^{32}PO_4$ (239) has indicated that the peak in transcriptional enhancement occurs very early (within 10–15 min of deflagellation), even though the peak in RNA accumulation occurs between 45 and 60 min. The brevity of the period during which tubulin RNA synthesis is maximal (the first 20 min) is as striking as the speed of activation. However, the peak transcription rate is estimated again to be only 4–7 fold above the initial rate even though the accumulation of tubulin RNAs reaches 10–14 fold above the initial levels. Careful consideration of these data mandates that in addition to an enhanced rate of transcription there must be a concomitant increase of at least 2–3 fold in tubulin mRNA stability. Moreover, redeflagellation of cells at times before the first flagellar regeneration is completed induces another burst in tubulin RNA synthesis which is identical to the first in magnitude and duration, indicating that the induction signal may act simply to reprogram the tubulin genes for a transient burst of maximal synthesis.

Cell Cycle Regulation of Tubulin Expression

Regulation of tubulin synthesis throughout the cell cycle has been addressed in several organisms. The tubulins are among the few polypeptides seen to vary a few fold in relative rate of synthesis during the cell cycle of cultured HeLa cells (241). Similarly, Howell and coworkers (242, 243) have investigated tubulin synthesis during the cell cycle of *Chlamydomonas*, which grows synchronously when maintained under alternating light/dark conditions. Using permeabilized cells, a two fold increase in tubulin transcription rate over the cell cycle was seen even though accumulated tubulin mRNAs fluctuate more than 10 fold (243). However, these cell cycle fluctuations reflect the superposition of true cell cycle requirements on top of the natural induction of flagellar proteins which occurs late in the dark phase of the cycle.

One particularly attractive system for study of tubulin requirements through the cell cycle is in the naturally synchronous cycle of the multinucleate plasmodium of *Physarum*. The tubulin polypeptides to be utilized for spindle microtubules [the only microtubules in the plasmodium (137)] have been found to be synthesized preferentially in the G2 phase of the cell cycle preceding entry into mitosis (244). Using cloned nucleic sequences, Schedl et al (245) have determined quantitatively that tubulin RNA levels rise at least 40 fold during this G2 period, followed by a rapid degradation of these RNAs concomitant with spindle disassembly. Although the mechanism leading to tubulin RNA accumulation during G2 is not known, the rapid loss of tubulin RNAs at the end of mitosis must arise from a major decrease in the mRNA stability (245).

Altered Tubulin Expression in the Life Form Switch of Leishmania

The life cycle of the parasitic protozoa *Leishmania* consists of two morphologically distinct forms, the first of which (the amastigote) resides inside macrophages of the mammalian host and the second of which (the promastigote) lives extracellularly and possess a flagellum. Although both forms contain intracellular microtubules, as might be anticipated, the flagellated form synthesizes tubulin at approximately three times the rate of the intracellular from (246, 247), presumably as required for assembly of the axonemal microtubules. On the basis of qualitative data from in vitro translations (248) and from tubulin RNA levels deduced from hybridization to cloned tubulin sequences (249), this difference in synthesis has been interpreted in *Leishmania mexicana* to arise from a surprising posttranslational control mechanism. This is not, however, the only interpretation of these data, and in light of the failure to find evidence for such a translational contribution in *Leishmania enrietti* (247), this intriguing possibility remains unproven.

NAGGING QUESTIONS: A PROSPECTUS FOR FUTURE DIRECTIONS

Tubulin has now been studied for eighteen years. But despite substantial progress, the determination of the ways in which multiple tubulin polypeptides participate in establishment of diverse microtubule function and the determination of the molecular pathways that regulate expression of multiple tubulin genes remain major questions to be addressed by future studies. Additional fundamental questions looming large on the horizon include what parameters dictate association of nontubulin proteins (MAPs) with individual microtubules and how microtubule polymerization from specific, but biochemically poorly defined, microtubule initiation sites within cells is regulated both spatially and temporally.

Given the many difficulties involved, how are these goals to be experimentally achieved? First, with the demonstration that antibodies of predetermined specificity may be raised against chemically synthesized peptide antigens (250), we suggest that it is all but certain that antibodies specific to individual tubulin polypeptides can and will establish how different tubulin subunits participate in various microtubule systems. Second, the ability to introduce exogenously engineered wild type or in vitro mutagenized tubulin genes into cells with DNA mediated transfection represents a most promising approach both for analysis of the contribution of particular tubulin genes to microtubule function and for dissecting pathways of tubulin gene regulation. Finally, in concert with more established genetic protocols, recently developed 'pseudo-genetic' approaches represent very powerful, promising new investigative tools. Methods already at hand include (a) reintroduction of in vitro engineered genes into the germ line of Drosophila (251) and (b) disruption of expression of a resident wild type tubulin gene by co-expression of the corresponding 'anti-sense' mRNA (252). Using these and related molecular approaches, we look forward to a continuing period of rapid progress in understanding the molecular biology and genetics of tubulin.

Literature Cited

1. Wilson, L., Friedkin, M. 1966. *Biochemistry* 5:2463–68
2. Borisy, G. G., Taylor, E. W. 1967. *J. Cell Biol.* 34:525–33
3. Mohri, H. 1968. *Nature* 217:1053–54
4. Renaud, F. L., Rowe, A. J., Gibbons, I. R. 1968. *J. Cell Biol.* 36:79–90
5. Shelanski, M. L., Taylor, E. W. 1967. *J. Cell Biol.* 34:549–54
6. Shelanski, M. L., Taylor, E. W. 1968. *J. Cell Biol.* 38:304–15
7. Weisenberg, R. C., Borisy, G. G., Taylor, E. W. 1968. *Biochemistry* 7: 4466–79
8. Weisenberg, R. C. 1972. *Science* 177:1104–5
9. Borisy, G. G., Olmsted, J. B. 1972. *Science* 177:1196–97
10. Snyder, J. A., McIntosh, J. R. 1976. *Ann. Rev. Biochem.* 45:699–720
11. Kirschner, M. W. 1978. *Int. Rev. Cytol.* 54:1–71
12. Timasheff, S. N., Gresham, L. M. 1981. *Ann. Rev. Biochem.* 49:565–91
13. McKeithan, T. W., Rosenbaum, J. L. 1984. In *Cell and Muscle Motility*, ed. J. W. Shay, 5:255–88. New York: Plenum
14. Dustin, P. 1978. *Microtubules.* Berlin: Springer-Verlag
15. Roberts, K., Hyams, J. S., eds. 1979. *Microtubules.* London: Academic, 595 pp.
16. Borisy, G. G., Cleveland, D. W., Murphy, D. G., eds. 1984. *Molecular Biology of the Cytoskeleton.* New York: Cold Spring Harbor Press. 512 pp.
17. Hyams, J. S., Stebbings, H. 1979. See Ref. 15, pp. 487–530
18. Berlin, R. D., Caron, J. M., Oliver, J. M. 1979. See Ref. 15, pp. 443–86
19. Heuser, J. E., Kirschner, M. W. 1980. *J. Cell Biol.* 86:212–34
20. Hirokawa, N., Heuser, J. E. 1981. *J. Cell Biol.* 399–409
21. Hirokawa, N. 1982. *J. Cell Biol.* 94: 129–42
22. Tucker, J. B. 1979. See Ref. 15, pp. 315–58

23. Behnke, O., Forer, A. 1967. *J. Cell Sci.* 2:169–92
24. Brinkley, B. R., Cartwright, J. 1975. *Ann. NY Acad. Sci.* 253:428–38
25. Olmsted, J. B., Witman, G. B., Carlson, K., Rosenbaum, J. L. 1971. *Proc. Natl. Acad. Sci. USA* 68:2273–77
26. Meza, I., Huang, B., Bryan, J. 1972. *Exp. Cell Res.* 74:535–40
27. Bryan, J. 1972. *Biochemistry* 11:2611–16
28. Fulton, C., Kane, R. E., Stephens, R. E. 1971. *J. Cell Biol.* 50:762–73
29. Dales, S. 1972. *J. Cell Biol.* 52:748–52
30. Wilson, L., Bryan, J. 1974. *Adv. Cell Mol. Biol.* 3:371–420
31. Wilson, L., Bamburg, J. R., Mizel, S. B., Grisham, L. M., Creswell, K. M. 1974. *Fed. Proc.* 33:158–66
32. Luduena, R. F., Woodward, D. O. 1973. *Proc. Natl. Acad. Sci. USA* 70:3594–98
33. Farrell, K., Morse, A., Wilson, L. 1978. *Biochemistry* 18:905–11
34. Raff, E. C. 1984. *J. Cell Biol.* 99:1–10
35. Bryan, J., Wilson, L. 1971. *Proc. Natl. Acad. Sci. USA* 68:1762–66
36. Feit, H., Slusarek, L., Shelanski, M. L. 1971. *Proc. Natl. Acad. Sci. USA* 68:2028–31
37. Fine, R. E. 1971. *Nature New Biol.* 233:283–85
38. Everhart, L. P. 1971. *J. Mol. Biol.* 61:745–60
39. Luduena, R. F., Shooter, E. M., Wilson, L. 1978. *J. Biol. Chem.* 252:7006–114
40. Amos, L. A. 1979. See Ref. 15, pp. 1–64
41. Crepeau, R. H., McEwen, B., Edelstein, S. J. 1978. *Proc. Natl. Acad. Sci. USA* 75:5006–10
41a. Luduena, R. F. 1979. See Ref. 15, pp. 65–116
42. Cleveland, D. W., Lopata, M. A., Sherline, P., Kirschner, M. W. 1981. *Cell* 25:537–46
43. Brunke, K., Young, E. E., Buchbinder, U., Weeks, D. P. 1982. *Nucleic Acids Res.* 10:1295–310
44. Witman, G. B., Carlson, K., Rosenbaum, J. L. 1972. *J. Cell Biol.* 54:540–56
45. Bibring, T., Baxandall, J., Denslow, S., Walker, B. 1976. *J. Cell Biol.* 69:301–12
46. Fulton, C., Simpson, P. A. 1976. In *Cell Motility,* ed. R. Goldman, T. Pollard, J. L. Rosenbaum, pp. 987–1005. New York: Cold Spring Harbor Press
47. Kowit, J. D., Fulton, C. 1974. *Proc. Natl. Acad. Sci. USA* 71:2877–81
48. Fulton, C., Simpson, P. A. 1979. See Ref. 15, pp. 117–74
49. Stephens, R. E. 1970. *J. Mol. Biol.* 47:353–63
50. Stephens, R. E. 1975. In *Molecules and Cell Movement,* ed. S. Inoue, R. E. Stephens. New York: Raven
51. Stephens, R. E. 1978. *Biochemistry* 14:2882–91
52. Linck, R. W., Langevin, G. L. 1982. *J. Cell Sci.* 58:1–22
53. Bibring, T., Baxandall, J. 1981. *Dev. Biol.* 83:122–26
54. Feit, H., Neudeck, U., Baskin, F. 1977. *J. Neurochem.* 28:697–706
55. Marotta, C. A., Harris, J. L., Gilbert, J. M. 1978. *J. Neurochem.* 30:1431–40
56. Gozes, I., Littauer, U. Z. 1978. *Nature* 276:411–13
57. Dahl, J. L., Weibel, V. J. 1979. *Biochem. Biophys. Res. Commun.* 86:822–28
58. Nelles, L. P., Bamburg, J. R. 1979. *J. Neurochem.* 32:477–89
59. Deleted in proof
60. George, H. J., Misra, L., Field, D. J., Lee, J. C. 1981. *Biochemistry* 20:2402–9
61. von Hungen, K., Chin, R. C., Baxter, C. F. 1981. *J. Neurochem.* 37:511–14
62. Wolff, A., Denoulet, P., Jeantet, C. 1982. *Neurosci. Lett.* 31:323–28
63. Deleted in proof
64. Sullivan, K. F., Wilson, L. 1984. *J. Neurochem.* 42:1363–71
65. Field, D. J., Collins, R. A., Lee, J. C. 1984. *Proc. Natl. Acad. Sci. USA* 81:4041–45
66. Denoulet, P., Edde, B., Jeantet, C., Gros, F. 1982. *Biochimie* 64:165–72
67. Gozes, I., Sweadner, K. J. 1981. *Nature* 294:477–80
68. Gozes, I., de Baetaelier, A., Littauer, U. Z. 1980. *Eur. J. Biochem.* 103:13–20
69. Strocchi, P., Marotta, C. A., Bonventre, J., Gilbert, J. M. 1981. *Brain Res.* 211:206–10
70. Gozes, I., Saya, D., Littauer, U. Z. 1979. *Brain Res.* 171:171–75
71. Moura Neto, V., Mallat, M., Junket, C., Prochaintz, A. 1983. *EMBO J.* 2:1243–48
72. Marotta, C. A., Strocchi, P., Gilbert, J. M. 1979. *J. Neurochem.* 33:231–46
73. Kemphues, K. J., Raff, E. C., Raff, R. A., Kaufman, T. C. 1979. *Proc. Natl. Acad. Sci. USA* 76:3991–95
74. Raff, E. C., Fuller, M. T., Kaufman, T. C., Kemphues, K. J., Raff, R. A. 1982. *Cell* 28:33–40
75. Macrae, T. H., Luduena, R. F. 1984. *Biochem. J.* 219:137–48
76. Burland, T. G., Gull, K., Schedl, T., Boston, R. S., Dove, W. T. 1983. *J. Cell Biol.* 97:1852–59
77. McKeithan, T. W., Lefebvre, P. A.,

Silflow, C. D., Rosenbaum, J. L. 1983. *J. Cell Biol.* 96:1056–63

78. McKeithan, T. W., Rosenbaum, J. L. 1981. *J. Cell Biol.* 91:352–60
79. Russell, K., Gull, K. 1984. *Mol. Cell. Biol.* 4:1182–85
80. Shier-Niess, G., Lai, M. H., Morris, N. R. 1978. *Cell* 15:638–47
81. Diez, J. C., Little, M., Avila, J. 1984. *Biochem. J.* 219:277–85
82. Little, M., Rohricht, C. O., Schroeder, D. 1983. *Exp. Cell Res.* 147:15–22
83. Deleted in proof
84. Murphy, D. B., Wallis, K. T. 1983. *J. Biol. Chem.* 258:7870–75
85. Murphy, D. B., Wallis, K. T. 1983. *J. Biol. Chem.* 258:8357–64
86. Deleted in proof
87. Cleveland, D. W., Lopata, M. A., MacDonald, R. J., Cowan, N. J., Rutter, W. J., Kirschner, M. W. 1980. *Cell* 20:95–105
88. Cleveland, D. W., Kirschner, M. W., Cowan, N. J. 1978. *Cell* 15:1021–31
89. Valenzuela, P., Quiroga, M., Zaldivar, J., Rutter, W. J., Kirschner, M. W., Cleveland, D. W. 1981. *Nature* 289:650–55
90. Sanchez, F., Natzle, J. E., Cleveland, D. W., Kirschner, M. W., McCarthy, B. J. 1980. *Cell* 22:845–54
91. Kalfayan, L., Wensink, P. C. 1981. *Cell* 24:97–106
92. Mischke, D., Pardue, M. L. 1982. *J. Mol. Biol.* 156:449–66
93. Silflow, C. D., Rosenbaum, J. L. 1981. *Cell* 24:81–88
94. Minami, S., Collis, P. S., Young, E. E., Weeks, D. P. 1981. *Cell* 24:837–45
95. Alexandraki, D., Ruderman, J. V. 1981. *Mol. Cell. Biol.* 1:1125–37
96. Lemischka, I. R., Farmer, S. R., Rancaniello, V. R., Sharp, P. A. 1981. *J. Mol. Biol.* 150:101–20
97. Ginzburg, I., Behar, L., Givol, D., Littauer, U. Z. 1981. *Nucleic Acids Res.* 9:2691–97
98. Cowan, N. J., Wilde, C. D., Chow, L. T., Wefald, F. C. 1981. *Proc. Natl. Acad. Sci. USA* 78:4877–81
99. Wilde, C. D., Chow, L. T., Wefald, F. C., Cowan, N. J. 1982. *Proc. Natl. Acad. Sci. USA* 79:96–100
100. Cowan, N. J., Dobner, P., Fuchs, E. V., Cleveland, D. W. 1983. *Mol. Cell. Biol.* 3:1738–45
101. Hall, J. L., Dudley, L., Dobner, P. R., Lewis, S. A., Cowan, N. J. 1983. *Mol. Cell. Biol.* 3:854–62
102. Gwo-Shu Lee, M., Lewis, S. A., Wilde, C. D., Cowan, N. J. 1983. *Cell* 33:477–87

103. Gwo-Shu Lee, M., Loomis, C., Cowan, N. J. 1984. *Nucleic Acids Res.* 12:5823–36
104. Lemischka, I. R., Sharp, P. A. 1982. *Nature* 300:330–35
105. Bond, J. F., Farmer, S. R. 1983. *Mol. Cell. Biol.* 3:1333–42
106. Bond, J. F., Robinson, G. S., Farmer, S. R. 1984. *Mol. Cell. Biol.* 4:1313–19
106a. Farmer, S. R., Bond, J. F., Robinson, G. S., Mbangkollo, D., Fenton, M. J., Berkowitz, E. M. 1984. See Ref. 16, pp. 333–42
107. Distel, R. J., Kleene, K. C., Hecht, N. B. 1984. *Science* 224:68–70
108. Thompson, M. A., Brinkley, B. R. 1983. *J. Cell Biol.* 97:215a
109. Cleveland, D. W., Hughes, S. H., Stubblefield, E., Kirschner, M. W., Varmus, H. E. 1981. *J. Biol. Chem.* 256:3130–34
110. Lopata, M. A., Havercroft, J. C., Chow, L. T., Cleveland, D. W. 1983. *Cell* 32:713–24
111. Havercroft, J. C., Cleveland, D. W. 1984. *J. Cell Biol.* 99:1927
112. Sullivan, K. F., Havercroft, J. C., Cleveland, D. W. 1984. See Ref. 16, pp. 321–332
113. Sullivan, K. F., Cleveland, D. W. 1984. *J. Cell Biol.* 99:1754–60
114. Alexandraki, D., Ruderman, J. V. 1983. *J. Mol. Evol.* 19:397–410
115. Alexandraki, D., Ruderman, J. V. 1985. *Dev. Biol.* In press
116. Kalfayan, L., Wensink, P. C. 1982. *Cell* 29:91–98
117. Natzle, J. E., McCarthy, B. J. 1984. *Dev. Biol.* 104:187–98
118. Thomashow, L. S., Milhausen, M., Rutter, W. J., Agabian, N. 1983. *Cell* 32:35–43
119. Landfear, S. M., McMahon-Pratt, D., Wirth, D. F. 1983. *Mol. Cell. Biol.* 3:1070–76
120. Huang, P. L., Roberts, B. E., Pratt, D. M., David, J. R., Miller, J. S. 1984. *Mol. Cell. Biol.* 4:1372–83
121. Landfear, S. M., Wirth, D. F. 1985. *Mol. Biochem. Parasitol.* In press
122. Burland, T. G., Schedl, T., Gull, K., Dove, W. F. 1984. *Genetics.* 108:123
123. Schedl, T., Owens, J., Dove, W. F., Burland, T. G. 1984. *Genetics.* 108:143
124. Roobol, A., Paul, E. C. A., Birkett, C. R., Foster, K. E., Gull, K., et al. 1984. See Ref. 16, pp. 223–34
125. Morris, N. R., Lai, M. H., Oakley, C. E. 1979. *Cell* 16:437–42
126. Weatherbee, J. A., Morris, N. R. 1984. *J. Biol. Chem.* 259:15452
127. Lai, E. Y., Remillard, S. P., Fulton, C. 1984. See Ref. 16, pp. 257–66

128. Calahan, R. C., Shalke, G., Gorovsky, M. 1984. *Cell* 36:441–45
129. Toda, T., Adachi, Y., Hiraoka, Y., Yanagida, M. 1984. *Cell* 37:233–42
130. Neff, N., Thomas, J. H., Grisafi, P., Botstein, D. 1983. *Cell* 33:211–19
131. Youngblom, J., Schloss, J. A., Silflow, C. D. 1984. *Mol. Cell. Biol.* 4:2686
132. Wilde, C. D., Crowther, C. E., Cripe, T. P., Lee, M. G.-S., Cowan, N. J. 1982. *Nature* 297:83–84
133. Ponstingl, H., Krauhs, E., Little, M., Kempf, T. 1981. *Proc. Natl. Acad. Sci. USA* 78:2757–61
134. Krauhs, E., Little, M., Kempf, T., Hofer-Warbinek, R., Ade, W., Ponstingl, H. 1981. *Proc. Natl. Acad. Sci. USA* 78:4156–60
135. Baum, H. J., Livneh, Y., Wensink, P. C. 1983. *Nucleic. Acids Res.* 11:5569–87
136. Schedl, T., Burland, T. G., Gull, K., Dove, W. F. 1984. *J. Cell Biol.* 99:155–65
137. Havercroft, J. C., Gull, K. 1983. *Eur. J. Cell Biol.* 32:67–74
138. Barra, H. S., Arce, C. A., Rodriguez, J. A., Caputto, R. 1973. *J. Neurochem.* 21:1241–51
139. Barra, H. S., Arce, C. A., Rodriguez, J. A., Caputto, R. 1974. *Biochem. Biophys. Res. Commun.* 60:1384–90
140. Arce, C. A., Rodriguez, J. A., Barra, H. S., Caputto, R. 1975. *Eur. J. Biochem.* 59:145–49
141. Raybin, D., Flavin, M. 1977. *Biochemistry* 16:2189–94
142. Murofushi, H. 1980. *J. Biochem.* 87:979–84
143. Deanin, G. G., Gordon, M. W. 1976. *Biochem. Biophys. Res. Commun.* 71:676–83
144. Raybin, D., Flavin, M. 1977. *J. Cell Biol.* 73:429–504
145. Preston, S. F., Deanin, G. G., Hanson, R. K., Gordon, M. W. 1979. *J. Mol. Evol.* 13:233–44
146. Deanin, G. G., Thompson, W. C., Gordon, M. W. 1977. *Dev. Biol.* 57:230–33
147. Rodriguez, J. A., Borisy, G. G. 1978. *Biochem. Biophys. Res. Commun.* 83:579–86
148. Argarana, C. E., Barra, H. S., Caputto, R. 1980. *Mol. Cell. Biochem.* 19:17–21
149. Argarana, C. E., Barra, H. S., Caputto, R. 1980. *J. Neurochem.* 34:64–73
150. Kumar, N., Flavin, M. 1981. *J. Biol. Chem.* 256:7678–86
151. Lu, P. C., Elzinga, M. 1978. *Biochim. Biophys. Acta* 537:320–28
152. Arce, C. A., Hallak, M. E., Rodriquez, J. A., Barra, H. S., Caputto, R. 1978. *J. Neurochem.* 31:205–10
153. Rodriquez, J. A., Borisy, G. G. 1979. *Biochem. Biophys. Res. Commun.* 89:893–99
154. Thompson, W. C. 1977. *FEBS. Lett.* 80:9–13
155. Nath, J., Flavin, M. 1979. *J. Biol. Chem.* 254:11505–10
156. Thompson, W. C., Deanin, G. C., Gordon, M. W. 1979. *Proc. Natl. Acad. Sci. USA* 76:1318–22
157. Kilmartin, J. V., Wright, B., Milstein, C. 1982. *J. Cell Biol.* 93:576–82
158. Wehland, J., Willingham, M. C., Sandoval, I. V. 1983. *J. Cell Biol.* 97:1467–75
159. Wehland, J., Schorder, H. C., Weber, C. 1984. *EMBO J.* 3:1295–1300
159a. Gundersen, G. G., Kalnoski, M. H., Bulinski, J. C. 1984. *Cell* 38:779–89
160. Wehland, J., Willingham, M. C. 1983. *J. Cell Biol.* 97:1476–90
161. Cumming, R., Burgoyne, R. D., Lytton, N. A. 1983. *Eur. J. Cell Biol.* 31:241–48
162. Cumming, R., Burgoyne, R. D., Lytton, N. A. 1984. *J. Cell Biol.* 98:347–51
163. Hoffman, P. N., Lasek, R. J. 1975. *J. Cell Biol.* 66:351–57
164. Brown, B. A., Nixon, R. A., Moratta, C. A. 1982. *J. Cell Biol.* 94:159–64
165. Goodrum, J. F., Morell, P. M. 1982. *J. Neurochem.* 39:443–51
166. Tashiro, T., Komiya, Y. 1983. *Neuroscience* 9:943–50
167. Brady, S. T., Tytell, M., Lasek, R. J. 1984. *J. Cell Biol.* 99:1716–24
168. Black, M. M., Cochran, J. M., Kurdyla, J. T. 1984. *Brain Res.* 295:255–63
169. Lefebvre, P. A., Silflow, C. D., Weiber, E. D., Rosenbaum, J. L. 1980. *Cell* 20:469–77
170. Brunke, K. J., Collis, P. S., Weeks, D. P. 1982. *Nature* 297:516–19
171. L'Hernault, S. W., Rosenbaum, J. L. 1983. *J. Cell Biol.* 97:258–63
172. Cumming, R., Williamson, J. 1984. *Cell Biol. Int. Rep.* 8:2
173. Eipper, B. A. 1972. *Proc. Natl. Acad. Sci. USA* 69:2283–87
174. Reddington, M. L., Tan, P., Lagnado, J. R. 1976. *J. Neurochem.* 27:1227–36
175. Goldenring, J. R., Gonzalez, B., McGuire, J. S., DeLorenzo, R. J. 1982. *J. Biol. Chem.* 258:12632–40
176. Bishop, J. M. 1983. *Ann. Rev. Biochem.* 52:301–54
177. Pallas, D., Solomon, F. S. 1983. *Cell* 30:407–14
178. Feit, H., Shelanski, M. L. 1975. *Biochem. Biophys. Res. Commun.* 66:920–27
179. Amir-Zaltsman, Y., Ezra, E., Scherson,

T., Zutra, A., Littauer, U. Z. 1982. *EMBO J.* 1:181–86

180. Oakley, B. R., Morris, N. R. 1980. *Cell* 19:255–62
181. Oakley, B. R., Morris, N. R. 1981. *Cell* 24:837–45
182. Gambino, J., Bergen, L. G., Morris, N. R. 1984. *J. Cell Biol.* 99:830–8
183. Morris, N. R., Weatherbee, J. A., Gambino, J., Bergen, L. G. 1984. See Ref. 16, pp. 211–22
184. Inoue, S. 1981. *J. Cell Biol.* 91:131S–47S
185. Water, R. D., Kleinsmith, L. J. 1976. *Biochem. Biophys. Res. Commun.* 70:704–8
186. Baum, P., Thorner, J., Honig, L. 1978. *Proc. Natl. Acad. Sci. USA* 76:4962–66
187. Kilmartin, J. 1981. *Biochemistry* 20:3629–33
188. Byers, B. 1981. In *Molecular Biology of the Yeast Saccharomyces,* ed. J. N. Strathern, E. W. Jones, J. R. Broach. New York: Cold Spring Harbor Press
189. King, S. M., Hyams, J. S., Luba, A. 1982. *J. Cell Biol.* 94:263–70
190. Peterson, J. B., Ris, H. 1976. *J. Cell Sci.* 22:219–42
191. Adams, A. E. M., Pringle, J. R. 1984. *J. Cell Biol.* 98:934–45
192. Kilmartin, J., Adams, A. E. M. 1984. *J. Cell Biol.* 98:922–33
193. Kilmartin, J. 1984. See Ref. 16, pp. 185–92
194. Pringle, J. R., Coleman, K., Adams, A., Little, S., Haarer, B., et al. 1984. See Ref. 16, pp. 193–209
195. Novick, P., Thomas, J. H., Botstein, D. 1984. In *Genetic Engineering Principles and Methods,* Vol. 6, ed. J. Setlow. New York: Plenum. In press
196. Kemphues, K. J., Raff, E. C., Raff, R. A., Kaufman, T. C. 1980. *Cell* 21:445–51
197. Kemphues, K. J., Kaufman, T. C., Raff, R. A., Raff, E. C. 1982. *Cell* 31:655–70
198. Kemphues, K. J., Raff, E. C., Kaufman, T. C. 1983. *Genetics* 105:345–56
199. Raff, E. C., Fuller, M. T. 1984. See Ref. 16, pp. 293–304
200. Ling, V., Aubin, J. E., Chase, A., Sarangi, F. 1979. *Cell* 18:423–30
201. Keates, R. A. B., Sarangi, F., Ling, V. 1981. *Proc. Natl. Acad. Sci. USA* 78:5638–42
202. Connolly, J. A., Kalnins, V. I., Ling, V. 1981. *Exp. Cell Res.* 132:147–55
203. Cabral, F., Sobel, M. E., Gottesman, M. M. 1980. *Cell* 20:29–36
204. Cabral, F., Abraham, I., Gottesman, M. M. 1981. *Proc. Natl. Acad. Sci. USA* 78:4388–91

205. Cabral, F., Abraham, I., Gottesman, M. M. 1982. *Mol. Cell. Biol.* 2:720–29
206. Cabral, F. 1983. *J. Cell Biol.* 97:22–29
207. Cabral, F., Wible, L., Brenner, S., Brinkley, B. R. 1983. *J. Cell Biol.* 97:30–39
208. Abraham, I., Marcus, M., Cabral, F., Gottesman, M. M. 1983. *J. Cell Biol.* 97:1055–61
209. Cabral, F., Schibler, M., Abraham, I., Whitfield, C., Kuriyama, R., et. al. 1984. See Ref. 16, pp. 305–17
210. Schiff, P. B., Faut, J., Horwitz, S. B. 1979. *Nature* 277:665–67
211. Schiff, P. B., Horwitz, S. B. 1980. *Proc. Natl. Acad. Sci. USA* 77:1561–65
212. Huang, B., Ramanis, Z., Luck, D. J. L. 1982. *Cell* 28:115–24
213. Luck, D. J. L. 1984. *J. Cell Biol.* 98:789–94
214. Chalfie, M., Thomson, J. N. 1979. *J. Cell Biol.* 82:278–89
215. Chalfie, M., Sulston, J. 1981. *Dev. Biol.* 82:358–70
216. Chalfie, M., Thomson, J. N. 1982. *J. Cell Biol.* 93:15–23
217. Ben Ze'ev, A., Farmer, S. R., Penman, S. 1978. *Cell* 17:319–25
218. Weber, K., Pollack, R., Bibring, T. 1975. *Proc. Natl. Acad. Sci. USA* 72:459–63
219. Hiller, G., Weber, K. 1978. *Cell* 14:795–804
220. Spiegelman, B. M., Penningroth, S. M., Kirschner, M. W. 1979. *Cell* 12:239–52
221. Bensch, K. B., Malawista, S. E. 1969. *J. Cell Biol.* 40:95–107
222. Bryan, J. 1972. *Biochemistry* 11:2611–16
223. De Brabander, M. J., Van De Veire, R. M. L., Aerts, F. E. M., Borgers, M., Janssen, P. A. J. 1976. *Cancer Res.* 36:905–16
224. Cleveland, D. W., Pittenger, M. F., Feramisco, J. R. 1983. *Nature* 305:738–40
225. Cleveland, D. W., Kirschner, M. W. 1982. *Cold Spring Harbor Symp. Quant. Biol.* 46:171–83
226. Cleveland, D. W., Havercroft, J. C. 1983. *J. Cell Biol.* 97:919–24
227. Stephens, R. E. 1977. *Dev. Biol.* 61:311–19
228. Guttman, S. D., Gorovsky, M. A. 1979. *Cell* 17:307–17
229. Brown, D. L., Rodgers, R. A. 1978. *Exp. Cell Res.* 117:313–24
230. Marcaud, L., Hayes, D. 1979. *Eur. J. Biochem.* 98:267–73
231. Lai, E. Y., Walsh, C., Wardell, D., Fulton, C. 1979. *Cell* 17:867–78
232. Merlino, G. T., Chamberlin, J. P.,

Kleinsmith, L. J. 1978. *J. Biol. Chem.* 253:7078–85

233. Rosenbaum, J. L., Moulder, J. E., Ringo, D. L. 1969. *J. Cell Biol.* 41:600–19

234. Weeks, D. P., Collis, P. S. 1976. *Cell* 9:15–27

235. Weeks, D. P., Collis, P. S., Gealt, M. A. 1977. *Nature* 268:667–68

236. Lefebvre, P. A., Nordstrom, S. A., Moulder, J. E., Rosenbaum, J. L. 1978. *J. Cell Biol.* 78:8–27

237. Schloss, J. A., Silflow, C. D., Rosenbaum, J. L. 1984. *Mol. Cell. Biol.* 4: 424–34

238. Keller, L. R., Schloss, J. A., Silflow, C. D., Rosenbaum, J. L. 1984. *J. Cell Biol.* 98:1138–43

239. Baker, E. J., Schloss, J. A., Rosenbaum, J. L. 1985. *J. Cell Biol.* 99:2074

240. Deleted in proof

241. Bravo, J., Celis, J. E. 1980. *J. Cell Biol.* 84:795–802

242. Ares, M., Howell, S. H. 1982. *Proc. Natl. Acad. Sci. USA* 79:5577–81

243. Dallman, T., Ares, M., Howell, S. H. 1983. *Mol. Cell. Biol.* 3:1537–39

244. Chang, M. T., Dove, W. F., Laffler, T. G. 1983. *J. Biol. Chem.* 258:1352–56

245. Schedl, T., Burland, T. G., Gull, K., Dove, W. F. 1984. *J. Cell Biol.* 99:155–65

246. Fong, D., Chang, K.-P. 1981. *Proc. Natl. Acad. Sci. USA* 78:7624–28

247. Landfear, S. M., Wirth, D. F. 1984. *Nature* 309:716–17

248. Wallach, M., Fong, D., Chang, K.-P. 1982. *Nature* 299:650–52

249. Fong, D., Wallach, M., Keithly, J., Melera, P. W., Chang, K.-P. 1984. *Proc. Natl. Acad. Sci. USA.* 81:5782–86

250. Lerner, R. A. 1982. *Nature* 299:592–96

251. Rubin, G. M., Spradling, A. C. 1983. *Nucleic Acids Res.* 11:6341–51

252. Izant, J. G., Weintraub, H. 1984. *Cell* 36:1007–15

Ann. Rev. Biochem. 1985. 54:367–402
Copyright © 1985 by Annual Reviews Inc. All rights reserved

NUCLEOSIDE PHOSPHOROTHIOATES

F. Eckstein

Max-Planck-Institut für experimentelle Medizin, Abteilung Chemie,
Hermann-Rein-Strasse 3, D-3400 Göttingen, West Germany

CONTENTS

0066-4154/85/0701-0367$02.00

INTRODUCTION

Analogs of naturally occurring substances have a firm place in the array of tools biochemists use to unravel enzyme functions and mechanisms. For example, substrate analogs have been constructed as reversible and irreversible inhibitors, transition state analogs, suicide substrates, and spectroscopic probes. These analogs all differ from the normal substrate in one or more properties that determine suitability for investigation of various aspects of enzyme action. Nucleoside phosphorothioates as analogs of nucleotides combine several properties that make them attractive. Firstly, the same modification, exchange of a nonbridging oxygen at a phosphate group for sulfur, can be introduced into almost all nucleoside phosphates and can reside, for example, on either the α-, β-, or γ-phosphate of a nucleoside triphosphate or at an internucleotidic phosphate linkage. Secondly, this modification often confers chirality on this particular phosphorus, and since nucleosides themselves are chiral, leads to the existence of a pair of diastereomers. Thirdly, as some of these phosphorothioates are substrates for many enzymes, this group can be transferred into the product of such a reaction; the most notable example is the synthesis of phosphorothioate-containing DNA from dNTPαS precursors.

Their versatility distinguishes nucleoside phosphorothioates from most other nucleotide analogs, and has made them attractive not only to enzymologists, but also, more recently, to molecular biologists. Several reviews (1–14) have previously dealt with particular aspects of the use of nucleoside phosphorothioates, often in a more thorough and detailed manner than is possible in the more general review given here.

SYNTHESIS AND STRUCTURE

Nucleoside 5'-Monophosphorothioates

The method originally devised for the synthesis of nucleoside 5'-phosphorothioates was the reaction of suitably protected ribo- and deoxyribonucleosides with triimidazolyl phosphinesulfide (15, 16). However, this procedure has been superseded by the direct thiophosphorylation of unprotected ribonucleosides (17–19) and of deoxynucleosides (20; 183) with $PSCl_3$ using triethyl or trimethyl phosphate as solvent. More recently, the four deoxynucleoside 5'-phosphorothioates have been prepared by phosphitylation of the unprotected deoxynucleosides and subsequent addition of sulfur, a method probably most suitable for the preparation of ^{35}S-labeled material (21).

Nucleoside 5'-Thiodi- and 5'-Thiotriphosphates

Nucleoside 5'-O-(1-thiodiphosphates) (Figure 1) are generally prepared by activation of nucleoside 5'-phosphorothioates with diphenylphosphorochloridate and subsequent reaction with phosphate (22, 23). Nucleoside 5'-O-(1-

thiotriphosphates) are synthesized in an analogous manner by substituting pyrophosphate for phosphate. These chemical syntheses are not stereospecific, and furnish a mixture of the Sp- and Rp-diastereomers of either nucleoside 5'-O-(1-thiodi) or triphosphate arising from the fact that the α-phosphorus now bears four non-identical substituents and is therefore chiral (Figure 2; for determination of absolute configuration see later in this section). For the study of stereochemical aspects of enzymatic and nonenzymatic reactions, pure diastereomers are essential; hence, considerable effort has been invested in their separation. Mixtures of diastereomers can be analyzed either by ^{31}P-NMR spectroscopy (19, 20, 24, 25) or reverse-phase HPLC (26, 19). Various methods based on stereoselective enzymatic phosphorylation exist for the preparation of pure diastereomers. The Sp-isomer of ATPαS can be stereospecifically synthesized from adenosine 5'-phosphorothioate by enzymatic phosphorylation with adenylate and pyruvate kinase (24, 25). However, since adenylate kinase is specific for the adenine nucleotides, this route is restricted to the synthesis of (Sp)-ATPαS, although (Sp)-dATPαS can also be obtained using large amounts of enzyme (27). The other enzymatic methods consist of the more or less stereospecific enzymatic phosphorylation of one of the two diastereomers of a nucleoside 5'-O-(1-thiodiphosphate) present in the mixture to the nucleoside 5'-O-(1-thio-

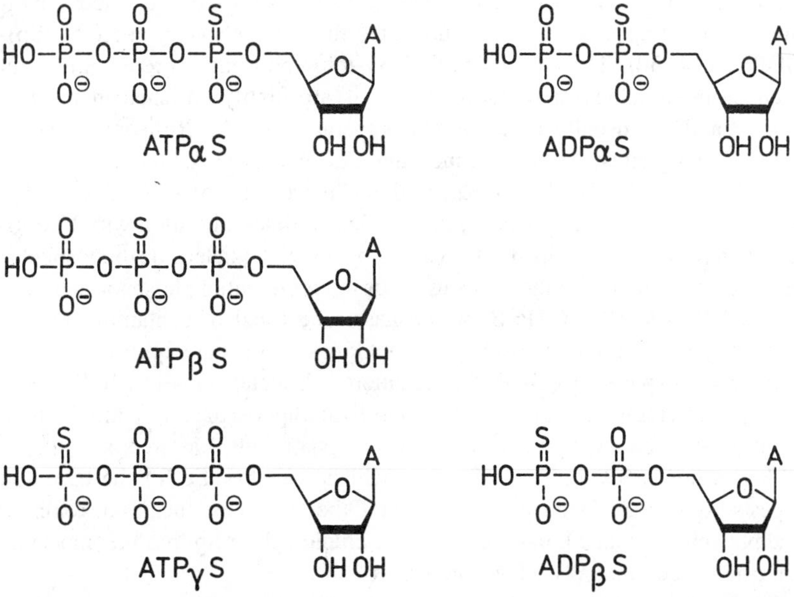

Figure 1   Structural formulae for the various adenosine 5'-phosphorothioates. For the analogs of nucleosides other than adenosine the formulae are corresponding.

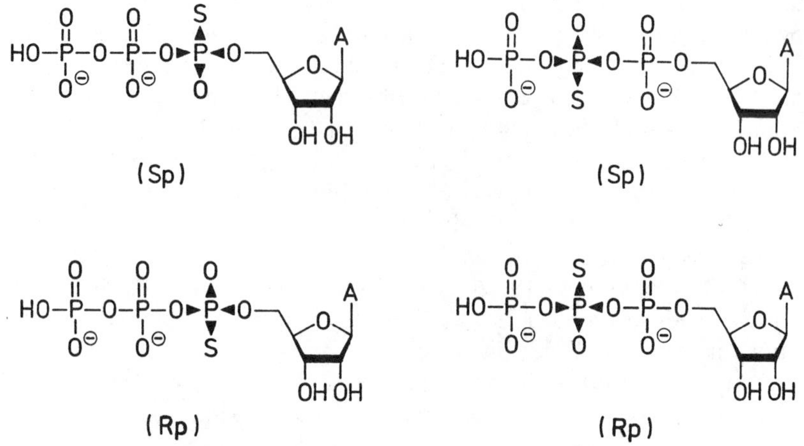

Figure 2 Absolute configuration of the diastereomers of adenosine 5'-O-(1-thiotriphosphates) and adenosine 5'-O-(2-thiotriphosphates).

triphosphate). Pyruvate kinase and phosphoglycerate kinase either preferentially or exclusively phosphorylate the Sp-isomers. Thus, pyruvate kinase preferentially phosphorylates (Sp)-ADPαS (23) and (Sp)-UDPαS (28), and exclusively (Sp)-dADPαS (183), to the corresponding α-thiotriphosphates. Phosphoglycerate kinase shows a higher specificity for (Sp)-ADPαS in the presence of Mg^{2+} than pyruvate kinase (29) and also is specific, under conditions of limiting amounts of cosubstrate, for the phosphorylation of (Sp)-GDPαS (19, 30). The nucleoside 5'-O-(1-thiodiphosphates) remaining from such preferential phosphorylation reactions are highly enriched in the Rp-isomer and are usually used for the preparation of the Rp-isomers of the α-thiotriphosphates. Thus, creatine kinase is employed for the synthesis of (Rp)-ATPαS (18, 23, 28) and (Rp)-UTPαS. Traces of (Sp)-ATPαS formed in this synthesis can be removed by stereoselective back digestion with hexokinase whereas traces of (Sp)-UTPαS are removed with nucleoside diphosphokinase. Succinyl-CoA synthetase is used for the preferential phosphorylation of (Rp)-GDPαS to (Rp)-GTPαS, which has to be freed of contaminating Sp-isomer by reaction with phosphoglycerate kinase. This nucleotide can also be obtained from pure (Rp)-GDPαS by reaction with acetate kinase (19). Pure Rp- and Sp-diastereomers of nucleoside 5'-O-(1-thiodiphosphates) are most conveniently obtained by hydrolysis of the corresponding diastereomers of the α-thiotriphosphates with myosin (23), which is unspecific for the base and digests both the Sp- and Rp-isomers (31). Other enzymes, such as hexokinase and phosphoglycerate kinase, can only be employed for Sp-diastereomers and are often predominantly adenosine specific.

The synthesis of nucleoside 5'-O-(2-thiotriphosphates) starts from the nucleoside 5'-O-(2-thiodiphosphates), whose synthesis is discussed below. Once again these methods rest on the observation that a number of kinases recognize the prochiral β-thiodiphosphate group stereospecifically producing preferentially or exclusively one of the two diastereomers of the nucleoside 5'-O-(2-thiotriphosphates) (23). Thus, pyruvate kinase preferentially produces the Sp-isomers of ATPβS and GTPβS, which can be freed of contaminating Rp-isomer by digestion with hexokinase (23, 28, 32) or glycerol kinase, respectively (19). The reaction of ADPβS with phosphoglycerate kinase yields diastereomerically pure (Sp)-ATPβS (29, 33). (Rp)-ATPβS and (Rp)-GTPβS are prepared using acetate kinase and by removing traces of contaminating Sp-isomer by digestion with myosin (19, 23, 28).

These enzymatic syntheses of the diastereomers of the nucleoside 5'-O-(1-thiotriphosphates) and -(2-thiotriphosphates) are obviously based on the specificities of the enzymes employed for these diastereomers. This specificity is given by V_{max}/K_m (34). General equations and useful graphs for the relationship of this discriminatory factor with percent conversion of substrate have been published (35). Unfortunately, this factor has only been determined for a few enzymes employed in the syntheses of the diastereomers of nucleoside phosphorothioates (Table 1). Since the K_m values generally differ only by factors of 2–5 for two diastereomers, the discrimination is mainly based on V_{max}.

Other methods of preparing the diastereomers of nucleoside phosphorothioates are based on nonstereoselective chemical synthesis followed by diastereomer separation. At least for small-scale preparations, such separation can be achieved by HPLC (26, 19). A chemical approach first developed for the stereospecific synthesis of [α-^{18}O]ADPαS, [β-^{18}O] ADPβS, and [γ-^{18}O]ATPγS (38–40) is particularly suitable for the preparation of the diastereomers of deoxynucleoside 5'-O-(1-thiodiphosphates) and 5'-O-(1-thiotriphosphates) (41). The diastereomers formed on condensation of the deoxynucleoside 5'-phosphorothioate with AMP are separable by chromatography on DEAE-Sephadex and the adenosine residue is removable by

Table 1 Stereospecificity of some kinases for adenosine thiotriphosphate diastereomers

Enzyme	Compound	$\dfrac{(V_{max}/K_m)_{S_p}}{(V_{max}/K_m)_{R_p}}$ [a]	Reference
Myosin	ATPβS	2770[b]	31
Creatine Kinase	ATPβS	0.02	36
Acetate Kinase	ATPβS	⩾4860[c]	37
Phosphoglycerate kinase	ATPαS	2940	29
Phosphoglycerate kinase	ATPβS	1320	29
Hexokinase	ATPαS	67	32
Hexokinase	ATPβS	0.0003	32

[a] All values obtained in the presence of Mg^{2+}.
[b] K_m for Rp-isomer not known; assumed to be the same as for Sp-isomer.
[c] Calculated with K_i-value and lower limit of detection for V_{max}-value for Sp-isomer.

oxidation with $NaJO_4$, yielding the separate diastereomers of the deoxynucleoside 5'-O-(1-thiodiphosphates), which can be enzymatically phosphorylated to the corresponding triphosphates. Nucleoside 5'-O-(3-thiotriphosphates) and 5'-O-(2-thiotriphosphates), such as ATPγS and ADPβS, are prepared by condensation of ADP or AMP, respectively, using β-cyanoethyl phosphorothioate and subsequently removing the β-cyanoethyl group (42). A more recent publication (43) describes an alternative synthesis of ADPβS based on that developed for $[\beta\text{-}^{18}O]$ADPβS described above (40): coupling AMPS with 2',3'-methoxymethylidene AMP, then removing the unprotected adenosine residue.

The absolute configurations of the diastereomers of ATPαS and ATPβS (Figure 2) have been determined by relating the difference in rates of enzymatic transformations of these diastereomers to compounds for which the configuration has been established by X-ray structural analysis. It had been observed that of the two diastereomers of ATPαS, essentially only one is hydrolyzed to AMPS by snake venom phosphodiesterase. This enzyme hydrolyzes the Sp-enantiomer of O-p-nitrophenyl phenylphosphonothioate (44) and the Rp-diastereomer of Up(S)A (45). Thus, the diastereomer of ATPαS that is hydrolyzed by snake venom phosphodiesterase has been assigned the Rp-configuration (44, 45). This has been confirmed in two ways: first, by determination of the structure of the diastereomer of the p-nitrophenylester of AMPS that is not hydrolyzed by snake venom phosphodiesterase as being of the Sp-configuration (46), and, second, by a stereospecific synthesis of $[\alpha\text{-}^{18}O]$ADPαS (47).

The diastereomers of ATPβS react with very different rates with hexokinase and glucose in the presence of Mg^{2+} (32, 48). This is also true for the diastereomers of β,γ-Co(NH$_3$)$_4$ ATP (49). The configuration of the diastereomer of the latter, which functions as substrate, was determined by X-ray structural analysis of the Co(NH$_3$)$_4$-triphosphate moiety, obtained from the nucleotide by removal of the adenosine residue, as being of the Λ screw sense (50). Thus, the diastereomer of Mg·ATPβS that is the better substrate must also have Λ screw sense that corresponds to the Sp-configuration (48).

Nucleoside 3',5'-Cyclic Phosphorothioates

The methods in use, at present, for the preparation of the diastereomers of nucleoside 3',5'-cyclic phosphorothioates are, unfortunately, not stereospecific (Figure 3). All procedures employed lead to a mixture of diastereomers that have to be separated either as precursors or as the final product. Stec and collaborators (3, 51) reported a synthesis of cAMPS in which N^6-dibenzoyl, 2'-O-benzoyl cAMP was converted into the mixture of diastereomeric phosphoroanilidates that could be separated chromatographically. Their configurational assignment rests on ^{31}P-NMR data, which has more recently been confirmed by an X-ray structure determination of one of the diastereomers of the anilidates of cdAMP (52). The phosphoroanilidates are subsequently reacted with potassium and CS_2, with retention of configuration at phosphorus to give the individual diastereomers of cAMPS. This publication represents the reference point for configurational assignment in this field. A relatively simple synthesis for the cyclic phosphorothioate derivatives of adenosine (53), uridine, and cytidine (F. Eckstein, unpublished) consists of the reaction of the nucleosides with bis(p-nitrophenyl)thiophosphorylchloride and subsequent cyclization with potassium tert-butoxide. This method yields approximately 5–10 times as much Sp- as Rp-isomer. (Sp)-cAMPS (53) and (Sp)-cUMPS can be obtained from the mixture of isomers by crystallization. The Rp-isomers can be purified by preparative reverse-phase HPLC. The report that the crystalline cAMPS represents a mixture of diastereomers (53) is erroneous, as was shown by a comparison with material obtained from the Stec group. The synthesis of cGMPS poses problems and, although very poor yields are observed, at present the reaction of GMPS with diphenylphosphorochloridate and subsequent cyclization represents the best method (30). The diastereomers of all of these cyclic phosphorothioates can be distinguished by HPLC and by ^{31}P-NMR spectroscopy.

Phosphorothioate-Containing Dinucleotides and Oligodeoxynucleotides

Dinucleoside phosphorothioates (Figure 4) can be prepared in a manner similar to that used with dinucleoside phosphates using both the phosphotriester and the phosphite approach. One of the earlier phosphotriester methods consisted of condensing β-cyanoethyl S-phosphorothioate with the 5'- and 3'-hydroxyl groups of the desired deoxynucleosides and subsequently removing this protecting group by treatment with base (54). Unfortunately, however, concurrently with the β-elimination reaction for the removal of the β-cyanoethyl group, direct nucleophilic attack of water on phosphorous led to sulfur loss in

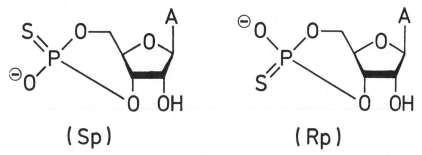

Figure 3   Absolute configurations of the two diastereomers of adenosine 3',5'-cyclic phosphorothioate.

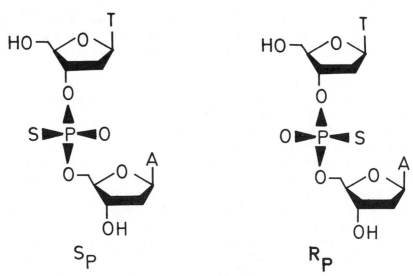

Figure 4   Absolute configurations of the diastereomers of d[Tp(S)A].

20–40% of the dinucleoside phosphates. Although Up(S)A (54), Tp(S)T (41, 55), d[Tp(S)A] (56), d[Ap(S)T] (57), and d[Cp(S)G] and d[Gp(S)C] (F. Eckstein, unpublished) were prepared by this route, the yields were poor and the separation of diastereomers could only be achieved by HPLC. Recently, a different variety of the phosphotriester method has been introduced for the synthesis of Tp(S)T; this method circumvents the difficulty of sulfur loss by using 2,5-dichlorophenyl phosphorodichloridothioate as phosphorylating agent (58). The diastereomers of the triester resulting from this reaction could be partially resolved by chromatography on SiO_2. Removal of the dichlorophenyl group by oxamate was stereospecific, but the stereochemical course of the reaction has yet to be determined.

The phosphite approach for the preparation of those compounds is attractive because it requires little deviation from the procedures established for the synthesis of dinucleoside phosphates; the only difference is the replacement of the iodine/H_2O oxidation step by the addition of sulfur to the phosphite triester. When this approach was first tried in the synthesis of Up(S)A with dichlorophenoxyphosphane as phosphitylating agent, the diastereomers obtained after addition of sulfur could not be separated (59). However, subsequent work using dichloro- or chloro(morpholino)methoxyphosphane for the preparation of Ap(S)A (60a), Up(S)U (60b), Tp(S)T (61); d[Gp(S)A] (62); d[Cp(S)T], d[Gp(S)T], and d[Gp(S)C] (R. Cosstick, F. Eckstein, unpublished) showed that the diastereomers can be separated by chromatography on silica gel type H. Removal of the protecting methyl group from the phosphorothioate using thiophenol proceeds with C-O bond cleavage and thus without any epimerization of configuration at phosphorus (63). The diastereomers of all the dinuc-leoside phosphorothioates can be distinguished by HPLC and by ^{31}P-NMR; their configuration can be determined by digestion with snake venom phospho-diesterase, which hydrolyzes only the Rp-diastereomers or nuclease P1, which digests the Sp-diastereomers (55, 57). Based on the data obtained with the dinucleoside phosphorothioates prepared above, it seems that the Rp-isomer resonates at lower field in the ^{31}P-NMR and has the shorter retention time in reverse-phase HPLC. In a project requiring a phosphorothioate group between dG and dA in the octamer d(GGAATTCC), it was demonstrated that the separated and partially protected diastereomers of d[Gp(S)A] obtained as described above could be incorporated as a block in the course of a polymer-supported phosphite synthesis, yielding the diastereomers of d(GGsAATTCC) (62). Moreover, the mixture of diastereomers could be synthesized by elonga-tion of the oligonucleotide on the polymer support in nearly the usual way, but replacing the normal iodine/H_2O oxidation step with addition of sulfur when reaching the appropriate position between dG and dA. Rather unexpectedly, the resulting mixture of diastereomers could be separated by HPLC after 5'-phosphorylation with polynucleotide kinase.

Exchange of Sulfur for Oxygen

Nucleotides containing phosphate groups that are chiral owing to the presence of the three stable isotopes of oxygen, ^{18}O, ^{17}O, and ^{16}O, are of considerable interest to enzymologists. They are useful for the elucidation of the stereo-chemical course of phosphoryl and nucleotidyl transfer reactions (see following section), and for ^{31}P- and ^{17}O-NMR studies (for reviews see 2, 7–9, 12, 64–67). One method for the stereospecific introduction of oxygen isotopes into nucleotides consists of the exchange of sulfur for oxygen in the diastereomers of nucleoside phosphorothioates, which links the phosphorothioate to the oxygen-isotope method for enzymological studies.

The most common methods involve the activation of sulfur in nucleoside phosphorothioates with iodine (62, 68), cyanogen bromide (69), N-bromosuccinimide (30, 70), or bromine (71) and the simultaneous replacement of the activated sulfur with ^{17}O or ^{18}O derived from $H_2{}^{17}$O or $H_2{}^{18}$O, respec-tively. Other methods include the methylation of sulfur with methyl iodide followed by displacement with $Na^{18}OH$ (72) and the direct reaction of phos-phorothioates with [^{18}O]chloral or [^{18}O]styreneoxide (73). The latter two methods have the disadvantage that the ^{18}O-labeled reagents must be prepared from commercially available $H_2{}^{18}$O. The reaction with iodine has only been performed with oligo- and polynucleotides. It is extremely mild. However, the stereochemical course of this reaction has yet to be determined.

The reaction of cyanogen bromide with β-cyanoethyl ADPαS and those of N-bromosuccinimide and bromine with ADPαS proceed with inversion of configuration at phosphorus. The pH value plays a decisive role in the outcome of these reactions. Those using N-bromosuccinimide and bromine are usually performed in unbuffered solutions that instantaneously become very acidic. In this case, ^{18}O is only incorporated in the α-phosphate with inversion of configuration. Performing these reactions in buffered solutions at neutral pH leads to incorporation of ^{18}O into both the α- and β-phosphate [(74); B. A. Connolly, F. Eckstein, unpublished]. The use of cyanogen bromide with ADPαS at acidic (74), neutral, or basic pH (75, 76) leads to ^{18}O incorporation into both the α- and the β-phosphate, and in this case, β-cyanoethyl ADPαS must be used. These results have been interpreted by the formation of a cyclo-diphosphate intermediate formed by nucleophilic attack of one of the β-phosphate oxygens at α-phosphorus (75, 76). Presumably cyclo-diphosphate formation is prevented by β-phosphate esterification in case of the cyanogen bromide reaction or protonation in the case of the N-bromosuccinimide and bromine reactions. None of these reactions can be used with ATPαS because the label is incorporated into both the γ- and the α-phosphate, probably via a six-membered ring intermediate (69–72).

The reaction of N-bromosuccinimide with cAMPS (30, 70) and cTMPS (71), in opposition to what has been claimed earlier (70), proceeds only with 80% inversion; the reactions with bromine or cyanogen bromide have not yet been tested on such cyclic phosphorothioates. Finally, N-bromosuccinimide has been shown to convert Tp(S)T to $[^{18}O]$TpT with inversion of configuration (55).

The reaction with methyl iodide followed by Na^{18}OH has been performed with cTMPS and Tp(S)T and proceeds with retention of configuration (72). As this method relies on the activation of sulfur by methylation, and since it is known that such S-methylation of ADPαS leads to decomposition (77), this method is unsuitable for α-thiodiphosphates. $[^{18}O]$Chloral and $[^{18}O]$styrene oxide have been used to convert a number of nucleoside phosphorothioates into the corresponding ^{18}O-labeled nucleotides with retention of configuration at phosphorus (73). When ATPαS is used as educt, ^{18}O-incorporation is observed only into the α- and not concomitantly into the γ-phosphate.

MECHANISTIC STUDIES WITH NUCLEOTIDYL AND PHOSPHORYL TRANSFERASES

Stereochemical Course of Reaction

The diastereomers of nucleoside phosphorothioates have been employed to determine the stereochemical course of a large number of enzymatic nucleotidyl and phosphoryl transfer reactions (Table 2). These studies give an answer to the question of whether a reaction proceeds with inversion or retention of configuration at phosphorous. Such a result represents the single

Table 2 Stereochemical course of enzymatic phosphoryl and nucleotidyl transfer reactions

Enzyme	Stereochemical course	Phosphorothioate method	Oxygen isotope method
A. Nucleotidyl transferases			
Ribonuclease A (transesterification step)	Inversion	80, 81	
Ribonuclease T$_1$	Inversion	82	
Ribonuclease T$_2$	Inversion	54	
Enterobacter aerogenes phosphohydrolase	Inversion	83	
Escherichia coli DNA-dependent RNA Polymerase			
Initiation	Inversion	28	
Elongation	Inversion	45	
DNA-dependent DNA Polymerase			
E. coli I	Inversion	84, 27	
Phage T4	Inversion	56	
Phage T7	Inversion	85	
Micrococcus Luteus	Inversion	86	
Reverse transcriptase	Inversion	41	
Polynucleotide phosphorylase			
Exchange	Retention	87	

Table 2 *(continued)*

Enzyme	Stereochemical course	Phosphorothioate method	Oxygen isotope method
Elongation	Inversion	68	
tRNA nucleotidyl transferase	Inversion	88	
RNA ligase, ligation step	Inversion	89	
UDP glucose pyrophosphorylase	Inversion	90	
Galactose-1-phosphate-uridyl transferase	Retention	90	
Adenylate Cyclase			
Bacterial	Inversion	91	92
Mammalian	Inversion	93	
Guanylate Cyclase	Inversion	30	30
Acetyl CoA-Synthetase (activation step)	Inversion	94, 95	
Tyr-tRNA synthetase (activation step)	Inversion	96	97
Met-tRNA synthetase (activation step)	Inversion	96	98
Phe-tRNA synthetase (activation step)	Inversion	79	
B. Nucleases			
Ribonuclease A, hydrolytic step	Inversion	99	
Cyclic phosphodiesterase	Inversion	100, 101	102, 103
Snake venom phosphodiesterase	Retention	44, 104	105, 106
Nuclease S1	Inversion	107	
Restriction endonuclease Eco R1	Inversion	108	
3' → 5' exonuclease of T4 DNA polymerase	Inversion	109	

C. Kinases			
Hexokinase	Inversion	110	111, 112
Glycerol kinase	Inversion	110, 113	111, 113
Pyruvate kinase	Inversion	110	111, 114
Polynucleotide kinase	Inversion	115	116
Nucleoside diphosphate kinase	Retention	90	
Adenylate kinase	Inversion	39, 90	
Adenosine kinase	Inversion	117	
Nucleoside phosphotransferase	Retention	118	
Phosphoglycerate kinase	Inversion	119, 120	
Ribulose phosphate kinase	Inversion	121	
D. ATPases and GTPases			
Myosin ATPase	Inversion	122	
Mitochondrial ATPase	Inversion	123	
ATPase from sacroplasmatic reticulum	Retention	124	
Elongation factor G GTPase	Inversion	125	
Elongation factor T GTPase	Inversion	126	
Snake venom 5'-nucleotidase	Inversion	119	
Thermophilic bacterium PS3 ATPase	Inversion	127	
Phosphoenolpyruvate carboxylase	Inversion	128	
Pyrophosphatase (measured as ATPase)	Inversion	129	
Adenylosuccinate synthetase	Inversion	130	
Phosphoenolpyruvate carboxykinase	Inversion	131	

most informative criterion indicating the absence or presence of a covalent enzyme intermediate. Several reviews deal with this subject by discussing the phosphorothioate or the oxygen-isotope method (1, 2, 4–9, 11–14).

The reactions studied by the phosphorothioate method can be divided into four classes, which are distinguished by the nature of the phosphoryl or nucleotidyl group transferred and the acceptor molecule involved. These in turn determine the degree of oxygen isotope usage necessary to preserve chirality in substrate and product. The four classes are: (A), nucleotidyl transfer by way of a transesterification where the substrate and the product are both diastereomeric phosphorothioate nucleotides (nucleotidyl transferases, no oxygen isotopes are required); (B), nucleotidyl transfer to water where the phosphorothioate product has to be made chiral by the use of $H_2^{18}O$ (nucleases); (C), phosphoryl transfer to acceptors other than water where the phosphorothioate group to be transferred has to be chiral due to the presence of ^{18}O (kinases); and (D), thiophosphoryl transfer to water where the phosphorothioate group of the donor has to be chiral due to the presence of two different isotopes of oxygen with the third isotope being introduced into the product from H_2O (ATPases and GTPases).

The nucleoside triphosphate polymerases, such as DNA or RNA polymerases, comprise the largest group of enzymes in class A. The analysis of the stereochemical courses of their reactions is straightforward. The diastereomeric substrates are easy to synthesize as described above. For product analysis, the polymer produced is in most cases enzymatically degraded to a dinucleoside phosphorothioate whose configuration can be determined by HPLC or ^{31}P-NMR spectroscopy by comparison with the two diastereomers obtained by chemical synthesis, or by further degradation with a nuclease stereospecific for one of the two diastereomers. A similar approach can be used for intramolecular transesterification as catalyzed by RNase A, T_1, and T_2 and by adenylyl and guanylyl cyclases, where the products are 2',3'- or 3',5'-cyclic phosphorothioates, respectively. Once again, the diastereomers of these products can also be distinguished by HPLC or ^{31}P-NMR and comparison with authentic material. A particular case is the reaction product of RNase A with Up(S)A, the uridine 2',3'-cyclic phosphorothioate (45). The Sp-(endo)-isomer of this cyclic phosphorothioate has been characterized by X-ray structural analysis and forms the basis for the phosphorothioate approach for stereochemical investigations (78).

The synthesis of the d[Np(S)A] substrates necessary to study the class B enzymes (the nucleases) has been described above. When these substrates are hydrolyzed in $H_2^{18}O$, an ^{18}O-containing AMPS or dAMPS is produced. The absolute configuration of these and hence the stereochemical course of the enzyme-catalyzed hydrolysis can be determined after stereospecific phosphorylation to $[^{18}O]ATP\alpha S$ or $[^{18}O]dATP\alpha S$ with adenylate and pyruvate kinase (24, 25, 27). Following this phosphorylation, the product analysis is reduced to the problem of determining whether the ^{18}O is in an α,β-phosphate bridging or an α-phosphate nonbridging position. This can be most easily achieved by ^{31}P-NMR spectroscopy taking advantage of the fact that ^{18}O directly bonded to phosphorus produces an upfield chemical shift of the ^{31} P-signal, the magnitude of the shift increasing with increasing bond order (see reviews 64–67). Thus α,β-bridging ^{18}O produces an upfield shift on both the α- and the β-phosphorus signals, whereas a nonbridging ^{18}O causes such a shift only at the α-phosphorus. This ^{18}O-induced shift is different for phosphate and phosphorothioate groups. Thus, $[\alpha,\beta$-$^{18}O]ATP\alpha S$ shows a shift of 1.3 Hz for the β-, and of 1.9 Hz for the α-phosphorus, although the bond order is the same (79). Greater chemical shift variation in phosphorothioate as compared to phosphate groups is also observed on metal binding and pH changes (25, 48, 132). The oxygen isotope position in $[^{18}O]ATP\alpha S$ can also be determined by mass spectrometry in the ribo series after removal of the adenosine residue. The thiotriphosphate moiety thus obtained is permethylated and hydrolyzed to a mixture of trimethyl phosphate and O,O,S-trimethyl phosphate. This in turn is subjected to mass spectroscopic analysis to determine the distribution of ^{18}O, which gives the ^{18}O position in the parent $[^{18}O]ATP\alpha S$ (38, 39). More recently, fast atom bombardment mass spectroscopy has been introduced for this analysis; its advantage is that the ATP or ATPαS molecule can be subjected to such analysis without prior chemical degradation (79).

The class C kinases were first studied by Knowles et al, who used $[\gamma$-$^{16}O,$ $^{18}O]ATP\gamma S$ of unknown configuration to show that hexokinase, pyruvate kinase, and glycerokinase all catalyze the transfer of the phosphorothioate group to their respective acceptors with the same unknown-at-the-time stereochemical course, either all with inversion or retention (110). This question was answered by determining the stereochemical course of the glycerokinase reaction following the development of a stereospecific synthesis for Rp- and Sp-$[\gamma$-$^{16}O,$ $^{18}O]ATP\gamma S$ (39, 40). It was shown that glycerokinase catalyzes the transfer with inversion of configuration as must therefore hexokinase and pyruvate kinase (113). The availability of Rp- and Sp-$[\gamma$-$^{16}O,$ $^{18}O]ATP\gamma S$ also facilitated the determination of the stereochemical course of the other kinases listed in Table 2. The reaction products were usually analyzed by ^{31}P-NMR spectroscopy or mass spectrometry.

The enzymes in group D of Table 2 are the ATPases and GTPases, which usually hydrolyze these nucleotides to the diphosphate and inorganic phosphate. Their stereochemical course can be determined using ATPγS or GTPγS stereospecifically labeled in the γ-position with ^{17}O. Performing the reaction in $H_2^{18}O$ leads to inorganic thiophosphate that is chiral owing to the presence of the three stable isotopes of oxygen. The determination of the absolute configuration of this enantiomeric molecule has been described by Webb &

Trentham (120) and by Tsai (119). The procedure is based on the incorporation of the inorganic, enantiomeric thiophosphate in a sequence of enzymatic reactions into ATPβS, which is analyzed by ^{31}P-NMR spectroscopy. The method has been reviewed in detail (6).

Before discussing the results obtained on the stereochemical courses of nucleotidyl and phosphoryl transfer reactions using phosphorothioate analogs, it is appropriate to discuss their relevance. Obviously these data are only of significance if it can be shown that the stereochemical course of an enzyme-catalyzed reaction is the same regardless of whether a phosphoryl or a thiophosphoryl group has been transferred. The only way to demonstrate this is by analyzing an enzymatic reaction using both a phosphorothioate substrate and a substrate where the phosphorus is chiral owing to the presence of the different isotopes of oxygen. This latter method has been developed by the groups of Knowles and of Lowe and has been the subject of various reviews (2, 5, 7, 8, 12, 64). The first direct comparison of the stereochemical course of an enzyme-catalyzed reaction as determined by the two methods was reported for glycerol kinase (113). Since then such a comparison has become available for another 10 enzymes, as indicated in the last column of Table 2. The satisfying result of all these comparisons is that an enzyme catalyzes its reaction by a mechanism that has the same stereochemical consequence independent of whether a phosphoryl or a thiophosphoryl group is transferred. This result could probably have been presumed since it is unlikely that the active site of an enzyme is constructed in such a way as to have two alternative reaction pathways available. However, it had been of some concern, as many enzymatic reactions proceed much more slowly with a phosphorothioate than with a phosphate substrate. It is interesting that for many reactions (e.g. those catalyzed by the DNA and RNA polymerases and myosin), a minimal difference in k_{cat} is observed between phosphates and phosphorothioates; thus it cannot be generally argued that the presence of a phosphorothioate group decelerates an enzymatic reaction. As a detailed kinetic study with myosin has demonstrated, a slowdown in the actual chemical step—which is expected because of the slower reactivity of phosphorothioates as compared to phosphates (133)—might not affect the rate-limiting step (134) and thus could lead to little or no change in the overall rate.

The stereochemical outcomes of enzyme-catalyzed reactions are summarized in Table 2, which shows that the majority proceed with inversion of configuration at phosphorus. This is also true for the additional 16 enzymes for which the stereochemical course of the reaction has been established solely by the oxygen isotope method and that have not been summarized here (2). Inversion of configuration can be the result of any odd number of nucleophilic substitution reactions at phosphorus (1, 2, 8, 9, 12). Because all of the inversion results can be explained by the smallest odd number of reactions, one, there is no compelling reason to postulate more complicated schemes. Thus, inversion of configuration would be the result of a single displacement reaction in an associative mechanism where the enzyme catalyzes the nucleophilic attack of one substrate on the other, resulting directly in product formation. The transition state of such a reaction would consist of a pentacoordinated phosphorus with trigonal bipyramidal geometry and the incoming and leaving groups in the apical positions. Retention of configuration can be the result either of two successive substitution reactions, each proceeding with inversion of configuration, or of one nucleophilic substitution reaction involving pseudorotation of the pentacoordinate intermediate (135). For most of the enzymes listed in Table 2 for which retention of configuration is observed, a covalent enyzme intermediate has been postulated either from kinetic arguments or from isolation. Thus, for these enzymes, retention can most easily be explained by two successive nucleophilic substitution reactions, the first by the enzyme on the first substrate resulting in the covalent intermediate, the second by the second substrate on this intermediate resulting in product. As both steps proceed with inversion of configuration, overall retention is observed. For other enzymes for which retention is observed but no independent indication for the existence of a covalent intermediate exists, the stereochemical analysis argues strongly for such an intermediate and represents the strongest single indication of it. To date pseudorotation does not seem to be involved in enzymatic phosphoryl and nucleotidyl transfer reactions. In addition, there is no indication of formation of mnetaphosphate in a dissociative mechanism where bond breaking precedes bond formation. In solution where the metaphosphate is free this would lead to epimerization. Since no such epimerization has been observed in any of the enzyme reactions studied so far, metaphosphate does not seem to occur as an intermediate that can dissociate from the enzyme or rotate at the active site. Strictly speaking, it is impossible to eliminate a mechanism involving a tightly bound metaphosphate that cannot rotate. However, the postulation of metaphosphate in an enzymatic reaction in which the second, the acceptor substrate, is not only present but is also juxtaposed so as to immediately react with the group to be transferred, is probably meaningless, since bond breaking and bond formation would occur simultaneously, thus approaching the situation of an associative mechanism (see 12).

Table 2 shows that retention of configuration is rarely observed and seems to be the exception rather than the rule for these reactions. Hypotheses about why nature has developed such exceptions have been put forward, but at present they are not fully satisfactory (8). For instance, it is unclear why the same type of reaction, hydrolysis of an internucleotidic linkage, should be catalyzed on the one hand by snake venom phosphodiesterase with retention of configuration, and on the other by nuclease S1 or P1 with inversion of configuration.

Structure of the Substrate Metal-ATP Complex

Probably all ATP-dependent enzymes use a Mg·ATP chelate as the substrate where the Mg^{2+} ion is coordinated to the oxygens of the triphosphate group. In principle it can form mono-, bi-, or tridentate chelates. Additionally, binding to the α- and the β-phosphate groups introduces new chiral centers and thus leads to the occurrence of various diastereomers (11). A detailed knowledge of the structure of the metal·ATP substrate, the coordination of the metal to the substrate during the reaction, and finally the metal·product complex is a prerequisite for the complete description of the mechanism of any enzyme utilizing ATP. At present, three methods are available to probe the coordination of the metal to the phosphate groups. Besides the approach based on the use of phosphorothioate analogs of ATP, which will be discussed here, there is the method employing exchange-inert Cr^{III}- and Co^{III}-ATP complexes (151–154), and that based on ^{17}O-containing phosphate groups in ATP and their interaction with Mn^{2+} as monitored by EPR (155–159). The use of the phosphorothioate analogs of ATP rests on the observation made by ^{31}P-NMR spectroscopy that Mg^{2+} as a hard metal ion coordinates preferentially to oxygen, and Cd^{2+} as a soft metal ion to sulfur (32, 48). Other cations, such as Co^{2+}, Mn^{2+}, and Zn^{2+}, are intermediate. The stability constants of Mg^{2+} and Cd^{2+} complexes of various phosphorothioate analogs of ATP and ADP have recently been determined and used to calculate the preference for O and S coordination (132). The preference for sulfur over oxygen coordination for Cd^{2+} is estimated to be 60 regardless of the position of the phosphorothioate group. Preference for oxygen over sulfur coordination for Mg^{2+} is calculated to be 31,000 for MgATPβS, 3,100–3,900 for CaATPβS, and 158–193 for MnATPβS.

This different coordination of Mg^{2+} and Cd^{2+} has the consequence that Sp-ATPαS (or ATPβS), when coordinated to Mg^{2+}, has the same screw sense [as defined in reference (151)] as the Cd^{2+} complex of Rp-ATPαS (or ATPβS) (Figure 5). In contrast, both the Mg^{2+} chelate of Rp-ATPαS (or ATPβS) and the Cd^{2+} chelate of Sp-ATPαS (or ATPβS) have the opposite screw sense to that formed above. With hexokinase it was shown that Rp-ATPβS was active in the presence of Mg^{2+}, whereas Sp-ATPβS showed activity in the presence of Cd^{2+}. This reversal of stereoselectivity is indicative of metal-coordination to the phosphorothioate group under study during the rate-limiting step of the reaction. No reversal was observed with the diastereomers of ATPαS (32, 48). This concept of change in stereoselectivity has been very fruitful and has been applied to the determination of the metal-ATP complexes in a number of enzyme reactions as summarized in Table 3. As elegant as this approach is, it has its pitfalls. As can be seen from Table 3, a change in stereoselectivity is most often observed for the β-phosphate group. Ideally, for a change in stereoselectivity, one would expect that $(V_{max}/K_m)_{Mg}/(V_{max}/K_m)_{Cd}$ is >1 for one diastereomer and <1 for the other. This is observed for many enzymes, e.g. hexokinase and myosin, and thus the interpretation is straightforward. However, for others, such as protein kinase (147) and RNA polymerase (148), this ratio is about 1×10^4 or 1×10^3, respectively, for the Sp-, but about unity for the Rp-ATPβS. Thus, in these cases no real reversal in stereoselectivity is observed, although it is obvious that the Sp-diastereomer is a much better substrate in the presence of Mg^{2+} than Cd^{2+}. However, the Rp-diastereomer exhibits about equal reactivity in the presence of both metal ions. The activity of the Sp-isomer with Mg^{2+} but not with Cd^{2+} suggests that the Δ-β,γ-bidentate complex is used, and in this case one expects to observe activity with Cd^{2+} and the Rp-isomer. The unexpected activity seen with Mg^{2+} and Rp-ATPβS, which to form the active Δ-screw sense isomer must involve unfavorable $Mg \rightarrow S$ coordination, was explained by assuming that the RNA polymerase formed an additional favorable interaction between an enzyme group and the noncoordinated β-phosphate oxygen. This favorable interaction, possibly a hydrogen bond, is thought to offset the $Mg \rightarrow S$ coordination. The same situation is observed for phe-tRNA synthetase (143), and Sp-ATPβS in the presence of Mg^{2+} and Cd^{2+}. These examples demonstrate that restraints exerted by the structure of the active site are able to counteract the most favorable metal-phosphorothioate interaction. Reversal at β-phosphorus is generally interpreted in terms of β,γ-bidentate-metal coordination, although coordination to the γ-phosphate cannot be directly demonstrated by this method. This assumption is based on the greater thermodynamic stability of such β,γ-six-membered chains over a β-monodentate complex (160). Reversal at the α- and the β-phosphate has been detected using ATPαS and ATPβS for creatine and arginine kinase, with additional reversal at the α-phosphate for ADPαS in the reverse reaction (36, 137). This latter result is most easily explained by postulation of an α,β-bidentate ADP complex for the product and is in agreement with the results obtained using the diastereomers of CrADP (154, 161). However, whether reversal at the α- and β-phosphate for the triphosphates indicates α-, β-, or β,γ-coordination of the substrate could not be decided. As it turns out, the answer is that this dual reversal is probably an indication of the involvement of an α,β,γ-metal tridentate complex in the reaction; this has been elegantly demonstrated by EPR measurements of a transition state complex consisting of ^{17}O-phosphate containing ADP, Mn^{2+}, nitrate, and enzyme (155, 156). More difficult are explanations for the lack of reversal of stereoselectivity found so often for the α-phosphate. It would argue against metal coordination to this phosphate, but on the other hand such a coordination might not occur at the rate-limiting step and therefore not manifest itself. It could also be due to the 10-fold lower preference of $Cd \rightarrow S$ interaction in ATPαS as compared to ATPβS and ADPαS or the higher tridentate content of this complex (132). In an attempt to shed some light on this question, ATPαS

Table 3 Determination of metal-ATP complex for substrates of some enzymes by change in stereoselectivity for diastereomers of ATPαS and ATPβS in presence of Mg^{2+} and Cd^{2+}

	$\dfrac{(V_{max}/K_m)_{Mg}}{(V_{max}/K_m)_{Cd}}$		$\dfrac{(V_{max}/K_m)_{Mg}}{(V_{max}/K_m)_{Cd}}$		β, γ-metal ATP-complex	Reference
	Sp-ATPαS	Rp-ATPαS	Sp-ATPβS	Rp-ATPβS		
Hexokinase, yeast	34.75	40.00	0.0024	1884	Λ	32
mam.[a]						136
Glucokinase[a]			1.8	165	Λ	136
Creatine kinase[a]	0.046	1266	2.28	238	Λ[i]	36
Arginine kinase	0.016	5.52	0.26	694	Λ[i]	137
Acetate kinase[b]	2.26	12.7	0.0228	83	Λ	37
Glutamine synthetase[j]	—	—	—	8.26	Λ	138
Mevalonate kinase	—	—	0.61	4.98	Λ	139
Quinolate phosphoribosyl transferase[j]	—	—	—	—	Λ[h]	140
Phosphoglycerate kinase[f]	2.13	0.008	4.5	3.12	—	29
Phe tRNA synthetase[a,m]	4	—	1.12	152	Λ	143
Met tRNA synthetase[c,f]	—	—	3.5	0.0001	Δ	144

Tryp tRNA synthetase	3.68[a]	4.2[a]	8.9	0.018	Δ[k]	145
Adenylate kinase[b]	1.65	3.8	12.43	0.03	Δ	141, 142[n]
Myosin[b]	—			0.19	Δ	31
Pyridoxal kinase		—	31.25		Δ	146
Protein kinase[a]			10,490	1.2	Δ	147
Thermophilic bacterium PS3 ATPase[a,b]	49	2.09	1.45	0.002	Δ	127
RNA polymerase[d,m]	18	—[j]	900	3.6	Δ	148
DNA polymerase[b,e]	1.09	0.02	9.0[o]	0.0048	Δ	84
Phosphoribosylpyrophosphate synthetase[a]	61			—	Δ[g]	149
EF-Tu-GTPase	0.039		211		Λ	150

[a] Only V_{max}-values taken into account because none or not all K_m values available.
[b] Some K_m values replaced by K_i values.
[c] Lower limits for K_m and V_{max} values for Mg ATPβS estimated since no activity observed.
[d] Lower limit for V_{max} value for Cd ATPβS estimated since no activity observed, K_m values replaced by K_i values, initial velocities for ATPαs used.
[e] Enzyme inactive in presence of Cd^{2+}, values in presence of Co^{2+} taken instead.
[f] Enzyme has low or no activity in presence of Cd^{2+}, values in presence of Zn^{2+} taken instead.
[g] Configuration at β,γ-position based on data obtained with $Co(NH_3)_4ATP$.
[h] Chelate complex of α,β-metal phosphoribosyldiphosphate.
[i] Δ for α,β-metal ADP product complex.
[j] Rp-dATPαS shows no activity with either metal ion.
[k] Rp-ATPβS not a substrate in presence of Mg^{2+}, Sp-ATPαS not a substrate in presence of Zn^{2+}.
[l] Data incomplete.
[m] Rp-ATPαS not a substrate in presence of either Mg^{2+} or Co^{2+} (or Cd^{2+}).
[n] On the basis of $[^{17}O]$ATP data, a β-monodentate metal-ATP complex is postulated by these authors.
[o] Lower limit for V_{max} value for Rp-MgdATPβS assumed since no activity observed.

has been methylated on sulfur so as to exclude metal coordination to this triester phosphate group (77). Two enzymes that show lack of reversal at the α-phosphate, hexokinase and acetate kinase, were able to use the Sp-ATPαSCH₃, confirming a lack of metal coordination to the α-phosphate. Phosphofructokinase, for which a β,γ-bidentate Mg·ATP complex had been postulated on the basis of experiments with Cr^{III} ATP, was able to use both diastereomers of ATPαSCH₃. In contrast, creatine kinase and phosphoglycerate kinase, for which metal coordination to the α-phosphate had been demonstrated, were unable to use ATPαSCH₃ as substrate. This suggests the necessity of metal coordination to the α-phosphate group, although the lack of activity seen with the methylated derivatives could be due simply to steric reasons.

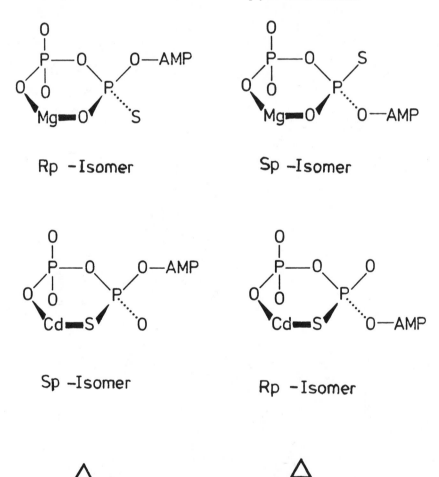

Figure 5   Structures of the Mg^{2+} and Cd^{2+} complexes of ATPβS.

A unique situation is represented by the observation of reversal solely at the α-phosphate in the phosphoglycerate kinase reaction (29). What prevents the coordination or detection of coordination to the β-phosphate that would be expected is not clear at present. It is worth mentioning in the context of these phosphoryl transfer reactions that the free energy difference between ATPβS and ADPβS is more exergonic by about 2.4 kcal/mol than between ATP and ADP. Thus, reactions involving phosphoryl transfer from ATP are shifted about 60 fold towards ADPβS formation relative to ADP formation when ATPβS is used as phosphoryl donor instead of ATP (10).

Summarizing the data in Table 3, it can be said that most enzymes investigated so far prefer a β,γ-bidentate metal chelate ATP complex with half the enzymes requiring the Λ, and the other the Δ screw sense.

INTERACTION OF NUCLEOSIDE 3',5'-CYCLIC PHOSPHOROTHIOATES WITH PROTEINS

cAMP is extremely important in the regulation of several biochemical systems and so the phosphorothioate analog cAMPS has been studied most intensely. Its synthesis was first reported by Eckstein et al (53). However, contrary to what was described in this report, the crystalline material obtained was not a mixture of diastereomers but, as has been shown by comparison with material from the Stec group (51), the pure Sp-isomer. Thus, this article's description of the very slow hydrolysis of this material by beef heart and rabbit brain phosphorodiesterases and of its binding to and activation of beef heart cAMP-dependent protein kinase refers to the pure Sp-isomer rather than the mixture. The Rp-isomer is not a substrate for the cyclic phosphodiesterase from beef heart (93, 101), but both isomers are hydrolyzed by the enzyme from baker's yeast (101). The interactions of the diastereomers of cAMPS with protein kinases have been studied in more detail. Thus, both type I and II kinases are activated by the Sp-isomer; but the Rp-isomer, although binding to these enzymes, does not cause activation providing that it is diastereomerically pure (162, 163). This isomer is not able to trigger release of the catalytic subunit for which a conformational change of the regulatory subunit is apparently required. A hydrogen bond or salt bridge interaction between the equatorial exocyclic negatively charged oxygen of cAMP—which is replaced in the Rp-cAMPS by sulfur—and the protein has been postulated. The small activation seen for the type II enzyme (163) was probably due to contamination of the Rp- by the Sp-isomer (162). Early experiments with cAMPS and rat parotid cells showed that this analog—i.e. the Sp-isomer, which at the time was thought to be a mixture of diastereomers—can elicit the same response, secretion of amylase, as seen on stimulation by isoproterenol (164). This result made it evident that the exchange of oxygen for sulfur enables the compound to penetrate the plasma membrane and to enter cells. This property coupled with the ability to activate protein kinases makes this analog an extremely useful tool for the study of the involvement of cAMP in biological systems. It actually has a certain advantage over the most commonly used analog, dibutyryl cAMP, since this releases butyric acid on hydrolysis, which can cause a number of reactions unrelated to the action of cAMP (165, 166) that might not be evident by applying butyric acid to the system as a control.

Rothermel et al (167) recently made the interesting observation that, in agreement with what has been found for isolated protein kinases, glycogenolysis in isolated rat hepatocytes is activated by Sp-cAMPS, whereas this activation as well as that induced by glucagon (168) can be inhibited by the Rp-isomer. Cyclic AMPS has also been applied to the study of various effects of cAMP on *Dictyostelium discoideum,* including chemotaxis, which will not be discussed in detail here [(169); reviews in (1, 13)].

Sp-cAMPS has also been found to bind to the *Escherichia coli* cAMP receptor protein (170). Rp-cAMPS is about 10 times more efficient in catabolite repression than Sp-cAMPS (171), which also inhibits the rate of triacylglycerol synthesis from glycerol in rat hepatocytes (172) and the rate of antigen-stimulated histamine release of mast cells (173). Thus, this analog elicits the effects expected of cAMP and should be a useful compound to probe more complex systems for the possible involvement of cAMP.

The diastereomers of cGMPS (30) still await application to a cGMP-regulated system. The light-activated cGMP phosphodiesterase of rod outer segments might be such a system where the resistance of cGMPS against phosphodiesterase could be usefully applied (174).

PHOSPHOROTHIOATE DNA

Hydrolysis by Nucleases

The slow rates of hydrolysis of phosphorothioate diesters by snake venom phosphodiesterase (41, 54, 84) suggested the possibility of introducing inter-nucleotidic linkages into DNA, which might be difficult to hydrolyze by some enzymes, but which at the same time would retain essential features of the normal phosphate group, such as a negative charge and a similar geometry, so as not to interfere gravely with the structure of DNA. The enzymatic incorporation of phosphorothioate groups into DNA can be achieved with surprising ease by all the DNA polymerases studied to date, employing any of the four deoxynucleoside 5'-O-(1-thiotriphosphates). All these enzymes accept only the Sp-diastereomer, and although V_{max}- and K_m-values have only been determined for *E. coli* DNA polymerase I and T4 DNA polymerase, and were found to be very close to those of the natural triphosphates (56, 84), the general experience is that all four dNTPαS nucleotides are incorporated by all polymerases quite efficiently (176). All polymerases for which the stereochemical course has been determined (see Table 2) catalyze the polymerization reaction with inversion of configuration at phosphorus, resulting in phosphorothioate internucleotidic groups of the Rp-configuration. The expectation that these might be hydrolyzed by nucleases rather slowly has in general been gratifyingly fulfilled. However, the degree of decrease in rates of hydrolysis in comparison to normal internucleotidic linkages differs between enzymes. For snake venom phosphodiesterase this factor is about 10 (41, 54, 84) for this isomer. For the Sp-isomer, which has so far not been available as part of a polymer, the rate difference for dinucleoside phosphates and phosphorothioates is several orders of magnitude (54). The opposite specificity is observed for nucleases S1 and P1, which hydrolyze dinucleoside phosphorothioates of the Sp- but not the Rp-configuration (55, 57, 107). Nuclease P1 hydrolyzes phosphorothioates without any of the partial desulfurization seen with snake venom phosphodiesterase and Nuclease S1 and is thus preferable for preparative or analytical purposes (57, 62, 108).

Very large differences are observed for the rates of hydrolysis catalyzed by the exonuclease activities of *E. coli* DNA polymerase I. Thus, the phosphorothioate internucleotidic linkage in a polynucleotide, i.e. of the Rp-configuration, is essentially resistant against the 5' → 3' and 3' → 5' exonuclease activities of this enzyme. The resistance of internucleotidic phosphorothioate linkages against the 5' → 3' exonuclease of pol I has been used to advantage in the hydrolysis of polydeoxynucleotides for configurational determination (27, 56, 84). However, contrary to an earlier report (176), the 3' → 5' exonuclease of T4 DNA polymerase cleaves such modified groups with similar rates as the normal substrates (177). The 3' → 5' exonuclease activities of bacterial and viral DNA polymerases fulfill a proofreading function during the polymerization reaction. Addition of a noncomplementary nucleotide to the growing DNA strand is prevented by excision of such a mismatched nucleotide by the 3' → 5' exonuclease before the polymerization continues (178). Obviously, if the internucleotidic linkage to the mismatched nucleotide cannot be hydrolyzed, one would expect the accuracy of replication to be decreased. This has been checked by using dCTPαS as phosphorothioate analog of dCTP and ΦX 174am3 single-stranded DNA as template (176). In this *amber* mutant a TGG triplet is changed to TAG in the plus strand so that incorporation during replication of deoxycytidine, leading to an A-C mismatch instead of an A-T base pair, leads to reversion, which can easily be quantitated by transfecting the copied DNA into *E. coli* spheroplasts and measuring the titer of the resultant progeny phage. In this system it has been found that the fidelity of *E. coli* DNA polymerase I is reduced 20 fold when dCTPαS is employed as a substrate rather than dCTP. This number should represent the contribution of the exonuclease activity towards fidelity, although it does not take into account the fact that mismatches might not be easily elongated. This interpretation is supported by the observation that mammalian DNA polymerases that lack this exonuclease activity show no decrease in fidelity when phosphorothioates are employed as substrates for the replication process. However, this interpretation is somewhat doubtful, since an approximately 500-fold reduction in fidelity was observed with T4 DNA polymerase, although it was later shown that the 3' → 5' exonuclease hydrolyzes phosphorothioates quite efficiently (177). At present this discrepancy remains unexplained. This decrease in fidelity can be exploited for the generation of mutants. Shortle et al have described the use of α-phosphorothioates in gap-misrepair mutagenesis (179), where a gap is introduced by a restriction enzyme in one strand, enlarged by a exonuclease, and then filled and ligated. When one of the four dNTP was replaced by a dNTPαS, the majority of the mutations recovered were single base substitutions of the type expected from the base of the nucleoside phosphorothioate used. Thus, it is expected that this phosphorothioate method might be another useful technique for mutagenesis. Exonuclease III is another enzyme that cannot cleave phosphorothioates. As a consequence, the incorporation of a phosphorothioate nucleotide at the 3'-end of one strand of a double-stranded DNA protects this strand against degradation whereas the other can be trimmed (180, 181).

Restriction endonucleases, in general, hydrolyze phosphorothioates more slowly than phosphates. This has first been shown in a hybrid system where the (+)strand of fd DNA contained all phosphate groups but the (−)strand carried phosphorothioate groups at the 5'-position of a particular nucleotide. Such hybrid DNA was cleaved slowly by various restriction endonucleases whenever a phosphorothioate was present directly at the cleavage site. The presence at other positions in the recognition sequence had little or no effect (182). More recent data obtained with M13 DNA show that it is possible to isolate the nicked intermediate of this reaction where one strand, presumably the all phosphate–containing (+)strand, is cleaved, but the other, most likely the phosphorothioate-substituted (−)strand, remains intact (183). This has been observed for the Bam H1- and Eco R1-catalyzed reactions. Such intermediates cannot be detected for cleavage of phosphorothioate-phosphate DNA hybrids by Sal I even at low enzyme concentration. The first reaction product seen is linear DNA. These results suggest that at least for restriction enzymes, which operate by a stepwise mechanism of strand cleavage as does Eco R1, and probably Bam H1 (184), cleavage can be directed towards one strand by phosphorothioate substitution.

The first study between a restriction enzyme and a chemically synthesized phosphorothioate oligonucleotide has been recently undertaken using the diastereomers of d(GGsAATTCC). This octanucleotide contains the recognition sequence for Eco R1 and has a phosphorothioate group at the site of cleavage between dG and dA (62). Only the Rp-isomer is cleaved at a rate about 15 times slower than the unmodified octamer, with inversion of configuration (Table 2). The Rp-isomer of the phosphorothioate is, of course, the one present when phosphorothioate groups are introduced enzymatically into DNA.

B → Z Transition

^{31}P-NMR studies on phosphorothioate-containing alternating polydeoxynucleotides such as poly [d(A-T)] and poly [d(G-C)] have led to an assignment of the phosphorus resonances in the spectra of these polynucleotides. This is possible because phosphorothioates resonate about 50 ppm downfield from phosphates. Thus, on replacing one of the phosphates by a phosphorothioate in such alternating polymers, the resonances can easily be assigned as long as the remaining phosphate resonance does not change. This is straightforward for poly [d(A-T)], where two resonances are observed at −4.2, and −4.4 ppm relative to trimethyl phosphate. The resonance at lower field could be assigned by this method to the d(ApT) group (185). The spectrum of poly [d(G-C)] in solutions of low salt concentrations shows only one broad resonance, but that spectrum taken at high concentrations of salt, conditions under which the Z-conformation is adopted, again shows two resonances. By synthesizing the phosphorothioate analog where dCMP is replaced by dCMPS, the low field resonance at high salt could be assigned to d(GpC) (86, 186). Thus, on going from the B- to the Z-conformation, this is the phosphate residue which experiences the greater conformational change. Such shifts to low field are usually due to a change in torsional angle of the ester groups from a gauche, gauche to a gauche, *trans*-conformation (187).

The experiments with the phosphorothioate analogs of poly [d(G-C)] revealed a surprising influence of the phosphorothioate group on the ability of the polymer to undergo the B → Z-conformational change. The poly [d(G-C)] analogs in which dCMP is replaced by dCMPS underwent this change much more easily than poly [d(G-C)] itself. The analog where dGMP was replaced of dGMPS, however, proved to be incapable of adopting the Z-conformation (86, 186, 188). At present these results are difficult to interpret. Further studies will have to show whether the prevention of the conformational change by a purine 5'-phosphorothioate, and its facilitation by a pyrimidine 5'-phosphorothioate, is a general phenomenon and whether possibly the configuration of the phosphorothioate group has an influence.

Phosphorothioate DNA has found a further, unexpected application in the sequencing of DNA. Biggin et al reported (189) that the use of [^{35}S]dATPαS instead of [α-^{32}P]dATP in the dideoxynucleotide sequencing technique increased the sharpness of the bands on a gel autioradiograph and so increased the resolution. This increase is due to the lower energy and hence the shorter path length of the β-particles emitted by ^{35}S in comparison to [32]P.

PHOSPHOROTHIOATE RNA

As Sp-nucleoside 5'-O-(1-thiotriphosphates) are substrates for *E. coli* DNA–dependent RNA polymerase, phosphorothioate groups can also be incorporated in polyribonucleotides (148, 190). As with the deoxy series, only Rp-phosphorothioate–containing polymers can be produced. However, such modified polymers have not been intensively studied. The notable exception is the use of the phosphorothioate analogs of poly[r(A-U)] and poly[r(I-C)] as inducers of interferon (191). On the other hand, the attachment of phosphorothioate groups at the 5'-end of RNA introduced during polymerization by employing either ATPγS or GTPγS, and more recently, ATPβS and GTPβS, as substrates (192, 193), has been found useful for the separation (using Hg-Sepharose columns) of newly synthesized RNA from endogenous RNA in isolated nuclei of mouse myeloma [for more examples see (1)]. Introduction at the 3'-end of RNA, in particular tRNA, can be achieved by the use of dpCp(S) as substrate for RNA ligase (194). As the sulfur in phosphorothioates is a good nucleophile for activated methylene groups, the terminal phosphorothioate groups can serve as functional groups for the attachment of reporter groups such as fluorescent labels. These can also be attached to dpCp(S) before the ligation reaction, as such compounds are also substrates for the ligase (194).

FORMATION OF STABLE PROTEIN-NUCLEOTIDE COMPLEXES

There exists the general problem in systems in which a particular reaction is coupled to the hydrolysis of ATP or GTP of deciding whether binding of the triphosphate is sufficient or whether hydrolysis is essential for this reaction to occur. Obviously, hydrolysis-resistant analogs should help to answer this question as long as they are similar enough to ATP or GTP to elicit the reaction in question. The imido analogs of ATP or GTP have been most useful in this respect since they are resistant to many hydrolases (195). However, the pKa-value of AMPPNP of 8.2 (196) is somewhat higher than that of 6.5–6.9 for ATP (25, 197), so that in many buffer systems it is not fully ionized. As thiophosphoric acid is more acidic than phosphoric acid (198), ATPγS also has a lower pKa than ATP. It has been measured to be 5.3 by ^{31}P (25), and 5.8 by ^{17}O NMR spectroscopy (199). Indeed, for some systems ATPγS and GTPγS show higher affinities than AMPPNP and GMPPNP, respectively (see below). Whether this is a reflection of the lower pKa or is owing to the higher hydrophobicity of sulfur or some other structural feature is not clear.

Of the many systems in which ATPγS or GTPγS have been used as hydrolysis-resistant analogs, only a few will be discussed here, and the reader is referred to one of the recent reviews (1) for more examples.

ATPγS·rec A Protein Complex

In the presence of ATP, rec A protein catalyzes the in vitro hybridization of DNA single-stranded regions with homologous regions in double-stranded DNA to form a D-loop which migrates until complete strand exchange has taken place (200, 201). In addition, this protein cleaves proteolytically various repressors such as λ, lex A, and P22 when DNA and ATP are present. Both of these activities reflect the various roles of rec A protein function in cellular processes that promote repair of damaged DNA. For both types of reactions, it is of interest to know whether hydrolysis of ATP is necessary. It was found that ATPγS binds to rec A protein but is not hydrolyzed (202). As a consequence, it became instrumental to investigate the role of ATP in these reactions. Such studies showed that D-loop formation does not require ATP hydrolysis but that branch migration does (203). The rec A protein DNA complex formed in the presence of ATPγS is quite stable. In this complex the DNA is partially unwound (204–207). Such a complex is thought to be one of the first steps in recombination for which the rec A protein is required. The protease activity of the rec A protein against repressors is also observed in the presence of ATPγS (208–210), leading to the conclusion that ATP hydrolysis is not required for this reaction whereas a conformational change introduced by the triphosphate probably is.

GTPγS·adenylate Cyclase Complex

Vertebrate hormonally regulated adenylate cyclase requires GTP for stimulation (211). The hydrolysis-resistant analogs GMPPNP and particularly GTPγS are much more potent than GTP, which is readily hydrolyzed in this system (212, 213). Because of this result, GTPγS became a most useful tool to help in the isolation and characterization of the protein N_s, which is responsible for the mediation of the stimulatory signal of the hormone from the receptor to the cyclase (214). It was thought that hydrolysis of GTP to GDP brings about cessation of activation (213). This idea gained support from the finding that GDPβS, a metabolically stable GDP analog, did indeed rapidly terminate activation (215, 216). This GDPβS-mediated turnoff has been used as an indicator for the involvement of GTP in stimulation of cyclase in a number of systems (217–219). More recently it was demonstrated that the effect of inhibitory hormones is also mediated by a guanine nucleotide-binding protein, N_i (220, 221). The use of GTPγS and GDPβS strengthened the argument for the GTP-protein complex to be responsible for this inhibition. The effect of GTP, as evidenced by the use of GTPγS, apparently is a dissociation of the two proteins N_s and N_i into their α- and β-subunits, the latter being identical for both. The α-subunit of the N_s-protein can be ADP-ribosylated by cholera toxin (222), that of the N_i-protein by Bordetella pertussis toxin (223). This second reaction was shown to be reduced when the N_i-protein was pretreated with GTPγS and thus dissociated. A model for the regulation of the cyclase activity by stimulatory and inhibitory hormones has been described (224). In the adenylate cyclase system GTPγS proved useful not only because of its resistance against GTPases but also because of its tight, quasi-irreversible binding to the proteins N_S and N_i.

GTPγS·transducin Complex

It is becoming more and more apparent that a system very similar to the hormonally regulated vertebrate adenylase cyclase exists for the coupling of photoexcitation of rhodopsin to the stimulation of a cGMP phosphodiesterase in rod outer segments (174, 225). A guanine nucleotide-binding protein, transducin, which also shows GTPase activity, is supposed to be the mediator in this process. It also consists of two subunits, of which the α-subunit is ADP-ribosylated by cholera toxin (226), as well as by pertussis toxin (227, 228). This latter reaction is greatly stimulated by the presence of GDPβS and inhibited by the presence of GTPγS indicating that it must be the GDP-form of transducin which is the substrate for this reaction. Strangely, however, the reaction with cholera toxin is inhibited by GTPγS, which is unexpected as this reaction is supposed to proceed wih the GTP-form as substrate.

The exchange of GTP for GDP bound to transducin catalyzed by photoexcited rhodopsin has also been investigated by light scattering employing GTPγS and GDPβS as stable analogs (229, 230). Interestingly, chemical excitation of Limulus photoreceptors can also be elicited by GTPγS, which also prolongs excitation evolved by dim light, suggesting that a GTP binding protein is also involved in the regulation of this system (231).

THIOPHOSPHORYLATION OF PROTEINS

Activation

In contrast to the poor substrate property of ATPγS towards phosphatases and most ATPases, it is a surprisingly good substrate for many kinases, in particular protein kinases. This dual behavior offers the possibility of transferring the phosphorothioate residue to proteins that are normally regulated by phosphorylation-dephosphorylation in the hope of making them resistant to the action of phosphatases. This was first verified for glycogen phosphorylase where phosphorothioate transfer led to phosphatase insensitive activation (1, 232). In addition, because of the large downfield shift of the ^{31}P-NMR resonance of the phosphorothioate, this group and that of the effector AMPS can easily be observed by NMR spectroscopy without interference from phosphate groups and can be used to monitor conformational changes (233). The role played by phosphorylation of the myosin light chains in smooth muscle contraction has also been investigated successfully by using ATPγS (1, 234–237). The metabolic stability of a thiophosphorylated protein has also been demonstrated for the receptor of epidermal growth factor (238, 239). Histones can be thiophosphorylated by ATPγS and protein kinase not only in vitro but also by application of inorganic thiophosphate to HeLa cells (240, 241).

Affinity Chromatography

The thiophosphorylation of proteins offers the possibility of isolating them on Hg-Sepharose and such a purification has been performed with thiophosphorylated histones (240). A purification of myosin light chain phosphatase has been described where the thiophosphorylated light chains have been immobilized and used for affinity chromatography separation (242). These systems have a clear similarity to that described for the isolation of 5'-thiophosphorylated RNA (see above).

MISCELLANEOUS PHOSPHOROTHIOATES

Phosphorothioate analogs have also been synthesized of compounds other than those described above and will be briefly mentioned here. Thus, a phosphorothioate analog of NAD has been prepared by chemical synthesis and separated into its diastereomers (243). Ribose 5-phosphorothioate has been synthesized and used for the enzymatic synthesis of 5-thiophosphoryibosyl-1-pyrophosphate (244). The enzymatic synthesis of 5-phosphoribosyl-1-O-(2-thiodiphosphate) and 1-O-(1-thiodiphosphate) has been described (245, 246). The phosphorothioate analog of pyridoxal phosphate has been prepared enzymatically (146). A new class of phosphorothioate analogs is represented by the phosphorothioate analogs of phospholipids. The two diastereomers of 1,2-dipalmitoyl-sn-glycero-3-thiophosphocholine and thiophosphoethanolamine show different susceptibilities toward phospholipases A$_2$ and C [(247–250); for more details see review (1)]. It is thus expected that they might become useful in similar ways as the nucleotide analogs described in this review.

CONCLUDING REMARKS

Phosphorothioate analogs of nucleotides have turned out to be extremely useful for a variety of studies. As the era of determining the stereochemical courses of enzymatic phosphoryl and nucleotidyl transfer reactions draws to a close, the applications of these analogs will probably shift to other areas. It is expected that, in particular, phosphorothioate-containing DNA will be increasingly useful in molecular biology. Also, phosphorothioate analogs of cyclic nucleotides, especially cAMPS, might eventually be used more frequently now that it has become commercially available.

ACKNOWLEDGMENTS

I would like to thank B. A. Connolly for many suggestions that considerably improved the manuscript and U. Rust for patiently typing various versions of it. Many colleagues helped by sending me material prior to publication for which I am thankful.

Literature Cited

1. Eckstein, F. 1983. *Angew. Chem.* 95: 431–47; *Agnew. Chem. Int. Ed. Engl.* 22:423–39
2. Gerlt, J. A., Coderre, J. A., Mehdi, S. 1983. *Adv. Enzymol.* 55:291–380
3. Stec, W. J. 1983. *Acc. Chem. Res.* 16:411–17
4. Eckstein, F., Romaniuk, P. J., Connolly, B. A. 1982. *Methods Enzymol.* 87:197–212
5. Frey, P. A., Richard, J. P., Ho, H.-T., Brody, R. S., Sammons, R. D., Sheu, K.-F. 1982. *Methods Enzymol.* 87:213–35
6. Webb, M. R. 1982. *Methods Enzymol.* 87:301–16
7. Buchwald, S. C., Hansen, D. E., Hassett, A., Knowles, J. R. 1982. *Methods Enzymol.* 87:279–301
8. Frey, P. A. 1982. *New Compr. Biochem.* 3:201–48
9. Frey, P. A. 1982. *Tetrahedron* 38:1541–67
10. Cohn, M. 1982. *Acc. Chem. Res.* 15: 326–32
11. Eckstein, F. 1980. *Trends Biochem. Sci.* 5:157–9
12. Knowles, J. R. 1980. *Ann. Rev. Biochem.* 49:877–919
13. Eckstein, F. 1979. *Acc. Chem. Res.* 12: 204–10
14. Eckstein, F. 1975. *Angew. Chem.* 87: 179–85; *Angew. Chem. Int. Ed. Engl.* 14:160–66
15. Eckstein, F. 1966. *J. Am. Chem. Soc.* 88:4292–94
16. Eckstein, F. 1970. *J. Am. Chem. Soc.* 92:4718–23
17. Murray, A. W., Atkinson, M. R. 1968. *Biochemistry* 7:4023–29
18. Sheu, K.-F., Richard, J. P., Frey, P. A. 1979. *Biochemistry* 18:5548–56
19. Connolly, B. A., Romaniuk, P. J., Eckstein, F., 1982. *Biochemistry* 21:1983–89
20. Brody, R. S., Frey, P. A. 1981. *Biochemistry* 20:1245–52
21. Chen, J.-T., Benkovic, S. J. 1983. *Nucleic Acids Res.* 11:3737–51
22. Eckstein, F., Gindl, H. 1967. *Biochim. Biophys. Acta* 149:35–40
23. Eckstein, F., Goody, R. S. 1976. *Biochemistry* 15:1685–91
24. Sheu, K. F., Frey, P. A. 1977. *J. Biol. Chem.* 252:4445–48
25. Jaffe, E. K., Cohn, M. 1978. *Biochemistry* 17:652–57
26. Bryant, F. R., Benkovic, S. J., Sammons, D., Frey, P. A. 1981. *J. Biol. Chem.* 256:5965–66
27. Brody, R. S., Frey, P. A. 1981. *Biochemistry* 20:1245–52
28. Yee, D., Armstrong, V. W., Eckstein, F. 1979. *Biochemistry* 18:4116–20
29. Jaffe, E. K., Nick, J., Cohn, M. 1982. *J. Biol. Chem.* 257:7650–56
30. Senter, P. D., Eckstein, F., Mülsch, A., Böhme, E. 1983. *J. Biol. Chem.* 258: 6741–45.
31. Connolly, B. A., Eckstein, F. 1981. *J. Biol. Chem.* 256:9450–56
32. Jaffe, E. K., Cohn, M. 1979. *J. Biol. Chem.* 254:10839–45
33. Stingelin, J., Boyle, D. W., Kaiser, E. T. 1980. *J. Biol. Chem.* 255:2022–25
34. Fersht, A. 1977. *Enzyme Structure and Mechanism*, p. 97. San Francisco: Freeman. 371 pp.
35. Chen, C.-S., Fujimoto, Y., Girdaukas, G., Sih, C. J. 1982. *J. Am. Chem. Soc.* 104:7294–99
36. Burgers, P. M. J., Eckstein, F. 1980. *J. Biol. Chem.* 255:8229–33
37. Romaniuk, P. J., Eckstein, F. 1981. *J. Biol. Chem.* 256:7322–28
38. Richard, J. P., Ho, H.-T., Frey, P. A. 1978. *J. Am. Chem. Soc.* 100:7756–57
39. Richard, J. P., Frey, P. A. 1978. *J. Am. Chem. Soc.* 100:7757–58
40. Richard, J. P., Frey, P. A. 1982. *J. Am. Chem. Soc.* 104:3476–81
41. Bartlett, P. A., Eckstein, F. 1982. *J. Biol. Chem.* 257:8879–84
42. Goody, R. S., Eckstein, F. 1971. *J. Am. Chem. Soc.* 93:6252–57
43. Ho, H.-T., Frey, P. A. 1984. *Biochemistry* 23:1978–83
44. Bryant, F. R., Benkovic, S. J. 1979. *Biochemistry* 18:2825–28
45. Burgers, P. M. J., Eckstein, F. 1978. *Proc. Natl. Acad. Sci. USA* 75:4798–800
46. Burgers, P. M. J., Sathyanarayana, B.

K., Saenger, W., Eckstein, F. 1979. *Eur. J. Biochem.* 100:585–91
47. Jarvest, R. L., Lowe, G. 1979. *J. Chem. Soc. Chem. Commun.* 1979:364–66
48. Jaffe, E. K., Cohn, M. 1978. *J. Biol. Chem.* 253:4823–25
49. Cornelius, R. D., Cleland, W. W. 1978. *Biochemistry* 17:3279–86
50. Merritt, E. A., Sundaralingam, M., Cornelius, R. D., Cleland, W. W. 1978. *Biochemistry* 17:3274–78
51. Baraniak, J., Kinas, R. W., Lesiak, K., Stec, W. J. 1979. *J. Chem. Soc. Chem. Commun.* 940–41
52. Lesnikowski, Z. J., Stec, W. J., Zielinski, W. S., Adamiak, D., Saenger, W. 1981. *J. Am. Chem. Soc.* 103:2862–63
53. Eckstein, F., Simonson, L. P., Baer, M. P. 1974. *Biochemistry* 13:3806–10
54. Burgers, P. M. J., Eckstein, F. 1978. *Biochemistry* 18:592–96
55. Potter, B. V. L., Connolly, B. A., Eckstein, F. 1983. *Biochemistry* 22:1369–77
56. Romaniuk, P. J., Eckstein, F. 1982. *J. Biol. Chem.* 257:7684–88
57. Spitzer, S. 1983. *Diplomarbeit. Fachbereich Chemie, Univ. Göttingen*
58. Kemal, Ö., Reese, C. B., Sarafinowska, H. T. 1983. *J. Chem. Soc. Chem. Commun.* 1983:591–93
59. Burgers, P. M. J., Eckstein, F. 1978. *Tetrahedron Lett.* 3835–38
60a. Marlier, J. F., Benkovic, S. J. 1980. *Tetrahedron Lett.* 21:1121–24
60b. Nemer, M. J., Ogilvie, K. K. 1980. *Tetrahedron Lett.* 21:4149–52
61. Uznanski, B., Nieviarowski, W., Stec, W. J. 1982. *Tetrahedron Lett.* 23:4289–92
62. Connolly, B. A., Potter, B. V. L., Eckstein, F., Pingoud, A., Grotjahn, L. 1984. *Biochemistry.* 23:3443–53
63. Daub, G. W., van Tamelen, E. E. 1976. *J. Am. Chem. Soc.* 99:3526–28
64. Lowe, G. 1983. *Acc. Chem. Res.* 16:224–51
65. Cohn, M. 1982. *Ann. Rev. Biophys. Bioeng.* 11:23–42
66. Tsai, M.-D., Bruzik, K. 1983. In *Biological Magnetic Resonance,* ed. L. J. Berliner, J. Reuben, 5:129–81. New York: Plenum
67. Tsai, M.-D. 1982. *Methods Enzymol.* 87:235–79
68. Burgers, P. M. J., Eckstein, F. 1979. *Biochemistry* 18:450–54
69. Sammons, R. D., Frey, P. A. 1982. *J. Biol. Chem.* 257:1138–41
70. Connolly, B. A., Füldner, H. H., Eckstein, F. 1982. *J. Biol. Chem.* 257:3382–84
71. Lowe, G., Tansley, G., Cullis, P. M. 1982. *J. Chem. Soc. Chem. Commun.* 1982:595–98

72. Cullis, P. M. 1983. *Tetrahedron Lett.* 24:5677–80
73. Guga, P., Okruszek, A. 1984. *Tetrahedron Lett.* 25:2897–900
74. Lowe, G., Sproat, B. S., Tansley, G., Cullis, P. M. 1983. *Biochemistry* 22:1229–36
75. Sammons, R. D., Ho, H.-T., Frey, P. A. 1982. *J. Am. Chem. Soc.* 104:5841–42
76. Iyengar, R., Ho, H.-T., Sammons, R. D., Frey, P. A. 1984. *J. Am. Chem. Soc.* 106:6038–49
77. Connolly, B. A., Eckstein, F. 1982. *Biochemistry* 21:6158–67
78. Saenger, W., Eckstein, F. 1970. *J. Am. Chem. Soc.* 92:4712–18
79. Connolly, B. A., Eckstein, F., Grotjahn, L. 1984. *Biochemistry* 23:2027–31
80. Saenger, W., Suck, D., Eckstein, F. 1974. *Eur. J. Biochem.* 46:559–67
81. Usher, D. A., Erenrich, E. S., Eckstein, F. 1972. *Proc. Natl. Acad. Sci. USA* 69:115–18
82. Eckstein, F., Schulz, H. H., Rüterjans, H., Haar, W., Maurer, W. 1972. *Biochemistry* 11:3507–12
83. Gerlt, J. A., Wan, H. Y. 1979. *Biochemistry* 18:4630–38
84. Burgers, P. M. J., Eckstein, F. 1979. *J. Biol. Chem.* 254:6889–93
85. Brody, R. S., Adler, S., Modrich, P., Stec, W. J., Leznikowski, Z. J., Frey, P. A. 1982. *Biochemistry* 21:2570–72.
86. Jovin, T., Eckstein, F., In preparation
87. Marlier, J. F., Bryant, F. R., Benkovic, S. J. 1981. *Biochemistry* 20:2212–19
88. Eckstein, F., Sternbach, H., von der Haar, F. 1977. *Biochemistry* 16:3429–32
89. Bryant, F. R., Benkovic, S. J. 1982. *Biochemistry* 21:5877–85
90. Sheu, K.-F. R., Richard, J. P., Frey, P. A. 1979. *Biochemistry* 18:5548–55
91. Gerlt, J. A., Coderre, J. A., Wolin, M. S. 1980. *J. Biol. Chem.* 255:331–34
92. Coderre, J. A., Gerlt, J. A. 1980. *J. Am. Chem. Soc.* 102:6594–97
93. Eckstein, F., Romaniuk, P. J., Heideman, W., Storm, D. R. 1981. *J. Biol. Chem.* 256:9118–20
94. Midelfort, C. F., Sarton-Miller, I. 1978. *J. Biol. Chem.* 253:7127–29
95. Tsai, M.-D. 1979. *Biochemistry* 18:1468–72
96. Langdon, S. P., Lowe, G. 1979. *Nature* 281:320–21
97. Lowe, G., Tansley, G. 1984. *Tetrahedron* 40:113–17
98. Lowe, G., Sproat, B. S., Tansley, G. 1983. *Eur. J. Biochem.* 130:341–45
99. Usher, D. A., Richardson, D. J., Eckstein, F. 1970. *Nature* 228:663–65
100. Burgers, P. M. J., Eckstein, F., Hunneman, D. H., Baraniak, J., Kinas, R. W. et al. 1979. *J. Biol. Chem.* 254:9959–61

101. Jarvest, R. L., Lowe, G., Baraniak, J., Stec, W. J. 1982. *Biochem. J.* 203:461–70
102. Cullis, P. M., Jarvest, R. L., Lowe, G., Potter, B. V. L. 1981. *J. Chem. Soc. Chem. Commun.* 1981:245–46
103. Coderre, J. A., Mehdi, S., Gerlt, J. A. 1981. *J. Am. Chem. Soc.* 103:1872–75
104. Burgers, P. M. J., Eckstein, F., Hunneman, D. H. 1979. *J. Biol. Chem.* 254:7476–78
105. Jarvest, R. L., Lowe, G. 1981. *Biochem. J.* 199:447–51
106. Mehdi, S., Gerlt, J. A. 1981. *J. Biol. Chem.* 256:12164–66
107. Potter, B. V. L., Romaniuk, P. J., Eckstein, F. 1983. *J. Biol. Chem.* 258:1758–60
108. Connolly, B. A., Eckstein, F., Pingoud, A. 1984. *J. Biol. Chem.* 259:10760–63
109. Gupta, A., De Brosse, C., Benkovic, S. J. 1982. *J. Biol Chem.* 257:7689–92
110. Orr, G. A., Simons, J., Jones, S. R., Chin, G. J., Knowles, J. R. 1978. *Proc. Natl. Acad. Sci. USA* 75:2230–33
111. Blättler, W. A., Knowles, J. R. 1979. *Biochemistry* 18:3927–33
112. Lowe, G., Potter, B. V. L. 1981. *Biochem. J.* 199:227–33
113. Pliura, D. H., Schomburg, D., Richard, J. P., Frey, P. A., Knowles, J. R. 1980. *Biochemistry* 19:325–29
114. Lowe, G., Cullis, P. M., Jarvest, R. L., Potter, B. V. L., Sproat, B. S. 1981. *Philos. Trans. R. Soc. London Ser. B* 293:75–92
115. Bryant, F. R., Benkovic, S. J., Sammons, D., Frey, P. A. 1981. *J. Biol. Chem.* 256:5965–66
116. Jarvest, R. L., Lowe, G. 1981. *Biochem. J.* 199:273–76
117. Richard, J. P., Carr, D. C., Ives, D. H., Frey, P. A. 1980. *Bichem. Biophys. Res. Commun.* 94:1052–56
118. Richard, J. P., Prasker, D. C., Ives, D. H., Frey, P. A. 1979. *J. Biol. Chem.* 254:4339–41
119. Tsai, M. D., Chang, T. T. 1980. *J. Am. Chem. Soc.* 102:5416–18
120. Webb, M. R., Trentham, D. R. 1980. *J. Biol. Chem.* 255:1775–79
121. Miziorko, H., Eckstein, F. 1984. *J. Biol. Chem.* 259:13037–40
122. Webb, M. R., Trentham, D. R. 1980. *J. Biol. Chem.* 255:8629–32
123. Webb, M. R., Grubmeyer, C., Penefsky, H. S., Trentham, D. R. 1980. *J. Biol. Chem.* 255:11637–39
124. Webb, M. R., Trentham, D. R. 1981. *J. Biol. Chem.* 256:4884–87
125. Webb, M. R., Eccleston, J. F. 1981. *J. Biol. Chem.* 256:7734–37
126. Eccleston, J. F., Webb, M. R. 1982. *J. Biol. Chem.* 257:5046–49
127. Senter, P. D., Eckstein, F., Kagawa, Y. 1983. *Biochemistry* 22:5514–18
128. Hansen, D. E., Knowles, J. R. 1982. *J. Biol. Chem.* 257:14795–98
129. Gonzalez, M. A., Webb, M. R., Welsh, K. M., Cooperman, B. S. 1984. *Biochemistry* 23:797–801
130. Webb, M. R., Reed, G. H., Cooper, B. F., Rudolph, F. B. 1984. *J. Biol. Chem.* 259:3044–46
131. Sheu, K.-F., Ho, H.-T., Nolan, L. D., Markovitz, P., Richard, J. P., et al. 1984. *Biochemistry* 23:1779–83
132. Pecoraro, V. L., Hermes, J. D., Cleland, W. W. 1984. *Biochemistry.* 23:5262–71
133. Ketelaar, J., Gersmann, H., Koopmanns, K. 1952. *Recl. Trav. Chim. Pays-Bas* 71:1253–58
134. Bagshaw, C. R., Eccleston, J. F., Trentham, D. R., Yates, D. R., Goody, R. S. 1972. *Cold Spring Harbor Symp. Quant. Biol.* 37:127–35
135. Westheimer, F. H. 1980. In *Rearrangements in Ground and Excited States*, ed. P. De Mayo, pp. 229–71. New York: Academic. 431 pp.
136. Darby, M. K., Trayer, I. P. 1982. *Eur. J. Biochem.* 129:555–60
137. Cohn, M., Shih, H., Nick, J. 1982. *J. Biol. Chem.* 257:7646–49
138. Pillai, R. P., Ranshel, F. M., Villafranca, J. J. 1980. *Arch. Biochem. Biophys.* 199:7–15
139. Lee, C. S., O'Sullivan, W. J. 1984. *Biochim. Biophys. Acta* 787:131–37
140. Kunjara, S., Lee, C. S., Smithers, G. W., O'Sullivan, W. J. 1983. *Fed. Asian Oceanic Biochem.* (Bangkok) 88–124
141. Tomasselli, A. G., Noda, L. H. 1983. *Eur. J. Biochem.* 132:109–15
142. Kalbitzer, H. R., Marquetant, R., Connolly, B. A., Goody, R. S. 1983. *Eur. J. Biochem.* 133:221–27
143. Connolly, B. A., von der Haar, F., Eckstein, F. 1980. *J. Biol. Chem.* 255:11301–7
144. Smith, L. T., Cohn, M. 1982. *Biochemistry* 21:1530–34
145. Piel, N., Freist, W., Cramer, F. 1983. *Bioorg. Chem.* 12:18–33
146. Churdrich, J. E., Wu, C. 1982. *J. Biol. Chem.* 257:12136–40
147. Bolen, D. W., Stingelin, J., Bramson, H. N., Kaiser, E. T. 1980. *Biochemistry* 19:1176–82
148. Armstrong, V. W., Yee, D., Eckstein, F. 1979. *Biochemistry* 18:4120–23
149. Gibson, K. J., Switzer, R. L. 1980. *J. Biol. Chem.* 255:694–96
150. Leupold, C. M., Goody, R. S., Witting-

hofer, A. 1983. *Eur. J. Biochem.* 135:237–41
151. Cornelius, R. D., Cleland, W. W. 1978. *Biochemistry* 17:3279–86
152. Dunaway-Mariano, D., Cleland, W. W., Gupta, R. K., Mildvan, A. S. 1979. *Biochemistry* 18:4347–54
153. Dunaway-Mariano, D., Cleland, W. W. 1980. *Biochemistry* 19:1496–1505
154. Dunaway-Mariano, D., Cleland, W. W. 1980. *Biochemistry* 19:1506–15
155. Reed, G. H., Leyh, T. S. 1980. *Biochemistry* 19:5472–80
156. Leyh, T. S., Sammons, R. D., Frey, P. A., Reed, G. H. 1982. *J. Biol. Chem.* 257:15047–53
157. Eccleston, J. F., Webb, M. R., Ash, D. E., Reed, G. H. 1981. *J. Biol. Chem.* 256:10774–77
158. Webb, M. R., Ash, D. E., Leyh, T. S., Trentham, D. R., Reed, G. H. 1982. *J. Biol. Chem.* 257:3068–72
159. Kalbitzer, H. R., Marquetant, R., Connolly, B. A., Goody, R. S. 1983. *Eur. J. Biochem.* 113:221–27
160. Huang, S. L., Tsai, M. D. 1982. *Biochemistry* 21:951–59
161. Pecoraro, V. L., Rawlings, J., Cleland, W. W. 1984. *Biochemistry* 23:153–58
162. De Wit, R. J. W., Hoppe, J., Stec, W. J., Baraniak, J., Jastorff, B. 1982. *Eur. J. Biochem.* 122:95–99
163. O'Brian, C. A., Roczniak, S. O., Bramson, H. N., Baraniak, J., Stec, W. J., Kaiser, E. T. 1982. *Biochemistry* 21:4371–76
164. Eckstein, F., Eimerl, S., Schramm, M. 1976. *FEBS Lett.* 64:92–94
165. Leder, A., Leder, P. 1975. *Cell* 5:319–22
166. Ito, F., Chou, J. Y. 1984. *J. Biol. Chem.* 259:2526–30
167. Rothermel, J. D., Stec, W. J., Baraniak, J., Jastorff, B., Botelho, L. H. P. 1983. *J. Biol. Chem.* 258:12125–28
168. Rothermel, J. D., Jastorff, B., Botelho, L. H. P. 1984. *J. Biol. Chem.* 259:8151–55
169. Van Haastert, P. J. M., Kloin, E. 1983. *J. Biol. Chem.* 258:9636–42
170. Gronenborn, A. M., Clore, G. M. 1982. *Biochemistry* 21:4040–48
171. Scholübbers, H.-G., Knippenberg, P. H., Baraniak, J., Stec, W. J., Morr, U., Jastorff, B. 1984. *Eur. J. Biochem.* 138:101–9
172. Pelech, S. L., Pritchard, P. H., Brindley, D. N., Vance, D. E., 1983. *Biochem. J.* 216:129–36
173. Eckstein, F., Foreman, J. C. 1978. *FEBS Lett.* 91:182–85
174. Fung, B. K.-K., Harley, J. B., Stryer, L.

1981. *Proc. Natl. Acad. Sci. USA* 78:152–56
175. Deleted in proof
176. Kunkel, T. A., Eckstein, F., Mildvan, A. S., Koplitz, R. M., Loeb, L. 1981. *Proc. Natl. Acad. Sci. USA* 78:6734–38
177. Gupta, A. P., Benkovic, P. A., Benkovic, S. J. 1983. *Nucleic Acids Res.* 12:5897–911
178. Kornberg, A. 1980. In *DNA Replication*, p. 127. San Francisco: Freeman. 724 pp.
179. Shortle, D., Grisafi, P., Benkovic, S. J., Botstein, D. 1982. *Proc. Natl. Acad. Sci. USA* 79:1588
180. Putney, S., Benkovic, S. J., Schimmel, P. 1981. *Proc. Natl. Acad. Sci. USA* 78:7350–54
181. Jasin, M., Regan, L., Schimmel, P. 1983. *Nature* 306:441–47
182. Vosberg, H. P., Eckstein, F. 1982. *J. Biol. Chem.* 257:6595–99
183. Potter, B. V. L., Eckstein, F. 1984. *J. Biol. Chem.* 259:14243–48
184. Halford, E. S. 1983. *Trends Biochem. Sci.* 8:455–61
185. Eckstein, F., Jovin, T. M. 1983. *Biochemistry* 22:4546–50
186. Jovin, T. M., van de Sande, J. H., Zarling, D. A., Arndt-Jovin, D. J., Eckstein, F., et al. 1983. *Cold Spring Harbor Symp. Quant. Biol.* 47:143–54
187. Gorenstein, D. 1981. *Ann. Rev. Biophys. Bioeng.* 10:355–86
188. Jovin, T. M., McIntosh, L. P., Arndt-Jovin, D. J., Zarling, D. A., Robert-Nicoud, M., et al. 1983. *J. Biomol. Struct. Dyn.* 1:21–57
189. Biggin, M. D., Gibson, T. J., Hong, G. F. 1983. *Proc. Natl. Acad. Sci. USA* 80:3936–65
190. Eckstein, F., Gindl, H. 1970. *Eur. J. Biochem.* 13:558–64
191. De Clercq, E., Eckstein, F., Sternbach, H., Merigan, T. C. 1970. *Virology* 42:421–28
192. Zhang, Z.-Y., Thompson, E. A., Stallcup, M. R. 1984. *Nucleic Acids Res.* 12:8115–28
193. Stallcup, M. R., Washington, L. D. 1983. *J. Biol. Chem.* 258:2802–7
194. Cosstick, R., McLaughlin, L. W., Eckstein, F. 1984. *Nucleic Acids Res.* 12:1791–1810
195. Yount, R. G. 1975. *Adv. Enzymol. Relat. Areas Mol. Biol.* 43:1–56
196. Reynolds, M. A., Gerlt, J. A., Demou, P. C., Oppenheimer, N. J., Kenyon, G. L. 1983. *J. Am. Chem. Soc.* 105:6475–81
197. Gerlt, J. A., Demou, P. C., Mehdi, S. 1982. *J. Am. Chem. Soc.* 104:2848–56
198. Peacock, C. J., Nickless, G. 1969. *Z. Naturforsch Teil A* 24:245–47

199. Gerlt, J. A., Reynolds, M. A., Demon, P. C., Kenyon, G. L. 1983. *J. Am. Chem. Soc.* 105:6469–74
200. Radding, C. M. 1981. *Cell* 25:3–4
201. Radding, C. M. 1982. *Ann. Rev. Genet.* 16:405–37
202. Weinstock, G. M., McEntee, K., Lehman, I. R. 1979. *Proc. Natl. Acad. Sci.* 76:126–30
203. Cox, M. M., Lehman, I. R. 1981. *Proc. Natl. Acad. Sci. USA* 78:3433–37
204. Stasiak, A., Di Capua, E. 1982. *Nature* 299:185–86
205. Di Capua, E., Engel, A., Stasiak, A., Koehler, T. 1982. *J. Mol. Biol.* 157:87–103
206. Chrysogelos, S., Register, J. C., Griffith, J. 1983. *J. Biol. Chem.* 258:12624–31
207. Dombrowski, D. F., Scraba, D. G., Bradley, R. D., Morgan, A. R. 1983. *Nucleic Acids Res.* 11:7487–504
208. Craig, N. L., Roberts, J. W. 1981. *J. Biol. Chem.* 256:8039–44
209. Weinstock, G. M., McEntee, K. 1981. *J. Biol. Chem.* 26:10883–88
210. Little, J. W., Mount, D. W., Yanisch-Perron, C. R. 1981. *Proc. Natl. Acad. Sci.* 78:4199–203
211. Ross, E. M., Gilman, A. G. 1980. *Ann. Rev. Biochem.* 49:533–64
212. Pfeuffer, T., Helmreich, E. J. M. 1975. *J. Biol Chem.* 250:867–76
213. Cassel, D., Selinger, Z. 1977. *Biochim. Biophys. Res. Commun.* 77:868–73
214. Sternweis, P. C., Northrup, J. K., Smigel, M. D., Gilman, A. G. 1981. *J. Biol. Chem.* 256:11517–26
215. Eckstein, F., Cassel, D., Lefkowitz, H., Lowe, M., Selinger, Z. 1979. *J. Biol. Chem.* 254:9829–34
216. Cassel, D., Eckstein, F., Lowe, M., Selinger, Z. 1979. *J. Biol. Chem.* 254:9835–38
217. Swoboda, M., Furnelle, J., Eckstein, F., Christophe, J. 1980. *FEBS Lett.* 109:275–79
218. Ezra, E., Salomon, Y. 1981. *J. Biol. Chem.* 256:5377–82
219. Casperson, G. F., Walker, N., Brazier, A. R., Bourne, H. R. 1983. *J. Biol. Chem.* 258:7911–14
220. Jakobs, K. H., Gehring, U., Gangler, B., Pfeuffer, T., Schultz, G. 1983. *Eur. J. Biochem.* 130:605–11
221. Jakobs, K. H., Aktories, K., Schultz, G. 1983. *Nature* 303:117–78
222. Cassel, D., Pfeuffer, T. 1978. *Proc. Natl. Acad. Sci. USA* 75:2669–73
223. Katada, T., Northrup, J. K., Bokoch, G.-M., Ui, M., Gilman, A. G. 1984. *J. Biol. Chem.* 259:3578–85
224. Katada, T., Bokoch, G. M., Smigel, M.

D., Ui, M., Gilman, A. G. 1984. *J. Biol. Chem.* 259:3586–95
225. Fung, B. K. K., Hurley, J. B., Stryer, L. 1981. *Proc. Natl. Acad. Sci. USA* 78:152–56
226. Abood, M. E., Hurley, J. B., Pappone, M.-C., Bourne, U. R., Stryer, L. 1982. *J. Biol. Chem.* 257:10540–43
227. Van Dop, C., Yamanaka, G., Steinberg, F., Sekura, R. D., Manclark, C. R., et al. 1984. *J. Biol. Chem.* 259:23–26
228. Watkins, R. A., Moss, J., Burns, D. L., Hewlett, E. L., Vaughan, M. 1984. *J. Biol Chem.* 259:1378–81
229. Emeis, D., Kühn, U., Reichert, J., Hofmann, K. P. 1982. *FEBS Lett.* 143:29–34
230. Bennett, N. 1982. *Eur. J. Biochem.* 123:133–39
231. Corson, D. W., Fein, A., Walthall, W. W. 1983. *J. Gen. Physiol.* 82:659–77
232. Gracos, D., Fischer, E. J. 1974. *Biochem. Biophys. Res. Commun.* 58:960–67
233. Withers, S. G., Madsen, N. B., Sykes, B. D. 1981. *Biochemistry* 20:1748–56
234. Hoar, P. E., Kerrick, W. G. L., Cassidy, P. S. 1979. *Science* 204:503–6
235. Kerrick, W. G. L., Hoar, P. E. 1981. *Nature* 292:253–55
236. Cassidy, P. S., Hoar, P. E., Kerrick, W. G. L. 1979. *J. Biol. Chem.* 254:11148–53
237. Sherry, J. M. F., Gorecka, A., Ahsoy, M. O., Dabrowska, R., Hartshorne, D. J. 1978. *Biochemistry* 17:4441
238. Cassel, D., Glaser, L. 1982. *Proc. Natl. Acad. Sci. USA* 79:2231–35
239. Cassel, D., Pike, L. J., Grant, G. A., Krebs, E. G., Glaser, L. 1983. *J. Biol. Chem.* 258:2945–50
240. Sun, I. Y.-C., Johnson, E. M., Allfrey, V. G. 1980. *J. Biol. Chem.* 255:742–47
241. Sun, I. Y.-C., Allfrey, V. G. 1982. *J. Biol. Chem.* 257:1347–53
242. Pato, M. D., Adelstein, R. S. 1980. *J. Biol. Chem.* 255:6535–38
243. Meyer, T., Wilckens, K., Hilz, H., Thien, J. 1984. *Eur. J. Biochem.* 140:531–37
244. Murray, A. W., Wong, P. C. L., Friedericks, B. 1969. *Biochem. J.* 112:741–46
245. Smithers, G. W., O'Sullivan, W. J. 1984. *Biochemistry* 23:4767–73
246. Smithers, G. W., O'Sullivan, W. J. 1984. *Biochemistry* 23:4773–78
247. Bruzik, K., Gupta, S. M., Tsai, M.-D. 1982. *J. Am. Chem. Soc.* 104:4682
248. Orr, G. A., Brewer, C. F., Heney, G. 1982. *Biochemistry* 21:3202–6
249. Bruzik, K., Jiang, R.-T., Tsai, M.-D. 1983. *Biochemistry* 22:2478–86
250. Jiang, R.-T., Shyy, Y.-J., Tsai, M.-D. 1984. *Biochemistry* 23:1661–67

Ann. Rev. Biochem. 1985. 54:403–23

GROWTH HORMONE RELEASING FACTORS

Nicholas Ling, Fusun Zeytin, Peter Böhlen, Fred Esch, Paul Brazeau, William B. Wehrenberg, Andrew Baird and Roger Guillemin

Laboratories for Neuroendocrinology, The Salk Institute for Biological Studies, La Jolla, California 92037

CONTENTS

PERSPECTIVES AND SUMMARY

Physiological studies led, forty years ago, to the concept that the secretion of all hormones from the anterior pituitary gland (adenohypophysis) would be controlled by specialized neurons of the hypothalamus. The controlling mechanism would be neurohumoral in nature, neuronal products reaching the pituitary cells through an extensive network of portal capillary vessels joining hypothalamus to adenohypophysis. The identification of such hypophysiotropic peptides for each and every known pituitary hormone was to take more than 30 years. Until recently four such regulatory peptides were known: TRF, a tripeptide-amide stimulating the secretion of thyrotropin and prolactin; GnRF,

403

a decapeptide-amide stimulating the secretion of both gonadotropins; somatostatin, an inhibitory factor for the secretion of growth hormone; and CRF, a 41-residue amidated peptide stimulating the secretion of adrenocorticotropin and β-endorphin.

GRF, growth hormone releasing factor, and several biologically active fragments were isolated and characterized less than two years ago from two rare tumors of the pancreas recognized as potential sources of GRF in two acromegalic patients with no pituitary tumor—the usual cause of acromegaly through (primary) hypersecretion of growth hormone. GRF is a 44-residue amidated peptide; only one of the tumors contained this mature form of the peptide, which has the highest potency in releasing growth hormone in in vitro tests (ED_{50} 0.5—1 × 10^{-11}M). That same tumor also contained two fragments of GRF corresponding to the sequences 1–37 and 1–40. The other tumor contained exclusively the form 1–40. The complete message for GRF has been cloned from one tumor; while a partial message lacking a portion of the putative signal sequence has been deduced from the other tumor, both messages contain a coding sequence corresponding to the 44-residue peptide, including evidence of an amidation signal (..Leu^{44}-Gly..). The proposed amino acid sequence for the precursor molecule of GRF also contains putative peptides outside the GRF sequence, one amino-terminal to GRF, probably composed of 7–9 residues, the other, carboxy-terminal to GRF, with 30 residues. A peptide with the latter amino acid sequence has been isolated and characterized from the tumor that generated human hypothalamic GRF (hGRF). The isolated peptide was shown to possess a pGLu amino-terminal and an amidated carboxy-terminal, thus implying biological significance. No biological activity has been found so far for this peptide.

GRF has now been isolated and characterized from the hypothalamus of six different species; in all cases but one, hypothalamic GRF is a 44-residue amidated peptide, the exception being rat GRF (characterized as a 43-residue peptide in the free-acid form). Recent knowledge of the structure of the precursor for rat GRF confirms the sequence as 1–43, with no evidence of an amidation signal for either position 43 or 44. Human hypothalamic GRF has been sequenced and shown to be identical to the tumor-derived 44-residue GRF.

A number of fragments and analogs of GRF have been synthesized by the solid-phase method. The active core of the molecule (intrinsic activity) resides in the amino-terminal region of the molecule, while carboxy-terminal sequences determine its potency (specific activity). Amidation of all the peptides with GRF activity increases the potency at least 2-fold. The primary structure of the GRF of all species shows major homologies with several peptides of the gastrointestinal tract and the pancreas, all of the glucagon-secretin-VIP (vasoactive intestinal peptide) family. Mechanisms of action of GRF, through membrane receptors not characterized at the moment, involve activation of the adenylate cyclase–cAMP system. GRF receptors would be both on the plasma membrane (of pituitary somatotrophs) and on the external membrane of secretory granules containing growth hormone, where GRF would stimulate exocytosis by activating a phosphorylating kinase. Release of growth hormone induced by GRF is antagonized in a noncompetitive manner by the hypothalamic peptide somatostatin and by somatomedin-C (insulin-like growth factor I, or IGF-I), a growth-promoting peptide originating from peripheral tissues under the regulation of growth hormone. This provides evidence of a negative feedback loop for the secretion of growth hormone. At the pituitary level, GRF stimulates the secretion of GH exclusively; there is no reliable evidence that GRF stimulates the secretion of any other pituitary hormone by normal pituitary cells. Extrapituitary actions of GRF have been shown: release of neurotensin and calcitonin by a cell-line of rat medullary thyroid carcinoma; also release of amylase and other pancreatic exocrine enzymes by the rat pancreas. In the latter case GRF seems to be acting through previously recognized VIP receptors, most probably because of the structural homologies between the amino-terminal region of GRF and VIP.

To complete this introduction with a subject not included elsewhere in this review, GRF has been shown to be a potent stimulator of the secretion of growth hormone in vivo in all species of vertebrates studied so far (see Wehrenberg et al in *Ann. Rev. Pharmacol. Toxicol.*, 1984), including man. Indeed, extensive clinical studies are in progress throughout the world, all supporting the proposal that synthetic human GRF should be a useful clinical medication in most cases of hypothalamic pituitary dwarfism as well as in other clinical circumstances in which one wishes to stimulate secretion of growth hormone.

ISOLATION AND CHARACTERIZATION OF GRFs FROM TUMOR AND HYPOTHALAMIC TISSUE

The existence of a growth hormone releasing factor in the hypothalamus was demonstrated by Deuben & Meites (1), who showed that hypothalamic extracts were capable of eliciting growth hormone release from pituitary tissue in vitro. Since then many efforts have been directed at the isolation of hypothalamic GRF. Earlier attempts produced materials that proved to be artifacts (2–4) or were of questionable purity and therefore never fully characterized (5–15). During the 1970s sufficient evidence was accumulated suggesting that certain rare tumors, such as pancreatic islet tumors and carcinoids of various origins, caused acromegaly, presumably owing to the presence in these tumors of substances capable of stimulating pituitary growth hormone secretion [see Leveston et al (16), Scheithauer et al (17) for reviews of such cases]. Frohman et al (18) showed that GRF activity from one such tumor was very similar chromatographically to that of hypothalamic GRF.

Tumor GRFs

Our first attempts at the isolation and chemical characterization of tumor peptides with GRF activity date back to 1981, when we received small pieces of a lung carcinoid (200 mg, provided by B. Scheithauer of the Mayo Clinic) and an islet cell tumor (400 mg, provided by M. Thorner, University of Virginia, the "Charlottesville tumor") that had caused acromegaly. Between 100 and 200 pmol of peptide with GRF activity were isolated from both tumors and their amino acid compositions were determined (19), but no sequencing was attempted owing to the small amounts of material isolated and the insensitive chemical characterization methodology then available to us. In April 1982 larger quantities of another pancreatic islet tumor from an acromegalic patient were provided to us by G. Sassolas, Lyon, France (the "Lyon tumor"). From this tissue we isolated three GRF peptides in May 1982 (20, 21) and determined their structures as follows: hpGRF-44, hpGRF-40, and hpGRF-37, consisting of 44, 40, and 37 amino acids, respectively. Human pancreas GRF-44 is the most potent peptide in the rat anterior pituitary monolayer culture assay, with hpGRF-40 and hpGRF-37 being somewhat less active.

Isolation of the three hpGRF peptides was accomplished by an efficient four-step procedure involving: (a) tissue extraction with 0.3 M HCl, (b) gel filtration on Sephadex G-75, and (c) two steps of reverse-phase HPLC using mobile phases of different solute selectivities (Figure 1). GRF activity was monitored with a sensitive in vitro bioassay that measures the capacity of the column fractions to stimulate the release of growth hormone from rat anterior pituitary cells in monolayer culture (22). Subsequently a peptide identical to hpGRF-40 was isolated by our group (21, 23) from larger quantities of the Charlottesville tumor, using the methodology outlined above, as well as by Rivier et al (24), who used similar procedures.

The isolation of several forms of hpGRF from the Lyon tumor, but only of one from the Charlottesville tumor, indicates that ectopic GRF production from a biosynthetic precursor and/or degradation may be regulated differently in different tumors. The two tumors also differed greatly with respect to GRF content. While the tumor from Lyon yielded 7.5 nmol total GRF/g tissue, only 0.45 nmol GRF/g was extracted from the other tumor. Dissimilarities also exist with regard to the expression of the somatostatin gene in the two tumors. Appreciable amounts of both somatostatin-14 and somatostatin-28 (21) were found in the tumor from Lyon, but none in the other tumor.

The primary structures of hpGRF-44, -40, and -37 were determined by a gas-phase protein sequencer (Applied Biosystems, Inc., Model 470A) in combination with a highly resolutive and sensitive HPLC system for the identification and quantitation of the phenylthiohydantoin amino acids (25). The strategy employed was as follows: Gas-phase sequence analyses of the intact peptides were performed at the low- or sub-nmol level to establish the bulk of the amino-terminal amino acid sequence. Confirmation of these data and determination of the remaining carboxy-terminal sequences were accomplished by cyanogen bromide digestions, reverse-phase HPLC isolation of the digestion fragments, and their subsequent structural characterization by amino acid analyses and microsequencing. Evidence for the nature of the carboxy-terminus of each of the hpGRF peptides was obtained from reverse-phase HPLC studies in which the native GRFs were co-chromatographed with synthetic replicates possessing either a free acid or an amidated carboxy-terminus.

The sequencing data showed that all the tumor-derived GRFs contained identical sequences from their amino-termini, with the structures of hpGRF-37 and hpGRF-40 being deduced as the 1–37 and 1–40 fragments of hpGRF-44. Only the 44-residue peptide possessed an amidated carboxy-terminus; the 40- and 37-amino-acid peptides contained free acids at their carboxy-termini. An independent report by Spiess et al (26) confirmed the structural analysis of hpGRF-40 from the Charlottesville tumor.

The structure of preproGRF was also established by molecular cloning and DNA sequence analysis (27). Poly(A)mRNA from the Lyon tumor was used to construct a cDNA library which was subsequently screened with two synthetic oligonucleotide probes directed against hpGRF-44(1–7) and hpGRF-44(35–39). Two preproGRF clones were isolated, each containing cDNA encoding a putative hydrophobic signal peptide (28) joined to a proGRF of 87 or 88 amino acids. The two precursors each contain the sequence of hpGRF-44 flanked by basic processing sites (29) and include a carboxy-terminal amidation signal (Leu^{44}-Gly) (30). The 107- and 108-amino-acid preproGRFs are identical except for the presence or absence of a serine at position 103 in the carboxy-terminal portions of the molecules. Mayo et al (31) confirmed these findings in a similar fashion using poly(A)mRNA from the Charlottesville tumor to generate a cDNA expression library. Screening with both synthetic oligonucleotide probes and an antisera raised against synthetic hpGRF-40 yielded a clone encoding preproGRF(6–108) that contained a serine at residue 103.

Besides the hpGRF-44 sequence, preproGRF also encodes a putative peptide of 7–9 amino acids at the amino-terminal end and a 30- or 31-residue peptide at the carboxy-terminal end of hpGRF. From the Lyon tumor a peptide corresponding to the predicated carboxy-terminal sequence of preproGRF-108 was isolated. Based on amino acid analysis and reverse-phase HPLC comparisons with synthetic replicates, the isolated peptide was shown to have 30 amino acids, with a pGlu at the amino-terminus and an amidated Gln at the carboxy-terminus (32).

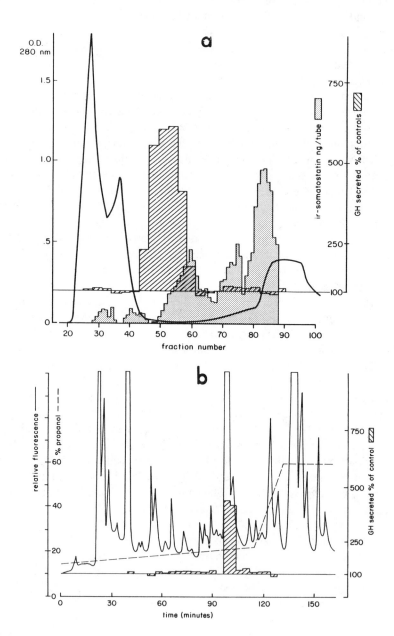

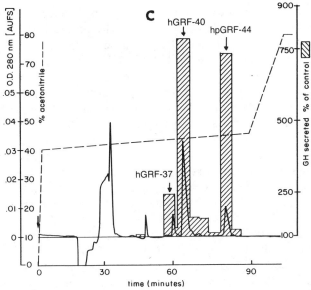

Figure 1 Isolation of GRF peptides from a pancreatic islet cell tumor (provided by G. Sassolas, Lyon, France). [This tumor also contained several forms of immunoreactive somatostatin (a), two of which were isolated and characterized as somatostatin-14 (fractions 78–88) and somatostatin-28 (fractions 52–65) (21).] (*a*) Gel filtration of acidic tissue extract (50 ml) of 7 g tumor on Sephadex G-75 (120 × 4.5 cm). Eluent 5 M acetic acid, flow rate 60 ml/hr, fraction volume 15 ml. The first and last absorbance peaks correspond to exclusion and salt volumes of the column, respectively. (*b*) Semipreparative reverse-phase HPLC of pool of bioactive gel filtration fractions 43–58 on C18 column using pyridine formate/n-propanol as mobile phase. Flow rate 0.8 ml/min. Bioassay was performed with pooled or individual aliquots (5–10 μl) of column fractions. To avoid losses due to lyophilization, pooled gel filtration fractions were applied directly to the reverse-phase column by pumping the sample onto the column prior to starting the elution gradient. (*c*) Reverse-phase HPLC of bioactive fractions from semipreparative chromatography (b) using analytical C18 column in conjunction with 0.2% (v/v) heptafluorobutyric acid/acetonitrile. Flow rate 0.6 ml/min. The sample was loaded (after 3-fold dilution) as described above. AUFS: Absorbance units full scale. Bioassay was performed with aliquots of individual or pooled column fractions that had been dried in the presence of 100 μg serum albumin.

Hypothalamic GRFs

The isolation of hpGRFs from tumor tissues raised the question as to whether hpGRF is identical in structure to the as yet uncharacterized hypothalamic GRF. In a first attempt to answer this question we analyzed extracts from eight human hypothalami by reverse-phase HPLC and two radioimmunoassays, using antibodies directed against different parts of hpGRF-44 and hpGRF-40 (33). One antibody that was directed against the central part of hpGRF recognized two major forms of GRF in human hypothalamic extracts coeluting with hpGRF-44 and hpGRF-40. The second antibody, which was directed against the amidated carboxy-terminus of hpGRF-44 (no cross-reactivity with the non-amidated form), recognized only the GRF form coeluting with hpGRF-44. These data provided strong evidence that human hypothalamic GRF is identical to tumor-derived GRF. This conclusion was also supported by findings of Spiess et al (34), who reported that HPLC peptide maps of human hypothalamic and tumor GRFs were indistinguishable.

Final proof of the identity of tumor-derived and human hypothalamic GRFs was provided just recently when sufficient quantities of the peptide were isolated from a few thousand human-pituitary-stalk/hypothalamic-median-eminence fragments to permit complete amino acid sequence determination (see Figure 2) (35). As expected, human hypothalamic GRF was found to be identical in structure to hpGRF-44. Meanwhile the structure of hypothalamic GRF from several other mammalian species has been characterized.

In 1983 Spiess et al (34) reported the isolation and characterization of rat hypothalamic GRF from acid extracts of 80,000 rat hypothalami, using preparative, semipreparative, and analytical reverse-phase HPLC. The structure of rat GRF is substantially different from that of human GRF as it consists of only 43 amino acid residues and possesses a free carboxy-terminus. It also differs from human GRF by 14 amino-acid substitutions (see Figure 3). Our own group devised an alternative approach to the isolation of hypothalamic GRFs that uses immunoaffinity chromatography with Affi-gel-bound IgG raised against hpGRF-40, gel filtration, and two steps of reverse-phase HPLC. This isolation strategy proved very effective and allowed us, within a few months, to isolate GRFs from hypothalamic extracts of the porcine (36), bovine (37), caprine, and ovine (37a) species. Improvements in reverse-phase HPLC analysis of the phenylthiohydantoin amino acid residues from the sequencer and in the hardware and software of the sequencer itself permitted the complete characterization of these hypothalamic GRFs to be accomplished with less than 1 nmol of each peptide. In most cases it was possible to ascertain the sequence of all but the carboxy-terminal amino acid by direct microsequencing of 500 pmol or less of the intact peptides. Comparison of the amino acid composition with the sequence data then permitted identification of the carboxy-terminal residue. Reverse-phase HPLC comparative studies with the appropriate synthetic replicates were carried out to establish the amidated nature of the carboxy-terminal amino acid. Figure 3 shows the structures of all the characterized hypothalamic GRFs and illustrates the high degree of homology that exists among all but the rat species. The consensus structure of the typical mammalian hypothalamic GRF is that of a 44-amino-acid peptide possessing an amidated carboxy-terminus.

Owing to the high efficiency of the isolation strategy based on immunoaffinity chromatography, nanomolar quantities of pure GRF peptides were isolated from a few hundred to a few thousand hypothalami (e.g. 1 nmol porcine GRF from 400 hypothalami). Other isolation schemes omitting the immunoaffinity

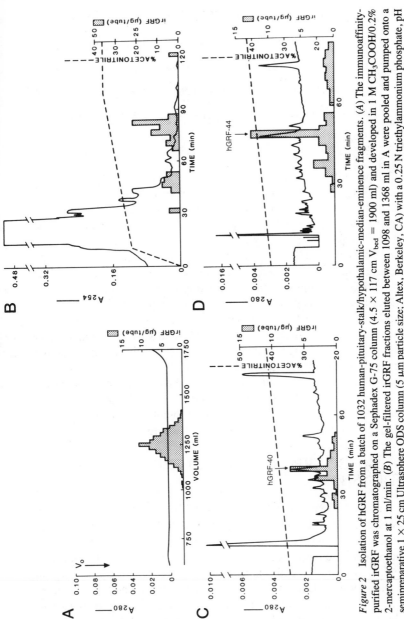

Figure 2 Isolation of hGRF from a batch of 1032 human-pituitary-stalk/hypothalamic-median-eminence fragments. (*A*) The immunoaffinity-purified irGRF was chromatographed on a Sephadex G-75 column (4.5 × 117 cm V_{bed} = 1900 ml) and developed in 1 M CH_3COOH/0.2% 2-mercaptoethanol at 1 ml/min. (*B*) The gel-filtered irGRF fractions eluted between 1098 and 1368 ml in *A* were pooled and pumped onto a semipreparative 1 × 25 cm Ultrasphere ODS column (5 μm particle size; Altex, Berkeley, CA) with a 0.25 N triethylammonium phosphate, pH 3.0/acetonitrile mobile phase. Fractions of 2.5 ml were collected at 1 ml/min. (*C*) The irGRF species eluted between 78 and 87 min in *B* were pooled and purified on a 0.46 × 25 cm Aquapore RP-300 column (7 μm particle size; Brownlee Labs) with a 0.2% heptafluorobutyric acid/acetonitrile solvent system. Fractions of 2.5 ml were collected at 1 ml/min. (*D*) The irGRF species eluted between 72 and 77 min in *B* were pooled and purified in the same manner as in *C*. At all chromatography steps, 1% aliquots of the column fractions were subjected to RIA after drying in a vacuum centrifuge (Savant) in the presence of 100 μg of serum albumin.

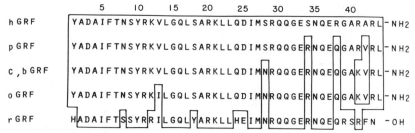

Figure 3   Amino acid sequences of mammalian hypothalamic GRFs. The letters h, p, c, b, o, and r designate the human, porcine, caprine, bovine, ovine, and rat species, respectively.

step yielded pure GRF as well, but with much lower yield (e.g. 80 pmol porcine GRF from 300 hypothalami). Recently we also completed the isolation of rat hypothalamic GRF using immunoaffinity chromatography with antibodies raised against synthetic rat GRF. Structural analysis of the isolated peptide confirmed the sequence previously determined by Spiess et al (34).

So far, GRFs have been characterized only from pancreas tumors and hypothalamic tissue. There is good evidence, based on reverse-phase HPLC, bioassay, amino acid composition, or detection by specific RIA, that other tumors that caused acromegaly, i.e. a lung carcinoid (19) and a liver metastatis [provided by W. Daughaday, St. Louis (16)] from a patient with a foregut carcinoid, contained GRF of structure identical to or at least very similar to that of hpGRF. Interestingly, Arimura et al (38) have reported the presence of other GRF-like substances in ovine brain and gut tissue. These authors have isolated from ovine brain and partially sequenced a peptide with a molecular weight of 4000–5000 that has a novel amino-terminal sequence 60–70% homologous to that of rat GRF and VIP. The possibility must therefore be considered that GRF-like peptides may also exist in nontumor, extrahypothalamic tissue.

SYNTHESIS OF GRFs AND STRUCTURE-ACTIVITY RELATIONSHIPS

Total synthesis of hGRF (20, 35), hGRF(1–40)OH (20, 24), hGRF(1–37)OH (20), as well as that of pGRF (36), bGRF (37), and rGRF (34), was achieved by solid-phase peptide synthesis methodology. Synthesis of hGRF by the classical fragment-condensation technique has also been reported (39). When the synthetic replicates of the hypothalamic GRFs were tested in the normal rat anterior pituitary monolayer culture system, rGRF was found to be the most potent, with the human, porcine, and bovine counterparts having about one half to one third the potency of rGRF (Ling et al, unpublished data).

Based on a potency ranking of 1 for hGRF, successive deletion of the carboxy-terminal residues of hGRF resulted in a gradual decreasing of bioactivity as shown in Table 1 (40). This finding differs from the in vivo data (41), in which hGRF, hGRF(1–40)OH and hGRF(1–37)OH were found to be equipotent on a molar basis. When the deletion reached the fragment (1–31)OH, however, the activity did not decrease any further. The next two deletion fragments, hGRF(1–30)OH and hGRF(1–29)OH, were almost equipotent to hGRF(1–31)OH. Moreover, amidation of the carboxy-terminal enhances the potency of these hGRF fragments to approximately twice the value of their corresponding free carboxylic-acid forms. Thus the amidated fragments of hGRF as small as 1–29, 1–30, and 1–31 possess about one half the potency of hGRF. These data are in agreement with the potency of hGRF analogs reported by Rivier et al (24).

Further deletion of the amino acids to yield fragments smaller than hGRF(1–27)NH$_2$ resulted in marked decreases in bioactivity. But despite the drastically decreased potencies, hGRF(1–23)NH$_2$, hGRF(1–22)NH$_2$ and hGRF(1–21)NH$_2$ gave parallel dose-response curves and reached the same maximal stimulation of GH release as that of hGRF (40). Only the fragment hGRF(1–19)NH$_2$ showed no activity up to 10^{-5}M.

In contrast, progressive deletion of amino-terminal amino acids resulted in a marked loss of bioactivity. Removal of the amino-terminal Tyr^1 residue caused the activity to drop 1000 fold (20). Further deletion of the Tyr^1-Ala^2 dipeptide

Table 1 Relative potencies of carboxy-terminal deleted analogs of hGRF[a]

Analogs	Potency	95% Confidence limits
hGRF(1–44)NH$_2$	1	
hGRF(1–44)OH	0.70	0.56–0.87
hGRF(1–40)NH$_2$	0.91	0.72–1.14
hGRF(1–40)OH	0.34	0.27–0.43
hGRF(1–37)NH$_2$	0.50	0.40–0.62
hGRF(1–37)OH	0.27	0.20–0.36
hGRF(1–34)OH	0.23	0.19–0.30
hGRF(1–31)NH$_2$	0.68	0.52–0.89
hGRF(1–31)OH	0.40	0.29–0.55
hGRF(1–30)NH$_2$	0.51	0.39–0.64
hGRF(1–30)OH	0.27	0.18–0.49
hGRF(1–29)NH$_2$	0.51	0.37–0.70
hGRF(1–29)OH	0.25	0.20–0.31
hGRF(1–27)NH$_2$	0.12	0.09–0.17
hGRF(1–24)OH	0.0002	—
hGRF(1–23)NH$_2$	0.0024	0.002–0.003
hGRF(1–22)NH$_2$	0.00001	—
hGRF(1–21)NH$_2$	0.000001	—
hGRF(1–19)NH$_2$	inactive up to 10^{-5} M	

[a] This table reprinted with permission of Academic Press.

decreased the activity even more, and deletion of the amino-terminal tripeptide Tyr^1-Ala^2Asp^3 yielded a fragment hGRF(4–44)NH_2 that had some GH-releasing activity at $10^{-5}M$, but the dose-response curve never reached the same maximum as that of the parent compound. Thus, together with the bioassay results obtained from the carboxy-terminally deleted fragments, the minimal biologically active core of hGRF possessing full intrinsic activity appears to comprise the fragment 3–21.

The importance of the tyrosine at the first amino-terminal position for endowing the hGRF molecule with high bioactivity was further demonstrated by its substitution with other amino acids or its modification as shown in Table 2 (42). Aside from the replacement with histidine, all other substitutions or modifications resulted in marked loss of bioactivity. Interestingly, replacement of the tyrosine-1 with its D-isomer decreased the activity of the resultant analog 50 fold, in comparison with the parent molecule, indicating that preserving the natural peptide-backbone conformation at that position is important for binding and activating the receptor. Blocking of the amino-terminal of hGRF with an acetyl group reduced its activity to 3% of that of the unblocked molecule. Methylation of the phenolic hydroxyl group of tyrosine-1 of hGRF or the N^τ-imidazole of [His^1]hGRF(1–44)NH_2 reduced the potency of the resulting compounds. However, the decrease in potency was not that pronounced when a 3-methyl group was added to the imidazole ring of histidine-1.

In contrast to these in vitro data, Lance et al (43) reported that [D-Tyr^1]hGRF(1–29)NH_2 and [(N-Ac)Tyr^1]hGRF(1–29)NH_2 were substantially more potent than hGRF(1–29)NH_2 in vivo. These differences, however, may simply reflect a decreased enzymatic degradation of the amino-terminal-modified analogs when administered to live animals.

Table 2 Relative potencies of hGRF analogs modified at position 1

Analogs	Potency	95% Confidence limits
[D-Tyr^1]hGRF(1–40)OH	0.022[a]	0.015–0.036
[Phe^1]hGRF(1–40)OH	0.038[a]	0.031–0.048
[Trp^1]hGRF(1–40)OH	0.003[a]	0.002–0.005
[His^1]hGRF(1–40)OH	0.351[a]	0.254–0.478
[Ala^1]hGRF(1–40)OH	0.010[a]	0.008–0.014
[(N-Ac)Tyr^1]hGRF(1–40)OH	0.032[a]	0.021–0.052
Arg^0-hGRF(1–40)OH	0.002[a]	0.001–0.002
Ala^0-hGRF(1–40)OH	0.007[a]	0.005–0.010
[(3-Me)His^1]hGRF(1–44)NH_2	0.132[b]	0.065–0.282
[(0-Me)Tyr^1]hGRF(1–44)NH_2	0.001	0.001–0.002

[a] relative to hGRF(1–40)OH = 1
[b] relative to hGRF(1–44)NH_2 = 1

In our original paper describing the isolation and characterization of GRF from a tumor of the pancreas (20), we had noted that the structure of GRF has considerable sequence homology with the glucagon-secreting family of brain-gut peptides, especially when the structure of GRF was compared to that of PHI and PHM. Effects of this family of peptides on GH-release in vivo have been reported by Murphy et al (44), who found that secretin, gastric inhibitory polypeptide, and glucagon significantly decreased plasma GH levels, while vasoactive intestinal peptide (VIP) had no effect. In addition, they also reported that secretin and VIP could alter the plasma GH response to hGRF(1–40)OH. A hybrid peptide composed of the 1–14 fragment of hGRF joined to the 15–27 fragment of PHI had been synthesized by Fujii et al (45) and tested in vivo. The synthetic hybrid was found to retain 10% of the activity of hGRF(1–44)NH$_2$.

Aside from the glucagon-secretin family of peptides and a hybrid of hGRF and PHI, a hexapeptide, His-D-Trp-Ala-Trp-D-Phe-Lys-NH$_2$, which has no sequence homology with hGRF, has been synthesized and reported to have GH-releasing activity both in vitro and in vivo (46). However, the potency of this hexapeptide is much less than that of hGRF.

Examination of the molecular models of various peptide hormones have lead Kaiser (47) and Mandell (48) to propose that the amphiphilic secondary structure of peptides (i.e. GRF, CRF) may play a significant role in the binding of these molecules to membrane surfaces. GRF has a hydrophobic sequence wavelength of ≈ 4 (48) (called π helix by Kaiser). According to Kaiser (47), hGRF(1–22) is likely to form a π helix in amphiphilic environments such as water and phospholipid surfaces. The application of the theories of Kaiser and Mandell to the design of GRF analogs may prove to be fruitful and could be of value in the design of superagonists or antagonists of GRF.

MECHANISM OF ACTION OF GRF IN VITRO

Effects of GRF at the Pituitary Level

In primary cultures of dispersed rat anterior pituitary cells, GRF stimulates secretion of GH (20, 22, 49–56). GRF has no effect upon the secretion of β-endorphin, follicle-stimulating hormone, luteinizing hormone, thyrotropin, or prolactin (22). The lack of effect of GRF upon PRL (prolactin) release was confirmed by Almeida (49) and Cronin (57, 58) using hGRF or culture medium in which a GRF-secreting tumor was placed after resection. At the high concentration of 10^{-6} M, hGRF(1–40)OH has been reported to cause release of PRL in incubated rat pituitary halves (59); since (a) hGRF elicits release of GH in vitro at concentrations as low as 10^{-13} (22), (b) the ED$_{50}$ of GRF for GH release in the rat pituitary ranges from 0.5 to 1×10^{-11} M, and (c) hGRF (1–40)OH has about 50% of the potency of hGRF, the physiological relevance of using micromolar concentrations of GRF to achieve PRL release is highly questionable.

In perifused rat anterior pituitary cells, a method whose particular advantage is the frequency with which release of hormone (i.e. GH) into the effluent is sampled and measured (quantitated), the rapid action (< 1 min) of GRF upon GH release has been shown (22, 49, 57, 60). For example, Brazeau et al (22) have shown a dose-dependent release of GH 30 sec following contact of rat pituitary cells with hGRF.

The specificity of the action of GRF has also been shown recently by Wehrenberg et al (61). In an elegant study in which a $2 \times 2 \times 2 \times 2$ factorial design was used, they found that there is no significant interaction on anterior pituitary hormone secretion between the four hypothalamic releasing factors, GnRF, CRF, TRF, and GRF, when rat anterior pituitary cells are simultaneously incubated with these agents. This study (in which the above-mentioned noninteractive effects of the releasing factors were also observed in vivo) suggests that the changes in pituitary hormone release under normal physiologic conditions are not caused by the interaction of hypothalamic releasing factors, each one of these peptides being highly specific in its effect upon the secretion of its target hormone(s).

It is noteworthy that there are 14 amino acid differences between rat hypothalamic GRF (rGRF) (34) and human hypothalamic GRF (hGRF) (35). In the first study on the biologic effect of rGRF, Spiess et al (34) reported that rGRF is slightly more potent (i.e. twofold) than hGRF in eliciting GH release in the normal rat anterior pituitary cell culture assay; these findings have been confirmed (62, 63).

The in vitro effect of GRF on GH release by human pituitary tumor tissues in primary culture has been reported (64–67). In one of these studies, the sensitivity of human GH-secreting adenoma cells to hGRF(1–40)OH was enhanced by preincubation with the synthetic glucocorticoid, dexamethasone (67). The effect of GRF upon GH secretion by normal human pituitary has not been reported.

Law et al (68) have shown that hGRF stimulates GH release in primary cultures of ovine pituitary cells. This group also reported a GRF-mediated PRL release, although the effect of GRF on PRL secretion was found to be only 20–25% greater than in controls.

In monolayer cultures of chicken pituitaries (69) and in dispersed chicken pituitary cells (70) hGRF stimulates release of GH. Leung et al (69) reported that hGRF is significantly less active than hGRF(1–40)OH in eliciting GH release. These findings (which are based on only three samples) differ markedly from the in vitro potencies of the GRFs as measured by the rat monolayer system.

Somatostatin (SS-14 or -28) inhibits GRF-stimulated pituitary GH release in vitro. In monolayer cultures of rat anterior pituitary cells SS-14 and SS-28 inhibit the GH response to GRF in a noncompetitive fashion (22, 56). SS-14 (22, 56) and SS-28 (22) induce lower GH secretion rates at maximally effective levels of GRF, although even at the highest concentrations of somatostatin used, some secretory response to GRF persists. SS-14 has been reported to inhibit the GRF-stimulated decrease in intracellular GH levels in rat anterior pituitary cells, while SS-28 is purported to block the release of GH without having a significant effect upon intracellular GH levels (63). Somatostatin has also been reported to inhibit GRF-stimulated GH release from human pituitary adenoma cells (64, 66) and from cultured ovine pituitary cells (68).

To ascertain whether one of the proposed peripheral regulators of pituitary GH secretion [somatomedin C (SmC) or insulin-like growth factor I (IGF-I)] affects GH release in vitro, rat pituitary cells in monolayer culture were incubated with these agents in the presence or absence of GRF. Highly purified SmC or IGF-I inhibited GRF-stimulated GH release (51). SmC is thought to block the release of GH without affecting the cellular content of GH (63).

The effect(s) of continuous or repeated stimulation of pituitary cells in vitro was tested. In rat pituitary cells in monolayer culture a 24-hr incubation with rGRF depleted the cellular content of GH and attenuated the response to a short-term incubation with rGRF thereafter (62). These findings do not enable us to determine whether the above data are indicative of a desensitization of pituitary somatotrophs to GRF.

In order to investigate whether the paradoxical stimulatory effect in acromegalic patients of TRF upon GH release is caused by the increased levels of GRF in these patients, Borges et al (60) carried out a study in which rat anterior pituitary cells were continuously perifused with hGRF(1–40)OH and treated with a 30-min pulse with TRF thereafter. They found that there was indeed a TRF-stimulated GH release. However, the findings of this study have not been confirmed by others nor has this model been tested in GH-secreting pituitary adenomas from acromegalic patients to ascertain whether GRF does sensitize the somatotrophs, thereby resulting in TRF stimulation of GH secretion.

Results of clinical studies indicate an attenuation of the acute GH secretion in response to hGRF with increasing age, there being a marked decrease after the fourth decade (71). In order to determine whether this observation could be replicated in vitro, pituitary slices from young (3–4 months) and old (19–21 months) rats were incubated with hGRF. An age-dependent difference in GH release was not found (72).

It has been shown that in late pregnancy (days 19 and 21 of gestation) the fetal rat pituitary is highly responsive to GRF, with a maximal effect of the peptide upon GH release being observed with 10^{-10} M GRF (72a). These results suggest that in the rat GRF may regulate GH levels in the fetal pituitary.

The effect of hGRF upon normal rat pituitary GH mRNA sequences was studied (73). The cytoplasmic dot hybridization methodology was used. Incubation of rat pituitary monolayer cultures with hGRF for 72 hr resulted in a 2- to 2.5-fold increase in GH mRNA levels, and maximal levels of stimulation were achieved at hGRF concentrations as low as 1 fM. hGRF did not stimulate prolactin release, nor did it affect specific prolactin mRNA levels in the pituitary cultures.

The transcriptional regulation of GH gene expression by hGRF(1–40)OH was also studied in rat anterior pituitary cells in culture (74). GRF increased the transcriptional rate of the GH gene approximately twofold; the rate of the prolactin gene transcription did not change in response to GRF. Moreover, the transcription rate of CHOB, a gene that is constitutively expressed in all mammalian cells and cell-lines thus far studied, was not altered in response to GRF. These findings suggest that GRF does not have an effect upon transcription in general but is specific for the GH gene. Moreover, our findings (73) and the report of Barinaga (74) suggest that there is little posttranscriptional regulation of the GH gene expression by GRF.

Thyroid hormone (T_3) and glucocorticoids increase the sensitivity of cultured rat pituitary cells to GRF (56). We have observed that following culture of normal rat pituitary cells in defined medium lacking T_3 and cortisol there is no significant stimulation of GH release into the medium following a 3-hr incubation with increasing concentrations of GRF; the normal response to GRF is achieved when the medium is supplemented with the synthetic glucocorticoid, dexamethasone.

In rat pituitary cells in monolayer culture hGRF causes release of stored pools of GH following labeling of cells for 12 hr to achieve a uniformly labeled pool of GH and does not affect the basal, tonic, "unregulated" release of GH (75).

Preliminary studies from several groups strongly suggest that extracellular calcium is required for GRF-stimulated GH release (50, 52). Incubation of normal rat pituitary cells in calcium-free buffer or medium or in the presence of calcium channel blockers such as cobalt chloride and verapamil attenuates the response of the cells' GH release to GRF. There is one report of a GRF-mediated increase in intracellular Ca^{2+} content in anterior pituitary cultures (76). As of August 1984 definitive studies upon calcium fluxes or calcium regulated GH release in the presence of GRF had not been carried out.

GRF causes both the accumulation and secretion of cAMP in pituitary cells. Although a pituitary cell culture consists of a heterogeneous population of cells, the great specificity of GRF in eliciting exclusively GH secretion allows one to interpret the findings described below as a (most likely) direct effect of GRF on somatotrophs. In rat pituitary cells in monolayer culture hGRF (52), GRF(1–40)-OH (50), and rGRF (62) increase cellular cAMP levels and stimulate cAMP efflux. One report (50) shows that ten times more GRF is required for half-maximal stimulation of cAMP production than for GH secretion. GRF-containing tumor medium also stimulates cAMP content of anterior pituitary cells (58).

Brazeau et al (52) reported that the release of GH in rat anterior pituitary cells in monolayer culture is stimulated by 8-bromo-cAMP, isobutyl methylxanthine (IBMX), cholera toxin, and forskolin, each of these substances having an identical maximal effect on hormone secretion. They also found that the addition of 8-bromo-cAMP, IBMX, cholera toxin, or forskolin to a maximally stimulating dose of GRF does not increase the response that remains at the E_{max} of GRF. Cronin et al (77) confirmed the above-mentioned findings with cholera toxin and forskolin.

A direct effect of hGRF(1–40)NH$_2$ (ED$_{50}$ 150 nM) upon adenylate cyclase levels was reported (78). In this study only a 1.4-fold stimulation of adenylate cyclase activity was found with a maximal concentration of GRF; thus the reported findings must be interpreted with great caution. A systematic study of the effect of GRF upon adenylate cyclase activation in normal rat pituitary cells has not been reported.

In a preliminary study in which there was only a 1.76-fold stimulation of GH release in response to hGRF(1–40)OH, the effect of GRF upon ^{32}P incorporation into phospholipids was studied using a rat anterior pituitary dispersed cell preparation (53). GRF appeared to stimulate ^{32}P incorporation into phosphatidylinositol in a dose-dependent manner but had no effect upon incorporation of ^{32}P into phosphatidylcholine and phosphatidylethanolamine.

In a preparation of hog anterior pituitary secretory granules that contain a cAMP-dependent protein kinase that catalyzes [γ^{32}P]-ATP histone phosphorylation (79, 80), hGRF stimulated the activity of the enzyme. Stimulation by hGRF was observed at concentrations as low as 0.3 pM and half-maximal effect was observed with 35 pM (80). hGRF did not significantly change the V_{max} of the reaction but produced a marked increase in the affinity of the enzyme for cAMP (the apparent K_m for the nucleotide decreased from 400 × 10^{-9}M in control, unstimulated conditions to 15 × 10^{-9} M in the presence of 100 pM hGRF). These findings suggest that GH secretion (exocytosis) is mediated via a phosphorylation mechanism involving a granular receptor coupled with a cAMP-dependent protein kinase.

The identification and characterization of pituitary GRF receptors has been difficult because structures of the human and rat GRF peptides have tyrosine or histidine residues at the amino terminal where the active core of the molecule is located and iodination of these molecules yields peptides with little or no biological activity. To circumvent this problem, several analogs of hGRF were synthesized, their effect upon GH release by normal rat pituitary cells were tested in vitro, and the hGRF analog(s) with full intrinsic biologic activity of native GRF were chosen. We have found that $[His^1, Phe^{10}, norVal^{27}, Tyr^{44}]hGRF(1-44)NH_2$ can be used to characterize GRF receptors in bovine pituitary membrane and granule preparations and in GH_3 cells (81).

A preliminary study reported characterization of GRF receptors in rat anterior pituitary homogenates using as a ligand ^{125}I-$[His^1, norleu^{27}]hGRF(1-32)NH_2$ (82). However, the above ligand cannot be used effectively in normal rat pituitary cells in primary culture, unless the cells are pretreated with dexamethasone.

A tumoral line of rat pituitary cells (GH_3) that has been used extensively to characterize GH and PRL gene expression, and that has receptors for somatostatin, was used to characterize the effects of hGRF upon GH and PRL mRNA levels and upon GH and/or PRL secretion (83). In GH_3 cells, hGRF does not significantly affect hormone release or specific GH or PRL mRNA levels. Most recently, however, we have found that rat GRF has much greater potency in GH_3 cells. There is approximately a twofold increase in GH release following incubation of GH_3 cells with rGRF. However, there is a marked (tenfold) stimulation of cAMP efflux, which is time and concentration dependent (84). Additionally rGRF stimulates adenylate cyclase levels in GH_3 cells, and this effect of rGRF is blocked by SS-14 and SS-28 in a noncompetitive manner (84). Studies with GH_3 cells indicate that rGRF elicits its effect on cAMP release and adenylate cyclase activation by interaction with VIP receptors (81, 84).

Extrapituitary Effects of GRF

In a line of rat neural crest–derived cells secreting calcitonin and neurotensin, rGRF stimulates release of these hormones with an ED_{50} of 10^{-11} and 10^{-9} M, respectively (85). In these cells rGRF markedly stimulates cAMP efflux. Somatostatin partially blocks peptide secretion and significantly inhibits cAMP efflux (85).

Laburthe et al (86) were the first to report that in rat intestinal membranes hGRF binds with low affinity to VIP receptors, and is a partial agonist of VIP-stimulated adenylate cyclase activity. The high concentrations (10^{-8} to 10^{-5} M) of hGRF used make it unlikely, however, that in the intestine hGRF elicits a physiological effect through its interaction with VIP receptors.

Most recently, rGRF has been shown to stimulate amylase release in a guinea-pig pancreas dispersed acini preparation (87). In contrast to rGRF, hGRF(1–40)OH or (1–44)NH$_2$ had potencies for amylase release that were less than 1% of that of rGRF. Moreover, these studies (87) suggest that in the guinea-pig exocrine pancreas GRF has a VIP-like effect and thus may stimulate digestive enzyme secretion by interacting with a "VIP-preferring" cellular receptor. It is obvious from the above-mentioned studies that extrapituitary effects of GRF should be studied with the caveat that marked differences of species specificity may be extant in response to GRF.

In summary, in vitro studies have shown that GRF causes the release of stored pools of GH and does not effect the basal, tonic release of GH; GRF specifically stimulates the secretion of pituitary GH. Somatostatin partially blocks GRF-stimulated GH release. Calcium appears to mediate (at least in part) the action of GRF. cAMP is released and accumulated of in response to GRF; this effect appears to be caused by the activation of adenylate cyclase. Glucocorticoids enhance GRF-mediated hormone release, and their presence appears to be a prerequisite for the characterization of GRF receptors in normal rat pituitary cells in culture. The peripheral modulator of GH release (SmC or IGF-I) inhibits GRF-stimulated GH release. GRF is specific in increasing GH mRNA in normal pituitary cells. The peptide also enhances the transcription rate of the GH gene without affecting that of the PRL gene.

ACKNOWLEDGMENTS

We wish to express our appreciation to D. Angeles, F. Castillo, T.-C. Chiang, M. K. Culkin, T. Durkin, C. Hoeman, G. Kleeman, R. Klepper, D. Lappi, D. Martineau, M. Mercado, D. Olshefsky, B. Phillips, R. Schroeder, G. Textor, K. Von Dessonneck, and R. Yates for their excellent technical assistance. Research supported by program grants from NIH (HD-09690 and AM-18811) and the Robert J. and Helen C. Kleberg Foundation.

Literature Cited

1. Deuben, R. R., Meites, J. 1964. *Endocrinology* 74:408–14
2. Schally, A. V., Baba, Y., Nair, R. M. G., Bennett, C. D. 1971. *J. Biol. Chem.* 246:6647–50
3. Veber, D. F., Bennett, C. D., Milkowski, J. D., Gal, G., Denkewalter, R. G., Hirschmann, R. 1971. *Biochem. Biophys. Res. Commun.* 45:235–39
4. Yudaev, N. A., Utesheva, Z. F., Novikova, T. E., Shvachkin, Y. P., Smirnova, A. P. 1972. *Dan. USSR* 210:731–32
5. Wilber, J. F., Nagel, T., White, W. F. 1971. *Endocrinology* 89:1419–24
6. Nair, R. M. G., de Villier, C., Barnes, M., Anatalis, J., Wilbur, D. L. 1978. *Endocrinology* 103:112–20
7. Sykes, J. E., Lowry, P. J. 1983 *Biochem. J.* 209:643–51
8. Dhariwal, A. P. S., Krulich, L., Katz, S.

H., McCann, S. M. 1965. *Endocrinology* 77:932–36
9. Schally, A. V., Sawano, S., Arimura, A., Barrett, J. F., Wakabayashi, I., Bowers, C. Y. 1969. *Endocrinology* 84:1493–1506
10. Malacara, J. M., Valverde, R. C., Reichlin, S., Bollinger, J. 1972. *Endocrinology* 91:1189–98
11. Stachura, M. E., Dhariwal, A. P. S., Frohman, L. A. 1972. *Endocrinology* 91:1071–78
12. Wilson, M. C., Steiner, A. L., Dhariwal, A. P. S., Peake, G. T. 1974. *Neuroendocrinology* 15:313–27
13. Johansson, K. N. G., Currie, B. L., Folkers, K., Bowers, C. Y. 1974. *Biochem. Biophys. Res. Commun.* 60:610–15
14. Currie, B. L., Johansson, K. N., Greibrokk, T., Folkers, K., Bowers, C. Y.

1974. *Biochem. Biophys. Res. Commun.* 60:605–9
15. Boyd, A. E., Sanchez-Franco, F., Spencer, E., Patel, Y. C., Jackson, I. M. D., Reichlin, S. 1978. *Endocrinology* 103:1075–83
16. Leveston, S. A., McKeel, D. W., Buckley, P. J., Deschryver, K., Greides, M. H., et al. 1981. *J. Clin. Endocrinol. Metab.* 53:682–89
17. Scheithauer, B. W., Carpenter, P. C., Bloch, B., Brazeau, P. 1984. *Am. J. Med.* 76:605–16
18. Frohman, L. A., Szabo, M., Berelowitz, M., Stachura, M. E. 1980. *J. Clin. Invest.* 65:43–54
19. Böhlen, P., Thorner, M., Cronin, M., Shively, J., Scheithauer, B. 1982. *Endocrinology* 110:A540
20. Guillemin, R., Brazeau, P., Böhlen, P., Esch, F., Ling, N., Wehrenberg, W. B. 1982. *Science* 218:585–87
21. Böhlen, P., Brazeau, P., Esch, F., Ling, N., Wehrenberg, W. B., Guillemin, R. 1983. *Reg. Peptides* 6:343–53
22. Brazeau, P., Ling, N., Böhlen, P., Esch, F., Ying, S. Y., Guillemin, R. 1982. *Proc. Natl. Acad. Sci. USA* 79:7909–13
23. Esch, F., Böhlen, P., Ling, N., Brazeau, P., Wehrenberg, W. B., et al. 1982. *Biochem. Biophys. Res. Commun.* 109:152–58
24. Rivier, J., Spiess, J., Thorner, M., Vale, W. 1982. *Nature* 300:276–78
25. Esch, F., Böhlen, P., Ling, N., Brazeau, P., Wehrenberg, W. B., Guillemin, R. 1983. *J. Biol. Chem.* 258:1806–12
26. Spiess, J., Rivier, J., Thorner, M., Vale, W. 1982. *Biochemistry* 21:6037–40
27. Gubler, U., Monahan, J., Lomedico, P., Blatt, P., Collier, K., et al. 1983. *Proc. Natl. Acad. Sci. USA* 80:4311–14
28. Blobel, G., Dobberstein, B. 1975. *J. Cell Biol.* 67:835–51
29. Steiner, D., Quinn, P., Chan, S., Marsh, J., Tager, H. 1980. *Ann. NY Acad. Sci.* 342:1–16
30. Bradbury, A., Finnie, M., Smyth, D. 1982. *Nature* 298:686–88
31. Mayo, K., Vale, W., Rivier, J., Rosenfeld, M., Evans, R. 1983. *Nature* 306:86–88
32. Baird, A., Böhlen, P., Esch, F., Ling, N. 1984. *7th Int. Endocrinol. Congr., Quebec City, Canada,* Abstr. 135
33. Böhlen, P., Brazeau, P., Bloch, B., Ling, N., Gaillard, R., Guillemin, R. 1983. *Biochem. Biophys. Res. Commun.* 114:930–36
34. Spiess, J., Rivier, J., Vale, W. 1983. *Nature* 303:532–35
35. Ling, N., Esch, F., Böhlen, P., Brazeau, P., Wehrenberg, W., Guillemin, R.

1984. *Proc. Natl. Acad. Sci. USA* 81:4302–6
36. Böhlen, P., Esch, F., Brazeau, P., Ling, N., Guillemin, R. 1983. *Biochem. Biophys. Res. Commun.* 116:726–34
37. Esch, F., Böhlen, P., Ling, N., Brazeau, P., Guillemin, R. 1983. *Biochem. Biophys. Res. Commun.* 117:772–79
37a. Brazeau, P., Böhlen, P., Esch, F., Ling, N., Wehrenberg, W. B., Guillemin, R. 1984. *Biochem. Biophys. Res. Commun.* 125:606–14
38. Arimura, A., Culler, M. D., Matsumoto, K., Kanda, M., Itoh, Z., et al. 1984. *Peptides* 5:41–44
39. Fijii, N., Shimokura, M., Nomizu, M., Yajima, H., Shono, F., et al. 1984. *Chem. Pharm. Bull.* 32:520–29
40. Ling, N., Baird, A., Wehrenberg, W. B., Ueno, N., Munegumi, T., Brazeau, P. 1984. *Biochem. Biophys. Res. Commun.* 123:854–61
41. Wehrenberg, W. B., Ling, N. 1983. *Biochem. Biophys. Res. Commun.* 115:525–30
42. Ling, N., Baird, A., Wehrenberg, W. B., Ueno, N., Munegumi, T., et al. 1984. *Biochem. Biophys. Res. Commun.* 122:304–10
43. Lance, V. A., Murphy, W. A., Sueiras-Diaz, J., Coy, D. H. 1984. *Biochem. Biophys. Res. Commun.* 119:265–72
44. Murphy, W. A., Lance, V. A., Sueiras-Diaz, J., Coy, D. H. 1983. *Biochem. Biophys. Res. Commun.* 112:469–74
45. Fujii, N., Lee, W., Shimokura, M., Yajima, H. 1984. *Chem. Pharmacol. Bull.* 32:739–43
46. Bowers, C. Y., Momany, F. A., Reynolds, G. A., Hong, A. 1984. *Endocrinology* 114:1537–45
47. Kaiser, E. T., Kezdy, F. J. 1984. *Science* 223:249–55
48. Mandell, A. 1983. *Synergetics of the Brain,* ed. E. Basar, H. Flohr, H. Haken, A. Mandell, pp. 365–76. New York: Springer-Verlag
49. Almeida, O. F. X., Schulte, H. M., Rittmaster, R. S., Chrousos, G. P., Loriaux, D. L., Merriam, G. R. 1984. *J. Clin. Endocrinol. Metab.* 58:309–12
50. Bilezikjian, L. M., Vale, W. W. 1983. *Endocrinology* 113:1726–31
51. Brazeau, P., Guillemin, R., Ling, N., van Wyk, J., Humbel, R. 1982. *C. R. Acad. Sci. III* 295:651–54
52. Brazeau, P., Ling, N., Esch, F., Böhlen, P., Mougin, C., Guillemin, R. 1982. *Biochem. Biophys. Res. Commun.* 109:588–94
53. Canonico, P. L., Cronin, M. J., Thorner, M. O., MacLeod, R. M. 1983. *Am. J. Physiol.* 245:E587–90

54. Guillemin, R., Zeytin, F., Ling, N., Böhlen, P., Esch, F., et al. 1984. *Proc. Soc. Exp. Biol. Med.* 175:407–13
55. Michel, D., Lefevre, G., Labrie, F. 1983. *Mol. Cell. Endocrinol.* 33:255–64
56. Vale, W., Vaughan, J., Yamamoto, G., Spiess, J., Rivier, J. 1983. *Endocrinology* 112:1553–55
57. Cronin, M. J., Rogol, A. D., MacLeod, R. M., Keefer, D. A., Login, I. S., et al. 1983. *Am. J. Physiol.* 244:E346–53
58. Cronin, M. J., Rogol, A. D., Dabney, L. G., Thorner, M. O. 1982. *J. Clin. Endocrinol. Metab.* 55:381–83
59. Arimura, A., Culler, M. D., Turkelson, C. M., Luciano, M. G., Thomas, C. R. 1983. *Peptides* 4:107–10
60. Borges, J. L. C., Uskavitch, D. R., Kaiser, D. L., Cronin, M. J., Evans, W. S., Thorner, M. O. 1983. *Endocrinology* 113:1519–21
61. Wehrenberg, W. B., Baird, A., Ying, S.-Y., Rivier, C., Ling, N., Guillemin, R. 1984. *Endocrinology* 114:1995–2001
62. Bilezikjian, L. M., Vale, W. W. 1984. *7th Int. Endocrinol. Congr., Quebec City, Canada,* Abstr. 401
63. Fukata, J., Martin, J. B. 1984. *7th Int. Endocrinol. Congr., Quebec City, Canada,* Abstr. 665
64. Adams, E. F., Winslow, C. L. J., Mashiter, K. 1983. *Lancet* 1:1100–1
65. Daniels, M., Turner, S. J., Cook, D. B., Mathias, D., Kendall-Taylor, P., et al. 1984. *7th Int. Endocrinol. Congr., Quebec City, Canada,* Abstr. 678
66. Lamberts, S. W. J., Verleun, T., Oosterom, R. 1984. *J. Clin. Endocrinol. Metab.* 58:250–54
67. Oosterom, R., Verleun, T., Lamberts, S. W. J. 1984. *J. Endocrinol.* 100:353–60
68. Law, G. J., Ray, K. P., Wallis, M. 1984. *FEBS Lett.* 166:189–93
69. Leung, F. C., Taylor, J. E. 1983. *Endocrinology* 113:1913–15
70. Scanes, C. G., Carsia, R. V., Lauterio, T. J., Huybrechts, L., Rivier, J., Vale, W. 1984. *Life Sci.* 34:1127–34
71. Shibasaki, T., Shizume, K., Nakahara, M., Masuda, A., Jibiki, K., et al. 1984. *J. Clin. Endocrinol. Metab.* 58:212–14

72. Sonntag, W. E., Hylka, V. W., Meites, J. 1983. *Endocrinology* 113:2305–7
72a. Baird, A., Wehrenberg, W. B., Ling, N. 1984. *Regul. Peptides.* 10:23–28
73. Gick, G., Zeytin, F., Brazeau, P., Ling, N., Esch, F., Bancroft, C. 1984. *Proc. Natl. Acad. Sci. USA* 81:1553–55
74. Barinaga, M., Yamonoto, G., Rivier, C., Vale, W., Evans, R., Rosenfeld, M. G. 1983. *Nature* 306:84–85
75. Zeytin, F., Ling, N., Esch, F. S., Wehrenberg, W. B., Baird, A., Guillemin, R. 1984. *7th Int. Endocrinol. Congr., Quebec City, Canada,* Abstr. 2628
76. Culler, M. D., Obara, N., Arimura, A. 1984. *7th Int. Endocrinol. Congr., Quebec City, Canada,* Abstr. 675
77. Cronin, M. J., Hewlett, E. L., Evans, W. S., Thorner, M. O., Rogol, A. D. 1984. *Endocrinology* 114:904–13
78. Labrie, F., Gagne, B., Lefèvre, G. 1983. *Life Sci.* 33:2229–33
79. Lewin, M. J. M., Reyl-Desmars, F., Ling, N. 1983. *C. R. Acad. Sci. III* 297:123–25
80. Lewin, M. J. M., Reyl-Desmars, F., Ling, N. 1983. *Proc. Natl. Acad. Sci. USA* 80:6538–41
81. Zeytin, F. N., Reyl-Desmars, F. 1985. *Biochem. Biophys. Res. Commun.* In press
82. Seifert, H., Bilezikjian, L., Rivier, J., Perrin, M., Vale, W. 1984. *7th Int. Endocrinol. Congr., Quebec City, Canada,* Abstr. 2320
83. Zeytin, F. N., Gick, G. G., Brazeau, P., Ling, N., McLaughlin, M., Bancroft, C. 1984. *Endocrinology* 114:2054–59
84. Reyl-Desmars, F., Zeytin, F. N. 1985. *Biochem. Biophys. Res. Commun.* In press
85. Zeytin, F., Brazeau, P. 1984. *Biochem. Biophys. Res. Commun.* 123:497–506
86. Laburthe, M., Amiranoff, B., Boige, N., Rouyer-Fessard, C., Tatemoto, K., Moroder, L. 1983. *FEBS Lett.* 159:89–92
87. Pandol, S. J., Seifert, H., Thomas, M. W., Rivier, J., Vale, W. 1984. *Science* 225:326–28

Ann. Rev. Biochem. 1985. 54:425–57

INDUCIBLE DNA REPAIR SYSTEMS

Graham C. Walker

Biology Department, Massachusetts Institute of Technology, Cambridge, Massachusetts 02139

CONTENTS

PERSPECTIVES AND SUMMARY

A great deal of research has been directed towards gaining an understanding of the mechanisms and regulation of DNA repair processes in *Escherichia coli*. Physiological studies of recovery from DNA damage have established the existence of DNA repair processes and offered insights into their mechanisms and regulation. Genetic studies have identified a large number of genes whose products participate in these repair processes and aided in the analysis of their regulation. Recombinant DNA techniques have allowed the cloning of the genes for a number of DNA repair proteins, thereby facilitating studies of the regulation and the biochemistry of their products. Biochemical studies

425

0066-4154/85/0701-0425$02.00

have helped define the mechanisms of action and physical properties of these DNA repair proteins. A number of reviews concerned with DNA repair have been published in recent years [(1–15) see ref. (16) for the proceedings of a recent meeting]. This review will stress recent advances in our understanding of the molecular mechanisms of inducible DNA repair systems in *E. coli*.

In *E. coli*, most of the inducible genes coding for DNA repair proteins are members of one of two major regulatory networks that can be induced by DNA damage: (*a*) the SOS network, which is controlled by the RecA and LexA proteins, and (*b*) the adaptive response network, which is controlled by the Ada protein. Between them, these two regulatory systems control the induction of more than 21 genes. The SOS network controls the expression of genes whose products are known to play roles in excision repair, daughter-strand gap repair, double-strand break repair, methyl-directed mismatch repair, and SOS processing. The adaptive response network controls the expression of proteins with roles in the direct removal of methyl and ethyl groups from DNA, in the excision repair of alkylated bases, and perhaps in other repair processes. In addition, studies showing that *E. coli* has an inducible system for repairing oxidizing damage have suggested that there may be at least one other regulatory circuit governing the expression of DNA repair proteins (17). Also, a number of DNA-damaging agents induce genes that are members of the heat shock regulatory network (18) but to date there is no evidence that any of the heat shock proteins are DNA repair proteins.

We often speak of DNA repair in terms of a number of distinct linear pathways that can help an organism recover from the introduction of lesions into its DNA. However, it is important to bear in mind that there are often several choices for how a given lesion or repair intermediate can be processed. Thus a representation of the possibilities for repairing a given lesion or repair intermediate can generate a branching tree of considerable complexity. Branches can arise if a cell contains enzymes with different activities that are capable of processing the same lesion or repair intermediate, thereby giving pathways differing in the logic of their repair of the lesion. Branches can also rise if a cell contains more than one enzyme that is capable of carrying out the same operation on a given lesion or repair intermediate, thereby giving different pathways that use the same logic to effect the repair. An awareness of this issue is particularly helpful when one is concerned with the inducibility of a repair system. As I will discuss, it has clearly been shown that a number of genes coding for DNA repair proteins can be induced by DNA damage. However, whether the inducibility of one of these genes is manifested at a physiological or biochemical level will depend on whether the induced protein, or a complex involving that protein, is normally rate-limiting for the process being examined.

REPAIR PROCESSES CONTROLLED BY THE SOS REGULATORY CIRCUIT

The SOS system was the first regulatory network to be characterized that is induced by DNA damage. It is the largest, most complex, and best understood DNA-damage inducible network. The existence of the SOS network was first clearly postulated by Defais et al (19) and this hypothesis was amplified and developed by Radman (20, 21). The earlier genetic and physiological evidence suggesting the existence of the SOS system was discussed in depth by Witkin (3) and more recent developments have been the subject of a number of reviews (11, 14, 22–25).

Regulation of the SOS Responses

Exposure of *E. coli* to agents that damage DNA or interfere with DNA replication results in the induction of a diverse set of physiological responses termed the SOS responses which include: (*a*) an increased capacity to reactivate UV-irradiated bacteriophage (Weigle-reactivation), (*b*) a capacity to mutate UV-irradiated bacteriophage (Weigle-mutagenesis), (*c*) the induction of functions that allow bacteria to be mutated by UV and a variety of agents, (*d*) filamentous growth, (*e*) an increased capacity to repair double-stranded breaks, (*f*) an alleviation of restriction, (*g*) a capacity to carry out long patch excision repair (3, 11, 14). As diagrammed in Figure 1, these physiological responses are due to the induction of more than seventeen genes which have often been referred to as *din* (*d*amage-*in*ducible) genes (26).

The expression of the genes in the SOS regulatory circuit is controlled by a complex circuitry involving the RecA and LexA gene products (3, 11, 14) (Figure 1). The LexA protein apparently serves as the direct repressor of every SOS gene that has been identified to date. Exposure of cells to DNA-damaging agents (e.g. UV or mitomycin C), or to treatments which interfere with DNA replication (e.g. shifting certain mutants that are temperature-sensitive for DNA replication to the restrictive temperature), generates an inducing signal which activates RecA molecules. When activated RecA interacts with a LexA monomer, an -ala-gly- bond in the LexA molecule is cleaved. As the LexA molecules in a cell are inactivated by this proteolytic cleavage, the various SOS genes are expressed at increased levels, and the SOS responses mediated by the products of these genes are observed. The repressors of bacteriophage such as lambda have homology to LexA (27) and are similarly cleaved at an -ala-gly- bond when they interact with activated RecA leading to prophage induction. As DNA repair helps the cells recover from the DNA-damaging treatment, the inducing signal disappears so that RecA molecules cease to be activated (28). LexA molecules then accumulate in the cells and repress the SOS genes.

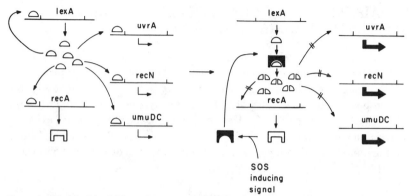

Figure 1   Model of the SOS regulatory system. LexA is the repressor of at least seventeen genes on the *E. coli* chromosome. The generation of an inducing signal as a consequence of DNA damage leads to an activation of RecA. The interaction of activated RecA with LexA results in the cleavage of LexA. As the LexA pools decrease the SOS genes are expressed at higher levels.

THE RecA PROTEIN   The RecA protein is a protein of 37,800 daltons that plays two roles in the cell. It is absolutely required for homologous recombination in *E. coli* and catalyzes synapsis and strand exchange between homologous DNA molecules [for a review see ref. (29)]. In addition, the RecA protein regulates the induction of the SOS responses. After activation RecA mediates the cleavage of the LexA protein; when it is in its activated state, RecA can also mediate the cleavage of the repressors of bacteriophage lambda, 434, and P22 (30–33). In vitro studies have shown that activation of RecA requires single-stranded DNA and a nucleoside triphosphate although hydrolysis of the triphosphate is not required for the cleavage of LexA or phage repressors (34–36). A variety of genetic and physiological studies are consistent with the in vivo inducing signal being single-stranded DNA or DNA containing a gap that is generated by processing of damaged DNA or by interference with replication fork movement (3, 14, 33, 37, 38). Several classes of *recA* mutations have been isolated (11, 14).

THE LexA PROTEIN   LexA is a 22,700-dalton protein that represses SOS genes by binding to sequences, often referred to as SOS boxes, that are located near their promoters (11, 14, 39–42). The consensus sequence for a LexA binding site is taCTGatata-a-aCAGta (14) and instances of cooperative binding of LexA to adjacent sites have been observed (42, 43). Most SOS genes are expressed at a significant basal level even in the absence of an SOS-inducing treatment; the uninduced level of expression of an SOS gene may be influenced by the physical relationship of the promoter to the LexA-binding sequence (44) or the presence of an additional unregulated promoter (45). The K_ds for LexA binding to various operator sites vary considerably (46) so that treatments which lead to a partial depletion of a cell's LexA pools only fully induce a subset of the SOS genes (28). Several classes of *lexA* mutations have been isolated (11, 14) including: (*a*) ones that interefere with the ability of LexA to bind to SOS boxes thereby causing constitutive expression of the SOS responses [*lexA*(Def)] and (*b*) ones that prevent the proteolytic cleavage of LexA thereby rendering the SOS responses noninducible [*lexA*(Ind⁻)].

CLEAVAGE OF THE LexA PROTEIN   The exact details of the molecular mechanism by which activated RecA mediates the proteolytic cleavage of LexA and phage repressors are still unclear. Little (47) has recently reported that specific cleavage of LexA and lambda repressors can occur in the absence of RecA. Incubation of highly purified LexA or lambda repressor under mild alkaline conditions in the presence of a divalent cation resulted in the cleavage of the same -ala-gly- bond that is normally cleaved under neutral conditions only when activated RecA is present. Little's experiments raise the possibility that RecA may not necessarily function itself as the direct catalytic agent in peptide bond cleavage. Rather the interaction of activated RecA with LexA may change the conformation of LexA in a way which increases the susceptibility of a specifically labile bond to simple hydrolysis by an exogenous water molecule. Alternatively, interaction with activated RecA may cause conformation changes in the LexA protein that facilitate its autodigestion; in such a model, an amino acid of LexA itself would participate directly as a catalyst in the cleavage of the -ala-gly- bond and the LexA-RecA complex would exert protease activity previously latent in the LexA protein. Whatever the exact mechanism of this cleavage reaction, the logic of the regulatory circuit remains the same: RecA must be activated in order to effect the cleavage of LexA and the resulting decrease in the pool of intact LexA protein in a cell leads to increased expression of the SOS genes.

Uvr⁺-Dependent Excision Repair

Excision repair is an important strategy for the accurate repair of DNA lesions and takes advantage of the fact that the information in a DNA molecule is present in two copies as a consequence of the complementary double-stranded structure of DNA. Lesions that only damage one strand of a DNA molecule can be accurately repaired by the following set of operations (*a*) the introduction of one or more incisions at or near the site of a lesion, (*b*) the excision of a fragment of DNA containing the damage, (*c*) the use of the complementary strand to direct the resynthesis of the DNA that was removed during the repair process, and (*d*) ligation (1, 8, 10, 12, 13).

Excision repair processes can be operationally separated into two categories on the basis of the mechanism used to effect the incision. (*a*) In excision repair processes in the first category, a glycosylase activity first removes the damaged or altered base to generate an apurinic or apyrimidinic site (AP site). An AP endonuclease (2) then breaks a phosphodiester bond at the AP site thereby initiating the process of excision repair. A number of glycosylases that recognize damaged bases have been described [see ref. (2) for a review]. In general, most of these glycosylases are highly specific and recognize only one damaged base. A broader-spectrum, inducible glycosylase that removes a variety of methylated bases is discussed in the section on the adaptive response regulatory network. Both T4 and *Micrococcus luteus* code for unusual enzymes having both glycosylase and AP endonuclease activities that initiate the excision repair of UV-induced pyrimidine dimers (Figure 1) (48–52). (*b*) In excision repair processes in the second category, a protein or protein complex recognizes a lesion in the DNA and then directly introduces one or more endonucleolytic cuts in the damaged DNA strand without first generating an AP site (1, 8, 10, 12, 13). At least in *E. coli,* this type of incision endonuclease is characterized by its broad specificity for many types of DNA lesions.

PROPERTIES OF *uvrA, uvrB, uvrC,* AND *uvrD* MUTANTS   The existence of excision repair in *E. coli* was first demonstrated by Boyce & Howard-Flanders (53) and by Setlow & Carrier (54) during studies of the repair of UV-induced pyrimidine dimers. Unlinked mutations termed *uvrA, uvrB,* and *uvrC* were then isolated which prevented the excision of pyrimidine dimers and made cells very sensitive to killing by UV irradiation (55). These *uvrA, uvrB,* and *uvrC* mutants were also very sensitive to killing by a wide variety of other DNA-damaging agents such as mitomycin C (56), 4-nitroquinoline-1-oxide (57), and psoralen plus near UV light (58).

Another gene, *uvrD,* was also identified (59–63) that interfered with the excision repair of pyrimidine dimers (64). Unlike *uvrA, uvrB,* and *uvrC* mutations, mutations at the *uvrD* locus can also cause a large increase in the spontaneous mutation rate. The reason for this additional phenotype is that the UvrD protein is also required for the process of methyl-directed mismatch repair which helps to correct mistakes introduced during DNA replication (65–67). In addition, *uvrD* mutations also have effects on the precise excision of transposons (68) and on recombination (69).

In vivo studies of the repair of UV-induced damage (70, 71), as well as studies using permeabilized whole cell systems (72, 73) showed that *uvrA* and *uvrB* mutants were unable to perform the incision step which initiates excision repair of pyrimidine dimers. These latter studies also indicated that the incision event required ATP hydrolysis. The most consistent view of a series of sometimes conflicting studies examining incision in *uvrC* mutants both in vivo (74–76) and in permeabilized whole cell systems (76, 77) seems to be that *uvrC* function is also required for efficient incision in vivo. $uvrA^+uvrB^+uvrC^+$-dependent incision of UV-irradiated DNA was demonstrated in vitro using a cell-free extract and the process was shown to require ATP (78). An in vitro complementation assay based on this system has subsequently been used to follow the purification of the UvrA, UvrB, and UvrC proteins (79–81).

CLONING AND REGULATION OF THE *uvr* GENES   The *uvrA* (82–87), *uvrB* (88–91), and *uvrC* (83, 92–96) genes have been cloned and their products shown to be 114,000 (84), 84,000 (89–96), and 70,000-dalton (83, 92–96) proteins, respectively. The *uvrD* has been cloned and its 75,000-dalton product has been shown to be helicase II (97–104). All four of the *uvr* genes have been shown to be inducible genes that are members of the SOS network. The inducibility and $recA^+lexA^+$-dependent regulation of the *uvrA* (26, 105), *uvrB* (105–107), *uvrC* (108, 109), and *uvrD* (104, 110–113) genes was initially demonstrated by the use of in vivo and in vitro operon fusion techniques.

The discovery of the inducibility of the *uvrA, uvrB,* and *uvrC* genes was initially surprising since, for a number of years, the *uvr* genes had been widely regarded as being constitutively expressed and not being under the control of the $recA^+lexA^+$-regulatory circuit (21). However, the inducibility of the *uvr* genes is consistent with a variety of studies that have suggested that *uvr*-dependent excision repair can be induced by DNA-damaging treatments (114–118; J. Hays, personal communication). Furthermore, the inducibility of these genes helps to explain certain earlier observations such as (*a*) inactivation of the LexA product increasing the UV-resistance of *recA* cells in a $uvrA^+$-dependent fashion (119) and (*b*) induced phage reactivation (W-reactivation) having a uvr^+-dependent component (4, 120, 121). The initial feeling that these genes were not inducible was due to the facts that they are expressed at significant basal levels in uninduced cells and that the basal level of expression is not affected very much by *recA*(Def) or *lexA*(Ind⁻) mutations.

In vitro footprinting studies have shown that LexA functions as the direct repressor of the *uvrA* gene (122); the LexA binding site for the *uvrA* gene overlaps with the *uvrA* promoter. Backendorf et al (44, 108) have presented evidence based on the use of operon fusions which suggests that this same LexA binding site also regulates the transcription of the divergently transcribed *ssb* gene. Other groups using other techniques have failed to observe either induction or LexA-control of *ssb* in vivo (123, 124) suggesting that perhaps *ssb* may be induced only modestly over its relatively high basal level of expression.

In vitro studies of the *uvrB* gene have shown that it has two adjacent promoters (45, 90) which initiate transcripts at +1 and −31 respectively. Both of these promoters have been shown to function in vivo with the downstream promoter being 10–20 fold more active than the upstream one (125). Footprinting experiments identified a LexA binding sequence that overlaps the upstream promoter (45, 46). Although LexA only inhibited transcription from the upstream promoter in vitro (45), S1 mapping studies have shown that LexA regulates the action of both promoters in vivo (125). It is not yet clear whether the differences between the in vitro and in vivo results are due to differences in the conformation of the templates between the two experiments or whether they are a reflection of additional complexities of *uvrB* regulation in vivo. A third promoter was located at -341 by in vitro experiments (45) but does not seem to function in vivo (125).

The regulation of the *uvrC* gene appears to be quite complex (109, 126). van Sluis et al (109) have used an operon fusion to demonstrate a modest induction of the *uvrC* gene that is controlled by the $recA^+ lexA^+$-regulatory circuit but found that the kinetics of its derepression were much slower than those reported for other SOS genes. By using S1 mapping they have obtained evidence that a regulated promoter is located approximately 220 bp ahead of the UvrC coding sequence. Sharma & Moses (126) have identified three putative promoters for *uvrC* by heparin-resistant RNA polymerase-DNA complex formation and suggest that a distal promoter, more than 1 kb upstream of the *uvrC* structural gene, is important for *uvrC* expression in vivo.

The promoter/regulatory region of the *uvrD* gene has been sequenced (127, 128). A LexA binding site located just downstream of the promoter has been shown to interfere with transcription (127). The presence of an apparent strong transcription terminator between the promoter and the start of the coding sequence has led to suggestions that *uvrD* is also subject to regulation by transcriptional attenuation (127, 128).

THE UvrABC ENDONUCLEASE   The cloning and overproduction of the *uvr* genes has greatly facilitated biochemical studies of the UvrABC endonuclease. The UvrA, UvrB, and UvrC proteins have all been purified to near homogeneity from overproducing strains (91, 129–131). Combining these three proteins in the presence of ATP and Mg^{2+} gave an activity that was capable of efficiently incising UV-irradiated DNA (91, 130). Omission of any one of the Uvr proteins or ATP or Mg^{2+} resulted in the loss of this incision activity. The optimal ratio of UvrA to UvrB to UvrC in these experiments was 1:1:1 (91).

A surprising result was obtained when the nature of the incisions made by the UvrABC endonuclease was investigated. When UV-irradiated DNA was used as the substrate, the UvrABC enzyme was shown to make incisions on both sides of pyrimidine dimers (Figure 2) (91, 129–131). The eighth phosphodiester bond to the 5' side of a pyrimidine dimer was hydrolyzed as well as the fourth or fifth phosphodiester bond to the 3' side, thus generating an oligonucleotide of 12–13 nucleotides in length. No incisions were found in the strand opposite the pyrimidine dimer (91). This mode of incision differs strikingly from that caused by the T4 and *M. luteus* pyrimidine-dimer-specific glycosylase/AP endonucleases mentioned above which introduce a single nick directly

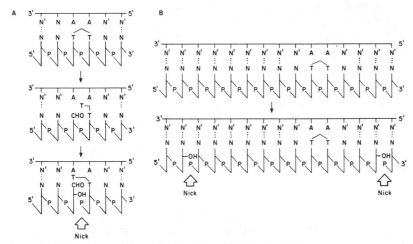

Figure 2   A comparison of the mode of incision of UV-irradiated DNA by (*A*) the pyrimidine dimer specific glycosylase/AP endonuclease activities of *M. luteus* and bacteriophage T4 and (*B*) the UvrABC endonuclease of *E. coli*.

at the site of the pyrimidine dimer via an intermediate containing an AP site (Figure 2).

The reconstituted UvrABC endonuclease also has been shown to recognize other types of DNA damage including the UV-induced (6-4) pyrimidine-pyrimidone photoproducts, guanosines modified with platinum (II) compounds, and psoralen adducts, and to make incisions on both sides of these lesions (130, 132); the (6-4) photoproducts appear to be at least one of the major classes of premutagenic lesions introduced by UV (13, 132–135). Partially purified UvrABC enzyme has been shown to incise DNA modified with 2-acetylaminofluorene derivatives (136) or benzo[*a*]pyrene diolepoxide (137). The ability of the UvrABC endonuclease to recognize a wide range of DNA lesions is consistent with the sensitivity of *uvrA, uvrB,* and *uvrC* mutants to killing by a wide variety of different DNA-damaging agents and suggests that the enzyme recognizes some distortion of the DNA that is common to the various substrates. The advantage of the displacement of the cutting sites from the damage sites is that bulky adducts, which might otherwise interfere with the incision, can be repaired. The capacity of the endonuclease to introduce two incisions flanking a lesion may also help with repair of cross-links.

Some insights have been gained into the function of the individual protein components of the enzyme. Purified UvrA protein has a DNA-independent ATPase and displays an ATP-dependent affinity for UV-irradiated duplex DNA (80, 138); it also binds significantly to unirradiated duplex DNA (91). Purified UvrB protein does not bind to UV-irradiated DNA by itself but stimulates the binding of UvrA to both UV-irradiated single- and double-stranded DNA in an ATP-dependent fashion (91, 139). Incision occurs when purified UvrC is added to a complex of UvrA and UvrB bound to UV-irradiated DNA (81, 91). The UvrC protein is capable of binding to single-strand DNA by itself (93). Although the repair of most lesions seems to require the three proteins, in vivo studies have raised the possibilities that the repair of a lesion introduced by N-hydroxy-aminofluorene may only require UvrC but not UvrA and UvrB (140) and that a resident enhanced repair of plasmids may require UvrA and UvrB but not UvrC (141). The relative complexity of UvrABC endonuclease action is probably related to the fact that it has a much broader substrate specificity than many of the other repair enzymes that have been studied.

POST-INCISION EVENTS IN EXCISION REPAIR   Studies of the action of the reconstituted UvrABC endonuclease have shown that, after the incision event, the UvrABC-DNA complex seems to persist (91). Thus, even though some release of the 12–13 long oligonucleotide has been observed (130), it seems possible that other proteins may participate in the disruption of the UvrABC complex in vivo and in the displacement of the oligonucleotide. A likely candidate for such a role is the UvrD protein (91, 104), which has been shown to be helicase II, although other proteins such as DNA polymerase I, single-stranded DNA binding protein (Ssb), or one of several exonucleases (12) are also possibilities. The suggestion of such a role for UvrD is consistent with studies of repair of UV-induced damage in *uvrD* mutants. These studies have demonstrated that, although *uvrD* mutants are proficient in incision, they exhibit a deficiency in the excision of pyrimidine dimers and a decrease in the rate and extent of rejoining of repaired DNA to parental DNA (64, 104, 142–148). In addition, recent studies of the *phr* gene have raised the possibility that the *phr* gene product plays a role in the $uvrA^+B^+C^+$-dependent repair of UV damage in the dark [(149–151b); J. Hays, personal communication].

The incision step of repair synthesis is followed by DNA synthesis which fills in the excision gaps (152). The repair patches that are produced are heterogeneous in size with 99% of the patches being approximately 20 nucleotides long and 1% being 1500 nucleotides or more in length (118, 153, 154). The relatively close correspondence between the size of the short patches (155–157) and the size of the oligonucleotides produced by the UvrABC double excision event suggests that in most cases a DNA polymerase fills in the gap left after the action of the UvrABC endonuclease with very little nick translation or exonucleolytic expansion of the gap. DNA Polymerase I appears to play a major role in the resynthesis step of short patch excision repair (1, 12, 114, 158); the inherent processivity of polymerase I may be a determinant of the patch size (159).

The process of long patch repair is inducible and is controlled by the SOS regulatory circuit (114, 160–163). The kinetics of the long patch repair synthesis differ from those of the short patch repair synthesis. Short patch synthesis begins immediately after UV irradiation and is virtually completed prior to the synthesis of the majority of the long patches (118). The nature of the induced function(s) that lead to long patch repair synthesis have not yet been determined; studies of repair synthesis in *ssb* mutants have suggested that the single-stranded DNA binding protein does not play a critical role in the determination of patch length (164). The fact that mutants deficient in DNA polymerase I predominantly exhibit long-patch excision repair had led to the suggestion that it is mediated by DNA polymerase II or III (114, 165–167). However, more recent studies have indicated that DNA polymerase I is responsible for the long-patch synthesis observed in wild-type cells (118).

Inducible Repair Involving Recombination Functions

To date, seven genes have been identified whose products play roles in homologous recombination—*recA, recB, recC, recF, recJ, recN*, and *ruv*. The expression of at least three of these genes—*recA, recN*, and *ruv*—is induced by DNA damage and is controlled by the SOS regulatory circuit. Mutation of any one of these seven genes results in an increased sensitivity to killing by agents that damage DNA, indicating that the products of these recombination genes play direct or indirect roles in the repair of DNA damage in *E. coli*. Two repair processes have been identified in which at least a subset of these recombination genes appear to act directly. These are: (*a*) daughter-strand gap repair, which repairs gaps formed in daughter DNA strands copied from damaged templates, and (*b*) double-strand break repair. The second role of the RecA protein in SOS processing will be discussed in a later section.

FUNCTIONS INVOLVED IN CONJUGAL RECOMBINATION   Genetic analysis of the process of conjugation recombination in *E. coli* has led to the identification of the seven recombination genes listed above. The product of the *recA* gene, described above, is absolutely required for homologous recombination and plays a central role in the regulation of the SOS network. The *recB* and *recC* genes encode subunits of the enzyme exonuclease V, a DNA-dependent ATPase that functions both as an exo- and endonuclease as well as a DNA helicase (168–170). However, unlike *recA*, the products of the *recB* and *recC* genes are not absolutely required for recombination since mutations in the *recB* and *recC* genes reduce recombination frequencies to about 1% of those seen in wild-type cells but do not have as severe as an effect on recombination as *recA* mutations. The recombination deficiencies of *recB* or *recC* mutants can be suppressed by either of two mutations, *sbcA* and *sbcB*, that affect enzymes involved in DNA metabolism. The pathway for conjugation recombination activated in *sbcA* mutants has been referred to as the RecE pathway and that activated in *sbcB* mutants as the RecF pathway (171). *sbcA* mutations lead to the activation of Exo VIII, an ATP-independent nuclease specified by *recE* gene of the Rac prophage, a cryptic prophage found in only some strains of *E. coli* K-12 (171–173) and will not be considered further in this review. *sbcB* is the structural gene for exonuclease I and mutations which inactivate *sbcB* restore conjugal recombination proficiency to *recB, recC,* or *recB recC* mutants (174, 175). This *recB recC*-independent pathway of recombination activated by *sbcB* mutations (the RecF pathway) has an additional requirement for the products of at least five different genes: *recA* (62), *recF* (62), *recJ* (176), *recN* (177), and *ruv* (178) and it seems likely that other genes will be shown to be required (171). The RecF pathway of conjugal recombination has been shown to be regulated by the products of the *recA* and *lexA* genes; consistent with this observation is the fact that at least three of the five genes shown to be required for the RecF pathway are inducible (177, 179–181).

Operon fusion techniques have been used to show that the *recN* (177) and *ruv* (182) genes are induced by DNA damage and regulated by the SOS regulatory circuit as is the *recA* gene. No evidence has been obtained to date that the *recF* gene is inducible and the regulation of *recJ* gene has not yet been examined (176).

The observation that *recF, recJ, recN,* and *ruv* mutations reduce conjugal recombination in *recB recC sbcB* strains but have very little effect on recombination in *recB⁺ recC⁺ sbcB⁺* strains has been explained by assuming that they affect recombination initiated independently of *recB recC* function but allow *recB⁺ recC⁺*-dependent recombination to proceed more or less normally (183, 184). This model assumes that the components required specifically for the *recF* pathway contribute very little to the conjugal recombination seen in a wild-type cell and it has been suggested that the major role of these recombination proteins is in DNA repair (181). However, three recent findings may force a reevaluation of the contribution of the RecF pathway in wild type cells: (*a*) *recB* or *recC* mutations reduce the expression of the *recN* gene, (*b*) *sbcB* mutations derepress *recN* expression, especially in a *recB recC* background, and (*c*) *lexA*(Ind⁻) mutations, in contrast to their effects on other SOS-regulated genes, cause a large decrease in the basal level of expression of recN (185, 186). Thus the recombination deficiencies of *recB* and *recC* mutants may not only be due to a loss of exonuclease V, but also to a decrease in recombination due to lowered expression of *recN*. Thus, since *recN* seems to be an important component of the RecF pathway of recombination, the restoration of recombination proficiency in a *recB recC* background that is caused by an *sbcB* mutation may not only be due to the loss of exonuclease I activity but also to a reestablishment of *recN* expression.

Detailed models have been proposed which attempt to assign specific roles in the processing of intermediates in homologous recombination to specific gene products required for recombination (171, 185). At least some of these proteins that can participate in conjugal recombination play roles in the repair of daughter-strand gaps and double-strand breaks. It would appear that these types of DNA damage are sufficiently like intermediates encountered in conjugal homologous recombination that certain recombination proteins can process both.

DAUGHTER-STRAND GAP REPAIR   When DNA synthesis occurs in cells that have been exposed to UV-irradiation, the newly synthesized DNA has a lower molecular weight than newly synthesized DNA from unirradiated cells (72). The low molecular weight is due to gaps or discontinuities in the nascent strand. These gaps apparently arise by replication being blocked at a pyrimidine dimer or other bulky lesion and then resuming at some site past the lesion, presumably at the next site for the initiation of an Okasaki fragment. *E. coli* has a strategy for repairing these gaps that has been referred to as "daughter-strand gap repair" or "postreplicational repair" [for reviews see refs. (1, 12)] (Figure 3). During this type of repair the gaps are filled and the discontinuous strands are joined into molecules of high molecular weight (72). The mechanism by which this occurs results in stretches of parental DNA becoming covalently attached to daughter strands, indicating that a recombinational strand exchange is involved (187). These strand exchanges occur at a frequency approaching one per daughter-strand gap (187). In excision-deficient cells, pyrimidine dimers remain in DNA during this process but become equally distributed between parental and progeny strands as a result of the strand exchanges (188, 189). Thus daughter-strand gap repair repairs gaps that are generated by the replication apparatus attempting to replicate past certain lesions but does not remove the original lesions themselves. It therefore can be viewed as a mechanism which allows cells to tolerate such lesions at least temporarily. However, once the strand exchange has occurred, both the lesions in the original strand as well as those transferred to the daughter strand can, at least in principle, be removed by an excision repair process.

Four classes of mutations appear to cause deficiencies in daughter-strand gap repair, *recA*(Def), *ruv*, *lexA*(Ind⁻), and *recF*. The *recA*(Def), *lexA*(Ind⁻), and *recF* mutations also block long-patch repair, but long-patch repair can be distinguished from daughter-strand gap repair because the former, but not the latter, requires the $uvrA^+B^+C^+$ genotype and can occur in the absence of normal replication (12).

recA(Def) mutants are defective in daughter-strand gap repair (190). Whatever the regulatory effects of RecA on daughter strand gap repair, some evidence has been obtained suggesting that the RecA protein does participate mechanistically in this recombinational repair process. Overproduction of

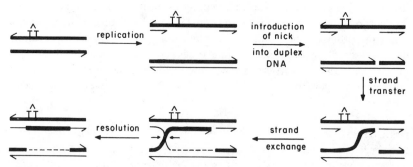

Figure 3   A model for daughter-strand gap repair. The model outlines the logic of some of the events that are thought to occur during daughter strand gap repair. The nicking of the duplex DNA does not necessarily precede strand invasion.

truncated *recA* proteins in *recA*⁺ cells results in a deficiency in homologous recombination and a sensitivity to killing by UV that correlates with a loss of recombinational repair; the ability of the cells to induce the SOS responses is largely unaffected (191–193). The inhibition of RecA function seen in such experiments presumably is a consequence of the formation of mixed multimers (191–195). A variety of in vitro studies have suggested that the RecA protein participates directly in certain of the steps of daughter-strand gap repair as modeled in Figure 3. If the duplex DNA that is homologous to the single-stranded gapped region contains a nick, RecA is able to initiate the strand exchange and then, in the presence of ATP, promote reciprocal exchanges in one direction starting at the site of the original crossover (196, 197). Furthermore, in the presence of SSB and ATP, RecA is able to promote such branch migration through a DNA strand containing pyrimidine dimers and other UV-induced photoproducts as required by the model for daughter-strand gap repair (198). The rate of branch migration is decreased by approximately 50-fold when a dimer is encountered but the reaction proceeds to completion even with heavily UV-irradiated templates. Furthermore, once dimers have been bypassed in the in vitro reaction they can be specifically incised by the T4 dimer-specific glycosylase/ATP endonuclease, suggesting that lesions can be removed by excision repair after the completion of daughter-strand gap repair (198).

Mutations in the *ruv* gene lead to sensitivity to radiation and increase filamentation after transient inhibition of DNA synthesis (178, 199, 200). Their increased sensitivity to UV-killing appears to arise not from the induced filamentation but rather from a defect in a repair process involving recombination functions (178). Since they are proficient in excision repair, SOS processing, and are proficient in SOS induction, it seems likely that their defect is in daughter-strand gap repair. Because *ruv* strains appear to convert low-molecular-weight DNA synthesized immediately after UV-irradiation into high-molecular-weight DNA (199), it has been suggested that the *ruv* gene product acts in some step in the repair process subsequent to strand rejoining, perhaps by participating in the resolution of the joint molecules into viable genomes (178). Since the *ruv* gene is inducible and regulated by the SOS circuit, it is a candidate for another inducible gene besides *recA* that is involved in daughter-strand gap repair.

The deficiency in daughter-strand gap repair seen in *lexA*(Ind⁻) mutant appears to be a consequence of the inability of such cells to induce proteins required for this type of recombinational repair: *recA,* probably *ruv,* and possibly others. The partial suppression of the UV sensitivity of *lexA*(Ind⁻) mutants by *recA* operator-constitutive mutations is consistent with there being a need to induce *recA* for efficient daughter strand-gap repair (201, 202). Similarly, the deficiency of *recF* mutants in daughter-strand gap repair (120, 189, 203–206) may also be a consequence of a deficiency in inducing SOS-controlled genes (162, 207–210). Since *recF* mutations cause a partial inhibition of SOS induction rather than a complete block, it is possible that the *recF* gene product affects the activation of *recA,* perhaps playing a role in the formation, stabilization, or utilization of the SOS-inducing signal. However, it is possible that the *recF* gene product also participates more directly in the mechanism of daughter-strand gap repair since a *lexA*(Def)::Tn5 mutation, which leads to complete derepression of the SOS functions, does not suppress the UV sensitivity of a *recF* mutant (211). The *recF* gene is located in the middle of a cluster of genes involved in DNA metabolism and codes for a 40,000-dalton protein but the biochemical function of the *recF* gene is still unknown (212, 213). The fact that two classes of *recA* mutations, *recA*(Tif) (211) and *recA*(SfrA) (214), can partially suppress the UV sensitivity of a *recF* mutant without inducing the SOS genes, raises the possibility that the RecF protein may physically interact with RecA thereby facilitating its role in recombinational repair. The effect of these mutations must be on the activity of the RecA protein rather than on its level of expression, since overproducing normal RecA, either by an operator constitutive mutation or by a multicopy plasmid, did not suppress the UV sensitivity of a *recF* mutant (202, 215).

No gene product has yet been shown to introduce the nick in the duplex DNA that is homologous to the gap (Figure 3), despite the fact that an activity ("cutting in trans") that is consistent with this function has been detected both in vivo (216) and in vitro (217). One way of rationalizing the observation that a *uvrB* mutation eliminates the residual daughter-strand gap repair seen in a *recF* mutant (204) would be to suggest that, by introducing a nick near a dimer in the duplex DNA homologous to a gap, the UvrABC endonuclease is able to provide the 3' end needed for reciprocal strand exchange. By such a mechanism it could substitute for an SOS-regulated endonuclease that is poorly induced in *recF* mutants.

DOUBLE-STRAND BREAK REPAIR   Double-strand breaks in DNA can be created by the action of various physical and chemical agents. One of the most widely studied of these agents is ionizing radiation. A variety of experiments have shown that bacteria have the capability to repair double-strand breaks (218). For *E. coli,* the capacity to repair double-strand breaks was shown to be inducible since pretreatment of cells with either X rays or UV resulted in an induced resistance to killing by X rays or gamma rays (219–225). This induced resistance was controlled by the *recA*$^+$*lexA*$^+$ regulatory circuit (219, 220, 226) and required de novo transcription and translation since it could be blocked by rifampicin or chloramphenicol (219, 220, 222, 225).

The inducible double-strand break repair seen in *E. coli* requires the presence of another DNA duplex that has the same base sequence as the broken double helix (221). As the number of copies of the genome within the same cell increases, the cells become increasingly radioresistant (224). *E. coli* cells growing normally in the lab can repair double-strand breaks since they contain approximately 4–5 genomes per cell. These latter observations suggest that the molecular mechanism for double-strand-break repair in *E. coli* may be similar to that described for yeast (227).

The inducible repair of double-strand breaks introduced by ionizing radiation or mitomycin C requires a functional *recA* gene (221, 225) and a functional *recN* gene (228). The *recN* gene codes for a 60,000-dalton protein (186) and its expression is regulated by the SOS control circuit (177). Recent in vitro experiments suggest that RecA also participates in the repair of double-strand breaks (229). Double-strand breaks are also introduced during the repair of UV-irradiated DNA and studies of various multiple mutants are consistent with these breaks being repaired by a *recB*-dependent process (205, 206, 230) that is different from the *recF*-dependent process involved in daughter-strand gap repair. The relationship between the *recN*-dependent repair of double-strand breaks and the *recB*-dependent repair of double-strand breaks in UV-irradiated DNA is not yet clear.

SOS Processing

In *E. coli,* mutagenesis by UV and a variety of chemical agents, such as 4-nitroquinoline-1-oxide and methyl methanesulfonate, is not a passive process but rather requires the intervention of an active cellular system that processes damaged DNA in such a way that mutations result. This processing has often been referred to as "error-prone repair," "SOS repair," or "misrepair." I will use the term SOS processing in an attempt to avoid implying any particular biochemical mechanism. SOS processing has been the subject of two intensive reviews (3, 14).

The existence of SOS processing was first deduced from physiological and genetic experiments that indicated that cellular functions had to be induced by DNA damage in order for mutations to arise as a consequence of DNA damage (3, 14, 19, 231–234). It requires the functions of at least three genes, *umuD*, *umuC*, and *recA*, all of which are inducible and regulated as part of the SOS network.

THE *umuD* AND *umuC* GENES   Mutations in the *umuD* and *umuC* genes (120, 235–237) abolish the ability of *E. coli* cells to be mutated by a wide variety of agents such as UV, 4-nitroquinoline-1-oxide (120, 235), methyl methanesulfonate (238), and neocarcinostatin (239). They do, however, retain the ability to be mutated by certain agents such as *N*-methyl-*N'*-nitro-*N*-nitrosoguanidine (MNNG) or ethyl methanesulfonate which introduce directly-mispairing lesions such as O^6-methylguanine (240, 241). *umuD* and *umuC* mutants are somewhat sensitive to killing by UV but are by no means as sensitive as *uvr* mutants. The *umuD* and *umuC* genes have been cloned (236, 237) and have been shown to code for proteins with molecular weights of 16,000 and 45,000, respectively. The two genes are located in an operon and sequencing studies have revealed that the final A of the UGA stop codon of *umuD* is the first A of the AUG codon that initiates the *umuC* gene (K. L. Perry, S. J. Elledge, G. C. Walker, unpublished results; T. Kato, personal communication). A variety of techniques including the use of operon and gene fusions have been used to show that the *umuDC* operon is inducible and regulated by the SOS control circuit (236, 237, 242).

A number of naturally occurring plasmids have the ability to make *E. coli* more mutable to UV and a variety of chemical agents and to alter its resistance to these agents (14, 243–251). Many of these plasmids appear to exert these effects by carrying analogs of the *umuD* and *umuC* genes (238, 252–256). The most intensively studied of these plasmids is pKM101 (14, 238, 257–264), an in vivo–derived deletion derivative of the clinically isolated plasmid R46 (260, 265, 266); pKM101 has played a major role in increasing the sensitivity to mutagenesis of the Ames *Salmonella typhimurium* strains used for detecting mutagens and carcinogens (267–269). pKM101 carries two genes *mucA* and *mucB* that can suppress the nonmutability of *umuD* and *umuC* strains. Like the *umuD* and *umuC* genes, the *mucA* and *mucB* genes are organized in an operon that is repressed by LexA, and code for products of 16,000 and 45,000 daltons respectively (262, 263). The reading frames for the two proteins overlap by 13 basepairs. The sequences of the *umu* proteins and the *muc* proteins have undergone considerable evolutionary divergence. The deduced amino acid sequences of the UmuD and MucA proteins are approximately 41% homologous and those of the UmuC and MucB proteins are approximately 55% homologous (K. L. Perry, S. J. Elledge, G. C. Walker, unpublished results). The phenotypes conferred by the *muc* genes are very similar to those conferred by the *umu* genes although some minor differences have been reported (14, 270–272). It is not yet clear whether these differences are due to differences in biochemical function or to differences in levels of expression.

Intriguingly, the UmuD and MucA proteins are about 30% homologous to the carboxy-terminal region of the LexA protein, including that portion containing the cleavage site, and to the repressors of lambda, 434, and P22. The significance of this homology, if any, is not yet clear. Possibilities include the UmuD and MucA proteins being proteolytically cleaved in a RecA-mediated fashion, the proteins interacting physically with the RecA protein, or the proteins existing as dimers as LexA does. Both UmuC and MucB are quite basic proteins. The phenotypes caused by the *umuDC* operon are highly dependent on its dosage and level of expression. *lexA*$^+$*uvrA* cells are sensitized to killing by UV by the presence of multicopy plasmid carrying the *umuDC* operon. *lexA*(Def) cells (which lack functional LexA) can only tolerate the presence of a *umuD*$^+$*umuC*$^+$ multicopy plasmid at 43°C; shifting such cells to 30°C leads to a cessation of DNA replication and cell death. Investigations of possible biochemical roles of the UmuD and UmuC proteins have been complicated by their relative instability both in vivo and during purification (K. L. Perry, S. J. Elledge, L. Marsh, L. Dodson, L. Vales, G. C. Walker, unpublished results).

THE *recA* GENE   Genetic evidence indicates that the RecA protein must play some other role in SOS processing besides mediating the cleavage of LexA that is necessary for the induction of the *umuD* and *umuC* genes. This can most clearly be seen by the fact that *lexA*(Def) *recA*(Def) strains are nonmutable with UV (11, 242, 273) despite the fact that the *umuD* and *umuC* genes are constitutively expressed at high levels in such backgrounds (242). Studies of other *recA* alleles have suggested that the ability of the RecA protein to be activated is important for this second role (14, 263, 273–277a). To date, no evidence has been obtained that would support the idea that the second role of RecA in mutagenesis is to induce a gene that is regulated by some RecA-cleavable repressor besides LexA (277b). Thus, the most likely possibilities would seem to be that (*a*) RecA is required for the cleavage of some protein which then participates in the actual biochemical process of mutagenesis and (*b*) that RecA participates mechanistically in SOS processing. If this latter possibility were the case, the characteristics of the RecA protein needed for this role must be the same as, or related to, the characteristics that are required for its activation to cleave LexA. As mentioned above, the fact that UmuD and MucA are homologous to LexA and several phage repressors raises the possibility that these proteins may interact with the RecA protein or even be cleaved by a RecA-mediated process.

IMPLICATIONS OF MUTATIONAL SEQUENCE CHANGES   The sequence changes of mutations resulting from treatment with SOS-processing-dependent mutagens such as UV, 4-nitroquinoline-1-oxide, and activated aflatoxin B1 have been extensively analyzed (278–295). These studies have indicated that different mutagens cause different spectra of mutations, both with regard to the sequence changes that are observed and also with regard to the location and relative frequency of the sequence changes. It thus appears that most mutagenesis resulting from SOS processing of damaged DNA templates is targeted (14, 296, 297). It cannot be due to the induction of some random mutator activity whose site of action is not influenced by lesions in the DNA (untargeted mutagenesis). As I have discussed previously (14), targeted mutagenesis may consist of two components: (a) locally targeted mutagenesis (the presence of the lesion leads to the introduction of mutations at its actual site) and (b) regionally targeted mutagenesis (the presence of a lesion leads to the introduction of mutations in its immediate vicinity but not directly at its site).

The majority of the events that arise from the action of the SOS processing system acting on damaged templates appear to be the result of a locally targeted process in which the nature of the premutagenic lesion influences the sequence change (14, 296–298). A considerable body of work has been reported suggesting that AP sites can serve as premutagenic lesions (299–301) and that the production of AP sites may be a significant factor in the mutagenic effects of a variety of chemical mutagens (285, 289, 295, 297, 302). Similarly, a variety of experiments are consistent with the suggestion that (6-4) pyrimidine-pyrimidone photoproducts are at least one of the significant premutagenic lesions introduced by UV irradiation (13, 133, 134, 293, 303). The minority of the mutagenic events resulting from SOS processing are targeted but random with respect to basepair change (298) and can be interpreted as being either the result of a regionally targeted process acting in the vicinity of lesions or of a locally targeted process acting at relatively rare noninformational lesions (14). It is interesting to note in passing that a regionally targeted mechanism could account for the clustering of point mutations found around rearranged antibody variable genes (304).

The induction of the SOS system leads to an increase in mutation frequency even with undamaged DNA templates; there appear to be both hot spots and cold spots for this type of mutagenesis (282, 284, 295, 298, 305–315). However, for the E. coli chromosome, the amount of mutagenesis resulting from this type of process is much less than that resulting from the processing of damaged templates; only a component of this type of mutagenesis appears to be

$umuD^+umuC^+$-dependent. The most likely explanation for such mutagenesis would seem to be that it is the result of the action both of untargeted mutational processes as well as targeted mutational processes acting at spontaneously occurring lesions. The observation of an increased SOS mutator effect in mutants deficient in mismatch repair has led to the suggestion that misincorporated bases introduced by untargeted processes can be corrected by methyl-directed mismatch repair (314). If this is the case, the amount of such untargeted mutagenesis that is finally observed may be an underestimate of the amount of misincorporation that actually occurred on the undamaged templates. Experiments in which phage DNA molecules are introduced into SOS-induced cells often indicate that a greater proportion of the final amount of mutagenesis observed is due to untargeted mutagenesis than seems to be the case with the E. coli chromosome. Such observations could be rationalized by the suggestions that (a) misincorporations from untargeted processes are more likely for phage DNAs since they undergo many rounds of replication in an SOS-induced cellular environment and (b) phage DNAs are less efficient substrates for methyl-directed mismatch repair than the E. coli chromosome (314).

POSSIBLE MECHANISMS FOR SOS PROCESSING   With regard to the possible mechanism of SOS processing, the most significant conclusion that can be drawn from these sequencing studies is that the nature of the premutagenic lesion can influence the nature of the sequence change. The simplest mechanism consistent with this conclusion would be that a polymerase with a relaxed template specificity introduces an incorrect base(s) opposite a lesion. If the lesion retained some capacity for base pairing it could influence the nature of the incoming base. If the lesion were truly noninformational and had no preferential base-pairing properties, the selection of the misincorporated base could be random or it could be influenced by some inherent preference of the polymerase for inserting a particular base.

Despite a great deal of work by many labs, the actual biochemical mechanism of SOS processing has not yet been elucidated. Radman's suggestion (20, 21) that the mutation fixation processes stimulated by SOS induction result from an induced infidelity of DNA replication was supported by the observation that pyrimidine dimers act as an absolute block for chain elongation by purified DNA polymerase I and polymerase III on UV-irradiated templates. This in turn led to the hypothesis that SOS mutagenesis might involve the inducible inhibition of the $3' \rightarrow 5'$ proofreading exonuclease of DNA polymerases (316–318). Subsequent studies of the action of purified polymerases on various damaged DNA templates have shown that lesions introduced by a number of different agents can block chain elongation (290, 300, 319–325). Polymerization usually terminates just before, or at, the site of the lesion. If the stringency of a polymerase is relaxed, for example by adding Mn^{2+}, additional nucleotides can be incorporated and model experiments have been carried out showing that polymerases are capable of incorporating nucleotides opposite AP sites. In general, mammalian polymerases, which as presently isolated lack proofreading activity, are able to bypass at least some lesions consistent with the suggestion that an inhibition of proofreading would contribute to misincorporation opposite lesions.

Despite the facts that the $3' \to 5'$ exonuclease activities of DNA polymerase I and polymerase III holoenzyme can be inhibited by deoxynucleoside monophosphates or by purified RecA protein (325–327) evidence has been presented suggesting that $3' \to 5'$ exonuclease may not be the only factor involved in elongation past lesions and that it is not required for termination at a lesion. The addition of Mn^{2+} facilitates incorporation opposite lesions by the Klenow fragment of polymerase I but has no effect on the $3' \to 5'$ exonuclease activity. Furthermore, Mn^{2+} has a similar effect on polymerase α which lacks a proofreading activity. The reduced fidelity of DNA polymerase I* (328a), an error-prone polymerase isolated from SOS-induced cells, appears to be attributable to reduced discrimination during dNTP selection rather than to a defect in proofreading or in template reading (328b). Although the appearance of polymerase I* is strictly correlated with the expression of the SOS functions there is, to date, no genetic evidence implicating it in SOS processing. Genetic studies have suggested that polymerase III may be required for SOS processing [(329, 330) A. Brotcorne-Lannoye, G. Maenhaut-Michel, M. Radman, personal communication]. The demonstration that the ϵ subunit [the $dnaQ(mutD)$ gene product] of DNA polymerase holoenzyme has a special role in defining the accuracy of DNA replication (331) has provided a precedent for how a protein could modulate the fidelity of a DNA polymerase. However, once again there is, to date, no evidence directly implicating polymerase III in SOS processing.

The roles of the UmuD and UmuC gene products in SOS processing are also unknown as of the writing of this review. These proteins could influence the misincorporation of bases opposite a lesion by such formal possibilities as (a) coding for a new polymerase activity themselves, (b) modifying the properties of an existing polymerase, or (c) regulating the induction of yet another protein which then participates directly in the biochemical mechanism. Alternatively, the UmuD and UmuC proteins could be involved in some step subsequent to the actual misincorporation event. Bridges & Woodgate (332) have recently made the intriguing suggestion that the UmuC protein is not required for misincorporation of bases opposite UV-induced pyrimidine dimers but rather for continued chain elongation past the misincorporated bases. This hypothesis is based on an experiment in which $uvrA$ $umuC$ cells, which are normally nonmutable with UV, were allowed to replicate their DNA for several hours after UV irradiation. If these cells were then exposed to visible light so that photoreactivation could occur, some mutagenesis was observed. This observation suggests a model in which misincorporation occurs opposite a pyrimidine dimer but elongation cannot continue because of the lack of *umuC* function; removal of the dimer by photoreactivation allows elongation to occur thereby fixing the mutation.

THE ADAPTIVE RESPONSE

The adaptive response regulatory network is the other system controlling the induction of repair processes in *E. coli* to have been extensively characterized. Physiological evidence suggesting the existence of this network was first obtained by Samson & Cairns (333). They observed that, if *E. coli* cells were first exposed to low concentrations of the methylating agent MNNG, they became resistant to the mutagenic and lethal effects of a subsequent challenge with a higher dose of this agent. This induced resistance is now known to be the result of a set of induced repair processes, collectively termed the adaptive response (334, 335), that repair DNA damage introduced by methylating and ethylating agents. The lesions repaired by these processes include purine bases alkylated at ring nitrogens or exocyclic oxygens, pyrimidine bases alkylated at exocyclic oxygens, and phosphotriesters. The regulation of the adaptive response is independent of the SOS regulatory network and is controlled by the product of the *ada* gene. Aspects of the biochemistry and regulation of the adaptive response have recently been reviewed (2, 14, 336, 337).

Induced Repair Activities

At least two different types of DNA repair enzymes have been purified that are induced during the adaptive response. One of these is a broad spectrum DNA glycosylase, termed 3-methyladenine-DNA glycosylase II (338, 339), which has a molecular weight of approximately 30,000 (340, 341). This enzyme is induced approximately 20-fold during the adaptive response and initiates excision repair of several methylated bases by removing 3-methyladenine, 3-methylguanine, O^2-methylcytosine, and O^2-methylthymine (338, 339, 342, 343). 3-Methyladenine-DNA glycosylase II is the product of the *alkA*$^+$ gene (339, 343, 344). Fusions derived by both in vivo [(345) M. R. Volkert, D. C. Nguyen, C. Beard, submitted] and in vitro (344) techniques have been used to show directly that expression of the *alkA* gene is inducible and regulated by the *ada* gene product. Since the methyl groups of 3-methyladenine, 3-methylguanine, O^2-methylcytosine, and O^2-methylthymine all protrude into the minor groove of the DNA double helix, McCarthy et al (343) have proposed that the AlkA glycosylase functions by patrolling the minor groove of the DNA double helix and removing potentially cytotoxic and mutagenic methyl groups.

The broad substrate specificity of the AlkA glycosylase differentiates it from most other DNA glycosylases which appear to be highly specific, each recognizing a single base lesion (2). In fact, *E. coli* codes for a second glycosylase (3-methyladenine-DNA glycosylase I), the *tag* gene product, that is only able to release 3-methyladenine (2, 338, 339, 346–348). The *tag* gene product accounts for 90–95% of the 3-methyladenine-DNA glycosylase activity in uninduced cells but, unlike the AlkA glycosylase, is not induced as part of the adaptive response. Increasing the dosage of the *tag* gene by cloning it on a multicopy plasmid suppresses most of the sensitivity to methyl methanesulfonate caused by an *alkA* mutation (341) suggesting that 3-methyladenine is the most important potentially lethal lesion caused by that agent. Glycolytic removal of the alkylated bases generates AP sites which are then cleaved by AP endonucleases. The excision repair process requires polymerase I and appears to play a major role in helping cells overcome the lethal effects of exposure to methylating and ethylating agents (349–352).

The second type of DNA repair activities induced as part of the adaptive response are methyltransferase functions that are able to directly remove particular methyl and ethyl groups from DNA molecules (336, 353–357). In contrast to many types of repair, no breakage or formation of phosphodiester bonds is required for this type of processing. The first protein purified with this type of activity was an 19,000-dalton protein termed O^6-alkylguanine-DNA alkyltransferase (2, 358–360). This protein is able to carry out a reaction in which the alkyl group of an O^6-methyl- or O^6-ethylguanine residue in DNA is transferred to a cysteine on the protein itself. This self-methylation results in a suicide inactivation of the methyltransferase so that each molecule is only able to act once. Although this repair strategy is somewhat costly to the cell, one molecule of protein being used for each methyl group removed, the repair of O^6-alkylguanine lesions is a matter of particular importance since they are premutagenic and give rise to mutations by mispairing during replication (361–364). The participation of the SOS processing system is not required for mutagenesis by O^6-methyl or O^6-ethylguanine lesions (365). Experiments using crude cell extracts have shown that O^4-methylthymine lesions are similarly repaired by a methyl transfer mechanism accompanied by the appearance of protein methyl groups (366, 367). The 19,000-dalton protein having the methyltransferase activity for O^6-alkylguanine lesions has also been shown to be capable of removing the methyl group from an O^4-methylthymine lesion and transferring it to one of its own cysteines (343). The action of the methyltransferase activity is somewhat complementary to that of the AlkA glycosylase since it is able to remove methyl groups bound to base oxygens from the major groove of the double helix (343). The identity of the gene coding for the O^6-alkylguanine/O^4-alkylthymine-DNA methyltransferase remained a mystery until recently when the 19,000-dalton protein was shown to be a derivative of the *ada* gene product which arose during purification (368).

Biochemical and Regulatory Roles of the Ada Protein

The *ada* gene product is a 39,000-dalton protein that plays multiple roles in the biochemistry and regulation of the adaptive response. This remarkable protein, which is about the size of the RecA protein, has some conceptual similarities to RecA in that (*a*) it participates mechanistically in the repair of damaged DNA and (*b*) it also acts positively to regulate the expression of a set of genes whose products are involved in various aspects of DNA repair. However the mechanisms by which Ada and RecA exert their effects are completely different.

The *ada* locus was first identified by mutations that block the induction of the adaptive response (369). Mutations (termed *adc*) causing constitutive expression of the adaptive response were then isolated (370) and shown to map at the *ada* locus (371). The *ada* gene was cloned by Sedgwick (372a) and was shown to code for 39,000-dalton protein (368, 372a, 372b). The gene adjacent to the *ada* gene that was cloned at the same time is *alkB* whose product plays some as yet undefined role in the repair of alkylated DNA (342, 373a, 373b). The observation that the levels of both the O^6-alkylguanine-DNA alkyltransferase and the AlkA glycosylase were increased by a multicopy plasmid carrying the ada^+ gene indicated that the Ada^+ protein is a positive regulator of the adaptive response (372a). This conclusion is supported by the observation that an insertion of a transposon in the *ada* locus results in a complete loss of the ability to induce the adaptive response (372b).

Teo et al (368) have recently shown that the 39,000-dalton Ada protein has not only regulatory properties but also alkyltransferase activities for the repair of O^6-methylguanine, O^4-methylthymine, and methyl phosphotriester lesions. The 19,000-dalton O^6-alkylguanine-DNA alkyltransferase that has been purified is derived from the carboxy-terminal region of the Ada protein (T. Lindahl, personal communication). Antibodies raised against the 19,000-dalton protein cross-react with the 19,000-dalton Ada protein. Furthermore, synthetic oligonucleotides corresponding to known partial amino acid sequences of the isolated enzyme were found to hybridize specifically to plasmid DNA containing the ada^+ gene. The 19,000-dalton protein apparently arises as the result of degradation or processing of the 39,000-dalton protein which occurs during purification (368). The 19,000-dalton protein and the 39,000-dalton protein appear to repair O^6-methylguanine and O^4-methylthymine lesions in DNA in an indistinguishable fashion. Furthermore, a methyltransferase activity for phosphotriesters, which is induced as part of the adaptive response (366), has been shown to reside in the Ada protein [T. McCarthy, T. Lindahl, unpublished data, quoted in ref. (368)].

The molecular mechanism by which the Ada protein exerts its regulatory effects has not yet been established. Operon fusions to the *ada* gene itself (372b) and to the *alkA* gene [(345) M. R. Volkert, D. C. Nguyen, C. Beard, submitted] are induced by methylating agents in an ada^+-dependent fashion suggesting that the Ada protein acts by increasing the level of transcription of the genes in the adaptive response network. Increasing the dosage of the ada^+ gene on a multicopy plasmid leads to an increase in the levels of both O^6-alkylguanine-DNA alkyltransferase activity and 3-methyladenine-DNA glycosylase-II (368) so that, even in the absence of exogenous DNA damage, high levels of Ada protein can exert a positively acting regulatory effect. Nevertheless, cells containing multiple copies of the *ada* gene can be further induced by exposure to a methylating agent (372a, 372b). Thus it seems likely that, in vivo, the Ada protein exerts its positive regulatory effects by becoming activated as a consequence of the exposure of the cell to a methylating or ethylating agent.

The nature of the inducing signal must be something that is specifically produced by methylating and ethylating agents since UV and many of the agents that induce the SOS response do not induce the adaptive response (334). Since the Ada protein has methyltransferase activities for the repair of three different lesions, the simplest model is that the Ada protein becomes activated by transferring a methyl group from one of these lesions to itself. Derivatives of the *ada* gene truncated at their 3'-ends that are supplied in trans cause an extremely high basal level of expression of an $ada'-lacZ^+$ operon fusion (372b). This indicates that the ability of the Ada protein to stimulate transcription resides in the amino terminus and suggests either that (*a*) the activation of the Ada protein might involve a conformation change which can be mimicked by truncating the carboxy terminus of the protein or (*b*) the activation process involves a proteolytic cleavage of the Ada protein that can be mimicked by truncating the carboxy terminus of the protein. The mechanism by which the adaptive response shuts off after induction is also unclear. However, if an activated form of Ada is required for the positively acting regulatory effects and that activated form of Ada is labile, the system would shut off when the inducing signal disappeared. A mathematical model for the mechanism of alkylation mutagenesis has recently been described (374).

The adaptive response network seems to be considerably smaller than the SOS network and presently is known to have four members: *ada, alkA, alkB* (which may be in an operon downstream of *ada*), and *aidB*. The *aidB* locus was identified by screening random Mu *dl*(Ap *lac*) insertions for *lac* operon fusions that were induced by methylating agents in an ada^+-dependent fashion (345).

CONCLUSIONS

E. coli has a major commitment to preserving the physical and informational integrity of its DNA. The products of a considerable number of genes are involved in DNA repair and complex control systems exist to regulate the expression of many of these genes. Although studies of the DNA repair systems of eukaryotic organisms such as yeast, *Drosophila,* and humans are still less advanced, it seems clear that the situation is at least as complex, and most likely more complex, than in *E. coli.* Nevertheless, Nature is a great reuser of strategies and it seems reasonable to suppose that many of the basic molecular processes that have been found to be involved in DNA repair in *E. coli* will have counterparts in higher cells. In addition, it is possible that some higher level strategies, for example the use of proteins that have dual regulatory and mechanistic roles in DNA repair, may also have counterparts. Some additional complexity will be necessitated by features of eukaryotic cells such as chromosome structure, the compartmentalization of nuclear, mitochondrial, and chloroplast DNAs, and temporal and developmental variations in DNA structure.

ACKNOWLEDGMENTS

I am deeply grateful to the members of my research group for their support and good spirits during the writing of this review. I would particularly like to thank L. Dodson, J. Kreuger, P. LeMotte, and L. Marsh for their insightful comments and M. White for her invaluable help in the preparation of the manuscript. Many people sent reprints, preprints, or shared unpublished results and I thank them all for their efforts. The comments of T. Lindahl and C. Walsh were very much appreciated. This work was supported by Public Health Service grants CA21615 from the National Cancer Institute and GM28988 from the National Institute of General Medical Sciences, and by grant NP-461A from the American Cancer Society.

Literature Cited

1. Hanawalt, P. C., Cooper, P. K., Ganesan, A. K., Smith, C. A. 1979. *Ann. Rev. Biochem.* 48:783–836
2. Lindahl, T. 1982. *Ann. Rev. Biochem.* 51:61–87
3. Witkin, E. M. 1976. *Bacteriol. Rev.* 40:869–907
4. Clark, A. J., Volkert, M. R. 1978. In *DNA Repair Mechanisms: ICN-UCLA Symp. Mol. Cell. Biol.,* ed. P. C. Hanawalt, E. C. Friedberg, C. F. Fox, 9:52–72. New York: Academic
5. Friedberg, E. C., Ehmann, U. K., Williams, J. I. 1979. *Adv. Radiat. Biol.* 8:85–174
6. Hartman, P. E. 1980. *Environ. Mutagen.* 2:3–16
7. Devoret, R. 1981. *Prog. Nucleic Acid Res. Mol. Biol.* 26:251–63
8. Grossman, L. 1981. *Arch. Biochem. Biophys.* 211:511–22
9. Hall, J. D., Mount, D. W. 1981. *Prog. Nucleic Acid Res. Mol. Biol.* 25:53–126
10. Schendel, P. F. 1981. *CRC Crit. Rev. Toxicol.* 8:311–62
11. Little, J. W., Mount, D. W. 1982. *Cell* 29:11–22
12. Ganesan, A. K., Cooper, P. K., Hanawalt, P. C., Smith, C. A. 1982. In *Progress in Mutation Research,* Vol. 4, ed. A. T. Natarajan, G. Obe, H. Altmann, pp. 313–23. Amsterdam: Elsevier
13. Haseltine, W. A. 1983. *Cell* 33:13–17
14. Walker, G. C. 1984. *Microbiol. Rev.* 48:60–93
15. Defais, M. J., Hanawalt, P. C., Sarasin, A. R. 1983. *Adv. Radiat. Biol.* 10:1–37
16. Friedberg, E. C., Bridges, B. A., eds. 1983. *Cellular Responses to DNA Damage,* Vol. 11. New York: Liss
17. Demple, B., Halbrook, J. 1983. *Nature* 304:466–68
18. Krueger, J. H., Walker, G. C. 1984. *Proc. Natl. Acad. Sci. USA* 81:1499–1503
19. Defais, M., Fauquet, P., Radman, M., Errera, M. 1971. *Virology* 43:495–503
20. Radman, M. 1974. In *Molecular and Environmental Aspects of Mutagenesis,* ed. L. Prakash, F. Sherman, M. Miller,

C. Lawrence, H. W. Tabor, pp. 128–42. Springfield, Ill: Thomas
21. Radman, M. 1975. In *Molecular Mechanisms for Repair of DNA*, Part A, ed. P. C. Hanawalt, R. B. Setlow, pp. 355–67. New York: Plenum
22. Gottesman, S. 1981. *Cell* 23:1–2
23. Echols, H. 1981. *Cell* 25:1–2
24. Witkin, E. M. 1982. *Biochimie* 64:549–55
25. Kenyon, C. J. 1983. *Trends Biochem. Sci.* 8:84–87
26. Kenyon, C. J., Walker, G. C. 1980. *Proc. Natl. Acad. Sci. USA* 77:2819–23
27. Sauer, R. T., Yocum, R. R., Doolittle, R. F., Lewis, M., Pabo, C. O. 1982. *Nature* 298:447–51
28. Little, J. W. 1983. *J. Mol. Biol.* 167:791–808
29. Radding, C. M. 1982. *Ann. Rev. Genet.* 16:405–37
30. Roberts, J. W., Roberts, C. W., Craig, N. L. 1978. *Proc. Natl. Acad. Sci. USA* 75:4714–18
31. Little, J. W., Edmiston, S. H., Pacelli, L. Z., Mount, D. W. 1980. *Proc. Natl. Acad. Sci. USA* 77:3225–29
32. Horii, T., Ogawa, T., Nakatani, T., Hase, T., Matsubara, H., Ogawa, H. 1981. *Cell* 27:515–22
33. Roberts, J. W., Devoret, R. 1983. In *Lambda II*, ed. R. W. Hendrix, J. W. Roberts, F. W. Stahl, R. A. Weisberg. New York: Cold Spring Harbor Lab.
34. Craig, N. L., Roberts, J. W. 1980. *Nature* 283:26–30
35. Craig, N. L., Roberts, J. W. 1981. *J. Biol. Chem.* 256:8039–44
36. Phizicky, E. M., Roberts, J. W. 1981. *Cell* 25:259–67
37. Quillardet, P., Huisman, O., D'Ari, R., Hofnung, M. 1982. *Proc. Natl. Acad. Sci. USA* 79:5971–75
38. D'Ari, R., Huisman, O. 1982. *Biochimie* 64:623–27
39. Little, J. W., Harper, J. E. 1979. *Proc. Natl. Acad. Sci. USA* 76:6147–51
40. Brent, R., Ptashne, M. 1980. *Proc. Natl. Acad. Sci. USA* 77:1932–36
41. Little, J. W., Mount, D. W., Yanisch-Perron, C. R. 1981. *Proc. Natl. Acad. Sci. USA* 78:4199–4203
42. Brent, R., Ptashne, M. 1981. *Proc. Natl. Acad. Sci. USA* 78:4204–8
43. Ebina, Y., Takahara, Y., Kishi, F., Nakazawa, A., Brent, R. 1983. *J. Biol. Chem.* 258:13258–69
44. Backendorf, C., Brandsma, J. A., Kartasova, T., van de Putte, P. 1983. *Nucleic Acids Res.* 11:5795–810
45. Sancar, G. B., Sancar, A., Little, J. W., Rupp, W. D. 1982. *Cell* 28:523–30
46. Brent, R. 1983. See Ref. 16, pp. 361–68.

47. Little, J. W. 1984. *Proc. Natl. Acad. Sci. USA* 81:1375–79
48. Haseltine, W. A., Gordon, L. K., Lindan, C. P., Grafstrom, R. H., Shaper, N. L., Grossman, L. 1980. *Nature* 285:240–44
49. Demple, B., Linn, S. 1980. *Nature* 287:203–7
50. Radany, E. H., Friedberg, E. C. 1980. *Nature* 286:182–85
51. Seawell, P. C., Smith, C. A., Ganesan, A. K. 1980. *J. Virol.* 35:790–97
52. Nakabeppu, Y., Sekiguchi, M. 1981. *Proc. Natl. Acad. Sci. USA* 78:2472–746
53. Boyce, R. P., Howard-Flanders, P. 1964. *Proc. Natl. Acad. Sci. USA* 51:293–300
54. Setlow, R. B., Carrier, W. L. 1964. *Proc. Natl. Acad. Sci. USA* 51:226–31
55. Howard-Flanders, P., Boyce, R. P., Theriot, L. 1964. *Genetics* 53:1119–36
56. Boyce, R. P., Howard-Flanders, P. 1964. *Z. Vererbungsl.* 95:345–50
57. Ikenaga, M., Ichikawa-Ryo, H., Kondo, S. 1975. *J. Mol. Biol.* 92:341–56
58. Cole, R. S., Levitan, D., Sinden, R. R. 1976. *J. Mol. Biol.* 103:39–59
59. Ogawa, H., Shimada, K., Tomizawa, J. 1968. *Mol. Gen. Genet.* 101:227–44
60. Siegel, E. C. 1973. *J. Bacteriol.* 113:145–60
61. Horii, Z. I., Clark, A. J. 1973. *J. Mol. Biol.* 80:327–44
62. Kushner, S. R., Shepherd, J., Edwards, G., Maples, V. F. 1978. See Ref. 4, pp. 251–54
63. Siegel, E. C. 1981. *Mol. Gen. Genet.* 184:526–30
64. Kuemmerle, N. B., Masker, W. E. 1980. *J. Bacteriol.* 142:535–46
65. Radman, M., Wagner, R. E. Jr., Glickman, B. W., Meselson, M. 1980. In *Progress in Environmental Mutagenesis*, ed. M. Alecevic, pp. 121–30. Amsterdam: Elsevier
66. Pukkila, P. J., Peterson, J., Herman, G., Modrich, P., Meselson, M. 1983. *Genetics* 104:571–82
67. Lu, A.-L., Clark, S., Modrich, P. 1983. *Proc. Natl. Acad. Sci. USA* 80:4639–43.
68. Kleckner, N. 1981. *Ann. Rev. Genet.* 15:341–404
69. Arthur, H. M., Lloyd, R. G. 1980. *Mol. Gen. Genet.* 180:185–91
70. Shimada, K., Ogawa, H., Tomizawa, J. 1968. *Mol. Gen. Genet.* 101:245–56
71. Rupp, W. D., Howard-Flanders, P. 1968. *J. Mol. Biol.* 31:291–304
72. Waldstein, E. A., Sharon, R., Ben-Ishai, R. 1974. *Proc. Natl. Acad. Sci. USA* 71:2651–54

73. Seeberg, E., Strike, P. 1976. *J. Bacteriol.* 125:787–95
74. Kato, T. 1972. *J. Bacteriol.* 112:1237–46
75. Seeberg, E., Johansen, I. 1973. *Mol. Gen. Genet.* 101:227–44
76. Seeberg, E., Rupp, W. D., Strike, P. 1980. *J. Bacteriol.* 144:97–104
77. Sharma, S., Moses, R. E. 1979. *J. Bacteriol.* 137:397–408
78. Seeberg, E., Nissen-Meyer, J., Strike, P. 1976. *Nature* 263:524–26
79. Seeberg, E. 1978. *Proc. Natl. Acad. Sci. USA* 75:2569–73
80. Seeberg, E., Steinum, A.-L. 1982. *Proc. Natl. Acad. Sci. USA* 79:988–92
81. Seeberg, E., Steinum, A.-L. 1983. See Ref. 16, pp. 39–49
82. Sancar, A., Rupp, W. D. 1979. *Biochem. Biophys. Res. Commun.* 90:123–29
83. Yoakum, G. H., Kushner, S. R., Grossman, L. 1980. *Gene* 12:243–48
84. Sancar, A., Wharton, R. P., Seltzer, S., Kacinski, B. M., Clarke, N. D., Rupp, W. D. 1981. *J. Mol. Biol.* 148:45–62
85. Brandsma, J. A., van Sluis, C. A., van de Putte, P. 1981. *J. Bacteriol.* 147:682–84
86. Brandsma, J. A., Stoorvogel, J., van Sluis, C. A., van de Putte, P. 1982. *Gene* 18:77–85
87. Yoakum, G. H., Yeung, A. T., Mattes, W. B., Grossman, L. 1982. *Proc. Natl. Acad. Sci. USA* 79:1766–70
88. Pannekoek, H., Noordermeer, I. A., van Sluis, C. A., van de Putte, P. 1978. *J. Bacteriol.* 133:884–96
89. Sancar, A., Clarke, N. D., Griswold, J., Kennedy, W. J., Rupp, W. D. 1981. *J. Mol. Biol.* 148:63–76
90. van den Berg, E., Zwetsloot, J., Noordermeer, I., Pannekoek, H., Dekker, B., et al. 1981. *Nucleic Acids Res.* 9:5623–43
91. Yeung, A. T., Mattes, W. B., Oh, E. Y., Yoakum, G. H., Grossman, L. 1983. *Proc. Natl. Acad. Sci. USA* 80:6157–61
92. Yoakum, G. H., Grossman, L. 1981. *Nature* 292:171–73
93. Sancar, A., Kacinski, B. M., Mott, D. L., Rupp, W. D. 1981. *Proc. Natl. Acad. Sci. USA* 78:5450–54
94. Sharma, S., Ohta, A., Dowhan, W., Moses, R. E. 1981. *Proc. Natl. Acad. Sci. USA* 78:6033–37
95. Auerbach, J. I., Howard-Flanders, P. 1981. *J. Bacteriol.* 146:713–17
96. van Sluis, C. A., Brandsma, J. A. 1981. In *Chromosome Damage and Repair*, ed. E. Seeberg, E. Kleppe, pp. 293–302. New York: Plenum
97. Oeda, K., Horiuchi, T., Sekiguchi, M. 1981. *Mol. Gen. Genet.* 184:191–99
98. Oeda, K., Horiuchi, T., Sekiguchi, M. 1982. *Nature* 298:98–100
99. Maples, V. F., Kushner, S. R. 1982. *Proc. Natl. Acad. Sci. USA* 79:5616–20
100. Arthur, H. M., Bramhill, D., Eastlake, P. B., Emmerson, P. T. 1982. *Gene* 19:285–95
101. Hickson, I. D., Arthur, H. M., Bramhill, D., Emmerson, P. T. 1983. *Mol. Gen. Genet.* 190:265–70
102. Taucher-Scholz, G., Hoffmann-Berling, H. 1983. *Eur. J. Biochem.* 137:573–80
103. Richet, E., Nishimura, Y., Hirota, Y., Kohiyama, M. 1983. *Mol. Gen. Genet.* 192:378–85
104. Kumura, K., Sekiguchi, M. 1984. *J. Biol. Chem.* 259:1560–65
105. Kenyon, C. J., Walker, G. C. 1981. *Nature* 289:808–10
106. Fogliano, M., Schendel, P. F. 1981. *Nature* 289:196–98
107. Schendel, P. F., Fogliano, M., Strausbaugh, L. D. 1982. *J. Bacteriol.* 150:676–85
108. Backendorf, C. M., van den Berg, E. A., Brandsma, J. A., Kartasova, T., van Sluis, C. A., van de Putte, P. 1983. See Ref. 16, pp. 161–71
109. van Sluis, C. A., Moolenaar, G. F., Backendorf, C. 1983. *EMBO J.* 2:2313–18
110. Siegel, E. C. 1983. *Mol. Gen. Genet.* 191:397–400
111. Nakayama, K., Irino, N., Nakayama, H. 1983. *Mol. Gen. Genet.* 192:391–94
112. Arthur, H. M., Eastlake, P. B. 1983. *Gene* 25:309–16
113. Pang, P. P., Walker, G. C. 1983. *J. Bacteriol.* 153:1172–79
114. Cooper, P. K., Hanawalt, P. C. 1972. *Proc. Natl. Acad. Sci. USA* 69:1156–60
115. Sedliakova, M., Slezarikova, V., Pirsel, M. 1978. *Mol. Gen. Genet.* 167:209–15
116. Sedliakova, M., Slezarikova, V., Brozmanova, J., Masek, F., Bayerova, V. 1980. *Mutat. Res.* 71:15–23
117. Silber, J. R., Achey, P. M. 1984. *Mutat. Res.* 131:1–10
118. Cooper, P. K. 1982. *Mol. Gen. Genet.* 185:189–97
119. Mount, D. W., Kosel, C., Walker, A. 1976. *Mol. Gen. Genet.* 146:37–42
120. Kato, T., Shinoura, Y. 1977. *Mol. Gen. Genet.* 156:121–31
121. Rothman, R. H., Margossian, L. J., Clark, A. J. 1979. *Mol. Gen. Genet.* 169:279–87
122. Sancar, A., Sancar, G. B., Rupp, W. D., Little, J. W., Mount, D. W. 1982. *Nature* 298:96–98

123. Salles, B., Paoletti, C., Villani, G. 1983. *Mol. Gen. Genet.* 189:175–77
124. Alazard, R. J. 1983. *Mutat. Res.* 109:155–68
125. van den Berg, E. A., Geerse, R. H., Pannekoek, H., van de Putte, P. 1983. *Nucleic Acids Res.* 11:4355–63
126. Sharma, S., Stark, T., Moses, R. E. 1984. *Nucleic Acids Res.* 12:5341–54
127. Easton, A. M., Kushner, S. R. 1983. *Nucleic Acids Res.* 11:8625–40
128. Finch, P., Emmerson, P. T. 1983. *Gene* 25:317–23
129. Rupp, W. D., Sancar, A., Sancar, G. B. 1982. *Biochimie* 64:595–98
130. Sancar, A., Rupp, W. D. 1983. *Cell* 33:249–60
131. Yeung, A. T., Mattes, W. B., Oh, E. Y., Grossman, L. 1983. See Ref. 16, pp. 77–86
132. Franklin, W. A., Haseltine, W. A. 1984. *Proc. Natl. Acad. Sci. USA* 81:3821–24
133. Franklin, W. A., Lo, K. M., Haseltine, W. A. 1982. *J. Biol. Chem.* 257:13535–43
134. Brash, D. E., Haseltine, W. A. 1982. *Nature* 298:189–92
135. Franklin, W. A., Haseltine, W. A. 1984. *Proc. Natl. Acad. Sci. USA.* 81:3821–24
136. Fuchs, R. P. P., Seeberg, E. 1984. *EMBO J.* 3:757–60
137. Seeberg, E., Steinum, A.-L., Nordenskjold, M., Soderhall, S., Jernstrom, B. 1983. *Mutat. Res.* 112:139–45
138. Kacinski, B. M., Sancar, A., Rupp, W. D. 1981. *Nucleic Acids Res.* 9:4495–508
139. Kacinski, B. M., Rupp, W. D. 1981. *Nature* 294:480–81
140. Tang, M-S., Lieberman, M. W. 1982. *Nature* 299:646–48
141. Strike, P., Roberts, R. J. 1982. *J. Bacteriol.* 150:385–88
142. Sinzinis, B. I., Smirnov, G. B., Saenko, A. A. 1973. *Biochem. Biophys. Res. Commun.* 53:309–16
143. van Sluis, C. A., Mattern, I. E., Paterson, M. C. 1974. *Mutat. Res.* 25:273–79
144. Rothman, R. H., Clark, A. J. 1977. *Mol. Gen. Genet.* 155:267–77
145. Rothman, R. H. 1978. *J. Bacteriol.* 136:444–48
146. Ben-Ishai, R., Sharon, R. 1981. See Ref. 96, pp. 147–151
147. Kuemmerle, N. B., Ley, R. D., Masker, W. E. 1982. *Mutat. Res.* 94:285–97
148. Kuemmerle, N. B., Masker, W. E. 1983. *Nucleic Acids Res.* 11:2193–204
149. Yamamoto, K., Satake, M., Shinagawa, H., Fujiwara, Y., 1983. *Mol. Gen. Genet.* 190:511–15
150. Yamamoto, K., Fujiwara, Y., Shinagawa, H. 1983. *Mol. Gen Genet.* 192:282–84
151a. Yamamoto, K., Satake, M., Shinagawa, H. 1984. *Mutat. Res.* 131:11–18
151b. Sancar, A., Franklin, K. A., Sancar, G. B. *Proc. Natl. Acad. Sci. USA* 81:In press
152. Pettijohn, D. E., Hanawalt, P. C. 1964. *J. Mol. Biol.* 9:395–410
153. Cooper, P. K., Hanawalt, P. C. 1972. *J. Mol. Biol.* 67:1–10
154. Kuemmerle, N., Ley, R., Masker, W. 1981. *J. Bacteriol.* 147:333–39
155. Ley, R. D., Setlow, R. B. 1972. *Biophys. J.* 12:420–31
156. Ben-Ishai, R., Sharon, R. 1978. *J. Mol. Biol.* 120:423–32
157. Masker, W. E. 1977. *J. Bacteriol.* 129:1415–23
158. Kelly, R. B., Atkinson, M. R., Huberman, J. A., Kornberg, A. 1969. *Nature* 224:495–501
159. Matson, S. W., Bambara, R. A. 1981. *J. Bacteriol.* 146:275–84
160. Youngs, D. A., van der Schuren, E., Smith, K. C. 1974. *J. Bacteriol.* 117:717–25
161. Cooper, P. K. 1981. See Ref. 96, pp. 139–46
162. Cooper, P. K., Ganesan, A. K. 1984. *Photochem. Photobiol.* 39(Suppl):82
163. Ganesan, A. K., Cooper, P. K., Hanawalt, P. C. 1984. See Ref. 162
164. Whittier, R. F., Chase, J. W., Masker, W. E. 1983. *Mutat. Res.* 112:275–86
165. Masker, W. E., Hanawalt, P., Shizuya, H. 1973. *Nature New Biol.* 244:242–43
166. Youngs, D. A., Smith, K. C. 1973. *Nature New Biol.* 244:240–41
167. Tait, R. C., Harris, A. L., Smith, D. W. 1974. *Proc. Natl. Acad. Sci. USA* 71:675–79
168. Goldmark, P. J., Linn, S. 1972. *J. Biol. Chem.* 247:1849–60
169. Taylor, A., Smith, G. R. 1980. *Cell* 22:447–57
170. Telander-Muskavitch, K. M., Linn, S. 1982. *J. Biol. Chem.* 257:2641–48
171. Clark, A. J., Sandler, S. J., Willis, K. D., Chu, C. C., Blanar, M. A., Lovett, S. T. 1984. *Cold Spring Harbor Symp. Quant. Biol.* 49:453–62
172. Barbour, S. D., Nagaishi, H., Templin, A., Clark, A. J. 1970. *Proc. Natl. Acad. Sci. USA* 67:128–35
173. Gillen, J. R., Willis, D. K., Clark, A. J. 1981. *J. Bacteriol.* 145:521–32
174. Kushner, S. R., Nagaishi, H., Templin, A., Clark, A. J. 1971. *Proc. Natl. Acad. Sci. USA* 68:824–27
175. Kushner, S. R., Nagaishi, H., Clark,

A. J. 1972. *Proc. Natl. Acad. Sci. USA* 69:1366–70
176. Lovett, S. T., Clark, A. J. 1984. *J. Bacteriol.* 157:190–96
177. Lloyd, R. G., Picksley, S. M., Prescott, C. 1983. *Mol. Gen. Genet.* 190:162–67
178. Lloyd, R. G., Benson, F. E., Shurvinton, C. E. 1984. *Mol. Gen. Genet.* 194:303–9
179. Armengod, M.-E. 1982. *Biochimie* 64:629–32
180. Clark, A. J. 1982. *Biochimie* 64:669–75
181. Lovett, S. T., Clark, A. J. 1983. *J. Bacteriol.* 153:1471–78
182. Shurvinton, C. E., Lloyd, R. G. 1982. *Mol. Gen. Genet.* 185:352–55
183. Clark, A. J. 1973. *Ann. Rev. Genet.* 7:67–86
184. Clark, A. J. 1980. In *Mechanistic Studies of DNA Replication and Genetic Recombination*, ed. B. Alberts, C. F. Fox, pp. 891–99. New York: Academic
185. Lloyd, R. G., Thomas, A. 1984. *Mol. Gen. Genet.* In press
186. Picksley, S. M., Lloyd, R. G., Buckman, C. 1984. *Cold Spring Harbor Symp. Quant. Biol.* 49:469–74
187. Rupp, W. D., Wilde, C. E. III., Reno, D. L., Howard-Flanders, P. 1971. *J. Mol. Biol.* 61:25–44
188. Ganesan, A. K. 1974. *J. Mol. Biol.* 87:103–19
189. Ganesan, A. K., Seawell, P. C. 1975. *Mol. Gen. Genet.* 141:189–205
190. Smith, K. C., Muen, D. H. C. 1970. *J. Mol. Biol.* 51:459–72
191. Sedgwick, S. G., Yarranton, G. T. 1982. *Mol. Gen. Genet.* 185:93–98
192. Yarranton, G. T., Sedgwick, S. G. 1982. *Mol. Gen. Genet.* 185:99–104
193. Sedgwick, S. G., Yarranton, G. T. 1983. See Ref. 16, pp. 437–45
194. Ogawa, T., Wabiko, H., Tsurimoto, T., Horii, T., Masukata, H., Ogawa, H. 1978. *Cold Spring Harbor Symp. Quant. Biol.* 43:909–15
195. Yancey, S. D., Porter, R. D. 1984. *Mol. Gen. Genet.* 193:53–57
196. West, S. C., Cassuto, E., Howard-Flanders, P. 1981. *Nature* 294:659–62
197. West, S. C., Cassuto, E., Howard-Flanders, P. 1982. *Mol. Gen. Genet.* 187:209–17
198. Livneh, Z., Lehman, I. R. 1982. *Proc. Natl. Acad. Sci. USA* 79:3171–75
199. Otsuji, N., Iyehara, H., Hideshima, Y. 1974. *J. Bacteriol.* 117:337–44
200. Shurvinton, C. E., Lloyd, R. G., Benson, F. E., Attfield, P. V. 1984. *Mol. Gen. Genet.* 194:322–29
201. Quillardet, P., Moreau, P. L., Ginsburg,

H., Mount, D. W., Devoret, R. 1982. *Mol. Gen. Genet.* 188:37–43
202. Clark, A. J., Volkert, M. R., Margossian, L. J., Nagaishi, H. 1982. *Mutat. Res.* 106:11–26
203. Kato, T. 1977. *Mol. Gen. Genet.* 156:115–20
204. Rothman, R. H., Clark, A. J. 1977. *Mol. Gen. Genet.* 155:279–86
205. Wang, T.-C. V., Smith, K. C. 1983. *J. Bacteriol.* 156:1093–98
206. Wang, T.-C. V., Smith, K. C. 1984. *J. Bacteriol.* 158:727–29
207. Armengod, M. E., Blanco, M. 1978. *Mutat. Res.* 52:37–47
208. Karu, A. E., Belk, E. D. 1982. *Mol. Gen. Genet.* 185:275–82
209. Salles, B., Paoletti, C. 1983. *Proc. Natl. Acad. Sci. USA* 80:65–69
210. Smith, C. L. 1983. *Proc. Natl. Acad. Sci. USA* 80:2510–13
211. Volkert, M. R., Margossian, L. J., Clark, A. J. 1984. *J. Bacteriol.* 160:702–5
212. Ream, L. W., Margossian, L., Clark, A. J., Hansen, F. G., von Meyenburg, K. 1980. *Mol. Gen. Genet.* 180:115–21
213. Blanar, M. A., Sandler, S. J., Armengod, M.-E., Ream, L. W., Clark, A. J. 1984. *Proc. Natl. Acad. Sci. USA.* 81:4622–26
214. Volkert, M. R., Hartke, M. A. 1984. *J. Bacteriol.* 157:498–506
215. Clark, A. J., Volkert, M. R., Margossian, L. J. 1978. *Cold Spring Harbor Symp. Quant. Biol.* 43:887–92
216. Ross, P., Howard-Flanders, P. 1977. *J. Mol. Biol.* 117:137–58
217. Cassuto, E., Mursalim, J., Howard-Flanders, P. 1978. *Proc. Natl. Acad. Sci. USA* 75:620–24
218. Hutchinson, F. 1978. See Ref. 4, pp. 457–64
219. Pollard, E. C., Achey, P. M. 1975. *Biophys. J.* 15:1141–54
220. Smith, K. C., Martignoni, K. D. 1976. *Photochem. Photobiol.* 24:515–23
221. Krasin, F., Hutchinson, F. 1977. *J. Mol. Biol.* 116:81–98
222. Pollard, E. C., Fugate, J. K. Jr. 1978. *Biophys. J.* 24:429–37
223. Brozmanova, J., Iljina, T. P., Saenko, A. S., Sedliakova, M. 1981. *Stud. Biophys.* 85:195–201
224. Pollard, E. C., Fluke, D. J., Kazanis, D. 1981. *Mol. Gen. Genet.* 184:421–29
225. Krasin, F., Hutchinson, F. 1981. *Proc. Natl. Acad. Sci. USA* 78:3450–53
226. Pollard, E. C., Person, S., Rader, M., Fluke, D. J. 1977. *Radiat. Res.* 72:519–32
227. Szostak, J. W., Orr-Weaver, T. L.,

Rothstein, R. J., Stahl, F. W. 1983. *Cell* 33:25–35

228. Picksley, S. M., Attfield, P. V., Lloyd, R. G. 1984. *Mol. Gen. Genet.* 195:267–74

229. West, S. C., Howard-Flanders, P. 1984. *Cell* 37:683–91

230. Wang, T.-C., Smith, K. C. 1981. *Mol. Gen. Genet.* 183:37–44

231. Weigle, J. J. 1953. *Proc. Natl. Acad. Sci. USA* 39:628–36

232. Bleichrodt, J. F., Verheij, W. S. D. 1974. *Mol. Gen. Genet.* 135:19–27

233. Yatagai, F., Kitayama, S., Matsuyama, A. 1981. *Mutat. Res.* 91:3–7

234. Defais, M. 1983. *Mol. Gen. Genet.* 192:509–11

235. Steinborn, G. 1978. *Mol. Gen. Genet.* 165:87–93

236. Elledge, S. J., Walker, G. C. 1983. *J. Mol. Biol.* 164:175–92

237. Shinagawa, H., Kato, T., Ise, T., Makino, K., Nakata, A. 1983. *Gene* 23:167–74

238. Walker, G. C., Dobson, P. P. 1979. *Mol. Gen. Genet.* 172:17–24

239. Eisenstadt, E., Wolf, M., Goldberg, I. H. 1980. *J. Bacteriol.* 144:656–60

240. Schendel, P. F., Defais, M. 1980. *Mol. Gen. Genet.* 177:661–65

241. Kato, T., Ise, T., Shinagawa, H. 1982. *Biochimie* 64:731–33

242. Bagg, A., Kenyon, C. J., Walker, G. C. 1981. *Proc. Natl. Acad. Sci. USA* 78:5749–53

243. Howarth, S. 1965. *J. Gen. Microbiol.* 40:43–55

244. Howarth, S. 1966. *Mutat. Res.* 3:129–34

245. Drabble, W. T., Stocker, B. A. D. 1968. *J. Gen. Microbiol.* 53:109–23

246. MacPhee, D. G. 1973. *Mutat. Res.* 19:356–59

247. Mortelmans, K. E., Stocker, B. A. D. 1976. *J. Bacteriol.* 128:271–82

248. Babudri, N., Monti-Bragadin, C. 1977. *Mol. Gen. Genet.* 155:287–90

249. Molina, A. M., Babudri, N., Tamaro, M., Venturini, S., Monti-Bragadin, C. 1979. *FEMS Microbiol. Lett.* 5:33–37

250. Pinney, R. J. 1980. *Mutat. Res.* 72:155–59

251. Chernin, L. S., Mikoyan, V. S. 1981. *Plasmid* 6:119–40

252. Walker, G. C. 1979. *Cold Spring Harbor Symp. Quant. Biol.* 43:893–96

253. Steinborn, G. 1979. *Mol. Gen. Genet.* 175:203–8

254. Upton, C., Pinney, R. J. 1983. *Mutat. Res.* 112:261–73

255. Dowden, S. B., Glazebrook, J. A., Strike, P. 1984. *Mol. Gen. Genet.* 193:316–21

256. Kopylov, V. M., Khmel, I. A., Vorobjeva, I. P., Lipasova, V. A., Kolot, M. N. 1984. *Mol. Gen. Genet.* 193:520–24

257. Walker, G. C. 1977. *Mol. Gen. Genet.* 152:93–103

258. Walker, G. C. 1978. *J. Bacteriol.* 133:1203–11

259. Walker, G. C. 1978. *J. Bacteriol.* 135:415–21

260. Mortelmans, K. E., Stocker, B. A. D. 1979. *Mol. Gen. Genet.* 167:317–28

261. Shanabruch, W. G., Walker, G. C. 1980. *Mol. Gen. Genet.* 179:289–97

262. Perry, K. L., Walker, G. C. 1982. *Nature* 300:278–81

263. Elledge, S. J., Walker, G. C. 1983. *J. Bacteriol.* 155:1306–15

264. Winans, S. C., Walker, G. C. 1984. *J. Bacteriol.* 161:402–10

265. Langer, P. J., Walker, G. C. 1981. *Mol. Gen. Genet.* 182:268–72

266. Brown, A. M. C., Willetts, N. S. 1981. *Plasmid* 5:188–201

267. McCann, J., Spingarn, N. E., Kobori, J., Ames, B. N. 1975. *Proc. Natl. Acad. Sci. USA* 72:979–83

268. McCann, J., Choi, E., Yamasaki, E., Ames, B. N. 1975. *Proc. Natl. Acad. Sci. USA* 72:5135–39

269. McCann, J., Ames, B. N. 1976. *Proc. Natl. Acad. Sci. USA* 73:950–54

270. Goze, A., Devoret, R. 1979. *Mutat. Res.* 61:163–79

271. Brouwer, J., Adhin, M. R., van de Putte, P. 1983. *J. Bacteriol.* 156:1275–81

272. Attfield, P. V., Pinney, R. J. 1984. *Mutat. Res.* 139:101–5

273. Blanco, M., Herrera, G., Collado, P., Rebollo, J., Botella, L. M. 1982. *Biochimie* 64:633–36

274. Blanco, M. A., Levine, A., Devoret, R., 1975. In *Molecular Mechanisms for the Repair of DNA*, ed. P. C. Hanawalt, R. B. Setlow, pp. 379–82. New York: Plenum

275. Glickman, B. W., Guijt, N., Morand, P. 1977. *Mol. Gen. Genet.* 157:83–89

276. Witkin, E. M., McCall, O. J., Volkert, M. R., Wermendsen, I. E. 1982. *Mol. Gen. Genet.* 185:43–50

277a. Mount, D. W., Wertmann, K. F., Ennis, D. G., Peterson, K. R., Fisher, B. L., Lyons, G. 1983. See Ref. 16, pp. 343–52

277b. Witkin, E. M., Kogoma, T. *Proc. Natl. Acad. Sci. USA.* 1985. In press

278. Coulondre, C., Miller, J. H. 1977. *J. Mol. Biol.* 117:577–606

279. Miller, J. H., Coulondre, C., Farabaugh, P. J. 1978. *Nature* 274:770–75

280. Brouwer, J., van de Putte, P., Fichtinger-

Schepman, J., Reedijk, J. 1981. *Proc. Natl. Acad. Sci. USA* 78:7010–14
281. Fuchs, R. P. P., Schwartz, N., Duane, M. P. 1981. *Nature* 294:657–59
282. Mizusawa, H., Lee, C.-H., Kafefuda, T., McKenney, K., Shimatake, H., Rosenberg, M. 1981. *Proc. Natl. Acad. Sci. USA* 78:6817–20
283. Brandenburger, A., Godson, G. N., Radman, M., Glickman, B. W., van Sluis, C. A., Doubleday, O. P. 1981. *Nature* 294:180–82
284. LeClerc, J. E., Istock, N. L. 1982. *Nature* 297:596–98
285. Eisenstadt, E., Warren, A. J., Porter, J., Atkins, D., Miller, J. H. 1982. *Proc. Natl. Acad. Sci. USA* 79:1945–49
286. Schaaper, R. M., Glickman, B. W. 1982. *Mol. Gen. Genet.* 185:404–7
287. Todd, P. A., Glickman, B. W. 1982. *Proc. Natl. Acad. Sci. USA* 79:4123–27
288. Foster, P. L., Eisenstadt, E. 1983. *J. Bacteriol.* 153:379–83
289. Foster, P. L., Eisenstadt, E., Miller, J. H. 1983. *Proc. Natl. Acad. Sci. USA* 80:2695–98
290. Schaaper, R. M., Kunkel, T. A., Loeb, L. A. 1983. *Proc. Natl. Acad. Sci. USA* 80:487–91
291. Livneh, Z. 1983. *Proc. Natl. Acad. Sci. USA* 80:237–41
292. Shinoura, Y., Ise, T., Kato, T., Glickman, B. W. 1983. *Mutat. Res.* 111:51–59
293. Wood, R. D., Skopek, T. R., Hutchinson, F. 1984. *J. Mol. Biol.* 173:273–91
294. Kunkel, T. A. 1984. *Proc. Natl. Acad. Sci. USA* 81:1494–98
295. Miller, J. H., Low, K. B. 1984. *Cell* 37:675–82
296. Miller, J. H. 1982. *Cell* 31:5–7
297. Miller, J. H. 1983. *Ann. Rev. Genet.* 12:215–38
298. Foster, P. L., Eisenstadt, E., Cairns, J. 1982. *Nature* 299:365–67
299. Schaaper, R. M., Loeb, L. A. 1981. *Proc. Natl. Acad. Sci. USA* 78:1773–77
300. Kunkel, T. A., Shearman, C. W., Loeb, L. A. 1981. *Nature* 291:349–51
301. Schaaper, R. M., Glickman, B. W., Loeb, L. A. 1982. *Mutat. Res.* 106:1–9
302. Schaaper, R. M., Glickman, B. W., Loeb, L. A. 1982. *Cancer Res.* 42:3480–85
303. Haseltine, W. A. 1983. See Ref. 16, pp. 3–22
304. Gearhart, P. J., Bogenhagen, D. F. 1983. *Proc. Natl. Acad. Sci. USA* 80:3439–43
305. Witkin, E. M. 1974. *Proc. Natl. Acad. Sci. USA* 71:1930–34
306. Witkin, E. M. 1975. *Mol. Gen. Genet.* 142:87–103
307. George, J., Castellazzi, M., Buttin, G. 1975. *Mol. Gen. Genet.* 140:309–32
308. Ichikawa-Ryo, H., Kondo, S. 1975. *J. Mol. Biol.* 97:77–92
309. Sargentini, N. J., Smith, K. C. 1981. *Carcinogenesis* 2:863–72
310. Ciesla, Z. 1982. *Mol. Gen. Genet.* 186:298–300
311. Quillardet, P., Devoret, R. 1982. *Biochimie* 64:789–96
312. Bridges, B. A., Southworth, M. 1982. *Biochimie* 64:655–59
313. Wood, R. D., Hutchinson, F. 1984. *J. Mol. Biol.* 173:293–305
314. Caillet-Fauquet, P., Maenhaut-Michel, G., Radman, M. 1984. *EMBO J.* 3:707–12
315. Maenhaut-Michel, G., Caillet-Fauquet, P. 1984. *J. Mol. Biol.* 177:181–88
316. Radman, M., Villani, G., Boiteux, S., Defais, M., Caillet-Fauquet, P., Spadari, S. 1977. In *Origins of Human Cancer*, ed. H. Hiatt, J. D. Watson, J. A. Winsten, Book B, pp. 903–22. New York: Cold Spring Harbor Lab.
317. Villani, G., Boiteux, S., Radman, M. 1978. *Proc. Natl. Acad. Sci. USA* 75:3037–41
318. Boiteux, S., Villani, G., Spadari, S., Zambrano, F., Radman, M. 1978. See Ref. 4, pp. 73–84
319. Moore, P. D., Bose, K. K., Rabkin, S. D., Strauss, B. S. 1981. *Proc. Natl. Acad. Sci. USA* 78:110–14
320. Moore, P. D., Rabkin, S. D., Osborn, A. L., King, C. M., Strauss, B. S. 1982. *Proc. Natl. Acad. Sci. USA* 79:7166–70
321. Strauss, B., Rabkin, S., Sagher, D., Moore, P. 1982. *Biochimie* 64:829–38
322. Philippe, M., Wang, T. S. F., Hanawalt, P. C., Korn, D. 1982. *Biochimie* 64:783–88
323. Kunkel, T. A., Schaaper, R. M., Loeb, L. A. 1983. *Biochemistry* 22:2378–84
324. Sagher, D., Strauss, B. 1983. *Biochemistry* 22:4518–26
325. Rabkin, S. D., Strauss, B. S. 1984. *J. Mol. Biol.* 178:569–94
326. Kunkel, T. A., Schaaper, R. M., Beckman, R., Loeb, L. A. 1981. *J. Biol. Chem.* 256:9883–89
327. Fersht, A. R., Knill-Jones, J. W. 1983. *J. Mol. Biol.* 165:669–82
328a. Lackey, D., Krauss, S. W., Linn, S. 1982. *Proc. Natl. Acad. Sci. USA* 79:330–34
328b. Lackey, D., Krauss, S. W., Linn, S. 1985. *J. Biol. Chem.* In press
329. Bridges, B. A., Mottershead, R. P.,

Sedgwick, S. G. 1976. *Mol. Gen. Genet.* 144:53–58
330. Bridges, B. A., Mottershead, R. P. 1978. *Mol. Gen. Genet.* 162:35–41
331. Scheuermann, R., Tam, S., Burgers, P. M. J., Lu, C., Echols, H. 1983. *Proc. Natl. Acad. Sci. USA* 80:7085–89
332. Bridges, B. A., Woodgate, R. 1984. *Mol. Gen. Genet.* 196:364–66
333. Samson, L., Cairns, J. 1977. *Nature* 267:281–82
334. Jeggo, P., Defais, M., Samson, L., Schendel, P. 1977. *Mol. Gen. Genet.* 157:1–9
335. Schendel, P. F., Defais, M., Jeggo, P., Samson, L., Cairns, J. 1978. *J. Bacteriol.* 135:466–75
336. Cairns, J., Robins, P., Sedgwick, B., Talmud, P. 1981. *Prog. Nucleic Acids Res. Mol. Biol.* 26:237–44
337. Lindahl, T., Rydberg, B., Hjelmgren, T., Olsson, M., Jacobsson, A. 1982. In *Molecular and Cellular Mechanisms of Mutagenesis,* ed. J. F. Lemontt, W. M. Generoso, pp. 89–102. New York: Plenum
338. Karran, P., Hjelmgren, T., Lindahl, T. 1982. *Nature* 296:770–73
339. Evensen, G., Seeberg, E. 1982. *Nature* 296:773–75
340. Thomas, L., Yang, C. H., Goldthwait, D. A., 1982. *Biochemistry* 21:1162–69
341. Nakabeppu, Y., Kondo, H., Sekiguchi, M. 1984. *J. Biol. Chem.* In press
342. Yamamoto, Y., Kataoka, H., Nakabeppu, Y., Tsuzuki, T., Sekiguchi, M. 1983. See Ref. 16, pp. 271–78
343. McCarthy, T. V., Karran, P., Lindahl, T. 1984. *EMBO J.* 3:545–50
344. Nakabeppu, Y., Miyata, T., Kondo, H., Iwanaga, S., Sekiguchi, M. 1984. *J. Biol. Chem.* In press
345. Volkert, M. R., Nguyen, D. C. 1984. *Proc. Natl. Acad. Sci. USA* 81:4110–14
346. Lindahl, T. 1976. *Nature* 259:64–66
347. Riazzudin, S., Lindahl, T. 1978. *Biochemistry* 17:2110–18
348. Karran, P., Lindahl, T., Ofsteng, I., Evensen, G. B., Seeberg, E. 1980. *J. Mol. Biol.* 140:101–27
349. Jeggo, P., Defais, M., Samson, L., Schendel, P. 1978. *Mol. Gen. Genet.* 162:299–305
350. Billen, D., Hellermann, G. R. 1979. *J. Bacteriol.* 137:1439–42
351. Yamamoto, Y., Sekiguchi, M. 1979. *Mol. Gen. Genet.* 171:251–59
352. Karran, P., Stevens, S., Sedgwick, B. 1982. *Mutat. Res.* 104:67–73
353. Schendel, P. F., Robins, P. E. 1978. *Proc. Natl. Acad. Sci. USA* 75:6017–20
354. Karran, P., Lindahl, T., Griffin, B. E. 1979. *Nature* 280:76–77
355. Cairns, J. 1980. *Nature* 286:176–78
356. Foote, R. S., Mitra, S., Pal, B. C. 1980. *Biochem. Biophys. Res. Commun.* 97:654–59
357. Olsson, M., Lindahl, T. 1980. *J. Biol. Chem.* 255:10569–71
358. Lindahl, T., Demple, B., Robins, P. 1982. *EMBO J.* 1:1359–63
359. Demple, B., Jacobsson, A., Olsson, M., Robins, P., Lindahl, T. 1982. *J. Biol. Chem.* 257:13776–80
360. Sedgwick, B., Lindahl, T. 1982. *J. Mol. Biol.* 154:169–75
361. Singer, B., Kusmierek, J. T., 1982. *Ann. Rev. Biochem.* 52:655–93
362. Dodson, L. A., Foote, R. S., Mitra, S., Masker, W. E. 1982. *Proc. Natl. Acad. Sci. USA* 79:7440–44
363. Eadie, J. S., Conrad, M., Toorchen, D., Topol, M. D. 1984. *Nature* 308:201–3
364. Loechler, E. L., Green, C. L., Essigmann, J. M. 1984. *Proc. Natl. Acad. Sci. USA* 81:6271–75
365. Todd, M. L., Schendel, P. F. 1983. *J. Bacteriol.* 156:6–12
366. McCarthy, J. G., Edington, B. V., Schendel, P. F. 1983. *Proc. Natl. Acad. Sci. USA* 80:7380–84
367. Ahmmed, Z., Laval, J. 1984. *Biochem. Biophys. Res. Commun.* 120:1–8
368. Teo, I., Sedgwick, B., Demple, B., Li, B., Lindahl, T. 1984. *EMBO J.* 3:2151–57
369. Jeggo, P. 1979. *J. Bacteriol.* 139:783–91
370. Sedgwick, B., Robins, P. 1980. *Mol. Gen. Genet.* 180:85–90
371. Sedgwick, B. 1982. *J. Bacteriol.* 150:984–88
372a. Sedgwick, B. 1983. *Mol. Gen. Genet.* 191:466–72
372b. LeMotte, P. K., Walker, G. C. 1985. *J. Bacteriol.* In press
373a. Kataoka, H., Yamamoto, Y., Sekiguchi, M. 1983. *J. Bacteriol.* 153:1301–7
373b. Kataoka, H., Sekiguchi, M. 1985. *Mol. Gen. Genet.* In press
374. Schendel, P. F., Michaeli, I. 1984. *Mutat. Res.* 125:1–14

Ann. Rev. Biochem. 1985. 54:459–77
Copyright © 1985 by Annual Reviews Inc. All rights reserved

VITAMIN K–DEPENDENT CARBOXYLASE

J. W. Suttie

Department of Biochemistry, College of Agricultural and Life Sciences, University of
Wisconsin-Madison, Madison, Wisconsin 53706

CONTENTS

PERSPECTIVES AND SUMMARY

Compounds with vitamin K activity are 2-methyl-1,4-naphthoquiones substituted at the 3-position with a polyisoprenoid. The vitamin K–dependent carboxylase is a microsomal enzyme catalyzing the posttranslational conversion of glutamyl residues in precursor proteins to γ-carboxyglutamyl (Gla) residues in mature proteins. This amino acid was first identified in prothrombin and subsequently in the other vitamin K–dependent plasma proteins, clotting Factor VII, IX, and X, and protein C, protein S, and protein Z. Gla residues are required for the calcium-dependent interaction between prothrombin and a negatively charged phospholipid surface which is essential for the activation of prothrombin to thrombin. Although less extensively studied, similar interac-

459

tions are important for the physiological action of the other vitamin K–dependent plasma proteins, and their structure and function have recently been reviewed (1). Proteins containing Gla residues were subsequently discovered in other tissues, most notably bone (2, 3), and activity of the enzyme has now been detected in numerous nonskeletal tissues and organs.

Status of research efforts in this field were last reviewed in this series in 1977 (4). At that time, studies leading up to the identification of Gla residues in prothrombin (5, 6) and the demonstration of a vitamin K–dependent carboxylase activity in rat liver microsomes (7) were reviewed. The enzyme was described as a reduced vitamin K (vitamin KH_2)– and O_2-dependent carboxylase, but specificity of the glutamyl or vitamin K binding site had not been probed, nor had hypotheses of the possible mechanism of action of the enzyme been formulated. This chapter will define current knowledge of the requirements and specificity of the enzyme, and review the data which has led to the generally accepted hypothesis that the formation of vitamin K-2,3-epoxide is an obligatory step in the carboxylation reaction. A number of recent reviews (8–13) of vitamin K action are available, and some of them contain peripheral information on vitamin K nutrition and metabolism, and properties of vitamin K–dependent proteins that will not be covered in this chapter.

SPECIES AND ORGAN DISTRIBUTION

Vitamin K–dependent carboxylase activity was first demonstrated (7) with a crude rat liver microsomal preparation, and most studies have utilized this source of enzyme. Early studies (14) demonstrated that the activity of the enzyme as assayed with a low-molecular-weight peptide substrate was increased 2–3 fold by a vitamin K deficiency or Warfarin treatment, and livers from these rats are used by most investigators. Although it has been reported (15) that the presence of endogenous precursors or decarboxylated plasma prothrombin will activate the enzyme, this observation has not been repeated by others (16). It is more likely that the increased prothrombin precursors resulting from the hypoprothrombinemic state induce the synthesis of the enzyme, alter its turnover rate, or stabilize the enzyme during preparation (17).

The activity of the enzyme has been measured in the liver of a number of species (12, 18), and in most cases an increased activity of the enzyme is observed when vitamin K action has been blocked by anticoagulant action. This appears to be less true in the case of the bovine liver preparation (19), and this source of the enzyme has been extensively used by some investigators in recent years. Significant activity has been demonstrated in human liver (20), and the enzyme from this source appears to have similar properties to that studied in other species. In only a few cases have direct comparisons of the optimal incubation conditions for preparations from different sources been made (21); and it is possible that if conditions were optimized, significantly higher activity of the enzyme than has previously been reported might be observed in some species.

Vitamin K–dependent proteins of unknown function are apparently produced in most tissues, and the enzyme has been detected in microsomal preparations from various organs. A protein containing Gla residues has been isolated from kidney (22), and the kidney microsomal preparation has been most extensively studied (23–28). Although there is a great deal of interest in the function of the vitamin K–dependent bone proteins, the difficulty of obtaining microsomes from this source has limited investigations to a detection of the activity (25–29). Activity has also been detected in spleen (25–27, 30), pancreas (25, 31), lung (25–27, 30, 32), placenta (20, 33), testes (25, 27, 30), and thyroid, thymus, cartilage, tendon, and uterus (25). The enzyme activity has also been demonstrated in cultured renal cells and fibroblasts (31) and in a number of tumor cells of various origin (34). As far as is known, there is nothing unique about the enzyme from any of these sources, and most studies of the properties of the enzyme have utilized the rat or bovine liver preparation.

ENZYME REQUIREMENTS AND SPECIFICITY

General Properties and Standard Conditions

Initial studies of the vitamin K–dependent carboxylase (7, 35–38) were carried out in washed rat liver microsomes, where the incorporation of $^{14}CO_2$ into endogenous precursor proteins was shown to require O_2, vitamin K, and $HCO_3{}^-$. The activity was stimulated by an energy source and factor(s) present in the postmicrosomal supernatant. Studies in this system demonstrated a requirement for NAD(P)H and/or a reduced pyridine nucleotide-generating system, and it was subsequently shown that the chemically reduced form of the vitamin, the hydroquinone (vitamin KH_2), could substitute for these cytoplasmic factors (36, 37, 39). Crude microsomal preparations contain several vitamin K reductase activities, and the carboxylase activity has been studied with [NADH + vitamin K] or with vitamin KH_2 as a vitamin source. The vitamin K–dependent carboxylase activity was soon solubilized (39, 40), and these preparations were shown to retain the basic requirement for reduced vitamin K and O_2 of the membrane-associated system. The elimination of ATP or the addition of an ATP analog did not inhibit the solubilized system, suggesting that the only energy needed to drive the carboxylation came from the reoxidation of reduced vitamin K. Biotin was shown (41) not to be involved in the reaction, and the available data (42) indicate that the active species in the carboxylation reaction is CO_2 rather than $HCO_3{}^-$.

The vitamin K–dependent carboxylase has usually been studied at a pH of 7.2–7.4, and the activity falls off sharply above pH 8 or below pH 7. Early studies were carried out at 37° C, but it is apparent that the enzyme is not very stable at this temperature and that much lower temperatures are more desirable. At temperatures below 20° C, extended linear rates of incorporation of $^{14}CO_2$ into exogenous peptide substrates can be observed (43). When endogenous microsomal proteins are used as substrates, the reaction is substrate limited; and at temperatures above 5–10° C the reaction is over in a few minutes. At low temperatures, linear rates of incorporation can be observed. The K_m for total $[CO_2/HCO_3{}^-]$ in the media has been reported to be on the order of 0.2–0.4 mM (42), but accurate measurements have been difficult to obtain and higher values have been observed (21, 44). Most investigators have not controlled CO_2 concentration, and most studies have probably been carried out with total CO_2 concentration of around 1 mM. There are relatively few measurements of the oxygen requirement of the enzyme available. A K_m of 10–15 μM has been reported for the in vitro conversion of microsomal precursors to active prothrombin (38) in intact microsomes, and a K_m of 45 μM has been reported (45) for the carboxylation reaction in a detergent-solubilized preparation.

Vitamin K Requirement and Specificity

The vitamin K requirement of the carboxylase appears to depend on whether intact microsomes or a detergent-solubilized system is studied. In general, the requirement appears to be increased in a detergent-solubilized preparation utilizing an exogenous peptide substrate compared to an intact microsomal preparation carboxylating endogenous precursors. The available data are difficult to interpret, as there is no indication of the extent to which the vitamin is free, lipoprotein bound, or micellar associated in the various preparations that have been used (46). Most investigators have utilized 50–500 μM vitamin KH_2, and it is likely the vitamin concentration was not initially limiting as other variables were studied. However, in most studies no attempt has been made to determine the rate at which the reduced vitamin is spontaneously oxidized during incubation, and in lengthy incubations it is likely that the reaction has been limited by decreasing concentration of reduced vitamin K.

The specificity of the enzyme for different forms of vitamin K has been studied in a number of systems. Menaquinones (MK) contain a polyunsaturated isoprenoid chain at the 3-position, and their activity has been assessed (36) utilizing the carboxylation of endogenous precursors by intact microsomes. Utilizing a fixed time point assay, MK-2 had 10 times the relative molar activity of vitamin K_1 (phylloquinone) and MK-3 had 80 times the activity. The cis isomer of vitamin K_1, the 2-desmethyl derivative of phylloquinone, MK-1, or menadione (2-methyl-1,4-naphthoquinone) had little or no activity. Inactive compounds were also assayed in the hydroquinone form to assure that it was not the specificity of a microsomal vitamin K reductase which was being assessed. In a study (38) of prothrombin production in postmitochondrial supernatants,

MK-2 through MK-7 were found to have similar activity to phylloquinone, although the lower homologs of the vitamin were relatively much better forms of the vitamin at low temperatures. A number of menaquinones have also been reported to have similar activity in detergent-solubilized microsomal preparations (47, 48). All of these studies measured $^{14}CO_2$ incorporation into the endogenous precursor proteins or production of biological prothrombin activity at a fixed time point. As the reaction may have reached completion during the incubation, the data obtained probably bear little or no relationship to rates of reaction. Responses to different menaquinones have now been reinvestigated (49) in a detergent-solubilized system utilizing a peptide substrate. When [vitamin K + NADH] was used, MK-3 was 2–4 times as active as phylloquinone or the higher MK homologs. Much of this difference was lost when the reduced form of the vitamin was used. These data suggested that the solubilized carboxylase shows little specificity for the chain length of the substituent at the 3-position of the reduced cofactor, but that generation of the hydroquinone in crude systems is dependent on a quinone reductase which is maximally active with MK-3.

Although a number of early studies demonstrated that menadione was not a substrate for the enzyme, it was subsequently shown (47) that, in the presence of dithiothreitol (DTT), menadione is derivatized at the 3-position by DTT to form an active vitamin. A number of other 3-thioethers, and 3-O-ethers of menadione have been prepared (50), and many have appreciable activity. In general, hydrophobic or noncharged hydrophilic thioethers are active, but derivatives with charged thioether side chains, such as a cysteinyl group, are inactive. Although their biological activity has been assessed (11), an extensive investigation of the activity of various ring substituted forms of vitamin K as substrates for the carboxylase has not been carried out. The limited amount of available data suggest, however, that little deviation from the native structure is allowed. Effective vitamin K activity at the active site, therefore, appears to be limited to a 2-methyl-1,4-naphthoquinone with some degree of nonpolar substitution on the 3-position.

Specificity of the Glu Binding Site

Inspection of the amino acid sequences of the Gla region of the known vitamin K–dependent clotting factors (1) fails to reveal any unique sequence that might be a recognition signal for this enzyme. The demonstration that a peptide comprising residues 5–9 of the bovine prothrombin precursor would serve as a substrate for the carboxylase (14) opened the possibility of assessing sequence requirements for substrate activity, and a large number of low-molecular-weight peptide substrates have now been synthesized (51–54). Although these low-molecular-weight peptide substrates have relatively high apparent K_m values, in the range of a few mM, they have provided some assessment of binding site specificity. Early studies demonstrated that only the first of two adjacent Glu residues in the peptide Phe-Leu-Glu-Glu-Leu (55) or Phe-Leu-Glu-Glu-Val (56) were carboxylated. More recent investigations (57, 58) have confirmed these data and have shown that a postulated second modification of the substrate (55, 56) is most likely due to proteolytic degradation of the substrate or product. Although carboxylation of the first of two adjacent Glu residues in a peptide substrate is now generally assumed, it has been directly demonstrated for only a few substrates. The available data (59) would suggest that bound endogenous precursor proteins are nearly completely carboxylated by the enzyme.

The vitamin K–dependent plasma proteins contain a number of Gla-Gla sequences (1); and, in general, peptides with Glu-Glu sequences are better substrates than those with single Glu residues. Although the enzyme will carboxylate a derivatized Glu residue (60), modifications of the residue preceding the Glu-Glu sequence of a short peptide do have an influence on substrate activity (52). Peptides containing Asp, D-Glu, Gln, or Homo-Glu residues were originally reported not to be carboxylated to any significant extent (52), but subsequent studies (61) suggested that the enzyme can form β-carboxyaspartic acid. A more detailed reinvestigation of this possibility (62) has shown that Boc-Asp-OBzl is carboxylated to about 10% the extent of Boc-Glu-OBzl, and the product has been unambiguously identified as β-carboxyaspartic acid. However, peptides containing Asp-Asp sequences are very poor substrates compared to those containing Glu residues, and in a peptide containing an Asp-Glu sequence, only the Glu residue is carboxylated. Knowledge of the Glu site specificity has been extended by the demonstration of Dubois et al (63) that it is the pro-S hydrogen at the γ-position of the glutamyl residue which is eliminated prior to carboxylation.

Most low-molecular-weight substrates have been reported to have K_m values in the range of a few mM, although a tripeptide linked to menadione has a K_m of about 50 μM (64). The basis for this interaction is not yet clear. The low apparent affinity of most peptide substrates suggests that some secondary or tertiary structure might be important in providing tight binding to the enzyme, and peptides homologous to prothrombin precursor sequence 18–23 whose structure is constrained by a disulfide loop and a Pro-residue have been synthesized (53, 54) to probe this possibility. These peptides were not good substrates, suggesting that the tertiary structure of the precursor or a spatially removed recognition sequence are more important in determining specific Glu sites in carboxylation. The peptide corresponding to prothrombin precursor residues 13–29, which has six Glu residues, has been isolated (65) from an enzymatic digest and reported to have an apparent K_m of only 1 μM, suggesting that longer peptides with either a multiple Glu-Glu sequence or a significant secondary structure would be good substrates. The decarboxylated form of the vitamin K–dependent bone protein, osteocalcin, has recently been shown (66) to be a substrate for the hepatic carboxylase. This 6800-molecular-weight protein which contains 3 Gla residues has a reported K_m of 25 μM and might be very useful in subsequent studies.

The natural substrates for the vitamin K–dependent carboxylase are high-molecular-weight proteins undergoing intracellular processing. Two forms of the prothrombin precursor have been isolated from rat liver (67, 68), and additional forms of this protein can be distinguished by isoelectric focusing or SDS-PAGE (69, 70). The structural features responsible for these variants have not been completely elucidated, but they may be to a large extent related to stages in carbohydrate processing (70). There is apparently a tight association between these precursor proteins and the membrane-associated carboxylase. At relatively low detergent concentrations, low-molecular-weight exogenous substrates are not utilized by the enzyme until the endogenous proteins are carboxylated (71–73); but, when rat liver preparations depleted of precursors are studied, carboxylation of low-molecular-weight substrates proceeds with no lag (72, 73).

In the human and bovine, anticoagulant treatment results in the excretion into the plasma of descarboxy and undercarboxylated forms of the vitamin K–dependent clotting factors (9, 10). Proteins containing Gla residues can also be decarboxylated by heating in the dry state (74–76). These preparations have been reported to have no substrate activity (15, 16) or to have significant activity only at relatively high concentrations (65, 77). Preparations that were the most undercarboxylated were the best substrates (77). Recent studies (16) indicate that purified rat liver prothrombin precursors added to an in vitro carboxylase preparation are carboxylated, but that decarboxylated vitamin K–dependent rat plasma proteins are not, suggesting that there are additional recognition sites on the material isolated from liver that are lacking in the plasma forms. It is also possible that the heat decarboxylation has introduced other alterations in the protein.

From 60–70% of the vitamin K–dependent carboxylase activity in bovine liver can be retained on a column of Sepharose-linked bovine Factor X antibody (78), and about 20% of the carboxylase activity in rat liver microsomes binds (59) to a Sepharose bound antibody directed against rat prothrombin. The interaction is presumably between the antibody associated precursor protein and the enzyme, and these solid-phase preparations will carboxylate both the bound precursors and added low-molecular-weight peptide substrates. The enzyme does not readily dissociate from these supports after carboxylation is initiated, suggesting some association between the enzyme and precursor proteins in addition to that directed to the Glu-binding site. Recent data obtained by sequencing the cDNA of vitamin K–dependent clotting factors (79, 80) makes it clear that the primary gene product of these proteins contains a short basic polypeptide sequence between the end of the signal peptide and the amino terminus of the plasma form of the protein. The intracellular prothrombin precursors also appear to contain this amino terminal extension (70), and it is possible that it may be an important enzyme recognition site.

Stimulation and Inhibition

Early studies (43) demonstrated that pyridoxal-5'-phosphate stimulated the carboxylation of exogenous peptide substrates but not endogenous proteins. Further studies of this response (81, 82) suggest that the stimulation is not due to formation of a Schiff-base with the free amino group of the substrate but that it is probably due to a modification of the enzyme. There is no data to suggest that this activation is of any physiological significance. The carboxylase can also be substantially stimulated by rather high concentrations of Mn^{2+} and to some extent other divalent cations (83). High concentrations of Mn^{2+} have been reported to significantly decrease (73) or to have no effect (21) on the apparent K_m of a peptide substrate for the enzyme. The large Mn^{2+} stimulation observed in some preparations (21) could not be explained by the alterations in K_m that have been observed (73), and it has also been suggested (84) that the efficiency of the coupling of carboxylation to epoxidation is increased by Mn^{2+}. Other studies (85) have observed a stimulation of both carboxylation and epoxidation by the addition of Mn^{2+}. Differences in the preparations used by different investigators may have accounted for these diverse results; and, at the present time, the nature of Mn^{2+} stimulation is unclear. Crude phospholipid has been added to partially purified carboxylase preparations to obtain maximal activity (86), and a specific requirement for mixed micelles of phosphatidyl choline and cholate has been clearly demonstrated (71). In some preparations, substantial stimulation by 1 M $(NH_4)_2SO_4$ (66) or by relatively high concentration of organic solvents such as dimethyl sulphoxide, ketones, or acetonitrile can be observed (84). In most preparations, the activity of the enzyme is stimulated by mM concentrations of DTT (36, 40, 43).

The vitamin K–dependent carboxylase system is inhibited by a number of vitamin K antagonists and nonspecific inhibitors. Chloro-K (2-Cl-3-phytyl-1,4-naphthoquinone), an effective in vivo antagonist of vitamin K (87), inhibits the enzyme in an apparent competitive fashion (39). The compound 2,3,5,6-tetrachloro-4-pyridinol (TCP) has anticoagulant action (88) and is an effective in vitro inhibitor of the carboxylase (37, 89). A study of the inhibitory properties of TCP and a related imidazopyridine (90) has revealed the complex nature of this effect but could not clearly describe its mechanism of action. Vitamin K–dependent carboxylation can also be blocked by mM concentrations of CN^- (91). Kinetic analysis indicates that CN^- is competitive against CO_2, but the data do not distinguish if a single or multiple sites were involved. A number of nonspecific inhibitors have also been shown to influence carboxylase activity (8, 46). These include some sulfhydryl reagents, particularly *p*-hydroxymercuribenzoate, some spin trapping agents, and high concentrations of ethanol. In an attempt to demonstrate the involvement of superoxide in the reaction mechanism, Cu^{2+} complexes have been shown to inhibit the reaction (92), and similar inhibitions have been observed with free Cu^{2+} and with hematin (93) and methemoglobin (86). More recent studies (94) suggest that the inhibition observed under these conditions is the result of a metal-catalyzed oxidation of the hydroquinone substrate to the inactive quinone.

Some specific inhibitors of the Glu binding site carboxylase have been synthesized. Most of the low-molecular-weight peptides synthesized as substrates for the carboxylase (52) have had varying degrees of substrate activity but have not been inhibitors of the reaction. The peptide Boc-Ser(OPO$_4$)-Ser(OPO$_4$)-Leu-OMe has been shown (95) to inhibit the enzyme in a manner that is apparently competitive with regard to other peptide substrates. A more potent inhibitor of this type has been produced (96) by incorporation of a L-threo-γ-methylglutamyl rather than a glutamyl residue at the third position of the peptide Phe-Leu-Glu-Glu-Leu to produce a Glu site inhibitor with a K_i of about 65 μM. A possible analog of the vitamin K binding site, 1-acetoxy-2-methyl-3-phytyl-4-hydroxynaphthalene, has been shown (97) to be an inhibitor of the carboxylase, but the nature of the inhibition was not determined. A derivative of 2-methyl-1,4-naphthoquinone with a peptide substrate linked to the 3-position has been shown (97) to be a substrate for the enzyme at low concentrations, but an inhibitor at higher concentrations. The degree of inhibition is strongly decreased as the concentration of vitamin K in the incubation is increased, but the detailed mechanism of the inhibition is not clear. A number of naphthoquinones have been reported to inhibit the vitamin K–dependent carboxylase by oxidation of the reduced vitamin (98), and this vitamin derivative may be acting in the same way.

FRACTIONATION AND PURIFICATION

The vitamin K–dependent carboxylase is a rough endoplasmic reticulum protein (99–101), and purification of the activity has been difficult. It requires appreciable concentrations of detergent to solubilize the enzyme (102) and is not removed from the membrane by techniques that are able to remove many peripheral or lumenal microsomal proteins. Various detergents have been used in solubilization and fractionation procedures; and, although CHAPS has been claimed to be particularly useful (103), other detergents have been more successful in other laboratories. Solubilization and ammonium sulfate fractionation, followed by gel and hydrophobic column chromatography, have been utilized to prepare a preparation that is claimed to be about 400-fold purified from crude microsomes (93, 104). Similar purification of carboxylase activity has been achieved (105) by methods that depend largely on the interaction of adsorbents with the enzyme-bound prothrombin precursor proteins, but the yield claimed in this method was somewhat lower. A different approach has been utilized (86) to obtain a purification of this enzyme through the selective removal of contaminating proteins by sequential extraction with octyl glucoside, sodium perchlorate, and sodium cholate. The resulting pellet was treated with ammonium sulfate to remove the cholate and additional protein to yield a lipoprotein preparation which was 100–150 fold increased in specific activity over the starting microsomal preparation. Much of the apparent purification resulted from the removal of a protein inhibitor of the carboxylase which is present in the crude microsomal preparation. This preparation could be solubilized in higher detergent concentrations, but attempts at further fractionation resulted in only small increases in specific activity (106). It is difficult to assess the state of purity of the preparations that have been described, and it is likely that substantial further purification will be required to produce a homogeneous preparation. None of these procedures has resulted in a preparation with the activity desired for a number of potentially useful mechanism studies. Crude liver microsomal preparations have also been used to obtain an active acetone powder preparation (107), but no purification has been reported from this starting material.

A more recent advance has been the demonstration (78) that an active carboxylase fraction can be obtained from detergent-solubilized bovine microsomes by adsorption onto a Sephadex-bound anti–Factor X antibody. The interaction is presumably through the associated precursors of Factor X, and it has not yet been possible to remove the enzyme from the support without loss of activity. This general approach has been used (59, 105) to immobilize an active rat liver carboxylase preparation onto Sepharose-bound anti-rat prothrombin. Although it is difficult to determine what degree of purification of the enzyme has been achieved by these immunoadsorptive procedures (108), it is likely that these immunocomplexes represent the most highly purified preparations of the carboxylase available for routine use. They at least have the advantage of being less contaminated with other microsomal activities than the crude rat liver microsomal preparation normally used.

MOLECULAR ROLE OF VITAMIN K

Speculations on the molecular action of vitamin K originally centered around two possibilities: that the vitamin was a cofactor utilized to labilize the hydrogen on the γ-position of the glutamyl residue to allow CO_2 attack at this position, or that it was involved as a CO_2 carrier. Evidence obtained in the last few years strongly supports the former hypothesis. The pentapeptide Phe-Leu-Glu-Glu-Leu, tritiated at the γ-carbon of each Glu residue, was synthesized and used to demonstrate (44) that the rat liver microsomal system catalyzed a vitamin KH_2 and oxygen-dependent but CO_2 -independent release of tritium from this substrate. These data shed no light on the mechanism of hydrogen abstraction but did offer final evidence that the role of the vitamin was not to activate or carry the CO_2 required in the reaction.

Current ideas of the mechanism of the carboxylation reaction stress its relationship to the microsomal vitamin K epoxidase activity (109), which converts vitamin K hydroquinone to the 2,3-epoxide of the vitamin. Over a period of time following the discovery of this internal mono-oxygenase, it became clear (110, 111) that this activity was closely associated with the vitamin K–dependent carboxylase activity. Both activities are enriched in the rough endoplasmic reticulum (101) and tend to fractionate together (106). Both utilize the reduced vitamin as a substrate (112), are located in the same tissues (113), and have similar activity with different homologs of the vitamin (114). With the exception of low CO_2 concentrations, which limit carboxylation but have no effect on epoxidase activity (112), incubation conditions which stimulate or inhibit one reaction have the same effect on the other (111). A more direct demonstration of the close association between the two reactions was the observation (115) that increasing the concentration of a peptide substrate for the carboxylase, which increases the number of carboxylation events in the system, also increases epoxidase activity. This observation would be difficult to explain if there were not a close relationship between the two activities. It was subsequently shown (85) that at saturating concentrations of CO_2 there is an apparent equivalent stoichiometry between epoxide formation and Gla formation, but as the CO_2 concentration was lowered, a large excess of vitamin K epoxide was produced. This relationship has been reinvestigated (116) utilizing a more accurate measurement of the amount of Gla formed in the system, and the one-to-one relationship between the two products has been confirmed. The degree to which these two reactions are coupled in routine incubations is to some extent dependent on incubation conditions. In addition to its apparent interaction at a CO_2 binding site, high concentrations of CN^- will uncouple epoxidation from carboxylation (85) and lead to high epoxide to Gla ratios. It seems significant that data to support a hypothesis that carboxylation can proceed in the absence of epoxidation has not been presented.

How epoxide formation and γ-hydrogen abstraction are coupled is not clear at this time, but one possibility would be through an oxygenated intermediate that would be on the pathway of epoxide formation. The inhibition of both vitamin K epoxidase and vitamin K–dependent carboxylase activity in detergent-solubilized microsomes by the addition of glutathione peroxidase has been reported (117) and interpreted as evidence that a hydroperoxide of the vitamin, a logical intermediate in epoxide formation, was involved in the carboxylation reaction. The same study reported that t-butyl-hydroperoxide was both an inhibitor of the carboxylase and would promote a low level of carboxylation in the absence of the vitamin. It has now been shown (118) that t-butyl-hydroperoxide will rapidly oxidize the reduced vitamin in the incubation which could account for the inhibition observed. Additional reports of the ability of t-butyl-hydroperoxide to act as a weak vitamin K agonist have appeared (84, 91, 119), and this aspect needs further clarification.

If an oxygenated intermediate of the vitamin is involved in the reaction, the available data relating to its possible formation are conflicting. The direct involvement of superoxide in the carboxylation reaction has been suggested (92, 120), but the evidence to support this conclusion has recently been questioned (94). The effects of specific scavengers of various active oxygen species either added to or generated in the incubation have been studied (121), and these data suggest that superoxide, singlet oxygen, and hydroxyl radical are not involved in the reaction. Scavengers that react with vitamin K semiquinone do, however, inhibit carboxylation (37, 121, 122). More recent data (122) obtained with laser flash photolysis have demonstrated that vitamin K–dependent carboxylase preparations are able to stabilize vitamin K semiquinone and that the degree of stabilization is dependent on concentrations of a peptide substrate. These data are consistent with an attack of a semiquinone form of the vitamin on oxygen to produce an activated oxygenated species rather than a superoxide attack on the vitamin. Peroxy radicals of vitamin K have also been produced by pulse radiolysis (123), but have not been shown to be related to the enzymatic reaction.

Direct evidence for the formation or presence of an oxygenated intermediate of the vitamin bound to the enzyme is still lacking, although perturbation of the system has been reported (119) to lead to the 3-hydroxy derivative of the vitamin rather than to the epoxide. Any detailed mechanism of how such an oxygenated form could be used to drive the carboxylation is at the present time speculative. Although most investigators have suggested that hydrogen removal is an abstraction of a proton to leave a formal carbanion on the glutamyl residue, a radical-mediated sequence of events (124) has also been considered. It has now been shown (125) that the enzyme will catalyze a vitamin KH_2– and oxygen-dependent exchange of ^3H from ^3H_2O into the γ-position of a Glu residue in the substrate Boc-Glu-Glu-Leu-OMe. Exchange of ^3H with the γ-carbon hydrogen is decreased as the concentration of HCO_3^- in the media is increased. This exchange reaction has been confirmed, and it has also been demonstrated (126) that the fate of the activated Glu residue in the absence of CO_2 is to protonate rather than form an adduct with some other component of the incubation that would result in an altered Glu residue. These data place severe constraints on a radical mechanism and are consistent with the model shown in Figure 1, which indicates that the role of vitamin K is to abstract the γ-hydrogen of a Glu residue as a proton, leaving a carbanion which is attacked

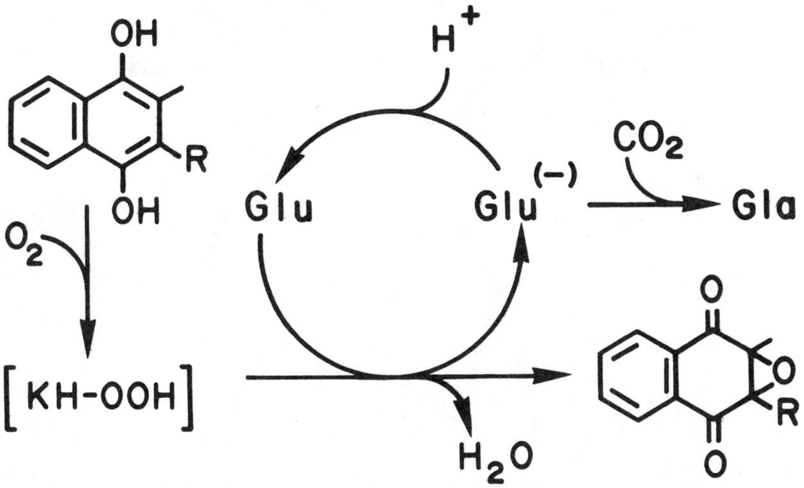

Figure 1 Proposed mechanism for the vitamin K–dependent carboxylase/epoxidase system. The available data strongly support the vitamin K–dependent formation of a carbanion followed by carboxylation in a step not involving the vitamin. Neither the chemical nature of the proposed oxygenated intermediate nor the mechanism by which hydrogen abstraction is linked to epoxide formation can be determined from the available data.

by free CO_2 to form γ-carboxyglutamic acid. Proof of this hypothesis will, however, require a much clearer understanding of the mechanism by which hydrogen abstraction is coupled to epoxide formation. Direct interaction of the vitamin with a heme center has also been considered (105), although evidence for the presence of iron in the enzyme is conflicting (93, 105). Indirect evidence (127) also suggests the lack of flavoprotein involvement.

LIVER VITAMIN K METABOLISM

If the coproduct of Gla formation is vitamin K epoxide, microsomes must contain efficient mechanisms for recycling this metabolite to the active form, the hydroquinone. The microsomal associated activities which have been identified as being involved in these interconversions of the liver vitamin K pool are shown in Figure 2 and include a vitamin K epoxide reductase and at least two and possibly other quinone reductases. A microsomal epoxide reductase activity was originally studied (128, 129) in intact microsomal preparations. The enzyme requires a dithiol rather than a reduced pyridine nucleotide for activity and is commonly studied with DTT as a reductant (130). A cytosolic protein which enhances the microsomal epoxide reductase activity has been purified (131), but the mechanism by which it stimulates activity has not been clarified. The epoxide reductase activity has been solubilized (132,

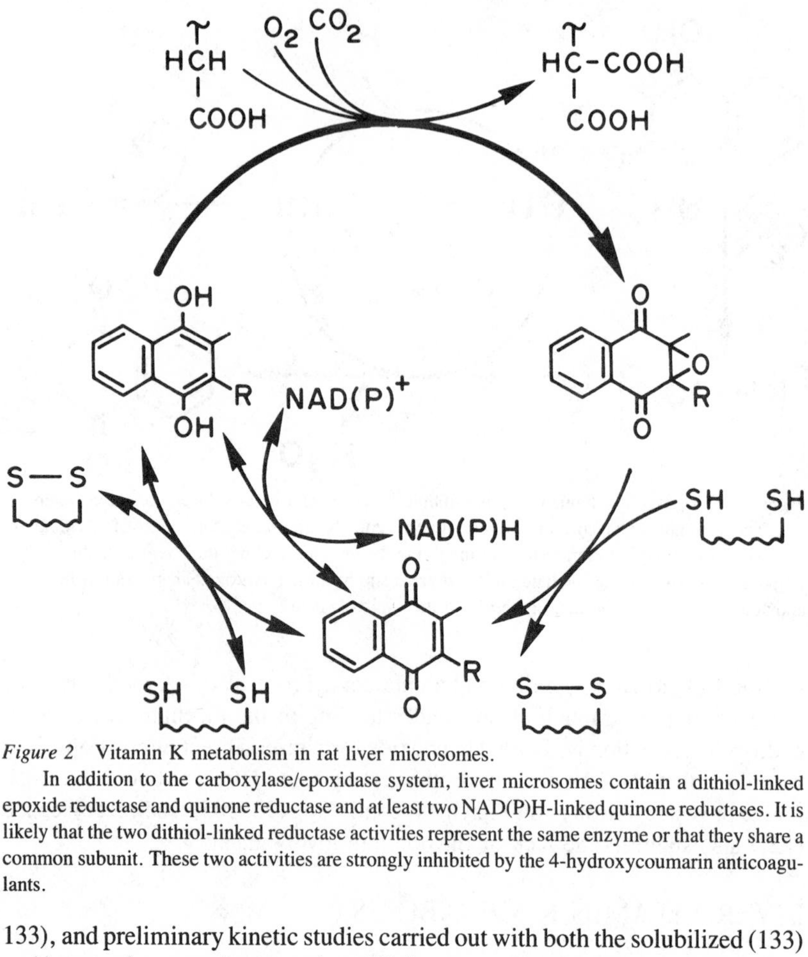

Figure 2 Vitamin K metabolism in rat liver microsomes.

In addition to the carboxylase/epoxidase system, liver microsomes contain a dithiol-linked epoxide reductase and quinone reductase and at least two NAD(P)H-linked quinone reductases. It is likely that the two dithiol-linked reductase activities represent the same enzyme or that they share a common subunit. These two activities are strongly inhibited by the 4-hydroxycoumarin anticoagulants.

133), and preliminary kinetic studies carried out with both the solubilized (133) and intact microsomal preparations (134) suggest that the reaction catalyzed by vitamin K epoxide reductase is between the epoxide and a reduced disulfide at the active site. These data and model studies (135, 136) are consistent with a mechanism where a free sulfhydryl on the enzyme attacks and opens the epoxide ring and forms a thioether adduct. It is suggested that this adduct is converted to an enzyme-bound enolate of 3-hydroxyvitamin K with the re-formation of the disulfide form of the enzyme. Elimination of H_2O from the 3-hydroxy derivative would then result in formation of the reduction product, the quinone. The hypothesis that this series of reactions represents the enzyme-catalyzed reaction has been strengthened by the observation (137) that microsomal preparations from a Warfarin-resistant strain of rats form a significant amount of hydroxyvitamin K as well as the normal quinone product.

The reactions in Figure 2 indicate that there are a number of ways in which the vitamin KH_2 needed for the carboxylase/epoxidase reaction can be generated. The relationship of the vitamin K reductase activity required for the microsomal vitamin K–dependent carboxylase system to the extensively studied DT-diaphorase of liver has been investigated by isolation of an oxidoreductase from both cytosol and microsomes that appeared to be the same as DT-diaphorase (89, 138). Detergent-solubilized-microsomal preparations lacking this activity did not demonstrate a vitamin K–dependent carboxylase activity unless purified reductase and NADH were added, suggesting that this enzyme is one of the physiologically important vitamin K activities in liver. DT-diaphorase has been shown to use vitamin K as a substrate (139), and vitamin KH_2 can be demonstrated (140) as a product of this reaction. Microsomes also contain two other NAD(P)H-linked reductase activities, cytochrome P-450 reductase, and cytochrome b_5 reductase which do not appear (141) to be involved in vitamin K metabolism. More recently it has been shown (142) that antibodies directed against DT-diaphorase will neutralize only part of the [vitamin K + NADH]-dependent carboxylase activity in detergent-solubilized microsomes which suggests that there is a second non–Warfarin sensitive NADH-linked dehydrogenase that can reduce vitamin K. Little is currently known about the properties of this enzyme.

In intact microsomes, DTT will also serve as a vitamin K reductant for the carboxylase (130), and it has been clearly demonstrated (143, 144) that intact microsomal preparations will catalyze a DTT-dependent formation of vitamin KH_2 from either vitamin K epoxide or vitamin K quinone. The kinetics of formation from vitamin K epoxide (145) have suggested that the quinone is an intermediate in this reaction.

A number of studies, which have been adequately reviewed (110), have suggested that the site of action of the widely used 4-hydroxycoumarin anticoagulants is at the epoxide reductase. Inhibition of this step would block the recycling of the liver vitamin K epoxide pool to the hydroquinone and limit action of the carboxylase. The sensitivity of the other vitamin K–dependent enzymes that might be involved in the response has recently been investigated (146), and these data indicated that the epoxide reductase was the most sensitive site and the most likely site of action of the drug. One activity which was not assayed was the DTT-dependent vitamin K quinone reductase which had been reported (145) to be Warfarin sensitive. It has now been shown (147) that this activity, like that of the epoxide reductase, is less sensitive to Warfarin when assayed in tissues of the Warfarin-resistant rat, and it is likely that the effects of 4-hydroxycoumarins involve both the reduction of vitamin K epoxide to the quinone and the reduction of the quinone to the hydroquinone.

Whether or not these two DTT-dependent activities are catalyzed by the same active site, two active sites on the same protein, or on two different proteins is not clear at the present time, and conflicting data have been presented (148, 149). The details of the interaction between Warfarin and the enzyme are still unsolved although kinetic analysis (134) has provided some information. The structural similarity between the postulated enzyme-bound hydroxyvitamin K intermediate (136, 137), and the 4-hydroxycoumarins and other anticoagulants (150) suggest that these anticoagulants are acting as an analog of this transition state. Other hypotheses have been suggested (151); and, if the site of coumarin action is not at the catalytic site but rather at a subunit involved in transferring electrons to the catalytic site, it is likely that this structural similarity is fortuitous.

CONCLUSIONS AND PROSPECTS

Studies of the microsomal vitamin K–dependent carboxylase in recent years have resulted in a proposed mechanism of action of the enzyme which is consistent with the available data but which lacks rigorous proof. The inability of investigators to purify this minor membrane protein has limited the approaches that can be used to study the mechanism and has precluded the use of information which could be gained by structural and functional group analysis of the purified protein. Unless the postulated oxygenated intermediate of the enzyme can be identified or better knowledge of the enzyme itself can be obtained, it is likely that final proof of the chemical course of this apparently unique reaction will be hard to obtain. The evidence that the vitamin K–dependent aspect of the reaction is removal of a proton to form a carbanion is fairly conclusive, and this opens the possibility that there are systems where the enzyme catalyzes a reaction other than carboxylation. At this time, none have been discovered. Studies of the associated microsomal vitamin K epoxide reductase and vitamin K quinone reductase in recent years have greatly advanced knowledge in this area. These studies have led to an appreciation of the efficiency by which this important cofactor is conserved and recycled, and have clarified the site of action of the clinically important 4-hydroxycoumarin anticoagulants.

Literature Cited

1. Jackson, C. M., Nemerson, Y. 1980. *Ann. Rev. Biochem.* 49:765–811
2. Gallop, P. M., Lian, J. B., Hauschka, P. V. 1980. *N. Engl. J. Med.* 302:1460–66
3. Price, P. 1983. In *Bone and Mineral Research*, ed. W. A. Peck, 157–90. Amsterdam: Excerpta Medica
4. Stenflo, J., Suttie, J. W. 1977. *Ann. Rev. Biochem.* 46:157–72
5. Stenflo, J., Fernlund, P., Egan, W., Roepstorff, P. 1974. *Proc. Natl. Acad. Sci. USA* 71:2730–33
6. Nelsestuen, G. L., Zytkovicz, T. H., Howard, J. B. 1974. *J. Biol. Chem.* 249:6347–50
7. Esmon, C. T., Sadowski, J. A., Suttie, J. W. 1975. *J. Biol. Chem.* 250:4744–48

8. Johnson, B. C. 1981. *Mol. Cell. Biochem.* 38:77–121
9. Liebman, H. A., Furie, B. C., Furie, B. 1982. *Hepatology* 2:488–94
10. Suttie, J. W. 1983. In *Plasma Protein Secretion by the Liver,* ed. H. Glaumann, T. Peters Jr., C. Redman, pp. 375–403. London: Academic
11. Suttie, J. W. 1984. In *Handbook of Vitamins: Nutritional, Biochemical, and Clinical Aspects,* ed. L. J. Machlin, pp. 147–98. New York: Dekker
12. Vermeer, C. 1984. *Mol. Cell. Biochem.* 61:17–35
13. Olson, R. E. 1984. *Ann. Rev. Nutr.* 4:281–337
14. Suttie, J. W., Hageman, J. M., Lehrman, S. R., Rich, D. H. 1976. *J. Biol. Chem.* 251:5827–30
15. Dubin, A., Suen, E. T., Delaney, R., Chiu, A., Johnson, B. C. 1980. *J. Biol. Chem.* 255:349–52
16. Shah, D. V., Swanson, J. C., Suttie, J. W. 1983. *Arch. Biochem. Biophys.* 222:216–21
17. Shah, D. V., Suttie, J. W. 1978. *Arch. Biochem. Biophys.* 191:571–77
18. Shah, D. V., Suttie, J. W. 1979. *Proc. Soc. Exp. Biol. Med.* 161:498–501
19. Vermeer, C., Soute, B. A. M., De Metz, M., Hemker, H. C. 1982. *Biochim. Biophys. Acta* 714:361–65
20. Soute, B. A. M., De Metz, M., Vermeer, C. 1982. *FEBS Lett.* 146:365–68
21. Uotila, L., Suttie, J. W. 1982. *Biochem. J.* 201:249–58
22. Griep, A. E., Friedman, P. A. 1980. In *Vitamin K Metabolism and Vitamin K Dependent Proteins,* ed. J. W. Suttie, pp. 307–10. Baltimore: Univ. Park Press
23. Hauschka, P. V., Friedman, P. A., Traverso, H. P., Gallop, P. M. 1976. *Biochem. Biophys. Res. Commun.* 71:1207–13
24. Buchthal, S. D., Bell, R. G. 1980. See Ref. 22, pp. 299–302
25. Vermeer, C., Hendrix, H., Daemen, M. 1982. *FEBS Lett.* 148:317–20
26. Vermeer, C., Ulrich, M. 1982. *Thromb. Res.* 28:171–77
27. Buchthal, S. D., Bell, R. G. 1983. *Biochemistry* 22:1077–82
28. Karl, P. I., Friedman, P. A. 1983. *J. Biol. Chem.* 258:6623–26
29. Lian, J. B., Friedman, P. A. 1978. *J. Biol. Chem.* 253:6623–26
30. Roncaglioni, M. C., Soute, B. A. M., de Boer-van den Berg, M. A. G., Vermeer, C. 1984. *Biochem. Biophys. Res. Commun.* 114:991–97
31. Traverso, H. P., Hauschka, P. V., Gallop, P. V. 1980. See Ref. 22, pp. 311–14
32. Bell, R. G. 1980. *Arch. Biochem. Biophys.* 203:58–64
33. Friedman, P. A., Hauschka, P. V., Shia, M. A., Wallace, J. K. 1979. *Biochim. Biophys. Acta* 583:261–65
34. Buchthal, S. D., McAllister, C. G., Laux, D. C., Bell, R. G. 1982. *Biochem. Biophys. Res. Commun.* 109:55–62
35. Sadowski, J. A., Esmon, C. T., Suttie, J. W. 1976. *J. Biol. Chem.* 251:2770–75
36. Friedman, P. A., Shia, M. 1976. *Biochem. Biophys. Res. Commun.* 70:647–54
37. Girardot, J.-M., Mack, D. O., Floyd, R. A., Johnson, B. C. 1976. *Biochem. Biophys. Res. Commun.* 70:655–62
38. Jones, J. P., Fausto, A., Houser, R. M., Gardner, E. J., Olson, R. E. 1976. *Biochem. Biophys. Res. Commun.* 72:589–97
39. Esmon, C. T., Suttie, J. W. 1976. *J. Biol. Chem.* 251:6238–43
40. Mack, D. O., Suen, E. T., Girardot, J.-M., Miller, J. A., Delaney, R., Johnson, B. C. 1976. *J. Biol. Chem.* 251:3269–76
41. Friedman, P. A., Shia, M. A. 1977. *Biochem. J.* 163:39–43
42. Jones, J. P., Gardner, E. J., Cooper, T. G., Olson, R. E. 1977. *J. Biol. Chem.* 252:7738–42
43. Suttie, J. W., Lehrman, S. R., Geweke, L. O., Hageman, J. M., Rich, D. H. 1979. *Biochem. Biophys. Res. Commun.* 86:500–7
44. Friedman, P. A., Shia, M. A., Gallop, P. M., Griep, A. E. 1979. *Proc. Natl. Acad. Sci. USA* 76:3126–29
45. Suttie, J. W., Preusch, P. C., McTigue, J. J. 1983. In *Posttranslational Covalent Modifications of Proteins,* ed. B. C. Johnson, pp. 253–79. New York: Academic
46. Suttie, J. W. 1980. *CRC Crit. Rev. Biochem.* 8:191–223
47. Gardner, E. J. 1977. *Fed. Proc.* 36:307 (Abstr.)
48. Yen, C. S., Mack, D. O. 1980. *Proc. Soc. Exp. Biol. Med.* 165:306–8
49. Newton-Nash, D. K., Sadowski, J. A. 1983. *Fed. Proc.* 42:923 (Abstr.)
50. Mack, D. O., Wolfensberger, M., Girardot, J.-M., Miller, J. A., Johnson, B. C. 1979. *J. Biol. Chem.* 254:2656–64
51. Houser, R. M., Carey, D. J., Dus, K. M., Marshall, G. R., Olson, R. E. 1977. *FEBS Lett.* 75:226–30
52. Rich, D. H., Lehrman, S. R., Kawai, M., Goodman, H. L., Suttie, J. W. 1981. *J. Med. Chem.* 24:706–11
53. Rich, D. H., Kawai, M., Goodman, H. L., Suttie, J. W. 1981. *Int. J. Pept. Protein Res.* 18:41–51

54. Rich, D. H., Kawai, M., Goodman, H. L., Suttie, J. W. 1983. *J. Med. Chem.* 26:910–16
55. Finnan, J. L., Goodman, H. L., Suttie, J. W. 1980. See Ref. 22, pp. 480–83
56. Decottignies-Le Marechal, P., Rikong-Adie, H., Azerad, R. 1979. *Biochem. Biophys. Res. Commun.* 90:700–7
57. Burgess, A. I., Esnouf, M. P., Rose, K., Offord, R. E. 1983. *Biochem. J.* 215:75–81
58. Decottignies-Le Marechal, P., Le Marechal, P., Azerad, R. 1984. *Biochem. Biophys. Res. Commun.* 119:836–40
59. Swanson, J. C., Suttie, J. W. 1982. *Biochemistry* 21:6011–18
60. Finnan, J. L., Suttie, J. W. 1980. See Ref. 22, pp. 509–17
61. Hamilton, S. E., Tesch, D., Zerner, B. 1982. *Biochem. Biophys. Res. Commun.* 107:246–49
62. McTigue, J. J., Dhaon, M. K., Rich, D. H., Suttie, J. W. 1984. *J. Biol. Chem.* 259:4272–78
63. Dubois, J., Gaudry, M., Bory, S., Azerad, R., Marquet, A. 1983. *J. Biol. Chem.* 258:7897–99
64. Dhaon, M. K., Lehrman, S. R., Rich, D. H., Engelke, J. A., Suttie, J. W. 1985. *J. Med. Chem.* In press
65. Soute, B. A. M., Vermeer, C., De Metz, M., Hemker, H. C., Lijnen, H. R. 1981. *Biochim. Biophys. Acta* 676:101–7
66. Vermeer, C., Soute, B. A. M., Hendrix, H., de Boer-van den Berg, M. A. G. 1984. *FEBS Lett.* 165:16–20
67. Esmon, C. T., Grant, G. A., Suttie, J. W. 1975. *Biochemistry* 14:1595–1600
68. Grant, G. A., Suttie, J. W. 1976. *Biochemistry* 15:5387–93
69. Graves, C. B., Grabau, G. G., Olson, R. E., Munns, T. W. 1980. *Biochemistry* 19:266–72
70. Swanson, J. C., Suttie, J. W. 1983. *Fed. Proc.* 42:1861 (Abstr.)
71. De Metz, M., Vermeer, C., Soute, B. A. M., Hemker, H. C. 1981. *J. Biol. Chem.* 256:10843–46
72. Shah, D. V., Suttie, J. W. 1983. *Proc. Soc. Exp. Biol. Med.* 173:148–52
73. Kappel, W. K., Olson, R. E. 1984. *Arch. Biochem. Biophys.* 230:294–99
74. Poser, J. W., Price, P. A. 1979. *J. Biol. Chem.* 254:431–36
75. Tuhy, P. M., Bloom, J. W., Mann, K. G. 1979. *Biochemistry* 18:5842–48
76. Hauschka, P. V. 1979. *Biochemistry* 18:4992–99
77. Malhotra, O. P. 1983. *Biochim. Biophys. Acta* 746:81–86
78. De Metz, M., Vermeer, C., Soute, B. A. M., Van Scharrenburg, G. J. M., Slotboom, A. J., Hemker, H. C. 1981. *FEBS Lett.* 123:215–18
79. Kurachi, K., Davie, E. W. 1982. *Proc. Natl. Acad. Sci. USA* 79:6461–64
80. Degen, S. J. F., MacGillivray, R. T. A., Davie, E. W. 1983. *Biochemistry* 22:2087–97
81. Suttie, J. W., Geweke, L. O., Finnan, J. L., Lehrman, S. R., Rich, D. H. 1980. See Ref. 22, pp. 450–54
82. Dubin, A., Suen, E. T., Delaney, R., Chiu, A., Johnson, B. C. 1979. *Biochem. Biophys. Res. Commun.* 88:1024–29
83. Larson, A. E., Suttie, J. W. 1980. *FEBS Lett.* 118:95–98
84. De Metz, M., Soute, B. A. M., Hemker, H. C., Vermeer, C. 1983. *Biochem. J.* 209:719–24
85. Larson, A. E., Friedman, P. A., Suttie, J. W. 1981. *J. Biol. Chem.* 256:11032–35
86. Canfield, L. M., Sinsky, T. A., Suttie, J. W. 1980. *Arch. Biochem. Biophys.* 202:515–24
87. Lowenthal, J., MacFarlane, J. A. 1967. *J. Pharmacol. Exp. Ther.* 157:672–80
88. Marshall, F. N. 1972. *Proc. Soc. Exp. Biol. Med.* 139:223–27
89. Wallin, R., Gebhardt, O., Prydz, H. 1978. *Biochem. J.* 169:95–101
90. Friedman, P. A., Griep, A. E. 1980. *Biochemistry* 19:3381–86
91. De Metz, M., Soute, B. A. M., Hemker, H. C., Vermeer, C. 1982. *FEBS Lett.* 137:253–56
92. Esnouf, M. P., Green, M. R., Hill, H. A. O., Walter, S. J. 1979. *FEBS Lett.* 107:146–50
93. Girardot, J.-M., Kanabus-Kaminska, J. M., Johnson, B. C. 1983. See Ref. 45, pp. 205–27
94. Kanabus-Kaminska, J. M., Girardot, J.-M. 1984. *Arch. Biochem. Biophys.* 228:646–52
95. Rich, D. H., Kawai, M., Goodman, H. L., Engelke, J., Suttie, J. W. 1983. *FEBS Lett.* 152:79–82
96. Gaudry, M., Bory, S., Dubois, J., Azerad, R., Marquet, A. 1983. *Biochem. Biophys. Res. Commun.* 113:454–61
97. Chander, K. S., Gaudry, M., Marquet, A., Rikong-Adie, H., Decottignies-Le Marechal, P., Azerad, R. 1981. *Biochim. Biophys. Acta* 673:157–62
98. Johnson, B. C., Mack, D. O., Delaney, R., Wolfensberger, M. R., Esmon, C., et al. 1980. See Ref. 22, pp. 455–60
99. Helgeland, L. 1977. *Biochim. Biophys. Acta* 499:181–93

100. Wallin, R., Prydz, H. 1979. *Thromb. Haemostas.* 41:529–36
101. Carlisle, T. L., Suttie, J. W. 1980. *Biochemistry* 19:1161–67
102. Brody, T., Suttie, J. W. 1984. *Methods. Enzymol.* 107:552–63
103. Girardot, J.-M., Johnson, B. C. 1982. *Anal. Biochem.* 121:315–20
104. Girardot, J.-M. 1982. *J. Biol. Chem.* 257:15008–11
105. Olson, R. E., Hall, A. L., Lee, F. C., Kappel, W. K., Meyer, R. G., Bettger, W. J. 1983. See Ref. 45, pp. 295–316
106. Wallin, R., Suttie, J. W. 1982. *Arch. Biochem. Biophys.* 214:155–63
107. Friedman, P. A., Shia, M. A. 1980. *Biochim. Biophys. Acta* 616:362–70
108. Vermeer, C., Soute, B. A. M., De Metz, M. 1983. See Ref. 45, pp. 231–50
109. Willingham, A. K., Matschiner, J. T. 1974. *Biochem. J.* 140:435–41
110. Bell, R. G. 1978. *Fed. Proc.* 37:2599–604
111. Suttie, J. W., Larson, A. E., Canfield, L. M., Carlisle, T. L. 1978. *Fed. Proc.* 37:2605–9
112. Sadowski, J. A., Schnoes, H. K., Suttie, J. W. 1977. *Biochemistry* 16:3856–63
113. Friedman, P. A., Smith, M. W. 1977. *Biochem. Pharmacol.* 26:804–5
114. Friedman, P. A., Smith, M. W. 1979. *Biochem. Pharmacol.* 28:937–38
115. Suttie, J. W., Geweke, L. O., Martin, S. L., Willingham, A. K. 1980. *FEBS Lett.* 109:267–70
116. Sadowski, J. A. 1983. *Fed. Proc.* 42:924
117. Larson, A. E., Suttie, J. W. 1978. *Proc. Natl. Acad. Sci. USA* 75:5413–16
118. Hall, A. L., Kloepper, R., Zee-Cheng, R. K.-Y., Chiu, Y. J. D., Lee, F. C., Olson, R. E. 1982. *Arch. Biochem. Biophys.* 214:45–50
119. De Metz, M., Soute, B. A. M., Hemker, H. C., Fokkens, R., Lugtenburg, J., Vermeer, C. 1982. *J. Biol. Chem.* 257:5326–29
120. Esnouf, M. P., Green, M. R., Hill, H. A. O., Irvine, G., Walter, S. J. 1978. *Biochem. J.* 174:345–48
121. Larson, A. E., McTigue, J. J., Suttie, J. W. 1980. See Ref. 22, pp. 413–21
122. Canfield, L. M., Ramelow, U. 1984. *Arch. Biochem. Biophys.* 230:389–99
123. Rao, P. S., Olson, R. E. 1982. *Biochem. Biophys. Res. Commun.* 105:231–35
124. Gallop, P. M., Friedman, P. A., Henson, E. 1980. See Ref. 22, pp. 408–12
125. McTigue, J. J., Suttie, J. W. 1983. *J. Biol. Chem.* 258:12129–31
126. Anton, D. L., Friedman, P. A. 1983. *J. Biol. Chem.* 258:14084–87
127. Preusch, P. C., Suttie, J. W. 1981. *J. Nutr.* 111:2087–97
128. Matschiner, J. T., Zimmerman, A., Bell, R. G. 1974. *Thromb. Diath. Haemorrh. Suppl.* 57:45–52
129. Zimmerman, A., Matschiner, J. T. 1974. *Biochem. Pharmacol.* 23:1033–40
130. Whitlon, D. S., Sadowski, J. A., Suttie, J. W. 1978. *Biochemistry* 17:1371–77
131. Siegfried, C. M. 1983. *Arch. Biochem. Biophys.* 223:129–39
132. Siegfried, C. M. 1978. *Biochem. Biophys. Res. Commun.* 83:1488–95
133. Hildebrandt, E. F., Preusch, P. C., Patterson, J. L., Suttie, J. W. 1984. *Arch. Biochem. Biophys.* 228:480–92
134. Fasco, M. J., Principe, L. M., Walsh, W. A., Friedman, P. A. 1983. *Biochemistry* 22:5655–60
135. Silverman, R. B. 1981. *J. Am. Chem. Soc.* 103:5939–41
136. Preusch, P. C., Suttie, J. W. 1983. *J. Org. Chem.* 48:3301–5
137. Fasco, M. J., Preusch, P. C., Hildebrandt, E., Suttie, J. W. 1983. *J. Biol. Chem.* 258:4372–80
138. Wallin, R. 1979. *Biochem. J.* 178:513–19
139. Martius, C., Ganser, R., Viviani, A. 1975. *FEBS Lett.* 59:13–14
140. Fasco, M. J., Principe, L. M. 1982. *Biochem. Biophys. Res. Commun.* 104:187–92
141. Wallin, R., Suttie, J. W. 1981. *Biochem. J.* 194:983–88
142. Wallin, R., Hutson, S. 1982. *J. Biol. Chem.* 257:1583–86
143. Fasco, M. J., Principe, L. M. 1980. *Biochem. Biophys. Res. Commun.* 97:1487–92
144. Sherman, P. A., Sander, E. G. 1981. *Biochem. Biophys. Res. Commun.* 103:997–1005
145. Fasco, M. J., Principe, L. M. 1982. *J. Biol. Chem.* 257:4894–901
146. Hildebrandt, E. F., Suttie, J. W. 1982. *Biochemistry* 21:2406–11
147. Fasco, M. J., Hildebrandt, E. F., Suttie, J. W. 1982. *J. Biol. Chem.* 257:11210–12
148. Lee, J. J., Fasco, M. J. 1984. *Biochemistry* 23:2246–52
149. Preusch, P. C., Suttie, J. W. 1984. *Biochim. Biophys. Acta* 798:141–43
150. Preusch, P. C., Suttie, J. W. 1984. *Arch. Biochem. Biophys.* 234:405–12
151. Silverman, R. B. 1980. *J. Am. Chem. Soc.* 102:5421–23

Ann. Rev. Biochem. 1985. 54:479–506
Copyright © 1985 by Annual Reviews Inc. All rights reserved

PROTEIN CARBOXYL METHYLTRANSFERASES: TWO DISTINCT CLASSES OF ENZYMES

Steven Clarke

Department of Chemistry and Biochemistry and the Molecular Biology Institute, University of California, Los Angeles, California 90024

CONTENTS

PERSPECTIVES AND SUMMARY

Posttranslational covalent modification can be an important element in protein function. One such modification is the S-adenosyl-L-methionine-dependent methylation of protein carboxyl groups. Enzymes that catalyze these reactions have been found in all cells and tissues examined to date. It now appears that there are at least two classes of these enzymes which play two very different physiological roles.

479

0066-4154/85/0701-0479$02.00

The first class of enzymes catalyzes the methylation of carboxylic acid residues of normal proteins. These enzymes function as part of systems to reversibly modulate the activities of methyl-accepting substrate proteins. The only example of these enzymes presently known are bacterial proteins that catalyze the formation of L-glutamyl γ-methyl esters on specific membrane chemoreceptors. Here, processing of sensory input appears to depend on the differential action of methylated and unmethylated chemoreceptors.

A second class of methyltransferases appears to recognize structurally altered carboxylic acids on a large variety of proteins. Enzymes from bovine brain and human red blood cells have been shown to form methyl esters on subpopulations of proteins that contain D-aspartyl and L-isoaspartyl residues, for example. These methylation reactions typically modify only a small fraction of a given polypeptide in intact cells and apparently do not function to regulate the activity of these proteins. The physiological role of these enzymes is still poorly understood, but it is tempting to speculate that these enzymes specifically recognize proteins that have suffered spontaneous chemical damage and that the methylation reaction may be the first step in the repair or further metabolism of these proteins. Table 1 compares the general properties of the two classes of protein carboxyl methyltransferases.

Methylation reactions that occur in a wide variety of cells and tissues result in methyl group incorporation into proteins in base-labile linkages characteristic of carboxyl methyl esters. Since both classes of methyltransferases can form such products, the identification of specific roles for these reactions will depend on recognizing which types of enzyme are active in a particular cell type. In this review, experimental criteria are suggested to distinguish between the two classes of methyltransferases and to emphasize what types of experiments are needed to better understand the function of these modification reactions.

Carboxyl methyl ester formation is a reversible process in cells. Proteins that catalyze the hydrolysis of the products of the first class of methyltransferases

Table 1 Two classes of protein methyltransferases[a]

	Class I	Class II
Residue Methylated	L-Glutamyl	D-Aspartyl
		L-Isoaspartyl
Methyl-Acceptor Specificity	Specific	Nonspecific
Stoichiometry of Methylation	Up to 4 methyl groups/poly-peptide	10^{-6} to 10^{-2} methyl groups/ polypeptide
Tissue Distribution	Chemotactic bacteria	Erythrocytes, brain
Function	Regulatory	Metabolism of altered proteins, possible repair

[a]Both classes of enzyme are currently included under E.C. 2.1.1.24, S-adenosylmethionine:protein carboxyl O-methyltransferase.

appear to be simple esterases and these enzymes have been isolated from chemotactic bacteria. On the other hand, more complicated mechanisms of demethylation and hydrolysis may occur for the products of the second class of methyltransferases.

Recent reviews on protein carboxyl methylation include those of Gagnon & Heisler (1), Springer et al (2), Borchardt (3), Paik & Kim (4), O'Dea et al (5), and Clarke & O'Connor (6).

PROTEIN METHYLATION AT L-GLUTAMYL RESIDUES

Chemotactic Bacteria

ESCHERICHIA COLI AND *SALMONELLA TYPHIMURIUM*   The role of a methylation reaction in the physiology of the closely related bacteria *E. coli* and *S. typhimurium* was first suggested by the requirement of methionine (7, 8) and S-adenosylmethionine (9, 10) for the chemotactic response of these cells. This reaction results in the formation of membrane-bound methylated proteins (11) and subsequent studies have identified three components of the methylation system. These include an S-adenosyl-L-methionine-dependent L-glutamyl protein methyltransferase, an L-glutamyl protein γ-methyl esterase, and a set of specific methyl-accepting protein substrates for these enzymes which includes membrane-bound chemoreceptors for serine and aspartate and a membrane-bound signal transduction component for ribose and galactose chemotaxis. The homology between the genes coding for these proteins in *E. coli* and *Salmonella* has been shown by interspecies complementation analysis (12, 13) as well as by direct DNA sequence analysis (see below). The exact role of this protein methylation system in chemotaxis is not yet clear but it does appear that changes in the level of methylation of these proteins, determined by the relative activity of the methyltransferase and methylesterase, modulates their signaling action and maximizes receptor function [for reviews see Springer et al. (2), Koshland (14), Boyd & Simon (15), Taylor & Panasenko (16)].

L-Glutamyl protein methyltransferases   A protein methyltransferase involved in chemotaxis was first identified in *S. typhimurium* (17). This cytosolic enzyme, the product of the *cheR* gene (17–19), catalyzes the S- adenosyl-L-methionine-dependent methylation of a class of 60,000-dalton membrane proteins from *S. typhimurium* and *E. coli* (17–20). The site of methylation was shown to be a glutamic acid residue from experiments in which radiolabeled glutamic acid γ-methyl ester was isolated from proteolytic digests of *Salmonella* membranes incubated with *Salmonella* cytosolic extracts and S-adenosyl-L-[methyl-^{14}C]methionine (21), or from similar digests of *E. coli* membranes isolated from intact cells incubated with L-[methyl-^{3}H]methionine as a precursor of S-adenosyl-L-[methyl-^{3}H]methionine (22). Although it has not been established directly, the stoichiometry of the reaction and the ease of proteolytic digestion suggest that the glutamate residue is in the usual L-configuration in these proteins (23).

This methyltransferase has been partially purified and characterized from cytosolic extracts of *S. typhimurium*. This enzyme is made up of a 31,000-dalton polypeptide chain as determined by dodecyl sulfate gel electrophoresis (13, 19). A native molecular weight of about 41,000 was determined for the enzyme in crude extracts using a measured $S_{20,w}$ of 3.5 S from sucrose velocity gradients and a value of the $D_{20,w}$ of 7.6×10^{-7} cm^{2}sec^{-1} from gel filtration chromatography (18). This methyltransferase has been purified to about 90% homogeneity and in this form exhibits a native molecular weight of about 30,000 [S. Simms, J. Stock, unpublished; cf. Ref. (19)]. The analogous enzyme from *E. coli* appears to be more unstable and/or less active than the *Salmonella* enzyme and has been less well characterized (19).

The enzyme appears to be completely specific for the methyl-accepting chemotaxis proteins. These proteins include four polypeptides of approximately 60,000 daltons that are integral proteins of the inner membrane of *E. coli* and *Salmonella* (24, 25). These proteins are generally referred to by their genetic designations: the tar protein functions as the aspartate chemoreceptor (26, 27); the tsr protein is the serine chemoreceptor (26–28); the trg protein is the binding site for the ribose and galactose periplasmic chemoreceptors (29); and the function of the tap protein is still uncertain (30, 31). Four specific methyl-accepting sites for the methyltransferase have recently been mapped on the tar polypeptide of *Salmonella typhimurium* (32) and the tsr polypeptide of *Escherichia coli* (33). These sites are depicted in Table 2. It is likely that the sites in the trg and tap proteins will be in similar positions because these proteins share extensive sequence homology, especially in the region of the methylated sites (24, 25, 34, 35). It is significant to note that the *Salmonella* methyltransferase does not catalyze the methylation of a variety of soluble proteins including ovalbumin, ribonuclease, and lysozyme (17, 18). These proteins are all methyl-accepting substrates for the eucaryotic methyltransferases characterized so far (see below).

L-glutamyl γ-methyl esterase   An enzyme capable of catalyzing the hydrolysis of membrane-bound methyl esters was found in cytoplasmic extracts of *Salmonella* and *E. coli* (36). This protein, the product of the *cheB* gene (36–38), has been purified from the soluble fraction of *S. typhimurium* and is composed of a single 37,000-dalton polypeptide with a K_m for the methylated substrate proteins estimated at 15μM (37). Competition studies indicated that the enzyme does not catalyze the hydrolysis of a variety of peptidyl and naphthyl esters (37), and this enzyme is apparently specific for the tar-, tsr-, trg-, and tap-methylated proteins. It has recently become clear, however, that the methylesterase can catalyze a second type of covalent modification of the membrane methylated proteins (39, 40). This reaction is a deamidation step in

Table 2 Methyl-accepting sequences for bacterial chemotaxis methyltransferases on the *Salmonella* tar protein (32) and on the *E. coli* tsr protein (33, 34). The position in the sequence of the first residue is indicated at the left, the methylation site is indicated by the arrow, and asterisked residues (Glu*) were originally present as glutaminyl residues.

289 ⇓⇓	
Leu Ser Ser Arg Thr Glu Gln Gln Ala Ser Ala	tar site 1
292	
Ser Ser Arg Thr Glu Glu*Gln Ala Ala Ser Leu	tsr site 1
297	
Ala Ser Ala Leu Glu Glu Thr Ala Ala Ser Met	tar site 2
299	
Ala Ala Ser Leu Glu Glu Thr Ala Ala Ser Met	tsr site 2
304,306	
Ala Ala Ser Met Glu Glu*Leu Thr Ala Thr Val	tar site 3, tsr site 3
486	
Ala Ser Leu Val Gln Glu Ser Ala Ala Ala Ala	tar site 4
488	
Ala Ala Leu Val Glu Glu Ser Ala Ala Ala Ala	tsr site 4
X Ser X X Hpl Glu X X Ala Ser Hpb	CONSENSUS (Hpl, hydrophilic;
Ala Thr	Hpb, hydrophobic)
Ala	

which a specific glutamine residue is hydrolyzed to form a glutamic acid residue which can be subsequently methylated [(32, 33), cf. Table 2]. Thus the L-glutamyl γ-methyl esterase can apparently recognize both the methyl ester and amide as substrates.

Role of the methylation reaction in receptor and transducer function   While it is clear that the level of methylation of the chemoreceptor proteins is a function of the chemoeffector concentration in the cell medium (18, 20, 41–43), it is unclear how conformational changes that may occur with chemoeffector binding or the occupancy of each methylated site are linked to changes in the flagellar rotation of the cell. It has been proposed that these proteins are arranged in the membrane with an N-terminal domain containing the chemoeffector binding site exposed to the periplasm, while a carboxyl terminal domain containing the methylation sites is exposed to the cytosol (24, 25).

Because mutants lacking methyltransferase and methylesterase activity are generally nonchemotactic, it is tempting to speculate that these proteins function in the central information processing reactions of chemotaxis. However, it is clear that many chemotaxis signals can be processed without methylation reactions (44), and chemotactic mutants that completely lack methyltransferase activity can be compensated by second mutations that abolish methylesterase function (45, 46). A model that is consistent with much of the physiological evidence presented so far is that the methylation reaction can effectively attenuate the signal generated from attractant binding or repellent dissociation at each of the methylated proteins. Furthermore, the demethylation reactions can attenuate the signal generated by repellent binding or attractant dissociation (2). Thus, the methylation and demethylation reactions may be largely responsible for the phenomenon of adaptation in which only changes in chemoeffector concentrations are effective in controlling the swimming behavior of these cells (2, 14). More generally, the methylation and demethylation reactions can be seen as regulatory controls that can either turn up or turn down chemoreceptor function. The loss of all chemotactic responses in cells lacking methyltransferase or methylesterase activity can be explained if the chemoreceptor input to the central processing system is overloaded when the methylation and demethylation reactions are out of balance. This type of function for methylation reactions is clearly analogous to functions described for protein phosphorylation reactions in which covalent modifications can modulate either the affinity for substrates or the maximal velocity of enzymatic reactions (47).

BACILLUS SUBTILUS AND OTHER CHEMOTACTIC BACTERIA   Protein methylation reactions have also been identified in the Gram positive chemotactic bacterium *B. subtilus*. Two distinct protein methyltransferases have been purified from these cells. One of these enzymes catalyzes the methylation of a group of *B. subtilus* membrane polypeptides of about 60,000 daltons whose level of methylation is regulated by the concentration of attractants and repellents in the medium (48, 49). This species, methyltransferase II, has been purified from the cytosol as a native monomer of a 30,000-dalton polypeptide (48) and the product of the enzymatic reaction has been shown to be an L-glutamyl γ-methyl ester (50). The activity is absent in at least one *B. subtilus* chemotaxis mutant (48) and appears to be functionally homologous to the *E. coli* cheR methyltransferase (51). The second enzyme, methyltransferase I (52), does not appear to be an L-glutamyl methyltransferase and its role in chemotaxis is uncertain (G. Ordal, personal communication).

A protein methylesterase has recently been purified from the cytosolic fraction of *B. subtilus* (53). This enzyme is a monomer of 41,000 daltons and it catalyzes the formation of methanol from [^{3}H]methylated *B. subtilus* membrane proteins. The K_m of this enzyme for these methylated proteins was determined to be approximately 10 nM (53).

The physiological role of this methylation system is also not established although it appears likely that it also participates in an adaptation process to sensory stimuli. It is significant to note that increases in the level of chemoattractants (such as aspartate) cause decreased levels of membrane protein methylation in *B. subtilus* (50). This is in contrast to the situation in *E. coli* (41) and *Salmonella* (18–20) where the level of methylation increases as a result of exposure to attractant.

Evidence for the participation of methylation reactions in the chemotaxis of a wide variety of bacteria has recently been reported. For example, chemoeffector-induced changes in the methylation of 55,000–82,000 dalton membrane proteins have been found in the widely divergent species *Halobacterium halobium* (54, 55), *Pseudomonas aeruginosa* (56), *Caulobacter crescentus* (57), and *Spirochaeta aurantia* (58).

Eucaryotic Cells?

The utilization of protein methylation reactions to regulate receptor action in chemotactic bacteria, as well as the need to regulate both receptor and other protein function in higher cells, has prompted efforts to discover analogous L-glutamyl protein methyltransferases in eucaryotes. While it is clear that protein carboxyl methylation does occur in these cells, no direct evidence has been found for methylation reactions at L-glutamyl residues. Although it had been suggested that a calf brain protein carboxyl methyltransferase could catalyze the methylation of a glutamyl residue at position 28 of adrenocorticotropic hormone (ACTH)(59), L-glutamic acid γ-methyl ester has not been isolated from this preparation, and more recent work suggests that the site of methylation of ACTH by the brain enzyme is likely to be instead at an L-isoaspartyl residue derived from the asparagine residue at position 25 (60). There have been no other reports of the formation of L-glutamyl γ-methyl esters in eucaryotic tissues, although an effort has been made to detect the products of putative L-glutamyl methyltransferases in human erythrocytes (23). Based on the base-lability characteristics of presently known eucaryotic protein methyl esters, the substoichiometric incorporation of methyl groups into these proteins, and the discovery of methylation sites at D-aspartyl and L-isoaspartyl residues, it seems likely that the bulk of the eucaryotic methylation reactions studied to date may not represent modification of normal L-glutamyl (or L-aspartyl) residues (6). If such reactions do occur in eucaryotic cells, they may be masked by the presence of the products of the nonspecific D-aspartyl/L-isoaspartyl methyltransferases which appear to be present in most cells (see below). Specific criteria that may be useful in identifying L-glutamyl methyltransferases in eucaryotic cells are discussed below.

PROTEIN METHYLATION AT
D-ASPARTYL/L-ISOASPARTYL RESIDUES

Erythrocytes

Mammalian red blood cells have proved to be a useful system for studying protein carboxyl methylation reactions. The simplicity of these cells, which contain a single plasma membrane, no internal organelles, and restricted macromolecular synthesis, facilitates the analysis of methylated proteins. The polypeptides of the membrane fraction are well-separated by dodecyl sulfate gel electrophoresis and most of these species have been functionally identified and characterized (61, 62). Much is known as well about the composition of the cytosolic proteins (63). Additionally, protein carboxyl methylation appears to be the major transmethylation reaction in the human erythrocyte. Methyl acceptors such as DNA and RNA are not present, little methyl group incorporation is seen in base-stable polypeptide linkages (such as at lysyl, arginyl, or histidyl residues)(64, 65), and the level of phospholipid methylation has been estimated at less than 2% of that of the methylated membrane proteins (64). Polyamine biosynthesis has not been detected in mature erythrocytes, and the major utilization of intracellular S-adenosyl-L-methionine may be in protein carboxyl methylation reactions (66). Finally, the lack of protein synthesis in these cells allows methyl groups in intact cells to be labeled with L-methionine, isotopically labeled in the methyl position, as a precursor to S-adenosyl-L-methionine without complications resulting from the incorporation of methionine into the polypeptide backbone or from the use of protein synthesis inhibitors. (S-Adenosyl-L-methionine is not permeable to most cells, including erythrocytes).

METHYLATION REACTIONS IN INTACT CELLS   Analysis by dodecyl sulfate gel electrophoresis at pH 7.4 of the methylated membrane polypeptides of human erythrocytes labeled by incubation of intact cells with L-[methyl-^{3}H]methionine has revealed at least five major carboxyl-methylated species (68). A subsequent study employing a pH 2.4 dodecyl sulfate gel system revealed the presence of other methylated polypeptides which may not have been originally detected due to hydrolysis in the more basic pH 7.4 gel system (64). While the methylated glutamyl residues on bacterial chemoreceptor proteins show a rather remarkable stability to base hydrolysis (21, 22), and these methylated proteins can be analyzed without significant loss even using the relatively alkaline Laemmli gel system [(69); cf. ref. (19)], erythrocyte and other eucaryotic methylated proteins are much less stable at alkaline pH (70, 71). In our studies we have employed a slightly modified version of the pH 2.4 dodecyl sulfate system of Fairbanks & Avruch (64, 65, 71, 72). Other low-pH denaturing gel systems employing cationic detergents such as N-cetylpyridinium chloride have also been described and include both continuous (73) and discontinuous (74, 75) buffer formulations. These latter systems may provide greater resolving power for individual polypeptide bands but probably do not stabilize protein methyl esters as well as anionic detergents, which may electrostatically repel hydroxyl ions from the detergent-protein complex [cf. (22, 76)].

From studies in which the identification of methylated polypeptides from dodecyl sulfate gels was confirmed by several independent criteria (such as by the degree of membrane attachment or by the proteolytic susceptibility), the major methylated membrane proteins in intact human erythrocytes were found to include the band 3–anion transport protein, the cytoskeletal proteins band 2.1 and 4.1, and several other as yet unidentified polypeptides (64, 65, 77). The stoichiometry of methylation of each of these proteins at steady state is very low. Even in the oldest population of human erythrocytes, which are the most heavily methylated (65, 77), less than 6 methyl groups are present per thousand band 3 polypeptides at steady state and less than 80 methyl groups are present per thousand band 2.1 polypeptides. When appropriately sensitive analytical methods are used, it appears that almost all erythrocyte membrane proteins can be methyl acceptors in intact cells (81).

The red cell cytosolic fraction also includes a large number of methyl acceptors (78, 79). Indeed, when intact human erythrocytes are incubated with L-[methyl-^{3}H]methionine and protein methyl esters with cellular half-lives of 150 min or less are quantitated, the number of cytosolic methyl esters (115,000 per cell) is actually greater than the number of membrane methyl esters (38,000 per cell) (78). Again these cytosolic methyl groups are present on a large number of polypeptide species, and the degree of methylation of each species is often very low. For example, hemoglobin is methylated to the extent of 3 methyl groups per million polypeptides, while carbonic anhydrase was found to contain 60 methyl groups per million polypeptides (78).

METHYLATION REACTIONS IN VITRO   The red blood cell protein carboxyl methylation system can be resolved into a cytosolic protein carboxyl methyltransferase (see below) and a set of cytosolic and membrane-bound methylacceptor proteins. The substoichiometric methylation of a variety of membrane proteins when incubated with methyltransferase preparations and S-adenosyl-L-methionine has been described (71, 76, 80, 81). A significant result of this work is that many proteins that are not methyl acceptors in intact cells do accept methyl groups in these broken cell preparations. For example, the transmembrane protein glycophorin is a major methyl group acceptor in both a "cracked-cell" system (71) and in membranes incubated with purified methyltransferase (80). However, no detectable methylation of this protein has been found in intact cell preparations (64). It is possible that many of the in vitro methylaccepting sites are located on the external portion of membrane proteins facing the extracellular medium and may not be accessible to the cytosolic methyltransferase in vivo (81). The physiological significance of these methylation reactions is unclear because no protein carboxyl methyltransferase activity has been detected in plasma (68, 82) and the plasma level of S-adenosyl-L-methionine is very low (83). On the other hand, some of the major methylaccepting polypeptides in vivo (such as bands 2.1 and 4.1) are poor methyl acceptors in vitro (64, 71). The methylation sites of these proteins may be saturated in intact cells since preincubation of cells in the presence of methylation inhibitors, or mild base treatment, can enhance the ability of these proteins to accept methyl groups in vitro (81). Thus, the results of in vitro methylation experiments must be interpreted cautiously.

A REGULATORY ROLE FOR PROTEIN METHYLATION? The identification of methyl-accepting proteins with known functions in intact cells allows one to test the hypothesis that the covalent modification of these proteins by methyl esterification can alter their physiological function, as in the case of the bacterial chemoreceptor proteins. The erythrocyte cytoskeletal proteins band 2.1 and 4.1 are important elements in the control of cell shape and deformability (61). Although only a small fraction of each polypeptide is methylated under steady-state conditions, these proteins represent major cellular methyl acceptors, and the methylation of these species could conceivably regulate cytoskeletal function (64). Here the modification of only a few proteins in a large network might have long-range consequences, especially since both band 2.1 and band 4.1 perform "linker" functions in maintaining the cytoskeletal network. Nevertheless, two experimental tests of this hypothesis have failed to indicate a direct role for carboxyl methylation reactions in the regulation of the cytoskeleton. In the first place, no change in cell shape, deformability, or osmotic fragility is seen when erythrocytes are incubated with methylation inhibitors under conditions where the steady-state level of methylation is estimated at less than 20% that of control cells (84). Secondly, when intact red cells are incubated under conditions where the cell shape is markedly altered, no direct change in protein methylation levels of cytoskeletal proteins is observed (85).

Because band 3 is also a major methyl acceptor, similar experiments have been done to examine a role for carboxyl methylation in the regulation of band 3–catalyzed anion transport in intact erythrocytes. Once again, incubation of erythrocytes with methylation inhibitors produced no direct change in the anion transport activity of these cells (L. L. Lou, S. Clarke, unpublished).

SITE OF METHYLATION IN ERYTHROCYTE PROTEINS   When [^{3}H]methylated erythrocyte proteins are exhaustively digested with proteolytic enzymes, aspartic acid β-[^{3}H]methyl ester can be recovered in yields of up to 15%, while the bulk of the remaining radioactivity is recovered as [^{3}H]methanol from demethylated esters (23, 78, 81, 87, 88). The assignment of the aspartic acid β-methyl ester structure to the proteolytic product was based on the behavior of this material in ion-exchange, thin-layer, and gel filtration chromatography as well as by its hydrolytic lability. The yield of aspartic acid β-methyl ester is highest when preparations of membranes from intact cells are protease-digested (23, 87, 88), although this methyl ester has also been isolated from digestions of fractions of cytosolic proteins labeled in intact cells (78), and from in vitro labeled, then digested membrane proteins (23, 81). No evidence for the presence of glutamic acid γ-methyl ester in these digests has been found (23).

If the sites of methylation were at typical L-aspartyl residues, one would expect that a homogenous protein would be capable of accepting one, two, or more methyl groups per polypeptide chain. The repeated failure to observe these stoichiometries led us to pursue the idea that the methylation site was at a modified aspartyl residue that is present in only a small subpopulation of a given protein. It had been well established that L-aspartic acid (and/or L-asparagine) residues are especially prone to spontaneous racemization reactions that result in the generation of D-aspartyl derivatives both in vitro (90–93) and in vivo (94, 95). When the stereoconfiguration of the aspartic acid β-[^{3}H]methyl ester isolated from proteolytic digests of erythrocyte membranes was determined by the peptide diastereomer method (96) and by utilizing L- and D-specific amino acid oxidases, it was found to be exclusively the uncommon D-stereoisomer (88). No evidence for the presence of L-aspartic acid β-[^{3}H]methyl ester was found. The same result was obtained for aspartic acid β-methyl ester isolated from erythrocyte cytosolic (78) and in vitro methylated membrane proteins [(81); cf. ref. (23)].

Thus it appeared that the red cell methylation system might be specific for the small subpopulation of proteins containing D-aspartyl residues. However, recent studies have shown that at least one other type of modified aspartyl residue may also be a substrate for erythrocyte methyltransferases. This modification involves the formation of an L-isoaspartyl residue in which the α-carboxyl group is free and the β-carboxyl group is linked in a peptide bond to the rest of the polypeptide chain. These isopeptides may be derived from errors in protein synthesis in which peptide bond formation occurs at the β- rather than the α-carboxyl group of aspartic acid residues (89). Alternatively, L-isoaspartyl residues may originate from the degradation of L-aspartyl and L-asparaginyl residues in proteins (Figure 1). The latter reactions probably involve succinimide intermediates (97) generated largely by deamidation of asparagine residues (98). In these studies, synthetic peptides corresponding to the sequence L-Val-L-Tyr-L-Pro-(Asp)-Gly-L-Ala were made containing the aspartyl residue either in the L- or the D- configuration and with either a normal α-peptide bond to the glycyl residue or an isopeptide bond from the side chain(β) carboxyl of the aspartyl residue to the glycyl residue. It was found that the hexapeptide containing the L-isoaspartyl residue was a stoichiometric, high-affinity, substrate for the enzyme in red cell cytosol (89). No methylation was detected with peptides containing L-aspartyl or D-isoaspartyl residues, or surprisingly, D-aspartyl residues.

RED CELL D-ASPARTYL/L-ISOASPARTYL METHYLTRANSFERASE   A protein carboxyl methyltransferase has been purified from human erythrocytes based on its ability to catalyze the transfer of methyl groups from S-adenosyl-L-

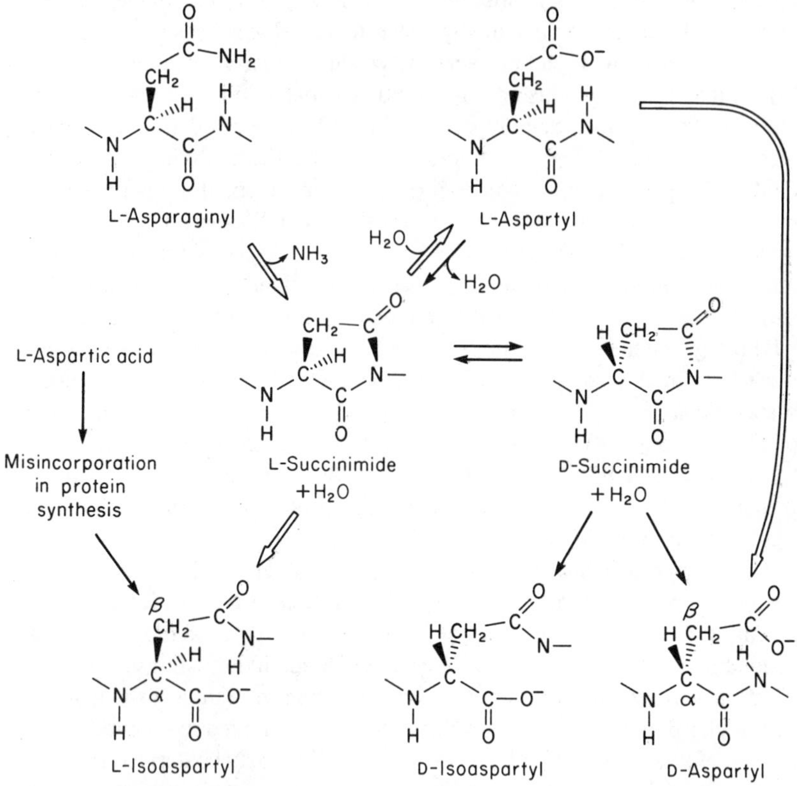

Figure 1   Chemical routes to the formation of D-aspartyl and L-isoaspartyl residues from naturally occurring L-aspartyl and L-asparaginyl residues in proteins. The heavy arrows indicate steps supported by experimental data in model systems; the light arrows indicate other feasible reactions.

methionine to γ-globulin (99). This cytosolic enzyme appears to consist of two similar monomeric 28,000-dalton isozymes which can be separated by DEAE-cellulose chromatography (I. Ota, S. Clarke, unpublished). The specificity of these enzymes for catalyzing the formation of D-aspartyl protein β-methyl esters is indicated by the fact that D-aspartic acid β-[³H]methyl ester can be isolated from erythrocyte membranes incubated with S-adenosyl-L-[methyl-³H]methionine and a partially purified enzyme preparation (81). These enzymes can also catalyze the methylation of the synthetic hexapeptide containing an L-isoaspartyl residue (89). It is possible that the same enzyme(s) can catalyze both methylation reactions because the stereochemical configuration of the free carboxyl group is similar in both modified sites (89). It is troubling, however, that the erythrocyte methyltransferase does not catalyze the methylation of the D-aspartyl residue of L-Val-L-Tyr-L-Pro-D-Asp-Gly-L-Ala, which is otherwise identical to the isoaspartyl substrate (89). It is possible that different sequences and/or three-dimensional features are required for the enzyme to catalyze each type of methylation reaction.

At this point, a working hypothesis can be made that the red cell methyltransferase(s) can catalyze the formation of both D-aspartyl β-methyl esters and L-isoaspartyl α-methyl esters. Although evidence for the former reaction has only been found in intact and in vitro labeled erythrocyte preparations, and evidence of the latter reaction only in in vitro studies with synthetic peptides, it is clear that both of these reactions represent methylation at modified L-aspartyl (or L-asparaginyl) residues. Chemical pathways to the formation of each of these modified residues are shown in Figure 1. It is still unclear whether other sites of methylation exist in erythrocytes. More work needs to be done with different synthetic peptides to definitively establish that the enzyme directly catalyzes the carboxyl methylation of D-aspartyl residues. It will also be important to find evidence for L-isoaspartyl α-methyl ester residues in intact cells, although this might be technically difficult (89).

DEMETHYLATION REACTIONS   Based on pulse-chase experiments, protein methyl esters turn over in intact erythrocytes with half times varying from 0.75 to 25 hours (65, 77, 84). However, the rate of this turnover is similar to the rate of the presumably spontaneous demethylation of isolated [^{3}H]methylated proteins (100). Based on a comparative analysis of the rates of hydrolysis of enzymatically methylated proteins and model peptide esters, it appears that a major mechanism for the demethylation reaction does not involve direct cleavage by hydroxide ion or water (71). Rather, an initial intramolecular reaction results in the demethylation of the ester and the formation of an anhydride or succinimide intermediate (Figure 2). This situation is very different from that seen in the bacterial chemotaxis methylation system where the relatively stable methyl esters are hydrolyzed by a specific enzymatic mechanism. No evidence to support a protein "esterase" activity has been found in erythrocytes to date (100).

FUNCTION OF D-ASPARTYL/L-ISOASPARTYL METHYLATION IN RED BLOOD CELLS   The physiological role of this enzymatic reaction is presently unknown. Based on the unusual nature of the substrate site and the negative results obtained in direct tests of the effect of methylation levels on cytoskeletal and anion transport activity (84, 85), it appears that a regulatory role for this protein covalent modification reaction is unlikely, at least for these proteins.

On the other hand, a role for this reaction in the metabolism of altered proteins does appear to be a reasonable possibility. Several specific roles have been postulated that assume that the methylation reaction is a recognition step that identifies already damaged proteins for one or more subsequent steps (88, 89)(Figure 2). In analogy to the ubiquitin-protein degradation system (101), this latter step may be the proteolysis of the methylated protein and would result

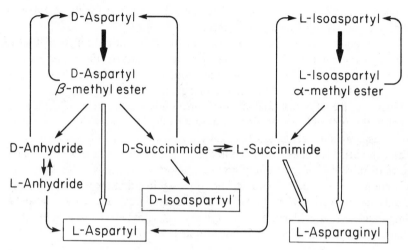

Figure 2 Possible routes of cellular metabolism of D-aspartyl β-methyl ester residues and L-isoaspartyl α-methyl ester residues. S-adenosyl-L-methionine dependent enzymatic methylation reactions are indicated by heavy filled-in arrows. Heavy open arrows indicate possible enzymatic repair reactions while the light arrows indicate nonenzymatic (or possibly enzymatic) steps. Anhydride refers a postulated 5-membered ring intermediate of aspartyl β-methyl ester demethylation. The ring contains the aspartyl α and β carbon atoms, the α carbonyl carbon and oxygen atoms, and the β carbonyl carbon atom. This intermediate may be formed in a reaction analogous to succinimide formation where the intramolecular attack is done by the α-carbonyl oxygen atom rather than the adjacent nitrogen atom (cf. Figure 1).

in the removal of the altered protein (65, 88). Although erythrocytes have been shown to rapidly degrade proteins damaged by oxidation (102), their overall rate of proteolysis in vivo is very low (103) and the normal rate of protein loss is much less than the rate of protein methylation.

A second possible fate of the protein methyl ester is to go though a reaction in which the hydrolysis of the methyl ester is coupled to the "repair" of the originally damaged aspartyl residue (88). For example, D-aspartic acid β-methyl ester residues can be converted to L-aspartyl residues via a coupled epimerization/hydrolysis reaction. Alternatively, L-isoaspartyl α-methyl ester residues can be converted to L-asparagine residues by a coupled transpeptidation/amidation reaction. Although these latter reactions may be enzymatically catalyzed, it is also possible to achieve partial repair by series of nonenzymatic steps [Figure 2; cf. ref. (88, 89)]. Although enzymatic repair pathways are well established for DNA molecules (104), less is known about the possibility of reversing spontaneous damage to proteins. However, widely distributed enzymes can reduce oxidized methionine residues in proteins (105). One can speculate that other enzymatically catalyzed protein modification reactions, such as the possible phosphorylation of D-serine, threonine, or tyrosine residues, may also be involved in protein repair reactions (88).

To test these specific hypotheses, it will be necessary to trace the metabolism of the methylated protein in the cell. These studies have been limited to date because rapid demethylation (100) results in the loss of the isotopically labeled methyl group that identifies the substrate protein. However, with the development of synthetic stoichiometric methyl-accepting peptide substrates, it should be possible to directly follow the metabolism of methylated peptides in cellular extracts, and efforts in this direction are in progress (60, 86, 89).

Another approach to understanding the function of this methylation reaction has been made by investigating the reaction in aberrant red cells. For example, the level of protein carboxyl methylation had been found to be altered in erythrocytes from patients with sickle cell anemia (106, 107). Although the issue of whether the methylation level is increased or decreased in specific subpopulations of these cells is still unresolved, perhaps due to methodological difficulties, it will be interesting to see if the pathophysiology of sickling is associated with changes in protein carboxyl methylation reactions.

Brain and Other Tissues

Two major isozymes of protein carboxyl methyltransferase have been purified to near homogeneity from a cytosolic extract of bovine brain (108). Both isozymes were found to be monomers of 24,300-molecular-weight polypeptides. In a previous study using calf brain, a single activity was purified, although several components were detected when the preparation was fractionated by native gel electrophoresis (109). On the other hand, fractionation of an ovalbumin methylating activity from ox brain resulted in the purification of a 35,000-dalton protein (110). The relation of this enzyme to the lower-molecular-weight species is not clear.

These enzymes catalyze the substoichiometric methylation of a variety of proteins, including calmodulin, ACTH, and a major 46,000-dalton endogenous brain protein (60, 74). However, if ACTH is deamidated under conditions that result in the partial formation of an L-isoaspartyl derivative, this peptide becomes a near-stoichiometric methyl group acceptor (60). Both brain isozymes catalyze the stoichiometric methylation of the synthetic peptide L-Val-L-Tyr-L-Pro-L-isoAsp-Gly-L-Ala, but do not catalyze the methylation of analogs containing L-aspartyl, D-aspartyl, or D-isoaspartyl groups (60, 89). In these respects, the brain isozymes and the erythrocyte methyltransferase(s) behave identically (60, 89). In a study comparing the human erythrocyte and cow brain enzymes, it is found that the methyl-acceptor specificity for erythrocyte membrane polypeptides was similar for both the red cell and brain methyltransferases, and D-aspartic acid β-[^{3}H]methyl ester could be isolated from erythrocyte membranes labeled with either the red cell or the brain enzymes (81, 111).

Thus it appears that both bovine brain methyltransferase isozymes are D-aspartyl/L-isoaspartyl methyltransferases and share similar catalytic and structural properties with the erythrocyte enzymes. It is still unclear how the brain isozymes differ from one another. Although these enzymes are readily separated by DEAE-celluose chromatography, their physical and kinetic properties are very similar (108). Their specificity towards purified methyl-accepting peptides and proteins also appears to be similar (60, 89, 108), but these enzymes can be differentiated by their relative activity with ovalbumin and red cell membranes as methyl acceptors (111). It will be interesting to see if the physiological role(s) of the brain enzymes differ(s) from that of the erythrocyte methyltransferases.

It appears that the bulk of the non-substrate-specific protein carboxyl methyltransferase activity in the brain is in the cytosolic fraction. Although some enzymatic activity is found in vesicular or membrane fractions (82, 109, 110, 112), this activity may represent cytosolic enzyme trapped during the homogenization procedure. The absence of specific membrane or organellar-specific protein carboxyl methyltransferases appears to be a general feature of most mammalian cells (see below).

The properties of protein carboxyl methyltransferases characterized from various tissues are summarized in Table 3. With the exception of the L-glutamyl methyltransferases involved in bacterial chemotaxis, it is apparent that enzymes from many procaryotic and eucaryotic cell types share many of the properties of the D-aspartyl/L-isoaspartyl methyltransferases from human erythrocytes and bovine brain described above. With few exceptions, cytosolic extracts of all tissues tested have been found to catalyze an S-adenosyl-L-methionine dependent methylation of proteins such as ovalbumin, γ-globulin, or ACTH. The eucaryotic enzymes all appear to be monomers of 25,000-dalton polypeptide chains with similar Michaelis constants for the substrate S-adenosyl-L-methionine and the product inhibitor S-adenosyl-L-homocysteine. Isozymes with various isoelectric points have been described in many of these cell types. The availability of L-Val-L-Tyr-L-Pro-L-isoAsp-Gly-L-Ala, a synthetic substrate for the erythrocyte and brain enzyme (89), will now allow extracts of various cell types to be assayed specifically for L-isoaspartyl methyltransferase activity. We have recently demonstrated such peptide-methylating activities in crude extracts of *Salmonella typhimurium*, *Xenopus laevis* oocytes, as well as in a lymphoma and an adrenal medulla (PC 12) cell line (C. M. O'Connor, S. Clarke submitted for publication).

REGULATORY PROTEIN CARBOXYL METHYLTRANSFERASES IN EUCARYOTIC CELLS?

The example of regulating protein function by reversible carboxyl methylation in the chemotactic bacteria has inspired the search for similar regulatory systems in higher cells. In fact, in most of the studies performed to date with eucaryotic cells and tissues, the assumption has been that the methylated protein may have different functional properties from those of the unmethylated species and that the distribution of proteins between these states can be controlled by the activity of specific methyltransferases and methylesterases. However, if the D-aspartyl/L-isoaspartyl enzyme is widely distributed in cells, it will be necessary to distinguish methylation reactions catalyzed by these non-substrate-specific enzymes from reactions catalyzed by methyltransferases specific for L-glutamyl or L-aspartyl residues. Only the latter reactions might be expected to have regulatory significance, although possible regulatory roles of L-isoaspartyl methylation have been suggested (60).

It appears that distinguishing these two types of methylation reactions can be very difficult, especially when considering data from studies performed before it was realized that D-aspartyl/L-isoaspartyl methyltransferases existed. It may now be useful to outline basic criteria for establishing a regulatory role for methylation in a given tissue. Such criteria can be derived from results obtained with the bacterial methylation system involved in chemotaxis and from other regulatory covalent modification systems such as those involving protein phosphorylation reactions.

Criteria for Identifying Regulatory Methylation Reactions

1. METHYLATION SHOULD OCCUR ON SPECIFIC PROTEINS IN INTACT CELLS   Many studies have been performed with eucaryotic cells in vitro using cell homogenates or mixtures of cell proteins with purified methyltransferases. The problem with such approaches is that methylation reactions can occur on extracellular sites and other sites that would not normally be exposed to the methyltransferase(s). Additionally, good methyl acceptors in intact cells may be poor methyl acceptors in lysed cell systems, especially if these proteins are already fully methylated. This problem will be aggravated by the presence of D-aspartyl/L-isoaspartyl methyltransferases, which are very active on extracellular (and presumably nonphysiological) sites [see for example ref. (81)]. Although the bulk of studies performed so far have been in broken cell preparations, several non-erythroid membrane-sealed systems have been used. For example, isotopically labeled methionine has been used as an S-adenosylmethionine precursor in intact platelets (124), isolated pituitary stalks (125), or rat hypothalamic synaptosome preparations (126). Similarly, brain proteins can be labeled in situ by the intraventricular injection of isotopic methionine (127). Even in these preparations, partial disruption of the plasma membrane can result in the leakage of both methyltransferase and isotopically labeled S-adenosyl-L-methionine from these cells or organelles and the subsequent labeling of extracellular sites. To control for such nonphysiological methylation reactions, as well as for methylation reactions that might occur during the work up of methylated cells, S-adenosyl-L-homocysteine can be added to the incubation medium (79). This methyltransferase inhibitor is not permeable to the plasma membrane (at least in erythrocytes) and should inhibit all methylation reactions that do not occur in the cell cytosol.

Table 3 Comparison of the properties of cytosolic protein carboxyl methyltransferases from eucaryotic and procaryotic cells

Source	Methyl-accepting substrates	Residue methylated	Native M_r	Polypeptide M_r	K_m AdoMet μM	K_i AdoHcy μM	pI	Ref.
Class I: Substrate-specific								
Salmonella typhimurium cheR[a]	60,000-dalton membrane chemoreceptors[b]	L-Glu	41,000	31,000	12		>9	(18,19,21)
Bacillus subtilus type II	60,000-dalton membrane polypeptides[b]	L-Glu	30,000	30,000	5	0.2		(48,50,51)
Class II: Non-substrate-specific								
Escherichia coli	IgG, ovalbumin>> BSA, Hb[c]		d		2	1.8		(113)
Salmonella typhimurium	ACTH>>ovalbumin, ribonuclease							(18)
Wheat germ	ACTH> gelatin, glucagon> IgG, ovalbumin, BSA		41,000		5	1.5		(114)
Calf thymus	IgG, ovalbumin>> BSA		35,000		1.0–1.6		4.85	(115, 116)
Bovine pituitary	Ovalbumin>>BSA				1.5			(117)

Source	Methyl acceptor	Substrate specificity	MW	MW			pH optimum	Reference
Rat testes	Gelatin		25,000		1.3		6.1	(118) 6.7 7.35
Bovine brain[e]	IgG>>BSA	D-Asp, L-isoAsp	27,000–28,000	24,300	1.0–2.0	0.65	5.6, 5.7, 6.5	(60,108, 111, 119)
Rat erythrocyte	IgG, ovalbumin>> BSA, Hb		25,000		2.0–3.1		4.9, 5.5, 6.0	(120)
Horse erythrocyte	BSA		25,400–26,000	24,500	1.1	0.7	5.6	(121)
Human erythrocyte[e]	IgG, ovalbumin	D-Asp, L-isoAsp	25,000	28,000	1.8	0.1–1.6	5.5[f], 6.6	(88, 89, 99, 114, 120, 122)
Bovine lens	Ovalbumin		27,000					(123)

[a]The E. coli cheR enzyme is probably very similar (12,19).
[b]Ovalbumin or other exogenous proteins do not serve as methyl acceptors.
[c]IgG, γ-globulin; BSA, bovine serum albumin; Hb, hemoglobin; ACTH, adrenocorticotropic hormone.
[d]Reversible association/dissociation behavior in gel filtration chromatography.
[e]Two isozymes can be separated by DEAE-celluose chromatography with nearly identical properties.
[f]T. Ota and S. Clarke, unpublished.

The identification of methyl acceptor proteins in one tissue using an in vitro assay with a purified methyltransferase from another tissue has the additional problem that the methyl acceptor specificity will reflect only that of the purified methyltransferase. For example, in studies where the red cell methyltransferase was used as a reagent to probe methyl acceptor sites (128–131), the methylated proteins identified presumably included those that contained D-aspartyl and/or L-isoaspartyl residues. To identify potential sites of regulation, it will be necessary to both use homologous methyltransferases and to exclude the participation of endogenous D-aspartyl/L-isoaspartyl specific enzymes. This latter problem may be at least partially solved by the use of the synthetic peptide L-Val-L-Tyr-L-Pro-L-isoAsp-Gly-L-Ala. This compound is a competitive inhibitor of the D-aspartyl/L-isoaspartyl methyltransferase (60, 89, 111), but would not be expected to inhibit a specific regulatory enzyme.

2. METHYLATION SHOULD BE STOICHIOMETRIC   Under the appropriate conditions, each homogeneous polypeptide chain should be capable of incorporating an integral number of methyl groups. In studies of intact cells, this stoichiometry determination cannot generally be made until the specific methyl acceptors are purified. In almost all cases where this has been examined so far in eucaryotic cells, the steady-state degree of modification is markedly substoichiometric and it appears that only a subpopulation of proteins, possibly those containing D-aspartyl or L-isoaspartyl residues, are methyl acceptors [see for example, ref. (74, 128, 132, 133)]. Of course, these latter results do not preclude the possible detection in these cells of other methyl-accepting proteins which may be present at low concentration and are in fact capable of stoichiometric methylation.

3. THE SITE OF METHYLATION SHOULD BE AT A SPECIFIC L-GLUTAMYL OR L-ASPARTYL RESIDUE   No L-glutamyl γ- or L-aspartyl β-methyl esters have yet been identified in eucaryotic proteins (see above). These modified amino acids can be readily identified from proteolytic digests of isotopically labeled proteins (19–23). It is especially important to identify the modified carboxylic acid not only to distinguish D-aspartyl/L-isoaspartyl linkages from L-glutamyl/L-aspartyl linkages, but also to directly demonstrate the presence of a methyl ester. Many assays presently used for "carboxyl methylation" rely simply on the release of radioactivity from cell components after mild base treatment. Although the product of such treatment has been demonstrated to be [^{3}H]methanol in many (but not all systems) examined so far, definitive proof for a methyl ester requires the isolation of the amino acid residue. It should be noted that methanol can also be formed from the base-treatment of certain lipids (134). Although the most likely sites for regulatory methylation reactions are on the side chains of aspartyl and glutamyl residues, it is certainly possible that modifications of other carboxylic acid groups present in stoichiometric

amounts can also modulate protein function. Such groups may include the C-terminal α-carboxyl group or one of the two carboxyl groups on γ-carboxyl glutamic acid.

4. METHYLATION SHOULD ALTER THE FUNCTIONAL PROPERTIES OF THE MODIFIED PROTEINS   In theory, the most precise way to determine this is to compare the properties of the fully methylated protein to that of the unmethylated protein, when both proteins are obtained from the same cell preparation and are chemically separated. This has not been accomplished to date, but its importance is underscored by the difficulties in interpreting results in systems where only a small fraction of the proteins have been modified, often by methyltransferases from other tissues (see above).

A simple and widely used technique to demonstrate the role of carboxyl methylation in specific cell types is to assess the physiological effects of inhibitors of methylation such as S-adenosyl-L-homocysteine and its derivatives. These inhibitors, however, are not specific for carboxyl methylation reactions but affect other S-adenosyl-L-methionine dependent methyltransferases as well (3). These inhibitors may also affect cAMP metabolism (135) and the synthesis of specific proteins (136). Other side effects of these inhibitors have been demonstrated and it appears that caution is required in interpreting data obtained with such compounds (135–139). It is of interest to note that some phosphodiesterase inhibitors may also inhibit methylation reactions (124).

5. SPECIFIC METHYLTRANSFERASES AND METHYLESTERASES MUST BE ISOLATED   Table 3, which compares the properties of protein carboxyl methyltransferases characterized to date, shows that no eucaryotic enzyme exists with substrate-specific properties similar to the regulatory enzymes involved in bacterial chemotaxis. Because it seems unlikely that substrate-specific methyltransferases would catalyze the methylation of proteins such as ovalbumin, ribonuclease, or γ-globulin, which have been used to assay the nonspecific D-aspartyl/L-isoaspartyl enzyme, the future identification of eucaryotic L-glutamyl and L-aspartyl methyltransferases would appear to depend upon the prior identification of specific methyl accepting substrates in intact cells so that specific in vitro assays can be developed. It is conceivable, however, that regulatory subunits that may confer methyl-acceptor specificity are lost during the isolation of eucaryotic methyltransferases. In this latter case, however, the site of methylation should be found at a normal L-carboxylic acid residue. Similarly, one would expect to find substrate-specific esterases analogous to those described in bacterial chemotaxis in eucaryotic regulatory methylation systems. An esterase activity has been detected in mammalian cells based on the ability of cell extracts to remove methyl esters formed on gelatin by action of the bovine erythrocyte carboxyl methyltransferase (140,

141, 141a). However, if the bovine erythrocyte methyltransferase is similar to the human erythrocyte enzyme, the gelatin substrate will probably contain L-isoaspartyl α-methyl esters and/or D-aspartyl β-methyl esters and this activity may represent the metabolism of altered proteins rather than the reversal of a regulatory covalent modification reaction.

Critical Appraisal of Postulated Regulatory Methylation Mechanisms

Protein carboxyl methylation reactions have been proposed to fulfill a variety of regulatory functions in eucaryotic cells. In the section below the evidence for such functions is considered in light of the criteria developed above. From these studies, it is clear that in no tissue has an unequivocal regulatory function been shown; more work is needed to identify specific methyl-accepting substrates in intact cells and specific methyltransferases and esterases.

LEUCOCYTE CHEMOTAXIS   The hypothesis that similar biochemical mechanisms, such as receptor methylation, are utilized in the chemotaxic behavior of both bacterial and eucaryotic cells was initially supported by data showing that the chemoattractant f-Met-Leu-Phe caused a rapid dose-dependent but transient increase in the overall level of protein carboxyl methylation in rabbit neutrophils (142). However, it was found that the attractant-induced changes in methylation were highly variable in these cells (143) and could not be demonstrated in human neutrophils (144, 145) or in chemotactic macrophage (144). Inhibitor studies have also been used to support a role of carboxyl methylation in leukocyte chemotaxis. Conditions that result in elevated intracellular levels of S-adenosyl-L-homocysteine were found to result in both inhibition of chemotaxis and protein carboxyl methylation in human monocytes (146). Methylation inhibitors reduced chemotaxis in human neutrophils and guinea pig macrophage as well (147, 148). It is not established, however, whether these physiological effects were due to changes in protein methylation or alterations in other methylation reactions such as those involving phospholipid acceptors (149). Furthermore, subsequent studies have shown that methylation inhibitors may have complex effects on cells. For example, the inhibition of chemotaxis of rabbit neutrophils by 3-deazaadenosine is reversed by incubation with adenosine, while the inhibition of protein carboxyl methylation is not (138). Chemotaxis was not affected by adenosine alone but protein methylation was inhibited (138).

PLATELETS   It was originally reported that one part of the response of platelets to thrombin, which includes both cellular aggregation reactions and the extracellular release of granule contents, was a 60% increase in the total protein carboxyl methylation levels of intact cells (150). Although two subsequent studies confirmed the presence of a protein carboxyl methylation system in these cells, both were unable to confirm any stimulation of methylation by thrombin or other platelet stimuli (124, 151).

SPERM   A protein carboxyl methyltransferase has been identified in sperm using gelatin as a methyl acceptor (130). The possible involvement of this enzyme in sperm function was suggested by the finding that the level of this enzyme was 4–7 times lower in immotile sperm from infertile men (152). Since gelatin is a substrate for the D-aspartyl/L-isoaspartyl protein carboxyl methyltransferase of erythrocyte and brain, the significance of these results is not clear. [See also Ref. (152a,b)]

NEURAL AND SECRETORY TISSUES   The possibility of regulating neural and secretory function by protein carboxyl methylation has been explored in several laboratories. Functions of methylation reactions have been proposed in controlling neurotransmitter release (153, 154), and specific modes of regulation by reversible carboxyl methylation have been suggested for benzodiazepine receptor maturation (127), acetylcholine receptor function (128, 129), calmodulin function (132, 155, 156), and phosphodiesterase activity (133). Such regulatory roles would presumably require substrate-specific methyltransferases. However, such enzymes have not been described so far. Cytosolic methyltransferases active with ribonuclease, γ-globulins, ovalbumin, luteinizing hormone, as well as endogeneous proteins as methyl accepting substrates have been found in all regions of bovine, rabbit, and rat brain examined (82, 112). The properties of these enzyme activities, especially in terms of the broad substrate specificity, are similar to those of the human erythrocyte and isolated bovine brain D-aspartyl/L-isoaspartyl methyltransferases. (See above.)

Based on the ability of the red blood cell protein carboxyl methyltransferase to catalyze the methylation of electric fish *(Torpedo californica)* acetylcholine receptor subunits, it was proposed that receptor activity could be modulated (128, 129). This conclusion is now in doubt based on our present knowledge of the specificity of the red cell enzyme for D-aspartyl and L-isoaspartyl residues. Furthermore, the extent of methylation in these experiments was highly substoichiometric [only about 3 in one hundred molecules could be methylated (128)] and many of these sites appeared to be on portions of the receptor exposed to the extracellular space where they would not be accessible to a cytosolic methyltransferase (129). Thus, it now appears that this methylation reaction may only recognize damaged acetylcholine receptors.

Evidence has been presented that protein carboxyl methylation can modulate the activity of calmodulin (132, 155, 156) or the cAMP phosphodiesterase (133, 157) in brain and retina but further studies will be necessary to verify these hypotheses.

Regulatory roles for protein carboxyl methylation reactions in secretory processes have also been suggested. Postulated physiological roles include inactivating anterior pituitary hormones (117), modulating the interaction between neurophysin and oxytocin/vasopressin (158–161), and facilitating the secretion of newly synthesized proteins (126, 162–164). Recent studies have focused on the latter possibility in the posterior pituitary. When isolated pituitary lobes are incubated with isotopically labeled L-methionine, base-labile methyl groups were found to be incorporated into a number of polypeptides, including some identified as neurophysins (125). The level of these methyl acceptors was altered in Brattleboro rats which lack the vasopressin-associated neurophysin (125). The physiological importance of these reactions in regulating neurosecretion is not clear, however, because it has been shown that similar changes in the methyl-accepting capacity of Brattleboro and control rat posterior pituitary proteins can be seen when the red blood-cell D-aspartyl/L-isoaspartyl protein carboxyl methyltransferase is used to probe the methylation sites (164).

Another function proposed for carboxyl methylation in secretory cells is the neutralization of negative charges on the surface proteins of the secretory granules which would allow their fusion with the plasma membrane and release of the granule contents from the cell (165). Several in vitro methyl-accepting polypeptides associated with the chromaffin granules have been identified (73, 166), but no direct tests of the hypothesis have been performed so far.

Finally, it has been reported that the β-adrenergic agent isoproterenol stimulates both the secretion of amylase and protein carboxyl methylation of rat parotid and pancreatic proteins (167, 168). However, a reinvestigation of this phenomenon has revealed that the apparent stimulus-induced changes in methylation specific activity could be accounted for by the loss of amylase protein from the cell; when methyl acceptor and protein carboxyl methyltransferase activity were normalized to cellular DNA rather than cellular protein content (which varies as amylase is secreted), these values actually decreased as a function of stimulation (169). Other methodological difficulties with the original studies have been brought up recently (170).

OTHER POSSIBLE SITES OF CELLULAR PROTEIN CARBOXYL METHYLATION

Enzymatic methylation at L-glutamyl, and at D-aspartyl and L-isoaspartyl residues have now been demonstrated in procaryotic and eucaryotic systems respectively. At the same time, there have been no direct demonstrations that methylation reactions occur at L-aspartyl residues. It will be interesting to see if enzymes are present in cells that may catalyze reactions at L-aspartyl residues, or at other carboxyl groups present in "normal" proteins. These other groups would include the α-carboxyl group at the carboxyl terminus of the protein as well as carboxyl groups on β-carboxyl aspartyl residues or γ-carboxy glutamyl residues (171). Finally, it will be interesting to examine whether sites on other types of structurally altered carboxylic acid residues can be enzymatically methylated. These sites would include D-isoaspartyl, D-glutamyl, D-isoglutamyl, as well as L-isoglutamyl residues.

ACKNOWLEDGMENTS

Work in the author's laboratory is supported by Grants GM 26020 and EY 04942 from the National Institutes of Health, by a Grant-in-Aid from the American Heart Association with funds contributed in part by the Greater Los Angeles Affiliate, and with funds supplied by the Alfred P. Sloan Foundation. Many thanks are due to my colleagues for their thoughtful comments on the manuscript.

Literature Cited

1. Gagnon, C., Heisler, S. 1979. *Life Sci.* 25:993–1000
2. Springer, M. S., Goy, M. F., Adler, J. 1979. *Nature* 280:279–84
3. Borchardt, R. T. 1980. *J. Med. Chem.* 23:347–57
4. Paik, W. K., Kim, S. 1980. *Protein Methylation,* pp. 202–31. New York: Wiley
5. O'Dea, R. F., Viveros, O. H., Diliberto, E. J. Jr. 1981. *Biochem. Pharmacol.* 30:1163–68
6. Clarke, S., O'Connor, C. M. 1983. *Trends Biochem. Sci.* 8:391–94
7. Adler, J., Dahl, M. M. 1967. *J. Gen. Microbiol.* 46:161–73
8. Aswad, D., Koshland, D. E. Jr. 1974. *J. Bacteriol.* 118:640–45
9. Armstrong, J. B. 1972. *Can. J. Microbiol.* 18:1695–1701
10. Aswad, D. W., Koshland, D. E. Jr. 1975. *J. Mol. Biol.* 97:207–23
11. Kort, E. N., Goy, M. F., Larsen, S. H., Adler, J. 1975. *Proc. Natl. Acad. Sci. USA* 72:3939–43
12. DeFranco, A. L., Parkinson, J. S., Koshland, D. E. Jr. 1979. *J. Bacteriol.* 139:107–14
13. DeFranco, A. L., Koshland, D. E. Jr. 1981. *J. Bacteriol.* 147:390–400
14. Koshland, D. E. Jr. 1981. *Ann. Rev. Biochem.* 50:765–82
15. Boyd, A., Simon, M. 1982. *Ann. Rev. Physiol.* 44:501–17
16. Taylor, B. L., Panasenko, S. M. 1984. In *Membranes and Sensory Transduction,* ed. G. Colombetti, F. Lenci. New York: Plenum
17. Springer, W. R., Koshland, D. E. Jr. 1977. *Proc. Natl. Acad. Sci. USA* 74:533–37
18. Clarke, S., Sparrow, K., Panasenko, S., Koshland, D. E. Jr. 1980. *J. Supramol. Struct.* 13:315–28
19. Stock, J. B., Clarke, S., Koshland, D. E. Jr. 1984. *Methods Enzymol.* 106:310–21
20. Kleene, S. J., Hobson, A. C., Adler, J. 1979. *Proc. Natl. Acad. Sci. USA* 76:6309–13
21. Van der Werf, P., Koshland, D. E. Jr. 1977. *J. Biol. Chem.* 252:2793–95
22. Kleene, S. J., Toews, M. L., Adler, J. 1977. *J. Biol. Chem.* 252:3214–18
23. Clarke, S., McFadden, P. N., O'Connor, C. M., Lou, L. L. 1984. *Methods Enzymol.* 106:330–44
24. Krikos, A., Mutoh, N., Boyd, A., Simon, M. I. 1983. *Cell* 33:615–22
25. Russo, A. F., Koshland, D. E. Jr. 1983. *Science* 220:1016–20
26. Clarke, S., Koshland, D. E. Jr. 1979. *J. Biol. Chem.* 254:9695–702
27. Wang, E. A., Koshland, D. E. Jr. 1980. *Proc. Natl. Acad. Sci. USA* 77:7157–61
28. Hedblom, M. L., Adler, J. 1983. *J. Bacterial.* 155:1463–66
29. Kondoh, H., Ball, C. B., Adler, J. 1979. *Proc. Natl. Acad. Sci. USA* 76:260–4
30. Slocum, M. K., Parkinson, J. S. 1983. *J. Bacteriol.* 155:565–77
31. Wang, E. A., Mowry, K. L., Clegg, D. O., Koshland, D. E. Jr. 1982. *J. Biol. Chem.* 257:4673–76
32. Terwilliger, T. C., Koshland, D. E. Jr. 1984. *J. Biol. Chem.* 259:7719–25
33. Kehry, M. R., Bond, M. W., Hunkapiller, M. W., Dahlquist, F. W. 1983. *Proc. Natl. Acad. Sci. USA* 80:3599–602
34. Boyd, A., Kendall, K., Simon, M. O. 1983. *Nature* 301:623–26
35. Bollinger, J., Park, C., Harayama, S.,

Hazelbauer, G. L. 1984. *Proc. Natl. Acad. Sci. USA* 81:3287–91
36. Stock, J. B., Koshland, D. E. Jr. 1978. *Proc. Natl. Acad. Sci. USA* 75:3659–63
37. Snyder, M. A., Stock, J. B., Koshland, D. E. Jr. 1984. *Methods Enzymol.* 106:321–30
38. Snyder, M. A., Koshland, D. E. Jr. 1981. *Biochimie* 63:113–17
39. Rollins, C., Dahlquist, F. W. 1981. *Cell* 25:333–40
40. Sherris, D., Parkinson, J. S. 1981. *Proc. Natl. Acad. Sci. USA* 78:6051–55
41. Goy, M. F., Springer, M. S., Adler, J. 1977. *Proc. Natl. Acad. Sci. USA* 74:4964–68
42. Stock, J. B., Koshland, D. E. Jr. 1981. *J. Biol. Chem.* 256:10826–33
43. Oosawa, K., Imae, Y. 1984. *J. Bacteriol.* 157:576–81
44. Niwano, M., Taylor, B. L. 1982. *Proc. Natl. Acad. Sci. USA* 79:11–15
45. Stock, J. B., Maderis, A. M., Koshland, D. E. Jr. 1981. *Cell* 27:37–44
46. Stock, J. B., Kersulis, G., Koshland, D. E. Jr. 1983. *Fed. Proc.* 42:2136
47. Krebs, E. G., Beavo, J. A. 1979. *Ann. Rev. Biochem.* 48:923–59
48. Burgess-Cassler, A., Ullah, A. H. J., Ordal, G. W. 1982. *J. Biol. Chem.* 257:8412–17
49. Goldman, D. J., Worobec, S. W., Siegel, R. B., Hecker, R. V., Ordal, G. W. 1982. *Biochemistry* 21:915–20
50. Ahlgren, J. A., Ordal, G. W. 1983. *Biochem. J.* 213:759–63
51. Burgess-Cassler, A., Ordal, G. W. 1982. *J. Biol. Chem.* 257:12835–38
52. Ullah, A. H. J., Ordal, G. W. 1981. *Biochem. J.* 199:795–805
53. Goldman, D. J., Nettleton, D. O., Ordal, G. W. 1984. *Biochemistry* 23:675–80
54. Schimz, A. 1981. *FEBS Lett.* 125:205–7
55. Bibikov, S. I., Baryshev, V. A., Glagolev, A. N. 1982. *FEBS Lett.* 146:255–58
56. Craven, R. C., Montie, T. C. 1983. *J. Bacteriol.* 154:780–86
57. Shaw, P., Gomes, S. L., Sweeney, K., Ely, B., Shapiro, L. 1983. *Proc. Natl. Acad. Sci. USA* 80:5261–65
58. Kathariou, S., Greenberg, E. P. 1983. *J. Bacteriol.* 156:95–100
59. Kim, S., Li, C. H. 1979. *Proc. Natl. Acad. Sci. USA* 76:4255–57
60. Aswad, D. W. 1984. *J. Biol. Chem.* 259:10714–21
61. Gratzer, W. B. 1981. *Biochem. J.* 198:1–8
62. Marchesi, V. T. 1979. *Semin. Hematol.* 16:3–20
63. Edwards, J. J., Anderson, N. G., Nance,

S. L., Anderson, N. L. 1979. *Blood* 53:1121–32
64. Freitag, C., Clarke, S. 1981. *J. Biol. Chem.* 256:6102–8
65. Barber, J. R., Clarke, S. 1983. *J. Biol. Chem.* 258:1189–96
66. Oden, K. L., Clarke, S. 1983. *Biochemistry* 22:2978–86
67. Deleted in proof
68. Kim, S., Galletti, P., Paik, W. K. 1980. *J. Biol. Chem.* 255:338–41
69. Laemmli, U. K. 1970. *Nature* 227:680–85
70. Kim, S., Paik, W. K. 1976. *Experientia* 32:982–84
71. Terwilliger, T. C., Clarke, S. 1981. *J. Biol. Chem.* 256:3067–76
72. Fairbanks, G., Avruch, J. 1972. *J. Supramol. Struct.* 1:66–75
73. Gagnon, C., Viveros, O. H., Diliberto, E. J. Jr., Axelrod, J. 1978. *J. Biol. Chem.* 253:3778–81
74. Aswad, D. W., Deight, E. A. 1983. *J. Neurochem.* 41:1702–9
75. Macfarlane, D. E. 1983. *Anal. Biochem.* 132:231–35
76. Galletti, P., Paik, W. K., Kim, S. 1978. *Biochemistry* 17:4272–76
77. Galletti, P., Ingrosso, D., Nappi, A., Gragnaniello, V., Iolascon, A., Pinto, L. 1983. *Eur. J. Biochem.* 135:25–31
78. O'Connor, C. M., Clarke, S. 1984. *J. Biol. Chem.* 259:2570–78
79. Runte, L., Jurgensmeier, H. L., Unger, C., Soling, H. D. 1982. *FEBS Lett.* 147:125–30
80. Galletti, P., Paik, W. K., Kim, S. 1979. *Eur. J. Biochem.* 97:221–27
81. O'Connor, C. M., Clarke, S. 1983. *J. Biol. Chem.* 258:8485–92
82. Kim, S., Wasserman, L., Lew, B., Paik, W. K. 1975. *J. Neurochem.* 24:625–29
83. Giulidori, P., Stramentinoli, G. 1984. *Anal. Biochem.* 137:217–20
84. Barber, J. R., Clarke, S. 1984. *J. Biol. Chem.* 259:7115–22
85. Barber, J. R., Clarke, S. 1984. *Biochem. Biophys. Res. Commun.* 123:133–40
86. Johnson, B. A., Aswod, D. W. 1985. *Biochemistry.* In press
87. Janson, C. A., Clarke, S. 1980. *J. Biol. Chem.* 255:11640–43
88. McFadden, P. N., Clarke, S. 1982. *Proc. Natl. Acad. Sci. USA* 79:2460–64
89. Murray, E. D. Jr., Clarke, S. 1984. *J. Biol. Chem.* 259:10722–32
90. Bada, J. L., Shou, M.-Y. 1980. In *Biogeochemistry of Amino Acids,* ed. P. E. Hare, pp. 234–55. New York: Wiley
91. Smith, G. G., Evans, R. C. 1980. See Ref. 90, pp. 257–82

92. Friedman, M., Masters, P. M. 1982. *J. Food Sci.* 47:760–64
93. Masters, P. M. 1983. *J. Am. Geriatr. Soc.* 31:426–34
94. Masters, P. M., Bada, J. L., Zigler, J. S. Jr. 1978. *Proc. Natl. Acad. Sci. USA* 75:1204–8
95. Bada, J. L., Brown, S. E. 1980. *Trends Biochem. Sci.* 5(9):R3–R5
96. Manning, J. M., Moore, S. 1968. *J. Biol. Chem.* 243:5591–97
97. Bornstein, P. 1970. *Biochemistry* 9:2408–21
98. Robinson, A. B., Rudd, C. J. 1974. *Curr. Top. Cell. Regul.* 8:247–95
99. Kim, S., Choi, J., Jun, G.-J. 1983. *J. Biochem. Biophys. Methods* 8:9–14
100. Barber, J. R. 1984. *The Function of Protein Carboxyl Methylation in the Human Erythrocyte* PhD thesis. UCLA, Chap. 8. 171 pp.
101. Ciechanover, A., Finley, D., Varshavsky, A. 1984. *J. Cell. Biochem.* 24:27–53
102. Goldberg, A. L., Boches, F. S. 1982. *Science* 215:1107–9
103. Cohen, N. S., Ekholm, J. E., Luthra, M. G., Hanahan, D. J. 1976. *Biochim. Biophys. Acta* 419:229–42
104. Gabrielli, F. 1983. *Life Sci.* 33:805–16
105. Brot, N., Weissbach, H. 1983. *Arch. Biochem. Biophys.* 223:271–81
106. Green, G. A., Sikka, S. C., Kalra, V. K. 1983. *J. Biol. Chem.* 258:12958–66
107. Ro, J.-Y., Neilan, B., Kim, S. 1983. *Biochem. Med.* 30:342–48
108. Aswad, D. W., Deight, E. A. 1983. *J. Neurochem.* 40:1718–26
109. Kim, S., Nochumson, S., Chin, W., Paik, W. K. 1978. *Anal. Biochem.* 84:415–22
110. Iqbal, M., Steenson, T. 1976. *J. Neurochem.* 27:605–8
111. O'Connor, C. M., Aswad, D. W., Clarke, S. 1984. *Proc. Natl. Acad. Sci. USA.* 81:7757–61
112. Diliberto, E. J. Jr., Axelrod, J. 1976. *J. Neurochem.* 26:1159–65
113. Kim, S., Lew, B., Chang, F. N. 1977. *J. Bacteriol.* 130:839–45
114. Trivedi, L., Gupta, A., Paik, W. K., Kim, S. 1982. *Eur. J. Biochem.* 128:349–54
115. Kim, S., Paik, W. K. 1970. *J. Biol. Chem.* 245:1806–13
116. Kim, S. 1973. *Arch. Biochem. Biophys.* 157:476–84
117. Diliberto, E. J. Jr., Axelrod, J. 1974. *Proc. Natl. Acad. Sci. USA* 71:1701–4
118. Cusan, L., Gordeladze, J. O., Andersen, D., Hansson, V. 1981. *Arch. Androl.* 7:263–74
119. Oliva, A., Galletti, P., Zappia, V., Paik, W. K., Kim, S. 1980. *Eur. J. Biochem.* 104:595–602
120. Kim, S. 1974. *Arch. Biochem. Biophys.* 161:652–57
121. Polastro, E. T., Deconinck, M. M., Devogel, M. R., Mailier, E. L., Looza, Y. B., et al. 1978. *Biochem. Biophys. Res. Commun.* 81:920–27
122. Gillet, L., Looze, Y., Deconinck, M., Leonis, J. 1979. *Experientia* 35:1007–9
123. McFadden, P. N., Horwitz, J., Clarke, S. 1983. *Biochem. Biophys. Res. Commun.* 113:418–24
124. Macfarlane, D. E. 1984. *J. Biol. Chem.* 259:1357–62
125. Kloog, Y., Saavedra, J. M. 1983. *J. Biol. Chem.* 258:7129–33
126. Eiden, L. E., Borchardt, R. T., Rutledge, C. O. 1982. *J. Neurochem.* 38:631–37
127. Gregor, P., Sellinger, O. Z. 1983. *Biochem. Biophys. Res. Commun.* 116:1056–63
128. Kloog, Y., Flynn, D., Hoffman, A. R., Axelrod, J. 1980. *Biochem. Biophys. Res. Commun.* 97:1474–80
129. Flynn, D. D., Kloog, Y., Potter, L. T., Axelrod, J. 1982. *J. Biol. Chem.* 257:9513–17
130. Bouchard, P., Gagnon, C., Phillips, D. M., Bardin, C. W. 1980. *J. Cell Biol.* 86:417–23
131. Kloog, Y., Axelrod, J., Spector, I. 1983. *J. Neurochem.* 40:522–29
132. Billingsley, M. L., Velletri, P. A., Roth, R. H., DeLorenzo, R. J. 1983. *J. Biol. Chem.* 258:5352–57
133. Billingsley, M., Kuhn, D., Velletri, P. A., Kincaid, R., Lovenberg, W. 1984. *J. Biol. Chem.* 259:6630–35
134. Zatz, M., Dudley, P. A., Kloog, Y., Markey, S. P. 1981. *J. Biol. Chem.* 256:10028–32
135. Zimmerman, T. P., Schmitges, C. J., Wolberg, G., Deeprose, R. D., Duncan, G. S., et al. 1980. *Proc. Natl. Acad. Sci. USA* 77:5639–43
136. Aksamit, R. R., Backlund, P. S. Jr., Cantoni, G. L. 1983. *J. Biol. Chem.* 258:20–23
137. Aksamit, R. R., Falk, W., Cantoni, G. L. 1982. *J. Biol. Chem.* 257:621–25
138. Garcia-Castro, I., Mato, J. M., Vasanthakumar, G., Wiesmann, W. P., Schiffmann, E., Chiang, P. K. 1983. *J. Biol. Chem.* 258:4345–49
139. Zimmerman, T. P., Iannore, M., Wolberg, G. 1984. *J. Biol. Chem.* 259:1122–26
140. Gagnon, C. 1979. *Biochem. Biophys. Res. Commun.* 88:847–53

506   CLARKE

141. Chene, L., Bourget, L., Vinay, P., Gagnon, C. 1982. *Arch. Biochem. Biophys.* 213:299–305
141a. Gagnon, C., Harbour, D., Camato, R. 1984. *J. Biol. Chem.* 259:10212–15
142. O'Dea, R. F., Viveros, O. H., Axelrod, J., Aswanikumar, S., Schiffmann, E., Corcoran, B. A. 1978. *Nature* 272:462–64
143. Venkatasubramanian, K., Hirata, F., Gagnon, C., Corcoran, B. A., O'Dea, R. F., et al. 1980. *Mol. Immunol.* 17:201–7
144. Schiffmann, E. 1982. *Ann. Rev. Physiol.* 44:553–68
145. Barber, J. R. 1984. See Ref. 100, Chap. 2
146. Pike, M. C., Kredich, N. M., Snyderman, R. 1978. *Proc. Natl. Acad. Sci. USA* 75:3928–32
147. Snyderman, R., Pike, M. C., Kredich, N. M. 1980. *Mol. Immunol.* 17:209–18
148. Pike, M. C., Snyderman, R. 1982. *Cell* 28:107–14
149. Pike, M. C., Snyderman, R. 1981. *J. Cell Biol.* 91:221–26
150. O'Dea, R. F., Viveros, O. H., Acheson, A., Gorman, C., Axelrod, J. 1978. *Biochem. Pharmacol.* 27:679–84
151. Shattil, S. J., McDonough, M., Burch, J. W. 1981. *Blood* 57:537–44
152. Gagnon, C., Sherins, R. J., Phillips, D. M., Bardin, C. W. 1982. *N. Engl. J. Med.* 306:821–25
152a. Hatch, R., Harvey, S. E., Williams-Ashman, H. G. 1985. *Fert. Steril.* In press
152b. Williams-Ashman, H. G. 1985. *Adv. Enzymol. Regul.* 23: In press
153. Gilad, G. M., Gagnon, C., Kopin, I. J. 1980. *Brain Res.* 183:393–402
154. Billingsley, M. L., Roth, R. H. 1982. *J. Pharmacol. Exp. Ther.* 223:681–88

155. Gagnon, C., Kelly, S., Manganiello, V., Vaughan, M., Odya, C., et al. 1981. *Nature* 291:515–17
156. Gagnon, C. 1983. *Can. J. Biochem. Cell Biol.* 61:921–26
157. Swanson, R. J., Applebury, M. L. 1983. *J. Biol. Chem.* 258:10599–605
158. Edgar, D. H., Hope, D. B. 1974. *FEBS Lett.* 49:145–48
159. Kim, S., Pearson, D., Paik, W. K. 1975. *Biochem. Biophys. Res. Commun.* 67:448–54
160. Edgar, D. H., Hope, D. B. 1976. *J. Neurochem.* 27:949–55
161. Diliberto, E. J. Jr., Axelrod, J., Chaiken, I. M. 1976. *Biochem. Biophys. Res. Commun.* 73:1063–67
162. Chen, J.-K., Liss, M. 1978. *Biochem. Biophys. Res. Commun.* 84:261–68
163. Gagnon, C., Axelrod, J., Brownstein, M. J. 1978. *Life Sci.* 22:2155–64
164. Saavedra, J. M., Kloog, Y., Chevillard, C., Fernandez-Pardal, J. 1983. *J. Neurochem.* 41:195–200
165. Diliberto, E. J. Jr., Viveros, O. H., Axelrod, J. 1976. *Proc. Natl. Acad. Sci. USA* 73:4050–54
166. Borchardt, R. T., Olsen, J., Eiden, L., Schowen, R. L., Rutledge, C. O. 1978. *Biochem. Biophys. Res. Commun.* 83:970–76
167. Strittmatter, W. J., Gagnon, C., Axelrod, J. 1978. *J. Pharmacol. Exp. Ther.* 207:419–31
168. Povilaitis, V., Gagnon, C., Heisler, S. 1981. *Am. J. Physiol.* 240:G199–205
169. Unger, C., Jahn, R., Soling, H. D. 1981. *FEBS Lett.* 23:211–14
170. Davison, J. S. 1982. *Am. J. Physiol.* 242:G76
171. McTigue, J. J., Dhaon, M. K., Rich, D. H., Suttie, J. W. 1984. *J. Biol. Chem.* 259:4272–78

Ann. Rev. Biochem. 1985. 54:507–30
Copyright © 1985 by Annual Reviews Inc. All rights reserved

EVOLVING RIBOSOME STRUCTURE: DOMAINS IN ARCHAEBACTERIA, EUBACTERIA, EOCYTES AND EUKARYOTES

James A. Lake

Molecular Biology Institute and Department of Biology, University of California, Los Angeles, California 90024

CONTENTS

507

PERSPECTIVES AND SUMMARY

Ribosomes are ubiquitous organelles found in all cells and are the site of protein synthesis. They are large, 25–30 nm in diameter, and are composed of two parts, the small and large subunits. Approximately two thirds of the mass of the small subunit is a single rRNA molecule (16S or 18S rRNA), while 20–40 small proteins contribute the remainder one third of the subunit's mass. In the large subunit, two (23S and 5S in prokaryotic subunits) or three (28S, 5.8S, and 5S in eukaryotic subunits) rRNAs again contribute approximately two thirds of the subunit mass while 30–50 proteins contribute the remaining one third. During protein synthesis the primary function of the small subunit is to hold the messenger RNA and the tRNAs, while the large subunit catalyzes the formation of peptide bonds in the proteins being synthesized. The three-dimensional structure of the ribosome is quite highly conserved, so that even the most distantly related organisms yet discovered have similar ribosomal structures. It is thought that this extreme conservation reflects the central role of the ribosome in cellular function.

Major advances have been made in recent years in our understanding of the structure, function, and evolution of ribosomes. Among these advances one notes that the overall three-dimensional structures of ribosomes and ribosomal subunits are known; the primary and secondary structures of the rRNAs are known for many diverse organisms [for a review see (1)]; the approximate locations of many ribosomal proteins (r-proteins) and of some sequences of the ribosomal RNAs (rRNA) are known; many aspects of ribosome function have been related to ribosome structure; and finally, comparative studies of ribosomes have revealed some of the early steps in the evolution of ribosomes and of the cells that contain them.

In this chapter, I emphasize the detailed structural and biochemical information available for the *Esherichia coli* ribosome, and then use it as a basis to interpret the less detailed structural and biochemical information available for archaebacterial, eocytic, and eukaryotic ribosomes. In this way, I hope to present a unified view of the structural and functional bases of protein synthesis and of the evolution of the ribosome.

STRUCTURE OF THE *E. COLI* RIBOSOME: THE CONSENSUS MODEL

There is now general agreement concerning the overall structure and morphology of the *E. coli* ribosome. The asymmetric models for the ribosome (2) and ribosomal subunits (2, 5, 6) have gradually become accepted. In this review I refer to the structures originally proposed as the consensus model (3), since their main features have been confirmed repeatedly and are generally accepted.

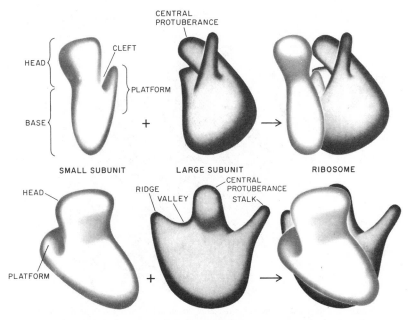

Figure 1 The three-dimensional consensus model of the ribosome (2, 3). This model gives an asymmetric shape to the two subunits of which a ribosome consists. The small subunit (left) includes a head, a base and a platform. The large subunit (second from left) includes a central protuberance, flanked by a ridge on one side and a stalk on the other. Two orientations of the model are shown. The length of a ribosome is about 250 Å.

The consensus structures of the *E. coli* ribosome and of its component subunits are illustrated in Figure 1. The smaller (30S) subunit is divided into two unequal parts by an indentation and a region of accumulated negative stain (2, 4–6). The parts are the head, or the upper third, and the base, or lower two thirds. A region of the subunit, called the platform, extends from the base of the small subunit and forms a cleft between it and the head (7). The 30S model derived from images of heavy metal–shadowed subunits (6) is similar to that shown in Figure 1 although it lacks the cleft. All other models (6, 8, 9) now used in immune mapping studies are in general agreement with the consensus structure (3).

The consensus model of the large subunit, like that of the small subunit, is asymmetric (2, 3). It consists of a central protuberance, or head, and protrusions inclined approximately 50° to either side of the central protuberance (lower central panel of Figure 1). One of these, the "L7/L12 stalk," is at the right and contains the only multiple copy proteins present in the *E. coli* ribosome. In a projection approximately orthogonal to this (shown in the upper central panel of Figure 1), the large subunit is characterized by a notch on the upper surface. All other large subunit models (10–13) are asymmetric and in

Figure 2   Three-dimensional structure of the large ribosomal subunit of *E. coli* at 35 Å resolution. This structure was determined by Fourier reconstruction of electron micrographs of in vitro grown crystals of large ribosomal subunits [(13, 14), Lake, J. A., Clark, M. W. Leonard, K. R., in preparation].

general agreement with the consensus model. In addition this structure is in accord with the three-dimensional structure determined by Fourier reconstruction from in vitro crystals of *E. coli* large subunits shown in Figure 2 [(14, 15) and J. A. Lake, M. W. Clark, K. M. Leonard, in preparation].

In the monomeric ribosome the small subunit is positioned asymmetrically on the large subunit (2, 3) as shown in Figure 1. This allows the platform of the small subunit to contact the large subunit, so that the partition between the head and body of the small subunit is approximately aligned with the notch of the large subunit. With one exception (16), all current models agree with this structure. Supporting experiments include a double labeling immune electron microscopy study from our laboratory (17), and others (18), three-dimensional analysis of ribosomes and ribosomal complexes (13, 19), and functional studies mapping bound ligands (20–22). Furthermore, this structure has been shown to apply to 70S ribosomes in several translational states (23).

Significant advances have come through using these structures to interpret the results of immune electron microscopy, neutron diffraction, chemical cross-linking, and other techniques [reviewed in (24)]. In the limited space available, I will emphasize immune electron microscopy at the expense of the others. This technique (5, 25–27), by combining immunology and electron microscopy, allows one to map specific ribosomal protein and RNA components and to determine their locations in three dimensions. Using these three-dimensional structures, locations of many ribosomal proteins have been mapped by the author in collaboration with L. Kahan (5, 7, 28, 29), W. Strycharz (30, 31), and M. Nomura (5, 30). The initial results of Stoffler and coworkers (32), interpreted using a symmetrical model, are not included; however, the new results and new interpretations of this group are included (18, 33–35). Small subunit proteins mapped by neutron diffraction (36–39) are also included in the following discussions.

THE TRANSLATIONAL DOMAIN

Ribosomes from all organisms are divided into two general functional regions. They are the translational domain and the exit, or secretory, domain (21). The translational and exit domains are found at opposite ends of the ribosome (see Figure 3). The translational domain includes the head and platform of the small subunit and the L7/L12 stalk, the central protuberance, and the L1 ridge of the large subunit.

In general, all of the proteins of the *E. coli* small ribosomal subunit that have been mapped are located in this domain (see Figure 4). In the large subunit, many ribosomal proteins are found in the translational domain, although at least one, L17, is located in the exit domain (see Figure 4). Within the translational domain itself several functional regions are found. These regions are discussed in the following sections.

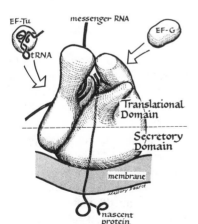

Figure 3 Diagrammatic representation of the exit and translational domains of the ribosome and their orientations with respect to the membrane binding site. Adapted from Bernabeu & Lake (21).

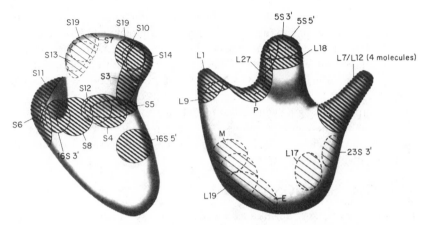

Figure 4   Summary map of protein, RNA, and functional sites on the ribosome. Details of these immune electron microscopy and neutron diffraction mappings are given in the text. Lightly shaded sites are located on the far side of the subunits. The letters P, M, and E represent the peptidyl transferase site, the membrane binding site, and the nascent protein exit site. The letters S and L refer to small and large subunit protein and 16S 5', 23S 3', etc refer to the 5' and 3' ends, respectively, of the ribosomal RNAs.

The Initial Site of tRNA Recognition (R site) is Very Likely on the External Surface of the Small Subunit.

The placement of the small subunit on the large subunit makes possible an operational distinction between 30S proteins located on the "exterior" of the subunit (i.e., the side away from the 50S subunit) and those at the "interface" between subunits. In the view of the 30S subunit characterized by a platform and a cleft shown in the upper left corner of Figure 1, exterior proteins are present on the side of the small subunit opposite the platform. Initially, proteins S3, S10, S14, and S19 provided a clue that the exterior surface of the small subunit functioned in tRNA binding (40). Several types of experiments suggested that proteins at the concave edge of the head of the subunit (see Figure 4) were involved in aspects of tRNA binding (41–43). Single-component omission experiments (41), for example, implied an important role for proteins S3, S10, S14, and S19 in poly U–dependent Phe-tRNA binding.

Adjacent to this group is a second region of small subunit proteins mapping on the external surface. It includes the proteins located at the constriction between the base and the head, i.e., S4, S5, S8, and S12. These proteins have all been implicated in tRNA recognition by drug-induced misreading studies (44–47). Hence these experiments have all argued, indirectly, that the recognition tRNA binding site is on the exterior surface of the small subunit (40).

Recently, the recognition site has been mapped directly by immune electron microscopy of the EF-Tu binding site [(48) and J. Langer, J. Lake, in prepara-

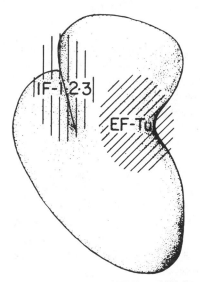
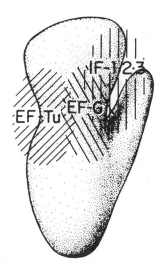

Figure 5   Locations of EF-Tu and EF-G and suggested location of IF-1, IF-2, and IF-3 on the small subunit. A view of the exterior surface of the small subunit is shown in the panel on the at left and a view down the cleft is shown in the panel on the right.

tion]. Although the EF-Tu.GTP.tRNA complex is only transiently bound to 70S ribosomes, the complex is stably bound to messenger RNA programmed 30S subunits. Immune mapping of the 30S.EF-Tu complex places EF-Tu at the concave edge of the external surface of the 30S subunit (as shown in Figure 5), adjacent to the EF-G site [the EF-G site is, however, located near the interface surface of the subunit, (20)].

In addition to beginning to appreciate the function of the external 30S surface, we are learning something about the tertiary structure of the rRNA in this region. Proteins S4 and S8, mapping at the exterior surface, for example, have further important roles in RNA binding. Proteins S8 is quite accessible and on the surface of the subunit (28) while S4 is "beneath" proteins S5 and S12 (29). Protein S4 is accessible for antibody binding only in ribosomal subunits lacking S5 and S12. Both S4 and S8 "protect" extensive regions of RNA from digestion by ribonuclease A and both proteins function to autogenously regulate the synthesis of ribosomal proteins [for a review see (49)]. For this reason they are useful markers of rRNA positions. In addition to these proteins, the $_m^7G$ located at position 526 in 16S rRNA has been mapped near the top of the base of the small subunit (50). Thus a number of RNA markers are known on this surface of the ribosome. These sites have recently been used to construct a model (shown in Figure 6) for the path of parts of the rRNA on the external surface of the ribosome (51).

The Platform is the Likely Site of the Codon-Anticodon Interaction.

One special group of small subunit proteins is found on the platform. Proteins S6, S11, S15, and S18 are in this group (see Figure 4). Because of the thinness (approximately 30 Å) of the platform, it is possible for a protein to be exposed on both the exterior and the interface surfaces of the platform.

Several lines of experimentation have linked these proteins to mRNA binding. Single-component omission experiments [(41); discussed in (52)] have suggested that S11 may be participating directly in the selection of the correct tRNA, i.e., in the codon-anticodon interaction. Protein S18 has been cross-linked to mRNA using a variety of affinity-labeling analogs [reviewed in (53)]. Also the platform has been implicated as the decoding site by affinity immune electron microscopy of the 3' end of a cross-linked tRNA (54).

Important rRNA markers are found on the platform. The 3' end of 16S RNA has been localized there (55, 56) and the two N6-dimethyladenosine nucleosides at positions 1518 and 1519 have also been placed there by immune electron microscopy (57). Recently the initiation codon of a Shine-Delgarno containing message has been mapped on the platform (58). In addition, proteins S6, S15, and S18, also located on the platform, protect specific regions of 16S rRNA from digestion by RNase A and provide important clues to the location of rRNA within the platform. Thus the platform represents one of the best mapped and understood regions of the ribosome (51, 59).

Initiation Factors May Bind Near the Cleft.

The locations of proteins S13 and S19, the locations of the platform proteins, and the locations of S12 taken together suggest the binding site for initiation factors IF-1, IF-2, and IF-3 (7). Consideration of the results of both cross-linking (60–64) and protein localization by immune electron microscopy suggests that IF-3, IF-2, and IF-1 are positioned across the cleft between S13, S19-II, and S12 on the head of the small subunit and S11 on the platform in the general region indicated in Figure 5. In eukaryotic small subunits, eIF3 binds in a similar region on the platform (65).

Recent experiments have also started to reveal some elements of the tertiary structure of rRNA in the vicinity of the cleft. Oakes et al (58) have mapped sequence 1392–1408 at the cleft and on the side of the head. By hybridizing a triester-synthesized DNA probe that was complementary to the 16S rRNA sequence, they determined the location shown on Figure 7. This mapping adds support to the proposal of Noller & Lake (51) for the existence of a helical structure, "the cleft anchor," that traverses the small subunit. Also found on the interface side of the head is protein S7 (37, 52). This protein, together with S9, S13, and S19, protects a significant fragment of rRNA from digestion by

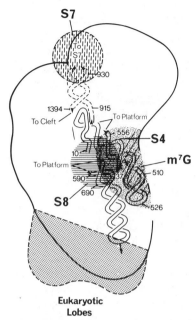

Eukaryotic Lobes

Figure 6 Model for the path of 16S rRNA on the external surface of the small subunit compared with experimentally mapped markers of specific rRNA sites. Protein S7, indicated by dashed lines, is on the opposite side of the subunit. All other sites are on the exterior 30S surface. Note the eukaryotic lobes, the presumed location of 18S rRNA inserts, are also shown. Adapted from Noller & Lake (51).

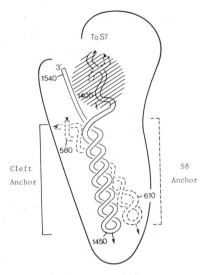

Figure 7 Diagram of the proposed locations of the "cleft anchor" and the "S8 anchor" within the 30S subunit. Also shown is the location mapped for 16S sequence 1392–1408 (58). Adapted from Noller & Lake (51).

RNase T1 (66). Thus a number of biochemical markers are available in the vicinity of the head near the initiation factor binding site.

Proteins L7 and L12—Implicated in GTP Hydrolysis—Form the Stalk of the Large Subunit.

Ribosomal proteins L7 and L12 have unique properties (67, 68). L12 is identical to protein L7 except that the amino terminus of L7 is acetylated. L7 and L12 are the only ribosomal proteins present in multiple copies, and the total number of L7 and L12 copies per ribosome is most likely four (69, 70). Both proteins are intimately involved in elongation factor Tu (EF-Tu), elongation factor G (EF-G), and IF-2–dependent GTP hydrolysis. The location of L7/L12 at the L7/L12 stalk (30) is illustrated in Figure 3.

The stalk is a constant feature of all ribosomes. Indeed, among eubacteria large subunits from gram negative bacteria, gram positive bacteria, cyanobacteria, and chloroplasts are nearly indistinguishable (71). More recent studies show the stalk is a relatively constant structural feature even across the archaebacterial, eubacterial, eocytic, and eukaryotic lineages (72). The L7/L12 proteins have been sequenced and characterized for several diverse organisms. The eubacterial L7/L12 sequence differs from the general form found in archaebacteria, eocytes, and eukaryotes in that it contains segments in a transposed order (73). This important landmark appears to be a universal ribosomal structure.

The Peptidyl Transferase and 5S rRNA are on the Central Protuberance of the Large Subunit.

The first evidence suggesting that the central protuberance was the site of the peptidyl transferase center was provided by immune electron microscopy. In particular, this was inferred from the location of the codon-anticodon site on the platform of the small subunit (2). More direct evidence was provided by a mapping of protein L27 on the side of the central protuberance opposite the L7/L12 stalk (14, 31), since L27 is consistently found among the proteins labeled by modified aminoacyl-tRNAs when they are bound to either the peptidyl site or to the aminoacyl site [for discussions, see (74, 75)]. Most recently the binding site of an analog of puromycin has been directly mapped on the central protuberance (76). Hence evidence for this site is quite strong.

The central protuberance was shown to also be the location of the 3' end of 5S rRNA in an important set of experiments by Shatsky et al (11). Subsequently, the location of the 3' end of 5S rRNA has been confirmed (77) and the 5' end of the 5S molecule has been mapped at the same region (78). This latter mapping is illustrated by electron micrographs of subunits that are coupled by covalently linked 3S and 5S rRNAs produced by *E. coli* RNase III mutant strain AB301–105 (Figure 8). As previously shown in Figure 4, the locations of the 3' end of the 23S rRNA (79), and of the rRNA binding proteins L1, L17 (31), and L18 (80) are also known. These sites provide important markers to follow the path of 23S rRNA through the large subunit. Using these markers, locations for parts of 23S rRNA have been tentatively proposed (51). These are summarized in Figure 9.

The Elongation Cycle.

During protein synthesis, amino acids are added to the growing polypeptide chain in a series of reactions, known as the elongation cycle. One of the major goals of studying ribosome structure is to learn what structural changes correspond to this cycle of additions. The biochemistry of the elongation cycle is quite well understood although the structural correlates of this process are only

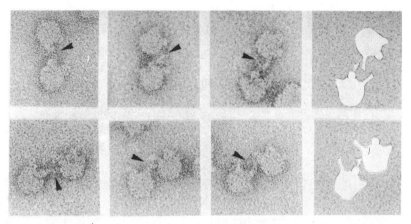

Figure 8 Pairs of large ribosomal subunits isolated from *Escherichia coli* rRNA processing mutant AB301–105. This strain is deficient in RNase III processing activity so that the primary rRNA transcript is inefficiently processed and 23S rRNA and 5S rRNA frequently remain attached. In the top row, the RNA strand leaves the lower subunit of the pair from the central protuberance (the 5' end of the 5S rRNA) and enters the other subunit just below the base of the L7/L12 stalk (the 3' end of the 23S rRNA). In the bottom row, the RNA strand leaves the right-most subunit just below the base of the L7/L12 stalk and enters the left-most subunit at its central protuberance. Adapted from Clark & Lake (78).

starting to become understood. Each iteration of the cycle consists of the participation of molecules called elongation factors (EFs) and also two molecules of GTP. At the beginning of the iteration a peptidyl-tRNA (a tRNA bearing the nascent chain) is bound to the ribosome at a site called the P site. Next an incoming aminoacyl-tRNA bearing the amino acid specified by the next unread codon on the mRNA binds to the R site (81, 82). It binds in complex with EF-Tu and GTP. The binding is controlled by the binding of a codon on the mRNA to three bases (the anticodon) on the tRNA. The tRNA is transferred to the A site, where its amino acid accepts the nascent chain. Finally, the P-site tRNA is ejected and the A-site tRNA takes its place. A cycle within a cycle prepares complexes of EF-Tu, GTP, and an aminoacyl-tRNA.

If I attempt to incorporate the data presented in this review on the locations of functional sites into the elongation cycle, then the models at the right of Figure 10 seem the most reasonable. The codon-anticodon site should almost certainly be placed on the platform near the cleft, and the bulk of data now suggest that the recognition site is on the external surface of the small subunit, as illustrated in Figure 10. The most likely locations for the A and P tRNA binding sites would seem to be in the cleft. Although the data support the choice of these sites, other alternatives are possible given the current tentative state of our knowledge [see for example (83–85)]. The locations in Figure 10 have been proposed previously (3, 81) and they are especially attractive because switch-

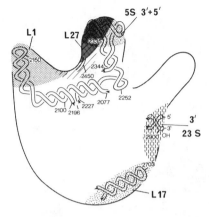

Figure 9 Hypothetical path of certain regions of 23S rRNA within the large ribosomal subunit. Locations of some of the ribosomal components that suggest this path have been indicated. Adapted from Noller & Lake (51).

ing of a tRNA from the R site to the A site can be accomplished by a conformational change of the anticodon loop from the 5' stacked configuration to the 3' stacked configuration (3, 81).

THE EXIT DOMAIN

In contrast with the detailed knowledge of the translational domain, relatively little was known until recently about the part of the ribosome now known as the ribosomal exit domain (21). In particular, all translational functions of the ribosome are found in the upper half of the small subunit and in the adjacent region of the large subunit. Information on the role of the other one half of the ribosomal surface is, by comparison, meager.

The Exit Domain is Opposite the Peptidyl Transferase.

In *E. coli,* information on the exit domain comes from immune mapping of the nascent chain as it emerges from the ribosome, a function associated with protein secretion rather than the translation steps just described. Using antibodies directed against the enzyme beta-galactosidase to map the exit site of the protein chain, it was found that the nascent chain exits from the ribosome at a single region, located on the large subunit (21). Electron micrographs of 70S ribosomes labeled with IgGs are shown in the upper half of Figure 11. In these micrographs the IgGs are bound at the end of the ribosome opposite the head of the small subunit. Hence, the nascent chain emerges from the large subunit nearly 150 Å from the peptidyl transferase located at the central protuberance of the large subunit. Both the peptidyl transferase (shown by the letter P) and the exit site (shown by the letter E) are indicated in Figure 4.

The Exit Domain is in the Same Relative Region in both Eukaryotic and Eubacterial Ribosomes.

As the structure of the *E. coli* ribosomes was studied more intensively, many of the newly discovered morphological features of the prokaryotic ribosome could

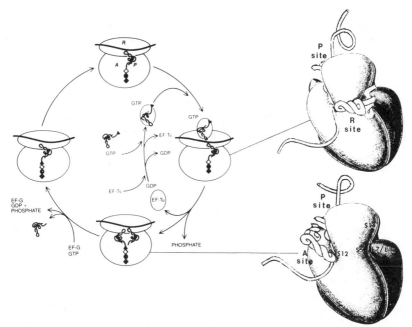

Figure 10 The elongation cycle of protein synthesis. At the right are shown the current "best guesses" for the locations of the tRNA binding sites during specific stages of the cycle (3).

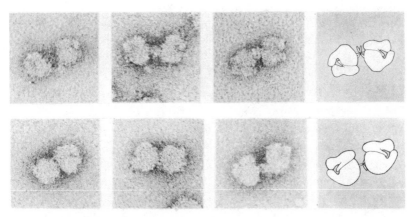

Figure 11 Electron micrographs of ribosomes linked by antibodies against nascent chains. *E. coli* ribosomes are in the top row and eukaryotic, cytoplasmic *(Lemna gibba)* ribosomes are in the bottom row. The orientation of the ribosomes is shown diagrammatically in the last frame of each row. Adapted from (94).

be related to the eukaryotic ribosome. When the location of eukaryotic initiation factor eIF-3 was mapped (65) on native small subunits from rabbit reticulocytes, for example, the existence of the platform in eukaryotic ribosomes was recognized. The main feature of the model developed for the eukaryotic small subunit at that time was that much of the additional mass in the large eukaryotic subunit was concentrated in the lower two thirds, or base. Subsequently, great progress has been made studying the structure of eukaryotic ribosomes. Three-dimensional models have been developed, immune mappings have been performed, and the structures have been analyzed, in projection and in three dimensions, by three-dimensional Fourier reconstruction and computer analysis (86–93).

Since many of the structural features present in the translational domain in prokaryotes are also present in eukaryotes, this suggested that the translation mechanisms might be functioning similarly in both. Even those translational functions that are quite different, such as initiation, seem to occur on comparable surfaces in ribosomes of both types [e.g., eIF-3, (65)]. For other aspects of ribosomal organization, particularly those involved with protein secretion and processing, it was thought that perhaps they could differ extensively in eukaryotic and prokaryotic ribosomes since the rough endoplasmic reticulum has no obvious counterpart in the prokaryotic cell. Thus we were interested in determining whether the exit domain had a similar arrangement in both prokaryotes and eukaryotes.

It now appears that the exit domain in both types of organisms has a similar organization. In particular, the exit site of the nascent protein chain in eukaryotic ribosomes (94), mapped using antibodies directed against the enzyme ribulose-1,5-bisphosphate carboxylase (i.e., Ribisco) emerges from the large subunit opposite the translational domain and in the same relative position as found in the *E. coli* ribosome. These micrographs are shown at the bottom of Figure 11 (94). Hence, in spite of the greater complexity of eukaryotic as compared with prokaryotic ribosomes, the overall organization of the exit domain on ribosomes, as reflected by the location mapped for the exit site, seems to be similar in both.

Eukaryotic Ribosomes Bind to the Rough Endoplasmic Reticulum through the Exit Domain.

Eukaryotic cells contain two populations of actively synthesizing ribosomes which are distinguished according to whether they are free in the cytoplasm or are bound to membranes. In general, the two populations synthesize different sets of proteins. In one population, the membrane-bound ribosomes of the rough endoplasmic reticulum (RER), each ribosome functions to synthesize proteins for export from the cell (95). During synthesis these proteins are vectorially discharged through the membrane (96). These ribosomes are attached to the RER by two types of interaction (97). One attachment can be released by treatment with the antibiotic puromycin, indicating that it occurs through an anchoring effect of the nascent chain. The other interaction is sensitive to high concentrations of monovalent salts and possibly involves integral membrane proteins (97–99) and a signal recognition particle (100).

The first clue to the existence of specialized ribosomal structures involved in protein secretion was provided by the work of Unwin (90). In the lizard *Lacerta sicula,* ordered arrays of membrane-bound ribosomes form in the ovarian follicles during its winter hibernation. Density maps calculated by three-dimensional reconstruction techniques from these arrays show a structure (RNA and/or protein) extending from the large ribosomal subunit into the rough endoplasmic reticulum (RER) linking the ribosome to the membrane. When the eukaryotic ribosome was compared with the *E. coli* ribosome (59), the attachment site of membrane-bound ribosomes was found to be near the exit site found in both eubacteria and eukaryotes.

Both sites are shown on the *E. coli* large subunit in Figure 4. The membrane site is indicated by the letter "M." By this criterion both the membrane binding site and the nascent chain exit sites are adjacent (within 60 Å of each other). Hence, in a general sense, protein synthesis appears to be organized into translational domains and exit domains in both prokaryotes and eukaryotes. Within each of these two major functional domains, the functional model illustrated in Figure 3 seems to be applicable to both prokaryotic and eukaryotic ribosomes.

EVOLUTION OF RIBOSOME STRUCTURE

In using ribosome structure to search for evolutionary relationships between eukaryotic and prokaryotic (eubacterial) ribosomes, one is immediately impressed by the apparent gulf between the two very different types of ribosome structures. Clearly, if one hopes to sort out evolutionary relationships between the two groups it would be helpful if other forms of ribosomes existed that could provide a baseline for judging which ribosome properties are ancestral and which are derived. One such group of organisms was recognized in 1977, when Fox & Woese proposed that archaebacteria represented a separate line of cellular descent (101). Another group of organisms, the eocytes, has been shown to have a unique ribosome structure (102, 103). Recently, these organisms, which have a sulfur based metabolism and thrive in thermal springs at temperatures above 90°C (104) have been proposed to represent a fourth aboriginal line of cellular descent (105).

These unusual groups of organisms provide a unique opportunity for studying the evolution of ribosome structure. Just as three-dimensional molecular structure has been used to measure bacterial evolution within lineages (106), the divergences among differing types of ribosomes (103) can be used to measure the evolution of cellular lineages.

Structure of Ribosomes from Archaebacteria, Eubacteria, Eocytes, and Eukaryotes.

Electron micrographs of small ribosomal subunits from eubacteria, from archaebacteria, from eocytes, and from eukaryotes (cytoplasmic ribosomes) (102, 103) are shown in the gallery in Figure 13. Because the ribosomal subunits are randomly oriented on the carbon support films used for electron microscopy, images vary according to their orientation. Hence, in Figure 13, images corresponding to the asymmetric projections of the *E. coli* small subunit and the quasisymmetric projection of the *E. coli* large subunit are displayed (2). Also differences between the eukaryotic, eocytic, archaebacterial, and eubacterial structures are most readily apparent in these projections. The eukaryotic structure and images corresponding to its characteristic projections (59, 65, 86, 87, 92, 103) were identified by analogy with the asymmetric projection of the eubacterial structure. The projections of the archaebacterial (103) and eocytic (102) subunits were similarly identified.

A field of eubacterial *(Escherichia coli)* small subunits is shown in Figure 12. Also present are several eukaryotic small subunits that have been included as size markers. They are quite apparent because of their larger size and different shape. (Four of them have been circled and a fifth is left unmarked for the reader to identify.) A striking feature of small subunits from eubacteria is the extreme constancy of the characteristic projections. Profiles, representing the extreme divisions of the eubacteria (103, 107) are nearly indistinguishable.

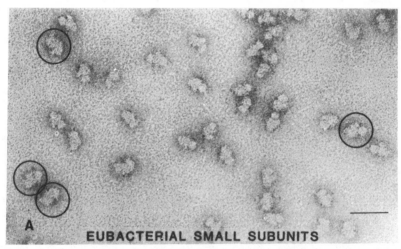

EUBACTERIAL SMALL SUBUNITS

Figure 12 Electron micrograph of a field of small ribosomal subunits from the eubacterium *Escherichia coli*. As a control four circled (and one not circled for the reader to find) eukaryotic subunits are present. They are generally larger and special features of them are described in the text. The scale bar represents 500 Å.

Figure 13 Electron micrographs of the ribosomes of representative eubacteria, archaebacteria, eocytes, and eukaryotes. From left to right, the organisms are *Synechocystis* 6701, a cyanobacterium (eubacteria); *Halobacterium cutirubrum,* an extreme halophile (archaebacteria); *Thermoproteus tenax,* a sulfur respiring anaerobe (eocyte): and *Saccharomyces cerevisiae,* a yeast (eukaryote). Small subunits are shown in row A, large subunits are shown in row C, and diagrams of their profiles are shown below each micrograph (in rows B and D).

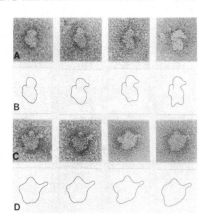

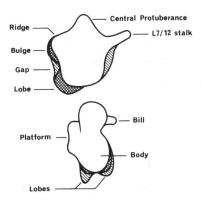

Figure 14 A summary of the structural features found in the four ribosomal types. Features found in eocytes (but absent in both archaebacteria or eubacteria) are indicated by diagonal (lower left to upper right) striping. Eukaryotic features (absent in both archaebacteria and eubacteria) are indicated by diagonal (lower right to upper left) striping. Crosshatched features are present in both eocytes and eukaryotes and lacking in both eubacteria and archaebacteria. Adapted from (102).

The structures of eukaryotic ribosomes, while somewhat more variable, are nevertheless also remarkably constant.

Representative electron micrographs of ribosomal subunits from eubacteria, archaebacteria, eocytes, and eukaryotes are shown in the first through fourth columns, respectively, of Figure 13. Small subunits are illustrated in row A and large subunits are in row C. These are interpreted in diagrams below each figure. Subunits from these four types of ribosomes contain several unusual structural features that are described below and shown in Figure 14.

Small subunits from archaebacteria, eocytes, and eukaryotes contain a structure that resembles a duck bill—the archaebacterial bill. Only a small (presumably vestigial) bill is present in eubacterial ribosomes. It extends from the head of the subunit and, in archaebacteria, is estimated from a comparison of its size with that of the L7/L12 stalk to have a molecular weight ($\pm$ SEM) of 44,000 $\pm$ 7,000. Biochemical and structural evidence, although indirect, suggests the bill may function in the factor-related steps of protein synthesis since the bill is located near the factor binding sites (103). The structure of the archaebacterial small subunit (shown schematically in Figures 13 and 14) is that of the eubacteria with the addition of the bill.

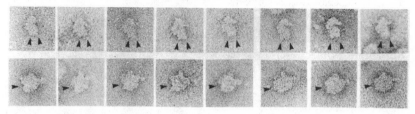

Figure 15   Electron micrographs of small subunits (top row) and large subunits (bottom row) of eocytes (left) and representative eukaryotes (right). The eocytic organisms, from left to right, are *Thermoproteus tenax, Sulfolobus acidocaldarius, Desulfurococcus mucosus, Thermofilum pendens* and *Thermococcus celer*. The eukaryotic organisms are, from left to right, *Tetrahymena thermophila* (a ciliate), *Saccharomyces cerevisiae* (a yeast), and *Triticum aestivum* (wheat). Adapted from (103).

Eocytic and eukaryotic small subunits, in addition to containing all the features of the eubacterial subunits, also contain the archaebacterial bill and possess additional structures at the bottom of the subunit called the eukaryotic lobes (see Figure 15). These lobes are absent in both eubacteria and archaebacteria, and are present in a reduced form in eocytes (102). The eukaryotic lobes are thought to be composed primarily of RNA and to contain the equivalent of about 300 nucleotides in eukaryotes. Three-dimensional reconstruction of the RNA and protein distributions in eukaryotic ribosomes has indicated that the region of the small subunit containing the lobes is predominantly RNA (91). Consistent with the base of the subunit being primarily composed of RNA, in *E. coli* no small subunit proteins have yet been mapped on the bottom of the subunit either by immune electron microscopy (28) or by neutron diffraction (39). The lobes are thought to correspond to eukaryotic "inserts" (variable blocks of 18S rRNA sequence inserted into the 16S sequences near positions 200, 580, and 1450 in the *E. coli* numbering system). Direct evidence for this is lacking, however. The function of the lobes is not known. The presence of the lobes at the base of the eukaryotic small subunit, however, provides a potential constraint on the folding of 18S rRNA (see Figure 6, for example).

The three-dimensional structures of eubacterial and archaebacterial large ribosomal subunits are similar. In particular, the halobacterial structure is nearly that found in eubacteria, while the structure of ribosomes from methanogens are somewhat more variable (103). Three structural features are present in the large ribosomal subunits of eocytes and/or eukaryotes that are lacking in eubacterial and archaebacterial large subunits (see Figures 14, 15). Lobes are present at the base of both eukaryotic and eocytic large subunits and a bulge near the L1 ridge (the "eocytic bulge") is also present in both. In eocytic large subunits these regions are separated by a gap and in eukaryotic large subunits they are not. Little is known about these regions; however, the eocytic bulge, judging from its location in the translational domain and its negative staining properties, probably contains a high protein content.

Both the eocytic lobe and the eocytic gap are in the exit domain near the nascent chain exit site. Nothing is known about their function or protein/RNA composition, however. In the following section we show how these features can be used to measure steps in the evolution of ribosome structure.

The Four Types of Ribosomes are Most Parsimoniously Fit by a Single Evolutionary Tree.

Within each of the four ribosomal types, three-dimensional ribosomal structure is relatively constant. Hence, the variations in structure between these four groups provide a phylogenetic basis for relating their evolution. If, as the constancy of ribosome structure within lineages at the resolution limit of electron microscopic images suggests, the individual structural features of each ribosomal type arose only once, then a parsimony analysis is appropriate. Of the three possible unrooted trees for connecting four taxa (108), only one dendrogram most parsimoniously fits the data in Table 1. This tree, shown in Figure 16, places the eukaryotes and eocytes on one side and the eubacteria and archaebacteria on the other. It is not the only possible interpretation of the data, but it is the simplest.

The unrooted tree shows the topology of the evolutionary divisions; it does not, however, show the direction of the flow of time. Thus, these data do not suggest which organisms are primitive. Other types of data, however, can be used to root the tree (109, 110). They tentatively suggest that the branching of eubacteria is that shown in Figure 17C. The rooted trees consistent with these

Table 1   Structural features of the four ribosomal types[a]

	Large subunit			Small subunit	
	Lobe	Filled gap	Bulge	Bill	Lobes
Eubacteria	−	−	−	−	−
Archaebacteria	−+	−	−	+	−
Eukaryotes	++	+	+	+	++
Eocytes	++	−	++	+	+
Thermofilum	++	−	++	+	+
Thermococcus	++	−	++	+	+
Desulforococcus	++	−	++	+	+
Sulfolobus	++	−	++	+	+
Thermoproteus	++	−	++	+	+
Thermoplasma	+	−	++	+	+

[a]Within eqcytes, the size of the 50S lobe varies from large (++ in *Thermoproteus tenax*) to intermediate (+ in *Thermoplasma acidophilus*). For this reason we regard *Thermoplasma* as representing a transitional organism. In archaebacteria, and to a lesser extent in eubacteria, a small 50S lobe is present.

Eubacteria   Eocytes

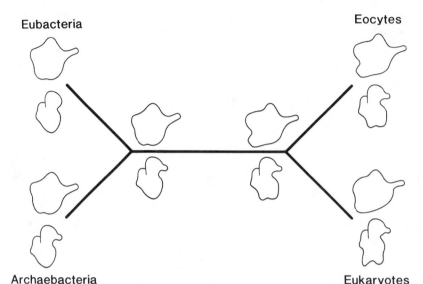

Archaebacteria   Eukaryotes

Figure 16   The unrooted dendrogram relating the steps in the evolution of taxa from the four lineages. This is the most parsimonious tree relating the four groups. Characters listed as "+ +" in Table 1 are assumed to represent ordered transitions from + + to + to −. The ribosomal subunits shown at the nodes correspond to the ribosomes present in the most parsimonious interpretation. In this unrooted tree, no assumptions have been made about the flow of time since our data do not directly provide this information.

constraints are the three trees in Figure 17*D-F*. If these arguments are correct, then certain ribosomal features, such as the bill, must be primitive. The three rooted trees also imply that eocytes and eukaryotes are "older" than (Figure 17*D* and *E*) or at least as old as (Figure 17*F*), the archaebacteria and eubacteria. It is interesting to note that all three rooted trees are consistent with the proposals that the ability to splice RNA is a primitive property (111, 112).

Eocytes have a Topologically Close Evolutionary Relationship to Eukaryotes.

The close, phylogenetic relationship of eocytes to eukaryotes, when compared with archaebacterial and eubacterial features, is reflected in the similarity of many molecular processes in eocytes and eukaryotes. Both the fundamental nature of these common properties and the variety of the processes involved supports the proposed phylogenic tree.

Many molecular properties of eocytes are more like those of eukaryotes than they are like those of archaebacteria and eubacteria. The DNA-dependent RNA polymerases of eocytes are composed of protein subunits with molecular weights and immunological properties that resemble those of eukaryotic polymerase A(I) more closely than the patterns found in archaebacteria and

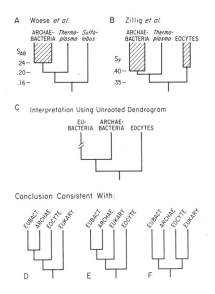

Figure 17 The three rooted trees that are consistent with data obtained from oligonucleotide catalogs and DNA-rRNA hybridization data. A S_{AB} of 1.0 indicates perfect oligonucleotide correspondence between 16S rRNAs of taxa A and B, and a S_{AB} of 0.0 indicates no common oligonucleotides. Similarly, S_F's of 1.0 and 0.0 represent 100% and 0% hybridization, respectively, between pairs of taxa. (*A*) Summary of the oligonucleotide catalog data using S_{AB} analysis (109). *Sulfolobus* is an eocyte. (*B*) Summary of the DNA.rRNA crosshybridization data using fractional stability, S_F, analysis (110). The sulfur-dependent bacteria are listed as eocytes. (*C*) The unique interpretation for the placement of eubacteria that is consistent with the unrooted dendrogram and with the data in *A* and *B*. (*D–F*) The three alternative trees that are consistent with the interpretation in *C*.

eubacteria (113). The secondary structures of 5S ribosomal RNA from *sulfolobus,* and from *Thermoplasma* to a lesser degree, while related to the eukaryotic 5S pattern, differ more from the archaebacterial and the eubacterial patterns than does even eukaryotic 5S rRNA (115). The initiator tRNAs of archaebacteria show a closeness to eubacteria (for example, the sequence of the initiator tRNA of *Halococcus* resembles those of eubacteria more than those of eukaryotes), while the sequence of *Sulfolobus* initiator tRNA is closer to those of eukaryotes, and that of *Thermoplasma* is intermediate (116). Significant amounts of long polyadenylylated sequences are found in *Sulfolobus* RNA that are similar to those in eukaryotic mRNAs, whereas only much lower amounts (1/30th) are found in eubacteria (117). The citric acid cycle enzymes, citrate synthase and succinate thiokinase, in *Sulfolobus* closely resemble the eukaryotic enzymes while the archaebacterial and eubacterial ones do not (118). Thus, it appears that many molecular properties support the close topological relationship of eocytes and eukaryotes.

In recent years, the view that "the concept of genealogical relationships is the starting point for any true understanding of microbial diversity (119)" has gained dominance. Even as early as 1946, Stanier & van Niel [cited in (119)] advocated a phylogenetic-based classification of organisms. "Even granting that the true course of evolution can never be known and that any phylogenetic system has to be based to some extent on hypothesis, there is good reason to prefer an admittedly imperfect natural system to a purely empirical one. A phylogenetic system has at least a rational basis, and can be altered and improved as new facts come to light; its very weaknesses will suggest the type of experimental work necessary for improvement (120)." Hence if the phylogenetic conclusions relating eocytes and eukaryotes, based in large part on ribosome structure, are borne out by subsequent molecular biological studies, then eocytes will require a taxonomic status equal to that of the other three kingdoms.

Thus, three-dimensional ribosome structure is providing information on the functioning, on the arrangement of components and of domains, on evolutionary diversity, and on the evolution of ribosomes and of the cells that contain them. Clearly these are exciting times for continuing investigations into the molecular biology of the ribosome.

Literature Cited

1. Noller, H. F. 1984. *Ann. Rev. Biochem.* 53:119–62
2. Lake, J. A. 1976. *J. Mol. Biol.* 105:131–59
3. Lake, J. A. 1981. *Sci. Am.* 245:84–97
4. Lake, J. A., Nonomura, Y., Sabatini, D. D. 1974. In *Ribosomes*, ed. M. Nomura, A. Tissiers, P. Lengyel, pp. 543–57. New York: Cold Spring Harbor Lab.
5. Lake, J. A., Pendergast, M., Kahan, L., Nomura, M. 1974. *Proc. Natl. Acad. Sci. USA* 71:4688–92
6. Vasiliev, V. D. 1974. *Acta Biol. Med. Ger.* 33:779–93
7. Lake, J. A., Kahan, L. 1975. *J. Mol. Biol.* 99:631–44
8. Boublik, M., Hellmann, W. 1978. *Proc. Natl. Acad. Sci. USA* 75:2829–33
9. Stoffler-Meilicke, M., Epe, B., Steinhauser, K. G., Woolley, P., Stoffler, G. 1983. *FEBS Lett.* 163:94–98
10. Boublik, M., Hellmann, W., Roth, H. E. 1976. *J. Mol. Biol.* 107:479–90
11. Shatsky, I. N., Evstafieva, A. G., Bystrova, T. F., Bogdanov, A. A., Vasiliev, V. D. 1980. *FEBS Lett.* 121:97–100
12. Dabbs, E. R., Ehrlich, R., Hasenbank, R., Schroeter, B.-H., Stoffler-Meilicke, M., Stoffler, G. 1981. *J. Mol. Biol.* 149:553–78
13. Vasiliev, V. D., Selivanova, O. M., Ryazantsev, S. N. 1983. *J. Mol. Biol.* 171:561–69
14. Lake, J. A. 1980. See Ref. 24, pp. 207–36
15. Clark, M. W., Leonard, K., Lake, J. A. 1982. *Science* 216:999–1000
16. Boublik, M., Hellmann, W., Roth, H. E. 1976. *J. Mol. Biol.* 107:479–90
17. Lake, J. A. 1982. *J. Mol. Biol.* 161:89–106
18. Stoffler-Meilicke, M., Stoffler, G. 1981. *Proc. Natl. Acad. Sci. USA* 78:6652–56
19. Bernabeu, C., Lake, J. A. 1982. *J. Mol. Biol.* 160:369–73
20. Girshovich, A. S., Kurtschaliov, T. V., Ovchinnikov, Yu, A., Vasiliev, V. D. 1981. *FEBS Lett.* 130:54–59
21. Bernabeu, C., Lake, J. A. 1982. *Proc. Natl. Acad. Sci. USA* 79:3111–15
22. Evstafieva, A. G., Shatsky, I. N., Bogdanov, A. A., Semenkov, Y. P., Vasiliev, V. D. 1983. *EMBO J.* 2:799–804
23. Vasiliev, V. D., Selivanova, O. M., Baranov, V. I., Spirin, A. S. 1983. *FEBS Lett.* 1557:167–72
24. Chambliss, G., Craven, G. R., Davies, J., Davis, K., Kahan, L., Nomura, M. eds. 1980. *Ribosomes: Structure, Function and Genetics.* Baltimore: Univ. Park 984 pp.
25. Wabl, M. R. 1974. *J. Mol. Biol.* 84:241–47
26. Lake, J. A., Pendergast, M., Kahan, L., Nomura, M. 1974. *J. Supramol. Struct.* 2:189–95
27. Tischendorf, G. W., Zeichardt, H., Stoffler, G. 1974. *Mol. Gen. Genet.* 134:187–208
28. Kahan, L., Winkelmann, D. A., Lake, J. A. 1981. *J. Mol. Biol.* 145:193–214
29. Winkelmann, D. A., Kahan, L., Lake, J. A. 1982. *Proc. Natl. Acad. Sci. USA* 79:5189–98
30. Strycharz, W. A., Nomura, M., Lake, J. A. 1978. *J. Mol. Biol.* 126:123–140
31. Lake, J. A., Strycharz, W. A. 1981. *J. Mol. Biol.* 153:979–92
32. Tischendorf, G. W., Zeichardt, H., Stoffler, G. 1975. *Proc. Natl. Acad. Sci. USA* 72:4820–24
33. Kastner, B., Stoffler-Meilicke, M., Stof-

fler, G. 1981. *Proc. Natl. Acad. Sci. USA* 78:6652–56
34. Lotti, M., Dabbs, E. R., Hasenbank, R., Stoffler-Meilicke, M., Stoffler, G. 1983. *Mol. Gen. Genet.* 192:295–380
35. Stoffler-Meilicke, M., Noah, M., Stoffler, G. 1983. *Proc. Natl. Acad. Sci. USA* 80:6780–87
36. Schindler, D. G., Langer, J. A., Engelman, D. M., Moore, P. B. 1979. *J. Mol. Biol.* 134:595–620
37. Moore, P. B., Engelman, D. M., Langer, J. A., Ramakrishnan, V. R., Schindler, D. G., et al. 1976. In *Brookhaven National Laboratories Neutron Symp.* Springfield, Va: Natl. Tech. Info. Serv., US Dept. Commerce
38. Chambliss, G., Craven, G. R., Davies, J., Davis, K., Kahan, L., Nomura, M., eds. 1980. See Ref. 24, pp. 111–34
39. Ramakrishnan, V. R., Yabuki, S., Sillers, I.-Y., Schindler, D. G., Engelman, D. M., Moore, P. B. 1981. *J. Mol. Biol.* 153:739–60
40. Lake, J. A. 1977. *Proc. Natl. Acad. Sci. USA* 7:1903–7
41. Nomura, M., Mizushima, S., Ozaki, M., Traub, P., Lowry, C. V. 1969. *Cold Spring Harbor Symp. Quant. Biol.* 34:49–61
42. Noller, H. F., Chang, C., Thomas, G., Aldridge, J. 1971. *J. Mol. Biol.* 61:669–79
43. Rummel, D. P., Noller, H. F. 1973. *Nature New Biol.* 245:72–75
44. Gorini, L. 1971. *Nature New Biol.* 234:261–65
45. Anderson, P., Davies, J., Davis, B. D. 1967. *J. Mol. Biol.* 29:203–15
46. Gorini, L. C. 1969. *Cold Spring Harbor Symp. Quant. Biol.* 34:101–11
47. Kuwano, M., Endo, H., Ohnishi, Y. 1969. *J. Bacteriol.* 97:940–43
48. Langer, J. A., Jurnak, F., Lake, J. A. 1984. *Biochemistry.* 23:6171–78
49. Nomura, M., Gourse, R., Baughman, G. 1974. *Ann. Rev. Biochem.* 53:75–117
50. Trempe, M. R., Ohgi, K., Glitz, D. G. 1982. *J. Biol. Chem.* 257:9822–29
51. Noller, H. F., Lake, J. A. 1984. In *Membrane Structure and Function*, ed. E. Bittar, 6:217–97. New York: Wiley
52. Lake, J. A. 1979. In *Transfer RNA: Structure, Properties and Recognition*, pp. 393–411. New York: Cold Spring Harbor Press
53. Cooperman, B. S. 1978. In *Bioorganic Chemistry. A Treatise to Supplement Bioorganic Chemistry, an International Journal*, ed. E. van Tamelen, 4:81–115. New York: Academic
54. Keren-Zur, M., Boublik, M., Ofengand,

J. 1979. *Proc. Natl. Acad. Sci. USA* 76:1054–58
55. Olson, H. M., Glitz, D. G. 1979. *Proc. Natl. Acad. Sci. USA* 76:3769–73
56. Shatsky, I. N., Mochalova, L. V., Kojouharova, M. S., Bogdanov, A. A., Vasiliev, V. D. 1979. *J. Mol. Biol.* 133:501–15
57. Politz, S. M., Glitz, D. G. 1977. *Proc. Natl. Acad. Sci. USA* 74:1468–72
58. Oakes, M., Clark, M., Henderson, E., Lake, J. A. 1984. *J. Cell. Biol.* (Abstr.) 99: no. 4, part 2, 13a
59. Lake, J. A. 1981. In *Electron Microscopy of Proteins*, ed. R. Harris, pp. 167–95. London: Academic
60. Heimark, R. L., Kahan, L., Johnston, K., Hershey, J. W. B., Traut, R. R. 1976. *J. Mol. Biol.* 105:219–30
61. Van Duin, J. J., Kurland, C. G., Dandon, J., Grunberg-Manago, M. 1975. *FEBS Lett.* 59:287–90
62. Bollen, A., Heimark, R. L., Cozzone, A., Traut, R. R., Hershey, J. W. B., Kahan, L. 1975. *J. Biol. Chem.* 250:4310–14
63. Langberg, S., Kahan, L., Traut, R. R., Hershey, J. W. B. 1977. *J. Mol. Biol.* 117:307–19
64. Schwartz, I., Kahan, L. Personal communication
65. Emanuilov, I., Sabatini, D. D., Lake, J. A., Freienstein, C. 1978. *Proc. Natl. Acad. Sci. USA* 75:1389–93
66. Yuki, A., Brimacombe, R. 1975. *Eur. J. Biochem.* 56:23–34
67. Moller, W. 1974. See Ref. 4, pp. 711–31
68. Weissbach, H., Pestka, S., eds., 1977. In *Molecular Mechanisms of Protein Biosynthesis*. New York: Academic
69. Hardy, S. J. S. 1975. *Mol. Gen. Genet.* 140:253–74
70. Subramanian, A. R. 1975. *J. Mol. Biol.* 95:1–8
71. Marquis, D. M., Fahnestock, S. R., Henderson, E., Woo, D., Schwinge, S. 1981. *J. Mol. Biol.* 150:121–32
72. Henderson, E., Oakes, M., Clark, M. W., Lake, J. A., Matheson, A. T., Zillig, W. 1984. *Science* 225:510–12
73. Matheson, A. T., Nazar, R. N., Willick, G. E., Yaguchii, M. 1980. In *Genetics and Evolution of RNA Polymerases, tRNA and Ribosomes*, ed. S. Osawa, H. Ozeki, H. Uchida, T. Yura, pp. 625–37. Tokyo: Tokyo Press
74. Traut, R. R., Heimark, R. L., Sun, T.-T., Hershey, J. W. B., Bollen, A. 1974. See Ref. 4, pp. 271–308
75. Cooperman, B. S. 1978. In *Bioorganic Chemistry. A Treatise to Supplement Bioorganic Chemistry. An International*

Journal, ed. E. van Tamelen, 4:81–115. New York: Academic

76. Olson, H. M., Grant, P. G., Cooperman, B. S., Glitz, D. H. 1982. *J. Biol. Chem.* 257:2649–56

77. Stoffler-Meilicke, M., Stoffler, G., Odom, O. W., Zinn, A., Kramer, G., Hardesty, B. 1981. *Proc. Natl. Acad. Sci. USA* 78:5538–42

78. Clark, M. W., Lake, J. A. 1984. *J. Bacteriol.* 157:971–74

79. Shatsky, I. N., Evstafieva, A. G., Bystrova, T. F., Bogdanov, A. A., Vasiliev, V. D. 1980. *FEBS Lett.* 121:97–100

80. Stoffler-Meilicke, M., Noah, M., Stoffler, G. 1983. *Proc. Natl. Acad. Sci. USA* 80:6780–84

81. Lake, J. A. 1977. *Proc. Natl. Acad. Sci. USA* 74:1903–7

82. Johnson, A. E., Fairclough, R. H., Cantor, C. R. 1977. In *Nucleic Acid-Protein Recognition,* ed. H. J. Vogel, pp. 469–90. New York: Academic

83. Woese, C. 1970. *Nature* 226:817–20

84. Ofengand, J., Gornicki, P., Nurse, K., Boublik, M. 1984. In *Alfred Benzon Symp. Proc.* In press

85. Spirin, A. S. 1983. *FEBS Lett.* 156:217–21

86. Boublik, M., Hellmann, W. 1978. *Proc. Natl. Acad. Sci. USA* 75:2829–33

87. Lutsch, G., Noll, F., Theise, H., Enzmann, G., Bielka, H. 1979. *Mol. Gen. Genet.* 176:281–91

88. Bommer, U.-A., Noll, F., Lutsch, G., Bielka, H. 1980. *FEBS Lett.* 111:171–74

89. Emanuilov, I., Sabatini, D. D. 1981. *Ultrastruct. Res.* 76:263–76

90. Unwin, P. N. T. 1979. *J. Mol. Biol.* 132:69–84

91. Kuhlbrandt, W., Unwin, P. N. T. 1982. *J. Mol. Biol.* 156:431–48

92. Frank, J., Verschoor, A., Boublik, M. 1981. *Science* 214:1353–55

93. Frank, J., Verschoor, A., Boublik, M. 1982. *J. Mol. Biol.* 161:107–37

94. Bernabeu, C., Tobin, E. M., Fowler, A., Zabin, I., Lake, J. A. 1983. *J. Cell Biol.* 96:1471–74

95. Siekevitz, P., Palade, G. E. 1960. *J. Biophys. Biochem. Cytol.* 7:619–44

96. Redman, C. M., Sabatini, D. D. 1966. *Proc. Natl. Acad. Sci. USA* 56:608–15

97. Adelman, M. R., Sabatini, D. D., Blobel, G. 1973. *J. Cell Biol.* 56:206–29

98. Sabatini, D. D., Kreibich, G. 1976. In *The Enzymes of Biological Membranes,* ed. A. Martonosi, 2:531–79. New York: Plenum

99. Chua, N. H., Blobel, G., Siekevitz, P., Palade, G. E. 1976. *J. Cell Biol.* 71:497–514

100. Walter, P., Blobel, G. 1982. *Nature* 299:691–98

101. Woese, C. R., Fox, G. E. 1977. *Proc. Natl. Acad. Sci. USA* 74:5088–90

102. Henderson, E., Oakes, M., Clark, M. W., Lake, J. A., Matheson, A. T., Zillig, W. 1984. *Science* 225:510–12

103. Lake, J. A., Henderson, E., Clark, M. W., Matheson, A. T. 1982. *Proc. Natl. Acad. Sci. USA* 79:5948–52

104. Zillig, W., Schnabel, R., Stetter, K. O. 1984. *Curr. Top. Microbiol.* In press

105. Lake, J. A., Henderson, E., Oakes, M., Clark, M. W. 1984. *Proc. Natl. Acad. Sci. USA* 81:3786–90

106. Dickerson, R. E. 1980. *Nature* 283:210–12

107. Henderson, E., Pierson, B. K., Lake, J. A. 1983. *J. Bacteriol.* 155:900–2

108. Fitch, W. M. 1971. *Syst. Zool.* 20:406–16

109. Woese, C. R. 1981. *Sci. Am.* 244:98–122

110. Tu, J., Prangishvilli, D., Huber, H., Wildgruber, G., Zillig, W., Stetter, K. O. 1982. *J. Mol. Evol.* 18:109–14

111. Darnell, J. E. 1978. *Science* 202:1257–60

112. Doolittle, W. F. 1978. *Nature* 272:581–82

113. Zillig, W., Stetter, K., Schnabel, R., Madon, J., Gierl, A. 1982. In *Archaebacteria,* ed. O. Kandler, Stuttgart: Fisher. 1983. C3:218–27

114. Woese, C. R., Kaine, B., Gupta, R. 1983. *Proc. Natl. Acad. Sci. USA* 80:3309–14

115. Fox, G. E., Luehrsen, K. R., Woese, C. R. 1982. See Ref. 113, pp. 330–45

116. Kuchino, Y., Ihara, M., Yabusaki, Y., Nishimura, S. 1982. *Nature* 298:684–85

117. Ohba, M., Oshima, T. 1982. See Ref. 113, p. 353

118. Danson, M. J. 1984. *16th Meet. FEBS,* p. 240 (Abstr.) Moscow, U.S.S.R.

119. Balch, W. E., Fox, G. E., Magrum, L. T., Woese, C. R., Wolfe, R. S. 1979. *Microbiol. Rev.* 43:260–96

120. van Niel, C. B. 1946. *Cold Spring Harbor Symp. Quant. Biol.* 117:285–301

Ann. Rev. Biochem. 1985. 54:531–64

VIROIDS

Detlev Riesner

Institut für Physikalische Biologie, Universität Düsseldorf, Universitätsstraβe 1, D-4000 Düsseldof 1, West Germany

Hans J. Gross

Institut für Biochemie, Bayerische Julius-Maximilians-Universität, Röntgenring 11, D-8700 Würzburg, West Germany

---

CONTENTS

531

0066-4154/85/0701-0531$02.00

INTRODUCTION

If one looks into modern textbooks of biochemistry, molecular biology, or microbiology, one can even nowadays find the notion that small RNA viruses are the smallest self-replicable structures at the lowest level in the hierarchy of life. It is the purpose of this review to make the biochemically oriented scientific community acquainted with the new and independent class of extremely small pathogens, the viroids. Viroids are an exciting and challenging problem, not only for molecular biology, but also for virology, plant pathology, biophysics, and other related areas. In this review we concentrate on the recent biochemical studies that led to our present knowledge of viroid structure, replication, and pathogenesis. The remarkable increase in our knowledge of viroids may be seen if one compares the present review with similar reports a few years ago (1, 2). The reader who is interested in the details of phytopathology and microbiology may refer to the extensive descriptions of Diener (3) and Sänger (4). Collections of up-to-date research reports on viroids were given in two symposium reports (5, 6).

Viroids are single-stranded, circular ribonucleic acids (RNAs) of a few hundred nucleotides, which occur as pathogens only in higher plants. Tomato plants infected with viroid strains of different virulence are shown in Figure 1. Viroid diseases of economically important plants have been observed since the first quarter of the twentieth century; however, they were initially thought to be caused by classical plant viruses. It was in fact the search for the presumable pathogenic viruses which led to the discovery of viroids. An extensive account of these discoveries is given in T. O. Diener's book (3).

The first pathogen to be recognized as a viroid was the agent causing the spindle tuber disease of potato, PSTV, which had long been thought to be a virus. The first evidence that this latter idea might be wrong was reported by Diener & Raymer (7), who showed that the pathogen is a free RNA and that conventional virus particles are not present in the infected plant. When the low molecular weight of this pathogen was realized (8, 9), the term "viroid" was proposed in order to differentiate these small, protein-free infectious RNAs from conventional viruses which have an encapsidated genome (8). The next plant pathogens to be recognized as viroids were the agent of the citrus exocortis disease CEV (10, 11), and the viroid of the chrysanthemum stunt disease CSV (12, 13). Until now, about ten different viroids have been detected, the majority of them causing diseases of economically important crop plants (cf Table 1). The cadang-cadang viroid of the coconut palm, for example, causes severe losses in the vast palm plantations and hence seriously threatens the economy of a whole country, the Philippines (14, 15). It is hoped that a detailed knowledge of viroid structure, replication, and viroid-host interaction might finally contribute to control their propagation and thus to decrease the enormous losses they cause in agriculture.

Figure 1   Tomato plants (Cultivar Rutgers) 8 weeks after infection with PSTV field isolates that differ in their virulence. From left to right infected with a lethal, severe, intermediate, and mild PSTV isolate, respectively, and the uninfected control plant. The figure is reproduced with permission from Sänger (4a).

In some respects, a phenomenon similar to viroids are the "virusoids." These are RNA molecules with viroid-like size and structure that are encapsidated together with a larger RNA in several plant viruses (15a, 16, 17). They do not replicate autonomously, but some behave as satellite RNAs of conventional viruses (18), and others can be an integral part of the replicative machinery of a virus (19). Virusoids will not be reviewed systematically in this article but will be taken into consideration for comparison.

PREPARATION AND DIAGNOSIS

Sources

Viroids have a fairly wide host range. This implies that viroid RNA may be prepared not only from the host in which the viroid was originally discovered, but also from other plants that may be more suitable for viroid preparation. In many cases transfer of viroids to other host plants does not lead to the expression of symptoms; therefore other diagnostic methods have to be used. A comprehensive list of species and families that serve as viroid hosts is given by Diener (3).

Table 1 Viroid diseases

Viroid disease	Abbreviation of viroid	Reference
Potato spindle tuber	PSTV	Diener, 1971 (8); Singh & Clark, 1971 (9)
Citrus exocortis	CEV	Sänger, 1972 (10); Semancik & Weathers, 1972 (11)
Chrysanthemum stunt	CSV	Hollings & Stone, 1973 (12); Diener & Lawson, 1973 (13)
Chrysanthemum chlorotic mottle	CCMV	Romaine & Horst, 1975 (20)
Cucumber pale fruit	CPFV	Van Dorst & Peters, 1974 (21); Sänger et al, 1976 (22)
Coconut cadang-cadang	CCCV	Randles, 1975 (23); Randles et al, 1976 (14)
Hop stunt	HSV	Sasaki & Shikata, 1977 (24)
Columnea erythrophae[a]		Owens et al, 1978 (25)
Avocado sun blotch	ASBV	Thomas & Mohamed, 1979 (26); Mohamed & Thomas, 1980 (27)
Tomato apical stunt[b]	TASV	Walter, 1981 (28)
Tomato planta macho	TPMV	Galindo et al, 1982 (29)
Burdock stunt	BSV	Chen et al, 1983 (30)

[a]The viroid infection in the ornamental plant *Columnea erythrophae* is symptomless.
[b]In earlier reports on the disease it was called tomato bunchy top disease.

Many species have been reported to be susceptible to infection with PSTV (cf 3), most of which belong to the family of Solanaceae. Tomato apical stunt viroid is similar to PSTV in respect to the host range. Citrus exocortis viroid replicates not only in citrus plants but also in Compositae, e.g., *Gynura aurantiaca,* and in Solanaceae. Chrysanthemum stunt viroid and chrysanthemum chlorotic mottle viroid could be transferred only to other species of the Compositae. Also the cucumber pale fruit viroid seems to be restricted mainly to the family of its original host, i.e. Cucurbitaceae; the only other host reported so far is tomato (22). Hop stunt viroid also infects Solanaceae, cucurbitaceae, and Moraceae (31). Infections with the avocado sunblotch viroid have been found only in avocado and cinnamon which are both members of the Lauraceae (32). The host range of the coconut cadang-cadang viroid is distinct from all other viroids, because it has been found only in coconut palms, and this viroid is the only viroid that infects monocotyledons.

The transfer of viroid infections to experimental host plants is important for biochemical studies, because the artificial hosts in some cases grow much faster, under controlled climate conditions, and their tissue is easier to handle for preparative isolations of the viroid. Tomato plants have proven particularly favorable experimental host plants. It has been shown that PSTV and CEV undergo only minor changes after transfer from tomato to *Gynura* (33).

As a general rule, expression of symptoms occurs in tomato plants 10–14 days after inoculation, if the plants are grown under high light intensity and at elevated temperature (30–35°C during the day, around 20°C during the night; cf 3). This period may be prolonged to 2–3 months under growth conditions of 20°C. It has been shown (34) that viroid replication and expression of symptoms take place nearly at the same time and that optimum yield of viroid RNA is obtained 3–6 weeks after inoculation.

As viroids infect the plant systemically, they may in principle be prepared from every part of the plant. If infected tomato plants were grown under optimum greenhouse conditions, a purification procedure combining a Cs_2SO_4-density-gradient centrifugation and a high performance liquid chromatography (HPLC) yielded the following amounts of PSTV: from 0.5-kg leaves: 500 μg PSTV; from 0.5-kg stems: 300 μg PSTV; and from 0.5-kg roots: 180 μg PSTV (35). Most preparation procedures start from leaves, but stems may be used as well if they are easy to homogenize as in the case of tomato plants.

Viroids can be grown continuously in cell suspensions, in protoplasts, and in callus cultures as shown experimentally in cultures from tomato (36–39) and from potato (36, 37). Up to now, however, these cultures have been used solely for mechanistic studies on viroids and not for isolating viroids.

Diagnosis

Viroid infections cannot be recognized unambiguously by their symptoms. Even after transfer of viroid infections to "diagnostic hosts" and judgment by an experienced phytopathologist, diagnosis by detection of symptoms is neither safe nor generally applicable. Viroids can also not be identified by serological methods because antibodies against viroid RNA could not be obtained in spite of several attempts (40).

Today, correct diagnosis is achieved by gel electrophoretic detection of a viroid band or by molecular hybridization. The gel electrophoretic methods include homogenization of one or a few grams of tissue, nucleic acid extraction, in some cases removal of high-molecular-weight nucleic acids by a LiCl precipitation, and an ethanol precipitation of the remaining nucleic acids. The resulting crude nucleic acid extract is analyzed on polyacrylamide slab gels. Either 5% polyacrylamide gels (41–43) for viroids of molecular weights up to 120,000 (e.g. PSTV) or 2.5–3% for larger viroids as cadang-cadang (23) were applied. Due to the addition of 8M urea, narrower bands were obtained (22). The native secondary structure of viroids is maintained. An appreciable increase in sensitivity and decrease in time needed for the electrophoretic diagnosis was achieved when a run under native conditions was combined with a consecutive run under denaturing conditions in a two-dimensional gel electrophoresis (44). This procedure resulted in a clear separation of circular viroid molecules from all cellular nucleic acids and may be applied to completely unfractionated crude extracts. Together with the silver staining method (45) as little as 60-ng viroid RNA/g tomato leaf tissue and 600 pg in a single band were detected; more recent values are 2-ng viroid/g tomato tissue and 80 pg in single band (J. Schumacher, unpublished). The "bidirectional" combination of the two electrophoretic runs facilitates a routine application to many tissue samples simultaneously (44).

Diagnostic tests for viroids that are based on hybridization of highly radioactive complementary DNA have been reported for several viroids. The assays were carried out by dot-spot hybridization with cloned cDNA for PSTV (46, 47), CSV (56), and HSV (48), or by liquid hybridization with cDNA probes for cadang-cadang RNA (49) and ASBV (50, 51). ASBV was also detected by self-hybridization with a (^{32}P)-labeled ASBV-RNA (52). The sensitivities reported for the hybridization assays were between 20 and 80 ng viroid per gram of infected tissue. More recent hybridization technique as with highly labeled M13 probes may exceed the published values probably by one order of magnitude. The gel electrophoretic test is faster, about 1 day compared to 4 days, but hybridization may handle more samples. Gel electrophoresis is sensitive to the specific viroid structure, i.e. the circularity, and cannot differentiate between different sequences, whereas hybridization merely detects specific sequences regardless of the structure. Therefore, both methods clearly complement each other's properties.

Molecular Cloning

At present, cloned viroid sequences find many applications in viroid research: (*a*) sequencing, (*b*) hybridization for viroid diagnosis, studies on intermediates in viroid replication, and cellular distribution of viroids, and (*c*) infectivity of cloned cDNA and site-directed mutagenesis.

Viroid cloning follows standard procedures which need not be outlined here. Table 2 summarizes the full length and oligomeric-sized clones reported together with their plasmids, sites and direction of insertion, and promoters. In addition, partial clones for CPFV (59), TPMV, and TASV (60) have been described.

Purification

Viroid purification proceeds mainly in two steps, (*a*) preparation of a low-molecular-weight RNA extract and (*b*) final purification of viroid RNA from this extract.

In most procedures (3, 4, 61), total nucleic acid is extracted from the homogenized tissue by a buffer/phenol system containing the detergent SDS and the RNase-inhibitor bentonite. Polysaccharides and DNA are separated from the RNA by a combination of two phase extraction systems and precipitations with cetyltrimethylammoniumbromide. High-molecular-weight RNA is removed by 2 M LiCl precipitation. An enrichment of the viroid content in the

Table 2 Viroid cDNA clones

Viroid	Plasmid	Promoter	Insertion site	Size and direction	Reference
PSTV	pGL 101 H	lac	HaeIII (146)	monomeric (+) and (−)	Cress et al, 1983 (53)
PSTV	pBR 322	tet	Bam H1 (87)	monomeric (+) and (−)	Cress at al, 1983 (53)
				dimeric (+) and (−)	
PSTV	pBR 322	—	Bam H1 (87)	monomeric (+)	van Wezenbeek et al, 1982 (54)
HSV	pGL 101	lac UV5	Eco R1 (296)	monomeric (+) and (−)	Ohno et al, 1983 (55)
				dimeric (+) and (−)	Ohno et al, 1983 (48)
				tetrameric (+) and (−)	
CEV	pBR 322	—	Bam H1 (87)	monomeric (+)	Visvander & Symons, 1983 (57)
ASBV	pBR 322	—			Murphy [cited in (58)]

low-molecular-weight RNA extract may be obtained by several chromatographic steps. More recently, very similar procedures have been applied for the
preparation of a crude extract from hop stunt viroid (62), TPMV and TASV
(60).

In a different procedure (35) the low-molecular-weight RNA extract was
obtained after a single Cs_2SO_4-density-gradient centrifugation; i.e. polysaccharides, DNA, high-molecular-weight RNA, and other components of the
tissue, could be removed in a single step. A combination of 2 M LiCl precipitation and subsequent precipitation from the supernatant with 0.5 volume of
ethanol yields an RNA extract without DNA and high-molecular-weight RNA
with little tRNA content (63). The precipitation of RNA from high salt solution
with low concentrations of ethanol could also be applied for the fast recovery of
low-molecular-weight RNA from the Cs_2SO_4 gradient if the corresponding
fractions were adjusted to a final density of 1.18 g/cm^3 (J. Schumacher,
personal communication).

In the second step, viroids are purified from the crude RNA extract to
homogeneity by polyacrylamide gel electrophoresis. Preferably, two subsequent runs are carried out, the first under native, the second under denaturing
conditions (4, 22, 64). The disadvantages of the preparative gel electrophoresis, i.e. low yield and loss of time, can be circumvented by the
application of a newly developed ion-exchanger resin for the application in
high-performance liquid chromatography (35, 65). From crude RNA extract,
prepared by Cs_2SO_4-density-gradient centrifugation, 200 μg homogenous
PSTV RNA have been obtained from a single chromatographic run. Both the
yield and the purity were over 95%.

STRUCTURE

Primary and Secondary Structure

The characterization of viroids as protein-free RNA (7), the evidence for the
small size (8–11), and the detection of their circularity (22, 66) were prereq-
uisites for studies of their nucleotide sequence and their complex secondary
structure. However, important details of the structure, as rod-like shape (67,
22) and single-strandedness (22), became known together with a secondary
structure model consisting of loops and helixes (68), before the nucleotide
sequence had been established.

For various reasons discussed earlier (2), only one method appeared to be
promising for successful elaboration of the complete nucleotide sequence of a
viroid: in vitro 5'-labeling of viroid fragments obtained by complete or partial
digestion with specific ribonucleases with 5'-polynucleotide kinase and (γ-^{32}P)
ATP. The first complete viroid structure, that of PSTV, was established by the
application of these procedures (69). This confirmed unambiguously the circu-

larity and revealed several features of the structure that turned out to be typical for all viroids. Modified nucleotides such as those present in tRNA were not detected.

The secondary structure of viroids was derived both from experimental results and from theoretical calculations. The physico-chemical approach is reviewed in the following section. The secondary structure of viroids in the form of an unbranched series of short double helixes and small internal loops became evident already from the original work on PSTV (69, 70) and has been confirmed by all sequences subsequently determined (71–74).

Later, a second sequencing strategy was applied successfully: transcription of linear viroid into cDNA, cloning, and sequencing of the cloned viroid cDNA by the established DNA sequencing techniques (48, 60, 57, 59, 75). Although direct sequencing by in vitro labeling had revealed that viroid populations contain sequence heterogeneities, sequencing of many cDNA clones of one viroid allows establishment of the individual sequences of variants even in those cases where the heterogeneity is not detected by direct RNA sequencing (57).

In Figure 2 all known viroid sequences are depicted in their secondary structures, which are taken from the original publications if not stated otherwise in the figure legend. Comparison of the sequences revealed that there are at least three groups of viroids:

1. The PSTV-group, which includes PSTV, TPMV, TASV, CEV, CSV, and to a lesser extent HSV and CPFV,
2. ASBV, and
3. CCCV.

In Table 3 some quantitative data on the sequences and the secondary structures are collected. The members of the PSTV group are characterized by a sequence homology of about 60–70% or more. Their G:C, A:U, and hence their purine:pyrimidine ratio are around one, and there are about twice as many G:C as A:U pairs. Between 4 and 16% of the base pairs are G:U pairs. Near the middle of the viroid secondary structure, there exists a region of high sequence homology shown as a boxed region in Figure 2. A long polypurine stretch is located near this conserved central region, and the sequence homology is higher in the left half of the secondary structure than in the right half. The left half is thermodynamically less stable than the right half (cf STRUCTURAL TRANSITIONS section). Obviously the secondary structure of these viroids can be divided into a left part which needs conserved sequences together with (or in order to maintain) a certain extent of structural flexibility due to a decreased thermodynamic stability, and a right half, for which stability is more important than the nucleotide sequence.

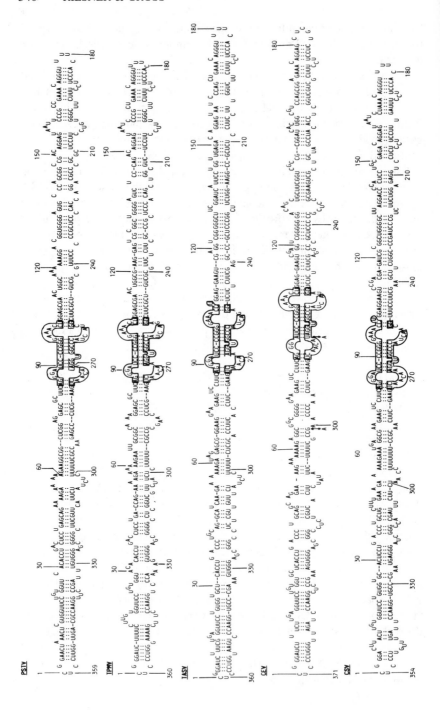

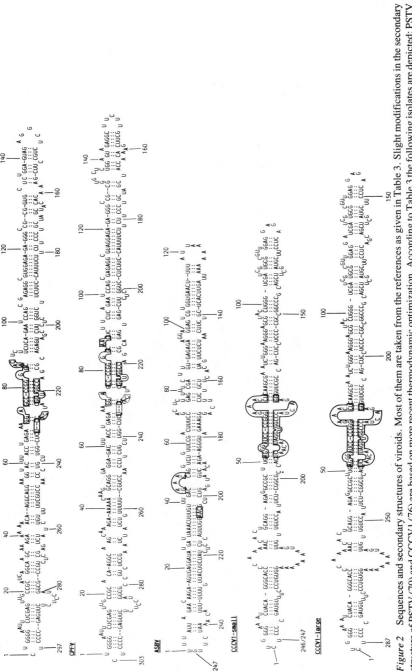

Figure 2 Sequences and secondary structures of viroids. Most of them are taken from the references as given in Table 3. Slight modifications in the secondary structure of PSTV (70) and CCCV1 (76) are based on more recent thermodynamic optimization. According to Table 3 the following isolates are depicted: PSTV "Type," CEV "C," CCCV1-small with 247 nucleotides (cf C*), CCCV1-large "Baao 54." The sequences in the boxed regions are the highly conservative regions.

Table 3 Primary and secondary structure of viroids[a]

	Nucleotides					Sequence homology		Numbers of nucleotides changed among isolates			Base pairs					References
	Total	A	U	G	C	with PSTV (%)	among isolates (%)	Exchange	Insertion	Deletion	Degree (%)	Total number	A:U (%)	G:C (%)	G:U (%)	
PSTV Type	359	73	77	101	108	100	100	—	—	—	70	126	29	58	13	69
Mild	359	70	80	101	108		99	2	1	1	71	128	30	57	13	71
Severe	359	74	77	100	108			4	0	0	69	124	30	58	12	4
TPMV	360	72	81	99	108	83	99	—	—	—	68	122	31	60	9	60
TASV	360	70	90	101	99	73		—	—		73	131	32	57	11	60
CEV A	371	72	75	112	112	73	100	—	—	—	69	128	28	56	16	75
C	371	73	76	110	112		99	4	0	0	69	128	28	56	16	73
AM	371	72	76	110	113		99	4	0	0	67	125	28	56	16	57
DE25	371	71	76	112	112		99	4	0	0	69	128	28	56	16	57
DE26	371	69	80	112	110		93	15	6	6	68	126	28	58	14	57

CSV E	354	74	93	90	97	73	100	—	—	—	69	122	34	56	10	73
A	356	75	93	89	99	69	97	6	4	2	70	124	35	52	13	72
HSV	297	61	69	79	88	55	100	—	—	—	67	100	29	64	7	48
CPFV[b]	303	64	70	81	88	55	97	8	7	1	69	105	31	65	4	59
ASBV	247	68	85	51	43	18		—	—	—	67	83	51	34	14	58
CCCV1																
small[c]	246/7	53	47	73	73/4	11	100	—	—	—	65	80	24	69	8	74
large[d] Baao 54	287	59	58	86	84		100	0	41	0	64	92	25	67	8	74
Ligao 14B	296	59	59	91	87		100	0	50	0	65	96	24	68	8	74
Ligao T1	301	60	59	93	89		100	0	55	0	64	97	24	68	8	74
San Nascisco	297	59	59	91	88		100	0	51	0	65	96	24	68	8	74

[a] The data are taken from the references as given, from Sänger, (4) from Sano et al (59), and from the secondary structures in Figure 2.

[b] CPFV is considered in respect to its sequence homology as a variation of HSV (cf 59).

[c] CCCV1 small is a mixture of variants, one containing C_{197} and one containing CC in that position. CCCV2 small and large are not contained in the Table because they are exact duplicates of the corresponding CCCV1 RNAs.

[d] All isolates of CCCV1 large differ from CCCV1 small by the duplication of a sequence at the right end of the secondary structure. Therefore, sequence homology is 100%. The duplicated region is counted under "insertion." The isolates Baao 54, Ligao 14B, and Ligao T1 are derived from CCCV1 small with 246 nucleotides; only San Nascisco is derived from CCCV1 small with 247 nucleotides.

For two viroids of the PSTV group, namely PSTV itself and CEV, different isolates ("mutants") are known which differ by their sequence and partially in their virulence i.e. symptom expression.

ASBV and CCCV differ from each other as much as they differ from the viroids of the PSTV group. For example, ASBV and CCCV do not have a polypurine sequence. In CCCV the conserved central region is found, whereas in ASBV only a common sequence GAAAC is found. GAAAC is also found in virusoids, in the central region of the secondary structure (cf Figure 4). Taking into account most of the data of Table 3, ASBV shows the lowest homology with other viroids.

In addition, CCCV differs from all other viroids because four CCCV variants of different length are associated wih the cadang-cadang disease of coconut (77). They are called "CCCV1-small" (246 nucleotides, cf Figure 2) and its exact duplicate "CCCV2-small" (492 nucleotides). They occur in the early stages of the disease, whereas two additional RNAs appear late, i.e. after years of infection ("CCCV1-large," 287 nucleotides, and "CCCV2-large," 574 nucleotides). CCCV1-large differs from CCCV1-small in a duplication of 41 (isolate "Baao 54") nucleotides at the right end of the secondary structure. The size of the duplicated region varies between 41 and 55 among different CCCV isolates (74).

Physical Properties of Viroids in Solution

The physical properties of several viroids have been outlined already in detail in review articles (2–4, 78). Therefore the main features will be repeated here only briefly and completed by a few more recent studies.

MOLECULAR WEIGHT   Molecular weights had been estimated originally from gel electrophoresis, sucrose gradients, and electron micrographs (cf 3). The first accurate molecular weights were obtained from equilibrium sedimentation runs in the analytical ultracentrifuge (22). Since at present the sequences of viroids are known, the most accurate molecular weights may be calculated from the sequences. Good values will be obtained with an average molecular weight of 333 per nucleotide including bound cations (22).

SHAPE   Viroids show up in electron micrographs as rod-like structures if prepared under native conditions (e.g. 0.1 M NaCl, pH 7.0). This result, long known with PSTV (67), was found also with CPFV (22), CCCV (79), CSV (64), and HSV (62). From velocity sedimentation studies on PSTV and the four forms of CCCV (80) it could be concluded that viroids also have a rod-like structure in solution. Furthermore, they show in solution some flexibility which may quantitatively be characterized by a unique persistence length (persistence length is roughly the radius of the circle to which the polymer may be bent) of

300 Å; for comparison, the persistence length of double-stranded DNA is 600 Å; that of double-stranded RNA has not yet been determined. From the analysis of the sedimentation coefficients it could also be derived (80) that even the duplicate forms of CCCV, the so-called RNA2, assume in solution the extended form, whereas Haseloff et al (74) concluded on the basis of the sequence that extended as well as cruciform structures were possible. Also ligand interaction studies (cf next section) are in accordance with a rod-like structure.

SECONDARY STRUCTURE   As mentioned before, the secondary structure of viroids was derived both from experimental results as well as from theoretical calculations. Experiments with chemical modification, dye binding, oligonucleotide binding, and evaluation of the phosphodiester bonds accessible to enzymatic attack [reviewed in (2)] showed clearly the presence of single-stranded as well as double-stranded regions. It was also derived that most of the molecule is accessible to ligand interaction and not covered by an additional tertiary structure leading to a globular shape. The fact, however, that the single-stranded regions and the double helixes were arranged in serial manner without bifurcation could only be concluded from the quantitative evaluation of thermal denaturation curves of viroids (68, 70).

Theoretical derivations of the secondary structure were performed at three levels. First, the extended rod shape of the molecule was taken from the experimental results and, by the trial and error method, base pairing schemes were optimized for maximum numbers of base pairs (e.g. 69). Second, at a higher level of refinement, a semi-empirical search for thermodynamically most stable structures was carried out by the method of Tinoco et al (81). Finally, the algorithm of Nussinov & Jacobson (82) and of Zuker & Stiegler (83) was applied for a nearly exact calculation of the structure of lowest free energy. Recently, the program of Zuker & Stiegler was modified slightly (84a) to account for circular strands and more recent values for the stabilities of base pairs and loops. In summary, the very simple method of optimizing for maximum numbers of base pairs and the most sophisticated algorithm resulted in nearly identical secondary structure; differences were found only in limited regions of the molecule. Very similar secondary structures were also obtained when different sets of elementary stability data were used. As an example, the secondary structures of PSTV published by Riesner et al (70) or Nussinov & Jacobson (82) agree mostly. It may safely be stated that viroid structure is an unambiguous structure, and therefore even rough estimations lead to nearly the right structure. In Table 4 the stabilities of the secondary structures of several viroids are characterized by the mean free energy of structure formation $\Delta G/N$.

STRUCTURAL TRANSITIONS   The principles of the structural transitions of viroids have been reviewed earlier (2, 78, 84). Besides a brief résumé of the

Table 4 Thermodynamic parameters of the structural transitions of viroids, virusoids, and random sequences[a].

	$\Delta G/N$ [kJ/mol]	$T_m[°C]$	$\Delta T_{1/2}[°C]$	Mechanism
Viroids				
PSTV	1.67	51	0.9	
CEV	1.62	51	1.0	formation
CSV	1.61	48.5	1.1	of stable
CCCV1-large	1.53	49.1	1.2	hairpins
CCCV1-small	1.53	49.1	1.4	
ASBV	1.13	37.5	1.5	not determined
Virusoids				
SNMV2	1.43	38	2.8	—
VTMoV2	1.32	38	2.0	—
Random sequences	1.27±0.1	36±5	>5	—

[a]For random sequences the number of the nucleotides, and the content of A, U, G, and C are taken from PSTV. The parameters listed except $\Delta G/N$ (difference in free energy per nucleotide during formation of the native structure from the completely coiled circle) and except the results on random sequences are from experiments. The values for random sequences are average values of 5 different sequences. T_{m_-} and $\Delta T_{1/2}$-values refer to an ionic strength of 0.011 Na$^+$, pH 6.8. Data are taken from Riesner et al (84) and G. Steger et al (84a).

results on PSTV, more recent studies carried out on other viroid species are described here. The structural transitions of viroids have been studied by thermal denaturation curves (68), by microcalorimetry (85), and by kinetic methods as the temperature-jump technique (86). The melting curve, whether recorded by UV-hypochromicity or by differential calorimetry, exhibits a narrow main transition at around 50°C in 0.011 Na$^+$, pH 6.8. The midpoint-temperature (T_m) is about 20°C lower than the T_m-value of double-stranded DNA and about 30°C lower than that of double-stranded RNA, both with comparable GC-content. The T_m-value of all viroid-species measured are listed in Table 4 together with the halfwidths of the transitions $\Delta T_{1/2}$ and the calculated mean free energies per nucleotide $\Delta G/N$. One or two transitions of much lower hypochromicity and larger $\Delta T_{1/2}$ are observed at temperatures 10–20°C higher than the main transition. A synopsis of the melting processes is given in Figure 3. The interpretation was given first for PSTV, CEV, and CSV (86, 70) and also experimentally confirmed for CCCV (76). In a highly cooperative main transition, all base pairs in the native structure dissociate and complementary segments located in distant portions of the native structure recombine to form new stable hairpins. At higher temperatures, these hairpins denature in separate thermal transitions. They have also been observed in electron micrographs (70). The stable hairpins found experimentally at higher temperatures are listed in Table 5. It should be noted that hairpin I is located in the highly conserved region. Hairpins I and II also have to be expected in TPMV and TASV because of the sequence homology, but have not been proven experimentally.

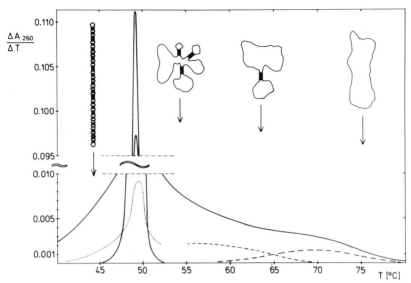

Figure 3 Synopsis of the structural transitions of PSTV. The differentiated melting curve is composed from equilibrium and kinetic measurements as described by Henco et al (86). The conditions differ slightly from those of Table 4. Equilibrium melting curve (——, upper curve), slow contribution denaturation process of the main transition (———-), fast contribution to the main transition (· · · ·), intermediate transition (· - · · -), high temperature transition (- - - -). The figure is reproduced with permission from Henco et al (86).

The process of the structural rearrangement from the extended native structure to a partially denatured structure with newly formed hairpins, as depicted in Figure 3, was first observed experimentally. Recently, the same mechanism could be derived theoretically, without a priori assumptions about the secondary structure. The Nussinov-Zuker (82, 83) algorithm was applied at different temperatures and showed that the obtained structure switches in a narrow transition range from the extended structure to a branched structure as depicted in Figure 3 (84a). Although the main transition is highly cooperative, the calculations showed that denaturation starts in the left half of the secondary structure (70). There are two regions where secondary structure is most labile: the polypurine stretch and the region neighboring the left side of the highly conservative region. These weak points in the structure are nearly identical in PSTV, CEV and CSV (78, 87).

COMPARISON OF VIROIDS, VIRUSOIDS, AND RANDOM SEQUENCES   The special features of viroids can best be seen in comparison with "viroid-like" RNA molecules, such as virusoids, and circular RNAs of similar size and GC content but random, i.e. computer-generated sequence. Figure 4 shows the nucleotide sequence (88) and secondary structures of two different virusoids, that from velvet tobacco mottle virus (VTMoV) and from *Solanum nodiflorum*

Table 5 Stable hairpins in viroids[a]

Viroids	-Hairpins-		
	I	II	III
PSTV	79 87 CGCUUCAGG :::::::: 14 GCGAGGUCC 110 102	227 236 CCCUCGCCCC :::::::::: 82 GGGAGCGGGG 328 319	127 135 CGGUGGGGA :::::::: 28 GCCGCCCUU 168 160
CEV	(possible but not studied)	239 251 CCCUCGCCCGGAG ::::::::::::: 79 GGGAGCGGGCCUC 339 327	not possible
CSV	(possible but not studied)	223 233 CCCUAGCCCGG ::::::::::: 76 GGGAUCGGGCC 322 312	not possible
CCCV1-large	46 54 CGCUUGAGG ::::::::: 14 GCGAACUCC 77 69	not possible	not possible

[a]The hairpins listed are formed during the main transition and dissociate at temperatures between 5 and 15°C higher than the main transition. They were derived from comparing the values for T_m, content of G:C base pairs, loop size, and number of base pairs, either calculated from the sequence or measured in thermodynamic and kinetic experiments (86,70,76,78). Only those hairpins that were studied experimentally are depicted. The numbers refer to the nucleotides of the based paired regions and to the size of the loops.

mottle virus (SNMV). The degree of base pairing is 56% in VTMoV and 57% in SNMV, and a mean value of 57% was calculated from optimizing the secondary structure of 5 different random sequences (78). This shows that the degree of base pairing of at least some viroids, e.g. 64% in CCCV, is not much higher than in virusoids or random sequences. The absence of bifurcations, however, is a distinguishing feature that is found only in viroids. A distinguishing feature can also be seen in thermodynamic stability, either expressed by $\Delta G/N$ or by the T_m-value. The cooperativity represented by $\Delta T_{1/2}$ is only high in viroid transitions, and the formation of stable hairpins during the highly cooperative transition occurs only in viroids.

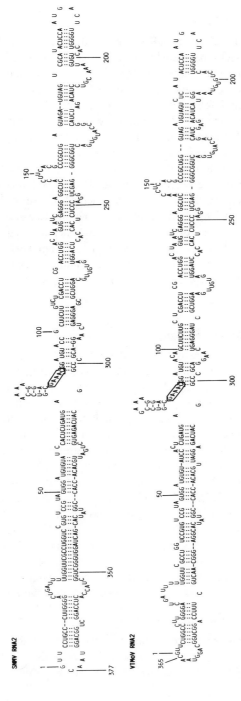

Figure 4 Sequence and secondary structures of two virusoids. Sequences are from Haseloff & Symons (88), the secondary structures according to the more recent thermodynamic optimization (78; 84a). The sequences in the boxed regions agree with the highly conservative region of viroids.

In Vivo Structure of Viroids

Purified viroids consist exclusively of RNA and yet display all of the biological properties of the pathogens that were known from the early studies with cruder preparations. The possibility, however, that, in situ, viroids are present in complexes with cellular constituents and are located in particular compartments or organelles of the cell, has been discussed several times (3, 4). Early studies based on infectivity tests suggested that viroids were associated primarily with nuclei (10, 89, 90) and/or membranes (91). In a more recent investigation (92), highly purified nuclei and chloroplasts from tomato leaf tissue were used, and the PSTV content was determined quantitatively by bidirectional gel electrophoresis (44). Ninety five percent of viroid RNA was found in nuclei. Further fractionation showed that it is not homogeneously distributed inside the nucleus but that it is associated with the nucleolar fraction. Because increasing ionic strength leads to a release of the viroids, it was concluded that viroids are associated with the nucleoli by nucleic acid–protein complexes. Depending upon the progress of the disease, average viroid copy numbers are between 200 and 10,000 per cell. In chloroplasts, practically no viroids were detected.

REPLICATION

Absence of Translation Products

Viroids are the only autonomously replicating system that does not code for its replicase or a subunit of its replicating enzyme. Although there is no final experimental proof that the genetic information of viroids is not expressed in a small protein, several points of negative evidence argue against it. The coding capacity of a viroid would result in a protein of not more than 120 amino acids. For PSTV e.g. the longest possible protein could result from more than two rounds of translation, provided that in this case the terminator codon UGA would be suppressed. The lack of ribosome binding sites, the absence of the "cap" structure, nearest neighbor sequences, the stable secondary structure, and the circularity disfavor mRNA activity (69, 93). Similar arguments hold for other viroids. In in vitro translation systems no specific products could be detected if PSTV (94) or CEV (95) were offered as mRNA. Careful comparison of the protein patterns from CEV-infected *Gynura aurantiaca* plants (96) and PSTV-infected tomato plants (97, 98) with the corresponding controls from healthy plants revealed no viroid-coded differences. In summary, one has to assume that viroid replication is dependent upon host polymerases.

Candidates of Host Polymerases

First indications of the mode of viroid replication came from inhibition studies. Actinomycin D inhibits viroid replication as well as synthesis of cellular RNA as was shown with leaf discs (99), nuclei (89), infiltrated foliar tissue or sprouts

(100), and tomato protoplasts (101). This inhibition has to be interpreted as a nonspecific secondary effect of DNA-dependent RNA synthesis on viroid replication since it became known that viroid replicative intermediates exist only as RNA and never as DNA (cf next section). It is only justified to exclude the involvement of preexisting RNA-dependent RNA replicases during viroid replication, which are known to be actinomycin D–resistant in plants infected with conventional plant viruses (102). More specific conclusions could be drawn from studies of Mühlbach & Sänger (101) on PSTV synthesis in protoplasts if α-amanitin is added as an inhibitor. At an intracellular α-amanitin concentration of 2×10^{-8} M, viroid synthesis was inhibited by 75% whereas total RNA synthesis was inhibited only by 10%. Because it is known that 10^{-8} M, α-amanitin inhibits specifically eukaryotic DNA-dependent RNA polymerase II, the finding strongly suggests that RNA polymerase II is involved, directly or indirectly, in viroid replication. Moreover, the dependence of CEV synthesis in nucleirich fractions of *Gynura aurantiaca* upon ion concentration such as Mn^{2+}, Mg^{2+}, and $(NH_4)_2SO_4$ indicated an RNA polymerase II activity (103). From a trace level of CEV synthesis even at high α-amanitin concentrations (10^{-8}–10^{-6} M), a participation of RNA polymerase I and/or III was considered.

The possibility of viroid replication by RNA polymerase II was supported by in vitro studies with RNA polymerase II purified from healthy tomato tissue or wheat germ (104; cf also 4). It was shown that in the presence of Mn^{2+}-ions the enzyme transcribes viroid RNA into linear negative strands of full length and that viroid is accepted as template with a higher activity compared to other natural or synthetic RNA. Furthermore, during the synthesis intermediates of defined length accumulate, indicating a defined position of initiation. It has been reported (105) that Mn^{2+}-ions which normally reduce the template specificity of polymerases are not needed, and that viroids, even if compared with virusoids, show an order of magnitude higher template activity. In analytical ultracentrifugation experiments (84, 105) the binding constant between polymerase II from wheat germ and PSTV was found to be about 10^7 M^{-1}. This number is low compared with polymerase-promoter binding but around an order of magnitude higher than binding of that polymerase to other natural RNAs including virusoids. In electron micrographs (105) it was seen that the RNA polymerase II from wheat germ may bind to both ends of the PSTV secondary structure. The fact that viroids possibly can also be replicated in vivo by the host enzyme polymerase II which normally accepts double-stranded DNA as template would suggest that viroids are mistaken in the cell for a piece of DNA. This may reflect the unique structural and dynamic features of viroids which had been described earlier as "DNA-like" (70).

In vitro transcription of PSTV RNA into full-length copies by RNA-dependent RNA polymerase from healthy leaf tissue has also been reported

(106). This enzyme is clearly a candidate for the replication of plant RNA viruses (107). In vitro viroid transcription is, however, not inhibited by α-amanitin (106). Therefore, these results would not explain the in vivo α-amanitin sensitivity of viroid synthesis observed in protoplasts (101).

So far, there are no reports on viroid synthesis with purified DNA-dependent RNA polymerase I or III. Therefore, one should also consider them as candidates for viroid replication. The location of the mature viroids in the nucleolus (cf *In Vivo Structure of Viroids* section) would argue in favor of polymerase I which normally carries out transcription of ribosomal RNA in the nucleolus. Indeed, sequence homologies between DNA promoter sequences from mouse with sequences at the right end of the secondary structure of PSTV, CSV, or CEV have been found (P. Palukaitis, personal communication; cf also 4a, 5). On the other hand, the location of the mature viroids in the nucleolus does not exclude their synthesis by polymerase II which is known to be present in the nucleoplasma or associated with the chromatin. This may be inferred (92) from comparison with the small nuclear RNA U3 which at least in animal cells is transcribed by polymerase II from the chromatin and afterwards becomes associated with the nucleolus.

Intermediates of Viroid Replication

As mentioned before, hybridization studies from several laboratories have established that host plants whether healthy or infected do not contain detectable amounts of viroid-specific DNA (108–110). When the RNA of the cell, however, was tested for viroid-specific sequences, viroid-complementary RNA was found for CEV (111), in PSTV (112–114), and in ASBV (115). PSTV-infected tissues contain RNA complementary to the complete viroid (116). At present, it is commonly accepted that only RNA intermediates are involved in viroid replication.

For further characterization of the viroid-specific RNA, hybridization probes specific for (−)strand sequences or for (+)strand sequences were used. The probes specific for (−)strands were ^{125}I-labeled viroids (113, 114), (+)strand deoxynucleotides (117), and M13mp93 clones containing the (+)strand sequence (115). (+)Strand sequences were tested with (−)strand deoxynucleotides (117), ^{32}P-labeled cDNA (118), or M13mp99 clones containing the (−)strand sequence (115).

Northern blotting analysis of minus strand viroid sequences (113, 114) revealed that the (−)strands are multimers of 2–5 times unit length. Larger than unit length sequences were also detected for CEV (119) and ASBV (115). (−)Strands of unit length could not be detected safely. However, whenever one member of a pair of complementary strands is present in large excess over the other, the minority strand is difficult, or impossible, to detect by Northern hybridization analysis (118).

Oligomeric (+)strands of viroids have also been detected. In PSTV-infected tissue low levels of dimers and trimers (117, 119a), and in ASBV-infected tissue high levels of oligomers up to eightmer (115), were identified. In ASBV, circular dimers were described (115) similar to those of cadang-cadang RNA2 (cf Table 3).

Splitting of Oligomer Viroid Precursors and Circularization of Viroids

The circular structure confers an enormous selective advantage to viroids, since the polymerase, once associated with the RNA, would be able to produce long multimeric copies which then could be cut to the monomer, i.e. unit length molecules. Linear monomers, at least the (+)strands, have to be circularized to form mature viroids.

It is not yet clear by what activity multimers are cleaved to unit length RNAs. The multimers might contain preformed secondary structures of the monomers, so that ribonuclease could release the monomers. In the course of the investigations on the litigation of viroids (see below), two types of linear molecules have been identified in preparations of circular PSTV, which differ by a nick between C_{181} and A_{182}, and between C_{348} and A_{349}, respectively (120). At least one of these "natural" linears was thought to derive from the cleavage of the multimer precursor. For thermodynamic reasons one has to expect a structure for the multimers like a series of the extended viroid secondary structures. In this series the nick between C_{181} and A_{182} would be in the linkage between the extended structures. C-A neighbors have also been identified in other RNAs as particularly weak points, and in a precursor RNA from bacteriophage T4 a self-cleavage at those points has been claimed (121). Besides ribonuclease digestion and self-cleavage, even a combination of self-cleavage and ligation, i.e. self-splicing to form the circular unit directly, as shown for a pre-rRNA of *Tetrahymena* (122) may be considered. On the other hand, i.e. if cleavage and ligation are separate reactions, an enzyme has been detected in plants (123, 124) that ligates 2',3'-cyclophosphate-terminated RNA to 5'-phosphorylated RNA to form 2'-phosphomonoester-3',5'-phosphodiester bonds. This enzyme has been shown to circularize covalently so-called "natural" linear as well as artificially linearized viroid molecules (120, 125).

Models of Replication

Models of viroid replication are derived mainly from the experimental findings of viroid-specific sequences of different length and polarity (see above). Although not definitely proven, it is assumed that the oligomeric (+)- and (−)strands are intermediates of replication rather than end products. They appear concomitantly with the progress of infection. A discussion of the different aspects of viroid replication was given recently by Branch & Robertson (118).

On the basis of the evidence for multimeric PSTV (−)strands and of the circularity of mature viroids, a rolling circle mechanism was proposed (113). Synthesis of the (−)strand starts from the infecting circular (+)strand (step 1 in Figure 5). The rolling circle mechanism leads to a multimeric (−)strand (step 2). The multimeric (−)strands then serve as a template for the production of multimeric (+)strands (step 3). These must be cleaved to give unit-length molecules with characteristic end groups (as in step 4) and circularized (step 5) to yield progeny circles. This hypothetical scheme relies exclusively on viroid-specific nucleic acid species that can be readily detected in PSTV-infected plants. Furthermore, as outlined in the previous paragraphs, the assumed enzymatic activities have been found in plant cells, although none of them has been proven definitively to take part in viroid replication in vivo as suggested in Figure 5.

From their finding of high levels of multimeric (+)strands of ASBV and very low levels of (−)sequences (see above) Bruening et al (115) derived a different rolling circle mechanism. The invading monomeric ASBV could be converted by host enzymes to a (−)circular molecule which then acts as a template for the rolling circle synthesis of a continuous (+)ASBV sequence which is processed to give full-length linear monomeric ASBV and after ligation the predominant monomer circular ASBV.

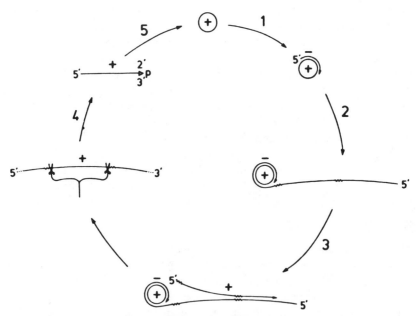

Figure 5   A hypothetical scheme for PSTV replication. The figure is reproduced with permission from Branch & Robertson (118).

A third mechanism (118) involves two rolling circles. Multimeric (−)strands are synthesized as in Figure 5 but processed to give a (−)strand monomer circle which serves as the template for a second rolling circle to produce (+)strand multimers. Cleavage and ligation leads to the progeny viroid.

PATHOGENESIS

Pathogenesis will be discussed in this paragraph exclusively under biochemical aspects. The reader who is interested in phytopathological aspects, cytopathic effects, influence of cross-protection, etc may see the reviews mentioned earlier (3, 4).

Disease Related Proteins and Other Compounds

As mentioned above (cf *Absence of Translation Products* section), viroid-coded proteins have not been found. However, in CEV-infected tissue from *Gynura aurantiaca*, tomato, and potato (96, 126, 127), two proteins in the molecular weight range 12,000–18,000 accumulate, and in PSTV-infected tomato tissue (97, 98) a 14,000-dalton protein was found in much higher concentrations as compared to healthy tissue. These proteins are not specific for viroid infections; they appear also after virus infections (97) and in naturally senescent plants (127). They may be regarded as a general pathophysiological response of the host plant. Changes after viroid infections have also been reported for the total concentration of DNA, RNA, and proteins (128), in elemental composition (129), and in the concentration of chlorogenic acid (130).

Sequence Homology with Small Nuclear RNAs

Between viroids and small nuclear RNAs (U1–U6) from eukaryotic organisms, some remarkable sequence homologies have been detected.

Most evident is the sequence homology between the 5'-end of U1RNA and the highly conservative region (lower strand in the secondary structure) of viroids of the PSTV class (73, 131). According to recent models, U1RNA is involved in the splicing process in which the introns are cut from the heteronuclear RNA, and the processed mRNA is formed. In these models the 5'-end of U1RNA forms base pairs with both ends of the intron (132). It was speculated that the viroid could replace U1RNA and interfere with the splicing process. It was also considered that the (−)strand viroid is interacting with and thereby blocking the U1RNA. Formation of stable hairpins (cf STRUCTURAL TRANSITIONS section) would favor this interaction (76, 78). In a different model (133) PSTV would interact by two nonadjacent regions with the splicing site of host mRNA. Similar but not identical regions are also present in a mild isolate of PSTV and CSV. Severeness of symptoms in chrysanthemum after infection with the three viroids mentioned is in parallel with the expected interaction of the viroids with the splicing site on the mRNAs.

However, the location of mature viroids in the nucleolus (92) and the presence of U1RNA in ribonucleoprotein particles disfavor the above models of pathogenesis. Only the small nuclear RNA U3B is known to be present in the nucleolus (134), and sequence homologies between PSTV and snRNA U3B from Novikoff hepatoma cells (135, 92) have actually been found. Detailed models on a splicing interference or another pathogenic mechanism were not derived from these homologies.

Sequence Variation

Sequences of viroid strains with differences in the symptom expression have been determined for PSTV (4, 4a, 69, 71) and for CEV (57, 73, 75). In CEV the symptoms of the different isolates could not be classified from mild up to necrotic, nor could they be interpreted in terms of a well-defined "virulence modulating region" on the viroid. The situation seems more clear with PSTV (4a). The symptoms may easily be classified from mild to necrotic (cf Figure 1), and the variation in the sequence, as shown in Figure 6, is restricted to a major part on the secondary structure formed by nucleotides 40–55 and 305–320 and to a minor part around nucleotides 115–125. Sänger and his colleagues argued that the major variation influences virulence by changing the thermodynamic stability of that region and that the minor variation compensates for the total number of nucleotides which is 359 for all strains. The region of major variation is located in the oligopurine stretch (see above). It was shown earlier (78, 87) that this region is one of the regions of lowest thermodynamic stability, i.e. which would undergo a premelting process. Estimation of the thermal stability of these premelting areas yields the correlation that decreasing stability, i.e. more expressed premelting, goes with increasing virulence (4a).

Infectious Viroid cDNA Clones

Construction of viroid cDNA clones was mentioned in the section *Molecular Cloning* (Table 2). Owens and his colleagues found that plasmid DNA containing PSTV cDNA dimers were infectious (53). RNA transcripts from these plasmids when containing the sequence of PSTV were also infectious. The sequences of the viroid progeny and the cloned DNA were identical. Infectivity could not be shown with plasmids containing PSTV cDNA monomers. Ohno et al (55) synthesized in vitro infectious RNA molecules from cloned HSV cDNA. The clones were tandemly repeated HSV cDNAs carrying the *lac* promoter (Table 2). Meshi et al (136) reported that even double-stranded cDNAs consisting of 1–3 units of HSV sequences without promoters were infectious. The infectivity of 1-unit sequences was low, but was found in cDNA that was derived from cuts in the viroid circle on different sites. The

PSTV

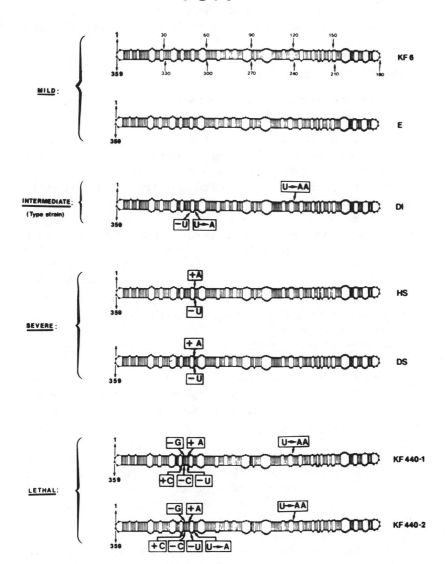

Figure 6 Location of the nucleotide exchanges in various PSTV isolates causing mild, intermediate, severe, and lethal symptoms in tomato as shown in Figure 1. The exchanges are related to the mild isolate as standard and indicated by boxes. It should be noted that the resulting influence on the secondary structure is not shown. The figure is reproduced with permission from Sänger (4a).

authors speculate, however, that the unit-length cDNA may be ligated to dimers in the infected plant before an infectious RNA is synthesized.

All authors working with infectious cloned viroid cDNA emphasized that viroid sequences may be mutated by site-directed mutagenesis methods and this may be used as a promising tool for studying the mechanism of pathogenesis and replication.

POSSIBLE ORIGIN OF VIROIDS

Many of the ideas about origin of viroids have already been summarized by Diener (3) and were discussed and extended by others (4).

1. One of the important observations with respect to the origin of viroids is related to their discovery and has been summarized by Diener (3), namely the fact that all viroid diseases have been detected during this century, some of them quite recently. This is in total contrast to plant diseases caused by conventional viruses, many of which were observed by the nineteenth century, the first being documented in the early seventeenth century on Dutch still-life paintings. Thus, the diseases caused by PSTV and CEV were first observed in the early 1920s, whereas those caused by HSV and CPFV were first described as late as 1963–1970 and 1974, respectively (3). In particular, it is well-documented that the disease caused by CPFV first appeared in a single greenhouse and spread from there, whereas the coconut cadang-cadang disease first occurred on one small island of the Philippines (3). Hence it appears that viroids may be of recent origin, and it was postulated that human activities, e.g., the introduction of vast monocultures, may have contributed to the generation and spreading of viroids and viroid diseases. Since viroids often do not cause symptoms in wild plants, it was proposed that they originate from any unknown wild host plant where they are not pathogenic. Transfer to sensible cultivated plants, e.g., by grafting or accidentally, would thus give rise to a pathogenic, replicative RNA, the viroid (3).
2. Alternatively it has been proposed that viroids have accidentally evolved from normal, regulatory nuclear RNAs in their natural cultivated hosts or in wild plants into structures of high stability and the capacity of self-replication and intercellular and intracellular mobility. Consequently, Diener (3) predicted that new viroid diseases of cultivated plants will continue to develop and to appear unexpectedly.
3. Viroids may have evolved from intervening sequences (73, 131, 133, 137–139). This hypothesis seems to be supported by a sequence homology between most sequenced viroids [exception: ASBV (139)] and the 5'-end of U1RNA, a small nuclear RNA believed to be involved in mRNA splicing

(132, 140). This observation was discussed already with respect to viroid pathogenesis (preceding paragraph). U1RNA has also been found in plant nuclei (141). During the splicing process, U1RNA is thought to interact with both ends of the intron to be excised. Circularization of an excised intron by plant RNA ligase (142) could create a viroid-like RNA with some properties of a viroid, i.e., replicability and intercellular and interorganismal mobility. The sequence homology between several viroids and the 5'-end of U1RNA would indicate that not the viroids themselves, but rather their complementary replication intermediates would derive from an "escaped intron."

4. Viroids may have evolved from "virusoids," small circular RNAs encapsidated in certain plant viruses with dipartite genome (15a, 17). The fundamental difference between a viroid and a virusoid seems to be that the former is replicated by a host polymerase, whereas the latter depends on a viral replicase for replication. A direct phylogenetic relation between viroids and the virusoids studied so far (88) seems unlikely for the following reasons:

(a) there is no significant sequence homology between viroids and virusoids;

(b) virusoids differ from viroids in their thermodynamic and hydrodynamic properties in that they closely resemble random sequences with the length, base-composition, and circularity of viroids (Table 4 and COMPARISON OF VIROIDS, VIRUSOIDS AND RANDOM SEQUENCES section).

5. Viroids and RNA viruses may have originated from a system that exchanges genetic information between eukaryotic cells and cell organelles. This highly speculative hypothesis (143) postulates that viroids are able to infect a plant systemically and can rely on host mechanisms for intercellular transportation and replication because normal plant cells contain structurally related RNAs which are also transportable and replicable in order to facilitate their amplification and extrachromosomal inheritance. It is evident that such a system, if existing, would be a source for the evolution of viroids and RNA viruses.

6. It has been proposed that viroids evolved from prokaryotic RNA through the infection of higher plants by prokaryotes (4). This hypothesis is based on the finding that not only eukaryotic RNA polymerases, but also RNA- and DNA-polymerases from *E. coli* are able to transcribe viroid RNA in vitro into complementary RNA and DNA copies, respectively (144). However, viroid-related sequences have not yet been found in prokaryotes.

7. Viroids may have derived from transposable genetic elements. Kiefer et al (60) have noted that viroid structures contain sequences that resemble those occurring at the ends of certain moveable genetic elements and retroviral

proviruses, including the presence of inverted repeats often ending with the dinucleotides U-G and C-A, flanking imperfect direct repeats, and a long stretch of purines present in many viroids, which resembles the putative reverse transcriptase primer sequence for (+)strand DNA synthesis of retroviral proviruses. Thus viroids would have evolved from transposable elements or retroviral proviruses by deletion of most of the interior regions as well as of intervening host sequences.

So far, the different hypotheses on viroid evolution have not been supported by direct experimental evidence. However, some conclusions should be possible. The fact remains that there are at least three families of viroids: (a) The PSTV group, which includes PSTV, TPMV, TASV, CSV, CEV, HSV, and CPFV, (b) ASBV, and (c) CCCV. The sequence homology of more than 70% among the viroids of the PSTV group may indicate their origin from a common ancestor, or, perhaps more likely, their evolution in geographically separated events from different, however related, RNA ancestors. At least the recent origin of viroids from intervening sequences or from other DNA-encoded RNA of their presently known host plants is rather unlikely, since viroid sequences do not occur in host DNA. It is also unlikely that viroids derive from extrachromosomally replicating RNA, if it exists at all, of their present hosts, since this would make it difficult to understand why most viroid diseases appeared only recently. Hence it might be more likely that viroids, because of the sequence homology with U1RNA and the structural features resembling transposable elements and retroviral proviruses, originate from nonpathogenic RNAs of unrelated eukaryotic species. Any unrelated plants, insects, fungi, protozoa, symbionts, or pathogens, whose close contact with the affected host plant is favored by human activities or mass propagation of plants in monoculture, could hence be the cryptic carrier of potential viroid RNA. The fact that viroids often do not cause symptoms in wild plants (3) might even allow speculation that wild plants exist that carry nonpathogenic, viroid-like RNA which can be transferred by mechanical contact into competing plants which then suffer from a viroid infection. Plants with such a defense mechanism, if they exist, would certainly have a selective advantage as soon as there is competition with other plants for space and light. These hypothetical viroid hosts might even contain viroid sequences in their DNA.

Another enigma is the fact that viroids appear to be restricted to higher plants. It seems reasonable that viroids do not occur in single-cellular algae, because the uncoated viroid RNA is too sensible to survive the extracellular phase of an infection cycle. If viroids or viroid-like pathogens do not occur at all in animals and *homo sapiens*, this should be an indication for a unique relationship between plants and viroids.

CONCLUDING REMARKS

This review has tried to emphasize the development in viroid research from a purely phytopathological problem to a present-day question of biochemistry, physical biochemistry, and molecular biology. A first approach towards an understanding of viroid function started from the elucidation of the in vitro structure. This is well understood today: viroids may be described as a class of molecules with extraordinary but homogeneous structural features, in that sense comparable to tRNA. Next to tRNA their structure is best understood. A further increase in knowledge of viroid structure is possible only from X-ray analysis. This, however, would result not only in structural details on the level of atomic coordinates of viroids themselves, but would clarify structural features of the different kinds of loops as hairpin, internal, and bulge loops, information which is not available at present. A prerequisite for those studies is now fulfilled due to the availability of mg amounts of purified viroids. Other progress in viroid structure research has to come from the in vivo structure, i.e. details of the native complexes of viroids and viroid intermediates in the nucleus.

The study of host polymerase candidates and replication intermediates led to the present models of the mode of replication. There is right now no alternative to a rolling circle model. Potential intermediates have been detected in the cell and host enzymes which are able to catalyze every single step are known. Future effort will concentrate on proving which intermediates and which enzymes are actually involved in vivo.

At present, little is known on the mechanism of pathogenesis. In PSTV, a virulence-modulating region could be localized. This is a starting point to search for the interacting counterparts. The difference in the expression of proteins seems right now not to be specific for viroid infections. Either viroid pathogenesis and virus pathogenesis converge after a particular step in the functional chain, or the detected differences in the protein pattern are late secondary effects.

The finding that some viroid cDNA clones are infectious offers a wide variety of techniques from molecular biology. It will be possible, mainly by site-directed mutagenesis, to identify the sequences in the viroid chain that are responsible for cellular organization, initiation of replication, splitting, ligation, initial steps of pathogenesis, and other steps.

The possible origin of viroids is still highly speculative and will be a challenging subject in the future.

ACKNOWLEDGMENTS

We thank the colleagues who sent us reprints and preprints for the preparation of this manuscript. The help of Ms. Gruber, Ms. Grigoleit, Dr. Colpan, and Dr.

Steger is gratefully acknowledged. The work of the authors was supported by grants from the Deutsche Forschungsgemeinschaft and the Fonds der Chemischen Industrie.

Literature Cited

1. Diener, T. O. 1979. *Science* 205:859–66
2. Gross, H. J., Riesner, D. 1980. *Angew. Chem. Int. Ed. Engl.* 19:231–43; *Angew. Chem.* 92:233–45
3. Diener, T. O. 1979. *Viroids and Viroid Diseases.* New York: Wiley
4. Sänger, H. L. 1982. In *Encyclopedia of Plant Physiology* (NS), ed. B. Parthier, D. Boulter, 14B:368–454. Berlin/Heidelberg: Springer-Verlag
4a. Sänger, H. L. 1984. In *The Microbe 1984: Part I Viruses,* ed. B. W. J. Mahy, J. R. Pattison, pp. 281–334. Cambridge Univ. Press
5. Robertson, H. D., Howell, S. H., Zaitlin, M., Malmberg, R. L., eds. 1983. *Current Communications in Molecular Biology.* New York: Cold Spring Harbor Lab.
6. Maramorosch, K., McKelvey, J. J., eds. 1984. *Subviral Pathogens of Plant and Animals: Viroids and Prions.* New York: Academic
7. Diener, T. O., Raymer, W. B. 1967. *Science* 158:378–81
8. Diener, T. O. 1971. *Virology* 45:411–28
9. Singh, R. P., Clark, M. C. 1971. *Biochem. Biophys. Res. Commun.* 44:1077–82
10. Sänger, H. L. 1972. *Adv. Biosci.* 8:103–116
11. Semacik, J. S., Weathers, L. G. 1972. *Nature New Biol.* 237:242–44
12. Hollings, M., Stone, O. M. 1973. *Ann. Appl. Biol.* 74:333–48
13. Diener, T. O., Lawson, R. H. 1973. *Virology* 51:94–101
14. Randles, J. W., Rillo, W. P., Diener, T. O. 1976. *Virology* 74:128–39
15. Zelasny, B., Randles, J. W., Boccardo, G., Imperial, J. S. 1982. *Sci. Filipinas* 2:46–63
15a. Randles, J. W., Davies, C., Hatta, T., Gould, A. R., Francki, R. I. B. 1981. *Virology* 108:111–22
16. Tien, P., Davies, C., Hatta, T., Francki, R. I. B. 1981. *FEBS Lett.* 132:353–56
17. Francki, R. I. B., Randles, J. W., Hatta, T., Davies, C., Chu, P. W. C., McLean, G. D. 1983. *Plant Pathol.* 32:47–59
18. Jones, A. T., Mayo, M. A., Duncan, G. H. 1983. *J. Gen. Virol.* 64:1167–73
19. Gould, A. R., Francki, R. I. B., Randles, J. W. 1981. *Virology* 110:420–26

20. Romaine, C. P., Horst, R. K. 1975. *Virology* 64:86–95
21. Van Dorst, H. J. M., Peters, D. 1974. *Neth. J. Plant. Pathol.* 80:85–96
22. Sänger, H. L., Klotz, G., Riesner, D., Gross, H. J., Kleinschmidt, A. K. 1976. *Proc. Natl. Acad. Sci. USA* 73:3852–56
23. Randles, J. W. 1975. *Phytopathology* 65:163–67
24. Sasaki, M., Shikata, E. 1977. *Proc. Jpn. Acad. Ser. B* 53:109–12
25. Owens, R. A., Smith, D. R., Diener, T. O. 1978. *Virology* 89:388–94
26. Thomas, W., Mohamed, N. A. 1979. *Aust. Plant Pathol. Soc. Newslett.* 8:1–2
27. Mohamed, N. A., Thomas, W. 1980. *J. Gen. Virol.* 46:157–67
28. Walter, B. 1981. *CR Acad. Sci. Paris III* 292:537–42
29. Galindo, J., Smith, D. R., Diener, T. O. 1982. *Phytopathology* 72:49–54
30. Chen, W., Tien, P., Zhu, Y. X., Liu, Y. 1983. *J. Gen. Virol.* 64:409–14
31. Sasaki, M., Shikata, E. 1980. *Rev. Plant Prot. Res.* 13:97–113
32. Da Graca, J. V., van Vuuren, S. P. 1980. *Plant Dis.* 64:475
33. Dickson, E., Diener, T. O., Robertson, H. D. 1978. *Proc. Natl. Acad. Sci. USA* 75:951–54
34. Sänger, H. L., Ramm, K. 1975. In *Modification of the Information Content of Plant Cells,* ed. R. Markham, D. R. Davies, D. A. Hopwood, R. W. Horne, pp. 229–52. Amsterdam: Elsevier
35. Colpan, M., Schumacher, J., Brüggemann, W., Sänger, H. L., Riesner, D. 1983. *Anal. Biochem.* 131:257–65
36. Mühlbach, H. P., Sänger, H. L. 1981. *Biosci. Rep.* 1:79–87
37. Mühlbach, H. P. 1982. *Curr. Top. Microbiol. Immunol.* 99:81–129
38. Zelcer, A., Vanadels, J., Leonhard, D. A., Zaitlin, M. 1981. *Virology* 109:314–22
39. Marton, L., Duran-Vila, N., Lin, J. J., Semancik, J. S. 1982. *Virology* 122:229–38
40. Stollar, B. D., Diener, T. O. 1971. *Virology* 46:168–70
41. Morris, T. J., Smith, E. M. 1977. *Phytopathology* 67:145–50
42. Pfannenstiel, M. A., Slack, S. A., Lane, L. S. 1980. *Phytopathology* 70:1015–18

43. Singh, R. P. 1982. *Can. Plant Dis. Surv.* 62:41–44
44. Schumacher, J., Randles, J. W., Riesner, D. 1983. *Anal. Biochem.* 135:288–95
45. Sammons, D. W., Adams, L. D., Nishizawa, E. E. 1981. *Electrophoresis* 2:135–41
46. Owens, R. A., Diener, T. O. 1981. *Science* 213:670–72
47. Salazar, L. F., Owens, R. A., Smith, D. R., Diener, T. O. 1983. *Am. Potato J.* 60:587–97
48. Ohno, T., Takamatsu, N., Meshi, T., Okada, Y. 1983. *Nucleic Acids Res.* 11:6185–97
49. Randles, J. W., Palukaitis, P. 1979. *J. Gen. Virol.* 43:649–62
50. Palukaitis, P., Rakowski, A. G., Alexander, D. M., Symons, R. H. 1981. *Ann. Appl. Biol.* 98:439–49
51. Allen, R. N. 1982. *Intervirology* 18:76–82
52. Rosner, A., Spiegel, S., Alper, M., Bar-Joseph, M. 1983. *Plant Mol. Biol.* 2:15–18
53. Cress, D. E., Kiefer, M. C., Owens, R. A. 1983. *Nucleic Acids Res.* 11:6821–35
54. Van Wezenbeek, P., Vos, P., van Boom, J., van Kammen, A. 1982. *Nucleic Acids Res.* 10:7947–57
55. Ohno, T., Ishikawa, M., Takamatsu, N., Meshi, T., Okada, Y., et al. 1983. *Proc. Jpn. Acad. Ser. B* 59:251–54
56. Palukaitis, P., Symons, R. H. 1978. *FEBS Lett.* 92:268–72
57. Visvader, J. E., Symons, R. H. 1983. *Virology* 130:232–37
58. Symons, R. H. 1981. *Nucleic Acids Res.* 9:6527–37
59. Sano, T., Uyeda, J., Shikata, E., Ohno, T., Okada, Y. 1984. *Nucleic Acids Res.* 12:3427–34
60. Kiefer, M. C., Owens, R. A., Diener, T. O. 1983. *Proc. Natl. Acad. Sci. USA* 80:6234–38
61. Dickson, E. 1979. In *Nucleic Acids in Plants,* Vol. II, ed. T. C. Hall, J. W. Davies. Boca Raton: Fla: CRC
62. Ohno, T., Akiya, J., Higuchi, M., Okada, Y., Yoshikawa, N., et al. 1982. *Virology* 118:54–63
63. Granell, A., Flores, R., Conejero, V. 1983. *Anal. Biochem.* 134:479–82
64. Palukaitis, P., Symons, R. H. 1980. *J. Gen. Virol.* 46:477–89
65. Colpan, M., Riesner, D. 1984. *J. Chromatogr.* 296:339–53
66. McClements, W. L., Kaesberg, P. 1977. *Virology* 76:477–84
67. Sogo, J. M., Koller, T., Diener, T. O. 1973. *Virology* 55:70–80
68. Langowski, J., Henco, K., Riesner, D., Sänger, H. L. 1978. *Nucleic Acids Res.* 5:1589–610
69. Gross, H. J., Domdey, H., Lossow, C., Jank, P., Raba, M., et al. 1978. *Nature* 273:203–8
70. Riesner, D., Henco, K., Rokohl, U., Klotz, G., Kleinschmidt, A. K., et al. 1979. *J. Mol. Biol.* 133:85–115
71. Gross, H. J., Liebl, U., Alberty, H., Krupp, G., Domdey, H., et al. 1981. *Biosci. Rep.* 1:235–241
72. Haseloff, J., Symons, R. H. 1981. *Nucleic Acids Res.* 9:2741–52
73. Gross, H. J., Krupp, G., Domdey, H., Raba, M., Alberty, H., et al. 1982. *Eur. J. Biochem.* 121:249–57
74. Haseloff, J., Mohamed, N. A., Symons, R. H. 1982. *Nature* 299:316–22
75. Visvader, J. E., Gould, A. R., Bruening, G. E., Symons, R. H. 1982. *FEBS Lett.* 137:288–92
76. Randles, J. W., Steger, G., Riesner, D. 1982. *Nucleic Acids Res.* 10:5569–86
77. Imperial, J. S., Rodriguez, J. B., Randles, J. W. 1981. *J. Gen. Virol.* 56:77–85
78. Riesner, D., Steger, G., Schumacher, J., Gross, H. J., Randles, J. W., Sänger, H. L. 1983. *Biophys. Struct. Mech.* 9:143–70
79. Randles, J. W., Hatta, T. 1979. *Virology* 96:47–53
80. Riesner, D., Kaper, J., Randles, J. W. 1982. *Nucleic Acids Res.* 10:5587–98
81. Tinoco, I. Jr., Uhlenbeck, O. C., Levine, M. D. 1971. *Nature* 230:362–67
82. Nussinov, R., Jacobson, A. 1980. *Proc. Natl. Acad. Sci. USA* 77:6309–13
83. Zuker, M., Stiegler, P. 1981. *Nucleic Acids Res.* 9:133–48
84. Riesner, D., Colpan, M., Goodman, Th. C., Nagel, L., Schumacher, J., et al. 1983. *J. Biomol. Struct. Dyn.* 1:669–88
84a. Steger, G., Hofmann, H., Förtsch, J., Gross, H. J., Randles, J. W., et al. 1984. *J. Biomol. Struct. Dyn.* 2:543–71
85. Klump, H., Riesner, D., Sänger, H. L. 1978. *Nucleic Acids Res.* 5:1581–87
86. Henco, K., Sänger, H. L., Riesner, D. 1979. *Nucleic Acids Res.* 6:3041–59
87. Gross, H. J., Krupp, G., Domdey, H., Steger, G., Riesner, D., Sänger, H. L. 1981. *Nucleic Acids Res. Symp. Ser.* 10:91–98
88. Haseloff, J., Symons, R. H. 1982. *Nucleic Acids Res.* 10:3681–91
89. Takahashi, T., Diener, T. O. 1975. *Virology* 64:106–14
90. Takahashi, T., Yaguchi, S., Oikawa, S., Kamita, T. 1982. *Phytopathol. Z.* 103:285–93

91. Semancik, J. S., Tsuruda, D., Zaner, L., Geelen, J. L. M. C., Weathers, J. G. 1976. *Virology* 69:669–76
92. Schumacher, J., Sänger, H. L., Riesner, D. 1983. *EMBO J.* 22:1549–55
93. Kozak, M. 1979. *Nature* 280:82–85
94. Davies, J. W., Kaesberg, P., Diener, T. O. 1974. *Virology* 61:281–86
95. Hall, T. C., Wepprich, R. K., Davies, J. W., Weathers, L. G., Semancik, J. S. 1974. *Virology* 61:486–92
96. Conejero, V., Semancik, J. S. 1977. *Virology* 77:221–32
97. Camacho Henriquez, A., Sänger, H. L. 1982. *Arch. Virol.* 74:181–96
98. Camacho Henriquez, A., Sänger, H. L. 1982. *Arch. Virol.* 74:167–80
99. Diener, T. O., Smith, D. R. 1975. *Virology* 63:421–27
100. Grill, L. K., Semancik, J. S. 1980. *Nature* 283:399–400
101. Mühlbach, H. P., Sänger, H. L. 1979. *Nature* 278:185–88
102. Sänger, H. L., Knight, C. A. 1963. *Biochem. Biophys. Res. Commun.* 13:455–61
103. Semancik, J. S., Harper, K. L. 1984. *Proc. Natl. Acad. Sci. USA.* 81:4429–33
104. Rackwitz, H. R., Rohde, W., Sänger, H. L. 1981. *Nature* 291:297–301
105. Goodman, T. C., Nagel, L., Rappold, W., Klotz, G., Riesner, D. 1984. *Nucleic Acids Res.* 12:6231–46
106. Boege, F., Rohde, W., Sänger, H. L. 1982. *Biosci. Rep.* 2:185–94
107. Fraenkel-Conrat, H. 1977. *Trends Biochem. Sci.* 4:184–86
108. Zaitlin, M., Niblett, C. L., Dickson, E., Goldberg, R. B. 1980. *Virology* 104:1–9
109. Branch, A. D., Dickson, E. 1980. *Virology* 104:10–26
110. Hadidi, A., Cress, D. E., Diener, T. O. 1981. *Proc. Natl. Acad. Sci. USA* 78:6932–35
111. Grill, L. K., Semancik, J. S. 1978. *Proc. Natl. Acad. Sci. USA* 75:896–900
112. Owens, R. A., Cress, D. E. 1980. *Proc. Natl. Acad. Sci. USA* 77:5302–6
113. Branch, A. D., Robertson, H. D., Dickson, E. 1981. *Proc. Natl. Acad. Sci. USA* 78:6381–85
114. Rohde, W., Sänger, H. L. 1981. *Biosci. Rep.* 1:327–36
115. Bruening, G., Gould, A. R., Murphy, P. J., Symons, R. H. 1982. *FEBS Lett.* 148:71–78
116. Zelcer, A., Zaitlin, M., Robertson, H. D., Dickson, E. 1982. *J. Gen. Virol.* 59:139–48
117. Spiesmacher, E., Mühlbach, H.-P., Schnölzer, M., Haas, B., Sänger, H. L. 1983. *Biosci. Rep.* 3:767–74
118. Branch, A. D., Robertson, H. D. 1984. *Science* 223:450–55

119. Grill, L. K., Negruk, V. I., Semancik, J. S. 1980. *Virology* 107:24–33
119a. Branch, A. D., Wills, K. K., Davatelis, G., Robertson, H. D. 1984. See Ref. 6
120. Kikuchi, Y., Tyc, K., Filipowicz, W., Sänger, H. L., Gross, H. J. 1982. *Nucleic Acids Res.* 10:7521–29
121. Watson, N., Gurevitz, M., Ford, J., Apirion, D. 1984. *J. Mol. Biol.* 172:301–23
122. Zaug, A. J., Grabowski, P. J., Cech, Th. R. 1983. *Nature* 301:578–83
123. Konarska, M., Filipowicz, W., Gross, H. J. 1982. *Proc. Natl. Acad. Sci. USA* 79:1474–78
124. Konarska, M., Filipowicz, W., Domdey, H., Gross, H. J. 1981. *Nature* 293:112–16
125. Branch, A. D., Robertson, H. D., Greer, Ch., Gegenheimer, P., Peebles, C., Abelson, J. 1982. *Science* 217:1147–49
126. Flores, R., Chroboczek, J., Semancik, J. S. 1978. *Physiol. Plant Pathol.* 13:193–201
127. Conejero, V., Picazo, I., Segado, P. 1979. *Virology* 97:454–56
128. Kaur, H., Cheema, S. S., Kapur, S. P. 1981. *J. Res. Punjab. Agric. Univ.* 18:140–43
129. Ter-Saakov, A. A., Kusinnyi, V. D. 1983. *Fisiol. Rast. (Moscow)* 30:184–86
130. De Fazio, G., Barradas, M. M. 1982. *Arg. Inst. Biol. Sao Paulo* 49:89–92
131. Diener, T. O. 1981. *Proc. Natl. Acad. Sci. USA* 78:5014–15
132. Lerner, M. R., Boyle, J. A., Mount, S. M., Wolin, S. L., Steitz, J. A. 1980. *Nature* 283:220–24
133. Dickson, E. 1981. *Virology* 115:216–21
134. Busch, H., Reddy, R., Rothblum, L., Choi, Y. C. 1982. *Ann. Rev. Biochem.* 51:617–54
135. Kiss, T., Posfai, J., Solymosy, F. 1983. *FEBS Lett.* 163:217–20
136. Meshi, T., Ishikawa, M., Ohno, T., Okada, Y., Sano, T., et al. 1984. *J. Biochem.* 95:1521–24
137. Roberts, R. J. 1978. *Nature* 247:530
138. Crick, F. 1979. *Science* 204:264–71
139. Kiss, T., Solymosy, F. 1982. *FEBS Lett.* 144:318–20
140. Rogers, J., Wall, R. 1980. *Proc. Natl. Acad. Sci. USA* 77:1877–79
141. Krol, A., Ebel, J.-P., Rinke, J., Lührmann, R. 1983. *Nucleic Acids Res.* 11:8583–94
142. Filipowicz, W., Gross, H. J. 1984. *Trends Biochem. Sci.* 9:68–71
143. Zimmern, D. 1982. *Trends Biochem. Sci.* 8:205–7
144. Rohde, W., Rackwitz, H. R., Boege, F., Sänger, H. L. 1982. *Biosci. Rep.* 2:929–39

Ann. Rev. Biochem. 1985. 54:565–95

THE CHEMICAL MODIFICATION OF ENZYMATIC SPECIFICITY

Emil Thomas Kaiser, David Scott Lawrence and Steven Edward Rokita

Laboratory of Bioorganic Chemistry and Biochemistry, The Rockefeller University, 1230 York Avenue, New York, NY 10021

CONTENTS

PERSPECTIVES AND SUMMARY

The chemical modification of enzymes has been a major tool in the elucidation of enzymatic properties. Chemical modification studies have been used to

565

0066-4154/85/0701-0565$02.00

pinpoint the nature of the active site residues and to differentiate between those amino acids that participate in the catalytic act and those that are important in substrate binding. These investigations have also substantiated the existence of effector sites that control the overall reactivity of the enzyme molecule. Other aspects of enzymatic catalysis that have been studied through chemical modification include reaction and substrate specificity, cooperativity, and the ionization behavior of functional groups.

Besides the use of chemical modification in these areas, a further use of this technique has been to alter enzyme specificity systematically. It is on the latter use of chemical modification that our chapter will focus. There are many ways in which enzymatic specificity can be altered by chemical modification. In particular, through chemical modification it has been possible to alter the pH optima for the action of enzymes, to alter the relative reactivity of an enzyme towards different types of substrates, to change patterns of substrate inhibition and activation and even, as will be discussed at length in this chapter, to alter the type of reaction catalyzed by an enzyme. The effects of chemical modification on enzymatic specificity can be expressed through changes in relative k_{cat} and/or K_m values for the action of the enzyme on different categories of substrates. In more radical modifications, new active functional groups can be introduced in the structure of an existing enzyme and the catalytic behavior of the enzyme can be altered completely. For example, by appropriate modification with a coenzyme analog it has been possible to generate an oxidoreductase at the active site of a naturally occurring proteolytic enzyme.

Once an enzyme molecule has been chemically altered, generally an analysis is performed to determine the efficacy with which the modified and native enzymes catalyze a particular reaction. The covalent transformation of amino acid side chains, particularly at the active site, is often manifested by dramatic changes in the specificity of the enzyme. Once this effect has been noted, it is then necessary to correlate cause with effect, i.e. to elucidate the identity of the amino acid that has been altered. The importance of this exercise cannot be overestimated since the ultimate point is the determination of those residues that play a pivotal role in enzyme specificity. Moreover, once the altered amino acid residue has been identified, a further series of chemical modifications is often carried out to change rationally the specificity of the enzyme.

Traditionally, chemical modification studies precede the crystallographic analysis of proteins. X-ray analysis, in turn, can be used to confirm and complement the data produced by chemical means. In the subsequent sections we will focus on various changes in enzyme specificity and correlate them, where possible, with the known tertiary structure of the enzyme molecule.

PART I: GENERAL CHEMICAL MODIFICATIONS OF ENZYMES

α-Chymotrypsin

One of the most studied amino acid residues in enzymology has been the methionine-192 in chymotrypsin. This residue has been oxidized by periodate (1), photooxidation (2), and hydrogen peroxide (3), and alkylated with iodoacetic acid (3), α-bromoketones (4), iodoacetylphenylalanine ester (5), and α-bromoacetanilide (6).

The oxidation of methionine-192 to the sulfoxide provides an enzyme derivative possessing 55% of the native esterase activity toward L-tyrosine ethyl ester. The hydrolytic rate constants (acylation and deacylation steps) for both nonspecific and specific substrates were nearly identical for both enzyme species. However, whereas the K_m for the nonspecific substrates remained the same, the K_m for the specific substrates increased two- to threefold for the methionine-192 sulfoxide form of chymotrypsin. Knowles has ascribed this to steric factors (1). The nonspecific substrates are by definition relatively small molecules (for example, p-nitrophenylacetate) which do not make extensive use of the binding pocket in the enzyme. Specific substrates generally have lower K_m values which may result from greater utilization of the noncovalent binding regions within the active site. Since the specific substrates of chymotrypsin are aromatic molecules (for example, N-acetyl-L-tryptophan ethyl ester), Knowles proposed that the methionine-192 sulfoxide is directed into the active site region where the substrates' aromatic side chains are bound. This hypothesis is supported by the alkylation experiments of Lawson & Schramm (4) in which methionine-192 is converted to S-(acetyl α-aminoisobutyric acid) sulfonium salt. In this case, the presence of the more sterically hindered methionine residue (that is sterically hindered relative to methionine sulfoxide) led to a 10-fold increase in K_m for specific substrates.

In 1965 Bender & Kézdy (7) proposed that more than one productive binding mode could occur with nonspecific substrates of chymotrypsin. In order to test this, they employed Lawson & Schramm's alkylated methionine-192 chymotrypsin since the alkyl group introduced at methionine-192 may interfere with only one of the many possible productive binding sites present in the enzyme (8). In order to elucidate the alkyl group's region of interference, the effect of modified methionine-chymotrypsin on kinetic parameters was related to distinct regions of the active site.

The standard kinetic scheme I for the formation of an acylenzyme intermediate is

$$S + E \underset{k_{-1}}{\overset{k_1}{\rightleftharpoons}} ES \xrightarrow{k_2} ES' \xrightarrow[\quad\quad]{\overset{k_3}{H_2O}} E + P_1 \quad\quad\quad \text{Scheme I}$$
$$\xrightarrow[CH_3OH]{k_4} E + P_2$$

where ES' represents the acylenzyme and k_3 is the rate constant for deacylation. The trapping of the acylenzyme intermediate with methanol was studied and the rate constant, k_4, was determined. It was found that the ratio k_4/k_3 was identical for both the modified and native enzymes. Therefore the authors concluded that, on the basis of the symmetry of the enzyme mechanism, the alkyl group on methionine-192 does not interact with the leaving group of the substrate. If such an interaction did exist, then the methanol molecule would have relatively less access to the acylenzyme intermediate than the smaller water molecule in the alkylated versus native enzyme. The constant value of k_4/k_3 for both enzymatic species belies such an interaction.

In contrast to other interpretations, these authors were able to rule out interference at the hydrophobic (aromatic) binding site by demonstrating that the K_i for an inhibitor, benzamide, was unaltered in both enzymatic species. Thus, by a process of elimination, it appears that the alkyl group on the methionine interacts with the acylamido group of the substrate.

Therefore, the nonspecific substrates (with small leaving groups) probably bind at the hydrophobic site and do not come in contact with the oxidized or alkylated methionine, thus accounting for their unaltered kinetic parameters with respect to the modified and native enzymatic species. Kézdy, Feder, and Bender conclude that the alkyl group on the methionine is acting as an intramolecular competitive inhibitor (8).

Finally, a major difference was noted between the oxidized and alkylated methionine-192 chymotrypsin derivatives. It was found that in the latter species k_3 was accelerated whereas in the former it remained unchanged. This has been interpreted to represent an independent phenomenon such as steric strain introduced by the bulky alkyl group.

The covalent modification of methionine-192 in these studies was used to map the active site of α-chymotrypsin. These newly created enzymes display activity toward specific and nonspecific substrates altered significantly from that of the native enzyme.

Trypsin

In 1949, Fraenkel-Conrat and coworkers published a paper reporting the chemical modification of trypsin upon which much of the subsequent work in this area was based (9). The initial purpose of this investigation was to elucidate the nature of the essential functional groups in trypsin. Although acetylation with acetic anhydride was demonstrated to have little effect upon the enzymatic activity of trypsin, the trypsin-inhibiting action of ovomucoid was completely impaired. This was attributed to the acylation of the trypsin amino groups which could then no longer interact with the carboxyl groups of ovomucoid. It was also observed that acetyltrypsin was not as susceptible to self-digestion as was the native enzyme. This is, undoubtedly, reflected by trypsin's specificity for positively charged lysine (and arginine) residues. On the other hand, Riordan et al showed that acylation of trypsin with acetylimidazole under nondenaturing conditions resulted in no change in enzymatic activity toward peptide (casein) or ester [benzoylarginine ethyl ester (BAEE)] substrates (10).

Subsequently Trenholm et al (11) studied both the acetic anhydride and N-acetylimidazole-acetylated trypsins and likewise found that both hydrolyzed BAEE at rates comparable to that of the native enzyme. However, with N-α-p-toluenesulfonyl-L-arginine methyl ester (TAME) as substrate, the specific enzymatic activity of acetyltrypsin shows greater substrate activation than does the native enzyme at high TAME concentrations. This phenomenon is evident but to a much lesser degree in the presence of BAEE. Therefore, the acylamido moiety is crucial in determining the differential activation of acetylated versus native trypsin. Acetylation of trypsin increases the enzyme's activity toward both p-toluenesulfonylarginine amide (TAA) and benzamidine by a factor of two (12). Furthermore, substrate activation does not appear to be an important factor in the hydrolysis of TAA. Therefore, the effects of p-toluenesulfonyl- and benzoyl moieties on the amide substrates appear to be similar. Trenholm et al (11) have demonstrated that the enhanced activity of acetyltrypsin relative to trypsin is due to a combination of a decrease in K_m and an increase in k_{cat} for the modified enzyme.

The enhanced activity of acetyltrypsin can be reversed by treatment with hydroxylamine and therefore Trenholm has proposed that tyrosines are the sites of modification. Chevallier et al (13) reported that trypsin acetylated at 4 tyrosine residues possessed similar properties to the species produced by Trenholm. Unfortunately, none of these reports identified which tyrosine residues were responsible for the modulation of enzymatic activity.

Elastase

Gertler et al have proposed that the α-amino function of the N-terminal valine residue plays an important role in maintaining the activity and conformation of elastase (14). Kaplan & Dugas tested this hypothesis by acetylating elastase with acetic anhydride (15). This treatment modified the ε-amino groups of lysine and the α-amino group of the N-terminal valine. The acetylated and native forms of elastase were shown to have the same activity toward the methyl ester of N-benzoyl-L-alanine. Therefore, it was concluded that the ion pair formed between the α-amino of Val-16 and the β-carboxyl of Asp-194 was not important for elastase activity. (Numbering of these residues is based upon the sequence of the zymogen.) However, the elastolytic activity of modified elastase is only 50% of that of the unmodified enzyme. Therefore, it may be that acetylation of the α-amino group can affect the substrate specificity.

In 1971 Gertler (16) proposed that the adsorption of elastase onto elastin, at low ionic strength and pH 5–10, is an electrostatic nonspecific process that occurs between the basic groups of elastase and the carboxyl groups of elastin. To test this hypothesis, Gertler maleylated the primary amino groups of elastase (17). As with acetylated elastase, the esterase activity of the maleylated enzyme was similar to elastase; however, the elastolytic activity dramatically decreased 40 fold. Futhermore, Gertler did perform an N-terminal analysis and noted that the α-amino terminus of the maleylated enzyme was unmodified. For the maleylated enzyme, the change in substrate specificity relative to the native enzyme appears to be due to a general electrostatic effect.

A modified elastase derivative containing two chemically altered arginine residues has been recently prepared. The reaction of elastase with 1,2-cyclohexanedione provided a species that possessed 15% of the activity of the native enzyme against elastin. However, the modified enzyme possessed 6–104% activity toward a series of synthetic substrates. In all of these cases, the K_m value increased whereas the k_{cat} value either increased or decreased depending upon the particular substrate. For anilide substrates Davril et al (18) proposed that the increase in k_{cat} for the modified enzyme may be a result of better positioning of the scissile anilide bond toward the serine hydroxyl of the catalytic site. This arginine modification also slows the formation of the α_1-antitrypsin-elastase complex fourfold.

Recently, a great deal of interest has been generated with respect to the chemical modification of elastase and inhibitors. The breakdown of elastin by elastase in polymorphonuclear leukocytes has been implicated as a major factor in emphysema. The increased activity of elastase in emphysema patients has been proposed to arise from cigarette smoke oxidation of a methionine residue in the inhibitor of elastase, α_1-antitrypsin (19, 20). Preliminary results indicate that the oxidized form of α_1-antitrypsin can no longer interact with elastase.

There have been a number of attempts to block the degradative ability of elastase by modifying its active site with various affinity labels (21a,b). In the context of this review, however, we note that a chemically induced alteration in elastase's specificity toward elastin could also achieve the same desired result. Indeed, as indicated earlier, modification of elastase's amino groups selectively abolishes activity toward elastin. Obviously, such a modification would be difficult to achieve in vivo, but it represents an alternative to traditional methods of enzyme inactivation. On the other hand, a chemical modification resulting in enhanced binding of elastase to its inhibitor (either the native form or its oxidized derivative) may prove effective.

Subtilisin

Subtilisin Carlsberg and subtilisin BPN' are serine-proteinases isolated from *Bacillus subtilis*. They differ at 84 positions out of 275 residues. However, X-ray analysis of the tertiary structure indicates that with one exception, the 84 differences occurred on the protein's exterior. Both enzymes have been the subject of extensive chemical modification investigations, the majority of which have been performed at the Carlsberg laboratories in Copenhagen.

Employing the procedure developed by Vallee and coworkers (22), Svendsen et al noted an increase in proteolytic activity of subtilisin Carlsberg upon nitration (23). Nitrated subtilisin Carlsberg hydrolyzed clupein six times more efficiently than the native enzyme. Iodination and succinylation of subtilisin resulted in a modified enzyme possessing a six- and sevenfold enhancement in this activity, respectively. The increased enzymatic activity displayed by these derivatives of subtilisin Carlsberg cannot be ascribed to the relief of marked substrate inhibition as it can be with modified carboxypeptidase A (see below). However, the activity of the iodinated, nitrated, and succinylated forms of subtilisin Carlsberg was found to be nearly equivalent to that of the native enzyme for the hydrolysis of the neutral substrates and *N*-benzoyltyrosine ethyl ester.

In effect, iodination, nitration, and succinylation of subtilisin Carlsberg serve to introduce negative charge which enhances the binding of clupein (a positively charged protein) without affecting the binding of casein of benzoyl tyrosine ethyl ester (neutral substrates). The negative charge results from the lowered pK_a of the iodo- or nitrotyrosine hydroxyl relative to that of the tyrosine hydroxyl, or from the presence of a free carboxyl group in succinylated tyrosine. Therefore, it is not surprising that the modified subtilisin's activity toward clupein returned to that of the native enzyme when the nitrotyrosine was reduced to an aminotyrosine (24). Thus, it was unexpected that the positively charged *p*-toluenesulfonyl-L-arginine methyl ester was hydrolyzed with equal efficiency by nitrated, iodinated, and native subtilisin Carlsberg. An explanation was suggested by a detailed investigation of succinyl and glutarylsubtilisin Carlsberg (25) in conjunction with known X-ray data (26). Succinylation (as well as nitration and iodination) of subtilisin Carlsberg can result in changes in secondary binding sites while leaving the primary binding site unchanged. This explanation provides a rationalization for the experimental observation that small substrates, such as *p*-toluenesulfonyl-L-arginine methyl ester, that bind only at the primary site, are hydrolyzed equally well by either the modified or the native enzyme. Larger substrates, such as clupein and gelatin, interact with the secondary sites, and their hydrolysis rates are therefore sensitive to the protein modification.

In 1974, Svendsen reported (27) that tyrosine-104 was the first of eight tyrosine residues to be modified by tetranitromethane. This, in conjunction with the fact that modification of between one and two tyrosine residues provided maximal change in activity toward clupein (28), led Svendsen to conclude that residue 104 is important to clupein binding. Therefore tyrosine-104 must be within the secondary binding site. This is corroborated by X-ray crystallographic analysis of the interaction of a chloroemethylketone inhibitor (29, 30) with subtilisin BPN'.

A detailed investigation of the reaction of clupein's three subunits (Y^I, Y^{II}, Z) with nitrated and native subtilisin Carlsberg revealed interesting differences in cleavage specificity (31). The major sites of hydrolysis in clupein Z are Ser_{21}-Arg_{22} and Ala_9-Ser_{10}. Svendsen demonstrated that nitrated subtilisin Carlsberg was more specific for the cleavage of the Ser-Arg bond than the native species. The nitrated enzyme also showed increased specificity for the loci Arg_5-Ser_6 and Ser_6-Ser_7 in clupein Y^I compared to both subtilisin Carlsberg and BPN'.

The introduction of negative charge at certain regions in the binding site of subtilisin produces an enzyme with increased hydrolytic activity toward large positively charged substrates. Furthermore, specific bonds within these substrates are preferentially subject to hydrolysis in the presence of the modified enzyme.

Papain

Papain is probably the most extensively studied of all the cysteine proteinases. Much of the attention has centered around the functional role of the tryptophan residues. By a variety of chemical modification studies, such as N-bromosuccinimide (NBS) oxidation (32), photooxidation (33, 34), and alkylation (35–38), these tryptophans have been implicated to play an important functional role in papain's catalytic action. The NBS oxidation was originally reported to inactivate papain (39). However, later studies (40, 41) reported that papain's activity toward benzoylarginine amide derivatives was enhanced upon modification. One of the later studies (41) also noted that oxidized papain had increased peptidase activity but decreased esterase (N-benzoylarginine ethyl ester) and proteinase (casein, protamine) activities relative to the native enzyme. Unfortunately, the modified residues were not identified. Therefore, it is possible that in all of these studies a mixture of several modified species was present.

For this reason Lowe & Whitwork undertook a systematic study of the NBS reaction (32). They protected the active site cysteine from oxidation by treatment of papain with a hydroxyethyldisulfide to provide papain-S-S-CH_2CH_2OH. The protected enzyme was then oxidized with six equivalents of NBS. Subsequently, three modified enzymes were isolated. Amino acid analysis indicated no change in the tyrosine or histidine content whereas for two of these species the tryptophan content dropped from five to three residues. The

Table 1   Apparent pK_a values of k_{cat} and k_{cat}/K_m for the enzyme-catalyzed hydrolysis of N-benzyloxycarbonylglycine p-nitrophenyl ester

	pK_a from the pH dependency of k_{cat}	pK_a from the pH dependency of k_{cat}/K_m
Native papain	3.50	4.30
Trp-67[0]-papain	3.45	4.70
Trp-67[0],Trp-177[0]-papain (species 1)	4.55	5.35
Trp-67[0],Trp-177[0]-papain (species 2)	4.25	5.45

tryptophan residues modified were determined to be Trp-69 and Trp-177 by peptide mapping. The third species was oxidized only at Trp-69. Subsequent to the oxidation, cysteine was used to regenerate the active site sulfhydryl from its derivatized form. The pK_a for k_{cat} and k_{cat}/K_m for the hydrolysis of N-benzyloxycarbonylglycine p-nitrophenyl ester was similar for both papain and the modified enzyme oxidized at Trp-69. However, for the enzymes with two oxidized tryptophan residues the pK_a for both k_{cat} and k_{cat}/K_m increased by about a full unit over that of the native enzyme (Table 1).

In the native protein, the relatively acidic pK_a values of k_{cat} and k_{cat}/K_m have been ascribed to the ionization of histidine. Oxidation of Trp-177 may then change the hydrophobicity of the active site and therefore alter somewhat the ionization behavior of this histidine residue. Thus, Trp-177 is at least partially responsible for maintaining the hydrophobic nature of the active site in papain.

The oxidized papains were also analyzed for their ability to hydrolyze N-benzyloxycarbonyl-Gly-Gly-NH$_2$ (Z-Gly-Gly-NH$_2$) and Z-Gly-Trp-NH$_2$. In these cases, k_{cat} represents k_2, the rate constant for the acylation step, and K_m is equivalent to K_s, the dissociation constant for the Michaelis complex (Equation 1). The results are shown in Table 2. In all cases there is an increase in K_m. The increase is clearly associated with Trp-69 since oxidation of Trp-177 does not significantly affect K_m. It is probable that the oxidation of Trp-69 decreases

Table 2   Kinetic parameters for peptide hydrolysis by native and oxidized papains at pH 5.0

	k_{cat} (s^{-1})	K_m(mM)	k_{cat}/K_m (M^{-1} s^{-1})
Z-Gly-Gly-NH$_2$			
papain	0.068	6.4	10.6
Trp-69[0],Trp-177[0] papain	0.040	34	1.2
Z-Gly-Trp-NH$_2$			
papain	0.60	0.9	670
Trp-69[0] papain	0.49	5.3	92
Trp-69[0],Trp-177[0] papain	0.064	5.2	12

the hydrophobicity of one of the subsites and thus weakens substrate binding. On the other hand, the oxidation of Trp-177 appears to have a dramatic effect on k_{cat} for the hydrolysis of Z-Gly-Trp-NH$_2$.

In a series of papers, Horton and coworkers demonstrated that Trp-177 can be alkylated when the active site cysteine sulfhydryl is unprotected and this results in an enzyme possessing enhanced catalytic activity (35–38). The design of the alkylating agent 1 (See Figure 1) is based upon the work of Schramm & Lawsson (4). Hydroxynitrobenzylation (HNB) of Trp-177 resulted in an enzyme that had a 24% increase in k_{cat}/K_m toward benzoylarginine ethyl ester (BzArgOEt) and a 240% increase in k_{cat}/K_m toward benzoylarginine nitroanalide (BzArgNA) relative to papain. The increased k_{cat} for the ester substrate is due to a 50% decrease in K_m (k_{cat}/K_m however is lowered by 28%). The K_m for the nitroanilide substrate is decreased by 60% while the k_{cat} is increased by 27%. A detailed examination of the hydrolysis of BzArgNA by HNB-papain revealed that K_s had increased from 7.9 to 14.9 mM and k_2 (acylation rate constant) had increased by sevenfold. No significant change was found for the deacylation rate constant, k_3.

$$E + S \underset{}{\overset{K_s}{\rightleftharpoons}} ES \xrightarrow{k_2} ES' \xrightarrow{k_3} E + P \qquad 1.$$

The increased rate of acylation in HNB-papain may be due to an increased nucleophilicity of the active site cysteine sulfhydryl. This has been supported by considering the nature of the active site of papain. Drenth and colleagues (38) have suggested that His-159 and Cys-25 exist in an imidazolium mercaptide ion-pair form in native papain. A charge-transfer interaction between Trp-177 and the imidazolium moiety of His-159 stabilizes this ion-pair form which, in turn, is responsible for the nucleophilicity of Cys-25. Horton (35) proposed that the p-nitrophenoxide ion attached to Trp-177 in HNB-papain is in a position to interact with and stabilize the His-159-Cys-25 ion pair even further, thereby enhancing the reactivity of Cys-25 over that observed in native papain.

In summary, selective changes in papain's substrate specificity have been introduced by the oxidation and alkylation of Trp-177. Oxidation results in a

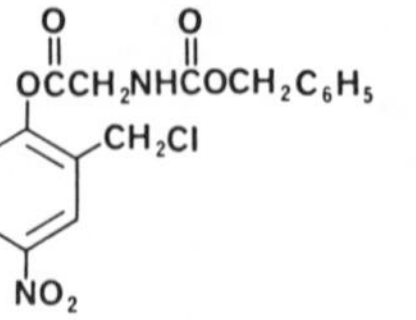

Figure 1   *Structure 1*. 2-Chloromethyl-4-nitrophenyl (*N*-carbobenzoxy)glycinate.

higher pK_a of the active site histidine whereas alkylation results in a higher nucleophilicity of the active site cysteine.

Pepsin

As early as 1925 chemical modification studies on pepsin were already under way (42). In 1934 Herriott & Northrop acetylated porcine pepsin with ketene. In this report, they describe the rationale behind chemical modification studies that is still invoked today (43). Their reasoning went as follows:

> Pepsin is known to have such free groups as carboxyl (COOH), amino (NH$_2$), and phenolic hydroxyl (OH). The effect on activity resulting from the modification of these groups should furnish information as to their relation to the molecular structure responsible for the activity. If the modification of these groups in native pepsin resulted in a complete loss in activity it would be probable that the active group, or one closely associated with the active group, had been altered by the treatment. If the modification of these groups failed to affect the activity it would follow that the modified groups are not closely related to the active part of the enzyme.

They produced three modified pepsins (each in crystalline form; the photographs of two of these are shown in the original article) which differed in their degree of acetylation. One of the derivatives contained 3 or 4 acetyl groups and could be isolated after brief exposure to ketene. It hydrolyzed hemoglobin as readily as the native enzyme. A second derivative was found to contain 6–11 acetyl groups and possessed nearly 60% of the activity of pepsin. The third modified enzyme had 20–30 of its residues modified and had 10% of the original activity of pepsin. Herriott & Northrop were able to demonstrate that the 60% active derivative, upon standing in strong acid solution, loses some of its acetyl groups and regains full activity. These authors concluded that acetylation of the first three or four primary amino groups does not change the activity of pepsin and that acetylation at other sites on the enzyme molecule must be responsible for the alteration in activity.

Thirty years later, G. E. Perlmann reinvestigated the acetylation of porcine pepsin and extended the acetylation procedure to pepsin's zymogen, pepsinogen (44). Perlmann employed N-acetylimidazole to produce an acetylated pepsin that was only 40% as active toward hemoglobin as was the native species. However, the relative specific activity of the modified enzyme toward the synthetic substrate, N-acetyl-DL-phenylalanyldiiodotyrosine, was enhanced. Furthermore, the activity of the acetylpepsin toward the oxidized and reduced B chain of insulin was approximately equal to that of native pepsin. Since insulin is intermediate in molecular weight between the synthetic substrate and hemoglobin, Perlman concluded that acetylation of pepsin enhances its activity toward small substrates while it inhibits its activity toward large substrates.

Neither Herriott & Northrop nor Perlmann identified which residues were acetylated in pepsin. However, based on chemical and spectral evidence,

Herriott (45) concluded that tyrosine acetylation is responsible for the activity modulation.

The reaction of pepsinogen with acetylimidazole resulted, after activation of the zymogen, in an enzymatic species that was 20–25% as active toward hemoglobin as is pepsin. Interestingly, the activity of this acetylpepsin toward the dipeptide substrate was also decreased. This is in marked contrast to the activity arising from the acetylation of native pepsin. The obvious conclusion is that differential acetylation occurred in the reaction of pepsin and pepsinogen with acetylimidazole.

An enhancement of the rate parameters of chicken pepsin–catalyzed hydrolysis of Z-Ala-Ala-Phe-Phe-OPhe was observed upon sulfenylation and alkylation of the single cysteinyl residue (46, 47). In contrast, these altered pepsins are less active than the native enzyme toward hemoglobin. This, of course, is reminiscent of the decreased catalytic efficiency displayed by acetylated pepsin. Additionally, the groups used to modify the cysteine residue were systematically varied with a concomitant modulation of pepsin activity. In all cases, the K_m for the hydrolysis of the tetrapeptide substrate decreased upon cysteinyl derivatization. The k_{cat} for this reaction varied from 30% (for pepsin-S-Hg-benzoic acid) to 500% (for pepsin-S-S-p-nitro-o-benzoic acid) of that of the native enzyme. The authors concluded that the -SH group does not participate in binding since its derivatization with bulky groups such as -(S-p-nitro-m-benzoyl-Phe-Phe) does not affect binding of the substrate. However, the rate of modification of the cysteinyl sulfhydryl group is dependent upon the nature of the modifying reagent. In general, the sulfhydryl shows a preference for aromatic derivatizing agents. This indicates that the reagent must penetrate a hydrophobic area to reach the cysteine residue.

Both chicken and porcine pepsin catalyze the hydrolysis of hemoglobin at similar rates. However, it has also been observed that chicken pepsin does not catalyze the hydrolysis of synthetic substrates as readily as the porcine species. The introduction of an aromatic molecule at the cysteinyl SH in chicken pepsin produces a species as active as porcine pepsin. These authors speculated that the porcine enzyme may, in fact, possess an aromatic group in this position whereas the chicken pepsin may not.

Carboxypeptidase A

In the early 1960s Vallee and coworkers began to report their work on the chemical modification of the zinc-containing proteinase, carboxypeptidase A (CPA). Their initial purpose was to identify those residues involved in the hydrolytic mechanism. They observed that acetylation of the enzyme with acetic anhydride or N-acetylimidazole modified 2 of the 19 tyrosyl residues (48). Interestingly, esterase activity was enhanced sixfold whereas peptidase activity was almost entirely abolished. Diazotization with 5-diazo-1H-tetrazole resulted in the covalent modification of only one tyrosine residue and a concomitant 1.8-fold increase in esterase efficiency without a corresponding change in peptidase activity (49). When a great excess of this reagent was employed a second tyrosine residue as well a histidine residue were modified. This derivatized CPA no longer possessed peptidase activity; however, esterase activity remained intact. Unfortunately, Vallee and coworkers did not identify the specific amino acid residues modified. Lipscomb was able to deduce the sites of chemical alteration based upon the crystal structure (50). Acetylation and diazotization resulted in the modification of tyrosine residues 198 and 248, though it was not clear which histidine residue was diazotized. Sokolovsky & Vallee have suggested that the elimination of peptidase activity corresponds directly to the modification of histidine (49). In the presence of the competitive inhibitor β-phenylpropionate, this modification is prevented. Therefore, it seems likely that the histidine residue affected lies at or near the enzyme's active site.

Arginine modification with butanedione in borate buffer (51) decreases CPA's peptidase activity while it increases its apparent esterase activity threefold over that of the native enzyme. Riordan proposed that the peptide substrates bind differently than ester substrates in CPA's active site. This series of investigations suggests that the specificity of CPA is readily changed by chemical modification and results in differential activity toward ester versus peptide substrates. However, a detailed kinetic analysis by Bender and coworkers with acetylcarboxypeptidase A indicated that acetylation of the enzyme results in the elimination of substrate inhibition (52). They pointed out that the substrate concentrations employed by Vallee and coworkers (53) were in the range where substrate inhibition occurs in native CPA. Thus, under these circumstances, the esterase activity appears to be greater in acetylated CPA than in the native enzyme. Actually, acetylated CPA shows decreased activity (in terms of k_{cat}/K_m) toward both ester and amide substrates. The analysis by Bender et al was corroborated by Hall et al who observed reduced activity in the action of acylated CPA toward the ester substrate O-(*trans*-β-cinnamoyl)-L-β-phenyllactate, an ester that does not show the complication of substrate inhibition (54).

In the case of acetylated CPA, and possibly with the other modified CPAs, the apparent increase in enzymatic activity seen for some ester substrates appears to be due primarily to the elimination of substrate inhibition encountered with the native enzyme at high substrate concentration.

Lysozyme

In general, chemical modification studies have predated crystallographic analyses of proteins. However, lysozyme is the exception since the three-dimensional structure, elucidated by Phillips and coworkers (55), served as a stimulus for the derivatization of lysozyme.

The catalytic activity of lysozyme is most commonly assayed with chitin and the bacterium *Micrococcus lysodeikticus*. These two substrates serve as the focal point for most of the studies involving the alteration of lysozyme's specificity.

A quintessential example of specific chemical modification is the conversion of Trp-62 in hen egg-white lysozyme to oxindolealanine-62 by NBS (56, 57). The importance of Trp-62 in substrate binding has been noted by X-ray crystallography (58). This has been corroborated by the chemical properties of Trp-62 oxidized lysozyme. This modified enzyme possesses decreased affinity for glycol chitin (57, 59, 60). Indeed, a threefold increase in K_m was found for the modified versus the native enzyme species. Furthermore, Trp-62-oxidized lysozyme displays a V_{max} for glycol chitin of less than 10% of that of the native enzyme. Interestingly, oxidation of the enzyme does not affect lytic activity (29%) as drastically as glycol chitin activity (10%). Imoto et al (57) explained this difference by considering the interaction between positively charged lysozyme and negatively charged bacterial cell wall components (61). This interaction would serve to align negatively charged substrates with the positively charged arginines (61, 73, 112, and 114) located at or near the active site of lysozyme. Since the positively charged arginines remain unmodified during the oxidation of Trp-62, the binding of negatively charged substrates (bacterial cell walls) should be relatively unaffected compared to that of neutral substrates (glycol chitin). However, if Trp-62 plays an important role in maintaining the hydrophobic nature of the binding site then it is understandable that the neutral species, glycol chitin, is a poor substrate when this residue is oxidized.

Imoto and coworkers next oxidized Trp-62 with ozone to form the *N*-formyl kynurenine derivative of lysozyme. This species possessed 40% of the native cell lysis ability. However, it exhibits only 15% of the native enzyme's activity toward glycol chitin. The *N*-formyl kynurenine lysozyme was then deformylated by acid hydrolysis to provide kynurenine-62 lysozyme. Deformylation resulted in a species with increased cell lytic activity (80% of native enzyme) and glycol chitin cleavage activity (30% of native enzyme). These authors suggested that the relatively high catalytic efficacy of Kyn-62 lysozyme may be due to the ability of kynurenine to form a coplanar structure similar to an indole ring.

Davies & Neuberger (62) specifically inhibited lysozyme's lytic activity toward *M. lysodeikticus* by acetylating the six lysines of the enzyme. Although this physiologically important reaction was inhibited, the lytic activity toward the tetramer of *N*-acetylglucosamine [(NAG)(4)] was unaltered. These two activities were differentially affected with other reagents as well. For example, the modification of seven of the eleven arginine residues with butanedione had a similar effect on the catalytic activity.

The change in specificity resulting from these modifications may be due to the elimination of the positive charge on the lysine amino group or to the introduction of increased steric bulk. In order to differentiate between these two possibilities the lysine residues were modified with ethyl acetamidate. This modification introduces an appendage of similar size to the acetyl group without affecting surface charge. Once again, the hydrolytic activity of this derivatized enzyme remains unaffected toward (NAG)$_4$. In contrast, the cell lysis activity is no longer inhibited, and, in fact, the activity is slightly increased relative to that of the native enzyme.

In this section, it has been shown that conversion of charged residues to uncharged derivatives selectively decreases the lytic activity of lysozyme toward *M. lysodeikticus,* whereas the chemical modification of the hydrophobic binding site disrupts the hydrolytic activity of lysozyme toward only small neutral substrates.

Pancreatic Ribonuclease

A vast literature exists on the chemical modification of pancreatic ribonuclease (RNase A) [for review see (63)]. RNase A is certainly one of the most, if not the most, extensively modified enzymes. It is a favorite target for the development of new modifying reagents, and therefore, investigations on this enzyme can often be regarded as indications of future trends in protein chemistry. For example, ribonuclease was reacted with chloroamineruthenium dichloride to provide the first pentaamineretheniumhistidine complex in a protein (64). Subsequently, Gray and others have made extensive use of the specificity shown by the ruthenium reagent for histidine residues and have created semi-synthetic proteins possessing unique properties (see below). Since RNase A appears to represent a popular testing ground for chemical modification, we will present in this section the more unusual and novel alterations achieved with this enzyme; those investigations that have employed standard chemical reagents have been previously reviewed (63).

A novel result achieved by chemical modification was reported in 1967 by Sela and coworkers (65). RNase A was treated with phosphorothionate under nondenaturing conditions to produce a species in which two out of the four disulfide bonds in the enzyme were cleaved and derivatized. The reaction between the disulfide bonds and the phosphorous reagent in the presence of oxygen provides the doubly phosphorylated species:

$$R\text{-}S\text{-}S\text{-}R' + HPO_3S^{2-} + \tfrac{1}{2}O_2 \rightarrow R\text{-}S\text{-}SPO_3^{2-} + R'\text{-}S\text{-}SPO_3^{2-} + H_2O \qquad 2.$$

This modified enzyme, denoted 4PS-RNase, was homogeneous. The reactive disulfide bonds were between half-cystines 65–72 and 58–110. Since disulfide bonds often play a dominant role in preserving the tertiary structure of an enzyme, it is surprising that 4PS-RNase is as active toward RNA as the native enzyme. The introduction of four doubly charged phosphate groups into an enzyme molecule would certainly have been predicted to result in disruption of the tertiary structure. In fact, not only is the ability to hydrolyze RNA unaffected in 4PS-RNase, but its activity toward cytidine 2', 3'-cyclic phosphate is increased to 220% of that of the native enzyme. In spite of what would seem to be a potentially disruptive covalent modification, spectrophotometric titrations and CD measurements (66) showed little difference between 4PS-RNase and the native enzyme.

Related studies on RNase include a report on the creation of cysteine-mercury-cysteine cross-links in place of disulfide bonds (67). This RNase-mercury derivative possesses 5% and 25% of the hydrolytic activity of the native enzyme toward RNA and cytidine 2', 3'-cyclic monophosphate, respectively. The insertion of mercury elongates the internal disulfide cross-links in RNase by 3Å and thus may alter the conformation necessary for optimal activity.

Watkins & Benz (68) reported that incubation of RNase with mercaptoethanol provided a number of modified RNases. Interestingly, the most active of these possessed four to five times the hydrolytic activity of the native enzyme toward cytidine-2',3'-cyclic monophosphate. This increase in activity is reflected by a tenfold decrease in the Michaelis constant. This new enzyme is possibly a mixed disulfide of mercaptoethanol and RNase; however, this has not been demonstrated. In fact, with or without exogenous thiol, this modified RNase reverts back to the native enzyme.

RNase has also been cross-linked by various reagents to provide changes in enzymatic specificity. Moore et al (69, 70) dimerized pancreatic RNase by treatment with the bifunctional reagent dimethylsuberimidate [CH$_3$OC(=NH)(CH$_2$)$_6$C(=NH)OCH$_3$]. RNase of bovine seminal plasma, unlike other RNase isozymes, is known to exist as a dimer (71). This dimer has the ability to hydrolyze double-stranded RNA at appreciable rates under conditions where the activity of the monomeric pancreatic enzyme is low. The synthetic dimeric species prepared by these authors contained 19 free amines. This would indicate that 3 amino groups per dimer had been imidylated—only one over the minimum necessary for duplex formation. The synthetic dimer contained 78 times the activity toward polyadenylate-polyuridylate and 440 times the activity toward another double-stranded RNA than the native monomer.

An interesting variation on the dimerization of RNase was reported by Wang (72) who cross-linked RNase with DNase. The eventual hope of this study was to develop a new enzymatic species that could disrupt the reproductive cycle of certain cancer viruses by hydrolyzing both RNA and DNA. The hybrid enzyme was prepared by treating (73) the amino groups of DNase with N-acetyl-DL-homo-cysteinethiolactone in the presence of 4,4'-dithiopyridine. This provided the mixed disulfide, DNase-S-S-pyridine. This same procedure with

RNase in the absence of 4,4-dithiopyridine provided the RNase-SH. Disulfide exchange between RNase-SH and DNase-S-S-pyridine gave the desired species in 25–33% yield. This hybrid enzyme possessed 75% of the activity of native DNase toward calf thymus DNA and 40% of the activity of native RNase toward yeast or transfer RNA. A novel result of this study was that a hybrid substrate, phage f_1 DNA•RNA was hydrolyzed by the hybrid enzyme, but not by RNase.

Recently, vigorous conditions were used to cross-link RNase into an unnatural conformation that possessed a new catalytic function, esterase activity (74). Native RNase displays no esterase activity. This activity was induced by cross-linking RNase with glutaraldehyde in the presence of a known competitive inhibitor of RNase, indolepropionic acid. The presence of this competitive inhibitor was not only important for the creation of esterase activity but also for the production of the subsequent substrate specificity. Both tryptophan ethyl ester and N-benzoyl-L-arginine ethyl ester were efficiently hydrolyzed by the modified RNase; glycine ethyl ester and lysine ethyl ester, however, were not. In addition, this new enzyme species is capable of amidase activity toward benzoyl-L-arginine-p-nitroanilide. Because of these encouraging preliminary results, future experiments with this type of modified enzyme are awaited with anticipation.

Fructose-1,6-Bisphosphatase

Fructose-1,6-bisphosphatase is an allosteric enzyme that can be inhibited by the effector, adenosine monophosphate (AMP). Many of the chemical modification studies on this catalyst have altered the AMP binding site. This in turn has led to an alteration of the enzyme's allosteric properties.

Han and coworkers have demonstrated that aspirin desensitizes the chicken enzyme toward inhibition by AMP (75). A similar phenomenon was noted when the porcine enzyme was treated with salicylate (76). However, in the first case aspirin appears to acetylate two distinct regions of the enzyme since both AMP inhibition and substrate inhibition are abolished. In the second case, salicylate is proposed to act as a competitor of AMP for the AMP binding site.

Modification of arginine residues in the porcine enzyme can lead to a loss of activation by monovalent cations and a loss of inhibition by AMP (77). Treatment of fructose-1,6-bisphosphatase with butanedione in the presence of fructose-1,6-bisphosphate and AMP resulted in the modification of one arginine residue without any apparent effect on the enzyme's ability to respond to effectors (77). Treatment of the enzyme with butanedione in the absence of substrate and AMP produces a catalyst with three modified arginine residues that cannot be inhibited or activated by effectors. The presence of substrate appears to induce a conformational change in the enzyme molecule which prevents modification of an arginine residue located away from the substrate binding site. This conclusion is supported by the fact that these modifications do not lead to a change in enzyme activity (only a change in response to effectors) and therefore do not appear to alter chemically the active site. Desensitization to AMP inhibition has also been achieved by acetylating four tyrosine residues with acetylimidazole (78) and modification of four lysine residues with pyridoxal phosphate (79).

Horecker and colleagues have published a series of papers in which fructose-1,6-bisphosphatase was activated by the modification of specific cysteine residues. These enzyme derivatives include mixed disulfides with mercaptoethanol, ethylmercaptan, N,N-diacetylcystamine, cystamine (80), CoA, and acyl carrier protein (81, 82). Modification with homocystine (83), Ellman's reagent (84), iodoacetamide fluorodinitrobenzene (85), and p-hydroxy-mercuribenzoate (86) also enhanced enzymatic activity. Activation ranged from 1.8-fold with p-hydroxymercuribenzoate to more than eightfold with homocystine. Interestingly, the inhibition response to AMP of fructose-1,6-bisphosphatase treated with p-chloromercuribenzoate (87) or cystamine (84) was greater than that of the native enzyme. Clearly, these studies indicate that one or more free cysteine residues may play an important role in fructose-1,6-bisphophatase activation. An X-ray crystallographic analysis of FDPase would provide a valuable basis for interpretation of these studies.

Allosteric Enzymes

The allosteric properties of a variety of enzymes other than fructose-1,6-bisphosphatase have also been characterized by covalent modification, but probably none to the extent illustrated by fructose-1,6-bisphosphatase. This section presents the results obtained from the modification of other allosteric enzymes and focuses specifically on the enzymes' allosteric properties. Allosteric interactions can be divided into two categories, homotropic and heterotropic, each of which can be modulated by covalent protein modification. Homotropic interactions produce the characteristic sigmoidal enzyme-substrate saturation curve of cooperative protein subunits, whereas heterotropic interactions result from the binding of catalytic effectors to enzymes at sites other than the active sites. The modifications of fructose-1,6-bisphosphatase described above would then be classified as affecting the protein's heterotropic properties.

When aspartate transcarbamylase was treated with tetranitromethane, both the catalytic activity and allosteric properties decreased over the course of the reaction (88). However, the kinetics of modification indicate that the enzyme's cooperativity was lost more quickly than its catalytic activity. This loss of cooperativity is concomitant with the nitration of only one tyrosine per each regulatory and catalytic subunit duplex. Extensive nitration of the enzyme in the presence of the transition state analog, N-(phosphonacetyl)-L-aspartate, eliminated both homotropic and heterotropic interactions. Under these conditions the catalytic activity was hardly affected. Hybrid enzymes containing unmodified regulatory subunits but singularly nitrated catalytic subunits displayed no homotropic interactions at pH = 8.3 (89). However, hybrids consisting of modified regulatory subunits and unmodified catalytic subunits retained their cooperativity, CTP inhibition, and ATP activation (88, 90).

The allosteric nature of aspartate transcarbamylase has also been studied using another type of modification procedure, protein cross-linking. Recently, protein cross-linking procedures have offered new techniques for investigating and controlling the allosteric properties of enzymes. Therefore, rather than modifying individual amino acids in hopes of altering a protein's allosteric properties, bifunctional reagents can often selectively stabilize one form of an allosteric protein. When aspartate transcarbamylase was cross-linked with the bifunctional reagent, tartaryl diazide, in the presence of substrate analogs, the enzyme was forced to maintain a relaxed state, displaying a characteristically low K_m for aspartate of 7 mM (91). If the cross-linking was performed in the absence of any enzyme ligands, the enzyme was forced into a taut form that exhibits a K_m for aspartate of 130 mM (92). The cross-linking, approximately one cross-link per enzyme subunit, in either case was sufficient to inhibit the potency of the catalytic effectors.

Phosphofructokinase, another allosteric enzyme, has been subjected to both individual amino acid modifications and general cross-linking. Since the results of the former technique have already been reviewed (93), only the recent cross-linking experiments are presented here. Treatment of phosphofructokinase with 1 mM glutaraldehyde abolished the sigmoidal binding properties of the substrate, fructose-6-phosphate, and produced an enzyme with Michaelian behavior (94). This procedure, however, did not eliminate its heterotropic interactions (the AMP stimulation and ATP inhibition of catalysis). When the enzyme was cross-linked with glutaraldehyde and immobilized in polymers formed in the added presence of lysine, the protein also displayed Michaelian behavior. The catalytic activity of phosphofructokinase only became insensitive to the heterotropic interactions of ATP and AMP when the immobilization was carried out in the presence of both ATP and AMP.

Finally, the cross-linking studies on dCMP-aminohydrolase presented by Nucci et al (95) and Raia et al (96) suggest that protein subunit cross-linking may become a rather general procedure to stabilize the high- and low-substrate affinity forms of an enzyme. dCMP aminohydrolase was cross-linked with glutaraldehyde in the presence of an allosteric activator, dCTP, and a competitive inhibitor, dAMP. This modification freezes the enzyme in an activated state by stabilizing the native, hexameric form of the enzyme. The activators could no longer boost enzyme activity. Conversely, the same enzyme was fixed in its taut, low-substrate affinity mode after it was cross-linked with glutaraldehyde in the presence of dTTP, an effector that stabilizes the taut form. Protein covalent cross-linking now appears to be a proven way to stabilize specific forms of allosteric proteins.

Coenzyme-dependent Enzymes

The covalent modification of some of the enzymes presented in this section (glyceraldehyde-3-phosphate dehydrogenase, xanthine dehydrogenase, lysine monooxygenase, and amine oxidase) was previously reviewed by Gorkin (97, 98) covering the literature through part of 1975. Many of the early studies included in those reviews are still valuable enough to deserve a recapitulation in the context of this presentation. However, more recent studies will be discussed in greater detail.

Glyceraldehyde-3-phosphate dehydrogenase catalyzes a series of reactions during its catalytic cycle. These reactions include (a) formation of a hemithioacetal between an enzyme thiol and the substrate, (b) subsequent oxidation of the hemithioacetal to a thiol acyl ester and (c) transfer of the acyl group to a variety of acceptors. Some of the earliest chemical studies on this enzyme relied on reagents that could modify specific amino acids and allow identification of the amino acids most crucial for the enzyme's function. In 1958, Park & Koshland (99) reported the modification of the active site thiol (Cys-149) of glyceraldehyde-3-phosphate dehydrogenase with iodoacetate. This new enzyme did not lose its ability to hydrolyze acetyl phosphate. However, the other activities of the enzyme were destroyed. Ehring & Colowick (100) later published "an optimal conversion of glyceraldehyde-3-phosphate dehydrogenase of muscle and yeast by oxidation with 2-iodosobenzoate to a highly active acyl phosphatase." Again, the active site thiol was modified, this time to a sulfenic acid, producing the phosphatase activity but inactivating the other enzyme functions. A variety of other oxidizing agents, iodine (100), H_2O_2 (101), iodine monochloride (102), and trinitroglycerin (103) were used to oxidize the active site cysteine to a stable sulfenic acid derivative. Although sulfenic acids are rarely stable in small organic molecules, a variety of proteins seem to form sulfenic acid derivatives under mild oxidizing, but nondenaturing, conditions (104).

Lactate dehydrogenase also contains an active site thiol group. Its specific modification produces contrasting results to that of glyceraldehyde-3-phosphate dehydrogenase because the thiol of lactate dehydrogenase is not an integral part of the catalytic apparatus. The active site cysteine, Cys-165, was modified with methanethiosulfonate (105, 106). Essentially the only change in the kinetic properties induced by this modification was an increase (300-fold and 100-fold, respectively) in the apparent K_m of pyruvate and lactate; the V_{max} of turnover and K_m for NADH and NAD were virtually unaffected. Therefore, the consequences of thiol modification as illustrated for the above two enzymes are very much determined by the thiol's role in catalysis.

Cysteine has been the most common target of covalent amino acid modification in the studies of the coenzyme-dependent proteins. Malic enzyme represents no exception; approximately ten thiol groups per unit (or 30 per native trimer) were reacted with 5,5'-dithiobis(2-nitrobenzoic acid)(DTNB) under nondenaturing conditions (107). A recent review by Hsu (108) on pigeon liver malic enzyme described a variety of other thiol specific reagents that derivatized this enzyme including N-ethylmaleimide (NEM), bromopyruvate, and p-chloromercuribenzoate. Each of these reagents effectively prohibited the carbon-carbon bond synthesis and cleavage reactions, thereby blocking the overall oxidative decarboxylase activity with malate and reductive carboxylase activity with pyruvate. Other reactions catalyzed by this multistep enzyme were not inhibited by the thiol modification. The oxalacetate reductase activity was unaffected by these reagents and the pyruvate reductase activity was actually stimulated between 1.5- and 8-fold depending on the reagent used.

Thiol specific reagents have also been used to study the role of cysteine residues in the active site of *Escherichia coli* tryptophan synthetase (protein B). From these studies (109), two distinct binding regions, one for indole and one for serine, were characterized. If apoprotein (devoid of its bound pyridoxal phosphate) was reacted with either NEM or DTNB, two thiols were modified and all enzyme activity was inhibited. However, if the enzyme was modified with these reagents in the presence of pyridoxal phosphate, only one thiol was modified. When DTNB was used in this second procedure, tryptophan synthesis and a minor reaction, serine deamination, were both inhibited although the deaminase activity was less sensitive to this modification. A smaller thiol modifying reagent, NEM, produced very different results with the holoenzyme. The rate of tryptophan synthesis catalyzed by the NEM-protein was inhibited by over 90%, whereas serine deaminase activity increased 20% over that of the native enzyme. However, it is still not clear whether the appendage on the thiol directly interferes with indole binding or causes a conformational change in the enzyme. This study, and the study of malic enzyme, suggest procedures in which certain minor enzymic reactions can be forced to become the dominant reactions.

Thiol modification of lysine monooxygenase, a flavin-dependent protein, serves as another example of how minor reactions catalyzed by coenzyme-dependent proteins can be activated concomitantly with inhibition of the physiologically important reaction. The sulfhydryl reagents, p-chloromercuribenzoic acid, mercuric chloride, and phenyl mercuric acetate, all converted lysine monooxygenase into a protein that primarily catalyzes a lysine oxidase reaction (110, 111).

Xanthine dehydrogenase, also a flavin-containing protein, can be converted to a xanthine oxidase when treated in a similar manner. 2-Iodosobenzoic acid, p-chloromercuribenzoate, DTNB, tetraethylthioperoxydicarbonic diamide (disulphiram) (112) and O_2/Cu(II) (113) transform the enzyme-catalyzed oxidation of xanthine from a NAD^+-dependent reaction to an O_2-dependent reaction. However, only the modification using disulphiram converted the enzyme into an oxidase without altering the oxidation rate of xanthine. The kinetic analysis of disulphiram's inactivation of the NAD^+-dependent activity is consistent with a model of a dehydrogenase to oxidase conversion requiring the modification of only one thiol.

In contrast to the previous case, Baumanis et al (114) demonstrated that the derivatization of a flavin-dependent protein can alter the substrate specificity without necessarily affecting the overall type of reaction catalyzed. Monoamine oxidase, once treated with the hydroxyethylhydrazide of cyanoacetic acid, retained very little of its native tyramine deaminase activity. Instead, this protein acquired new deaminase activity toward the monoamine, histamine, as well as with some di- and polyamines such as putrescine, cadaverine, spermine, and spermidine. Any interpretation comparing the change in substrate specificity with protein modification must now await the identification of the site of modification.

A survey such as this is not complete without demonstrating that covalent modification can systematically affect the most subtle characteristics of protein catalysis as well as those described above. For example, the action of firefly luciferase is very sensitive to the protonation state of the excited, oxidized derivative of luciferin. The normal yellow-green emission of luciferase was substituted by a red emission when an amino acid base in the active site was ethylated by the affinity reagent, ethyl 2-benzothiazolesulfonate (115). Thus, even a simple proton transfer step can be rationally and selectively blocked by covalent modification.

Covalent modification of proteins also has a demonstrable utility in characterizing the flow of electrons between electron transferases. The protein binding properties of the coenzyme-dependent NADPH-adrenodoxin reductase was altered after one lysine, probably the one located near the sugar region of this glycoprotein, was modified with pyridoxal phosphate (116). This modification was sufficient to inhibit the flow of electrons from adrenodoxin reductase to adrenodoxin. However, the modified adrenodoxin reductase still could accept electrons from NADPH. Only the specific protein-protein interactions necessary for electron transfer were disrupted; the electron transfer itself was not prohibited since the derivatized adrenodoxin reductase could still catalyze the transfer of electrons from NADPH to ferricyanide. This change in catalytic specificity seems to affect the enzyme's surface interactions only; no conformational change was evident from CD spectroscopy.

Covalent modification of an enzyme surface was also used by Margoliash et al (117) to trace the flow of electrons in and out of cytochrome c. A different set of covalent modifications were successfully employed to limit specifically either the reduction of cytochrome c by succinate-cytochrome c reductase or the oxidation of cytochrome c by cytochrome c oxidase. Monoiodination of Tyr-74, mononitration of Tyr-67, carboxymethylation of Met-80 or oxidation of Trp-59 to formyltryptophan in cytochrome c inhibited enzyme reduction by the reductase, yet enzyme oxidation by the oxidase was hardly affected. Conversely, modification of Lys-13 with bis-phenylglyoxal in *Clostridium krusei* cytochrome c inhibited the cytochrome from passing its electron to the oxidase. Its ability to accept an electron from the reductase was unimpaired. Therefore, selective control of electron flow between proteins is now quite feasible. Limiting the interactions between small oxidants or reductants and redox active proteins is a more difficult task and must be left for the future.

PART II: CHANGES IN ENZYME SPECIFICITY CAUSED BY AMINO ACID SUBSTITUTION

The previous sections have dealt with changes in enzymatic specificity resulting from the covalent modification of amino acid residues. These chemically altered enzymes contain appendages (e.g. alkyl, acyl, oxo) not present in their native counterparts. However, the introduction of auxiliary molecular components onto the framework of the protein molecule is not the only way to modify enzymatic properties. An alternative approach is the replacement of one amino acid residue with a new one. This has been accomplished by both chemical and genetic methodologies. The chemical approach was first developed and reported almost simultaneously by Koshland & Neet (118) and Bender & Polgar (119). Theoretically, amino acid substitution can be effected chemically by (a) specific hydrolytic removal of an amino acid residue and insertion of its replacement or (b) transformation of an amino acid residue by modification of its side chain component. Clearly, organic chemistry has not yet reached the level of sophistication required to accomplish the complex operation in (a). Although this procedure has been achieved enzymatically, its application, at the present time, is far from general (120a,b).

The chemical conversion of one amino acid side chain to a new one has been termed "chemical mutagenesis" (121). It is hampered by the limited ability to modify specifically one amino acid residue in an entire protein. It is also restricted by the present state of the art of organic chemistry (e.g. the mild conversion of a tyrosine to a phenylalanine residue might be useful, but is currently not possible by chemical means).

Nevertheless, Bender & Koshland successfully converted the active site serine residue in subtilisin into a cysteine moiety (Scheme II).

$$E - CH_2OH + PMSE \rightarrow E - CH_2-O-PMS$$

$$E-CH_2-O-PMS + K^+ \; {}^-SC-CH_3 \overset{O}{\underset{\|}{}} \rightarrow E-CH_2-S-\overset{O}{\underset{\|}{C}}-CH_3$$

$$E-CH_2-S-\overset{O}{\underset{\|}{C}}-CH_3 \rightarrow E-CH_2-SH$$

Scheme II

This serine's hydroxyl group specifically reacts with phenylmethane sulfonyl fluoride (PMSF) to provide an activated species which is then displaced by the potassium salt of thioacetic acid. The product formed is equivalent to an acyl enzyme intermediate and therefore spontaneously deacetylates to provide thiol-subtilisin. This new enzyme does not possess any hydrolytic activity toward peptides or proteins. However, it is capable of hydrolyzing highly activated substrates, like nitrophenyl esters. Other enzymes chemically mutated include trypsin (cysteine substituted for a serine residue) (122) and papain (serine substituted for a cysteine residue) (123). However, for the reasons cited earlier, the chemical mutation approach is limited to the conversion of a few amino acids in a handful of enzymes. Polgar & Bender recognized in 1970 the importance of developing a general method that would permit amino acid substitution anywhere in the protein molecule (121).

> Amino acid substitutions induced in proteins by point mutations are widely known. The fortuitous mutation, as occurs in nature, has not yet been controlled by the biochemist. He cannot generally produce by mutation a protein that differs from another protein in a particular amino acid. Nevertheless, such progress in molecular biology is most desirable since it would greatly contribute to the understanding of the conformation as well as the function of biologically active proteins.

In the context of this review, a general method of such power would also provide the biochemist with a way to control enzyme specificity. Site-specific mutagenesis has been developed from relatively recent advances in recombinant DNA technology and has already been employed to convert the active site serine of β-lactamase to a cysteine (124) and a threonine residue (125). It is possible to speculate that site-specific mutagenesis could eventually provide a fructose-1,6-bisphosphatase enzyme which is no longer subject to allosteric regulation by AMP, a trypsin enzyme which is no longer inhibited by bovine pancreatic trypsin inhibitor, an elastase enzyme which is inhibited by both the native and oxidized forms of α_1-antitrypsin, and in general many altered enzymes possessing properties that are deemed desirable.

PART III: MODIFICATION OF ENZYMES WITH COENZYME ANALOGS AND OTHER COFACTORS

In contrast to the studies presented thus far, this section deals with the chemical modification of enzymes with reagents possessing their own catalytic activities. The modification of coenzyme-dependent enzymes with reagents structurally and catalytically similar to the essential coenzyme serves to maintain the catalytic action of the parent enzyme. In particular, reactive analogs of NAD^+ have been used as substrates of and affinity labels for NAD^+-requiring dehydrogenases. Alcohol dehydrogenase was modified with a catalytically competent NAD^+ derivative, nicotinamide-5-bromoacetyl-4-methyl-imidazole dinucleotide. Subsequent analysis of the covalently modified enzyme did yield data on the spatial arrangement of the active site but resulted in an inactive enzyme (126). Later work by Mansson et al (127) produced a derivative of alcohol dehydrogenase that had an NAD^+ analog, N^6-[N-(6-aminohexyl)carbamoylmethyl]-NAD, covalently bound at the active site in a catalytically active conformation. Therefore, this enzyme, once requiring the addition of a freely diffusing redox cofactor for activity, was converted into an enzyme that no longer required an exogenous cofactor. The authors of this study suggested that the technique of linking a coenzyme to a coenzyme-dependent enzyme may be useful in industrial enzymology. This procedure could eliminate the need for large amounts of expensive coenzymes in bulk processes. Mansson et al (128) later demonstrated that the nicotinamide-linked alcohol dehydrogenase can be efficiently recycled to the oxidized form by adding lactate dehydrogenase and pyruvate. In this way, once the modified alcohol dehydrogenase is reduced by ethanol, the lactate dehydrogenase could then reoxidize the modified enzyme and subsequently reduce pyruvate to lactate.

An alternative method for linking NAD^+ analogs to dehydrogenases has been presented by Woenckhaus et al (129). This method utilizes a diazoniumaryl derivative of the adenosine in NAD^+. Both alcohol dehydrogenase and lactate dehydrogenase covalently modified by this derivative appear extremely redox active in the absence of added NAD^+. Although the covalent modification of dehydrogenases with reactive analogs of NAD^+ does not produce much change in enzyme specificity, it does stabilize enzyme-substrate interactions and change the chemical requirements for complete enzyme turnover.

Additionally, flavin analogs have been used to modify flavin-dependent enzymes covalently. The primary goals of these studies were to probe the nature of the flavoprotein's active site (130) and, therefore, they will not be covered here.

In a major departure from previous studies, the modification of an enzyme's active site has been used to introduce a catalytic activity different than that of the native enzyme. This has been accomplished by the reaction of coenzyme analogs with amino acid residues in the active site of the hydrolytic enzyme, papain. Flavopapain, as the name implies, is an enzyme built from papain but transformed into a flavoprotein by alkylation of the active site cysteine with a flavin derivative. Consequently, the hydrolytic activity of papain is converted into an oxidoreductase activity of flavopapain.

Papain was chosen as the protein template for the initial work based on its ready availability and the structural information known for it from both X-ray investigations and solution studies. A long groove capable of accommodating extended substrates exists near the active site residue cysteine-25, and model building indicated that considerable room would exist for the binding of potential substrates to the modified enzyme if coenzyme analogs were attached covalently to this residue. Flavins were chosen as the modifying agents for the preparation of the first "semisynthetic" enzymes because of their known catalytic versatility and because it appeared from the literature that the high catalytic activity of a number of naturally occurring flavoenzymes did not depend on the specific involvement of functional groups from the protein.

The most effective flavopapain produced to date is species 2. This semi-synthetic enzyme is produced from the reaction between papain's Cys-25 and the bromomethylketone 3. The activity exhibited by flavopapain toward dihydronicotinamides is comparable to naturally occurring flavoenzymes. For example, it oxidizes N-hexyldihydronicotinamide with a k_{cat}/K_M of 5.7×10^5 $M^{-1}s^{-1}$ (131). This value compares favorably with old yellow enzyme (6×10^2 $M^{-1}s^{-1}$), and NADPH-specific FMN oxidoreductase (8.5×10^5 $M^{-1}s^{-1}$).

Other flavopapains synthesized and characterized include 4 and 5 produced in a manner comparable to that of 2 [for a recent review on flavopapains, see reference (132a)]. Both of these flavoenzymes exhibit oxidoreductase activity with N-alkyldihydronicotinamides, albeit with lower efficiency than semi-synthetic enzyme 2(132b). These flavopapains also have other activities associated with flavoproteins such as the oxidation of mercaptans to disulfides (133) and the ability to distinguish between the prochiral hydrogens of dihydronicotinamides (134).

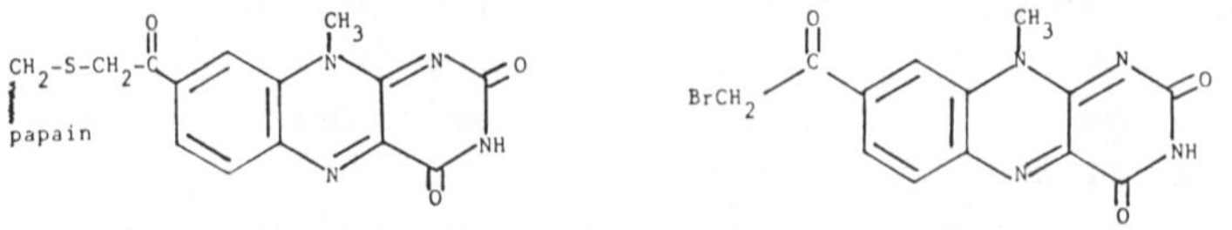

Figure 2   Structure 2. The product of alkylation of papain by Structure 3.
        Structure 3. 8α-bromoacetyl-10-methylisoalloxazine.

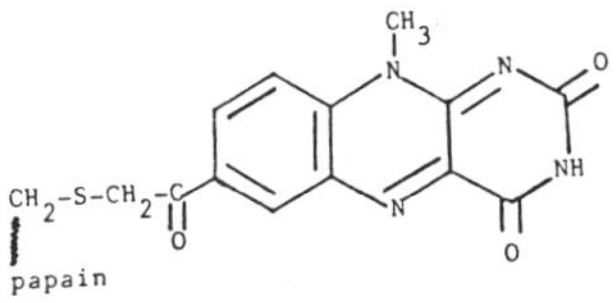

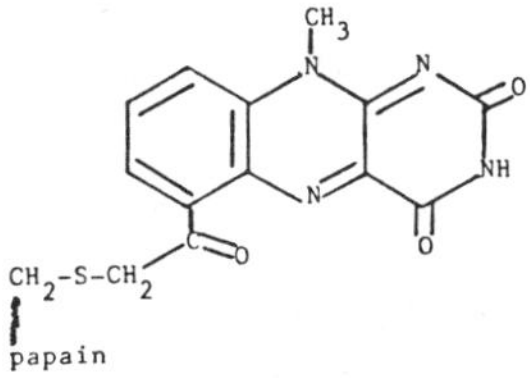

Figure 3   *Structure 4.* The product of alkylation of papain by 7α-bromoacetyl-10-methylisoalloxazine.
Structure 5. The product of alkylation of papain by 6α-bromoacetyl-10-methylisoalloxazine.

Flavopapains demonstrate the possibility of introducing a new catalytic activity in a protein quite distinct from the natural activity of the native enzyme. An early example relevant to this type of work was published over 25 years ago. In this report, the apo form of glyceraldehyde-3-phosphate dehydrogenase was converted into a diaphorase by derivatizing the enzyme's active site cysteine with the dye, 2,6-dichlorophenolindophenol (DCIP) (135). Only much later, 1973, was this hybrid enzyme fully characterized (136). Apo-glyceraldehyde-3-phosphate dehydrogenase reacted with DCIP to form a yellow enzyme complex. This covalent complex was then oxidized in the presence of air to a blue species and reversibly reduced with NADH reforming the yellow species. Although more than one dye molecule derivatized each protein subunit, 80% of the diaphorase activity was attributed to the addition of only one dye molecule/subunit. Presumably, the DCIP that modifies the active site cysteine exhibits the redox activity. This redox catalyst behaves as a typical enzyme; it exhibits Ping-Pong kinetics and the pH profile of NADH oxidation and DCIP reduction is not equivalent to the nonenzymatic reaction.

Along somewhat different lines, redox active semisynthetic enzymes can also be constructed by derivatizing proteins with inorganic complexes (137). The laboratory of Gray has modified sperm whale myoglobin with three molecules of a pentacoordinated ruthenium complex producing a new enzyme capable of oxidizing ascorbate, durohydroquinone, and hydroquinone with Michaelis-Menten type kinetics. This group has also recently produced a pentaaminoruthenium (III) derivative of ferricytochrome at His-33 to study the process of electron transfer between multiple sites in a protein (138). With the successes mentioned here and earlier in this section, the potential for creating new catalytic species from commonplace proteins is clearly growing. The synthesis of semisynthetic enzymes has been limited by the need to modify specifically particular amino acid residues favorably located near binding sites in existing protein molecules. Therefore, site-specific mutagenesis, which has the potential to introduce appropriate amino acids available for chemical modification in combination with classical chemical methodology, should provide the means to generate a vast array of new enzymes.

SUMMARY

The chemical modification of enzymes has played and will continue to play an important role in probing the mechanism of enzyme activity. This technique can be utilized for identification of those individual amino acid residues responsible for the catalytic properties of the entire protein. In chemical modification experiments, changes in enzymatic specificity have been noted, but often not predicted. However, in recent years, rational approaches for the alteration of enzymatic properties have become feasible by means of site-specific mutagenesis and chemical methodology. In the first method, one amino acid can be replaced by a new one; in the second method, not only can new amino acid residues be introduced but also new catalytic entities (such as flavins) can be affixed to the protein molecule. Both methodologies are in their infancy, yet they represent a potentially powerful approach toward the design and synthesis of enzymes possessing new specificities.

Literature Cited

1. Knowles, J. R. 1965. *Biochem. J.* 95:180–90
2. Ray, W. J., Latham, H. G., Katsoulis, M., Koshland, D. E. Jr. 1960. *J. Am. Chem. Soc.* 82:4743–44
3. Koshland, D. E. Jr., Strumeyer, D. H., Ray, W. J. 1962. *Brookhaven Symp. Biol.* 15:101–33
4. Lawson, W. B., Schramm, H.-J. 1965. *Biochemistry* 4:377–86
5. Gundlach, G., Turba, F. 1962. *Biochem. Z.* 335:573–81
6. Schramm, H.-J., Lawson, W. B. 1963. *Z. Physiol. Chem.* 332:97–100
7. Bender, M. L., Kézdy, F. J. 1965. *Ann. Rev. Biochem.* 34:49–76
8. Kézdy, F. J., Feder, J., Bender, M. L. 1967. *J. Am. Chem. Soc.* 89:1009–16
9. Fraenkel-Conrat, H., Bean, R. S., Lineweaver, H. 1949. *J. Biol. Chem.* 385–403
10. Riordan, J. F., Wacker, W. E. C., Vallee, B. L. 1965. *Nature* 208:1209–11
11. Trenholm, H. L., Spomer, W. E., Wootton J. F. 1966. *J. Am. Chem. Soc.* 88:4281–82
12. Trenholm, H. L., Spomer, W. E., Wootton, J. F. 1969. *Biochemistry* 8:1741–47
13. Chevallier, J., Yon, J., Spotorno, G., Labouesse, J. 1968. *Eur. J. Biochem.* 5:480–86
14. Gertler, A., Hofmann, T. 1967. *J. Biol. Chem.* 242:2522–27
15. Kaplan, H., Dugas, H. 1969. *Biochem. Biophys. Res. Commun.* 34:681–85
16. Gertler, A. 1971. *Eur. J. Biochem.* 20:541–46
17. Gertler, A. 1971. *Eur. J. Biochem.* 23:36–40
18. Davril, M., Jung, M.-L., Duportail, G., Lohez, M., Han, K. K., Bieth, J. G. 1984. *J. Biol. Chem.* 259:3851–57
19. Beatty, K., Bieth, J., Travis, J. 1980. *J. Biol. Chem.* 255:3931–34
20. Johnson, D., Travis, J. 1979. *J. Biol. Chem.* 254:4022–26
21a. Yoshimura, T., Barker, L. N., Powers, J. C. 1982. *J. Biol. Chem.* 257:5077–84
21b. Teshima, T., Griffin, J. C., Powers, J. C. 1982. *J. Biol. Chem.* 257:5085–91
22. Sokolovsky, M., Riordan, J. F., Vallee, B. L. 1966. *Biochemistry* 5:3582–89
23. Johansen, J. T., Ottesen, M., Svendsen, I. 1967. *Biochim. Biophys. Acta* 139:211–14
24. Svendsen, I. 1967. *C. R. Trav. Lab. Carlsberg* 36:235–46
25. Johansen, J. T. 1969. *C. R. Trav. Lab. Carlsberg* 37:145–77
26. Drenth, J., Hol, W. G. J. 1967. *J. Mol. Biol.* 28:543–44

27. Svendsen, I. 1974. *C. R. Trav. Lab. Carlsberg* 39:375–98
28. Svendsen, I. 1967. *C. R. Trav. Lab. Carlsberg* 36:347–63
29. Robertus, J. D., Alden, R. A., Birktoft, J. J., Kraut, J., Powers, C., Wilcox, P. E. 1972. *Biochemistry* 11:2439–49
30. Robertus, J. D., Kraut, J., Alden, R. A., Birktoft, J. J. 1972. *Biochemistry* 11:4293–303
31. Svendsen, I. 1976. *Carlsberg Res. Commun.* 41:131–42
32. Lowe, G., Whitworth, A. S. 1974. *Biochem. J.* 141:503–15
33. Jori, G., Galiazzo, G. 1971. *Photochem. Photobiol.* 14:607–19
34. Jori, G., Gennari, G., Toniolo, C., Scoffone, E. 1971. *J. Mol. Biol.* 59:151–68
35. Mole, J. E., Horton, H. R. 1973. *Biochemistry* 12:5278–89
36. Mole, J. E., Horton, H. R. 1973. *Biochemistry* 12:5285–89
37. Chang, S.-M., Horton, H. R. 1979. *Biochemistry* 18:1559–63
38. Evans, B. L. B., Knopp, J. A., Horton, H. R. 1981. *Arch. Biochem. Biophys.* 206:362–71
39. Yu-Kun, S., Chen-Lu, T. 1963. *Sci. Sin.* 12:879–84
40. Kirschenbaum, D. M. 1971. *Biochim. Biophys. Acta* 235:159–63
41. Loffler, H. G., Schneider, F. 1972. *Z. Naturforsch.* 27:1490–97
42. Hugounenq, L., Loiseleur, J. 1925. *C. R.* 181:149–51
43. Herriott, R. M., Northrop, J. H. 1934. *J. Gen. Physiol.* 18:35–67
44. Perlmann, G. E. 1966. *J. Biol. Chem.* 241:153–57
45. Herriott, R. M. 1935. *J. Gen. Physiol.* 19:283–99
46. Shechter, Y., Rubinstein, M., Becker, R., Bohak, Z. 1975. *Eur. J. Biochem.* 58:123–31
47. Becker, R., Shechter, Y., Bohak, Z. 1973. *FEBS Lett.* 36:49–52
48. Simpson, R. T., Riordan, J. F., Vallee, B. L. 1963. *Biochemistry* 2:616–22
49. Sokolovsky, M., Vallee, B. L. 1967. *Biochemistry* 6:700–8
50. Hartsuck, J. A., Lipscomb, W. N. 1971. *Enzymes* 3:1–56
51. Riordan, J. F. 1973. *Biochemistry* 12:3915–23
52. Bender, M. L., Whitaker, J. R., Menger, F. 1965. *Proc. Natl. Acad. Sci. USA* 53:711–16
53. Whitaker, J. R., Menger, F., Bender, M. L. 1966. *Biochemistry* 5:386–92
54. Hall, F. L., Kaiser, E. T. 1967. *Biochem. Biophys. Res. Commun.* 29:205–10
55. Imoto, T., Johnson, L. N., North, A. C. T., Phillips, D. C., Rupley, J. A. 1972. *Enzymes* 7:665–868
56. Hayashi, K., Imoto, T., Funatsu, G., Funatsu, M. 1965. *J. Biochem.* 58:227–35
57. Imoto, T., Fujimoto, M., Yagishita, K. 1974. *J. Biochem.* 76:745–53
58. Blake, C. C. F., Johnson, L. N., Mair, G. A., North, A. C. T., Phillips, D. C., Sarma, V. R. 1967. *Proc. R. Soc. London Ser. B* 167:378–88
59. Imoto, T., Hayashi, K., Funatsu, M. 1968. *J. Biochem.* 64:387–92
60. Imoto, T., Yagishita, K. 1973. *Agric. Biol. Chem.* 37:465–70
61. Davies, R. C., Neuberger, A., Wilson, B. M. 1969. *Biochim. Biophys. Acta* 178:294–305
62. Davies, R. C., Neuberger, A. 1969. *Biochim. Biophys. Acta* 178:306–17
63. Blackburn, P., Moore, S. 1982. *Enzymes* 15:317–433
64. Matthews, C. R., Erickson, P. M., Van Vliet, D. L., Petersheim, M. 1978. *J. Am. Chem. Soc.* 100:2260–62
65. Neumann, H., Steinberg, I. Z., Brown, J. R., Goldberger, R. F., Sela, M. 1967. *Eur. J. Biochem.* 3:171–82
66. Tamburro, A. M., Boccu, E., Celotti, L. 1970. *Int. J. Protein Res.* 11:157–64
67. Sperling, R., Steinberg, I. Z. 1971. *J. Biol. Chem.* 246:715–18
68. Watkins, J. B., Benz, F. W. 1978. *Science* 199:1084–87
69. Bartholeyns, J., Moore, S. 1974. *Science* 186:444–45
70. Wang, D., Wilson, G., Moore, S. 1976. *Biochemistry* 15:660–65
71. D'Alessio, G. D., Malorni, M. C., Parente, A. 1975. *Biochemistry* 14:1116–22
72. Wang, D. 1979. *Biochemistry* 18:4449–52
73. King, T. P., Li, Y., Kochoumian, L. 1978. *Biochemistry* 17:1499–506
74. Saraswathi, S., Keyes, M. H. 1984. *Enzyme Microb. Technol.* 6:98–100
75. Han, P. F., Han, G. Y., McBay, H. C., Johnson, L. Jr. 1978. *Biochem. Biophys. Res. Commun.* 85:747–55
76. Marcus, F. 1976. *FEBS Lett.* 70:159–62
77. Marcus, F. 1976. *Biochemistry* 15:3505–9
78. Pontremoli, S., Grazi, E., Accorsi, A. 1966. *Biochemistry* 5:3568–74
79. Colombo, G., Hubert, E., Marcus, F. 1972. *Biochemistry* 11:1798–803
80. Pontremoli, S., Traniello, S., Enser, M., Shapiro, S., Horecker, B. L. 1967. *Proc. Natl. Acad. Sci. USA* 58:286–93
81. Nakashima, K., Pontremoli, S., Horecker, B. L. 1969. *Proc. Natl. Acad. Sci. USA* 64:947–51

82. Nakashima, K., Horecker, B. L., Traniello, S., Pontremoli, S. 1970. *Arch. Biochem. Biophys.* 139:190–99
83. Nakashima, K., Horecker, B. L., Pontremoli, S. 1970. *Arch. Biochem. Biophys.* 141:579–87
84. Pontremoli, S., Traniello, S., Enser, M., Shapiro, S., Horecker, B. L. 1967. *Proc. Natl. Acad. Sci. USA* 58:286–93
85. Pontremoli, S., Luppis, B., Wood, W. A., Traniello, S., Horecker, B. L. 1965. *J. Biol. Chem.* 240:3464–68
86. Pontremoli, S., Luppis, B., Traniello, S., Rippa, M., Horecker, B. L. 1965. *Arch. Biochem. Biophys.* 112:7–15
87. Little, C., Sanner, T., Pihl, A. 1969. *Eur. J. Biochem.* 8:229–36
88. Landfear, S. M., Lipscomb, W. N. 1978. *J. Biol. Chem.* 253:3988–96
89. Landfear, S. M., Evans, D. R., Lipscomb, W. N. 1978. *Proc. Natl. Acad. Sci. USA* 75:2654–58
90. Wang, C.-M., Yang, Y. R., Hu, C. Y., Schachman, H. K. 1981. *J. Biol. Chem.* 256:7028–34
91. Enns, C. A., Chan, W. W.-C. 1978. *J. Biol. Chem.* 253:2511–13
92. Enns, C. A., Chan, W. W.-C. 1979. *J. Biol. Chem.* 254:6180–86
93. Uyeda, K. 1979. *Adv. Enzymol.* 48:194–237
94. Cambou, B., Laurent, M., Hervagault, J.-F., Thomas, D. 1981. *Eur. J. Biochem.* 121:99–104
95. Nucci, R., Raia, C. A., Vaccaro, C., Sepe, S., Scarano, E., Rossi, M. 1978. *J. Mol. Biol.* 124:133–45
96. Raia, C. A., Nucci, R., Vaccaro, C., Sepe, S., Rella, R. 1982. *J. Mol. Biol.* 157:557–70
97. Gorkin, V. Z. 1976. *Mol. Biol.* 10:589–604
98. Gorkin, V. Z. 1977. *Horizons Biochem. Biophys.* 3:1–35
99. Park, J. H., Koshland, D. E. Jr. 1958. *J. Biol. Chem.* 233:986–90
100. Ehring, R., Colowick, S. P. 1969. *J. Biol. Chem.* 244:4589–99
101. Little, C., O'Brien, P. J. 1969. *Eur. J. Biochem.* 10:533–38
102. Parker, D. J., Allison, W. S. 1969. *J. Biol. Chem.* 244:180–89
103. You, K.-S., Benitez, L. V., McConachie, W. A., Allison, W. S. 1975. *Biochim. Biophys. Acta* 384:317–30
104. Allison, W. S. 1976. *Acc. Chem. Res.* 9:293–99
105. Bloxham, D. P., Wilton, D. C. 1977. *Biochem. J.* 161:643–51
106. Bloxham, D. P., Sharma, R. P., Wilton, D. C. 1979. *Biochem. J.* 177:769–80
107. Tang, C. L., Hsu, R. Y. 1974. *J. Biol. Chem.* 249:3916–22
108. Hsu, R. Y. 1982. *Mol. Cell. Biochem.* 43:3–26
109. Miles, E. W. 1970. *J. Biol. Chem.* 245:6016–25
110. Yamauchi, T., Yamamoto, S., Hayaishi, O. 1973. *J. Biol. Chem.* 248:3750–52
111. Yamauchi, T., Yamamoto, S., Hayaishi, O. 1975. *J. Biol. Chem.* 250:7127–33
112. Kaminski, Z. W., Jezewska, M. M. 1982. *Biochem. J.* 207:341–46
113. Kaminski, Z. W., Jezewska, M. M. 1979. *Biochem. J.* 181:177–82
114. Baumanis, E. A., Kalnina, I. E., Moskvitina, T. A., Gorkin, V. Z. 1977. *Biochem. Pharmacol.* 26:1059–63
115. White, E. H., Branchini, B. R. 1974. *J. Am. Chem. Soc.* 97:1243–45
116. Hamamoto, I., Ichikawa, Y. 1984. *Biochim. Biophys. Acta* 786:32–41
117. Margoliash, E., Ferguson-Miller, S., Tulloss, J., Kang, C. H., Feinberg, B. A., et al. 1973. *Proc. Natl. Acad. Sci. USA* 70:3245–49
118. Neet, K. E., Koshland, D. E. Jr. 1966. *Proc. Natl. Acad. Sci. USA* 56:1606–11
119. Polgar, L., Bender, M. L. 1966. *J. Am. Chem. Soc.* 88:3153–54
120a. Kowalski, D., Laskowski, M. 1976. *Biochemistry* 15:1300–9
120b. Kowalski, D., Laskowski, M. 1976. *Biochemistry* 15:1309–15
121. Polgar, L., Bender, M. L. 1970. *Adv. Enzymol.* 33:381–400
122. Yokosawa, H., Ojima, S., Ishii, S.-I. 1977. *J. Biochem.* 82:869–76
123. Clark, P. I., Lowe, G. 1978. *Eur. J. Biochem.* 84:293–99
124. Sigal, I. S., Harwood, B. G., Arentzen, R. 1982. *Proc. Natl. Acad. Sci. USA* 79:7157–60
125. Dalbadie-McFarland, G., Cohen, L. W., Riggs, A. D., Morin, C., Itakura, K., Richards, J. H. 1982. *Proc. Natl. Acad. Sci. USA* 79:6409–13
126. Jornvall, H., Woenckhaus, C., Johnscher, G. 1975. *Eur. J. Biochem.* 53:71–81
127. Mansson, M.-O., Larsson, P.-O., Mosbach, K. 1978. *Eur. J. Biochem.* 86:455–63
128. Mansson, M.-O., Larsson, P.-O., Mosbach, K. 1979. *FEBS Lett.* 98:309–13
129. Woenckhaus, C., Koob, R., Burkhard, A., Schaefer, H.-G. 1983. *Bioorg. Chem.* 12:45–47
130. Massey, V., Claiborne, A., Biemann, M., Ghisla, S. 1984. *J. Biol. Chem.* 259:9667–78 and references cited therein
131. Slama, J. T., Oruganti, S. R., Kaiser, E.

T. 1981. *J. Am. Chem. Soc.* 103:6211–13

132a. Kaiser, E. T., Lawrence, D. S. 1984. *Science* 226:505–11

132b. Slama, J. T., Radziejewski, C., Oruganti, S.-R., Kaiser, E. T. 1984. *J. Am. Chem. Soc.* 106:6778–85

133. Fried, H. E., Kaiser, E. T. 1981. *J. Am. Chem. Soc.* 103:182–84

134. Levine, H. L., Kaiser, E. T. 1980. *J. Am. Chem. Soc.* 102:343–45

135. Rafter, G. W., Colowick, S. P. 1957. *Arch. Biochem. Biophys.* 66:190–207

136. Benitez, L. V., Allison, W. S. 1973. *Arch. Biochem. Biophys.* 159:89–96

137. Margalit, R., Pecht, I., Gray, H. B. 1983. *J. Am. Chem. Soc.* 105:301–2

138. Yocom, K. M., Winkler, J. R., Nocera, D. G., Bordignon, E., Gray, H. B. 1983. *Chem. Scr.* 21:29–33

Ann. Rev. Biochem. 1985. 54:597–629

EFFECTS OF SITE-SPECIFIC AMINO ACID MODIFICATION ON PROTEIN INTERACTIONS AND BIOLOGICAL FUNCTION

Gary K. Ackers and Francine R. Smith

Department of Biology, The Johns Hopkins University, Baltimore, Maryland 21218

CONTENTS

PERSPECTIVES AND SUMMARY

Currently, there is intense interest in site-directed mutagenic approaches to the study of fundamental structure-function problems in proteins. The ability to manipulate protein structures by altering individual amino acid residues at desired locations through genetic engineering has stimulated much new interest in classical problems such as the mechanisms of protein folding, protein-protein interaction, nucleated polymerization, cooperative protein-ligand interactions, and the bases of enzymatic activity. It is now possible to design new proteins that have desired functional properties, either by modifying existing structures or by creating entirely new ones.

An overriding issue in structure-function problems is that of understanding the roles played by the local parts of a molecular structure in the overall functional processes of interest. Which amino acid side chains and specific residue locations are crucial to function? How do functional effects or their control arise from particular domains, local binding sites, or interfaces? By what pathways within a molecular structure are the local changes coupled to each other? In this article we review recent results and rationale for studying these issues through local-site alterations of amino acid residues, which may be accomplished through techniques of site-directed mutagenesis. The structural alterations may, of course, be generated by other methods, such as selection of naturally occurring genetic variants, or by chemical modification of the native protein species. To date, the most extensive studies relating structure and function to single-site modifications have been carried out using proteins obtained by the latter approaches. In these cases the investigations have essentially started at the point where the corresponding project by recombinant DNA methods would have successfully provided the starting materials.

In this article we do not review the methods for making proteins by genetic engineering, which have been discussed extensively [cf (1, 2)]. We are concerned rather with the question of what can be learned about structure-function problems using proteins modified at single sites, as exemplified by recent studies. The lessons learned from results obtained to date may provide guidelines for studying the plethora of modified proteins currently on the horizon. The strategies and systems we discuss are restricted to functional processes where the protein molecules retain nearly the same tertiary structures throughout. Thus, the structural alterations brought about by single-site modifications and by functional change are both in the realm of *perturbations*. These functional processes include subunit assembly, ligand binding, conformational changes, and steady-state enzymatic activities. Strategies are discussed for studying molecular mechanisms of these processes using two general approaches. The first, and classical approach, relies on the interpretation of functional perturbation (i.e. deviation from normal behavior) in terms of the *local stereochemistry* of the modified site. The second approach *(mapping by structure-function perturbation)* is based upon the patterns of functional perturbation arising from proteins modified at many sites distributed throughout a protein structure. The magnitudes of functional perturbations versus structural locations of the modified sites provide a "map" from which inferences may be drawn regarding (*a*) the structural locations of functional events and (*b*) structural pathways for coupling and transmission of functional events within the molecule. The combination of these two approaches provides an especially powerful tool in systems where long-range coupling effects play important biological roles. We review recent results obtained with six different protein systems that illustrate the types of effects observed with single-site modifications and the concepts currently used in their interpretation.

DEFINITIONS AND ISSUES

Local versus System Properties

Throughout this review we will use the term "function" to denote any process or combination of processes involving intermolecular interactions or intramolecular change that is singled out for study. In the literature to be reviewed, these processes include interactions between protein subunits to form complexes of biological interest, enzymatic activities, conformational changes, ligand binding, and the modulation of these interactions and activities. Each of these "functions" is a "system property" reflecting observable behavior of the protein as a whole. The goal of structure-function studies is to learn how the roles played by the local parts of the protein contribute to its net functional behavior. For example, when the function of interest is the association between two protein molecules to form a noncovalent complex, the desired information may include the determination of which amino acid groups are involved in the interactions and how each contributes to the energetics of stabilization for the complex. Do the groups at the subunit interface act independently or are there cooperative effects between them? Are the pairwise intersubunit interactions at the interface independent of the remainder of the molecule or are they modulated by changes in conformation and intramolecular interaction that accompany complex formation? If so, which residues are involved in the conformational changes and in their coupling to the interfacial interactions? In these senses then, is the whole a sum of its parts? These questions illustrate the types of information sought. A range of more specific questions will be discussed in conjunction with the experimental studies to be reviewed. It should be noted at this point that amino acid substitutions or deletions are also of great use in probing the roles of domains or other large regions of a protein, in addition to their use in dissecting the roles of individual residues themselves.

Structural Modifications and Functional Perturbations

The functional processes mentioned above fall into a general class where the protein molecules each retain a similar structure throughout as regards the way the polypeptide chain is folded. These functional processes therefore involve only a "perturbation" of structure which entails a corresponding perturbation in energetic state. Since the overall structures of the proteins are similar before and after the reaction (e.g. subunit assembly) the change in Gibbs free energy is very small compared to the absolute free energies of the various forms assumed by the molecules during the functional process. We thus will utilize the concept of a *pathway of functional perturbation* within a protein to denote that region of the molecule where local changes in structure and interaction energy occur during the normal functional process. Likewise, the substitution or alteration of a single amino acid residue generally preserves the protein's folded tertiary structure. Validity of the strategies and interpretations of the work reviewed in this article will be limited for the most part to systems where both the structural modifications and functional processes merely represent variations on the same structural theme. This similarity makes it possible in many cases to use structural modifications as models for functional perturbation. The problem of protein folding, which will not be reviewed here, appears to represent a rather different class of problems which involve extreme states of the protein structure.

The functional interactions to be discussed may be mediated by the following types of effects:

1. *Local Effects*—in the neighborhood of an amino acid residue, short-range steric and noncovalent bonding interactions, including ionic bonds (salt bridges), H-bonds, hydrophobic and van der Waals interactions, as well as local conformational changes
2. *Global Effects*—overall changes of conformation or solvation that may affect, or be propagated throughout, the molecule as a whole
3. *Specific Long-Range Effects*—these might include specific pathways whereby structural modifications at a site are transmitted to another site.

Consider the following scheme depicting a (native, or wild-type) protein, P, which undergoes a functional process, $(P \rightarrow P')$ for which we can obtain a quantitative measure of its magnitude, F.

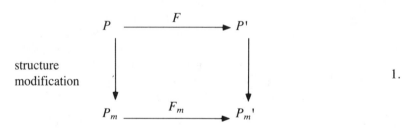

structure
modification

1.

The quantity F might represent a Gibbs free energy of protein association, or an appropriate function of the rate constant for enzymatic reaction, etc. The lower part of the diagram depicts a modified form of the protein P_m, which differs structurally from P at a single amino acid residue. The protein P_m may be capable of the same functional process as P but with a quantitatively altered magnitude, F_m. There is thus, as a result of the structural modification, a perturbation δF_m in the functional property F:

$$\delta F_m = F_m - F \qquad\qquad 2.$$

The possible ways in which a protein may respond to a single-site modification to yield the observed functional perturbation δF_m has been discussed extensively (3). These possibilities may be summarized as follows.

RESPONSES TO A SINGLE-SITE MODIFICATION   In general, the system's response to a given structural modification may fall into one of three classes (3):

1. *Function is completely eliminated by structural modification* ($\delta F_m = -F$). This could result from any of the following causes:
   a. Local structures directly required for function are eliminated (e.g., at an active site).
   b. The modification precludes conformations that would permit function to occur, either through local steric effects or long-range "global" effects.
   c. Alterations of noncovalent interactions are "coincidentally" of such magnitude that $\delta F_m = -F$.
2. *Function is unperturbed by structural modifications* ($\delta F_m = 0$):
   a. The modification site is not involved in function F, i.e., does not lie on the pathway of functional perturbation.
   b. The modification site is within the pathway of functional perturbation, but $\delta F_m = 0$ as a result of compensating effects or because the magnitude of the perturbation resulting from the structure modification does not exceed a minimal threshold.
3. *Function is perturbed by structural modification* ($\delta F_m \neq 0$, $\delta F_m \neq -F$):
   a. The site of modification is in the pathway of functional perturbation.
   b. The modification is at a site remote from the functional pathway but exerts an effect through long-range coupling either via a specific pathway, or via "global" conformational effects.

These possibilities define the nature of the problem encountered by the investigator in using single site modification as a tool. What are the best strategies for design of experiments and interpretation of results?

It is obvious from the possibilities listed above that simplistic interpretations can be treacherous and that interpretation can be greatly aided if much is already known about the mechanism of the functional process. Usually the functional response F_m to a single amino acid modification can only be interpreted in a speculative fashion. When a residue is altered and it is observed that the functional property of interest has also been changed, it is hazardous to interpret the alteration solely in terms of the local interactions of the altered residue site. This is because the observed functional property is a "system property," and may thus reflect contributions from any part of the molecule that is concomitantly altered during the overall molecular process. A ligand binding constant, for example, may reflect more than just the free energy of local interactions at the binding site if perturbations at that site are coupled to structural alterations of other parts of the molecule. Undoubtedly there are some binding sites or enzyme active sites where the interaction effects are not propagated significantly outside of the local site. For other cases, particularly with allosteric systems, the opposite is true so that the extent and pathway for transmission of interactions becomes a major issue in understanding the molecular mechanism of biological function and regulation.

EXPERIMENTAL INFORMATION REQUIRED   In order to use structurally modified proteins as tools for studying structure-function issues, it is desirable to have considerable information on both the structural and functional aspects of the system. The amount and level of resolution of structural information required depends greatly on the types of component parts whose roles are under investigation. For cases where the components are the residues themselves, an atomic level X-ray structure of at least one form of the protein is generally the minimal requirement. Structures representing several stages in the functional process are highly desirable, as is additional structural information obtainable by techniques such as NMR, fluorescence, or vibrational spectroscopy. Structural results obtained on the modified protein are of course highly desired.

Functional characterization of the "system" also needs to be exacting both in qualitative and quantitative terms. If, for example, a subunit polymerization process (the "function" of interest) is accompanied by an absorption of protons, it is desirable to have a quantitative characterization, or at least experimental control, over each of these coupled reactions. It is highly desirable to have as much mechanistic information as possible. Such information may be obtained, for example, from thermodynamic, kinetic, and spectroscopic analyses of the native and modified proteins in various states of functional behavior.

STEREOCHEMISTRY OF LOCAL EFFECTS   An extremely important tool in protein structure-function studies is the analysis of local noncovalent interactions in the immediate environment of an amino acid residue that is believed to play a functional role, or that has been altered by structural modification. Useful theories can be developed using (*a*) principles of stereochemistry, (*b*) assessment or postulation of roles played by bonds such as ionic, H-bonds, hydrophobic, and van der Waals interactions, and (*c*) calculated interactions based on electrostatic theory, empirical potential energy functions for pairwise interactions, and estimated solvent exposure effects. This type of analysis is the classical approach to detailed structure-function problems in protein chemistry, and has yielded much useful insight into the possible roles of individual amino acid side chains in mechanistic processes. In the following section we review recent work on the study of enzyme active sites using the combination of this approach with site-specific amino acid modifications.

ENZYMATIC ACTIVITIES

Tyrosyl-tRNA Synthetase

In an elegant series of studies by Fersht and associates, this enzyme from *Bacillus Stearothermophilus* has been subjected to detailed analyses of the relationship between structure and activity using techniques of "protein engineering" [cf (4) for review]. By means of oligonucleotide mutagenesis (2, 5, 6), over 70 mutant forms of this protein have been generated involving single mutations, double mutations, and deletions (A. Fersht, personal communication). These mutants have been designed and utilized to study interactions of the enzyme with substrates. To date, only a handful of mutants have been studied in detail, but the effects observed so far are very illuminating and indicate much promise for the future of enzyme active site modification.

The *B. Stearothermophilus* enzyme consists of a dimer of two identical 47.5 kd subunits and exhibits "half-of-sites reactivity" so that only one site per dimer catalyzes the following two-step reaction:

$$E + Tyr + ATP \rightleftarrows E \cdot Tyr - AMP + PP_i \qquad 3.$$

$$E \cdot Tyr - AMP + tRNA^{Tyr} \rightleftarrows Tyr - tRNA^{Tyr} + AMP + E \qquad 4.$$

Crystallographic structure determinations have been published at a resolution of 3.0 Å (7) and the crystal structure of the enzyme-bound tyrosyl-adenylate complex has also been determined (8), providing comparative data for the active site in the presence and absence of bound substrate.

BINDING ACTIVATED SUBSTRATES   The strategy employed for studying interactions at the active site has been based on the relationship between the overall rate constant, k_{cat}/K_M, of the Michaelis-Menten equation and the free energy of binding between enzyme and activated substrate, ΔG_s (9).

$$RT\ln (k_{cat}/K_M) = RT\ln (kT/h) - \Delta G^{\neq} - \Delta G_s \qquad 5.$$

In this eqn. k_{cat} and K_M are the catalytic and Michaelis-Menten constants, respectively (6). $\Delta G^{\neq}$ is the energy of activation for the catalytic steps of bond making and breaking, k and h are respectively the Boltzmann and Planck constants, R the gas constant and T the temperature (degrees Kelvin).

For reactions with two structurally different enzymes, determination of the k_{cat}/K_M values provides a measure of the difference in binding free energy ΔG_s between the enzyme and activated substrate:

$$\Delta G_s{}' - \Delta G_s = RT\ln \left[\frac{(k_{cat}/K_M)'}{(k_{cat}/K_M)} \right] \qquad 6.$$

where the primed and unprimed symbols refer, for example, to modified and native enzyme species. It should be noted that this relationship is only valid when values of the activation energy $\Delta G^{\neq}$ for the bond making and bond breaking steps are identical for both reactions—a criterion that appears to be satisfied in the studies reviewed here. By altering through mutagenesis the side chains of amino acid residues that interact with the substrate, Fersht and his associates were able to alter the binding energy ΔG_s. The difference $\Delta G_s{}' - \Delta G_s$ then measures the magnitude of the functional perturbation δF_m for each modified enzyme. These studies have concentrated on exploring the roles of hydrogen bonding between the substrate and the enzyme's active site. The principal interacting groups of this active site are diagrammed in Figure 1.

An initial experiment (5) involved the mutation of Cys35 to Ser35, altering the possible hydrogen bond with the 3'-hydroxyl of the ribose group of Tyr-AMP (Figure 1). It was found that the wild-type enzyme (i.e. with Cys) had a higher affinity for substrate by 1.1 kcal/mol. This difference can be

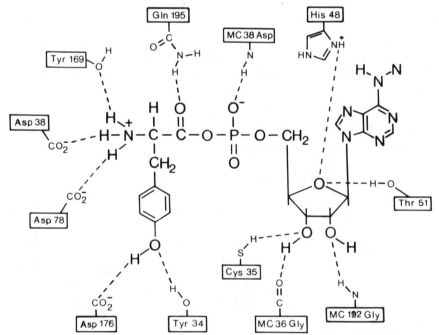

Figure 1 Active site of tyrosyl-tRNA synthetase with bound tyrosyl adenylate. Hydrogen bonds are indicated by dashed lines. Reproduced with permission from (4).

rationalized in terms of the local hydrogen-bonding strengths (with the active site versus competing solvent) assuming that the effect of the mutation is not propagated to regions of the molecular structure beyond the local site of the altered residue. The observed order of affinities is consistent with the structural observation that geometrical constraints on hydrogen bond formation exist within the, presumably rigid, active site so that the O-O distance in the OH–O bond is 5Å longer than the S-O distance in SH–O. These results, and others, suggest that the presence of a poor hydrogen bond in an enzyme-substrate complex can weaken the binding and lower k_{cat}/K_m (5, 6).

ENGINEERING INCREASED ENZYME-SUBSTRATE AFFINITY   Mutants were constructed (10, 11) to alter the hydrogen bond between Thr51 and substrate (see Figure 1) in two different ways: First Thr51 was changed to Ala51, leading to a reduced K_M for ATP by a factor of two without significantly changing k_{cat}. Thus k_{cat}/K_M was increased twofold. In a separate construction Thr51 was changed to Pro51, which resulted in a 25-fold increase in k_{cat}/K_M (10). The enzymatic activity is thus increased dramatically by a single point mutation. These results have shown for the first time that an enzymatic activity can be improved by genetic engineering of the active site.

To explore the structural basis of this enhancement of activity, it was surmised that the role of proline was to distort the helix structure in the region of residues 47–50 (11). A mutagenesis experiment modifying His48 to glycine, Figure 2a, yielded a sixfold reduction of k_{cat}/K_M corresponding to 1.1 kcal in binding energy (which the authors assigned to the H-bond between His48 and the substrate's ribose moiety). Working from the hypothesis that the effect of introducing Pro51 into the helix was to distort the polypeptide backbone such that His48 would interact more favorably with the substrate, a double mutant was constructed (His48,Thr51→Gly48,Pro51). The relationships between perturbations in binding free energy resulting from these modifications are diagrammed in Figure 2 along with the k_{cat}/K_M of each species. It was found that k_{cat}/K_M for the double mutant was approximately the same as that of the single (His48→Gly48) mutant. The dramatic effect of changing Thr51 to Pro51 was only observed when His48 was unaltered. These results clearly show that structural modification at the two sites are coupled: the free energy values representing functional perturbations for the single mutations are $\delta F_{Gly} = 1.1$ kcal, and $\delta F_{Pro} = -1.9$ kcal which sum to -0.8 kcal. The energetic perturbation of the double mutation $\delta F_{Gly-Pro}$ was 1.1 kcal (Figure 2a). A coupling free energy of 1.9 kcal is thus found between the two perturbations. In order to assess further the nature of the activity enhancement generated by the mutation Thr51→Pro51, we may consider results from two additional double mutations involving the positions 48, 51, and 35, as shown in Figure 2b and 2c. Results from an additional double mutation, Gly35,Pro51, and the constituent single mutations are shown in Figure 2b. It was found that mutation at position 51 from Thr (wild-type) to Pro results in an enhancement of k_{cat}/K_M by a factor of about 50 fold if the residue at position 35 was a glycine rather than the native cysteine (Figure 2b). However, the final activity resulting from the double mutation was still much lower than that from the single mutation Thr51→Pro51 since the effect of eliminating the side chain Cys35 is to lower the affinity substantially. The overall coupling free energy in this set of mutations is 0.4 kcal, which is considerably less than the 1.9 kcal found for the double mutation Gly48, Pro51. Results for the double mutation Gly35,Gly48, which eliminates both amino acid side chains at these positions, are shown in Figure 2c. This leads to a decrease in activated substrate binding free energy of 2.3 kcal. This value equals the sum of free energy perturbations found for the separate mutations at each of the two sites. Thus there is no coupling between the functional perturbations arising from modifications at these two sites.

It was concluded from these studies that the enhancement of activity due to Pro51 is entirely dependent upon the interactions between His48 and the substrate, and that distorting the helix leads to a more favorable interaction at that site (11).

These studies provide considerable insight into the types of effects that can result from specific modification at the active site of an enzyme and of the possible mechanistic sources of the effects. Future studies may shed light on the nature of long-range coupling effects in the enzyme, in particular as regards the origins of the interactions responsible for the half-of-sites reactivity in this system. Studies of long-range coupling in the aminoacyl tRNA synthetases will be summarized in a later section.

Dihydrofolate Reductase

Dihydrofolate reductase (DHFR) catalyzes the NADPH-dependent reduction of 7,8-dihydrofolate to 5,6,7,8-tetrahydrofolate. The enzyme is present in all organisms and is especially abundant in rapidly dividing cell lines. Crystal structures of DHFR from three species [*Escherichia coli* (12, 13), *Lactobacillus casei* (13, 14), and chicken (15)] have been determined to high resolution and indicate a highly conserved backbone structure. Recently, Kraut and his associates have constructed three mutant forms of the *E. Coli* enzyme, using the technique of oligonucleotide-directed mutagenesis of the cloned gene (16). Mutations were introduced at position 27 (Asp → Asn) in the enzyme active site, and at residues 39 (Pro → Cys), and 95 (Gly → Ala).

Based on the structures of the complexes of the enzyme with the cofactor NADPH and inhibitors, the carboxyl group of Asp27, located in the active site of the enzyme, is believed to stabilize the protonated transition state complex. Substitution of the aspartate residue by asparagine, which removes this poten-

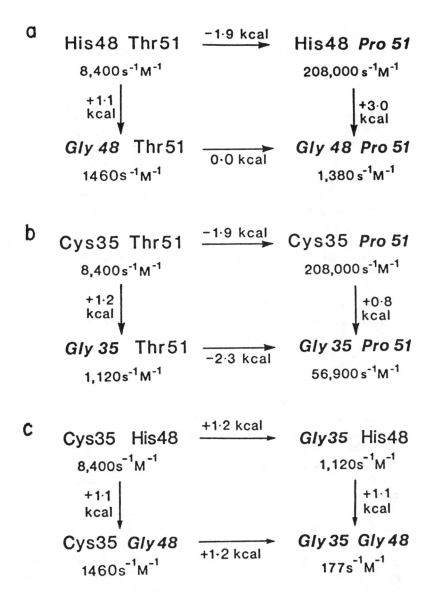

Figure 2 Schemes for double modifications of tyrosyl tRNA synthetase showing relationships between k_{cat}/K_M values and corresponding free energies of activated substrate binding. Adapted with permission from (11).

tial source of stabilization, was found to reduce catalytic activity to 0.1 percent of wild-type. This result supports the hypothesized role of Asp27 in catalysis, and also indicates that other factors may be involved in stabilization of the transition state.

In the wild-type *E. Coli* enzyme, Gly95 and Gly96 are linked by a *cis* peptide bond. This structural feature is also observed in the *L. casei* and chicken enzymes, and is thought to play a role in the stabilization of a local conformation of the molecule required in the catalytic mechanism. Examination of the contacts in the wild-type structure indicates that mutation at residue 95 should change the conformation of this region of the molecule, possibly leading to a change in enzyme activity. It was found that mutation at position 95 from glycine to alanine resulted in complete loss of enzymatic activity. Preliminary characterization of the modified protein also indicated that a small change in conformation had occurred.

The substitution of cysteine for proline at position 39 was designed to examine questions of protein folding and molecular dynamics in DHFR. This substitution had no effect on enzyme activity in the fully reduced form of the protein. However, activity was diminished upon oxidation of the protein, which is believed to lead to the formation of a disulfide bond between Cys39 and Cys85. The formation of a new disulfide bond may result in a loss of flexibility in the enzyme and the concomitant reduction in activity.

β-*Lactamase*

Techniques of "protein engineering" have been used to construct active-site mutants of *E. coli* β-lactamase. Using restriction enzyme methods, the active site serine (position 70) was replaced with cysteine, producing thiol β-lactamase (17). This modification resulted in an apparent decrease in enzymatic activity, although the authors note that a quantitative comparison of the activity of thiol-β-lactamase and the native enzyme requires a knowledge of the effect of an additional cysteine residue on the folding, stability, and processing of the enzyme.

The active site serine and adjacent threonine (position 71) residues were altered, using techniques of oligonucleotide-directed mutagenesis (18). Three mutants in this region were constructed; Ser 70 → Thr, which is catalytically inactive, Thr 71 → Ser, which shows a low level of activity, and a double mutation, Ser 70 → Thr, Thr 71 → Ser, which is catalytically inactive. Results of these studies provide support for the catalytic mechanism of β-lactamase, which is thought to proceed through an acyl intermediate involving serine 70 (19). The authors point out that such studies may allow delineation of the roles of active site residues in the binding of substrate as well as in catalysis.

Site-directed mutagenesis has also been used to construct other mutants of β-lactamase in the structural gene for the protein (20), although the catalytic properties of these mutants have not been determined. Additionally, mutants of β-lactamase with altered catalytic properties have been isolated by selection for increased resistance to cephalosporin C (21). Recently, directed mutagenesis of the β-lactamase signal sequence has been used to investigate secretion of proteins across membranes (22, 23).

MAPPING BY STRUCTURE-FUNCTION PERTURBATION

The strategy we review next provides a different approach, based not on the behavior of any single residue nor even a small group of residues but upon the patterns of functional response that arise from numerous site-specific structure modifications distributed throughout the protein's molecular structure (24–28). The perturbations of functional behavior δF_m, versus the structural location of the modified sites m provide a "map" from which inferences regarding structure-function relationships may be drawn. In this approach, the structural modifications are treated as "arbitrary" perturbations of structure at specific locations. This strategy has been found useful for (a) localizing functional events within a macromolecular structure i.e. determining the pathway of functional perturbations (24–28) and (b) detecting and mapping the pathways of coupling between local structural perturbations (3, 8, 28, 29). When combined with the more classical individual-site interaction analysis described above this approach provides an especially powerful combination. The two methods are capable of providing highly complementary (and supplementary) types of information.

Rationale

Noting again that both the structural and functional changes that the protein undergoes are only small perturbations of state, the rationale for this strategy is based upon studying the effects of one type of perturbation (covalent structure modification) upon perturbations of a second type—i.e., interactions representing the functional processes of interest. The method may be viewed crudely by thinking of the protein as a network of coupled springs to which various weights (atoms and groups of atoms) are attached. We now attach a "test weight" at a specific site on the network and measure the displacement of this weight when acted upon by the gravitational force and resisted by the entire system of springs and weights. Next we modify one of the weights and the adjacent springs by which it is coupled to its neighbors. The "test weight" is attached again to the same site as in the previous experiment and its net displacement measured again. If we see a significant difference in the response (i.e., net displacement) upon structural modification of the local site, we may infer that the structural location of the altered weights is a region of the system that participates significantly in generating the response first observed with the unmodified system. This inference, however, may be greatly strengthened or weakened by a systematic series of modifications in which many sites (weight and accompanying springs) are modified in several different ways and the entire pattern of resulting effects is studied. A dynamic analogy of this static model would arise if we plucked one of the springs at a particular site and measured the effects of modifying the other sites upon the sounds produced.

These mechanical analogies portray the general concept of the method, even though the actual processes represented differ somewhat from those in proteins. The power of this approach will be maximized when the system property F is chosen to reflect the energy states of the molecule as a whole (e.g. the Gibbs free energy) and is at the same time directly related to its biological function.

The Functional Filter

Structural modification of a given amino acid residue may generate local structural changes at other sites that may not be involved in function. Even if these local structure changes are monitored (e.g. spectroscopically), there remains the problem of determining which ones are relevant to function. This problem is bypassed when the system property F is directly related to function, since it automatically selects contributions only from the sites where structural perturbation affects functional behavior (3).

Interpreting the Map

As noted earlier, the functional response to a single alteration of an amino acid can frequently be interpreted in only a speculative fashion except in cases where a great deal is already known about the detailed roles of the specific groups. However, the development of a pattern of behavior over an extensive set of single-site modifications distributed throughout the molecular structure may provide a strong case for inferences regarding the localization and coupling of functional effects.

Considering the three classes of response to a single site modification discussed earlier, we see that a procedure based on grouping the sites according to the system's response may be highly misleading and is not a recommended procedure. It is necessary, rather, to consider the entire map to see whether well-defined patterns of effects are present. In general the sites m should be grouped according to structural regions of the molecule and the corresponding values of δF_m tabulated. If a striking pattern exists such that the only sites where modifications produce significant values of δF_m lie in a certain region of the molecular structure, then it may be reasonably inferred that these sites are altered during the functional process in the native molecule. In such a case it is desirable to explore the observed patterns by further modifications and by multiple modifications as described in the next section. If the patterns are not striking (i.e., the magnitudes of differences small), it is necessary to obtain a more extensive data base before conclusions may be drawn (depending on what one already knows).

Functional Coupling Between Sites

Once a region or *functional pathway* has been identified, it is of great value to explore the coupling of effects within this region of structure through the interplay of multiple modifications. The simplest way to accomplish this is to first construct two separate proteins P_1 and P_2 which are modified at different locations. The corresponding functional perturbations (equation 2.) will be δF_1 and δF_2. Next construct a molecular species P_{12} that has both modifications, and determine the resulting perturbation of function δF_{12} for the doubly modified molecule. If the quantity F is proportional to a conserved property, say the Gibbs free energy, we can test for the independence of perturbations resulting from modifications at sites "1" and "2" by the relationship:

$$\delta F_{12} = \delta F_1 + \delta F_2 \qquad\qquad 7.$$

If this equality holds, this provides evidence (barring coincidence) that the two sites are not directly coupled in generating the functional response F. When equation 7. fails, it is useful to assess the magnitude of "functional coupling" by noting that P_{12} could be constructed along two routes depending on the order of introducing the modifications as shown in scheme 8.

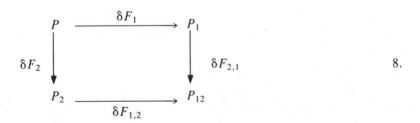

In this scheme $\delta F_{2,1}$ is the functional perturbation arising from modification at site "2," given that modification at site "1" has already been carried out. The analogous perturbation resulting from structural modification at site "1" for a molecule previously modified at site "2" is given by $\delta F_{1,2}$. We define a coupling constant δF_c by the relationships

$$\delta F_c = \delta F_{2,1} - \delta F_2 = \delta F_{1,2} - \delta F_1 \qquad\qquad 9.$$

If there is no site-site coupling so that $\delta F_c = 0$, then the relationship of equation 7. follows, since $\delta F_{12} = \delta F_1 + \delta F_{2,1} = \delta F_2 + \delta F_{1,2}$. This type of coupling is illustrated by the double mutant studies with aminoacyl Tyr-tRNA synthetase of *B. Stearothermophilus* shown in Figure 2c.

These coupling relationships for effects of covalent modification are analogous to those for "linked functions" involving noncovalent interactions (30–32). That one can mix the two types of perturbation in many combinations provides great versatility in devising experimental strategies. Suppose, for example, that a second molecular species Q (e.g. another protein, a small ligand, or a nucleic acid) associates with P to form a complex PQ; then we can write the following scheme in strict analogy to scheme 8.

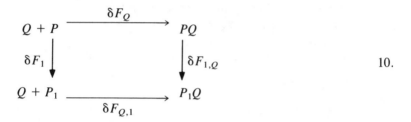

10.

Here the coupling between an amino acid substitution at site "1" of protein P and the noncovalent interaction with species Q is given by

$$\delta F_c = \delta F_{1,Q} - \delta F_1 = \delta F_{Q,1} - \delta F_Q \qquad 11.$$

where each of these terms has entirely analogous significance to the corresponding term of equation 9.

Indirect Coupling and Triangulation

Even if $\delta F_c = 0$ for two site-specific structure modifications (equation 9.), it is entirely possible for both sites to lie within the functional pathway, if, for example, they are both "driven" by changes at a third site. Directional, or anisotropic effects in proteins are well known (e.g. actin-myosin complexes) and can arise either through stereospecific pathways or through widespread conformational changes that may be "triggered" by alterations at some sites but not others. We consider two sites, say "1" and "2," each of which is coupled to a third site "3" as in equation 9. But sites "1" and "2," when modified in the absence of modification at site "3," generate perturbations that satisfy equation 7. Then it follows that modification of site "3" will simultaneously affect perturbations at sites "1" and "2," i.e. we have the simultaneous relationships

$$\delta F_{12} = \delta F_1 + \delta F_2$$

$$\delta F_{12,3} = \delta F_{1,3} + \delta F_{2,3} \qquad\qquad 12.$$

$$\delta F_{1,3} \neq \delta F_1 \; ; \; \delta F_{2,3} \neq \delta F_2$$

where the subscripts denote the respective combinations of modified sites. By exploring all sites in the protein system, it is possible, in principle, to map all such couplings by this triangulation procedure. Additional combinations of modifications can be usefully generated and studied in a hierarchical fashion, limited only by exhaustion levels of the investigators.

SUBUNIT INTERACTIONS AND LIGAND BINDING

Interactions of Serine Proteinases with Ovomucoid Inhibitors

Serine proteinases, including tyrpsin, chymotrypsin, elastase, and subtilisin are inhibited by ovomucoids, which comprise a major component of avian egg white. These proteins consist of three homologous domains, each of which is capable of binding to a serine proteinase molecule, and thereby inhibiting its activity [cf (33) for recent review]. The "third domain," consisting of 56 amino acids, can be proteolytically cleaved from the other two, yielding a compact, heterogeneously glycosylated, globular structure with full inhibitory capabilities.

An extensive series of quantitative interaction studies have been carried out by Laskowski and associates on the effects of single and multiple amino acid modifications on the interactions of ovomucoid third domains with serine proteinases (25, 29). The long-term goal of these studies is to develop a "sequence to reactivity algorithm" which would permit a quantitative prediction of the interactions between the serine proteinases and an ovomucoid of any specified sequence. To date, a series of exceptionally accurate and reliable equilibrium association constants, K_a, have been determined for a large series of ovomucoid third domains which differ from each other in contiguous sequences of single amino acid substitutions (25). It has also been possible to construct "hybrid" molecules by a procedure involving proteolytic cleavage of two ovomucoid species, refolding of the cleaved polypeptides in a hybrid form, and finally enzymatic ligation at the original cleaved site (29). The effects of possible coupling between perturbations at multiple sites could thus be investigated. Crystallographic structures have been obtained from ovomucoids of the turkey (34) and Japanese quail (35) and the thermodynamic findings have been correlated with these structural results. Studies by NMR spectroscopy (36) have provided additional insights into the structural and functional aspects of the molecular interactions.

Table 1 gives a summary of results of the effect of single and multiple-site modifications on the association equilibrium constants for three different serine proteinases. From these and other similar results the following generalizations have been drawn. (*a*) Amino acid alterations of the ovomucoids within the region of intersubunit contact exhibit large effects on the association constants as well as dramatic differential effects toward the various proteinase enzymes. By contrast, the amino acid substitutions at other regions of the ovomucoid have essentially no effect on the association constants. Thus the "map of structure-function perturbation" shows the subunit interface to be the sole region or pathway of functional perturbation in this system. (*b*) The equilibrium constant ratios of Table 1 provide determinations of the net free energies of transfer of side chains from the surface of the inhibitor to the enzyme-inhibitor interface. These free energies of transfer generally follow the expected trend for transfer of side chains from aqueous to nonaqueous environments (37). Although exceptions exist, this trend is consistent with the generally hydrophobic character of the protein-protein interface. These results and the deviations from them can be interpreted to a large extent by recourse to the crystallographic data, using models of local stereochemical effects. (*c*) In many cases where double modifications (hybrid molecules) have been studied, the observed perturbations in free energy of association are equal to the sum of perturbations which result from each of the modifications when carried out separately. Thus, to within limits of experimental accuracy, the effects of local site modification do not appear to be coupled, i.e., $\delta F_c = 0$ in equation 9.

Local Effects versus Long-Range Coupling in Human Hemoglobins

Human hemoglobin has served for many years as a vehicle for attempts to understand how a macromolecule can regulate its own functions in response to "signals" it receives from the local environment [cf (38) for historical review]. In the case of hemoglobin, the signals take the form of changes in the environmental concentrations of small regulatory molecules including oxygen, protons, carbon dioxide, 2,3-diphosphoglycerate, and chloride (cf 39–43). An ensemble of hemoglobin tetramers, such as that within a red cell, is able to "estimate" the concentration of each regulatory species in the environment. The hemoglobin tetramers consequently undergo molecular adjustments that alter their affinity for each of the small regulatory species. These adjustments entail both changes of structure (44, 45) and altered free energies of intramolecular interaction (45). How much of this physiologically-important control is exerted

Table 1 Equilibrium constant factors associated with amino acid substitutions[a]

position of substitution(s)[d]	substitution A$_1$ $\rightarrow$ A$_2$		Ratio of equilibrium constants[b] (K_2/K_1) chymotrypsin	elastase	subtilisin
15	Asp	$\rightarrow$ Ala	1.2	160	2800
18	Leu	$\rightarrow$ Met	0.56	0.21	1.9
20	His	$\rightarrow$ Tyr	25	44	130
28	Asn	$\rightarrow$ Ser	0.72	0.86	0.97
			0.53	0.68	0.53
32	Gly	$\rightarrow$ Ala	2.3	1.9	0.11
			3.0	2.4	0.20
32	Asp	$\rightarrow$ Gly	450	86	150
33	Ser	$\rightarrow$ Asn	1.5	19	0.65
45	Asn	$\rightarrow$ Asn[c]	0.94	1.3	0.67
			1.1	0.77	0.79
51	Ser	$\rightarrow$ Asn	0.67	1.0	0.30
52	His	$\rightarrow$ Arg ⎱	1.9	0.92	0.34
55	Lys	$\rightarrow$ Glu ⎰			
51	Ser	$\rightarrow$ Asn ⎱			
52	His	$\rightarrow$ Arg ⎬	0.41	1.0	0.12
55	Lys	$\rightarrow$ Gln ⎰			
10	Glu	$\rightarrow$ Gly ⎱			
21	Arg	$\rightarrow$ Met ⎬			
43	Glu	$\rightarrow$ Asp ⎬	0.08	15	0.39
55	Lys	$\rightarrow$ Glu ⎰			
10	Glu	$\rightarrow$ Asp ⎱			
11	Tyr	$\rightarrow$ His ⎪			
15	Ala	$\rightarrow$ Val ⎪			
17	Thr	$\rightarrow$ Ser ⎬			
21	Arg	$\rightarrow$ Met ⎪	10^{-4}	0.39	0.22
28	Asn	$\rightarrow$ Ser ⎪			
32	Gly	$\rightarrow$ Ser ⎪			
36	Asn	$\rightarrow$ Asp ⎰			
10	Glu	$\rightarrow$ Asp ⎱			
18	Met	$\rightarrow$ Val ⎪			
21	Arg	$\rightarrow$ Met ⎬	10^{-5}	14	0.02
43	Glu	$\rightarrow$ Asp ⎰			

[a] Adapted with permission from (25).
[b] K_2 is the equilibrium constant for an ovomucoid third domain designated A$_2$ binding to each of the three proteinases. K_1 is the corresponding equilibrium constant for an ovomucoid designated A$_1$.
[c] Glycosylated
[d] Numbering of the inhibitor residue is based on (34) and (35).

within each individual molecule, as opposed to merely being an average property over many molecules, is a fundamental issue of hemoglobin research (47–49). The use of site-specific structural modifications in multiple combinations affords a means of mapping the intramolecular "communication" in terms of functional and energetic responses. It may be expected that many of the presently unresolved questions regarding the mechanism of free energy coupling and its relationship to biological function and regulation will be answered using these techniques. Some first steps in this direction have recently been taken through studies of regulatory subunit-subunit interactions in single-site mutants and chemically modified hemoglobins (24, 3).

The hemoglobin molecule as depicted in Figure 3 is a tetrameric structure comprised of four subunits (two α and two β chains), each of which contains a heme group where oxygen is bound. The subunits are designated α^1, α^2, β^1, and β^2 denoting their relative locations within the tetramer. The molecule may

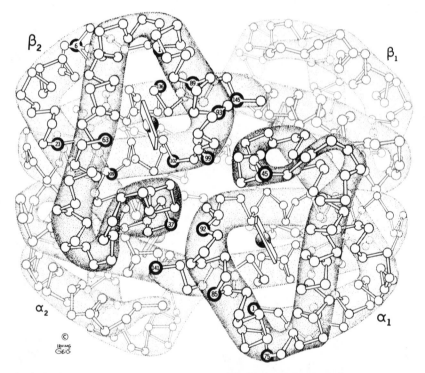

Figure 3   Tetrameric structure of human hemoglobin showing locations of altered sites for mutant and chemically modified hemoglobins. Circled α carbons indicate residue numbers of modification sites for data in Table 2, except for positions β94, β101 and β95 (not shown). Adapted with permission from (11); illustration copyright, Irvin Geis.

be viewed as a system of interacting dimers $\alpha^1\beta^1$ and $\alpha^2\beta^2$. When oxygen is bound, major changes occur at the interface separating the two dimer pairs— the "$\alpha^1\beta^2$ intersubunit contact region" (44, 45). This region contains three pairwise contacts $\alpha^1\beta^2$, $\alpha^2\beta^1$, and $\alpha^1\alpha^2$ (β chains don't touch). The pairwise intersubunit contacts consist of hydrogen bonds, ion pairs (salt bridges), hydrophobic interactions and van der Waals interactions. The $\alpha^1\beta^1$ and $\alpha^2\beta^2$ contacts do not undergo appreciable alterations upon oxygenation. The binding of oxygen at the heme of a given subunit leads to structure changes at the heme itself, tertiary changes within the oxygenated subunit, and quaternary changes affecting the subunit contacts mentioned above [cf (43)]. An elegant series of NMR studies on these processes using mutant hemoglobins has been reviewed recently (28).

A central issue in hemoglobin research has always been the mechanism of cooperative oxygenation as reflected in the sigmoidal shape of the oxygen binding isotherm. The sigmoid shape of this curve is a direct manifestation of the differences in Gibbs free energies for the successive oxygenation steps (50). Cooperativity in oxygen binding is, in essence, a thermodynamic phenomenon. Consequently, all meaningful efforts to define the structural and energetic bases of cooperativity must be directed toward the understanding of this energetic effect.

A meaningful measure of site-site interaction is the *regulatory free energy*, ΔG_r. This quantity relates the binding free energies for successive steps to the intrinsic free energies of the constituent subunits (24). An experimental determination of ΔG_r measures the amount of energy invested by the molecule over all four oxygen binding steps to "pay" for cooperativity. Understanding cooperativity in molecular terms therefore depends upon understanding the origins of the regulatory free energy. Effective methods for accurately determining ΔG_r have been developed using subunit dissociation (24).

SUBUNIT DISSOCIATION AS A PROBE OF REGULATORY INTER-ACTIONS   Since the self-regulation of hemoglobin oxygen binding is mediated through interactions at the intersubunit contacts of the tetrameric molecule, subunit dissociation has proven to be a powerful quantitative tool for probing the regulatory energy changes that accompany the functional cycle of oxygenation-deoxygenation (42). The rationale for this approach is as follows (51, 52): (*a*) Interactions within the tetrameric molecule that are responsible for cooperativity can be decoupled by dissociation of the tetramers into noncooperative ($\alpha^1\beta^1$) dimers. (*b*) The difference between energies of dimer-tetramer assembly at the extreme stages of oxygenation (deoxy and oxy) provides a measure of how much the interaction energy within the tetramer is changed as a result of the oxygen binding. Dissociation equilibrium constants for unligated and fully oxygenated molecules differ by as much as five orders of magnitude, providing an exquisitely sensitive probe of the regulatory interactions (42, 46, 52). The regulatory free energy has been shown to equal the difference between Gibbs free energies of dimer-tetramer assembly, i.e. oxyhemoglobin minus deoxyhemoglobin (24). Under a standard set of conditions used for all comparisons reviewed here (42), these assembly free energies are: $^{0}\Delta G_2 = -14.3$ kcal (deoxy) and $^{4}\Delta G_2 = -8.0$ kcal (oxy).

LOCALIZATION OF REGULATORY FREE ENERGY   Recent studies have determined the values of $^{0}\Delta G_2$ and $^{4}\Delta G_2$ for a series of mutant and chemically modified hemoglobins in which each tetrameric molecule bears two structural modifications, i.e. in the α chains or β chains (24, 3). Table 2 lists values of functional perturbations δF_i and δF^{oxy}_i for 27 human hemoglobins that have these site-specific structural modifications. δF_i is the deviation in $^{0}\Delta G_2$ due to the structural modifications at site i of the appropriate subunits, and δF^{oxy}_i are the corresponding deviations for the fully oxygenated species. Among these hemoglobins there are 21 sites of modification, 18 of which are shown in Figure 3. With few exceptions (Hb S and C), the variants of Table 2 are found in minute populations and in heterozygous form, and they do not appear to be subject to any strong selection pressures. Of the more than 450 known mutant human hemoglobins, the great majority may be viewed as "genetic noise." The hemoglobins are listed in Table 2 in two series of entries. The first series (A) includes modification in the $\alpha^1\beta^2$ intersubunit contact region of the molecule (53); the second (B) lists modifications at other regions including the heme pocket, the $\alpha^1\beta^1$ intersubunit contact, the external surface, the interior of the molecule, and the "central cavity" [cf (43, 53)]. The combination of Table 2 and a structural model of the hemoglobin tetramer in its deoxy and oxy forms constitutes a "map of structure-function perturbation" from which inferences have been drawn regarding the location of the residue sites responsible for cooperativity in oxygen binding.

A striking pattern is seen in Table 2 for the effects of structure modifications on stabilizing interactions of the deoxy tetramer. For nearly all of the sites in the $\alpha^1\beta^2$ contact region the values of δF_i are large and correspond to changes in the equilibrium constant $^{0}K_2$ by factors of 10^2–10^5. By contrast, the values of δF_i for sites outside of this region are very small. Modifications denoted by asterisks at eleven sites in the $\alpha^1\beta^2$ contact region destabilize the molecule by 3.6 ± 2.3 kcal (mean and standard deviation). This corresponds to a 470-fold less stable molecule on the average. Modifications at the 9 sites not in the interface region destabilize by 0.3 ± 0.7 kcal corresponding to a 1.7-fold less stable molecule. On the average then, a modification in the $\alpha^1\beta^2$ intersubunit contact region provides a molecule that is less stable than one modified elsewhere by a factor of 276 fold. Stability of the deoxy molecule is thus seen to be hypersensitive to structural perturbations within the $\alpha^1\beta^2$ region but very

Table 2   Effects of site specific structure modifications on subunit interactions in human hemoglobins[a]

Hb	Modification	Perturbations of interaction free energies		
		δF_i	δF_i^{oxy}	$(\delta F_i^{oxy} - \delta F_i)$
A.	Modified sites in the $\alpha^1\beta^2$ intersubunit contact region			
* Hotel Dieu	(β99 Asp $\rightarrow$ Gly)	6.1	0.0	−6.1
Kempsey	(β99 Asp $\rightarrow$ Asn)	5.9	− .7	−6.6
Yakima	(β99 Asp $\rightarrow$ His)	4.5	−1.5	−6.0
* Osler	(β145 Tyr $\rightarrow$ Asp)	5.5	0.6	−4.9
* Creteil	(β89 Ser $\rightarrow$ Asn)	5.8	0.3	−5.5
des Arg	(α141 deleted)	4.2	−1.0	−5.2
* Legnano	(α141 Arg $\rightarrow$ Leu)	4.8	0.1	−4.9
* Chesapeake	(α92 Arg $\rightarrow$ Leu)	2.6	−1.6	−4.2
* G–Georgia	(α95 Pro $\rightarrow$ Leu)	4.0	0.0	−4.0
* NES	(β93 modified)	2.8	−0.5	−3.3
Kansas	(β102 Asn $\rightarrow$ Thr)	0.7	2.2	1.5
* Saint Mandé	(β102 Asn $\rightarrow$ Tyr)	−0.6	1.6	2.2
Rush	(β101 Glu $\rightarrow$ Gln)	−0.5	0.2	0.7
* British Columbia	(β101 Glu $\rightarrow$ Lys)	2.8	2.1	−0.7
* Hirose	(β37 Trp $\rightarrow$ Ser)	5.6	—	—
* $\alpha^c \beta$	(α1 carbamylated)	−0.4	−0.5	−0.1
$\alpha^c \beta^c$	(α1β1 carbamylated)	−0.1	−1.4	−1.3
B.	Modification sites not in the $\alpha^1\beta^2$ contact region			
* Fort de France	(α45 His $\rightarrow$ Arg)	0.7	0.0	−0.7
* Winnipeg	(α75 Asp $\rightarrow$ Tyr)	−0.2	0.1	0.3
* G Norfolk	(α85 Asp $\rightarrow$ Asn)	0.1	−0.3	−0.4
S$_o$	(β6 Glu $\rightarrow$ Val)	0.0	0.0	0.0
* C	(β6 Glu $\rightarrow$ Lys)	−0.4	−0.2	0.2
* Strasbourg	(β23 Val $\rightarrow$ Asp)	0.7	−0.3	−1.0
* Zurich	(β63 His $\rightarrow$ Arg)	1.4	1.1	−0.3
* San Diego	(β109 Val $\rightarrow$ Met)	0.9	0.4	−0.5
* Hope	(β136 Gly $\rightarrow$ Asp)	−0.8	0.4	1.2
* Barcelona	(β94 Asp $\rightarrow$ His)	0.5	0.0	−0.5

[a] Values calculated from data given in (24) and (3).
* Values used in statistical calculations (see text).

insensitive to modifications of amino acid residues elsewhere in the molecule. These results strongly suggest that the interactions responsible for stability of the deoxy tetramer are localized at the interface itself and are not the result of an assembly mechanism that depends on a change in conformation of the $\alpha^1\beta^1$ dimers between their dissociated and assembled deoxy states.

There is one dramatic exception to this insensitivity of the intersubunit interactions to perturbations at sites remote from the intersubunit contact region. The exception is for perturbations at the heme brought about by the binding of oxygen. This perturbation (i.e. complete oxygenation at all four sites) destabilizes the dimer-dimer interactions by 6.3 kcal corresponding to a factor of over 10^5 in the equilibrium constant (46). Furthermore, these heme site perturbations bring about large and simultaneous alterations in the perturbations δF_i due to the site modifications within the $\alpha^1\beta^2$ interface. The perturbation values for oxyhemoglobin are listed in Table 2 (δF^{oxy}_i values) as well as the magnitudes of the shifts ($\delta F^{oxy}_i - \delta F_i$) brought about by oxygenation. It should be noted that the differences ($\delta F^{oxy}_i - \delta F_i$) are just the perturbations in regulatory free energy resulting from the structure modification. In sharp contrast to the sites within the $\alpha^1\beta^2$ interface, oxygen binding has very little effect on values of δF_i for mutations outside of the $\alpha^1\beta^2$ intersubunit contact region, i.e. the perturbations in regulatory energies for these sites are very small. The average shift in ($\delta F^{oxy}_i - \delta F_i$) for $\alpha^1\beta^2$ modifications is -3.1 ± 2.5 kcal, whereas for sites elsewhere it is only -0.2 ± 0.6 kcal. This is a striking example of both the indirect coupling and triangulation relationships.

THE PATHWAY OF REGULATORY FREE ENERGY COUPLING   From the studies just reviewed it was inferred that the $\alpha^1\beta^2$ interface region is the location of most regulatory free energy change within the tetrameric molecule and that this energy change is controlled by ligand binding perturbations at the heme sites. These results raise the important question of how much coupling exists between the residue sites that lie within the $\alpha^1\beta^2$ contact region. Figure 4 shows a diagram of one of the identical halves of this region and many of the pairwise contacts across the interface are indicated. The experimental results summarized alone are highly suggestive of coupling within this interface. (*a*) Structural perturbations at widely separated regions of the interface lead to similar alterations of regulatory energy (e.g. Hotel Dieu and des-Arg). Magnitudes of the largest free energy perturbation due to modified sites at different locations are strikingly similar (e.g., Kempsey, Creteil, des-Arg, and Osler). Furthermore, the perturbations in regulatory energy never exceed the magnitude of the regulatory energy for normal hemoglobin. These observations suggest a common set of delicately balanced interactions which act in an interdependent way. (*b*) Substitutions of different amino acid residues at the same position lead to identical changes in regulatory energy. At the $\beta99$ position, hemoglobins Kempsey, Yakima, and Hotel Dieu have amino acid side chains of widely different steric and chemical natures. Yet they show nearly identical perturbations in regulatory energy (Table 2). (*c*) Structural studies (54) have shown hemoglobins Osler and Creteil to have similar deoxy structures with identically altered COOH-terminal tetrapeptides. Thermodynamic results (24) show identical energetics for those hemoglobins (Table 2). Thus the structural and energetic changes that result from altering local interactions at the $\beta89$ and $\beta145$ positions are found to be highly interdependent.

The possible pairwise coupling between modified residue sites within the

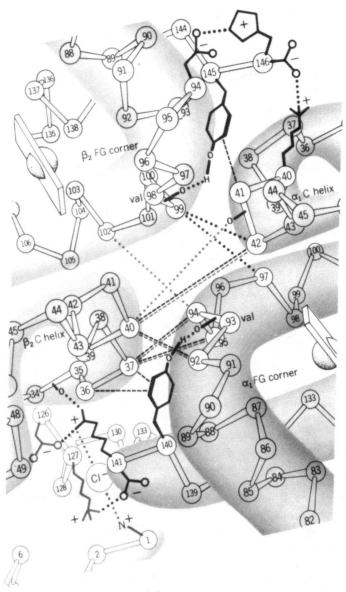

Figure 4 Enlarged front view of the $\alpha^1\beta^2$ contact region. Each numbered circle gives the locations of one of the two symmetry-related amino acid positions that occur in each tetramer: (···) and (- - -), which illustrate H-bond and nonbonded packing contact, respectively. Reproduced with permission from (43).

$\alpha^1\beta^2$ interface has been explored by constructing a series of hybrid tetramers (Smith, Ackers, manuscript in preparation). These hybrids consist of one $\alpha^1\beta^1$ dimer from each of the parent molecules, as shown topographically in the following diagrams.

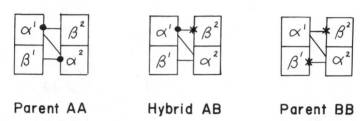

Parent AA   **Hybrid AB**   **Parent BB**

Here the three pairwise contacts $\alpha^1\beta^2$, $\alpha^2\beta^1$, and $\alpha^1\alpha^2$ are indicated by lines connecting the respective subunits. The filled circles and crosses indicate mutation sites. These molecules have been studied through the development of techniques that resolve properties of the hybrid in the presence of both parent molecules with which it is in equilibrium (Smith, Ackers, manuscript in preparation). The perturbations in the free energy of formation of the deoxy hybrid species are given in Table 3. Values of δF_{hybrid} listed are the differences between $^0\Delta G_2$ values for the hybrid molecules and normal hemoglobin A_o, which bears no modified sites. The first set of values are for hybrids between normal hemoglobin A_o and a mutant or chemically modified species, so that each tetramer carries only a single modified site. For comparison, values are listed as $\delta F_{i/2}$ for the corresponding parent molecules carrying the two identical modifications. The second group of hybrids (Table 3B) provides data on cases of two different modifications within the same tetramers. The values listed as $(\delta F_{\text{AA}} + \delta F_{\text{BB}})/2$ are the free energy perturbations expected if the effects of the two modifications are independent (i.e $\delta F_c = 0$ in eqn. 9). These values were calculated from the values of δF_i for the parent molecules (Table 2). Also listed for both types of hybrids are the distances of closest approach between the residues at the altered sites (55).

It was found in these studies that, to within limits of experimental accuracy, the individual-site perturbations appear to be "additive" and hence to exhibit pairwise independence over all the combinations tested. There thus appears to be no direct long-range coupling within the interface itself. Additional studies will be required with multiple modifications in closer proximity in order to locate the ranges of coupling within the regulatory interface.

Of major significance regarding the nature and pathway of functional perturbation in the hemoglobin system is the observation that long-range coupling does exist between the heme site and each of the sites represented in Table 3. Even though these sites do not appear to affect each other, they are all simultaneously affected by oxygen binding at the heme sites, as shown in Table 2. We thus have *indirect coupling* among these sites through perturbations at

Table 3   Free energy perturbations of dimer-tetramer assembly for hybrids of mutant hemoglobins

A. Hybrids with normal hemoglobin A_o			distance between sites
Hybrid (site)	δF_{hybrid}	$\delta F_{i/2}$	
A_o-Yakima (β99)	2.5	2.3	23 Å
A_o-Kempsey (β99)	3.4	3.0	23 Å
A_o-NES (β93)	1.1	1.4	35 Å
A_o-Chesapeake (β92)	0.9	1.3	33 Å
A_o-des Arg (α141)	2.4	2.1	18 Å
A_o-St. Mandé (β102)	-0.2	-0.3	26 Å

B. Hybrids with double modifications at the Regulatory interface			distance between sites
Hybrid	δF_{hybrid}	$(\delta F_{AA} + \delta F_{BB})/2$	
St. Mande-Kempsey (β102–β99)	2.9	2.7	24 Å
Chesapeake-Kempsey (α92–β99)	3.7	4.3	8 Å
des Arg-Kempsey (α141–β99)	5.3	5.1	16 Å
des Arg-NES (α141–β93)	3.5	3.5	24 Å

the heme, as described by the "triangulation relations" equation 12. The triangulation pathways need to be explored further by additional combinations of modifications. From crystallographic studies it is known which residues undergo the largest structural alteration within a subunit upon oxygenation (56). Some of these are included in the modifications studied to date, but further explorations along these lines will be necessary to resolve details of the coupling pathways in energetic terms.

SINGLE-SITE ELECTROSTATIC CONTRIBUTIONS TO THE FREE ENERGY   In an elegant and extensive series of studies, Gurd and associates have developed an effective model for calculating the electrostatic free energies arising from charge group interactions in a great variety of proteins, including myoglobins and hemoglobins [see (41) for a recent review]. The net electrostatic free energies of stabilization in tetramers versus dimers (57, 41) are given in Table 4 for deoxy and oxy hemoglobins, respectively. The net values for deoxy hemoglobin are listed as $^0\Delta G_{2,el}$ and the oxyhemoglobin entries as $^4\Delta G_{2,el}$. The differences between these values and the overall experimentally determined free energies (24, 3) are listed as $^0\Delta G_{2,non\text{-}el}$ and $^4\Delta G_{2,non\text{-}el}$, respectively.

The net electrostatic stabilizations due to charged group interactions in normal hemoglobin A_o are given in the first row of Table 4. It is seen that the electrostatic interactions account for roughly half of the stabilization energy in both unliganded and oxy molecules, whereas the actual stabilization due to electrostatic interactions is considerably greater in the deoxy species.

The effects of local site modifications on these electrostatic interactions have been explored in detail in terms of the local (and global) charge array present in each of the modified hemoglobins (41). Only the net results are shown in Table 4 for a representative series of mutants where the modified sites are within the molecule's regulatory interface region. The perturbations in electrostatic free energy for sites modified at locations not in this interface region (not shown) are found to be very minor in comparison with those given in Table 4. Particularly striking cases are the modifications of deoxy hemoglobin at sites β99 (hemoglobins Kempsey, Yakima) and β101 (British Columbia), whereas the effects of these modifications on the oxygenated tetramer are much smaller. The results cited here illustrate in an overall way a few of the many findings contributed to date by these investigators on the roles of local charged amino acid residues in the overall system properties of proteins.

Functional Domains and Long-Range Coupling in Aminoacyl-tRNA Synthetase

Members of the 20 aminoacyl tRNA synthetases exhibit widely varying sizes and quaternary structures, even within a single organism [see (58) for recent

Table 4 Electrostatic and non-electrostatic components of free energy of dimer-tetramer assembly[a]

Hemoglobin	Modification			$^0\Delta G_{2,el}$	$^0\Delta G_{2,non-el}$	$^4\Delta G_{2,el}$	$^4\Delta G_{2,non-el}$
A_o				−7.9	−6.4	−4.4	−3.6
Kempsey	β99	Asp	→ Asn	−9.3	+0.9	−4.5	−4.2
Yakima	β99	Asp	→ His	−9.8	0.0	−4.4	−5.1
Rush	β101	Glu	→ Gln	−8.7	−6.3	−3.3	−4.5
British Columbia	β101	Glu	→ Lys	−11.7	+0.3	−3.5	−2.4
Legnano	α141	Arg	→ Leu	−2.8	−7.3	−0.5	−8.5
Chesapeake	α92	Arg	→ Leu	−7.3	−4.4	−4.2	−5.2
Creteil	β89	Ser	→ Asn	−7.9	−0.6	−4.4	−3.3
Kansas	β102	Asn	→ Thr	−7.6	−6.0	−4.3	−1.5

[a] Reproduced with permission from (41).

review]. The component subunits range in size between 330 amino acids to over 1000 amino acids and are found in states of assembly denoted by α, α_2, $\alpha_2\beta_2$, and α_4 (59–62). Recent studies on the *E. coli* Ala-tRNA synthetase by Schimmel and associates have provided a mapping of this enzyme's 875 amino acid sequence into functional domains, and have established long-range integrative coupling between them (63–68).

The alanine-specific enzyme, which is the largest of the aminoacyl tRNA synthetases characterized to date, participates in the following functions: (*a*) formation of aminoacyl adenylate (the activation reaction given as equation 3 in the previous section), (*b*) attachment of the activated amino acid to tRNA^{Ala} (equation 4.) (*c*) subunit association to form tetramers. In the tetrameric form the enzyme may repress its own gene transcription by binding to a site adjacent to the controlling promoter (63). Using recombinant DNA techniques (66), a series of truncated polypeptide fragments were created of different lengths, starting at the amino terminal end of the 875 amino acid chain. These fragments were made from in vitro gene deletions on recombinant plasmids which contained the Ala-tRNA synthetase gene. The resected plasmids were then introduced into *E. Coli* via transformation. The nested set of truncated polypeptides that were stable when isolated for study ranged in length from 257 amino acids to 852 amino acids. The smallest protein (257 amino acids) had no enzymatic activity, whereas a molecule of 385 residues was found to have adenylate synthesis activity with a specific activity equal to half that of native enzyme. Among twelve proteins studied, all fragments that extended beyond residue 385 also had adenylate synthesis activity. It was concluded that residues beyond 385 are neither required for, nor do they interfere with, adenylate synthesis. The minimum length residue required for complete aminoacylation of tRNA^{Ala} (combined reactions 3 and 4) was found to contain residues 1–461 (a fragment of 404 residues did not suffice). Therefore only about half of the native protein is sufficient to provide catalytic activity. The specific activity of the 461 residue protein was only 20% of that for native protein. However the amino acid specificity was found to be essentially that of native enzyme.

The ability of the truncated proteins to self-associate into tetrameric complexes was studied by gel chromatography using calibrated Sephadex G-200 columns (66). It was found that sites essential for tetrameric assembly lie between residues 699 and 808 (i.e. the 699 fragment runs as a monomer and the 808 fragment as a tetramer on the column). The tetramerization is believed not to be required for protein synthesis in vivo. It is surmised rather to be necessary for the gene regulation mediated by this enzyme in binding to the DNA site mentioned above. Thus each of the three activities appears to require specific domains arranged linearly along the polypeptide. It was postulated (66) that the diverse sizes of aminoacyl tRNA synthetase may be accounted for in terms of a fundamental catalytic core onto which various fusions of extra polypeptide domains carrying additional functions may have occurred.

The existence of separate functional domains in this large protein system evokes an intriguing replay of the question regarding overall system properties versus local part properties. Is the whole a sum of its parts when the parts are triple-domain 875-residue subunits of a tetrameric complex? Or is there intersubunit coupling between the functional behavior of the various domains? Recent studies (68) indicate the latter to be the case. A series of 8 proteins were constructed containing deletions of various lengths within the amino-terminal region of the Ala-tRNA synthetase molecule. The largest deletion removed 442 amino acids and the resulting protein consisted of the first 7 N-terminal amino acids fused to the C-terminal half of the molecule. The smallest deletion removed just amino acids 149–210. Subunits of each of these proteins were combined, through in vivo complementation, with that of a temperature-sensitive point mutation, (AlaS5) in the C-terminal domain. Under conditions where the AlaS5 protein alone has no enzymatic activity, the hybridized tetramers (AlaS5 protein + fragment) exhibited approximately half the activity of wild-type enzyme (Figure 5). Thus the inactive synthetase fragments, which have deletions in their catalytic domains, are each able to activate the catalytic domains of AlaS5 subunit. Several lines of evidence indicate that this reversal of the effect of mutation in one subunit by interaction with a second subunit (itself incapable of activity) occurs through subunit contact sites in the C-terminal domains of both subunits. Furthermore, these contact sites are believed to be remote from the catalytic site that becomes activated (68). There thus appears to be a long-range coupling between the C-terminal contact and the activated catalytic site. It is possible that this functional coupling is mediated through a specific pathway within the protein subunits. An alternative explanation is that the functional coupling is mediated by "global" effects, i.e. a conformational form of the AlaS5 subunit having catalytic activity is stabilized by interaction with protein fragments.

ACKNOWLEDGMENTS

A number of the concepts discussed in this article have evolved over the last several years through discussions among members of our research group; in particular we wish to acknowledge discussions with Donald W. Pettigrew and Benjamin W. Turner. We gratefully acknowledge the many colleagues who provided us with unpublished results. We thank Joan Sarkin for help in preparing this manuscript. The work described in this review was supported by NSF grant PCM 80-14533 and NIH grant GM 24486.

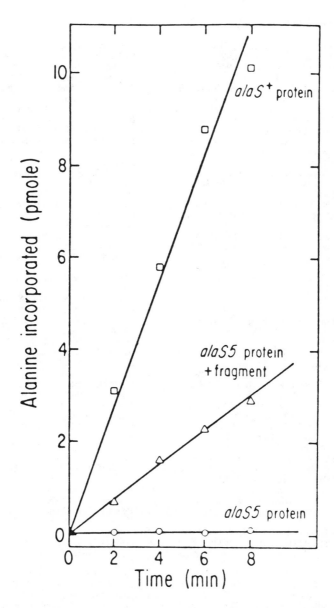

Figure 5   Aminoacylation activities of protein extracts from wild-type cells (alaS⁺), mutant cells (alaS5), and mutant cells with alaS5 and a fragment containing the C-terminal domain. Reproduced with permission from (68).

Literature Cited

1. Shortle, D., DiMaio, D., Nathans, D. 1981. *Ann. Rev. Genet.* 15:265–94
2. Zoller, M. J., Smith, M. 1983. *Methods Enzymol.* 100(B):468–500
3. Ackers, G. K., Smith, F. R., Turner, B. W. 1985. *Biophys. J.* In press
4. Fersht, A. R., Shi, J. P., Wilkinson, A. J., Blow, D. M., Carter, P., et al. 1984. *Angew. Chem.* 96:455–62
5. Winter, G., Fersht, A. R., Wilkinson, A. J., Zoller, M., Smith, M. 1982. *Nature* 299:756–58
6. Wilkinson, A. J., Fersht, A. R., Blow, D. M., Winter, G. 1983. *Biochemistry* 22:3581–86
7. Bhat, T. N., Blow, D. M., Brick, P., Nyborg, J. 1982. *J. Mol. Biol.* 122:407–19
8. Bhat, T. N., Blow, D. M., Brick, P., Nyborg, J. 1982. *J. Mol. Biol.* 158:699–709
9. Fersht, A. R. 1977. *Enzyme Structure and Mechanism.* San Francisco: Freeman
10. Wilkinson, A. J., Fersht, A. R., Blow, D. M., Carter, P., Winter, G. 1984. *Nature* 307:187–88
11. Carter, P. J., Winter, G., Wilkinson, A. J., Fersht, A. R. 1984. *Cell.* In press
12. Matthews, D. A., Alden, R. A., Bolin, J. T., Freer, S. T., Hamlin, R., et al. 1977. *Science* 197:452–55
13. Bolin, J. T., Filman, D. J., Matthews, D. A., Hamlin, R. C., Kraut, J. 1982. *J. Biol. Chem.* 257:13650–62
14. Matthews, D. A., Alden, R. A., Bolin, J. T., Filman, D. J., Freer, S. T., et al. 1978. *J. Biol. Chem.* 253:6946–54
15. Volz, K. W., Matthews, D. A., Alden, R. A., Freer, S. T., Hansch, C., et al. 1982. *J. Biol. Chem.* 257:2528–36
16. Villafranca, J. E., Howell, E. E., Voet, D. H., Strobel, M. S., Ogden, R. C., et al. 1983. *Science* 222:782–88
17. Sigal, I. S., Harwood, B. G., Arentzen, R. 1982. *Proc. Natl. Acad. Sci. USA* 79:7157–60
18. Dalbadie-McFarland, G., Cohen, L. W., Riggs, A. D., Morin, C., Itakura, K., et al. 1982. *Proc. Natl. Acad. Sci. USA* 79:6409–13
19. Fisher, J., Charnas, R. L., Bradley, S. M., Knowles, J. R. 1980. *Biochemistry* 20:2726–31
20. Shortle, D., Koshland, D., Weinstock, G. M., Botstein, D. 1980. *Proc. Natl. Acad. Sci. USA* 77:5375–79
21. Hall, A., Knowles, J. R. 1976. *Nature* 264:803–4
22. Charles, A. D., Gautier, A. E., Edge, M. D., Knowles, J. R. 1982. *J. Biol. Chem.* 257:7930–32
23. Kadogna, J. T., Gautier, A. E., Straus, D. R., Charles, A. D., Edge, M. D., et al. 1984. *J. Biol. Chem.* 259:2149–54
24. Pettigrew, D. W., Romeo, P. H., Tsapis, A., Thillet, J., Smith, M. L., et al. 1982. *Proc. Natl. Acad. Sci. USA* 79:1849–53
25. Empie, M. W., Laskowski, M. 1982. *Biochemistry* 21:2274–84
26. Nelson, H. C. M., Hecht, M. H., Sauer, R. T. 1983. *Cold Spring Harbor Symp. Quant. Biol.* 47:441–49
27. Hecht, M. H., Nelson, H. C. M., Sauer, R. T. 1983. *Proc. Natl. Acad. Sci. USA* 80:2676–80
28. Ho, C., Russu, I. M. 1985. In *New Methodologies in Studies of Protein Configuration,* ed. T. T. Wu. New York: Van Nostrand. In press
29. Wieczorek, M., Laskowski, M. 1983. *Biochemistry* 22:2630–36
30. Wyman, J. 1964. *Adv. Protein Chem.* 18:223–86
31. Weber, G. 1975. *Adv. Protein Chem.* 29:1–83
32. Ackers, G. K., Shea, M. A., Smith, F. R. 1983. *J. Mol. Biol.* 170:223–42
33. Laskowski, M., Kato, I. 1980. *Ann. Rev. Biochem.* 49:593–626
34. Fujinaga, M., Read, R. J., Sielecki, A., Ardelt, W., Laskowski, M., et al. 1982. *Proc. Natl. Acad. Sci. USA* 79:4868–72
35. Papamokos, E., Weber, E., Bode, W., Huber, R., Empie, M. W., et al. 1982. *J. Mol. Biol.* 158:515–37
36. Ogino, T., Croll, D. H., Kato, I., Markley, J. L. 1982. *Biochemistry* 21:2452–60
37. Nozaki, Y., Tanford, C. 1971. *J. Biol. Chem.* 246:2211–17
38. Edsall, J. T. 1972. *J. Hist. Biol.* 5:205–57
39. Baldwin, J. M. 1975. *Prog. Biophys. Mol. Biol.* 29:225–320
40. Benesch, R. E., Benesch, R. 1974. *Adv. Protein Chem.* 28:211–37
41. Matthew, J. B., Flanagan, M. A., Garcia-Moreno, E. B., March, K., Shire, S. J., et al. 1985. *CRC Crit. Rev. Biochem.* In press
42. Chu, A. H., Turner, B. W., Ackers, G. K. 1984. *Biochemistry* 23:604–17
43. Dickerson, R., Geis, I. 1982. *Hemoglobin,* pp. 43, 146. Menlo Park: Benjamin/Cummings
44. Perutz, M. F. 1970. *Nature* 228:726–33
45. Perutz, M. F. 1970. *Nature* 228:734–41
46. Ip, S. H. C., Johnson, M. L., Ackers, G. K. 1976. *Biochemistry* 15:654–60
47. Monod, J., Wyman, J., Changeux, J. P. 1965. *J. Mol. Biol.* 12:88–118

48. Koshland, D. E., Nemethy, G., Filmer, D. 1966. *Biochemistry* 5:364–85
49. Johnson, M. L., Turner, B. W., Ackers, G. K. 1984. *Proc. Natl. Acad. Sci. USA* 81:1093–97
50. Roughton, F. J. W., Lyster, R. L. J. 1965. *Hvalradets Skr.* 48:185–97
51. Ackers, G. K., Halvorson, H. R. 1974. *Proc. Natl. Acad. Sci. USA* 71:4312–15
52. Mills, F. C., Johnson, M. L., Ackers, G. K. 1976. *Biochemistry* 15:5350–62
53. Sack, J., Andrews, L., Magnus, K., Hanson, J., Rubin, J., et al. 1978. *Hemoglobin* 2:153–69
54. Arnone, A., Thillet, J., Rosa, J. 1981. *J. Biol. Chem.* 256:8545–52
55. Perutz, M. F., Fermi, G. 1981. *Hemoglobin and Myoglobin.* New York: Oxford Univ. Press
56. Baldwin, J., Chothia, C. 1979. *J. Mol. Biol.* 129:175–220
57. Flanagan, M. A., Ackers, G. K., Matthew, J. B., Hanania, G. I. H., Gurd, F. R. N. 1981. *Biochemistry* 20: 7439–49
58. Schimmel, P. R., Soll, D. 1979. *Ann. Rev. Biochem.* 48:601–48
59. Ofengand, J. 1977. In *Molecular Mechanism of Protein Biosynthesis,* ed. H. Weisback, S. Pestka. New York: Academic
60. Hall, C. V., van Cleemput, M., Muench, K. H., Yanofsky, C. 1982. *J. Biol. Chem.* 257:6132–36
61. Baldwin, A. N., Berg, P. 1966. *J. Biol. Chem.* 241:831–38
62. Yaniv, M., Gros, F. 1969. *J. Mol. Biol.* 44:1–15
63. Putney, S. D., Schimmel, P. R. 1981. *Nature* 291:632–35
64. Putney, S. D., Sauer, R. T., Schimmel, P. R. 1981. *J. Biol. Chem.* 256:198–204
65. Putney, S. D., Benkovic, S. J., Schimmel, P. R. 1981. *Proc. Natl. Acad. Sci. USA* 78:7350–54
66. Jasin, M., Regan, L., Schimmel, P. 1983. *Nature* 306:441–47
67. Jasin, M., Schimmel, P. R. 1984. In *Alfred Benzon Symp. 19,* ed. B. F. C. Clark, H. U. Petersen, pp. 223–35. Copenhagen: Munksgaard
68. Jasin, M., Regan, L., Schimmel, P. R. 1984. *Cell* 36:1089–95

Ann. Rev. Biochem. 1985. 54:631–64

ASSEMBLY OF ASPARAGINE-LINKED OLIGOSACCHARIDES

Rosalind Kornfeld and Stuart Kornfeld

Departments of Internal Medicine and Biochemisty, Washington University School of Medicine, St. Louis, Missouri 63110

CONTENTS

PERSPECTIVES AND SUMMARY

The subject of asparagine-linked oligosaccharide synthesis and processing was last reviewed in this series in 1981 by Hubbard & Ivatt (1). These authors concluded their excellent chapter with the following statement: "Although the main outlines of this pathway are now clear, several areas, including the detailed enzymology of the synthesizing and processing enzymes, the organization of these enzymes in the cell, the regulation of the pathway, and the

631

0066-4154/85/0701-0631$02.00

factors that direct the terminal processing reactions are only beginning to be understood."

During the past few years considerable progress has been made in each of these areas. In addition, much new information has accumulated on the occurrence of oligosaccharide processing in the animal and plant kingdoms and on the repertoire of asparagine-linked oligosaccharides that are synthesized in nature. In this review we have attempted to focus on this new information, particularly the factors that serve to control the extent of processing and thereby determine the final oligosaccharide structures that are assembled. We will begin with a brief review of asparagine-linked oligosaccharide structure, including examples of recently deduced structures. Due to space limitations we are not able to review the important advances that have been made in the area of oligosaccharide structural analysis and the determination of oligosaccharide conformation. The interested reader is referred to several recent reviews of these topics (2–5). In addition, the reader may wish to consult other recent reviews on the synthesis of asparagine-linked oligosaccharides (6–9a).

STRUCTURES OF ASPARAGINE-LINKED OLIGOSACCHARIDES

The advances in techniques for oligosaccharide structural analysis, including the use of 360 and 500 MHz ^{1}H-NMR, have made it possible to deduce the complete structure of hundreds of asparagine-linked oligosaccharides from a variety of plant and animal sources. When these structures are examined, it is evident that they fall into three main categories termed high mannose, hybrid, and complex. Typical examples of these oligosaccharides are shown in Figure 1. They all share the common core structure Manα1–3(Manα1–6) Manβ1–4GlcNAcβ1–4 GlcNAc-Asn, contained within the boxed area in Figure 1, but differ in their outer branches. The high mannose-type oligosaccharides typically have two to six additional mannose residues linked to the pentasaccharide core. The hybrid molecules are so named because they have features of both high-mannose and complex-type oligosaccharides. Most hybrid molecules contain a "bisecting" N-acetylglucosamine linked β1,4 to the β-linked mannose residue, although some exceptions exist (10–12). The complex-type structure shown in Figure 1 contains two outer branches with the typical sialyl lactosamine sequence and shows two other commonly found substituents, namely a fucose in α1,6 linkage to the innermost N-acetylglucosamine residue, and a "bisecting" N-acetylglucosamine linked β1,4 to the β-linked mannose residue. This complex-type structure may be modified both by the addition of extra branches on the α mannose residues or by the addition of extra sugar residues that elongate the outer chains. The majority of complex-type oligosaccharides contain two, three, or four outer branches, but units with five

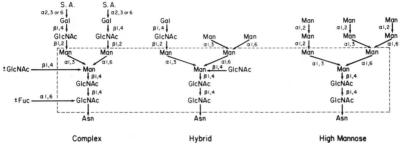

Complex   Hybrid   High Mannose

Figure 1   Structures of the major types of asparagine-linked oligosaccharides. The boxed area encloses the pentasaccharide core common to all N-linked structures.

outer branches have been found on avian ovomucoids (13–16). Figure 2 shows the variations in the substituents which have been found attached to outer chain N-acetylglucosamine residues of complex oligosaccharides. The basic Galβ1-4GlcNAc or lactosamine sequence may contain a fucose in α1,3 linkage to the GlcNAc, as for example in human lactotransferrin (17, 18), or a sialic acid in α2,6 linkage to galactose, as in human transferrin (19), but it cannot contain both substituents (20, 21). Alternatively the galactose may be substituted with a fucose residue in α1,2 or α1,6 linkage (2) or with another galactose residue in α1,3 linkage (22–24). An unusual polysialosyl sequence (NeuAcα2–8)$_n$ NeuAcα2–3Gal→ has been found in the complex glycopeptides of developing rat brain by Finne (25). The number of sialic acid residues varied from 8 to 12 in glycopeptides having 3 or 4 outer branches. Rothbard et al (26) have shown that the developmentally regulated neural cell adhesion molecule (N-CAM) from chicken brain is a glycoprotein containing 30% sialic acid in the embryonic form and only 10% in the adult form. This sialic acid probably occurs in polysialosyl chains since Finne et al (27) have shown that the analogous mouse brain cell surface protein (BSP-2) contains polysialosyl units of α2,8-linked sialic acid residues which decrease in both amount and chain length during embryonic to adult development. Polysialosyl glycopeptides have also been isolated from a neural tumor cell line (28). Another unusual outer chain sequence, shown in Figure 2, is the β1,3-linked Gal→GlcNAc disaccharide in which the GlcNAc bears a sialic acid substituent. This sequence has another sialic acid attached to the galactose through an α2,4 linkage in bovine serum fibronectin (29) and through an α2,3 linkage in the bovine coagulation factors IX, X, and prothrombin (30–32). It is obvious that a vast number of different oligosaccharide structures can be constructed by varying the number and type of outer chains, but in fact the three outer branches to the left in Figure 2 have been found to occur very frequently and those to the right more rarely.

Another outer branch structure consists of repeating lactosamine disaccharides (Galβ1–4GlcNAcβ1–3)$_n$. First discovered on erythrocyte membrane

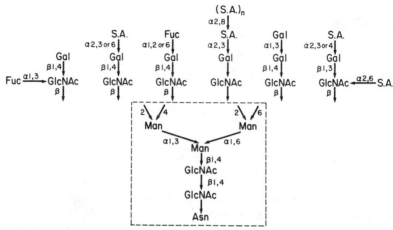

Figure 2   Various outer chain sequences found in complex-type oligosaccharides. The boxed area encloses the core region to which four outer chains may be attached.

glycoproteins (33, 34) and called erythroglycans, these polylactosamine oligosaccharides may be substituted in various ways and have been shown to carry ABH and Ii blood group antigens (33, 35). Fukuda et al. (36, 37) have deduced the complete structure of the biantennary lactosaminoglycans derived from the Band 3 glycoprotein of fetal and adult erythrocytes and shown that the adult form (I) contains chains with multiple branches of the type

$$\text{Gal}\beta 1\text{–}4\text{GlcNAc}\beta 1$$
$$(\text{Gal}\beta 1\text{–}4\text{GlcNAc}\beta 1\text{–}3)_n^6 \, \text{Gal}\beta 1\text{–}4\text{GlcNAc}\beta 1\text{–}3$$

whereas the fetal form (i) has unbranched chains. Spooncer et al (38) have isolated a tetraantennary neutral lactosaminoglycan from human granulocyte glycoproteins in which the *N*-acetylglucosamine residues of the lactosamine chain are substituted with fucose to form the antigenic determinant Galβ1–4(Fucα1–3)GlcNAc expressed on granulocytes (39). This determinant is also expressed on other cell types of human and murine origin, including the murine preimplantation embryo where it has been detected as a stage specific embryonic antigen (SSEA-1) (39–41). Galβ1–4GlcNAcβ1–3, sulfated on C-6 of the GlcNAc and some of the Gal residues, is also the repeating disaccharide of corneal keratan sulfate, the only proteoglycan known to be derived from a complex type of N-linked oligosaccharide precursor (42–47).

All of the asparagine-linked oligosaccharide structures have the common pentasaccharide core structure (see Figure 1) because they all arise from the same biosynthetic precursor lipid-linked oligosaccharide which is transferred to nascent peptide chains and then processed to form these various structures.

ASSEMBLY AND TRANSFER OF THE LIPID-LINKED OLIGOSACCHARIDE

Assembly

The steps in the synthesis of the lipid-linked oligosaccharide precursor

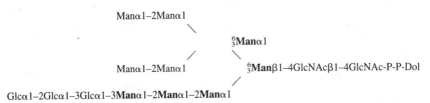

have been reviewed (1, 6, 9). The oligosaccharide is assembled in the endoplasmic reticulum on the lipid carrier dolichol phosphate. The sugars are added in a stepwise fashion with the first seven sugars (two GlcNAc and five Man residues) derived from the nucleotide sugars, UDP-GlcNAc and GDP-Man (**Man** residues in bold type) whereas the next seven sugars (four Man and three Glc residues) are derived from the lipid intermediates, Dol-P-Man and Dol-P-Glc. While it has been established that protein glycosylation occurs in the lumen of the rough endoplasmic reticulum (48–53), the site of lipid-linked oligosaccharide assembly is not fully understood. The problem centers around the fact that the nucleotide sugar donors are unable to enter the lumen of the RER (54, 55). Several studies have attempted to define the membrane orientation of the lipid-linked oligosaccharide intermediates and the enzymes involved in their synthesis. Snider and coworkers (56, 57) used the lectin Con A to probe the orientation of the lipid-linked oligosaccharide precursors in ER-derived vesicles. Their data indicate that the Man$_{3-5}$GlcNAc$_2$ species face the cytoplasm whereas the Man$_{6-9}$GlcNAc$_2$ and Glc$_{1-3}$Man$_9$GlcNAc$_2$ species face the lumen of the ER. This suggests that the Man$_5$GlcNAc$_2$ species is assembled on the cytoplasmic side of the ER membrane and then is translocated to the lumenal side where assembly is completed and transfer to protein occurs. Since dolichol derivates do not translocate spontaneously across the lipid bilayer (58), it is likely that the translocation is protein mediated. This pathway fits nicely with the fact that the Man$_5$GlcNAc$_2$ intermediate is assembled directly from nucleotide sugars. However, this scheme also requires that Dol-P-Man and Dol-P-Glc face the lumen of the ER even though these intermediates are synthesized from cytoplasmic nucleotide sugars by the reaction: Dol-P + XDP-Sugar → Dol-P-Sugar + XDP. Haselbeck & Tanner (59) have proposed a solution to this problem. They have shown that yeast Dol-P-Man synthetase, when incorporated into an artificial membrane, can catalyze the transmembrane movement of Dol-P-Man. In this way the lipid-linked monosaccharides could be synthesized from nucleotide sugars on the cytoplasmic face and then translocated to the lumenal face where they could serve as donors for the elongation of the $Man_5GlcNAc_2$ intermediate. Many of the enzymes involved in both the early and late stages of lipid-linked oligosaccharide assembly have protease-sensitive sites on the cytoplasmic side of microsomal membranes (55, 60–63), although the $Man_5GlcNAc_2$ translocation mechanism predicts that early enzymes of this pathway would be protease-sensitive but late enzymes, e.g., the glucosyltransferases, would not be. This discrepancy can be explained by proposing that the biosynthetic enzymes are in a transmembrane orientation (55). Also at variance with the model is the observation by Hanover & Lennarz (55, 64) that the intermediate $GlcNAc_2$-P-P-Dol in sealed hen oviduct microsomes is inaccessible to a soluble galactosyltransferase. In contrast, exogenous $GlcNAc_2$-P-P-Dol added to the microsomes was accessible to the galactosyltransferase and did not become inaccessible as a function of time. Possibly the endogenous intermediate is localized within the membrane in a manner which makes it inaccessible to the probe, but still capable of being elongated at the cytoplasmic face by an intramembraneous mannosyltransferase.

Transfer-Oligosaccharide Structural Requirements

Several studies (65–72) have shown that the glucose residues on the lipid-linked oligosaccharide facilitate the in vitro transfer of the oligosaccharide to protein, but the presence of glucose residues is not an absolute requirement for transfer. Robbins and coworkers have shown that yeast mutants defective in various steps in the synthesis of the lipid-linked oligosaccharide transfer nonglucosylated oligosaccharides ranging in size from $Man_9GlcNAc_2$ to $Man_{1-2}GlcNAc_2$ to protein (73, 74). The transfer of $Man_{1-2}GlcNAc_2$ is consistent with previous in vitro studies (71, 72, 75–78). The protozoa *Trypanosoma cruzi* (79–81), *Crithidia fasciculata* (82, 83), and *Leishmania mexicana* (84) synthesize nonglucosylated lipid-linked oligosaccharides which are transferred directly to protein. Interestingly, in *Trypanosoma cruzi,* a transient glucosylation occurs after the oligosaccharide is bound to protein (80, 81). It is not known if the oligosaccharyltranferase of higher organisms will transfer nonglucosylated lipid-linked oligosaccharides. A Con A–resistant cell line that cannot glucosylate lipid-linked oligosaccharides has been isolated, but transfer of this oligosaccharide to protein has not been demonstrated (85).

Hoflack et al (86) have proposed that glucosylation protects the lipid-linked oligosaccharides from degradation. They studied intact rat spleen lymphocytes which incorporate nucleotide sugars into lipid-linked oligosaccharides and found that the nonglucosylated lipid-linked oligosaccharides were selectively degraded by a phosphodiesterase.

While glucose residues facilitate oligosaccharide transfer to protein in some cell types, the number of mannose residues has little effect on protein glycosylation. In protozoa the completed lipid-linked oligosaccharides contain nine mannoses (79), seven mannoses (82, 83) and six mannoses (84). The green flagellate *Volvox carteri* synthesizes a lipid-linked oligosaccharide with the composition $Glc_1Man_5GlcNAc_2$ which is transferred to protein (87, 88). The oligosaccharyltransferase of higher organisms also appears to be indifferent to the number of mannose residues on the lipid-linked oligosaccharide precursor. Removal of mannose residues from the $Glc_3Man_9GlcNAc_2$ precursor with α-mannosidase does not impair transfer to protein (66, 69). Mutant cell lines that synthesize truncated lipid-linked oligosaccharides with five or seven mannoses transfer these species to protein (89–93). Cells starved of glucose (94–96) or treated with CCCP, an uncoupler of oxidative phosphorylation (97), synthesize lipid-linked $Glc_3Man_5GlcNAc_2$ which is transferred to protein. The effects seen with glucose starvation or energy depletion vary with the cell type and cell density. Spiro et al (98) found that energy deprivation of thyroid slices resulted in depletion of $Glc_3Man_9GlcNAc_2$-P-P-Dol, an accumulation of the $Man_9GlcNAc_2$ lipid-linked species and a concomitant decrease in protein glycosylation. In this system, the $Man_9GlcNAc_2$-P-P-Dol was stable.

Transfer-Role of Peptide Acceptor

Glycosylated asparagines almost always occur in the sequence Asn-X-Ser/Thr where X can be any amino acid except possibly proline and aspartic acid (99, 100). There is one report of glycosylation at an Asn-Ala-Cys sequence (100a). In vitro glycosylation with synthetic peptides has shown that the Asn-X-Ser/Thr sequence is sufficient for oligosaccharide transfer although glycosylation is not efficient unless both the α-amino group of asparagine and the α-carboxyl moiety of the hydroxyl amino acid are blocked (71, 72, 101–105). Increasing the acceptor chain length to a hexa- or octapeptide greatly enhances the rate of oligosaccharide transfer, possibly by allowing the formation of a higher order structure which appears to facilitate the reaction (106, 107).

Studies by Bause & Legler on the role of the hydroxyl amino acids in the glycosylation reaction have provided important insights into the mechanism of the reaction and the requirement of a favorable conformation (108). Using a series of hexapeptides derived from Tyr-Asn-Gly-X-Ser-Val in which X was varied, they found that threonine-, serine- and cysteine-containing derivatives could be glycosylated, but at very different rates (threonine>serine>cysteine) whereas valine and O-methylthreonine analogs were inactive. They concluded that there is an absolute requirement in the glycosylation reaction mechanism for formation of a hydrogen bond in the side chain of the hydroxyl amino acid. A model was proposed in which a hydrogen-bond interaction between the amide of asparagine (the hydrogen-bond donor) and the oxygen of the hydroxyl group of the hydroxyl amino acid (the hydrogen acceptor) increased the nucleophilicity of the amide electron pair, resulting in a higher reactivity toward the glycosyl donor. By analyzing space-filling models, Bause found that β-turns or loops represent spatial arrangements of the peptide chain that favor the required hydrogen-bonded contacts (109). Proline-containing peptide analogs that did not serve as acceptors were unable to achieve a conformation that allowed the necessary hydrogen bonding. The relevance of these findings to the in vivo situation is supported by statistical studies of naturally occurring glycoproteins which indicate that most N-glycosylated asparagines are located in peptide segments that favor the formation of β-turns (110, 111).

Lau et al (112) examined the effect of synthetic peptides on the transfer of oligosaccharide to nascent polypeptides during coupled translation/glycosylation in a reticulocyte lysate, dog pancreas microsome system. They found that peptides that were acceptors acted as competitive inhibitors of nascent chain glycosylation, but did not affect translocation of the nascent chains. Nonacceptor peptides had no effect, indicating that the inhibition was a result of competition for the active site of the oligosaccharyltransferase.

An examination of protein sequences has revealed that only about one third of the potential Asn-X-Ser/Thr sites in proteins are actually glycosylated (113). In addition, while glycosylation in general is highly efficient, the process is incomplete at some Asn-X-Ser/Thr sites (114–116). It would appear that the ability of the tripeptide signal to achieve a favorable conformation in the presence of short-range interactions with neighboring amino acids may have a significant effect on the rate and extent of glycosylation. Such favorable conformations would be needed to provide the correct hydrogen bonding as well as adequate accessibility to the oligosaccharyltransferase. Since glycosylation occurs cotranslationally, the asparagine that is to be glycosylated is part of a growing peptide chain that is in the process of folding. Consequently, the period of time during which glycosylation can occur may be quite brief. Once the protein has folded, potential glycosylation sites are no longer accessible to the oligosaccharyltransferase (117). Another factor in determining the level of glycosylation is the availability of the lipid-linked oligosaccharide donor. Carson et al found that added Dol-P increased the glycosylation of secreted RNase from 12% to 90% in bovine pancreas tissue slices (118). Thus, Dol-P availability may regulate asparagine-linked glycosylation under some circumstances. In other instances, such as the regenerating rat liver (118a), and hormone treated thyroid (118b) and oviduct tissue (118c), the level of oligosaccharyltransferase activity may control the extent of protein glycosylation. The oligosaccharyltransferase has been partially purified by two groups (119, 120).

In summary, the efficient glycosylation of proteins is dependent on a sufficient pool of completely assembled and glucosylated lipid-linked oligosaccharide donor, an adequate activity of oligosaccharyltransferase, and a properly oriented and accessible Asn-X-Ser(Thr) sequence in the acceptor.

OLIGOSACCHARIDE PROCESSING

Sequence of Processing

The enzymatic pathway of oligosaccharide processing and the subcellular location of the various reactions in that sequence are schematically depicted in Figure 3. The precursor $Glc_3Man_9GlcNAc_2$ is transferred from the lipid donor to an asparagine in a nascent polypeptide during its vectorial transport across the membrane of the RER (reaction 1). Processing is initiated by the removal of the terminal glucose residue by a specific $\alpha 1,2$ glucosidase I (reaction 2). The two inner glucose residues are then removed (reaction 3) by a single $\alpha 1,3$ specific glucosidase II. These glucosidases are located in the membranes of the RER as is a specific α-mannosidase which catalyzes the removal of at least one $\alpha 1,2$-linked mannose residue (reaction 4). These processing events have been shown by Atkinson & Lee (121) to occur cotranslationally (i.e., on polysome-bound nascent polypeptides) for the viral membrane glycoprotein (G protein) of vesicular stomatitis virus-infected HeLa cells. Both integral membrane and secreted glycoproteins undergo this same sequence of events in the RER, but with various glycoproteins some or all of the steps, including the glycosylation event, may occur after the polypeptide chain is completely synthesized (122–123a). Resident ER glycoproteins such as HMG CoA reductase (124) and the ribophorins (121) and glycoproteins which exit the ER only after a long lag time (125, 126) may undergo further mannose trimming to give rise to $Man_6GlcNAc_2$ species. Parodi and his coworkers have shown that protein-bound nonglucosylated high mannose units may be transiently reglucosylated by an ER glucosyltransferase (127). This reaction has been demonstrated in mammalian, avian, protozoan, and plant cells but is not detectable in yeast (81, 127, 128). While no function for this reglucosylation has been demonstrated, Parodi has speculated that it could serve to protect protein-linked oligosaccharides from extensive degradation by the ER mannosidase. With the exception of glycoproteins that are permanent residents of the ER membrane, the newly synthesized glycoproteins are next transported to the cis Golgi cisternae by means of vesicles which are believed to bud from the RER and then fuse with the Golgi membrane (129). The rate at which these proteins exit the RER varies widely depending on the specific glycoprotein and the tissue examined (126, 130, 131). This variation in the rate of exit from the ER of different glycoproteins in the same cell suggests that the vesicular transport may involve a receptor-mediated process rather than simple pinching off of bulk phase lumenal contents. 1-Deoxynojirimycin, a specific inhibitor of glucosidases I and II, delays the exit of some, but not all, glycoproteins from the RER (132–134). On this basis it has been postulated that glucose trimming is necessary for efficient movement from the RER to the Golgi, possibly because the deglucosylated oligosaccharides form part of a recognition site for a transport receptor for certain secretory proteins (132). An alternative explanation for these findings is that glucose removal is necessary for the glycoprotein to mature to a correct functional conformation (135).

When the glycoproteins arrive in the Golgi, they traverse the stack from the

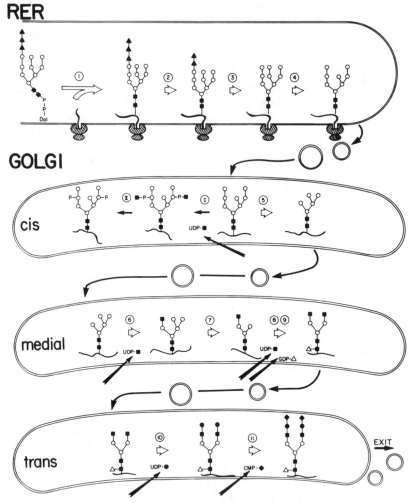

Figure 3 Schematic pathway of oligosaccharide processing on newly synthesized glycoproteins. The reactions are catalyzed by the following enzymes: (1) oligosaccharyltransferase, (2) α-glucosidase I, (3) α-glucosidase II, (4) ER α1,2-mannosidase, (I) *N*-acetylglucosaminylphosphotransferase, (II) *N*-acetylglucosamine-1-phosphodiester α-*N*-acetylglucosaminidase, (5) Golgi α-mannosidase I, (6) *N*-acetylglucosaminyltransferase I, (7) Golgi α-mannosidase II, (8) *N*-acetylglucosaminyltransferase II, (9) fucosyltransferase, (10) galactosyltransferase, (11) sialyltransferase. The symbols represent: ■, *N*-acetylglucosamine; ○, mannose; ▲, glucose; △, fucose; ●, galactose; ◆, sialic acid.

cis through medial to trans cisternae by vesicular transport (136–138). A special subset of glycoproteins, the lysosomal enzymes, undergo a highly specific mannose phosphorylation catalyzed by two enzymes, N-acetylglucosaminylphosphotransferase (reaction I) and N-acetylglucosamine-1-phosphodiester α-N-acetylglucosaminidase (reaction II), which have been localized to a dense Golgi membrane subfraction presumed to correspond to the cis Golgi cisternae. The high mannose oligosaccharides on nonlysosomal glycoproteins can be further trimmed by the Golgi α1,2 mannosidase (Golgi mannosidase I) to yield a $Man_5GlcNAc_2$ structure (reaction 5). This reaction is believed to occur in either the cis or the medial cisternae (137, 138). In the medial cisternae, those oligosaccharide chains destined to become complex-type structures are further processed by addition of a N-acetylglucosamine residue catalyzed by N-acetylglucosaminyltransferase I (reaction 6), followed by the removal of two mannose residues by Golgi α-mannosidase II (reaction 7) and the subsequent addition of another outer chain N-acetylglucosamine residue catalyzed by N-acetylglucosaminyltransferase II (reaction 8). At this stage fucosyltransferase may act to transfer a fucose residue to the innermost GlcNAc residue on the oligosaccharide (reaction 9). The final steps of complex oligosaccharide synthesis occur in the trans Golgi cisternae and consist of addition of outer chain galactose and sialic acid residues catalyzed by galactosyl- and sialyl-transferases in reactions 10 and 11. Other terminal sugar additions (Figure 2) must also occur at this late stage. The newly synthesized glycoproteins then exit the Golgi and are transported to their final destination.

The specific localization of the processing enzymes to the RER or the cis, medial, or trans Golgi cisternae provides a mechanism for controlling their sequential action on newly synthesized glycoproteins in both space and time. The final oligosaccharide structure assembled on a glycoprotein is dictated to a large extent by the order in which that glycoprotein encounters the processing glycosidases and glycosyltransferases, and their specificity. These factors are reviewed in more detail below.

Subcellular Localization of Processing Enzymes

A variety of techniques have been employed to localize oligosaccharide processing reactions to specific intracellular organelles. Subcellular fractionation by density gradient centrifugation provides separation of the RER membranes from the Golgi complex membranes and on this basis many of the processing enzymes have been localized to one or the other compartment. Finer mapping of some of the Golgi-associated enzymes to specific regions of the Golgi stack has been achieved more recently. The most definitive method to emerge involves immunocytochemical labeling of ultrathin sections to localize specific proteins within the cell at the electron microscope level. Using colloidal gold particles coated with Protein A as the marker, Roth & Berger (139) showed that galactosyltransferase was localized in only two or three trans cisternae of the Golgi stacks in thin sections of HeLa cells pretreated with antibody to galactosyltransferase. Slot & Geuze (140), using immunogold labeling of ultrathin cryosections, also observed that galactosyltransferase was confined to the trans-most cisternae in human hepatoma cells. Dunphy & Rothman (141), using horseradish peroxidase coupled to Protein A as the marker to detect bound antibody on thin sections, have localized N-acetylglucosaminyltransferase I to the medial Golgi cisternae. Novikoff et al (142), using a similar method, found Golgi α-mannosidase II distributed throughout the Golgi stack and in portions of the ER in rat liver. Other approaches have provided convincing, albeit less definitive, evidence for the intracellular localization of a number of these enzymes. Dunphy et al (143) devised a sucrose density gradient centrifugation method that gave partial separation of Chinese hamster ovary cell membranes containing the early-acting α-mannosidase I from the late-acting galactosyl- and sialyltransferases, and proposed that the enzymes were in separate compartments of the Golgi. Dunphy & Rothman (144) extended this study to show that N-acetylglucosaminyltransferases I and II and α-mannosidase II cofractionate with α-mannosidase I. Using a modification of these workers' sucrose gradient technique to fractionate the membranes of cultured mouse lymphoma cells, Goldberg & Kornfeld (145) observed that Golgi associated enzyme activities could be partially separated into four regions on the gradient. Going from most to least dense these regions contained (1) N-acetylglucosaminylphosphotransferase, (2) N-acetylglucosamine-1-phosphodiester α-N-acetylglucosaminidase (3) N-acetylglucos aminyltransferases I and V, α-mannosidase II, and fucosyltransferase, and (4) galactosyltransferase. Deutscher et al (146) were able to separate rat liver Golgi membranes containing N-acetylglucosamine-1-phosphotransferase from those enriched in N-acetylglucosamine-1-phosphodiester α-N-acetyl glucosaminidase.

Other investigators have mapped the subcellular distribution of glycoprotein oligosaccharides by using lectins that react with specific carbohydrate structures. For example, using the immunogold technique, Griffiths et al (147) found that the galactose binding lectin RCA-I only labeled Golgi cisternae in the middle and on the trans side of the stack in frozen thin sections of virus-infected cells. They concluded that the viral membrane proteins acquire galactose in the trans Golgi cisternae. Using lightly fixed saponinpermeabilized, IgM-secreting myeloma cells, Tartakoff & Vassalli (148) found that peroxidase-conjugated Concanavalin A (Con A), which preferentially binds to high mannose oligosaccharides, stained the nuclear envelope, the RER, and the proximal (or cis) Golgi cisternae but not the cell surface. In contrast, peroxidase-conjugated WGA, which binds to sialic acid and outer chain N-acetylglucosamine residues of complex-type oligosaccharides, stained the more distal (or trans) Golgi cisternae and associated vesicles and the cell surface, but not the nuclear envelope, RER, or cis Golgi cisternae. Although such lectin studies only indirectly localize the processing enzymes by mapping their oligosaccharide products, they are entirely consistent with the idea that these enzymes have a restricted distribution in the Golgi cisternae that fits the sequential processing of glycoproteins moving from the cis to trans face of the stack. Based on studies employing a combination of mapping techniques in conjunction with the use of monensin to inhibit glycoprotein transport through the Golgi, Griffiths et al (149) concluded that the Golgi stack can be divided into three functionally distinct compartments, the cis, medial, and trans. Rothman and his coworkers have provided evidence for this Golgi compartmentation using a different approach. They followed the transport of the VSV G protein from one Golgi compartment to another by monitoring its acquisition of outer chain sugar residues. The initial in vitro studies (150, 151) employed Golgi membranes from VSV-infected mutant cells lacking N-acetylglucosaminyl transferase I, mixed with Golgi membranes from uninfected wild-type cells and showed that the G protein acquired outer chain N-acetylglucosamine presumably due to transfer of the G protein between Golgi subcompartments. In subsequent in vivo studies (137) the same result occurred when pulse-labeled VSV-infected mutant cells were fused with wild-type cells. In similar experiments (138) the G protein from VSV infected cells deficient in galactosylation or in sialylation were shown to acquire galactose or sialic acid residues after cell fusion with uninfected cells having that capacity. These intercompartment transfers were very efficient. Importantly, the transport of a cohort of labeled G protein from one compartment to another in fused cells only occurred during a brief interval after the pulse. If fusion was delayed so that transport had already occurred to the next compartment in the host cell, G protein could not acquire the later sugars in the Golgi of the fusion partner. The authors concluded that transfer is a dissociative process most probably mediated by vesicles budding from the rims of the donor Golgi cisterna and moving to and fusing with the target cisterna. These experiments further support the idea that there are three functionally distinct Golgi compartments, the cis, medial, and trans, as shown in Figure 3.

Processing in Lower Organisms

Plant cells synthesize lipid-linked $Glc_3Man_9GlcNAc_2$ which is transferred to protein (69, 152, 153). In soybean cells, the protein-bound oligosaccharide is processed by the removal of the glucose residues and up to four mannose residues (154). In the case of soybean agglutinin only glucose is removed to form the $Man_9GlcNAc_2$ oligosaccharide found on the mature agglutinin (155). Vitale & Chrispeels have studied the biosynthesis of phytohemagglutinin (PHA) in cotyledons of the common bean (*Phaseolus vulgaris* L.) (156). The mature protein contains two Asn-linked oligosaccharides, one with the composition $Man_{8-9}GlcNAc_2$ and the other with the composition $Xyl_{0.5}Fuc_{0.6}Man_{3.8}GlcNAc_2$. This latter oligosaccharide is derived from a high mannose unit which becomes resistant to endo H after the protein moves from the ER to the Golgi, indicating that the late stage processing events occur in this organelle. The oligosaccharide also acquires one or more outer GlcNAc residues which are subsequently removed when the PHA arrives at its destination in the protein body, a lysosome-like organelle. Pineapple stem bromelein also has an N-linked oligosaccharide-containing xylose, fucose, mannose, and N-acetylglucosamine which presumably is formed by processing of a high mannose precursor (157).

In yeast, the precursor $Glc_3Man_9GlcNAc_2$ is trimmed to $Man_8GlcNAc_2$ by α-glucosidases I and II (158) and an $\alpha1,2$ mannosidase localized in the ER (152, 159–161a). In the Golgi body the $Man_8GlcNAc_2$ species is then elongated by the action of a series of mannosyltranferases to form the large mannan oligosaccharides found on yeast mannoproteins. These structures usually contain 50–150 mannose residues and some of them have Man-P-Man sequences (162, 163). Mosquito cells trim the $Glc_3Man_9GlcNAc_2$ precursor all the way down to a $Man_3GlcNAc_2$ structure (164), but are unable to form complex-type oligosaccharides (164, 165). *Trypanosoma cruzi* cells trim the $Glc_3Man_9GlcNAc_2$ precursor to a mixture of high mannose species, the smallest of which is $Man_6GlcNAc_2$ (81).

Specificity of Processing Enzymes

EARLY ACTING ENZYMES   It is apparent that many thousands of different oligosaccharide structures could be assembled from the combined and sequential action of the processing enzymes. Yet in fact the synthesis of only a limited number of structures is observed. This is due to the rigid substrate specificity of these enzymes. Since the earlier reviews (1, 7, 8) include detailed discussions of prior studies on the specificity of the various processing enzymes, this account will emphasize the more recent findings.

Glucosidase I has been purified from calf liver (166) while glucosidase II has been purified from rat liver and kidney (167, 168). Evidence that glucosidase II removes both $\alpha1,3$ glucose residues from $Glc_2Man_9GlcNAc_2$ has come from studies with the purified enzyme and from the finding that extracts of a glucosidase II–deficient mouse lymphoma cell line are unable to remove glucose from either $Glc_1Man_9GlcNAc_2$ or $Glc_2Man_9GlcNAc_2$ (169). Burns & Touster (167) showed that glucosidase II action on the artificial substrate p-nitrophenyl-α-glucoside was activated by 2-deoxyglucose and mannose whereas those compounds inhibited its action on $Glc_{1-2}Man_9GlcNAc$. Based on their observation and the earlier findings of others (170–172) that glucosidase II splits $Glc_{1-2}Man_9GlcNAc$ much better than $Glc_{1-2}Man_7GlcNAc$ or

$Glc_{1-2}Man_4GlcNAc$, they suggested that glucosidase II has, in addition to its catalytic site, a binding site for one or more $\alpha 1,2$ mannose residues. Such a site would provide the basis for very high affinity binding to the physiological substrate. The finding that a mutant cell line that synthesizes a truncated lipid-linked oligosaccharide with the structure $Glc_3Man_5GlcNAc_2$ is able to process this oligosaccharide to typical complex-type species demonstrates that glucosidase II can act on such structures in vivo (173).

The other processing enzyme in the RER is an α-mannosidase which removes $\alpha 1,2$ mannose residues from $Man_9GlcNAc$ and less well from $Man_{6-8}GlcNAc$ (174). This enzyme has been solubilized from rat liver ER and differs from the two Golgi processing α-mannosidases in its substrate specificity and its failure to bind to Con A-Sepharose. However, its properties are virtually identical to the "cytosolic" α-mannosidase (175), suggesting that the latter enzyme may be derived by proteolytic cleavage of the ER α-mannosidase. Yeast cell extracts contain a very specific $\alpha 1,2$ mannosidase that has similarities to mammalian cell ER $\alpha 1,2$ mannosidase (159). The yeast enzyme removes a single mannose residue from the middle branch of $Man_9GlcNAc$ to produce the $Man_8GlcNAc$ isomer shown in Figure 3, which is the precursor to the elongated polymannose chains of yeast mannoprotein (163). High field proton NMR studies, which indicated that the $\alpha 1,6$-linked branch has a different conformation in this $Man_8GlcNAc$ than in $Man_9GlcNAc$, led Byrd et al (159) to propose that the altered conformation could facilitate the subsequent addition of mannose to the $\alpha 1,6$ branch. Interestingly, rat liver Golgi α-mannosidase I, although capable of removing all four $\alpha 1,2$ mannose residues from $Man_9GlcNAc$ in vitro (176, 177), works better on the $Man_8GlcNAc$ shown in Figure 3 and its pattern of mannose removal from $Man_9GlcNAc$ generates $Man_8GlcNAc$, $Man_7GlcNAc$, and $Man_6GlcNAc$ isomers that retain the $\alpha 1,2$ mannose residue on the middle branch (176).

The particular isomers of the Man_8-, Man_7-, and $Man_6GlcNAc_2$ oligosaccharides found on mature glycoproteins is a reflection of the pattern of mannose removal they underwent during processing. That pattern, being determined by the specificities of the α-mannosidases responsible for the trimming, should reflect the extent of mannose removal that occurred in each processing compartment. In general, the smallest high mannose oligosaccharide found on mature glycoproteins is the $Man_5GlcNAc_2$ from which all $\alpha 1,2$ mannose residues have been removed. This oligosaccharide is the preferred substrate in vitro for N-acetylglucosaminyltransferase I purified from bovine colostrum (178) and rabbit liver (179) but $Man\alpha 1-3(Man\alpha 1-6)Man\beta 1-4$ $GlcNAc_2$-Asn will accept as will the trisaccharide $Man\alpha 1-3Man\beta 1-4GlcNAc$ at much higher concentrations (180). The first committed step in complex oligosaccharide synthesis is catalyzed by Golgi α-mannosidase II which removes the terminal $\alpha 1,3$ and $\alpha 1,6$ linked mannose residues from $GlcNAcMan_5$

GlcNAc$_2$-Asn (177, 181–183). This enzyme has a very strict substrate specificity and cannot remove those same mannose residues from Man$_5$GlcNAc or from GlcNAc(GlcNAc)Man$_5$GlcNAc$_2$-Asn in which a "bisecting" GlcNAc residue has been attached to the core β-mannose residue. This latter reaction is catalyzed by N-acetylglucosaminyltransferase III and represents a committment to hybrid oligosaccharide synthesis. Harpaz & Schachter (182) suggested that the relative abundance of α mannosidase II and N-acetylglucosaminyltransferase III in a given tissue could determine whether it synthesized complex or hybrid oligosaccharides. In a latter study (184) it was shown that membranes from hen oviduct, which produce ovalbumin containing hybrid oligosaccharides, catalyzed the in vitro conversion of GlcNAcMan$_5$GlcNAc$_2$-Asn to hybrid oligosaccharides containing the "bisecting" GlcNAc residue. The plant alkaloid swainsonine, which blocks the synthesis of complex oligosaccharides in vivo (185–189) causing accumulation of GlcNAcMan$_5$-GlcNAc$_2$, inhibits purified rat liver Golgi α-mannosidase II in vitro (183).

CHAIN ELONGATING ENZYMES   The product of α mannosidase II action, GlcNAcMan$_3$GlcNAc$_2$-Asn, serves as the substrate for N-acetylglucosaminyltransferase II catalyzed addition of an N-acetylglucosamine residue in β1,2 linkage to the terminal Manα1,6 residue (178, 190, 191). This transferase cannot act on Man$_3$GlcNAc or Man$_5$GlcNAc. Its product, GlcNAc$_2$-Man$_3$GlcNAc$_2$-Asn, is the precursor to complex oligosaccharides having two, three, or four outer branches. If N-acetylglucosaminyltransferase III acts on the product to attach a "bisecting" GlcNAc residue to the β mannose then the oligosaccharide cannot be further branched (192). Campbell & Stanley (193) have described a lectin resistant CHO cell line that has induced expression of N-acetylglucosaminyltransferase III, an enzyme activity undetectable in the parental or a revertant cell line. A high proportion of the oligosaccharide chains on VSV G protein synthesized by the mutant were biantennary complex-type structures containing a "bisecting" GlcNAc residue, whereas no such structures were found on G protein synthesized in parental cells.

In the absence of prior action by N-acetylglucosaminyltransferase III on GlcNAc$_2$Man$_3$GlcNAc$_2$-Asn, other N-acetylglucosaminyltransferases can act to add N-acetylglucosamine in β1,4 linkage to the α1,3 mannose (transferase IV) or in β1,6 linkage to the α1,6 mannose (transferase V) to produce structures with three or four outer branch points. Schachter et al (7) have reviewed the substrate specificities known for a number of these N-acetylglucosaminyltransferases and point out that they often compete for a common substrate and the order in which specific GlcNAc residues are attached can determine the subsequent route of synthesis. This occurs because transfer of a GlcNAc to a particular linkage in an oligosaccharide can convert it from being a good substrate for a subsequent transferase to being a poor substrate and vice versa. This situation also obtains for the action of the fucosyl and galactosyl-transferases. N-acetylglucosaminyltransferase IV from hen oviduct (194) can transfer GlcNAc to $GlcNAc_2Man_3GlcNAc_2$-Asn in vitro to give

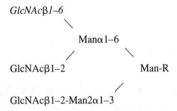

when a Gal residue is on the GlcNAcβ1–2Manα1–6 arm but not when the Gal residue is on the GlcNAcβ1–2Manα1–3 arm. N-acetylglucosaminyltransferase V from mouse lymphoma cells (195) acts in vitro on $GlcNAc_2Man_3GlcNAc_2$-Asn to give

$$
\begin{array}{c}
GlcNAc\beta1{-}6 \\
\searrow \\
Man\alpha1{-}6 \\
\diagup \qquad \searrow \\
GlcNAc\beta1{-}2 \qquad\qquad Man\text{-}R \\
\diagup \\
GlcNAc\beta1{-}2{-}Man2\alpha1{-}3
\end{array}
$$

and cannot act on $Gal_2GlcNAc_2Man_3GlcNAc_2$-Asn. Since galactosyltrans-ferase is localized to a later Golgi compartment than the N-acetylglucosaminyl-transferases, galactosylated oligosaccharides are probably not encountered by N-acetylglucosaminyltransferases in vivo.

The fucosyltransferase that transfers fucose in α1,6 linkage to the N-acetylglucosamine residue attached to asparagine acts on oligosaccharides with the structure

$$
\begin{array}{c}
R\text{-}Man\alpha1{-}6 \\
\searrow \\
Man\beta1{-}4GlcNAc\beta1{-}4GlcNAc\text{-}Asn \\
\diagup \\
GlcNAc\beta1{-}2Man\alpha1{-}3
\end{array}
$$

where R is H, GlcNAcβ1–2, or Manα1–6(Manα1–3), but not on substrates where the β Man carries a "bisecting" N-acetylglucosamine residue or when GlcNAcβ1–2 on both arms is substituted with galactose (196). The require-ment of the fucosyltransferase for the GlcNAcβ1–2Manα1–3 sequence ex-plains why most high mannose oligosaccharides do not have fucose linked to the core *N*-acetylglucosamine residue. The two exceptions that have been reported are high mannose oligosaccharides on β-glucuronidase (197) and cathepsin D (198).

Schachter et al (7) has noted that a number of processing enzymes appear to interact with the GlcNAcβ1–2Manα1–3 sequence which is present in both

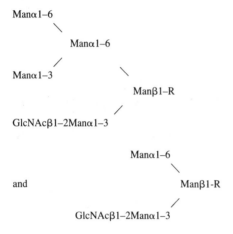

(where R = GlcNAcβ1–4GlcNAc-Asn) although their catalytic attack occurs on very different parts of the structures. These enzymes include those that can act on both substrates (*N*-acetylglucosaminyltranferases III and IV and the α1,6 fucosyltransferase); an enzyme that acts only on GlcNAcMan₅GlcNAc₂-Asn (α-mannosidase II); and an enzyme that acts only on the GlcNAcMan₃GlcNAc₂-Asn (*N*-acetylglucosaminyltransferase II). However, introduction of a "bisecting" GlcNAc residue into either structure by *N*-acetylglucosaminyltransferase III converts that structure to a nonacceptor for the other enzymes. An explanation for these observations comes from the three-dimensional conformation of these oligosaccharide structures determined by Brisson & Carver (199). The Manα1–3 arm in these N-linked oligosaccharides maintains a fixed orientation in relation to the core Manβ1–4GlcNAcβ1–4GlcNAc segment even when the "bisecting" GlcNAc residue is attached to C-4 of the Manβ1–4 residue. But the added GlcNAc residue blocks access to one face of the GlcNAcβ1–2Manα1–3 segment, with which the other enzymes must presumably interact.

The galactosyltransferase that transfers galactose residues in β1,4 linkage to terminal *N*-acetylglucosamine residues on N-linked oligosaccharides has been purified to homogeneity from a variety of tissues and secretions and the extensive earlier studies on this transferase have been reviewed by Beyer et al (8). Rao & Mendicino (200) first observed that this enzyme acted best on a biantennary complex oligosaccharide with two terminal *N*-acetylglucosamine residues (K_m 0.25mM), less well on the monogalactosylated derivative (K_m = 2mM) and very poorly on a substrate with a S.A.→Gal on one branch (K_m 10mM).

More recently two groups have established that β1,4 galactosyltransferase displays branch specificity in its action. Paquet et al (201) using purified rat liver Golgi galactosyltransferase with the substrates

GlcNAcβ1–2Manα1–6
$\diagdown$

        Man-R   (I)

$\diagup$

GlcNAcβ1–2Manα1–3

      GlcNAcβ1–2Manα1–6
            $\diagdown$

              Man-R   (II)

            $\diagup$

      Galβ1–4GlcNAcβ1–2Manα1–3

and

      Galβ1–4GlcNAcβ1–2Manα1–6
            $\diagdown$

              Man-R   (III) where R = GlcNAcβ1,4GlcNAc-Asn

            $\diagup$

      GlcNAcβ1–2Manα1–3

found that their apparent K_m's as acceptors were 0.13, 0.43, and 6.28 mM respectively. By monitoring the products formed by galactosyltransferase acting on 0.18 mM structure (I) these workers showed that structure (II) was formed rapidly, structure (III) more slowly, and the digalactosylated product gradually accumulated. The rate constants derived from this study showed that galactose transfer to the GlcNAcβ1–2Manα1–3 branch occurred about 5 times faster than transfer to the GlcNAcβ1–2Manα1–6 branch. Using a double-radio-labeling technique, Blanken et al (202) have shown that the galactosyltransferases from bovine colostrum and calf thymus also preferentially transfer galactose to the 1,3 branch of an acceptor oligosaccharide with structure (I) with a preference ratio of about 4 : 1 over transfer to the 1,6 branch. These specificity studies were carried out in vitro with glycopeptide or oligosaccharide substrates. However, in vivo the biantennary complex oligosaccharide of human and bovine IgG is monogalactosylated on the Manα1–6 arm (203, 204). This seeming paradox suggests that the relative accessibility of the Manα1–3 arm and the Manα1–6 arm to galactosyltransferase is altered when the oligosaccharide is covalently attached to a protein with which it can interact. In fact, X-ray crystallographic studies of the F_c fragment of human IgG have shown that the oligosaccharide is disposed in a fixed conformation stabilized by carbohydrate-protein interactions (205, 206). Similar observations have been made on crystals of the F_c fragment of rabbit IgG (207).

LATE ACTING ENZYMES   The β-galactoside α2–6 sialyltransferase that is responsible for sialylating N-linked oligosaccharides has been purified from bovine colstrum and its substrate specificity extensively studied as reviewed by Beyer et al (8). Recently van den Eijnden et al (208) showed that this enzyme displays "branch specificity", preferring to sialylate the galactose on the Galβ1–4GlcNAcβ1–2Manα1–3 arm of a biantennary glycopeptide. In the triantennary glycopeptide both branches arising from Manα1–3 were sialylated whereas the Galβ1–4GlcNAcβ1–2Manαβ1–6 branch was not. Interestingly, the core GlcNAc residue to which the β-mannose residue is attached in biantennary oligosaccharides is essential for this branch specificity (209). Reduction of this residue to N-acetylglucosaminitol led to decreased branch specificity as well as decreased sialic acid transfer and its removal led to random attachment of sialic acid to the galactose residues in both branches. The authors proposed that interaction of the sialyltransferase with this N-acetylglucosamine residue in acceptor oligosaccharides positions the enzyme on the substrate in such a way that transfer to galactose on the 1–3 branch is favored.

Paulson et al (20) and Beyer et al (21) have studied the in vitro glycosylation of asialotransferrin, which contains biantennary complex oligosaccharides, using purified β-galactoside α2,6 sialyltransferase and N-acetylglucosaminide α1,3 fucosyltransferase. The latter enzyme adds fucose to C-3 of the N-acetylglucosamine residue in the sequence Galβ1–4GlcNAcβ1–2Man and the product of its action Galβ1–4(Fucα1–3)GlcNAcβ1–2Man will not serve as a substrate for the α2,6 sialyltransferase. Similarly, prior sialylation of the asialotransferrin prevented subsequent fucosylation of the N-acetylglucosamine residues. The mutually exclusive action of these two transferases on the same oligosaccharide substrate accounts for the fact that the sequence S.A.α2–6Galβ1–4(Fucα1–3) GlcNAc- has not been found on glycoprotein oligosaccharides. An analogous situation appears to account for the observations of Finne et al (210, 211) that a variant mouse melanoma cell line selected for resistance to the sialic acid binding lectin wheat germ agglutinin has increased sensitivity to the fucose binding lectin from *Lotus tetragonolobus* and a 70-fold increase in N-acetylglucosaminide α1,3 fucosyltransferase activity compared to the parent melanoma cell line. The oligosaccharides synthesized by the parent cells contained α2,3-linked sialic acid which was reduced in the variant cells concomitant with the appearance of high amounts of Fucα1–3GlcNAc. Thus prior action by the 1,3 fucosyltransferase may also hinder the action of α2,3 sialyltransferase on the same oligosaccharide and vice versa, although structures containing both substituents have been reported, e.g., in rat brain by Krusius & Finne (212). Campbell & Stanley (213) have also described two mutant Chinese hamster ovary cell lines that are resistant to wheat germ agglutinin and express N-acetylglucosaminide $\alpha 1,3$ fucosyltransferase activity not present in the parental cell line. The mutant cells also express the stage-specific embryonic antigen SSEA-1, Galβ1–4(Fucα1–3)GlcNAc, (40, 41) on their surfaces and the parental cells do not.

The $\alpha 1$–3 fucosyltransferase activity that is found in tissues from many species (including human) appears to be a different enzyme than the human Lewis blood group specified $\alpha 1$–3/4 fucosyltransferase which forms the Le^a antigenic structure Galβ1–3(Fucα1–4)GlcNAc (214, 215). This latter enzyme, isolated from human milk by Prieels et al (216), was shown to catalyze fucose transfer to N-acetylglucosamine in $\alpha 1$–3 linkage using Galβ1–4GlcNAc as acceptor and in $\alpha 1$–4 linkage using Galβ1–3GlcNAc as acceptor. The two activities occurred in a constant ratio throughout a 500,000-fold purification indicating that they are catalyzed by a single enzyme. This enzyme is thus an exception to the dictum: one linkage—one enzyme.

The $\alpha 2,3$ sialyltransferase that sialylates Galβ1–3(4)GlcNAc sequences in N-linked oligosaccharides has been purified to homogeneity from rat liver by Weinstein et al (217, 218) and its substrate specificity for oligosaccharides and glycoproteins compared with that of the Galβ1–4GlcNAc $\alpha 2$–6 sialyltransferase purified to homogeneity from the same source. The best acceptors for the $\alpha 2,3$ sialyltransferase contained the sequence Galβ1–3GlcNAc-R which was inactive with the $\alpha 2,6$ transferase. The product, SA$\alpha 2$–3Galβ1–3 GlcNAc-R, has only been reported to occur in glycoproteins containing the disialylated structure SA$\alpha 2$–3Galβ1–3(SA$\alpha 2$–6)GlcNAc-(30–32). The $\alpha 2,3$ sialyl transferase also acted on acceptors with the sequence Galβ1–4GlcNAc-R to form SA$\alpha 2$–3Galβ1–4GlcNAc-R. These acceptors were even more active with the $\alpha 2,6$ sialyltransferase which formed SA$\alpha 2$–6Galβ1–4GlcNAc-R. This is in accord with the fact that the N-linked oligosaccharides of many glycoproteins contain one or both of these sialylated sequences in their outer branches. Both enzymes showed higher affinity for glycoprotein acceptors than for the analogous oligosaccharide acceptors. Asialo α-1 acid glycoprotein could be almost quantitatively sialylated by either enzyme even though in its isolated form it contains sialic acid in $\alpha 2,6$ linkage. These investigators, on the basis of transfer experiments using mixtures of the $\alpha 2,3$ and $\alpha 2,6$ enzymes, suggest that the relative proportions of these enzymes in a tissue will not totally account for the relative proportion of $\alpha 2,3$ vs $\alpha 2,6$ sialylation on glycoproteins and that the two enzymes probably have complementary but overlapping branch specificity. Interestingly, the $\alpha 2,3$ sialyltransferase could not act on the fucosylated pentasaccharide Galβ1–3(Fucα1–4)GlcNAcβ1–3Galβ1–4Glc. The isomeric structure containing Galβ1–4(Fucα1–3)GlcNAc-R was not tested as an acceptor but its α2,3-sialylated form has been found in oligosaccharides (212).

Paulson et al (219) have identified an N-acetylglucosaminide α2,6 sialyltransferase in rat liver Golgi which, acting with α2,3 sialyltransferase, can form the disialylated sequence SAα2–3Galβ1–3(SAα2–6)GlcNAc found in bovine prothrombin (32). SAα2–3Galβ1–3GlcNAc was the preferred substrate of the GlcNAc α2,6 sialyltransferase. Using desialyzed acceptors (either oligosaccharide or glycoprotein) and both α2,3 sialyltransferase and GlcNAc α2,6 sialyltransferase it was shown that the former acted first and the latter second to form the disialylated product. However the α2,3 sialyltransferase can act on the milk oligosaccharide Galβ1–3 (SAα2–6)GlcNAcβ1–3Galβ1–3Glc (218).

The repeating disaccharides (Galβ1–4GlcNAcβ1–3) of polylactosamine chains are thought to be assembled by the alternating action of β1,4 galactosyltransferase and β1,3 N-acetylglucosaminyltransferase, and the branches formed by the action of a β1,6 N-acetylglucosaminyltransferase. Three groups (220–222) have reported that normal human serum contains a β1,3 N-acetylglucosaminyltransferase activity that acts on lactose to form GlcNAcβ1–3Galβ1–4Glc. Each group found that Galβ1–4GlcNAc was a better acceptor than lactose while Yates & Watkins (221) and Piller & Cartron (222) found that Galβ1–3GlcNAc was a poor acceptor, but a variety of structures with terminal Galβ1–4GlcNAc- sequences were good acceptors. Interestingly the latter group found that the branched tetrasaccharide Galβ1–4GlcNAcβ1–3(GlcNAcβ1–6)Gal had poor acceptor activity but the branched pentasaccharide Galβ1–4GlcNAcβ1–3(Galβ1–4GlcNAcβ1–6)Gal was very active. Piller et al have identified the β1,6 N-acetylglucosaminyltransferase using GlcNAcβ1–3Galβ1–4Glcβ-OMe as substrate (223). This enzyme has no activity with acceptors having terminal Gal residues. These results indicate that branching of polylactosamine structures only occurs during elongation of the chains and that at branching points two GlcNAc residues are incorporated in a strictly ordered manner into terminal nonreducing galactose. The addition of the Gal residues is also ordered, with the GlcNAc-linked β1–6 being substituted first (224). Using a different approach van den Eijnden et al (225) tested asialo α-1 acid glycoprotein labeled in its terminal galactose residues as an acceptor for N-acetylglucosaminyltransferases in crude homogenates of Novikoff ascites tumor cells. They observed that N-acetyl–glucosamine was transferred in both β1,3 and β1,6 linkage to terminal galactose residues as revealed by methylation analysis of the product. However, they did not find evidence for branch formation on these terminal galactoses. The serum β1,3 N-acetylglucosaminyltransferase activity (221, 222) was more active on asialo α-1 acid glycoprotein than on asialo fetuin which fits with the fact that the oligosaccharides of the former glycoprotein have been shown to have the repeating lactosamine sequence on some of their

GlcNAcβ1–6 Manα1–6 arms whereas fetuin oligosaccharides do not have the repeating sequence. Structural analysis of mouse lymphoma cells (226) and BHK (227) glycoproteins have shown that the repeating lactosamine sequence is present on a high proportion of the tetra and triantennary N-linked oligosaccharides, but rarely on the biantennary oligosaccharides. A mutant cell line deficient in α mannoside β1,6 N-acetylglucosaminyltransferase (N-acetylglucosaminyltransferase V) has a greatly reduced content of repeating lactosamine sequences and those present are found predominantly on triantennary chains (226). Taken together these observations suggest that the β galactoside β1,3 N-acetylglucosaminyltransferase may have a preference for elongating oligosaccharide chains linked to C-6 of the Manα1-6 residue.

Some of the polylactosamine chains in the mouse lymphoma oligosaccharides referred to above were terminated by an additional Galα1–3 residue (226). This was also observed by Eckhardt & Goldstein (22) in the case of the polylactosamine chains on tetraantennary oligosaccharides in Ehrlich ascites tumor cells. Blake & Goldstein (228) obtained a partially purified α1,3 galactosyltransferase activity from these cells and showed that it catalyzed transfer of galactose from UDP-Gal to lactosamine to form Galα1–3Galβ1–4GlcNAc. The transferase preferred the Galβ1–4GlcNAc-R sequence but also acted on Galβ1–3GlcNAc-R. The enzyme was distinguished from the blood group B α-galactosyltransferase by its lack of reaction with Fucα1–2Galβ1–4GlcNAc-R, the preferred substrate for that transferase. Van Halbeek et al (229) and van den Eijnden et al (230) have studied an α-galactosyltransferase activity in calf thymus that transfers galactose in α1,3 linkage to Galβ1–4GlcNAc and to the Galβ1–4GlcNAc-termini in asialo α-1 acid glycoprotein. This enzyme is different from other galactosyltranferases in calf thymus that act on GalNAc-R in ovine submaxillary asialo mucin and on GlcNAc. During the analysis of the products of the α1,3 galactosyltransferase, it was noted that the added Gal residue was cleaved by some β galactosidases due to their contamination with α galactosidase activity, often barely detectable against p-NO$_2$-phenyl-α-galactoside (230). This prompted the authors to suggest that the Galβ1–3Galβ1–4 GlcNAc- sequences previously reported to occur on calf thymus membrane glycoproteins (231, 232), in which the terminal linkage had been assigned as β on the basis of its susceptibility to cleavage by β galactosidase, should be reevaluated.

OTHER POSTTRANSLATIONAL MODIFICATIONS

Asparagine-linked oligosaccharides may undergo further posttranslational modifications, including phosphorylation of mannose residues, sulfation of mannose and N-acetylhexosamine residues, and O-acetylation of sialic acid residues. Lysosomal hydrolases acquire phosphomannosyl residues which serve as the essential component of a recognition marker which allows binding to a specific receptor (the Man 6-P receptor) in the Golgi and subsequent translocation to lysosomes (233). This recognition marker is generated by the sequential action of two Golgi enzymes (Figure 3). First, N-acetylglucosaminylphosphotransferase transfers N-acetylglucosamine 1-phosphate from UDP-GlcNAc to selected mannose residues of high mannose oligosaccharides and then N-acetylglucosamine 1-phosphodiester α-N-acetylglucosaminidase removes the N-acetylglucosamine residue to expose the phosphomannosyl group (234). The phosphates may be linked to five different mannose residues on the oligosaccharide and individual molecules may contain one or two phosphates (234). While oligosaccharides with phosphomonoesters are the physiologic ligands for binding to the receptor (11, 235–237), lysosomal enzymes containing several oligosaccharides with phosphodiesters have a weak interaction with the receptor (238). Partially purified N-acetylglucosaminylphosphotransferase phosphorylates lysosomal enzymes at least 100-fold better than nonlysosomal glycoproteins which contain identical high mannose–type oligosaccharides (239, 240). Isolated high mannose oligosaccharides and glycopeptides are extremely poor substrates for the enzyme, indicating that the high affinity interaction between the transferase and lysosomal enzymes is mediated primarily by protein-protein interactions. This has been shown directly by demonstrating that deglycosylated lysosomal enzymes are potent inhibitors of the phosphorylation of intact lysosomal enzymes (241). Based on these data it has been proposed that the N-acetylglucosaminylphosphotransferase recognizes a protein domain that is common to all lysosomal enzymes, but is absent in nonlysosomal glycoproteins (239–241); however, the identity of this common protein domain is unknown. Fibroblasts from patients with the lysosomal enzyme storage diseases, I-cell disease (mucolipidosis II), and pseudo-Hurler polydystrophy (mucolipidosis III) are severely deficient in this enzyme activity (233, 234). Their inability to generate phosphomannosyl residues prevents the receptor-mediated targeting of newly synthesized acid hydrolases to lysosomes, and consequently the enzymes are secreted into the extracellular milieu. Cell fusion studies indicate that there are three complementation groups among patients with mucolipidosis III (241). Fibroblast extracts of all three groups have very low N-acetylglucosaminylphosphotransferase activity toward lysosomal enzyme acceptors, but in one group there is nearly normal activity toward the substrate analog, α-methylmannoside (242, 243). To explain this result, it was proposed that the N-acetylglucosaminylphosphotransferase contains a binding site, distinct from the catalytic site, which is involved in the specific recognition of lysosomal enzymes, and that this site is selectively altered by the mutation (243). High mannose units with Man-6-P are also present on lysosomal enzymes of the slime mold *Dictyostelium discoideum* (244, 245), but almost all of these residues are in the form of an acid-stable diester which has been identified as methylphosphomannose (246). The pathway by which this diester is synthesized is unknown.

Another recognition system involving Man-6-P has been detected in the embryonic chick-neural retina where a family of glycoproteins bind to the cell-surface baseplate protein, called ligatin (247). These glycoproteins contain high mannose oligosaccharides with Man-6-P residues in diester linkage to the 1-carbon of terminal glucose residues (248). The binding to ligatin is mediated by the diester rather than the monoester as in the case of the Man-6-P receptor. A UDP-glucose : glycoprotein glucose 1-phosphotransferase has been detected in homogenates of embryonic chicken neural retina (249). Thus this diester is synthesized in a fashion analogous to the transfer of GlcNAc-P to mannose residues of lysosomal enzymes. Presumably in this system the glucose is not removed since the specificity of the ligatin is for the phosphodiester (248). Hen oviduct contains an enzyme which transfers Gal-1-P from UDP-Gal to UDP-GlcNAc to form UDP-GlcNAc-6-P-Gal (250). The latter compound had previously been isolated from hen oviduct, but its physiologic function is not known. The enzyme was also detected in the liver, ovary, and uterus of the laying hen.

A variety of secretory and membrane glycoproteins have been shown to contain sulfate residues on their N-linked oligosaccharides (12, 245, 251–260). In calf thyroid plasma membrane glycoproteins, the asparagine-linked GlcNAc is 6-O-sulfated and the outer GlcNAc residues are 4-O-sulfated (256). The glycoproteins of paramyxovirus SV5 also contain 6-O-sulfate residues on their innermost GlcNAc residues (255). A small fraction (3%) of ovalbumin contains an unusual hybrid-type species in which one of the outer mannose residues is 4-O-sulfated (12). Several pituitary glycoproteins, including bovine lutropin (LH) and thyroid stimulating hormone (TSH), contain, on their α-subunits, biantennary complex-type oligosaccharides with 4-O-sulfated GlcNAc and GalNAc residues at the two nonreducing termini (257–259). By contrast, placental hCG, which shares the common α subunit, lacks SO_4 and has terminal sialic acid residues indicating that the sulfation reaction may be tissue specific (261, 262). The sulfation of the pituitary glycoprotein hormones also differs from most of the other examples in that the oligosaccharides have a full residue of SO_4 per hexosamine whereas in other glycoproteins the substitution is incomplete (253, 255, 256). The high mannose units of the *D. discoideum* lysosomal enzymes contain SO_4 in addition to methylphosphate residues (245). The SO_4, which appears to be present on both the *N*-acetylglucosamine and mannose residues, has been identified as the "common" antigen of these proteins (263).

CONTROL OF OLIGOSACCHARIDE PROCESSING

An analysis of oligosaccharide structures on the same protein from different species and even different tissues reveals that major variations frequently exist (24, 29, 32, 264–270). It is evident from the preceding discussion that a key factor in determining the synthesis of particular N-linked oligosaccharides is the level of expression of the various glycosyltransferases. Differences in the relative activity of these enzymes among species and tissues can account for many of the variations in oligosaccharide structures that occur. But beyond this another level of control must be operative in individual cells. Although most cells have the capacity to process the common protein-bound oligosaccharide precursor to a great number of structures ranging from simple high mannose units to extended complex-type units and potentially could produce such an array at any particular asparagine that is glycosylated, this type of gross heterogeneity is not common. Rather, individual glycosylation sites tend to have characteristic oligosaccharides, and when heterogeneity is encountered, it usually consists of a family of closely related oligosaccharides which may differ from oligosaccharides at another site on the same protein (271). The factors which control this level of oligosaccharide processing are beginning to be understood. One determinant of oligosaccharide processing is the location of the protein in the cell. Thus resident ER glycoproteins, not having been exposed to the Golgi processing enzymes, would be expected to have only high mannose units. In fact, HMG-CoA reductase has $Man_8GlcNAc_2$ and $Man_6GlcNAc_2$ units (124). Oligosaccharide processing of proteins that exit the ER is controlled by a number of factors, including the conformation of the polypeptide backbone. This was demonstrated by analyzing the types of oligosaccharides present on closely related viral glycoproteins isolated from the same host cells. Since the virus utilizes the host cell processing enzymes, any difference in oligosaccharide structure should reflect differences in the viral glycoproteins. Hunt et al (10) found that the G protein of the Indiana serotype of vesicular stomatitis virus (VSV) contains sialylated complex-type oligosaccharides at both its glycosylation sites when grown in baby hamster kidney cells. In contrast, the Hazelhurst subtype of the New Jersey serotype of VSV contained a similar unit at one site, but a mixture of hybrid and high mannose (predominantly $Man_{5-6}GlcNAc_2$) species at the other site. Similar differences in oligosaccharide processing were observed when closely related murine leukemia viruses were grown in a number of hosts under various growth conditions (272). The simplest explanation for these findings is that differences in the conformation of these closely related viral glycoproteins influences the extent of processing. Further evidence that protein structure influences oligosaccharide processing has been obtained by incorporating the amino acid analogs β-hydroxyleucine (leucine analog) and 4-thioisoleucine (isoleucine analog) into the immunoglobulin light chain synthesized by MOPC-46B cells (273). The incorporation of these analogs prevented the normal processing of the high mannose precursor to a complex-type unit without significantly decreasing the secretion of the protein.

Some insight into one mechanism by which polypeptide structure influences oligosaccharide processing has come from studies using endo H to probe the relative accessibility of high mannose units on native and denatured glycoproteins. Since this enzyme cleaves the chitobiose unit of the inner core, it can be used to probe the exposure of an oligosaccharide. Three different studies have shown that the extent of oligosaccharide processing correlates with the susceptibility to endo H cleavage. Hsieh et al grew Sindbis virus in a clone of Chinese hamster ovary cells that is deficient in N-acetylglucosaminyltransferase I (274). In this cell line the formation of complex-type units is blocked at the $Man_5GlcNAc_2$ stage so that all the N-linked oligosaccharides remain endo H-sensitive. For both Sindbis glycoproteins, endo H preferentially cleaved the oligosaccharides located at the sites that normally contain complex-type units. The selectivity of the endo H cleavage was lost when viruses were digested with pronase or incubated with detergent. Similarly, treatment of yeast carboxypeptidase Y and invertase with endo H released the highly processed large, phosphorylated oligosaccharides ($Man_{11-18}GlcNAc$) whereas the smaller, less processed species ($Man_{8-12}GlcNAc$ with no phosphate) were inaccessible to endo H until the proteins were denatured (275). Finally, treatment of human spleen β-glucuronidase with endo H released the most processed oligosaccharides (the phosphorylated high mannose species and $Man_{5-7}GlcNAc$) whereas the $Man_9GlcNAc_2$ units were only released after the protein was denatured (276). These findings indicate that physical accessibility of the oligosaccharide to processing enzymes can control its processing. The interaction of protein subunits may also influence oligosaccharide processing by this same mechanism. For example, the association of the two heavy chains of IgG places the two oligosaccharide units in direct contact with each other, leading to carbohydrate-carbohydrate interactions as well as protein-carbohydrate interactions (207). It has been postulated that the specific pairing of these oligosaccharides dictates the extent of terminal glycosylation.

There is also evidence that host-dependent factors influence the extent of oligosaccharide processing. Hsieh et al (277) examined the Asn-linked oligosaccharides at the individual glycosylation sites of the two Sindbis virus glycoproteins in virus grown in three different host cells. They found that one of the two glycosylation sites in E1 had exclusively complex-type oligosaccharides regardless of the host cell type. However the other site had a high mannose unit on virus grown in chick embryo fibroblasts, a complex-type unit in virus from baby hamster kidney cells, and both complex and high mannose-type units in virus from Chinese hamster ovary cells. A significant amount of the high mannose units were processed to the $Man_5GlcNAc_2$ species, suggesting that the next processing step, the addition of GlcNAc by N-acetylglucosaminyltransferase I, occurs more slowly at some glycosylation sites than at others. Similar host-dependent variation in oligosaccharide processing of viral glycoproteins has been noted by others (278–281). Hsieh et al pointed out that these differences in processing among cell types could be due to intrinsic differences in the cellular processing enzymes, differences in the duration of intracellular transit of glycoproteins, or differences in the physical accessibility of oligosaccharides during processing (277). Williams & Lennarz have obtained evidence that indicates that the substrate specificities of the processing enzymes can differ between tissues and/or species (282). They found that rat liver Golgi membranes could process the high mannose units of native bovine RNase B (mainly $Man_5GlcNAc_2$) to complex-type units whereas bovine pancreas Golgi membranes failed to do so unless the RNase B was denatured. Presumably one or more of the processing enzymes of the bovine pancreas (presumably N-acetylglucosaminyltransferase I and/or α-mannosidase II) are constrained by some aspect of the native conformation of RNase B that is not inhibitory to the same processing enzymes of the rat liver Golgi. These findings indicate that the control of processing involves more sophisticated mechanisms than simple exposure of oligosaccharide chains.

It may be that interaction of protein signals with certain glycosyltransferases which mediate the terminal steps in oligosaccharide assembly determines which of the many possible structures the mature oligosaccharide will have. This would be analogous to the specific recognition of a protein determinant on lysosomal enzymes by N-acetylglucosaminylphosphotransferase. Alternatively, differences in the interaction of the oligosaccharide with the underlying peptide may determine the nature of the final structure. In this case accessibility rather than a specific signal would be the critical factor.

Pollack & Atkinson noted that the location of glycosylation sites in the polypeptide chain correlates with processing (283). Their survey of glycoproteins revealed that glycosylation sites in the first 100 amino acid residues were enriched in complex-type units whereas high mannose units predominated at sites at amino acid residue 200 or higher. The mechanism whereby the location of an oligosaccharide along the linear polypeptide chain influences processing is unclear since, in most instances, the polypeptide will have folded by the time it enters the Golgi.

Literature Cited

1. Hubbard, S. C., Ivatt, R. J. 1981. *Ann. Rev. Biochem.* 50:555–83
2. Kobata, A. 1984. In *Biology of Carbohydrates*, ed. V. Ginsburg, P. W. Robbins, 2:87–161. New York: Wiley-Intersciences
3. Vliegenthart, J. F. G., Dorland, L., van Halbeek, H. 1983. *Adv. Carbohydr. Chem. Biochem.* 41:209–374
4. Carver, J. P., Brisson, J. R. 1984. See Ref. 2, pp. 289–331
5. Montreuil, J. 1980. See Ref. 3, 37:157–223
6. Snider, M. D. 1984. See Ref. 2, pp. 163–98

7. Schachter, H., Narasimhan, S., Gleeson, P., Vella, G. 1983. *Can. J. Biochem. Cell Biol.* 61:1049–66
8. Beyer, T. A., Sadler, J. E., Rearick, J. I., Paulson, J. C., Hill, R. L. 1981. *Adv. Enzymol. Relat. Areas Mol. Biol.* 52:23–175
9. Staneloni, R. J., Leloir, L. F. 1982. *Crit. Rev. Biochem.* 12:289–326
9a. Presper, K., Heath, E. C. 1983. In *The Enzymes*, ed. P. D. Boyer, 16:449. Academic
10. Hunt, L. A., Davidson, S. K., Golemboski, D. B. 1983. *Arch. Biochem. Biophys.* 226:347–56

11. Varki, A., Kornfeld, S. 1983. *J. Biol. Chem.* 258:2808–18
12. Yamashita, K., Veda, I., Kobata, A. 1983. *J. Biol. Chem.* 258:14144–47
13. Paz-Parente, J., Wieruszeski, J. M., Strecker, G., Montreuil, J., Fournet, B., et. al. 1982. *J. Biol. Chem.* 257:13173–76
14. Yamashita, K., Kammerling, J. P., Kobata, A. 1983. *J. Biol. Chem.* 258:3099–106
15. Paz-Parente, J., Strecker, G., Leroy, Y., Montreuil, J., Fournet, B., et. al. 1983. *FEBS Lett.* 152:145–52
16. Francois-Gerard, C., Pierce-Cretel, A., Andre, A., Dorland, L., van Halbeek, H., et. al. 1983. In *Glycoconjugates, Proc. 7th Int. Symp. Glycoconjugates,* ed. M. A. Chester, D. Heinegard, A. Lundblad, S. Svensson, pp. 169–70. Sweden: Lund-Ronneby
17. Spik, G., Strecker, G., Fournet, B., Bouquelet, S., Montreuil, J., et. al. 1982. *Eur. J. Biochem.* 121:413–19
18. Matsumoto, A., Yoshima, J., Takasaki, S., Kobata, A. 1982. *J. Biochem.* 91:143–55
19. Dorland, L., Haverkamp, J., Schut, B., Vliegenthart, J. F. G., Spik, G., et. al. 1977. *FEBS Lett.* 77:15–20
20. Paulson, J. C., Prieels, J.-P., Glasgow, L. R., Hill, R. L. 1978. *J. Biol. Chem.* 253:5617–24
21. Beyer, T. A., Rearick, J. I., Paulson, J. C., Prieels, J.-P., Sadler, J. E., Hill, R. L. 1979. *J. Biol. Chem.* 254:12531–41
22. Eckhardt, A. E., Goldstein, I. J. 1983. *Biochemistry* 22:5290–97
23. Dorland, L., van Halbeek, H., Vliegenthart, J. F. G. 1984. *Biochem. Biophys. Res. Commun.* 122:859–66
24. Spiro, R. G., Bhoyroo, V. D. 1984. *J. Biol. Chem.* 259:9858–66
25. Finne, J. 1982. *J. Biol. Chem.* 257:11966–70
26. Rothbard, J. B., Brackenbury, R., Cunningham, B. A., Edelman, G. M. 1982. *J. Biol. Chem.* 257:11064–69
27. Finne, J., Finne, U., Deagostini-Bazin, H., Goridis, C. 1983. *Biochem. Biophys. Res. Commun.* 112:482–87
28. Margolis, R. K., Margolis, R. V. 1983. *Biochem. Biophys. Res. Commun.* 116:889–94
29. Takasaki, S., Yamashita, K., Suzuki, K., Iwanaga, S., Kobata, A. 1979. *J. Biol. Chem.* 254:8548–53
30. Mizuochi, T., Taniguchi, T., Fujikawa, K., Titani, K., Kobata, A. 1983. *J. Biol. Chem.* 258:6020–24
31. Mizuochi, T., Yamashita, K., Fujikawa, K., Titani, K., Kobata, A. 1980. *J. Biol. Chem.* 255:3526–31
32. Mizuochi, T., Yamashita, K., Fujikawa, K., Kisiel, W., Kobata, A. 1979. *J. Biol. Chem.* 254:6419–25
33. Krusius, T., Finne, J., Rauvala, H. 1978. *Eur. J. Biochem.* 92:289–300
34. Jarnefelt, J., Rush, J., Li, Y.-T., Laine, R. A. 1978. *J. Biol. Chem.* 253:8006–9
35. Childs, R. A., Feizi, T., Fukuda, M., Hakomori, S. I. 1978. *Biochem. J.* 173:333–36
36. Fukuda, M., Dell, A., Fukuda, M. N. 1984. *J. Biol. Chem.* 259:4782–91
37. Fukuda, M., Dell, A., Oates, J. E., Fukuda, M. N. 1984. *J. Biol. Chem.* 259:8260–73
38. Spooncer, E., Fukuda, M., Klock, J. C., Oates, J. E., Dell, A. 1984. *J. Biol. Chem.* 259:4792–801
39. Huang, L. C., Civin, C. I., Magnani, J. L., Shaper, J. H., Ginsburg, V. 1983. *Blood* 61:1020–23
40. Gooi, H. C., Feizi, T., Kapadia, A., Knowles, B. B., Solter, D., Evans, M. J. 1981. *Nature* 292:156–58
41. Kannagi, R., Nudelman, E., Levery, S. B., Hakomori, S. 1982. *J. Biol. Chem.* 257:14865–74
42. Nilsson, B., Nakazawa, K., Hassell, J. R., Newsome, D. A., Hascall, V. C. 1983. *J. Biol. Chem.* 258:6056–63
43. Nakazawa, K., Newsome, D. A., Nilsson, B., Hascall, V. C., Hassell, J. R. 1983. *J. Biol. Chem.* 258:6051–55
44. Hassell, J. R., Newsome, D. A., Hascall, V. C. 1979. *J. Biol. Chem.* 254:12346–54
45. Roden, L. 1980. In *The Biochemistry of Glycoproteins and Proteoglycans.* ed. W. J. Lennarz, pp. 267–371. New York: Plenum
46. Keller, R., Stein, T., Stuhlsatz, H. W., Greiling, H., Ohst, E., Muller, E., Scharf, H.-D. 1981. *Hoppe-Seyler's Z. Physiol. Chem.* 362:327–36
47. Stein, T., Keller, R., Stuhlsatz, H. W., Greiling, H., Ohst, E., Muller, E., Scharf, H.-D. 1982. *Hoppe-Seyler's Z. Physiol. Chem.* 363:825–33
48. Bergman, L. W., Kuehl, W. M. 1978. *Biochemistry* 17:5174–80
49. Roberts, J. L., Phillips, M., Rosa, P. A., Herbert, E. 1978. *Biochemistry* 17:3609–18
50. Ronnett, G. O., Lane, M. D. 1981. *J. Biol. Chem.* 256:4704–7
51. Katz, F. W., Rothman, J. E., Lingappa, V. R., Blobel, G., Lodish, H. F. 1977. *Proc. Natl. Acad. Sci. USA* 74:3278–82
52. Lingappa, V. R., Lingappa, J. R., Prasad, R., Ebner, K. E., Blobel, G. 1978. *Proc. Natl. Acad. Sci. USA* 75:2338–42
53. Hanover, J. A., Lennarz, W. J. 1980. *J. Biol. Chem.* 255:3600–4

54. Carey, D. J., Sommers, L. W., Hirschberg, C. B. 1980. *Cell* 19:597–605
55. Hanover, J. A., Lennarz, W. J. 1982. *J. Biol. Chem.* 257:2787–94
56. Snider, M. D., Rogers, O. C. 1984. *Cell* 36:753–61
57. Snider, M. D., Robbins, P. W. 1982. *J. Biol. Chem.* 257:6796–801
58. McCloskey, M. A., Troy, F. A. 1980. *Biochemistry* 19:2061–66
59. Haselbeck, A., Tanner, W. 1982. *Proc. Natl. Acad. Sci. USA* 79:1520–24
60. Snider, M. D., Sultzman, L. A., Robbins, P. W. 1980. *Cell* 21:385–92
61. Nilsson, O. S., DeTomas, M. E., Peterson, E., Bergman, A., Dallner, G., Hemming, F. W. 1978. *Eur. J. Biochem.* 89:619–28
62. Bergman, A., Dallner, G. 1978. *Biochem. Biophys. Acta* 512:123–35
63. Eggens, I., Dallner, G. 1980. *FEBS Lett.* 122:247–50
64. Hanover, J. A., Lennarz, W. J. 1979. *J. Biol. Chem.* 254:9237–46
65. Turco, S. J., Stetson, B., Robbins, P. W. 1977. *Proc. Natl. Acad. Sci. USA* 74:4411–14
66. Spiro, M. J., Spiro, R. G., Bhoyroo, V. D. 1979. *J. Biol. Chem.* 254:7668–74
67. Staneloni, R. J., Ugalde, R. A., Leloir, L. F. 1980. *Eur. J. Biochem.* 105:275–78
68. Murphy, L. A., Spiro, R. G. 1981. *J. Biol. Chem.* 256:7487–94
69. Staneloni, R. J., Tolmasky, M. E., Petriella, C., Leloir, L. F. 1981. *Plant Physiol.* 68:1175–79
70. Trimble, R. B., Byrd, J. C., Maley, F. 1980. *J. Biol. Chem.* 255:11892–95
71. Sharma, C. B., Lehle, L., Tanner, W. 1981. *Eur. J. Biochem.* 116:101–8
72. Lehle, L., Bause, E. 1984. *Biochem. Biophys. Acta* 799:246–51
73. Huffaker, T. C., Robbins, P. W. 1983. *Proc. Natl. Acad. Sci. USA* 80:7466–70
74. Runge, K. W., Huffaker, T. C., Robbins, P. W. 1984. *J. Biol. Chem.* 259:412–17
75. Chen, W. W., Lennarz, W. J. 1977. *J. Biol. Chem.* 252:3473–79
76. Harford, J. B., Waechter, C. J. 1979. *Arch. Biochem. Biophys.* 197:424–35
77. Nakayama, K., Araki, Y., Ito, E. 1976. *FEBS Lett.* 72:287–90
78. Hoflack, B., Debeire, P., Cacan, R., Montreuil, J., Verbert, A. 1982. *Eur. J. Biochem.* 124:527–31
79. Parodi, A. J., Quesada-Allue, L. A. 1982. *J. Biol. Chem.* 257:7637–40
80. Parodi, A. J., Cazzulo, J. J. 1982. *J. Biol. Chem.* 257:7641–45
81. Parodi, A. J., Lederkremer, G. Z., Mendelzon, D. H. 1983. *J. Biol. Chem.* 258:5589–95
82. Parodi, A. J., Quesada-Allue, L. A., Cazzulo, J. J. 1981. *Proc. Natl. Acad. Sci. USA* 78:6201–5
83. Katial, A., Prakash, C., Vijay, I. K. 1984. *Eur. J. Biochem.* 141:521–26
84. Parodi, A., Martin-Barrientos, J. 1984. *Biochem. Biophys. Res. Commun.* 118:1–7
85. Krag, S. S. 1979. *J. Biol. Chem.* 254:9167–77
86. Hoflack, B., Cacan, R., Verbert, A. 1981. *Eur. J. Biochem.* 117:285–90
87. Bause, E., Muller, T., Jaenicke, L. 1983. *Arch. Biochem. Biophys.* 220:200–7
88. Muller, T., Bause, E., Jaenicke, L. 1984. *Eur. J. Biochem.* 138:153–59
89. Trowbridge, I. S., Hyman, R. 1979. *Cell* 17:503–8
90. Chapman, A., Trowbridge, I. S., Hyman, R., Kornfeld, S. 1979. *Cell* 17:509–15
91. Stoll, J., Robbins, A. R., Krag, S. S. 1982. *Proc. Natl. Acad. Sci. USA* 79:2296–300
92. Chapman, A., Li, E., Kornfeld, S. 1979. *J. Biol. Chem.* 254:10243–49
93. Hunt, L. A. 1980. *Cell* 21:407–15
94. Gershman, H., Robbins, P. W. 1981. *J. Biol. Chem.* 256:7774–80
95. Rearick, J. I., Chapman, A., Kornfeld, S. 1981. *J. Biol. Chem.* 256:6255–61
96. Turco, S. J. 1980. *Arch. Biochem. Biophys.* 205:330–39
97. Datema, R., Schwarz, R. T. 1981. *J. Biol. Chem.* 256:11191–98
98. Spiro, R. G., Spiro, M. J., Bhoyroo, V. D. 1983. *J. Biol. Chem.* 258:9469–76
99. Marshall, R. D. 1972. *Ann. Rev. Biochem.* 41:673–702
100. Marshall, R. D. 1974. *Biochem. Soc. Symp.* 40:17–26
100a. Stenflo, J., Fernlund, P. 1982. *J. Biol. Chem.* 257:12180–90
101. Struck, D. K., Lennarz, W. J., Brew, K. 1978. *J. Biol. Chem.* 253:5786–94
102. Hart, G. W., Brew, K., Grant, G. A., Bradshaw, R. A., Lennarz, W. J. 1979. *J. Biol. Chem.* 254:9747–53
103. Bause, E., Lehle, L. 1979. *Eur. J. Biochem.* 101:531–40
104. Ronin, C., Granier, C., Caseti, C., Bouchilloux, S., van Rietschoten, J. 1981. *Eur. J. Biochem.* 118:159–64
105. Welply, J. K., Shenbagamurthi, P., Lennarz, W. J., Naider, F. 1983. *J. Biol. Chem.* 258:11856–63
106. Aubert, J. P., Helbecque, N., Loucheux-Lefebvre, M. H. 1981. *Arch. Biochem. Biophys.* 208:20–29

107. Ronin, C., Aubert, J. P. 1982. *Biochem. Biophys. Res. Commun.* 105:909–15
108. Bause, E., Legler, G. 1981. *Biochem. J.* 195:639–44
109. Bause, E. 1983. *Biochem. J.* 209:331–36
110. Aubert, J. P., Biserte, G., Loucheux-Lefebvre, M. H. 1976. *Arch. Biochem. Biophys.* 175:410–18
111. Beeley, J. G. 1977. *Biochem. Biophys. Res. Commun.* 76:1051–55
112. Lau, J. T. Y., Welply, J. K., Shenbagamurthi, P., Naider, F., Lennarz, W. J. 1983. *J. Biol. Chem.* 258:15255–60
113. Kronquist, K. E., Lennarz, W. J. 1978. *J. Supramol. Struct.* 8:51–65
114. Plummer, T. H., Hirs, C. H. W. 1964. *J. Biol. Chem.* 239:2530–38
115. Anderson, D. R., Samaraweera, P., Grimes, W. J. 1983. *Biochem. Biophys. Res. Commun.* 116:771–76
116. McCune, J., Fu, S., Kunkel, H., Blobel, G. 1981. *Proc. Natl. Acad. Sci. USA* 78:5127–31
117. Pless, D. D., Lennarz, W. J. 1977. *Proc. Natl. Acad. Sci. USA* 74:134–38
118. Carson, D. D., Earles, B. J., Lennarz, W. J. 1981. *J. Biol. Chem.* 256:11552–57
118a. Oda-Tamai, S., Kato, S., Hara, S., Akamatsu, N. 1985. *J. Biol. Chem.* 260:57–63
118b. Franc, J.-L., Hovsepian, S., Fayet, G., Bouchilloux, S. 1984. *Biochem. Biophys. Res. Commun.* 118:910–15
118c. Singh, B. N., Lucas, J. J. 1981. *J. Biol. Chem.* 256:12018–22
119. Aubert, J. P., Chiroutre, M., Kerckaert, J. P., Helbecque, N., Loucheux-Lefebvre, M. H. 1982. *Biochem. Biophys. Res. Commun.* 104:1550–59
120. Das, R. C., Heath, E. C. 1980. *Proc. Natl. Acad. Sci. USA* 77:3811–15
121. Atkinson, P. H., Lee, J. T. 1984. *J. Cell Biol.* 98:2245–49
122. Hubbard, S. C., Robbins, P. W. 1979. *J. Biol. Chem.* 254:4568–76
123. Bergman, L. W., Kuehl, W. M. 1982. In *The Glycoconjugates,* ed. M. I. Horowitz, III:81–98. New York: Academic
123a. Compton, T., Courtney, R. 1984. *J. Virology* 52:630–37
124. Liscum, L., Cummings, R. D., Anderson, R. G. W., DeMartino, G. N., Goldstein, J. L., Brown, M. S. 1983. *Proc. Natl. Acad. Sci. USA* 80:7165–69
125. Hickman, S., Theodorakis, J. L., Greco, J. M., Brown, P. H. 1984. *J. Cell Biol.* 98:407–16
126. Gabel, C., Kornfeld, S. 1984. *J. Cell Biol.* 99:296–305
127. Parodi, A. J., Mendelzon, D. H., Lederkremer, G. Z., Martin-Barrientos, J. 1984. *J. Biol. Chem.* 259:6351–57
128. Parodi, A. J., Mendelzon, D. H., Lederkremer, G. Z. 1983. *J. Biol. Chem.* 258:8260–65
129. Jamieson, J. D., Palade, G. E. 1968. *J. Cell Biol.* 39:589–603
130. Lodish, H. F., Kong, N., Snider, M., Strous, G. J. A. M. 1983. *Nature* 304:80–83
131. Fries, E., Gustafsson, L., Peterson, P. A. 1984. *EMBO J.* 3:147–52
132. Lodish, H. F., Kong, N. 1984. *J. Cell Biol.* 98:1720–29
133. Gross, V., Andus, T., Tran-Thi, T.-A., Schwarz, R. T., Decker, K., Heinrich, P. C. 1983. *J. Biol. Chem.* 258:203–9
134. Lemansky, P., Gieselmann, V., Hasilik, A., von Figura, K. 1984. *J. Biol. Chem.* 259:10129–35
135. Schlesinger, S., Malfer, C., Schlesinger, M. J. 1984. *J. Biol. Chem.* 259:7597–601
136. Bergmann, J. E., Singer, S. J. 1983. *J. Cell Biol.* 97:1777–87
137. Rothman, J. E., Urbani, L. J., Brands, R. 1984. *J. Cell Biol.* 99:248–59
138. Rothman, J. E., Miller, R. L., Urbani, L. J. 1984. *J. Cell Biol.* 99:260–71
139. Roth, J., Berger, E. G. 1982. *J. Cell Biol.* 93:223–29
140. Slot, J. W., Geuze, H. J. 1983. *J. Histochem. Cytochem.* 31:1049–56
141. Dunphy, W. G., Rothman, J. E. 1985. *Cell.* 40:463–72
142. Novikoff, P. M., Tulsiani, D. R. P., Touster, O., Yam, A., Novikoff, A. B. 1983. *Proc. Natl. Acad. Sci. USA* 80:4364–68
143. Dunphy, W. G., Fries, E., Urbani, L. J., Rothman, J. E. 1981. *Proc. Natl. Acad. Sci. USA* 78:7453–57
144. Dunphy, W. G., Rothman, J. E. 1983. *J. Cell Biol.* 97:270–75
145. Goldberg, D., Kornfeld, S. 1983. *J. Biol. Chem.* 258:3159–65
146. Deutscher, S. L., Creek, K. E., Merion, M., Hirschberg, C. B. 1983. *Proc. Natl. Acad. Sci. USA* 80:3938–42
147. Griffiths, G., Brands, R., Burke, B., Louvard, D., Warren, G. 1982. *J. Cell Biol.* 95:781–92
148. Tartakoff, A. M., Vassalli, P. 1983. *J. Cell Biol.* 97:1243–48
149. Griffiths, G., Quinn, P., Warren, G. 1983. *J. Cell Biol.* 96:835–50
150. Fries, E., Rothman, J. E. 1981. *J. Cell Biol.* 90:697–704
151. Fries, E., Rothman, J. E. 1980. *Proc. Natl. Acad. Sci. USA* 77:3870–74
152. Lehle, L. 1981. *FEBS Lett.* 123:63–66
153. Hori, H., James, D. W. Jr., Elbein, A. D. 1982. *Arch. Biochem. Biophys.* 215:12–21

154. Hori, H., Elbein, A. D. 1983. *Arch. Biochem. Biophys.* 220:415–25
155. Dorland, L., van Halbeek, H., Vliegenthart, J. F. G., Lis, H., Sharon, N. 1981. *J. Biol. Chem.* 256:7708–11
156. Vitale, A., Chrispeels, M. J. 1984. *J. Cell Biol.* 99:133–40
157. Ishihara, H., Takahashi, N., Oguir, S., Tejima, S. 1979. *J. Biol. Chem.* 254:10715–19
158. Kilker, R. D. Jr., Saunier, B., Tkacz, J. S., Herscovics, A. 1981. *J. Biol. Chem.* 256:5299–303
159. Byrd, J. C., Tarentino, A. L., Maley, F., Atkinson, P. H., Trimble, R. B. 1982. *J. Biol. Chem.* 257:14657–66
160. Parodi, A. J. 1981. *Arch. Biochem. Biophys.* 210:372–82
161. Esmon, B., Esmon, P. C., Schekman, R. 1984. *J. Biol. Chem.* 259:10322–27
161a. Tsai, P.-K., Ballou, L., Esmon, B., Schekman, R., Ballou, C. E. 1984. *Proc. Natl. Acad. Sci. USA* 81:6340–43
162. Ballou, C. E. 1980. In *Fungal Polysaccharides, ACS Symp. Ser.*, ed. P. A. Sandford, D. Matsuda, 126:1–14. Washington: Am. Chem. Soc., 284 pp.
163. Tsai, P. K., Frevert, J., Ballou, C. E. 1984. *J. Biol. Chem.* 259:3805–11
164. Hsieh, P., Robbins, P. W. 1984. *J. Biol. Chem.* 259:2375–82
165. Butters, T. D., Hughes, R. C. 1981. *Biochem. Biophys. Acta* 640:655–71
166. Hettkamp, H., Legler, G., Bause, E. 1984. *Eur. J. Biochem.* 142:85–90
167. Burns, D. M., Touster, O. 1982. *J. Biol. Chem.* 257:9991–10,000
168. Brada, D., Dubach, U. 1984. *Eur. J. Biochem.* 141:149–56
169. Reitman, M. L., Trowbridge, I. S., Kornfeld, S. 1982. *J. Biol. Chem.* 257:10357–63
170. Grinna, L. S., Robbins, P. W. 1980. *J. Biol. Chem.* 255:2255–58
171. Michael, J. M., Kornfeld, S. 1980. *Arch. Biochem. Biophys.* 199:249–58
172. Spiro, R. G., Spiro, M. J., Bhoyroo, V. D. 1979. *J. Biol. Chem.* 254:7659–67
173. Kornfeld, S., Gregory, W., Chapman, A. 1979. *J. Biol. Chem.* 254:11649–54
174. Bischoff, J., Kornfeld, R. 1983. *J. Biol. Chem.* 258:7907–10
175. Shoup, V. A., Touster, O. 1976. *J. Biol. Chem.* 251:3845–52
176. Tabas, I., Kornfeld, S. 1979. *J. Biol. Chem.* 254:11655–63
177. Tulsiani, D. R. P., Hubbard, S. C., Robbins, P. W., Touster, O. 1982. *J. Biol. Chem.* 257:3660–68
178. Harpaz, N., Schachter, H. 1980. *J. Biol. Chem.* 255:4885–93
179. Oppenheimer, C. L., Hill, R. L. 1981. *J. Biol. Chem.* 256:799–804
180. Vella, G. J., Paulsen, H., Schachter, H. 1984. *Can. J. Biochem. Cell Biol.* 62:409–17
181. Tabas, I., Kornfeld, S. 1978. *J. Biol. Chem.* 253:7779–86
182. Harpaz, N., Schachter, H. 1980. *J. Biol. Chem.* 255:4894–902
183. Tulsiani, D. R. P., Touster, O. 1983. *J. Biol. Chem.* 258:7578–85
184. Allen, S. D., Tsai, D., Schachter, H. 1984. *J. Biol. Chem.* 259:6984–90
185. Elbein, A. D., Solf, R., Dorling, P. R., Vosbeck, K. 1981. *Proc. Natl. Acad. Sci. USA* 78:7393–97
186. Elbein, A. D., Dorling, P. R., Vosbeck, K., Horisberger, M. 1982. *J. Biol. Chem.* 257:1573–76
187. Gross, V., Tran-Thi, T.-A., Vosbeck, K., Heinrich, P. C. 1983. *J. Biol. Chem.* 258:4032–36
188. Kang, M. S., Elbein, A. D. 1983. *J. Virol.* 46:60–69
189. Tulsiani, D. R. P., Harris, T. M., Touster, O. 1982. *J. Biol. Chem.* 257:7936–39
190. Oppenheimer, C. L., Eckhardt, A. E., Hill, R. L. 1981. *J. Biol. Chem.* 256:11477–82
191. Mendicino, J., Chandrasekaran, E. V., Anumula, K. R., Davila, M. 1981. *Biochemistry* 20:967–76
192. Narasimhan, S. 1982. *J. Biol. Chem.* 257:10235–42
193. Campbell, C., Stanley, P. 1984. *J. Biol. Chem.* 259:13370–78
194. Gleeson, P., Schachter, H. 1983. *J. Biol. Chem.* 258:6162–73
195. Cummings, R. D., Trowbridge, I. S., Kornfeld, S. 1982. *J. Biol. Chem.* 257:13421–27
196. Longmore, G. D., Schachter, H. 1982. *Carbohydr. Res.* 100:365–92
197. Howard, D. R., Natowicz, M., Baenziger, J. U. 1982. *J. Biol. Chem.* 257:10861–68
198. Takahashi, T., Schmidt, P. G., Tang, J. 1983. *J. Biol. Chem.* 258:2819–30
199. Brisson, J. R., Carver, J. P. 1983. *Can. J. Biochem. Cell Biol.* 61:1067–78
200. Rao, A. K., Mendicino, J. 1978. *Biochemistry* 17:5632–38
201. Paquet, M. R., Narasimhan, S., Schachter, H., Moscarello, M. A. 1984. *J. Biol. Chem.* 259:4716–21
202. Blanken, W. M., van Vliet, A., van den Eijnden, D. H. 1984. *J. Biol. Chem.* 259:15131–35
203. Kornfeld, R., Kornfeld, S. 1980. In *The Biochemistry of Glycoproteins and Proteoglycans.*, ed. W. J. Lennarz, pp. 1–34. New York: Plenum
204. Tai, T., Ito, S., Yamashita, K., Mura-

matsu, T., Kobata, A. 1975. *Biochem. Biophys. Res. Commun.* 65:968–74
205. Deisenhofer, J. 1981. *Biochemistry* 20: 2361–70
206. Deisenhofer, J., Colman, P. M., Epp, O., Huber, R. 1976. *Hoppe-Seyler's Z. Physiol. Chem.* 357:1421–34
207. Sutton, B. J., Phillips, D. C. 1983. *Biochem. Soc. Trans.* 11:130–32
208. van den Eijnden, D. H., Joziasse, D. H., Dorland, L., van Halbeek, H., Vliegenthart, J. F. G., Schmid, K. 1980. *Biochem. Biophys. Res. Commun.* 92: 839–45
209. Joziasse, D. H., Schiphorst, W. E. C. M., van den Eijnden, D. H., van Kuik, J. A., van Halbeek, H., Vliegenthart, J. F. G. 1985. *J. Biol. Chem.* 260:714–19
210. Finne, J., Tao, T. W., Burger, M. M. 1980. *Cancer Res.* 40:2580–87
211. Finne, J., Burger, M. M., Prieels, J.-P. 1982. *J. Cell Biol.* 92:277–82
212. Krusius, T., Finne, J. 1978. *Eur. J. Biochem.* 84:395–403
213. Campbell, C., Stanley, P. 1983. *Cell* 35:303–9
214. Johnson, P. H., Yates, A. D., Watkins, W. M. 1981. *Biochem. Biophys. Res. Commun.* 100:1611–18
215. Johnson, P. H., Watkins, W. M. 1982. *Biochem. Soc. Trans.* 10:445–46
216. Prieels, J.-P., Monnom, D., Dolmans, M., Beyer, T. A., Hill, R. L. 1981. *J. Biol. Chem.* 256:10456–63
217. Weinstein, J., de Souza-e-Silva, U., Paulson, J. C. 1982. *J. Biol. Chem.* 257:13835–44
218. Weinstein, J., de Souza-e-Silva, U., Paulson, J. C. 1982. *J. Biol. Chem.* 257:13845–53
219. Paulson, J. C., Weinstein, J., de Souza-e-Silva, U. 1984. *Eur. J. Biochem.* 140:523–30
220. Zielenski, J., Koscielak, J. 1983. *FEBS Lett.* 158:164–68
221. Yates, A. D., Watkins, W. M. 1983. *Carbohydr. Res.* 120:251–68
222. Piller, F., Cartron, J.-P. 1983. *J. Biol. Chem.* 258:12293–99
223. Piller, F., Cartron, J.-P., Maranduba, A., Veyrieres, A., Leroy, Y., Fournet, B. 1984. *J. Biol. Chem.* 259:13385–90
224. Blanken, W. M., Hooghwinkel, G. J. M., van den Eijnden, D. H. 1982. *Eur. J. Biochem.* 127:547–52
225. van den Eijnden, D. H., Winterwerp, H., Smeeman, P., Schiphorst, W. E. C. M. 1983. *J. Biol. Chem.* 258:3435–37
226. Cummings, R. D., Kornfeld, S. 1984. *J. Biol. Chem.* 259:6253–60
227. Yamashita, K., Ohkura, T., Tachibana, Y., Takasaki, S., Kobata, A. 1984. *J. Biol. Chem.* 259:10834–40

228. Blake, D. A., Goldstein, I. J. 1981. *J. Biol. Chem.* 256:5387–93
229. van Halbeek, H., Vliegenthart, J. F. G., Winterwerp, H., Blanken, W. M., van den Eijnden, D. H. 1983. *Biochem. Biophys. Res. Commun.* 110:124–31
230. van den Eijnden, D. H., Blanken, W. M., Winterwerp, H., Schiphorst, W. E. C. M. 1983. *Eur. J. Biochem.* 134:523–30
231. Kornfeld, R. 1978. *Biochemistry* 17: 1415–23
232. Yoshima, H., Takasaki, S., Kobata, A., 1980. *J. Biol. Chem.* 255:10793–804
233. Creek, K. E., Sly, W. S. 1984. In *Lysosomes in Biology and Pathology,* ed. J. T. Dingle, R. T. Dean, W. S. Sly, pp. 63–82. New York: Elsevier
234. Goldberg, D., Gabel, C., Kornfeld, S. 1984. See Ref. 233, pp. 45–62
235. Gabel, C. A., Goldberg, D. E., Kornfeld, S. 1982. *J. Cell Biol.* 95:536–42
236. Fischer, H. D., Creek, K. E., Sly, W. S. 1982. *J. Biol. Chem.* 257:9938–43
237. Natowicz, M., Hallett, D. W., Frier, C., Chi, M., Schlesinger, P. H., Baenziger, J. U. 1983. *J. Cell Biol.* 96:915–19
238. Talkad, V., Sly, W. S. 1983. *J. Biol. Chem.* 258:7345–51
239. Reitman, M. L., Kornfeld, S. 1981. *J. Biol. Chem.* 256:11977–80
240. Waheed, A., Hasilik, A., von Figura, K. 1982. *J. Biol. Chem.* 257:12322–31
241. Lang, L., Reitman, M., Tang, J., Roberts, R. M., Kornfeld, S. 1984. *J. Biol. Chem.* 259:14663–71
242. Mueller, O. T., Honey, N. K., Little, L. E., Miller, A. L., Shows, T. B. 1983. *J. Clin. Invest.* 72:1016–23
243. Varki, A. P., Reitman, M. L., Kornfeld, S. 1981. *Proc. Natl. Acad. Sci. USA* 78:7773–77
244. Freeze, H. H., Miller, A. L., Kaplan, A. 1980. *J. Biol. Chem.* 255:11081–84
245. Freeze, H. H., Yeh, R., Miller, A. L., Kornfeld, S. 1983. *J. Biol. Chem.* 258:14874–79
246. Gabel, C. A., Costello, C. E., Reinhold, V. N., Kurz, L., Kornfeld, S. 1984. *J. Biol. Chem.* 259:13762–69
247. Jakoi, E. R., Marchase, R. B. 1979. *J. Cell Biol.* 80:642–50
248. Marchase, R. B., Koro, L. A., Kelly, C. M., McClay, D. R. 1982. *Cell* 28:813–20
249. Koro, L. A., Marchase, R. B. 1982. *Cell* 31:739–48
250. Nakanishi, Y., Otsu, K., Suzuki, S. 1983. *FEBS Lett.* 151:15–18
251. Heifetz, A., Kinsey, W. H., Lennarz, W. J. 1980. *J. Biol. Chem.* 255:4528–34
252. Heifetz, A., Watson, C., Johnson, A.

R., Roberts, M. K. 1982. *J. Biol. Chem.* 257:13581–86

253. Cummings, R. D., Kornfeld, S., Schneider, W. J., Hobgood, K. K., Tolleshaug, H., et. al. 1983. *J. Biol. Chem.* 258:15261–73

254. Nakamura, K., Compans, R. W. 1978. *Virology* 84:303–19

255. Prehm, P., Scheid, A., Choppin, P. W. 1979. *J. Biol. Chem.* 254:9669–77

256. Edge, A. S., Spiro, R. G. 1984. *J. Biol. Chem.* 259:4710–13

257. Parsons, T. F., Pierce, J. G. 1980. *Proc. Natl. Acad. Sci. USA* 77:7089–93

258. Bedi, G. S., French, W. C., Bahl, O. P. 1982. *J. Biol. Chem.* 257:4345–55

259. Anumula, K. R., Bahl, O. P. 1983. *Arch. Biochem. Biophys.* 220:645–51

260. Hsu, C. H., Kingsbury, D. W. 1982. *J. Biol. Chem.* 257:9035–38

261. Mizuochi, T., Kobata, A. 1980. *Biochem. Biophys. Res. Commun.* 97:772–78

262. Kessler, M. J., Reddy, M. S., Shah, R. H., Bahl, O. P. 1979. *J. Biol. Chem.* 254:7901–8

263. Freeze, H. H., Mierendorf, R. C., Wunderlich, R., Dimond, R. L. 1984. *J. Biol. Chem.* 259:10641–43

264. Yamashita, K., Hitoi, A., Tateishi, N., Higashi, T., Sakamoto, Y., Kobata, A. 1983. *Arch. Biochem. Biophys.* 225:993–96

265. Mizuochi, T., Fujii, J., Kisiel, W., Kobata, A. 1981. *J. Biochem.* 90:1023–31

266. Yoshima, H., Matsumoto, A., Mizuochi, T., Kawasaki, T., Kobata, A. 1981. *J. Biol. Chem.* 256:8476–84

267. Mizuochi, T., Nishimura, R., Derappe, C., Taniguchi, T., Hamamoto, T., Tamotsu, H., et. al. 1983. *J. Biol. Chem.* 258:14126–29

268. Takasaki, S., Yamashita, K., Suzuki, K., Kobata, A. 1980. *J. Biochem.* 88:1587–94

269. Fukuda, M., Levery, S. B., Hakomori, S. 1982. *J. Biol. Chem.* 257:6856–60

270. Zhu, B. C-R., Fisher, S. F., Pande, H., Calaycay, J., Shively, J. E., Laine, R. H. 1984. *J. Biol. Chem.* 259:3962–70

271. Anderson, D. R., Grimes, W. J. 1982. *J. Biol. Chem.* 257:14858–64

272. Rosner, M. R., Grinna, L. S., Robbins, P. W. 1980. *Proc. Natl. Acad. Sci. USA* 77:67–71

273. Green, M. 1982. *J. Biol. Chem.* 257:9039–42

274. Hsieh, P., Rosner, M. R., Robbins, P. W. 1983. *J. Biol. Chem.* 258:2555–61

275. Trimble, R. B., Maley, F., Chu, F. K. 1983. *J. Biol. Chem.* 258:2562–67

276. Natowicz, M., Baenziger, J. U., Sly, W. S. 1982. *J. Biol. Chem.* 257:4412–20

277. Hsieh, P., Rosner, M. R., Robbins, P. W. 1983. *J. Biol. Chem.* 258:2548–54

278. Keegstra, K., Sefton, B., Burke, D. J. 1975. *J. Virol.* 16:613–20

279. Burke, D. J., Keegstra, K. 1976. *J. Virol.* 20:676–86

280. Burke, D. J., Keegstra, K. 1979. *J. Virol.* 29:546–54

281. Hunt, L. A. 1981. *Virology* 113:534–43

282. Williams, D. B., Lennarz, W. J. 1984. *J. Biol. Chem.* 259:5105–14

283. Pollack, L., Atkinson, P. H. 1983. *J. Cell Biol.* 97:293–300

Ann. Rev. Biochem. 1985. 54:665–97
Copyright © 1985 by Annual Reviews Inc. All rights reserved

DNA TOPOISOMERASES

James C. Wang

Department of Biochemistry and Molecular Biology, Harvard University, Cambridge, Massachusetts 02138

CONTENTS

PERSPECTIVES AND SUMMARY

The double helix structure of DNA imposes certain problems of a topological nature that the cellular apparatus must solve in order to propagate the molecule as the genetic material. A well-known example is the untwining of the parental

665

strands during semiconservative replication; the topological restriction of separating two intertwined strands is particularly evident when the DNA is the form of a double-stranded ring. In some cases, it appears that the cellular apparatus also takes advantage of the unique topology of two intertwined DNA strands—in prokaryotes at least, the maintenance of duplex DNA in a negatively supercoiled or underwound state is important for a number of cellular processes.

DNA topoisomerases are enzymes that control and modify the topological states of DNA. By transiently breaking a DNA strand and passing another strand through the transient break (type I topoisomerases), or by transiently breaking a pair of complementary strands and passing another double-stranded segment (type II topoisomerases), these enzymes can catalyze many types of interconversions between DNA topological isomers (topoisomers). Examples are catenation and decatenation, and knotting and unknotting. DNA topoisomerases have been found to affect a number of vital biological functions, including the replication of DNA, and have been implicated in others. A number of reviews and commentaries on this class of enzymes have been published (1–16).

This review summarizes studies carried out in roughly a four-year period since the publication of the last comprehensive review in these volumes (5). During this period much progress has been made. In the bacterium *Escherichia coli,* a new type I topoisomerase, DNA topoisomerase III, has been found. This enzyme, similar to the other type I enzymes, DNA topoisomerase I (known in earlier years as the ω protein), relaxes negatively supercoiled DNA. The only known type II topoisomerase in *E. coli,* DNA gyrase (DNA topoisomerase II), catalyzes the negative supercoiling (linking number reduction) of DNA at the expense of ATP hydrolysis. In a thermophilic archebacterium *Sulfolobus acidocaldarius,* a topoisomerase termed "reverse gyrase," appears to use ATP hydrolysis to drive the positive supercoiling (linking number increment) of DNA. There are several other topoisomerases in the *Sulfolobus,* including a negative supercoiling DNA gyrase activity. The existence of multiple topoisomerases with diametrically opposing activities in both eubacteria and archebacteria raises interesting questions on their regulation. In *E. coli,* there is evidence that they are regulated homeostatically through the degree of supercoiling of DNA.

The purification of the topoisomerases has been much improved. For the bacterial enzymes, the cloning of the genes has simplified the preparation of large amounts of the enzymes. Improvements in procedures for the eukaryotic topoisomerases have provided nearly homogeneous preparations in high yields; thus putting enzymological studies of the eukaryotic enzymes on equal footing with studies of the prokaryotic ones. Mechanistic studies of the enzymes have been aided by the use of unique DNA substrates, and by the development of electron microscopic methods that permit the detailed characterization of DNA topology. The transient cleavage of DNA accompanied by the simultaneous formation of an enzyme-DNA covalent link, a hallmark of topoisomerases, has been established and in several cases, the link has been shown to be a phosphotyrosine bond. A number of antitumor drugs are found to induce eukaryotic DNA topoisomerase II–mediated cleavage of DNA in vitro, raising the possibility that their pharmacological target in vivo might be the topoisomerase.

Much progress has been made in the genetic and biochemical dissection of the functional roles of the topoisomerases. In *E. coli,* the gene *topA* encoding DNA topoisomerase I has been identified, cloned, and sequenced. The genes *gyrA* and *gyrB* encoding the two subunits of gyrase have been cloned. A number of temperature-sensitive gyrase mutants have been isolated. In the lower eukaryote yeast, mutants in both the type I and the type II enzyme have been isolated and the gene *top2* encoding DNA topoisomerase II has been cloned. These mutants and cloned genes have provided powerful tools in deducing the biological functions of the enzymes. The essentiality of the type II topoisomerase in all organisms is now well established; one important function of the enzyme is the resolution of a pair of intertwined double-stranded progeny DNA molecules in the terminal stage of replication. In *E. coli,* genetic analysis shows that DNA topoisomerase I normally serves an essential cellular function, though its absence can be tolerated in strains carrying compensatory mutations in other genes.

Biochemical studies of the enzymes have complemented genetic analysis of their functions and regulation. Studies on poly-ADP-ribosylation and phosphorylation of topoisomerases have been initiated. The presence of a tightly associated or intrinsic protein kinase activity in eukaryotic topoisomerase II raises interesting questions. Studies on the cellular locations of the enzymes have begun. Immunostaining of *Drosophila* polytene chromosomes suggests that eukaryotic topoisomerase I is enriched in actively transcribed regions, including the nucleolus where transcription of rRNA by RNA polymerase I occurs. Eukaryotic DNA topoisomerase II has been identified as a major scaffold protein of mitotic chromosomes and is present in the nuclear matrix.

It is clear that through their actions on DNA topology and the breakage and rejoining of DNA strands, the topoisomerases are involved in a diverse number of vital cellular processes. Their study offers a rare opportunity to combine chemical topology, enzymology, and biochemical and molecular genetics.

INTRODUCTION

The early research of a DNA topoisomerase was inspired by the problem of untwining the two parental strands of DNA by semiconservative replication (17). With the discovery of circular DNA, it became clear that a "swivel" must be introduced into a circular DNA composed of two intertwined single-stranded rings to permit strand separation (18, 19). It was reported in 1969 that an activity in *E. coli* extracts is capable of relaxing supercoiled DNA (20). Initially, it was thought that the activity was an endonuclease, and that the relaxation of the DNA occurred by a cycle of endonucleolytic nicking and resealing of the nick by DNA ligase. Subsequent purification of the enzyme, however, shows that a single enzyme is capable of relaxing negatively super-coiled DNA (21). This enzyme, originally designated as the ω protein (21), has been renamed *E. coli* DNA topoisomerase I (3). Its lack of a requirement for a high-energy cofactor, such as ATP or NAD, has led to the postulate that the enzyme breaks transiently a DNA backbone bond and forms simultaneously a covalent enzyme-DNA intermediate; this transient breakage of the DNA back-bone bond permits changes in the linking number (termed the topological winding number in the earlier literature) between the strands, and the dissocia-tion of the enzyme from the DNA is accompanied by the reformation of the DNA backbone bond (21). An activity from mouse cell extract capable of relaxing both positively and negatively supercoiled DNA in the absence of a high-energy cofactor was found in 1972, and the same mechanism was pos-tulated for the eukaryotic enzyme (22). It turns out that both the bacterial and the eukaryotic enzyme belong to a class of topoisomerases classified as the type I enzymes (23). Mechanistically, they are characterized by the transient break-age of one strand at a time to permit the passage of another strand through the break. The two enzymes differ, however, in the covalent intermediates they form with the DNA; the eukaryotic enzyme is linked to a DNA 3' phosphoryl group (24–26), whereas the bacterial enzyme is linked to a DNA 5' phosphoryl group (27).

In 1976, an entirely new DNA topoisomerase, DNA gyrase, was discovered (28). This enzyme negatively supercoils the DNA, that is, reduces the linking number of a double-stranded DNA ring in an ATP-dependent fashion. The enzyme achieves the topological change by breaking transiently a pair of complementary strands in concert to pass another duplex DNA (3–5, 7, 9, 11, 13, 15, 29–31). An interesting consequence of this double-stranded breakage, passage, and rejoining is that the linking number changes by 2 in each cycle (23, 29, 31). Topoisomerases that catalyze double-stranded breakage, passage, and rejoining are now classified as type II enzymes (4, 5, 23). Another unique type II enzyme, the DNA topoisomerase of phage T4, was found in 1979 (32). It is ATP-dependent; in contrast to DNA gyrase, it catalyzes the relaxation of negatively or positively supercoiled DNA but not DNA supercoiling (32, 33). Remarkably, the T4 enzyme is similar in this respect to the type II DNA topoisomerase that has been subsequently isolated from eukaryotic organisms (34, 35). All type II DNA topoisomerases, in addition to their changing the supercoiled state of DNA, have been found to catalyze the catenation/decatena-tion and knotting/unknotting of duplex rings, (4, 5, 7, 10). As will be described later, the passage of duplex DNA molecules through each other is of key biological importance.

The discovery of DNA gyrase also brings into focus the concept that the cellular apparatus not only has to solve the topological problem of DNA that comes with its double-helix structure, but also appears to take advantage of that unique topological property. Although DNA purified from various natural sources has been known to be negatively supercoiled (deficient in linking number) relative to purified DNA (2), before the discovery of gyrase it was generally thought that the linking number deficiency was due to passive mechanisms such as stoichiometric binding of proteins. The existence of gyrase points to the active maintenance of the negatively supercoiled state of DNA, at least in prokaryotes, presumably for the purpose of facilitating certain cellular processes.

This review covers mainly recent studies on the type I and type II enzymes exemplified by the prokaryotic and eukaryotic enzymes mentioned above, which will be referred to in some passages below as the "archetype" topoisomerases. It should be pointed out, as was done in previous reviews (3, 5), that under certain conditions a topoisomerase may break a strand and transfer the enzyme-attached strand to a receiving group on a different strand. In such a case, the topoisomerase functions as a DNA strand-transferase. There are also classes of enzymes that normally function in DNA strand-transfer but under proper conditions can catalyze DNA topoisomerization as well. The most extensively studied examples include a class of enzymes involved in the initiation of DNA replication, such as the gene A protein of phage ϕX174 and the gene 2 protein of phage fd, and a class of enzymes involved in site-specific recombination, such as the phage λ *int* gene product and the resolvase of transposons Tn3 and $\gamma\delta$. Other enzymes involved in site-specific recombination, such as the phage P1 *cre* gene product and the yeast 2 micron plasmid *flp* gene product, share the basic characteristics of DNA topoisomerases in their reversible breakage and rejoining of DNA strands. All these DNA strand-transferases can be considered as special topoisomerases and they offer fascinating and diverse examples of DNA topoisomerization (3, 5, 36–54); their established roles in strand transfer in vivo also raise questions regarding whether the archetype topoisomerases might have similar roles. The page limitation, however, makes it impossible to cover these special topoisomerases aside from a few cursory comments.

NEW ENZYMES AND METHODS

New Enzymes

In *E. coli*, a new type I enzyme, DNA topoisomerase III, has been identified (55, 56). Similar to DNA topoisomerase I, the enzyme relaxes negatively (55–57) but not positively (57) supercoiled DNA, and is strongly inhibited by single-stranded DNA (57). If a single DNA topoisomer of a fixed linking number is used in the relaxation reaction, the linking numbers of the products of

DNA topoisomerase III differ from the initial linking number by both odd and even integers, which is diagnostic of the enzyme as a type I topoisomerase (23, 29–31). The generally similar properties between this enzyme and *E. coli* topoisomerase I suggest that the two might have parallel functions.

A type I topoisomerase has been partially purified from extracts of spinach chloroplasts (58). The enzyme requires Mg(II) ions and relaxes negatively but not positively supercoiled DNA. In these respects the enzyme resembles prokaryotic type I topoisomerases more than the eukaryotic type I enzyme, including the type I enzyme from plants (59–61). The similarity between the choloroplast and bacterial enzymes favors the theory of origination of chloroplasts by prokaryotic invasion (58).

A presumably type I intramitochondrial topoisomerase activity from rat liver, mouse L cells, and *Xenopus laevis* oocytes has been reported (62–64). It has yet to be resolved whether the activity is distinct from the nuclear enzyme (62–65).

In addition to topoisomerases from eubacteria and eukaryotes, enzymes from an archebacterium (66) *S. acidocaldarius,* a species found in acidic hot springs, have been examined (67). Four chromatographically distinct activities have been partially purified. One of these, Fraction 4, appears to cause the positive supercoiling of DNA in the presence of ATP and Mg (II). At its reaction temperature of 75°C, the specific linking difference [or superhelical density; for the definition of the quantity, see (5, 68–70)] of the DNA product is estimated to be about +0.03. This positive supercoiling activity has been termed "reverse gyrase" (67). Further purification of the enzyme is needed to determine whether the linking number increment of the DNA substrate is the result of an active process, or is due to the stoichiometric binding of the enzyme or contaminants in the reaction medium. In *E. coli,* positively supercoiled plasmid has been isolated from cells following inhibition of DNA gyrase by coumermycin (72). This linking number increment can be explained, however, by the stoichiometric binding of the drug-inactivated gyrase (16, 72), which is expected to increase the linking number from in vitro measurements with purified gyrase (73). Treatment of a covalently closed DNA with *E. coli* topoisomerase II', which consists of the gyrase A subunit and a 50,000-dalton fragment derived from gyrase B subunit, has also been shown to yield a positively supercoiled DNA product because of the stoichiometric binding of the enzyme (78). In an independent study, an ATP-dependent topoisomerase from *S. acidocaldarius* that relaxes negatively but not positively supercoiled DNA has been reported (71).

An activity resembling the archetype eukaryotic topoisomerase II has been partially purified from the trypanosomatid *Crithidia faciculata* (74), a flagellate protozoan with an unusual kinetoplast DNA structure in its single mitochondrion. The kinetoplast DNA consists of a catenated network of thousands of small duplex rings and a few larger rings [for a review, see (75)].

The trypanosomatid activity, similar to other type II eukaryotic enzymes, can catenate or decatenate covalently closed double-stranded DNA rings depending upon the ionic media. With singly nicked DNA as the substrate, however, catenation occurs but not decatenation (74). Catenated network of covalently closed rings after relaxation by a type I topoisomerase activity is readily decatenated; thus the enzyme appears to distinguish the relaxed and nicked substrates (74). It has been postulated that the small rings are released from the network prior to their replication and the replicated progeny rings are re-attached to the network (75). Newly replicated small rings of kinetoplast DNA contain nicks (75). Thus the discrimination of nicked circles in the decatenation reaction might be functionally significant (74).

Because both catenation and decatenation involve the same strand passage events, it appears paradoxical that the enzyme can catenate nicked rings but not decatenate them. The catenation and decatenation reactions in vitro are, however, carried out under different conditions. Thus one possibility is that in the decatenation assay but not the catenation assay nicks are inhibitory; another possibility is that an additional factor is involved in substrate discrimination. It is unknown whether the trypanosoamatid activity is of nuclear or mitochondrial origin.

A presumably type II topoisomerase activity from *Trypanosoma cruzi* has been reported (76, 77). It can catenate covalently closed DNA rings in the absence of ATP. It has been shown previously that a fragment of the *E. coli* *gyrB* protein can complement the *gyrA* protein to give an ATP-independent DNA relaxation activity termed DNA topoisomerase II' (78, 79); thus it is plausible that the *T. cruzi* catenation activity is a partially proteolyzed type II topoisomerase.

In addition to the isolation of the novel activities described above, progress has been made in the purification of the other topoisomerases (59, 65, 80–85b). Examples are the enzyme from HeLa cells (100,000 daltons), mouse cells (102,000 daltons), *Drosophila melanogaster* embryo (135,000 daltons), wheat germ (110,000 daltons), yeast (90,000 daltons), and avian erythrocyte nuclei (105,000 daltons). The circa 100,000–dalton type I enzyme from these organisms appears to be monomeric in solution, relaxes both negatively and positively supercoiled DNA, requires no ATP, and is stimulated by Mg(II). Smaller proteolytic products of the protein can apparently retain full relaxation activity of the enzyme (81, 86).

A type I topoisomerase with a polypeptide weight of 67,000 has been purified from mature ovaries and from nuclei of stage 6 oocytes of the frog *Xenopus laevis* (87). The ovaries activity is active in 150 mM monovalent salt and is stimulated by Mg(II). The oocyte nuclei activity, however, appears to be inactive in the absence of Mg(II) (87); this strict Mg(II) dependence has not been observed for the type I enzyme from other eukaryotic sources (87).

It is unknown whether the protein mass not required for the relaxation activity might serve additional functions, such as interacting with other cellular components. Eukaryotic DNA topoisomerase I has been shown to interact with nucleosomes and several chromosomal proteins (88, 89, 271), and its relaxation activity is stimulated by histone H1 (90) and the high mobility group (HMG) proteins (88, 90). The existence of a nuclear (91) and cytoplasmic (92) assembly of enzymes containing a topoisomerase activity, and a possible association between a type I topoisomerase activity and ribonucleoprotein particles (93), have been reported.

Eukaryotic type I topoisomerase with molecular weights higher than 200,000 have also been reported for the lower eukaryote *Ustilago maydis* (90) and the plant *Cauliflower influorescence* (60, 61), based on gel filtration of partially purified preparations. The *Ustilago* activity is inhibited by ATP and its nonhydrolyzable analogs (90). The inhibitory effect of ATP has similarly been observed in the case of the type I topoisomerase from chick erythrocytes (113). This inhibitory effect is surprising, in view of the lack of such an effect for eukaryotic topoisomerase I from a number of other organisms [see for example, (65)]. The type I topoisomerase from vaccinia virus core particles is stimulated by ATP, inhibited by ADP, β, γ-imido ATP, or coumermycin, although the enzyme does not hydrolyze ATP (93). Two general comments can be made about these observations. First, they show that the effects of ATP, coumermycin, and their analogs are insufficient for the identification of a type II topoisomerase activity. Second, ATP could affect topoisomerase activity indirectly through ion chelation and protein phosophorylation (see a later section), particularly when impure enzyme preparations are used.

The archetype type II eukaryotic topoisomerase has been purified to near homogeneity from HeLa (94), *D. melanogaster* (95, 95a, 96), and yeast (65, 83). The purified enzyme has a single subunit, and preparations from different species give peptide molecular weights in the range 150,000–180,000. The cloning of the structure gene *top2* encoding the yeast enzyme and the determination of the size of the coding sequence provide an independent estimate of the size of the subunit (97), which agrees with that measured for the purified protein (65, 83).

For the *E. coli* topoisomerases, the cloning of the genes *topA, gyrA,* and *gyrB* on phage or multicopy plasmid vectors has allowed overproduction of the enzymes they code in cells (98, 99). Genes encoding the subunits of *Bacillus subtilis* have also been cloned (100). In addition to the use of more conventional fractionation steps, affinity chromatography on novobiocin-Sepharose has been used in the purification of gyrase from *E. coli* (101) and *B. subtilis* (102). Novobiocin-Sepharose has also been used in the purification of *E. coli* DNA topoisomerase III (57), to which the drug is not an inhibitor (55–57). Recent purification procedures of phage T4 DNA topoisomerase yield enzyme containing gene 39, gene 52, and gene 60 products with peptide weights of 56,500,

48,000, and 18,000, respectively (102a, 103), confirming previous genetic and biochemical evidence that the enzyme has three subunits (32, 33).

Assays of DNA Topoisomerases

The earlier assays of DNA topoisomerases have been reviewed previously (8). Gel electrophoresis techniques continue to occupy the center stage. The use of unknotting or decatenation reaction to assay specifically type II topoisomerase activities has been developed (30, 31, 94, 104). Tailless capsids of phage P2 and P4 provide a convenient source of knotted DNA (104, 105), and the catenated network of trypanosomatid kinetoplast DNA is particularly convenient for decatenation assays (94, 106). The large size of the network also makes it sedimentable when spun in a benchtop centrifuge, and thus the dissolution of the network by a topoisomerase can be followed by measuring the amount of DNA in the supernatant (107). Catenated networks of DNA can also be prepared in high yield by the use of an appropriate topoisomerase and spermidine (182), histone H1 (35), or HMG17 protein (108).

A rapid assay based on the retention of catenated DNA networks on nitrocellulose filter has been developed (115). This assay, first used in studying the catenation of gapped DNA rings by *E. coli* DNA topoisomerase I (115), should be generally applicable in catenation as well as decatenation reactions.

The sodium dodecyl sulfate-induced formation of the covalent topoisomerase-DNA product has been used to assay the enzymes. The addition of K^+ to the mixture containing sodium dodecyl sulfate selectively and quantitatively precipitates DNA covalently linked to protein (111–113). When end-labeled DNA substrates are used, an activity such as the eukaryotic type I topoisomerase that links covalently to the 3' DNA phosphoryl group (24–26) is readily distinguishable from topoisomerases that link to the 5' phosphoryl group in their covalent complexes. Topoisomerases such as *E. coli* DNA topoisomerase I and III, both of which are linked to the 5' phosphoryl end when cleaving DNA, can be distinguished and assayed by their different sequence specificities (113a); each enzyme generates a distinctive family of DNA fragments when a restriction fragment labeled at one appropriate end is cleaved by the enzyme under DNA-excess conditions.

Unique DNA Substrates and the Detailed Characterization of DNA Topology

Covalently closed duplex DNA relaxed in the presence of the drug netropsin has been found to have a higher linking number than DNA relaxed in the absence of the drug; this provides a method of preparing moderately positively supercoiled DNA (114). DNA rings containing nicks, small gaps (115, 116), closed duplex rings with mismatched single-stranded regions (117), and oligonucleotides of defined lengths (118) have been used to test various topoisomerization reactions. Cyclization and covalent closure of restriction fragments shorter than 600 bp usually yield rings of a single linking number (119, 120), which are useful for classifying topoisomerases into type I and type II enzymes.

The use of various two-dimensional gel electrophoresis techniques has helped the resolution of topoisomers of different linking numbers (121, 122). Two-dimensional gel electrophoresis has also been used ingeniously in the identification of different classes of multiply intertwined catenated dimers (123).

The coating of DNA with *E. coli recA* protein or a phage T4 protein *uvsX* increases greatly the diameter of the fiber, and electron microscopic procedures have been developed to determine, whenever there is a crossover of DNA segments, which segment is over and which segment is under (124–126). The use of these specific DNA binding proteins has improved the earlier procedures (127, 128), and has permitted the detailed characterization of the topology of DNA rings. The introduction of the Schubert description of curves has also facilitated the classification of the various stereoisomers of catenanes and knots (50, 129).

MECHANISTIC ASPECTS

The Topoisomerase-DNA Covalent Complex and Strand Transfer

It is now established that the transient breakage of a DNA backbone bond by a topoisomerase is accompanied by the formation of a covalent enzyme-DNA intermediate. The covalent linkage is found to be an O^4-phosphotyrosine bond for all archetype DNA topoisomerases examined to date, including *E. coli* DNA topoisomerase I (130), rat liver DNA topoisomerase I (131), bacterial gyrase (DNA linked to subunit A; 130), and T4 DNA topoisomerase (DNA linked to the gene 52 protein subunit; 132). The ϕX174 gene A protein also forms a phosphotyrosine linkage with DNA (37, 133). In the case of the transposon $\gamma\delta$ resolvase, the covalent enzyme-DNA link has been shown to be a phosphoserine bond (134, 135).

Direct demonstration that the covalent topoisomerase-DNA complex is an active intermediate in DNA topoisomerization has been achieved. Treatment of single-stranded DNA with rat liver type I topoisomerase in the absence of a protein denaturant yields a covalent complex in which the native protein is attached to a 3' terminal phosphate; cyclization of the DNA attached to the protein to give a single-stranded DNA ring, or the transfer of the attached strand to a 5' OH-receiving group on a different strand, has been demonstrated (136). Similar experiments have been carried out with HeLa DNA topoisomerase I (137). When bacterial DNA topoisomerase I is incubated with a short oligonucleotide, the enzyme cleaves the oligonucleotide to give a covalent complex in which the native protein is attached to a 5' phosphoryl group (118). The enzyme-linked oligonucleotide can be transferred to a 3' OH acceptor (138). These experiments with the eukaryotic and prokaryotic topoisomerases are important not only in establishing the functionality of the covalent intermediate, but also in showing that under particular circumstances the topoisomerases can be DNA-strand transferases.

The Cleavage Reaction

Because of the ease of mapping the sites of cleavage, either in purified systems or in vivo, extensive studies have been carried out on the cleavage of DNA by topoisomerases. Strong DNA binding sites that show little cleavage upon addition of protein denaturants to the enzyme-DNA complex are known (5, 139); thus strong binding sites are not necessarily strong cleavage sites, and by inference strong cleavage sites may not be strong binding sites or preferential sites in the catalysis of DNA breakage and rejoining. Nevertheless, a cleavage site is necessarily a binding site and it seems reasonable to assume that the sequence of the site and the position of cleavage within the site reflect the way a DNA topoisomerase interacts with its substrate.

DNA CLEAVAGE BY TYPE I DNA TOPOISOMERASES   As shown by studies with *E. coli* and *Microccocus luteus* DNA topoisomerse I, relaxed double-stranded DNA is refractory to cleavage by bacterial DNA topoisomerase I (27) and single-stranded DNA is cleaved at the position indicated by the arrow in the sequence 5' - - - -CNNN $\downarrow$ - - - - (N represents any base) (130, 55). The presence of a C four nucleotides away on the 5' side of the break (position -4) is observed in most of the sites mapped, but is neither sufficient nor necessary for the cleavage reaction (27, 118, 130). The oligomers dA_7 and dT_8 are the smallest oligomers that can be cleaved; in each case, cleavage occurs between the fourth and fifth nucleotide from the 3' end of the starting material (118). With longer oligomers, a cluster of cleavage sites in the middle of the oligomer extending to 5 nucleotides away from each end were observed; cleavage of oligo dG and oligo dC occurs less readily compared with cleavage of oligo dA and dT (118).

DNA substrates containing single-stranded gaps or loops have been used to determine the structural determinants of the cleavage sites generated by the addition of OH^- to the enzyme-DNA complex (116). Cleavage studies with these substrates indicate that bacterial DNA topoisomerase I preferentially interacts with a segment of DNA composed of both a single- and double-stranded region, and that the site of cleavage falls within the single-stranded region. Presumably, when the enzyme binds to a negative supercoiled DNA it unpairs a short region of the duplex DNA, which was first postulated to account for the enzyme's specificity for negatively supercoiled DNA and its inhibition in its relaxation reaction by single-stranded DNA (21). The sites of cleavage of a negatively supercoiled DNA by *Haemophilus gallinarium* DNA topoisomerase I appear to correlate with the sites of cleavage by the single-strand specific nuclease S1 (140).

Cleavage of single-stranded DNA by the type I enzyme *E. coli* DNA topoisomerase III has also been studied (55, 56). The sequences of the cleavage sites differ from those of topoisomerase I.

Cleavage of DNA by eukaryotic type I DNA topoisomerase has been studied extensively. In addition to its linkage to a 3' rather than 5' phosphoryl group when cleaving the DNA (24–26), the eukaryotic enzyme differs from bacterial DNA topoisomerase I in a number of ways in the cleavage reaction. First, whereas relaxed DNA is refractory to cleavage by the bacterial enzyme (27), it is readily cleaved by the eukaryotic enzyme (24, 141–143), though at a slower rate than the rate of cleavage of supercoiled DNA (144). Second, as mentioned earlier, the physical disjoining of a long DNA strand by the bacterial enzyme is observable only when a protein denaturant is added (27), whereas disjoining of a single-stranded DNA by the eukaryotic enzyme can occur in the absence of a protein denaturant (136, 137, 142). Third, the sequence specificities of the two enzymes differ. With the eukaryotic type I enzyme, there is significant bias in nucleotide composition for the four nucleotides 5' to the break site (−4 to −1 positions), and perhaps the nucleotide immediately 3' to the break site (+1 position) (141, 143). Furthermore, with single-stranded DNA, it has been shown that cleavage by the eukaryotic enzyme occurs preferentially at sites with sequences that can pair with complementary sequences on the same strand (145, 146). This indicates that cleavage by the eukaryotic enzymes occurs preferentially in short duplex regions, in contrast to the case with the prokaryotic enzyme described above. Within the eukaryotic kingdom, the type I enzymes isolated from different species including yeast, *Drosophila*, wheat, and human show common cleavage sites, although the cleavage patterns of a given DNA sequence differ considerably for enzymes from phylogenetically distant sources (116, 142, 143).

The results summarized above for DNA cleavage by eukaryotic DNA topoisomerase I suggest that the enzyme interacts with its substrate over a region covering at least four base pairs 5' to the point of cleavage, with the DNA in the double-stranded form. There is probably little interaction in the region 3' to the point of breakage.

DNA CLEAVAGE BY TYPE II TOPOISOMERASES   Earlier studies of the gyrase-DNA complex and the cleavage of the DNA by the enzyme have been reviewed (4, 5, 7, 10; see also 147–156). The sequences of a number of cleavage sites generated in vitro have been determined; they show sequence preferences but the determinants of sequence specificity are less apparent (139, 150, 151, 154).

Many gyrase cleavage sites on an intracellular plasmid in *E. coli* in the presence of oxolinic acid have been mapped (157, 158). The in vivo sites share the nucleotide preferences observed in vitro (157).

The cleavage of DNA by the type II phage T4 DNA topoisomerase has been analyzed (159). Similar to *E. coli* gyrase, the T4 enzyme introduces the same type of staggered breaks accompanied by the covalent linkage of the gene 52 protein subunit (132) to the 5' ends, and the cleavage is strongly stimulated by oxolonic acid (159). The T4 enzyme shows little specificity on cytosine-containing T4 DNA but higher specificity on the native T4 DNA containing glucosylated hydroxymethylcytosine (159). ATP stimulates cleavage of the glucosylated native T4 DNA but has little effect on ordinary C-containing DNA (159); the stimulatory effect of ATP on cleavage by *E. coli* gyrase (160) and by *Drosophila* type II topoisomerase (161) has also been reported. In contrast to bacterial gyrase, which does not cleave single-stranded DNA (152, 162), the T4 DNA topoisomerase is shown to bind tightly and to cleave single-stranded DNA (162). Denaturation of the enzyme is not necessary to disjoin the broken single strand (162). Surprisingly, whereas oxolinic acid stimulates cleavage of duplex DNA by 100-fold, it is inhibitory to cleavage of single-stranded DNA (162). The sequences of several of the cleavage sites in single-stranded DNA suggest that the T4 enzyme might cleave near the base of a hairpinned structure. Single-stranded DNA coated by the T4 gene 32 protein blocks the topoisomerase cleavage reaction (162).

The eukaryotic type II topoisomerase, which is a homodimer in solution (94, 96, 65), yields identical cleavage site structure as the other type II enzymes described above (95). The sequences of cleavage sites of the *Drosophila* enzyme have been analyzed (163).

ANTITUMOR DRUGS THAT STIMULATE EUKARYOTIC DNA TOPOISOMERASE II MEDIATED DNA BREAKAGE   A number of antitumor drugs are known to induce DNA breakage in mammalian cells. In several cases it has been shown that the DNA fragments contain tightly bound protein or blocked ends (164–171), and the possible involvement of a topoisomerase has been proposed (165, 167, 168, 170, 172–174).

Direct evidence that some of these drugs enhance DNA cleavage by purified eukaryotic DNA topoisomerase II has been reported recently (175–178). The intercalative acridine derivative 4'-(9-acridinylamino)methanesulfon-*m*-anisidide (*m*-AMSA), stimulates the breakage of DNA by mammalian DNA topoisomerase II upon treatment with sodium dodecyl sulfate. Interestingly, T4 DNA topoisomerase–mediated DNA cleavage is strongly stimulated by the drug, and the possibility that T4 topoisomerase may be a specific target of the drug in vivo has been raised (132). The ortho-isomer of the drug, *o*-AMSA, is not an effective antitumor drug and is less effective in enhancing topoisomerase

II–mediated DNA cleavage in vitro (175). Preliminary studies suggest that a number of other intercalative and nonintercalative antitumor drugs including adriamycin, 5-iminodaunorubicin, ellipticine, 2-methyl-9-hydroxyellipticine, and epipodophyllotoxins, VP-16 and VM-26, function similarly in vitro (175, 177–178b). The drug m-AMSA is also inhibitory to mammalian DNA topoisomerase II at a concentration of 20 μg/ml (175). A number of intercalative drugs have also been tested for their inhibition of *T. cruzi* topoisomerase activities (76, 179). One of the drugs, 2-6-dimethyl-9-hydroxyellipticinium, is a potent inhibitor of the trypanosome type I topoisomerase but not mammalian type I topoisomerase (76).

In simian virus 40 (SV40)–infected cells, the cleavage sites in the SV40 chromosome upon addition of sodium dodecyl sulfate to m-AMSA treated cells have been mapped (180, 181). The major site is in the regulatory region of the viral chromosome; this cleavage is not observed in cells treated with o-AMSA (180, 181). The possibility that the target of m-AMSA in mammalian cells is DNA topoisomerase II is supported by examining DNA fragments produced in cells treated with the drug (180, 181).

DNA Binding and the Catalysis of DNA Topoisomerization

The topoisomerization reactions catalyzed by the type I and type II DNA topoisomerases have been reviewed (3–10). Basically, the type I enzymes can catalyze the relaxation of supercoiled DNA (including the intertwining of single-stranded rings of complementary sequences, which can be considered as relaxing a duplex ring of linking number zero), knotting/unknotting of single-stranded rings, and knotting/unknotting or catenating/decatenating of double-stranded rings containing single-stranded scissions; the type II enzymes can change the linking number of a covalently closed DNA ring in steps of 2, and can catenate/decatenate or knot/unknot double-stranded rings with or without nicks.

It has been somewhat puzzling why the bacterial type I enzyme relaxes negatively but not positively supercoiled DNA (21). The notion that the relaxation reaction requires the disruption of a short helical segment by the enzyme, which has been described earlier, is further supported by two recent experiments. In one, it is found that duplex DNA rings with short single-stranded gaps are rapidly linked into a catenated network by bacterial topoisomerase I in the presence of a hydrophilic polymer such as polyvinyl alcohol (115). The effect of the polymer is probably to favor DNA condensation; spermidine (182), HMG17 (108), H1 (35), or a yeast-DNA-binding protein DBP1 (83) have been found in other systems to favor catenation by bringing DNA molecules together. In a second experiment, a covalently closed double-stranded DNA ring containing a short single-stranded loop is used as the substrate for the bacterial enzyme. It is shown that the presence of the single-stranded loop allows the enzyme to relax the duplex ring even when it is positively supercoiled (117).

The knotted products produced by treatment of nicked double-stranded rings with *E. coli* topoisomerase I have been analyzed by electron microscopy, using the *recA* protein coating procedure (183). All isomers of knots that are theoretically possible are observed; formation of complex knots requires a high temperature and a large excess of the topoisomerase. It has been proposed that excess enzyme might be needed in contorting the DNA into a form that yields a knot upon strand passage.

The kinetics of DNA relaxation by a homogeneous preparation of *D. melanogaster* DNA topoisomerase II has been studied extensively (161). The enzyme interacts preferentially with negatively supercoiled DNA over relaxed or nicked DNA, from kinetic measurements of DNA relaxation and ATP-hydrolysis rates (161), and from substrate binding studies (184). The enzyme relaxes positively and negatively supercoiled DNA, however, at a comparable rate (161). In this respect, *E. coli* gyrase binds preferentially to relaxed DNA over negatively supercoiled DNA by an order of magnitude (185).

The knotting of DNA by *Drosophila* topoisomerase II has also been studied (186). Excess enzyme is required. In contrast to T4 topoisomerase, which is the first type II enzyme known to knot DNA (21), knotting with the *Drosophila* enzyme does not require the DNA to be supercoiled. Furthermore, the reaction is stimulated by ATP, and the knotted DNA appears to be a stable product rather than a transient intermediate.

Strand Passage

The topology of DNA topoisomerization dictates the passage of DNA strands; the linking of two single-stranded rings, for example, cannot occur without one strand passing through the other. Experiments have established that there are at least two distinctive modes of strand passage: the transient breakage of one strand to permit the passage of another, and the transient breakage of a pair of complementary strands in a duplex DNA to permit the passage of another duplex DNA segment. These modes form the basis of classifying the topoisomerases into the type I and the type II enzymes (3–11, 23).

The molecular aspects of the strand passage events for the various topoisomerases are largely unknown. As described earlier, the specificity of bacterial topoisomerase I for relaxing negatively supercoiled DNA can be attributed to the requirement of disrupting a short duplex DNA segment, and the enzyme appears to interact simultaneously with both double- and single-stranded regions. The cleavage specificity suggests that the enzyme is interacting with both sides of the point it cleaves. Thus it is likely that the enzyme bridges the two ends of the broken strands, through covalent and noncovalent interactions with one side and noncovalent interactions with the other. In this sense, the bacterial type I enzyme might be similar to the type II enzymes in that the transiently broken DNA strands are bridged by the enzymes (3, 5, 10, 23, 29). With the eukaryotic type I topoisomerase, however, the cleavage specificity does not suggest enzyme bridging; the transient break appears to occur, however within a double-stranded region (145). Thus the modes of actions of both the prokaryotic and eukaryotic type I topoisomerase suggest the need to keep the transiently broken DNA ends from coming apart. In the case of the prokaryotic enzyme, the breakage appears to occur in a single-stranded region, but the enzyme probably bridges the two ends; in the case of the eukaryotic enzyme, enzyme-bridging is probably not involved, but the break appears to occur in a double-stranded region and thus the ends are held together by the complementary strand.

One potential way of testing the enzyme bridging model for a type I enzyme is to determine whether for a type I enzyme each cycle of breakage and rejoining changes the linking number by unity. If the enzyme is not bridging the break, multiple strand passage events can occur in between the breakage and rejoining events, and the linking number change can be greater than one. It has been shown that in a type I toposiomerase catalyzed relaxation reaction the linking number changes by integral numbers odd or even (187, 188). Furthermore, for *E. coli* DNA topoisomerase I, relaxation of negatively supercoiled DNA does not follow a one-hit mechanism: the linking number of each DNA increases gradually (21, 188a), and the increment can be as small as one when the reaction is slowed down sufficiently (188). The lack of an experimental method to determine the number of breakage and rejoining cycles, however, makes it impossible to determine the linkage change per cycle under a wide range of experimental conditions.

BIOLOGICAL ASPECTS

Cellular Locations

The presence of topoisomerase activity in mitochondrion and chloroplast has been mentioned earlier. The bulk of the topoisomerase activity in a eukaryotic cell is present in the nucleus. Immunofluorescent staining of polytene chromosomes of *Drosophila* third instar larval salivary glands with affinity-purified antibodies against *Drosophila* DNA topoisomerase I indicates that the enzyme is preferentially associated with the transcriptionally active loci (189).

For *Drosophila* DNA topoisomerase II, immunocytochemical studies performed with permeabilized whole cells from third instar larval salivary glands indicate that the enzyme is also associated with the polytene chromosome;

some banding pattern is evident, but the distribution of the enzyme along the polytene chromosome is not particularly uneven (190). During mitosis, the enzyme appears to distribute diffusely throughout the cell (190).

An interesting feature of the organization of mitotic chromosomes is the folding of the nucleoprotein fiber into a looped structure, and "scaffolding" proteins that are present in folded chromosomes isolated from eukaryotic cells have been postulated to interact with the base of the loops (191). One of the major scaffolding proteins of chick mitotic chromosomes, CSc-1, has been identified as the eukaryotic DNA topoisomerase II (192). The enzyme has also been identified as a major polypeptide of *Drosophila* nuclear matrix-pore complex-lamina fraction (190), an entity operationally defined as the insoluble residue of nuclease-digested nuclei after extraction with detergent and concentrated salt solutions (193). The association of DNA topoisomerase II with chromosomal scaffolding and nuclear matrix in eukaryotes may be related to the earlier finding in *E. coli* that the products of several genes of phage T4, which have been shown subsequently to encode T4 DNA topoisomerase, appear to be membrane-associated (194, 195). The association of an ATP-dependent DNA nicking activity in epidermal growth factor receptor has also been reported (195a). The possibility that this activity might also have a DNA rejoining activity similar to a topoisomerse has been pointed out (195a).

Modifications of Topoisomerases

DNA topoisomerase I purified from Novikoff ascites cell extracts is found to be phosphorylated (196, 197) at one major site (198). The partial purification of a protein kinase from human cells that phosphorylates eukaryotic topoisomerase I has been reported (199). In vitro, the kinase phosphorylates three serine and threonine residues at major sites, one of which coincides with the in vivo site of phosphorylation (198). The relaxation activity of the topoisomerase is enhanced by phosphorylation and is decreased by treatment with phosphatase (197).

In vitro, it has been demonstrated that retrovirus and cell-encoded tyrosyl protein kinases phosphorylate eukaryotic and bacterial DNA topoisomerase I (200). The formation of the DNA phosphoryl-protein tyrosyl covalent link by pre-incubation of *E. coli* DNA topoisomerase I with oligo dA or oligo dT blocks tyrosine phosphorylation by the kinases, suggesting that the site of kinase phosphorylation might be the essential tyrosine involved in DNA covalent linkage or one that is very close to it (200). Chemical modification of the tyrosine residues in bacterial topoisomerase I and gyrase with tetranitromethane also leads to inactivation of the enzymes; the binding of DNA to the enzymes retards this inactivation (201).

A protein kinase tightly associated with *Drosophila* DNA topoisomerase II has been found (202). Serine and threonine residues are modified (202). Coinactivation of the kinase and the topoisomerase by heating or *N*-ethyl maleimide treatment has been observed, thus raising the possibility that the kinase might be an intrinsic activity of the topoisomerase.

Recently it has been found that purified poly(ADP-ribose) synthetase, a DNA strand breakage–activated enzyme present in most eukaryotic nuclei (203), modifies and inactivates eukaryotic DNA topoisomerase I (204–207). Although the high turnover rate of poly-ADP-ribosylation prevents the demonstration of this reaction in vivo, the finding that DNA topoisomerase I is the best exogenous acceptor of poly(ADP-ribose) known to date (206) hints that the modification of the enzyme might be biologically significant.

Roles of Topoisomerases in the Initiation of Replication

Earlier studies of bacteria have strongly implicated DNA gyrase and T4 DNA topoisomerase in the initiation of replication [reviewed in (3–5)]. In vitro systems containing soluble enzyme fractions capable of initiating bidirectional replication from the unique E. coli oriC chromosomal origin cloned on plasmids have been developed, and gyrase is required in these systems (207a–207c). In a system containing purified proteins, both gyrase subunits are necessary for replicating the oriC plasmid; furthermore, gyrase, dnaA protein, dnaB protein, and dnaC protein are needed in addition to RNA polymerase to give a functional intermediate that can support DNA replication by DNA polymerase III holoenzyme and other components without further RNA synthesis (207c). In a system containing purified proteins and a high concentration of polyvinyl alcohol, the replication of oriC plasmids requires gyrB, though the gyrA requirement is less certain (207c). The role of gyrase in these systems remains to be elucidated; the dependence on gyrase for replication of linearized oriC plasmids is as strong as that of the circular plasmids (207c).

In the in vitro studies described above, it has also been observed that there is a class of proteins that suppress initiations on plasmids lacking the oriC sequence and that do not depend on dnaA gene protein (207b). DNA topoisomerase I is one of this class of "specificity factors" (207b). The role of E. coli DNA topoisomerase I does not appear to be simply adjusting the degree of supercoiling of the DNA; eukaryotic DNA topoisomerase I cannot substitute the bacterial enzyme in this system (207b).

Resolution of Intertwined Pairs of Newly Replicated DNA by Type II DNA Topoisomerases

A crucial role of a type II DNA topoisomerase in the terminal stage of replication in both prokaryotes and eukaryotes has been implicated. The production of DNA catenanes by replication has been known for some time (207d, 207e). In eukaryotic cells, there is now strong evidence that a pair of newly replicated progeny DNA molecules are multiply intertwined (123, 208, 209). Strong support for a crucial role of a type II DNA topoisomerase in the resolution of intertwined progeny molecules comes from studies with mutants of yeast DNA topoisomerase II. The yeast *top2* gene is essential for viability (97, 211, 212). Using a temperature-sensitive *top2* mutant, it has been shown that at a nonpermissive temperature yeast plasmids accumulate as multiply intertwined catenated dimers after one round of DNA synthesis (211). This experiment shows that at least in yeast, the known DNA topoisomerase II is the enzyme responsible for the resolution of intertwined progeny molecules. Although eukaryotic chromosomes are linear topologically, their great length, looped structure, and the presence of multiple replicons appear to require the action of DNA topoisomerase II to resolve pairs of intertwined pairs of progeny chromosomes. Analysis of synchronized populations of replicating *Saccharomyces cerevisiae* cells carrying *top2* ts mutations shows that cells are arrested at medial nuclear division (211). With a *top2* ts mutant of the fission yeast *Schizosaccharomyces pombe,* it has been shown that blockage of DNA topoisomerase II does not prevent cell division; segregation of the DNA into the progenies no longer occurs normally, however. Light microscopy of dividing cells stained with a DNA binding fluorescent dye shows that some DNA is pulled out of a pair of newly divided cells at a nonpermissible temperature (212). The requirement of a type II DNA topoisomerase in resolving a pair of newly divided catenated chromosomes also appears true in bacteria (213).

An essential role of eukaryotic DNA topoisomerase II in replication is also consistent with the observation that the level of this enzyme, in rat liver nuclei, but not that of DNA topoisomerase I, is significantly increased during liver regeneration after partial hepatectomy (210). The stimulation of toposiomerase II in human and mouse fibroblasts by epidermal growth factor has also been reported (92).

Untwining of the Parental Strands During DNA Replication

The results summarized in the above section establish the involvement of the known type II DNA topoisomerase in the resolution of intertwined pairs of newly replicated DNA molecules. At the same time, because the blockage of the yeast type II topoisomerase seems to block only the unlinking of the terminally replicated DNA molecules, the results suggest that another enzyme or enzymes might be responsible for solving the topological problem of untwining the parental strands during DNA replication in eukaryotes. It cannot be ruled out, however, that the type II topoisomerase itself is the "swivelase" of replication, and that the decatenation defect observed in yeast *top2* mutants reflects only leakiness of the mutants and a more stringent requirement of the enzyme in decatenation.

In *E. coli,* the type II DNA topoisomerase is probably involved in the propagation step, although there has been some diversity in interpretation. Mutants of phage T4 topoisomerase appear to require *E. coli* gyrase for initiation but not propagation (214). Gyrase inhibitors stop *E. coli* DNA replication more rapidly, suggesting blockage of the elongation step; this might be related to, however, the properties of the gyrase-inhibitor-DNA ternary complex (215–219). Several ts mutants of gyrase have been shown to block the initiation and termination (see the preceding section) step, but does not appear to affect the elongation step (213–216, 220, 221, 247; see also 222, 223). One gyrase ts mutant, however, has been reported to stop replication rapidly at a nonpermissive temperature and is implicated in blocking both initiation and chain elongation (224).

It is probable that other topoisomerases are also partially responsible for solving the swivel problem of replication. A protein isolated from nuclear extracts of uninfected HeLa cells has been found to be required for the elongation of replicating intermediates of adenovirus DNA to full length in an in vitro system (225). Highly purified preparations of this protein, which has a molecular weight of 25,000–45,000 and is designated nuclear factor II, contain a topoisomerase activity (225). Furthermore, purified eukaryotic DNA topoisomerase I, but not *E. Coli* DNA topoisomerase I, can substitute for this factor (225). In yeast, the lack of a replicative phenotype in *top1* mutants (212, 226) suggests that the type I topoisomerase cannot be the only enzyme responsible for the untwinement of parental strands. For *S. pombe,* although *top1* mutation lacks a phenotype, *top1 top2* double mutants differ from *top2* mutants in their being blocked at an earlier replicative stage (212).

In *E. coli,* mutants are known in which the *topA* gene is deleted (227); thus *E. coli* DNA topoisomerase I cannot be the sole enzyme that provides the "swivel" in the unraveling of the strands of a replicating DNA. Studies on the physiological roles of the enzyme are complicated by the finding that mutations in *topA* are often accompanied by mutations elsewhere to compensate the physiological consequences of topoisomerase I dysfunction (228–230). To avoid such complication, a strain in which the only copy of *topA* gene is expressed from a regulated *lac* promoter has been constructed; maintenance of such a strain in the presence of the *lac* operon-inducer IPTG avoids the acquisition of compensatory mutations (69, 116). It is found that a reduction in the intracellular level of the enzyme leads to a reduction in the growth rate (234). The use of a strain carrying a nonsense mutation in *topA* and a temperature-sensitive suppressor has led to the same conclusion (232). The involvement of DNA topoisomerase I in modulating the supercoiling of intracellular DNA (see the section below), which affects a multitude of cellular processes, makes it difficult to sort out the pleiotropic effects of dysfunction of this enzyme. The presence of another type I topoisomerase in *E. coli,* DNA topoisomerase III (55–57), which may complement or supplement the functions of topoisomerase I, complicates the situation further.

In relation to the involvement of topoisomerases in DNA replication, the presence of topoisomerase activities in the multienzyme complexes is of interest. The presence of a topoisomerase activity in a nuclear and a cytoplasmic assembly (90, 91) has been mentioned earlier. A large complex from extracts of yeast cells capable of replicating the 2-micron plasmid in vitro has been reported to contain DNA topoisomerase II (235). In the case of T4 topoisomerase, an essential role of the gene 39 subunit in deoxyribonucleotide synthesis (236) may be related to its plausible presence in a multienzyme replication complex (237). A cautionary note might be interjected, however, regarding genetic studies on the biological roles of multisubunit topoisomerases when mutants of a single subunit were employed. Strictly speaking, such studies pertain to the particular subunit only. As an example, in the synthesis of the viral strand of ϕX174, it appears that the *gyrA* subunit but not the *gyrB* subunit is required (238).

Regulation of DNA Supercoiling by Multiple Topoisomerases

The strongest genetic evidence in favor of a dynamic interplay among the topoisomerases in the maintenance of the degree of supercoiling of intracellular *E. coli* DNA comes from studies of mutants that compensate the otherwise lethal effect of deleting the *topA* gene; some of the compensatory mutations have been mapped in *gyrA* or *gyrB* (228–230). Direct measurements of the linking number of plasmid DNA isolated from partially defective *topA* mutants without compensatory mutations show that the linking number is lower (i.e., the DNA is more negatively supercoiled) in *E. coli* (227, 229) as well as in *S. typhimurian* (238a). A net decrease in the negative superhelicity of intracellular DNA in some of the strains carrying *gyrA* or *gyrB* compensatory mutations has been observed (228, 229, 239), indicating that the *gyrA* or *gyrB* compensatory mutations are down-mutations that reduce the intracellular activity of gyrase. The use of strains in which *topA* is expressed from the regulated *lac* promoter further confirms that switching off *topA* indeed increases the negative superhelicity of intracellular DNA (69, 234).

It appears that the syntheses of both DNA gyrase subunits are themselves regulated by the degree of DNA supercoiling (240). In vivo, the rates of synthesis of both the A and B subunits of gyrase are increased up to 10 fold by blocking gyrase activity with either coumermycin or novobiocin, or by shifting to a nonpermissive temperature using a *gyrB* ts mutant or *amber* mutant carrying a ts suppressor (240). The induction of gyrase synthesis by coumermycin occurs more rapidly in a *topA*$^+$ strain than a *topA*$^-$ control, suggesting that DNA superhelicity rather than a repressor role of gyrase itself is regulating the synthesis of gyrase synthesis (240). In a cell-free system, the addition of purified *E. coli* or HeLa DNA topoisomerase I stimulates the synthesis of gyrase subunit proteins in this system; furthermore, when plasmids carrying *gyrA* or *gyrB* genes are used as templates the highest rate of synthesis is found when relaxed rather than negatively supercoiled plasmid DNA is used and novobiocin is added to the cell extract to prevent the supercoiling of the added

DNA (240). These results have led to the postulate of homeostatic control of DNA supercoiling in prokaryotes (240). Such a homeostatic control would presumably be exerted at the transcription level. For the *gyrB* transcript, it is found that the amount of this transcript is much greater in cells that have been treated with novobiocin (242).

Effects of Prokaryotic DNA Topoisomerases on Transcription: Template Supercoiling

Earlier studies on the effects of DNA supercoiling on transcription in vivo have been discussed in a number of reviews (3–7, 14). The bulk of the earlier studies make use of bacterial DNA gyrase inhibitors and a spectrum of response to the inhibitors, ranging from reduction to increment of the levels of expression, are observed depending on the particular genes. Some of the complications due to the use of inhibitors, in particular nalidixic or oxolinic acid, have been discussed (243).

Several lines of genetic evidence indicate that topoisomerases affect transcription through their effects on DNA superhelicity. The mutation *supX* in *Salmonella typhimurian,* which suppresses a mutation *leu500* in the leucine operon, has been subject to extensive studies genetically (244). The identification of *topA* gene and the establishment of *topA* and *supX* being the same gene (227, 245), immediately led to the suggestion that *topA* affects transcription through its effect on DNA supercoiling (227, 245); it is known that *leu500* is a promoter down-mutation (246). Studies of compensatory mutations in *topA* mutants have led to the finding that the *bgl* locus, a normally inactive operon for the utilization of β-glucosides, is turned on in *topA* strains carrying compensatory mutations in gyrase (228). It has also been shown that the Bgl$^+$ phenotype parallels the temperature sensitivity phenotype of a *gyrB* ts mutant (247). *GyrA* mutants have also been reported to express a normally silent gene encoding adenylate synthetase as well as the *bgl* operon (248).

Several mutants in the structural gene *rpoB* encoding the B subunit of *E. coli* RNA polymerase have been shown to affect the sensitivity of cells to gyrase inhibitors (224, 250, 251). A *rpoB* mutant has also been found to suppress the hypersensitivity of a dnaA46 mutant to coumermycin (251). Because of the involvement of *dnaA* in the initiation of DNA replication, it is suggested that gyrase might affect the synthesis of the initiator RNA for replication through its affect on supercoiling (251). The addition of a low level (6–8 µg/ml) of rifampcin, an inhibitor of RNA polymerase, has been reported to improve the viability of a *gyrB* ts mutant at its nonpermissive temperature (224).

In gyrase ts mutants, shifting to the nonpermissive temperature shows a larger effect on RNA synthesis than the gyr^+ isogenic controls, although the magnitude of the effects are usually not as high as those elicited by gyrase inhibitors (249). Again, different operons respond differently to inhibition by mutations in gyrase. Expression of the *gyrA, gyrB,* and *bgl* appears to increase when gyrase is inactivated, as discussed earlier, whereas the rates of synthesis of total or stable RNA appear to decrease. There is considerable spread in the observed magnitudes of the effects. For rRNA synthesis, as an example, one group reports that gyrase inhibitors or mutations strongly inhibits the synthesis (252–254) whereas another reports much smaller differences with a different *gyrB* ts mutant (255).

An increase in the rate of induction of the catabolic-sensitive enzymes including tryptophanase and β-galactosidase has been observed in a partial *topA* mutant (227). In phage T4–infected cells, transcription of T4 late genes is affected by mutations in the phage topoisomerase genes (257).

From a mechanistic point of view, the effects of topoisomerase levels on transcription is likely to be complex. Even if the topoisomerases are assumed to exert their effects solely through supercoiling and that the pleiotropic effects of supercoiling on other cellular processes are assumed not to influence transcription, the effect of supercoiling on the rate-limiting step would still be dependent on the particular promoter sequence, the intracellular RNA polymerase concentration, and regulatory elements. Supercoiling has been shown, for example, to affect *lac* repressor binding by a factor of ten in vitro (258), and measurements of the temperature-dependence of induced and uninduced levels of β-galactocidase synthesis in a *gyrB* ts mutant are qualitatively consistent with less tight repressor binding when gyrase is inhibited (249).

Possible Roles of DNA Topoisomerases in Transcription in Eukaryotes: The Question of a "Swivel"

The possible requirement of a swivel in DNA transcription, so that the rotation of a DNA segment around its helical axis could be carried out rather than the rotation of the entire nascent RNA chain around the DNA, was first pointed out in 1966 (259); the discovery of DNA topoisomerase naturally led to discussions on its possible involvement in transcription as well as in replication (188a). Several lines of indirect evidence are suggestive that in eukaryotes in particular, the type I enzyme might have such a function.

Earlier hints of a possible role of eukaryotic DNA topoisomerase I in transcription came from the presence of an activity in the core particle of vaccinia virus (260), which also contains a number of enzymes required for transcription and transcript modification, and the presence of DNA topoisomerase I in amplified nucleoli from immature oocytes of *X. laevis* (261), where DNA replication activity is not detectable but transcription of ribosomal DNA by RNA polymerase I is very active.

More recently (262, 263), it is found that treatment of macronuclear chromatin from *Tetrahymena thermophila* with sodium dodecyl sulfate causes specific cleavages in the extra-chromosomal rDNA. Proteins are found to attach covalently to the 3' ends generated by these cleavages, suggesting that they are due to the action of DNA topoisomerase I. The cleavages are located at 1000, 600, and 150 bp upstream to the transcription initiation point; these and a fourth cleavage site are located close to sites that are hypersensitive to attack by pancreatic DNase I. Because of the correlation between DNase I hypersensitive and plausible sites of transcriptional regulation (264–266), the presence of DNA topoisomerase I near the DNase I hypersensitive sites suggests its possible involvement in transcription.

More direct indication of the presence of eukaryotic DNA topoisomerase I at locations of active transcription comes from immunofluorescent staining of *Drosophila* polytene chromosomes with affinity-purified rabbit antibodies specific to *Drosophila* DNA topoisomerase I (189). Some of these results have been described in the section on the cellular location of the enzyme. In addition, when a mutant fly BER-1 carrying a deletion in the control region of a gene Sgs-4 at position 3C11-12 of the X-chromosome is compared with its wild-type control, it is found that the locus 3C11-12 is much less strained by the antibodies in the mutant (189), where transcription of Sgs-4 is abolished because of the deletion. On the other hand, the lack of a transcriptional phenotype of yeast mutants deficient in DNA topoisomerase I or II (211, 212, 226) indicates that if a eukaryotic DNA topoisomerase is required for transcription, one or the other of the two known topoisomerases would probably suffice.

Possible Roles of DNA Topoisomerases in Transcription in Eukaryotes: The Question of Template Supercoiling

Although folded chromosomes from both bacterial and eukaryotic sources contain negatively supercoiled loops of DNA after deproteinization, in the eukaryotic case the deficiency in linking number relative to pure DNA could be attributed entirely to the coiling of the DNA around histone cores in nucleosomes (267, 268). Studies of psoralen adduct formation in vivo with intact and γ-ray severed DNA show that DNA breakage in eukaryotic cells does not affect the binding of the intercalator (269), in contrast to the case in *E. coli* (269, 270); this result suggests that the bulk of intracellular DNA in eukaryotes is not under strain imposed by topological constraints. It is plausible, however, that a fraction of intracellular DNA or some regions of the chromosomes might be in a negatively supercoiled state over and above the linking number reduction due to nucleosome formation; furthermore, this supercoiled state might influence gene expression in eukaryotes. This possibility is suggested by several recent findings, in particular by three types of experiments summarized below [see (271–275) for additional examples]. First, nuclease-hypersensitive sites, which are present in active genes, appear to correlate with sites of attack by reagents that react preferentially to single-stranded and negatively supercoiled DNA over relaxed double-stranded DNA (276, 277). Second, transcription in

Xenopus oocytes of genes on plasmids injected into the oocytes requires circularity of the template; linearization of the injected plasmid by injecting a restriction enzyme into the oocyte abolishes the transcription of the genes (278). Third, there appears to be a transcriptionally active subpopulation of simian virus 40 minichromosomes with its DNA in a negatively supercoiled state; treatment of purified SV40 minichromosomes with DNA topoisomerase I increases the linking number of this subpopulation (279, 280).

Injection of DNA topoisomerase I into *Xenopus* oocytes also increases the linking number of a substantial fraction of plasmid DNA that was injected earlier into oocyte nuclei; the relaxation of this subpopulation can be accomplished by the injection of DNase I as well, because of the presence of excess cellular DNA ligase. Injection of novobiocin induces this relaxation, and the target of novobiocin is thought to be DNA topoisomerase II. It has been suggested that this subpopulation might be actively kept in the negatively supercoiled form by DNA topoisomerase II (281). The time course of transcription of 5S RNA gene present on the injected plasmid and that of linking number reduction of the plasmid suggest that the subpopulation contains transcriptionally active templates. Studies of nucleosome assembly in *Xenopus* oocyte extracts indicate that ATP amd Mg(II) are required, and that the linking number reduction of the plasmid is inhibited by novobiocin (282). These studies led to the postulation that chromatin assembly is an active, ATP-driven process involving DNA topoisomerase II.

Other Plausible Biological Roles

The exquisite studies of specific DNA topoisomerases that catalyze site-specific recombination (38–54) will not be reviewed here due to page limitations. For the archetype topoisomerases, three of their biochemical characteristics hint that they might be involved in recombination. First, the type I topoisomerases permit intertwining of complementary strands without ends (283, 284). In the presence of a type I topoisomerase and either the *recA* gene protein of *E. coli* (285, 286), which is essential for recombination, or a similar protein *rec1* in the lower eukaryote *Ustilago maydis* (287), circular single strands can pair plectonemically with its complementary strand in a duplex ring. Extensive unwinding of a duplex DNA ring by *recA* has also been shown under certain conditions, provided that a topoisomerase is present to permit the reduction in linking number (288). These experiments suggest a plausible role of a topoisomerase in the pairing of strands undergoing recombination. Second, the ability of a topoisomerase to break and rejoin DNA strands, and to carry out intermolecular strand transfer in some cases, suggests that the archetype topoisomerases might also catalyze strand breakage and rejoining in DNA recombination, much like the special topoisomerases of site-specific recombinations. Third, in prokaryotes at least, the role of topoisomerases in maintaining the supercoiled state of DNA may in turn affect certain recombination pathways. Examples are the site-specific integration catalyzed by phage λ *int* protein (38, 39) and Tn5 transposition (see below).

Studies of multiplicity reactivation of damaged phage T4 has implicated a role of phage T4 DNA topoisomerase in recombination (289). The increase in the spontaneous level of recombination in T4 topoisomerase mutants (290), and the requirement of a functional T4 topoisomerase in recombinational initiation of DNA replication (291), have been reported. Illegitimate recombination in *E. coli* extracts mediated by DNA gyrase has been implicated; the reaction is stimulated by nalidixic and oxolinic acid, but the stimulation is abolished by coumermycin. These results have been interpreted as involving transient double-stranded DNA cleavage by gyrase and intermolecular exchange of gyrase subunits with their attached DNA (292, 293). Sequence analysis of *recA* independent phage-plasmid recombinants also suggests a possible role of gyrase in illegitimate recombination (293a, 293b).

In eukaryotes, the possible involvement of eukaryotic DNA topoisomerase I in the excision of integrated viral sequences has been raised based on the nucleotide sequences of junctions created by the excision events (294). It is mentioned earlier that a number of antitumor drugs appear to stimulate eukaryotic DNA topoisomerase II–mediated cleavage of DNA. The same drugs are known to increase the frequency of sister-chromatid exchange in vivo, thus a possible role of DNA topoisomerase II in sister-chromatid exchange is implicated (170, 295). The possible involvement of DNA topoisomerase II in sister-chromatid exchange has also been raised based on its enzymological properties (296, 297).

Blockage of *E. coli* gyrase with coumermycin, or using a gyrase temperature-sensitive *gyrA* mutant at nonpermissive temperature, inhibits the transposition of Tn5 inserted in the *E. coli* chromosome to a nonreplicating intracellular DNA; deletion of *topA* in the *gyrA* mutant, however, increases transposition frequency (298). Thus the effect of gyrase on transposition is likely to be through its effect on supercoiling. It has been observed earlier, using a different assay of transposition frequency, that the deletion of *topA* severely represses Tn5 transposition (227). The interpretation of this observation is unclear, and the large reduction in transposition frequency may be due to the particular assay used.

Genetic studies show that *E. coli topA* deletion mutants are recombination proficient (227). Plasmid recombination, however, is much lowered in Δ*topA* strains at 30°, but not significantly at 42° (299). This temperature-sensitive reduction in plasmid recombination in Δ*topA* strain is eliminated by the introduction of a cloned copy of *topA* gene on plasmid. One interesting possibility is that a type I topoisomerase is involved in plasmid recombination and that at the higher temperature DNA topoisomerase III is sufficiently active to compensate the deficiency in topoisomerase I. Genetic evidence in favor of a functional relation between gyrase and the *recF* gene product (300), and between *gyrB* and a phage mu *lig* gene product (301), has been reported. *TopA* mutants of *Salmonella typhimurian*, which seems to be more tolerant to the mutation than *E. coli*, have been found to increase the sensitivity of killing by ultraviolet (uv) irradiation, and to abolish mutation induction by uv irradiation and direct and indirect mutagensis by alkalyting agents (244). The involvement of *E. coli* gyrase in the repair of uv-damaged DNA has been reported (302–305). In contrast to the effects of *topA* on sensitivity to uv and mutagensis, a mutation in *E. coli gyrA* is found to enhance resistance to killing by nitrosoguanidine (304). Mutation in *E. coli gyrB* has also been shown to give an apparent enhancement of survival after uv irradiation in strains possessing only excision repair and a reduction in strains possessing only postreplicative repair (305). In vitro, uv irradiation of DNA strongly inhibits *M. luteus* and *E. coli* DNA topoisomerse I; sensitivity of *E. coli topA* mutants with and without *recA* mutation has also been studied (306, 307). The Poly-ADP-ribosylation of eukaryotic topoisomerase in vitro has raised questions regarding its possible involvement in repair (204–207). A possible role of a type II DNA topoisomerase in short-patch alkylation repair in human cells has also been raised through studies on the inhibitory effect of novobiocin on uv and methyl methanesulfonate-induced damage (308). The role of gyrase in the maintenance of plasmid and phage genome has also been studied (310–313).

ACKNOWLEDGMENTS

I am most grateful to many colleagues for generously providing manuscripts prior to publication and to Nick Cozzarelli, Marty Gellert, Leroy Liu, and Charles Richardson for their critical reading of this review; I am solely responsible, of course, for any errors in it.

Literature Cited

1. Champoux, J. J. 1978. *Ann. Rev. Biochem.* 47:449–79
2. Bauer, W. R. 1978. *Ann. Rev. Biophys. Bioeng.* 7:287–313
3. Wang, J. C., Liu, L. F. 1979. In *Molecular Genetics*, ed. J. H. Taylor, pp. 65–88. New York: Academic
4. Cozzarelli, N. R. 1980. *Science* 207: 953–60
5. Gellert, M. 1981. *Ann. Rev. Biochem.* 50:879–910
6. Wang, J. C. 1981. *Enzymes* 14:331–44
7. Gellert, M. 1981. *Enzymes* 14:345–66
8. Wang, J. C., Kirkegaard, K. 1981. In *Gene Amplification and Analysis,* ed. J. G. Chirikjian, T. S. Papas, 2:456–73. New York: Elsevier/North-Holland
9. Wang, J. C. 1982. *Sci. Am.* 247:94–109
10. Wang, J. C. 1982. In *Nucleases,* ed. R. Roberts, S. Linn, pp. 41–57. New York: Cold Spring Harbor Lab.
11. Liu, L. F. 1983. CRC *Crit. Rev. Biochem.* 15:1–24
12. Denhardt, D. T. 1979. *Nature* 280:196–98
13. Cozzarelli, N. R. 1980. *Cell* 22:327–28
14. Smith, G. R. 1981. *Cell* 24:599–600
15. Fisher, L. M. 1984. *Nature* 307:686–87
16. Wang, J. C. 1984. *Nature* 309:669–70
17. Delbruck, M., Stent, G. 1957. In *The Chemical Basis of Heredity,* ed. W. D. McElroy, B. Glass, pp. 699–736. Baltimore: The Johns Hopkins Press
18. Cairns, J. 1963. *J. Mol. Biol.* 6:208–13

19. Cairns, J. 1963. *Cold Spring Harbor Symp. Quant. Biol.* 28:43–45
20. Wang, J. C. 1969. *J. Mol. Biol.* 43:263–72
21. Wang, J. C. 1971. *J. Mol. Biol.* 55:523–33
22. Champoux, J. J., Dulbecco, R. 1972. *Proc. Natl. Acad. Sci. USA* 69:143–46
23. Liu, L. F., Liu, C.-C., Alberts, B. M. 1980. *Cell* 19:697–708
24. Champoux, J. J. 1977. *Proc. Natl. Acad. Sci. USA* 74:3800–4
25. Champoux, J. J. 1978. *J. Mol. Biol* 118:441–46
26. Prell, B., Vosberg, H.-P. 1980. *Eur. J. Biochem.* 108:389–98
27. Depew, R. E., Liu, L. F., Wang, J. C. 1978. *J. Biol. Chem.* 253:511–18
28. Gellert, M., Mizuuchi, K., O'Dea, M. H., Nash, H. A. 1976. *Proc. Natl. Acad. Sci. USA* 73:3872–76
29. Brown, P. O., Cozzarelli, N. R. 1979. *Science* 206:1081–83
30. Kreuzer, K. N., Cozzarelli, N. R. 1980. *Cell* 20:245–54
31. Mizuuchi, K., Fisher, L. M., O'Dea, M. H., Gellert, M. 1980. *Proc. Natl. Acad. Sci. USA* 77:1847–51
32. Liu, L. F., Liu, C.-C., Alberts, B. M. 1979. *Nature* 281:456–61
33. Stetler, G. L., King, G. J., Huang, W. M. 1979. *Proc. Natl. Acad. Sci. USA* 76:3737–41
34. Baldi, M. I., Benedetti, P., Mattoccia, E., Tocchini-Valentini, G. P. 1980. *Cell* 20:461–67
35. Hsieh, T.-S., Brutlag, D. 1980. *Cell* 21:115–25
36. Brown, D. R., Roth, M. J., Reinberg, D., Hurwitz, J. 1984. *J. Biol. Chem.* 259:10545–55
37. Roth, M. J., Brown, D. R., Hurwitz, J. 1984. *J. Biol. Chem.* 259:10556–68
38. Nash, H. A. 1981. *Ann. Rev. Genet.* 15:143–66
39. Weisberg, R., Landy, A. 1983. In *Lambda II*, ed. R. Hendrix, J. Roberts, F. Stahl, R. Weisberg, pp. 211–50. New York: Cold Spring Harbor Lab.
40. Nash, H. A., Pollock, T. J. 1983. *J. Mol. Biol.* 170:19–38
41. Pollock, T. J., Nash, H. A. 1983. *J. Mol. Biol.* 170:1–18
42. Craig, N. L., Nash, H. A. 1983. *Cell* 35:795–803
43. Craig, N. L., Nash, H. A. 1983. In *Mechanism of DNA Replication and Recombination*, ed. N. R. Cozzarelli, pp. 617–36. New York: Alan Liss
44. Abremski, K., Hoess, R., Sternberg, N. 1983. *Cell* 32:1301–11
45. Reed, R. R. 1981. *Cell* 25:713–19
46. Reed, R. R., Grindley, N. D. F. 1981. *Cell* 25:721–28
47. Grindley, N. D. F. 1983. *Cell* 32:3–5
48. Kostriken, R., Morita, C., Heffron, F. 1981. *Proc. Natl. Acad. Sci. USA* 78:4041–45
49. Benjamin, H. W., Matzuk, M. M., Krasnow, M. A., Cozzarelli, N. R. 1985. *Cell.* 40:147–58
50. Cozzarelli, N. R., Krasnow, M. A., Gerrard, S. P., White, J. H. 1984. *Cold Spring Harbor Symp. Quant. Biol.* 49: In press
51. Spengler, S. J., Stasiak, A., Cozzarelli, N. R. 1984. *Cold Spring Harbor Quant. Biol.* 49: In press
52. Echols, H., Dodson, M., Better, M., Roberts, J. D., McMacken, R. 1984. *Cold Spring Harbor Symp. Quant. Biol.* 49: In press
53. Cox, M. M. 1983. *Proc. Natl. Acad. Sci. USA* 80:4223–27
54. Vetter, D., Andrews, B. J., Roberts-Beatty, L., Sadowski, P. D. 1983. *Proc. Natl. Acad. Sci. USA* 80:7284–88
55. Dean, F., Krasnow, M. A., Otter, R., Matzuk, M. M., Spengler, S. J., et al. 1982. *Cold Spring Harbor Symp. Quant. Biol.* 47:769–77
56. Pastorcic, M. 1982. *Purification and Characterization of a New Type I Topoisomerase in E. coli*. PhD thesis. Univ. Chicago, Ill.
57. Srivenugopal, K. S., Lockshon, D., Morris, D. R. 1984. *Biochemistry* 23:1899–1906
58. Siedlecki, J., Zimmerman, W., Weissbach, A. 1983. *Nucleic Acids Res.* 11:1523–36
59. Dynan, W. S., Jendrisak, J. J., Hager, D. A., Burgess, R. R. 1981. *J. Biol. Chem.* 256:5860–65
60. Fukata, H., Fukasawa, H. 1982. *J. Biochem.* 91:1337–42
61. Fukata, H., Fukasawa, H. 1983. *Jpn. J. Genet.* 58:425–32
62. Fairfield, F. R., Bauer, W. R., Simpson, M. V. 1979. *J. Biol. Chem.* 254:9352–54
63. Fairfield, F. R., Bauer, W. R., Simpson, M. V. 1985. *Biophys. Biochim. Acta* 824:48–57
64. Brun, G., Vannier, P., Scovassi, I., Callen, J.-C. 1981. *Eur. J. Biochem.* 118:407–15
65. Goto, T., Laipis, P., Wang, J. C. 1984. *J. Biol. Chem.* 16:10422–29
66. Woese, C. R. 1981. *Sci. Am.* 244:98–122
67. Kikuchi, A., Asai, K. 1984. *Nature* 309:677–81
68. Wang, J. C. 1980. *Trends Biochem. Sci.* 5:219–21
69. Wang, J. C., Peck, L. J., Becherer, K. 1983. *Cold Spring Harbor Symp. Quant. Biol.* 47:85–92
70. Wang, J. C. 1983. In *Nucleic Acid Re-*

search: Future Development, ed. K. Mizobuchi, I. Watanabe, J. D. Watson, pp. 549–66. New York: Academic
71. Mirambeau, G., Duguet, M., Forterre, P. 1984. *J. Mol. Biol.* 179:559–63
72. Lockshon, D., Morris, D. R. 1983. *Nucleic Acids Res.* 11:2999–3017
73. Liu, L. F., Wang, J. C. 1978. *Proc. Natl. Acad. Sci. USA* 75:2098–102
74. Shlomai, J., Zadok, A. 1983. *Nucleic Acids Res.* 11:4019–34
75. Englund, P. T., Hajduk, S. L., Marini, J. C. 1982. *Ann. Rev. Biochem.* 51:695–726
76. Douc-Rasy, S., Kayser, A., Riou, G. F. 1984. *EMBO J.* 3:11–16
77. Riou, G. F., Gabillot, M., Douc-Rasy, S., Kayser, A., Barrois, M. 1983. *Eur. J. Biochem.* 134:479–84
78. Brown, P. O., Peebles, C. L., Cozzarelli, N. R. 1979. *Proc. Natl. Acad. Sci. USA* 76:6110–14
79. Gellert, M., Fisher, L. M., O'Dea, M. H. 1979. *Proc. Natl. Acad. Sci. USA* 76:6289–93
80. Liu, L. F. 1980. In *ICN-UCLA Symp. Mol. Cell. Biol.*, ed. B. Albert, F. Fox, 19:817–31. New York: Academic
81. Liu, L. F., Miller, K. G. 1981. *Proc. Natl. Acad. Sci. USA* 78:3487–91
82. Javaherian, K., Tse, Y.-C., Vega, J. 1982. *Nucleic Acids. Res.* 10:6945–55
83. Goto, T. G., Wang, J. C. 1982. *J. Biol. Chem.* 257:5866–72
84. Ishii, K., Hasegawa, T., Fujisawa, K., Andoh, T. 1983. *J. Biol. Chem.* 258:12728–32
85. Tricoli, J. V., Kowalski, D. 1983. *Biochemistry* 22:2025–31
85a. Liu, L. F., 1983. In *Methods in Enzymology*, pp. 171–80
85b. Martin, S. R., McCoubrey, W. K. Jr., McConaughgy, B. L., Young, L. S., Been, M. D., Brewer, B. J., Champoux, J. J. 1983. *Methods in Enzymology*, pp. 137–44
86. Pulleyblank, D. E., Ellison, J. J. 1982. *Biochemistry* 21:1155–61
87. Attardi, D. G., Paolis, A. D., Tocchini-Valentini, G. P. 1981. *J. Biol. Chem.* 256:3654–61
88. Javaherian, K., Liu, L. F. 1983. *Nucleic Acids Res.* 11:461–72
89. Muller, M. T. 1983. *Biochem. Biophys. Res. Commun.* 114:99–106
90. Rowe, T. C., Rusche, J. R., Brougham, M. J., Holloman, W. K. 1981. *J. Biol. Chem.* 256:10354–61
91. Reddy, G. P. V., Pardee, A. B. 1983. *Nature* 304:86–88
92. Miskimins, R., Miskimins, W. K., Bernstein, H., Shimizu, N. 1983. *Exp. Cell Res.* 146:53–62
93. Tsubota, Y., Waqar, M. A., Burke, J.

F., Milavetz, B. I., Evans, M. J., et al. 1979. *Cold Spring Harbor Symp. Quant. Biol.* 43:693–704
94. Miller, K. G., Liu, L. F., Englund, P. T. 1981. *J. Biol. Chem.* 256:9334–39
95. Sander, M., Hsieh, T. 1983. *J. Biol. Chem.* 258:8421–28
95a. Hsieh, T.-S. 1983. *Methods Enzymol.* 100:144–60
96. Shelton, E. R., Osheroff, N., Brutlag, D. L. 1983. *J. Biol. Chem.* 258:9530–35
97. Goto, T., Wang, J. C. 1984. *Cell* 36:1073–80
98. Wang, J. C., Becherer, K. 1983. *Nucleic Acids Res.* 11:1773–90
99. Mizuuchi, K., Mizuuchi, M., O'Dea, M. H., Gellert, M. 1984. *J. Biol. Chem.* 259:9199–201
100. Lampe, M. F., Bott, K. F. 1984. *Nucleic Acids Res.* 12:6307–23
101. Staudenbauer, W. L., Orr, E. 1981. *Nucleic Acids. Res.* 9:3589–603
102. Orr, E., Staudenbauer, W. L. 1982. *J. Bacteriol.* 151:524–27
102a. Kreuzer, K. N., Jongeneel, C. V. 1983. *Methods in Enzymology*, 144–160
103. Seasholtz, A. F., Greenberg, G. R. 1983. *J. Biol. Chem.* 258:1221–26
104. Liu, L. F., Davis, J. L., Calendar, R. 1981. *Nucleic Acids Res.* 9:3979–89
105. Liu, L. F., Perkocha, L., Calendar, R., Wang, J. C. 1981. *Proc. Natl. Acad. Sci. USA* 78:5498–502
106. Kayser, A., Douc-Rasy, S., Riou, G. 1982. *Biochimie* 64:285–88
107. Sahai, B. M., Kaplan, J. G. 1984. *Fed. Proc.* 43:1542 (Abstr.)
108. Tse, Y.-C., Javaherian, K., Wang, J. C. 1984. *Arch. Biochem. Biophys.* 231: 169–74
109. Tse, Y.-C., Wang, J. C. 1980. *Cell* 22:269–76
110. Brown, P. O., Cozzarelli, N. R. 1981. *Proc. Natl. Acad. Sci. USA* 78:843–47
111. Liu, L. F., Rowe, T. C., Yang, L., Tewey, K. M., Chen, G. L. 1983. *J. Biol. Chem.* 258:15365–70
112. Muller, M. T. 1983. *Biochem. Biophys. Res. Commun.* 114:99–105
113. Trask, D. K., DiDonato, J. A., Muller, M. T. 1984. *EMBO J.* 3:671–76
113a. Dean, F., Cozzarelli, N. R. 1985. *J. Biol. Chem.* In press
114. Malcolm, A. D. B., Snounou, G. 1983. *Cold Spring Harbor Symp. Quant. Biol.* 47:323–26
115. Low, R. L., Kaguni, J. M., Kornberg, A. 1984. *J. Biol. Chem.* 259:4576–81
116. Kirkegaard, K., Pflugfelder, G., Wang, J. C. 1984. *Cold Spring Harbor Symp. Quant. Biol.* 49: In press
117. Kirkegaard, K. 1983. *Mechanistic Studies of DNA Topoisomerases*. PhD thesis. Harvard Univ. Cambridge, Mass.

694   WANG

118. Tse-Dinh, Y.-C., McCarron, B. G. H., Arentzen, R., Chowdhry, V. 1983. *Nucleic Acids Res.* 11:8691–701
119. Horowitz, D., Wang, J. C. 1984. *J. Mol. Biol.* 173:75–91
120. Shore, D., Baldwin, R. L. 1983. *J. Mol. Biol.* 170:983–1008
121. Lee, C.-H., Mizusawa, H., Kakefuda, T. 1981. *Proc. Natl. Acad. Sci. USA* 78:2838–42
122. Peck, L. J., Wang, J. C. 1983. *Proc. Natl. Acad. Sci. USA* 80:6206–10
123. Sundin, O., Varshavsky, A. 1980. *Cell* 21:103–14
124. Stasiak, A., DiCapua, E., Koller, Th. 1983. *Cold Spring Harbor Symp. Quant. Biol.* 47:811–20
125. Krasnow, M. A., Stasiak, A., Spengler, S. J., Dean, F., Koller, T., et al. 1983. *Nature* 304:559–60
126. Griffith, J., Nash, H. 1985. *Proc. Natl. Acad. Sci. USA.* In press
127. Hudson, B., Vinograd, J. 1967. *Nature* 216:647–51
128. Iwamoto, S., Hsu, M.-T. 1983. *Nature* 305:70–72
129. White, J. H., Cozzarelli, N. R. 1984. *Proc. Natl. Acad. Sci. USA* 81:3322–26
130. Tse, Y.-C., Kirkegaard, K., Wang, J. C. 1980. *J. Biol. Chem.* 255:5560–65
131. Champoux, J. J. 1981. *J. Biol. Chem.* 256:4805–9
132. Rowe, T. C., Tewey, K. M., Liu, L. F. 1984. *J. Biol. Chem.* 259:9177–81
133. Sanhueza, S., Eisenberg, S. 1984. *Proc. Natl. Acad. Sci. USA* 81:4285–89
134. Reed, R. R., Moser, C. D. 1984. *Cold Spring Harbor Symp. Quant. Biol.* 49: In press
135. Newman, B. J., Grindley, N. D. F. 1984. *Cell.* 38:463–69
136. Been, M. D., Champoux, J. J. 1980. *Proc. Natl. Acad. Sci. USA* 78:2883–87
137. Halligan, B. D., Davis, J. L., Edwards, K. A., Liu, L. F. 1982. *J. Biol. Chem.* 257:3995–4000
138. Deleted in proof
139. Kirkegaard, K., Wang, J. C. 1981. *Cell* 23:721–29
140. Shishido, K., Noguchi, N., Ando, T. 1983. *Biochim. Biophys. Acta* 740:108–17
141. Edwards, K. A., Halligan, B. D., Davis, J. L., Nivera, N. L., Liu, L. F. 1982. *Nucleic Acids Res.* 10:2565–76
142. Been, M. D., Burgess, R. R., Champoux, J. J. 1984. *Biochim. Biophys. Acta* 782:304–12
143. Been, M. D., Burgess, R. R., Champoux, J. J. 1984. *Nucleic Acids Res.* 12:3097–114
144. Deleted in proof
145. Been, M. D., Champoux, J. J. 1984. *J. Mol. Biol.* In press

146. Champoux, J. J., McCoubrey, W. K. Jr., Been, M. D. 1984. *Cold Spring Harbor Symp. Quant. Biol.* 49: In press
147. Klevan, L., Wang, J. C. 1980. *Biochemistry* 19:5229–34
148. Lother, H., Lurz, R., Orr, E. 1984. *Nucleic Acids Res.* 12:901–14
149. Liu, L. F., Wang, J. C. 1978. *Cell* 15:979–84
150. Fisher, L. M., Mizuuchi, K., O'Dea, M. H., Ohmori, H., Gellert, M. 1981. *Proc. Natl. Acad. Sci. USA* 78:4165–69
151. Morrison, A., Cozzarelli, N. R. 1981. *Proc. Natl. Acad. Sci. USA* 78:1416–20
152. Sugino, A., Peebles, C. L., Kreuzer, K. N., Cozzarelli, N. R. 1977. *Proc. Natl. Acad. Sci. USA* 74:4767–71
153. Gellert, M., Mizuuchi, K., O'Dea, M. H., Itoh, T., Tomizawa, J. 1977. *Proc. natl. Acad. Sci. USA* 74:4772–76
154. Morrison, A., Cozzarelli, N. R. 1979. *Cell* 17:175–84
155. Sugino, A., Higgins, N. P., Cozzarelli, N. R. 1980. *Nucleic Acids Res.* 8:3865–74
156. Snyder, M., Drlica, K. 1979. *J. Mol. Biol.* 131:287–302
157. Lockshon, D., Morris, D. R. 1985. *J. Mol. Biol.* 181:63–74
158. O'Connor, M. B., Malamy, M. H. 1984. *Cold Spring Harbor Symp. Quant. Biol.* 49: In press
159. Kreuzer, K. N., Alberts, B. M. 1984. *J. Biol. Chem.* 259:5339–46
160. Morrison, A., Higgins, N. P., Cozzarelli, N. R. 1980. *J. Biol. Chem.* 255:2211–19
161. Osheroff, N., Shelton, E. R., Brutlag, D. L. 1983. *J. Biol. Chem.* 258:9536–43
162. Kreuzer, K. N. 1984. *J. Biol. Chem.* 259:5347–54
163. Sander, M., Hsieh, T.-S. 1984. *Fed. Proc.* 43:1543 (Abstr.)
164. Ross, W. E., Glaubiger, D. L., Kohn, K. W. 1978. *Biochim. Biophys. Acta* 519:23–30
165. Ross, W. E., Glaubiger, D. L., Kohn, K. W. 1979. *Biochim. Biophys. Acta* 562:41–50
166. Paoletti, C., Lesca, C., Cros, S., Malvy, S., Auclair, C. 1979. *Biochem. Pharmacol.* 28:345–50
167. Ross, W. E., Bradley, M. O. 1981. *Biochim. Biophys. Acta* 654:129–34
168. Zwelling, L. A., Michaels, S., Erickson, L. C., Ungerleider, R. S., Nichols, M., Kohn, K. W. 1981. *Biochemistry* 20:6553–63
169. Zwelling, L. A., Michaels, S., Kerrigan, D., Pommier, Y., Kohn, K. W. 1982. *Biochem. Pharmacol.* 31:3261–67
170. Marshall, B., Ralph, R. K., Hancock, R. 1983. *Nucleic Acids Res.* 11:4251–56

171. Filipski, J., Kohn, K. W. 1983. *Biochim. Biophys. Acta* 698:280–86
172. Pommier, Y., Mattern, M. R. Schwartz, R. E., Zwelling, L. A. 1984. *Biochemistry* 23:2922–27
173. Pommier, Y., Mattern, M. R., Schwartz, R. E., Zwelling, L. A., Kohn, K. W. 1984. *Biochemistry* 23:2927–32
174. Pommier, Y., Schwartz, R. E., Kohn, K. W., Zwelling, L. A. 1984. *Biochemistry* 23:3194–201
175. Nelson, E. M., Tewey, K. M., Liu, L. F. 1984. *Proc. Natl. Acad. Sci. USA* 81:1361–65
176. Tewey, K. M., Chen, G. L., Nelson, E. M., Liu, L. F. 1984. *J. Biol. Chem.* 259:9182–87
177. Ross, W., Rowe, T., Yalowich, J., Glisson, B., Liu, L. F. 1984. *Cancer Res.* 44:5857–60
178. Minocha, A., Long, B. H. 1984. *Fed. Proc.* 43:1542 (Abstr.)
178a. Tewey, K. M., Rowe, T. C., Yang, L., Halligan, B. B., Liu, L. F. 1984. *Science.* 226:466–8
178b. Chen, G. L., Yang, L., Rowe, T. C., Halligan, B. B., Tewey, K. M., Liu, L. F. 1984. *J. Biol. Chem.* 259:13560–6
179. Douc-Rasy, S., Kayser, A., Riou, G. 1983. *Biochem. Biophys. Res. Commun.* 117:1–5
180. Yang, L., Rowe, T. C., Nelson, E. M., Liu, L. F. 1985. *Cell.* In press
181. Nelson, E. M. 1984. *Interactions between Mammalian DNA Topoisomerase II and Chromatin.* PhD thesis. Johns Hopkins Univ., Baltimore, Md.
182. Krasnow, M. A., Cozzarelli, N. R. 1982. *J. Biol. Chem.* 257:2687–93
183. Dean, F. B., Stasiak, A., Koller, T., Cozzarelli, N. R. 1985. *J. Biol. Chem.* In press
184. Osheroff, N., Brutlag, D. L. 1983. See Ref. 43, pp. 55–64
185. Higgins, N. R., Cozzarelli, N. R. 1982. *Nucleic Acids Res.* 10:6833–47
186. Hsieh, T.-S. 1983. *J. Biol. Chem.* 258:8413–20
187. Pulleyblank, D. E., Shure, M., Tang, D., Vinograd, J., Vosberg, H.-P. 1975. *Proc. Natl. Acad. Sci. USA* 72:4280–84
188. Brown, P. O., Cozzarelli, N. R. *Proc. Natl. Acad. Sci. USA* 78:843–7
188a. Wang, J. C. 1973. In *DNA Synthesis in Vitro,* ed. R. D. Wells, R. B. Inman, pp. 163–74. Baltimore: Univ. Park Press
189. Fleischmann, G., Pflugfelder, G., Steiner, E. K., Javaherian, K., Howard, G. C., et al. 1984. *Proc. Natl. Acad. Sci. USA* 81:6958–62
190. Berrios, M., Osheroff, N., Fisher, P. A. 1985. *Proc. Natl. Acad. Sci. USA.* In press

191. Lewis, C. D., Laemmli, U. K. 1982. *Cell* 17:849–58
192. Earnshaw, W. C., Halligan, B., Cooke, C. A., Liu, L. F. 1985. *J. Cell Biol.* In press
193. Fisher, P. A., Berrios, M., Blobel, G. 1982. *J. Cell. Biol.* 92:674–86
194. Huang, W. M. 1975. *Virology* 66:508–21
195. Takacs, B. J., Rosenbusch, J. P. 1975. *J. Biol. Chem.* 250:2339–50
195a. Mroczykowski, B., Mosig, G., Cohen, S. 1984. *Nature.* 309:270–73
196. Durban, E., Roll, D., Beckner, G., Busch, H. 1981. *Cancer Res.* 41:537–45
197. Durban, E., Mills, J. S., Roll, D., Busch, H. 1983. *Biochem. Biophys. Res. Commun.* 111:897–905
198. Durban, E., Goodenough, M., Busch, H. 1984. *Fed. Proc.* 43:1542 (Abstr.)
199. Mills, J. S., Busch, H., Durban, E. 1982. *Biochem. Biophys. Res. Commun.* 109:1222–27
200. Tse-Dinh, Y.-C., Wong, T. W., Goldberg, A. R. 1984. *Nature.* 312:785–6
201. Klevan, L., Tse, Y.-C. 1983. *Biochim. Biophys. Acta* 745:175–80
202. Sander, M., Nolan, J., Hsieh, T.-S. 1984. *Proc. Natl. Acad. Sci. USA* 81:6938–42
203. Hayaishi, O., Ueda, K., eds. 1983. *ADP-Ribosylation Reactions, Biology and Medicine.* New York: Academic
204. Ferro, A. M., Higgins, N. P., Olivera, B. M. 1983. *J. Biol. Chem.* 258:6000–3
205. Jongstra-Bilen, J., Ittel, M.-E., Niedergang, C., Vosberg, H.-P., Mandel, P. 1983. *Eur. J. Biochem.* 136:391–96
206. Ferro, A. M., Olivera, B. M. 1984. *J. Biol. Chem.* 259:547–54
207. Ferro, A. M., McElwain, M. C., Olivera, B. M. 1984. *Cold Spring Harbor Symp. Quant. Biol.* 49: In press
207a. Fuller, R. S., Kaguni, J. M., Kornberg, A. 1981. *Proc. Natl. Acad. Sci. USA* 12:7370–74
207b. Kaguni, J. M., Kornberg, A. 1984. *J. Biol. Chem.* 259:8578–83
207c. Kaguni, J. M., Kornberg, A. 1984. *Cell* 38:183–90
207d. Sogo, J. M., Greenstein, M., Skalka, A. 1976. *J. Mol. Biol.* 103:537–62
207e. Gefter, M. L. 1975. *Ann. Rev. Biochem.* 44:45–78
208. Sundin, O., Varshavsky, A. 1980. *Cell* 25:659–69
209. Varshavsky, A., Sundin, O., Ozgaynak, E., Pan, R., Soloman, M., et al. 1983. See Ref. 43, pp. 463–94
210. Duguet, M., Lavenot, C., Harper, F., Mirambeau, G., DeRecondo, A.-M. 1983. *Nucleic Acids Res.* 11:1059–75

211. DiNardo, S., Voelkel, K., Sternglanz, R. 1984. *Proc. Natl. Acad. Sci. USA* 81:2616–20
212. Uemura, T., Yanagida, M. 1984. *EMBO J.* 3:1737–44
213. Steck, T. R., Drlica, K. 1984. *Cell* 36:1081–88
214. McCarthy, D. 1979. *J. Mol. Biol.* 127:265–83
215. Kreuzer, K. N., Cozzarelli, N. R. 1979. *J. Bacteriol.* 140:424–35
216. Orr, E., Fairweather, N. F., Holland, I. B., Pritchard, R. H. 1979. *Mol. Gen. Genet.* 177:103–12
217. Kano, Y., Miyashita, T., Nakamura, H., Kuroki, K., Nagata, A., Imamoto, F. 1981. *Gene* 13:178–84
218. Engle, E. C., Mans, S. H., Drlica, K. 1982. *J. Bacteriol.* 149:92–98
219. Drlica, K., Engle, E. C., Manes, S. H. 1980. *Proc. Natl. Acad. Sci. USA* 77:6879–83
220. Orr, E., Staudenbauer, W. 1981. *Mol. Gen. Genet.* 181:52–56
221. Orr, E., Lother, H., Lurz, R., Wahle, E. 1984. In *Proteins Involved in DNA Replication*, ed. U. Hubscher, S. Spadari, pp. 395–408. New York: Plenum
222. Ogasawara, N., Seiki, M., Yoshikawa, H. 1981. *Mol. Gen. Genet.* 181:332–37
223. Hecker, M. 1982. *Z. Allg. Mikrobiol.* 22:529–34
224. Filutowicz, M., Jonczyk, P. 1983. *Mol. Gen. Genet.* 191:282–87
225. Nagata, K., Guggenheimer, R. A., Hurwitz, J. 1983. *Proc. Natl. Acad. Sci. USA* 80:4266–70
226. Thrash, C., Voelkel, K., DiNardo, S., Sternglanz, R. 1984. *J. Biol. Chem.* 259:1375–77
227. Sternglanz, R., DiNardo, S., Voelkel, K. A., Nishimura, Y., Hirota, Y., et al. 1981. *Proc. Natl. Acad. Sci. USA* 78:2747–51
228. DiNardo, S., Voelkel, K. A., Sternglanz, R., Reynolds, A. E., Wright, A. 1982. *Cell* 31:43–51
229. Pruss, G. J., Manes, S. H., Drlica, K. 1982. *Cell* 31:35–42
230. Laufer, C. S., Depew, R. E. 1984. *Fed. Proc.* 43:1542 (Abstr.)
231. Deleted in proof
232. Stamkiewicz, A. F., Depew, R. E. 1983. *Abstr. Ann. Meeting Am. Soc. Microbiol.* p. 122, Abstr. No. H102
233. Deleted in proof
234. Wang, J. C. 1984. *J. Cell. Sci.*, Suppl. 1. In press
235. Jazwinski, S. M., Edelman, G. M. 1984. *J. Biol. Chem.* 259:6852–57
236. Wirak, D. O., Greenberg, G. R. 1980. *J. Biol. Chem.* 255:1896–904
237. Allen, J. R., Lasser, G. W., Goldman, D. A., Booth, J. W., Mathews, C. K. 1983. *J. Biol. Chem.* 258:5746–53
238. Hamatake, R. K., Mukai, R., Hayashi, M. 1981. *Proc. Natl. Acad. Sci. USA* 78:1532–36
238a. Richardson, S. M. H., Higgins, C. F., Lilley, D. M. J. 1984. *EMBO J.* 3:1745–52
239. Gellert, M., Menzel, R., Mizuuchi, K., O'Dea, M. H., Friedman, D. I. 1983. *Cold Spring Harbor Symp. Quant. Biol.* 47:763–67
240. Menzel, R., Gellert, M. 1983. *Cell* 34:105–13
241. Deleted in proof
242. Adachi, T., Mizuuchi, K., Menzel, R., Gellert, M. 1984. *Nucleic Acids Res.* 12:6389–95
243. Manes, S. H., Pruss, G. J., Drlica, K. 1983. *J. Bacteriol.* 155:4120–423
244. Overbye, K. M., Basu, S. K., Margolin, P. 1983. *Cold Spring Harbor Symp. Quant. Biol.* 47:785–91
245. Trucksis, M., Depew, R. E. 1981. *Proc. Natl. Acad. Sci. USA* 78:2164–68
246. Gemmil, G., Calvo, J. 1979. Cited in Ref. 244
247. Mirkin, S. M., Shmerling, Zh. G. 1982. *Mol. Gen. Genet.* 188:91–95
248. Ephrati-Elizur, E., Chronis-Anner, B. 1984. In *Proteins Involved in DNA Replication*, ed. W. Hubsher, S. Spadari, New York: Plenum
249. Wahle, E., Mueller, K., Orr, E. 1984. *EMBO J.* 3:315–20
250. Mirkin, S. M., Bogdanova, E. S., Gorlenko, S. M., Gragerov, A. I., Larionov, O. A. 1979. *Mol. Gen. Genet.* 177:169–75
251. Filutowicz, M., Jonczyk, P. 1981. *Mol. Gen. Genet.* 183:134–38
252. Oostra, B. A., Ab, G., Gruber, M. 1980. *Nucleic Acids. Res.* 8:4235–46
253. Oostra, B. A., Ab, G., Gruber, M. 1980. *Mol. Gen. Genet.* 177:291–95
254. Oostra, B. A., van Vliet, A. D., Ab, G., Gruber, M. 1981. *J. Bacteriol.* 148:782–87
255. Wahle, E., Mueller, K. 1980. *Mol. Gen. Genet.* 179:661–67
256. Deleted in proof
257. Mosig, G., MacDonald, P., Lin, G., Levin, M., Seaby, R. 1983. See Ref. 43, pp. 173–86
258. Wang, J. C., Barkley, M. D., Bourgeois, S. 1974. *Nature* 251:247–49
259. Maaløe, O., Kjeldgaard, N. O. 1966. In *Control of Macromolecular Synthesis*. New York: Benjamin
260. Bauer, W. R., Ressner, E. C., Kates, J., Patzke, J. V. 1977. *Proc. Natl. Acad. Sci. USA* 74:1841–45
261. Higashinakagawa, T., Wahn, H., Reeder, R. H. 1977. *Dev. Biol.* 55:375–86

262. Bonven, B., Westergaard, O. 1982. *Nucleic Acids Res.* 10:7593–608
263. Gocke, E., Bonven, B. J., Westergaard, O. 1983. *Nucleic Acids Res.* 11:7661–78
264. Wu, C. 1980. *Nature* 286:854–60
265. Weintraub, H. 1983. *Cell* 32:1191–203
266. Elgin, S. C. R. 1984. *Nature* 309:213–14
267. Germond, J. E., Hirt, B., Oudet, P., Gross-Bellard, M., Chambon, P. 1975. *Proc. Natl. Acad. Sci. USA* 72:1843–47
268. Benyajati, C., Worcel, A. 1976. *Cell* 9:393–407
269. Sinden, R. R., Carlson, J. O., Pettijohn, D. E. 1980. *Cell* 21:773–83
270. Pettijohn, D. E., Pfenninger, O. 1980. *Proc. Natl. Acad. Sci. USA* 77:1331–34
271. Weisbrod, S. 1982. *Nature* 297:289–95
272. Lilley, D. M. J. 1983. *Nature* 305:276–77
273. Rowe, T. C., Liu, L. F. 1984. In *Comments Mol. Cell. Biophys: Comments Modern Biol.*, Part A, 2:267–83
274. Schon, E., Evans, T., Welsh, J., Efstratiadis, A. 1983. *Cell* 35:837–48
275. Selleck, S. B., Elgin, S. C. R., Cartwright, I. I. 1984. *J. Mol. Biol.* 178:17–33
276. Larsen, A., Weintraub, H. 1982. *Cell* 29:609–22
277. Kohwi-Shigematsu, T., Gelinas, R., Weintraub, H. 1983. *Proc. Natl. Acad. Sci. USA* 80:4389–93
278. Harland, R. M., Weintraub, H., McKnight, S. L. 1983. *Nature* 302:38–43
279. Luchnik, A. N., Bakayev, V. V., Zbarsky, I. B., Georgiev, G. P. 1982. *EMBO J.* 1:1353–58
280. Luchnik, A. N., Bakayev, V. V., Glaser, V. M. 1982. *Cold Spring Harbor Symp. Quant. Biol.* 47:793–801
281. Ryoji, M., Worcel, A. 1984. *Cell* 37:21–32
282. Glikin, G. C., Ruberti, I., Worcel, A. 1984. *Cell* 37:33–41
283. Champoux, J. J. 1977. *Proc. Natl. Acad. Sci. USA* 74:5328–31
284. Kirkegaard, K., Wang, J. C. 1978. *Nucleic Acids Res.* 5:3811–20
285. Cunningham, R. P., Wu, A. M., Shibata, T., Das Gupta, C., Radding, C. M. 1981. *Cell* 24:213–33
286. Wu, A. M., Bianchi, M., Das Gupta, C., Radding, C. M. 1983. *Proc. Natl. Acad. Sci. USA* 80:1256–60
287. Kmiec, E. B., Kroeger, P. E., Brougham, M. J., Holloman, W. K. 1983. *Cell* 34:919–29
288. Iwabuchi, M.-A., Shibata, T., Ohtani, T., Natori, M., Ando, T. 1983. *J. Biol. Chem.* 258:12394–404
289. Miskimins, R., Schneider, S., Johns, V.,

Bernstein, H. 1982. *Genetics* 101:157–77
290. Mufti, S., Bernstein, H. 1974. *J. Virol.* 14:860–71
291. Mosig, G. 1984. *Cold Spring Harbor Symp. Quant. Biol.* 49: In press
292. Ikeda, H., Aoki, K., Naito, A. 1982. *Proc. Natl. Acad. Sci. USA* 79:3724–28
293. Ikeda, H., Miura, A., Shiozaki, M. 1984. *Cold Spring Harbor Symp. Quant. Biol.* 49: In press
293a. King, S. R., Krolewski, M. A., Marvo, S. L., Lipson, P. J., Pogue-Geile, K. L. et al. 1982. *Mol. Gen. Genet.* 186:548–57
293b. Marvo, S. L., King, S. R., Jaskunas, S. R. 1983. *Proc. Natl. Acad. Sci. USA* 80:2452–56
294. Bullock, P., Forrester, W., Botchan, M. 1984. *J. Mol. Biol.* 174:55–84
295. Deaven, L. L., Oka, M. S., Tobey, R. A. 1978. *J. Natl. Cancer Inst.* 60:1155–68
296. Painter, R. B. 1980. *Mut. Res.* 70:337–41
297. Taylor, J. H. 1982. *Sister Chromatid Exchange*, ed. S. Wolff, pp. 1–16. New York: Wiley
298. Isberg, R. R., Syvanen, M. 1982. *Cell* 30:9–18
299. Fishel, R. A., Kolodner, R. 1984. *J. Bacteriol.* 160:1168–70
300. Smith, C. L. 1983. *Proc. Natl. Acad. Sci. USA* 80:2510–13
301. Ghelardini, P., Liebart, J. C., Marchelli, C., Pedrini, A. M., Paolozzi, L. 1984. *J. Bacteriol.* 157:665–68
302. Hays, J. B., Boehmer, S. 1978. *Proc. Natl. Acad. Sci. USA* 75:4125–29
303. Crumplin, G. C. 1981. *Carcinogenesis* 2:157–60
304. Chao, L., Tillman, D. M. 1982. *J. Bacteriol.* 151:764–70
305. Wright, A., Bridges, B. A. 1981. *J. Bacteriol.* 146:18–23
306. Pedrini, A. M., Ciarrochi, G. 1983. *Proc. Natl. Acad. Sci. USA* 80:1787–81
307. Pedrini, A. M. 1984. In *Proteins Involved in DNA Replication*, ed. W. Hubsher, S. Spadari. New York: Plenum. pp. 449–54
308. Snyder, R. D., van Houten, F., Regan, J. D. 1982. *Nucleic Acids Res.* 10:6207–19
309. Deleted in proof
310. Wolfson, J. S., Hooper, D. C., Swartz, M. N., Swartz, M. D., McHush, G. L. 1983. *J. Bacteriol.* 156:1165–70
311. Wolfson, J. S., Hooper, D. C., Swartz, M. N., McHush, G. L. 1982. *J. Bacteriol.* 152:338–44
312. Friedman, D. I., Plantefaber, L. C., Olson, E. J., Carver, D., O'Dea, M. H., et al. 1984. *J. Bacteriol.* 157:490–97
313. Alonso, J. C., Sarachu, A. N., Grau, O. 1981. *J. Virol.* 39:855–60

Ann. Rev. Biochem. 1985. 54:699–727
Copyright © 1985 by Annual Reviews Inc. All rights reserved

HUMAN APOLIPOPROTEIN MOLECULAR BIOLOGY AND GENETIC VARIATION

Jan L. Breslow

The Rockefeller University, New York, NY 10021

CONTENTS

0066-4154/85/0701-0699$02.00

PERSPECTIVES AND SUMMARY

Apolipoproteins are important structural constituents of lipoprotein particles and have been shown to participate in lipoprotein synthesis, secretion, processing, and catabolism. These subjects have been recently reviewed (1, 2). In the last 15 years, advances in protein chemistry techniques have allowed the identification, isolation, and characterization of at least eight apolipoproteins (A-I, A-II, A-IV, B, CI, CII, CIII, E). Protein sequencing techniques have allowed the derivation of the primary amino acid sequence of the plasma form of six of these polypeptides. In the last two years, cDNA and in some cases genomic clones have been derived for seven of the apolipoproteins. The DNA sequences combined with cell-free synthesis and tissue and organ culture studies have revealed the presence of apolipoprotein precursors containing NH_2-terminal extensions, including prepeptides and in some cases propeptides. In addition, most of the apolipoprotein genes have been mapped in the human genome. Finally, human mutations in the apolipoprotein genes have been identified at both the amino acid and DNA level. Some of these have profound effects on lipoprotein metabolism and are associated with premature atherosclerosis. This chapter will review current knowledge of apolipoprotein gene structure, function, and genetic variation.

APO A-I

Apo A-I is the major protein constituent of high density lipoproteins (HDL). HDL particles are about 50% protein and 50% lipid, and apo A-I is 70% of HDL protein (2). HDL levels are inversely correlated with susceptibility to coronary artery disease and recently the same association has been demonstrated for apo A-I (3). Apo A-I is abundant in plasma with a concentration of 1.0–1.2 mg/ml. Apo A-I is thought to participate by two mechanisms in the reverse transport of cholesterol from tissues to the liver for excretion. Apo A-I can promote cholesterol efflux from tissues (4), perhaps through a recently described receptor that is up-regulated by cholesterol loading (5). Apo A-I also displays cofactor activity for the lecithin cholesterol acyltransferase (LCAT) enzyme, which is responsible for almost all plasma cholesterol esterification (6). This reaction is thought to play a role in transforming nascent HDL to mature HDL particles. In mammals, apo A-I synthesis is approximately equally divided between liver and small intestine, whereas in avians other major sites of synthesis have been identified [for references, see review listed as reference (1)].

Apo A-I cDNA

Apo A-I cDNA clones have been obtained by several laboratories and their DNA sequences derived (7–13). From this information, apo A-I mRNA is thought to be 893 bp in length and includes a 5' untranslated region of 35 bp, a region coding for 267 amino acids of 801 bp, a termination codon, TGA, and a 3' untranslated region of 54 bp followed by a poly A tail. This is compatible with apo A-I mRNA size of 950 bp determined by Northern blotting analysis of human liver RNA.

The cDNA sequence and NH_2-terminal microsequencing of the primary translation product of apo A-I mRNA in cell-free synthesis experiments indicates translation initiation at the methionine 24 amino acids upstream of the NH_2 terminus of the mature protein. The NH_2-terminal 18 amino acids can be cotranslationally cleaved by microsomal membranes; they represent the apo A-I prepeptide. The 6 amino acids adjacent to the NH_2 terminus of the mature apo A-I is the propeptide and has the rather unusual sequence Arg-His-Phe-Trp-Gln-Gln (14–16). The propeptide is not cleaved intracellularly, but rather is present in secreted apo A-I (15–19). Thus, it is necessary to postulate the existence of a previously unsuspected protease activity in lymph and/or plasma required to cleave this hexapeptide and generate mature apo A-I. This converting protease presumably plays a role in apo A-I processing and may be an important determinant of apo A-I and, thereby, HDL metabolism. Recently, protease activity, which is inhibited by EDTA, has been demonstrated in human serum and on the surface of human endothelial cells and hepatoma cells (20, 21). The reported cDNA sequences specify an amino acid sequence for mature apo A-I of 243 amino acids, which is very similar to that derived previously by protein sequencing methods (22). The only difference is at residue 34 where the protein sequence specified Gln and the cDNA sequence indicates Glu.

It has been previously noted that the apo A-I amino acid sequence of residues 99–230 was composed of six tandem 22 amino acid repeats, and five of the six repeats begin with proline (23, 24). Examination of the DNA sequence in this region confirms this and shows a tandemly repeated DNA structure 66 bp in length (11). This finding suggests that this portion of the apo A-I gene arose by intragenic duplications. When the six 66-bp repeats are aligned and a consensus nucleotide at each position of the repeat derived, the consensus sequence is 64–80% homologous with each of the repeats (11). Translation of the consensus sequence reveals an interesting underlying protein structure for this region of apo A-I. As noted from the protein structure, proline, an alpha-helix breaker, occurs every 22 amino acids. The intervening amino acids, when placed in an Edmundsen wheel diagram (25), specify an alpha-helix with a nonpolar and a polar face. This is the general character of the amphipathic alpha-helical configuration which is a common feature of the apolipoproteins (26). It is thought that the nonpolar face interacts with the hydrophobic lipid core of the lipoprotein particle, whereas the polar face interacts with the aqueous plasma environment. In addition, the positively charged residues tend to cluster between the nonpolar and polar faces. The latter has been shown to be important in stabilizing the lipid protein association (27). A recent, more sophisticated computer analysis derived to specifically look for DNA repeats in apo A-I reveals that the basic structure is a 33-bp (11-codon) repeat and a model based on gene duplication and unequal crossing-over events has been derived (28).

Apo A-I Gene

The apo A-I gene has been isolated and its DNA sequence derived (11–13; Figure 1). The gene is 1863 bp in length. A comparison of the sequence of the apo A-I gene with the cDNA reveals 3 introns (IVS) in the apo A-I gene. IVS-1 is 197 bp long and occurs in the 5' untranslated region between bases 20 and 21 upstream of the codon for Met that initiates translation. IVS-2 is 186 bp long and interrupts the codon specifying amino acid -10 which is in the apo A-I prepeptide. IVS-3 is 588 bp long and interrupts the codon specifying amino acid 43 of the mature protein. The intron locations indicate that apo A-I exons may code for functionally distinct regions of apo A-I. For instance, exon 2 contains most of the apo A-I prepeptide, exon 3 contains the propeptide and the NH_2-terminal sequences, whereas exon 4 contains codons for the 200 amino acids which comprise the COOH-terminal portion of the molecule. The latter includes the 66-bp tandem DNA repeats.

The apo A-I gene transcription initiation site has been designated based on the length of several apo A-I cDNA clones (13). Further experiments have

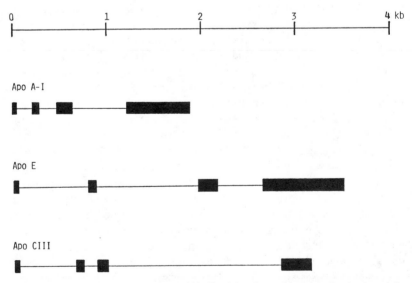

Figure 1 The exon (thick boxes), intron (thin lines) lengths, and locations for the three fully characterized apolipoprotein genes are indicated.

failed to find sequences upstream of this site occurring commonly in liver-derived apo A-I mRNA. Upstream of the proposed apo A-I transcription initiation site is a 7-bp-long AT-rich region which may be the apo A-I promoter, "TATA box" (11–13).

Apo A-I Gene Location

Somatic cell hybrids and DNA probes have been used to map the gene for human apo A-I (29–31, Figure 2). In all studies, the apo A-I gene appears to be at a single locus and cosegregates with human chromosome 11. Some of the hybrids examined contained only a portion of chromosome 11, and apo A-I cosegregated with p11—qter (29), p11—q13 (30), and q13—qter (31) in three different studies. In the mouse, apo A-I has been mapped to chromosome 9 (32) and has been shown to be 1.3±0.7 centimorgans from uroporphyrinogen I synthase (33). The latter has been mapped in humans to chromosome region

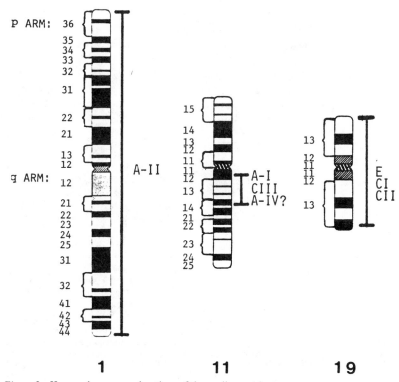

Figure 2   Human chromosome locations of the apolipoprotein genes.

11q13—qter (34). These data strongly suggest that the human apo A-I gene is located on the long arm of chromosome 11 in the vicinity of q13 (35).

Apo A-I Mutations

Apo A-I is the principal structural protein in HDL. Because of the importance of HDL levels in predicting atherosclerosis susceptibility, attention has been paid to apo A-I both by screening for variants in populations, principally by isoelectric focusing, as well as by studying apo A-I in individuals with altered HDL levels. Thus far, these studies have resulted in the discovery of eight proven apo A-I genetic variants (Table 1). Seven of these variants have been shown in people who appear to be heterozygotes for one normal apo A-I structural allele and another allele that specifies a gene product that is either one charge unit more acidic or one charge unit more basic than wild type. The acidic alleles have been designated A-I_{Milano} (36, 37), A-$I_{Marburg}$ (38, 39) and A-$I_{Munster2}$ (40). The former mutation results from a substitution of Cys for Arg at residue 173 (37), whereas both of the latter two mutations appear to be due to a deletion of Lys at residue 107 (41). The basic alleles have been designated A-$I_{Giessen}$ (38), A-$I_{Munster3A}$, A-$I_{Munster3B}$ and A-$I_{Munster3C}$ (40). These mutations result from substitutions of Arg for Pro at residue 143 (41), Asn for Asp at residue 103, Arg for Pro at residue 4, and His for Pro at residue 3 (42), respectively. A-I_{Milano} and A-$I_{Marburg}$, but not the other structural variants,

Table 1   Apo A-I genetic variants

Name[a]	Charge difference[b]	Defect[c]	HDL level[d]
A-I_{Milano}	−1	$Arg_{173} \rightarrow Cys$	↓
A-$I_{Marburg}$	−1	$Lys_{107} \rightarrow 0$	↓
A-$I_{Munster2}$	−1	$Lys_{107} \rightarrow 0$	NL
A-$I_{Giessen}$	+1	$Pro_{143} \rightarrow Arg$	NL
A-$I_{Munster3A}$	+1	$Asp_{103} \rightarrow Asn$	NL
A-$I_{Munster3B}$	+1	$Pro_4 \rightarrow Arg$	NL
A-$I_{Munster3C}$	+1	$Pro_3 \rightarrow His$	NL
A-I-CIII Deficiency	NA	A-I gene insert	↓ ↓

[a]The apo A-I genetic variants, except for A-I-CIII deficiency, have been named for the cities in which they were discovered. A-I-CIII deficiency was discovered in Detroit by Norum et al and designates a clinical syndrome of severe premature atherosclerosis associated with low HDL levels and plasma deficiency of two apolipoproteins, apo A-I and apo CIII.

[b]The major normal human apo A-I isoprotein has an isoelectric point of pH 5.64. Individuals with apo A-I charge variants are heterozygotes with one normal major apo A-I isoprotein and one major apo A-I isoprotein with an isoelectric point of either pH 5.52 (−1) or pH 5.74 (+1).

[c]For each of the charge variants, the altered protein has been isolated, protein sequence determined, and the altered amino acid residue identified. For two of the variants, it appears that Lys_{107} has been deleted. In apo A-I-CIII deficiency, a DNA insert of apo CIII sequences in the apo A-I gene has rendered both genes dysfunctional.

[d]HDL levels in probands has been determined and found to be either normal (NL), mildly reduced (↓), or severely reduced (↓ ↓) as indicated.

have been associated with reduced HDL levels (36, 39). Normal apo A-I activates lecithin cholesterol acyl transferase and A-I$_{Giessen}$ has been reported to be defective in this regard (41). The apo A-I genetic variant designated apo A-I apo CIII deficiency appears to be due to a DNA insertion in the coding region of the apo A-I gene and will be discussed in a subsequent section (43–46).

Another disorder, called Tangier disease, characterized by an autosomal recessive form of inheritance, very low HDL levels, and cholesteryl ester accumulation in reticuloendothelial cells (2), may be associated with an apo A-I abnormality (47). Normal plasma apo A-I has a ratio of propeptide to mature peptide of .01 to .02, whereas in Tangier disease this ratio is 1 to 1.5 (47). It was hypothesized that this observation could be due to a lack of Tangier converting protease activity, a defective substrate precluding normal conversion, or a very unstable conversion product (47). Each of these possibilities has been studied. Conversion of proapo A-I to apo A-I appears to be normal (48) and NH$_2$-terminal sequencing of Tangier proapo A-I reveals a normal amino acid sequence around the cleavage point suggesting a normal substrate for conversion (49). There is some suggestion of instability of mature Tangier apo A-I. Altered lipid binding (50) as well as enhanced in vivo catabolism (51) have been demonstrated for the isolated material. However, the Tangier apo A-I gene has recently been cloned (V. I. Zannis, J. L. Breslow, unpublished observations). Thus far, the DNA sequence of the one allele from one patient analyzed has shown that the coding region appears to specify a normal amino acid sequence. If this is replicated in studies of other alleles from other patients, it would indicate that a structural apo A-I abnormality does not underlie Tangier disease. Other explanations would have to be found for the altered lipid binding, accelerated in vivo catabolism, and increased ratio of proapo A-I to apo A-I. Apo A-I is not a glycoprotein, therefore, altered posttranslational modification with sugars cannot be involved as a mechanism.

APO A-II

Apo A-II is the second most abundant protein in HDL comprising approximately 20% of its protein (2). Plasma apo A-II concentrations are 0.3–0.5 mg/ml. In vitro apo A-II has been shown to displace apo A-I from HDL particles (52) as well as both activate hepatic lipase (53) and inhibit LCAT (6). However, the physiological role of apo A-II has not been determined, and primary qualitative or quantitative abnormalities of human apo A-II have not been reported. Apo A-II is made in the liver and intestine (54).

Apo A-II cDNA

Apo A-II cDNA clones have been isolated and the longest sequence reported is 437 bp (13). The transcription initiation site has not been identified but 22 bp of

5' untranslated sequence have been reported. The rest of the DNA sequence specifies a region coding for 100 amino acids of 300 bp, a termination codon TGA, and a 3' untranslated region of 112 bp.

The cDNA sequence (13) and NH_2-terminal microsequencing of the primary translation product of apo A-II mRNA in cell-free synthesis experiments (55), indicates translation initiation at the methionine 23 amino acids upstream of the NH_2-terminus of the mature protein. Cotranslational cleavage of the primary translation product by microsomal membranes removes the 18 NH_2-terminal amino acids which presumably represent the apo A-II prepeptide. The remaining 5 amino acids adjacent to the NH_2-terminus of mature apo A-II is a propeptide with the sequence Ala-Leu-Val-Arg-Arg. The occurrence of two basic amino acids in the propeptide adjacent to the NH_2 terminus of the mature protein is rather typical of propeptides and quite different from the apo A-I propeptide (14–16). Therefore, apo A-II should be cleaved intracellularly, in contrast to apo A-I. Studies with the human hepatoma cell line, HepG2 (55), indicates that proapo A-II can be cleaved intracellularly but only approximately half is cleaved before export. The explanation for this may be of interest physiologically, or may be a function of this particular in vitro system.

The cDNA sequence specifies an amino acid sequence for mature apo A-II of 77 amino acids, very similar to the previously reported amino acid sequence determined by protein sequencing methods (56). The difference was at residue 35, where the DNA sequence predicts Glu, and the protein derived sequence specified Gln. In human plasma, apo A-II exists as a dimer, the monomers are connected by a disulfide bridge between the only cysteine in the mature protein which resides at amino acid residue 6 (56). At this time, apo A-II gene structure has not been reported. Recently, apo A-II has been mapped to human chromosome 1 using a cDNA probe and a panel of somatic cell hybrids (57, Figure 2).

APO A-IV

Apo A-IV was originally discovered as a major constituent of rat HDL (58). In humans, it is a relatively minor HDL protein with most apo A-IV in the nonlipoprotein plasma fraction (59). Human plasma apo A-IV concentrations are approximately 0.16 mg/ml. In rats, apo A-IV synthesis occurs equally in intestine and liver (54) and in this animal fat feeding doubles intestinal apo A-IV synthesis (60). On this basis, it is hypothesized that apo A-IV plays a role in the biogenesis and/or secretion of intestinal triglyceride–rich lipoproteins, but direct proof of this or any other functional role for apo A-IV is lacking.

Apo A-IV cDNA

Human apo A-IV cDNA clones have not yet been reported; however, some relevant information is available. RNA from human intestinal mucosa has been isolated and translated in wheat germ lysates (61). The NH_2-terminal sequence of the primary translation product has been analyzed by microsequencing techniques and compared to the NH_2-terminal sequence of the mature protein. In this manner, it was determined that human apo A-IV contains a prepeptide sequence 20 amino acids in length. Unlike apo A-I and apo A-II, apo A-IV does not contain a propeptide. The apo A-IV prepeptide is conserved between rats (62) and humans with at least a 55% sequence homology at the amino acid level. The rat apo A-IV prepeptide is very similar to the rat apo A-I prepeptide (14) with 67% homology at the amino acid level.

The rat apo A-IV cDNA sequence has been recently reported (63). From this information, rat apo A-IV mRNA is thought to be 1422 bp in length, and includes a 5' untranslated region of 90 bp, a region coding for 391 amino acids of 1173 bp, a termination codon, TGA, and a 3' untranslated region of 156 bp followed by a poly A tail. The deduced protein sequence was analyzed and 13 repetitions of segments approximately 22 amino acids in length with amphipathic alpha-helical character were identified. In analogy with the previous findings for human apo A-I (11), eight of the apo A-IV repeats, from amino acids 95 through 270, occur as tandem repetitions of exactly 22 amino acids. Seven of these eight repeats begin with proline. When the DNA segments coding for these repeats are aligned and a consensus nucleotide at each position of the repeat derived, the consensus sequence is 64–83% homologous with each of the rat apo A-IV repeats and 49% homologous to a similarly derived consensus sequence for the six human apo A-I DNA repeats that code for apo A-I residues 99–230 (11). In an Edmundson wheel diagram, the consensus codes for an alpha-helix with a polar and nonpolar face as shown for apo A-I. In the case of rat apo A-IV, and human apo A-I, repetitive amino acid and DNA sequences may exist outside of the region of amino acids 95–270, and 99–230, respectively, but the requirement for exact tandem repetition of 22-amino-acid segments has been lost. This presumably represents relaxation of a biological requirement outside of these regions, but future work will be necessary to validate this hypothesis.

Apo A-IV Gene and Genetic Variation

The genomic structure has not been reported for either the human or rat apo A-IV genes. However, genetic variation in human apo A-IV has been demonstrated (38, 40). In most people, isoelectric focusing of plasma apo A-IV results in a single major isoprotein of pH 5.50; other people have this isoprotein plus another one charge unit more basic; and a few people have just the more basic isoprotein. In one large German study, the frequencies of these patterns in a normal population were 85.6%, 13.8%, and 0.6%, respectively (40). Genetic studies were consistent with a single genetic locus two allele model (64). From these data, the major allele frequency, specifying the more acidic gene product, was 92.5%, and the minor allele frequency, specifying the more basic gene product, was 7.5%. Neither heterozygosity nor homozygosity for the minor allele have been associated with plasma lipoprotein abnormalities or athero-sclerosis susceptibility. The apo A-IV protein polymorphism is potentially useful in studying linkage relationships of the apo A-IV gene to other genes. In family studies, a preliminary result showing cosegregation of the apo A-IV protein variant with an apo A-I protein variant has been reported (Lod = 2.7 at θ = 0.0) (35). If this is confirmed, it would mean that the apo A-IV gene resides on human chromosome 11 in the region of q13 (Figure 2).

APO B

Apo B is the major protein constituent of low density lipoproteins (LDL) but is also found in chylomicrons and very low density lipoproteins (VLDL) (2). LDL particles are approximately 25% protein and 75% lipid and virtually all of the protein is apo B. LDL levels are directly correlated with coronary artery disease susceptibility and recently the same association has been demonstrated for apo B levels (3). Apo B is abundant in plasma with a concentration of 0.7–1.0 mg/ml. Apo B is thought to be required for the secretion into plasma of intestinal and hepatic triglyceride-rich lipoproteins (2). Apo B is also recognized by specific high affinity receptors that mediate clearance of LDL particles from plasma (65).

Human apo B is a glycoprotein which occurs in two forms, designated B-100 and B-48 (66). B-100 is thought to be a single polypeptide of $M_r \sim 400,000$ produced primarily in liver, whereas B-48 is approximately half that molecular weight and is produced primarily in the small intestine (66). Very little information about the primary structure of apo B is available. Two problems have been encountered. The protein becomes quite insoluble after delipidation and standard methods of proteolytic digestion have not resulted in a high enough yield of unique peptides for structural studies (66). It has been suggested that the insolubility problem is due to an abnormal sensitivity of the protein to oxidation after delipidation, and recently, with proper precautions, soluble preparations of delipidated apo B have been obtained (67). In addition, the use of bacterial proteases has improved the yield of unique apo B peptides (68). A recent report provides the partial amino acid sequence of two peptides and more structural data should be forthcoming in the near future (68).

The paucity of data on apo B structure, using standard protein chemistry techniques, has lead to immunochemical studies to gain relevant information about this protein. In this regard, several groups have developed monoclonal antibodies to human apo B that recognize distinct epitopes and some interesting information has been derived. In one study, the immunoreactivity of apo B in VLDL changed significantly after in vitro lipolysis, suggesting that apo B conformation might change at different stages of lipoprotein metabolism (69). In another study, epitopes of apo B-100 have been mapped in a linear non-repetitive array. For a subset of these epitopes, the monoclonal antibodies disrupt apo B binding to the LDL receptor, but do not bind to B-48. For another subset of epitopes, the monoclonal antibodies bind B-48, but do not disrupt receptor binding (70). Thus, B-48 and B-100 are antigenically related and it has been suggested that B-48 represents approximately one half of the B-100 protein and that this part of apo B is not involved in receptor binding (66, 70).

Apo B Genetic Variation

Genetically determined variation in both the quality and quantity of apo B has been demonstrated. Antisera from multiply transfused patients have been used to define two series of allelic variants called Lp and Ag. Both systems appear to have marginal effects on plasma cholesterol levels and risk of atherosclerosis [see reference (71) for a recent review]. Recently, in some individuals reduced binding to each of three anti apo B monoclonal antibodies has been demonstrated (72). Three phenotypes of strong, weak, and intermediate binding have been identified and family studies suggest a genetic basis for this consistent with a single genetic locus with two alleles specifying strong and weak binding forms of apo B. Thus, the intermediate binding pattern is the result of heterozygosity for the strong and weak binding alleles, whereas the other two patterns represent homozygosity for their respective alleles. In a study of the phenotype frequencies in a small number of unrelated individuals, the major allele frequency was found to be 70% (72). It was suggested that this apo B antigenic variation resulted from an alteration in the amino acid sequence of the apo B polypeptide affecting the configuration of a single domain in which the epitopes recognized by the three monoclonal antibodies reside. The effect of this recently described genetic variation in apo B on plasma lipoprotein levels has yet to be determined.

Inherited disorders of lipoprotein metabolism associated with diminished plasma levels of apo B have been described. The most striking of these is abetalipoproteinemia (2). In this disorder, individuals suffer from fat malabsorption and lack apo B–containing lipoproteins in their plasma, including chylomicrons, VLDL, and LDL. Parents of affected individuals have normal lipoprotein and apo B levels, however, siblings with this disorder have been described and inheritance is assumed to be autosomal recessive. Immunologically detectable apo B is absent from both the plasma and tissues of these individuals and the disorder is thought to be a genetic defect in apo B synthesis. A phenotypically similar disorder has been described, termed homozygous hypobetalipoproteinemia, in which parents of affected individuals have half-normal levels of LDL cholesterol and apo B (2). This condition may also involve a genetic defect in apo B synthesis, but by a different mechanism.

Finally, individuals have been described with normal fat absorption and the ability to produce chylomicrons, but low to absent levels of LDL cholesterol (73, 74). Apparently, these individuals can produce the intestinal form of apo B, B-48, but not the hepatic form, B-100. The existence of this disorder suggests separate genetic control of B-48 and B-100 synthesis. However, at this time, it cannot be determined whether these two gene products are the result of distinct genetic loci or represent differential splicing of the transcript produced from a single genetic locus.

Increased apo B levels are associated with atherosclerosis susceptibility (3). In familial hypercholesterolemia (FH), an autosomal dominant disorder characterized by premature atherosclerosis, apo B and cholesterol in LDL are elevated and the genetic lesion is a defect in the apo B/E receptor (65). A substantial fraction of nonFH individuals with coronary disease have been shown to have increased apo B, but normal cholesterol in LDL. This phenotype has been called hyperapobetalipoproteinemia (75). A subset of these individuals may have the autosomal dominant disorder associated with premature atherosclerosis, familial combined hyperlipidemia (FCHL) (76). Plasma from these individuals consistently contains elevated plasma apo B levels but only occasionally is the LDL cholesterol elevated (77). One possible genetic explanation for the hyperapobetalipoproteinemia phenotype could be a primary overproduction of apo B (78, 79). Further efforts aimed at studying the regulation of apo B synthesis are hampered by the absence of cDNA and genomic clones for apo B.

APO CI

Apo CI is a constituent of VLDL and HDL (2). VLDL particles are approximately 10% protein and 90% lipid and apo CI is 10% of VLDL protein. As previously noted, HDL particles are approximately 50% protein and 50% lipid and apo CI is 2% of HDL protein. Human plasma apo CI concentrations are in the range of 0.04–0.06 mg/ml. In vitro apo CI has been shown to activate LCAT, but not as efficiently as apo A-I (80). The physiological role(s) of apo CI has not been defined. Synthesis mainly occurs in liver and to a minor degree in intestine, but has not been evaluated in other organs (2). Primary qualitative or quantitative abnormalities of human apo CI have not been reported.

Apo CI cDNA

Apo CI cDNA clones between 400 and 480 bp in length have recently been isolated (81). Northern blotting of adult human liver RNA with a probe made from apo CI cDNA indicated two species of apo CI mRNA approximately 580 and 560 bp in length. This meant both that there could be heterogeneity in the transcription initiation of apo CI mRNA and that the inserts in the cDNA clones were not full length. The inserts appeared to lack sequences corresponding to the 5' end of apo CI mRNA. Primer extension on a template of human liver mRNA indicated that apo CI mRNAs have 63 and 40 bp 5' untranslated regions. The DNA sequence provided includes 56 bp in the 5' untranslated region. The rest of the DNA sequence specifies a region coding for 83 amino acids of 249 bp, a termination codon, TGA, and a 3' untranslated region of 111 bp.

The DNA-derived amino acid sequence contains the 57 residues of mature apo CI and agrees entirely with the results derived by protein sequencing techniques (82, 83). The DNA sequence also specifies a 26-amino-acid NH_2-terminal extension (81). This amino acid sequence is compatible with the entire 26 amino acids being the apo CI prepeptide. However, the possible existence of an apo CI propeptide with an unusual sequence and two-step processing of the apo CI primary translation product must be formally excluded.

Apo CI gene structure has not yet been reported. Recently, utilizing somatic cell hybrids and a cDNA probe, the apo CI gene has been mapped to human chromosome 19 (84) (Figure 2). Utilizing cloned DNA, an apo CI gene has been located approximately 4 kb 3' to the apo E gene (Das, Breslow, unpublished observations)

APO CII

Apo CII is a constituent of VLDL and HDL and comprises 10% of VLDL protein and approximately 1% of HDL protein (2). Human plasma apo CII concentrations are in the range of 0.03–0.05 mg/ml. Purified apo CII has cofactor activity for the enzyme lipoprotein lipase, which catalyzes the hydrolysis of triglycerides in chylomicrons and VLDL. The physiological importance of apo CII in activating lipoprotein lipase has been established by the finding of patients with inherited apo CII deficiency, who are severely hypertriglyceridemic and have functional lipoprotein lipase deficiency [see reference (85) for review]. In such cases, both the hypertriglyceridemia and the lipase deficiency can be relieved by an exogenous source of apo CII (86). Studies using tryptic fragments of apo CII have shown that the COOH-terminal amino acids 55–79 are necessary for maximal lipoprotein lipase activation (87, 88). Synthesis of apo CII is mainly in liver and to a minor degree in intestine.

Apo CII cDNA

Apo CII cDNA clones have been obtained by several laboratories (13, 89, 90). DNA sequence information is not yet available on the 5' untranslated region, but beginning with the codon that initiates translation 450 bp of sequence has been specified. Northern blotting analysis of liver and intestine RNA indicates that apo CII mRNA is approximately 500 bp in length (90). This suggests a short 5' untranslated region of at the most 50 bp. The cDNA sequence indicates a coding region for 101 amino acids of 303 bp, a termination codon, TAA, and a 3' untranslated region of 144 bp, followed by a poly A tail.

The DNA-derived amino acid sequence contains the 79 residues of mature apo CII. This is in agreement with a recent result derived by protein sequencing techniques (91) but differs somewhat from the previously derived amino acid sequence (92). In the latter case, a polypeptide of 78 amino acids was reported. In addition, the DNA sequence indicates that residues 2 and 17 are glutamine not glutamic acid, amino acid 27 is glutamic acid not glutamine, and amino acids 20–26 are Glu-Ser-Leu-Ser-Ser-Tyr-Trp not the reported Glu-Trp-Leu-Ser-Ser-Tyr. The DNA sequence also specifies a 22-amino-acid NH_2-terminal extension compatible with the existence of an apo CII prepeptide (13, 89, 90).

There is a striking homology between the NH_2-terminal regions of both apo A-I and apo CII at the amino acid level. Apo A-I amino acids -2, -1, $+1$, $+2$ are identical to apo CII amino acids 5, 6, 7, and 8 and, in each case, specify Gln-Gln-Asp-Glu. This region spans the site of cleavage of the protease that converts proapo A-I to mature apo A-I. This suggests that apo CII should be a substrate for the apo A-I converting protease. However, in plasma, mature apo CII is found to be the full 79 amino acids in length and is fully active as a lipoprotein lipase activator in this form. Thus, apo CII is not a physiological substrate for the apo A-I converting protease for reasons that remain to be elucidated. Preliminary analysis of the apo A-I and apo CII cDNA sequences reveals extensive homology at the DNA level extending between apo A-I bp 102–285 (amino acids -2 to 59) and apo CII bp 79–261 (amino acids 5–65). Alignment of the DNA sequences in these regions by metric analysis indicates 56% overall matching with runs of matches 11, 8, and 10 bp in length (B. Erickson, J. L. Breslow, unpublished observations). These observations suggest a close phylogenetic relationship between apo A-I and apo CII.

Apo CII Gene Structure, Mapping, and Genetic Variation

The apo CII gene has been isolated and a restriction map reported (89, 93). Preliminary sequencing information reveals introns interrupting the codons specifying amino acids -3 and 50. There are introns in similar locations in the other apolipoprotein genes where this has been studied. This is discussed in the sections about the apo A-I, apo CIII, and apo E genes. In addition, these other genes all have introns interrupting the 5' noncoding region of their cDNAs (Figure 1). Therefore, it is expected that apo CII will also have an intron in this location, but this information is not yet available. The two known apo CII introns divide the DNA sequence coding for the signal peptide from that coding for the mature protein, and the DNA sequence coding for the NH_2 from that coding for the COOH-terminal portions of the mature protein, respectively.

Somatic cell hybrids and a probe made from an apo CII cDNA clone have been used to map the gene for human apo CII to chomosome 19 (89, 93, Figure 2). A discordancy was noted with a somatic cell hybrid that contained a rearranged chromosome 19 and lacked two markers for the distal end of the long arm of this chromosome. Apo CII can therefore be localized to 19pter—q13 (89, 93).

Southern blot analysis of human DNA after digestion with the restriction endonuclease TaqI reveals a common polymorphism in the vicinity of the apo CII gene (94, 95). The major allele frequency in normal individuals is 0.60 (95). As will be discussed in the section on apo E, the human apo E gene also resides on chromosome 19 and a common polymorphism has been demonstrated. The exact relationship between the apo CII and apo E genes has not been defined. However, family studies done by two different groups show cosegregation of the apo CII TaqI polymorphism and the apo E protein polymorphism (in each case, Lod > 4.0 at θ = 0.0) (94, 95). In addition, linkage disequilibrium has been demonstrated between the alleles of both genes (95). The combined data from the two studies indicate that no recombinations have occurred in 53 observed meioses. Therefore, these two genes appear to be less than 2 centimorgans apart. The genomic clone for apo CII does not hybridize strongly with the cDNA clone for apo E and vice versa. From the location of these genes on their respective genomic clones, one can conclude that they are no closer than 5–7 kb (J. L. Breslow, unpublished data).

Several patients have been described who lack apo CII in their plasma. They are functionally lipoprotein lipase deficient and severely hypertriglyceridemic. Apo CII deficiency appears to be a recessive genetic disorder with obligate heterozygotes having half-normal apo CII levels, but normal triglyceride concentrations (85). Southern blot analysis with a probe made from an apo CII cDNA clone reveals that at least a subset of these patients possess the apo CII gene and that it is grossly intact (96). Two independent families of probands with this disorder have been studied using the TaqI polymorphism to assess disease linkage with the apo CII gene locus. In each family, cosegregation was observed, which is consistent with the defect in this condition being within or near the apo CII gene (96).

A protein polymorphism of apo CII has also been described (97). In most individuals, plasma apo CII is a single spot on two-dimensional gels and a single band on one-dimensional size or charge separation gels. Recently, three hypertriglyceridemic individuals have been described who have a normal apo CII component and an apo CII isoprotein 1 charge unit more acidic than normal (97). It has been determined that this is due to a substitution of glutamine for lysine at residue 55. The apo CII DNA sequence at this residue determined for cDNA clone pCII-711 is AAA which specifies lysine. It is possible to explain the occurrence of glutamine at this residue by a single base substitution in the first base of this codon, substituting C for A. Although this mutant form of apo

CII was isolated from hypertriglyceridemic patients, it appears to activate lipoprotein lipase normally. Thus, the relationship between the amino acid substitution and the hypertriglyceridemia probably is not cause and effect.

APO CIII

Apo CIII is a constituent of VLDL and HDL and comprises about 50% of VLDL protein and 2% of HDL protein (2). Human plasma apo CIII concentrations are in the range of 0.12–0.14 mg/ml. Apo CIII is a glycoprotein containing 1 mole each of galactose, galactosamine, and either 0, 1, or 2 moles of sialic acid (98). The three resultant isoproteins recognizable by isoelectric focusing are designated CIII-0, CIII-1, and CIII-2 and comprise 14%, 59%, and 27% of plasma apo CIII, respectively (99). In vitro, apo CIII has been shown to inhibit the activities of both lipoprotein lipase (100, 101) and hepatic lipase (102). Apo CIII has also been shown to decrease the uptake of lymph chylomicrons by the perfused rat liver (103–106). These in vitro studies suggest that apo CIII might delay catabolism of triglyceride-rich particles. Recently, patients with combined apo A-I, apo CIII deficiency were shown to have low plasma triglyceride levels (43) and in vivo studies showed that they rapidly convert VLDL to LDL (107). In vitro lipolysis of their VLDL was inhibited by added apo CIII (107). Thus, it appears that primary abnormalities in the quantity or quality of apo CIII may affect plasma triglyceride levels and the physiological role of apo CIII may be in the regulation of the catabolism of triglyceride-rich lipoproteins. Functional domains of apo CIII have been demonstrated. The NH_2-terminal 40 amino acids do not bind phospholipid, whereas the COOH-terminal 39 amino acids do (108). Synthesis of apo CIII is mainly in liver and to a lesser degree in intestine (2).

Apo CIII cDNA

Apo CIII cDNA sequence has been reported (13, 45, 109). The complete sequence of the apo CIII 5' untranslated region is not currently known but 20 bp of sequence information is available. These data indicate that apo CIII mRNA is at least 507 bp in length and includes a coding region for 99 amino acids of 297 bp, a termination codon, TGA, and a 3' untranslated region of 187 bp followed by a poly A tail. This is compatible with the size of apo CIII mRNA of about 550 bp determined by Northern blotting analysis of human liver mRNA (13).

The DNA-derived apo CIII amino acid sequence (13, 109) differs from the previously reported protein-derived apo CIII amino acid sequence (98) at residues 32, 33, 37, and 39. At these locations, the DNA sequence predicts Glu, Ser, Gln, Ala, respectively, whereas the previously reported protein-derived sequence specified Ser, Gln, Ala, Gln, respectively. Three cDNA clones from two separate cDNA libraries (13, 109) all have shown the same DNA-derived amino acid sequence. In addition, the DNA sequence coding for residues 32 and 33, GAGTCC, includes a recognition site for the restriction endonuclease HinfI (GANTC), whereas the DNA sequence required to code for the corresponding residues in the protein-derived sequence could not possibly contain a HinfI recognition site (109). Southern blotting analysis of genomic DNA from four normal and two hypertriglyceridemic individuals identifies a HinfI site in this region, also compatible with the DNA-derived amino acid sequence (109). Thus, the protein sequence, unless it was derived from a person homozygous for a rare apo CIII allele, is probably in error and should be revised.

The DNA-derived amino acid sequence indicates a 20-amino-acid NH_2-terminal extension for the primary translation product of apo CIII (13, 109). The sequence is compatible with previously reported prepeptide sequences. Cell-free synthesis experiments using mRNA from rat liver and intestine indicates rat apo CIII is made with a 20-amino-acid NH_2-terminal extension (110). This can be cotranslationally cleaved by signal peptidase to yield a product with the same NH_2-terminus as the mature protein (110). Therefore, apo CIII is made as a preprotein and does not contain a propeptide sequence.

Apo CIII Gene Structure, Mapping, and Genetic Variation

Apo CIII cDNA clones were used to identify the apo CIII gene on human genomic DNA cloned in lambda phage which contained the apo A-I gene (45). Mapping of the apo CIII gene reveals that it is about 2500 bp from the 3' end of the apo A-I gene (45). Further mapping and DNA sequence analysis revealed that these genes are coded for by opposite DNA strands (45, 109). The 3' end of the apo CIII gene is located closest to the 3' end of the apo A-I gene, and the 5' end of the apo CIII gene, containing the apo CIII promoter, is furthest away from the apo A-I gene. Thus, these two genes are convergently transcribed. It is not known where their primary transcripts end, or whether there is any functional significance to this unusual configuration.

The apo CIII gene is approximately 3000 bp in length and contains 3 introns (109; Figure 1). IVS-1 is approximately 600 bp long and occurs in the 5' untranslated region between bases 13 and 14 upstream of the codon for Met that initiates translation. IVS-2 is approximately 125 bp long and interrupts the codon specifying amino acid −2, which is in the apo CIII prepeptide. IVS-3 is about 1800 bp long and interrupts the codon specifying amino acid 40 of the mature protein. Thus, as for apo A-I and apo CII, the intron locations indicate that apo CIII exons may code for functionally important domains of the protein. For instance, intron 2 separates the prepeptide from the mature protein, and intron 3 seems to separate the phospholipid binding domain from the nonbinding domain.

Apo CIII cDNA clones and somatic cell hybrids have been used to map the apo CIII gene to a single locus on human chromosome 11 (29, Figure 2). This is compatible with the close linkage demonstrated by finding both genes on a single cloned genomic fragment of DNA (45). Thus the human apo A-I, apo CIII, and probably apo A-IV (35) genes all reside in the same region of the genome in the vicinity of 11q13.

Human genetic variation in apo CIII has been documented. Southern blotting of human DNA after digestion with SstI or its isoschizomer, SacI, with an apo A-I probe revealed a DNA polymorphism flanking the apo A-I gene (111). This was subsequently shown to be due to a single base substitution in the 3' untranslated region of the neighboring apo CIII gene (45). A clinical study of 28 hypertriglyceridemic subjects found ten heterozygotes and two homozygotes for this polymorphism, whereas in 70 controls, only three were found to be heterozygotes (111). Thus, an association was suggested between hypertriglyceridemia and genetic variation in the apo CIII gene (111). This is compatible with the proposed physiological role of apo CIII in regulating the metabolism of triglyceride-rich lipoproteins. However, it will be necessary to confirm this association by studying other hypertriglyceridemic as well as control populations. It also remains to be determined how a DNA base substitution in the 3' untranslated region could affect the quantity or quality of the apo CIII protein in a person's plasma. To this end, three apo CIII cDNA clones containing the entire coding region, one with and two without the SacI site, have been sequenced (13, 109). Three sites of variation, including the polymorphic SacI site, have been identified. One site of variation is in the coding region affecting the third base of a codon-specifying amino acid residue 14, but in each case, the codon specifies glycine. The other two sites of variation are 31 bp apart in the 3' untranslated region. Thus far, the SacI polymorphism cannot be associated with another DNA variation affecting the primary amino acid sequence and thereby the quality of apo CIII. Whether the bp variations identified in the 3' untranslated region can affect the quantity of apo CIII remains to be determined.

Apo A-I Apo CIII Gene Rearrangement

A family has been described in which two sisters had very low HDL but normal LDL levels, xanthomas, severe premature atherosclerosis, and absent plasma apo A-I and apo CIII (43). First degree relatives of these individuals had half-normal plasma levels of HDL, apo A-I, and apo CIII. Southern blotting of genomic DNA from the probands, after digestion with EcoRI, with an apo A-I cDNA probe revealed a single band of 6.5 kb, whereas normals showed a single 13-kb band (44). First degree relatives, including the mother and father, of the probands showed one normal band and one abnormal band (44). Therefore, they appeared to be carriers of a mutant allele associated with the apo A-I gene,

APO A-I APO CIII DEFICIENCY
PROBANDS AND FIRST-DEGREE RELATIVES
SOUTHERN BLOT ANALYSIS
APO A-I GENE

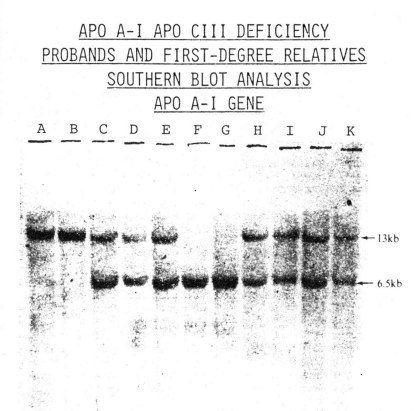

Figure 3   Southern blot of EcoRI-digested DNA from a normal individual (A) and from the apo A-I-deficient probands (F) and (G). Blots are also shown for the maternal grandfather (B), father (C), mother (D) and brother (E) of the probands. In addition, the son (H) and daughter (I) of proband (G), and the son (J) and daughter (K) of proband (F) are also shown.

and the probands appeared to be homozygous for this mutant allele (Figure 3). Southern blotting of probands' DNA, after digestion with other restriction endonucleases with an apo A-I cDNA probe, consistently revealed differences from wild type (46). This suggested that the genetic lesion was not a single bp substitution, but rather a more major DNA alteration. Southern blotting with other probes derived from the region of the apo A-I gene indicated that the fourth exon of the apo A-I gene was interrupted at approximately the codon specifying residue 80 of the mature protein and this may explain the lack of apo A-I in the plasma of these patients. Recently, the insertion sequences have been cloned and found to correspond to apo CIII sequences. Thus a DNA rearrangement at the apo A-I apo CIII genetic locus has occurred in these patients (S. K. Karathanis, V. I. Zannis, J. L. Breslow, unpublished data).

APO E

Apo E in normal plasma is equally divided between VLDL and HDL. It comprises about 10–20% of VLDL protein and 1–2% of HDL protein. Apo E occurs in a metabolically distinct subfraction of HDL particles where it is a larger fraction of the protein (2). Human plasma apo E concentrations are in the range of 0.025–0.050 mg/ml. Two-dimensional gel electrophoresis of human plasma apo E has shown it to consist of several isoproteins that differ in size and/or charge (112, 113). This is the result of both common genetic variation of apo E in the population (discussed below) and posttranslational modification of apo E with carbohydrate chains containing sialic acid (112–116). Apo E is synthesized and secreted as sialo apo E and subsequently desialated in plasma (18, 19, 117), but the physiological significance of this process is unknown. Apo E can be recognized by high affinity receptors and mediate the binding, internalization, and catabolism of lipoprotein particles. Apo E can serve as a ligand for the LDL (apo B/E) receptor present on hepatic as well as extrahepatic tissues (118–120). Hepatic tissues also possess a high affinity receptor that recognizes particles that contain apo E, but not apo B (121–123). This receptor is genetically distinct from the LDL receptor and has been called the chylomicron remnant or apo E receptor. Mature apo E is a 299-amino-acid polypeptide (124) and the receptor binding region has been localized to the middle portion of the polypeptide chain between residues 140 and 150 with residue 158 being important for the conformation of the binding domain (125–127). Structural mutations in apo E affect receptor recognition and are believed to underlie type III hyperlipoproteinemia (HLP) (discussed below), a condition associated with increased plasma levels of cholesterol and triglycerides, xanthomas, and premature atherosclerosis [for reviews, see references (128, 129)].

Apo E synthesis occurs in liver and to a minor extent in intestine. However, in contrast to the other apolipoproteins, synthesis has been documented in a wide variety of other tissues including kidney, adrenal gland, and reticuloendothelial cells (130, 131).

Apo E cDNA

Apo E cDNA sequences have been reported (117, 132–135). From the proposed transcription initiation point (136), these data indicate that apo E mRNA is 1163 bp in length and includes a 5' untranslated region of 67 bp, a region coding for 317 amino acids of 951 bp, a termination codon, TGA, and a 3' untranslated region of 142 bp. This is compatible with apo E mRNA size of 1150 bp determined by Northern blotting analysis of human liver mRNA (134).

The cDNA sequence and NH_2-terminal microsequencing of the primary translation product of apo E mRNA in cell-free synthesis experiments indicates translation initiation at the methionine 18 amino acids upstream of the mature protein (117). The NH_2-terminal 18 amino acids can be cotranslationally cleaved by microsomal membranes and represent the apo E signal peptide (117). There is no propeptide.

In analogy with both apo A-I and apo A-IV, human apo E contains eight tandem repetitions of exactly 22 amino acids from residues 62 to 237 (136). Only one of these repeats actually begins with proline. However, the sequence of charges of the amino acids in each repeat is strikingly similar. For example, two consecutive acidic amino acids occur in the same position in six of the eight repeats. When the DNA segments coding for these repeats are aligned and a consensus nucleotide at each position of the repeat derived, the consensus sequence is 51–75% homologous with each of the apo E repeats and 72% homologous to a similarly derived consensus sequence for the six human apo A-I DNA repeats that code for apo A-I residues 99–230. Thus, extreme similarity exists with respect to the 66 bp repeats in apo E, apo A-I, and apo A-IV, suggestive of a common ancestral origin of this portion of these three apolipoprotein genes. The consensus amino acid sequence when placed in an Edmundson wheel diagram specifies an amphipathic alpha-helix.

Apo E Gene Structure and Mapping

The apo E gene has been isolated and sequenced (136; Figure 1). The gene is about 3.7 kb in length, and contains four exons and three introns (IVS). IVS-1 is about 700 bp long and occurs in the 5' untranslated region between bases 23 and 24 upstream of the codon for Met that initiates translation. IVS-2 is about 1100 bp long and interrupts the codon specifying amino acid -4 which is in the apo E prepeptide. IVS-3 is about 600 bp long and interrupts the codon specifying amino acid 61 of the mature protein. The intron locations are strikingly similar to those identified for the apo A-I, apo CII, and apo CIII genes and, as previously suggested, may indicate that each exon codes for a functionally distinct region of apo E. Similar intron locations of the apolipoprotein genes may be in favor of a common ancestral origin of this gene family as previously suggested (23). The apo E transcription initiation site has been tentatively assigned to the A 44 bp upstream of the GT that begins the first intron based on S1 nuclease protection experiments with human liver mRNA (136). The sequence TATAATT occurs beginning 33 bp upstream of the proposed transcription initiation site and is the putative apo E promoter.

Family studies were used to show that the inheritance of the apo E protein polymorphism cosegregated with the protein polymorphism for the third component of complement (137). Since the latter has been mapped using a DNA probe and somatic cell hybrids to human chromosome 19 (138), the apo E gene was also assigned to this chromosome (137). Apo E cDNA probes and somatic cell hybrids have now been used to confirm this assignment (136). As previously noted, the apo CI (84) and apo CII (89) genes have also been mapped to chromosome 19. The apo E and apo CII genes are within 2 centimorgans (94, 95), but their exact relationship has not been defined. The apo E gene lies adjacent to the apo CI gene (Das, Breslow, unpublished data). The disorder familial hypercholesterolemia involves a defect in the LDL (apo B/E) receptor and the receptor gene has been assigned to chromosome 19 (139). It is of interest that the gene for a receptor and one of its ligands both reside on the same chromosome. However, preliminary data suggest that they are not closely linked (140).

Apo E Mutations

One-dimensional isoelectric focusing of human plasma apo E reveals several bands whose relative concentrations vary between different individuals (114, 116). Utilizing two-dimensional gel electrophoresis, it was possible to determine that some of these bands were due to sialo apo E isoproteins and others due to variations in the isoelectric point of the major asialo apo E isoprotein(s) (112, 113). Studies of large numbers of individuals revealed six common apo E phenotypes in the population (112, 113). Family studies showed that these phenotypes were the result of a single apo E gene locus with three common alleles (112, 113). The alleles have been designated $\epsilon4$, $\epsilon3$, and $\epsilon2$ and their gene products from basic to acidic are E4, E3, and E2, respectively. There are three homozygous phenotypes, E4/4, E3/3, and E2/2, and three heterozygous phenotypes, E4/3, E3/2, and E4/2 (141; Figure 4). Five relatively large studies of apo E phenotype prevalence have been reported (129, 142–145). These have been done in diverse geographical areas but primarily in Caucasians. The range of allele frequencies were $\epsilon4$ 11–15%, $\epsilon3$ 74–78%, and $\epsilon2$ 8–13%. Large studies assessing the frequencies of the apo E alleles in other racial groups have not been reported. In Caucasians, it appears that apo E allele frequencies are similar worldwide.

This common apo E polymorphism has been found to play a role in type III HLP (112–114, 146). This disorder is characterized by elevated cholesterol and triglyceride levels, as a result of delayed chylomicron remnant clearance, xanthomas, and premature coronary as well as peripheral vascular disease (128, 129). Over 90% of individuals with type III HLP have the E2/2 phenotype (129, 147), whereas this occurs in only 0.5–1.4% of normal individuals (129, 142–145). In addition, when E2 is isolated and studied in vitro, it does not bind as well as E3 or E4 to high affinity lipoprotein receptors (148–150). It has been suggested that chylomicron remnants with E2 on their surface are recognized poorly by receptors, cleared slowly from plasma, and accumulated in plasma. Chylomicron remnants are quite potent stimulators of macrophage cholesteryl ester accumulation in vitro and high plasma concentrations of these particles may be involved in the atherogenic process in vivo [reviewed in references (128, 129)]. These data all suggest that homozygosity for the $\epsilon2$

APO E ALLELES

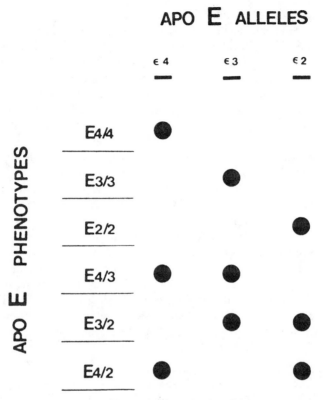

Figure 4   Schematic presentation of the three-allele model of apo E inheritance and nomenclature of the apo E alleles and phenotypes. The closed circles represent the major asialo apo E isoproteins.

allele may be the underlying cause of type III HLP. However, the disease frequency is such that only 1–2% of people with the E2/2 phenotype actually express the disease. It is known that other hormonal and environmental factors are necessary for disease expression. However, the current belief is that type III HLP is the result of two gene defects. One of these is in the apo E structural gene and the other in a gene that influences chylomicron remnant synthesis or catabolism in a synergistic fashion. The second gene product has yet to be defined (128, 129).

In addition to the striking involvement of the E2/2 phenotype in type III HLP, it appears that the apo E gene locus may be one of the factors influencing lipid levels in the general population. A recent review of five studies in the literature suggests that the ε2 allele exerts a stepwise gene dosage effect on lowering LDL cholesterol levels (129). The ε2 allele also appeared to influence the VLDL fraction and results in an increase in VLDL cholesterol and tri-glyceride levels in a similar stepwise manner (129). Recent studies have found

E2 more frequent in patients with hypertriglyceridemia, E4 more frequent in hypercholesterolemia, and E2 and E4 more frequent in mixed hyperlipidemia, and suggested a specific effect of E4 on blood lipid values (142, 143). Future studies will have to determine the exact nature of the effect of the apo E phenotype on plasma lipid levels in the general population.

Amino acid sequence analysis established that the two common variants of apo E, E4 and E2, differ from E3 by single amino acid substitutions (124). E4 differs from E3, at residue 112, because of an arginine for cysteine substitution, and E2 differs at residue 158, because of a cysteine for arginine substitution. Isoelectric focusing and amino acid and DNA sequencing have identified other rare apo E alleles. In all, nine alleles are known and these are listed in Table 2. In all but one of these alleles, E5 (151), the amino acid substitution underlying the variation has been identified. Alleles E3** (127), E2 (124), E2* (150), E2** (152), and E1 (153) all involve amino acid substitutions replacing positively charged with neutral amino acids in the apo E receptor binding region. Where information is available, these gene products have been shown to be defective in receptor binding and/or isolated from individuals with the type III HLP phenotype. This emphasizes the importance of the positively charged amino acid residues in the receptor binding region. Allele E3* was determined from the DNA sequence (135), but the substitution of proline for alanine at residue 152 might have a significant effect on the conformation of the

Table 2 Human apo E protein polymorphism

Name[a]	Charge difference[a]	Defect[b]
E5	+2	?
E4	+1	$Cys_{112} \rightarrow Arg$
E3	0	—
E3*	0	$Ala_{99} \rightarrow Thr$, $Ala_{152} \rightarrow Pro$
E3**	0	$Cys_{112} \rightarrow Arg$, $Arg_{142} \rightarrow Cys$
E2	−1	$Arg_{158} \rightarrow Cys$
E2*	−1	$Arg_{145} \rightarrow Cys$
E2**	−1	$Lys_{146} \rightarrow Gln$
E1	−1	$Gly_{127} \rightarrow Asp$, $Arg_{158} \rightarrow Cys$

[a]Nomenclature for the apo E allele gene products recognized by the isoelectric focusing position of their major asialo apo E isoprotein as specified by Zannis et al (141). The most common allele gene product E3 has an isoelectric point of pH 6.02. Alleles specifying gene products E4 and E5 are 1 and 2 charge units, respectively, more basic, and E2 and E1 are 1 and 2 charge units, respectively, more acidic than wild type.

[b]The amino acid sequence of the most common allele gene product, E3, has been specified by protein and DNA sequencing. Further protein sequencing studies of apo E derived from different individuals has revealed amino acid substitutions as indicated. The nature of the protein defect responsible for E5 has yet to be determined.

*, **The stars indicate rare apo E variants recently discovered with the same isoelectric focusing pattern as E3 and E2.

receptor binding region. Functional studies of this gene product have not yet been reported. Finally, the alteration responsible for the E4 allele is not in the receptor binding region, and this gene product is fully functional in receptor binding studies (148–150).

Further information about apo E genetic variation has been forthcoming from DNA sequencing. Four cDNA clones and one genomic clone have been sequenced and seven sites of variation identified (117, 132, 133, 135 Table 3). A cDNA clone pE-368 varies at bp 9 (117); cDNA clones pHAE-112 and pHAE-178 vary at bp 865 (135). cDNA clone pHAE-813 differs at 4 bp, 376, 416, 575, and 790 (135). Two of the bp substitutions actually change the coding sequence and result in the gene product E3* discussed above and presented in Table 3. The lambda apo E #1 clone of genomic DNA differs at bp 455 and appears to specify the E4 gene product (136). Thus in addition to the nine apo E alleles specified in Table 3, DNA sequence analysis reveals the presence of at least two more alleles. In addition to the large number of mutations in the apo E exonic regions, other significant variations may exist in the apo E gene, as suggested by the recent report of an individual with absence of plasma apo E and the type III HLP phenotype (154).

ACKNOWLEDGMENTS

This work was supported by grants from the National Institutes of Health (HL32354, HL32435, AGO4727). Dr. Jan L. Breslow is an Established Investigator of the American Heart Association. I would like to express my sincere appreciation to Miss Lorraine Duda, Mr. Jeffrey Levine, and Mr. Alexander Pertsemlidis for their extraordinary and expert assistance in preparing this review.

Table 3 Human apo E genetic variation

Clone[b]	Base pair[a]							Gene product[c]
	9	376	416	455	575	790	865	
pE-368	G	G	G	T	G	C	G	E3
λ apo E #1	C	G	G	C	G	C	G	E4
pHAE-112	C	G	G	T	G	C	A	E3
pHAE-178	C	G	G	T	G	C	A	E3
pHAE-813	C	A	A	T	C	T	G	E3*

[a]The bp's are numbered to correspond to those in apo E in RNA beginning at the cap site [see Ref. (136)]
[b]cDNA and genomic clones were sequenced in two different laboratories. pE-368, pHAE-112, pHAE-178, and pHAE-813 are cDNA clones. λ apo E #1 is a genomic clone.
[c]Gene product specified by that allele. E3* refers to a new allele identified by DNA sequencing. This allele has the same isoelectric focusing pattern as the E3 allele but differs in two uncharged amino acids and four bp's.

Literature Cited

1. Zannis, V. I., Breslow, J. L. 1984. *Adv. Hum Genet.* In press
2. Herbert, P. N., Assmann, G., Gotto, A. M. Jr., Fredrickson, D. S. 1982. *The Metabolic Basis of Inherited Disease,* pp. 589–651. New York: McGraw-Hill
3. Heiss, G., Tyroler, H. A. 1982. *Proc. Workshop on Apolipoprotein Quantification,* pp. 7–24. Bethesda, MD: US Dept. of Health and Human Services, Natl. Inst. Health. NIH Publ. 83-1266
4. Stein, O., Stein, Y. 1973. *Biochim. Biophys. Acta* 326:232–44
5. Brinton, E. A., Bierman, E. L. 1983. *J. Clin. Invest.* 72:1611–21
6. Fielding, C. J., Shore, V. G., Fielding, P. E. 1973. *Biochim. Biophys. Acta* 270:513–18
7. Breslow, J. L., Ross, D., McPherson, J., Williams, H., Kurnit, D., et al. 1982. *Proc. Natl. Acad. Sci. USA* 79:6861–65
8. Shoulders, C. C., Baralle, F. E. 1982. *Nucleic Acids Res.* 10:4873–82
9. Cheung, P., Chan, L. 1983. *Nucleic Acids Res.* 11:3703–15
10. Law, S. W., Brewer, H. B. Jr. 1984. *Proc. Natl. Acad. Sci. USA* 81:66–70
11. Karathanasis, S. K., Zannis, V. I., Breslow, J. L. 1983. *Proc. Natl. Acad. Sci. USA* 80:6147–51
12. Shoulders, C. C., Kornblihtt, A. R., Munro, B. S., Baralle, F. E. 1983. *Nucleic Acids Res.* 11:2827–37
13. Sharpe, C. R., Sidoli, A., Shelley, C. S., Lucero, M. A., Shoulders, C. C., et al. 1984. *Nucleic Acids Res.* 12:3917–32
14. Gordon, J. I., Smith, D. P., Andy, R., Alpers, D. H., Schonfeld, G., et al. 1982. *J. Biol. Chem.* 257:971–78
15. Zannis, V. I., Karathanasis, S. K., Keutmann, H., Goldberger, G., Breslow, J. L. 1983. *Proc. Natl. Acad. Sci. USA* 80:2574–78
16. Gordon, J. I., Sims, H. F., Lentz, S. R., Edelstein, C., Scanu, A. M., et al. 1983. *J. Biol. Chem.* 258:4037–44
17. Zannis, V. I., Breslow, J. L., Katz, A. J. 1980. *J. Biol. Chem.* 255:8612–17
18. Zannis, V. I., Breslow, J. L., SanGiacomo, T. R., Aden, D. P., Knowles, B. B. 1981. *Biochemistry* 20:7089–96
19. Zannis, V. I., Kurnit, D. M., Breslow, J. L. 1982. *J. Biol. Chem.* 257:536–44
20. Edelstein, C., Gordon, J. I., Toscas, K., Sims, H. F., Strauss, A. W. et al. 1983. *J. Biol. Chem.* 258:11430–33
21. Kooistra, T., VanHinsbergh, V.,

Havekes, L., Kempen, H. J. 1984. *FEBS Lett.* 170:109–13
22. Brewer, H. B. Jr., Fairwell, T., LaRue, A., Ronan, R., Houser, A., et al. 1978. *Biochem. Biophys. Res. Commun.* 80:623–30
23. Barker, W. C., Dayhoff, M. O. 1977. *Comp. Biochem. Physiol. B* 57:309–15
24. Fitch, W. M. 1977. *Genetics* 86:623–44
25. Schiffer, M., Edmundson, A. B. 1967. *Biophys. J.* 7:121–35
26. Segrest, J. P., Jackson, R. L., Morrisett, J. D., Gotto, A. M. 1974. *FEBS Lett.* 38:247–53
27. Segrest, J. P., Chung, B. H., Brouillette, C. G., Kanellis, P., McGahan, R. 1983. *J. Biol. Chem.* 258:2290–95
28. Fitch, W. M., Smith, T., Breslow, J. L. 1984. *Methods Enzymol.* In press
29. Bruns, G. A. P., Karathanasis, S. K., Breslow, J. L. 1984. *Arteriosclerosis* 4:97–102
30. Law, S. W., Gray, G., Brewer, H. B. Jr., Sakaguchi, A. Y., Naylor, S. L. 1984. *Biochem. Biophys. Res. Commun.* 118:934–42
31. Cheung, P., Kao, F. T., Law, M. L., Jones, C., Puck, T. T., et al. 1984. *Proc. Natl. Acad. Sci. USA* 81:508–11
32. Lusis, A. J., Taylor, B. A., Wangenstein, R. W., LeBoeuf, R. L. 1983. *J. Biol. Chem.* 258:5071–78
33. Antonucci, T. K., VonDeimling, O. H., Rosenblum, B. B., Skow, L. C., Meisler, M. H. 1984. *Genetics* 107:463–75
34. Meisler, M. H., Wanner, L., Kao, F. T., Jones, C. 1981. *Cytogenet. Cell Genet.* 31:124–28
35. Gerald, P. S., Grzeschik, K. H. 1984. *Cytogenet. Cell Genet.* 37:103–26
36. Franceschini, G., Sirtori, C. R., Capurso, A., Weisgraber, K. H., Mahley, R. W. 1980. *J. Clin. Invest.* 66:892–900
37. Weisgraber, K. H., Rall, S. C., Bersot, T. P., Mahley, R. W., Franceschini, G., et al. 1983. *J. Biol. Chem.* 258:2508–13
38. Utermann, G., Feussner, G., Franceschini, G., Haas, J., Steinmetz, A. 1982. *J. Biol. Chem.* 257:501–7
39. Utermann, G., Steinmetz, A., Paetzold, R., Wilk, J., Feussner, G., et al. 1982. *Hum. Genet.* 61:329–37
40. Menzel, H. J., Kladetzky, L., Assmann, G. 1982. *J. Lipid Res.* 23:915–22
41. Rall, S. C., Menzel, H. J., Assmann, G., Utermann, G., Haas, J., et al. 1983. *Arteriosclerosis* 3:515a

42. Menzel, H. J., Assmann, G., Rall, S. C., Weisgraber, K. H., Mahley, R. W. 1984. *J. Biol. Chem.* 259:3070–76
43. Norum, R. A., Lakier, J. B., Goldstein, S., Angel, A., Goldberg, R. B., et al. 1982. *N. Engl. J. Med.* 306:1513–19
44. Karathanasis, S. K., Norum, R. A., Zannis, V. I., Breslow, J. L. 1983. *Nature* 301:718–20
45. Karathanasis, S. K., McPherson, J., Zannis, V. I., Breslow, J. L. 1983. *Nature* 304:371–73
46. Karathanasis, S. K., Zannis, V. I., Breslow, J. L. 1983. *Nature* 305:823–25
47. Zannis, V. I., Lees, A. M., Lees, R. S., Breslow, J. L. 1982. *J. Biol. Chem.* 257:4978–86
48. Bojanovski, D., Gregg, R. E., Zech, L. A., Meng, M. S., Ronan, R., et al. 1984. *Clin. Res.* 32:390a
49. Brewer, H. B. Jr., Fairwell, T., Meng, M., Kay, L., Ronan, R. 1983. *Biochem. Biophys. Res. Commun.* 113:934–40
50. Rosseneu, M., Assmann, G., Taveirne, M. J., Schmitz, G. 1984. *J. Lipid Res.* 25:111–20
51. Schaefer, E. J., Kay, L. L., Zech, L. A., Brewer, H. B. Jr. 1982. *J. Clin. Invest.* 70:934–45
52. Logocki, P. A., Scanu, A. M. 1980. *J. Biol. Chem.* 255:3701–6
53. Jahn, C. E., Osborne, J. C. Jr., Schaefer, E. J., Brewer, H. B. Jr. 1983. *Eur. J. Biochem.* 131:25–29
54. Wu, A. L., Windmueller, H. G. 1979. *J. Biol. Chem.* 254:7316–22
55. Gordon, J. I., Budelier, K. A., Sims, H. E., Edelstein, C., Scanu, A. M., et al. 1983. *J. Biol. Chem.* 258:14054–59
56. Brewer, H. B. Jr., Lux, S. E., Ronan, R., John, K. M. 1972. *Proc. Natl. Acad. Sci. USA* 69:1304–8
57. Moore, M. N., Kao, F. T., Tsao, Y. K., Chan, L. 1984. *Biochem. Biophys. Res. Commun.* 123:1–7
58. Swaney, J. B., Reese, H., Eder, H. A. 1974. *Biochem. Biophys. Res. Commun.* 59:513–18
59. Weisgraber, K. H., Bersot, T. P., Mahley, R. W. 1978. *Biochem. Biophys. Res. Commun.* 85:287–92
60. Gordon, J. I., Smith, D. P., Alpers, D. H., Strauss, A. W. 1982. *Biochemistry* 21:5424–30
61. Gordon, J. I., Bisgaier, C. L., Sims, H. F., Sachdev, O. P., Glickman, R. M., et al. 1984. *J. Biol. Chem.* 259:468–74
62. Gordon, J. I., Smith, D. P., Alpers, D. H., Strauss, A. W. 1982. *J. Biol. Chem.* 257:8418–23
63. Boguski, M. S., Elshourbagy, N., Taylor, J. M., Gordon, J. I. 1984. *Proc. Natl. Acad. Sci. USA.* 81:5021–25
64. Menzel, H. J., Kovary, P. M., Assmann, G. 1982. *Hum. Genet.* 62:349–52
65. Goldstein, J. L., Brown, M. S. 1982. *The Metabolic Basis of Inherited Disease,* pp. 672–712. New York: McGraw-Hill
66. Kane, J. P. 1983. *Ann. Rev. Physiol.* 45:637–50
67. Lee, D. M., Valente, A. J., Kuo, W. H., Maeda, H. 1981. *Biochim. Biophys. Acta* 666:133–46
68. LeBoeuf, R. C., Miller, C., Shively, J. E., Schumaker, V. N., Balla, M. A., et al. 1984. *FEBS Lett.* 170:105–8
69. Schonfeld, G., Patsch, W., Pfleger, B., Witztum, J. L., Weidman, S. W. 1979. *J. Clin. Invest.* 64:1288–97
70. Marcel, Y. L., Hogue, M., Theolis, R., Milne, R. W. 1982. *J. Biol. Chem.* 257:13165–68
71. Berg, K. 1983. *Progress in Medical Genetics,* pp. 35–90. Philadelphia, PA: Saunders
72. Schumaker, V. N., Robinson, M. T., Curtiss, L. K., Butler, R., Sparkes, R. S. 1984. *J. Biol. Chem.* 259:6423–30
73. Malloy, M. J., Kane, J. P., Hardman, D. A., Hamilton, R. L., Dolal, K. B. 1981. *J. Clin. Invest.* 67:1441–50
74. Hyams, J., Herbert, P., Bernier, D., Saritelli, A., Lynch, K., Berman, M. 1984. *Clin. Res.* 32:399a
75. Sniderman, A. D., Shapiro, S., Marpole, D., Skinner, B., Teng, B., et al. 1980. *Proc. Natl. Acad. Sci. USA* 77:604–8
76. Goldstein, J. L., Schrott, H. G., Hazzard, W. R., Bierman, E. L., Motulsky, A. R. 1973. *J. Clin. Invest.* 52:1544–68
77. Brunzell, J. D., Albers, J. J., Chait, A., Grundy, S. M., Groszek, E., et al. 1983. *J. Lipid Res.* 24:147–55
78. Chait, A., Albers, J. J., Brunzell, J. D. 1980. *Eur. J. Clin. Invest.* 10:17–22
79. Janus, E. D., Nicoll, A. M., Turner, P. R., Magill, P., Lewis, B. 1980. *Eur. J. Clin. Invest.* 10:161–72
80. Soutar, A. K., Garner, C. W., Baker, H. N., Sparrow, J. J., Jackson, R. L. 1975. *Biochemistry* 14:3057–64
81. Knott, T. J., Robertson, M. E., Priestley, L. M., Urdea, M., Wallis, S., et al. 1984. *Nucleic Acids. Res.* 12:3909–15
82. Jackson, R. L., Sparrow, J. T., Baker, H. N., Morrisett, J., Taunton, O. D., et al. 1974. *J. Biol. Chem.* 249:5308–13
83. Shulman, R. S., Herbert, P. N., Wehrly, K., Fredrickson, D. S. 1975. *J. Biol. Chem.* 280:182–90
84. Tata, F., Henri, I., Markham, A., Weil, D., Williamson, R., et al. 1984. *Hum. Genet.* In press
85. Nikkila, E. A. 1983. *The Metabolic*

Basis of Inherited Disease, pp. 622–42. New York: McGraw-Hill

86. Breckenridge, W. C., Little, J. A., Steiner, G., Chow, A., Poapst, M. 1978. *N. Engl. J. Med.* 298:1265–73

87. Musliner, T. A., Church, E. C., Herbert, P. N., Kingston, M. J., Shulman, R. S. 1977. *Proc. Natl. Acad. Sci. USA* 74:5358–62

88. Kinnunen, P. K. J., Jackson, R. L., Smith, L. C., Gotto, A. M., Sparrow, J. T. 1977. *Proc. Natl. Acad. Sci. USA* 74:4848–51

89. Jackson, C. L., Bruns, G. A. P., Breslow, J. L. 1984. *Proc. Natl. Acad. Sci. USA* 81:2945–49

90. Myklebost, O., Williamson, B., Markham, A. F., Myklebost, S. R., Rogers, J., et al. 1984. *J. Biol. Chem.* 259:4401–4

91. Hospattankar, A. V., Fairwell, T., Ronan, R., Brewer, H. B. Jr. 1984. *J. Biol. Chem.* 259:318–22

92. Jackson, R. L., Baker, H. N., Gilliam, E. B., Gotto, A. M. 1977. *Proc. Natl. Acad. Sci. USA* 74:1942–45

93. Jackson, C. L., Bruns, G. A. P., Breslow, J. L. 1984. *Methods Enzymol.*, In press

94. Myklebost, O., Rogne, S., Olaisen, B., Gedde-Dahl, T. Jr., Prydz, H. 1984. *Hum. Genet.* 67:309–12

95. Humphries, S. E., Borresen, A. L., Gill, L., Cumming, A. M., Robertson, F. W., et al. 1984. *Clin. Genet.* In press

96. Humphries, S. E., Williams, L., Myklebost, O., Stalenhoef, A. F. H., Demacker, P. N. M., et al. 1984. *Hum. Genet.* 67:151–55

97. Havel, R. J., Kotite, L., Kane, J. P. 1979. *Biochem. Med.* 21:121–28

98. Brewer, H. B., Shulman, R., Herbert, P., Ronan, R., Wehrly, K. 1974. *J. Biol. Chem.* 249:4975–84

99. Kashyap, M. L., Srivastava, L. S., Hynd, B. A., Gartside, P. S., Perisutti, G. 1981. *J. Lipid Res.* 22:800–10

100. Brown, W. V., Baginsky, M. L. 1972. *Biochem. Biophys. Res. Commun.* 46:375–82

101. Krauss, R. M., Herbert, P. N., Levy, R. I., Fredrickson, D. S. 1973. *Circ. Res.* 33:403–11

102. Kinnunen, P. K. J., Ehnholm, C. 1976. *FEBS Lett.* 65:354–57

103. Windler, E., Chao, Y., Havel, R. J. 1980. *J. Biol. Chem.* 255:5475–80

104. Shelburne, F., Hanks, J., Meyers, W., Quarfordt, S. 1980. *J. Clin. Invest.* 65:652–58

105. Windler, E., Chao, Y., Havel, R. J. 1980. *J. Biol. Chem.* 255:8303–7

106. Quarfordt, S. H., Michalopoulos, G.,

Schirmer, B. 1982. *J. Biol. Chem.* 257:14642–47

107. Ginsberg, H., Le, N. A., Norum, R. A., Gibson, J., Brown, W. V. 1984. *Clin. Res.* 32:396a

108. Sparrow, J. T., Pownall, H. J., Hsu, F. J., Blumenthal, L. E., Culwell, A. R., et al. 1977. *Biochemistry* 16:5427–31

109. Karathanasis, S. K., Zannis, V. I., Breslow, J. L. 1984. *J. Lipid Res.* In press

110. Blaufuss, M. C., Gordon, J. I., Schonfeld, G., Strauss, A. W., Alpers, D. H. 1984. *J. Biol. Chem.* 259:2452–56

111. Rees, A., Shoulders, C. C., Stocks, J., Galton, D. J. 1983. *Lancet* 1(8322):444–46

112. Zannis, V. I., Just, P. W., Breslow, J. L. 1981. *Am. J. Hum. Genet.* 33:11–24

113. Zannis, V. I., Breslow, J. L. 1981. *Biochemistry* 21:1033–41

114. Utermann, G., Jaeschke, M., Mangel, J. 1975. *FEBS Lett.* 56:352–55

115. Jain, R. S., Quarfordt, S. H. 1979. *Life Sci.* 25:1315–23

116. Utermann, G., Langenback, U., Beisiegel, U., Weber, W. 1980. *Am. J. Hum. Genet.* 32:339–47

117. Zannis, V. I., McPherson, J., Goldberger, G., Karathanasis, S. K., Breslow, J. L. 1984. *J. Biol. Chem.* 259:5495–99

118. Pitas, E., Innerarity, T. L., Arnold, K. S., Mahley, R. W. 1981. *Proc. Natl. Acad. Sci. USA* 76:2311–15

119. Bersot, T. P., Mahley, R. W., Brown, M. S., Goldstein, J. L. 1976. *J. Biol. Chem.* 251:2395–98

120. Innerarity, T. L., Mahley, R. W. 1978. *Biochemistry* 17:1440–47

121. Hui, D. Y., Innerarity, T. L., Mahley, R. W. 1981. *J. Biol. Chem.* 256:5646–55

122. Carrela, M., Cooper, A. D. 1979. *Proc. Natl. Acad. Sci. USA* 76:338–42

123. Sherrill, B. C., Innerarity, T. L., Mahley, R. W. 1980. *J. Biol. Chem.* 255:1804–7

124. Rall, S. C., Weisgraber, K. H., Mahley, R. W. 1981. *J. Biol. Chem.* 257:4171–78

125. Innerarity, T. L., Friedlander, E. J., Rall, S. C. Jr., Weisgraber, K. H., Mahley, R. W. 1983. *J. Biol. Chem.* 258:12341–47

126. Weisgraber, K. H., Innerarity, T. L., Harder, K. J., Mahley, R. W., Milne, R. W., et al. 1983. *J. Biol. Chem.* 258:12348–54

127. Innerarity, T. L., Weisgraber, K. H., Arnold, K. S., Rall, S. C. Jr., Mahley, R. W. 1984. *J. Biol. Chem.* 259:7261–67

128. Mahley, R. W., Angelin, B. 1984. *Adv. Intern. Med.* 29:385–411

129. Breslow, J. L., Zannis, V. I. 1984. *Arteriosclerosis Reviews.* New York: Raven. In press

130. Blue, M. L., Williams, D. L., Zucker, S., Khan, S. A., Blum, C. B. 1983. *Proc. Natl. Acad. Sci. USA* 80:283–87
131. Basu, S. K., Brown, M. S., Ho, Y. K., Havel, R. J., Goldstein, J. L. 1981. *Proc. Natl. Acad. Sci. USA* 78:7545–49
132. Breslow, J. L., McPherson, J., Nussbaum, A. L., Williams, H. W., Lofquist-Kahl, F., et al. 1982. *J. Biol. Chem.* 257:14639–41
133. Breslow, J. L., McPherson, J., Nussbaum, A. L., Williams, H. W., Lofquist-Kahl, F., et al. 1983. *J. Biol. Chem.* 258:11422
134. Wallis, S. C., Rogne, S., Gill, L., Markham, A., Edge, M., et al. 1983. *EMBO J.* 2:2369–73
135. McLean, J. W., Elshourbagy, N. A., Chang, D. J., Mahley, R. W., Taylor, J. M. 1984. *J. Biol. Chem.* 259:6498–504
136. Das, H. K., McPherson, J., Bruns, G. A. P., Karathanasis, S. K., Breslow, J. L. 1985. *J. Biol. Chem.* In press
137. Olaisen, B., Teisberg, P., Gedde-Dahl, T. 1982. *Hum. Genet.* 62:233–36
138. Whitehead, A. S., Bruns, G. A. P., Markham, A. P., Colten, H. R., Woods, D. E. 1983. *Science* 221:69–71
139. Francke, U., Brown, M. S., Goldstein, J. L. 1984. *Proc. Natl. Acad. Sci. USA* 81:2826–30
140. Berg, K., Borresen, A. L., Heiberg, A., Maartmann-Moe, K. 1984. *Cytogenet. Cell Genet.* 37:418a
141. Zannis, V. I., Breslow, J. L., Utermann, G., Mahley, R. W., Weisgraber, K. H., et al. 1982. *J. Lipid Res.* 23:911–14
142. Utermann, G., Kindermann, I., Kaffar-nik, H., Steinmetz, A. 1984. *Hum. Genet.* 65:232–36
143. Assmann, G., Schmitz, G., Menzel, H. J., Schulte, H. 1984. *Clin. Chem.* 30/5:641–43
144. Wardell, M. R., Suckling, P. A., Janus, E. D. 1982. *J. Lipid Res.* 23:1174–82
145. Cumming, A. M., Robertson, F. W. 1984. *Clin. Genet.* 25:310–13
146. Zannis, V. I., Breslow, J. L. 1980. *J. Biol. Chem.* 255:1759–62
147. Breslow, J. L., Zannis, V. I., SanGiacomo, T. R., Third, J. L. H. C., Tracy, T., et al. 1982. *J. Lipid Res.* 23:1224–35
148. Schneider, W. J., Kovanen, P. T., Brown, M. S., Goldstein, J. L., Utermann, G., et al. 1981. *J. Clin. Invest.* 68:1075–85
149. Weisgraber, K. H., Innerarity, T. L., Mahley, R. W. 1982. *J. Biol. Chem.* 257:2518–21
150. Rall, S. C. Jr., Weisgraber, K. H., Innerarity, T. L., Mahley, R. W. 1982. *Proc. Natl. Acad. Sci. USA* 79:4696–700
151. Yamamura, T., Yamamoto, A., Hiramori, K., Nambu, S. 1984. *Atherosclerosis* 50:159–72
152. Rall, S. C. Jr., Weisgraber, K. H., Innerarity, T. L., Bersot, T. P., Mahley, R. W., et al. 1983. *J. Clin. Invest.* 72:1288–97
153. Weisgraber, K. H., Rall, S. C. Jr., Innerarity, T. L., Mahley, R. W., Kuusi, W., et al. 1984. *J. Clin. Invest.* 73:1024–33
154. Ghiselli, G., Schaefer, E. J., Gascon, P., Brewer, H. B. Jr. 1981. *Science* 214:1239–41

Ann. Rev. Biochem. 1985. 54:729–64
Copyright © 1985 by Annual Reviews Inc. All rights reserved

BIOSYNTHESIS AND METABOLISM OF TETRAHYDROBIOPTERIN AND MOLYBDOPTERIN[1]

Charles A. Nichol, Gary K. Smith and David S. Duch

Department of Medicinal Biochemistry, The Wellcome Research Laboratories, 3030 Cornwallis Road, Research Triangle Park, North Carolina 27709

CONTENTS

[1]Abbreviations used: ACTH, adrenocorticotropic hormone; biopterin, 6-(L-erythro-1',2'-dihydroxypropyl)pterin; 8-Br cAMP, 8-Bromocyclic AMP; Bt$_2$ cAMP, dibutyrylcyclic AMP; CHO, Chinese hamster ovary; CSF, cerebrospinal fluid; DHFR, 5,6,7,8-tetrahydrofolate-NADP$^+$ oxido-reductase, EC 1.5.1.3 (dihydrofolate reductase); DHPR, quinonoid dihydropteridine reductase, EC 1.6.99.10; DOPA, 3,4-dihydroxyphenylalanine; GTP-CH, GTP cyclohydrolase I, EC 3.5.4.16; HPLC, high performance liquid chromatography; H$_2$, dihydro; H$_4$, tetrahydro; H$_2$neopterin-PPP, 6-(D-erythro-1',2',3'-trihydroxypropyl)-7,8-dihydropterin-3'-triphosphate; H$_4$biopterin, 6R-(L-erythro-1',2'-dihydroxypropyl)-5,6,7,8-tetrahydropterin (tetrahydrobiopterin); MTX, methotrexate; NAS, N-acetyl serotonin; PAH, phenylalanine-4-monooxygenase; PKU, phenylketonuria; TH, tyrosine-3-monooxygenase, EC 1.14.16.2.

729

0066-4154/85/0701-0729$02.00

PERSPECTIVES AND SUMMARY

Pterins include those compounds that have in common the 2-amino-4-hydroxypteridine structure and that occur in tetrahydro, dihydro and fully oxidized forms (Figure 1). The folate cofactors are called "conjugated" pteridines, referring to the linkage of p-aminobenzoylglutamate(s) to the pterin, whereas other pterins, including biopterin, molybdopterin, and pterin pigments, are referred to as unconjugated pteridines. Although the folate and biopterin cofactors have such unrelated functions, namely one-carbon transfer reactions and utilization of molecular oxygen, respectively, there are a number of interrelationships involving the biosynthesis and metabolism of these cofactors. GTP is a common precursor of both H_4folate and H_4biopterin in cells which have the capacity for their biosynthesis. Regeneration of these cofactors is catalyzed in each case by reduced pyridine nucleotide-dependent reductases. Some folate enzymes can accept unconjugated pteridines (1–5). Methotrexate, which is best known as an inhibitor of DHFR, at higher concentrations can also inhibit DHPR, the enzyme responsible for the regeneration of H_4biopterin from quinonoid H_2biopterin (6), but has no effect on the de novo biosynthesis of H_4biopterin from GTP (7). During biochemical evolution, the ability to conjugate p-aminobenzoylglutamate to the pterin moiety was lost by vertebrates and other organisms that have a nutritional requirement for folates, whereas the capacity for the de novo biosynthesis of H_4biopterin from GTP has been retained in the cells and tissues where it is required. Our extensive knowledge of folate metabolism stands in contrast to the limited information on biopterin metabolism. Now pharmacological and physiological roles of H_4biopterin are receiving more attention. H_4biopterin is known to be essential for the biosynthesis of catecholamines and serotonin, serving as the cofactor for tyrosine and tryptophan hydroxylases, respectively. Recent evidence that H_4biopterin is formed by an inducible enzyme system which is subject to neuronal and

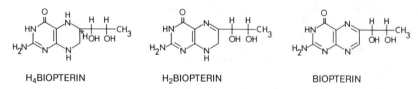

H_4BIOPTERIN H_2BIOPTERIN   BIOPTERIN

Figure 1   Redox states of biopterin (see abbreviations for chemical names).

hormonal control directs attention to the role of this cofactor in modulating neurotransmitter synthesis (7).

The stable urinary metabolites derived from H$_4$biopterin and molybdopterin were identified before their relationships to these cofactors was established. Biopterin, the growth factor for *Crithidia fasciculata* isolated from human urine in 1955 (8), was initially thought to be related to the folate family of pterins. A sulfur-containing pterin derived from molybdopterin by alkaline oxidation (9) was found in 1980 to have the same properties as an oxidation product of urothione, a compound isolated from human urine in 1943 (10). Both molybdopterin and the biopterin cofactor are reduced pterins, a class of compounds well known for their instability and ease of oxidation. Because of this instability, the structures of intermediates in the biosynthesis of H$_4$biopterin have not been fully characterized and the structure of molybdopterin is inferred from stable oxidation products.

As yet, there are no animal models of H$_4$biopterin or molybdopterin deficiencies that would enable study of related metabolic dysfunctions. However, several human inborn errors of metabolism have been discovered that cause defects in the biosynthesis of H$_4$biopterin and molybdopterin and have become an important resource to elucidate the consequences of these cofactor deficiencies. The development of specific assays of sufficient sensitivity to measure pterins in plasma, cerebrospinal fluid, and urine has facilitated the acquisition of data on changes in the levels of biopterin and related compounds during the course of various diseases. Within the past several years, H$_4$biopterin has been used in the treatment of atypical PKU. Also, clinical trials using H$_4$biopterin in the treatment of motor dysfunctions related to dopamine deficiency diseases, such as parkinsonism, have given added impetus to investigations on the role of the biopterin cofactor in neurotransmitter synthesis.

Although there are many intriguing aspects of the comparative biochemistry of pterin cofactors and pigments in bacteria, insects, amphibia, and reptiles, the main focus of this review will be on mammalian metabolism. Evidence for the biosynthesis of the biopterin cofactor in mammals was presented in 1974 (11) whereas the presence of molybdenum bound to a pterin was discovered more recently (9). For both cofactors, the biosynthetic intermediates have not been fully defined. A number of monographs and reviews have dealt in detail with the chemistry and biology of pteridines (12–17) and with the role of the aromatic amino acid hydroxylases in the synthesis of monoamine neurotransmitters (18–20) as well as with the defects in phenylalanine metabolism in PKU (21–24).

MOLYBDOPTERIN

The molybdenum cofactor is a pterin found in a wide variety of molybdenum-containing enzymes such as xanthine oxidase, nitrate reductase, sulfite oxidase, formate dehydrogenase, and aldehyde oxidase. These enzymes are widely distributed in bacteria, fungi, plants, and animals (25). The distinction between the molybdenum-pterin cofactor for these enzymes and the molybdenum-iron cofactor for nitrogenase was discussed in a previous review (26). Evidence for the existence of an organic prosthetic group associated with molybdenum as well as for a common genetic determinant for different molybdoenzymes was first presented by Pateman et al (27). Further evidence for the existence of a cofactor common to several molybdoenzymes was obtained in studies with the nit-1 mutant of *Neurospora crassa* (28–30). These studies demonstrated the in vitro activation of inactive nitrate reductase from this mutant by the addition of acid extracts of wild-type and other non-allelic nitrate reductase *Neurospora* mutants, as well as by extracts of other acid-treated molybdenum-containing enzymes from diverse phylogenetic sources. No activation was observed by the addition of acid extracts of proteins devoid of molybdenum, molybdate anion, or model molybdenum-sulfur amino acid complexes; demolybdosulfite oxidase purified from livers of tungsten-treated rats can be activated by extracts of the same cofactor sources active in the *Neurospora* nit-1 system (31). This activation has served as the basis of the assay for the molybdenum cofactor.

This cofactor does not exist stably in a free and soluble state (31). In *Neurospora* and in liver, the cofactor is associated with a particulate fraction and in rat liver is associated with the outer mitochondrial membrane. The extreme lability of the cofactor following its release from proteins makes it difficult to isolate and characterize this cofactor. Rajagopalan and coworkers (9, 32–34) chose to isolate and identify the stable inactive forms of the cofactor in order to deduce the structure of the active parent compound. An oxidized fluorescent derivative of the cofactor (Form A) was obtained by denaturation of molybdoenzymes from several sources in the presence of KI and I_2. (9, 32). The absorption and fluorescence spectra of Form A after purification were similar but not identical to those obtained for biopterin, suggesting the presence of a pteridine ring in the cofactor molecule. Form A, as isolated, is a phosphate ester which can be dephosphorylated by alkaline phosphatase and contains one phosphate group per pterin moiety (32, 33). Periodate cleaved only the dephosphorylated form indicating the presence of vicinal hydroxyl groups which are blocked by the attached phosphate (32). Alkaline permanganate converted Form A to pterin-6-carboxylic acid (9). Results of chemical, mass spectral, and NMR studies (34) indicated a side-chain formulation of $-C \equiv C-CHOHCH_2OPO_3{}^{2-}$ (Figure 2).

A second fluorescent derivative of molybdopterin (Form B) can be isolated from molybdoenzymes when they are denatured in the absence of KI and I_2 (35). Form B is also a phosphate ester and like Form A is susceptible to periodate cleavage only after dephosphorylation. Oxidation of Form B with

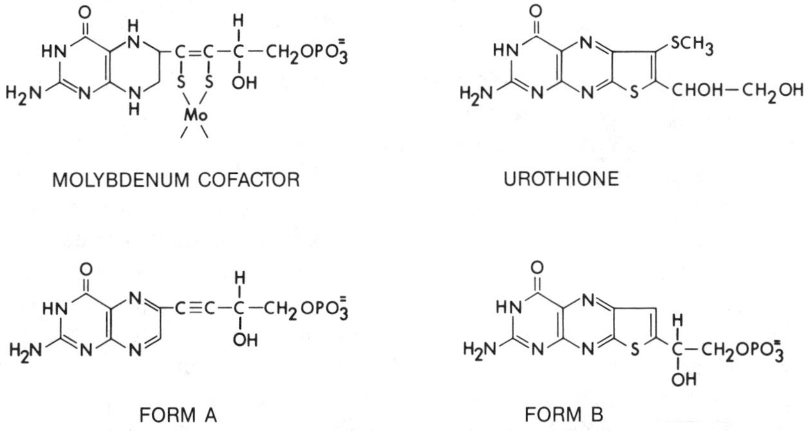

MOLYBDENUM COFACTOR   UROTHIONE

FORM A   FORM B

Figure 2   Proposed structure of the molybdenum cofactor (molybdopterin) and its oxidative decomposition products.

alkaline permanganate yielded a product with chemical and spectral properties of pterin-6-carboxylic-7-sulfonic acid. This product was identical to that obtained by the permanganate oxidation of urothione, a compound isolated from urine (10, 36) and characterized as a sulfur-containing pterin (37, 38). Chemical, spectral, and NMR studies on Form B (34) indicated that it is a phosphorylated analog of urothione lacking the 3-methylthiol group. Moreover, dephosphorylated Form B can be synthesized from urothione by desulfuration with Raney nickel and oxidation with SeO_2. The structural similarities between urothione and Form B of molybdopterin led to the suggestion (34, 35, 39) that urothione might be the urinary excretion product of this cofactor. Other studies on the structure of the molybdenum cofactor for *E. coli* nitrate reductase and xanthine oxidase indicated that the oxidation product of the molybdenum cofactor is a thienopterin derivative with an unidentified side chain in the 2'-position (40). The purification of reduced molybdenum cofactor using HPLC has been reported (41); however, no evidence was presented indicating the purity of the cofactor. The structures for the molybdenum cofactor as well as the inactive Forms A and B proposed by Rajagopalan and coworkers are illustrated in Figure 2.

Inborn errors of metabolism in which there is a combined deficiency of sulfite oxidase and xanthine dehydrogenase have been recently described (42–44). The metabolic defect responsible for loss of both enzyme activities was determined to be due to the lack of the molybdenum cofactor. Analysis of urine from patients with the cofactor deficiency (36, 41) indicated that urothione was totally absent in these patients, thus supporting the suggestion that urothione was derived from molybdopterin.

TETRAHYDROBIOPTERIN: COFACTOR ROLE AND REACTION MECHANISM

Since the biochemical role of H_4biopterin and the enzyme mechanisms involving the cofactor have been reviewed (18, 45–50), only a brief discussion will be given. Four mammalian enzymes are known to have a requirement for an H_4pterin cofactor. These are PAH (51, 52), TH (53, 54), tryptophan hydroxylase (18, 55), and alkylglycerol monooxygenase (56, 57), which convert Phe to Tyr, Tyr to DOPA (the precursor of dopamine, norepinephrine, and epinephrine), Trp to 5-hydroxytryptophan (the precursor of serotonin and melatonin), and catalyze the cleavage of alkylglycerol ethers, respectively. The natural cofactor is assumed to be H_4biopterin; however, for all four enzymes, synthetic H_4pterins, such as 6-methyl-H_4pterin and 6,7-dimethyl-H_4pterin, were often used as cofactors (18, 45, 51–59). With Phe, Tyr, and Trp hydroxylases, the natural cofactor endows the enzymes with properties not observed with the synthetic cofactors; K_m^{pterin}, $K_m^{substrate}$, or regulatory properties may change (18, 58, 59).

The cofactor function of H_4biopterin appears to be related to its ability to reduce molecular oxygen. Thus, in the overall enzyme reaction shown in Figure 3, H_4pterin provides electrons to reduce O_2 and is in turn oxidized to the quinonoid H_2pterin. Only the mechanism of PAH has been examined in detail; however, it is generally assumed that the mechanisms of the other enzymes will be similar (18, 45–48, 56, 60, 61). The H_4pterin (I) is thought to activate molecular oxygen by formation of a 4a-hydroperoxy-H_4pterin (II) or a similar species also involving an iron atom in the active site (18, 45–48, 60, 61). Although a 4a-hydroperoxide has been demonstrated in the case of flavin-dependent oxygenases, it has not been directly observed with any H_4pterin-dependent enzyme (60). As expected for this type of hydroxylating intermediate, one atom of oxygen from molecular oxygen is incorporated into the product (62, 63). Upon hydroxylation of Phe, the first pterin product released from the enzyme is the 4a-carbinolamine-H_4pterin (III) (64). This compound then spontaneously dehydrates or is enzymatically dehydrated by 4a-carbinolamine dehydratase to form the quinonoid-H_2pterin (IV) (61, 65, 66). DHPR completes the catalytic cycle by reducing the quinonoid-H_2pterin back to the H_4pterin (6) (Figure 3). The quinonoid (IV) can also rearrange to the 7,8-H_2pterin (V), which can only be reduced by DHFR (2–5, 18).

Cofactor Availability for Pterin-dependent Monooxygenases

TH is the rate-limiting enzyme in the biosynthesis of the neurotransmitters dopamine, norepinephrine, and epinephrine (54). The content of H_4biopterin in the brain correlates with regions containing tyrosine and tryptophan hydroxylases (67, 68). There is considerable evidence for both short-term and long-

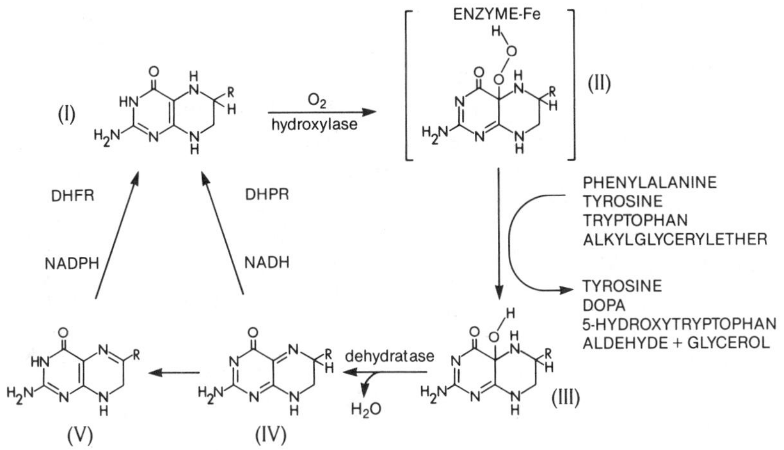

Figure 3 H₄Pterin cofactor role in H₄pterin-dependent monooxygenases.

term regulation of Tyr hydroxylation and catecholamine biosynthesis (20). Long-term regulation involves increased synthesis of TH, dopamine-β-hydroxylase, and catecholamine storage vesicles (69, 70). Short-term regulation of TH appears to be due to a decrease in the K_m of TH for H₄biopterin as a result of phosphorylation of the enzyme (71, 72). K_m values for H₄biopterin of 10–30 μM have been reported for the activated high-affinity form of the enzyme while K_m values of 100–600 μM have been reported for the low-affinity form (73–75). Estimates of H₄biopterin levels have indicated concentrations of 11–14 μM in bovine adrenal medullary chromaffin cells in culture (76), 1–10 μM in adrenal medulla (77, 78), and 100 μM in rat striatum (74). The major portion of the TH in rat striatum and adrenal medulla is in the low-affinity high K_m form and as a result H₄biopterin levels are subsaturating (74, 79). In studies using guinea pig vas deferens with attached hypogastric nerves, both high-affinity and low-affinity forms of TH have been observed (80); however, following electrical stimulation of the nerves only the high-affinity form of the enzyme was present, indicating that electrical stimulation resulted in conversion of the less active to the more active form. Similar results were also obtained by Morgenroth et al (81).

Other evidence suggests that levels of H₄biopterin are subsaturating in monoaminergic tissues. The intraventricular administration of H₄biopterin increased the hydroxylation of Tyr in rat striatum (82) and the addition of biopterin to cultures of chick embryo sympathetic ganglia resulted in increased synthesis of catecholamines (83). Addition of H₄biopterin to rat brain slices (84) or of 6-methyl-H₄pterin to rat brain striatal synaptosomes (85) or guinea pig vas deferens preparations (86, 87) resulted in stimulation of DOPA and catecholamine synthesis. In synaptosomes prepared from rat hypothalamus in

which TH had been induced with reserpine, there was no enhanced hydroxylation of Tyr relative to that seen in synaptosomes prepared from hypothalamus
of control rats unless exogenous H_4biopterin was added to the incubation
medium (88, 89). There is feedback inhibition of TH by catecholamines which
are competitive inhibitors of the enzyme with respect to the reduced pteridine
cofactor (90, 91). Thus, the reduction in the K_m value for H_4biopterin following
activation of TH would bring the K_m values closer to the concentration of this
cofactor in tissues and may also free the enzyme from feedback inhibition by
catecholamines. Similar studies have indicated that activation of tryptophan
hydroxylase using phosphorylating conditions also results in a decrease of the
K_m for the reduced pterin (92–96); however, in these activation studies, a
synthetic H_4pterin rather than H_4biopterin, the natural cofactor, has been used
and thus the effect of activation on the affinity of the natural cofactor has not yet
been determined. In contrast, activation of PAH by phosphorylation causes an
increase in the V_{max} with little or no change in the K_m for either Phe or
H_4biopterin (96a).

ANALYSIS OF PTERINS IN TISSUES
AND BODY FLUIDS

A microbial assay for the determination of unconjugated pteridines was based
on the initial observations of Cowperthwaite et al (97, 98) who showed that the
trypanosomatid insect parasite *Crithidia fasciculata* had a nutritional requirement for an unconjugated pteridine, subsequently identified and named biopterin by Patterson et al (8). The method was developed by Dewey & Kidder (99)
and Guttman (100) and was further refined by Baker et al (101, 102). Because
several pterins serve as growth factors for this organism, the content in tissues
and body fluids could only be expressed as biopterin equivalents. However, the
Crithidia assay can distinguish between the D- and L-erythro side chain configuration of the pterins much more readily than any other technique since the
L-erythro pterins accelerate growth at much lower concentrations than the
D-erythro isomers (103). Pteridines have also been separated using paper and
ion-exchange column chromatography (104–106), but the large amount of
sample needed as well as the losses incurred limited the usefulness of these
procedures as an analytical tool. The use of gas chromatography–mass spectrometry for the analysis of trimethylsilylated pteridines has been described
(107–110). PAH obtained from rat liver and *Pseudomonas* has been used
(111–114) for the determination of H_4biopterin in tissues and body fluids. This
assay is specific for H_4pterins, and unlike other assays, measures total hydroxylase cofactor activity. A comparison of the PAH, *Crithidia*, and HPLC assays
for several tissues and body fluids indicated that similar values are obtained
with the *Crithidia* and HPLC assays whereas the PAH assay gave much higher

values (115). This may be due to the fact that H_4pterins other than H_4biopterin can serve as substrates or that racemic mixtures of H_4biopterin are used as standards. It has been shown that the naturally occuring 6-R-isomer of H_4biopterin has a V_{max} four times greater than that of the unnatural 6-S-isomer (116). Radioimmunoassays have been described for the determination of L-erythro-biopterin (117–119), D-erythro-neopterin (103, 119, 120) and 6,7-dimethylpterin (117). The radioimmunoassay techniques allow the determination of pterin levels in a large number of samples, but an inherent disadvantage is that antibodies must be prepared for each pterin of interest and determination of total pterin content requires oxidation of the reduced forms.

By far the largest body of work on the analysis of pterins in tissues and body fluids has made use of HPLC coupled with either fluorescent or electrochemical detection. The HPLC methodology has utilized either ion-exchange (115, 121–123) or reverse-phase (115, 124–130) chromatography. In several of these methods (122–124) the chromatographic profiles are markedly affected by pH, ionic strength, or concentration of cation in the mobile phase. In both the *Crithidia* and HPLC assays, various methods of sample preparation have been used. In some studies (102, 124), samples were autoclaved at 100° at pH 4.5, a procedure known to convert H_4biopterin to other pterins (131, 132), and oxidized biopterin was used to determine recoveries. Under such conditions, a large fraction of H_4biopterin, the predominant form of the cofactor, is converted to pterins which will not support the growth of *Crithidia* (131, 132), but these losses can be prevented by iodine oxidation of tissue samples under conditions where H_4biopterin is quantitatively converted to the fully oxidized form prior to autoclaving (132). Detailed studies (115, 123, 133) on the use of iodine oxidation and sample preparations have resulted in procedures for the selective oxidation of reduced pterins to stable fluorescent products which allows the determination of total as well as H_4pterins. An improved chromatographic analysis based on this methodology has been developed (125). Except for the assays based on the use of PAH, none of the above methods allows for the direct determination of the reduced forms of the pterins. The use of HPLC coupled with electrochemical detection (126–130) has made this possible. Additionally, LC/EC using a dual-electrode detector in a parallel adjacent configuration allows the determination of both oxidized and reduced pterins in biological samples.

Tissue Distribution

Unconjugated pteridines are widely distributed in plants, insects, bacteria, and mammals (134, 135). In tissues and body fluids of mammalian species, H_4biopterin is the predominant unconjugated pterin, while neopterin, xanthopterin, isoxanthopterin, pterin-6-carboxylic acid, 6-hydroxymethylpterin, and the lumazines have also been demonstrated (106, 115, 118, 123, 125,

136–143). With the exception of certain cell cultures which have no detectable biopterins (141), all mammalian tissues and body fluids which have been examined contain biopterin. Studies in several mammalian species (mouse, rat, dog, monkey) demonstrated that tissues with the highest levels were pineal (0.4–12.5 μg/g), liver (0.3–1.6 μg/g), bone marrow (0.4–1.2 μg/g), spleen (.04–.35 μg/g), pituitary (0.18–0.44 μg/g), and adrenal gland (0.45–0.95 μg/g). H_4biopterin is present in both the adrenal medulla and cortex. In the rat adrenal gland, the concentration of H_4biopterin in the medulla is 10-fold higher than that in the cortex (76). Plasma, erythrocytes, and cells in the buffy coat all contain biopterin, and as was observed with other tissues, there appears to be a species variation in content. Biopterin was not evenly distributed in brain; highest levels were found in the hypothalamus and corpus striatum (68, 115, 141, 142). In tissues, almost all of the biopterin is present as the tetrahydro form (115, 142).

The presence of H_4biopterin has also been demonstrated in a wide variety of mammalian cells in culture. Initial observations (11) indicated that CHO and C-1300 neuroblastoma produced a *Crithidia*-active biopterin-like substance. High levels of *Crithidia*-active material, presumed to be biopterin, were found in several human adrenergic neuroblastomas, while markedly lower levels were observed in other cell lines (144). High levels of H_4biopterin and GTP-CH were found in the N1E-115 neuroblastoma; lower levels were found in Y-1 adrenal cortical tumor cells, CHO cells, and PC-12 cells while in some cell lines no H_4biopterin could be detected (141). In another study (145) high levels of H_4biopterin were measured in the N1E-115 and N2A neuroblastomas and PC-12 pheochromocytoma.

High levels of biopterin are also excreted in the urine. In various mammalian species biopterin levels of 0.5–10.7 μg/mg creatinine have been reported, while in humans, levels ranged from 0.5–2.2 μg/mg creatinine (118, 123, 125, 142). Tissues and body fluids of humans and monkeys also contain substantial amounts of neopterin (106, 115, 123, 125, 141, 142). Much lower levels have been found in dog liver and urine (142). This pterin has not been demonstrated in significant amounts in any other species. The origin of neopterin is unknown, although it is presumed to be derived from D-erythro-H_2neopterin-PPP, the product of the first enzymic reaction in the biosynthesis of H_4biopterin. The demonstration that the neopterin had the D-erythro configuration supports this view (103). Studies of normal individuals indicated that the urinary excretion of pterins, when based on creatinine content, was constant over a 24-hour period and that there was little variation in total pterins excreted daily over a ten day period (123).

Ontogenetic studies on H_4biopterin indicated a differential developmental pattern of H_4biopterin in rat brain and pineal gland (146). In the pineal, H_4biopterin levels were low prior to birth and then increased to adult levels five days after birth. In contrast, brain levels were highest two days before birth and subsequently declined to adult levels. In rat midbrain, biopterin levels were found to peak at 12 days after birth and then decline (147). Studies in humans (148–151) demonstrated that neopterin and biopterin excretion in urine was highest during the early years of life and declined with age.

BIOSYNTHESIS OF TETRAHYDROBIOPTERIN

Early work on the biosynthesis of H_4biopterin has been reviewed by several authors (45, 152, 153). Many organisms (11, 154–159) are able to convert [^{14}C]guanosine to [^{14}C]biopterin. In contrast, no [^{14}C]biopterin was synthesized from [2-^{14}C]folic acid (11). GTP was shown to be the true precursor (160–162), with all of the carbon atoms in biopterin being derived from this source (163, 164).

GTP-CH catalyzes the first reaction on the biosynthetic pathway for H_4biopterin, and its activity correlates well with H_4biopterin levels in a variety of mammalian cells and tissues (76, 141, 142, 165, 166). The enzyme has no apparent cofactor requirement (162, 165, 167, 168) and has not been purified to homogeneity from any mammalian source, but its molecular weight has been reported to be 125,000 (162). The products of the reaction are D-erythro-H_2neopterin-PPP and formate (161, 162, 167). The mechanism of the reaction has been proposed to be imidazole ring opening, followed by loss of C-8 as formate, an Amadori type rearrangement to place a keto function on the 2' carbon, and finally forming the pyrazine ring by formation of the N^5–C^6 double bond to yield 7,8-H_2neopterin-PPP directly (153, 169). A different series of reactions and very different enzyme properties have been reported (152), but these results have not been reproducible (76, 141, 153, 161, 162, 165–168, 175). In those organisms that can synthesize folate, H_2neopterin-PPP is also the first intermediate in the biosynthesis of H_2folate (153). Only D-erythro-H_2neopterin-PPP, and not H_2neopterin, serves as the precursor for H_4biopterin (164, 170). Thus the mechanism of phosphate removal is most likely an elimination reaction (163), probably involving the acidic 2'hydrogen (Figure 4).

Concept of the Dihydropterin Pathway

Initially the pathway for the biosynthesis of H_4biopterin from H_2neopterin-PPP appeared to proceed through H_2pterin intermediates (Figure 5). When [^{14}C]guanosine was administered to bullfrog tadpoles, [^{14}C]sepiapterin as well as [^{14}C]biopterin were produced (155). Also, partially purified extracts of *Drosophila, Ascaris,* chicken kidney, and several mammalian tissues in the presence of Mg^{2+} and NADPH produced sepiapterin from H_2neopterin-PPP (158, 170–174). Further, sepiapterin is converted to H_2biopterin by a variety of

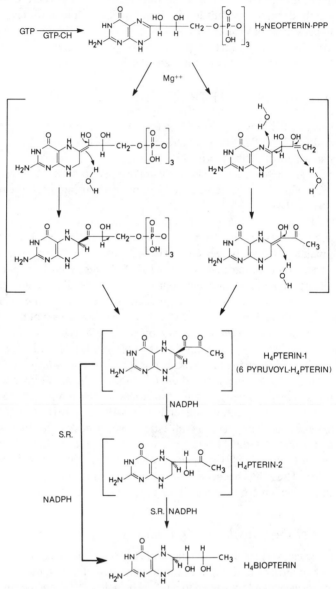

Figure 4   The de novo tetrahydropterin pathway for H₄biopterin biosynthesis. It is not known whether electron transfer from the side chain to the pteridine moiety to produce the H₄pterin moiety precedes or follows triphosphate elimination; thus, both alternatives have been included in the reaction proceeding from H₂neopterin-PPP. S.R. is sepiapterin reductase (H₄biopterin synthase). Side chains of the structures in brackets have not been fully characterized.

tissues and organisms in vivo and in vitro (11, 155, 158, 163, 171–176). Such studies led to proposals that sepiapterin is an intermediate on the biosynthetic pathway for H_4biopterin (11, 155, 158, 163, 171–174). These proposals were supported by the evidence that the product of sepiapterin reductase is L-erythro-H_2biopterin (177, 178), and that the enzyme is found in all of the tissues that produce H_4biopterin (141, 179).

More recently, it was observed that in the presence of Mg^{2+} but the absence of NADPH, partially purified enzyme fractions from avian and mammalian tissues produced an unidentified compound from H_2neopterin-PPP (171, 173, 174, 180). Tanaka et al (171) found that the "compound X" fluoresces under UV light, does not contain phosphate, and reacts with dinitrophenylhydrazine, yielding pterin and pyruvate phenylhydrazone, and proposed that the structure of "X" was 6-(1,2-dioxopropyl)-H_2pterin (171). It was found that "X" could be converted to sepiapterin by the biosynthetic enzymes and NADPH; thus, this compound was proposed to be the immediate precursor of sepiapterin on the biosynthetic pathway (171). Krivi & Brown (163), using partially purified *Drosophila* extracts in the absence of NADPH, also observed the production of "a labile intermediate" from H_2neopterin-PPP which was not characterized.

The enzymes proposed to convert H_2neopterin-PPP to sepiapterin have been partially purified from several species (158, 163, 164, 171, 174, 180). The characteristics of the enzymes in all of the systems except *Ascaris* were found to be very similar. The enzymes were precipitated from the soluble fraction in 40–65% saturated ammonium sulfate, and eluted from an Ultrogel AcA-34 (164, 171, 180) or AcA-44 (163) in the included volume. Similarly, the reactions all require Mg^{2+} and NADPH (163, 164, 170, 171, 180) for the production of sepiapterin from H_2neopterin-PPP. In the case of the hamster kidney preparation, the combination of NADPH plus NADH was reported to be most active (164).

The enzymes from chicken kidney and *Drosophila* were purified further (163, 171). Two enzymes were required for the conversion of H_2neopterin-PPP

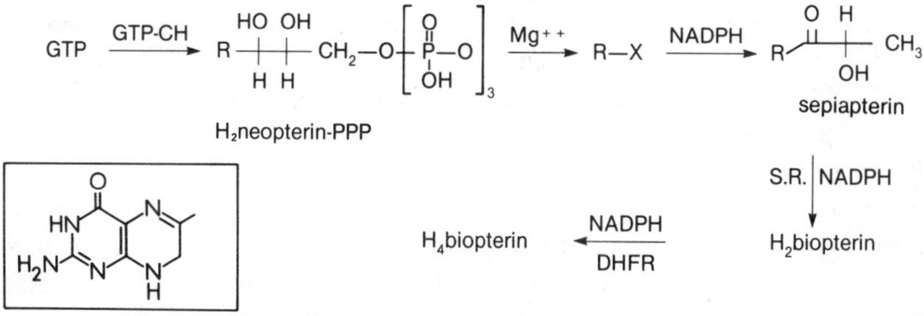

Figure 5   The dihydropterin pathway for H_4biopterin biosynthesis. "R-X" is an intermediate for which two structures were proposed (163, 171). S.R. is sepiapterin reductase.

to sepiapterin by either tissue. They had molecular weights of 77,000 and 30,000 in the chicken (171) and 82,000 and 36,000 in *Drosophila* (163, 170). In both cases, the higher-molecular-weight enzyme catalyzed the Mg^{2+} dependent release of the phosphoryl groups (163). The second enzyme plus NADPH converted the product of the first into sepiapterin (Figure 5). There is one report that the rat brain enzyme system precipitates in the 0–35% ammonium sulfate pellet and does not require NADPH or Mg^{2+} (181), but this result has not been found to be reproducible in other preparations from brain (172, 174) and these cofactors are clearly required for the biosynthesis in other tissues (163, 164, 170, 171, 180, 182).

SEPIAPTERIN REDUCTASE   This enzyme has been purified to homogeneity from horse liver, rat erythrocytes, and brain (183–185) and partially purified from other sources (171, 175, 177, 180, 186, 187). The reported properties of the avian and mammalian enzymes are quite similar. The enzyme was found in the soluble homogenate fraction, precipitated with 40–60% saturated ammonium sulfate and had a molecular weight in the range of 40,000–60,000. The rat erythrocyte enzyme, which has been best characterized, was found to be a dimer of 27,500-dalton subunits (184). The levels of the enzyme have been determined in several mammalian cell lines and tissues (141, 179). It was found that H_4biopterin levels correlate with GTP-CH levels, but not with sepiapterin reductase activity (141).

The best known reaction catalyzed by sepiapterin reductase is the reduction of L-sepiapterin to L-erythro-H_2biopterin (178, 179). The kinetic mechanism of the reaction has been reported to be ordered bi-bi for the rat erythrocyte enzyme and the K_m values for substrate and cofactor are 15.4 μM and 1.7 μM, respectively, for that enzyme (184). It had been thought that the substrate specificity of the enzyme was very strict (179). Recently, however, the homogenous rat erythrocyte enzyme was reported to reduce α-diketones at rates comparable to or greater than sepiapterin (188), and it was also found that the α-diketone biacetyl binds to the enzyme 30 times tighter than its α-hydroxyketone analog, acetoin. The *Drosophila* enzyme also reduces "oxidized sepiapterin" (6-lactoyl pterin) to biopterin (187) and 6-lactoyl-H_4pterin as well as a compound proposed to be 6-pyruvoyl-H_4pterin to H_4biopterin (189). NAS is a competitive inhibitor of sepiapterin reductase, having K_i values of 0.20 μM and 0.17 μM for the rat brain and erythrocyte enzymes, respectively (190). Based on inhibition by NAS in bovine adrenal medulla enzyme preparations, 6-lactoyl-H_4pterin and another H_4pterin (H_4pterin II) appear to serve as substrates for sepiapterin reductase (191, 192). Recognition of rather broad substrate specificity of sepiapterin reductase has caused speculation that the function of the enzyme in vivo may not be solely for the reduction of sepiapterin to H_2biopterin (182, 188, 189, 191, 192).

DIHYDROFOLATE REDUCTASE   The ability of DHFR to reduce 7,8-H$_2$biopterin led to proposals that DHFR catalyzed the terminal reduction in the H$_2$pterin biosynthetic pathway (2, 4, 193). The role of DHFR in brain as well as in H$_4$biopterin biosynthesis has been the subject of some speculation. In contrast to early studies (194) in which no activity could be detected in the brains of several mammalian species, subsequent studies have reported the presence of DHFR in brain (193, 195–197). Developmental studies (197–200) have shown that embryonic tissues have a relatively high content of the enzyme. In the chick embryo and human fetus, reductase levels were highest in the earliest stages of development (198, 199). In rabbit brain the activity was highest in fetal brain, but remained high throughout life (200). Studies using rat brain and liver (197) showed that in brain, enzyme levels were highest before birth and declined to low levels after birth. In liver, activity was highest ten days after birth and then declined to adult levels by 24 days of age.

The substrate specificity of DHFR is quite broad (201–204). In addition to the conjugated pteridines, the enzyme can also reduce unconjugated pteridines. H$_2$biopterin can be reduced to H$_4$biopterin, although at a rate less than 10% and a K_m 50–100 times greater than that for H$_2$folate (2–5, 205). Inhibition of the reduction of H$_2$biopterin to H$_4$biopterin by MTX indicates that DHFR is the enzyme that catalyzes this reduction when H$_2$biopterin is administered to animals or to cells in culture (141, 175, 176, 205, 206). Even though it is implausible that a single enzyme would be essential for the formation of both H$_4$folate and H$_4$biopterin, concern nonetheless has been expressed that the chemotherapy of cancer with folate antagonists would cause interference with the formation of monoamine neurotransmitters and might be a factor in the CNS toxicity of MTX (207).

De Novo Synthesis of Tetrahydrobiopterin Independent of Dihydrofolate Reductase

Evidence against the "dihydropterin pathway" for H$_4$biopterin biosynthesis has accumulated. Stone (3) found that although trimethoprim is an effective inhibitor of the reduction of H$_2$biopterin to H$_4$biopterin by rat DHFR in vitro ($K_i + = 0.285$ μM), administration of large doses of trimethoprim to rats had little effect upon H$_4$biopterin levels in liver. From these results, it was postulated that either trimethoprim did not reach the site of DHFR action, which is unlikely, or that the biosynthesis of H$_4$biopterin in rat did not involve reduction of H$_2$biopterin. Gál et al (208) reported that MTX had no effect on H$_4$biopterin biosynthesis either in vivo or in vitro and suggested that DHFR was not involved in the biosynthetic pathway. However, these experiments were questionable because the properties reported for H$_2$biopterin, quinonoid-H$_2$biopterin, and H$_4$biopterin were inconsistent with the known chemical properties of the compounds (115, 209, 210). Duch et al (211) found that total biopterin was not changed by MTX treatment in any tissue studied except the liver, and the percent of the total H_4biopterin remained the same as in the control tissue. Also, the lipophilic DHFR inhibitor, metoprine, which enters the brain better than MTX, did not reduce H_4biopterin levels in brain or adrenals (211). The neuroblastoma clone, N1E-115, can grow well in the presence of sufficient MTX to totally inhibit DHFR if thymidine and hypoxanthine are added to the medium to relieve the H_4folate requirement. In such cultures, total biopterin increased and its percent as H_4biopterin remained the same as controls (211). Since new H_4biopterin synthesis occurred, the experiment showed that inhibition of DHFR does not inhibit H_4biopterin biosynthesis de novo. Similar results were also found using CHO cells in culture. DUKX-B11, a mutant line of CHO, has no detectable DHFR, yet formation of H_4biopterin de novo is comparable to that in the parent line which does possess DHFR (176).

Other in vitro experiments support the conclusion that DHFR is not involved in de novo H_4biopterin biosynthesis (175, 176, 192). Crude extracts of bovine adrenal medulla convert both GTP and H_2neopterin-PPP to H_4biopterin at the same rate in the presence or absence of up to 100 μM MTX (175, 176). Also, partially purified adrenal medulla enzyme fractions which had been freed of DHFR convert H_2neopterin-PPP to H_4biopterin (175).

In contrast to the results on de novo biosynthesis from GTP and H_2neopterin-PPP, DHFR is required for the conversion of sepiapterin and H_2biopterin to H_4biopterin (175, 176, 192). Thus, CHO cells can efficiently convert exogenous sepiapterin to H_4biopterin, but the DHFR-deficient DUKX-B11 mutant cell line converts sepiapterin only to H_2biopterin (176). Similarly, intracisternal injection of sepiapterin or H_2biopterin resulted in increased levels of H_4biopterin in rat brain, but administration of MTX prior to the injection of sepiapterin or H_2biopterin prevented conversion to H_4biopterin (176). Crude extracts of bovine adrenal medulla can also convert sepiapterin and H_2biopterin to H_4biopterin. MTX prevents this conversion, and purified fractions lacking DHFR do not produce H_4biopterin from either sepiapterin or H_2biopterin (175, 176, 192). Further, in the presence of MTX or the absence of DHFR, unlabeled sepiapterin does not dilute radioactivity in [^{14}C]H_4biopterin produced from [^{14}C]H_2neopterin-PPP by adrenal medulla extracts. Thus, sepiapterin cannot enter the running de novo biosynthetic pathway (192, 212). Since the overall de novo biosynthetic pathway from GTP and H_2neopterin-PPP does not require DHFR, but H_4biopterin synthesis from sepiapterin and H_2biopterin does, the latter compounds are not de novo biosynthetic intermediates. It has been suggested that the conversion of these H_2pterins to H_4biopterin may represent a pterin salvage pathway (176).

The Tetrahydropterin De Novo Pathway

The evidence that neither sepiapterin nor H_2biopterin are de novo biosynthetic intermediates required a new hypothesis. Several groups have reported the isolation of sepiapterin from biosynthetic reactions (155, 158, 163, 170–174, 180, 191). Thus, it appears to be most likely that these compounds, and especially sepiapterin, are decomposition products of the true intermediates which are unstable under the conditions used for incubation or isolation (182, 189, 191, 192, 212, 213). The structures of these unstable intermediates were then proposed to be H_4pterins (182, 189, 191, 192, 213). H_4pterins are known to be very easily oxidized and to decompose to H_2pterins (2, 131, 214, 215); thus, production of sepiapterin upon oxidation of some H_4pterin intermediate is reasonable. Also, a biosynthetic pathway with H_4pterin intermediates would obviate the requirement for DHFR as the terminal enzyme on the pathway.

The mechanism proposed for the conversion of H_2neopterin-PPP to an H_4pterin intermediate is based on the Lobry de Bruyn-Alberda van Ekenstein reaction, well known in carbohydrate chemistry (216), and also on the mechanism proposed by Pfleiderer (131) for the conversion of H_2biopterin to 6-lactoyl-H_4pterin. The proposed mechanism, shown in the top of Figure 4, involves transfer of electrons from the 1'hydroxyl of the side chain to the pterin moiety. The mechanism does not require the involvement of any NADPH-dependent reductase such as DHFR. The overall proposed H_4pterin pathway is shown in Figure 4.

Recently, using anaerobic techniques to prevent the oxidation of H_4pterins to H_2pterins, direct evidence has been provided for the presence of H_4pterin intermediates (189, 192, 217). Switchenko et al (189) partially purified three enzymes from *Drosophila* which together convert H_2neopterin-PPP to H_4biopterin. They found that under anaerobic conditions, the first enzyme (A) required only Mg^{2+} for the conversion of H_2neopterin-PPP to a compound with a UV spectrum characteristic of H_4pterins, λ_{max} at pH 7 = 300 nm, and with no 330 or 420 nm absorbance band characteristic of H_2pterins (218). The compound was proposed to be 6-pyruvoyl-H_4pterin (189). When the second enzyme (B) plus NADPH were added to the reaction anaerobically, no further spectral change occurred, but under aerobic conditions the compound made by enzyme B was nonenzymatically converted to sepiapterin. The authors proposed that enzyme B converted 6-pyruvoyl-H_4pterin to 6-lactoyl-H_4pterin, which oxidized to sepiapterin in air. When the products of enzyme A or B were anaerobically exposed to the third enzyme (biopterin synthase or sepiapterin reductase) both compounds were converted to H_4biopterin. However, since the yield of H_4biopterin from the product of enzyme A was 3.4 times the yield from the product of enzyme B, the authors concluded that the product of enzyme B is not a kinetically competent intermediate. Based on these data, it was proposed that only enzyme A and biopterin synthase are required for the conversion of H_2neopterin-PPP to H_4biopterin, and that biopterin synthase catalyzes the reduction of both carbonyls of 6-pyruvoyl-H_4pterin (189) (Figure 4).

From experiments using partially purified extracts from bovine adrenal medulla, Smith & Nichol (192, 217) came to similar, though not identical conclusions. Anaerobic HPLC conditions coupled with electrochemical detection were used to look for H_4pterin intermediates. It was found that H_2neopterin-PPP, in the absence of NADPH but the presence of Mg^{2+}, was converted to a compound (referred to as H_4pterin-1) that oxidized on the glassy carbon electrode at 200 mV and had a UV spectrum at pH 3 with $\lambda_{max} = 265$ nm. These are properties of H_4pterins and not H_2pterins (128, 218). H_4Pterin-1 may be the same compound observed by Switchenko et al (189), that is, 6-pyruvoyl-H_4pterin. H_2neopterin-PPP was converted to a second H_4pterin in the presence of Mg^{2+}, NADPH, and 1–10 μM of the sepiapterin reductase inhibitor NAS. The compound also accumulated in preparations low in sepiapterin reductase. This compound, referred to as H_4pterin-2, exhibited oxidation and UV spectral characteristics of an H_4pterin (192, 217). The HPLC retention times of H_4pterin-1 and H_4pterin-2 indicated that neither one is 6-lactoyl-H_4pterin. Concentrations of NAS 10–100 times higher than that required for >95% inhibition of H_4biopterin synthesis (10 μM) can cause some accumulation (2–10% of the optimal H_4pterin-2 production) of 6-lactoyl-H_4pterin (G. Smith, unpublished). This may simply reflect a side reaction producing 6-lactoyl-H_4pterin as was implied in the results of Switchenko et al (189).

When either H_4pterin-1, H_4pterin-2, or 6-lactoyl-H_4pterin is exposed to NADPH and the biosynthetic enzymes, 6R-H_4biopterin is produced at a rate faster than that for the conversion of H_2neopterin-PPP to H_4biopterin; thus, the reaction rates indicate that these compounds are kinetically competent with the overall biosynthesis from H_2neopterin-PPP (191, 192). These data obtained with adrenal medulla extracts indicate that H_4pterin-1 and H_4pterin-2 are intermediates in the biosynthesis of H_4biopterin, whereas it is not yet certain whether 6-lactoyl-H_4pterin is an intermediate. The misnomer "tetrahydrosepiapterin" (which denotes the addition of four hydrogen atoms to a H_2pterin) has been used in some reports instead of the correct name 6-lactoyl-H_4pterin. The mechanism for the conversion of H_2neopterin-PPP to the first H_4pterin intermediate predicts that the proton at C-6 of H_4biopterin must derive from H_2O (Figure 4). It was found that when the conversion of H_2neopterin-PPP to H_4biopterin was run in D_2O, deuterium was incorporated at C-6 (G. Smith, unpublished). Thus, the evidence that the C-6 proton is derived from water rather than NADPH supports this general mechanism.

Milstien & Kaufman (191) have shown that NAS is an inhibitor of

H_4biopterin biosynthesis. Since NAS is an inhibitor of the conversion of H_2neopterin-PPP to H_4biopterin, the results suggest that sepiapterin reductase is on the pathway. The results of Smith & Nichol (192) suggest that the substrate for the enzyme on the pathway is H_4pterin-2 since the latter accumulates when the enzyme is absent or inhibited by NAS. The interpretation that sepiapterin reductase catalyzes the last reaction(s) on the pathway agrees with the results of Switchenko et al (189). The observation (189) that the enzyme catalyzes two steps on the pathway, reduction of 1'- and 2'-carbonyls, has not yet been verified, but the report (188) that rat erythrocyte sepiapterin reductase can catalyze the reduction of diketones such as biacetyl supports this idea. Since sepiapterin is clearly not the natural substrate, the enzyme involved can be preferably designated as "H_4biopterin synthase."

The bovine adrenal medullary enzymes that catalyze the DHFR-independent conversion of H_2neopterin-PPP to H_4biopterin have been partially purified (175, 192). The enzymes were found in the soluble homogenate supernatant, precipitate in the 40–60% ammonium sulfate fraction, elute from a column of Ultrogel AcA-34 in the 60–70,000 dalton range, and bind to DEAE-Sephacel. Further, the biosynthesis of H_4biopterin by the partially purified fractions require NADPH and Mg^{2+}. These properties are very similar to those of the enzymes on the "dihydropterin pathway" (163, 164, 170, 171, 180) and indicate that the enzymes on the "dihydropterin pathway" and the H_4pterin pathway are probably the same.

There is a logical progression from the initial concept of a pathway with H_2pterin intermediates to the perception of a H_4pterin pathway. Thus, the current views can be summarized as follows: (*a*) although sepiapterin and H_2biopterin are not intermediates on the de novo pathway, both compounds can be converted to H_4biopterin by enzymes that may have a salvage function, that is, by sepiapterin reductase and DHFR; (*b*) "compound X," which was reported to be 6-pyruvoyl-H_2pterin, appears to be an oxidation product of 6-pyruvoyl-H_4pterin; (*c*) there is a lack of dependence on DHFR since the first H_4pterin intermediate is formed without any requirement for NADPH; (*d*) an enzyme with properties similar to sepiapterin reductase, H_4biopterin synthase, appears to be on the biosynthetic pathway but its natural substrate appears to be H_4pterin-2 and/or 6-pyruvoyl-H_4pterin; (*e*) the enzymes on the H_4pterin pathway appear, in general, to be the same as those originally described for the H_2pterin pathway, except that the true substrates are H_4pterins. However, the side chain structures of the H_4pterin intermediates and the actual sequence of reactions are yet to be verified. There is now general agreement that the biosynthesis of H_4biopterin proceeds by a H_4pterin pathway in organisms as different as bovine and *Drosophila* (Figure 4).

Regulation of Tetrahydrobiopterin Biosynthesis

The biosynthesis of H_4biopterin appears to be under dynamic regulation. There is evidence that the aromatic amino acid precursors of the monoamines have an effect on the regulation of H_4biopterin biosynthesis. Following intraperitoneal administration of Tyr to rats, the total biopterin content of striatum, adrenal gland, and serum was increased 1.5- to 3-fold (219). In the same study, it was reported that administration of Tyr to normal human subjects resulted in a 3- to 7-fold increase in serum biopterin levels. Administration of Phe to normal human subjects (220) or rats (221) also resulted in increased levels of biopterin in blood. An oral Phe dose increased the incorporation of ^{14}C-guanosine into biopterin in rats (221), thus providing additional evidence for the increased synthesis of biopterin under these conditions. In contrast to the above results, however, other experiments indicated that Tyr and Trp loading have no effect on biopterin levels in the tissues examined (220, 221). The role of H_4biopterin in physiological responses is beginning to be studied. Biopterin levels are decreased in the murine neurological mutant "rolling" and can be increased following the administration of thyrotropin releasing hormone (222). In the adrenal medulla, reserpine acts by blocking the uptake of catecholamines into storage vesicles and can also act indirectly like insulin by increasing splanchnic nerve discharge (69, 223–225). In addition to causing a depletion of catecholamines, reserpine treatment or insulin-induced hypoglycemia in rats caused significant increases in both GTP-CH and H_4biopterin in the adrenal medulla (76, 165, 226, 227). Following reserpine treatment, catecholamines were maximally depleted at 24 hr and then returned to control levels by 72 hr. During this period, there was an increase in GTP-CH as early as 4 hr after reserpine which preceded the increase in H_4biopterin. Both enzyme and cofactor levels were highest 24 hr after reserpine. These changes were blocked by adrenal denervation or cycloheximide, indicating that enzyme and cofactor synthesis in the medulla are regulated by enchanced splanchnic-adrenal medullary discharge. Immobilization, a milder form of stress than either reserpine or insulin, also produced significant increases in medullary GTP-CH and H_4biopterin (226, 228).

Reserpine and insulin treatment produced similar changes in GTP-CH and H_4biopterin in the adrenal cortex of rats, and cycloheximide prevented these drug-induced rises in H_4biopterin biosynthesis (76, 165, 166, 226, 229). Unlike the adrenal medulla, splanchnic denervation of the adrenal gland had no effect on these responses. The increases appeared to be specific for the adrenal gland since H_4biopterin levels were unchanged in liver, kidney, and corpus striatum and only slightly decreased in the pineal (76). Immobilization stress also resulted in induction of H_4biopterin biosynthesis in the adrenal cortex. Twenty-four hr after a 2.5-hr period of immobilization, GTP-CH and H_4biopterin levels were elevated to the same extent as was observed with insulin and reserpine (166, 229). Reserpine, insulin, and immobilization stress all stimulate the adrenal cortex by increasing the release of ACTH from the pituitary (230–233). While hypophysectomy had no effect on the reserpine or insulin-induced increases in GTP-CH and H_4biopterin in the adrenal medulla, it did prevent their rise in the adrenal cortex (76, 165, 166). Administration of purified porcine ACTH or synthetic $ACTH_{1-24}$ to intact or hypophysectomized rats elevated cortical levels of GTP-CH and H_4biopterin (166). Using Y-1 mouse adrenal cortical cells in culture, the addition of ACTH to the culture medium also produced an induction of GTP-CH and H_4biopterin (234). Maximal stimulation of H_4biopterin biosynthesis occurred at the same concentration of ACTH that produced maximal stimulation of steroidogenesis. Thus, in the adrenal medulla H_4biopterin metabolism appears to be under neural control while in the adrenal cortex regulation appears to be dependent on hormones secreted from the pituitary.

GTP-CH activity in primary cultures of adrenal medullary chromaffin cells appears to be regulated in a manner similar to that observed in vivo (228). Treatment of cultures with reserpine or tetrabenazine, which results in the cellular depletion of catecholamines, caused increases in the cellular levels of GTP-CH. Moreover, agents which elevate cAMP levels, such as 8Br cAMP or Bt_2 cAMP also elevated GTP-CH in these cells. In cultures of neuroblastoma cells, 8Br cAMP, Bt_2 cAMP, and phosphodiesterase inhibitors all produced significant stimulation of the levels of GTP-CH and H_4biopterin (235). cAMP elevation caused a decrease in the biopterin content of pineal glands in organ culture through inhibition of biopterin biosynthesis (236). These results suggest that regulation of H_4biopterin biosynthesis occurs by a cAMP-dependent mechanism.

PTERIN CATABOLISM

A number of pterins in addition to biopterin and neopterin have been identified in the urine of mammals including xanthopterin, dihydroxanthopterin, isoxanthopterin, 3'-hydroxysepiapterin, pterin, pterin-6-carboxylic acid, 6-hydroxymethylpterin, and the lumazines (106, 123, 125, 237, 238, 244). Pterin-6-carboxylic acid, 6-hydroxymethylpterin and probably some fraction of the pterin present in urine appear to be derived from folates rather than from unconjugated pteridines (11, 239). Following administration of $[^{14}C]H_4$biopterin or $[^{14}C]H_4$neopterin to rats, Rembold et al (239) found that there was no conversion of either pterin to $[^{14}C]$6-hydroxymethylpterin. In cell culture studies (11) the addition of ^{14}C-folic acid to the medium resulted in label being found in pterin, pterin-6-carboxylic acid and 6-hydroxymethylpterin. When $[2^{14}C]$guanine was added to the medium, significant amounts of label were found in pterin and biopterin, but none in pterin-6-carboxylic acid and 6-hydroxymethylpterin.

Based on studies in vitro with rat liver homogenates and in vivo in rats, Rembold has proposed a metabolic sequence to account for the pterin metabolites derived from 6-hydroxyalkyltetrahydropterins (239–242). The metabolic scheme is based on the presence of quinonoid intermediates, although none could be detected in these studies. In the proposed pathway the initial reaction is a nonenzymatic oxidative cleavage of the polyhydroxyalkyl side chain of the H_4pterin, resulting in the production of 7,8-H_2pterin. Detailed studies on the aerobic nonenzymatic oxidation of H_4biopterin at physiological pH by Armarego et al (210) added further support to the scheme proposed by Rembold (242). The existence of an enzyme in *Drosophila* that releases the side chain from H_2neopterin-PPP has also been described (243). H_2pterin produced by cleavage of the side chain can then be oxidized by xanthine oxidase, resulting in the formation of xanthopterin, isoxanthopterin, or leucopterin or can serve as the substrate for pterin deaminase. The isolation and characterization of a pterin deaminase from rat liver, which appears to be specific for H_2- and H_4pterin, has been described (241). The action of pterin deaminase alone or in combination with xanthine oxidase results in the formation of lumazine, 6-hydroxylumazine, 7-hydroxylumazine, or 6,7-dihydroxylumazine. Xanthopterin, lumazine, and 6-hydroxylumazine can also be excreted as the dihydro derivatives. High levels of urinary dihydroxanthopterin have been found in PKU (238, 245, 246) and it has been proposed that measurement of urinary levels of this pterin may be useful in the diagnosis of these diseases. The presence of 3'-hydroxysepiapterin in urine has also been reported (244), although the origin and significance of this pterin are unknown at the present time. Changes in biopterin and neopterin levels in urine apparently reflect changes in rates of biosynthesis.

PTERIN METABOLISM IN DISEASE

The evidence that biopterin metabolism is altered in disease states has grown during the past ten years since the observation by Baker et al (101) that levels of biopterin were elevated in the blood from patients with gout, uremia, and alcoholic liver disease. Altered biopterin levels have been found in blood and urine in patients with several diseases (137, 138, 247).

Cancer

Several studies of unconjugated pterin metabolism in disease states have centered on the changes which occur in neoplastic disease. Increased levels of pterin-6-aldehyde and 6-hydroxymethylpterin have been reported in the urines of patients with malignant diseases (248–250) although the source of these pterins was probably folate rather than biopterin. Elevated levels of biopterin in the blood of patients with a wide variety of malignant diseases have been reported (143, 251). The effects of neoplastic disease on the urinary excretion of a variety of unconjugated pterins were carried out by Rokos et al (252). Elevated neopterin levels were observed in 70% of the cancer patients while biopterin levels were elevated in 22% of the patients. The amount of xanthopterin, which in most cases was excreted as the dihydro derivative, was generally increased when neopterin and/or biopterin excretion was elevated. In general neopterin/biopterin ratios were higher in cancer patients than in the control group. Pterin levels were lower than controls in most patients and 6-hydroxymethylpterin, pterin-6-carboxylic acid, and isoxanthopterin levels were similar in both groups (252). In another study of patients with cancer, Stea et al (253) found a significant increase in the mean urinary excretion levels of xanthopterin, neopterin, and pterin and a significant decrease in isoxanthopterin. Excretion of biopterin, pterin-6-carboxylic acid and 6-hydroxymethylpterin did not differ from controls. In both studies, the neopterin/biopterin ratios appeared to be the important factor in the assessment of altered pterin metabolism. In general, the neopterin/biopterin ratio was higher in cancer patients than in controls. In some cases it was also found that normal pterin levels correlated with a stable clinical status (252, 253).

In a series of studies, Wachter and coworkers (254–256) have assessed changes in pteridine metabolism in cancer patients by the direct measurement of urinary neopterin levels. With the exception of patients with Stage I multiple myeloma, patients with active hematological disease had significantly higher mean neopterin levels than the controls. Patients with disease in remission or with benign hematological diseases had neopterin levels within the normal range. However, only that fraction of neopterin which is present in the urine as the fully oxidized form was measured in these studies. Since there is evidence that more than half of the neopterin present in urine is in reduced forms (115, 257), iodine oxidation during sample preparation could give a measure of total neopterin excretion. The tissue(s) of origin and the reasons for the changes in neopterin and biopterin excretion in cancer patients are at present unknown.

Immune Disorders

Changes in pteridine metabolism have also been investigated as a marker of T-lymphocyte activation in vitro and in vivo. Increased pterin levels in blood and urine occur in acquired immunodeficiency syndrome, generalized lymphadenopathy, coeliac disease, allograft rejection, viral infections, and bone marrow transplantation (258–267). Pterins may be involved in the control of lymphocyte stimulation and lymphoblast proliferation (268). Xanthopterin and isoxanthopterin, catabolic products derived from H_4biopterin, inhibited the proliferation of Conconavalin-A stimulated lymphocytes and the growth of the

LS-2 lymphoid cell line in culture. Conversely, the addition of H_4biopterin, H_2biopterin, or sepiapterin stimulated the activation of lymphocytes produced by suboptimal or supraoptimal concentrations of Conconavalin A. Additionally, the isolation and characterization of a pteridine-binding α_1-acid glycoprotein have been reported (269, 270).

Hyperphenylalaninemia Due to Tetrahydrobiopterin Deficiency

Phenylketonuria is now known to be caused by genetic defects resulting in either a deficiency of PAH apoenzyme, referred to as "classical" PKU, or of the H_4pterin cofactor, referred to as "atypical" PKU or malignant hyperphenylalaninemia (271). The term "H_4biopterin deficiency" is preferable to "atypical PKU" since defective neurotransmitter synthesis is part of the syndrome (272). Screening at birth for abnormally high Phe levels in blood (273) is mandatory in most countries since prompt treatment may avoid progressive mental retardation (274). Diagnosis of H_4biopterin deficiency is based on a decrease in serum Phe within 2–4 hr following a small oral dose of H_4biopterin (275–277). The incidence of H_4biopterin deficiency is low (1–2% of PKU patients) (277, 278) and one case of PKU unresponsive to low Phe diet and without evidence of H_4biopterin deficiency has been reported (279). An international registry compiled by Dhondt (24) lists the diagnosis, treatment, and clinical course of 50 cases of H_4biopterin deficiency. For such individuals, H_4biopterin is an essential nutrient but it does not serve as a satisfactory vitamin due to instability, poor bioavailability, and limited entry into brain. Consequently, attention has been directed to the use of H_4biopterin precursors (276) or lipophilic H_4pterins (280) which can replace the natural cofactor.

A subclass of PKU was recognized about ten years ago when infants with hyperphenylalaninemia failed to respond to low Phe diets and developed severe neurological dysfunction which responded to treatment with L-DOPA and 5-hydroxytryptophan (281–283). This indicated a defect in the synthesis or availability of the cofactor for the aromatic amino acid hydroxylases. Two different metabolic lesions, a deficiency of DHPR (284) and defective synthesis of H_4biopterin (285–288) were soon identified. More recently, a third type has been attributed to the lack of GTP-CH (289). The locus of the enzyme defect in each type can be indicated by the urinary pterin profile (150, 151, 290–294). Much clinical variability is observed with both classical and atypical PKU and there are variations in the degree of enzyme deficiency (23, 295, 296). Heterogeneity of gene mutations was observed in patients with DHPR deficiency (296).

GTP CYCLOHYDROLASE DEFICIENCY   During screening for H_4biopterin deficiency, an unusual urinary pterin profile was observed in one case of PKU, which was characterized by very low levels of neopterin, biopterin, and pterin

(289, 297). CSF levels of neopterin, biopterin, homovanillic acid, and 5-hydroxyindoleacetic acid were also extremely low. The lack of neopterin, in contrast to other types of H_4biopterin deficiency, indicated the likelihood of a defect in GTP-CH. GTP-CH activity in two liver biopsies was less than 4% of control levels. Also, phytohemagglutinin-stimulated lymphocytes from this patient had no measurable GTP-CH activity in contrast to 32 and 49% of normal activity in lymphocytes from the father and mother, respectively, indicating that the patient's lymphocytes were unable to form neopterin (289). If molybdopterin was formed from H_2neopterin triphosphate, abnormalities in the activity of xanthine oxidase and sulfite oxidase would be expected, but the metabolite excretion pattern indicated normal function of these enzymes. Also, a detailed evaluation of the immune system in this patient was carried out to examine the putative role of neopterin metabolism in the immune response. There were no gross abnormalities in the patient's immune system (289, 297, 298).

DEFICIENCY IN COFACTOR BIOSYNTHESIS FROM DIHYDRONEOPTERIN TRIPHOSPHATE (206, 278, 280, 285, 286, 288, 299–303)   A high level of neopterin and low level of biopterin, giving a high neopterin:biopterin ratio, characterizes the urinary pterin profile in patients with this metabolic defect. Among the 27 reported cases with this type of H_4biopterin deficiency, some variations or subtypes have been observed and have been described as "partial forms," in which dietary Phe tolerance develops more rapidly (24, 301), or "peripheral forms," having normal CSF biopterin and neurotransmitter metabolite levels along with elevated plasma Phe responsive to H_4biopterin treatment (206, 299). In many instances there was substantial clinical improvement following treatment with H_4biopterin but in most cases low CSF neurotransmitter metabolites indicated the need for therapy with L-DOPA and 5-hydroxytryptophan. Kaufman et al (280, 304) have cited the advantages of cofactor replacement over neurotransmitter replacement. In order to localize the biosynthesis defect, Yoshioka and coworkers (305) developed enzyme assays for the conversion of H_2neopterin-PPP to "compound X" and its conversion to sepiapterin and presented evidence for a 50–60% decrease in the formation of "compound X" in peripheral erythrocytes of two children with this form of H_4biopterin deficiency (305).

DIHYDROPTERIDINE REDUCTASE DEFICIENCY (277, 278, 284, 286, 296, 306–314)   Since this metabolic defect involves the regeneration of the active cofactor rather than the biosynthesis, quinonoid-H_2biopterin (see Figure 3) accumulates and appears as H_2biopterin in blood and urine. The lower than normal neopterin:biopterin ratio is not in itself adequate for diagnosis (150, 151, 290–294). Although diagnosed as having PKU, some patients do not develop neurological signs until approximately four months of age. Breast milk contains significant quantities of H_4biopterin (139) and might supply sufficient amounts of cofactor to lower the neonatal Phe levels of these patients. As a result, the correct diagnosis may be masked by the mild hyperphenylalaninemia during the period of breast feeding. Consequently, procedures have been developed to measure DHPR deficiency in peripheral blood cells (310), cultured fibroblasts and lymphoid cells (311), in the blood spots on PKU screening cards (312), and in liver biopsies (277, 284). Clinical improvement, in cases of DHPR deficiency, has been quite limited and less responsive to cofactor replacement than in patients with H_4biopterin biosynthesis defects.

Neurologic and Psychiatric Diseases

Levels of H_4biopterin in the brain and CSF of patients affected with these diseases have been examined. In CSF, biopterin levels of 17.5–20.5 pmol/ml have been reported in healthy individuals (77, 114, 118, 315–322) and H_4biopterin and total biopterin have been found to decrease slightly with age (315, 321, 323, 324). In patients with Parkinson's disease, the CSF cofactor concentration was approximately 50% that of age matched controls, and there was a positive correlation between cofactor level and the dopamine metabolite homovanillic acid (114, 317). The total biopterin content in the caudate nucleus was 70% lower in the post-mortem brains of patients with Parkinson's disease than in controls (77, 118). Also, serum biopterin levels were slightly lower than those of controls (219), while urinary levels of biopterin were not different from controls (118).

Similarly, in a family with dystonias, CSF hydroxylase cofactor levels were shown to be lowered approximately 85% in four sisters with generalized dystonia and 70% in other family members with focal dystonia (316). In another study, total CSF biopterin was decreased an average of 40% in eight patients with dystonia (322); however, a third study (317) reported no difference from controls in patients with adult onset focal dystonia. This same study reported that the CSF hydroxylase cofactor was decreased to 50% of control levels in patients with Steele-Richardson syndrome, essential tremor, Huntington's disease and pre-senile dementia, to 25% of control levels in patients with Shy-Drager syndrome, and to 40% in patients with senile dementia of the Alzheimer type (317). In another study in patients with Alzheimer's disease, total CSF biopterin levels were also found to be 40% of control (318). Decreased serum biopterin levels have been reported in patients with senile dementia (325) and with dementia of the Alzheimer type (326). Individuals with Down's syndrome were found to have elevated serum H_2biopterin levels, while levels in mentally retarded individuals without Down's syndrome were found to be depressed (327).

In two reports, CSF hydroxylase cofactor levels in depressed patients were not significantly different from controls (319, 320), but in another report total

CSF biopterin was found to be decreased in a single patient with endogenous depression (328). Patients with depression were found to have normal biopterin levels in serum, whereas elevated levels were found in patients receiving tricyclic antidepressants (329). Other studies demonstrated an increased urinary excretion of biopterin as well as neopterin in patients with unipolar depression (330). No significant differences were found in CSF or urinary levels of biopterin in schizophrenia or bipolar depression (321, 330). The majority of CSF biopterin almost surely derives from the brain; thus, CSF biopterin can provide a measure of changes in brain biopterin levels (315, 331). However, the source of serum and urinary biopterin is unknown, but is most likely derived from a variety of peripheral tissues.

L-DOPA therapy for replacement of deficient dopamine in Parkinsonian patients has been used successfully for many years (332). Since H_4biopterin is required for DOPA synthesis in the brain, the low brain and CSF levels of H_4biopterin, due at least in part to the neuronal loss associated with Parkinson's disease and other central nervous system diseases (317), led to the suggestion that exogenous H_4biopterin might accelerate dopamine synthesis. Oral administration of a single dose of H_4biopterin (10–15 mg/kg) to Parkinsonian patients gave partial relief of symptoms for several hours (333, 334). In another study, however, no effect was observed in two patients given 2.5–10 mg/kg i.v. (324). It was suggested that better responses in Parkinson's disease would be obtained with patients in an early stage of the disease since neuronal loss continues as the disease progresses (324, 333, 334) and at later stages, insufficient neurons exist for H_4biopterin to affect a significant increase in striatal dopamine synthesis. In this regard, a study reported best results with patients having a short case history (334). One group, which reported no effect upon Parkinsonian patients (324), observed substantial improvement in three dystonia patients given a single dose of 1.5–5 mg/kg H_4biopterin (322). A positive effect in three of six depressed patients who received approximately 10 mg/kg H_4biopterin was also reported (328, 333).

Efficacy in terms of clinical status has not been evaluated since there were very few patients in each study and treatment was with only a single dose in all but one case. H_4biopterin has poor oral bioavailability and does not cross the blood-brain barrier well, so treatment with this compound in particular is fraught with problems (331, 335–337). 6-Methyl-H_4pterin, on the other hand, which is also a good cofactor for all three hydroxylases, has been shown to enter the brain of rats ten times better than H_4biopterin (331) and to have some effects in atypical PKU (302, 303). Thus, more lipophilic H_4pterins, like 6-methyl-H_4pterin, might be better candidates than H_4biopterin for treatment. Further, it has been proposed that the amounts of H_4biopterin administered in the studies of neurological and psychiatric diseases were not sufficient to have a substantial effect on neuronal H_4biopterin concentration (74, 338). Clearly, more studies, and especially more long-term studies like the one of Curtius et al

(328), are needed to determine the feasibility of H_4pterin treatment for these diseases. In addition, there is a need for more bioavailability and brain penetration studies. The use of animal models, such as the 6-hydroxydopamine-lesioned rat model (74, 339) and the 1-methyl-4-phenyl-1,2,5,6-tetrahydropyridine-lesioned monkey and mouse models (340, 341) of Parkinson's disease, should also be of use in a preliminary evaluation of the efficacy of H_4pterins.

CONCLUDING REMARKS

In this review, we have attempted to show the growing momentum in the area of H_4biopterin research which stems in part from the recent development of new and sensitive assay methods, the recognition of the importance of the cofactor in neurotransmitter synthesis, and the recognition of the tetrahydropterin pathway for H_4biopterin biosynthesis.

The early hypothesis that H_4biopterin is required for mitochondrial electron transport (342, 343) should be reexamined in view of the observations that cells lacking both GTP-CH and H_4biopterin can grow at normal rates (141). It is apparent that H_4biopterin has specialized functions in cells and tissues containing the aromatic amino acid hydroxylases. However, the cofactor and its biosynthetic enzymes are also present in tissues lacking these enzymes. Neither the substrate(s) nor the enzyme(s) requiring H_4biopterin in tissues such as adrenal cortex, bone marrow, or spleen are known, yet regulation of synthesis, at least in the adrenal cortex, suggests an important role for this cofactor. The role of H_4biopterin in the glycerol ether cleavage reaction (56) also deserves further attention. The idea "to develop antimetabolites that specifically affect biopterin-requiring enzymes" put forward in 1964 (56) is still relevant since this may be useful in determining other functions.

The temporary correction of the loss of muscle control characteristic of dystonia and Parkinson's disease following treatment with H_4biopterin indicates some degree of cofactor deficiency in diseases involving such motor dysfunctions. The clinical use of the natural cofactor and analogs in the treatment of hyperphenylalaninemia due to H_4biopterin deficiency has already indicated some of the limitations of cofactor replacement therapy in restoring normal neurotransmitter synthesis. These clinical investigations, however, and the evidence of marked changes in pterin metabolism associated with different disease states, have given real impetus to studies on the regulation of the synthesis and the additional biochemical functions of H_4biopterin.

In a symposium on oxygenases in 1982, a perceptive comment by Bloch on monooxygenase reactions is pertinent to unanswered questions: "Whether in a given instance electrons travel from NADPH to O_2 via FAD or biopterin and thence by way of cytochromes, ferredoxins or some other none-heme iron protein has not yet been rationalized" (344).

ACKNOWLEDGMENTS

We gratefully acknowledge the help of Ms. Tonya H. Beasley in the preparation of the manuscript and thank Mr. Jeffrey H. Woolf for his expert technical assistance.

Literature Cited

1. Matthews, R. G., Kaufman, S. 1980. *J. Biol. Chem.* 255:6014–17
2. Kaufman, S. 1967. *J. Biol. Chem.* 242:3934–43
3. Stone, K. J. 1976. *Biochem. J.* 157:105–9
4. Abelson, H. T., Spector, R., Gorka, C., Fosburg, M. 1978. *Biochem. J.* 171:267–68
5. Spector, R., Fosburg, M., Levy, P., Abelson, H. T. 1978. *J. Neurochem.* 30:899–901
6. Craine, J. E., Hall, E. S., Kaufman, S. 1972. *J. Biol. Chem.* 247:6082–91
7. Nichol, C. A., Viveros, O. H., Duch, D. S., Abou-Donia, M. M., Smith, G. K. 1983. See Ref. 12, pp. 131–51
8. Patterson, E. L., Broquist, H. P., Albrecht, A. M., von Saltza, M. H., Stokstad, E. L. R. 1955. *J. Am. Chem. Soc.* 77:3167–68
9. Johnson, J. L., Hainline, B. E., Rajagopalan, K. V. 1980. *J. Biol. Chem.* 255:1783–86
10. Koschara, W. 1943. *Hoppe-Seyler's Z. Physiol. Chem.* 277:284–87
11. Fukushima, T., Shiota, T. 1974. *J. Biol. Chem.* 249:4445–51
12. Blair, J. A., ed. 1983. *Chemistry and Biology of Pteridines.* Berlin: de Gruyter. 1070 pp.
13. Kisliuk, R. L., Brown, G. M., eds. 1979. *Chemistry and Biology of Pteridines.* New York: Elsevier. 713 pp.
14. Pfleiderer, W., ed. 1975. *Chemistry and Biology of Pteridines.* Berlin: de Gruyter. 949 pp.
15. Iwai, K., Akino, M., Goto, M., Iwanami, Y., eds. 1970. *Chemistry and Biology of Pteridines.* Tokyo: Int. Acad. Print. 481 pp.
16. Wachter, H., Curtius, H.-Ch., Pfleiderer, W., eds. 1982. *Biochemical and Clinical Aspects of Pteridines,* Vol. 1. Berlin: de Gruyter. 373 pp.
17. Curtius, H.-Ch., Pfleiderer, W., Wachter, H., eds. 1983. *Biochemical and Clinical Aspects of Pteridines,* Vol. 2. Berlin: de Gruyter. 435 pp.
18. Kaufman, S., Fisher, D. B. 1974. In *Molecular Mechanisms of Oxygen Activation,* ed. O. Hayaishi, pp. 285–369. New York: Academic 678 pp.
19. Mandell, A. J. 1978. *Ann. Rev. Pharmacol. Toxicol.* 18:461–93
20. Usdin, E., Weiner, N., Youdim, M. B. H., eds. 1981. *Function and Regulation of Monoamine Enzymes.* London: Macmillan. 961 pp.
21. Blaskovics, M. E. 1974. *Clin. Endocrinol. Metab.* 3:87–105
22. Cotton, R. G. H. 1977. *Int. J. Biochem.* 8:333–41
23. Kaufman, S. 1983. *Adv. Hum. Genet.* 13:217–97
24. Dhondt, J. L. 1984. *J. Pediatr.* 104:501–8
25. Johnson, J. L. 1980. *Molybdenum and Molybdenum-Containing Enzymes,* ed. M. P. Coughlan, pp. 345–83. Oxford: Pergamon. 577 pp.
26. Shah, V. K., Ugalde, R. A., Imperial, J., Brill, W. J. 1984. *Ann. Rev. Biochem.* 53:231–57
27. Pateman, J. A., Cove, D. J., Rever, B. M., Roberts, D. B. 1964. *Nature* 201:58–60
28. Nason, A., Antoine, A. D., Ketchum, P. A., Frazier, W. A. III, Lee, D. K. 1970. *Proc. Natl. Acad. Sci. USA* 65:137–44
29. Ketchum, P. A., Cambier, H. Y., Frazier, W. A. III, Madansky, C. H., Nason, A. 1970. *Proc. Natl. Acad. Sci. USA* 66:1016–23
30. Nason, A., Lee, K. Y., Pan, S. S., Ketchum, P. A., Lamberti, A., DeVries, J. 1971. *Proc. Natl. Acad. Sci. USA* 68:3242–46
31. Johnson, J. L., Jones, H. P., Rajagopalan, K. V. 1977. *J. Biol. Chem.* 252:4994–5003
32. Rajagopalan, K. V., Johnson, J. L., Hainline, B. E. 1982. *Fed. Proc.* 41:2608–12
33. Rajagopalan, K. V., Johnson, J. L., Hainline, B. E. 1980. *Fed. Proc.* 40:1652 (Abstr.)
34. Johnson, J. L., Hainline, B. E., Rajago-

palan, K. V., Arison, B. H. 1984. *J. Biol. Chem.* 259:5414–22
35. Johnson, J. L., Rajagopalan, K. V. 1982. *Proc. Natl. Acad. Sci. USA* 79:6856–60
36. Koschara, W. 1943. *Hoppe-Seyler's Z. Physiol. Chem.* 279:44–52
37. Goto, M., Sakurai, A., Ohta, K., Yamakami, H. 1967. *Tetrahedron Lett.* 45:4507–11
38. Goto, M., Sakurai, A., Ohta, K., Yamakami, H. 1969. *J. Biochem.* 65:611–20
39. Johnson, J. L., Hainline, B. E., Rajagopalan, K. V. 1983. See Ref. 12, pp. 1043–47
40. Ishizuka, M., Ushio, K., Toraya, T., Fukui, S. 1983. *Biochem. Biophys. Res. Commun.* 111:537–43
41. Claassen, V. P., Oltmann, L. F., Van't Riet, J., Brinkman, U. A. Th., Stouthamer, A. H. 1982. *FEBS Lett.* 142:133–37
42. Johnson, J. L., Waud, W. R., Rajagopalan, K. V., Duran, M., Beemer, F. A., Wadman, S. K. 1980. *Proc. Natl. Acad. Sci. USA* 77:3715–19
43. Ogier, H., Saudubray, J. M., Charpentier, C., Munnich, A., Perignon, J. L., et al. 1982. *Ann. Med. Int.* 133:594–96
44. Ogier, H., Wadman, S. K., Johnson, J. L., Saudubray, J. M., Duran, M., et al. 1983. *Lancet* 2:1363–64
45. Kaufman, S. 1967. *Ann. Rev. Biochem.* 36:171–84
46. Massey, V., Hemmerich, P. 1975. *Enzymes* 12:191–252
47. Benkovic, S. J. 1980. *Ann. Rev. Biochem.* 49:227–51
48. Ayling, J. E., Bailey, S. W. 1982. In *Oxygenases and Oxygen Metabolism*, ed. M. Nozaki, S. Yamamoto, Y. Ishimura, M. J. Coon, L. Ernster, R. W. Estabrook, pp. 267–79. New York: Academic. 664 pp.
49. Kaufman, S. 1971. *Adv. Enzymol.* 35:245–319
50. Roth, R. H. 1980. In *Neurobiology of Dopamine*, ed. A. S. Horn, J. Korf, B. H. C. Westerbaink, pp. 101–22. London: Academic. 723 pp.
51. Kaufman, S. 1959. *J. Biol. Chem.* 234:2677–82
52. Kaufman, S., Levenberg, B. 1959. *J. Biol. Chem.* 234:2683–88
53. Brenneman, A. R., Kaufman, S. 1964. *Biochem. Biophys. Res. Commun.* 17:177–83
54. Nagatsu, T., Levitt, M., Udenfriend, S. 1964. *J. Biol. Chem.* 239:2910–17
55. Friedman, P. A., Kappelman, A. H., Kaufman, S. 1972. *J. Biol. Chem.* 247:4165–73
56. Tietz, A., Lindberg, M., Kennedy, E. P. 1964. *J. Biol. Chem.* 239:4081–90
57. Soodsma, J. F., Piantadosi, C., Snyder, F. 1972. *J. Biol. Chem.* 247:3923–29
58. Oka, K., Kato, T., Sugimoto, T., Matsuura, S., Nagatsu, T. 1981. *Biochim. Biophys. Acta* 661:45–53
59. Kato, T., Yamaguchi, T., Nagatsu, T., Sugimoto, T., Matsuura, S. 1980. *Biochim. Biophys. Acta* 611:241–50
60. Wessiak, A., Bruice, T. C. 1983. *J. Am. Chem. Soc.* 105:4809–25
61. Kaufman, S. 1975. See Ref. 14, pp. 291–301
62. Kaufman, S., Bridgers, W. F., Eisenberg, F., Friedman, S. 1962. *Biochem. Biophys. Res. Commun.* 9:497–502
63. Daly, J. W., Levitt, M., Guroff, G., Udenfriend, S. 1968. *Arch. Biochem. Biophys.* 126:593–98
64. Lazarus, R. A., De Brosse, C. W., Benkovic, S. J. 1982. *J. Am. Chem. Soc.* 104:6869–71
65. Lazarus, R. A., Dietrich, R. F., Wallick, D. E., Benkovic, S. J. 1981. *Biochemistry* 20:6834–41
66. Lazarus, R. A., Benkovic, S. J., Kaufman, S. 1983. *J. Biol. Chem.* 258:10960–62
67. Bullard, W. P., Guthrie, P. B., Russo, P. V., Mandell, A. J. 1978. *J. Pharmacol. Exp. Ther.* 206:4–20
68. Levine, R. A., Kuhn, D. M., Lovenberg, W. 1979. *J. Neurochem.* 32:1575–78
69. Viveros, O. H., Arqueros, L., Connett, R. J., Kirshner, N. 1969. *Mol. Pharmacol.* 5:69–82
70. Mueller, R. A., Thoenen, H., Axelrod, J. 1969. *J. Pharmacol. Exp. Ther.* 169:74–79
71. Lovenberg, W., Bruckwick, E. A., Hanbauer, I. 1975. *Proc. Natl. Acad. Sci. USA* 72:2955–58
72. Morgenroth, V. H. III, Hegstrand, L. R., Roth, R. H., Greengard, P. 1975. *J. Biol. Chem.* 250:1946–48
73. Wilson, S. P., Abou-Donia, M. M., Chang, K.-J., Viveros, O. H. 1981. *Neuroscience* 6:71–79
74. Levine, R. A., Miller, L. P., Lovenberg, W. 1981. *Science* 214:919–21
75. Oka, K., Ashiba, G., Sugimoto, T., Matsuura, S., Nagatsu, T. 1982. *Biochim. Biophys. Acta* 706:188–96
76. Abou-Donia, M. M., Viveros, O. H. 1981. *Proc. Natl. Acad. Sci. USA* 78:2703–6
77. Nagatsu, T. 1981. *Trends Pharmacol. Sci.* 2:276–79
78. Nagatsu, T. 1983. *Neurochem. Int.* 5:27–38

79. Masserano, J. M., Weiner, N. 1979. *Mol. Pharmacol.* 16:513–28
80. Weiner, N., Lee, F.-L., Dreyer, E., Barnes, E. 1978. *Life Sci.* 22:1197–216
81. Morgenroth, V. H. III, Boadle-Biber, M., Roth, R. H. 1974. *Proc. Natl. Acad. Sci. USA* 71:4283–87
82. Kettler, R., Bartholini, G., Pletscher, A. 1974. *Nature* 249:476–78
83. Coté, L. J., Benitez, H. H., Murray, M. R. 1975. *J. Neurobiol.* 6:233–43
84. Hirata, Y., Togari, A., Nagatsu, T. 1983. *J. Neurochem.* 40:1585–89
85. Patrick, R. L., Barchas, J. D. 1976. *J. Phamarcol. Exp. Ther.* 197:97–104
86. Boadle-Biber, M. C., Roth, R. H. 1972. *Br. J. Pharmacol.* 46:696–707
87. Cloutier, G., Weiner, N. 1973. *J. Pharmacol. Exp. Ther.* 186:75–85
88. Boarder, M. R., Fillenz, M. 1979. *Biochem. Pharmacol.* 28:1675–77
89. Boarder, M. R., Fillenz, M. 1980. *J. Neurochem.* 34:1016–18
90. Udenfriend, S., Zaltzman-Nirenberg, P., Nagatsu, T. 1965. *Biochem. Pharmacol.* 14:837–45
91. Nagatsu, T., Mizutani, K., Nagatsu, I., Matsuura, S., Sugimoto, T. 1972. *Biochem. Pharmacol.* 21:1945–53
92. Lovenberg, W., Jequier, E., Sjoerdsma, A. 1967. *Science* 155:217–19
93. Kuhn, D. M., Vogel, R. L., Lovenberg, W. 1978. *Biochem. Biophys. Res. Commun.* 82:759–66
94. Hamon, M., Bourgoin, S., Hery, F., Simonnet, G. 1978. *Mol. Pharmacol.* 14:99–110
95. Yamauchi, T., Fujisawa, H. 1979. *Arch. Biochem. Biophys.* 198:219–26
96. Boadle-Biber, M. C. 1980. *Biochem. Pharmacol.* 29:669–72
96a. Kaufman, S. 1981. See Ref. 20, pp. 165–73
97. Cowperthwaite, J., Weber, M. M., Packer, L., Hutner, S. H. 1953. *Ann. NY Acad. Sci.* 56:972–81
98. Nathan, H. A., Cowperthwaite, J. 1955. *J. Protozool.* 2:37–42
99. Dewey, V. C., Kidder, G. W. 1971. *Methods Enzymol.* 18b:618–24
100. Guttman, H. N. 1972. *Anal. Microbiol.* 2:457–77
101. Baker, H., Frank, O., Bacchi, C. J., Hutner, S. H. 1974. *Am. J. Clin. Nutr.* 27:1247–53
102. Baker, H., Frank, O., Shapiro, A., Hutner, S. H. 1980. *Methods Enzymol.* 66:490–500
103. Rokos, H., Rokos, K. 1983. See Ref. 12, pp. 815–19
104. Rembold, V. H., Metzger, H. 1967.

105. Rembold, H. 1971. *Methods Enzymol.* 18b:652–60
106. Fukushima, T., Shiota, T. 1972. *J. Biol. Chem.* 247:4549–56
107. Kobayashi, K., Goto, M. 1970. See Ref. 15, pp. 57–70
108. Lloyd, T., Markey, S., Weiner, N. 1971. *Anal. Biochem.* 42:108–12
109. Röthler, F., Karobath, M. 1976. *Clin. Chim. Acta* 69:457–62
110. Kuster, T., Niederwieser, A. 1983. *J. Chromatogr.* 278:245–54
111. Kaufman, S. 1958. *J. Biol. Chem.* 230:931–39
112. Guroff, G., Rhoads, C. A., Abramowitz, A. 1967. *Anal. Biochem.* 21:273–78
113. Guroff, G. 1971. *Methods Enzymol.* 18b:600–5
114. Lovenberg, W., Levine, R. A., Robinson, D. S., Ebert, M., Williams, A. C., et al. 1979. *Science* 204:624–26
115. Fukushima, T., Nixon, J. C. 1980. *Anal. Biochem.* 102:176–88
116. Bailey, S. W., Ayling, J. E. 1978. *J. Biol. Chem.* 253:1598–1605
117. Nagatsu, T., Yamaguchi, T., Kato, T., Sugimoto, T., Matsuura, S., et al. 1979. *Proc. Jpn Acad. Ser. B* 55:317–22
118. Nagatsu, T., Yamaguchi, T., Kato, T., Sugimoto, T., Matsuura, S., et al. 1981. *Anal. Biochem.* 110:182–89
119. Sugimoto, T., Matsuura, S., Yamaguchi, T., Sawada, M., Nagatsu, T. 1983. See Ref. 17, pp. 53–62
120. Nagatsu, T., Sawada, M., Yamaguchi, T., Sugimoto, T., Matsuura, S., et al. 1983. See Ref. 12, pp. 821–25
121. Gál, E. M., Sherman, A. D. 1977. *Prep. Biochem.* 7:155–64
122. Stea, B., Halpern, R. M., Smith, R. A. 1979. *J. Chromatogr.* 168:385–93
123. Stea, B., Halpern, R. M., Halpern, B. C., Smith, R. A. 1980. *J. Chromatogr.* 188:363–75
124. Andondonskaja-Renz, B., Zeitler, H.-J. 1983. *Anal. Biochem.* 133:68–78
125. Woolf, J. H., Nichol, C. A., Duch, D. S. 1983. *J. Chromatogr.* 274:398–402
126. Bräutigam, M., Dreesen, R., Herken, H. 1982. *Hoppe-Seyler's Z. Physiol. Chem.* 363:341–43
127. Bräutigam, M., Dreesen, R. 1982. *Hoppe-Seyler's Z. Physiol. Chem.* 363:1203–7
128. Lunte, C. E., Kissinger, P. T. 1983. *Anal. Chem.* 55:1458–62
129. Lunte, C. E., Kissinger, P. T. 1983. *Anal. Biochem.* 129:377–86
130. Lunte, C. E., Kissinger, P. T. 1983. *J. Liq. Chromatogr.* 6:1863–72

Hoppe-Seyler's Z. Physiol. Chem. 348:194–98

131. Pfleiderer, W. 1975. See Ref. 14, pp. 941–49
132. Milstien, S. 1983. See Ref. 17, pp. 65–70
133. Fukushima, T., Kobayashi, K.-I., Eto, I., Shiota, T. 1978. *Anal. Biochem.* 89:71–79
134. Rembold, H., Gyure, W. L. 1972. *Angew. Chem. Int. Ed. Engl.* 11:1061–72
135. Wachter, H., Hausen, A., Reider, E., Schweiger, M. 1980. *Naturwissenschaften* 67:610–11
136. Gál, E. M., Hanson, G., Sherman, A. 1976. *Neurochem. Res.* 1:511–23
137. Leeming, R. J., Blair, J. A., Melikian, V., O'Gorman, D. J. 1976. *J. Clin. Pathol.* 29:444–51
138. Leeming, R. J., Blair, J. A. 1980. *Clin. Chim. Acta* 108:103–11
139. Dhondt, J.-L., Delcroix, M., Farriaux, J.-P. 1982. *Clin. Chim. Acta* 121:33–35
140. Hennings, G., Rembold, H. 1982. *Int. J. Vitam. Nutr. Res.* 52:36–43
141. Duch, D. S., Nichol, C. A. 1983. See Ref. 12, pp. 839–43
142. Duch, D. S., Bowers, S. W., Woolf, J. H., Nichol, C. A. 1984. *Life Sci.* 35:1895–901
143. Ziegler, I., Fink, M., Wilmanns, W. 1982. *Blut* 44:231–40
144. Albrecht, A. M., Biedler, J. L., Baker, H., Frank, O., Hutner, S. H. 1978. *Res. Commun. Chem. Pathol. Pharmacol.* 19:377–80
145. Bräutigam, M., Dreesen, R., Herken, H. 1984. *J. Neurochem.* 42:390–96
146. Kapatos, G., Kaufman, S., Weller, J. L., Klein, D. C. 1983. *Brain Res.* 258:351–55
147. Kato, T., Yamaguchi, T., Togari, A., Nagatsu, T., Yajima, T., et al. 1982. *J. Neurochem.* 38:896–901
148. Niederwieser, A., Curtius, H.-Ch., Gitzelmann, R., Otten, A., Baerlocher, K., et al. 1980. *Helv. Paediatr. Acta* 35:335–42
149. Shintaku, H., Isshiki, G., Hase, Y., Tsuruhara, T., Oura, T. 1982. *J. Inher. Metab. Dis.* 5:241–42
150. Matalon, R., Michals, K., Lee, C.-L., Nixon, J. C. 1982. *Ann. Clin. Lab Sci.* 12:411–14
151. Dhondt, J.-L., Ardouin, P., Hayte, J.-M., Farriaux, J.-P. 1981. *Clin. Chim. Acta* 116:143–52
152. Gál, E. M. 1982. *Adv. Neurochem.* 4:83–148
153. Brown, G. M. 1971. *Adv. Enzymol.* 35:35–77
154. Levy, C. C. 1964. *J. Biol. Chem.* 239:560–66
155. Fukushima, T. 1970. *Arch. Biochem. Biophys.* 139:361–69
156. Sugiura, K., Goto, M. 1973. *Experientia* 29:1481–82
157. Gál, E. M., Sherman, A. D. 1976. *Neurochem. Res.* 1:627–39
158. Otsuka, H., Sugiura, K., Goto, M. 1980. *Biochim. Biophys. Acta* 629:69–76
159. Buff, K., Dairman, W. 1975. *Mol. Pharmacol.* 11:87–93
160. Fukushima, K., Eto, I., Saliba, D., Shiota, T. 1975. *Biochem. Biophys. Res. Commun.* 65:644–51
161. Fan, C. L., Brown, G. M. 1976. *Biochem. Genet.* 14:371–89
162. Fukushima, K., Richter, W. E. Jr., Shiota, T. 1977. *J. Biol. Chem.* 252:5750–55
163. Krivi, G. G., Brown, G. M. 1979. *Biochem. Genet.* 17:371–90
164. Eto, I., Fukushima, K., Shiota, T. 1976. *J. Biol. Chem.* 251:6505–12
165. Viveros, O. H., Lee, C.-L., Abou-Donia, M. M., Nixon, J. C., Nichol, C. A. 1981. *Science* 213:349–50
166. Abou-Donia, M. M., Duch, D. S., Nichol, C. A., Viveros, O. H. 1983. *Endocrinology* 112:2088–94
167. Blau, N., Niederwieser, A. 1983. *Anal. Biochem.* 128:446–52
168. Bellahsene, Z., Dhondt, J.-L., Farriaux, J.-P. 1984. *Biochem. J.* 217:59–65
169. Yim, J. J., Brown, G. M. 1976. *J. Biol. Chem.* 251:5087–94
170. Fan, C. L., Krivi, G. G., Brown, G. M. 1975. *Biochem. Biophys. Res. Commun.* 67:1047–54
171. Tanaka, K., Akino, M., Hagi, Y., Doi, M., Shiota, T. 1981. *J. Biol. Chem.* 256:2963–72
172. Kapatos, G., Katoh, S., Kaufman, S. 1982. *J. Neurochem.* 39:1152–62
173. Curtius, H.-Ch., Häusermann, M., Niederwieser, A., Ghisla, S. 1982. *Biochemical and Clinical Aspects of Pteridines* 1:27–50
174. Yoshioka, S.-I., Masada, M., Yoshida, T., Inoue, K., Mizokami, T., et al. 1983. *Biochim. Biophys. Acta* 756:279–85
175. Smith, G. K., Nichol, C. A. 1983. *Arch. Biochem. Biophys.* 227:272–78
176. Nichol, C. A., Lee, C.-L., Edelstein, M. P., Chao, J. Y., Duch, D. S. 1983. *Proc. Natl. Acad. Sci. USA* 80:1546–50
177. Matsubara, M., Katoh, S., Akino, M., Kaufman, S. 1966. *Biochim. Biophys. Acta* 122:202–12
178. Nagai, M. 1968. *Arch. Biochem. Biophys.* 126:426–35
179. Katoh, S., Nagai, M., Nagai, Y., Fukushima, T., Akino, M. 1970. See Ref. 15, pp. 225–34

180. Häusermann, M., Ghisla, S., Nieder-wieser, A., Curtius, H.-Ch. 1981. *FEBS Lett.* 131:275–78
181. Gál, E. M., Nelson, J. M., Sherman, A. D. 1978. *Neurochem. Res.* 3:69–88
182. Heintel, D., Ghisla, S., Curtius, H.-Ch., Niederwieser, A., Levine, R. A. 1984. *Neurochem. Int.* 6:141–55
183. Katoh, S. 1971. *Arch. Biochem. Biophys.* 146:202–14
184. Sueoka, T., Katoh, S. 1982. *Biochim. Biophys. Acta* 717:265–71
185. Katoh, S., Sueoka, T., Yamada, S. 1983. See Ref. 12, pp. 789–93
186. Katoh, S., Sueoka, T. 1982. *Comp. Biochem. Physiol.* B71:33–39
187. Fan, C. L., Brown, G. M. 1979. *Biochem. Genet.* 17:351–69
188. Katoh, S., Sueoka, T. 1984. *Biochem. Biophys. Res. Commun.* 118:859–66
189. Switchenko, A. C., Primus, J. P., Brown, G. M. 1984. *Biochem. Biophys. Res. Commun.* 120:754–60
190. Katoh, S., Sueoka, T., Yamada, S. 1982. *Biochem. Biophys. Res. Commun.* 105:75–81
191. Milstien, S., Kaufman, S. 1983. *Biochem. Biophys. Res. Commun.* 115:888–93
192. Smith, G. K., Nichol, C. A. 1984. *Biochem. Biophys. Res. Commun.* 120:761–66
193. Spector, R., Levy, P., Abelson, H. T. 1977. *Biochem. Pharmacol.* 26:1507–11
194. Makulu, D. R., Smith, E. F., Bertino, J. R. 1973. *J. Neurochem.* 21:241–45
195. Lynn, R., Rueter, M. E., Guynn, R. W. 1977. *J. Neurochem.* 29:1147–49
196. Pollock, R. J., Kaufman, S. 1978. *J. Neurochem.* 30:253–56
197. Duch, D. S., Bigner, D. D., Bowers, S. W., Nichol, C. A. 1979. *Cancer Res.* 39:487–91
198. Silber, R., Huennekens, F. M., Gabrio, B. W. 1962. *Arch. Biochem. Biophys.* 99:328–33
199. Roberts, D., Hall, T. C. 1965. *Cancer Res.* 25:1894–98
200. Spector, R., Levy, P., Abelson, H. T. 1977. *J. Neurochem.* 29:919–21
201. Zakrzewski, S. F. 1960. *J. Biol. Chem.* 235:1780–84
202. Nath, R., Greenberg, D. M. 1962. *Biochemistry* 1:435–41
203. Morales, D. R., Greenberg, D. M. 1964. *Biochim. Biophys. Acta* 85:360–76
204. Greenberg, D. M., Tam, B.-D., Jenny, E., Payes, B. 1966. *Biochim. Biophys. Acta* 122:423–35
205. Reinhard, J. F., Chao, J. Y., Smith, G. K., Duch, D. S., Nichol, C. A. 1984. *Anal. Biochem.* 140:548–52

206. Milstien, S., Kaufman, S. 1983. See Ref. 17, pp. 133–38
207. Abelson, H. T. 1978. *Cancer Treat. Rep.* 62:1999–2001
208. Gál, E. M., Bybee, J. A., Sherman, A. D. 1979. *J. Neurochem.* 32:179–86
209. Bublitz, C. 1977. *Biochem. Med.* 17:13–19
210. Armarego, W. L., Randles, D., Taguchi, H. 1983. *Eur. J. Biochem.* 135:393–403
211. Duch, D. S., Lee, C.-L., Edelstein, M. P., Nichol, C. A. 1983. *Mol. Pharmacol.* 24:103–8
212. Smith, G. K., Nichol, C. A. 1983. *Fed. Proc.* 42:2104 (Abstr.)
213. Smith, G. K., Nichol, C. A. 1983. See Ref. 17, pp. 123–31
214. Blakley, R. L. 1969. *The Biochemistry of Folic Acid and Related Pteridines.* Amsterdam: North Holland. 569 pp.
215. Fukushima, T., Nixon, J. C. 1979. See Ref. 13, pp. 31–34
216. Speck, J. C. Jr. 1958. *Adv. Carbohydr. Chem.* 13:63–103
217. Smith, G. K., Nichol, C. A. 1984. *Fed. Proc.* 43:1998 (Abstr.)
218. Pfleiderer, W. 1982. See Ref. 16, pp. 3–26
219. Yamaguchi, T., Nagatsu, T., Sugimoto, T., Matsuura, S., Kondo, T., et al. 1983. *Science* 219:75–77
220. Leeming, R. J., Blair, J. A., Green, A., Raine, D. N. 1976. *Arch. Dis. Child.* 51:771–77
221. Milstien, S., Kaufman, S. 1983. See Ref. 12, pp. 753–57
222. Yamaguchi, T., Hirata, Y., Nagatsu, T., Oda, S., Sugimoto, T., et al. 1982. *Neurochem. Int.* 4:491–94
223. Viveros, O. H., Arqueros, L., Kirshner, N. 1971. *Mol. Pharmacol.* 7:434–43
224. Viveros, O. H., Arqueros, L., Kirshner, N. 1971. *Mol. Pharmacol.* 7:444–54
225. Mueller, R. A., Thoenen, H., Axelrod, J. 1969. *Mol. Pharmacol.* 5:463–69
226. Abou-Donia, M. M., Nichol, C. A., Diliberto, E. Jr., Viveros, O. H. 1982. In *Advances in the Biosciences,* ed. F. Izumi, 36:187–93. Oxford/New York: Pergamon
227. Viveros, O. H., Abou-Donia, M. M., Lee, C. L., Wilson, S. P., Nichol, C. A. 1981. See Ref. 20, pp. 241–51
228. Abou-Donia, M. M., Daniels, A. J., Wilson, S. P., Nichol, C. A., Viveros, O. H. 1983. See Ref. 12, pp. 777–81
229. Abou-Donia, M. M., Daniels, A. J., Nichol, C. A., Viveros, O. H. 1983. See Ref. 12, pp. 783–87
230. Munson, P. L. 1961. *Ann. Rev. Pharmacol.* 1:315–50

231. Gaunt, R., Chart, J. J., Renzi, A. A. 1963. *Ann. Rev. Pharmacol.* 3:109–28
232. Maickel, R. P., Westermann, E. O., Brodie, B. B. 1961. *J. Pharmacol. Exp. Ther.* 134:167–75
233. Sun, C. L., Thoa, N. B., Kopin, I. J. 1979. *Endocrinology* 105:306–11
234. Duch, D. S., Woolf, J. H., Edelstein, M. P., Nichol, C. A. 1984. *7th Int. Congr. Endocrinol.*, p. 542 (Abstr.)
235. Duch, D. S., Woolf, J. H., Edelstein, M. P., Nichol, C. A. 1983. *J. Neurochem.* 41(Suppl.):41(Abstr.)
236. Kapatos, G., Kaufman, S., Weller, J. L., Klein, D. C. 1981. *Science* 213:1129–31
237. Wachter, H., Grassmayr, K., Hausen, A. 1979. *Cancer Lett.* 6:61–66
238. Watson, B. M., Schlesinger, P., Cotton, R. G. H. 1977. *Clin. Chim. Acta* 78:417–23
239. Rembold, H., Chandrashekar, V., Sudershan, P. 1971. *Biochim. Biophys. Acta* 237:365–68
240. Rembold, H., Metzger, H., Sudershan, P., Gutensohn, W. 1969. *Biochim. Biophys. Acta* 184:386–96
241. Rembold, H., Simmersbach, F. 1969. *Biochim. Biophys. Acta* 184:589–96
242. Rembold, H., Metzger, H., Gutensohn, W. 1971. *Biochim. Biophys. Acta* 230:117–26
243. Yim, J. J., Jacobson, K. B., Crummett, D. C. 1981. *Insect Biochem.* 11:363–70
244. Niederwieser, A., Matasovic, A., Curtius, H.-Ch., Endres, W., Schaub, J. 1980. *FEBS Lett.* 118:299–302
245. Schlesinger, P., Watson, B. M., Cotton, R. G. H., Danks, D. M. 1979. *Clin. Chim. Acta* 92:187–95
246. Watson, B. M., Armarego, W. L. F., Schlesinger, P., Cotton, R. G. H., Danks, D. M. 1979. See Ref. 13, pp. 159–64
247. Ziegler, I., Fink, M. 1982. *Biochem. Med.* 27:401–4
248. Halpern, R., Halpern, B. C., Stea, B., Dunlap, A., Conklin, K., et al. 1977. *Proc. Natl. Acad. Sci. USA* 74:587–91
249. Rao, K. N., Trehan, S., Noronha, J. M. 1981. *Cancer* 48:1656–63
250. Trehan, S., Rao, K. N., Shetty, P. A., Noronha, J. M. 1982. *Cancer* 50:114–17
251. Kokolis, N., Ziegler, I. 1977. *Cancer Biochem. Biophys.* 2:79–85
252. Rokos, H., Rokos, K., Frisius, H., Kirstaeder, H. J. 1980. *Clin. Chim. Acta* 105:275–86
253. Stea, B., Halpern, R. M., Halpern, B. C., Smith, R. A. 1981. *Clin. Chim. Acta* 113:231–42
254. Hausen, A., Fuchs, D., Grünewald, K., Huber, H., König, K., Wachter, H. 1981. *Clin. Chim. Acta* 117:297–305
255. Hausen, A., Fuchs, D., Grünewald, K., Huber, H., König, K., Wachter, H. 1982. *Clin. Biochem.* 15:34–37
256. Bichler, A., Fuchs, D., Hausen, A., Hetzel, H., König, K., Wachter, H. 1982. *Clin. Biochem.* 15:38–40
257. Rokos, K., Rokos, H., Frisius, H., Hüfner, M. 1983. See Ref. 12, pp. 153–57
258. Wachter, H., Hausen, A., Grassmayr, K. 1979. *Hoppe-Seyler's Z. Physiol. Chem.* 360:1957–60
259. Fuchs, D., Granditsch, G., Hausen, A., Reibnegger, G., Wachter, H. 1983. *Lancet* 2:463–64
260. Wachter, H., Fuchs, D., Hausen, A., Huber, C., Knosp, O., et al. 1983. *Hoppe-Seyler's Z. Physiol. Chem.* 364:1345–46
261. Fuchs, D., Hausen, A., Huber, C., Margreiter, R., Reibnegger, G., et al. 1982. *Hoppe-Seyler's Z. Physiol. Chem.* 363:661–64
262. Niederwieser, D., Huber, C., Wachter, H. 1983. *Wien. Klin. Wochenschr.* 95:161–64
263. Huber, C., Niederwieser, D., Schönitzer, D., Fuchs, D., Hausen, A., et al. 1983. See Ref. 17, pp. 177–83
264. Huber, C., Fuchs, D., Hausen, A., Margreiter, R., Reibnegger, G., et al. 1983. *J. Immunol.* 130:1047–50
265. Niederwieser, D., Huber, C., Gratwohl, A., Bannerth, P., Speck, D., et al. 1983. See Ref. 17, pp. 195–202
266. Ziegler, I., Kolb, H. J., Bedenberger, U., Wilmanns, W. 1982. *Blut* 44:261–70
267. Fink, M., Jehn, U., Wilmanns, W., Rokos, H., Ziegler, I. 1983. See Ref. 17, pp. 223–34
268. Ziegler, I., Hamm, U., Berndt, J. 1983. *Cancer Res.* 43:5356–59
269. Ziegler, I., Maier, K., Fink, M. 1982. *Cancer Res.* 42:1567–73
270. Fink, M., Ziegler, I., Maier, K., Wilmanns, W. 1982. *Cancer Res.* 42:1574–78
271. Wellner, D., Meister, A. 1981. *Ann. Rev. Biochem.* 50:911–68
272. Curtius, H.-Ch., Niederwieser, A., Viscontini, M., Leimbacher, W., Wegman, H., et al. 1981. In *Serotonin. Current Aspects of Neurochemistry and Function,* ed. B. Haber, S. Gabay, M. Issidorides, S. Alivisatos, pp. 277–91. New York: Plenum 824 pp.
273. Guthrie, R., Susi, A. 1963. *Pediatrics* 32:338–43
274. Tourian, A. Y., Sidbury, J. B. 1978. In *The Metabolic Basis of Inherited Diseases,* ed. J. B. Stanbury, J. B. Wyngaarden, D. S. Fredrickson. New York: McGraw-Hill. 1986 pp.
275. Danks, D. M., Cotton, R. G. H., Schle-

singer, P. 1979. *Arch. Dis. Child.* 54: 329–30
276. Curtius, H.-Ch., Niederwieser, A., Viscontini, M., Otten, A., Schaub, J., et al 1979. *Clin. Chim. Acta* 93:251–362
277. Danks, D. M., Schlesinger, P., Firgaira, F., Cotton, R. G. H., Watson, B., Rembold, H., Hennings, G. 1979. *Pediatr. Res.* 13:1150–55
278. Niederwieser, A., Curtius, H.-Ch., Wang, M., Leupold, D. 1982. *Eur. J. Pediatr.* 138:110–12
279. Westwood, A., Barr, D. G. D. 1982. *Acta Pediatr. Scand.* 71:859–61
280. Kaufman, S., Kapatos, G., Rizzo, W. B., Schulman, J. D., Tamarkin, L., Van Loon, G. R. 1983. *Ann. Neurol.* 14:308–15
281. Bartholomé, K. 1974. *Lancet* 2:1580
282. Bartholomé, K., Byrd, D. J. 1975. *Lancet* 2:1042–43
283. Smith, I., Clayton, B. E., Wolff, O. H. 1975. *Lancet* 1:1108–11
284. Kaufman, S., Holtzman, N. A., Milstien, S., Butler, I. J., Krumholz, A. 1975. *N. Engl. J. Med.* 293:785–90
285. Rey, F., Blandin-Savoja, F., Rey, J. 1976. *N. Engl. J. Med.* 295:1138–39
286. Rey, F., Harpey, J. P., Leeming, R. J., Blair, J. A., Aicardi, J., Rey, J. 1977. *Arch. Fr. Pediatr.* 34:109–20
287. Bartholomé, K., Byrd, D. J., Kaufman, S., Milstien, S. 1977. *Pediatrics* 59:757–61
288. Niederwieser, A., Curtius, H.-Ch., Bettoni, O., Bieri, J., Schircks, B., et al. 1979. *Lancet* 1:131–33
289. Niederwieser, A., Blau, N., Wang, M., Joller, P., Atares, M., Cardesa-Garcia, J. 1984. *Eur. J. Pediatr.* 141:208–14
290. Nixon, J. C., Lee, C.-L., Milstien, S., Kaufman, S., Bartholomé, K. 1980. *J. Neurochem.* 35:898–904
291. Dhondt, J.-L., Largilliere, C., Ardouin, P., Farriaux, J.-P., Dautrevaux, M. 1981. *Clin. Chim. Acta* 110:205–14
292. Dhondt, J.-L., Farriaux, J.-P., Largilliere, C., Dautrevaux, M., Ardouin, P. 1981. *J. Inher. Metab. Dis.* 4:47–48
293. Matalon, R. 1984. *J. Pediatr.* 104(4):579–81
294. Dhondt, J.-L., Farriaux, J.-P. 1981. *Arch. Fr. Pediatr.* 38:573–78
295. Firgaira, F. A., Choo, K. H., Cotton, R. G. H., Danks, D. M. 1981. *Biochem. J.* 197:45–53
296. Firgaira, F. A., Choo, K. H., Cotton, R. G. H., Danks, D. M. 1981. *Biochem. J.* 198:677–82
297. Niederwieser, A., Staudenmann, W., Wang, M., Curtius, H.-C., Atares, M., Cardesa-Garcia, J. 1983. See Ref. 12, pp. 183–87
298. Joller, P. W., Blau, N., Atares, M., Niederwieser, A., Cardesa-Garcia, J. 1983. See Ref. 17, pp. 167–76
299. Hreidarsson, S., Valle, D., Holtzman, N., Coyle, J., Singer, H., et al. 1982. *Pediatr. Res.* 16:192A
300. Tanaka, T., Aihara, K., Iwai, K., Kohashi, M., Tomita, K., et al. 1981. *Eur. J. Pediatr.* 136:275–80
301. Dhondt, J.-L., Leroux, B., Farriaux, J.-P., Largilliere, C., Leeming, R. J. 1983. *Eur. J. Pediatr.* 141:92–95
302. Endres, W., Niederwieser, A., Curtius, H.-Ch., Wang, M., Ohrt, B., Schaub, J. 1982. *Helv. Paediat. Acta* 37:489–98
303. Kaufman, S., Kapatos, G., McInnes, R. R., Schulman, J. D., Rizzo, W. B. 1982. *Pediatrics* 70:376–80
304. McInnes, R. R., Kaufman, S., Warsh, J. J., Van Loon, G. R., Milstien, S., et al. 1984. *J. Clin. Invest.* 73:458–69
305. Yoshioka, S., Masada, M., Yoshida, T., Mizokami, T., Akino, M., Matsui, N. 1984. *Zoolog. Sci.* 1:74–81
306. Kaufman, S., Max, E. E., Kang, E. S. 1975. *Pediatr. Res.* 9:632–34
307. Milstien, S., Kaufman, S. 1975. *Biochem. Biophys. Res. Commun.* 66:475–81
308. Brewster, T. G., Abroms, I. F., Kaufman, S., Breslow, J. L., Moskowitz, M. A., et al. 1976. *Pediatr. Res.* 10:446
309. Narisawa, K., Arai, N., Ishizawa, S., Ogasawara, Y., Onuma, A., et al. 1980. *Clin. Chim. Acta* 105:335–42
310. Firgaira, F. A., Cotton, R. G. H., Danks, D. M. 1979. *Lancet* 2:1260–63
311. Firgaira, F. A., Cotton, R. G. H., Danks, D. M. 1979. *Clin. Chim. Acta* 95:47–59
312. Leeming, R. J., Barford, P. A., Blair, A., Smith, I. 1984. *Arch. Dis. Child.* 59:58–61
313. Kaufman, S. 1975. *Lancet* 2:767
314. Danks, D. M., Cotton, R. G. H., Schlesinger, P. 1976. *Lancet* 1:1236–37
315. Williams, A., Ballenger, J., Levine, R., Lovenberg, W., Calne, D. 1980. *Neurology* 30:1244–46
316. Williams, A., Eldridge, R., Levine, R., Lovenberg, W., Paulson, G. 1979. *Lancet* 2:410–11
317. Williams, A. C., Levine, R. A., Chase, T. N., Lovenberg, W., Calne, D. B. 1980. *J. Neurol. Neurosurg. Psychiatry.* 43:735–38
318. Morar, C., Whitburn, S., Blair, J. A., Leeming, R. J., Wilcock, G. K. 1983. *J. Neurol. Neurosurg. Psychiatry.* 46:582
319. Van Kammen, D. P., Levine, R., Sternberg, D., Ballenger, J., Marder, S., et al. 1978. *Psychopharmacol. Bull.* 14:51–52
320. Kellner, C. H., Rakita, R. M., Rubinow,

D. A., Gold, P. W., Ballenger, J. C., Post, R. M. 1983. *Lancet* 2:55–56
321. Garbutt, J. C., van Kammen, D. P., Levine, R. A., Sternberg, D. E., Murphy, D. L., et al. 1982. *Psychiatry Res.* 6:145–51
322. Le Witt, P. A., Newman, R. P., Miller, L. P., Lovenberg, W., Eldridge, R. 1983. *N. Engl. J. Med.* 308:157–58
323. Levine, R. A., Williams, A. C., Robinson, D. S., Calne, D. B., Lovenberg, W. 1979. *Adv. Neurol.* 24:303–07
324. Le Witt, P. A., Miller, L. P., Newman, R. P., Burns, R. S., Insel, T., et al. 1984. *Adv. Neurol.* 40:459–62
325. Leeming, R. J., Blair, J. A., Melikian, V. 1979. *Lancet* 1:215
326. Young, J. H., Kelly, B., Clayton, B. E. 1982. *J. Clin. Exp. Gerontol.* 4:389–402
327. Aziz, A. A., Blair, J. A., Leeming, R. J., Sylvester, P. E. 1982. *J. Ment. Defic. Res.* 26:67–71
328. Curtius, H.-Ch., Niederwieser, A., Levine, R. A., Lovenberg, W., Woggon, B., Angst, J. 1983. *Lancet* 1:657–58
329. Leeming, R. J., Blair, J. A., Walters, J. 1982. *Psych. Med.* 12:191–92
330. Duch, D. S., Woolf, J. H., Nichol, C. A., Davidson, J. R., Garbutt, J. C. 1984. *Psychiatry Res.* 11:83–89

331. Kapatos, G., Kaufman, S. 1981. *Science* 212:955–56
332. Cotzias, G. C., Van Woert, M. H., Schiffer, L. M. 1967. *N. Engl. J. Med.* 276:374–79
333. Curtius, H.-Ch., Muldner, H., Niederwieser, A. 1982. *J. Neural Transm.* 55:301–8
334. Narabayashi, H., Kondo, T., Nagatsu, T., Sugimoto, T., Matsuura, S. 1982. *Proc. Jpn Acad. Ser.* B 58:283–87
335. Rembold, H., Metzger, H. 1967. *Z. Naturforsch.* 226:827–30
336. Rembold, H. 1982. See Ref. 16, pp. 51–63
337. Leeming, R. J., Blair, J. A., Melikian, V. 1983. *Biochem. Med.* 30:328–32
338. Levine, R. A., Lovenberg, W., Curtius, H.-C., Niederwieser, A. 1983. See Ref. 12, pp. 833–37
339. Pycock, C. J. 1980. *Neuroscience* 5:461–514
340. Burns, R. S., Chiueh, C. C., Markey, S. P., Ebert, M. H., Jacobowitz, D. M., Kopin, I. J. 1983. *Proc. Natl. Acad. Sci. USA* 80:4546–50
341. Heikkila, R. E., Hess, A., Duvoisin, R. C. 1984. *Science* 224:1451–53
342. Rembold, H., Buff, K. 1972. *Eur. J. Biochem.* 28:579–85
343. Rembold, H., Buff, K. 1972. *Eur. J. Biochem.* 28:586–91
344. Bloch, K. 1982. See Ref. 48, pp. 645–53

Ann. Rev. Biochem. 1985. 54:765–801
Copyright © 1985 by Annual Reviews Inc. All rights reserved

STRUCTURAL ANALYSIS OF GLYCOCONJUGATES BY MASS SPECTROMETRY AND NUCLEAR MAGNETIC RESONANCE SPECTROSCOPY

Charles C. Sweeley

Department of Biochemistry, Michigan State University, East Lansing, Michigan 48824

Hernan A. Nunez

Michigan Department of Public Health, Blood Derivatives Section, Lansing, Michigan 48909

CONTENTS

PERSPECTIVES AND SUMMARY

It has long been believed that the carbohydrate chains of glycoconjugates such as glycoproteins, glycolipids, and proteoglycans confer biological specificity at the cell surface, where they may be involved in adhesion, communication,

765

0066-4154/85/0701-0765$02.00

recognition, modulation of protein receptors, antigenic specificity, and regulation of cell growth. The far greater potential of oligosaccharides for structural diversity, as compared with polypeptides and oligonucleotides, provides a rich "vocabulary" which could accommodate the specificity inherent in these biological processes.

New methods for the isolation and structural analysis of complex carbohydrates, developed over the past two decades, have provided the opportunity to correlate biological function and structural diversity of the glycoconjugates. Our present understanding of the structures of glycoconjugates may be rudimentary, as compared with proteins, but enormous progress has been made since 1975 and well over 400 glycolipids, glycoproteins, and complex polysaccharides have been completely characterized in terms of the primary structures of their carbohydrate moieties. Three-dimensional structures of a few substances have been determined. It appears that primary structure will dictate the three-dimensional form of a complex carbohydrate, which in turn is an important aspect of substrate specificity in biosynthetic pathways.

The complete structural characterization of a complex carbohydrate typically requires purification of the glycoconjugate to homogeneity, liberation of the carbohydrate moiety from the remaining aglycone in some instances, and determination of the following structural aspects of the carbohydrate: 1. sugar composition, 2. anomeric configuration of each of the sugar residues, 3. conformation of the sugar rings, 4. the sequence or arrangement of the sugars, 5. the positions of glycosidic linkages between the sugars, 6. the conformation of the intersugar residue linkages, 7. the identity, points of attachment, and stereochemistry (if appropriate) of nonsugar substituents, and 8. the secondary structure or three-dimensional orientation of the carbohydrate chain.

This review will focus on two instrumental methods that have revolutionized work on the chemistry of complex carbohydrates, mass spectrometry (MS) and nuclear magnetic resonance spectroscopy (NMR). It is meant to serve as an overview for nonexperts about how these instruments have been employed to determine carbohydrate structures and will highlight recent advances, sensitivity and limitations of current methods, and some important applications. The review covers a period from about 1979 through the middle of 1984, with a few references to important historical advances and valuable reviews.

MASS SPECTROMETRY

Methylation Analysis

SIMPLE OLIGOSACCHARIDES   Methylation analysis is the traditional procedure used to establish the positions of glycosidic linkages and the ring size of sugar substituents (1–3). Free hydroxyl groups in the glycoconjugate are converted to methoxyl groups, acetamide residues are N-methylated, and free carboxyl groups are converted to methyl esters. The permethylation reaction is carried out in dimethylsulfoxide containing methyl iodide and sodium methylsulfinylmethanide (4, 5). The permethylated product is hydrolyzed in 80–95% acetic acid containing 0.5 N sulfuric acid (6, 7), yielding a mixture of partially methylated monosaccharides. The mixture is reduced with sodium borohydride or sodium borodeuteride and the partially methylated alditols are acetylated with acetic anhydride (Figure 1). The products contain O-acetyl groups at positions of substitution in the glycoconjugate and on carbon atoms attached to the ring oxygen. These products are readily analyzed by comparison of their retention behavior on several different GC columns (8) and by their characteristic fragmentation patterns, obtained by combined gas chromatography–mass spectrometry (GC-MS). For example, the electron impact (EI) mass spectrum of the partially methylated alditol acetate (A, Figure 1) obtained from the fucose at the nonreducing end of the trisaccharide would be expected to have fragment ions at m/z 117, 131, 161, and 175 because fission occurs predominantly between carbon atoms that are methoxylated. The internal N-acetylglucosamine residue yields a product (B, Figure 1) that can be deduced to have been derived from a 3-substituted pyranose form of GlcNAc by the characteristic fragment ions at m/z 45, 161, and 274 which arise by fission between carbon atoms containing one methoxyl group and one acetoxyl group with charge retention on the fragment with the methoxyl group (1). Mass spectra of products of N-acetylhexosamines also have a characteristic ion at m/z 158, derived from C-1 and C-2 of the partially methylated alditol acetate, and a companion ion at m/z 116, derived from m/z 158 by the elimination of ketene (6, 9, 10). Substituents with acetamido groups at unusual positions can be conveniently analyzed by methylation analysis (11).

Determination of which end of the partially methylated alditol acetate is C-1 is usually apparent. Ambiguities are readily resolved by reduction of the partially methylated monosaccharides with sodium borodeuteride, which places a deuterium atom on C-1 and increases the m/z values of fragment ions containing C-1 by one mass unit (C, Figure 1).

Methylation analysis of glycopeptides is complicated by the appearance on GC recordings of peaks derived from noncarbohydrate material. To circumvent this problem, the partially methylated hexitol acetates and N-acetylhexosaminitol acetates can be purified on a silica gel G column (12). Alternatively, high resolution fused silica capillary GC columns give excellent resolution of the partially methylated hexitol acetates and N-acetylhexosaminitol acetates. The retention behavior of 30 products commonly encountered by methylation analysis of oligosaccharides from N-linked glycoproteins are reported on several capillary columns (13). In another report the capillary GC profile and relative retention times are given of 14 partially methylated alditol acetates from the oligosaccharides of blood group H active human ovarian cyst mucin (14).

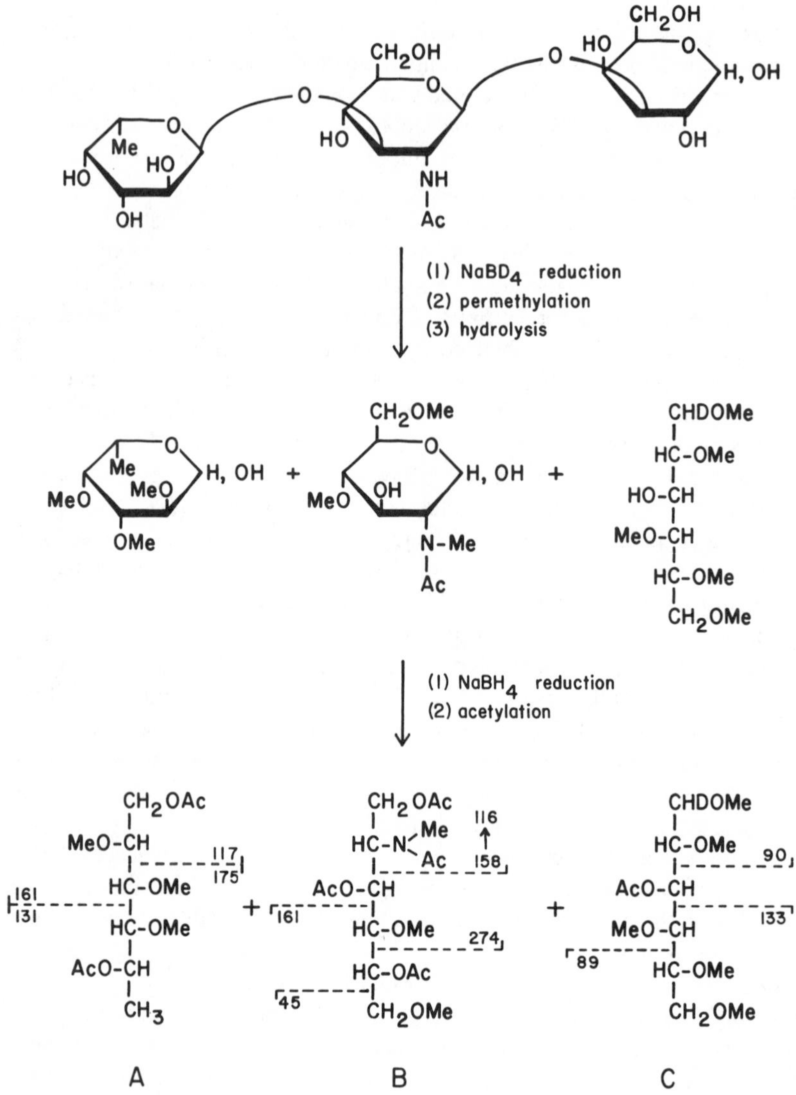

Figure 1   Stepwise conversion of Fuc(α1–3)GlcNAc(β1–3)Gal to partially methylated alditol acetates, as described in (3).

The gas chromatographic retention times and mass spectra of several partially *O*-methylated derivatives of the methyl ester methyl ketoside of *N*-acetyl-*N*-methylneuraminic acid have been reported (15), extending the application of methylation analysis to sialic acid residues of glycoconjugates. In this case, the permethylated glycoconjugate is degraded by methanolysis to partially methylated methyl glycosides, which are analyzed directly by GC-MS as *O*-trimethylsilyl or *O*-acetyl derivatives. The methane chemical ionization mass spectrum of the *O*-trimethylsilyl derivative has an intense ion at m/z 610 (MH-16)[+], which has been used to determine sialic acid content in the 10–1000 ng range by GC-MS (16). Use of partially methylated methyl glycosides is also advocated for the GC-MS analysis of permethylated oligosaccharides from glycoproteins, using the GC retention times and mass spectra of 52 reference compounds of galactose, mannose, glucose, and *N*-acetylglucosamine in making structural assignments (17). Multiple peaks can be obtained from α and β anomers in this procedure, which can be both an advantage (more than one characteristic retention time for each monosaccharide) and a disadvantage (complex GC patterns). Selected ion monitoring by chemical ionization (ammonia reagent gas) GC-MS has also been used with a micromethylation procedure and high resolution GC on glass capillary columns to improve the sensitivity of methylation analyses of *N*-linked glycoprotein oligosaccharides (18).

STRATEGIES FOR COMPLEX POLYSACCHARIDES   The traditional procedure for methylation analysis (see above) fails to distinguish between 4-linked aldohexopyranose residues and 5-linked aldohexofuranose residues, both of which would yield 4,5-di-*O*-acetyl hexitols, and the mixtures become very complex with higher-molecular-weight polysaccharides. Two alternative approaches have been described and have significant advantages over the standard procedures of methylation analysis. Reductive cleavage of permethyl-ated glycans with boron trifluoride etherate (BF_3-Et_2O) and trimethylsilyl trifluoromethanesulfonate as the catalyst in the presence of triethylsilane yields stable 1,5-anhydroalditols, as shown in Figure 2 (19). The ring forms of the monosaccharide residues are preserved and linkage positions are readily estab-lished by GC-MS after acetylation of the free hydroxyl group(s) in the pro-ducts. It also appears possible that partial reductive cleavage of permethylated polysaccharides will give mixtures of partially methylated 1,5-anhydroalditols

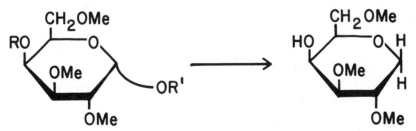

Figure 2   Reductive cleavage of oligosaccharide with boron trifluoride etherate and trimethylsilyl trifluoromethanesulfonate.

of di-, tri- and tetra-saccharides that can be directly analyzed by GC-MS. The full scope of this interesting new approach will be dependent upon access by investigators to a bibliography of retention behavior of reference compounds.

The other approach takes advantage of HPLC to separate mixtures of partially methylated oligosaccharides derived from a parent complex polysaccharide (20–22). The strategy here is to degrade the polysaccharide into somewhat simpler oligosaccharides by controlled partial hydrolysis, separate them, and analyze each fragment by the standard procedure of methylation analysis. To distinguish between free hydroxyl groups generated by the initial partial hydrolysis from those formed when individual oligosaccharides are further degraded to partially methylated alditols, the intermediate partially methylated oligosaccharides are ethylated. This can be understood in a conceptual way by examination of a hypothetical trisaccharide alditol (Figure 3) formed from a parent methylated polysaccharide by partial hydrolysis, borodeuteride reduction, and ethylation. The presence of an O-ethyl group at C-4 on the partially methylated glucose at the nonreducing end of the trisaccharide derivative indicates that this glycose unit was substituted at C-4 in the parent polysaccharide. The O-ethyl group at C-4 of the substituted glucitol residue of the trisaccharide derivative indicates the presence of a glycosidic linkage to this carbon atom in the parent polysaccharide, whereas the O-ethyl groups at C-1 and C-5 of the glucitol are indicative of a 1,5-pyranose ring before reduction of the partially methylated trisaccharide. After HPLC purification of the methylated and ethylated trisaccharide derivative, complete hydrolysis yields partially methylated, partially ethylated alditols that can be acetylated and analyzed in the usual manner by GC-MS. An important further advantage of this approach is that the partially methylated partially ethylated trisaccharide alditol can also be analyzed directly by GC-MS to establish the number of glycose units and their arrangement (20, 21). Higher-molecular-weight intermediate peralkylated oligosaccharide alditols may be analyzed by direct probe mass spectrometry (22).

APPLICATIONS   There are numerous examples in the recent literature where methylation analysis was employed to determine the positions of glycosidic linkages in glycolipids, polysaccharides, glycoproteins, and other glycoconjugates. The references given here are illustrative of the breadth of applications but are by no means inclusive of all reports. Applications are reported for glycolipids, glycoproteins, and miscellaneous carbohydrates in Table 1, arranged according to the types of parent compounds that were investigated and their biological sources.

Mass Spectrometry of Oligosaccharides and Glycoconjugates

Improvements in the mass range of commercial mass spectrometers have encouraged research on the direct analysis of intact glycoconjugates and

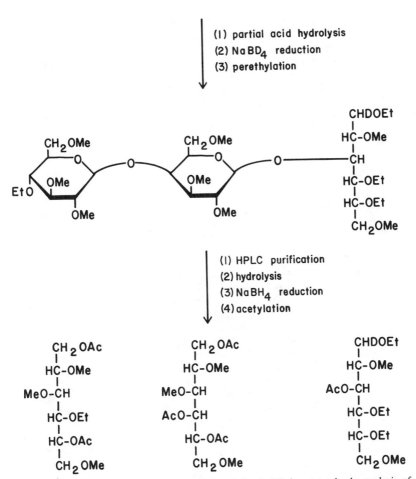

Figure 3 Use of partially methylated, partially ethylated alditol acetates in the analysis of complex polysaccharides, as described in (20–22).

oligosaccharides, in the underivatized state or as peralkylated or trimethylsily-lated derivatives. Two of the techniques, traditional electron impact ionization (EI) and chemical ionization (CI) of samples introduced by GC or via a direct probe inlet, require volatilization as a prerequisite of ionization. These two methods of ionization are therefore restricted to samples that can successfully be volatilized. As a rule, oligosaccharides must be derivatized, and those with greater than about 12 monosaccharide residues will probably have vapor pressures that are too low, even in the derivatized form.

Several new methods of direct ionization have been employed to study the

Table 1   Methylation analysis of various glycoconjugates

Substances analyzed	References
Glycosphingolipids	
Glycosphingolipids of rat intestine	23–28
Fucoganglioside of rat hepatoma	29
Sulfated gangliotriaosylceramide of rat kidney	30
9-O-Acetyl-tetrasialoganglioside of mouse brain	31
9-O-Acetyl-trisialoganglioside of mouse brain	32
Gangliosides of rabbit thymus	33
I-Active Gly_{10} Cer from rabbit erythrocytes	34
VIII^3-α-NeuAc-Lc_8Cer from rabbit skeletal muscle	35
Isoglobotriaosylceramide of canine intestine	36
Disialoganglioside of pig skeletal muscle	37
Disialogangliosides of bovine adrenal medulla	38
Fucosyl ganglioside of bovine thyroid gland	39
Neutral glycosphingolipids and gangliosides of bovine thyroid gland	40
2-O-Acyl-galactosylceramide of whale brain	41
Pentaglycosylceramide of green monkey kidney	42
H-Active pentaglycosylceramide of human pancreas	43
B-Active difucosylheptaglycosylceramide of human intestine	44
Lewis-active fucolipids of human and canine intestine	45
Glycosphingolipids of human meconium	46
Lewis- and H-active glycosphingolipids of human plasma	47, 48
A-Active glycosphingolipids of human erythrocytes	49
Globo-type glycosphingolipids of human teratocarcinoma cells	50
Gangliosides of chronic myelogenous leukemia cells	51
Neutral glycolipids of human myeloid leukemias	52
Glycosylceramides from fresh-water bivalve, *Hyriopsis schlegelii*	53–56
Glycosylceramides of fresh-water bivalve, *Corbicula sandai*	57
Glycolipids of star fish, *Asteriua pictinifera*	58
Glycolipids of *Lucilia caesar* larvae	59, 60
Glycosyl-inositolphosphoceramide from tobacco leaves	61
Glycoproteins	
Polysialoglycoproteins of Pacific salmon eggs	62
Fetuin	63
Bovine rhodopsin	64
Bovine plasma cold-insoluble globulin	65
Band 3 glycoprotein of human erythrocytes	66
Human cervical mucin	67
Human thrombin	68
Miscellaneous Polysaccharides	
Polysaccharide Type 18C of pneumococcal capsule	69
Polysaccharide of *Klebsiella* K33 capsule	70
Lipoligosaccharides of *Mycobacterium kansasii*	71
Konjac glucomannan of *Amorphophallus konjac*	72
Heptasaccharide from pectic rhamnogalacturonan II	73
N- and O-Linked oligoaccharides of Swarm rat chondrosarcoma proteoglycans	74
Urinary oligosaccharides from mannosidosis patients	75–77
Urinary oligosaccharides	78

underivatized glycoconjugates and oligosaccharides. The important feature of these methods is the desorption of samples in an ionized form directly from the condensed phase (solid or liquid). Thus, the term desorption ionization is generally employed to describe the mechanisms by which ionization occurs. The several methods of desorption ionization used to characterize glycoconjugates and other biomolecules of intermediate molecular weight are secondary ion mass spectrometry (SIMS), fast atom bombardment (FAB), plasma desorption (PD), field desorption (FD), electrohydrodynamic ionization (EHMS), and thermal desorption (TD). A recent report (79) on the plasma desorption mass spectrometry of bovine insulin [calculated mass (CM) = 5733.5], a cobra neurotoxin (CM=7821), porcine proinsulin (CM=9082), cytochrome C (CM=12384), bovine ribonuclease (CM=13682), and porcine phospholipase A_2 (CM=13980) dramatically illustrates the potential of these techniques for high-molecular-weight samples. Successful mass spectrometry of oligosaccharides of such high molecular weight has not been reported yet, but recently reported FAB mass spectra indicate a broad range of applications and a similar molecular weight capability. Recent reviews describe several methods of desorption ionization in detail and discuss important applications of these methods (80–85).

ELECTRON IONIZATION MASS SPECTROMETRY   Much of the structural work involving direct analysis of oligosaccharides and glycosphingolipids has been carried out by electron ionization (EI) of volatile derivatives obtained by methylation, acetylation, or trimethylsilylation. Permethylation is the most laborious of the methods of derivatization and requires rigorously purified solvents to avoid artifactual contributions to the mass spectra, but is the preferred approach because methyl groups contribute the least to the molecular weight. Oligosaccharides containing up to five or six monosaccharide residues can be analyzed by GC-MS of the permethylated reduced (alditol) derivatives (73, 77, 78, 86–91). Difficulties are encountered with small oligosaccharides containing N-acetylhexosamine residues that increase the retention times on GC. Two methods to overcome this problem are transamidation of the glycopeptides (from N-linked glycoproteins), glycosphingolipids, or oligosaccharides with a mixture of trifluoroacetic acid and trifluoroacetic anhydride (1:100) at 100°C, which gives N,O-pertrifluoroacetylated oligosaccharides in high yields (90–92). O-Trifluoroacetyl groups are easily removed with methanol and the further conversion to N-trifluoroacetyl permethyl derivatives gives products ready for GC-MS analysis. The N-trifluoroacetyl group stabilizes positively charged fragment ions and enhances the value of EI mass spectrometry for structural analyses of oligosaccharides (88). Alternatively, the amide linkages in hexosamine-containing glycoconjugates can be removed by N-deacetylation in hot, alkaline dimethyl sulfoxide (93) followed by deamination with nitrous acid (87).

When higher-molecular-weight oligosaccharides cannot be separated by GC, TLC (89) or HPLC (22) can be used to purify individual permethylated oligosaccharide alditols prior to direct probe EI mass spectrometry. The highest masses observed so far by direct probe EI mass spectrometry of permethylated derivatives have been in the mass spectra of a permethylated and reduced blood group B–active dodecaglycosylceramide (m/z 2835 to 2947) (94) and the permethylated N-linked high mannose oligosaccharide alditol, Man₉GlcNAc-ol (m/z 2099) (95, 96). Still larger molecules may be studied by EI-MS although ions in the molecular region may not be obtained. For example, the terminal 13 sugars of a 25-sugar blood group B-active glycolipid were successfully sequenced by this method (97). About 100 μg of sample must be used with EI-MS of the permethylated oligosaccharides and permethylated glycosphingolipids because the fragment ions at high mass are not very intense in these spectra.

The sequence of monosaccharide units of a permethylated oligosaccharide or oligosaccharide alditol can be directly deduced from the EI mass spectrum. The most valuable fragment ions (A series) are formed by random homolytic cleavage of glycosidic bonds with positive charge retention on the ring oxygen, as shown in Figure 4. The fragment ion at m/z 219 indicates that the nonreducing terminal monosaccharide is a hexose. Oligosaccharides with terminal fucose and other deoxyhexoses have intense ions at m/z 189 and those with terminal N-acetylhexosamines have intense ions at m/z 260. Similarly, cleavage of the internal glycosidic linkage gives m/z 423 for the nonreducing terminal disaccharide residue. These cleavages can be expected for all glycosidic linkages in most oligosaccharides, allowing deduction of their sequences. Figure 4 shows a second series of ions at m/z 276 and 480, resulting from homolytic cleavage on the other side of the glycosidic oxygen atoms; this series also gives sequence information.

Very little additional structural information can be gleaned from the EI mass spectra of permethylated derivatives. In a few instances the positions of

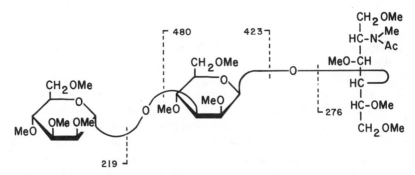

Figure 4   Electron impact ionization of permethylated oligosaccharide alditol yields fragment ions that are predominantly formed by cleavages at glycosidic linkages.

specific glycosidic linkages can be inferred from unique fragment ions (such as so-called J ions) or the intensities of ions resulting from the loss of methanol from the A series. The fragmentation processes and structures of these ions have been discussed in some detail (84, 87, 88, 90, 91, 95, 96, 98, 99).

The sequence information obtained by EI-MS of permethylated derivatives has been used in support of structural studies of oligosaccharides derived from glycopeptides by hydrazinolysis and deamination (100), mannose-containing oligosaccharides in urine from patients with mannosidosis (77, 95, 96), and the carbohydrate chains of prothrombin (101) and transferrin (102). The partial structures of a family of inositol- and glycerol-containing oligosaccharides of human urine were derived from EI mass spectra of methylated derivatives (103). By far, the greatest number of applications of EI mass spectrometry relate to sequence analysis of glycosphingolipids. The mass spectra of methylated and lithium aluminum hydride-reduced and methylated neutral glycosphingolipids and gangliosides provide carbohydrate sequence information as well as the molecular composition of the ceramide group in terms of their sphingoid bases and fatty acids (104, 105). Applications include structural studies of the glycosphingolipids of human meconium (106), a triantennary blood group I-active pentadeca-glycosylceramide from rabbit erythrocytes (107), a human tumor–associated glycolipid antigen (108), blood group ABH and Lewis antigens of human kidney (109), human lymphocyte and neutrophil gangliosides (110), glycolipids of spleen and liver from patients with Gaucher's disease (111), a tetrasialoganglioside from human brain (112), and rat small intestinal glycosphingolipids (23–26, 28, 113–115). The sequences were determined by EI-MS of oligosaccharides derived from neutral glycosphingolipids of murine B cell hybridomas by trifluoroacetolysis and permethylation (90).

CHEMICAL IONIZATION MASS SPECTROMETRY   Chemical ionization (CI) mass spectrometry generally affords higher intensities of molecular ions and fragment ions in the high mass range, as compared with electron impact mass spectrometry. This technique, like EI, requires volatilization prior to ionization and therefore has been applied primarily to the analysis of derivatized oligosaccharides and glycolipids. Exceptions are the CI mass spectra recorded for underivatized menthyl glycosides (116), glycosphingolipids with one to four sugar residues that were directly introduced from HPLC in a combined HPLC-MS system (117), some glucuronides introduced with a special thermospray interface from HPLC (118) and also by moving belt LC-MS with ammonia as reagent gas (119), and a synthetic lactosylceramide analyzed by direct insertion of a polyimide-coated wire containing the sample into a chemical ionization reagent gas plasma (120). This method, known as direct chemical ionization (DCI), has been discussed in some detail in a recent review (84).

Most investigations of CI mass spectra of higher molecular weight carbohydrates have employed permethylated derivatives. The DCI mass spectrum of a methylated glucose tetrasaccharide alditol, obtained with ammonia as the reagent gas, exhibited an intense molecular adduct ion at m/z 897 $(M+NH_4)^+$ along with fragment ions of lower intensity that could be used to determine the sugar sequence (84). Permethylated and permethylated-reduced glycosphingolipids and gangliosides have been analyzed for their sugar sequences by direct inlet CI mass spectrometry (121–124). Sialo-oligosaccharides liberated from gangliosides by ozonolysis have also been analyzed as permethylated alditols by CI mass spectrometry (125). Permethylated oligosaccharides coupled with a fluorophore at the reducing end have been separated by HPLC and then sequenced by DCI mass spectrometry (126). This approach, when coupled with the partial reductive cleavage of polysaccharides with boron trifluoride etherate and triethylsilane (19), affords an excellent approach for the purification and sequence analysis of high-molecular-weight polysaccharides. Exposure of a permethylated β-cyclodextrin with seven monosaccharide residues to reductive cleavage for four minutes gave a mixture of the intact heptamer and the di-, tri-, tetra-, penta-, and hexa-saccharide derivatives which were completely separated by HPLC and gave molecular adduct ions $(M+NH_4)^+$ by ammonia CI-MS (126).

FIELD DESORPTION MASS SPECTROMETRY   Field desorption (FD) mass spectrometry does not require prior derivatization, since ionization occurs concurrently as the sample is desorbed from an activated emitter wire in a high electric field (127). The FD mass spectrum of underivatized β-cyclodextrin has an intense cluster ion $(M+Na)^+$ at m/z 1157 for the intact cyclic heptasaccharide and two fragment ions for loss of one sugar residue $(M+Na-162)^+$ at m/z 995 and loss of three sugar residues $(M+Na-486)^+$ at m/z 671 (128). A laser beam was used in this case as an alternative to direct heating of the emitter wire. The FD mass spectrum of a partially methylated linear mannose oligomer of ten sugar residues had a cluster ion $(M+Na)^+$ at m/z 1801 and a cluster ion was observed at m/z 2506 in the FD spectrum of a mycobacterial methylmannose polysaccharide with 13 sugar residues (129). Underivatized glycosphingolipids with up to five sugar residues, including neutral glycosphingolipids, gangliosides, and sulfoglycosphingolipids, have given molecular cluster ions $(M+Na)^+$ by FD mass spectrometry (130–132). The FD technique is somewhat difficult in the sense that ions are generally formed for brief periods at a precise temperature of the emitter wire, called the best anode temperature, and recording of the molecular ion region during this transient high intensity ion current is difficult. Recent reviews have covered the theory and earlier applications of FD mass spectrometry in greater detail (83, 84, 133).

SECONDARY ION MASS SPECTROMETRY   Unlike field desorption mass spectrometry, secondary ion mass spectrometry (SIMS) does not require heating of the sample for its desorption and ionization. Accelerated ion beams (argon, xenon, or cesium) bombard the sample, which is placed onto a metal with or without other molecules such as glycerol or diethanolamine that provide a matrix for the sample. This is an increasingly popular method for the desorption ionization of nonvolatile biomolecules without prior derivatization. When matrices are not included with the sample, the intense production (sputtering) of Ni, Cu, or Ag ions from the bombardment of their salts with $(Ar)^+$ or $(Xe)^+$ causes desorption ionization of the substance of interest, primarily as cluster ions containing the metal substrate [e.g., $(M+Ag)^+$ and $(M+Na)^+$]. On the other hand, the presence of an organic matrix generally intensifies the protonated molecular ion $(M+H)^+$ and may result in the appearance of cluster ions containing the matrix molecule $(M+H+glycerol)^+$.

Molecular SIMS mass spectra of several kanamycins gave molecular weight information but nothing about the sequence of the carbohydrate moiety (134). Similarly, the SIMS mass spectrum of γ-cyclodextrin, recorded using a diethanolamine matrix, had intense cluster ions at m/z 1402 $(M+H+die-thanolamine)^+$ and 1319 $(M+Na)^+$ but it was not clear that other structural information could be obtained from the mass spectrum (135). On the other hand, the SIMS mass spectrum of the tetrasaccharide stachyose in admixture with sodium chloride (1:1) gave m/z 365 $(M+Na)^+$ and a series of cationized fragment ions (136). The SIMS mass spectra of some underivatized neutral glycosphingolipids have recently been reported (137), and the technique has been used to obtain cationized molecular ion peaks of sialo-oligosaccharides of fish egg polysialoglycoproteins (138).

FAST ATOM BOMBARDMENT MASS SPECTROMETRY   Fast atom bombardment (FAB) mass spectrometry involves the transfer of kinetic energy from a beam of highly energetic atoms such as argon or xenon to a matrix such as glycerol and then to the sample. This technique is therefore closely related to SIMS, which uses a beam of ions rather than atoms. FAB has gained wide popularity and is now used in many applications for the structural analysis of complex carbohydrates. Like the other kinds of desorption ionization, it has the advantage that nonvolatile polar substances usually give intense peaks in the molecular ion region. Another advantage, especially as compared with field desorption (FD), is that a high intensity of ions from the sample can be maintained for prolonged periods of time, even with rather small amounts of sample. Fragment ions indicative of structure may be observed under appropriate conditions. The degree of fragmentation appears to be dependent upon structural parameters that are not entirely understood. FAB spectra have many of the same features as SIMS, and duration of a sample ion beam is about the same in both techniques. Typical sample sizes for FAB are about one nmol dissolved in one μliter of matrix. Glycerol and other substances have been used as the matrix material (80–82, 84).

The negative ion FAB mass spectrum of underivatized rabbit blood group B-active pentaglycosylceramide is shown in Figure 5. The molecular weight is apparent from the intense ion at m/z 1497 (M-H)⁻ and this m/z value can be used to deduce the size of the carbohydrate chain [four hexose (4×162) and one N-acetylhexosamine (1×203) added to 646 (ceramide) totals 1497]. Random fragmentation at glycosidic bonds with negative charge retention on the oligoglycosylceramide residue gives a series of fragment ions at m/z 1335, 1173, 970, 808, and 646 which can be used to deduce the sequence of glycose units in the glycolipid.

Comparison of the EI and FAB mass spectra of permethylated gangliotetraosylceramide from ox brain indicates that FAB spectra may yield ions in the molecular weight region that are as much as 50 times more intense than those recorded by the EI technique (139). A comparison has been made of the positive- and negative-ion FAB and EI spectra of a tetrasaccharide (Galβ1-4GlcNAcβ1-3Galβ1-4GlcNAc) methyl glycoside, the permethylated and permethylated reduced forms, and the peracetylated reduced form (140). The FAB spectra had the advantage of being applicable to the nonderivatized sample and

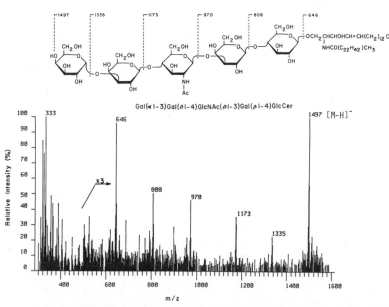

Figure 5   Negative ion FAB mass spectrum of a blood group B-active pentaglycosylceramide, run using triethanolamine as matrix. The mass scale was computer-calibrated using cesium iodide in glycerol. The mass spectrometer-computer system was a Kratos MS50/DS55 operating at 6 kV accelerating potential, with a fast atom bean of xenon produced using a saddle field ion source (Ion Tech Ltd.) operating with a tube current of 1.5 mA at an energy of 7 KeV. This bar graph plot (average of five scans) was obtained in 1983 by Ian Jardine, Department of Pharmacology, May Clinic, and is reproduced with his permission and copyright 1982 by the AAAS.

gave good molecular weight information but the abundance of fragment ions providing structural information varied from sample to sample. Measurement conditions, especially proton affinity of the matrix, were important in their effects on the FAB spectra. The more suitable mode (+ or −) for sequence analysis of glycolipids depends on the relative proton affinity of the saccharide moiety to that of the aglycone. Thus, negative-ion FAB is preferred for the sequence analysis of gangliosides, in which negatively charged fragment ions always involve the sialo group(s). The positive mode may work better, however, with other kinds of glycoplipids.

It was reported that FAB spectra were not easily obtained with a series of peracetylated L-fucose-containing milk oligosaccharide alditols unless the target mixture of sample and glycerol matrix was heated (139). Permethylated derivatives were recommended for best results in sequence analysis by FAB mass spectrometry. The addition of a salt to a matrix-sample mixture (dosing or doping) sometimes improved the formation of molecular ion signals in the positive ion and negative ion modes (141). These authors reported a positive ion FAB spectrum of a mixture of reduced permethylated glucose polymers (50–100 pmol of each component) and glycerol, dosed with thioglycerol and ammonium thiocyanate; ions between 5000 and 6700 were observed for components with 25–32 glucose residues.

The difference between failure and success in FAB mass spectral studies of carbohydrates may depend upon the choice of the matrix material as well as dosing and appropriate choice of positive or negative ion mode. The positive ion FAB mass spectrum of underivatized GM1 ganglioside in glycerol contained no high mass ions whereas use of triethanolamine as a matrix with a small amount of tetramethylurea gave a series of relatively intense cationized species such as $(M+Na)^+$ and $(M+2Na-H)^+$ (142). The disialoganglioside, GD1a, similarly gave peaks at m/z 1859–1909 for various cationized molecular ion species. Only two µg of a ganglioside were required for these analyses. The negative ion FAB spectrum of GM1 ganglioside appears to be more suitable, both in terms of the intensity of the molecular ion $(M-H)^-$ and the appearance of sequence ions (143). The best negative ion FAB spectra of gangliosides were obtained with the mixture of triethanolamine and tetramethylurea as matrix, which gave molecular ions of the trisialogangliosides, GT1a and GT1b, at m/z 2126 and 2154 for two forms with sphingosine (C-18) and eicosasphingosine (C-20) as the long-chain bases of the ceramide moiety (144). Unique fragment ions allowed these two isomers to be distinguished from their FAB spectra. FAB mass spectrometry was used to study the distribution of gangliosides in various rat tissues (145) and the structures of I/i glycosphingolipids of rabbit erythrocyte membranes (146).

The N-glycosidically linked oligosaccharides derived from hen ovomucoid were analyzed by FAB mass spectrometry, which confirmed the structures deduced from permethylation analysis and 500 MHz proton NMR spectroscopy (147). Four minor components up to an undecasaccharide were uniquely found by FAB MS of the reduced permethylated mixture; the highest mass recorded was at m/z 2861. The fragmentation processes that occur in positive- and negative-ion FAB mass spectrometry of underivatized oligosaccharides and glycopeptides from various glycoproteins have been investigated (148). Positive-ion FAB mass spectrometry was used to establish the molecular weights of urinary oligosaccharides obtained from guinea pigs, rats, and sheep that had been treated with swainsonine (149).

A detailed investigation was made of the negative-ion FAB mass spectral data obtained with samples of a mycobacterial O-methyl-D-glucose polysaccharide (MGP) and a mixture of acylated O-methyl-D-glucose lipopolysaccharides (MGLP) (150). A molecular ion peak for MGP at m/z 3513 confirmed that MGP contains 8 glucose, 12 O-methylglucose, and 1 glyceric acid unit. Four species of MGLP gave molecular ions between m/z 3794 and m/z 4036, confirming the distribution of neutral and acidic acyl groups and information about their specific locations. An earlier report described positive-ion FAB spectra (thioglycerol matrix) and negative-ion FAB spectra (glycerol matrix) of an enzymatic digestion product of MGP with four fewer sugar residues (151).

Lipid A obtained from the lipopolysaccharide of *Salmonella typhimurium* gave negative-ion FAB spectra with (M-H)$^-$ ions at m/z 1716 and 1730 for two species of monophosphorylated lipid A with the molecular formulae of $C_{94}H_{177}O_{22}H_2P$ and $C_{95}H_{179}O_{22}N_2P$, respectively (152). The repeating oligosaccharide of pneumococcal capsular polysaccharide (type 18C) was obtained by aqueous HF treatment; FAB mass spectrometry in a glycerol matrix (doped with sodium and potassium chlorides) yielded cationized molecular ions of a pentaglycosylglycerol monoacetate (69).

The molecular weight of an elicitor from soybean cell walls and citrus pectin was established by negative-ion FAB mass spectrometry; an (M-H)$^-$ molecular ion at m/z 2519 corresponded to a di-pentafluorobenzyl derivative of a dodecasaccharide composed of galacturonosyl residues (153). The cationized molecular ion at m/z 1371 from a positive ion FAB mass spectrum was used in the structural characterization of a complex heptasaccharide derived from pectic rhamnogalacturonan II of suspension-cultured sycamore cells (73). FAB mass spectrometry and ^{13}C NMR spectroscopy were used to unambiguously show that β-D-(1-2)-linked D-glucans secreted by *Rhizobium leguminosarum* and *Agrobacterium tumefaciens* C58-C1 are unbranched, cyclic structures with 17–24 sugar residues (154). Oligosaccharides isolated from human milk (155) and urine (156) have been analyzed by FAB mass spectrometry.

One of the few papers on the structural analysis of glycosaminoglycans by mass spectrometry has positive and negative ion FAB spectra of sulfated di-, tetra-, and hexa-saccharides from chondroitin sulfate (157). Polymers of 4-O-

sulfate and 6-O-sulfate isomers could be distinguished, and the presence of products containing more than one sulfate residue per repeating unit of GlcU-GalNAc could be detected. This study indicates that FAB mass spectrometry may be uniquely appropriate for detailed structural studies of sulfated glycosaminoglycans.

NUCLEAR MAGNETIC RESONANCE SPECTROSCOPY

Analysis of Simple Sugars

Carbohydrate molecules have built-in probes for studying their structure by NMR spectroscopy; i.e., their normal elemental compositions of ^1H, ^{13}C and occasionally ^{31}P or other nuclei provide NMR signals that encode the structure. To simplify the structural problem, the vast majority of glycoconjugates contain sugar residues in the pyranosidic form, which adopts a very rigid and stable chair conformation (158, 159). In addition, the glycosidic linkages between sugar residues, for steric and electronic reasons, allow the existence of a few, perhaps only one, predominant conformer under conditions normal to biological systems (158–160). For example, most β1-4(D) glycosidic linkages between two glycopyranosidic residues appear to exist in aqueous solutions as depicted in Figure 6, despite the infinite number of other possible three-dimensional structures. This situation, which seems to hold in more complex carbohydrates (161), appears to be the reason for the extraordinary biological specificity of glycoconjugates (162). It is, at the same time, one of the basic reasons that NMR spectroscopy is so useful in determining the structures of carbohydrates; i.e., each nucleus is in a relatively fixed electronic environment and in a relatively fixed spatial relationship to other nuclei. As will be discussed in the next few paragraphs, this is revealed in the most useful NMR parameters: chemical shifts, areas under the peaks, coupling constants, nuclear Overhauser enhancement (NOE), and spin-lattice relaxation times (T_1).

CHEMICAL SHIFTS   The chemical shift is a representation of the frequency at which a given nucleus, placed in an external magnetic field, resonates. In an NMR spectrometer, the external magnetic field is meant to be equal (homogeneous) for all the nuclei in the sample. The actual magnetic field that a given nucleus sees, however, is the external magnetic field modified by that created by the electron density ("shielding") surrounding that nucleus (163). This is the reason why chemical shift spectra are highly informative about the chemical structures as well as inter- and intra-molecular interactions (164). As the chemical shifts and coupling constants of the common sugars and their derivatives have been essential for using NMR spectroscopy for the structural determination of more complex carbohydrates, great emphasis has been placed on those assignments during the last decade. The efforts have been highly

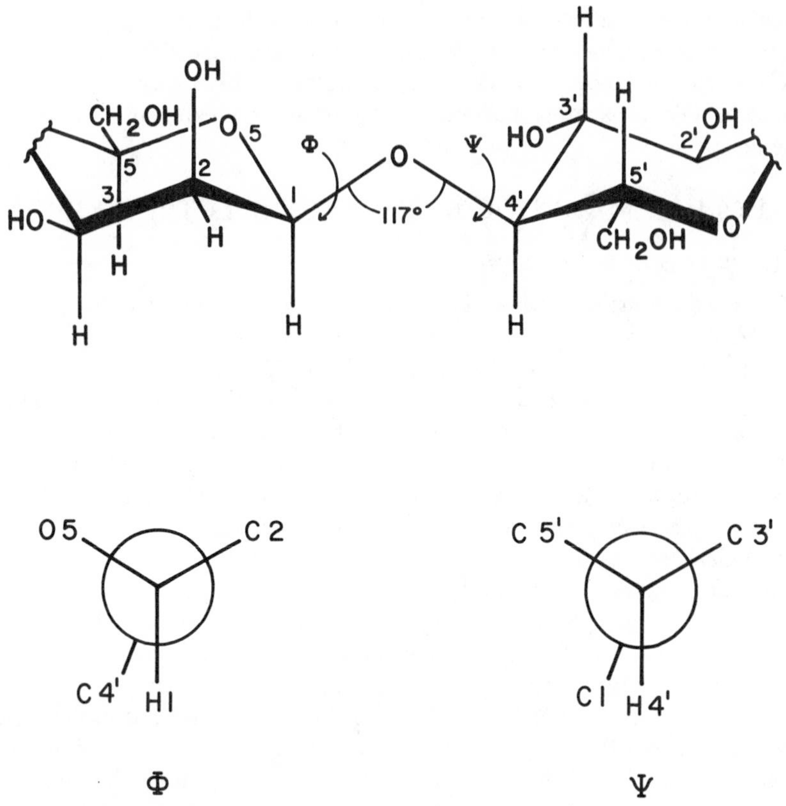

Figure 6   The most abundant conformers of hypothetical, β-(1–4)-linked D-glycopyranosides in the $^4C_1$ chair conformation.

successful; as summarized in two recent reviews (165, 166), the assignments are quite comprehensive. Although NMR spectra are unique for each compound, this "library" of spectra for known compounds is enormously useful for determining and corroborating structures of unknown but related compounds (167, 168). The library, however, is not a panacea in relatively crowded spectra. Nor is it essential in those cases in which the connectivity among, and the configuration of every nucleus in the molecule can be determined by NMR spectroscopy (see below).

COUPLING CONSTANTS   Coupling constants are the result of spin-spin coupling or splitting between two or more nuclei possessing magnetic moments. The magnitude of the splitting is the coupling constant, measured in Hz. The coupling constant is independent of the magnetic field. As the spin-spin coupling is thought to be mediated by the bonding electrons, the magnitude of the coupling constant is dependent on the distance, the chemical bond type, and the bond angle between the nuclei in question, in addition to the nature of the nuclei. The one- and two-bond ^1H-^1H coupling constants have been measured in numerous carbohydrates (165). The one- and two-bond ^{13}C-^1H coupling constants have been measured in numerous carbohydrates (165). The one- and two-bond ^{13}C-^1H and ^{13}C-^{13}C, coupling constants, however, have been measured in only a few mono- and oligosaccharides, mainly because of the low natural abundance (1.1%) of ^{13}C. To solve this problem, ^{13}C-enriched carbohydrates have been used (169). Several values of the two-bond ^{31}P-^{13}C coupling constant are available (170). Three- and four-bond couplings are especially informative about the spacial relationships between two nuclei. Well-established is the Karplus-type (171) angular dependence for the three-bond coupling in which there is a minimum value at 90° and maximum values at 0° and 180°. The Karplus relationship has been shown to be applicable to ^1H-^1H, ^{13}C-^{13}C, and ^1H-^{13}C three-bond coupling constants in carbohydrates (158, 159, 169). Since glycoconjugates are overwhelmingly formed by sugar residues in the pyranosidic form, the three-bond coupling constants are valuable for establishing both pyranosidic ring configuration and ring conformation. In the pyranosyl system (Figure 6) the ^3H axial-H axial (180°) and the ^3H axial-^3H equatorial (60°) have values of approximately 8 Hz and 2.5 Hz, respectively. Using this information, the anomeric conformation and the pyranosidic structure can be unambiguously assigned (172). The Karplus relationship for three-bond ^{13}C-^{13}C and ^{13}C-^1H coupling constants is also applicable to carbohydrates (169). Apparently, the homo- and hetero-nuclear four-bond coupling also has a stereo dependence, being of maximum magnitude when all the bonds are in, or close to, the same plane. The ^{31}P-^1H four-bond coupling through the glycosidic linkage (173) and the couplings equivalent to that between H1 and H4' through the glycosidic linkage in Figure 6 of several di- and tri-saccharides have been reported (174). Couplings across the glycosidic linkage have great value for establishing the linkage conformation and for establishing unambiguously the connectivity between the sugar residues in oligosaccharides. Up to now only NOE has been used for the latter purpose, as discussed in the next paragraph.

NUCLEAR OVERHAUSER ENHANCEMENT (NOE) AND SPIN-LATTICE RELAXATION TIME (T_1)   The population of nuclei in a given energy level can be changed by saturating a nearby nucleus. The change in population is reflected in the corresponding peak size, that is, the NOE. The effect is transmitted through the space (through the "lattice") and is a function of the distance between the nuclei involved (163, 164). Protons within an approximate distance of 3 Å will show measurable enhancements (maximum 15%) and the NOE will be observable in both inter- and intra-sugar residue protons. The

NOE is especially useful for determining sugar residue sequences and glycosidic linkage conformations since the effect produced by irradiating, for example, the anomeric proton will be observed across the glycosidic linkage only if a proton or protons of the aglycone fall within range. Although in the structure depicted in Figure 6 there are many combinations of ϕ and ψ in which all the aglycone protons are outside the NOE range, in all the carbohydrates studied to date, at least one proton, usually that proton attached to the linked sugar aglycone carbon, falls within the NOE range. Per se, these NOE data narrow the possible conformers to a few and assist in establishing the connectivity between sugar residues.

T_1 can be interpreted as the rate at which the NMR signal (peak) intensity decreases as a function of the time lapse between the sample irradiation and the beginning of the collection or storage of the free induction decay (FID) in a spectrometer computer. It is a function of the nearby environment of the observed nucleus. It gives, therefore, an indication of the spatial location of one nucleus with respect to other nuclei, usually within the same molecule (163, 164).

Traditionally, the assignments of chemical shifts and coupling constants in the most crowded regions of both ^{1}H and ^{13}C NMR spectra have been a major problem, even for simplest carbohydrates. To solve it, one or more techniques like homo- and heteronuclear decoupling (175) or isotope (176, 177) enrichment have been mandatory. The introduction of high magnetic field magnets (for example, 11.4 Tesla; 500 MHz for ^{1}H and 125 MHz for ^{13}C) and the application of two-dimensional (2-D) NMR techniques to carbohydrates have decreased the difficulties to the extent that the deduction of complete structures can be made solely from 2-D NMR spectral information. This is because encoded in the 2-D NMR spectra is the connectivity information (see below) which allows the determination of the skeletal structure of sugar residues without additional information (178).

Several classes of 2-D NMR have been devised. The J-resolved (179), the J-correlated (180), and the 2-D NOE NMR (181) have been shown to be most useful for structural elucidation of carbohydrates. In the J-resolved type, one frequency axis (F_1) contains the coupling constant information and the other frequency axis (F_2) contains the chemical shifts. In the correlated 2-D NMR both frequency axes contain the chemical shifts and the coupling information either from the same (for example, ^{1}H) or from different (for example, ^{1}H and ^{13}C) nuclei. This gives rise to the diagonal peaks. Due to the J (scalar) coupling there are also nondiagonal (cross) peaks that indicate J coupling (connectivity) between the various diagonal peaks. Thus, 2-D homo- or heterocorrelated NMR can replace the tedious and sometimes impossible homo- or heterodecoupling experiments (182).

The 2-D SECSY (spin-echo correlated) and J-resolved NMR spectroscopy have been used quite successfully in complex carbohydrate structure analysis as reviewed in the section on applications of NMR to complex carbohydrates (see below). For simpler carbohydrates, these techniques are indeed able to provide completely resolved chemical shifts and coupling constants and unambiguous assignments, a situation that could only have been achieved a few years ago with multiple isotopic substitutions. Thus, for example, a complete set of ^1H and ^{13}C chemical shifts and the ^{13}C-^1H, ^1H-^1H and hydroxyl hydrogen ^1H-^1H coupling for α and β glucopyranose in dimethylsulfoxide (183) and in D_2O (184), and the complete, self-consistent ^1H and ^{13}C spectra assignments for cellobiononitrile-octaacetate (185) have been published. Other 2-D NMR spectroscopic studies include ^1H and ^{13}C NMR spectra of ^2H and ^{13}C-enriched monosaccharides (186), and ^1H NMR of methyl α-glucopyranoside (187), fluorinated monosaccharides (188), and a few polysaccharides in dimethylsulfoxide (189).

As stated above, the ^{13}C-^{13}C couplings provide valuable structural information (159, 169), but because of the low natural abundance of ^{13}C (1.1%) and its low sensitivity (1/64 of that of ^1H) it is difficult to obtain. Two types of pulsed NMR techniques have been used to obtain this information using natural abundance ^{13}C NMR spectroscopy: polarization transfer (190) and double-quantum transfer (191). Applied to carbohydrates, these techniques have allowed the assignments of all ^1H and ^{13}C chemical shifts and the ^{13}C-^{13}C connectivities within each of the sugar residues of four different per-acetylated trisaccharides of glucose (192) and cellobiose (193).

Nuclear Magnetic Resonance Spectroscopy of Complex Carbohydrates

The following examples, although not inclusive of all the reports, are illustrative of the structural information the NMR parameters and techniques discussed in the previous section can provide for glycoconjugates. The minimal amount of oligosaccharides necessary for obtaining these data falls approximately in the range 20–100 μg for ^1H– and 0.5–1 mg for ^{13}C 1-D NMR; 1 mg for J-resolved, 2 mg for NOE, 5–10 mg for ^{13}C-^1H connectivity, and 500 mg for ^{13}C-^{13}C connectivity measurements by 2-D NMR spectroscopy.

GLYCOLIPIDS   In one of the most recent series of studies, use was made of the homonuclear 2-D, J-correlated and 2-D, NOE ^1H NMR to obtain sugar-residue composition, anomeric configuration, aglycone structure, sugar-residue sequence, and intersugar linkage sites of several gangliosides with minimal assistance from chemical analyses of sugar composition or a chemical shift-coupling constant library from model compounds (194–196). The carbohydrate structure of G_{M1} is shown in Figure 7. To apply the 2-D-SECSY NMR technique, it was essential to know the chemical shift assignment of at least one peak per sugar residue in the 1-D or 2-D NMR spectrum. By using the connectivity properties of the contour density map (182), the peaks associated

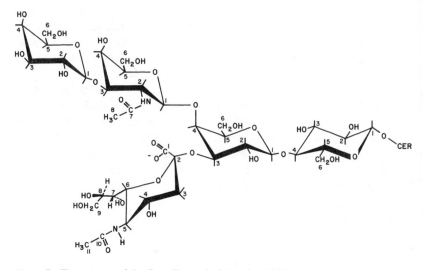

Figure 7   The structure of the G_{M1} oligosaccharide moiety (194).

with a given sugar residue could be assigned unambiguously within each one of the sugar residues. Because the J-connectivity across the glycosidic linkage was either small or lacking, the sugar residue sequence could not be obtained using the 2-D-SECSY NMR experiment. This information was obtained by 2-D NOE NMR spectroscopy by measuring the effect on the peak intensity of a specific proton of one sugar residue while "saturating" the anomeric proton belonging to another sugar residue. Each of these experiments invariably affected the proton of the linked aglycone carbon, thus establishing the sequence in an unambiguous manner. The sialic acid residue, which lacks the anomeric proton, could not be sequenced from information obtained by the NOE experiment, but glycosidation shift information could be used.

In a related study, the distances between the anomeric proton and two protons of the corresponding aglycone of globoside (GalNAcβ1-3Galα1-4Galβ1-4-Glcβ1-Ceramide) were determined by NOE techniques (196). The observed distances were 2.5 Å for GalNAcβH1-GalαH3, 2.8 Å for GalNAcβH1-GalαH4, 2.5 Å for GalαH1-GalβH4, and 3.2 Å for GalαH1-GalβH5. These distances can be achieved from many combinations of ϕ and ψ (Figure 6) and consequently are not sufficient for establishing the glycosidic linkage conformation. The study, however, unambiguously determines the primary structure. The primary structure of ceramide pentadecasaccharide from rabbit erythrocyte membranes was partially determined by 2-D, J-resolved, and J-correlated ^{1}H NMR spectroscopy (197). The connectivity information could not be established for all the nuclei because of the complexity (overlap) of the contour plot, but comparison with spectra of related compounds established that in all probability the pentadecasaccharide has a tri-antennary structure (198). The primary structure determination was later completed using glycosylation shifts (199). Many of the glycolipid structure determinations have involved a systematic use of a library of spectra of structurally related compounds, from which spectral lines can be assigned. Thus, the ^{1}H and the ^{13}C NMR spectra of several glycolipids, as a series of related compounds, have been interpreted in terms of their primary structure (200–203). As a result, it is now possible to identify several members of the series from their ^{13}C NMR spectra ("fingerprint") alone. It has even been possible to predict the ^{13}C NMR spectrum of ganglioside G_{M1b}, a positional isomer of G_{M1} (204). Other glycolipid primary structures determined with assistance of NMR spectroscopy include glycosphingolipids of rat intestine (23, 24), sulfated gangliotriaosylceramide of rat kidney (30), pentaglycosyl-ceramide of green monkey kidney (42), globo-type glycosphingolipids of human teratocarcinoma cells (50), glycosylceramides from fresh-water bivalve *Hyriopsis schlegelli* (53, 55), and globotetraosylceramide from human urinary sediment epithelial cells (205).

GLYCOPROTEINS   Determination of the detailed structures of six major classes of carbohydrate chains that are *N*-linked to proteins of cell membranes and serum represents an elegant application of the 1-D ^{1}H-NMR techniques (206–208). The data obtained and the rationale of the approach were recently summarized (209). As shown in Figure 8, all of these carbohydrates have a common core of Manα1-6(Manα1-3)Manβ1-4GlcNAcβ1-4 (±Fucα1-6)GlcNAcβ1-Asn. Because of this structural similarity it was convenient to study the core NMR spectrum first to assign chemical shifts, coupling constants and the most stable conformations of the glycosidic linkages. This was accomplished using (*a*) the angular dependence of the three-bond ^{13}C-H coupling constants across the linkage with a 1-^{13}C enriched sugar, (*b*) the distance dependence of the NOE for already assigned protons at strategic positions, (*c*) the spin-lattice relaxation times (T_1) of specific protons, and (*d*) an extensive application of hard sphere calculations, which included the anomeric effect, to find the best combinations of φ and ψ (Figure 6) that fit all the experimental data. Then the more complex structures were sequentially studied in the same fashion, as more sugar residues were added to the carbohydrate core.

Several conclusions with important biological implications were drawn from these studies (209): 1. these compounds have well-defined three-dimensional structures with relatively few degrees of freedom; 2. the degrees of freedom that do exist are associated with the orientation of the Manα1-6Manβ1-arm, which in some classes of glycoproteins may have two orientations but in others is fixed; 3. the orientation of the Manα1-6Manβ1-arm may be stabilized by specific interactions between parts of the molecules remote in the primary structure; and 4. the hard sphere calculation approach to obtain the minimum

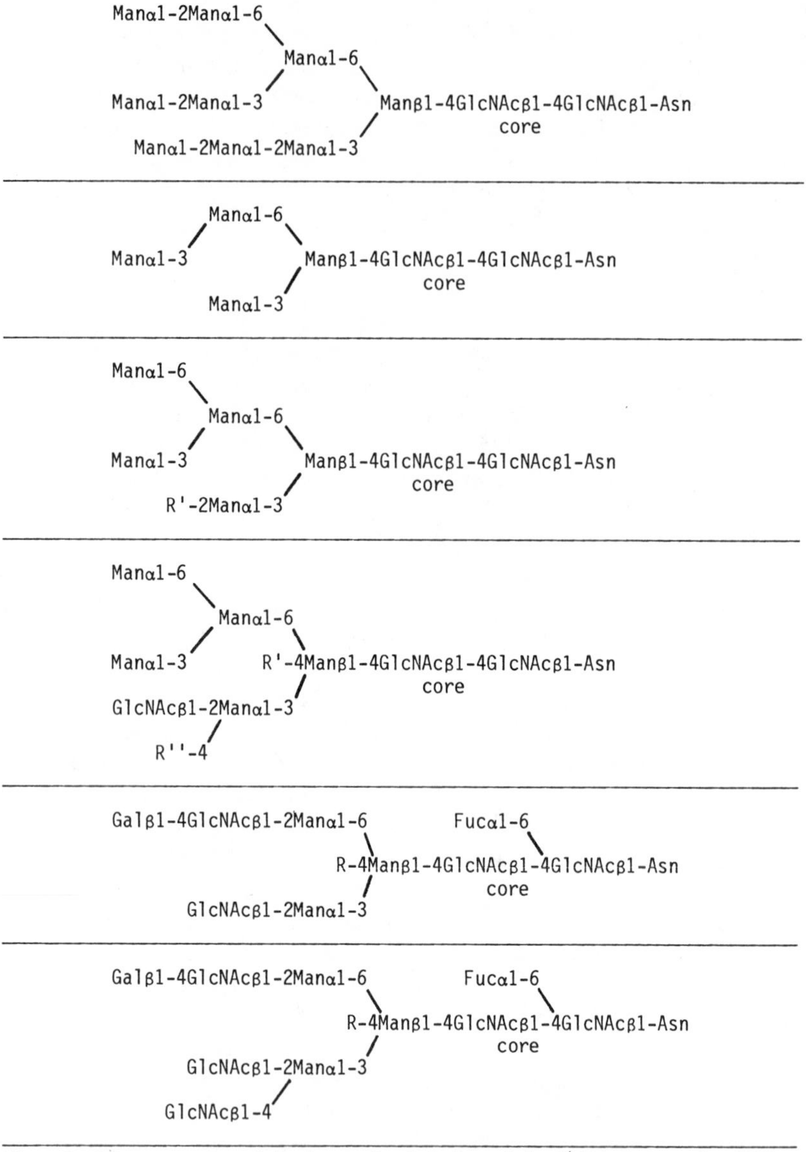

```
Manα1-2Manα1-6
             \
              Manα1-6
             /        \
Manα1-2Manα1-3         Manβ1-4GlcNAcβ1-4GlcNAcβ1-Asn
                       core
Manα1-2Manα1-2Manα1-3
```

```
        Manα1-6
       /       \
Manα1-3         Manβ1-4GlcNAcβ1-4GlcNAcβ1-Asn
       \       /        core
        Manα1-3
```

```
Manα1-6
       \
        Manα1-6
       /       \
Manα1-3         Manβ1-4GlcNAcβ1-4GlcNAcβ1-Asn
       \       /        core
R'-2Manα1-3
```

```
Manα1-6
       \
        Manα1-6
       /       \
Manα1-3         R'-4Manβ1-4GlcNAcβ1-4GlcNAcβ1-Asn
       \       /        core
GlcNAcβ1-2Manα1-3
       \
        R''-4
```

```
Galβ1-4GlcNAcβ1-2Manα1-6          Fucα1-6
                        \               \
                         R-4Manβ1-4GlcNAcβ1-4GlcNAcβ1-Asn
                        /               core
       GlcNAcβ1-2Manα1-3
```

```
Galβ1-4GlcNAcβ1-2Manα1-6          Fucα1-6
                        \               \
                         R-4Manβ1-4GlcNAcβ1-4GlcNAcβ1-Asn
                        /               core
       GlcNAcβ1-2Manα1-3
              \
               GlcNAcβ1-4
```

Figure 8 Primary structure of six classes of carbohydrate chains N-linked to proteins (209).

energy conformation gives estimates of the ranges of allowed values of Φ and Ψ (Figure 6) that overlap with those deduced from NMR data. These and other studies strongly suggest that the basis for the different biological activities of the N-linked oligosaccharides depends on their primary as well as their three-dimensional structures, determined by the orientation of the Manα1-6Manβ1-

arm and the protein chain structure (210). As far as enzymatic specificity is concerned there is NMR evidence indicating that the oligosaccharide structure controls bond-making (211, 212) as well as the bond-breaking specificity (213, 214).

A detailed ^{13}C NMR study of the N-linked, high mannose glycopeptides of hen ovalbumin was recently published (215). The main goal of the study was to demonstrate that use of a large library of ^{13}C chemical shifts, one-bond ^{1}H-^{13}C coupling constants, and T_1 data of related compounds, was sufficient to determine the structures of these types of carbohydrates, in this case containing up to nine sugar residues. As shown before for Manα1-3Manβ1-4GlcNAc (216), the method is based on a systematic positioning, testing, and rejecting of certain sugar composition and structural possibilities until the best fit is found.

The ^{1}H NMR spectra of 72 known oligosaccharide structures derived from glycoproteins was published recently (217). The spectra catalog contains coupling constant and chemical shift assignments for the anomeric protons and the "structural reporter groups" (168) of all these compounds. As a means to facilitate the use of a chemical shift library, the spectral position of the anomeric and the H2 protons of 63 different glycopeptides and oligosaccharides of known primary structure were determined under controlled conditions and tabulated (218). The results showed that anomeric proton resonances fall in 41 clusters. The clusters are useful for deducing the sequence and branching pattern of most classes of saccharides found in glycoproteins (218). In a conceptually similar approach, 2-D, J-resolved NMR spectroscopy was applied to the anomeric protons of ovalbumin glycopeptides to assist in their primary structure determination (219). Proton and ^{13}C NMR spectroscopy have also assisted in the primary structure determination of carbohydrates from human γ-immunoglobulin G (Tem) myeloma protein (212), ceruloplasmin (220), ovalbumin (221), human fibrinogen (222), human plasma galactoprotein (223), polysialoglycoproteins from rainbow trout eggs (224), human platelet glycocalicin (225), glycoproteins from alveoli of patients with alveolar proteinosis (226), glycoprotein of vesicular stomatitis virus (227), yeast invertase (228), lysosomal cathepsin D from porcine spleen (229), urine from patients with mannosidosis (230), glucoamylase from *Aspergillus niger* (231), human milk (232), hen ovalbumin (233), ovarian cyst mucins (234), polysialoglycoproteins from salmon eggs (62), urine from patients with Gaucher's disease (235), and urine of patients with inositol- and glycerol-containing oligosaccharides (103).

Using NOE and relaxation times, the motion of the carbohydrate moieties of hen ovalbumin was studied using [1-^{13}C] Gal enzymatically attached to the glycoprotein (236). In a similar type of study, the solution structures and mobility of the sialo- and asialo ovine submaxillary mucin were investigated using natural abundance ^{13}C NMR spectroscopy (237). In this latter study, it was found that removing the terminal sialic acid does not alter the dynamics of the peptide core but increases the conformational mobility of the oligosaccharide chains, probably as a result of net loss of hydrogen-bonding. Also by natural abundance ^{13}C NMR spectroscopy the structures of the oligosaccharides of bovine pancreatic ribonuclease B were studied. The results suggested that carbohydrate chains have negligible effects on the conformation of the whole glycoprotein (238).

BLOOD GROUP SUBSTANCES   Both ^{1}H and ^{13}C NMR spectroscopy have been extensively used to determine the three-dimensional structures of blood group substances. In one series of studies (239, 240) the anomeric carbons of Gal and Fuc residues of the nonreducing terminal carbohydrates of Type 2-0 blood substance (Fucα1-2Galβ1-4GlcNAc) were enriched with 90% ^{13}C. This isotopic enrichment allowed determination of the most abundant conformers about the fucopyranosyl- and the galactopyranosyl glycosidic linkages by using three-bond ^{1}H-^{13}C and ^{13}C coupling constants across the linkage in both glycosidic bonds. In addition, the detailed analysis of the homo- and heteronuclear, intraresidue coupling pattern demonstrated the rigidity of the sugar ring chair conformation and its conservation after derivatization (240). In another series of studies of Type 2 (Galβ1-4GlcNAc) and Type 1 (Galβ1-3GlcNAc) as well as the corresponding O(H), A and B determinants (241, 242), ^{1}H NMR spectroscopy and the use of NOE and T_1 measurements led to the conclusion that, although several conformational rotamers of the glycosidic bonds are allowed by hard sphere calculations, only one of them is corroborated by the ^{1}H NMR experiment. Interestingly, these studies and those of related carbohydrates confirm that the most abundant conformation about the glycosidic bond is one in which the derivatized aglycone carbon lies, when viewed through a C_1-O_1 bond projection, between O5 and H1 of the glycopyranosyl residue, as depicted in Figures 6 and 9. The exo-anomeric effect appears to be large and dominant in stabilizing these conformations and its inclusion in the hard sphere calculations seems highly appropriate (242).

The intact Lewis blood-group-active glycosphingolipids, as well as the Type 2 glycolipids, lactotetraosylceramide and neolactotetraosylceramide, were studied by ^{1}H-NMR in [^{2}H]-dimethyl-sulfoxide (243). The common carbohydrate core in these structures is GlcNAcβ1-3Galβ1-4Glcβ1-1'Cer. The sequence was established by NOE experiments. Single and double substitutions of the core GlcNAc residue and other substitutions along the oligosaccharide chain do not affect the chemical shifts of the lactosyl portion of the core. This finding is interpreted as a reflection of the conservation of the three-dimensional structure of the core as more sugar residues are added to it.

Other blood group determinants studied by ^{1}H NMR spectroscopy include A- and H-type from rat small intestine (25, 244) A and H active glycosphingoli-

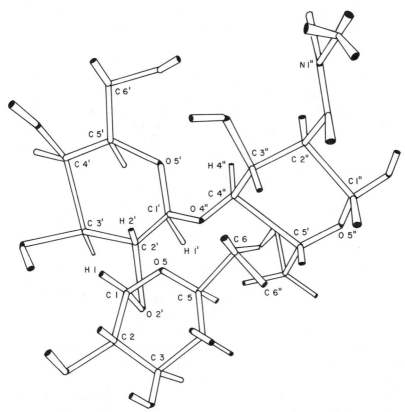

Figure 9   The structure of Type 2 blood group (H) oligosaccharide (L-Fucα1-2-D-Galβ1–4-D-GlcNAc) (240).

pids (26), B-active tetraglycosylceramide (27), H-active triglycosylceramide (28), I-active decaglycosylceramide from rabbit erythrocytes (24), H-active pentaglycosylceramide from human pancreas (33), B-active difucosylhepta-glycosylceramide of human intestine (34), Lewis-active fucolipids of human and canine intestine (45), and glycosphingolipids of human meconium (46).

MICROBIAL ANTIGENIC STRUCTURES   A review (245) and a monograph (246) have addressed the structure determination by ^{13}C NMR spectroscopy of capsular polysaccharide antigens from *Haemophilus influenzae* and *Neisseria meningitidis* and the capsular polysaccharides used as human vaccines, respectively. In some pathogenic bacteria (for example, *Salmonella* and *Shigella*) the capsular polysaccharide is replaced by the *O*-chain polysaccharide of their lipopolysaccharides. A recent study used ^{1}H and ^{13}C NMR spectroscopy to investigate the solution conformation of the complex, linear repeating oligosaccharide units responsible for the sero group Y, O-antigenic properties of *Shigella flexneri* (247). This polysaccharide contains the basic repeating unit [2-L-Rhamα1-2-L-Rhamα1-3-LRhamα1-3-D-GlcNAcβ1]$_n$ upon which α-D—glucopyranosyl and O-acetyl substituents build up the more complex polysaccharide chains that define all the *S. Flexneri* sero groups (247). In order to infer the O-antigenic three-dimensional structure, ten model compounds were used to build a library of NMR parameters: ^1H and ^{13}C chemical shifts, ^1H-^1H and ^{13}C-^1H coupling constants, and T_1 and NOE values. Per se, these data indicated that the L-rhamnopyranosyl residue is in the $^1C_4$ (L) chain conformation whereas the D-N-acetylglucosaminopyranosyl residue is in the $^4C_1$ (D) conformation. These data also indicated the anomeric configurations of each sugar residue. For determining the glycosidic linkage conformation at least the following factors were considered: (*a*) NOE, for establishing the distance between an anomeric and the corresponding aglycone interresidual proton(s); (*b*) three-bond coupling constant across the glycosidic linkage (equivalent to $^3J_{C1-H4'}$ in Figure 6), to establish, using the Karplus equation, the most probable φ and ψ, (*c*) hard sphere calculations that include the exo-anomeric effect, to find the theoretically most stable conformers; and (*d*) model building, to estimate the most stable conformers from steric considerations. Altogether the analysis of this information reduced the number of conformers, compatible with the NMR data, to one with the smallest degrees of freedom for φ and ψ between each one of the sugar residues present in any given oligosaccharide structure. A comparison between the O-antigen NMR parameters and those of the library of model compounds indicated that both the sugar residue rings and the glycosidic linkage conformation found in the model compounds prevail also in the O-antigen structure. With this information, a computer generated hard-sphere drawing showed that two different surfaces, each one containing two tetrasaccharide repeating units, can be visualized in the linear polysaccharide. Because of the polymer inflections some portions of the polymer must deviate from the highly organized three-dimensional arrangement, but at any given time a large portion of the chain would have the inferred structures. It is not known if a direct relationship exists between these two three-dimensional structures and the antigenic properties. Interestingly, however, the polysaccharide generates in vivo the O-factor 3 and the O-factor 4 antibody specificities. Using the information from these studies it is suggested that one part of the antigenic determinant might be related to the acetamido group and to the methyl group of the rhamnose residue. The latter region could provide initial hydrophobic binding to the antibody-combining site followed by polar interactions with the acetamido group. Hydrophobic binding has been proposed as the basis of recognition for certain carbohydrate antibody interactions (248–250) in which 6-deoxy sugars of bacterial polysaccharides may play an important role.

In other recent lipopolysaccharide studies, the solution conformation of *Salmonella* O-antigenic oligosaccharides of sero groups A, B, and D_1 was also determined by ^{1}H and ^{13}C NMR spectroscopy (251, 252). The approach was similar to that used for *Shigella* (247). An axial configuration for the glycosidically linked phosphate groups in *Salmonella* lipopolysaccharides was established by selective decoupling and polarization transfer using ^{1}H and ^{31}P NMR spectroscopy (253). Using ^{13}C NMR spectroscopy, the structures of specific lipopolysaccharides from *Shigella boydii* (254), *Escherichia coli* (255, 256), *Salmonella minnesota* (255), and *Lactobacillus fermentum* (255) have been investigated. The repeating unit structures of a free extracellular polysaccharide from *Mycobacterium salivarus 76* (257) and *Mycobacterium album B-88* (258) were determined by ^{13}C NMR spectroscopy. The *E. coli* O-antigen 25 structure was studied by ^{1}H and ^{13}C NMR spectroscopy (259). Proton-, ^{13}C- and ^{31}P NMR spectroscopy were used for the structural characterization of lipopolysaccharide antigens from *Pseudomonas aeruginosa* (260). The overall temperature dependence of the rotational mobility of the ^{31}P nuclei in *E. coli* lipopolysaccharide was studied by ^{1}H-^{31}P nuclear magnetic double resonance spectroscopy (261). The study provided a numerical estimation of the percentage of lipopolysaccharide that is motionally restricted at temperatures between 24 and 46°C (261).

With respect to capsular polysaccharides (245), the structure of one from *Klebsiella K23* was studied by 2-D ^{1}H-NMR spectroscopy (262). Proton and ^{13}C NMR were used to study capsular polysaccharides from *Klebsiella* (263–266) and *Streptococcus pneumoniae* (267). *Klebsiella* K50 has a heptasaccharide repeating unit and is the only known five-plus-two repeating unit (264). Spin-lattice relaxation rates determined by ^{13}C NMR were used to study the branched-chain polysaccharide dynamics of *Klebsiella* K18 and K41 (265). This branched polysaccharide contains hexa- and heptasaccharide repeating units, respectively, to which trisaccharide units are attached. The study indicated that the least mobile carbon nuclei are those at the branching point, a feature that can be used for sequencing complex polysaccharides (265). The structure of the capsular polysaccharide from *Haemophilus influenzae* Type D was studied by ^{13}C NMR (268).

GLYCOSAMINOGLYCANS   A review on ^{13}C NMR spectroscopy of polysaccharides was published recently (269). From the results of a series of ^{1}H-NMR studies on the secondary structure of oligosaccharides of hyaluronate, chondroitin sulfate, dermatan sulfate, and keratan sulfate, an interesting structure-function relationship has been proposed for these glycosaminoglycans (270). The secondary structure was deduced by studying the hydrogen-bonding between intersugar residues rather than the glycosidic linkage conformations. The glycosaminoglycans were dissolved in [^{2}H$_6$] dimethylsulfoxide. This approach has extensively been used for determining the three-dimensional structures of other biopolymers such as proteins, polypeptides (271), and nucleic acids (272). In the case of hyaluronate (270), the NMR data are consistent with the existence of several hydrogen-bonds between contiguous sugar residues, as depicted in Figure 10. Similar hydrogen bonds in chondroitin 4- and 6-sulfates can be formed, whereas a lesser number of hydrogen-bonds can be formed in dermatan sulfate and none in keratan sulfate. From these results it was suggested that the "stiffness" of the most hydrogen-bonded glycosaminoglycans makes them especially adapted for counteracting the mechanical forces that try to compress the tissue compartments where these glycosaminoglycans are localized. The least hydrogen-bonded glycosaminoglycans could function mostly in recognition by biological receptors. Hydrogen-bonds have been observed as well by X-ray diffraction studies on oriented fibers (273) and have been postulated as the reason for the slow periodate oxidation of glucuronic acid in hyaluronate and condroitin sulfate (274). It is usually assumed that aqueous solutions are more relevant than dimethylsulfoxide solutions to study the biological environment where these biopolymers operate. It is, therefore, interesting that the hydrogen-bond between the amide proton and the carboxyl oxygen of hyaluronate was not observed at pH 2.5 or 5.5 when sought by ^{1}H NMR in aqueous solution (275), but the changes of the N-acetyl group relaxation time as a function of temperature was interpreted as being caused by the hydrogen-bond disruption (276).

Heparin has been extensively investigated by ^{1}H- and ^{13}C-NMR spectrometry for structural characterization. The polysaccharide chain does not seem to contain a unique repeating unit although a disaccharide containing L-idopyranosyluronate-2-O-sulfate linked α1-4 to 2-deoxy-2-sulfamino—D-glucopyranosyl (L-IdUA(2S)α1-4GlcNS) is highly represented (277). The biological significance of this diversity is unknown but heparin has been shown to play a role in platelet activation (278), lipoprotein-lipase liberation (279), fibronectin binding (280), and most of all as an inhibitor of the proteolytic activity of thrombin and Factor Xa mediated by antithrombin III (281). It is clear, however, that a high affinity for antithrombin III is expressed in an oligosaccharide sequence present in approximately one third of the molecular

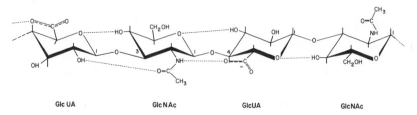

Figure 10 The structure of a tetrasaccharide of hyaluronate showing hydrogen bonds (270).

species present in heparin preparations (282). This sequence is different from that present in the rest of the heparin molecule and provides the basis for the therapeutic use of heparin as an anticoagulant drug (283). Most NMR studies have been focused on this special sugar residue sequence. As a result, it seems that a pentasaccharide, GlcNAc(6S)α1-4GlcUAβ1-4GlcNS(3,6—diS)α1-4-IdUA(2S)α1-4GlcNS(6S), is, or forms part of, the required sequence (277, 284). Other heparin oligosaccharides representing major sequences of the heparin molecule have also been characterized by NMR spectroscopy (284, 285). The primary structure is different from that associated with AT-III binding capacity and indeed could be responsible for other so far unknown biological functions of heparin.

ACKNOWLEDGMENTS

Preparation of this review was supported by research grants (AM12434 and RR00480) from the National Institutes of Health. We are grateful to Robert Barker, Klaas Hallenga and Yoko Ohashi for their generous efforts in reading and critiquing a draft of the review.

Literature Cited

1. Bjorndal, H., Hellerquist, C. G., Lindberg, B., Svensson, S. 1970. *Angew. Chem. Int. Ed. Eng.* 9:610–19
2. Lindberg, B. 1972. *Methods Enzymol.* 28:178–95
3. Lindberg, B., Lonngren, J. 1978. *Methods Enzymol.* 50:3–33
4. Hakomori, S. 1964. *J. Biochem.* 55:205–8
5. Berger, E. G., Buddecke, E., Kamerling, J. P., Kobata, A., Paulson, J. C., Vliegenthart, J. F. G. 1982. *Experientia* 38:1129–258
6. Stellner, K., Saito, H., Hakomori, S. I. 1973. *Arch. Biochem. Biophys.* 155: 464–72
7. Fukuda, M., Hakomori, S. 1979. *J. Biol. Chem.* 254:5451–56
8. Jansson, P.-E., Kenne, L., Liedgren, H., Lindberg, B., Lonngren, J. 1976. *Chem. Commun. Univ. Stockholm*, pp. 2–75
9. Stoffel, W., Hanfland, P. 1973. *Hoppe-Seyler's Z. Physiol. Chem.* 354:21–31
10. Schwarzmann, G. O. H., Jeanloz, R. W. 1974. *Carbohydr. Res.* 34:161–68
11. Banoub, J. H. 1982. *Carbohydr. Res.* 100:C17–23
12. Lowe, M. E., Nilsson, B. 1984. *Anal. Biochem.* 136:187–91
13. Geyer, R., Geyer, H., Kuhnhardt, S., Mink, W., Stirm, S. 1982. *Anal. Biochem.* 121:263–74
14. Lawson, A. M., Hounsell, E. F., Feizi, T. 1983. *Int. J. Mass Spectrom. Ion Phys.* 48:149–52
15. van Halbeek, H., Haverkamp, J., Kamerling, J. P., Vliegenthart, J. F. G. 1978. *Carbohydr. Res.* 60:51–62
16. Ashraf, J., Butterfield, D. A., Jarnefelt, J., Laine, R. A. 1980. *J. Lipid Res.* 21:1137–41
17. Fournet, B., Strecker, G., Leroy, Y., Montreuil, J. 1981. *Anal. Biochem.* 116:489–502
18. Geyer, R., Geyer, H., Kuhnhardt, S., Mink, W., Stirm, S. 1983. *Anal. Biochem.* 133:197–207
19. Rolf, D., Gray, G. R. 1982. *J. Am. Chem. Soc.* 104:3539–41
20. Valent, B. S., Darvill, A. G., McNeil, M., Robertsen, B. K., Albersheim, P. 1980. *Carbohydr. Res.* 79:165–92
21. Robertsen, B. K., Åman, P., Darvill, A. G., McNeil, M., Albersheim, P. 1981. *Plant Physiol.* 67:389–400
22. McNeil, M., Darvill, A. G., Åman, P., Franzen, L.-E., Albersheim, P. 1982. *Methods Enzymol.* 83:3–45
23. Ångstrom, J., Breimer, M. E., Falk, K.-E., Hansson, G. C., Karlsson, K.-A., et al. 1982. *Arch. Biochem. Biophys.* 213:708–25
24. Ångstrom, J., Breimer, M. E., Falk, K.-E., Hansson, G. C., Karlsson, K.-A., Leffler, H. 1982. *J. Biol. Chem.* 257:682–88

25. Breimer, M. E., Falk, K.-E., Hansson, G. C., Karlsson, K.-A. 1982. *J. Biol. Chem.* 257:50–59
26. Breimer, M. E., Hansson, G. C., Karlsson, K.-A., Leffler, H. 1982. *J. Biol. Chem.* 257:906–12
27. Hansson, G. C., Karlsson, K.-A., Thurin, J. 1980. *Biochim. Biophys. Acta* 620:270–80
28. Breimer, M. E., Hansson, G. C., Karlsson, K.-A., Leffler, H. 1980. *Biochim. Biophys. Acta* 617:85–96
29. Holmes, E. H., Hakomori, S. 1982. *J. Biol. Chem.* 257:7698–703
30. Tadano, K., Ishizuka, I. 1982. *J. Biol. Chem.* 257:1482–90
31. Chigorno, V., Sonnino, S., Ghidoni, R., Tettamanti, G. 1982. *Neurochem. Int.* 4:531–39
32. Ghidoni, R., Sonnino, S., Tettamanti, G., Baumann, N., Renter, G., Schauer, R. 1980. *J. Biol. Chem.* 255:6990–95
33. Iwamori, M., Nagai, Y. 1981. *Biochim. Biophys. Acta* 665:205–13
34. Hanfland, P., Egge, H., Dabrowski, U., Kuhn, S., Roelcke, D., Dabrowski, J. 1981. *Biochemistry* 20:5310–19
35. Iwamori, M., Nagai, Y. 1981. *J. Biochem.* 89:1253–64
36. Sung, S.-S. J., Sweeley, C. C. 1979. *Biochim. Biophys. Acta* 575:295–98
37. Ariga, T., Sekine, M., Nakamura, K., Igarashi, M., Nagashima, M., et al. 1983. *J. Biochem.* 93:889–93
38. Ariga, T., Sekine, M., Yu, R. K., Miyatake, T. 1982. *J. Biol. Chem.* 257:2230–35
39. Macher, B. A., Pacuszka, T., Mullin, B. R., Sweeley, C. C., Brady, R. O., Fishman, P. H. 1979. *Biochim. Biophys. Acta* 588:35–43
40. Iwamori, M., Sawada, K., Hara, Y., Nishio, M., Fujisawa, T., et al. 1982. *J. Biochem.* 91:1875–87
41. Yasugi, E., Saito, E., Kasama, T., Kojima, H., Yamakawa, T. 1982. *J. Biochem.* 91:1121–27
42. Blomberg, J., Breimer, M. E., Karlsson, K.-A. 1982. *Biochim. Biophys. Acta* 711:466–77
43. Breimer, M. E., Karlsson, K.-A., Samuelsson, B. E. 1981. *J. Biol. Chem.* 256:3810–16
44. Breimer, M. E., Karlsson, K.-A., Samuelsson, B. E. 1982. *J. Biol. Chem.* 257:1079–85
45. McKibbin, J. M., Spencer, W. A., Smith, E. L., Mansson, J.-E., Karlsson, K.-A., et al. 1982. *J. Biol. Chem.* 257:755–60
46. Karlsson, K.-A., Larson, G. 1981. *J. Biol. Chem.* 256:3512–24

47. Egge, H., Hanfland, P. 1981. *Arch. Biochem. Biophys.* 210:396–404
48. Hanfland, P., Graham, H. A. 1981. *Arch. Biochem. Biophys.* 210:383–95
49. Fukuda, M. N., Hakomori, S. 1982. *J. Biol. Chem.* 257:446–55
50. Kannagi, R., Levery, S. B., Ishigami, F., Hakomori, S., Shevinsky, L. H., et al. 1983. *J. Biol. Chem.* 258:8934–42
51. Westrick, M. A., Lee, W. M. F., Macher, B. A. 1983. *Cancer Res.* 43:5890–94
52. Klock, J. C., D'Angona, J. L., Macher, B. A. 1981. *J. Lipid Res.* 22:1079–83
53. Hori, T., Sugita, M., Ando, S., Tsukuda, K., Shiota, K., et al. 1983. *J. Biol. Chem.* 258:2239–45
54. Itasaka, O., Sugita, M., Kataoka, H., Hori, T. 1983. *Biochim. Biophys. Acta* 751:8–13
55. Hori, T., Sugita, M., Ando, S., Kuwahara, M., Kumauchi, K., et al. 1981. *J. Biol. Chem.* 256:10979–85
56. Sugita, M., Yamamoto, T., Masuda, S., Itasaka, O., Hori, T. 1981. *J. Biochem.* 90:1529–35
57. Itasaka, O., Kosuga, M., Okayama, M., Hori, T. 1983. *Biochim. Biophys. Acta* 750:440–46
58. Sugita, M. 1979. *J. Biochem.* 86:289–300
59. Sugita, M., Nishida, M., Hori, T. 1982. *J. Biochem.* 92:327–34
60. Sugita, M., Iwasaki, Y., Hori, T. 1982. *J. Biochem.* 92:881–87
61. Hsieh, T. C.-Y., Kaul, K., Laine, R. A., Lester, R. L. 1978. *Biochemistry* 17:3575–81
62. Shimamura, M., Endo, T., Inoue, Y., Inoue, S. 1983. *Biochemistry* 22:959–63
63. Nilsson, B., Norden, N. E., Svensson, S. 1979. *J. Biol. Chem.* 254:4545–53
64. Liang, C.-J., Yamashita, K., Muellenberg, C. G., Shichi, H., Kobata, A. 1979. *J. Biol. Chem.* 254:6414–18
65. Takasaki, S., Yamashita, K., Suzuki, K., Iwanaga, S., Kobata, A. 1979. *J. Biol. Chem.* 254:8548–53
66. Tsuji, T., Irimura, T., Osawa, T. 1981. *J. Biol. Chem.* 256:10497–502
67. Yurewicz, E. C., Matsuura, F., Moghissi, K. S. 1982. *J. Biol. Chem.* 257:2314–22
68. Nilsson, B., Horne, M. K., Gralnick, H. R. 1983. *Arch. Biochem. Biophys.* 224:127–33
69. Phillips, L. R., Nishimura, O., Fraser, B. A. 1983. *Carbohydr. Res.* 121:243–55
70. Rao, A. S., Kabat, E. A., Whittaker, N. F., Nilsson, B., Zopf, D. A., Nimmich, W. 1983. *Carbohydr. Res.* 116:271–76

71. Hunter, S. W., Murphy, R. C., Clay, K., Goren, M. B., Brennan, P. J. 1983. *J. Biol. Chem.* 258:10481–87
72. Maeda, M., Shimahara, H., Sugiyama, N. 1980. *Agric. Biol. Chem.* 44:245–52
73. Spellman, M. W., McNeil, M., Darvill, A. G., Albersheim, P., Dell, A. 1983. *Carbohydr. Res.* 122:131–53
74. Nilsson, B., DeLuca, S., Lohmander, S., Hascall, V. C. 1982. *J. Biol. Chem.* 257:10920–27
75. Strecker, G., Fournet, B., Bouquelet, S., Montreuil, J., Dhondt, J. L., Farriaux, J. P. 1976. *Biochimie* 58:579–86
76. Yamashita, K., Tachibana, Y., Mihara, K., Okada, S., Yabuuchi, H., Kobata, A. 1980. *J. Biol. Chem.* 255:5126–33
77. Matsuura, F., Nunez, H. A., Grabowski, G. A., Sweeley, C. C. 1981. *Arch. Biochem. Biophys.* 207:337–52
78. Hallgren, P., Lundblad, A. 1977. *J. Biol. Chem.* 252:1014–22
79. Sundqvist, B., Kamensky, I., Håkansson, P., Kjellberg, J., Salehpour, M., et al. 1984. *Biomed. Mass Spectrom.* 11:242–57
80. Busch, K. L., Cooks, R. G. 1982. *Science* 218:247–54
81. Rinehart, K. L. 1982. *Science* 218:254–60
82. Fenselau, C. 1982. *Anal. Chem.* 54A:105–16
83. Wood, G. W. 1982. *Mass Spectrom. Rev.* 1:63–102
84. Reinhold, V. N., Carr, S. A. 1983. *Mass. Spectrom. Rev.* 2:153–221
85. Burlingame, A. L., Whitney, J. O., Russell, D. H. 1984. *Anal. Chem.* 56:R417–67
86. Strecker, G., Pierce-Cretel, A., Fournet, B., Spik, G., Montreuil, J. 1981. *Anal. Biochem.* 111:17–26
87. Mononen, I. 1982. *Carbohydr. Res.* 104:1–9
88. Nilsson, B., Zopf, D. 1983. *Arch. Biochem. Biophys.* 222:628–48
89. Wang, W.-T., Matsuura, F., Sweeley, C. C. 1983. *Anal. Biochem.* 134:398–405
90. Ugorski, M., Nilsson, B., Schroer, K., Cashel, J. A., Zopf, D. 1984. *J. Biol. Chem.* 259:481–86
91. Nilsson, B., Zopf, D. 1982. *Methods Enzymol.* 83:46–58
92. Nilsson, B., Svensson, S. 1979. *Carbohydr. Res.* 69:292–96
93. Erbing, C., Granath, K., Kenne, L., Lindberg, B. 1976. *Carbohydr. Res.* 47:C5–7
94. Breimer, M. E., Hansson, G. C., Karlsson, K.-A., Leffler, H., Pimlott, W.,

Samuelsson, B. E. 1981. *FEBS Lett.* 124:299–303
95. Egge, H., Michalski, J. C., Strecker, G. 1982. *Arch. Biochem. Biophys.* 213:318–26
96. Jardine, I., Matsuura, F., Sweeley, C. C. 1984. *Biomed. Mass Spectrom.* 11:562–68
97. Breimer, M. E., Hansson, G. C., Karlsson, K.-A., Larson, G., Leffler, H., et al. 1983. *Int. J. Mass Spectrom. Ion Phys.* 48:113–16
98. Rauvala, H., Finne, J., Krusius, T., Karkkainen, J. 1981. *Adv. Carbohydr. Chem. Biochem.* 38:389–416
99. DeJone, E. G., Heerma, W., Dijkstra, G. 1980. *Biomed. Mass Spectrom.* 7:127–31
100. Strecker, G., Pierce-Cretel, A., Fournet, B., Spik, G., Montreuil, J. 1981. *Anal. Biochem.* 111:17–26
101. Taylor, G. W., Morris, H. R., Petersen, T. E., Magnusson, S. 1979. *Adv. Mass Spectrom.* 8:1090–96
102. Karlsson, K.-A., Pascher, I., Samuelsson, B. E., Finne, J., Krusius, T., Rauvala, H. 1978. *FEBS Lett.* 94:413–17
103. Wang, W.-T., Matsuura, F., Nunez, H., LeDonne, N., Baltzer, B., Sweeley, C. C. 1984. *Glycoconjugate J.* 1:17–35
104. Egge, H. 1978. *Chem. Phys. Lipids* 21:349–60
105. Karlsson, K.-A. 1978. *Prog. Chem. Fats Other Lipids* 16:207–30
106. Ångstrom, J., Falk, K.-E., Karlsson, K.-A., Larson, G. 1982. *Biochim. Biophys. Acta* 712:274–82
107. Hanfland, P., Kordowicz, M. 1984. *J. Biol. Chem.* In press
108. Falk, K.-E., Karlsson, K.-A., Larson, G., Thurin, J., Blaszczyk, M., et al. 1983. *Biochem. Biophys. Res. Commun.* 110:383–91
109. Breimer, M. E., Karlsson, K.-A. 1983. *Biochim. Biophys. Acta* 755:170–77
110. Macher, B. A., Klock, J. C., Fukuda, M. N., Fukuda, M. 1981. *J. Biol. Chem.* 256:1968–74
111. Nilsson, O., Månsson, J.-E., Håkansson, G., Svennerholm, L. 1982. *Biochim. Biophys. Acta* 712:453–63
112. Fredman, P., Månsson, J.-E., Svennerholm, L., Karlsson, K.-A., Pascher, I., Samuelsson, B. E. 1980. *FEBS Lett.* 110:80–84
113. Breimer, M. E., Hansson, G. C., Karlsson, K.-A., Leffler, H. 1982. *J. Biol. Chem.* 257:557–68
114. Breimer, M. E., Hansson, G. C., Karlsson, K.-A., Leffler, H. 1982. *Biochim. Biophys. Acta* 710:415–27
115. Breimer, M. E., Hansson, G. C., Karls-

son, K.-A., Leffler, H. 1983. *J. Biochem.* 93:1473–85
116. Takeda, N., Harada, K., Suzuki, M., Tatematsu, A., Sakata, I. 1983. *Biomed. Mass Spectrom.* 10:608–13
117. Evans, J. E., McCluer, R. H., Kadowaki, H. 1983. *31st Ann. Conf. Mass. Spectr. Allied Top.*, pp. 793–94
118. Liberato, D. J., Fenselau, C. C., Vestal, M. L., Yergey, A. L. 1983. *Anal. Chem.* 55:1741–44
119. Cairns, T., Siegmund, E. G. 1982. *Anal. Chem.* 54:2456–61
120. Reinhold, V. N., Carr, S. A. 1982. *Anal. Chem.* 54:499–503
121. Carr, S. A., Reinhold, V. N. 1984. *Biomed. Mass Spectrom.* 11:633–42
122. Ariga, T., Yu, R. K., Suzuki, M., Ando, S., Miyatake, T. 1982. *J. Lipid Res.* 23:437–42
123. Ariga, T., Murata, T., Oshima, M., Maezawa, M., Miyatake, T. 1980. *J. Lipid Res.* 21:879–87
124. Ariga, T., Yu, R. K., Miyatake, T. 1984. *J. Lipid Res.* 25:1096–1101
125. Tanaka, Y., Yu, R. K., Ando, S., Ariga, T., Itoh, T. 1984. *Carbohydr. Res.* 126:1–14
126. Reinhold, V. N., Coles, E., Carr, S. A. 1983. *J. Carbohydr. Chem.* 2:1–18
127. Beckey, H. 1969. *Int. J. Mass Spectrom. Ion Phys.* 2:500–3
128. Schulten, H.-R., Komori, T., Fujita, K., Shinoda, A., Imoto, T., Kawasaki, T. 1982. *Carbohydr. Res.* 107:177–86
129. Linscheid, M., D'Angona, J., Burlingame, A. L., Dell, A., Ballou, C. E. 1981. *Proc. Natl. Acad. Sci. USA* 78:1471–75
130. Costello, C. E., Wilson, B. W., Biemann, K., Reinhold, V. N. 1980. In *Cell Surface Glycolipids*, ed. C. C. Sweeley, pp. 35–54. Washington:Am. Chem. Soc.
131. Kushi, Y., Handa, S. 1982. *J. Biochem.* 91:923–31
132. Kushi, Y., Handa, S., Kambara, H., Shizukuishi, K. 1983. *J. Biochem.* 94:1841–50
133. Schulten, H.-R. 1979. *Int. J. Mass Spectrom. Ion Phys.* 32:97–283
134. Harada, K.-I., Suzuki, M., Takeda, N., Tatematsu, A., Kambara, H. 1982. *J. Antibiot.* 35:102–5
135. Harada, K.-I., Suzuki, M., Kambara, H. 1982. *Org. Mass Spectrom.* 17:386–91
136. Kambara, H. 1982. *Org. Mass Spectrom.* 17:29–33
137. Handa, S., Kushi, Y., Kambara, H., Shizukuishi, K. 1983. *J. Biochem.* 93:315–18
138. Shimamura, M., Endo, T., Inoue, Y.,

Inoue, S., Kambara, H. 1984. *Biochemistry* 23:317–22
139. Dell, A., Morris, H. R., Egge, H., von Nicolai, H., Strecker, G. 1983. *Carbohydr. Res.* 115:41–52
140. Hounsell, E. F., Madigan, M. J., Lawson, A. M. 1984. *Biochem. J.* 219:947–52
141. Dell, A., Oates, J. E., Morris, H. R., Egge, H. 1983. *Int. J. Mass Spectrom. Ion Phys.* 46:415–18
142. Arita, M., Iwamori, M., Higuchi, T., Nagai, Y. 1983. *J. Biochem.* 93:319–22
143. Arita, M., Iwamori, M., Higuchi, T., Nagai, Y. 1983. *J. Biochem.* 94:249–56
144. Arita, M., Iwamori, M., Higuchi, T., Nagai, Y. 1984. *J. Biochem.* 95:971–81
145. Iwamori, M., Shimomura, J., Tsuyuhara, S., Nagai, Y. 1984. *J. Biochem.* 95:761–70
146. Egge, H., Peter-Katalinic, J., Kordowicz, M., Hanfland, P. 1984. In press
147. Egge, H., Peter-Katalinic, J., Paz-Parente, J., Strecker, G., Montreuil, J., Fournet, B. 1983. *FEBS Lett.* 156:357–62
148. Kamerling, J. P., Heerma, W., Vliegenthart, J. F. G., Green, B. N., Lewis, I. A. S., et al. 1983. *Biomed. Mass Spectrom.* 10:420–25
149. Abraham, D. J., Sidebothom, R., Winchester, B. G., Dorling, P. R., Dell, A. 1983. *FEBS Lett.* 163:110–13
150. Dell, A., Ballou, C. E. 1983. *Carbohydr. Res.* 120:95–111
151. Dell, A., Ballou, C. E. 1983. *Biomed. Mass Spectrom.* 10:50–56
152. Qureshi, N., Takayama, K., Ribi, E. 1982. *J. Biol. Chem.* 257:11808–15
153. Nothnagel, E. A., McNeil, M., Albersheim, P., Dell, A. 1983. *Plant Physiol.* 71:916–26
154. Dell, A., York, W. S., McNeil, M., Darvill, A. G., Albersheim, P. 1983. *Carbohydr. Res.* 117:185–200
155. Egge, H., Dell, A., von Nicolai, H. 1983. *Arch. Biochem. Biophys.* 224:235–53
156. Wang, W.-T., LeDonne, N. C., Ackermann, B., Sweeley, C. C. 1984. *Anal. Biochem.* 141:366–81
157. Carr, S. A., Reinhold, V. N. 1984. *J. Carbohydr. Chem.* 3:381–401
158. Lemieux, R. U. 1978. *Chem. Soc. Rev.* 7:423–52
159. Barker, R., Nunez, H. A., Rosevear, P., Serianni, A. S. 1982. *Methods Enzymol.* 83:58–69
160. Lemieux, R. U., Koto, S., Voisin, D. 1979. *Am. Chem. Soc. Symp. Ser.* 87:17–29

161. Bock, K. 1983. *Pure Appl. Chem.* 55:605–22
162. Lemieux, R. U. 1982. In *28th IUPAC Congr.*, ed. K. J. Laidler, pp. 3–24. New York:Pergamon
163. Harris, R. K. 1983. *Nuclear Magnetic Resonance Spectroscopy, A Physicochemical View.* Marshfield, Mass: Pitman
164. Gadian, D. G. 1982. *Nuclear Magnetic Resonance and its Applications to Living Systems.* pp. 23–42. New York:Oxford Univ. Press.
165. Bock, K., Thogersen, H. 1982. *Ann. Rep. NMR Spectrosc.* 13:1–57
166. Bock, K., Pedersen, C. 1983. *Adv. Carbohydr. Chem. Biochem.* 41:27–66
167. Dill, K., Allerhand, A. 1979. *J. Biol. Chem.* 254:4524–31
168. Vliegenthart, J. F. G., van Halbeek, H., Dorland, L. 1980. In *27th Int. Congr. Pure and Applied Chemistry*, ed. A. Varmavuori, pp. 253–62. New York: Pergamon
169. Hayes, M. L., Serianni, A. S., Barker, R. 1982. *Carbohydr. Res.* 100:87–107
170. Ewing, D. F. 1983. *Nucl. Magn. Reson.* 12:68–95
171. Karplus, M. 1963. *J. Am. Chem. Soc.* 85:2870–71
172. Altona, C., Haasnoot, C. A. G. 1980. *Org. Magn. Reson.* 13:417–29
173. O'Connor, J. V., Nunez, H. A., Barker, R. 1979. *Biochemistry* 18:500–7
174. Batta, G., Liptak, A. 1984. *J. Am. Chem. Soc.* 106:248–50
175. McFarland, W., Rycroft, D. S. 1983. *Nucl. Magn. Reson.* 12:158–87
176. Barker, R., Clark, E. L., Nunez, H. A., Pierce, J., Rosevear, P., Serianni, A. S. 1982. *Anal. Chem. Symp. Ser.* 11 (Stable Isotopes):719–30
177. Christofides, J. C., Davies, D. B. 1983. *J. Am. Chem. Sox.* 105:5099–105
178. States, D. J., Haberkorn, R. A., Ruben, D. J. 1982. *J. Magn. Reson.* 48:286–92
179. Bodenhausen, G., Freeman, R., Turner, D. L. 1976. *J. Chem. Phys.* 65:839–40
180. Aue, W. P., Bartholdi, E., Ernst, R. R. 1976. *J. Chem. Phys.* 64:2229–46
181. Macura, S., Huang, Y., Suter, D., Ernst, R. R. 1981. *J. Magn. Reson.* 43:259–81
182. Benn, R., Gunther, H. 1983. *Angew. Chem. Int. Ed. Engl.* 22:350–80
183. Coxon, B. 1983. *Anal. Chem.* 55:2361–66
184. Curatolo, W., Neuringer, L. J., Ruben, D., Haberkorn, R. 1983. *Carbohydr. Res.* 112:297–300
185. Szilagyi, L. 1983. *Carbohydr. Res.* 118:269–75
186. Taravel, F. R., Vignon, M. R. 1982. *Nouv. J. Chim.* 6:37–42
187. Bernstein, M. A., Hall, L. D., Sukumar, S. 1982. *Carbohydr. Res.* 103:C1–6
188. Card, P. T., Reddy, G. S. 1982. *J. Org. Chem.* 48:4734–43
189. Taravel, F. R., Vignon, M. R. 1982. *Polym. Bull.* 7:153–57
190. Chalmers, A. A., Pachler, K. G. R., Wessels, P. L. 1974. *Org. Magn. Reson.* 6:445–47
191. Bolton, P. 1982. *J. Magn. Reson.* 48:336–40
192. Bigler, P., Ammann, W., Richarz, R. 1984. *Org. Magn. Reson.* 22:109–13
193. Patt, S. L., Sauriol, F., Perlin, A. S. 1982. *Carbohydr. Res.* 107:C1–4
194. Koerner, T. A. W. Jr., Prestegard, J. H., Demou, P. C., Yu, R. K. 1983. *Biochemistry* 22:2676–87
195. Koerner, T. A. W. Jr., Prestegard, J. H., Demou, P. C., Yu, R. K. 1983. *Biochemistry* 22:2687–90
196. Yu, R. K., Koerner, T. A. W. Jr., Demou, P. C., Scarsdale, J. N., Prestegard, J. H. 1984. In *Advances in Experimental Medicine and Biology*, 174:87–102
197. Dabrowski, J., Hanfland, P. 1982. *FEBS Lett.* 142:138–42
198. Dabrowski, J., Hanfland, P., Egge, H. 1982. *Methods Enzymol.* 83:69–86
199. Dabrowski, U., Hanfland, P., Egge, H., Kuhn, S., Dabrowski, J. 1984. *J. Biol. Chem.* 259:7648–51
200. Egge, H., Dabrowski, J., Hanfland, P., Dell, A., Dabrowski, U. 1982. *Adv. Exp. Med. Biol.* 152:33–40
201. Dabrowski, J., Hanfland, P., Egge, H. 1980. *Biochemistry* 19:5622–58
202. Yu, R. K., Sillerud, L. O. 1982. *Adv. Exp. Med. Biol.* 152:41–46
203. Sillerud, L. O., Yu, R. K., Schafer, D. E. 1982. *Biochemistry* 21:1260–71
204. Sillerud, L. O., Yu, R. K. 1983. *Carbohydr. Res.* 113:173–88
205. Falk, K. E., Jovall, P. A., Winyard, P. 1982. *Acta Chem. Scand. Ser. B* 36:558–60
206. Brisson, J.-R., Carver, J. P. 1983. *Biochemistry* 22:1362–68
207. Brisson, J.-R., Carver, J. P. 1983. *Biochemistry* 22:3671–80
208. Brisson, J.-R., Carver, J. P. 1983. *Biochemistry* 22:3680–83
209. Carver, J. P., Brisson, J.-R. 1984. In *Biology of Carbohydrates*, ed. V. Ginsburg, P. W. Robbins, 2:289–331. New York: Wiley
210. Shimizu, A., Honzawa, M., Ito, S., Miyazaki, T., Matsumoto, H., Nakamura, H., Michaelsen, T. E., Arata, Y. 1983. *Mol. Immunol.* 20:141–48
211. Longmore, G. D., Schachter, H. 1982. *Carbohydr. Res.* 100:365–92
212. Grey, A. A., Narasimhan, S., Brisson, J.

R., Schachter, H., Carver, J. P. 1982. *Can. J. Biochem.* 60:1123–31
213. Paulson, J. C., Weinstein, J., Dorland, L., Van Halbeek, H., Vliegenthart, J. F. G. 1982. *J. Biol. Chem.* 257:12734–38
214. Berman, E., Allerhand, A. 1981. *J. Biol. Chem.* 256:6657–62
215. Allerhand, A., Berman, E. 1984. *J. Am. Chem. Soc.* 106:2400–12
216. Nunez, H. A., Matsuura, F., Sweeley, C. C. 1981. *Arch. Biochem. Biophys.* 212:638–43
217. Vliegenthart, J. F. G., Dorland, L., Van Halbeek, H. 1984. *Adv. Carbohydr. Chem. Biochem.* 41:209–374
218. Carver, J. P., Grey, A. A. 1981. *Biochemistry* 20:6607–16
219. Bruch, R. C., Bruch, M. D. 1982. *J. Biol. Chem.* 257:3409–13
220. Endo, M., Suzuki, K., Schmid, K., Fournet, B., Karamanos, Y., et al. 1982. *J. Biol. Chem.* 257:8755–60
221. Ceccarini, C., Lorenzoni, P., Atkinson, P. H. 1983. *Biochim. Biophys. Acta* 759:214–21
222. Townsend, R. R., Hilliber, E., Li, Y. T., Loune, R. A., Bell, W. R., Lee, Y. C. 1982. *J. Biol. Chem.* 257:9704–10
223. Akiyama, K., Simons, E. R., Bernasconi, R., Schmid, K., Van Halbeek, H., et al. 1984. *J. Biol. Chem.* 259:7151–54
224. Iwasaki, M., Inoue, S., Kitajima, K., Nomoto, H., Inoue, Y. 1984. *Biochemistry* 23:305–22
225. Korrel, S. A. M., Clemetson, K. J., Van Halbeek, H., Kamerling, J. P., Sixma, J. J., Vliegenthart, J. F. G. 1984. *FEBS Lett.* 571–76
226. Bhattacharyya, S. N., Lynn, W. S., Dabrowski, J., Trauner, K., Hull, W. E. 1984. *Arch. Biochem. Biophys.* 231:72–85
227. Stanley, P., Vivona, G., Atkinson, P. H. 1984. *Arch. Biochem. Biophys.* 230:363–74
228. Lehle, L., Cohen, R. E., Ballou, C. E. 1979. *J. Biol. Chem.* 254:12209–18
229. Takahashi, T., Schmidt, P. G., Tang, J. 1983. *J. Biol. Chem.* 258:2819–30
230. Van Halbeek, H., Dorland, L., Veldink, G. A., Vliegenthart, J. F. G., Strecker, G., et al. 1980. *FEBS Lett.* 121:71–77
231. Dill, K., Allerhand, A. 1979. *J. Biol. Chem.* 254:4524–31
232. Bush, C. A., Panitch, M. M., Dua, V. K., Rohr, T. E. 1985. *Anal. Biochem.* 145:124–36
233. Dua, V. K., Bush, C. A. 1984. *Anal. Biochem.* 137:33–40
234. Dua, V. K., Dube, V. E., Bush, C. A.

1984. *Biochim. Biophys. Acta.* 802:29–40
235. Van Halbeek, H., Dorland, L., Veldink, G. A., Vliegenthart, J. F. G., Michalski, J. C., et al. 1980. *FEBS Lett.* 121:65–70
236. Goux, W. J., Perry, C., James, T. L. 1982. *J. Biol. Chem.* 257:1829–35
237. Gerken, T. A., Dearborn, D. G. 1984. *Biochemistry* 23:1485–97
238. Berman, E., Walters, D. E., Allerhand, A. 1981. *J. Biol. Chem.* 256:3853–57
239. Nunez, H. A., Barker, R. 1980. *Biochemistry* 19:489–95
240. Rosevear, P. R., Nunez, H. A., Barker, R. 1982. *Biochemistry* 21:1421–31
241. Lemieux, R. U., Bock, K., Selbaere, T. J., Koto, S., Rao, V. S. 1980. *Can. J. Chem.* 58:631–53
242. Thøgersen, H., Lemieux, R. U., Bock, K., Meyer, B. 1982. *Can. J. Chem.* 60:44–57
243. Dabrowski, J., Hanfland, P., Egge, H., Dabrowski, U. 1981. *Arch. Biochem. Biophys.* 210:405–10
244. Hansson, G. C. 1983. *J. Biol. Chem.* 258:9612–15
245. Egan, W. 1980. In *Magnetic Resonance in Biology*, ed. J. S. Cohen, pp. 197–258. New York: Wiley
246. Jennings, H. J. 1983. *Adv. Carbohydr. Chem. Biochem.* 41:155–207
247. Bock, K., Josephson, S., Bundle, D. R. 1982. *J. Chem. Soc. Perkin Trans.* 2:59–70
248. Schalch, W., Wright, J. K., Rodkey, S., Braun, J. 1979. *Exp. Med.* 149:923–37
249. Lemieux, R. U., Boullanger, P. H., Bundle, D. R., Baker, D. A., Nagpurkar, A., Venot, A. 1978. *Nouv. J. Chim.* 2:321–29
250. DiFabio, J., Dutton, G. G. S. 1981. *Carbohydr. Res.* 92:287–98
251. Bock, K., Meldal, M. 1984. *Carbohydr. Res.* 130:35–53
252. Bock, K., Meldal, M., Bundle, D. R., Iversen, T., Garegg, P. J. 1984. *Carbohydr. Res.* 130:23–34
253. Batley, M., Packer, N., Redmond, J. W. 1982. *Biochemistry* 21:6580–86
254. Lvov, V. L., Tokhtamysheva, N. V., Shashkov, A. S., Dmitriev, B. A., Kochetkov, N. K. 1983. *Bioorg. Khim.* 9:60–73
255. Batley, M., Packer, N., Redmond, J. 1981. *Proc. Conf. Chem. Biol. Act. Bact. Surf. Amphiphiles*, pp. 125–36 New York: Academic
256. Vasiliev, V. N., Zakharova, I. Y., Shashkov, A. S. 1982. *Bioorg. Khim.* 8:120–25

257. Sviridov, A. F., Shashkov, A. S., Kochetkov, N. K., Botvinko, I. V., Egorov, N. A. 1982. *Bioorg. Khim.* 8:1242–51
258. Shashkov, A. S., Sviridov, A. F. 1982. *Bioorg. Khim.* 8:1252–55
259. Kenne, L., Lindberg, B., Madden, J. K. 1983. *Carbohydr. Res.* 122:249–56
260. Horton, D., Riley, D. A., Samreth, S., Schweitzer, M. G. 1983. In *Bacterial Lipopolysaccharides Structure, Synthesis, and Biological Activity, ACS Symp. Ser.* 231, Chap 2, pp. 21–47 Washington, DC:American Chemical Society
261. Egan, W., Leive, L. 1982. *Biochim. Biophys. Acta* 692:165–67
262. Gagnaire, D. Y., Taravel, F. R., Vignon, M. R. 1982. *Macromolecules* 15:126–29
263. Dutton, G. G. S., Karunaratne, D. N. 1983. *Carbohydr. Res.* 119:157–69
264. Altman, E., Dutton, G. G. S. 1983. *Carbohydr. Res.* 118:183–94
265. Vignon, M., Michon, F., Joseleau, J. P., Bock, K. 1983. *Macromolecules* 16:835–38
266. DiFabio, J., Dutton, G. G. S. 1981. *Carbohydr. Res.* 92:287–98
267. Jansson, P. E., Lindberg, B., Lindquist, U. 1981. *Carbohydr. Res.* 95:73–80
268. Branefors-Helander, P., Kenne, L., Lindberg, B., Peterson, K., Unger, P. 1981. *Carbohydr. Res.* 97:285–91
269. Gorin, P. A. J. 1981. *Adv. Carbohydr. Chem. Biochem.* 38:13–104
270. Scott, J. E., Heatley, F., Hull, W. E. 1984. *Biochem. J.* 220:197–205

271. Krishna, N. R., Huang, D.-H., Vaughn, J. B., Heavner, G. A., Goldstein, G. 1981. *Biochemistry* 20:3933–40
272. Reid, B. R. 1981. In *Topics in Nucleic Acid Structure*, ed. S. Neidle, pp. 113–39. New York: Wiley
273. Atkins, E. D. T., Meader, D., Scott, J. E. 1980. *Int. J. Biol. Macromol.* 2:318–19
274. Scott, J. E., Tigwell, M. J. 1978. *Biochem. J.* 173:103–14
275. Cowman, M. K., Cozart, D., Nakanishi, D., Balazs, E. A. 1984. *Arch. Biochem. Biophys.* 230:203–12
276. Hofman, H., Schmut, O. 1983. *Int. J. Biol. Macromol.* 5:229–32
277. Casu, B., Oreste, P., Torri, G., Zoppetti, G., Choay, J., et al. 1981. *Biochem. J.* 197:599–609
278. Mohammad, S. F., Anderson, W. H., Smith, J. B., Chuang, Y. K., Mason, R. G. 1981. *Am. J. Pathol.* 104:64–73
279. Korn, E. D. 1955. *J. Biol. Chem.* 215:1–14
280. Smith, D. E., Furch, L. T. 1982. *J. Biol. Chem.* 257:6518–23
281. Nesheim, E. M. 1983. *J. Biol. Chem.* 258:14708–17
282. Jordan, R., Beeler, D., Rosenberg, R. 1979. *J. Biol. Chem.* 254:2902–13
283. Lindahl, U., Backstrom, G., Thunberg, L., Leder, I. G. 1980. *Proc. Natl. Acad. Sci. USA* 77:6551–55
284. Ototani, N., Kikuchi, M., Yosizawa, Z. 1982. *Biochem. J.* 205:23–30
285. Linker, A., Horringh, P. 1984. *Carbohydr. Res.* 127:75–94

Ann. Rev. Biochem. 1985. 54:803–30
Copyright © 1985 by Annual Reviews Inc. All rights reserved

ORIGIN OF IMMUNE DIVERSITY: GENETIC VARIATION AND SELECTION

Tasuku Honjo

Department of Medical Chemistry, Faculty of Medicine, Kyoto University, Sakyo-Ku, Kyoto 606, Japan

Sonoko Habu

Department of Cell Biology, Tokai University School of Medicine, Isehara, Kanagawa 259–11, Japan

CONTENTS

PERSPECTIVES AND SUMMARY

The immune system comprises a variety of cells with diverse functions, including B and T lymphocytes, macrophages, granulocytes, and mast cells,

803

0066-4154/85/0701-0803$02.00

most of which differentiate from stem cells in the bone marrow. Among these only B and T lymphocytes are responsible for antigen recognition. B cells, recognizing antigens, are activated to produce immunoglobulins specific to the antigens. Helper and suppressor T cells regulate the immunoglobulin synthesis by B cells in an antigen-specific manner. Cytotoxic T cells kill the host cells carrying antigens (e.g. virus-infected cells) but not the normal host cells. Macrophages are involved in antigen presentation to lymphocytes. Granulocytes and mast cells serve as effectors of the immune reaction. Concerted function of these cells is essential to express and maintain the immune diversity.

It is the function of the immune system to recognize and neutralize numerous antigens invading animals. The more diverse the antigen recognition of the organism, the more efficient is its defense system. There are two types of antigen recognition molecules in the immune system; the immunoglobulin produced by B cells and the T cell antigen receptor. These molecules have variable (V) regions which recognize antigens and constant (C) regions which mediate physiological functions. The latter provide additional diversity to the immune system. For example, the same heavy chain (H) V regions recognizing influenza hemoagglutinin are expressed as the μ, γ, ϵ, and α chains, each of which constitutes a different class of the immunoglobulin.

Since the V and C regions are encoded by separate sets of DNA segments, it is the genetic variability of the V gene that determines the antigen recognition diversity. Recent molecular genetic studies on the immunoglobulin gene and the T cell antigen receptor gene indicate that the V region genes of both molecules produce somatic as well as evolutionary variations. Somatic variations include joining of two or three germ-line segments by site-specific recombinations and somatic base replacements. Evolutionary variations include gene duplication, segments transfer (gene conversion), mutational drifts, and so on. Since genetic events, be they somatic or evolutionary, tend to be random, not only useful but also harmful immunoglobulins are produced. Therefore, appropriate selection, be it positive or negative, is a prerequisite for the proper function of the genetic diversity of the immune system as the defense mechanism. Combination of genetic variability and appropriate selection is the basic idea to explain biological diversity since Darwin and Wallace, and was applied to the immune system by Jerne and Burnet as the clonal selection theory which still remains valid in essence.

The basic questions to be asked are: (*a*) How are genetic variants of the immunoglobulin and T cell receptor genes created? and (*b*) How are the variants selected? First, we will summarize recent knowledge of the origin of genetic variations of the immunoglobulin and T cell antigen receptor genes. We tried to avoid extensive overlap with previous reviews (1, 2) and readers are advised to consult them about details of earlier findings. Then, we wish to describe our personal view of the selection mechanism which remains a matter of controversy. The other molecules contributing to immune diversity are products of the major histocompatibility complex (MHC) which play important roles for antigen recognition by T cells. Because of the limited space allowed this article, we are unable to discuss the origin of diversity of MHC molecules, which show extensive polymorphic variations but no somatic changes.

VARIABILITY OF IMMUNOGLOBULIN GENES

The V regions of immunoglobulins contain three hypervariable segments which make contact with various antigen determinants (3). These segments are called the complementarity determining region (CDR), and the rest of the V region constitutes the framework segments (FR1, FR2, FR3, and FR4). Somatic changes of immunoglobulin gene sequences contribute enormously to amplification of the V region diversity (1, 2). The somatic mechanisms involve (a) randomly paired recombination of V, diversity (D), and joining (J) segments to complete an expressed V gene, (b) variability at a joining site between the recombined segments, and (c) base replacements in the completed V gene (somatic mutation). Although hypervariability of the third CDR is thus explained by the reassortment of D and J segments, the genetic basis of hypervariability in the first and second CDRs is less understood and probably depends on both evolutionary processes and somatic base replacements.

Organization and Number of V Gene Segments

The number and organization of the immunoglobulin V genes, which are determined by the evolutionary process, provide the basis of immunoglobulin diversity. The complete V region gene comprises two or three germ-line segments; V and J for the light (L) chain and V, D, and J for the heavy (H) chain (Figure 1). Three clusters (L_K, L_λ, and H) of the V and C genes are located on different chromosomes in man and mouse (4–10). Among them the mouse L_λ system has the simplest organization: each of two V_λ segments is linked to two pairs of the J_λ and C_λ segments (11, 12, 12a). The human L_λ system has many more V_λ and C_λ segments, although the characterization of the system remains incomplete (13). At least six C_λ genes have been identified in the human genome but only three C_λ genes sequenced and correspond to three known nonallelic λ chain isotypes.

The L_K gene systems of man and mouse consist of many V_K segments, five J_K segments, and one C_K gene (14–16). One of the mouse J_K segments is inactive, probably because of a base replacement in the splice donor site and divergence in the recombination signal sequence. The total number of V_K segments was estimated by hybridization with randomly selected V_K probes. Zeelon et al (17) estimated that the total number of mouse V_K segment is 2000

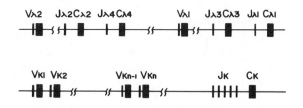

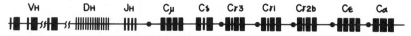

Figure 1   Organization of mouse immunoglobulin genes. Three linkage groups (L_λ, L_K, and H) were schematically represented. Wide rectangles indicate exons. Closed circles show S regions.

using saturation hybridization kinetics. On the other hand, the estimated number of V_K segments based on the number of restriction fragments hybridizing to 10 different V_K probes fell in the range 90–320 (18).

Nishi, Kataoka, and Honjo (submitted for publication) took a completely different approach to estimate the total number of the mouse V_K segments. They have found that about one third of the known V_K segments contains a KpnI site in the V_K exon. They have digested mouse spleen DNA with KpnI and BstEII which cleaves at the 3' flanking region of the J_{K5} segment. Since KpnI does not cleave the J_K region, the KpnI-BstEII cleavage should yield a series of J_K segments rearranged to those V_K segments containing the KpnI site (V_K-KpnI family). Since the V_K-J_K rearrangement in spleen B cells yields a number of heterogenous V_K genes, most restriction enzymes produce smeary rather than discrete bands of rearranged J_L fragments. The KpnI-BstEII digestion of spleen DNA is unique in producing discrete rearranged J_K fragments corresponding to J_{K1}, J_{K2}, J_{K4}, and J_{K5} segments (Figure 2). KpnI-BstEII fragments containing rearranged J_{K1} segments were cloned and sequenced. Eight rearranged V_K-J_K fragments were obtained and they were different from each other except for two clones which share the identical germ-line V_K segment but contain different base insertions at the V_K-J_K junction. Assuming that the members of the V_K-KpnI family associate randomly with the J_{K1} segment, the total number of the V_K-KpnI family is estimated to be 12–96 by statistical calculation. Since the portion of the KpnI site containing V_K segment is about 36%, the total number of the V_K segments that can rearrange is between 36 and 300 in general agreement with previous estimations.

Bently & Rabbitts (19) estimated that the number of human V_K segments is only 15–20 on the basis of the number of hybridized bands with three murine V_K probes. Bently (20) extended the study to analysis of human spleen V_K mRNA and showed that a family of about 25 human germ-line V_K genes encodes about 53–61% of K mRNA in spleen or peripheral blood lymphocytes

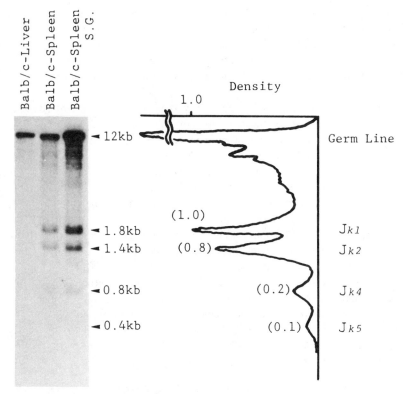

Figure 2 Unequal utilization of the J_K segments. Balb/c liver and spleen DNAs were digested with KpnI/BstEII and electrophoresed in an agarose gel. DNA was transferred to a nitrocellulose filter and hybridized with mouse J_{K4-5} probe. Autoradiogram was traced by a densitometer. Four bands corresponding to V_K-J_{K1} (1.8 kb), V_K-J_{K2} (1.4 kb), V_K-J_{K4} (0.8 kb), and V_K-J_{K5} (0.4 kb) were seen in addition to the germ-line J_K fragment. Balb/c spleen S.G. DNA was enriched for V_K-J_K fragment by sucrose gradient centrifugation after KpnI digestion.

of two individuals. This study simply indicates that a major portion of expressed L_K mRNA in two individuals is derived from a limited number of the V_K segments, which is not a direct answer to the question how many V_K germ-line segments are present in the human genome.

A more direct approach to elucidating the organization of the human V_K gene family is to clone all the human V_K genes with cosmid vectors, which is under way in Zachau's laboratory (21, 22). Among 15 different V_{KI} segments sequenced so far are 6 pseudogenes and 9 potentially functional genes. The longest linkage group extends over a 80-kb region containing 6 V_K segments, two of which are pseudogene segments. It is worth noting that one V_{KII} pseudogene segment is located among a cluster of V_{KI} segments. Furthermore, all the V_K segments of the cluster are orientated in the same direction.

The H chain gene system of the mouse is more complex and comprises many V_H, about 15 D_H, and five J_H segments, in addition to eight C_H genes (22a, Figure 1). The human H chain gene system is basically similar to the mouse system except that the numbers of the J_H segments and the C_H genes are 6 and 11, respectively (22b). The total number of the human D_H segments is not clear.

The number of the murine V_H segments was estimated to be 160 by counting the restriction DNA fragments hybridizing with three murine V_H probes (23). Brodeur and Riblet (submitted for publication) have carried out more extensive studies on 18 inbred strains using 24 V_H probes. They have classified V_H genes into seven groups each containing 2–60 members. The total number was estimated to be 100. The effective number of V genes seems to be considerably smaller than that estimated by hybridization because of abundant pseudogenes. Up to 40% of the V_H genes are pseudogenes according to one estimation (24).

The organization of the human V_H segments was also examined directly by cloning genomic V_H fragments using cosmid vectors (M. Kodaira, I. Umemura, T. Honjo, unpublished data). They have cloned 60 V_H segments dispersed among a 900-kb region. The clustering of homologous V_H segments does not appear to be absolute because a clone contains four V_H segments: two subgroup III members, and one member of each of subgroups II and III. Johnson et al (25) determined that a V_H probe (pCE-1) belonging to subgroup II (26) was 4 centi-Morgan (about 4000 kb) away from the C_μ gene by LOD score analysis of the restriction fragment length polymorphism.

Immunoglobulin Gene Variation During Evolution

The genetic variability of the immunoglobulin gene is determined by two factors: the germ-line gene diversity (number and sequence) and somatic variations. The former is the result of evolutionary processes which include genetic variations by repeated recombinations and drifts, and natural selection. It is obvious that all the multigene families were created by repeated duplication events which lend the organism a new dimension of genetic capacity through diversification of the members of each family. The increment of the number of the V, D, and J segments would serve as a selective advantage up to a certain level and the decrease would work in an opposite manner. In addition, the initial duplication increases the chance of the next duplication by the increase of the recombination target size. Once duplication of a gene begins, members of the gene family cannot avoid divergence by drift. Selection pressure operates to maintain minimal homology among the members. It is not obvious on which structure the selection pressure of the immunoglobulin V segments works as most of them are quite different from each other. Framework regions and recombination signals may be essential for V gene function, though we do not know whether selection of these structures alone is strong enough to conserve the immunoglobulin V gene family.

Comparison of nucleotide sequences of the murine $C_{\gamma 1}$ and $C_{\gamma 2b}$ genes first demonstrated segmental homology between members of the multigene family (27). Similar observations were reported in other C_H genes (28–30), V_K segments (21, 31), and V_H segments (24, 32, 33). Although observations were often interpreted as resulting from gene conversion which unilaterally corrects one sequence by another, it is almost impossible to prove gene conversion events in mammals and distinguish them from unequal double crossing-over. We would like to call these phenomena *segment transfer* because gene conversion is a well-defined genetic term in lower organisms and has not been illustrated in mammals. Segment transfer between family members, be it unilateral or reciprocal, contributes to homogenize diverged sequences within the family. The frequency of segment transfer in the immunoglobulin V gene may be high enough to maintain framework homology, yet low enough to allow divergence in CDR regions. Extensive homology in flanking regions of V genes (34–37) would be the result of segment transfer and at the same time facilitate future recombination.

The high percentage of germ-line pseudogenes of the immunoglobulin V segments might be another source of diversity (24, 26, 33, 38–41). All 14 immunoglobulin V_H pseudogenes characterized to date except one (26) differ from most pseudogenes of other multigene families in two aspects: (*a*) they carry only a few deletion mutations and (*b*) they show no evidence of increased divergence from intact V_H genes (32). The results appear to indicate that these pseudogenes evolved only recently from active genes. However, the high number of pseudogenes that did not diverge beyond one or two crippling mutations suggests that these pseudogenes are being continuously corrected and may be revived by recombination such as segment transfer.

Recent reports on sequence homologies between D_H segments and CDRs of V_H segments (42, 43) revived a modified concept of minigenes (44) as possible substrates for segment transfer events. Deletion of CDR from V_H segments, which are capable of $V_H D_H J_H$ recombination, may implicate translocation of the minigene segment during evolution (26). In these cases, the segment transfer would help to increase diversity by distributing drifts fixed in other regions.

In summary, evolutionary genetic events include two mechanisms: divergence by drift and convergence by recombination (or segment transfer). In addition, segment transfer may help to increase immunoglobulin diversity by scattering drifts into other genes and reviving pseudogenes.

V(D)J Recombination

There are two somatic mechanisms that amplify the immunoglobulin V gene diversity inherited by an individual; V(D)J recombination that randomly assorts two or three sets of germ-line segments (V, D, and J), and somatic mutation that introduces random base replacements into the completed V gene.

SITE-SPECIFICITY AND FLEXIBILITY   V(D)J recombination is a sort of site-specific recombination recognizing the conserved nonamer and heptamer sequences flanking the V, D, and J segments (45, 46). The nonamer and heptamer sequences are separated by either 12 ± 1-bp or 23 ± 1-bp spacers and recombination takes place between a pair of segments containing the signal sequences with different spacer lengths. Although V(D)J recombination is relatively site-specific, the specificity is not absolute but flexible at the nucleotide level as long as recombination does not result in a frame-shift mutation. In the case of the V_H gene there are many insertions or deletions at junctions with D segments. The inserted bases rich in G are called N regions and proposed to be due to the activity of terminal transferase (47). This flexibility at the junction endows organisms with another power to augment the diversity of V genes (14, 15, 47).

RANDOMNESS   The most important principle of V(D)J recombination to guarantee the amplification of the V gene diversity is random pairing of different sets of segments. There is no doubt about the fact that each J is able to associate with many different V and D segments. The question is how randomly the V segments are selected. Recent studies described below suggest that V(D)J recombination is reasonably random but not completely random. There might be physical and chemical restrictions for V(D)J pairing. The maximum repertoire of immunoglobulins could not be calculated by single multiplication of the numbers of the germ-line V, D, J segments. A preferential utilization of the most D-proximal V_H family was reported in the pre-B cell population transformed by Abelson murine leukemia virus (A-MuLV) (48).

There are several examples of the same V segment assorting with different J segments (49). Does one particular V segment recombine to the different J segments with equal probability? Nishi, Kataoka, and Honjo (submitted for publication) demonstrated that V_K segments of the V_K-KpnI family utilize the J_{K1} and J_{K2} segments more efficiently than the J_{K4} and J_{K5} segments. The J_{K1} was utilized by the V_K-KpnI family significantly more frequently than the J_{K5} (Figure 2). They have also shown that the J_{K3} segment is hardly, if ever, used, probably because of divergence in the recombination signal sequence. The results indicate that the recombination efficiency of the murine J_K segments declines towards the distal direction from the V_K cluster for unknown reason(s).

Another feature of randomness is association of the L and H chains. It is clear that any H chain has some preference for certain L chains (50). At the same time, it seems to be true that the H chain can pair with a fairly large variety of L_K chain families. Subclones derived from pre-B cell clones, producing an identical μ chain, rearranged different V_K segments during culture (51). The results indicate that the formation of the V_H sequence does not determine which V_K segment is to be rearranged in a particular lymphocyte.

MOLECULAR MECHANISMS

Order of Rearrangement   During ontogeny the V_H gene rearranges before the V_L gene. Pre-B cell was defined as a stage of the B cell maturation which contains the μ chain in the cytoplasm but no L chains. In other words, the pre-B cell has accomplished successful V_H-D_H-J_H recombination but has not succeeded in V_L-J_L recombination. Thus the pre-B cell contains the μ chain in the cytoplasm but not on the cell surface. Subsequent to the $V_H D_H J_H$ rearrangement, the V_K gene seems to rearrange before the V_λ gene (52–54).

V_H-D_H-J_H recombination proceeds in a definite order; formation of the D_H-J_H complex and then the completion of the V_H-$D_H J_H$ complex (55, 56). AMuLV-transformed fetal liver cells often have finished $D_H J_H$ rearrangements at both J_H loci (47). A similar $D_H J_H$-containing line was also established from BALB/c spleen (57). These lines continued DNA rearrangement in culture, in most cases by joining a V_H gene segment to an existing $D_H J_H$ complex with the simultaneous deletion of intervening DNA sequences (55, 56). None of these lines or their progenies showed evidence of V_H-D_H or D_H-D_H rearrangements. No evidence of $V_H D_H$ or $D_H D_H$ rearrangement was obtained in pre-B cells or normal B-lymphocytes. These results support an ordered pathway of the V_H gene formation in which D_H-to-D_H rearrangement generally occurs first on both chromosomes, followed by V_H-to-$D_H J_H$ rearrangement. In one case a pre-existing $D_H J_{H3}$ complex was completely replaced by a different $D_H J_{H4}$, which then recombined with a V_H segment to complete a V_H gene (55). The $V_H D_H J_H$ rearrangement resulted in the complete deletion of the D_H segments. Recently, an earlier stage of the B cell line was established using Epstein-Barr virus transformation (58). This cell, which we propose to call pre-pro-B cell, has no rearrangement in either the V_H or V_L locus. Consequently, the B cell stage in ontogeny that has the $D_H J_H$ rearrangement on either homolog should be termed pro-B-cell. The maturation steps of the B cell lineage can be classified by the rearrangement profile of the immunoglobulin genes and proceed as shown in Figure 3; pre-pro-B cell (no rearrangement) → pro-B cell ($D_H J_H$) → pre-B cell ($V_H D_H J_H$) → B cell ($V_H D_H J_H$ + $V_L J_L$) → plasma cell (class-switch recombination) as shown in Figure 3.

Inversion Versus Deletion Model   Three general models for antibody gene recombination have been postulated. These models invoke either (*a*) DNA excision (15, 59–61), (*b*) sister chromatid exchange (62), or (*c*) DNA inversion (47, 63, 64) to account for V-J joining. Early evidence (15, 59–61) favored a simple "excision-deletion" model, in which the DNA between V and J segments was excised and lost from the genome. The V_H locus rearrangement seems to operate by an excision deletion mechanism (55) as originally proposed in the class-switch recombination (60). More recently, in conflict with this excision-deletion model, several reports have indicated that the DNA initially

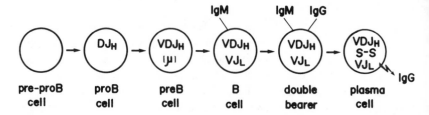

Figure 3   Maturation steps of B cells. DNA rearrangement types shown in the circle can define stages of B cell maturation. In the case of IgM-secreting plasma cells, no S-S recombination takes place.

separating V_K and J_K genes in the germ line is not always absent from the genomes of antibody-producing cells (47, 62, 64, 65).

Selsing et al (66) carried out extensive analyses of "remnant" DNA which is initially located between the V_K and J_K segments and unexpectedly remains in the genome when a simple excision-deletion model would predict its absence. Selsing et al have examined active V_K genes and remnant DNA in twenty cell lines, and have found sixteen remnant DNAs in plasmacytomas or hybridomas, all of which involved recombination at the J_{K1} recognition site. No remnant DNAs directly reciprocal to antibody genes recombined at J_{K2}, J_{K3}, J_{K4}, or J_{K5} were found. The results indicate that the remnant DNA is not necessarily the reciprocal product of the V_K-J_K joining. However, in myelomas (CH2 and T) and an A-MuLV-transformed cell (ABE8), the V_K segments in the active gene and in the remnant DNAs are derived from the same V_K family, suggesting some correlation between remnant DNA and V_K-J_K joining. Selsing et al argue that all three models are compatible with their data assuming that subsequent multistep recombination will follow. The most critical test for the inversion model is to show the orientation of the V_K segment relative to the J_K segment. So far all the V_K segments linked physically are oriented in the same direction with each other (21).

In Vitro Studies On Immunoglobulin Gene Rearrangements   Recently, Lewis et al (67) demonstrated that inversion of the V_K segment took place to join with the J_K segment inserted in a retroviral vector which was introduced into an A-MuLV-transformed pre-B cell. They have employed an elegant selection method for the recombinant. They constructed a defective retrovirus of Moloney sarcoma virus containing V_{K21-C} and J_K segments which are orientated in an opposite direction. The retrovirus also contains between the V_{K21-C} and J_K segments the *Escherichia coli* xanthine-guanine phosphoribosyl transferanse (*gpt*) gene in inverted orientation relative to the promotor, LTR. The *gpt* gene can be transcribed only when inversion took place. The retrovirus was introduced into A-MuLV-transformed lymphoid cells, which were then cultured in selective medium containing mycophenolic acid. Recombinant DNAs were cloned from mycophenolic acid–resistant cells and their nucleotide sequences were determined. One recombinant $V_K J_K$ recombined at -3 and -2 bp away from the heptamer sequences of the V_{K21-C} and J_K segments, respectively. This crossing-over point on the V_{K21-C} segment is identical to that observed in a myeloma MOPC321. Similarly, the cross-over point on J_{K1} is consistent with the site where J_{K1} recombined in another myeloma, MOPC41. The reading frames of the fused V_K and J_K elements do not match. These results not only provide strong support for the inversion model but also imply that the joining reaction is not dependent on long range chromatin or DNA structure, nor even the presence of the C_K gene.

Blackwell & Alt (68) constructed a recombinant plasmid containing the herpes simplex virus thymidine kinase (*tk*) gene, flanked on one side by two murine D_{Q52} segments and on the other by two murine J_H (J_{H3} and J_{H4}) segments. The plasmid was introduced into a tk^- variant of an AMuLV-transformed pre-B cell line 38B9. The four possible site-specific joining events between D_{Q52} and J_H segments occurred frequently during passage of the cloned line under nonselective conditions, and deletion of the internal *tk* gene as a result of these joining events was, by far, the predominant mechanism of resistance to BudR within this line. Nucleotide sequences surrounding recombination sites were close to the ends of the coding region but not exactly precise as often seen in the in vivo product. All of them had 1–10 bp deletion in the coding region but no additional bases corresponding to the N region which was claimed to be due to terminal deoxynucleotidyl transferase. The 38B9 tk^- cell line has very low terminal deoxynucleotidyl transferase levels relative to certain other A-MuLV transformants.

Kataoka et al (69) constructed a recombinant plasmid containing a pair of recombining segments D_{Q52} and J_H, and a genetic marker inserted between them. The genetic marker and host bacteria were chosen so that only the plasmid that lost the marker allows proliferation of the host bacteria in selection drugs. Incubation of the plasmid DNA with lymphocyte extracts and subsequent selection of recombinant plasmids in *E. coli* allowed them to isolate a number of candidates for successful recombinants. After careful analyses of the candidates only a few were shown to have recombined at regions close to those found in B-lymphocytes. The assay system that depends on deletion of the genetic marker tends to pick up many background recombinants due to deleterious recombinations within the marker gene. Nucleotide sequence determination of a recombinant revealed that recombination took place within the J_{H1} coding region. The results can be interpreted in several ways. Since the recombination occurred in the 11-bp homology regions of the germ-line D_{Q52} and J_{H1} segments, the reaction might be ascribed to general homologous recombination. The recombination site was located relatively close (28–29 bp)

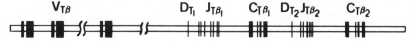

Figure 4 Organization of the T_β genes. Wide closed rectangles indicate exons.

to the 5' end of the J_{H1} segments. This recombination could be due to an aberrant D_H-J_H recombination as the D_H-J_H recombination sites so far determined in B cells are rarely found at the precise end of the coding region (55, 56).

Endonuclease J Attempts to isolate enzymes involved in V-J recombination succeeded in partial purification of an enzyme that introduces double-stranded cleavages near recombinational signal sequences of the immunoglobulin gene (70–72). The enzyme, termed endonuclease J, preferentially cleaves G clusters located slightly off the 5' ends of J_K segments cloned in pBR322. The endonuclease J activity was found not only in lymphoid tissues such as the chicken bursa of *Fabricius* but also in nonlymphoid cells like HeLa cells and mouse liver. In addition, the enzyme cleaved several G clusters in pBR322, which carries V_K and J_K segments. These results make it uncertain whether the enzyme is involved in V-J recombination. However, endonuclease J does not cleave pBR322 per se nor C_H genes (S. Kondo, unpublished data; H. Muriald, personal communication). The results suggest that the recombination signal sequence is required as an entry site of endonuclease J. Deletion of the nonamer sequence of the J_{K4} segment drastically reduced the endonuclease J cleavage activity of the J_{K4} segment, suggesting that the nonamer sequence may constitute at least part of the recognition signal of endonuclase J. Desiderio & Baltimore (72) did not observe the cleavage near the J_{K3} where the heptamer and nonamer signal sequences deviate one base each from the concensus structures. However, the Japanese group did see the J_{K3} cleavage weakly. Endonuclease J should be characterized more extensively before its physiological function can be fully elucidated.

Allelic Exclusion The efficient selection of functionally appropriate genotypes among diverse genotypes is essential if the immune system is to serve as defense mechanism. Selection should ideally operate at the genotype level but in fact it works at the cellular level. Efficient selection of a single phenotype (or clonotype) must coincide with selection of a single genotype, which can be accomplished by allelic exclusion allowing only one set of the H and L chain genes to be expressed in any given lymphocyte. Most data are consistent with the stochastic model that explains allelic exclusion by random competition for successful rearrangements among the different immunoglobulin loci, i.e. homologs and isotypes (K and λ), with a single winner as reviewed before (1,

73). However, regulation does not seem to be strict and may allow leakage in the expression of the other V_K allele, at a maximal level estimated at 8% in rat (74).

Stochastic competition may also exist between immunoglobulin and T cell receptor loci (73, 75). Striking homology of the signal sequence between immunoglobulin genes and T cell receptor genes supports this assumption. The V_H and V_T loci may use a common recombinational apparatus and the successful gene may determine the fate of the lymphocyte, to become a T or B cell.

Somatic Mutation

A number of V_L and V_H gene sequences have clearly demonstrated that somatic mutation takes place not only in CDRs but also in other regions including FR, introns, and flanking regions (38, 46, 76–82). But the C gene does not have any mutations (80). One of the important questions about somatic mutation is again how random it is. It is, however, extremely difficult to answer this question. First of all, most V gene sequences so far available were derived from antibody-producing B cells which had probably been selected by antigen stimulation. This is especially true for well-characterized systems like anti-phosphorylcholine and anti-nitrophosphate antibodies (33, 76, 83). Besides, antigen stimulation, which is inevitably accompanied by selective clonal expansion, seems to enhance the somatic mutation rate. Secondly, there are usually few somatically substituted sites in a single V gene. One should accumulate about one hundred rearranged V genes derived from a single germ-line V segment to evaluate the distribution of mutations. Though these data are not available at this moment, the distribution of base-substitution seems reasonably random within a V gene except for higher incidence in CDRs.

Somatic mutation seems to take place only in the rearranged V gene (84, 85), suggesting that the J region might be required for somatic mutation. The frequency of somatic mutation was estimated to be 10^{-3}/base/generation (M. Weigert, personal communication).

Studies in two systems have suggested that somatic mutation is linked, in some manner, to class switching. Immunoglobulins that have switched from IgM to other classes frequently, but not always, express somatic mutations whereas no mutations have been observed in IgM molecules (16, 32, 53). Other groups showed that somatic mutation was likely to be a continuous process occurring throughout the ontogeny of a B-cell line committed to antibody production and probably not associated with class switching (86, 87). Apparent association of higher mutation rate with non-IgM antibodies suggests that cell division or DNA synthesis, which may not necessarily be accompanied by class switching, may be involved in somatic mutation.

Amino acid substitutions in phosphorylcholine-binding antibodies were

Table 1   Amino acid substitutions in phosphorylcholine-binding antibodies classified according to site and probable underlying mechanism[a]

	Heavy chain		Light chain
	HV3[b]	Total	Total
1. Different D segments	8	8	—
2. Recombination site shift	5	5	—
3. Somatic mutation	3	26	20
4. N region diversity	10	10	—
5. Somatic Mutation or N region	6	6	—
Totals	32	55	20
Total number of substitutions = 107			

[a] Includes 19 complete V_H sequences, 23 N-terminal V_H sequences, 6 complete V_L sequences and 36 N-terminal V_L sequences. Taken from Perlmutter et al (83).
[b] Third hypervariable region.

classified according to the underlying genetic mechanisms, which clearly indicate that somatic mutation plays significant roles in immunoglobulin diversification (Table 1). A similar study on nitrophosphate-binding antibodies has also demonstrated that somatic mutations play important roles for immune diversity (33).

Class Switching

B lymphocytes that have completed the correct V_H-D-J_H and V_L-J_L recombinations produce IgM because the C_μ gene is the closest to the J_H segments. The progeny of a single lymphocyte retains essentially the same V_H region sequence, except for base replacements due to somatic mutation. However, the Ig produced among the progeny often changes from IgM to another class. Paradoxically, the constant region (C_μ) of the immunoglobulin is replaced by another C_H region while the variable region (V_H) remains constant. This phenomenon, known as class switching, is mediated by another DNA rearrangement, called S-S recombination.

Molecular genetic aspects of class switching were recently reviewed elsewhere (75).

DIVERSITY OF THE T CELL ANTIGEN RECEPTOR GENE

Until recently, the nature of the T cell antigen receptor was controversial (108, 109). The generation of antibodies specific for individual T cell clones (clonotypic antibodies) has permitted the first biochemical characterization of the T cell antigen receptor (110–113). The T cell receptor consists of two heterodimeric polypeptide chains, α and β, with molecular weights of 40,000 to

50,000. Peptide map analyses of α and β chains indicate that each chain is composed of constant and variable peptides.

Recent cloning of cDNAs encoding the β chains of mouse and human T cell receptors by subtraction and differential screening techniques (114–116) have triggered the extensive studies on organization and structure of the β chain genes of the T cell receptor. The same technique was employed for cloning of another chain cDNA (117). This field is now rapidly opening up and the results summarized here (at the end of July) are not sufficient to answer many fundamental questions. However, what emerged from these early studies on the T cell receptor genes is strikingly similar to the immunoglobulin genes in structure, organization, and rearrangement. It is, therefore, reasonable to assume that random genetic variability and selection play essential roles in eliciting diversity of the T cell antigen receptor as well.

Structure and Organization

β CHAIN GENES   The structure of the β chain cDNA clone (115, 116, 118) has revealed that the β chain is comprised of the V region (109 residues) and the C region (173 residues). There are two distinct C region genes separated by 6 kb with the same transcriptional orientation in the mouse genome. Each is associated with a cluster of J segments located immediately downstream and the whole complex is located within a 20-kb region (118, 119). The two J clusters contain seven elements each, twelve of which may be functional (Figure 4). The two C_β genes are arranged in four exons, i.e. each encoding the external domain (125 residues), the hinge-like region (6 residues), the transmembrane region (36 residues), and the cytoplasmic tail (6 residues) plus the 3' untranslated region. Introns with similar sizes interrupt both genes at exactly the same positions. The two C_β genes are strikingly similar over their entire coding region sequence: only five amino acid substitutions in 173 residues. This is in marked contrast to the almost complete divergence in the 3' untranslated regions of the two genes (45% homology). There must be strong functional constraints on this gene, much more than seen in the immunoglobulin gene. This assumption is consistent with the close nucleotide and amino acid homology of the C_β gene between mouse and man (114, 116). The amino acid sequence deduced from a human β chain cDNA coincided with that of the T cell receptor (114, 114a). The β chain genes are located on chromosome 6 in mice and chromosome 7 in humans (120, 121).

Comparison of the expressed and the germ-line forms of the β chain genes has clearly demonstrated that the V region gene of the β chain is composed of three distinct germ-line segments; V_β, D_β, and J_β segments (118a, 122, 123). Germ-line D_β segments were found approximately 650 and 580 bp 5' to the $J_{\beta 1}$ and $J_{\beta 2}$ clusters, respectively (122, 124). The number of restriction fragments hybridizing to cloned V_β probes was relatively smaller (a few bands by a V_β probe) than that of the immunoglobulin V segments. However, we have recently cloned four different human V_β clones and estimated the number of the germ-line V_β gene by Southern blot hybridization. One of the V_β probes lighted up about 8 fragments. The minimum number we obtained was 16 (K. Ikuta, A. Shimizu, T. Ogura, T. Honjo, unpublished data).

Recombination

Analyses of the germ-line V_β, D_β, and J_β segments indicate that these segments are flanked by conserved sequences similar to the heptamer and nonamer sequences characteristic of the immunoglobulin gene, separated by either 12 bp or 23 bp (118, 119). As shown in Table 2, the spacer lengths of the signal sequences located 3' to the D segment and 5' to the J_β segment are inverted as compared to the D_H-J_H pairs of the immunoglobulin H chain gene. This arrangement raises the possibility that the D_β-D_β and V_β-J_β rearrangements take place because both segmental pairs contain a pair of signal sequences with 12-bp and 23-bp spacers.

The striking homology in the nonamer and heptamer sequences and their spacer lengths suggests that the immunoglobulin gene and the T cell receptor gene might share at least a part of the recombination machinery during differentiation. Comparison of nucleotide sequences of the D_β-J_β recombination site and their germ-line counterparts indicates that D_β-J_β recombination is imprecise with one-base insertion and four-base deletion at the junction (118, 122). The results indicate that the recombination of the T cell receptor gene is

Table 2 Nucleotide sequence comparison of T-cell receptor and immunoglobulin genes[a]

3' Flanking	Heptamer	Spacer (bp)	Nonamer
V_K	CACAGTG	12 ± 1	GGTTTTTGT
Vλ	CACAATG	23	GGTTTTTGC
V_H	CACAGTG	23 ± 1	NGTTTTTGT
$V_{T\beta}$	CACAGCA	23	AGTTTTTGT
D_H (3' side)	CACAGTG	12	GATTTTTGT
D_T (3' side)	CACGGTG	23	CTTTTTTGT

5' Flanking	Nonamer	Spacer (bp)	Heptamer
J_K	ACAAAAACC	23 ± 1	CACTGTG
Jλ1	ACAAGAACA	12	CACAGTG
J_H	ACAAAAACC	23 ± 1	NACTGTG
$J_{T\beta}$	GCATAAACC	12 ± 2	NGCTGTG
D_H (5' side)	ACAAAAACC	12	TACTGTG
D_T (5' side)	ACAAAAACC	12	CATTGTG

[a] Taken from Kavaler et al (124).

also flexible, which contributes to augment the genetic diversity of the T cell receptor gene. Since there are little, if any, structural data about the V region of the T cell receptor, we are unable to convince readers about the extent of flexibility and randomness of V_β-D_β-J_β recombination of the T cell receptor gene. However, the striking similarity in the recombination signal sequence is certainly suggestive that the high level of randomness established in the immunoglobulin gene prevails in the T cell receptor gene.

An expressed murine V_β gene did not contain any somatic mutations (123). Although human V_β cDNA sequence was shown to contain ten base changes as compared with a putative germ-line counterpart (118a), more recent analysis revealed that there was the other germ-line V_β gene matching completely with the cDNA sequence (L. Hood, personal communication).

SELECTION MECHANISM

Random genetic events, on the one hand, contribute to amplification of limited germ-line diversity of the immunoglobulin and T cell receptor genes but, on the other hand, take a chance to produce B and T cell antigen receptors that recognize and potentially react with self-components. Without proper selection mechanisms, the immune system will become a troop shooting its own people. Autoimmune diseases are in a way unwelcome by-products of the sophisticated defense system. Selection within the immune system takes place at the clonal level, which requires that the genotype match with the clonotype.

Our understanding of the mechanism involved in clonal selection is controversial or more precisely little, if at all, known on a molecular basis although there are several consensus facts as described below. It is impossible for anyone working in the immune system to discuss these problems in an unbiased manner. No one has attempted to construct a general model that describes the whole story of clonal selection of T and B cells because of conflicting views presented for each aspect. Therefore, we will first summarize the basic phenomena and then propose a model explaining most of the known facts.

Experimental Facts

SELF-MHC RESTRICTION   The MHC contains a series of genes, the products of which can be classified into three general categories designated classes I, II, and III on the basis of both functional and structural characteristics (125). The class I molecules, which are expressed on most cells, are membrane glycoproteins composed of $\beta2$ microgobulin and a heavy chain. The class II molecules, which are expressed predominantly on B cells, macrophages, and thymus epithelial cells, consist of α and β chains. The class III molecules are components of the complement system.

MHC molecules are highly polymorphic and T cells are generally inert to antigens presented by cells expressing MHC molecules of a different type from the T cells. This phenomenon is termed self-MHC restriction. In general, helper and inducer T cells recognize antigens on the antigen-presenting cells which express class II molecules identical to those present on the functional T cells. Cytotoxic T cells recognize antigens such as viral glycoprotein and/or membrane proteins of other cells with class I molecules and lyse the target cells of the appropriate MHC type only.

MHC restriction is a direct consequence of the dual specificity of the T cell antigen receptor for both antigen and an MHC-encoded molecule (127). One of the hottest arguments about the T cell receptor molecule is whether or not it has two recognition units, one for MHC and the other for the antigen (126). Accumulating evidence indicates that T cells are clonally specific and recognize the foreign antigens in association with one allelic MHC gene product, usually a self-MHC molecule (127). The α and β chain cDNA sequences of the T cell antigen receptor have revealed that both chains have two domains, the V region and the C region. The proposed structure is similar to MHC molecules (114–117a, 118a) (Figure 5). The clonotypic antibody precipitates only the α and β chains (110–113), indicating that only these two chains are physically linked in agreement with the one recognition unit theory rather than the two recognition unit theory. However, we cannot exclude the possibility that T cells contain another set of receptors specific to the MHC molecules.

On the other hand, it is supposed that B cells are not directly restricted by the MHC. In other words, B cells recognize antigen without association with MHC molecules, though B cells require activation of antigen-specific helper T cells

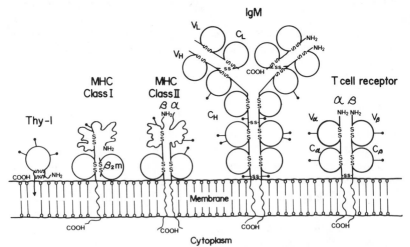

Figure 5   Structure of the T cell receptor and related surface proteins.

which are restricted by the MHC (127a). Some antigens induce immunoglobulin synthesis without the T cell help.

ROLE OF THE THYMUS IN MHC RESTRICTION   The self-MHC restriction of T cells is considered not to be determined by the genotypic constraints on the potential specificities of the T cell receptors but by the allelic MHC products expressed in the thymus. This concept was derived from the experimental results of Zinkernagel et al (128–130), Bevan (131), Bevan & Fink (131a), and many others. T cell–depleted bone marrow cells from a strain A were injected into a lethally irradiated F_1 strain (A×B) recipient. T cells from the bone marrow transplantation chimera can now recognize antigens in association with the MHC of both A and B strains. Lymphocytes in the thymus are selected to recognize the MHC gene products expressed in the thymus as self. It should be emphasized that the selection of T cells in the thymus takes place prior to any encounter with foreign antigens.

ANTIGEN INDUCES PROLIFERATION OF CLONES SPECIFIC TO THE ANTIGEN   It is well known that both peripheral T and B cells are stimulated by specific antigens to proliferate and consequently to expand their progenies. Once the T cell antigen receptor interacts with an antigen under MHC restriction, the T cell proliferation is regulated by the lymphokine interleukin 2 (IL-2) through its receptor. Both IL-2 and IL-2 receptor genes are expressed only in activated T cells at the early stage after the antigen stimulation (132–134). A resting B cell is activated with an antigen in collaboration with class II MHC-compatible helper T cells or can be activated without helper T cells by T cell–independent antigens such as bacterial polysaccharide, ficol, and lipopolysaccharide. These specific activation steps for B cells are supposed to induce receptors for growth-promoting lymphokines, generally termed B cell growth factors (BCGF) which are produced by helper T cells upon antigen stimulation (135). B cell proliferation is triggered by these lymphokine/receptor interactions. Molecular properties and identity of BCGF from different laboratories are not completely determined. The maturation step that leads B cells to secretion of the immunoglobulin also depends on separate lymphokine/receptor system(s) (136).

Antigen-dependent proliferation of T and B cells expands the antigen-specific clone selectively. It is worth noting that the only activation step is in general antigen-specific and subsequent proliferation is mediated by the lymphokine/receptor system in a polyclonal manner. The regulatory system for expression of the lymphokine receptor on both T and B cells appears to be a safeguard against a catastrophic spread of lymphocyte proliferation by an immunogenic stimulus.

TOLERANCE   Tolerance is defined as a specific depression of the immune response induced by previous exposure to the antigen. Induction of tolerance depends on several factors: dose of antigen, properties of the antigens, and physiological state of organism. Detailed dose-response studies by Mitchison (137) indicated that tolerance to a protein antigen is induced in the adult over two distinct ranges of dosage: above and below the immunizing range, which became known as high and low zone tolerance. The low zone tolerance is due to unresponsiveness of the helper T cell population, but is not universal to antigens (138). The high zone tolerance is ascribed to unresponsiveness of both the B cell and the helper T cell populations. Generally, it is very easy to induce tolerance with nonmetabolizable antigens consisting of repeating units, such as polysaccharide. The results suggest that maintenance of a high concentration of the antigen and multivalency may be important for the tolerance induction (139). In this respect, it might be an attractive idea that tolerance is due to continuous down-regulation of antigen receptors by persistent antigens. Self-tolerance, however, is not necessarily absolute, but rather is leaky because autoantibodies are occasionally found in serum of normal individuals as well as of autoimmune patients (140, 141).

Positive Affinity Selection Model

This model is proposed to explain clonal selection mechanisms of T and B cells in molecular terms depending on many speculations. The model postulates two positive selection steps in the maturation pathway of T cells; the first selection in the thymus and the second selection in the periphery. The selection mechanism depends on the affinity of the antigen receptor to the recognizing molecule. For B cells only the second selective mechanism operates during the maturation pathway. We have adopted several principles of the selection mechanism based on the experimental facts described above. (a) The lymphocyte clone is positively and specifically selected by proliferation triggered by signals through the antigen receptor. (b) Selection depends on the total intensity of the signals transmitted upon interaction with ligands through the antigen receptor expressed on the lymphocyte. (c) The T cell receptor consisting of the α and β chains constitutes one recognition unit which recognizes both the antigen and the MHC molecule. This includes two possibilities; simultaneous recognition of the antigen-MHC product complex and separate recognition of the antigen and the MHC product by cross-specificity of a single receptor. The outline of this model is schematically shown in Figure 6.

SELECTION IN THE THYMUS   T cells are selected in the thymus so that self-MHC restriction is imprinted on each T cell clone. This selection could be most easily explained and accomplished by some interaction between the T cell receptor and MHC molecules in the thymus because it is shown that class II

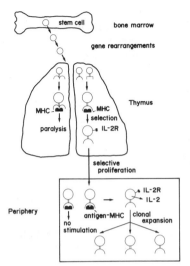

Figure 6 Schematic representation of T cell selection by the model. The scheme was drawn according to the model that the T cell receptor recognizes the antigen-MHC complex. IL-2R, IL-2 receptor.

molecules first appear in the subcortical area of the thymus in ontogeny (142). T cells are, therefore, required to have completed rearrangements of the receptor genes by the time or soon after they arrive at the thymus so that T cells in the thymus establish their clonality for antigens including their own MHC by random rearrangements of the T cell receptor gene (142a).

The criterion of the selection is based on the intensity of the signals transmitted to the T cell cytoplasm through the T cell receptor molecules. The intensity of the signal is defined by k_A × [T cell receptor concentration] × [MHC product concentration] where k_A represents the affinity constant of the T cell receptor. We assume that only moderate interaction with the T cell antigen receptor could give rise to a signal that induces proliferation and maturation of the T cell. Strong (or excessive) interaction with the receptor would result in paralysis of the proliferation signal transfer and failure to receive a positive selection signal. Thus, those T cells whose receptors have strong affinity to self-MHC products would be unable to mature and eventually die. This assumption is supported by the report that an excessive epidermal growth factor receptor signal causes the growth inhibition of A431 cells which express 20-fold more epidermal growth factor receptor (143, 144). If the affinity of a T cell receptor to self-MHC products is lower than a certain limit, the T cell may be unable to proliferate and mature.

What is the nature of the maturation signal induced by the moderate interaction with MHC molecules? We assume that the T cells receiving the signal in the thymus acquire inducibility of the IL-2 receptor. In other words, the IL-2 receptor gene becomes ready to accept induction signals by antigen binding through the T cell receptor, leading to the expression of the IL-2 receptor on the surface. The thymus may provide a special environment that helps change the inducibility of the IL-2 receptor gene: an appropriate concentration of MHC molecules, and/or specific lymphokines. The stimulation of the T cell receptor may be coupled with another lymphokine system in the thymus and combination of the two signals may result in activation of the IL-2 receptor gene. The appearance of the IL-2 receptor in thymocytes during early stages of ontogeny supports that the IL-2 receptor is involved in T cell maturation (145). The positive selection mechanism allows only those T cells that can interact moderately with their own MHC molecules to mature one step further and to prepare for the future clonal expansion upon encountering the antigen and MHC product.

The MHC molecules are evolutionarily related to the T cell receptor. The T cell receptor may have a potential affinity to MHC molecules by analogy to the receptor for transepithelial transport of IgA and IgM which binds immunoglobulins through its immunoglobulin-like domains (146). The T cell receptor may have evolved to recognize self-MHC products by like-like interaction. The T cell receptor repertoire, however, may contain clones that are capable of interacting with self-components other than MHC molecules. A similar selection in the thymus, therefore, could occur on the basis of the affinity to other self-components expressed on thymic lymphocytes and stromal cells. If this happens, T cell clones reacting strongly with self-components in the thymus would not be allowed to expand. The T cell selection in the thymus may thus explain the self-tolerance at least in part.

SELECTION IN THE PERIPHERY   We assume that the threshold of the signal intensity required for the activation of the peripheral T cell is different from that for the thymic T cell. Let us assume that the peripheral T cell requires a stronger affinity of the T cell receptor with a ligand. This could be because the lymphokines that helped IL-2 receptor expression in the thymus are no longer available in the periphery. One can also imagine that the type of the interaction with the T cell receptor in the thymus would be different from that in the periphery. Assuming that the T cell receptor is monovalent (Figure 6), the interactions of the T cell receptor with MHC molecules alone in the thymus would not cause clustering of the receptor. However, antigens may be processed and presented as a polyvalent structure in association with MHC molecules on the macrophage. Such antigen-MHC product complexes with polyvalent nature may induce clustering of the T cell receptor in the periphery.

Inversely, the peripheral T cell may require a weaker affinity of the T cell receptor for a ligand. The concentrations of the MHC molecules and the T cell receptor might be higher in the periphery (147). The induction mechanism of the IL-2/IL-2 receptor genes might become more sensitive after selection in the thymus; the second hit could be easier than the first one.

In either case, the T cell clone positively selected in the thymus is, thus, unable to be stimulated by its own MHC molecules alone in the periphery. A proper combination of antigen and MHC product as complex may be able to interact with the MHC-selected T cell receptor with a stronger (or weaker) affinity. A T cell clone which happens to interact properly with an antigen-MHC product complex is activated and induced to express the IL-2/IL-2 receptor system, resulting in extensive proliferation of the clone.

An important question is whether or not somatic mutation in the T cell receptor gene is as frequent as in the immunoglobulin genes. The absence of somatic mutation makes it easy to avoid the possibility that the T cell clones once selected with moderate affinity receptors to their own MHC may turn into clones with receptors of a stronger or weaker affinity to their own MHC, which will destroy the selection mechanism. If so, another important question arises whether the T cell receptor repertoire is large enough to encounter numerous antigens without the somatic mutation mechanism. It is hard to answer this question at this stage. However, the presence of 16 restriction fragments containing the V_β segments with 4 V_β probes in the human genome suggests that the germ-line V_β repertoire is comparable to or slightly smaller than that of the immunoglobulin V segments. Somatic mutation in the immune system may be more effective to increase the affinity of the receptor to an antigen than to change the specificity of the receptor. Anyway, we will probably know the frequency of somatic mutation in the T cell receptor gene by the time this review is printed. A B cell is selected only in the periphery by direct interaction of the surface immunoglobulin with an antigen. The selection is also positive and depends on the intensity of the signal through the antigen receptor. Only those B cells interacting with antigens with proper signal intensity are activated and then proliferate as described above.

A too-intense signal through the antigen receptor of T and B cells could cause paralysis of the proliferation signal (tolerance) in the periphery as well. Abundant self-antigens, when associated with their own MHC molecules, would normally induce and maintain tolerance.

It should be noted that the selection mechanism proposed is not absolute because the positive affinity selection model does not impose the active elimination mechanism of any clones. Thus, the occasional finding of autoreactive clones is not unexpected. In addition, the T cell receptor molecule has had a broad spectrum of cross-specificity (152). Some clones probably escape MHC-restriction by the cross-specificity of the receptor (148–151).

Experimental Tests of Positive Affinity Selection Model

Experimental tests of the positive affinity selection model will be carried out from several aspects. The essential evidence required for our model is that prethymic T cells must be stimulated with self-MHC molecules alone without foreign antigens. This can be tested by isolation of prethymic T cells expressing the T cell receptor and culturing them with thymus epithelial cells expressing the MHC molecules. Thymic epithelial cells may be replaced by L cells transfected with an MHC gene or by artificial membranes carrying MHC products. Soluble MHC products synthesized by recombinant DNA may also help such studies.

The model predicts that the threshold of the signal intensity through the T cell receptor changes between pre- and postthymic T cells. A monoclonal antibody against the constant portion of the T cell receptor may be used to trigger T cell activation as already shown by a clonotypic antibody (111, 154). Careful evaluation of the amount of the anti T cell receptor monoclonal antibody required for the activation of the pre- and postthymic T cell clones would reveal the difference of the signal intensity threshold between the pre- and postthymic T cells.

A direct test for one binding unit theory would be available by DNA transfection and in vitro mutagenesis of the T cell receptor gene. If the change in the V_β or V_α sequence results in a change of both antigen and MHC specificities, the α and β chains must be responsible for recognition of both antigen and MHC products.

One of the important claims of the theory is that postthymic T cells acquire inducibility of the IL-2/IL-2 receptor system. It is feasible to see whether or not the IL-2 receptor is inducible in prethymic T cells. Recently, cDNAs encoding the IL-2 receptor were cloned from adult T cell leukemia cells (155, 156). The genomic clone of this receptor was also isolated and characterized, which enabled us to study the regulatory mechanism of the IL-2 receptor gene and to test whether the expression of this gene changes after selection in the thymus. If the IL-2 receptor induction is involved in the selection mechanism in the thymus, one can test whether the self-MHC restriction is destroyed by introducing an unregulated recombinant form of the IL-2 receptor gene in stem cells.

ACKNOWLEDGMENTS

We are grateful to Drs. M. Mishina, J. Yodoi, S. Narumiya, A. Shimizu, and Y. Yaoita for stimulating discussions and reading of the manuscripts, and to Miss K. Kawabata for preparation of the manuscript. This investigation was supported by grants from the Ministry of Science, Culture and Education of Japan.

Literature Cited

1. Honjo, T. 1983. *Ann. Rev. Immunol.* 1:499–528
2. Tonegawa, S. 1983. *Nature* 302:575–81
3. Wu, T. T., Kabat, E. A. 1970. *J. Exp. Med.* 132:211–50
4. Swan, D., D'Eustachio, P., Leinwand, L., Seidman, J., Keithley, D., Ruddle, F. H. 1979. *Proc. Natl. Acad. Sci. USA* 76:2735–39
5. D'Eustachio, P., Bothwell, A. L. M., Takaro, T. K., Baltimore, D., Ruddle, F. H. 1981. *J. Exp. Med.* 153:793–800
6. D'Eustachio, P., Pravtcheva, D., Marcu, K., Ruddle, F. H. 1980. *J. Exp. Med.* 151:1545–50
7. Ellison, J., Hood, L. 1982. *Proc. Natl. Acad. Sci. USA* 79:1984–88
8. McBride, O. W., Hieter, P. A., Hollis, G. F., Swan, D., Otey, M. C., Leder, P. 1982. *J. Exp. Med.* 155:1480–90
9. Erikson, J., Martinis, J., Croce, C. M. 1981. *Nature* 294:173–75
10. Kirsch, I. R., Morton, C. C., Nakahara, K., Leder, P. 1982. *Science* 216:301–3
11. Blomberg, B., Tonegawa, S. 1982. *Proc. Natl. Acad. Sci. USA* 79:530–33
12. Miller, J., Selsing, E., Storb, U. 1982. *Nature* 295:428–30
12a. Elliott, B. W. Jr., Eisen, H. N., Steiner, L. A. 1982. *Nature* 299:559–61
13. Hieter, P. A., Hollis, G. F., Korsmeyer, S. J., Waldmann, T. A., Leder, P. 1981. *Nature* 294:536–40
14. Max, E. E., Seidman, J. G., Leder, P. 1979. *Proc. Natl. Acad. Sci. USA* 76:3450–54
15. Sakano, H., Huppi, K., Heinrich, G., Tonegawa, S. 1979. *Nature* 280:288–94
16. Hieter, P. A., Maizel, J. V., Leder, P. 1982. *J. Biol. Chem.* 257:1516–22
17. Zeelon, E. P., Bothwell, A. L. M., Kantor, F., Schechter, I. 1981. *Nucleic Acids Res.* 9:3809–20
18. Cory, S., Tyler, B. M., Adams, J. M. 1981. *J. Mol. Appl. Genet.* 1:103–16
19. Bentley, D. L., Rabbitts, T. H. 1980. *Nature* 288:730–33
20. Bentley, D. L. 1984. *Nature* 307:77–80
21. Pech, M., Jaenichen, H.-R., Pohlenz, H.-D., Neumaier, P. S., Klobeck, H.-G., Zachau, H. G. 1984. *J. Mol. Biol.* 176:189–204
22. Straubinger, B., Pech, M., Mühlebach, K., Jaenichen, H.-R., Bauer, H.-G., Zachau, H. G. 1984. *Nucleic Acids Res.* 12:5265–75
22a. Shimizu, A., Takahashi, N., Yaoita, Y., Honjo, T. 1982. *Cell* 28:499–506
22b. Flanagan, J. G., Rabbitts, T. H. 1982. *Nature* 300:709–13

23. Kemp, D. J., Tyler, B., Bernard, O., Gough, N., Gerondakis, S., et al. 1981. *J. Mol. Appl. Genet.* 1:245–61
24. Rechavi, G., Ram, D., Glazer, L., Zakut, R., Givol, D. 1983. *Proc. Natl. Acad. Sci. USA* 80:855–59
25. Johnson, M. J., Natali, A. M., Cann, H. M., Honjo, T., Cavalli-Sforza, L. L. 1984. *Proc. Natl. Acad. Sci. USA* In press
26. Takahashi, N., Noma, T., Honjo, T. 1984. *Proc. Natl. Acad. Sci. USA* 81:5194–98
27. Miyata, T., Yasunaga, T., Yamawaki-Kataoka, Y., Obata, M., Honjo, T. 1980. *Proc. Natl. Acad. Sci. USA* 77:2143–47
28. Ollo, R., Auffray, C., Morchamps, C., Rougeon, F. 1981. *Proc. Natl. Acad. Sci. USA* 78:2442–46
29. Ollo, R., Rougeon, F. 1983. *Cell* 32:515–23
30. Schreier, P. H., Bothwell, A. L. M., Mueller-Hill, B., Baltimore, D. 1981. *Proc. Natl. Acad. Sci. USA* 78:4495–99
31. Jaenichen, H.-R., Pech, M., Lindenmaier, W., Wildgruber, N., Zachau, H. G. 1984. *Nucleic Acid. Res.* In press
32. Cohen, J. B., Givol, D. 1983. *EMBO J.* 2:2013–18
33. Loh, D. Y., Bothwell, A. L. M., White-Scharf, M. E., Imanishi-Kari, T., Baltimore, D. 1983. *Cell* 33:85–93
34. Seidman, J. G., Max, E. E., Leder, P. 1979. *Nature* 280:370–75
35. Cohen, J. B., Effron, K., Rechavi, G., Ben-Neriah, Y., Zakut, R., Givol, D. 1982. *Nucleic Acids Res.* 10:3353–70
36. Gebhard, W., Zachau, H. G. 1983. *J. Mol. Biol.* 170:255–70
37. Gebhard, W., Zachau, H. G. 1983. *J. Mol. Biol.* 170:567–73
38. Bothwell, A. L. M., Paskind, M., Reth, M., Imanishi-Kari, T., Rajewsky, K., Baltimore, D. 1981. *Cell* 24:625–37
39. Givol, D., Zakut, R., Effron, K., Rechavi, G., Ram, D., Cohen, J. B. 1981. *Nature* 292:426–30
40. Rechavi, G., Bienz, B., Ram, D., Ben-Neriah, Y., Cohen, J. B., et al. 1982. *Proc. Natl. Acad. Sci. USA* 79:4405–9
41. Huang, H., Crews, S., Hood, L. 1981. *J. Mol. Appl. Genet.* 1:93–101
42. Wu, T. T., Kabat, W. A. 1982. *Proc. Natl. Acad. Sci. USA* 79:5031–32
43. Bernstein, K. E., Reddy, E. P., Alexander, C. B., Mage, R. G. 1982. *Nature* 300:74–76

44. Kabat, E. A., Wu, T. T., Bilofsky, H. 1978. *Proc. Natl. Acad. Sci. USA* 75:2429–33
45. Early, P., Huang, H., Davis, M., Calame, K., Hood, L. 1980. *Cell* 19:981–92
46. Sakano, H., Hüppi, K., Heinrich, G., Tonegawa, S. 1979. *Nature* 280:288–94
47. Alt, F. W., Baltimore, D. 1982. *Proc. Natl. Acad. Sci. USA* 79:4118–22
48. Blackwell, T. K., Yancopoulos, G. D., Alt, F. W. 1984. *UCLA Symp. Mol. Cell Biol.* (NS) 19: In press
49. Rocca-Serra, J., Tonnelle, C., Fougereau, M. 1983. *Nature* 304:353–55
50. Kwan, S-P., Max, E. E., Seidman, J. G., Leder, P., Scharff, M. D. 1981. *Cell* 26:57–66
51. Ziegler, S. F., Treiman, L. J., Witte, O. N. 1984. *Proc. Natl. Acad. Sci. USA* 81:1529–33
52. Korsmeyer, S. J., Hieter, P. A., Sharrow, S. O., Goldman, C. K., Leder, P., Waldmann, T. A. 1982. *J. Exp. Med.* 156:975–85
53. Hieter, P. A., Korsmeyer, S. J., Waldmann, T. A., Leder, P. 1981. *Nature* 290:368–72
54. Coleclough, C., Perry, R. P., Karjalainen, K., Weigert, M. 1981. *Nature* 290:372–78
55. Yaoita, Y., Matsunami, N., Choi, C. Y., Sugiyama, H., Kishimoto, T., Honjo, T. 1983. *Nucleic Acids Res.* 11:7303–16
56. Alt, F. W., Yancopoulos, G. D., Blackwell, T. K., Wood, C., Thomas, E., et al. 1984. *EMBO J.* 3:1209–19
57. Sugiyama, H., Akira, S., Kikutani, H., Kishimoto, S., Yamamura, Y., Kishimoto, T. 1983. *Nature* 303:812–15
58. Katamine, S., Otsu, M., Tada, K., Tsuchiya, S., Sato, T., et al. 1984. *Nature* 309:369–72
59. Cory, S., Adams, J. M. 1980. *Cell* 19:37–51
60. Honjo, T., Kataoka, T. 1978. *Proc. Natl. Acad. Sci. USA* 75:2140–44
61. Seidman, J. G., Nau, M. M., Norman, B., Kwan, S. P., Scharff, M., Leder, P. 1980. *Proc. Natl. Acad. Sci. USA* 77:6022–26
62. Steinmetz, M., Altenburger, W., Zachau, H. G. 1980. *Nucleic Acids Res.* 8:1709–20
63. Lewis, S., Rosenberg, N., Alt, F., Baltimore, D. 1982. *Cell* 30:807–16
64. Hochtl, J., Zachau, H. G. 1983. *Nature* 302:260–63
65. Van Ness, B. G., Coleclough, C., Perry,

R. P. Weigert, M. 1982. *Proc. Natl. Acad. Sci. USA* 79:262–66
66. Selsing, E., Voss, J., Storb, U. 1984. *Nucleic Acids Res.* 12:4229–46
67. Lewis, S., Gifford, A., Baltimore, D. 1984. *Nature* 308:425–28
68. Blackwell, T. K., Alt, F. W. 1984. *Cell* 37:105–12
69. Kataoka, T., Nishi, M., Kondo, S., Takeda, S., Honjo, T. 1983. *Progress in Immunology V*, ed. Y. Yamamura, T. Tada, 123–33. New York: Academic
70. Kataoka, T., Kondo, S., Nishi, M., Kodaira, M., Honjo, T. 1984. *Nucleic Acids Res.* 12:5995–6010
71. Kondo, S., Kataoka, T., Nishi, M., Kodaira, M., Takeda, S., Honjo, T. 1984. *Cold Spring Harbor Symp. Quant. Biol.* 49: In press
72. Desiderio, S., Baltimore, D. 1984. *Nature* 308:860–62
73. Coleclough, C. 1983. *Nature* 303:23–26
74. Tsukamoto, A., Weissman, I. L., Hunt, S. V. 1984. *EMBO J.* 3:975–81
75. Shimizu, A., Honjo, T. 1984. *Cell* 36:801–3
76. Crews, S., Griffin, J., Huang, H., Calame, K., Hood, L. 1981. *Cell* 25:59–66
77. Gershenfeld, H. K., Tsukamoto, A., Weissman, I. L., Joho, R. 1981. *Proc. Natl. Acad. Sci. USA* 78:7674–78
78. Hozumi, N., Wu, G. E., Murialdo, H., Roberts, L., Vetter, D., et al. 1981. *Proc. Natl. Acad. Sci. USA* 78:7019–23
79. Kataoka, T., Nikaido, T., Miyata, T., Moriwaki, K., Honjo, T. 1982. *J. Biol. Chem.* 10:277–85
80. Kim, S., Davis, M., Sinn, E., Patten, P., Hood, L. 1981. *Cell* 27:573–81
81. Pech, M., Hochtl, J., Schnell, H., Zachau, H. G. 1981. *Nature* 291:668–70
82. Selsing, E., Storb, U. 1981. *Cell* 25:47–58
83. Perlmutter, R. M., Crews, S. T., Douglas, R., Sorensen, G., Johnson, N., et al. 1984. *Adv. Immunol.* In press
84. Gorski, J., Rollini, P., Mach, B. 1983. *Science* 220:1179–81
85. Gearhart, P. J., Bogenhagen, D. G. 1983. *Proc. Natl. Acad. Sci. USA* 80:3439–43
86. Rudikoff, S., Pawlita, M., Pumphrey, J., Heller, M. 1984. *Proc. Natl. Acad. Sci. USA* 81:2162–66
87. Owen, J. A., Sigal, N. H., Klinman, R. 1982. *Nature* 295:347–48
88 –107 Deleted in proof
108. Nakanishi, K., Sugimura, K., Yaoita,

Y., Maeda, K., Kashiwamura, S., et al. 1982. *Proc. Natl. Acad. Sci. USA* 79:6984–88

109. Kronenberg, M., Kraig, E., Hood, L. 1983. *Cell* 34:327–29

110. Acuto, O., Hussey, R. E., Fitzgerald, K. A., Protentis, J. P., Meuer, S. C., et al. 1983. *Cell* 34:717–26

111. Kappler, J., Kubo, R., Haskins, K., White, J., Marrack, P. 1983. *Cell* 34:727–37

112. McIntyre, B., Allison, J. 1983. *Cell* 34:739–46

113. Samelson, L., Germain, R., Schwartz, R. 1983. *Proc. Natl. Acad. Sci. USA* 80: 6972–76

114. Yanagi, Y., Yoshikai, Y., Leggett, K., Clark, S. P., Aleksander, I., Mak, T. W. 1984. *Nature* 308:145–49

114a. Acuto, O., Fabbi, M., Smart, J., Poole, C., Protentis, J., et al. 1984. *Proc. Natl. Acad. Sci. USA.* 81:3851–55

115. Hedrick, S. M., Cohen, D. I., Nielsen, E. A., Davis, M. M. 1984. *Nature* 308:149–53

116. Hedrick, S. M., Nielsen, E. A., Kavaler, J., Cohen, D. I., Davis, M. M. 1984. *Nature* 308:153–58

117. Chien, Y., Becker, D. M., Lindsten, T., Okamura, M., Cohen, D. I., Davis, M. M. 1984. *Nature* 312:31–35

117a. Saito, H., Kranz, D. M., Takagaki, Y., Hayday, A. C., Eisen, H. N., Tonegawa, S. 1984. *Nature* 312:36–40

118. Gascoigne, N. R. J., Chien, Y. H., Becker, D. M., Kavaler, J., Davis, M. M. 1984. *Nature* 310:387–91

118a. Siu, G., Clark, S. P., Yoshikai, Y., Malissen, M., Yanagi, Y., et al. 1984. *Cell* 37:393–401

119. Malissen, M., Minard, K., Mjolsness, S., Kronenberg, M., Goverman, J., et al. 1984. *Cell* 37:1101–10

120. Caccia, N., Kronenberg, M., Saxe, D., Haars, R., Bruns, G. A. P., et al. 1984. *Cell* 37:1091–99

121. Lee, N. E., D'Eustachio, P., Pravtcheva, D., Ruddle, F. H., Hedrick, S. M., Davis, M. M. 1984. *J. Exp. Med.* 160:905–13

122. Siu, G., Kronenberg, M., Strauss, E., Haars, R., Mak, T. W., Hood, L. 1984. *Nature.* 311:344–50

123. Chien, Y., Gascoigne, N. R. J., Kavaler, J., Lee, N. E., Davis, M. M. 1984. *Nature* 309:322–26

124. Kavaler, J., Davis, M. M., Chien, Y. 1984. *Nature* 310:421–23

125. Klein, J., Figueroa, F., Nagy, Z. A. 1983. *Ann. Rev. Immunol.* 1:119–42

126. Cohn, M. 1983. *Cell* 33:657–69

127. Schwartz, R. H. 1984. *Fundamental Immunology*, ed. W. E. Paul, pp. 379–438 New York: Raven Press

127a. Singer, A., Asano, Y., Shigeta, M., Hathcoca, K. S., Ahmed, A., Fathman, C. G., Hodes, R. J. 1982. *Immunol. Rev.* 64:137–60

128. Zinkernagel, R. M., Klein, P. A., Klein, J. 1978. *Immunogenetics* 7:73–77

129. Zinkernagel, R. M., Callahan, G. N., Althage, A., Cooper, S., Klein, P. A., Klein, J. 1978. *J. Exp. Med.* 147:882–96

130. Zinkernagel, R. M., Callahan, G. N., Klein, J., Dennert, G. 1978. *Nature* 271:251–53

131. Bevan, M. J. 1977. *Nature* 269:417–19

131a. Bevan, M. J., Fink, P. J. 1978. *Immunol. Rev.* 42:3–19

132. Hemler, M. E., Brenner, M. B., Mclean, J. M., Strominger, J. L. 1984. *Proc. Natl. Acad. Sci. USA* 81:2172–75

133. Yachie, A., Miyawaki, T., Uwadana, N., Ohzeki, S., Taniguchi, N. 1983. *J. Immunol.* 131:731–35

134. Cantrell, D. A., Smith, K. A. 1984. *Science* 224:1312–16

135. Melchers, F., Anderson, J. 1980. *Cell* 37:715–20

136. Howard, M., Paul, W. E. 1983. *Ann. Rev. Immunol.* 1:307–33

137. Mitchison, N. A. 1968. *Proc. R. Soc. London Ser. B* 161:275–92

138. Weigle, W. O. 1971. *Clin. Exp. Immunol.* 9:437–47

139. Nossal, G. J. V. 1983. *Ann. Rev. Immunol.* 1:33–62

140. Feizi, T., Wernet, P., Kunkel, H. G., Douglas, S. D. 1973. *Blood* 42:753–62

141. Bankhurst, A. D., Torrigiani, G., Allison, A. C. 1973. *Lancet* 1:226–29

142. Natali, P. G., Nicotra, M. R., Giacomini, P., Pellegrino, M. A., Ferrone, S. 1981. *Immunogenetics* 14:359–65

142a. Royer, H. D., Acuto, O., Fabbi, M., Tizard, R., Ramachandran, K., et al. 1984. *Cell* 261–66

143. Gill, G. N., Lazar, C. S. 1981. *Nature* 293:305–7

144. Barnes, D. W. 1982. *J. Cell Biol.* 93:1–4

145. Habu, S., Okumura, K., Diamanstein, T., Shevach, E. M. 1984. *Eur. J. Immunol.* In press

146. Mostov, K. E., Friedlander, M., Blobel, G. 1984. *Nature* 308:37–43

147. Roehm, N., Herron, L., Cambier, J., DiGuisto, D., Hanskins, K., et al. 1984. *Cell* 38:577–84

148. Carel, S., Bron, C., Corradin, G. 1983. *Proc. Natl. Acad. Sci. USA* 80:4832–36

149. Rao, A., Ko, W. W.-P., Faas, S. J., Cantor, H. 1984. *Cell* 36:879–88
150. Rao, A., Faas, S. J., Cantor, H. 1984. *Cell* 36:889–906
151. Mitsuya, H., Guo, H.-G., Cossman, J., Megson, M., Reitz, M. S. Jr., Broder, S. 1984. *Science* 225:1484–86
152. Hunig, T. R., Bevan, M. J. 1982. *J. Exp. Med.* 155:111–25
153. Fathman, C. G., Hengartner, H. 1979. *Proc. Natl. Acad. Sci. USA* 76:5863–66

154. Meuer, S. C., Hussey, R. E., Cantrell, D. A., Hodgdon, J. C., Schlossman, S. F., et al. 1984. *Proc. Natl. Acad. Sci. USA* 81:1509–13
155. Nikaido, T., Shimizu, A., Ishida, N., Sabe, H., Teshigawara, K., et al. 1984. *Nature.* 311:631–35
156. Leonard, W. J., Depper, J. M., Crabtree, G. R., Rudikoff, S., Pumphrey, J., et al. 1984. *Nature.* 311:626–31

Ann. Rev. Biochem. 1985. 54:831–62

THE CREATINE-CREATINE PHOSPHATE ENERGY SHUTTLE

Samuel P. Bessman

Department of Pharmacology and Nutrition, University of Southern California, School of Medicine, Los Angeles, California 90033

Christopher L. Carpenter

University of Texas, Health Sciences Center, Dallas, Texas 75235

CONTENTS

831

0066-4154/85/0701-0831$02.00

PERSPECTIVES AND SUMMARY

The endergonic functions of the animal organism are fueled ultimately by ATP which is generated primarily through oxidative metabolism in the mitochondrion and through anaerobic glycolysis. A conventional view is that consumption of energy causes the formation of ADP which returns to the mitochondrion, stimulating it (acceptor effect-respiratory control) to consume oxygen and rephosphorylate the ADP to ATP, which diffuses back to the sites of utilization. That this view is insufficient is shown by several observations.

Studies on muscle contraction have shown little relation between myofibrillar activity energized by myosin ATPase and the availability of ATP. There appears to be insufficient ADP formation during muscle activity to cause significant release from respiratory control even though muscle activity does produce a marked parallel rise in oxygen uptake. On the other hand, there is a close parallel between skeletal and heart muscle activity, the concentration of creatine phosphate, and the activity of creatine phosphokinase. The creatine phosphate shuttle was proposed to explain why when a diabetic individual exercised, the blood glucose level and general metabolism were altered exactly as if a dose of insulin had been given. In all regimes for diabetics the amount and timing of exercise is taken into account in determining dosage of insulin.

It was proposed in 1960 (1) that insulin acted to stimulate all endergonic reactions by attaching hexokinase to mitochondria. This could provide for more efficient respiratory control and availability of ATP. Figure 1 shows the proposal as depicted in 1966 (2). Exercise causes a liberation of creatine from contracting muscle fibers. This creatine moves to the mitochondria where nascent ATP would produce creatine phosphate and immediately return ADP, stimulating oxygen uptake, at the same time increasing the flux of creatine phosphate to the muscle fiber. (Figure 2).

In the original proposal there were three parts to the creatine phosphate system of respiratory control and energy delivery.

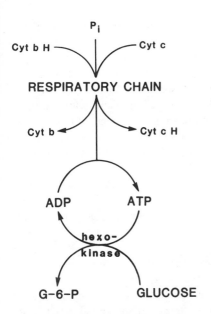

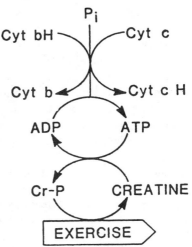

Figure 1 Respiratory control is made more efficient when insulin connects hexokinase to the mitochondrion causing glucose phosphorylation to replace ADP as rapidly as it is formed at certain sites on the mitochondrion.

Figure 2 The role of exercise, accelerating oxygen uptake, mimics the effect of insulin in attaching hexokinase to the mitochondrion. This occurs because creatine is liberated by the contracting myofibril, moves to the mitochondrion and there consumes ATP and replaces ADP in situ. The creatine phosphate returns to the myofibril to fuel further contractions.

1. The mitochondrial end which received the creatine "signal," was stimulated by it, and returned the activated (phosphorylated) creatine.
2. The intervening space of diffusion of creatine and creatine phosphate between mitochondrion and every utilization point.
3. The peripheral utilization points where bound CPK would rephosphorylate nascent ADP and allow return of free creatine to the mitochondria.

The concept of the creatine phosphate energy shuttle brings together and sets in context the excellent observations by many workers that were unexplainable by conventional muscle energetics. Its novelty is attested to by the fact that to date no textbook or monograph on muscle contraction refers to the creatine system other than as a reservoir of energy.

CREATINE KINASE

Lohmann (3) recognized the creatine kinase reaction while studying the chemistry of muscle contraction. He found that liberation of creatine during muscle contraction required ADP as a cofactor, and proposed that creatine phosphate reacted with ADP to produce ATP and creatine. Lehmann (4, 5) showed that the reaction was reversible and subsequently determined the equilibrium values for the reaction. The enzyme was progressively purified (6–8) and Kuby et al crystallized it from rabbit muscle (9).

In 1964 Burger et al (10) described three isozymes of creatine kinase, separable on agar gel electrophoresis. Type I isozyme was found in brain, Type III isozyme was found in skeletal muscle, and Type II isozyme, with intermediate electrophoretic properties, was found in smooth muscle and heart. In the same year Jacobs et al (11) discovered a fourth electrophoretically distinct isozyme of creatine kinase present in mitochondria isolated from brain, heart, and skeletal muscle.

Dance & Watts (12) analyzed the amino acid content of muscle creatine kinase. Half of the expected number of peptides were present after tryptic digestion, suggesting that creatine kinase was a dimer composed of two similar subunits. Dawson et al (13), using chicken brain, muscle, and heart, showed that creatine kinase was a dimer. They found that the three isozymes of creatine kinase described by Burger et al were composed of combinations of two subunits: the B subunit, found in brain, and the M subunit, found in muscle.

The brain isozyme of creatine kinase (Burger's type I) was found to be composed of two B subunits and the muscle enzyme (Burger's type III) of two M subunits. The third isozyme (Burger's type II) was composed of one M and one B subunit.

Eppenberger et al (14) showed that the two subunits were immunologically distinct, that their peptide maps and amino acid composition were distinct, and that the isozymes were kinetically different. They also showed that the isozymes were of similar molecular weight and that the MB form could occur in vitro by association of M and B subunits (15).

Creatine kinase is present in the cytosol and mitochondria of heart, skeletal muscle, and brain of vertebrates. Creatine kinase activity is also found in tumors (16), adipose tissue (17), white blood cells (18), and smooth muscle (19). Minimal activity has been found in many other tissues (20). Studies of the function of creatine kinase, the focus of this review, have been done primarily on heart and skeletal muscle.

In general, the mitochondrial isozyme is found in mitochondria isolated from vertebrate heart, skeletal muscle, and brain. The only other isozyme found in brain is the BB isozyme. In normal skeletal muscle the only additional isozyme is the MM form. Heart contains primarily the MM form but in most species, significant amounts of both the MB and BB forms are also present. The chicken heart is an exception; it contains only the mitochondrial and BB forms.

Invertebrate tissues contain an analogous enzyme: argininephosphokinase. Arginine phosphate, also a guanidino phosphate, seems to serve in invertebrates as creatine phosphate does in vertebrates (21). Arginine kinase which catalyzes the transphosphorylation reaction between arginine phosphate and ATP is also reversible (22).

CREATINE AND CREATINE PHOSPHATE

Creatine is synthesized in two sequential reactions; guanidinoacetic acid is formed from the transguanidination of arginine and glycine, and then N-methylated to form creatine. These reactions occur primarily in liver, pancreas, and kidney. The amount of each enzyme present in various tissue differs with the species. Neither enzyme has been found in skeletal muscle, heart, or brain, the tissues which contain the highest concentrations of creatine and creatine phosphate. Both creatine and creatine phosphate react nonenzymatically to form creatinine which is not reconvertible to creatine in significant amounts. These tissues must therefore replenish their creatine pools from the plasma in which it is normally about 0.40 millimolar. The mean muscle concentration of creatine at rest is approximately 5.0 millimolar and the concentration in red blood cells is about 0.5 millimolar.

Thunberg (23) was the first to observe that creatine was involved in muscle metabolism. In 1911 he showed that creatine added to a muscle mince stimulated oxygen consumption. Eggleton & Eggleton (24) found that muscle contained a highly acid-labile organic phosphate. Fiske & Subbarow (25) discovered the compound at about the same time and proved that it was a phosphorylated derivative of creatine. The interest in creatine phosphate, or "phosphagen" focused on its role in energy metabolism.

Meyerhof & Lohmann (26) showed that creatine phosphate had a large free energy of hydrolysis (12 kcal/mol), suggesting it might serve as the source of energy for muscle contraction. Lipmann & Meyerhof (27) established that creatine was liberated during muscle contraction, indicating that the hydrolysis of creatine phosphate was indeed coupled to muscle contraction. Lundsgaard (28) showed that creatine phosphate could be synthesized aerobically.

Almost thirty years after Thunberg's report (23), Belitzer & Tsybakova (29) showed that creatine phosphate synthesis in a muscle mince was coupled to oxygen consumption in the oxidation of 3 and 4 carbon compounds. Thus oxidative phosphorylation was first discovered as a function of creatine.

ATP was discovered in 1929 by Fiske & Subbarow (30). Much evidence accumulated indicating that hydrolysis of ATP provided the ultimate source of energy for muscle contraction. Myosin was found to be an ATPase (31) and it was shown that ATP would cause actomyosin fibrils to contract in solution (32). No one could demonstrate, however, that ATP was hydrolyzed by intact muscle during contraction. This led A. V. Hill (33) to challenge biochemists to prove that ATP truly supplied the energy for muscle contraction. A decade later

Cain & Davies (34) demonstrated a decrease in ATP levels in contracting muscle that had been treated with 2,4-dinitrofluorobenzene to inhibit creatine kinase completely. Although this experiment established the fact that ATP hydrolysis is the immediate source of energy for muscle contraction, it also emphasized the close association of creatine kinase and creatine phosphate with the contractile process, a fact that was not considered in subsequent discussions of the energetics of muscle contraction.

COMPARTMENTATION OF CELLULAR ENERGY METABOLISM

The creatine phosphate energy transfer system is an example of metabolic compartmentation: the localization of enzymes and intermediates resulting in enhanced efficiency. There are several recent symposia devoted to metabolic compartmentation (35, 36). Compartments restrict or facilitate the access of substrates to enzymes. By compartmentation an enzyme's environment can be optimized for pH, concentrations of ions, cofactors, substrates, and location of precursor enzymes. Compartmentation allows local variation in these factors to regulate enzyme activity.

Compartments are created when substrate diffusion is restricted either by binding or membrane barriers. The most obvious examples of compartments are those that are membrane delimited, such as mitochondria, nuclei, and lysosomes. Nevertheless the proximity or binding of sequential enzymes in the same metabolic pathway may also limit diffusion because a nascent intermediate is delivered primarily to the active site of the next enzyme in the pathway, such as in the fatty acid synthetase complex. Binding of intermediates also results in compartmentation, as in the glyceraldehyde phosphate dehydrogenase reaction (37). The acyl intermediate is bound to the enzyme at a tenfold higher concentration than the free compound.

THE CREATINE PHOSPHATE SHUTTLE

The creatine phosphate shuttle explains the role of creatine, creatine phosphate, and creatine kinase in facilitating energy distribution and responding to energy demand. The concept arose from studies of insulin action. Insulin has no effect on the turnover of creatine phosphate or on muscle contraction, but when a diabetic exercises, the blood glucose falls and other chemical changes occur that are the same as those occurring after a dose of insulin. Bessman (38) proposed that insulin acts by causing the binding of hexokinase at strategic mitochondrial sites where it could phosphorylate glucose with nascent ATP and resupply ADP efficiently for respiratory control (Figure 1). From this foundation, and based on subsequent work to be reviewed, Bessman proposed that creatine from the contracting muscle provided the stimulus for oxygen uptake (Figure 2). This was further refined (39) in 1972. Creatine phosphate is synthesized in mitochondria. The creatine phosphate diffuses to the myofibrils, where the MM isozyme of creatine kinase is bound. As contraction generates ADP, creatine kinase catalyzes the resynthesis of ATP to allow continued contraction. The creatine formed at the myofibril is thus rephosphorylated by mitochondria. This process was named the "Creatine Phosphate Shuttle" in 1978 (39a) and reviewed in detail in 1980 (40).

Gudbjarnason et al (41) proposed that ATP and creatine phosphate are compartmented at sites of utilization, based on the observation that contraction in anoxic heart stops when ATP levels are only minimally decreased, but creatine phosphate is depleted. Apparently unaware of experimental evidence for creatine as the respiratory control mediator in muscle (42) they proposed that adenine nucleotides are compartmented into mitochondrial and myofibrillar pools and creatine phosphate mediates the transfer of energy. Only a portion of total cellular ATP was thought to have access to myofibrils.

The creatine phosphate shuttle is based on the view that adenine nucleotides and creatine kinase are indeed compartmented. Adenine nucleotides are thought to be located primarily in mitochondria and near or bound to peripheral ATPases, e.g. myosin ATPase. Creatine kinase, bound as various isozymes to mitochondria and myofibrils, is thought to have the same peripheral distribution as the ATPases and also to synthesize creatine phosphate in mitochondria from nascent ATP. There are structural and functional data to indicate that this view is correct. There is also evidence that other ATPases in muscle, heart, and brain may derive the ATP they use by the creatine phosphate shuttle using adjacent isozymes of creatine kinase.

There are several reviews of the creatine phosphate shuttle (40, 43) and it has been the subject of symposia in 1979 and 1984 (43, 44).

COMPARTMENTATION OF ENERGY METABOLITES IN HEART AND SKELETAL MUSCLE ADENINE NUCLEOTIDES

Perry (45) long ago proposed that adenine nucleotides might be compartmented. He thought they were localized to myofibrils and that "such a system would have the advantage of maintaining ATP precisely where it is needed for contraction." In subsequent years a great deal of evidence has accumulated that suggests that adenine nucleotides are indeed compartmented. Mommaerts (46), using contracting frog muscle, showed there was no detectable change in ATP or ADP during contraction, and that changes in creatine phosphate alone accounted for the work.

Seraydarian et al (47) have shown that ADP binds to actin. Veech et al (48) measured concentrations of intermediates of reactions that are presumed to be in equilibrium and in which ADP is a substrate. The free ADP concentration in skeletal muscle, brain, and liver was estimated to be 20-fold lower than the total ADP concentration. They proposed that sequestration in mitochondria may account for low free ADP levels in brain and liver. Provision of ATP formed by ATPases to mitochondria by diffusion must be restricted by the low concentration of ADP, which would limit respiratory control by ATP availability, as pointed out originally by Chance (49).

D. K. Hill (50) used ^{3}H-labeled adenine to study the distribution of adenine nucleotides in frog muscle. 50–80% of the adenine nucleotides were located at the A-band of the myofibrils. Ottaway & Mowbray (51) suggested that the possibility of artifact limits the conclusions that can be drawn from this study. Lanthanum was used in the fixative. It catalyzes the hydrolysis of ATP and seems to be limited to the extramyofibrillar space.

Gudbjarnason et al (41) studied the effect of ischemia on the concentrations of high energy phosphates in dog heart. Ischemic muscle stopped contracting when 75% of the creatine phosphate had been depleted even though 80% of the ATP remained. A large proportion of the ATP appears not to have access to the myofibrillar ATPase. Dhalla et al (52) and Neeley (53) also found that decreasing cardiac output in ischemic hearts correlated well with creatine phosphate levels and only minimally with ATP levels.

Studies of muscle that have been stimulated to fatigue have shown that a large proportion of the ATP remains and that depletion of creatine phosphate correlates with fatigue (54–56).

Nassar-Gentina et al (57) proposed that a defect in excitation-contraction coupling during ischemia and fatigue would prevent utilization of ATP and might explain conservation of ATP in ischemic and fatigued muscle and heart. After a muscle has been stimulated to fatigue, addition of 5 mM caffeine allowed an additional tetanic contraction to occur during which most of the remaining ATP was hydrolyzed. They suggested that caffeine releases calcium from the sarcoplasmic reticulum, overcoming the defect in excitation contraction coupling. Caffeine is well known to disrupt muscle ultrastructure (59) and probably releases calcium by disrupting the membrane. Since the effect of caffeine is apparently the result of destruction of compartments, the compartmentation of ATP could also have been affected, allowing access of previously compartmented ATP to the myofibrillar ATPase.

Studies of muscle and heart in which creatine kinase has been inhibited with 2,4-dinitrofluorobenzene (FDNB) also suggest that ATP is compartmented. Infante & Davies (60) studied contractions in frog sartorius muscle in which creatine kdinase had been completely inhibited by preincubation with FDNB to prevent regeneration of ATP from creatine phosphate. The muscle could be stimulated to contract three or four times with a 50% reduction in ATP. After ten minutes the muscle would again contract fully and after another ten minutes the muscle would contract weakly with a further decrease in ATP. Although there was no synthesis of ATP during the ten-minute rest periods, some of the normally sequestered ATP became available for contraction, suggesting that ATP compartmentation is not absolute. Over the ten-minute period sufficient ATP must have diffused to the myofibrillar ATPase sites to support additional contraction. Gercken & Schlette (61) measured ATP levels in rat heart perfused with FDNB. At the time of failure 85% of the ATP remained, as did most of the creatine phosphate.

In frog heart, which is permeable to phosphorylated compounds, contractile force decreases with creatine phosphate depletion, even though ATP levels remain at about 75% of the control value (62, 63). Reperfusion with creatine leads to a parallel increase in creatine phosphate and contractile force. ATP increases to 90% of the control value. These results were confirmed by Vassort & Ventura-Clapier (64). These data emphasize the dependence of contractile function on creatine phosphate and the limited access of the total ATP pool to the contractile apparatus, in amphibia as well as mammalia.

McClellan et al (65) used rat heart, made hyperpermeable by EDTA treatment, to study energy transport. An increase in tension of the heart muscle strips resulted from a low energy state. In these nucleotide-depleted tissues, relaxation occurred with creatine phosphate perfusion, confirming that sufficient ADP must have been present at the myofibrils to be rephosphorylated. Tissues perfused with mitochondrial substrates and inorganic phosphate did not relax, indicating that the ADP produced by the myofibrils was not available to the mitochondria for rephosphorylation [cf Chance (49)]. Cells did relax if exogenous ADP was supplied, which presumably did have access to mitochondria.

^{31}P NMR has been used to investigate ATP compartmentation on the basis of the fact that a difference in pH of the environment causes differing NMR signals. Nunnally & Hollis (66) found two ATP pools, whereas Busby et al (67) concluded there was only one pool. If ATP compartments were not distinguished by a difference in pH, or if the concentrations of ATP in some compartments were small in relation to other compartments, they would not be detected by the presently insensitive [31]p NMR methods.

COMPARTMENTATION OF CREATINE AND CREATINE PHOSPHATE

Hill (68) also studied the distribution of tritium-labeled creatine phosphate in frog muscle. He found that almost all of the creatine phosphate was localized in narrow bands along the edges of the I-band of the myofibrils and estimated the level of concentration there to be about 150 mM.

Lee & Visscher (69) studied the distribution of ^{14}C-labeled creatine in perfused rabbit hearts. The specific activity (SA) of creatine phosphate rises more rapidly than that of the "free" creatine pool. On washout the SA of the free creatine drops at three times the rate of the creatine phosphate pool. The data are consistent with the existence of two pools of free creatine. Lee & Visscher proposed that a portion of the free creatine might be bound intracellularly. Savabi & Bessman (70) have shown that creatine is not lost from hearts subjected to anoxia, in which there is almost complete loss of creatine phosphate, for when they are reoxygenated there is regeneration of creatine phosphate even to higher than control levels. We have found that beating atria brought to a standstill in anoxia in which the creatine phosphate is almost completely depleted, do not lose any of the intracellular creatine to the medium over a 2-hour period (Savabi, 70) for on reoxygeneration 100% or more of the original creatine phosphate reappeared. This also suggests that when creatine phosphate is hydrolyzed the liberated creatine may be bound. It appears likely that the chemical activity of intracellular creatine may be considerably lower than the measured concentration of nonphosphorylated creatine.

MITOCHONDRIAL CREATINE KINASE AND THE SYNTHESIS OF CREATINE PHOSPHATE

Isolation and Characterization of Mitochondrial Creatine Kinase

Mitochondrial creatine kinase was first described by Jacobs et al (11) in 1964. They discovered an isozyme of creatine kinase electrophoretically distinct from the cytosolic forms of creatine kinase, in mitochondria of heart, brain, and skeletal muscle. Sobel et al (71) confirmed the findings of Jacobs using electrophoretic methods that were more sensitive and resolved the four isozymes of creatine kinase. Vial et al (72) were the first to study the kinetics of the enzyme, using intact mitochondria.

Farrell et al (73) extracted the enzyme from bovine heart mitochondria with inorganic phosphate. Scholte et al (74) localized the mitochondrial enzyme to the outside of the inner mitochondrial membrane. They found that a fraction of sonicated mitochondrial particles lose their creatine kinase activity, presumably those that are inside out. The activity returned with detergent treatment, which they concluded dissolved the membrane and released the creatine kinase.

Jacobus & Lehninger (20) showed that creatine phosphate synthesis from creatine and ATP formed from oxidative phosphorylation was inhibited by atractyloside, an inhibitor of the adenine nucleotide translocase. They showed that creatine kinase did not have access to matrix nucleotides and confirmed the fact (42) that it was an excellent respiratory control signal and that it was located on the outside of the inner membrane of the mitochondrion.

Hall, Addis, and DeLuca (75, 76) have purified mitochondrial creatine kinase from beef heart and characterized the pure enzyme. They have studied its kinetics, determined the effects of ions and pH on the rate, and determined the amino acid composition. Roberts & Grace (77) purified the mitochondrial isozyme from dog heart and showed it to be immunologically pure. Recently Grace et al (78) have purified human mitochondrial creatine kinase to a specific activity greater than 400 IU/mg.

The mitochondrial isozyme, like the cytosolic isozymes, is a dimer of about 80,000 daltons. The mitochondrial isozyme is distinguished on the basis of kinetic constants, and its migration toward the anode on cellulose acetate membrane electrophoresis in Tris barbital at pH 8.8.

FUNCTION OF THE MITOCHONDRIAL ISOZYME

The creatine phosphate shuttle theory proposed that the function of the mitochondrial isozyme of creatine kinase is to synthesize creatine phosphate from creatine and ATP generated de novo at the same time returning ADP to the respiratory system thereby stimulating oxidative phosphorylation. Bessman & Fonyo (42) showed, shortly after the discovery of mitochondrial creatine kinase, that creatine stimulated mitochondrial respiration, suggesting that the proposed relation between exercise and mitochondrial energy generation was correct (1). Creatine, acting as an acceptor for the gamma phosphate of ATP, provided ADP for the mitochondrial membrane to stimulate oxidative phosphorylation. Since then, both kinetic and direct labeling studies have elucidated this functional relationship between oxidative phosphorylation and the mitochondrial creatine kinase reaction.

KINETIC SUPPORT FOR THE MITOCHONDRIAL END OF THE CREATINE PHOSPHATE SHUTTLE

Jacobus & Lehninger (20) confirmed the original findings (42) that creatine functioned in respiratory control and showed that rat heart mitochondria contained enough bound creatine kinase activity to permit them to use all of the ATP generated at maximal respiration to synthesize creatine phosphate. Saks et al (79) confirmed their findings. It is clear that heart mitochondria contain enough creatine kinase to produce sufficient creatine phosphate exclusively to deliver the high energy phosphoryl group to myosin ATPase; there is no enzymatic requirement to rely on diffusion of ATP from mitochondrion to myofibril. They showed that the forward mitochondrial reaction (synthesis of creatine phosphate) was kinetically favored (80). They calculated that mitochondrially generated ATP had preferred access to mitochondrial creatine kinase for creatine phosphate synthesis compared to extramitochondrial ATP

(81). Using oligomycin, they studied ATP transport by the adenine nucleotide translocase as measured by oxygen consumption. Creatine phosphate synthesis increased with increasing transport of ATP (oxygen consumption) even if the concentration of ATP in the incubation medium was increased, suggesting that the de novo synthesized ATP had preferred access to creatine kinase. They concluded that these data demonstrated an association between the adenine nucleotide translocase and creatine kinase. They did not rule out the possibility that increasing concentrations of ATP might have produced more ADP which would stimulate oxygen consumption, however.

A different kinetic (82) approach showed that the apparent K_m of mitochondrial creatine kinase for ATP, when generated by mitochondria, was 37 μM. When ATP was provided by extramitochondrial phosphoenolpyruvate (PEP) and pyruvate kinase, the apparent K_m was 200 μM. The apparent K_m was also 200 μM for the solubilized mitochondrial enzyme supplied with ATP by PEP and pyruvate kinase. The apparent K_m for the solubilized mitochondrial was 145 μM. Product inhibition by creatine phosphate was compared in respiring heart mitochondria to liver mitochondria incubated with solubilized mitochondrial creatine kinase. The apparent K_m for ATP was 5–7-fold lower at each creatine phosphate concentration for the heart mitochondria with bound enzyme. The conclusion drawn from these experiments is that the bound creatine kinase of heart mitochondria is exposed to a higher effective concentration of ATP. It is also protected from added creatine phosphate so that there is a lower effective inhibitory concentration of creatine phosphate near the enzyme. These data do not distinguish between substrate compartmentation or close association of nucleotide translocase with creatine kinase.

The effect of the source of ATP on the rate of creatine phosphate synthesis was also studied (83). The rate of mitochondrial creatine phosphate synthesis was higher when ATP was supplied by oxidative phosphorylation than by PEP and pyruvate kinase incubation with mitochondria inhibited by oligomycin, again supporting the concept of the mitochondrial compartment.

Altschuld & Brierly (84) had earlier reported that the rate of creatine phosphate synthesis was unaffected by the source of ATP. They compared respiring mitochondria with mitochondria inhibited with atractyloside and incubated with PEP and pyruvate kinase. The rate of oxygen consumption, and thus ATP synthesis by the mitochondria, was low. Limited ATP synthesis probably decreased the rate of creatine phosphate synthesis in their respiring beef heart mitochondria.

Moreadith & Jacobus (85) took a similar approach in studying the apparent K_m of ADP for respiration. They found that ADP provided by the mitochondrial creatine kinase reaction had an apparent K_m for respiration of 2–4 μM. ADP provided by extramitochondrial glucose, ATP, and hexokinase had an apparent K_m of 19–20 μM. To further test if ADP generated by the mitochondrial creatine kinase reaction results in a higher concentration of ADP near the translocase, they compared the effect of the source of ADP on a competitive and noncompetitive inhibitor of the adenine nucleotide translocase. They found that atractyloside (a competitive inhibitor) was less effective in inhibiting respiration when the ADP was provided by the creatine kinase reaction, confirming that the effective concentration of ADP near the translocase is higher in this system. To try to identify whether the outer mitochondrial membrane or an association between creatine kinase and the adenine nucleotide translocase was responsible for the compartmentation, they compared ADP generated in the intermembrane space by nucleoside diphosphokinase (NUDI-KI) to that generated by creatine kinase. Liver mitochondria contain NUDIKI in the intermembrane space. UDP was added to act as a phosphate acceptor and regenerate ADP to stimulate respiration. The ability of atractyloside to inhibit respiration under these conditions was compared to its inhibition when ADP was generated by the extramitochondrial hexokinase reaction. Atractyloside inhibited respiration equally for ADP generated by NUDIKI or hexokinase. In heart mitochondria, atractyloside was less effective in inhibiting respiration stimulated by ADP provided by the creatine kinase reaction. They concluded that the compartmentation is conferred by an association between the adenine nucleotide translocase and creatine kinase and not the mitochondrial outer membrane, since ADP generated in the inner membrane by NUDIKI was less effective at competing with atractyloside. A close examination of the data reveals that this was not a fair comparison. In the absence of atractyloside the maximum rate for NUDIKI-stimulated respiration was only 67% of the maximal rate when ADP was added directly to stimulate respiration. ADP generation limited respiration in the experiment with NUDIKI, whereas in heart mitochondria there was a minimal difference in the rates of respiration stimulated by creatine and that stimulated by added ADP. Since NUDIKI is unable to generate sufficient ADP to drive mitochondria to their maximal respiratory rate it is not valid to compare its ability to generate locally high concentrations of ADP to that of creatine kinase, which under the same experimental conditions is able to drive mitochondria to their maximal respiratory rate.

Gellerich & Saks (86) also studied ADP provision for mitochondrial respiration. They used PEP and pyruvate kinase to consume extramitochondrial ADP and showed that respiration was not increased when ADP was added directly or through the hexokinase reaction. When provided by ATP and creatine through the creatine kinase reaction, ADP stimulated respiration. This experiment emphasizes the fact of compartmentation of adenine nucleotides in mitochondria but does not define the nature of the compartment as a direct association between the translocase and creatine kinase. Jacobus & Saks (87) found that ADP supplied to bound mitochondrial creatine kinase by oxidative phosphorylation resulted in a 2–3-fold decrease in the K_a of ATP for the ternary complex

E·MgATP, but a tenfold decrease in the K_a of ATP for the ternary complex E·MgATP creatine. Provision of ATP through oxidative phosphorylation stabilizes the ternary complex, compared to ATP from PEP and pyruvate kinase. If compartmentation were due to a localized increase in the concentration of ATP, the decrease in the K_a of ATP for the binary and ternary complexes should be equal. This suggests that the bound enzyme kinetics are not rapid-equilibrium random binding, as is found with the solubilized enzyme. Applying the fact that phosphate ion solubilizes mitochondrial creatine kinase (87a) Hall & DeLuca (87b) showed that increases in inorganic phosphate in the physiological range diminished the preferential use of mitochondrial ATP in synthesis of creatine phosphate.

DIRECT ANALYSIS OF MITOCHONDRIAL COMPARTMENT LABELING

Labeling studies have used either $^{32}P_i$ inorganic phosphate or gamma $^{32}P_i$-labeled ATP directly to assess the contribution of ATP synthesized de novo to mitochondrial creatine phosphate synthesis. These studies have been done primarily in our laboratory.

Yang et al (88) used gamma $^{32}P_i$-labeled ATP incubated with respiring rabbit heart mitochondria and measured the amount and specific activities of ATP and creatine phosphate at 5 and 10 sec. The specific activity of creatine phosphate was less than half that of ATP indicating that a large amount of the creatine phosphate had been synthesized from mitochondrial ATP. When mitochondrial ATP synthesis was inhibited by the uncoupler carbonyl cyanide chlorophenylhydrazone (CCCH) or atractyloside the specific activity of the creatine phosphate was equal to that of the added ATP, indicating no mitochondrial contribution.

Erickson-Viitanen et al (89) studied in more detail the contribution of mitochondrial ATP to the synthesis of creatine phosphate in isolated mitochondria. Analysis of these experiments rests on consideration of two distinct ATP pools: that present in the medium and that synthesized by mitochondria. Since there was insignificant breakdown of creatine phosphate during the short time course of these experiments and the rate of creatine phosphate synthesis is linear, its specific activity was the algebraic sum of the gamma phosphate contributions from the medium ATP and mitochondrially synthesized ATP. Because of the rapid rate of phosphate transport into mitochondria (90) and its incorporation into newly synthesized ATP, the specific activity of the newly synthesized ATP is closely approximated by the specific activity of the inorganic phosphate, in experiments in which P_i was used as the label. The mitochondrial contribution to creatine phosphate synthesis is calculated using the following equation:

$$\text{fractional mitochondrial contribution} = \frac{(SA_{CP}) - (\overline{SA}_{ATP})}{(SA_{P_i}) - (SA_{ATP})} \qquad 1.$$

SA_{CP} is the specific activity of creatine phosphate at the end of the incubation interval, SA_{P_i} is the specific activity of inorganic phosphate (which does not change over the short incubation periods used), and $\overline{SA}_{ATP}$ is the mean specific activity of ATP during the incubation interval.

These experiments revealed that demonstration of a significant mitochondrial component of creatine phosphate synthesis depended on the concentration of ATP added to the medium. At ATP concentrations below 0.2 mM a significant mitochondrial contribution can be seen. Experiments done with gamma-labeled ATP, rather than labeled P_i, give the same results. The ATP concentration at which compartmentation is evident is low; several factors may account for this. The isolation procedures may damage the mitochondria making the outer membrane more permeable to ATP or, in vivo, since most of the ATP is nondiffusible, the activity of ATP in the vicinity of the mitochondria may be less than 0.2 mM. [See ref. (39a) for more data on the inaccessibility of the large pool of ATP to the mitochondrial pool.]

The specific activity of creatine phosphate was higher than the mean specific activity of ATP between 5 and 10 seconds indicating preferred access of mitochondrial ATP to creatine kinase. The specific activity of glucose-6-phosphate synthesized by added hexokinase is equal to the mean specific activity of ATP. There is no preferred access of newly synthesized ATP to the extramitochondrial hexokinase reaction. These results are both qualitatively and quantitatively consistent with the kinetic calculations of the previous section.

Erickson-Viitanen et al (91) used gentle digitonin treatment to remove the outer mitochondrial membrane to determine whether the outer membrane was responsible for compartmentation. The mitochondria retained all but about 10% of their creatine kinase activity and respired normally suggesting no major change in the translocase. It was not possible, however, to demonstrate preferred access of newly synthesized ATP to creatine kinase in mitochondria without an outer membrane. Compartmentation of the mitochondrial creatine kinase reaction is apparently due to the outer mitochondrial membrane, which presumably restricts diffusion, resulting in a high concentration of ATP in the intermembrane space. Although it is true that the outer mitochondrial membrane is permeable to adenine nucleotides this does not mean that it does not limit loss of nucleotide at all. The limiting effects of this membrane are worthy of much further study to give insight into the dimensions of this form of compartmentation. These experiments do not support the hypothesis that there is close association between the translocase and creatine kinase. If there were such an association, a mitochondrial contribution to creatine phosphate synthesis should still have been detected in the mitochondria stripped of their outer membrane.

Recent experiments by Barbour et al (91a) using ^{3}H-labeled nucleotides confirm the role of creatine in respiratory control (42) and agree very well with the report of Erickson-Viitanen et al (91).

When Altschuld & Brierly (84) studied the synthesis of creatine phosphate by beef heart mitochondria using labeled inorganic phosphate they found no evidence for compartmentation. However, they used 3 mM ATP in the incubation medium, which, as shown above, masks the mitochondrial contribution. They also conducted their experiments for a minimum of 1 min. In our hands creatine phosphate synthesis is not linear over this long a time period.

Lipskaya et al (92) have also used labeled inorganic phosphate to study mitochondrial creatine kinase. In their first experiment, the specific activity of creatine phosphate lagged behind that of ATP. This is as expected, since the mean ATP specific activity from which accumulating creatine phosphate is formed must be compared to the creatine phosphate activity found to find the apparent mitochondrial contribution to creatine phosphate synthesis. In a second experiment they found that by 2 min the specific activity of creatine phosphate exceeded that of ATP. We have found that by 5 min the specific activity of creatine phosphate approaches that of ATP, but never exceeds it. If creatine phosphate accumulates as ATP turns over we do not see how such a result could be achieved. If both creatine phosphate and ATP turned over equally then by 5 min both should be equally labeled. The only way that creatine phosphate activity could exceed ATP activity is if ATP did not turn over as fast as CP.

Both kinetic and direct labeling experiments demonstrate that mitochondrial creatine kinase uses newly synthesized mitochondrial ATP to synthesize creatine phosphate. This compartmentation is due to the outer mitochondrial membrane.

CREATINE PHOSPHATE SYNTHESIS AND GLYCOLYSIS

Early work on creatine phosphate revealed that it is synthesized under anaerobic conditions, presumably through glycolysis (93). When creatine was added to a cytosolic fraction from rat hearts that contained the glycolytic enzymes, glycolytic cofactors and substrates, the production of lactate was accompanied by the synthesis of creatine phosphate. When fructose-1,6-biphosphate was used as the substrate, the ratio of creatine phosphate to lactate formed was about 2, the expected ratio if all of the ATP formed were converted to creatine phosphate (94). Higher concentrations of creatine phosphate inhibited glycolysis, presumably because ADP was converted to ATP as the reactions went to equilibrium. These results can be accounted for by the contamination of the fraction with creatine kinase. No one has yet shown an association of creatine kinase with the glycolytic enzymes.

THE USE OF CREATINE PHOSPHATE TO REGENERATE ATP

The creatine phosphate shuttle proposes that the cytosolic isozymes of creatine kinase are located very near the ATPases and catalyze the resynthesis of ATP in situ for use by the ATPases. Three major cell functions that require ATP have been shown to be terminals for the creatine phosphate shuttle: contraction, macromolecular synthesis, and maintenance of ion gradients.

CONTRACTION AND CREATINE PHOSPHATE

Binding of Creatine Kinase to Myofibrils

Botts & Stone (95) discovered that myosin ATPase activity was lower in the presence of creatine kinase and proposed that there was an interaction between the two enzymes. Yagi & Mase (96) described noncompetitive inhibition of myosin ATPase by creatine kinase, suggesting interaction of the two enzymes. Ottaway (97) isolated creatine kinase from ox heart myofibrils, indicating that in vivo creatine kinase was bound to myofibrils. Scholte et al (74) fractionated rat heart and skeletal muscle and also found creatine kinase activity in myofibrils.

The site of the binding of CK remains incompletely explained. Turner et al (98) showed that the M-line protein of myofibrils was in part composed of the MM isozyme of creatine kinase. Immunologic studies of chicken heart and skeletal muscle showed that creatine kinase was localized to the M-line in these tissues (99, 100). Herasymowych et al (101) eluted the MM isozyme of creatine kinase from bovine heart myofibrils at low ionic strength and showed rebinding of the M-line. Mani & Kay (102) isolated a 165,000-dalton M-line protein that is a competitive inhibitor of creatine kinase and binds creatine kinase in a one to one ratio.

Houk & Putnam (103) used fluorescence polarization measurement of tagged creatine kinase and found interaction with the rod portion of myosin. Botts et al (104), using more sensitive techniques, found interaction of creatine kinase with intact myosin, heavy meromyosin, and subfragment one. These results were confirmed by Mani & Kay (105) using circular dichroism. They also found interaction between creatine kinase and myosin, heavy meromyosin, and subfragment one. Neither Botts et al nor Mani & Kay found an interaction of creatine kinase with the rod portion of myosin. The evidence indicates that creatine kinase is bound to the M-line of myofibrils, perhaps through binding to the 165,000-dalton M-line protein, and also to subfragment one of heavy meromyosin, in the vicinity of the ATPase.

FUNCTIONAL STUDIES OF MYOFIBRILLAR CREATINE KINASE

Saks et al (106) studied the kinetics of myofibrillar creatine kinase—the MM isozyme. In contrast to mitochondrial creatine kinase, the kinetics of the myofibrillar isozyme at the concentrations of the substrates normally found in cells favor the synthesis of ATP from creatine phosphate and ADP.

Bessman et al (107) used ATP labeled with ^{32}P in the gamma position to study the functional relationship between myofibrillar creatine kinase and myosin ATPase. Myofibrils were incubated with creatine phosphate and ATP. The specific activity of the inorganic phosphate formed by the myofibrillar ATPase was compared to the mean specific activity of ATP over the incubation period. When the reaction was carried out in the presence of 1.6 mM ATP and 1.6 mM creatine phosphate the specific activity of the inorganic phosphate was less than the mean specific activity of the ATP, indicating that ATP regenerated from creatine phosphate had preferred access to the myosin ATPase. During a one-minute incubation, 30% of the inorganic phosphate came from creatine phosphate.

Further proof of the reliance of myofibrils on creatine phosphate to provide ATP for normal contraction was obtained by studying the contractile response of glycerinated muscle fibers to ATP and creatine phosphate (108, 109). In the presence of 250 μM ADP, the physiologic concentration of creatine phosphate (10 mM) produced faster and stronger contraction and faster and more complete relaxation than ATP at concentrations up to 10 mM. The apparent K_m for ADP added to fibers preincubated with 10 mM creatine phosphate was 1.2 mM. If the fibers were preincubated with ADP and contraction was initiated by the addition of 10 mM creatine phosphate, the apparent K_m of ADP for contraction was 0.076 mM. The apparent K_m of ADP for ATP synthesis by soluble creatine kinase was more than twice this value (0.15 mM), further indicating that when myofibrillar creatine kinase is compartmented and is associated with the myofibrillar ATPase, the juxtaposition of the two enzymes allows for more efficient provision of ATP by rephosphorylation of ADP than can occur in solution. These experiments were done in the presence of Ap5A, an inhibitor of adenylate kinase, to reduce any effects this enzyme would have on nucleotide concentrations. A slight difference in the diffusion coefficients of creatine phosphate and ATP has been suggested to play some role in explaining the greater efficiency of creatine phosphate and ADP than ATP alone in providing the energy for contraction, but this is not a complete explanation. The observation that 4 μM ADP and 10 mM creatine phosphate gives a greater contraction than 100 μM ATP, a gradient that should overcome any differences in the diffusion coefficients, suggests that the enzymes are functionally linked (108, 109).

PROTEIN AND LIPID SYNTHESIS

Kleine (110) showed in 1965 that creatine kinase was associated with the microsomal fraction isolated from brain, skeletal muscle, and heart. Baba et al (111) showed that the MM isozyme was bound to microsomes in heart and skeletal muscle. Oganro et al (112) also described the presence of creatine kinase in the microsomal fraction of guinea pig heart.

We have shown that inhibition of creatine kinase with 2,4-fluoro-dinitrobenzene (FDNB) results in a parallel inhibition of lipid and protein synthesis in rat diaphragm (113, 114). FDNB at the concentrations used in diaphragm had minimal effects on protein and lipid synthesis in hepatocytes. At higher concentrations of FDNB there was a decrease in synthesis in hepatocytes, which correlated with a decrease in ATP levels, but ATP levels were not affected in diaphragm in FDNB concentrations that inhibited both lipid and protein synthesis. These experiments strongly suggest that the effects of FDNB on protein and lipid synthesis are due to inhibition of creatine kinase and not to some other effect of FDNB, since hepatocytes, which lack creatine kinase, and therefore must not utilize the creatine phosphate shuttle, are spared.

We have further shown that polysomes isolated from rat skeletal muscle contain endogenous creatine kinase (115). In the presence of creatine phosphate, ATP, and GTP, higher rates of protein synthesis are observed than with an ATP generating system of phosphoenolpyruvate and pyruvate kinase plus ATP and GTP. These data suggest that as with myofibrils, creatine phosphate provision of nucleoside triphosphates by rephosphorylation is more efficient than diffusion of nucleotides. When ^{33}P-creatine phosphate was incubated with the protein synthesizing system, the labeled phosphate was transferred to GTP. Presumably creatine phosphate regenerated ATP through the creatine kinase reaction. This ATP was used by nucleoside diphosphokinase, also present in the polysomes, to regenerate GTP. Creatine phosphate can therefore provide both ATP and GTP for protein synthesis in skeletal muscle.

Further confirmation of the role of the creatine phosphate shuttle in providing energy for protein synthesis is provided by work done by Bessman & Pal (116, 117). Creatine phosphate levels decrease in brain in a model of hepatic coma in which rats are injected with ammonium chloride. There is a smaller decrease in GTP and ATP levels. Accompanying the decrease in creatine phosphate is a decrease in protein synthesis in brains of animals treated with ammonium chloride. The inability of the mitochondria of ammonia-poisoned brain to regenerate creatine phosphate also prevents protein synthesis. Preliminary experiments with FDNB show that protein synthesis in brain slices is even more sensitive to creatine kinase inhibition than in muscle tissue.

MAINTENANCE OF ION GRADIENTS

Sarcoplasmic Reticulum

Sarcoplasmic reticulum has been isolated in the microsomal fraction from muscle and heart, together with polypolysomes, making fractionation studies difficult to interpret. Baskin & Deamer (118) isolated microsomes from rabbit skeletal muscle and found creatine kinase activity. They proposed that the creatine kinase was associated with sarcoplasmic reticulum fragments and provided energy for calcium transport. Histochemical studies confirm the localization of creatine kinase in the sarcoplasmic reticulum, among other sites (111, 119).

Levitsky et al (120) isolated the MM isozyme from carefully fractionated sarcoplasmic reticulum. Creatine phosphate maintained a faster rate of calcium uptake by sarcoplasmic reticulum vesicles than an exogenous ATP-generating system of PEP and pyruvate kinase. This suggests that there is a functional coupling between ATP-dependent calcium transport and creatine kinase.

PLASMA MEMBRANE

Histochemical studies localized creatine kinase to the plasma membrane (119, 120) and Saks et al (121) found creatine kinase activity in isolated plasma membrane vesicles.

Several studies suggest that the creatine kinase associated with the plasma membrane functions to regenerate ATP for the sodium-potassium ATPase and ATP-dependent calcium transport. In 1960 Caldwell et al (122) showed that arginine phosphate injected into squid giant axon in which mitochondrial ATP production had been inhibited with cyanide maintained the sodium gradient. More recently Saks et al (121) have shown that the MM isozyme is associated with the plasma membrane in heart. Activity for the ATPase in a vesicle preparation is higher when ATP is regenerated by creatine phosphate rather than PEP and pyruvate kinase. Hydrolysis of creatine phosphate is inhibited by ouabain, an inhibitor of the sodium-potassium ATPase. Grosse et al (123) have shown that in the presence of creatine phosphate only 0.3 mM ATP is required to support maximal transport. These data again suggest functional coupling between creatine kinase and ATP requiring enzymes.

Spitzer et al (124) have shown, in similar experiments, that calcium transport by plasma membrane vesicles is faster when ATP is provided by creatine phosphate than the PEP, pyruvate kinase system.

Histochemical studies have identified creatine kinase in the nuclear membrane (119) and Erashova et al (125) have found creatine kinase activity in nuclei isolated from heart. The function of creatine kinase in nuclei of heart and skeletal muscle cells has not been investigated. We found that RNA synthesis was not inhibited by the creatine kinase inhibitor FDNB (114). The creatine phosphate shuttle functions to regenerate ATP and in the case of nucleic acid synthesis, most of the nucleotides would be consumed and not available for rephosphorylation. Furthermore, the major free nucleotide products would be mononucleotides, which are inactive in the creatine kinase reaction. It is possible that some linkage with adenylate kinase, as in the "comparticle" (see below) might be active in the nucleus.

CREATINE KINASE IN OTHER TISSUES

Creatine kinase is found in other tissues than in muscle, heart, and brain. It seems generally to be associated with contractile proteins and cell movement. Creatine kinase is induced by estrogen in rat uterus (126, 127). The sequence of events suggests that it could play a role in providing the energy for protein synthesis. After estrogen stimulation creatine kinase activity increases for the next 6 hours. Thereafter there is an increase in protein synthesis by the uterus.

Creatine kinase is also found in leukocytes (128) and macrophages (129). In macrophages a large increase in the rate of turnover of creatine phosphate occurs with phagocytosis, suggesting that creatine phosphate provided the energy required.

AN INTEGRATED VIEW OF THE CREATINE PHOSPHATE SHUTTLE

The studies discussed have detailed the synthesis of creatine phosphate by mitochondria, and its use by ATP requiring reactions throughout the cell. The use of creatine phosphate by these reactions involves a close association of creatine kinase and the local ATPase which allows creatine phosphate to regenerate ATP at or very near the active site of the ATPase. This allows the effective concentration of ATP near the active site to be very high so that the reaction is not limited by the availability of energy. As a corollary of this theory, adenine nucleotides are considered to be compartmented at strategic sites with a high concentration of ATP only in the intramitochondrial space and of ADP at the peripheral ATPase.

The presence of creatine kinase, creatine, and creatine phosphate in high concentrations in only three tissues, heart, skeletal muscle, and brain, may provide a clue to the evolutionary origin of the creatine phosphate shuttle. Unique to skeletal muscle and heart is their contractile function. Creatine phosphate may have served as an energy buffer, for it is far more basic than

ATP and would remain electrostatically in the region of the acidic myosin. Release of an even more basic free creatine would tend to displace phosphocreatine providing more energy for further contraction. An advantage may have been conferred when creatine kinase became closely associated with the myosin ATPase. ATP would be more efficiently regenerated and could better support contractile work. Development of the mitochondrial creatine kinase compartment resulted in more efficient synthesis of creatine phosphate. The more rapid diffusion of creatine phosphate than ATP provided a more efficient energy supply. Creatine phosphate no longer was simply an energy buffer but evolved into the means of energy transport.

In parallel with these developments creatine kinase may have become associated with other ATPases. This is a necessary development to assure ATP provision in the presence of adenine nucleotide compartmentation.

Brain tissue presumably has no major variation in the demand for energy as does the muscle and would not require creatine to buffer this demand. Continued energy supply may be so crucial for brain that a creatine phosphate shuttle system developed not to buffer changes in energy demand but to ensure an adequate energy supply under extreme conditions such as hypoxia.

Our view of the creatine phosphate shuttle is that it serves two distinct functions in the heart, muscle, and brain. The first is to deliver creatine as the signal or stimulus of oxygen uptake mediated by respiratory control, and the second is to provide creatine phosphate, the form of energy moving from mitochondrion to utilization site.

Mainwood & Rakusan (130) presented a model of intracellular energy transport in which mitochondria cluster around capillaries and the creatine kinase reaction is a near-equilibrium reaction at all rates of ATP consumption. According to this model individual molecules of creatine phosphate do not diffuse from mitochondria to ATPases in the cytsosol, rather the reaction is maintained in near-equilibrium throughout the cell. Deviations from equilibrium in a part of the cell are quickly corrected by distribution throughout the cell. Meyer et al (131) have termed this model "facilitated diffusion." Evidence suggests that the assumptions on which the model is based are not valid. The substrates of the creatine kinase reaction and the enzyme itself are compartmented, not distributed homogeneously throughout the cell. About half of the creatine kinase in muscle cells is bound to myofibrils and mitochondria. The state of the rest of the creatine kinase in the intact cell is not known. Based on evidence that creatine kinase is bound to myofibrils and in association with the ATP- and GTP-requiring reactions of protein synthesis and to the enzymes of lipid synthesis, as well as to the ATP-dependent ion pumps, creatine kinase may not be free in the cytosol. It may be bound near these and other ATPases, but released by the fractionation procedures making it appear to be a soluble enzyme.

The question of whether the creatine kinase reaction is in equilibrium has been addressed directly using ^{31}P-NMR techniques. As would be expected the reaction appears to be in equilibrium in resting skeletal muscle (132). However, in contracting skeletal muscle and heart, the reaction does not appear to be in equilibrium. Gadian et al (133) concluded that the creatine kinase reaction was in equilibrium in resting muscle if a correction was made for a presumed underestimation of flux from ATP to creatine phosphate. In contracting muscle the reaction was not in equilibrium. Fossel et al (134) found that in contracting heart creatine phosphate and ATP levels decreased by 15% between diastole and systole. This is compatible neither with the energetics of the process nor the direct results of any other investigators. Mathews et al (135) found that the rate of ATP synthesis in perfused rat heart is five times less than the flux through the creatine kinase reaction and concluded that the creatine kinase reaction could be maintained near equilibrium. They point out, however, that the oxygen consumption, and thus ATP synthesis, in a working heart is five times that of the Langendorf preparation used in their experiment (136). It is likely that the reaction is not in equilibrium in in vivo working heart and contracting muscle in vivo. Since creatine phosphate can only be converted to ATP, flux should be equal for creatine phosphate and ATP. The discrepancies above indicate the serious technical problems that beset NMR measurements to date.

STUDIES OF THE CREATINE PHOSPHATE SHUTTLE USING CREATINE ANALOGS

Several analogs of creatine have been used to study the creatine phosphate shuttle in intact tissue. The analogs were fed to animals (mice, rats, or chickens) over a period of weeks. Creatine uptake was inhibited by the analogs, which are themselves taken up and phosphorylated, resulting in a depletion of creatine phosphate and creatine and an accumulation of the analog and its phosphorylated derivative (137). The analogs and the phospho-derivatives are poor substrates for creatine kinase (138).

Shields et al (139) found that type II fibers (white) from rat gastrocnemius muscle were smaller in animals treated with B-guanidinopropionic acid (B-GPA). Petrofsky & Fitch (140) found that the plantaris muscle in rats treated with B-GPA was 30% lighter (wet weight) than in control animals. There was no difference in the soleus muscle. Laskowski et al (141) studied chickens fed with B-GPA and found growth retardation and weakness. Microscopically there was loss of thin and thick filaments, disruption of the Z band, and dilatation of mitochondria and sarcoplasmic reticulum.

All investigators have found marked depletion of creatine and creatine phosphate, usually 80–90%, and accumulation of the analogs. There is also a depletion in ATP levels of 40–80%. Shields et al (139) also described a decrease in creatine kinase activity. Petrofsky & Fitch (140) found little difference in the contractile characteristics of plantaris muscle from creatine depleted and normal rats. There were marked differences in the soleus muscle, however: a decrease in the rate and amplitude of contraction and in the rate of relaxation. Mainwood et al (142, 143) found decreased response to electrical stimulation in diaphragm from rats treated with B-GPA. Diaphragms from treated animals would contract only 4 times compared to 15 times for the control. They also found a decreased maximum tension and a decrease in the rate of tension development and relaxation.

Fitch et al (144) described a prolongation of isometric contraction in soleus, but not plantaris muscle from treated rats. Annesley & Walker (145) found that treatment with cyclocreatine led to a delay in the development of rigor.

Although chickens are severely affected by creatine depletion, rats and mice are grossly relatively little affected. An increase in the flux of the creatine phosphate that remains may be sufficient to supply energy for animals under normal, unstressed circumstances. We must again consider the possible major difference between measured concentrations and biological activity of creatine and creatine phosphate.

DEVELOPMENT OF THE CREATINE PHOSPHATE SHUTTLE

Only small concentrations of creatine and creatine phosphate are necessary to transfer energy under ordinary circumstances. The remainder of the creatine phosphate may be a reservoir, bound to myosin. The creatine phosphate shuttle does not seem to exist in fetal animals' heart, yet the heart contracts, albeit slowly. The creatine phosphate shuttle develops around the time of birth, both the enzymes and the substrates increasing roughly in parallel to the activity of the heart.

Hall & DeLuca (146) have shown that there is no mitochondrial creatine kinase in neonatal rat hearts, so there could be no compartmented synthesis of creatine phosphate coupled to oxidative phosphorylation. Ingwall and coworkers (147, 148) have shown that from about gestational day 14 through birth there is an 8–9-fold increase in total creatine kinase activity and a 13–15-fold increase in the concentration of total creatine (creatine phosphate plus creatine phosphate). Roberts & Bessman (149) have shown that there is about a 7-fold increase in the concentration of creatine phosphate from gestation day 14 to the newborn period.

Pette (149a) has conducted a series of studies on muscle function and enzyme behavior that relate enzyme composition to type of stimulation. The whole muscle fibers that are fast contain relatively few mitochondria and are primarily glycolytic whereas the slow, red fibers contain large amounts of mitochondria, and are primarily oxidative. When the fast muscles are continuously stimulated at the low rates normal for slow muscles, or when the innervation of fast muscles is changed to the innervation of slow muscles there are marked changes in the enzyme content of the muscle fibers. The enzymes of the glycolytic pathway, except for hexokinase, all fall, and the enzymes of the oxidative pathways increase, with the increase of mitochondria. Hexokinase also increases, which has led Pette (149a) to conclude that hexokinase activity is more a function of respiratory control than of glycolysis (cf the theory of insulin action above). Creatine kinase total activity drops by 30–40%.

There is no information on the isozyme distribution as affected by the change in stimulation, but if the pattern shown by all of the glycolytic and oxidative enzymes tested also applies to creatine kinase one would expect a large increase in mitochondrial creatine kinase and a drop in the MM function. Studies of the isozymal distribution under these circumstances might give very interesting insights into the actual quantity of "free" creatine kinase, for that portion should be related to the soluble glycolytic system. Gonner, Loike et al (149b) have shown a relation between differentiation of macrophages in culture and the appearance of creatine kinase (BB) and creatine phosphate.

ADENYLATE KINASE AND THE CREATINE PHOSPHATE SHUTTLE

Evidence has been obtained for a direct role of adenylate kinase in both the mitochondrial and myofibrillar ends of the shuttle.

Yang et al (88), in their early tracer studies of the mitochondrial synthesis of creatine phosphate from nascent ATP, found that specific activities of creatine phosphate measured in 5 and 10 sec incubations showed that about one-third of the phosphoryl group of creatine phosphate phosphorus could have come directly from oxidative phosphorylation and the rest must have come from other sources of ATP including the gamma ^{31}P-labeled ATP added as tracer. They concluded that two adenylate kinase molecules might be interposed between the oxidative phosphorylation site (translocase) and the mitochondrial creatine kinase. (Figure 3). The two adenylate kinase molecules acting in sequence in opposite directions on a common pool of ADP could produce a distribution of isotope which would match the experimental data. A particle or compartment containing two molecules of adenylate kinase and one of creatine kinase all clustered in a tetrahedral pattern (Fig. 4) around the translocase was designated the "comparticle" (150).

In studies of the histochemical distribution of ATPase in myofibrils using

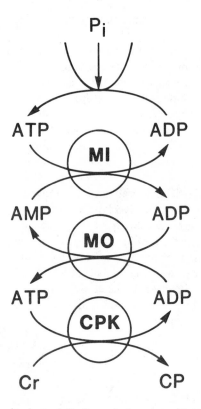

Figure 3 The sequential operation of two adenylate kinase enzymes, designated MI and MO, which tend to maintain direction and appropriate concentration for the conversion of creatine to creatine phosphate in the mitochondrion.

lead precipitation of phosphate liberated, Tice & Barnett (151) showed that both ATP and, to a lesser extent, ADP, caused the formation of free phosphate throughout the M-band as well as the Z-line. They concluded that adenylate kinase was located near the ATPase activities. Savabi et al (152) observed that glycerinated rabbit muscle fibers that contracted on the addition of ATP would relax spontaneously to a great extent without the addition of relaxing solutions. This relaxation could be abolished by the addition of P^1,P^5-di (adenosine 5')pentaphosphate Ap5A a specific inhibitor of adenylate kinase. Furthermore, in the presence of Ap5A it was possible to show that any concentration of creatine phosphate with 200 μM ADP was more effective in causing both contraction and relaxation than an equivalent concentration of ATP. This suggests that the "comparticle" structure may also reside in the myofibril. The first enzymes to contact creatine phosphate would be a pair of adenylate kinase molecules followed by myosin ATPase (Figure 5). There would be an ADP pool in the central region as in the mitochondrial "comparticle" (Figure 6).

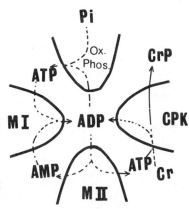

Figure 4 The arrangement of the three enzymes, the two adenylate kinase molecules and one creatine kinase around the site of oxidative phosphorylation to produce a common pool of ADP and sites of entry for ATP, creatine-creatine phosphate, and AMP.

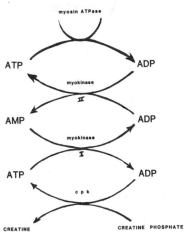

Figure 5 The sequential arrangement of adenylate kinase molecules to maintain substrate flow between creatine kinase and myosin ATPase.

POSSIBLE PATHOLOGICAL SIGNIFICANCE OF THE CREATINE PHOSPHATE SHUTTLE

Two general areas of concern relevant to the creatine phosphate shuttle and disease are, first, those diseases in which there would be a deficiency of creatine kinase, and second, those diseases that might arise from a deficiency of creatine.

In the first case it was shown in our laboratory (153) that as muscular dystrophy develops in the chicken model there is a progressive fall in mitochondrial creatine kinase activity. DeLuca's laboratory reports (154) an extensive analysis of this phenomenon, showing a strong correlation between function and the ratio of mitochondrial creatine kinase and succinate-iodotetrazolium reductase, a mitochondrial marker. The dystrophic chicks had a greater resistance to loss of creatine kinase from mitochondria caused by phosphate. The relevance of these findings to the human disease is discussed and several parallels are drawn. It is possible from their data to propose at least two forms of human muscular dystrophy—one deficient in the myofibrillar enzyme and one deficient in the mitochondrial isozyme. It would be necessary in any assay, in view of the foregoing evidence, to test for these possible lesions by the forward synthetic reaction in the mitochondria and the backward (ATP synthesis) reaction in myofibrils.

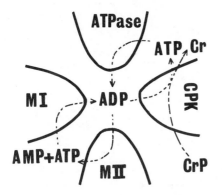

Figure 6 A tetrahedral arrangement of adenylate kinase molecules with creatine kinase around the myosin ATPase, in such a way that there is a common pool of ADP and entrance sites for ATP, AMP, creatine, and creatine phosphate. In this diagram the principal pathway for energy to be made available for the complex is from creatine phosphate to ATP.

PHOSPHATE DEPLETION

Work by Brautbar et al (155, 156) suggests that the heart and muscle disease that result from phosphate depletion may be a result of a defect in the creatine phosphate shuttle. Rats maintained on a phosphate deficient diet became deficient in myocardial inorganic phosphate, creatine phosphate, and adenine nucleotides. Total and mitochondrial creatine kinase levels were also decreased in both heart and skeletal muscle as was the mitochondrial respiratory rate. A decreased ability to produce, transport, and use creatine phosphate may explain the myopathy and cardiomyopathy of phosphate depletion.

The source of tissue creatine is a complex series of reactions involving, in the human, the liver, kidney, and pancreas. The liver is of particular importance, for the precursors of creatine, arginine and glycine, are synthesized there. One could expect serious curtailment of supply of creatine in liver disease. Perhaps this is one mechanism by which some beneficial effects have been reported for use of large amounts of arginine in treatment of hepatic coma (157). It has been estimated that at least 87 g of arginine are synthesized and broken down daily by the normal adult (158).

The liver is also the site of methylation of the guanidoacetic acid formed from glycine and arginine in the kidney. Since the methylation requires methionine, deficiency of this amino acid might be responsible for weakness, and, in the brain, of functional problems due to creatine deficiency. Mental retardation could be related to deficiency of creatine in utero caused by any lesion of the arginine cycle (improperly called the urea cycle). The fetus would make little arginine and would receive less than normal amounts from the heterozygous mother (159). This could be the basis of mental retardation associated with inability to synthesize glycine as well.

Gyrate atrophy of the retina has been shown to be associated with ornithinuria (160–162). It is possible that creatine synthesis from arginine is impaired and reveals itself as a failure in energy generation in the retina. The arginine cycle is a function of the prenatal brain (158) and probably the retina as well.

EXERCISE AND MUSCLE GROWTH

The muscle hypertrophy of exercise is accompanied by the increased activity of the shuttle caused by muscle contraction. The delivery of creatine phosphate to protein synthesizing sites will be increased by exercise and at least 70% of protein synthesis is dependent on this source of energy (113–115).

Hypertensive cardiac hypertrophy could occur by a similar mechanism. The increased vascular resistance in hypertension would stimulate increased cardiac contraction and this increased contractile activity would stimulate protein synthesis resulting in the enlarged myocardium found in chronic hypertension.

Literature Cited

1. Bessman, S. P. 1960. *J. Pediatr.* 56: 191
2. Bessman, S. P. 1966. *Am. J. Med.* 40:740
3. Lohmann, K. 1934. *Naturwissenschaften* 22:409–13
4. Lehmann, H. 1935. *Biochem. Z.* 281: 271–91
5. Lehmann, H. 1936. *Biochem. Z.* 286: 336–43
6. Banga, D. 1943. *Stud. Inst. Med. Chem. Szeged* 3:59
7. Soreni, E. T., Pegtyen, R. U. 1948. *Ukrain. Biochem. J.* 20:34
8. Askonas, B. A. 1951. *Biochem. J.* 48: 42–48
9. Kuby, S. A., Neda, L., Lardy, H. A. 1954 *J. Biol. Chem.* 209:191–201
10. Burger, A., Richterich, P., Aebi, H. 1964. *Biochem. Z.* 339:305–14
11. Jacobs, H., Heldt, H. W., Klingenberg, M. 1964. *Biochem. Biophys. Res. Commun.* 16:516–21
12. Dance, N., Watts, D. C. 1962. *Biochem. J.* 84:114P–15P
13. Dawson, D. M., Eppenberger, H. M., Kaplan, N. O. 1965. *Biochem. Biophys. Res. Commun.* 21:346–53
14. Eppenberger, H. M., Dawson, D. M., Kaplan, N. O. 1967. *J. Biol. Chem.* 242:204–9
15. Dawson, D. M., Eppenberger, H. M., Kaplan, N. O. 1967. *J. Biol. Chem.* 242:210–17
16. Shattor, J. B., Morris, H. P., Weinhause, S. 1979. *Cancer Res.* 39: 492–501
17. Berlet, H. H., Bonsmann, I., Birringer, H. 1976. *Biochim. Biophys. Acta* 437: 166–74
18. Tranielo, S., D'Aloya, M. A., Grazi, E. 1975. *Experientia* 31:278–80
19. Reiss, N. A., Kaye, A. M. 1981. *J. Biol. Chem.* 256:5741–49
20. Jacobus, W. E., Lehninger, A. L. 1973. *J. Biol. Chem.* 248:4803–10
21. Storey, K. B. 1977. *Arch. Biochem. Biophys.* 179:518–26
22. Walker, J. B. 1979. *Adv. Enzymol.* 50:177–242
23. Thunberg, T. 1911. *Zentralbl. Physiol.* 25:915
24. Eggleton, P., Eggleton, G. P. 1927. *Biochem. J.* 21:190–95
25. Fiske, C. H., Subbarow, Y. 1927. *Science* 65:401–3
26. Meyerhof, O., Lohmann, K. 1932. *Biochem. Z.* 252:431–61
27. Lipmann, F., Meyerhof, O. 1930. *Biochem. Z.* 227:84–109
28. Lundsgaard, E. 1932. *Dan. Hospit.* 75: 84–95
29. Belitzer, V. A., Tsybakova, E. T. 1939. *Biokhimiya* 4:516–34
30. Fiske, C. H., Subbarow, Y. 1929. *Science* 70:381–83
31. Engelhardt, V. A., Lyubimova, M. N. 1939. *Nature* 144:668–69
32. Banga, I., Szent-Gyorgyi, A. 1941–1942. *Stud. Inst. Med. Chem. Univ. Szeged* 1:5–00
33. Hill, A. V. 1950. *Biochim. Biophys. Acta* 41:4–11
34. Cain, D. F., Davies, R. E. 1962. *Biochem. Biophys. Res. Commun.* 8: 361–66

35. Srere, P. H., Estabrook, R. W., eds. 1978. *Microennvironments and Metabolic Compartmentation.* New York: Academic
36. Sers, H., ed. 1982. *Metabolic Compartmentation.* New York: Academic
37. Bloch, W., Macquarrie, R. A., Bernhard, S. A. 1971. *J. Biol. Chem.* 246:780–90
38. Bessman, S. P. 1954. *In Fat Metabolism,* ed. V. Najjar, p. 133. Baltimore: Johns Hopkins Press
39. Bessman, S. P. 1972. *Isr. J. Med. Sci.* 8:344–51
40. Bessman, S. P., Geiger, P. J. 1981. *Science* 211:448–52
41. Gudbjarnason, S., Mathes, P., Ravens, K. A. 1970. *J. Mol. Cell. Cardiol.* 1:325–39
42. Bessman, S. P., Fonyo, A. 1966. *Biochem. Biophys. Res. Commun.* 22:597
43. Jacobus, W. E., Ingwall, J. S., eds. 1980. *Heart Creatine Kinase, the Integration of Isozymes for Energy Distribution.* Baltimore: Williams & Wilkins
44. Brautbar, N. 1984. Symposium
45. Perry, S. V. 1954. *Biochem. J.* 57:427–34
46. Mommaerts, W. F. H. M. 1954. *Nature* 174:1083
47. Seraydarian, M., Mommaerts, W. F. H. M., Wallner, A. 1962. *Biochim. Biophys. Acta* 65:443–60
48. Veech, R. L., Lawson, J. W. R., Cornell, N. W., Krebs, H. A. 1979. *J. Biol. Chem.* 254:6538–47
49. Chance, B., Maumello, G., Maubert, X. 1960. *2nd Int. Res. Conf. Muscle Tissue,* ed. A. Szent Gyorgy, pp. 128–45
50. Hill, D. K. 1960. *J. Physiol.* 153:433–46
51. Ottaway, J. H., Mowbray, J. 1977. *Curr. Top. Cell. Regul.* 12:107–208
52. Dhalla, N. S., Yates, J. C., Walz, D. A., McDonald, V. A., Olson, R. E. 1972. *Can. J. Physiol. Pharmacol.* 50:333–45
53. Neeley, R., Rovetto, M. J., Whitmer, J. I., Morgan, H. E. 1973. *Am. J. Physiol.* 225:651–58
54. Bergstrom, J., Harris, R. C., Hultman, E., Nordejso, L. O. 1971. *Adv. Exp. Med. Biol.* 11:341–55
55. Meyer, R. A., Terjung, R. L. 1972. *Am. J. Physiol.* 237C:111–18
56. Nassar-Gentina, V., Passonneau, J. V., Vergara, J. L., Rapoport, S. I. 1978. *J. Gen. Physiol.* 72:593–606
57. Nassar-Gentina, V., Passonneau, J. V., Rapoport, S. I. 1981. *Am. J. Physiol.* 241C:160–66

58. Conway, C., Sakai, T. 1960. *Proc. Natl. Acad. Sci. USA* 46:897–903
59. Nassar-Gentina, V., Passonneau, J. V., Rapoport, S. I. 1977. *Biophys. J.* 17:173a (Abstr.)
60. Infante, A. A., Davies, R. E. 1965. *J. Biol. Chem.* 240:3996–4001
61. Gercken, A., Schlette, U. 1968. *Experientia* 24:17–19
62. Saks, V. A., Rosenshtraukh, L. V., Undrovinas, A. I., Smirnov, V. N., Chazov, E. I. 1976. *Biochem. Med.* 16:21–36
63. Rosenshtraukh, L. V., Saks, V. A., Undrovinas, A. I., Chazov, E. I., Smirnov, V. N., Sharov, V. G. 1978. *Biochem. Med.* 19:148–64
64. Vassort, G., Ventura-Clapier, R. 1977. *J. Physiol.* 269P:86–87
65. McClellan, G., Weisberg, A., Winegard, S. 1983. *Am. J. Physiol.* 245C:423–27
66. Nunnally, R. L., Hollis, D. P. 1979. *Biochemistry* 18:3642–46
67. Busby, S. J. W., Gadian, D. G., Radda, G. K., Richards, R. E., Seeley, D. J. 1978. *Biochem. J.* 170:103–14
68. Hill, D. R. 1961. *J. Physiol.* 164:31–50
69. Lee, Y. C. P., Visscher, M. B. 1961. *Proc. Natl. Acad. Sci USA* 47:1510–15
70. Savabi, F., Bessman, S. P. 1984. *J. Mol. Cell. Cardiol.* Submitted for publication
71. Sobel, B. E., Shell, W. E., Klein, M. S. 1972. *J. Mol. Cell. Cardiol.* 4:367–80
72. Vial, C., Godinot, C., Gautheron, D. 1972. *Biochimie* 54:843–52
73. Farrell, E. C., Baba, N., Brierley, G. P., Grumer, H. D. 1972. *Lab. Invest.* 27:209–13
74. Scholte, H. R., Weijers, P. J., Wit-Peeters, E. M. 1973. *Biochim. Biophys. Acta* 291:764–73
75. Hall, N., Addis, P., DeLuca, M. 1977. *Biochem. Biophys. Res. Commun.* 76:950–56
76. Hall, N., Addis, P., DeLuca, M. 1979. *Biochemistry* 18:1745–51
77. Roberts, R., Grace, A. M. 1980. *J. Biol. Chem.* 255:2870–77
78. Grace, A. M., Perryman, M. B., Roberts, R. L. 1983. *J. Biol. Chem.* 258:15346–54
79. Saks, V. A., Chernousova, G. B., Voronkov, I. I., Smirnov, V. N., Chazov, E. I. 1974. *Circ. Res. Suppl. III* 34/35(3):138–49
80. Saks, V. A., Chernousova, G. B., Gukovsky, D. E., Smirnov, V. N., Chazov, E. I. 1975. *Eur. J. Biochem.* 57:273–90
81. Saks, V. A., Kupriyanov, V. V., Eliza-

rova, G., Jacobus, W. E. 1980. *J. Biol. Chem.* 255:755–63

82. Saks, V. A., Lipina, N. V., Smirnov, V. N., Chazov, E. I. 1976. *Arch. Biochem. Biophys.* 173:34–41

83. Saks, V. A., Seppet, E. K., Smirnov, V. N. 1979. *J. Mol. Cell. Cardiol.* 11:1265–73

84. Altschuld, R. A., Brierly, G. 1977. *J. Mol. Cell. Cardiol.* 9:875–96

85. Moreadith, R. N., Jacobus, W. E. 1982. *J. Biol. Chem.* 257:889–905

86. Gellerich, F., Saks, V. A. 1982. *Biochem. Biophys. Res. Commun.* 105:1473–81

87. Jacobus, W. E., Saks, V. A. 1982. *Arch. Biochem. Biophys.* 219:167–78

87a. Vial, C., Font, B., Goldschmidt, D., Gautheron, D. C. 1979. *Biochem. Biophys. Res. Commun.* 88:1352–59

87b. Hall, N., DeLuca, M. 1984. Submitted for publication

88. Yang, W. C. T., Geiger, P. J., Bessman, S. P., Borrebaek, B. 1977. *Biochem. Biophys. Res. Commun.* 76:882–87

89. Erickson-Viitanen, S., Viitanen, P., Geiger, P. J., Yang, W. C. T., Bessman, S. P. 1982. *J. Biol. Chem.* 257:14395–404

90. Coty, W. A., Pedersen, P. L. 1974. *J. Biol. Chem.* 249:2593–98

91. Erickson-Viitanen, S., Geiger, P. J., Viitanen, P., Bessman, S. P. 1982. *J. Biol. Chem.* 257:14405–91

91a. Barbour, R. L., Ribaudo, J., Chan, S. H. P. 1984. *J. Biol. Chem.* 259(13):8246–51

92. Lipskaya, T. V., Temple, V. J., Belousova, L. V., Molokova, E. V. 1980. *Biokhimiya* 45:1347–51

93. Meyerhof, O., Lohmann, K. 1928. *Biochim. Z.* 196:22–48

94. Kupriyanov, V. V., Seppet, E. K., Emelin, I. V., Saks, V. A. 1980. *Biochim. Biophys. Acta* 592:197–210

95. Botts, J., Stone, M. J. 1960. *Fed. Proc.* 19:256

96. Yagi, K., Mase, R. 1962. *J. Biol. Chem.* 237:397–403

97. Ottaway, J. H. 1967. *Nature* 215:521–22

98. Turner, D. C., Walliman, T., Eppenberger, H. M. 1973. *Proc. Natl. Acad. Sci. USA* 70:702–5

99. Walliman, T., Turner, D. C., Eppenberger, H. M. 1977. *J. Cell. Biol.* 75:297–317

100. Walliman, T., Kuhn, H. J., Pelloni, G., Turner, D. C., Eppenberger, H. M. 1977. *J. Cell. Biol.* 75:318–25

101. Herasymowych, O. S., Mani, R. S., Kay, C. M., Bradley, R. D., Scraba, D. G. 1980. *J. Mol. Biol.* 136:193–98

102. Mani, R. S., Kay, C. M. 1978. *Biochim. Biophys. Acta* 533:248–56

103. Houk, T. W., Putnam, S. V. 1973. *Biochem. Biophys. Res. Commun.* 55:1271–77

104. Botts, J., Stone, B., Wang, A., Mendelson, R. 1975. *J. Supramol. Struct.* 3:141–45

105. Mani, R. S., Kay, C. M. 1976. *Biochim. Biophys. Acta* 453:391–99

106. Saks, V. A., Chernousova, G. B., Vetter, R., Smirnov, V. N., Chazov, E. I. 1976. *FEBS Lett.* 62:293–96

107. Bessman, S. P., Yang, W. C. T., Geiger, P. J., Erickson-Viitanen, S. 1980. *Biochem. Biophys. Res. Commun.* 96:1414–20

108. Savabi, F., Geiger, P. J., Bessman, S. P. 1983. *Biochem. Biophys. Res. Commun.* 114:785–90

109. Savabi, F., Geiger, P. J., Bessman, S. P. 1984. *Am. J. Physiol.* In press

110. Kleine, T. O. 1965. *Nature* 207:1393–94

111. Baba, N., Kim, S., Farrell, E. C. 1976. *J. Mol. Cell. Cardiol.* 8:599–617

112. Oganro, E. A., Peters, T. J., Hearse, D. J. 1977. *Cardiovasc. Res.* 11:250–59

113. Carpenter, C. L., Mohan, C., Bessman, S. P. 1983. *Biochem. Biophys. Res. Commun.* 111:884–89

114. Carpenter, C. L., Mohan, C., Bessman, S. P. 1984. Submitted for publication

115. Carpenter, C. L., Savabi, F., Bessman, S. P. 1984. Submitted for publication

116. Bessman, S. P., Pal, N. 1976. In *The Krebs Cycle Depletion Theory of Hepatic Coma in The Urea Cycle*, ed. S. Grisolia, R. Baguena, F. Mayor, pp. 83–89. New York: Wiley

117. Bessman, S. P., Pal, N. 1982. *Isr. Med. Sci.* 18:171–75

118. Baskin, R. J., Deamer, D. N. 1970. *J. Biol. Chem.* 245:1345–47

119. Sharov, V. G., Saks, V. A., Smirnov, V. N., Chazov, E. I. 1977. *Biochim. Biophys. Acta* 468:495–501

120. Levitsky, D. O., Levchenko, T. S., Saks, V. A., Sharov, V. G., Smirnov, V. I. 1978. *Membr. Biochem.* 2:81–86

121. Saks, V. A., Lipina, N. V., Sharov, V. G., Smirnov, V. I., Chazov, E., Grosse, R. 1977. *Biochim. Biophys. Acta* 465:550–58

122. Caldwell, P. C., Hodgkin, A. L., Keynes, R. D., Shaw, T. I. 1960. *J. Physiol.* 152:561–90

123. Grosse, R., Spitzer, E., Kupriyanov, V. V., Saks, V. A., Repke, K. R. H. 1980. *Biochim. Biophys. Acta* 603:142–56

124. Spitzer, E., Grosse, R., Kuprijanov, V., Preobrazhensky, A. 1981. *Acta Biol. Med. Germ.* 40:1111–22

125. Erashova, S., Saks, V. A., Sharov, V. G., Lyzlova, S. N. 1979. *Biokhimiya* 44:372–78

126. Reiss, N. A., Kaye, A. M. 1981. *J. Biol. Chem.* 256:5741–49

127. Abstract presented at 1984 CK conferences (not printed)

128. Traniello, S., D'Aloya, M. A., Grazi, E. 1975. *Experientia* 31:278–80

129. Loike, J. D., Kozler, V. F., Siverstein, S. C. 1979. *J. Biol. Chem.* 254:9558–64

130. Mainwood, G. W., Rakusan, K. 1982. *Can. J. Physiol. Pharmacol.* 60:98–102

131. Meyer, R. A., Sweeney, H. L., Kushmerick, M. J. 1984. *Am. J. Physiol.* In press

132. Meyer, R. A., Kushmerick, M. J., Brown, T. R. 1982. *Am. J. Physiol.* 242C:1–11

133. Gadian, D. G., Radda, G. K., Brown, T. R., Chance, E. M., Dawson, M. J., Wilkie, D. R. 1981. *Biochem. J* 194:215–28

134. Fossel, E. T., Morgan, H. E., Ingwall, J. S. 1980. *Proc. Natl. Acad. Sci. USA* 77:3654–58

135. Mathews, P. M., Bland, J. L., Gadian, D. G., Radda, G. K. 1981. *Biochem. Biophys. Res. Commun.* 103:1052–59

136. Taegtmeyer, H., Hems, R., Krebs, H. A. 1980. *Biochem. J.* 186:701–11

137. Fitch, C. D., Jellinek, M., Mueller, E. J. 1974. *J. Biol. Chem.* 249:1060–63

138. Fitch, C. D., Chevli, R. 1980. *Metabolism* 29:686–90

139. Shields, R. P., Whitehair, C. K., Carrow, R. E., Heusner, W. W., Van Huss, W. D. 1975. *Lab. Invest.* 33:151–58

140. Petrofsky, J. S., Fitch, C. D. 1980. *Pfluegers Arch.* 384:123–29

141. Laskowski, M. B., Chevli, R., Fitch, C. D. 1981. *Metabolism* 30:1080–85

142. Mainwood, G. W., Alward, M., Eiself, B. 1982. *Can. J. Physiol. Pharmacol.* 60:114–19

143. Mainwood, G. W., Alward, M., Eiself, B. 1982. *Can. J. Physiol. Pharmacol.* 60:120–27

144. Fitch, C. D., Chevli, R., Petrofsky, J. S., Kopp, S. J. 1978. *Life Sci.* 23:1285–92

145. Annesley, T. M., Walker, J. B. 1980. *J. Biol. Chem.* 255:3924–30

146. Hall, N., DeLuca, M. 1975. *Biochem. Biophys. Res. Commun.* 66:988–94

147. Ingwall, J. S., Kaufman, I. A., Mayer, S. E. 1976. *J. Cell Biol.* 70:239a

148. Ingwall, J. S., Wildenthal, K. 1978. In *Recent Advances in Studies on Cardiac Structure and Metabolism*, 12:621–33. Baltimore: Univ. Park Press

149. Roberts, C. M., Bessman, S. P. 1980. *Biochem. Biophys. Res. Commun.* 93:617–24

149a. Green, H. J., Reichman, H., Pette, D. 1983. *Pflügers Arch.* 399:216–22

149b. Loike, J. D., Kozler, V. F., Silverstein, S. C. 1984. *J. Exp. Med.* In press

150. Bessman, S. P., Borrebaek, B., Geiger, P. J., Ben-Or, S. 1978. In *Microenvironments and Cellular Compartmentation*, ed. P. Srere, R. W. Estabrook, p. 111. New York: Academic

151. Tice, L. W., Barnett, R. J. 1962. *J. Cell Biol.* 15:401–16

152. Savabi, F., Geiger, P. J., Bessman, S. P. 1984. *Biochem. Biophys. Res. Commun.* In press

153. Mahler, M. 1979. *Biochem. Biophys. Res. Commun.* 88:895–906

154. Bennett, V. D., Hall, N., DeLuca, M., Suelter, C. H. 1984. JBC In press

155. Brautbar, N., Baczynski, R., Carpenter, C., Moser, S., Geiger, P., Finander, P., Massry, S. G. 1982. *Am. J. Physiol.* 242:F699–704

156. Brautbar, N., Carpenter, C., Baczynski, R., Kohan, R., Massry, S. G. 1983. *Kidney Int.* 24:53–57

157. Najarian, J. S., Harper, H. 1956. *Proc. Soc. Exp. Biol. Med.* 92:558–60

158. Bessman, S. P. 1972. *J. Pediatr.* 81:834

159. Bessman, S. P. 1979. *Nutr. Rev.* 37:7:209–20

160. Sipila, I. 1980. *Biochim. Biophys. Acta* 613:79–84

161. Sipila, I., Simell, O., Arjomaa, P. 1980. *J. Clin. Invest.* 66:684–87

162. Sipila, I., Rapola, J., Simell, O., Vannas, A. 1981. *N. Engl. J. Med.* 304:867–70

Ann. Rev. Biochem. 1985. 54:863–96

TRANSPOSITIONAL RECOMBINATION IN PROKARYOTES

Nigel D. F. Grindley

Department of Molecular Biophysics and Biochemistry, Yale University Medical School, New Haven, Connecticut 06510

Randall R. Reed

Howard Hughes Medical Institute, Department of Molecular Biology and Genetics, The Johns Hopkins University, School of Medicine, Baltimore, Maryland 21205

CONTENTS

0066-4154/85/0701-0863$02.00

PERSPECTIVES AND SUMMARY

Two specialized forms of recombination have been recognized as playing fundamental roles in restructuring chromosomes, either by reassorting the genetic information or by adding to it or deleting from it. Site-specific recombination is one restructuring mechanism. In essence this involves a single reciprocal crossover by a specific breakage-reunion reaction between two very short regions of homology. The enzymes involved are a specialized form of sequence specific topoisomerase. Normally no DNA synthesis is required and, hence, the process can be described as a conservative reaction. Site-specific recombination results either in excision or inversion of a DNA segment if the reaction is intramolecular, or in reciprocal translocations or integration of a circular replicon if the reaction is intermolecular. The best known site-specific recombination systems are integration and excision of bacteriophage λ (1), control of flagellar phase variation in *Salmonella* by an inversion reaction (2, 3), and resolution of cointegrates mediated by the transposons Tn*3* or γδ (4, 5).

Transpositional recombination, the primary subject of this review, is a second chromosome restructuring mechanism. The vectors of transpositional recombination, called transposable elements or transposons, are specific DNA segments that can move from one genetic location to another. Although transposons have been studied most extensively in bacteria, they are found in virtually all organisms and were discovered through the elegant genetic and cytogenetic studies of McClintock in maize (6, 7). Like site-specific recombination, transpositional recombination is distinct from, and independent of, the mechanisms of generalized homologous recombination. The distinction between site-specific and transpositional recombination is a difficult one to define, however, partly because there may be several forms of the latter and our knowledge of the biochemical events is at a much more primitive level. Two features of transpositional recombination that so far appear to be characteristic are its nonreciprocal nature and the mandatory involvement of at least some DNA replication. As will be seen below, in all cases repair synthesis is necessary to generate duplications of a short sequence at the integration site, and in some cases the entire transposable segment may be replicated.

Much of our knowledge that pertains to the mechanisms of DNA transposition has come from the intensive studies of certain transposons in enterobacteria, so the bulk of this review will focus on these studies. More general and comprehensive reviews of transposable DNA (both prokaryotic and eukaryotic) include a recent collection edited by Shapiro (8) and a review by Kleckner (9). In addition transposition mechanisms were the subject of a brief review by Bukhari (10). Despite the appearance of this article in *Annual Review of Biochemistry*, most of our insights into transpositional mechanisms have come from genetic studies; biochemical studies, severely hampered by the very low frequency of most transposition events, have only started to yield informative results in the last year or so. Because of this, all mechanisms to be discussed are of a speculative nature and await biochemical proof.

GENERAL FEATURES OF BACTERIAL TRANSPOSONS

Genetic Rearrangements

In addition to promoting the movement of a transposon from one genomic location to another, transpositional recombination can give rise to a variety of genetic rearrangements. All the rearrangements involve joining transposon ends to new sites in inter- or intra-molecular events. Intermolecular events are of two types: simple insertions and replicon fusions (Figure 1A). In the former, the element alone is inserted into a target replicon; in the latter, the entire donor replicon is inserted into the target and the transposon itself is duplicated—the resulting structure is commonly called a cointegrate. Throughout this review we shall make a distinction between the transpositional processes that directly generate a simple insertion (the simple insertion pathway) and those that generate a cointegrate (the cointegrate pathway). Transpositions involving an intramolecular target result in either adjacent deletions or inversion-insertions (also called duplicative inversions) (Figure 1B). As is shown diagramatically in Figure 2, both deletions and inversion-insertions are the intramolecular equivalents of cointegrates, not of simple insertions. If a transposon interacts with an intramolecular target and follows the pathway to cointegrate formation, then, depending only on the orientation of the target site relative to the transposon ends, either a deletion (giving two reciprocal products as shown in Figure 1B) or an inversion-insertion results. An alternative pathway for deletion formation would be simply to join one end of the transposon to an adjacent target site; in this case the deleted segment would not be excised as a circular product with a transposon copy.

In essence one can imagine all transposon-promoted rearrangements as resulting from one of only two types of transposon-target interaction. In the simple insertion pathway both ends of a single copy of the transposon are joined to the target sequences. By contrast, in the cointegrate pathway two copies of the element are formed and each copy retains a parental joint at one end while the other end is attached to new target sequences. The cointegrate pathway must be a replicative process since two copies of the transposon are produced. The simple insertion pathway, however, could be either a replicative process in which a copy of the transposon is retained at the donor site, or a conservative process in which the transposon is excised from the donor site without replication.

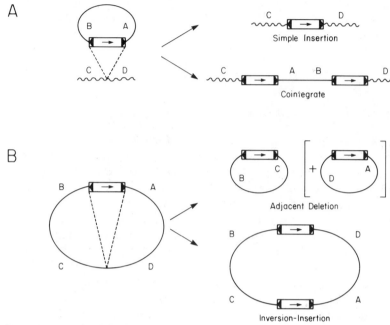

Figure 1   Rearrangements Promoted by Transposons.
A. Intermolecular Transposition. Interaction with an unlinked target site can produce simple insertions or cointegrates.
B. Intramolecular Transposition. Interaction with a genetically linked target site may lead to adjacent deletions or inversion-insertion.

Duplication of a Target Sequence

When a transposon inserts into a new target site, the ends of the element are joined precisely to target DNA, and a short segment of the target (from 3 bp to 13 bp) is duplicated, with one copy at each end of the inserted transposon. This is true for all prokaryotic (and eukaryotic) transposons investigated so far with the single exception of Tn*554* (9, 12). The size of the duplicated segment is characteristic of the transposon, however, a few transposons generate duplications of variable size. The available evidence indicates that the duplication is not a preexisting sequence that is brought in as part of the transposed DNA [as occurs with the integration of bacteriophage λ; ref (1)] but, rather, is a consequence of the mechanism of insertion (13, 14). The flanking repeat is not required for the subsequent transposition of an element (although it may be involved in the rare precise excision of a transposon—an activity that is not usually associated with transpositional recombination).

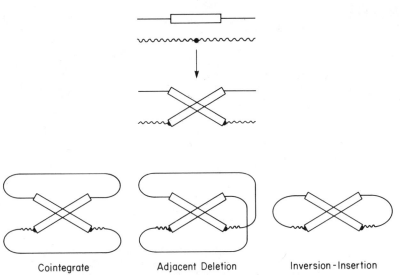

Cointegrate   Adjacent Deletion   Inversion-Insertion

Figure 2   Cointegrates, Adjacent Deltions, and Inversion-Insertions are Mechanistically Equivalent. These three different rearrangements all result from transposition by a replicative, cointegrate pathway; the outcome depends only on the linkage between the donor and target DNA. For cointegrates, donor and target are unlinked (*left*); for adjacent deletions and inversion-insertions, donor and target are linked but the targets are in opposite orientations (*center* and *right*). Adapted with permission from (11).

Transposon Ends

The ends of transposons are the specific sites used for transpositional recombination. It is not surprising, therefore, that the DNA sequences at each end of a transposon are identical or nearly so; these are called terminal inverted repeats (IRs). The only exceptions to this are the transposing bacteriophage Mu (and D108) and Tn*554* (12, 15). The terminal IRs of different transposons vary over a wide range both in length and in their degree of sequence conservation. They range from 8 bp to 40 bp and may show perfect homology (38/38 for Tn*3*) or imperfect homology (12/18 for IS*50*). These IRs almost certainly provide the recognition sites for the element's transposase protein. Transpositional recombination is generally dependent on the presence and integrity of both transposon ends even for rearrangements that, formally, only join a single end to target DNA. Both ends must be correctly oriented and must lie on the same DNA molecule, that is, they must define a transposing segment. A few exceptions to this general rule will be discussed below.

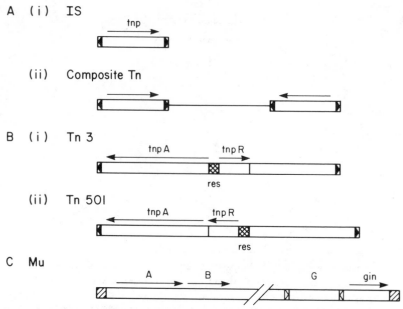

Figure 3 Three Classes of Bacterial Transposons.

A. ISs and Composite Transposons. ISs range in size from about 750 bp to 1600 bp. They generally have a single transposase gene (*tnp*). Composite transposons consist of a DNA segment (often with a phenotypic determinant such as a drug-resistance gene) flanked by two copies of an IS; the IS copies may be in the same or inverted orientations.

B. Tn3-family Transposons. *tnpA* and *tnpR* are the genes for transposase and the cointegrate resolvase respectively, *res* is the site of cointegrate resolution. Two subgroups have been distinguished and are exemplified by (i) Tn*3* and (ii) Tn*501*. In these subgroups the DNA segment that contains *tnpR* and its *res* site has adopted opposite orientations. The *tnpA*-(*res*-*tnpR*) segment is about 3800 bp in size. The remainder of the transposon may carry additional genetic markers (e.g. ampicillin resistance in Tn*3*, mercury resistance in Tn*501*).

C. Mu. The whole genome is about 38 kb. The A and B genes, necessary for efficient transposition and replication, lie near the left (immunity) end, *attL*. Near the right end lies a 3-kb invertible DNA segment (the G segment) and the *gin* gene; *gin* is related to the *tnpR* genes of Tn*3* and Tn*501* and encodes a site-specific recombinase that inverts the G segment.

GENERAL CLASSES OF BACTERIAL TRANSPOSONS

Although it is an oversimplification, it has become customary to divide bacterial transposons into three classes. These are (*a*) the Insertion Sequences (IS) together with the composite transposons which have IS constituents, (*b*) the Tn3-family of transposons, and (*c*) the transposing bacteriophage Mu and related phages (Figure 3). An increasingly large number of bacterial elements clearly do not fall into any of these groups. Among these "unclassified" transposons are Tn7 (16), Tn*916* (17), and Tn*554* (12). Because information concerning the transpositional processes of these transposons is limited we will not dwell on them further.

IS Elements And Composite Transposons

The IS elements are short segments of DNA (750 bp–1600 bp) that encode only one or two genes involved in transpositional recombination (9, 18). ISs are found singly as natural constituents of bacterial chromosomes, or as pairs (in the same or opposed orientations) as the driving force of a composite transposon. Although many ISs are known, only a few have been subjected to mechanistic studies. In this review we will focus on the properties of just four examples of this class: IS1, IS10, IS50, and IS903. These ISs are, respectively, constituents of the composite transposons Tn9, Tn10, Tn5, and Tn903. Although these ISs are similar in their recombinational behavior and genetic organization, there is no indication that they are related to one another; generalizations, therefore, have to be treated with caution.

Genetic evidence and DNA sequence analyses show that the ISs encode a protein, transposase, that is required for transpositional recombination. IS10, IS50, and IS903 each appear to encode a single transposase polypeptide; and in each case the gene occupies close to 90% of the total coding capacity of the transposon (19–21). In addition IS50 encodes a second product from an alternative initiation codon within the transposase reading frame; this product appears to function as an inhibitor of transposition (22–24). In the case of IS1, expression of two polypeptides encoded by adjacent genes is necessary for transposition (25–27).

A remarkable property of the IS transposases is that they act efficiently only in cis, that is, on transposon ends on the same DNA molecule as the transposase gene and with strong preference for closely linked ends (26, 28–30). Complementation in trans of a transposon with a defective transposase occurs with an efficiency of about 1% or less. An attractive explanation for the strong cis-preference is that the transposases (all very basic proteins; ref. 9) bind nonspecifically to DNA sequences close to the transposase gene immediately upon synthesis (or even during translation) and then search for specific DNA binding sites (transposon ends) by one-dimensional diffusion along the DNA (31).

The IS elements discussed here promote the various rearrangements shown in Figure 1 with varying efficiencies. In all cases, simple insertions occur considerably more frequently then replicon fusions; in fact, IS10 and IS50 appear to make cointegrates extremely rarely if at all (<0.1% of the frequency of simple insertions) (32, 33). On the other hand, both IS1 and IS903 do make cointegrates at between 1% and 5% of the simple insertion frequency (21, 34). IS1, IS903 and IS10 all promote deletion formation (IS50 has not been assayed for this activity), however, IS10 does so only with low efficiency (35–39). Few inversion-insertions have been detected with any of these IS elements (one with IS10, two with IS1) and since all were seen in recA+ strains, there is some doubt as to whether they arose directly from a single transpositional event (39, 40). For IS903, the only IS whose intramolecular rearrangements have been

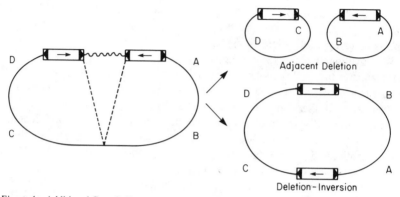

Figure 4   Additional Genetic Rearrangements Promoted by a Composite Transposon. Interaction of the "inside" ends of the two ISs with a linked target site generates an adjacent deletion or a deletion-inversion depending on target orientation. In both cases the internal, non-IS, DNA of the composite transposon is lost (41).

systematically studied, it is clear that adjacent deletions occur at least 100-fold more frequently than inversion-insertions[1] (34).

A segment of DNA flanked by two copies of an IS is itself a transposable element and is called a composite transposon. The ISs may be in the same or inverted orientation (see Figure 3). Composite transposons have four IS ends that can be used in transpositional recombination; the interaction of the two "inside ends" with a target allows a wider repertoire of genetic rearrangements (Figure 4). Intermolecular transposition (by the simple insertion pathway) results in insertion of the "inverse transposon," that is, all sequences of the donor replicon with the exception of the markers contained between the ISs of the conventional composite transposon. Of more interest, intramolecular transpositions (again by way of the simple insertion pathway) result in either deletions or deletion-inversions; in both cases the DNA segment between the inside ends is deleted, and the DNA between the inside end and the target site is either excised as a circle or inverted depending on the orientation of the target. Both these rearrangements are promoted at quite high frequencies by Tn*10* (41), and deletions (but not deletion-inversions) by Tn*903* (37).

---

[1]The intramolecular cointegrate process would be expected to generate equal numbers of deletions and inversion-insertions. Since IS*903* forms adjacent deletions by a cointegrate process (see below) the absence of inversion-insertions suggests that there is a strong bias for a particular orientation of the target site. We proposed that the transposase traces the path of the DNA between the transposon and the target site (37). The absence of deletion-inversions promoted by Tn*903* is consistent with this explanation.

IS TRANSPOSITIONAL RECOMBINATION: FREQUENTLY CONSERVATIVE, OCCASIONALLY REPLICATIVE   A question that is central to understanding the mechanism of transpositional recombination is whether genetic rearrangements are accompanied by duplication of the transposon that promotes them. The most frequent transpositional event seen with an IS is a simple insertion. Events that obviously involve transposon duplication—cointegrate formation and inversion-insertion—are observed much less frequently than simple insertions (or not at all). In the case of Tn3 and related transposons, there is compelling evidence that most, perhaps all, transpositional recombination is replicative, proceeding through a cointegrate intermediate (see below). By contrast, several observations argue against the involvement of cointegrates as an intermediate in simple insertion of an IS. First, cointegrates are formed infrequently relative to simple insertions, yet, once formed, are stable in a $recA^-$ strain. Second, the stability of cointegrates and the limited coding capacity of ISs argue against a separate site-specific cointegrate resolution system like that found in Tn3-related transposons. Third, all sites essential for the formation of simple insertions are found close to the termini of the ISs (within 27 bp for IS10, within 20 bp for IS903) (36, 38); this suggests that an internal cointegrate resolution site is not required although it does not preclude a site at the transposon end. Fourth, although some IS elements have been subjected to extensive random mutagenesis, no mutants have been isolated that give rise only to cointegrates (analogous to the $tnpR^-$ and res^- mutants of Tn3). Although none of these observations by itself is compelling, together they strongly suggest that simple insertions of an IS are formed directly, not via a cointegrate intermediate, while cointegrates are an alternative (and minor) final product of IS transpositional recombination.

Two questions are raised by this conclusion. First, are the rare cointegrates truly replicative; second, are IS simple insertions formed by a replicative or a conservative process?

It has been noted for both IS50 and IS10 that the formation of cointegrates (but not of simple insertions) is strongly stimulated in a $recA^+$ host strain (33, 42). From this, Berg (42) has proposed that these cointegrates are actually derived by a simple insertion process from dimers of the donor replicon. Both IS1 and IS903 promote replicon fusion in $recA^-$ strains, suggesting that these two ISs both have a replicative cointegrate pathway of transpositional recombination (21, 34). Perhaps the most convincing evidence for IS903 duplication has come from the analysis of IS903-promoted adjacent deletions (36). Using a plasmid that contained two origins of replication, Weinert et al were able to recover two circular products from a single deletion-forming recombinational event in a $recA^-$ strain (see Figure 1B). Since each product contained a copy of the IS903 that had promoted the deletion, this showed that the IS was duplicated during the process. Moreover, it showed that adjacent deletions were formed by the cointegrate pathway; not surprisingly, therefore, both IS ends were required for the event. Additional support for this came from a similar analysis of IS10, which gives simple insertions efficiently but fails to promote cointegrates. IS10 was found to generate adjacent deletions only at a very low frequency (37). Thus for the both IS903 and IS10 the ability to promote adjacent deletions correlated well with their ability to form cointegrates rather than with their ability to give simple insertions.

Recently Biel & Berg (43) have presented convincing evidence that IS1 also has a true cointegrate pathway. They analyzed cointegrate-like molecules formed between an F factor and either monomers or dimers of a pBR322::Tn9 plasmid in a $recA^-$ host, following selection for transfer of resistance to both chloramphenicol (carried by Tn9) and ampicillin (carried by pBR322). When derived from the monomer-containing donor, most recombinant molecules had 3 copies of IS1 (i.e. the structure of IS1-mediated cointegrates). However, when derived from the dimer-containing donor, most recombinants had 4 IS1s (simple insertions using the two inside ends of a Tn9—the closest IS1 ends available). These results showed that the formation of cointegrates from the pBR322::Tn9 monomers did not first require dimerization of the plasmid, since the predominant final products are different.

The existence of a duplicative, cointegrate pathway for at least some IS elements does not, of course, address the question of whether simple insertions also involve IS duplication. In an elegant experiment, Kleckner et al have shown that simple insertions of Tn10 do not involve extensive replication of the transposon (44). The experiment involved construction of a λ::Tn10 derivative that carried the lacZ gene within Tn10. Heteroduplex DNA molecules were made that carried $lacZ^+$ information on one strand and $lacZ^-$ information on the other. The heteroduplexes (and homoduplexes) were packaged into λ virions in vitro, Escherichia coli was infected with the λ phages (under conditions in which λ could not replicate), and tetracycline-resistant transductants (i.e. Tn10 transpositions) were selected. A high proportion of Tc^R colonies showed Lac^+/Lac^- sectoring; analysis showed the sectors of a single colony resulted from a single simple insertion of the heteroduplex Tn10 derivative. The simplest interpretation was that both parental strands of the Tn10 were integrated with little or no replication of the transposon until after insertion.

A corollary of simple insertion of unreplicated transposon DNA is that the element must have been excised from the donor site, causing destruction of the donor replicon.[2] Indirect evidence that simple insertions do cause destruction

---

[2]In principle, excision of an element either could destroy the donor replicon or could be accompanied by repair of the donor. The latter possibility is ruled out by observations made with several transposons (9). First, "precise excision" (as measured by regaining function of an inactivated gene) does not require transposase function, is not accompanied by reinsertion of the transposon, and occurs at a lower frequency than transposition. Second, in strains in which a transposition event has occurred, the transposon is retained at the donor site; this is compatible either with a replicative transposition or with a destructive excision process (from a second copy of the replicon within the same cell).

of their donor has been obtained with both IS*10* (for which there is no convincing evidence for a replicative mode of transposition) and IS*903* (which does appear to promote replicative transposition). As noted above, Weinert et al found that adjacent deletion formation promoted by an IS equated with the ability of the IS to produce cointegrates (36). However, one important exception was observed: deletions (and deletion-inversions) promoted by the inverse transposon (see Figure 4) occurred at a relatively high frequency, compatible with their formation by the more efficient simple insertion process (37). From this result, Weinert et al argued that viable replicons could only be recovered from intramolecular transposition events either if the events occurred by the cointegrate pathway or if the target site lay within the transposon sequences, in which case they could occur by the simple insertion pathway. Since simple insertions presumably occurred to sites outside the transposon (or IS), the failure to recover viable products from these events implied that the integrity of the donor/target replicon was destroyed (see Figure 5).

The similarity in behavior of the four IS elements suggests that all are likely to share a common transpositional mechanism. On the basis of the results discussed above, it appears that the most frequent transpositional event, the simple insertion, involves excision of the IS from the donor (causing destruction of the donor replicon) and its integration into a target site with little or no replication of the element; some repair synthesis is necessary to generate the target site duplications. Much less frequently (about 1–5% for IS*1* and IS*903*, virtually undetectable for IS*10* and IS*50*) a truly replicative, cointegrate process occurs that preserves both donor junctions while creating two new target junctions.[3]

The Tn3 Family Of Transposons

Tn*3* and its relatives form a completely distinct class of transposons (4, 5). Unlike the IS elements, transposons of this class are all related to one another. They share a common two-step transposition mechanism. They encode two gene products, both of which play a role in the formation of the final products of transposition: the *tnpA* gene encodes a transposase of about 120,000 daltons; the *tnpR* gene encodes resolvase (21,000 daltons), a site-specific recombinase. Both gene products are *trans* acting, in contrast to the IS transposases. These transposons all have long (35–40 bp) terminal inverted repeats of related sequence and duplicate a 5-bp target sequence upon transposition. In most cases the genetic markers of the transposon (e.g. antibiotic resistance) are not segregated from the transposition functions and sites (as is the case in the

---

[3]The observation that there is a binding site for the *dnaA* protein (45) near the end of IS*50*, within sequences that are essential for transposition, prompted Johnson & Reznikoff (46) to propose the involvement of this replication initiation factor in IS*50* transposition. The failure to detect any transposition events mediated by IS*50* that are unambiguously replicative is at odds with this proposal. However, the *dnaA* protein might play a regulatory role.

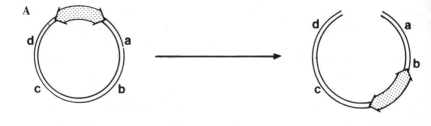

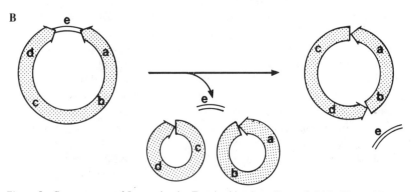

Figure 5   Consequences of Intramolecular Transposition by a Donor Suicide (Cut-and-Paste) Process.
A. Target site outside the transposon. An IS element is excised from the donor site and inserted into the target site; as a result, the replicon is linearized and thereby destroyed.
B. With an inverse transposon, the target lies within the element (the stippled segment defines the transposon). The ends of the element are joined to the target in one of two possible orientations, generating either deletions or deletion-inversions. Used with permission from (37).

composite transposons) but, rather, are contained within the body of the basic transposing segment.

Based on their genetic organization and on the ability of recombination functions to complement one another, transposons of the Tn3 family can be divided into 2 subgroups (see Figure 3B). In one group, represented by Tn3 itself and γδ, the *tnpA* and *tnpR* genes are divergently transcribed from a shared regulatory region (47). The *tnpR* functions, but not the *tnpA*, are interchangeable (48). The site of cointegrate resolution, *res*, lies within the intergenic region (49, 50) and the *tnpR* gene product is a transcriptional repressor (of both *tnpA* and *tnpR*) as well as a site-specific recombinase (51–53). In the Tn501 subgroup the region that contains *res* and *tnpR* is inverted relative to its orientation in Tn*3* (see Figure 3). The best characterized members of this subgroup are Tn*501*, Tn*21*, and Tn*1721*; the *tnpA* and *tnpR* functions are interchangeable amongst these transposons (54–56).

TN3 TRANSPOSITION—A REPLICATIVE TWO STEP PROCESS   Intermolecular transposition of Tn*3* and related transposons proceeds in two distinct stages (Figure 6) (4, 5). First, through the action of the *tnpA* gene product, donor and target replicons fuse to form a cointegrate. The cointegrate is a transient intermediate and, in a site-specific recombination, is rapidly broken down by the action of resolvase at the two *res* sites. This yields the final transposition product—a simple insertion of the transposon in the target replicon—and regenerates the donor replicon. In the absence of a functional transposase, no transpositional recombination is observed. In the absence of either *tnpR* function or active *res* sites the process stops with the formation of a cointegrate.

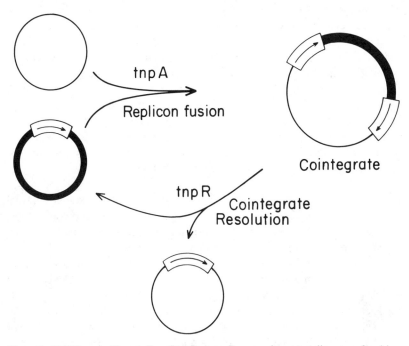

Figure 6   Tn*3* Transposition: A Two Step Process. Donor and target replicons are fused in a replicative process by the action of transposase (the *tnpA* gene product) to form a cointegrate. Site-specific recombination between the two transposon copies (mediated by the action of resolvase at the two *res* sites) breaks down the cointegrate into its two constituent replicons: the target with inserted transposon and the regenerated donor. Used with permission from (4).

Subsequent provision of active TnpR protein efficiently resolves a preformed cointegrate (in vivo or in vitro) showing that there is no obligatory coupling of transposition and resolution processes. Because a cointegrate, with its two transposon copies, is an intermediate in transposition, it is clear that the process normally involves transposon duplication. Bennett et al (57) have noted that, in the absence of *tnpR*, simple insertions are occasionally formed (at about 5% of the frequency of cointegrates). Whether this represents a low efficiency direct pathway to simple insertions or a low efficiency, *tnpR*-independent resolution of cointegrate intermediates is not known.

Intramolecular transposition involves just the first (cointegrate) step of the intermolecular process. It yields either adjacent deletions or inversion-insertions depending on the orientation of the target relative to the transposon ends. Resolvase plays no role in the formation of these products.

It appears that the 35–40 bp terminal inverted repeats are both necessary and sufficient sites for normal transpositional activity (i.e. cointegrate formation). A mutation that removes all but the terminal 14 bp of the Tn3 right IR transposes at about 0.1% of the normal frequency (58). Yet IS*101*, a 209-bp defective element that is activated by the γδ *tnpA* gene product, has homology to γδ only within the 35 bp IRs (59). Recently it has been observed that Tn*21* and Tn*1721* continue to transpose, although at a substantially reduced frequency, even when one end is completely deleted (60, 61). The product of this aberrant transposition is analogous to a cointegrate; one joint between the two replicons is at the IR and a segment, that includes the IR and a variable portion of the donor DNA, is duplicated at the other junction. The structures of these molecules were interpreted as supporting an asymmetric transposition process (60, 61); however, we believe that they are equally compatible with symmetric mechanisms (see *Transposition Mechanisms* below).

TRANSPOSITION IMMUNITY   Tn3 and related transposons show an unusual property that is unique to this family. A replicon containing a copy of Tn3 acts as a target for additional copies at a substantially reduced frequency (less than 5% of the nonimmune). Transposition immunity, as this phenomenon is called, is a *cis*-acting process—other replicons in the same cell have unimpaired target properties (62)—and it appears to affect only intermolecular transposition events (63). Just one terminal IR of the homologous type appears to be sufficient to confer immunity to a target replicon in at least some cases (64), although additional sequences of the transposon may increase the magnitude of the effect (65). An active *tnpR* gene is not necessary for immunity.

The specificity of the immunity process appears to follow the specificity of the transposases (5). Since the terminal IR is the only necessary sequence, it seems likely that immunity involves the recognition of an homologous IR in the target replicon by the transposase (or transposase-DNA complex) of the transposing element. Recognition of an IR anywhere within the target DNA implies that the transposase (or a pretransposition complex) scans the target sequences before choosing an integration site. (This is consistent with the observations of preferred sites for integration.) It is not known whether simply the recognition of an IR within the target by a pretransposition complex is sufficient to block transposition, or whether transposition is initiated but is aborted through the interference of the third IR. The nonimmunity of intramolecular transpositions is consistent with the notion that both transposon ends are complexed with transposase before the search for a target site (which would then not contain any free IR). This suggests that a third IR on a replicon might inhibit intramolecular transpositions; this has recently been observed (66).

Bacteriophage Mu

The genome of phage Mu (and related phages) is integrated into host sequences at all stages in its life cycle, including when packaged into mature virions [for a recent review, see ref. (67)]. The phage DNA is a linear molecule of about 38 kb consisting of 36 kb of Mu DNA with about 1.5 kb of host DNA at one end and 50–150 bp of host DNA at the other. The attached host sequences reflect the life style of the phage—to replicate, it undergoes repeated cycles of transposition in the cell (about 100 events per cell during the lytic cycle), and is finally packaged along with adjacent host sequences into virus particles. As with other transposons, Mu integration results in duplication of a small target sequence (5 bp in the case of Mu). However, in contrast to other transposons, the ends of Mu are not inverted repeats (15), even though they are clearly the sites of recombination and are both bound by purified Mu transposase (the A protein) in vitro (68).

Two gene products are involved in the transposition process. The A gene product, a polypeptide of 70,000 daltons, is the transposase; the B gene product (33,000 daltons) is apparently an accessory protein necessary for replication and full transpositional activity (see Figure 3). Although Mu fails to replicate in the absence of B (69), integration of infecting Mu is unaffected (as measured directly or by lysogen formation) (70, 71) and other Mu-mediated rearrangements occur at about 1%–10% of the wild-type frequency (72). Deficiencies in A or B are complemented in trans quite efficiently. An unusual property of the Mu A protein is that it appears to act stoichiometrically in vivo (73). If DNA replication was blocked at the time of induction of a Mu lysogen, then, upon simultaneous release of the block and inhibition of further protein synthesis, only a single round of Mu replication occurred. Further rounds of Mu replication required new synthesis of A protein.

TRANSPOSITIONAL PROCESSES AND GENETIC REARRANGEMENTS IN VIVO   During the lytic cycle, the stage at which Mu transpositional activity is greatest, Mu transposes predominantly by a replicative cointegrate process and simple insertions are rarely, if ever, seen (74, 75). As with cointegrates mediated by IS elements, Mu cointegrates are stable—there is no evidence for a Mu cointegrate resolvase. Recently M. Pato and B. Waggoner (personal communication) have been able to look at the consequences of a single synchronized round of Mu replication in vivo following prophage induction. Although both of the original Mu-host junctions are separately conserved [as was first shown by Ljungquist & Bukhari; (76)], the original prophage location is not retained intact. The results show that replication of Mu has generated two copies—one copy retains the original host-Mu *attL* junction and gains a new junction with random host sequences at the right end, the other copy retains the original host-*attR* junction and gains a new host junction at the left end. Induction of the Mu replication cycle has therefore resulted in transposition by a cointegrate process (to generate either inversion-insertions, reciprocal adjacent deletions, or intermolecular cointegrates) and not by a replicative simple insertion process.

Despite the predominance of cointegrates during Mu growth, simple insertions of Mu do occur under specific conditions. First, when Mu forms a lysogen, the end product is a simple insertion of Mu. Since lysogen formation is an inefficient process one might argue that simple insertions are formed either by occasional failures of the cointegrate pathway or from rare resolutions of a cointegrate intermediate. However, recent data indicate that infecting Mu first integrates to form a simple insertion even when the lytic cycle will ensue (77). Possible reasons for the predominance of simple insertions following phage infection include: (*a*) the linear, nonsupercoiled nature of infecting DNA (although it is actually formed into a circular (but noncovalent) protein-bound molecule after injection; refs (78, 79); (*b*) injection of specific Mu proteins with the DNA; (*c*) differential expression of Mu proteins immediately after infection relative to the expression upon prophage induction. In an entirely different situation, induction of a mini-Mu prophage, Harshey (80) has found that the choice between forming cointegrates and simple insertions may be determined by the B protein: in the absence of B or in the presence of overexpressed A, simple insertions were preferentially formed.

While cointegrates must be generated by a replicative process, the same is not true of simple insertions. In fact both genetic and biochemical data show that the initial postinfection insertions of Mu involve little or no replication of the transposon (81–83). In vitro experiments (84) support this conclusion (see below). In a situation where simple insertions occur from a viable circular replicon, they also appear to destroy the donor replicon as is expected for a nonreplicative process (85). Thus simple insertions of Mu, like those of IS elements, appear to be generated by an essentially conservative mechanism.

Several analyses of Mu replication and transposition in vivo have indicated that the process may be an asymmetric one. As with other transposons, two Mu ends are required for transpositional activity and phage replication. In the case of Mu it appears that one left end and one right end is required suggesting that the different ends play a distinct role in the transposition process. Although a DNA segment that is flanked by two Mu copies can be transposed (a structure analogous to the IS composite transposons), all the observed examples involve copies of Mu in the same orientation so that the whole composite has one left end and one right end of Mu at its termini (86). Segments flanked by inverted copies have never been seen to transpose even though electron microscope studies show that such structures exist (87). Biochemical evidence for asymmetry has come from an analysis of the strand-specificity of Okazaki fragments from a pulse-labeled, induced-Mu lysogen. Data obtained in vivo suggested that most replication was initiated at the left (immunity) end of Mu and ran from left to right (88); a similar result has been obtained in vitro [(89) see below]. Electron microscope studies of replicating intermediates also suggested asymmetric replication (90, 91); however, in contrast to the biochemical data, initiation of replication appeared to take place at either end (and in some cases at both ends). The electron microscope analyses suggested not only that replication usually started at one end, but also that attachment is generally limited to a single end. However, any tendency of DNA to break at single-stranded regions (e.g. a replication fork) due to shear forces during DNA preparation and spreading would bias the data in favor of molecules with a single fork.

Molecular details of transposition intermediates are starting to be revealed. Making use of the ability to synchronize the Mu transposition process and accummulate transpositional potential by blocking DNA synthesis (73), it has recently been shown that DNA restriction fragments containing the left and right Mu-host junctions are altered in electrophoretic mobility (M. Pato, C. Reich, personal communication). The altered fragments have been analyzed by 2-dimensional agarose gel electrophoresis with a denaturing second dimension to detect interrupted strands and strand-specific probes to distinguish the two strands of Mu at the immunity end (92). The data suggested that the 5' end of *attL* remained attached to the parental DNA whereas the 3' end was joined to the target (see also in vitro studies below).

MU REPLICATION IN VITRO   A considerable advance towards understanding transposition mechanisms has been the development of in vitro systems for studying Mu transposition both by Higgins et al (89) using a cell lysate on a cellophane disc, and by Mizuuchi (84, 93) using a more conventional cell-free soluble extract to which exogenous DNA substrates can be added.

Using the cellophane disc lysate method it has been possible to show that the Mu-specific replication is essentially confined to the Mu sequences and does not extend into adjacent host sequences. One strand of Mu is synthesized continuously, the other discontinuously; replication was initiated predominantly at the left end. New initiations of Mu replication are not seen in the cellophane disc lysate procedure consistent with the need for new synthesis of the Mu A protein that is observed in vivo (73).

Of greater use in analyzing the molecular features of Mu transposition is the fully cell-free system (84). Using replication proficient cell extracts from *E. coli* supplemented with extracts from strains carrying an overproducer of the Mu A and B gene products, efficient intermolecular (and probably intramolecular) transposition was detected (close to 10% of the input molecules were converted to products). Reactions were totally dependent on Mu A protein and on the substrate having two correctly oriented ends. Even using a gel assay, which would detect intermediates or products that might not yield viable final products, no detectable reaction was observed with a substrate with incorrectly oriented ends. Under optimum conditions about 2/3 of the intermolecular products were cointegrates, the rest were simple insertions. Treatments that interfered with DNA synthesis (novobiocin, dideoxynucleotides) reduced the cointegrate : simple insertion ratio.

Further insights into the Mu transposition mechanism have come from analysis of incorporation and distribution of labeled nucleotides into the products of the in vitro reaction (93). The amount of label incorporated into simple insertions showed that a substantial portion of the Mu sequence was unreplicated parental DNA, indicative of a conservative (or partially conservative) process. In cointegrate molecules, each copy of the transposon had one newly synthesized strand and one parental strand showing semiconservative replication (and segregation) of the copies. The newly synthesized strands were those joined to the 3' ends of the target DNA (see Figure 7). This showed that 3' ends of the parental Mu DNA are joined to the 5' ends of the target and that, to generate the 5-bp target duplication, the target must be cut to give 5' single strand extensions.

TRANSPOSITION MECHANISMS

The actual mechanism(s) of transposition must explain the following features of prokaryotic transposons.

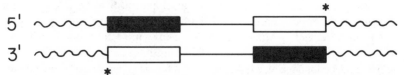

Figure 7 Distribution of Labeled DNA in Cointegrates Formed in vitro (93). Solid bars show the newly synthesized labeled DNA, open bars show the parental strands of the transposon. Asterisks mark the site of attachment of parental transposon DNA to the cleaved target DNA to create the new junctions.

1. Simple Insertions of IS elements and of Mu do not involve extensive replication of the transposon. These are the most frequent events for IS elements, the least frequent for Mu. For transposons of the Tn*3* family, simple insertions may only be aberrant products.

2. Cointegrates (and the intramolecular equivalents, deletions and inversion-insertions) clearly do involve replication of the entire transposon. Cointegrates are the major product of transpositional recombination of Mu (during the lytic cycle) and of Tn*3*-like transposons. IS elements only make cointegrates rarely, if at all.

A common feature of all the models is the way they account for the flanking duplications of a short target sequence (Figure 8). It is proposed that the target DNA is cleaved with staggered cuts; the extended single-stranded ends are then joined to transposon DNA and are filled in by repair synthesis (94). Because the size of the duplicated segment is characteristic of the transposon, it seems likely that the transposase itself either cleaves, or directs the cleavage, of the target DNA.

Conservative Mechanisms

The simplest way to imagine insertion of an excised, unreplicated transposon is a cut-and-paste process (95). Double-stranded cuts are made by the action of transposase at the two transposon ends. The excised duplex segment is integrated into the target site by ligating the extended single strands of the target DNA to the appropriate strands at the transposon ends. Excision of the transposon leads to linearization and consequent destruction of the donor replicon. The excision and insertion steps could be independent or coupled processes. It could be that the excised transposon, complexed with transposase, attacks a target site. Alternatively, the cleavages necessary for excision and integration might occur in a complex that contains the target site and both transposon ends (still attached to the adjacent DNA), all held together by the transposase.

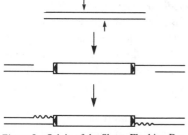

Figure 8   Origin of the Short, Flanking Duplications of Target DNA. The target is cleaved with staggered cuts and the extended single strands of the target are then joined to the transposon termini. Repair of the single-stranded gaps (94) completes the duplication.

Figure 9   Tn*10* Terminal Positions following Transposon Excision-Circularization (96). The sequence shows the 10 terminal base pairs of Tn*10* with dashes representing the adjacent vector sequences. Arrows show the sites of cleavage that produce free 5' ends (top line) and a free 3' end (bottom line). For further details see the text.

Although it has become increasingly clear that simple insertions in many cases occur without transposon duplication, there is very little evidence that addresses the actual mechanism. Evidence for a transposase-mediated excision activity has been obtained with both Tn*10* (96) and Mu (85). Morisato & Kleckner (96) have observed that induction of a high level of expression of the IS*10* transposase results in excision of a Tn*10* analog from its plasmid vector. The remaining vector "backbone" sequences were detected as an unstable linear molecule that is rapidly degraded. By contrast, the transposon sequences were not degraded; rather, they were found as a circular form with one uninterrupted and one broken strand. The breaks, which have 5' phosphates and 3' hydroxyls, are located at specific sites relative to the transposon ends. 5' ends were found at positions 1 and 3 of the transposon, 3' ends were found at position 5 (see Figure 9); additional breaks at various sites within the excised segment were also observed at lower frequency. Although these positions were determined by measurements of just one of the two IS ends, the identity of the two transposon ends suggests that the same break points would be observed at the other end. Whether the predominant 5' and 3' ends represent primary cleavages by transposase, or transposase-protected ends that have resulted from cleavage (by transposase) followed by nonspecific degradation, is not known. The new joint across the uninterrupted strand has not yet been characterized—it could result either from a ligation between the 3' end and one or the other of the 5' ends that are observed in the broken strands, or from an alternative ligation event (e.g. position 1 joined to position 1).

A somewhat different consequence of inducing a transposase, but also consistent with double strand cuts, has been observed with phage Mu by Harshey (85). Starting with a 25-kb plasmid consisting of a 16-kb mini-Mu inserted into a 9-kb vector, induction of the Mu A protein resulted in formation of 25-kb and 16-kb linear forms with a smear of heterogeneous DNA in between (but not extending below the 16-kb band). Harshey's interpretation is that the Mu has undergone a destructive excision and intramolecular insertion to generate the 25-kb linear species (as in Figure 5A); degradation of these linears gives the smear; protection of the ends of Mu by binding of transposase terminates the heterogeneous smear at the 16-kb Mu fragment.

A mechanism that uses transposase-generated double-stranded cuts accounts satisfactorily for an excision-insertion process that destroys the donor replicon. However, a limitation of such a mechanism is that it is not compatible with any replicative process that retains the original host-transposon junctions. We shall describe below a mechanism that uses a common intermediate to generate either conservative simple insertions or replicative cointegrates.

Replicative Mechanisms

Replicative transfer of genetic information is best achieved by transferring a single strand of the DNA segment; complementary strands are then synthesized in donor and recipient. The conjugational transfer of bacterial DNA occurs by such a mechanism (97) and, together with the rolling circle mechanism of replication (98) has provided the basis for models of replicative transposition. Following these precedents, models for transposition propose that the transposase makes specific single-strand cleavages at the transposon termini and transfers the ends to a target site. Although conjugational transfer and rolling circle replication are asymmetric processes, the symmetry of most transposons (i.e. the terminal inverted repeats) suggests that transposition might be a symmetric process. Consequently, the models can be divided into two classes—the symmetric and the asymmetric—depending on whether events are initiated at both or just one of the transposon ends.

ASYMMETRIC MODELS   A number of asymmetric mechanisms have been proposed starting with the first molecular model for transposition in 1978 (94, 99, 100). These models are all very similar, differing only in small details. They propose that transposition is initiated by cutting a strand at one end of the transposon and transferring it to the cleaved target site (Figure 10). The joint creates a replication fork which travels through the transposon carrying the transposase. When the opposite end of the transposon is reached (and replicated) transposase is presented with two alternative substrates that can be used to terminate the transposition: either the other parental strand or the newly synthesized strand (Figure 10, III, pathways 2 and 1 respectively). Joining of the former to the remaining target strand generates a cointegrate; joining of the latter gives an intermediate that, by nonspecific single stranded breaks, can be processed into a simple insertion. The overall process can be summarized as initiate-replicate-terminate.

Two alternative polarities of strand transfer have been proposed: either the 5' end of the transposon can be joined to the 3' end of the target [(100); Figure 10A] or the 3' end of the transposon can be joined to the 5' end of the target [(94, 99); Figure 10B]. The polarity determines whether the primer site for continuous DNA synthesis is in the donor DNA or the target (Figure 10, II). In addition, if the 5' end of the transposon is transferred, the second strand of the target does not need to be cut until the termination event.

Asymmetric replicative models now appear to be unattractive, at least for the ISs and Mu, for a variety of reasons:

1. They predicted that both simple insertions and cointegrates are formed by replicative processes, and that the donor is preserved during formation of a simple insertion. The available data, however, suggest that simple insertions destroy the donor and do not involve substantial transposon replication.

2. Asymmetric models allow for the formation of simple insertions in both intermolecular and intramolecular transpositions. However, intramolecular

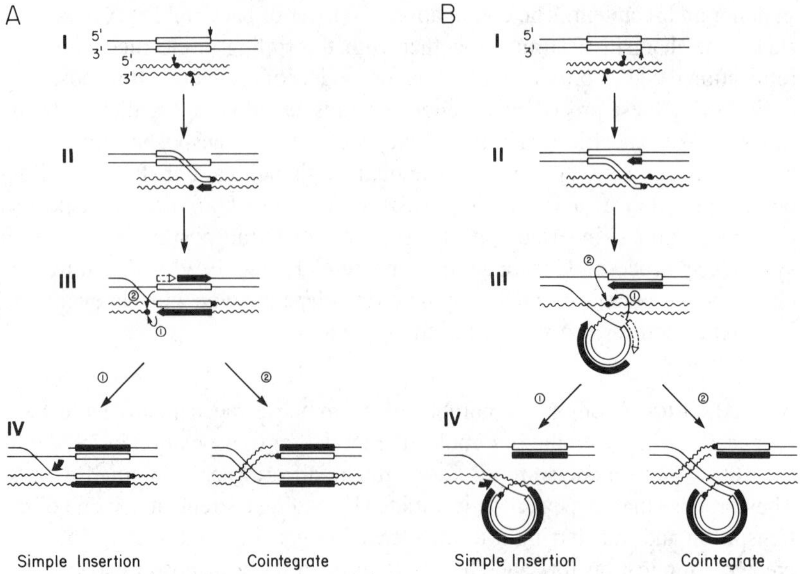

Figure 10 Asymmetric Transposition Models.
A. Transfer of the 3' end of the transposon to the 5' end of the target DNA.
B. Transfer of the 5' end of the transposon to the 3' end of the target DNA. Parental transposon DNA is shown as open bars, newly synthesized DNA is shown as solid bars with arrow heads to indicate the direction of synthesis. The open dashed arrows in A and B III are to indicate the most 5' Okazaki fragments. (1) and (2) show termination options for joining transposon end to the target after completion of transposon replication. Option (1) which joins the newly replicated transposon end gives a precursor to a simple insertion. Option (2) which joins the parental end gives a cointegrate. In B III the most 5' Okazaki fragment (open dashed arrow) must be made before termination option (1) is possible. The arrows in A and B IV show the single-stranded connections between donor and target replicons that must be broken to produce a simple insertion.

simple insertions of IS elements are rarely, if ever, seen even though simple insertions predominate in intermolecular events.

3. They proposed that the transposase is transported from one end of the transposon to the other by the replication process. Support for this appeared to come from the affect of length of IS-composite transposons on their transposition frequency; it was suggested that transposon replication might abort in a length-dependent manner (30, 101). Various experiments now suggest that length dependence is a property not of the length of the segment to be transposed but, rather, of the distance between the two ends of the transposon by any path (37, 43, 102). An extreme example of this has been observed in studies of transposition with a circular λ::Tn*9* donor molecule (102). The 3-kb Tn*9* transposon (IS*1*-CM^r-IS*1*) and the 45-kb IS*1*-λ-IS*1* segment both transposed at approximately the same frequency. Length-dependency is probably best explained by a requirement that the two

transposon ends are brought together in a complex with transposase before any recombination events are initiated; the efficiency of formation of such a complex could well be determined by the shortest distance between the two transposon ends. The requirements for two ends of Mu for initiation of replication (103), and for two ends of IS903 or the inverse Tn903 for formation of adjacent deletions (36, 37) is consistent with this. Evidence that the transposase does not find the terminating end by a replication process that follows formation of the initiating joint, comes from the intramolecular rearrangements mediated by the inverse transposons of Tn10 and Tn903 (37). Since the target site in these events lies within the transposing segment, cleavage of the target DNA interrupts the path of the DNA between the transposon ends that would be replicated in the asymmetric transposition mechanism.

4. Using the transposon excision-circularization reaction of Tn10, Morisato & Kleckner (96) have shown that the transposase chooses the genetically most active pair of ends available in a way that actually suppresses interactions with a less active end, even if the latter is interposed between the other two. If a processive replicative mechanism were used to find the second end, the weaker site should be used with an efficiency that is not influenced by the presence of the stronger but distal site. The result suggests, therefore, that the entire vector (or at least a substantial segment of DNA) is scanned by transposase in its search for a correctly oriented pair of transposon termini. Lack of influence of interposed ends is also seen in the analysis of transpositions from dimers of a pBR322::Tn9 plasmid (43). When transposition to an F plasmid of resistances to both ampicillin and chloramphenicol was selected, products with four IS1s were favored over products with three; a processive mechanism predicts that products with the lowest numbers of an IS would be preferentially recovered.

A SYMMETRIC MODEL   In 1979 Shapiro (104) and subsequently Arthur & Sherratt (53) proposed a transposition mechanism that used the same single-strand transfer process as the asymmetric mechanism, but initiated at both transposon ends simultaneously (Figure 11). As with the asymmetric models, the transfers could involve either the 5' or the 3' ends of the transposon segments; we favor the latter (as shown in Figure 11) at least for IS elements and Mu (see below). Attachment of both ends of the transposon to the target immediately forms two replication forks (Figure 11, II). The original proposals then involved assembly of a replisome at each joint and replication of the complete transposon (Figure 11, III–IV). Final sealing of the replicated DNA to flanking sequences at each end generates a cointegrate (Figure 11, IV). It was proposed that a cointegrate was the intermediate in all intermolecular transposition events, and that simple insertions were formed from the cointegrate by site-specific recombination (cointegrate resolution) between the two transposon copies (Figure 11, V). This is clearly true for Tn*3*-like transposons but does not appear to be so for ISs or Mu. In any case, such resolution generates a replicative simple insertion in which the donor replicon is regenerated—a situation that does not pertain to ISs and Mu.

In 1980, at a time when it was becoming increasingly clear that simple insertions of an IS were formed directly, Ohtsubo et al (105) proposed a modification of the Shapiro model that generated simple insertions as an alternative to cointegrates. If repair synthesis were initiated at the primer termini in the target DNA and the displaced single strands attaching the transposon to the donor replicon were broken [perhaps by the 5′ exonuclease of

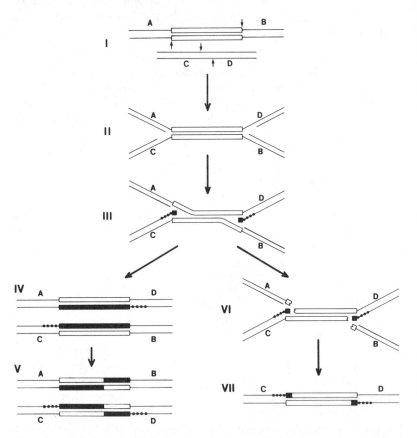

Figure 11 A Symmetric Transposition Model. The diagram shows the Shapiro model for replication transposition (104) and a modification that allows the simple insertion of unreplicated transposon DNA (105). Open bars show parental transposon DNA, solid bars show newly synthesized transposon DNA. The heavy dots indicate replication of the target sequences. For more details see text. Adapted with permission from (37).

DNA polymerase I; (106)], then a simple insertion would result (Figure 11, VI–VII). Because it predicted nonreplicative simple insertions, this proposal was not accorded a great deal of notice at the time; however, the very same prediction has now made it an attractive hypothesis as has been noted by several groups recently (37, 43, 44, 93).

Although the polarity of strand transfer makes very little difference in the fully replicative mechanism proposed by Shapiro (Figure 11, I–IV), it might well affect the outcome in the modified branch shown in Figure 11, VI–VII. If the primer for repair synthesis is in the target DNA (as is shown) the donor DNA will be displaced and become susceptible to breakage. However, if the polarity was reversed, that is, the 5' ends of the transposon were transferred to the 3' ends of the target DNA, then the primer would be in the donor DNA. In that case, repair replication would displace the transposon strands attached to the target; breakage of these strands would destroy the target and no transposition would be detected (9). Since both Mu and the ISs give conservative simple insertions, we propose that the transfer joins the 3' ends of the transposon to the 5' ends of the target. The studies on Mu described above show that this is indeed the polarity used (93). However, the cleavages detected with Tn10 do not include one at the precise 3' end (although they do include one at the 5' end) (96).

As we have seen, the ratio of simple insertions to cointegrates varies widely from one transposon to another. In the symmetric mechanism, the cointegrate pathway is distinguished from the simple insertion process by its requirement for a complete replisome at the terminal joints. It seems likely then, that it is the efficiency of assembly of the replisome that determines whether a cointegrate process occurs. In the case of Mu, the B protein, which appears to influence the choice of pathways (80), may act as a factor that promotes replisome formation. For the ISs the transposase itself may influence the decision: perhaps some transposases, those of IS1 and IS903 for example, have weak affinity for a component of the replisome, and others, those of IS10 and IS50 for example, have no such affinity. The fact that Tn3 follows the cointegrate pathway most, perhaps all, of the time could be explained in one of two ways: either there is a strong affinity between its transposase and replication factors, or the polarity of strand transfer is reversed so that single strand breakage would result in loss of the target molecule.

The Models—A Summing Up

IS ELEMENTS   The conservative formation of an IS simple insertion and the concomitant destruction of the donor replicon are explained either by the cut-and-paste conservative process or by the modified Shapiro mechanism. Only the Shapiro model, however, also allows for the formation of cointegrates as seen with IS1 and IS903. If the results of Morisato & Kleckner (96) with

Tn*10* really indicate that the transposase itself makes double-stranded cuts, then they support the cut-and-paste process and explain why IS*10* never seems to make cointegrates. However, their data are also compatible with the modified Shapiro mechanism: the vector backbone sequences could represent the combined effects of transposase-mediated single strand cleavage and a subsequent nonspecific breakage of the single-strand connections between transposon and donor. Isolation and characterization of intermediates in the transposition process are clearly necessary before the mechanism can be fully elucidated.

THE TN3 FAMILY   The symmetric mechanism appears to be the more satisfactory model for the replicative transposition of members of the Tn*3* family although more data are needed and both symmetric and asymmetric models have been advocated in recent papers. Heritage et al (65) have examined the transposition of some Tn*802* (Tn*3*) derivatives in which three IRs, defining two alternative and separately marked transposons, were available. As with the Tn*10* studies of Morisato & Kleckner (96), they found that ends were used with a hierarchy that was more or less independent of their order. The left end of Tn*802* was preferred over the right end despite the fact that the two sequences have identical 38-bp terminal IRs[4]. For example, a segment with two left ends transposed with a 7-fold preference over one with one left and one right end, even though the right end lay between the two left ends and defined a shorter transposon. More dramatically, a transposon with one left and one right end transposed more than 100-fold more frequently than one with two right ends. From these and other results, Heritage et al (65) proposed a mechanism in which the Tn*802* transposase first bound to one IR (with strong preference for a left end) and then searched for a second IR, again with a preference for the left end.

As mentioned above, derivatives of Tn*1721* and Tn*21* with only a single IR can fuse replicons (60, 61). The structures observed are of two types: cointegrates with a duplication of a small part of the transposon (just the intact IR) at each junction, and "simple insertions" of a segment that runs from the intact IR to one of a number of sites within the transposon or just beyond it in the donor vector sequences. Although interpreted as supporting an asymmetric process, all the observed structures are exactly as predicted by the Shapiro symmetric mechanism. If transposase binds to the intact IR and fails to find a second one, it might find a secondary site which it can use in transpositional recombination at a reduced frequency. Symmetric attachment and complete replication (as is

---

[4]This observation is analogous to the increased transposition immunity shown by the left end of Tn*802* and Tn*3*. These two results suggest that the left end may interact more effectively with the transposase, presumably as a result of adjacent sequences with the transposon (65).

expected with transposons of the Tn*3* family) will produce a "cointegrate" with a duplication of the segment that separates the IR from the secondary site. If this segment contains *res*, the cointegrate resolution system will resolve the cointegrate to generate a simple insertion of the segment—this is the second class of products seen. If, however, the segment does not contain *res*, the cointegrate will be stable—this is the first class of products seen. The preponderance of cointegrates with just the IR region duplicated may result from the way transposase binds to the transposon termini. If the transposase forms a dimer and each monomer contains a single IR binding site, then with a DNA segment that contains only one IR, one half of the transposase dimer will be unoccupied. The nearest piece of DNA to which this unoccupied monomer can bind is, therefore, the segment adjacent to the IR.

MU   As with the IS elements, the modified Shapiro mechanism has the virtue of explaining both cointegrates and conservative simple insertions. However, the apparent (and somewhat controversial) asymmetry of Mu replication (see above) does not readily fit the Shapiro model and was the foundation for one of the proposals for an asymmetric model (99); unfortunately the asymmetric model cannot give rise to conservative simple insertions. There are at least two ways around these contradictions. First, perhaps the symmetric process is used but subsequent replication is asymmetric. One way this might occur would be if the B protein, participating as a promoter of replication, only acts at one end of Mu—since the two ends of Mu are remarkably different, such a scenario is plausible. Replication, repair, or processing at the other end could all be blocked by the bound transposase. Second, the mechanisms of simple insertion and cointegrate formation might be different and controlled by the B protein. A symmetric process (either cut-and-paste or modified Shapiro) might be used for the simple insertions, an asymmetric mechanism for cointegrates (which, however, would always have to terminate by the cointegrate option; Figure 10). Although it has been shown that the junction between the left end of Mu and target DNA is made in the absence of replication (92), it is not yet known whether this is true for the right end.

COINTEGRATE RESOLUTION

Transposons of the Tn*3* family all encode a site-specific recombination system whose role is to break down the cointegrates, formed during intermolecular transposition, into their two constituent replicons. Cointegrate resolution is not a form of transpositional recombination, nor is it required for the transpositional recombination of Tn*3*-like elements. However, it does play a role in the normal formation of the final products of intermolecular transposition and, as such, we will briefly discuss it here. We will not attempt to cover the whole field of site-specific recombination but will confine ourselves to the cointegrate resolvases.

Based on complementation studies, two groups of resolvases have been identified—those of the Tn3 subgroup and those of the Tn501 subgroup. Within each group, pairs of resolvase proteins are very closely related (>80% identical amino acids); between groups, the homology between pairs drops to around 30%. It is remarkable that about the same degree of homology is seen between the resolvases and a group of site-specific DNA-invertases that include Gin, Cin, Hin, and Pin (3). Among all four resolvases of known sequence (Tn3, γδ, Tn501 and Tn21) and the DNA invertases, 20% of the amino acid residues are fully conserved and another 10% or so are highly homologous [see ref (107)].

Cointegrate Resolution: The Res Site

The res sites of γδ and Tn3 have been extensively mapped and characterized (108–110). They consist of three separate resolvase binding sites, spanning a total of about 120 bp (Figure 12). Site I, the most tnpR-distal site, contains the recombinational crossover point (49, 50, 108). In the case of γδ the entire 120-bp segment has been shown to be both necessary and sufficient for efficient cointegrate resolution in vivo and in vitro (110). Very recent results show that the res sites of the Tn501 subgroup have the same organization (S. Halford, R. Schmitt, personal communication).

Each of the three binding sites consists of inverted copies of a conserved "resolvase binding sequence" flanking a variable-sized spacer segment. Assuming that each site is bound by a dimer of resolvase, one consequence of spacer variability is that the relative positions of the monomeric units at each site must differ both in separation and in angular rotation about the DNA helix, implying a remarkable flexibility in the structure of the (presumed) resolvase dimers. The overall geometry of the res site is clearly important although not yet fully understood. For example, it appears that resolvase can accomodate additions that approximate to one complete helical turn between Sites I and II but does not tolerate additions of a partial turn (C. Kline, R. Wells, N. Grindley, unpublished observations).

Site I, the recombination site, differs from the other two both in the size of the spacer between the consensus-related binding sequences and in the sequence around the crossover point. Correct spacing alone is not sufficient for determining the recombination site since changes in the primary sequence across the crossover point are inhibitory (110). This suggests that there is specific recognition between the enzymatic portion of resolvase and the crossover region. Additional evidence for a Site I-specific resolvase-DNA interaction has come from an analysis of some resolution-deficient mutants of tnpR [(111) see below].

Figure 12 The *res* Site of γδ. Sites I, II, and III are binding sites for the resolvase protein. The site cleaved by resolvase in the minus Mg^{2+} reaction is marked. The horizontal half arrows show the "binding sequences" that are thought to be recognized by the C-terminal domain of resolvase. Stars mark the G and A bases within Site I that, when methylated, inhibit cointegrate resolution in vitro (107). The −35 and −10 regions of the *tnpA* and *tnpR* promoters are indicated.

The Resolvase Protein

Studies on the γδ resolvase, the most extensively characterized of the resolvases, show that these proteins consist of two distinct domains (112). Resolvase can be cleaved by mild proteolysis into a 140-amino-acid N-terminal fragment and a 45-amino-acid C-terminal fragment. The small C-terminal fragment determines the DNA binding specificity of resolvase, while the large N-terminal domain mediates protein-protein interactions and carries the enzymatic activity necessary for recombination. The C-terminal domain bound independently to each half of a binding sequence, apparently recognizing the conserved binding sequence. The cooperative binding of the intact resolvase to the two halves of a site, and the protection of the crossover region at the center of Site I both clearly depend on the N-terminal domain. Methylation interference experiments showed that the small domain interacts with bases in the major groove of the binding sequence whereas the large domain appeared to interact with the minor groove in the crossover region [(107) E. Falvey, N. Grindley, unpublished observations]. Crystallographic studies imply that the two domains are joined by a flexible hinge; this may account for the ability of resolvase dimers to bind cooperatively to individual binding sites with different geometric configurations (112). Characterization of recombination-deficient but DNA binding-proficient mutants of resolvase also showed that the enzymatic activity is located in the N-terminal domain, and suggested that the extreme N-terminus (residues 1–20) of resolvase may interact with the DNA at the crossover point (111).

The Recombinational Process

Cointegrate resolution has been studied in vitro and a number of biochemical details have been elucidated (113–115). Resolvase is the only protein required for the complete recombination reaction. The only other requirements are for a superhelical cointegrate substrate, a suitable buffer, and Mg^{2+}; no high energy cofactor is needed. Resolvase acts stoichiometrically and probably 12 resolvase monomers are required per cointegrate substrate. The form of the substrate has a strong influence on the recombination: loss of superhelicity decreases the recombination efficiency as does inversion of a *res* site. Only about 1% of a substrate with inverted *res* sites is converted to recombinant products in vitro. Intermolecular recombination is also not detected in vitro.

The product of cointegrate resolution is a catenated pair of circles (113, 115). That all the catenanes are singly interlinked, regardless of the degree of substrate superhelicity or the separation of the *res* sites, is a strong indication that resolvase detects that path of the DNA between the recombining sites (115). This has led to the proposal that two *res* sites are brought into a synaptic complex by resolvase that is bound to one *res* site, sliding along the DNA in a one-dimensional walk until a second, correctly oriented *res* is located. This would exclude DNA interwraps that would otherwise result in formation of more complex catenated products, and could also account for the intramolecular nature of the reaction and for *res* orientation specificity. Although resolvase shows a strong preference for adjacent *res* sites, other tests of the sliding model have not turned out to be very conclusive (116). However, the model remains the most attractive way of explaining the highly specific form of the recombinant products. It should be noted that complexes of resolvase and two *res* sites are formed with apparently the same efficiency whether the *res* sites are inverted or in the same orientation (J. Salvo, N. Grindley, unpublished observations). Therefore, the block to the inversion reaction may occur at a step in the process beyond synapsis.

Resolvase clearly acts as a site-specific topoisomerase that normally forms recombinant molecules. Omission of Mg^{2+} from (or addition of glycerol to) the resolution reaction inhibits recombination but still allows formation of a synaptic complex. If the complex was fixed and spread for electron microscopy, figure-8 molecules were observed with paired *res* sites covered with a protein blob (J. Salvo, N. Grindley, unpublished observations). If, however, a $-Mg^{2+}$ reaction was either proteolysed or spread for electron microscopy without fixation, linear DNA molecules were seen that had resulted from cleavage of both the *res* sites of each substrate molecule [(114) J. Salvo, unpublished observations]. Since superhelical molecules with only a single *res* or with inverted *res* sites are not linearized under the same conditions, the cleavage reaction is dependent on the formation of a potentially active synaptic complex. The resolvase-induced cleavage occurred at the center of binding site I symmetrically within the palindromic sequence 5'-TTATAA to produce 3' single-strand extensions of two bases (114) (Figure 12). The 3' ends were free hydroxyls; the 5' ends were blocked. Recently Reed & Moser (117) have shown that cleavage results in formation of a phosphoserine linkage between the DNA and resolvase. We presume that formation of this linkage provides the means to conserve the energy of the broken phosphodiester bonds and accounts for the lack of a requirement for a high energy nucleotide cofactor. Other nonspecific topoisomerases make use of transient covalent linkages between protein and DNA although a phosphotyrosine linkage has been identified in all cases examined so far (118). Krasnow & Cozzarelli (115) have shown directly that resolvase has a type 1 topoisomerase activity that, like the $-Mg^{2+}$ cleavage activity, is dependent on a proper cointegrate substrate. Clearly the recombinational potential of resolvase is only activated upon formation of a correct synaptic complex with two *res* sites. During the recombination process we do not yet know whether resolvase nicks, exchanges, and rejoins DNA strands one at a time or whether both strands (of both partners) are cut, partners are exchanged and then resealed. That resolvase acts as a type 1 topoisomerase does not prove that it is a nicking-closing (rather than a double-strand cutting) enzyme. The observation that resolvase-induced substrate cleavage generally involves both *res* sites in a single substrate molecule, rather than just one, suggests to us that all the necessary cleavages may be coupled.

ACKNOWLEDGMENTS

We thank our many colleagues for helpful discussions and unpublished information. Work in our laboratories was supported by grants to Nigel D. F. Grindley from the NIH (GM 28470) and to Randall R. Reed from the Howard Hughes Medical Institute.

Literature Cited

1. Nash, H. A. 1981. *Ann. Rev. Genet.* 15:143–167
2. Simon, M. I., Silverman, M. 1983. In *Gene Function in Prokaryotes*, ed. J. Beckwith, pp. 211–27. Cold Spring Harbor, NY: Cold Spring Harbor Laboratory
3. Silverman, M., Simon, M. I. 1983. See Ref. 8, pp. 537–57
4. Grindley, N. D. F. 1983. *Cell* 32:3–5
5. Heffron, F. 1983. See Ref. 8, pp. 223–60
6. McClintock, B. 1965. *Brookhaven Symp. Biol.* 18:162–84
7. Fedoroff, N. 1983. See Ref. 8, pp. 1–63
8. Shapiro, J. A., ed. 1983. *Mobile Genetic Elements*. New York: Academic
9. Kleckner, N. 1981. *Ann. Rev. Genet.* 15:341–404
10. Bukhari, A. I. 1981. *Trends. Biol. Sci.* 6:56–60
11. Kleckner, N. 1983. See Ref. 8, pp. 261–98
12. Murphy, E., Lofdahl, S. 1984. *Nature* 307:292–94
13. Kleckner, N. 1979. *Cell* 16:711–20
14. Reed, R. R., Young, R. A., Steitz, J. A., Grindley, N. D. F., Guyer, M. S. 1979. *Proc. Natl. Acad. Sci. USA* 76:4882–86
15. Kahmann, R., Kamp, D. 1979. *Nature* 280:247–50
16. Hauer, B., Shapiro, J. A. 1984. *Mol. Gen. Genet.* 194:149–58
17. Gawron-Burke, C., Clewell, D. B. 1982. *Nature* 300:281–84
18. Iida, S., Meyer, J., Arber, W. 1983. See Ref. 8, pp. 159–221
19. Halling, S. M., Simons, R. W., Way, J. C., Walsh, R. B., Kleckner, N. 1982. *Proc. Natl. Acad. Sci.* 79:2608–12
20. Auerswald, E.-A., Ludwig, G., Schaller, H. 1981. *Cold Spring Harbor Symp. Quant. Biol.* 45:107–13
21. Grindley, N. D. F., Joyce, C. M. 1980. *Proc. Natl. Acad. Sci. USA* 77:7176–80

22. Johnson, R. C., Reznikoff, W. S. 1981. *Nucleic Acid Res.* 9:1873–83
23. Johnson, R. C., Yin, J. C. P., Reznikoff, W. S. 1982. *Cell* 30:873–82
24. Isberg, R. R., Lazaar, A. L., Syvanen, M. 1982. *Cell* 30:883–92
25. Ohtsubo, H., Nyman, K., Doroszkiewicz, W., Ohtsubo, E. 1981. *Nature* 292:640–43
26. Machida, Y., Machida, C., Ohtsubo, H., Ohtsubo, E. 1982. *Proc. Natl. Acad. Sci. USA* 79:277–81
27. Machida, Y., Machida, C., Ohtsubo, E. 1984. *J. Mol. Biol.* 177:229–45
28. Grindley, N. D. F., Joyce, C. M. 1981. *Cold Spring Harbor Symp. Quant. Biol.* 45:125–33
29. Isberg, R. R., Syvanen, M. 1981. *J. Mol. Biol.* 150:15–32
30. Morisato, D., Way, J. C., Kim, H.-J., Kleckner, N. 1983. *Cell* 32:799–807
31. Berg, O. G., Winter, R. B., von Hippel, P. H. 1981. *Biochemistry* 20:6929–48
32. Harayama, S., Oguchi, T., Iino, T. 1984. *Mol. Gen. Genet.* 194:444–50
33. Hirschel, B. J., Galas, D. J., Chandler, M. 1982. *Proc. Natl. Acad. Sci. USA* 79:4530–34
34. Ohtsubo, E., Zenilman, M., Ohtsubo, H. 1980. *Proc. Natl. Acad. Sci. USA* 77:750–54
35. Reif, H. J., Saedler, H. 1975. *Mol. Gen. Genet.* 137:17–28
36. Weinert, T. A., Schaus, N. A., Grindley, N. D. F. 1983. *Science* 222:755–65
37. Weinert, T. A., Derbyshire, K., Hughson, F. M., Grindley, N. D. F. 1984. *Cold Spring Harbor Symp. Quant. Biol.* 49:251–60
38. Way, J. C., Kleckner, N. 1984. *Proc. Natl. Acad. Sci. USA* 81:3452–56
39. Kleckner, N., Ross, D. G. 1980. *J. Mol. Biol.* 144:215–21

40. Cornelis, G., Saedler, H. 1980. *Mol. Gen. Genet.* 178:367–74
41. Ross, D. G., Swan, J., Kleckner, N. 1979. *Cell* 16:721–31
42. Berg, D. E. 1983. *Proc. Natl. Acad. Sci. USA* 80:792–96
43. Biel, S. W., Berg, D. E. 1984. *Genetics* 108:319–30
44. Kleckner, N., Morisato, D., Roberts, D., Bender, J. 1984. *Cold Spring Harbor Symp. Quant. Biol.* 49:235–44
45. Fuller, R. S., Kornberg, A. 1983. *Proc. Natl. Acad. Sci. USA* 80:5817–21
46. Johnson, R. C., Reznikoff, W. S. 1984. *Nature* 304:280–82
47. Heffron, F., McCarthy, B. J., Ohtsubo, H., Ohtsubo, E. 1979. *Cell* 18:1153–63
48. Kitts, P., Symington, L., Burke, M., Reed, R., Sherratt, D. 1982. *Proc. Natl. Acad. Sci. USA* 79:46–50
49. Reed, R. R. 1981. *Proc. Natl. Acad. Sci. USA* 78:3428–32
50. Kostriken, R., Morita, C., Heffron, F. 1981. *Proc. Natl. Acad. Sci. USA* 78:4041–45
51. Chou, J., Casadaban, M. J., Lemaux, P. G., Cohen, S. N. 1979. *Proc. Natl. Acad. Sci. USA* 76:4020–24
52. Gill, R. E., Heffron, F., Falkow, S. 1979. *Nature* 282:797–801
53. Arthur, A., Sherratt, D. J. 1979. *Mol. Gen. Genet.* 15:267–74
54. Diver, W. P., Grinsted, J., Fritzinger, D. C., Brown, N. L., Altenbuchner, J., et al. 1983. *Mol. Gen. Genet.* 191:189–93
55. Rogowsky, P., Schmitt, R. 1984. *Mol. Gen. Genet.* 193:162–66
56. Grinsted, J., de la Cruz, F., Altenbuchner, J., Schmitt, R. 1982. *Plasmid* 8:276–86
57. Bennett, P. M., de la Cruz, F., Grinsted, J. 1983. *Nature* 305:743–44
58. McCormick, M., Wishart, W., Ohtsubo, H., Heffron, F., Ohtsubo, E. 1981. *Gene* 15:103–18
59. Fischhoff, D. A., Vovis, G. F., Zinder, N. D. 1980. *J. Mol. Biol.* 144:247–65
60. Avila, P., de la Cruz, F., Ward, E., Grinsted, J. 1984. *Mol. Gen. Genet.* 195:288–93
61. Motsch, S., Schmitt, R. 1984. *Mol. Gen. Genet.* 195:281–87
62. Robinson, M. K., Bennett, P. M., Richmond, M. H. 1977. *J. Bacteriol.* 129:407–14
63. Chiang, S.-J., Clowes, R. C. 1980. *J. Bacteriol.* 142:668–82
64. Lee, C.-H., Bhagwat, A., Heffron, F. 1983. *Proc. Natl. Acad. Sci. USA* 80:6765–69

65. Heritage, J., Perry, A. C. F., Bennett, P. M. 1984. Submitted for publication
66. Bishop, R., Sherratt, D. 1984. *Mol. Gen. Genet.* 196:117–22
67. Toussaint, A., Resibois, A. 1983. See Ref. 8, pp. 105–58
68. Craigie, R., Mizuuchi, M., Mizuuchi, K. 1984. *Cell* 39:387–94
69. Wijffelman, C., Lotterman, B. 1977. *Mol. Gen. Genet.* 151:169–74
70. Chaconas, G., Gloor, G., Miller, J. L., Kennedy, D. L., Giddens, E. B., Nagainis, C. R. 1984. *Cold Spring Harbor Symp. Quant. Biol.* 49:279–84
71. O'Day, K. J., Schultz, D. W., Howe, M. M. 1978. In *Microbiology 1978*, ed. D. Schlessinger, pp. 48–51. Washington, DC: Am. Soc. Microbiol.
72. Faelen, M., Resibois, A., Toussaint, A. 1979. *Cold Spring Harbor Symp. Quant. Biol.* 43:1169–77
73. Pato, M. L., Reich, C. 1984. *Cell* 36:197–202
74. Chaconas, G., Harshey, R. M., Sarvetnick, N., Bukhari, A. I. 1981. *J. Mol. Biol.* 150:341–59
75. Howe, M. M., Schumm, J. W. 1981. *Cold Spring Harbor Symp. Quant. Biol.* 45:337–46
76. Ljungquist, E., Bukhari, A. I. 1977. *Proc. Natl. Acad. Sci. USA* 74:3143–47
77. Chaconas, G., Kennedy, D. L., Evans, D. 1983. *Virology* 128:48–59
78. Harshey, R. M., Bukhari, A. I. 1983. *J. Mol. Biol.* 167:427–41
79. Puspurs, A. H., Trun, N. J., Reeve, J. N. 1983. *EMBO J.* 2:345–52
80. Harshey, R. M. 1983. *Proc. Natl. Acad. Sci. USA* 80:2012–16
81. Liebart, J. C., Ghelardini, P., Paolozzi, L. 1982. *Proc. Natl. Acad. Sci. USA* 79:4362–66
82. Akroyd, J. E., Symonds, N. 1983. *Nature* 303:84–86
83. Harshey, R. M. 1984. *Nature* 311:580–81
84. Mizuuchi, K. 1983. *Cell* 35:785–94
85. Harshey, R. 1984. *Cold Spring Harbor Symp. Quant. Biol.* 49:273–78
86. Toussaint, A., Faelen, M. 1973. *Nature New Biol.* 242:1–4
87. Resibois, A., Toussaint, A., van Gijsegem, F., Faelen, M. 1981. *Gene* 14:103–13
88. Goosen, T. 1977. In *DNA Synthesis: Present and Future*, ed. I. Molyneux, M. Kohiyama, pp. 121–26. New York: Plenum
89. Higgins, N. P., Moncecchi, D., Manlapez-Ramos, P., Olivera, B. 1981. *J. Biol. Chem.* 58:4293–97

90. Harshey, R. M., McKay, R., Bukhari, A. I. 1982. *Cell* 29:561–71
91. Resibois, A., Colet, M., Toussaint, A. 1982. *EMBO J.* 1:965–69
92. Krause, H. M., Higgins, N. P. 1984. *Cold Spring Harbor Symp. Quant. Biol.* 49:827–34
93. Mizuuchi, K. 1984. *Cell* 39:395–404
94. Grindley, N. D. F., Sherratt, D. J. 1979. *Cold Spring Harbor Symp. Quant. Biol.* 43:1257–61
95. Berg, D. E. 1977. In *DNA Insertion Elements, Plasmids and Episomes*, ed. A. I. Bukhari, J. A. Shapiro, S. L. Adhya, pp. 205–12. New York: Cold Spring Harbor Press
96. Morisato, D., Kleckner, N. 1984. *Cell* 39:181–90
97. Willetts, N., Wilkins, B. 1984. *Microbiol. Rev.* 48:24–41
98. Kornberg, A. 1980. *DNA Replication.* San Francisco: Freeman
99. Harshey, R. M., Bukhari, A. I. 1981. *Proc. Natl. Acad. Sci. USA* 78:1090–94
100. Galas, D. J., Chandler, M. 1981. *Proc. Natl. Acad. Sci. USA* 78:4858–62
101. Chandler, M., Clerget, M., Galas, D. J. 1982. *J. Mol. Biol.* 154:229–43
102. Rosner, J. L., Guyer, M. S. 1980. *Mol. Gen. Genet.* 178:111–20
103. Waggoner, B., Pato, M., Toussaint, A., Faelen, M. 1981. *Virology* 113:379–87
104. Shapiro, J. A. 1979. *Proc. Natl. Acad. Sci. USA* 76:1933–37
105. Ohtsubo, E., Zenilman, M., Ohtsubo, H., McCormick, M., Machida, C.,

Machida, Y. 1981. *Cold Spring Harbor Symp. Quant. Biol.* 45:283–95
106. Liu, L. F., Wang, J. C. 1975. In *DNA Synthesis and its Regulation*, ed. M. Goulian, P. Hanawalt, pp. 38–63. Menlo Park, Calif: Benjamin
107. Grindley, N. D. F., Newman, B. J., Wiater, L. A., Falvey, E. 1984. In *Genome Rearrangements*, ed. I. Herskowitz, M. Simon. UCLA Symp. Mol. Cell. Biol. New York: Liss. In press.
108. Grindley, N. D. F., Lauth, M. R., Wells, R. G., Wityk, R. J., Salvo, J. J. 1982. *Cell* 30:19–27
109. Kitts, P. A., Symington, L. S., Dyson, P., Sherratt, D. J. 1983. *EMBO J.* 2:1055–60
110. Wells, R. G., Grindley, N. D. F. 1984. *J. Mol. Biol.* 179:667–81
111. Newman, B. J., Grindley, N. D. F. 1984. *Cell* 38:463–69
112. Abdel-Meguid, S. S., Grindley, N. D. F., Templeton, N. S., Steitz, T. A. 1984. *Proc. Natl. Acad. Sci. USA* 81:2001–5
113. Reed, R. R. 1981. *Cell* 25:713–19
114. Reed, R. R., Grindley, N. D. F. 1981. *Cell* 25:721–28
115. Krasnow, M. A., Cozzarelli, N. R. 1983. *Cell* 32:1313–24
116. Benjamin, H. W., Matzuk, M. M., Krasnow, M. A., Cozzarelli, N. R. 1984. *Cell* In press
117. Reed, R. R., Moser, C. D. 1984. *Cold Spring Harbor Symp. Quant. Biol.* 49:45–49
118. Tse, Y.-C., Kirkegaard, K., Wang, J. C. 1980. *J. Biol. Chem.* 255:5560–65

Note added in proof.   Recent results indicate an essential role for the Mu B protein in Mu transposition (G. Chaconas, personal communication). Although *amber* mutations that map late in B affect only replication of Mu and not its nonreplicative integration, early *ambers* block both activities. It is unlikely, therefore, that differential expression of B determines the choice between simple insertions and cointegrates.

Ann. Rev. Biochem. 1985. 54:897–930

PROTEIN-TYROSINE KINASES

Tony Hunter and Jonathan A. Cooper[1]

Molecular Biology and Virology Laboratory, The Salk Institute, Post Office Box 85800, San Diego, California 92138

CONTENTS

PERSPECTIVES AND SUMMARY

Phosphate tightly associated with protein has been known since the late nineteenth century. Early hints that this phosphate might be covalently linked were gleaned in 1906, but the first phosphoamino acid was not isolated from protein until 1933. Since then a variety of covalent linkages of phosphate to proteins have been found. The most common involve esterification of phosphate to serine and threonine, with smaller amounts being linked to lysine, arginine, histidine, aspartic acid, glutamic acid, and cysteine. Even though the existence of phosphodiester bonds linking tyrosine to mono- or polyribonu-

[1]Present address: Fred Hutchinson Cancer Research Center, 1124 Columbia Street, Seattle, Washington 98104

cleotides had set a precedent, reports of orthophosphate esterified to tyrosine did not emerge until 1979 (1). Phosphotyrosine (P.Tyr) was first detected in hydrolysates of viral transforming proteins labeled by incubation of immuno-precipitates with radioactive ATP (1–3). Initially it was suggested that this might either be an aberrant in vitro reaction or represent a phosphoenzyme intermediate, but when suitable separation techniques for P.Tyr were developed it became clear that all animal cells do indeed contain low levels of P.Tyr in protein (2, 4). The previous failure to detect P.Tyr upon conventional analysis of protein hydrolysates had been due to its masking by the much more abundant phosphothreonine (P.Thr).

The occurrence of phosphorylated tyrosine as a stable yet reversible modification of proteins implies the existence of one or more protein kinases capable of phosphorylating tyrosine and also of protein phosphatases capable of hydrolyzing P. Tyr. The last five years have seen a burgeoning of systems in which tyrosine phosphorylation has been implicated, including viral transformation and growth control. Already six cellular genes have been identified that encode tyrosine-specific protein kinases, and there are indications of several more. Some of these protein-tyrosine kinases (PTK) have been purified and extensively characterized.

The first PTK activities to be detected were associated with viral transforming proteins, particularly those of acutely oncogenic retroviruses. Although virally coded, the provenance of these transforming proteins is ultimately cellular, since the transforming genes of these viruses are recognizable as coopted cellular genes [for review see (5)]. There are at least five cellular genes of this kind, and they all appear to encode proteins with PTK activity. The second type of PTK activity was detected in association with growth factor receptors (6). The receptor PTKs are stimulated by their respective ligands, and there are at least four enzymes of this sort.

At the outset there was some concern whether the PTK activities of the two types of protein were intrinsic or due to associated enzymes. It is now clear, however, that these proteins do indeed all have a catalytic domain capable of transferring phosphate to proteins. The most persuasive evidence comes in those cases where primary amino acid sequences have been predicted from molecularly cloned genes. A stretch of about 260 amino acids within each of these proteins has obvious homology with the sequence of the catalytic subunit of the serine-specific cyclic AMP–dependent protein kinase (A-kinase). Despite this homology the PTKs all show an exquisitely strict specificity for tyrosine in protein.

Genes corresponding to the PTK genes are readily recognizable in *Drosophila* (7, 8). From the fact that PTK activity can be detected in insects (8) and in even simpler eukaryotic species (9), it seems likely that these genes have the same function as in higher eukaryotes. The ancient evolutionary origin and the high degree of conservation of these genes suggest that tyrosine phosphorylation plays an important role in eukaryotic cellular physiology.

By analogy with the documented effects of phosphorylation on serine and threonine, the reversible phosphorylation of proteins on tyrosine would be expected to modulate protein function. Protein phosphatases specific for P. Tyr have been reported [for review see (10)] fulfilling a critical requirement for a regulatory system based on tyrosine phosphorylation. Nevertheless, although some progress has been made in identifying intracellular substrates for PTKs, in no case has the phosphorylation been of physiological significance.

While there is little doubt that the enzymes in question can and do phosphorylate proteins on tyrosine, there are recent suggestions that they may also phosphorylate nonprotein substrates, such as phosphatidylinositol (12, 13). If these activities prove to be intrinsic, the true function of PTKs will have to be reassessed.

The aim of this review is to describe our current knowledge of PTKs with emphasis on the enzymes themselves rather than their substrates.

TWO TYPES OF PROTEIN-TYROSINE KINASE

Initially it appeared that one could classify the PTKs based on the distinction between those whose genes have become part of retroviruses and those associated with growth factor receptors. Recently, however, it has become apparent that there is overlap between these two types (11). Nevertheless it is still useful to group the growth factor receptor PTKs separately because their regulation is a distinct feature (Table 1). For the purposes of this review we have selected a representative of each type for detailed consideration; these are pp60^{v-src} and the epidermal growth factor (EGF) receptor respectively. pp60^{v-src}, the 60,000-dalton product of the Rous sarcoma virus (RSV) src gene, is the prototypic retroviral PTK, while the EGF receptor is by far the best understood growth factor receptor PTK. The growing number of other PTKs that do not fit naturally into either category will be discussed separately.

RETROVIRAL PROTEIN-TYROSINE KINASES AND THEIR CELLULAR HOMOLOGS

The products of the v-src (2, 14, 15), v-yes (16), v-fgr (17), v-fps (18–20), v-fes (21), v-abl (3, 22), and v-ros (23) retroviral oncogenes (v-onc genes) all have PTK activities (Table 1). The v-src, v-yes, v-fps, and v-ros genes are the oncogenes of different chicken sarcoma viruses, the v-fgr and v-fes genes are the oncogenes of cat sarcoma viruses, and the v-abl gene is the oncogene of a mouse lymphoma virus. These v-onc genes originated from cellular genes (c-onc genes) of the species in question. The nucleotide sequences of all the

Table 1   Protein-tyrosine kinases[a]

Gene	Protein	Intracellular location	Gene	Protein	Intracellular location
Virus-related					
c-*src*	pp60[c-*src*]	Membrane	v-*src*	pp60[v-*src*]	Membrane/cyto-skeleton
c-*yes*	—	—	v-*yes*	P90[gag-yes]	Membrane
c-*fgr*	—	—	v-*fgr*	P70[gag-fgr]	Cytoplasm
c-*fps*	p98[c-*fps*]	Soluble	v-*fps*	P140[gag-fps]	Membrane/soluble
c-*fes*	p92[c-*fes*]	—	v-*fes*	P85[gag-fes]	Cytoplasm
c-*abl*	p150[c-*abl*]	Cytoplasm	v-*abl*	P120[gag-abl]	Membrane
c-*ros*	—	—	v-*ros*	P68[gag-ros]	Membrane
Growth factor receptors					
c-*erb*-B	EGF receptor	Plasma membrane	v-*erb*-B	gp68/74[v-*erb*-B]	Membrane
—	PDGF receptor	Plasma membrane	—		
—	Insulin receptor	Plasma membrane	—		
—	IGF-1 receptor	Plasma membrane	—		
Others					
—	p56[lstra]	Membrane	—		
—	p75	Cytoplasm	—		

[a] The genetic origins (references in text) and subcellular localizations [(57-63,200,202,219,221, reviewed in 218)] of known cellular (left columns) and their corresponding viral (right columns) protein-tyrosine kinases are listed.

viral genes and some of their cellular counterparts have been determined from molecular clones (24–34). Several points emerge from the predicted protein sequences. From their very high degree of conservation (31, 32) and other evidence (35), it appears likely that the c-*fps* and c-*fes* genes are homologous chicken and cat genes respectively, leaving us with six distinct cellular genes encoding PTKs of this sort. Although the complete sequences of most of the cellular gene products are not available, we can deduce a great deal from the sequences of the viral proteins. The different proteins are all related over a continuous stretch of about 260 amino acids, but are more or less divergent over the rest of the protein (Figure 1). For pp60[v-*src*] this region comprises its COOH-terminal half, the part of the protein shown to be sufficient for phosphate transfer. This is also the region of pp60[v-*src*] which is homologous to the A-kinase catalytic subunit (36, 37). A comparable 'catalytic domain' is found in various positions within the other viral transforming proteins. The sequences outside the catalytic domain are presumed to determine in large part the unique properties of the individual members of this gene family.

The products of the v-*src*, v-*yes*, v-*fgr*, v-*fps*, v-*fes*, v-*abl*, and v-*ros* genes were all shown to have PTK activity (2, 14–23) by assay in immunoprecipitates (38, 39). In such 'solid state' reactions either the heavy chain of the combining antibody or else the transforming protein itself is phosphorylated. While this technique serves as a rapid diagnosis for whether a given oncogene product is likely to be a PTK, it is not very useful for studying the protein's enzymatic properties. This requires conventional purification of the protein in a soluble form. Unfortunately, in contrast to other protein kinases, which are usually isolated from tissues, it is hard to obtain large amounts of virally transformed cells, and none of the viral transforming proteins is very abundant. An alternate and potentially copious source of protein would be bacteria expressing oncogenes from recombinant plasmids. The v-*src*, v-*abl*, and v-*fps* proteins have all been expressed in bacteria and shown to have PTK activity, providing strong support for the idea that this activity is innate (40–42). It has been hard, however, to obtain the expressed proteins in a soluble state, although progress has been made in purifying an enzymatically active fragment of the v-*abl* protein synthesized in *E. coli* (43).

Despite the frequency of oncogenes encoding PTKs, it should be stressed that not all retroviral oncogene products have this activity. Indeed the predicted sequences of the v-*sis*, v-*ras*, v-*myc*, v-*myb*, v-*fos*, v-*ski*, v-*rel*, v-*erb*-A, and v-*ets* oncogenes have little or no relationship to those of the PTK family, nor do their products have detectable protein kinase activity. Intriguingly, however, the products of the v-*mil* (or v-*mht*) (44), v-*raf* (45), v-*fms* (46), v-*mos* (47), and v-*erb*-B (48) oncogenes have sequence homology with the catalytic domain of the protein kinases, and yet these proteins do not have detectable PTK activity in immunoprecipitates. Nevertheless, since the v-*erb*-B gene product has turned out to be a fragment of the EGF receptor PTK (11), one should be cautious in dismissing the possibility that the v-*mil*, v-*raf*, v-*fms*, and v-*mos* proteins are PTKs with covert activity or phosphotransferases with another specificity. We may simply lack the appropriate assay for demonstrating such activities.

pp60$^{v\text{-}src}$

STRUCTURE AND SYNTHESIS OF pp60$^{v\text{-}src}$ The complete sequences of the v-*src* genes from two strains of RSV and the chicken c-*src* gene are known (24–28). pp60$^{v\text{-}src}$ is a protein of 526 amino acids which is synthesized on soluble ribosomes, but which rapidly associates with membranes following synthesis (49). The mature protein is indistinguishable in size from the primary translation product (50, 51), which implies that association of pp60$^{v\text{-}src}$ with

```
                 (1)TRDAALPGSHSTHGFYENYEPREIIGRGVSSVVRRCIHKEPTCKE.....YAVKIIDVTGGSFSAEEVQELREATLKVDILRKVSGHPNIIQLKDT
PhK-γ
cGPK  318/DSFKHLIGGLDDVSNKAYEDAEAKAYKEAEAAFFANLKLSDFNIIDTGVGGRGRVELVQLKSEESKT.....PAMKILKKRHI.....VDTRQQEHIRSEKQIMQGAH.SDFIVRLYRT
cAPK    (1)GNAAAKKGSEQESVKEFLAEAKAKEDFLKKWENPAQNTAHLDQFERIKTLGTGSFGRVMLVKHMETGNH.....YAMKILDKQKV.....VKLQIEHTLNEKRILQAVN.PFFLVKLEFS
                                    LG G  V           *G G RG V G   VA*K  *         E   E I    ** L  *

src   226/LVAYYSKHADGLCHRLANVCPTSKPGTQGLAKDAWEIPRESLRLEAKLGQGCFGEVWMGTWNDTTR.......VAIKTLKPGTM.......SPEAFLQEAQVMKKLRHEKLVQLYAVV.SEE
c-src 226/................-T-................-V----.....-G---.....................VAIKTLKPGTM.......
yes   510/LVKHYREHADGLCHLLTVCPTVKPGTQGLAKDAWEIPRESLRLEVKLGQGCFGEVWMGTWNGTTK........VAIKTLKPGTM.......MPEAFLQEAQIMKKLRHDKLVPLYAVV.SEE
fgr   363/LVQHYVEVNDGLCHLLTAACTTMKPGTWGLAKDAWEISRSSITLQRRIGTGCFGDVWLGMWNGSTK.......VAVKTLKPGTM.......SPKASLEEAQIMKKLRHDKLVQLYAVV.PEE
abl   323/HSTVADGLITTLHYPAPKRNKPTILYGVSPNYDKWEMERTDITMKHKLGGGQYGEVYEGVWKKYSLT......VAVKTLKEDTM.......EVEEFLKEAAVMKEIKHPNIVQLLGVCTREP
fps   880/LPTIPLLIDHLLQSQRPITRKSGIVLTRAVLKDKWVLNHEDVLLGERIGRGNFGEVFSGRLRADNTP......VAVKSCRE.TL.......PPELKAKFLQEARILKQCNHPNIVRLIGVCT
fes   656/FASIPLLVDHLLRSQQPLTKKSGIVLNRAVPKDKWVLNHEDLVIGEQIGRGNFGEVFSGRLRADNTL......VAVKSCRE.TL.......PPDIKAKFLQEAKILKQYSHPNIVRLIGVCT
ros   217/KELAQLRGMAEFVGLANACYAVSTLPSQAEIESLPAFPPRDKLNLHKLLGSSAFGEVYEGTALDILADGSGESRVAVKTLKRGAT.......DQEKSEFLKEAHLMSKFDHPHILKLLGVCL
erb-B  98/HIVRKRTLRRLLQERELVEPLTPSGEAPNQAHLRIKETEFKKV.KVLGSGAFGTIYKGLWIPEGEK..VTIPVAIKELREATS.........PKANKEILDEAYVMASVDNPHVCRLLGICL
EGF.R 649/...................-.........-L------I---..-V-------..........................


PhK-γ    YETNTFFLVFDLMKKGELFDYLTEKVTL.......SEKETRKIMRALLEVICALHKLNIVHRDLKPENILLDDDM.......NIKLTDFGFSCQLDPGEKLREV...CGT...PS.YLAPEIIECS
cGPK     FKDSKYLYMLMEACLGGELWTILRDRGSF.......EDSTTRFYTACVVEAFAYLHSKGIIYRDLKPENLLLDHRG......YAKLVDRGFAKKIGFGKKTWTF..CGT...PE.YLAPEIILNK
cAPK     FKDNSNLYMVMEYVPGGEMFSHLRRIGRF.......SEPHARFYAAQIVLTFEYLHSLDLIYRDLKPENLLIDQQG......YIQVTDFGFAKR..VKGRTWTL..CGT...P  YLAPEII
              *  *  GE   L         R    LH        GMYE  *HRDL A N LV      RDLKPEN***D              K* DFG  R    Y   G    P  *W APE

src   PIIVIEYMS........KGSLLDFLKEGEGKYL.......RLPQLVDMAAQIAGGMAYVERMNYVHRDLRAANILVGENL.......VCKVADFGLARL..IEDNEYTARQGAKF.PIKWTAPEAALY
c-src .....-T-.................................................................................
yes   PIIVIREPMT.......KGSLLDFLKEGEGKFL.......KLPQLVDMAAQVAGGMAYVERMNYIHRDLRAANILVGENL.......VCKIADFGLARL..IEDNEYTARQGAKF.PIKWTAPEAALY
fgr   PFYIVREPMT.......KGSLLEFLKDQGGQDL.......TLPQLVDMAAQVABGMAYMERMDYIHRDLRAANILVGERL.......VCKIADFGLARL..IEDNEYNPRQGAKF.PIKWTAPEAALF
abl   QPFYIIREPMT......YGNLLDYLRECNRQEV.......SAVVLLYMATQISSAMEYLEKKNFIHRDLAARNCLVGENH.......LVKVADFGLSRL..MTGDTYTAHAGAKF.PIKWTAPESLAY
fps   QKQPIYIVME.......LVQGGDFLSFLRSKGPRL.....KMKKLIKMMENAAAGMEYLESKHCIHRDLAARNCLVTEKN.......TLKISDFGMSRQ..EEDGVYASTGGMKQ.PIKWTAPEALNY
fes   QKQPIYIVME.......LVQGGDFLTFLRTEGARL.....RMKTLQMVGDAAAGMEYLESKCCIHRDLAARNCLVTEKN.......VLKISDFGMSRE..EADGVYAASGGLRLVPVKWTAPEALNY
ros   LNEPQYIILE.......LMBGGDLLSYLRGARKQKPQ.SPLLTLTDLLDICLDICKGCVYLEKMPFIHRDLAARNCLVSEKQYGSCSRVVKIGDFGLARD..IYKNDYYRKRGBGLLPVRWMAPESLID
erb-B TSTVQ.LITQ.......LMPYGCLLDYIREHKDNI....GSQYLLNWCVQIAKGMNYLEERRIVHRDLAARNVLIKTPQ........HVKITDFGLAKL.LGADEKEYHABGG.KV.PIKWMALESILH
EGF.R ........-...........-F----V-......-D-...........................................-D------.........-E-.......


                   DVWSFG*L* E  G   PY          EG R  P  CP    M CW     RP F             *
PhK-γ MNDNEHGYGKEVDMWSTGVIMYTLLA.GSPPFWHRKQMLMLRMIMSGNYQFGSPEWDDYSDFVKDLVSRFLVVQPQKRYTAEEALAHPFFQQYVVEVRHFSPRGKFKVICLTVLASVRIY/318
cGPK  ......GHDISADYWSLGILMYELLT.GSPPFFSGPDPMKTYNIILRGIDMIEPPKKIA..KNAANLIKKLCRDNPSERLGNLKNGVKDIQKHKWFEGFNWEGLRRGTLTPPIIPSVASPTD/644
cAPK  .....GYNKAVDWWALGVLIYEMAA.GYPPFPAQPIQIYEKIVSG..KVRFPSHFS..SD.LKDLLRNLLQVDLTKRFGNLKDGVNDIKNHKWFATTDWIAIYQRKVEAPFIPKFKGPGD/323
           G  D W G** Y  G  PFF                I  G    P    L       R

src   ........GRFTIKSDVWSFGILLTELVTKGRVPYPGMVNREVLDQVERGYRMPCPQGCPES....LHDLMOQCWRKDPEERPTFKYLQAQLLPACVLEVAB(526)
c-src ............-T-.................................-R----.-E----FLEDYFTSTEPQYQPGENL(530)
yes   ........GRFTIKSDVWSFGILLTELISKGRVPYPGMNNREVLEQVEHGYHMPCPPGCPAS....LYEAMEQTWRLDPEERPTFEYLQSPLEDYFTAAEPSGY(812)
fgr   ........GRFTIKSDVWSFGILLTELISKGRVPYPGMVNREVLEQVERGYRMPCPQGCPAS....VVELMRACWQWNPSDRPSFAEIHQAPETMQLSSISDEVEKELGKRGTRGGAGSMLQAP/647
abl   ........NKFSIKSDVWAFGVLLWEIATYGMSPYPGIDLSQVYELLEKDYRMERPEGCPEK....VYELMRQCWQRNPDDRPSFAEIHQAFETMFQESSISDEVEKELGKRGTRGGAGSMLQAP/647
fps   ........GWYSEESDVWSFGILLWEAFSLGAVPYANLSNQQTREAIEQGVRLEPPEQCPED....VYRLMRQCWEYDPHRRPSRGAVHQDLIAIRKRHR(1182)
fes   ........GRYSSESDVWSFGILLWETFSLGASMPWNLSNQQTREVERGRLPCELPECDDA.....VFRLMEQCWAYEPGQRPSAIYELQSIRKRHR(958)
ros   ........GVFTHHSDVWAFGVLVWETLTLGQQPYPGLSNIEVLHHVRSGGRLESPNNCPDD....IRDLMTRCWAQDPHNRPTFYIQHKLQEIRHSPLCFSYFPLGDKESVAPLRIQTAPPQPL(562)
erb-B .........RIYTHQSDVWSYGVTWELMTFGSKPYDGIPASEISSIVLEKGERLPQPPICTID....VYMIMVKCWMIDADSRPKFRELIAEPSRMARDPPRYLVIGDERMHLPSPTDSKFYRTL/427
EGF.R ........-.....................................................-I----.....N---A-/978
```

membranes is not mediated by a cleaved signal sequence. The transport of newly synthesized pp60^{v-src} may involve the formation of a complex with two cellular proteins, pp50 and pp89 (52, 53). In this complex pp89 is phosphorylated on serine, whereas pp50 is phosphorylated on serine and tyrosine (2, 52). Thus it seems likely that p50 is a substrate for pp60^{v-src}, but this has not been proven directly. In normal cells, pp50 has been detected in a complex with pp89 and contains only phosphoserine (P. Ser) (54). Two other viral PTKs, P140$^{gag-fps}$ and P90$^{gag-yes}$, each form a similar complex with pp50 and pp89 (55, 56), but this is not a universal feature of PTKs. For instance pp60^{c-src} is not found in this state, which is of interest because pp60^{c-src} has a different COOH terminus from pp60^{v-src}.

The best evidence suggests that pp60^{v-src} is not an integral membrane protein, but rather associates peripherally with the cytoplasmic face of several sorts of cytoplasmic membranes including the plasma membrane and perinuclear membranes (57–63). Because it is membrane-associated, purification of pp60^{v-src} requires detergent solubilization. With this caveat the bulk of pp60^{v-src} appears monomeric, and it is not found stably associated with proteins apart from pp50 and pp89.

POSTTRANSLATIONAL MODIFICATION OF pp60^{v-src} pp60^{v-src} is known to undergo two types of posttranslational modification. Following removal of the initiating methionine residue, the α amino group of the NH$_2$-terminal glycine is modified by the covalent attachment of a myristyl group (64–66). This acylation occurs either cotranslationally or else very rapidly following release of the nascent chain, such that the pp60^{v-src} molecules complexed with pp50 and pp89 contain myristate (67). The signal for myristylation probably lies close to the NH$_2$ terminus, since deletions or substitutions in the first 10 amino acids of pp60^{v-src}, created by site-directed mutagenesis, prevent addition of myristate (68). The myristyl group seems to have the same half-life as the protein

←_____

Figure 1   The amino acid sequences for three protein-serine/threonine protein kinases, phosphorylase kinase γ subunit (PhK-γ) (105), G-kinase (cGPK) (104), and A-kinase (cAPK) (36), are aligned with the predicted amino acid sequences of ten protein-tyrosine kinases: pp60^{v-src} (*src*) (24, 27), pp60^{c-src} (*c-src*) (26, 28), P90$^{gag-yes}$ (*yes*) (29), P70$^{gag-fgr}$ (*fgr*) (30), P120$^{gag-abl}$ (*abl*) (33), P140$^{gag-fps}$ (*fps*) (31), P85$^{gag-fes}$ (*fes*) (32), P68$^{gag-ros}$ (*ros*) (34), gp68/74$^{v-erb-B}$ (*erb*-B) (48), and the EGF receptor (EGF.R) (114, 148, 149). The sequences were aligned by eye for maximum homology. The single letter amino acid code is used with . indicating a gap and - in the pp60^{c-src} and EGF receptor sequences indicating identity with the cognate viral protein. Residue numbers are indicated: x/, starting number; /x, finishing number; (x), terminus. Between the sequences of the protein-serine/threonine kinases and the protein-tyrosine kinases invariant residues are shown, and highly conserved residues are indicated (*). The lysine (K) that is modified by the ATP analog FSBA is in bold type. The tyrosine (Y) that is phosphorylated in autophosphorylation reactions of the viral proteins is in bold type. The COOH termini of pp60^{v-src}, P90$^{gag-yes}$, and P70$^{gag-fgr}$ are in bold type where they diverge from the corresponding cellular sequences.

implying that it does not turn over (67). The NH$_2$-terminal 8 kDa of pp60$^{v\text{-}src}$ does not contain an extended hydrophobic region but is known to be important for membrane association (69). A direct involvement of the myristyl group in this association is suggested by the fact that the mutant pp60$^{v\text{-}src}$ molecules which lack myristate fail to become membrane associated (68).

pp60$^{v\text{-}src}$ belongs to a small group of proteins, including the murine leukemia virus p15^{gag} protein (70), the catalytic subunit of A-kinase (71), and the B subunit of the protein phosphatase calcineurin (72), which are modified in this fashion. Membrane association is not an inevitable consequence of this modification since the latter two proteins are soluble. It is curious that pp60$^{v\text{-}src}$ and the catalytic subunit of A-kinase share this modification, since, despite their extensive homology, their NH$_2$ termini, apart from the terminal glycine, are unrelated. Not surprisingly, pp60$^{c\text{-}src}$, which has the same NH$_2$-terminal sequence as pp60$^{v\text{-}src}$, is both myristylated (66) and membrane-bound (58). One other putative cellular PTK, p56^{lstra}, has been found to contain myristate, but may contain palmitate as well (74). Three other viral PTKs are myristylated, P120$^{gag\text{-}abl}$, P70$^{gag\text{-}fgr}$, and the gag-fes proteins (74), but here the myristyl group is attached to the p15^{gag} sequences present at their NH$_2$ termini. These proteins are associated with either membranes and/or the cytoskeleton, but it is unclear whether the myristyl group is involved. Whether the corresponding c-*abl*, c-*fgr*, and c-*fes* proteins are myristylated is unknown. P90$^{gag\text{-}yes}$, P140$^{gag\text{-}fps}$, and P68$^{gag\text{-}ros}$, three other viral PTKs, are presumably NH$_2$-terminally acetylated rather than myristylated, because they have the avian p19^{gag} NH$_2$ terminus which is normally acetylated. Thus the modification of PTKs by myristylation is apparently unimportant for enzymatic activity per se. The fact that the mutant pp60$^{v\text{-}src}$ molecules lacking myristate are active as PTKs, but do not transform (68) suggests that transformation requires the proper subcellular localization of the viral PTK, which may in turn dictate its substrate specificity.

In addition to myristylation, pp60$^{v\text{-}src}$ also undergoes phosphorylation. Two major sites of phosphorylation have been identified, Ser 17 and Tyr 416 (2, 75–77). There are also minor sites of both serine and tyrosine phosphorylation which have not been mapped in detail, but all of which are in the NH$_2$-terminal half of pp60$^{v\text{-}src}$. In view of the regulatory effects of phosphorylation on enzyme function, the phosphorylation of pp60$^{v\text{-}src}$ is of obvious interest.

PHOSPHORYLATION OF pp60$^{v\text{-}src}$ AT TYROSINE 416   Tyr 416 lies in the catalytic domain of pp60$^{v\text{-}src}$. At steady state it has been estimated that 10–30% of pp60$^{v\text{-}src}$ molecules are phosphorylated at this site (78). This phosphate turns over rapidly, but is apparently not transferred to other proteins. An equivalent tyrosine is found in the catalytic domain of each of the viral PTKs, although the sequence surrounding it is not as highly conserved as some other parts of the catalytic domain (Figure 1). This tyrosine has been shown to be the major phosphorylated tyrosine in P90$^{gag\text{-}yes}$ (79), P105$^{gag\text{-}fps}$ (79), P85$^{gag\text{-}fes}$ (77), and P120$^{gag\text{-}abl}$ (77) isolated from cells, although most of these proteins have other sites of tyrosine phosphorylation as well. P68$^{gag\text{-}ros}$ does not contain detectable P. Tyr when isolated from cells (23), although a single tyrosine in P68$^{gag\text{-}ros}$ is phosphorylated in vitro. The properties of the phosphopeptide (80), however, are not readily reconciled with that expected to contain Tyr 428. pp60$^{c\text{-}src}$ also contains P. Tyr in its COOH-terminal half, but although Tyr 416 and the sequence surrounding it are identical to pp60$^{v\text{-}src}$ this tyrosine is not phosphorylated (75, 81). It is not known which tyrosine is phosphorylated in pp60$^{c\text{-}src}$, but one might speculate that it is near the COOH terminus of the protein where its structure diverges from that of pp60$^{v\text{-}src}$.

There are two paramount questions concerning Tyr 416. What is the nature of the PTK(s) which phosphorylate it, and does occupancy of this site affect the function of pp60$^{v\text{-}src}$? There is reason to believe that this residue can be phosphorylated by pp60$^{v\text{-}src}$ itself in a so-called 'autophosphorylation' reaction. Purified preparations of pp60$^{v\text{-}src}$ can autophosphorylate at Tyr 416 and this is the only tyrosine phosphorylated either at low ATP concentrations or in the presence of GTP (82–84). At high concentrations of ATP one or more tyrosines in the NH$_2$-terminal half of the protein also accept phosphate (85, 86). It has been difficult to determine whether the phosphorylation at Tyr 416 is an inter- or intramolecular reaction because pp60$^{v\text{-}src}$ tends to aggregate. Whether other PTKs can phosphorylate Tyr 416 in pp60$^{v\text{-}src}$ has not been adequately tested, but it should be noted that synthetic peptides corresponding to the sequence around Tyr 416 can be phosphorylated by most PTKs examined. In pp60$^{v\text{-}src}$, however, Tyr 416 might be inaccessible to exogenous PTKs. Mutants of pp60$^{v\text{-}src}$ thermosensitive for transformation are also thermosensitive for PTK activity (38, 87). At the restrictive temperature, pp60$^{v\text{-}src}$ is found to be hypophosphorylated at Tyr 416 (52), which suggests that this modification is normally carried out by pp60$^{v\text{-}src}$ itself. Phosphorylation of the other viral PTKs in vitro also occurs at their major sites of in vivo tyrosine phosphorylation, although additional tyrosines may also be phosphorylated.

Does the conservation of an autophosphorylation site in the different viral PTKs indicate that it serves a function? Attempts to determine the significance of pp60$^{v\text{-}src}$ autophosphorylation have been made in two ways. Conversion of Tyr 416 to Phe by site-directed mutagenesis does not noticeably reduce the ability of pp60$^{v\text{-}src}$ to act as a PTK either in vitro or in the cell, nor does it affect its transforming potential for mouse fibroblasts (88). Although this might imply that phosphorylation at Tyr 416 is dispensible for pp60$^{v\text{-}src}$ function, the Phe 416 mutant is defective at inducing tumor formation (89). Also replacement of the equivalent Tyr 1073 in P140$^{gag\text{-}fps}$ by Phe has a detrimental effect on both PTK activity and fibroblast transformation (90).

Amino acid substitution by site-directed mutagenesis is not an entirely satisfactory way to answer this question since one cannot readily distinguish the effects of the amino acid replacement itself from the inability of the residue to be phosphorylated. A more direct approach is to compare the properties of a pp60$^{v\text{-}src}$ preparation completely dephosphorylated at Tyr 416 with one fully modified at this position. Most purified preparations of pp60$^{v\text{-}src}$ contain very little phosphate at Tyr 416. A comparison of the casein phosphorylating activity of such a preparation with one that has been fully phosphorylated at Tyr 416 in the presence of GTP shows that phosphorylation leads to a 2–3 fold increase in activity (84). This evidence is consistent with a regulatory role for Tyr 416 autophosphorylation.

The more minor sites of tyrosine phosphorylation which lie in the NH$_2$-terminal half of pp60$^{v\text{-}src}$ may also be important. Phosphorylation at these sites is observed upon incubating pp60$^{v\text{-}src}$ at high ATP concentrations (85, 86). Their occupancy correlates with a large elevation in activity in vitro (86). When cells are incubated with the phosphatase inhibitor sodium vanadate, pp60$^{v\text{-}src}$ phosphorylated at apparently the same tyrosine(s) accumulates (91, 92). These NH$_2$-terminal phosphorylations could regulate pp60$^{v\text{-}src}$ PTK activity in a manner akin to the tyrosine phosphorylations that may activate the EGF receptor PTK which occur likewise at sites outside the catalytic domain.

There are also intimations that phosphorylation of tyrosines near the NH$_2$ terminus of pp60$^{c\text{-}src}$ can regulate its activity. Part of the pp60$^{c\text{-}src}$ population in cells transformed or infected by polyoma virus is complexed with the middle-sized T-antigen (mT) (93). The mT protein is the major substrate for tyrosine phosphorylation in this complex (1). pp60$^{c\text{-}src}$ complexed with mT has a greater PTK activity when assayed in vitro with peptide or protein substrates than free pp60$^{c\text{-}src}$ (94). As a result the total pp60$^{c\text{-}src}$ PTK activity of polyoma-transformed cells is greater than that of normal cells. The pp60$^{c\text{-}src}$ molecules complexed with mT have a lower gel mobility, akin to the mobility shift seen with hyperphosphorylated pp60$^{v\text{-}src}$. The NH$_2$-terminal half of complexed pp60$^{c\text{-}src}$ is more highly phosphorylated than that of free pp60$^{c\text{-}src}$ (W. Yonemoto, J. S. Brugge, personal communication). A similar form of pp60$^{c\text{-}src}$ is induced by PDGF treatment of cells (R. Ralston, J. M. Bishop, personal communication). Tyrosine phosphorylation in the NH$_2$-terminal half of pp60$^{c\text{-}src}$ correlates in each case with increased PTK activity.

PHOSPHORYLATION OF pp60$^{v\text{-}src}$ AT SERINE 17   In RSV-transformed chick cells about 60% of pp60$^{v\text{-}src}$ molecules are phosphorylated on serine at steady state (78) and the majority of this phosphate appears to be located at Ser 17 (76, 77, 95). The sequence preceding Ser 17 has the hallmarks of an A-kinase phosphorylation site and this enzyme will phosphorylate pp60$^{v\text{-}src}$ in vitro on Ser 17 (76). Treatment of certain lines of RSV-transformed cells with agents that elevate the intracellular concentrations of cAMP leads to increased serine phosphorylation of pp60$^{v\text{-}src}$ (96), probably at Ser 17. It is not clear, however, whether A-kinase is the only protein kinase that can phosphorylate Ser 17, since several other protein kinases have rather similar primary sequence requirements for phosphorylation of their substrates, including the cGMP-dependent protein kinase (G-kinase), the multifunctional calmodulin-dependent protein kinase, and the Ca^{2+}/phospholipid-dependent protein kinase (C-kinase).

The effect of phosphorylation of Ser 17 on the activity of pp60$^{v\text{-}src}$ is also difficult to ascertain. No precise substitution of Ser 17 has yet been made but a series of small deletions covering Ser 17 have little effect on either PTK activity or tumorigenicity (95). This indicates that phosphorylation at this site is not crucial for the biological activity of pp60$^{v\text{-}src}$, but does not adequately address the question of whether phosphorylation at this site modifies enzymatic activity. There is some evidence that in vitro phosphorylation of pp60$^{v\text{-}src}$ by A-kinase increases its PTK activity (R. L. Erikson, personal communication). Moreover, treatment of RSV-transformed CHO hamster cells with cholera toxin or 8Br-cAMP increased the phosphorylation of Ig heavy chain in the immune complex kinase assay of pp60$^{v\text{-}src}$ (96). The latter result, however, is not seen with all RSV-transformed cell types, and could depend on examining cells in which the phosphorylation level of pp60$^{v\text{-}src}$ at Ser 17 is low. It is worth pointing out that Ser 17 is very close to the membrane attachment site of pp60$^{v\text{-}src}$ and that therefore phosphorylation of Ser 17 could affect properties of pp60$^{v\text{-}src}$, such as its disposition on the membrane (e.g. via repulsion from phospholipid head groups) and its interaction with other proteins and potential substrates in the cortical skeleton.

THE CATALYTIC DOMAIN OF pp60$^{v\text{-}src}$ There are several reasons for believing that the catalytic domain of pp60$^{v\text{-}src}$ is contained wholly within the COOH-terminal 30 kDa. Mild trypsin treatment of pp60$^{v\text{-}src}$ generates a 30-kDa fragment which retains PTK activity (69, 97). Similar protease-resistant catalytically active domains can be obtained from P90$^{gag\text{-}yes}$ (97) and P140$^{gag\text{-}fps}$ (98), as well as from the EGF receptor. This may be a general feature of proteins containing a catalytic domain recognizably related to those of pp60$^{v\text{-}src}$ and A-kinase. The COOH-terminal fragment of pp60$^{v\text{-}src}$ appears to have greater specific activity than the intact molecule (96), suggesting that the NH$_2$-terminal domain may modulate the PTK activity. A second line of evidence for the location of the catalytic activity within pp60$^{v\text{-}src}$ comes from mutant molecules. There is a phenotypically wild-type revertant of a nontransforming mutant of RSV which expresses a pp60$^{v\text{-}src}$-related protein consisting only of the COOH-terminal 45 kDa of pp60$^{v\text{-}src}$ (99). Some of the extensive deletions created in the NH$_2$-terminal half of pp60$^{v\text{-}src}$ have no effect on PTK activity, whereas other deletions render it thermolabile (95, 100, 101). The latter observation, coupled with the fact that all the other known mutations that render pp60^{v-src} thermosensitive for both transformation and PTK activity have NH$_2$-terminal locations, lends further credence to the idea that this region of pp60^{v-src} has a regulatory rather than a catalytic function.

Little is known about the precise catalytic mechanism of pp60^{v-src}. Studies have been hampered by a shortage of pure protein. Intact pp60^{v-src} has been purified only in microgram amounts (82), while somewhat larger quantities of natural proteolytic fragments have been isolated e.g. p52^{src} (15) and p54^{src} (102). Several divalent metal ions will support phosphorylation, although Mg^{2+} or Mn^{2+} are preferred. Measurement of kinetic parameters is problematic in the absence of sufficient quantities of authentic substrates. Synthetic tyrosine-containing peptides have been used, although their K_m's are much higher than those of protein substrates. With a preparation of p54^{src} the apparent K_m for ATP was 30 μM and the turnover number using casein as a substrate was about 2 μmol/min/μmol (102). This is 10–100 fold lower than the values determined for A-kinase (103), but experiments with another related PTK, the bacterially expressed fragment of the v-abl protein (43), using angiotensin as a substrate, show that the catalytic activity of PTKs can approach that of A-kinase.

Although the three-dimensional structure of pp60^{v-src} is not known, it is possible to say something about regions in the catalytic domain that are important for catalysis. A comparison of the primary amino acid sequence of the catalytic domain of pp60^{v-src} with those of the other PTKs, the A-kinase catalytic subunit (36), G-kinase (104), and the phosphorylase kinase γ subunit (105) is revealing (Figure 1). The PTKs are closely related to one another over a region of about 260 amino acids corresponding to residues 255–516 in pp60^{v-src}. Recognizable sequence homology between A-kinase and pp60^{v-src} starts at about residue 40 in the former and 265 in the latter and extends to about residue 280 and 505 respectively. A number of residues are conserved in all the protein kinases. Among these is Lys 295 in pp60^{v-src} which corresponds to Lys 71 in the catalytic subunit of the A-kinase. Through the use of the ATP affinity analog 5'-p-fluorosulphonylbenzoyladenosine (FSBA), Lys 71 has been shown to form part of the ATP binding site (106). Similar experiments with pp60^{v-src} have shown that Lys 295 becomes covalently modified upon treatment with FSBA (107). In both cases the lysine is about 15 residues to the COOH-terminal side of a sequence Gly.X.Gly.X.X.Gly. A similar sequence is observed in a number of nucleotide binding proteins (108) and in the cases where a three-dimensional structure is available, this sequence is found to form an elbow making contact with the ribose ring. A glycine-rich region and following lysine are found in all the protein kinases. Site-directed mutagenesis of Lys 295 in pp60^{v-src} is underway. Preliminary indications are that any alteration at this site abolishes transforming potential. Other highly conserved sequences in all the protein kinases are Arg.Asp.Leu (residues 385–387 in pp60^{v-src}), Asp.Phe.Gly (residues 404–406) and Ala.Pro.Glu (residues 430–432) which are found with similar spacing in every case. It seems likely that these residues are important for catalysis. Independent mutation of Ala 430 to Val, Pro 431 to Ser, Glu 432 to Lys, or Ala 433 to Thr each abrogates transforming ability and drastically diminishes the PTK activity of pp60^{v-src} (101, 109, 110).

Sequences to the COOH-terminal side of Ala 433 also seem to be critical for phosphate transfer by pp60^{v-src}. Deletion of residues 504–506 abolishes both the PTK and transforming activities of pp60^{v-src} (101). Antibodies against a synthetic peptide corresponding to residues 498–512 inhibit the PTK activity of pp60^{v-src} (111). In contrast, antibodies to residues 521–526, the very COOH terminus of pp60^{v-src}, do not affect PTK activity (112). Deletions of more than 11 residues at the COOH terminus (i.e. extending beyond Leu 516) decrease both PTK activity and transforming potential (101). This is of interest because the sequence of pp60^{v-src} diverges from that of its presumed progenitor pp60^{c-src} at residue 514, although residue 516 is Leu in both cases. Replacement of residues 518–526 with 9 totally unrelated amino acids has no phenotypic effect on pp60^{v-src} (113). This evidence places the boundary of the sequences involved in catalysis close to the COOH termini of pp60^{v-src} and pp60^{c-src}, with the same being true for the products of the v-*yes*, v-*fgr*, v-*fps*, and v-*fes* genes. A terminal location for the catalytic sequence, however, is not a sine qua non for catalysis since this domain is near the NH$_2$ terminus of P120$^{gag-abl}$ (33) and 20 kDa from the COOH terminus of the EGF receptor (114).

Some regions in the NH$_2$-terminal half of pp60^{v-src} are clearly dispensable for catalytic activity. For instance many small deletions [e.g. those around Ser 17 (95)] have little or no effect on either PTK or transforming activity. Larger deletions may affect transformed cell morphology [e.g. deletion of residues 16–82, 16–138 (95) or 135–236 (115)] or render pp60^{v-src} thermosensitive for transformation [e.g. deletion of residues 169–225 (100)] without a great effect on PTK activity. The phenotype of cells infected with the latter mutants suggests that their pp60^{v-src}'s have subtle changes in subcellular location or substrate specificity.

DOES pp60^{v-src} HAVE OTHER CATALYTIC ACTIVITIES?   In the presence of ADP, pp60^{v-src} will dephosphorylate tyrosine-phosphorylated Ig heavy chain, forming labeled ATP (116). The bacterially synthesized fragment of the v-*abl* gene has a similar property (43). In this sense phosphate esterified to tyrosine in protein is 'energy-rich' in distinction to serine phosphate. For a number of reasons, however, it seems unlikely that this activity will contribute significantly to removal of phosphate from proteins in the cell. In the absence of a peptide substrate A-kinase can act as an ATPase. pp60$^{v\text{-}src}$ has not been tested for this activity to our knowledge. The phosphorylation of nonprotein substrates by pp60$^{v\text{-}src}$ will be discussed below.

SUBSTRATES OF pp60$^{v\text{-}src}$ In general PTKs are rather nonspecific in vitro, so physiological substrates for pp60$^{v\text{-}src}$ which might be critical for transformation have been sought by analysis of phosphoproteins from RSV-transformed cells. A number of proteins containing P.Tyr have been detected specifically in cells transformed by RSV [for review see (117)]. These proteins are not phosphorylated by the cognate cellular PTK, pp60$^{c\text{-}src}$, even when it is expressed at high level, although they are phosphorylated in cells transformed by other viruses whose oncogenes encode PTKs. The striking correlation between tyrosine phosphorylation of these substrates and transformation leads us to anticipate that they may be involved in transformation but their functions are in many cases unknown. Three of the proteins containing P.Tyr in transformed cells, p36, p81, and vinculin, are found together in a submembraneous cortical skeleton where they are in a position to be phosphorylated directly by the viral PTKs, many of which are membrane-associated. Soluble proteins are also phosphorylated in RSV-transformed cells; among these are enolase, phosphoglycerate mutase, and lactate dehydrogenase (LDH). The steady state stoichiometry of phosphorylation of most of these proteins is rather low (in the 1–10% range). In no case is there yet evidence for a functional consequence of tyrosine phosphorylation. At the moment it is difficult to decide whether phosphorylation of any of the identified substrates is important to the transformed phenotype or whether they are simply modified gratituitously. Most of the substrates are rather abundant cellular proteins and it may well be that the crucial targets are rare proteins that await identification.

GROWTH FACTOR RECEPTOR–ASSOCIATED PROTEIN-TYROSINE KINASES

An EGF-activated protein kinase activity was first detected in membranes made from A431 cells (118). This protein kinase was stimulated 2–4 fold by added EGF, but had a significant basal activity without EGF (118). Because A431 cells, derived from a human epidermoid carcinoma, are a rich source of EGF receptors, much of the subsequent information obtained about the EGF receptor PTK has come from these cells. A431 cells, however, are unusual in their growth response to EGF; instead of being stimulated by EGF, A431 cell growth is inhibited (119, 120). Although this might suggest that A431 cell EGF receptors are atypical in their structure or enzymatic properties, their study seems justified since so far they have properties indistinguishable from the EGF receptors obtained from other cell lines (80, 121, 122), liver (123), or other tissues. The recent identification of the EGF receptor gene will now enable the direct comparison of A431 cell and germ-line human EGF receptor sequences.

In A431 cell membranes EGF stimulated the phosphorylation of added proteins such as casein, histones, antibodies to pp60^{v-src}, and a variety of peptides (124–127), as well as a number of endogenous membrane proteins, the most prominent being 170 kDa (128). The residue phosphorylated in all cases is now known to be tyrosine (6). Since then overwhelming evidence has been obtained that the EGF-stimulated PTK activity is intrinsic to the receptor. The PTK activity copurifies with EGF binding activity and the 170-kDa substrate protein (124, 129). Cross-linking of radioiodinated EGF to cell surfaces also results in labeling of a 170-kDa protein (130). The ATP analog FSBA labels the purified EGF-binding protein with concomitant loss of PTK activity (131). Finally, the recent cDNA cloning of the EGF receptor shows that it has sequences homologous to the catalytic domains of other protein kinases (11, 114).

Following the reports of EGF receptor PTK activity, other growth factor receptors were examined for a similar ligand-stimulated property. The platelet-derived growth factor (PDGF) (132, 133), insulin (134–137), and insulin-like growth factor 1 (IGF-1) (138, 139) receptors all prove to have ligand-stimulated autophosphorylating activities specific for tyrosine. Like the EGF receptor, these receptors can phosphorylate exogenous substrates in vitro. The evidence that PTK activity is intrinsic to these other receptors is less complete. In its size and disposition in the membrane the glycosylated single chain 180-kDa PDGF receptor molecule is similar to the EGF receptor (140–142). The structures of the insulin and IGF-1 receptors, however, are quite different from the EGF and PDGF receptors. In each case a single precursor polypeptide chain is cleaved to form two subunits that remain joined by disulphide bridges in an $\alpha_2 \beta_2$ structure (143, 145). Both subunits are glycosylated and exposed on the surface. In the insulin receptor, the 135-kDa α subunit binds insulin whereas the 90-kDa β subunit is autophosphorylated (134–137). Highly purified insulin receptor retains PTK activity (145) and the β subunit can be labeled by ATP affinity analogs (137, 146, 147). Verification of the intrinsic nature of the PTK activities of these proteins, however, awaits their cloning and sequencing.

Not all growth factor receptors have ligand-stimulated PTK activities. For instance the interleukin 2, nerve growth factor, and IGF-2 receptors do not have detectable protein kinase activity. Nevertheless it is likely that other growth factor receptors will prove to have this property. Among the candidates are the as yet uncharacterized receptors for fibroblast growth factor and endothelial cell growth factor, as well as those for several hematopoietic growth factors.

The EGF receptor

STRUCTURE AND SYNTHESIS OF THE EGF RECEPTOR   The complete amino acid sequence of the human EGF receptor has now been deduced from the nucleotide sequences of a series of overlapping cDNA clones (114, 148, 149). This matches exactly with the sequences of a total of 19 tryptic and cyanogen bromide fragments derived from purified human EGF receptors (11, 114), and with an extended NH_2-terminal sequence of the mature protein (150). In all over 250 residues of the protein have been sequenced and they fit that predicted from the cDNA clones precisely.

The primary amino acid sequence suggests the following picture for the synthesis and structure of the EGF receptor (114). The protein commences with a hydrophobic signal peptide of 24 amino acids. The known NH_2-terminal sequence of the mature EGF receptor starts at position 25 of the predicted sequence, indicating that the signal peptide is removed. The remainder of the 1186-residue protein contains only one long stretch of preponderantly hydrophobic amino acids. This run of 26 amino acids divides the receptor into an NH_2-terminal region of 621 amino acids and a COOH-terminal domain of 542 amino acids. By analogy with other membrane proteins, the signal peptide would direct polysomes translating EGF receptor mRNA to the endoplasmic reticulum where the NH_2 terminus is inserted into the membrane. The signal peptide would be cleaved cotranslationally, and the polypeptide chain then extruded through the membrane until the hydrophobic stretch is reached. Nine of the 15 amino acids after the putative 26-amino-acid-transmembrane domain are basic. This is common in sequences on the cytoplasmic side of transmembrane domains; the basic residues may prevent further translocation by interacting with phospholipid head groups. The NH_2 terminus of the newly synthesized EGF receptor is thus within the lumen of the endoplasmic reticulum, and is destined to be exposed outside the cell while the COOH terminus remains inside the cell. The NH_2-terminal region is rich in cysteine, and disulphide bonds may stabilize the extracellular domain. This includes the EGF binding site, which binds only a single EGF molecule (151). There are 12 potential N-linked glycosylation sites in the external domain. Metabolic labeling and immunoprecipitation confirms that the protein undergoes glycosylation, first into an endoglycosidase H-sensitive mannose-rich form in a tunicamycin-sensitive reaction and subsequently into a form with complex carbohydrates (152–156). Protein analysis confirms that there are at least 11 carbohydrate chains (154). These may be important for EGF binding since the immature receptors from tunicamycin-treated cells do not bind EGF (G. Carpenter, personal communication). In addition certain lectins interfere with EGF binding (157).

Beginning at residue 557 the sequence of the EGF receptor shows nearly complete identity with the predicted product of the v-*erb*-B oncogene (48, 114,

148, 149). Indeed it seems very likely that the EGF receptor gene is the c-*erb*-B gene, particularly since the human EGF receptor (158, 159) and a human v-*erb*-B related sequence (160) have been mapped independently to chromosome 7. Starting about 50 residues from the transmembrane segment, the COOH-terminal cytoplasmic region of the EGF receptor shows very strong homology to the catalytic domains of the other PTKs in the *src* gene family. Presumably this part of the EGF receptor accounts for its PTK activity. The final 240 amino acids at the COOH terminus of the EGF receptor show no homology with known proteins and form an additional domain whose function may be regulatory.

There is considerable experimental support for the proposed topology of the EGF receptor in the plasma membrane. The sizes of fragments obtained by proteolysis of native EGF receptor agree well with those of the domains predicted from the primary sequence, implying that the domain boundaries are exposed on the surface of the protein. Treatment of intact cells with trypsin releases a glycosylated fragment of about 115 kDa, corresponding to the 621-amino-acid external domain with its 11 or 12 carbohydrate chains (154). Proteases can also act on the cytoplasmic half of the protein. If cells are lysed in the presence of Ca^{2+}, a fragment of 15–20 kDa is clipped from the cytoplasmic domain of the EGF receptor leaving a 150-kDa fragment which retains EGF-stimulated PTK activity (129, 161–163). The 15–20-kDa fragment presumably corresponds to the region extending beyond the PTK domain. Several proteases, including trypsin, elastase (164), and a Ca^{2+}-activated neutral protease purified from A431 cells (165) cleave purified EGF receptor in a similar manner. More extensive proteolysis of the purified receptor gives rise to 30–40 kDa- and 110–125-kDa fragments (154, 161, 166, 167). The larger pieces contain the EGF-binding site and the glycosyl side chains (167). The smaller fragments may correspond to the PTK domain, minus the COOH-terminal addendum.

POSTTRANSLATIONAL MODIFICATION OF THE EGF RECEPTOR   Although the EGF receptor is not modified by covalent attachment of a lipid moiety, it is phosphorylated. When isolated from A431 cells, the receptor contains 1–1.5 mol phosphate/mol linked to serine and threonine residues (168, 169). Tryptic phosphopeptide analysis suggests that there are several phosphorylation sites (168, 170). It appears that no one serine or threonine is phosphorylated in all receptor molecules in the cell. Partial digestion experiments indicate that most of the sites are within the 30–40-kDa tryptic fragments (167) that contain the PTK domain. The effects of these modifications on receptor function are unknown and may be difficult to ascertain until the protein kinases responsible have been identified. It is clear, however, that the total content of phosphate in A431 cell EGF receptors doubles after exposure of the cells to EGF (168, 169).

Most of this increase in phosphorylation occurs at the same serines and threonines that are phosphorylated constitutively. Significantly, up to 4 major additional phosphopeptides have been detected in EGF receptors from EGF-treated A431 cells (168, 170). One of these contains P.Tyr, and can also be labeled in vitro in a presumptive autophosphorylation reaction (80). The others contain P.Thr and are thought to be phosphorylated by C-kinase (169, 170).

PHOSPHORYLATION OF THE EGF RECEPTOR AT TYROSINE   While there is a single major site of tyrosine phosphorylation in EGF receptors isolated from EGF-treated cells (168, 171), incubation of crude or purified EGF receptor preparations with ATP results in 'autophosphorylation' of the receptor at several tyrosines (80, 172). Three of the 4 major sites are contained in the 15–20-kDa tail appended to the protein kinase domain (166, 171). The precise location of the fourth site is not known. The main site of tyrosine phosphorylation in vivo is residue 1173, one of the three sites in the COOH-terminal region (171). The two other in vitro phosphorylation sites in the COOH terminus, which are only weakly phosphorylated in the cell in response to EGF, have been mapped to residues 1068 and 1148 (171). Unlike $pp60^{v-src}$ and the other PTKs described above, the EGF receptor appears not to be phosphorylated at the tyrosine Tyr 845 in the heart of the PTK domain corresponding to Tyr 416 in $pp60^{v-src}$. Perhaps the environment of Tyr 845 precludes phosphorylation. It differs from Tyr 416 of $pp60^{v-src}$ in having a basic amino acid amongst the acidic residues to its NH_2-terminal side.

Autophosphorylation proceeds rapidly at ATP concentrations lower than are required for phosphorylation of added substrates. The K_m for autophosphorylation is 0.2–0.3 μM, whereas the K_m for phosphorylation of added proteins is 2–3 μM [(151) D. Garbers, personal communication]. The initial rate of the autophosphorylation reaction is independent of receptor concentration and proceeds to greater than one mol phosphate/mol receptor (151). From this one might conclude that the phosphorylation is intramolecular such that the PTK domain phosphorylates one or more tyrosines in the COOH-terminal tail of the same receptor molecule. The reaction, however, could still be intermolecular if receptors form stable multimers. In this case the K_d for the interaction would have to be lower than 0.5 nM, the lowest concentration tested in the assay.

Since phosphorylation of Tyr 1173 both in vitro and in vivo depends on the binding of EGF, occupation of this site correlates well with increased phosphorylation of exogenous substrates in vitro or cellular proteins in vivo. It is not clear, however, whether autophosphorylation causes or results from activation of the receptor protein kinase or both. The intramolecular nature of the autophosphorylation suggests that phosphorylation of Tyr 1173 could lead to an unhinging of the COOH-terminal domain from a resting configuration in which it prevents access of exogenous substrates to the active site. Autophosphorylation is not obligatory for activation in vitro, since the 150-kDa form of the receptor, which lacks at least three autophosphorylation sites, is still stimulated by EGF (129). It is not known whether autophosphorylated EGF receptor molecules remain active after EGF is removed. If this were the case then stimulation of PTK activity would need only transient binding of EGF and the duration of stimulation would depend on the turnover rate of phosphate at Tyr 1173. It is worth noting that autophosphorylation of the insulin receptor renders it independent of insulin for full PTK activity (173).

SERINE AND THREONINE PHOSPHORYLATION OF THE EGF RECEPTOR   The multiplicity of phosphorylated peptides in the EGF receptor means that several serine/threonine protein kinases might regulate its activity, although, other than the overall increase in phosphorylation at these sites seen with EGF and tumor promoters, conditions that cause specific modulation of phosphorylation at these sites have not yet been reported. The EGF receptor can be phosphorylated in vitro by A-kinase but the sites phosphorylated do not appear to be extensively phosphorylated in intact cells (174, 175). There is good evidence, however, that β-adrenergic agonists, which increase cAMP levels, reduce EGF binding to adipocytes.

Addition of tumor promoting agents, including phorbol esters and indole alkaloids, to A431 cells causes a rapid doubling of phosphorylation of EGF receptors, but no P.Tyr is detected (169, 170). Tryptic mapping of the EGF receptor from tumor promoter-treated cells shows three new P.Thr-containing peptides in addition to those that are constitutively phosphorylated (169). Tumor promoters can activate C-kinase, which is specific for serine and threonine residues (176, 179). Indeed this protein kinase is found to be the quantitatively major binding protein for phorbol esters in the cell. It seems probable that C-kinase is responsible for these new phosphorylations, since purified C-kinase phosphorylates purified or crude EGF receptor preparations at the same three peptides as are phosphorylated in vivo (169). It transpires that all three peptides contain the same phosphorylated threonine, Thr 654, which lies in a highly basic region that causes incomplete tryptic digestion (180). Since active C-kinase is membrane-associated, Thr 654 is in a favorable site to be phosphorylated in the cell. It has been estimated that up to 30% of the EGF receptor molecules in A431 cells are phosphorylated at Thr 654 following tumor promoter treatment (169). The EGF receptor of normal human fibroblasts is also phosphorylated at Thr 654 upon treatment with tumor promoters (181). It is not known whether other serine/threonine-protein kinases can phosphorylate Thr 654.

Phosphorylation of EGF receptors by C-kinase appears to regulate their function. A431 cells exposed to tumor promoters and then to EGF exhibit lower levels of P.Tyr in their EGF receptors and other cell proteins than control cells exposed to EGF alone (169, 182). Purified EGF receptors have decreased EGF-stimulated PTK activity if first phosphorylated by purified C-kinase. Thus phosphorylation of Thr 654 appears to compromise stimulation of EGF receptor PTK activity by EGF, either by affecting the EGF-binding domain or the PTK domain, or perhaps by altering receptor:receptor interactions. Thr 654 is nine amino acids from the membrane on the cytoplasmic side, in a position from which conformational changes may be propagated either out through the membrane to the EGF-binding region or inward to the PTK domain. Tumor promoters decrease EGF binding in many cell types, either reducing receptor affinity or number (183–186). In A431 cells most of the EGF binding sites have a K_d of 5–10 nM but about 0.2–2% of binding sites appear to have a 1000-fold higher K_d (182). The latter small population of binding sites is absent in A431 cells treated with tumor promoters (182). This change in EGF binding may be attributable to C-kinase phosphorylation, but perhaps the more significant effect on A431 cell EGF receptors is to block the activation of the kinase domain by EGF.

Tumor promoters appear to mimic a physiological activator of C-kinase, diacylglycerol (176). One effect of EGF on A431 cells is to stimulate phospholipid turnover and production of diacylglycerol (187). This might in turn activate C-kinase; indeed, Thr 654 is also phosphorylated upon EGF treatment under certain conditions (170). This phosphorylation may be a physiological mechanism that can partly explain down-modulation of EGF binding by EGF itself and by other extracellular factors. For instance PDGF induces phospholipid breakdown (188), and reduces EGF binding to cells that bear both PDGF and EGF receptors (189, 190). Consistent with this idea PDGF stimulates Thr 654 phosphorylation (191). The functions of other receptors may also be regulated through phosphorylation by C-kinase, since changes in binding of insulin (192) and transferrin (193) have been observed in tumor promoter treated cells.

THE CATALYTIC DOMAIN OF THE EGF RECEPTOR   Evidence that the catalytic domain of the EGF receptor is contained within the predicted 30-kDa region has been obtained by analysis of proteolytic fragments. Mild trypsinolysis of purified receptor in solution generates first a 150-kDa fragment and the 15–20 kDa tail. Upon further digestion the 150-kDa fragment releases a 40-kDa fragment which can autophosphorylate and has about 30% of the activity of the intact receptor towards exogenous substrates (166). The PTK activity of this fragment is not stimulated by EGF and is inhibited by the ATP analog FSBA concomitant with covalent attachment of FSBA to the 40-kDa fragment. By analogy with pp60^{v-src} this would be expected to be Lys 721 which is 21 residues downstream of a Gly.X.Gly.X.X.Gly sequence. Most of the other highly conserved sequence features of the *src* gene family catalytic domain are found in the EGF receptor, except that the residue equivalent to Pro 431 is Leu. Although replacement of Pro 431 by Ser abolishes pp60$^{v\text{-}src}$ function, substitution with Leu has not been tested.

Some details of the catalytic mechanism of the EGF receptor have been reported. A kinetic analysis conducted using a synthetic tyrosine-containing peptide as a substrate suggests that the enzyme catalyzes a reaction in which peptide binds before ATP and phosphopeptide is released before ADP (194). Such an ordered Bi Bi mechanism does not require a phosphoenzyme intermediate. The K_m for ATP is about 5 μM. The major effect of EGF is to increase V_{max}, although there is a small increase in K_m for peptide. The availability of cDNA clones for the EGF receptor will allow a study of the importance of particular regions of the catalytic domain in phosphate transfer by site-directed mutagenesis.

How does EGF binding increase V_{max}? In particular, can a protein chain that only traverses the plasma membrane once transmit an allosteric signal from outside to inside? Oligomerization of the receptor in the membrane could be involved, although one has to explain how EGF stimulates the PTK activity of the EGF receptor in solution. Another consideration is that gp68/74$^{v\text{-}erb\text{-}B}$, which lacks the majority of the external domain, has only very weak PTK activity. Thus simple removal of the regulatory domain apparently does not cause constitutive activation of the catalytic domain. The inhibitory effect of Thr 654 phosphorylation on EGF stimulation suggests that this region of the protein is important in signal transmission. The phosphate on Thr 654 nine residues from the membrane might provide a charge acceptor alternative to the phospholipid head groups for its neighboring basic residues and thus prevent translocation of this region closer to the membrane. A comparative study of the structures of the other receptors with inducible PTK activities, such as the PDGF and insulin receptors, may give us clues to the mechanism of signal transduction.

DOES THE EGF RECEPTOR HAVE OTHER CATALYTIC ACTIVITIES?   The ATP-dependent DNA nicking activity reported to copurify with the EGF receptor (195) can apparently be separated upon more stringent purification. Because EGF induces increased turnover of phosphatidylinositol, an important question is whether the receptor can phosphorylate phosphatidylinositol, a step which is thought to be the first in its breakdown. So far there have been no positive reports.

SUBSTRATES FOR THE EGF RECEPTOR   The substrates of the growth factor receptor PTKs are of obvious interest with regard to mechanisms of growth control. In addition to the EGF receptor itself, several proteins rapidly phosphorylated on tyrosine have been detected in cells treated with EGF (168, 196,

197). In A431 cells these proteins include p36 and p81 also identified as substrates for the retroviral PTKs. A 35-kDa protein, which is phosphorylated in vitro by the EGF receptor, has been isolated on the basis of Ca^{2+}-dependent binding to A431 cell membranes (198). In mitogenically responsive fibroblasts, however, proteins of about 40 kDa are the major tyrosine phosphorylated proteins (196). Among these p42 is the most prominent. The unphosphorylated form of p42 is a minor soluble cytoplasmic protein (0.002% of total cell protein), which is highly conserved among vertebrates. Greater than 50% of p42 molecules are phosphorylated in response to EGF treatment, but no function has yet been ascribed to this protein. Its phosphorylation is also induced by PDGF (196). Some cell types that respond to PDGF but not to EGF as a mitogen show equal p42 phosphorylation in response to either factor, suggesting that tyrosine phosphorylation of p42 is not sufficient for mitogenesis. Whether tyrosine phosphorylation in general is necessary for a mitogenic response has not yet been established.

OTHER PROTEIN-TYROSINE KINASES

A number of other tyrosine phosphorylating activities have been reported. Among these are $p56^{lstra}$ (73, 199–201) and PTKs obtained from normal tissues (202–204). A soluble cytoplasmic PTK has been partially purified from rat liver (202). It has a native molecular weight of about 70 kDa. Based on the size of an endogenous substrate in this preparation it has been called p75. Several proteins are phosphorylated on tyrosine in membrane preparations from lymphoid cells (203, 204). B cells contain polypeptides of 61 and 55 kDa, and T cells proteins of 64 and 58 kDa (204). Since LSTRA cells are of thymic origin, it is likely that $p56^{lstra}$ is closely related to the 58-kDa protein seen in normal T cells. Based on a series of structural studies $p56^{lstra}$ appears to be distinct from all of the viral PTKs and the EGF receptor (73).

Another good candidate PTK is the v-*fms* gene product. The v-*fms* glycoprotein autophosphorylates on tyrosine in vitro (205). Moreover, it has sequence homology with the catalytic domain of the *src* gene family including a tyrosine homologous to Tyr 416 (46). The rest of the predicted v-*fms* protein structure is reminiscent of that of the EGF receptor, and the c-*fms* protein could well be a surface receptor for an unknown ligand. There are still doubts about the v-*fms* protein PTK activity because cells transformed by the v-*fms* gene do not show any alterations in the pattern of tyrosine phosphorylation unlike those transformed by $pp60^{v-src}$ (205). Parts of the v-*raf*, v-*mil*, and v-*mos* proteins have homology with the catalytic domain of the *src* family, but preliminary indications are that these proteins may phosphorylate serine and threonine rather than tyrosine (206, 207).

GENERAL PROPERTIES OF PROTEIN-TYROSINE KINASES

Given the close relationship between their catalytic domains one would anticipate that the PTKs share many properties. They all exhibit a tight specificity for tyrosine, and do not phosphorylate serine, threonine, or hydroxyproline in synthetic peptide substrates (208–210). Many, but not all, of the enzymes show an apparent preference for Mn^{2+} over Mg^{2+} for phosphorylation of both protein and peptide substrates, but this could be due indirectly to Mn^{2+} inhibition of phosphatases. $pp60^{v-src}$ and $P120^{gag-abl}$ will use other divalent cations such as Co^{2+} and Zn^{2+}. Ca^{2+}, however, is inactive in all cases (3, 211, 212). ATP is the preferred nucleotide. In some instances GTP can be utilized, but it always has a higher K_m than ATP. Turnover numbers range from 2–3 $\mu mol/min/\mu mol$ for $pp60^{v-src}$ (213) to 170 $\mu mol/min/\mu mol$ for the bacterially synthesized v-abl fragment (43), with the latter value being close to that for A-kinase. Except for the insulin and IGF-1 receptors, the PTKs all have a single polypeptide chain which is active in the monomeric state.

Very little is known about catalytic mechanism, but given the relatedness, the PTKs seem likely to be similar to A-kinase. Unfortunately, even though large quantities of enzyme are available, mechanistic investigations with A-kinase have not proceeded very far. Some details of the conformation required for the peptide substrate, the chirality of ATP, and its metal ion binding have been reported for A-kinase, but it is not yet ruled out that an enzyme-phosphate intermediate exists transiently although this seems unlikely. Progress would be facilitated by the knowledge of a three-dimensional structure, but it has only just become possible to grow suitable crystals for X-ray diffraction studies.

The viral PTKs all autophosphorylate at a homologous tyrosine in the heart of the catalytic domain. In every case this tyrosine is preceded by one or more acidic residues, which, based on studies of model synthetic peptide substrates, seem likely to be involved in recognition of target tyrosine residues (127, 203, 208, 214). A-kinase is phosphorylated at Thr 196 in the corresponding position (36). Thr 196 is preceded by basic residues, a feature common to A-kinase phosphorylation sites (103). An attractive speculation is that protein kinases may all contain a canonical phosphorylation site sequence at a corresponding position and that the type of hydroxyamino acid in this site could indicate the amino acid specificity of the protein kinase in question. For example, the predicted v-fms protein has a tyrosine preceded by acidic residues in this region consistent with its reported tyrosine autophosphorylating activity. In contrast the predicted v-mos and v-raf/mil gene products lack both tyrosine and acidic residues in this region, but have serines preceded by basic residues at this location instead. One might predict therefore that they are serine protein kinases, which is consistent with preliminary reports of such an activity associated with immunoprecipitates of these proteins (206, 207).

With the viral PTKs there are indications that autophosphorylation augments catalytic activity (84). Phosphorylation of the EGF receptor accompanies activation, but in this case the major autophosphorylation sites lie outside the catalytic domain (171). The activity of other PTKs may also be increased by phosphorylation of tyrosines outside the catalytic region. For example pp60^{v-src} is activated by phosphorylation of tyrosines near its NH_2 terminus. All of the PTKs are also modified by at least one serine kinase. Such serine phosphorylations could act as positive or negative modulators of activity.

Apart from the growth factors, physiological regulators of PTKs are not known. None of these enzymes displays dependence on cyclic nucleotides, nor on Ca^{2+} or calmodulin. In this regard p150^{c-abl}, the product of the c-abl gene, is of interest. It does not display detectable PTK activity in vitro (215), yet an altered form of this protein, p210^{c-abl}, autophosphorylates on tyrosine (216, 217). The difference between p210^{c-abl} and p150^{c-abl} has been placed tentatively at their NH_2 termini (216). Given the fact that P120$^{gag-abl}$ also differs from p150^{c-abl} at its NH_2 terminus, this suggests that p150^{c-abl} may have latent PTK activity which is regulated by an NH_2-terminal domain.

All of the PTKs found to date are located in the cytoplasm [for review see (218)]. P.Tyr, however, can be detected in proteins which purify with nuclei, and there could be nuclear PTKs. Many of the enzymes are associated with cytoplasmic membranes, in particular with the plasma membrane, and/or with the cytoskeleton. Of these some are integral membrane proteins [e.g. the growth factor receptors and P120$^{gag-abl}$ (219)], while others are peripherally associated with cytoplasmic membranes [e.g. pp60^{v-src} (57–63), P140$^{gag-fps}$ (220), and p56^{lstra} (199)]. Yet other PTKs appear to be soluble [e.g. p75 (202) and the c-fps protein (221)]. These varying subcellular locations may well dictate that the different enzymes have access to different sets of substrates.

Substrate Selection

All of the serine/threonine-protein kinases so far examined select serine or threonine residues in their substrates based at least in part on the primary sequence surrounding the target residue. In every case the canonical recognition sequence contains charged residues in the vicinity of the hydroxy amino acid. What can we say about the sequences surrounding phosphorylated tyrosines? The autophosphorylation sites in the viral PTKs have as common features a basic amino acid 7 residues to the NH_2-terminal side of the tyrosine and one or more acidic amino acids amongst the intervening amino acids (Figure 1). The similarity of these sites of phosphorylation probably reflects both the specificity of the relevant kinase and the fact that they are located within highly homologous regions of these proteins.

The requirements for specifically located charged residues within a phosphorylation site have usually been tested using synthetic peptides. Synthetic peptides based on the sequence surrounding Tyr 416 in $pp60^{v-src}$ can be phosphorylated by both viral and growth factor PTKs (127, 199, 203, 208). Comparison of the efficiency of phosphorylation of peptides in which the positions and number of acidic residues on the NH_2-terminal side of the tyrosine were varied suggests that acidic residues play some role in recognition. The lowest K_m values for peptides based on the sequence surrounding Tyr 416, however, are about 1 mM, which is approximately 100-fold higher than that of the optimal peptide substrates for the A-kinase.

One might argue that the appropriate peptide substrate would not be one based on the sequence of an autophosphorylation site but rather one based on that of a substrate phosphorylation site. To date the sequences of only two natural substrate phosphorylation sites are known, those in enolase and LDH (222). That in LDH is similar to the PTK autophosphorylation sites. The enolase phosphorylation site, however, differs significantly in having no basic or acidic amino acids within 10 residues on the NH_2-terminal side of the tyrosine. Instead there are glutamic acids at positions 1 and 4 on the COOH-terminal side. A synthetic peptide based on this sequence has a K_m of about 150 μM (222). This is a reasonably low value, but is still 100-fold worse than that for enolase itself. The two best peptide substrates for any PTK found to date are gastrin (223), with five consecutive glutamic acids upstream of the tyrosine and a peptide based on the NH_2 terminus of the red cell anion transport protein (Band 3) which also has multiple acidic residues upstream of the tyrosine (224). Both have K_m's in the 100 μM range. Acidic residues, however, are not essential for phosphorylation. For example [val^5]-angiotensin II, which has no acidic amino acids, has a K_m of about 0.9 mM (210), and both tyramine and N-acetyl-tyrosine can be phosphorylated albeit only when used at very high concentrations (209). In conclusion, while acidic residues in the vicinity of target tyrosines probably play some role, it seems likely that factors in addition to primary sequence are important for recognition.

Rather little kinetic analysis has been done using natural substrates. The K_m for enolase is about 1 μM (222). Two substrates for the EGF receptor PTK have K_m's about 100 nM (225). One of these is the progesterone receptor and the other a 94-kDa protein isolated from placenta. It is not clear whether either of these is a physiological substrate, but in both cases the stoichiometry of phosphorylation is good. The same is true for the phosphorylation of myosin light chain 1 (226), which does not appear to be a natural substrate.

From the different phenotypes induced by proteins with PTK activity one would surmise that the various enzymes have different substrate specificities. For instance this could explain why $pp60^{v-src}$, but not $pp60^{c-src}$, induces transformation, and why activation of the growth factor receptor PTKs correlates with mitogenesis and not transformation. So far, however, the most striking feature of the substrate specificities of the PTKs is their similarity rather than their difference, both in the cell and in vitro. It may not be possible to determine to what extent the various enzymes differ in substrate specificity until physiologically relevant substrates are identified.

Inhibitors of Protein-Tyrosine Kinases

A number of inhibitors of PTKs have been reported. The ATP analog, FSBA, irreversibly inactivates all PTKs tested so far through covalent modification of the ATP binding site (106, 107). Several halomethylektones (227), including tosyl-lysyl chloromethylketone (TLCK) (228) are also irreversible, but rather nonspecific inhibitors. Quercetin, a quinone that inhibits a number of ATPases, inhibits PTKs and G-kinase but not A-kinase (229, 230). Amiloride, a drug that blocks passive Na^+ uptake by cells, also inhibits PTKs. Its K_i for the EGF receptor kinase is 300 μM, and it acts competitively with ATP (231). The regulatory molecule Ap4A is a potent inhibitor of $pp60^{v\text{-}src}$, being less active against $pp60^{c\text{-}src}$ (9, 232). The dipeptide Tyr.Arg (kyotorphin) inhibits the EGF receptor PTK (233). Site-specific antibodies directed against parts of the catalytic domains of $pp60^{v\text{-}src}$ and $P120^{gag\text{-}abl}$ inhibit their PTK activities (111, 234). Several of the synthetic peptide substrates compete for phosphorylation of protein substrates (235). So far no suicide peptide substrates have been designed for the PTKs, although one has been devised for A-kinase. Random copolymers containing Ala, Glu, and Tyr serve as good substrates for most PTKs, whereas the ordered polymer $(Tyr.Glu.Ala.Gly)_n$ is a strong inhibitor (233).

LIPID PHOSPHORYLATION BY PROTEIN-TYROSINE KINASES

Preparations of two of the viral PTKs, $pp60^{v\text{-}src}$ and $P68^{gag\text{-}ros}$, have recently been shown to be capable of phosphorylating hydroxyl groups in glycerol (213, 236) and certain lipids, including phosphatidylinositol, 4-phosphophosphatidylinositol, and diacylglycerol (12, 13). With $pp60^{v\text{-}src}$, phosphorylation of glycerol has a K_m of greater than 100 mM (236), but the apparent K_m's for phosphorylation of phosphatidylinositol and diacylglycerol are 75 and 60 μM respectively (12). This ability may not be a feature unique to the PTKs, however, since A-kinase can also phosphorylate 4-phosphophosphatidylinositol (12) and phosphorylase kinase can phosphorylate phosphatidylinositol (237). The lipid phosphorylating activity is intriguing, but absolute proof that it is an intrinsic rather than adventitious property of the protein kinases in question is still lacking. Many of these enzymes contain covalently bound lipid and have to be worked with in detergent solution, thus increasing the risk of contaminating lipid kinase activities. It also is hard to envisage how the active site of the PTKs can specifically discriminate against serine and threonine in favor of the planar tyrosine ring and at the same time recognize the nonplanar inositol ring. Perhaps the best test of would be to ask whether PTKs synthesized in *E. coli,* which lack phosphatidylinositol, can phosphorylate lipids. Preliminary results with the bacterially synthesized fragment of the v-*abl* protein are negative in this regard (M. Whitman, G. Foulkes, personal communication), although it should be noted that this protein unlike P120$^{gag-abl}$ lacks an NH_2-terminal myristyl group.

If intrinsic to the PTKs, the significance of such a nonprotein phosphorylating activity could be manifold. Phosphorylation of phosphatidylinositol appears to be obligatory for its turnover, which produces diacylglycerol and triphosphoinositide. Both these molecules serve second messenger functions. Diacylglycerol activates C-kinase (238), while triphosphoinositide is believed to mobilize intracellular Ca^{2+} stores (239). Phosphorylation of proteins by C-kinase and events activated by increased Ca^{2+} could play important roles in both growth control and viral transformation. Where measured, however, the V_{max}'s for lipid phosphorylation by PTKs are very low, and it is not clear whether there would be sufficient activity in transformed cells to account for alterations in cellular phospholipid turnover.

CONCLUSIONS AND PROSPECTS

Although first identified in animal cells, tyrosine phosphorylation appears to be of ancient evolutionary origin. Evidence is accumulating for the existence in protozoans and other simple eukaryotes both of genes homologous to known PTK genes in higher organisms and of PTKs themselves. Whether PTKs occur in prokaryotes is moot, the issue being complicated by the presence of nucleotidyl tyrosine linkages. In general there are rather few phosphoproteins in prokaryotes and likewise little protein kinase activity, although protein phosphorylation has been shown to serve a regulatory function in at least one case. Protein phosphorylation seems to have become more common in parallel with the increase in complexity of eukaryotic organisms. Probably the first protein kinase originated from a nucleotide binding protein and had specificity for serine and/or threonine. Subsequently a PTK evolved from a protein-serine kinase. The expansion in the number of PTK genes may have coincided with the evolution of multicellular differentiated organisms. It is interesting that most PTKs are associated with the plasma membrane and thus are in a position to transduce signals from the outside of the cell. A clear prerequisite for a multicellular organism is a means of cell-cell signaling and the PTK nature of many of the growth factor receptors may be pertinent here.

The existence of PTKs in lower organisms opens up the possibility that one will be able to use genetics to analyze the roles of tyrosine phosphorylation in development and differentiation as well as for defining critical substrates. A stellar beginning has been made with *Drosophila* where several PTK genes have been cloned. For example the D*src* gene, which is homologous to the vertebrate c-*src* gene, has a complex transcription pattern that is regulated throughout development. Attempts are being made to determine whether the deletion of the D*src* gene is a lethal event. Rather little is known about the products of such *Drosophila* genes, but since their sequences are available we can anticipate rapid progress in this area. Nematodes and *Dictyostelium* are other simple organisms where genes related to PTKs have been detected.

Given the large number of PTK genes, the paucity of P.Tyr in cellular protein is perhaps surprising. Partly this stems from the rapid turnover of phosphate on tyrosine. In addition most of the PTKs are scarce proteins and not all of the PTK genes are expressed in the same cell at the same time. The rarity of tyrosine phosphorylation may reflect the fact that the substrates for the PTKs are themselves nonabundant proteins whose functions are regulatory. Indeed tyrosine phosphorylation could be reserved for the regulation of control pathways rather than of metabolic processes per se. The sensitivity of control systems relying on protein phosphorylation depends heavily on the turnover rate of the modulating phosphate groups. Tyrosine phosphorylation systems could have a high degree of sensitivity, since phosphates on tyrosine turn over rapidly, possibly in part because of the 'high-energy' nature of the phospho-diester bond. The role of phosphotyrosine phosphatases in the functions of tyrosine phosphorylation is therefore likely to be crucial. A beginning has been made in this area, but we need to know how many such enzymes there are and how are they regulated.

The cloning of further PTK genes will be a high priority, and the genes for the PDGF, insulin, and IGF-1 receptors are likely to be among the first obtained. The cloning of PTK genes should be assisted by the high degree of homology between the family members within the catalytic domain. We can also expect progress in the cloning of protein-serine and protein-threonine kinases. Indeed there are already cDNA clones for the catalytic subunit of A-kinase. The availability of cloned PTK genes will enable an extensive analysis of both catalytic and regulatory mechanisms by site-directed mutagenesis.

The purification and crystallization of PTKs should proceed apace particularly if the technical problems with bacterially synthesized proteins can be solved. A true understanding of catalytic mechanism and regulation ultimately requires a knowledge of the three-dimensional structure of these enzymes. Such information may also reveal the principles by which the different protein kinases select their substrates particularly with regard to the target amino acid. This ought to allow the rational design of protein kinase-specific suicide substrates. One may also be able to determine whether the catalytic center can recognize nonprotein substrates as well as peptide substrates.

The identification of crucial substrates for the PTKs and demonstration of functional effects of tyrosine phosphorylation will also receive considerable attention, especially in the areas of viral transformation and growth control. It will be important to find out whether the different PTKs differ absolutely in substrate specificity or whether this is dictated by subcellular location and differential regulation. The possible existence of regulatory molecules or other systems for regulation by covalent modification deserves consideration.

Our knowledge of tyrosine phosphorylation has come a long way in just five years. The next five years will be crucial in establishing whether or not tyrosine phosphorylation plays the central role in cellular physiology which the results so far have suggested.

ACKNOWLEDGMENTS

We thank our colleagues for their comments and for their generosity in providing unpublished material. This review is indirectly attributable to the laziness of one of us (T. H.) in not making up fresh electrophoresis buffer.

Literature Cited

1. Eckhart, W., Hutchinson, M. A., Hunter, T. 1979. *Cell* 18:925–33
2. Hunter, T., Sefton, B. M. 1980. *Proc. Natl. Acad. Sci. USA* 77:1311–15
3. Witte, O. N., Dasgupta, A., Baltimore, D. 1980. *Nature* 283:826–31
4. Sefton, B. M., Hunter, T., Beemon, K., Eckhart, W. 1980. *Cell* 20:807–16
5. Bishop, J. M. 1983. *Ann. Rev. Biochem.* 52:301–54
6. Ushiro, H., Cohen, S. 1980. *J. Biol. Chem.* 255:8363–65
7. Shilo, B.-Z., Weinberg, R. A. 1981. *Proc. Natl. Acad. Sci. USA* 78:6789–92
8. Simon, M. A., Kornberg, T., Bishop, J. M. 1983. *Nature* 302:837–39
9. Schartl, M., Barnekow, A. 1982. *Differentiation* 23:109–14
10. Foulkes, J. G. 1983. *Curr. Top. Microbiol. Immunol.* 107:163–80
11. Downward, J., Yarden, Y., Mayes, E., Scrace, G., Totty, N., et al. 1984. *Nature* 307:521–27
12. Sugimoto, Y., Whitman, M., Cantley, L. C., Erikson, R. L. 1984. *Proc. Natl. Acad. Sci. USA* 81:2117–21
13. Macara, I. G., Marinetti, G. V., Balduzzi, P. C. 1984. *Proc. Natl. Acad. Sci. USA* 81:2728–32
14. Collett, M. S., Purchio, A. F., Erikson, R. L. 1980. *Nature* 285:167–69

15. Levinson, A. D., Oppermann, H., Varmus, H. E., Bishop, J. M. 1980. *J. Biol. Chem.* 255:11973–80
16. Kawai, S., Yoshida, M., Segawa, K., Sugiyama, H., Ishizaki, R., Toyoshima, K. 1980. *Proc. Natl. Acad. Sci. USA* 77:6199–203
17. Naharro, G., Dunn, C. Y., Robbins, K. C. 1983. *Virology* 125:502–7
18. Neil, J. C., Ghysdael, J., Vogt, P. K. 1981. *Virology* 109:223–28
19. Feldman, R. A., Hanafusa, T., Hanafusa, H. 1980. *Cell* 22:757–65
20. Pawson, T., Guyden, J., Kung, T.-H., Radke, K., Gilmore, T., Martin, G. S. 1980. *Cell* 22:767–76
21. Barbacid, M., Lauver, A. V. 1981. *J. Virol.* 40:812–21
22. Van de Ven, W. J. M., Reynolds, F. H., Stephenson, J. R. 1980. *Virology* 101:185–97
23. Feldman, R. A., Wang, L.-H., Hanafusa, H., Balduzzi, P. C. 1982. *J. Virol.* 42:228–36
24. Czernilofsky, A. P., Levinson, A. D., Varmus, H. E., Bishop, J. M., Tischler, E., Goodman, H. M. 1980. *Nature* 287:193–203
25. Czernilofsky, A. P., Levinson, A. D., Varmus, H. E., Bishop, J. M., Tischler, E., Goodman, H. M. 1983. *Nature* 301:736–38

26. Takeya, T., Hanafusa, H. 1983. *Cell* 32:881–90
27. Schwartz, D., Tizard, R., Gilbert, W. 1983. *Cell* 32:853–69
28. Swanstrom, R., Parker, R. C., Varmus, H. E., Bishop, J. M. 1983. *Proc. Natl. Acad. Sci. USA* 80:2519–23
29. Kitamura, N., Kitamura, A., Toyoshima, K., Hirayama, Y., Yoshida, M. 1982. *Nature* 297:205–8
30. Naharro, G., Robbins, K. C., Reddy, E. P. 1984. *Science* 223:63–66
31. Shibuya, M., Hanafusa, H. 1982. *Cell* 30:787–95
32. Hampe, A., Laprevotte, I., Galibert, F., Fedele, L. A., Sherr, C. J. 1982. *Cell* 30:775–85
33. Reddy, E. P., Smith, M. J., Srinivasan, A. 1983. *Proc. Natl. Acad. Sci. USA* 80:3623–27
34. Neckameyer, W., Wang, L.-H. 1985. *J. Virol.* 53:879–84
35. Groffen, J., Heisterkamp, N., Shibuya, M., Hanafusa, H., Stephenson, J. R. 1983. *Virology* 125:480–86
36. Shoji, S., Parmelee, D. C., Wade, R. D., Kumar, S., Ericsson, L. H., et al. 1981. *Proc. Natl. Acad. Sci. USA* 78:848–51
37. Barker, W. C., Dayhoff, M. O. 1982. *Proc. Natl. Acad. Sci. USA* 79:2836–39
38. Collett, M. S., Erikson, R. L. 1978. *Proc. Natl. Acad. Sci. USA* 75:2021–24
39. Levinson, A. D., Oppermann, H., Levintow, L., Varmus, H. E., Bishop, J. M. 1978. *Cell* 15:561–72
40. Gilmer, T. M., Erikson, R. L. 1981. *Nature* 294:771–73
41. McGrath, J. P., Levinson, A. D. 1982. *Nature* 295:423–25
42. Wang, J. Y. J., Queen, C., Baltimore, D. 1982. *J. Biol. Chem.* 257:13181–84
43. Foulkes, J. G., Chow, M., Gorka, C., Frackelton, A. R., Baltimore, D. 1985. *J. Biol. Chem.* 260:In press
44. Kan, N. C., Flordellis, C. S., Mark, G. E., Duesberg, P. H., Papas, T. S. 1984. *Science* 223:813–16
45. Mark, G. E., Rapp, U. R. 1984. *Science* 224:285–89
46. Hampe, A., Gobet, M., Sherr, C. J., Galibert, F. 1984. *Proc. Natl. Acad. Sci. USA* 81:85–89
47. Van Beveren, C., Galleshaw, J. A., Jonas, V., Berns, A. J. M., Doolittle, R. F., et al. 1981. *Nature* 289:258–62
48. Yamamoto, T., Nishida, T., Miyajima, N., Kawai, S., Ooi, T., Toyoshima, K. 1983. *Cell* 35:71–78
49. Courtneidge, S. A., Bishop, J. M. 1982. *Proc. Natl. Acad. Sci. USA* 79:7117–21
50. Purchio, A. F., Erikson, E., Brugge, J. S., Erikson, R. L. 1978. *Proc. Natl. Acad. Sci. USA* 75:1567–71
51. Sefton, B. M., Beemon, K., Hunter, T. 1978. *J. Virol.* 28:957–71
52. Oppermann, H., Levinson, A. D., Levintow, L., Varmus, H. E., Bishop, J. M., Kawai, S. 1981. *Virology* 113:736–51
53. Brugge, J., Erikson, E., Erikson, R. L. 1981. *Cell* 25:363–72
54. Brugge, J., Darrow, D. 1982. *Nature* 295:250–53
55. Adkins, B., Hunter, T., Sefton, B. M. 1982. *J. Virol.* 43:448–55
56. Lipsich, L. A., Cutt, J. R., Brugge, J. S. 1982. *Mol. Cell. Biol.* 2:875–80
57. Willingham, M. C., Jay, G., Pastan, I. 1979. *Cell* 18:125–34
58. Courtneidge, S. A., Levinson, A. D., Bishop, J. M. 1980. *Proc. Natl. Acad. Sci. USA* 77:3783–87
59. Rohrschneider, L. R. 1980. *Proc. Natl. Acad. Sci. USA* 77:3514–18
60. Kreuger, J. G., Wang, E., Goldberg, A. R. 1980. *Virology* 101:25–40
61. Nigg, E. A., Sefton, B. M., Hunter, T., Walter, G., Singer, S. J. 1982. *Proc. Natl. Acad. Sci. USA* 79:5322–26
62. Parsons, S. J., McCarley, D. J., Ely, C. M., Benjamin, D. C., Parsons, J. T. 1984. *J. Virol.* 51:272–82
63. Resh, M. D., Erikson, R. L. 1985. *J. Cell Biol.* 100:409–17
64. Sefton, B. M., Trowbridge, I. S., Cooper, J. A., Scolnick, E. M. 1982. *Cell* 31:465–74
65. Garber, E. A., Krueger, J. G., Hanafusa, H., Goldberg, A. R. 1983. *Nature* 302:161–63
66. Buss, J. E., Sefton, B. M. 1985. *J. Virol.* 53:7–12
67. Buss, J. E., Kamps, M. P., Sefton, B. M. 1984. *Mol. Cell. Biol.* 4:2679–704
68. Cross, F. R., Garber, E. A., Pellman, D., Hanafusa, H. 1984. *Mol. Cell. Biol.* 4:1834–42
69. Levinson, A. D., Courtneidge, S. A., Bishop, J. M. 1981. *Proc. Natl. Acad. Sci. USA* 78:1624–28
70. Henderson, L. E., Krutzsch, H. C., Oroszlan, S. 1983. *Proc. Natl. Acad. Sci. USA* 80:339–43
71. Carr, S. A., Biemann, K., Shoji, S., Parmelee, D. C., Titani, K. 1982. *Proc. Natl. Acad. Sci. USA* 79:6128–31
72. Aitken, A., Cohen, P., Santikarn, S., Williams, D. H., Calder, A. G., et al. 1982. *FEBS Lett.* 150:314–18
73. Voronova, A. F., Buss, J. E., Patschinsky, T., Hunter, T., Sefton, B. M. 1984. *Mol. Cell. Biol.* 4:2705–13

74. Schultz, A., Oroszlan, S. 1984. *Virology* 133:431–37
75. Smart, J. E., Oppermann, H., Czernilofsky, A. P., Purchio, A. F., Erikson, R. L., Bishop, J. M. 1981. *Proc. Natl. Acad. Sci. USA* 78:6013–17
76. Collett, M. S., Erikson, E., Erikson, R. L. 1979. *J. Virol.* 29:770–81
77. Patschinsky, T., Hunter, T., Esch, F. S., Cooper, J. A., Sefton, B. M. 1982. *Proc. Natl. Acad. Sci. USA* 79:973–77
78. Sefton, B. M., Patschinsky, T., Berdot, C., Hunter, T., Elliott, T. 1982. *J. Virol.* 41:813–20
79. Neil, J. C., Ghysdael, J., Vogt, P. K., Smart, J. E. 1981. *Nature* 291:675–77
80. Hunter, T., Cooper, J. A. 1983. In *Evolution of Hormone/Receptor Systems. UCLA Symp. Mol. Cell. Biol.*, ed. R. A. Bradshaw, G. N. Gill, 6:369–82. New York: Liss
81. Karess, R. E., Hanafusa, H. 1981. *Cell* 24:155–64
82. Erikson, R. L., Collett, M. S., Erikson, E., Purchio, A. F. 1979. *Proc. Natl. Acad. Sci. USA* 76:6260–64
83. Purchio, A. F. 1982. *J. Virol.* 41:1–7
84. Graziani, Y., Erikson, E., Erikson, R. L. 1983. *J. Biol. Chem.* 258:6344–51
85. Collett, M. S., Wells, S. K., Purchio, A. F. 1983. *Virology* 128:285–97
86. Purchio, A. F., Wells, S. K., Collett, M. S. 1983. *Mol. Cell. Biol.* 3:1589–97
87. Sefton, B. M., Hunter, T., Beemon, K. 1980. *J. Virol.* 33:220–29
88. Snyder, M. A., Bishop, J. M., Colby, W. W., Levinson, A. D. 1983. *Cell* 32:891–901
89. Snyder, M. A., Bishop, J. M. 1984. *Virology* 136:375–86
90. Weinmaster, G., Zoller, M. J., Smith, M., Hinze, E., Pawson, T. 1984. *Cell* 37:559–68
91. Brown, D. J., Gordon, J. A. 1984. *J. Biol. Chem.* 259:9580–86
92. Collett, M. S., Belzer, S. K., Purchio, A. F. 1984. *Mol. Cell. Biol.* 4:1213–20
93. Courtneidge, S. A., Smith, A. E. 1983. *Nature* 303:435–39
94. Bolen, J. B., Thiele, C. J., Israel, M. A., Yonemoto, W., Brugge, J. S. 1984. *Cell* 38:767–77
95. Cross, F. R., Hanafusa, H. 1983. *Cell* 34:597–607
96. Roth, C. W., Richert, N. D., Pastan, I., Gottesman, M. M. 1983. *J. Biol. Chem.* 258:10768–73
97. Brugge, J. S., Darrow, D. 1984. *J. Biol. Chem.* 259:4550–57
98. Weinmaster, G., Hinze, E., Pawson, T. 1983. *J. Virol.* 46:29–41
99. Oppermann, H., Levinson, A. D., Varmus, H. E. 1981. *Virology* 108:47–70
100. Bryant, D., Parsons, J. T. 1982. *J. Virol.* 44:683–91
101. Parsons, J. T., Bryant, D., Wilkerson, V., Gilmartin, G., Parsons, S. J. 1984. In *Cancer Cells*, ed. G. F. Vande Woude, A. J. Levine, W. C. Topp, J. D. Watson, 2:37–42. New York: Cold Spring Harbor
102. Blithe, D. L., Richert, N. D., Pastan, I. H. 1982. *J. Biol. Chem.* 257:7135–42
103. Kemp, B. E., Graves, D. J., Benjamini, E., Krebs, E. G. 1977. *J. Biol. Chem.* 252:4888–94
104. Takio, K., Wade, R. D., Smith, S. B., Krebs, E. G., Walsh, K. A., Titani, K. 1984. *Biochemistry* 23:4207–18
105. Reimann, E. M., Titani, K., Ericsson, L. H., Wade, R. D., Fischer, E. H., Walsh, K. A. 1984. *Biochemistry* 23:4185–92
106. Zoller, M. J., Nelson, N. C., Taylor, S. S. 1981. *J. Biol. Chem.* 256:10837–42
107. Kamps, M. P., Taylor, S. S., Sefton, B. M. 1984. *Nature* 310:589–92
108. Wierenga, R., Hol, W. 1983. *Nature* 302:258–62
109. Bryant, D. L., Parsons, J. T. 1984. *Mol. Cell. Biol.* 4:862–66
110. Bryant, D., Parsons, J. T. 1983. *J. Virol.* 45:1211–16
111. Gentry, L. E., Rohrschneider, L. R., Casnellie, J. E., Krebs, E. G. 1983. *J. Biol. Chem.* 258:11219–28
112. Sefton, B. M., Walter, G. 1982. *J. Virol.* 44:467–74
113. Shalloway, D., Coussens, P. M., Yaciuck, Y. P. 1984. In *Cancer Cells*, ed. G. F. Vande Woude, A. J. Levine, W. C. Topp, J. D. Watson, 2:9–18. New York: Cold Spring Harbor
114. Ullrich, A., Coussens, L., Hayflick, J. S., Dull, T. J., Gray, A., et al. 1984. *Nature* 309:418–25
115. Kitamura, N., Yoshida, M. 1983. *J. Virol.* 46:985–92
116. Fukami, Y., Lipmann, F. 1983. *Proc. Natl. Acad. Sci. USA* 80:1872–76
117. Cooper, J. A., Hunter, T. 1984. *Curr. Top. Microbiol. Immunol.* 107:125–162
118. Carpenter, G., King, L., Cohen, S. 1978. *Nature* 276:409–10
119. Barnes, D. 1982. *J. Cell. Biol.* 93:1–4
120. Gill, G. N., Lazar, C. S. 1981. *Nature* 293:305–7
121. Hapgood, J., Libermann, T. A., Lax, I., Yarden, Y., Schreiber, A. B., et al. 1983. *Proc. Natl. Acad. Sci. USA* 80:6451–55
122. Fernandez-Pol, J. A. 1981. *Biochemistry* 20:3907–12

928   HUNTER & COOPER

123. Cohen, S., Fava, R. A., Sawyer, S. T. 1982. *Proc. Natl. Acad. Sci. USA* 79:6237–41
124. Cohen, S., Carpenter, G., King, L. E. Jr. 1980. *J. Biol. Chem.* 255:4834–42
125. Kudlow, J. E., Buss, J. E., Gill, G. N. 1981. *Nature* 290:519–21
126. Chinkers, M., Cohen, S. 1981. *Nature* 290:516–19
127. Pike, L. J., Gallis, B., Casnellie, J. E., Bornstein, P., Krebs, E. G. 1982. *Proc. Natl. Acad. Sci. USA* 79:1443–47
128. King, L. E. Jr., Carpenter, G., Cohen, S. 1980. *Biochemistry* 19:1524–28
129. Cohen, S., Ushiro, H., Stoscheck, C., Chinkers, M. 1982. *J. Biol. Chem.* 257:1523–31
130. Das, M., Miyakawa, T., Fox, C. F., Pruss, R. M., Aharonov, A., Herschman, H. R. 1977. *Proc. Natl. Acad. Sci. USA* 74:2790–94
131. Buhrow, S. A., Cohen, S., Staros, J. V. 1982. *J. Biol. Chem.* 257:4019–22
132. Ek, B., Westermark, B., Wasteson, A., Heldin, C.-H. 1982. *Nature* 295:419–20
133. Nishimura, J., Huang, J. S., Deuel, T. F. 1982. *Proc. Natl. Acad. Sci. USA* 79:4303–7
134. Kasuga, M., Karlsson, F. A., Kahn, C. R. 1981. *Science* 215:185–87
135. Petruzzelli, L. M., Ganguly, S., Smith, C. J., Cobb, M. H., Rubin, C. S., Rosen, O. M. 1982. *Proc. Natl. Acad. Sci. USA* 79:6792–96
136. Avruch, J., Nemenoff, R. A., Blackshear, P. J., Pierce, M. W., Osathanondh, R. 1982. *J. Biol. Chem.* 257:15162–66
137. Shia, M. A., Pilch, P. F. 1983. *Biochemistry* 22:717–21
138. Jacobs, S., Kull, F. C., Earp, H. S., Svoboda, M. E., Van Wyk, J. J., Cuatrecasas, P. 1983. *J. Biol. Chem.* 258:9581–84
139. Rubin, J. B., Shia, M. A., Pilch, P. F. 1983. *Nature* 305:438–40
140. Heldin, C.-H., Ek, B., Rönnstrand, L. 1983. *J. Biol. Chem.* 258:10054–161
141. Frackelton, A. R., Tremble, P. M., Williams, L. T. 1984. *J. Biol. Chem.* 259:7909–15
142. Glenn, K., Bowen-Pope, D. F., Ross, R. 1982. *J. Biol. Chem.* 257:5172–76
143. Pilch, P. F., Czech, M. P. 1980. *J. Biol. Chem.* 255:1722–31
144. Massague, J., Guillette, B. J., Czech, M. P. 1981. *J. Biol. Chem.* 256:2122–25
145. Kasuga, M., Fujita-Yamaguchi, Y., Blithe, D. L., Kahn, C. R. 1983. *Proc. Natl. Acad. Sci. USA* 80:2137–41
146. van Obberghen, E., Rossi, B., Kowalski, A., Gazzano, H., Ponzio, G. 1983. *Proc. Natl. Acad. Sci. USA* 80:945–49
147. Roth, R. A., Cassell, D. J. 1983. *Science* 219:299–301
148. Xu, Y.-H., Ishii, S., Clark, A. J. L., Sullivan, M., Wilson, R. K., et al. 1984. *Nature* 309:806–10
149. Lin, C. R., Chen, W. S., Kruijer, W., Stolarsky, L. S., Weber, W., et al. 1984. *Science* 224:843–48
150. Weber, W., Gill, G. N., Speiss, J. 1984. *Science* 224:294–97
151. Weber, W., Bertics, P. J., Gill, G. N. 1984. *J. Biol. Chem.* 259:14631–36
152. Carlin, C. R., Knowles, B. B. 1984. *J. Biol. Chem.* 259:7902–8
153. Decker, S. J. 1984. *Mol. Cell. Biol.* 4:571–75
154. Mayes, E. L. V., Waterfield, M. D. 1984. *EMBO J.* 3:531–37
155. Cooper, J. A., Scolnick, E. M., Ozanne, B., Hunter, T. 1983. *J. Virol.* 48:752–64
156. Stoscheck, C. M., Carpenter, G. 1984. *J. Cell Biol.* 98:1048–53
157. Carpenter, G., Cohen, S. 1977. *Biochem. Biophys. Res. Commun.* 79:545–52
158. Shimizu, N., Behazdian, M. A., Shimizu, Y. 1980. *Proc. Natl. Acad. Sci. USA* 77:3600–4
159. Davis, R. L., Grosse, V. A., Kucherlapati, R., Bothwell, M. 1980. *Proc. Natl. Acad. Sci. USA* 77:4188–92
160. Spurr, N. K., Solomon, E., Jansson, M., Sheer, D., Goodfellow, P. N., et al. 1984. *EMBO J.* 3:159–63
161. Linsley, P. S., Fox, C. F. 1980. *J. Supramol. Struct.* 14:461–71
162. Cassel, D., Glaser, L. 1982. *J. Biol. Chem.* 257:9845–48
163. Gates, R. E., King, L. E. Jr. 1982. *Mol. Cell. Endocrinol.* 27:263–76
164. King, L. E. Jr., Gates, R. E. 1985. *Biochemistry* 24:In press
165. Yeaton, R. W., Lipari, M. T., Fox, C. F. 1983. *J. Biol. Chem.* 258:9254–61
166. Basu, M., Biswas, R., Das, M. 1984. *Nature* 311:477–80
167. Chinkers, M., Brugge, J. S. 1984. *J. Biol. Chem.* 259:11534–42
168. Hunter, T., Cooper, J. A. 1981. *Cell* 24:741–52
169. Cochet, C., Gill, G. N., Meisenhelder, J., Cooper, J. A., Hunter, T. 1984. *J. Biol. Chem.* 259:2553–58
170. Iwashita, S., Fox, C. F. 1984. *J. Biol. Chem.* 259:2559–67
171. Downward, J., Parker, P., Waterfield, M. D. 1984. *Nature* 311:483–85
172. Gates, R. E., King, L. E. 1982. *Biochem. Biophys. Res. Commun.* 105:57–66

173. Rosen, O. M., Herrera, R., Olowe, Y., Petruzzelli, L. M., Cobb, M. H. 1983. *Proc. Natl. Acad. Sci. USA* 80:3237–40
174. Ghosh-Dastidar, P., Fox, C. F. 1984. *J. Biol. Chem.* 259:3864–69
175. Rackoff, W. R., Rubin, R. A., Earp, H. S. 1984. *Mol. Cell. Endocrinol.* 34:113–19
176. Castagna, M., Takai, Y., Kaibuchi, K., Sano, K., Kikkawa, U., Nishizuka, Y. 1982. *J. Biol. Chem.* 257:7847–51
177. Niedel, J. E., Kuhn, L. J., Vandenbark, G. R. 1983. *Proc. Natl. Acad. Sci. USA* 80:36–40
178. Leach, K. L., James, M. L., Blumberg, P. M. 1983. *Proc. Natl. Acad. Sci. USA* 80:4208–12
179. Parker, P. J., Stabel, S., Waterfield, M. D. 1984. *EMBO J.* 3:953–59
180. Hunter, T., Ling, N., Cooper, J. A. 1984. *Nature* 311:480–83
181. Decker, S. J. 1984. *Mol. Cell. Biol.* 4:1718–24
182. Friedman, B., Frackelton, A. R., Ross, A. H., Connors, J. M., Fujiki, H., et al. 1984. *Proc. Natl. Acad. Sci. USA* 81:3034–38
183. Lee, L.-S., Weinstein, I. B. 1978. *Science* 202:313–15
184. Shoyab, M., DeLarco, J. E., Todaro, G. J. 1979. *Nature* 279:387–91
185. Salomon, D. S. 1981. *J. Biol. Chem.* 256:7958–66
186. King, A. C., Cuatrecasas, P. 1982. *J. Biol. Chem.* 257:3053–60
187. Sawyer, S. T., Cohen, S. 1981. *Biochemistry* 20:6280–86
188. Habenicht, A. J. R., Glomset, J. A., King, W. C., Nist, C., Mitchell, C. D., Ross, R. 1981. *J. Biol. Chem.* 256:12329–35
189. Wharton, W., Leof, E., Pledger, W. J., O'Keefe, E. J. 1982. *Proc. Natl. Acad. Sci. USA* 79:5567–71
190. Bowen-Pope, D., Dicorleto, P. E., Ross, R. 1983. *J. Cell Biol.* 96:679–83
191. Davis, R. J., Czech, M. P. 1984. *Proc. Natl. Acad. Sci. USA* 81:7797–801
192. Takeyama, S., White, M. F., Lauris, V., Kahn, C. R. 1985. *J. Cell. Biochem.* In press
193. May, W. S., Jacobs, S., Cuatrecasas, P. 1984. *Proc. Natl. Acad. Sci. USA* 81:2016–20
194. Erneux, C., Cohen, S., Garbers, D. L. 1983. *J. Biol. Chem.* 258:4137–42
195. Mroczykowski, B., Mosig, G., Cohen, S. 1984. *Nature* 309:270–73
196. Cooper, J. A., Bowen-Pope, D., Raines, E., Ross, R., Hunter, T. 1982. *Cell* 31:263–73
197. Frackelton, A. R., Ross, A. H., Eisen, H. N. 1983. *Mol. Cell. Biol.* 3:1343–52
198. Fava, R., Cohen, S. 1984. *J. Biol. Chem.* 259:2636–45
199. Casnellie, J. E., Harrison, M. L., Pike, L. J., Hellstrom, K. E., Krebs, E. G. 1982. *Proc. Natl. Acad. Sci. USA* 79:1443–47
200. Casnellie, J. E., Harrison, M. L., Hellstrom, K. E., Krebs, E. G. 1983. *J. Biol. Chem.* 258:10738–42
201. Gacon, G., Gisselbrecht, S., Piau, J. P., Boissel, J. P., Tolle, J., Fisher, S. 1982. *EMBO J.* 1:1579–82
202. Wong, T.-W., Goldberg, A. R. 1983. *Proc. Natl. Acad. Sci. USA* 80:2529–33
203. Swarup, G., Dasgupta, J. D., Garbers, D. L. 1983. *J. Biol. Chem.* 258:10341–47
204. Earp, H. S., Austin, K. S., Buessow, S. C., Dy, R., Gillespie, G. Y. 1984. *Proc. Natl. Acad. Sci. USA* 81:2347–51
205. Barbacid, M., Beemon, K., Devare, S. G. 1980. *Proc. Natl. Acad. Sci. USA* 77:5158–62
206. Kloetzer, W. S., Maxwell, S. A., Arlinghaus, R. B. 1983. *Proc. Natl. Acad. Sci. USA* 80:412–16
207. Moelling, K., Heimann, B., Beimling, P., Rapp, U. R., Sander, T. 1984. *Nature* 312:558–61
208. Hunter, T. 1982. *J. Biol. Chem.* 257:4843–48
209. Braun, S., Abdel Ghany, M., Racker, E. 1983. *Anal. Biochem.* 135:369–78
210. Wong, T.-W., Goldberg, A. R. 1983. *J. Biol. Chem.* 258:1022–25
211. Richert, N. D., Davies, P. J. A., Jay, G., Pastan, I. H. 1979. *J. Virol.* 31:695–706
212. Wong, T.-W., Goldberg, A. R. 1984. *J. Biol. Chem.* 259:8505–12
213. Richert, N. D., Blithe, D. L., Pastan, I. H. 1982. *J. Biol. Chem.* 257:7143–50
214. House, C., Baldwin, G. S., Kemp, B. E. 1984. *Eur. J. Biochem.* 140:363–67
215. Ponticelli, A. S., Whitlock, C. A., Rosenberg, N., Witte, O. N. 1982. *Cell* 29:953–60
216. Konopka, J. B., Watanabe, S. M., Witte, O. N. 1984. *Cell* 37:1035–42
217. Kloetzer, W., Kurzrock, R., Smith, L., Talpaz, M., Spiller, M., et al. 1985. *Virology* 140:230–38
218. Gentry, L., Rohrschneider, L. R. 1984. *Adv. Viral Oncology* 4:269–306
219. Schiff-Maker, L., Konopka, J. B., Davis, R. L., Watanabe, S. M., Ponticelli, A. S., et al. 1985. *J. Virol.* In press
220. Moss, P., Radke, K., Carter, C., Young, J., Gilmore, T., Martin, G. S. 1984. *J. Virol.* 52:557–65

221. Young, J. C., Martin, G. S. 1984. *J. Virol.* 52:913–18
222. Cooper, J. A., Esch, F., Taylor, S. S., Hunter, T. 1984. *J. Biol. Chem.* 259: 7835–41
223. Baldwin, G. S., Knesel, J., Monckton, J. M. 1983. *Nature* 301:435–37
224. Dekowski, S. A., Rybicki, A., Drickamer, K. 1983. *J. Biol. Chem.* 258:2750–53
225. Ghosh-Dastidar, P., Coty, W. A., Griest, R. E., Woo, D. D. L., Fox, C. F. 1984. *Proc. Natl. Acad. Sci. USA* 81: 1654–58
226. Gallis, B., Edelman, A. M., Casnellie, J. E., Krebs, E. G. 1983. *J. Biol. Chem.* 258:13089–93
227. Navarro, J., Abdel Ghany, M., Racker, E. 1982. *Biochemistry* 21:6138–44
228. Richert, N., Davies, P. J. A., Jay, G., Pastan, I. H. 1979. *Cell* 18:369–74
229. Graziani, Y., Erikson, E., Erikson, R. L. 1983. *Eur. J. Biochem.* 135:583–89
230. Cochet, C., Feige, J. J., Pirollet, F.,

Keramidas, M., Chambaz, E. M. 1982. *Biochem. Pharmacol.* 31:1357–61
231. Davis, R. J., Czech, M. P. 1985. *J. Biol. Chem.* 260:2543–51
232. Maness, P. F., Perry, M. E., Levy, B. T. 1983. *J. Biol. Chem.* 258:4055–58
233. Braun, S., Raymond, W. E., Racker, E. 1984. *J. Biol. Chem.* 259:2051–54
234. Konopka, J. B., Davis, R. L., Watanabe, S. M., Ponticelli, A. S., Schiff-Maker, L., et al. 1984. *J. Virol.* 51:223–32
235. Wong, T.-W., Goldberg, A. R. 1981. *Proc. Natl. Acad. Sci. USA* 78:7412–16
236. Graziani, Y., Erikson, E., Erikson, R. L. 1983. *J. Biol. Chem.* 258:2126–29
237. Varsanyi, M., Tölle, H.-G., Heilmeyer, L. M. G., Dawson, R. M. C., Irvine, R. F. 1983. *EMBO J.* 2:1543–48
238. Takai, Y., Kishimoto, A., Iwasa, Y., Kawahara, Y., Mori, M., Nishizuka, Y. 1979. *J. Biol. Chem.* 254:3692–95
239. Streb, H., Irvine, R. F., Berridge, M. J., Schulz, I. 1983. *Nature* 306:67–69

Ann. Rev. Biochem. 1985. 54:931–76

PROTEIN KINASES IN THE BRAIN

Angus C. Nairn, Hugh C. Hemmings, Jr., and Paul Greengard

Laboratory of Molecular and Cellular Neuroscience, The Rockefeller University, 1230 York Ave., New York, NY 10021

CONTENTS

PERSPECTIVES AND SUMMARY

The molecular mechanisms by which cells communicate with each other are of central importance to the biology of multicellular organisms. Work by Sutherland and his colleagues in the late 1950s on the hormonal control of glycogen metabolism in the mammalian liver revealed that epinephrine and glucagon stimulated glycogenolysis by increasing the synthesis of the intracellular second messenger cyclic AMP (cAMP) (1–5). Subsequently, Krebs and his colleagues discovered a protein kinase in skeletal muscle that was activated by

931

0066-4154/85/0701-0931$02.00

physiological increases in the levels of cAMP, and demonstrated that epinephrine stimulated glycogenolysis through activation of this protein kinase (6). The mechanisms of action of a number of additional extracellular signals have since been found to involve regulation of the formation of cAMP and thereby control of its target enzyme cAMP-dependent protein kinase (7). Other second messengers, including cyclic GMP (cGMP) (8, 9) and calcium ion (Ca^{2+}) (10–12), and some extracellular signals themselves (13–16), have also been found to regulate the activity of distinct classes of protein kinases. Protein kinases can be divided into two classes, those that are regulated by known second messengers (e.g. cAMP, cGMP, and Ca^{2+}) and those that are not (the Ca^{2+} - and cyclic nucleotide–independent protein kinases). The regulation of the state of phosphorylation of specific substrates by a variety of protein kinases appears to be a general mechanism by which many hormones, neurotransmitters, and other extracellular signals produce their physiological responses in specific target cells (7, 17, 18).

The brain is made up of many distinct cell types that are both anatomically and biochemically specialized for intercellular communication. The biochemical specialization includes a variety of hormone and neurotransmitter regulated protein phosphorylation systems containing high concentrations of adenylate cyclases (19, 20), guanylate cyclases (21, 22), phosphodiesterases (20, 23), protein kinases (24), protein kinase substrates (18), and phosphoprotein phosphatases (25). Many of these protein phosphorylation systems appear to be primarily neuronal, some being highly concentrated in synaptic junctions, the part of the nerve cell anatomically specialized for intercellular communication. Extensive evidence has implicated protein phosphorylation as the molecular mechanism by which many extracellular signals regulate intracellular functions in the brain including intermediary metabolism, neuronal excitability, neurotransmitter biosynthesis and release, and neuronal growth, differentiation, and morphology [for review see (18)].

The study of the physiological role and regulation of protein kinases in the brain, and the identification and characterization of their specific substrates, are active areas of research, and have already provided valuable information on the molecular mechanisms involved in the mediation or modulation of many neurophysiological processes (18). Although the properties of the protein kinases of the brain are generally similar to those of non-nervous tissues, there are important differences, especially in their relative concentrations and in their cellular and subcellular distributions. Furthermore, the brain appears to contain a greater diversity of substrates for most protein kinases than do non-nervous tissues (18, 26, 27). Since the general properties of protein kinases have been reviewed (9, 12, 17, 28, 29), the focus of this review will be on those aspects of protein kinases that are relevant to neuronal function. A discussion of substrate proteins for brain protein kinases will be included, as will a brief discussion of protein phosphatases in the brain.

CYCLIC NUCLEOTIDE–DEPENDENT PROTEIN KINASES

Mammalian brain contains two distinct cyclic nucleotide–dependent protein kinases, namely cAMP-dependent protein kinase and cGMP-dependent protein kinase [for review see (30)]. Using photoaffinity analogs of cAMP and cGMP, these enzymes have been shown to be the principal intracellular receptor proteins for these second messengers in the brain and other tissues (31, 32). Most, if not all, of the actions of both cAMP and cGMP are believed to be mediated through the activation of these protein kinases and the concomitant phosphorylation of specific substrate protein effectors (7). Although structural homologies between cAMP-dependent protein kinase and cGMP-dependent protein kinase suggest their evolution from a common ancestral protein kinase (33, 34), there are major differences in their substrate specificities, sensitivities to and mechanisms of activation by cyclic nucleotides, sensitivities to modulatory proteins, immunochemical cross-reactivity, and tissue distributions (see below). These differences support the hypothesis that these two protein kinases are involved in the regulation of distinct physiological processes.

Cyclic AMP–Dependent Protein Kinase

Following the discovery of cAMP-dependent kinase in rabbit skeletal muscle (6), the widespread distribution of this protein kinase throughout various tissues and phyla of the animal kingdom was established (35). This led to the hypothesis (35) that all of the diverse effects of cAMP on cell function are mediated through the activation of this enzyme, which has been shown to be the principal intracellular receptor for cAMP (31–33).

SUBUNIT STRUCTURE   cAMP-dependent protein kinase exists as a tetramer composed of two types of dissimilar subunits, the regulatory (R) subunit $(M_r = 49,000-55,000)$[1] and the catalytic (C) subunit $(M_r = 40,000)$. In the absence of cAMP, the inactive holoenzyme tetramer consists of two R subunits joined by disulfide bonds, bound to two C subunits (R_2C_2) (36–43). The binding of cAMP to the R subunits of the inactive holoenzyme lowers their affinity for the C subunits and leads to the dissociation from the holoenzyme of the two free C subunits expressing phosphotransferase activity (44, 45). Each R subunit contains two binding sites for cAMP, which activate the kinase synergistically and exhibit positively cooperative cAMP binding (44, 46–51).

[1]Throughout this article M_r is used to define the relative molecular weight determined by SDS-polyacrylamide gel electrophoresis. The actual molecular weights of the R and C subunits of cAMP-dependent protein kinase (67–70) and of cGMP-dependent protein kinase (34) have been determined by amino acid sequencing.

POLYMORPHISM   Two isozymes of cAMP-dependent protein kinase exist in brain and other tissues that share identical C subunits but differ in their R subunits (37, 52–57). These isozymes are classified as type I and type II cAMP-dependent protein kinases, and are distinguished by containing either regulatory subunit R_I (M_r=49,000) or R_{II} (M_r=52,000–55,000), respectively. Type I and type II cAMP-dependent protein kinases differ in several biochemical properties due to differences between R_I and R_{II} (36, 54), including their elution from DEAE-cellulose (52, 53, 58), affinities for cAMP, cAMP analogs, and Mg^{2+}ATP (49, 54, 58a), interactions with C subunit (58b), apparent molecular weights (54–56), peptide maps (56, 59), spectroscopic characteristics (60), ability to be autophosphorylated (39–43), antigenic determinants (61–66), tissue distributions (31, 53, 62), and amino acid sequences (52, 56, 67–70).

There is evidence for differences between the type II regulatory subunits from different tissues. Although brain type II cAMP-dependent protein kinase resembles cardiac muscle type II cAMP-dependent protein kinase in many of its physico-chemical properties (71), antibodies prepared against R_{II} from each of these tissues do not exhibit complete cross-reactivity with the kinase from the other tissue. These immunochemical differences between the R subunits of brain and cardiac muscle type II cAMP-dependent protein kinase are present in each of several mammalian species (61, 71–74). Further evidence for a distinct brain-specific form of R_{II} has been obtained by comparative tryptic peptide mapping studies of R_{II} purified from brain, skeletal muscle, and cardiac muscle (75, 76). Furthermore, the brain and muscle forms of type II cyclic AMP–dependent protein kinase may be functionally distinct, since the brain form of R_{II} appears to interact differently with the C subunit than does the muscle form of R_{II} (58b). Brain R_{II} also differs from muscle R_{II} in its electrophoretic mobility in SDS-polyacrylamide gels following autophosphorylation (76). These differences between brain and muscle type II cAMP-dependent protein kinase may reflect adaptations of the brain enzyme for neuron-specific processes (7, 79). No tissue differences have been reported for type I cyclic AMP-dependent protein kinase.

There is evidence for heterogeneity of the regulatory subunits of cAMP-dependent protein kinase within brain. Labeling of mammalian brain cAMP-binding proteins with the photoaffinity label 8-N_3-cAMP followed by isoelectric focusing indicates that both type I and type II cAMP-dependent protein kinases exhibit isoelectric variants in their R subunits (71, 78, 80). In addition, more than one form of brain R_{II} is observed following SDS-polyacrylamide gel electrophoresis. Such heterogeneity may result from the existence of distinct isozymic forms of R_I and R_{II} or from posttranslational modification, for example, phosphorylation. The heterogeneity in R_I and R_{II} in brain may reflect the diversity of cell types and specialized neuronal functions involving cAMP-regulated protein phosphorylation in the nervous system.

CELLULAR AND SUBCELLULAR LOCALIZATION   The brain contains relatively high levels of cAMP-dependent protein kinase (24), which is present at similar levels in all regions of the brain examined (81–83). Rat brain contains both the type I and type II forms of cAMP-dependent protein kinase, although there is three- to fourfold more of the type II enzyme (53, 54). The type II cAMP-dependent protein kinase appears to be predominantly of neuronal origin and the type I cAMP-dependent protein kinase predominantly non-neuronal since injection of the neurotoxin kainic acid into rat striatum causes a substantial reduction in type II cAMP-dependent protein kinase while only moderately reducing type I cAMP-dependent protein kinase (84). A neuronal localization for type II cAMP-dependent protein kinase is supported by im-munocytochemical evidence obtained using antibodies directed against R_{II}: R_{II} was found to be highly concentrated in neurons, but was present at different concentrations and in different subcellular compartments in different types of neurons (85, 85a). A non-neuronal localization of type I cAMP-dependent protein kinase is supported by photoaffinity labeling experiments that show that myelin contains predominantly R_I (85b).

Although cAMP-dependent protein kinase is largely soluble in most tissues (62), the brain contains approximately equal amounts of both soluble and particulate enzyme (24, 62, 71, 86–88). The particulate and cytosolic forms of the type II enzyme exhibit virtually identical biochemical and immunological properties (66, 71, 72), suggesting either that the same enzyme can exist in both subcellular compartments or that the partitioning is due to a minor, as yet undetected difference. The association of cAMP-dependent protein kinase with the particulate fraction appears to be determined by the properties of the R subunits and not those of the C subunit since cAMP can release C subunit, but not R subunit, from the particulate fraction (89, 90). cAMP-dependent protein kinase is present in all primary subcellular fractions of rat brain, with the highest specific activities being in the cytosol and synaptic membrane-enriched fractions (81). It is enriched in highly purified synaptic structures, including synaptic junctions, synaptic plasma membranes, and postsynaptic densities (91–93), suggesting that cAMP-dependent protein kinase plays an important role in synaptic function (86).

In several tissues, including bovine brain (86), endogenous cAMP-dependent protein kinase can be detected in the nuclear fraction. In addition, several hormones and neurotransmitters that elevate cellular cAMP levels have been reported to induce the translocation of the C subunit of cAMP-dependent protein kinase from the cytoplasm to the nucleus [for review see (94)]. Furthermore, the β-adrenergic receptor agonist isoproterenol, which appears to induce the translocation of C subunit into the nucleus of C6 glioma cells (95, 96), alters the phosphorylation of histones (96, 97) and induces the de novo synthesis of several proteins in these cells (97, 98). There is also evidence for the translocation of the R_{II} subunit of cAMP-dependent protein kinase into the nucleus (99, 100). These observations suggest that some of the effects of cAMP-dependent protein kinase on the regulation of gene expression may be mediated by the translocation of C and/or R subunits into the nucleus.

MACROMOLECULAR INTERACTIONS   Recent findings have demonstrated the high affinity binding of type II cAMP-dependent protein kinase to specific cytoplasmic proteins mediated by its regulatory subunit R_{II}. Apparently specific interactions have been detected between R_{II} and neuronal microtubule-associated protein 2 (MAP-2) (101, 102), calcineurin (103), and several other proteins with unknown functions (104, 105). Some of these interactions appear to be specific for R_{II} from brain as opposed to cardiac muscle (104, 105), indicating that the tissue-specific differences between brain and cardiac muscle R_{II} are also evident in their interactions with other proteins. The interactions of R_{II} with other proteins may be involved either in concentrating type II cAMP-dependent protein kinase near specific substrates and subcellular compartments where it plays important physiological roles, or in regulating the functions of other proteins in addition to the C subunit. The evidence for the interaction between R_{II} and MAP-2 will be discussed further as it has been most thoroughly studied.

A significant fraction ($\approx$30%) of cAMP-dependent protein kinase activity found in brain cytosol cofractionates with MAP-2 (101, 102, 105), a high molecular weight protein ($M_r \approx 270,000$) which copurifies with brain microtubules and which is a major substrate for cAMP-dependent protein kinase (106, 107). The high fraction of cAMP-dependent protein kinase associated with MAP-2 may be an overestimate of the amount found in situ resulting from posthomogenization interaction. The interaction between type II cAMP-dependent protein kinase and MAP-2 is mediated through R_{II} since C subunit can be released from the complex by cAMP. R_{II} is associated with the projection portion of MAP-2 (101), a large domain that projects from the surface of microtubules and is probably involved in the interactions of microtubules with other components of the cell. Both the projection portion and the assembly portion [a smaller domain responsible for the microtubule assembly promoting activity of MAP-2 (108)] are phosphorylated to a high stoichiometry by the associated type II cAMP-dependent protein kinase. In addition, MAP-2 is phosphorylated by both Ca^{2+}-dependent and Ca^{2+}- and cyclic nucleotide–independent protein kinases [(83, 107, 109, 110) and see below]. Phosphorylation of MAP-2 by cAMP-dependent protein kinase has been shown to inhibit its microtubule assembly promoting activity (111–113), and its interaction with actin (114, 115) and neurofilaments (116). It is possible that only a specific subclass of R_{II} binds to MAP-2 since there appear to be free binding sites on MAP-2 isolated from brain in spite of excess type II cAMP-dependent protein kinase (102, 117). The observed heterogeneity in R_{II} (see above) may be due to the presence of specific binding sites on R_{II} for MAP-2 and/or other proteins. The localization of MAP-2 and cAMP-dependent protein kinase in neurons, and within neurons their selective concentration in dendrites (117–122), suggest that the phosphorylation of MAP-2 may play an important role in neuronal function.

NEURONAL SUBSTRATES   The phosphorylation of specific substrates brought about by cAMP-dependent protein kinase represents the next step in the molecular pathway by which cAMP produces its biological responses. It follows that the identification and characterization of the specific substrate(s) phosphorylated in response to cAMP are important goals in the study of agents whose actions are mediated by cAMP. A large number of specific substrates for cAMP-dependent protein kinase exist in the mammalian brain (18, 82, 83). Two of the most prominent of these substrates are MAP-2 (discussed above) and the autophosphorylated form of R_{II}. Two more of the most prominent substrates, synapsin I and Protein III, and other neuronal proteins phosphorylated by cAMP-dependent protein kinase, will be discussed in this section.

Synapsin I   Synapsin I is a major substrate for cAMP-dependent protein kinase and is present in particulate fractions from brain (123, 124). It has been purified and characterized (125, 126), and consists of two closely related proteins, synapsin Ia and synapsin Ib (M_r=86,000 and 80,000, respectively), in a 1:2 molar ratio. A variety of studies indicate that synapsin I is neuron-specific (127–130), and that it is concentrated within virtually all nerve terminals, where it is associated primarily with the outer surface of synaptic vesicles (93, 127–132a). It represents about 6% of the total protein present in highly purified preparations of synaptic vesicles (132). A variety of neurotransmitters are capable of stimulating the phosphorylation of synapsin I in slices prepared from appropriate brain regions (18, 133–136). Phosphorylation of synapsin I by cAMP-dependent protein kinase occurs on a serine residue (site 1) which is also phosphorylated by Ca^{2+}/calmodulin-dependent protein kinase I, while two other serine residues (collectively termed site II) are phosphorylated by Ca^{2+}/calmodulin-dependent protein kinase II [(126, 137, 138) and see below]. Although the precise function of synapsin I is unknown, its presence at high concentrations on synaptic vesicles in virtually all nerve terminals suggests that it is involved in regulating some aspect of neurotransmitter release. This interpretation has recently received direct experimental support from microinjection experiments in the squid giant synapse (see below).

Protein III   Protein III consists of Protein IIIa and Protein IIIb ($M_r=74,000$ and 55,000, respectively), which appear to be closely related by several criteria (18, 139). Protein III is a major substrate for cAMP-dependent protein kinase in the nervous system, and appears to be a synaptic vesicle-associated protein that shares several properties with synapsin I (18, 139). It can be phosphorylated by several neurotransmitters acting through cAMP and Ca^{2+} (probably by $Ca^{2+}/$ calmodulin-dependent protein kinase I, see below) (18, 136, 140, 141), and may play an important role in synaptic function.

In addition to synapsin I and Protein III, a variety of other phosphoproteins have been identified as endogenous substrates for cAMP-dependent protein kinase in synaptic membrane and synaptic vesicle fractions, some of which are discussed in a recent review (142).

DARPP-32   DARPP-32 (*d*opamine- and cAMP-*r*egulated *p*hospho*p*rotein, $M_r=32,000$) is an acid-soluble and heat-stable cytosolic phosphoprotein which is phosphorylated in intact nerve cells in response to dopamine and to cAMP, and in broken cell preparations by endogenous cAMP-dependent protein kinase (143–146). DARPP-32 is phosphorylated on a single threonine residue (146, 147) surrounded by an amino acid sequence (148) very similar to those that surround the phosphorylatable threonine residues of phosphatase inhibitor-1 (149, 150) and of G-substrate [(147, 148) and see below]. Phosphorylated, but not dephosphorylated, DARPP-32, like phosphatase inhibitor-1, functions as a potent inhibitor of protein phosphatase-1[(151) and see below]. The regional distribution of DARPP-32 in rat brain, determined by both biochemical (145) and immunochemical [(143) and H. C. Hemmings, Jr. and P. Greengard, unpublished results] methods, has been found to closely parallel that of dopamine receptors coupled to adenylate cyclase [D_1-dopamine receptors, (152)]. Thus, DARPP-32 may be involved in mediating some of the transsynaptic effects of dopamine on dopaminoceptive cells possessing D_1-dopamine receptors by regulating the activity of protein phosphatase-1.

Voltage-Sensitive Sodium Channel   The voltage-sensitive Na^+ channel is responsible for the Na^+ current associated with the action potential in excitable membranes (153). Analysis of the polypeptides present in highly purified preparations of the Na^+ channel from rat brain (154–156) revealed an α subunit ($M_r \approx 260,000$), a $\beta 1$ subunit ($M_r \approx 39,000$), and a $\beta 2$ subunit ($M_r \approx 37,000$) in a molar ratio of $1:1:1$. The α subunit, which appears to play an essential role in Na^+ channel function (157), is phosphorylated by endogenous or exogenous cAMP-dependent protein kinase using either a partially purified preparation of the Na^+ channel (158) or a synaptosomal preparation (159) from rat brain. Phosphorylation appears to result in a reduction in neurotoxin-activated $^{22}Na^+$ influx in the synaptosomal preparation (159). Further studies will be necessary to establish the physiological significance of this phosphorylation.

Potassium Channels   Although none of the several classes of K^+ channels have yet been purified and biochemically characterized, considerable evidence obtained using a variety of invertebrate preparations suggests that they can be regulated by cAMP-dependent phosphorylation (160–162). Microinjection of the C subunit of cAMP-dependent protein kinase into *Aplysia* bag cell neurons mimics the effects of synaptic activation and of exogenous cAMP on these cells, apparently by decreasing the conductances of up to three distinct voltage-dependent K^+ channels (161). Similarly, microinjection of the C subunit of cAMP-dependent protein kinase mimics the effects of synaptic activation and of exogenous serotonin and cAMP in facilitating neurotransmitter release in response to nerve impulses in *Aplysia* sensory neurons; this effect is achieved by decreasing the conductance of serotonin-regulated K^+ channels (162, 163). Further evidence that cAMP-dependent protein kinase plays an essential role in these physiological processes is provided by the finding that microinjection of the cAMP-dependent protein kinase inhibitor abolishes the physiological response to synaptic activation, serotonin, or cAMP (164). Microinjection of the C subunit of cAMP-dependent protein kinase has also been found to increase the conductance of Ca^{2+}-dependent (165) K^+ channels in *Helix* neurons, (165) and to decrease the conductance of both early (I_A) and late (I_B) voltage-dependent K^+ channels in *Hermissenda* photoreceptor cells (166).

These findings provide convincing evidence that activation of cAMP-dependent protein kinase is both a necessary and sufficient step in the process by which K^+ channels are regulated by synaptic activation or neurotransmitter stimulation. It is not known whether the cAMP-regulation of K^+ channel function is mediated by phosphorylation of the channels themselves, or by phosphorylation of modulatory proteins which then modify the K^+ channels indirectly. Indirect regulation of Ca^{2+} transport by cAMP-regulated phosphorylation of phospholamban has been found in cardiac muscle sarcoplasmic reticulum [for review see (167)].

Nicotinic Acetylcholine Receptor   The nicotinic acetylcholine receptor is a neurotransmitter-regulated ion channel which mediates synaptic transmission at nicotinic cholinoceptive neurons, the neuromuscular junction, and in electric organs of various fishes [for reviews see (168–170)]. The purified receptor from *Torpedo* electric organ (M_r=250,000) consists of four types of subunit, α

(M_r=40,000), β (M_r=50,000), γ (M_r=60,000), and δ (M_r=65,000), in a molar ratio of $\alpha_2\beta\gamma\delta$. Both the γ and δ subunits are rapidly and stoichiometrically phosphorylated by cAMP-dependent protein kinase (171); however, the physiological effects of this phosphorylation are not known.

β-*Adrenergic Receptor*   The β-adrenergic receptor mediates the stimulatory effects of catecholamines on many tissues through activation of adenylate cyclase. The receptor binding subunit has been purified to apparent homogeneity from several sources (172, 173) and used to restore adrenergic responsiveness to receptor-deficient cells (174). Prolonged exposure of turkey erythrocytes to β-adrenergic agonists results in the desensitization of adenylate cyclase activity, which is correlated with phosphorylation of the β-adrenergic receptor (175, 176). Both desensitization and phosphorylation can be partially reproduced by exogenous cAMP, suggesting that the reaction is catalyzed by endogenous cAMP-dependent protein kinase. Recently, the tumor-promoting phorbol esters have also been found to induce phosphorylation and desensitization of the β-adrenergic receptor in turkey erythrocytes (177), which is probably mediated by Ca^{2+}/phospholipid-dependent protein kinase (protein kinase C).

Tyrosine Hydroxylase   Tyrosine hydroxylase (M_r=60,000), the first and rate-limiting enzyme in the biosynthesis of catecholamines (178), is phosphorylated and activated by cAMP-dependent protein kinase (179–184). This phosphorylation reaction may be part of the mechanism for the accelerated catecholamine synthesis observed in response to nerve impulse conduction or neurotransmitter stimulation in nervous tissue in vivo (185–187). Ca^{2+}-dependent phosphorylation also appears to be involved in the regulation of tyrosine hydroxylase activity (see below).

Neurofilaments   Neurofilaments, a class of intermediate filaments specific to neurons (188), are composed of three polypeptides, NF1 (M_r≈70,000), NF2 (M_r≈150,000), and NF3 (M_r≈200,000). cAMP-dependent protein kinase phosphorylates neurofilaments, predominantly on NF2 (189), and may thereby regulate the interaction between neurofilaments and the cytoskeleton.

Myelin Basic Protein   Myelin basic protein (M_r=14,000–18,000) is one of the major protein constituents of myelin [for review see (190)]. Although myelin basic protein is not phosphorylated by endogenous cAMP-dependent protein kinase in myelin fractions (191), purified myelin basic protein is phosphorylated by purified cAMP-dependent protein kinase (192, 192a, 192b). This discrepancy suggests that myelin basic protein may not be a substrate for cAMP-dependent protein kinase in vivo, possibly due to a lack of accessibility of the kinase to the substrate in the myelin membrane.

Guanylate Cyclase   Guanylate cyclase, the enzyme catalyzing the formation of cGMP, has been shown to be regulated by a variety of hormones and neurotransmitters (see below). cAMP has been shown to increase cGMP levels in several cell types (193), suggesting the involvement of cAMP-dependent phosphorylation. Purified cAMP-dependent protein kinase has been found to phosphorylate and activate purified rat brain guanylate cyclase (193), although in light of a contradictory finding employing impure guanylate cyclase (194), these results need to be confirmed.

Cyclic Nucleotide Phosphodiesterase   Purified Ca^{2+}/calmodulin-dependent cyclic nucleotide phosphodiesterase from bovine brain, an enzyme involved in the hydrolysis of cyclic nucleotides, has been shown to be phosphorylated by purified cAMP-dependent protein kinase (195). Phosphorylation did not appear to affect the activity of the enzyme; however, experiments employing liver membranes indicated that cAMP phosphodiesterase was activated by cAMP plus insulin, presumably mediated by cAMP-dependent phosphorylation (196).

GABA-Modulin   GABA-modulin ($M_r = 16,500$) has been reported to decrease the number of high affinity binding sites for the inhibitory neurotransmitter γ-aminobutyric acid (GABA), and to inhibit the ability of GABA to stimulate diazepam binding, in synaptic membrane preparations (197). Purified GABA-modulin from rat brain has been shown to be phosphorylated by purified cAMP-dependent protein kinase (198). Phosphorylation by cAMP-dependent protein kinase has been reported to decrease the ability of GABA-modulin to alter GABA binding sites in synaptic membranes (198).

REGULATION   A variety of hormones and neurotransmitters have been found to regulate the levels of cAMP in mammalian brain (18, 26), including dopamine (199), histamine (200), and serotonin (201). Adenylate cyclase, the membrane-bound enzyme which catalyzes the formation of cAMP, is controlled by receptor-mediated stimulation and inhibition [for review see (202)]. Stimulatory hormones and neurotransmitters act through a stimulatory GTP-binding regulatory protein (G_s), while inhibitory hormones and neurotransmitters act through an inhibitory GTP-binding protein (G_i), to regulate the activity of the catalytic subunit of adenylate cyclase. The brain also contains a form of adenylate cyclase which is regulated by Ca^{2+}/calmodulin (203, 204).

As discussed above, activation of cAMP-dependent protein kinase occurs when cAMP binds to the R subunits of the inactive holoenzyme, causing dissociation and release of free active C subunits. The primary mechanism of protein kinase inactivation involves reassociation of the R and C subunits following the hydrolysis of cAMP by cyclic nucleotide phosphodiesterase (205). The autophosphorylation of R_{II} decreases the rate of reassociation of type II cAMP-dependent protein kinase (38), enhancing its response to cAMP. Steroid hormones have been shown to alter the state of phosphorylation of R_{II}, an effect which may mediate some of the actions of these hormones (206, 207).

Although all normal tissues thus far examined contain equal concentrations of R and C subunits (62), some cultured neuronal cell lines contain excess R subunits (208, 209). Excess R subunits can be induced by treatment with cAMP derivatives or with agents which elevate cAMP levels, procedures which can also induce differentiation of these cells. Thus, changes in the total amount of R_I, an example of chronic regulation of type I cAMP-dependent protein kinase, have been observed in neuroblastoma and neuroblastoma × glioma hybrid cells following exposure to cAMP derivatives (208, 209). The functional role of excess R subunits is unknown, but may be to serve as a buffer for excess levels of cAMP in order to maintain low basal protein kinase activity.

Many tissues contain a heat-stable protein inhibitor of cAMP-dependent protein kinase, with that from rabbit skeletal muscle being the most extensively studied (210). The function of this inhibitor remains uncertain, although it may act to inhibit any free catalytic subunit present under basal conditions (211). Changes in the concentration of the cAMP-dependent protein kinase inhibitor, associated with drug treatments or with alterations in hormonal status, have been observed in several tissues including brain. Such changes may play a role in the chronic regulation of cAMP-dependent protein kinase activity (212–216).

Cyclic GMP–Dependent Protein Kinase

cGMP-dependent protein kinase, which was initially identified in invertebrate tissue (8), has been shown to be present in a number of mammalian tissues (217, 218). Significant differences between cGMP-dependent protein kinase and cAMP-dependent protein kinase have been described [(29, 30, 33, 219, 220) and see above] which indicate distinct physiological roles for these two enzymes.

SUBUNIT STRUCTURE   cGMP-dependent protein kinase, purified from bovine lung (221, 222) or heart (223), consists of a dimer of identical subunits, the amino acid sequence of which has recently been determined (34). Each subunit (M_r=74,000) contains a cGMP-binding domain, with two binding sites for cGMP (224, 225), and a catalytic domain. Upon binding of cGMP, a conformational change occurs in the enzyme which exposes the active catalytic domain (65, 226). cGMP-dependent protein kinase undergoes autophosphorylation which is stimulated by cGMP (224–227). Autophosphorylation of the enzyme has recently been found to increase the V_{max} of the phosphotransferase reaction (224, 225) in contrast to the results of earlier studies (226, 227). The enzyme has been partially purified from bovine cerebellum (228) and found to have properties similar to those of the kinase from other tissues. Radioimmunoassay (229, 230) and photoaffinity labeling (32, 229, 231) techniques have also demonstrated that the kinase from brain and other tissues is similar in all species examined.

CELLULAR AND SUBCELLULAR LOCALIZATION   In contrast to cAMP-dependent protein kinase, which is present in similar concentrations in most mammalian tissues (31, 62, 81), cGMP-dependent protein kinase has an uneven tissue distribution. Relatively high concentrations of the enzyme are found in lung, heart, smooth muscle, and intestine (218, 230, 232–234). Enzyme activity measurements (218), photoaffinity labeling (231), and radioimmunoassay (230, 232, 235) have been used to study the distribution of cGMP-dependent protein kinase in the mammalian brain. The highest concentration has been found in the cerebellum, with very low levels found in other major brain regions. cGMP-dependent protein kinase appears to be predominantly cytosolic in most tissues (236) including cerebellum (231, 237). Enzyme activity measurements (238) and photoaffinity labeling (231) have suggested that within the cerebellum cGMP-dependent protein kinase is concentrated in the Purkinje cells. A comprehensive immunocytochemical study of the distribution of cyclic GMP-dependent protein kinase in rat brain has allowed a detailed analysis of the architecture and projections of cerebellar Purkinje cells (235, 239). Immunoreactivity was found in the cytoplasm, dendritic trees, axons and axon terminals, but not in the nucleus, of all Purkinje cells.

The low concentrations of cGMP-dependent protein kinase found in regions of the brain other than the cerebellum are partly attributable to its presence in smooth muscle cells of blood vessels (231). Some of the enzyme found may also be due to its presence in neurons in which cGMP appears to play a role as a second messenger (240–242). Low levels of immunoreactivity have been suggested to be present in medium-sized spiny neurons of the caudatoputamen (243); however, cGMP-dependent protein kinase has not been conclusively detected in any neurons in the brain other than Purkinje cells.

NEURONAL SUBSTRATES   In vitro, cGMP-dependent and cAMP-dependent protein kinases show similar substrate specificities with respect to histones and synthetic peptides (29, 33, 244–246). Although many physiological substrates for cAMP-dependent protein kinase have been identified (see above), only a few specific physiological substrates for cGMP-dependent protein kinase have been found. These include proteins from nonvascular and vascular smooth muscle (233, 247) and from the intestinal brush border (234). In mammalian brain only one specific substrate, a protein termed G-substrate, has been described (237). Although the R subunit of type I cAMP-dependent protein kinase has been shown to be phosphorylated by cGMP-dependent protein kinase (248), it is not believed to be a physiological substrate.

G-substrate   G-substrate was detected in the cytosol of rabbit cerebellum as an endogenous substrate for cGMP-dependent, but not cAMP-dependent, protein kinase (237). The protein has been purified to homogeneity from cerebellum and extensively characterized (249–251). G-substrate is an acid-soluble and heat-stable protein with $M_r = 23,000$. cGMP-dependent protein kinase phosphorylates two threonine residues in G-substrate. The amino acid sequences surrounding these two phosphorylation sites exhibit a great degree of homology suggesting an internal gene duplication (251).

Endogenous protein phosphorylation (231), immunoprecipitation, peptide mapping, and radioimmunoassay (252) have been used to study the regional and cellular distribution of G-substrate in the mammalian brain. G-substrate was found to be highly concentrated in the cerebellum, where it was specifically localized to Purkinje cells (252). In addition, direct evidence for a Purkinje cell localization has been obtained by immunocytochemical studies using antibodies to G-substrate (P. Miller, C. C. Ouimet, A. C. Nairn, P. Greengard, unpublished results).

The phosphorylated form of G-substrate has been found to inhibit a protein phosphatase isolated from cerebellum (P. Simonelli, H.-C. Li, A. C. Nairn, P. Greengard, unpublished results). In addition, the phosphorylated form of G-substrate has been found to inhibit both protein phosphatase-1 and protein phosphatase-2A purified from rabbit skeletal muscle (A. C. Nairn, H. C. Hemmings, Jr., P. Greengard, unpublished results). The ability to act as a protein phosphatase inhibitor is one of several properties that G-substrate has in common with phosphatase inhibitor-1 (251, 252) and DARPP-32 (147, 148). G-substrate may therefore function as a Purkinje cell-specific protein phosphatase inhibitor.

REGULATION   cGMP-dependent protein kinase is activated by increases in intracellular cGMP, the formation of which is catalyzed by guanylate cyclase (21). A variety of hormones and neurotransmitters have been shown to increase cGMP (21, 22). In many cases, activation of guanylate cyclase appears to be interrelated with phosphatidylinositol turnover and Ca^{2+} mobilization (253–

259). The primary mechanism of inactivation of cGMP-dependent protein kinase results from hydrolysis of cGMP by cyclic nucleotide phosphodiesterase, of which there is a form specific for cGMP (259a). Considerable evidence suggests that cGMP plays an important role in the nervous system (18, 21, 22, 30, 220, 240–242). Studies of retina (260), superior cervical ganglion (261, 262), cerebellum (30, 220, 241, 242), and other brain regions (240–242) all implicate cGMP in neuronal function. The concentration of cGMP is altered by a variety of neurotransmitters, drugs, and other conditions that affect neuronal function (22, 242, 263). Within the mammalian brain, cGMP has been found in highest concentration in the Purkinje cells of the cerebellum (22). In addition guanylate cyclase (264), cGMP-dependent protein kinase (231, 235, 239), and G-substrate (252) are all highly enriched in this cell type. The localization of several elements of the cGMP-dependent protein phosphorylation system in Purkinje cells suggests an important and selective role for cGMP-dependent protein phosphorylation in this type of neuron. All excitatory neurotransmitters have been found to raise cGMP levels in Purkinje cells (22, 242, 265), and cGMP applied locally to Purkinje cells mimics the action of excitatory neurotransmitters (B. E. Hoffer, personal communication). Purkinje cells, which receive both excitatory and inhibitory inputs from several neuronal pathways (266), represent the only efferent pathway from the cerebellum. It is possible that cGMP-dependent protein phosphorylation in Purkinje cells is related to the integrative properties of these cells.

Intracellular injection of cAMP-dependent protein kinase resulted in increased input resistance of specific neurons of the mammalian motor cortex, mimicking the actions on these cells of extracellularly applied acetylcholine and intracellularly applied cGMP (T. Bartsai, C. D. Woody, E. Gruen, A. C. Nairn, T. Greengard, manuscript in preparation). These results provide direct evidence for the involvement of this kinase in mediating certain of the neurotransmitter actions of acetylcholine.

CALCIUM-DEPENDENT PROTEIN KINASES

Ca^{2+}-dependent activation of protein kinases was originally observed for phosphorylase kinase (267, 268) and myosin light chain kinase (269). The discovery that Ca^{2+}-dependent protein phosphorylation in brain synaptosomal membranes (270–273) required the ubiquitous multifunctional Ca^{2+}-binding protein calmodulin (10, 11) [for review of calmodulin, see (274)] led to the suggestion that Ca^{2+}/calmodulin-dependent protein phosphorylation would be widespread in various tissues (11, 12, 275). This hypothesis was supported by evidence that Ca^{2+}-dependent activation of myosin light chain kinase, from both smooth muscle (11a, 277) and skeletal muscle (278, 279), and of phosphorylase kinase (11b) was also dependent on calmodulin.

Following the observation that synaptosomal membranes contained Ca^{2+}/calmodulin-dependent protein kinase activity (10, 11), it was shown that a major substrate for Ca^{2+}/calmodulin-dependent protein phosphorylation in brain was synapsin I (137), previously known to be a substrate for both cAMP-dependent and Ca^{2+}-dependent protein kinases (124–126, 281, 282). A number of studies have established that two distinct Ca^{2+}/calmodulin-dependent protein kinases phosphorylate synapsin I (138, 283–286). Ca^{2+}/calmodulin-dependent protein kinase I (Ca^{2+}/calmodulin kinase I) specifically phosphorylates site I of synapsin I and Ca^{2+}/calmodulin-dependent protein kinase II (Ca^{2+}/calmodulin kinase II) specifically phosphorylates site II. These two enzymes have been purified and characterized, and have been shown to be distinct from myosin light chain kinase (103, 137, 287) and phosphorylase kinase (137, 288, 289), two well-characterized Ca^{2+}/calmodulin-dependent protein kinases which are also present in brain.

Ca^{2+}/calmodulin kinases I and II, myosin light chain kinase, and phosphorylase kinase appear to be responsible for much of the Ca^{2+}/calmodulin-dependent protein kinase activity present in mammalian brain. It was originally suspected that all Ca^{2+}-dependent protein phosphorylation might be regulated by calmodulin (290). However, a second type of Ca^{2+}-dependent protein kinase, the Ca^{2+}/phospholipid-dependent protein kinase, has been discovered in brain and other tissues (291–294). The brain contains significantly higher concentrations of Ca^{2+}-dependent protein kinases than do most non-neuronal tissues (11, 82, 83, 137, 291, 294a). In contrast to the cyclic nucleotide-dependent protein kinases, which have been purified and characterized primarily from non-nervous tissues, much of the work on Ca^{2+}-dependent protein kinases has been carried out on the enzymes from brain. This section will review the properties of the Ca^{2+}/calmodulin-dependent and Ca^{2+}/phospholipid-dependent protein kinases, which have been well-characterized in the mammalian brain.

Calcium/Calmodulin-Dependent Protein Kinase I

SUBUNIT STRUCTURE Ca^{2+}/calmodulin kinase I has been purified to apparent homogeneity from bovine brain, using synapsin I as substrate (284, 285). The purified enzyme consists of two major polypeptides in approximately equal amounts (M_r=37,000 and 39,000), and a less prominent polypeptide (M_r=42,000). In the presence of Ca^{2+} plus calmodulin, all three polypeptides bound calmodulin, were autophosphorylated on threonine residues at a slow rate relative to the rate of phosphorylation of synapsin I, and were labeled by the photoaffinity label 8-N$_3$-ATP. Peptide mapping studies of the three autophosphorylated polypeptides showed them to be very similar. While the purified kinase was found to contain polypeptides of M_r=37,000, 39,000, and 42,000 on SDS-polyacrylamide gels, a single symmetrical peak of activity was obtained on gel filtration with a M_r=49,000, suggesting that the kinase is a monomer. These (284, 285) and other studies (A. C. Nairn, P. Greengard, unpublished results) suggest that enzyme activity is associated with each of the three polypeptides. It is possible that the lower molecular weight polypeptides are derived from the polypeptide of M_r=42,000 or a polypeptide of higher molecular weight by proteolysis; alternatively, these polypeptides may represent isozymes of Ca^{2+}/calmodulin kinase I.

CELLULAR AND SUBCELLULAR DISTRIBUTION Ca^{2+}/calmodulin kinase I has a widespread species and tissue distribution. High concentrations of the enzyme have been detected in the brains of a number of mammalian species (137, 285, 295) and in the *Aplysia* nervous system (296). The kinase is widely distributed in the mammalian brain. It appears to be present in all parts of the neuron and may be enriched in the synaptic terminal (A. C. Nairn, P. Greengard, unpublished results). Ca^{2+}/calmodulin kinase I activity has been found to be predominantly cytosolic in all rat tissues examined (137, 285). Furthermore, Ca^{2+}/calmodulin kinase I partially purified from bovine heart has been found to have properties similar to those of the enzyme prepared from brain (285).

NEURONAL SUBSTRATES Ca^{2+}/calmodulin kinase I has been found to phosphorylate the neuronal proteins synapsin I and Protein III. Ca^{2+}/calmodulin kinase I also phosphorylates smooth muscle myosin light chain, but at a relatively slow rate. The enzyme does not phosphorylate histone, phosphorylase b, glycogen synthase, tubulin, or MAP-2 (285), all of which are substrates for other Ca^{2+}-dependent protein kinases (12). The widespread neuronal and tissue distribution of Ca^{2+}/calmodulin kinase I suggests that physiological substrates, in addition to synapsin I and Protein III, exist for this enzyme in both neuronal and non-neuronal mammalian tissues. It appears, however, that Ca^{2+}/calmodulin kinase I, like cGMP-dependent protein kinase, is an enzyme with a limited substrate specificity.

Synapsin I   Synapsin I (site I) has been found to be the best substrate tested for Ca^{2+}/calmodulin kinase I. The K_m value of Ca^{2+}/calmodulin kinase I for synapsin I (site I) was similar to that of cAMP-dependent protein kinase (285).

Protein III Ca^{2+}/calmodulin kinase I has been found to phosphorylate both Protein IIIa and Protein IIIb [collectively known as Protein III, (141)]. The kinase phosphorylated the same site(s) on Protein III as did cAMP-dependent protein kinase, and the two enzymes exhibited similar K_m values for Protein III [(285) and A. C. Nairn, M. D. Browning, P. Greengard, unpublished results].

REGULATION   Most calmodulin-regulated enzymes, including Ca^{2+}/calmodulin kinase I, appear to be activated by a similar mechanism, with the exception of phosphorylase kinase which contains calmodulin as an integral subunit (11b, 298). Calmodulin does not bind to the enzyme in the absence of Ca^{2+}; however, in the presence of micromolar concentrations of Ca^{2+}, calmodulin undergoes a marked conformational change exposing hydrophobic binding sites which interact with a calmodulin-binding domain on the enzyme (274, 299).

A role for Ca^{2+}-dependent protein phosphorylation in nervous tissue was first suggested by studies using isolated nerve terminals (synaptosomes) from cerebral cortex (271). In these and other experiments, depolarization of synaptosomes led to the Ca^{2+}-dependent phosphorylation of synapsin I (both site I and site II) (271, 281). In a variety of related studies, the phosphorylation of synapsin I (site I and site II) (18, 136, 140, 300, 301) and Protein III (18, 136, 140) has been found to be stimulated in intact nervous tissue by manipulations that increase Ca^{2+} concentration in neurons. These experiments suggest that Ca^{2+}/calmodulin kinase I is activated in vivo by depolarization-dependent Ca^{2+} influx into neurons.

Synapsin I is phosphorylated at a single site (site I) by both Ca^{2+}/calmodulin kinase I and cAMP-dependent protein kinase. Protein III is also phosphorylated at a single site by these two enzymes. Both Ca^{2+} (302, 303) and cAMP (160, 241, 304–306) are known to potentiate neurotransmitter release. Since Ca^{2+} and cAMP can each stimulate the phosphorylation of synapsin I and Protein III, one or both of these proteins might be involved in the facilitation of neurotransmitter release.

Calcium/Calmodulin-Dependent Protein Kinase II

Ca^{2+}/calmodulin kinase II has been purified from rat brain to apparent homogeneity using synapsin I as substrate (138, 283, 307, 308). The purified enzyme was found to have a relatively broad substrate specificity (see below), with synapsin I (site II) being the best substrate tested (307–309). Several other Ca^{2+}/calmodulin-dependent protein kinases, which have physico-chemical properties very similar to those of Ca^{2+}/calmodulin kinase II, have also been prepared from rat forebrain. These include enzymes in which the substrate used for purification was MAP-2 (310, 311), tryptophan hydroxylase and tyrosine hydroxylase (312), tubulin (313, 314, 314a), or myelin basic protein (311). Based on the similar properties of these enzyme preparations (307–310, 315, 316), they appear to be identical to Ca^{2+}/calmodulin kinase II. Differences reported in the substrate specificities of these preparations presumably reflect variations in the assay procedures (for example, substrate preparations used), rather than in the properties of the kinase.

SUBUNIT STRUCTURE Ca^{2+}/calmodulin kinase II, purified from rat forebrain (307, 308), has a native $M_r=550,000-650,000$, and contains a major polypeptide of $M_r=50,000$ and less prominent polypeptides of $M_r=58,000$ and 60,000. Peptide mapping studies have shown that the polypeptides of $M_r=50,000$ and 60,000 are distinct. These studies have also shown that the polypeptides of $M_r=58,000$ and 60,000 are very similar, suggesting that the polypeptide of $M_r=58,000$ may have been generated from the polypeptide of $M_r=60,000$. The polypeptides of $M_r=50,000$ and 58,000/60,000 were found to be present in approximately a 3:1 ratio. Monoclonal antibodies prepared against the $M_r=50,000$ polypeptide recognize all three subunits suggesting that the three polypeptides are immunologically related, although not identical (308). Each polypeptide was found to be autophosphorylated in a Ca^{2+}/calmodulin-dependent manner (307, 308). In addition, each polypeptide was shown to bind calmodulin in the presence of Ca^{2+} (307, 308) and to be labeled by the photoaffinity label 8-N_3-ATP (Y. Lai, P. Greengard, unpublished results). These results suggest that protein kinase activity is associated with all three subunits.

POLYMORPHISM   Several studies have suggested that Ca^{2+}/calmodulin kinase II from rat forebrain is an isozyme of a widespread Ca^{2+}/calmodulin-dependent protein kinase. Ca^{2+}/calmodulin kinase II, purified from rat cerebellum, has purification characteristics, physical properties, and substrate specificity similar to those of the enzyme purified from rat forebrain (308, 317, 318). The cerebellar kinase, however, consists of the polypeptides of $M_r=50,000$ and 58,000/60,000 in the approximate ratio of 1:4. The polypeptides of corresponding molecular weights in the two preparations were found to be identical by peptide mapping studies (308, 317, 318). Rat forebrain Ca^{2+}/calmodulin kinase II has similar properties to those of rabbit skeletal muscle Ca^{2+}/calmodulin-dependent glycogen synthase kinase (309). The enzyme from skeletal muscle has a substrate specificity, peptide maps, physical properties and immunological properties similar to those of the brain enzyme, but consists only of a polypeptide doublet of $M_r=58,000/60,000$. Ca^{2+}/calmodulin-dependent protein kinases have been prepared from various other tissues which exhibit properties similar to those of Ca^{2+}/calmodulin kinase II; namely, they have similar substrate specificities, have native $M_r=500,000-800,000$, and contain autophosphorylatable subunits with $M_r=50,000-60,000$. These include protein kinases prepared from *Torpedo* electric organ (319), *Aplysia* nervous system (296), turkey erythrocytes (320), bovine heart (321), rat pancreas (322), and rabbit liver (323, 324). The similar physical and immunologicalproperties and broad substrate specificities of these different Ca^{2+}/ calmodulin-dependent protein kinases strongly suggest that they and Ca^{2+}/ calmodulin kinase II are isozymes of a multifunctional Ca^{2+}/calmodulin-dependent protein kinase present in many animal species and tissues.

CELLULAR AND SUBCELLULAR DISTRIBUTION Ca^{2+}/calmodulin kinase II is present in very high concentrations in both soluble and particulate fractions from brain, perhaps comprising as much as 0.4% of total brain protein (307, 308). In a regional survey of Ca^{2+}/calmodulin kinase II activity in rat brain, the enzyme was found to have a widespread distribution (82, 83). High activity was found in cortical regions, particularly in the hippocampus, and in most subcortical forebrain regions, while relatively low activity was found in the cerebellum, diencephalon, mesencephalon, pons/medulla, and spinal cord.

The distribution of Ca^{2+}/calmodulin kinase II has also been studied immunocytochemically, using an antibody that recognizes all three subunits of the enzyme (286, 325). The enzyme was observed only in neurons and was present throughout the brain. The regional distribution of the kinase determined immunocytochemically closely paralleled that determined by enzyme activity measurements [(82, 83) and see above]. With respect to its subcellular distribution, immunoreactivity was found throughout the cell and was particularly enriched in neuronal somata and dendrites. Weaker immunoreactivity was observed in axons, nerve terminals, and postsynaptic densities. However, no immunoreactivity appeared to be associated with the nucleus. Biochemical studies have confirmed the presence of Ca^{2+}/calmodulin kinase II in the nerve terminal. Thus, Ca^{2+}/calmodulin kinase II activity has been found to be associated with purified synaptic vesicles (R. Jahn, W. Scheibler, P. Greengard, unpublished results). In addition, the enzyme has been found to be present in purified postsynaptic densities, where the "major PSD protein" (326–328) has been shown to be identical to the $M_r=50,000$ subunit of Ca^{2+}/calmodulin kinase II (329–331). In contrast to the immunocytochemical results, recent results have suggested that an enzyme with properties similar to those of Ca^{2+}/calmodulin kinase II is present in the neuronal nuclear matrix (332, 333).

In a detailed study of the regional distribution of protein phosphorylation systems in the rat brain (82, 83), marked differences were found in the relative amounts of three prominent substrates for Ca^{2+}/calmodulin-dependent protein kinase (82, 83). These substrates have subsequently been shown to be the autophosphorylated subunits of Ca^{2+}/calmodulin kinase II of $M_r=50,000$, 58,000, and 60,000 (S. I. Walaas, T. L. McGuinness, P. Greengard, unpublished results). The polypeptide of $M_r=50,000$ was found to be concentrated in forebrain regions, less prominent in the thalamus and hypothalamus, and present in very low levels in cerebellum, pons/medulla, and spinal cord. In contrast, the polypeptides of $M_r = 58,000$ and 60,000 were present in relatively similar concentrations throughout the brain. This differential distribution of the enzyme subunits suggests that a variety of isozymes of the kinase, which differ in their subunit compositions, might exist, not only in forebrain and cerebellum (see above), but also in other brain regions. It is possible that such isozymes may be responsible for determining the subcellular distribution of Ca^{2+}/calmodulin kinase II in brain, although it is not known if different isozymes are present in different subcellular compartments.

NEURONAL SUBSTRATES   In contrast to other Ca^{2+}/calmodulin kinases, for example phosphorylase kinase or myosin light chain kinase, the isozymes of Ca^{2+}/calmodulin kinase II from brain and other tissues exhibit a relatively broad substrate specificity. Good substrates for this enzyme include synapsin I, MAP-2, glycogen synthase, smooth muscle myosin light chain (110, 307–311), tau-protein (310, 334), tyrosine hydroxylase, tryptophan hydroxylase (312, 316, 335), myelin basic protein (311, 314), ribosomal protein S6 (322), and Ca^{2+}/calmodulin-sensitive cyclic nucleotide phosphodiesterase (336). With the exception of glycogen synthase, myosin light chain, and ribosomal protein S6, these proteins are predominantly neuronal. Tubulin has been used to purify Ca^{2+}/calmodulin kinase II from brain (313, 314, 314a), but appears to be a relatively poor substrate for the enzyme (307–310, 314a). Given the widespread tissue and species distribution of isozymes of this protein kinase, it is likely that additional neuronal and non-neuronal substrates exist. We will briefly discuss several of the well-characterized neuronal substrates for Ca^{2+}/calmodulin kinase II.

Synapsin I   Synapsin I (site II) has been found to be the best substrate tested for brain Ca^{2+}/calmodulin kinase II and isozymes of the protein kinase from non-nervous tissues. Synapsin I is bound through its tail region (which contains site II) at a specific high affinity saturable site on synaptic vesicles (132, 337). Phosphorylation of the tail region on site II by Ca^{2+}/calmodulin kinase II has recently been found to reduce this binding (337). The specific attachment of synapsin I to synaptic vesicles, and the regulation of this attachment by Ca^{2+}/calmodulin kinase II, suggest the involvement of this substrate and enzyme in synaptic vesicle function, possibly in the release of neurotransmitters. In support of this interpretation, recent studies have shown that micro-injection of synapsin I or Ca^{2+}/calmodulin kinase II into presynaptic nerve terminals of the squid giant synapse modifies neurotransmitter release (338).

MAP-2   Recent studies have shown that MAP-2 is phosphorylated by Ca^{2+}/calmodulin kinase II (110, 307–311) and protein kinase C (82, 83) in addition to cAMP-dependent protein kinase [(106, 107) and see above]. Phosphorylation of purified MAP-2 by Ca^{2+}/calmodulin kinase II was found to occur at several sites, which appeared to be distinct from those phosphorylated by cAMP-dependent protein kinase (110, 310, 314). The phosphorylation of MAP-2 by Ca^{2+}/calmodulin kinase II, like that by cAMP-dependent protein kinase, has been found to induce disassembly of microtubules (339). Ca^{2+}-dependent phosphorylation of MAP-2 may, therefore, represent an important mechanism for the control of microtubule assembly.

Tyrosine hydroxylase and tryptophan hydroxylase   Tyrosine hydroxylase, the enzyme that catalyzes the first step in the biosynthesis of the catecholamines, has been found to be phosphorylated by Ca^{2+}/calmodulin kinase II (335, 340, 341). Phosphorylation did not change the enzyme properties directly, but allowed an "activator protein" to activate the enzyme (341, 342). This "activator protein," which has been purified and characterized, is not specifically localized to catecholamine-producing cells, but is an abundant protein found throughout brain and in non-nervous tissues (342).

Recent evidence suggests that tryptophan hydroxylase, the enzyme that catalyzes the first step in the biosynthesis of serotonin (343), is also phosphorylated by Ca^{2+}/calmodulin kinase II (312, 316, 342). Phosphorylation of tryptophan hydroxylase resulted in a twofold activation of the enzyme in the presence of the same "activator protein" used with tyrosine hydroxylase (316, 342). Ca^{2+}/calmodulin-dependent phosphorylation of neurotransmitter-synthesizing enzymes may, therefore, represent a common regulatory mechanism.

REGULATION   As described above for Ca^{2+}/calmodulin kinase I, Ca^{2+}/calmodulin kinase II is activated by Ca^{2+} plus calmodulin. Phosphorylation of site II of synapsin I has been observed in intact nerve tissue, suggesting that the kinase is activated in vivo by increases in the intracellular concentration of Ca^{2+} (18, 136, 281, 300, 301). Ca^{2+}/calmodulin kinase II does not appear to be phosphorylated by other protein kinases. However, autophosphorylation of the enzyme appears to inhibit its activity (T. L. McGuinness, Y. Lai, P. Greengard, unpublished results) and has been reported to increase its affinity for calmodulin (343a). Carboxymethylation of a Ca^{2+}/calmodulin-dependent protein kinase, which appears to be identical to Ca^{2+}/calmodulin kinase II, has also been shown to inhibit kinase activity (344).

The cellular and subcellular distribution of Ca^{2+}/calmodulin kinase II is important with respect to its regulation, substrate specificity, and physiological function. The distribution of the kinase in rat brain is strikingly similar to that of sodium-independent glutamate binding sites (S. Halpain, C. C. Ouimet, T. C. Rainbow, P. Greengard, unpublished results). Since these binding sites probably represent glutamate receptors, and since glutamate has been shown to stimulate Ca^{2+} influx into neurons (345), Ca^{2+}/calmodulin kinase II may be involved in mediating certain of the effects of glutamate. Interestingly, the enzyme is present in postsynaptic densities (329–331), where it may have a structural role (346). While substrates for the enzyme have not been identified in postsynaptic densities, it is probable that they exist, and are involved in mediating some of the postsynaptic effects of Ca^{2+}. The kinase has also been found in dendrites, where it may be closely associated with MAP-2, as well as on the outer surface of synaptic vesicles, where it may be closely associated with synapsin I (286, 325).

It is not known what factors regulate the macromolecular interactions of the kinase. Enzyme activity is higher in particulate fractions in cerebellum than in the forebrain, and it has been suggested that the different isozymes of Ca^{2+}/ calmodulin kinase II found in these two brain regions may be responsible for this difference (308, 317, 318). Autophosphorylation has also been suggested to regulate subcellular distribution (307). In addition it has been found that application of serotonin to certain neurons in *Aplysia* results in changes in the subcellular distribution and activity of Ca^{2+}/calmodulin kinase II (347).

Myosin Light Chain Kinase and Phosphorylase Kinase

MYOSIN LIGHT CHAIN KINASE   Myosin light chain kinase has been purified from forebrain and found to have properties similar to those of the kinase found in smooth muscle (103, 287). The purified brain kinase ($M_r = 130,000$) appears to be a highly specific enzyme which preferentially phosphorylates smooth muscle myosin light chain (103). A number of other studies have reported partial purifications from brain of protein kinases that phosphorylate myosin light chain (137, 348, 349). Since both Ca^{2+}/calmodulin kinases I and II (284, 285, 307–309) and several other protein kinases (350, 351) phosphorylate myosin light chain in vitro, it is not clear if these partially purified preparations represent myosin light chain kinase.

Although myosin has been isolated from brain, its precise function in this tissue is unknown (352). Phosphorylation of smooth muscle or nonmuscle myosin light chain is believed to be a prerequisite for interaction between myosin and actin (353, 354). Brain myosin has been found to be phosphorylated and activated by myosin light chain kinase prepared from smooth muscle (355). Since troponin does not appear to be present in brain (356), myosin light chain kinase probably plays the same role in brain as it does in smooth muscle and nonmuscle cells.

PHOSPHORYLASE KINASE   Phosphorylase kinase has been identified in, but not purified from, brain (288, 289). The brain enzyme, which appears to be similar to phosphorylase kinase from skeletal muscle, is activated both by Ca^{2+} and by phosphorylation by cAMP-dependent protein kinase (289). However, the brain kinase is only partly cross-reactive with antibody prepared against the enzyme from skeletal muscle, indicating that it is a distinct phosphorylase kinase isozyme (289). Low concentrations of glycogen and glycogen-metabolizing enzymes are found throughout the brain, and are present in both glia and neurons (357). Electrical stimulation enhances glycogen breakdown in the nervous system (358), presumably mediated by Ca^{2+}-dependent activation of phosphorylase kinase.

Calcium/Phospholipid-Dependent Protein Kinase

Ca^{2+}/phospholipid-dependent protein kinase (also known as protein kinase C) was first purified from cerebellum as a cyclic nucleotide–independent protein kinase which could be activated by a Ca^{2+}-dependent protease also present in brain (359, 360). The holoenzyme was subsequently found to be activated by the addition of Ca^{2+}, diacylglycerol, and membrane phospholipid (361–364). Protein kinase C has recently been shown to be the intracellular receptor for, and to be activated by, the tumor-promoting phorbol esters (365–367). Therefore, the enzyme may mediate some of the tumor-promoting effects of these compounds. The role of protein kinase C in cellular regulation has been the subject of a number of reviews (291–294).

SUBUNIT STRUCTURE   Protein kinase C has been purified from brain (368, 369) and other tissues (370–372), and the properties of the enzyme from different sources have been found to be very similar. The enzyme from brain is a monomer of $M_r=80,000-87,000$ (368, 369). Partial proteolysis of the holoenzyme produces a fragment of $M_r=51,000$ that is catalytically active in the absence of Ca^{2+}, diacylglycerol, and phospholipid (368). It has been suggested that the holoenzyme contains a hydrophobic domain which interacts with Ca^{2+} and phospholipid, and a hydrophilic domain which is catalytically active (368).

CELLULAR AND SUBCELLULAR LOCALIZATION   Protein kinase C, which has a broad species, tissue, and cellular distribution (373, 374), is highly concentrated and widely distributed in brain (82, 83). In a regional survey of protein kinase C activity in rat brain, activity was highest in cortical regions and the cerebellum, intermediate in subcortical telencephalic regions, and lowest in the diencephalon, mesencephalon, rhombencephalon (except the cerebellum) and spinal cord (82, 83). Although a detailed immunocytochemical study of its distribution has not been performed, it is probable that in brain the enzyme is predominantly neuronal (374a, 368). In addition, the ontogeny of protein kinase C in rat brain parallels that of synaptogenesis and myelinogenesis (375).

In most non-nervous tissues protein kinase C is predominantly soluble (368, 373, 376). In contrast, a significant proportion of the enzyme in brain is particulate (368, 373, 376). The particulate kinase differs from the soluble enzyme in that it requires the addition of a non-ionic detergent in order to phosphorylate exogenously added substrates (368, 376). Soluble and particulate protein kinase C from brain are biochemically indistinguishable (368).

NEURONAL SUBSTRATES   Protein kinase C has a broad substrate specificity, which differs from those of both cyclic nucleotide–dependent and $Ca^{2+}/$ calmodulin-dependent protein kinases. The enzyme phosphorylates epidermal growth factor receptor (377), phospholamban (378, 379), histone H1 (368, 371), ribosomal protein S6 (380), vinculin (381), eukaryotic initiation factor eIF-2 (382), troponin T (383), smooth muscle myosin light chain (384, 385), the β-adrenergic receptor (176, 177), glycogen synthase (386, 387), the nicotinic acetylcholine receptor (388), and erythrocyte band 4.1 (389), among others. In addition, a variety of previously uncharacterized substrates have been found (371, 390–393).

In the brain tyrosine hydroxylase (340), GABA-modulin (198), myelin basic protein (371, 394), MAP-2 (82, 83), the "87 kDa" protein (82, 83, 374a, 395), the B-50 protein (396), and several uncharacterized substrates (82, 83, 390) have been identified as possible physiological substrates for the enzyme. Synapsin I, phosphorylated by protein kinase C in vitro (A. C. Nairn, P. Greengard, R. Minakuchi, Y. Takai, Y. Nishizuka, unpublished results), is not phosphorylated by tumor-promoting phorbol esters in synaptosomes (J. Wang, S. I. Walaas, P. Greengard, unpublished results), suggesting that it is not a physiological substrate for this enzyme. Several properties of the better-characterized neuronal substrates will be briefly discussed.

Tyrosine hydroxylase   Protein kinase C has recently been found to phosphorylate and activate tyrosine hydroxylase prepared from PC-12 pheochromocytoma cells (340). The kinetic changes and the phosphorylation site specificity observed with protein kinase C suggest that protein kinase C and cAMP-dependent protein kinase phosphorylate tyrosine hydroxylase at the same site and activate the enzyme by the same mechanism (340). Protein kinase C may be responsible for the Ca^{2+}-dependent activation of tyrosine hydroxylase observed in situ (187, 397, 398). However, a role for Ca^{2+}/calmodulin kinase II has also been suggested (335, 357).

Myelin basic protein   Protein kinase C, present in myelin, phosphorylates myelin basic protein in intact membranes, in detergent extracts, and in a purified preparation of myelin basic protein (394). In contrast to previous studies (399, 400), no evidence for Ca^{2+}/calmodulin-dependent phosphorylation of myelin basic protein was found. The state of phosphorylation of myelin basic protein is increased in vivo by K^+-induced depolarization (401) or by impulse conduction (402). Recently, a specific site phosphorylated in myelin basic protein in vivo (402a) has been shown to correspond to a site phosphorylated by protein kinase C in vitro (402b).

87-kDa Protein   Studies of intact synaptosome preparations prelabeled with $^{32}PO_4$ have shown that depolarization-induced calcium influx results in the phosphorylation of a protein of $M_r=87,000$ (the 87-kDa protein) (374a). Incubation of synaptosomes with tumor-promoting phorbol ester also resulted in phosphorylation of this protein (J. Wang, S. I. Walaas, P. Greengard, unpublished results). Further studies using crude cytosol or membrane preparations have shown that the 87-kDa protein could be phosphorylated by the addition of Ca^{2+} plus phosphatidylserine, but not Ca^{2+} plus calmodulin (82, 83, 374a).

The 87-kDa protein, which has been purified to homogeneity (395), appears to be widely distributed throughout the brain (82, 83). Subcellular fractionation studies of rat brain have shown that it is enriched in the crude synaptosomal fraction, and that it is present in both cytosol and membrane fractions [(82, 83, 374a) and S. I. Walaas, unpublished results].

Extensive studies in platelets (292, 293), adrenal chromaffin cells (403, 404), and pancreatic islets (405) have indicated that activation of protein kinase C may be involved in the regulation of hormone release. By analogy, Ca^{2+}/phospholipid-dependent phosphorylation of the 87-kDa protein in nerve terminals may be involved in Ca^{2+}-dependent mediation or modulation of neurotransmitter release.

B-50 Protein   B-50 protein is the name given to a brain membrane protein ($M_r=48,000$), the phosphorylation of which was originally found to be inhibited by high concentrations of ACTH (406, 407). Recent evidence has shown that protein kinase C is responsible for the phosphorylation of the B-50 protein (396, 408). Immunocytochemical studies have suggested that the B-50 protein is associated with presynaptic plasma membranes, possibly as an integral membrane protein (409–411). Partially purified B-50 protein has been reported to have phosphatidylinositol 4-phosphate kinase activity (412). In addition, both ACTH and an anti-B-50 protein antibody have been reported to increase the formation of phosphatidylinositol diphosphate (412, 413). Therefore, phosphatidylinositol-linked signal transmission in the brain (257, 258, 414, 415) may be regulated by Ca^{2+}-dependent phosphorylation of the B-50 protein.

REGULATION   Protein kinase C is activated by micromolar concentrations of Ca^{2+} and membrane phospholipids of which phosphatidylserine is the most active (361). Addition of low concentrations of diacylglycerol decreases the K_a for Ca^{2+} from 7×10^{-5} M to 5×10^{-6} M. Diacylglycerols containing side chains with one or two unsaturated fatty acids are most effective, whereas monoacylglycerols, triacylglycerols, and free fatty acids are ineffective (362, 363, 416).

Based on a series of studies in platelets, a model has been proposed in which protein kinase C is activated by an increase in the concentration of diacylglycerol produced by receptor-stimulated phosphatidylinositol turnover (291–293, 417). The activation of the kinase by diacylglycerol, although dependent on micromolar concentrations of Ca^{2+}, does not appear to be dependent on increases in intracellular Ca^{2+}. In contrast, the phosphorylation of the 87-kDa protein seen in response to depolarization in synaptosomes, and presumably mediated by protein kinase C, requires extracellular Ca^{2+} (374a).

Tumor-promoting phorbol esters, which appear to substitute for diacylglycerol in the activation of protein kinase C (365, 366, 369), have been employed in a number of studies on the regulation of the enzyme, both in vivo and in vitro (293, 365, 404, 418). While such studies have provided important evidence for a role of protein kinase C in secretion (418), they do not clarify the roles of Ca^{2+} and diacylglycerol as second messengers in the activation of this enzyme. The hydrolysis of phosphatidylinositol diphosphate produces diacylglycerol and inositol triphosphate, and the latter compound is believed to mobilize Ca^{2+} in cells (257–259, 418a). It seems likely, therefore, that activation of protein kinase C results from the synergistic actions of increases in the intracellular concentrations of both Ca^{2+} and diacylglycerol. The contributions of each second messenger may vary, however, depending on the cell type or receptor-mediated event.

The molecular mechanisms involved in the regulation of protein kinase C by Ca^{2+}, diacylglycerol, and phosphatidylserine are unknown. Reconstituted vesicles containing phosphatidylserine and diacylglycerol have been used as the membrane source in studies of the purified enzyme. Ca^{2+} does not appear to bind to the kinase directly; however, it is known that Ca^{2+} will bind to, and cause fusion of, phosphatidylserine vesicles (419, 420). Activation of the enzyme appears to require the simultaneous interaction of Ca^{2+}, a hydrophobic domain of the enzyme, and the phospholipid vesicle surface (368). Protein kinase C does not contain calmodulin (368); however, calmodulin and other Ca^{2+}-binding proteins have been found to markedly inhibit its activity (395). It is not known if this inhibition is related to the mechanism of activation of the enzyme or if it is a distinct regulatory device. In addition, a variety of compounds, including phenothiazines, inhibit protein kinase C (294, 421, 422).

Activation of protein kinase C in vivo (423, 424), or addition of Ca^{2+} to the enzyme in the presence of membrane in vitro (368, 369), appears to lead to its association with the particulate fraction. Since many of the substrates for protein kinase C are cytosolic (82, 83, 390), the exact mechanism by which phosphorylation occurs in situ is not known. A study of the association of the enzyme and its substrates with various cellular components would add greatly to our understanding of the function of protein kinase C in the brain and other tissues.

It is likely that protein kinase C is involved in many functions in the nervous system. The enzyme is activated by depolarization-dependent Ca^{2+} influx into synaptosomes (374a). In addition, a variety of neurotransmitters, as well as K^+-induced depolarization and electrical stimulation, stimulate phosphatidyl-inositol turnover [for reviews see (414, 415)]. Finally, either application of tumor-promoting phorbol esters or intracellular injection of protein kinase C enhances voltage-sensitive Ca^{2+} current in the bag cell neurons of *Aplysia* (425). The latter results provide direct evidence for the involvement of protein kinase C in the regulation of neuronal function.

CALCIUM- AND CYCLIC NUCLEOTIDE–INDEPENDENT PROTEIN KINASES

In addition to the protein kinases regulated by cAMP, cGMP, and Ca^{2+}, several other classes of protein kinases have been identified (17, 29). Those which have been identified in nervous tissue include: (*a*) the "independent" protein kinases, a group of enzymes not activated by cAMP, cGMP, Ca^{2+}, or any other known intracellular messenger, and which are usually named according to the substrates on which they act; (*b*) the tyrosine-specific protein kinases, originally identified as being associated with the transforming proteins of a number of retroviruses, and subsequently found to be associated with various cell surface growth factor receptors and within normal cells; and (*c*) other protein kinases which appear to be regulated by intracellular messengers other than cAMP, cGMP, and Ca^{2+}.

Independent Protein Kinases

The casein kinases are a class of independent protein kinases that preferentially phosphorylate acidic proteins, including casein, the substrate used in their purification and from which they take their names (426). Two distinct casein kinases, designated casein kinase I and casein kinase II, have been identified in many tissues and cell types, including brain [(427–429) for review see (426)]. Casein kinase I (monomeric, $M_r=37,000$) and casein kinase II ($\alpha_2\beta_2$; α $M_r=43,000$, β $M_r=24,000$) are distinguished from each other, not only by their subunit structures, but also by their physico-chemical properties, substrate specificities, and sensitivities to various inhibitors. Although it has been difficult to identify endogenous substrates for these protein kinases (426), phosphorylation by one or both of these protein kinases of eukaryotic initiation factors (430), RNA polymerases I and II (431), spectrin (432), myosin (433), glycogen synthase (434, 435), the R subunit of type II cAMP-dependent protein kinase (436), calsequestrin (437), and phosphatase inhibitor-1 (437) has been reported. Many of these substrate proteins are present in brain. Myelin fractions of brain contain a protein kinase activity independent of Ca^{2+} and cyclic nucleotides which is very active toward myelin basic protein (191, 438, 439). The relationship of this protein kinase to the casein kinases or to active proteolytic fragments of protein kinase C or other protein kinases is not known. Given the high activity of protein kinase C toward myelin basic protein and its high concentration in myelin (394), it will be important to rule out this kinase or an active proteolytic fragment as the source of this independent kinase activity. Another independent protein kinase which is capable of phosphorylating myelin basic protein has been partially purified from brain particulate fractions (440), and has several properties in common with casein kinase I.

Coated vesicles are vesicles enclosed by a polyhedral protein lattice. These vesicles appear to be involved in many cellular processes involving repeated membrane fusion and fission, for example, the recycling of synaptic vesicles [(441), for review see (442)]. A Ca^{2+}- and cyclic nucleotide–independent protein kinase activity has been identified in brain coated vesicle preparations that is capable of phosphorylating several coated vesicle proteins (443, 444); however, this kinase may be artifactually derived from a Ca^{2+}/calmodulin-dependent protein kinase which has been identified in brain coated vesicles (445).

Preparations of neurofilaments from the brains of several mammalian species contain tightly associated cyclic nucleotide–independent protein kinase activity that phosphorylates all three neurofilament subunits (446–448). Exogenous MAP-2 increases the rate of phosphorylation of neurofilaments and is itself extensively phosphorylated by the neurofilament-associated protein kinase (446), suggesting the possible involvement of this kinase in regulating the interaction between neurofilaments and microtubules in vivo. The identity of the endogenous protein kinase(s) that phosphorylates neurofilaments is unknown. It was found to be Ca^{2+}- and cAMP-independent, was insensitive to the cAMP-dependent protein kinase inhibitor, and could be separated from cAMP-dependent protein kinase (446). In another study, two protein kinases were extracted from neurofilament preparations by high salt, one of which did not phosphorylate neurofilaments and appeared to be casein kinase I, and another of which appeared to phosphorylate neurofilaments specifically (449).

Glycogen synthase kinase-3, a Ca^{2+}- and cyclic nucleotide–independent protein kinase, is one of five protein kinases that has been identified in skeletal muscle capable of phosphorylating glycogen synthase (435). In addition to glycogen synthase, it has also been found to phosphorylate the R subunit of type II cAMP-dependent protein kinase (450) and phosphatase inhibitor-2

(451). The phosphorylation of phosphatase inhibitor-2 is involved in the activation of the Mg^{2+}ATP-dependent protein phosphatase (452, 452a). The involvement of glycogen synthase kinase-3 and the Mg^{2+}ATP-dependent protein phosphatase (see below) in the regulation of glycogen metabolism makes it likely that this kinase and phosphatase inhibitor-2 are also present in brain, although this has not yet been established.

The independent protein kinases are characterized by their lack of sensitivity to various known physiological regulators. It is possible that the state of phosphorylation of substrates for these protein kinases is regulated in vivo by inhibitory modulators of protein kinase activity, by substrate-directed regulators, by regulation of the absolute levels of the protein kinases or their substrates, by regulation of protein kinase activity by other protein kinases, or by regulation of protein phosphatase activity. Finally, it is possible that some protein kinases that have been classified as "independent" are in fact catalytically active proteolytic fragments of regulated protein kinases, for example, protein kinase C (see above), produced during extraction procedures.

Tyrosine-Specific Protein Kinases

The viral gene products mediating transformation by a number of retroviruses specifically phosphorylate substrate proteins on tyrosine residues rather than on the more commonly phosphorylated serine or threonine residues [for review see (453)]. The *src* gene of Rous sarcoma virus encodes a phosphoprotein of $M_r=60,000$ (pp60[v-src]) with tyrosine kinase activity (454), which is homologous to a normal cellular protein (pp60[c-src]) present at low levels in a variety of normal vertebrate cells (455, 456). The level of pp60[c-src] in vertebrate brain is higher than that in any other nontransformed tissue tested (457–459). Tyrosine kinase activity has been found in brain and other normal nonproliferating tissues (460), suggesting that tyrosine-specific phosphorylation may be involved in processes other than stimulating cell proliferation. Possibly this type of activity may be involved in the control of cell-cell interactions underlying synaptic plasticity in the brain.

The plasma membrane receptors for epidermal growth factor (EGF) (13), platelet-derived growth factor (PDGF) (14), insulin (15), and somatomedin-C (16) from various tissues are closely associated with tyrosine-specific protein kinases which appear to reside within the receptor molecules themselves. The protein kinase activities are specifically stimulated by their respective ligands, resulting in the autophosphorylation of the receptor/protein kinase and, at least in the case of the insulin receptor, its concomitant activation (461, 462). The substrate specificities of the EGF receptor and insulin receptor protein kinases are similar to those of the viral-associated protein kinases (15, 463–465). The EGF receptor protein kinase shares certain immunological determinants (466, 467) and amino acid sequence homology (468) with certain retroviral transforming proteins. These findings suggest that the growth factor receptor-associated protein kinases are related to the retroviral protein kinases. The brain contains both insulin (468a) and insulin receptors coupled to tyrosine-specific kinase activity (469, 469a, 470), and may contain EGF receptors since EGF cross-reacting material has been found (471).

Although the roles of these tyrosine protein kinase-receptors in the brain are unknown, they may be involved in neuron-specific processes that are mediated through the phosphorylation of specific intracellular substrate proteins. The nicotinic acetylcholine receptor from *Torpedo* electric organ is phosphorylated by a tyrosine-specific protein kinase (472); however, it remains to be determined whether the receptor itself is the protein kinase. Future studies should reveal whether additional receptors for peptides and neurotransmitters in the brain are phosphorylated on tyrosine and/or possess intrinsic protein kinase activity.

Miscellaneous Regulated Protein Kinases

Several protein kinases that are sensitive to various regulators other than Ca^{2+} and cyclic nucleotides have been identified in brain. Although some of these are involved in the regulation of generalized cellular functions, it is possible that some also subserve neuron-specific functions. Double-stranded RNA-dependent protein kinase has been identified in mouse brain (473). This enzyme inhibits protein synthesis in reticulocyte lysates by phosphorylating the α-subunit of eukaryotic initiation factor eIF-2 (474), and its levels are increased in brain and other tissues by interferon. Two protease-activated protein kinases, designated protease-activated kinases I and II, have been identified in rabbit reticulocytes (475, 475a, 476). The enzymes are present in an inactive form and are activated by limited proteolysis with trypsin, chymotrypsin, or an endogenous Ca^{2+}-dependent protease. Both enzymes are present in a number of tissues, but only protease-activated kinase I has so far been found in brain (476).

Several protein kinases, the activities of which appear to be regulated by effects on their substrates, have been identified in nervous tissue. The phosphorylation of rhodopsin by rhodopsin kinase is stimulated by light through a direct effect on rhodopsin (477, 478). This kinase is distinct from cyclic nucleotide–dependent, Ca^{2+}-dependent, and casein protein kinases, and is specific for rhodopsin, although its role in retinal physiology remains unknown. Brain contains pyruvate dehydrogenase kinase, a mitochondrial enzyme that specifically phosphorylates the α subunit of pyruvate dehydrogenase, thereby inhibiting the enzyme (479, 480). Phosphorylation is stimulated by acetyl CoA and NADH and inhibited by pyruvate, coenzyme A, and NAD^+, by affecting pyruvate dehydrogenase as a substrate (479). The state of phosphorylation of pyruvate dehydrogenase is altered by tetanic electrical stimulation in hippocampal slices (481, 482), probably as a result of changes in energy metabolism caused by the electrical stimulation (483). Finally, another protein phosphorylation reaction in brain appears to be regulated at the substrate level by the Ca^{2+}-binding protein S–100 in a Ca^{2+}-independent manner (484, 485).

PHOSPHOPROTEIN PHOSPHATASES

The state of phosphorylation of specific substrates depends on the relative activities of phosphoprotein phosphatases (protein phosphatases) in addition to the activities of protein kinases. The protein phosphatases involved in the dephosphorylation of most of the known proteins phosphorylated on serine or threonine residues can be accounted for by four distinct enzymes (486–489). These enzymes are grouped into two classes, type 1 protein phosphatase (protein phosphatase-1) and type 2 protein phosphatases (protein phosphatase-2A, -2B, and -2C). Type 1 protein phosphatase selectively dephosphorylates the β-subunit of phosphorylase kinase and is inhibited by nanomolar concentrations of phosphatase inhibitor-1 or phosphatase inhibitor-2, while type 2 protein phosphatases selectively dephosphorylate the α-subunit of phosphorylase kinase and are insensitive to these inhibitors. The type 2 protein phosphatases are further distinguished by their requirements for divalent cations. Recently, multiple forms of phosphotyrosyl-protein phosphatases have been demonstrated in several tissues which appear to be distinct from the phosphoseryl/phosphothreonyl-protein phosphatases (490–494).

Protein phosphatase-1, -2A, -2B, -2C, (25, 495), the phosphotyrosyl-protein phosphatases (490–494), a mitochondrial protein phosphatase specific for pyruvate dehydrogenase (496) and phosphatase inhibitor-1(252) have been demonstrated in brain. Protein phosphatase-2A and -2B are present in brain at higher concentrations than in other tissues (25, 495). Protein phosphatase activity from nervous tissue has been detected in myelin (497–499), and in partially purified preparations of rhodopsin (500), nicotinic acetylcholine receptor (501, 502), and MAP-2 (503). A significant fraction of brain protein phosphatase activity appears to be associated with particulate fractions (504), suggesting a possible role for protein phosphatases in synaptic processes.

Protein phosphatases-1, -2A, and -2C all exhibit relatively broad substrate specificities, while that of protein phosphatase-2B (also known as calcineurin) appears to be more restricted (489, 505). Although the substrate specificity of protein phosphatase-2B is limited, several neuronal phosphoproteins are particularly good substrates for this enzyme (151, 506). These include DARPP-32, a substrate for cAMP-dependent protein kinase, and G-substrate, a substrate for cGMP-dependent protein kinase (see above). Both DARPP-32 (143) and protein phosphatase-2B (495, 507) are highly concentrated in the neostriatum, most likely in the medium-sized spiny neurons, suggesting that the state of phosphorylation of DARPP-32 may be regulated by both Ca^{2+} and cAMP in this brain region. It is becoming increasingly clear that protein phosphatases, like protein kinases, are under physiological regulation. Protein phosphatase-2B, a prominent calmodulin-binding protein in brain (508), is activated by Ca^{2+} plus calmodulin (505). cAMP regulates protein phosphatase-1 by stimulating the phosphorylation and activation of DARPP-32 and phosphatase inhibitor-1 (150, 151, 487). This provides a positive feedback mechanism for amplifying the effects of cAMP, and a mechanism for cAMP to modulate the phosphorylation state of substrate proteins for protein kinases other than cAMP-dependent protein kinases. Protein phosphatase-1 is also regulated by its interaction with phosphatase inhibitor-2. A complex of protein phosphatase-1 with phosphatase inhibitor-2 (together known as the Mg^{2+}ATP-dependent protein phosphatase) is inactive, but is activated by incubation with glycogen synthase kinase-3 plus Mg^{2+}ATP (452, 452a). In addition to calmodulin, DARPP-32, G-substrate, phosphatase inhibitor-1, and phosphatase inhibitor-2, several other modulatory proteins have been identified that appear to be involved in regulating various protein phosphatase activities (486, 509–511).

The regulation of protein phosphatase activity appears to be particularly important in brain. Protein phosphatase-1, protein phosphatase-2B, and phosphatase inhibitor-1 are all present in brain (25, 252). Furthermore, two neuronal protein phosphatase inhibitors that are regulated by phosphorylation (DARPP-32 and G-substrate, see above) have been identified in brain. These findings suggest that neurotransmitters produce some of their physiological effects by regulating protein phosphatase activity, in some cases involving molecular mechanisms apparently specific to nervous tissue.

CONCLUDING REMARKS

Protein phosphorylation is involved in the regulation of many aspects of neuronal function. A variety of regulatory agents, including neurotransmitters, hormones, and the nerve impulse itself, are capable of regulating the phosphorylation of many specific substrate proteins in brain. The great diversity of physiological effects elicited by these regulatory agents and mediated by protein phosphorylation in brain is reflected in the diversity and specificity of its protein phosphorylation systems.

The substrate specificities of protein kinases and protein phosphatases fall into two general categories. cGMP-dependent protein kinase, Ca^{2+}/calmodulin kinase I, myosin light chain kinase, phosphorylase kinase, and protein phosphatase-2B exhibit narrow substrate specificities and are presumably involved in specialized aspects of neuronal function. cAMP-dependent protein kinase, Ca^{2+}/calmodulin kinase II, protein kinase C, and several protein phosphatases exhibit broad substrate specificities and appear to be involved in many aspects of neuronal function.

A number of protein kinases and substrates, for example, cAMP-dependent protein kinase and one of its substrates synapsin I, have been found to be widely and fairly evenly distributed throughout the brain, and appear to be present in virtually all neurons. Such a distribution suggests their involvement in functions common to all neurons. In contrast, other protein kinases and substrates have been found to be enriched in particular classes of neurons where they may mediate specific intracellular actions of certain neurotransmitters; for example, Ca^{2+}/calmodulin kinase II may be regulated by glutamate and DARPP-32, a substrate for cAMP-dependent protein kinase, may be regulated by dopamine. Other protein kinases and substrates, for example, cGMP-dependent protein kinase and one of its substrates, G-substrate, are highly enriched in one region of the brain, apparently within a single type of neuron, where they are presumably involved in some specialized function. In addition to these differences observed in cellular distribution in brain, many protein kinases and substrates are concentrated in specific subcellular compartments, for example, Ca^{2+}/calmodulin kinase II, MAP-2, and synapsin I, suggesting their localization to those parts of the neuron where they play important functional roles.

Given the numerous molecular mechanisms involved in the control of protein phosphorylation in brain, it is not surprising that many interactions exist, both between separate components within a single molecular pathway and between various molecular pathways. For example, both the formation and degradation of cAMP and cGMP can be regulated by Ca^{2+}/calmodulin-dependent enzymes; numerous substrate proteins, including synapsin I, MAP-2, tyrosine hydroxylase, and the nicotinic acetylcholine receptor, are phosphorylated at separate sites or at the same site by both Ca^{2+}-dependent, cyclic nucleotide-dependent, and Ca^{2+}- and cyclic nucleotide–independent protein kinases; several substrates phosphorylated by cAMP-dependent protein kinase are substrates for the Ca^{2+}/calmodulin-dependent protein phosphatase, protein phosphatase-2B; and several substrates phosphorylated by Ca^{2+}-dependent protein kinases are substrates for protein phosphatase-1, an enzyme which can be regulated by the cAMP-dependent phosphorylation of phosphatase inhibitor-1 or DARPP-32, and by the Ca^{2+}- and cyclic nucleotide–independent phosphorylation of phosphatase inhibitor-2. These various types of interaction provide potential mechanisms for positive and negative feedback within molecular pathways and for either synergistic or antagonistic interactions between separate molecular pathways controlling protein phosphorylation.

The study of protein kinases, protein phosphatases, and their substrate proteins in the brain has provided considerable information concerning their roles in neuron-specific processes. In addition, important information has been obtained concerning their general biochemical and physiological properties which is also relevant to non-nervous tissues. The observations that certain protein kinases and their specific substrates exhibit limited regional distributions in the brain have made possible a novel approach to anatomical, developmental, and clinical studies of the brain. The study of protein kinases, protein phosphatases, and their specific substrates in brain should continue to provide important information relevant to the molecular mechanisms involved in the regulation of neuronal function.

Literature Cited

1. Rall, T. W., Sutherland, E. W., Berthet, J. 1957. *J. Biol. Chem.* 224:463–75
2. Rall, T. W., Sutherland, E. W. 1958. *J. Biol. Chem.* 232:1065–76
3. Sutherland, E. W., Rall, T. W. 1958. *J. Biol. Chem.* 232:1077–91
4. Robison, G. A., Butcher, R. W., Sutherland, E. W. 1968. *Ann. Rev. Biochem.* 37:149–74
5. Sutherland, E. W., Rall, T. W. 1960. *Pharmacol. Rev.* 12:265–86
6. Walsh, D. A., Perkins, J. P., Krebs, E. G. 1968. *J. Biol. Chem.* 243:3763–74
7. Greengard, P. 1978. *Science* 199:146–52
8. Kuo, J. F., Greengard, P. 1970. *J. Biol. Chem.* 245:2493–98
9. Kuo, J. F., Shoji, M. 1982. *Handb. Exp. Pharmacol.* 58(Pt. I):393–424
10. Schulman, H., Greengard, P. 1978. *Nature* 271:478–79
11. Schulman, H., Greengard, P. 1978. *Proc. Natl. Acad. Sci. USA* 75:5432–36
11a. Dabrowska, R., Aromatorio, D., Sherry, J. M. F., Hartshorne, D. J. 1977. *Biochem. Biophys. Res. Commun.* 78:1263–71
11b. Cohen, P., Burchell, A., Foulkes, J. G., Cohen, P. T. W., Vanaman, T. C., Nairn, A. C. 1978. *FEBS Lett.* 92:287–93
12. Schulman, H. 1982. *Handb. Exp. Pharmacol.* 58(Pt. I):425–78
13. Cohen, S., Carpenter, G., King, L. 1980. *J. Biol. Chem.* 255:4834–42
14. Heldin, C. H., Ek, B., Rönnstrand, L. 1983. *J. Biol. Chem.* 258:10054–61
15. Kasuga, M., Fujita-Yamaguchi, Y., Blithe, D. L., Kahn, C. R. 1983. *Proc. Natl. Acad. Sci. USA* 80:2137–41
16. Jacobs, S., Kull, F. C. Jr., Earp, H. S., Svoboda, M. E., Van Wyk, J. J., Cuatrecasas, P. 1983. *J. Biol. Chem.* 258:9581–84
17. Krebs, E. G., Beavo, J. A. 1979. *Ann. Rev. Biochem.* 48:923–59
18. Nestler, E. J., Greengard, P. 1984. *Protein Phosphorylation In The Nervous System.* New York: Wiley
19. Sutherland, E. W., Rall, T. W., Menon, T. 1962. *J. Biol. Chem.* 237:1220–27
20. De Robertis, E., Rodriguez de Lores Arnaiz, G. R., Alberici, M., Butcher, R. W., Sutherland, E. W. 1967. *J. Biol. Chem.* 242:3487–93
21. Goldberg, N. D., Haddox, M. K. 1977. *Ann. Rev. Biochem.* 46:823–96
22. Ferrendelli, J. A. 1978. *Adv. Cyclic Nucleotide Res.* 9:453–64
23. Butcher, R. W., Sutherland, E. W. 1962. *J. Biol. Chem.* 237:1244–50
24. Miyamoto, E., Kuo, J. F., Greengard, P. 1969. *J. Biol. Chem.* 244:6395–402
25. Ingebritsen, T. S., Stewart, A. A., Cohen, P. 1983. *Eur. J. Biochem.* 132:297–307
26. Nestler, E. J., Walaas, S. I., Greengard, P. 1984. *Science* 225:1357–64
27. Greengard, P. 1981. *Harvey Lect.* 1981 (75):277–331
28. Beavo, J. A., Mumby, M. C. 1982. *Handb. Exp. Pharmacol.* 58(Pt. I):363–92
29. Flockhart, D. A., Corbin, J. D. 1982. *CRC Crit. Rev. Biochem.* 12:133–86
30. Walter, U., Greengard, P. 1981. *Curr. Top. Cell. Regul.* 19:219–56
31. Walter, U., Uno, I., Liu, A. Y.-C., Greengard, P. 1977. *J. Biol. Chem.* 252:6494–500
32. Casnellie, J. E., Schlichter, D. J., Walter, U., Greengard, P. 1978. *J. Biol. Chem.* 253:4771–76
33. Glass, D. B., Krebs, E. G. 1980. *Ann. Rev. Pharmacol. Toxicol.* 20:363–88
34. Takio, K., Wade, R. D., Smith, S. B., Krebs, E. G., Walsh, K. A., Titani, K. 1984. *Biochemistry* 23:4207–18
35. Kuo, J. F., Greengard, P. 1969. *Proc. Natl. Acad. Sci. USA* 64:1349–55
36. Nimmo, H. G., Cohen, P. 1977. *Adv. Cyclic Nucleotide Res.* 8:145–266

37. Walsh, D. A., Krebs, E. G. 1973. *The Enzymes* 7:555–81
38. Rubin, C. S., Rosen, O. M. 1975. *Ann. Rev. Biochem.* 44:831–87
39. Erlichman, J., Rosenfeld, R., Rosen, O. M. 1974. *J. Biol. Chem.* 249:5000–3
40. Rosen, O. M., Erlichman, J. 1975. *J. Biol. Chem.* 250:7788–94
41. Maeno, H., Reyes, P. L., Ueda, T., Rudolph, S. A., Greengard, P. 1974. *Arch. Biochem. Biophys.* 164:551–59
42. Beavo, J. A., Bechtel, P. J., Krebs, E. G. 1975. *Adv. Cyclic Nucleotide Res.* 6:241–51
43. Rosen, O. M., Erlichman, J., Rubin, C. S. 1975. *Adv. Cyclic Nucleotide Res.* 5:253–63
44. Corbin, J. D., Sugden, P. H., West, L., Flockhart, D. A., Lincoln, T. M., McCarthy, D. 1978. *J. Biol. Chem.* 253:3997–4003
45. Hofmann, F. 1980. *J. Biol. Chem.* 255:1559–64
46. Weber, W., Hilz, M. 1979. *Biochem. Biophys. Res. Commun.* 90:1073–81
47. Weber, W., Vogel, C. W., Hilz, H. 1978. *FEBS Lett.* 99:62–66
48. Rannels, S. R., Corbin, J. D. 1980. *J. Biol. Chem.* 255:7085–88
49. Builder, S. E., Beavo, J. A., Krebs, E. G. 1980. *J. Biol. Chem.* 255:2350–54
50. Øgreid, D., Døskeland, S. O., Miller, J. P. 1983. *J. Biol. Chem.* 258:1041–49
51. Robinson-Steiner, A. M., Corbin, J. D. 1983. *J. Biol. Chem.* 258:1032–40
52. Reimann, E. M., Walsh, D. A., Krebs, E. G. 1971. *J. Biol. Chem.* 246:1986–95
53. Corbin, J. D., Kelly, S. L., Park, C. R. 1975. *J. Biol. Chem.* 250:218–25
54. Hofmann, F., Beavo, J. A., Bechtel, P. J., Krebs, E. G. 1975. *J. Biol. Chem.* 250:7795–801
55. Robinson-Steiner, A. M., Beebe, S. J., Rannels, S. R., Corbin, J. D. 1984. *J. Biol. Chem.* 259:10596–605
56. Zoller, M. J., Kerlavage, A. R., Taylor, S. S. 1979. *J. Biol. Chem.* 254:2408–12
57. Potter, R. L., Taylor, S. S. 1979. *J. Biol. Chem.* 254:2413–18
58. Corbin, J. D., Kelly, S. L., Soderling, T. R., Park, C. R. 1975. *Adv. Cyclic Nucleotide Res.* 5:265–79
58a. Corbin, J. D., Rannels, S. R., Flockhart, D. A., Robinson-Steiner, A. M., Tigani, M. C., et al. 1982. *Eur. J. Biochem.* 125:259–66
58b. Hartl, F. T., Roskoski, R. Jr. 1982. *Biochemistry* 21:5175–83
59. Rangel-Aldao, R., Kupiec, J. W., Rosen, O. M. 1979. *J. Biol. Chem.* 254:2499–508
60. LaPorte, D. C., Builder, S. E., Storm, D. R. 1980. *J. Biol. Chem.* 255:2343–49
61. Fleischer, N., Rosen, O. M., Reichlin, M. 1976. *Proc. Natl. Acad. Sci. USA* 73:54–58
62. Hofmann, F., Bechtel, P. J., Krebs, E. G. 1977. *J. Biol. Chem.* 252:1441–47
63. Rubin, C. S., Fleischer, N., Sarkar, D., Erlichman, J. 1981. *Cold Spring Harbor Conf. Cell Prolif.* 8:1333–46
64. Kapoor, C., Beavo, J. A., Steiner, A. 1979. *J. Biol. Chem.* 254:12427–32
65. Gill, G. N., McCune, R. W. 1979. *Curr. Top. Cell. Regul.* 15:1–45
66. Lohmann, S. M., Walter, U., Greengard, P. 1980. *J. Biol. Chem.* 255:9985–92
67. Takio, K., Smith, S. B., Krebs, E. G., Walsh, K. A., Titani, K. 1982. *Proc. Natl. Acad. Sci. USA* 79:2544–48
68. Hashimoto, E., Takio, K., Krebs, E. G. 1981. *J. Biol. Chem.* 256:5604–7
69. Titani, K., Sasagawa, T., Ericsson, L. H., Kumar, S., Smith, S. B., et al. 1984. *Biochemistry* 23:4193–99
70. Takio, K., Smith, S. B., Krebs, E. G., Walsh, K. A., Titani, K. 1984. *Biochemistry* 23:4200–6
71. Rubin, C. S., Rangel-Aldao, R., Sarkar, D., Erlichman, J., Fleischer, N. 1979. *J. Biol. Chem.* 254:3797–805
72. Rubin, C. S., Fleischer, N., Sarkar, D., Erlichman, J. 1981. *Cold Spring Harbor Conf. Cell Prolif.* 8:1333–46
73. Mumby, M., Beavo, J. A. 1981. *Cold Spring Harbor Conf. Cell Prolif.* 8:105–25
74. Erlichman, J., Sarkar, D., Fleischer, N., Rubin, C. S. 1980. *J. Biol. Chem.* 255:8179–84
75. Stein, J. C., Sarkar, D., Rubin, C. S. 1984. *J. Neurochem.* 42:547–53
76. Hartl, F. T., Roskoski, R. Jr. 1983. *J. Biol. Chem.* 258:3950–55
77. Deleted in proof
78. Strocchi, P., Sapirstein, V. S., Rubin, C. S., Gilbert, J. M. 1984. *J. Neurochem.* 43:466–71
79. Greengard, P. 1979. *Fed. Proc.* 38:2208–17
80. Panter, S. S., Butley, M. S., Malkinson, A. M. 1981. *J. Neurochem.* 36:2081–85
81. Walter, U., Kanof, P., Schulman, H., Greengard, P. 1978. *J. Biol. Chem.* 253:6275–80
82. Walaas, S. I., Nairn, A. C., Greengard, P. 1983. *J. Neurosci.* 3:291–301
83. Walaas, S. I., Nairn, A. C., Greengard, P. 1983. *J. Neurosci.* 3:302–311
84. Walter, U., Lohmann, S. M., Sieghart, W., Greengard, P. 1979. *J. Biol. Chem.* 254:12235–39

85. De Camilli, P. 1984. *Adv. Cyclic Nucleotide Protein Phosphorylation Res.* 17:489–99
85a. De Camilli, P., Moretti, M., Navone, F., Lohmann, S. M., Walter, U. 1984. *Adv. Cyclic Nucleotide Protein Phosphorylation Res.* 17A:140
85b. Bradbury, J. M., Thompson, R. J. 1984. *Biochem. J.* 221:361–68
86. Maeno, H., Johnson, E. M., Greengard, P. 1971. *J. Biol. Chem.* 246:134–42
87. Uno, I., Ueda, T., Greengard, P. 1976. *J. Biol. Chem.* 251:2192–95
88. Uno, I., Ueda, T., Greengard, P. 1977. *J. Biol. Chem.* 252:5164–74
89. Rubin, C. S., Erlichman, J., Rosen, O. M. 1972. *J. Biol. Chem.* 247:6135–39
90. Corbin, J. D., Sugden, P. H., Lincoln, T. M., Keely, S. L. 1977. *J. Biol. Chem.* 252:3854–61
91. Weller, M., Morgan, I. G. 1976. *Biochem. Biophys. Acta* 433:223–28
92. Kelly, P. T., Cotman, C. W., Largen, M. 1979. *J. Biol. Chem.* 254:1564–75
93. Ueda, T., Greengard, P., Berzins, K., Cohen, R. S., Blomberg, F., Grab, D. J., Siekevitz, P. 1979. *J. Cell Biol.* 83:308–19
94. Johnson, E. M. 1982. *Handb. Exp. Pharmacol.* 58(Pt. I):507–33
95. Schwartz, J. P., Costa, E. 1980. *J. Biol. Chem.* 255:2943–48
96. Jungmann, R. A., Lakes, M. S., Harrison, J. J., Siter, P., Jones, C. E. 1981. *Cold Spring Harbor Conf. Cell Prolif.* 8:1109–25
97. Jungmann, R. A., Harrison, J. J., Milkowski, D., Lee, S.-K., Schweppe, J. S., Miles, M. F. 1982. *Prog. Brain Res.* 56:163–78
98. Browning, E. T., Niklas, J. 1982. *J. Neurochem.* 39:336–41
99. Cho-Chung, Y. S., Archibald, D., Clair, T. 1979. *Science* 205:1390–92
100. Cho-Chung, Y. S. 1980. *J. Cyclic Nucleotide Res.* 6:163–77
101. Vallee, R. B., DiBartolomeis, M. J., Theurkauf, W. E. 1981. *J. Cell Biol.* 90:568–76
102. Theurkauf, W. E., Vallee, R. B. 1982. *J. Biol. Chem.* 257:3284–90
103. Hathaway, D. R., Adelstein, R. S., Klee, C. B. 1981. *J. Biol. Chem.* 256:8183–89
104. Sarkar, D., Erlichman, J., Rubin, C. S. 1984. *J. Biol. Chem.* 259:9840–46
105. Lohmann, S. M., DeCamilli, P., Einig, I., Walter, U. 1984. *Proc. Natl. Acad. Sci. USA* 81:6723–27
106. Sloboda, R. D., Rudolf, S. A., Rosenbaum, J. L., Greengard, P. 1975. *Proc. Natl. Acad. Sci. USA* 72:177–81

107. Vallee, R. 1980. *Proc. Natl. Acad. Sci. USA* 77:3206–10
108. Kim, H., Binder, L., Rosenbaum, J. L. 1979. *J. Cell Biol.* 80:266–76
109. Theurkauf, W. E., Vallee, R. B. 1983. *J. Biol. Chem.* 258:7883–86
110. Yamauchi, T., Fujisawa, H. 1982. *Biochem. Biophys. Res. Commun.* 109:975–81
111. Jameson, L., Caplow, M. 1981. *Proc. Natl. Acad. Sci. USA* 78:3413–17
112. Jameson, L., Frey, T., Zeeberg, B., Dalldorf, F., Caplow, M. 1980. *Biochemistry* 19:2472–78
113. Murthy, A. S. N., Flann, M. 1983. *Eur. J. Biochem.* 137:37–46
114. Nishida, E., Kuwaki, T., Sakai, H. 1981. *J. Biochem.* 90:575–78
115. Selden, S. C., Pollard, T. D. 1983. *J. Biol. Chem.* 258:7064–71
116. Aamodt, E., Williams, R. C. Jr. 1983. *Biophys. J.* 41:86a
117. Miller, P., Walter, U., Theurkauf, W. E., Vallee, R. B., De Camilli, P. 1982. *Proc. Natl. Acad. Sci. USA* 79:5562–66
118. Matus, A., Bernhardt, R., Hugh-Jones, T. 1981. *Proc. Natl. Acad. Sci. USA* 78:3010–14
119. Wiche, G., Briones, E., Hirt, H., Krepler, R., Artlieb, U., Denk, H. 1983. *EMBO J.* 2:1915–20
120. De Camilli, P., Miller, P. E., Navone, F., Theurkauf, W. E., Vallee, R. B. 1984. *Neuroscience* 11:819–46
121. Huber, G., Matus, A. 1984. *J. Neurosci.* 4:151–60
122. Caceres, A., Binder, L. I., Payne, M. R., Bender, P., Rebhun, L., Steward, O. 1984. *J. Neurosci.* 4:394–410
123. Johnson, E. M., Ueda, T., Maeno, H., Greengard, P. 1972. *J. Biol. Chem.* 247:5650–62
124. Ueda, T., Maeno, H., Greengard, P. 1973. *J. Biol. Chem.* 248:8295–305
125. Ueda, T., Greengard, P. 1977. *J. Biol. Chem.* 252:5155–63
126. Huttner, W. B., DeGennaro, L. J., Greengard, P. 1981. *J. Biol. Chem.* 256:1482–88
127. Sieghart, W., Forn, J., Schwarcz, R., Coyle, J. T., Greengard, P. 1978. *Brain Res.* 156:345–50
128. De Camilli, P., Ueda, T., Bloom, F. E., Battenberg, E., Greengard, P. 1979. *Proc. Natl. Acad. Sci. USA* 76:5977–81
129. De Camilli, P., Cameron, R., Greengard, P. 1983. *J. Cell Biol.* 96:1337–54
130. Fried, G., Nestler, E. J., DeCamilli, P., Stjärne, L., Olson, L., et al. 1982. *Proc. Natl. Acad. Sci. USA* 79:2717–21
131. De Camilli, P., Harris, S. M., Huttner,

W. B., Greengard, P. 1983. *J. Cell Biol.* 96:1355–73

132. Huttner, W. B., Scheibler, W., Greengard, P., De Camilli, P. 1983. *J. Cell Biol.* 96:1374–88

132a. Navone, F., Greengard, P., De Camilli, P. 1984. *Science* 226:1209–11

133. Dolphin, A. C., Greengard, P. 1981. *Nature* 289:76–79

134. Dolphin, A. C., Greengard, P. 1981. *J. Neurosci.* 1:192–203

135. Nestler, E. J., Greengard, P. 1980. *Proc. Natl. Acad. Sci. USA* 77:7479–83

136. Tsou, K., Greengard, P. 1982. *Proc. Natl. Acad. Sci. USA* 79:6075–79

137. Kennedy, M. B., Greengard, P. 1981. *Proc. Natl. Acad. Sci. USA* 78:1293–97

138. Kennedy, M. B., McGuinness, T., Greengard, P. 1983. *J. Neurosci.* 3:818–31

139. Browning, M. D., Huang, C.-K., Greengard, P. 1982. *Soc. Neurosci. Abstr.* 8:794

140. Forn, J., Greengard, P. 1978. *Proc. Natl. Acad. Sci. USA* 75:5195–99

141. Huang, C.-K., Browning, M. D., Greengard, P. 1982. *J. Biol. Chem.* 257:6524–28

142. Rodnight, R. 1982. *Prog. Brain Res.* 56:1–25

143. Ouimet, C. C., Miller, P. E., Hemmings, H. C. Jr., Walaas, S. I., Greengard, P. 1984. *J. Neurosci.* 4:111–24

144. Walaas, S. I., Aswad, D. W., Greengard, P. 1983. *Nature* 301:69–71

145. Walaas, S. I., Greengard, P. 1984. *J. Neurosci.* 4:84–98

146. Hemmings, H. C. Jr., Nairn, A. C., Aswad, D. W., Greengard, P. 1984. *J. Neurosci.* 4:99–110

147. Hemmings, H. C. Jr., Nairn, A. C., Greengard, P. 1984. *J. Biol. Chem.* 259:14491–97

148. Hemmings, H. C. Jr., Williams, K. R., Konigsberg, W. H., Greengard, P. 1984. *J. Biol. Chem.* 259:14486–90

149. Huang, F. L., Glinsmann, W. H. 1976. *Eur. J. Biochem.* 70:419–26

150. Nimmo, G. A., Cohen, P. 1978. *Eur. J. Biochem.* 87:341–51

151. Hemmings, H. C. Jr., Greengard, P., Tung, H. Y. L., Cohen, P. 1984. *Nature* 310:503–5

152. Kebabian, J. W., Calne, D. B. 1979. *Nature* 277:93–96

153. Ritchie, J. M. 1979. *Ann. Rev. Neurosci.* 2:341–62

154. Hartshorne, R. P., Catterall, W. A. 1981. *Proc. Natl. Acad. Sci. USA* 78:4620–24

155. Hartshorne, R. P., Messner, D. J., Cop-

persmith, J. C., Catterall, W. A. 1982. *J. Biol. Chem.* 257:3888–91

156. Hartshorne, R. P., Catterall, W. A. 1984. *J. Biol. Chem.* 259:1667–75

157. Costa, M. R. C., Catterall, W. A. 1982. *Mol. Pharmacol.* 22:196–203

158. Costa, M. R. C., Casnellie, J. E., Catterall, W. A. 1982. *J. Biol. Chem.* 257: 7918–21

159. Costa, M. R. C., Catterall, W. A. 1984. *J. Biol. Chem.* 259:8210–18

160. Kandel, E. R., Schwartz, J. H. 1982. *Science* 218:433–42

161. Kaczmarek, L. K., Jennings, K. R., Strumwasser, F., Nairn, A. C., Walter, U., et al. 1980. *Proc. Natl. Acad. Sci. USA* 77:7487–91

162. Castellucci, V. F., Kandel, E. R., Schwartz, J. H., Wilson, F. D., Nairn, A. C., Greengard, P. 1980. *Proc. Natl. Acad. Sci. USA* 77:7492–96

163. Siegelbaum, S. A., Camardo, J. S., Kandel, E. R. 1982. *Nature* 299:413–17

164. Castellucci, V. F., Nairn, A. C., Greengard, P., Schwartz, J. H., Kandel, E. R. 1982. *J. Neurosci.* 2:1673–81

165. dePeyer, J. E., Cachelin, A. B., Levitan, I. B., Reuter, H. 1982. *Proc. Natl. Acad. Sci. USA* 79:4207–11

166. Alkon, D. L., Acosta-Urguidi, J., Olds, J., Kuzma, G., Neary, J. T. 1983. *Science* 219:303–6

167. Demaille, J. G., Pechere, J.-F. 1983. *Adv. Cyclic Nucleotide Res.* 15:337–71

168. Fambrough, D. M. 1979. *Physiol. Rev.* 59:165–227

169. Conti-Tronconi, B. M., Raftery, M. A. 1982. *Ann. Rev. Biochem.* 51:491–530

170. Changeux, J.-P. 1981. *Harvey Lect.* 1981 (75):85–254

171. Huganir, R. L., Greengard, P. 1983. *Proc. Natl. Acad. Sci. USA* 80:1130–34

172. Shorr, R. G. L., Heald, S. L., Jeffs, P. W., Lavin, T. N., Strohsacker, M. W., et al. 1982. *Proc. Natl. Acad. Sci. USA* 79:2778–82

173. Shorr, R. G. L., Strohsacker, M. W., Lavin, T. N., Lefkowitz, R. J., Caron, M. G. 1982. *J. Biol. Chem.* 257:12341–50

174. Cerione, R. A., Strulovici, B., Benovic, J. L., Strader, C. D., Caron, M. G., Lefkowitz, R. J. 1983. *Proc. Natl. Acad. Sci. USA* 80:4899–903

175. Stadel, J. M., Nambi, P., Shorr, R. G. L., Sawyer, D. F., Caron, M. G., Lefkowitz, R. J. 1983. *Proc. Natl. Acad. Sci. USA* 80:3173–77

176. Sibley, D. R., Peters, J. R., Nambi, P., Caron, M. G., Lefkowitz, R. J. 1984. *J. Biol. Chem.* 259:9742–49

177. Kelleher, D. J., Pessin, J. E., Ruoho, A.

E., Johnson, G. L. 1984. *J. Biol. Chem.* 81:4316–20
178. Nagatsu, T., Levitt, M., Udenfriend, S. 1964. *J. Biol. Chem.* 239:2910–17
179. Morgenroth, V. A. III, Hegstrand, L. R., Roth, R. H., Greengard, P. 1975. *J. Biol. Chem.* 250:1946–48
180. Joh, T. H., Park, D. H., Reis, D. J. 1978. *Proc. Natl. Acad. Sci. USA* 75:4744–48
181. Yamauchi, T., Fugisawa, H. 1979. *J. Biol. Chem.* 254:503–7
182. Vulliet, P. R., Langan, T. A., Weiner, N. 1980. *Proc. Natl. Acad. Sci. USA* 77:92–96
183. Markey, K. A., Kondo, S., Shenkman, L., Goldstein, M. 1980. *Mol. Pharmacol.* 17:79–85
184. Edelman, A. M., Raese, J. D., Lazar, M. A., Barchas, J. D. 1981. *J. Pharmacol. Exp. Ther.* 216:647–53
185. Weiner, N. 1980. *Monogr. Neural Sci.* 7:146–60
186. Salzman, P. M., Roth, R. H. 1978. In *Neuropsychopharmacology*, ed. P. Deniker, C. Radouco-Thomas, A. Villeneuve, 2:1439–55. New York: Pergamon
187. Haycock, J. W., Meligeni, J. A., Bennett, W. F., Waymire, J. C. 1982. *J. Biol. Chem.* 257:12641–48
188. Lazarides, E. 1980. *Nature* 283:249–56
189. Leterrier, J.-F., Liem, R. K. H., Shelanski, M. L. 1981. *J. Cell Biol.* 90:755–60
190. Carnegie, P. R., Moore, W. J. 1980. In *Proteins of the Nervous System*, ed. R. A. Bradshaw, D. M. Schneider, pp. 119–43. New York: Raven
191. Miyamoto, E. 1976. *J. Neurochem.* 26:573–77
192. Miyamoto, E., Kakiuchi, S. 1974. *J. Biol. Chem.* 249:2769–77
192a. Carnegie, P. R., Kemp, B. E., Dunkley, P. R., Murray, A. W. 1973. *Biochem. J.* 135:569–72
192b. Steck, A. J., Appel, S. H. 1974. *J. Biol. Chem.* 249:5416–20
193. Zwiller, J., Revel, M.-O., Basset, P. 1981. *Biochem. Biophys. Res. Commun.* 101:1381–87
194. Kumakura, K., Battaini, F., Hofmann, M., Spano, P. F., Trabucchi, M. 1978. *FEBS Lett.* 93:231–34
195. Sharma, R. K., Wang, T. H., Wirch, E., Wang, J. H. 1980. *J. Biol. Chem.* 255:5916–23
196. Marchmont, R. J., Houslay, M. D. 1980. *Nature* 286:904–6
197. Guidotti, A., Konkel, D. R., Ebstein, B., Corda, M. G., Wise, B. C., et al. 1982. *Proc. Natl. Acad. Sci. USA* 79:6084–88
198. Wise, B. C., Guidotti, A., Costa, E. 1983. *Proc. Natl. Acad. Sci. USA* 80:886–90
199. Kebabian, J. W., Petzold, G. L., Greengard, P. 1972. *Proc. Natl. Acad. Sci. USA* 69:2145–49
200. Hegstrand, L. R., Kanof, P. D., Greengard, P. 1976. *Nature* 260:163–65
201. von Hungen, K., Roberts, S., Hill, D. F. 1975. *Brain Res.* 84:257–67
202. Smigel, M. D., Ross, E. M., Gilman, A. G. 1984. In *Cell Membranes, Methods and Reviews*, ed. E. L. Elson, W. A. Frazier, L. Glaser, 2:247–94. New York: Plenum
203. Brostrom, C. O., Huang, Y. C., Breckenridge, B. M., Wolff, D. J. 1975. *Proc. Natl. Acad. Sci. USA* 72:64–68
204. Cheung, W. Y., Bradham, L. S., Lynch, T. J., Lin, Y. M., Tallant, E. A. 1975. *Biochem. Biophys. Res. Commun.* 66:1055–62
205. Rangel-Aldao, R., Rosen, O. M. 1976. *J. Biol. Chem.* 251:3375–80
206. Liu, A. Y.-C., Greengard, P. 1976. *Proc. Natl. Acad. Sci. USA* 73:568–72
207. Liu, A. Y.-C., Walter, U., Greengard, P. 1981. *Eur. J. Biochem.* 114:539–48
208. Walter, U., Costa, M. R. C., Breakefield, X. O., Greengard, P. 1979. *Proc. Natl. Acad. Sci. USA* 76:3251–55
209. Prashad, N., Rosenberg, R. N., Wischmeyer, B., Ulrich, C., Sparkman, D. 1979. *Biochemistry* 18:2717–25
210. Walsh, D. A., Ashby, C. D., Gonzalez, C., Calkins, D., Fischer, E. H., Krebs, E. G. 1971. *J. Biol. Chem.* 246:1977–85
211. Beavo, J. A., Bechtel, P. J., Krebs, E. G. 1974. *Proc. Natl. Acad. Sci. USA* 71:3580–83
212. Kuo, J. F. 1975. *Biochem. Biophys. Res. Commun.* 65:1214–20
213. Costa, M. 1977. *Biochem. Biophys. Res. Commun.* 78:1311–18
214. Kuo, W. N., Williams, J. L., Floyd-Jones, T., Duggans, C. F., Boone, D. L., et al. 1979. *Expertania* 35:997–98
215. Szmigielski, A. 1981. *Arch. Int. Pharmacodyn.* 249:64–71
216. Szmigielski, A., Guidotti, A. 1979. *Neurochem. Res.* 4:189–99
217. Kuo, J. F. 1974. *Proc. Natl. Acad. Sci. USA* 71:4037–41
218. Lincoln, T. M., Hall, C. L., Park, C. R., Corbin, J. D. 1976. *Proc. Natl. Acad. Sci. USA* 73:2559–63
219. Lincoln, T. M., Corbin, J. D. 1983. *Adv. Cyclic Nucleotide Res.* 15:139–92
220. Nairn, A. C., Greengard, P. 1983. *Fed. Proc.* 42:3107–13
221. Gill, G. N., Walton, G. M., Sperry, P. J. 1977. *J. Biol. Chem.* 252:6443–49

222. Lincoln, T. M., Dills, W. L., Corbin, J. D. 1977. *J. Biol. Chem.* 252:4269–75
223. Flockerzi, V., Speichermann, N., Hofmann, F. 1978. *J. Biol. Chem.* 253: 3395–99
224. Hofmann, F., Flockerzi, V. 1983. *Eur. J. Biochem.* 130:599–603
225. Hofmann, F., Sensheimer, H.-P., Gobel, C. 1983. *FEBS Lett.* 164:350–54
226. Lincoln, T. M., Flockhart, D. A., Corbin, J. D. 1978. *J. Biol. Chem.* 253: 6002–9
227. Foster, J. L., Guttmann, J., Rosen, O. M. 1981. *J. Biol. Chem.* 256:5029–36
228. Takai, Y., Nishiyama, U., Yamamura, H., Nishizuka, Y. 1975. *J. Biol. Chem.* 250:4690–95
229. Walter, U., Miller, P., Wilson, F., Menkes, D., Greengard, P. 1980. *J. Biol. Chem.* 255:3757–62
230. Walter, U. 1981. *Eur. J. Biochem.* 118:339–46
231. Schlichter, D. J., Detre, J. A., Aswad, D. W., Chehrazi, B., Greengard, P. 1980. *Proc. Natl. Acad. Sci. USA* 77:5537–41
232. Walter, U., De Camilli, P., Lohmann, S. M., Miller, P., Greengard, P. 1981. *Cold Spring Harbor Conf. Cell Prolif.* 8:141–57
233. Casnellie, J. E., Greengard, P. 1974. *Proc. Natl. Acad. Sci. USA* 71:1891–95
234. de Jonge, H. R. 1976. *Nature* 262:590–92
235. Lohmann, S. M., Walter, U., Miller, P. E., Greengard, P. 1981. *Proc. Natl. Acad. Sci. USA* 78:653–57
236. Kuo, J. F. 1975. *J. Cyclic Nucleotide Res.* 1:151–57
237. Schlichter, D. J., Casnellie, J. E., Greengard, P. 1978. *Nature* 273:61–62
238. Bandle, E., Guidotti, A. 1978. *Brain Res.* 156:412–16
239. De Camilli, P., Miller, P. E., Levitt, P., Walter, U., Greengard, P. 1984. *Neuroscience* 11:761–817
240. Stone, T. W., Taylor, D. A. 1977. *J. Physiol.* 266:523–43
241. Dunwiddie, T. V., Hoffer, B. J. 1982. *Handb. Exp. Pharmacol.* 58(Pt. II):389–463
242. Drummond, G. I. 1983. *Adv. Cyclic Nucleotide Res.* 15:373–494
243. Ariano, M. A. 1983. *Neuroscience* 10: 707–23
244. Lincoln, T. M., Corbin, J. D. 1977. *Proc. Natl. Acad. Sci. USA* 74:3239–43
245. Glass, D. B., Krebs, E. G. 1979. *J. Biol. Chem.* 254:9728–38
246. Glass, D. B., Krebs, E. G. 1982. *J. Biol. Chem.* 257:1196–200
247. Casnellie, J. E., Ives, H. E., Jamieson, J.

D., Greengard, P. 1980. *J. Biol. Chem.* 255:3770–76
248. Geahlen, R. L., Krebs, E. G. 1980. *J. Biol. Chem.* 255:1164–69
249. Aswad, D. W., Greengard, P. 1981. *J. Biol. Chem.* 256:3487–93
250. Aswad, D. W., Greengard, P. 1981. *J. Biol. Chem.* 256:3494–500
251. Aitken, A., Bilham, D., Cohen, P., Aswad, D. W., Greengard, P. 1981. *J. Biol. Chem.* 256:3501–6
252. Detre, J. A., Nairn, A. C., Aswad, D. W., Greengard, P. 1984. *J. Neurosci.* 4:2843–49
253. Schultz, G., Hardman, J. G., Schultz, K., Baird, C. E., Sutherland, E. W. 1973. *Proc. Natl. Acad. Sci. USA* 70:3889–93
254. Berridge, M. J. 1975. *Adv. Cyclic Nucleotide Res.* 6:1–98
255. Berridge, M. J. 1980. *Trends Pharmacol. Sci.* 1:419–24
256. Ohsako, S., DeGuchi, T. 1981. *J. Biol. Chem.* 256:10945–48
257. Berridge, M. J. 1982. *Handb. Exp. Pharmacol.* 58(Pt. II):227–70
258. Michell, R. H. 1975. *Biochem. Biophys. Acta* 415:81–147
259. Michell, R. H. 1979. *Trends Biochem. Sci.* 4:128–31
259a. Strada, S. J., Martin, M. W., Thompson, W. J. 1984. *Adv. Cyclic Nucleotide Protein Phosphorylation Res.* 16:13–29
260. Ferrendelli, J. A., DeVries, G. 1982. *Fed. Proc.* 42:3103–6
261. Greengard, P., Kebabian, J. W. 1974. *Fed. Proc.* 33:1059–67
262. Volle, R. L., Quenzer, L. F. 1981. *Fed. Proc.* 42:3099–102
263. Ferrendelli, J. A., Kinscherf, D. A., Chang, M. M. 1973. *Mol. Pharmacol.* 9:445–54
264. Ariano, M. A., Lewicki, J. A., Brandwein, H. J., Murad, F. 1982. *Proc. Natl. Acad. Sci. USA* 79:1316–20
265. Biggio, E., Brodie, B. B., Costa, E., Guidotti, A. 1977. *Proc. Natl. Acad. Sci. USA* 74:3592–96
266. Llinas, R. R. 1975. *Sci. Am.* 232:56–57
267. Brostrom, C. O., Hunkeler, F. L., Krebs, E. G. 1971. *J. Biol. Chem.* 246:1961–67
268. Cohen, P. 1973. *Eur. J. Biochem.* 34:1–4
269. Pires, E. M. V., Perry, S. V. 1977. *Biochem. J.* 167:137–46
270. Krueger, B. K., Forn, J., Greengard, P. 1976. *Soc. Neurosci. Abstr.* 2:1007
271. Krueger, B. K., Forn, J., Greengard, P. 1977. *J. Biol. Chem.* 252:2764–73
272. DeLorenzo, R. J. 1976. *Biochem. Biophys. Res. Commun.* 71:590–97

273. DeLorenzo, R. J., Freedman, S. D. 1977. *Biochem. Biophys. Res. Commun.* 77:1036–43
274. Cheung, W. Y. 1980. *Science* 207:19–27
275. Schulman, H., Huttner, W. B., Greengard, P. 1980. In *Calcium and Cell Function*, ed. W. Y. Cheung, 1:219–52. New York: Academic
276. Deleted in proof
277. Dabrowska, R., Sherry, J. M. F., Aromatorio, D. K., Hartshorne, D. J. 1978. *Biochemistry* 17:253–58
278. Perry, S. V., Cole, H. A., Frearson, H., Moir, A. J. G., Nairn, A. C., Solaro, R. J. 1978. *Proc. FEBS 12th Meet.* 54:147–59
279. Yagi, K., Yazawa, M., Kakiuchi, S., Ohshima, M., Uenishi, K. 1978. *J. Biol. Chem.* 253:1338–40
280. Deleted in proof
281. Huttner, W. B., Greengard, P. 1979. *Proc. Natl. Acad. Sci. USA* 76:5402–6
282. Sieghart, W., Forn, J., Greengard, P. 1979. *Proc. Natl. Acad. Sci. USA* 76:2475–79
283. Lai, Y., McGuinness, T. L., Greengard, P. 1983. *Soc. Neurosci. Abstr.* 9:1029
284. Nairn, A. C., Greengard, P. 1983. *Soc. Neurosci. Abstr.* 9:1029
285. Nairn, A. C., Greengard, P. 1985. *J. Biol. Chem.* Submitted
286. McGuinness, T. L., Lai, Y., Ouimet, C. C., Greengard, P. 1984. In *Calcium in Biological Systems*, ed. R. P. Rubin, J. W. Putney, E. Weiss, pp. 291–305. New York: Plenum
287. Dabrowska, R., Hartshorne, D. J. 1978. *Biochem. Biophys. Res. Commun.* 85: 1352–59
288. Ozawa, E. 1973. *J. Neurochem.* 20: 1487–88
289. Taira, T., Kii, R., Sakai, K., Tabuchi, H., Takimoto, S., et al. 1982. *J. Biochem.* 91:883–88
290. Yamauchi, T., Fujisawa, H. 1979. *Biochem. Biophys. Res. Commun.* 90: 1172–78
291. Nishizuka, Y. 1980. *Mol. Biol. Biochem. Biophys.* 32:113–35
292. Nishizuka, Y. 1983. *Trends Biochem. Sci.* 8:13–16
293. Nishizuka, Y. 1984. *Science* 225:1365–70
294. Kuo, J. F., Schatzman, R. C., Turner, S., Mazzei, G. J. 1984. *Mol. Cell. Endocrinol.* 35:65–73
294a. Walaas, S. I., Nairn, A. C., 1985. In *Calcium Physiology*, ed. D. Marme, pp. 238–64. Springer-Verlag
295. Nairn, A. C., Greengard, P. 1984. *Fed. Proc.* 43:1467
296. DeReimer, S. A., Kaczmarek, L. K.,

297. Lai, Y., McGuinness, T. L., Greengard, P. 1984. *J. Neurosci.* 4:1618–25
297. Deleted in proof
298. Shenolikar, S., Cohen, P. R. W., Cohen, P., Nairn, A. C., Perry, S. V. 1979. *Eur. J. Biochem.* 100:329–37
299. Klee, C. B., Vanaman, T. C. 1982. *Adv. Protein Chem.* 35:213–321
300. Nestler, E. J., Greengard, P. 1982. *Nature* 296:452–54
301. Nestler, E. J., Greengard, P. 1982. *J. Neurosci.* 2:1011–23
302. Rosenthal, J. 1969. *J. Physiol.* 203:121–33
303. Zengel, J. E., Magleby, K. L., Horn, J. P., McAfee, D. A., Yarowsky, P. J. 1980. *J. Gen. Physiol.* 76:213–31
304. Miyamoto, M. D., Breckenridge, B. M. 1974. *J. Gen. Physiol.* 63:609–24
305. Reubi, J.-C., Iversen, L. L., Jessell, T. M. 1977. *Nature* 268:652–54
306. Kuba, K., Kato, E., Kumamoto, E., Koketsu, K., Hirai, K. 1981. *Nature* 291:654–56
307. Bennett, M. K., Erondu, N. E., Kennedy, M. B. 1983. *J. Biol. Chem.* 258:12735–44
308. McGuinness, T. L., Lai, Y., Greengard, P. 1985. *J. Biol. Chem.* 260:1696–1704
309. McGuinness, T. L., Lai, Y., Greengard, P., Woodgett, J. R., Cohen, P. 1983. *FEBS Lett.* 163:329–34
310. Schulman, H. 1984. *J. Cell Biol.* 1984. 99:11–19
311. Fukunaga, K., Yamamoto, H., Matsui, K., Higashi, K., Miyamoto, E. 1982. *J. Neurochem.* 39:1607–17
312. Yamauchi, T., Fujisawa, H. 1983. *Eur. J. Biochem.* 132:15–20
313. Goldenring, J. R., Gonzalez, B., DeLorenzo, R. J. 1982. *Biochem. Biophys. Res. Commun.* 108:421–28
314. Goldenring, J. R., Gonzalez, B., McGuire, J. S. Jr., DeLorenzo, R. J. 1983. *J. Biol. Chem.* 258:12632–40
314a. Yamauchi, T., Fujisawa, H. 1984. *Arch. Biochem. Biophys.* 234:89–96
315. Iwasa, T., Fukunaga, K., Yamamoto, N., Tanaka, E., Miyamoto, E. 1983. *FEBS Lett.* 161:28–32
316. Woodgett, J. R., Cohen, P., Yamauchi, T., Fujisawa, H. 1984. *FEBS Lett.* 170:49–54
317. McGuinness, T. L., Lai, Y., Greengard, P. 1984. *Soc. Neurosci. Abstr.* 10:919
318. Miller, S. G., Kennedy, M. B. 1984. *Soc. Neurosci. Abstr.* 10:544
319. Palfrey, H. C., Rothlein, J. E., Greengard, P. 1983. *J. Biol. Chem.* 258:9496–503
320. Palfrey, H. C., Lai, Y., Greengard, P. 1984. In *Proc. 6th Int. Conf. Red Cell*

Structure and Metabolism, ed. G. Brewer, pp. 291–308. New York: Liss

321. Palfrey, H. C. 1984. *Fed. Proc.* 43:1466
322. Gorelick, F. S., Cohn, J. A., Freedman, S. D., Delahunt, N. G., Gershoni, J. M., Jamieson, J. D. 1983. *J. Cell Biol.* 97:1294–98
323. Ahmad, Z., De Paoli-Roach, A. A., Roach, P. J. 1982. *J. Biol. Chem.* 257:8348–55
324. Payne, M. E., Schworer, C. M., Soderling, T. R. 1983. *J. Biol. Chem.* 258:2376–82
325. Ouimet, C. C., McGuinness, T. L., Greengard, P. 1984. *Proc. Natl. Acad. Sci. USA.* 81:5604–8
326. Banker, G., Churchill, L., Cotman, C. W. 1974. *J. Cell Biol.* 63:456–65
327. Blombers, F., Cohen, R. S., Siekevitz, P. 1977. *J. Cell Biol.* 74:204–25
328. Kelly, P. T., Montgomery, P. R. 1982. *Brain Res.* 233:265–86
329. Kennedy, M. B., Bennett, M. K., Erondu, N. E. 1983. *Proc. Natl. Acad. Sci. USA* 80:7359–61
330. Kelly, P. T., McGuinness, T. L., Greengard, P. 1984. *Proc. Natl. Acad. Sci. USA* 81:945–49
331. Goldenring, J. R., McGuire, J. S., DeLorenzo, R. J. 1984. *J. Neurochem.* 42:1077–84
332. Sahyoun, N., Levine, H., Cuatrecasas, P. 1984. *Proc. Natl. Acad. Sci. USA* 81:4311–15
333. Sahyoun, N., Levine, H., Bronson, D., Cuatrecasas, P. 1984. *J. Biol. Chem.* 259:9341–44
334. Yamamoto, H., Fukunaga, K., Tanaka, E., Miyamoto, E. 1983. *J. Neurochem.* 41:1119–25
335. Vulliet, P. R., Woodgett, J. R., Cohen, P. 1984. *J. Biol. Chem.* 259:13680–83
336. Fukunaga, K., Yamamoto, H., Tanaka, E., Iwasa, T., Miyamoto, E. 1984. *Life Sci.* 35:493–500
337. Schiebler, W., Rothlein, J., Jahn, R., Doucet, J. P., Greengard, P. 1983. *Soc. Neurosci. Abstr.* 9:882
338. Llinas, R., McGuinness, T. L., Leonard, C. S., Sugimori, M., Greengard, P. 1985. *Proc. Natl. Acad. Sci. USA* In press
339. Yamauchi, T., Fujisawa, H. 1983. *Biochem. Biophys. Res. Commun.* 110:287–91
340. Albert, K. A., Helmer-Matyjek, E., Nairn, A. C., Müller, T. H., Haycock, J. W., et al. 1984. *Proc. Natl. Acad. Sci. USA.* 84:7713–17
341. Yamauchi, T., Fujisawa, H. 1981. *Biochem. Biophys. Res. Commun.* 100:807–13
342. Yamauchi, T., Nakata, H., Fujisawa, H. 1981. *J. Biol. Chem.* 256:5404–9
343. Cooper, J. R., Bloom, F. E., Roth, R. H. 1982. *The Biochemical Basis of Neuropharmacology.* New York/Oxford: Oxford Univ. Press
343a. Shields, S. M., Vernon, P. J., Kelly, P. T. 1984. *J. Neurochem.* 43:1599–1609
344. Billingsley, M. L., Velletri, P. S., Kuhn, D. M., Levine, R. A., Lovenberg, W. 1983. *Soc. Neurosci. Abstr.* 9:596
345. Sonnhof, V., Buhrle, C. 1981. *Adv. Biochem. Psychopharmacol.* 27:195–98
346. Shaw, G. 1984. *Nature* 308:496
347. Saitoh, T., Schwartz, J. H. 1985. *J. Cell Biol.* 100:835–42
348. Yamauchi, T., Fujisawa, H. 1980. *FEBS Lett.* 116:141–44
349. Miyamoto, E., Fukunaga, M. K., Iwasa, Y. 1981. *J. Neurochem.* 37:1324–30
350. Matsumura, S., Murakami, N., Yasuda, S., Kumon, A. 1982. *Biochem. Biophys. Res. Commun.* 109:683–88
351. Singh, T. J., Akatsuka, A., Huang, K.-P. 1983. *FEBS Lett.* 159:217–20
352. Trifaro, J. 1978. *Neuroscience* 3:1–24
353. Adelstein, R. S. 1980. *Fed. Proc.* 39:1544–46
354. Mrwa, U., Hartshorne, D. J. 1980. *Fed. Proc.* 39:1564–68
355. Barylko, B., Sobieszek, A. 1983. *EMBO J.* 2:369–74
356. Anthony, F. A., Babitch, J. H., Sharma, S. U. 1984. *J. Neurochem.* 42:1343–49
357. Knull, H. R., Khandelwal, R. L. 1982. *Neurochem. Res.* 7:1307–17
358. King, L. J., Lowry, O. H., Passonneau, J. V., Venson, V. 1967. *J. Neurochem.* 14:599–611
359. Takai, Y., Kishimoto, A., Inoue, M., Nishizuka, Y. 1977. *J. Biol. Chem.* 252:7603–9
360. Inoue, M., Kishimoto, A., Takai, Y., Nishizuka, Y. 1977. *J. Biol. Chem.* 252:7610–16
361. Takai, Y., Kishimoto, A., Kawahara, Y., Mori, T., Nishizuka, Y. 1979. *J. Biol. Chem.* 254:3692–95
362. Takai, Y., Kishimoto, A., Kikkawa, U., Mori, T., Nishizuka, Y. 1979. *Biochem. Biophys. Res. Commun.* 91:1218–24
363. Kishimoto, A., Takai, Y., Mori, T., Kikkawa, U., Nishizuka, Y. 1980. *J. Biol. Chem.* 255:2273–78
364. Kaibuchi, K., Takai, Y., Nishizuka, Y. 1981. *J. Biol. Chem.* 256:7146–49
365. Castagna, M., Takai, Y., Kaibuchi, K., Sano, K., Kikkawa, U., Nishizuka, Y. 1982. *J. Biol. Chem.* 257:7847–51
366. Kikkawa, U., Takai, Y., Tanaka, Y., Miyake, R., Nishizuka, Y. 1983. *J. Biol. Chem.* 258:1144–45

367. Neidel, J. E., Kuhn, L. J., Vandenbark, G. R. 1983. *Proc. Natl. Acad. Sci. USA* 80:36–40

368. Kikkawa, U., Takai, Y., Minakuchi, R., Inohara, S., Nishizuka, Y. 1982. *J. Biol. Chem.* 257:13341–48

369. Parker, P. J., Stabel, S., Waterfield, M. D. 1984. *EMBO J.* 3:953–59

370. Wise, B. C., Raynor, R. L., Kuo, J. F. 1982. *J. Biol. Chem.* 257:8481–88

371. Wise, B. C., Glass, D. B., Chou, C.-H. J., Raynor, R. L., Katoh, N., et al. 1982. *J. Biol. Chem.* 257:8489–95

372. Schatzman, R. C., Raynor, R. L., Fritz, R. B., Kuo, J. F. 1983. *Biochem. J.* 209:435–43

373. Kuo, J. F., Andersson, R. G. G., Wise, B. C., Mackerlova, L., Salomonsson, I., et al. 1980. *Proc. Natl. Acad. Sci. USA* 77:7039–43

374. Minakuchi, R., Takai, Y., Yu, B., Nishizuka, Y. 1981. *J. Biochem.* 89: 1651–54

374a. Wu, W. C.-S., Walaas, S. I., Nairn, A. C., Greengard, P. 1982. *Proc. Natl. Acad. Sci. USA* 79:5249–53

375. Turner, R. S., Raynor, R. Y., Mazzei, G. J., Girard, P. R., Kuo, J. F. 1984. *Proc. Natl. Acad. Sci. USA* 81:3143–47

376. Katoh, N., Kuo, J. F. 1982. *Biochem. Biophys. Res. Commun.* 106:590–95

377. Cochet, C., Gill, G. N., Meisenhelder, J., Cooper, J. A., Hunter, T. 1984. *J. Biol. Chem.* 259:2553–58

378. Iwasa, Y., Hosey, M. M. 1984. *J. Biol. Chem.* 259:536–40

379. Mousesian, M. A., Nishikawa, M., Adelstein, R. S. 1984. *J. Biol. Chem.* 259:8029–32

380. LePeuch, C. J., Ballester, R., Rosen, O. M. 1983. *Proc. Natl. Acad. Sci. USA* 80:6858–62

381. Werth, D. K., Niedel, J. E., Pastan, I. 1983. *J. Biol. Chem.* 258:11423–26

382. Schatzman, R. C., Grifo, J. A., Merrick, W. C., Kuo, J. F. 1983. *FEBS Lett.* 159:167–70

383. Katoh, N., Wise, B. C., Kuo, J. F. 1983. *Biochem. J.* 209:189–95

384. Endo, T., Naka, M., Hidaka, H. 1982. *Biochem. Biophys. Res. Commun.* 105: 942–48

385. Nishikawa, M., Hidaka, H., Adelstein, R. S. 1983. *J. Biol. Chem.* 258:14069–72

386. Roach, P. J., Goldman, M. 1983. *Proc. Natl. Acad. Sci. USA* 80:7170–73

387. Ahmad, Z., Lee, F.-T., De Paoli-Roach, A., Roach, P. J. 1984. *J. Biol. Chem.* 259:8743–47

388. Huganir, R. L., Albert, K. A., Green-

389. Ling, E., Sapirstein, V. 1984. *Biochem. Biophys. Res. Commun.* 120:291–98

390. Wrenn, R. W., Katoh, N., Wise, B. C., Kuo, J. F. 1980. *J. Biol. Chem.* 255:12042–46

391. Katoh, N., Wrenn, R. W., Wise, B. C., Shoji, M., Kuo, J. F. 1981. *Proc. Natl. Acad. Sci. USA* 78:4813–17

392. Wrenn, R. W., Katoh, N., Kuo, J. F. 1981. *Biochem. Biophys. Acta* 676:266–69

393. Wrenn, R. W., Wooten, M. W. 1984. *Life Sci.* 35:267–76

394. Turner, R. S., Chou, C.-H. J., Kibler, R. F., Kuo, J. F. 1982. *J. Neurochem.* 39:1397–404

395. Albert, K. A., Wu, W. C.-S., Nairn, A. C., Greengard, P. 1984. *Proc. Natl. Acad. Sci. USA* 81:3622–25

396. Aloyo, V. J., Zwiers, H., Gispen, W. H. 1983. *J. Neurochem.* 44:649–53

397. Haycock, J. W., Bennett, W. F., George, R. J., Waymire, J. C. 1982. *J. Biol. Chem.* 257:13699–703

398. El Mestikawy, S., Glowinsky, J., Hamon, M. 1983. *Nature* 302:830–32

399. Endo, T., Hidaka, H. 1980. *Biochem. Biophys. Res. Commun.* 97:553–58

400. Sulakhe, P. V., Petrali, E. H., Davis, E. R., Thiessen, B. J. 1980. *Biochemistry* 19:5363–71

401. Murray, N., Steck, A. J. 1983. *J. Neurochem.* 41:543–48

402. Murray, N., Steck, A. J. 1984. *J. Neurochem.* 43:243–48

402a. Martenson, R. E., Law, M. J., Deibler, G. E. 1983. *J. Biol. Chem.* 258:930–37

402b. Turner, R. S., Chou, C.-H. J., Mazzei, G. J., Dembure, P., Kuo, J. F. 1984. *J. Neurochem.* 43:1257–64

403. Knight, D. E., Baker, P. F. 1983. *FEBS Lett.* 160:98–100

404. Baker, P. F., Knight, D. E. 1984. *Trends Neurosci.* 7:120–26

405. Tanigawa, K., Kuzuya, H., Imura, H., Taniguchi, H., Baba, S. et al. 1982. *FEBS Lett.* 138:183–86

406. Zwiers, H., Veldhuis, H. D., Schotman, P., Gispen, W. H. 1976. *Neurochem. Res.* 1:669–77

407. Zwiers, H., Wiegant, V. M., Schotman, P., Gispen, W. H. 1978. *Neurochem. Res.* 3:455–63

408. Zwiers, H., Schotman, P., Gispen, W. H. 1980. *J. Neurochem.* 34:1689–99

409. Oestreicher, A. B., Zwiers, H., Schotman, P., Gispen, W. H. 1981. *Brain Res. Bull.* 6:145–53

410. Sorenson, R. G., Kleine, L. P., Mahler, H. R. 1981. *Brain Res. Bull.* 7:57–61

411. Kristjansson, G. I., Zwiers, H., Oestreicher, A. B., Gispen, W. H. 1982. *J. Neurochem.* 39:371–78
412. Jolles, J., van Dongen, C. J., Schotman, P., Wirtz, K. W. A., Gispen, W. H. 1980. *Nature* 286:623–25
413. Oestreicher, A. B., van Dongen, C. J., Zwiers, H., Gispen, W. H. 1983. *J. Neurochem.* 41:331–40
414. Hawthorne, J. N., Pickard, M. R. 1979. *J. Neurochem.* 32:5–14
415. Downes, C. P. 1983. *Trends Neurosci.* 6:313–16
416. Mori, T., Takai, Y., Yu, B., Takahashi, J., Nishizuka, Y., Fujikura, T. 1982. *J. Biochem.* 91:427–31
417. Kawahara, Y., Takai, Y., Minakuchi, R., Sano, K., Nishizuka, Y. 1980. *Biochem. Biophys. Res. Commun.* 97:309–17
418. Baker, P. 1984. *Nature* 310:629–30
418a. Berridge, M. J., Irvine, R. F. 1984. *Nature* 312:315–21
419. Papahadjopoulous, D., Poste, G., Vail, W. J. 1979. *Methods Membr. Biol.* 10:1–21
420. Ekerdt, R., Papahadjopoulous, D. 1982. *Proc. Natl. Acad. Sci. USA* 79:2273–77
421. Mori, T., Takai, Y., Minakuchi, R., Yu, B., Nishizuka, Y. 1980. *J. Biol. Chem.* 255:8378–80
422. Schatzman, R. C., Wise, B. C., Kuo, J. F. 1981. *Biochem. Biophys. Res. Commun.* 98:669–76
423. Kraft, A. S., Anderson, W. B., Cooper, H. L., Sando, J. J. 1982. *J. Biol. Chem.* 257:13193–96
424. Kraft, A. S., Anderson, W. B. 1983. *Nature* 301:621–23
425. DeReimer, S. A., Strong, J. A., Albert, K. A., Greengard, P., Kaczmarek, L. K. 1985. *Nature.* 313:313–16
426. Hathaway, G. M., Traugh, J. A. 1982. *Curr. Top. Cell. Regul.* 21:101–27
427. Rabinowitz, M., Lipmann, F. 1960. *J. Biol. Chem.* 235:1043–50
428. Rodnight, R., Lavin, B. E. 1964. *Biochem. J.* 93:84–91
429. Wålinder, O. 1973. *Biochim. Biophys. Acta* 293:140–49
430. Hathaway, G. M., Lundak, T. S., Tahara, S. M., Traugh, J. A. 1979. *Methods Enzymol.* 60:495–511
431. Dahmus, M. E. 1981. *J. Biol. Chem.* 256:3332–39
432. Hosey, M. M., Tao, M. 1977. *Biochemistry* 16:4578–83
433. Murakami, N., Matsumura, S., Kumon, A. 1984. *J. Biochem.* 95:651–60
434. DePaoli-Roach, A. A., Roach, P. J.,

Larner, J. 1979. *J. Biol. Chem.* 254:12062–68
435. Cohen, P., Yellowlees, D., Aitken, A., Donella-Deana, A., Hemmings, B. A., Parker, P. J. 1982. *Eur. J. Biochem.* 124:21–35
436. Hemmings, B. A., Aitken, A., Cohen, P., Rymond, M., Hofmann, F. 1982. *Eur. J. Biochem.* 127:473–81
437. Meggio, F., Donella-Deana, A., Pinna, L. A. 1981. *J. Biol. Chem.* 256:11958–61
438. Miyamoto, E. 1975. *J. Neurochem.* 24:503–12
439. Miyamoto, E., Miyzaki, K., Hirose, R., Kashiba, A. 1978. *J. Neurochem.* 31:269–75
440. Chan, K.-F. J., Romano, C., Greengard, P. 1983. *Fed. Proc.* 42:2248
441. Heuser, J. E., Reese, T. S. 1979. In *The Neurosciences, Fourth Study Program,* ed. F. O. Schmitt, F. G. Worden, pp. 573–600. Cambridge: MIT
442. Pearse, B. M. F., Bretscher, M. S. 1981. *Ann. Rev. Biochem.* 50:85–101
443. Pauloin, A., Bernier, I., Jolles, P. 1982. *Nature* 298:574–76
444. Kadota, K., Usami, M., Takahashi, A. 1982. *Biomed. Res.* 3:575–78
445. Moskowitz, N., Glassman, A., Ores, C., Schook, W., Puszkin, S. 1983. *J. Neurochem.* 40:711–18
446. Runge, M. S., El-Maghrabi, M. R., Claus, T. S., Pilkis, S. J., Williams, R. C. Jr. 1981. *Biochemistry* 20:175–80
447. Shecket, G., Lasek, R. J. 1982. *J. Biol. Chem.* 257:4788–95
448. Julien, J.-P., Smoluk, G. D., Mushynski, W. E. 1983. *Biochim. Biophys. Acta* 755:25–31
449. Toru-Delbauffe, D., Pierre, M. 1983. *FEBS Lett.* 162:230–34
450. Hemmings, B. A., Aitken, A., Cohen, P., Rymond, M., Hofmann, F. 1982. *Eur. J. Biochem.* 127:473–81
451. Hemmings, B. A., Resink, T. J., Cohen, P. 1982. *FEBS Lett.* 150:319–24
452. Hemmings, B. A., Yellowlees, D., Kernohan, J. C., Cohen, P. 1981. *Eur. J. Biochem.* 119:443–51
452a. Resink, T. J., Hemmings, B. A., Tung, H. Y. L., Cohen, P. 1983. *Eur. J. Biochem.* 133:455–61
453. Hunter, T., Sefton, B. M. 1982. *Mol. Aspects Cell. Regul.* 2:337–70
454. Hunter, T., Sefton, B. M. 1980. *Proc. Natl. Acad. Sci. USA* 77:1311–15
455. Oppermann, H., Levinson, A. D., Varmus, H. E., Levintow, L., Bishop, J. M. 1979. *Proc. Natl. Acad. Sci. USA* 76:1804–8

456. Sefton, B. M., Hunter, T., Beeman, K. 1980. *Proc. Natl. Acad. Sci. USA* 77:2059–63
457. Barnekow, A., Schartl, M., Anders, F., Bauer, H. 1982. *Cancer Res.* 42:2429–33
458. Cotton, P. C., Brugge, J. S. 1983. *Mol. Cell. Biol.* 3:1157–62
459. Jacobs, C., Rübsamen, H. 1983. *Cancer Res.* 43:1696–702
460. Tuy, F. P. D., Henry, J., Rosenfeld, C., Kahn, A. 1983. *Nature* 305:435–38
461. Zick, Y., Whittaker, J., Roth, J. 1983. *J. Biol. Chem.* 258:3431–34
462. Rosen, O. M., Herrera, R., Olowe, Y., Petruzzelli, L. M., Cobb, M. H. 1983. *Proc. Natl. Acad. Sci. USA* 80:3237–40
463. Cohen, S., Fava, R. A., Sawyer, S. T. 1982. *Proc. Natl. Acad. Sci. USA* 79:6237–41
464. Avruch, J., Nemenoff, R. A., Blackshear, P. J., Pierce, M. W., Osathanondh, R. 1982. *J. Biol. Chem.* 257:15162–66
465. Erikson, E., Shealy, D. J., Erikson, R. L. 1981. *J. Biol. Chem.* 256:11381–84
466. Chinkers, M., Cohen, S. 1981. *Nature* 290:516–19
467. Kudlow, J. E., Buss, J. E., Gill, G. N. 1981. *Nature* 290:519–21
468. Downward, J., Yarden, Y., Mayes, E., Scrace, G., Totty, N., et al. 1984. *Nature* 307:521–27
468a. Havrankova, J., Schmechel, D., Roth, J., Brownstein, M. 1978. *Proc. Natl. Acad. Sci. USA* 75:5737–41
469. Havrankova, J., Roth, J., Brownstein, M. 1978. *Nature* 272:827–29
469a. Pacold, S. T., Blackard, W. G. 1979. *Endocrinology* 105:1452–57
470. Rees-Jones, R. W., Hendricks, S. A., Quarum, M., Roth, J. 1984. *J. Biol. Chem.* 259:3470–74
471. Fallon, J., Seroogy, K. B., Loughlin, S. E., Morrison, R. S., Bradshaw, R. A., et al. 1984. *Science* 224:1107–9
472. Huganir, R. L., Miles, K., Greengard, P. 1984. *Proc. Natl. Acad. Sci. USA.* 81:6968–72
473. Krust, B., Riviere, Y., Hovanessian, A. G. 1982. *Virology* 120:240–46
474. Levin, D. H., London, I. M. 1978. *Proc. Natl. Acad. Sci. USA* 75:1121–25
475. Tahara, S. M., Traugh, J. A. 1981. *J. Biol. Chem.* 256:11558–64
475a. Lubben, T. H., Traugh, J. A. 1983. *J. Biol. Chem.* 258:13992–97
476. Tuazon, P. T., Traugh, J. A. 1984. *J. Biol. Chem.* 259:541–46

477. Shichi, H., Somers, R. L. 1978. *J. Biol. Chem.* 253:7040–46
478. Lee, R. H., Brown, B. M., Lolley, R. N. 1981. *Biochemistry* 20:7532–38
479. Randle, P. S. 1981. *Curr. Top. Cell. Regul.* 18:107–29
480. Sheu, K.-F. R., Lai, J. C. K., Blass, J. P. 1984. *J. Neurochem.* 42:230–36
481. Browning, M., Dunwiddie, T., Bennett, W., Gispen, W., Lynch, G. 1979. *Science* 203:60–62
482. Lynch, G., Browning, M., Bennett, W. F. 1979. *Fed. Proc.* 38:2117–32
483. Magilein, G., Gordon, A., Au, A., Diamond, I. 1981. *J. Neurochem.* 36:1861–64
484. Qi, D.-F., Turner, R. S., Kuo, J. F. 1984. *J. Neurochem.* 42:458–65
485. Qi, D.-F., Kuo, J. F. 1984. *J. Neurochem.* 43:256–60
486. Li, H.-C. 1982. *Curr. Top. Cell. Regul.* 21:129–74
487. Cohen, P. 1982. *Nature* 296:613–20
488. Ingebritsen, T. S., Cohen, P. 1983. *Science* 221:331–38
489. Ingebritsen, T. S., Cohen, P. 1983. *Eur. J. Biochem.* 132:255–61
490. Brautigan, D. L., Bornstein, P., Gallis, B. 1981. *J. Biol. Chem.* 256:6519–22
491. Foulkes, J. G., Howard, R. F., Ziemiecki, A. 1981. *FEBS Lett.* 130:197–200
492. Hörlein, D., Gallis, B., Brautigan, D. L., Bornstein, P. 1982. *Biochemistry* 21:5577–84
493. Foulkes, J. G., Erikson, E., Erikson, R. L. 1983. *J. Biol. Chem.* 258:431–38
494. Chernoff, J., Li, H.-C. 1983. *Arch. Biochem. Biophys.* 226:517–30
495. Wallace, R. W., Tallant, E. A., Cheung, W. Y. 1980. *Biochemistry* 19:1831–37
496. Sheu, K.-F. R., Lai, J. C. K., Blass, J. P. 1983. *J. Neurochem.* 40:1366–72
497. Miyamoto, E., Kakiuchi, S. 1975. *Biochim. Biophys. Acta* 384:458–65
498. McNamara, J. O., Appel, S. H. 1977. *J. Neurochem.* 29:27–35
499. Wu, N. C., Martinez, J. J., Ahmad, F. 1980. *FEBS Lett.* 116:157–60
500. Goridis, C., Weller, M. 1976. *Adv. Biochem. Psychopharmacol.* 15:391–412
501. Teichberg, V. I., Sobel, A., Changeux, J.-P. 1977. *Nature* 267:540–42
502. Gordon, A. S., Milfay, D., Davis, C. G., Diamond, I. 1979. *Biochem. Biophys. Res. Commun.* 87:876–83
503. Coughlin, B. A., White, H. D., Purich,

D. L. 1980. *Biochem. Res. Commun.* 92:89–94

504. Maeno, H., Greengard, P. 1972. *J. Biol. Chem.* 247:3269–77

505. Stewart, A. A., Ingebritsen, T. S., Manalan, A., Klee, C. B., Cohen, P. 1982. *FEBS Lett.* 137:80–84

506. King, M. M., Huang, C. Y., Chock, P. B., Nairn, A. C., Hemmings, H. C. Jr., et al. 1984. *J. Biol. Chem.* 259:8080–83

507. Wood, J. G., Wallace, R. W., Whitaker, J. N., Cheung, W. Y. 1980. *J. Cell Biol.* 84:66–76

508. Klee, C. B., Crouch, T. H., Krinks, M. H. 1979. *Proc. Natl. Acad. Sci. USA* 76:6270–73

509. Khandelwal, R. L., Sloan, S. K., Craw, D. J. 1981. *Cold Spring Harbor Conf. Cell Prolif.* 8:473–81

510. Grankowski, N., Lehmusvirta, D., Stearns, G. B., Kramer, G., Hardesty, B. 1980. *J. Biol. Chem.* 255:5755–62

511. Defreyn, G., Goris, J., Merlevede, W. 1977. *FEBS Lett.* 79:125–28

Ann. Rev. Biochem. 1985. 54:977–1014

CHEMICAL PROBES OF THE MITOCHONDRIAL ATP SYNTHESIS AND TRANSLOCATION[1]

Pierre V. Vignais and Joël Lunardi

Laboratoire de Biochimie, Département de Recherche Fondamentale, Centre d'Etudes Nucléaires, 85 X, 38041 Grenoble, France

CONTENTS

[1] Abbreviations used: AdN, Adenine nucleotide; ATP synthase, ATPase complex, or F_0-F_1 complex; F_1, soluble sector of the ATP synthase; F_0, membrane sector of the ATP synthase; OSCP, oligomycin sensitivity conferring protein; dansyl-ADP(ATP), 2'-(5-dimethylamino naphtalene-1-sulfonyl)amino-2'-desoxy ADP(ATP); 8-Br-ADP(ATP), 8-bromo ADP(ATP); CATR, carboxyatractyloside; ATR, atractyloside; BA, bongkrekic acid; isoBA, isobongkrekic acid; EEDQ, N-ethoxy-carbonyl-2-ethoxy-1,2-dihydroquinoline; DCCD, dicyclohexylcarbodiimide; Nbf, 4-chloro-7-nitrobenzofurazan; NEM, N-ethylmaleimide; EMA, eosine 5-maleimide; EDAC, ethyl-3[(dimethylamino)propyl]carbodiimide; MABI, methyl 4-azidobenzimidate; HNB, 2-hydroxy-5-nitrobenzylbromide; ANPP, 4-azido-2-nitrophenyl phosphate. Other abbreviations are given in Table 1.

977

0066-4154/85/0701-0977$02.00

PERSPECTIVES AND SUMMARY

The last ten years have witnessed considerable developments in the use of chemical probes in the field of bioenergetics. The purpose of the present review is to discuss the utilization of these probes for the study of the structure and the functioning of the mitochondrial ATPase complex and AdN carrier. In eukaryotic cells, most of the ATP required to drive energy-consuming reactions, such as the synthesis of biomolecules, active transport, and mechanical work in muscle tissue, is synthesized in mitochondria via the so-called ADP/ATP cycle. The bulk of ATP is synthesized from ADP within the mitochondrial matrix by the enzyme ATP synthase, via the process known as oxidative phosphorylation. ATP is then exported from the mitochondria in exchange for cytosolic ADP, in a 1:1 stoichiometry. This exchange transport of ATP and ADP is catalyzed by an enzyme known as the adenine nucleotide (AdN) carrier. The intense activity of the ADP/ATP cycle, and of its key enzymes, the ATP synthase and the AdN carrier, is illustrated by the fact that an active human adult uses and resynthesizes about its own weight in ATP every day.

Chemical probes can be divided into two groups, depending on whether their binding is covalent or noncovalent. The noncovalent probes that have been applied to ATP-synthase and the AdN carrier comprise nucleotide analogs and specific inhibitors, and include fluorescent, spin-labeled and radiolabeled molecules. These probes have been used to determine the binding parameters of strategic sites and to monitor conformational changes in both the membrane-bound and isolated forms of ATP synthase and the AdN carrier. Covalent probes can be subdivided into group-directed reagents, cross-linking reagents, and active site–directed reagents. Group-directed reagents are relatively selective chemical modifiers of amino acid side chains; the identification of the modified amino acid residues together with the correlation between the extent of binding of a modifier and the resulting inactivation have provided interesting information about the active site and the catalytic mechanism of both the mitochondrial ATP synthase and the AdN carrier. Cross-linkers, which can also be considered as group-directed reagents, have been applied to the ATP synthase to investigate the arrangement of its subunits. Active site–directed reagents comprise affinity and photoaffinity labels, and have been used to map specific sites in ATP synthase and the AdN carrier.

Whereas some topographical and mechanistic aspects of the ATP synthase and AdN carrier have been explored successfully with chemical probes, others remain to be elucidated. Chemical probes have proved useful tools for determining the stoichiometry and the arrangement of subunits and for mapping nucleotide binding sites in the F_1 sector of the mitochondrial ATP synthase. However the structural organization of the mitochondrial F_0 sector remains unclear. As a result of chemical studies, the concept of catalytic cooperativity between subunits in the F_1 sector has gained wide acceptance, and its detailed mechanism is currently being investigated. Chemical studies of the AdN carrier have resulted in the demonstration of ADP- and ATP-dependent conformational changes, and a specific inhibitor binding site has been mapped. However, the mechanism of cooperativity between the nucleotide binding sites of the carrier, and the relationship of this cooperativity to the vectorial step of AdN carrier transport, are still poorly understood.

INTRODUCTION

In the present article, the application of noncovalent and covalent probes to the study of the mitochondrial ATP synthase and the AdN carrier will be discussed in turn. The experimental data will then be integrated around selected topics bearing on topography and mechanism. We have not covered the use of immunochemical probes, nor that of intrinsic probes. As a basis for discussion of the utilization of chemical probes, we shall now outline the essential structural features of the mitochondrial ATP synthase and AdN carrier.

The mitochondrial ATP synthase, also called the ATPase complex, catalyzes both ATP synthesis and hydrolysis. It consists of two structurally distinct components, a membrane-bound sector, F_0, which functions as a proton-translocating channel (1), and an extramembrane sector, F_1, which has catalytic activity (2). There are similarities in structure and mechanism between the mitochondrial, bacterial, and chloroplast ATP synthases, all of which are H^+-dependent. Mitochondrial F_1 ($M_r = 360,000-347,000$) (3, 4) is composed of five types of subunit called α, β, γ, δ, and ϵ; the stoichiometry of the subunits is $\alpha_3\beta_3\gamma\delta\epsilon$ (5, 6). The catalytic site is most likely on the β subunit. A small peptide of M_r 9,578 (7, 8), known as the natural ATPase inhibitor (IF_1) (9), is loosely attached to the β subunit of the mitochondrial F_1; this protein is involved in the regulation of the catalytic activity of F_1. Although the F_1 and F_0 sectors are likely in contact in certain areas, additional links are provided by two polypeptides of low molecular weight, the oligomycin-sensitivity conferring protein (OSCP) (10) of M_r 20,967 (11) and F_6 of M_r about 8,000 (12). These proteins are thought to compose the stalk which in electron micrographs appears to join F_1 and F_0. Whereas the structural organization of the subunits of mitochondrial F_1 is virtually elucidated, that of the subunits of the mitochondrial F_0 sector is far from being understood, and even their number and molecular weights are still debated (13–21).

The AdN carrier is a typical intrinsic membrane protein, which spans the entire width of the inner mitochondrial membrane. Exploration of the structure and functioning of the AdN carrier has been considerably aided by the use of two classes of specific inhibitors, the atractylosides, namely atractyloside (ATR) and carboxyatractyloside (CATR), which bind to the cytosolic face of the AdN carrier in the inner membrane of mitochondria, and the bongkrekic acids, namely bongkrekic acid (BA) and isobongkrekic acid (isoBA), which bind to the matrix face of the AdN carrier (22). The minimal molecular weight of the AdN carrier is about 32,000. When isolated in the presence of the specific and quasi-irreversible inhibitor, CATR, the AdN carrier behaves as a dimer (23, 24); its degree of oligomerization in the membrane is, however, unknown.

NONCOVALENT PROBES

Adenine Nucleotide Analogs

Many analogs exist of ATP and ADP, the substrates of ATP synthase and the AdN carrier. These analogs can be divided into three classes, according to which of the three structural moieties—phosphate chain, ribose sugar, or adenine base—is modified. Since the true substrate for ATP synthase is ATP or ADP complexed with Mg^{2+}, rather than the free nucleotide, it is possible to consider a fourth class of analogs in which Mg^{2+} is replaced by another cation (metallo-derivatives of ADP and ATP).

MODIFIED PHOSPHATE CHAIN   The length of the phosphate chain is critical in nucleotide recognition by mitochondrial ATPase and by the AdN carrier; among the natural nucleotides only di- and triphosphates are recognized. The oxygen bridge between the β and γ phosphorus atoms in ATP can be replaced by a $-CH_2$ group or by an $-NH_2$ group, resulting in adenylyl β,γ methylene diphosphate (AMPPCP) or adenylyl β,γ imido-diphosphate (AMPPNP), respectively (25, 26). These analogs are not hydrolyzed by mitochondrial F_1; however they are tiansported by the AdN carrier (26, 27). Similarly, the α,β methylene analog of ADP and the hypophosphate of ADP (no α,β oxygen bridge) are transported into mitochondria, but are not phosphorylated (27, 28). Although not hydrolyzed by F_1, AMPPNP binds to the enzyme with high affinity, making this analog an interesting probe of the nucleotide binding sites of F_1. Binding studies with (^{3}H)AMPPNP have revealed the presence of six nucleotide binding sites on beef heart F_1; nucleotides are freely exchangeable at three sites, and tightly bound at the others (29).

Nucleotide phosphorothioates have been synthesized by replacing a non-bridging oxygen atom of a phosphate group of ADP or ATP with a sulfur atom (30, 31). A number of these analogs are transported by the mitochondrial AdN carrier, although the rate of transport is lower than that of the unmodified nucleotides, presumably because of the larger size and the lower electronegativity of sulfur compared to oxygen (32). One thiophosphate analog of ATP, adenosine 5'-O-(3 thio-triphosphate) (ATPγS), labeled stereospecifically with ^{18}O in the γ position, has provided useful information on the stereochemical course of phosphoric residue transfer catalyzed by beef heart mitochondrial ATPase. The results showed that the most likely mechanism for ATP hydrolysis by mitochondrial F_1 is direct, in-line displacement of ADP by a water oxygen in a single step (33).

Arsenate, being isosteric and isoelectronic with phosphate, can replace phosphate in a number of enzymatic reactions. Thus, beef heart submitochondrial particles oxidizing succinate showed a transient synthesis of ADP-arsenate from ADP and arsenate (34). The V_{max} and K_m values for the synthesis of ADP-arsenate from ADP and arsenate were similar to those for the synthesis of ATP from ADP and P_i (35).

MODIFIED RIBOSE MOIETY   The stringent steric requirement of the nucleotide site of the AdN carrier for the ribose moiety and the purine ring of the substrate contrasts with the correspondingly loose stereospecificity of mitochondrial F_1. The structural features of the sugar moiety which are essential for AdN transport have been extensively discussed (36). Among the ribose-modified analogs, the 3'-derivatives are of particular interest, since modification at the C3' position does not significantly alter recognition of the analogs by mitochondrial ATPase and the AdN carrier. The 3'-esters of ATP act as substrates for mitochondrial F_1-ATPase, whereas the 3'-esters of ADP are powerful and highly specific inhibitors, this inhibition being competitive with respect to ADP (37, 38). In addition to the photoreactive 3'-analogs which will be discussed later (37–43), a number of fluorescent molecules such as 3'-naphthoyl ADP (N-ADP), 3'-dimethyl-aminonaphthoyl ADP (DMAN-ADP) (44–48), and 2',3'-O(2,4,6)trinitrophenyl ATP (TNP-ATP) or its ADP homolog (TNP-ADP) (49, 50) have been used to explore the nucleotide binding sites of mitochondrial F_1.

The thermodynamic binding parameters of ADP, N-ADP, and DMAN-ADP to nucleotide-depleted F_1 have been compared; the results were interpreted in terms of anticooperative interactions between the nucleotide sites (45). Titrations of the mitochondrial F_1 with TNP-ATP and TNP-ADP revealed two binding sites; one site had a very high affinity which was beyond the measurable range; the K_d of the other site was in the range of 20–80 nM (49, 50). Occupancy of the very high affinity site by trace amounts of the radioactive analog, TNP-(γ-^{32}P)ATP, was accompanied by a low rate of hydrolysis of this analog. Further addition of nonradioactive TNP-ATP, at a concentration sufficient to fill the second catalytic site, accelerated the rate of hydrolysis of the previously bound TNP-(γ-^{32}P)ATP. A number of hydrolyzable nucleotides, including ATP, GTP, and ITP, also promoted hydrolysis of the previously bound TNP-(γ-^{32}P)ATP. These data afforded strong support for the existence of catalytic site cooperativity. The observation that ATP promotes hydrolysis of 2'-(5-dimethylamino-naphthalene-1-sulfonyl)amino-2'-desoxy ATP (dansyl ATP) by beef heart F_1 (51) can be explained in the same way. Differential effects of DMAN-ADP and TNP-ADP (or TNP-ATP) on the synthesis and hydrolysis of ATP in submitochondrial particles have been reported and interpreted in terms of a two-state model of bound F_1, one directed to ATP synthesis and the other to ATP hydrolysis (52).

N-ADP binds to the mitochondrial AdN carrier, but is not transported (46). In the case of the membrane-bound carrier, the sites titrated by N-ADP were those located on the outer face of the carrier (53). A single type of sites was demonstrated for the binding of N-ADP in contrast to two types of interacting sites for the transportable nucleotides, ADP, ATP, and AMPPCP. It is possible that only transportable nucleotides are able to induce, upon binding to the carrier, a conformational change that acts as a triggering signal for initiation of the vectorial reaction of transport. In the case of the isolated AdN carrier in detergent, the binding parameters for N-ADP were different from those of the membrane-bound carrier; instead of one type of sites, two types of sites ($K_d <$ 10 nM and 0.4 μM) were revealed (54).

N-ADP has also been used to probe the interactions of the two specific inhibitors of AdN transport, CATR and BA (47). The data indicated the existence of two distinct conformational states of the membrane-bound carrier, recognized by CATR and BA. The transition between the two conformations, which was rather slow in the absence of added nucleotides, was accelerated by transportable nucleotides such as ADP and ATP. Evidence for an ADP/ATP-dependent acceleration of the transition between the CATR and BA conformations has also been obtained by exploration of the changes in the intrinsic fluorescence of the isolated AdN carrier in detergent (55).

MODIFIED PURINE RING   Two-dimensional fluorescent probes of nucleotide binding sites with a stretched structure have been extensively studied. They are the N^1,N^6-etheno (ϵ) nucleotides, ϵ-ADP or ϵ-ATP (56), and the 8-amino-3(β-D-ribo-furanosyl)-imidazo[4–5]quinazoline-5'-nucleotides (*lin*-benzo ADP and ATP) (57). ϵ-ADP binds to a high affinity binding site of mitochondrial F_1; the rate of binding is slow, however, suggesting that a rate-limiting conformational change of F_1 is required for binding (58). ϵ-ATP is not transported by the AdN carrier of rat liver mitochondria, but it is hydrolyzed by the exposed F_1 of inverted submitochondrial particles, at a rate six times lower than that of ATP (59, 60). The K_m value for hydrolysis of ϵ-ATP is similar to that of ATP. *Lin*-benzo ADP and ATP differ from ADP and ATP by the insertion of a benzene molecule into the purine ring which makes these derivatives 2.4 Å wider than the natural nucleotides (57). *Lin*-benzo ADP is a substrate for oxidative phosphorylation in submitochondrial particles, while *lin*-benzo ATP is hydrolyzed by soluble F_1 (61). The binding affinity of P_i to F_1 is not altered by *lin*-benzo ADP, suggesting that the lateral extension of the purine ring by

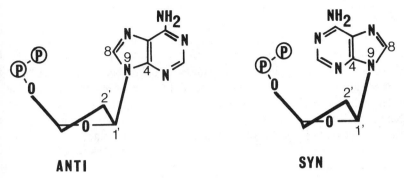

ANTI   **SYN**

Figure 1   Anti and syn conformations of ADP

2.4 Å does not obstruct the P_i binding site. The facile recognition of *lin*-benzo ADP and ATP by mitochondrial F_1 contrasts with their inability to be transported by the AdN carrier (61).

Nucleotides can adopt two conformations, *anti* and *syn*, depending on the orientation of the planar purine base with respect to the quasi planar ribose (Figure 1). For ADP or ATP analogs, this will depend on the substituted groups at the purine ring. In solution, ADP and ATP are preferentially in the *anti* conformation (62). To be transported by the AdN carrier, a nucleotide must have a non-fixed *anti* conformation; additionally, transport requires a C6 positioned amino group and an unsubstituted C2 atom (36, 63). For example, 8-Br ADP (64, 65) that is blocked in the *syn* conformation, and the derived 8-azido ADP in the dark (66) bind to the AdN carrier, but are not transported. The 2-azido ADP analog presumably has the favorable *anti* conformation (67–70). However, substitution in position 2 prevents transport (36). On the other hand, formycin di- and triphosphate (FDP and FTP) (63, 71), tubercidin di- and triphosphate (TuDP and TuTP) (63, 72), and l-*N* oxide-ADP and -ATP (32, 73) that have the *anti* conformation, and no steric restriction at difference with 2-azido ADP, bind to the AdN carrier and are transported.

The binding parameters of the transportable nucleotide, FTP, with respect to the isolated AdN carrier in detergent, have been assessed (74), and compared to those of the non-transportable nucleotide N-ATP (cf section on Modified Ribose Moiety). Instead of two classes of sites found for N-ATP, four distinct sites were titrated with FTP; one site was characterized by a much higher affinity for FTP ($K_d < 10$ nM) than the others (K_d 0.5–2 μM). This difference in the binding parameters of the solubilized AdN carrier for FTP and N-ATP most probably reflects a conformational difference between the resting (binding without transport) and functioning (binding with transport) carrier.

As already mentioned, the conformation requirement is less stringent for the nucleotide sites of F_1 than for those of the AdN carrier. For example, 8-azido

ADP, which is not transported, is able to bind reversibly in the dark to mitochondrial F_1 and to compete with ATP hydrolysis by F_1 (75). Covalent photolabeling by the 8- and 2-azido nucleotides will be discussed later (68, 69, 75–83). On the other hand, TuTP (72) and 1-N oxide ATP, which are transported by the AdN carrier, are good substrates for mitochondrial F_1 (84).

The study of the interactions of GDP, GTP, IDP, and ITP with mitochondrial F_1 led to the conclusion that F_1 possesses both catalytic and regulatory sites and that the nucleotide specificity for the regulatory sites is more strict than for the catalytic sites (85, 86). A factor of complexity, however, is the presence of the so-called tight nucleotide sites in mitochondrial F_1, i.e. sites occupied by non-exchangeable ATP and ADP (87). The tight sites have a stringent specificity for a restricted number of nucleotides (88, 89). For example, it was possible to deplete beef heart mitochondrial F_1 from its tightly bound nucleotides and to replace them by the analogs desoxy-ATP, FTP, and iso-GTP; this was not the case for GTP, ITP, and ϵ-ATP, although the latter nucleotides were readily hydrolyzed by F_1 (88).

The different noncovalent analogs of ADP and ATP which have been reviewed above are listed in Table 1. For convenience, Table 1 also includes ADP and ATP analogs used as affinity and photoaffinity labels, to be examined later in this article.

METALLO-DERIVATIVES OF ADP AND ATP   A large number of nucleotide-dependent enzymes have a requirement for divalent cations such as Mg^{2+}, Ca^{2+}, and Zn^{2+}. For efficient ATP hydrolysis by mitochondrial F_1-ATPase to occur, the medium must be supplemented with Mg^{2+} (90). On the other hand, ADP and ATP are transported as free nucleotides by the mitochondrial AdN carrier (55). Very little information is available concerning the function of Mg^{2+} during ATP hydrolysis or synthesis, and the particular form of Mg-ATP used by F_1. The difficulty resides in the fact that magnesium exchanges oxygen ligands very quickly and that consequently the Mg-ATP complex can rapidly assume many possible chelate structures; it can be overcome by the use of stable chromium-nucleotide complexes. In fact, chromium (III) ligands exchange 10^{10}–10^{13} times more slowly than those of magnesium (II) (91–96). Mono-, bi- and tridentate complexes of metal nucleotides are now available, and the geometrical and optical isomers can be readily separated, allowing more accurate exploration of the binding sites. Among Cr-ATP complexes, only the mono and tridentate complexes acted as inhibitors of mitochondrial F_1 (97). On the other hand, monodentate Cr-ADP behaved as a substrate for soluble mitochondrial F_1; in the presence of P_i, the condensation product P_i-Cr(III)ADP was formed (98, 99). The analog P_i-Cr(III)ADP transiently accumulated upon incubation of F_1 with Cr(III)ATP, in accordance with the view that the reaction pathway for ATP hydrolysis is the same as for ATP synthesis.

^{18}O from $H^{18}OH$ as Probe of Elementary Steps of the F_1-Catalyzed Reaction

The extent of the F_1-catalyzed incorporation of ^{18}O from $H^{18}OH$ into P_i and ATP has been used to probe the reaction rates of F_1-bound intermediates (100). The premise was that incorporation of ^{18}O occurs by exchange and that this exchange reflects the reversible cleavage of the bound ATP at the catalytic site into ADP and P_i. The systematic application of this technique to the study of F_1 mechanism stemmed from the unexpected finding that the $P_i \rightleftharpoons H^{18}OH$ exchange that occurs during net ATP synthesis in coupled submitochondrial particles is insensitive to uncouplers (101, 102). An important conclusion was that in the course of ATP synthesis, energy is required essentially to change ATP binding from tight to loose, and, in contrast, the P_i and ADP binding from loose to tight. A mechanism of catalytic cooperativity between alternating catalytic sites was proposed. This alternating site mechanism was further substantiated by ^{18}O exchange experiments carried out with soluble mitochondrial F_1. During the course of the F_1-catalyzed intermediate $P_i \rightleftharpoons H^{18}OH$ exchange, where P_i comes from ATP hydrolysis, decreasing the concentration of ATP resulted in extensive incorporation of ^{18}O into P_i released (103). Just the opposite occurred when the concentration of ATP was raised. The given explanation was that, at low concentration of ATP, the products ADP and P_i accumulated at a catalytic site in a tightly bound form inaccessible to pyruvate kinase (104) and that $P_i \rightleftharpoons H^{18}OH$ exchange occurred at their level through the reaction $ATP + H^{18}OH \rightleftharpoons ADP + P_i$. At higher concentrations of ATP, all three catalytic sites were filled, and the bound P_i was readily released, resulting in incorporation of slightly more than one ^{18}O atom into each molecule of P_i released. In complementary experiments, $(\gamma\text{-}^{18}O)ATP$ was exposed to soluble F_1, and the distribution of species with 3, 2, 1, or zero ^{18}O atoms per molecule of P_i formed was measured by mass spectrometry (105, 106). The observed distribution ranged from a limit of about 1 at saturating ATP concentrations to a limit close to 4 at very low ATP concentrations; it was entirely consistent with the theoretical pattern predicted by the alternating site mechanism.

Inhibitors of the Mitochondrial ATPase Complex

The reversible inhibitors of the mitochondrial ATPase complex can be divided into the following groups: specific inhibitors of the F_1 sector, which comprise the aurovertins and efrapeptin; nonspecific inhibitors of the F_1 sector, which include quercetin, bathophenanthroline and azide; and finally the oligomycins and venturicidins which are fairly selective inhibitors of the F_0 sector. The structure and properties of these inhibitors have been reviewed in detail (107), and only new data on the specific inhibitors aurovertin D and efrapeptin will be reported.

Table 1 Most commonly used covalent and noncovalent analogs of adenine nucleotides

Name	Abbreviation	Formula	Reference(s)
Phosphate Chain Modified Nucleotides			
Adenylyl β,γ,imidodiphosphate	AMPPNP	Ado–O–(P)–O–(P)–N–(P)	25, 26, 29
Adenylyl β,γ,methylene diphosphate	AMPPCP	Ado–O–(P)–O–(P)–C–(P)	26, 27
Adenosine 5'-O-thiodiphosphate thiotriphosphate	ADP(αor βS) ATP(α,β or γS)	Ado–O–(P)–O–(P)–O–(P) (S)α (S)β (S)γ	30–33
Mixed anhydride of ADP and mestiylenecarboxylic acid	ADP-MC	Ado–O–(P)–O–(P)–O–CO– (CH₃ CH₃ CH₃)	222
p-Fluorosulfonyl-benzoyl-5'-adenosine	FSBA	Ado–O–CO–⬡–SO₂F	223
Ribose Modified Nucleotides			
2' or 3' deoxy ADP or ATP	d-ADP or ATP	Ad O CH₂–O–R OH (2') ; Ad O CH₂–O–R OH (3')	27, 84
2',3' dialdehyde ADP or ATP	dial ADP or ATP	Ad O CH₂–O–R CHO CHO	217–220

2',3'-O-(2,4,6-trinitro phenyl) ADP or ATP	TNP-ADP or ATP	49, 50
3'-esters of ADP or ATP		
3'-O-naphtoyl ADP or ATP	N-ADP or ATP	44, 47
3'-O-(5 dimethylamino naphtoyl-1) ADP	DMAN-ADP	45, 48
3'-O-[(N-(4-azido-2-nitro phenyl)amino)propionyl or butyryl]-ADP or ATP or AMPPNP	3'-arylazido ADP or ATP or AMPPNP	37–41, 224, 225
3'-O-(4-benzoyl)benzoyl ATP	Bz-ATP	42

Table 1 (*continued*)

Name	Abbreviation	Formula	Reference(s)
Purine Modified Nucleotides			
7-deaza ADP or ATP Tubercidin di or triphosphate	Tu-DP or Tu-TP		63, 72
1-N oxide ADP or ATP			32, 73
Formycin di or triphosphate	FDP or FTP		63, 71, 74
N^1,N^6-etheno ADP or ATP	ε-ADP or ε-ATP		56, 58–60
8-amino-3(β-D-ribofuranosyl) imidazo(4-5)quinazoline-5'- di or triphosphate	lin-benzo ADP or ATP		57, 61
8-azido ADP or ATP	8-N$_3$ ADP or ATP		75–81

			References
2-azido ADP			67–70, 227, 228
Mixed Modified Nucleotides			
8-azido-N^1,N^6-etheno ATP	8-N$_3$-ε-ATP		82, 226
2',3' dialdehyde-N^1,N^6-etheno ATP	dial-ε-ATP		221
8-azido-3'-O-[(N-(4-azido-2 nitrophenyl)aminopropionyl] ATP	3'-arylazido-8-N$_3$ ATP		83

Aurovertin D (108) binds to both mitochondrial and *E. coli* F_1 at the level of the β-subunit (109, 110). However, chloroplast F_1 and TF_1 from the thermophilic bacteria *PS3* are insensitive to and fail to bind aurovertin (111, 112). The fluorescence of aurovertin D is greatly enhanced when it binds to mitochondrial and bacterial F_1 (113–116). Moreover, the fluorescence of the F_1-aurovertin complex is markedly altered by the addition of nucleotides, P_i and Mg^{2+}. Quantitative measurements have been made of the fluorescence changes occurring upon binding of aurovertin D, ADP, and ATP (117). After binding of one mol aurovertin D to one mol mitochondrial F_1, the additional binding of one mol ADP was sufficient to produce maximal increase of fluorescence, whereas binding of one mol ATP per mol F_1 promoted maximal quenching.

(^{14}C)aurovertin D has been prepared by deacetylation followed by reacetylation with (^{14}C)acetic anhydride (109). On the basis of the incorporation of one (^{14}C)acetyl group per aurovertin D, a molar extinction coefficient (ϵ_{368nm}) of 36,100 $M^{-1}cm^{-1}$ was calculated (109). Direct titration with (^{14}C) aurovertin D, by means of equilibrium dialysis, demonstrated unambiguously that mitochondrial F_1 possesses three aurovertin binding sites of different affinities, all located on the β subunits (110). One of the sites bound (^{14}C)aurovertin D with high affinity (K_d 0.3 μM), whereas the other two exhibited an affinity one order of magnitude lower (K_d 3–5 μM). The isolated β subunit is capable of binding (^{14}C)aurovertin D with a K_d of 1 μM and a stoichiometry of 1 mol aurovertin per mol β subunit. As each mol F_1 bound three mol (^{14}C)aurovertin, it is clear that each F_1 contains three β subunits. Upon addition of ATP, two high affinity aurovertin binding sites (K_d 0.7 μM) were observed and only one low affinity binding site (K_d ~5 μM) remained. As clearly demonstrated for *E. coli* F_1 (109), aurovertin D inhibits the ATPase activity primarily by binding to the high affinity site of F_1.

While aurovertin D increased the binding affinity of beef heart F_1 for P_i (118) and that of rat liver F_1 for ADP in the presence of Mg^{2+} (119), it facilitated the dissociation of IF_1 from F_1, induced by the functioning of the respiratory chain (120). Conversely, IF_1 decreased the binding affinity of (^{14}C)aurovertin D for the three aurovertin binding sites of mitochondrial F_1 (110).

Efrapeptin is considered as a reliable probe of the catalytic site of mitochondrial F_1, on the basis of the following experimental data (*a*) it prevents P_i binding to mitochondrial F_1 (118); (*b*) it competes with both ADP and P_i during ATP synthesis by submitochondrial particles (121); (*c*) it prevents modification by phenylglyoxal of an essential arginyl residue in mitochondrial F_1 (122).

Inhibitors of the AdN Carrier

The specific reversible inhibitors of AdN transport in mitochondria fall into two classes. The first class comprises ATR and CATR, two heterosides extracted from the thistle *Atractylis gummifera*. The second class of inhibitors consists of

BA and isoBA, two fermentation products of *Pseudomonas cocovenenans,* which are long unsaturated tricarboxylic acids with pK values ranging from 4 to 6. These inhibitors have been used as probes of the orientation and conformation of the AdN carrier (22, 123).

It has been shown that radiolabeled ATR and CATR bind to the cytosolic face of the AdN carrier in mitochondria (123, 124), whereas radiolabeled BA and isoBA bind to the matrix face of the carrier (126, 127). To inhibit AdN transport, BA and isoBA have to penetrate through the lipid core of the mitochondrial membrane. They do so after protonation, which requires a slightly acidic pH. The reverse situation holds for inside-out submitochondrial particles. These results clearly indicate the membrane sidedness of the ATR(CATR) and BA(isoBA) binding sites of the AdN carrier. Radiolabeled BA, ATR, and CATR have also been used to determine the density of carrier sites in mitochondria from different sources: heart, liver, yeast, and plants (128). The proportion of the AdN carrier to the F_1-ATPase of about 2 in liver mitochondria and between 4 and 5 in heart mitochondria is in accordance with the fact that, in the cellular ADP/ATP cycle, AdN transport is less rate-limiting in heart than in liver (129).

The primary alcohol group of the glucose disulfate residue in ATR or CATR can be esterified without loss of biological activity (130). Esterified ATR derivatives have been prepared for affinity chromatography (131), ESR motion studies of the membrane-bound carrier (132), photoaffinity labeling (133), and fluorescent labeling (134) (Table 2).

That CATR and BA cannot bind to the same AdN transport unit at the same time was demonstrated in a double labeling experiment in which mitochondria were preincubated with (^{14}C)acetyl CATR; the prebound (^{14}C)acetyl CATR was released upon binding of (^{3}H)BA (130). Since the binding of CATR and that of BA are mutually exclusive, it is likely that two distinct conformations of the AdN carrier exist that are in equilibrium and that bind CATR (or ATR), and BA (or isoBA), respectively. As mentioned in a preceding chapter and as supported by current studies conducted with the fluorescent dansyl ATR, dansylaminobutyryl ATR, and naphthoyl ATR (135), the transition between the CATR and BA conformations is markedly accelerated by micromolar concentrations of ADP or ATP.

In the spin-labeled acyl ATR, the non-permeant ATR moiety interacts with the cytosolic side of the carrier, whereas the spin-labeled acyl chain is buried in the core of the membrane (132). Spin-labeled palmityl ATR was partially immobilized after incubation with mitochondria; upon addition of ADP, ATP, or inhibitory ligands, the spin label regained its mobility. This can be explained if, in the absence of added ligand, the spin-labeled acyl chain is in contact with the AdN carrier, and upon addition of a specific ligand, the spin-labeled chain is released to the lipid core. Consequently the AdN carrier and the lipid core are

Table 2 Noncovalent and covalent derivatives of atractyloside

	R_1	R_2	References
atractyloside (ATR)	–H	–H	123
6'-O-succinyl-ATR	–H	$-CO-(CH_2)_2-COOH$	131
6'-O-azidonitrophenyl-ATR	–H	$-CO-(CH_2)_n-$ (nitroazidophenyl)	133
6'-O-azidobenzoyl-ATR	–H	$-CO-$ (azidophenyl)	245

6'-O-spin labeled acyl-ATR	–H	$-CO-(CH_2)_n-C-(CH_2)_m-CH_3$ (N–O)	132
6'-O-dansyl-ATR	–H	$-SO_2$– naphthyl–$N-(CH_3)_2$	134
6'-O-naphtoyl-ATR	–H	$-CO$– naphthyl	134
carboxy-ATR	–COOH	–H	124
6'-O-acetyl-carboxy-ATR	–COOH	$-CO-CH_3$	130

in direct contact. Investigation of the degree of immobilization of the spin label as a function of its position in the acyl chain led to the conclusion that the carrier protein has a more globular shape at the surface of the membrane than in the middle of the core.

Palmityl CoA is a nonspecific inhibitor of the AdN carrier; however it binds to the AdN carrier with high affinity (136). Spin-labeled palmityl CoA has also been used to probe the membrane-bound AdN carrier. Essentially the same results were obtained as with spin-labeled palmityl ATR (137).

COVALENT PROBES

Group-Directed Reagents

Group-directed reagents are chemical modifiers that bind covalently with some degree of specificity to side-chains of amino acid residues in protein. Their binding, especially if it occurs in or close to the active site of an enzyme, leads to inactivation. Generally, partial inactivation is reflected in a decrease in V_{max}, with no change in K_m. Due to their general electrophilic nature, chemical modifiers preferentially react with nucleophiles at the active site (138). The reagent specificity with respect to a given amino acid is never absolute, and modification of an amino acid residue at distance from the active site may result in modification of the geometry of the site by a propagated change of conformation, so that the activity of the enzyme is lost. Protection against inactivation by substrates added prior to the chemical reagent is indicative that modification may have occurred at the active site. The kinetic determination of the stoichiometry of the inactivation reaction is advantageously complemented by chemical determination of the number of bound reagent molecules by means of radiolabeled reagents. The use of radiolabeled reagents also enables the exact position of the modified amino acid residue to be determined.

Cross-linking reagents comprise a class of modifiers which, by virtue of their bifunctional reactivity, can bind to two amino acid side chains (139). By varying the length of the cross-linker, it is possible to explore the distance between residues located in different subunits in oligomeric enzymes such as the F_1-F_0 complex. However the absence of cross-links does not rule out the vicinity of subunits, since, to be achieved, cross-linking requires not only contact between subunits, but also the accessibility of reactive amino acid residues.

CHEMICAL MODIFIERS OF ATP SYNTHASE

Carboxyl Group-Directed Reagents   Studies with the two carboxyl reagents, N-ethoxy-carbonyl-2-ethoxy-1,2-dihydroquinoline (EEDQ) (140) and dicyclohexylcarbodiimide (DCCD) (141), have led to the finding that the active site of mitochondrial F_1 contains a carboxyl group with an important role in catalysis. Addition of EEDQ to soluble mitochondrial F_1 resulted in a decrease in the rate of ATP hydrolysis, which was correlated with the amount of EEDQ bound; the binding stoichiometry was 1 mol EEDQ per mol active site, both for isolated F_1 and for membrane-bound F_1 in submitochondrial particles (142). EEDQ was most effective at neutral or slightly acidic pH; the half inactivation occurred at pH 7.3–7.5, indicating that the inactivated carboxyl had an unusually high pK value. ATP and divalent cations such as Mg^{2+}, Mn^{2+}, and Ca^{2+} protected against EEDQ inactivation, suggesting that the EEDQ-reactive carboxyl group was at or close to the active site, and that it might act as a ligand for divalent cations. *Escherichia coli* F_1 was also inactivated by EEDQ (143). By means of (^{14}C)glycyl ethyl ester, a nucleophile which binds to the activated carboxyl group by an amide bond after displacement of EEDQ, the EEDQ reactive carboxyl of *E. coli* F_1 was located in the β subunit. Most interestingly, inactivation of one β subunit out of three was sufficient to block ATPase activity completely (143).

At alkaline pH, DCCD binds to the F_0 sector of the mitochondrial ATPase complex at the level of a hydrophobic subunit known as the DCCD-binding protein; this results in the total inhibition of ATP synthesis (144, 145). At neutral or slightly acid pH, however, DCCD binds to the F_1 sector as well as to F_0 (146, 147). A single glutamic acid residue of the DCCD-binding protein of the F_0 sector was found to react with DCCD (148, 149). Although the DCCD-binding protein of the mitochondrial F_0 sector behaved as an hexamer (149), the binding of 1 mol DCCD per mol ATPase complex was sufficient to fully inhibit F_1-linked functions, such as ATP hydrolysis and ATP-P_i exchange (150). Covalent binding of (^{14}C)DCCD to the F_0 sector was inhibited by prior treatment of the ATPase complex with rutamycin and venturicidin; this suggested that the two antibiotics and DCCD might bind to the same protein of the F_0 sector, namely the DCCD-binding protein (151). Through the use of a spin-labeled derivative of DCCD applied to submitochondrial particles, the DCCD binding site of the F_0 sector was located at about 20 Å from an ATP binding site probed by Mn-ATP in F_1 (152).

As previously mentioned, DCCD also reacts with the F_1 sector of mitochondrial ATPase: the binding of two mol DCCD per mol F_1 resulted in full inactivation, and bound DCCD was located in the β subunit (146, 147). One of the two bound DCCD was readily displaced by glycyl ethyl ester, demonstrating that the displaced DCCD was initially bound to a carboxyl group. This carboxyl belongs to a glutamic acid residue (153) located at position 199 in the amino acid sequence of the mitochondrial β subunit (81). Inactivation by DCCD was decreased by Mg^{2+}, suggesting a role of the DCCD-reactive carboxyl in the binding of Mg^{2+}. Inactivation of F_1 by the modification of a carboxyl-group located in the β subunit has also been demonstrated for chloroplast F_1 (154) and for F_1 from *E. coli* (155, 156), *Rhodospirillum rubrum* (157), and the thermophilic bacterium PS3 (158). The binding stoichiometry for full inactivation of *E.coli* F_1 was of 1 mol DCCD bound per mol BF_1 (155).

In summary, the reaction of one EEDQ or DCCD with one β subunit among the three β present in F_1 results in full inactivation. Since the β subunit contains the catalytic site, this strongly suggests a ⅓ of the site reactivity. Under the conditions used for inactivation, the DCCD-induced cross-linking of subunits in F_1 is negligible (146, 155); however, a slight degree of cross-linking was observed with EEDQ (143).

Arginyl Group-Directed Reagents   Arginyl residues are commonly found at the active site of nucleotide processing enzymes. The positively charged guanidinium group in arginyl residues is ideally suited for interaction with nucleotides because of its planar structure and its ability to form hydrogen bonds with the phosphate chain of nucleotides (159). Butanedione, in the presence of borate, and phenylglyoxal have been used as reagents to modify arginyl residues in mitochondrial F_1.

In a study of beef heart and rat liver F_1 (160), the reaction order with respect to butanedione was found to be one, as was also the case for the inactivation of beef heart F_1 by phenylglyoxal. Binding studies with (^{14}C)phenylglyoxal clearly indicated that the preferential modification of a single arginine residue is responsible for inactivation. Likewise, inhibition of the ATPase and ATP-P_i exchange activities of an ATP synthase preparation (complex V) from beef heart mitochondria by phenylglyoxal and butanedione appeared to involve the binding of one molecule of inhibitor per active site (161).

Amino Group-Directed Reagents   The ε amino group of lysine is able to react with pyridoxal phosphate to form a Schiff base complex, which can be stabilized by borohydride; the reaction is best achieved at alkaline pH to ensure partial deprotonation of this amino group. In contrast to the other group-reagents mentioned above, substitution of a large number of lysine residues (about ten) was required for inactivation to occur in F_1 from mitochondria (162, 163), bacteria (164), and chloroplast (165). However, the fact that pyridoxal phosphate decreased the number of exchangeable nucleotide binding sites by one, and also the fact that ADP and ATP protected against inactivation, suggested that a positively charged lysine residue is present at the catalytic site.

The dial ADP and ATP, which are affinity labels, and 4-chloro-7-nitrobenzofurazan (Nbf) at alkaline pH react with lysine residues of mitochondrial F_1 with concomitant inactivation, as to be detailed later.

Labeling of a protein by a group-directed reagent may leave intact the biological activity of the protein. This is the case for the limited labeling of the lysine residues in the natural ATPase inhibitor, IF_1, by (^{14}C)phenyl isothiocyanate, [(^{14}C)PITC] (166), and by the hetero-bifunctional reagent, (^{14}C)methyl 4-azidobenzimidate, [(^{14}C)MABI], in the dark (167). With radiolabeling by (^{14}C)PITC or (^{14}C)MABI being restricted to less than three out of the ten lysine residues present in the IF_1 molecule, the inhibitory activity of IF_1 remained intact. The (^{14}C) labeled IF_1 was found to bind specifically to the β subunit; the binding of one mole IF_1 was per mol F_1 sufficient to fully inhibit F_1-ATPase (166, 167). Similar results have been obtained with radioiodinated IF_1 (168, 169).

Tyrosyl Group-Directed Reagents Mitochondrial F_1-ATPase is inhibited by treatment at neutral pH with Nbf; inhibition results from the reversible modification of a single tyrosine residue in one out of the three β subunits, another example of the $\frac{1}{3}$ of the site reactivity in F_1 (170). The reaction probably occurred by nucleophilic attack of the oxygen atom of the reactive tyrosine on the C-4 atom of Nbf leading to an O-Nbf-F_1 complex (171); it was accompanied by the development of an absorption peak at 385 nm (170–172). At pH 9, the tyrosine-bound Nbf was transferred intramolecularly to a nitrogen group in the β subunit to form a stable derivative, N-Nbf-F_1 (173). Nbf also reacted with a β subunit in yeast mitochondrial F_1 (174), and in F_1 from $E. coli$ (175), $R. rubrum$ (176), and chloroplasts (177). ATPase activity was restored by the addition of sulfhydryl compounds, such as N-acetylcysteine, glutathione, and dithiothreitol to O-Nbf-F_1; the absorption at 385 nm decreased concomitantly and that at 425 nm increased, owing to the displacement of the Nbf residue from the phenolic group of tyrosine and the formation of a Nbf-thiol derivative with the added thiols. The side-chain of free tyrosine did not react with Nbf at neutral pH (172); its ability to react when inserted in a protein like F_1 therefore suggests an unusual microenvironment for this residue (171). Reaction of mitochondrial F_1 with Nbf did not prevent nitration by tetranitromethane of other tyrosine residue(s) (170).

To study the localization of the Nbf-modified tyrosine in mitochondrial F_1, the bond between radiolabeled Nbf and tyrosine was stabilized by reduction either with Zn^{2+} in the presence of 4-4'dipyridyl (178), or with dithionite (179). Accordingly, the Nbf-reactive tyrosine was found in position 197 (178) close to the essential Glu 199 modified by DCCD or in position 311 (179) close to a region that is photolabeled by 8-azido ATP. The question of whether the Nbf-reactive tyrosine is located at the catalytic site of mitochondrial F_1 is a debatable one (170, 175–181). The affinity label, p-fluorosulfonyl-benzoyl-5'-adenosine (FSBA), (see section on Active-Site Directed Reagents) which has a relatively broad specificity towards amino acid residues, was reported to bind covalently to a tyrosine residue in the β subunit of F_1, resulting in inactivation of the enzyme (182); the modified residue Tyr 368 (81, 183) was different from the Nbf-reactive tyrosine residues.

Tyrosine residues in the F_0 sector of the ATPase complex with a critical function in the energy linked proton translocation have been revealed through binding studies with tetranitromethane (184).

Thiol Group-Directed Reagents   The ATPase activity of native mitochondrial F_1 is insensitive to -SH reagents, although the enzyme contains reactive cysteinyl residues. These residues, which are located in the α, γ, and ϵ subunits, are readily accessible to N-ethylmaleimide (NEM) (185). The ATPase activity of the nucleotide-depleted mitochondrial F_1 is, however, inhibited by NEM (186). Early studies on NEM inactivation of Factor B, a subunit of the F_0 sector of the mitochondrial ATPase complex (187, 188), have been recently extended with the demonstration of the inhibitory effect of two other thiol reagents, copper o-phenanthroline and iodosobenzoate, which oxidize vicinal -SH groups with the formation of disulfide bridges (189). The ATP-P_i exchange activity of an ATP-synthase preparation from heart mitochondria (complex V) was also found to be sensitive to -SH reagents such as carboxypyridine disulfide and p-chloro-mercuri-benzoate. The -SH groups reacting with these reagents were located in a small unidentified protein of M_r 8000, and in the α subunit of the F_1 sector (190).

It is noteworthy that binding of NEM to the unique cysteinyl residue of OSCP (Cys 118) does not alter the biological activity of this peptide (191). This property has been utilized to prepare a radiolabeled, biologically active, (^{14}C)NEM-OSCP; the study of its binding to F_1 by cross-linking revealed interactions of OSCP with the α and β subunits of mitochondrial F_1 (192).

In summary, the amino acid residues of mitochondrial F_1 whose modification results in loss of ATPase activity are as follows: two or more distinct tyrosine residues reactive with Nbf at neutral pH, FSBA, or tetranitromethane, one lysine residue labeled by Nbf at alkaline pH, one arginine residue modified by phenylglyoxal or butanedione, and at least one carboxyl group modified by DCCD or EEDQ. Since in most of these cases inactivation was prevented by substrates, it is probable that the reactive amino acid residues are located at the active site of F_1. The selective recognition of these residues and their unusually high reactivity to modifiers at neutral pH implies that, due to their environment, their pKs are shifted to values that make them more reactive to the modifiers.

The hydrolytic breakdown of ATP into ADP and P_i, and its synthesis from ADP and P_i, are probably catalyzed at the same active site of the ATPase complex. Under appropriate conditions of incubation, soluble mitochondrial F_1 was found to be able not only to hydrolyze ATP, but also to synthesize a minute amount of ATP that remains bound to the enzyme (193). Along this line, puzzling differential effects of chemical modifiers on hydrolytic and synthetic activities of F_1 were reported. For example, upon addition of DCCD, the ATPase activity of F_1 was lost, but full capacity to synthesize enzyme-bound

ATP was retained (193). Similar observations concern the functioning of the F_1-F_0 complexes reconstituted with Nbf-modified F_1 (194) and DCCD-modified F_1 (195).

CHEMICAL MODIFIERS OF THE AdN CARRIER   The chemical modification technique has been applied to the AdN carrier in order to determine whether one or several sites are recognized by ATR (or CATR), BA, and the substrates ADP and ATP. As previously discussed, the AdN carrier is characterized by a binding asymmetry for ATR or CATR on one hand, and for BA or isoBA on the other. CATR and ATR bind to the cytosolic face of the AdN carrier in the CATR conformation, whereas BA and isoBA bind to the matrix face of the AdN carrier in the BA conformation. Our studies on chemical modifications support the conclusion that distinct sets of amino acids are used for binding of CATR or ATR and for binding of BA or isoBA respectively (197, 198) in contrast with the view that a common binding site would be used for all ligands (199). In the latter concept, the CATR and BA conformations are referred to as the C and M states respectively, based on the assumption that, when the site of the membrane-bound carrier is opened to the cytosol, it binds CATR (C state) and when it is turned to the matrix space, it binds BA (M state).

In addition to standard chemical modifiers, like butanedione, phenylglyoxal, NEM, and EEDQ, another reagent, 2-hydroxyl-5-nitrobenzyl bromide (HNB), which is rather selective for tryptophan, has been applied to the AdN carrier, which is relatively rich in tryptophanyl residues (five residues per subunit of AdN carrier) (197). Addition of HNB to mitochondria resulted in the inhibition of ATR binding, but had no effect on BA binding. The binding of two mol HNB per mol carrier subunit was required to fully inhibit ATR binding; protection against inhibition was afforded by ADP, ATP, or ATR. In contrast to HNB which is a permeant reagent, a nonpermeant sulfonium derivative of HNB had no effect on ATR binding, indicating that the tryptophanyl residues modified by HBN were either buried in the hydrophobic core of the membrane or located on the matrix side.

NEM and EEDQ, like HNB, inhibited ATR binding, but not BA binding, whereas butanedione and phenylglyoxal inhibited the binding of both ATP and BA (197). BA or isoBA, and also ADP or ATP, potentiated the inhibition of ATR binding by NEM. The NEM inactivation data can be explained by assuming that, upon binding of NEM to a critical amino acid residue, the equilibrium between the CATR and BA conformations is shifted towards the BA conformation (which makes the CATR conformation unavailable for ATR binding) and that the transition between the two conformations is accelerated by ADP or ATP (198). A single amino acid residue of the AdN carrier, Cys 56, is modified by NEM (200). Cys 56 is probably in a hydrophobic environment, since fuscin, a penetrant SH-reagent like NEM, inhibited AdN transport and ATR binding whereas non-penetrant SH-reagents including mersalyl, dithiobis-nitrobenzoic acid, dithiobis-nicotinic acid, and diamide were ineffective (201).

In the case of inactivation by phenylglyoxal, the reaction order was one for inhibition of ATR binding and two for inhibition of BA binding. This was in good agreement with the results of binding studies, which showed that two mol bound (^{14}C)phenylglyoxal per mol carrier dimer were required to inactivate BA binding, compared with one mol reagent in the case of ATR binding (197). Preincubation with ATR protected against inactivation of ATR binding by phenylglyoxal and butanedione, and conversely preincubation with BA protected against inactivation of BA binding. Further, ADP and ATP protected both ATR binding and BA binding against butanedione inactivation. These data taken together are consistent with the idea that there is more than one binding site per carrier unit for the inhibitors ATR and BA and for the substrates ADP and ATP; it is, however, plausible that these sites are located in the same region of the carrier molecule (active center), possibly at the interface between the subunits, each subunit containing an active center.

The maleimide analog, eosine-5-maleimide (EMA), a covalent triplet probe, has been used to probe the motion of the carrier in the inner mitochondrial membrane (202). From the induced absorption anisotropy of EMA, it was concluded that the carrier rotates with a time constant of 2×10^{-4} s at 5°C and 10^{-4} s at 37°C, and that a significant proportion of the carrier molecules is immobile at low temperature.

CROSS-LINKING REAGENTS   In oligomeric proteins, cross-linking molecules of several Å length, with dimethylimido ester groups at both ends, bind covalently to amino groups of suitably positioned lysine residues of adjacent subunits (139). Zero length cross-linkers, such as ethyl-3[(dimethyl amino)-propyl] carbodiimide (EDAC) and EEDQ, function by activating carboxyl groups; when the activated carboxyl group in one subunit is close to an amino group in an adjacent subunit, cross-linking occurs.

In the first experiments with beef heart F_1, dimethylsuberimidate, a 12 Å length cross-linker, was used (203). The predominant cross-linked products were found to be αα and αβ; two minor products were later identified as αγ and βγ (166). No cross-linking was detected between the β subunits. These results suggest that close contact occurs between the α subunits and between each α and β subunit. In addition to the cross-linked products mentioned above, the following dimers γγ, γε, and δε have been identified in cross-linked beef heart F_1 (204). This was done using a refined technique involving two-dimensional sodium dodecyl sulfate polyacrylamide gel electrophoresis and two cross-linking reagents: dimethyl-3,3'-dithio-bis-propionimidate which forms an 11 Å-bridge, and cupric phenanthrolinate which cross-links SH groups at the subunit interface. The same strategy applied to rat liver F_1 yielded the following products: αα, αβ, αγ, βγ, and γε (205).

The arrangement of subunits in the ATPase complex of the yeast *Saccharomyces cerevisiae* has been explored with methyl-4-mercapto-butyrimidate (206) and with the cleavable cross-linker, dithiobis-(succinimidyl) propionate (207). Besides the dimers already reported in cross-linked beef heart and rat liver F_1, self association occurred for subunit 6 of M_r 21,900 and subunit 9 of M_r about 8,000 (DCCD-binding protein), both of which belong to the F_0 sector. Cross-linking experiments have also allowed identification of the subunit(s) of beef heart F_1 that react(s) with the natural ATPase inhibitor IF_1 and OSCP. The strategy of identification was based on the use of chemically radiolabeled, biologically active, IF_1 and OSCP. (^{14}C)PITC-IF_1 was cross-linked with F_1 by means of EDAC or EEDQ (166). On the other hand, (^{14}C)MABI-IF_1 was photoirradiated in the presence of F_1 in order to photogenerate a reactive nitrene which bound covalently to F_1 (167). In both cases, the β subunit of F_1 reacted with IF_1. In the case of (^{14}C)NEM-OSCP, cross-linking with F_1 by means of EDAC and EEDQ showed interactions of OSCP with the α and β subunits of F_1 (192).

Treatment of electron transport particles from beef heart mitochondria with the reversible oxidizing reagent, copper-o-phenantroline, has been shown to result in the formation of inter- and intramolecular disulfide bonds in a 29,000-dalton protein identified as the AdN carrier (207a).

Active Site–Directed Reagents

Active site–directed reagents comprise affinity labels and photoaffinity labels. An affinity label is characterized by two moieties; one of these has a structure closely related to that of the substrate (or a specific inhibitor), and therefore complementary to the active site; the other is a group, generally of an electrophilic nature, which reacts with a nucleophilic amino acid residue to form a covalent bond. Photoaffinity labels differ from affinity labels in that the modifying group is photoreactive; inert in the dark, this group is activated by light to generate a highly reactive intermediate. This intermediate then binds covalently to virtually any neighboring group. For details on the principle of the affinity and photoaffinity labeling technique and its application to biological systems, the reader is referred to recent reviews (208, 209).

AFFINITY LABELS   The extensively used affinity label FSBA behaves as an analog of ADP and ATP (210): in addition to the adenine and ribose moieties, it contains a carbonyl group adjacent to the 5' position of ribose, which mimics the α phosphoryl group of ADP and ATP. The sulfonyl fluoride moiety displays high reactivity and broad specificity towards lysine, tyrosine, histidine, and serine residues. Labeling by radioactive FSBA has been used to explore the active site of mitochondrial F_1 from beef heart (182, 183), pig heart (211, 212), and yeast (213). In the case of beef heart F_1, (^{14}C)FSBA was incorporated into the two largest subunits, α and β, with marked preference for β (182). FSBA-modified Tyr of beef heart F_1 is Tyr 368; it must be recalled that a different Tyr residue is labeled by Nbf (see section on Group-Directed Reagents). The amino acid sequence around Tyr 368 is strikingly homologous not only with that of *S. cerevisiae* F_1 (213), but also with that of F_1 from *E. coli* (214), and from spinach and maize chloroplasts (215, 216).

The dialdehyde derivatives of ADP and ATP (dial or *o*-ADP and dial or *o*-ATP) have been used as affinity labels of mitochondrial F_1 (217–220). Dial ATP forms a stable Schiff base with amino groups of F_1, presumably ε-NH_2 groups of lysine residues; because of this stability, reduction with borohydride is not required for full inhibition to occur. (^{14}C)Dial ATP binds to both α and β subunits; 50% inhibition of ATPase activity was attained upon binding of 1 mol dial-ATP per mol F_1 (217). Careful chemical determinations showed that dial ATP has a tendency to lose its triphosphate group and to generate an adenine-containing compound with a highly reactive conjugated aldehyde group, explaining the formation of the stable conjugated Schiff base with F_1 (218, 219). With (^{14}C) dial ADP, both α and β subunits were also labeled, and full inactivation was attained with 2–3 mol reagent bound per mol F_1 (217, 220). In the presence of Mg^{2+}, dial ATP was hydrolyzed to give dial-ADP, indicating that dial ATP binds at the active site of F_1. Unexpectedly dial AMP was also able to bind to and inactivate F_1, possibly because the cleaved ribose ring increases the length of the molecule (220). An etheno derivative of dial ATP (dial ε-ATP) has been reported to inactivate beef liver F_1, with full inactivation being attained for 1 mol reagent bound per mol F_1 at the level of the β subunit (221).

Two other affinity labels have been used with mitochondrial F_1, namely the mixed anhydride formed by condensation of ATP and mesitylene carboxylic acid (222) and the ATP-γ-4(*N*-2-chloroethyl-*N*-methylamino) benzoylamidate (223). These analogs bound to the β subunit of mitochondrial F_1, full inhibition being attained upon the binding of 1–2 mol inhibitor per mol F_1. ATP protected efficiently against inactivation, suggesting that the reagents were indeed probing a nucleotide site.

PHOTOAFFINITY LABELS   In early experiments, photolabels derived from ADP and ATP were used to locate nucleotide binding sites in subunits of mitochondrial F_1, and also to identify the AdN carrier by the characteristic inhibitory effect of CATR on its photolabeling. The ultimate aim of covalent photolabeling is, however, to map the nucleotide binding site(s) in these two proteins. The derivatives used contained photoreactive groups placed in the 2 or 8 position of the adenine ring or in the 3' position of the ribose moiety. Direct attachment of the azido group to the 2 or 8 position of the adenine ring is expected to insure specific binding of the photogenerated nitrene at the nucleotide site of the protein studies. As previously mentioned, the 2 and 8 substitutions have different consequences on the conformation of the modified nucleotide. Substitution at the 8-position shifts the nucleotide conformation about the N-glycosidic linkage from *anti* to *syn,* and introduces obvious complications in the recognition of the nucleotide binding sites, since ADP and ATP have a non-fixed *anti* conformation (62). On the other hand, the 2-azido nucleotides is supposed to take up the favorable *anti* conformation (68). It must be kept in mind that 2-azido nucleotides spontaneously isomerize to form tetrazole rings (67) with 50% of the nucleotide being transformed at equilibrium (70).

In the 3'-arylazido ADP or ATP derivatives, the azido group is attached to an aromatic ring which itself is linked to the 3' carbon atom of the ribose moiety by a chain whose length can be varied. Their purine ring is left unmodified which explains that 3'-arylazido ADP and ATP, in the dark, bind reversibly to F_1 and the AdN carrier with the same affinity as ADP and ATP. The first photolabel of this series to be synthesized was 3'-O-[(N-(4 azido-2-nitrophenyl) amino)pro-pionyl]ATP, (3'-arylazido ATP), which has been applied to myosin-ATPase (224) and mitochondrial F_1 (37). Propionyl and butyryl chains have been extensively used, to provide a long and flexible arm, of approximately 10 Å, that is free to move about within and around the binding site. Because of this large distance, covalent labeling data obtained with these photolabels may not reflect the exact location of the nucleotide site. This difficulty has been overcome by shortening the length of the arm. For example, in 3'-O-(4-azidobenzoyl)ATP or ADP, the benzoyl ring is directly linked to the ribose moiety (225). After this brief outline on the structural features of currently used azido nucleotides, we shall compare their binding properties, first with respect to the mitochondrial F_1 and second to the AdN carrier.

Irrespective of which photoactivable derivative of ADP or ATP is used, photolabeling of mitochondrial F_1 is always restricted either to the α and β subunit, or to the β subunit alone. Upon photoirradiation in the absence of Mg^{2+}, 8-azido ATP bound specifically to the β subunit of mitochondrial F_1, full inactivation being attained upon binding of two mol 8-azido ATP per mol F_1 (75). Whereas Mg^{2+} did not alter the binding stoichiometry, it markedly modified the distribution of the photolabel in the α and β subunits. In the absence of Mg^{2+}, 8-azido ADP bound in almost equal quantities to the α and β subunits of F_1 whereas in the presence of Mg^{2+}, the α subunit was predominantly labeled (76). A total number of six nucleotide binding sites in mitochondrial F_1 has been determined by sequential photolabeling with 8-azido ATP and 8-azido ADP (77) or with 3'-O-[(N-(azido-2-nitro-phenyl)amino)butyryl]AMPPNP, (3'-arylazido AMPPNP) (39).

An interesting derivative of 8-azido ATP, 8-azido-N^1,N^6-etheno ATP (8-azido-ϵ-ATP), which combines the site-directed specificity with the power of fluorescence probes, has been applied to F_1 from the yeast *S. cerevisiae* (82, 226). Contrary to ϵ-ATP, 8-azido-ϵ-ATP does not fluoresce, due to the quenching of the etheno fluorescence by the azido group; fluorescence develops after photolysis. By this means, the fluorescent label was localized in the β subunit of yeast F_1.

In spite of the fact that 8-azido ATP has a nonfavorable fixed *syn* conformation, and may give rise to unspecific labeling, the regions of the β subunit of beef heart F_1 photolabeled with 8-azido(^{3}H)ATP have been explored (81). Two regions were labeled. The first corresponded to residues 1–12, a region apparently not essential for catalysis. In the second, radioactivity was localized at the level of Lys 301, Ile 304, and Tyr 311; these residues were believed to belong to the active site of F_1, since photolabeling by 8-azido(^{3}H)ATP was prevented by preincubation with ATP.

Due to their presumed *anti* conformation, the 2-azido nucleotides appear to be promising tools for exploration of nucleotide binding sites in F_1. Recent data from this laboratory indicate that 2-azido ADP either in the presence or absence of Mg^{2+}, binds to the β subunit of mitochondrial F_1 (227). A similar result was reported in the case of chloroplast F_1 together with the interesting finding that 2-azido-ADP remains tightly bound to chloroplast F_1 in the dark and that in thylakoid membranes the tightly bound 2-azido-ADP is photophosphorylated into 2-azido-ATP (68–70, 228).

Photoirradiation of mitochondrial F_1 in the presence of 3'-aryl azido butyryl ADP, either with or without $MgCl_2$, resulted in the labeling of both the α and β subunits; the covalent binding of two mol probe per mol F_1, one on α and the other on β, resulted in full inactivation of the enzyme. Photolabeling was prevented by preincubation with ADP, ATP, and AMPPNP indicating specific binding at nucleotides sites (38). The reversible binding of Nbf at neutral pH prevented the covalent photolabeling of the β subunit, but not that of the α subunit (229). This α-photolabeled F_1 was devoid of ATPase activity. In beef heart submitochondrial particles, both the α and β subunits of the membrane-bound ATPase were photolabeled by 3'-arylazido propionyl ATP (41) and by 3'-arylazido butyryl ADP (40). Upon photoirradiation, a photoreactive analog of 3'-O-naphthoyl-ADP, 3'-O-[5-azido-naphthoyl]ADP, was also found to bind to the α and β subunits of mitochondrial F_1, total inactivation occurring at a covalent occupancy of 2 mol/mol F_1; it was proposed that each nucleotide site is made of two domains belonging to α and β juxtaposed subunits (230).

A difficulty encountered with azido derivatives of ATP is that they are hydrolyzed by F_1, especially when the medium is supplemented with $MgCl_2$. However AMPPNP, a nonhydrolyzable derivative of ATP, can be substituted for ATP. The corresponding photolabel (3'-arylazido AMPPNP) was found to bind in the presence of $MgCl_2$ to both the α and β subunits of mitochondrial F_1, with full inactivation occurring upon binding of 2 mol label per mol F_1 (39). A careful study of the distribution of bound arylazido AMPPNP indicated that at very low concentrations the label was preferentially located on the α subunit, whereas at higher concentrations it was equally distributed between the α and β subunits. However, the extent of inactivation was proportional to the extent of photolabeling at both low and high concentrations of arylazido AMPPNP. In other words, photolabeling of the α subunit inactivated F_1 to the same extent as photolabeling of the β subunit did, although the catalytic site is most probably located on β.

3'-O-(4-benzoyl)benzoyl ATP (Bz-ATP) is another photoprobe with an unmodified adenine ring and a photoreactive benzophenone group branched on ribose (42). Bz-ATP behaved in the dark as a substrate for the hydrolysis reaction catalyzed by free and bound F_1, with a K_m close to that found for ATP and a V_{max} 10 times lower. Under light, in the presence of Mg^{2+} and Bz-ATP, both the α and β subunits were labeled. In the absence of Mg^{2+}, only β was labeled. This was interpreted in terms of an interfacial domain between α and β, accessible to Mg-ATP, H_2O, and Mg-ADP. However, there were no data on the relationship between photolabeling and inactivation of F_1-ATPase.

An azido derivative of P_i, 4-azido 2-nitrophenyl phosphate (ANPP), was found to bind covalently to the β subunit of mitochondrial F_1 after photoirradiation. Full inactivation occurred after binding of 1 mol ANPP per mol F_1, and protection was afforded by preincubation with P_i (231). Similar results were obtained in photolabeling assays carried out with F_1 from *E. coli* and the thermophilic bacterium *PS3* (232) and with chloroplast F_1 (233). The specific localization of the P_i binding site in the β subunit of mitochondrial, bacterial, and chloroplast F_1 can be considered as a major criterion to ascribe a catalytic role to the β subunit.

Examination of the relationship between the binding of a number of azido derivatives to mitochondrial F_1, and the resulting inactivation, is consistent with a partial site reactivity of F_1, since the covalent labeling of one β, or one β plus one α, out of the 3 α and 3 β present in F_1, results in complete inactivation. The same observation holds for the affinity labels and several chemical modifiers (Table 3).

Azido derivatives of uncouplers have been synthesized in an attempt to identify an uncoupler binding protein (234–241). A typical example is that of 2-azido-4-nitrophenyl, a structural analog of the classical uncoupler 2,4-dinitrophenol (234). Upon photoirradiation in the presence of submitochondrial particles, 2-azido-4-nitrophenyl was reported to inactivate oxidative phosphorylation and to become associated with proteins of M_r 30,000 and 50,000 (235), and also with proteins of M_r 9,000 and 20,000 (236). The 50,000 M_r protein was believed to be the α subunit of F_1 (235) and the 9,000 M_r protein the DCCD-binding protein of the F_0 sector (236). It remains to be explained, however, why spin-labeled 2-4 dinitrophenol, when added to mitochondria, is

Table 3 Partial site reactivity of mitochondrial F_1

Probes	Subunit labeled	Stoichiometry (number of mol reagent bound/ mol F_1) for full inactivation	References
DCCD	β	2 (1 displaced by glycyl ethyl ester)	146
Nbf	β	1	170
Phenylglyoxal	α, β	1	161
FSBA	β	3	182
Bz amidate ATP $(Mg^{2+})^a$	β	1–2	223
ATP mixed anhydride $(EDTA)^b$	β	1	222
8-azido ATP (EDTA)	β	2	75
8-azido ATP (Mg^{2+})	α + β	2	76
8-azido ADP (EDTA)	α + β	2	76
8-azido ADP (Mg^{2+})	α > β	2	76
3'-arylazido ADP	α + β	2	38
3'-arylazido AMPPNP	α + β	2	39
2-azido ADP	β	2	(This laboratory unpublished)
ANPP	β	1	231

[a] Bz amidate ATP = ATP- -4(N-2-chloroethyl-N-methylamino)benzylamidate
[b] ATP mixed anhydride = mixed anhydride of ATP and mesitylenecarboxylic acid

distributed mainly in the phospholipid core of the mitochondrial membrane (242).

Azido derivatives of ADP have been used to characterize the AdN carrier in situ, i.e. bound to the mitochondrial membrane. When assayed in the dark, the affinity of the AdN carrier for 8-azido ADP was relatively poor (K_i for inhibition of ADP transport of 400 μM) (79) compared to that of 3'-arylazido bytyryl ADP (K_i 10 μM) (40) and 2-azido ADP (K_i 15–20 μM) (227). 3'-Arylazido ADP has been especially useful in the characterization of the AdN carrier in beef heart (40), yeast (40), and plant mitochondria (243). In all these cases, the specific photolabeling of the AdN carrier in mitochondria was ascertained by the protection afforded by CATR. In inside-out submitochondrial particles, both the α and β subunits of F_1 and the AdN carrier were labeled; preincubation with BA decreased specifically the photolabeling of the AdN carrier.

Radiolabeled 6'-0- [N-(4-azido-2-nitrophenyl)amino)butyryl]ATR was designed to characterize the membrane-bound AdN carrier in mitochondria and to map the ATR site in the polypeptide chain of the carrier. In heart, yeast, and plant mitochondria (40, 243), a single protein corresponding to the AdN carrier was photolabeled by this probe. The mapping of the ATR binding site in the photolabeled membrane-bound AdN carrier was recently reported (244, 245). In the sequence of 297 amino acids of the AdN carrier (246), the ATR binding site was found to be located in a segment extending from Cys 159 to Met 200 (245). The same result was obtained with the short-arm derivative, 6'-O-*p*-azidobenzoyl ATR (247). Since ATR and its photoactivable derivatives do not penetrate the mitochondrial membrane, it was concluded that the segment Cys 159-Met 200 is accessible to ATR from the cytosol.

CONCLUDING REMARKS

The experimental data obtained with noncovalent and covalent probes for the mitochondrial ATPase complex and AdN carrier will now be related to selected aspects of the structure and function of these two proteins, namely the stoichiometry and arrangement of subunits, the mapping and functioning of sites, and the role of subunit interactions and conformational changes.

For many years, much controversy has surrounded the stoichiometry of subunits in F_1. However, the $\alpha_3\beta_3$ stoichiometry is now well established, aided considerably by the use of chemical probes, as briefly summarized in the case of mitochondrial F_1. 1. Upon extensive photolabeling of the α and β subunits of F_1 with 8-azido ADP and 8-azido ATP (77) or with 3'-arylazido AMPPNP (39), followed by determination of the remaining tightly bound ADP or ATP, up to six nucleotide binding sites per F_1 are found. 2. Three mol (^{14}C)aurovertin D bind per mol F_1 at the level of the β subunit, each β subunit being able to bind one molecule of (^{14}C)aurovertin D (110). 3. The existence of three β subunits per F_1 has also been demonstrated by affinity labeling with FSBA (182).

Cross-linking data from different research groups (166, 203–205) indicate that in mitochondrondrial F_1 close contacts exist between the three α subunits, and between α and β subunits, but not between the β subunits; the γ subunit appears to be in contact with both α and β. This suggests that the α and β subunits are arranged in alternating sequence, with the three α and the three β subunits on two levels, the β subunits being eccentric with respect to α. The cross-linking technique has also revealed that the natural ATPase inhibitor IF_1 binds to the β subunit of mitochondrial F_1 (166, 167) and that OSCP is in contact with both the α and β subunits (192).

Affinity labels, photoaffinity labels, and chemical modifiers have been applied to the mitochondrial F_1 to map the DCCD-reactive glutamic residue (153), the Nbf- and FSBA-reactive tyrosyl residues (178, 179, 183) and the nucleotide binding site in the β subunit (81). The correlation of ATPase inactivation with the binding of a large number of chemical modifiers, and affinity and photoaffinity labels, to the β subunit of the mitochondrial F_1 (Table 3) strongly suggests that the β subunit contains the catalytic site.

Different chemical approaches have been used to investigate the arrangement of the peptide chain of the AdN carrier in the mitochondrial membrane. Covalent labeling by photoprobes derived from ATR has shown that the ATR binding site is located in the middle of the peptide chain of the AdN carrier (245). A single cystein residue, Cys 56, has been shown to be unmasked in the membrane-bound carrier upon addition of ADP or ATP. A direct contact of the AdN carrier with the lipid core of the mitochondrial membrane has been demonstrated by the use of spin-labeled long chain acyl ATR and acyl CoAs (132, 136). Structural mapping of mitochondrial F_1 and the AdN carrier may benefit from the technique of fluorescence resonance transfer which has been extensively used in the case of chloroplast F_1 (248 and ref. therein).

Chemical probes have been beneficial not only to our understanding of the topography of the ATPase complex and the AdN carrier, but also for allowing choices between different possible mechanisms in the functioning of these proteins. The existence of catalytic cooperativity between alternating sites in F_1, with binding of substrate at one site promoting catalytic events and release of products at other sites (102), is supported by the following lines of evidence. 1. The distribution pattern of the $(^{18}O)P_i$ species formed during $(\gamma\text{-}^{18}O)ATP$ hydrolysis by F_1 over a large range of ATP concentrations conforms best to the mechanism of alternating catalytic sites (106). 2. TNP-ATP added at a very low concentration, just sufficient to saturate one catalytic site of F_1, firmly binds to this site, and is not rapidly hydrolyzed; when more TNP-ATP is added to saturate the other two sites, the rate of TNP-ATP hydrolysis is markedly enhanced (49, 50). 3. AMPPNP firmly bound to F_1 is released by addition of an excess ADP and P_i (249). 4. Covalent modification of only one β subunit per F_1 by chemical modifiers causes full inactivation of F_1, in spite of the fact that there are three β subunits per F_1; similar results were obtained with covalent labeling of F_1 by photoactivable derivatives of ADP, ATP, and P_i, although in this case labeling may occur on both the α and the β subunits (Table 3).

Although the mechanism of ADP/ATP transport is far from being elucidated, a number of new findings concerning the conformation of the AdN carrier protein and its binding sites have recently arisen from the use of chemical probes. It has been ascertained that during the course of ADP/ATP transport, the AdN carrier assumes two distinct conformations that are recognized by CATR (CATR conformation) and BA (BA conformation). The transition between the two conformations is markedly accelerated by ADP or ATP, as demonstrated in experiments measuring either the intrinsic fluorescence of the isolated AdN carrier (54), or the extrinsic fluorescence of N-ADP, added to mitochondria as a conformational probe (47). Chemical modification data (197, 198) clearly indicate that CATR (or ATR) and BA (or isoBA) bind to two different sets of amino acids in the CATR and BA conformations, respectively. This, however, does not preclude the possibility that these sets of amino acids are situated in the same specialized region of the carrier molecule. Comparison of the binding parameters of nontransportable and transportable nucleotide analogs has led to the conclusion that the binding of nontransportable nucleotides occurs without site-site interactions, whereas negative cooperativity is exhibited for the binding of transportable nucleotides (53). This intriguing behavior leads us to question whether binding followed by transport is not accompanied, at the binding step, by a conformational change of the carrier protein which would be mandatory for transport to occur.

ACKNOWLEDGMENTS

We would like to acknowledge the contributions of M. Block, M. Bof, F. Boulay, G. Brandolin, P. Dalbon, A. C. Dianoux, J. Doussière, A. Dupuis, J. P. Issartel, G. Klein, G. Lauquin, R. Pougeois, and M. Satre to the research work from our laboratory during the past ten years. We thank J. Willison for the critical reading of the manuscript. We are indebted to J. Bournet and R. Césarini for their excellent assistance in the preparation of the manuscript. The studies from the authors' laboratories were supported by grants from the French National Center of Scientific Research, The French National Institute for Medical Research, and the Medical Faculty, Scientific University of Grenoble.

Literature Cited

1. Mitchell, P. 1973. *FEBS Lett.* 33:267–72
2. Pullman, M. E., Penefsky, H. S., Datta, A., Racker, E. 1960. *J. Biol. Chem.* 235:3322–28
3. Lambeth, D. O., Lardy, H. A., Senior, A. E., Brooks, J. C. 1971. *FEBS Lett.* 17:330–32
4. Knowles, A. F., Penefsky, H. S. 1972. *J. Biol. Chem.* 247:6624–30
5. Amzel, L. M., Pedersen, P. L. 1983. *Ann. Rev. Biochem.* 52:801–24
6. Vignais, P. V., Satre, M. 1984. *Mol. Cell. Biochem.* 60:33–70
7. Frangione, B., Rosenwasser, E., Penefsky, H. S., Pullman, M. E. 1981. *Proc. Natl. Acad. Sci. USA* 78:7403–7
8. Dianoux, A. C., Tsugita, A., Przybylski, M. 1984. *FEBS Lett.* 174:151–56
9. Pullman, M. E., Monroy, G. C. 1963. *J. Biol. Chem.* 238:3762–69
10. McLennan, D. H., Tzagoloff, A. 1968. *Biochemistry* 7:1603–10
11. Ovchinnikov, Y. A., Modyanov, N. N., Grinkevich, V. A., Aldanova, N. A., Trubetskaya, O. E., et al. 1983. *FEBS Lett.* 166:19–22
12. Fang, J. K., Jacobs, J. W., Kanner, B. I., Racker, E., Bradshaw, R. A. 1984. *Proc. Natl. Acad. Sci. USA* 81:6603–7
13. Serrano, R., Kanner, B. I., Racker, E. 1976. *J. Biol. Chem.* 251:2453–61
14. Berden, J. A., Voor-Brouwer, M. M. 1978. *Biochim. Biophys. Acta* 501:424–39
15. Galante, Y. M., Wong, S. Y., Hatefi, Y. 1979. *J. Biol. Chem.* 254:12372–78
16. Ludwig, B., Prochaska, L., Capaldi, R. A. 1980. *Biochemistry* 19:1516–23
17. Hughes, J., Joshi, S., Torok, K., Sanadi, D. R. 1982. *J. Bioenerg. Biomembr.* 14:287–95
18. Soper, J. W., Decker, G. L., Pedersen, P. L. 1979. *J. Biol. Chem.* 254:11170–76
19. Tzagoloff, A., Meagher, P. 1971. *J. Biol. Chem.* 246:7328–33
20. Ryrie, I. J., Gallagher, A. 1979. *Biochim. Biophys. Acta* 545:1–14
21. Todd, R. D., Griesenbeck, T. A., Douglas, M. G. 1980. *J. Biol. Chem.* 255:5461–67
22. Vignais, P. V., Block, M. R., Boulay, F., Brandolin, G., Lauquin, G. J.-M. 1982. In *Membranes and Transport* ed. A. N. Martonosi, 1:405–13. New York: Plenum
23. Hackenberg, H., Klingenberg, M. 1980. *Biochemistry* 19:548–55
24. Block, M. R., Zaccaï, G., Lauquin, G. J.-M., Vignais, P. V. 1982. *Biochem. Biophys. Res. Commun.* 109:471–77
25. Yount, R. G., Babcock, D., Ballantyne, W., Ojala, D. 1971. *Biochemistry* 10:2484–89
26. Yount, R. G. 1975. *Adv. Enzymol.* 43:1–56
27. Duée, E., Vignais, P. V. 1969. *J. Biol. Chem.* 244:3920–31

1010   VIGNAIS & LUNARDI

28. Vignais, P. V., Setondji, J., Ebel, J. P. 1971. *Biochimie* 53:127–29
29. Cross, R. L., Nalin, C. M. 1982. *J. Biol. Chem.* 257:2874–81
30. Eckstein, F. 1983. *Angew. Chem. Int. Ed. Engl.* 22:423–39
31. Knowles, J. R. 1980. *Ann. Rev. Biochem.* 49:877–919
32. Schlimme, E., Lamprecht, W., Eckstein, F., Goody, R. S. 1973. *Eur. J. Biochem.* 40:485–91
33. Webb, M. R., Grubmeyer, C., Penefsky, H. S., Trentham, D. R. 1980. *J. Biol. Chem.* 255:11637–39
34. Gresser, M. J. 1981. *J. Biol. Chem.* 256:5981–83
35. Moore, S. A., Moennich, D. M. C., Gresser, M. J. 1983. *J. Biol. Chem.* 258:6266–71
36. Boos, K. S., Schlimme, E. 1979. *Biochemistry* 18:5304–9
37. Russell, J., Jeng, S. J., Guillory, R. J. 1976. *Biochem. Biophys. Res. Commun.* 70:1225–34
38. Lunardi, J., Lauquin, G. J.-M., Vignais, P. V. 1977. *FEBS Lett.* 80:317–23
39. Lunardi, J., Vignais, P. V. 1982. *Biochim. Biophys. Acta* 682:124–34
40. Lauquin, G. J.-M., Brandolin, G., Lunardi, J., Vignais, P. V. 1978. *Biochim. Biophys. Acta* 501:10–19
41. Cosson, J. J., Guillory, R. J. 1979. *J. Biol. Chem.* 254:2946–55
42. Williams, N., Coleman, P. S. 1982. *J. Biol. Chem.* 257:2834–41
43. Lübben, M., Lücken, U., Weber, J., Schäfer, G. 1984. *Eur. J. Biochem.* 143:483–90
44. Schäfer, G., Onur, G. 1979. *Eur. J. Biochem.* 97:415–24
45. Tiedge, H., Lücken, U., Weber, J., Schäfer, G. 1982. *Eur. J. Biochem.* 127:291–99
46. Block, M. R., Lauquin, G. J.-M., Vignais, P. V. 1982. *Biochemistry* 21:5451–57
47. Block, M. R., Lauquin, G. J.-M., Vignais, P. V. 1983. *Biochemistry* 22:2202–8
48. Schäfer, G., Onur, G. 1980. *FEBS Lett.* 109:197–201
49. Grubmeyer, C., Penefsky, H. S. 1981. *J. Biol. Chem.* 256:3718–27
50. Grubmeyer, C., Penefsky, H. S. 1981. *J. Biol. Chem.* 256:3728–34
51. Matsuoka, I., Watanabe, T., Tonomura, Y. 1981. *J. Biochem.* 90:967–89
52. Schäfer, G. 1982. *FEBS Lett.* 139:271–75
53. Block, M. R., Vignais, P. V. 1984. *Biochim. Biophys. Acta* 767:369–76
54. Brandolin, G., Dupont, Y., Vignais, P.

V. 1981. *Biochem. Biophys. Res. Commun.* 98:28–35
55. Dupont, Y., Brandolin, G., Vignais, P. V. 1982. *Biochemistry* 21:6343–47
56. Secrist, J. A. III, Barrio, J. R., Leonard, N. J., Weber, G. 1972. *Biochemistry* 11:3499–506
57. Leonard, N. J. 1984. *Crit. Rev. Biochem.* 15:125–99
58. Tondre, C., Hammes, G. G. 1973. *Biochim. Biophys. Acta* 314:245–48
59. Barzu, O., Kiss, L., Bojan, O., Niac, G., Mantsch, H. H. 1976. *Biochem. Biophys. Res. Commun.* 73:894–902
60. Kaplan, R. S., Coleman, P. S. 1978. *Biochim. Biophys. Acta* 501:269–74
61. Kauffman, R. F., Lardy, H. A., Barrio, J. R., Barrio, M. C. G., Leonard, N. J. 1978. *Biochemistry* 17:3686–92
62. Davies, D. B., Danyluk, S. S. 1974. *Biochemistry* 13:4417–34
63. Schlimme, E., Boos, K. S., De Groot, E. J. 1980. *Biochemistry* 19:5569–74
64. Muneyama, K., Bauer, R. J., Shuman, D. A., Robins, R. K., Simon, N. 1971. *Biochemistry* 10:2390–95
65. Schlimme, E., Stahl, K. W. 1974. *Hoppe-Seyler's Z. Physiol. Chem.* 335:1139–42
66. Schäfer, G., Penades, S. 1977. *Biochem. Biophys. Res. Commun.* 78:811–18
67. Temple, C., Thorpe, M. C., Coburn, W. C., Montgomery, J. A. 1966. *J. Am. Chem. Soc.* 31:935–58
68. Czarnecki, J. J., Abbott, M. S., Selman, B. R. 1982. *Proc. Natl. Acad. Sci. USA* 79:7744–48
69. Czarnecki, J. J., Abbott, M. S., Selman, B. R. 1983. *Eur. J. Biochem.* 136:19–24
70. Czarnecki, J. J. 1984. *Biochim. Biophys. Acta* 800:41–51
71. Graüe, C., Klingenberg, M. 1979. *Biochim. Biophys. Acta* 546:539–50
72. Petrescu, I., Lascu, I., Goia, I., Markert, M., Schmidt, F. H., et al. 1982. *Biochemistry* 21:886–93
73. Schlimme, E., Schäfer, G. 1972. *FEBS Lett.* 20:359–63
74. Brandolin, G., Dupont, Y., Vignais, P. V. 1982. *Biochemistry* 24:6348–53
75. Wagenvoord, R. J., Van der Kraan, I., Kemp, A. 1977. *Biochim. Biophys. Acta* 460:17–24
76. Wagenvoord, R. J., Van der Kraan, I., Kemp, A. 1979. *Biochim. Biophys. Acta* 548:85–95
77. Wagenvoord, R. J., Kemp, A., Slater, E. C. 1980. *Biochim. Biophys. Acta* 593:204–11
78. Wagenvoord, R. J., Verschoor, G. J., Kemp, A. 1981. *Biochim. Biophys. Acta* 634:229–36
79. Schäfer, G., Schrader, E., Rowohl-

Quisthoudt, G., Rimpler, M., Penades, S. 1976. FEBS Lett. 64:185–89
80. Boos, K. S., Bridenbaugh, R., Ronald, R., Yount, R. G. 1978. FEBS Lett. 91:285–87
81. Hollemans, M., Runswick, M. J., Fearnley, J. M., Walker, J. E. 1983. J. Biol. Chem. 258:9307–13
82. Gregory, R., Recktenwald, D., Hess, B., Schäfer, H. J., Scheurich, P., Dose, K. 1979. FEBS Lett. 108:253–56
83. Schäfer, H. J., Scheurich, P., Rathgeber, G., Dose, K., Mayer, A., Klingenberg, M. 1980. Biochem. Biophys. Res. Commun. 95:562–68
84. Harris, D. A., Dall-Larsen, T., Klungsøyr, L. 1981. Biochim. Biophys. Acta 635:412–28
85. Schuster, S. M., Ebel, R. E., Lardy, H. A. 1975. Arch. Biochem. Biophys. 171:656–61
86. Pedersen, P. L. 1976. J. Biol. Chem. 251:934–40
87. Harris, D. A. 1978. Biochim. Biophys. Acta 463:245–73
88. Harris, D. A., Gomez-Fernandez, J. C., Klungsøyr, L., Radda, G. K. 1978. Biochim. Biophys. Acta 504:364–83
89. Sakamoto, J. 1984. J. Biochem. 96:475–82
90. Pougeois, R. 1983. FEBS Lett. 154:47–50
91. De Pamphilis, M. L., Cleland, W. W. 1973. Biochemistry 12:3714–24
92. Danenberg, K. D., Cleland, W. W. 1975. Biochemistry 14:28–34
93. Cornelius, R. D., Cleland, W. W. 1978. Biochemistry 17:3279–86
94. Cleland, W. W., Mildvan, A. S. 1979. Adv. Inorg. Biochem. 1:163–91
95. Dunaway-Mariano, D., Cleland, W. W. 1980. Biochemistry 19:1506–15
96. Cleland, W. W. 1982. Methods Enzymol. 87:159–79
97. Bossard, M. J., Schuster, S. M. 1981. J. Biol. Chem. 256:6617–22
98. Bossard, M. J., Vik, T. A., Schuster, S. M. 1980. J. Biol. Chem. 255:5342–46
99. Bossard, M. J., Schuster, S. M. 1981. J. Biol. Chem. 256:1518–21
100. Boyer, P. D., Gresser, M., Vinkler, C., Hackney, D., Choate, G. L. 1977. In Structure and Function of Energy Transducing Membranes, ed. K. Van Dam, B. F. Van Gelder, pp. 261–74. Amsterdam: Elsevier/North Holland Biomedical Press
101. Boyer, P. D., Cross, R. L., Momsen, W. 1973. Proc. Natl. Acad. Sci. USA 70:2837–39
102. Kayalar, C., Rosing, J., Boyer, P. D. 1977. J. Biol. Chem. 252:2486–91
103. Choate, G. L., Hutton, R. L., Boyer, P. D. 1979. J. Biol. Chem. 254:286–90
104. Gresser, M. J., Myers, J., Boyer, P. D. 1982. J. Biol. Chem. 257:12030–38
105. Hutton, R. L., Boyer, P. D. 1979. J. Biol. Chem. 254:9990–93
106. O'Neal, C. C., Boyer, P. D. 1984. J. Biol. Chem. 259:5761–67
107. Linnett, P. E., Beechey, R. B. 1979. Methods Enzymol. 55:472–518
108. Lardy, H. A., Connelly, J. L., Johnson, D. 1964. Biochemistry 3:1961–68
109. Issartel, J.-P., Klein, G., Satre, M., Vignais, P. V. 1983. Biochemistry 22:3485–92
110. Issartel, J.-P., Klein, G., Satre, M., Vignais, P. V. 1983. Biochemistry 22:3492–97
111. Avron, M., Shavit, N. 1965. Biochim. Biophys. Acta 109:317–31
112. Kagawa, Y., Nukiwa, N. 1981. Biochem. Biophys. Res. Commun. 100:1370–76
113. Chang, T. M., Penefsky, H. S. 1973. J. Biol. Chem. 248:2746–54
114. Chang, T. M., Penefsky, H. S. 1974. J. Biol. Chem. 249:1090–98
115. Satre, M., Klein, G., Vignais, P. V. 1978. J. Bacteriol. 134:17–23
116. Satre, M., Bof, M., Vignais, P. V. 1980. J. Bacteriol. 142:768–76
117. Lunardi, J., Klein, G., Vignais, P. V. 1984. In H+-ATPase (ATP synthase): Structure, Function, Biogenesis. The F_0F_1 Complex of Coupling Membranes, ed. S. Papa, K. Altendorf, L. Ernster, L. Packer, pp. 229–30. Bari: Adriatica Editrice
118. Kasahara, M., Penefsky, H. S. 1977. See Ref. 100, pp. 295–305
119. Catterall, W. A., Pedersen, P. L. 1972. J. Biol. Chem. 247:7969–76
120. Harris, D. A., Von Tscharner, V., Radda, G. K. 1979. Biochim. Biophys. Acta 548:72–84
121. Kohlbrenner, W. E., Cross, R. L. 1979. Arch. Biochem. Biophys. 198:598–607
122. Kohlbrenner, W. E., Cross, R. L. 1978. J. Biol. Chem. 253:7609–11
123. Vignais, P. V. 1976. Biochim. Biophys. Acta 456:1–38
124. Vignais, P. V., Vignais, P. M., Defaye, G. 1973. Biochemistry 12:1508–19
125. Lauquin, G. J.-M., Villiers, C., Michejda, J., Hryniewiecka, L. V., Vignais, P. V. 1977. Biochim. Biophys. Acta 460:331–45
126. Lauquin, G. J.-M., Vignais, P. V. 1976. Biochemistry 15:2316–22
127. Lauquin, G. J.-M., Duplaa, A. M., Klein, G., Rousseau, A., Vignais, P. V. 1976. Biochemistry 15:2323–27
128. Vignais, P. M., Brandolin, G., Lauquin, G. J.-M., Chabert, J. 1979. Methods Enzymol. 55:518–32

129. Doussière, J., Ligeti, E., Brandolin, G., Vignais, P. V. 1984. *Biochim. Biophys. Acta* 766:492–500
130. Block, M. R., Pougeois, R., Vignais, P. V. 1980. *FEBS Lett.* 117:335–40
131. Brandolin, G., Meyer, C., Defaye, G., Vignais, P. M., Vignais, P. V. 1974. *FEBS Lett.* 46:149–53
132. Lauquin, G. J.-M., Bienvenüe, A., Villiers, C., Vignais, P. V. 1977. *Biochemistry* 16:1202–8
133. Lauquin, G. J.-M., Brandolin, G., Vignais, P. V. 1976. *FEBS Lett.* 67:306–11
134. Boulay, F., Brandolin, G., Lauquin, G. J.-M., Vignais, P. V. 1983. *Anal. Biochem.* 128:323–30
135. Boulay, F. 1983. *Etude topographique de la proteine de transport ADP/ATP.* PhD thesis. Univ. de Grenoble, France. 153 pp. (In French)
136. Morel, F., Lauquin, G., Lunardi, J., Duszynski, J., Vignais, P. V. 1974. *FEBS Lett.* 38:133–37
137. Devaux, P., Bienvenüe, A., Lauquin, G. J.-M., Brisson, A., Vignais, P. M., Vignais, P. V. 1975. *Biochemistry* 14:1272–80
138. Vallee, B., Riordan, J. F. 1969. *Ann. Rev. Biochem.* 38:733–94
139. Ji, T. H. 1983. *Methods Enzymol.* 91:580–609
140. Belleau, B., Martel, R., Lacasse, G., Menard, M., Weinberg, N. L., Perron, Y. G. 1968. *J. Am. Chem. Soc.* 90:823–24
141. Khorana, H. G. 1953. *Chem. Rev.* 53:145–66
142. Pougeois, R., Satre, M., Vignais, P. V. 1978. *Biochemistry* 17:3018–23
143. Satre, M., Dupuis, A., Bof, M., Vignais, P. V. 1983. *Biochem. Biophys. Res. Commun.* 114:684–89
144. Beechey, R. B., Holloway, C. T., Knight, I. G., Robertson, A. M. 1966. *Biochem. Biophys. Res. Commun.* 23:75–80
145. Beechey, R. B., Robertson, A. M., Holloway, C. T., Knight, I. G. 1967. *Biochemistry* 6:3867–79
146. Pougeois, R., Satre, M., Vignais, P. V. 1979. *Biochemistry* 18:1408–13
147. Pougeois, R., Satre, M., Vignais, P. V. 1980. *FEBS Lett.* 117:344–48
148. Sebald, W., Wachter, E. 1978. In *Energy Conservation in Biological Membranes*, ed. G. Schäfer, M. Klingenberg, pp. 228–36. Berlin: Springer-Verlag
149. Sebald, W., Graf, T. R., Lukins, H. B. 1979. *Eur. J. Biochem.* 93:587–99
150. Kopecky, J., Dedina, J., Votruba, J., Svoboda, P., Houstek, J., Babitch, S., Drahota, Z. 1982. *Biochim. Biophys. Acta* 680:80–87

151. Kiehl, R., Hatefi, Y. 1980. *Biochemistry* 19:541–48
152. Azzi, A., Bragadin, M. A., Tamburro, A. M., Santato, M. 1973. *J. Biol. Chem.* 248:5520–26
153. Esch, F. S., Bohlen, P., Otsuka, A. S., Yoshida, M., Allison, W. S. 1981. *J. Biol. Chem.* 256:9084–89
154. Shoshan, V., Selman, B. R. 1980. *J. Biol. Chem.* 255:384–89
155. Satre, M., Lunardi, J., Pougeois, R., Vignais, P. V. 1979. *Biochemistry* 18:3134–40
156. Satre, M., Bof, M., Issartel, J.-P., Vignais, P. V. 1982. *Biochemistry* 21:4772–76
157. Khananshvili, D., Gromet-Elhanan, Z. 1983. *J. Biol. Chem.* 258:3720–25
158. Yoshida, M., Poser, J. W., Allison, W. S., Esch, F. S. 1981. *J. Biol. Chem.* 256:148–53
159. Riordan, J. F., McElvany, K. D., Borders, C. L. 1977. *Science* 195:884–86
160. Marcus, F., Schuster, S. M., Lardy, H. A. 1976. *J. Biol. Chem.* 251:1775–80
161. Frigeri, L., Galante, Y. M., Hanstein, W. G., Hatefi, Y. 1977. *J. Biol. Chem.* 252:3147–52
162. Godinot, C., Penin, F., Gautheron, D. C. 1979. *Arch. Biochem. Biophys.* 192:225–34
163. Koga, P. C., Cross, R. L. 1982. *Biochim. Biophys. Acta* 679:269–78
164. Peters, H., Risi, S., Dose, K. 1980. *Biochem. Biophys. Res. Commun.* 97:1215–19
165. Sugiyama, Y., Mukohata, Y. 1979. *FEBS Lett.* 98:276–80
166. Klein, G., Satre, M., Dianoux, A. C., Vignais, P. V. 1980. *Biochemistry* 19:2919–25
167. Klein, G., Satre, M., Dianoux, A. C., Vignais, P. V. 1981. *Biochemistry* 20:1339–44
168. Wong, S. Y., Galante, Y. M., Hatefi, Y. 1982. *Biochemistry* 21:5781–87
169. Power, J., Cross, R. L., Harris, D. A. 1983. *Biochim. Biophys. Acta* 724:128–41
170. Ferguson, S. J., Lloyd, W. J., Lyons, M. H., Radda, G. K. 1975. *Eur. J. Biochem.* 54:117–26
171. Sutton, R., Ferguson, S. J. 1984. *Eur. J. Biochem.* 142:387–92
172. Aboderin, A. A., Boedefeld, E., Luisi, P. L. 1973. *Biochim. Biophys. Acta* 328:20–30
173. Ferguson, S. J., Lloyd, W. J., Radda, G. K. 1975. *Eur. J. Biochem.* 54:127–33
174. Gregory, R., Recktenwald, D., Hess, B. 1981. *Biochim. Biophys. Acta* 635:284–94

175. Lunardi, J., Satre, M., Bof, M., Vignais, P. V. 1979. *Biochemistry* 24:5310–16
176. Khananshvili, D., Gromet-Elhanan, Z. 1983. *J. Biol. Chem.* 258:3714–19
177. Deters, D. W., Racker, E., Nelson, N., Nelson, H. 1975. *J. Biol. Chem.* 250:1041–47
178. Ho, J. W., Wang, J. H. 1983. *Biochem. Biophys. Res. Commun.* 116:599–604
179. Andrews, W. W., Hill, F. C., Allison, W. S. 1984. *J. Biol. Chem.* 259:8219–25
180. Ting, L. P., Wang, J. H. 1980. *Biochemistry* 19:5665–70
181. Ferguson, S. J., Lloyd, W. J., Radda, G. K., Slater, E. C. 1976. *Biochim. Biophys. Acta* 430:189–93
182. Esch, F. S., Allison, W. S. 1979. *J. Biol. Chem.* 254:10740–46
183. Esch, F. S., Allison, W. S. 1978. *J. Biol. Chem.* 253:6100–6
184. Guerrieri, F., Yagi, A., Yagi, T., Papa, S. 1984. *J. Bioenerg. Biomembr.* 16:251–62
185. Senior, A. E. 1973. *Biochemistry* 12:3622–27
186. Tamura, J. K., Wang, J. H. 1983. *Biochemistry* 22:1947–54
187. Lam, K. W., Warshaw, J. B., Sanadi, D. R. 1967. *Arch. Biochem. Biophys.* 199:477–84
188. Hughes, J., Joshi, S., Sanadi, D. R. 1983. *FEBS Lett.* 153:441–46
189. Laskshni Kantham, B. C., Hughes, J. B., Pringle, M. I., Sanadi, D. R. 1984. *J. Biol. Chem.* 259:10627–32
190. Godinot, C., Gautheron, D., Galante, Y. M., Hatefi, Y. 1981. *J. Biol. Chem.* 256:6776–82
191. Dupuis, A., Issartel, J.-P., Lunardi, J., Satre, M., Vignais, P. V. 1984. *Biochemistry.* 24:728–33
192. Dupuis, A., Lunardi, J., Issartel, J.-P., Vignais, P. V. 1984. *Biochemistry.* 24:734–39
193. Sakamoto, J., Tonomura, Y. 1982. *J. Biochem.* 93:1601–14
194. Steinmeier, R. C., Wang, J. H. 1979. *Biochemistry* 19:5665–70
195. Matsuno-Yagi, A., Hatefi, Y. 1984. *Biochemistry* 23:3508–14
196. Song, K. S., Wang, J. H. 1984. *Biochemistry* 23:136–41
197. Block, M. R., Lauquin, G. J.-M., Vignais, P. V. 1981. *Biochemistry* 20:2692–99
198. Block, M. R., Lauquin, G. J.-M., Vignais, P. V. 1981. *FEBS Lett.* 131:213–18
199. Klingenberg, M., Appel, M. 1980. *FEBS Lett.* 119:195–99
200. Boulay, F., Vignais, P. V. 1984. *Biochemistry* 23:4807–12
201. Vignais, P. M., Chabert, J., Vignais, P. V. 1975. In *Biomembrane, Structure and*
Function, ed. G. Gárdos, J. Szász, pp. 307–13. Amsterdam: North Holland
202. Mülher, M., Krebs, J. R. J., Cherry, R. J., Kawato, S. 1982. *J. Biol. Chem.* 257:1117–20
203. Satre, M., Klein, G., Vignais, P. V. 1976. *Biochim. Biophys. Acta* 453:111–20
204. Baird, B. A., Hammes, G. G. 1977. *J. Biol. Chem.* 252:4743–48
205. Bragg, P. D., Hou, C. 1982. *Biochem. Int.* 4:31–38
206. Enns, R. K., Criddle, R. S. 1977. *Arch. Biochem. Biophys.* 182:587–600
207. Todd, R. D., Douglas, M. G. 1981. *J. Biol. Chem.* 256:6984–89
207a. Joshi, S., Torok, K. 1984. *J. Biol. Chem.* 259:12742–48
208. Plapp, B. V. 1982. *Methods Enzymol.* 87:469–99
209. Guillory, R. J. 1979. *Curr. Top. Bioenerg.* 9:267–414
210. Pal, P. K., Wechter, W. J., Colman, R. F. 1975. *Biochemistry* 14:705–15
211. Di Pietro, A., Godinot, C., Martin, J. C., Gautheron, D. C. 1979. *Biochemistry* 18:1738–45
212. Di Pietro, A., Godinot, C., Gautheron, D. C. 1981. *Biochemistry* 20:6312–18
213. Bitar, K. G. 1982. *Biochem. Biophys. Res. Commun.* 109:30–35
214. Saraste, M., Gay, N. J., Eberle, A., Runswick, M. J., Walker, J. E. 1981. *Nucleic Acids Res.* 9:5287–96
215. Krebbers, E. T., Larrinna, I. M., McIntosh, L., Bogorad, L. 1982. *Nucleic Acids Res.* 10:4945–5002
216. Zurawaski, G., Sottomley, W., Whitfield, P. R. 1982. *Proc. Natl. Acad. Sci. USA* 79:6260–64
217. Lowe, P. N., Beechey, R. B. 1982. *Biochemistry* 21:4073–82
218. Lowe, P. N., Baum, H., Beechey, R. B. 1979. *Biochem. Soc. Trans.* 7:1133–36
219. Lowe, P. N., Beechey, R. B. 1982. *Bioorg. Chem.* 11:55–71
220. De Melo, D. F., Satre, M., Vignais, P. V. 1984. *FEBS Lett.* 169:101–6
221. Wakagi, T., Ohta, T. 1982. *J. Biochem.* 92:1403–12
222. Drutsa, V. I., Kozlov, I. A., Milgrom, Y. M., Shabarova, Z. A., Sokolova, N. I. 1979. *Biochem. J.* 182:617–19
223. Budker, V. G., Kozlov, I. A., Kurbatov, V. A., Milgrom, Y. M. 1977. *FEBS Lett.* 83:11–14
224. Jeng, S. J., Guillory, R. J. 1975. *J. Supramol. Struct.* 3:448–68
225. Lunardi, J., Satre, M., Vignais, P. V. 1979. *Biochemistry* 20:473–80
226. Schäfer, H. J., Scheurich, P., Rathgeber, G., Dose, K. 1978. *Nucleic Acids Res.* 5:1345–51

227. Dalbon, P., Boulay, F., Vignais, P. V. 1985. *FEBS Lett.* 180:212–18
228. Abbott, M. S., Czarnecki, J. J., Selman, B. R. 1984. *J. Biol. Chem.* 259:12271–78
229. Lunardi, J., Vignais, P. V. 1979. *FEBS Lett.* 102:23–28
230. Lübben, M., Lücken, U., Weber, J., Schäfer, G. 1984. *Eur. J. Biochem.* 143:483–90
231. Lauquin, G. J.-M., Pougeois, R., Vignais, P. V. 1980. *Biochemistry* 19:4620–26
232. Pougeois, R., Lauquin, G. J.-M., Vignais, P. V. 1983. *FEBS Lett.* 153:65–70
233. Pougeois, R., Lauquin, G. J.-M., Vignais, P. V. 1983. *Biochemistry* 22:1241–45
234. Hanstein, W. G., Hatefi, Y. 1974. *J. Biol. Chem.* 249:1356–62
235. Hanstein, W. G., Hatefi, Y. 1979. *Biochemistry* 18:1019–25
236. Kurup, K. R., Sanadi, D. R. 1977. *J. Bioenerg. Biomembr.* 9:1–15
237. Katre, N. V., Wilson, D. B. 1977. *Arch. Biochem. Biophys.* 184:578–85
238. Katre, N. V., Wilson, D. B. 1978. *Arch. Biochem. Biophys.* 191:647–56
239. Partis, M. D., Griffiths, D. E., Beechey, R. B. 1984. *Arch. Biochem. Biophys.* 232:610–15
240. Blondin, G. A. 1980. *Biochem. Biophys. Res. Commun.* 96:1587–94
241. Senior, A. E., Tomesko, A. M. 1975. I *Electron Transfer Chain and Oxidativ Phosphorylation,* ed. E. Quagliariello S. Papa, F. Palmieri, E. Slater, N. Sili prandi, pp. 155–70. Amsterdam: Nort Holland
242. Hsia, J. C., Chen, W., Long, R. A. Wong, L. T., Kalow, W. 1972. *Proc Natl. Acad. Sci. USA* 69:3412–15
243. Brandolin, G., Lauquin, G. J.-M., Silva Lima, M., Vignais, P. V. 1979 *Biochim. Biophys. Acta* 548:30–37
244. Boulay, F., Lauquin, G. J.-M., Vignais P. V. 1979. *FEBS Lett.* 108:390–94
245. Boulay, F., Lauquin, G. J.-M., Tsugita A., Vignais, P. V. 1983. *Biochemistr* 22:477–84
246. Aquila, H., Misra, D., Eulitz, M. Klingenberg, M. 1982. *Hoppe-Seyler' Z. Physiol. Chem.* 363:345–49
247. Boulay, F., Lauquin, G. J.-M., Vignais P. V. 1982. *FEBS Lett.* 143:268–72
248. Snyder, B., Hammes, G. G. 1984 *Biochemistry* 23:5787–95

nn. Rev. Biochem. 1985. 54:1015–69

THE MITOCHONDRIAL ELECTRON TRANSPORT AND OXIDATIVE PHOSPHORYLATION SYSTEM[1]

Youssef Hatefi

Division of Biochemistry, Department of Basic and Clinical Research, Scripps Clinic and Research Foundation, La Jolla, California 92037

CONTENTS

[1]Abbreviations used: M_r, apparent molecular weight; SMP, submitochondrial particles; mtDNA, mitochondrial DNA; PMS, phenazine methosulfate; TMPD, N,N,N',N'-tetramethyl-*p*-phenylenediamine; HQNO, 2-n-heptyl-4-hydroxyquinoline N-oxide; UHDBT, 5-*n*-undecyl-6-hydroxy-4,7-dioxobenzothiazole; DCCD, N,N'-dicyclohexylcarbodiimide; and EDTA, ethylenediaminetetraacetic acid. Other abbreviations are indicated in the text.

0066-4154/85/0701-1015$02.00

PERSPECTIVES AND SUMMARY

It is now well established that the mitochondrial energy-yielding and energy-consuming reactions are linked by way of protonic circuits (1). Electron transfer from respiratory chain substrates to molecular oxygen results in the vectorial translocation of protons, and this protonic energy is utilized for ATP synthesis, ion translocation, and protein importation. The transduction of oxidative energy to a protonmotive force and the utilization of protonic energy for ATP synthesis are catalyzed by discrete multisubunit enzyme complexes located in the mitochondrial inner membrane (2). Consistent with the principle of energy transduction through vectorial proton translocation, all energy transducing enzyme systems of mitochondria traverse the inner membrane and catalyze proton translocation coupled to their scalar reactions (e.g., electron transfer or ATP synthesis/hydrolysis) in the same direction, depending on whether the scalar reaction yields or consumes energy. In mitochondria, the energy-yielding reactions result in outward proton translocation, and the energy-consuming reactions are coupled to inward movement of protons. The same basic mechanistic principles of energy transduction and the same basic structural design for vectorial catalysis are involved in oxidative and photosynthetic phosphorylation in bacteria and plants (1, 3). Particularly remarkable is the conservation in structure and composition of two energy-transducing enzyme complexes, the ATP synthase and the quinol-cytochrome c (plastocyanin) oxidoreductase, which occur in mitochondria, prokaryotic plasma membranes, bacterial chromatophores, and chloroplast thylakoid membranes (2, 4–6).

Recent volumes of *Annual Review of Biochemistry* contain elegant articles on selected structural and bioenergetic aspects of mitochondrial and bacterial oxidative phosphorylation enzyme systems (7–16). This review will attempt to offer a brief account of current knowledge of the structure and function of the entire mammalian mitochondrial oxidative phosphorylation system. Where necessary, information accrued from prokaryotic as well as other eukaryotic systems will also be included to make the picture more complete. However, space limitation will necessitate selection and reference to review articles for greater detail.

INTRODUCTION

The oxidative phosphorylation system of eukaryotes is contained in the mitochondrial inner membrane, which is composed of approximately 70% protein and 30% lipid. The lipids are predominantly phosphotidylcholine (~40%), phosphatidylethanolamine (~35%), and cardiolipin (~15%) (17). In bovine heart mitochondria, approximately 50% of the inner membrane protein is associated with the enzyme systems that catalyze oxidative phosphorylation.

The remainder includes, among other things, various transport proteins and enzymes which act as electron tributaries of the respiratory chain. Functionally, the mitochondrial oxidative phosphorylation system is composed of five protein-lipid enzyme complexes. They are:

NADH:ubiquinone oxidoreductase (complex I),
succinate:ubiquinone oxidoreductase (complex II),
ubiquinol:ferricytochrome c oxidoreductase (complex III),
ferrocytochrome c:oxygen oxidoreductase (complex IV), and
ATP synthase (complex V).

Table 1 contains information regarding their molecular weights, polypeptide composition, prosthetic groups, and approximate relative ratios in bovine heart mitochondria. Of the sixty or so unlike polypeptides that are contained in these five enzyme complexes, only six are known to be encoded by the mammalian mitochondrial DNA and synthesized within the mitochondria (Table 1) (18). The remainder are cytoribosomal products and are imported.

Functionally, the five enzyme complexes of the mitochondrial oxidative phosphorylation system interact as shown in Figure 1. Complexes I, II, III, and IV plus ubiquinone (Q) and cytochrome c make up the respiratory chain. The electron carriers of the respiratory chain are quinoid structures (FMN, FAD, Q) and transition metal complexes (iron-sulfur clusters, hemes, protein-bound copper), with quinoid structures and iron-sulfur clusters predominating in the region of $E_m < 0$, and hemes and copper in the region of $E_m > 0$. As shown on the E_m scale atop Figure 1, the electron carriers of the respiratory chain appear to comprise three quasi-equipotential regions in each of which energy drop is less than 100 mV [the E_h values calculated for representative electron carriers of mitochondria in the resting state are very close to their E_m values, ref. (19)].

Table 1 Molecular weights, polypeptides, prosthetic groups, and relative abundance in the mitochondrial inner membrane of complexes I, II, III, IV, and V

Complex	$M_r \times 10^6$ (monomer)[a]	Poly-peptides	mtDNA Encoded polypeptides	Prosthetic groups	Ratio in mito.[b]
I	0.7–0.9	25	—	FMN, Fe-S clusters	1
II	0.14	4–5	—	FAD, Fe-S clusters, b_{560} heme	2
III	0.25	9–10	1	b_{562}, b_{566}, c_1 hemes, [2Fe-2S] cluster	3
IV	0.16–0.17[c]	8[c]	3	aa_3 hemes, Cu_a Cu_{a_3}	6–7
V	0.5	12–14	2	adenine nucleotides, Mg^{2+}	3–5

[a] Protein only.
[b] Based on bovine SMP content of FMN (I), covalently-bound FAD (II), cyt. c_1 (III), cyt. aa_3 (IV), and F_1 (V), corrected to nearest integer relative to I.
[c] From (261).

There also appear to be intervals (represented by the breaks in the E_m scale of Figure 1) where energy drop between respiratory chain components appears to be greater than 100 mV. The significance of these features relative to the mechanism of energy transduction by the respiratory chain is not known, except that in the quasi-equipotential regions there would be minimum energy loss and maximum energy conservation.

At three stages along the respiratory chain, represented by complexes I, III, and IV, oxidative energy is conserved via coupled vectorial proton translocation and creation of a membrane electrochemical potential of protons ($\Delta\bar{\mu}_{H^+}$). The protonic energy so created then drives ATP synthesis. The system up to the level of cytochrome c (i.e., the third equipotential region shown in Figure 1) is reversible (20). However, the final step of electron transfer from cytochrome a_3 to oxygen is not reversible, which is important, because it displaces the

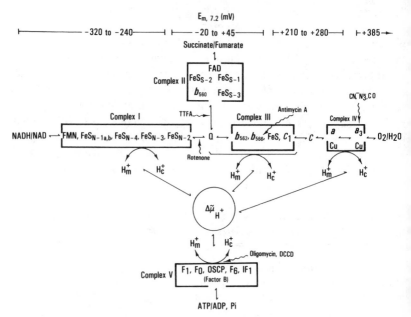

Figure 1 Profile of the mitochondrial electron transport-oxidative phosphorylation system, showing the well-characterized components of complexes I, II, III, IV, and V, and energy communication by way of $\Delta\bar{\mu}_{H^+}$ among the energy transducing complexes I, III, IV, and V. FeS, iron-sulfur cluster [only the ESR-detectable clusters are shown; FeS subscripts denote individual FeS clusters according to the nomenclature of Ohnishi, ref. (65)]; a, b, and c, cytochromes a, b, and c, respectively (subscripts on b cytochromes denote position of their α peaks at ambient temperatures); H_m^+ and H_c^+, protons on the matrix and cystolic sides of the mitochondrial inner membrane; TTFA, thenoyltrifluoroacetone. For other notations, see abbreviations list and text. The E_m scale is applicable to all components of the respiratory chain, except FeS_{N-1a} (apparent $E_m \leqslant -400$ mV), FeS_{S-2} (apparent $E_m \sim -400$ mV), and b_{560} (apparent $E_m < -100$ mV).

quilibrium of the system in the direction of ATP synthesis. Isolated complexes , III, and IV have each been shown to catalyze proton translocation coupled to lectron transfer when embedded in phospholipid vesicles (11, 21–24), and oincorporation of complex I or IV with ATP synthase in liposomes has been hown to reconstitute oxidative phosphorylation (25–27).

NADH:UBIQUINONE OXIDOREDUCTASE (COMPLEX I)

This enzyme complex was first isolated in 1961 (2, 28, 29) from bovine heart mitochondria. The preparation catalyzes the rotenone-sensitive reduction of Q analogs by NADH, and vectorial proton translocation coupled to electron transfer (2, 21, 29, 30). Functionally similar enzymes appear to be present in all mitochondria possessing coupling site 1 of oxidative phosphorylation (e.g., 31–34), as well as in *Paracoccus denitrificans* membranes (35–38). However, the isolation of a complex I-like enzyme with retention of the above physiological properties has not yet been achieved from another source, or by an alternative procedure from the same source (39–42).

Composition of Complex I

The nonprotein components of complex I are FMN, nonheme iron (Fe), acid-labile sulfide (S^{2-}), Q, and phospholipids. The ratio of FMN : Fe : S^{2-} is 1 : 22–24 : 22–24, despite variations in the FMN content (1.0 to 1.4 nmol/mg of protein) of preparations from different laboratories (30, 43–45). However, as with preparations of complexes II and III, the Q content is variable. In complex I, as well as in complexes II to V, the nature and the relative ratios of the three classes of phospholipids are the same as in the mitochondrial inner membrane.

Preparations of complex I contain approximately 25 unlike polypeptides (46–49), all but one of which (a 42,000 M_r polypeptide) can be immunoprecipitated from detergent-solubilized mitochondria by antisera directed against a limited number of complex I polypeptides (50). On the basis of Coomassie blue stain intensity, most high M_r polypeptides appear to be present in complex I at concentrations equimolar with FMN, and the sum of the M_r of one copy of all the polypeptides is close to the M_r of complex I as estimated from its FMN content (51).

The polypeptides of complex I can be divided into three groups. When treated with chaotropes (e.g., 0.5 M $NaClO_4$ at pH 8.0 and 30°C), complex I resolves into a water-soluble (about 30% of the total protein), and a water-insoluble fraction (30, 52, 53). The former can be further divided by ammonium sulfate into two distinct fractions, a brown, water-soluble iron-sulfur protein fraction (IP) and an amber, water-soluble iron-sulfur flavoprotein fraction (FP) (30, 54). The composition of IP, FP, and the insoluble hydrophobic fraction (HP) in terms of number of polypeptides and content of FMN, Fe, and S^{2-}, is shown in Table 2. HP contains the bulk of the phospholipids of complex I. Its polypeptides are extremely hydrophobic and, on the whole, smaller and with more alkaline isoelectric points than the polypeptides of IP and FP (2, 47). The M_r values of the polypeptides of IP are 75, 49, 30, 18, 15, and 13×10^3, and those of FP are 51, 24, and 9×10^3. The polypeptides of IP (except the 15,000 M_r polypeptide) can be immunoprecipitated together by antiserum specific for any one of the M_r 75, 49, 30, and 13×10^3 polypeptides (2), and the three polypeptides of FP elute together in a narrow band from Sephadex or agarose columns (54), indicating the physical association of each set of polypeptides in solutions of IP or FP. The isolated FP is a diaphorase type NADH dehydrogenase, with a wide specificity for quinones and ferric complexes (including ubiquinone analogs and cytochrome c) as electron acceptors (30, 54, 55). It also catalyzes transhydrogenation from NADH to NAD analogs, and exhibits a low NADPH dehydrogenase activity at pH ≤ 6.0 (54). IP and HP have no demonstrable catalytic properties, although it has been claimed recently that a 42,000 M_r polypeptide of complex I has a distinct NADH $\rightarrow$ NAD transhydrogenase activity, which is sensitive to treatment of complex I with 0.5 M NaClO$_4$ (56). There appear to be five binuclear and three tetranuclear iron-sulfur (FeS) clusters in complex I, which are considered to be associated with the following fractions: one binuclear [2Fe$-$2S] and one tetranuclear [4Fe$-$4S] cluster each in FP and HP, and three binuclear and one tetranuclear clusters in IP (for further details, see ref. 2, 57–59).

Light Absorption and ESR Spectral Properties of Complex I

Early studies showed that treatment of complex I with NADH results in considerable bleaching in the absorption spectrum of the enzyme between 350 and 500 nm, only about 50% of which at 450 nm could be attributed to flavin reduction. The extra bleaching was suggested to be due to reduction of nonheme iron (29, 60, 61). The resolution of complex I into FP, IP, and HP, and the finding of Fe and S^{2-} in each fraction (62) suggested the presence in

Table 2 Composition of principal fractions from chaotropic resolution of complex I[a]

Fraction	Polypeptides	FMN	Fe	Fe/complex I FMN
		(nmol/mg)	(nmol/mg)	(mol/mol)
Complex I	~25	0.98	22.1	22.6
HP	~16	0.04	7.2	6.6
IP	6	0.52	48.2	9.3
FP	3	12.5	77.9	6.2

[a] Columns under FMN and Fe give measured values (59); the last column gives the number of complex I iron atoms accounted for by each fragment after correction for losses and cross contamination. In all fractions, the S^{2-} content was the same as the Fe content. In (54), average values per mg FP are 13.5 nmol FMN, 74 nmol Fe, and 72 nmol S^{2-}. From (2).

complex I of multiple forms of FeS clusters, which were subsequently demonstrated by ESR examination of NADH-treated complex I at $\leq$ 13 K (63, 64). Preparations of complex I contain four confirmed and two controversial ESR-detectable FeS clusters (65, 66). Two additional, and apparently ESR-silent, clusters are also considered to be present (2). The characteristics of the ESR-detectable clusters are summarized in Table 3. The controversial clusters are N1a and N5. The former has been detected only by Ohnishi and coworkers (57, 65) and occurs at a spin concentration of at most 0.5/FMN in complex I. Its apparent $E_{m,7}$ is about -380 mV in SMP and considerably lower (< -500 mV) in complex I (57). The latter, N5, has been detected in complex I in variable low amounts [less than 0.25/FMN (67–69)], and not at all in *Candida utilis* and plant mitochondria (68, 70). The established clusters are N1b, N2, N3, and N4 according to the nomenclature of Ohnishi (65) [respectively, clusters 1, 2, 4, and 3 according to the nomenclature of Albracht, ref. (66)]. These four clusters appear to be present in complex I at concentrations comparable to FMN (Table 3). The apparent $E_{m,7}$ values given in Table 3 are from potentiometric titration studies (57, 65, 69, 71). However, there is reason to believe that the reduction potentials of clusters N3 and N4 are not similar. The low temperature (13 K) ESR spectrum of complex I treated with NADPH at neutral pH, shows little or no signals due to reduced clusters N1b and N4 (72, 73). This is because at pH $\geq$ 7, NADPH-treated complex I is in an equilibrium state of partial reduction owing to the fact that it is very slowly reduced by NADPH and very slowly oxidized by molecular oxygen. However, when pH is lowered below pH 7 to increase the rate of electron transfer from NADPH to complex I, and rotenone is added to diminish the autoxidation rate of complex I, then signals due to clusters N1b and N4 appear [(72); for additional details, see (73)]. These results together with other data suggest a reduction potential order of N2 $\geq$ N3 > N4 $\geq$ N1b. This order could explain the downfield shift of the g = 2.1 signal,

Table 3 Properties of complex I FeS clusters[a]

Cluster[b]	Field postions	Max. conc.	$E_{m,7}$ (apparent)	Cluster structure
	$g_{x,y,z}$	(e$^-$/FMN)	(mV)	
N1a	1.91, 1.95, 2.03	0.5	-370	binuclear
N1b	1.92, 1.94, 2.02	0.8	-245	binuclear
N2	1.92, 1.92, 2.05	0.9	$- 20$	tetranuclear
N3	1.86, 1.93, 2.04	} 1.8	-245	tetranuclear
N4	1.87, 1.93, 2.10		-245	tetranuclear
N5	2.11, 1.93, 1.88	0.25	-270	—

[a] From ref. (2), (65), (66) and references therein.
[b] Nomenclature of Ohnishi (69). According to Albracht's designations, N1b is 1 (or 1a plus 1b), N2 is 2, N3 is 4 and N4 is 3 (66).

observed by Orme-Johnson et al (63, 64, 66), as it develops during the redox titration of complex I, because at $g = 2.1$ the g_z signals of N3 and N4 overlap, but the combined g_z signal of N3 and N4 is about 3 Gauss downfield as compared to the g_z of N3 (72). The kinetic sequence of complex I electron carriers (including FMN) is not known. All the ESR-visible FeS clusters N1b, N2, N3, and N4 were reduced with NADH at 20°C within 10 msec, and attempts at discerning a kinetic sequence resulted in a thermodynamic order of appearance of the ESR signals of the above clusters (64). However, cluster N2 is considered to be the direct electron donor to Q (65, 74). Also, cluster N3 is considered to undergo spin-spin interaction with FMN (69, 75); hence it may be the cluster contained in the 51,000 M_r polypeptide that presumably carries the flavin (2, 59).

Structure of Complex I

Chemical probes used to explore the nature of complex I polypeptides in contact with the lipid bilayer (e.g., arylazido derivatives of phospholipids and iodonaphthylazide) showed that several polypeptides of the HP fraction, but none of IP or FP, were labeled (48, 76). Other probes used for modifying polypeptides in contact with the aqueous phase (e.g., lactoperoxidase-catalyzed iodination and diazobenzenesulfonate) labeled several polypeptides of HP and IP, but none of FP. Among the IP polypeptides, the 75, 49, 30, and 13 × 10³ M_r polypeptides were labeled, and the three largest have been shown to be transmembranous (2, 50, 77).

Thus, it appears that FP and IP are surrounded by HP polypeptides (Figure 2). Neither is in contact with the membrane, and FP polypeptides are shielded from the aqueous phases on opposite sides of the membrane as well (2). However, arylazido-β-alanyl-NAD has been shown to label the 51,000 M_r polypeptide of the FP fraction (78), which carries a tetranuclear FeS cluster (N3?) and possibly FMN. Therefore, there must be an access from the matrix side of the membrane to this polypeptide, allowing entry of NAD(H) and certain electron acceptors (e.g., ferricyanide) (Figure 2). It has been suggested that complex I is a dimer (73, 79) or tetramer (80), but data in support of these possibilities are considered inadequate (2, 66).

Catalytic Properties and Mechanism of Action of Complex I

Complex I catalyzes electron transfer from NADH (also at a slow rate at pH ≤ 6 from NADPH) to ubiquinone homologs, ferricyanide, and NAD (30, 55, 81). The reduction of Q, but not of ferricyanide, is coupled to vectorial proton translocation, and is inhibited by rotenone, piericidin A, barbiturates, Demerol, and mercurials. These inhibitors appear to act in the same region; they all inhibit the reduction of Q and oxidation of all the ESR-detectable FeS clusters. The apparent K_m's for NADH and Q_1 are, respectively, about 14 and

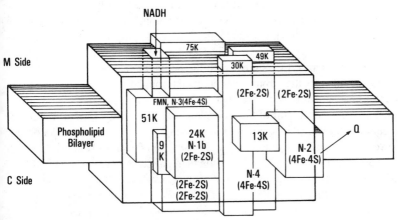

Figure 2 Schematic representation of the arrangement of complex I subunits. The 51, 24, and 9 K (kilodalton) blocks are the polypeptide components of FP; and the 75, 49, 30, and 13 K blocks are the polypeptide components of IP. The large block containing the FP and IP polypeptides is the HP fraction. The relative sizes and placement of the protein blocks are not precise. For details see text and ref. (2).

40 μM, and V_{max}^{app} for Q_1 reduction at 38°C is about 25 μmol min^{-1} mg^{-1} of protein (30, 55). The ferricyanide reductase reaction mechanism appears to be ping pong bi bi with double substrate inhibition, i.e., NADH inhibiting the interaction of the reduced enzyme with ferricyanide, and ferricyanide inhibiting the interaction of the oxidized enzyme with NADH (82). These results imply that NADH and ferricyanide react at the same or overlapping sites on the enzyme.

In contrast to ferricyanide, little is known about the mechanism of Q reduction by complex I, except that the effects of the above inhibitors as well as data from systems deficient in one or more FeS clusters implicate these redox centers in the pathway of electron transfer from NADH to Q (31, 32, 36, 37, 83). The finding of bound ubisemiquinone radicals in NADH-treated complex I has suggested single electron reduction of Q by the enzyme (84). However, whether ubisemiquinone is released from complex I before complete reduction, and is the species that transports electrons from complex I to complex III is not known.

The H$^+$/e$^-$ stoichiometry of proton translocation coupled to electron transfer by complex I has been estimated to be 1, 1.5, or 2 (33, 85–91). In the absence of precise information regarding H$^+$/e$^-$ and q$^+$/e$^-$ (i.e., number of charges translocated per electron transferred), it would be premature to speculate on the mechanism of redox-linked proton translocation. In principle, two types of proton translocation mechanisms may be conceived. The redox loop mechanism, originally proposed by Mitchell (85, 92), suggests that electron carriers such as flavin and Q take up and release protons vectorially across the membrane during their cyclic reduction and oxidation. This type of mechanism would be restricted to complexes I and III where appropriate electron carriers such as FMN and Q are found. The proton pump mechanism, originally suggested by Chance by analogy with the hemoglobin Bohr effect [(93), see also (94)], suggests that protons are pumped through the membrane via redox-coupled pK changes of appropriate prototropic protein residues. Several proton translocation mechanisms of the redox-loop type, involving FMN (89, 95, 96) or ubisemiquinone (84) as the proton-carrying vehicle, have been proposed. However, they are all hypothetical at this time and not entirely consistent with the available information (2, 66). While there are no data regarding the role of FMN and Q in proton translocation, there is evidence that FeS clusters are necessary for this function. In *Candida utilis*, rotenone and piericidin A sensitivity and site 1 coupling seem to be associated with the presence of ESR signals due to FeS clusters 1 and 2 (presumably N1b and N2, respectively) [(31, 32); see, however, (97–99)]. Also, *Paracoccus denitrificans* cells grown under sulfate-limited conditions have been shown to lack rotenone sensitivity, ESR signals due to FeS cluster 2 (N2), and site 1 coupling (37). The latter two characteristics were also absent in a rotenone-insensitive mutant of this organism (36). Furthermore, the E_m of cluster N2 (and presumably N1a) is considered to be pH-dependent ($\Delta E_m/\Delta pH = 60$ mV) (71), and the redox poise of N2 (and N4) has been shown to be affected by ATP-induced membrane energization (71, 100–102). The above results suggest that the subunit carrying cluster N2 may be involved in proton translocation. However, any mechanistic speculation should consider that complex I appears to contain two ESR-silent clusters whose redox properties are unknown.

SUCCINATE: UBIQUINONE OXIDOREDUCTASE (COMPLEX II)

The major component of this enzyme complex is succinate dehydrogenase (SDH), which occurs in all aerobic organisms as a membrane-bound enzyme of the citric acid cycle. SDH was discovered in 1909 (103), first solubilized in 1950 from horse heart mitochondria [(104), see also (105) and (106)], and first purified in 1971 from bovine heart mitochondria (107). While this enzyme has been the subject of intensive study in a number of laboratories [for reviews, see (2, 30, 108, 109)], it is difficult to consider SDH as a discrete and complete enzyme. Like the FP fragment of complex I, the catalytic activity of SDH is assayable only in the presence of artificial electron acceptors, and only in association with other mitochondrial polypeptides can SDH express a physiological activity. Thus, the physiological unit of function is complex II, which catalyzes electron transfer from succinate to ubiquinone (2, 30).

Composition and Structure of Complex II

Mammalian complex II appears to be composed of four polypeptides of M_r 70, 27, 15.5, and 13.5 $\times$ 10^3, as has been demonstrated by reconstitution of purified components (110). When the enzyme is directly isolated from mitochondria or complex II-III preparations, minor impurities due to complex III polypeptides are also present (46). Treatment of complex II with chaotropes (e.g., 0.5 to 1 M NaClO$_4$) results in the selective extraction of SDH, which is a water-soluble protein composed of two subunits of M_r 70 and 27 $\times$ 10^3 (30, 53, 107). SDH contains per mol one mol of covalently bound FAD, 7–8 g atoms of Fe and 7–8 mol of acid-labile sulfide (S^{2-}) (107). Separation of the two subunits under nondenaturing conditions has shown that the larger polypeptide contains the covalently bound FAD plus 4 equivalents each of Fe and S^{2-}, and that the smaller polypeptide contains 3 Fe and 3 S^{2-} per mol (107). This basic structure has been found in all the prokaryotic and eukaryotic SDH preparations that have been examined, even in *Vibrio succinogenes* fumarate reductase (Table 4). They all contain a ferroflavoprotein (FP) subunit of M_r 60–70 $\times$ 10^3 and an ironprotein (IP) subunit of M_r 25–30 $\times$ 10^3. It might also be added that the nature of the FAD linkage in the FP subunits of mammalian SDH and in *V. succinogenes* and *E. coli* fumarate reductases is the same, i.e., 8α[N(3)histidyl]FAD (117, 121), and that in the *E. coli* FP a sequence of 9 amino acids as determined from the gene sequence is the same as that which in the bovine FP contains the histidyl-FAD (120).

Preparations of complex II also contain a *b*-type cytochrome equimolar to FAD. Spectroscopic studies indicated that this cytochrome is distinct from the *b* cytochromes of complex III (122). Subsequently, it was shown in *Neurospora crassa* that the *b* cytochrome of complex II is a cytoribosomal product, while the *b* cytochrome of complex III (presumably one polypeptide, two hemes) is encoded by the mitochrondrial DNA (111). The cytochrome *b* of

Table 4 Composition of succinate dehydrogenases, succinate-ubiquinone oxidoreductases, and fumarate reductases from various sources

Source	Electron transfer activity	Polypeptides $M_r \times 10^3$	λ_{max} (α) of reduced cytochrome *b*	Ref.
Bovis	succ. → Q,dyes	70(FP),27(IP),15.5,13.5	560	110
N. crassa	succ. → Q,dyes	72(FP),28(IP),14(*b*)	559	111
B. subtilis	succ. → dyes	65(FP),28(IP),19(*b*)	558	112
M. luteus	succ. → dyes	72(FP),30(IP),17,15	556[a]	113, 114
R. rubum	succ. → dyes	60(FP),25(IP)	—	115, 116
V. succinogenes	menaquinol → fum.	79(FP),31(IP),25(*b*)[b]	560, 563[c]	117, 118
E. coli	dyes → fum.	66(FP),24	—	119, 120

[a] At 77 K

[b] The molar ratios of polypeptides in the enzyme complex are, respectively, 1 : 1 : 2.

[c] Two *b* cytochromes with E_m of -20 and -200 mV, respectively.

bovine complex II has been purified in a preparation composed of two polypeptides of M_r 15.5 and 13.5 × 10³ (110). These two polypeptides will be referred to as C_{II-3} and C_{II-4} in accordance with the designation of Ackrell et al (123). The *Neurospora* complex II, isolated from mitochondria of cells grown in the presence of chloramphenicol to inhibit mitochondrial protein synthesis, contains only three polypeptides of M_r 72, 28, and 14 × 10³, of which the first two are presumably the FP and IP subunits of SDH, and the last is a *b* cytochrome (111). A *b*-type cytochrome is present in *Bacillus subtilis* (112) and *Micrococcus luteus* (113) SDH complex preparations (Table 4), and has been found associated with SDH in other organisms as well [for reviews, see (30, 112)]. In *B. subtilis*, the genes encoding the SDH subunits and cytochrome b_{558} (Table 4) appear to be located in the same operon, with the order *b*-FP-IP (124, 125). Cytochrome b_{558} of *B. subtilis* appears to be a membrane attachment site for SDH. In cytochrome-deficient mutants SDH is elaborated, but does not become membrane-bound (126, 127). In the bovine system, the hydrophobic polypeptide C_{II-3}, which is required for the Q reductase activity of complex II and probably carries the b_{560} heme, appears to traverse the mitochondrial inner membrane as it is accessible to membrane-impermeable protein modifiers from the cytosolic side (128). Thus, by analogy to the *B. subtilis* system, C_{II-3} of bovine complex II could be a membrane attachment site for SDH. The release of SDH from membranes effected by chaotropes, the reversal of this process by introduction of water structure-forming ions (e.g., sulfate or phosphate), and the effect of D_2O versus H_2O on this resolution-reconstitution equilibrium have indicated that the interaction of SDH with membranes is largely hydrophobic (53, 129), although ionic interaction involving lysyl amino groups has also been suggested (130). This membrane interaction probably occurs more, or exclusively, through the IP subunit, since arylazidophospholipids were shown to label C_{II-3}, C_{II-4}, and IP, but not FP, in complex II incorporated in egg lecithin vesicles (128).

Light Absorption and ESR Spectral Properties of Complex II and SDH

The chromophores of complex II are cytochrome b_{560}, FAD, and three FeS clusters. Reduced b_{560} exhibits absorption maxima at 560 (α), 526 (β), and 424 (Soret) nm at room temperature (respectively, 557, 523 and 422 nm at 77 K) (110). The oxidized FAD and FeS clusters, as observed in SDH, absorb visible light between 400 and 550 nm with maxima at 450 nm due to FAD and about 420 nm due to FeS clusters (30).

The ESR studies of Ohnishi and coworkers have suggested that SDH contains two binuclear and one tetranuclear FeS clusters, designated respectively as S-1, S-2, and S-3 (65, 66, 69). The results of core extrusion and interprotein core transfer agree with this conclusion (131). Cluster S-1 exhibits an ESR spectrum centered at $g = 1.93$ and an apparent $E_m \sim 0$ mV. S-3 is a HiPIP-type cluster. It is paramagnetic in the oxidized state, has an ESR spectrum centered at $g = 2.014$, and exhibits an apparent E_m of $+65$ mV in complex II and $+120$ mV in intact mitochrondria (65). Cluster S-3 is stable in membrane-bound SDH, but in soluble SDH it is sensitive to oxygen and its ESR signal deteriorates in parallel with the loss of reconstitution and low-K_m ferricyanide reductase activities of the enzyme (see below). Recent applications of continuous wave ESR and study of the magnetic field dependence of g value shifts caused by an externally applied electric field (linear electric field effect) have indicated that in air-oxidized complex II S-3 is a 3-iron cluster, possibly of the [3Fe-4S] type (132). However, whether in intact mitochondria S-3 might be a [4Fe-4S] cluster or whether the redox cycle of SDH might involve an interconversion between [3Fe-4S] and [4Fe-4S] are not known (132). Clusters S-1 and S-3 are present in SDH in amounts equimolar to FAD (65, 66).

Cluster S-2, originally detected by Ohnishi et al, is considered to be a binuclear cluster with an apparent E_m of about -400 mV in isolated SDH and about -260 mV in particles (65). This cluster has a shorter spin relaxation time than S-1, and an ESR spectrum very similar to that of S-1. It is observed as an enhancement of the S-1 ESR signals ($g_{x,y,z}$ at 1.91, 1.93, 2.03) when, instead of succinate, SDH is reduced with dithionite (133). The intensity of the signal induced by dithionite (presumably due to S-1 plus S-2) amounts to less than twice the concentrations of either FAD or S-3, hence the suggestion that S-1 and S-2 are spin-coupled (65, 66, 132). Also, while S-2 is seen in complex II and isolated (especially reconstitutionally inactive) SDH, it is not detected in submitochondrial particles (66, 134). Therefore, not every worker in the field is agreed on the existence of S-2 [(134), see also (132)], even though the FAD : Fe : S^{2-} ratio of 1 : 7–8 : 7–8 and the results of core extrusion studies (see above) favor the existence in SDH of more than one binuclear and one tetranuclear (or [3Fe-4S]) clusters [for further details, see (65, 66)].

It has been proposed that the FP subunit of SDH contains the two binuclear clusters S-1 and S-2, and that the IP subunit contains S-3 (65). However, there is no direct evidence for this distribution in SDH. In *V. succinogenes* fumarate reductase, the HiPIP cluster appeared to be contained in FP (albeit the isolated FP had retained only a marginal amount of this cluster), and the single binuclear cluster detected (118) was found in the fraction containing the smaller subunit plus cytochrome b_{560} (117).

Catalytic Properties and Mechanism of Action of Complex II and SDH

COMPLEX II   Complex II catalyzes electron transfer from succinate to ubiquinone and dyes such as ferricyanide and phenazine methosulfate (PMS). The

Q reductase activity of complex II is inhibited by 2-thenoyltrifluoroacetone [(30), see also ref. (135) regarding the inhibitory effects of carboxins]. Isolated SDH also reacts with the above dyes, but not with Q.

When purified SDH is admixed with a cytochrome b_{560} preparation obtained from complex II, a highly active succinate-Q reductase complex is reconstituted, which is composed of equimolar amounts of the two SDH subunits and the two polypeptides (C_{II-3} and C_{II-4}) of the cytochrome b_{560} preparation (2, 110). C_{II-3}, which probably carries the b_{560} heme, has been shown to be required for expression of Q reductase activity (123). Also, an excellent correlation has emerged between the Q reductase activities of reconstituted complex II systems and the b_{560} heme content of the C_{II-3}/C_{II-4} preparations employed by different laboratories [(110, 123, 136, 137), see also Table XIII in ref. (2)]. It has been shown that dithionite-reduced b_{560} of complex II is rapidly oxidized by fumarate via SDH, and that the addition of succinate together with fumarate lowers the rate and the extent of b_{560} oxidation (110). These experiments have allowed calculation of an $E_{h,7}$ for complex II b_{560} of less than -80 mV. Reduced b_{560} is also oxidized by added Q and, in keeping with its low reduction potential, the oxidized cytochrome is reduced to a small extent by succinate (110). These results indicate, therefore, that cytochrome b_{560} of complex II is capable of electron transfer from SDH to Q. The only problem in considering b_{560} as an obligatory electron carrier in complex II is its low reduction potential, which might pose a thermodynamic barrier against succinate oxidation, especially as fumarate accumulates. However, it is possible that the reduction potential of b_{560} in intact mitochondria is more positive than that in complex II.

Another view is that the cytochrome b of complex II is an impurity (see, however, Table 4 and above), and that the smallest polypeptide of complex II (corresponding to C_{II-4}) is a specific ubiquinone apoprotein (designated QP_s-I), which is required for Q binding and conferral of Q reductase activity to SDH (136–138). Preparations of QP_s-I (136) have a low cytochrome b content (< 1 nmol/mg), and when combined with SDH reconstitute a marginal Q reductase activity [about 4% as compared to the Q reductase activity obtained with the use of a more intact cytochrome b_{560} preparation, ref. (110)]. In general, the evidence regarding the existence in mitochondria of specific Q-binding apoproteins (i.e., as distinct from cytochromes and FeS proteins which might bind and reduce or oxidize Q) is circumstantial, and attempts to identify such apoproteins with the use of photoreactive ubiquinone analogs have resulted in labeling of hydrophobic polypeptides (e.g., both C_{II-3} and C_{II-4} of complex II, cytochrome b of complex III) (139, 140), which does not allow a distinction between partitioning of the Q analog in the hydrophobic environment of these complexes and its specific interaction with a Q-binding apoprotein [for further discussion of this issue, see ref. (2)]. It should also be mentioned that the

Neurospora complex II contains only three polypeptides (the two SDH subunits and cytochrome b_{559}), and lacks a polypeptide corresponding to QP_s-I (111). Unfortunately, however, the Q reductase activity of the isolated *Neurospora* complex II is only about 1% of that of the bovine enzyme (110, 111), and the possibility of the presence of marginal amounts of a QP_s-I-like polypeptide in the former preparation cannot be ruled out.

SDH   Preparations of SDH exhibit two kinds of catalytic properties. One is succinate-dye reductase activity; the other is succinate-X reductase activity, where X (cytochrome b_{560}?) is the natural electron acceptor of SDH in the respiratory chain. The former activity is much more stable than the latter, and is exhibited even by SDH preparations containing <50% of their Fe and S^{2-} content (141). The dyes commonly used for assay of this activity are ferricyanide or PMS, the latter in the presence of 2,6-dichlorophenol indophenol as electron acceptor from the easily air-oxidized PMS. SDH is also capable of slow electron transfer from a suitable dye to fumarate (108, 109, 142). The activity of SDH to transmit electrons to its natural acceptor is stable when the enzyme is membrane-bound, but in isolated SDH it deteriorates rapidly under aerobic conditions (30, 143). The deterioration of reconstitution activity of SDH is paralleled by modification and disappearance of the ESR signal of FeS cluster S-3 (144, 145). It can also be monitored by the loss of a low-K_m (0.25 mM) ferricyanide reductase activity which is exhibited by soluble SDH (146, 147). The high-K_m (3 mM) ferricyanide reductase activity exhibited by both membrane-bound and soluble SDH does not appear to be affected by deterioration of the reconstitution activity and the ESR signal of cluster S-3. Reconstitutionally inactive SDH can be repaired in this respect by treatment with sodium sulfide and ferrous ions (148, 149) or thiosulfate in the presence of rhodanese (150). However, the difference between reconstitutionally active and inactive SDH cannot be detected by assay of their Fe and S^{2-} content (148, 151). Thus, it is possible that conversion to reconstitutionally inactive SDH involves displacement, but not loss, of an S-3 component, such as a sulfur moiety.

The pH/activity profile of SDH both in the direction of succinate oxidation and fumarate reduction has suggested participation in catalysis of a group with a pK ~ 7.0 (142). It has also been shown that SDH is inhibited by ethoxyformic anhydride (142) and dansyl chloride at pH 6.0 (L. Hederstedt, Y. Hatefi, unpublished), and that these inhibitory effects are prevented in the presence of substrates and competitive inhibitors. Together the above results have suggested the possible participation of an active site histidyl residue in catalysis, with the unprotonated species being involved in succinate oxidation and the protonated species in fumarate reduction (142). In addition, SDH has been known since 1938 to contain an essential thiol (152) at or near the active site (153, 154), and recently the presence of active site arginyl residue(s), presumably involved in electrostatic interaction with substrate carboxyl groups, has been reported (155).

UBIQUINOL:CYTOCHROME c OXIDOREDUCTASE (COMPLEX III)

Complex III was discovered and purified from bovine heart mitochondria in 1961 (28, 156). The enzyme catalyzes electron transfer from dihydroubiquinone (QH_2) to cytochrome c, and this reaction is coupled to transmembrane proton translocation (157). Similar or analogous enzyme complexes, which catalyze proton translocation-coupled electron transfer from a reduced quinone to an acceptor, have been isolated from *N. crassa* (111), *S. cerevisiae* (158, 159), *Rhodopseudomonas sphaeroides* (160–162), *Anabaena variabilis* (163), and chloroplasts (164, 165), and discussed in excellent recent reviews (5, 6, 9, 11, 166–169). They all contain two b-type and one c-type cytochromes and an iron-sulfur protein. The electron acceptor is cytochrome c for the mitochondrial complex III, cytochrome c_2 for the *R. sphaeroides* enzyme, and plastocyanin for the analogous complexes from *A. variabilis* and chloroplasts.

Composition and Structure of Complex III

The polypeptide composition of complex III-like preparations from several sources is shown in Table 5. The mammalian enzyme is composed of 9–10 unlike polypeptides, three of which are associated with redox centers. These centers are b_{562}, b_{566}, and c_1 hemes and a [2Fe-2S] cluster. In addition, two ubisemiquinone species are considered to be present in two separate domains of complex III (175–177).

The two b hemes are considered to be associated with a single hydrophobic polypeptide of M_r 42,540 [based on the bovine cytochrome b gene sequence, ref. (178)], which is encoded by the mitochondrial DNA, synthesized within the mitochondrion, and incorporated into the inner membrane, presumably without further processing (111, 178–180). The occurrence of 2 b hemes on a single polypeptide has not been demonstrated, however, and the possible presence in complex III of two apocytochrome b polypeptides each containing one b heme agrees with the polypeptide stoichiometries shown in Table 5. The cytochrome b of complex III has been purified from bovine (181, 182), *Neurospora* (183, 184), and yeast mitochondria (185, 186). The estimated heme contents of the purified preparations (30 nmol/mg of protein for the cytochrome b purified from bovine heart mitochondria) are considerably lower than that expected from two hemes bound to a polypeptide of 42,000 M_r (178). The highest is the heme content of 37 nmol/mg of protein reported for a preparation of cytochrome b from yeast complex III (186). Also, while in mitochondria and complex III the two b cytochromes (b_{562} and b_{566}) exhibit

Table 5 Polypeptide composition and molecular weights of quinol-cytochrome c (plastocyanin) oxidoreductases (apparent M_r in thousands)

Polypeptide	Bovine heart Ref: 166, 170	Rat liver 171	N. crassa 111, 172	S. cerevisiae 159, 173	Rps. sphaeroides 5	Spinach chloroplast 5	A. variabilis 5
1 (core protein 1)	45.5(1)[a,b]	50	50(1)[c]	49	—	—	—
2 (core protein 2)	44 (1)	46	45(2)	39	—	—	—
3 (cyt. b)	42 (2)	33	35(2)	30	40[d]	23 (1)[e]	22.5 (2)[f]
4 (cyt. c_1 or f)	31 (1)	25	31(1)	29	34	33/34(1)	31/38(1)
5 (FeS protein)	24.5(1)	12.5	25(1)	22.4	25	20 (1)	22
6	15 (2)	10	14(1)	13.4	—	17 (1)	16 (1)
7	9 (2)	—	12(1)	11.1	—	—	—
8	7.8(2)	5.6	11(1)	10	—	—	8
9	4.8		8	9	6	5	—
10	3						

			Molecular Weights of the Enzyme Complexes				
Monomer	250	250	290	238, 253	250	142, 185	≤220
Dimer	440	250	550	450			

[a] Numbers in parentheses indicate relative stoichiometry of the polypeptides

[b] Based on Coomassie blue stain intensity of protein bands on dodecyl sulfate gels (174)

[c] Based on [^{3}H]leucine labeling (111)

[d] Respectively, 48, 30, 24 and 12 according to (162)

[e] Based on Coomassie blue and amido black stain intensity of protein bands on dodecyl sulfate gel slabs (165)

[f] Based on Coomassie blue stain intensity of protein bands on dodecyl sulfate gels (163)

different spectral properties (187, 188), the purified preparations have a single cytochrome absorption spectrum characteristic of b_{562} (182–186). Conversion of the b_{566} absorption spectrum to that of b_{562} has been observed by perturbation of complex III structure with detergents or chaotropes (187, 189). In various complex III preparations, the apparent E_m values of b_{562} and b_{566} differ by 100 ± 20 mV (5, 190). A dimeric cytochrome b purified from bovine mitochondria has also been reported to exhibit two $E_{m,7}$ values of -5 mV and -85 mV, each about 100 mV lower than the $E_{m,7}$ values measured by the same authors in the parent complex III (182). However, the yeast cytochrome b preparation referred to earlier exhibited only a single $E_{m,7.4} = -44$ mV with n $= 1$ (186).

It has been shown that the amino acid sequences of b cytochromes from human, bovine, murine, yeast, and *Aspergillus nidulans* mitochondria, and from spinach chloroplasts (cytochrome b_6) have a high degree of homology, including conserved histidine residue pairs His-82—His-197/198 and His-96—His-183 which are considered to be the ligand pairs for the two b hemes (191). Hydropathy analyses have indicated that the 42,000 M_r protein could form 8–9 membrane-spanning hydrophobic domains (191). The histidine residues 82 and 96 occur in domain II, and their corresponding residues 197/198 and 183 in domain V. The two hemes appear to be located perpendicular to the membrane, with b_{566} close to the cytosolic side, and b_{562} close to the matrix side (191, 192). Most interesting is the fact that the chloroplast b_6 contains only 210 residues ($M_r = 23,000$) and can form only five membrane-spanning hydrophobic domains, corresponding to the first five heme-bearing domains of the mitochondrial cytochrome (190, 191). However, the chloroplast b_6-f complex also contains a 17,000 M_r polypeptide which has several regions homologous to the C-terminal end of the mitochondrial cytochrome b (191). This polypeptide, which seems to cofractionate with b_6, is encoded by a gene adjacent to that of b_6 on the chloroplast genome (5).

Cytochrome c_1 is water-soluble in the purified state, and appears to be composed of two unlike polypeptides (193, 194). The larger polypeptide, first isolated in 1960 (195), has been purified and sequenced (196). It is composed of 241 amino acid residues, the heme is attached covalently near the amino terminus to cysteinyl residues 37 and 40, and the calculated molecular weight (protein plus heme) is 27,924. The protein contains a hydrophobic cluster near the carboxy terminus, which is considered to be its membrane attachment domain. The smaller polypeptide has also been sequenced [(197), see also (198)]. It is composed of 78 amino acids (M_r 9,175), with a high content (27%) of glutamic acid and glutamine and no methionine, isoleucine, tryptophan, and tyrosine. There are eight consecutive glutamic acid residues in positions 5–12. This polypeptide, designated hinge protein, is required for interaction of the heme-bearing polypeptide with cytochrome c (199). Whether the eight consecutive glutamic acid residues play a role in c_1-c interaction is not known, but probable considering the basic nature of cytochrome c (pI = 10.3) (200).

The bovine FeS protein was first isolated in its succinylated form by Rieske (201) after whom this type of FeS protein is often named. More recently, the protein was purified in unmodified and reconstitutable state (167, 202). The FeS protein of complex III is water-soluble, has an apparent M_r of 24,500, a binuclear FeS cluster with a characteristic ESR signal centered at $g = 1.90$, and an $E_{m,7} = +280$ mV (n = 1) (65, 66, 167). An FeS protein with similar ESR spectrum and E_m is present in complex III and analogous preparations from other sources (Table 5) (5, 6, 66). The FeS protein of *Neurospora* complex III (M_r 25,000) is considered to be an amphiphilic molecule with a hydrophilic domain of 16,000 M_r which is cleavable from the remaining hydrophobic domain by proteolysis (203). The former has been isolated and shown to contain the FeS cluster with unaltered ESR characteristics, and the hydrophobic domain is thought to be the region that anchors the exposed hydrophilic domain to the membrane (203).

Low-resolution electron microscopic studies of two-dimensional crystals of *Neurospora* complex III, the mode of cleavage of the subunits by mild procedures, chemical labeling, and cross-linking data, as well as considerations of the solubility properties of complex III subunits, have suggested the following with regard to the structure of the *Neurospora* enzyme (172, 203). Purified *Neurospora* complex III is dimeric. In phospholipid bilayers, the dimer spans the bilayer (about 5 nm thick), extending out 7 nm from one side and 3 nm from the other. The locations of the b and c_1 cytochromes and the FeS protein are considered to be as shown in Figure 3. The two largest polypeptides of complex III, misleadingly referred to as core proteins 1 and 2, extend out into the matrix space. This picture agrees with the results of chemical labeling and cross-linking studies on bovine and yeast complex III as well as with information accrued from the use of antibodies (204–209), except that in the case of yeast complex III the core proteins appeared to be almost completely buried in the membrane (159, 210). Preparations of *R. sphaeroides* complex III and of cytochrome b_6-f complexes from spinach chloroplasts and *A. variabilis* do not contain polypeptides corresponding to the core proteins (Table 5), and the role of these polypeptides in mitochondrial complex III is not known. From *Neurospora* complex III, a subcomplex containing b and c_1 cytochromes plus the four small subunits of the complex, and lacking core proteins 1 and 2 and the FeS protein, has been isolated (203, 211). This subcomplex plus the FeS protein could be considered analogous to the non-mitochondrial enzyme complexes. However, whether the bc_1 subcomplex in the absence or presence of added FeS protein has any catalytic properties has not been reported. Except apocytochrome b, all the other subunits of *Neurospora* complex III have been synthesized as precursors in a reticulocyte lysate (212). All the precursors were

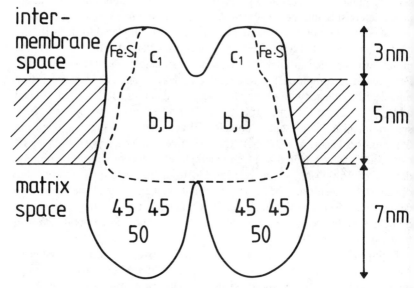

Figure 3 Schematic representation of the gross structure of *N. crassa* complex III. Dashed line shows the bc_1 subcomplex. Letters refer to the redox centers, and numbers to the M_r values of "core" proteins 1 and 2. From (203).

synthesized as larger polypeptides, except a subunit of $14,000\,M_r$. The processing and transfer of several subunits into isolated mitochondria were shown to be dependent on the mitochondrial membrane potential (212).

Catalytic Properties and Mechanism of Action of Complex III

Complex III catalyzes electron transfer from ubiquinol to cytochrome c. The original preparations isolated with the use of deoxycholate and cholate exhibited maximal turnover rates of 2.5 to $4 \times 10^3\ \mathrm{s}^{-1}$ at 38°C and pH 8.0 in the presence of chromatographically pure ubiquinol-2 as substrate (156, 166). More recent procedures for the isolation of complex III use Triton X-100, either alone or in combination with cholate and/or octylglucoside. These preparations exhibit extremely low activities [for review, see (5)], which is in part because of the deleterious effect of Triton X-100. Among other things, this detergent destabilizes the structure of the complex and results in the release of the FeS protein and changes in the spectral properties of the b cytochromes (213–215). As discussed above and elsewhere (2), Triton X-100 also damages the cytochrome b_{560} of complex II, resulting in preparations with low heme content. The natural electron donor and acceptor for b_6-f complexes are, respectively, plastoquinol and plastocyanin (also cytochrome c-553 for the *A. variabilis* enzyme), but they also exhibit oxidoreductase activity with ubiquinol and cytochrome c (5). Similarly, the mammalian complex III can oxidize duroquinol and menaquinol (216, 217), and reduce various c type cytochromes, including cytochrome f (218). Mammalian complex III recombines stoichiometrically with complexes I, II, and IV (in the presence of added cytochrome c) to reconstitute the entire respiratory chain (2, 28, 219, 220). It also reconstitutes in the presence of cytochrome c with a detergent-solubilized reaction center preparation from $R.$ $sphaeroides$ to reconstitute a light-activated cyclic electron transfer system [(221), see also (222)].

Electron transfer through complex III is coupled to transmembrane proton translocation with a stoichiometry of $H^+/e^- = 2$ and a net positive charge translocation outward $q^+/e^- = 1$ (11). The mechanisms of electron transfer and proton translocation by complex III are not clear. A possible mechanism considered best to fit the available information is that known as the Q-cycle, which was originally proposed by Mitchell (223). The central feature of this and other mechanisms advanced by Wikström [the b-cycle, ref. (11)] and Papa et al (224) is a branched electron transfer pathway first proposed by Wikström & Berden (225) to explain the phenomenon of the oxidant-induced extra reduction of cytochromes b (see below). In general, complex III may be thought of as a system of two high-potential (n = 1) and two low-potential (n = 1) redox centers (5, 6). The former are the FeS protein ($E_{m,7}$ = +280 mV) and cytochrome c_1 ($E_{m,7}$ = +230 mV), and the latter are cytochromes b_{566} ($E_{m,7}$ = -30 ± 10 mV) and b_{562} [E_{m7} = $+30 \pm 10$ mV or +93 mV, ref. (226)] (227). Ubiquinol donates one electron to the high-potential centers ($\xleftrightarrow{e}$ FeS protein $\xleftrightarrow{e}$ c_1) and one electron to the low potential centers ($\xleftrightarrow{e}$ b_{566} $\xleftrightarrow{e}$ b_{562}). The reduction of the high-potential centers is inhibited by myxothiazol (228) or by 2,3-dimercaptopropanol (BAL, British anti-Lewisite) which destroys the FeS center (229). According to the Q-cycle, (Figure 4A), the withdrawal of one electron from QH_2 (at a location near the C side of the membrane, center o) is accompanied by vectorial release of two protons to the C side and formation of a ubisemiquinone anion, a powerful reductant for the low-potential centers ($E_{m,7}$ for $Q^{\div}/Q \sim -240$ mV) (5, 6). The b-cycle (Figure 4B) relegates proton translocation to cytochromes b by a redox-coupled proton pump mechanism, which could be mechanistically analogous to the hemoglobin Bohr effect (93, 94, 224). To complete the cycle, reduced b_{562} then donates an electron back to ubiquinone in a reaction which is inhibited by antimycin or HQNO. According to the Q-cycle, the recipient of electron from b_{562} is a species of ubisemiquinone anion (or ubiquinone, Figure 4A) located near the M side of the membrane (center i), which would also take up $2H^+$ from the M side to reform QH_2 (169). According to the b-cycle, the electron recipient is the protonated species $QH\cdot$ (11). The net result in either scheme is the scalar transfer of one electron from ubiquinol to cytochrome c coupled to the vectorial translocation of protons from the M to the C side. In the Q-cycle the vehicle for proton translocation is Q and the proton stoichiometry is necessarily restricted to $2H^+/e^-$ [however, see (230)]. In the b-cycle, the instrument for proton trans-

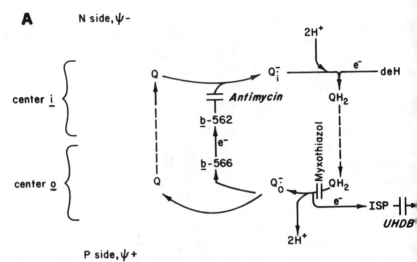

A   N side, ψ –

P side, ψ +

Figure 4A   The proposed protonmotive Q-cycle of complex III. The dashed arrows designate the limited mobilities of ubiquinone and ubiquinol. The subscripts i and o designate specific species of ubisemiquinone anion, which react at the electronegative N (matrix) side (center i) or electropositive P (cytosolic) side (center o) of the inner mitochondrial membrane. ISP, iron-sulfur protein; deH, dehydrogenase. From (167) and (236).

location is the transmembranous cytochrome *b* molecule and the proton stoichiometry would in principle depend on the details of the mechanism of proton translocation (e.g., redox-coupled pK_a changes of the prototropic residues on the C and M sides of the protein).

Both the Q-cycle and the *b*-cycle hypotheses agree with the following important features of the system: (*a*) removal of the FeS protein or its in situ modification by BAL results in inhibition of substrate-induced *b* reduction in antimycin-treated preparations (167, 169, 202, 231–233); (*b*) the phenomenon

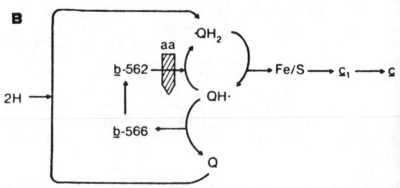

Figure 4B   The proposed *b*-cycle of complex III. 2H indicates input of reducing equivalents from the ubiquinone pool. Q, QH· and QH$_2$ represent protein-bound ubiquinone, ubisemiquinone, and ubihydroquinone, respectively. aa, antimycin. From (258).

of oxidant-induced extra reduction of cytochrome b (i.e., oxidation of the high-potential centers results in increased reduction of the low-potential centers), which is particularly prominent in antimycin-treated systems, and the slow reduction of b in antimycin-treated and c_1 prereduced preparations (5, 6, 13, 187); and (c) the fact that inhibition of electron flow through the high-potential centers by myxothiazol, UHDBT, or BAL does not inhibit cytochrome b reduction, but will do so when antimycin is also added, presumably to block electron backflow to cytochrome b (e.g., from center i) (169, 233–236). In addition, the observation that oxidized Q affects the ESR line-shape of the FeS cluster is considered to be in agreement with Q being capable of interaction with the FeS protein (237). The Q-cycle is supported by ESR evidence for the existence in complex III of two types of stable $Q\overline{\cdot}$ radicals, one sensitive to antimycin and the other sensitive to BAL but not to antimycin (169, 175–177). The b cycle is strengthened, on the other hand, by the pH dependence of the apparent E_m of both b cytochromes (the E_m's of the FeS protein and c_1 also show pH dependency, but only at pH values > 8) (5, 227, 238). Furthermore, in yeast complex III, DCCD reacts specifically with cytochrome b [which also appears to bind antimycin, ref. (239)] and inhibits proton translocation but not electron transfer (173, 240, 241). DCCD also inhibits proton translocation in bovine complex III, but it is considered to react with several polypeptides with or without appreciable binding to cytochromes b [(242–247); see, however, (248)].

There are also many observations that do not fully agree with the formulations depicted in Figure 4. For example, both the Q- and the b-cycles predict that single-turnover reduction of c_1 by substrates should be antimycin-insensitive, and evidence in agreement with this prediction has been presented (249). However, others have argued that the former experiments involved multiple turnovers and slow reduction of c_1 via electron leak through the antimycin block, and have published data indicating that rapid c_1 reduction by Q_1H_2 is antimycin-sensitive (250). Also, it has been shown that in the absence of antimycin, b_{562} is reduced before the FeS cluster, which does not agree with the proposition that the FeS cluster is the first (or concomitant) recipient of electrons from QH_2 (237). Another finding which may require modification of the schemes of Figure 4 concerns the fact that in these formulations b_{566} and b_{562} are placed in kinetic sequence for which there is no experimental evidence. The finding is that in antimycin-treated SMP, reduced b_{566} and b_{562} are oxidized through the apparent leak in the antimycin block at two different rates, the rate of b_{566} oxidation being 10 times faster than that of b_{562} (251). These results suggest that in antimycin-treated systems b_{566} and b_{562} are not in kinetic sequence. Another explanation could be that b_{562} is oxidized slowly through the antimycin block, while b_{566} is oxidized faster in the opposite direction (e.g., via center o in Figure 4A). However, this would require an extremely slow redox equilibration between b_{562} and b_{566}, which is not in agreement with published results (252). Other observations not easily rationalized by the schemes of Figure 4 have prompted Slater and coworkers to propose a complicated double Q-cycle, involving a complex III dimer with two kinetically different species of FeS clusters, four different b cytochromes, and four species of Q^- (169, 237). Another possible mechanism which combines the redox-loop feature of the Q-cycle and the proton-pump attribute of the b-cycle should be mentioned. It has been shown in antimycin-treated rat liver mitochondria that QH_2 oxidation via a TMPD bypass results in the extrusion of $2H^+/2e^-$, regardless of whether QH_2 was generated by succinate oxidation on the matrix side or by glycerol phosphate oxidation on the cytosolic side of the inner membrane (253). When the electrons were allowed to pass through the bc_1 segment in the absence of antimycin and TMPD, then the proton stoichiometry for QH_2 oxidation via complex III was $4H^+/2e^-$. These results indicate that $2H^+$ emanate from QH_2 oxidation as proposed by the Q-cycle, and $2H^+$ are translocated as a result of electron flow through complex III.

The basic mechanistic features discussed above also pertain to other complex III and analogous structures referred to above (Table 5). They are all composed of two low-potential and two high-potential redox centers, they all exhibit oxidant-induced b reduction, and are all inhibited by HQNO and UHDBT (5, 6). The b_6-f systems are not inhibited by stoichiometric amounts of antimycin (5, 6), which may be related to the structural difference between b_6 and the mitochondrial cytochrome b. Also, the chloroplast b_6-f complex is not inhibited by myxothiazol (6). Other characteristics of the prokaryotic and plant quinol-cytochrome c (plastocyanin) oxidoreductases are summarized in recent comprehensive reviews (5, 6, 168, 254).

FERROCYTOCHROME c: OXYGEN OXIDOREDUCTASE (COMPLEX IV)

Der Atmungsferment of Warburg, designated cytochrome oxidase by Keilin (255), is a multisubunit enzyme complex that catalyzes the terminal act of respiration by delivery of electrons derived from the stepwise oxidation of foodstuffs to molecular oxygen. According to Malmström (13), there are more than 200 enzymes known for which molecular oxygen is a substrate. However, the present section will be confined to a review of the mitochondrial aa_3 type cytochrome oxidase. Excellent recent reviews are available on various oxidases and oxygenases (13), on bacterial cytochrome oxidases (256), as well as on the structure, function, and biogenesis of mitochondrial cytochrome oxidase (24, 257–265).

Polypeptide Composition and Structure of Cytochrome c Oxidase

Cytochrome oxidase preparations from bovine heart, rat liver, human placenta, *S. cerevisiae,* and *N. crassa* have in common seven or eight unlike polypeptides of roughly comparable size (260). When subjected to dodecyl sulfate gel electrophoresis, most preparations of the bovine enzyme exhibit a minimum of eight polypeptides in equimolar concentrations, plus significant amounts of at least three other polypeptides (a, b, and c, Table 6). An additional polypeptide comigrating with subunit VII-Ile (Table 6) and, unlike VII-Ile, apparently containing cysteine residues is also thought to be present (261). However, Capaldi and coworkers consider a, b, and c as impurities, because these polypeptides are not present in column-purified oxidase, are not precipitated together with the major polypeptides by antibody raised to a highly purified preparation, and can be digested by proteases without affecting the electron transfer and proton translocation activities of the enzyme (261). The three largest polypeptides of cytochrome oxidase are encoded by the mitochondrial DNA, synthesized within the mitochondrion, and inserted into the membrane from the matrix side (8, 264). The mammalian subunits I, II, and III do not appear to undergo posttranslational processing (260, 266, 267), whereas the corresponding subunits of the yeast and *Neurospora* enzymes seem to be produced with N-terminal leader sequences that are subsequently deleted (268). The amino acid sequences of the eight subunits of the bovine cytochrome oxidase have been determined (I, II, and III from the gene sequence; the remainder as well as II by direct sequence analysis of the polypeptides) [(178, 260), (268–276), see also (277)]. The calculated molecular weight of the eight-subunit enzyme monomer, based on amino acid sequence data (Table 6), is 162,000.

Electron microscopy and image reconstitution of two-dimensional crystals of monomeric and dimeric cytochrome oxidase have shown that the enzyme is

Table 6   Molecular weights of the polypeptides of bovine heart cytochrome c oxidase[a]

Polypeptide	Molecular weight
I	56,065[b]
II	26,021[b]
III	29,918[b]
IV	17,153
V	12,436
VI	10,026
VII-Ser	5,541[c]
VII-Ile	4,962[c]
a	14,000
b	11,000
c	8,480

[a] From Capaldi, R. A. 1982. *Biochim. Biophys. Acta* 694:291–306.
[b] Calculated from mtDNA sequence data.
[c] The N-terminal sequences are Ser-His-Tyr-Glu for VII-Ser, and Ile-Thr-Ala-Lys for VII-Ile.

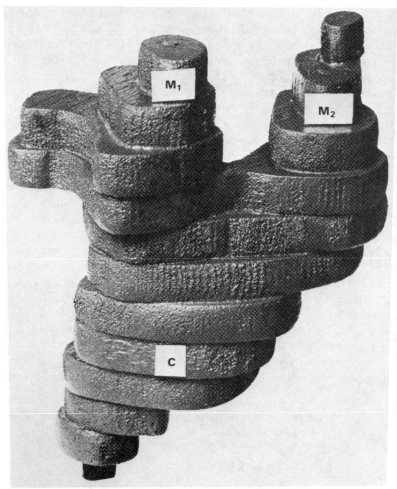

Figure 5 A model of cytochrome *c* oxidase monomer deduced from electron microscopic analysis of two-dimensional crystals. M_1 and M_2 are the two matrix domains, and C is the cytosolic domain. From (261) and ref. (16) therein.

Y-shaped (Figure 5) (170, 261). The stem of the Y extends 5.5 nm from the C side, and the two arms of the Y (M_1 and M_2) span the inner membrane and protrude at most about 2 nm into the matrix space. The M to C length of the molecule is about 12 nm and the widest cross section, which is within the membrane, is about 10 nm across (262). M_1 is larger than M_2, and their center to center distance is about 4 nm (261). In dimeric cytochrome oxidase, the two monomers make contact in the C domain (Figure 6).

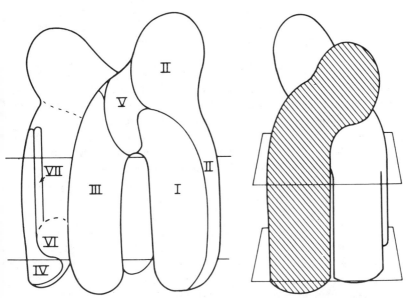

Figure 6   Arrangement of subunits in cytochrome *c* oxidase according to Capaldi et al (261). The figure on the right shows the juxtapositioning of monomers in the oxidase dimer. From (261).

The arrangement of subunits within the oxidase molecule has been investigated with the use of membrane-impermeable and membrane-permeable protein modifying reagents (278–281), subunit-specific antibodies (282, 283), arylazidophospholipids (284, 285), and protein cross-linking reagents (286–290). The results have led Capaldi and coworkers (261) to construct the model shown in Figure 6. Subunits I, II, III, IV, VII-Ser, and VII-Ile are transmembranous, VI is considered to be exposed on the M side, and V is placed by these investigators on the C side on the basis of cross-linking results (291, 292). This position does not agree, however, with the antibody recognition results of others, which suggest that subunit V is exposed on the M side (282). It has been pointed out that subunits I, II, III, IV, VII-Ser, and VII-Ile contain, respectively, 10, 2, 6, 1, 1, and 1 long stretches of hydrophobic amino acids, which agrees with their transmembranous positioning in Figure 6 (170, 261).

Cytochrome oxidases of the aa_3 type are also found in a variety of microorganisms (e.g., *Thiobacillus, Nitrobacter, Rhodopseudomonas,* cyanobacteria, and thermophiles), and have been purified from *P. denitrificans, N. agilis, R. sphaeroides, Thermus thermophilus,* and the thermophilic bacterium PS3 [for review, see (256)]. The enzymes from *P. denitrificans* and *N. agilis* have two subunits each with M_r 45 and 28×10^3, and 51 and 31×10^3, respectively. The others have been isolated as three-subunit enzymes, with those from the thermophiles being cytochrome aa_3-c complexes. The preparations from thermophilic bacterium PS3, *N. agilis,* and *P. denitrificans* have been shown to contain copper equimolar to heme aa_3, and the *Paracoccus* enzyme also contains stoichiometric amounts of tightly-bound manganese (293).

Prosthetic Groups of Cytochrome c Oxidase

The cytochrome oxidase molecule contains two identical hemes *a* in two different environments, each of which is associated with an atom of copper. The two hemes are designated *a* and a_3 and the associated coppers are denoted, respectively, as Cu_a (or Cu_A) and Cu_{a_3} (or Cu_B). The orientations of the hemes in the mitochondrial inner membrane are nearly normal to the plane of the membrane. The iron of heme *a* (Fe_a) is hexacoordinate, low-spin (g = 1.45, 2.21, 3.0), and is liganded in the axial positions to nitrogens of two neutral imidazoles (262, 294). It has been shown that among a number of low-spin heme compounds only the g values of neutral *bis*-imidazole complex of heme *a* are the same as those of heme *a* in cytochrome oxidase (294). Most of the absorbance of reduced cytochrome oxidase at 605 nm (α band) and half of the absorbance at 455 nm (Soret band) are considered to be due to *a,* and the remainder due to a_3 (24). Cu_a displays an intense and temperature-dependent ESR spectrum with g values of 1.99, 2.03, and 2.18 (38, 295), and is responsible for the wide and relatively flat absorption centered at about 830 nm (13, 262, 296, 297). The results of X-ray absorption studies (298) and extended X-ray absorption fine structure (EXAFS) measurements (299) together with comparison of ESR and electron nuclear double resonance (ENDOR) spectra from native yeast cytochrome oxidase and preparations containing [^{15}N] histidine or ^{2}H at the β-methylene of cysteine have prompted Chan and coworkers (262, 300) to propose that Cu_a in the reduced state is liganded to two cysteine sulfurs (one a cysteinate and the other a sulfur radical) and one imidazole nitrogen (Figure 7). The fourth ligand is also thought to be a nitrogen. Further, ESR studies have suggested to these investigators that there is a dipolar interaction between Cu_a and Fe_a, indicating that these centers are not more than 1.5 nm apart. Similarly, the absence of an effect of the Cu_{a_3}-Fe_{a_3} site on the relaxation behavior of Cu_a has suggested that this site is relatively distant from the location of Cu_a (262). Based in part on the previous suggestions of others (266, 301), Capaldi et al (261) have proposed the structure shown in Figure 8 for the position of Cu_a in subunit II. This structure incorporates the findings of Chan and coworkers (262). It might be added for comparison that in blue oxidases containing type 1 Cu^{2+}, such as azurin and plastocyanin, the Cu ligands, as determined from X-ray crystal structure analysis to 2.7 Å, are two imidazole nitrogens and one cysteine and one methionine sulfur (302, 303).

Heme a_3 and Cu_{a_3} comprise an S = 2 ESR silent center in both reduced and oxidized cytochrome oxidase. In the oxidized enzyme, Fe_{a_3} and Cu_{a_3} are considered to be antiferromagnetically coupled, thus resulting in an ESR silent

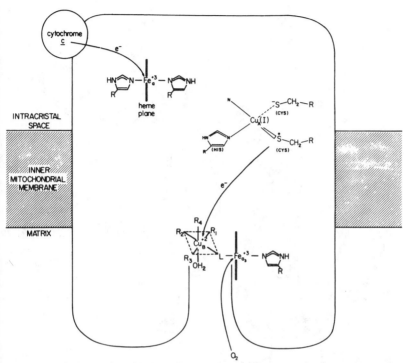

Figure 7 The ligand structure of cytochromes a and a_3, Cu_a and Cu_{a_3} in oxidized cytochrome c oxidase. From (262).

pair. Such strong antiferromagnetic coupling is also seen between the two Cu^{2+} atoms in tyrosinase and hemocyanin (304). The coupling between Fe_{a_3} and Cu_{a_3} can be broken by addition of NO to reduced cytochrome oxidase, and by partial reduction of the enzyme (305, 306). In the latter case, a high-spin heme signal at $g = 6$, ascribed to $a_3 \cdot H_2O$, is observed at neutral pH, which is converted at alkaline pH values to a low spin signal ($g = 2.6$), presumably due to $a_3 \cdot OH^-$ (13, 307). Heme a_3 is considered to have an imidazole nitrogen as an axial ligand in a position distal to Cu_{a_3}, and the second axial ligand is thought to be shared with Cu_{a_3} (Figure 7). EXAFS studies have suggested that in the oxidized enzyme the shared ligand is a cysteine sulfur (308). Other ligands of Cu_{a_3} are not known, but the possibility of three nitrogens has been mentioned (262).

There is evidence that the location of the prosthetic groups of cytochrome oxidase is confined to subunits I and II (309), which agrees with the occurrence of all four prosthetic groups (i.e., hemes a, a_3, Cu_a, and Cu_{a_3}) in the two-subunit cytochrome c oxidases purified from *P. denitrificans* (310) and *N. agilis* (311). It is also known that subunit III of the mitochondrial enzyme does

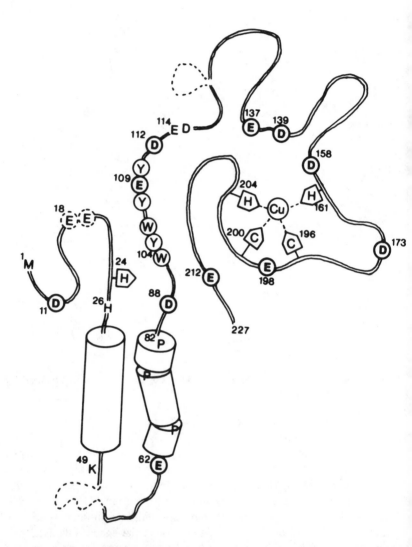

Figure 8   Schematic diagram of the possible folding pattern of subunit II of cytochrome *c* oxidase. The hydrophobic sequences 27–48 and 63–82 are shown as α-helices traversing the phospholipid bilayer, placing residues 1–26 and 83–227 on the cytoplasmic side of the membrane. All aspartic acid, glutamic acid, histidine, and cysteine residues fully conserved in subunit II from bovine, human, mouse, yeast, and maize cytochrome *c* oxidase are circled. Conserved residues in the sequence 104–115 are shown and the proposed copper-binding residues are enclosed by a pentagon. The dashed lines indicate extra residues in the sequence of subunit II from yeast. From (261) and ref. (123) therein.

not carry any prosthetic groups, because oxidase preparations lacking subunit III have been shown to contain the expected amount of heme and to exhibit unaltered spectral and redox properties (312). When the subunits of the bovine enzyme were separated under carefully controlled conditions, 90% of the Cu was found on subunit II, and heme was associated in equal amounts with subunits I and II (309). However, another view is that only Cu_a is in subunit II (Figure 8), and that hemes a and a_3 and Cu_{a_3} are all associated with subunit I (R. A. Capaldi, private communication).

Catalytic Properties and Mechanism of Action of Cytochrome c Oxidase

ELECTRON TRANSFER   Cytochrome oxidase catalyzes the 4-electron reduction of molecular oxygen by the single-electron donor ferrocytochrome c. The enzyme may be thought of as being composed of two redox centers differing in apparent E_m by about 100 mV. The first center is cytochrome a Cu_a which is the recipient of electrons from ferrocytochrome c, and the second center is cytochrome a_3 Cu_{a_3} which is the site of oxygen reduction. Kinetic studies have provided strong evidence that cytochrome a is the first recipient of electrons from cytochrome c (313–316). However, the possibility that Cu_a may fulfill this role has been suggested in view of the binding site of cytochrome c on the oxidase (261). Cytochrome oxidase has a high-affinity and a low-affinity cytochrome c binding site per monomer (317–320). Arylazidocytochrome c modified at Lys-13 near the heme cleft binds at the high-affinity site of the oxidase and when photoactivated cross-links to subunit II at or near His-161 (Figure 8), which is considered to provide a ligand for Cu_a (288, 321, 322). Yeast cytochrome c derivatized with thionitrobenzoate at its Cys-107 on the back side also binds to the high-affinity site of bovine cytochrome oxidase, and forms a disulfide linkage with subunit III of the bovine or the yeast enzyme (286, 289, 290, 323). Thus, it has been suggested that the high-affinity binding site of cytochrome c might be the cleft formed on the C side of the oxidase dimer between subunit II of one monomer and subunit III of the other (261). The front side of cytochrome c would face subunit II, and several conserved aspartic and glutamic acid residues of subunit II which are circled in Figure 8 could interact electrostatically with the conserved lysine residues surrounding the heme cleft of cytochrome c (261). Not much is known about the low-affinity site, except that cardiolipin forms a part of this site and is involved in cytochrome c binding (288, 324). Also, the same domain on the surface of cytochrome c that binds to the high-affinity site is considered to bind to the low-affinity site (325). A regulatory role for cytochrome c at the low-affinity site has been proposed (326).

The mechanism of oxygen reduction by reduced or "mixed-valence" (a Cu_a

oxidized, $a_3 Cu_{a_3}$ reduced) cytochrome oxidase has been studied by Chance and coworkers (327, 328), using a low-temperature trapping technique to stabilize possible intermediates for static spectroscopic analysis. They were thus able to identify three intermediates in the reaction of "mixed-valence" cytochrome oxidase with oxygen. Subsequently, Clore et al (329–331) confirmed these findings and gave further definition to the redox states of the intermediates by optical and ESR measurements at 173 K. Intermediate I is formed from the reaction of dioxygen with either the fully reduced or "mixed valence" cytochrome oxidase to yield a peroxide complex:

$$a_3^{2+} \ Cu_{a_3}^{1+} + O_2 \rightarrow [\ a_3^{3+}\diagup \overset{-O}{\underset{O_-}{\diagdown}}\diagdown_{O_-}\diagup Cu_{a_3}^{2+}].$$

Intermediate II is marked by a $Cu_{a_3}^{2+}$ ESR signal, which appears to be influenced by a nearby paramagnetic center (332). This center was first proposed by Brudvig et al (333) to be a ferryl ion. Malmström and colleagues (13, 331) believe that intermediate I receives $1e^-$ and $2H^+$ to yield one molecule of water and $[Cu_{a_3}^{2+} \ O = a_3^{4+}]$ as intermediate II. Chan and coworkers (262) utilize $1e^-$ and $1H^+$ to give $[Cu_{a_3}^{2+}—^-OH \ O = a_3^{4+}]$ as intermediate II. Intermediate III is a further 1-electron reduced state, with the structure $[Cu_{a_3}^{2+} \ HO^-—a_3^{3+}]$ according to Malmström (13), and $[Cu_{a_3}^{2+}—^-OH \ HO^-—a_3^{3+}]$ according to Chan (262). Other details regarding the stages of H_2O release and additional possible intermediates, including dead-end species with unique ESR signals (e.g., at $g = 2.6$ and $g = 12$), are given in the recent reviews by Malmström (13) and Blair et al (262), and will not be repeated here. Although the two proposed mechanisms differ in certain details, their basic features are the same. Upon binding, dioxygen is reduced by the $a_3^{2+} \ Cu_{a_3}^{1+}$ pair to the peroxide level. Aquisition of a third electron in a subsequent step results in cleavage of the oxygen-oxygen bond and reduction of one atom of oxygen to the level of water. Finally, a fourth electron is received in another step and the second oxygen is further reduced.

Whether the minimal active unit of cytochrome oxidase is a monomer or a dimer has been a subject of long debate. Isolated bovine cytochrome oxidase is dimeric at neutral pH and in the presence of moderate concentrations of salts and detergents. Whether the enzyme in the mitochondrial inner membrane is also dimeric is not known. Monomeric bovine cytochrome oxidase has been prepared and shown to retain high activity (334). In addition, cytochrome oxidases from shark and camel hearts (335), and *P. denitrificans* (336) are all monomeric in the isolated state. However, the covalent binding of a single molecule of appropriately derivatized cytochrome *c* (see above) to either subunit II or III of purified dimeric cytochrome oxidase completely blocks electron transfer from added ferrocytochrome *c* to the dimer [(286, 288, 289,

(21, 322), see, however, (337) regarding the lack of inhibition of monomeric cytochrome oxidase by cytochrome c covalently bound to its subunit III]. The possible half-of-the-sites reactivity or site-site "anticooperativity" suggested by these results has been regarded by Wikström et al (11, 24) as an important clue to the mechanism of proton translocation by cytochrome oxidase.

PROTON TRANSLOCATION   Despite reports to the contrary (338–342), there is considerable evidence that cytochrome oxidase comprises an energy-transducing site in the respiratory chain, and proteoliposomes prepared from the isolated enzyme catalyze proton translocation coupled to electron transfer [for reviews, see (11, 24, 259)]. The H^+/e^- stoichiometry is uncertain, and values of $H^+/e^- = 1$ or 2 have been claimed for the mammalian enzyme [see (259)].

An important finding in support of energy transduction by cytochrome oxidase is that DCCD inhibits proton translocation by the mitochondrial enzyme with little effect on its electron transfer activity [(344–346), see also (347) for the oxidase from thermophilic bacterium PS3]. Similar results have been obtained upon treatment of bovine and yeast complex III with DCCD (see above). The primary target of DCCD in bovine cytochrome oxidase appears to be Glu-90 in subunit III (345, 346), the sequence around which is similar to that around the DCCD-reactive glutamic acid residue of the DCCD-binding proteolipid of the bovine ATP synthase complex (346). Thus, by analogy with the latter, subunit III of the bovine cytochrome oxidase is considered to form part of the proton channel of the enzyme. This conclusion agrees with the claim that oxidase preparations deficient in subunit III are incapable of proton translocation [(348); see, however, (349)]. It has been reported that cytochrome oxidases purified from the thermophilic bacterium PS3 (347, 350, 351), $T.$ $thermophilus$ (352), and $P.$ $denitrificans$ (353, 354) pump protons, while the $N.$ $agilis$ enzyme does not (352). However, the proton pumping activity of the $P.$ $denitrificans$ oxidase [$H^+/e^- \leqslant 0.6$, ref. (353)], which lacks a subunit analogous to the mitochondrial subunit III, is apparently insensitive to DCCD (354).

In spite of the rather precise information now available on the details of dioxygen reduction by cytochrome oxidase, clues regarding the mechanism of proton translocation by this enzyme are scarce. Because of the pH-dependence of their apparent E_m, cytochromes a and a_3 have been implicated in proton translocation by the mitochondrial oxidase [(11, 24); see, however, (257) regarding the uncertainty in assigning measured E_m values to one or another redox center], and mechanisms based almost entirely on theoretical considerations have been proposed (11, 24). Perhaps future studies on the relatively simple, two-subunit oxidase of $P.$ $denitrificans$ will provide clues to the mechanism of proton translocation, if proton pumping by this enzyme is indeed a firm finding.

ATP SYNTHASE (COMPLEX V)

This enzyme complex (also referred to as the proton-ATPase complex or F_o-F_1) is responsible for ATP synthesis from ADP and inorganic phosphate (P_i) at the expense of protonic energy derived from the operation of the respiratory complexes I, III, and IV. Complex V is also capable of ATP hydrolysis linked to proton translocation from the matrix to the cytosolic side of the mitochondrial inner membrane. In this sense, the ATP synthase complex may be classed among a large number of membrane-bound enzymes whose scalar catalytic function (i.e., ATP synthesis or hydrolysis) is mechanistically coupled to transmembrane translocation of cations, such as H^+, K^+, Na^+, and Ca^{2+} (2, 15, 355–361). However, the ATP synthases have a much more complex structure than cation-transport ATPases, including the proton-transport ATPases found in fungal plasma membranes (356).

Composition and Structure of ATP Synthase

The mitochondrial ATP synthase is composed of two principal sectors, F_1 and F_o. In electron micrographs of negatively stained preparations, a stalk is also seen between F_1 and F_o (Figure 9). F_1 is the catalytic sector, which can be isolated as a highly active ATPase (turnover ~ 600 s^{-1}), and F_o is the membrane sector and is concerned with proton translocation. The mitochondrial ATP synthase has a more complex structure and composition than its prokaryotic counterpart, and is not as well characterized especially as regards

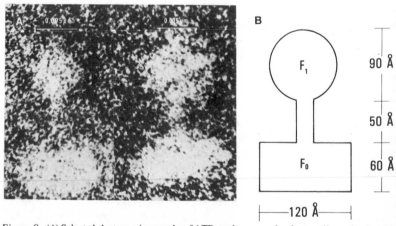

Figure 9 (A) Selected electron micrographs of ATP synthase complex from rat liver mitochondria × 4,800,000 (from Soper, J. W., Decker, G. L., Pedersen, P. L. 1979. *J. Biol. Chem.* 254:11170–76. Courtesy of Dr. P. L. Pedersen). (B) An idealized drawing of the ATP synthase complex showing the approximate dimensions of F_1, F_o and the stalk.

the composition of F_o. For clarity it may be useful, therefore, to start with a brief description of the prokaryotic ATP synthase. The purified ATP synthases of *Escherichia coli* and the thermophilic bacterium PS3 are each composed of eight unlike subunits, five in F_1 and three in F_o [for reviews, see (4, 7, 10, 362–365)]. They do not have a sector corresponding to the stalk of the mitochondrial enzyme (Figure 9). The chloroplast ATP synthase has also been claimed to have eight or nine subunits [(367), see also (368)]. However, the system discussed here will be the *E. coli* enzyme regarding which the structural information is much more precise.

The subunits of the *E. coli* ATP synthase are encoded by eight genes which occur in an operon, designated the *unc* (also *atp, pap,* or *bcf*) operon (4, 363, 364, 369). Also phages containing all the F_o-F_1 genes or plasmids containing all or segments of the *unc* operon have been isolated (363, 364). The *unc* operon maps at approximately 83 min on the *E. coli* linkage map, and is composed of nine genes, designated in gene order *unc* I, B, E, F, H, A, G, D, C (4, 363, 364, 369). The role of *unc* I is not known. Others encode, respectively, the subunits *a, c,* and *b* of F_o, and δ, α, γ, β, and ε of F_1 (*a, b,* and *c* are also referred to as χ, ψ, and ω, respectively) (4, 363, 364). All the genes of the *unc* operon have been sequenced, and the primary structure of the corresponding F_o-F_1 subunits are known from the gene sequence [for reviews, see (363, 364)]. Their molecular masses in kilodaltons are α(55.2), β(50.1), γ(31.4), δ(19.4), ε(14.9), *a*(30.3), *b*(17.2), and *c*(8.2). The subunit stoichiometry of F_1 from *E. coli* as well as from several other prokaryotic and eukaryotic sources is $\alpha_3\beta_3\gamma\delta\epsilon$ [see (368) and (370) for the subunit stoichiometry of the chloroplast F_1], and the subunit stoichiometry of F_o appears to be $a_1b_2c_{10\text{-}12}$ in *E. coli* (371, 372), and $a_1b_2c_{3\text{-}6}$ in the thermophilic bacterium PS3 (373). In mitochondria and chloroplasts the subunit corresponding to *c* (the DCCD-binding proteolipid) appears to be present as six copies per F_o (374, 375). Other similarities in the structure and composition of F_o-F_1 from various sources have been discussed extensively in recent reviews (4, 363, 364) and will be considered in part below.

F_1 from rat liver mitochondria and thermophilic bacterium PS3 has been crystallized (16, 365, 376). X-ray diffraction of single crystals of rat liver F_1 to 9 Å resolution has indicated the presence of two equivalent halves, each composed of three masses of approximately equal size (16, 376). Two pairs of masses appear to be equivalent and are considered to contain two α and two β subunits, while the third pair is thought to contain one α and one β in association with the smaller subunits [for details, see (16)]. Electron microscopic studies of bovine heart F_1 have also demonstrated the existence of six globular masses (presumably corresponding to three α and three β subunits) arrayed around a seventh central mass, which is considered to contain γ (377). This picture agrees with the pattern of reversible cold-resolution (at 4°C) of rat liver F_1 which has shown the presence of αγ and βγ complexes in the dissoci-

ated enzyme (378). It also agrees with the fact that in *E. coli* and the thermophilic bacterium PS3 isolated α, β, and γ form a complex with ATPase activity
(362, 379, 380).

As mentioned above, the mitochondrial ATP synthases differ from their
prokaryotic counterparts in possessing two polypeptides which are thought to
form the stalk and in the composition of their F_o. The former are oligomycin
sensitivity-conferring protein (OSCP) and coupling factor 6 (F_6). OSCP from
bovine heart mitochondria has been sequenced. It is composed of 190 amino
acids and has a molecular weight of 20,967 (381). The protein is an elongated
molecule, with an axial ratio >3, and has about 40% α helical structure (382).
The central region of the molecule is hydrophobic while the two terminal
segments have a high concentration of charged residues (381). F_6 has a
molecular weight of about 8,000, and is required for the binding of F_1 to F_o
(383). In what manner OSCP and F_6 participate in proton translocation from F_o
to F_1 remains to be determined.

The number of unlike polypeptides in the mitochondrial F_o is not known.
Highly purified preparations of ATP synthase with 9–10 polypeptides from rat
liver (384), 11–12 polypeptides from fungal and mammalian sources (385–
388), and 14 polypeptides from bovine heart (389) have been reported. However, the minimum number of polypeptides in the mammalian ATP synthase is
probably 12. These are α, β, γ, δ, ϵ, OSCP, and F_6, as mentioned above, plus
the ATPase inhibitor protein (IF_1) (390), coupling factor B (F_B) (391), the
mtDNA-encoded subunits 6 and 8 (178, 392–394), and the DCCD-binding
proteolipid which in *N. crassa, S. cerevisiae,* and possibly maize is also
encoded by mtDNA and synthesized within the mitochondrion (392, 395–397).
In addition, it has been shown that subunit 6 in yeast (398, 399) and bovine
(389) ATP synthases is composed of two electrophoretically separable components, which in yeast appear to be structurally related but not identical (399). If
this is not an isolation artifact, then the minimum number of ATP synthase
subunits might be 13, at least for the bovine heart enzyme. The roles of subunits
6 and 8 are obscure, but IF_1, F_B, and the DCCD-binding proteolipid are
involved in ATP synthesis and hydrolysis.

ATPase INHIBITOR PROTEIN IF_1 from bovine heart (400) and yeast (401) has
been sequenced [for a review on IF_1, see (402)]. The former is composed of 84
amino acids, and its molecular weight is 9,578 (400). IF_1 binds to the β subunit
of F_1 (~ 1 mol/mol F_1) (403), and inhibits ATP hydrolysis as well as the initial,
but not the steady-state, phase of ATP synthesis (404–406). The binding of IF_1
to F_1 and inhibition of ATPase activity require the presence of a hydrolyzable
substrate plus Mg^{2+}, and an active state of F_1 (407, 408). Presumably, IF_1
binds to a transient conformational state of F_1 produced during catalysis (407).

The binding is accelerated at pH <7.0, and the dissociation constant, as determined in the presence of complex V and MgATP at pH 6.7, is about 0.75 μM [(408, 409), see also (410)]. In the absence of MgATP, enzyme-bound IF_1 can be released at pH > 7.0 (409) or by respiration-induced membrane energization (406). The latter has suggested a regulatory role for IF_1. The isolated bacterial ATPase complexes do not contain an ATPase inhibitor protein. In *E. coli* (411) and chloroplast (412), the ε subunit has been shown to cause partial inhibition of ATPase activity when added to F_1 or ε-deficient F_1 preparations. However, this may be mechanistically analogous to the decrease in the ATPase activity of F_1 when it is bound to F_o. It has been reported that the interaction between yeast IF_1 and membrane-bound, but not isolated, F_1 is facilitated and stabilized by two protein factors (413). One of these has been sequenced and found to have 83 amino acids and a molecular weight of 9,450 (414). This polypeptide has significant homology with yeast IF_1 (414).

FACTOR B F_B is a water-soluble protein with an apparent M_r of $11–12 \times 10^3$ (415) or 15×10^3 (391). Preparations with F_B activity and apparent M_r of 29,200 and 47,000 have also been reported (416, 417). F_B contains an essential vicinal dithiol (418, 419), and appears to be required for energy transfer to and from F_1. SMP extracted with ammonia-EDTA is considered to be F_B-deficient [(391); see, however, (420)]. These particles have low activities for oxidative phosphorylation, ATP-$^{32}P_i$ exchange, and ATP-induced reverse electron transfer from succinate to NAD and transhydrogenation from NADH to NADP. Addition of F_B restores these activities (418). F_B has been purified only from bovine heart mitochondria (415), but the presence of F_B-like antigen has been claimed in several other sources, even *E. coli* (391). It has been shown recently that bovine heart F_o contains a moiety whose modification by monothiol and vicinal dithiol reagents (e.g., *N*-ethylmaleimide, mercurials, diamide, Cd^{2+}, phenylarsene oxide, or Cu^{2+} plus orthophenanthroline) is energy-potentiated and results in uncoupling, which is reversible when a modifier other than *N*-ethylmaleimide is used (420). There are indications that the energy-potentiated uncoupling effected by the above reagents is not due to modification of F_B (420). An energy-potentiated uncoupling by a dithiol reagent (*o*-phenylenedimaleimide) has also been reported in chloroplasts (421). However, in this case, the uncoupling is the result of internal cross-linking in the γ subunit of F_1 (421). Mono- and dithiols as well as disulfide-dithiol interchange have been implicated in membrane energy transduction and transfer (422–426). Whether the above dithiols play a mechanistic role in the bovine F_o and chloroplast F_1 is not known. However, it might be of interest to point out that *E. coli* F_o contains only a single cysteine residue (Cys-21), which is in subunit *b* (364, 427).

DCCD-BINDING PROTEOLIPID   The literature on the DCCD-binding proteolipid (subunit 9 of mammalian and yeast ATPase, subunit c of $E.$ $coli$ F_o) has
been reviewed in recent comprehensive articles (4, 10, 363, 364, 396, 428).
Thus, this section will contain only a brief description of the basic features of
this polypeptide. The sequence of the DCCD-binding proteolipid from ten
different prokaryotic and eukaryotic sources has been determined (364). These
proteolipids are composed of 70–82 amino acids of which about 80% are
nonpolar. This high degree of hydrophobicity is probably responsible for the
fact that the protein from various sources can be extracted into chloroform/
methanol (364). All DCCD-binding proteins are made up of two hydrophobic
stretches joined to a central small hydrophilic segment with β-turn capability.
Prediction of secondary structure from the amino acid sequence has suggested a
"hairpin" structure with the two hydrophobic stretches traversing the membrane, possibly as antiparallel α helices, and the hydrophilic middle segment
forming the bend of the pin and protruding from the F_1 side of the membrane (4,
364, 396, 429). There is an invariant glutamic acid residue (Glu-58 in the
bovine proteolipid) near the center of one hydrophobic stretch (except in $E.$ $coli$
in which it is Asp-61) which reacts with DCCD and is required for proton
translocation through F_o (4, 363, 364, 396). $E.$ $coli$ mutants in which Asp-61
has been replaced by Gly or Asn are incapable of ATP-induced proton translocation (364, 430, 431). There are also residues in the vicinity of Asp-61 or on
the second hydrophobic arm of the "hairpin," but presumably near Asp-61
(e.g., Ile-28), whose mutational exchange results in a defective and DCCD-
resistant F_o (364). These results have suggested that the environment around
the reactive carboxyl group provides a tight fit for DCCD and three other
hydrophobic carbodiimides which exhibit high inhibitor potency (364). In
addition, however, there are point mutations resulting in defective F_o which are
considered to straighten a bend in one arm of the "hairpin" or introduce a bend
in the other arm and thereby interfere with the stacking of the proteolipid
molecules (432). Whatever the arrangement of the multiple proteolipid copies
in the membrane, modification of only one copy by DCCD appears to be
sufficient for complete inhibition (433). Whether the DCCD-reactive carboxyl
group is directly involved in proton translocation is not clear. It has been
reported that incorporation of the yeast or chloroplast proteolipid into lipid
bilayers increases the bilayer proton permeability (434–438). However, studies
in $E.$ $coli$ have shown that subunit c alone is not sufficient for efficient proton
conduction through F_o, and that subunits a and b are also required for this
function (364).

In addition to the homology (especially in the polarity profiles) between
various DCCD-binding proteins (4, 364, 396), segmental sequence and polarity homologies have been shown to exist between the mitochondrial subunits α,
β, γ, δ, OSCP, and ATPase-6 and the $E.$ $coli$ α, β, γ, ε, δ, and $a,$ respectively

(363, 364). Noteworthy is a sequence homology of about 70% among the β subunits of bovine, *E. coli,* maize, spinach, and *S. cerevisiae* F_1 (363). Segmental sequence homologies have also been detected between α and β, suggesting an evolutionary relationship (363, 364). In addition, segmental sequence homologies have been shown to exist between α and β and a number of ATP-binding proteins [for review, see (16)], as well as between β and the human bladder carcinoma oncogene product p21, which suggested that p21 may have nucleotidase activity (439). This latter point has been subsequently confirmed. On the other hand, mitochondrial ε, F_6, subunit 8, and IF_1 do not have homologous counterparts in *E. coli,* nor does the *E. coli* subunit *b* have a compositional counterpart in mitochondria. Thus, in structural and mechanistic details, it may be misleading to extrapolate freely from prokaryotic F_1 and F_o to eukaryotic F_1 and F_o. There are many comparable features, but there are also significant disparities.

Catalytic Properties of ATP Synthase

Preparations of the ATP synthase complex catalyze DCCD-sensitive ATP hydrolysis and DCCD- and uncoupler-sensitive ATP-$^{32}P_i$ exchange. These activities of the mitochondrial enzyme are also sensitive to oligomycin. Most preparations of ATP synthase exhibit either no or various low levels of ATP-$^{32}P_i$ exchange activity, which suggests damage during purification. Only two preparations isolated from bovine and porcine heart mitochondria with the use of lysolecithin as detergent have been shown to exhibit ATP-$^{32}P_i$ exchange activities ($\geqslant 1$ μmol·min^{-1}·mg^{-1}) commensurate with their degree of purification (440, 441). However, these preparations are not free from respiratory chain components, and attempts at further purification of the porcine preparation resulted in the loss of exchange activity (441). The mitochondrial ATP synthase also uses GTP and ITP as substrate. At V_{max}, the rates of GTP and ITP hydrolysis are comparable to that of ATP, although the K_m^{app} of GTP and ITP are considerably higher (442). In addition, substitution of purine with pyrimidine for either NTP hydrolysis or synthesis, and modifications of the ribose moiety are tolerated by the enzyme (2, 443–446), indicating a lack of absolute specificity at the nucleoside level. This agrees with the results of inhibitory chemical modifications of F_1, which seem to suggest that the essential active site residues are located mainly in the binding domain for the phosphoryl groups of ATP (447–456, for a review of the chemical probes of ATPase, see chapter by P. Vignais and J. Lunardi in this volume).

ATP-$^{32}P_i$ exchange appears to be the result of two separate reactions catalyzed by two ATP synthase molecules. It has been shown that the porcine enzyme can catalyze an NTP-$^{32}P_i$ exchange reaction in which ATP or GTP hydrolysis supplies the required energy for the phosphorylation, respectively, of GDP or ADP by $^{32}P_i$ to form [γ-^{32}P]GTP or ATP (441). Studies of ATP-$^{32}P_i$

exchange by chloroplast thylakoids have also suggested a similar mechanism, namely ATP hydrolysis by one ATP synthase complex to energize the system, and [γ-^{32}P]ATP synthesis from ADP and ^{32}P$_i$ by a second ATP synthase complex embedded in the same membrane or phospholipid vesicle (457). In SMP, the ATP synthase complex also catalyzes ATP-ADP exchange and P$_i$-H$_2$O oxygen exchange, of which the latter property has been explored to great advantage in discerning the mechanism of ATP synthesis (see below).

F$_1$ binds six mol of adenine nucleotides per mol (4, 12, 458). These binding sites are located on the α and β subunits. Covalent modification of F$_1$ with inhibitory ATP analogs has indicated that the β subunit bears the catalytic site (459–461). Photolabeling of F$_1$ with an inhibitory analog of P$_i$ (4-azido-2-nitrophenylphosphate) also occurs on β (462). In addition, several ATPase inhibitors (447–456), including IF$_1$ (403, 410), efrapeptin (458, 463) and aurovertin (464), bind specifically to the β subunit. However, there are also suggestions that the active site of F$_1$ is composed of apposing segments of both the α and the β subunits [(4, 465), see also (466)]. The nucleotide binding sites on the α subunit are specific for adenine nucleotides, bind adenine nucleotides very tightly, and are considered to have a regulatory function (467–469). This is, however, uncertain (470), and F$_1$ from the thermophilic bacterium PS3 appears to lack tightly bound nucleotides on the α subunits (471).

Mechanism of ATP Hydrolysis and Synthesis

The recent literature on this subject is vast and far beyond the limited space of this article for a thorough survey [for recent reviews, see (2, 4, 12, 15, 16, 472–475)]. However, the major advances in our understanding of the mechanism of ATP synthesis and hydrolysis by the mitochondrial ATP synthase are relatively recent, of which the principal features will be discussed below.

Not only in structure, but also in mechanistic details, the ATP synthases are distinct from the cation-transport ATPases. The mechanism of ATP synthesis and hydrolysis by the transport ATPases involves a phosphoenzyme (acyl phosphate) intermediate in a high-energy (E $\sim$ P) and low-energy (E'-P) state, of which the former can react with ADP to form ATP (15, 361, 476, 477). By contrast, ATP synthases do not appear to involve an X $\sim$ P intermediate. It has been shown that the bovine ATPase-catalyzed hydrolysis of ATP stereospecifically labeled at the oxygens of the γ-phosphoryl group results in inversion at the γ-phosphorus (478) [similar results have been obtained with the myosin ATPase, ref. (479)]. This suggests a one-step (or an odd number of steps) mechanism for ATP hydrolysis. A two-step mechanism, where an X-P intermediate is formed, would involve two inversions, resulting in P$_i$ with retention of chirality.

This and other key features of the mechanism of ATP hydrolysis and synthesis by the mitochondrial enzyme were, in fact, predicted by Boyer and coworkers. They showed in 1973 that SMP catalyzes two types of P_i-H_2O oxygen exchange reaction: an uncoupler-insensitive exchange, designated "intermediate exchange," which was considered to involve oxygen exchange at P_i formed from ATP hydrolysis and prior to its release from the catalytic site, and an uncoupler-sensistive exchange, designated "medium exchange," which involved P_i from the reaction medium (480, 481). Boyer and his colleagues rationalized that "intermediate exchange" indicates that enzyme-bound ATP undergoes hydrolysis and resynthesis at the catalytic site, thus suggesting that the formation of enzyme-bound ATP from enzyme-bound ADP and P_i does not require energy. They also interpreted the uncoupler sensitivity of the "medium exchange" reaction to mean that substrate binding to the enzyme is an energy-promoted process. Thus, on the basis of these and subsequent results, they concluded that the synthesis of enzyme-bound ATP from enzyme-bound ADP and P_i does not require energy, and that the energy-requiring steps in oxidative phosphorylation are substrate binding and product releasing (473–475). It has since been shown that isolated F_1-ATPase from chloroplasts (also F_1 on uncoupled thylakoid membranes) (482, 483), bovine heart [(484), see also (485)], or the thermophilic bacterium PS3 (486) is capable of F_1-bound ATP synthesis from ADP and P_i, and the calculated equilibrium constants have suggested that the formation of enzyme-bound ATP from enzyme-bound ADP and P_i is essentially an isoenergetic process involving negligible free energy change. Further, it has been demonstrated that mitochondrial membrane energization increases the affinities of the mitochondrial energy-linked enzymes for their respective substrates [(487), see also (488)]. The systems tested were respiration-driven ATP synthesis, and ATP-driven transhydrogenation from NADH to NADP and uphill electron transfer from succinate to NAD. The basis for the enzyme-substrate affinity increase is considered to be energy-induced enzyme conformation change, and it has been shown in SMP that the energy-induced fluorescence increase of F_1-bound aurovertin (a presumed reporter of F_1 conformation change) decreases in parallel with the membrane potential and V_{max}^{app} for ATP synthesis when the particles are treated with incremental amounts of an uncoupler (489). These findings suggest that, as in the case of transport ATPases, ligand(proton?)-induced enzyme conformation change and the consequent change in the binding energies of substrates and products are fundamental mechanistic features of the mitochondrial energy-linked reactions, including ATP synthesis (15, 474–476, 487). Thus, a minimum of two enzyme conformations may be involved in ATP synthesis such that in conformation E_1 enzyme-bound ADP, P_i, and ATP would each be in equilibrium with its counterpart species in the medium, and in conformation E_2 enzyme-bound ADP and P_i would be in equilibrium with enzyme-bound ATP (475, 487, 490).

Another important mechanistic feature of the ATP synthases is the cooperativity of the catalytic sites (three per F_1 molecule). In isolated F_1, this is displayed as negative cooperativity with respect to substrate concentration

(MgATP), and as positive catalytic cooperativity in the sense that substrate binding to the second and third sites greatly increases enzyme turnover. The negative cooperativity with respect to [MgATP], which was first discovered by Lardy and coworkers (467, 491), has been further studied recently (450, 492, 493). Analysis of the kinetics of ATP hydrolysis by isolated F_1 from bovine heart and thermophilic bacterium PS3 over a wide range of MgATP concentration has indicated three K_m^{app} values of the order of $\leq 10^{-6}$, 10^{-4}, and 10^{-3} M (450). When one β subunit per F_1 was inactivated by modification with DCCD, 95% of the activity was lost and the high K_m^{app} for MgATP was eliminated, but the other two K_m^{app} values were retained, thus indicating negative cooperativity even between two active catalytic sites (450). The loss of 95% ATPase activity upon modification of one catalytic site per F_1 is an expression of the positive catalytic cooperativity of the enzyme, which has been more elegantly demonstrated in the experiments of Penefsky and coworkers (444, 445, 485, 493). They have shown that uni-site ATP hydrolysis (i.e., at an F_1 : ATP molar ratio of 3) is a very slow process ($V = 10^{-4} s^{-1}$), which is increased 10^6-fold to $V = 600 s^{-1}$ when ATP concentration is increased (485, 493). Their results further indicate that this dramatic rate increase is primarily due to a 10^6-fold increase in the rate of product release from one site upon substrate binding to additional sites (493). The catalytic cooperativity of the mammalian and chloroplast F_1 has also been extensively investigated by Boyer and his colleagues (492, 494–496) and Nalin & Cross (497). These and other observations have, therefore, led Boyer and coworkers to suggest a sequential binding change mechanism for ATP synthesis and hydrolysis involving three equivalent catalytic sites. It is proposed (*a*) that at any one time during catalysis, one site contains tightly-bound reactants undergoing interconversion (ATP $\xleftarrow{\quad H_2O \quad}$ ADP + P_i) while the other two sites contain loosely bound reactants and products at different stages of the catalytic process, and (*b*) that the binding of ADP and P_i to one site promotes the release of preformed and tightly-bound ATP from another site (474, 475, 492, 496). It is not known whether protons, which are conducted through F_o, participate directly in the energy-requiring steps during ATP synthesis (i.e., substrate binding and product release). Studies on the reconstitution of thermophilic bacterium F_o-F_1 complex have suggested that the γ subunit of F_1 in the presence of δ and ϵ seals the proton permeability of F_o-containing membranes (498). Also, as mentioned earlier, internal cross-linking of the thiol groups of the γ subunit of chloroplast F_1 results in uncoupling due to an increase in proton permeability of the thylakoid membranes (421). Therefore, it is possible that protons are experienced at least by the γ subunit of F_1. However, on the basis of theoretical considerations, especially as pertains to cation-transport ATPases, it is felt that the vectorial and the scalar reactions follow separate paths (15).

An important issue of current controversy is whether, as suggested by the chemiosmotic hypothesis, energy communication in transducing membranes takes place via the medium that surrounds the membrane and is, therefore, delocalized over the entire membrane, or whether such energy transfer is more localized involving direct free energy transfer between the energy-yielding and the energy consuming units (14, 499–504). Hybrid reconstitution experiments in which light-driven ATP synthesis has been demonstrated by coincorporation into the same phospholipid vesicle of bacteriorhodopsin and ATP synthase from different sources strongly suggest that a transmembrane electrochemical potential of protons is capable of driving ATP synthesis [(505–507), see also (366)]. However, the low ATP synthase turnover rates achieved in these experiments ($\leq 1\%$ as compared to the intact systems) have raised questions as to whether bulk-phase $\Delta\bar{\mu}_H+$ is the principal mode of energy communication in the native membranes. Indeed, the results of experiments designed to probe this question more directly have led a number of investigators to suggest that energy transfer is localized and direct (504, 508–510). For example, it has been shown in oxidative phosphorylation and ATP-driven electron transfer from succinate to NAD that partial (up to about 60%) inhibition of the energy-yielding reaction results in a parallel inhibition of the energy-driven reaction with little or no change in the magnitude of the membrane potential or total $\Delta\bar{\mu}_H+$ (489, 511). It has also been demonstrated that the same degree of depression of $\Delta\bar{\mu}_H+$ by uncouplers versus valinomycin $+ K^+$ produces vastly different effects in the rate of ATP synthesis (508). On the other hand, more recent studies have revealed that partial uncoupling increases the K_m^{app} of the substrates of the energy-driven reactions of mitochondria, while partial inhibition of the rates of the energy-yielding reactions decreases these K_m^{app} values (487, 489). Thus, it was shown that regardless of whether the perturbant was an uncoupler or an inhibitor, there was a good correlation between membrane potential changes and V_{max}/K_m of the energy-driven reactions (487, 489).

Another type of experiment which is considered to discriminate between localized and delocalized energy transfer is double-inhibition studies (512–515). In principle, double inhibition experiments involve the following. In an electron transport-driven phosphorylation system, the ATP synthase is partially (e.g., 50%) inhibited by a specific covalent modifier (e.g., DCCD). Then the rate of electron transfer is modulated by a second specific inhibitor (e.g., antimycin), and the effect of incremental inhibition of electron transfer is studied on the rate of ATP synthesis. It is argued that if energy transfer between the energy-yielding and energy-consuming units is delocalized, then partial inhibition of the former would be expected to have a relatively less effect on the rate of ATP synthesis by the system containing partially inhibited ATP synthase than by the control containing uninhibited ATP synthase. If, on the other hand, energy transfer is localized, then only the paired exergonic/endergonic units containing uninhibited ATP synthase would function, and the effect of incremental inhibition of electron transfer on the rate of ATP synthesis would be parallel to that of the control system containing unmodified ATP synthase.

The results of such experiments on the photophosphorylation system of bacterial chromatophores have followed the second pattern, leading the authors to suggest that energy transfer between the chromatophore electron transport and ATP synthase complexes is strictly localized (509, 510, 512–515). In addition, dual inhibition experiments in which the rate of energy transfer to the ATP synthase was modulated by an uncoupler, instead of a respiratory chain inhibitor, have further suggested to these investigators that the action of uncouplers is also localized (512–515). Particularly noteworthy, however, are the results of recent studies on mutations on the DCCD-binding subunit c of $E.$ $coli,$ which seem to agree with an intramembranous mechanism of energy communication as originally proposed by Williams [(499), see also (501, 502)]. It has been shown that Leu-31 → Phe substitution on subunit c in one strain and Pro-64 → Leu and Ala-20 → Pro substitution in another strain result in membranes capable of oxidative phosphorylation at high rates (432, 516). However, in the former strain ATPase activity was resistant to inhibition by DCCD and F_1-depleted membranes were proton-impermeable, and in the latter strain ATP-dependent proton translocation could not be demonstrated and the ATPase activity of isolated F_1 was inhibited when it was bound to the membranes. These results have led the authors to the conclusion that "apparently proton-impermeable membranes are capable of carrying out oxidative phosphorylation" (432, 516).

ACKNOWLEDGMENTS

The author is grateful to many colleagues whose outstanding contributions are cited for providing him with reprints and preprints of their recent work. He also wishes to thank Professors P. D. Boyer and F. M. Huennekens and Drs. T. Yagi and A. Matsuno-Yagi for their critical examination of the article, and Ms. P. Sarkar for typing of the manuscript. The work of the author's laboratory was supported by United States Public Health Service grants AM08126 and GM24887. This is publication No. 3607-BCR from the Research Institute of Scripps Clinic, La Jolla, California.

NOTE ADDED IN PROOF: The amino acid sequence of bovine coupling factor 6 (76 amino acids, calculated M_r, 9006) has been determined (517). Also, the nucleotide sequences of the genes encoding the FP and the IP subunits of $E.$ $coli$ succinate dehydrogenase have been published (518, 519). Striking amino acid sequence similarities, as deduced from the nucleotide sequences of the respective genes, have been noted between the corresponding FP and IP subunits of $E.$ $coli$ succinate dehydrogenase and fumarate reductase. The FP subunits of the two enzymes contain only one conserved cysteine residue each. The others (total of 10 cysteines in fumarate reductase and 11 cysteines in succinate dehydrogenase) are scattered (518). The IP subunits contain 11 cysteine residues each, of which 10 are in identical positions and form three clusters similar to those of bacterial ferredoxins (519).

Literature Cited

1. Skulachev, V. P. 1981. *Chemiosmotic Proton Circuits in Biological Membranes*, ed. V. P. Skulachev, P. C. Hinkle, pp. 3–46. Reading, Mass: Addison-Wesley

2. Hatefi, Y., Ragan, C. I., Galante, Y. M. 1985. *The Enzymes of Biological Membranes*, Vol. 4, ed. A. Martonosi, pp. 1–70. New York: Plenum. 2nd ed.

3. Skulachev, V. P. 1984. *Trends Biochem. Sci.* 9:182–85

4. Senior, A. E., Wise, J. G. 1983. *J. Membr. Biol.* 73:105–24

5. Hauska, G., Hurt, E., Gabellini, N., Lockau, W. 1983. *Biochim. Biophys. Acta* 726:97–133

6. Rich, P. R. 1984. *Biochim. Biophys. Acta* 768:53–79

7. Downie, J. A., Gibson, F., Cox, G. B. 1979. *Ann. Rev. Biochem.* 48:103–31

8. Tzagoloff, A., Macino, G., Sebald, W. 1979. *Ann Rev. Biochem.* 48:419–41

9. von Jagow, G., Sebald, W. 1980. *Ann. Rev. Biochem.* 49:281–314

10. Fillingame, R. H. 1980. *Ann. Rev. Biochem.* 49:1079–113

11. Wikström, M., Krab, K., Saraste, M. 1981. *Ann. Rev. Biochem.* 50:623–55

12. Cross, R. L. 1981. *Ann. Rev. Biochem.* 50:681–714

13. Malmström, B. G. 1982. *Ann. Rev. Biochem.* 51:21–59

14. Ferguson, S. J., Sorgato, M. C. 1982. *Ann. Rev. Biochem.* 51:185–217

15. Tanford, C. 1983. *Ann. Rev. Biochem.* 52:379–409

16. Amzel, L. M., Pedersen, P. L. 1983. *Ann. Rev. Biochem.* 52:801–24

17. Fleischer, S., Brierley, G., Klouwen, H. 1961. *J. Biol. Chem.* 236:2936–41

18. Borst, P., Grivell, L. A., Groot, G. S. P. 1984. *Trends Biochem. Sci.* 9:128–30

19. Wilson, D. F., Erecinska, M., Owen, C. S., Mela, L. 1974. *Dynamics of Energy Transducing Membranes*, ed. L. Ernster et al, pp. 221–31. Amsterdam: Elsevier

20. Forman, N. G., Wilson, D. F. 1982. *J. Biol. Chem.* 257:12908–15

21. Ragan, C. I., Hinkle, P. C. 1975. *J. Biol. Chem.* 250:8472–76

22. Hinkle, P. C., Leung, K. H. 1974. *Membrane Proteins in Transport and Phosphorylation*, ed. G. F. Azzone et al, pp. 73–78. New York: Elsevier

23. Wikström, M. K. F. 1977. *Nature* 266:271–73

24. Wikström, M., Krab, K., Saraste, M. 1981. *Cytochrome Oxidase, A Synthesis*. New York: Academic

25. Racker, E., Kandrach, A. 1971. *J. Biol. Chem.* 246:7069–71

26. Racker, E., Kandrach, A. 1973. *J. Biol. Chem.* 248:5841–47

27. Ragan, C. I., Racker, E. 1973. *J. Biol. Chem.* 248:2563–69

28. Hatefi, Y., Haavik, A. G., Griffiths, D. E. 1961. *Biochem. Biophys. Res. Commun.* 4:441–46, 447–53

29. Hatefi, Y., Haavik, A. G., Griffiths, D. E. 1962. *J. Biol. Chem.* 237:1676–80

30. Hatefi, Y., Stiggall, D. L. 1976. *The Enzymes*, ed. P. D. Boyer, pp. 175–297. New York: Academic. 3rd ed.

31. Grossman, S., Cobley, J. G., Singer, T. P., Beinert, H. 1974. *J. Biol. Chem.* 249:3819–26

32. Cobley, J. G., Grossman, S., Singer, T. P., Beinert, H. 1975. *J. Biol. Chem.* 250:211–17

33. DiVirgilio, F., Azzone, G. F. 1982. *J. Biol. Chem.* 257:4106–13

34. Granger, D. L., Lehninger, A. L. 1982. *J. Cell Biol.* 95:527–35

35. Stouthamer, A. H. 1980. *Trends. Biochem. Sci.* 5:164–66

36. Meijer, E. M., Schuitenmaker, M. G., Boogerd, F. C., Wever, R., Stouthamer, A. H. 1978. *Arch. Microbiol.* 119:119–27

37. Meijer, E. M., Wever, R., Stouthamer, A. H. 1977. *Eur. J. Biochem.* 81:267–75

38. Albracht, S. P. J., van Verseveld, H. W., Hagen, W. R., Kalkman, M. L. 1980. *Biochim. Biophys. Acta* 593:173–86

39. Tottmar, S. O. C., Ragan, C. I. 1971. *Biochem. J.* 124:853–65

40. Singer, T. P., Gutman, M. 1971. *Adv. Enzymol.* 34:79–153

41. Baugh, R. F., King, T. E. 1972. *Biochem. Biophys. Res. Commun.* 49:1165–73

42. Huang, P. C., Pharo, R. L. 1971. *Biochim. Biophys. Acta* 245:240–44

43. Ragan, C. I., Racker, E. 1973. *J. Biol. Chem.* 248:6876–84

44. Fry, M., Green, D. E. 1981. *J. Biol. Chem.* 256:1874–80

45. Ragan, C. I., Galante, Y. M., Hatefi, Y. 1982. *Biochemistry* 21:2518–24

46. Hatefi, Y., Galante, Y. M., Stiggall, D. L., Ragan, C. I. 1979. *Methods Enzymol.* 56:577–602

47. Heron, C., Smith, S., Ragan, C. I. 1979. *Biochem. J.* 181:435–43

48. Earley, F. G. P., Ragan, C. I. 1981. *FEBS Lett.* 127:45–47

49. Hare, J. F., Hodges, R. 1982. *Biochem. Biophys. Res. Commun.* 105:1250–56

50. Smith, S., Ragan, C. I. 1980. *Biochem. J.* 185:315–26

51. Ragan, C. I. 1980. *Subcellular Bio-*

chemistry, ed. D. B. Roodyn, 7:267–307. New York: Plenum

52. Davis, K. A., Hatefi, Y. 1969. *Biochemistry* 9:3355–61

53. Hatefi, Y., Hanstein, W. G. 1974. *Methods Enzymol.* 31:770–90

54. Galante, Y. M., Hatefi, Y. 1979. *Arch. Biochem. Biophys.* 192:559–68

55. Hatefi, Y., Stempel, K. E. 1969. *J. Biol. Chem.* 244:2350–57

56. Chen, S., Guillory, R. J. 1984. *J. Biol. Chem.* 259:5124–31

57. Ohnishi, T., Blum, H., Galante, Y. M., Hatefi, Y. 1981. *J. Biol. Chem.* 256:9216–20

58. Ragan, C. I., Galante, Y. M., Hatefi, Y. 1982. *Biochemistry* 21:2518–24

59. Ragan, C. I., Galante, Y. M., Hatefi, Y., Ohnishi, T. 1982. *Biochemistry* 21:590–94

60. Hatefi, Y. 1963. *The Enzymes,* ed. P. D. Boyer et al, 7:495–515. New York: Academic 2nd ed.

61. Hatefi, Y., Jurtshuk, P., Haavik, A. G. 1961. *Biochim. Biophys. Acta* 52:119–29

62. Hatefi, Y., Stempel, K. E. 1967. *Biochem. Biophys. Res. Commun.* 26:301–8

63. Orme-Johnson, N. R., Hansen, R. E., Beinert, H. 1974. *J. Biol. Chem.* 249:1922–27

64. Orme-Johnson, N. R., Orme-Johnson, W. H., Hansen, R. E., Beinert, H., Hatefi, Y. 1971. *Biochem. Biophys. Res. Commun.* 44:446–52

65. Ohnishi, T., Salerno, J. C. 1982. *Iron-Sulfur Proteins,* ed. T. G. Spiro, 4:285–327. New York: Wiley

66. Beinert, H., Albracht, S. P. J. 1982. *Biochim. Biophys. Acta* 683:245–77

67. Beinert, H., Ruzicka, F. J. 1975. *Electron Transfer Chains and Oxidative Phosphorylation,* ed. S. Papa et al, pp. 37–42. Amsterdam: North-Holland

68. Albracht, S. P. J., Dooijewaard, G., Leeuwerik F. J., Van Swol, B. 1977. *Biochim. Biophys. Acta* 459:300–17

69. Ohnishi, T. 1979. *Membrane Proteins in Energy Transduction,* ed. R. A. Capaldi, pp. 1–87. New York: Dekker

70. Rich, P., Bonner, W. D. 1978. *Functions of Alternative Respiratory Oxidases,* ed. D. Lloyd et al, pp. 61–68. New York: Pergamon

71. Ingledew, W. J., Ohnishi, T. 1980. *Biochem. J.* 186:111–17

72. Hatefi, Y., Bearden, A. J. 1976. *Biochem. Biophys. Res. Commun.* 69:1032–38

73. Albracht, S. P. J. 1982. *Flavins and Flavoproteins,* ed. V. Massey, C. H. Williams, pp. 759–62. Amsterdam: Elsevier/North-Holland

74. Ohnishi, T., Leigh, J. S., Ragan, C. I., Racker, E. 1974. *Biochem. Biophys. Res. Commun.* 56:775–82

75. Salerno, J. C., Ohnishi, T., Lim, J., Widger, W. R., King, T. E. 1977. *Biochem. Biophys. Res. Commun.* 75:618–24

76. Earley, F. G. P., Ragan, C. I. 1980. *Biochem. J.* 191:429–36

77. Ragan, C. I. 1976. *Biochem. J.* 154:295–305

78. Chen, S., Guillory, R. J. 1981. *J. Biol. Chem.* 256:8318–23

79. Dooijewaard, G., De Bruin, G. J. M., Van Dijk, P. J., Slater, E. C. 1978. *Biochim. Biophys. Acta* 501:458–69

80. Boekema, E. J., Van Breenen, J. F. L., Keegstra, W., Van Bruggen, E. F. J., Albracht, S. P. J. 1982. *Biochim. Biophys. Acta* 679:7–11

81. Hatefi, Y., Galante, Y. M. 1978. *The Molecular Biology of Membranes,* ed. S. Fleischer et al, pp. 163–78. New York: Plenum

82. Dooijewaard, G., Slater, E. C. 1976. *Biochim. Biophys. Acta* 440:1–15

83. Moreadith, R. W., Batshaw, M. L., Ohnishi, T., Kerr, D., Knox, B., et al. 1984. *J. Clin. Invest.* 74:685–97

84. Suzuki, H., King, T. E. 1983. *J. Biol. Chem.* 258:352–58

85. Mitchell, P. 1966. *Chemiosmotic Coupling in Oxidative and Photosynthetic Phosphorylation.* Bodmin, England: Glynn Res.

86. Reynafarje, B., Brand, M. D., Lehninger, A. L. 1976. *J. Biol. Chem.* 251:7442–51

87. Rottenberg, H., Gutman, M. 1977. *Biochemistry* 16:3220–27

88. Al-Shawi, M. K., Brand, M. D. 1981 *Biochem. J.* 200:539–46

89. Hinkle, P. C. 1981. See Ref. 1, pp. 49–58

90. De Jonge, P. C., Westerhoff, H. V. 1982. *Biochem. J.* 204:515–23

91. Wikström, M. 1984. *FEBS Lett.* 169:300–4

92. Mitchell, P. 1978. *Les Prix Nobel en 1978,* pp. 135–72. Stockholm: The Nobel Found.

93. Chance, B. 1972. *FEBS Lett.* 23:3–20

94. Papa, S. 1976. *Biochim. Biophys. Acta* 456:39–84

95. Lawford, H. G., Garland, P. B. 1972. *Biochem. J.* 130:1029–44

96. Ragan, C. I. 1984. *Coenzyme Q,* ed. G. Lenaz. Chichester, England: Wiley. In press

97. Light, P. A., Garland, P. B. 1971. *Biochem. J.* 124:123–34

98. Clegg, R. A., Garland, P. B. 1971. *Biochem. J.* 124:135–54

99. Haddock, B. A., Garland, P. B. 1971. *Biochem. J.* 124:155–70
100. Gutman, M., Singer, T. P., Beinert, H. 1972. *Biochemistry* 11:556–62
101. Ohnishi, T. 1973. *Biochim. Biophys. Acta* 301:105–28
102. Ohnishi, T. 1976. *Eur. J. Biochem.* 64:91–103
103. Thunberg, T. 1909. *Skand. Arch. Physiol.* 22:430–36
104. Morton, R. K. 1950. *Nature* 166:1092–95
105. Hogeboom, G. H. 1946. *J. Biol. Chem.* 162:739–40
106. Singer, T. P., Kearney, E. B. 1954. *Biochim. Biophys. Acta* 15:151–53
107. Davis, K. A., Hatefi, Y. 1971. *Biochemistry* 10:2509–16
108. Singer, T. P., Kearney, E. B., Kenney, W. C. 1973. *Adv. Enzymol.* 37:189–272
109. Singer, T. P., Gutman, M., Massey, V. 1973. *Iron-Sulfur Proteins*, ed. W. Lovenberg, 1:225–300. New York: Academic
110. Hatefi, Y., Galante, Y. M. 1980. *J. Biol. Chem.* 255:5530–37
111. Weiss, H., Kolb, H. J. 1979. *Eur. J. Biochem.* 99:139–49
112. Hederstedt, L., Rutberg, L. 1981. *Microbiol. Rev.* 45:542–55
113. Cammack, R., Rowe, B., Owen, P. 1982. *Biochem. Soc. Trans.* 10:261–62
114. Crowe, B. A., Owen, P., Patil, D. S., Cammack, R. 1983. *Eur. J. Biochem.* 137:191–96
115. Hatefi, Y., Davis, K. A., Baltscheffsky, H., Baltscheffsky, M., Johansson, B. C. 1972. *Arch. Biochem. Biophys.* 152:613–18
116. Davis, K. A., Hatefi, Y., Crawford, I. P., Baltscheffsky, H. 1977. *Arch. Biochem. Biophys.* 180:459–64
117. Unden, G., Hackenberg, H., Kröger, A. 1980. *Biochim. Biophys. Acta* 591:275–88
118. Albracht, S. P. J., Unden, G., Kröger, A. 1981. *Biochim. Biophys. Acta* 661:295–302
119. Dickie, P., Weiner, J. H. 1979. *Can. J. Biochem.* 57:813–21
120. Cole, S. T. 1982. *Eur. J. Biochem.* 122:479–84
121. Weiner, J. H., Dickie, P. 1979. *J. Biol. Chem.* 254:8590–93
122. Davis, K. A., Hatefi, Y., Poff, K. L., Butler, W. L. 1973. *Biochim. Biophys. Acta* 325:341–56
123. Ackrell, B. A. C., Ball, M. B., Kearney, E. B. 1980. *J. Biol. Chem.* 255:2761–69
124. Hederstedt, L., Magnusson, K., Rutberg, L. 1982. *J. Bacteriol.* 152:157–65
125. Magnusson, K., Rutberg, B., Hederstedt, L., Rutberg, L. 1983. *J. Gen. Microbiol.* 129:917–22
126. Hederstedt, L., Holmgren, E., Rutberg, L. 1979. *J. Bacteriol.* 138:370–76
127. Holmgren, E., Hederstedt, L., Rutberg, L. 1979. *J. Bacteriol.* 138:377–82
128. Girdlestone, J., Bisson, R., Capaldi, R. A. 1981. *Biochemistry* 20:152–56
129. Davis, K. A., Hatefi, Y. 1972. *Arch. Biochem. Biophys.* 149:505–12
130. Yu, L., Yu, C. A. 1981. *Biochim. Biophys. Acta* 637:383–86
131. Coles, C. J., Holm, R. H., Kurtz, D. M. Jr., Orme-Johnson, W. H., Rawlings, J., et al. 1979. *Proc. Natl. Acad. Sci. USA* 76:3805–8
132. Ackrell, B. A. C., Kearney, E. B., Mims, W. B., Peisach, J., Beinert, H. 1984. *J. Biol. Chem.* 259:4015–18
133. Salerno, J. C., Lim, J., King, T. E., Blum, H., Ohnishi, T. 1979. *J. Biol. Chem.* 254:4828–35
134. Albracht, S. P. J. 1980. *Biochim. Biophys. Acta* 612:11–28
135. Mowery, P. C., Steenkamp, D. J., Ackrell, B. A. C., Singer, T. P., White, G. A. 1977. *Arch. Biochem Biophys.* 178:495–506
136. Yu, C. A., Yu, L., King, T. E. 1977. *Biochem. Biophys. Res. Commun.* 78:259–65
137. Yu, C. A., Yu, L. 1980. *Biochemistry* 19:3579–85
138. King, T. E. 1979. *Structure and Function of Biomembranes*, ed. K. Yagi, pp. 149–65. Tokyo: Jpn. Sci. Soc.
139. Yu, C. A., Yu, L. 1982. *J. Biol. Chem.* 257:6127–31
140. Yu, L., Yu, C. A. 1982. *J. Biol. Chem.* 257:10215–21
141. Singer, T. P. 1966. *Comprehensive Biochemistry*, ed. M. Florkin, E. H. Stotz, 14:127–98. Amsterdam: Elsevier
142. Vik, S. B., Hatefi, Y. 1981. *Proc. Natl. Acad. Sci. USA* 78:6749–53
143. Ackrell, B. A. C., Kearney, E. B., Coles, C. J. 1977. *J. Biol. Chem.* 252:6963–65
144. Ohnishi, T., Lim, J., Winter, D. B., King, T. E. 1976. *J. Biol. Chem.* 251:2105–9
145. Beinert, H., Ackrell, B. A. C., Vinogradov, A. D., Kearney, E. B., Singer, T. P. 1977. *Arch. Biochem. Biophys.* 182:95–106
146. Vinogradov, A. D., Garrikova, E. V., Goloveshkina, V. G. 1975. *Biochem. Biophys. Res. Commun.* 65:1264–69
147. Vinogradov, A. D., Ackrell, B. A. C., Singer, T. P. 1975. *Biochem. Biophys. Res. Commun.* 67:803–9
148. Baginsky, M. L., Hatefi, Y. 1969. *J. Biol. Chem.* 244:5313–19

149. King, T. E., Winter, D., Steele, W. 1972. *Oxidation Reduction Enzymes*, ed. A. Akeson, A. Ehrenberg, pp. 519–32. New York: Pergamon

150. Bonomi, F., Pagani, S., Cerletti, P., Cannella, C. 1977. *Eur. J. Biochem.* 72:17–24

151. King, T. E. 1964. *Biochem. Biophys. Res. Commun.* 16:511–15

152. Hopkins, F. G., Morgan, E., Lutwak-Mann, C. 1938. *Biochem. J.* 32:1829–48

153. Vinogradov, A. D., Winter, D. W., King, T. E. 1972. *Biochem. Biophys. Res. Commun.* 49:441–44

154. Kenney, W. C. 1975. *J. Biol. Chem.* 250:3089–94

155. Kotlyar, A. B., Vinogradov, A. D. 1984. *Biochem. Int.* 8:545–52

156. Hatefi, Y., Haavik, A. G., Griffiths, D. E. 1962. *J. Biol. Chem.* 237:1681–85

157. Leung, K. H., Hinkle, P. C. 1975. *J. Biol. Chem.* 250:8467–71

158. Siedoro, J. N., Pomer, S., De La Rosa, F. F., Palmer, G. 1978. *J. Biol. Chem.* 253:2392–99

159. Sidhu, A., Beattie, D. S. 1982. *J. Biol. Chem.* 257:7879–86

160. Gabellini, N., Bowyer, J. R., Hurt, E., Melandri, B. A., Hauska, G. 1982. *Eur. J. Biochem.* 126:105–11

161. Yu, L., Yu, C. A. 1982. *Biochem. Biophys. Res. Commun.* 108:1285–92

162. Yu, L., Mei, Q.-C., Yu, C. A. 1984. *J. Biol. Chem.* 259:5752–60

163. Krinner, M., Hauska, G., Hurt, E., Lockau, W. 1982. *Biochim. Biophys. Acta* 681:110–17

164. Hurt, E., Hauska, G. 1981. *Eur. J. Biochem.* 117:591–99

165. Hurt, E., Hauska, G. 1982. *J. Bioenerg. Biomembr.* 14:405–24

166. Rieske, J. S. 1976. *Biochim. Biophys. Acta* 456:195–247

167. Trumpower, B. L. 1981. *Biochim. Biophys. Acta* 639:129–55

168. Crofts, A. R., Wraight, C. A. 1983. *Biochim. Biophys. Acta* 726:149–85

169. Slater, E. C. 1983. *Trends Biochem. Sci.* 8:239–42

170. Capaldi, R. A. 1982. *Biochim. Biophys. Acta* 694:291–306

171. Gellerfors, P., Johnsson, T., Nelson, B. D. 1981. *Eur. J. Biochem.* 115:275–78

172. Leonard, K., Hovmöller, S., Karlsson, B., Wingfield, P., Li, Y., et al 1981. *Vectorial Reactions in Electron and Ion Transport in Mitochondria and Bacteria*, ed. F. Palmieri et al, pp. 127–37. Amsterdam: Elsevier/North-Holland

173. Beattie, D. S., Clejan, L., Bosch, C. G. 1984. *J. Biol. Chem.* 259:10526–32

174. Marres, C. A. M., Slater, E. C. 1977. *Biochim. Biophys. Acta* 462:531–48

175. Ohnishi, T., Trumpower, B. L. 1980. *J. Biol. Chem.* 255:3278–84

176. De Vries, S., Berden, J. A., Slater, E. C. 1980. *FEBS Lett.* 122:143–48

177. De Vries, S., Albracht, S. P. J., Berden, J. A., Slater, E. C. 1981. *J. Biol. Chem.* 256:11996–98

178. Anderson, S., De Bruijn, M. H. L., Coulson, A. R., Eperon, I. C., Sanger, F., Young, I. G. 1982. *J. Mol. Biol.* 156:683–717

179. Nobrega, F. G., Tzagoloff, A. 1980. *J. Biol. Chem.* 255:9828–37

180. von Jagow, G., Engel, W. D., Schägger, H. 1981. See Ref. 172, pp. 149–61

181. Goldberger, R., Green, D. E. 1963. See Ref. 60, 8:81–95

182. Jagow, G., Schägger, H., Engel, W. D., Hackenberg, H., Kolb, H. J. 1978. *Energy Conservation in Biological Membranes*, ed. G. Schafer, M. Klingenberg, pp. 43–52. Berlin: Springer-Verlag

183. Weiss, H., Ziganke, B. 1974. *Eur. J. Biochem.* 41:63–71

184. Weiss, H. 1976. *Biochim. Biophys. Acta* 456:291–313

185. Lin, L.-F. H., Beattie, D. S. 1978. *J. Biol. Chem.* 253:2412–18

186. Tsai, A.-L., Palmer, G. 1982. *Biochim. Biophys. Acta* 681:484–95

187. Davis, K. A., Hatefi, Y., Poff, K. L., Butler, W. L. 1973. *Biochim. Biophys. Acta* 325:341–56

188. Reddy, K. V. S., Hendler, R. W. 1983. *J. Biol. Chem.* 258:8568–81

189. Erecinska, M., Oshino, R., Oshino, N., Chance, B. 1973. *Arch. Biochem. Biophys.* 157:431–45

190. Barber, J. 1984. *Trends Biochem. Sci.* 11:209–11

191. Widger, W. R., Cramer, W. A., Herrmann, R. G., Trebst, A. 1984. *Proc. Natl. Acad. Sci. USA* 81:674–78

192. Erecinska, M., Wilson, D. F. 1979. *Arch. Biochem. Biophys* 192:80–85

193. King, T. E. 1978. *Methods Enzymol.* 53:181–91

194. King, T. E. 1983. *Adv. Enzymol.* 54:267–366

195. Bomstein, R., Goldberger, R., Tisdale, H. 1961. *Biochim. Biophys. Acta* 50:527–43

196. Wakabayashi, S., Matsubara, H., Kim, C. H., King, T. E. 1982. *J. Biol. Chem.* 257:9335–44

197. Wakabayashi, S., Takeda, H., Matsubara, H., Kim, C. H., King, T. E. 1982. *J. Biochem.* 91:2077–85

198. Schägger, H., von Jagow, G. 1983. *Hoppe-Seyler's Z. Physiol. Chem.* 364:307–11

199. King, T. E., Kim, C. H. 1983. *J. Biol. Chem.* 258:13543–51

200. Margoliash, E., Schejter, A. 1966. *Adv. Protein Chem.* 21:113–286
201. Rieske, J. S. 1965. *Non-Heme Iron Proteins*, ed. A. San Pietro, pp. 461–68. Yellow Springs: Antioch
202. Trumpower, B. L., Edwards, C. A. 1979. *J. Biol. Chem.* 254:8697–706
203. Li, Y., De Vries, S., Leonard, K., Weiss, H. 1981. *FEBS Lett.* 135:277–80
204. Beattie, D. S., Clejan, L., Chen, Y.-S., Lin, C.-I. P., Sidhu, A. 1981. *J. Bioenerg. Biomembr.* 13:357–73
205. Bell, R. L., Sweetland, J., Ludwig, B., Capaldi, R. A. 1979. *Proc. Natl. Acad. Sci. USA* 76:741–45
206. Gutweniger, H., Bisson, R., Montecucco, C. 1981. *J. Biol. Chem.* 256:11132–36
207. Gellerfors, P., Nelson, B. D. 1977. *Eur. J. Biochem.* 80:275–82
208. Mendel-Hartvig, I., Nelson, B. D. 1978. *FEBS Lett.* 92:36–40
209. Mendel-Hartvig, I., Nelson, B. D. 1981. *Biochim. Biophys. Acta* 636:91–97
210. Beattie, D. S., Sidhu, A., Clejan, L. 1982. *Proc. 2nd Eur. Bioenerg. Conf.*, pp. 203–4. Lyon: Univ. Claude Bernard
211. Hovmöller, S., Leonard, K., Weiss, H. 1981. *FEBS Lett.* 123:118–22
212. Teintze, M., Slaughter, M., Weiss, H., Neupert, W. 1982. *J. Biol. Chem.* 257:10364–71
213. Engel, W. D., Schägger, H., von Jagow, G. 1980. *Biochim. Biophys. Acta* 592:211–22
214. von Jagow, G., Schägger, H., Engel, W. D., Riccio, P., Kolb, H. J., Klingenberg, M. 1978. *Methods. Enzymol.* 53:92–98
215. Engel, W. D., Michalski, C., von Jagow, G. 1983. *Eur. J. Biochem.* 132:395–402
216. Kröger, A., Klingenberg, M. 1973. *Eur. J. Biochem.* 39:313–23
217. Zhu, Q. S., Berden, J. A., De Vries, S., Slater, E. C. 1982. *Biochim. Biophys. Acta* 680:69–79
218. Davis, K. A., Hatefi, Y., Salemme, F. R., Kamen, M. D. 1972. *Biochem. Biophys. Res. Commun.* 49:1329–35
219. Hatefi, Y., Haavik, A. G., Fowler, L. R., Griffiths, D. E. 1962. *J. Biol. Chem.* 237:2661–69
220. Hatefi, Y. 1966. See Ref. 141, pp. 199–231
221. Packham, N. K., Tiede, D. M., Mueller, P., Dutton, P. L. 1980. *Proc. Natl. Acad. Sci. USA* 77:6339–43
222. Prince, R. C., Matsuura, K., Hurt, E., Hauska, G., Dutton, P. L. 1982. *J. Biol. Chem.* 257:3379–81
223. Mitchell, P. 1976. *J. Theor. Biol.* 62:327–67
224. Papa, S., Guerrieri, F., Lorusso, M., Izzo, G., Boffoli, D., Maida, I. 1981. See Ref. 172, pp. 57–69
225. Wikström, M. K. F., Berden, J. A. 1972. *Biochim. Biophys. Acta* 283:403–20
226. Nelson, B. D., Gellerfors, P. 1976. *Biochim. Biophys. Acta* 357:358–64
227. Lindsay, J. G., Dutton, P. L., Wilson, D. F. 1972. *Biochemistry* 11:1937–42
228. Becker, W. F., von Jagow, G., Anke, T., Steglich, W. 1981. *FEBS Lett.* 132:329–33
229. Slater, E. C., De Vries, S. 1980. *Nature* 288:717–18
230. Mitchell, P., Moyle, J. 1982. *Functions of Quinones in Energy-Conserving Systems*, ed. B. L. Trumpower, pp. 553–75. New York: Academic
231. Trumpower, B. L. 1976. *Biochem. Biophys. Res. Commun.* 70:73–80
232. Trumpower, B. L., Edwards, C. A. 1979. *FEBS Lett.* 100:13–16
233. Slater, E. C. 1981. See Ref. 1, pp. 69–104
234. von Jagow, G., Engel, W. D. 1981. *FEBS Lett.* 136:19–24
235. Bowyer, J. R., Edwards, C. A., Ohnishi, T., Trumpower, B. L. 1982. *J. Biol. Chem.* 257:8321–30
236. von Jagow, G., Ljungdahl, P. O., Graf, P., Ohnishi, T., Trumpower, B. L. 1984. *J. Biol. Chem.* 259:6318–26
237. De Vries, S., Albracht, S. P. J., Berden, J. A., Slater, E. C. 1982. *Biochim. Biophys. Acta* 681:41–53
238. Prince, R. C., Dutton, P. L. 1976. *FEBS Lett.* 65:117–19
239. Roberts, H., Smith, S. C., Marzuki, S., Linnane, A. W. 1980. *Arch. Biochem. Biophys.* 200:387–95
240. Beattie, D. S., Villalobo, A. 1982. *J. Biol. Chem.* 257:14745–52
241. Beattie, D. S., Clejan, L. 1982. *FEBS Lett.* 149:245–48
242. Esposti, M. D., Saus, J. B., Timoneda, J., Bertoli, E., Lenaz, G. 1982. *FEBS Lett.* 147:101–5
243. Esposti, M. D., Meier, E. M. M., Timoneda, J., Lenaz, G. 1983. *Biochim. Biophys. Acta* 725:349–60
244. Price, B. D., Brand, M. D. 1982. *Biochem. J.* 206:419–21
245. Price, B. D., Brand, M. D. 1983. *Eur. J. Biochem.* 132:595–601
246. Nalecz, M. J., Casey, R. P., Azzi, A. 1983. *Biochim. Biophys. Acta* 724:75–82
247. Lorusso, M., Gatti, D., Boffoli, D., Bellomo, E., Papa, S. 1983. *Eur. J. Biochem.* 137:413–20
248. Clejan, L., Bosch, C. G., Beattie, D. S. 1984. *J. Biol. Chem.* 259:11169–72
249. Bowyer, J. R., Trumpower, B. L. 1981. *J. Biol. Chem.* 256:2245–51

250. Esposti, M. D., Lenaz, G. 1982. *FEBS Lett.* 142:49–53
251. Hatefi, Y., Yagi, T. 1982. *Biochemistry* 21:6614–18
252. Eisenbach, M., Gutman, M. 1975. *Eur. J. Biochem.* 52:107–116
253. Alexandre, A., Galiazzo, F., Lehninger, A. L. 1980. *J. Biol. Chem.* 255:10721–30
254. Crofts, A. R. 1985. *The Enzymes of Biological Membranes,* ed. A. Martonosi, Vol. 4, pp. 347–82. New York:Plenum. 2nd ed.
255. Keilin, D. 1966. *The History of Cell Respiration and Cytochrome.* Cambridge: Cambridge Univ. Press
256. Poole, R. K. 1983. *Biochim. Biophys. Acta* 726:205–43
257. Malmström, B. G. 1979. *Biochim. Biophys. Acta* 549:281–303
258. Wikström, M., Krab, K. 1980. *Curr. Top. Bioenerg.* 10:51–101
259. Wikström, M., Krab, K. 1979. *Biochim. Biophys. Acta* 549:177–222
260. Azzi, A. 1980. *Biochim. Biophys. Acta* 594:231–52
261. Capaldi, R. A., Malatesta, F., Darley-Usmar, V. M. 1983. *Biochim. Biophys. Acta* 726:135–48
262. Blair, D. F., Martin, C. T., Gelles, J., Wang, H., Brudvig, G. W., et al. 1983. *Chemica Scripta* 21:43–53
263. Schatz, G., Mason, T. L. 1974. *Ann. Rev. Biochem.* 50:51–87
264. Tzagoloff, A., Foury, F., Macino, G. 1978. *Biochemistry and Genetics of Yeast: Pure and Applied Aspects,* ed. M. Bacila et al, pp. 477–88. New York: Academic
265. Tzagoloff, A. 1982. *Mitochondria.* New York: Plenum
266. Buse, G., Steffens, G. J., Steffens, G. C. M. 1978. *Hoppe-Seyler's Z. Physiol. Chem.* 359:1011–13
267. Darley-Usmar, V. M., Fuller, S. D. 1981. *FEBS Lett.* 135:164–66
268. Machleidt, W., Werner, S. 1979. *FEBS Lett.* 107:327–30
269. Anderson, S., Bankier, A. T., Barrell, B. G., De Bruijn, M. H. L., Coulson, A. R., et al. 1981. *Nature* 290:457–65
270. Thalenfeld, B. E., Tzagoloff, A. 1980. *J. Biol. Chem.* 255:6173–80
271. Bonitz, S. G., Coruzzi, G., Li, M., Macino, G., Nobrega, F. G., et al. 1980. *J. Biol. Chem.* 255:11927–41
272. Steffens, G. J., Buse, G. 1979. *Hoppe-Seyler's Z. Physiol. Chem.* 360:613–19
273. Yasumoku, K., Tanaka, M., Wei, Y.-H., King, T. E. 1980. *J. Sci. Industr. Res.* 39:796–801
274. Fox, T. D. 1979. *Proc. Natl. Acad. Sci. USA* 76:6534–38
275. Barrell, B. G., Bankier, A. T., Drouin, J. 1979. *Nature* 282:189–94
276. Bibb, M. I., Van Etten, G. M., Wright, C. T., Walberg, M. W., Clayton, D. A. 1981. *Cell* 26:167–80
277. Steffens, G. C. M., Buse, G., Oppliger, W., Ludwig, B. 1983. *Biochem. Biophys. Res. Commun.* 116:335–40
278. Eytan, G. D., Carroll, R. C., Schatz, G., Racker, E. 1975. *J. Biol. Chem.* 250:8598–603
279. Eytan, G. D., Broza, R. 1978. *J. Biol. Chem.* 253:3196–202
280. Ludwig, B., Downer, N. W., Capaldi, R. A. 1979. *Biochemistry* 18:1401–7
281. Cerletti, N., Schatz, G. 1979. *J. Biol. Chem.* 254:7746–51
282. Chan, S. H. P., Tracy, R. P. 1979. *Eur. J. Biochem.* 89:595–605
283. Frey, T. G., Chan, S. H. P., Schatz, G. 1978. *J. Biol. Chem.* 253:4389–95
284. Bisson, R., Montecucco, C., Gutweniger, H., Azzi, A. 1979. *J. Biol. Chem.* 254:9962–65
285. Prochaska, L., Bisson, R., Capaldi, R. A. 1980. *Biochemistry* 19:3174–79
286. Birchmeier, W., Kohler, C. E., Schatz, G. 1976. *Proc. Natl. Acad. Sci. USA* 73:4334–38
287. Bisson, R., Azzi, A., Gutweniger, H., Colonna, R., Montecucco, C., Zanotti, A. 1978. *J. Biol. Chem.* 253:1874–80
288. Bisson, R., Jacobs, B., Capaldi, R. A. 1980. *Biochemistry* 19:4173–78
289. Moreland, R. N., Docktor, E. 1981. *Biochem. Biophys. Res. Commun.* 99:339–46
290. Fuller, S. D., Darley-Usmar, V. M., Capaldi, R. A. 1981. *Biochemistry* 20:7046–53
291. Briggs, M. M., Capaldi, R. A. 1977. *Biochemistry* 16:73–77
292. Briggs, M. M., Capaldi, R. A. 1978. *Biochem. Biophys. Res. Commun.* 80:553–59
293. Seelig, A., Ludwig, B., Seelig, J., Schatz, G. 1981. *Biochim. Biophys. Acta* 636:162–67
294. Peisach, J. 1978. *Frontiers of Biological Energetics,* ed. P. L. Dutton et al, 2:873–81. New York: Academic
295. Greenaway, F. T., Chan, S. H. P., Vincow, G. 1977. *Biochim. Biophys. Acta* 490:62–78
296. Griffiths, D. E., Wharton, D. C. 1961. *J. Biol. Chem.* 236:1850–56
297. Wharton, D. C., Tzagoloff, A. 1964. *J. Biol. Chem.* 239:2036–41
298. Hu, V. W., Chan, S. I., Brown, G. S. 1977. *Proc. Natl. Acad. Sci. USA* 74:3821–25
299. Scott, R. A., Cramer, S. P., Shaw, R.

W., Beinert, H., Gray, H. B. 1981. *Proc. Natl. Acad. Sci. USA* 78:664–67

300. Stevens, T. H., Martin, C. T., Wang, H., Brudvig, G. W., Scholes, C. P., Chan, S. I. 1982. *J. Biol. Chem.* 257:12106–13

301. Buse, G. 1981. *Biological Chemistry of Organelle Formation,* ed. T. H. Bucher et al, pp. 59–70. Berlin: Springer-Verlag

302. Colman, P. M., Freeman, H. C., Guss, J. M., Murata, M., Norris, V. A., et al. 1978. *Nature* 272:319–24

303. Adman, E. T., Jensen, L. H. 1981. *Isr. J. Chem.* 21:8–12

304. Mason, H. S. 1976. *Iron and Copper Proteins,* ed. K. T. Yasunobu et al, pp. 464–69. New York: Plenum

305. Aasa, R., Albracht, S. P. J., Falk, K.-E., Lanne, B., Vanngard, T. 1976. *Biochim. Biophys. Acta* 422:260–72

306. Beinert, H., Shaw, R. W. 1977. *Biochim. Biophys. Acta* 462:121–30

307. Hartzell, C. R., Beinert, H. 1974. *Biochim. Biophys. Acta* 368:318–38

308. Powers, L., Chance, B., Ching, Y., Angiolillo, P. 1981. *Biophys. J.* 34:465–98

309. Winter, D. B., Bruyninckx, W. J., Foulke, F. G., Grinich, N. P., Mason, H. S. 1980. *J. Biol. Chem.* 255:11408–14

310. Ludwig, B., Schatz, G. 1980. *Proc. Natl. Acad. Sci. USA* 77:196–200

311. Yamanaka, T., Kamita, Y., Fukumori, Y. 1981. *J. Biochem.* 89:265–73

312. Saraste, M., Penttilä, T., Wikström, M. 1980. *FEBS Lett.* 114:35–38

313. Brunori, M., Antonini, E., Wilson, M. T. 1981. *Metal Ions in Biological Systems,* ed. M. Sigel, pp. 115–42. New York: Dekker

314. Wilson, M. T., Greenwood, C., Brunori, M., Antonini, E. 1975. *Biochem. J.* 147:145–53

315. Greenwood, C., Brittain, T., Wilson, M. T., Brunori, M. 1976. *Biochem. J.* 157:591–98

316. Beinert, H., Hansen, R. E., Hartzell, C. R. 1976. *Biochim. Biophys. Acta* 423:339–55

317. Ferguson-Miller, S., Brautigan, D. L., Margoliash, E. 1976. *J. Biol. Chem.* 251:1104–15

318. Ferguson-Miller, S., Brautigan, D. L., Margoliash, E. 1978. *J. Biol. Chem.* 253:149–59

319. Osheroff, N., Brautigan, D. L., Margoliash, E. 1980. *J. Biol. Chem.* 255:8245–51

320. Osheroff, N., Speck, S. H., Margoliash, E., Veerman, E. C. I., Wilms, J., et al. 1983. *J. Biol. Chem.* 257:5731–38

321. Bisson, R., Azzi, A., Gutweniger, H.,

Colonna, R., Montecucco, C., Zanotti, A. 1978. *J. Biol. Chem.* 253:1874–80

322. Bisson, R., Steffens, G. C. M., Capaldi, R. A., Buse, G. 1982. *FEBS Lett.* 144:359–63

323. Malatesta, F., Capaldi, R. A. 1982. *Biochem. Biophys. Res. Commun.* 109:1180–85

324. Vik, S. B., Georgevich, G., Capaldi, R. A. 1981. *Proc. Natl. Acad. Sci. USA* 78:1456–60

325. Veerman, E. C. I., Wilms, J., Dekker, H. L., Muijsers, A. O., van Buuren, K. J. H., et al. 1983. *J. Biol. Chem.* 257:5739–45

326. Speck, S. H., Dye, D., Margoliash, E. 1984. *Proc. Natl. Acad. Sci. USA* 81:347–51

327. Chance, B., Saronio, C., Leigh, J. S. Jr. 1975. *J. Biol. Chem.* 250:9226–37

328. Chance, B., Saronio, C., Leigh, J. S. Jr., Ingledew, W. J., King, T. E. 1978. *Biochem. J.* 171:787–98

329. Clore, G. M., Chance, E. M. 1978. *Biochem. J.* 173:799–810

330. Clore, G. M., Andreasson, L.-E., Karlsson, B., Aasa, R., Malmström, B. G. 1980. *Biochem. J.* 185:139–54

331. Clore, G. M., Andreasson, L.-E., Karlsson, B., Aasa, R., Malmström, B. G. 1978. *Biochem. J.* 185:155–67

332. Karlsson, B., Aasa, R., Vänngård, T., Malmström, B. G. 1981. *FEBS Lett.* 131:186–88

333. Brudvig, G. W., Stevens, T. H., Chan, S. I. 1980. *Biochemistry* 19:5275–85

334. Georgevich, G., Darley-Usmar, V. M., Malatesta, F., Capaldi, R. A. 1983. *Biochemistry* 22:1317–22

335. Darley-Usmar, V. M., Alizai, N., Ayashi, A. I., Jones, G. D., Sharpe, A., Wilson, M. T. 1981. *Comp. Biochem. Physiol. B* 68:445–56

336. Ludwig, B., Grabo, M., Gregor, I., Lustig, A., Regenass, M., Rosenbusch, J. P. 1982. *J. Biol. Chem.* 257:5576–78

337. Darley-Usmar, V. M., Georgevich, G., Capaldi, R. A. 1984. *FEBS Lett.* 166:131–35

338. Mitchell, P., Moyle, J. 1967. *Biochem. J.* 105:1147–62

339. Moyle, J., Mitchell, P. 1978. *FEBS Lett.* 88:268–72

340. Moyle, J., Mitchell, P. 1978. *FEBS Lett.* 90:361–65

341. Lorusso, M., Capuano, F., Boffoli, D., Stefanelli, R., Papa, S. 1979. *Biochem. J.* 182:133–47

342. Papa, S., Guerrieri, F., Izzo, G. 1983. *Biochem. J.* 216:259–72

343. Deleted in proof

344. Casey, R. P., Thelen, M., Azzi, A.

1066   HATEFI

1979. *Biochem. Biophys. Res. Commun.*
87:1044–51
345. Casey, R. P., Thelen, M., Azzi, A.
1980. *J. Biol. Chem.* 255:3994–4000
346. Prochaska, L. J., Bisson, R., Capaldi, R.
A., Steffens, G. C. M., Buse, G. 1981.
Biochim. Biophys. Acta 637:360–73
347. Sone, N., Hinkle, P. C. 1982. *J. Biol.
Chem.* 257:12600–4
348. Saraste, M., Penttila, T., Wikström, M.
1981. *Eur. J. Biochem.* 115:261–68
349. Thompson, D. A., Ferguson-Miller, S.
1983. *Biochemistry* 22:3178–87
350. Sone, N., Yanagita, Y. 1982. *Biochim.
Biophys. Acta* 682:216–26
351. Sone, N., Yanagita, Y. 1984. *J. Biol.
Chem.* 259:1405–8
352. Sone, N., Yanagita, Y., Hon-Nami, K.,
Fukumori, Y., Yamanaka, T. 1983.
FEBS Lett. 155:150–54
353. Solioz, M., Carafoli, E., Ludwig, B.
1982. *J. Biol. Chem.* 257:1579–82
354. Püttner, I., Solioz, M., Carafoli, E.,
Ludwig, B. 1983. *Eur. J. Biochem.*
134:33–37
355. Pedersen, P. L. 1982. *Ann. NY Acad.
Sci.* 402:1–20
356. Goffeau, A., Slayman, C. W. 1981.
Biochim. Biophys. Acta 639:197–223
357. Skou, J. C., Nørby, J. G., eds. 1979.
Na, K-ATPase. New York: Academic.
549 pp.
358. Cantley, L. C. 1981. *Curr. Top. Bioenerg.* 11:201–37
359. Jørgensen, P. L. 1982. *Biochim. Biophys. Acta* 694:27–68
360. de Meis, L. 1981. *The Sarcoplasmic Reticulum.* New York:Wiley. 163 pp.
361. Sacks, G., Wallmark, B., Saccomani,
G., Rabon, E., Stewart, H. B., et al.
1982. *Curr. Top. Membr. Transp.*
16:135–60
362. Dunn, S. D., Heppel, L. A. 1981. *Arch.
Biochem. Biophys.* 210:421–36
363. Futai, M., Kanazawa, H. 1983. *Microbiol. Rev.* 47:285–312
364. Hoppe, J., Sebald, W. 1984. *Biochim.
Biophys. Acta* 768:1–27
365. Kagawa, Y., Sone, N., Hirata, H.,
Yoshida, M. 1979. *J. Bioenerg.
Biomembr.* 11:39–78
366. Kagawa, Y. 1982. *Curr. Top. Membr.
Transp.* 16:195–213
367. Nelson, N. 1981. *Curr. Top. Bioenerg.*
11:1–33
368. McCarty, R. E., Carmeli, C. 1982.
*Photosynthesis: Energy Conversion by
Plants and Bacteria,* ed. Govindjee.
1:647–95. New York:Academic
369. Gibson, F. 1983. *Biochem. Soc. Trans.*
11:229–40
370. Moroney, J. V., Lopresti, L., McEwen,

B. F., McCarty, R. E., Hammes, G. G.
1983. *FEBS Lett.* 158:58–62
371. Foster, D. L., Fillingame, R. H. 1982. *J.
Biol. Chem.* 257:2009–15
372. Von Meyenburg, K., Nielsen, J., Jørgensen, B. B., Michaelsen, O., Hansen, F.
1983. *Tokai J. Exp. Clin. Med. Special
Symp.* 7:23–31
373. Kagawa, Y. 1982. *Transport and
Bioenergetics in Biomembranes,* ed. R.
Sato, Y. Kagawa, pp. 37–56. Tokyo:
Jpn. Sci. Soc.
374. Sebald, W., Graf, T., Lukins, H. B.
1979. *Eur. J. Biochem.* 93:587–99
375. Sigrist-Nelson, K., Sigrist, H., Azzi, A.
1978. *Eur. J. Biochem.* 92:9–14
376. Amzel, L. M., McKinney, M.,
Narayanan, P., Pedersen, P. L. 1982.
Proc. Natl. Acad. Sci. USA 79:5852–56
377. Tiedge, H., Schäffer, G., Mayer, F.
1983. *Eur. J. Biochem.* 132:37–45
378. Williams, N., Hullihen, J. M., Pedersen,
P. L. 1984. *Biochemistry* 23:780–85
379. Futai, M. 1977. *Biochem. Biophys. Res.
Commun.* 79:1231–37
380. Kagawa, Y., Nukiwa, N. 1981. *Biochem. Biophys. Res. Commun.* 100:
1370–76
381. Ovchinnikov, Y. A., Modyanov, N. N.,
Grinkevich, V. A., Aldanova, N. A.,
Trubetskaya, O. E., et al. 1984. *FEBS
Lett.* 166:19–22
382. Dupuis, A., Zaccai, G., Satre, M. 1983.
Biochemistry 22:5951–56
383. Knowles, A. F., Guillory, R., Racker, E.
1971. *J. Biol. Chem.* 246:2672–79
384. McEnery, M. W., Buhle, E. L. Jr., Aebi,
U., Pedersen, P. L. 1984. *J. Biol. Chem.*
259:4642–51
385. Ryrie, I. J. 1977. *Arch. Biochem. Biophys.* 184:464–75
386. Tzagoloff, A. 1979. *Methods Enzymol.*
55:351–58
387. Sebald, W., Wild, G. 1979. *Methods Enzymol.* 55:344–51
388. Serrano, R., Kanner, B. I., Racker, E.
1976. *J. Biol. Chem.* 251:2453–61
389. Galante, Y. M., Wong, S.-Y., Hatefi, Y.
1979. *J. Biol. Chem.* 254:12372–78
390. Pullman, M. E., Monroy, G. C. 1963. *J.
Biol. Chem.* 238:3762–69
391. Sanadi, D. R. 1982. *Biochim. Biophys.
Acta* 683:39–56
392. Macino, G., Tzagoloff, A. 1980. *Cell*
20:507–17
393. Macreadie, I. G., Novitski, C. E., Maxwell, R. J., John, U., Ooi, B.-G., et al.
1983. *Nucleic Acids Res.* 11:4435–51
394. Chomyn, A., Mariottini, P., Gonzalez-
Cadavid, N., Attardi, G., Strong, D. D.,
et al. 1983. *Proc. Natl. Acad. Sci. USA*
80:5535–39

395. Sebald, W., Graf, T., Lukins, H. B. 1979. *Eur. J. Biochem.* 93:587–99
396. Sebald, W., Hoppe, J. 1981. *Curr. Top. Bioenerg.* 12:1–64
397. Borst, P., Grivell, L. A., Groot, G. S. P. 1984. *Trends Biochem. Sci.* 9:128–30
398. Stephenson, G., Marzuki, S., Linnane, A. W. 1980. *Biochim. Biophys. Acta* 609:329–40
399. Todd, R. D., Douglas, M. G. 1983. *Arch. Biochem. Biophys.* 227:106–10
400. Frangione, B., Rosenwasser, E., Penefsky, H. S., Pullman, M. E. 1981. *Proc. Natl. Acad. Sci. USA* 78:7403–7
401. Matsubara, H., Hase, T., Hashimoto, T., Tagawa, K. 1981. *J. Biochem.* 90:1159–65
402. Pedersen, P. L., Schwerzmann, K., Cintron, N. 1981. *Curr. Top. Bioenerg.* 11:149–99
403. Klein, G., Satre, M., Dianoux, A.-C., Vignais, P. V. 1981. *Biochemistry* 20:1339–44
404. Gomez-Puyou, A., Gomez-Puyou, M. T., Ernster, L. 1979. *Biochim. Biophys. Acta* 547:252–57
405. Harris, D. A., Von Tscharner, V., Radda, G. K. 1979. *Biochim. Biophys. Acta* 548:72–84
406. Schwerzmann, K., Pedersen, P. L. 1981. *Biochemistry* 20:6305–11
407. Gomez-Fernandez, J. C., Harris, D. A. 1978. *Biochem. J.* 176:967–75
408. Wong, S.-Y., Galante, Y. M., Hatefi, Y. 1982. *Biochemistry* 21:5781–87
409. Galante, Y. M., Wong, S.-Y., Hatefi, Y. 1981. *Biochemistry* 20:2671–78
410. Schwerzmann, K., Hillihen, J., Pedersen, P. L. 1982. *J. Biol. Chem.* 257:9555–60
411. Dreyfus, G., Satre, M. 1984. *Arch. Biochem. Biophys.* 229:212–19
412. Finel, M., Rubinstein, M., Pick, U. 1984. *FEBS Lett.* 166:85–89
413. Hashimoto, T., Yoshida, Y., Tagawa, K. 1984. *J. Biochem.* 95:131–36
414. Yosida, Y., Wakabayashi, S., Matsubara, H., Hashimoto, T., Tagawa, K. 1984. *FEBS Lett.* 170:135–38
415. You, K.-S., Hatefi, Y. 1976. *Biochim. Biophys. Acta* 423:398–412
416. Lam, K. W., Swann, D., Elzinga, M. 1969. *Arch. Biochem. Biophys.* 130:175–82
417. Shankaran, R., Sani, B. P., Sanadi, D. R. 1975. *Arch. Biochem. Biophys.* 168:394–402
418. Stiggall, D. L., Galante, Y. M., Kiehl, R., Hatefi, Y. 1979. *Arch. Biochem. Biophys* 196:638–44
419. Joshi, S., Hughes, J. B. 1981. *J. Biol. Chem.* 256:11112–16
420. Yagi, T., Hatefi, Y. 1984. *Biochemistry* 23:2449–55
421. Weiss, M. A., McCarty, R. E. 1977. *J. Biol. Chem.* 252:8007–12
422. Abou-Khalil, S., Sabadi-Pialoux, N., Gautheron, D. 1975. *Biochem. Pharmacol.* 24:49–56
423. Conn, D. E., Kaczorowski, G. J., Kaback, H. R. 1981. *Biochemistry* 20:3308–13
424. Siliprandi, D., Scutari, G., Zoccarato, F., Siliprandi, N. 1974. *FEBS Lett.* 42:197–99
425. Moroney, J. V., Warncke, K., McCarty, R. E. 1982. *J. Bioenerg. Biomembr.* 14:347–59
426. Robillard, G. T., Konings, W. N. 1982. *Eur. J. Biochem.* 127:597–604
427. Nielsen, J., Hansen, F. G., Hoppe, J., Friedl, P., von Meyenburg, K. 1981. *Mol. Gen. Genet.* 184:33–39
428. Solioz, M. 1984. *Trends Biochem. Sci.* 9:309–12
429. Schairer, H. U., Hoppe, J., Sebald, W., Friedl, P. 1982. *Biosci. Rep.* 2:631–39
430. Hoppe, J., Schairer, H. U., Sebald, W. 1980. *FEBS Lett.* 109:107–11
431. Wachter, E., Schmid, R., Deckers, G., Altendorf, K. 1980. *FEBS Lett.* 113:265–70
432. Fimmel, A. L., Jans, D. A., Langman, L., James, L. B., Ash, G. R., et al. 1983. *Biochem. J.* 213:451–58
433. Kopecky, J., Glaser, E., Norling, B., Ernster, L. 1981. *FEBS Lett.* 131:208–12
434. Nelson, N., Eytan, E., Notsani, B. E., Sigrist, H., Sigrist-Nelson, K., Gitler, C. 1977. *Proc. Natl. Acad. Sci. USA* 74:2375–78
435. Nelson, N. 1982. *Methods in Chloroplast Molecular Biology*, ed. Edelman et al, pp. 899–905. Amsterdam: Elsevier
436. Celis, H. 1980. *Biochem. Biophys. Res Commun.* 92:26–31
437. Sigrist-Nelson, K., Azzi, A. 1980. *J. Biol. Chem.* 255:10638–43
438. Schindler, H., Nelson, N. 1982. *Biochemistry* 21:5787–94
439. Gay, N. J., Walker, J. E. 1982. *Nature* 301:262–64
440. Hughes, J., Joshi, S., Torok, K., Sanadi, D. R. 1982. *J. Bioenerg. Biomembr.* 14:287–95
441. Penin, F., Godinot, C., Comte, J., Gautheron, D. C. 1982. *Biochim. Biophys. Acta* 679:198–209
442. Stiggall, D. L., Galante, Y. M., Hatefi, Y. 1978. *J. Biol. Chem.* 253:956–64
443. Russell, J., Jeng, S. J., Guillory, R. J. 1976. *Biochem. Biophys. Res. Commun.* 70:1225–34

444. Grubmeyer, C., Penefsky, H. S. 1981. *J. Biol. Chem.* 256:3718–27
445. Grubmeyer, C., Penefsky, H. S. 1981. *J. Biol. Chem.* 256:3728–34
446. Lowe, P. N., Beechey, R. B. 1982. *Biochemistry* 21:4073–82
447. Ferguson, S. J., Lloyd, W. J., Lyons, M. H., Radda, G. K. 1975. *Eur. J. Biochem.* 54:117–26
448. Lunardi, J., Vignais, P. V. 1979. *FEBS Lett.* 102:23–28
449. Pougeois, R., Satre, M., Vignais, P. V. 1979. *Biochemistry* 18:1408–13
450. Wong, S.-Y., Matsuno-Yagi, A., Hatefi, Y. 1984. *Biochemistry.* 23:5004–9
451. Marcus, F., Schuster, S., Lardy, H. A. 1976. *J. Biol. Chem.* 251:1775–80
452. Frigeri, L., Galante, Y. M., Hanstein, W. G., Hatefi, Y. 1977. *J. Biol. Chem.* 252:3147–52
453. Ting, L. P., Wang, J. H. 1980. *Biochemistry* 19:5665–70
454. Ting, L. P., Wang, J. H. 1980. *J. Bioenerg. Biomembr.* 12:79–93
455. Ting, L. P., Wang, J. H. 1981. *Biochem. Biophys. Res. Commun.* 101:934–38
456. Ting, L. P., Wang, J. H. 1982. *Biochemistry* 21:269–75
457. Davenport, J. W., McCarty, R. E. 1981. *J. Biol. Chem.* 256:8947–54
458. Cross, R. L., Nalin, C. M. 1982. *J. Biol. Chem.* 257:2874–81
459. Budker, V. G., Kozlov, I. A., Krubatov, V. A., Milgrom, Y. M. 1977. *FEBS Lett.* 83:11–14
460. Esch, F. S., Allison, W. S. 1978. *J. Biol. Chem.* 253:6100–6
461. Drutsa, V. L., Kozlov, I. A., Milgrom, Y. M., Shabarova, Z. A., Sokolova, N. I. 1979. *Biochem. J.* 182:617–19
462. Lauquin, G., Pougeois, R., Vignais, P. V. 1980. *Biochemistry* 19:4620–26
463. Cross, R. L., Kohlbrenner, W. E. 1978. *J. Biol. Chem.* 253:4865–73
464. Chang, T.-M., Penefsky, H. S. 1973. *J. Biol. Chem.* 248:2746–54
465. Williams, N., Coleman, P. S. 1982. *J. Biol. Chem.* 257:2834–41
466. Matsuoka, I., Takeda, K., Futai, M., Tonomura, Y. 1982. *J. Biochem.* 92:1383–98
467. Schuster, S., Ebel, R. E., Lardy, H. A. 1975. *J. Biol. Chem.* 250:7848–53
468. Ohta, S. Tsuboi, M., Yoshida, M., Kagawa, Y. 1980. *Biochemistry* 19:2160–65
469. Tamura, J. K., Wang, J. H. 1983. *Biochemistry* 22:1947–54
470. Myers, J. A., Boyer, P. D. 1983. *FEBS Lett.* 162:277–81
471. Yoshida, M. 1983. *Biochem. Biophys. Res. Commun.* 114:907–12
472. Penefsky, H. S. 1979. *Adv. Enzymol.* 49:223–80
473. Boyer, P. D., Chance, B., Ernster, L., Mitchell, P., Racker, E., Slater, E. C. 1977. *Ann. Rev. Biochem.* 46:957–1026
474. Boyer, P. D. 1979. *Membrane Bioenergetics,* ed. C. P. Lee et al, pp. 461–79. Mass: Addison-Wesley
475. Boyer, P. D., Kohlbrenner, W. E., McIntosh, D. B., Smith, L. T., O'Neal, C. C. 1982. *Ann. NY Acad. Sci.* 402:65–83
476. Jencks, W. P. 1980. *Adv. Enzymol.* 51:75–106
477. Taniguchi, K., Post, R. L. 1975. *J. Biol. Chem.* 250:3010–18
478. Webb, M. R., Grubmeyer, C., Penefsky, H. S., Trentham, D. R. 1980. *J. Biol. Chem.* 255:11637–39
479. Webb, M. R., Trentham, D. R. 1980. *J. Biol. Chem.* 255:8629–32
480. Boyer, P. D., Cross, R. L., Momsen, W. 1973. *Proc. Natl. Acad. Sci. USA* 70: 2837–39
481. Rosing, J., Kayalar, C., Boyer, P. D. 1976. *The Structural Basis of Membrane Function,* ed. Y. Hatefi, L. Djavadi-Ohaniance, pp. 189–204. New York: Academic
482. Feldman, R. I., Sigman, D. S. 1982. *J. Biol. Chem.* 257:1676–83
483. Feldman, R. I., Sigman, D. S. 1983. *J. Biol. Chem.* 258:12178–83
484. Sakamoto, J., Tonomura, Y. 1983. *J. Biochem.* 93:1601–14
485. Grubmeyer, C., Cross, R. L., Penefsky, H. S. 1982. *J. Biol. Chem.* 257:12092–100
486. Yosida, M. 1983. *Biochem. Biophys. Res. Commun.* 114:907–12
487. Hatefi, Y., Yagi, T., Phelps, D. C., Wong, S.-Y., Vik, S. B., Galante, Y. M. 1982. *Proc. Natl. Acad. Sci. USA* 79:1756–60
488. Kayalar, C., Rosing, J., Boyer, P. D. 1976. *Biochem. Biophys. Res. Commun.* 72:1153–59
489. Yagi, T., Matsuno-Yagi, A., Vik, S. B., Hatefi, Y. 1984. *Biochemistry* 23:1029–36
490. Hammes, G. G. 1982. *Proc. Natl. Acad. Sci. USA* 79:6881–84
491. Ebel, R. E., Lardy, H. A. 1975. *J. Biol. Chem.* 250:191–96
492. Gresser, M. J., Myers, J. A., Boyer, P. D. 1982. *J. Biol. Chem.* 257:12030–38
493. Cross, R. L., Grubmeyer, C., Penefsky, H. S. 1982. *J. Biol. Chem.* 257:12101–5
494. Hackney, D. D., Boyer, P. D. 1978. *J. Biol. Chem.* 253:3164–70
495. Kohlbrenner, W. E., Boyer, P. D. 1983. *J. Biol. Chem.* 258:10881–86

496. O'Neal, C. C., Boyer, P. D. 1984. *J. Biol. Chem.* 259:5761–67
497. Nalin, C. M., Cross, R. L. 1982. *J. Biol. Chem.* 257:8055–60
498. Yoshida, M., Okamoto, H., Sone, N., Hirata, H., Kagawa, Y. 1977. *Proc. Natl. Acad. Sci. USA* 74:936–40
499. Williams, R. J. P. 1978. *Biochim. Biophys. Acta* 505:1–44
500. Williams, R. J. P. 1983. *Trends Biochem. Sci.* 8:48
501. Kell, D. B. 1979. *Biochim. Biophys. Acta* 549:59–99
502. Rottenberg, H. 1979. *Biochim. Biophys. Acta* 549:225–53
503. Slater, E. C. 1980. *Trends Biochem. Sci.* 5:10–11
504. See Ref. 172, pp. 331–426
505. Racker, E., Stoeckenius, W. 1974. *J. Biol. Chem.* 249:662–63
506. Dewey, T. G., Hammes, G. G. 1981. *J. Biol. Chem.* 256:8941–46
507. Takabe, T., Hammes, G. G. 1981. *Biochemistry* 20:6859–64
508. Zoratti, M., Pietrobon, D., Azzone, G. F. 1982. *Eur. J. Biochem.* 126:443–51
509. Mandolino, G., DeSantis, A., Melandri, B. A. 1982. *Biochim. Biophys. Acta* 723:428–39
510. Westerhoff, H. V., Melandri, B. A., Venturoli, G., Azzone, G. F., Kell, D. B. 1984. *FEBS Lett.* 165:1–5
511. Sorgato, M. C., Branca, D., Ferguson, S. J. 1980. *Biochem. J.* 188:945–48
512. Hitchens, G. D., Kell, D. B. 1982. *Biochem. J.* 206:351–57
513. Hitchens, G. D., Kell, D. B. 1982. *Biosci. Rep.* 2:743–49
514. Hitchens, G. D., Kell, D. B. 1983. *Biochem. J.* 212:25–30
515. Hitchens, G. D., Kell, D. B. 1983. *Biochim. Biophys. Acta* 723:308–16
516. Cox, G. B., Jans, D. A., Gibson, F., Langman, L., Senior, A. E., Fimmel, A. L. 1983. *Biochem. J.* 216:143–50
517. Fang, J.-K., Jacobs, J. W., Kanner, B. I., Racker, E., Bradshaw, R. A. 1984. *Proc. Natl. Acad. Sci. USA* 81:6603–7
518. Wood, D., Darlison, M. G., Wilde, J. R., Guest, J. R. 1984. *Biochem. J.* 222:519–34
519. Darlison, M. G., Guest, J. R. 1984. *Biochem. J.* 223:507–17

Ann. Rev. Biochem. 1985. 54:1071–108

DEVELOPMENTAL REGULATION OF HUMAN GLOBIN GENES[1]

Stefan Karlsson

Arthur W. Nienhuis

Clinical Hematology Branch, National Heart, Lung, and Blood Institute, National Institutes of Health, Bethesda, Maryland

CONTENTS

PERSPECTIVES AND SUMMARY

Hemoglobin switching during ontogeny reflects the sequential expression of the individual genes within the globin gene clusters. These phenomena have been of interest to investigators for several reasons. First, hemoglobin switching is prototypical for the problem that is central to developmental biology, namely the mechanism(s) by which specific genes are expressed in particular cells. Second, the perinatal switch from fetal (HbF=$\alpha_2\gamma_2$) to adult (HbA =$\alpha_2\beta_2$) hemoglobin precedes the clinical onset of the severe hemoglobinopathies and thalassemias. Thereafter, the level of fetal hemoglobin in individual patients is a major modulator of disease severity. Third, there are many well-characterized mutations that affect the production of fetal hemoglobin during adult life. Study of thalassemia mutations has provided many lessons about the DNA sequences required for efficient gene expression whereas analyses of mutations that increase HbF synthesis should provide insights into the mechanisms of gene control. Fourth, there has been recent success in manipulating HbF synthesis with drugs in adult individuals with sickle cell anemia or severe β thalassemia.

We have chosen to focus our review on hemoglobin switching with its many implications. The interested reader should also consult several excellent recent reviews on globin gene structure and function and the thalassemias (1–7).

From an analysis of the ontogeny of hemoglobin switching has come the observation that changes in hemoglobin phenotype are related to developmental age of the organism and appear to be intrinsic to erythroid stem and progenitor cells. Molecular studies of the globin gene cluster have provided a detailed framework on which to consider the regulatory mechanisms that operate in cells in defining the pattern of gene expression. Recent discovery of point mutations that enhance γ gene expression, and therefore HbF synthesis in adults, provides strong evidence that DNA sequences within the promoter influence developmental expression of individual genes. The role of putative cis-acting enhancer elements, either within or flanking genes, is far more conjectural. Many lines of evidence point to the existence of trans-acting factors (proteins) that influence gene expression by interacting with specific DNA sequences. DNA binding protein(s) capable of reconstituting a nucleosome-free site on the promoter region of the chicken β-globin gene is (are) the best studied of these potential trans-acting factors.

The cell biology of hemoglobin switching has been characterized in vitro by clonogenic assay of erythroid progenitor cells. From such studies has come the observation that the pattern of globin gene expression at different developmental stages is already defined in the earliest detectable erythroid progenitors. Erythroid progenitors from adult bone marrow and peripheral blood give rise to colonies containing cells with only HbA or HbF and HbA. The latter are the analogs of peripheral blood cells called F-cells. F-cell production is determined during differentiation within the erythroid progenitor and precursor compartment.

More than 20 deletion mutants that remove all or parts of the beta-globin gene cluster have been characterized. These variably influence the pattern of globin gene expression in adults. Of great interest is the apparent polarity of the clusters; deletions that remove the upstream or 5' end (with reference to globin genes) totally inactivate the gene cluster, whereas mutations that remove the downstream or 3' end enhance expression of the remaining γ-globin gene(s). Several different mutations that influence HbF synthesis in adults but leave the globin gene clusters intact also have been characterized by clinical and genetic studies. Three of these are point mutations in one of the γ gene promoters.

Administration of certain drugs to adults with sickle cell anemia or severe β thalassemia results in increased HbF production. The first drug tested, 5-azacytidine, demethylates DNA and also activates genes in cultured cells, but two other drugs—hydroxyurea and cytosine arabinoside—that have no direct effects on DNA methylation, also enhance HbF synthesis. The mechanism by which these drugs act is currently controversial but most evidence is consistent with a direct inductive action on erythroid progenitor and precursor cells.

THE ONTOGENY OF HEMOGLOBIN SYNTHESIS

In man, there are two developmental switches in hemoglobin phenotype, the embryonic to fetal switch that occurs very early in gestation, and the fetal to adult hemoglobin switch that occurs predominantly around the time of birth (1–7). The identity of the hemoglobins found in red cells at the various developmental stages, their globin subunit compositions, and the organization of the globin genes on chromosomes 11 and 16 are shown in Figure 1.

The Embryonic to Fetal Switch

The embryonic hemoglobins are found in relatively large (200 cubic microns) nucleated red cells formed in the placental yolk sac. These cells, called megaloblasts, circulate until approximately the twentieth week of gestation (Figure 2). With the onset of erythropoiesis in the liver, fetal red cells are formed. These macrocytic cells lack a nucleus and thus resemble adult red cells, although they are considerably larger (125 compared to 80 cubic microns) (7).

Early studies suggested that primitive (yolk sac) and definitive (liver and bone marrow) red cells contained only embryonic or fetal hemoglobin, respectively, and therefore might be derived from separate stem cell populations, thereby accounting for their differences in hemoglobin phenotype (8, 9). More recently, sensitive fluorescent assays (10) and analysis of chick-quail chimeras (11)

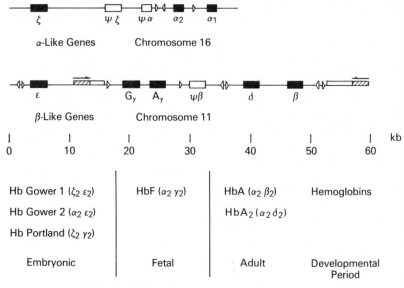

Figure 1 Organization of the globin genes on chromosomes 11 and 16. The composition of the various embryonic, fetal, and adult hemoglobins is also indicated. Closed boxes indicate active genes and open boxes pseudogenes. The triangles (▷) indicate Alu repetitive sequences and their orientation. The two open boxes partially hatched above the line in the β-like gene cluster indicate Kpn repeat sequences and their respective orientation. The Kpn sequence between the ε and $^G\gamma$ genes in fact consists of two tandemly linked Kpn repeats (see text).

have clearly shown that fetal or adult-type globin chains may be found in yolk sac derived erythroblasts and that embryonic chains may be found in enucleated red cells of both avian and murine origin. Similarly hamster embryonic and adult globin genes were found to be co-expressed in both yolk-sac erythroid cells and neonatal erythroid cells of liver and spleen (12).

The human embryonic to fetal hemoglobin switch does not coincide with the transition from yolk sac to fetal erythropoiesis (13, 14). Rather, α-globin production can be detected in megaloblastic cells indicating that the ζ to α switch begins in the yolk sac. Megaloblastic cells also contain γ globin; the ε to γ globin switch appears to occur slightly later than the globin gene switch in the α cluster (15, 16). Macrocytic cells of fetal liver origin formed at about the same time contain both ζ and ε in addition to α and γ globin. Thus, the embryonic to fetal hemoglobin switch occurs at a specific time in gestation and appears to be unrelated to the exact site of erythropoiesis.

The Fetal to Adult Switch

Synthesis of β chain begins at the very earliest stages in erythropoiesis. Erythrocytes present in 8–12 week old embryos have been shown to contain

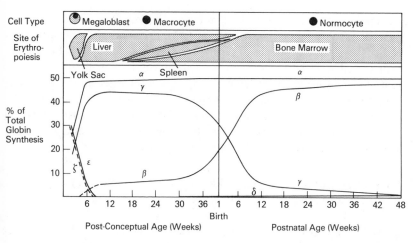

Figure 2 A representation of the synthesis of different globin chains at various stages of embryonic and fetal development. The site of erythropoiesis is also indicated. (From Weatherall and Clegg, 1981, with permission).

this globin (17, 18). In fact, clonal erythroid colonies derived from single progenitors have been shown in culture to contain ϵ, γ, and β globin chains (13, 14). There is a steady increase in β chain synthesis throughout gestation so that just before birth it accounts for 15–20% of the non-α chain synthesis (18, 19). The onset of erythropoiesis in the spleen or bone marrow (Figure 2) does not coincide with any significant alteration in the proportions of γ- and β-globin synthesis (20). Beta globin is uniformly distributed among fetal macrocytic red cells in contrast to the situation in adult red cells in which the small residual amounts of HbF that are produced are found in a subpopulation of red cells (see below) (21).

The major switch from HbF ($\alpha_2\gamma_2$) to HbA ($\alpha_2\beta_2$) synthesis occurs about the time of birth (22–24). Red cells formed during the switching period, either in vivo or in clonal erythroid cultures in vitro, contain varying proportions of γ and β globin (24) with continuous distribution. Some have argued that red cells derived from 'fetal' or 'adult' stem or progenitor cells can be distinguished by virtue of their relative amounts of the two γ globins ($^{G}\gamma$ or $^{A}\gamma$) rather than the amount of γ and β globin that they contain (25), but this idea has not gained wide acceptance.

Control of the fetal to adult switch appears to be related to the developmental time of the organism (23, 26). The switch is not influenced by the exact timing of birth nor do major perturbations in the hormonal balance in fetuses of experimental animals have a major influence on its progression (20). Transplantation of fetal hematopoietic cells into irradiated recipient sheep has been used to characterize the switching mechanism (27–29). Cells obtained during the latter stages of gestation complete the switch in the adult bone marrow with a temporal pattern that resembles that by which these cells would have completed the switch in situ in the developing fetus. In contrast, cells obtained from earlier sheep fetuses continue to generate erythrocytes containing HbF in adult recipients. The switch in these transplanted animals is gradual and incomplete. Recently transplantation of human early fetal liver cells into a child with leukemia provided the opportunity to make analogous observations (30, 31). The fetal stem (and/or progenitor cells) maintained their fetal characteristic in postnatal bone marrow. These observations are consistent with regulation of switching by operation of a "developmental clock."

ORGANIZATION, STRUCTURE, AND FUNCTION OF THE HUMAN GLOBIN GENES

Several features of the globin gene clusters are pertinent to a detailed consideration of the hemoglobin switches that occur during development (1).

Arrangement of the Globin Gene Clusters on Chromosomes 11 and 16

The eight functional globin genes and three globin pseudogenes are arranged in two gene clusters, the α- and β-globin gene clusters (Figure 1). The α gene is located on chromosome 16 (32) and the β gene is on chromosome 11 (33). The α globin gene has been mapped to the subband region pter-p12 on the short arm of chromosome 16 (34, 35), but the precise location of the β-globin gene cluster on chromosome 11 is more controversial (36–39). Other genes have also been localized on 11p, including insulin (40), the c-Ha-Ras I oncogene (41), the gene for parathyroid hormone (42), and the insulin-like growth factor II gene (43, 44). The gene order beginning from the centromere, parathyroid hormone-β-globin-insulin, was suggested by a recent linkage study (45).

The β-globin gene cluster spans over 60 kb of DNA (Figure 1) (46–52), all of which has been sequenced; the sequence data is included in a recent review (2). The pseudogene, $\psi\beta$, lies between the $^A\gamma$ and the δ gene. Another β-like pseudogene was originally described (1) but recent DNA sequence analysis has failed to confirm the existence of $\psi\beta_2$ (53). The duplicated fetal globin genes encode almost identical polypeptides which differ by only one amino acid, glycine ($^G\gamma$) or alanine ($^A\gamma$), at codon 136 while the sequence divergence of the adult δ and β genes is more extensive. The α-globin gene cluster spans 30 kb and contains three functional genes, the embryonic ζ gene and two adult α genes, $\alpha1$ and $\alpha2$, that encode identical polypeptides (Figure 1). Two pseudogenes, $\psi\zeta$ and $\psi\alpha$ are situated between the ζ and $\alpha2$ genes (54–57). Although the duplicated α genes encode identical polypeptide chains, their 3' untranslated regions are quite divergent (58, 59).

Repetitive Sequence Families Within and Flanking the Globin Clusters

Approximately 20–30% of human DNA consists of repetitive DNA sequences (60–62). Moderately repetitive sequences are interspersed with single copy sequences; these repeats are not necessarily internally repetitive and rarely occur in tandem arrays. Such sequences are divided into 'families' based on the sequence homology of their individual members. Individual copies of moderately repetitive sequence families may be short (100–300 bp—short interspersed repeats or SINES) or long (>6 kb—long interspersed repeats or LINES). Members of two moderately repetitive DNA families, one from each class, have been described around or within the globin gene clusters (Figure 1).

THE ALU FAMILY   This repetitive family, consisting of 3–5×10^5 interspersed copies, derives its name from a common Alu I enzyme cleavage site (63) present in 60% of the copies. The consensus sequence of the Alu family is generally conserved in primate species but on average Alu members are only 60% homologous to one another (64, 65). The human Alu sequence is an imprecise dimer formed of two directly repeated, approximately 130-bp monomers with a 31-bp insertion in the second monomer (65–68). Many human Alu family members are flanked by direct repeats of 7–20 bp (66, 69, 70) which, by analogy to bacterial transposons, may indicate that Alu sequences are mobile DNA elements.

Alu sequences may be found within introns or in 3' untranslated regions and may therefore be transcribed into RNA. Alu family members share sequence homology with double-stranded heterogenous nuclear RNA (71, 72). In addition, a sequence in the first monomer of human Alu is likely to be an internal 'control' region for RNA polymerase III (73). Due to this internal promoter region RNA polymerase III transcripts could function as intermediates in the proposed dispersal of Alu sequences into new chromosomal sites (61).

The α-globin cluster includes three Alu repeats (74), and the β-globin gene cluster contains eight Alu family members (49, 75) (Figure 1). Several of these sequences are transcribed in vitro and in vivo (76–80) as discussed in more detail below.

THE KPN FAMILY   The second major family of interspersed repetitive DNA sequences in the human genome is the Kpn family (more recently called the LINE 1 or L1 family) which derives its name from Kpn I restriction endonuclease cleavage sites located at analogous positions in many copies of the sequences [reviewed in ref. (62)]. In primates, there are 10^4 or more members of the Kpn family that are up to 6.0 kb long and are polymorphic in sequence (81–89). The shorter members of the Kpn family usually miss sequences, and are homologous to the 5' end of a Kpn consensus sequence deduced from available sequence information for several Kpn members of both human and other primate origin (62). Several Kpn family members in monkeys have polyadenylation signals followed by an (A)-rich stretch (90). It has been suggested that transposition of Kpn-like sequences may occur by way of a polyadenylated RNA intermediate (88, 91, 92).

Nine segments of DNA, varying in length from 70 bp to several kb, having homology to the Kpn family, have been found in and around the beta-globin gene cluster (62). There is a complete family member (reported as 6.4 kb in length) immediately downstream from the β-globin gene in a position analogous to a long repeat sequence downstream from the mouse β globin gene (81). Between the ε and ^{G}γ globin genes are two family members of 3 and 4 kb in length in a tandem array and in an inverted orientation compared to the complete family member downstream from the β globin gene. Both are homologous mainly to the 3' end of the consensus sequence. An LTR-like structure initially reported as being part of this repeat sequence (87) is in fact several hundred bp downstream (J. Skowronski, M. Singer, personal communication). There are 70-bp segments of DNA, homologous to the 3' end of the consensus Kpn sequence, at the ends of the γ-globin gene duplication unit (52). One such segment is found at the 5' end of the duplication unit, one is between, and the third at the 3' end (J. Skowronski, M. Singer, personal communication).

Heterogenous nuclear RNA and cytoplasmic RNA has been reported to be complementary to the human Kpn I repeat sequence (92, 93). Potter (94) has recently detected a 650-bp long open reading frame within a 3000-bp human Kpn sequence. An open reading frame that can potentially code for 326 amino acids has also been detected in an L1 repeat family (previously called the Bam HI family) within the β-globin cluster of mice (95). To date however there is no direct evidence for the transcription of the Kpn families in the human β globin cluster and their role remains to be elucidated.

Structure of the Prototypical Globin Gene

All human globin genes have three coding blocks (exons) separated by two intervening sequences (introns), probably reflecting their common origin from a single ancesteral globin gene (46–52, 54–57, 96). The length of the introns varies between the α- and β-like genes.

Conserved sequences important for gene function are found in and around the globin genes. These include sequences in the promoter region, at the exon/intron boundaries, and at the 3' end of the mRNA coding sequences (Figure 3). The available evidence indicates that the major modulation of globin gene expression occurs at the transcriptional level although posttranscriptional differences in RNA processing efficiency, mRNA stability, or the rate of mRNA translation may have minor but significant modulating influences on the final proportions of the individual globins found in red cells.

STRUCTURE OF THE HUMAN GLOBIN GENES

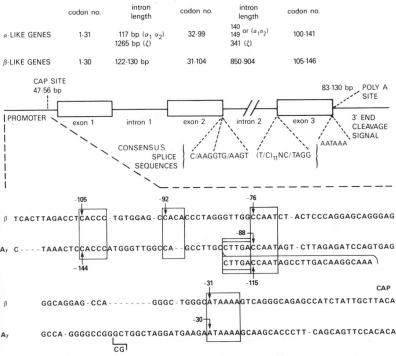

Figure 3 Structure of a prototypical globin gene showing conserved sequences at the exon-intron boundaries (splice sequences) and at the 3' end of the gene (polyadenylation cleavage signal) The length of each exon (in codons) and introns (in base pairs) is also shown. Below the sequences of the human ^Aγ and β promoters are shown. Indicated are the important conserved sequences, TATA box, CAT box (duplicated in the ^Aγ promoter) and CACA box (duplicated in the β promoter) Also shown (▤) is the area just upstream from the CAT boxes in the ^Aγ promoter that is conserved in globin genes expressed during fetal life. A mutation in this area has been detected at position −117. This mutation results in increased γ gene expression (see text).

RNA Processing

During and following transcription, the globin mRNA precursors are modified by capping, splicing, and polyadenylation to form mature mRNA molecules.

RNA CAPPING   'Capping' refers to the addition of a 7-methyl-guanosine triphosphate residue to the 5' end of the mRNA in a 5' to 5' linkage (97, 98) and subsequent methylation of the nitrogenous bases and ribose residues of the first two or three nucleotides. The Cap structure promotes binding of globin mRNA to ribosomes during the initiation of protein synthesis (99) and capped RNA precursors are spliced more efficiently than uncapped molecules and produce fewer aberrant RNA species in vitro (100). Capping is not absolutely necessary for splicing or protein synthesis initiation, however; these processes are just less efficient when RNA is uncapped.

RNA SPLICING   Comparison of exon/intron boundary sequences from more than 100 genes identified consensus 5' and 3' splice sequences (101) (Figure 3). The obligatory 'GT' in the 5' splice sequence and 'AG' in the 3' splice sequence are required for splicing while the remaining consensus nucleotides contribute to splice junction selection and efficiency (102, 103). The 5' terminus of U1 RNA, a small nuclear riboprotein ('snurp'), displays complementarity to the consensus 5' and 3' splice sequences (101–104). This suggested that U1 RNA could base-pair with 5' and 3' splice junction sequences and thereby facilitate accurate ligation of the two exons involved (105–107). The U1 snurp has now been shown to bind selectively to a 5' splice site of an unspliced mouse β-globin transcript (108) and antibodies directed against U1 snurps can be demonstrated to inhibit the in vitro splicing of adenovirus late RNA (109).

Recently a detailed model of the RNA splicing mechanism has been proposed (110–113). The first step appears to be cleavage of the 5' exon-intron boundary; this step is either directly coupled or closely followed by formation of a lariat structure. The 5' end of the intron is involved in an atypical 2'-5' phosphodiester linkage to an adenosine residue that is part of a consensus sequence near the 3' end of intron I. Subsequently, the 3' intron-exon boundary site is cleaved and the two exons are ligated to one another.

Specific internal sequences are not required for efficient splicing of sequences transcribed from intron II of the β-globin gene although a minimum intron length is mandatory (114); the sequence requirements for formation of the 2'-5' phosphodiester bond may not be very stringent. Several attempts have been made to determine which end of the intron first becomes associated with the splicing components, but the available data do not conclusively support either a 5' to 3' or 3' to 5' scanning model (115, 116).

POLYADENYLATION AND TERMINATION OF RNA TRANSCRIPTION   Transcription of the mouse β-globin gene continues beyond the gene. A specific termination site has been described but this was based on an experimental error (117, 118). The 3' untranslated regions of globin and other RNAs contain a highly conserved sequence, 'AAUAAA,' originally called the poly A addition signal (119). This signal now appears to be primarily involved in endonucleolytic cleavage of the RNA transcript followed by polyadenylation (120–124). A recently identified mutation in the 3' end cleavage sequence of the human α2 thalassemic gene (AAUAAA→AAUAAG) produces α-globin mRNA with 3' ends extending far beyond the normal polyadenylation site. Inhibition of the endonucleolytic cleavage reaction apparently results in unstable transcripts (123). Similarly a β-thalassemia gene has been described with a mutation in the 3' end cleavage signal (AATAAA→AACAAA) (124). Recently sequences beyond the 3' end of mature mRNA were shown to be required for cleavage of the RNA transcript and generation of the poly A addition site (125–128). Proudfoot and colleagues have shown that more than 15 but no more than 51 nucleotides downstream from the AATAAA of the human α globin gene are required for the cleavage polyadenylation reaction (128).

The role of poly A track is to increase the stability of mRNA in the cytoplasm. Poly A⁻ mRNA is readily transported out of the nucleus, and it can be translated in the cytoplasm, but its half-life in the cytoplasm is much shorter than that of poly A⁺ mRNA (129, 130). Polyadenylation is not believed to play a major role in differential expression of the globin genes.

mRNA Translation

Translation of globin mRNA into the protein product involves the interaction of numerous protein initiation, elongation, and termination factors in addition to ribosomes and a complement of tRNA molecules. The rate of initiation of protein synthesis is defined by several regulatory mechanisms some of which are controlled by the concentration of hemin but there is no evidence that these mechanisms are involved in hemoglobin switching. Several excellent reviews of protein synthesis have recently been prepared (131–133), and therefore this process will not be considered in detail here.

MOLECULAR MECHANISMS OF HEMOGLOBIN SWITCHING AND HbF SYNTHESIS

The structure of the promoter region, including its sequence, methylation, and chromatin structure, may be involved in modulating gene expression. Cis-acting modulating sequences that enhance promoter function have been described in viruses and eukaryotic genes. Trans-acting molecules that interact with promoter regions thereby possibly regulating tissue-specific expression of genes have also been demonstrated recently. We will discuss these structural and molecular interactions in the context of the control of hemoglobin switching.

Promoter Region

The promoter includes those DNA sequences necessary for binding of RNA polymerase to effect efficient and accurate initiation of transcription (134). Several strategies have been used to deduce those sequences required for promoter function (Figure 3). First, comparison of the sequences of many globin genes identified several conserved blocks of nucleotides (boxes) (46–

52, 54–57). Second, mutations involving replacement of one nucleotide, several nucleotides (linker scan), or deletions have been studied both by in vitro transcription reactions or after introduction of the mutated genes into HeLa, monkey kidney, or mouse cells (135–139). Third, five point mutations have been identified within the promoter region of β-globin genes isolated from thalassemic individuals; these point mutations are thought to be responsible for decreased β-globin gene expression and provide direct evidence for the importance of specific conserved 'boxes' for efficient gene expression in vivo (5). Fourth, a limited number of studies have now been performed to characterize the function of promoter regions of human globin genes during erythroid maturation of mouse erythroleukemia cells into which these genes have been introduced by DNA transfection (140–144). Figure 3 contains the sequences of the γ and β promoters and shows the location of the conserved sequences in each.

A structural and functional comparison of the γ and β gene promoters may be relevant to their differential expression during development (see Figure 3). There are indeed very distinct structural differences. For example, a 16-nucleotide segment that includes the CCAAT box is tandemly duplicated in the human γ gene promoters; the two CCAAT boxes are 27 nucleotides apart. In contrast, the 'CACA' sequence appears only once and is located between 141 and 148 nucleotides (nt) upstream from the start site of transcription. A deletion mutation that removes sequences down to within 131 nucleotides of the start site of transcription functions at only 25% of the level of the normal γ gene promoter with 260 nt of 5' flanking sequence in a transient assay in HeLa cells (N. Anagnou, A. Moulton, and A. W. Nienhuis, unpublished observations). Further truncation to remove the distal 'CCAAT' box reduces function to 10% of normal. In contrast, linker scan or deletion mutations that remove the proximal CCAAT box but leave the sequences further upstream intact, result in a 2–4-fold increase in promoter function in HeLa cells.

Of considerable interest are three recently discovered point mutations within the γ-gene promoter region that appear to increase expression of the affected γ gene in adult individuals. Substitution of A for G at position −117 of the ^{A}γ gene promoter has been found in two individuals with a type of heridCitary persistence of fetal hemoglobin (HPFH) characterized by an increase in ^{A}γ production only (145, 145a) (see below). This position is only two nt upstream from the conserved distal 'CCAAT' box. The six nucleotides immediately upstream from this 'CCAAT' box are conserved in globin genes expressed during the embryonic and fetal development periods of man, goat, rabbit, and mouse; the conserved block has the sequence YYTTGA where Y is a pyrimidine (2). This sequence is not conserved in genes expressed during the adult developmental period. An interesting corollary of the −117 substitution in the ^{A}γ gene is the fact that there is decreased expression of the β-globin gene and apparent increased expression of the ^{G}γ gene on the same chromosome.

Two other mutations have been found upstream of the CAP site in γ genes that are over-expressed in vivo. Substitution of G for C at −202 has been reported in one ^{G}γ gene of six black individuals with increased HbF containing ^{G}γ chains; this substitution was not found in 170 ^{G}γ genes from individuals with normal levels of HbF (146, 146a). Recently a second mutation in this region has been described in a Southern Italian with the ^{A}γ type of HPFH. Substitution of T for C was found at position −196 of the ^{A}γ-globin gene promoter (147). Of interest in this context is recent evidence that DNA binding proteins may react with globin promoters as far as 300 bp upstream from the Cap site (148). Other genes have also been shown to have transcriptional regulatory sequences in this region (149, 150).

There is a switch from fetal to adult hemoglobin in sheep and goats that is quite analogous to that seen in man. The major transition from γ to β synthesis occurs during the perinatal period. Comparison of the promoter sequences of the γ and β genes from sheep (151) and goats (152, 153) provides an interesting contrast to that seen for the human genes. The sequences within 150 bp upstream from the start site of transcription are virtually identical although two nt differences are located in the conserved segment immediately 5' to the single CCAAT box. The large introns of the goat and sheep globin genes contain inserted elements that differ in sequence from one another (152). Whether small differences in promoter structure or difference in intron or coding sequences (see later discussion) are more important in developmental regulation of the sheep and goat globin genes remains to be determined.

Although the vast majority of globin gene transcripts initiate at the CAP site, recent observations indicate that a minor fraction of β-globin mRNA molecules are initiated at several sites (mainly at −172) upstream (154, 155). Discrete upstream initiation sites have also been identified for the human ε and γ-globin genes at least one of which is utilized by polymerase III from an Alu sequence (78, 156). There is no clear evidence as to the special function or regulatory significance of the various transcription initiation sites in the β-globin clusters.

Sequences 3' to the CAP Site May Be Involved in Globin Gene Regulation

Study of promoter function in cell extracts or in intact HeLa cells or fibroblasts cannot be expected to lead to identification of all sequences that function in the developmental regulation of genes. To date, no truly adequate assay system has been devised although use of mouse erythroleukemia cells has yielded interesting data. Mouse erythroleukemia (MEL or Friend) cells are transformed cells that appear to be arrested at the pro-erythroblast stage (156a). Culture of these cells in one of several chemical inducers leads to induction of erythroid maturation. Human globin genes have been introduced into such cells by two methods, chromosome transfer via cell fusion and by DNA transfection.

Transfer of either chromosome 16 (α globin genes) or chromosome 11 (β-globin genes) into mouse erythroleukemia cells from fibroblasts (157) or bone marrow cells (158, 159) by cell fusion leads to regulated expression of the human adult globin genes. In cell hybrids containing chromosome 16, the α-globin genes are expressed at low levels in noninduced cells and show a several-fold increment in expression upon chemically induced erythroid maturation (34). The embryonic ζ gene is not expressed even when the human chromosome number 16 in the MEL cell hybrid is derived from a human erythroleukemia cell line (K562) in which the ζ gene is active (160). In cell hybrids containing human chromosome 11 derived from primary normal adult or tissue culture cells, the β-globin gene is expressed constitutively at low levels and its expression increases several-fold on induction. In contrast, the ε and γ genes are expressed at very low levels both before and after induction (157–160). These results suggest that there are trans-acting factors (either negative, positive, or both) that modulate globin gene expression in cell hybrids. Fusion of MEL cells to fetal liver erythroid cells has yielded cell hybrids containing chromosome 11 that expresses the γ gene preferentially (161) suggesting that the origin of the chromosome is also relevant with respect to gene expression.

The method of DNA-mediated transfer provides the opportunity to define sequences required for tissue specific expression although the regulatory influences of chromatin structure and DNA methylation are not reproduced by this experimental approach. Study of the human globin genes introduced into MEL cells by DNA transfection has yielded the following results. The human α-globin gene is expressed at a high level in many but not all transformants but shows no significant increase in expression upon induction. The human β-globin gene is expressed at a relatively low level in uninduced cells but exhibits a 5–20-fold increase in the level of expression upon induction. The human ε and γ genes are expressed at low levels in MEL cells and fail to show significant induction in the vast majority of transformed cell clones (140–142).

One of the first clues that some of these sequences might lie to the 3' side of the CAP site was data showing that the rabbit β-globin gene promoter could be truncated to 58 nt from the CAP site without loss of induction although the level of expression in both induced and uninduced cells was significantly reduced compared to that of a gene with an intact promoter (143). Hybrid genes have been constructed in which the human α or γ promoter was linked to the β coding sequences (143, 144). In both cases, the α promoter–β coding sequence and γ promoter–β coding sequence hybrid genes showed a fairly low level of expression in most uninduced clones and a significant increase in expression upon induced erythroid maturation. A β promoter–α coding sequence hybrid gene was expressed constitutively at a fairly high level and showed no significant increase in expression upon induction. A β promoter–γ coding sequence hybrid gene also showed significant induction in several transformed clones, and so did a hybrid gene consisting of a β promoter and the noninducible murine H-2K^{bm1} class 1 MHC gene, suggesting that either the β promoter or β coding sequence could confer the property of inducibility. Analysis of nuclear run-off transcripts showed that the induction results at least in part from transcriptional activation (143). The high constitutive level of the α-globin gene expression in MEL cell clones appears to be a property of its coding sequences.

Enhancer Elements

Enhancer elements were first described in the DNA tumor virus, simian virus 40 (SV40). The SV40 enhancer is a 72-bp repeat that acts in cis to activate SV40 early genes in vivo (162, 163). The SV40 and other viral enhancers can also activate heterologous genes and can act independent of orientation and relatively independent of distance (164) with respect to the activated promoter [reviewed by Gruss (165) and Khoury & Gruss (166)]. Enhancers function only in cis to specific promoters and may only act on the nearest strong promoter but not on more distal promoter elements (167). Recently viral enhancers have been shown to demonstrate host specificity which could partially determine the host range of a particular virus (169).

Function of the γ and β but not of the α promoter in vector molecules 48 hours after introduction into nonerythroid cells depends upon the presence of an enhancer within the vector. The amount of correctly initiated rabbit β-globin mRNA was increased 200-fold by the SV40 72-bp repeat (135). The human β promoter has also been shown to be nearly totally dependent on the SV40 enhancer in both monkey kidney and HeLa cells (138). When α and β-globin genes are put in the same vector molecule, only the α but not the β promoter functions to generate correctly initiated RNA molecules. These data seemed to suggest that the α-globin gene promoter works independent of an enhancer but now these results can be explained by virtue of the fact that a specific enhancer may only act on one promoter within a vector molecule (167). Indeed, recently an α-promoter–β coding sequence hybrid gene has been shown to be enhancer dependent in a transient assay in HeLa cells whereas the β promoter–α coding sequence hybrid gene yields correctly initiated transcripts from the β promoter independent of the presence of a viral enhancer (T. Maniatis, personal communication).

As yet there is no evidence linking an enhancer element to tissue specific expression of the γ and β genes. Clearly, the sequence within the coding portion of the β gene that confers the property of inducibility in MEL cells does not 'enhance' β promoter function in heterologous cells in the transient assay system. This control element may not have the properties of a classical enhancer or it may require a tissue specific trans-acting factor for its function.

There are precedents however, for supposing that enhancer-like elements may control tissue specific gene expression. For example, the large introns of the constant region of the immunoglobulin heavy and kappa light chain genes contain an enhancer sequence whose activity is orientation independent and exerted over relatively long distances (170–173). Tissue specific enhancer-like elements have also been detected in the 5' flanking regions, approximately 100–300 bp upstream from the CAP site of the insulin and chymotrypsin genes (150). The transcription enhancing effects of these elements is independent of position and orientation and they also act in a tissue specific manner when they are in reverse orientation (M. Walker, W. Rutter, personal communication). Similarly, a cell type–specific enhancer element has been reported upstream from the mouse MHC gene Eβ (174).

How enhancers work is still a mystery. Much evidence is consistent with their acting as an entry point for RNA polymerase II although no direct data support this mechanism (164). Alternatively, enhancers may interact with specific protein molecules that modulate their activity or that may facilitate binding of specific regions of the genome to strutural components of the cell nucleus. Scholer & Gruss (175) in an elegant study, demonstrated that both the SV40 and MSV enhancers interact with the same class of cellular molecules as demonstrated in a competition assay.

Inhibitory Sequences

Recombinant plasmid molecules containing the mouse β major globin promoter or the human ε globin promoter linked to the herpes simplex virus (HSV-1) thymidine kinase (tk) gene have been constructed and used to transform TK⁻ fibroblast cells (176). When a DNA fragment (H1) containing sequences between 344 and 1413 bp upstream from the mouse β major gene was inserted in either orientation upstream from a globin promoter, transformation was almost completely abolished. The H1 fragment had no effect on transformation frequency when the HSV-1 TK promoter was used instead of the globin promoters. Similar results were also obtained in transient expression assays. An alternating purine pyrimidine sequence found in the H1 fragment might predispose this region to adapt the Z-form of DNA. It has been suggested that a switch from B to Z conformation might be involved in gene regulation (177).

Recently a strong DNAse I hypersensitive site (site I) has been identified upstream from the nonrearranged germline c-myc gene in Burkitt lymphoma cells. This gene is not expressed. Other DNAse I hypersensitive sites upstream from and around the c-myc promoter which are found when the c-myc gene is actively expressed are not detected in the germline c-myc gene in Burkitts lymphoma. This led the authors to propose that the DNAse I hypersensitive site, called site I, mediate negative transcriptional control through binding of a hypothetical trans-acting repressor molecule (149).

DNA Methylation

In mammalian DNA, 5-methylcytosine (^{m}C) is the only well-characterized modified nucleotide. Methylated cytosine residues are predominantly in the sequence 5' CpG3' (178, 179). In their analysis of the β-globin gene cluster, Van der Ploeg & Flavell concluded that there was an inverse correlation between DNA methylation and gene expression (180). Seventeen cleavage sites for restriction endonucleases, sensitive to ^{m}C residues, were used to demonstrate that the globin gene region was fully methylated in sperm DNA, that it exhibited a variable but relatively high degree of methylation in tissues not expressing globin, and that in general there was relative undermethylation of specific sites around globin genes in tissues in which these genes were expressed. Recently, the human ε- and γ-gene promoter regions were shown to exhibit a reciprocal relationship between methylation and expression of those genes in erythroid cells from very early embryos (181).

The human γ-globin gene has been methylated in vitro (182) and introduced into L-cells, and stable transformants were analyzed for γ-globin gene expression. Methylation in the promoter region inhibited transcription, but methylation of the coding parts of the gene had no effect (183). These and other data (184) suggest that methylation of specific residues may be relevant to gene expression whereas methylation of other sites may not be important.

Doubt about the significance of DNA methylation has arisen because certain genes may be actively transcribed despite extensive methylation (185, 186). Estrogen induction of the chicken vitellogenin leads to progressive demethylation of a HpaII site 600 bp upstream from the transcriptional start site but this decrease in methylation follows rather than precedes the peak of vitellogenin gene transcription (184). A general conclusion from these disparate observations is that there is no obligatory relationship between overall methylation and gene expression. The methylation status of specific sites within control regions of genes may be relevant, however, at least in some genes such as the globin genes. Although methylation may not be the primary causal factor in leading to gene expression, once specific sites become hypomethylated, the DNA conformation might favor interaction with specific regulatory proteins. Indeed, methylation appears to favor the B-form of DNA whereas non-B forms of DNA may occur in control regions of transcriptionally active genes (187).

Culture of cells in 5-azacytidine leads to incorporation of this compound into DNA and inhibition of methyltransferase activity. The result is global demethylation of DNA (188, 189). This is often associated with activation of specific genes (179) or assumption of a new phenotype by cells, e.g., induction of chrondroblast, myoblast, or adipocyte formation by mouse 3T3 cells (190). This property of 5-azacytidine formed the basis for the initial pharmacological attempts to increase HbF synthesis in vivo in experimental animals and man (discussed later) (191–194).

Activation of the human γ-globin gene has also been demonstrated in cultured MEL cells in vitro. Somatic cell hybrids formed by fusion of MEL cells and a human fibroblast line in which a portion of the X chromosome had been translocated to chromosome 11, allowed selective retention of the β-globin gene cluster on the translocated chromosome. Induction of these hybrid cells leads to expression of the β- but not the γ-globin genes as discussed earlier (195). When these cells were cultured for 48 hours in 5-azacytidine and then induced with another chemical, γ gene expression increased more than 20-fold over baseline (196). Demethylation of the ε gene was also observed, although its expression was minimal. Hypomethylation of specific genes may contribute to their expression but is insufficient to lead to full gene activation. Remethylation of transcriptionally inactive sequences was also demonstrated during the 48 hours of induction, although the transcriptionally active γ gene promoters remained hypomethylated.

Chromatin Structure

Nuclease sensitivity of specific DNA sequences within isolated nuclei has been useful for defining those features of chromatin structure associated with transcriptional activity (197). General sensitivity of rather large blocks of DNA has led to the concept of "domains." Within these large domains, transcribed genes have a greater sensitivity; this enhanced sensitivity has been associated with the binding of the nonhistone proteins, HMG14 and HMG17 (198–200). Another type of sensitivity exhibited by DNA sequences in the promoter region and occasionally in other areas of the gene is referred to as hypersensitivity (200a). These regions are thought to be relatively nucleosome-free as a consequence of their interaction with proteins involved in gene transcription. The chromatin structure of globin genes in primitive yolk sac-derived erythroblasts of developing chick embryos has been shown to have the features of transcriptionally active chromatin prior to the onset of gene transcription (201).

DNA sequences between 270 and 70 nucleotides upstream from the CAP site of the transcriptionally active chicken adult β^A globin gene are hypersensitive to nuclease digestion (202). Digestion of nuclear chromatin with MspI released a short DNA fragment from this region to which is bound a protein(s). An assay was devised to test for activity of a protein involved in forming the hypersensitive region (148). Nucleosomes were deposited on plasmid molecules containing the chicken β^A globin gene with extracts of Xenopus nuclei. In the presence of proteins from adult chicken erythroid nuclei, a hypersensitive region formed at the β^A gene promoter. Proteins from nuclei isolated from chicken brain or oviduct cells did not confer hypersensitivity to the histone/plasmid complex in the promoter region. Furthermore, extracts from the nuclei of 5-day-old chicken erythrocytes, which do not express the β^A globin, were unable to induce hypersensitivity. In a filter binding assay several proteins were found to bind specifically to DNA fragments that are adjacent to, and within the hypersensitive region. The 5' boundary of the DNA fragment was found to lie between −303 and −268 and the 3' boundary between −164 and −40 (148). This localization is similar to the hypersensitive region in erythrocyte chromatin extending from sequences −270 to −70 of the promoter.

Two studies report an increase in DNAse hypersensitivity in a small region near the 5' terminus of the β major globin gene following induction. Hexamethylene bis-acetamide (HMBA) induced a 6–10-fold increase in the primary hypersensitive site just upstream from the β major gene but did not alter the methylation status of the promoter region (203). Dimethyl sulfoxide (DMSO) mediated induction increased the DNAse I hypersensitivity at this site 2–4 fold (204). Following DMSO induction, changes in chromatin conformation could be detected within 12 h whereas β-globin mRNA were not detectable until after 18–20 h of induction. This indicates that changes in chromatin conformation of the globin genes precedes their transcription following chemical induction.

The human β-globin gene cluster in erythroid cells is generally more sensitive to DNAse I than it is in leukocytes. Hypersensitive sites were found within 200 bp 5' to the CAP sites of the γ, δ, and β genes in fetal liver cells whereas adult bone marrow cells only exhibited hypersensitive sites upstream from the δ and β genes (205). The β gene is expressed in fetal liver although at a much lower level than the γ genes.

The β-globin gene in the erythroleukemia cell line, K562, is structually normal and transcriptionally active when cloned and tested in HeLa cells (205a). Nonetheless, this gene is not expressed in intact cells. The ε, $^G\gamma$, $^A\gamma$, and ζ gene promoters exhibit hypersensitive sites in K562 cells but the β promoter lacks a hypersensitive region (206, 207). In contrast, chromatin in the human erythroleukemia cell line (HEL) has hypersensitive sites at the promoter regions of all five genes despite lack of β-gene expression in these cells also (205).

Trans-acting Factors

In contrast to cis-acting elements that influence activity of promoters on the same DNA strand, putative trans-acting factors are encoded by distant genes. One likely candidate for such a trans-acting factor is the DNA binding protein described above (148). This binding protein is found in nuclei of the definitive chick erythroid series where the β-globin gene is expressed but it is absent from nuclei of primitive erythroblasts or brain cells.

The E1A protein encoded by the early region of adenovirus substitutes for a cis-acting enhancer element when the human β-globin gene is assayed for expression 48 hours after introduction into cells by DNA transfection. Only 36 bp of DNA upstream from the CAP site are required for this action of the E1A protein as opposed to the 100 bp of flanking sequence required for promoter function when a cis-acting enhancer element is included in for the vector (208, 209).

As discussed in detail in an earlier section, results obtained by introduction of human globin genes into MEL cells by chromosome transfer or DNA transfection are compatible with the existence of regulatory factors that interact with these genes in a sequence specific manner. Several other observations provide circumstantial evidence for the operation of trans-acting factors. The structurally intact but transcriptionally inactive β globin gene in K562 cells can be activated during culture in 5-azacytidine or hydroxyurea at concentrations that do not affect the cell growth rate. As the β gene was substantially hypomethylated prior to treatment, demethylation does not appear to explain the results nor does cell selection as the growth rate was not affected (210). 5-azacytidine has also been found to activate the human ζ gene in K562 × MEL hybrids (211). The transcriptionally inactive β globin gene in HEL cells is activated in MEL × HEL hybrids, perhaps due to the action of trans-acting regulatory molecules (212). The ζ globin gene promoter functions after microinjection into oocytes but not after transfection into HeLa or Cos cells, suggesting that there might be transcriptional factors specific for embryonic genes (213).

CELL BIOLOGY OF HEMOGLOBIN SWITCHING

Release of enucleated red cells from the bone marrow or other erythroid organs reflects the culmination of cell differentiation, proliferation, and maturation that begins with a multipotential stem cell and ends with a fully hemoglobinized erythrocyte (Figure 4) (214, 215). "Early" and "late" progenitor cells with significant but declining proliferative potential are classified as Burst Forming Units-Erythroid (BFU-E) or Colony Forming Units-Erythroid (CFU-E), respectively, based on the size and morphology of the erythroblast colonies formed in semi-solid medium in vitro. Hemoglobin accumulation begins within the precursor compartment at approximately the pro-erythroblast stage (216).

Erythroid progenitors from the fetal developmental stage form erythroid colonies containing predominantly HbF whereas progenitor from the adult bone marrow form colonies containing predominantly HbA and also significant amounts of HbF. Progenitors present in fetal cord blood or hematopoietic tissues during the perinatal period form colonies containing HbF or HbA in varying proportions (24, 217). Thus, control appears to take the form of a "developmental clock" (24, 28) although the hemoglobin content of colonies derived from adult progenitors may be modulated by exogenous factors (218, 219). Significant amounts of HbF may be found in colonies derived from individual adult progenitors. While HbA is found in all erythroblasts, HbF is found in only a fraction and these are clustered within sectors of the colony

MODULATION OF HbF SYNTHESIS DURING ERYTHROPOIESIS

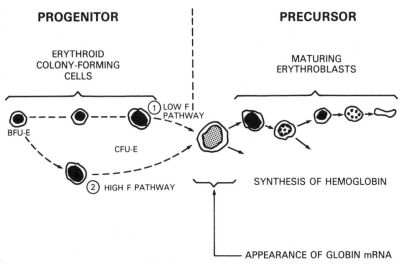

Figure 4 A scheme showing the differentiation pathway from erythroid progenitor cells, the burst forming units-erythroid (BFU-E) and the colony forming units-erythroid (CFU-E) to erythroblast precursors and ending in mature red cells. See text for discussion of the two pathways in the progenitor compartment that may result in red cells containing different amounts of HbF.

(220). Thus a single progenitor has the capability of giving rise to erythroblasts containing either HbA or HbA + HbF and the regulatory event apparently occurs during erythroid progenitor differentiation (221).

The amount of HbF was reported as being greater in colonies derived from earlier progenitors compared to those derived from late progenitors (220) although not all studies support these early observations (222–224). The property of HbF production appears to be acquired with a certain probability during each cell division in the early stages of colony formation (221). A "switching factor" present in fetal sheep serum reduces HbF synthesis, compared to that observed with fetal calf serum, in colonies formed by adult erythroid progenitors (218). Of interest is that progenitors obtained from individuals in whom the β-globin gene cluster is disrupted by several types of deletion mutations that increase HbF synthesis in vivo (see below), nonetheless respond to this "switching factor" (227a). Enhanced production of HbF is observed during bone marrow regeneration following transplantation (225) or recovery from cytotoxic drugs (226). This has been attributed to recruitment of early progenitors that have a higher intrinsic capacity for forming erythroblasts making HbF (Figure 4) (220, 227).

Asynchronous synthesis of HbF and HbA during erythroblast maturation has been demonstrated in clonal erythroid cultures (228), in adult bone marrow (229), and during the perinatal switch in experimental animals (230). Most γ globin is synthesized in pro- and basophilic erythroblasts whereas β-globin synthesis may begin slightly later. Peripheral blood reticulocytes do make both globins, however. This "hemoglobin switch" within the erythroblast compartment recapitulates the perinatal switch from HbF to HbA although its mechanism is unknown. One might suppose that the chromatin structure of the β cluster in early adult erythroblasts resembles that found in fetal erythroblasts (205) and that as cellular maturation proceeds, the hypersensitive site 5' to the genes are closed. This model suggests that regulation is primarily by suppression of γ gene expression.

HbF PRODUCTION IN ADULTS

Genetic Control of HbF Synthesis in Normal Adults

Small amounts of HbF is found in the red cells of normal adults. Most individuals have less than 1% HbF although the range is 0.3–1.2%. Only 0.2–7% of red cells contain HbF; these are called F-cells. When HbF is less than 5%, the concentration of HbF in F-cells is constant and the relative amount of HbF is directly proportional to the number of F-cells (7). Serial studies have shown that the F-cell level remains constant and reproducible in adults (231–233). Strong genetic component(s) determine the relative number of F-cells in normal adults as shown by family studies (234).

Production of HbF in Hematological Diseases

The pathophysiology of severe β thalassemia reflects the imbalance in α and non-α globin synthesis. Deficient or absent production of β chains results in a marked relative excess of α globin, which forms highly insoluble aggregates that precipitate in the red cells (4–7). Individuals with severe β thalassemia have an increased amount of HbF in their red cells compared to normal individuals. Gamma globin is synthesized in a small proportion of red cells in both thalassemic and normal individuals. In patients with thalassemia, those erythroblasts that synthesize even small amounts of γ globin have less severe chain imbalance and therefore possess a selective advantage favoring their release from the bone marrow and survival in the circulation. By this mechanism alone, it has been estimated that the intrinsic capacity for globin synthesis (usually less than 1% of β) may be amplified 30–40 fold (7). Among 331 β thalassemia heterozygotes the HbF distribution was continuous between 0% and 4%. A few individuals had values of more than 5%. These values are on the average three times higher than those of normal adults (7).

The pathophysiology of sickle cell anemia is based on the propensity of HbS

:o form structured polymers upon deoxygenation (235–237). HbF inhibits polymerization by reducing the concentration of HbF and by forming mixed :etramers ($\alpha_2\beta^S\gamma$) that do not participate in, and may inhibit, the polymerization reaction. The HbF concentration is elevated in most patients with sickle :ell anemia compared to controls (238, 239). HbF may influence the severity of sickle cell anemia, although the levels of HbF found in most patients appear not :o have a profound effect on their clinical course (240). Individuals with sickle :ell anemia in Saudi Arabia have HbF levels of 30–40% (241, 242); recent studies suggest that these high levels have a genetic basis (see below).

Interaction of the β^S or β-thalassemia point mutations with various deletion and nondeletion mutations that increase HbF is evident in many well-characterized pedigrees. The genetic propensity for HbF synthesis that is conferred by these determinants has a modulating effect on disease severity that is roughly proportional to the level of HbF production achieved. The complexity of these interactions reflects the potential genetic heterogeneity in the several determinants that may define the level of HbF synthesis in normal individuals and those with a genetic propensity to increased HbF production.

Mutations that Increase HbF Synthesis

There are two general classes of mutations that increase HbF synthesis, those that reflect deletions within the β-globin gene cluster and those in which the β-globin gene cluster remains structurally intact. These are referred to as 'deletion' and 'nondeletion' mutations respectively (243).

DELETION MUTATIONS   These mutations may be further classified with respect to the pattern of distribution of HbF in heterozygotes. A heterocellular (uneven) distribution of HbF among red cells is characteristic of those mutations that also result in red cell abnormalities characteristic of thalassemia. Microcytosis and decreased red cell content of hemoglobin is also characteristic of these deletions reflecting an overall decrease in non-α globin synthesis. As summarized in Table 1, the amount of HbF found in heterozygotes with these mutations is quite variable but usually less than 20%. Generally, between 10 and 50% of red cells of individuals with these deletions contain HbF.

Those deletions that are found in individuals with morphologically normal red cells are classified as Hereditary Persistence of Fetal Hemoglobin (HPFH) rather than thalassemia mutations. The red cells of heterozygous individuals with these deletions contain 25–30% of HbF in a pancellular distribution. There is minimal or no significant imbalance in α and non-α globin synthesis. Homozygotes for HPFH mutations do have mild microcytosis and a slight decrease in red cell hemoglobin content. Thus the distinction between HPFH and the thalassemia-like deletion mutations is relative and not absolute.

One of the main reasons for analyzing these deletion mutations in detail is an

Table 1 Deletions in the β-gene cluster

Structural Coordinates diagram with markers: ε (0), Gγ Aγ (10-20), ψβ1 (30), δ (40), β (50), (60 Kb)

	ETHNIC GROUP	DELETION SIZE	DELETION COORDINATES	HbF LEVEL IN HETERO-ZYGOTES	OTHER INFORMATION	REFERENCE
A. β - Thalassemias (heterocellular)						
1.	Indian	619 bp	48.4 → 49 5' = β IVS 2 3' = 209 bp into 3' flanking region	Unavailable	β°-thalassemia	244-248
2.	Black (USA)	1.3 kb	46.7 → 48	Unavailable	β°-thalassemia	249
3.	Dutch	10 kb	46 → 56	4-11%	β°-thalassemia	250
4.	Greek Italian Iranian Yugoslavian Papuan	7 kb	41 → 48	1-5%	Hb Lepore δ-β fusion	251-253
B. (δβ)° - Thalassemias (heterocellular)						
1.	Mediteranean	6.5 kb	33.5 → 40	0.5-5%		254
2.	Sicilian	10 kb	40.5 → 50.5	5-12%		48, 255, 256
3.	Spanish	> 70 kb	36 → Unknown	5-12%		257
4.	Black (USA)	12 kb	37 → 49	25%*	Pancellular in a sickle cell anemia patient	258
5.	Japanese	> 25 kb	26.5 → Unknown	5-7%		259
C. (Aγδβ)° - Thalassemias (heterocellular)						
1.	Indian Iranian Kuwaiti	8.5 kb deleted 15.5 kb are inverted	24 → 25 40.5 → 48 25 → 40.5 is inverted	10-15%		260, 265
2.	Black (USA)	34 kb	26 → 70	8-14%		261
3.	Turkish Black (USA)	34 kb	21.5 → 55.5	10-13%		D. Mager[a]
4.	Turkish	40 kb	22 → 62	10-15%		48, 245, 262
5.	Malaysian	Unknown	22 → Unknown	Not Known		263
6.	Chinese	Unknown	25 → Unknown	10-15%		264
7.	German	> 60 kb	23 → Unknown	10-15%		265
D. (γδβ)° - Thalassemias (heterocellular)						
1.	English	Large	Unknown → 19 5' site upstream from ε	0%		266
2.	Dutch	> 60 kb	Unknown → 44	0%		267
3.	Anglo-Saxon	Large	Unknown → 47.5	0%		268, 269
4.	Mexican	> 105 kb	Unknown → Unknown the whole β cluster is deleted	0%		270
5.	Scotch-Irish	> 150 kb	Unknown → Unknown	0%		271
E. Hereditary Persistence of Fetal Hemoglobin (pancellular)						
1.	Black (U.S.A.)	> 70 kb	36.5 → Unknown	25-30%	HPFH-1	48, 262, 272, 273, 274
2.	Black (Ghana)	> 70 kb	32 → Unknown	25-30%	HPFH-2	262, 272
3.	Italian	15-37 kb	35.5 → Unknown	17-30%		G. Saglio[a]
4.	Indian	> 30 kb	27 → Unknown			P. Henthorn[a] T.H.A. Huisman, O. Smithies
5.	Kenyan	22.5 kb	24.5 → 47	6-7%	Hb Kenya Aγ-β fusion	275, 276

[a]Personal Communication

effort to understand how specific mutations influence expression of the γ-globin genes. Inspection of Table 1 will show that in general the level of HbF synthesis in heterozygotes increases with the size of the deletion. A sufficient number of exceptions exist, however to make this correlation less than perfect. For example the Spanish $\delta\beta$ thalassemia deletion (B3) is greater than 70 kb in length and a recently described $(^A\gamma\delta\beta)^0$ thalassemia deletion is also large and similar in size to two deletions that produce the HPFH phenotype (277, 278).

A decade ago Huisman and coworkers proposed that a sequence between the $^A\gamma$ and δ genes acted as a cis-acting negative regulatory locus for the γ-globin genes (279). The clearest evidence that this might be so comes from comparison of the Lepore and Kenya mutations. Both result from unequal crossover events that generate fusion genes, either the δ-β gene with the Lepore mutation or the $^A\gamma$-β gene with the Kenya mutation (Table 1). The Lepore deletion is associated with a very small increase in HbF. This is heterogeneously distributed among the red cells of heterozygous individuals. In contrast, although the Kenya deletion results in increase in HbF of similar magnitude from the remaining intact $^G\gamma$ locus, this HbF has a pancellular distribution.

The discovery of the Alu sequences upstream from the δ gene focused attention on them as potential negative regulatory elements for γ gene expression in adult life. Indeed, the Alu sequence closest to the δ globin gene functions as a transcriptional template for RNA polymerase III in vitro and sequences immediately downstream from this element are transcribed in erythroid cells (80). There is no experimental evidence for this extension of Huisman's original theory and recently direct evidence against it was obtained. A Mediterranean individual with 100% HbF was found to be homozygous for a 6.5-kb deletion that starts 3.5 kb downstream from the $\psi\beta$ gene and extends to the 3' end of the δ gene, effectively deleting the two Alu repeats (B1, Table 1). Ten heterozygotes in his family have HbF values between 0.5 and 5% or similar values to those of β-thalassemia heterozygotes. The red cell morphology is also characteristic of thalassemia trait. The β gene is intact in this individual but not expressed. It has not been sequenced to rule out a second mutation in the β gene but the deletion mutation is currently classified as a $\delta\beta$ thalassemia mutation. Similarly a recently described Japanese $\delta\beta$ thalassemia mutation (B5, Table 1) deletes both Alu repeats upstream from the δ gene, and preserves both γ genes but does not cause an HPFH phenotype. Two deletion mutations (B3 and E3, Table 1) have similar 5' endpoints and conserve the distal Alu repeat upstream from the δ gene. The proximal Alu is deleted (E3) or partially deleted (B3) in these mutations. Despite this similarity these mutations have different phenotypes, B3 giving rise to a $\delta\beta$-thalassemia and E3 causing an HPFH phenotype. These results collectively indicate that the loss of a region bearing the two Alu repeats upstream from the δ gene is not sufficient to achieve a high level of expression of the γ genes.

Perhaps the DNA that is brought into the vicinity of the γ-globin genes by the HPFH deletions contains sequences that increase γ-gene expression (262). A 0.9-kb fragment from the very end of the 3′ breakpoint of the HPFH-1 mutation was found to increase CAT expression in plasmid vectors containing the SV40 early promoter linked to the CAT gene. This fragment could only increase expression when situated upstream from the promoter and unlike viral enhancers (see above) could only act in one orientation (280). Recently Mager & Henthorn (278) identified a new 5–6 kb retrovirus-like repetitive element in human DNA, the hs VL-H family. In the Chinese $(^A\gamma\delta\beta)^0$ thalassemia deletion an hs VL-H element is found juxtaposed to the γ genes whereas this element has been deleted from the HPFH deletions. The authors propose that this retrovirus-like element reduces the expression of the γ-globin gene in the $(^A\gamma\delta\beta)^0$ thalassemias.

Another very important general feature that these mutations reveals about function of the β-globin gene cluster is its polarity. In both the α and β clusters, the individual genes are expressed during development in the same order that they appear on the chromosome (from left to right as commonly written) (Figure 1). That this polarity has an important functional basis is suggested by the differing effects of deletions from the left and right side of the β cluster. The $(\delta\beta)^0$, $(A\gamma\delta\beta)^0$, and HPFH mutations are all characterized by increased expression of the remaining intact globin gene(s). In contrast, two $(\gamma\delta\beta)^0$ thalassemia mutations (Table 1, D1 and D2) remove the left portion of the cluster along with large amounts of flanking sequence but leave either the $^A\gamma$, δ, and β genes (D1) or only the β gene (D2) intact. None of these remaining genes appear to be expressed.

The most thoroughly studied is the Dutch (γδβ) mutation (D2) in which the β gene and 3 kb of flanking sequences are preserved. This gene has been cloned and found to have a normal DNA sequence. Furthermore, the gene functions normally in a heterologous transient expression assay despite its lack of function in vivo (281). The features of the chromatin structure of this gene are those of an inactive locus but the β-globin gene on the normal chromosome has the structural features of a transcriptionally active gene.

NONDELETION MUTATIONS   There are several distinct mutations or genetic traits that are characterized by increased production of HbF in clinically normal adults with intact β globin gene clusters (Table 2). This list in Table 2 suggests considerable genetic heterogenity as judged by the variable level of HbF, variable $^G\gamma/^A\gamma$ ratios, and whether each syndrome is characterized by a pancellular or a heterocellular HbF distribution in the red cells. Only 4 of the 10 nondeletion HPFH syndromes have been characterized at the molecular level. They were all found to have point mutations within the promoter region of a γ gene that shows increased expression (three of them were discussed above).

Table 2 Nondeletion HPFH Syndromes

Reference	Type	HbF in heterozygotes	$^G\gamma/^A\delta$ ratio	Distribution of HbF	Molecular defect
1. 146	Negroγ β^+	15–20%	$^G\gamma$ only	Pancellular	C → G at −202
2. 147	Italianγ β^+	16%	95%$^A\gamma$, 5%$^G\gamma$	Pancellular	C → T at −196
3. 145	Greekγ β^+	10–20%	90%$^A\gamma$, 10%$^G\gamma$	Pancellular	C → A at −117
4. 282	Chinese	10–14%	predominantly$^A\gamma$	Pancellular?[a]	—
5. 283	German Associated with a translocation involving chromosome 11	5–8%	60%^{A} γ, 40%^{G} γ	Heterocellular	—
6. 284, 285	British	4–12%	90%^{A} γ, 10%^{G} γ	Heterocellular	—
7. 285, 286	Georgia	2.6–6%	Mainly^G γ but ^{A} γ also detected	Heterocellular	—
8. 287	Swiss	0.8–3.4%	both ^{A} γ and ^{G} γ ratio variable in different families	Heterocellular	—
9. 288, 289	Atlanta	2.3–3.8%	^{G} γ only	Heterocellular	C → T at −158
10. 290	Seattle	3.7–7.8%	equal porportions of ^{G} γ and ^{A} γ	Heterocellular	—

[a]Heterozygotes contain 70–80% F cells.

The remaining six HPFH syndromes were reported in unrelated families and are probably distinct mutations but need not be.

The ratio of $^G\gamma$ to $^A\gamma$ chains in the newborn (7/3) changes with age to become about 2/3 in the adult (291). The cause of this shift in ratio is unknown. Patients with β-thalassemia and sickle cell anemia generally have 40% $^G\gamma$ but some patients have a higher value or 60–70% $^G\gamma$ (292, 293). Some of these high $^G\gamma$ individuals have now been shown to have a C→T mutation at position −158. The relative $^G\gamma$ value is 54–66% in heterozygotes and 65–74% in homozygotes for this mutation. The Atlanta nondeletion HPFH (no. 9, Table 2) has also been found to have this −158 mutation upstream from both γ gene loci on the same chromosome (294). In fact both loci produce $^G\gamma$ chains so the chromosome therefore has two $^G\gamma$ genes, a 5' $^G\gamma$ and a 3' one, both having a C→T mutation 158 bp upstream from their Cap site. Chromosomes with two $^G\gamma$ genes ($^G\gamma$-$^G\gamma$) or two $^A\gamma$ genes ($^A\gamma$-$^A\gamma$) have been described recently in individuals with normal HbF (295). It was proposed that these unusual γ gene arrangements arose by a point mutation or an intrachromosomal gene conversion event involving only a short segment of DNA.

Recently an interesting form of nondeletion HPFH has been inferred in sickle cell anemia patients from Saudi Arabia. Only the patients but not their parents or normal siblings have a raised HbF (241, 242). When bone marrow cells from the patients' parents have been cultured in vitro increased production of HbF compared to controls has been detected in at least one of the parents of each patient (296). This indicates that there is a genetic predisposition towards increased HbF production during hematopoietic stress in these families which is not detectable when the individual is not anemic.

Several linkage studies have been attempted to genetically map the various "HPFH determinants" other than those that reflect point mutations. Only one of them provides convincing data however. This is the study of a British nondeletion HPFH which showed linkage between the "HPFH determinant" and a particular polymorphic restriction endonuclease site within the β globin gene cluster (297). In this family the HPFH individuals have high HbF values (the range for the British type is 4–12%) and are easily distinguished from normal individuals. The family studied was a three generation family without any accompanying conditions like a β thalassemia trait that increase HbF indirectly. Furthermore there is evidence that two individuals in the family are homozygotes for the mutation.

Three recent studies have analyzed the linkage between the Swiss type "HPFH determinant" and polymorphic restriction sites in the β-globin cluster (297–299). The Swiss type of HPFH is characterized by a very slight increase in HbF distributed in a heterocellular fashion. Two studies indicate linkage but the third showed no linkage between the β-globin cluster and the HPFH determinant. As the families studied were not related (four in one paper and one n each of the other two) they need not have the same genetic defect although hey were all classified as the Swiss type of HPFH. The major problem here owever is that the Swiss type has relatively low values of HbF (0.8–3.4%) and herefore there is overlap between high normal HbF values and values taken to ndicate the presence of the HPFH determinant. Secondly, all the families analyzed have individuals with β-thalassemia trait, a condition which can raise he HbF level. Classification of individuals as normal or HPFH therefore presents a problem. The conflicting reports of linked or nonlinked Swiss type HPFH might therefore reflect this classification difficulty.

PHARMACOLOGICAL MANIPULATION OF HbF SYNTHESIS

Stimulation of HbF synthesis in patients with sickle cell anemia and β thalassemia is a highly desirable therapeutic goal. An increase in HbF production could modulate disease severity whereas a complete reversion to HbF synthesis might eliminate all disease manifestations. The initial efforts to achieve this therapeutic goal with drugs followed studies in animals by DeSimone, Heller, and their colleagues (191). Their use of 5-azacytidine was based on the fact that this agent induced hypomethylation in tissue culture cells and activated specific genes. Administration of 5-azacytidine to anemic baboons resulted in a very marked increase in HbF synthesis and in some animals there was complete reversion to synthesis of HbF (191).

Clinical Effects of 5-Azacytidine

The animal experiments provided the rationale for testing 5-azacytidine clinically, although the physiology of HbF production in human adults is quite different from that in baboons (7,300). 5-Azacytidine was shown to produce small but reproducible increases in γ-globin gene expression in humans with severe β thalassemia or sickle cell anemia (192–194, 301). The increase in γ-globin gene expression was associated with a significant reduction in methylation frequency of several residues within the β-globin gene cluster including the CpG that is within the γ-globin gene promoter region. The increase in HbF synthesis ranged 4–7-fold over the baseline. In patients with sickle cell amemia the maximum γ-globin synthesis achieved was 15–20% of β globin production (302). Limited experience with cyclic administration of 5-azacytidine over two months in two patients with severe β thalassemia failed to eliminate the transfusion requirement (303) but prolonged administration of the compound by parental and oral routes in three patients with sickle cell anemia led to an apparent reduction in crises frequency and correction of anemia (302).

Mechanism of Action of 5-Azacytidine

Surveys of several genes in the bone marrow DNA of patients receiving 5-azacytidine, revealed that the hypomethylating effects were rather general (192, 304). For example, the ϵ-globin gene was undermethylated to the same degree as the γ gene following 5-azacytidine administration. Nonetheless, no more than 1–2 copies/cell of ϵ mRNA were detected compared to the several hundred copies of γ mRNA that appeared on drug treatment. This apparent specific effect was difficult to reconcile with a simple hypothesis that demethylation activated specific genes. These facts prompted an immediate consideration of alternate mechanisms by which 5-azacytidine might increase HbF synthesis (192). The drug is known to kill cells; indeed, this cytotoxic property makes it useful for treatment for leukemia. The fraction of late erythroid progenitors (CFU-E) that are engaged in DNA synthesis and therefore susceptible to 5-azacytidine is much higher than for the early erythroid progenitors (BFU-E) (214, 215). Hence, the drug might be more cytotoxic to late progenitors allowing repopulation of the bone marrow by erythroblasts derived from early progenitors with a higher propensity to HbF synthesis (305).

Two experimental strategies were devised to investigate the mechanism by which 5-azacytidine affected γ gene expression. The first involved the use of mouse erythroleukemia cells into which human chromosome 11 had been introduced by cell fusion (196). As described above exposure of such cells to 5-azacytidine resulted in selective activation of the quiescent γ-globin gene despite a global reduction in DNA methylation. The second strategy designed to define the mechanism of 5-azacytidine action involved careful analysis of its effect on erythroid progenitors and precursors (301–305). In experimental animals, doses higher than those employed clinically were shown to result in a reduction in colonies formed by late progenitors whereas the effects on early progenitor colony formation were much less striking (305). These perturbations in the erythroid progenitor pool were interpreted as supporting the hypothesis that 5-azacytidine worked primarily by a cytotoxic effect on frequently cycling late progenitor cells, followed by their regeneration from an earlier progenitor pool (Figure 4). In humans given low doses, 5-azacytidine increased F-reticulocyte production and increased the number and HbF content of CFU-E derived colonies without significant cytotoxicity (301). These results seemed more consistent with an inductive rather than a cytotoxic effect on erythroid progenitors and precursors (Figure 4).

The Effect of Other Drugs on HbF Synthesis

Arabinosylcytosine and hydroxurea, S-phase specific cytoxic drugs, have been shown to induce HbF synthesis in experimental animals and man (306–308). Some studies suggest that the amount of HbF produced in response to these agents is similar to that observed with 5-azacytidine (306) whereas other studies have consistently shown that hydroxurea and arabinosylcytosine are less potent than the demethylating agent (T. Ley, A. Nienhuis, unpublished observations). Both drugs, in the doses employed, decrease reticulocyte production and perturb erythroid progenitor kinetics in a manner analogous to that seen with high dose 5-azacytidine (306–308). A reduction in methylation frequency of a CpG residue in the γ-globin gene promoter has been shown with hydroxurea, although the overall methylation of DNA is unchanged (308). These studies certainly show that the direct hypomethylating effect of 5-azacytidine is not essential to achieve an increase in HbF synthesis. In support of the interpretation that specific drugs may act by different mechanisms is the observation that 5-azacytidine but not hydroxurea, produces a significant increase in γ-globin synthesis in the bone marrow of patients with β thalassemia (303). Further studies are required to evaluate the mechanism by which the S-phase drugs act and to evaluate their potential clinical usefulness.

ACKNOWLEDGMENTS

We are indebted to our colleagues for allowing us to quote their unpublished data. We would also like to thank Drs. N. Anagnou, T. Ley, and T. Rutherford for critical reading of the manuscript and Mrs. E. Murray for excellent editorial assistance.

Literature Cited

1. Maniatis, T., Fritsch, E. F., Lauer, J., Lawn, R. M. 1980. *Ann. Rev. Genet.* 14:145–78
2. Collins, F. S., Weissman, S. M. 1984. *Prog. Nucleic Acid Res. Mol. Biol.* 31:315–438
3. Higgs, D. R., Weatherall, D. J. 1983. *Curr. Top. Hematol.* 4:37–97
4. Nienhuis, A. W., Anagnou, N. P., Ley, T. J. 1984. *Blood* 63:738–58
5. Orkin, S. M., Kazazian, H. H. Jr. 1984. *Ann. Rev. Genet.* 18:131–71
6. Spritz, R. A., Forget, B. G. 1983. *Am. J. Hum. Genet.* 35:333–61
7. Weatherall, D. J., Clegg, J. B. 1981. *The Thalassemia Syndromes*. Boston: Blackwell
8. Ingram, V. M. 1972. *Nature* 235:338–39
9. Ingram, V. M. 1981. In *Hemoglobins in Development and Differentiation*, ed. G. Stamatoyannopoulos, A. W. Nienhuis, pp. 147–60. New York: Liss
10. Tobin, A. J., Chapman, B. S., Hansen, D. A., Lasky, L., Selvig, S. E. 1979. In *Cellular and Molecular Regulation of Hemoglobin Switching*, ed. G. Stamatoyannopoulos, A. W. Nienhuis, pp. 205–12. New York: Grune & Stratton
11. Beaupain, D., Martin, C., DieterleuLievre, F. 1981. See Ref. 9, pp. 161–69
12. Boussios, T., Bertles, J. F., Clegg, J. B. 1982. *Science* 218:1225–27
13. Peschle, C., Migliaccio, A. R., Migliaccio, G., Russo, G., Mastroberardino, G., et al. 1983. In *Globin Gene Expression and Hemopoietic Differentiation*, ed. G. Stamatoyannopoulos, A. W. Nienhuis, pp. 411–19, New York: Liss
14. Peschle, C., Migliaccio, A. R., Migliaccio, G., Petrini, M., Caladrini, M. 1984. *Proc. Natl. Acad. Sci. USA* 81:2416–20
15. Peschle, C., Mavilio, F., Carè, A., Migliaccio, G., Migliaccio, A. R. et al. 1985. *Nature* 313:235–38
16. Peschle, C., Mavilio, F., Migliaccio, G., Migliaccio, A. R., Russo, G. et al. 1985. In *Experimental Approaches for Study of Hemaglobin Switching*, eds. G. Stamatoyannopoulos, A. W. Nienhuis. In press
17. Pataryas, H. A., Stamatoyannopoulos, G. 1972. *Blood* 39:688–96
18. Kazazian, H. H. Jr., Woodhead, A. P. 1973. *N. Engl. J. Med.* 289:58–62
19. Kazazian, H. H. Jr., Woodhead, A. P. 1974. *Ann. NY Acad. Sci.* 241:691–98
20. Wood, W. G., Nash, J., Weatherall, D. J., Robinson, J. S., Harrison, F. A. 1979. In *Cellular and Molecular Regulation of Hemoglobin Switching*, ed. G. Sta-

matoyannopoulos, A. W. Nienhuis, pp. 153–67. New York: Grune & Stratton
21. Papayannopoulou, T., Shepard, T. H., Stamatoyannopoulos, G. 1983. See Ref. 13, pp. 421–30
22. Huehns, E. R., Dance, N., Beaven, G. H., Hecht, F., Motulsky, A. G. 1964. *Cold Spring Harbor Symp. Quant. Biol.* 29:327–31
23. Bard, H. D., Makowski, E. L., Meschia, G., Battaglia, F. C. 1970. *Pediatrics* 45:766–72
24. Stamatoyannopoulos, G., Papayannopoulou, T., Brice, M., Jurachi, S., Nakamoto, B., et al. 1981. See Ref. 9, pp. 287–305
25. Weinberg, R. S., Goldberg, J. P., Schofield, J. M., Lenes, A. L., Styczynski, R., et al. 1983. *J. Clin. Invest.* 71:785–94
26. Wood, W. G., Clegg, J. B., Weatherall, D. J. 1977. In *Progress in Hematology,* ed. E. B. Brown, 10:43–90. New York: Grune & Stratton
27. Zanjani, E. D., McGlave, P. B., Bhakthavathsalan, A., Stamatoyannopoulos, G. 1979. *Nature* 280:495–96
28. Wood, W. G., Bunch, C. 1983. See Ref. 13, pp. 511–21
29. Wood, W. G., Bunch, C., Kelly, S., Gunn, Y., Breckon, G. 1985. *Nature.* 313:320–23
30. Delfini, C., Saglio, G., Mazza, U., Muretto, P., Filippetti, A., Lucarelli, G. 1983. *Br. J. Haematol.* 55:609–14
31. Papayannopoulou, T., Nakamoto, B., Agostinelli, F., Lucarelli, G., Stamatoyannopoulos, G. 1984. *Blood* 64:63a
32. Deisseroth, A., Nienhuis, A., Turner, P., Velez, R., Anderson, W. F., et al. 1977. *Cell* 12:205–18
33. Deisseroth, A., Nienhuis, A., Lawrence, J., Giles, R., Turner, P., Ruddle, F. H. 1978. *Proc. Natl. Acad. Sci. USA* 75:1456–60
34. Deisseroth, A., Hendrick, D. 1978. *Cell* 15:55–63
35. Barton, P., Malcolm, S., Murphy, C., Ferguson-Smith, M. A. 1982. *J. Mol. Biol.* 156:269–78
36. Gusella, J., Varsanyi-Breiner, A., Kao, F. T., Jones, C., Puck, T. T., et al. 1979. *Proc. Natl. Acad. Sci. USA* 76:5239–43
37. de Martinville, B., Francke, U. 1983. *Nature* 305:641–43
38. Huerre, C., Despoisse, S., Gilgenkrantz, S., Lenoir, G. M., Junien, C. 1983. *Nature* 305:638–41
39. Morton, C. C., Kirsch, I. R., Taub, R. A., Orkin, S. M., Brown, J. A. 1984. *Am. J. Hum. Genet.* 36:576–85
40. Harper, M. E., Ullrich, A., Saunders, G.

A. 1981. *Proc. Natl. Acad. Sci. USA* 78:4458–60
41. de Martinville, B., Giacalone, J., Shih, C., Weinberg, R., Francke, U. 1983. *Science* 219:638–41
42. Gerhard, D. S., Kidd, K. K., Housman, D., Gusella, J. F., Kidd, J. R. 1984. *Cytogenet. Cell Genet.* 37:478
43. Brissenden, J. E., Ullrich, A., Francke, U. 1984. *Nature* 310:781–84
44. Tricoli, J. V., Rall, L. B., Scott, J., Bell, G. I., Shows, T. B. 1984. *Nature* 310:784–86
45. Antonarakis, S. E., Phillips, J. A. III, Mallonee, R. L., Kazazian, H. H. Jr., Fearon, E. R., et al. 1983. *Proc. Natl. Acad. Sci. USA* 80:6615–19
46. Baralle, F. E., Shoulders, C. C., Proudfoot, N. J. 1980. *Cell* 21:621–26
47. Efstradiatis, A., Posakony, J. W., Maniatis, T., Lawn, R. M., O'Connell, C., et al. 1980. *Cell* 21:653–68
48. Fritsch, E. F., Lawn, R. M., Maniatis, T. 1979. *Nature* 279:598–603
49. Fritsch, E. F., Lawn, R. M., Maniatis, T. 1980. *Cell* 19:959–72
50. Kaufman, R. E., Kretschmer, P. J., Adams, J. W., Coon, H. C., Anderson, W. F., et al. 1980. *Proc. Natl. Acad. Sci. USA* 77:4229–33
51. Poncz, M., Schwartz, E., Ballantine, M., Surrey, S. 1983. *J. Biol. Chem.* 258:11599–609
52. Slightom, J. L., Blechl, A. E., Smithies, O. 1980. *Cell* 21:627–38
53. Shen, S. H., Smithies, O. 1982. *Nucleic Acids Res.* 10:7809–18.
54. Lauer, J., Shen, C.-K. J., Maniatis, T. 1980. *Cell* 20:119–30
55. Liebhaber, S. A., Goossens, M., Kan, Y. W. 1981. *Nature* 290:26–29
56. Proudfoot, N. J., Gil, A., Maniatis, T. 1982. *Cell* 31:553–63
57. Proudfoot, N. J., Maniatis, T. 1980. *Cell* 21:537–44
58. Orkin, S. M., Goff, S. C. 1981. *Cell* 24:345–51
59. Liebhaber, S. A., Kan, Y. W. 1982. *J. Biol. Chem.* 257:11852–55
60. Davidson, E. M., Posakony, J. W. 1982. *Nature* 297:633–35
61. Schmid, C. W., Jelinek, W. R. 1982. *Science* 216:1065–70
62. Singer, M. F., Skowronski, J. 1985. *Trends Biochem. Sci.* 10:119–21
63. Houck, C. M., Rinehart, F. P., Schmid, C. W. 1979. *J. Mol. Biol.* 132:289–306
64. Rubin, C. M., Houck, C. M., Deininger, P. L., Friedman, T., Schmid, C. W. 1980. *Nature* 284:372–75
65. Deininger, P. L., Jolly, D. J., Rubin, C.

M., Friedman, T., Schmid, C. W. 1981. *J. Mol. Biol.* 151:17–33

66. Bell, G. I., Pictet, R., Rutter, W. J. 1980. *Nucleic Acids Res.* 8:4091–109

67. Haynes, S. R., Toomey, T. P., Leinwand, L., Jelinek, W. R. 1981. *Mol. Cell. Biol.* 1:573–83

68. Pan, J., Elder, J. T., Duncan, C., Weissman, S. M. 1981. *Nucleic Acids Res.* 9:1151–70

69. Baralle, F. E., Shoulders, C. C., Goodbourne, S., Jeffreys, A., Proudfoot, N. J. 1980. *Nucleic Acids Res.* 8:4393–404

70. Duncan, C. H., Jagadeeswaran, P., Wang, R. R. C., Weissman, S. M. 1981. *Gene* 13:185–96

71. Jelinek, W. R., Toomey, T. P., Leinwand, L., Duncan, C. H., Choudary, P. V., et al. 1980. *Proc. Natl. Acad. Sci. USA* 77:1398–402

72. Fritsch, E. F., Shen, C. K. J., Lawn, R. M., Maniatis, T. 1981. *Cold Spring Harbor Symp. Quant. Biol.* 45:761–75

73. Elder, J. T., Pan, J., Duncan, C. H., Weissman, S. M. 1981. *Nucleic Acids Res.* 9:1171–89

74. Hess, J. F., Fox, M., Schmid, C., Shen, C.-K. J. 1983. *Proc. Natl. Acad. Sci. USA* 80:5970–74

75. Coggins, L. W., Grindlay, G. J., Vass, J. K., Slater, A. A., Montague, P., et al. 1980. *Nucleic Acids Res.* 8:3319–34

76. Duncan, C. H., Biro, P. A., Choudary, P. V., Elder, J. T., Wang, R. R. C., et al. 1979. *Proc. Natl. Acad. Sci. USA* 76:5095–99

77. Manley, J. L., Colozzo, M. T. 1982. *Nature* 300:376–79

78. Allan, M., Lanyon, W. G., Paul, J. 1983. *Cell* 35:187–97

79. Allan, M., Paul, J. 1984. *Nucleic Acids Res.* 12:1193–99

80. Zavodny, P. J., Roginski, R. S., Skoultchi, A. I. See Ref. 13, pp. 53–62

81. Adams, J. W., Kaufman, R. E., Kretschmer, P. J., Harrison, M., Nienhuis, A. W. 1980. *Nucleic Acids Res.* 8:6113–28

82. Manuelidis, L., Biro, P. 1982. *Nucleic Acids Res.* 10:3221–39

83. Manuelidis, L. 1982. *Nucleic Acids Res.* 10:3211–19

84. Shafit-Zagardo, B., Maio, J. J., Brown, F. L. 1982. *Nucleic Acids Res.* 10:3175–93

85. Shafit-Zagardo, B., Brown, F. L., Maio, J. J., Adams, J.W. 1982. *Gene* 20:397–407

86. Girmaldi, G., Singer, M. F. 1983. *Nucleic Acids Res.* 11:321–38

87. Forget, B. G., Tuan, D., Biro, P. A., Jagadeeswaran, P., Weissman, S. M.

88. Lerman, M. I., Thayer, R. E., Singer, M. F. 1983. *Proc. Natl. Acad. Sci. USA* 80:3966–70

89. Thayer, R. E., Singer, M. F. 1983. *Mol. Cell. Biol.* 3:967–73

90. Grimaldi, G., Skowronski, J., Singer, M. F. 1984. *EMBO J.* 3:1753–59

91. DiGiovanni, L., Haynes, S. R., Misra, R., Jelinek, W. R. 1983. *Proc. Natl. Acad. Sci USA* 80:6533–37

92. Kole, L. B., Haynes, S. R., Jelinek, W. R. 1983. *J. Mol. Biol.* 165:257–86

93. Shafit-Zagardo, B., Brown, F. L., Zavodny, P. J., Maio, J. J. 1983. *Nature* 304:277–80

94. Potter, S. S. 1984. *Proc. Natl. Acad. Sci. USA* 81:1012–16

95. Martin, S. L., Voliva, C. F., Burton, F. H., Edgell, M. H., Hutchison, C. A. 1984. *Proc. Natl. Acad. Sci. USA* 81: 2308–12

96. Leder, P., Hansen, J. N., Konkel, D., Leder, A., Nishioka, Y., et al. 1980. *Science* 209:1336–42

97. Furuichi, Y., Morgan, M., Muthukrishnan, S., Shatkin, A. J. 1975. *Proc. Natl. Acad. Sci. USA* 72:362–66

98. Wei, C. M., Moss, B. 1974. *Proc. Natl. Acad. Sci. USA* 71:3014–18

99. Zanowalczewska, M., Bretner, M., Sierakowska, H., Szczesna, E., Filipowicz, W., Shatkin, A. J. 1977. *Nucleic Acids Res.* 4:3065–81

100. Krainer, A. R., Maniatis, T., Ruskin, B., Green, M. R. 1984. *Cell* 36:993–1005

101. Mount, S. M. 1982. *Nucleic Acids Res.* 10:459–72

102. Breathnach, R., Chambon, P. 1981. *Ann. Rev. Biochem.* 50:349–83

103. Sharp, P. A. 1981. *Cell* 23:643–46

104. Breathnach, R., Benoist, C., O'Hare, K., Gannon, F., Chambon, P. 1978. *Proc. Natl. Acad. Sci. USA* 75:4853–57

105. Lerner, M. R., Boyle, J. A., Mount, S. M., Wolin, S., Steitz, J. A. 1980. *Nature* 283:220–24

106. Rogers, J., Wall, R. 1980. *Proc. Natl. Acad. Sci. USA* 77:1877–79

107. Yang, V. W., Lerner, M. R., Steitz, J. A., Flint, S. J. 1981. *Proc. Natl. Acad. Sci. USA* 78:1371–75

108. Mount, S. M., Pettersson, I., Hinterberger, M., Karmas, A., Steitz, J. A. 1983. *Cell* 33:509–18

109. Padgett, R. A., Mount, S. M., Steitz, J. A., Sharp, P. A. 1983. *Cell* 35:101–7

110. Weissmann, C. 1984. *Nature* 311:103–4

111. Grabowski, P. J., Padgett, R. A., Sharp, P. A. 1984. *Cell* 37:415

112. Padgett, R. A., Konarska, M. M., Grabowski, P. J., Hardy, S. F., Sharp, P. A. 1984. *Science* 225:898
113. Ruskin, B., Krainer, A. R., Maniatis, T., Green, M. R. 1984. *Cell* 38:317–31
114. Wieringa, B., Hofer, E., Weissmann, C. 1984. *Cell* 37:915–25
115. Lang, K. M., Spritz, R. A. 1983. *Science* 220:1351–55
116. Kühne, T., Wieringa, B., Reiser, J., Weissmann, C. 1983. *EMBO J.* 2:727–33
117. Salditt-Georgieff, M., Darnell, J. E. Jr. 1983. *Proc. Natl. Acad. Sci. USA* 80:4694–98
118. Salditt-Georgieff, M., Darnell, J. E. Jr. 1984. *Proc. Natl. Acad. Sci. USA* 81·2274
119. Proudfoot, N. J., Brownlee, G. G. 1976. *Nature* 263:211–14
120. Proudfoot, N. J. 1984. *Nature* 307:412–13
121. Fitzgerald, M., Shenk, T. 1981. *Cell* 24:251–60
122. Montell, C., Fisher, E. F., Caruthers, M. H., Berk, A. J. 1983. *Nature* 305:600–5
123. Higgs, D. R., Goodbourne, S. E. Y., Lamb, J., Clegg, J. B., Weatherall, D. J. 1983. *Nature* 306:398–400
124. Orkin, S. H., Cheng, T. C., Antonarakis, S. E., Kazazian, H. H. Jr. 1985. *EMBO J.* 4:453–56
125. McDevitt, M. A., Imperiate, M. J., Ali, H., Nevins, J. R. 1984. *Cell* 37:1993–99
126. Sadofsky, M., Alwine, J. C. 1984. *Mol. Cell. Biol.* 4:1460–68
127. Woychik, R. P., Lyons, R. H., Post, L., Rottman, F. M. 1984. *Proc. Natl. Acad. Sci. USA* 81:3944–48
128. Gil, A., Proudfoot, N. J. 1984. *Nature.* 312:473–74
129. Darnell, J. E. Jr. 1982. *Nature* 297:365–71
130. Zeevi, M., Nevins, J. R., Darnell, J. E. Jr. 1982. *Mol. Cell. Biol.* 2:517–25
131. Hershey, J. W. B. 1980. In *Cell Biology: A Comprehensive Treatise*, ed. D. M. Prescott, L. Goldstein, 4:1–68. New York: Academic
132. Jagus, R., Anderson, W. F., Safer, B. 1981. *Prog. Nucleic Acid Res. Mol. Biol.* 25:127–85
133. Maitra, U., Stringer, E. A., Chaudhuri, A. 1982. *Ann. Rev. Biochem.* 51:869–900
134. McKnight, S. L., Kingsbury, R. 1982. *Science* 217:316–24
135. Banerji, J., Rusconi, S., Schaffner, W. 1981. *Cell* 27:299–308
136. Mellon, P., Parker, V., Gluzman, Y., Maniatis, T. 1981. *Cell* 27:279–88
137. Grosveld, G. C., deBoer, E., Shewmaker, C. K., Flavell, R. A. 1982. *Nature* 295:120–26
138. Humphries, R. K., Ley, T., Turner, P., Moulton, A. D., Nienhuis, A. W. 1982. *Cell* 30:173–83
139. Dierks, P., van Ooyen, A., Cochran, M. D., Dobkin, C., Reiser, J., Weissmann, C. 1983. *Cell* 32:695–706
140. Spandidos, D. A., Paul, Z. 1982. *EMBO J.* 1:15–20
141. Chao, M. V., Mellon, P., Charnay, P., Maniatis, T., Axel, R. 1983. *Cell* 32:483–93
142. Wright, S., deBoer, E., Grosveld, F. G., Flavell, R. A. 1983. *Nature* 305:333–36
143. Wright, S., Rosenthal, A., Flavell, R., Grosveld, F. 1984. *Cell* 38:265–73
144. Charnay, P., Treisman, R., Mellon, P., Chao, M., Axel, R., et al. 1984. *Cell* 38:251–63
145. Collins, F. S., Metherall, J. E., Yamakawa, M., Pan, J., Weissman, S. M., Forget, B. G. 1985. *Nature* 313:325–26
145a. Gelinas, R., Endlich, B., Pfeiffer, C., Yagi, M., Stamatoyannopoulos, G. 1985. *Nature* 313:323–25
146. Collins, F. S., Stoeckert, C. J. Jr., Serjeant, G. R., Forget, B. G., Weissman, S. M. 1984. *Proc. Natl. Acad. Sci. USA* 81:4894–98
146a. Collins, F. S., Boehm, C.D., Waber, P. G., Stoeckert, C. J. Jr., Weissman, S. H. et al. *Blood* 64:1292–96
147. Giglioni, B., Casini, C., Mautovani, R., Merli, S., Comi, P., et al. 1984. *EMBO J.* 3:2641–45
148. Emerson, B. M., Felsenfeld, G. 1984. *Proc. Natl. Acad. Sci. USA* 81:95–99
149. Siebenlist, U., Hennighausen, L., Battey, J., Leder, P. 1984. *Cell* 37:381–91
150. Walker, M. D., Edlund, T., Boulet, A. M., Rutter, W. J. 1983. *Nature* 306:557–61
151. Kretschmer, P. J., Coon, H. C., Davis, A., Harrison, M., Nienhuis, A. W. 1981. *J. Biol. Chem.* 256:1975–82
152. Schon, E. A., Cleary, M. L., Haynes, J. R., Lingrel, J. B. 1981. *Cell* 27:359–69
153. Townes, T. M., Fitzgerald, M. C., Lingrel, J. B. 1984. *Proc. Natl. Acad. Sci. USA* 81:6589–93
154. Ley, T. J., Nienhuis, A. W. 1983. *Biochem. Biophys. Res. Commun.* 112:1041–48
155. Carlson, D. P., Ross, J. 1983. *Cell* 34:857–64
156. Grindlay, G. J., Lanyon, W. G., Allan, M., Paul, J. 1984. *Nucleic Acids Res.* 12:1811–20
156a. Marks, P. A., Rifkind, R. A. 1978. *Ann. Rev. Biochem.* 47:419–26
157. Willing, M. C., Nienhuis, A. W.,

Anderson, W. F. 1979. *Nature* 277:534–38

158. Deisseroth, A., Barker, J., Anderson, W. F., Nienhuis, A. W. 1975. *Proc. Natl. Acad. Sci. USA* 72:2682–86

159. Deisseroth, A., Burk, R., Picciano, D., Minna, J., Anderson, W. F., et al. 1975. *Proc. Natl. Acad. Sci. USA* 72:1102–6

160. Anagnou, N. P., Lim, E., Yuan, T. Y., Helder, J., Wieder, S., et al. 1985. *Blood.* 65:705–12

161. Papayannopoulou, Th., Brice, M., Kurachi, S., Hemenway, T., Stamatoyannopoulos, G. 1984. *Blood* 64:62a

162. Benoist, C., Chambon, P. 1981. *Nature* 290:304–10

163. Gruss, P., Dhar, R., Khoury, G. 1981. *Proc. Natl. Acad. Sci. USA* 78:943–47

164. Moreau, P., Hen, R., Wasylyk, B., Everett, R., Gaub, M. P., Chambon, P. 1981. *Nucleic Acids Res.* 9:6047–68

165. Gruss, P. 1984. *DNA* 3:1–5

166. Khoury, G., Gruss, P. 1983. *Cell* 33:313–14

167. Wasylyk, B., Wasylyk, C., Chambon, P. 1984. *Nucleic Acids Res.* 12:5589–608

168. Deleted in proof

169. Lamins, L., Khoury, G., Gorman, C., Howard, B., Gruss, P. 1982. *Proc. Natl. Acad. Sci. USA* 79:6453–57

170. Banerji, J., Olson, L., Schaffner, W. 1983. *Cell* 33:729–40

171. Gillies, S. D., Morrison, S. L., Oi, V. T., Tonegawa, S. 1983. *Cell* 33:717–28

172. Neuberger, M. S. 1983. *EMBO J.* 2:1373–78

173. Picard, D., Schaffner, W. 1984. *Nature* 307:80–82

174. Gillies, S. D., Folsom, V., Tonegawa, S. 1984. *Nature* 310:594–97

175. Scholer, H. R., Gruss, P. 1984. *Cell* 36:403–11

176. Gilmour, R. S., Spandidos, D.A., Vass, J. K., Gow, J. W., Paul, J. 1984. *EMBO J.* 3:1263–72

177. Rich, A. 1983. *Cold Spring Harbor Symp. Quant. Biol.* 47:1–12

178. Riggs, A. D., Jones, P. A. 1983. *Adv. Cancer Res.* 40:1–30

179. Cooper, D. N. 1983. *Hum. Genet.* 64:315–33

180. Van der Ploeg, L. H. T., Flavell, R. A. 1980. *Cell* 19:947–58

181. Mavilio, F., Giampaolo, A., Care, A., Migliaccio, G., Calandrini, M., et al. 1983. *Proc. Natl. Acad. Sci. USA* 80:6907–11

182. Stein, R., Gruenbaum, Y., Pollack, Y., Razin, A., Cedar, H. 1982. *Proc. Natl. Acad. Sci. USA* 79:61–65

183. Busslinger, M., Hurst, J., Flavell, R. A. 1983. *Cell* 34:197–206

184. Bird, A. P. 1984. *Nature* 307:503–4

185. Kunnath, L., Locker, J. 1983. *EMBO J.* 2:317–24

186. Tanaka, K., Appella, E., Jay, G. 1983. *Cell* 35:457–65

187. Vardimon, L., Rich, A. 1984. *Proc. Natl. Acad. Sci. USA* 81:3268–72

188. Creusot, F., Acs, G., Christman, J. K. 1982. *J. Biol. Chem.* 257:2041–48

189. Taylor, S. M., Jones, P. A. 1982. *J. Mol. Biol.* 162:679–92

190. Constantinides, P. G., Jones, P. A., Gevers, W. 1977. *Nature* 267:364–66

191. DeSimone, J., Heller, P., Hall, L., Zwiers, D. 1982. *Proc. Natl. Acad. Sci. USA* 79:4428–31

192. Ley, T. J., DeSimone, J., Anagnou, N. P., Keller, G. H., Humphries, R. K., et al. 1982. *N. Engl. J. Med.* 307:1469–75

193. Ley, T. J., DeSimone, J., Noguchi, C. T., Turner, P. H., Schechter, A. N., et al. 1983. *Blood* 62:370–80

194. Charache, S., Dover, G., Smith, K., Talbot, C. C. Jr., Moyer, M., Boyer, S. 1983. *Proc. Natl. Acad. Sci. USA* 80:4842–46

195. Chiang, Y. L., Ley, T. J., Sanders-Haigh, L., Anderson, W. F. 1984. *Somatic Cell Mol. Genet.* 10:399–407

196. Ley, T. J., Chiang, Y. L., Haidaris, D., Anagnou, N. P., Wilson, V. L., et al. 1984. *Proc. Natl. Acad. Sci. USA* 81:6618–22

197. Weisbrod, S. 1982. *Nature* 297:289–95

198. Weisbrod, S., Weintraub, H. 1979. *Proc. Natl. Acad. Sci. USA* 76:630–34

199. Weisbrod, S., Groudine, M., Weintraub, H. 1980. *Cell* 14:289–301

200. Sandeen, G., Wood, W. I., Felsenfeld, G. 1980. *Nucleic Acids Res.* 8:3757–78

200a. Elgin, S. C. R. 1981. *Cell* 27:413–15

201. Groudine, M., Weintraub, H. 1981. *Cell* 24:393–401

202. McGhee, J. D., Wood, W. I., Dolan, M., Engel, J. D., Felsenfeld, G. 1981. *Cell* 27:45–55

203. Shefferey, M., Rifkind, R. A., Marks, P. A. 1982. *Proc. Natl. Acad. Sci. USA* 79:1180–84

204. Balcarek, J. M., McMorris, F. A. 1983. *J. Biol. Chem.* 258:10622–28

205. Groudine, M., Kohwi-Shigematsu, T., Gelinas, R., Stamatoyannopoulos, G., Papayannopoulou, T. 1983. *Proc. Natl. Acad. Sci. USA* 80:7551–55

205a. Fordis, C. M., Anagnou, N. P., Dean, A., Nienhuis, A. W., Schechter, A. N. 1984. *Proc. Natl. Acad. Sci.* 81:4485–89

206. Lachman, H. M., Mears, J. G. 1983. *Nucleic Acids Res.* 11:6065–77
207. Tuan, D., London, I. M. 1984. *Proc. Natl. Acad. Sci. USA* 81:2718–22
208. Treisman, R., Green, M. R., Maniatis, T. 1983. *Proc. Natl. Acad. Sci. USA* 80:7428–32
209. Green, M. R., Treisman, R., Maniatis, T. 1983. *Cell* 35:137–48
210. Dean, A., Wu, Y., Ley, T., Fordis, C. M., Schechter, A. N. 1985. See Ref. 16. In press
211. Yuan, T. Y., Deisseroth, A. 1984. *Clin. Res.* 32:501A
212. Papayannopoulou, Th., Lindsley, D., Kurachi, S., Lewison, K., Hemenway, T. et al. 1985. *Proc. Natl. Acad. Sci USA* 82:780–84
213. Proudfoot, N. J., Rutherford, T. R., Partington, G. A. 1984. *EMBO J.* 3:1533–40
214. Gregory, C. J., Eaves, A. C. 1977. *Blood* 49:855–64
215. Quesenberry, P., Levitt, L. 1979. *N. Engl. J. Med.* 301:819–23
216. Nienhuis, A. W., Benz, E. J. Jr. 1977. *N. Engl. J. Med.* 297:1318–28, 1371–81, 1430–36
217. Papayannopoulou, T., Nakamoto, B., Kurachi, S., Kurnit, D., Stamatoyannopoulos, G. 1981. See Ref. 9, pp. 307–20
218. Papayannopoulou, T., Kurachi, S., Nakamoto, B., Zanjani, E. D., Stamatoyannopoulos, G. 1982. *Proc. Natl. Acad. Sci. USA* 79:6579–83
219. Stamatoyannopoulos, G., Nakamoto, B., Kurachi, S., Papayannopoulou, T. 1983. *Proc. Natl. Acad. Sci. USA* 80: 5650–54
220. Papayannopoulou, T., Brice, M., Stamatoyannopoulos, G. 1977. *Proc. Natl. Acad. Sci. USA* 74:2923–27
221. Stamatoyannopoulos, G., Kurnit, D. M., Papayannopoulou, T. 1981. *Proc. Natl. Acad. Sci. USA* 78:7005–9
222. Dover, G., Ogawa, M. 1980. *J. Clin. Invest.* 66:1175–78
223. Kidoguchi, K., Ogawa, M., Karam, J. D. 1979. *J. Clin. Invest.* 63:804–6
224. Weinberg, R. S., Goldberg, S. D., Alter, B. P. 1981. *Blood* 58:682
225. Alter, B. P., Rappeport, J. M., Huisman, T. H. J., Schroeder, W. A., Nathan, D. G. 1976. *Blood* 48:843–53
226. Dover, G. J., Boyer, S. H., Zinkbaum, W. H. 1979. *J. Clin. Invest.* 63:173
227. Nathan, D. G. 1983. See Ref. 13, pp. 365–78
227a. Papayannopoulou, T., Tatsis, B., Kurachi, S., Nakamoto, B., Stamatoyannopoulos, G. 1984. *Nature* 309:71–73
228. Papayannopoulou, T., Kalmantis, T., Stamatoyannopoulos, G. 1979. *Proc. Natl. Acad. Sci. USA* 76:6420–24
229. Dover, G. J., Boyer, S. H. 1980. *Blood* 56:1082–91
230. Wood, W. G., Jones, R. W. 1981. See Ref. 9, pp. 147–60
231. Boyer, S. H., Belding, T. K., Margolet, L., Noyes, A. N., Burke, P. J., Bell, W. R. 1975. *Johns Hopkins Med. J.* 137:105–15
232. Boyer, S. H., Belding, T. K., Margolet, L., Noyes, A. N. 1975. *Science* 188: 361–63
233. Wood, W. G., Stamatoyannopoulos, G., Lim, G., Nute, P. E. 1975. *Blood* 46: 671–82
234. Zago, H. A., Wood, W. G., Clegg, J. B., Weatherall, D. J., O'Sullivan, M., et al. 1979. *Blood* 53:977–86
235. Dean, J., Schechter, A. N. 1978. *N. Engl. J. Med.* 299:752–63, 804–11, 863–70
236. Eaton, W. A., Hofrichter, J., Ross, P. D. 1976. *Blood* 275:621–27
237. Nagel, R. L., Bookchin, R. M., Johnson, J., Labie, D., Wajcman, H., Matsutomo, K. 1979. *Proc. Natl. Acad. Sci. USA.* 76:670–72
238. Boyer, S. H., Dover, G. J. 1979. In *Cellular and Molecular Regulation of Hemoglobin Switching,* ed. G. Stamatoyannopoulos, A. Nienhuis, pp. 47–73. New York: Grune & Stratton
239. Brittinghan, G. N., Schechter, A. N., Noguchi, C. 1985. *Blood.* 65:183–89
240. Powers, D. R., Weiss, J. N., Cham, L. S., Schroeder, W. A. 1984. *Blood* 63: 921–26
241. Wood, W. G., Pembrey, M. E., Serjeant, G., Perrine, R. P., Weatherall, D. J. 1980. *Br. J. Haematol.* 45:431–45
242. Weatherall, D. J., Wood, W. G., Clegg, J. B. 1979. See Ref. 238, pp. 3–25
243. Wood, W. G., Clegg, J. B., Weatherall, D. J. 1979. *Br. J. Haematol.* 43:509–20
244. Flavell, R. A., Bernards, R., Kooter, J. M., deBoer, E., Little, P. F. R., et al. 1979. *Nucleic Acids Res.* 6:2749–60
245. Orkin, S. H., Old, J. M., Weatherall, D. J., Nathan, D. G. 1979. *Proc. Natl. Acad. Sci. USA* 76:2400–4
246. Orkin, S. H., Kolodner, R., Michelson, A. M., Husson, R. 1980. *Proc. Natl. Acad. Sci. USA* 77:3558–62
247. Spritz, R. A., Orkin, S. H. 1982. *Nucleic Acids Res.* 10:8025–929
248. Kazazian, H. H. Jr., Orkin, S. H., Antonarakis, S. E., Sexton, J. P., Boehm, C. D., et al. 1984. *EMBO J.* 3:593–96
249. Boehm, C. D., Dowling, C. E. D., Kazazian, H. H. Jr. 1984. *Am. J. Hum. Genet.* 36:1335

250. Gilman, J. G., Huisman, T. H. J., Abels, J. 1984. *Br. J. Haematol.* 56:339–48
251. Mears, J. G., Ramirez, F., Leibowitz, D., Nakamura, F., Bloom, A. 1978. *Proc. Natl. Acad. Sci. USA* 75:1222–26
252. Flavell, R. A., Kooter, J. M., deBoer, E., Little, P. F. R., Williamson, R., et al. 1978. *Cell* 15:25–41
253. Baird, M., Driscoll, C., Schreiner, H., Sciarrata, G. V., Sansone, G., et al. 1981. *Proc. Natl. Acad. Sci. USA* 78: 4218–21
254. Wainscoat, J. S., Thein, S. L., Wood, W. G., Weatherall, D. J., Tzotos, S. et al. 1985. *Ann. NY Acad. Sci.* In press
255. Ottolenghi, S., Giglioni, B., Comi, P., Gianni, A. M., Polli, E. 1979. *Nature* 278:654–56
256. Bernards, R., Kooter, J. M., Flavell, R. A. 1979. *Gene* 6:265–80
257. Ottolenghi, S., Giglioni, B. 1982. *Nature* 300:770–71
258. Anagnou, N. P., Papayannopoulou, T., Stamatoyannopoulos, G., Nienhuis, A. W. 1985. *Blood.* 65:1245–51
259. Matsunaga, E., Kimura, A. Yamada, H., Fukumaki, Y., Takagi, Y. 1985. *Biochem. Biophys. Res. Commun.* 126:185–91
260. Jones, R. W., Old, J. M., Trent, R. J., Clegg, J. B., Weatherall, D. J. 1981. *Nature* 291:39–44
261. Henthorn, P. S., Smithies, O., Nakatsuji, T., Felice, A. E., Gardiner, M. B., et al. 1985. *Br. J. Haematol.* 59:343–56
262. Tuan, D., Feingold, E., Newman, M., Weissman, S. M., Forget, B. G. 1983. *Proc. Natl. Acad. Sci. USA* 80:6937–41
263. Trent, R. J., Jones, R. W., Clegg, J. B., Weatherall, D. J., Davidson, R. 1984. *Br. J. Haematol.* 57:279–89
264. Jones, R. W., Old, J. M., Trent, R. J., Clegg, J. B., Weatherall, D. J. 1981. *Nucleic Acids Res.* 9:6813–25
265. Anagnou, N. P., Papayannopoulou, T., Stamatoyannopoulos, G., Nienhuis, A. W. 1985. *Blood.* 64:61a
266. Pirastu, M., Curtin, P., Kan, Y. W., Gobert-Jones, J. A., Stephens, A. D., Lehmann, H. 1984. *Clin. Res.* 32:493A
267. Van der Ploeg, L. H. T., Konings, A., Oort, M., Roos, D., Bernini, L., Flavell, R. A. 1980. *Nature* 283:637–42
268. Kan, Y. W., Forget, B. G., Nathan, D. G. 1972. *N. Engl. J. Med.* 286:1290–134
269. Orkin, S. H., Goff, S. C., Nathan, D. G. 1981. *J. Clin. Invest.* 67:878–84
270. Fearon, E. R., Kazazian, H. H. Jr., Waber, P. G., Lee, J. I., Antonarakis, S. E., et al. 1984. *Blood* 61:1273–78
271. Pirastu, M., Kan, Y. W., Lin, C. C., Baine, R. M., Holbrook, C. T. 1983. *J. Clin. Invest.* 72:602–9

272. Tuan, D., Murnane, M. J., de Riel, J.K., Forget, B. G. 1980. *Nature* 285:335–37
273. Bernards, R., Flavell, R. A. 1980. *Nucleic Acids Res.* 9:1521–34
274. Jagadeeswaran, P., Tuan, D., Forget, B. G., Weissman, S. M. 1982. *Nature* 296:469–70
275. Huisman, T. H. J., Wrightstone, R. W., Wilson, J. B., Schroeder, W. A., Kendall, W. A. 1972. *Arch. Biochem. Biophys.* 152:850–55
276. Ojwang, P. J., Nakatsuji, T., Gardiner, M. B., Reese, A. L., Gilman, J. G., Huisman, T. H. J. 1983. *Hemoglobin* 7:115–23
277. Vanin, E. F., Henthorn, P. S., Kioussis, D., Grosveld, F., Smithies, O. 1983. *Cell* 35:701–9
278. Mager, D. L., Henthorn, P. S. 1984. *Proc. Natl. Acad. Sci. USA.* 81:7510–14
279. Huisman, T. H. J., Schroeder, W. A., Efremov, G. D., Duma, H., Mladenovsky, B., et al. 1974. *Ann. NY Acad. Sci.* 232:107–24
280. Feingold, E., Collins, F. S., Metherall, J. E., Stoeckert, C. J. Jr., Weissman, F. M., Forget, B. 1985. *Ann. NY Acad. Sci.* In press
281. Kioussis, D., Vanin, E., deLange, T., Flavell, R. A., Grosveld, F. G. 1983. *Nature* 306:662–66
282. Farquhar, M., Gelinas, R., Tatsis, B., Murray, J., Yagi, M., et al. 1983. *Am. J. Hum. Genet.* 35:611–20
283. Jensen, M., Wintz, A., Walther, J.-U., Schemken, E. M., Laryea, M. D., et al. 1984. *Br. J. Haematol.* 56:87–94
284. Weatherall, D. J., Cartner, R., Clegg, J. B., Wood, W. G., Macrae, I. A., et al. 1975. *Br. J. Haematol.* 29:205–19
285. Boyer, S. H., Margolet, L., Boyer, M. L., Huisman, T. H. J., Schroeder, W. A., et al. 1977. *Am. J. Hum. Genet.* 29:256–71
286. Sukumaran, P. K., Huisman, T. H. J., Schroeder, W. A., McCurdy, P. R., Freehafer, J. T., et al. 1972. *Br. J. Haematol.* 23:403–17
287. Marti, H. R. 1963. *Normale und Abnormale Menschliche Hämoglobine.* Berlin: Springer-Verlag
288. Altay, C., Huisman, T. H. J., Schroeder, W. A. 1976. *Hemoglobin* 1:125–33
289. Altay, C., Schroeder, W. A., Huisman, T. H. J. 1977. *Am. J. Hematol.* 3:1–14
290. Stamatoyannopoulos, G., Wood, W. G., Papayannopoulou, T., Nute, P. E. 1975. *Blood* 46:683–92
291. Schroeder, W. A., Huisman, T. H. J., Brown, A. K., Uy, R., Bouver, N. G., et al. 1971. *Pediatr. Res.* 5:493–99

292. Gilman, J. G., Huisman, T. H. J. 1984. *Blood* 64:452–57
293. Harano, T., Reese, A. L., Ryan, R., Abraham, B. L., Huisman, T. H. J. 1985. *Br. J. Haematol.* 59:333–42
294. Gilman, J. G., Huisman, T. H. J. 1984. *Blood*:64:62a
295. Powers, P. A., Altay, C., Huisman, T. H. J., Smithies, O. 1984. *Nucleic Acids Res.* 12:7023–34
296. Miller, B. A., Salameh, M., Ahmed, M., Antognetti, G., Weatherall, D. J., Nathan, D. G. 1984. *Blood.* 64:51a
297. Old, J. M., Ayyub, H., Wood, W. G., Clegg, J. B., Weatherall, D. J. 1982. *Science* 215:981–82
298. Gianni, A. M., Bregni, M., Cappellini, M. D., Fiorelli, G., Taramelli, R., et al. 1983. *EMBO J.* 2:921–25
299. Giampaolo, A., Mavilio, F., Sposi, N. M., Care, A., Massa, A., et al. 1984. *Hum. Genet.* 66:151–56
300. DeSimone, J., Biel, S. I., Heller, P. 1978. *Proc. Natl. Acad. Sci. USA* 75:2937–40
301. Humphries, R. K., Dover, G., Young, N. S., Moore, J. G., Charache, S., Ley T., Nienhuis, A. W. 1985. *J. Clin. Invest.* 75:547–57
302. Dover, G., Charache, S., Vogelsang, G 1983. *Blood* 62(Suppl. 1):55a
303. Nienhuis, A. W., Ley, T. J., Humphries R. K., Young, N. S., Dover, G. 1985 *Ann. NY Acad. Sci.* In press
304. Ley, T. J., Anagnou, N. P., Noguchi, C T., Schechter, A. N., DeSimone, J 1983. See Ref. 13, pp. 457–74
305. Torrealba-de Ron, A. T., Papayannopoulou, T., Napp, M. S., Feng-Tuen Fu M., Nitter, J., Stamatoyannopolos, G 1984. *Blood* 63:201–10
306. Papayannopoulou, T., Torrealba-de Ron, A. T., Veith, R., Knitter, G., Stamatoyannopoulos, G. 1984. *Science* 224:617–19
307. Letvin, N. L., Linch, D. C., Beardsley, G. P., McIntyre, K. W., Nathan, D. G 1984. *N. Engl. J. Med.* 310:869–73
308. Platt, O. S., Orkin, S. H., Dover, G., Beardsley, G. P., Miller, B., Nathan, D. G. 1984. *J. Clin. Invest.* 74:652–56

nn. Rev. Biochem. 1985. 54:1109–49

EUKARYOTIC PROTEIN SYNTHESIS

Kivie Moldave[1]

Department of Biological Chemistry, University of California, Irvine, California 92717

CONTENTS

PERSPECTIVES AND SUMMARY

The last 10 years have seen the emergence of a picture of eukaryotic protein synthesis that reveals the sequence of reactions involved and the large number of components, in particular initiation factors, that participate in this complicated process. Detailed information regarding the structure of translational components such as ribosomes, initiation, elongation and termination factors, mRNA, etc, and their role in various intermediary reactions has become available. More recent studies are elucidating some of the regulatory mecha-

[1]Current address: University of California, Santa Cruz, Santa Cruz, California 95064

0066-4154/85/0701-1109$02.00

nisms that affect translation directly, the genes that encode specific mRNAs and the sequences that control their expression, as well as posttranscriptiona and posttranslational events that profoundly affect gene expression.

A decade ago, the mechanism of protein synthesis was better understood i prokaryotes than in eukaryotes. The three initiation factors (IF-1, IF-2, an IF-3), the two elongation factors (EF-Tu·Ts and EF-G), and the three termina tion factors (RF-1, RF-2, and RF-3) in bacteria had been identified an extensively characterized as to their structure and function. Some aspects o protein synthesis in eukaryotes have turned out to be quite different from thos in prokaryotes. This is particularly true in initiation, where at least 10 eukaryot ic factors are required, as compared to 3 for prokaryotes, and the sequence c interactions between the small ribosomal subunit, initiator tRNA, and mRN/ are dissimilar. The basic mRNA structure in the two systems is also quit different; prokaryotic mRNAs are predominantly polycistronic while mos eukaryotic mRNAs appear to be monocistronic and contain a unique "cap structure at the 5'-end and a poly(A) sequence at the 3'-terminus. On the othe hand, although the structures of the ribosomal particles, initiator tRNAs, an termination factors are somewhat different in the two systems, their overal roles in translation are quite similar, and this is particularly true in the case o polypeptide chain elongation. Two comprehensive reviews dealing with th initiation of protein synthesis in prokaryotic and eukaryotic systems hav appeared recently (1, 2). Several other reviews dealing with various aspects o eukaryotic translation and its components or regulation have appeared (3–8) and the reader's attention is called to the proceedings of the NATO Advance Study Institute on Protein Biosynthesis in Eukaryotes (9) and of the Fogart International Conference on Translational/Transcriptional Regulation of Gen Expression (10).

Studies with highly purified translational factors and with cell-free system that translate a variety of mRNAs are providing insights into the exact roles o translational factors, the regulatory mechanisms that affect protein synthesis a the level of these factors, and into essential features of mRNAs that affect thei interaction with the translational machinery. This review covers some struc ture-function aspects of initiation, elongation, and termination components i eukaryotic systems.

OVERVIEW OF TRANSLATION

The decoding of the codons in mRNA occurs in three distinct phases, eac utilizing different soluble factors that catalyze specific interactions among th large number of components that participate in this process. The three phase are initiation, elongation, and termination. Peptide chain initiation serves t decode the initiation codon, AUG, which codes for methionine at the beginnin f the cistron. Of the full complement of aminoacyl-tRNAs produced by the ell, only one, Met-tRNA$_f$ (or Met-tRNA$_i$) is used for initiation. The initiation rocess involves GTP, ATP, a number of initiation factors (eIFs), mRNA, and ibosomal subunits, to form an 80S initiation complex. The 80S initiation omplex, containing mRNA with its initiation codon base-paired to the anti-odon of initiator Met-tRNA$_f$, free of initiation factors, is then ready to decode ll of the meaningful internal codons in the mRNA cistron.

Peptide chain elongation serves to translate all of the internal codons, etween the initiation and termination triplets. It represents a series of reactions n 80S ribosomes that are repeated as each codon is translated sequentially, xtending the nascent polypeptide chain from the N-terminal to the C-terminal esidue. All of the aminoacyl-tRNAs, other than Met-tRNA$_f$, are used in this process which also requires GTP and elongation factors (EFs). The elongation eactions, binding of the appropriate aminoacyl-tRNA to the ribosomal A site, ranspeptidation to form a peptide bond between the incoming aminoacyl noiety and methionyl (or peptidyl) residue at the P site, and movement of the ibosome on mRNA, are repeated until a termination codon enters the A site.

The termination phase of protein synthesis, when the A (decoding) site on he ribosome contains a termination codon, requires GTP and a specific protein actor (RF). The GTP-dependent, codon-specific binding of the termination actor to the ribosomal A site results in the hydrolysis of the peptidyl-tRNA ster at the P site and release of the completed polypeptide chain.

PEPTIDE CHAIN INITIATION

A number of distinct intermediary steps are involved in the assembly of the 80S nitiation complex: 1. Ribosome dissociation and accumulation of ribosomal ubunits; 2. Formation of a ternary complex containing eIF-2, GTP, and Met-tRNA$_f$; 3. Binding of the ternary complex to 40S ribosomal subunits to orm the 40S preinitiation complex; 4. Binding of mRNA to the 40S preinitia-ion complex; and 5. Joining of 60S ribosomal subunits to the 40S intermediary complex to form the 80S initiation complex. The initial and most extensive solation of eukaryotic initiation factors has been carried out with reticulocyte preparations (11–27); however, most or some of the factors have also been solated from wheat germ (28–34), ascites cells (35, 36), mammalian liver 39–41), HeLa cells (37, 38), etc. In most cases, the structural and biological characteristics of individual factors from various sources appear to be quite similar. Indeed, antisera raised in goats against several rabbit reticulocyte and HeLa cell initiation factors cross-reacted with the cognate factors from both species, and evidence of some degree of conservation in some of the factors has been obtained (37).

A large number of initiation factors are currently recognized as playing direct role in specific reactions in the pathway. Multiple roles for some of these factors, for example, eIF-2, eIF-3, and eIF-4C, as discussed below, have no been ruled out. Other proteins have been reported that may be required for factor recycling, stabilization of intermediates or ancillary functions, such a (e)RF, GEF, eIF-2B, Co-eIF-2A, Co-eIF-2B, Co-eIF-2C, etc (8, 9, 36, 42 46), and some have been reported to which no specific function has bee assigned, such as eIF-4D (19), cIF-1, or IF-M1, etc (12, 29, 30, 40, 47–51 One molecule of GTP and one of ATP are hydrolyzed in the initiation proces and the factors presumably cycle between the particles and the cytoso although the precise points at which they are released have not been establishe for all of them. The 80S initiation complex does not contain any initiatio factors.

Ribosome Dissociation (Anti-association)

Ribosomal subunits are produced as a consequence of the ribosome cycle i protein synthesis, in which ribosomes are released from polysomes on termina tion; ribosomal subunits then accumulate as the result of active dissociation 80S ribosomes or because they are prevented from reassociating, and an reused in initiation to form polysomal ribosomes (52–55). Unless otherwis specified, the terms dissociation and anti-association are used here interchange ably. Ribosomal subunits, essential participants in the initiation process, an usually found in association with a number of nonribosomal proteins (39, 5 56–59). In the case of the small (40S) ribosomal subunit, a somewhat faste sedimenting (43S–46S) particle has been isolated from a variety of cells and ha been reconstituted from derived subunits, free of nonribosomal proteins, an initiation factors in vitro (53, 56–60).

Initiation factor eIF-3 (M_r 500,000–700,000; 7–10 polypeptides) has bee isolated from native 40S subunits (16, 34, 39, 56, 61–67), and purified eIF-binds to purified 40S subunits to form a 43S–46S particle (34, 39, 61, 65 Several laboratories (16, 19, 34, 39, 61, 68–72) have reported that this hig molecular weight factor, eIF-3, functions as a ribosome dissociation factor; th 40S·eIF-3 complex that is formed does not directly associate with 60S riboson al subunits. However, eIF-3 preparations have been described that are devoi of ribosome dissociation activity (73, 74), and the existence of low-molecula weight proteins with dissociation factor activity has been reported (32, 34, 41 63, 69, 71, 75–79).

Whether dissociation factor(s) react directly with ribosomes to carry o dissociation, or with subunits formed as a consequence of spontaneous dis sociation preventing their reassociation into ribosomes, has been the subject c considerable discussion (54, 55, 60). Most of the studies cited above, with variety of dissociation preparations, are consistent with an anti-associatio nction; however, evidence has been presented for active dissociation with a gh-molecular-weight factor from rat liver (39, 71) and with a low-molecular-eight factor from wheat germ and rat liver (34, 71). The complexity of the action(s) responsible for the generation of ribosomal subunits is further nderscored by the following observations: two distinct ribosome dissociating ctivities have been detected in reticulocytes (77), rat liver (71) and wheat germ 4); in rat liver, eIF-3 appears to have both dissociation and anti-association ctivities (39); in wheat germ, eIF-3 prevents reassociation only, and a low-(M_r 00,000)-molecular-weight factor actively dissociates 80S ribosomes (34); me wheat germ eIF-3 preparations seem to be devoid of any dissociation ctivity (73, 74); and in a variety of cells, ribosome dissociation activity has een assigned to components with molecular weights of 500,000–700,000 (16, 9, 34, 39, 62, 68–72), 100,000–300,000 (34, 68, 70, 74), or 23,000–25,000 2, 41, 78, 79). Further, another initiation factor, eIF-4C, which stimulates lobin mRNA translation and the formation of initiation intermediates in the resence of other initiation factors (11, 13, 15–17, 19, 34, 63, 72, 80) has been ported to act as an accessory to eIF-3 in ribosome dissociation (72). Initiation ctor eIF-4C is a single polypeptide chain with a molecular weight of 17,500–2,500 (13, 17, 34, 80), it has been isolated from native 40S subunits, and it inds in vitro to 40S subunits to prevent association of the particle with 60S ubunits to form 80S ribosomes or with other 40S subunits to form dimers (72). oth the high (eIF-3_H)- and the low-(eIF-3_L)-molecular-weight dissociation actors from rat liver also exhibit dissociation activity with dimers of ribosomal ubunits, resulting in the formation of monomers (39, 71).

Most eIF-3 preparations have sedimentation coefficients of 13S–16S, nolecular weights of 500,000–700,000, 7–10 polypeptide chains with molecu-ar weights of 30,000–170,000 (for example: 36K, 40K, 44K, 47K, 66K, 15K, and 170K) on SDS polyacrylamide gel electrophoresis, and appear to eact specifically with 40S subunits. Analysis of 40S·eIF-3 complexes by hemical cross-linking indicates that eIF-3 becomes covalently bound to 18S RNA (81) and to ribosomal proteins S3a, S4, S6, S7, S8, S9, S10, S23/24, and 27 (82) or proteins S2, S3, S3a, S4, S5, S6, S8, S9, S11, S12, S13, S14, S15, 16, S19, S24, S25, and S26 (83). In this latter study, proteins were cross-inked directly with eIF-3 or indirectly through one bridging protein. The nteraction between 18S RNA and eIF-3 is highly specific in that only the 6,000-dalton subunit of the initiation factor is involved in the cross-linking 81). The binding of eIF-3 appears to be at the ribosomal interphase (84) and losely adjacent to the attachment sites for eIF-2, mRNA, and 60S subunits.

Recent reports have described a low-(M_r 23,000–25,000)-molecular-weight issociation factor from wheat germ and calf liver (32, 41, 77, 79), designated IF-6, that reacts with 60S rather than 40S subunits and prevents subunit eassociation. These studies questioned the role of eIF-3 in ribosome dissociation and suggest that the results may be due to contamination of eIF-3 with eIF-6 (1, 2, 41); the absence of polypeptide chains smaller than 28,000–35,000 daltons in the reticulocyte (59, 65), rat liver (39), and wheat germ (34) eIF-preparations, however, would argue against this explanation.

The possibility exists that both eIF-3 and eIF-6 possess dissociation (or anti-association) activity, although one of the two factors may be quantitatively more significant. A model consistent with many of the findings on ribosome dissociation suggests that: Most of the dissociation activity can be accounted for by an anti-association mechanism; eIF-6 reacts with 60S subunits and prevents their association with 40S subunits; concomitant with the 60S-eIF-6 reaction, or immediately afterwards, eIF-3 and possibly eIF-4C bind to 40S subunits to form a 40S·eIF-3·-eIF-4C complex which is also unable to react directly with 60S or 60S·eIF-6 particles; and the reactions between these factors and the subunits do not require any other initiation components or nucleotides. Two distinct dissociation activities have been reported in reticulocytes (77) and rat liver (39, 71); one is specific for 40S subunits and the other for 60S subunits. As described below, the interaction of eIF-3 with 40S subunits is an essential prerequisite to the subsequent reactions between the 40S particle and other initiation components such as Met-tRNA$_f$·eIF-2·GTP ternary complex and mRNA; a similar role for eIF-4C, in the interaction between 40S subunits and ternary complex or in the stabilization of the resulting intermediate, has also been suggested (19, 34, 59).

Formation of the Ternary Complex

When aminoacyl-tRNAs react with ribosomal particles under physiological conditions, in initiation or elongation, the reactions are catalyzed by a specific protein factor, GTP is required, and a ternary complex is formed between the aminoacyl-tRNA, the binding factor, and GTP. In chain initiation, the complex is formed between initiator Met-tRNA$_f$, eIF-2, and GTP (11, 14, 40, 61, 85–96). The interactions between the aminoacyl-tRNA binding factors and other components in translation are highly specific. For example, eIF-2 reacts with Met-tRNA$_f$ and ribosomal 40S subunits, but does not recognize any of the other aminoacyl-tRNAs or the 80S ribosome. By contrast, the aminoacyl-tRNA binding factor in chain elongation (EF-1) described below, reacts with GTP, all of the aminoacyl-tRNAs other than the initiator Met-tRNA$_f$, and 80S ribosomes but not 40S subunits.

A number of studies have been carried out to identify features in the structures of tRNAs that would distinguish the initiator molecules from those used in chain elongation. tRNA$_f^{Met}$ molecules from a variety of prokaryotic and eukaryotic sources appear to have 4 G-C base pairs in identical positions; two of them are in the stem of the anticodon (I) loop, one is in the stem of the dihydrouridylate synthetase recognition (II) loop, and the other is in the amino acid acceptor stem, near the 5' and 3' termini of tRNA (97). A unique

:onformation in the anticodon loop of initiator tRNA and some specific :hanges in nucleotides at otherwise unvariant positions in noninitiator tRNAs, ₁ave also been identified (98, 99). A more detailed description of the structural eatures of tRNA that may relate to their specific interaction with other ₁anslational components is included in the review by Kozak (2). It would :ertainly be important to establish the exact structural basis for the specificity in he interaction between the tRNAs and their respective binding factors, in ₁articular. Such specificity would not extend to the interaction between tRNAs ₁nd their cognate activating enzymes; a single Met-tRNA ligase is responsible or the aminoacylation of both $tRNA_f^{Met}$ which is used for initiation, and ₁RNA$_m^{Met}$ which is used for elongation. However, although selectivity appears ₁o occur primarily at the level of the binding factors, ribosomal particles may ₁lso participate in the aminoacyl-tRNA recognition reaction contributing a :ertain degree of specificity (100). In the absence of GTP and binding factors ₁IF-2 or EF-1, high Mg^{2+}-dependent binding of Met-tRNA$_f$ to purified rat liver ₁0S subunits was significantly greater than that of Met-tRNA$_m$, binding of Met-tRNA$_f$ to 40S subunits was greater than to 80S ribosomes, and 80S ribosomes bound Met-tRNA$_m$ more efficiently than Met-tRNA$_f$ (100).

Initiation factor eIF-2 has been purified in a large number of laboratories, from a variety of cells such as reticulocytes, *Artemia salina,* wheat germ, liver, cultured cells, yeast, etc (8, 11, 15–17, 28, 33, 35, 37, 40, 85, 87, 89, 90, 92, 93, 95, 96, 101–108). Values for the native molecular weight of about 145,000 have been obtained with many of these eIF-2 preparations, and SDS gel electrophoretic analyses indicate that it is a polymeric protein composed of 3 nonidentical subunits (8, 9, 11, 15–17, 29, 40, 43, 102, 109–111), referred to as α, β, and γ. Apparent subunit molecular weights of about 38,000, 50,000, and 56,000 have been obtained by SDS gel electrophoresis but, with the exception of the α subunit (M_r 38,000), assignment of molecular weights to the β and γ subunits has been confusing. This is particularly true for the β subunit which migrates faster, slower, or the same as the γ subunit depending on the analytical electrophoretic procedure (27, 102, 106, 113). It has been suggested that the aberrant behavior of the β subunit in gel electrophoresis may be due to the high content of basic amino acids (112). For the purpose of this review, α is used to denote the 35,000–38,000 dalton polypeptide and γ is used for the 48,000–52,000 dalton chain; the anomalous β subunit is frequently assigned a molecular weight of 52,000–56,000.

There are a number of parameters that distinguish the subunits of eIF-2, that are more useful for their characterization than molecular weights based on SDS gels. The α subunit is phosphorylated by a kinase (eIF-2α kinase) induced by heme deficiency in reticulocytes, and by double-stranded (ds)RNA, or interferon treatment (114–126), which affect the function of eIF-2 in initiation; its isoelectric point is estimated to be pH 5.1–5.8 (43, 109, 112); and it binds GDP with high affinity (109). The β subunit has a pI of 5.2–6.6, and is phosphorylated by casein kinase II (17, 103, 106), which has no effect on the function of eIF-2 (106, 107, 121, 127); in addition, as discussed below, this subunit is missing in some of the eIF-2 preparations that have been described (8, 41, 95 96, 107, 128). The γ subunit has a pI of 8.9, is not phosphorylated by either of the kinases that modify the α or β subunit, and binds Met-tRNA$_f$ and mRNA (109).

Some preparations with many of the functional characteristics of eIF-2, but which have a lower (85,000–115,000) molecular weight and contain only two of the polypeptide chains (M_r 38,000 and M_r 50,000), have been purified (8 14, 42, 95, 96, 107, 128, 129). It has been suggested that detection of only two components may be due to the aberrant behavior of the β subunit in some gel electrophoretic procedures that fail to resolve it from the γ chain (17, 27, 104 112, 113), the loss of the β subunit from eIF-2 which may be due to proteolysis of this sensitive polypeptide during purification (9, 107, 130, 131), that the active factor may indeed consist of only two subunits (95, 96), or that one of the subunits dissociates as a consequence of its catalytic function in initiation (8 42).

Initiation factor eIF-2 has a much greater affinity for GDP than for GTP, i binds GDP strongly, and formation of ternary complex is inhibited by low concentrations of GDP (109, 132–135). Some of these studies (132–134) relate the adenylate energy charge in the cell to the guanine nucleotide pool and to the regulation of protein synthesis at the level of initiation. GDP plays a central role in the initiation process; it is a product of the hydrolysis of GTP in the ternary complex, it is released in a complex with eIF-2 (eIF·GDP) which is inactive and must be replaced with GTP in order to generate functional eIF-2 (9, 43, 44 136–140).

Ternary complexes can also be prepared with nonhydrolyzable analogs of GTP, such as GMP-P(NH)P and GMP-P(CH$_2$)P (14, 85, 102, 137, 141). The resulting complexes appear to participate properly in some of the subsequent steps in initiation, such as the binding to 40S subunits, but are inactive in promoting formation of the final initiation complex or protein synthesis. The results obtained on the formation of ternary complex and binding to 40S subunits in initiation, with GTP analogs, are reminiscent of those on the aminoacyl-tRNA binding reaction in chain elongation (142–144). The formation of the ternary complex containing noninitiator aminoacyl-tRNAs, elongation factor 1, and guanine nucleotide, as well as the binding of this ternary complex to 80S ribosomes, occur when GMP-P(CH$_2$)P is used instead of GTP however, elongation reactions beyond binding do not occur without GTP hydrolysis. These findings suggest that although a guanine nucleotide is required for the interaction between the binding factor (eIF-2 or EF-1), the appropriate aminoacyl-tRNA (Met-tRNA$_f$ or noninitiator aminoacyl-tRNAs) and the appropriate ribosomal particle (40S subunit or 80S ribosome), GTP hydrolysis is not essential; it has been suggested that the hydrolytic role of GTP is related to a conformational effect on the binding factor (144).

A number of proteins have been described that affect the synthesis, stability, or utilization of eIF-2·GTP·Met-tRNA$_f$ ternary complex. In addition, some of these proteins relieve the inhibition in protein synthesis and initiation in heme-deficient reticulocyte lysates (42–45, 145, 146). These proteins have been referred to as ancillary cofactors Co-eIF-2A, Co-eIF-2B, and Co-eIF-2C (23, 24, 45, 147–154), restoring factor, recycling factor, inhibition reversal factor, stabilizing factor, RF, SF, or SRF (3, 9, 18, 43, 45, 155–161), ESP, for eIF-2 stimulating protein (162–166), GEF, for guanine nucleotide exchange factor (9, 36, 167–169), anti-inhibitor, anti-HRI or eRF, for regulatory initiation factor (8, 21, 42), eIF-2B (46, 170), etc. These new factors have been isolated from a variety of cells such as reticulocytes, ascites tumor cells, *A. salina,* liver, etc. Examination of the physical, chemical, and biological properties reported for these modulating proteins reveals a great deal of similarity, and suggests that most of them may indeed be identical. With the exception of Co-eIF-2A and some protein cofactors from wheat germ (129, 171, 172) which also affect eIF-2 activity, all of the preparations are between 250,000 and 300,000 daltons, are found in complexes with eIF-2, and contain five polypeptide subunits with molecular weights of 80K–85K, 65K–67K, 52K–58K, 37K–45K, and 21K–40K (9, 36, 38, 43–46, 153, 170); one preparation has been described that contains only 4 subunits and appears to be missing the 65K–67K polypeptide (8, 21, 42). This new polymeric factor stimulates the formation of ternary complex and overcomes the inhibitory effect of Mg^{2+}, catalyzes guanine nucleotide (GDP/GTP) exchange on eIF-2, stimulates the catalytic binding of initiator Met-tRNA$_f$ to initiation complexes, and restores translation in heme-deficient lysates. For the purpose of this review, the term eIF-2B (46, 170) is used for this factor, consistent with the convention for naming eukaryotic initiation factors (173), particularly those that affect eIF-2.

It now seems clear that the role of eIF-2B is to recycle eIF-2, converting the inactive factor (eIF-2·GDP) produced in the course of initiation to the active form (eIF-2·GTP) capable of participating in ternary complex synthesis. A number of models have been proposed that account for the recycling of eIF-2, that are consistent with much of the data available. One of the models (A) for the catalytic role of eIF-2B in eIF-2 recycling (9, 44) is based on the mechanism previously proposed for the recycling of eukaryotic elongation factor 1α (174–177) and of prokaryotic elongation factor EF-Tu (178, 179):

$$eIF\text{-}2\cdot GDP + eIF\text{-}2B \rightleftharpoons eIF\text{-}2B\cdot eIF\text{-}2 + GDP$$

$$eIF\text{-}2B\cdot eIF\text{-}2 + GTP \rightleftharpoons eIF\text{-}2\cdot GTP + eIF\text{-}2B$$

$$eIF\text{-}2\cdot GTP + Met\text{-}tRNA_f \rightleftharpoons eIF\text{-}2\cdot GTP\cdot Met\text{-}tRNA_f$$

According to this scheme, the eIF-2·GDP complex, released as such in the process of initiation (136, 137, 138) or formed from GDP and released eIF-2 (133), particularly in the presence of Mg^{2+}, reacts with eIF-2B to form a binary complex containing eIF-2 and eIF-2B, with the release of GDP. GTP then displaces eIF-2B from that complex, with the formation of eIF-2·GTP, the active form of the binding factor, which can now react with Met-tRNA$_f$ and produce the ternary complex. Thus, the first two steps represent an exchange reaction between GDP and GTP, catalyzed by eIF-2B: eIF-2·GDP + GTP $\rightleftharpoons$ eIF-2·GTP + GDP.

Another model (B) suggests that a GDP/GTP exchange occurs at the level of the eIF-2B·eIF-2 complex (43, 46):

eIF-2·GDP + eIF-2B·eIF-2·GTP $\rightleftharpoons$ eIF-2·GTP + eIF-2B·eIF-2·GDP

eIF-2B·eIF-2·GDP + GTP $\rightleftharpoons$ eIF-2B·eIF-2·GTP + GDP

eIF-2·GTP + Met-tRNA$_f$ $\rightleftharpoons$ eIF-2·GTP·Met-tRNA$_f$

This scheme proposes that eIF-2·GDP exchanges with the eIF-2·GTP component of the eIF-2B·eIF-2·GTP complex, releasing active eIF-2·GTP to react with Met-tRNA$_f$. The GDP in the resulting eIF-2B·eIF-2·GDP complex then exchanges with free GTP because of the differential affinity of the two-factor complex for GDP and GTP; thus, GDP/GTP exchange does not occur on eIF-2 but on the eIF-2B·eIF-2 complex.

A third model (C) has been proposed that does not require factor-associated GDP/GTP exchange (8, 42). According to this scheme, eIF-2 containing only the α and γ subunits binds to the 40S subunit (as a ternary complex with GTP and Met-tRNA$_f$), is released on initiation, and reacts with eIF-2B and eIF-2(β). This latter reaction reconstitutes the binding factor eIF-2(αβγ), either prior to or at the same time that it forms a complex with the cycling factor eIF-2B, and is now capable of reacting with GTP and Met-tRNA$_f$. When the quaternary complex reacts with 40S subunits, only eIF-2(αγ)·GTP·Met-tRNA are transferred to it. Eventually, GTP is hydrolyzed, and eIF-2(αγ), GDP, and P$_i$ are released.

The existence of some of the crucial intermediates, such as eIF-2B·eIF-2·GDP, 40S·eIF-2(αγ), and eIF-2B·eIF-2(β), etc, remains to be established (9, 42, 141). It does not appear that the eIF-2B catalyzed nucleotide exchange proposed in Model A can be explained simply on the basis of increasing affinities (GDP < eIF-2B < GTP) for the same site on eIF-2α, without the contribution of energy. One possible explanation is that the binding site for GTP on eIF-2α, induced by the interaction of eIF-2B with eIF-2, is different from the one for GDP. Another possibility is that nucleotide exchange occurs not at the level of eIF-2 but of the eIF-2B·eIF-2 complex, as proposed in Model

B; thus, as compared to GDP, the affinity of GTP would be lower for free eIF-2 but higher for eIF-2 associated with eIF-2B. It should be noted that the same concerns regarding nucleotide exchange here pertain as well to the recycling-associated nucleotide exchange reaction involving prokaryotic EF-T and eukaryotic EF-1 in chain elongation.

The formation of binary complexes from free eIF-2 and GTP (14, 180) or Met-tRNA$_f$ (14) has been reported, and the formation of a transient intermediate (eIF-2·GDP), which reacts with Met-tRNA$_f$ to form the ternary complex, has been suggested (14). In view of the information regarding the recycling of eIF-2, the formation or release of eIF-2·GDP complex, and the guanine nucleotide exchange catalyzed by eIF-2B, it appears unlikely that free eIF-2 exists to any great extent in the cell. The formation of ternary complex via the reaction of free eIF-2 with GTP and Met-tRNA$_f$, as is commonly described, must be a rare event indeed, except in the case of newly synthesized factor. Most likely, the quantitatively major cellular precursor of the ternary complex is eIF-2·GDP, and the formation of the eIF-2B·eIF-2 complex, with or without associated nucleotides, must be an obligatory intermediate in this process.

The recycling of eIF-2, with its associated GDP/GTP exchange catalyzed by eIF-2B, provides an explanation for the inhibition of initiation obtained in heme deficiency or with double stranded RNA (114–125). In these cases, protein kinases are activated that phosphorylate specifically the α subunit of eIF-2 in the presence of ATP. According to two of the models described above (A and B), eIF-2(αP) reacts with eIF-2B but forms such a tight complex that it cannot be dissociated, thus sequestering the small amount of eIF-2B available for recycling (9, 46, 161). According to the third model (C), eIF-2 requires eIF-2B for maximal activity, but eIF-2(αP) is incapable of reacting or forming a complex with eIF-2B (8, 42). The sequestering of eIF-2B by eIF-2(αP) explains why translation is almost completely inhibited when only 20–30% of eIF-2 is phosphorylated (49, 181) and why eIF-2(αP) is as active as eIF-2 in assays that measure stoichiometric activity only (49). Studies with interferon-treated mouse L-cell extracts indicate that phosphorylation of eIF-2(α) is not sufficient to cause inhibition of protein synthesis. Inhibition in these extracts induced by dsRNA, correlates better with activation of (2'–5')-oligoadenylate synthetase than with activation of eIF-2α kinase (182).

The Interaction of Ternary Complex with 40S Ribosomal Subunits

The binding of ternary complex to 40S subunits, as determined by the eIF-2 and GTP-dependent binding of radioactive Met-tRNA$_f$ to the small ribosomal subunit, has been demonstrated in many laboratories, with preparations obtained from a variety of cells such as reticulocytes, *A. salina,* wheat germ, liver, cultured cells, yeast, etc (11, 38, 40, 61, 85–90, 92, 93, 96, 102–104,

112, 136, 141). In many of these experiments, the [eIF-2·Met-tRNA$_f$·GTP] ternary complex was bound to purified 40S subunits devoid of any other initiation factors, in the absence of mRNA or exogenous initiation factors other than the eIF-2. However, it seems evident from studies with unfractionated cell-free systems that Met-tRNA$_f$ binds to native 40S (43S–46S) subunits that represent a complex of 40S particles and nonribosomal proteins, that contain no mRNA, and are active in protein synthesis (39, 52, 56–58, 60, 64, 183–189). Buoyant density analysis by cesium chloride gradient centrifugation indicated values of 1.41 g/cm^3 for native subunits as compared to 1.51 g/cm^3 for purified 40S subunits (35, 39, 53, 190). Calculations based on these buoyant density values suggested that about 700,000 daltons of nonribosomal protein are complexed to 40S subunits. The nonribosomal proteins can be dissociated from the native 40S subunits by extraction with high salt (0.5 M KCl)-containing solutions, and have been shown to contain eIF-3 (39, 56, 64), as well as other initiation factors.

Although binding of Met-tRNA$_f$ to 40S subunits in vitro can be detected with GTP and eIF-2, in the absence of eIF-3, the reaction is markedly stimulated by or requires eIF-3 (19, 34, 68, 102, 141, 191). This effect appears to be directly on the binding reaction rather than as a consequence of any ribosome dissociation activity of eIF-3 (191). Also, studies with a temperature sensitive yeast mutant (*ts* 187) with an altered initiation factor (192) suggest that the defective component is eIF-3. The normal (wild-type) factor converts 40S subunits to 43S particles, and formation of this [40S·eIF-3] complex is obligatory for the interaction with ternary complex (193); when the mutant eIF-3 is inactivated at the nonpermissive temperature, the 43S complex is not formed and Met-tRNA$_f$ does not bind to the particle.

These data, on the binding of ternary complex to native 40S subunits or to reconstituted [40S·eIF-3], suggest an important role for eIF-3 either in the formation or the stabilization of the 40S-preinitiation complex containing eIF-3 and [eIF-2·Met-tRNA$_f$·GTP]. The three components of the ternary complex (eIF-2, Met-tRNA$_f$, and GTP) are associated with the ribosomal subunit in a 1:1:1 ratio (136, 141), as are all of the polypeptide components of eIF-2 and eIF-3 (141, 191); however, there are reports that eIF-3 had no effect on this binding reaction (74, 102) or that the β subunit of eIF-2 is not transferred with the ternary complex to 40S subunits (8, 42).

In addition to eIF-3, other initiation factors such as eIF-4C (19, 34, 59, 72), eIF-2B, or eRF (8, 19, 89, 148), and eIF-1 (16, 194), have been implicated in the 40S-ternary complex interaction. Factor eIF-4C, as in the case of eIF-3 discussed above, may play a role in ribosome dissociation and in the binding of the ternary complex. The eIF-2B is the eIF-2 recycling or guanine nucleotide exchange factor, according to two of the schemes presented above (Models A and B), but is clearly an essential component in the transfer of Met-tRNA$_f$ to the

40S subunit; eIF-1 is a single polypeptide of about 15,000 molecular weight, which appears to stabilize various initiation complexes, including, possibly, the 40S-preinitiation intermediate. The requirement for mRNA or the initiation codon AUG, in the binding of ternary complex, has been reported (23, 85, 96, 188, 195); however, the data described above, indicating that Met-tRNA$_f$ binds to native 40S subunits devoid of mRNA, and the observations in many laboratories that templates have no effect on the binding reaction (15, 18, 61, 88, 196) indicate that the binding of the ternary complex precedes that of mRNA in the initiation process.

The currently accepted model calls for the binding of eIF-3 and then eIF-4C to 40S subunits, either during dissociation or immediately thereafter, to form the obligatory 43S–45S acceptor for the [eIF-2·Met-tRNA$_f$·GTP] ternary complex; the resulting complex is referred to as the preinitiation complex.

Binding of mRNA to the 40S-Preinitiation Complex

The ordered sequence for the formation of intermediates in initiation suggests that binding of eIF-3 and eIF-4C to the 40S subunit preceeds that of ternary complex and that mRNA then binds to the complete preinitiation complex containing 40S subunits, eIF-2, eIF-3, eIF-4C, Met-tRNA$_f$, and GTP (15, 18, 19, 88, 89, 183, 197). A number of factors have been implicated in the interaction between mRNA and the preinitiation complex: eIF-4A (15–19, 198, 199), eIF-4B (15–19, 198–205), cap binding protein (CBP) I (20, 204, 206–212) and CBP II or eIF-4F (199, 204, 205, 213–215); in addition, eIF-1 (18, 216), eIF-2 (217–224), eIF-3 (18, 19, 213, 225–227), and ATP (15, 17–19, 203, 228–230) also appear to play a role in this reaction. Some of these initiation factors have been shown to be required in or to stimulate fractionated systems, form complexes with or cross-link to mRNA, or to directly enhance the binding of mRNA to the 40S-preinitiation intermediate.

Eukaryotic cellular mRNAs seem to be exclusively monocistronic. Each polynucleotide chain coding for a protein has a single initiation codon, which is AUG only (231–235), and a single termination codon (UAA). The primary, direct product of translation is a single polypeptide sequence per mRNA molecule. It should be emphasized, however, that considerable posttranslational processing occurs, frequently giving rise to more than one active protein from the primary product [see, for example, ref. (236)]. Cellular mRNAs contain untranslated, noncoding regions at both the 5' and the 3' ends of the coding sequence, which commonly range between 40 and 150 nucleotides in length (235); these noncoding sequences tend to be less conserved than the coding sequences [see ref. (2)]. This latter observation is of particular interest since there are suggestions of two conserved positions close or adjacent to the initiation codon, a conserved region several nucleotides 5' to the AUG, and possibly another very near the 5' terminus (2); it is possible that these nucleotides or sequences may play an important role in the accurate and efficient binding and/or positioning of mRNA to 40S ribosomal subunits. Translation is almost invariably initiated at the first AUG from the 5' end of the mRNA, and there is no convincing evidence for two or more nonoverlapping (polycistronic) coding sequences, although some viral RNAs contain overlapping coding regions (237–241); in the latter case, translation can be initiated both at the 5'-proximal AUG and at an initiation codon other than the 5'-proximal one, the remaining internal codons are read in frame, both polypeptides are terminated at the single termination codon, and two structurally related polypeptides are produced.

The extreme termini of eukaryotic mRNAs contain unique structures which represent two of a number of events that modify mRNA. The 5' terminus consists of a 7-methyl guanosine linked by a 5'–5' triphosphate bridge to the penultimate nucleotide, which is methylated at the 2'-hydroxyl position, m^7-GpppNm. . . . (242–245). The m^7-GTP cap increases the binding of mRNA to ribosomal 40S subunits (204, 245–247) and its stability (204, 248–252). Indeed, the importance of the cap structure is underscored by the observations that cap analogs such as m^7-GMP, m^7-GDP, m^7-GTP, etc, severely inhibit the interaction of mRNA with various initiation factors and 40S subunits, as well as in vitro translation (200, 221, 253–264). Also, the enzymatic addition of the cap structure to a normally uncapped (prokaryotic) mRNA allows translation in a eukaryotic cell-free system (265) while removal of the cap from a normally capped (eukaryotic) mRNA decreases its translational efficiency in such systems (266, 267). The posttranscriptional modification at the 3' end is the addition of a polyadenylate "tail," usually 50–150 nucleotides in length, to the end of the 3' noncoding sequence (268–271). The addition of poly(A) to the primary mRNA transcript, as well as other facets in the synthesis, processing and metabolism, have been reviewed recently (7, 272). An interesting, highly conserved, homologous sequence, AAUAAA, has been reported in the 3' noncoding region of eukaryotic mRNAs, some 10–30 nucleotides from the start of the poly(A) tail (273). This sequence may be involved in the addition of poly(A), possibly as a recognition site for cleavage or for poly(A) polymerase. The role of poly(A) in eukaryotic mRNAs remains to be established. It has been suggested that poly(A) enhances the stability of mRNA. Although some mRNAs are more stable in the polyadenylated form (274–280), the stability of other messengers, lacking poly(A), is not affected (281) and some viral poly(A)$^+$ mRNAs decay rapidly under some conditions and slowly under others (274, 282–284). Poly(A) is not essential for protein synthesis, since poly(A)$^-$ mRNA is readily translated (272, 280, 281), and is not required for splicing of intervening sequences from the mRNA transcript since several early adenovirus RNAs are correctly spliced even when polyadenylation is blocked with cordycepin (285).

The involvement of initiation factors eIF-1, eIF-2, and eIF-3, in the formation of the 40S-preinitiation complex has been described above. Factor eIF-1, which stabilizes the 40S-preinitiation intermediate, is assigned a similar role with respect to the new 40S complex containing [eIF-2·Met-tRNA$_f$·GTP] and mRNA (216). Factor eIF-2, the initiator Met-tRNA$_f$-binding factor, also forms complexes with free eukaryotic mRNA, with high affinity, and this reaction is inhibited by 5'-cap analogs such as m^7-GMP (217–224). These observations have suggested that eIF-2 possesses a binding site in addition to those for Met-tRNA$_f$, GTP, and native 40S subunits, that reacts with and stabilizes mRNA when it becomes associated with the 40S preinitiation complex (219). Factor eIF-3, which is involved in ribosome dissociation and binding of ternary complex to 40S subunits, is an essential component of the particle to which mRNA binds, and is also capable of binding to isolated mRNA (200). Several possibilities exist for the role of eIF-3 in mRNA binding: In integrating with 40S subunits, eIF-3 affects the conformation of the particle, generating a binding site for mRNA; the 40S-bound eIF-3 is a part of the active site and participates directly in the recognition and binding of mRNA (225); and/or eIF-3 relaxes and unwinds the extensive secondary structure in mRNA (286, 287).

The other protein factors required for mRNA binding, eIF-4A, eIF-4B, and CBP (I and II) play a more direct role in this reaction. Factor eIF-4A is a single polypeptide chain with a molecular weight of 44,000–55,000 (22, 25, 27, 203, 288, 289), and pI of 6.1 (203); eIF-4B is also a single polypeptide with a molecular weight of about 80,000 (18, 19, 25), and CBP I has a molecular weight of approximately 24,000 (24, 206, 208, 218). Complexes containing eIF-4A and eIF-4B (16, 65) and CBP I plus eIF-3 or eIF-4B (203, 206, 290), etc have been reported. However, more recent evidence suggests that some of these factors, and at least one other polypeptide, exist as a high (7S–10S) molecular weight complex, referred to as CBP II or eIF-4F (199, 204, 205, 213, 215, 289, 291). This polymeric protein, with a molecular weight of about 300,000, contains 3 subunits of M_r 24,000 (CBP I), 50,000 (eIF-4A), and 200,000–220,000 (199), although preparations containing an additional subunit, with a molecular weight of 73,000 (205) or of 55,000 (289), have been reported. It should be emphasized that many of the early studies on the mRNA-binding factors were carried out with less than pure preparations, containing other activities; for example, eIF-4B preparations were frequently contaminated with CBP I (206, 208), or possibly eIF-4F (205).

Factors eIF-4A and eIF-4B stimulate the binding of mRNA to 40S subunits, bind to and form complexes with free mRNA, and cross-link specifically to the oxidized 5' cap structure of mRNA in the presence of ATP (18, 19, 200, 202–206, 214, 292, 293). Most of these reactions are inhibited by cap analogs (200, 202, 203, 213, 291). CBP I also cross-links specifically to oxidized mRNA, but ATP is not required; in addition, cross-linking of the 28-, 32-, 50-, 80-, and 240- kilodalton polypeptides has been reported (214, 294). In the presence of eIF-4B and ATP, the high-molecular-weight cap binding complex eIF-4F exhibits the same activities as free eIF-4A and CBP I. Indeed, whereas cross-linking of CBP I to oxidized mRNA is ATP-independent, the cross-linking of CBP I in the high-molecular-weight complex is ATP-dependent (213). It seems most likely that the 24,000- and the 50,000-dalton polypeptides in the 7S complex are probably CBP I and eIF-4A, respectively; they also seem to be more active in the complex than in the free form (204). Whether the 73,000–80,000 dalton polypeptide detected in eIF-4F is related to eIF-4B (204) and whether it is an essential part of the complex, remains to be firmly established, as is the significance of the other polypeptides that react with the 5' cap structure of oxidized mRNA.

It would be tempting to assume that eIF-4F/CBP II contains all of the initiation factors required for the ATP-dependent binding of capped mRNAs to the 40S preinitiation complex, and that no other components are necessary. However, experiments suggest that the ATP-dependent, cap-specific binding and cross-linking of the 24-, 50-, and 80-kilodalton polypeptides require not only the high molecular weight cap binding protein complex, but eIF-4A and eIF-4B as well (204, 289). The possibility that this step in initiation involves a cycling of these mRNA-binding factors between free and complexed forms, and that some structural modifications of the polypeptides may also be involved in this partitioning, must be considered. In this respect, it is of interest that in the process of mRNA binding to 40S ribosomal subunits, eIF-4A and eIF-4B are released and are not found in the ribosomal complex that is formed (18, 19). Whether the free and the complexed forms of the factors are indeed identical, even though they react with an antibody against cap-binding protein, will require additional studies with pure initiation factors and a more detailed analysis of their structures.

Additional evidence for the role of cap binding proteins in the recognition and binding of mRNA has been obtained from studies on cells in which host protein is shut off after infection with some viruses, such as the picornaviruses and some plant viruses (295–297). In contrast to most eukaryotic mRNAs, these viruses do not contain the cap structure m^7-GpppNm at the 5' terminus (298–304). When mammalian cells such as HeLa cells are infected with picornaviruses such as poliovirus, host protein synthesis is inhibited, host polysomes are disaggregated although the mRNA is not degraded (305–307), and capping, methylation, and polyadenylation of host mRNA appears to be unchanged (306). Most capped cellular and viral mRNAs are not translated in poliovirus-infected cells (308–310) but mRNAs isolated from infected cells are fully functional; that is, they are translated normally in heterologous (wheat germ) cell-free systems (311). These observations suggest that in infected cells, host protein synthesis is blocked at the level of chain initiation, capped mRNAs are not translated but uncapped viral mRNAs are, and that a cellular component required for the translation of capped mRNAs is inactivated by viral infection.

Studies using cell-free extracts and initiation factor-containing fractions from poliovirus-infected HeLa cells showed that uncapped mRNAs were translated in vitro but capped mRNAs were not (214, 310). Earlier evidence suggested that eIF-3 or eIF-4B was inactivated in the infected cells (308–310, 312); however, later studies failed to detect any alterations in eIF-2, eIF-3, eIF-4A, or eIF-4B (313). Subsequently, the 24 kilodalton CBP I, which is frequently co-purified with eIF-3 and eIF-4B, was thought to restore the capacity of extracts from infected cells to translate capped mRNAs (290). More recent evidence assigns this restoring activity, perhaps exclusively, to the high-molecular-weight cap binding protein complex eIF-4F/CBP II (199, 204, 205, 214, 294); in fact, it has also been shown that eIF-4F stimulates the cell-free translation of globin mRNA whereas CBP I does so only poorly (204, 205). After poliovirus infection, cells contain significantly lower levels of all of the cap binding proteins, as determined by cross-linking analysis (214, 294), although the effect on the 24 kilodalton CBP I level is uncertain (210, 212). An activity is also present in infected cells that causes a decrease in the levels of cap binding proteins in control extracts and prevents the cap binding protein-dependent restoration of capped mRNA translation (294, 314). Thus, it seems that eIF-4F/CBP II, which contains the 24 kilodalton CBP I, 50 kilodalton eIF-4A, and possibly the 80 kilodalton eIF-4B, may be the essential cap binding protein complex that is inactivated in poliovirus infection. The polypeptides of molecular weight higher than the 24 K subunit are therefore also essential for the translation of capped mRNAs. Experiments with monoclonal antibodies have revealed that the various polypeptides in the complex share some structural features (214, 215). A viral gene product, a protein, may be responsible for the shut off of host protein synthesis by disrupting the complex or hydrolyzing one or more of its polypeptides (294, 309, 314–317). The high-molecular-weight cap binding protein(s) could conceivably be the precursor(s) of some of the lower-molecular-weight polypeptides, such as the 24 kilodalton CBP I (213, 215, 291). Cleavage of the largest (M_r 220,000) polypeptide in eIF-4F/CBP II during poliovirus infection has been correlated with the inhibition in protein synthesis (318).

Although the studies with virus-infected cells emphasize the critical functions that these mRNA-binding initiation factors play in the translation of capped mRNAs, the exact role of the cap binding proteins and of the 5' cap structure itself must be carefully considered, since uncapped messengers are properly translated and do not require the cap binding proteins. It has been suggested (203, 204, 213, 214, 291, 295) that cap binding proteins and ATP are necessary to destabilize the secondary structure of capped mRNAs, which facilitates the binding of mRNA to the 40S preinitiation complex, and that there is a relationship between the degree of secondary structure and the extent to which mRNA binding is dependent on the cap structure, cap binding proteins, and ATP. Consistent with these suggestions are the findings that some mRNAs with less secondary structure such as capped synthetic inosine-substituted reovirus, capped AMV-4, and naturally uncapped cowpea mosaic virus and EMC virus, etc, are less dependent on the cap structure, on some initiation factors, or ATP (4, 230, 293, 318–320); also, a monoclonal antibody against cap binding protein inhibited the binding of the preinitiation complex to native mRNA but not to inosine-substituted capped reovirus mRNA or to uncapped mRNA (209, 291).

There may be regulatory mechanisms, other than the effect on cap binding protein, that operate in the selective translation of mRNAs. Experiments with mouse L-cells showed that infection with vesicular stomatitis virus (VSV) effected the selective translation of VSV RNA, which has the 5' cap structure (321). Translation of L-cell mRNA was markedly lower in extracts prepared from infected cells, inhibition was at the level of initiation and, although the eIF-2 levels appeared to be similar in both control and infected extracts, optimal recovery of translational activity was obtained with a partially-purified preparation of eIF-2 from noninfected cells. However, whether the results are due to the presence of other initiation factors in the partially-purified eIF-2 or whether eIF-2 modulation directs the selective translation of VSV mRNA in infected cells remains to be shown. Other studies, with mengovirus RNA, also implicate eIF-2 in regulation; it should also be borne in mind that eIF-2 forms specific complexes with mRNA, as discussed previously. Under conditions that support protein synthesis in vitro, ribosomes recognize and bind to several sequences in mengovirus RNA that are far removed from the 5' end, and are detected in 80S initiation complexes (322); eIF-2 also binds to some of these sequences. In addition, mengovirus RNA is translated much more effectively than globin mRNA in reticulocyte lysates, and addition of purified eIF-2, which does not stimulate overall protein synthesis, shifts translation in favor of globin mRNA (323); the ability of these two RNAs to compete in translation correlates with their ability to bind to eIF-2. Other evidence of discrimination among capped mRNAs (reovirus versus globin mRNA) indicates that eIF-4A and/or the cap binding protein complex is involved, and suggests that some feature other than the 5' cap structure is recognized by the discriminatory factor (324).

The requirement for ATP in eukaryotic protein synthesis has been localized to the interaction between capped mRNAs and the 40S preinitiation complex (18, 19, 203, 204, 213, 289, 291, 293, 319–326). Hydrolysis of ATP is required as evidenced by the failure of nonhydrolyzable analogs to support this reaction (18, 203, 208, 325). Evidence has been obtained with initiation factors from wheat germ that the primary interaction of ATP is with eIF-4A (326). A number of possibilities occur as to the site of action and function of ATP (and of mRNA-binding factors) in this process, but the exact molecular mechanism is not at all clear. The nucleotide may be required for actively melting the 5'-proximal secondary structure of mRNA to facilitate the subsequent binding of cap recognition factors; it may be required directly for the binding of the cap recognition factors to mRNA; ATP may participate with the mRNA-binding factors to melt the 5'-terminal secondary structure to facilitate binding of the 40S preinitiation complex to the appropriate position at the 5' end of mRNA, after the interaction of the mRNA with the cap recognition factors; it may be required for the "scanning" or migration (2, 230, 235, 327–329) of the 40S subunit from an entry site at the 5' terminus to the translation initiation site at the 5' proximal AUG codon. The findings that mRNAs with diminished secondary structure have a lower dependence on ATP for the formation of initiation complexes are consistent with the hypothesis that ATP hydrolysis is required for the melting of the 5' proximal structure (230, 291, 293, 319). However, studies on the requirement of ATP for the formation of complexes or for cross-linking between initiation factors and mRNA, as a function of RNA secondary structure, are at variance. In one study, the cross-linking of the 50- and 80-kilodalton polypeptides to mRNA was found to be ATP-independent when the secondary structure of the mRNA was less ordered (293). The results suggested that the secondary structure at the 5'-end determines the accessibility of the cap structure to the mRNA-binding proteins, and that ATP is required for the melting of that region as a prerequisite to the interaction between certain cap recognition factors and the cap structure. In other studies (204, 325), the interaction of eIF-4A and eIF-4B with mRNA was found to be ATP-dependent regardless of the extent of secondary structure. These results suggested that ATP hydrolysis is required for some process other than melting the 5'-proximal structure, that the attachment of initiation factors to mRNA occurs prior to denaturation of mRNA, and that denaturation occurs before or at the same time as the attachment to and scanning of 40S subunits on the template. Another interesting observation made in the latter study (325) was that the complexes that were formed between 40S subunits and native, denatured, or uncapped mRNA in the absence of ATP were not normal; that is, in contrast to complexes formed with ATP, the 5' capped sequence in these complexes was not protected by the 40S subunits against RNase digestion, and their formation was not inhibited by cap analogs. Thus, the possibility that ATP hydrolysis is related not to the denaturation of mRNA but to some other event is strongly suggested.

The scanning mechanism in translation initiation is an interesting concept that calls for the migration of the 40S preinitiation complex along the mRNA strand, from its entry or binding site at the 5' end to the initiator AUG codon (230, 327). This hypothesis has been reviewed recently (2). Although in most eukaryotic mRNAs translation is initiated at the first AUG from the 5' terminus, in a few cases initiation occurs at an AUG other than the one that is closest to the 5' end (230, 329, 330). Recognition of the appropriate AUG triplet for initiation appears to depend on the flanking sequence, and ^A_CNNAUGG has been proposed as the optimal sequence context for initiation (234). If the sequence adjacent to the 5' proximal AUG is suboptimal, all or some of the 40S subunits may bypass that site and initiate at the next potential site farther downstream. If both AUGs are in frame, it is possible that initiation could also occur at either site, which would result in the synthesis of two closely related proteins on a single monocistronic mRNA.

The Joining of 60S Subunits to Form the 80S Initiation Complex

The reaction of mRNA with the 40S preinitiation complex results in the formation of a new preinitiation intermediate containing 40S subunits, eIF-2, eIF-3, GTP, Met-tRNA$_f$, mRNA, and possibly eIF-4C. The fate of some of the other initiation factors is not clear, although the available evidence suggests that eIF-1, eIF-4A, eIF-4B, and eIF-4C are not present in the mRNA-containing 40S complex (19). One would suspect, on the basis of the ribosome dissociation role proposed for eIF-4C (72), that this factor is still in the complex at this stage. The anticodon of Met-tRNA$_f$ is base-paired to the initiator AUG codon in mRNA at this time, and the GTP has not been hydrolyzed yet.

The only additional initiation factor required for the next step, the joining of the large (60S) ribosomal subunit to the new 40S preinitiation intermediate, is eIF-5. This factor is a single polypeptide chain, and estimates of the molecular weight vary between 100,000 and 160,000, with most grouping at the higher value (13, 17–19, 34, 80, 137, 141, 224). When eIF-5 reacts with the 40S complex, GTP is hydrolyzed to GDP and P$_i$, eIF-2·GDP and eIF-3 are released, and the 40S·Met-tRNA$_f$·mRNA complex can react with free 60S subunits to form the 80S initiation complex. The eIF-2·GDP complex, as described above, reacts with eIF-2B and GTP in a recycling reaction that generates eIF-2·GTP, picks up a molecule of Met-tRNA$_f$ to form the ternary complex [eIF-2·GTP·Met-tRNA$_f$], and is ready to participate in another series of initiation reactions. The hydrolysis of GTP, triggered by the reaction with eIF-5, is a critical intermediary step. Factor eIF-5 catalyzes the hydrolysis of 40S-bound GTP (137, 331), and the release of eIF-2 and eIF-3 (136, 137, 141, 191) in the absence of 60 subunits. When the reactions that precede this step are carried out with nonhydrolyzable analogs of GTP, binding of Met-tRNA$_f$ and mRNA to 40S subunits occurs (14, 15, 18, 191, 332) but the subsequent release of eIF-2 and eIF-3 (130, 168), and the joining of 60S subunits (14, 15, 18, 191) does not take place. It has also been reported that dsRNA-induced phosphorylation of eIF-2 allows the formation of a 48S preinitiation complex containing Met-tRNA$_f$, guanine nucleotide, and eIF-2(αP) which can bind mRNA, but the ensuing complex is unable to react with 60S subunits (125). It would appear that the release of the high-molecular-weight ribosome dissociation factor eIF-3 (and possibly eIF-4C) would be sufficient to allow free 60S subunits to join the 40S complex; however, in view of the recent reports on the low-molecular-weight dissociation factor eIF-6, and the formation of 60S·eIF-6 complexes (32, 39, 41, 71, 77–79), an additional explanation is necessary. Either eIF-6 is released from 60S subunits by some mechanism prior to the joining step or the factor is released as a consequence of the reaction between the two ribosomal subunit complexes. The product of these reactions is the formation of an 80S initiation complex that is ready to participate in chain elongation reactions.

Additional Factors

A factor, eIF-4D, has been identified (19, 333), that does not affect translation of natural mRNA or any of the initiation reactions but stimulates the synthesis of methionyl-puromycin in incubations containing puromycin and the components necessary to form the 80S initiation complex. Factor eIF-4D is a single 14,000–19,000 dalton polypeptide chain with a pI of about 5.1 (19, 333, 334). A number of possible roles for eIF-4D have been proposed which would explain its effect on the reactivity of ribosome-bound Met-tRNA$_f$ with puromycin: It stabilizes intermediate complexes in initiation (59); it is responsible for the correct alignment of Met-tRNA$_f$ on the ribosome (19); it stimulates the activity of eIF-5 and thus, the formation of the 80S initiation complex (18); or it may act directly in the reactions involved in the formation of the first peptide bond in chain elongation, discussed below. If the latter explanation is correct, this factor would more properly be described as an "interphase" factor whose role bridges the initiation and elongation phases of translation. Factor eIF-4D is also the cellular component that is uniquely modified posttranslationally at a lysine, in growth-stimulated lymphocytes, forming hypusine (334); it has been suggested that this modification may play a role in the regulation of protein synthesis during lymphocyte growth stimulation.

In addition to eIF-2, another protein factor referred to here as eIF-2A (also known as IF-M1, EIF-1, cIF-1, or eIF-1B) has been shown to catalyze the binding of initiator Met-tRNA$_f$ to ribosomal particles (12, 29, 30, 40, 47–51, 335–341). This factor differs in many respects from eIF-2; it does not affect the translation of natural mRNA but promotes the AUG codon-dependent, GTP-independent binding of Met-tRNA$_f$ to 40S subunits. Factor eIF-2A has been isolated from the cytosol and/or ribosomal high salt extracts from a variety of cells such as *Artemia salina* (47, 336), liver (46, 48, 51, 337–339), reticulocytes (12, 49, 340), hen oviduct (50), wheat germ (29, 30), and human tonsils (341). The physical properties for different purified preparations vary considerably. The factor from *A. salina* has a molecular weight of 145,000 and is composed of two 74,000-dalton subunits (47); the one from rat liver is 4? kilodaltons and has two unequal subunits of M_r 18,000 and 28,500 (40); one report on reticulocyte eIF-2A estimates a molecular weight of 50,000 and two subunits of 20 kilodaltons and 30 kilodaltons (49), while another describe reticulocyte eIF-2A as a single polypeptide chain of 65,000 daltons (12) Experiments with isotopically labeled rat liver factor revealed that both polypeptide chains bind to 40S subunits to form a 40S·eIF-2A complex which then reacts with Met-tRNA$_f$ and AUG to form an eIF-2A·40S·Met-tRNA$_f$·AUC complex (51). When 60S subunits are added to the quaternary complex, the factor is released from the particles, and a new complex is formed [80S·Met tRNA$_f$·AUG], which is capable of reacting with puromycin to yield methionyl puromycin (51). However, the template-dependent binding of Met-tRNA$_f$ or the formation of a puromycin-sensitive "80S initiation" complex has not been obtained with eIF-2A when natural mRNA is used as the template (51, 196) Recent studies (51) have also shown that eIF-2A catalyzes the removal of deacylated tRNA from 40S subunits generated as a consequence of chain termination, a prerequisite to the binding of Met-tRNA$_f$. Thus, this factor may function by participating in various interactions at the level of the ribosoma subunit pool in the ribosome cycle.

PEPTIDE CHAIN ELONGATION

The initiation series of reactions are finalized with the formation of an 80S initiation complex containing a ribosome, initiator Met-tRNA$_f$, and mRNA but no initiation factors. The Met-tRNA$_f$ is bound to a ribosomal site referred to as the P (or donor) site, and its anticodon is base-paired at that site with the initiation codon AUG in mRNA. Initiation is therefore responsible for the translation of the appropriate AUG as methionine. The next codon to be translated, that is, the first internal codon, is in an open ribosomal position adjacent to the P site, referred to as the A (or acceptor) site. The appropriate aminoacyl-tRNA, depending on the codon available at the A site, then binds to that site in a reaction catalyzed by GTP and a binding factor for aminoacyl-tRNA that is specific for elongation. When the two ribosomal binding sites are filled, the P site with methionyl-tRNA and the A site with the newly bound aminoacyl-tRNA, a peptide bond is formed between the methionine residue and the incoming aminoacyl moiety. Peptide bond formation results in the synthesis of a dipeptide attached to tRNA (peptidyl-tRNA) at the A site, while the P site is left with a deacylated tRNA. This reaction does not require any soluble (nonribosomal) protein factors or nucleotides. In order to translate the next codon in mRNA, that triplet must be moved to the A site, the decoding site for chain elongation. Whereas translation of the initiation (AUG) codon takes place at a position that eventually becomes the ribosomal P site, translation of ll of the internal mRNA codons is carried out at the ribosomal A site. Movement of the next codon to the A site is accomplished by a translocation factor and GTP, which catalyze the movement of the ribosome and of mRNA in relation to each other, by the equivalent of one codon. In this process, the peptidyl-tRNA at the A site is translocated to the ribosomal P site (along with the codon to which it is base-paired), a new codon is shifted to the A site, and the deacylated tRNA at the P site is released. The ribosome, with peptidyl-tRNA at the P site and a new codon at the open A site, is now ready to repeat the series of reactions involving binding of aminoacyl-tRNA, peptide bond formation and translocation, until all of the codons that are translatable into amino acids have been decoded. In this manner, amino acids from the aminoacyl-tRNA pool are added one at a time to the nascent peptidyl chain, and growth occurs from the N-terminal residue toward the C-terminal residue. The two soluble translational factors that are required for chain elongation were first discovered in mammalian preparations (342–346), and then found in all cells that synthesize protein.

The Binding of Aminoacyl-tRNA to the Ribosomal A Site

The eukaryotic binding factor for aminoacyl-tRNA in chain elongation, EF-1, has been isolated from a variety of cells including liver (347–353), reticulocytes (354–358), brain (359, 360), wheat germ (33, 361–365), A. salina (174, 366–368), yeast (369–372), ascites (373, 374), silk glands (375, 376), etc. In most cell extracts, EF-1 activity is detected in multiple forms ranging in molecular weight from 140,000 to 500,000 or higher (347, 349, 358, 359, 377, 378). The higher-molecular-weight forms, referred to as EF-1_H, appear to be aggregates of lower ones. The major component of EF-1, regardless of molecular weight or tissue of origin, is a polypeptide of 50,000–60,000 daltons (174, 350, 352, 357, 359, 365, 367, 368, 373, 374, 378–381) which is referred to as EF-1_α, EF-1_L or eEF-Tu. Analyses of calf liver EF-1_H and EF-1_α indicated that their amino acid compositions were quite similar (382). Molecular genetic analyses of A. salina with a plasmid carrying a cDNA sequence coding for EF-1_α (383, 384) revealed between one and four genes coding for that polypeptide, the presence of intervening sequences, and an mRNA about one-third longer than the coding sequence for EF-1_α (384). Another component, designated as EF-1_β or eEF-Ts, M_r 30,000, is commonly found in association with EF-1_α in a variety of EF-1 preparations (174, 176, 350–352, 355, 357, 364, 367, 368, 374–376, 378, 385–387), and plays an important role in the reactions catalyzed by EF-1. The presence of variable amounts of other polypeptides in EF-1 preparations has also been reported (176, 351, 352, 358, 365, 385, 388, 389); one of these (EF-1_γ, M_r 55,000) has been found in a complex with EF-1_β but does not seem to be required for activity (376, 386, 390).

Binding of the aminoacyl-tRNAs (other than the initiator Met-tRNA$_f$) to the ribosomal particle in eukaryotic chain elongation requires the intermediary formation of a ternary complex between the aminoacyl-tRNA, GTP, and the binding factor, in this case EF-1. These steps were suggested by findings that EF-1 is more stable at 37°C in the presence of aminoacyl-tRNA and GTP, but not tRNA or GDP (391), and that the resulting complex reacts with and becomes associated with the ribosome (142). EF-1$_\alpha$ is the subunit that participates in the formation of the ternary complex (355, 364, 392–394), via the formation of a binary complex with GTP (367, 377, 393–396). This subunit has a somewhat greater affinity for GTP than does the aggregate form EF-1$_H$ (367, 377), and it binds GTP better than GDP (377, 393, 395); further, EF-1$_H$ reacts poorly with aminoacyl-tRNAs (392, 394), and it is converted to EF-1$_\alpha$ on incubation with GTP and aminoacyl-tRNA (355, 364) resulting in the formation of an EF-1$_\alpha$·GTP·aminoacyl-tRNA ternary complex.

The mechanism of action of EF-1$_\alpha$ is similar to that of eukaryotic initiation factor eIF-2 and prokaryotic elongation factor EF-Tu, described above. It binds GTP and the resulting binary complex then reacts with aminoacyl-tRNAs to form a ternary complex; the ternary complex binds to and delivers the aminoacyl-tRNA to the ribosomal particle (144, 391, 397, 398), GTP is hydrolyzed (394, 397, 399), EF-1·GDP is formed (397), and the factor is released from the ribosome (398). The function of EF-1$_\beta$ appears to be analogous to that of eukaryotic eIF-2B and prokaryotic EF-Ts, also described above. It catalyzes GDP/GTP exchange on EF-1$_\alpha$ (376, 386, 387, 390, 400). Thus, according to one model, EF-1$_\beta$ reacts with EF-1$_\alpha$·GDP, displacing the GDP and forming an EF-1$_\alpha$·EF-1$_\beta$ binary complex; GTP then displaces EF-1$_\beta$ resulting in the formation of EF-1$_\alpha$·GTP which can bind another aminoacyl-tRNA. Evidence for the formation of an EF-1$_{\alpha,\beta}$ binary complex and its dissociation by GTP has been presented (386). Another attractive possibility is that EF-1$_\alpha$·GDP exchanges with EF-1$_\alpha$·GTP in EF-1$_{\alpha\beta}$·GTP, and that nucleotide exchange occurs at the level of this ternary complex: EF-1$_{\alpha\beta}$·GDP $\xrightarrow{\text{GTP}}$ EF-1$_{\alpha\beta}$·GTP + GDP. The high-molecular-weight EF-1$_H$ probably consists of aggregates of EF-1$_\alpha$ plus varying amounts of EF-1$_\beta$ and EF-1$_\alpha$·EF-1$_\beta$ and possibly EF-1$_\gamma$; failures to detect EF-1$_\beta$ may be due to the relatively small amounts of this polypeptide some EF-1 preparations.

EF-1-dependent binding of aminoacyl-tRNA to ribosomes also occurs when non-hydrolyzable analogs of GTP are used instead of GTP (143, 144, 398) however, the aminoacyl-tRNA bound to the ribosome does not react in the next step in chain elongation, peptide bond formation, unless GTP is present and hydrolysis occurs. One possible interpretation of these results is that analog-containing ternary complexes can bind to the ribosomal A site in a codon-dependent reaction that does not require GTP hydrolysis, but that an intermediate step must take place, which requires GTP hydrolysis and positions the aminoacyl-tRNAs appropriately on the A site allowing them to participate in peptide bond formation (143, 144). The finding that ribosome-bound EF-1 is not released from the ribosome when the binding reaction is carried out with a nonhydrolyzable GTP analog (398) suggests an alternate explanation; that is, peptide bond formation cannot occur until EF-1 is released from the ribosome, which requires GTP hydrolysis.

Several studies suggest that EF-1 plays a role in the quantitative regulation of translation. For example, decreases in protein synthesis concomitant with decreases in EF-1 activity have been observed in rat liver following thyroidectomy (401), in aging rats (402) and Drosophila (403), and as a consequence of transition from the exponential to the stationary phase of growth in cultured cells by serum deprivation, high cell density, etc (404–408). Increases in EF-1 activity in tissues of cold-adapting fish (409) and in response to immune challenge in mice (410) have been reported. Thus, it is possible that the rate of protein synthesis in a given tissue may be affected at the level of chain elongation by the changes in EF-1 activity. Analyses in control and SV40-transformed 3T3B cells revealed that the EF-1$_\alpha$ from 3T3B/SV40 cells was methylated to a higher extent, and that the growth properties of the two cell types correlated positively with the extent of methylation (411).

Peptide Bond Formation

When the incoming aminoacyl-tRNA has been bound and appropriately positioned or aligned at the ribosomal A site adjacent to the peptidyl-tRNA (or Met-tRNA$_f$ in the case of the first elongation cycle) at the P site, a ribosome-catalyzed reaction (peptidyltransferase) occurs. The ester carboxyl group of the peptidyl-tRNA is released and becomes bound through a peptide linkage to the amino group of the aminoacyl-tRNA. This reaction, which does not require any soluble nonribosomal translational factors or nucleotides, results in the transfer of the peptidyl moiety from the tRNA at the P site to the amino group of the aminoacyl-tRNA at the A site. A new peptide bond is formed, a deacylated tRNA is left at the P site, and the peptidyl-tRNA which is one amino acid longer is attached to the A site through the tRNA that brought in the aminoacyl-tRNA.

A great deal of work has been carried out on the peptidyltransferase activity in prokaryotic ribosomes; the structural organization of the peptidyltransferase center, the substrate specificity, the mechanism of action, etc, have been thoroughly reviewed (412, 413). Although studies with eukaryotic ribosomes have not been as extensive, the results, in general, are similar to those obtained with prokaryotes. For example, formation of a peptide bond in model systems containing peptidyl-tRNA or N-blocked aminoacyl-tRNA "donors" (P site) and aminoacyl-tRNA or puromycin "acceptors" (A site) has also been demonstrated with eukaryotic ribosomes and with isolated 60S ribosomal subunits from a variety of tissues (143, 144, 414–428).

The monovalent cation (K^+, NH_4^+) requirement for many in vitro translation systems is about 100 mM ($\pm$ 50mM), and the divalent cation (Mg^{2+}) requirement is below 2 mM. In model peptidyltransferase systems, in which the reactions are carried out with isolated ribosomes or subunits and analogs of P and A site substrates, the optimal monovalent cation concentration is considerably higher, near 0.3 mM (143, 144, 420–422); this requirement appears to be related directly to the catalytic step rather than to the binding of substrates to the particle.

Analogs of P site substrates that have been used in prokaryotic and eukaryotic peptidyltransferase assays include a variety of acylaminoacyl-tRNAs such as AcetylPhe-tRNA, and acetylaminoacyl-oligonucleotides such as CACCA(Ac-Phe) prepared from tRNA derivatives by nuclease digestion; the most common A site substrate analog is puromycin, although aminoacyl-oligonucleotides and aminoacyl-tRNAs have also been used (412, 413, 420–422, 429–443). In the experiments with isolated 60S ribosomal subunits, in the absence of template and the small ribosomal subunit, high (30–50%) concentrations of alcohol are required for activity, and this requirement is related to both the binding of donor substrates to the P site and to the peptidyltransferase catalytic step (422). Binding of acylaminoacylated oligonucleotides (3' fragments of tRNA) require, in addition, high (55 mM) concentrations of magnesium ion (421). The results obtained indicate that the model 60S transpeptidation reaction between substrates bound to the A and P sites stringently require high concentrations of both alcohol and monovalent cations.

The ability to carry out peptide bond synthesis with isolated 60S subunits indicates that the peptidyltransferase center is a property of the large ribosomal subunit, although in the 80S ribosome, the small subunit may also influence or be involved in the formation of the catalytic center. The latter could be the case if the 60S peptidyltransferase center is localized at or very near the interface between the two subunits. A number of ribosomal proteins in eukaryotic preparations have been tentatively identified as participating in the formation or function of the peptidyltransferase cener; proteins L7, L11, L21, L24, L26, L27, L28, L29, and/or L36, as well as S4, S13, S18, S20, and S21 have been implicated (413, 423, 424, 440, 444). It is clear from studies with very low-molecular-weight analogs such as puromycin, acetyl- or formyl-aminoacyl-oligonucleotides, formylmethionyladenosine, etc, that the peptidyltransferase catalytic center is limited in size and probably covers only a very small portion of the surface of the large ribosomal subunit. Apparently, an adenine nucleoside attached to an amino acid (at the A site) and to an N-blocked amino acid (at the P site) is sufficient to trigger transpeptidation. Thus, it would seem that the appropriate binding and alignment of the peptidyl-tRNA at the P site and of the aminoacyl-tRNA at the A site serve to place the 3'-termini of the tRNAs with their esterified moieties within the peptidyltransferase center. One possibility for the mechanism of transpeptidation is that the responsible pro-eins on the surface of the 60S subunit allow the reaction components to bind in he appropriate stereochemical configuration to allow for the spontaneous synthesis of the peptide bond. A more attractive hypothesis suggests nucleophilic or acid-base catalysis by the functional groups of the peptidyltransferase center, and a nucleophilic attack by the α-amino nitrogen of the incoming aminoacyl-tRNA on the carbonyl carbon atom of the peptidyl-tRNA. It should be noted, as is apparent from studies on the mechanism of action of peptidyltransferase that at each cycle, the energy for the formation of the peptide bond is derived not from the ester linkage in the aminoacyl-tRNA at the A site, but from the hydrolysis of the ester bond in the peptidyl-tRNA at the P site.

Translocation

As the result of the peptidyltransferase reaction, a ribosome is generated in which the P site is occupied by a deacylated tRNA and the A site contains the newly-formed peptidyl-tRNA. The tRNA at the A site is attached to the peptidyl moiety through the original ester linkage to its cognate amino acid and is base-paired to the codon in mRNA which it decoded. Before the next codon can be translated, it must be moved into the A (decoding) site, which must otherwise be unoccupied and available for binding of the next, appropriate, aminoacyl-tRNA. Thus, concomitant with the movement of the next codon to the A site, the peptidyl-tRNA and the translated codon must be shifted to the P site. This reaction, translocation, involves the movement of the ribosome and/or mRNA in relation to each other, by precisely one triplet, from the 5'- to the 3'-end of mRNA.

The translocation reaction is catalyzed by EF-2 and GTP (143, 144, 414, 415). Following the binding of EF-2 and GTP to pretranslocated ribosomes, a number of reactions have been shown to occur: Peptidyl-tRNA and its corresponding codon are shifted from the A to the P site; mRNA moves by one triplet placing the next codon to be translated at the open A site; deacylated tRNA is released from the P site; GTP is hydrolyzed to GDP and P_i; and EF-2 is released from the posttranslocated ribosome. In contrast to pretranslocated ribosomes, posttranslocated ribosomes are capable of binding aminoacyl-tRNA to the A site and reacting with puromycin to form peptidyl-puromycin (144, 414–416, 445, 446) but do not support the ribosome-dependent hydrolysis of GTP catalyzed by EF-2 (447–451). EF-2 has been purified from a variety of sources including rat liver (348, 447, 452, 453), pig liver (352, 448), reticuloyctes (356, 358), silk glands (449), calf brain (452), wheat germ (362, 365, 454), ascites cells (374), yeast (371, 372, 455), sea urchin embryos (456), *A. salina* (457), etc. It is a single polypeptide chain, and most preparations have a molecular weight between 100,000 and 110,000 (348, 356, 365, 371, 374, 448, 450, 453, 456).

Elongation factor 2 contains a number of sulfhydryl groups that are essential for activity (458, 459), and a single posttranslational derivative of histidine,

2-[3-carboxyamido-3-(trimethylammonio)propyl]histidine, referred to as diphthamide (460–466). Inhibition of eukaryotic protein synthesis by diphtheria toxin is related to the ADP-ribosylation and inactivation of EF-2 (450, 461, 467–471); the toxin catalyzes the transfer of ADP-ribose from the coenzyme NAD$^+$ to the N-1 nitrogen of the histidine imidazole ring of the diphthamide residue (460, 463–466, 472, 473). The ADP-ribosylated EF-2 is able to form a binary complex with GTP (450, 451, 471), but is unable to catalyze protein synthesis, poly(U)-dependent synthesis of polyphenylalanine, translocation reactions, or GTP hydrolysis (356, 447, 450, 451, 471, 472). Prokaryotic and mitochondrial translocation factors (EF-G), which do not contain this unusual amino acid, are not ADP-ribosylated or inhibited by diphtheria toxin-NAD$^+$. Mutants of Chinese hamster ovary, polyoma virus-transformed baby hamster kidney, and human fibroblast cells with altered forms of EF-2 that are resistant to ADP-ribosylation and inactivation by the toxin, have been isolated (466, 474–477). Some of the mutants appear to contain an altered structural gene coding for EF-2, which could involve the histidine position that is modified or some other residue required for the modification. Other mutants possess alterations that affect the posttranslational conversion of histidine to diphthamide, which seems to require three distinct enzymatic reactions (463, 466, 475). Although diphthamide has been found in practically all of the eukaryotic EF-2 examined, the ability of toxin-resistant mutants to synthesize protein suggests that modification of this histidine residue to diphthamide is not essential for translocation or protein synthesis.

The mechanism of action of EF-2 in translocation involves the formation of two intermediary complexes. The first is a binary complex with GTP (447, 450, 451, 478–481), and the second is a ternary complex containing the factor, GTP, and the ribosome (144, 414, 450, 451, 478–483). The differential sensitivities of rat liver EF-2 to sulfhydryl or amino group reactive reagents suggest that the binding sites involved in the binary as compared to the ternary complex correspond to different conformational states or possibly distinct physical entities (484). The observations that these complexes can be formed with nonhydrolyzable analogs of GTP indicate that GTP hydrolysis is not required for the nucleotide-dependent binding of EF-2 to ribosomes. However, when EF-2·GTP reacts with pretranslocated ribosomes, a ternary complex containing GTP is formed (450, 451, 478–480, 483). Thus, GTP hydrolysis occurs in a step subsequent to the binding of EF-2 to the ribosome. One possibility is that energy from GTP hydrolysis is used for translocation; that is, the movement of the ribosome on mRNA resulting in the shift of the peptidyl-tRNA to the P site and of a new codon into the A site (414, 415). Another possibility is that hydrolysis of GTP is required for the release and recycling of EF-2, and that translocation can occur with ribosome-bound EF-2 in the absence of hydrolysis, such as in the presence of nonhydrolyzable analogs (398, 485). Following translocation, EF-2·GDP is released from the ribosome and reacts with free GTP to regenerate the active EF-2·GTP binary complex (480). Additional evidence has been obtained in support of two binding states for EF-2 on 80S ribosomes, with different affinities (481); one is a high affinity pretranslocation state specific for EF-2·GTP, and the other is a low affinity posttranslocation state in which EF-2·GDP is bound in a less stable and specific manner. The molecular mechanism of eukaryotic or prokaryotic ribosomal translocation and the role of the translocation factor in this process are not well understood, although a number of ideas have been proposed (486–489). It would seem obvious that translocation would involve energy-dependent changes in ribosomal conformation. Whether such changes involve GTP energy, the emptying of a ribosomal site (for example, an EF-2-dependent release of tRNA from the P site) inducing the ribosome to seek a more stable conformation, etc, remains to be elucidated.

Other Elongation Factors and Sites

Occasionally, evidence for additional factors that may play a role in eukaryotic chain elongation has been reported. For example, a protein that stimulates poly(U) binding and the activity of EF-1 in polyphenylalanine biosynthesis has been purified from wheat germ embryos (490). In yeast, considerable evidence has accumulated for an additional elongation factor, designated as EF-3 (371, 372, 455, 491, 492). Whereas rat liver, HeLa cell, or *Artemia salina* ribosomes carry out the poly(U)-dependent synthesis of polyphenylalanine with EF-1, EF-2, and GTP, (as do practically all eukaryotic ribosomes), yeast ribosomes also require EF-3. Studies with EF-3 purified by ion exchange (371, 372) or immunoaffinity (455) chromatography reveal that it is a single polypeptide chain, approximately 125,000 in molecular weight. The factor exhibits strong GTP-ase and ATPase activities (371, 372), but its exact site and mechanism of action in the elongation series of reactions remains to be established. Recent studies with a monoclonal antibody specific against EF-3 have shown that this factor is also essential for the translation of natural mRNA, and that it is required in every cycle of chain elongation (455). The in vitro translation of natural mRNA, after the initiation reactions had taken place, were immediately and completely inhibited by the addition of the antibody to incubations containing yeast ribosomes; protein synthesis in incubations containing rat liver ribosomes and yeast elongation factors were not inhibited by the anti-EF-3 antibody.

Most of the information on chain elongation summarized above has been interpreted on the basis of a ribosomal two-site model (493) applicable to both eukaryotes and prokaryotes. Results consistent with two nonoverlapping tRNA-binding sites on *E. coli* ribosomes have been obtained using centrifugation techniques to isolate intermediates (494) and with matrix-bound poly(U) in which individual steps in chain elongation were synchronized (495); some of these data, however, would not exclude overlapping or low-affinity sites.

Indeed, a number of reports suggest the existence of more than two tRNA-binding sites on ribosomes (143, 144, 487, 496–505). Some have proposed an entry or recognition site before the A site, and others an exit site after the P site. The possibility that some of these additional sites may be functionally but not necessarily physically distinct from the two accepted ribosomal sites, must be considered. Most of the data regarding multiple sites has been obtained with prokaryotic systems; proof of such a model would clearly have similar implications on the mechanism of action of eukaryotic ribosomes.

PEPTIDE CHAIN TERMINATION

When all of the codons in a cistron are translated, formation of the last peptide bond results in a polypeptide chain attached to tRNA at the A site through its C-terminal amino acid residue. Translocation with EF-2 and GTP then shifts the polypeptidyl-tRNA and its complementary mRNA codon to the P site, bringing a termination codon into the empty A site. Three termination codons exist (506), UAA, UAG, and UGA, for which there are no known normal tRNAs with complementary anticodons that can effect their translation; however, mutant yeast suppressor tRNAs have been obtained that are capable of translating termination codons such as UAA and UAG (507, 508). When the peptidyl-tRNA is at the ribosomal P site and a termination codon is at the A site, a protein factor (RF) binds to the A site in the presence of GTP and catalyzes the termination reaction. This reaction involves the hydrolysis of the peptidyl-tRNA ester bond, the hydrolysis of GTP, and the release of the completed polypeptide chain, the deacylated tRNA, and the ribosome from mRNA.

Whereas termination in prokaryotes requires 3 release factors and there is some specificity with respect to the codons that are recognized, only one release factor is required in the eukaryotic systems that have been investigated (506, 509–515). Most of the results reported on reticulocyte RF are based on the release of formylmethionine from fMet-tRNA·ribosome·oligo-nucleotide complexes (506, 508–511, 514, 516), in a model assay that requires very high Mg^{2+} concentrations, tetranucleotide (UAAA, UAGA, or UGAA) templates, and/or ethanol. The molecular weight of reticulocyte RF is about 110,000 and is composed of 55,000-dalton subunits (516), although higher values were reported for the native factor in earlier studies (514, 515).

The codon-dependent binding of RF to terminating ribosomes requires GTP but does not involve hydrolysis of the nucleotide at this stage since binding can also be obtained with nonhydrolyzable analogs of GTP (510); GDP is ineffective. Reticulocyte release factor exhibits a ribosome-dependent GTPase activity (506), but it seems unlikely that this hydrolysis is directly linked to peptidyl-tRNA hydrolysis because it occurs with ribosomes not carrying peptidyl-tRNA, and it is not inhibited by antibiotics that interfere with the hydrolyis of peptidyl-tRNA. Most probably, GTP hydrolysis occurs at a later step in termination and is required for the dissociation of the release factor from the ribosome, as has been suggested for a number of other ribosome-bound translational factor·GTP complexes.

Antibiotics that inhibit peptidyltransferase, such as sparsomycin and trichodermin (509, 511, 515, 517), also inhibit the termination reaction, suggesting that the activity that normally is responsible for the formation of peptide bonds in chain elongation is also, in some way, involved in chain termination. A plausible model for termination is that following the formation of the last peptide bond, the peptidyl-tRNA is translocated to the P site and a termination codon moves into the A site. Release factor and GTP then bind to the open ribosomal A site. When the P site is occupied by peptidyl-tRNA and the A site is occupied by RF, ribosomal peptidyltransferase catalyzes the hydrolysis of the peptidyl-tRNA ester. Peptidyltransferase seems to be triggered when the P and A sites are occupied by the appropriate substrates. However, whereas in chain elongation peptidyltransferase catalyzes the transfer of the peptidyl group to the amino group of the incoming aminoacyl-tRNA, in chain termination the transfer involves the elements of water resulting in the release of the completed polypeptide chain from the tRNA and the ribosome. Subsequent to these reactions, GTP is hydrolyzed to GDP and P_i, RF is released from the ribosome, and the terminal tRNA as well as the ribosome are released from mRNA. Subunits from released ribosomes enter the subunit pool and are available for another round of protein synthesis on mRNA by participating in the chain initiation series of reactions with initiator Met-tRNA$_f$, GTP, ATP, and the various initiation factors required in this process.

ACKNOWLEDGMENTS

Work from the author's laboratory cited in this review was supported in part by research grants AM 15156, GM 27999, and AG 00538 from the National Institutes of Health, USA, and NP-88 from the American Cancer Society.

Literature Cited

1. Maitra, U., Stringer, E. A., Chaudhuri, A. 1982. *Ann. Rev. Biochem.* 51:869–900
2. Kozak, M. 1983. *Microbiol. Rev.* 47:1–45
3. Austin, S. A., Kay, J. E. 1982. *Essays Biochem.* 18:79–120
4. Abraham, A. K. 1983. *Progr. Nucleic Acid Res. Mol. Biol.* 28:82–100
5. Lake, J. A. 1983. *Progr. Nucleic Acid Res. Mol. Biol.* 30:164–94
6. Nevins, J. R. 1983. *Ann. Rev. Biochem.* 52:441–66
7. Voorma, H. O., Goumans, H., Amesz, H., Benne, R. 1983. *Curr. Top. Cell. Regul.* 22:51–70
8. Ochoa, S. 1983. *Arch. Biochem. Biophys.* 223:325–49
9. Perez-Bercoff, R., ed. 1980. *Protein Biosynthesis in Eukaryotes, A NATO Adv. Study Inst. Ser., Ser. A: Life Sci.* New York: Plenum
10. Grunberg-Manago, M., Safer, B., eds. 1982. *Interaction of Translational and Transcriptional Controls in the Regulation of Gene Expression.* New York: Elsevier Biomedical
11. Safer, B., Anderson, W. F., Merrick,

W. C. 1975. *J. Biol. Chem.* 250:9067–75

12. Merrick, W. C., Anderson, W. F. 1975. *J. Biol. Chem.* 250:1197–1206

13. Merrick, W. C., Kemper, W. M., Anderson, W. F. 1975. *J. Biol. Chem.* 250:5556–62

14. Safer, B., Adams, S. L., Anderson, W. F., Merrick, W. C. 1975. *J. Biol. Chem.* 250:9076–82

15. Staehelin, T., Trachsel, H., Erni, B., Boschetti, A., Schrier, M. H. 1975. *FEBS Proc. Meet.* 10:309–23

16. Safer, B., Adams, S. L., Kemper, W. M., Berry, K. W., Lloyd, M., Merrick, W. C. 1976. *Proc. Natl. Acad. Sci. USA* 73:2584–86

17. Schrier, M. H., Erni, B., Staehelin, T. 1977. *J. Mol. Biol.* 116:727–54

18. Trachsel, H., Erni, B., Schrier, M. H., Staehelin, T. 1977. *J. Mol. Biol.* 116:755–67

19. Benne, R., Hershey, J. W. B. 1978. *J. Biol. Chem.* 253:3078–87

20. Sonenberg, N., Rupprecht, K. M., Hecht, S. M., Shatkin, A. J. 1979. *Proc. Natl. Acad. Sci. USA* 76:4345–49

21. Amesz, H., Goumans, H., Hanbrich-Morree, T., Voorma, H. O., Benne, R. 1979. *Eur. J. Biochem.* 98:513–20

22. Benne, R., Brown-Luedi, M. L., Hershey, J. W. B. 1979. *Methods Enzymol.* 60:15–35

23. Majumdar, A., Dasgupta, A., Chatterjee, B., Das, H. K., Gupta, N. K. 1979. *Methods Enzymol.* 60:35–52

24. Dasgupta, A., Das, A., Roy, R., Ralston, R., Majumdar, A., Gupta, N. K. 1979. *Methods Enzymol.* 60:53–61

25. Merrick, W. C. 1979. *Methods Enzymol.* 60:101–8

26. Voorma, H. O., Thomas, W., Goumans, H., Amesz, H., van der Mast, C. 1979. *Methods Enzymol.* 60:124–35

27. Staehelin, T., Erni, B., Schrier, M. H. 1979. *Methods Enzymol.* 60:136–65

28. Giesen, M., Roman, R., Seal, S. N., Marcus, A. 1976. *J. Biol. Chem.* 251:6075–81

29. Spermulli, L. L., Walthall, B. J., Lax, S. R., Ravel, J. M. 1977. *Arch. Biochem. Biophys.* 178:565–75

30. Treadwell, B. V., Mauser, L., Robinson, W. G. 1979. *Methods Enzymol.* 60:181–93

31. Walthall, B. J., Spermulli, L. L., Lax, S. R., Ravel, J. M. 1979. *Methods Enzymol.* 60:193–204

32. Russell, D. W., Spermulli, L. L. 1980. *Arch. Biochem. Biophys.* 201:518–26

33. Seal, S. N., Schmidt, A., Marcus, A. 1983. *J. Biol. Chem.* 258:859–65

34. Seal, S. N., Schmidt, A., Marcus, A. 1983. *J. Biol. Chem.* 258:866–71

35. Smith, K. E., Henshaw, E. C. 1975. *Biochemistry* 14:1060–67

36. Panniers, R., Henshaw, E. C. 1983. *J. Biol. Chem.* 258:7928–34

37. Brown-Luedi, M. L., Meyer, L. J., Milburn, S. C., Yan, P. M., Corbett, S. Hershey, J. W. B. 1982. *Biochemistry* 21:4202–6

38. Meyer, L. J., Milburn, S. C., Hershey, J. W. B. 1982. *Biochemistry* 21:4206–12

39. Thompson, H. A., Sadnik, I., Scheinbuks, J., Moldave, K. 1977. *Biochemistry* 16:2221–30

40. Herrera, F., Sadnik, I., Gough, G., Moldave, K. 1977. *Biochemistry* 16:4664–7

41. Valenzuela, D. M., Chaudhuri, A. Maitra, U. 1982. *J. Biol. Chem.* 257:7712–19

42. Voorma, H. O., Amesz, H. 1982. See Ref. 10, pp. 297–309

43. Safer, B., Jagus, R., Konieczny, A. Crouch, D. 1982. See Ref. 10, pp. 311 25

44. Siekierka, J. J., Mauser, L., Ochoa, S. 1982. See Ref. 10, pp. 327–37

45. Gupta, N. K., Grace, M., Banerjee, A C., Bagchi, M. 1982. See Ref. 10, pp 339–58

46. Safer, B. 1983. *Cell* 33:7–8

47. Zasloff, M., Ochoa, S. 1973. *J. Mol. Biol.* 73:391–402

48. Grummt, F. 1974. *Eur. J. Biochem.* 43:337–42

49. Cimadevilla, J. M., Hardesty, B. 1975 *J. Biol. Chem.* 250:4389–97

50. Hejtmancik, J. F., Comstock, J. P. 1976 *Biochemistry* 15:3804–12

51. Herring, S., Sadnik, I., Moldave, K 1982. *J. Biol. Chem.* 257:4882–87

52. Henshaw, E. C., Guiney, D. G., Hirsch C. A. 1973. *J. Biol. Chem.* 248:4367 76

53. Hirsch, C. A., Cox, M. A., Van Venrooij, W. J., Henshaw, E. C. 1973. *J. Biol. Chem.* 248:4377–85

54. Kaempfer, R. 1974. In *Ribosomes*, ed M. Nomura, A. Tissieres, P. Lengyel pp. 679–704. New York: Cold Spring Harbor Lab.

55. Davis, B. D. 1974. See Ref. 54, pp 705–10

56. Freienstein, C., Blobel, G. 1975. *Proc Natl. Acad. Sci. USA* 72:3392–96

57. Sadnik, I., Herrera, F., McCuiston, J. Thompson, H. A., Moldave, K. 1975 *Biochemistry* 14:5328–35

58. Van Venrooij, W. J., Janssen, A. P. M. Hoeymakers, J. H., DeMan, B. M 1976. *Eur. J. Biochem.* 64:429–35

59. Safer, B., Anderson, W. F. 1978. *CRC Critical Rev. Biochem.* 5:261–90

60. Ayuso-Parilla, M., Henshaw, E. C. Hirsch, C. A. 1973. *J. Biol. Chem.* 248:4386–93

61. Schrier, M. H., Staehelin, T. 1973. *Nature New Biol.* 242:35–38
62. Schrier, M. H., Staehelin, T. 1973. *J. Mol. Biol.* 73:329–49
63. Prichard, P. M., Anderson, W. F. 1974. *Methods Enzymol.* 30:136–41
64. Sundkvist, I. C., Staehelin, T. 1975. *J. Mol. Biol.* 99:401–18
65. Benne, R., Hershey, J. W. B. 1976. *Proc. Natl. Acad. Sci. USA* 73:3005–9
66. Moldave, K., Thompson, H. A., Sadnik, I. 1979. *Methods Enzymol.* 60:290–97
67. Nygard, O., Westermann, P. 1982. *Biochim. Biophys. Acta* 697:263–69
68. Trachsel, H., Staehelin, T. 1979. *Biochim. Biophys. Acta* 565:305–14
69. Nakaya, K., Ranu, R., Wool, I. 1973. *Biochim. Biophys. Res. Commun.* 54:246–55
70. Merrick, W. C., Lubsen, N. H., Anderson, W. F. 1973. *Proc. Natl. Acad. Sci. USA* 70:2220–23
71. Jones, R. L., Sadnik, I., Thompson, H. A., Moldave, K. 1980. *Arch. Biochem. Biophys.* 199:277–85
72. Goumans, H., Thomas, A., Verhoeven, A., Voorma, H. O., Benne, R. 1980. *Biochim. Biophys. Acta* 608:39–46
73. Ceglarz, E., Goumans, H., Thomas, A., Benne, R. 1980. *Biochim. Biophys. Acta* 610:181–88
74. Checkley, J. W., Cooley, L., Ravel, J. M. 1981. *J. Biol. Chem.* 256:1582–86
75. Lawford, G. R., Kaiser, J., Hey, W. D. 1971. *Can. J. Biochem.*, 49:1301–6
76. Mizuno, S., Rabinowitz, M. 1973. *Proc. Natl. Acad. Sci. USA* 70:787–91
77. Lubsen, N. H., Davis, B. D. 1974. *Biochim. Biophys. Acta* 335:196–200
78. Russell, D. W., Spermulli, L. L. 1978. *J. Biol. Chem.* 253:6647–49
79. Russell, D. W., Spermulli, L. L. 1979. *J. Biol. Chem.* 254:8796–800
80. Benne, R., Brown-Luedi, M. L., Hershey, J. W. B. 1978. *J. Biol. Chem.* 253:3070–77
81. Nygard, O., Westermann, P. 1982. *Nucleic Acids Res.* 10:1327–34
82. Westermann, P., Nygard, O. 1983. *Biochim. Biophys. Acta* 741:103–8
83. Tolan, D. R., Hershey, J. W. B., Traut, R. T. 1983. *Biochimie* 65:427–36
84. Emanuilov, I., Sabatini, D. D., Lake, J. A., Freienstein, C. 1978. *Proc. Natl. Acad. USA* 75:1389–93
85. Dittman, G. L., Stanley, W. M. 1972. *Biochim. Biophys. Acta* 287:124–33
86. Chen, Y. C., Woodley, C. L., Bose, K. K., Gupta, N. K. 1972. *Biochem. Biophys. Res. Commun.* 48:1–9
87. Gupta, N. K., Woodley, C. L., Chen, Y. C., Bose, K. K. 1973. *J. Biol. Chem.* 248:4500–11
88. Levin, D. H., Kyner, D., Acs, G. 1973. *J. Biol. Chem.* 248:6416–25
89. Levin, D. H., Kyner, D., Acs, G. 1973. *Proc. Natl. Acad. Sci. USA* 70:41–45
90. Cashion, L. M., Stanley, W. M. 1974. *Proc. Natl. Acad. Sci. USA* 71:7675–81
91. Elson, N. A., Adams, S. L., Merrick, W. C., Safer, B., Anderson, W. F. 1975. *J. Biol. Chem.* 250:3074–79
92. Treadwell, B. V., Robinson, W. G. 1975. *Biochem. Biophys. Res. Commun.* 65:176–83
93. Ranu, R. S., Wool, I. G. 1976. *J. Biol. Chem.* 251:1926–35
94. Benne, R., Amesz, H., Hershey, J. W. B., Voorma, H. 1979. *J. Biol. Chem.* 254:3201–5
95. Stringer, E. A., Chaudhuri, A., Maitra, U. 1979. *J. Biol. Chem.* 254:6845–48
96. Stringer, E. A., Chaudhuri, A., Valenzuela, D., Maitra, U. 1980. *Proc. Natl. Acad. Sci. USA* 77:3356–59
97. Calagan, J. L., Pirtle, R., Pirtle, I., Kashdan, M., Vreman, H., Dudock, B. 1980. *J. Biol. Chem.* 255:9981–84
98. Wrede, P., Woo, N., Rich, A. 1979. *Proc. Natl. Acad. Sci. USA* 76:3289–93
99. Woo, N., Roe, B., Rich, A. 1980. *Nature* 286:346–51
100. Schroer, R. A., Moldave, K. 1973. *Arch. Biochem. Biophys.* 154:422–30
101. Filipowicz, W., Sierra, J., Ochoa, S. 1975. *Proc. Natl. Acad. Sci. USA* 72:3947–51
102. Benne, R., Wong, C., Luedi, M., Hershey, J. W. B. 1976. *J. Biol. Chem.* 251:7675–81
103. Issinger, O.-G., Benne, R., Hershey, J. W., Traut, R. R. 1976. *J. Biol. Chem.* 251:6471–74
104. Harbitz, I., Hauge, J. G. 1976. *Arch. Biochem. Biophys.* 176:766–78
105. Traugh, J. A., Tahara, S. M., Sharp, S. B., Safer, B., Merrick, W. C. 1976. *Nature* 263:163–65
106. Tahara, S. M., Traugh, J. A., Sharp, S. B., Lundak, T. S., Safer, B., Merrick, W. C. 1978. *Proc. Natl. Acad. Sci. USA* 75:789–93
107. Mitsui, K.-I., Datta, A., Ochoa, S. 1981. *Proc. Natl. Acad. Sci. USA* 78:4128–32
108. Mehta, H. B., Woodley, C. L., Wahba, A. J. 1983. *J. Biol. Chem.* 258:3438–41
109. Barrieux, A., Rosenfeld, M. 1977. *J. Biol. Chem.* 252:3843–47
110. Barrieux, A., Rosenfeld, M. 1978. *J. Biol. Chem.* 253:6311–14
111. Baan, R. A., Keller, P. B., Dahlberg, A. E. 1981. *J. Biol. Chem.* 256:1063–66
112. Lloyd, M. A., Osborne, J. C., Safer, B., Powell, G. M., Merrick, W. C. 1980. *J. Biol. Chem.* 255:1189–94

113. Meyer, L. J., Brown-Luedi, M., Corbett, S., Tolan, D. R., Hershey, J. W. B. 1981. *J. Biol. Chem.* 256:351–56
114. Kramer, G., Cimadevilla, J. M., Hardesty, B. 1976. *Proc. Natl. Acad. Sci. USA* 73:3078–82
115. Levin, D. H., Ranu, R. S., Ernst, V., London, I. M. 1976. *Proc. Natl. Acad. Sci. USA* 73:3112–16
116. Ranu, R. S., London, I. M. 1976. *Proc. Natl. Acad. Sci. USA* 73:4349–53
117. Farrell, P. J., Balkow, K., Hunt, T., Jackson, R. J., Trachsel, H. 1977. *Cell* 11:182–200
118. Gross, M., Mendelewski, J. 1977. *Biochem. Biophys. Res. Commun.* 74:559–69
119. Levin, D. H., London, I. M. 1978. *Proc. Natl. Acad. Sci. USA* 75:1121–25
120. Ernst, V., Levin, D. H., Leroux, A., London, I. M. 1980. *Proc. Natl. Acad. Sci. USA* 77:1286–90
121. Benne, R., Edman, J., Traut, R. R., Hershey, J. W. B. 1978. *Proc. Natl. Acad. Sci. USA* 75:108–12
122. Sen, G. C., Taira, H., Lengyel, P. 1978. *J. Biol. Chem.* 253:5915–21
123. Kimchi, A., Zilberstein, A., Schmidt, A., Shulman, L., Revel, M. 1979. *J. Biol. Chem.* 254:9846–53
124. West, D. K., Baglioni, C. 1979. *Eur. J. Biochem.* 101:461–68
125. Petryshyn, R., Levin, D. H., London, I. M. 1982. *Proc. Natl. Acad. Sci. USA* 79:6512–16
126. DeBenedetti, A., Baglioni, C. 1983. *J. Biol. Chem.* 258:14556–62
127. DeHaro, C., Ochoa, S. 1979. *Proc. Natl. Acad. Sci. USA* 76:1741–45
128. Harbitz, I., Hauge, J. G. 1979. *Methods Enzymol.* 60:240–46
129. Seal, S. N., Schmidt, A., Marcus, A. 1983. *J. Biol. Chem.* 258:10573–76
130. Das, A., Bagchi, M. K., Gosh-Dastidar, P., Gupta, N. K. 1982. *J. Biol. Chem.* 257:1282–88
131. Zardeneta, G., Kramer, G., Hardesty, B. 1982. *Proc. Natl. Acad. Sci. USA* 79:3158–61
132. Walton, G. M., Gill, G. N. 1975. *Biochim. Biophys. Acta* 390:231–41
133. Walton, G. M., Gill, G. N. 1976. *Biochim. Biophys. Acta* 418:195–203
134. Walton, G. M., Gill, G. N. 1976. *Biochim. Biophys. Acta* 447:11–19
135. Stringer, E. A., Chaudhuri, A., Maitra, U. 1977. *Biochem. Biophys. Res. Commun.* 76:586–92
136. Trachsel, H., Staehelin, T. 1978. *Proc. Natl. Acad. Sci. USA* 75:204–8
137. Peterson, D. T., Safer, B., Merrick, W. C. 1979. *J. Biol. Chem.* 254:7730–35
138. Merrick, W. C. 1979. *J. Biol. Chem.* 254:3708–10
139. Siekierka, J., Mauser, L., Ochoa, 1982. *Proc. Natl. Acad. Sci. USA* 7 2537–40
140. Siekierka, J., Manne, V., Mauser, I Ochoa, S. 1983. *Proc. Natl. Acad. S. USA* 80:1232–35
141. Peterson, D. T., Merrick, W. C., Saf B. 1979. *J. Biol. Chem.* 254:2509–1(
142. Ibuki, F., Moldave, K. 1968. *J. Bi Chem.* 243:791–98
143. Skogerson, L., Moldave, K. 1968. *Arc Biochem. Biophys.* 125:497–505
144. Moldave, K., Ibuki, F., Rao, Schneir, M., Skogerson, L., Sutter, P. 1968. In *Some Regulatory Mecha isms of Protein Synthesis in Mammali Cells,* ed. A. San Pietro, M. R. Lambo: F. T. Kenney, pp. 191–229. New Yo: Academic
145. London, I. M., Clemens, M. J., Rar R. S., Levin, D. H., Cherbas, L. I Ernst, V. 1976. *Fed. Proc.* 35:2218–
146. Grace, M., Ralston, R. O., Banerjee, C., Gupta, N. K. 1982. *Proc. Na Acad. Sci. USA* 79:6517–21
147. Dasgupta, A., Majumdar, A., Georg A. B., Gupta, N. K. 1976. *Bioche Biophys. Res. Commun.* 71:1234–41
148. Majumdar, A., Roy, R., Das, A., D gupta, A., Gupta, N. K. 1977. *Bioche Biophys. Res. Commun.* 78:161–69
149. Dasgupta, A., Das, A., Roy, R., R ston, R., Majumdar, A., Gupta, N. 1978. *J. Biol. Chem.* 253:6054–59
150. Das, A., Ralston, R. O., Grace, M Roy, R., Gosh-Dastidar, P., et al. 197 *Proc. Natl. Acad. Sci. USA* 76:507 79
151. Gupta, N. K. 1982. In *Protein Biosyn esis in Eukaryotes,* ed. R. Bercoff, r 419–40. New York: Plenum
152. Gupta, N. K. 1982. *Curr. Top. Ce Regul.* 21:1–33
153. Das, A., Bagchi, M., Roy, R., Gos Dastidar, P., Gupta, N. K. 198 *Biochem. Biophys. Res. Commu 104:89–98
154. Bagchi, M. K., Banerjee, A. C., Rc R., Chakrabarty, I., Gupta, N. K. 198 *Nucleic Acids Res.* 10:6501–10
155. Gross, M. 1975. *Biochem. Biophys. R Commun.* 67:1507–15
156. Gross, M. 1976. *Biochem. Biophys. A(447:445–49
157. Ranu, R. S., London, I. M. 1979. *Pr(Natl. Acad. Sci. USA* 76:1079–83
158. Ralston, R. O., Das, A., Grace, M Das, H., Gupta, N. K. 1979. *Proc. Na Acad. Sci. USA* 76:5490–94
159. Ochoa, S. 1981. *Eur. J. Cell Bi(26:212–16

160. Ranu, R. S. 1982. *Biochem. Biophys. Res. Commun.* 107:828–33
161. Matts, R. L., Levin, D. H., London, I. M. 1983. *Proc. Natl. Acad. Sci. USA* 80:2559–63
162. DeHaro, C., Datta, A., Ochoa, S. 1978. *Proc. Natl. Acad. Sci. USA* 75:243–47
163. DeHaro, C., Ochoa, S. 1978. *Proc. Natl. Acad. Sci. USA* 75:2713–16
164. Malathi, V. G., Mazumder, R. 1978. *FEBS Lett.* 86:155–59
165. Siekierka, J., Mitsui, K. I., Ochoa, S. 1981. *Proc. Natl. Acad. Sci. USA* 78:220–23
166. Siekerka, J., Datta, A., Mauser, L., Ochoa, S. 1982. *J. Biol. Chem.* 257:162–65
167. Clemens, M. J., Pain, V. M., Wong, S.-T., Henshaw, E. C. 1982. *Nature* 296:93–95
168. Proud, C. G., Clemens, M. J., Pain, V. M. 1982. *FEBS Lett.* 148:214–20
169. Pain, V. M., Clemens, M. J. 1983. *Biochemistry* 22:726–32
170. Komenczny, A., Safer, B. 1983. *J. Biol. Chem.* 258:3402–8
171. Lax, S. R., Osterhout, J. J., Ravel, J. M. 1982. *J. Biol. Chem.* 257:8233–37
172. Osterhout, J. J., Lax, S. R., Ravel, J. M. 1983. *J. Biol. Chem.* 258:8285–89
173. Anderson, W. F., Bosch, L., Cohn, W. E., Lodish, H., Merrick, W. C., et al. 1977. *FEBS Lett.* 76:1–10
174. Slobin, L. I., Moller, W. 1975. *Nature* 258:452–54
175. Nagata, S., Iwasaki, K., Kaziro, Y. 1976. *Arch. Biochem. Biophys.* 172:168–77
176. Nagata, S., Motoyoshi, K., Iwasaki, K. 1976. *Biochem. Biophys. Res. Commun.* 71:933–38
177. Nagata, S., Motoyoshi, K., Iwasaki, K. 1978. *J. Biochem.* 83:423–29
178. Miller, O. L., Weissbach, H. 1977. In *Molecular Mechanisms of Protein Biosynthesis*, ed. H. Weissbach, S. Pestka, pp. 323–73. New York: Academic
179. Kaziro, Y. 1978. *Biochim. Biophys. Acta* 505:95–127
180. DeHaro, C., Ochoa, S. 1979. *Proc. Natl. Acad. Sci. USA* 76:2163–64
181. Leroux, A., London, I. M. 1982. *Proc. Natl. Acad. Sci. USA* 79:2147–51
182. Jacobsen, H., Epstein, D. A., Friedman, R. M., Safer, B., Torrence, P. F. 1983. *Proc. Natl. Acad. Sci. USA* 80:41–45
183. Darnbrough, C., Legon, S., Hunt, T., Jackson, R. J. 1973. *J. Mol. Biol.* 76:379–403
184. Sundkvist, I. C., McKeehan, W. L., Schrier, M. H., Staehelin, T. 1974. *J. Biol. Chem.* 249:6512–16
185. Freienstein, C., Blobel, G. 1974. *Proc. Natl. Acad. Sci. USA* 71:3435–39
186. Smith, K. E., Henshaw, E. C. 1975. *J. Biol. Chem.* 250:6880–84
187. Howard, G. A., Herbert, E. 1975. *Eur. J. Biochem.* 54:75–80
188. Smith, K. E., Richards, A. C., Arnstein, H. R. V. 1976. *Eur. J. Biochem.* 62:243–55
189. Safer, B., Kemper, W., Jagus, R. 1978. *J. Biol. Chem.* 253:3384–86
190. Wool, I. G., Stoffler, G. 1974. See Ref. 54, pp. 417–60
191. Safer, B., Peterson, D., Merrick, W. C. 1977. In *Translation of Natural and Synthetic Polynucleotides*, ed. A. B. Legocki, pp. 24–31. Poznan: Univ. Agric. Poznan
192. Feinberg, B., McLaughlin, C. S., Moldave, K. 1982. *J. Biol. Chem.* 257:10846–51
193. Feinberg, B., Moldave, K. Unpublished results
194. Seal, S. N., Schmidt, A., Marcus, A. 1982. *J. Biol. Chem.* 257:8634–37
195. Gupta, N. K., Chatterjee, B., Chen, Y. C., Majumdar, A. 1975. *J. Biol. Chem.* 250:853–62
196. Adams, S. L., Safer, B., Anderson, W. F., Merrick, W. C. 1975. *J. Biol. Chem.* 250:9083–89
197. Hunter, A. R., Jackson, R. J., Hunt, T. 1977. *Eur. J. Biochem.* 75:159–70
198. Benne, R., Leudi, M., Hershey, J. W. B. 1977. *J. Biol. Chem.* 252:5798–803
199. Tahara, S. M., Morgan, M. A., Shatkin, A. J. 1981. *J. Biol. Chem.* 256:7691–94
200. Shafritz, D., Weinstein, J., Safer, B., Merrick, W. C., Weber, L., et al. 1976. *Nature* 261:291–94
201. Kabat, D., Chappell, M. R. 1977. *J. Biol. Chem.* 252:2684–90
202. Padilla, M., Canaani, D., Groner, Y., Weinstein, J. A., Bar-Joseph, M., et al. 1978. *J. Biol. Chem.* 253:5939–45
203. Grifo, J. A., Tahara, S. M., Leis, J. P., Morgan, M. A., Shatkin, A. J., Merrick, W. C. 1982. *J. Biol. Chem.* 257:5246–52
204. Tahara, S. M., Morgan, M. A., Grifo, J. A., Merrick, W. C., Shatkin, A. J. 1982. See Ref. 10, pp. 359–72
205. Grifo, J. A., Tahara, S. M., Morgan, M. A., Shatkin, A. J., Merrick, W. C. 1983. *J. Biol. Chem.* 258:5804–10
206. Sonenberg, N., Morgan, M., Merrick, W. C., Shatkin, A. J. 1978. *Proc. Natl. Acad. Sci. USA* 75:4843–47
207. Sonenberg, N., Morgan, M. A., Testa, D., Colonno, R. J., Shatkin, A. J. 1979. *Nucleic Acids Res.* 7:15–29
208. Sonenberg, N., Trachsel, H., Hecht, S., Shatkin, A. J. 1980. *Nature* 285:331–33
209. Sonenberg, N., Skup, D., Trachsel, H.,

Millward, S. 1981. *J. Biol. Chem.* 256:4138–41

210. Hansen, J., Ehrenfeld, E. 1981. *J. Virol.* 38:438–45

211. Rupprecht, K. M., Sonenberg, N., Shatkin, A. J., Hecht, S. M. 1981. *Biochemistry* 20:6570–77

212. Hansen, J., Etchison, D., Hershey, J. W. B., Ehrenfeld, E. 1982. *J. Virol.* 42:200–7

213. Sonenberg, N. 1981. *Nucleic Acids Res.* 9:1643–56

214. Sonenberg, N., Lee, K. A. W. 1982. See Ref. 10, pp. 373–88

215. Sonenberg, N., Trachsel, H. 1982. *Curr. Top. Cell. Regul.* 21:65–88

216. Thomas, A., Spaan, W., van Steeg, H., Voorma, H. O., Benne, R. 1980. *FEBS Lett.* 116:67–71

217. Hellerman, J. G., Shafritz, D. A. 1975. *Proc. Natl. Acad. Sci. USA* 72:1021–25

218. Sonenberg, N., Shatkin, A. J. 1978. *J. Biol. Chem.* 253:6630–32

219. Kaempfer, R., Hollender, R., Abrams, W., Israeli, R. 1978. *Proc. Natl. Acad. Sci. USA* 75:209–31

220. Kaempfer, R., Rosen, H., Israeli, R. 1978. *Proc. Natl. Acad. Sci. USA* 75:650–54

221. Gasior, E., Herrera, F., McLaughlin, C. S., Moldave, K. 1979. *J. Biol. Chem.* 254:3970–76

222. DiSegni, G. D., Rosen, H., Kaempfer, R. 1979. *Biochemistry* 18:2847–54

223. Kaempfer, R., van Eurmelo, J., Fiers, W. 1981. *Proc. Natl. Acad. Sci. USA* 78:1542–46

224. Chaudhuri, A., Stringer, E. A., Valenzuela, D., Maitra, U. 1981. *J. Biol. Chem.* 256:3988–94

225. Ilan, J., Ilan, J. 1976. *J. Biol. Chem.* 251:5718–25

226. Heywood, S. M., Kennedy, D. S. 1979. *Arch. Biochem. Biophys.* 192:270–81

227. Vlasik, T. N., Domogatsky, S. P., Bezlepkina, T. A., Ovchinnikov, L. P. 1980. *FEBS Lett.* 116:8–10

228. Marcus, A. 1970. *J. Biol. Chem.* 245:955–61

229. Marcus, A. 1970. *J. Biol. Chem.* 245:962–66

230. Kozak, M. 1980. *Cell* 22:459–67

231. Brown, J. C., Smith, A. E. 1970. *Nature* 226:610–12

232. Stewart, J. W., Sherman, F. 1971. *J. Biol. Chem.* 246:7429–45

233. Sherman, F., McKnight, G., Stewart, J. W. 1980. *Biochim. Biophys. Acta* 609:343–46

234. Kozak, M. 1981. *Nucleic Acids Res.* 9:5233–52

235. Kozak, M. 1981. *Curr. Top. Microbiol. Immunol.* 93:81–123

236. Zimmerman, M., Mumford, R. A., Steiner, D. F., eds. 1980. *Precursor Processing in the Biosynthesis of Proteins,* Ann. NY Acad. Sci., Vol. 343. New York: NY Acad. Sci.

237. Pitak, M., Ghosh, P. K., Reddy, V. B., Lebowitz, P., Weissman, S. M. 1979. In *ICN-UCLA Symp. Mol. Cell. Biol.* 15:199–215. New York: Academic

238. Preston, C. M., McGeoch, D. J. 1981. *J. Virol.* 38:593–605

239. Bos, J. L., Polder, L., Bernards, R., Schrier, P., van den Elsen, P., et al. 1981. *Cell* 27:121–31

240. Kozak, M. 1982. *J. Mol. Biol.* 156:807–20

241. Bishop, D., Gould, K., Akashi, H., Clerx-van Haaster, C. 1982. *Nucleic Acids Res.* 10:3703–13

242. Both, G. W., Banerjee, A. K., Shatkin, A. J. 1975. *Proc. Natl. Acad. Sci. USA* 72:1189–93

243. Muthukrishnan, S., Both, G. W., Furuichi, Y., Shatkin, A. J. 1975. *Nature* 225:33–37

244. Furuichi, Y., Tomasz, J., Shatkin, A. J. 1976. In *Progress in Nucleic Acid Research and Molecular Biology,* ed. W. E. Cohen, 19:3–20. New York: Academic

245. Shatkin, A. J. 1976. *Cell* 9:645–53

246. Muthukrishnan, S., Moss, B., Cooper, J. A., Maxwell, E. S. 1978. *J. Biol. Chem.* 253:1710–15

247. Banerjee, A. K. 1980. *Microbiol. Rev.* 44:175–205

248. Furuichi, Y., La Fiandra, A., Shatkin, A. J. 1977. *Nature* 266:235–39

249. Shimotohno, K., Kodama, Y., Hashimoto, J., Miura, K-I. 1977. *Proc. Natl. Acad. Sci. USA* 74:2734–38

250. Lockard, R. E., Lane, C. 1978. *Nucleic Acid Res.* 5:3237–47

251. Gedamu, L., Dixon, G. H. 1978 *Biochem. Biophys. Res. Commun.* 85 114–24

252. Mayer, S. A. 1981. *Virology* 112:157–68

253. Hickey, E. D., Weber, L. A., Baglioni C. 1976. *Proc. Natl. Acad. Sci. US* 73:19–23

254. Filipowicz, W., Furuichi, Y., Sierra, J. M., Muthukrishnan, S., Shatkin, A. J. Ochoa, S. 1976. *Proc. Natl. Acad. Sci USA* 73:1559–63

255. Weber, L. A., Fernan, E. R., Hickey, E D., Williams, M. C., Baglioni, C. 1976 *J. Biol. Chem.* 251:5657–62

256. Roman, R. J. D., Booker, S. N., Marcus, A. 1976. *Nature* 260:359–60

257. Canaani, D., Revel, M., Groner, Y 1976. *FEBS Lett.* 64:326–31

258. Groner, Y., Grosfeld, H., Littauer, U. Z 1976. *Eur. J. Biochem.* 71:281–93

259. Suzuki, H. 1977. *FEBS Lett.* 79:11–1

60. Weber, L. A., Hickey, E. D., Baglioni, C. 1978. *J. Biol. Chem.* 253:178–83

61. Asselbergs, F. A., Peters, W., Van Venrooij, W. J., Bloemendal, H. 1978. *Eur. J. Biochem.* 88:483–88

62. Willems, M., Wieringa, B. E., Mulder, J., Ab, G., Gruber, M. 1979. *Eur. J. Biochem.* 93:469–79

63. Nakashima, K., Darzynklewicz, E., Shatkin, A. J. 1980. *Nature* 286:226–30

64. Wieringa, B. E., van der Swaag-Gerritsen, J., Mulder, J., Ab, G., Gruber, M. 1981. *Eur. J. Biochem.* 114:635–41

65. Paterson, B. M., Rosenberg, M. 1979. *Nature* 279:692–96

66. Lodish, H. F., Rose, J. K. 1977. *J. Biol. Chem.* 252:1181–88

67. Wodnar-Filipowicz, A., Szezesna, M., Zan-Kowalczewska, S., Muthukrishnan, S., Szybiak, U., et al. 1978. *Eur. J. Biochem.* 92:69–80

68. Edmonds, M., Caramela, M. G. 1969. *J. Biol. Chem.* 244:1314–24

69. Kates, J. 1970. *Cold Spring Harbor Symp. Quant. Biol.* 35:743–52

70. Darnell, J. E., Wall, R., Tushinski, R. J. 1971. *Proc. Natl. Acad. Sci. USA* 68:1321–25

71. Edmonds, M., Vaughan, M. H. Jr., Nakazato, H. 1971. *Proc. Natl. Acad. Sci. USA* 68:1336–40

72. Brawerman, G. 1981. *CRC Crit. Rev. Biochem.* 10:1–38

73. Proudfoot, N. J., Brownlee, G. G. 1976. *Nature* 263:211–14

74. Huez, G., Marbaix, G., Hubert, E., Leclercq, M., Nudel, U., et al. 1974. *Proc. Natl. Acad. Sci. USA* 71:3143–46

75. Marbaix, G., Huez, G., Burny, A., Cleuter, Y., Hubert, E., et al. 1975. *Proc. Natl. Acad. Sci. USA* 72:3065–67

76. Huez, G., Marbaix, G., Hubert, E., Cleuter, Y., Leclercq, M., et al. 1975. *Eur. J. Biochem.* 59:589–92

77. Nudel, U., Soreq, H., Littauer, U. Z., Marbaix, G., Huez, G., et al. 1976. *Eur. J. Biochem.* 64:115–21

78. Huez, G., Marbaix, G., Gallwitz, D., Weinberg, E., Devos, R., et al. 1978. *Nature* 271:572–73

79. Huez, G., Bruck, C., Cleuter, Y. 1981. *Proc. Natl. Acad. Sci. USA* 78:908–11

80. Zeevi, M., Nevins, J. R., Darnell, J. E. 1982. *Mol. Cell. Biol.* 2:517–25

81. Soreq, H., Sagar, A. D., Sehgal, P. B. 1981. *Proc. Natl. Acad. Sci. USA* 78:1741–45

82. Wilson, M. C., Nevins, J. R., Blanchard, J.-M., Ginsberg, H. S., Darnell, J. E. 1980. *Cold Spring Harbor Symp. Quant. Biol.* 44:447–55

83. Chung, S., Landfear, S. M., Blumberg, D. D., Cohen, N. S., Lodish, H. F. 1981. *Cell* 24:785–97

284. Babich, A., Nevins, J. R. 1981. *Cell* 26:371–79

285. Zeevi, M., Nevins, J. R., Darnell, J. E. 1981. *Cell* 26:39–46

286. Ilan, J., Ilan, J. 1976. *Proc. Natl. Acad. Sci. USA* 74:2325–29

287. Szer, W., Thomas, J. O., Feienstein, C., Kolb, A. 1977. See Ref. 191, pp. 70–78

288. Van Der Mast, C., Voorma, H. O. 1980. *Biochim. Biophys. Acta* 607:512–19

289. Edry, I., Humbelin, M., Darveau, A., Lee, K. A. W., Milburn, S., et al. 1983. *J. Biol. Chem.* 258:11398–403

290. Trachsel, H., Sonenberg, N., Shatkin, A. J., Rose, J. K., Leong, K., et al. 1980. *Proc. Natl. Acad. Sci. USA* 77:770–74

291. Sonenberg, N., Guertin, D., Cleveland, D., Trachsel, H. 1981. *Cell* 27:563–72

292. Jagus, R. W., Anderson, W. F., Safer, B. 1981. *Prog. Nucleic Acid Res. Mol. Biol.* 25:127–85

293. Lee, K. A. W., Guertin, D., Sonenberg, N. 1983. *J. Biol. Chem.* 258:707–10

294. Lee, K. A. W., Sonenberg, N. 1982. *Proc. Natl. Acad. Sci. USA* 79:3447–51

295. Baltimore, D. 1969. In *The Biochemistry of Viruses*, ed. H. B. Levy, pp. 101–76. New York: Dekker

296. Balbanian, R. 1975. In *Progress in Medical Virology*, ed. J. L. Melnick, pp. 40–83. Basel: Karger

297. Lucas-Lenard, J. M. 1979. In *The Molecular Biology of Picornaviruses*, ed. R. Perez-Bercoff, pp. 78–99. New York: Plenum

298. Horst, J., Fraenkel-Conrat, H., Mandeles, S. 1971. *Biochemistry*, 10:4748–52

299. Frisby, D., Eaton, M., Fellner, P. 1976. *Nucleic Acids Res.* 3:2771–87

300. Nomoto, A., Lee, Y. F., Wimmer, E. 1976. *Proc. Natl. Acad. Sci. USA* 73:375–80

301. Hewlett, M. J., Rose, J. K., Baltimore, D. 1976. *Proc. Natl. Acad. Sci. USA* 73:327–30

302. Sangar, D. V., Rowlands, D. J., Harris, T. J. R., Brown, F. 1977. *Nature* 268:648–50

303. Klootwijk, J., Klein, I., Zabel, P., Van Kammen, A. 1977. *Cell* 11:73–82

304. Ghosh, A., Dasgupta, R., Salerno-Rife, T., Rutgers, T., Kalsberg, P. 1979. *Nucleic Acids Res.*, 7:2137–46

305. Leibowitz, R., Penman, S. 1971. *J. Virol.* 8:661–68

306. Fernandez-Munoz, R., Darnell, J. E. 1976. *J. Virol.* 126:719–26

307. Kaufman, Y., Goldstein, E., Penman, S. 1976. *Proc. Natl. Acad. Sci. USA* 73:1834–38

308. Doyle, M., Holland, J. J. 1972. *J. Virol.* 9:22–28
309. Rose, J. K., Trachsel, H., Leong, K., Baltimore, D. 1978. *Proc. Natl. Acad. Sci. USA* 73:2732–36
310. Helentjaris, T., Ehrenfeld, E. 1978. *J. Virol.* 26:510–21
311. Ehrenfeld, E., Lund, H. 1977. *Virology* 80:279–308
312. Helentjaris, T., Ehrenfeld, E., Brown-Luedi, M. L., Hershey, J. W. B. 1979. *J. Biol. Chem.* 254:10973–78
313. Duncan, R., Etchison, D., Hershey, J. W. B. 1983. *J. Biol. Chem.* 258:7236–39
314. Brown, B. A., Ehrenfeld, E. 1980. *Virology* 103:327–39
315. Penman, S., Summers, D. 1965. *Virology* 27:614–20
316. Steiner-Pryor, A., Cooper, P. D. 1973. *J. Gen. Virol.* 21:215–25
317. Helentjaris, T., Ehrenfeld, E. 1977. *J. Virol.* 21:259–67
318. Etchison, D., Milburn, S. C., Edry, I., Sonenberg, N., Hershey, J. W. B. 1982. *J. Biol. Chem.* 257:14806–10
319. Morgan, M. A., Shatkin, A. J. 1980. *Biochemistry* 19:5960–66
320. Kozak, M. 1980. *Cell* 19:79–90
321. Centrella, M., Lucas-Lenard, J. 1982. *J. Virol.* 41:781–91
322. Perez-Bercoff, R., Kaempfer, R. 1982. *J. Virol.* 41:30–41
323. Rosen, H., DiSegni, G., Kaempfer, R. 1982. *J. Biol. Chem.* 257:946–52
324. Ray, B. K., Brendler, T. G., Adya, S., Daniels-McQueen, S., Miller, J. K., et al. 1983. *Proc. Natl. Acad. Sci. USA* 80:663–67
325. Tahara, S. M., Morgan, M. A., Shatkin, A. J. 1983. *J. Biol. Chem.* 258:11350–53
326. Seal, S. N., Schmidt, A., Marcus, A. 1983. *Proc. Natl. Acad. Sci. USA* 80:6562–65
327. Kozak, M., Shatkin, A. J. 1978. *J. Biol. Chem.* 253:6568–77
328. Kozak, M. 1978. *Cell* 15:1109–23
329. Kozak, M. 1980. *Cell* 22:7–8
330. Kozak, M. 1982. *Biochem. Soc. Symp.* 47:113–28
331. Odom, O. W., Kramer, G., Henderson, A. B., Pinphanichakarn, P., Hardesty, B. 1978. *J. Biol. Chem.* 253:1807–16
332. Nombela, C., Nombela, N. A., Ochoa, S., Safer, B., Anderson, W. F., Merrick, W. C. 1976. *Proc. Natl. Acad. Sci. USA* 73:298–301
333. Kemper, W. M., Berry, K. W., Merrick, W. C. 1976. *J. Biol. Chem.* 251:5551–57
334. Cooper, H. L., Park, M. H., Folk, J. E., Safer, B., Braverman, R. 1983. *Proc. Natl. Acad. Sci. USA* 80:1854–57
335. Gasior, E., Rao, P., Moldave, K. 1971. *Biochim. Biophys. Acta* 254:331–40
336. Zasloff, M., Ochoa, S. 1971. *Proc. Na. Acad. Sci. USA* 68:3059–63
337. Gasior, E., Moldave, K. 1972. *J. M. Biol.* 66:391–402
338. Leader, D. P., Wool, I. G. 19⁷ *Biochim. Biophys. Acta* 262:360–70
339. McCuiston, J., Parker, R., Moldave, 1976. *Arch. Biochem. Biophys.* 1⁷ 387–98
340. Picciano, D. J., Prichard, P. M., M rick, W. C., Shafritz, D. A., Graf, H. al. 1973. *J. Biol. Chem.* 248:204–14
341. Tiryaki, D., Bermek, E. 1976. *Hopp Seyler's Z. Physiol. Chem.* 357:721–
342. Fessenden, J. M., Moldave, K. 196 *Biochem. Biophys. Res. Commu* 6:232–35
343. Bishop, J., Schweet, R. 1961. *Biochi Biophys. Acta* 54:617–19
344. Fessenden, J. M., Moldave, K. 196 *Biochemistry* 1:485–90
345. Fessenden, J. M., Moldave, K. 1963. *Biol. Chem.* 238:1479–84
346. Arlinghaus, R., Favelukes, G., Schm R. 1963. *Biochem. Biophys. Res. Co mun.* 11:92–96
347. Schneir, M., Moldave, K. 196 *Biochim. Biophys. Acta* 166:58–67
348. Raeburn, S., Collins, J. F., Moon, M., Maxwell, E. S. 1971. *J. Biol. Che* 246:1041–48
349. Collins, J. F., Moon, H.-M., Maxwe E. S. 1972. *Biochemistry* 11:4187–9⁴
350. Iwasaki, K., Nagata, S., Mizumoto, Kaziro, Y. 1974. *J. Biol. Che* 249:5008–10
351. Iwasaki, K., Motoyoshi, K., Nagata, Kaziro, Y. 1976. *J. Biol. Chem.* 2⁵ 1843–45
352. Iwasaki, K., Kaziro, Y. 1979. *Meth Enzymol.* 60:657–76
353. Coppard, N. J., Cramer, F., Clark, B. 1982. *FEBS Lett.* 145:332–36
354. McKeehan, W. L., Hardesty, B. 1969. *Biol. Chem.* 244:4330–39
355. Prather, N., Ravel, J. M., Hardesty, Shive, W. 1974. *Biochem. Biophys. R Commun.* 57:578–83
356. Merrick, W. C., Kemper, W. M., Ka tor, J. A., Anderson, W. F. 1975. *Biol. Chem.* 250:2620–25
357. Kemper, W. M., Merrick, W. C., Re field, B., Liu, C., Weissbach, H. 19⁷ *Arch. Biochem. Biophys.* 174:603–12
358. Kemper, W. M., Merrick, W. C. 19⁷ *Methods Enzymol.* 60:638–48
359. Moon, H.-M., Redfield, B., Millard, Vane, F., Weissbach, H. 1973. *Pr Natl. Acad. Sci. USA* 70:3282–86
360. Murakami, K., Miyamoto, K. 1982. *Neurochem.* 38:1315–22
361. Golinska, B., Legocki, A. B. 19⁷ *Biochim. Biophys. Acta* 324:156–70

62. Tivardowski, T., Legocki, A. B. 1973. *Biochim. Biophys. Acta* 324:171–83
63. Lanzani, G. A., Bollini, R., Soffientini, A. N. 1974. *Biochim. Biophys. Acta* 335:275–83
64. Bollini, R., Soffientini, A. N., Bertani, A., Lanzani, G. A. 1974. *Biochemistry* 13:5431–45
65. Lauer, S. J., Burks, E., Irvin, J. D., Ravel, J. M. 1984. *J. Biol. Chem.* 259:1644–48
66. Nombela, C., Redfield, B., Ochoa, S., Weissbach, H. 1976. *Eur. J. Biochem* 65:395–402
67. Slobin, L. I., Moller, W. 1976. *Eur. J. Biochem.* 69:351–66
68. Slobin, L. I., Moller, W. 1979. *Methods Enzymol.* 60:685–703
69. Richter, D., Lipmann, F. 1970. *Biochemistry* 9:5065–70
70. Spermulli, L. L., Ravel, J. M. 1976. *Arch. Biochem. Biophys.* 172:261–69
71. Skogerson, L. 1979. *Methods Enzymol.* 60:676–85
72. Dasmahaparta, B., Chakraburtty, K. 1981. *J. Biol. Chem.* 256:9999–10004
73. Drews, J., Bednarik, K., Grasmuk, H. 1974. *Eur. J. Biochem.* 41:217–27
74. Nolan, R. D., Grasmuk, H., Drews, J. 1979. *Methods Enzymol.* 60:649–57
75. Ejiri, S., Taira, H., Shimura, K. 1973. *J. Biochem.* 74:195–97
76. Ejiri, S., Murakami, K., Katsumoto, T. 1977. *FEBS Lett.* 82:111–14
77. Legocki, A. B., Redfield, B., Weissbach, H. 1974. *Arch. Biochem. Biophys.* 161:709–12
78. Nagata, S., Iwasaki, K., Kaziro, Y. 1976. *J. Biochem.* 80:73–77
79. Grasmuk, H., Nolan, R. D., Drews, J. 1976. *Eur. J. Biochem.* 67:421–31
80. Slobin, L. I., Clark, R. V., Olson, M. O. 1983. *Biochemistry* 22:1911–17
81. Slobin, L. I. 1983. *J. Biol. Chem.* 258:4895–900
82. Liu, C. K., Legocki, A. B., Weissbach, H. 1974. In *Lipmann Symposium: Energy, Biosynthesis and Regulation in Molecular Biology,* ed. D. Richter, pp. 384–98. Berlin/New York: de Gruyter
83. Van Hemert, F. J., Van Ormondt, H., Moller, W. 1983. *FEBS Lett.* 157:289–93
84. Van Hemert, F. J., Lenstra, J. A., Moller, W. 1983. *FEBS Lett.* 157:295–99
85. Iwasaki, K., Mizumoto, K., Tanaka, M., Kaziro, Y. 1973. *J. Biochem.* 74:849–52
86. Slobin, L. I., Moller, W. 1978. *Eur. J. Biochem.* 84:69–77
87. Grasmuk, H., Nolan, R. D., Drews, J. 1978. *Eur. J. Biochem.* 92:479–90

388. Hattori, S., Iwasaki, K. 1983. *J. Biochem.* 94:79–85
389. Ejiri, S., Ebata, N., Kawamura, R., Katsumoto, T. 1983. *J. Biochem.* 94:319–22
390. Motoyoshi, K., Iwasaki, K. 1977. *J. Biochem.* 82:703–8
391. Ibuki, F., Moldave, K. 1968. *J. Biol. Chem.* 243:44–50
392. Moon, H.-M., Redfield, B., Weissbach, H. 1972. *Proc. Natl. Acad. Sci. USA* 69:1249–52
393. Nagata, S., Iwasaki, K., Kaziro, Y. 1976. *Arch. Biochem. Biophys.* 172:168–77
394. Slobin, L. I., Moller, W. 1976. *Eur. J. Biochem.* 69:367–75
395. Moon, H.-M., Weissbach, H. 1972. *Biochem. Biophys. Res. Commun.* 46:254–62
396. Nolan, R. D., Grasmuk, H., Hogenauer, G., Drews, J. 1974. *Eur. J. Biochem.* 45:601–9
397. Weissbach, H., Redfield, B., Moon, H.-M. 1973. *Arch. Biochem. Biophys.* 156:267–75
398. Nolan, R. D., Grasmuk, H., Drews, J. 1975. *Eur. J. Biochem.* 50:391–402
399. Lin, S. Y., McKeehan, W. L., Culp, W., Hardesty, B. 1969. *J. Biol. Chem.* 244:4340–50
400. Motoyoshi, K., Iwasaki, K., Kaziro, Y. 1977. *J. Biochem.* 82:145–55
401. Nielson, J. B. K., Plant, P. W., Haschemeyer, A. E. V. 1976. *Nature* 264:804–6
402. Moldave, K., Harris, J., Sabo, W., Sadnik, I. 1979. *Fed. Proc.* 38:1979–83
403. Webster, G. C., Webster, S. L. 1983. *Mech. Aging Dev.* 22:121–28
404. Engelhardt, D. L., Sarnoski, J. 1975. *J. Cell Physiol.* 86:15–30
405. Hassell, J. A., Engelhardt, D. L. 1976. *Biochemistry* 15:1375–80
406. Fischer, I., Moldave, K. 1980. *Biochemistry* 19:1417–25
407. Moldave, K., David, E. T., Hutchison, J. S., Laidlaw, S. A., Fischer, I. 1982. See Ref. 10, pp. 455–71
408. Nielsen, P. J., McConkey, E. H. 1980. *J. Cell Physiol.* 104:269–81
409. Haschemeyer, A. E. V. 1969. *Proc. Natl. Acad. Sci. USA* 62:128–35
410. Willis, D. B., Starr, J. L. 1971. *J. Biol. Chem.* 246:2828–34
411. Coppard, N. J., Clark, B. F., Cramer, F. 1983. *FEBS Lett.* 164:330–34
412. Pestka, S. 1977. In *Molecular Mechanisms of Protein Biosynthesis,* ed. H. Weissbach, S. Pestka, pp. 467–53. New York: Academic
413. Krayevsky, A., Kukhanova, M. K. 1979. In *Progress in Nucleic Acid Re-*

search and Molecular Biology, ed. W. E. Cohn, pp. 1–51. New York: Academic

414. Skogerson, L., Moldave, K. 1968. *J. Biol. Chem.* 243:5354–60
415. Skogerson, L., Moldave, K. 1968. *J. Biol. Chem.* 243:5361–67
416. Siler, J., Moldave, K. 1969. *Biochim. Biophys. Acta* 195:130–37
417. Vazquez, D., Battaner, E., Neth, R., Heller, G., Monro, R. E. 1969. *Cold Spring Harbor Symp. Quant. Biol.* 34:369–75
418. Falvey, A. K., Staehelin, T. 1970. *J. Mol. Biol.* 53:1–19
419. Pestka, S., Goorha, R., Rosenfeld, H., Neurath, C., Hintikka, H. 1972. *J. Biol. Chem.* 247:4258–63
420. Thompson, H. A., Moldave, K. 1974. *Biochemistry* 13:1348–53
421. Edens, B., Thompson, H. A., Moldave, K. 1974. See Ref. 382, pp. 179–91
422. Edens, B., Thompson, H. A., Moldave, K. 1975. *Biochemistry* 14:54–60
423. Reyes, R., Vazquez, D., Ballesta, J. P. G. 1976. *Biochim. Biophys. Acta* 435:317–32
424. Reyes, R., Vazquez, D., Ballesta, J. P. G. 1977. *Eur. J. Biochem.* 73:25–31
425. Neth, R., Monro, R. E., Heller, G., Battaner, E., Vazquez, D. 1970. *FEBS Lett.* 6:198–202
426. Battaner, E., Vazquez, D. 1971. *Biochim. Biophys. Acta* 254:316–30
427. Carrasco, L., Vazquez, D. 1975. *Eur. J. Biochem.* 50:317–23
428. Sikorski, M. M., Cerna, J., Rychlik, I., Legocki, A. B. 1977. *Biochim. Biophys. Acta* 475:123–30
429. Monro, R. E., Marcker, K. A. 1967. *J. Mol. Biol.* 25:347–50
430. Miskin, R., Zamir, A., Elson, D. 1968. *Biochem. Biophys. Res. Commun.* 33:551–57
431. Monro, R. E., Staehelin, T., Celma, M. L., Vazquez, D. 1969. *Cold Spring Harbor Symp. Quant. Biol.* 34:357–68
432. Pestka, S. 1970. *Arch. Biochem. Biophys.* 136:80–88
433. Pestka, S., Hishizawa, T., Lessard, J. L. 1970. *J. Biol. Chem.* 245:6208–19
434. Cerna, J. 1971. *FEBS Lett.* 15:101–4
435. Monro, R. E. 1971. *Methods Enzymol.* 20:472–81
436. Watanabe, S. 1972. *J. Mol. Biol.* 67:443–57
437. Hussain, L., Ofengand, J. 1972. *Biochem. Biophys. Res. Commun.* 25:233–38
438. Lessard, J. L., Pestka, S. 1972. *J. Biol. Chem.* 247:6909–12
439. Mercer, J. F. B., Symons, R. H. 1972. *Eur. J. Biochem.* 28:38–45
440. Nierhaus, K. H., Montejo, V. 1973. *Proc. Natl. Acad. Sci. USA* 70:1931–35

441. Cerna, J., Rychlik, I., Krayevsky, A. A., Gottikh, B. P. 1973. *FEBS Lett.* 37:188–91
442. Chladek, S., Ringer, D., Zemlicka, J. 1973. *Biochemistry* 12:5135–38
443. Chladek, S., Ringer, D., Quiggle, K. 1974. *Biochemistry* 13:2727–35
444. Westermann, P., Gross, B., Haumann, W. 1974. *Acta Biol. Med. Ger.* 33:699–707
445. Siler, J., Moldave, K. 1969. *Biochim. Biophys. Acta* 195:138–44
446. Moldave, K., Galasinski, W., Rao, P., Siler, J. 1969. *Cold Spring Harbor Symp. Quant. Biol.* 34:347–56
447. Henriksen, O., Robinson, E. A., Maxwell, E. S. 1975. *J. Biol. Chem.* 250:720–24
448. Muzumoto, K., Iwasaki, K., Kaziro, Y., Nojiri, C., Yamada, Y. 1974. *J. Biochem.* 75:1057–62
449. Taira, H., Ejiri, S., Shimura, K. 1972. *J. Biochem.* 72:1527–35
450. Chuang, D., Weissbach, H. 1972. *Arch. Biochem. Biophys.* 152:114–24
451. Bermek, E., Matthaei, H. 1971. *Biochemistry* 10:4906–12
452. Galasinski, W., Moldave, K. 1969. *J. Biol. Chem.* 244:6527–32
453. Collins, J. F., Raeburn, S., Maxwell, E. S. 1971. *J. Biol. Chem.* 246:1049–54
454. Legocki, A. 1979. *Methods Enzymol.* 60:703–12
455. Hutchison, J. S., Feinberg, B., Rothwell, T. C., Moldave, K. 1984. *Biochemistry* 23:3055–64
456. Yablonka-Reuveni, Z., Hille, M. B. 1983. *Biochemistry* 22:5205–12
457. Yablonka-Reuveni, Z., Fontaine, J. J., Warner, A. H. 1983. *Can. J. Biochem. Cell. Biol.* 61:833–39
458. Sutter, R. P., Moldave, K. 1966. *J. Biol. Chem.* 241:1698–1704
459. Robinson, E. A., Maxwell, E. S. 1972. *J. Biol. Chem.* 247:7023–28
460. Robinson, E. A., Henriksen, O., Maxwell, E. S. 1974. *J. Biol. Chem.* 249:5088–93
461. Van Ness, B. G., Howard, J. B., Bodley, J. W. 1978. *J. Biol. Chem.* 253:8687–90
462. Brown, B. A., Bodley, J. W. 1979. *FEBS Lett.* 103:253–55
463. Van Ness, B. G., Howard, J. B., Bodley, J. W. 1980. *J. Biol. Chem.* 255:10710–16
464. Van Ness, B. G., Howard, J. B., Bodley, J. W. 1980. *J. Biol. Chem.* 255:10717–20
465. Bodley, J. W., Dunlop, P. C., Van Ness, B. G. 1984. *Methods Enzymol.* 106:378–87
466. Moehring, J. M., Moehring, T. J. 1984. *Methods Enzymol.* 106:388–95

467. Collier, R. J., Pappenheimer, A. M. Jr. 1964. *J. Exp. Med.* 120:1019–39
468. Collier, R. J. 1967. *J. Mol. Biol.* 25:83–98
469. Goor, R. S., Pappenheimer, A. M. Jr. 1967. *J. Exp. Med.* 126:899–912
470. Goor, R. S., Pappenheimer, A. M. Jr. 1967. *J. Exp. Med.* 126:913–21
471. Raeburn, S., Goor, R. S., Schneider, J. A., Maxwell, E. S. 1968. *Proc. Natl. Acad. Sci. USA* 61:1428–34
472. Honjo, T., Nishizuka, Y., Hayaishi, O., Kato, I. 1968. *J. Biol. Chem.* 243:3553–55
473. Oppenheimer, N. J., Bodley, J. W. 1981. *J. Biol. Chem.* 256:8579–81
474. Moehring, T. J., Danley, D. E., Moehring, J. M. 1979. *Somatic Cell Genet.* 5:469–80
475. Moehring, J. M., Moehring, T. J., Danley, D. E. 1980. *Proc. Natl. Acad. Sci. USA* 77:1010–14
476. Iglewski, W. J., Lee, H. 1983. *Eur. J. Biochem.* 134:237–40
477. Aust, A. E., Drinkwater, N. R., Debien, K., Maher, V. M., McCormick, J. J. 1984. *Mutat. Res.* 125:95–104
478. Bodley, J. W., Lin, L. 1970. *Nature* 227:60–62
479. Baliga, B. S., Munro, H. N. 1972. *Biochim. Biophys. Acta* 277:368–83
480. Mizumoto, K., Iwasaki, K., Kaziro, Y. 1974. *J. Biochem.* 76:1269–80
481. Nygard, O., Nilsson, L. 1984. *Eur. J. Biochem.* 140:93–96
482. Skogerson, L., Moldave, K. 1967. *Biochem. Biophys. Res. Commun.* 27:568–72
483. Henriksen, O., Robinson, E. A., Maxwell, E. S. 1975. *J. Biol. Chem.* 250:725–30
484. Nurten, R., Aktar, N. B., Bermek, E. 1983. *FEBS Lett.* 154:391–94
485. Tanaka, M., Iwasaki, K., Kaziro, Y. 1977. *J. Biochem.* 82:1035–43
486. Spirin, A. 1969. *Cold Spring Harbor Symp. Quant. Biol.* 34:197–207
487. Hardesty, B., Culp, W., McKeehan, W. 1969. *Cold Spring Harbor Symp. Quant. Biol.* 34:331–45
488. Gupta, S. L., Waterson, J., Sopari, M. L., Weissman, S. M., Lengyel, P. 1971. *Biochemistry* 10:4410–21
489. Schrier, M. H., Noll, H. 1971. *Proc. Natl. Acad. Sci. USA* 68:805–9
490. Caldiroli, E., Zocchi, G., Cocucci, S. 1983. *Eur. J. Biochem.* 131:255–59
491. Skogerson, L., Wakatama, E. 1976. *Proc. Natl. Acad. Sci. USA* 73:73–76
492. Skogerson, L., Engelhardt, D. 1977. *J. Biol. Chem.* 252:1471–75
493. Watson, J. D. 1964. *Bull. Soc. Chim. Biol.* 46:1399–425
494. Schmitt, M., Neugebauer, U., Berg-mann, C., Gassen, H. G., Riesner, D. 1982. *Eur. J. Biochem.* 127:525–29
495. Spirin, A. S. 1984. *FEBS Lett.* 165:280–84
496. Wettstein, F. O., Noll, H. 1965. *J. Mol. Biol.* 11:35–53
497. Haenni, A.-L., Lucas-Lenard, J. 1968. *Proc. Natl. Acad. Sci. USA* 61:1363–69
498. Swan, D., Sander, G., Bermek, E., Kramer, W., Kreuzer, T., et al. 1969. *Cold Spring Harbor Symp. Quant. Biol.* 34:179–96
499. Lake, J. A. 1980. In *Ribosomes: Structure, Function and Genetics*, ed. G. Chambliss, G. R. Craven, J. Davies, K. Davis, L. Kahan, M. Nomura, pp. 207–36. Baltimore: University Park Press
500. Rheinberger, H.-J., Sternbach, H., Nierhaus, K. H. 1981. *Proc. Natl. Acad. Sci. USA* 78:5310–14
501. Grajevskaja, R. A., Ivanov, Y. V., Saminsky, E. M. 1982. *Eur. J. Biochem.* 128:47–52
502. Rheinberger, H.-J., Schilling, S., Nierhaus, K. H. 1983. *Eur. J. Biochem.* 134:421–28
503. Kirillov, S. V., Makarov, E. M., Semenkov, Y. P. 1983. *FEBS Lett.* 157:91–94
504. Robbins, D., Hardesty, B. 1983. *Biochemistry* 22:5675–79
505. Rheinberger, H.-J., Nierhaus, K. H. 1983. *Proc. Natl. Acad. Sci. USA* 80:4213–17
506. Beaudet, A. L., Caskey, C. T. 1971. *Proc. Natl. Acad. Sci. USA* 68:619–24
507. Capecchi, M. R., Hughes, S. H., Wahl, G. M. 1975. *Cell* 6:269–77
508. Gesteland, R. F., Wolfner, M., Grisafi, P., Fink, G., Bottstein, D., Roth, J. R. 1976. *Cell* 7:381–90
509. Goldstein, J. L., Beaudet, A. L., Caskey, C. T. 1970. *Proc. Natl. Acad. Sci. USA* 67:99–105
510. Tate, W. P., Beaudet, A. L., Caskey, C. T. 1973. *Proc. Natl. Acad. Sci. USA* 70:2350–52
511. Tate, W. P., Caskey, C. T. 1973. *J. Biol. Chem.* 248:7970–72
512. Ilan, J. 1973. *J. Mol. Biol.* 77:437–48
513. Innanen, V. T., Nicholls, D. M. 1973. *Biochim. Biophys. Acta* 324:533–44
514. Caskey, C. T., Beaudet, A. L., Tate, W. P. 1974. *Methods Enzymol.* 30:293–303
515. Tate, W. P., Caskey, C. T. 1974. *Enzymes* 10:87–118
516. Konecki, D. S., Aune, K. C., Tate, W., Caskey, C. T. 1977. *J. Biol. Chem.* 252:4514–20
517. Wei, C.-M., Hansen, B. S., Vaughan, M. H. Jr., McLaughlin, C. S. 1974. *Proc. Natl. Acad. Sci. USA* 71:713–17

nn. Rev. Biochem. 1985. 54:1151–93
Copyright © 1985 by Annual Reviews Inc. All rights reserved.

PSORALENS AS PHOTOACTIVE PROBES OF NUCLEIC ACID STRUCTURE AND FUNCTION: ORGANIC CHEMISTRY, PHOTOCHEMISTRY, AND BIOCHEMISTRY

George D. Cimino, Howard B. Gamper, Stephen T. Isaacs, and John E. Hearst

Department of Chemistry, University of California, Berkeley, California 94720

CONTENTS

1151

0066-4154/85/0701-1151$2.00

PERSPECTIVES AND SUMMARY

Psoralens comprise the most important class of photochemical reagents for the investigation of nucleic acid structure and function. They have been used for determining the structure of both DNA and RNA in viral, bacterial, and mammalian systems, and also for studying functional questions, such as the role of the small nuclear RNAs in processing heteronuclear RNA. A list of some of the major applications of these compounds during the last ten years is presented in Table 1.

Psoralens are unique in their ability to freeze helical regions of nucleic acid. Psoralens react with DNA and RNA by a two-step mechanism. First, the planar psoralen molecule intercalates within a double helical region of nucleic acid. Covalent addition of the psoralen is effected by controlled irradiation into an absorption band of the psoralen molecule. Stable, but photoreversible, covalent adducts form with pyrimidine bases at one or both ends of the psoralen molecule. By forming covalent crosslinks with base-paired structures, psoralens can probe both static and dynamic structural features. Psoralens can trap long-range interactions which are in dynamic equilibrium. This allows both the occurrence of the interaction to be established and its position within the structure to be mapped. Psoralens can also be used temporally, such as in following the fate of short-lived nucleic acid species in vivo.

The details of the interaction between psoralens and nucleic acid are well understood at the molecular level. The structure of the psoralen adducts formed with DNA have been determined, the polarity of the reaction which converts monoadduct to crosslink established, and methods for the exclusive formation of monoaddition products worked out. This advanced state of chemical control makes the psoralens extremely versatile reagents. As more information is compiled about structure-activity parameters, a fine tuning of the reaction of psoralens with nucleic acid will be realized.

Future application of psoralens for investigating nucleic acid structure and function will be aided by the following developments. The preparation of hybridization probes which carry psoralen monoadducts is currently under way. These probes will be used to form covalent hybrids for locating particular sequences and also for site-specific placement of psoralen monoadducts in nucleic acid structures via photochemical transfer of the psoralen. The transferred monoadduct will be used for fixation of "dynamic" base paired intrastructural conformations by crosslink formation. Chemical schemes for the site specific cleavage of DNA and RNA at the position of psoralen addition are also being developed. These procedures will allow for the direct mapping of secondary structure at the position of crosslink formation. Finally, many new psoralen derivatives are being synthesized for specific applications such as site-directed crosslinking of DNA and protein-nucleic acid crosslinking. Psor-

Table 1 Publications employing psoralen as a probe for nucleic acid structure and function

Subject Study	Author	Year	Ref.
Chromatin Structure			
Drosophila melanogaster nuclei	Hanson et al	1976	34
Mouse liver nuclei	Cech & Pardue	1977	143
Drosophila melanogaster nuclei	Wiesehahn et al	1977	39
Mouse liver nuclei	Cech et al	1977	144
Simian virus 40 minichromosomes	Hallick et al	1978	145
Main-band and satellite DNAs in *Drosophila melanogaster* nuclei	Shen & Hearst	1978	146
Ribosomal RNA genes in *Tetrahymena thermophila* cells	Cech & Karrer	1980	151
Mitochondrial DNA in *Drosophila melanogaster* cells	Potter et al	1980	40
Escherichia coli DNA	Hallick et al	1980	153
Mitochondrial DNA in HeLa cells	De Francesco & Attardi	1981	152
Simian virus 40 DNA in lytically infected CV-1 cells	Carlson et al	1982	148
Simian virus 40 DNA in lytically infected CV-1 cells	Robinson & Hallick	1982	149
Simian virus 40 DNA in the virion	Kondoleon et al	1983	150
Calf thymus chromatin	Mitra et al	1984	191
Cruciforms			
Mouse tissue culture cells	Cech & Pardue	1976	154
Ribosomal RNA genes in *Tetrahymena thermophila* cells	Cech & Karrer	1980	151
Lac operator DNA in *Escherichia coli* cells	Sinden et al	1983	155
Torsional Tension			
Prokaryotic and eukaryotic cells	Sinden et al	1980	156
Domains of supercoiling	Sinden & Pettijohn	1981	192
Secondary Structure in Single-Stranded DNA			
Bacteriophage fd DNA in solution	Shen & Hearst	1976	157
Denatured simian virus 40 DNA in solution	Shen & Hearst	1977	160
Bacteriophage fd DNA in the virion	Shen et al	1979	158
Denatured simian virus 40 DNA in solution	Shen & Hearst	1979	161
Chimeric phage M13Gori1 in the virion	Ikoku & Hearst	1981	159
Secondary Structure in Ribosomal RNA			
Drosophila melanogaster 18S and 26S RNA in solution	Wollenzien et al	1978	35
E. coli 16S RNA in solution	Wollenzien & Hearst	1979	162

Table 1 *(continued)*

Subject Study	Author	Year	Ref.
E. coli 16S RNA within the 30S subunit	Thammana et al	1979	163
Polarity of *E. coli* crosslinked 16S RNA	Wollenzien et al	1979	164
E. coli 5S RNA in solution	Rabin & Crothers	1979	33
Sequence analysis of *E. coli* crosslinked 16S RNA	Cantor et al	1980	165
Drosophila melanogaster 5S RNA in solution	Thompson et al	1981	168
Electrophoretic separation of *E. coli* crosslinked 16S RNA	Wollenzien & Cantor	1982	166
E. coli 16S RNA in solution	Turner et al	1982	29
Comparison of *E. coli* 16S RNA in active and inactive 30S subunits	Chu et al	1983	167
E. coli 16S RNA in solution	Thompson & Hearst	1983	30
Functional significance of long range crosslinks in *E. coli* 16S RNA	Thompson & Hearst	1983	171
E. coli 23S RNA in solution	Turner & Noller	1983	193
Secondary Structure in Transfer RNA			
E. coli phenylalanine tRNA in solution	Bachellerie & Hearst	1982	172
Yeast phenylalanine tRNA in solution	Garrett-Wheeler et al	1984	37
Secondary Structure in Heterogeneous and Small Nuclear RNAs			
Heterogeneous nuclear RNA in HeLa cell nuclei	Calvet & Pederson	1979	173
Heterogeneous nuclear RNA in HeLa cells	Calvet & Pederson	1979	174
Complex between U1 RNA and heterogeneous nuclear RNA in HeLa cells	Calvet & Pederson	1981	176
Complex between U2 RNA and heterogeneous nuclear RNA in HeLa cells	Calvet et al	1982	177
Neurospora mitochondrial 35S precursor rRNA within ribonucleoprotein particles	Wollenzien et al	1983	175
Ribonucleoprotein organization of the U1 RNA complex with heterogeneous nuclear RNA in HeLa cells	Setyono & Pederson	1984	178
Secondary Structure in Viral RNA Genomes			
Reovirus RNA genome within the virion	Nakashima & Shatkin	1978	194
Rous sarcoma RNA genome within the virion	Swanstrom et al	1981	195
Fixation of Nucleic Acid Complexes			
Ternary transcription complexes	Shen & Hearst	1978	180
R-loops	Kaback et al	1979	182
R-loops	Wittig & Wittig	1979	32

Table 1 *(continued)*

Subject Study	Author	Year	Ref.
D-loops	DeFrancesco & Attardi	1981	152
Adenovirus 5 replicative intermediates	Revet & Benichou	1981	181
R-loops	Chatterjee & Cantor	1982	183
M13 hybridization probes	Brown et al	1982	185
Determination of the polarity of single-stranded RNA in the electron microscope	Wollenzien & Cantor	1982	186
Tertiary Interactions in Nucleoprotein Complexes			
DNA-DNA interactions in λ bacteriophage	Haas et al	1982	87
DNA-DNA and DNA-protein interactions in bacteriophage λ	Schwartz et al	1983	86

alen analogs which will crosslink purine to pyrimidine and purine to purine are also being considered.

It is not the intent of this review to catalog the properties and applications of every psoralen derivative known. Rather we try to show how a basic understanding of the organic chemistry, photochemistry, and biochemistry of these compounds has produced a versatile molecular tool for the elucidation of nucleic acid structure and function. The use of psoralens for the determination of nucleic acid secondary structure will be emphasized here. Recent reviews include coverage of other aspects of psoralens including clinical applications (1, 2), mutagenesis, toxicity and repair (3), and photochemistry and photobiology (4–6).

INTRODUCTION

Psoralens are bifunctional photoreagents which form covalent bonds with the pyrimidine bases of nucleic acids (4, 7). Structurally, psoralens are tricyclic compounds formed by the linear fusion of a furan ring with a coumarin. An angular fusion of the two-ring systems forms an isopsoralen, which is also known as angelicin. The structures of psoralen, isopsoralen, and several psoralen derivatives are shown in Figure 1. Both psoralens and isopsoralens can intercalate into double-stranded nucleic acid end undergo covalent photocycloaddition at either the furan or coumarin and (8, 9). The photoaddition to nucleic acid occurs with incident light of wavelength of 320–400 nm, a region of the electromagnetic spectrum to which nucleic acids are transparent. Binding of both ends of a psoralen to opposite strands of a nucleic acid helix results inthe formation of a covalent interstrand crosslink (10). Isopsoralens, due to their angular geometry, cannot form crosslinks with DNA in dilute aqueous solutions; however, they are reported to crosslink certain types of folded DNA (11–14). Psoralens also react with single-stranded nucleic acid, but to a much smaller extent than with double-stranded structures (15). Psoralens have some reactivity with proteins (16–20) and lipid membranes (21); however, these reactivities are minor compared to the reaction with nucleic acid.

Psoralen reacts primarily with thymidine in DNA and uridine in RNA, although a minor reaction with cytosine also occurs. There is a single report in the literature of 8-methoxypsoralen reacting with adenosine (22), but this is a very minor reaction compared to the reaction with the pyrimidines. Aside from the alteration which occurs at the site of the chemical modification, there is no additional degradation of either RNA or DNA associated with the photochemistry provided that it is carried out under anoxygenic conditions. If singlet oxygen is produced during the photochemistry it will bring about degradation of all biological molecules with which it comes in contact (23, 24).

Isaacs et al (25) proposed a mechanism for the reaction of the psoralens with nucleic acid helices:

$$P + S \rightleftharpoons PS$$

$$PS + h\nu \rightarrow A$$

$$A + h\nu \rightarrow X$$

$$P + h\nu \rightarrow B$$

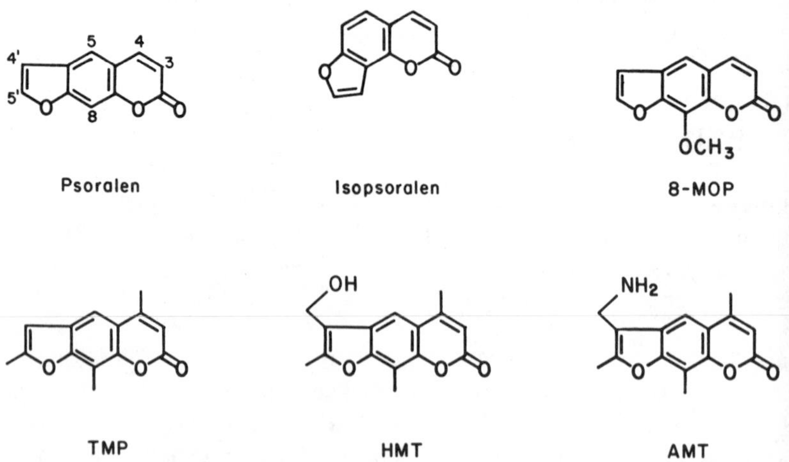

Figure 1 The structure and numbering system used for psoralen, isopsoralen (angelicin), 8-methoxypsoralen (8-MOP), 4,5',8-trimethylpsoralen (TMP), 4'-hydroxymethyl-4,5',8-trimethylpsoralen (HMT) and 4'aminomethyl-4,5',8-trimethylpsoralen (AMT).

where P is the psoralen derivative in question, S is a psoralen intercalation site in a nucleic acid helix, PS is the noncovalent intercalation complex between psoralen and the DNA or RNA site, A refers to the covalent monoadduct of the psoralen to the nucleic acid, X refers to the covalent crosslink in the nucleic acid helix, B represents photobreakdown products of psoralen, and hν is a photon of light.

The versatility of psoralen photochemistry allows for control of the degree of reaction by light dose, control of the ratio of monoaddition to crosslinkage by either selection of suitable wavelengths to induce the chemistry (26, 27), or by controlled timing of the delivery of the light (28). Furthermore, the monoadducts and diadducts (crosslinks) formed between psoralen and pyrimidines can be reversed with short wavelength ultraviolet light. This property is used to great advantage in the determination of nucleic acid secondary structure using psoralens (29, 30).

The monoadduct and crosslink are chemically stable, allowing for analysis under varied chemical conditions. The photochemistry can be carried out under a wide variety of conditions including broad ranges of temperature and ionic strength, in the presence or absence of divalent cations (31), and even in some organic solvents (32). The positions of the covalent crosslinks can therefore be enzymatically mapped (33). They can also be mapped by visualization in the electron microscope when the DNA or RNA is spread under denaturing conditions (34, 35). Enzymatic methods for locating monoadducts in nucleic acids sequences have been developed (36, 37), and chemical cleavage methods are sure to follow shortly (38). Since proteins that bind DNA such as nucleosomes protect the DNA from reaction with psoralen, these mapping techniques provide information about protein–nucleic acid interactions as well (39, 40). In such experiments a very high level of chemical substitution of the DNA with psoralen can be achieved.

Psoralens are used in a number of diverse ways. Clinically, psoralens are used for the treatment of psoraisis (41), vitiligo (42), and other skin disorders (43). Extensive studies on the metabolism of these compounds have been made (44). As irreversible nucleic acid specific crosslinking reagents, psoralens are used to inactivate viruses and other pathogens with little disruption of proteinacious antigenic structures. Many viruses that cannot be otherwise inactivated for vaccine production are effectively inactivated by psoralen photochemistry (45, 46). As ligands that bind stereospecifically to nucleic acid, psoralens form a limited set of known and characterized adducts, which make them extremely useful as model compounds for mutagenesis and repair studies (47–51). The selective reaction of psoralens with nucleic acids, which is most efficient in helical regions, allows one to probe both static and dynamic nucleic acid structure both in vitro and in vivo. The psoralens are able to penetrate most biological structures and are not highly toxic to cells in the absence of actinic light, so they are ideal probes for nucleic acid structure wherever long wave ultraviolet light can be delivered. Both DNA and RNA structure have been probed by psoralen crosslinking (see Table 1).

Control of the crosslinking reaction at each step in the psoralen-nucleic acid interaction is possible by molecular architecture. Since the properties of a given psoralen derivative are directly a function of its structure, control at the level of chemical synthesis is possible by judicious choice of substituent, with respect to both type and placement on the tricyclic ring system (52, 53). As more information is compiled about structure-activity parameters for both the dark binding and photoaddition of psoralens to nucleic acid, a fine tuning of the interaction for the study of nucleic acid will become possible.

ORGANIC AND STRUCTURAL CHEMISTRY OF PSORALEN AND ITS ADDUCTS

Naturally Occurring Psoralens

Both psoralens and isopsoralens are natural products found in plants. They are most abundant in the *Umbelliferae, Rutacea,* and *Leguminosae* families (54). Several dozen psoralens have been characterized from natural sources (55); some of these are thought to act as natural insecticides (56). Psoralens have also been isolated from microorganisms including fungi. The two most widely used psoralens are 8-methoxypsoralen, isolated from Ammi Majus, and 4,5',8-trimethylpsoralen, derived from a fungus which grows on diseased celery (57). Both of these compounds are readily prepared by chemical synthesis (58–60).

Synthesis and Radiolabeling of Psoralens

The synthetic chemistry which has been developed falls into two areas: (*a*) preparation of the psoralen tricyclic ring system which is built from resorcinol, and (*b*) exocyclic modification of the intact ring system. The combined synthetic procedures provide different psoralen derivatives with particular characteristics for different applications.

Ring synthesis of both psoralen and isopsoralen (angelicin) proceeds by converting resorcinol to either a coumarin or benzofuran then adding the furan or pyrone ring respectively. The final substituents at the 5 and 8 positions of the psoralen or isopsoralen are those present on the resorcinol used in the initial step. When proceeding from the coumarin, the presence or absence of substituents at the 6 and 8 position of the coumarin determine if a psoralen, isopsoralen, or a mixture of the two is obtained (61–63). Substituents at the 8 position of the coumarin direct furan ring closure to the 6 position giving a psoralen, while substituents at the 6 position of the coumarin have just the opposite effect and are precursors of isopsoralens. When neither the 6 or 8 position of the coumarin is substituted, a mixture of psoralen and isopsoralen usually results.

Proceeding from benzofuran, the key synthetic intermediate is 6-acetoxycoumaran (64–67). Psoralens that contain a substituent on the 3 but not the 4 carbon, such as 3-carbethoxypsoralen (68) and 3-methylpsoralen (69) are prepared by this method. The more common procedure, however, is to first synthesize the coumarin and then add the furan ring. Here resorcinol is treated with a β-ketoester giving a coumarin via the von Pechmann condensation. The substituents at the 3 and 4 positions are controlled by which β-ketoester is used for the reaction. Psoralens such as 4,5',8-trimethylpsoralen (60) and 3-n-butyl-4,5',8-trimethylpsoralen (70) are synthesized by this procedure. A number of methods have been developed for the addition of the furan ring to the coumarin. Psoralens and isopsoralens containing alkyl groups at the 4' or 5' positions (60, 63, 71) or with an unsubstituted furan ring have been prepared (72, 73).

Exocyclic modification of the intact psoralen ring system is used to fine tune the molecular characteristics of the molecule. Psoralens undergo a variety of substitution and addition reactions such as nitration, halogenation, quinone formation, catalytic reduction, chlorosulfonation, and chloromethylation (25, 52, 73–81). Psoralen derivatives substituted at the 8 position with aminomethyl, aminoethyl, hydroxymethyl, and hydroxyethyl substituents have been prepared by manipulation of a family of 8-acetylpsoralens (63).

Chloromethylation has been a singularly valuable reaction for providing a synthetic handle for the preparation of a whole family of new psoralen derivatives. Among these are the highly water soluble derivatives 4'-aminomethyl-4,5',8-trimethylpsoralen (25), 5-aminomethyl-8-methoxypsoralen (82), 5'-aminomethyl-4,4',8-trimethylpsoralen (83), and 4'-aminomethyl-4,5'-dimethylisopsoralen (84). Specialized psoralens for site directed crosslinking of DNA (85), for protein-nucleic acid crosslinking (86), and for investigating the packaging of phage DNA (87) have also been synthesized starting with 4'-chloromethyl-4,5',8-trimethylpsoralen.

The use of psoralens for determining nucleic acid structure often requires that the psoralen be radiolabeled. Both tritium and carbon-14 labeled psoralens have been synthesized by routine synthetic manipulation to obtain radiolabeled derivatives. Methods used include exchange reactions with tritium gas (88, 89) and tritiated water (25), catalytic reduction (80, 81) and hydrogenolysis with tritium gas (90), alkylation with tritium (80, 81) or carbon-14 labeled alkyl halides, borotritide reductions of aldehydes and ketones (80), and various ring syntheses with carbon-14 labeled precursors (47). The exchange procedures typically give low specific activities (50–300 mCi/mmol) and do not site-specifically label the molecule. Higher specific activities (1–20 Ci/mmol) are obtained with site-specific reduction or alkylation. The position of the label is often important, such as in metabolism studies where the structure is sequentially degraded.

Monofunctional Psoralens

There are a variety of uses for psoralen derivatives which only form monoadducts with helical nucleic acid (91). The debate over the relative mutagenicity and/or carcinogenicity of monoadducts vs crosslinks has resulted in the clinical community looking closely at psoralens which form only monoadducts for use in photochemotherapy (92). Investigators studying DNA repair have used monofunctional vs bifunctional psoralens extensively in order to address cellular response to the two different kinds of damage (47–50). The approach has been to use compounds that due to their geometry cannot form crosslinks, such as the isopsoralens, or to use psoralens which have been modified by substituents that limit the reactivity to one end of the molecule. In this second approach, substituents that exert either steric or electronic effects have been used.

Isopsoralens are reported not to crosslink double-stranded B form nucleic acid due to their angular structure (93, 94). The geometry of the isopsoralen is such that once a monoadduct has been formed, subsequent reaction of the remaining double bond with a pyrimidine on the adjacent nucleic acid strand is not possible due to a misalignment of the two reactive double bonds. The absence of crosslinking is usually demonstrated by the inability of DNA to "snap back" rapidly after denaturation following irradiation in the presence of these compounds (95–97). Absolute lack of crosslinking by the isopsoralens has not been confirmed by total characterization of the adducts formed by HPLC and NMR analysis. Several investigators have reported that some crosslinks are formed by isopsoralen with phage lambda DNA (11–14). The explanation proposed is that the packaged DNA is amenable to crosslinking by isopsoralen due to its special folded structure within the phage head.

Monofunctional psoralens have been prepared by chemical synthesis for the most part. An exception is 3-dimethylallylpsoralen, which is a natural product. 3-dimethylallylpsoralen is reported to form only 2.5% crosslink based on the denaturation assay (98). This presumably results from the bulky dimethylallyl side chain sterically preventing the efficient conversion of the monoaddition product to crosslink. Among the synthetic derivatives, the approach has been to attach strongly electron withdrawing or donating groups to one of the two reactive double bonds. 3-Carbethoxypsoralen is reported to form little if any crosslink due to the electron withdrawing effect of the 3-carbethoxy group (47); however, this compound binds very poorly to DNA in vitro, having virtually no detectable dark association and approximately 3–5% of the photoreactivity of

3-methoxypsoralen with DNA (81). Another unreactive psoralen with an electron withdrawing group at the 3 position is 3-cyanopsoralen (99). Psoralens with electron donating substituents at the 3 position have also been prepared as nonofunctional reagents, such as 3-amino and 3-methoxypsoralen (100). The 3-amino compound reportedly formed only monoadducts and reacted faster than 8-methoxypsoralen, but the total amount of the compound bound to the DNA was much less than with 8-MOP. With the 3-methoxy compound, little binding was detected and there was a significant amount of crosslink formed by the bound compound. In an attempt to inhibit the reactivity of the furan 4',5' double bond, 5-carbomethoxy-4,8-dimethylpsoralen was prepared and surprisingly was found to be highly reactive with DNA forming a significant amount of crosslink. The isomeric 3-carbomethoxy-4',8-dimethylpsoralen, however, was found to be essentially unreactive (101). There is one report of a dihydropsoralen, reduced at the furan end, being slightly reactive with DNA (91).

Relative Reactivity of Different Psoralens with Nucleic Acids

The substituents of a psoralen effect each step in its interaction with nucleic acid. The position, the steric, and the electronic characteristics of each group on the psoralen ring system determine its ability to dark bind and photoreact with DNA and RNA, as well as the distribution of adducts formed. Many correlations as to what the effect of various substituent groups are have been made (25, 52, 53, 63, 102, 103). Some of the general trends that are known are the following. The photochemistry of methylated psoralens and isopsoralens is relatively fast. Methylation of a psoralen or isopsoralen increases the dark binding affinity, the quantum yield of photoaddition, and the quantum yield of photobreakdown of the compound (52, 102, 103). 4,5',8-trimethylpsoralen and 4,4',6-trimethylisopsoralen are therefore much more reactive than unsubstituted psoralen or isopsoralen respectively (63). A methoxy groups at the 8 position slows the photochemistry, with 8-methoxypsoralen adding much more slowly to DNA and also being much longer lived in solution than 4,5',8-trimethylpsoralen (52).

Strong electron withdrawing or donating substituents such as hydroxy, amino, and nitro groups either drastically reduce or completely eliminate the ability of the psoralen to undergo photocycloaddition with nucleic acids. 5-Nitro-8-methoxypsoralen and 5-amino-8-methoxypsoralen have no reactivity with DNA (52). Substitution at the 3 position of the psoralen nucleus with electronically active substituents is particularily unfavorable for high reactivity with nucleic acid (47, 99–101). However, relatively bulky groups which are positively charged placed at the 4' and 5 positions of the psoralen ring system form compounds which have both high dark association constants and high photoreactivity with DNA (52, 104).

The presence or absence of a methyl group at the 4 position has been shown to have a major role in controlling the amount of pyrone-side monoadduct formed with DNA (53). Psoralens that contain a 4 methyl group, such as 4,5',8-trimethylpsoralen and 4'-hydroxymethyl-4,5',8-trimethylpsoralen, form less than 2% pyrone side monoadduct. This has been attributed to steric interference between the 4 methyl group of the psoralen and the 5 methyl group of thymidine with which the psoralen predominately reacts. Psoralens that do not contain a 4 methyl group, such as 8-methoxypsoralen and psoralen, form up to 20% pyrone-side monoadduct with DNA (10). With RNA, 4'-hydroxy-methyl-4,5',8-trimethylpsoralen forms up to 20% pyrone side monoadduct (105). This result is consistent with the DNA result since the reactive base in RNA is uracil which does not contain a methyl group at the 5 position.

Psoralen Monoadducts and Diadducts Formed with DNA and RNA

The photochemical addition products formed between psoralen and the pyrimidine bases of DNA and RNA have been characterized extensively. The detailed structure of both monoadducts and diadducts formed with several psoralen derivatives have been reported (8–10, 106), including the crystal structure of the furan-side 8-methoxypsoralen-thymine monoadduct (107–109). If the reaction occurs with intact, double-stranded nucleic acid, the number of adducts is limited and their stereochemistry controlled by the geometry of the intercalation complex. A wider variety of adducts are formed when psoralen is reacted with monomeric pyrimidine bases free in solution, in a frozen matrix or as a thin film.

Irradiation of 8-methoxypsoralen and double-stranded DNA starting with one 8-MOP per four base pairs will covalently bind 25–30% of the 8-MOP to the DNA. The distribution of adducts formed is approximately 25% thymidine-8-MOP-thymidine diadduct (crosslink), 45% furan-side thymidine-8-MOP monoadduct, and 20% pyrone-side thymidine-8-MOP monoadduct. A small amount of furan-side deoxyuridine-8-MOP monoadduct (ca 2%), derived from reaction with cytosine followed by hydrolytic deamination, is also formed. 8-MOP preferentially binds to 5'-TpA crosslinkable sites in the DNA (27). If the initial 8-MOP:base pair ratio is low, this preference becomes apparent, with a larger proportion of the total adduct being crosslink (60–70%), assuming a sufficient dose of light has been provided. As the amount of psoralen available for photoaddition is increased, crosslinkable sites become filled and the ratio of crosslink to monoadduct correspondingly decreases.

The structures of the monoadducts and diadducts formed between 8-MOP and DNA are shown in Figure 2. The eight possible configurations for psoralen-thymidine monoadducts are shown schematically in Figure 3. The stereochemistry of all the 8-MOP DNA adducts shown in Figure 2 is *cis-syn*. *Syn* for the furan-side describes the structure having the furan 1'-oxygen and the pyrimidine 1-nitrogen on adjacent corners of the cyclobutane ring. *Syn* for the pyrone side is defined having the 2-carbon of the psoralen and the 1-nitrogen of the pyrimidine on adjacent corners of the cyclobutane ring. The formation of only one particular set of configurational isomers reflects the stringent restrictions on modes of psoralen intercalation imposed by the double helical DNA conformation. A complete discussion of the possible stereochem-istries for psoralen-pyrimidine monoadducts and diadducts has been presented elsewhere (8–10).

As shown in Figure 4, 8-MOP can react from either the 3' or the 5' face of thymine, depending on whether the 8-MOP is positioned on top of or under-neath the plane of the base within the binding site. Reaction within either a 5'-TpX (3' face) or 3'-TpX (5' face) sequence results in a pair of enantiomeric thymine monoadducts being formed. In the case of nucleoside monoadducts, a pair of diasteromers is formed since the chirality of the deoxyribose is the same

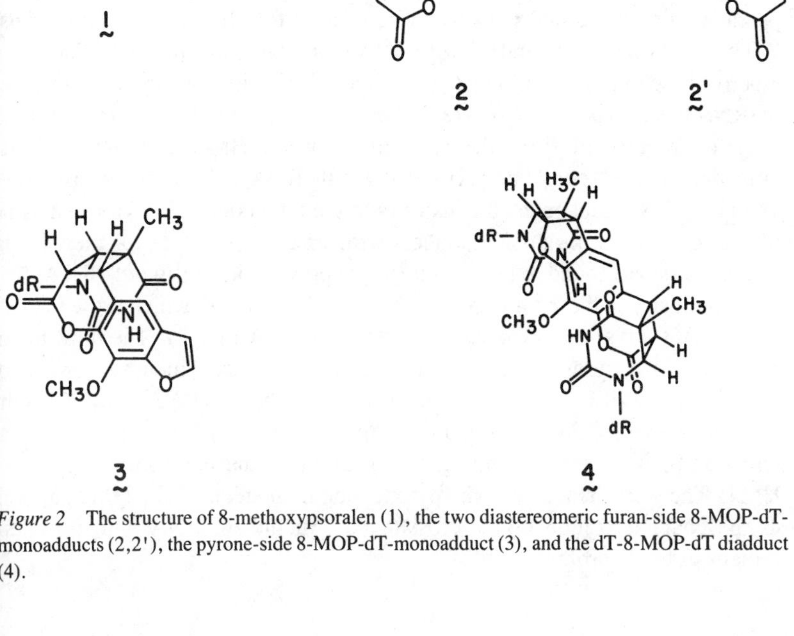

Figure 2   The structure of 8-methoxypsoralen (1), the two diastereomeric furan-side 8-MOP-dT-monoadducts (2,2'), the pyrone-side 8-MOP-dT-monoadduct (3), and the dT-8-MOP-dT diadduct (4).

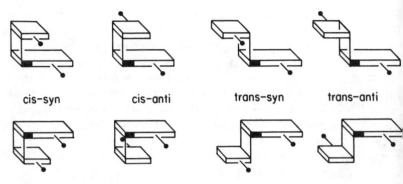

cis-syn   cis-anti   trans-syn   trans-anti

Figure 3 Nomenclature and schematic representation of the 8 possible configurational isomer for the 8-methoxypsoralen-thymine monoadduct. The large slabs represent 8-methoxypsoralen with the 5-membered furan ring at the left as indicated by the blackened markers. Handle projecting from these slabs represent the methoxy group at C-8 of the psoralen ring. The smalle slabs are for thymine bases with the handles representing TN-1-TH-1. The isomers on the top row are mirror images of the isomers on the bottom row.

in both cases. Two furan-side monoadducts between 8-MOP and thymidine (Figure 2, structures 2 and 2') have been isolated from DNA as nucleoside adducts while only one pyrone side nucleoside diastereomer (structure 3) has been identified (9). The thymidine-8-MOP-thymidine diadduct (structure 4 also occurs as a pair of diastereomers with *cis-syn* stereochemistry. The *cis-syn* configuration for both mono and diadducts has been found with four differen psoralen derivatives and is likely to be general for all linear furocoumarins

The reaction of the psoralens with RNA is similar to that of DNA. The majo reactive base in RNA is uridine (110), although reaction with cytosine residue: in tRNA has also been reported (37). Three major monoaddition products have been characterized from the reaction between HMT and polyuridylic polyadenylic acid (polyU·polyA) and with bulk RNA isolated from yeast (105) With both RNA substrates, the majority of the total bound HMT formed a pai of diastereometric furan-side adducts with uridine. The ratio of the two di astereomers was found to be 1 : 10 in polyU·polyA and 1 : 1 in bulk RNA. The sterochemistry of these furan side monoadducts is *cis-syn,* which is the same a in the HMT-thymine monoadducts formed in DNA (8, 10). The third majo product was a pyrone-side HMT-uridine monoadduct which accounted fo approximately 20% of the total adduct in polyU·polyA. In DNA, only 2% of the analogous pyrone-side monoadduct is formed, presumably due to steric in teraction between the 5-methyl group of thymine and the 4-methyl group o HMT. The sterochemistry of the pyrone-side monoadduct with HMT, as wel as the characterization of the crosslinked structure in RNA with psoralen remain to be determined.

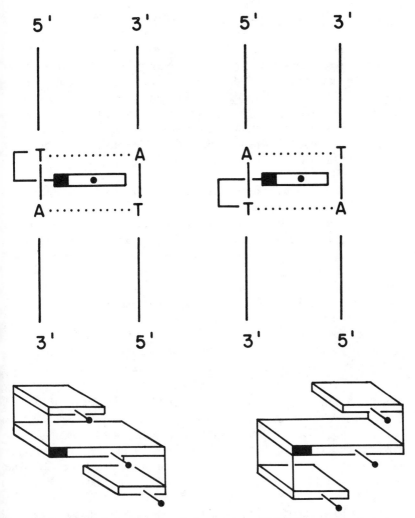

Figure 4   *Top:* Schematic drawing to show the intercalation and 8-MOP-T monoadduct formation at 5' . . . T-A . . . 3' and 5' . . . A-T . . . 3' sequences. *Bottom:* Schematic drawing of 2 diadduct enantiomers of *cis-syn* configuration. The representations are the same as in Figure 3.

Irradiation of psoralen with pyrimidine bases or nucleosides in solution, as frozen matrices or as thin films, also produce psoralen-pyrimidine mono- and diadducts. There are numerous reports in the literature which describe various adducts with stereochemistries different than those formed within helical nucleic acid (111–115). A greater variety of adducts can be formed under these conditions since the constraints of intercalation are not present to limit the stereochemical outcome of the reaction.

PHOTOCHEMICAL PROPERTIES OF PSORALENS

The photophysical and photochemical properties of psoralen and its derivatives have been extensively studied since the early 1970s. The excited states of psoralen have been identified. Interactions of these excited states of psoralen with bases, nucleosides, DNA, and RNA have been observed. Although much is known about psoralen excited states and their interactions, there is still some controversy about which excited state is photoactive in reactions with polynucleotides. Extensive reviews of this literature can be found by Parsons (5) and also Song & Tapeley (4). Only a brief account will be presented here.

The reactivity of any excited molecule will be influenced by several factors. The lifetime of the excited molecule must be long enough to encounter a second molecule in order to effect a reaction. During an encounter, both steric and electronic properties contribute to the overall reactivity. The relative influence of these factors on the cycloaddition of psoralen to a pyrimidine base is further determined by the local environment of the base. When psoralen is irradiated with free nucleoside or base in solution, the reaction will be diffusion controlled, and hence, the lifetime of the excited has a greater influence on the type and distribution of adducts. Since the reaction of psoralen with polynucleotides involves intercalation of psoralen into a double helical region, the lifetime of the excited state has less influence on the reactivity and the steric and electronic effects assume larger roles.

Spectroscopic Properties of Psoralen Molecules

The synthetic psoralen derivative AMT has a strong binding constant (25) for polynucleotides (K_d DNA = 6.6×10^{-6}), thus permitting the characterization of spectroscopic properties of both free and intercalated psoralen (25). The absorption and fluorescence properties of AMT (116) are shown in Figures 5 and 6. These spectra are representative of most psoralen derivatives. The absorption spectrum has maxima at about 250 nm, 300 nm, and 340 nm, and extends out to 400 nm. Fluorescence emission is observed from ca. 380 nm to 600 nm. The lifetime of the fluorescence state is small and varies with the type of psoralen derivative and the solvent conditions (4, 23, 117–119). Typical values range from 1 to 5 ns (2.0 ns for free AMT in aqueous solutions). The quantum yield for fluorescence is also very small, ranging from 0.01 to 0.02 (120). Fluorescence excitation of AMT is maximized by the band centered at 340 nm. This band is also the photoactive band for photocycloaddition. Emission from the triplet state of psoralens is significantly greater than that of the singlet due to efficient intersystem crossing (23, 120, 121). Phosphorescence is observed from ca. 450 nm to 600 nm. The lifetime of the triplet state

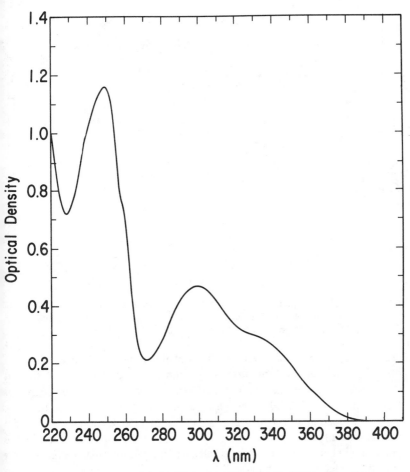

Figure 5   Absorption spectra of AMT (46.6μm) in 10mM Tris/1mEDTA, pH = 7.0.

also varies with the type of psoralen derivative, ranging from 1 μs to 1 s (100 μs for AMT) (117, 119–121).

Psoralen and its derivatives have two transitions in the 320 nm to 400 nm range: an $n \rightarrow \pi^*$ transition resulting from the excitation of a non-bonding electron on the C-2 carbonyl group to the π^* orbital, and a $\pi \rightarrow \pi^*$ transition occuring when a π electron in the psoralen ring system is excited to the π^* orbital. Mantulin & Song (120) have measured several spectroscopic parameters of coumarins and psoralens. By using substituted derivatives, they were able to assign the energy levels of the singlet and triplet states of psoralen. The relative orientation of the energy levels of these states are shown in the Jablonski diagram of Figure 7. The lowest singlet (120, 123) and triplet states

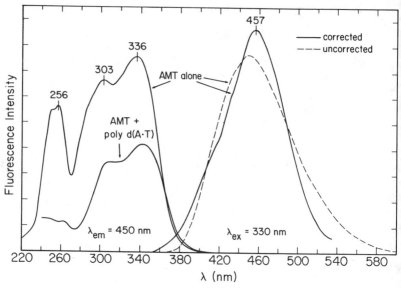

Figure 6 Fluorescence spectra of AMT. Corrected excitation *(left)* and emission *(right)* spectra of AMT alone (27.8 μM in 10 mM Tris/1 mM EDTA, pH = 7.0), and corrected excitation spectrum for AMT (25.9 μM) with poly d(A·T) [0.152 mM, 1 AMT/5.7 base pairs].

(120) are the (π,π^*) states. The reactivity of either the $1(\pi,P^*)$ or the $3(\pi,\pi^*)$ toward pyrimidine cycloaddition is determined by kinetic (i.e. lifetimes), steric, and electronic factors. The electronic factors consist of the degree of local excitation and the electron density at the 3,4 and 4',5' groups (122). By analogy with coumarin, it has been established that the 3,4 double bond of the pyrone is locally excited in the triplet state. Local excitation of the 4',5' double bond has not been observed with psoralen itself. Intramolecular charge transfer from the π-electron system to the 3,4 or 4',5' double bonds affects the electron density in these reactive regions. Theoretical calculations of the electron density in these excited states of psoralen show that the 3,4 double bond has more charge transfer character than the 4',5' double bond (120, 122, 125–127). Based on these results, it has been suggested that the triplet photoreactivity resides with the pyrone moiety rather than with the furyl group (4, 120, 122, 124). Reactivity of the 4',5' double bond therefore appears to be determined by steric and kinetic factors rather than electronic factors. However, solvent perturbation of the triplets of 8-MOP, 5-MOP, and 3-carbethoxypsoralen has recently been explained by redistribution of charge at both the 3,4 and 4',5' positions (128).

The separation of the energy levels in Figure 7 affects the kinetics of psoralen addition by altering the lifetimes and distribution of the singlet and triplet

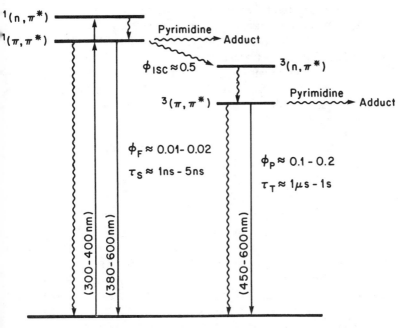

Figure 7   Jablonski diagram showing the relative energy levels of the excited states of Psoralen.

states. Lai et al (118) have recently measured several luminescent properties of the excited states of 8-MOP and TMP in different solvents. They conclude that the energy gap between the $1(\pi,\pi^*)$ and the $1(n,\pi^*)$ states determine the extent to which rapid, radiationless transitions occur between the $1(\pi,\pi^*)$ state and the ground state. The fluorescence quantum yield and fluorescence lifetime of 8-MOP were observed to increase upon going from a hydrophobic to a polar solvent. The quantum yield and lifetime of the triplet state changed in parallel with the fluorescence changes. It was suggested that the energy level of the $1(n,\pi^*)$ state is raised in polar solvents, thereby increasing the energy gap between the two singlet states. The increased energy gap between these two states reduces the interactions between the states and therefore reduces the radiationless transition probability. It has been suggested by Song (122) that this same effect upon the lifetime of the excited states of psoralen can also be induced by substitutients on the psoralen ring system. Electron donating groups, such as methyl and methoxy groups, would further raise the $1(n,\pi^*)$ and enhance the photocycloaddition by way of the singlet state. Lai, Lim and Lim (118) suggest that the greater fluorescence observed with TMP (30 times greater than 8-MOP) is due to a larger separation of the $1(n,\pi^*)$ and $1(\pi,\pi^*)$ states in TMP compared to 8-MOP.

Triplet-triplet absorption has been measured for several psoralen derivatives (117, 121, 128–133). Absorption occurs between 350 nm and 600 nm, with maximum extinction coefficients in the range of 3,700 to 49,500. AMT has a triplet-triplet extinction coefficient of 24,200 (117). Quantitation of the extinction coefficients of these transitions has facilitated the measurement of intersystem crossing quantum yields, ϕ_t, for several psoralen derivatives. Values of ϕ_t have been observed in the range of 0.1 to 0.4 (121). 5-methoxypsoralen is an exception, with $\phi_t < 0.01$.

Reactive States for Monoadduct and Crosslink Formation in Polynucleotides

The identity of the reactive state that gives rise to monoadduct and crosslink formation has been a topic of debate for many years. It was predicted by Song et al (124) that the triplet excited state is more reactive than the singlet excited state. Since the excitation of the triplet state is localized in the 3,4 carbon-carbon double bond, it was expected that the 3,4 monoadduct would be predominant in the reaction with pyrimidines (120). Several 3,4 psoralen monoadducts have been isolated and characterized from reactions between psoralens and thymine (11, 112, 115) and psoralens and DNA (9, 10, 53). The distribution of adducts observed from reactions with DNA is very different than the distribution of adducts seen upon reacting free thymine with psoralen. Formation of adducts from different excited states might contribute to this observation. Involvement of the triplet state of free psoralen in solution has been confirmed by observing triplet quenching upon addition of nucleic acid bases and amino acids (121, 134, 135). O_2 and paramagnetic ions were found to quench the photoreaction between psoralen and thymine (136), and also quench 8-MOP inactivation of bacteriophage lambda (137). This latter observation is not in complete agreement with the work of Goyal & Grossweiner (138), who showed that intercalation of 8-MOP with calf thymus DNA inhibited the accessibility to O_2.

It is now clearly established that psoralen intercalation into a DNA helix determines the stereospecificity and distribution of adduct products. The 4',5' psoralen-adduct is the predominant monoadduct formed in reactions with DNA and is the precursor to crosslink formation (10). This result is not surprising, since the 4',5' monoadduct retains absorption above 320 nm. The wavelength dependance for AMT crosslinking of pBR322 DNA (139) and the action spectra for the photo-inactivation of phage with 8-MOP (140) both parallel the absorption spectrum of 4',5' monoadducts (Figure 8). Beaumont et al (134) have suggested that the 4',5' monoadducts are formed via a singlet excited psoralen upon absorption of the first photon. Crosslink formation would then be formed via the triplet state upon absorption of a second photon. Indirect evidence for the singlet excited state formation of the 4',5' adduct has been reported by Salet et al (117) and by Beaumont et al (119). Both groups were unable to detect the triplet excited state of AMT intercalated in calf thymus DNA. The quantum yield of AMT fluorescence when bound to DNA was also reduced compared to free AMT. The fluorescence lifetime of free AMT was found to be 2.0 ns, while intercalated AMT had a lifetime of 0.03 ns (119). Involvement of the triplet excited state in the formation of a diadduct from a 4',5' monoadduct has not been established. However, the triplet excited states of 4',5' dihydropsoralen (129) and the 4',5' photoadduct of psoralen and thymine (130) are both quenched by thymine.

The major adducts formed with 8-MOP and DNA are shown in Figure 2 and the absorption properties of these adducts are presented in Figure 8. The 4',5' monoadducts have an absorption peak at about 340 nm. Irradiation into this band converts 4',5' monoadducts in DNA to crosslinks (10, 26, 27). The presence of 4',5' monoadducts and the conversion of these adducts to cross-links in polynucleotides can be observed spectroscopically since these adducts have a distinct fluorescence (116). Both the 3,4 monoadduct and the diadduct

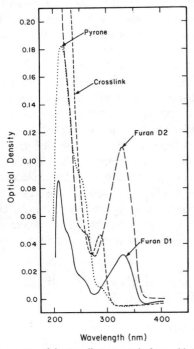

Figure 8   The absorption spectra of the two diastereomeric furan-side 8-MOP-dT monoadducts (D1, —, 1.0 μg/ml; D2, — — —, 2.7 μg/ml), the pyrone-side 8-MOP-dT monoadduct (. . . , 2.7 μg/ml), and the dT-8-MOP-dT diadduct(- - -, 11.5 μg/ml). The spectra were determined in 30% methanol : 70% 20 mM KH_2PO_4, pH 2.2.

have little absorption above 320 nm. The capability of a 3,4 monoadduct to form a diadduct has not been established. A low lying triplet state of the 3,4 monoadduct exists (141). Ben-Hur & Song (3) propose that this state is unreactive with respect to a second photocyoladdition because of its rapid relaxation to ground state. Further characterization of the properties of 3,4 monoadducts is needed.

Controlled Production of Monoadducts and Diadducts in Double-Stranded Nucleic Acids

At present there are three methods of generating monoadducts in DNA that are capable of subsequent crosslinkage. The first exploits the structural properties of a psoralen adduct in DNA. The dihedral angle in a psoralen monoadduct between the plane of a psoralen molecule and the plane of a pyrimidine base requires a conformational change in the DNA backbone to accommodate a diadduct (108). Lown & Sim (11) first observed this phenomenon by irradiating PM2 DNA in the presence of different psoralen derivatives. Monoadducts capable of subsequent crosslinkage were greatly enhanced by irradiating in frozen solutions. Johnston et al (28) subsequently showed that the controlled production of monoadducts could be achieved by irradiating a solution of AMT and DNA with a single laser pulse of short duration (15 ns) sufficient for monoadduct formation but too short for a conformational change which occurs in the DNA and facilitates diadduct formation. A second method of generating monoadducts was reported by Chatterjee & Cantor (26). 4',5' monoadducts in DNA are greatly enhanced by irradiating with wavelengths above 380 nm. This selectivity is a result of the absorption properties of 4',5' monoadducts. Tessman et al (27) have recently used this procedure to show that the quantum yield for monoadduct formation (0.008) is smaller than the quantum yield for diadduct formation from a monoadduct (0.028). The last method of generating monoadducts in DNA is applicable to short oligonucleotides containing a single crosslinked site. Controlled photoreversal of the crosslinked molecule permits the isolation of monoadducted oligonucleotides (142). The quantum yield for photoreversal and the wavelength selectivity for photoreversal (furan vs pyrone monoadducted oligonucleotide) are areas that need further research.

THE USE OF PSORALENS AS PROBES FOR NUCLEIC ACID STRUCTURE AND FUNCTION

Chromatin Structure

Hanson et al (34) first proposed the use of psoralens as probes for chromatin structure based upon the results of an electron microscopic analysis of denatured DNA isolated from TMP photoreacted *Drosophila* embryo nuclei. This

DNA exhibited an alternating pattern of loops (single strands) and bridges (apparent double strands). Interstrand DNA crosslinks were confined to the bridged regions. A weight averaged histogram analysis of the loops indicated that their length peaked at 160–200 base pairs and at intervals thereof. Approximately 60% of the total DNA existed as single-stranded loops. These loops were interpreted to represent regions on the DNA protected from TMP crosslinkage by associated histones. In agreement with this interpretation purified Drosophila DNA similarly photoreacted with TMP and analyzed by denaturation electron microscopy was almost completely double-stranded. The similarity in periodicity between the nucleosome subunit of chromatin and the looped pattern exhibited by in vivo crosslinked and formaldehyde denatured DNA led to the proposal that TMP crosslinking is restricted to the DNA between nucleosomes.

The selective photoreaction of TMP with internucleosomal DNA was independently confirmed by two groups (39, 143). Their studies with TMP photoreacted nuclei showed that micrococcal nuclease, itself a probe for chromatin structure which preferentially digests internucleosomal DNA, releases TMP at a rate much greater than bulk DNA when digesting the modified chromatin. In the limit digest, 92% of the TMP was released when 45% of the DNA had been rendered acid soluble. Control experiments demonstrated that TMP crosslinkage did not significantly alter the kinetics of micrococcal nuclease digestion or the size distribution of enzyme resistant DNA fragments. It was concluded that both TMP and micrococcal nuclease react with the same subset of DNA in chromatin. TMP saturation experiments conducted by Wiesehahn et al (39) support this conclusion. These experiments showed that DNA in chromatin was 90% protected from TMP photoreaction relative to naked DNA. Saturation levels of binding for the two DNAs were, respectively, 25 psoralens or 250 psoralens per 1000 base pairs.

TMP crosslinkage has been used as an in situ probe for chromatin structure in several systems. In each case the crosslinked DNA was purified and analyzed by denaturation electron microscopy. An alternating pattern of loops and bridges characteristic of chromatin structure was observed for mouse liver DNA (144), for SV40 DNA derived from both intracellular and viral nucleoprotein complexes (145), and for the main band and satellite I, II, and III DNAs from Drosophila nuclei (146). Meaningful results were not obtained for the satellite IV fraction since this DNA consists of short A-T containing segments separated by 250 base pair tracts of noncrosslinkable polypyrimidine/polypurine sequences (147).

Sequence specific positioning of nucleosomes on DNA has been addressed by Carlson et al (148) utilizing TMP photoaddition together with Hind III restriction analysis. Hind III was used in this study because preliminary experiments with TMP modified plasmid DNA showed that its recognition sequence was 15-fold more susceptible to psoralen inactivation than othe restriction sites examined. TMP mono- or diadducts at or near the *Hind* II recognition site are postulated to prevent cleavage of one or both strands by th enzyme. This phenomenon was used to probe the positioning of nucleosome on the SV40 minichromosome. SV40 DNA from TMP photoreacted monke cells was restricted with *Hind* III and electrophoresed on an agarose gel t resolve both partial and complete digestion fragments. Partial fragments con tained one or more intact *Hind* III sites resistant to the enzyme. Since TM photoreaction is limited to the internucleosomal DNA of chromatin, that DN/ associated with the nucleosomal core will remain unmodified and restrictabl by *Hind* III. A careful analysis of the restriction pattern indicated that no on *Hind* III site was either completely resistant or completely sensitive to inactiva tion by the psoralen probe. The evidence described here supports earlier in vitr studies which concluded that nucleosomes on the SV40 minichromosom occupy a random or nearly random set of positions on the underlying DN/ sequence.

A different conclusion regarding the distribution of nucleosomes on SV4 minichromosomes has been reached by Robinson & Hallick (149). Thes investigators photoreacted intracellular SV40 chromatin with ^{3}H-labele HMT. The viral DNA was then isolated, multiply restricted, and the fragment electrophoresed on a polyacrylamide gel. Specific activity measurements of th fragments indicated that HMT had preferentially reacted with a 400 base pa region located between 0.65 and 0.73 on the SV40 physical map. This regio encompasses the origin of replication and the promoters for early and lat mRNA synthesis. The accessibility of this region to HMT addition is consister with other evidence indicating that it is nucleosome-free. In contrast, howevei the same group (150) could not detect preferential reaction of HMT with th origin region of viral chromatin implying that a complete randomization c nucleoprotein structure occurs upon encapsidation of the minichromosome.

Cech & Karrer (151) have applied TMP crosslinkage to the in vivo study c transcriptionally active rDNA of *tetrahymena thermophila*. Unique propertie make this DNA an ideal substrate for crosslinkage analysis. Within the nucleu it exists as a separate extranucleosomal entity which can be isolated free of bul DNA by isopycnic centrifugation. Additionally, it has a palindromic structur which means that each half of the molecule is structurally as well as functiona ly equivalent to the other half. Thus in the electron microscope each molecul can by systematically divided into three nontranscribable spacer regions an two coding regions. Log phase *tetrahymena* cultures were irradiated with lon wave ultraviolet light in the presence of TMP to fix the transcriptionally activ rDNA. This DNA was then isolated and examined by denaturation electro microscopy. Representative molecules were scored for crosslinks (double stranded bridged regions) and bubbles (single-stranded loops). Each structur was assigned its position on the rDNA map which in turn was divided int

nown coding and spacer regions. As a control, rDNA photoreacted in vitro
ith TMP was similarly analyzed. The results clearly showed that the frequen-
y of crosslinkage in vivo was greater in the coding regions than in the spacer
egions of the rDNA. Furthermore, the peak loop sizes measured in the coding
gions were 215, 335, and 460 base pairs as compared to 200 and 400 base
airs for the peak loop sizes of main band DNA. These observations indicate
at the chromatin structure of transcriptionally active rDNA possesses an
tered conformation rendering it more open and accessible. One attractive
odel proposed by the authors is that such chromatin exists as a series of
alf-nucleosomes" exhibiting a repeat unit length of approximately 110 base
airs.

The in situ structures of three non-nuclear DNAs have been probed by
soralen crosslinkage followed by denaturation electron microscopy. Uniform
rosslinkage was observed over at least 90% of the contour length of
itochondrial DNA from *Drosophila* embryos and HeLa cell cultures (40, 152)
nd over the entire contour length of *E. coli* DNA (153) thus confirming a
on-nucleosomal structure for each. However, 6–10% of the contour length
as looped out at the origin of replication in each mitochondrial DNA prepara-
on. Since these sequences were crosslinked in vitro, the in situ protection was
tributed to origin specific nucleoprotein complexes.

ruciforms

egions in DNA with two fold symmetry, commonly referred to as inverted
peats or palindromic sequences, can exist in two alternate configurations, a
near form and a cruciform. Methods exist for the detection of cruciforms in
itro. Psoralen photochemistry, however, is unique in providing a way to probe
or the existence of such structures within the cell. By crosslinking either arm of
cruciform with psoralen, its rearrangement to the linear form is prevented.
imilarly, crosslinkage of the linear form precludes its conversion to cruci-
orm. Thus DNA crosslinked in vivo can be assayed for cruciform structure in
itro without altering the original distribution. Cech & Pardue (154) have used
is approach to search for cruciforms in the DNA of intact mouse tissue culture
ells. After photofixing the DNA in situ with TMP crosslinks, it was purified,
enatured in the presence of glyoxal, and examined by electron microscopy. Of
e inverted repeats in mouse DNA long enough to be detected by electron
icroscopy, essentially none were found to be in the cruciform state. Using a
imilar protocol, Cech & Karrer (151) have probed the in situ configuration of
trahymena thermophila rDNA. This long 20,400 base pair palindromic DNA
as crosslinked in log phase cells and examined by electron microscopy under
ondenaturing conditions. Again, no cruciforms were detected.

More recently Sinden et al (155) have employed psoralen crosslinkage of
NA as part of an assay for cruciform structure in a plasmid containing the

palindromic *E. coli* lac operator sequence. This DNA sequence can exist as either a 66 base pair long linear fragment or as a cruciform with two 33 base pair long hairpins. After TMP fixation and EcoR1 restriction, the conformation of the crosslinked lac operator containing restriction fragments was determined electrophoretically. Results of the assay showed that formation of the cruciform was dependent upon the presence of torsional tension in the molecule. In highly supercoiled DNA all of the lac operator sequence was in the cruciform state while in nicked circular DNA the same sequence existed in the linear state. Interconversion between the two forms was rapid at 37°C and slow at 0°C. When the assay was performed on plasmid DNA in bacterial cells, no cruciforms were detected. This indicates that in vivo the DNA lacks sufficient torsional tension to induce cruciform formation. In contrast to the S1 nuclease digestion method, this approach, which utilizes a TMP fixation step, permits an accurate determination of the relative amount of cruciform present in an in vitro or an in vivo sample without perturbing the equilibrium between the two forms.

Torsional Tension

Sinden et al (156) have described an assay for the determination of DNA torsional strain in both prokaryotic and eukaryotic cells based upon the rate of photobinding of labeled TMP to DNA. Like other intercalating agents, TMP has an enhanced affinity for negatively supercoiled DNA. In vitro, the rate of photobinding to covalently closed circular plasmid DNA is proportional to its superhelical density. This dependence provides an indirect assay for torsional tension. In the assay living cells were irradiated with near ultraviolet light in the presence of TMP to induce approximately one psoralen adduct per 10^3 base pairs. The reaction conditions were chosen to minimize photochemical nicking and perturbation of chromosome structure due to drug binding. After photoreaction, DNA and rRNA were purified and the extent of modification to each determined. The photoreaction was normally expressed as the ratio of TMP binding to DNA versus rRNA. In this manner rRNA served as an internal standard to correct for experimental variations in the absolute rate of binding.

The above assay was used to monitor the effects of ionizing radiation and coumermycin treatment on the rate of TMP photoaddition to *E. coli* DNA. In prokaryotes both agents relax supercoiled DNA by either direct nicking or by inhibiting DNA gyrase. With either treatment the initial rate of TMP photobinding to DNA was reduced by 40%. Thus the *E. coli* chromosome contains unrestrained torsional tension equivalent to that expected for a purified DNA duplex with a superhelical density of -0.05 ± 0.01. The same assay was extended to *Drosophila* cells and HeLa cells. However, the initial rate of TMP photoreaction in these cells was unaffected by treatment with ionizing radiation. This supports the general belief that superhelical DNA turns in the eukaryotic chromosome are restrained in nucleosome structures and that such DNA is not subject to torsional tension.

Secondary Structure in Single-Stranded Nucleic Acid

The use of psoralen derivatives to photocrosslink double-stranded regions in single-stranded nucleic acid is without parallel as a technique for determination of secondary structure. Both localized hairpins and long range interactions can be fixed in vitro or in situ by psoralen crosslinkage. In early studies, pioneered by Hearst, Cantor, and coworkers (35, 157–167), denaturation electron microscopy was used to map crosslinked secondary structures in both single-stranded DNA and RNA. With this technique structures could be mapped to a resolution of 50 base pairs. Recent advances in the analysis of crosslinked molecules now permit secondary structures to be mapped to sequence resolution (29, 33, 168). In this approach, which has been applied to 16S ribosomal RNA by Thompson & Hearst (30), small crosslinked fragments isolated from the photoreacted molecule are sequenced after photoreversal and purification of component strands, if necessary. This section will survey the ways in which psoralen crosslinkage has been used to probe structure in single-stranded fd and SV40 DNAs and in transfer, small nuclear, ribosomal, and heterogeneous nuclear RNAs.

SINGLE-STRANDED DNA   The psoralen crosslinkage technique was first applied by Shen & Hearst (157) to the single-stranded DNA genome of bacteriophage fd. Both circular and full length linear DNA were photoreacted with TMP at several different ionic strengths and then examined by electron microscopy. The single-stranded linear DNA had been prepared from replicative form by restriction and transient alkali denaturation. When crosslinkage was performed at 15°C in water no secondary structures were discernable. However, when the NaCl concentration was raised to 20 mM, TMP crosslinkage stabilized a single 200 ± 50 base pair hairpin. This hairpin was uniquely positioned near one end of the linear DNA at a site believed to correspond to the origin of replication. Crosslinkage at higher salt concentrations fixed other less stable hairpins. A comparison of the histogram map of these hairpins with known promoter sites of fd suggested a possible correlation.

In a subsequent study Shen et al (158) investigated the structure and orientation of single-stranded circular fd DNA inside its filamentous virion. For this analysis purified virions were photoreacted with TMP. The DNA was then deproteinized and analyzed by electron microscopy. Most of the DNA molecules were visualized as extended circles exhibiting a single prominent hairpin at one end together with occasional internal crosslinks. The end hairpin was mapped relative to a unique restriction endonuclease cut introduced into the crosslinked DNA after hybridizing it to a complementary linear single strand. Analysis showed that the end hairpin was identical to the stable hairpin seen in naked fd DNA. From these results it was concluded that the stable hairpin is located at one of the ends of the fd virus and that the DNA genome occupies an extended, fixed, and nonrandom orientation within the viral capsid.

Employing similar protocols, Ikoku & Hearst (159) have determined th
orientation of the single-stranded DNA genome in the filamentous chimeri
phase M13gori1, and Shen & Hearst (160, 161) have determined the secondar
structure of single-stranded SV40 DNA. In the latter studies after linearizin;
and denaturing form I DNA, helical regions on single strands were fixed b
TMP photoreaction and visualized by electron microscopy. Most of the shor
hairpins on SV40 mapped at initiation and termination sites for replication o
transcription while the larger hairpin loops bracketed previously identifie
intervening sequences.

RIBOSOMAL RNA   Our understanding of the secondary and tertiary structur
of rRNAs has been advanced by several psoralen crosslinkage studies. Th
initial application of this technology utilized electron microscopy to ma
psoralen crosslinks in *Drosophila melanogaster* 18S and 26S ribosomal RN/
(35). HMT crosslinkage of purified rRNA stabilized a centrally located hairpi
on 26S rRNA and an open loop structure near one end of the 18S rRNA.

Application of the psoralen-crosslinking technology to *E. coli* 16S ribosoma
RNA has been extensive (162–165). Electron microscopic visualization o
isolated 16S rRNA crosslinked by HMT revealed eleven large open loo
structures (162). The frequency of occurrences of these structures ranged fron
2% to 30%. The loop size and stem length of each secondary structure wer
measured and all mapped relative to one another on the 16S molecule. Th
polarity of the map was determined by hybridizing DNA complimentary t
specific regions of the 16S rRNA and visualizing the hybrids by electro
microscopy (164). 16S rRNA in the 30S subunit has also been probed b
psoralen crosslinkage and analyzed by electron microscopy (163). Both activ
and inactive conformations of the 30S subunit were photoreacted with HMT
Analysis of 16S rRNA crosslinked in the inactive subunit showed that i
contained seven of the eleven open loop structures previously characterized fo
the same molecule crosslinked in solution. Of those seven structures, six wer
present in 16S rRNA when crosslinked in the active subunit. A computer searcl
of the 16S rRNA sequence yielded energetically stable duplex structure
corresponding in positions to each of the observed psoralen crosslinking site
(165). All of the crosslinked features can occur simultaneously in the 16:
molecule because the base paired sequences are non-overlapping.

Two variations on the formation, processing, and electron microscopi
detection of psoralen crosslinks in *E. coli* 16S rRNA have been explored. A ge
electrophoretic technique for the isolation of 16S rRNA fractions containin;
unique psoralen crosslinks has been developed by Wollenzein & Cantor (166)
Purified 16S rRNA containing ca. 10 adducts per RNA molecule was furthe
resolved by electrophoresis through a gradient polyacrylamide gel made i
formamide and low salt to promote denaturation. This gel system resolve

rosslinked 16S rRNA into several discrete bands. Each band was analyzed eparately by electron microscopy and found to contain 16S rRNA molecules ith long range crosslinks at different positions within the 16S molecule. Six of ie previously identified long range interactions were found by this technique. ecause of the reduced heterogeneity of products, it is expected that this chnique will be invaluable for the analysis of crosslinks at sequence resolu- on.

In the second variation of the psoralen–crosslinking technique applied to 16S RNA, Chu et al (167) utilized a two-step irradiation protocol to compare the soralen stabilized secondary structure of 16S rRNA in active and inactive 30S bosomal subunits. The first photoreaction, carried out with 390 nm light, was sed to generate AMT monoadducts in inactive 30S particles. The modified articles were freed of unreacted AMT and, if desired, converted into active ubunits. AMT monoadducts in the 30S particles were chased into crosslinks y reirradiation at 360 nm. The locations of crosslinks were determined by lectron microscopy of the purified 16S RNA. Since this approach inserts AMT ato the RNA prior to conformational changes induced by activation of the bosome, any decrease in the frequency of a crosslinked feature may be ttributed to a loss of base paired structure in the vicinity of an AMT monoad- uct. When the secondary structure of 16S rRNA from active and inactive ubunits was examined, the overall patterns were similar except for the de- reased frequency of a 3' end loop in RNA isolated from active subunits. This nplies that activation of the ribosome is accompanied by a change in the econdary and possibly tertiary structure of the 3' terminus of 16S rRNA.

Sequence resolution of psoralen-crosslinked fragments was first reported by abin & Crothers in 1979 (33). Using AMT crosslinked *E. coli* 5S rRNA, they vere able to isolate, photoreverse with 250 nm light, and sequence both strands f a crosslinked fragment of 5S rRNA. The sequence analysis of each strand onfirmed the existence of a stem structure in 5S RNA formed between esidues 1–10 and 110–119. This work demonstrates the utility of psoralen hotochemistry in the unequivocal characterization of secondary structure eatures in RNA.

The approach initiated by Rabin & Crothers has been extended by Thompson t al (168). These authors photoreacted uniformly ^{32}P-labeled *Drosophila nelanogaster* 5S RNA in solution with HMT. After photoreaction the RNA vas digested to completion with RNase T1 and the fragments were separated on 20% polyacrylamide gel. Since eukaryotic 5S RNA is only 120 nucleotides ong, the straightforward digestion pattern permitted the separation of cross- inked fragments from one another and from unmodified oligonucleotides. ach crosslinked band was extracted from the gel and photoreversed, and the equence of the fragments determined. This information permitted the authors o unequivocally identify the location of two crosslinks and to tentatively identify the location of three others. The secondary structure model for 5S rRNA predicted by these crosslinks is very similar to ones proposed on the basis of evolutionary and enzymatic digestion data.

The first application of sequencing analysis to psoralen crosslinked 16S rRNA was reported by Turner et al (29). After photocrosslinking uniformly ^{32}P-labeled 16S RNA with HMT, the RNA was partially digested with RNase T1. The fragments were resolved by a two-dimensional electrophoresis system. Crosslinked fragments were completely photoreversed between dimensions. In this system, crosslinked fragments migrate as a pair of off diagonal spots. A hairpin localized at positions 434–497 was identified in this study. The putative location of the HMT crosslink was deduced to be an intercalation site formed between base pairs $U_{458} \cdot G_{474}$ and $A_{459} \cdot U_{473}$. Based on the known reactivity of psoralens, this site should be a reaction hotspot for HMT since it contains a pair of crosslinkable uridines in a loose helix as defined by a G·U base pair directly abutting a run of three A·U base pairs.

Thompson & Hearst (30) have refined the mapping of psoralen crosslinks and in so doing have characterized 13 unique secondary structures in *E. coli* 16S RNA. The crosslinked fragments were isolated by modification of a two-dimensional gel electrophoretic technique originated by Zweib & Brimacombe (169). The isolation protocol is outlined in Figure 9. Off diagonal spots containing crosslinked fragments were excised and photoreversed. Component strands were purified and sequenced by standard enzymatic methods. In Figure 10 the positions of the characterized crosslinks are superimposed on the secondary structure model of Noller & Woese (170). Several long range interactions elucidated by HMT crosslinkage appear to exist in dynamic equilibrium with local secondary interactions. These equilibria may reflect conformational changes which occur in 16S RNA during functioning of the ribosome (171).

TRANSFER RNA   The photoreaction of HMT with phenylalanine tRNA has been addressed by two groups. In each case the reaction pattern was determined to sequence resolution and the results were consistent with the three dimensional structure determined by X-ray crystallography. In the earlier study, conducted by Bachellerie & Hearst (172), the sites of HMT monoaddition on *E. coli* tRNA^{Phe} were mapped by a combination of techniques including chemical and enzymatic digestions and electrophoretic separations on gel and paper. By far the most reactive site was ^{51}U, a residue base paired to ^{63}G in the T-stem, although every other uracil in the molecule save one exhibited some residual activity with HMT. In the more recent study, Garrett-Wheeler et al (37) photocrosslinked yeast tRNA^{Phe} in solution with HMT. A partial T1 digest of the modified tRNA was ^{32}P end-labeled and fractionated by a two dimensional electrophoresis scheme similar to that devised by Turner et al (29)

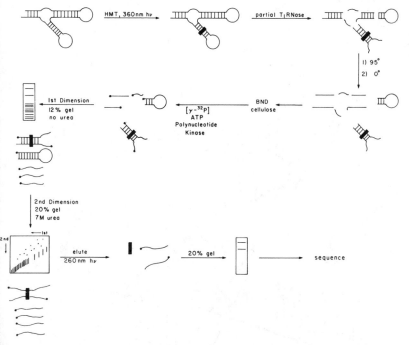

Figure 9   Schematic diagram of the methodology used by Thompson & Hearst (30) to analyze crosslinks in 16S RNA.

This scheme, however, incorporated a photoreversal step which was reduced in both time and intensity to minimize loss of monoadducts. Oligonucleotides formerly involved in a crosslink were isolated and sequenced enzymatically. Additionally, monoadducted pyrimidines were mapped by electrophoresing purine specific U_2 and cytidine specific CL3 ribonuclease digests of the oligonucleotides. These enzymes were observed to anomalously cleave RNA adjacent to psoralen monoadducts. The exact positions of five crosslinks were determined. Four of the crosslinks occurred in stems and one bridged a tertiary interaction between two trans oriented pyrimidines. Surprisingly, the crosslinks involved one C-C interaction, three C-U interactions, and only one classical U-U interaction. The ability to enzymatically map psoralen crosslinks should prove very useful in probing the tertiary structure of other RNA molecules.

HETEROGENEOUS AND SMALL NUCLEAR RNAs   In both prokaryotes and eukaryotes high molecular weight primary transcripts are processed into mature tRNA, rRNA, and mRNA. An important facet of RNA processing, not clearly understood, is the enzymatic removal of extraneous intervening sequences or

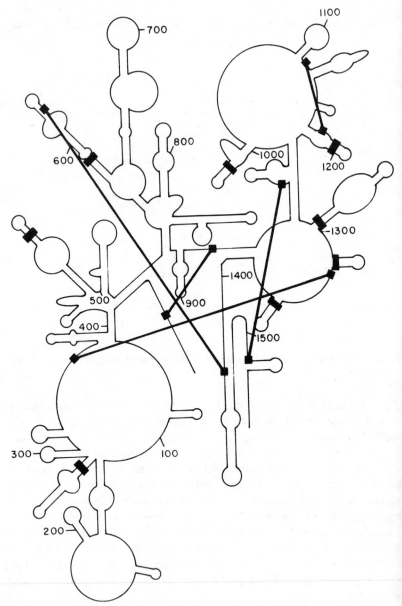

Figure 10   Position of observed crosslinks in *E. coli* 16S ribosomal RNA.

introns. The coupled cleavage-ligation reaction referred to as splicing is site specific and often involves regions of RNA secondary structure in the primary transcript. In eukaryotes heterogeneous nuclear RNA (hnRNA) is the nuclear precursor to cytoplasmic mRNA. This high molecular weight RNA possesses n ordered ribonucleoprotein structure (hnRNP) and, unlike mRNA, contains RNase resistant double-stranded regions when deproteinized. These regions, whose existence in vivo may be regulated by specific proteins, probably play a role in RNA processing. Calvet & Pederson (173, 174), utilizing AMT cross-linkage in vivo, have verified the existence of double-stranded regions in hnRNP complexes. They digested hnRNP from control or AMT photoreacted HeLa cells with proteinase K, pancreatic DNase, and the single-strand specific enzymes pancreatic RNase and RNase T1. The enzyme resistant double-stranded RNA was purified on a Cs_2SO_4 gradient and analyzed for rapidly renaturing or "snap back" RNA. The RNA which had been crosslinked with AMT in vivo showed a several fold enhancement in the rapidly renaturing fraction relative to the control sample. These results demonstrated the existence in vivo of double-stranded regions in hnRNA accessible to AMT intercalation and crosslinkage.

Evidence for the potential involvement of RNA secondary structure in splicing has been obtained by Wollenzein et al (175) in an electron microscopic analysis of AMT crosslinked 35S RNA from *Neurospora* mitochondria. Mature 25S rRNA is formed from this precursor upon removal of a 2.2 kilobase intron. The 35S RNA used in this study had been photoreacted with AMT in isolated RNP particles and then spread for electron microscopy after deproteinization and denaturation in the presence of formaldehyde. Two mutually exclusive structures were identified: a central hairpin 105 base pairs long and a large loop 2.2 kilobases in length. When crosslinkage was conducted in RNP particles at low temperature in the presence of $MgCl_2$ and KCl, the loop structure predominated. However, when the same RNA, previously monoadducted with AMT in RNP particles, was photoreacted at 365 nm in the absence of protein, the central hairpin was more frequent. The location and size of the loop structure indicated that it was in fact the intron sequence and that the corresponding crosslink bridged the two splice sites. Taken together, the results suggested that one or more RNP proteins stabilize a psoralen accessible base paired region formed between two complementary sequences on opposite splice sites.

The small nuclear RNAs (snRNAs) are a class of abundant RNA molecules 100–300 nucleotides in length which are postulated to play a role in RNA processing. Two members of this class, U1 and U2, have potential sequence complementarity with intron-exon junctions in messenger RNA precursors. Pederson and coworkers (176, 177) have investigated the occurrence of these interactions in vivo by psoralen crosslinkage. Very simply, they isolated hnRNA under denaturing conditions from AMT photoreacted HeLa cells. The RNA was then photoreversed and electrophoresed on a 10% polyacrylamide gel made in formamide. Gel blots with U1 and U2 specific probes lit up bands for the respective snRNAs. When the photoreversed step was omitted, the probes hybridized to the high molecular weight hnRNA band at the top of the

gel. This electrophoresis pattern confirmed the existence of base paired com
plexes between U1 and U2 with hnRNA.

In a continuation of their work, Setyono & Pederson (178) have examined
the ribonucleoprotein organization of the U1 RNA fraction base paired in vivo
to hnRNA. Previous studies had shown that U1 RNA within the cell exists as a
ribonucleoprotein complex and it was expected that this was the species bound
to hnRNA. For this analysis isolated HeLa cell nuclei were pulse-labeled with
^{3}H-uridine and photoreacted with AMT. Nuclear hnRNA-ribonucleoprotein
particles were purified on a sucrose gradient containing 50% formamide. In this
concentration of formamide noncovalently bound U1 RNA was lost but ribonucleoprotein structure was retained. The purified RNP, labeled with ^{3}H in
hnRNA, was passed through an anti-U1 RNP immuno-affinity column. In the
absence of photoreversal, a small fraction of the hnRNA in hnRNP particles
reproducibly bound to the column. These binding results support the notion that
both U1 RNA and hnRNA interact with one another as ribonucleoprotein
complexes.

Stabilization of Nucleic Acid Complexes

The psoralen derivatives TMP, HMT, and AMT have been shown to crosslink
DNA·DNA, DNA·RNA, and RNA·RNA double strands in the presence of near
ultraviolet light (25, 179). This photoreaction has been used to stabilize a
variety of nucleic acid structures including transcriptional complexes, replicative intermediates, R-loops, and D-loops. Psoralen crosslinkage has also been
used as a tool in various hybridization, electron microscopic, and cloning
procedures. This section will survey these uses.

The ability of psoralen derivatives to crosslink DNA·RNA hybrids has been
utilized by Shen & Hearst (180) to study the *E. coli* RNA polymerase ternary
complex. ^{14}C-labeled SV40 DNA was transcribed by the *E. coli* enzyme in the
presence of ^{3}H-labeled ribonucleoside triphosphates. Ternary complexes were
photoreacted with AMT and spun through an SDS-sucrose gradient. Only in
the crosslinked samples did the nascent RNA and viral DNA cosediment.
These results indicate that AMT can intercalate into and crosslink the short
DNA·RNA hybrid formed within the ternary complex between nascent RNA
and the coding DNA strand. The authors suggest that in vivo crosslinkage of
ternary complexes could facilitate the isolation of transcriptionally active DNA
via isopycnic gradient ultracentrifugation.

Revet & Benichou (181) have utilized in vivo TMP crosslinkage to fix
adenovirus 5 replicative form intermediates. The replicating viral DNA molecules were purified by ultracentrifugation through consecutive sucrose and
CsCl gradients and examined by electron microscopy under both native and
denaturing conditions. The original distribution of replication forks on each
molecule was preserved during workup and analysis by the presence of TMP

crosslinks. These crosslinks prevented branch migration. The authors concluded that in vivo TMP crosslinkage could be extended to the study of other labile double-stranded structures such as Okasaki fragments and recombination intermediates.

When an RNA strand hybridizes to a complementary section of double-stranded DNA, the RNA displaces the homologous strand in the duplex forming a DNA·RNA hybrid with a looped out single-stranded DNA. This structure, referred to as an R-loop, is unstable under physiological conditions. Three protocols, utilizing psoralen crosslinkage to stabilize the R-loop structure have been published. In a procedure developed by Davidson and coworkers (182), the DNA was lightly crosslinked with TMP prior to R-loop formation. During hybridization the presence of crosslinks stabilized the DNA double helix by preventing complete strand separation and by reducing the probability of fragmentation due to denaturation between widely spaced nicks on opposite strands. After R-loop formation the single-stranded DNA was modified at hydrogen bonding positions with glyoxal to prevent branch migration. Together, the two treatments stabilized R-loop structures for subsequent electron microscopic analysis.

Two other protocols for stabilizing R-loop structures with psoralen crosslinks have been published. Wittig & Wittig (32) have reported that R-loops in DNA can be fixed by crosslinking with TMP and near ultraviolet light directly in the R-loop formation buffer (70% formamide at 4°C) and Chatterjee & Cantor (183) have demonstrated that R-loops formed with AMT monoadducted RNA can be stabilized by subsequent irradiation at 360 nm. The ultraviolet treatment converts many of the monoadducts to crosslinks. In both of these studies the R-loop structures were successfully cloned after removal of the RNA strand by alkali treatment or the DNA loop by S1 nuclease digestion.

The approaches described above can also be applied to the stabilization of D-loops. A D-loop is typically formed when a single-stranded DNA fragment hybridizes to a complementary double-stranded DNA and displaces the homologous strand. D-loops are even less stable than R-loops and are usually formed on negatively supercoiled DNA. DeFrancesco & Attardi (152) have used HMT crosslinkage to stabilize in situ a unique D-loop present on HeLa cell mitochondrial DNA. In their analysis the mtDNA was isolated from HMT photoreacted mitochondria, linearized with a single-hit restriction endonuclease, and spread for electron microscopy under nondenaturing conditions. Approximately 25% of the molecules contained a 560 ± 70 base pair D-loop located at the origin of replication. This loop was linked to the in vivo synthesis of a 7S DNA fragment complementary to the L-strand of the molecule. A previous study (184) had shown that this particular D-loop in the absence of crosslinkage was rapidly lost upon restriction of the mtDNA.

Two applications of psoralen crosslinkage deserve mention here. In 1982

Brown et al (185) described the preparation and use of unique M13 hybridiza
tion probes labeled to a high specific activity with a TMP crosslinked and
enzymatically extended primer. The strand specific probes are ideally suited for
use in Northern hybridization experiments and for use in M13 plaque hybridi
zation assays. Also, in 1982 Wollenzein & Cantor (186) described a procedure
for marking the polarity of single-stranded DNA or RNA to be analyzed by
electron microscopy. In this procedure a DNA restriction fragment com
plementary to an asymmetric site on the molecule of interest is monoadducted
with AMT and then hybridized and crosslinked to the target nucleic acid. In a
denaturing spread, the marker DNA strand serves as a point of reference.

FUTURE DIRECTIONS

From our perspective, three avenues of ongoing research hold the potential for
dramatically increasing the versatility of psoralens for use as photochemical
probes of nucleic acid secondary structure and function. The three areas of
research are, respectively, the synthesis of modified psoralens exhibiting
unique crosslinking specificities or containing reporter molecules, the develop
ment of a direct mapping protocol for psoralen monoadducts and crosslinks in
both RNA and DNA, and the refinement of photochemical as well as chemical
procedures for the site-specific placement of crosslinkable psoralen monoad
ducts onto DNA oligonucleotides. In this section the advances in each area will
be reviewed and the likely applications of each technology will be discussed.

Cantor and coworkers (86) have synthesized three modified psoralen deriva
tives capable of forming intermolecular crosslinks in nucleoprotein complexes.
One derivative, a bis psoralen compound, has been photoreacted with bacterio
phage lambda (87). DNA helices in close proximity should be crosslinked to
one another by this compound. Nondenaturing electron microscopic analysis of
a Bgl II digest of the modified lambda DNA has verified the formation of
interhelical crosslinks through the detection of X-shaped molecules. The cross
linking frequencies between all six possible pairs of the four largest Bgl I
fragments of lambda DNA were determined. The results ruled out purely
random packaging of the DNA within the virion but at the same time were
inconsistent with two previously suggested models for intraphage DNA pack
ing. A second psoralen derivative, also applied to the structure analysis of
lambda bacteriophage, was an AMT molecule linked at the 4' position to a
reactive succinate group by a long alkyl chain (86). In the lambda model
system, this compound crosslinked viral capsid to the DNA genome. Protein
linkage occurred chemically via the succinate group while DNA linkage took
place photochemically via AMT. The third psoralen derivative synthesized by
the Cantor group was another AMT molecule containing a 4' linker arm
terminating in a sulfhydryl group (85). The reagent was selectively dark bound

o mercurated patches in pBR322 DNA via the formation of a Hg-S linkage. After removal of excess reagent and upon irradiation with 365 nm light the ite-specifically bound AMT formed intramolecular as well as intermolecular DNA crosslinks. All three compounds described above have the potential to lucidate tertiary interactions in condensed nucleoprotein complexes.

A second class of modified psoralens with obvious application to structure determination studies are furocoumarins which contain an attached reporter group. An example of such a group is the biotin moiety. We have synthesized AMT molecules with the group attached to the 4' position by various linker chains. Preliminary studies indicate that these AMT derivatives can intercalate nto and crosslink double-stranded nucleic acid. The adducts can then be detected colorimetrically or fluorescently by standard methods which are based on the avidin-biotin interaction. Our expectation is that this same interaction can be used to isolate by affinity chromatography monoadducted as well as crosslinked oligonucleotides. In structure determination analyses the oligonucleotides would be derived from partial enzymatic digests of psoralen photoreacted single-stranded RNA.

Another potential breakthrough in psoralen based secondary structure analysis would be the development of a direct chemical mapping protocol for the localization of psoralen modified bases in DNA and RNA. Two enzymatic approaches exist but each has disadvantages. In 1982 Youvan & Hearst (36) published a protocol for mapping HMT monoadducts in RNA based on the sequencing of reverse transcriptase cDNA transcripts initiated from a unique primer on a psoralen modified RNA. This study demonstrated that reverse transcriptase terminates synthesis opposite HMT uridine monoadducts in RNA. A major drawback of the approach is the need for biologically prepared or chemically synthesized DNA primers. In 1984 a second enzymatic mapping protocol was published by Garrett-Wheeler et al (37) who reported that ribonucleases U^2 and CL3 anomalously cleave RNA after HMT monoadducted pyrimidines. A disadvantage of this approach is that the psoralen specific cleavages are superimposed upon the naturally occurring cuts dictated by the nucleotide specificities of the enzymes.

The development of a chemical cleavage method for mapping psoralen monoadducts and crosslinks in nucleic acid is premised upon the reducibility of pyrimidines containing a saturated 5,6-double bond (187). Psoralen photocycloaddition to pyrimidines occurs at this site and the resultant cyclobutane ring is saturated. Reduction of modified pyrimidines with excess $NaBH_4$ should ing open the base (187–189) thereby permitting acid catalyzed depyrimidination followed by base catalyzed β-elimination and strand scission. This protocol should be straightforward for psoralen modified thymidine and uridine. Application to cytosine, however, must follow deamination of the 4 position on the base converting it to a carbonyl group. This reaction occurs spontaneously upon saturation of the 5,6-double bond (190). Becker & Wang (38) have reported that psoralen binding to DNA can be detected by chemical cleavage induced by $NaBH_4$ reduction of modified pyrimidines followed by acidic aniline treatment. Perfection of this technique together with the inclusion of a crosslink photoreversal step if necessary and the use of gel electrophoresis to resolve the cleavage products should permit chemical mapping of psoralen adducts in both DNA and RNA.

A third area of activity is the development of a straightforward chemistry for the synthesis of DNA oligonucleotides containing site-specifically placed crosslinkable psoralen monoadducts. If successful, this technology could not only have a significant impact on the use of psoralens as probes for nucleic acid structure and function but also extend the advantages of psoralen based crosslinking to several biochemical techniques. Gamper et al (142) have described a photochemical protocol for the generation of psoralen monoadducted oligonucleotides. They observed that the TpA sequence in double-stranded DNA oligomers is a preferred site for psoralen photoaddition. When the self-complementary Kpn I linker CGGTACCG was photoreacted with HMT, the central TpA sequence was crosslinked. Controlled photoreversal of the crosslink generated a furan side HMT monoadduct. The photochemical approach outlined above is limited to relatively short oligonucleotides which contain a single TpA sequence. We are therefore developing a direct chemical approach to the synthesis of psoralen modified DNA oligonucleotides. Preliminary studies indicate that large quantities of the furan side 8-MOP monoadduct to thymidine can be enzymatically prepared from photoreacted DNA. This adduct, after appropriate protection and activation steps, should substitute for thymidine in the standard oligonucleotide synthesis schemes.

The availability of chemically synthesized photoactive DNA oligonucleotides provides a vehicle for the sequence specific targeting of psoralen crosslinks into both single and double-stranded nucleic acid. By hybridizing and crosslinking judiciously selected psoralen monoadducted oligonucleotides to single-stranded DNA or RNA, the role of the complementary regions in secondary structure and function can be probed under reconstitution conditions. Ligation and crosslinkage of such oligonucleotides into appropriately gapped double-stranded DNA molecules will provide unique substrates with site-specific blocks to replication and transcription. Characterization of the structure and response of the stalled enzymatic complexes should be quite informative.

Advances in our understanding of psoralen photochemistry may permit actual strand transfer of the psoralen from the modified oligonucleotide to the complementary region in the receptor nucleic acid. This could be accomplished by selectively photoreversing the crosslink to give pyrone side monoadduct now attached to the complementary strand. After removal of the targeting oligonucleotide and renaturation of the single-stranded DNA or RNA, secondary structure in the vicinity of the monoadduct could be probed by irradiating it back to crosslink. This technique will permit site-specific positioning of a psoralen monoadduct within a large DNA or RNA molecule and will therefore expand the potential reactive sites, reduce the complexity of psoralen crosslink analysis, and ultimately provide structure-function information not obtainable by conventional means. Psoralen strand transfer should also be applicable to superhelical DNA through the formation of an intermediate D-loop with the modified oligonucleotide. Success of the strand transfer concept hinges upon the development of an effective way to crosslink pyrone side monoadducts.

The use of psoralen monoadducted oligonucleotides as hybridization probes could have a significant and far reaching impact on hybridization technology and clinical diagnostics (K. Yabusaki, H. Gamper, S. Isaacs, patent pending). By crosslinking the probe-target mixture under stringent conditions, a complex is formed that is not only hydrogen bonded but also covalently linked. This in turn permits the use of rigorous, even denaturing, wash conditions thereby reducing background and increasing sensitivity. The use of short crosslinkable probes should also reduce hybridization times, favor simple solution assays, and permit the use of techniques such as gel filtration to separate probe-target complex from excess oligomer under denaturing conditions.

ACKNOWLEDGMENTS

We are grateful to the National Institutes of Health for their long term support of psoralen chemistry research in our laboratory (GM 11180, GM 25151), and to many colleagues around the world who generously provided us with current information about the psoralen activities in their laboratories.

Literature Cited

1. Scott, B. R., Pathak, M. A., Mohn, G. R. 1976. *Mutat. Res.* 39:29–74
2. Rodighiero, G., Dall'Acqua, F. 1976. *Photochem. Photobiol.* 24:647–53
3. Ben-Hur, E., Song, P. S. 1984. *Adv. Radiat. Biol.* 11:131–71
4. Song, P. S., Tapley, K. J. 1979. *Photochem. Photobiol.* 29:1177–97
5. Parsons, B. J. 1980. *Photochem. Photobiol.* 32:813–21
6. Averbeck, D. 1984. *Proc. Jpn. Invest. Dermatol.* 8:52–73
7. Hearst, J. E., Isaacs, S. T., Kanne, D., Rapoport, H., Straub, K. 1984. *Q. Rev. Biophys.* 17:1–44
8. Straub, K., Kanne, D., Hearst, J. E., Rapoport, H. 1981. *J. Am. Chem. Soc.* 103:2347–55
9. Kanne, D., Straub, K., Rapoport, H.,

Hearst, J. E. 1982. *Biochemistry* 21:861–71
10. Kanne, D., Straub, K., Hearst, J. E., Rapoport, H. 1982. *J. Am. Chem. Soc.* 104:6754–64
11. Lown, J. W., Sim, S.-K. 1978. *Bioorganic Chem.* 7:85–95
12. Kittler, L., Hradecna, Z., Suhnel, J. 1980. *Biochim. Biophys. Acta* 607:215–20
13. Kittler, L. 1982. *Chem. Abstr.* 97:51833j
14. Kittler, L., Lober, G. 1983. *Studia Biophys.* 97:61–67
15. Thompson, J. F., Bachellerie, J.-P., Hall, K., Hearst, J. E. 1982. *Biochemistry* 21:1363–68
16. Yoshikawa, K., Mori, N., Sakakibara, N., Mizuno, N., Song, P. S. 1979. *Photochem. Photobiol.* 29:1127–33

17. Lerman, S., Megaw, J., Willis, I. 1980. *Photochem. Photobiol.* 31:233–42
18. Veronese, F. M., Schiavon, O., Bevilacqua, R., Bordin, F., Rodighiero, G. 1981. *Photochem. Photobiol.* 34:351–54
19. Veronese, F. M., Schiavon, O., Bevilacqua, R., Bordin, F., Rodighiero, G. 1982. *Photochem. Photobiol.* 36:25–30
20. Granger, M., Toulme, F., Helene, C. 1982. *Photochem. Photobiol.* 36:175–80
21. Kittler, L., Midden, W. R., Wang, S. Y. 1983. *Photochem. Photobiol.* Suppl. 37:S16
22. Ou, C. N., Song, P. S. 1978. *Biochemistry* 17:1054–59
23. Poppe, W., Grossweiner, L. I. 1975. *Photochem. Photobiol.* 22:217
24. Singh, H., Vadasz, J. A. 1978. *Photochem. Photobiol.* 28:539
25. Isaacs, S. T., Shen, C.-K. J., Hearst, J. E., Rapoport, H. 1977. *Biochemistry* 16:1058–64
26. Chatterjee, P. K., Cantor, C. R. 1978. *Nucleic Acids Res.* 5:3619–33
27. Tessman, J. W., Isaacs, S., Hearst, J. E. 1985. *Biochemistry*. In press
28. Johnston, B. H., Johnson, M. A., Moore, C. B., Hearst, J. E. 1977. *Science* 197:906–8
29. Turner, S., Thompson, J. F., Hearst, J. E., Noller, H. F. 1982. *Nucleic Acids Res.* 10:2839–49
30. Thompson, J. F., Hearst, J. E. 1983. *Cell* 32:1355–65
31. Hyde, J. E., Hearst, J. E. 1978. *Biochemistry* 17:1251–57
32. Wittig, B., Wittig, S. 1979. *Biochem. Biophys. Res. Commun.* 91:554–62
33. Rabin, D., Crothers, D. M. 1979. *Nucleic Acids Res.* 7:689–703
34. Hanson, C. V., Shen, C.-K. J., Hearst, J. E. 1976. *Science* 193:62–64
35. Wollenzien, P. L., Youvan, D. C., Hearst, J. E. 1978. *Proc. Natl. Acad. Sci. USA* 75:1642–46
36. Youvan, D. C., Hearst, J. E. 1982. *Anal. Biochem.* 119:86–89
37. Garrett-Wheeler, E., Lockard, R. E., Kumar, A. 1984. *Nucleic Acids Res.* 12:3405–23
38. Becker, M. M., Wang, J. C. 1984. *Nature* 309:682–87
39. Wiesehahn, G. P., Hyde, J. E., Hearst, J. E. 1977. *Biochemistry* 16:925–32
40. Potter, D. A., Fostel, J. M., Berninger, M., Pardue, M. L., Cech, T. R. 1980. *Proc. Natl. Acad. Sci. USA* 77:4118–22
41. Parrish, J. A., Fitzpatrick, T. B., Tanenbaum, L., Pathak, M. A. 1974. *N. Engl. J. Med.* 291:1207–1211
42. Perone, V. B. 1972. *Microbial Toxins* 8:71–81
43. Honigsmann, A. K., Konrad, K., Schnait, F. G., Wolf, K. 1967. Presented at the 7th *Int. Congr. Photobiol., Rome*. Book of Abstracts
44. Kolis, S. J., Williams, T. H., Postma, E. J., Sasso, G. J., Confalone, P. N., Schwartz, M. A. 1979. *Drug Metab. Disposit.* 7:220–25
45. Hearst, J. E., Thiry, L. 1977. *Nucleic Acids Res.* 4:1339–47
46. Hanson, C. V., Riggs, J. L., Lennette, E. H. 1978. *J. Gen. Virol.* 40:345–58
47. Averbeck, D., Moustacchi, E., Bisagni, E. 1978. *Biochim. Biophys. Acta* 518:464–81
48. Hanawalt, P. C., Kaye, J., Smith, C. A., Zolan, M. 1981. In *Psoralens in Cosmetics and Dermatology*, ed. J. Chan, pp. 133–42. Paris: Pergamon
49. Zolan, M. E., Cortopassi, G. A., Smith, C. A., Hanawalt, P. C. 1982. *Cell* 28:613–19
50. Zolan, M. E., Smith, C. A., Hanawalt, P. C. 1984. *Biochemistry* 23:63–69
51. Piette, J. G., Hearst, J. E. 1983. *Proc. Natl. Acad. Sci. USA* 80:5540–45
52. Isaacs, S. T., Chun, C., Hyde, J. E., Rapoport, H., Hearst, J. E. 1982. *Trends in Photobiology*, ed. C. Helene, M. Charlier, Th. Montenay-Garestier, G. Laustriat, pp. 279–94. New York: Plenum
53. Kanne, D., Rapoport, H., Hearst, J. E. 1984. *J. Med. Chem.* 27:531–34
54. Mustafa, A. 1967. The Chemistry of Heterocyclic Compounds, Vol. 23. In *Furopyrans and Furopyrones*, p. 15. New York: Wiley
55. Mustafa, A. 1967. See Ref. 54, p. 14
56. Barenbaum, M. 1978. *Science* 201:532–34
57. Scheel, L. D., Perone, V. B., Larkin, R. L., Kupel, R. E. 1963. *Biochemistry* 2:1127–31
58. Confalone, P. N., Lollar, E. D., Pizzolato, G., Uskokovic, M. 1978. *US Patent No. 4,130,568*
59. Kaufman, K. D. 1961. *J. Org. Chem.* 26:117–21
60. Bender, D. R., Hearst, J. E., Rapoport, H. 1979. *J. Org. Chem.* 44:2176–81
61. Guiotto, A., Rodighiero, P., Pastorini, G., Manzini, P., Bordin, F., Dall'Acqua, F., et al. 1981. *Med. Biol. Environ.* 9:313–19
62. Bender, D. R., Kanne, D., Frazier, J. D., Rapoport, H. 1983. *J. Org. Chem.* 48:2709–19
63. Guiotto, A., Rodighiero, P., Manzini, P., Pastorini, G., Bordin, F., et al. 1984. *J. Med. Chem.* 27:959–66
64. Horning, E. C., Reisner, D. B. 1948. *J. Am. Chem. Soc.* 70:3619–20

65. Lagercrantz, C. 1956. *Acta Chem. Scand.* 10:647–54
66. Chatterjee, D. K., Chatterjee, R. B., Sen, K. 1964. *J. Org. Chem.* 29:2467–69
67. Chatterjee, D. K., Sen, K. 1969. *Tetrahedron Lett.* 59:5223–24
68. Queval, P., Bisagni, E. 1974. *Eur. J. Med. Chem.—Chim. Ther.* 9:335–40
69. Worden, L. R., Kaufman, K. D., Weis, J. A., Schaff, T. K. 1969. *J. Org. Chem.* 34:2311–13
70. Kaufman, K. D., Gaiser, F. J., Leth, T. D., Worden, L. R. 1961. *J. Org. Chem.* 26:2443–46
71. Bender, D. R., Hearst, J. E., Rapoport, H. 1983. *US Patent No. 4,398,031*
72. Spath, E., Pesta, O. 1934. *Chem. Ber.* 67:1212
73. Raizada, K. S., Sarin, P. S., Seshadri, T. R. 1960. *J. Sci. Ind. Res.* 19B:76
74. Kaufman, K. D., Worden, L. R., Lode, E. T., Strong, K. E., Reitz, N. C. 1970. *J. Org. Chem.* 35:157–60
75. Brokke, M. E., Christensen, B. E. 1958. *J. Org. Chem.* 23:589–96
76. Asker, W., Shalaby, A. F. A. M., Zayed, S. M. A. D. 1958. *J. Org. Chem.* 23:1781–83
77. Brokke, M. E., Christensen, B. E. 1959. *J. Org. Chem.* 24:523–26
78. Aboulezz, A. F., El-Attar, A. A., El-Sockary, M. A. 1973. *Acta Chim. Acad. Sci. Hung.* 77:205–10
79. Loutfy, M. A., Abu-Shady, H. A. 1977. *Pharmazie* 32:240
80. Isaacs, S. T., Rapoport, H., Hearst, J. E. 1982. *J. Labelled Compds. Radiopharm.* 19:345–56
81. Isaacs, S. T., Wiesehahn, G. P., Hallick, L. M. 1984. *Natl. Cancer Inst., Monogr.* No. 66, pp. 21–30
82. Hansen, J. B., Buchardt, O. 1981. *Tetrahedron Lett.* 22:187–1848
83. Kaufman, K. D. 1981. *US Patent No. 4,294,822*
84. Dall'Acqua, F., Vedaldi, D., Caffieri, S., Guiotto, A., Rodighiero, P., et al. 1981. *J. Med. Chem.* 24:1748–84
85. Saffran, W. A., Goldenberg, M., Cantor, C. R. 1982. *Proc. Natl. Acad. Sci. USA* 79:4594–98
86. Schwartz, D. C., Saffran, W., Welsh, J., Haas, R., Goldenberg, M., Cantor, C. R. 1983. *Cold Spring Harbor Symp. Quant. Biol.* 47:189–95
87. Haas, R., Murphy, R. F., Cantor, C. R. 1982. *J. Mol. Biol.* 159:71–92
88. Rodighiero, G., Chandra, P., Wacker, A. 1970. *FEBS Lett.* 10:29–32
89. Rodighiero, G., Musajo, L., Dall'Acqua, F., Marciani, S., Caporale, G., Ciavatta, L. 1970. *Biochim. Biophys. Acta* 217:40–49
90. Liebman, A. A., Delaney, C. W. 1981. *J. Labelled Compds. Radiopharm.* 18:1167–72
91. Rodighiero, G., Dall'Acqua, F. 1984. In *Topics in Photomedicine*, ed. K. Smith, pp. 319–97. New York: Plenum
92. Gange, R. W., Levins, P., Murray, J., Anderson, R. R., Parrish, J. A. 1984. *J. Invest. Derm.* 82:219–22
93. Bordin, F., Carlassare, F., Baccichetti, F., Guiotto, A., Rodighiero, P., et al. 1979. *Photochem. Photobiol.* 29:1063
94. Bordin, F., Carlassare, F., Baccichetti, F., Guiotto, A., Rodighiero, P. 1979. *Experientia* 35:1567
95. Dall'Acqua, F., Marciani, S., Ciavatta, L., Rodighiero, G. 1971. *Z. Naturforsch. Teil B* 26:561
96. Dall'Acqua, F., Marciani, S., Rodighiero, G. 1970. *FEBS Lett.* 9:121
97. Lowry, P. D., Brookes, P. 1967. *J. Mol. Biol.* 25:143
98. Vedaldi, D., Dall'Acqua, F., Caffieri, S., Rodighiero, G. 1979. *Photochem. Photobiol.* 29:277
99. Averbeck, D., Bisagni, E., Marquet, J. P., Vigny, P., Gaboriau, F. 1980. *Photochem. Photobiol.* 30:547
100. Baccichetti, F., Bordin, F., Carlassare, F., Rodighiero, P., Guiotto, A., et al. 1981. *Farmaco—Ed. Sci.* 36:585–97
101. Magno, S. M., Rodighiero, P., Gia, O., Bordin, F., Baccichetti, F., Guiotto, A. 1981. *Farmaco—Ed. Sci.* 36:629–47
102. Guiotto, A., Rodighiero, P., Pastorini, G., Manzini, P., Bordin, F., et al. 1981. *Eur. J. Med. Chem.* 16:489–94
103. Dall'Acqua, F., Vedaldi, D., Bordin, F., Baccichetti, F., Carlassare, F., et al. 1983. *J. Med. Chem.* 26:870–76
104. Hearst, J. E., Rapoport, H., Isaacs, S., Shen, C.-K. J. 1979. *US Patent No. 4,169,204*
105. Isaacs, S. T., Thompson, J. F., Hearst, J. E. 1983. *Photochem. Photobiol. Suppl.* 37:S100
106. Dall'Acqua, F., Caffieri, S., Vedaldi, D., Guiotto, A., Rodighiero, G. 1981. *Photochem. Photobiol.* 33:261–64
107. Land, E. J., Rushton, F. A. P., Beddoes, R. L., Bruce, J. M., Cernik, R. J., et al. 1982. *J. Chem. Soc. Chem. Commun.* 1982:22–23
108. Peckler, S., Graves, B., Kanne, D., Rapoport, H., Hearst, J. E., Kim, S.-H. 1982. *J. Mol. Biol.* 162:157–72
109. Kim, S.-H., Peckler, S., Graves, B., Kanne, D., Rapoport, H., Hearst, J. E. 1983. *Cold Spring Harbor Symp. Quant. Biol.* 47:361–65
110. Bachellerie, J.-P., Thompson, J. F., Wegnez, M. R., Hearst, J. E. 1981. *Nucleic Acids Res.* 9:2207–22
111. Hahn, B. S., Joshi, P. C., Kan, L. S.,

Wang, S. Y. 1981. *Photobiochem. Photobiophys.* 3:113–24

112. Shim, S. C., Kim, Y. Z. 1983. *Photochem. Photobiol.* 38:265–71

113. Dall'Acqua, F., Caffieri, S., Vedaldi, D., Guiotto, A., Rodighiero, P. 1983. *Photochem. Photobiol.* 37:373–79

114. Caffieri, S., Dall'Acqua, F., Rodighiero, P. 1983. *Med. Biol. Envior.* 11:387–91

115. Cadet, J., Voituriez, L., Gaboriau, F., Vigny, P., DellaNegra, S. 1983. *Photochem. Photobiol.* 37:363–71

116. Johnston, B. H., Hearst, J. E. 1981. *Photochem. Photobiol.* 33:785–92

117. Salet, C., Macedo De Sa E Melo, T., Bensasson, R. V., Land, E. J. 1980. *Biochim. Biophys. Acta* 607:379–83

118. Lai, T., Lim, B. T., Lim, E. C. 1982. *J. Am. Chem. Soc.* 104:7631–35

119. Beaumont, P. C., Parsons, B. J., Navratnam, S., Phillips, G. O. 1983. *Photochem. Photobiol.* 5:359–64

120. Mantulin, W. W., Song, P.-S. 1973. *J. Am. Chem. Soc.* 95:5122–29

121. Bensasson, R. V., Land, E. J., Salet, C. 1978. *Photochem. Photobiol.* 27:273–80

122. Song, P.-S. 1984. *Natl. Cancer Inst. Monogr.* No. 66

123. Song, P.-S., Chae, Q. 1976. *J. Luminesc.* 12/13:831–37

124. Song, P.-S., Harter, M. L., Moore, T. A., Herndon, W. C. 1971. *Photochem. Photobiol.* 14:521–30

125. Song, P.-S., Chin, C. A., Yamazaki, I., Baba, H. 1975. *Int. J. Quant. Chem., Quantum Biol. Symp.* No. 2:1–8

126. Harrigan, E. T., Charkrabarti, A., Hirota, N. 1976. *J. Am. Chem. Soc.* 98:3460–65

127. Moore, T. A., Montgomery, A. B., Kwiram, A. L. 1976. *Photochem. Photobiol.* 24:83–86

128. Bensasson, R. V., Chalvet, O., Land, E. J., Ronfard-Haret, J. C. 1984. *Photochem. Photobiol.* 39:287–91

129. Land, E. J., Truscott, T. G. 1979. *Photochem. Photobiol.* 29:861–66

130. Bensasson, R. V., Salet, C., Land, E. J., Rushton, F. P. 1980. *Photochem. Photobiol.* 31:129–33

131. Melo, M. T. S. E., Averbeck, D., Bensasson, R. V., Land, E. J., Salet, C. 1979. *Photochem. Photobiol.* 30:645–51

132. Ronfard-Haret, J. C., Averbeck, D., Bensasson, R. V., Bisagni, E., Land, E. J. 1982. *Photochem. Photobiol.* 35:479–89

133. Bensasson, R. V., Land, E. J., Salet, C., Sloper, R. W., Truscott, T. G. 1979. In *Radiation Biology and Chemistry: Research Developments*, ed. H. E. Edwards, S. Navaratnam, B. J. Parsons, G.

O. Phillips, pp. 431–39. Amsterdam New York: Elsevier

134. Beaumont, P. C., Parsons, B. J., Phillips, G. O., Allen, J. C. 1979. *Biochim Biophys. Acta* 562:214–21

135. Beaumont, P. C., Parsons, B. J., Navaratnam, S., Phillips, G. O., Allen, J. C. 1979. See Ref. 133, pp. 441–51

136. Bevilacqua, R., Bordin, F. 1973. *Photochem. Photobiol.* 17:191–94

137. Fujita, H., Kitakami, M. 1977. *Photochem. Photobiol.* 26:647–49

138. Goyal, G. C., Grossweiner, L. I. 1979 *Photochem. Photobiol.* 29:847–50

139. Gasparro, F. P., Saffran, W. A., Cantor C. R., Edelson, R. L. 1984. *Photochem Photobiol.* 40:215–19

140. Suzuki, K., Fujita, H., Yanagisawa, F. Kitakami, M. 1977. *Photochem. Photobiol.* 26:49–51

141. Ou, C. N., Tsai, C. H., Song, P.-S 1977. In *Research in Photobiology*, ed A. Castellani, pp. 257–65. New York Plenum

142. Gamper, H., Piette, J., Hearst, J. E 1984. *Photochem. Photobiol.* 40:29–34

143. Cech, T., Pardue, M. L. 1977. *Cell* 11:631–40

144. Cech, T., Potter, D., Pardue, M. L 1977. *Biochemistry* 16:5313–21

145. Hallick, L. M., Yokota, H. A., Bartholomew, J. C., Hearst, J. E. 1978. *J. Virol* 27:127–35

146. Shen, C.-K. J., Hearst, J. E. 1978. *Cold Spring Harbor Symp. Quant. Biol* 47:179–89

147. Shen, C.-K. J., Hearst, J. E. 1977. *J Mol. Biol.* 112:495–507

148. Carlson, J. O., Pfenninger, O., Sinden R. R., Lehman, J. M., Pettijohn, D. E 1982. *Nucleic Acids Res.* 10:2043–63

149. Robinson, G. W., Hallick, L. M. 1982 *J. Virol.* 41:78–87

150. Kondoleon, S. K., Robinson, G. W. Hallick, L. M. 1983. *Virology* 129:261–73

151. Cech, T. R., Karrer, K. M. 1980. *J. Mol Biol.* 136:395–416

152. DeFrancesco, L., Attardi, G. 1981. *Nucleic Acids Res.* 9:6017–30

153. Hallick, L. M., Hanson, C. V., Cacciapuoti, J. O., Hearst, J. E. 1980. *Nucleic Acids Res.* 8:611–22

154. Cech, T. R., Pardue, M. L. 1976. *Proc. Natl. Acad. Sci. USA* 73:2644–48

155. Sinden, R. R., Broyles, S. S., Pettijohn, D. E. 1983. *Proc. Natl. Acad. Sci. USA* 80:1797–801

156. Sinden, R. R., Carlson, J. O., Pettijohn, D. E. 1980. *Cell* 21:773–83

157. Shen, C.-K. J., Hearst, J. E. 1976. *Proc. Natl. Acad. Sci. USA* 73:2649–53

158. Shen, C.-K. J., Ikoku, A., Hearst, J. E. 1979. *J. Mol. Biol.* 127:163–75
159. Ikoku, A. S., Hearst, J. E. 1981. *J. Mol. Biol.* 151:245–59
160. Shen, C.-K. J., Hearst, J. E. 1977. *Proc. Natl. Acad. Sci. USA* 74:1363–67
161. Shen, C.-K. J., Hearst, J. E. 1979. *Anal. Biochem.* 95:108–16
162. Wollenzien, P., Hearst, J. E., Thammana, P., Cantor, C. R. 1979. *J. Mol. Biol.* 135:255–69
163. Thammana, P., Cantor, C. R., Wollenzien, P. L., Hearst, J. E. 1979. *J. Mol. Biol.* 135:271–83
164. Wollenzien, P. L., Hearst, J. E., Squires, C., Squires, C. 1979. *J. Mol. Biol.* 135:285–91
165. Cantor, C. R., Wollenzien, P. L., Hearst, J. E. 1980. *Nucleic Acids Res.* 8:1855–72
166. Wollenzien, P. L., Cantor, C. R. 1982. *J. Mol. Biol.* 159:151–66
167. Chu, Y. G., Wollenzien, P. L., Cantor, C. R. 1983. *J. Biomolecular Structure Dynamics* 1:647–56
168. Thompson, J. F., Wegnez, M. R., Hearst, J. E. 1981. *J. Mol. Biol.* 147:417–36
169. Zwieb, C., Brimacombe, R. 1980. *Nucleic Acids Res.* 8:2397–411
170. Noller, H. F., Woese, C. R. 1981. *Science* 212:403–11
171. Thompson, J. F., Hearst, J. E. 1983. *Cell* 33:19–24
172. Bachellerie, J.-P., Hearst, J. E. 1982. *Biochemistry* 21:1357–63
173. Calvet, J. P., Pederson, T. 1979. *Nucleic Acids Res.* 6:1993–2001
174. Calvet, J. P., Pederson, T. 1979. *Proc. Natl. Acad. Sci. USA* 76:755–59
175. Wollenzien, P. L., Cantor, C. R., Grant, D. M., Lambowitz, A. M. 1983. *Cell* 32:397–407
176. Calvet, J. P., Pederson, T. 1981. *Cell* 26:363–70
177. Calvet, J. P., Meyer, L. M., Pederson, T. 1982. *Science* 217:456–58
178. Setyono, B., Pederson, T. 1984. *J. Mol. Biol.* 174:285–95
179. Shen, C.-K. J., Hsieh, T.-S., Wang, J. C., Hearst, J. E. 1977. *J. Mol. Biol.* 116:661–79
180. Shen, C.-K. J., Hearst, J. E. 1978. *Nucleic Acids Res.* 5:1429–41
181. Revet, B., Benichou, D. 1981. *Virol.* 114:60–70
182. Kaback, D. B., Angerer, L. M., Davidson, N. 1979. *Nucleic Acids Res.* 6:2499–517
183. Chatterjee, P. K., Cantor, C. R. 1982. *J. Biol. Chem.* 257:9173–80
184. Brown, W. M., Vinograd, J. 1974. *Proc. Natl. Acad. Sci. USA* 71:4617–21
185. Brown, D. M., Frampton, J., Goelet, P., Karn, J. 1982. *Gene* 20:139–44
186. Wollenzien, P. L., Cantor, C. R. 1982. *Proc. Natl. Acad. Sci. USA* 79:3940–44
187. Cerutti, P., Miller, N. 1967. *J. Mol. Biol.* 26:55–66
188. Miller, N., Cerutti, P. 1968. *Proc. Natl. Acad. Sci. USA* 59:34–38
189. Kunieda, T., Witkop, B. 1971. *J. Am. Chem. Soc.* 93:3493–99
190. Green, M., Cohen, S. S. 1957. *J. Biol. Chem.* 225:397–407
191. Mitra, S., Sen, D., Crothers, D. M. 1984. *Nature* 308:247–50
192. Sinden, R. R., Pettijohn, D. E. 1981. *Proc. Natl. Acad. Sci. USA* 78:224–28
193. Turner, S., Noller, H. F. 1983. *Biochemistry* 22:4159–64
194. Nakashima, K., Shatkin, A. J. 1978. *J. Biol. Chem.* 253:8680–82
195. Swanstrom, R., Hallick, L. M., Jackson, J., Hearst, J. E., Bishop, J. E. 1981. *Virology* 113:613–22

Ann. Rev. Biochem. 1985. 54:1195–227

THE APPLICATION OF NEUTRON CRYSTALLOGRAPHY TO THE STUDY OF DYNAMIC AND HYDRATION PROPERTIES OF PROTEINS

Anthony A. Kossiakoff

Genentech, Inc., 460 Point San Bruno Boulevard, South San Francisco, California 94080 and Department of Pharmaceutical Chemistry, School of Pharmacy, University of California, San Francisco, California 94143

CONTENTS

0066-4154/85/0701-1195$02.00

PERSPECTIVES AND SUMMARY

X-ray and neutron diffraction techniques are the most direct and powerful methods available to study biological macromolecules at the atomic level. In many respects, these two techniques are similar in both their experimental methodologies and in the resulting informational content. However, the most significant attribute of neutron crystallography, setting it apart from its X-ray counterpart, is its ability to locate experimentally hydrogen (or deuterium) atoms in large molecules. This is particularly significant in the study of protein behavior because typically half the total number of atoms are hydrogen and these dominate much of the chemistry and physical structure of proteins. In X-ray crystallography hydrogen positions cannot be observed directly and can only be inferred from the stereochemistry of the heavier atoms to which they are bound, making it difficult to define accurately the chemical interactions taking place among groups in the protein. This limitation makes it likely that any unusual or unexpected stereochemistry for the hydrogens will be overlooked. This class of hydrogens, of course, is just the one that is of most interest to providing new structural information.

Until recently, the use of neutrons to study biological systems has been mainly limited to solution scattering studies of systems existing in partially ordered arrays like membranes, muscle, and collagen [representative studies found in Ref. (1)]. The advantages of neutrons over X rays to study these types of systems are well documented (2). Similar progress in high resolution neutron crystallography had been slowed largely because of experimental difficulties (3–5) caused by the intrinsically low beam flux of neutron sources. However, advances in several important areas, such as improved data acquisition by array detectors (6–11) and sophisticated data processing techniques (12–14) have made neutron diffraction capable of providing resolution comparable to that of X-ray diffraction.

The application of the neutron diffraction technique to assign H atom positions in proteins and to differentiate between H and D atoms has been mainly focused on structural issues in three research areas: 1. protein reaction mechanisms, 2. protein dynamics, and 3. protein-water interactions. Although several important findings concerning protein mechanism have been reported (15–21), this review will emphasize the areas of protein dynamics and hydration, in which most of the recent work in the neutron field has been concentrated. This work seeks to understand the effects of protein folding on the extent and nature of the protein's dynamic motions, the interaction of the protein with its envelope of solvent, and the way the solvent is involved in the phenomenon of dynamic mobility—all areas of prime importance to the understanding of biological processes. These questions are currently subjects of intensely active investigations by other techniques as well, with neutron diffraction being a major contributor of interpretable data.

In the following sections the application of neutron data to identify the character of a protein's conformational variability and solvent interaction will be discussed. Although certain technical aspects of the neutron technique are mentioned in detail, most of the discussion is centered on the interpretation of the observations and their relationship to structured issues. In the interest of making this review as current as possible, material is included that has been submitted for publication or is in press and is so designated in the text.

INTRODUCTION

Neutron Density Maps—Informational Content

Some of the types of structural information that can be derived from neutron density maps are illustrated in Figure 1. This figure is of the density of a well-ordered tyrosine ring in the 1.4 Å structure of the protein crambin (22). It is clear that a high resolution neutron map contains a striking degree of structural detail. One reason for this is that neutrons scatter off the nucleus of an atom, which acts virtually as a point source, whereas X rays scatter from the considerably more diffuse electron cloud. As a result, all other things being equal, the neutron map will be considerably sharper than an electron density map at the same resolution.

It can be seen from the figure that the ring hydrogen atom locations are in positions of negative density. These negative peaks arise from the fact that hydrogen is one of very few atom types whose nucleus induces a phase change of 180° in the scattered neutrons (23). This difference in the sign of the density peak is an important feature because it allows hydrogens to be distinguished readily from all other atom types in the protein. Scattering from deuterium atoms however, does not involve this phase change and gives rise to positive peaks in the map. As will be discussed, this capability to differentiate between H and D atoms can be exploited in a powerful way to gain insight into the dynamic properties of a protein.

The figure also illustrates that the intensity peaks marking the hydrogen and deuterium atoms are of similar magnitude as the peaks due to the ring carbons and polypeptide chain atoms, resulting from the fact that neutron scattering by an atom is independent of its atomic number (23, 24). (Scattering potentials of atom types common to proteins are listed in the caption of Figure 1). X-ray scattering, on the other hand, is directly proportional to atomic number, and for this reason hydrogen and deuterium peaks are considerably smaller than the other protein atom types (C,N,O), and cannot be detected in electron density maps.

The failure of X-ray diffraction to locate hydrogens directly is of little significance in cases where the hydrogen is bound to a rigid portion of the

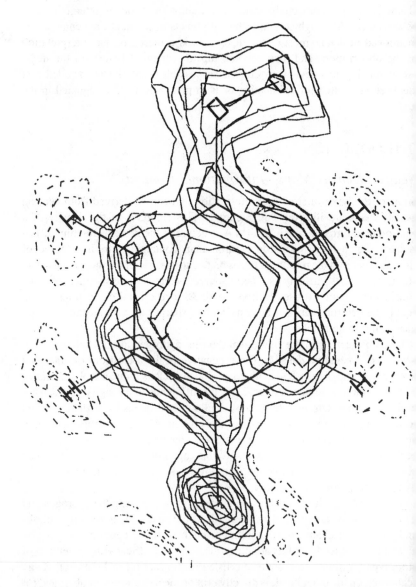

Figure 1   Neutron scattering density of a well-ordered tyrosine ring in the 1.4 Å refined structural model of crambin (22). Hydrogen atoms appear as negative peaks because of a phase change of 180° induced by a resonance level of their nucleus. The relevant scattering potentials for protein atoms types are H=−3.8, D=6.7, C=6.6, N=9.4, O=5.8, S=2.8 (×10⁻¹³ cm).

molecule (such as the ring in the case of the tyrosine in Figure 1) and their position can be inferred with high accuracy. In the case of the hydroxyl group, however, the clearly resolved deuterium peak in the neutron map of Figure 1 would be completely invisible in an X-ray electron density map. The inherent ability of neutron diffraction to make definitive conformational assignments of key hydrogen sites enables this method to provide unambiguous data on important chemical and structural interactions not available from any other source.

In the field of study of biological macromolecules, knowledge of the conformation or position of certain protons (or deuterons) can be important. For example in structure/function studies, the detailed determination of hydrogen bonding and protonation states is critical for the understanding of the mechanism of the system. Recent reviews have summarized the contributions made by neutron diffraction in these determinations (15–21), so there will be no attempt to review these studies further here. However, in the interest of showing the power of the method and the importance of assigning certain proton positions, two examples are cited below.

The neutron analysis of trypsin was initiated in an attempt to resolve a much debated issue concerning the identification of the catalytic group in the reaction mechanism of the serine protease family of enzymes (15, 16). A number of biophysical techniques had previously been brought to bear on this question with conflicting findings, some implicating His-57 and others Asp-102 as the catalytic group (25–33). The answer depended on the identification of which of these groups is protonated during the transition state of the reaction.

The approach to resolving this question used by Kossiakoff & Spencer was to study a derivative form of trypsin which had been inhibited by a transition state analog (16). Therefore, the atomic arrangement of the active site closely mimicked that of the true transition state form of the enzyme. The results of the study demonstrated conclusively that the catalytic group was His-57, not Asp-102. Figure 2a shows how directly this assignment could be made from the neutron density map.

In another example, Phillips recently reported the structure of the oxygen binding site in oxy-myoglobin (19, 34). Here again there was a controversy concerning the protonation state of a histidine group. From Figure 2b it is evident that the NE_2 nitrogen is protonated to form a H-bond to the O_2 ligand. This finding suggests that the imidazole functions as a control on oxygen affinity in myoglobin (and by analogy, hemoglobin).

In both examples cited above the assignments of the protonated positions were definitive from the density maps. These findings represent examples of the direct fashion by which neutron density maps can be interpreted to give data unobtainable by other experimental approaches.

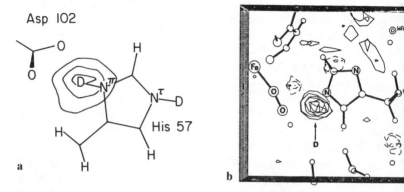

Figure 2 (*a*) A difference map [(F$_o$-F$_c$) exp iΦ_α] calculated with only the deuterium between the His-57 and Asp-102 side chains left out of the phases. The difference peak is approximately 5.5σ above the background level and shows the deuterium bound to the imidazole nitrogen. (*b*) Difference density map (as above) centered on the plane of the E7 imidazole ring of oxymyoglobin. A strong positive peak indicates the presence of the deuterium bound to NE2. Reprinted with permission (34).

Experimental Limitations of the Method

In the preceding section, the attributes of the neutron crystallography approach to studying protein structure have been presented in such a manner that one might ask why the method has not been more generally applied. Several experimental difficulties have been responsible for limiting its application. Most serious of these difficulties is of course that the method requires the use of a high flux beam neutron reactor of which there are only 4 or 5 worldwide. Second, even the most powerful of these reactors produce neutron beams with fluxes 4–5 orders of magnitude less than X-ray fluxes from a standard rotating anode generator. Low neutron flux normally necessitates data collection times in the range of one or two months. It also makes the use of very large crystals (> 2 mm^3) a requirement.

Another data limitation is that the scattering of neutrons by biological samples has an inherently large background component because of the substantial incoherent scattering component of neutrons by hydrogen atoms (23). This effect can be reduced significantly, but not eliminated, by soaking the crystal in D$_2$O, thereby exchanging all waters of crystallization and labile protons by deuterium atoms, which have considerably lower incoherent scattering potential than hydrogens. This soaking procedure is standard practice in most neutron studies of proteins and is why most labile protons appear as deuterium rather than hydrogen atoms in the density maps.

These data acquisition problems are being dealt with by the development of improved instrumentation and data handling procedures (6–14). In addition each neutron facility has a user program that provides investigators from other nstitutions the use of the equipment and instruction and advice from the n-house staff. Although the technique will never have the same availability as s possible with X-rays, there is reason to believe that with increasing interest he number of studies undertaken will grow significantly.

PROTEIN DYNAMICS AND FOLDING

Three-dimensional analysis has shown that the folding pattern of a typical protein is characterized by an architecture consisting of sets of structural units oined together by linking sections of polypeptide chain, forming highly ordered structures. Some of these units are compact, folding their interacting groups into relatively close proximity to one another; others, however, have extended structures. It has also been shown by a number of structural and chemical investigations that these component segments can exhibit substantial fluctuations from their lowest energy conformation (35–40), even under physiological conditions. These deformations are transient in nature and differ in degree over a large spectrum of structural variation, ranging from individual atomic vibration to significant tertiary unfolding or "denaturation."

A majority of these conformational fluctuations appear to involve localized deformation of small segments of structure resulting from the natural equilibrium between protein-protein and protein-solvent interactions. Although most of these fluctuations may have little effect on functional activity of the protein, there is evidence to indicate that conformational change plays an important role in a significant number of biological processes and therefore warrants investigation (41–45).

In the following discussion of the use of neutron data to study conformational dynamics, the subject matter is divided into two categories, differing in the relative scale of deformation that is involved. The first category concerns substantial structural deformations which usually result from the breaking of several hydrogen bonds in the secondary structure. These excursions produce deformations extensive enough to solvate groups in the protein interior which in the normal lowest energy configuration are well buried (46–50). This category will be referred to by the term "regional melting," denoting localized unfolding or "denaturation."

The second form of fluctuation will be referred to as protein "breathing," and is defined here as a set of motions involving a transient local redistribution of the packing density in the interior or at the surface of the protein. This type of structural deformation differs from the first in that it does not involve significant denaturation of the protein structure. An important point to emphasize at the beginning of this discussion is that it has not been established definitively whether these two classes of conformational fluctuation occur through the same mechanism, differing only in extent, or whether they are the result of different phenomena.

The following subsections discuss experimental findings from a series of neutron diffraction studies and indicate briefly the experimental approaches that were used to obtain them. Two distinct types of neutron data are used to study the individual classes of conformational variability. The larger conformational fluctuations described by the term "regional melting" are measured by employing the H/D exchange method, the data being derived through the ability of neutron diffraction to discriminate between H and D atoms. In the case of protein "breathing," the conformation and relative degree of order of side chain methyl groups are used as a probe for local motion.

Regional Denaturation Investigated by H/D Exchange

The term regional denaturation (or regional melting) describes a class of molecular fluctuations involving transient denaturation of limited regions of a protein's structure. Such a process usually entails the breaking of some number of hydrogen bonds and thus represents as a class conformational fluctuation variations larger than those associated with protein breathing as described above.

Clarification of the details of the nature of protein fluctuations requires an experimental technique that can identify those segments of the polypeptide chain involved in transient motions. Since its introduction by Linderstrom-Lang and his colleagues in the 1950s, the H/D exchange method has been recognized to be a powerful probe for protein conformational change (35, 46–50). This is because the rate of exchange of a peptide hydrogen in the interior of a protein molecule is a direct function of the energy required to distort the polypeptide chain to provide access for the deuterated solvent. Indeed, the H to D exchange rates of interior sites are found to differ by several orders of magnitude in an intact protein, providing a sensitive measure of the extent of shielding from the solvent provided at a given site by the protein's secondary and tertiary structure (47–49).

H/D exchange has advantages over other labeling techniques in that a deuteron has a negligible space requirement and has nearly equivalent chemical properties to the proton it replaces in the structure. Furthermore, potentially labile sites are distributed fairly uniformly throughout the molecule and therefore probe the structural variability of the whole molecule.

Unfortunately, many previous H/D experiments have not yielded readily interpretable results because of their inability to relate exchange rates to specific groups or even regions of the polypeptide chain. This limitation has been recognized and recently a series of NMR (51–54) and chemical analyses have begun to provide data on exchange characteristics at specific sites.

A more definitive approach to the problem is offered by neutron diffraction, which determines the location of each exchange site of the protein and permits the unambiguous characterization of its exchange status by direct examination of the neutron density map (48, 55, 56). As can be seen in Figure 3, the task of assigning a site as having either H or D character is straightforward because the amplitude of H and D are of opposite signs. The practicality of using neutron diffraction for H/D analysis was first shown by Schoenborn and his colleagues (57, 58), and subsequently verified by other investigators (see discussion below).

The use of a crystallographic technique to study protein dynamics might appear at first thought to be inappropriate because it is by its nature a time-averaged sampling of the low energy conformation of the molecule. However, it turns out in practice that no contradiction is involved. This is because protein crystals typically contain 50% solvent (59) and are best described as an ordered, but open array of molecules held together by a relatively small number of intermolecular contact points. Thus, proteins in crystals are highly solvated and because of the extent of unoccupied volume (solvent) in the crystal,

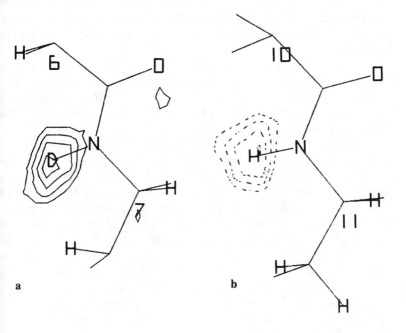

Figure 3 Sections of a neutron difference map taken in the plane of the peptide group. (*a*) Fully exchanged site; (*b*) unexchanged site. In cases where partial exchange has taken place, the expected density at the H/D site is near zero. This is because with H and D having scattering amplitudes of opposite sign, the composite scattering density tends to cancel in the Fourier summation.

substrates and ions can diffuse through the solvent channels to interact with th protein surface. A number of studies support the point that crystalline an solvated proteins are similar by showing that in many instances the system exhibit the same chemical and physical properties (60–64). Investigations hav also shown that, except possibly at the lattice contact points, moderate confo mational mobility of the protein is not greatly impeded in its crystalline for (50, 68).

In the following subsections the observed pattern of exchange reported fo several proteins will be discussed with emphasis on the systematic relationshij between exchangeable sites and the protein's structural and chemical pro erties.

Interpretation of H/D Exchange Data and Its Implication to Analysis of Conformational Fluctuations

As of this writing H/D studies on six proteins have been published: myoglobi (21, 34, 65), ribonuclease (20, 66), trypsin (50, 67), lysozyme (18, 68 crambin (22), and pancreatic trypsin inhibitor (PTI) (69). The results of thes studies, when viewed together, suggest several general relationships governin conformational variability in protein structure, although, as might be expecte for molecules of such complexity, there are some findings that deviate from th general pattern.

A consistent finding is a strong correlation of the H to D exchange rates o the location of the site with respect to the principal secondary structural units c the protein, notably the β-sheet and α-helix. The exchange characteristics these two types of structure have been analyzed for several of the abov proteins (18, 50, 65–67, 69).

The exchange pattern of trypsin as a function of its structure is represente schematically in Figure 4a (50). The individual peptide sites shown as ope circles are fully exchanged amide peptide groups; half-filled circles represe partially exchanged sites, and filled circles unexchanged sites. From inspectio of the figure it is evident that the unexchanged sites form a distinctive patteri whose most prominent feature is the clustering of unexchanged sites in regior corresponding to the β-sheet structures of the protein. There are two suc structures in trypsin, called "beta barrels," each consisting of six antiparall strands laced together by a network of hydrogen bonds. In fact, about 85% (4 of 54) of the unexchanged hydrogens are found in the β-sheet structures. It interesting to note that only 11 of the peptide groups involved in β-she hydrogen bonding are completely exchanged, and all of these are located edges of the sheet structures.

Similar patterns of exchange of β-sheet structures have been observed fo other proteins (50, 66, 67, 69). Figure 4b shows the exchange pattern of PT (black circles are unexchanged groups) (69). These exchange assignments we

und to be in good qualitative agreement with assignments determined by
MR (51–53), giving an independent verification of the accuracy of H/D
signments by the neutron method. (PTI is one of the very few proteins that
s an appreciable percentage of its amide peptide protons assigned in the NMR
ectrum). As in trypsin, the majority of the unexchanged groups are located in
e sheet structure. Several of these amide protons are surprisingly resistant to
change considering the small degree of shielding that a protein the size of PTI
n provide (52).

The exchange characteristics of α-helix units appear to be different from
ose observed for sheet structures. In the two short segments of the α-helix in
ypsin (Figure 4a), the peptide groups are almost fully exchanged (50). In the
terminal helix the only fully unexchanged sites, I-238 and I-242, are located
ong the interface between the helix and two strands of β-sheet. A similar
change pattern was observed by Wlodawer & Sjölin for two of the helixes in
onuclease (66). In addition, Phillips has reported that the unexchanged sites
oxymyoglobin are located predominantly on the inside, less accessible
rfaces of the helixes (34). The finding that exchange seems to be impeded
ly at points of intermolecular contact suggests that steric shielding plays an
portant role in the exchange properties of α-helixes.

The pattern of exchange for helixes described above, although consistent
th most observations, has some exceptions. Mason and his colleagues report
at in lysozyme there are several stretches of α-helix that remain fully unex-
anged yet are solvent accessible (18). A similar situation is found for crambin
2). However, an important factor that may affect the exchange of these two
otein systems is that their crystalline lattice is unusually tightly packed
ving relatively little solvent content (≅30%).

orrelation of Exchange with Packing Effects

e above discussion, reporting the general pattern of exchange for several
oteins as determined by the neutron method, makes a qualitative argument
r the premise that the exchange properties of a group have a fundamental
pendence on the degree of its local secondary structure. The development of
nore quantitative model for clarifying exchange properties of different sites
quires the understanding of the influences of each structural feature within the
cal vicinity. Definitive information of this type is not presently available.
wever, for a general understanding of the correspondence between exchange
d readily computed indices of local structure, three simple parameters, 1.
omic packing density, 2. solvent accessibility, and 3. local thermal motion,
ve been found to provide further insight into the relationship between
change and structure. The results of these analyses are of sufficient general
terest to warrant a brief discussion here.

PACKING DENSITY   It has been shown that accurate assessments of seconda
structure can be derived from computed atomic packing densities when radii
6–9 Å are used (70). The packing density is computed by placing the origin of
sphere of a chosen radius at an amide peptide site and counting the number
enclosed atoms; in trypsin a radius of 7 Å was found to give good results (6
To gain additional information on the effect of structure, the number
neighboring H-bonds can also be computed for each amide peptide nitrogen in
similar way.

The heavier line in Figure 5 traces the packing density of the pepti
nitrogens in trypsin as a function of their residue number; the lighter line t
number of neighboring H-bonds in the same sphere. The unexchanged pepti
groups are marked at the top of the figure. Comparison of this curve with t
schematic in Figure 4a shows that the regions of highest packing densi
correspond to the β-sheet segments of the protein. Conversely, points of l

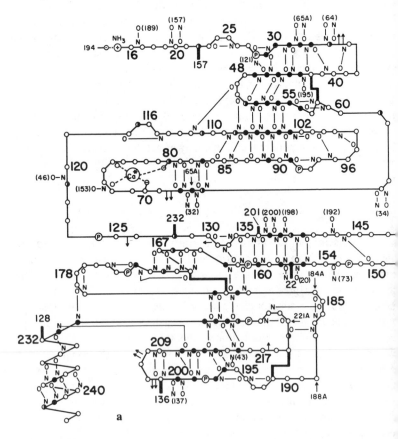

a

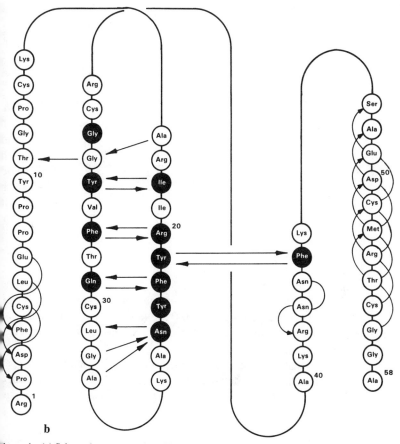

b

Figure 4 (a) Schematic representation of H/D exchange at each amide peptide site in the trypsin molecule. Key: Full exchange ○, partial exchange ◐, unexchanged ●, proline ℗, sequence insertions ↓, deletions ↑ (based on the chymotrypsinogen numbering scheme), carboxylate side chains ⊖. Peptide NH and carbonyl oxygens are shown when H-bonded. Disulfide bridges are indicated by broad lines. Because of the effects of thermal motion or local disorder, the H/D exchange information for residues 60–62, 110–111, and 147–150 is not reliable. (b) H/D exchange pattern in pancreatic trypsin inhibitor (PTI). Filled circles correspond to protected hydrogens. Reprinted with permission (69).

density are seen to be usually associated with segments of the polypeptide chain which are in turns and loops. These findings are what one would generally expect.

An interesting feature of this trypsin analysis was the observation that there existed a surprisingly distinct partitioning of the two exchange categories (exchanged/unexchanged) as a function of packing density of the protein (67). The average packing density was 117 atoms/7 Å radius for the exchanged

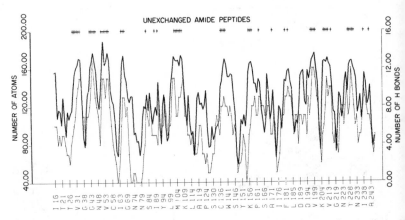

Figure 5 Packing density profile of trypsin, showing the number of atoms (heavier line) and main-chain hydrogen bonds (lighter line) contained within a 7 Å sphere around each amide peptide proton of the molecule. Unexchanged groups are designated as crosses at the top of the figure.

groups and 161 for the unexchanged sites. The packing density has an upper bound of about 180, and for two thirds of the unexchanged sites it ranges from 160 to 175. When a threshold value of 160 is chosen, it was found that about 80% of the unexchanged sites and only 5% of the exchanged sites have packing densities greater than 160.

In summary, what these packing parameters show when applied to trypsin structure is that exchange characteristics follow a path parallel to the amount of local structure, a conclusion that is consistent with most proposed mechanisms of exchange.

PROXIMITY TO SOLVENT   As a first approximation, it might be expected that the degree of exchange of sites located near the surface of the molecule to be greater than those in the interior. This concept suggests that a protein's size would be a principal factor in determining the percent of groups protected from exchange. It has been pointed out, however, that the overall degree of exchange seems to be independent of the protein's size (69). For instance in trypsin (M_r 25,000) about 25% of its groups are resistant to exchange. This compares closely with the observations from smaller proteins—PTI 20% ($M_r \sim 6,500$) ribonuclease 22.5% ($M_r \sim 13,500$), and lysozyme 43% ($M_r \sim 14,000$).

Quantitative data on solvent accessibility/exchange correlations have been reported for the β-sheet rich protein, trypsin (50, 57), and for myoglobin which is mainly α-helical (34). A plot of the variation of group accessibility versus exchange for oxy-myoglobin is shown graphically in Figure 6 (34). Phillips reported that all unexchanged groups have very little or no accessibility; however, there are a significant number of other peptides that are equally

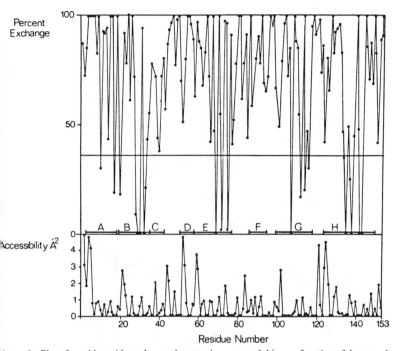

Percent
Exchange

Accessibility Å²

Residue Number

igure 6 Plot of peptide amide exchange character in oxymyoglobin as a function of the group's
·egree of static accessibility. Helix assignments are shown as bars on the horizontal axis (A–H).
ower portion of the plot shows static accessibility of the groups as calculated from the X-ray
·ordinates. Reprinted with permission (34).

·ell buried yet are fully exchanged indicating that in myoglobin low accessibil-
·y alone is not sufficient to impede exchange.

The situation is more complicated in β-sheet proteins. This is because the
ffects of hydrogen bonding vs depth are not easily separated since β-structures
·re on average more deeply buried than most other types of structure. For
·ypsin, the average distance between peptide sites and the solvent interface
·as determined to be 3.8 Å and, as anticipated, the great majority of the
·nexchanged sites (~90%) were found to be located at least 4 Å from the
·iterface (50). In the β-sheet structure, the average depth of the 11 fully
·xchanged sites was found to be 4.0 Å. This corresponded closely to the
·verage depth (3.8 Å) of the total β-sheet population, implying that solvent
·ccessibility is not a primary factor in imposing exchange characteristics on
·-sheet structures.

Several additional examples from the trypsin analysis provide direct data
·elevant to the relative importance of accessibility vs H-bonding. Illustrated in
·igure 7 is a region of beta structure containing an occluded water molecule

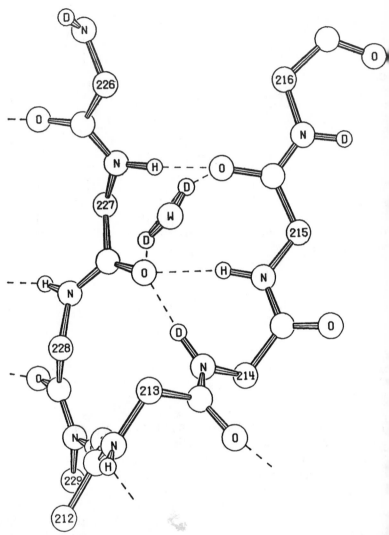

Figure 7 A segment of β-sheet structure showing the H-bond interaction between 227 HN and 215 O, and between 227 O and 215 NH. An ordered D_2O molecule shares this H-bond, yet the peptide NH's are unexchanged. The locations of the D's on the D_2O molecule were determined from the neutron density map to indicate directly how the D_2O was oriented relative to the coordination ligands (215 O and 227 O). Note that the 214 amide is fully exchanged. Calculations indicate that the 214 and 215 amide protons are equally close to the solvent interface. The respective hydrogen bond strengths are quite different. The 227 O to 215 N hydrogen bond is classified as normal (2.9 Å), whereas the 227 O to 214 N distance is quite long (3.5 Å), indicating a weak interaction.

tween peptides 215 and 217. This highly ordered water molecule is perpendicular to the plane of the β sheet and H-bonds to both carbonyl oxygens and to r 190 OG (not shown). (A similar arrangement of β sheet H-bonding to included waters is observed in five other places in the trypsin molecule). The finement results indicate that both the 215 and 227 amide peptides are drogenated, not deuterated. Clearly, solvent has access to these sites, as monstrated by the observation that the adjacent 214 NH is fully exchanged. hese observations make a very strong argument for the fact that structure fects dominate over solvent accessibility effects, although exceptions certainexist.

There are also instances where NH groups not involved in H-bonding vertheless remain unexchanged even when the group's packing and accessility parameters do not suggest any additional shielding when compared to the ass of NHs that are H-bonded and yet do exchange (22, 67). Modeling studies ill have to be applied in order to sort out the structural features important in otecting non-H-bonded groups from exchange.

HERMAL MOTION   Several investigations have looked at correlations of a oup's H/D exchange behavior and the corresponding B factors of atoms in its cal vicinity (34, 50). It was found that unexchanged groups were in environments of low overall B values, indicating that they are predominantly located in terior, highly structured regions of the protein, at least in the time-averaged nse, but must become exposed to solvent as a function of the general eathing modes of the protein. Because many exchanged interior sites have omparable B factor values with those of sites that remain unexchanged, no stinct correlation exists between exchange of an interior site and its observed ermal motion.

terpretative Limitations of the Method

he detail to which H/D exchange data can be interpreted is a direct function of e accuracy of the densities contained in the neutron Fourier map. To quantify e H/D exchange ratios for amide peptides, several different refinement hemes have been employed (4, 5). In all cases, because of the general noise atures observed in the neutron density maps, the obtained ratio values had to viewed as being more qualitative than quantitative. Consequently, the change ratios are most accurately subdivided into three major categories: 1. nexchanged, 2. partially exchanged, and 3. fully exchanged; finer gradation f these categories is of questionable significance at resolutions >1.8 Å. By miting the number of categories in this manner, one retains the essential formational content of the H/D data while guarding against possible overterpretation of the relevance of small changes in exchange ratio.

The primary obstacle to obtaining accurate H/D occupancies is that, at th resolution of the data used in most protein analyses, any errors in the position or thermal refinement parameters of the parent amide nitrogen are very likely t affect the density characteristics of the peptide proton. Furthermore, exper ence has shown that useful H/D information can be obtained only fro well-ordered segments of the polypeptide chain (50). A substantial fraction c the groups classified as partially exchanged on the basis of their zero density ar usually found to be located in "loosely ordered" regions of the molecule. Thi finding leaves the interpretation of the exchange status of these groups open t question.

A basic limitation of the neutron approach to H/D experiments is that it ca distinguish only between sites that exchange quickly and those that exchang slowly relative to the possible time frame over which the reaction is allowed t proceed. In theory, higher sensitivity is possible, and quantitative exchang rates can be obtained for the partially exchanged sites. However, in practic the precise pH in the crystal and the effective period of soaking are difficult t estimate. The reason for this is that when the crystal is mounted in the capillar for data collection (a process of several months) the "mother liquor" is remove from direct contact with it, and although the crystal remains in an environme of high relative humidity, one cannot assume that exchange proceeds at th same rate as in solution.

Mason et al (18) have suggested a set of experiments with lysozyme usin low temperature crystallography which could decrease the degree of unknow variability in the exchange conditions. They observed that lysozyme crysta could be taken to a temperature of 260°K without suffering damage. The propose that a nondeuterated crystal of the protein could be mounted in th conventional manner in a low temperature device. The capillary would have "plug" of frozen D_2O in its upper end removed from the crystal. Data of th undeuterated crystal would be taken. After data collection, the H/D exchang process could be initiated by raising the temperature to melt the D_2O "plug. After a suitable time period the temperature could again be lowered refreezin the plug and thereby quenching the exchange. A second set of data would b collected and the process repeated until the full range of exchange times we explored. Most likely a majority of the available information could be obtaine in three sets of data.

An alternative to temperature variation is the use of pH change. From neutr pH, each change of a pH unit enhances (if the pH is raised) or impedes (lowered) exchange by about an order of magnitude. Effects of exchange durin the period of data collection could be eliminated by always collecting data at th same pH. How well these methods work, or whether they work at all i practice, must wait for experimental verification.

SMALL-SCALE STRUCTURAL FLUCTUATIONS: PROTEIN BREATHING

In a number of crystallographic analyses, attempts have been made to gain insights into small transient fluctuations in proteins using temperature (or B) factor data [reviewed in Ref. (45)]. The attraction of this method is that it identifies specific groups in the protein structure whose locations show positional variation due to thermal motion. Typically, B factor correlations describe small atomic displacements on the order of 0.5–1.5 Å. Note that in order to obtain accurate B factor data it is crucial that the analysis be done at high resolution (<2.0 Å) and that the structure be highly refined because B factor values are inherently sensitive to data and refinement errors. As a result of the difficulty in assigning accurate individual B factor values, local trends of group B values offer a considerably more reliable measure of motion; therefore, during the refinement procedure, B factors are usually restrained to be similar to those of their near neighbors.

The general picture that emerges from these thermal factor analyses is that, as expected, protein molecules appear to have considerably more flexibility at their surfaces than in their interior (37, 38, 45). However, the fact that groups in the interior look well-ordered in the time-averaged structure does not exclude the possibility that they constantly undergo rapid reorientations, too rapid to be observed in a diffraction experiment. Indeed, the mobility of the protein structure was confirmed several years ago by NMR spectroscopy, which showed that in many proteins, bulky ring structures (i.e., PHE and TYR) can undergo ring "flips" (180° rotations around CB-CG bonds) even when these groups are buried (71, 72).

NMR studies also provide data on the mobility of another important class of side chain groups—terminal methyl groups (73–75). It is observed that these groups rotate rapidly (1–15 ps) around their rotor axes (36, 75). These findings suggest that there operates in proteins a set of cooperative forces permitting rapid redistribution of the packing density in the molecule's interior region. These motions represent as a class, fluctuations larger than those described by temperature factors. It would appear, therefore, that detailed evaluations of the rotational properties of interior side chains can be employed as a probe of protein "breathing" motions and of the effects of local structural environment on a group's conformation.

While NMR data provide important information about the time frame of rotational disorder, it does not directly address questions related to structure, such as whether these groups have preferred torsional conformations, and if so, what those conformations are. Such structural information would provide valuable insights into the relative magnitude of forces involved in secondary and tertiary interactions and how these forces compete with the torsional force of side-chain groups to define the protein's low energy conformation.

Exactly this type of information is provided by studying the configuration o terminal methyl groups by neutron diffraction (16, 76, 77). Terminal methy groups are widely distributed throughout the protein structure. In the free stat they possess a well-defined rotational energy barrier of about 3 kcal, which i within the range of expected energies associated with secondary and tertiar interactions in the protein. Thus, an observed departure of a given methyl grou from its normally preferred staggered orientation would mean that the loca structural forces were comparable to or exceeded the normal restoring force o the rotor. A neutron density map allowing identification of hydrogen aton positions provides a means for directly visualizing the rotational orientation o each well-ordered methyl group in the molecule, as related to its surroundin structure.

Assignment of Methyl Rotor Conformations

The rapid rotatory motions of the methyl groups would not result in a dis ordered pattern in a neutron density map so long as the group had a well-define minimum energy orientation in which it resided most of the time. Thus, onl those groups for which external forces coincidentally cancelled the intrinsi rotational barrier would show no preferred orientation. The others would b expected to result in an interpretable set of peaks in the neutron density map regardless of the time frame of reorientation.

Previous neutron studies determined that high resolution data and a well refined phasing model are prerequisite to any type of detailed methyl roto analysis (16, 77). For this reason, Kossiakoff & Shteyn selected crambin, small protein of 46 amino acid residues (22, 78, 79), with 27 terminal methy groups as an ideal candidate for this type of study (76).

To analyze rotor conformations a method was employed that minimized th effects of phasing errors, and that allowed for interpretation independent of preconceived model of the rotor orientation (76, 77). This was done in a Fourie difference map by subtracting out the scattering contribution of the methy carbon and by not incorporating the methyl hydrogens in the phasing mode The result of this treatment effectively reduces the overall peak heights of th excluded methyl H's by about ½, but does not perturb their peak positions c overall shape of their scattering profile.

Examples of the methyl group neutron densities from the crambin analysis 1.4 Å resolution are shown in Figure 8. It is seen that in three of these, th positions of all three H atoms are sharply resolved with peaks spaced 12C apart. The fourth example (8d) is considerably less sharp, but appears as well t show a preferred orientation. These examples are representative of the othe methyls in crambin, with most profiles exhibiting a high degree of order.

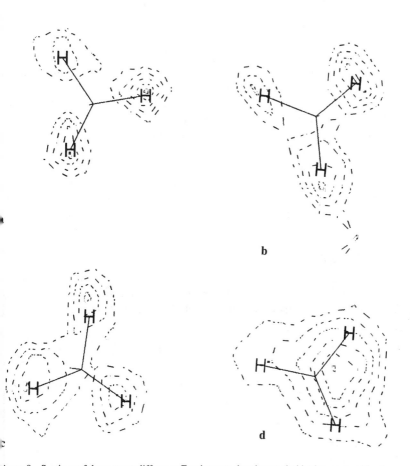

Figure 8 Sections of the neutron difference Fourier map showing methyl hydrogen densities for several representative methyl groups. Because of the rather unique scattering characteristics of neutrons by hydrogen atoms, hydrogen atom positions appear in neutron density maps as negative peaks. No phasing information about the methyl hydrogens was included in the model; therefore, hydrogens should appear in the difference map at their true positions but at reduced density ($\simeq \frac{1}{2}$ weight). This crambin analysis represents the first time that methyl hydrogens have ever been observed at atomic resolution in a Fourier density map. The groups shown are: (*a*) ala 24, (*b*) thr 21, (*c*) thr 28 and (*d*) ala 45.

In order to obtain a more quantitative measure of the preferred orientations of the methyl groups, the neutron density plots were sampled every 10° around the rotor axis at the stereochemically appropriate location for a possible methyl hydrogen atom. These plots confirmed the finding that most methyl groups possess a high degree of order, exhibiting a clearly defined pattern with maxima and minima spaced at 120° intervals. Since these maps are unbiased

towards the character of the density distribution of the rotor hydrogens, th
occurrence of this pattern by chance is highly unlikely.

A further extension of the density analysis was made by applying a threefol
averaging procedure, which reduces local perturbations and increases th
accuracy of locating the preferred orientation (76). This analysis was applied t
every methyl group in the crambin molecule. Figure 9 shows this type o
averaging using as a representative example the alanine groups. This pl
clearly suggests that the preferred conformation of the alanines is staggere
(refer to the figure caption for the torsion angle convention). The trend towar
staggered conformations was observed for the other groups (VAL, THR, ILI
LEU) in crambin (76). Similar findings were observed in two other analyse
trypsin (A. A. Kossiakoff, unpublished results), and PTI (69), but becaus
those studies were done at lower resolution (1.8 Å), the interpretation of th
density plots was less definitive.

Factors Affecting Interpretation of Methyl Rotor Plots

Before discussing the implications of the findings for rotor conformation:
several limitations for the interpretability of these plots are mentioned. The
are three primary factors which could potentially attenuate the scatterin

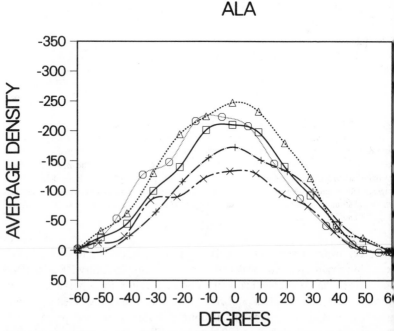

Figure 9   Rotationally averaged rotor plots for the alanine groups in crambin. The position of th
density maximum defines the groups torsional angle. The staggered conformation is defined as 0

density of rotor hydrogens (77). Two of the factors, 1. general rotational disorder, and 2. vibrational disorder of the parent methyl carbon, result directly from the physical characteristics of the protein motion. The third factor, incorrect phasing, is associated with refinement errors. Minimization of these phasing errors is best achieved by using the type of difference Fourier synthesis discussed above.

Of the first two factors, vibrational disorder of the parent methyl carbon probably represents the more prominent effect. This effect can usually be identified in a refined structure by a large temperature factor for the methyl carbon. In point of fact, in the crambin structure there seems to be definitive correlation between side chain length and methyl hydrogen density. What is observed is that rotors having the highest order by density criteria are those associated with alanines where side chain motion is strongly coupled to the main chain (76).

Structural Implications of Methyl Rotor Orientation

The experimental observations indicating that a majority of methyl rotors have their preferred orientations in the staggered conformation address an important question about protein packing. Examination of space filling models of proteins (a time-averaged structure representation) shows that their interior cores are usually extremely close packed, with constituent groups having little room to maneuver. Therefore, it is important to assess whether in the tightly packed core regions there exist mutual effects established through van der Waals interactions which perturb rotors towards orientations of higher torsional energy.

In the systems that have been analyzed in detail [trypsin (A. A. Kossiakoff, unpublished results), crambin (76, 77), and PTI (69)], cases in which rotors are pushed out of the intrinsic low energy conformer by packing effects are rare. One of the few clear examples of a rotor having a perturbed conformation was reported for crambin (76). The explanation for this occurrence was that the methyl group makes close van der Waals contact with a portion of a symmetry related molecule in the crystal packing lattice. On the other hand, at other similar intermolecular contact regions (of which there are several), methyl groups were observed to have the normal staggered conformation; so from these results alone there is no general rule which can be applied to these types of interactions.

In summary, the finding that the staggered conformation is highly preferred strongly suggests that the principal determinant in conferring orientation is the group's intrinsic barrier to rotation rather than local structure environment. Presumably, the inherent "breathing" of the protein and its small but numerous packing defects are important in minimizing packing effects of neighboring groups.

Experimental observations from NMR (showing rapid rotatory motions) together with the neutron results (indicating highly preferred conformations) depict methyl rotations as being "quantized" in 120° steps about a position of highest stability. The lifetime in any particular orientation is short in absolute terms; however, it is a relatively much longer time than times spent in rotational transition from one threefold energy minimum to the next.

PROTEIN HYDRATION

A protein molecule in water solutions is surrounded by a layer of solvent which mediates its functional conformation as well as its chemical characteristics. The interaction of water molecules with the protein surface depends on the surface' specific local character; waters are held tightly by hydrophilic sites, while those contacting hydrophobic surface regions are poorly ordered and have similar properties of solvent in the bulk phase. The intimate interaction between the protein and solvent is a dominant factor in the processes of protein folding and chemical activity, and hence an understanding of its structure is of considerable importance to the interpretation of protein behavior.

As mentioned in the section discussing regional denaturation, the solvent environment of protein molecules in the crystalline state has been shown to be very similar to that in solution. This makes it possible to apply the structural analysis power of crystallographic techniques to determine just how the protein interacts with the solvent as a function of specific physical and chemical properties of its surface features, and thus develop an understanding of the structural principles that govern this interaction. [For details refer to recent reviews of water structure (80–83)].

Assigning Water Molecule Positions—Advantages of Neutrons

The standard method of identifying potential water molecule sites by crystallography is to pick out those density peaks observed at proper H-bonding locations at the surface of the protein (82). While tightly bound water molecules can be generally located accurately by this procedure, because their peaks are significantly above the signal/noise threshold, partially ordered waters are often difficult to characterize with any degree of confidence. In this instance the neutron diffraction technique is inherently better suited than X rays to study the water structure of proteins because it has a threefold greater relative scattering of ordered water (D_2O), providing a correspondingly greater signal noise ratio in assigning their location. Another advantage of the neutron technique is that a full set of data can be collected from a single crystal because unlike X rays, neutrons cause no radiation damage to the sample. In X-ray studies it is normally necessary to use several crystals to obtain high resolution data. Scaling errors due to the collation of partial sets of data are notorious for enerating the types of systematic errors that introduce artificial features into he resulting density map. The presence of such features in the protein portion f the map is a bothersome problem; in the case of solvent regions, where rdered water molecules are less highly constrained by specific stereochemical ules, this problem can be fatal to accurate interpretation if not recognized and lealt with properly.

Methods to Improve Density Map Interpretations

The first detailed analysis of water structure by neutrons was performed on nyoglobin by Schoenborn and his colleagues (84). Since then, detailed studies ave been made of oxy-myoglobin (34), trypsin (16), crambin (22), ribonuc-ease (66), and PTI (69). These latter investigations reported that about 80% of he water located in the neutron map were also found in essentially equivalent ositions in the X-ray map.

These comparisons suggest that given a high quality phasing model, the vell-ordered water molecules can be located accurately by either neutron or X-ray methods using the conventional method for water position assignment. The advantage of the neutron method (for the reasons discussed above) is in the dentification of the location of partially ordered waters. New methods to mprove the accuracy of discriminating between density features produced by partially ordered water and those due to noise are currently being developed in everal laboratories. (65, 77).

BULK SOLVENT DENSITY MODIFICATION   Raghavan & Schoenborn are in-vestigating metmyoglobin using a method that improves the treatment of the bulk solvent scattering component of the phasing model (65). This was done by applying a correction to the data during refinement to model the average density scatter of the disordered solvent. They report the following observations: 1. For ow angle data the contributions from the protein and the bulk solvent are of the ame magnitude, and 2. the phase of the bulk is always different from that of the protein. These observations in a direct way show why the fitting of low angle data using the protein model alone has never been successful (86).

H_2O/D_2O SOLVENT DIFFERENCE MAPS   The large difference in the scattering characteristics of neutrons by H_2O compared to D_2O has been used to advan-age through density matching and exchange labeling for several years in small angle neutron scattering experiments (2). This difference can likewise be exploited in neutron protein crystallography to determine the detailed structural characteristics of protein hydration through the calculation of solvent differ-ence maps (A. A. Kossiakoff, unpublished results). In practice, such maps are obtained by comparing the changes in diffracted intensities between two sets of data, one obtained from a crystal having H_2O as the major solvent constituent and a second where D_2O is the solvent medium. To a good approximation, the protein atom contributions to the scattering intensities in both data sets are equal and cancel, but since H_2O and D_2O have very different scattering properties, their differences are accentuated to reveal an accurate and nearly unbiased representation of the solvent structure.

The features of a solvent difference map of this type are not as affected by errors in the phasing model as conventional difference Fourier maps. In addition, there are refinement procedures that can be applied to them that lead to significant enhancement in signal/noise discrimination (87–90). Without going into detail, the basic feature of the method is a set of density modification steps based on the fact that a considerable amount of information about the density distribution of the crystallographic unit cell is known. For instance, it is known that the region of the unit cell occupied by protein atoms should be featureless in solvent maps. It can also be assumed that, as an approximation, the solvent regions further than 4 Å from the protein surface have bulk solvent characteristics and can be treated as a constant density region. Combining these two regions gives about 50–60% of the total volume of the unit cell.

Knowledge of the density content of such a large percentage of the unit cell places a strong constraint on the overall character of the Fourier transform, a fact that can be used to improve the quality of the experimentally determined phases (87–89).

In such a refinement procedure, it is also important to minimize the effects of series termination errors which are inherent in the Fourier synthesis. Although these density effects are not large compared to the densities of the well-ordered waters, they are of the same magnitude as the peaks of partially ordered waters and must be eliminated (or at least identified) for high confidence to be placed in partial water assignments.

An example of the improvement in the quality of the neutron solvent map of trypsin using this method can be seen by comparing the representative section in Figure 10. Fig. 10a is the initial D_2O-H_2O solvent map; 10b is the same area after nine cycles of density modification. (The algorithms used in the refinement are relatively simple and completion of a cycle takes only several minutes on a computer of moderate speed.) It is clear that there is a distinct difference in the quality of the two maps. In the unrefined map, a significant percentage of peaks due to real waters have similar amplitudes to those due to noise. Also, there were several waters whose peak heights are below the contouring threshold in this example.

The situation is quite different for the refined case. Here most well-ordered waters are at least at the 10σ level. It should be emphasized here that the refinement method makes no a priori assumptions about where water peaks are located except requiring that they be closer than 4 Å from the protein surface. As will be discussed in the next section, one measure of confidence that the

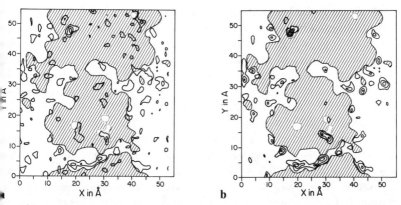

igure 10   Section of D_2O-H_2O solvent difference map of trypsin; (*a*) initial map (no refinement), *b*) same map after 9 cycles of refinement. Dotted areas of map represent regions of the unit cell where the trypsin protein is located.

efined maps are accurate is that all the well-ordered waters determined in the X-ray structure coincide with large peaks in the solvent maps. In fact, a series of model studies have shown that in the refined solvent maps, water molecules scattering at 10% occupancy can be located readily.

Solvent Structure Characteristics Determined by Solvent Difference Maps

As seen above, using the solvent difference map approach, a high degree of accuracy has been achieved in determining the solvent binding characteristics in the protein trypsin. The trypsin solvent system provides a good comparative test for the approach because the solvent structure model from the X-ray structure of trypsin is of extremely high quality. In the X-ray model (refined at 1.5 Å resolution), about 150 water positions were assigned (J. Finer-Moore, unpublished results). Finer-Moore and her colleagues compared these positions with the solvent density determined in the neutron solvent difference map. They found an extremely good correspondence; in addition, they also observed that the neutron map contained a number of additional density peaks adjacent to the primary sites. These density peaks presumably are waters which likely form the second hydration layer at particular points along the protein surface. Assignment of solvent sites to this additional density resulted in a total of about 350 unique waters surrounding the trypsin molecule.

An obvious question to ask is, given this improved model of solvent-protein interaction, are there new structure data in trypsin which heretofore have not been observed? Such situations have been identified for partially ordered waters in interior protein cavities.

M. Connelly and I. Kuntz (unpublished results) have surveyed a number o proteins (trypsin included) to determine the number of interior cavities they contained large enough to fit at least one water molecule. Their calculation: indicated that most proteins of any significant size contained a number of these interior cavities. A further comparison with the reported water structure sites o these proteins indicated that most of these cavities actually contained boun water. However, there were some which, although large enough to contai several waters, were reported to contain none. The question remained un answered as to whether these cavities are actually unoccupied or whether the contain solvent that was so poorly ordered that the resulting scattering was to low to be normally unobservable.

From the X-ray results there are several such cases of unoccupied cavities i trypsin. In the neutron solvent difference maps, however, there existed inter pretable neutron density in these cavity regions. An example of this is shown i Figure 11, where two isolated solvent molecules are bound in internal pocket: of the molecule. It is noteworthy that neither water has the usual types o hydrogen bonding ligands usually associated with protein-solvent interactions Analysis of the groups that form the interface of these water cavities indicate that the waters resided in a predominantly hydrophobic environment. Thi example is one of several instances observed where partially ordered water: are associated with hydrophobic environments. An understanding of the ener getic compromises that favor water/hydrophobic interactions in certain situ ations will be an important step in determining the rules governing protei folding.

ORIENTATION OF WATER MOLECULES   For the reasons described in th previous sections, neutron solvent difference maps are a powerful method fo locating weakly scattering waters. The application of these solvent maps car also provide information on the orientation of highly ordered waters (D_2O). Ir the case of a highly ordered water in a D_2O-H_2O difference map, the oxyger scattering components will cancel (having identical locations and scattering potential), but because H and D have very different scattering properties (H = -3.8, D = $+6.7$) large peaks will be found in the map at the D-H positions Consequently, these ordered waters usually can be oriented with reasonabl accuracy.

An example of this is shown in Figure 12. This water is located in an interna cavity and makes several strong H-bonds to hydrophilic ligands. The position of the oxygen (as determined in the X-ray structure) is marked by Ow. It i evident that the density peak is asymmetric with respect to the oxygen position a consequence of the scattering of the D-H sites alone. From the trypsir analysis the D-H sites for 5–7 waters can be assigned (A. A. Kossiakoff

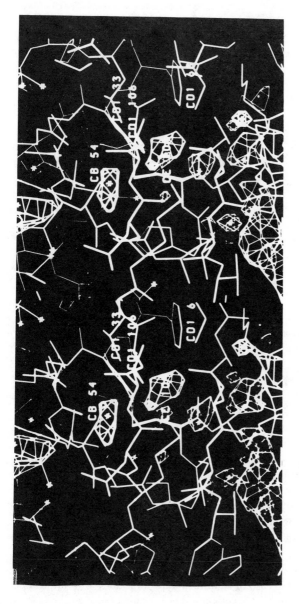

Figure 11   Stereo view of internal waters in two hydrophobic cavities in trypsin.

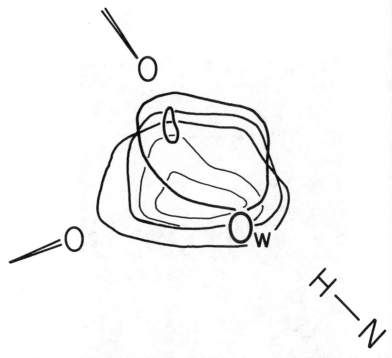

Figure 12   Orientation of water molecules. D_2O-H_2O solvent map; Ow is the position of the oxygen as determined by the X-ray analysis. Resulting density is due to scattering of the D-H positions alone.

unpublished results). In crambin, where the resolution is higher (1.4 vs 1.8 Å), about 20 waters can be oriented [(85), M. M. Teeter, unpublished results].

The importance of the ability to assign orientations of water in gaining insights into the electrostatic forces at the protein's surface cannot be overstated. In order to survey electrostatic forces in a protein, a probe strongly influenced by these forces is needed. As such, waters bound in interior cavities of proteins provide an almost perfect model. Interior waters are segregated from outside solvent forces; their orientation determined predominantly by the influence of their coordinating ligands. Computational methods exist to model electrostatic surface forces as a function of the protein's local charge environment (91–95). Methods to predict the experimentally observed orientation of waters will provide a crucial test for the accuracy of the currently used potential functions, and in case of significant deviations they will furnish experimental data for investigators developing these computational methods to "fine tune" their algorithms.

FUTURE APPLICATIONS

The impact of any technique on a field of science ultimately is measured by the amount of new information it brings to bear on issues not previously resolved by other methods. The goal of this review has been to demonstrate that by this criterion neutron diffraction has made a major impact in several important areas. The future of the technique looks promising as well. For instance, new neutron diffraction data are becoming available for the widely studied protein systems, crambin and pancreatic trypsin inhibitor (PTI).

In the case of the crambin study, there exists the opportunity to study by neutron diffraction a protein at true atomic resolution. Some important findings have been reported from the 1.4 Å analysis (as for example, the methyl rotor analysis described in the dynamics section of this review). However, data have now been collected by M. M. Teeter and her colleagues (personal communication) to 0.9 Å resolution. Location of hydrogen atoms at the very high resolution of the available data should settle a fundamental issue in protein structure concerning the degree of deviation from ideal bonding geometry present in biomacromolecules.

PTI is the most highly studied protein by biophysical methods. The neutron work by Wlodawer, Sjölin, and their coworkers offers the opportunity to compare findings between neutron diffraction and NMR and other techniques. This will be the first time in a protein system that this type of direct comparison has been made.

With the above two studies added to the existing neutron data discussed in the review, it is clear that neutron diffraction will establish itself as an equal partner of X-ray diffraction in the effort to unravel the details of protein structure.

ACKNOWLEDGMENTS

I would like to thank J. Shpungin and S. Shteyn for technical assistance and to acknowledge the comments of Drs. M. Teeter, R. Levy, J. Hanson, and G. Rose. Dr. Alex Kossiakoff assisted in the preparation of the manuscript. Work was supported by NIH grant no. GM 29616.

Literature Cited

1. Schoenborn, B. P., ed. 1984. In *Neutrons in Biology, Basic Life Sciences*, Vol. 27. New York: Plenum 462 pp.
2. Schoenborn, B. P. 1975. *Brookhaven Symp. Biol.* 27(1):1–10
3. Schoenborn, B. P., Nunes, A. C. 1972. *Ann. Rev. Biophys. Bioeng.* 1:529–52
4. Kossiakoff, A. A. 1983. *Ann. Rev. Biophys. Bioeng.* 12:159–82

5. Wlodawer, A. 1982. *Prog. Biophys. Mol. Biol.* 40:115–59
6. Alberi, J. L. 1975. *Brookhaven Symp. Biol.* 27(8):24–42
7. Alberi, J. L., Fischer, J., Radeka, V., Rogers, L. C., Schoenborn, B. P. 1975. *IEEE Trans. Nucl. Sci.* (NS) 22(1):255–68
8. Arndt, U. W., Gilmore, D. J. 1975. *Brookhaven Symp. Biol.* 27(8):16–23

9. Bartunik, H. D., Jacobe, V. 1975. In *Proc. Neutron Diffraction Conf., Petten*, pp. 430–38. Petten: Reactor Centrum Nederland
10. Caine, J. E., Norvell, J. C., Schoenborn, B. P. 1976. *Brookhaven Symp. Biol.* 27(8):43–50
11. Davidson, J. B. 1975. *Brookhaven Symp. Biol.* 27(8):3–15
12. Spencer, S. A., Kossiakoff, A. A. 1980. *J. Appl. Crystallogr.* 13:563
13. Sjölin, L., Wlodawer, A. 1981. *Acta Crystallogr.* A 37:594–604
14. Schoenborn, B. P. 1984. See Ref. 1, pp. 261–79
15. Kossiakoff, A. A., Spencer, S. A. 1980. *Nature* 288:414–16
16. Kossiakoff, A. A., Spencer, S. A. 1981. *Biochemistry* 20:6462–74
17. Bentley, G. A., Duee, E. D., Mason, S. A., Nunes, A. C. 1979. *J. Chim. Phys.* 76:817–21
18. Mason, S. A., Bentley, G. A., McIntyre, G. J. 1984. See Ref. 1, pp. 323–34
19. Phillips, S. E. V., Schoenborn, B. P. 1981. *Nature* 292:81
20. Wlodawer, A., Sjölin, L. 1981. *Proc. Natl. Acad. Sci. USA* 78:2853
21. Hanson, J. C., Schoenborn, B. P. 1981. *J. Mol. Biol.* 153:117–46
22. Teeter, M. M., Kossiakoff, A. A. 1984. See Ref. 1, pp. 335–48
23. Bacon, G. E. 1975. In *Neutron Diffraction*, pp. 155–88. Oxford: Clarendon
24. Schoenborn, B. P. 1976. *Brookhaven Symp. Biol.* 27(1):10–17
25. Robillard, G., Shulman, R. G. 1974. *J. Mol. Biol.* 86:541–58
26. Markley, J. L., Ibanez, I. B. 1978. *Biochemistry* 17:4627–40
27. Bachovchin, W. W., Roberts, J. D. 1978. *J. Am. Chem. Soc.* 100:8041–47
28. Hunkapiller, M. W., Smallcombe, S. H., Whitaker, D. R., Richards, J. H. 1973. *Biochemistry* 12:4732
29. Hunkapiller, M. W., Forgac, M. D., Richards, J. H. 1976. *Biochemistry* 15:5581–88
30. Koeppe, R. E., Stroud, R. M. 1976. *Biochemistry* 15:3450–58
31. Scheiner, S., Lipscomb, W. H. 1976. *Proc. Natl. Acad. Sci. USA* 73:432–36
32. Kollman, P. A., Hayes, D. M. 1981. *J. Am. Chem. Soc.* 103:2955–61
33. Nakagawa, S., Umeyama, H., Kudo, T. 1980. *Chem. Pharm. Bull.* 28:1342–44
34. Phillips, S. E. V. 1984. See Ref. 1, pp. 305–22
35. Linderstrom-Lang, K. U., Schellman, J. A. 1959. In *The Enzymes*, ed. D. Boyer, H. Lardy, K. Myrback, Vol. 1. New York: Academic. 443 pp. 2nd ed.
36. Gurd, F. N., Rothgeb, T. M. 1979. *Adv. Protein Chem.* 33:73–165
37. Sternberg, M. J., Grace, D. E., Phillips D. C. 1979. *J. Mol. Biol.* 130:231–53
38. Frauenfelder, H., Petsko, G., Tsernoglou, D. 1979. *Nature* 280:558–63
39. Debrunner, P. G., Frauenfelder, H. 1982. *Ann. Rev. Phys. Chem.* 33:283–99
40. Karplus, M., McCammon, J. A. 1981. *CRC Crit. Rev. Biochem.* 9:293–349
41. Lumry, R., Biltonen, R. 1969. In *Structure and Stability of Biological Macromolecules*. New York: Dekker
42. Baldwin, R. L. 1975. *Ann. Rev. Biochem.* 44:453–75
43. Blumenfeld, L. A. 1976. *J. Theor. Biol.* 58:269–84
44. Perutz, M. F. 1979. *Ann. Rev. Biochem.* 48:327–86
45. Petsko, G. A., Ringe, D. 1984. *Ann. Rev. Biophys. Bioeng.* 13:331–71
46. Hvidt, A., Nielson, S. O. 1966. *Adv. Protein Chem.* 21:287–386
47. Englander, S. W., Downer, N. W., Teitelbaum, H. 1972. *Ann. Rev. Biochem.* 41:903–24
48. Englander, S. W., Calhoun, D. B., Englander, J. J., Kallenbach, N. R., Liem, R. K. H., et al. 1980. *Biophys. J.* 32:557–89
49. Woodward, C. K., Hilton, B. D. 1979. *Ann. Rev. Biophys. Bioeng.* 8:99–127
50. Kossiakoff, A. A. 1982. *Nature* 296:713–21
51. Wüthrich, K., Wagner, G. 1979. *J. Mol. Biol.* 130:1–18
52. Wagner, G., Wüthrich, K. 1979. *J. Mol. Biol.* 134:75–94
53. Hilton, B. D., Woodward, C. K. 1978. *Biochemistry* 17:3325–32
54. Kuwajima, K., Baldwin, R. L. 1983. *J. Mol. Biol.* 169:281–97
55. Schreier, A. A., Baldwin, R. L. 1976. *J. Mol. Biol.* 105:409–26
56. Rosa, J. J., Richards, F. M. 1979. *J. Mol. Biol.* 133:399–416
57. Schoenborn, B. P. 1971. *Cold Spring Harbor Symp. Quant. Biol.* 36:569–75
58. Schoenborn, B. P. 1972. In *Structure and Function of Oxidation Reduction Enzymes*, ed. A. Akeson, A. Ehreberg, pp. 109–16. Oxford/New York: Pergamon
59. Matthews, B. W. 1968. *J. Mol. Biol.* 33:491–96
60. Rupley, J. A. 1969. See Ref. 41, pp. 291
61. Haggis, G. H. 1957. *Biochim. Biophys. Acta* 23:494
62. Praissman, M., Rupley, J. A. 1968. *Biochemistry* 7:2431
63. Tuchsen, E., Hvidt, A., Ottesen, M. 1980. *Biochimie* 62:563–66
64. Makinen, M. W., Fink, A. L. 1977. *Ann. Rev. Biophys. Bioeng.* 6:301–43

65. Raghavan, N. V., Schoenborn, B. P. 1984. See Ref. 1, pp. 247–59
66. Wlodawer, A., Sjölin, L. 1983. *Biochemistry* 22:2720–28
67. Kossiakoff, A. A. 1984. See Ref. 1, pp. 281–304
68. Bentley, G. A., Delepierre, M., Dobson, C. M., Wedin, R. E., Mason, S. A., Poulsen, F. M. 1983. *J. Mol. Biol.* 170:243–47
69. Wlodawer, A., Walter, J., Huber, R., Sjölin, L. 1984. *J. Mol. Biol.* In press
70. Rose, G. D., Roy, S. 1980. *Proc. Natl. Acad. Sci. USA* 77:4643–47
71. Campbell, I. D., Dobson, C. M., Williams, R. J. P. 1975. *Proc. R. Soc. London Ser. B* 189:503–9
72. McCammon, J. A., Karplus, M. 1980. *Biopolymers* 19:1375
73. Oldfield, E., Norton, R. S., Allerhand, A. 1975. *J. Biol. Chem.* 250:6368
74. Wittebort, R. J., Rothgeb, T. M., Szabo, A., Gurd, F. R. 1979. *Proc. Natl. Acad. Sci. USA* 76:1059
75. Lipari, G., Szabo, A., Levy, R. M. 1982. *Nature* 300:197–98
76. Kossiakoff, A. A., Shetyn, S. 1984. *Nature.* 311:582–83
77. Kossiakoff, A. A. 1984. *Methods Enzymol.* In press
78. Teeter, M. M., Mazer, J. A., L'Italien, J. J. 1981. *Biochemistry* 20:5437–43
79. Hendrickson, W. A., Teeter, M. M. 1981. *Nature* 290:107–13
80. Kuntz, I. D., Kauzmann, W. 1974. *Adv. Protein Chem.* 28:239
81. Cooke, R., Kuntz, I. D. 1974. *Ann. Rev. Biophys. Bioeng.* 3:95–126
82. Finney, J. L. 1979. In *Water: A Comprehensive Treatise,* ed. F. Franks, 6:47. New York: Plenum
83. Edsall, J. T., McKenzie, A. A. 1983. *Adv. Biophys.* 16:53–183
84. Schoenborn, B. P., Hanson, J. C. 1980. In *Water in Polymers,* ed. S. P. Rowland, ACS Symp. 127, pp. 215–24. Washington, DC: Am. Chem. Soc.
85. Teeter, M. M. 1984. *Proc. Natl. Acad. Sci. USA.* 81:6014–18
86. Chambers, J. L., Stroud, R. M. 1977. *Acta Crystallogr. B* 33:1824
87. Raghavan, N. V., Tulinsky, A. 1979. *Acta Crystallogr. B* 35:1776–85
88. Bhat, T. N., Blow, D. M. 1982. *Acta Crystallogr. A* 38:21–29
89. Schevitz, R. W., Podjarny, A. D., Zwick, M., Hughes, J. J., Sigler, P. B. 1981. *Acta Crystallogr. A* 37:669–77
90. Agard, D. A., Stroud, R. M. 1982. *Biophys. J.* 37:589–602
91. Hermans, J., Vacatello, M. 1980. See Ref. 84, pp. 199–214
92. Hagler, A. T., Moult, J. 1978. *Nature* 272:222–26
93. Clementi, E., Corongiu, G., Jonsson, B., Romano, S. 1979. *FEBS Lett.* 100:313–17
94. Stillinger, F. H., Rahman, A. J. 1974. *J. Chem. Phys.* 60:1545
95. Karplus, M., Rossky, P. J. 1980. See Ref. 84, pp. 23–42

AUTHOR INDEX

1234 AUTHOR INDEX

1292   AUTHOR INDEX

Vik, S. B., 1029, 1045, 1055, 1057
Vik, T. A., 984
Vilgrain, I., 226
Villafranca, J. E., 606
Villafranca, J. J., 386
Villalobo, A., 1037
Villalobos-Molina, R., 215
Villani, G., 431, 444
Villarroya, H., 124
Villiers, C., 991, 993, 1008
Vina, J., 318
Vinay, P., 500
Vincow, G., 1042
Vinkler, C., 985
Vinograd, J., 674, 680, 1185
Vinogradov, A. D., 1029, 1030
Virtanen, I., 295
Viscontini, M., 752
Visscher, M. B., 840
Visser, A. J. W. G., 55, 65
Visvader, J. E., 537, 539, 542, 556
Vitale, A., 643
Vitti, P., 76, 79, 80
Viveros, O. H., 481, 486, 500, 502, 730, 731, 735, 738, 739, 748, 749
Viviani, A., 473
Vivona, G., 789
Vladimirov, Yu. A., 44
Vlasik, T. N., 1121
Vlasuk, G. P., 124, 125
Vliegenthart, J. F. G., 632, 633, 643, 650, 653, 767, 768, 780, 782, 789
Voelkel, K., 683, 684, 688
Voelkel, K. A., 684-86, 690
Voet, D. H., 606
Voetman, A. A., 317
Vogel, C. W., 933
Vogel, R. H., 48, 66
Vogel, R. L., 736
Vogelsang, G., 1099, 1100
Vogt, K., 263
Vogt, P. K., 899, 901, 905
Vohra, M. M., 216
Voisin, D., 781
Voituriez, L., 1165, 1170
Voliva, C. F., 1078
Volkert, M. R., 439, 442, 446, 448, 449
Volle, R. L., 945
Volpi, M., 216, 226
Volwerk, J. J., 55
Volz, K. W., 606
von Bahr-Lindström, H., 242-44, 249, 266
Von Deimling, O. H., 703
von der Haar, F., 378, 385, 386
von Figura, K., 639, 654
von Gabain, M., 181

von Hippel, P. H., 173, 180, 182, 185, 196, 869
von Hungen, K., 336, 941
von Jagow, G., 1016, 1030, 1032, 1034-37
Von Meyenburg, K., 439, 1049, 1051
von Nicolai, H., 778-80
von Saltza, M. H., 731, 736
Von Tscharner, V., 990, 1050
Voor-Brouwer, M. M., 979
Voorma, H. O., 1110-14, 1116-23, 1128
Vorob'eva, L. I., 27
Vorobjeva, I. P., 441
Voronkov, I. I., 841
Voronova, A. F., 918
Vos, P., 537
Vosbeck, K., 646
Vosberg, H.-P., 83, 392, 668, 673, 676, 680, 682, 691
Voss, J., 812
Votruba, J., 995
Vovis, G. F., 876
Vreman, H., 1114
Vulliet, P. R., 940, 951, 952, 955
Vurek, G. G., 53, 58

W

Waber, P. G., 1083, 1094
Wabiko, H., 438
Wabl, M. R., 511
Wachter, E., 995, 1052
Wachter, H., 731, 737, 749, 751
Wacker, A., 1159
Wacker, W. E. C., 569
Wade, R. D., 900, 903, 908, 919, 933, 943
Wadman, S. K., 733
Waechter, C. J., 636
Wagenvoord, R. J., 984, 988, 1003, 1006, 1007
Waggoner, B., 885
Wagner, G., 1202, 1205
Wagner, P., 295
Wagner, R. E. Jr., 430
Wagstaff, W., 277
Wahba, A. J., 1115
Waheed, A., 654
Wahl, G. M., 1138
Wahl, P., 47, 49, 52-57, 62, 64, 65, 67
Wahle, E., 684, 686, 687
Wahllander, A., 308
Wahn, H., 687
Wainscoat, J. S., 277, 1094
Waite, M., 28, 29
Wajcman, H., 1093
Wakabayashi, S., 1032, 1051
Wakagi, T., 989, 1002

Wakatama, E., 1137
Wakil, S. J., 28, 29
Walaas, O., 76, 79
Walaas, S. I., 932, 935-38, 941, 946, 950, 951, 954-58, 962
Walberg, M. W., 1039
Walbridge, D. G., 54-56
Waldeck, D. H., 46
Walder, J. A., 277
Waldmann, T. A., 805, 811, 815
Waldstein, E. A., 430, 437
Wålinder, O., 958
Walker, A., 431
Walker, B., 335
WALKER, G. C., 425-57; 199, 431, 441, 442, 448, 449
Walker, J. B., 835, 854
Walker, J. E., 984, 988, 995, 997, 1002, 1004, 1007, 1053
Walker, M. D., 1083, 1086
Walker, N., 395
Wall, R., 559, 1080, 1122
Wallace, B. J., 242-44, 249, 266
Wallace, J. K., 461
Wallace, M., 215
Wallace, M. A., 216
Wallace, R. W., 962
Wallach, D. F. H., 212
Wallach, M., 359
Walliman, T., 847
Wallin, K. T., 280, 466-69, 473
Wallis, K. T., 336, 340, 345
Wallis, M., 416, 417
Wallis, S., 710, 711
Wallis, S. C., 718
Wallmark, B., 1048, 1054
Wallner, A., 837
Walsh, C., 357
Walsh, D. A., 932-34, 942
Walsh, K. A., 903, 908, 933, 934, 943
Walsh, R. B., 869
Walsh, W. A., 472, 474
Walter, B., 534
Walter, G., 172, 900, 903, 909, 920
Walter, J., 1204, 1207, 1208, 1216, 1217, 1219
Walter, P., 102, 103, 126, 128, 521
Walter, S. J., 467, 470
Walter, U., 933-37, 939, 942, 943, 945
Walters, D. E., 790
Walters, J., 755
Walthall, B. J., 1111, 1112, 1115, 1129
Walthall, W. W., 396
Walther, J.-U., 1097

Wingfield, P., 1031, 1033
Winkelmann, D. A., 511, 513, 524
Winkler, J. R., 591
Winslow, C. L. J., 416, 417
Winter, D. B., 1029, 1043, 1045
Winter, G., 603, 605-7, 616
Winter, R. B., 869
Winterhalter, K. H., 76, 81, 82
Winterwerp, H., 652, 653
Wintz, A., 1097
Winyard, P., 787
Wirak, D. O., 685
Wirch, E., 941
Wirth, D. F., 338, 339, 359
Wirtz, K. W. A., 55, 216, 218, 956
Wischmeyer, B., 942
Wise, B. C., 941, 954, 955, 957
Wise, J. G., 1016, 1049, 1052, 1054
Wishart, W., 876
Withers, S. G., 396
Withnell, M. R., 225
Witkin, E. M., 442
Witkop, B., 1187
Witman, G. B., 335
Wit-Peeters, E. M., 840, 847
Witte, O. N., 810, 898, 899, 919, 920
Wittebort, R. J., 1213
Wittekind, W., 105, 108, 117
Wittig, B., 1154, 1157, 1185
Wittig, S., 1154, 1157, 1185
Wittinghofer, A., 387
Wityk, R. J., 890
Witztum, J. L., 709
Wlodawer, A., 1196, 1199, 1200, 1204, 1205, 1207, 1208, 1211, 1216, 1217, 1219
Wodak, S., 49
Wodnar-Filipowicz, A., 1122
Woenckhaus, C., 589
Woese, C. R., 517, 521, 525, 527, 670, 1180
Wofsy, L., 284
Woggon, B., 755, 756
Wolberg, G., 499
Wold, F., 74
Wolf, K., 1157
Wolf, M., 441
Wolfe, L. C., 278, 290
Wolfe, L. S., 224
Wolfe, P. B., 125
Wolfe, R. S., 527
Wolfensberger, M., 463
Wolfensberger, M. R., 467
Wolff, A., 336
Wolff, D. J., 941
Wolff, J., 296

Wolff, O. H., 752
Wolfner, M., 1138
Wolfson, J. S., 691
Wolin, M. S., 378
Wolin, S. L., 555, 559, 1080
Wollenzien, P. L., 1153-55, 1157, 1177-79, 1183, 1186
Wollheim, C. B., 221
Wollny, E., 316
Wolosiuk, R. A., 239, 241, 261-63, 266, 321
Wolynes, P. G., 63
Wong, A. L., 281, 289
Wong, C., 1115, 1116, 1119, 1120
Wong, L. T., 1006
Wong, M., 86, 92
Wong, P. C. L., 397
Wong, S.-L., 174
Wong, S.-T., 1117
Wong, S.-Y., 979, 997, 1050, 1051, 1053-57
Wong, T.-W., 681, 900, 918-22
Woo, D., 516
Woo, D. D. L., 921
Woo, N., 1115
Wood, C., 811, 814
Wood, D. C., 188, 1058
Wood, G. W., 773, 776
WOOD, H. G., 1-41; 6-11, 14, 15, 17, 19-21, 23-32, 35
Wood, J. G., 962
Wood, R. D., 443
Wood, W. A., 582
Wood, W. G., 1075, 1090, 1092-94, 1097, 1098
Wood, W. I., 1088
Woodgate, R., 445
Woodgett, J. R., 948, 949, 951-53, 955
Woodhead, A. P., 1075
Woodley, C. L., 1114, 1115, 1119
Woods, D. E., 719
Woodward, C. K., 1201, 1202, 1205
Woodward, D. O., 334, 341
Woody, A.-Y. M., 184
Woody, R. W., 184
Wool, I. G., 1112-15, 1119, 1120, 1129
Woolf, J. H., 737-39, 749, 755
Woolfe, G. J., 46, 48, 51, 52
Woolford, C., 177
Woolley, D. E., 317
Woolley, P., 509
Wooten, M. W., 955
Wootton, J. F., 569
Worcel, A., 688, 689
Worden, L. R., 1159
Worobec, S. W., 484
Wortham, K. A., 141, 144
Woychik, R. P., 1081

Wraight, C. A., 1030, 1038
Wrede, P., 1115
Wrenn, R. W., 955, 957
Wright, A., 109, 684-86, 691
Wright, B., 347
Wright, C. T., 1039
Wright, J. K., 792
Wright, R. R., 309, 314
Wright, S., 1082, 1084, 1085
Wrightstone, R. W., 1094
Wu, A. L., 705, 706
Wu, A. M., 689
Wu, C.-W., 65, 173, 174, 183, 387, 397, 688
Wu, D., 307
Wu, E. S., 274
Wu, F. Y.-H., 65, 174, 183
Wu, G. E., 815
Wu, G-Y., 262
Wu, H. C., 103-5, 113, 122, 125, 126
Wu, H. C.-P., 103
Wu, H.-M., 178
Wu, J. M., 316
Wu, N. C., 962
Wu, T.-H., 188
Wu, T. T., 805, 809
Wu, W. C.-S., 954-58
Wu, X-Y., 262
Wu, Y., 1090
Wühr, B., 36
Wulff, D. L., 191, 192
Wunderlich, R., 655
Wusteman, M. M., 216, 217
Wüthrich, K., 1202, 1205
Wyatt, I., 309, 314
Wyatt, J., 275
Wyche, A., 224
Wyman, J., 612, 616
Wyss, S. R., 323

X

Xu, Y.-H., 903, 912, 913

Y

Yablonka-Reuveni, Z., 1135
Yabuki, S., 511, 524
Yabusaki, Y., 527
Yabuuchi, H., 772
Yachie, A., 821
Yaciuck, Y. P., 909
Yagi, A., 998
Yagi, K., 847, 946
Yagi, M., 1082, 1097
Yagi, T., 998, 1037, 1051, 1055, 1057
Yagihara, Y., 215
Yagishita, K., 578
Yaguchi, S., 550
Yaguchii, M., 516
Yajima, H., 412, 415

SUBJECT INDEX

CUMULATIVE INDEXES

CONTRIBUTING AUTHORS, VOLUMES 50–54

CHAPTER TITLES, VOLUMES 50–54

Annual Reviews Inc. | *ORDER FORM*

NONPROFIT SCIENTIFIC PUBLISHER

139 El Camino Way, Palo Alto, CA 94306-9981, USA • (415) 493-4400

ews Inc. publications are available directly from our office by mail or telephone (paid by credit card or der), through booksellers and subscription agents, worldwide, and through participating professional ices subject to change without notice.

ls: Prepayment required on new accounts by check or money order (in U.S. dollars, check drawn on) or charge to credit card — American Express, VISA, MasterCard.

nal buyers: Please include purchase order number.

: $10.00 discount from retail price, per volume. Prepayment required. Proof of student status must be photocopy of student I.D. or signature of department secretary is acceptable). Students must send ct to Annual Reviews. Orders received through bookstores and institutions requesting student rates urned.

nal Society Members: Members of professional societies that have a contractual arrangement with eviews may order books through their society at a reduced rate. Check with your society for infor-

ders: Please list the volumes you wish to order by volume number.

rders: New volume in the series will be sent to you automatically each year upon publication. Cancel-e made at any time. Please indicate volume number to begin standing order.

ion orders: Volumes not yet published will be shipped in month and year indicated.

orders: Add applicable sales tax.

id (4th class bookrate/surface mail) **by Annual Reviews Inc.** Airmail postage extra.

L REVIEWS SERIES		Prices Postpaid per volume USA/elsewhere	Regular Order Please send:	Standing Order Begin with:
			Vol. number	Vol. number

w of ANTHROPOLOGY (Prices of Volumes in brackets effective until 12/31/85)

10	(1972-1981)	$20.00/$21.00]		
	(1982)	$22.00/$25.00]		
-14	(1983-1985)	$27.00/$30.00]		
14	(1972-1985)	$27.00/$30.00		
	(avail. Oct. 1986)	$31.00/$34.00	Vol(s). _____	Vol. _____

w of ASTRONOMY AND ASTROPHYSICS (Prices of Volumes in brackets effective until 12/31/85)

2, 4-19	(1963-1964; 1966-1981)	$20.00/$21.00]		
	(1982)	$22.00/$25.00]		
-23	(1983-1985)	$44.00/$47.00]		
2, 4-20	(1963-1964; 1966-1982)	$27.00/$30.00		
-23	(1983-1985)	$44.00/$47.00		
	(avail. Sept. 1986)	$44.00/$47.00	Vol(s). _____	Vol. _____

w of BIOCHEMISTRY (Prices of Volumes in brackets effective until 12/31/85)

-34, 36-50	(1961-1965; 1967-1981)	$21.00/$22.00]		
	(1982)	$23.00/$26.00]		
-54	(1983-1985)	$29.00/$32.00]		
-34, 36-54	(1961-1965; 1967-1985)	$29.00/$32.00		
	(avail. July 1986)	$33.00/$36.00	Vol(s). _____	Vol. _____

w of BIOPHYSICS AND BIOPHYSICAL CHEMISTRY (Prices of Vols. in brackets effective until 12/31/85)
nnual Review of Biophysics and Bioengineering)

0	(1972-1981)	$20.00/$21.00]		
	(1982)	$22.00/$25.00]		
14	(1983-1985)	$47.00/$50.00]		
1	(1972-1982)	$27.00/$30.00		
14	(1983-1985)	$47.00/$50.00		
	(avail. June 1986)	$47.00/$50.00	Vol(s). _____	Vol. _____

of CELL BIOLOGY

	(1985)	$27.00/$30.00		
	(avail. Nov. 1986)	$31.00/$34.00	Vol(s). _____	Vol. _____

w of COMPUTER SCIENCE

	(avail. late 1986)	Price not yet established	Vol. _____	Vol. _____

of EARTH AND PLANETARY SCIENCES (Prices of Volumes in brackets effective until 12/31/85)

	(1973-1981)	$20.00/$21.00]		
	(1982)	$22.00/$25.00]		
13	(1983-1985)	$44.00/$47.00]		
0	(1973-1982)	$27.00/$30.00		
13	(1983-1985)	$44.00/$47.00		
	(avail. May 1986)	$44.00/$47.00	Vol(s). _____	Vol. _____